This volume is dedicated to

Robert J. Dowsett and Françoise Dowsett-Lemaire

ornithologists, sound-recordists, environmentalists, explorers, bibliographers and publishers

in recognition of their remarkable achievements in furthering

knowledge of African birds

Recommended citations for reference to this volume in scientific and general publications are:
 Fry, C. H., Keith, S. and Urban, E. K. (Eds) (2000). *The Birds of Africa* Vol. VI. Academic Press, London.

or, for example:

 Irwin, M. P. S. (2000). *Hedydipna pallidigastra*, in C. H. Fry, S. Keith and E. K. Urban (Eds), *The Birds of Africa* Vol. VI. Academic Press, London.

The Birds of Africa
Volume VI

C. HILARY FRY   STUART KEITH

ADRIAN J. F. K. CRAIG   LLEWELLYN GRIMES
MICHAEL P. S. IRWIN   DAVID J. PEARSON
EMIL K. URBAN   DAVID WIGGINS   ROGER WILKINSON

Edited by

C. HILARY FRY
EXECUTIVE EDITOR
*Department of Zoology
University of Aberdeen
Aberdeen, UK*

STUART KEITH
*Department of Ornithology
American Museum of Natural History
New York
New York, USA*

EMIL K. URBAN
*Department of Biology
Augusta State University
Augusta, Georgia, USA*

Colour Plates by Martin Woodcock

Line Drawings by Ian Willis
Discography by Claude Chappuis

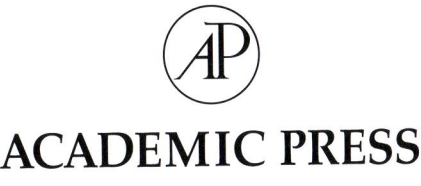

ACADEMIC PRESS
A Harcourt Science and Technology Company

San Diego · San Francisco · New York · Boston · London · Sydney · Tokyo

This book is printed on acid-free paper.

Copyright © 2000 by ACADEMIC PRESS

All Rights Reserved.
No part of this publication may be reproduced or transmitted in any form or by any means, electronic or mechanical, including photocopying, recording, or any information storage and retrieval system, without permission in writing from the publisher.

Academic Press
A Harcourt Science and Technology Company
Harcourt Place, 32 Jamestown Road, London NW1 7BY, UK
http://www.academicpress.com

Academic Press
A Harcourt Science and Technology Company
525 B Street, Suite 1900, San Diego, California 92101-4495, USA
http://www.academicpress.com

ISBN 0-12-137306-1

A catalogue record for this book is available from the British Library

Editorial and production services by Gray Publishing, Tunbridge Wells, Kent, UK
Typeset by Gray Publishing, Tunbridge Wells, Kent, UK
Colour plates printed by Bright Arts, Hong Kong
Printed and bound in Singapore by Imago

00 01 02 03 04 05 IM 9 8 7 6 5 4 3 2 1

CONTENTS

PLATES	vi
AUTHORSHIP	vii
ACKNOWLEDGEMENTS	viii
INTRODUCTION	xi

ORDER PASSERIFORMES (continued from Volume V)

Picathartidae, picathartes	1
Timaliidae, babblers	7
Aegithalidae, long-tailed tits	76
Paridae, tits	77
Remizidae, penduline tits	106
Sittidae, nuthatches and Wallcreeper	123
Salpornithidae, Spotted Creeper	128
Certhiidae, tree-creepers	132
Nectariniidae, sunbirds	135
Zosteropidae, white-eyes	305
Promeropidae, sugarbirds	326
Laniidae, true shrikes	339
Malaconotidae, bush-shrikes	383
Pycnonotidae, bulbuls	476
Prionopidae, helmet-shrikes	486
Oriolidae, orioles	502
Dicruridae, drongos	521
Corvidae, crows	531
Sturnidae, starlings	572
Buphagidae, oxpeckers	662

BIBLIOGRAPHY

General and Regional References	669
Family References	679
Acoustic References	707
INDEX	711

PLATES

Plate		Facing Page
1	Picathartes, babblers, rockjumpers	14
2	Babblers (*Illadopsis*)	15
3	Babblers (*Turdoides*)	30
4	Babblers (*Turdoides*)	31
5	Tits	78
6	Tits, penduline tits	79
7	Tits, nuthatches and others	94
8	Sunbirds (*Anthreptes*)	95
9	Sunbirds (*Anabathmis, Hedydipna*)	142
10	Sunbirds (*Cyanomitra*)	143
11	Sunbirds (*Chalcomitra*)	158
12	Sunbirds (*Nectarinia*)	159
13	Double-collared sunbirds	206
14	Sunbirds (*Cinnyris*)	207
15	Sunbirds (*Cinnyris*)	222
16	Sunbirds (*Cinnyris*)	223
17	White-eyes	270
18	Shrikes, fiscals (*Lanius*)	271
19	Shrikes (*Lanius*)	286
20	Shrikes, sugarbirds	287
21	Bush-Shrikes (*Malaconotus*)	334
22	Bush-Shrikes (*Malaconotus*)	335
23	Bush-Shrikes (*Telophorus*), nicators	350
24	Tchagras	351
25	Puffbacks	398
26	Boubous, gonoleks (*Laniarius*)	399
27	Bush-Shrikes, boubous (*Laniarius*)	414
28	Helmet-Shrikes	415
29	Orioles	462
30	Drongos, choughs and others	463
31	Piapiac, crows, ravens	478
32	Starlings (*Poeoptera, Onychognathus*)	479
33	Starlings (*Cinnyricinclus, Onychognathus*)	526
34	Starlings (*Lamprotornis*)	527
35	Starlings (*Lamprotornis, Spreo*)	542
36	Starlings (*Sturnus* and others), oxpeckers	543

AUTHORSHIP

C. H. Fry: Picathartidae, *Chaetops frenatus, Ptyrticus, Turdoides, Phyllanthus, Lioptilus, Kupeornis, Horizorhinus, Panurus,* Aegithalidae, *Parus fasciiventer, P. major, P. fringillinus, P. funereus, P. albiventris, P. leuconotus, P. leucomelas, P. cristatus, P. ater, P. caeruleus,* Remizidae, Sittidae, Salpornithidae, Certhiidae, *Anabathmis, Dreptes, Anthobaphes, Cyanomitra, Chalcomitra, Nectarinia, Hedydipna collaris, H. platura, H. metallica, Cinnyris,* Zosteropidae, Promeropidae, *Lanius collaris, L. newtoni, L. dorsalis, L. somalicus, L. cabanisi, L. mackinnoni, L. excubitoroides, L. gubernator, L. souzae, Corvina, Urolestes, Eurocephalus, Malaconotus, Telophorus zeylonus, T. cruentus, Dryoscopus, Laniarius, Nilaus,* Oriolidae (except *O. oriolus*), *Ptilostomus, Pica, Garrulus, Nucifraga, Pyrrhocorax, Poeoptera, Onychognathus* (except *O. morio, O. neumanni* and *O. nabouroup*), *Lamprotornis cupreocauda, L. purpureiceps, L. purpureus, L. chalcurus, L. chalybaeus, L. chloropterus, L. acuticaudus, L. ornatus, L. splendidus, L. caudatus, L. purpuropterus, L. mevesii, Speculipastor, Grafisia, Neocichla, Creatophora, Zavattariornis, Sturnus, Acridotheres,* Buphagidae.

S. Keith: **Field Characters** and **Voice Sections**; and *Illadopsis, Kakamega, Pseudalcippe, Telophorus viridis, T. dohertyi, Neolestes, Nicator, Corvus frugilegus, C. monedula.*

D. Pearson: most **Description Sections**; and *Lanius nubicus, L. minor, L. meridionalis, L. isabellinus, L. collurio, L. senator, Antichromus, Tchagra, Oriolus oriolus, Dicrurus, Pholia, Cinnyricinclus.*

A. J. F. Craig: *Onychognathus morio, O. neumanni, O. nabouroup, Lamprotornis nitens, L. corruscus, L. australis, Spreo bicolor.*

L. Grimes: *Chaetops pycnopygius.*

M. P. S. Irwin: *Anthreptes, Hedydipna pallidigastra.*

D. Wiggins: *Parus griseiventris, P. cinerascens, P. afer, P. thruppi, P. rufiventris, P. carpi, P. niger.*

R. Wilkinson: *Lamprotornis iris, L. unicolor, L. regius, L. shelleyi, L. hildebrandti, L. superbus, L. fischeri, Spreo albicapillus.*

E. K. Urban: *Prionops, Corvus* (remaining species).

ACKNOWLEDGEMENTS

A work as large and complex as even a single one of this series of volumes could not be contemplated, and certainly not brought to a timely conclusion, without its many contributors drawing upon the help and expertise of a small army of specialists. The editorial process is essentially one of constructive criticism and adaptation. It says a great deal about those specialists, colleagues, biologists, birders, authors, artists, sound-recordists and other colleagues and friends, that they have responded with enthusiasm and goodwill to the editors' demands on their patience in researching a vast literature and writing multiple drafts. For the skills, co-operation, tolerance and kindliness of all of the other members of the team, the editors and the Publisher are sincerely indebted.

David Pearson has become 'our man at the museum' and has taken descriptions and measurements of most species and subspecies in this volume, working on the collections of the Natural History Museum (formerly the British Museum of Natural History), Department of Ornithology, in Tring, UK. For access to specimen and library holdings and other facilities, and for the ready help and guidance of administrators, curators and librarians of this and other museums and libraries, we are most grateful to: the Trustees of the American Museum of Natural History (Department of Ornithology), Durban Natural History Museum, Field Museum of Natural History, Josselyn Van Tyne/Wilson Ornithological Society Library, Kendal Natural History Museum, Liverpool Museum of Natural History, Los Angeles County Museum (Bird Division), Maidstone Museum, Musée National d'Histoire Naturelle (Paris), Musée Royale de l'Afrique Centrale (Tervuren), Namibian Scientific Society, National Museums of Kenya (Nairobi), National Museum of Zimbabwe (Bulawayo), National Museums of Scotland (Royal Museum of Scotland, Edinburgh), North of England Zoological Society, Transvaal Museum (Bird Division, Pretoria), US National Museum of Natural History, Western Foundation of Vertebrate Zoology, and Zoological Society of London. The editors are further indebted for facilities at, and the support of, their respective institutions: Aberdeen University (Department of Zoology), American Museum of Natural History (Department of Ornithology) and Augusta State University (Department of Biology). Authors David Wiggins and Roger Wilkinson are similarly indebted respectively to Uppsala University (Department of Ecology) and North of England Zoological Society (Chester).

Martin Woodcock, almost as much at the museum as in his studio, whose commitment to the project has done so much to enrich it, is most grateful to Robert Prŷs-Jones, Mark Adams and Frank Steinheimer at the Natural History Museum and to Ed Jarzembowski at Maidstone Museum.

Linda Birch and Mike Wilson of the Alexander Library of the Edward Grey Institute of Field Ornithology, Oxford University, have become indispensable to the project in supplying literature not otherwise readily available to the executive editor.

With much generosity Neil Baker provided painstakingly detailed, point-plotted maps and distributional notes on nearly all Tanzanian species in this volume. Robert Dowsett and Françoise Dowsett-Lemaire have taken a particular interest in the entire project for almost as long as have the editors, and during the preparation of Volume VI have helped in numerous ways, from tape-recordings to hospitality and from refereeing to biology. At our behest Michael Irwin made a comprehensive review of the taxonomy of sunbirds (Nectariniidae) and we have adopted his arrangement. As work on sunbirds progressed, many draft accounts were exchanged with Robert Cheke, who is monographing the family. Robert Dawson contributed new observations on the biology of *Tchagra senegala*. Tony Harris was to have written the tchagra shrikes but circumstances prevented him from doing so; instead, he kindly provided us with copious notes and abundant data about them. Louis A. Hansen provided voluminous moult, mensural and weight data, from Tanzania, and so did Michael King of the Gambia Ringing Project. Tom Huels gave his unpublished research data on *Lamprotornis superbus*, and David Bygott and Geoffrey Field did likewise with *L. unicolor* and *Cinnyricinclus leucogaster*.

For refereeing species accounts we thank Chris Bowden (picathartes, several Cameroon bush-shrikes and sunbirds), Peter Davidson (Socotra birds), Richard Dean (some babblers), Françoise Dowsett-Lemaire (several babblers, *Nicator*, *Neolestes*, many bush-shrikes and *Onychognathus* starlings and numerous sunbirds), Colin Jackson (Tacazze Sunbird), Morné du Plessis (some babblers), David Pearson (all babblers and most sunbirds) and Roger Wilkinson (long-tailed starlings). For critiques of the colour illustrations we and Martin Woodcock are most grateful to John Ash, Alec Forbes-Watson and John Miskell.

Besides the above-named people who have particular areas of expertise, there are many who gave their time and invaluable help in answering a host of questions, or who willingly provided taxonomic and other advice, tape-recordings, ringing and nest-record data, access to museum material, information on particular species or regions, museum specimen data printouts, weights, longevities, reprints, translations, contacts, hospitality and field assistance. Editors, authors and artists alike take pleasure in recording their heartfelt thanks to: Gary Allport, Phil Angle, John Ash, John Atkins, Graeme Backhurst, G. Balança, M.-N. Balança, Neil Baker, Richard Banks, Clive Barlow, Phoebe Barnard, George Barrowclough, Kimbo Beakbane, Bruce M. Beehler, Leon Bennun, Aldo Berruti, Denise Brinkrow, Peter Britton, the late Richard Brooke, T. Brooks, Chris Brown, Tom M. Butynski, Tamar Cassidy, J. de Castro, M. de Castro, F. Cezilly, Robin Chancellor, Claude Chappuis, Robert Cheke, Patrice Christy, Philip Clancey, William Clarke, Peter Colston, Adrian Craig, Peter Davidson, Richard Dean, Bob Dickerman, Robert J. Dowsett, Françoise Dowsett-Lemaire, Jens Eriksen, Hanne Eriksen, J. H. Fanshawe, R. Farmer, Francine Faucher, Kate Ferguson, Geoffrey Field, Clem Fisher, Lincoln D. C. Fishpool, Jon Fjeldså, Roger Fotso, K. Garrett, Llewellyn Grimes, Mrs Dale B. Hanmer, J. Heigham, I. Henrichsen, C. F. C. Hesse, Chris Hines, Janet Hinshaw, Philip Hockey, Kit Hustler, Michael Irwin, Colin Jackson, Des Jackson, R. W. James, Michael C. Jennings, Peter Jones, Joël Joubert, Martine Joubert, Jan Kalina, Alan Kemp, Lloyd F. Kiff, Karen King-Sharp, A. Kitchener, Joris Komen, Peter Lack, Wesley Lanyon, Bruno Lamarche, Mary LeCroy, Peter Leonard, Manny Levine, Jeremy Lindsell, Lynda Lotz, Michel Louette, Linda R. Macaulay, Charles MacInnes, Yves-Michel de Martin de Viviès, Robert McGowan, John Mendelsohn, Gérard Morel, David Moyer, Eli Mulungu, René de Naurois, Gerhardt Nikolaus, Rick Nuttall, Chris G. Petrow, Storrs L. Olson, Joseph Oyugi, Robert B. Payne, Chris M. Perrins, Richard Porter, Robert Prŷs-Jones, S. P. Rodwell, Stephen J. R. Rumsey, C. Ryall, Antoni Sala, V. Salewski, Herbert Schifter, Jack Schmidt, Mary Lou Schmidt, Karl Schuchmann, Phil Shaw, Rob Simmons, Neville Skinner, Peter Slater, Richard Sloss, E. F. G. Smith, Walter Sontag, Jack Spearpoint, L. Svensson, Paul Sweet, Ann Sylph, Warwick Tarboton, Paul Taylor, M. Thévenot, Y. Thonnérieux, S. Tonge, Don Turner, Alan Tye, Stephanie Tyler, Staffan Ulfstrand, Anne Vale, Dirk Vanackere, J. P. Vande weghe, Carl J. Vernon, J. D. R. Vernon, T. Wacher, Stephanie Wehnelt, Hilary Welch, G. R. Welch, Michael Walters, L. M. Wilkinson, David Willard, T. Williams, Mary Wilson, Trevor Wilson, M. C. Wimer, the late J. M. Winterbottom, Malcolm Wright and Richard Zusi.

In preparing Voice accounts Stuart Keith listened to as many tapes as possible. Beyond the published tapes and discs, there is much unpublished material in institutional and private collections. Institutions generously provided tapes without charge, and many people took the time and trouble to make copies of their recordings; all of them should know that they made a significant contribution to the text of this volume. We wish to express our gratitude and appreciation to: Richard Ranft and the National Sound Archive, London (formerly British Library of Wildlife Sounds); Greg Budney and the Library of Natural Sounds, Cornell University; Arnoud van den Berg, Hazel Britton, Tom Butynski, Louis Hansen, Peter Jones, Linda Macaulay, David Moyer, Jens Otto Svendsen, Tony Walker, Roger Wilkinson, Martin Woodcock, and most especially Françoise Dowsett-Lemaire and Claude Chappuis.

As in earlier volumes, Martin Woodcock painted the colour plates. We are particularly grateful both to him and to Ian Willis, who drew the great majority of figures in the text, for their skill and good-natured collaboration. Line drawings in the text, mainly by I. Willis, also by C. H. Fry and R. de Naurois, are from material in the Royal Museum of Scotland (Edinburgh) and Natural History Museum (Tring) or are taken from or based upon published or unpublished source material as follows: PICATHARTIDAE – plate in P. R. Lowe, *Ibis* **1938**; photo by BirdLife International in *Bull. Afr. Bird Club* **1**, 1994, 17; photo by D. Halleux in *Malimbus* **16**, 1994, 21; photo in P. R. Colston and K. Curry-Lindahl, *The Birds of Mount Nimba, Liberia*, 1986, 78; drawing by M. O'Brien in *Bull. Br. Orn. Club* **116**, 1996, 16; photos by A. Devez in A. Brosset and C. Erard, *Les Oiseaux des Régions Forestières du Nord-est du Gabon*, 1986, 205; photos in L. G. Grimes and N. Gardiner, *Nigerian Field* **28**, 1963; TIMALIIDAE – photos by A. Devez in *Alauda* **42**, 1974, 385–396; photo by P. Blasdale in *Malimbus* **10**, 1988, 168; drawings by Caroline Ash in *Bull. Br. Orn. Club* **101**, 1981, 400; photos by P. Shaw (*T. gymnogenys*); photo by G. Selfe in *Afr. Birds and Birding* **4** (4), 1999, 29; field drawing by R. Green (*Panurus biarmicus*); PARIDAE – drawings by D. Nurney in S. Cramp and C. M. Perrins (Eds), *The Birds of the Western Palearctic*, VII, 1993, 267–268; REMIZIDAE – photos by W. Massyn in *Bokmakierie* **27**, 1975, 8–9; photo by C. J. Skead, *Ostrich* Suppl. **3**, 1959, 279; photo by C. J. Uys in *Bokmakierie* **18**, 1966, 80–82; photos in P. Steyn, *Nesting Birds*, 1996, 154; photos by D. and J. Bartlett in *National Geographic* **163**, 3, 1983, 344–385; drawing by A. Seidel after F. C. Holman in J. P. Chapin, *The Birds of the Belgian Congo*, part 4, 1954, 460; SALPORNITHIDAE – photos in P. Steyn, *op. cit.*, 1996, 155; CERTHIIDAE – photo by H. Schouten in *Brit. Birds* **86**, 1993, 370; NECTARINIIDAE – photos and drawings in C. J. Skead, *Sunbirds of Southern Africa, also the Sugarbirds, the White-eyes and the Spotted Creeper*, 1967, 111, 116, 118, 160, opp. 161, 172, 212, 219; photos by P. J. Ginn, C. Laubscher and N. Myburgh in P. J. Ginn, W. G. McIlleron and P. le S. Milstein *The Complete Book of Southern African Birds* 1989, 630, 632, 634, 644; drawing in B. Quantrill and R. Quantrill, *Malimbus* **20**, 1998, 4; drawings by R. de

Naurois (*Anabathmis newtonii*, *Dreptes thomensis*); photos in P. Steyn, *op. cit.*, 1996, 194, 196, 199; photo by G. J. Broekhuysen in *Ostrich* **34**, 1963, 226; photos by A. Devez in *Terre et Vie* **28**, 1974, 585 and in A. Brosset and C. Erard, *op. cit.*, 1986, 254; drawing in W. W. Howells, *Ostrich* **42**, 1971, 104; ZOSTEROPIDAE – drawings in P. Kunkel, *Zool. Afr.* **10**, 1975; field drawing by C. H. Fry (*Zosterops griseovirescens*); photo in M. J. S. Harrison, *Malimbus* **11**, 1990, 140; photos and drawings in C. J. Skead, *op. cit.*, 1967, 302, 305; PROMEROPIDAE – photos and drawings in C. J. Skead, *op. cit.*, 1967, 242, 247, 259, 265, opp. 272, 273; LANIIDAE – photo by J. and H. Eriksen in C. H. Fry, *Oman Birds in Focus* 1990, 91; photo by P. Steyn, cover of *Promerops* 210, 1993; photo by W. Tarboton in *Afr. Birds and Birding* **3** (5), 1998, 56; photos by M. Wright (*Lanius cabanisi*); photos in P. Steyn, *Nesting Birds*, 1996, 185, 187; drawing by G. Arnott in A. Harris and G. Arnott, *Shrikes of Southern Africa*, 1988, 34; photo by C. Haagner in P. A. R. Hockey, *Birds of Southern Africa*, 1991, 75; photo by P. J. Ginn in P. J. Ginn, W. G. McIlleron and P. le S. Milstein *op. cit.*, 1989, 610; MALACONOTIDAE – photo on plate 1 in J. P. Chapin, *op. cit.*, 1954; field drawing by M. Willis (*Telophorus cruentus*); drawings by G. Arnott in A. Harris and G. Arnott *op. cit.*, 1988, 127, 136; photos in P. Steyn *op. cit.*, 1996, 186, 187; photo by C. H. Fry of mounted specimen in Percy FitzPatrick Institute of Southern African Ornithology (*Dryoscopus cubla*); photo by A. Devez in A. Brosset and C. Erard *op. cit.*, 1986, 152; photos by E. F. G. Smith (*Laniarius liberatus*); photos by W. Tarboton in *Ostrich* **42**, 1971, 279 and *Afr. Birds and Birding* **3** (2), 1998, 30, 32; photos by R. M. Bloomfield and J. L. Viljoen in P. J. Ginn, W. G. McIlleron and P. le S. Milstein *op. cit.*, 1989, 599; PYCNONOTIDAE – drawing by A. Seidel and J. P. Chapin, *The Birds of the Belgian Congo*, part 3, 1953, 184; PRIONOPIDAE – photos by J. Arnett in W. Tarboton *Bokmakierie* **15**, 1963, 2; photo by C. Laubscher in P. J. Ginn, W. G. McIlleron and P. le S. Milstein *op. cit.*, 1989, 608; ORIOLIDAE – photo in V. G. L. van Someren, *Days with Birds*, 1956, 415; photo in P. Steyn *op. cit.*, 1996, 150; photos by P. J. Ginn in P. J. Ginn, W. G. McIlleron and P. le S. Milstein *op. cit.*, 1989, 452, 453; CORVIDAE – drawings by D. Nurney and D. Quinn in S. Cramp and C. M. Perrins (Eds), *The Birds of the Western Palearctic*, VIII, 1994, pp 18, 19, 201, 214, 215; field drawings of *Pica pica mauritanica* by I. Willis; drawings by C. J. F. Coombs in C. J. F. Coombs *The Crows*, 1978, pp 27, 28, 34; photos by D. and J. Bartlett in *National Geographic* **163**, 1983, 365, 366; photos by D. Waters in C. H. Fry, *Oman Birds in Focus*, 1990, 93, 94; STURNIDAE – photo in P. Steyn *op. cit.*, 1996, 191; drawings by A. Seidel in J. P. Chapin, *op. cit*, 1954, 150; photos by D. Bygott (*Lamprotornis unicolor*); photos by R. Wilkinson (*L. regius*, *L. pulcher*); drawing by R. Restall in *Avicult. Mag.* **74**, 1986; drawing in P. A. Clancey, *The Birds of Natal and Zululand*, 1964, 419; photos by C. J. Uys in *Bokmakierie* **29**, 1977, 87 and in P. J. Ginn, W. G. McIlleron and P. le S. Milstein *op. cit.*, 1989, 614; photos by W. A. Sontag in *Malimbus* **7**, 1985, 132, 133; field drawing by C.W. Benson (*Zavattariornis stresemanni* nest); photos by H. Shirihai in J. Francis and H. Shirihai, *Ethiopia in Search of Endemic Birds*, 1999; drawings by G. Wilson in C. Feare, *The Starling*, 1984, 6; photo by M. Danegger in *Birds*, **16**, 1998, 84; BUPHAGIDAE – photo by W. Paton/NHPA in *BBC Wildlife*, **15** (4), 1977, 21; photos in P. Johnson, *As Free as a Bird*, 1976, 161. In Fig. 2 in the Introduction, map **D** is redrawn from R. A. Cheke and J. F. Walsh (1996) *The Birds of Togo* with permission of the British Ornithologists' Union and map **C** from W. Gatter (1997) *Birds of Liberia* with permission of Pica Press.

Our wives Kathie Fry, Sallyann Keith and Lois Urban, and doubtless Nancy Irwin, Maggie Pearson, Lynn Wilkinson, Diane Willis, Barbara Woodcock and the wives and families of our other main contributors, have had their lives variously affected by the demands of this project, and we thank them for their forbearance. Sallyann shouldered a particular burden, standing by whilst her husband underwent dangerous surgery.

Twelve years have passed since it fell to C. H. Fry to act as *primus inter pares* or principal co-ordinator of one of the volumes, III, in this work. In the present volume he has also taken on the roles of main author and executive editor, cartographer and bibliographer, and he wishes to record here his sincere thanks to his patient wife Kathie for her continual support, understanding and encouragement in the first years of what might otherwise have been a relaxed retirement. Amongst ourselves we have long called *The Birds of Africa* simply *BoA*. We will leave to the final volume any reference to *BoA constrictor*! Last, we give special thanks to Andrew Richford, Commissioning Editor at Academic Press, for continuing to give direction to the project and every encouragement to its team, as well as for contributing to this volume his particular knowledge of *Corvus monedula*.

January 2000

C. Hilary Fry
Stuart Keith
Emil K. Urban

INTRODUCTION

This is the last-but-one volume in a series the planning of which began a quarter-century ago. It was inspired by Leslie Brown, who died whilst the first volume was in preparation; a history of the project appeared in *The Birdwatcher's Yearbook and Diary 1998* (Fry 1997) and will be updated in the seventh and final volume.

Our objectives are as before: to gather and integrate material from diverse sources and to produce a definitive handbook of the avifauna of Africa and its islands. To quote from a recent review of Vol. V: 'More and more people are watching birds in Africa, but the majority are doing surveys or making lists. Few are taking the places of the great African naturalists that fill the references, van Someren, Chapin, Brosset, Erard, Lynes, Moreau and Vincent. Hopefully this volume and the series as a whole will encourage a band of dedicated indigenous replacements' (Kemp 1997). That succinctly expresses our wishes too.

For various reasons, intervals between the appearance of successive volumes have become more and more protracted, and there were no less than nine years between Vols III (1988) and V (1997). Not surprisingly, the Publishers have put their corporate foot down and insisted that operations be accelerated. This has necessitated a radical review of procedures, with a change in the apportionment of authorship and of editorial responsibilities. We have employed sectional or 'horizontal' as well as systematic or 'vertical' research and authorship – Stuart Keith has written nearly all of the VOICE and FIELD CHARACTERS sections and David Pearson most of the DESCRIPTION sections. Hilary Fry has written the remaining sections (RANGE AND STATUS, GENERAL HABITS, FOOD and BREEDING HABITS) of nearly three quarters of the 324 species. RANGE AND STATUS sections and mapping have improved in recent volumes and in the present one VOICE sections have also been expanded, with numerous transcriptions attempting to convey the sounds of all major vocalizations.

Literature

Whilst we have endeavoured to keep abreast of contemporary field studies, most information in this volume is culled from journals and recent books. Knowledge about the biology, distribution and identification of African birds is increasing exponentially. We have set down everything judged to be of salient biological interest, and for many well-known species we are now having to *summarize* the available information – especially observations that are readily quantifiable. Many other species remain so poorly known that their accounts amount to only a few hundred words, whilst the brevity of some sections in lengthy accounts indicates areas where knowledge about otherwise well-known species remains superficial.

To a greater extent than ever before, we have benefited from a wealth of information in recently published books. We would like to mention in particular the southern African atlas (Harrison *et al.* 1997), books on the birds of The Gambia (Barlow *et al.* 1997), Liberia (Gatter 1997), Togo (Cheke and Walsh 1996), Gabon (Christy and Clarke 1994), São Tomé and Príncipe islands (Christy and Clarke 1998), Somalia (Ash and Miskell 1998) and Kenya and northern Tanzania (Zimmerman *et al.* 1996), and treatises on tits and their relatives (Harrap and Quinn 1996), laniid shrikes (Lefranc and Worfolk 1997) and starlings (Feare and Craig 1998). Whilst only marginally African, Shirihai's birds of Israel (1996) contains much of relevance. We are aware of comprehensive works in progress on the birds of Morocco, the Gulf of Guinea and Angola, the last close to publication as we write late in 1999, and of avian atlases in preparation for Ethiopia, Uganda, Kenya, Zambia and Malaŵi. We received the atlas of Sul do Save, Mozambique (Parker 1999) too late for inclusion of most of its wealth of new information, but have made an exception in the case of the special Neergaard's Sunbird *Cinnyris neergaardi*, a near-endemic to Sul do Save, and have added population density values for several species. Of older works, Skead's monograph on sunbirds, sugarbirds and white-eyes (1967) and Harris and Arnott's on mainly malaconotid shrikes (1988) have proved invaluable. Other sources have already been identified in our ACKNOWLEDGEMENTS; an outstanding one that well merits repetition is Neil Baker's Tanzanian atlas.

In related fields, we take common and scientific names of mammals from Kingdon (1997) and plant names usually from Mabberley (1993); these are standard works embracing the whole continent.

The literature has been researched comprehensively up to the time of manuscript delivery near the beginning of 1999 and updated to about November 1999. In a few cases an important new distributional record in the text is not matched by an addition to the species map, since maps were finalized earlier than the texts.

As in earlier volumes, a short list of principal references is given at the end of most species accounts. They have all been used in compiling accounts, although they are not necessarily cited therein. All references cited in the main text, together with some that may not have been cited, are given in the complementary GENERAL AND REGIONAL REFERENCES, SYSTEMATIC REFERENCES and ACOUSTIC REFERENCES following the text.

Place Names and some Terms

350 names of national parks, faunal reserves, forests and highlands are listed, with regional maps, in Vol. V pp xiv–

xix. For place names, we generally follow *The Times Atlas of the World* (1990) but depart from it in several cases where prevalence of current local usage seems so to demand. The Democratic Republic of Congo has officially changed its name several times and for brevity, as well as to distinguish it from Congo (Brazzaville), we continue to call it Zaïre. Its south-eastern province Shaba reverts to its old name Katanga. We use both Free State and Orange Free State, Natal and KwaZulu/Natal, Impenetrable and Bwindi Forest, Mbini and Equatorial Guinea, and Murchison and Kabalega Falls; other inconsistencies occur but we have tried to minimize them. According to *The Times Atlas* (1990), Gebel Elba lies astride the Political Boundary between Egypt and Sudan but is well south of the Administrative Boundary; in past volumes it has been attributed to both countries. Further remarks about name changes in Africa have been made in the Introductions of earlier volumes.

As in previous volumes, linear *measurements* are in millimetres unless otherwise specified and *weights* are in grams. Under *Laying Dates* A, B, C, D and E refer to East African climatic zones, shown on p. xvii; months not in parentheses are those in which eggs have been found, and they are followed parenthetically by times of other breeding indications.

Having replaced 'Downy Young' with NESTLING three volumes ago we retain the latter as a subheading in Description sections, although what we actually mean is hatchling or pre-feathered nestling. In its first plumage of true feathers the bird is a *juvenile*, and in subsequent plumages it is *immature* for months or years until moulting into *adult* plumage at sexual maturity. A bird moulting from immature to adult plumages is sometimes called a *subadult*.

Age classes and terms for the various feather generations have been defined by Jenni and Winkler (1994) and others. Nomenclature is complicated by different species having different moult regimes and plumage sequences; by complete, partial and interrupted moults; and by some birds breeding in immature plumage. Our descriptions of age classes are often greatly hampered by the poor state of knowledge about immature moult schedules and plumage sequences in numerous species of African birds. We retain IMMATURE as a descriptive heading, and distinguish juveniles from immatures where we can. In some cases where IMMATURE and NESTLING stages have never been described we simply omit the headings.

For plumage and bare parts terminology the reader is referred to *Topography of African Passerine Birds* in Vol. IV, pp xiv–xv. In describing feathers we use *vane* and *web* interchangeably. By *flight feathers* we mean *primaries and secondaries* (some authorities include tail feathers too), which we often refer to also as *remiges* (sing. remex), in distinction from tail feathers or *rectrices* (sing. rectrix).

A *sedentary* species, population or pair of birds is one that remains in the same locality for years on end, individuals roaming no more than a few kilometres at the most. A species or population is *resident* when it occurs year-round in the same area and breeds there, even though the same individuals may not be in the same area all year round. A species resident throughout most of its range is a *partial migrant* if it occurs regularly towards the edges of its range only at given seasons. If that part of its range where it is resident is small in comparison with adjacent parts where it appears only seasonally, the species can be described as an *intra-African migrant* rather than a partial migrant; a species that breeds in one part of Africa and 'winters' in quite a different one is unquestionably an intra-African migrant. Species that breed in Europe or Palearctic Asia and winter in Africa are *Palearctic migrants*, whether or not they also breed in Palearctic North Africa.

Mapping

With so much recent emphasis upon national and regional atlassing, we are able to map most species continent-wide in much greater detail than has ever previously been possible.

Agricultural and industrial development and burgeoning human populations are affecting much of Africa. Woodland, wetland and other habitats have shrunk and been degraded, diminishing birds' ranges and putting some montane forest endemics at risk of extinction. Conversely, many farmland and garden species have expanded their ranges. We generally remark such range shifts in the RANGE AND STATUS sections, but have made little attempt to map them; rather, maps purport to show contemporary ranges. However, ranges remain sketchy in some parts where war, famine and civil unrest have precluded field studies, and for such areas our maps are often based on little more than the point-plotted distributions of Hall and Moreau (1970).

Figure 1 shows the colour and shading conventions used in species maps. X's indicate extralimital records. In practice it is often impossible to know exactly where *range*

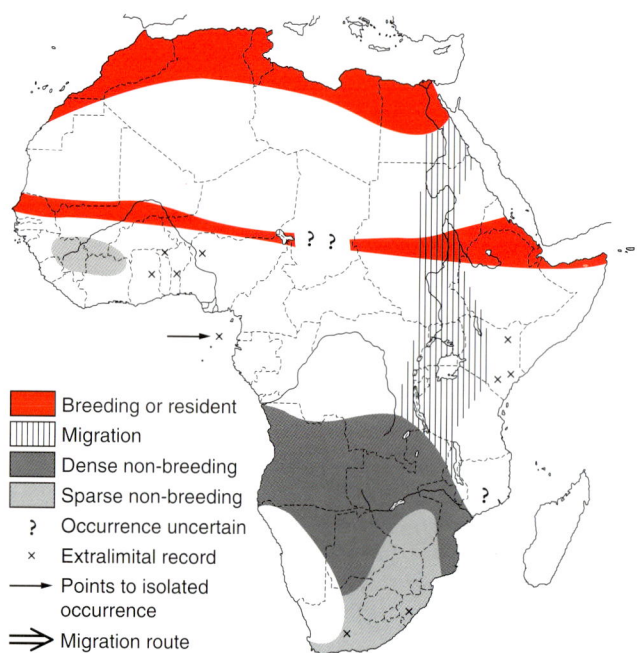

Fig. 1. Shading, symbols and arrows used on maps.

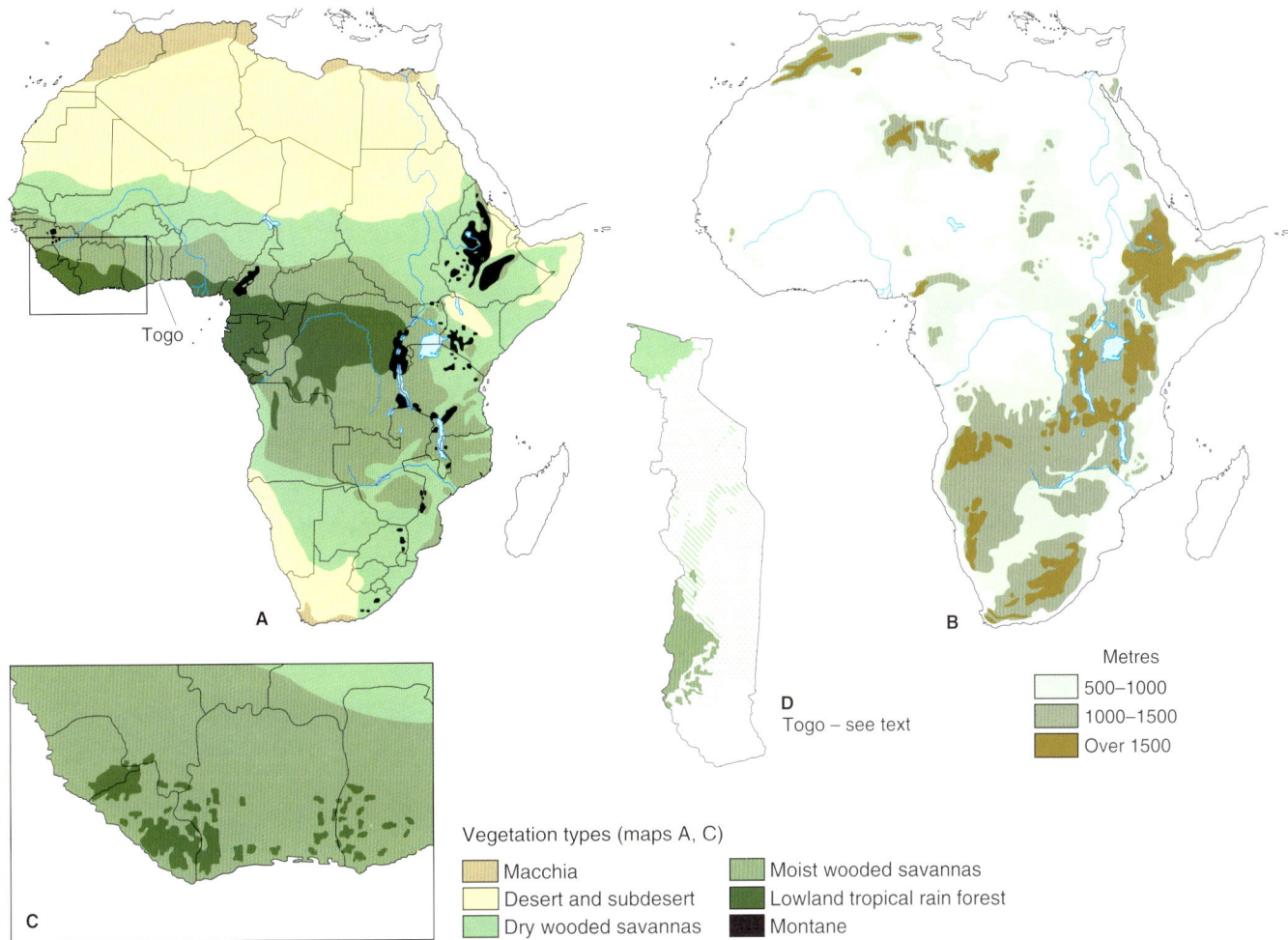

Fig. 2. **A**: Notional boundaries of major African vegetation zones. **B**: Topographic features. **C**: Enlargement of West Africa showing recent fragmentation of rain forest (adapted from Gatter 1997). **D**: Enlargement of Togo (adapted from Cheke and Walsh 1996).

limits should be drawn, and further research may show that some X's should have been shown as red dots, indicating a small resident population.

The political map of Africa is shown on p. xvii.

Vegetation

Africa's vegetation was discussed in Vol. I, pp 4–11 (Brown *et al.* 1982), with accompanying essays placing it in the context of the geological past, climate, topography, and bird habitats. Map **A** in Figure 2 here reproduces in colour the map in Vol. I, after Keay (1959); the principal lowland vegetation types are, from high to equatorial latitudes, *macchia*, *desert* and *subdesert*, *dry wooded savannas*, *moist wooded savannas*, and *tropical rain forest*. The subdesert southern fringe of the Sahara is termed the *Sahel*, with *sahelian* savannas; in West and N-central Africa, dry wooded savannas comprise the *soudanian* zone and, further south, the *northern guinean* zone, and moist savanna woodlands comprise *southern guinean* and *derived savanna* zones, the latter bordering coastal rain forest. We use these terms commonly. The map is the same size as species-distribution maps and superspecies maps throughout this book so that, by comparison, the degree to which numerous species have ranges determined by vegetation can readily be appreciated. Map **B** in Figure 2 shows the principal montane regions and rivers. Montane forest and other mountain vegetation types above the treeline provide wildlife habitats as distinctive as the main lowland types; also important are aquatic habitats – lakesides, rivers, marshes, estuaries, mangroves – and man-made habitats like towns and gardens. At this scale each vegetation zone appears to be a vast, continuous swath. That is not actually the case, however, as enlargement map **D** shows for one small country, Togo; satellite imagery shows Togo to have 15 eco-floristic zones, described in Cheke and Walsh (1996). Here we simplify them to: Dry wooded savannas, Moist wooded savannas, Upland wooded savannas (hatched), Degraded woodlands (dense stipple) and Coastal derived savannas (sparse stipple).

Vegetation boundaries in map **A** are now more theoretical than real. Map **C** shows the extent of closed-canopy rain forest in West Africa today, and even that may be optimistic (Gatter 1997).

Evolution and Biogeography

Figure 3 illustrates principles of evolution and biogeography, using the genus *Telophorus* as an example. These are semi-terrestrial bush-shrikes forming a parapatric series, from Red Sea coasts to Angola and South Africa, as in the map. The geographically central one, Doherty's Bush-Shrike *T. dohertyi* (which has a rare yellow-throated form), is montane, and separates differing populations of the lowland Gorgeous Bush-Shrike *T. viridis* to the west, east and south. The two lowland populations have often been treated as separate species, *T. viridis* (1) and *T. quadricolor* (2, 3, 4). Being closely related and apparently of immediate common descent, *T. dohertyi* and *T. viridis*

Fig. 3. Range and differentiation of *Telophorus* bush-shrikes: see the text. All subspecies are mapped but not all are illustrated. **1–4 Gorgeous Bush-Shrike *T. viridis*,** 1 *T. v. viridis*, 2 *T. v. nigricauda*, 3 *T. v. quartus*, 4 *T. v. quadricolor*; **5 Doherty's Bush-Shrike *T. dohertyi*** *T. dohertyi* (5r the normal red phase, 5y the rare yellow phase); **6–9 Bokmakierie *T. zeylonus*,** 6 *T. z. phanus*, 7 *T. z. thermophilus*, 8 *T. z. zeylonus*, 9 *T. z. restrictus*; **10–13 Rosy-patched Shrike *T. cruentus*,** 10 *T. c. kordofanicus*, 11 *T. c. cruentus*, 12 *T. c. hilgerti*, 13 *T. c. cathemagmenus*. The sexes of *T. dohertyi* and of *T. zeylonus* are alike. Regions of evident overlap are shown purple.

compose a superspecies (of three species, should '*T. quadricolor*' be regarded as a single one). The distinctive Bokmakierie *T. zeylonus* inhabits dry bushlands in the south; and the even more distinct Rosy-patched Shrike *T. cruentus* occupies semi-arid bushlands to the north. The two are similarly large, with white in the tail.

T. cruentus is the only bird in this complex that has no green or yellow in its plumage; it also has striking sexual dichromatism, and the male *cathmagmenus* possesses the female character of a black gorget. Loss of yellow-green colour is commonplace in sub-desert birds; all the same, the colour differences of *T. cruentus* have obscured what are now thought to be its true affinities. Hitherto, *T. cruentus* has generally been placed in *Tchagra* or in its own genus, *Rhodophoneus* (although relationship to the geographically distant *Telophorus zeylonus* has been hinted at: see *Telophorus* generic diagnosis, p. 408).

Some general principles are (i) Ecological and morphological differentiation tends to proceed outwards from the ancestral range, as populations slowly adapt to new or slightly different habitats; (ii) Closely allied species in a genus tend to be parapatric (i.e. with abutting ranges) in the same or similar habitats; (iii) Less closely related ones are often allopatric (with separated, distant ranges); (iv) The closest relative of a particular species is generally allopatric with it; (v) When new habitats are invaded – in our example arid bushland, from the ancestral mesic bushy woodlands – adaptation can produce novel plumage colours or patterns, proportions, bill and leg morphology, or behaviour; and (vi) Well differentiated congeners in differing habitats may ultimately become sympatric (not demonstrated well in this example: only by the purple patches). Differentiation rates vary greatly and are high when climatic and environmental changes proceed rapidly in the evolutionary time-scale.

Taxonomy

Hitherto, the traditional evolutionary taxonomy with its 'biological species concept' or BSC, championed by Mayr (1942), has been used widely by ornithologists. During the life of *The Birds of Africa*, evolutionary taxonomy has been challenged by alternative classification methodologies (such as phenetic taxonomy, distance-based evolutionary taxonomy, Hennigian cladistics and pattern cladistics: Greenwood 1997), and the BSC by alternative species concepts (such as evolutionary, recognition and phylogenetic ones). In particular, the phylogenetic species concept or PSC has received considerable advocacy, for instance by Hazevoet (1994, 1995) and Zink (1997). The PSC makes the species the smallest irreducible unit, without subspecies, and in effect upgrades well-defined subspecies, or most kinds of them, to the status of full species. Snow (1997) and Collar (1997) have pointed out difficulties with the PSC, from theoretical and practical standpoints respectively, and conclude that the BSC appeals better to common sense, serves ornithology best, and is likely to remain the most valuable and popular species concept among field ornithologists for a good many years to come. For a contrary opinion, see Preddy (1999).

We continue to use *subspecies* in the conventional manner. We take note of the numerous *species* definitions, mostly 'biological' rather than 'phylogenetic' ones, listed by Zink (1997). In the Introductions to previous volumes we have given short statements regarding our interpretation of the concept of *superspecies*. We still find it a useful taxon (level of classification) and we continue to employ it in this volume, applying the same criteria as before (see Vols II, xi–xii, III, xiii and IV, x–xi).

Systematic treatment is based largely upon the works of White (1962, 1963), Sibley and Monroe (1990), Dowsett and Forbes-Watson (1993) and M. P. S. Irwin (Nectariniidae, 1999 and pers. comm.). We continue to experience difficulties in ranking taxa, particularly allopatric ones. The problems are well illustrated for an Oriental warbler which under one interpretation of the PSC is four species and under another, and under the BSC, is but a single species (*Seicercus burkii*, Alström and Olsson 1999). If such a complex is deemed to comprise more than one species, it is the ideal example of a superspecies. Despite the extensive morphometric and plumage studies of taxonomists such as P. A. Clancey, few African taxa have received such thorough investigation as the *S. burkii* complex or, nearer home, the chiffchaff *Phylloscopus collybita* complex, with its bioacoustic and molecular evidences (Helbig *et al.* 1996). Until they do, doubts about taxonomic relationships and ranking of African birds will remain and controversies flourish.

References

Alström, P. and Olsson, U. (1999). The Golden-spectacled Warbler: a complex of sibling species, including a previously undescribed species. *Ibis* **141**, 545–568.

Ash, J. S. and Miskell, J. E. (1998). 'Birds of Somalia'. Pica Press, Robertsbridge, UK.

Barlow, C., Wacher, T. and Disley, T. (1997). 'A Field Guide to Birds of Gambia and Senegal'. Pica Press, Robertsbridge, UK.

Brown, L. H., Urban, E. K. and Newman, K. (1982). 'The Birds of Africa', Vol I. Academic Press, London.

Cheke, R. A. and Walsh, J. F. (1996). 'The Birds of Togo'. B.O.U. Check-list 14. Brit. Ornithologists' Union, Tring, UK.

Christy, P. and Clarke, W. V. (1994). 'Guide des Oiseaux de la Réserve de la Lopé'. ECOFAC, Libreville, Gabon.

Christy, P. and Clarke, W. V. (1998). 'Guide des Oiseaux de São Tomé et Príncipe'. ECOFAC, Libreville, Gabon.

Collar, N. J. (1997). Taxonomy and conservation: chicken and egg. *Bull. Br. Orn. Club* **117**, 122–136.

Dowsett, R.J. and Forbes-Watson, A.D. (1993). 'Checklist of Birds of the Afrotropical and Malagasy Regions.' Tauraco Press, Liège, Belgium.

Feare, C. and Craig, A. (1998). 'Starlings and Mynahs'. Christopher Helm/A. & C. Black, London.

Fry, C. H. (1997). The birds of Africa. Pp. 249–252 in J. E. Pemberton (Ed.) 'The Birdwatcher's Yearbook and Diary 1998'. Buckingham Press, Buckingham, UK.

Gatter, W. (1997). 'Birds of Liberia'. Pica Press, Robertsbridge, UK.

Greenwood, J. J. D. (1997). Introduction: the diversity of taxonomies. *Bull. Br. Orn. Club* **117**, 85–96.

Harrap, S. and Quinn, D. (1996). 'Tits, Nuthatches and Treecreepers'. Christopher Helm/A. & C. Black, London.

Harris, T. and Arnott, G. (1988), 'Shrikes of Southern Africa'. Struik Winchester, Cape Town.

Harrison, J. A., Allan, D. G., Underhill, L. G., Herremans, M., Tree, A. J., Parker, V. and Brown, C. J. (1997). 'The Atlas of Southern African Birds', Vol. 2: Passerines. BirdLife South Africa, Johannesburg.

Hazevoet, C. J. (1994). Species concepts and systematics. *Dutch Birding* **16**, 111–116.

Hazevoet, C. J. (1995) 'The Birds of the Cape Verde Islands'. B.O.U. Check-list 13. Brit. Ornithologists' Union, Tring, UK.

Helbig, A.J., Martens, J., Seibold, I., Henning, F., Schottler, B. and Wink, M. (1996). Phylogeny and species limits in the Palaearctic chiffchaff *Phylloscopus collybita* complex: mitochondrial genetic differentiation and bioacoustic evidence. *Ibis* **138**, 650–666.

Irwin, M.P.S. (1999). The genus *Nectarinia* and the evolution and diversification of sunbirds: an Afrotropical perspective. *Honeyguide* **45**, 45–58.

Jenni, L. and Winkler, R. (1994). 'Moult and Ageing of European Passerines'. Academic Press, London.

Keay, R. W. J. (1959). 'Vegetation Map of Africa South of the Tropic of Cancer. Explanatory Notes.' Oxford Univ. Press, Oxford.

Kemp, A. (1997). Review of *The Birds of Africa, Vol. 5. Ostrich* **68**, 85.

Kingdon, J. (1997). 'The Kingdon Field Guide to African Mammals.' Academic Press, London.

Lefranc, N. and Worfolk, T. (1997). 'Shrikes'. Pica Press, Robertsbridge, UK.

Mabberley, D. J. (1993). 'The Plant-Book. A Portable Dictionary of the Higher Plants.' Cambridge Univ. Press, Cambridge, UK.

Mayr, E. (1942). 'Systematics and the Origin of Species'. Columbia Univ. Press, New York.

Parker, V. (1999). 'The Atlas of the Birds of Sul do Save, Southern Mozambique'. Avian Demography Unit and Endangered Wildlife Trust, Johannesburg.

Preddy, S. (1999). The species-concepts debate. *Brit. Birds* **92**, 261–262.

Shirihai, H. (1996). 'The Birds of Israel'. Academic Press, London.

Sibley, C.G. and Monroe, B.L. (1990). 'Distribution and Taxonomy of Birds of the World.' Yale University Press, New Haven.

Skead, C.J. (1967). 'Sunbirds of Southern Africa, also the Sugarbirds, the White-eyes and the Spotted Creeper'. South African Bird Book Fund, A. A. Balkema, Cape Town.

Snow, D.W. (1997). Should the biological be superseded by the phylogenetic species concept? *Bull. Br. Orn. Club* **117**, 110–121.

'The Times Atlas of the World' Comprehensive Edition (1990) (ed. J. C. Bartholomew *et al.*). Times Books, London.

Zimmerman, D. A., Turner, D. A. and Pearson, D. J. (1996). 'Birds of Kenya and Northern Tanzania'. Princeton University Press, Princeton.

Zink, R. M. (1997). Species concepts. *Bull. Br. Orn. Club* **117**, 97–109.

White, C.M.N. (1962). 'A Revised Check List of African Shrikes, Orioles, Drongos, Starlings, Crows, Waxwings, Cuckoo-Shrikes, Bulbuls, Accentors, Thrushes and Babblers.' Government Printer, Lusaka.

White, C.M.N. (1963). 'A Revised Check List of African Flycatchers, Tits, Tree Creepers, Sunbirds, White-eyes, Honey Eaters, Buntings, Finches, Weavers and Waxbills.' Government Printer, Lusaka.

INTRODUCTION xvii

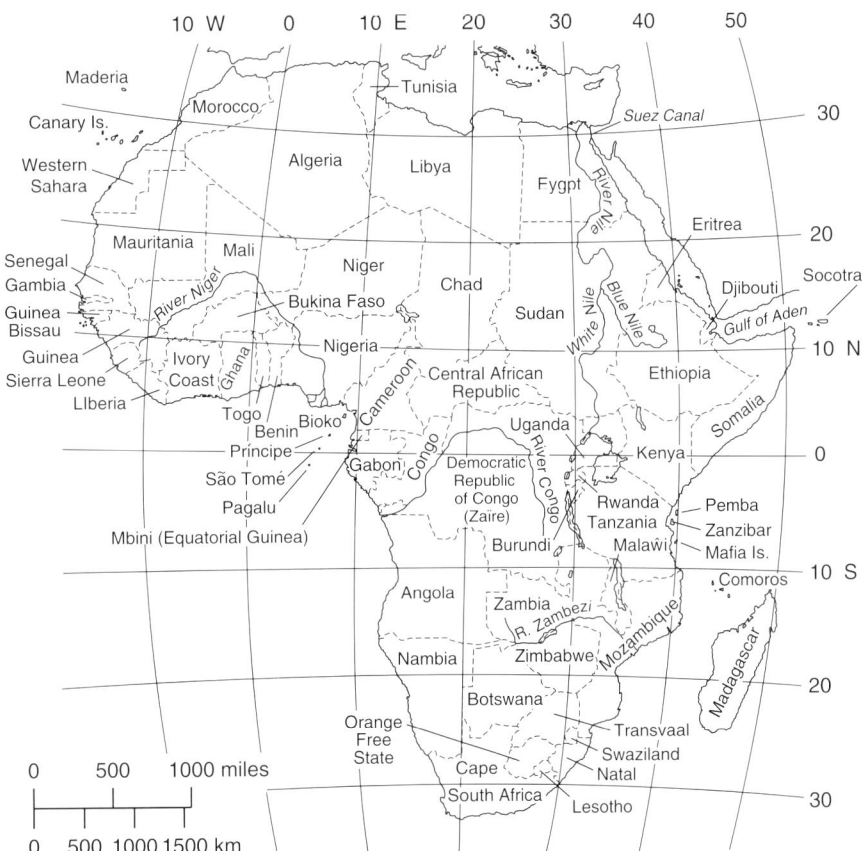

Fig. 4
Political map.

Fig. 5
East African climatic zones.

Family PICATHARTIDAE: picathartes

A family and genus of 2 distinctive West African birds of uncertain affinities. Large, dark above, white or cream below, with brightly coloured bare skin on head (black and yellow in *P. gymnocephalus*, **A**, blue and carmine in *P. oreas*, **B**). Rather small head, with erectile crest of tiny semiplumes intermediate between contour feathers and down (Tye 1986), and wispy ruff on hindneck; large eyes, robust bill, rather long and slender neck, lax plumage, long tail, muscular thighs, long tibia and long, strong tarsus. Semi-terrestrial; in rain forest, keeping near boulders and nesting in overhanging cliffs. Earthworms are major part of food given to nestlings.

Picathartes has been placed in crows Corvidae, starlings Sturnidae, and babblers Timaliidae (reviews by Delacour and Amadon 1951 and Sibley 1973). Anatomy of lachrymals, vomer and maxillo-palatine, myology and pterylosis, all indicate that *Picathartes* is not a crow Corvidae *sens str.*, but that it has some affinity with starlings (Lowe 1938). Delacour and Amadon concluded that it is a timaliid, but Serle (1952a) found that breeding biology is crow-like. Most recent authorities treat it as a timaliid, although Sibley and Monroe (1990) place it as a parvorder *incertae sedis* in their greatly expanded Family Corvidae, distant from other babblers in their Family Sylviidae. In any event the closest relative of *Picathartes* may be the Malaysian Rail-Babbler *Eupetes macrocerus* (Serle 1952a, Olson 1979). Within the Timaliidae, Sibley (1973) thought *Picathartes* to be closest to *Turdoides*, but there is little or no evidence behind the supposition (Olson 1979).

The 2 species were treated by Hall and Moreau (1970) and Dowsett and Forbes-Watson (1993) as composing a superspecies. They have about the same degree of differentiation – striking colour and pattern differences on the head – as allopatric species pairs that we have treated as a superspecies (e.g. the cranes *Balearica pavonina* and *B. regulorum*, Vol. II Plate 6, and the turacos *Musophaga violacea* and *M. rossae*, Vol III Plate 3) and as ones that we have treated as independent species (e.g. the ground hornbills *Bucorvus abyssinicus* and *B. cafer*, Vol. III Plate 24). Here we rank them as independent species.

Endemic. Single genus with 2 species, one in Upper and one in Lower Guinea rain forests.

A　　　　　　B

Genus *Picathartes* Lesson

Picathartes gymnocephalus (Temminck). White-necked Picathartes. Picatharte à cou blanc.

Plate 1 (Opp. p. 14)

Corvus gymnocephalus Temminck, 1825. Planch. Col. livr. 55, pl. 327; Guinea coast.

Range and Status. Endemic resident, from Guinea to Ghana. Guinea: in W, rare in remnant forest on Kounounkan Massif near Sierra Leone border (Hayman *et al.* 1995); in SE common and widespread in Macenta Préfecture with 9 nest sites found in Ziama Forest, and occurs just to north of Macenta (Halleux 1994). In Sierra Leone widespread except in N and NW; known from 30 colonies (Grimes and Darku 1968), and recently found in 6 out of 7 forest reserves: in Gola Forest, 37 active colonies with 190 nests, and in the other reserves together *c.* 63 colonies with *c.* 125 nests (Allport *et al.* 1989, Thompson and Fotso 1995); main centres in Loma and Tingi Hills in NE plateau, and in Peninsular Forest near Freetown (where 8 colonies with 18 nests, Allport 1991). Liberia: widespread in N and NE, less so in SW and S, with breeding colonies or other records at Zorzor, Geatown, Grebo, Krahn-Bassa, Wonegizi Mts, Fasswalazu, Wainjama, Zowolo, Mt Nimba, Tchien and Bong Mts (Collar and Stuart 1985); large numbers in Belle Nat. Forest (Allport 1991); numbers increase towards N highlands, 116 breeding sites reported and 500–1000 estimated (Gatter 1997). Ivory Coast: small colonies at foot of Mt Nimba and 6 single nests on E side of Mt Nienokoué (Gartshore *et al.* 1995); one colony of at least 20 birds with 15 new and old nests at Lamto (Demey and Fishpool 1991) was destroyed when ground cleared for banana plantation (Wood 1995);

Picathartes gymnocephalus

16 known colonies and >100 estimated in NW from Blolekin, Danané and Sipilou to Man (Gatter 1997); Mt Peko (*Bull. Afr. Bird Cl.* 6, 1999, 10). Ghana: uncommon and localized; breeding colonies on S scarp of Volta basin, with 8 sites in Mpraeso area and several others in Kwahu Tafo forests east of Mpraeso; south of scarp at Bodweseanwo, Fumso, and *c.* 80 km north of Cape Coast; northwest of Kumasi; nest found in 1894 near Togo border at Awatotse (= Apototsi, 06°57′N, 00°29′E: Cheke 1986).

Vulnerable – population highly fragmented and threatened. Still widespread in Sierra Leone and S Guinea and locally fairly common even if difficult to find, but the species has almost certainly declined and its range contracted: 'the rate of forest destruction in Africa west of the Dahomey Gap is so severe that any bird species endemic to primary forest in this region must now be considered gravely at risk' (Collar and Stuart 1985, p. 502: *P. gymnocephalus*). In Macenta Préfecture, Guinea, forests have been severely encroached upon in the last two decades for upland rice cultivation, and encroachment is growing worse because of farmers immigrating from the country's dry north (Halleux 1994). High incidence of abandoned nests in Gola Forest, Sierra Leone (38%) indicates pressure upon the population there (Thompson and Fotso 1995); all colonies outside the retreating forest edge have been abandoned (Allport *et al.*1989). Collecting *Picathartes* for zoos was formerly a widespread practice, and in Liberia animal dealers have destroyed many colonies. Many live-collected adults and juveniles die within 24 h; up to 70 birds listed in world zoos in 1970s; although species has bred in captivity a few times, attempts to establish stable captive breeding programmes have failed. Trapping for export has recently died down. Traditional hunting methods with traps are or were widespread (e.g. Ivory Coast: Collar and Stuart 1985), and in Guinea (Ziama) hunters trap rodents and hyraxes at *Picathartes* colonies and sometimes catch birds on their nests at night. In Liberia they are caught at night using torchlights (Gatter 1997). Birds are also caught by bat-catchers (Ivory Coast). In native lore in Sierra Leone, *Picathartes* is supposed to be guardian of the bizarre-looking forest rock formations where they nest, and a residual respect for the birds persists (Thompson and Fotso 1995). Chicks sometimes taken for food by migrant Liberian hunters (Allport 1991). Ghanaian population thought to be 200–300 pairs in 1970s.

During the 1990s this and the next species have become icons of bird ecotourism and symbols of rain forest conservation.

Description. ADULT ♂: head completely bare except for chin and throat, nape and neck, and for thin white fuzz on forehead. Bare skin bright yellow, but large circular black patch including ear behind eye, separated from its fellow by 2 mm of yellow skin in middle of hindcrown; black patches have raised edge and look as if ironed on. Eyelid black, forming thin black line around eye. Yellow skin sharply demarcated from soft black skin at base of upper and lower mandible. Bare skin between rami of lower mandible yellow. Chin and throat with white feathering, thin in centre of throat. Nape white, feathers short, thin, silky, fur-like, pointing inward and upward from hind edge of naked black 'ear' patch to form wispy ruff; hindneck almost naked, skin orange-yellow. Upper mantle matt black, merging into grey-black lower mantle and bluish grey back, scapulars, rump and uppertail-coverts. Plumage smooth, lax and fluffy, feathers long and on rump dense. Tail uniform dark brown, graduated (T6 35–65 shorter than T1), feather shafts light brown, tail tented or strongly decurved at sides, and outer vane of T6 almost curling below and in. Wings uniform dark brown above, the same but glossy below, primaries decurved, P9–7 longest, P6 5, P5 11, P4 30, P3 44 and P2 62 shorter. Underparts creamy white, creamiest on upper breast, silky, feathers with parallel barbs well separated. Bill black; eyes dark brown; legs and feet pale blue-grey or blue; claws horn-coloured in skins. Sexes alike. SIZE: wing, ♂ (n = 4) 150–158 (156), ♀ (n = 3) 147–154 (151); tail, ♂♀ (n = 9) 171–185 (178); bill ♂♀ (n = 3) 33–34 (33·3); tarsus, ♂♀ (n = 6) 63–70 (66·5). WEIGHT: (2 ♀♀) 166, 226.

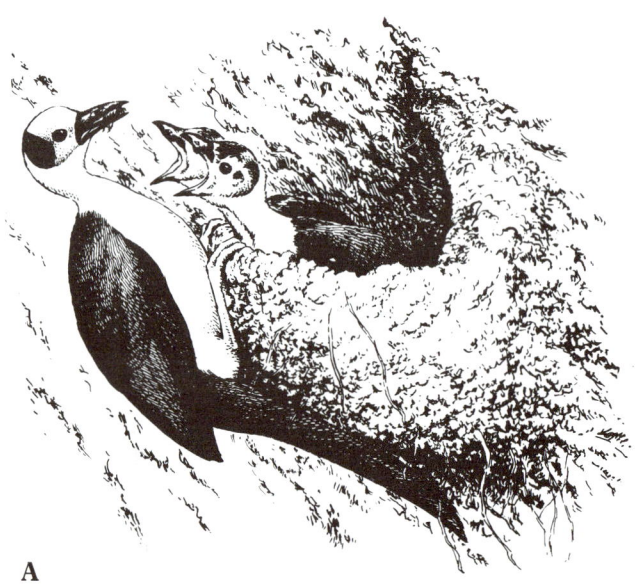

A

IMMATURE: like adult. Juvenile at age 4 weeks (tarsus 59, wing 113, tail 54) very like adult, but throat and neck thinly feathered, remaining underparts fluffy, silky, creamier than adult, especially lower breast and belly; yellow on head slightly paler than in adult and dark-mottled.

NESTLING: hatches naked with dark skin, gape orange-red changing after a few days to bright yellow-orange; edges of mandibles conspicuously dull yellow, at least until young ready to leave nest, when nestling plumage like adults' (**A**). Head skin all yellow, without black patches, until a week before leaving nest (Grimes 1964), but yellow colour said to develop to varying extent at fledging (Glanville 1954).

Field Characters. Length 38–41 cm. A shy bird of forest undergrowth, with curious springy, hopping gait; tail rather long. Unmistakable, with bald yellow head and large black ear-patch. In dim light body appears mainly black above and white below, but in better light upperparts grey-brown, sometimes even with a silvery sheen, underparts tinged with lemon.

Voice. Tape-recorded (B, BBC, ATT). Call (song?) a series of evenly-spaced chicken-like clucks, 'chuk-chuk-chuk ...' or 'choop-choop-choop ...', lasting a min or more, given at rates varying from 4 per 3 s to 8 per 5 s. Adult and fledgling infrequently makes weird, long-drawn 'owooh', 'kaaa' or 'raaa' note (sonagram in Grimes and Darku 1968). Contact call of fledgling in and out of nest a loud, melodious, quavering whistle lasting *c.* 1 s. Alarm, given readily, is continuous, guttural chatter, a low-pitched note repeated at rate of one per 0·8 s, probably same as continuous 'ow, ow, ow' of Bannerman (1948). Also described are a harsh 'chirr' and a longer 'kaaa'. Begging call of nestlings a rasp or squawk.

General Habits. Inhabits steep slopes in primary and mature secondary, rock-strewn forest in hilly lowland districts, up to 800 m. Keeps low in vegetation and on ground, near nesting site of cliffs, large boulders and towering rocks with overhangs and shallow caves; also inhabits huge fallen hollow trees (twice in *Entandrophragma candollei*, Liberia, Gatter 1997). Forages amongst mossy boulders with creepers and trees festooned with lianas and hanging moss, often on slopes, sometimes near stream; hops across streamside sand. Seldom seen away from breeding sites. Occasionally occurs in partly cleared, disturbed forest, so a bird may habitually range quite widely. Tolerates some disturbance: a colony in Ghana survived for at least 13 years after much of surrounding forest had been clear-felled, and another was completely surrounded by cocoa plantation. Some colonies in Sierra Leone survive in tiny fragments of rock-strewn or cliffy forest. A nest in Ivory Coast was on rock with only slight overhang in brightly lit, quite open area; and 7 nest sites there were dry, bare, rock faces looking NNE to SSE (Gartshore 1989).

Beautiful, strange, graceful birds, generally solitary or in pairs, sometimes in groups of 3–5. Rather silent. Secretive only in that preferred habitat is difficult for observer access; self-effacing – quick and silently evasive, but inquisitive and can approach near to observer to eye him. 10 birds in colony in streamside cliffs, Sierra Leone, used small sand beach, lianas and low branches to pause whilst moving to nests, and displayed, chased, preened and loafed at the site or rustled enigmatically in undergrowth (Wood 1995). Sometimes colony roosts in nests outside breeding season. Moves on ground, on low branches and lianas and in caves by hopping and bounding, with long, springing hops followed by statuesque pause, then more hops. Often disappears into mass of creepers and rocks. Can leap 6 m from floor of cave up to nest, partly using wings. Progresses by short, low-level flights between vines, lianas, buttress roots and trees (Thompson and Fotso 1995). Uses tail to maintain balance when hopping through undergrowth. Rarely flies far. Peers into gloom by stretching out its long neck. Forages on forest floor or in vegetation within 1 m of it; uses bill to turn leaves over and toss them aside. Often feeds at army ant swarm, in company of alethes, bristlebills and Finsch's Flycatcher-Thrushes *Neocossyphus finschii*, preying on flushed insects ahead of or among the ants. At swarm of *Dorylus* ants in Sierra Leone, solitary Picathartes bounded rapidly on ground and along logs or flew rapidly with sudden flutters from limb to limb up to 4 m above ants, sometimes with sideways twitch of tail; it foraged quietly by waiting on ground or stout horizontal perch or by ricochetting rapidly over the ground and low lianas, buttress roots or slender, vertical stems ahead of ant column; of 21 captures, one was a quick hop to pick prey off leaf 70 cm overhead, all others were on ground – 13 by pecking or rushing, 6 by tossing leaves and one a sally (Willis 1983). In Ivory Coast 7–8 birds gathered by nesting rocks, an hour before dusk, and one displayed (**B**) – possibly a pre-group-roosting intention-display: the birds perched on rocks, vines and branches 2–3 m above forest floor, when one leant forwards and downwards with legs flexed and wings raised and half-opened and neck arched down so that bill protruded behind the legs (Mudd and Martins 1996). Behaviour was repeated twice more in 30 min.

Food. Mainly insects: larval cockroaches, tettigonid grasshoppers, earwigs, click-beetles *Psephus*, histerid and anthribid beetles, ants *Dorylus* and *Bothroponera*, termites; millipedes, centipedes, snails, earthworms, small crus-

B

taceans, and small frogs and lizards. Prey taken near *Dorylus* ant column mostly tiny but included a large tettigonid swallowed nearly whole (Willis 1983). Food given to nestlings mainly worms (Grimes 1963, and photo in Colston and Curry-Lindahl 1986) and small frogs and lizards (Thompson and Fotso 1995).

Breeding Habits. Solitary or colonial nester in caves; evidently monogamous with birds in 'pairs', but may yet prove to breed co-operatively. 2 breeding seasons per year. Breeding sites usually deserted, but increasing use by group of birds is first sign of breeding. Group of 6–8 birds interact by chasing each other in circles, sometimes through the treetops, accompanied by peculiar display involving head and wings (Thompson and Fotso 1995). 31 colonies, Liberia, had av. 3·8 nests (Gatter 1997). Pair said to use 2 nests, one for breeding, one for roosting (Allport *et al.* 1989). Courtship and territorial behaviour not known.

NEST: a thick-walled, strong, deep cup made of mud, about size of half of a football, with varying amounts of vegetable fibres mixed in, lined with fibres and rootlets, placed 2–4 m above floor on walls or roof of cave or on cliff or large boulder (**A**). Nests once reported on bank of watercourse and on fallen treetrunk (Allport 1991). Many nests have short beard of bents hanging from their sides. Fibres are phloem fibres from monocotyledonous and dicotyledonous plant stems and leaves. How bird begins construction on smooth ceiling of cave is not known; cylindrical wasps' nests 2–3 cm long are common in nesting caves and are often embedded in mud of birds' nests; they may form the construction nucleus. Nests on open rock faces (not in caves) are always well clear of adjacent vegetation. Nests usually separated by *c.* 1 m, but one colony had 6 nests joined in pairs and 3 separate nests. One nest was partly built in one year and completed in the next, but was not lined and was never used for breeding. Another nest was complete but unlined in Mar, had lining added then removed in Oct, was relined next Mar and used for night roosting until eggs were laid in Oct. At a third nest young were reared in Sept and then the nest was not used until 3 years later when a clutch was laid in Sept. Nest height above ground (n = 34) 1–8 m (2·6 m) (Allport *et al.* 1989).

EGGS: 1–2, usually 2. Said to be almost exactly the same as eggs of *P. oreas*, *q.v.* SIZE: (n = 2) 40 × 25, 42 × 25. Most pairs double-brooded, with intervals of 102–147 (127) days between nestings in same year (n = 7); second clutch laid in same nest as the first.

LAYING DATES: Guinea, (breeding season July–Jan: Halleux 1994); Sierra Leone, Nov–Feb, Apr–Oct; Liberia, Sept, Oct–Dec (mainly Oct), Mar, July (and ♀ had recently laid in May) (Gatter 1997); Ghana, Mar–June and Sept–Nov. In W, laying commences both before and after the wet season (Allport 1991).

INCUBATION: period between 23 and 28 days.

DEVELOPMENT AND CARE OF YOUNG: eyes open between days 9 and 11; rectrices begin to emerge at day 7. Primaries, tail and mantle and back feathers grow earlier than rest of plumage. At first food brought to nest 4 times per h. Young excrete freely over side of nest from *c.* 10 days. When feeding young, adult clings with fluttering wings to side of nest; long tail may be used as a prop underneath it; feeding visits last 2–3 min. Nestling period between 23 and 27 days. Young just out of nest resemble short-tailed adults, moving on forest floor in long bounds; they return to cave to roost on nest with parents.

References
Allport, G. *et al.* (1989).
Gartshore, M.E. *et al.* (1995).
Grimes, L.G. (1963, 1964).
Grimes, L.G. and Darku, K. (1968).
Olson, S.L. (1979).
Thompson, H.S. (1993).
Thompson, H.S. and Fotso, R. (1995).

Plate 1 (Opp. p. 14)   *Picathartes oreas* Reichenow. Grey-necked Picathartes. Picatharte à cou gris.

Picathartes oreas Reichenow, 1899. Orn. Monats., p. 40; Victoria, Cameroon.

Range and Status. Endemic, resident, SE Nigeria to NE Gabon. Nigeria: locally common in undisturbed forest in SE, close to Cameroon border; 94 breeding sites with estimated 500–1000 birds (in 1987), mostly between villages of Olum, Kanyang, Bamba and Bashu, in area of *c.* 20 × 50 km centred on 06°15′N, 09°05′E (Ash 1991), and 42 further sites nearby in Oban Hills/Obudu region (Anon. 1995). Cameroon: widespread in SW, south of 06°N and west of 13°E, mainly in coastal lowland rain forest, up to 1800 m on Mt Cameroon and 1250 m on Mt Nlonako; 13 breeding sites in area of 3 × 5 km just northwest of Mundemba in Korup Nat. Park, one with 30 nests (in 1993: Green 1995); near Korup, known from Rumpi Hills and mountains near Usukutang and Nguti (Rodewald *et al.* 1994); in S, a large colony with 47 nests, at least 40 active, in Dja Forest Res. (Thompson and Fotso 1995). Bioko, 10 sightings in almost inaccessible terrain in SW, on 7 km stretch of Rio Olé at 500–750 m near 03°19′N, 08°29′E and in Gran Caldera, in 1986, 1989 and 1990 (Butynski and Koster 1989, Butynski *et al.* 1996); and strongly suspected to occur in Rio Muni, mainland Equatorial Guinea (T. M. Butynski and S.H. Koster, pers. comm.). Gabon: 7 colonies each of 5–15 birds known in Bélinga, Dibakouélé and Bengoué (Brosset and Erard 1986); several colonies of 2–3 nests in Réserve de la Lopé, near airport and on R. Mbay (Christy and Clarke 1994); bird caught alive, Mouila (01°50′S, 11°02′E), 1981, may have been an escape, but in 1994 pair seen not far north, in limestone cave east of Lastoursville (00°50′S, 12°43′E) (Y.-M. de Martin de Viviès, pers. comm.).

Not endangered, although population is thought to have contracted in area west of Mt Kupé, with the destruction of

Picathartes oreas

forest by road building and timber extraction east and north of Kumba (Moore 1997), and around Yaoundé due to habitat destruction (R. Fotso, pers. comm.).

Description. ADULT ♂: upper mandible behind nostril, forehead and forecrown powder blue, bare; lores, cheeks and ear region naked, skin black, with a few short thin bristles, and *c.* 4 bristlelike feathers 3–4 mm long in row across crown; hindcrown and nape bare, skin carmine, bare, 2 mm black band across crown between blue forecrown and carmine hindcrown. Hindneck skin carmine, feathers sparse, *c.* 10 long, grey, slanting inwards from sides of hindneck to midline where feather tips form a short ruff separating bare nape from bare skin above mantle; mantle, scapulars, back, rump and uppertail-coverts grey; parallel barbs of mantle and back feathers well separated, rump feathers long, dense, soft and silky. Tail grey. Chin and throat pale grey, bare skin between rami of lower mandible dull blackish; upper breast and sides of neck grey; lower breast, belly, flanks, thighs, vent and undertail-coverts rich lemon-buff (in skins soon fading to cream-white with slight greyish tinge). Flanks sometimes greyish (feather bases showing through where overlying cream feathers are sparse-barbed). Wing coverts and tertials grey, remiges black above and below. Upper mandible behind nostril pale blue, rest of bill black; eyes dark brown; legs and feet grey. Sexes alike. SIZE: wing, ♂ (n = 6) 143–158 (151), ♀ (n = 4) 143–152 (149); tail, ♂ (n = 6) 130–149 (142.5), ♀ (n = 4) 137–143 (140); bill, 2 ♂♂ 38, 39, 2 ♀♀ 35, 36; tarsus, ♂ (n = 5) 63–66 (65.0), ♀ (n = 5) 59–63 (61.2). WEIGHT: (n = 6) 202–239 (226.5) (Bowden 1986).

IMMATURE: like adult but bare patch on back of head golden yellow not carmine (Moore 1974).

NESTLING: hatches naked except for tiny primary quills 1 mm long and fine down 5–10 long in lines along spine, humerus, forearm and femur; skin dark pink, with black patches on upperside (extent of black varies considerably) (Tye 1987). Gape yellow. Late nestling has plumage like adult, except for white flecks on wing coverts; and bare skin of head is black or dark brown on forecrown and dark reddish brown on hindcrown.

Field Characters. Length 33–38 cm. A shy bird of forest undergrowth, with uncanny, springy hop/walk; tail rather long. Unmistakable, with bald blue, red and black head; neck and upperparts grey, underparts creamy lemon-yellow. At rest, primaries form distinctive black band separating yellowish vent from grey back (Thompson and Fotso 1995). Readily attracted by imitation of its call (Christy and Clarke 1994).

Voice. Tape-recorded (B, 104, ERA). Rather silent, but sometimes gives low, quiet, drawn-out rasping or hissing 'wheet' call, 1–2 s long, repeated several times every 4 s or so, with beak open and throat inflated (Butynski *et al.* 1996, Dolton 1995); easily imitated; perhaps the same as call 'like the sound of heavy furniture being pushed across a gritty wooden floor' (Thomas 1991). Adult approaching nest with food gives single or double 'peep' (Tye 1987). On arrival at nest, utters a regularly repeated low 'ga-a-a', half snore and half sigh (Moore 1974).

General Habits. Inhabits rugged terrain in closed-canopy, undisturbed rain forest; thick forest where trees interspersed with huge boulders, slabs of rock, caves and deep rocky gorges with bare, vertical or overhanging gneiss walls on high-rainfall hillsides, the forest understorey with many mosses, ferns, lianas and epiphytes but with sparse undergrowth and a fairly open floor. Habitat in SW Bioko Island is diabolical terrain, low forest with 10 m annual rainfall, dense undergrowth, vertical-sided gorges near caldera with 1200 m wall. Altitudinal range 450–2100 m, lower on Bioko.

Usually singly or in pairs but often in small flocks of 3–10 birds. After being disturbed, flock of 9 birds returned to colony site, keeping close together; 3 went straight onto nests and 6 hopped on floor of cave and over boulders at base of adjacent cliff face, boldly approaching observer (Moore 1974). Moves by curious series of runs and long,

A

springing hops, a small flock on ground almost in unison (Moore 1974), but also described as hopping but never walking or running (Butynski *et al.* 1996). Progresses among low branches, on rock ledges and on ground. Stands with legs flexed, tail down, body and small-looking head up (**A**). Flies fast, manoeuvring well through trees and rocks. Not shy; eyes observer curiously from relatively open situation, hopping back and forth on a few woody perches and sometimes giving 'wheet' call. When suspicious, raises 'crest' on crown and thin ruff on hindneck (**B**), and utters muffled groans. Study birds in Cameroon remained all year within 300 m of nest; during much of day, from 10h30 to 15h00, perches inactively in liana-clad tangles of herbage or inside cave some distance from nest site (Fotso 1993).

Forages in leaf litter, on dead tree trunks lying on ground, and leaps up to take prey from overhead foliage. Searches for invertebrates by standing in one place and scanning, but usually forages more actively, tossing leaf litter aside with bill then pouncing. Forages singly or in groups. Crushes snail shells in bill; beats struggling prey against ground. Has once been seen feeding on invertebrates disturbed by swarming army ants (Butynski *et al.* 1996). Rather inactive during midday hours; in Gabon, returns to roost in nest around midday and resumes foraging later in afternoon. In Cameroon several pairs gather near nest site in hour before going to roost (Fotso 1993). At night roosts sociably at cliff nesting site, arriving at dusk in pairs or small parties from surrounding forest and dispersing at daybreak (Serle 1952b). Bathes in small pools near colony site. Colony roosts together in nests, just before breeding season and perhaps at other seasons, but evidently birds do not always roost communally (Tye 1987).

Food. Invertebrates and small vertebrates. Items seen carried to nest were identified as beetles (weevils, rove beetles), lepidopterans, orthopterans, blattids, frogs and earthworms, also ?earwigs, a 95 mm long ?caterpillar, an ?ant-lion, ?silverfish, and sometimes beakfuls of ?ants (Tye 1987). Most of the larger items are worms. Regurgitated pellets, *c.* 20 long, contained ants, a crab *Potamon* with carapace 10–20 wide, and fragments of snail shell, grit, moss and a leaf. Stomach contents: nymphal roaches, tettigonid grasshoppers, earwigs, click beetles *Psephus*, histerid and anthribid beetles, ants *Dorylus* and *Bothroponera*, and fragments of small crabs and frogs (Lowe 1938). Diet mainly insects, with some lizards, small frogs, snails and slugs (Fotso 1993). At one Gabon site, thought to depend on arthropods feeding on bat guano and detritus near cave (Brosset 1965b).

Breeding Habits. Breeds solitarily or colonially. Not territorial nor aggessive. Seemingly monogamous; possibly a cooperative breeder (Brosset 1965a,b, Tye 1987), but probably not so (Fotso 1993). No courtship behaviour known. Within one colony timing of egg laying is not well synchronized.

NEST: half-cup of dried mud, dry grass fibres and dead leaves, thick-walled (usually 30–40 thick but once 140 – see below), somewhat uneven, roughly plastered to rock face, or a retaining wall of mud built across opening of small rock fissure (**C**). Mud sets into a very hard, stone-like structure, with many fibres sticking out from inside and outside surfaces. Occasionally green leaves incorporated. Untidily lined with rootlets and thin dry-grasslike vegetable strips. Nest also sited in mouth of cave, well into dark cavern, on vertical sides of gorges, either bare or moderately vegetated with moss, ferns and small herbs but well clear of woody branches, on cliffs beneath dry 'waterfalls', and once well into large log burnt hollow on forest floor (Lowe 1938). On Mt Cameroon nests were on cliffs, under 30–50 cm overhangs, but none in caves (Tye 1987). Often near small pools. Nest exterior *c.* 400 long, 290 wide, walls 140 thick, weight *c.* 3 kg (also said to be 15 kg); interior of cup usually *c.* 60 × 200 (Ash 1991). Height of nest above ground (n = 54) 1·2–5·2 (3·1) m. Nest built equally by ♂ and ♀, sometimes taking 2–3 months, sometimes over a year.

EGGS: 1–3 (av. *c.* 2·3); in W Cameroon study, 16 clutches of 2 and only one of 1. Second egg laid 48 h after first (or 24 h: R. Fotso, pers. comm.). Rather nightjar-like: dark fawn with dark brown blotches, creamy white with chocolate brown and grey blotches, or pale grey with brown

B

C

mottling. SIZE: (n = 4) 39·6–41·7 × 27·0–28·0 (40·5 × 27·5). WEIGHT: (n = 4) 14·7–16·0 (15·2).

LAYING DATES: Nigeria, breeding season given as Aug–Nov (Ash 1991); W Cameroon, Mar–Nov, with peaks in June–July and Oct (Fotso 1993); S Cameroon, main breeding season Oct–Dec, secondary one Apr–May; Gabon, Nov–Apr (21 clutches).

INCUBATION: begins at completion of clutch. Duration of bouts varies greatly, mainly <5 mins, but up to 80 min; intervals between stints the same, up to 110 min; or bouts regular, of 40–45 min (Fotso 1993). Both sexes incubate, and changeover is sometimes so rapid that incoming bird seems to chase incubating one off nest. Bird remains motionless whilst incubating; or restless, looking towards passing squirrels, monkeys and sunbirds. Arriving bird calls, prompting incubating bird to quit nest. Both birds leave and arrive following same route. Eggs incubated for 40–71% of day, young nestlings for 61%. 2 days before eggs due to hatch, adult often comes to nest and makes prodding movements, as if feeding young; food was brought twice but swallowed by adult (Tye 1987). Period: 21–24 days.

DEVELOPMENT AND CARE OF YOUNG: eggs hatch at intervals of 24 h. Nestling growth curves sigmoidal, from *c.* 12 g at hatching to 40 g at 13–14 days and 55 g at 15 days, and (another study) from 31 g at day 2–3 to 150 g at day 16–17. Weight of one chick declined a little in the 2 days before it left nest, from 162 g. At day 5, primary quills 5 long. From 5th day tail starts to grow and eyes to open. Adult removes eggshell fragments and drops them 5–10 m from nest. For first 10 days adult generally eats faecal sacs; thereafter they are carried away, but parents cease to remove them for last 2 days before chicks leave nest (Fotso 1993). At first food brought to nest 3 times per h. Prey caught away from nest but sometimes in vegetation and leaf litter below it. For first 10 days, one parent broods chicks or stays near nest to guard them, and other searches for food. Nestling period: about 24 days.

BREEDING SUCCESS/SURVIVAL: in brood of 2, 2nd-hatched young often fails to gain weight and dies or disappears; dead nestling sometimes removed from nest by adult and evidently cannibalized. Sometimes both nestlings disappear, the evidence again pointing to cannibalism (Tye 1987). Many nests destroyed for unknown reasons (Brosset 1965a). Snake predation, if it occurs at all, is rare. One nest with 2 eggs destroyed by Chimpanzee *Pan troglodytes*; built and destroyed again; built and destroyed with 2 eggs a third time by Chimpanzee *Pan troglodytes* (Rodewald *et al.* 1994). Adult birds roosting in nests sometimes killed by hunters who shelter in caves.

References
Ash, J.S. (1991).
Brosset, A. and Erard, C. (1986).
Fotso, R.C. (1993).
Green, A.A. (1995).
Thompson, H.S. and Fotso, R.C. (1995).
Tye, H. (1987).

Family TIMALIIDAE: babblers

A large, heterogeneous, mainly Asiatic family with a complex taxonomic history and fluctuating composition – many taxa have been moved between Timaliidae and Sylviidae or Turdidae. Small to quite large passerines, robust, strong-legged, often pale-eyed, adults of many species rufescent brown above and dark-streaked white below, sexes usually alike, rather sombre, but a few Asiatic species strongly coloured and patterned; juveniles unspotted; plumage rather hard to soft and lax, and back and rump feathering usually soft and profuse. Bills more robust than those of Sylviidae *sens str.* (probably the closest relatives of babblers), narrow, the culmen often decurved; many Asiatic genera with whole bill elongated and strongly decurved, and some with granivores' short and sturdy bill; bill sometimes notched. Rictal bristles, often conspicuous. Wing short and round-tipped, 10 primaries, P10 30–60% shorter than longest; outer primaries often markedly decurved; tail generally moderately to strongly graduated, or rounded but with T6 much shorter than T1–5. Legs and toes sturdy, legs long in some terrestrial species; tarsus scutellate. Characteristic egg-white protein profile (Sibley 1970). Syrinx lacks 'turdine thumb' which is so characteristic of Turdidae and Muscicapidae (Ames 1975, Olson 1984).

Essentially Paleotropical; arboreal and semi-terrestrial, in deciduous and evergreen woodland, most species inhabiting woody, shrubby understorey within 2–4 m of the ground, and many keeping near woodland boulders and caves; some inhabit dry scrubland, bamboos, or reedbeds; others forage mainly on ground, flicking aside leaves and debris like a thrush; most species grasp and clamp food item in foot. Mostly insectivorous, many partly frugivorous, some granivorous (and store large seeds), omnivorous (and catch small vertebrates) or nectarivorous. Flight weak – short flights low in dense vegetation, or of 50 m very low over open ground between clumps of shrubs. Gait a jump in vegetation or a bounding hop on ground; some also walk and run. Most scratch heads directly, a few genera indirectly. Numerous babblers stay in small family parties year-round and are known or likely co-operative breeders; individuals of many species clump or huddle tightly together when loafing and roosting. Noisy, with babbling, chuckling and chattering notes, often reciprocated or delivered by flock in unison; or monotonous repetition of single note, or fluty whistles. Almost completely sedentary. Nest a cup of coarse plant material, thickly lined with fine matter; eggs grey-blue, pale blue or whitish, immaculate or scantily marked, or strongly marked with red-brown.

About 50 genera and 232 species (Sibley and Monroe 1990), or 260–280 (Cramp and Perrins 1993); in Africa, Madagascar, Middle East, S and SE Asia, with a few in temperate Europe and Asia. 5 genera and 8 species endemic to Malagasy Region (Irwin 1983). The American *Chamaea fasciata* is sometimes included in Timaliidae. Australian babblers *Pomatostomus* belong to a different family.

37 species in Africa in 12 genera, including *Chaetops* (Olson 1998), all endemic except *Turdoides* and *Panurus*.

3 additional genera and species of putative timaliids have already been dealt with in previous volumes. They are *Modulatrix stictigula* and *Arcanator orostruthus* (Vol. IV, pp. 458–461 and Plate 25) (see Dowsett and Dowsett-Lemaire 1993a, p. 365, for a recent opinion that these species are babblers), and *Sphenoeacus afer* (Vol. V, p. 87 and Plate 6) (see following *Chaetops* generic diagnosis and footnote).

Genus *Chaetops* Swainson

2 species of petrophilous birds in SW and South Africa, with white eyebrows, dark-streaked brown crowns and backs, rufous rumps, white-tipped dark tails, white-tipped remicles, streaky breasts at least in ♀♀, and dark rufous bellies. Plumage of flanks and belly, lower back and (particularly) rump lax and fluffy in adults, more so in juveniles. Thighs long, legs long in *C. frenatus* (44% of wing length), much less so in *C. pycnopygius* (37% of wing length). Tail often cocked. Nest an untidy cup, neatly lined, on ground next to rock or in grass tuft. Eggs plain white, or pink with small spots. Songs sound very different, *C. pycnopygius* with a rich warble and *C. frenatus* tuneless (S. Keith, pers. comm.), but have a similar structure ('tip tip tip tootle tootle tootle' and 'teep teep teep prree prree prreee' respectively: Maclean 1993).

When first described, *pycnopygius* was placed in the grass-warbler genus *Sphenoeacus*, with *S. afer**. It was unwittingly redescribed in *Chaetops* in 1869 and was then generally made congeneric with *C. frenatus* until separated into *Achaetops* by Roberts (1922). Thereafter *Achaetops pycnopygius* and *Chaetops frenatus* became increasingly dissociated by taxonomists, culminating in Hall and Moreau (1970) making the former but not the latter consuperspecific with *S. afer* (and *Melocichla mentalis*: see Vol. V, p. 85). Olson (1998) finds no osteological differences apart from size by which '*A*'. *pycnopygius* can be distinguished generically from *C. frenatus*; since we are impressed by plumage resemblances and shared petrophilous and locomotory habits as well as by the osteological similarities, we return *pycnopygius* to *Chaetops*, notwithstanding the considerable difference in tarsal lengths.

Western and eastern populations of *C. frenatus* have generally been treated as separate species, *C. frenatus* and *C. aurantius*. However, tape-recorded song of *aurantius* elicited strong reaction when played back to *frenatus* (details and sonagrams in Dowsett and Dowsett-Lemaire 1993a) so the taxa are surely conspecific.

Endemic. 2 species.

*Songs of *S. afer* and *C. pycnopygius* are similar; nests of *afer*, *pycnopygius* and *frenatus* are alike; and eggshells form a linear series from *frenatus* through *afer* to *pycnopygius* (Clinning and Tarboton 1972). Some skull features of *afer* are more babbler-like than warbler-like (Olson 1998), and *Sphenoeacus* may yet prove to be timaliid, not sylviid (*cf.* Vol. V, p. 87).

Plate 1 (Opp. p. 14)

Chaetops pycnopygius (Sclater and Strickland). Rockrunner; Damara Rockjumper. Chétopse à flancs roux.

Sphenoeacus pycnopygius Sclater and Strickland, 1852. L.A. Co. Mus. Contrib. Orn., p. 148; Erongo Mts and Omaruru R., Damaraland.

Range and Status. Endemic resident. Locally not uncommon Namibia (N Great Namaqualand and Damaraland, Brandberg Mts, Waterberg Mts, Naukluft Mts, north to Opuwo, Owambo and Kaokoveld), and drier coastal areas in SW Angola (Namibe and Benguela and along coast from Capangombe to Catumbella, Iona Nat. Park, also Bocoio, NW Huila). Common in Waterberg Plateau Park, Namibia; Namibian population estimated at *c*. 96,000 birds, with *c*. 7500 in 3 protected areas (Harrison *et al.* 1997, Jarvis and Robertson 1997, *Afr. Birds & Birding* 4 (2) 1999, 56).

Description. *C. p. pycnopygius* (Strickland and Sclater): Namibia and coastal Angola. ADULT ♂: forehead, crown, nape and

Chaetops pycnopygius

scapulars dusky brown, each feather edged cinnamon; mantle and back paler, mantle feathers edged cinnamon; rump and uppertail-coverts tawny. Tail uniform dark brown; when freshly moulted, feather tips are paler. Supercilium white, lores blackish; ear-coverts streaked dusky brown and white; malar stripe white, bordered with black. Chin, throat and breast white, breast profusely spotted with black at sides, less so towards centre; upper belly white, merging into bright tawny on lower belly, flanks and undertail-coverts. Primaries and secondaries dark reddish brown, secondaries edged cinnamon, primaries edged cinnamon on outer webs only. Bill black, ivory white at base of lower mandible; eyes dark brown; legs and feet dark grey to blackish brown. ADULT ♀: like ♂ but tawny on lower belly more extensive. SIZE (10 ♂♂, 10 ♀♀): wing, ♂ 65–69 (67.6), ♀ 64–69 (65.9); tail, ♂ 71–82 (75.7), ♀ 69–78 (73.5); bill to feathers, ♂ 15–17 (16.2), ♀ 15–17 (16.1); tarsus, ♂ 23–25 (24.2), ♀ 23–25 (24.1). WEIGHT: ♂ (n = 2, Oct–Nov) 24, 28; ♀ (n = 2) 28, 30; 1 unsexed 26.5 (Friedmann and Northern 1975).

IMMATURE: like adult but less distinctly marked.

NESTLING: well-grown nestling brown, with black and white streaks on neck and mantle (photo in Steyn 1996). Bill brown, gape pale yellow, eyes brown, legs light brown.

C. p. spadix Clancey: escarpment zone of moist SW highlands of Angola (Huila and adjacent Namibe) at 1150–1900 m. Top of head, hindneck and upper mantle darker than in nominate form, shaft streaks deeper black and feather edges more vinaceous, less buffy; mantle and scapulars deeper, more vinous, brown, with heavier and blacker streaks; rump much darker brown, and innermost tail feathers blacker than in *pycnopygius*. SIZE: wing, ♂ (n = 8) 67.5–70.5 (68.7). WEIGHT: ♂ (n = 7) 26–34 (av. ?), 1 ♀ 30 (Clancey 1972).

Field Characters. Length 23–27 cm. A large, striking warbler-like bird, with streaked head and back, black and white face marks, spotted breast and tawny rump and belly. Tail relatively short, usually cocked. Perches on rocks but more comfortable in cover, where cryptic colouration makes it difficult to see. Restricted to rocky areas in N Namibia and SW Angola. Does not resemble any other bird in its range.

Voice. Tape-recorded (35, 88, 91–99, F, GIB, GIL, WAT). Song a beautiful, clear, liquid, bubbling warble, rich and throaty, preceded by some low notes in an undertone; usually rendered 'tip tip tootle tootle ti tootle tootle too', although that underestimates the number of notes both in the introduction and in the main song. Songs of one bird lasted 1.6–3.7 s (av. of 29 songs, 2.2 s); those of another lasted 2.9–3.6 (av. of 4, 3.2 s). Pauses between consecutive songs range from 5.7 to 12.0 (av. of 16 pauses, 7.6) s. Contact call or one of mild alarm 'hoo-boy' (a soft version of 'tootle too'); alarm, a harsh 'cheerrrrrrrrrr'. Said to mimic other birds (Clinning and Tarboton 1972).

General Habits. Inhabits dry rocky slopes of hills and mountains with scattered thorn bush, particularly along dry watercourses. Usually solitary or in pairs. Easily located by its silhouette when perched on rocky outcrop or exposed branch of shrub, particularly at dawn and dusk. Both sexes sing from call posts in early morning and at dusk, mainly in breeding season. Although not shy when singing, it is easily disturbed, and then either skulks in crevices or flies from boulder to boulder; when disturbed near nest, calls from prominent perch. Does not jump, but runs, mouselike, over rocks and along ledges (Clancey 1966). Often occurs alongside Hartlaub's Francolin *Francolinus hartlaubi*. Sedentary.

Food. Insects: beetles; green caterpillars and grasshoppers given to nestlings; bird seen with what appeared to be a small scorpion in its bill.

Breeding Habits. Solitary nester; territorial.

NEST: a thick, large, untidy walled structure made almost entirely of grass; outer rim made of coarse dry grass blades thinning towards the cup, which is lined with fine soft grass; some birds also add fine rootlets. Entrance to nest well matted to form 'verandah' (Clinning and Tarboton 1972). Nest well concealed and placed close to ground in centre of large tuft of grass *Digitaria dinteri*, or in low large-leaved shrub. Int. diam. 63–76, ext. diam. 89–140, height 89, depth of cup 57, length of 'verandah' 51.

EGGS: 2–3; pale buffy pink, with dark red-brown to light brown spots and underlying slaty blotches, mainly at large end. SIZE: (n = 14) 20.9–22.7 × 15.0–16.3 (21.4 × 15.5).

LAYING DATES: Namibia, Dec–Mar (♂ with large gonads, Okosongomingo, late Nov; Damaraland, ♂♂ in breeding condition, Dec; Brandberg Mts, nestlings mid Apr; Waterberg Mts, nests Mar); Angola (♂♂ in breeding condition, Dec).

DEVELOPMENT AND CARE OF YOUNG: when feeding nestlings, adult approaches and leaves nest along a preferred route, using the same perches on each visit. Young leave nest at a relatively early age and hide and move among rocks and grass like mice.

References
Clancey, P.A. (1972).
Clinning, C.F. and Tarboton, W.R. (1972).
Maclean, G.L. (1993).

Plate.1 (Opp. p. 14)

Chaetops frenatus (Temminck). Rockjumper. Chétopse bridé.

Malurus frenatus Temminck, 1826. Planches Col., livr. 65, pl. 385; River Zonde.

Chaetops frenatus

Range and Status. Endemic resident, South Africa and Lesotho. Rocky slopes and mountain ridges, in S Cape Prov. from Cape Hangklip (where down to sea level) and Somerset West, north through Hottentots Holland to the Cedarberg, east to Knysna, Kougaberge, Sneeuberg, E Griqualand, all Lesotho, W Natal in Drakensberg Range and outliers (where ranges only above 1000 m); Orange Free State, records only in Golden Gate Highlands Nat. Park and at Phuthadijhaba (Witsieshoek). (Records in Transvaal near Carolina and at Loskop Dam require confirmation.) Considerable gap between ranges of the 2 races, centred on 33°S, 29°E (Craig 1991).

Uncommon and local in SW Cape (commonest between Hottentots Holland and Cedarberg), elsewhere fairly common and in Lesotho common, mainly above 2250 m; occurs over 3000 m. One of the commonest birds in Natal alpine zone. Density in Lesotho of 4 birds per km^2 in summer and 1·6 per km^2 in winter; population there estimated to be in range of 10,000–100,000 birds, perhaps one-third of global population (Osborne and Tigar 1990). Density in Katse Basin, Lesotho, of about 1 pair or group per 30 ha (Harrison *et al.* 1997).

Description. *C. f. aurantius* Layard: E Cape from near Graaff-Reinet to Lesotho, Natal and Orange Free State. ADULT ♂ (breeding): forehead, crown, nape, hindneck, sides of neck, mantle, scapulars and back brownish grey, streaked black, the feathers from nape to back rather loose and fluffy, each one black with radiating, separated barbs at the sides grey; forehead to nape a browner grey (almost olivaceous) and finely black-streaked, mantle to back and sides of neck purer grey with black streaks coarser, in acute triangles pointing backwards. Sides of forehead and crown paler, silvery grey with very fine black streaks, forming long superciliary stripe. Lores, thin line over eye, and cheeks black. Ear-coverts dark grey, finely streaked black, merging into black of cheeks. Rump and uppertail-coverts bright foxy orange-brown. Tail above and below glossy black, well rounded, with white 'corners': T1 with tiny white dot or line at tip <1 mm deep (occasionally up to 9 mm deep), which can abrade away; T2 with round 2–4 mm white midline spot near tip or with white end 9 mm deep; white tips 16–20 deep on T3, 18–24 on T4, 23–28 on T5 and 26–30 on T6. T6 19–21 shorter than T1. Chin and throat glossy black, long malar stripe white; moustachial stripe black; breast orange-rufous, shade merging to bright buff on belly and flanks; hind part of flanks dark olivaceous grey; thighs and vent blackish with some overlying buff barbs (all feathers of underparts except chin and throat are lax and fluffy); undertail-coverts buff with black showing through. Overall shade of breast to belly and flanks varies from quite deep orange to rather pale orange-buff. Primaries, secondaries and tertials brownish black, outer primaries and inner secondaries with narrow white tips, tertials with narrow grey outer edge and broad rufous sides towards tips; primaries with concealed large white patch on base of inner vane; P6–7 longest, P8 10 and P9 30 shorter; upperwing-coverts black, alula and greater primary-coverts with white tips up to 10 deep, greater coverts with small white tips, greater primary, median and lesser coverts with barbs radiating, forming white crescents. Underside of remiges grey, paler at base where upperside white; underwing-coverts and axillaries glossy black. Bill black, eyes red-hazel, legs and feet black. Toes long, black, with strong claws.

Lesotho birds have underparts much paler tawny rufous and appear to represent an undescribed race.

ADULT ♂ (non-breeding): grey fringes to feathers from forehead to back wear away, so in worn plumage mantle and back are brown-black, mottled grey, and forehead and crown can be almost uniform black-brown. ADULT ♀: like ♂ but feather fringes from forehead to back olivaceous brown (not grey), tail brownish black, not glossy; lores, cheeks, ear-coverts, moustachial stripe and sides of neck grey-brown, mottled buff; malar stripe pale buff, chin buff, faintly dark-mottled, throat pale grey or buffy or greyish light orange, grey feather bases showing through, predominant throat colour either merging with orange-rufous upper breast (paler than in ♂) or forming gorget well demarcated from buff breast; lower throat, flanks and belly warm buff, paler than in ♂, feathers very fluffy, hind flanks, lower belly, vent, thighs and undertail-coverts duskier than in ♂; wings dark brown where ♂ black. SIZE: wing, ♂ (n = 5) 89–93 (90·2), ♀ (n = 5) 84–89 (85·8); tail, ♂ (n = 5) 86–93 (89·2), ♀ (n = 5) 79–86 (82·0); bill, unsexed (n = 8) 18–23·5; tarsus, ♂ (n = 5) 39–43 (40·6), ♀ (n = 5) 38–41 (39·75). WEIGHT: unknown.

IMMATURE: like adult but less strongly marked.

NESTLING: hatchling with pink skin and tufts of grey down; gape pale yellow, inside mouth bright yellow; later, whole upperside densely covered with slate grey down (photo in Steyn 1996).

C. f. frenatus (Temminck): SW Cape east to about Graaff-Reinet. Like *aurantius* but breast to belly uniform deep cinnamon in ♂, deep orange-cinnamon in ♀ with sparse small black arrowhead marks on breast. ♂ undertail-coverts black with 2 mm whitish tips. Eyes orange-red (♂) or hazel-red (♀). White marks at tip of T6 24 deep, T5 23, T4 19, T3 13, T2 3 deep; T1 all black. SIZE: (♂♀) wing 90–96, tail 100–115, bill 20–21, tarsus 38–40.

Field Characters. Length 21–22 cm in E of range, and 23–25 cm in W. ♂ has grey head, black throat and prominent white malar stripe; underparts and rump dark red in nominate race, orange in *aurantius*. ♀ lacks all striking colours except orange rump, has streaky greyish head and back, pale supercilium and malar stripe; breast to belly deep orange-cinnamon in nominate race, warm buff in *aurantius*. Rock-hopping behaviour, striking plumage of ♂ and white-tipped black tail of ♀ ensure they cannot be mistaken for any other species. Rock-thrushes *Monticola* in same habitat have short orange-and-brown tails and unstriped heads (blue in ♂♂), perch upright on rocks.

Voice. Tape-recorded (88, 91–99, F, GIB, GIL, LEM). Song a series of piping notes, quite lengthy and variable; bird takes a phrase and repeats it 3–6 times, 'pseeu-pseeu-psut', 'pew-wee-tsertsertser', a faster 'weeweeweewee ...', a more liquid 'plee-pleeu-pli', or thin 'tsi-tsi-tsi tirrrr', the last part with a watch-winding quality. Another song (call?) is a 5-s series of high, tuneless, partly trilled or rolling notes slightly descending the scale and decelerating, 'psi-psi-psi-psi-psee-psee-psee-psee-pseew-pseew-pseew-pseew' or 'fiurr, fiurr, fiurr', "like alarm clock running down" (Maclean 1993); motifs sometimes disyllabic, 'peeurree peeurree peeurree' (sonagrams in Dowsett and Dowsett-Lemaire 1993a). Birds also sing in chorus, rather tuneless and chattery repeated phrases, 'tsi-tsi-tsuwi', 'psuwee-psuwee-psuwee', 'tirree-tirree-tirree'. Voices of both races very similar. Said to have 'variety of loud piping alarm, anxiety or contact calls, 'peeurip-tri tri-trip chi-treee-prip', etc.' (Maclean 1993).

General Habits. A bird of high rainfall, fynbos-covered mountains with plenty of bare rock; in Lesotho inhabits mainly steep, rocky slopes, low escarpments, screes, rocky gullies and passes with grass and small shrubs; tends to avoid upland macchia areas lacking boulders, and to shun flat ground. Commonest in mountain fynbos near mountaintops; ranges below 1000 m only in SW Cape (where breeds down to the coast at Pringle Bay) but in Lesotho mainly above 2250 m; in June occurs up to >3000 m, when night temperatures fall to −12°C, streams freeze and there is 5% snow cover. In Natal alpine belt regularly occurs down to 2440 m in summer and to 2075 m in winter (Brown and Barnes 1984); in Lesotho probably moves to lower altitudes when snow cover is thick (Osborne and Tigar 1990).

Lives in pairs or small family parties and forms small flocks in winter (R. Prŷs-Jones, pers. comm.). Flock size smaller at lower altitude (Hockey *et al.* 1989). Perches on top of boulder, tail often a little cocked; hops, bounds agilely or flies from rock to rock, examining fissures and crannies for food; often disappears between rocks to come into the open again some distance away. Forages on ground, pecking and scratching soil surface and vegetable debris, digging out insects and their larvae and pupae (W. R. Tarboton in Ginn *et al.* 1989). Wary; vocal when disturbed, and when near nest or with young. Calls are far-carrying, given frequently; when bird excited, calls accompanied by tail fanning. Runs quickly over rock surface and ground, sometimes with tail half cocked. Flies with alternating fluttering and gliding; when disturbed flight low and heavy, but flies and glides well; takes cover among boulders, peering out at intruder.

Food. Insects, larvae and pupae; includes grasshoppers, moths and caterpillars; many grasshoppers fed to young (Tait 1948); occasionally lizards including a gecko (photo P. Steyn, 1994, *Promerops* 213, 8).

Breeding Habits. Solitary nester, monogamous, but ♀ often accompanied at the nest by 2 ♂♂ (Martin 1964, Craig 1991 and R. Prŷs-Jones, pers. comm.).

NEST: an untidy cup or bowl of coarse grass, rushes, moss, lichens and twigs, neatly lined with fine grass, roots, reddish brown *Protea* seeds and sheep and goat hair. Sited on ground, usually under boulder or detached slab or against a rock, occasionally in clump of palmiet vegetation, generally concealed under dense grass and low shrubs.

EGGS: 2–4, usually 2. Plain white. SIZE: (*C. f. frenatus*, n = 4) 25·4–26·4 × 19·6–19·8 (26·1 × 19·7), (*aurantius*, n = 10) 24·2–27·9 × 18·6–20·2 (26·4 × 19·4) and (n = 4) *c*. 29 × 20.

LAYING DATES: Cape, Sept–Nov; Natal, Oct–Dec; Lesotho, May (and nestlings and fledglings Dec and food-carrying Jan).

DEVELOPMENT AND CARE OF YOUNG: both sexes feed the young and remove faecal sacs from nest. Parents go to and from nest by a regular route and are confiding to observer (Steyn 1996).

References
Ginn, P.J. *et al.* (1989).
Maclean, G.L. (1993).
Osborne, P.E. and Tigar, B.J. (1990).
Tait, I.C. (1948).

Genus *Illadopsis* Heine

Small, shy babblers of forest undergrowth, with dull, dark plumage and semi-terrestrial habits. Bill variable, from small and weakly hooked to rather robust and well hooked; rictal bristles weakly to moderately well developed; tail relatively long. Plumage thick and fluffy; upperparts dull to rufous-brown or olive, one species with contrasting black cap; breast and flanks washed heavily with grey or olive, often contrasting with whitish throat. Nest a loose cup of dead leaves placed on or near the ground. Vocalizations vary from duets and group chorusing (whistled song accompanied by group chatter) in *I. rufipennis*, *I. fulvescens* and *I. pyrrhoptera*, and (though chatter much less marked) *I. puveli*, to unaccompanied ringing whistles (*I. cleaveri*/*I. albipectus*) and a single endlessly repeated phrase (*I. rufescens*).

Endemic, 7 species. *Illadopsis* was submerged in *Malacocincla* by Delacour (1946) and in *Trichastoma* by Mayr and Paynter (1964). However, the distribution of shared characters between *Illadopsis* and Asian *Trichastoma* is irregular, and similarities may be due to convergence (Ripley and Beehler 1985), and we follow these authors in considering *Illadopsis* distinct. *I. cleaveri* and *I. albipectus* form a superspecies; they are parapatric in Zaïre, and their voices are so similar that Chappuis (1975) thought they might even be conspecific. We see no reason to place *I. rufipennis* and *I. pyrrhoptera* in a superspecies, as was done by Hall and Moreau (1970); their plumage and voices are far more divergent than are those of *I. rufipennis* and *I. fulvescens*, which are closely sympatric.

Illadopsis albipectus superspecies

1 *I. cleaveri*
2 *I. albipectus*

Plate 2
(Opp. p. 15)

Illadopsis rufipennis (Sharpe). Pale-breasted Illadopsis. Akalat à poitrine blanche.

Trichastoma rufipennis Sharpe, 1872. Ann. Mag. Nat. Hist., ser.4, 10, p. 451; Gabon.

Range and Status. Endemic resident. Sierra Leone (fairly common), SE Guinea (Nzérékoré), Liberia (common), Ivory Coast north to Taï Nat. Park (common), Nimba and Yapo Forest (uncommon); Ghana, not uncommon from coast north to Kakum Nat. Park, Goaso, Mampong and Tafo; Nigeria, not uncommon in forest zone and absent north of it; S Cameroon, north and west to Mt Cameroon (to 850 m), Rumpi Hills (to 1050 m) and Korup Nat. Park (where one of the commonest birds in lowland forest understorey, uncommon above 1000 m: Rodewald et al. 1994); Bioko; Mbini; Gabon, common; Congo, common in Odzala and Nouabalé-Ndoki Nat. Parks, rare in S (Mayombe, Île M'Bamou near Brazzaville); Angola, Cabinda (only); birds from Canzele, shown in Hall and Moreau (1970), are referable to *I. albipectus* (Traylor 1962); SW and SE Central African Republic (Dzanga Res., Ouossi R.); Zaïre, known from Lukolela and lowland forests of E half of Congo basin but not yet from in between; extends to bases of Albertine Rift Mts, up to 1550 m west of L. Edward (Lutunguru), 1800 m in Itombwe (Ibachilo); not in Sudan (Nikolaus 1987); forests of W and S Uganda, from Budongo to Kalinzu, Kasyoha-Kitomi, Lugalambo and Malabigambo and east to Mabira and south into NW Tanzania at Bukoba; W Kenya in Kakamega and S Nandi forests, Trans-Mara Forest at 2240 m, and on Tanzania border at El Doinyo Orok; (single record from Cheranganis at 2300 m is probably referable to Mountain Illadopsis: Zimmerman et al. 1996). Tanzania in Usambaras (to 1200 m), Pugu Hills (rare), Ngurus, Ulugurus at 300–900 m, and Udzungwas; Zanzibar. Generally common, locally uncommon; race *T. r. puguensis* 'cannot be far from extinction' (Stuart and Van der Willigen 1979). Density in Gabon 12 pairs per km^2.

Illadopsis rufipennis

Description. *I. r. rufipennis* (Sharpe) (including '*bocagei*'): Nigeria to Angola, Kenya (except S) and NW Tanzania. ADULT ♂: top of head dark greyish olive, darker feather edges giving a scaled effect; rest of upperparts olive-brown with rufous tinge, strongest on rump and uppertail-coverts. Tail dark brown with slight russet tinge. Lores grey-brown, superciliary stripes faint, greyish; cheeks and ear-coverts olive-grey with paler streaks, sides of neck pale olive-brown. Chin and throat whitish, upper breast pale buffish brown, lower breast and belly whiter, sides of breast, flanks, thighs and undertail-coverts brown. Primaries, secondaries and primary coverts rich brown on outer webs, dark brown on inner webs; tertials and upperwing-coverts rich brown, like upperparts. Underwing-coverts and axillaries pale buffish-brown. Bill blackish above, grey, pale horn or blue-grey below; eyes light brown to russet orange or rufous, almost buff around pupil (Zimmerman 1972); legs brownish grey, blue-grey, dull purplish blue, or purplish grey. Sexes alike. SIZE (12 ♂♂, 10 ♀♀): wing, ♂ 71–78 (74·3), ♀ 68–74 (69·7); tail, ♂ 54–61 (56·6), ♀ 49–57 (52·7); bill, ♂ 17–20 (18·6), ♀ 16–18 (17·4); tarsus, ♂ 25–26 (25·5), ♀ 24–26 (24·4). WEIGHT: Central African Republic, ♂ (n = 3) 26–30 (28); Uganda, unsexed (Budongo, n = 25) av. 24 (J. Lindsell, pers. comm.); Kenya, ♂ (n = 3) 18–21 (19·7), ♀ (n = 4) 18–23 (20·0), unsexed (n = 21) 18–27 (22·2); unsexed (Kakamega, n = 24) 22·36 ± 3·51. Can change weight daily up to 18% (av. of 5 birds 14·3) (Mann 1985).

IMMATURE: see *I. r. distans*.

I. r. extrema (Bates): Sierra Leone to Ghana. Somewhat more rufous (less olive) than nominate race, top and sides of head rich brown, not grey, concolorous with mantle. WEIGHT: Sierra Leone, unsexed (n = 14) 20·8–30 (24·9); Liberia, ♂ (n = 13) 22–28 (25·0), (n = 16) 19–25 (22·0).

I. r. distans (Friedmann): S Kenya (Trans-Mara Forest, Ol Doinyo Orok), eastern arc mts of Tanzania, and Zanzibar. Like nominate race but breast-band pale grey rather than brownish; more olive above, top of head dull brown. Bill larger. WEIGHT: ♂ (n = 4) 27–28 (27·3), ♀ (n = 6) 25–29 (26·6).

JUVENILE: more rufous-brown above than adult, mantle and scapular feathers with long creamy-buff shaft streaks; creamy buff spots on sides of crown; downy feathering below: throat whitish, pale rufous brown on breast and flanks, large dusky brown spots across breast and diffuse dark blotches on flanks; centre of lower breast and belly creamy-white; creamy-buff streaks on lesser and median coverts. Base of lower mandible and gape yellow to orange, eyes dark brown to pale olive-brown or greyish tan, legs yellowish white to whitish flesh.

I. r. puguensis Grant and Praed: Pugu Hills, E Tanzania. Like *distans* but paler brown.

Field Characters. Length 12·5–14 cm. The least distinctive and most widespread illadopsis, partly sympatric with all the others. Brown above with lighter brown flanks and breast-band; white throat more contrasting than Brown Illadopsis *I. fulvescens* and more often puffed out (Uganda: J. Lindsell, pers. comm.). Belly white (often hard to see, especially in forest undergrowth), and in side view bird looks almost identical to Brown Illadopsis, which has brown belly. Pale-breasted is smaller and shorter-tailed, and head often has domed appearance whereas Brown is flatter crowned (J. Lindsell, pers. comm.); there is greater contrast between upper and undersides than in Brown (Upper Guinea: Allport *et al.* 1996). In the hand, combination of weight, short dark tarsus, dark bases to throat feathers and long fine rictal bristles make this species distinctive (J. Lindsell, pers. comm.). Immature Pale-breasted shows suggestion of dark malar streak and rusty brown wings, strongly suggesting Brown Illadopsis, and immature Brown has some white on belly (Zimmerman 1972). Also very similar to those Scaly-breasted Illadopsises *I. albipectus* that lack scaling on underparts; pale brown breast-band more sharply defined than pale grey breast of Scaly-breasted, legs usually darker and shorter. In W Africa overlaps with Puvel's and Rufous-winged Illadopsises *I. puveli* and *I. rufescens*, which are larger, with whiter underparts contrasting more with upperparts, longer and stronger legs. Mountain Illadopsis *I. pyrrhoptera* is dark grey below.

Voice. Tape-recorded (53, 104, B, C, KEI, LEM, PER, STJ, SVEN, WOOD, ZIM). Song given singly or as a duet or trio. Very variable, and easily confused with songs of Scaly-breasted (Chapin 1953), Black-capped *I. cleaveri* and Brown Illadopsises. In Gabon, 1 bird perched 1–2 m above ground and gave single long pure whistle on one pitch, not unlike that of Brown Illadopsis but less powerful, while 2 others perched within 1 m of each other a few cm above ground giving a continuous low chatter which frequently broke into louder notes ascending the scale, 'tuk-tuk-tuk-wik-wik-wik-week-week-week-week' (Chappuis 1975). In Nigeria, singer gave 1 short and 1 or 2 long whistles ascending the scale one tone apart, the third after a slight pause, 'twip ... tyooo ... weep', sometimes preceded by a brief 'wip': 'wip ... tyoo ... tweee'; accompanying bird(s) gave low 'tuktuktuktuk ...'. Whistles of one solo singer were more slurred, with slight pause after first, 'tyooo ... toowee ... toowee'. In Liberia, 2–3 notes run together into a single long whistle, preceded by chip, 'chip ... heeeee' or 'chip ... heeeewi'; second bird gives low 'charrr'. Whistles of Tanzanian birds richer, sweeter and more thrush-like, usually ending with upslurred whistle, 'tyooeep ... woy-whee', 'wee-tyoo-woy-twee', 'chip-taw-whee ... woy-whee', 'tyu-too-tooy'; described as a human-like whistle, a slow,

Plate 1

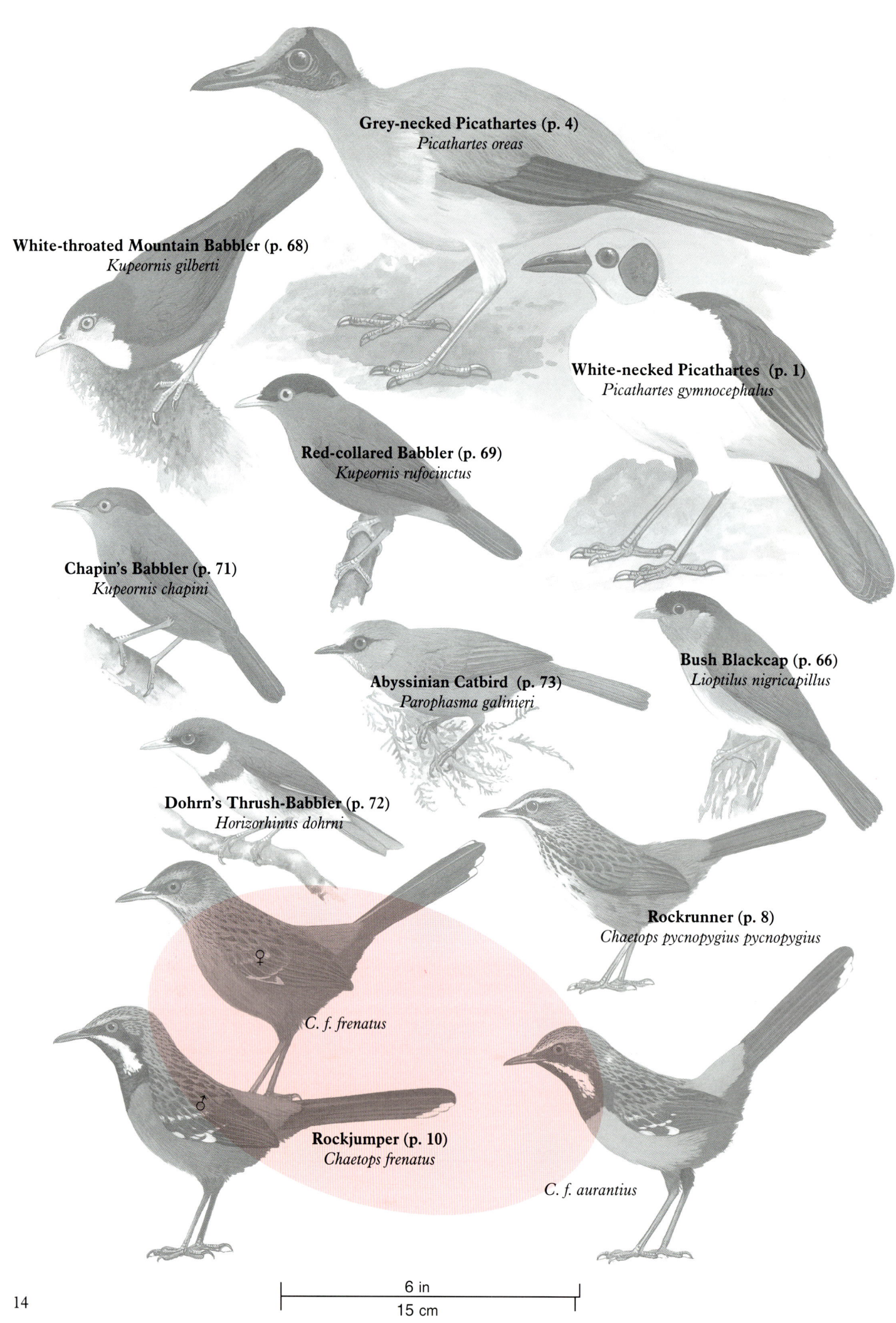

Plate 2

Grey-chested Babbler (p. 28)
Kakamega poliothorax

African Hill Babbler (p. 32)
Pseudoalcippe abyssinica

P. a. abyssinica
P. a. stierlingi
P. a. atriceps
P. a. monacha

Mountain Illadopsis (p. 20)
Illadopsis pyrrhoptera

Spotted Thrush-Babbler (p. 34)
Ptyrticus turdinus harterti

Brown Illadopsis (p. 17)
Illadopsis fulvescens

I. f. iboensis
I. f. ugandae
I. f. fulvescens

Blackcap Illadopsis (p. 23)
Illadopsis cleaveri

I. c. batesi
I. c. cleaveri

Puvel's Illadopsis (p. 26)
Illadopsis puveli puveli

Rufous-winged Illadopsis (p. 25)
Illadopsis rufescens

Scaly-breasted Illadopsis (p. 21)
Illadopsis albipectus

Well marked form (Kakamega)
Poorly marked form

Pale-breasted Illadopsis (p. 12)
Illadopsis rufipennis

I. r. puguensis
I. r. rufipennis

6 in
15 cm

meditative 'hoooit-hooooee', which may be answered in a similar tone by another bird in some dense patch of undergrowth (Sclater and Moreau 1932). Differences suggest these populations have been isolated for a long time (Chappuis 1975) and may even be different species (Zimmerman *et al.* 1996); however, accompanying birds of both groups give similar low, grating chatter (Chappuis 1975). Birds in foraging flock gave low, nasal 'tyank-tyank-tyank ...' (J.O. Svendsen and L. Hansen, pers. comm.) and much conversational churring. Alarm a grating 'ka-a-a-a' followed by throaty 'kwo-kwo' (*distans*: Sclater and Moreau 1932).

General Habits. Inhabits lowland, transition and montane primary and old secondary forest; thickets in secondary forest along roads and paths, dense bush in clearings in lowland and montane forest up to 800 m, also logged forest, farmbush and forestry plantations (Sierra Leone), swamp and dryland forest under closed canopy (Congo, Odzala Nat. Park: Dowsett-Lemaire 1997a). In Liberia, forest patches in grassland mosaic, often in damp places with large-leaved Marantaceae and Zingiberaceae (Gatter 1997). Bioko, moss forest, up to 1000 m; in Gabon, micro-habitat almost identical to that of Brown Illadopsis; if anything prefers even denser vegetation.

Lives in pairs or small family parties; noisy parties of 3–7 regular in monospecific groups or within mixed bird parties (Liberia: Gatter 1997). Usually remains <5 m above ground, occasionally up to 12 m. Forages on branches of trees, trunks of saplings, among lianas and Marantaceae, looking for dead leaves, small branches and other vegetable debris piled up among living leaves and branches and forming a kind of compost, into which it pokes its head; also searches through bunches of dead leaves still in place and in litter in epiphytes. More prone than Brown Illadopsis to forage among rotting vegetation on ground, turning over debris. Members of pair maintain contact with repeated nasal calls. Follows army ant invasions; often in company of squirrel *Funisciurus lemniscatus*. More sedentary than other illadopsis spp.: 4 ringed birds recaptured at same spots after 2, 7 and 14 months; 2 individuals netted together were recaptured together 12 months later (Gabon: Brosset and Erard 1986). In Gabon sings all year, somewhat less during long dry season. Puffs out white throat when alarmed or inquisitive (Allport *et al.* 1996).

Food. Invertebrates, mainly insects: moths, caterpillars, beetles, Hymenoptera, pentatomid bugs, elaterid larvae, ants, crickets, insect eggs, woodlice, earwigs, centipedes, spiders and small snails.

Breeding Habits. Territorial; defends territory by singing. Perhaps a co-operative breeder. In aggressive reaction, e.g. when 2 groups meet or after playback of song, several birds may sing together, so social group may include several adults (Brosset and Erard 1986).

NEST: 2 quite different types described (both of nominate race): (a) 9 nests (Gabon) were large and loose, looking like clump of dead leaves, placed on thick base of dead leaves, some skeletonized, rolled up to form a rather deep and narrow cup (**A**), lined with rootlets, dead leaves and strands of *Marasmius* fungus; ext. diam. 90, int. diam. 50, depth of cup 45; somewhat smaller, more compact and deeper than sympatric congeners, and placed somewhat higher, 0·8–2·5 m above ground in leafiest part of isolated bush on forest floor; one was 1·4 m up in base of large unopened leaves of a fleshy plant in a periodically flooded area (Brosset and Erard 1974); (b) 2 nests (Itombwe, Zaïre) described as rudimentary, flat cup of dark rootlets, twigs and dry leaves with an outer layer of moss, ext. diam. 60–80, ext. depth 30, placed 0·3 m above ground (Prigogine 1971, 1984). The first of these 2 nests (b) was said to resemble that of Green-tailed Bristlebill *Bleda eximia* (Brosset and Erard 1974), the second was more certainly identified (♂ taken on nest). Nest of *I. r. distans* said to be deep cup of dead leaves, ext. diam. 80, ext. depth 70, lined with *Marasmius* fungus, well hidden in leafy stump 1 m high (Moreau and Moreau 1937).

A

EGGS: 2, laid at 1–2 day intervals. In Gabon, bright blue-green, with strongly contrasting small spots of red-brown or blackish brown. SIZE: av. 20 × 16. In Zaïre (Itombwe) eggs beige, heavily spotted with brown and red-brown, especially around large end; SIZE: 22·8–23·1 × 16·4–17·0. Eggs of race *distans* greyish white speckled or blotched with red-brown; SIZE: (n = 2) 23·2–23·6 × 15·8–16·0 (Sclater and Moreau 1932).

LAYING DATES: Liberia, June, Aug (breeding condition Mar–Apr, July, Sept); Ghana, Mar–Apr; Nigeria (breeding condition June, Aug, Oct–Nov); Cameroon (breeding condition Mar–Aug, Oct–Nov); Gabon, mainly Dec–Feb, during short dry season; also July–Aug; Central African Republic (breeding condition June); Zaïre, records in nearly every month (Chapin 1953, Prigogine 1971); E Africa: Kenya (breeding condition Mar); Tanzania, Dec, Region B Feb, May, Dec, mainly in the rains.

INCUBATION: begins with second egg; by ♀ only. Period: 14 days. When disturbed, incubating bird may give 'rodent-run' distraction display, falling to the ground, running back and forth with hunched back, rustling the leaves (Brosset and Erard 1974).

DEVELOPMENT AND CARE OF YOUNG: young fed by both adults and one or two immatures, both in nest and after fledging. 2 adults fed young 14 times between 10h25 and 14h25, mainly with caterpillars, also other larvae and small crickets (A. Devez, *in* Brosset and Erard 1986). Young remain in nest 10 days; after leaving nest, brood defended by 'rodent-run' distraction display.

BREEDING SUCCESS/SURVIVAL: 7 out of 9 nests destroyed by predators; one parasitized by cuckoo, probably Olive Long-tailed Cuckoo *Cercococcyx olivinus*. 1 bird lived for 2 years 5 months.

Reference
Brosset, A. and Erard, C. (1974, 1986).

Illadopsis fulvescens (Cassin). Brown Illadopsis. Akalat brun.

Turdirostris fulvescens Cassin, 1859. Proc. Acad. Nat. Sci. Philadelphia, 11, p. 54; Cama River, Gabon.

Plate 2
(Opp. p. 15)

Range and Status. Endemic resident. Senegal (Basse Casamance), Guinea-Bissau (Contabane, 11°32′N, 14°43′W: Rodwell 1996); Guinea in SW (Kounounkan) and SE (Ziama, common); Mali (rare, valley of upper Bafing R.); Sierra Leone, Liberia (common throughout), Ivory Coast north to Bouaké, Comoé and Taï (but not in Taï Nat. Park: Gartshore *et al.* 1995), common in Yapo Forest; Ghana in forest zone (not uncommon) and coastal thicket (more common); S Togo (not uncommon: Cheke and Walsh 1996), S Benin, S Nigeria north to Ibadan and Enugu, also Kagoro; S Cameroon north to *c.* 05°14N, and in NW to Korup Nat. Park, Rumpi Hills and Mt Kupé; Mbini, Gabon (common to abundant), Congo, widespread and common from coast north to Brazzaville, and common in N at Odzala and Nouabalé-Ndoki Nat. Parks (Dowsett-Lemaire 1997a,b). NW and NE Angola; Central African Republic, many localities in SW (Germain and Cornet 1994) including Dzangha Res. (*contra* Green and Carroll 1991; recent specimens in American Museum of Natural History), also in SE at Rafai and Owossi R. Zaïre, in forests of Congo basin, where it is the commonest illadopsis (Chapin 1953), from northern borders south at least to Kasai and Maniema, east to base of Albertine Rift mts, to 1550 m at Lutunguru (east of L. Edward), 1450 m at Mt Nyombe in Itombwe; Burundi (Musigati: Gaugris *et al.* 1981); extreme S Sudan; common in forests of Uganda, 700–1800 m, where it is the most widespread illadopsis, from Budongo, Bugoma and Bwamba to Bwindi (Impenetrable) and Malabigambo, east to Mabira, Buvuma I. and Bukedi near Kenya border; W Tanzania in Kabogo Forest and further north on shore of L. Tanganyika, also in extreme NW; W Kenya in Kakamega and Malaba forests (below 1600 m). Density in Gabon, 10–12 pairs per km².

Description. *I. f. fulvescens* (*Cassin*): Cameroon (Mt Cameroon and east) to Congo and W Zaïre; intergrades with *iboensis*. ADULT ♂: upperparts olive-brown, forehead to nape tinged greyish, and with dark grey feather fringes producing faint scaly appearance, back and rump tinged rufous, uppertail-coverts more strongly rufous. Tail brown with rufous tinge. Lores and ear-coverts dark brown; superciliary stripe indistinct, greyish; cheeks dark brownish grey; sides of neck pale olive-brown. Chin and throat buffish white with a few diffuse dark grey streaks; rest of underparts buffish brown, paler on centre of breast and belly, darker on sides of breast, flanks and undertail-coverts; thighs grey. Primaries, secondaries and primary coverts olive-brown on outer webs, dark brown on inner webs; tertials and rest of upperwing-coverts olive-brown like upperparts. Underwing-coverts and axillaries pale brownish buff. Bill blackish brown above, pale grey or bluish grey below; eyes pale brown to reddish brown; legs dusky grey to pale blue-grey. Sexes alike. SIZE (10 ♂♂, 6 ♀♀): wing, ♂ 74–78 (76·1), ♀ 65–71 (68·5); tail, ♂ 65–68 (67·5), ♀ 58–64 (60·6); bill, ♂ 20–23 (21·1), ♀ 19–21 (19·9); tarsus, ♂ 24–26 (25·0), ♀ 22–24 (23·2). WEIGHT: Gabon, unsexed (n = 6) 23–29 (24·8).

IMMATURE: top of head browner than adult, less grey; upperparts mainly tawny, olive only on upper mantle; edges of wing-coverts, tertials and secondaries darker tawny, almost chestnut, much redder in tone than olive-brown of adult; tawny wash on breast, flanks and undertail-coverts.

JUVENILE: similar to immature but almost rufous on back and wings; downy feathering below; eyes grey; bill yellowish brown; legs whitish. 2 very young, almost tail-less birds (mainly feathered above, downy below) have top of head uniform dull dark grey-brown, no dark feather edges, upperparts and tail dull chestnut, rump tawny, paler than back; upperwing-coverts, secondaries and tertials dark chestnut edged brighter chestnut. Cheeks grey-brown with pale shaft-streaks, partly-feathered throat whitish, rest of underparts covered with dull brown down, somewhat brighter on lower flanks and undertail-coverts; centre of lower breast and belly whitish. Bill blackish with light yellow

Illadopsis fulvescens

at tip, gape yellow, eyes greyish brown, legs rather dark grey, with olive tinge.

I. f. ugandae (van Someren): central and E Zaïre to Sudan, Kenya and Tanzania. Chin and throat plain whitish, unstreaked or almost so, contrasting more strongly with dark grey cheeks; centre of lower breast and belly whiter. Eyes bright brown or orange-brown. WEIGHT: Central African Republic, ♂ (n = 4) 31–38 (34·7), ♀ (n = 4) 28–30·5 (29); Uganda, unsexed (Budongo Forest, n = 60) 20–37 (27·9); Kenya, unsexed (n = 13) 31·45 ± 2·73.

I. f. dilutior (White): Angola. Like *ugandae* but whitish throat faintly streaked; upper breast and flanks brighter buff.

I. f. gularis (Sharpe): Senegal to W Ghana. Underparts slightly more rufous than in nominate race; sides and top of head browner; little or no streaking on throat (like *ugandae*). WEIGHT: Liberia, ♂ (n = 10) 29–37 (33·2), ♀ (n = 5) 28–33 (30·2).

I. f. maloneyana (Sharpe): E Ghana and Togo. Throat brown like *iboensis* but upperparts more rufous, more tawny-rufous below.

I. f. iboensis (Herbert): Nigeria, intergrading with nominate race in Cameroon west of Mt Cameroon. Chin to breast tawny-ochre with no streaking, grading to tawny-brown on rest of underparts; upperparts like *gularis*. WEIGHT: Nigeria, unsexed (n = 6) 22·6–32·1 (27·9).

Field Characters. Length 14–15 cm. Extremely similar to Pale-breasted Illadopsis *I. rufipennis* with which it frequently occurs, even in same flock. Brown distinguished by warmer, tawny tone of underparts, especially belly, contrast between grey face and brown crown, flatter, less rounded crown and longer bill (J. Lindsell, pers. comm.). Contrasting white throat common to both species, except throat brown in races *I. f. iboensis* and *moloneyana*. Whistled songs are similar and can only be distinguished with experience, but 'dict-a-phone' call of Brown is characteristic, not given by Pale-breasted. Dumpier shape, smaller size and shorter and weaker-looking legs distinguish them from Rufous-winged and Puvel's Illadopsises, which have entire underparts whitish, contrasting with upperparts (Allport *et al.* 1996). In Kenya said to be longer-tailed and larger-billed than its relatives (Zimmerman *et al.* 1996) and to have dusky lower edge of grey cheeks and ear-coverts, forming weak malar stripe; lack of scales on breast separates it from local Scaly-breasted Illadopsis *I. albipectus*. In the hand, bases of throat feathers of Brown Illadopsis white (dark in Pale-breasted)

Voice. Tape-recorded (32, 53, 104, B, C, F, GAU, GRI, KEI, MOR, MOY, ZIM). Song usually a duet or trio, although one bird may sing alone. Principal singer gives a short, upslurred 'weep' followed by a longer, drawn-out, ringing whistle, usually at a lower (sometimes higher) pitch, '*weep ... hooooo*' or '*wheat ... germ*'; other versions are 'tip ... tyop ... waaaaa' (Ghana); 'tip ... tyop ... waaaah ... tip ... wooooo ... tyop ... waaah' (Gabon); 'chop ... weeeeh ... chop ... waaaah', 'chip ... chop ... hoooo', or '*weep ... weep-chi-waaaah*', sometimes followed by 'heee ... hawww' (Uganda); a single, rather weak 'waaaah', 'cha-waaaah' or 'wik ... chaaaah' (Kenya). Slow and deliberate, notes well separated; 2-note version lasts 2 s, 3-note version 3 s. Intervals between songs 2–3 s. Singers from Gabon to Kenya are accompanied by 2 or more birds with nasal 'jip ... jap*paaangh*', the last note twangy and very characteristic; originally rendered 'dict-a-phone' by Chapin (1953); this accompaniment now known (from tape-recorded and collected birds) to be made by 2 birds, one giving the 'dict' and the other the 'taphone' (Zimmerman *et al.* 1996); these authors suggest the second bird, perched nearby, is the presumed mate, but in that case who is giving the whistle which goes with the 'dict-a-phone'? At least 3 birds take part in this performance, and the whistler at least may be presumed to be a ♂. Singers in W Africa (Liberia, Ghana, Benin, races *gularis* and *moloneyana*) are accompanied by a rapid chatter on one pitch, 'gih-gih-ger-ger-ger-ger ...', rather than by 'dict-a-phone'. Song similar to Pale-breasted but louder, and whistles change in pitch, alternating ups and downs (J. Lindsell, pers. comm.). Contact calls from foraging group 'chip', 'chup', 'chippa', 'chrrr', 'jhreee-jajajaja'; around ants gives a short 'chiii' or 'cheu' (Willis 1983); young bird still in soft juvenile plumage repeated a triple nasal note louder than that of any adult (Chapin 1953). Alarm, a nasal 'tchaa' or 'chwaa' (Zimmerman *et al.* 1996).

General Habits. Inhabits thickets and vine tangles in lowland primary and secondary forest; also transition forest (Itombwe), farmbush, forestry plantation and logged forest (Sierra Leone: Gola), gallery forest in savanna (Liberia), coffee plantations (Guinea: Ziama), understorey of dryland forest, even in pure Marantaceae (Congo, Odzala Nat. Park: Dowsett-Lemaire 1997a); Zaïre, forest only, not clearings, gardens or banana groves (Chapin 1953). In Gabon, abundant around recent tree falls, among tangle of branches fallen from crowns of trees, and equally around older falls where gaps and clearings are in the process of filling in; also frequents parts of forest regularly battered by storms, and liana tangles in undergrowth and dense brush beside streams (Brosset and Erard 1986). Lower and middle levels, from 0·5 to 18 m above ground, typically 4–12 m, rarely higher except in long dry season when it reaches vine tangles in the canopy (Gabon: Brosset and Erard 1986). When in mixed flock with Pale-breasted Illadopsis, keeps higher than it, at 10–16 m. Lowlands, up to 1300 m in Cameroon, 1550 m in Zaïre, 1800 m in Uganda.

Crepuscular, singing and moving about mainly at sunrise and sunset in small groups, typically 2 adults and young of previous brood, sometimes parties of 4–6. Group members remain several m apart when travelling, sometimes more, but keep in contact with frequent rolling and grating calls. Regular member of mixed-species flocks; in Uganda (Budongo) commonly with White-throated Greenbul *Phyllastrephus albigularis*, Olive-green Camaroptera *Camaroptera chloronota* and Red-cheeked Wattle-eye *Dyaphorophyia blissetti jamesoni* (J. Lindsell, pers. comm.); often in company of squirrel *Funisciurus lemniscatus* (Gabon; Brosset and Erard 1974). Joins groups of insectivores around ant swarms, but diet is not restricted to ants. Pairs or small families wandered near 4 raiding parties of *Dorylus wilverthi* colony at Makokou (Gabon) for up to 75 min, mostly hopping on logs and low rank herbs 0·1–2 m up and peering in dense debris, dead or green leaves, stubs and treefalls. Resembles wrens Troglodytidae in pecking rather than sallying for prey, and is therefore restricted to dense foliage or litter (Willis 1983). Habitually works its way up liana tangles around tree trunks to *c.* 8 m, then flies down to next tangle (Uganda, Budongo: J. Lindsell, pers. comm.); gleans lianas, searching among interlaced stems and dead leaves still in place; strongly attracted to live brown leaves remaining among green leaves (Brosset and Erard 1986). Pries into bunches of dead leaves, especially those piled up at bases of large leaves of ground vegetation, but does not get as far into them as Pale-breasted Illadopsis. Inserts its beak, and sometimes the whole head, into rolled-up dead leaves. Extremely worn outer primaries in old plumage may indicate habitual use of forest floor in dense vegetation (Gatter 1997); however, never seen to forage on ground in Budongo Forest (J. Lindsell, pers. comm.), and said to hop and flit from bush to bush rather than feeding right on ground (Chapin 1953). Ringed birds seldom recaptured at same spot (of 27 ringed, 3 retaken after 24 h, 1 after 7 months, 1 after 11 months). Inquisitive as well as secretive and suspicious; joins birds mobbing snakes. When excited, puffs out throat feathers and displays white moustaches, like Yellow-whiskered Greenbul *Andropadus latirostris* (Brosset and Erard 1986). In Gabon sings all year, even at height of long dry season. In Zaïre, first bird to call at dawn (Chapin 1953).

Food. Arthropods: small beetles, Hemiptera, Orthoptera, moths, caterpillars, ants, termites (mostly winged), insect eggs; spiders, millipedes, centipedes up to 7 cm long. Also a few seeds.

Breeding Habits. Territorial; chases rivals along branches and through foliage. Singing male regularly gave food to his mate during breeding season (Brosset and Erard 1986).

NEST: in Gabon, large and loosely put together, making it look like a bunch of dead leaves; built upon thick base of dry leaves, composed of dry leaves, some skeletonized, bent round to form a cup (**A**), lined with rootlets, fine stems, decayed leaves and strands of fungus *Marasmius*; blends perfectly with its surroundings; ext. diam. 100, int. diam. 80, depth of cup 40; placed among bits of dead wood and dead leaves in dense bush, 0·4–1·4 m above ground; once 3·5 m up at base of low branch in small tree where there was an accumulation of dead leaves; another was situated at base of tree trunk, on hump in ground which raised it 0·5 m above surrounding young shoots (Brosset and Erard 1974, 1986).

EGGS: always 2; long ovals to short blunt ovals; somewhat glossy; somewhat smaller and rounder than those of Blackcap Illadopsis *I. cleaveri*; colour and markings very variable, pinkish white, creamy or pure white; a few rather large spots and blotches of bright maroon and purplish grey rather sparingly scattered all over the shell; or the entire shell densely mottled and spotted with rather fine markings of the same colour; or markings browner and less bright. One specimen has large blotches of dull purplish grey underlying the small mottlings, and 2 very blunt eggs have distinct zone or cap of deep chestnut-maroon surrounding large end (Ogilvie-Grant *in* Bannerman 1936). SIZE: (n = 10, Gabon) 18 × 16; (n = ?, Cameroon) 20–23·5 × 15–16·5.

LAYING DATES: Mali (breeding condition May); Sierra Leone (breeding condition June); Liberia (juvs Sept, fledglings July–Sept, breeding condition June–Jan); Ivory Coast (juvs Oct, Dec); Nigeria (juvs Aug, breeding condition Oct); Cameroon July, Sept–Oct (fledgling Mar, ♀ with brood patch May); Gabon, 2 seasons: second half of long dry season and beginning of the following rains (July–Oct), and short dry season and beginning of the following rains (Dec–Mar); Central African Republic (breeding condition June, Oct); Zaïre, Feb (juvs Dec–Mar, May, Sept–Oct, birds in breeding condition May, July, Dec; 'eggs must be laid almost throughout the year' (Chapin 1953)); Angola (breeding condition Sept); Sudan, Aug; E Africa: Region B, May.

INCUBATION: begins on day second egg laid; eggs laid at 1–2 day intervals; incubated by only one member of pair, probably the ♀; period at least 12 days. Bird surprised on nest allows very close approach, otherwise quickly leaves nest when it sees observer and silently disappears among vegetation.

DEVELOPMENT AND CARE OF YOUNG: young remain in nest 14 days (1 observation).

BREEDING SUCCESS/SURVIVAL: parasitized by *Chrysococcyx* cuckoos. Of 12 nests, 8 destroyed by predators and 4 produced only 4 fledged young. Ringed bird was retrapped 3 years 3 months later.

Reference
Brosset, A. and Erard, C. (1974, 1986).

A

Plate 2
(Opp. p. 15)

Illadopsis pyrrhoptera (Reichenow and Neumann). Mountain Illadopsis. Akalat montagnard.

Callene pyrrhoptera Reichenow and Neumann, 1895. Orn. Monatsb., 3, p. 75; Mau, Kenya.

Range and Status. Endemic resident. Highlands of Albertine Rift, in E Zaïre at 1140–2480 m, from Lendu Plateau and Rwenzoris through N and S Kivu to volcano region, Itombwe and Mt Kabobo; in W Uganda in Rwenzoris, Kigezi and Ankole, including riparian forest strips along Akagera R. at Merama Hill on Rwandan border (Vande weghe 1992); W Rwanda (very common), W Burundi (Buranga Lulenga, Teza, Ijenda, Bururi, also in E at Kigamba (1450–1550 m) near Tanzanian border: Gaugris *et al.* 1981); W Tanzania (Mahari Mt and Mugombazi, down to 1300 m). Highlands of W Kenya, 1550–2800 m, from Mt Elgon, Kapenguria and Cheranganis to N Nandi, Mau and Trans-Mara Forests; reported from Aberdares by Britton (1980) but not by Zimmerman *et al.* (1996); range entirely within humid, 1000+ mm rainfall areas. Highlands of N Malaŵi, 1600–2300 m, on E scarp of Nyika Plateau, N Viphya Mts (Uzumara, Chimaliro) and Nyankhowa Mt (Rumphi District); common at Uzumara, in every streambed on NE slopes, more scattered on E Nyika, local in less luxuriant forest of Chimaliro (Dowsett-Lemaire 1989). Uncommon to very common.

Illadopsis pyrrhoptera

Description. *I. p. pyrrhoptera* (Reichenow and Neumann): Zaïre to Kenya and Tanzania. ADULT ♂: top of head brownish grey, dark grey feather fringes producing faint scaly appearance; rest of upperparts brown, more olive on mantle, tinged rufous on back, rump and uppertail-coverts. Tail brown with rufous tinge. Lores grey, superciliary stripes indistinct, pale grey; ear-coverts, cheeks and sides of neck grey. Chin and throat greyish white, breast dark grey, belly paler, tinged brown; flanks and undertail-coverts rich brown; thighs grey. Primaries, secondaries and primary coverts rich brown on outer web, dark brown on inner web; tertials and rest of upperwing-coverts rufous-brown like back, lesser coverts more olive-brown. Primary underwing-coverts brown; rest of underwing-coverts pale greyish brown; axillaries grey. Bill dark brown or blackish above, pale grey or bluish grey below; eyes light brown or reddish brown; legs grey-brown to dark blue-grey, soles yellowish. Sexes alike. SIZE (10 ♂♂, 8 ♀♀): wing, ♂ 69–75 (71·6), ♀ 68–73 (69·8); tail, ♂ 52–60 (55·7), ♀ 47–55 (53·6); bill, ♂ 17–19 (18·0), ♀ 16–19 (17·5); tarsus, ♂ 24–25 (24·9), ♀ 24–25 (24·7). WEIGHT: Kenya, ♂ (n = 13) 22–30 (24·8), ♀ (n = 8) 20–26 (22·1), unsexed (n = 66) 20–30 (24·8).

IMMATURE: juvenile more reddish brown than adult on upperparts (including top of head) and flanks; downy feathering below, breast with a few rufous-brown spots; eyes dull brown; bill blackish brown above, dull yellowish below, gape yellow; legs greyish olive. An older bird (wing 68, tail 52) was unspotted and had dark grey chin, dark brown breast and belly, dark brown upperparts (Prigogine 1971).

I. p. nyasae (Benson) N Malaŵi: top of head brown (not grey); breast paler grey.

Field Characters. Length 12–13·4 cm. The only montane illadopsis, but at lower altitudes may overlap with other species. Small, with dark slightly reddish brown upperparts and *mainly uniform grey underparts*; pale eye-ring contrasts with dark iris (Dowsett-Lemaire and Dowsett 1983). Uppertail-coverts dull rufous-brown, contrasting with dark brown tail as bird darts into cover (Zimmerman *et al.* 1996). Larger Grey-breasted Babbler *Kakamega poliothorax* has grey breast-band but white belly and throat, and bright rufous upperparts; Scaly-breasted Illadopsis *I. albipectus* has whitish underparts often scaly, with greyish wash on breast; Brown Illadopsis *I. fulvescens* has tawny underparts with contrasting white throat, Pale-breasted Illadopsis *I. rufipennis* looks very pale below with brownish or greyish wash on breast and white belly. Voice distinctive.

Voice. Tape-recorded (32, 53, 104, B, C, F, KE1, LEM, McVIC). Song a duet; presumed ♂ gives 3 pure whistles descending the scale, well spaced and interspersed with brief 'tata' notes, '*wheee ... tatawhaaa ... whoah*', '*wheee ... tatatawhaaa ... tawhoah*'; sometimes just the first note, sometimes whistles without any 'tata' – '*wheee-hoo-whay-haw-where*'. Second bird or birds keep up constant unsynchronized chatter, '*chut-chut*', '*chip-chip-chip*', '*prrrrrt*' (Uganda: Keith and Gunn 1971). In recording from Rwanda (Nyungwe: LEM), ♂ often has 2 whistles, '*titiwheee ... wheeyoo*' and in addition to the chattering accompaniment, a third bird joins in with brief, high-pitched piping notes, making 3 different song types in this chorus. Song in Kenya said to be different, penetrating descending semitones, '*tweek twe tyew tu-wer tu-wer*', mingled with much low chucking and chattering (Zimmerman *et al.* 1996). Contact/alarm calls, '*tchek tchek*' and a rolled, nasal '*prret*', similar to those used in duet.

General Habits. Inhabits montane and transition (Zaïre) forest, both primary and secondary, and gallery forest;

enters lower edge of bamboo (Rwenzoris); keeps to deep shade, in dense shrubby undergrowth and on ground, also up to mid-levels. In Bwindi (Impenetrable) Forest, Uganda, dominant trees in forest around nest site at 2100 m were *Strombosia scheffleri*, *Chrysophyllum gorgungosanum*, *Tabbernaemontana holstii*, *Symphonia globulifera*, *Myrianthus arboreus* and *Fagara* sp. (Butynski and Kalina l989); tree canopy reached to *c.* 38 m, undergrowth included *Mimulopsis solmsii* and *Sericostachys tomentosa*. In Malaŵi, impenetrable thickets 2–4 m high often lining streams (Acanthaceae *Anisotes*, *Mimulopsis*) or on slopes (*Alchornea*, *Dracaena fragrans*) (Dowsett-Lemaire 1989).

Usually in small parties of 4–5, sometimes 2–3. Joins mixed-species flocks. Probes leaf clusters and twigs close to ground and up to 3–4 m; turns over dead leaves and other vegetable debris for small arthropods, on ground and 0·2–2 m above it. Forages wren-like through tangled fallen branches, recalling Evergreen-Forest Warbler *Bradypterus lopezi* in same habitat (Zimmerman *et al.* 1996). Spends little time in more open situations (e.g. *Sericostachys* thickets), where it moves through rather quickly; readily crosses gaps or roads; attracted to ant swarms and feeds over them (Nyungwe: Dowsett-Lemaire 1990). Flicks wings rapidly when alarmed. Stance rather horizontal.

Sedentary, although birds appearing rarely and irregularly in Kakamega Forest may be wanderers from higher elevations, e.g. Nandi Forest (Zimmerman 1972, Zimmerman *et al.* 1996).

Food. Insects, including ants and small crickets; also small snails; some berries and seeds.

Breeding Habits.

NEST (only 1 described: Butynski and Kalina 1989): cup-shaped, embedded 5 cm into dead, dry leaves which overlaid *c.* 10 cm of damp, decaying leaves on top of a rock ledge; built of green moss (*c.* 50%), interwoven with dry tree leaves and dry leaves and stems of herbs, thinly lined with dry rootlets and moss; ext. diam. 100, ext. depth 40, int. diam. 50, int. depth 20; overhung with fern leaves and dead tree leaves caught in base of fern; extremely well hidden, camouflaged, and protected from hail and rain; placed 1·4 m above ground on ledge in heavy shade on steep slope *c.* 15 m above small stream.

EGGS: 2; very pale blue-green with brown and purplish brown splashing covering *c.* 10% of surface, densest at large end. SIZE: (n = 2) 17 × 20, 17 × 23.

LAYING DATES: Zaïre, Jan–June (and possibly most of the year: Chapin 1953); Rwanda (increase in singing Dec, ♀♀ with eggs in oviduct Feb); Uganda, Feb (breeding condition and immature May); Kenya (breeding condition Mar); Malaŵi, Sept.

Reference
Butynski, T.M. and Kalina, J. (1989).

Illadopsis albipectus (Reichenow). Scaly-breasted Illadopsis. Akalat à poitrine écaillée. Plate 2

Turdinus albipectus Reichenow, 1887. J. Orn. 35, p. 307; Stanley Falls, Belgian Congo. (Opp. p. 15)

Forms a superspecies with *I. cleaveri*.

Range and Status. Endemic resident. NW Angola (Canzele, N. Cuanza Norte); SE Central African Republic (Ouossi R., common); Zaïre in lowland forest of Congo basin from Kunungu, Bolobo and Lukolela to Uele and Ituri (abundant), south to Kasai (Luebo), Maniema and S Kivu. Extreme SE Sudan, uncommon in foothills of Imatong Mts below 1500 m. Widespread and locally common in forests of W and S Uganda, 700–1500 m, from Budongo, Bwamba and Kibale east to Mabira, south to Bwindi (Impenetrable) and Malabigambo, and just into extreme NW Tanzania (Minziro Forest). W Kenya at 1600–2100 m in Kakamega (abundant) and Nandi Forests; listed for Elgon by Jackson and Sclater but omitted by Zimmerman *et al.* (1996). Uncommon to abundant.

Description. ADULT ♂: top of head dark greyish olive, with darker feather edges producing scaling; rest of upperparts olive-brown with rufous tinge, strongest on rump and uppertail-coverts, and a few narrow dark grey fringes to feathers of mantle, scapulars and back. Tail dark brown, tinged russet. Lores grey, superciliary stripes paler grey, indistinct; ear-coverts olive-grey, cheeks grey with whitish mottling, sides of neck pale olive-grey. Chin and throat greyish white, upper breast pale grey with slight olive tinge in centre, lower breast and belly whitish; sides of breast, flanks and thighs olive-grey; in some birds greyish feather

Illadopsis albipectus

fringes on throat, breast and flanks, giving scaly effect; in other birds barely visible, almost absent. Undertail-coverts olive-brown. Primaries, secondaries and primary coverts rich olive-brown on outer web, dark brown on inner web; tertials and rest of upperwing-coverts rich olive-brown like mantle; pale buff tips to median and lesser coverts. Underwing-coverts and axillaries pale buffish brown. Bill blackish above, pale grey or pale horn below; eyes brown; legs whitish grey or greyish pink with faint purple tinge, sometimes pale brown. Sexes alike. SIZE: wing, ♂ (n = 10) 73–79 (75·8), ♀ (n = 8) 68–75 (71·3); tail, ♂ (n = 9) 56–61 (58·1), ♀ (n = 8) 50–56 (52·5); bill, ♂ (n = 11) 15–19 (17·6), ♀ (n = 8) 15–18 (16·8); tarsus, ♂ (n = 10) 27–29 (27·8), ♀ (n = 8) 26–28 (27·5). WEIGHT: Central African Republic, ♂ (n = 4) 31–38 (35·2), ♀ (n = 3) 30–36 (32·3); Uganda, unsexed (n = 18) 26–35 (29·4); elsewhere, ♂ (n = 14) 27–34 (30·6), ♀ (n = 9) 25–32 (28·2), unsexed (n = 34) 23–36 (30·6). Can change daily weight by up to 18% (Mann 1985).

IMMATURE: like adult but more rufous on upperparts, flanks and undertail-coverts, top of head browner; tawny-buff spots on median and lesser coverts more pronounced; breast buffy brown (sometimes with faint yellow wash), with a few pale tawny-buff spots; conspicuous rufous supraorbital and eyelid feathers, this colour extending to the forehead and above the ear-coverts in some birds; some tawny-buff feathers are retained by immatures in first adult plumage (Zimmerman et al. 1996). Upper mandible black, lower mandible dusky brown with dull pink or yellowish base, gape yellow; eyes brown; tarsi and toes whitish or pinkish grey.

NESTLING: lightly covered with dark grey down when primaries starting to appear.

Field Characters. Length 13–15·5 cm. One of the more common, tame and vocal forest species in its range, but drab plumage and dense habitat make it very hard to observe (Butynski 1989). Best told from congeners by song. Extremely similar to Pale-breasted Illadopsis *I. rufipennis*, which often occurs in same forests. Individuals with well-marked scaling on breast, flanks and belly (commonest in Kenya) are distinguishable, but scaling elsewhere often vestigial or absent, especially in W of range. Pale grey breast less sharply defined than pale brownish breast-band of Pale-breasted Illadopsis, legs longer and usually paler; in the hand, rictal bristles half the length of Pale-breasted, contra Zimmerman et al. (1996) (J. Lindsell, pers. comm.), dark bases to pale throat feathers, tarsus length 26–30 (Zimmerman et al. 1996), also pale shaft streaks and pale feather tips of lesser coverts (J. Lindsell, pers. comm.). Juvenile has rufous feathers above eye, base of lower mandible dull pink or yellowish (bright yellow or orange in Pale-breasted). Brown Illadopsis *I. fulvescens* has uniform brownish underparts and contrasting white throat.

Voice. Tape-recorded (32, 53, 104, B, C, F, HUG, KEI, LEM, ZIM). Typical song in Kakamega, Kenya, 3, sometimes 2, ringing whistles in ascending semitones, with an introductory low 'tip' and a shorter 't' attached to the beginning of the first whistle, 'tip, t'weee-weee-weee'; duration, 2–2·5 s. Commonest song in Kibale Forest, Uganda the same – 'twit-twittweee-tweee-tweeeee', but can vary: 1–3 soft, squeaky chirps ('twit') followed by 1–4 loud, lively, high-pitched whistles ('tweee'); first whistle seems to begin as a chirp but immediately becomes a whistle ('twittweee'). Occasionally a soft 'wheet' is given about 2 s prior to the chirps; calls with 'wheets', multiple chirps and/or 1–4 whistles are uncommon; shortest variation 'twit-twittweee', longest and most complex is 'wheet ... twir ... wheet ... twit-twittwee-tweee-tweeeee', latter lasting 7–8 s' (Butynski 1989). Last note is longest and loudest. Chirps audible up to 50 m, last whistle up to 200 m. Bird often repeats song every 5–35 s (usually 6–16 s) for more than an hour, usually while foraging, but one bird sang for 5 min while preening on top of 10-cm high mound (Butynski 1989). 5 or more birds may sing at the same time, and often seem to be counter-singing. Song in Bwamba, Uganda (LEM), has a longer introduction and the 3 whistles are rather shorter: 'chup-chiweechi-pu-weee-weee-weee'.

Call given by foraging, possibly immature, birds, 2 loud, liquid, high-pitched drawn out whistles, 'peeeeep-peeeeep', audible up to 100 m. High-pitched, warbler-like 'see-u, see-u, see-u' preceded by a soft twittering 'tititititititi', given by full-sized immature in response to squeaking (Zimmerman 1972). Sings at any time of day, most often in early morning.

General Habits. Inhabits lowland primary forest, in S Kivu (Itombwe) also transitional forest to 1460 m, exceptionally temperate forest (Mt Nyombe, 1400 m); seasonal swamp forest (Uganda: Malabigambo). Kibale Forest, Uganda, has fairly open understorey and thick upper levels, creating dense shade on ground; commonest tree species are *Celtis durandii*, *Olea welwitschii* and *Strombosia scheffleri*, and there is dense ground cover of ferns and large plants (e.g. *Palisota schweinfurthii*, *Pollea condensata*, *Piper capensis*, *Aframomum* sp.); birds equally common in valley bottoms, hillsides or ridge tops (Butynski 1989).

Singly or in pairs or threes, once 5. One group of 3 contained 2 adults and an immature, suggesting groups are family parties. Sometimes near, but perhaps not part of, mixed-species flocks (Zimmerman et al. 1996). Forages mainly on ground, among leaf litter; sometimes in undergrowth, up to 2 m; overturns leaves with bill, sometimes scratches ground with feet. More terrestrial than Pale-breasted Illadopsis (Chapin 1953). Approximate duration of primary moult (n = 1), 170 days.

Food. Insects: small beetles, ants, caterpillars, many worker and soldier termites (Chapin 1953); insect eggs; also millipedes and small snails.

Breeding Habits. Monogamous; territorial. Sings in territory or home range all year. Singing confined to area of 0·8–1·2 ha. Singers may be within 5 m of one another. Site fidelity pronounced: birds were singing from same sites when observer revisited them 3 years later (Butynski 1989).

NEST (only 1 described: Butynski 1989): loose, shallow structure of brown, damp, rotting tree leaves (mostly of *Teclea nobilis* and *Bosqueia phoberos*) lined with dead leaves of *Celtis durandii*, with a few rootlets at bottom of nest cup; placed under small fern in slight indentation on damp (but not wet) ground; a large fallen leaf of *Palisota schweinfurthii* prevented access to the nest from one side besides providing some shelter from rain; extremely well concealed.

EGGS: 2; white, with light to dark brown blotches densest toward broad end.

LAYING DATES: Central African Republic (breeding condition June); Zaïre, Sept (breeding condition Jan–Feb, Apr–July, Oct–Nov, juv. Nov, i.e. probably breeds all year); Sudan, Aug–Nov; Uganda Nov (breeding condition Apr–May, Nov, juv. May, juv. out of nest Dec, full-grown immatures July).

INCUBATION: incubating bird allowed observer to approach to within 1 m and place hand to within 15 cm; then it suddenly flitted from the nest, landed 1 m away and moved off along the ground; when observer retreated, bird returned within a few min; it never flew to the nest but walked in from a few m away (Butynski 1989).

DEVELOPMENT AND CARE OF YOUNG: fledged young observed for over an hour moving with 2 foraging, calling adults (Butynski 1989).

BREEDING SUCCESS/SURVIVAL: at only known nest, both nestlings disappeared, presumably taken by predator, while nest remained intact. 2 birds lived 6 years 8 months and 8 years 3 months (Kakamega, Kenya: Mann 1985).

Reference
Butynski, T.M. (1989).

Illadopsis cleaveri (Shelley). Blackcap Illadopsis. Akalat à tête noire.

Plate 2
(Opp. p. 15)

Drymocataphus cleaveri Shelley, 1874. Ibis, p. 89; Fanti region, Gold Coast.

Forms a superspecies with *I. albipectus*.

Range and Status. Endemic resident. E Sierra Leone (Gola Forest, Daru), SE Guinea (Macenta), Liberia (not uncommon to common), Ivory Coast north to Mt Nimba and Bouaké, including Taï Nat. Park and Yapo Forest; SW Ghana north to Bia Nat. Park and Kumasi, uncommon but probably overlooked (Grimes 1987); common in Kakum Nat. Park. S Nigeria, uncommon, also in SE on Obudu Plateau. S Cameroon north to Korup Nat. Park, Rumpi Hills, Mt Kupé and Mt Nlonako; Bioko; SW Central African Republic (Dzanga Reserves, uncommon; La Maboké); Mbini; Gabon (common); Congo, common in Odzala and Nouabalé-Ndoki Nat. Parks and in S (Mouilou basin and Mayombe). Density in Gabon 12–14 pairs per km^2.

Illadopsis cleaveri

Description. *I. c. batesi* (Sharpe): SE Nigeria (Obudu Plateau) to Central African Republic and Congo. ADULT ♂: top of head blackish, hindneck grey, rest of upperparts rufous-brown. Tail dark russet brown. Lores blackish, superciliary stripe pale grey, broad and prominent and extending to behind ear-coverts; cheeks and ear-coverts blackish, the former with some pale grey mottling, sides of neck grey. Underparts from chin to belly whitish, upper breast tinged tawny-brown; sides of breast, flanks, thighs and undertail-coverts tawny-brown. Primaries, secondaries and primary coverts rufous-brown on outer web, dark brown on inner web; tertials and rest of upperwing-coverts rufous-brown. Primary underwing-coverts brownish; rest of underwing-coverts and axillaries rufous-buff. Upper mandible blackish, lower mandible light grey or whitish; eyes grey-brown, brown or reddish brown; legs pale, grey or blue-grey to fleshy white. Sexes alike. SIZE (10 ♂♂, 10 ♀♀): wing, ♂ 75–83 (78·9) ♀ 70–77 (73·1); tail, ♂ 54–63 (59·8), ♀ 54–59 (56·2); bill, ♂ 17–20 (18·1), ♀ 17–19 (18·3); tarsus, ♂ 27–29 (28·0), ♀ 25–27 (25·9). WEIGHT: Gabon, unsexed (n = 4) 29–36 (32·2); Congo, unsexed (n = 15) 28–34 (30·8 ± 1·8).

IMMATURE: more rufous on breast and flanks; some dusky mottling on underparts.

I. c. johnsoni (Büttikofer): Sierra Leone and Liberia (also Ivory Coast?). Top of head and ear-coverts blacker than in *batesi*, supercilium whiter and more contrasting; grey wash on breast becoming olive-grey at sides, flanks same tawny brown as *batesi*. WEIGHT: Liberia, ♂ (n = 14) 23–32 (26·5); ♀ (n = 5) 22–29 (26·4); Ivory Coast, unsexed (n = 5) 26–30 (26·5).

I. c. cleaveri (Shelley): Ghana. Upperparts, sides of head, flanks and undertail-coverts brighter rufous than previous two races; superciliary stripe whiter; centre of breast whitish without rufous wash.

I. c. marchanti Serle: S Nigeria. Crown dark olive-grey (not blackish); front of superciliary stripe buffish and less distinct; breast washed brown as in *batesi*.

I. c. poensis Bannerman: Bioko. Like *johnsoni* but upperparts darker with blackish barring on mantle and scapulars, sides of breast darker olive, flanks darker brown.

Field Characters. Length 15–17 cm. Readily distinguished from congeners by black cap (extending to nape) and contrasting well-defined long, whitish or grey supercilium; dark grey ear-coverts contrast with pale grey moustachial stripe which grades into white throat; underparts dirty white with rich brown flanks (difficult to see) and diffuse breast-band grey (*johnsoni*) or brown (*batesi*). Most resembles sympatric Brown-chested Alethe *Alethe poliocephala*; best distinguished by *broad grey supercilium* continuing *around ear-coverts* and down onto *sides of neck* (in the alethe, pale supercilium ends just behind eye, and ear-coverts and sides of neck are concolorous); also told by whitish lores, pale grey moustachial stripe; diffuse breast-band grey (not brown), flanks rich mid-brown (not pale grey-brown), upperparts olive-brown (not chestnut-brown) (Allport *et al.* 1996). Vocal distinctions very marked: characteristic ringing song of Blackcap Illadopsis very different from softer mournful whistles of Brown-chested Alethe.

Voice. Tape-recorded (53, 104, B, ERA, KEI, LEM, MAC). Song extremely similar to that of Scaly-breasted Illadopsis *I. albipectus*, one of the most characteristic forest sounds: 2–3 loud, ringing, penetrating whistles in ascending semitones, preceded by 1–3 low introductory notes, 'chup', 'chip' or 'weet'. The whistled section always lasts *c.* 1 s, whether it is one long note, 2 leisurely ones or 3 fast ones. In some 3-whistled phrases notes are partly run together, sounding at a distance more like a single note. Each of the 3 whistles is louder than preceding one; first whistle carries for 150 m, second for 300 m and third for 450 m (Brosset and Erard 1986). 2 disputing birds following army ants flitted wings, gave series of short calls, high and low 'weet', 'whit', 'chirp', 'eu', 'he', 'hi', 'hih' and a buzzy 'trrr' (Willis 1983).

General Habits. Inhabits lowland primary and old secondary forest, *Raphia* swamp forest (Congo: Dowsett-Lemaire and Dowsett 1991), logged forest and forest-shrub mosaic (Liberia). Frequents both dense undergrowth and more open understorey of tall forest (Gabon); avoids pure Marantaceae (Congo: Dowsett-Lemaire 1997b). Usually on forest floor or in lowest level of vegetation, seldom above 2 m except during territorial conflict or in response to playback of its song. Up to 1500 m in Liberia (Mt Nimba) and Nigeria (Obudu Plateau), 1850 m in Cameroon (Mt Cameroon).

Lives in pairs (pair members are not always close to one another), sometimes singly; small family groups rare (Gabon); parties of 6–8 reported in Ghana (Holman *in* Bannerman 1936) may have been mixed-species flocks. Readily joins bands of insectivores and mixed-species flocks around army ants. Individual birds briefly or irregularly walked and hopped on the ground ahead of 7 large raids of *Dorylus wilverthi*, often pecking small prey (Willis 1983). Turns over leaf litter; searches through masses of decomposing vegetation around bases of trees, beside fallen logs, and especially between stems of broad-leaved grasses and marantaceous plants. Bathes in small forest streams in afternoon. Skulking but tame: approaches observer to inspect him, moving constantly but managing to keep concealed behind foliage (Holman *in* Bannerman 1936). Sings shortly after dawn, rarely thereafter (Yapo, Ivory Coast: Demey and Fishpool 1994) but up to at least 0900 h in Kakum Nat. Park, Ghana and in Liberia (pers. obs.); degree of singing may vary with season. Sings all year; in Gabon, less in the long dry season, June–Sept. Also a form of ♂ song, 'wee-tawee-tyooo' or 'wee-tawilly-tyooo', mellower and less ringing, is sexual. Singer readily responds to human whistles. ♂ approached to within a few m, flitted about at ground level giving little rolling calls, then, in full view 8–10 m high, sang normal song while quivering, flapping wings and spasmodically raising tail (Brosset and Erard 1986).

Food. Small beetles, orthopterans, spiders, millipedes, small snails.

Breeding Habits. Monogamous; territorial. Territory advertised and defended by singing.

NEST: large and loosely constructed, built of dead leaves, some skeletal, rolled up to form a cup, placed on a thick base of dead leaves on or near ground; lined with hair-like strands of fungus *Marasmius* (**A**). Ext. diam. 120–130, int. diam. 60–75, int. depth 40–50. Resembles the bunches of dead leaves that accumulate in forks of bushes; looks like that of Brown Illadopsis *I. fulvescens* (Bates 1930) but better constructed and rim more flared, very like nest of Nightingale *Luscinia megarhynchos* (Brosset and Erard 1986); not in a depression in the ground but raised slightly above it, its edges supported by low branches whose leaves conceal the nest; once 15 cm up in a mass of dead leaves in a large clump of marantaceous plants; situated in open areas beneath shrubs whose branches do not touch the ground; 4 were very close to edge of a path, one on a sloping bank (Brosset and Erard 1996).

EGGS: always 2. Rather large and elongated; white or pale rose, densely covered with clearly-marked and sharply contrasting dark red-brown spots. SIZE: (n = 10) 24–25 × 15–16 (Gabon); av. 24 × 17 (Cameroon).

LAYING DATES: Liberia, Aug; Ivory Coast (bird carrying food Jan); Nigeria, (juvs June, Aug); Cameroon, (birds in breeding condition Mar (6), Apr (1), May (3), Aug (1), Sept (1), Dec (1), indicating breeding all year: Rodewald *et al.* 1994); Bioko (fledgling Feb); Gabon, mainly during short dry season (Jan–Feb), also Mar, Nov.

A

INCUBATION: begins with second egg; by ♀ only. Period unknown, but one bird incubated 2 infertile eggs for an entire month (Brosset and Erard 1974). If incubating bird sees observer coming it quickly leaves the nest and disappears into vegetation; when surprised it freezes, allowing observer to get very close and almost touch it. Occasionally uses 'rodent run' diversionary tactic: quickly leaves nest and at a distance of 5–6 m, with hunched back, runs back and forth along the ground, rustling the dead leaves (Brosset and Erard 1974).

BREEDING SUCCESS/SURVIVAL: 1 nest contained a cuckoo egg, possibly of Red-chested *Cuculus solitarius* (Brosset and Erard 1986: the cuckoo's egg did not hatch, and young illadopsis was raised normally). Out of 7 clutches, only 1 young fledged; 1 clutch was infertile, the others probably taken by predators.

Reference
Brosset, A. and Erard, C. (1974, 1986).

Illadopsis rufescens (Reichenow). Rufous-winged Illadopsis. Akalat à ailes rousses.

Plate 2
(Opp. p. 15)

Turdirostris rufescens Reichenow, 1878. J. Orn. 26, p. 209; Liberia.

Range and Status. Resident, endemic to Upper Guinea forests. Senegal (Basse-Casamance), SW Guinea, Sierra Leone (York Pass, Bintumane Peak, Gola Forest); Liberia (locally common throughout forest region; the commonest illadopsis at Mt Nimba: pers. obs.); Ivory Coast north to Taï Nat. Park, Mt Nimba, Azaguy and Marahoué Nat. Parks, Lamto (Thiollay 1985) and Yapo Forest but never recorded at Lamto by L.D.C. Fishpool (pers. comm.), who found Puvel's Illadopsis *I. puveli* common there. SW Ghana (not uncommon); one record Togo (Agbossomou Kope, 08°06'N, 00°37'E); specimen from Atakpamé (de Roo 1970) is *I. puveli* (M. Louette *in* Cheke and Walsh 1996). Common to uncommon. Listed as near-threatened by Collar and Stuart (1985) but probably overlooked and in no serious danger; only recently discovered in Guinea (Demey 1995). Density in primary forest, Sierra Leone, 16 pairs per km^2, in logged forest 4 pairs per km^2, giving total number of birds in Gola Forest 9604 (Allport *et al.* 1989); density in Liberia 4–8 pairs per km^2 (Gatter 1997).

Description. ADULT ♂: top of head, upperparts and tail uniform rufous-brown, with slight darker mottling on top of head. Lores and sides of head paler brown, some whitish streaks on cheeks; indistinct buffy superciliary stripe. Chin to belly whitish, tinged pale brown on centre of breast; sides of breast, flanks and undertail-coverts brownish grey. Primaries, secondaries and primary-coverts rufous-brown on outer webs, dark brown on inner webs; tertials and rest of upperwing-coverts rufous-brown. Primary underwing-coverts greyish brown; rest of underwing-coverts and axillaries greyish white. Bill black above, pale to pinkish horn or pale greyish below; eyes brown or tawny-brown; legs pale whitish flesh. Sexes alike. SIZE (10 ♂♂, 7 ♀♀): wing, ♂ 76–84 (79·0), ♀ 74–81 (75·8); tail, ♂ 64–74 (70·1), ♀ 58–69 (63·6); bill, ♂ 18–20 (18·9), ♀ 18–19 (18·6); tarsus, ♂ 27–31 (28·7), ♀ (27–30 (28·0). WEIGHT: (Liberia), ♂ (n = 9) 32–36 (35·0), ♀ (n = 6) 32–37 (35·0); Ivory Coast, unsexed (n = 6) 34–41 (37).
IMMATURE: ♀, Liberia, with partly ossified skull, is in plumage identical to adult.
NESTLING: unknown.

Field Characters. Length 18 cm. Larger than Pale-breasted Illadopsis *I. rufipennis* and Brown Illadopsis *I. fulvescens*, with large, round, dark eye; clean white underparts have only ill-defined breast-band, and contrast with upperparts, even in poor light; throat not markedly contrasting with rest of underparts; legs and feet pale,

Illadopsis rufescens

longer and stronger than in Pale-breasted and Brown Illadopsises (Allport *et al.* 1996). Very similar to Puvel's Illadopsis; best separated by voice and habitat (forest vs secondary scrub and gallery forest). On horizontal perch both species have upright stance, giving thrush-like appearance. Best plumage character is colour of underparts: diffuse breast-band brownish grey, flanks clean mid-grey (warm peachy buff in Puvel's: Allport *et al.* 1996); throat and lower breast white with no creamy tinge, inner third to half of lower mandible pale yellowish (all yellow in Puvel's).

Voice. Tape-recorded (53, 104, B, CHA, KEI). Song unvarying, on one pitch, lasting *c*. 1·5 s: 2–3 dry notes, a short whistle and 2 longer whistles, 'chup-chup-chup-he-heee-heee'; repeated for long periods with 1 s intervals between songs. Last 2 notes far-carrying, with ringing quality like that of Blackcap Illadopsis *I. cleaveri*. Sings alone without accompaniment.

General Habits. Occurs in lowland forest undergrowth. Habitat variously reported as 'closed-canopy forest only' (Allport *et al.* 1996), and this was also the principal habitat at Mt Nimba (pers. obs.), but in Sierra Leone (Gola Forest) also occurs in logged areas (Allport *et al.* 1989); Liberia, often prefers quite open ground in mature forest, also in logged and old secondary forest (Gatter 1997); mainly in forest at Taï Nat. Park, Ivory Coast, rarely in plantation and logged forest (Gartshore *et al.* 1995); in Ghana frequents forest edge and the coastal thicket zone, occasionally forest outliers, occurring in more open thicket than congeners (Grimes 1987). In pairs or small groups; sometimes joins mixed-species flocks. Active, terrestrial, scratching on ground with feet; also perches in saplings, especially when disturbed. Sings either from ground or high perch, often while moving rapidly around its territory (Allport *et al.* 1996).

Food. Beetles, grasshoppers, small amphibians.

Breeding Habits. Nothing known.
BREEDING INDICATIONS: Sierra Leone (breeding condition Apr, feeding young Dec); Liberia (independent young Nov–Jan; breeding condition all months); Ivory Coast (birds carrying food Feb).

Plate 2
(Opp. p. 15)

Illadopsis puveli (Salvadori). Puvel's Illadopsis. Akalat de Puvel.

Turdinus puveli Salvadori, 1901. Ann. Mus. Civ. Genova, ser. 2, 20 (1899–1901) p. 767; Rio Cacine, Portuguese Guinea.

Range and Status. Endemic resident. SW Senegal (Oussouye, Basse Casamance), Guinea-Bissau, W Guinea, Mali (valley of upper R. Bafing), Sierra Leone. N Liberia in Lofa and Nimba counties, and in SE at Zwedru. Ivory Coast north of evergreen forest zone in semideciduous forest and southern guinea savanna, from Soubré to Danané, Lamto, Bouaké, Boron and Comoé Nat. Park where common (Salewski 1997), perhaps to Korhogo: one of the most widespread timaliids (Thiollay 1985), but apparently absent from Taï and other southern forests. Ghana, definitely known only from Abokobi but probably occurs elsewhere in coastal region (Grimes 1987). Togo (Atakpamé, Misahöhe); rare, probably overlooked (Cheke and Walsh 1996). Nigeria, uncommon in undergrowth of forest zone from Lagos to Mamu, north to Ibadan and Ife; also Kagoro. In Cameroon restricted to forest/savanna contact zone, e.g. Yaoundé, Bafia, Ngoumé, Knounden (Louette 1981), also in Kimbi R. Game Res., 06°36′N, 10°23′E (F. Dowsett-Lemaire, pers. comm.). NE Zaïre (Uele) and extreme SW Sudan (Bengengai Forest, fairly common). Uganda, recently discovered in Budongo Forest (Plumptre and Owiunji 1997), where locally common (J. Lindsell, pers. comm.).

Description. *I. p. puveli* (Salvadori): Senegal to Togo. ADULT ♂: top of head brown, tinged rufous, with slight darker mottling, rest of upperparts uniform rufous-brown. Tail dark rufous-brown. Lores brown; sides of head warm brown, cheeks with whitish streaks; indistinct buffy superciliary stripe. Chin to belly whitish, tinged rufous-buff across breast; sides of breast brown, flanks and undertail-coverts rufous-buff. Primaries, secondaries and primary coverts rufous-brown on outer webs, dark brown on inner webs; tertials and rest of upperwing-coverts rufous-brown. Underwing-coverts and axillaries pale rufous-buff. Bill black or dark brown above, pale horn or pale greyish below; eyes brown; legs pale, pinkish grey to whitish flesh. Sexes alike. SIZE (7 ♂♂, 5 ♀♀): wing, ♂ 89–91 (90·0), 80–84 (81·2); tail, ♂ 72–75 (73·5), ♀ 63–69 (65·0); bill, ♂ 19–22 (20·1), ♀ 19–20 (19·4); tarsus, ♂ 31–33 (32·3), ♀ 29–31 (30·0). WEIGHT: Liberia, ♂ (n = 1) 43, ♀ (n = 2) 40, 38; Nigeria, unsexed (n = 7) 37·7–48·0 (43·1).
IMMATURE: juvenile similar to adult but deeper rufous, more extensively rufous-brown on breast.

Illadopsis puveli

NESTLING: featherless, eyes closed (Salewski 1997).
I. p. strenuipes (Bannerman): Nigeria to Sudan and Uganda. Upperparts darker than in nominate race, breast and flanks browner. SIZE (12 ♂♀, Uganda): wing 81–92 (87·1); tail 62–79 (70·9); tarsus 31–36 (35·1). WEIGHT: (12 ♂♀, Uganda) 41–52·5 (48·4).

Field Characters. Length 17 cm. A strikingly robust illadopsis with strong legs, rather long tail, large, dark eye and rufous-brown upperparts contrasting with creamy white underparts. In nominate race breast-band ill-defined and only visible with good view of underparts, warm peachy buff appearing almost orange in good light (Allport *et al.* 1996); in E race *strenuipes* white throat stands out against buff-brown breast-band, and in the hand is sometimes puffed out as in Red-tailed Greenbul *Criniger calurus* (Plumptre and Owiunji 1997); face pattern distinctive, brown with contrasting grey patch around eye (J. Lindsell, pers. comm.). In Upper Guinea forest overlaps with Rufous-winged Illadopsis *I. rufescens*; for differences see that species. Other sympatric congeners are smaller, with duller brown upperparts and different voices, and usually travel in groups.

Voice. Tape-recorded (53, 104, B, CHA, GRI, LEM, MOR). Song starts with 3–6 brief tuneless notes which lead into 2–3 descending whistles, 'chip-chippity-*hee-hee-heeer*', 'pit-chip-per-chip-per-chip-*hur-heeer*'; last 2 whistles may be repeated, 'chip-chipper-*hee-hee-heeer-hee-heeer*'. Whistles have a muted quality, lack ringing tone of smaller species. Aggressive call of ♀ a repeated down-slurred 'peeoo'. Song of *I. p. strenuipes* in Budongo Forest (Uganda) described as strong and greenbul-like in tone and chirped rather than whistled; an introductory phrase like that of Scaly-breasted Illadopsis followed by usually 5 notes, first on same pitch as third, second same as fifth, and fourth falling in between; also has distinctive call, a powerful down-slurred whistle 'peeeeyaaar' lasting 1–1.5 s (J. Lindsell, pers. comm.). Call of both juvenile and adult given as they sat in collecting bags described as '3–4 notes, rising and falling', to which free adult nearby replied with a churring call (Plumptre and Owiunji 1997).

General Habits. Inhabits lowland forest undergrowth, gallery forest, ravines, farmbush and second growth, thickets, secondary scrub. In Liberia, forest–grassland mosaic and savanna at 500–800 m, secondary vegetation (after burning) between 1100 and 1500 m on Mt Nimba, primary ridge forest at 1400 m, also 3–7-year-old farmland and *Harungana* scrub around airfield, where nearly always on ground (Gatter 1997). In Budongo Forest, Uganda, occurs in understorey of bushy shrub *Rinoria ilicifolia* below dense canopy of *Cynometra alexandri* (Plumptre and Owiunji 1997).

Singly, in pairs, or in groups of 3–6. Terrestrial; forages on ground, running from one spot to another, hopping and scratching for food among dry leaves. Often jumps towards a leaf, turns it over with bill, then jumps backwards and investigates what is underneath (Salewski 1997). Sometimes hops up onto low branches, where upright stance gives it a rather thrush-like appearance.

Food. Insects, including ants, and spiders.

Breeding Habits. Little known. Only 1 nest found (Salewski 1997).

NEST (**A**): a bowl of dry leaves, pieces of bark and some grass on a base of twigs up to 5 mm in diam., in hollow scratched in ground; ext. diam. of bowl 170, int. diam. 80, int. depth 25; situated under a fallen branch in an open place in otherwise dense forest. The whole construction was very loose and fell apart when collected (Salewski 1997).

EGGS: not described. Only known nest contained 2 young.

LAYING DATES: Liberia (breeding condition Aug); Ivory Coast, Nov; Zaïre (breeding condition Nov); Sudan, Mar.

INCUBATION: adult flushed off nest ran away in a quail-like manner (Salewski 1997).

DEVELOPMENT AND CARE OF YOUNG: 2 featherless young with closed eyes weighed 5.0 and 5.1 g; 3 days later, when remiges were pins, weights were 14.0 and 14.1 g; after another 3 days eyes were open and weights 26.5 and 27.3.

References
Plumptre, A.J. and Owiunji, I. (1997).
Salewski, V. (1997).

Genus *Kakamega* Mann, Burton and Lennerstedt

A single, somewhat anomalous babbler, with some similarities to *Illadopsis* and Asian *Trichastoma*, but differing as follows: rictal bristles much reduced; tongue longer relative to bill size, and with U-shaped cross-section; width/length ratio of skull greater, and skull more tapering; angle in dorsal anterior quadrant of intersection of post-orbital ligament and jugal bar considerably greater (112 compared to 87–95); in structure of jaw muscles *M. adductor mandibulae externus* and *M. pterygoideus*; in the structure of foot pads II:2, III:2, III:3, and IV:2; in lengths of basal phalanges in anterior digits (and in the pattern of pads and folds, it also differs from all members of Sylviinae and Turdinae studied by Mann *et al.* 1978); feathering finer and closer. Song pure, loud thrush-like whistles, without 'chorusing' or chuckling notes typical of *Illadopsis*. First described as an *Alethe* by Reichenow, it was not included in the Timaliinae by Delacour (1946). Chapin (1953) noted that 'in its small bill and long feet it differs noticeably from the other species of *Malacocincla* (=*Illadopsis*)', but he nevertheless believed it closer to the latter than to *Alethe*. Hall and Moreau (1970) noted some turdine similarities but retained it with some hesitation in *Trichastoma*. Ripley and Beehler (1985) accepted *Kakamega* as a monotypic genus, but noted that 'clear phenetic links between this species and (*Illadopsis*) *pyrrhoptera* leave little doubt in our minds that *poliothorax* is a member of the Pellorneini, and that *Kakamega* might better be considered a subgenus of *Illadopsis sens. lat.*'. Whatever its subgeneric status, we follow Dowsett and Forbes-Watson (1993) in considering *Kakamega* a valid genus.

Endemic; 1 species.

Plate 2 (Opp. p. 15)

Kakamega poliothorax (Reichenow). Grey-chested Babbler; Grey-chested Illadopsis. Akalat à poitrine grise.

Alethe poliothorax Reichenow, 1900. Orn. Monatsb., 8, p. 6; Bangwa, Cameroon.

Kakamega poliothorax

Range and Status. Endemic resident, in 2 well-separated areas. Extreme SE Nigeria, common on Gangirwal and Chappal Waddi (Gotel Mts) at 1900–2300 m, and on Obudu Plateau at 1600 m (Ash *et al.* 1989); widespread but generally uncommon in mountains of W Cameroon, 700–2200 m, from Bamenda Highlands (rare; absent Mt Oku), Sabga Pass, Bafut-Ngemba Forest Reserve and L. Bambulue to Foto near Dschang, Mts Manenguba, Nlonako and Kupé, Rumpi Hills (Dikume-Balue) and Mt Cameroon; Bioko. Highlands of E Zaïre at 1190–2450 m from Lendu Plateau to Rwenzoris, N and S Kivu, including Kahuzi-Biega Forest and volcanoes, to Itombwe, and Mt Kabobo; uncommon to rather rare. SW Rwanda (Nyungwe Forest, to 2650 m); Burundi (Teza, Bururi Forest); W Uganda (Rwenzoris, and Bwindi (Impenetrable) Forest at 1540–2150 m); Kenya, scarce in Kakamega and N and S Nandi Forests, 1550–2000 m (formerly also Mt Elgon and Lerundo).

Description. ADULT ♂: forehead and crown dark rufous-brown, paler feather tips giving slightly mottled appearance; rest of upperparts bright rufous-brown, deeper rufous on uppertail-coverts. Tail dark russet-brown. Chin and throat greyish white, barred or mottled darker grey. Upper breast and flanks dark grey; lower breast paler grey, grading to whitish on centre of belly; thighs and undertail-coverts grey. Primaries, secondaries, outer tertials and alula rufous-brown on outer web, dark brown on inner web; primary coverts dark brown with rufous-brown outer edge and tip; innermost tertial, median and lesser coverts bright rufous. Primary underwing-coverts greyish brown; rest of underwing-coverts grey with whitish fringes; axillaries pale grey. Bill blackish or dark grey above, light grey or bluish grey below; eyes brown or red-brown; legs grey or purplish grey. Sexes alike. SIZE (10 ♂♂, 10♀♀): wing, ♂ 81–91 (86·1), ♀ 77–82 (80·2); tail, ♂ 67–78 (73·3), ♀ 64–73 (69·7); bill, ♂ 20–22 (20·7), ♀ 19–22 (20·4); tarsus, ♂ 30–34 (31·7), ♀ 29–32 (30·4). WEIGHT: Cameroon, unsexed (n = 19) 31·5–42·1 (37·0); Uganda, 1 ♂ 35; Kenya, 1 ♂ 35, unsexed (n = 14) 35–40 (36·1); Kenya (Kakamega), unsexed (n = 49) 36·38 ± 1·2. Changes weight daily, av. of 7 birds 15%, all more than 10%.

IMMATURE: like adult but face, rump and lower back less chestnut and more gingery, olive wash and rufous-brown tips to grey breast feathers, eyes brown (red-brown in adult), soft yellow or orange-yellow gape (Mann *et al.* 1978).

JUVENILE (description from Holyoak and Seddon 1990): forehead and crown blackish grey with brown feather tips; nape, mantle and back rufous-brown; uppertail-coverts bright rufous-brown; wing-coverts and remiges blackish grey, with rufous-brown feather fringes; rectrices dark rufous-brown with rufous fringes; lores dull grey-brown; supercilium cheeks and ear-coverts warm brown; chin and upper throat pale grey (whitish); lower throat and upper breast light grey, with olive-brown feather tips forming breast-band which is browner and more obvious at sides; belly and vent white; flanks grey with dull, light olive feather tips; underwing-coverts black; undertail-coverts grey with dull, light olive tips. Upper mandible black, lower mandible orange-yellow with extensive blackish tip; tongue and gape orange; irides brown; tarsi, toes and claws light blue-grey (legs and feet notably strong). Wing 81 (feather bases still in sheath); tail 65 (feather bases still in sheath); bill from skull 17; tarsus 32. WEIGHT: 40.

NESTLING: unknown.

Field Characters. Length 16·5 cm. A large montane forest babbler, thrush-like and rather long-legged, with bright reddish upperparts, white throat and belly and contrasting grey breastband and flanks. Mountain Illadopsis *Illadopsis pyrrhoptera* is much smaller, duller and more uniformly coloured, travels in babbling groups. In mountains of Albertine Rift might be mistaken, especially in rear view, for Red-throated Alethe *Alethe poliophrys*, which has similar bright red upperparts and tail, but has black crown and face separated by broad grey eyebrow, and red throat (often hard to see). Elsewhere, Brown-chested Alethe *A. poliocephala* has red-brown back but black tail and pale eyebrow, brownish breast and flanks, and both alethes have single long whistled calls. Very shy, seldom seen but easily mist-netted (Mann *et al.* 1978).

Voice. Tape-recorded (53, 104, B, BUT, GREG, HAZB, LEM, McVIC, STJ). Song, repeated phrases of 4–5 pure, loud, liquid whistles, not rich and fluty like ground-thrush *Zoothera*, nor soft and mellow like an oriole, but with a rather melancholy quality, 'pyoo-pyoo-tseewo', 'pyaw-pyaw-pyoo-pyi', 'toplee-toto-tooyoo'; soft and loud notes often alternate: 'hee-*hwee*-hu-*hway*-ho', '*pee*-pat*yee*-pat*yoo*', 'pee*yaw*-ti-*day*wa'; sometimes with bubbling quality, 'pyi-pyoo-wolly-weelu', sometimes longer, 6–9 notes, some of which may be run together, 'pi-pyoo, pyee-pyaw, pyooty', 'pi-tyoo-widdlyoolyoo-pityay'; one song went up the scale, with emphasis on second syllable of each double-note: 'pu-*tyaw*, pu-*tyaw*, pu-*tyoo*, tee-*lee*'. Variation almost endless. Singer repeats a phrase several times before switching to another. Intervals between phrases 5–10 s. Sings alone, or individuals sing within a group, but no 'chorusing' as in *Illadopsis*, nor is song interspersed with babbling chuckles. Group singing is not the norm (see Mann *et al.* 1978), and duetting unrecorded. Short songs like those of African Hill Babbler *Pseudoalcippe abyssinica* given by birds following ant swarm (Willis 1983). Other calls include loud, explosive 'chichichichi ...' and a liquid 'tuituituituitui'. Alarm, a persistent 'ker-ker-ker', 'kertaker' or 'kertaker-taker'.

General Habits. Inhabits undergrowth and floor of mature montane forest, especially near streams; in Kivu also high-altitude gallery forest and transitional forest down to 1200 m. In Rwanda (Nyungwe) occurs only in closed, deep-shaded forest, favouring humid hollows, streamsides, and shrubby slopes with Acanthaceae and tree ferns (Dowsett-Lemaire 1990). In Kenya, occurs within 1000 + mm rainfall area.

Usually forages singly or in pairs; hops along ground, often turning over dead leaves. Shy and skulking, but sometimes comes into more open areas when following swarms of army ants *Dorylus*. Joins other species around army ants; birds following ants hopped from one low perch to another, fled rapidly on human approach (Willis 1983). Song given low down, 1–2 m above ground, but carries a long way (Dowsett-Lemaire 1990). One bird sang every morning repeatedly, stopped singing by midday (Willis 1983).

Duration of primary moult very variable in Kakamega Forest, Kenya: (n = 8) 39–283 (196) days (Mann 1985).

Bird at exceptionally low altitude of 580 m on Mt Cameroon may have been altitudinal migrant (Stuart 1986).

Food. Variety of insects taken when flushed by army ants.

Breeding Habits. Presumably monogamous. Very faithful to its territory; ringed birds frequently retrapped in same spot (Mann 1980). Nest and eggs unknown.

LAYING DATES: Cameroon (breeding condition Apr, Nov); Zaïre (birds in breeding condition Jan–Feb, Apr (Itombwe), Nov (Mt Kabobo)); Rwanda (increase in song intensity, and pair giving persistent alarm call, Oct, suggestive of nesting).

References
Holyoak, D.T. and Seddon, M.B. (1990).
Mann, C.F. *et al.* (1978).
Stuart, S.N. (Ed.) (1986).

Genus *Pseudoalcippe* Bannerman

In plumage and other characters similar to *Illadopsis*, q.v., but with more racial differentiation (one race with black head, others with streaked throats). Placed in *Illadopsis* by Ripley and Beehler (1985), but there are major behavioural and vocal differences: *P. abyssinica* is solitary, forages like a warbler, at middle levels, not in the undergrowth, and has beautiful mellow song unlike that of any *Illadopsis*. Placed in Asian genus *Alcippe* by Deignan (*in* Mayr and Paynter 1964) at one end of his timaliid sequence and *Illadopsis* (or rather, *Trichastoma*) almost at the other.

Endemic, 1 species. The distinctive race *atriceps* has at times been considered a separate species (Mayr and Paynter 1964, Sibley and Monroe 1990), but behaviourally and vocally it is identical to other races, and we agree with Dowsett and Dowsett-Lemaire (1993a) that it belongs in *P. abyssinica*.

Plate 3

Plate 4

Plate 2
(Opp. p. 15)

Pseudoalcippe abyssinica (Rüppell). African Hill Babbler. Akalat à tête sombre.

Drymophila abyssinica Rüppell, 1840. Neue Wirbelt., Vögel, 13, p. 108; Simen Mts, Begemdir Province, Ethiopia.

Range and Status. Endemic resident. Extreme SE Nigeria in Gotel Mts (very common Gangirwal, fairly common Chappal Waddi, 1800–2300 m). Cameroon, widespread and common in Bamenda Highlands, 2000–2900 m, from Mt Oku, Ndu and Banso Mts to Bafut-Ngemba Forest Reserve, Mt Lefo, Babadjou and Bamboutos Mts; commonest at highest altitudes, and apparently absent from slightly lower forest patches (1700–1800 m) just west of Bamenda Highlands (Stuart 1986); outlying populations to the northeast at Tchabal Mbaba (07°16′N, 12°10′E) and to the south on Mt Cameroon, 910–2440 m. Bioko, at 1100–1800 m in S highlands, 1200–2900 m on Pico Basilé; common above 1600 m. W-central Angola (Mt Moco and Mt Soque above 2000 m, Bailundu highlands). Ethiopia in W Highlands north to Simien Mts, including Tana Basin (Bahar Dar, Addis Zemen: Olson 1976), SE Highlands and S Ethiopia, 1800–3200 m (down to 900 m in dense vegetation in W Highlands), frequent. Common in extreme SE Sudan (Imatong, Dongotona and Didinga Mts), in forest above 1800 m. Common and widespread in mountains on both sides of Albertine Rift, 1350–3000 m, from Lendu Plateau, Rwenzoris and Kibale Forest (Uganda) to N and S Kivu (to Itombwe), Bwindi (Impenetrable) and Mafuga Plantation Forests, volcano region, W Rwanda, W Burundi south to Bururi Forest, and further south on Mt Kabobo and Marungu Plateau (Zaïre) and Mahari Mt and Ufipa Plateau (W Tanzania); small population in SE Zaïre (Shaba) at summit of Kibara Mts, 1750 m. Extreme NE Uganda (Mts Lonyili and Morongole); Mt Elgon (Kenya/Uganda); widespread and common in highlands of W and central Kenya, 1500–3000 m, east to Nyambenis, and present on forest islands of Mts Loima, Kulal, Nyiru, Ndotos and Maralal in N and Chyulu Hills in S; once at Lolgorien (01°14′S, 34°48′E: B.W. Finch *in* Turner 1993). Nguruman Hills (S Kenya), Loliondo and crater highlands of N Tanzania south to Oldeani and Mbulu, through Arusha Nat. Park and Mt Meru to Kilimanjaro (1600–2500 m), N Pare Mts (1500–2100 m), S Pares and W Usambaras; recently discovered in S Usambaras at 1250–1350 m on Mt Nilo (Cordeiro 1998); E highlands of Tanzania from Ukagurus (common) and Ulugurus (1200–2500 m) to Udzungwas (1400–1900 m), Mt Rungwe, Umalila, Njombe and Songea. Nyika Plateau (Zambia/Malaŵi), and south in Malaŵi to 11°51′S in S Viphya Mts, and on Namizimu Hill and Mt Mangoche; its absence from suitable forest on Misuku Hills (between populations on Nyika Plateau and Isoko Hills, S Tanzania) is anomalous (Dowsett-Lemaire 1989). Altitudinal range in Malaŵi 1550–2400 m. N Mozambique (Njesi Plateau). In Malaŵi, locally reaches densities of 4–5 pairs per 10 ha.

The population of *P. a. atriceps* in Bamenda Highlands, Cameroon, though still numerous, is at risk from forest clearance; *P. a. monacha* on Mt Cameroon is under no immediate threat (Stuart 1986).

Description. *P. a. abyssinica* (Rüppell): Ethiopia and Sudan to E Uganda (Mt Elgon), Kenya, N Tanzania (including Usambaras), W and SW Tanzania, SE Zaïre (Shaba) and Angola. ADULT ♂: top and

Pseudoalcippe abyssinica

side of head uniform grey. Mantle and scapulars to uppertail-coverts rich brown. Tail feathers dark brown, outer webs fringed rich brown. Chin and throat greyish white with darker grey streaking. Upper breast grey, grading to grey-white on belly and undertail-coverts; sides grey apart from brown patch on lower flanks. Primaries and secondaries dark brown, outer webs broadly fringed rich brown; primary coverts dark brown with rich brown outer edges and tips; tertials and rest of upperwing-coverts rich brown like upperparts. Underwing-coverts and axillaries greyish white with brown tinge. Bill blackish, dark grey or dark horn above, pale horn to whitish grey below; eyes brown or dark red; legs dark grey, grey-brown or pale grey. Sexes alike. SIZE (10 ♂♂, 10 ♀♀: wing, ♂ 64–73 (69.2), ♀ 66–73 (69.0); tail, ♂ 56–66 (61.3), ♀ 59–65 (61.9); bill, ♂ 15–17 (15.8), ♀ 15–17 (16.0); tarsus, ♂ 22–24 (22.7), ♀ 21–24 (23.6). WEIGHT: Kenya, ♂ (n = 4) 16–19 (17.3), ♂ (n = 23) 14–25 (18.8), ♀ (n = 11) 15–25 (18.7), ♀ (n = 3) 16–20 (18.0).

IMMATURE: like adult, but top and sides of head brownish.

P. a. monacha (Reichenow): Mt Cameroon. Upperparts more rufous than in nominate race but grey of head extends onto mantle; top of head with darker grey streaking; chin and throat prominently streaked.

P. a. claudei (Alexander): Bioko. Like nominate race, but whole mantle and inner scapulars grey.

P. a. atriceps (Sharpe): Nigeria, Cameroon except Mt Cameroon; E Zaïre to Rwanda and W Uganda (Rwenzori). Top and sides of head and chin black; throat, breast and sides darker grey than in nominate race, upperparts more rufous. Pure albino adult ♂ known from Zaïre. WEIGHT: (Zaïre) 2 ♂♂ 20, 20; (Uganda, ♂ (n = 14) 17–25 (21.9), ♀ (n = 7) 16–21 (19.8).

P. a. stierlingi (Reichenow): E and S Tanzania (Kilosa and Uluguru to Njombe, Songea and Rungwe). Forehead and sides of head blacker than in nominate race; chin and throat grey with bold black and whitish streaking; upperparts as nominate race. Immature has sides of head paler than adult, only an indication of streaks on throat, some brown on breast. WEIGHT: 1 ♀ 16.

P. a. stictigula (Shelley): Malaŵi, N Mozambique. Like *stierlingi*, but upperparts more rufous-brown.

Field Characters. Length 15 cm. A montane forest babbler usually first discovered by its sweet song. Back red-brown, breast grey and belly whitish throughout range, but head and neck either grey (*abyssinica*, *monacha*, *claudei*), black (*atriceps*) or grey with blackish forehead and face, black and white stripes on throat (*stierlingi*, *stictigula*). Best distinguished from Mountain Illadopsis *I. pyrrhoptera* by behaviour (forages in pairs at mid-levels, not in flocks in undergrowth) and single-bird song (not chattery group chorus). Grey or black head sharply cut off from red-brown back (in Mountain Illadopsis, brownish grey head merges into brown nape and back, and belly grey, not white). Less of a skulker than illadopsises. Sitting bird's pale bill very noticeable in gloom of forest (Moreau and Moreau 1939).

Voice. Tape-recorded (32, 53, 86, 104, 109, B, C, F, LEM, SVEN). Song a sweet, rich warble of 5–20 notes (usually 8–13) lasting 1·5–3·5 s; mellow whistled notes are mixed with louder upslurred 'whee' and low conversational 'toto'. Singer often repeats previous song with slight variation, or joins 2 different songs together in a continuum (Keith and Gunn 1971); very inventive. No transcription could do justice to this beautiful song, considered one of the finest in Africa (Keith and Gunn 1971); 'the finest master-singer I have heard among African birds' (H. Granvik *in* Jackson and Sclater 1938). Songs very variable throughout range but variation not correlated with race; typical songs of mellow notes are indistinguishable, whether from Mt Kenya (nominate *abyssinica*), Udzungwa Mts, Tanzania (*stierlingi*) or Impenetrable Forest, Uganda (*atriceps*). Song of *abyssinica* from Cherangani Mts (Kenya) is somewhat faster, sharper and less musical. One bird usually sings alone, but duet recorded from Mt Cameroon (*monacha*: Chappuis 1975): song of ♂ 'toto-*sweeya*-*sweeya*-toto-pwee-hoo' accompanied by low dry chatter from ♀. The '*sweeya*' notes of ♂ are shrill and almost tuneless, and only the final 'pwee-hoo' has the characteristic sweetness of singers elsewhere. Unlike illadopsises, does not chorus. Sings all year in Kenya (Mau-Narok) but most strongly in Apr–May and Nov–Dec (Sessions 1966). Both sexes said to sing all year (Cameroon: Serle 1950), but this must refer to ♂ only, since ♀ merely gives a chatter during duets. Said to sing after dark but sings mainly during the day.

General Habits. Inhabits montane primary and secondary forest, especially in glades, clearings and more open areas, and at forest edge; dense thickets in second-growth; gallery forest and riparian forest strips; also bracken-briar (Marungu) and bamboo (Rwenzoris). On Bioko, moss forest, lichen forest and plantations; in Ethiopia, Olive–*Podocarpus*–Juniper forest, 1800–2400 m, Juniper–*Podocarpus* forest, 2400–3200 m, possibly also *Hagenia* forest, 2900–3200 m. Reported from a eucalyptus plantation above Buea, Mt Cameroon (Grimes *in* Stuart 1986). Frequents understorey, sometimes within 1–2 m of ground and up to mid levels and occasionally canopy. In Malaŵi, mostly at 4–16 m in mid-stratum thickets within forest with hanging masses of creepers, especially where there is a break in the canopy (Dowsett-Lemaire 1989).

Occurs singly or in pairs, less often in small groups of 3–5; sometimes joins mixed-species flocks. Gleans insects from twigs and leaves and takes some small fruit; sometimes hovers, and once seen attending an ant swarm, catching small prey in foliage above ants (Malaŵi: Dowsett-Lemaire 1989). Readily ventures into clearings to visit *Sericostachys* thickets and low tangles around trees (Nyungwe: Dowsett-Lemaire 1990).

Said to have 'definite seasonal movements' (Mackworth-Praed and Grant 1960) but present all year at 3000 m around Mau Narok, Kenya (Sessions 1966), and no evidence of altitudinal migration in Malaŵi.

Food. Insects, including caterpillars and bark beetles; small molluscs; also seeds, berries and small fruits, including *Urera*, *Galiniera*, *Rutidea*, *Jasminum abyssinicum* and *Ochna holstii*.

Breeding Habits.

NEST: small basket of dry twigs woven with fine dry stalks, lined with thin plant fibres, 4 m above ground (*monacha*: Eisentraut 1963); neatly made, light but strong, regular cup with outer layer of plant stems, roots and gossamer, lined with fine grasses; ext. diam. 95, int. diam. 60, ext. depth 65, int. depth 40; slung between 2 upright shoots in undergrowth 1·2 m above ground (*atriceps*: Serle 1950); outer layer of moss and leaves, lined with very fine plant threads (*atriceps*: Eisentraut 1973); small cup composed of moss, beard lichen and dead flower spikes of a forest *Salvia*, lined with stiff fibres, sometimes with cobwebs aiding in support, slung between twigs 2 m above ground in a shrub (*abyssinica*: Moreau and Moreau 1939); sometimes horsehair in lining (Mackworth-Praed and Grant 1963); shallow, rather flimsy but neat open cup of spiralling tendrils and fine twigs, lined with very fine, hair-like, chestnut-coloured fibres, ext. diam. 100, int. diam. 60, ext. depth 40, int. depth 20 (*stierlingi*).

EGGS: 2; rounded ovate; smooth, without gloss; creamy or buffy cream, whiter at narrow end, variably spotted and speckled with dark brown or pale reddish brown over lilac, in belt near broad end. SIZE: (n = 2, Cameroon) 20·8 × 15·7, 20·4 × 15·5; (n = 2, Tanzania) 20·1 × 15·1, 19·4 × 15·1.

LAYING DATES: Nigeria, Mar; Cameroon: Mt Cameroon, Nov (juvs Feb–Mar, carrying nest material Dec, bird in breeding condition Feb); Bamenda, May, elsewhere, breeding indications Feb–May, Sept; Ethiopia Apr (breeding condition June, immature Dec); Sudan, Nov–Jan; Zaïre, (juvs Mar, Aug, birds in breeding condition Mar, May–June, Nov; probably breeds all year, including months with little rainfall (Prigogine 1971)). Rwanda (increase in singing output from December); Uganda (♀ with egg in oviduct and ♂ in breeding condition May); Kenya (♀ with egg in oviduct Apr, birds in breeding condition Mar); Tanzania, Oct–Feb (juvs and fresh brood patch July); Malaŵi, Oct (5), Nov (2).

INCUBATION: by both sexes. Sitting birds allowed observer to watch at distance of 1 m; when flushed from nest they gave no alarm call but sang loudly (Moreau and Moreau 1939).

Reference

Stuart, S.N. (ed.) (1986).

Genus *Ptyrticus* Hartlaub

A quite large babbler, rufous above, white below, with spotted breast, rather long hook-tipped bill with pale lower mandible, short wings, rather long, strong tarsus and strong feet. ♂ considerably larger than ♀. The only babbler in Africa with white underparts and spotted breast. In respect of plain brown upperparts and rufous undertail-coverts, long bill with pale lower mandible, and strong, pale legs and feet, *Ptyrticus* is apparently closely related to *Illadopsis*, particularly to *I. puveli*. It differs in its greater size (especially in N tropics) and colour and pattern of underparts, and voice is completely different.

Endemic; 1 polytypic species.

Plate 2
(Opp. p. 15)

Ptyrticus turdinus **Hartlaub. Spotted Thrush-Babbler. Akalat à dos roux.**

Ptyrticus turdinus Hartlaub, 1883. J. Orn., p. 425; Tamaja, Uelle.

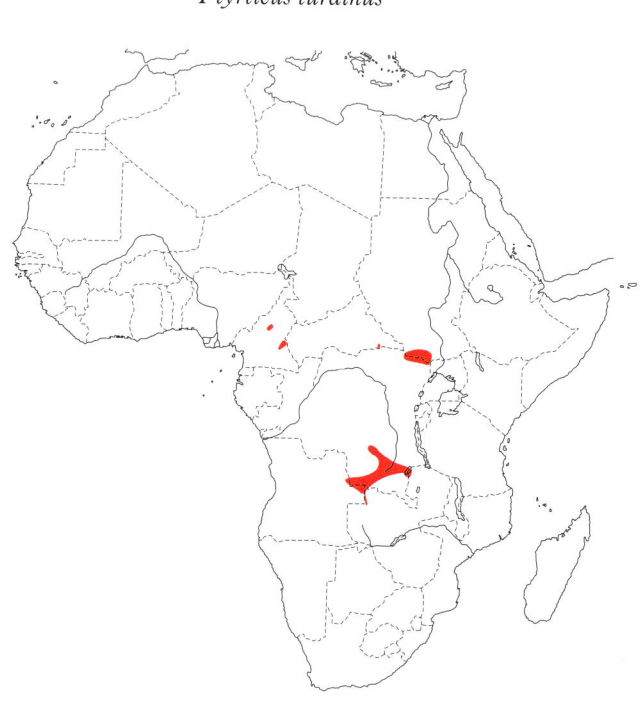

Ptyrticus turdinus

Range and Status. Endemic, resident. 3 separate populations in forest outliers in Cameroon, near Sudan/Zaïre border, and S Zaïre, and just across borders into Angola and NW and NE Zambia. Cameroon: Adamawa Plateau from southwest of Ngaoundéré (Galim, 07°06′N, 12°29′E) to Tibati and Ngaoundaba (Anon. 1995), and south flank of Adamawa Plateau near border with Central African Republic, between Bertoua (04°35′N, 13°41′E), Garoua Boulai (05°53′N, 14°33′E) and upper Kadei R. southeast of Bétaré Oya. Uncommon. Central African Republic: only record, a pair collected at 05°20′N, 24°20′E (Friedmann 1978); probably in Yade Hills, an extension of Cameroon's Adamawa Plateau in W. Sudan: uncommon, gallery and bowl forests from Bengengai (04°49′N, 27°43′E), Mt Baginze (04°29′N, 29°00′E) and Yambio (04°34′N, 28°23′E) to Tomaya (04°38′N, 29°50′E) and Yei (04°05′N, 30°40′E), north to about Mvolo (06°02′N, 29°53′E); other localities: Li Rangu, Sakure, Aza Forest. NE Zaïre: Kibali R. (headwaters of Uele R.), Niangara (03°42′N, 27°53′), Dungu (03°37′N, 28°33′E), Nzoro (03°17′N, 29°26′E), and between Faradje and Aba; S Zaïre: from Kabinda (06°08′S, 24°27′E: Lomami District) to Upemba Nat. Park at 1400–1600 m and Kibara Plateau (50 km northwest of L. Mweru), and southwest to Kasaji (10°23′S, 23°28′E: Lulua District) and Angola border. In Upemba Nat. Park known from Muye, Kanonga, Shinkulu and Kabwe, at 800–1630 m. Angola: uncommon resident along Luachimo R., N Lunda Norte. NW Zambia: N Mwinilunga at Salujinga (10°58′S, 24°07′E), Nyakaseya (11°09′S, 24°09′E, on Chinyashi R.) and on Lisombo (Isombu) Stream (about 11°15′S, 24°05′ E), and occurs southwest to Minyanya, Balovale (13°05′S, 22°23′E); NE Zambia: 2 parties seen on Luao R. in Nchelenge District (about 09°20′S, 28°44′E) (Aspinwall 1978).

Description. *P. t. turdinus* Hartlaub: S Sudan, NE Zaïre. ADULT ♂: forehead rufous in centre, whitish at sides, crown, nape, hindneck, sides of neck and mantle rufous, scapulars and back rufescent brown, rump rufous, uppertail-coverts bright rufous, long. Tail brown, outer vanes of feathers fringed rufous on proximal halves. Lores white, cheeks and ear-coverts mottled white and rufous. Chin and throat white, with small rufous teardrops at sides of throat; breast white, with a few large rufous triangles, many at sides of breast coalescing into large rufous patch; belly white, flanks white with obscure rufous smudges, thighs white inside, brownish outside, vent white, undertail-coverts white with rufous blotches. Wings rufous-brown, inner vanes of primaries and secondaries blackish, underside of remiges grey, underwing-coverts and axillaries white. Bill: upper mandible dark brown with pale bluish grey at extreme tip, and lower mandible light blue-grey; eyes bright brown or reddish brown; legs and feet very pale flesh or pinkish grey. Sexes alike. SIZE: wing, ♂ (n = 3) 104–111 (107), ♀ (n = 3) 90·5–97 (93·5), 1 unsexed 96; tail, ♂ (n = 3) 75–80·5 (78), ♀ (n = 3) 68–70 (69), 2 unsexed 71, 75; bill to feathers, 2 ♂♂ 22, 23, ♀ (n = 3) 19–23 (21·3); bill to skull, ♂ (n = 3) 23–26 (24·7), 1 ♀ 23, 2 unsexed 24, 25; tarsus, 1 ♂ 38, ♀ (n = 3) 32–35 (33·3), 1 unsexed 35. WEIGHT: 2 ♂♂ 66, 72·5, 1 ♀ 59.

IMMATURE: like adult but bill brownish black, with dull yellowish green along edges of both mandibles and at base of lower mandible, and eyes very dark brownish grey.

P. t. harterti Grote: Cameroon. Same as nominate race except that tail is rufous. SIZE (2 ♂♂): wing, 93·5, 108; tail 74, 82·5; bill to feathers 19, 21·5; bill to skull 23, 26; tarsus 34, 35.

P. t. upembae Verheyen: S Zaïre, NE Angola, NW Zambia. Adult like nominate race but hindneck to rump more olivaceous than rufescent; spots on breast brown (not rufous), more numerous but not forming patch at sides of breast, and tiny spots extend up onto throat; immature distinctly rufous. Smaller. SIZE (unsexed, n = 10): wing 89–100 (94·8), tail 69–75·5 (70·1), bill 21·2–23·3 (22·0), tarsus 28·4–31·2 (30·25) (W.R.J. Dean, pers. comm.); (2 ♂♂) wing 96, 104; tail 74, 77; bill to feathers 18, 18; bill to skull 22, 23; tarsus 32·5, 35.

Field Characters. Length 16·5 cm. A lowland undergrowth bird, superficially like a large illadopsis but with a very different song. Rufous-brown above, brighter rufous on rump, sides of neck and hind-collar; below white with *large brown breast-spots*. Grey-chested Illadopsis *Kakamega poliothorax* is same size, with rufous upperparts, but is a montane bird with unspotted grey breast and flanks. Other *Illadopsis* babblers are smaller, with duller upperparts, often travel in chorusing groups.

Voice. Tape-recorded (53, 102, 104, B, F, ASP, CART, CHA). Song by single bird, 3 loud whistles evenly spaced over 1·5 s, 'whoh ... tyaw ... tyaw', last 2 downslurred, first a little more clipped. Said by Chapin (1953) to be given by ♂ only, but recordings by Chappuis (1975) clearly show that birds also duet. In precisely-timed duet, second bird echoes each of partner's notes immediately after it, 'whoh/whoh, tyaw/tyaw, tyaw/tyaw'. Notes are pure and strong but lack any thrush-like sweetness; synchrony, quality and timbre more reminiscent of bush-shrike *Laniarius* than of any African babbler. They bear no resemblance to the hoarse chorusing of Leaf-love *Pyrrhurus scandens* (as was suggested by Chapin 1953; but see Keith *et al.* 1992). Said to give 'low clucking calls' (Chapin 1953), a 'shrike-like 'chuck' note' (Oatley 1969) and a loud cackling or chattering noise.

General Habits. Frequents moist evergreen riparian forest (Zambia), lowland swampy woodland and thickets (on grassy river-plain in E Cameroon) and damp undergrowth in gallery, outlier and bowl forests; often near streams, and seems generally to inhabit riverine forest. Altitudinal range about 800–1650 m. Habitat is same as that of much better known Leaf-love *Pyrrhurus scandens*, and habits of the 2 species said to be similar (Chapin 1953). Very wary. Keeps in parties of *c.* 6 birds. Whistles mainly at break of day, sometimes toward sundown, seldom in middle of day. Entire party of birds joins in the chattering. Responds strongly to tape-playback of voice (Aspinwall 1978, Anon. 1995). Calls loudly throughout year. A bird 'called up' in swamp forest in NW Zambia approached observer low down, uttering 'chuck' notes and spasmodically jerking tail up; on spotting observer it flew into thick mass of tangled lianas above a stream and 'danced up and down in a most excited fashion, pouring forth an extraordinary and continuous babble' (Oatley 1969). Its strong, quite long legs, and fact that it can be snared on ground, suggest that it forages mainly on ground, although only observation to that effect is 'found on the floor of the riparian evergreen forest' (Benson 1958).

Food. Insects, including beetles, small ants, cicadas, caterpillars, grasshoppers and winged termites. Birds in Cameroon caught in snares baited with termites (Chapin 1953). Once a spider and once a small frog.

Breeding Habits. Nothing known, but inferred to breed in NE Zaïre in rainy season, June–Nov (Chapin 1953), in S Zaïre in Oct–Dec (Lippens and Wille 1976), and in Angola about Feb when birds in breeding condition (Ripley and Heinrich 1966); juv. with unossified skull, Zambia, Oct.

Reference
Chapin, J. (1953).

Genus *Turdoides* Cretzschmar

Quite large, rather compact and robustly built babblers with curved culmen and almost decurved bill, white, yellow or red eyes, short rather curved wings, quite long, somewhat graduated and rounded tails, long in many Oriental species, and moderately strong legs. Bill and legs black (bill yellow in one Oriental species), or yellowish in slimmer, rufescent, desert-dwelling species of distinctive subgenus *Argya* (chatterers). Plumage grey-brown, stripey dark brown, or plain and rufescent, belly pale or underparts brown-stripey or white, feathers of head, breast and often mantle rounded with pale crescentic tips, or pointed with V-shaped pale tips, or very pointed, either with dark centres or with white arrow-marked tip or both, the white tips standing proud of breast feathers in one species (in subgenus *Turdoides*, 'babblers' proper); tail blackish, tail and tertials always with fine structural bars, rectrices narrower in juvenile than adult, 1 or 2 species with bare patches on sides of head. Plumage subject to fine structural banding, particularly on feathers of mantle, tertials and tail.

Highly sociable, year-round in small flocks; many species, perhaps all, are co-operative breeders. Sedentary. Inhabit thickets in savanna woodland and desert steppe, keeping low in woody vegetation, foraging, mainly on ground, with one bird typically posted on top of bush as a sentry. Can move rapidly on ground where gait a crouching hop. Noisy; chattering voices. Mainly insectivorous. Nest a large, loosely built open bowl. Eggs long ovals, immaculate blue except in African *T. gymnogenys* and Asiatic *T. altirostris*.

29 species, Asia and Africa; 17 in Africa, all endemic. 3 are rufous chatterers, related to 4 dry-country S Asiatic 'replacement' species, possibly superspecifically (Hall and Moreau 1970, Harrison 1982, Lees-Smith 1986). 2 African chatterers, the Saharan *T. fulvus* and the virtually parapatric E African *T. rubiginosus*, are alike in plumage and size and we regard them as composing a superspecies. We keep the smaller E African *T. aylmeri* independent.

The remaining 14 African species fall into 2 groups of parapatric species, so that most savanna woodland regions have 2 species only, one from each group. Their evolutionary differentiation is complex but, in the absence of DNA research and field studies at parapatric interfaces or in zones of marginal sympatry, remains poorly understood. No two taxonomists agree about close relationships. Hall and Moreau (1970) placed all 14 species in 2 superspecies; Sibley and Ahlquist (1990) used 4 superspecies of 2, 2, 3 and 3 species; and Dowsett and Forbes-Watson (1993) recognized only one (*T. plebejus/T. leucocephalus*). White (1962) admitted 12 full species, but Clancey (1984) separated *T. hartlaubii* from White's *T. leucopygius* and Clancey (1986) affirmed Hall and Moreau's separation of *T. sharpei* from White's *T. melanops*, making 14. Clancey (1984) argued relationships in and of the contentious *leucopygius/hartlaubii* group of populations at length, evolutionarily and persuasively. Short *et al.* (1990) disagreed with some of Clancey's conclusions; they think that *leucopygius* and *hartlaubii* are megasubspecies of one species, and their field experience indicates that *sharpei* and *melanops* are conspecific.

Considering those opinions and studying skins anew, we classify the numerous taxa (apart from the chatterers) into 14 species, being 3 superspecies and 6 independent species. The 2 widest-ranging species, N-tropical *T. plebejus* and S-tropical *T. jardineii*, are a superspecies, closely related to *T. leucocephalus* (all are silvery-cheeked; Sibley and Monroe 1990 place all 3 in a superspecies). North of the tropical rain forest, *T. reinwardtii* and *T. tenebrosus* are each other's nearest relative and form a superspecies, perhaps related to *T. melanops*. They are all large, large-billed, brown, mottle-throated, with long black tails and broad rectrices. *T. leucopygius* (Ethiopia), *T. sharpei*, *T. hartlaubii*, *T. melanops* (Caculovar valley, Angola, to Okavango R., Botswana) and *T. bicolor* (southern Africa) form a chain of nearly parapatric taxa. Although leapfrogging, *leucopygius* and *hartlaubii* clearly comprise a superspecies and so do *sharpei* and *melanops*. In our view *sharpei* and *hartlaubii*, exactly parapatric, are also very closely allied, and so we relate all 4 forms as a superspecies. Short *et al.* (1990) regard the 2 ends of the 5-species chain, *leucopygius* and *bicolor*, as a superspecies; but, because of extensive sympatry between *T. bicolor* and *T. hartlaubii* in N Botswana (Penry 1994) we exclude *bicolor* from its superspecies, whilst recognizing its close alliance. Each of 3 E African babblers, *T. hindei*, *T. hypoleucus* and *T. squamulatus*, is somewhat enigmatic; *T. hindei* of central Kenya is more closely related to *T. leucopygius* than to any other congener (Plumb 1979), and the other 2 remain of uncertain affinities. Last, the distinctive white-headed, bare-cheeked *T. gymnogenys* of arid SW Angola and N Namibia, may be a relation, distant temporally and spatially, of the white-headed, barish-lored *T. leucocephalus*.

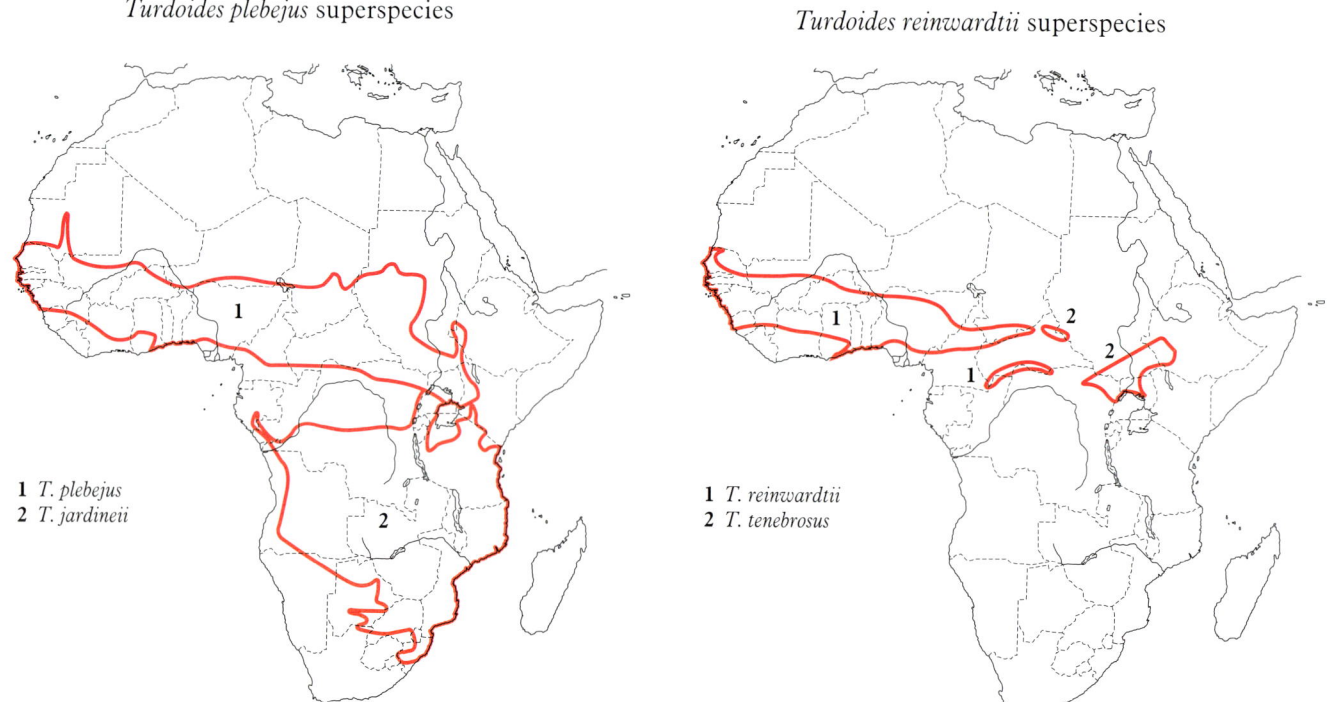

Turdoides plebejus superspecies

1 *T. plebejus*
2 *T. jardineii*

Turdoides reinwardtii superspecies

1 *T. reinwardtii*
2 *T. tenebrosus*

Turdoides hartlaubii superspecies

Turdoides fulvus superspecies

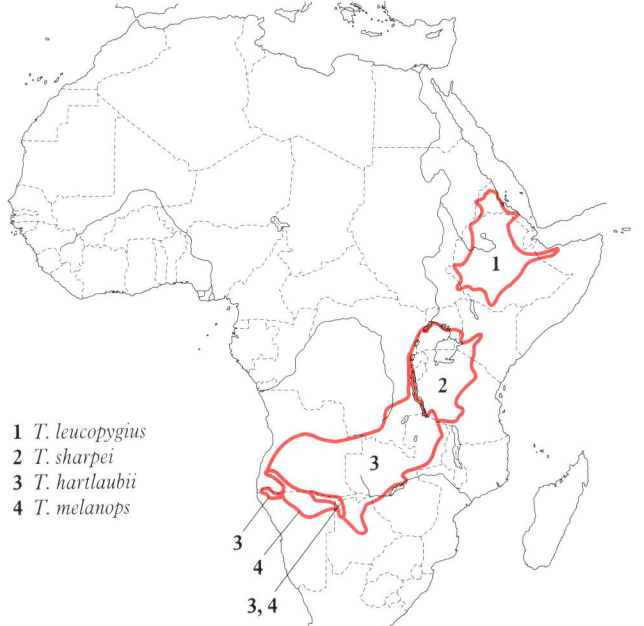
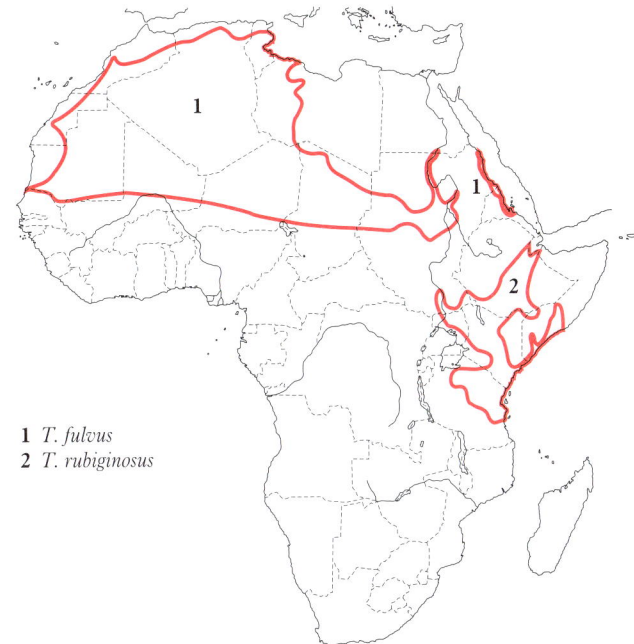

1 *T. leucopygius*
2 *T. sharpei*
3 *T. hartlaubii*
4 *T. melanops*

1 *T. fulvus*
2 *T. rubiginosus*

Turdoides plebejus (Cretzschmar). Brown Babbler. Cratérope brun.

Ixos plebejus Cretzschmar, 1828. In Rüppell, Atlas Vög, p. 35, pl. 23; Kordofan Province, Sudan.

**Plate 3
(Opp. p. 30)**

Forms a superspecies with *T. jardineii*.

Range and Status. Endemic resident, S Mauritania and Guinea to SW Ethiopia and W Kenya. Mauritania, isolated groups in woodland in Senegal R. valley and in massifs (Assaba, Tagant) north to 20°15′N in Adrar hills. Senegal, widespread, locally common, but in many areas scarce. Gambia, common throughout. Sierra Leone, records in Outamba area in NW (Happel 1985), Kamasigi and Karina district. Mali, common north to about 15° including Parc Nat. Boucle du Baoulé, and occurs on Dogon Plateau; and quite common in sahel zone in extreme E. Burkina Faso, widespread north at least to Ouagadougou area. Ivory Coast, widespread south to Bouaké. Ghana, widespread but local (Yendi, Tumu, Tamale, Pwalagu, Bolgatanga) and not common, south to woods around inselbergs in Accra Plains and to coastal areas west of Accra; once Cape Coast. Togo and Benin, common throughout, south to coastal scrub. Nigeria, frequent in coastal scrub at Badagri in SW and common in all types of savanna north of rainforest zone (south to Enugu in SE), particularly north of great rivers, and seemingly commonest towards Niger border. Niger, widespread in extreme SW (Parc du 'W', Tapoa, Gaya, Korogoungou); known from Maradi area in S and may be common there and elsewhere near Nigerian border (Giraudoux *et al.* 1988). Cameroon and Central African Republic, throughout savannas north of forest. Chad, widespread north to about 13°N; recorded in most southern wadis in Ouadi Rime – Ouadi Achim Faunal Res., north nearly to Wadi Hattat (Haddat), 14°32′N.

Turdoides plebejus

Sudan, widespread in W (Ennedi, Darfur), SW and S, common towards borders with Central African Republic and Zaïre; less common (or at least fewer records) northeast to drainage basin of Wadi Magrur (northeast of El Fasher) and to Kordofan; east of Nile occurs in E Equatoria north to about Mongalla and Kapoeta, but also records just north and south of Ilubabor salient – part of the Ilubabor (Ethiopia) population. Ethiopia, frequent to locally common in SW. N and NE Zaïre, from between Libenge (Zaïre) and Bangui (Central African Republic) to Uele grasslands especially north of Uele R., southwest to Buta and southeast to Semliki valley. Uganda, common, southwest to Mengo and Hoima. Kenya, locally common at 600–1500 m from Turkana and Pokot to Kerio valley, Kisumu and Muhoroni.

Description. *T. p. cinereus* (Heuglin) (includes '*gularis*'): SE Nigeria to S Sudan, SW Ethiopia and W Kenya. ADULT ♂: upperparts greyish brown, a little paler and greyer on rump; feathers of forehead to hindneck with dark brown centres forming distinct short streaks, accentuated on forehead and crown by shafts being black, shiny and quite wide; mantle, scapulars and back with diffuse, dark streaks, often barely discernible. Tail dark brown above, blackish brown below. Lores and ill-defined, short, narrow superciliary stripe mottled greyish buff; ear-coverts, cheeks and sides of neck greyish brown; moustachial area mottled grey-brown and buffish white. Chin pale greyish brown; throat and breast deeper brown, all feathers pale-fringed, with small, sharp buffish white spots at tip and sometimes with dark shaft-spot or streak behind tip. Belly whitish grey-buff, flanks, thighs and undertail-coverts buffish brown, plain or with ill-defined, diffuse darker brown streaks. Remiges, alula and upperwing-coverts brown, nearly as dark as tail, but with median and lesser coverts paler and greyer. Primary underwing-coverts brown; rest of underwing-coverts and axillaries pale buff; tawny-buff area on underside of flight feathers formed by broad pale inner borders. Eyes bright chrome yellow, sometimes pale straw, or white tinged olive (collectors' labels); bill blackish or dark horn; legs dark slate grey, blackish or blackish brown. Sexes alike. SIZE: (10 ♂♂, 10 ♀♀) wing, ♂ 98–113 (103), ♀ 99–106 (101); tail, ♂ 94–107 (100), ♀ 95–106 (98·9); bill, ♂ 24–27 (25·4), ♀ 23–25 (24·0); tarsus, ♂ 32–36 (33·7), ♀ 31–35 (33·6). WEIGHT: (Ghana, unsexed, n = 8) 61–71 (65·3), (Kenya, unsexed, n = 25) 52–70 (62·5).

IMMATURE: like adult but flanks and belly markedly buffier and eyes dull yellow. Juvenile not known, but see below.

NESTLING: not known. Newly fledged young has dark greyish brown eyes.

T. p. plebejus (Cretzschmar): N Nigeria (west to Kaduna) and L. Chad to W and central Sudan. Slightly paler above and considerably paler below than *cinereus*; lores and chin whiter, belly whitish. Throat and breast feathers with broader whitish fringe and rather more conspicuous dark spot behind it. Juvenile paler and plainer than adult, browns more tawny; lacks spots and streaks; underparts buffy white, with brown band across upper breast. Larger; wing (10 ♂♂) 106–117 (111·5).

T. p. platycircus (Swainson): Senegal to W Nigeria (Sokoto, Ilorin, Abeokuta). Like *cinereus* but top of head darker brown, feathers blackish brown with narrow grey-brown edges giving scaly rather than streaked effect; pale greyish rump contrasts more with brown upperparts; ear-coverts, cheeks and around eye plainer, often forming pale grey area contrasting with adjacent parts; throat and breast darker brown than in *cinereus*, feathers with whitish crescents and penultimate dark bands; underparts behind breast unstreaked. Large; wing (10 ♂♂) 106–115 (110). WEIGHT: (Ghana, unsexed, n = 4) 61–71 (66·4).

NESTLING: not described. Newly fledged young fluffy, with greyish brown eyes.

Field Characters. Length 21·5–22 cm. Rather uniform grey-brown, with pale lores and chin, yellow eye set in dark eye-ring and grey face; breast feathers somewhat darker, with pointed white tips. Immature has brown eye. Arrow-marked Babbler *T. jardineii* has darker look to head, with black lores, brown cheeks and throat, latter with pale streaks (spotted brown in Brown Babbler), and white feather tips on breast larger and more sharply defined. Their ranges come close in W Kenya but do not overlap (Zimmerman *et al.* 1996); both occur in Kerio Valley (Wilson and Wilson 1994) where, however, they appear to be altitudinally separated at about 1400 m. Immature uniform brown all over, somewhat paler below, with just a hint of dark streaking on breast (no pale tips) and top of head; smaller and paler than immature Arrow-marked Babbler. Overlaps in W Africa with Blackcap Babbler *T. reinwardtii*, which is easily distinguished by black crown and face; immature Blackcap similar but head darker, some black coming in on front part of head and ring of spots on breast; bill much longer. Dusky Babbler *T. tenebrosus*, confined to NE Zaïre/S Sudan/NW Uganda, is dark brown with dark mottling and pale scales on breast, jet black lores and around eye.

Voice. Tape-recorded (53, 104, 108, B, BRU, GREG, GRI, McVIC, STJ). Song in chorus, a variety of harsh, grating notes and phrases, which may be repeated once or twice, 'jeeya-jwor, jeeya-jwor, jeeya-jwor', 'gworrrr, gworrrr', 'jeeew, jeeew, jeeew', 'rraaakakakak'. Choruses typically in bursts of 5–10 s, separated either by silence or by 1–2 birds giving low 'kukukuk' or 'kekekek'. In a different interaction, individual repeated a harsh 'ch'kaaaaa', later others answered with chattering 'chakakakaka ...'. In mild anxiety, a short, dry 'tjuk'; alarm, a hissing 'tsa-tsa-tsa', often followed by a low 'tjuk'. Another scold 'a repeated buzzy *chay*-o, given with partly spread wings and fanned tail' (Zimmerman *et al.* 1996).

General Habits. Inhabits a wide variety of wooded savannas, parkland and farmland, often rocky; large gardens in rural areas, thin belts of fringing forest, woods and dense scrub around feet of inselbergs, thicket-clad banks of rivers, streams, dry watercourses in guinean and soudanian woodlands and bushy parts of wadis in sahel zone and sub-desert steppe regions of Mauritania and central Sudan. In Kenya in cold, bleak habitats up to 2300 m in *e.g.* Eldama Ravine and in scorching plains with flat-topped acacias at 850 m. Lives year-round in flocks of 4–12, usually 6–8. Retiring, disappearing quickly from view if disturbed in or near dense vegetation, but can often be detected by their frequent noisy chattering. Inquisitive, flock clumping in bushes to peer at observer, silently or 1–2 birds scolding a little from time to time. Skulks, creeping about rat-like in thicket, birds flying in ones and twos (often unseen) from far side. Perches with feet well apart (**A**). Forages on ground, creeping with short scuffling hops, often in grass-free open areas but never far from cover. Searches ground beneath thickets, rummaging amongst leaf litter and then working way onto open ground. One or more members generally perch well up in trees above or near foraging flock on ground, ready to sound the alarm;

Food. Insects, including beetles and ants. Perhaps dates (see above).

Breeding Habits. Co-operative breeder with solitary nests. Mating system unknown, but strong inference that 2 ♀♀ regularly lay in one nest (Serle 1977). Bird seen pecking cloaca of another, which had lifted tail and protruded vent (Cheke and Walsh 1996).

NEST: rather large, shallow bowl of small rootlets loosely woven with straw, sticks and leaves, lined with fine material, mainly grass; or sometimes nest made of grass and lined with fine rootlets. Placed in low shrub, acacia bush heavy with creepers, clump of withered bamboos, fork of small tree or (once) in deserted nest of Senegal Coucal *Centropus senegalensis*; at about 1–5.5 m above ground

EGGS: 2–4, av. (n = 9 clutches, Nigeria) 3·0. Slightly glossy; normally pale blue or deep turquoise-blue, but in Nigeria blue, grey-blue, mauve, lilac, pale pink or bright salmon pink, colour varying within and between clutches. Within-clutch variation (n = 3 clutches) inferred to arise from 2 ♀♀ laying in one nest. Abnormally-coloured eggs known from only small part of N Nigeria, between Kafanchan, Serakin Pawa, Kaduna and Kano; pink eggs almost exactly matched by eggs of African Striped Cuckoo *Oxylophus levaillantii* (Serle 1977). SIZE: (n = 25) 23·5–26·4 × 16·4–18·8 (24·8 × 17·7) and (n = 4) 25·6–25·9 × 18·5–18·7 (25·7 × 18·5). Shell weight: (n = 18) 251–332 (176) mg [shell weight of 2 African Striped Cuckoo eggs 452, 473 mg].

LAYING DATES: Mauritania, Apr, Aug; Senegal, (Richard-Toll) Mar–June, Aug, (Basse-Casamance) May–June; Gambia, June, Aug–Oct, Dec–Jan; Ghana, Apr–May, July, Nov; Nigeria, Sept–May in N, Dec, Jan and Apr in S, or 'almost all months' (Elgood *et al.* 1994); Niger, June–July; Sudan (Kordofan), Sept; Zaïre (fledglings Dec, Jan, Mar).

DEVELOPMENT AND CARE OF YOUNG: young leave nest prematurely and are tended by parents (Serle 1957).

BREEDING SUCCESS/SURVIVAL: known to be parasitized by African Striped Cuckoo (Serle 1977) and, with Blackcap and Dusky Babblers *T. reinwardtii* and *T. tenebrosus*, Brown is likely to be only host of this cuckoo west of Nile. However, any effect of cuckoo on breeding success of Brown Babbler is unknown.

References
Serle, W. (1977).
Shuel, R. (1938).

foraging birds also alert, often raising heads to scan around. May obtain some food in trees: once seen in date palm *Phoenix dactylifera* apparently eating dates. Drinks during day and before going to roost; flock approaches drinking pool cautiously, under cover of thick, woody vegetation, then bird drinks, clinging to thin overhanging branch, or flock comes to ground and drinks at edge of water, sitting in a row. Noisy, particularly in early morning and when retiring to roost in evening; but group calls at any time of day. Sometimes flaps wings when calling, keeping them stiff and bowed down. Flights short, seldom more than a few m, low down between thickets and trees, about 1 m above bare ground, sand or grass sward, strong but laboured, rather whirring (wings being short) with long, horizontal glide, often ascending at end; bird often seems to sail further than it flaps. Birds of flock keep within sight and sound of each other nearly all the time, but tend to fly singly or in twos and threes, following each other a few s apart.

Resident, but in central Burkina Faso noted more often in wet season than dry season.

Turdoides jardineii (Smith). Arrow-marked Babbler. Cratérope fléché.

Plate 3 (Opp. p. 30)

Cratopus Jardineii Smith, 1836. Rep. Exped. Centr. Afr., p. 45; banks of river beyond Kurrichane, northwest Transvaal.

Forms a superspecies with *T. plebejus*.

Range and Status. Endemic resident, southern tropics from S Uganda and Kenya to Cabinda, Angola, NE Namibia, N and E Botswana and NE South Africa. Gabon, several records in S, from Moabi, Ngounie near Mouila, and Tchibanga (Y.-M. de Martin de Viviès, pers. comm.). SW Uganda, north and east to Toro, Kampala and Jinja.

Turdoides jardineii

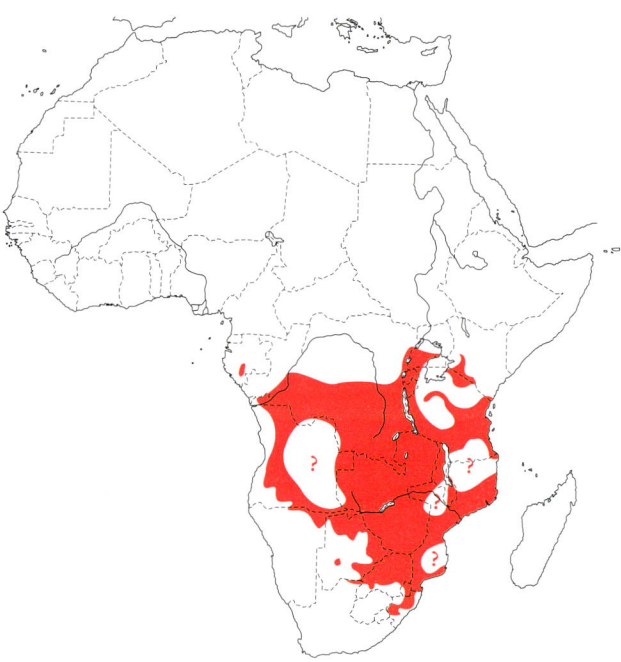

Kenya, locally common and widespread in SW, up to 2200 m (Mara Game Res., Narok district, Rift Valley north to Nakuru and Rongai); an old record from Taveta; also a record from Lamu. Rwanda, Burundi, Tanzania, ranges widely but somewhat patchily; sea level up to 2100 m (near Behungi, Rwanda). Zaïre, widespread but often uncommon in savannas south of Congo basin rainforest block, north on Congo R. to about 03°S. Cabinda and throughout Angola except in SW where replaced in dry coastal plain by Bare-cheeked Babbler *T. gymnogenys*. Namibia, Caprivi Strip and Kavango west to about Okipoko drainage and southwest nearly to Grootfontein. Zambia, common and widespread, a little less so in Northern and North Western Provs and in Kawambwa area (Luapula) and Mongu area (Barotse). Malaŵi, common below 1700 m; occurs on Nyika Plateau at 1850 m. Mozambique, status in N ill-known, but common and widespread in Sul do Save, over Manica Platform and in Tete district, uncommon north of Beira. Botswana, quite common, locally very common, in N, E and SE south to Lobatse but not west of Kanye R., and in N range stops abruptly at Mopipi/Boteti R. (Penry 1994). Zimbabwe, common and widespread, but only up to 1300 m in E highlands and scarce on seaward-facing slopes of mountains, e.g. near Honde valley and Lusitu-Haroni confluence. South Africa, common and widespread in Transvaal north of 26°S, but absent from Escarpment and highveld except as vagrant to Witwatersrand, Potchefstroom and Barberspan; in Orange Free State common to uncommon east and south of Harrismith, and 4 records of vagrants in W; in Natal, common in Drakensberg foothills, and uncommon in NE towards littoral plain and in Zululand, becoming commoner in Swaziland. Not in Lesotho.

Density of 89–101 birds in 18 flock-territories in 600 ha in central Transvaal, i.e. 0·13–0·15 birds per ha (Monadjem *et al.* 1995).

Description. *T. j. jardineii* (Smith) (including '*convergens*'): N and E South Africa, S Mozambique, Zimbabwe and central and NW Zambia. ADULT ♂: upperparts dark brown, rump slightly paler and greyish; browns with a vinaceous cast (more apparent in artificial than in natural light); forehead feathers pointed, blackish brown, with narrow brown edges; crown feathers less pointed, brown with blackish shaft-streaks; hindneck nearly uniform brown; mantle, scapulars and back brown with fine dark shaft streaks, and in fresh plumage buffy feathers form sparse, indistinctly pale spots (also a few indistinct pale feather-tips on rump and uppertail-coverts). Mantle feathers each with *c.* 5 narrow, dark, structural bands. Tail very dark brown above and below, with 17–20 exposed, narrow, structural bands. Lores very dark brown, ear-coverts silky brown with fine whitish streaks; cheeks and superciliary area with fine V-shaped whitish spots; sides of neck brown with fine whitish spots. Underparts brown, dark on chin, throat and breast, paler and buffier on belly, flanks and undertail-coverts; chin, throat and breast greyish brown, feathers pointed, barbs at tip not radiating but forming prominent, coherent, V-shaped white mark, 2 mm long on throat and 3 mm long on lower breast; rest of underparts with fine buffish white shaft streaks. Wings brown, median and lesser coverts with tiny pale spots; tertials with 8–9 exposed, narrow, dark, structural bands. Underwing-coverts and axillaries cinnamon-buff; rufous-cinnamon area on underside of flight feathers formed by broad inner borders. Eyes yellow to orange-yellow: outer part of iris bright red but inner part, next to pupil, bright yellow or orange; bill black; legs dark grey to dark brownish. Sexes alike. SIZE: (10 ♂♂, 10 ♀♀) wing, ♂ 101–111 (105), ♀ 103–111 (106); tail, ♂ 96–109 (101), ♀ 98–112 (102); bill, ♂ 25–28 (26·4), ♀ 24–27 (25·3); tarsus, ♂ 33–35 (33·8), ♀ 33–35 (34·3). WEIGHT: ♂ (n = 4) 70–73 (71·2), 1 ♀ 56·3, 10 unsexed 61·3–81·5 (72·2), 5 unsexed av. 77·4 ± 3·2.

IMMATURE: juvenile tawnier above than adult, head streaks indistinct; chin to upper breast buff, mottled with dark brown (lacks pointed, white-tipped feathers); eyes brown.

NESTLING: hatches blind, naked, skin pink-orange, gape yellow.

T. j. tamalakanae de Schauensee: N Botswana, SW Zambia (Barotseland) and S Angola. Upperparts paler and greyer than in nominate race, rump lighter grey; underparts colder brown.

T. j. kirkii (Sharpe): C and N Mozambique, Malaŵi, E Zambia, E Tanzania and SE Kenya; intergrades with nominate race. Flanks and belly paler than in nominate race. Mantle feathers, tertials and tail feathers usually with structural bands, as in nominate race. Smaller; wing (7 ♂♂) 98–105 (102). WEIGHT: Mozambique, 1 ♂ 67.

T. j. emini (Neumann): NW Tanzania, S and W Kenya, S Uganda, Rwanda, Burundi and adjacent Zaïre border areas. Head, neck and upperparts slightly darker and blotchier than in nominate race, and belly perhaps paler; white spots on underparts slightly smaller. Small; wing (10 ♂♂) 99–106 (103). WEIGHT: Uganda, 2 ♂♂ 65, 72; Kenya, unsexed (n = 27) 55–71 (65·6).

T. j. tanganjicae (Reichenow): E Angola to N Zambia (Mwinilunga and Copperbelt to Kasama) and SE Zaïre (Shaba to W shore of L. Tanganyika). Like *emini*, but forehead, crown, hindneck, cheeks and ear-coverts black with glossy feather shafts, the black sharply demarcated from mantle (or upper mantle sometimes blotched with black), less sharply from throat. Chin and throat blackish; chin to breast with spiky white V-spots or diamonds.

T. j. hypostictus (Cabanis and Reichenow): W Angola to W and S Zaïre (lower and middle Congo R., Kasai). Like nominate race but white spots small and less spiky; also quite like *T. plebejus cinereus*. Small; wing (3 ♂♂) 96–98 (96·7), (4 ♀♀) 93–103 (97·5).

Field Characters. Length 21·5–25 cm. Grey-brown, with *orange eye* (red-rimmed yellow eye in Southern Africa, yellow eye in immature), blackish lores (in southern Africa, whole preorbital region and eye surround black), and *narrow white tips* ('arrow-marks') to throat and breast feathers; crown and nape feathers have smaller and less distinct white tips. Dark lores, brown face and throat and more conspicuous white tips to breast feathers distinguish it from closely-related Brown Babbler *T. plebejus* in W Kenya. Overlaps in E and central Africa with Black-lored Babbler *T. sharpei*, which has white eye and mottled underparts without white streaking on breast, and in SW Africa with Black-faced Babbler *T. melanops* (yellow eye, scalloped white breast feathers). Immature has brown eye, plain head with black lores and a few white streaks on throat; larger and darker than immature Brown Babbler.

Voice. Tape-recorded (53, 86, 88, 91–99, 104, B, F, PAY). Group chorus a series of fast, excited, hoarse notes, 'jaa-jaa-jaa ...' or 'gaa-gaa-gaa ...', given in bursts of 5–12 s; in between bursts, individual birds give low, conversational 'jigga', 'jugga', 'kuk', 'kukukuk', so that whole performance is a wave-like continuum. Intensity varies; some smaller groups less vocal, give shorter bursts, 'zaa-zaa-zaa', with intervals of silence or the odd 'jijiji' or 'kik'. Typical note 'gaa' or 'jaa', sometimes with a small hiccup at start, 't'gaa', but may also be a raucous 'gaw', 'gew' or 'gyup'. Individual 'jaa' notes tend to be shorter than in Brown Babbler, but overall vocal similarity shows close relationship of the 2 species ('appartiennent sans aucun doute à une même super-espèce': Chappuis 1975).

General Habits. Frequents dense vegetation and thickets in high-rainfall bush and tree savanna, thick, low growth with long, rank grass in regenerating cover on edges of abandoned cultivation in woodland, and open woodland, edges of thickets, drier riparian forest, riverine reedbeds, gum, pine and other plantations, thick bush and dense Kalahari woodland. Habitat in central Transvaal dominated by tamboti tree *Spirostachys africana* on sparse grass cover, or by *Acacia karroo*, *A. nilotica*, *A. tortillis* and *A. burkei* on extensive grass cover (Monadjem *et al.* 1995). In S Zimbabwe absent from pure thorn-veld and occurs in open soft woodland only where there are bushes, shrubs and thick secondary growth particularly along watercourses; also in rough orchards, on wide lawns (foraging), and cultivated areas with *Parinarium mobola* trees which flock favours as a retreat (Vincent 1947).

Highly sociable; lives year-round in cohesive flocks of *c.* 4–8 and up to 15 birds, each flock remaining in its own territory; in Zimbabwe 28 flocks were of 3–13 (6) birds (Vernon 1976), and in central Transvaal mean group size was 5·9 (Oct) and 5·6 (next Mar) (Monadjem *et al.* 1995). Territorial boundaries defended by all members of flock. Group members, detecting a group in adjacent territory, move onto highest branches of tree and call continuously. A flock crossing boundary is quickly chased back into its own territory; if 2 groups are calling on adjacent treetops each in its own territory, neither group chases nor retreats, tempo of calling decreases after 1–2 min and the 2 groups soon disperse. But when adjacent group sizes are disparate (e.g. flocks of 9 and 3 birds), large flock may cross boundary and chase small flock deep into its territory. Flock of 6 birds, 5 of which had been mist-netted and ringed, subsequently avoided spot where they had been caught and growled in alarm whenever they encountered ringer (C.J. Vernon *in* Ginn *et al.* 1989).

Flock forages on or near the ground, creeping, clambering and jumping through undergrowth; most food seems to be taken on ground, birds in scattered flock hopping about and scuffling in leaf litter. Sometimes forages with Hartlaub's Babbler *T. hartlaubii*; and in S Mozambique often associates with Terrestrial Brownbul *Phyllastrephus terrestris* (Clancey 1971). 3 birds eating fallen fruit of loquat *Eriobotrya japonica* pecked vigorously at fruit held 'in either foot', the leg securing the fruit lowered so that whole tarsus rested on ground (Cooper 1970). Flock will come to garden bird-table, apparently eating seeds. One of 3 birds twice seen lying on ground on its back, other 2 pecked at its chest and belly (Gargett 1969). One of 3 birds fluttered 1 m along ground using its wings, then when approached by 2 others, turned onto its back with feet in air, was perfunctorily pecked by the 2, then all 3 flew to bush and called (Backshall 1993). Flock at bird-table behaves peacefully, though 6 birds once resolutely expelled an immature one; flock retreated when chased by Hartlaub's Babblers (Herremans and Herremans-Tonnoeyr 1995). From time to time flock calls whirring 'ra-ra-ra-ra-ra' in chorus. Flies short distances, flight straight, low over ground or grass and between bushes, with rather shallow, fluttering flaps and long horizontal glides, members of flock following each other one by one. At end of flight often dives out of sight in cover. Flock roosts communally. Flock or part of it feeds nestlings together; once 5 adults watched for 2 days feeding 2 young African Striped Cuckoos *Oxylophus levaillantii* in their nest; each cuckoo received av. 3·9 feeds per h; feeding visits to nest were (n = 87 intervals) 1–44 (8) min apart (P.G.H. Frost *in* Vernon 1976). Flock of babblers pursues African Striped Cuckoos at egg-laying time (Mortimer 1975) but ignores them near nest towards end of breeding season. 2 birds attacked metre-long Black Mamba on ground, one tugging at its tail and the other pecking top of its head and neck (Bradford 1966). Sedentary, though evidently given to vagrancy.

Food. Mainly insects (38 in one gizzard), including 5 species of long-horned grasshoppers, 3 of wingless termites, moths, caterpillars, flies, ants and beetles; also a solifugid (sun-spider), round seeds 3·2 mm across (Fry and Hosken 1983), fruits up to size of loquat, and nectar (Maclean 1993). Insect body lengths 3–35 mm. Fibres of palm-nut *Elaeis guineensis* in 3 stomachs (out of 4: Chapin 1953). Food brought to nestlings mainly insects, particularly grasshoppers and caterpillars (28% and 20% of 148 feeds); also spiders, lizards and snails. Adult birds eat nestlings' faecal sacs and members of flock seem to compete for them (Vernon 1976).

Breeding Habits. Social co-operative breeder, the pair, helped by other adults, building nest in group territory. Mating system not certainly known (for this or any other African *Turdoides* species, but in Indian Jungle Babbler *T. striatus* and Arabian Babbler *T. squamiceps* only one ♀ lays in each nest: Gaston 1977, 1978, Zahavi 1990). 18 group territories abutted onto each other, occupying all available habitat (Monadjem *et al.* 1995); mean size (14 territories)

37 ± 16·5 ha; territory boundaries remain fixed for at least several months. Size of territory is in proportion to number of birds in flock. All groups make 2–3 nesting attempts per breeding season, each new nest built elsewhere in territory.

NEST: bulky, untidy brown or yellowish mass of dry, coarse, broad-bladed straw-grass, stubbly plant stems with large leaves still attached, leaves, twigs and skeleton leaves, with substantial, neat, bowl-shaped hollow in middle lined with rootlets, fine fibres, dry grass and wiry creeper tendrils (sometimes dark reddish). One nest built and lined entirely with tendrils. Nest generally far more grassy than twiggy. Well hidden in dense vegetation. Often close to ground, but in Transvaal sited 1·8–7·0 (n = 13, av. 3·5) m up in *Acacia nilotica*, *Boscia foetida*, *Dichrostachys cinerea*, *Diospyros lycioides*, *Rhus pyroides* and *Spirostachys africana* trees; elsewhere placed in pile of driftwood, mass of dead material caught 1·5 m up in bush, thick-foliaged mimosa thorn, central fork of large bush (e.g. *Leucadendron rubrum* in *Protea* plantation), on tree-stump enclosed by densely-leaved saplings, in dense reeds (one nest only 30 cm above water), solid tangle of stems, *Bougainvillea* bush, and cavity in tree. SIZE: 165–205 wide, 95–115 deep, cup 75 across and 50 deep. Built co-operatively by flock members.

EGGS: 2–5, av. (25 clutches, Malaŵi) 2·9 and (13 clutches, Transvaal) 3·5. Laid daily at *c*. 07h00 and 08h00. Long ovals, slightly glossy, immaculate Cambridge blue, pale blue-green or turquoise. SIZE: (n = 75) 22·0–28·1 × 17·3–19·7 (24·9 × 18·6).

LAYING DATES: E Africa, Region C, Feb, May, Sept–Dec, Region D, Mar, May, Region E, Apr; Angola, May, Aug, Nov, Dec (and nestlings Sept); Zaïre (Manda) Aug, (Katapena) Mar, (Lubumbashi) Apr; Malaŵi, Sept–June, mainly Sept; Zambia, Sept–May; Botswana, Nov–Jan; Zimbabwe, all months, mainly Sept–Nov (56% of 265 nests: Irwin 1981); Mozambique, Sept–Nov; Transvaal, all months (62% of 53 nests in Sept–Nov), but in one intensive study eggs laid late Oct to Feb, mainly Nov, none Dec, with next peak in Jan–Feb (Monadjem *et al.* 1995); Natal, Oct–Mar.

INCUBATION: by all flock members, juveniles, subadults and adult 'helpers' as well as parents. Incubation commences immediately first egg is laid, bird sitting very tightly and leaving only when observer touched bush. Period: 13–14·5 days (Jones 1992); period measured as time lapsed from laying of last egg to hatching of last chick, (n = 6 nests) 16–17 (16·3) days (Monadjem *et al.* 1994).

DEVELOPMENT AND CARE OF YOUNG: when nestlings small, one adult stays at nest and broods them; chicks left uncovered from day 6–7, but one adult always remains perched in branches immediately above nest. That adult seems to be a sentry; if it sounds the alarm, rest of flock returns to scold intruder (Ginn 1993). Nestlings fed by all members of flock, up to 8 birds arriving together, though some adults do more work than others (3 adults gave 74% of feeds and other 3 adults in group gave 26%) and feedings by any one particular adult varies with time. Members of flock do not necessarily act in concert: when flock is foraging one or a few birds go away to feed nestlings and later a second subgroup will do so. Adults shuttle to and from nestlings and tend them constantly for short time (once 5 adults delivered 10 feeds in 17 min), then nest may be ignored for long period, e.g. intervals between feeds (n = 172) 1–84 (13) min (Vernon 1976). As group of adults (part of the flock) approaches nest one adult calls briefly, but not after nestlings reach 7 days old. Each nestling receives 1·3–4·4 (av. 2·4) feeds per h. After a feed adult waits for nestling to excrete faecal sac which it then eats, and often a second adult is ready to take faecal sac if food-giving bird hesitates. Nestlings not heard to importune for food until 5 days old; from 5 to 7 days they do so by hissing, and at 7 days they change to monotonous and repetitive call 'kek ... kek ... kek' like adult's. From day 11 nestlings make begging calls much like adult's, and from day 13 they use adult's contact call even when no adult is arriving (Ginn 1993). Fledglings beg with a similar call. Nestling period: (n = 3 nestlings) 11, 13 and 13 days (Ginn 1993); measured as time lapsed from hatching of last chick to departure of last chick from nest, (n = 3 nests) 18–21 (19·7) days (Monadjem *et al.* 1994). Once several adults observed trying to entice chicks to leave nest; they perched with food in bills in adjacent bushes and kept up constant chattering. On leaving nest, young unable to fly, but adept at clambering about in low branches, from where they flutter down to ground; active on ground, hopping about strongly. Can move up to 70 m away from nest on first day out. Young remain well hidden; located by food-carrying adults which keep up continuous chattering to which young respond (Jones 1992). Single surviving nestling left cuckoo-parasitized nest at only 14 days old; although unable to fly it moved 50 m away from nest in 2 days and was attended in clump of grass by adults (Jones 1985).

BREEDING SUCCESS/SURVIVAL: bird ringed as adult re-trapped after 4 years, 2 months. Of 13 nests only 3 successful; and 18% of 44 eggs resulted in fledged young (Monadjem *et al.* 1995). Parasitized by African Striped Cuckoo in E Africa, heavily so in southern Africa; indeed this cuckoo parasitizes *Turdoides* babblers almost exclusively, and in southern Africa north to S Zaïre Arrow-marked Babbler is practically sole host (Irwin 1988). Possibly parasitized by Jacobin Cuckoo *O. jacobinus* in Malaŵi and Zambia (Irwin 1988). African Striped Cuckoo (which has clutch of 4 eggs) may lay 2 in one Arrow-marked Babbler nest, several days apart (Vincent 1947, Vernon 1976, Jones 1985). In one nest with 3 babbler eggs 2 were found badly dented after cuckoo's visit; one dented egg later hatched and one embryo died. 4 young birds grew up together, 2 babblers and 2 cuckoos, fed only by babblers although a pair of cuckoos often looked on. One young cuckoo soon became much larger than young babblers, one of which died. At least 3 adult babblers fed the 3 surviving chicks; the 2 cuckoos left nest together and the young babbler left prematurely 1–2 days later (Jones 1985). Group of 3 babblers (including incubating bird) successfully defended their clutch by frenziedly mobbing 2 African Striped Cuckoos for 10–20 min at a time on 3 days; the clutch hatched successfully (Clark and Clarke 1985).

References
Jones, J.M.B. (1985, 1992).
Monadjem, A. *et al.* (1994, 1995).
Vernon, C.J. (1976).
Vincent, A.W. (1947).

Turdoides leucocephalus Cretzschmar. White-headed Babbler. Cratérope à tête blanche. **Plate 3**

Turdoides leucocephala [sic] Cretzschmar, 1826. *In* Rüppell, Atlas Vög., p. 6, pl. 4; 'Welled Medina', Sennar (= Blue Nile Prov., Sudan). **(Opp. p. 30)**

Range and Status. Endemic resident, E Sudan, NW Ethiopia and N Eritrea. Sudan: fairly common to uncommon in dry acacia woodland from Khartoum south along White Nile to about Renk and along Blue Nile and its tributaries to Ethiopian border, also in Atbara R. valley from just east of Atbara to Ethiopia, with record near Jebel Hamoyet north of Eritrea. Ethiopia and Eritrea, frequent, from Abay Wenz (Blue Nile) at Didesa confluence, northeast to N Eritrea, where ranges from east of Asmera north to Falkat R.

Turdoides leucocephalus

Description. ADULT ♂: head, including nape, ear-coverts and chin white, sharply demarcated from neck and throat. Amount of white on chin varies, from a few feathers between rami of lower mandible to patch *c.* 8 mm deep, sometimes with thin brownish malar stripe at sides (**A**). Hindneck to uppertail-coverts plain greyish brown with olivaceous cast, paler on rump. Tail dark brown, underside glossy very dark brown. Nearly bare skin from base of mandibles, on lores, around and behind eye black (at least in skins); a few wispy 2-mm white featherlets on lore and 0·5 mm ones on lower eye rim (**A**). Ear-coverts thin, making white feathers over black skin appear greyish. Sides of neck grey-brown. Chin white, throat and upper breast brown or greyish brown, with scaly buffish white feather fringes and more or less distinct tiny spots formed by whiter feather tips; but width of brown on throat varies (geographically?), or upper throat may be white in centre, sometimes with brown malar stripe at sides. Lower breast, belly, flanks, thighs and undertail-coverts paler buffish brown. Wings brown, with median and lesser coverts paler and slightly more greyish. Underwing-coverts and axillaries buff; tawny-buff area on underside of flight feathers formed by pale inner borders. Eyes yellow; bill black; legs greyish. Sexes alike. Albinistic examples occur; one has mantle white and throat mainly white with odd feathers brown. SIZE: (10 ♂♂, 8 ♀♀) wing, ♂ 104–113 (107), ♀ 104–113 (110); tail, ♂ 107–111 (108), ♀ 99–110 (105); bill, ♂ 24–27 (25·4), ♀ 24–27 (25·6); tarsus, ♂ 33–35 (33·8), ♀ 33–35 (34·2). WEIGHT: no data.

IMMATURE: top and sides of head brownish white, slightly paler than mantle. Cheeks and chin white; flanks and belly browner than in adult.

NESTLING: not known.

Field Characters. Length 25 cm. A pale desert babbler with white head, pale brown upperparts, throat and breast, buffy flanks and belly; tips of throat and breast feathers paler, giving lightly scaled effect; wings and tail medium brown. Approaches but does not meet Brown Babbler *T. plebejus* in central Sudan; latter is darker and smaller, uniform grey-brown with white tips to breast feathers.

Differs from white-headed races of White-rumped Babbler *T. l. leucopygius* and *T. l. limbatus* by pale body, lack of white rump.

Voice. Tape-recorded (MAC). Single bird calls 'wok'; in group, a conversational grating 'kwik' or 'kwik-u' periodically erupts into a loud 'waak-waak-waak ...'.

General Habits. Inhabits dense scrub beside wadis and arid acacia woodland, mainly below 1000 m, but up to *c.* 1500 m in NW Ethiopia. Lives in parties of about 5–6 and up to 12 birds. Noisy; inquisitive, to the point of being annoying to person, persistently following him and uttering a continuous string of rattling 'churr' alarm notes, a series lasting for 30 s or so; birds in a flock call in company, sitting side-by-side on a branch, touching one another (Witherby 1901). Sedentary.

Food. Insects.

Breeding Habits. Not known. Testes 'in breeding condition' May, despite birds being in moult.

A

Plate 3
(Opp. p. 30)

Turdoides reinwardtii (Swainson). Blackcap Babbler. Cratérope à tête noire.

Crateropus Reinwardii [sic] Swainson, 1831. Zool. Ill., ser. 2, 2, no. 17, pl. 80; 'Indian Islands', corrected to Senegal (Swainson, 1837, Birds W. Afr., 1, p. 276).

Forms a superspecies with *T. tenebrosus*.

Range and Status. Endemic resident, northern tropics from Gambia and Guinea to Central African Republic and just into Zaïre. Senegal, once at Richard-Toll and once at Louga in N, occurs south of 15°N in W, and quite common and widespread in S and in Gambia, Guinea-Bissau and Guinea except south of about Beyla. Sierra Leone, records south to Makeni, Waterloo and near Segbwema. Mali, quite common south of 14°N; widespread in Boucle du Baoulé and to southeast. Burkina Faso, frequent in Ouagadougou region; uncommon in Arli Nat. Park. Ivory Coast, uncommon, south to Lamto, but frequent between Katiola and Ouangolodougou and in N Comoé Nat. Park (Demey and Fishpool 1991). Ghana, widespread and not uncommon, south to Elmina and Legon, but not yet found in some localities where it is to be expected, namely Volta Region in SE or Tamale, Tumu and Bolgatanga in far N. Togo, common from N to coast. Benin, rare in Bétérou area. Nigeria, frequent and widespread between 08° and 10°N, uncommon and local further south and north; occurs south to coast west of Lagos and to Abeokuta, Enugu, Obudu (town: 06°28'N, 09°05'E) and Serti, and north to fringing forest patches (e.g. Danbagudu, Anara, Dunbi, Falgore and Gubuchi, 10°19' to 11°32'N, Fry 1975) and formerly Kano at about 12°N (Bannerman 1936). Cameroon, probably widespread in dense savanna woodland on Adamawa Plateau between 07° and 09°N (records from Ngaoundéré, Poli, Galim, Tello and Koubadjé), south to Banyo and Kimbe R. (06°38'N) in SW. Chad, quite common south of 09°N; possibly but not certainly north to 10°N. Central African Republic: 2 records north of Bangoran R. (Hall and Moreau 1970); reported from Lobaye Préfecture, perhaps in error (Germain 1992), though recorded also further south in Dzanga Res. (Green and Carroll 1991) and further east, a few km on Zaïre side of border – 2 birds at Yakoma (04°06'N, 22°23'E) in 1911. Brown Babbler *T. plebejus* occupies entire range of Blackcap Babbler, which is generally scarcer than Brown but reported as plentiful, replacing the Brown, south of 09°15'N in Niger Prov., Nigeria.

Description. *T. r. stictilaemus* (Alexander): Ghana and extreme S Mali to Cameroon, S Chad, Central African Republic and N Zaïre. ADULT ♂: forehead and crown brownish black, hindneck blackish brown, either sharply demarcated from olive-brown mantle, or merging through lower hindneck and upper mantle into olive-brown lower mantle and back. Rump and uppertail-coverts olive-brown. Tail brown above, greyish brown below. Lores and ear-coverts brownish black, cheeks mottled dark brown and pale greyish, sides of neck olive-brown. Chin to upper breast mottled, feathers dusky with 2–3 mm broad buff fringes; lower breast greyish buff, slightly mottled; rest of underparts pale buffy brown, often whiter on belly and rufescent on flanks. Primaries, secondaries, primary coverts and alula dark brown, tertials and rest of upperwing-coverts and axillaries pale buff. Eyes white or creamy white; bill black; legs brownish grey. Sexes alike. SIZE: (10 ♂♂, 10 ♀♀) wing, ♂ 112–128 (118), ♀ 104–119

Turdoides reinwardtii

(113); tail, ♂ 99–120 (111), ♀ 98–115 (108); bill, ♂ 26–28 (27·5), ♀ 26–28 (26·8); tarsus, ♂ 35–38 (36·6), ♀ 34–39 (35·7). WEIGHT: (Ghana, unsexed, n = 12) 69–91 (78·75).

IMMATURE: (paucity of museum skins suggests that juvenile plumage is ephemeral, or that species is long-lived with only a few juveniles surviving.) Juvenile has forehead and face dark brown, not sharply demarcated from mantle or throat; but is otherwise like adult except that mottling on underparts restricted to breast. Bill pale-tipped.

T. r. reinwardtii (Swainson): Senegal to Sierra Leone and Mali (except extreme south). Blackish top of head always sharply demarcated from olive-brown mantle. Cheeks blackish, contrasting with almost plain buffy-white chin and throat. Underparts paler buff than in *stictilaemus*, with dark mottling practically confined to upper breast.

Field Characters. Length 24–25 cm. A large W African babbler, readily distinguished by *solid black cap* extending to face and hind neck, cut off sharply from brown back and white throat; dark mottling on breast. Overlaps with Brown Babbler *T. plebejus*, which is smaller and uniform grey-brown except for pointy white feather tips to breast and dark streaks on head (not forming cap). Eyes ivory white or (Senegal to Sierra Leone and SW Mali) yellow. Immature similar to immature Brown but bill longer and head darker, especially forehead and face; some dark spots on breast.

Voice. Tape-recorded (53, 104, 108, B (BBC), BRU, CHA, LEM, MOR). Individual gives short, harsh, grating notes, singly or in series of 4–8, 'dzwit', 'jwiu', 'jijiji ...', 'jut', 'jui-jui', 'dzwuk-dzwuk-dzwuk-dzwuk ...'; another call is a loud braying or crowing, 4–5 notes going down the scale, 'waaa ... haa-haa-haa-haa'; contact, a continuous grating 'ja-ja-ja-...' or low, hurried 'jaajagagagagagaga', almost a trill. In chorus, short notes 'gaa-gaa-gaa ...' become louder double notes, '*gwaaa*-ga' or '*gwaya*-ga', or a single long scraping, tearing 'jaaaaaaaa' or 'jaaaaawi'.

General Habits. Frequents wooded ravines, gallery forest, dense growth around bases of inselbergs, bamboo groves, riverine thickets, and thick scrub in well-wooded savanna and on coast. Unlike Brown Babbler keeps to thick, woody vegetation, though flock once in a grove of ancient mango trees without ground cover. From sea level to about 1000 m. Habitat (and voice – a harsh chatter delivered from the depth of dense foliage) like that of Yellow-throated Leaf-love *Chlorocichla flavicollis.* Lives in parties of *c.* 6 but occasionally up to 10 birds. Shy and difficult to watch. Forages on ground and progresses through vegetation above it, sometimes almost at observer's walking pace, or spends several min at one place before moving on (Serle 1949). Forages thrush-like, a party of babblers fossicking quietly in leaf litter under tangled woody vegetation, crouching rather close to soil, and bills vigorously swiping at leaves and pecking at exposed earth surface. Climbs up through leafy bush to view observer, then abruptly disappears from view, sometimes calling. Flock drinks together from small pool with vegetation overhanging sides. Roosts socially, and vocal before going to roost.

Food. 5 stomachs contained insects and 2 contained berries.

Breeding Habits.
NEST: quite large, roughly-built structure with deep cup in middle, made of dry leaves and stems and lined with rootlets. One was sited amongst frond stumps of dead palm.
EGGS: 2 (–3?). Rather long and pointed; highly glossed; dark sky blue. Length *c.* 22–25 × 19.
LAYING DATES: Gambia, Nov–Dec; Senegal, Jan; Mali, Nov–July; Ghana (young out of nest being fed by adults, Jan); Nigeria, Dec, June.
INCUBATION: incubating bird, disturbed from eggs, reappeared some m away to chatter noisily; it later returned to nest stealthily and silently or (once) scolded observer from palm-stump nest site.
Nothing further known.

References
Bannerman, D.A. (1936).
Mackworth-Praed, C.W. and Grant, C.H.B. (1973).

Turdoides tenebrosus (Hartlaub). Dusky Babbler. Cratérope ombré; Cratérope fuligineux. Plate 3

Crateropus tenebrosus Hartlaub, 1883. J. f. Orn., 31, p. 425; Kudurma, Bahr-el-Ghazal, Sudan. (Opp. p. 30)

Forms a superspecies with *T. reinwardtii.*

Range and Status. Endemic resident, NE Central African Republic to NW Uganda and SW Ethiopia. Central African Republic, uncommon in Manovo-Gounda-Saint Floris Nat. Park (Carroll 1988). Sudan, rare or local and uncommon, records in SW on Boro R. near border with Central African Republic at 08°27′N, 24°47′E, in S from several localities close to Zaïre and Uganda borders including Li Rangu, Mt Bangenze (Baginzi) and Kajo Kaji, also Jebel Lado and Rejaf near Juba, and in SE from N Boma Hills to south of Naita (but not yet recorded across the border in Kenya, in Lokwanamoru Range for instance). Zaïre, from Bomokani R. at Poko (03°09′N, 26°53′E) in upper Uele District eastward, to Faradje (where not uncommon) and borders, and to Mahagi Port and Kasenye on W shores of L. Albert. Uganda, from Mt Kei White Rhino Sanctuary and West Nile, Gulu, Kitgum and SE Acholi, southeast to Lwampanga and Serere. SW Ethiopia, frequent from border country east to Gofa and 07°N on Omo R. and north to E Illubabor district.

Description. ADULT ♂: forehead silvery grey-brown, feathers at sides with narrow black shafts, crown olive-brown, greyish above eye; rest of upperparts plain brown, tinged tawny on back, rump and uppertail-coverts. Tail dark brown, above and below. Lores black; ear-coverts, cheeks and sides of neck olive-brown. Feathers

Turdoides tenebrosus

of chin, throat and upper breast blackish brown with broad (1–2 mm) pale greyish fringes, producing prominent scaly pattern; lower breast and centre of belly grey or greyish brown; sides of belly, thighs and undertail-coverts rich tawny brown. Remiges and upperwing-coverts dark brown; inner underwing-coverts paler tawny brown; axillaries brown. Bill black; eyes white; legs and feet black. Sexes alike. SIZE: (7 ♂♂, 7 ♀♀) wing, ♂ 110–121 (117), ♀ 109–120 (114); tail, ♂ 107–117 (112), ♀ 106–117 (113); bill, ♂ 25–28 (26·0), ♀ 22–27 (24·7); tarsus, ♂ 35–37 (35·9), ♀ 32–36 (33·7).

IMMATURE: young bird plainer than adult; forehead, crown, neck and mantle paler brown, warmer and less grey; back and rump rufous-brown, uppertail-coverts dark rufous; tail and wings quite dark rufous or rufescent brown; chin pale grey, throat grey, plain or faintly mottled, breast greyish brown with diffuse throat streaks but no scaling, belly, flanks, thighs and vent rufous, and undertail-coverts dark rufous.

Field Characters. Length 23–24 cm. *Dark brown* with blackish wings and tail, head lighter and forehead pale; *jet black* preorbital region and *eye surround* contrast with *white eye* and *brown face* (Brown Babbler *T. plebejus* has yellow eye, no black on face). Throat and breast with heavy dark mottling, feathers edged pale brown, giving scaly effect, but not white-tipped like Brown Babbler. Immature has richer brown upperparts, especially rump, paler head, less black on face, browner wings and tail, white chin, more pronounced pale breast scaling and tawny wash on flanks and belly. Adult and immature Brown Babbler are plain grey-brown. Range abuts that of Black-lored Babbler *T. sharpei* in NE Zaïre and NW Uganda; Black-lored is also much paler than Dusky, uniform light grey-brown, throat and breast lightly streaked, without heavy dark mottling, no black around eye.

Voice. Not tape-recorded. Gives an occasional hoarse 'chow'; from time to time ♂♂ give louder, more protracted sound, a repeated nasal 'what-cow' (Chapin 1953).

General Habits. Inhabits dense undergrowth near water in well wooded, hilly country at 600–1200 m. Shy, secretive and poorly known; lives in pairs and small parties. Difficult to see, but presence given away by hoarse 'chow' and repeated, nasal 'what-cow' calls. Immature bird moults remiges in May, at age probably *c.* 4 months. Sedentary.

Food. Insects in 5 out of 6 stomachs, including beetle larva; also small snails and millipedes and fruit (Chapin 1953).

Breeding Habits. Solitary nester; probably monogamous and with helpers at the nest.

NEST: rough open bowl, like that of Black-lored Babbler *T. sharpei* but somewhat more lightly built, made of dry leaves and grasses with fairly deep cup lined with rootlets, on foundation of dead leaves. Sited in thick cover with creepers, 1–2 m above ground.

EGGS: 2–3. Pale blue, very glossy. SIZE: (n = 8) 23·6–29·8 × 17·0–20·3 (25·85 × 18·7).

LAYING DATES: Sudan, Apr, July; Zaïre, Dec; Uganda, Feb–Apr, June.

Nothing further known.

Reference
Chapin, J.P. (1953).

Plate 3 *Turdoides leucopygius* (Rüppell). **White-rumped Babbler. Cratérope à croupion blanc.**
(Opp. p. 30)
Ixos leucopygius Rüppell, 1840. Neue Wirbelth., Vög., p. 82, pl. 30, fig. 1; coast of Eritrea from Massaua to Adigrat.

Forms a superspecies with *T. sharpei, T. hartlaubii* and *T. melanops*.

Range and Status. Endemic resident, Ethiopia and Eritrea, extending into Sudan and Somalia. Sudan: in E uncommon and local close to Eritrean border in Kassala district, and in SE fairly common near Ethiopian border in Boma highlands. Eritrea, from Anseba valley to 'Dancalia' (Smith 1957, i.e. Danakil country or Dangil, 14°40′N, 40°20′E) and throughout the intervening plateau, but absent from coastal plains. Ethiopia, frequent to common throughout highlands, but rare (or rarely encountered) in broad corridor between highlands and Sudan border and absent from hot lowlands north of Dire Dawa and from Ogaden.

Description. *T. l. smithii* (Sharpe): NW Somalia to E and SE Ethiopia. ADULT ♂: top of head blackish brown with prominent scaly greyish white feather edging; mantle, scapulars and back dark brown with a few scaly buffish markings; rump and shorter uppertail-coverts white; longer uppertail-coverts brown with paler fringes. Tail dark brown above and below. Greyish white facial patch includes lores, narrow stripe above eye, cheeks and chin; ear-coverts streaked brown and whitish. Sides of neck, throat and breast dark brown, feathers with scaly white fringes, broadest on lower breast; flanks, thighs, belly and undertail-coverts paler, plainer buffish brown. Wings dark brown, alula and median and lesser coverts narrowly edged buff. Underwing-coverts and axillaries tawny-buff; broad inner borders form tawny area on underside of flight feathers. Bill black; eyes orange-yellow or chestnut-brown; legs dark slate. Sexes alike. SIZE: (10 ♂♂, 10 ♀♀) wing, ♂ 112–121 (116), ♀ 110–120 (115); tail, ♂ 109–120 (113), ♀ 106–116 (112); bill, ♂ 25–28 (26·3), ♀ 25–27 (26·2); tarsus, 35–37 (36·0), ♀ 35–37 (35·5). WEIGHT: no data (but see below).

IMMATURE: juvenile plainer than adult and lacks scaly feather edging; browner above, top of head brown (not blackish) with faint mottling, wings warmer brown; sides of head and chin pale brown; throat to breast slightly mottled, flanks pale buff, belly whitish. Eyes pale brown.

NESTLING: not known.

T. l. leucopygius (Rüppell): plateau and E Eritrea (Danakil coast) to N Ethiopia at Adigrat. Differs from *smithii* in having whole head white, sharply demarcated from plumage behind; mantle, back and wings uniform dark brown; scaly markings on breast and neck narrow and inconspicuous.

Turdoides leucopygius

T. l. limbatus (Rüppell): W Eritrea and NW Ethiopia. Like *smithii* but forehead also white. Breast and neck scaling as in nominate race. WEIGHT: (n = 3, unsexed) 75–93 (82·4) (J.S. Ash, pers. comm.).

T. l. lacuum (Neumann): central Ethiopian Rift Valley (L. Zwai to L. Abassi). Like *smithii* but whitish or silvery face patch confined to ear-coverts and rear of cheek; lores, most of cheeks, chin and upper throat dark greyish with narrow white scaling. Breast prominently scaled, as in *smithii*..

T. l. omoensis (Neumann) (including '*clarkei*' of Baro R. district): S and SW Ethiopia (Sidamo and Omo drainage) to SE Sudan. Differs from *lacuum* in having lores, cheeks, chin and upper throat black; the black extends to upper breast where feathers sharply scaled white. WEIGHT: (n = 16, unsexed) 63–80 (70·6) (J.S. Ash, pers. comm.).

Field Characters. Length 25–27·5 cm. A large babbler of the highlands of Ethiopia and Somalia. Plumage very variable, but all races distinguished by *white rump* contrasting with dark upperparts, wings, and tail. White-headed races *leucopygius* and *limbatus* distinguished from White-headed Babbler *T. leucopygius* by mainly *dark underparts* contrasting with white head and belly. Other races told from Brown Babbler *T. plebejus* by variable amounts of white on face, chin and throat, and dark underparts with *white scaling*.

Voice. Tape-recorded (109, MAC). Harsh scolding calls similar to those of other babblers, often given in noisy chattering chorus.

General Habits. Inhabits elevated, open, stony ground with light woodland cover or scattered clumps of bushes and hedges, near ravines or scrub-clad depressions with a little water; eucalyptus woods (Eritrea); thick bushes and woodland at higher elevations (Sudan); near thick scrub on juniper-clad hills around Yavello and common in dense secondary growth on edge of evergreen forest at Alghe (Ethiopia). Keeps under cover, near bottom of hedgerows, thickets and bushes, where flock gives harsh, chattering, scolding call in chorus. Altitudinal range of *smithii* 1385–2000 m; other races about the same, but nominate *leucopygius* ranges much lower in Eritrea. Lives year-round in highly sociable parties of 6–8 birds. Not shy, but vigilant, and flock immediately sets up voluble and excited chatter when disturbed by person, retreating by flying in ones and twos to thicker cover 20–25 m ahead. Flock forages on ground, evidently much like congeners, though not reported to post a sentry in tree overhead (Archer and Godman 1961).

Breeding Habits. Only 2 nests found (Benson 1946b).

NEST: a slovenly-built cup, composed of rather coarse rootlets inside and coarse grass outside. Ext. diam. 150, int. diam. 90, ext. depth 75, depth of cup *c*. 63. One nest sited *c*. 1 m above ground in middle of thick, leafy bush.

EGGS: 2–5. Immaculate, rich turquoise-blue. SIZE: (n = 2) 27·0–27·5 × 20.

LAYING DATES: Ethiopia, Nov, Feb.

References
Archer, G. and Godman, E.M. (1961).
Clancey, P.A. (1984).

Turdoides sharpei (Reichenow). Black-lored Babbler. Cratérope de Sharpe.

Plate 3 (Opp. p. 30)

Crateropus sharpei Reichenow, 1891. J. f. Orn., 39, p. 432; Kakoma, Tabora, Tanganyika.

Forms a superspecies with *T. leucopygius*, *T. hartlaubii* and *T. melanops*.

Range and Status. Endemic resident, L. Turkana to W Rift Valley, south to L. Rukwa, Tanzania. Kenya, locally common at 1500–1900 m, down to 1000 m and exceptionally up to 2200 m, in and west of Rift Valley, east to Mosiro, Nanyuki, base of Mt Kenya and Laikipia Plateau, with old records from Kikuyu and north to Mt Nyiru. Uganda, widespread from Ankole north to L. Kyoga and Kabalega Falls Nat. Park and east to base of Mt Elgon. Zaïre, locally common in open country at lower elevations, up to 2430 m, along W side of Western Rift, from Kasenye (SW corner of L. Albert), S flank of Mt Rwenzori and Rutshuru Valley south to N end of L. Tanganyika. Rwanda and Burundi, widespread in open country. Tanzania, widespread west of *c*. 33°E but in L. Victoria basin east to Tarime and (old records) Serengeti Nat. Park and Mt Hanang, and south to L. Rukwa.

Turdoides sharpei

Description. *T. s. sharpei* (Reichenow): range of species except Nanyuki area, central Kenya. ADULT ♂: upperparts greyish brown; forehead and crown to mantle and scapulars mottled, feathers with pale grey tips and fringes and dark centres; back, rump and uppertail-coverts plain. Tail dark brown, underside silvery grey-brown. Lores and area around front half of eye blackish brown; sides of head plain greyish brown or pale-mottled dark brown, sides of neck mottled greyish brown. Chin and throat greyish brown, feathers with pale grey fringes and brown shaft streaks; upper breast greyish brown with prominent scaly greyish white feather fringes and some dark shaft streaks; rest of underparts pale greyish brown with faint darker streaking. Remiges and upperwing-coverts greyish brown, median and lesser coverts with paler, greyer fringes. Underwing-coverts and axillaries pale greyish brown. Bill dark horn to blackish; eyes white or ivory; legs dark brownish grey or blackish. Sexes alike. SIZE: (10 ♂♂, 10 ♀♀) wing, ♂ 111–115 (114), ♀ 111–114 (113); tail, ♂ 105–114 (111), ♀ 99–113 (109); bill, ♂ 22–25 (24·1), ♀ 22–25 (23·7); tarsus, ♂ 36–38 (36·7), ♀ 36–38 (36·8). WEIGHT: Uganda, ♂ (n = 3) 79–86 (82), ♀ (n = 3) 76–91 (82·3); Kenya, unsexed (n = 12) 66–84 (79·3).

IMMATURE: juvenile plumage soft and fluffy, more uniform than adult and lacks pale mottling and scaling; browner above; lores dark brown; chin and throat whitish, upper breast with dark brown band, sides of breast and flanks warm brown, rest of underparts greyish brown.

NESTLING: not described.

T. s. vepres (Sharpe): Nanyuki area, central Kenya. Very variable. Upperparts browner and more uniform than in nominate race, with pale flecking confined to top and sides of head; chin and throat either mottled brown and whitish or entirely white. Breast and flanks darker brown and more uniform than in nominate race; in white-throated birds, lower belly and undertail-coverts usually also whitish, and flanks streaked white; in dark-throated birds, lower belly and undertail-coverts usually buff. SIZE: (5 ♂♂, 2 ♀♀) wing, ♂ 110–115 (112), ♀ 110, 110; tail, ♂ 111–114 (113), ♀ 105, 108; bill, ♂ 21–22 (21·4), ♀ 21, 22; tarsus, ♂ 33–36 (35·6), ♀ 34, 35.

TAXONOMIC NOTE: *T. sharpei* is often considered conspecific with *T. melanops* (White 1962, Short *et al.* 1990, on the basis of their field experience, and Dowsett and Forbes-Watson 1993). However it is smaller, with a shorter, less decurved bill, and more mottling and scaly patterning in the plumage, and we follow Hall and Moreau (1970) and Clancey (1986) in separating them. Sibley and Monroe (1990) ally *T. sharpei* with *T. hypoleucus*; individuals with whitest underparts resemble *hypoleucus*, which is parapatric with *T. s. vepres*. The last-named race is highly variable, 'possibly reflecting past hybridization' with *T. hypoleucus* (Zimmerman *et al.* 1996).

Field Characters. Length 22–23 cm. Mottled grey-brown, with *black lores* and *white eye;* immature has dark eye. Plumage variable; birds around Naivasha (Kenya) have paler throats and darker wings and tail and may appear quite frosty-headed (Zimmerman *et al.* 1996). Race *vepres*, confined to area around Nanyuki, Kenya, even more variable, often with white throat and dark brown upperparts and underparts, and looks like a different species (Zimmerman *et al.* 1996). In head-only view, showing just white throat, black lores and white eye, very like Northern Pied Babbler *T. hypoleucus*, with which it may have hybridized in the past; in full view, latter shows white underparts with brown patch at sides of breast. In E Africa overlaps with Brown Babbler *T. plebejus* and Arrow-marked Babbler *T. jardinei*, which have orange or yellow eyes and white tips to breast feathers; Brown also has pale lores.

Voice. Tape-recorded (91–99, 104, B, C, F, CART, GREG, HOR). Contact call a high-pitched grating 'ska', 'skaaa' or prolonged 'skaaayaaa'; longer calls are down-slurred. In chorus other birds join in with deeper 'choo*kaah*-choo*kaah*-choo*kaah* ...'. Variable; for other renditions, see Zimmerman *et al.* (1996). Alarm, a cat-like 'nyaaa' (species used to be called Cat-bird in Uganda).

General Habits. Inhabits wooded plains, forest edge, dense bush, acacia thickets and grassy hillside savanna at middle altitude. In parties of 4–5, year-round. Flock forages in shrubbery and tall grass, often giving babbling chorus, particularly in early morning and late afternoon. Restless; suspicious of man, a flock takes cover in thickets or elephant grass, but on the way a bird may alight for a moment to assess danger from some small tree.

Food. Insects, but one stomach contained a small lizard and small fruits.

Breeding Habits.

NEST: a roughly-built open bowl, made of roots, thin creeper stems and strawlike grass, and lined with fine grass roots and finer creeper stems. Sited in large, isolated bush covered with dead creepers.

EGGS: 2–5. Uniform immaculate dark blue.

LAYING DATES: E Africa, Region A, Mar; Region B, Feb, Apr–Sept, Oct (mainly Apr–May); Region C, Mar–Apr; Region D May, July.

DEVELOPMENT AND CARE OF YOUNG: young attended by flock of adults – once 6 adults, all carrying food, were in bushes next to nest with week-old young, waiting to feed them (Reynolds 1965).

Turdoides hartlaubii (Barboza du Bocage). Hartlaub's Babbler; White-rumped Babbler. Cratérope de Hartlaub. **Plate 3 (Opp. p. 30)**

Crateropus Hartlaubii Barboza du Bocage, 1868. J. Sci. Math. Phys. Nat. Lisboa, 2, p. 48; Biballa, Mossamedes, Angola.

Forms a superspecies with *T. leucopygius*, *T. sharpei* and *T. melanops*.

Range and Status. Endemic resident, from E Zaïre and Rwanda to Zambia, N Botswana and Angola. Zaïre, E and SE, north to about Rutshuru, rather uncommon. Rwanda, common around Bugarama in extreme SW. Burundi, abundant in Ruzizi Plain and along shores of L. Tanganyika. Tanzania, locally common in SW on Ufipa Plateau. Zambia, almost throughout, but in NE (Isoka) only at Ikomba, not east of Muchinga Escarpment nor southeast of Serenje, and in S absent from southeast of line Livingstone-Kalomo-Choma-Mazabuka. Zimbabwe, extreme NW, west of Katombora Rapids: small population on Zambezi floodplain at Kazungula (Hustler 1997). Botswana, common to very common in Okavango basin and Nhabe, N Boteti, Linyanti and Chobe R. valleys; may extend south in rains to S Boteti R. and L. Ngami. Namibia, common in Caprivi Strip and recently found on lower Cunene valley in NW (Harrison *et al.* 1997). Angola, S Bihe, Huila, E Benguela, central plateau, central Malanje, east to Zambian border.

Description. *T. h. hartlaubii* (Bocage) (including '*ater*'): range of species except *griseosquamatus* – see below. ADULT ♂: top of head and neck dark brown with scaly greyish white feather fringes, fine fringing producing pale hoary effect on forehead, above eye and on side of neck; mantle, scapulars and back dark brown with a few pale scaly markings; lower rump whitish, and shorter uppertail-coverts whitish with dusky central streaks; longer uppertail-coverts grey-brown with dark brown central streaks. Tail dark brown above, dark grey-brown below. Lores plain blackish; ear-coverts dark brown with a few buff flecks. Cheeks and chin dark brown, with broad greyish feather fringes; throat and upper breast brown with a few short darker brown streaks; lower breast and flanks dark brown, with broad whitish feather fringes giving streaked rather than scaled appearance; belly, thighs and undertail-coverts buffish white. Wings dark brown. Underwing-coverts and axillaries cinnamon-buff; pale inner borders form buffish area on underside of flight feathers. Bill black; eyes crimson to deep orange-red, or red with yellow inner rim; legs brownish grey, dull slate or dark brown. Sexes alike. SIZE: (8 ♂♂, 9 ♀♀) wing, ♂ 105–118 (111), ♀ 105–119 (113); tail, ♂ 102–115 (110), ♀ 104–117 (112); bill, ♂ 22–25 (23.6), ♀ 22–25 (23.1); tarsus, 35–37 (36.0), ♀ 34–37 (35.2). A cline of increasing size from N to S: N end L. Tanganyika (unsexed, n = 5) wing 105–111, tail 97–106; NW Botswana (unsexed, n = ?) wing, 112–120, tail 113–116. WEIGHT: Zambia, ♂ (n = 3) 72–83 (77.3); Angola, 1 ♂ 70; also (unsexed, n = 3) 83–92 (Maclean 1993).
IMMATURE: like adult but paler, particularly on throat and breast, and has dark brown (not red) eyes.
NESTLING: hatches naked; egg tooth still present after a day or two; later, nestling has dark eyes.
T. l. griseosquamatus Clancey: N Botswana (south and east of Okavango Swamp, Maun, L. Ngami, Botletle R. to Chobe R.), just into Zimbabwe (Zambezi valley above Katombora Rapids) and Zambia (Mambova). Like *hartlaubii* but mantle, back and rump paler, and forehead and crown to hindneck and sides of neck hoary, with feathers edged greyish white. Larger. SIZE: (10 ♂♂, 10 ♀♀) wing, ♂ 115.5–122.5 (119), ♀ 113–124 (118); tail, ♂ 106–119 (113), ♀ 108.5–119 (114).

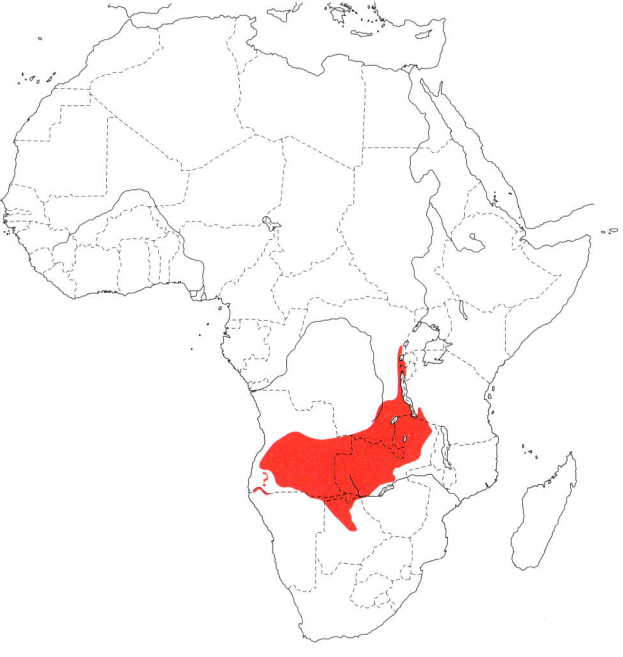

Turdoides hartlaubii

TAXONOMIC NOTE: for discussion of racial variation, including *ater*, see Clancey (1974). Formerly regarded as conspecific with *T. leucopygius*, but shown by Clancey (1984) to be specifically distinct.

Field Characters. Length 26 cm. A grey-brown babbler with extensive *white on rump* and uppertail-coverts conspicuous in flight. Immature paler than adult, especially on throat and breast. Perched bird with black lores and *orange eye* somewhat like Arrow-marked Babbler *T. jardineii*, but has *blotchy dark underparts* with *pale scales* on throat and breast, and *white belly;* Arrow-marked is uniform grey-brown below except for white streaks on breast. Black-faced Babbler *T. melanops* is plain-looking, with pale yellow eye and brown rump.

Voice. Tape-recorded (86, 88, 91–99, 104, B, F, LEM, MOY). Variety of harsh, grating calls, 'chaaa', 'cheeya', 'chyaa', a short 'jik' or longer 'chiwaaaya'; birds in chorus give double 'cheeka-cheeka-cheeka-cheeka ...'; some calls have a particularly nasal, petulant quality, 'jeeeya' or 'jeeeyeee'.

General Habits. Inhabits tall, dense vegetation in marshy valleys with clumps of bushes and high grass, up to 1300 m in most of SE Zaïre but to at least 1650 m in Marungu Highlands (Lumono), and at 1550–2150 m on Ufipa Plateau (Tanzania); borders of galleries of *Acacia poly-*

acantha (Rwanda); edges of riparian forest, bushy clumps around termite mounds, thickets on flood plains and in perennial river valleys (in Botswana keeping between open river and closed woodland); occasionally reed-beds and papyrus swamps.

Lives year-round in parties of 5–15 birds, often about 8, up to 20 recorded. Extremely noisy; moderately approachable. Flock maintains permanent territory, and when 2 flocks meet at territorial border they chorus raucously. Forages on ground, in the open, but spends plenty of time in dense woody growth and trees. Often mixes with flock of Arrow-marked Babblers *T. jardineii*, but 3 Hartlaub's once persistently and laboriously chased flock of 6 Arrow-markeds away from bird-table (Herremans and Herremans-Tonnoeyr 1995). Flight direct, low down, wingbeats rapid, rather shallow and fluttering, then a long, quite fast glide, followed by more flapping. Roosts communally, repeatedly in same tree or bush, also in reedbeds. Flock approaches roost calling, suddenly falls silent, and birds peer about, then move to sleeping branch, jostling and jumping over each other before settling down for the night, clumped in a row, plumage fluffed out (C. J. Vernon *in* Ginn *et al.* 1989).

Food. Not known.

Breeding Habits. Solitary nester; breeds co-operatively (i.e. young in and out of nest fed by several adults).

NEST: an untidy mass of grass and twigs with deep cup on top, the outside made of leaves and inside lined with fine grass and rootlets (Moyer 1982); nest also described as rough but neat (Hustler 1997). Sited in dense clump of vegetation such as flood debris caught up in bush or tree, and in reedbeds. One was 2·5 m up in a *Piliostigma thonningii* tree densely covered with vines which formed a shady canopy above nest; others, in and near reedbed in Zimbabwe, well concealed in dense tufts of reeds *Phragmites* and sedges, in *Croton megalobotrys* bush and *Acacia hebeclada* tree (Hustler 1997). One nest in reeds became completely flattened with growth of chicks. Cup diam. 93 × 95 and cup depth 70.

EGGS: 2–5. Glossy, rather deep blue or greenish blue or turquoise. SIZE: (n = 4) 25·0–28·0 × 18·4–20·5 (26·0 × 19·1); (n = 2) 27·8–28·2 × 20·1–20·3. WEIGHT (n = 2, well incubated): 6·0, 6·2.

LAYING DATES: Tanzania (Ufipa Plateau), Mar; Angola, July–Sept, Oct; Botswana, Apr, Aug, Oct, Dec; Zambia, Oct–Feb, Apr–May; Zaïre, (enlarged gonads Feb); Zimbabwe, Oct, Dec, Feb–Apr (clutch Mar; nestlings Mar; fledglings Jan, Mar, May; adults carrying material Oct, Dec, Feb: Hustler 1997).

DEVELOPMENT AND CARE OF YOUNG: in one nest, young fed regularly by 4 adults, which sometimes flew in line abreast, all with food in bill, and used same perches to get to the nest, each waiting until previous adult had delivered food; in 20 min the 4 brought at least 20 food itms to nest (Hustler 1997). Newly-fledged young follow adults whilst continuously vibrating their wings and soliciting food; fed by all members of flock; one brood still with flock 5 months after fledging.

BREEDING SUCCESS/SURVIVAL: occasionally parasitized by African Striped Cuckoo *Oxylophus levaillantii* and thought to be regularly parasitized by it in Kazungula reedbed population, Zimbabwe (Hustler 1997).

References
Clancey, P.A. (1974, 1984).
Ginn, P.J. *et al.* (1989).
Hustler, K. (1997).

Plate 3
(Opp. p. 30)

Turdoides melanops **(Hartlaub). Black-faced Babbler; Black-lored Babbler. Cratérope masqué.**

Crateropus melanops Hartlaub, 1867. Proc. Zool. Soc. London, pt 3 (1866), p. 435, pl. 37; Damaraland.

Forms a superspecies with *T. leucopygius*, *T. sharpei* and *T. hartlaubii*.

Range and Status. Endemic resident, SW Angola to NW Botswana. Angola: SW Huila Prov. from Jau and Humpata, west towards Namibe, and southeast along Caculovar R. valley to confluence with Cunene R., and common in woodland along short section of Cunene R.; also along Cubango R. at Namibian border, east of about 20°E. Namibia, extreme NW, and in NE in Okavango Delta and E Caprivi (Kwando, Linyanti, Chobe, Zambezi: Harrison *et al.* 1997). Botswana, uncommon to locally common in NW, west of Okavango delta, east to Gomare and Nokaneng. Extreme W Zimbabwe (near Katambora Rapids).

Description. *T. m. melanops* Hartlaub: SW Angola and N Namibia, west of 20°E. ADULT ♂: forehead, forecrown and superciliary area frosted or silvery greyish brown with fine white shaft streaks; hindcrown and nape brown in centre, greyish brown at sides; hindneck brown with paler feather fringes; mantle and scapulars brown, faintly mottled when fresh; back, rump and uppertail-coverts plain brown. Tail blackish brown above and below. Lores and crescent around front of eye blackish brown; cheeks and ear-coverts silvery greyish brown, finely streaked whitish; sides of neck brown with faint greyish buff scaling. Chin and throat almost plain brownish grey with silvery wash; upper breast brown with buff feather fringes; rest of underparts pale ochreous buff with diffuse darker brown streaking, sometimes barely discernible. Wing brown, darkest on remiges. Primary underwing-coverts dark brown; rest of underwing-coverts and axillaries paler buffish brown. Bill black; eyes yellow to greenish yellow; legs slaty grey to black. SIZE (7 ♂♂, 4 ♀♀): wing, 108–116 (114), ♀ 112–118 (114); tail, ♂ 112–125 (119), ♀ 119–125 (122); bill to feathers, 24–27 (25·4), ♀ 23–25 (24·4), bill to skull 29·5–31 (30·2); tarsus, ♂ 32–36 (34·9), ♀ 35–36 (35·8). WEIGHT: ♂ (n = 2) 76, 78.

IMMATURE: juvenile has brown head-top with narrow white streaks on forehead and in broad line over eye; rest of upperparts

Turdoides melanops

warm brown; tail black; lores dark grey, cheeks and ear-coverts grey-brown with fine pale streaks; underparts pale brown, whiter on throat and upper belly; breast with indistinct dark streaking; flanks and undertail-coverts with pale fulvous wash; narrow pale tips to some wing-coverts; eyes brown.

T. m. querulus Clancey: NW Botswana, NE Namibia (Caprivi Strip) and adjacently in Cuando-Cubango, SE Angola. Overall a greyer, cooler, darker brown than in *melanops*; cheeks and throat purer grey; more black around eye; bill shorter and less decurved. Smaller. SIZE: wing, ♂ (n = ?) 115–118·5, ♀ (n = 5) 109–116·5; bill to skull 26–28 (Clancey 1989).

Field Characters. Length 28 cm. A grey-brown babbler lacking white rump and belly of Hartlaub's *T. hartlaubii*. Distinguished from Arrow-marked Babbler *T. jardineii* by *pale yellow eye* and *more extensive black area between eye and base of bill* (black crescent around front of eye as well as black lores); colour more uniform, pale streaking confined to top of head, neck and breast lightly scaled buff, not streaked white. Perched Hartlaub's Babbler has orange eye. Underwing looks blackish in flight (tawny to buff in Arrow-marked and Hartlaub's Babblers: Maclean 1993). Immature has brown eye.

Voice. Tape-recorded (F, VDB, WALK). Typical *Turdoides* grating chorus, sometimes beginning with sharp 'jik', 'jik ... jaaa-jaaa-jaaa-jaaa-jaaa', 'jeeu-jeeu-jeeu ...', or nasal 'wha-wha-wha ...'. Other calls include higher-pitched 'jizi, jizi, jizi ...', fast chattering 'kakakakaka ...', cat-like 'waaaaaa' and low, conversational 'juk' or 'jerk'.

General Habits. In Angola occurs in any woodland but mainly acacia, towards edges of the Cunene and Caculovar valleys rather than on their floodplains; in Botswana inhabits thickets, creepers and secondary growth in understorey of broadleaved *Commiphora* and mixed acacia woodland with underlying long grass (Penry 1994). Lives in groups of 3–7 and up to 15 birds; av. group size (n = 28) 5·7 birds, including av. 1·0 fledglings/juveniles (Etosha Nat. Park: P. Shaw, pers. comm.). Keeps mainly to dense cover; shy, furtive but inquisitive; difficult to observe. Forages on ground and low down in grass, vigorously tossing leaf litter aside with bill. Flock watched for 5 h kept in dense bush in area *c.* 50 m wide; adult bird acts as a sentinel, perching on outer or top branch of tree or bush, above flock with juveniles foraging on ground; foraging flock consisting entirely of adults seems not to post a sentinel (P. Shaw, pers. comm.). Sentinel gives quiet call, 'cull, cull...' every 2–3 s, audible 10–15 m away. One flock left ground and entered a tree, where 3 birds sat up against each other; one ruffled its plumage, raised bill, closed its eyes, and had its throat preened by middle bird, which then turned and preened the third bird (C.J. Vernon in Ginn *et al.* 1989). Flock suddenly became alert, flew to distant tree, where birds called crescendo in chorus, fluffing plumage, lowering wings and fanning tail; another flock responded in kind, presumably across a territorial border. A fledgling and an adult once seen lying on ground with wings closed, facing each other 2–3 cm apart; the adult pushed fledgling with its feet, and after 5–10 s swivelled around and hopped onto it; fledgling then moved away, and adult lay on its back for a few s, tugging at dry grass blade with its feet (P. Shaw, pers. comm.).

Food. Insects, also small reptiles and fruits.

Breeding Habits.

NEST: quite large bowl, roughly made of grass, lined with fine grass and fibres. One nest *c.* 5 m above ground, wedged in upper and outer branches of *Commiphora* tree (Maclean 1993).

EGGS: 2–5. Deep turquoise, with fine nodules scattered over surface. SIZE: (n = 2) 26·5–28·5 × 19·0–19·5.

LAYING DATES: Angola, Nov (Humbe), (and nest Dec at Jau); Namibia (Etosha), Dec.

DEVELOPMENT AND CARE OF YOUNG: up to 4 helpers at nest. Fledgling has dark grey eyes (appearing black at a distance); bill dark grey or black; gape flange cream or pale yellow. Fledgling spends little time on ground, but perches just above foraging adults, almost always within 1–2 m of them (P. Shaw, pers. comm.). Fledgling sometimes fed by a juvenile.

BREEDING SUCCESS/SURVIVAL: one of the very few *Turdoides* species in Africa not parasitized by African Striped Cuckoo *Oxylophus levaillantii*.

References
Clancey, P.A. (1979).
Ginn, P.J. *et al.* (1989).

Plate 4
(Opp. p. 31)

Turdoides bicolor **(Jardine). Southern Pied Babbler; Pied Babbler. Cratérope bicolore.**

Crateropus bicolor Jardine, 1831. Edinburgh J. Nat. Geogr. Sci., 3, p. 97, pl. 3; Kuruman, South Africa.

Range and Status. Endemic resident, Namibia to Zimbabwe, south to N Cape Prov. No certain record in Angola, though occurs very close to border on Namibian side. Namibia, widespread in N, from Angolan border north of Etosha, and south to Windhoek and Gobabis. Botswana, common to very common throughout, except in N for Okavango and area to north of it and for Chobe region north of Nxai Pan, and absent from extreme SW corner and from woodlands in NE. N Cape Prov., locally common north of Orange R. and Vaal R. Zimbabwe, widespread in S half of Hwange Nat. Park north to line from Tamafupa Pans to Kennedy, but elsewhere localized, especially above 1200 m, northeast to Chiriza, east of Hwange nearly to Gwelo, and in S east along Limpopo valley to Mateke Hills (31°E). Transvaal, fairly common in NW, east to Bandolierskop and, on Limpopo, to Nwanedzi confluence (30°30′E), and 2 sight records in lowveld. Range boundaries shown are from Harrison *et al.* (1997). Not admitted to Mozambique (Clancey 1971, Dowsett and Dowsett-Lemaire 1993a), but reported from Incomati region 90 years ago (da Rosa Pinto and Lamm 1953) and point-plotted also in W Mozambique north of Limpopo by Hall and Moreau (1970). Orange Free State, 5 records, attributed to vagrants.

Turdoides bicolor

Description. ADULT ♂: whole head and body white. Tail brownish black above and below. Remiges, alula, primary and outer greater coverts brownish black, innermost greater coverts, median and lesser coverts white. Underwing-coverts and axillaries white. Bill black; eyes yellow, yellowish brown, yellowish orange, or orange; legs black or brown-black. Sexes alike. SIZE: (10 ♂♂, 10 ♀♀) wing, ♂ 107–116 (114), ♀ 108–116 (112); tail, ♂ 108–117 (112), ♀ 107–114 (111); bill, ♂ 28–31 (28·6), ♀ 27–32 (29·2); tarsus, ♂ 34–38 (36·3), ♀ 34–37 (35·0). WEIGHT: Namibia, ♂ (n = 3) 74–82 (78·7), ♀ (n = 3) 72–77 (73·7), 1 unsexed 77; also (n = 2, unsexed, ?locality) 69·2, 82·4; immature probably 4–6 months old, 53·5.
IMMATURE: juvenile has head and body brown, belly pale, wings dark brown with some rufescent coverts, tail blackish. Eyes dark brown. Immature acquires white plumage first on underparts and neck, then on head and back. Typically has white head, mixed brown and white feathers on mantle, brown wings, blackish tail, and white underparts with irregularly distributed brown feathers and large brown patches on flanks, belly and lower thighs. Attains full adult plumage in 4–6 months.
NESTLING: hatchling not described; late nestling feathered dark brown above, grey-brown below, some feathers with structural bars, gape yellow and inside mouth bright orange; eyes dark brown.

Field Characters. Length 26 cm. White head, body and wing-coverts, rest of wings and tail black; bill black, eye yellow or orange. Adult can be confused only with Bare-cheeked Babbler *T. gymnogenys*, which has mainly white head, rump and underparts but brown back, wings and tail, patches of bare black skin on face, tawny patch on sides of neck. Juvenile said to resemble immature Arrow-marked Babbler *T. jardineii* (Maclean 1993), mainly dull brown, paler on lower belly, with darker wings and tail; gradually whitens with age, first on underparts and neck, then on head and back, typically appearing mottled brown and white; adult plumage attained at 4–6 months.

Voice. Tape-recorded (58, 88, 91–99, B, F, CHA, LUT, STJ). Calls harsh, grating, rather high-pitched, often startling with repeated single note and becoming double note, 'jajajajijijijeeyi-jeeyi-jeeyi-jeeyi ...' or 'rak-rak-rak-rak-rak-jeeya-jeeya-jeeya ...', or a faster, sharper 'jijijijaaa-jaaa-jaaa ...'; individual bird may start calling like this and then be joined by others forming group chorus, which adds different notes including a double 'ka-*wack*-ka-*wack*-ka-*wack* ...'; chorus sometimes rises to a crescendo before dying away (Ginn *et al.* 1989), but varies greatly in length and intensity.

General Habits. Preferred habitat is semi-arid savanna woodlands with plenty of *Commiphora* and acacia and sparse grass cover (however, in Zimbabwe absent from watershed acacia savanna between Plumtree and Bulawayo); often in *Acacia mellifera* bushes (W Botswana) and camelthorns *A. erioloba* (N Cape). Keeps mainly to denser stands of thorny, woody growth, but also occurs in open areas with scattered clumps of bushes, for instance around termite hills.

Highly social and group-territorial year-round. Lives in groups of 3–15 birds, usually 6–10 birds including 2 or 3 immature birds. Noisy, vocalizing constantly; raucous flock chorus produced by some birds making cackling rattle and others a purring note (C.J. Vernon *in* Ginn *et al.* 1989); chorus rises to crescendo then dies away. 2 groups interact at territorial boundary by calling, with wings lowered and quivering and tails fanned. Flock frequently huddles together during day, and up to 15 adult birds scramble over each other whilst calling loudly (Lindeque and Kapner 1993). Spends much time loafing, birds of a group huddled together and preening each other. Bird closes eyes, points bill skywards, ruffles head feathers, and is preened by birds on either side of it; later it may preen them. Occasionally bird occurs solitarily. Flock largely ignores other bird species; Pied Babbler does not join mixed foraging parties nor responds to other species' alarm calls (C.J. Vernon *loc. cit.*). Often occurs alongside Crimson-breasted Shrike *Laniarius atrococcineus*, but this seems adventitious, not symbiotic. However, said to form mixed parties with Crimson-breasted Shrikes and Red-billed Buffalo Weavers *Bubalornis niger* (Maclean 1993). Once 5 Pied Babblers vigorously attacked an under-weight Gabar Goshawk *Micronisus gabar* which had seized a nearly full-grown immature babbler. After 10 min of sustained attack the hawk released its captive and flew 15 m away where it was again pinned face down with outspread wings and attacked for 5 min by babblers scrambling around and on top of it, calling loudly, and vigorously and sustainedly pecking its head and neck. The hawk died 36 h later and 75 m away, and had puncture wounds, haemorrhages, a fractured neck joint and spinal cord damage (Lindeque and Kapner 1993).

Forages mainly or entirely on ground, fossicking amongst leaf litter and other vegetable debris, lifting leaves like a thrush *Turdus*. Locomotion on ground a scuffling, short hop. Clambers purposefully and strongly amongst thorny twigs and branches, and creeps through tangled growth. Flies low down and direct, with fast-flapping alternating with gliding, in loose formation or birds following each other one by one along same route, whole flock taking 15–20 s to quit a tree. Flock roosts communally, birds huddling on branch in a row, and is strongly attached to favourite roost site in dense tree.

Sedentary, but evidently given to vagrancy.

Food. Said to be insects and other small animals.

Breeding Habits. Group-territorial, co-operative, ?monogamous breeder.

NEST: open bowl, not very compact and with edges sometimes loose and flimsy (like dove's nest), rather roughly made of dry, long grass stems, 1 mm thick climbing plant stems and very thin woody twigs, with quite deep cup in centre, neatly lined with hair, fine rootlets and soft fibres; lip of cup sometimes with grass inflorescences worked in. Sited on multiple fork in middle of leafy *Acacia*, *Ziziphus* or other thorn tree, but not necessarily in deepest shade. Nest diam. *c.* 150 × 175, cup diam. *c.* 70.

EGGS: 2–5, av. (n = 10) 3·3. Long ovals; pale blue, sometimes very pale, with fine nodules scattered over surface except at ends, which are smooth. SIZE: (n = 23) 23·5–27·5 × 18·0–20·1 (26·0 × 19·0).

LAYING DATES: Namibia, Oct–Apr; Botswana, Sept–Feb, mainly Oct–Dec (Skinner 1995); Zimbabwe, Aug, Oct–Jan (n = 27 clutches), mainly Nov (n = 12); Transvaal, Oct–Jan, Mar.

DEVELOPMENT AND CARE OF YOUNG: nest attended by up to 12 birds; one nest had 3 adults always present (but not known whether they shared incubation: Barbour 1972). Fledglings beg with tails fanned and wings lowered and quivering, and may perch huddled when begging.

BREEDING SUCCESS/SURVIVAL: one of the very few *Turdoides* species in Africa not parasitized by African Striped Cuckoo *Oxylophus levaillantii*. Gabar Goshawk thought to be a major predator (Lindeque and Kapner 1993).

References
Ginn, P.J. *et al.* (1989).
Lindeque, M. and Kapner, J. (1993).
Maclean, G.L. (1993).

Turdoides hindei (Sharpe). Hinde's Babbler; Hinde's Pied Babbler. Cratérope pie de Hinde.

Crateropus hindei Sharpe, 1900. Bull. Br. Orn. Club 11, p. 29; Athi River, Kenya.

Plate 4
(Opp. p. 31)

Range and Status. Endemic resident, Kenya. Scarce and very local in Meru, Embu, Karatina, Murang'a, Thika and Machakos districts and in Mwea Nat. Res. (Zimmerman *et al.* 1996). Range has contracted substantially, coincident with agricultural development: formerly occurred at Athi River and in Chuka, Mwingi and Kitui Districts. Range in 1940s covered 17,500 km^2 but by late 1970s thought to be common only within area of 1050 km^2 centred on Embu and headwaters of streams tributary to Tana R. (Plumb 1979). However, recently found in Kitui again and discovered in Mukurweini, Nyeri District, on W side of Mt Kenya. Density: 23 flocks with 87 birds found, by using tape-recorder playback of voices, in area of 27 km^2 at Mukurweini; much lower density in Muvuti Location and Kathekakai Ranch, Machakos District, where <40 birds located in 42 km^2 (Njoroge 1994); 4 flocks with 15+ birds (including 2 juveniles) in 1 km^2 at Kianyaga, Embu, but none found at all along 33 km of apparently suitable river-valley habitat (Shaw 1996).

Species could possibly number <1000 birds (Turner 1992). Thought not to be 'outcompeted' by the much commoner Northern Pied Babbler *T. hypoleucus*; but survival threatened by hunting and disappearance of suitable habitat (Anon. 1994). Many people in Kianyaga eat (or used to eat) birds, and babblers (Hinde's and Northern Pied) are favoured delicacies; *Lantana* patches (see General Habits) there are likely to be cleared in near future, which would be disastrous for Hinde's Babbler (Njoroge 1994).

Turdoides hindei

Description. ADULT ♂: mottled blackish brown, with rufous rump and uppertail-coverts and creamy belly, all dark parts extensively but irregularly scaled and blotched with whitish; white-tipped crown feathers often much longer than all-dark ones, but length of white tip or fringe seems unrelated to wear; some birds resemble semi-albinos, with e.g. an all-white remex in one wing, 2 or 3 partly white ones in the other, and a large white patch in tertials on one side only; rectrices and occasional contour feathers sometimes partly rusty. Top of head with dense white spotting or scale-like barring; hindneck densely barred, in some birds mainly whitish; broad whitish bars on mantle, back and scapulars, and a few on rump and uppertail-coverts. Tail dark brown, underside tinged tawny. Side of head and chin dark brown with extensive whitish mottling; sides of neck and throat dark brown with much white barring. Breast blotched dark brown and buffy white, in some birds mainly whitish; belly buffy grey blotched with brown; thighs dark brown, barred white; flanks and undertail-coverts pale tawny, marked with dark brown and white. Remiges dark brown, some feathers extensively white-tipped especially on inner web; alula brown, feather tips sometimes whitish; upperwing-coverts dark brown; in some birds many primary coverts with broad whitish tip and outer edge, and most greater coverts with large whitish tips; median and lesser coverts fringed whitish. Underwing-coverts tawny buff; axillaries tawny; tawny area on underside of flight feathers formed by inner borders. Bill dark slate, eyes rich orange-red, legs and feet dull umber. Sexes alike. SIZE: (9 ♂♂, 10 ♀♀) wing, ♂ 92–108 (100·5), ♀ 89–105 (97·8); tail, ♂ 96–109 (102), ♀ 95–107 (101); bill, ♂ 22–25 (22·9), ♀ 21–24 (22·4); tarsus, ♂ 30–35 (32·2), ♀ 32–34 (33·1); also (5 ♂♂, 6 ♀♀) wing, ♂ 100–108 (103); ♀ 97–105 (100); tail, ♂ 98–109 (104), ♀ 100–107 (104); bill, ♂ 22–25 (22·8), ♀ 21–24 (22·3); tarsus, ♂ 32–35 (33·1), ♀ 32–34 (33·3). WEIGHT: 1 ♂ 70, unsexed (n = 11) 58–77 (67·6).

IMMATURE: like adult but eyes grey not red.

NESTLING: half-grown nestling slate grey with bright yellow gape (Plumb 1979).

Field Characters. Length 20–23 cm. A highly variable babbler (no two individuals quite alike and species has been thought to be a hybrid: Plumb 1979); 1 or 2 birds in a flock often partially albinistic, either very white below, or almost white on mantle; odd flight feathers in wing and sides of tail are sometimes wholly white (but not necessarily symmetrically so). Blackish brown head, neck, mantle, lesser wing coverts and breast generally thickly covered with white crescents; rump rufous, flanks tawny, belly creamy. Orange-red eye striking in the field (eye yellow in Arrow-marked Babbler *T. jardineii*, creamy in Northern Pied Babbler *T. hypoleucus*). Juvenile has grey eye.

Voice. Tape-recorded (B, C, F, CART, KEI, NOR). Single bird gives low, conversational 'ruka'; group often breaks into noisy chatter, 'rukakakakakaka ...', faster than Northern Pied Babbler, almost like Green Wood-Hoopoe *Phoeniculus purpureus*. Also frequent is a grating, double note, 'ta-jaaa', lengthened during chorus into long screeching 'd'jaaaaaay'. Very different in quality is a repeated loud, down-slurred braying call, '*dayoo*'. Calls also rendered as a rambling 'chare-chare-chare-chare' (softer and more conversational than similar call of Northern Pied Babbler), and a chattering 'chirrr-chirrr-'; also, 'cherak-chwak-chakchakchakchak' or 'kwerak-chk-chk-chk' (Zimmerman *et al.* 1996). Alarm, a hiccough-like double note (Plumb 1979).

General Habits. Inhabits fringes of cultivation with scattered trees, open areas but plenty of cover, in small valleys and open bushland in semi-arid country at 915–1700 m (Plumb 1979); river valleys, particularly coffee plantations, food-crop farms and areas where alien shrub *Lantana* is well established (Britton 1980, Njoroge *et al.* 1998); also acacia woodland and dense vegetation, sometimes lush and green, bordering gulleys, streams and rivers with sand, boulders and thorn scrub. Original habitat thought to have been rocky hillsides and valleys dominated by *Combretum, Terminalia, Croton, Cussonia, Cassia* and *Commiphora* (Collar and Stuart 1985). Where it occurs in same valley as Northern Pied Babbler, Hinde's occupies lower slopes; sometimes interacts with Northern Pied, when disputes are always won by Hinde's (Njoroge 1994). Lives in parties of 6–8 year-round, including when breeding. Skulks in thickets and extremely difficult to detect except by calling-up using taped voices. Bubbling calls given mainly when flock on the move; otherwise flock can be silent for long periods, but tends to call when disturbed by observer (Lewis 1984). Frequents dense cover but flock forages on open ground, with 1–2 birds posted as sentinels on exposed perches in nearby treetop for minutes at a time. When disturbed, flock noisy at first but birds fly up into trees then become silent. In 3 h, foraging flock moves a few hundred m through habitat. Pair indulges in mutual preening, each bird in turn stretching its neck up with the other preening its neck feathers (Plumb 1979). Highly sedentary, flock keeping within small area, often part of a single valley. Group territory size (n = 5) 3·75 ± 1·6 ha (Njoroge *et al.* 1998).

Food. Not known. Nestlings fed on 'dark glutinous mass collected from below bushes' (Plumb 1979).

Breeding Habits. Solitary nester; co-operative breeder – parents at one nest accompanied by 2–3 other adults (Plumb 1979).
  NEST: open bowl of coarse grass lined with finer grass.
  EGGS: 2–3. Light blue, glossy. SIZE: about 26 × 18.
  LAYING DATES: Apr, Sept.
  DEVELOPMENT AND CARE OF YOUNG: young evidently attended by several adults.
  BREEDING SUCCESS/SURVIVAL: only one of 5 nesting attempts successful, 4 falling victim to predators or human disturbance (Anon. 1994). Parasitized by Jacobin Cuckoo *Oxylophus jacobinus*.

References
Njoroge, P. *et al.* (1998).
Plumb, W.J. (1979).
Shaw, P. (1996).

Turdoides hypoleucus (Cabanis). Northern Pied Babbler. Cratérope bigarré; Cratérope pie. Plate 4
Crateropus hypoleucus Cabanis, 1878. J. f. Orn., 26, p. 205; Kitui, Kenya. (Opp. p. 31)

Range and Status. Endemic resident, Kenya and N Tanzania. Kenya, locally common from centre to Tanzanian border, up to 1800 m, mainly above 1000 m; mainly east of Rift valley; north to Nyeri and Murang'a and east to Meru and Yatta Plateau (Kitui). Tanzania, 3 populations, one (confluent with Kenyan population) from about Loliondo to N Arusha, N Kilimanjaro and L. Manyara (50 localities: N. Baker, pers. comm.), one fairly common in Kilosa (23 localities), and one from Usambaras to near Tanga.

Description. *T. h. hypoleucus* (Cabanis): central and S Kenya to N Tanzania (Kilimanjaro and Arusha). ADULT ♂: upperparts including wings and tail uniform dark brown, except for pale flecks on forehead, forecrown and superciliary area produced by buffy brown feather tips and edges, and narrow white tips to median and lesser coverts when fresh. Lores, cheeks, sides of neck and ear-coverts dark brown, the last faintly white-streaked, sharply demarcated from white chin and throat. Upper breast brown at sides, forming prominent patches, and white in centre; lower breast to undertail-coverts white; flanks and thighs dark brown; disarray of feathers makes flanks patchily brown and white. Extent of brown on breast varies: brown patches sometimes coalesce to form breast band; belly sometimes dingy greybrown or patchily brown (Plumb 1979). Outer lesser underwingcoverts brown; rest of underwing-coverts and axillaries whitish. Bill black; eyes clear bluish white; legs dark greyish or greenish grey. Sexes alike. SIZE: (10 ♂♂, 10 ♀♀) wing, ♂ 104–118 (109), ♀ 105–113 (109); tail, ♂ 103–115 (108), ♀ 99–111 (107); bill, ♂ 23–24 (23·5), ♀ 23–25 (23·7); tarsus, ♂ 34–38 (35·9), ♀ 34–37 (35·6). WEIGHT: (Kenya, unsexed, n = 3) 67–80 (74).
  IMMATURE: juvenile like adult, but warmer brown; undertailcoverts browner; eyes dark.
  NESTLING: not known.
  T. h. rufuensis (Neumann): N and NE Tanzania (Mbulumbulu, L. Manyara Nat. Park, Tarangire, E Usambaras at Amani, and Kondoa, to Kibedya, Kilosa). Paler and greyer above than nominate race, feathers of forehead and crown with narrower buff edges, those of hindneck, mantle and back with broad greyish fringes, and rump and uppertail-coverts mainly grey. Sides of head and neck less dark than in *hypoleucos*, hence not as strongly contrasted with white throat. Centre of breast browner so that breast band nearly complete. Thighs greyish brown.

Field Characters. Length 20–23 cm. A truly *pied* babbler, brown above and white below, with brown flanks and brown patch on sides of breast. Plumage somewhat variable: brown breast patches may join to form breastband, or belly may be grey-brown or patchily brown;

Turdoides hypoleucus

underparts occasionally stained reddish from the red clay soil in their habitat (Jackson and Sclater 1938). Distinguishable from any partially-albino Hinde's Pied Babbler *T. hindei* by *white eye*; Hinde's also very different-looking; in addition to orange-red eye it has scaly head and neck. When head and neck only are visible, Northern Pied could be confused with white-throated variant of race *vepres* of Black-lored Babbler *T. sharpei*, which also has white eye and blackish lores, but breast and belly usually dark brown (or a little white streaking on upper belly); white-throated birds may also have white lower belly and undertail-coverts, but not entirely white underparts like Northern Pied Babbler. Immature has dark eye.

Voice. Tape-recorded (10, B, C, CHA, GREG, HOR, KEI, NOR, ZIM). Typical single-note call a nasal, complaining 'taaaa' or 'tyaaaa', slightly downslurred and fading somewhat at the end, rather conversational in tone, probably a simple contact call; more forceful calls in groups, often running into chorus, loud 'teeya-teeya-teeya' or longer, run-together 'teeeyayayayayaya', in addition to louder 'tyaaaa'. Voice also described as 'variety of loud raucous chattering, churring and chuckling calls' (Zimmerman *et al*. 1996).

General Habits. Occupies bushland at medium elevations, and edges of dry evergreen forest; attracted to vicinity of human habitation and now a common bird of Nairobi gardens. Always near thick cover. Occurs in same valleys as Hinde's Babbler *T. hindei* but extends away from them into areas of human habitation (Njoroge 1994). Keeps in territorial flocks of 3–12 (often 10–12) birds, av. (n = 3) 8·3; territory size (n = 3) 6·6 ± 1·8 ha (Njoroge *et al*. 1998). Noisy, restless and inquisitive. Becomes tame and confiding in Nairobi gardens. Forages on ground, under or very close to bushy cover, turning over dead leaves and flicking them aside like thrush *Turdus* sp. Flock moving from one tree to another flies in single file, and often choruses when or just after alighting. Departs from roost site at dawn; spends most of morning foraging; around midday, birds concentrate in *Lantana* thickets; in evening, feeds close to roost site; goes to roost at dusk (Njoroge *et al*. 1998).

Food. Unknown

Breeding Habits.
NEST: roughly constructed open bowl made of coarse dry grasses on foundation of small twigs, with deep cup lined with fine grasses and rootlets; one with *Bougainvillea* flowers and dog hair incorporated into nest fabric. Well hidden in isolated thorn tree or bush, 1·5–3·5 m above ground.
EGGS: 3. Not glossy. Uniform immaculate dark blue, or blue with a few dark brown spots at large end. SIZE: (n = ?) 25·5–26·0 × 18·5–19·0.
LAYING DATES: Kenya and Tanzania, Region D, all months, peaking in rains in Apr–May and Nov: Feb (2 clutches), Mar (2), Apr (4), May (4), July (1), Sept (3), Oct (2), Nov (3), Dec (1 clutch).
DEVELOPMENT AND CARE OF YOUNG: fledgling accompanies adults, begs by wing-shivering, and is fed by them; inside mouth bright vermilion red.

References
Jackson, F.J. and Sclater, W.L. (1938).
Njoroge, P. *et al*. (1998).

Plate 4
(Opp. p. 31)

Turdoides squamulatus (Shelley). Scaly Babbler. Cratérope maillé.

Crateropus squamulatus Shelley, 1884. Ibis, p. 45; Mombasa, Kenya.

Range and Status. Endemic resident, river valleys and near coasts in SE Ethiopia, Somalia and Kenya. 5 or 6 separate populations: (a) uncommon in R. Webi Shabeelle valley (Ethiopia and Somalia), in Ethiopia from Imi (42°10′E) to Gode and K'elafo (Callafo) and in Somalia near Muodisho (Mogadishu), from Jawhar (Jiohar) to Balcad (Balad) (Ash 1981); (b) very common in lower Jubba R. valley from Webi Shebeelle confluence north to Baardheere, and upstream known in Jubba valley from Luuq (Lugh; Somalia) to Dolo (Ethiopia); (c) an unnamed population resident in valley of Daua R. (which defines Ethiopian/Kenyan border between Mari and Mandera, and meets Jubba R. at Dolo), of quite white-headed birds, like those of distant population (a), known from Rhamu Dimtu and Ramu (Rhamu) (Lewis and Pomeroy 1989); (d) an unnamed population, perhaps the same as (c) though *c.* 180 km away and at head of a different drainage system (the Lak/Lagh Bor), known from Ethiopian side of border *c.* 50 km east of Moyale (03°31′N, 39°04′E); (e) an unnamed population in N Bale Prov., Ethiopia, in upper reaches of Webi Gestro at 06°59′N, 40°44′E near Ginir (Ghinnir) (Ash 1981); and (f) locally common, Kenya, in Tana R. valley north at least to Garissa and entering Somalia north to 30 km south of Kolbio at 01°22′S, 41°23′E, and in coastal lowlands south to Mombasa.

Turdoides squamulatus

Description. *T. s. squamulatus* (Shelley): Kenya coast and Tana R., and extreme S Somalia (see (f) above). ADULT ♂: forehead, forecrown and sides of hindcrown blackish brown with scaly greyish buff feather fringes; hindneck and sides of neck dark grey-brown with pale grey scaling; rest of upperparts plain greyish brown with olivaceous tinge. Tail dark brown above and below. Lores, cheeks and ear-coverts blackish brown. Chin to upper breast dark grey-brown, with prominent narrow whitish

crescents, most numerous on sides of chin (**A**); rest of underparts plain grey-brown, but flanks and undertail-coverts warm brown. Wings mid to dark brown, outer edges of outer primaries warm brown. Underwing-coverts and axillaries tawny-buff. Bill black; eyes orange-yellow; legs blackish or dark grey. Sexes alike. SIZE: wing, ♂ (n = 1) 105, ♀ (n = 3) 101–103 (102); tail, ♂ (n = 1) 105, ♀ (n = 2) 97, 104; bill, ♂ (n = 1) 24, ♀ (n = 2) 22, 23; tarsus, ♂ (n = 1) 32, ♀ (n = 3) 33–34 (33·3). WEIGHT: ♂ (n = 7) 55–85 (72·7), ♀ (n = 6) 60–78 (66·2), unsexed (n = 4) 60–81 (69·3).

IMMATURE: juvenile lacks scaly markings; tawny-brown above; tawny buff below, chin to upper breast streaked whitish and dark brown; eyes brown.

NESTLING: not known.

T. s. carolinae Ash: Webi Shabeelle R., S Somalia and SE Ethiopia (see (a) above). Differs from *jubaensis* in having forehead, lores and above eye to ear-coverts, cheeks, chin and upper throat white (**B**); dark head feathering usually restricted to crown and to region between base of mandible and eye. Slightly paler grey below than *jubaensis*. WEIGHT: (unsexed, n = 2) 62, 65 (J.S. Ash, pers. comm.).

T. s. jubaensis van Someren: Jubba R., S Somalia and SE Ethiopia (see (b) above). Paler and greyer above than nominate race, ear-coverts less dusky; centre of chin and upper throat often white (**C**); white head feathering occasionally more extensive. Wing (unsexed, n = 5) 101–105 (103). WEIGHT: unsexed (n = 5) 65–76 (69).

T. s. subsp.?: Lak Bor watercourse and Daua R., Ethiopian/ Kenyan border (see (c) and (d) above). Extent of white on head of (c) intermediate between *jubaensis* and *carolinae* (Lewis and Pomeroy 1989).

T. s. subsp.?: Webi Gestro R., Ethiopia (see (e) above). Upperparts uniform brown, underparts brownish white, underwing-coverts pale (?yellowish brown), eyes red (Ash 1981).

Field Characters. Length 22–23 cm. A rather shy and quiet babbler, *dark*, with pale scales on head, neck and breast, and *orange eye* in large dark face-patch. Does not overlap any other *Turdoides* babbler in Somalia or Kenya, but in Ethiopia the population in N Bale Prov. lies within the range of White-rumped Babbler *T. leucopygius*; differs in lack of white rump and belly, *uniform dark underparts* with scaling confined to neck and breast; dark face further distinguishes it from pale-faced *T. l. smithii*, scaly throat from black-throated *T. l. omoensis*, which also has broad scaling on head, back and underparts and pale ear-coverts.

Voice. Tape-recorded (B, F, CART, FIS, GREG, McVIC, STJ). Harsh single notes, 'chak', 'chuduk', 'chik' and a more explosive 'splik' or 'splidik'; in chorus a grating 'gagagagagaga ...'; in response to pishing, 'cheedada' or 'cheeda', and to playback, a single, repeated 'chwik'. Calls also described as "a distinctive, throaty 'wuk-a-ha, wuk-a-ha ...' and 'ch'wuk, ch'wuk', most reminiscent of Black-lored Babbler" (Zimmerman *et al.* 1996).

General Habits. Inhabits riverine thickets, forest and woodland undergrowth. Lives in flocks of about 5 birds. Silent and unobtrusive, though flock occasionally choruses loudly; but less vocal than most congeners. Flock skulks in very dense cover, though may emerge into the open when chorusing. Moulting of primaries and secondaries irregular (Wood 1989).

Food. Unknown.

Breeding Habits. One nest found but not described.

LAYING DATES: Somalia (*carolinae*) (slightly enlarged gonads, Sept; bird carrying nest material, Oct); Kenya (*squamulatus*), Apr.

Reference
Ash, J.S. (1981).

Turdoides gymnogenys (Hartlaub). Bare-cheeked Babbler. Cratérope à joues nues.

Plate 4
(Opp. p. 31)

Crateropus gymnogenys Hartlaub, 1865. Proc. Zool. Soc. London, p. 86, Benguella Prov., Angola.

Range and Status. Endemic resident, SW Angola and NW Namibia. Angola, coastal plain from Sumbe (Novo Redondo) to Capangombe, Caculovar R. from Humpata to Cahama, and east of Caculovar to about Quipungo. Namibia, locally common from Ruacana Falls and elsewhere on Cunene R. south to about Unjab/Achab R. and east to Etosha Pan, Namutoni and about Otjiwarongo.

Density, Etosha Nat. Park, Namibia, 7·3 birds in 1·2 groups per km^2 in riverine habitats and 0·7 birds in 0·1 groups per km^2 in other habitats (P. Shaw, pers. comm.).

Turdoides gymnogenys

Abundance in a given year is related with rainfall, at least in riverine habitats (Jarvis and Robertson 1997).

Description. ADULT ♂: forehead and forecrown greyish white, merging through hindcrown into mottled grey-brown nape; hindneck tawny-brown; mantle, back and scapulars dark brown with broad pale tawny-buff or greyish feather edging; rump white; uppertail-coverts grey-brown. Tail dark brown above, shiny blackish below with c. 18 structural bars. Bare skin from bill under eye to ear-coverts and another small bare patch around ear black; feathering on side of head, including narrow line under eye, white or greyish white. Sides of neck tawny in patch extending towards throat; flanks creamy, tinged rufous; underparts otherwise white; breast sometimes faintly dappled with grey. Wings dark brown, alula and median and lesser coverts tipped tawny-brown. Underwing-coverts and axillaries tawny-buff; tawny area on underside of flight feathers formed by broad inner borders. Bill black; eyes pale yellow; tarsus dark grey. Sexes alike. SIZE: (5 ♂♂, 7 ♀♀): wing, ♂ 107–116 (111), ♀ 104–114 (111); tail, ♂ 99–107 (102), ♀ 96–102 (98·4); bill, ♂ 25–28 (25·8), ♀ 25–28 (26·7); tarsus, ♂ 33–35 (34·0), ♀ 33–35 (34·1). WEIGHT: ♂ (n = 3, Namibia) 65–84 (75·3), ♀ (n = 2) 72, 80 (W.R.J. Dean, pers. comm.).

IMMATURE: recently fledged juvenile has top of head pale greyish brown, hindneck warm brown, upperwing-coverts and alula prominently edged tawny, secondaries and tertials with narrow rufous tips and inner edges, no rufous patch on side of neck, and pure white, fluffy underparts; eyes dark grey or blackish, becoming reddish brown (P. Shaw, pers. comm.).

NESTLING: not known.

TAXONOMIC NOTE: Namibian birds tend to be less white-headed and to have darker backs and paler sides than Angolan ones, and have been treated as a separate race, *kaokoensis* (including '*tsumebensis*') (Traylor 1963, Clancey 1989).

Field Characters. Length 24 cm. Head, rump and underparts white, flanks tawny, back and wings dark brown, tail blackish; bare black cheeks diagnostic but often crossed by thin stripes of white feathers, giving the bird a bar-cheeked rather than bare-cheeked appearance (Ginn *et al.* 1989); tawny patch on sides of neck extending round onto hindneck. Only bird looking at all similar is immature Southern Pied Babbler *T. bicolor*, which starts brown and becomes generally whitish with black wings and tail and irregular amounts of brown mottling over head and body; it lacks black face patches and tawny neck. Immature Bare-cheeked overall much darker than Pied, especially on back and nape (Sinclair *et al.* 1993). The 2 are ecologically separated, Bare-cheeked frequenting hilly mopane country, Pied in thornveld (Ginn *et al.* 1989).

Voice. Tape-recorded (91–99, B, F, GIB). Harsh 'jeee-jeee-jeee ...' or 'jaaa-jaaa-jaaa ...', often in chorus, when a raucous, crowing 'kaaa' or 'kaaay' also given. Also described as continuous low grating 'chuk-chuk' (Maclean 1993) or 'kerrakerra-kek-kek-kek' (Sinclair *et al.* 1993). Contact call, a low clucking (Ginn *et al.* 1989).

General Habits. Frequents bare, rocky ground among thick tangles of woody vegetation along dry watercourses; rocky, wooded hillsides; and middle and lower strata of well-developed woodland on open plains. Lives year round in cohesive, territorial flock. Group size (n = 47) 2–11 (6·0), consisting mainly of adults, with up to 4 young or very young birds (P. Shaw, pers. comm.). Vocal; especially noisy during territorial displays, social conflicts and just before going to roost (P. Shaw, pers. comm.); at other times unobtrusive. Birds in flock often give contact call, which is often taken up by all flock members at same time and becomes a sustained, loud babbling chatter lasting several s. Forages on ground, bird crouching low and appearing short-legged when deep in dry vegetable litter; may also forage in low scrub. Proceeds on ground by short, shuffling hops; turns over leaf litter and probes into base of grass clump (P. Shaw, pers. comm.). Flight low down, rather heavy and whirring, with quite long glides. Flock does not fly all at once but birds follow each other on same course, one by one. Adults take it in turn to act as sentinel, perching 3–4 m high on top of or on outer branches of shrub. Sentinel adult calls quietly, 'lull, lull, lull', at rate of 2–3 notes every 2–3 s. Foraging adult sometimes clambers up into bush to feed nestling, then takes over from the sentinel (P. Shaw, pers. comm). Rather conspicuous activities often attract other species, so that babbler flock can become nucleus of mixed foraging bird party (C.F. Clinning *in* Ginn *et al.* 1989). Mixes with other *Turdoides* babblers. During heat of day flock takes to leafy tree and loafs for long time, birds sitting close and preening themselves and each other (**A**). Roosts near middle or top of *Colophospermum* or acacia tree at height of 3–5 m; juveniles arrive first, adults appearing 5–10 min later; birds pack tightly together in a row, with juveniles in the middle (P. Shaw, pers. comm.).

Food. Insects. Mantids, beetles and small mopane worms (caterpillars) fed to young.

Breeding Habits. Nests are solitary; co-operative breeder.
NEST: bulky, deep bowl, made of dry grass and herb stems, lined with finer grass; sited in compound fork with

A

multiple upright branches in middle of *Spirostachys* or small *Terminalia* tree (C.F. Clinning *in* Ginn *et al.* 1989, Maclean 1993). Diam. 125, diam. of cup 80, depth of cup 50.

EGGS: 2 (n = 1 clutch). Glossy, smooth, turquoise-green. SIZE: (n = 2) 26·4 × 20·2, 26·6 × 20·2.

LAYING DATES: Namibia, Nov, (Etosha) Dec; (Kaokoland: fledglings being fed, May); breeding season July–Jan (Harrison *et al.* 1997), mainly Nov–Jan.

DEVELOPMENT AND CARE OF YOUNG: several adults in attendance at the one nest found; 2 fledglings attended and fed by 4 adults or by elder sibling. Plumage of newly-fledged bird loose and flight very weak (P. Shaw, pers. comm.). Fledgling perches 1–2 m above foraging adult. Adult finding food flutters its wings to attract attention of fledgling, which descends briefly to ground to be fed. Adult may try to lead fledgling on ground in the open to cover, by drooping and quivering its wings in 0·5 s bouts repeated for up to 3 min (P. Shaw, pers. comm.).

BREEDING SUCCESS/SURVIVAL: fledgling African Striped Cuckoo *Oxylophus levaillantii* being tended by flock of Bare-cheeked Babblers, Etosha, Dec, shows that the cuckoo parasitizes this species.

References
Ginn, P.J. *et al.* (1989).
Maclean, G.L. (1993).

Turdoides fulvus (Desfontaines). Fulvous Babbler; Fulvous Chatterer. Cratérope fauve.

Plate 4
(Opp. p. 31)

Turdus fulvus Desfontaines, 1789. Hist. Acad. Roy. Sci. (1787), p. 498, pl. 11; Gafsa, Tunisia.

Forms a superspecies with *T. rubiginosus*.

Range and Status. Endemic resident and short-distance migrant. Frequent and locally common, widespread in Maghreb from Western Sahara to Atlas foothills in S Morocco and adjacently in NW and N Algeria, Tunisia and NW Libya; throughout Tassili N'Ajjer and Hoggar in SE Algeria and Tibesti (N Chad), and across S Sahara from Mauritania and N Senegal to Eritrea and NE Ethiopia. Formerly recorded north of Atlas Mts in Essaouira, Marrakech and once near Meknes; and was common breeding resident in S Egypt in Nile Valley north to Aswan until Aswan High Dam constructed in mid 1960s, when the species became extinct there and has not returned. However, common near Gebel Elba at Egyptian/Sudanese border near Red Sea coast (Goodman and Meininger 1989).

Mauritania, frequent along Senegal R. valley, and isolated populations at massifs of Assaba, Tagant and Adrar. Senegal, rare, mouth of river and at Richard-Toll, where 2 small flocks seen in 30 years, Jan and Apr. Mali,

Turdoides fulvus

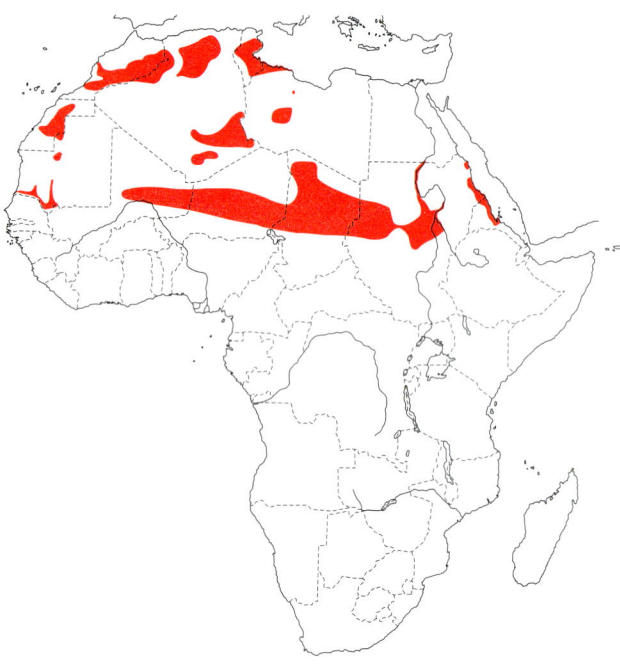

frequent, about 16–18°N. Niger, frequent if localized between 15° and 18°N, commoner in Aïr north to 19°N. Chad, characteristic of dense scrub in wadis in Ouadi Rime – Ouadi Achim Res., frequent and widespread in Tibesti, but few records elsewhere. Sudan, common in middle latitudes east to Blue Nile, frequent in Red Sea hills. Eritrea, E plains below 310 m, south to Buri Peninsula.

Description. *T. f. acaciae* (Lichtenstein): N Chad and W, central and coastal Sudan to N Eritrea. ADULT ♂: top of head greyish sandy-brown with fine dark brown streaks; rest of upperparts rufescent sandy-brown, mantle and back with diffuse darker brown streaks. Tail long and strongly graduated, feathers rather narrow, rufescent sandy-brown above, pale greyish rufous below. Lores to cheeks and ear-coverts dusky greyish brown; sides of neck greyish sandy-brown with fine, diffuse dark streaks. Chin whitish; throat, upper breast, flanks, thighs and undertail-coverts bright sandy buff or pale rufous, upper breast with very fine buffish white shaft streaks; lower breast and belly creamy buff. Remiges pale brown, outer webs paler sandy-buff; upperwing-coverts sandy brown. Underwing-coverts and axillaries pale tawny buff; inner borders of flight feather undersides sandy buff. Bill greenish or yellowish with dark brown tip; eyes bright brown or reddish brown; legs olive or greenish. Sexes alike. SIZE: (10 ♂♂, 10 ♀♀) wing, ♂ 93–100 (97.0), ♀ 91–99 (96.3); tail, ♂ 122–138 (131), ♀ 121–133 (129); bill, ♂ 22–26 (24.6), ♀ 22–25 (23.5); tarsus, ♂ 30–35 (32.7), ♀ 31–34 (32.1). WEIGHT: N Chad, ♂ (n = 2) 46, 54, ♀ (n = 1) 50; N Algeria (n = 6, Nov), ♂♂ 64–70, ♀♀ 62–63.

IMMATURE: juvenile similar to adult but lacks crown streaks; paler, more yellowish buff above; feathering shorter and looser, more fluffy below; eyes grey, bill blackish, legs grey. Can be told from adult in field by yellow gape and mouth lining (Goodman and Atta 1987).

NESTLING: no information.

T. f. buchanani (Hartert): SE Algeria (Ahaggar Mts) and sahelian zone of Mali to Niger and central Chad, and (probably this race) west to N Senegal and Mauritania. Slightly paler and sandier above than *acaciae*, top of head less grey and streaking less sharp; paler buff on breast and flanks, belly whiter; bill darker, more slender. WEIGHT: S Algeria and N Niger, ♂ (n = 4) 53–60 (55.5), ♀ (n = 3) 52–56 (53.5).

T. f. fulvus (Desfontaines): N Algeria (south to about Ouargla) to Tunisia and NW Libya. Like *acaciae*, but top of head tawny brown (rather than greyish brown), and streaking less sharp; underparts slightly paler, white throat a trifle more contrasting; bill dark brown or blackish.

T. f. maroccanus Lynes (including '*billypayni*'): S Morocco and adjacent NW Algerian Sahara, east to Béchar (and birds in Tassili N'Ajjer, S Algeria, and Fezzan, SW Libya perhaps of this race: Cramp and Perrins 1993). Like nominate *fulvus*, but upperparts slightly darker, more tawny brown; breast and flanks darker, more extensively tawny buff; bill dark, eyes pale grey.

Field Characters. Length 25 cm. Warm sandy brown above, paler below, rather uniform buff, including underwing, somewhat brighter on flanks and whiter on throat. Colours vary racially: *acaciae* has greyer crown with evident streaking; *maroccanus* has almost orange wash on breast and flanks. Tail long and graduated. The only babbler in its range except for White-headed *T. leucocephalus*, which it meets south of Khartoum and in N Eritrea; latter easily distinguished by white head sharply demarcated from brown back and breast, plain brown plumage (no buff), shorter, non-graduated dark tail.

Voice. Tape-recorded (53, 105, B, CHA, JOHN, ROC, TOMB, TUCK, WHI). Principal and most striking call, either contact call (Cramp and Perrins 1993) or song (Hollom *et al.* 1988), a series of 6–10 loud, clear, descending whistles, lasting 2–3 s, first one longer and more drawn-out, 'peeeeooo, peeoo-peeoo-peeoo-peeoo-peeoo-peeoo' (sonagram II in Cramp and Perrins 1993). A very different utterance, 'complex sequence of squeaky sounds and chirrups', called 'song' by J.-C. Roché, is more like a subsong, and resembles friendly-calls of roosting Arabian Babbler *T. squameiceps* (Cramp and Perrins 1993). Another frequent call, perhaps of mild alarm, given by single bird or more often by group, is shimmering metallic trill which rises and falls somewhat both in pitch and intensity (sonagram VII in Cramp and Perrins 1993), reminiscent of trill of Hunter's Cisticola *Cisticola hunteri*, or of higher-pitched and faster version of trill of Black Crake *Amaurornis flavirostris*. Same or related call, described as high-pitched, wavering tremolo, given by ♀ on nest when fed by another group member (sonagram VIII in Cramp and Perrins 1993). High intensity alarm, sharp 'pwit'. Flight calls, short 'pip' or clear 'pee', with twangy, pinging quality of Bearded Reedling *Panurus biarmicus*. Foraging call, quiet 'cheep' like chick of domestic fowl. For variations on these and other calls, see text and sonagrams in Cramp and Perrins (1993).

General Habits. Inhabits thorny, woody, desert thickets, scrub, open bushes and gnarled trees on sandy, stony, grassy and dead-vegetation-littered land, spending most of time in, under or near shady thorn-trees such as *Acacia*, *Ziziphus* and *Capparis* in wadis, and date palms *Phoenix* on watercourses and around oases. Sometimes uses plantations (*Eucalyptus*), and thorn brakes around melon field. Will fly into head of tall palm, but much prefers to stay in the thickest woody tangle it can find within 1–2 m of ground.

Rarely occurs solitarily, sometimes in pairs, generally in flock of 4–5 birds, occasionally up to 10–12, once 26; average (n = 33 flocks) 5·7 birds (Gebel Elba) and (n = 83 flocks) 5 (Morocco). Generally rather shy, but can become very tame; every day birds came to within 1 m of people to eat scraps left out for them (Goodman and Meininger 1989). Behaviourally characterful: 'comic' or 'silly'. Vocal, though thin 'peeoo' piping seems to fit a much smaller bird. Skulks in thickets. Flies low down, flock silent or giving subdued flight calls, birds leaving bush one by one, whole flock moving from one tangle to another over a few s, or in rapid succession if alarmed. Flight faintly gamebird-like, with spasmodic burst of wing beats followed by a quite fast glide, or a slow and uncertain one, with wings and tail spread, and an untidy landing in leafy vegetation, like a diminutive coucal *Centropus* sp. Flock tends to pitch at foot of bush, birds creep silently to top, then descend cautiously to ground to forage, leaving one member perched out in open on topmost twig, as a sentinel. Vocal. Loafs in thick cover during heat of day. Forages mainly on ground, probing earth with bill, or probes amongst fibrous bases of green palm fronds near foot of trunk. Sometimes utilizes head of tall palm. Moves through twigs, woody creepers and branches with strongly-clinging, creeping or bounding gait, and on ground by hopping, creeping or leaping amongst leafy and twiggy litter, with tail raised; also said to run. Flock forages over wide area every day. Flock roosts together in thickly-foliaged tree, leaving in morning at first light.

Food. Invertebrates and berries. Grasshoppers, caterpillars, flies, beetles (including *Pimelia, Erodius*), ants and termites; 5–50 mm long. Small beetles and grasshoppers are commonest prey. One stomach contained 6 beetles and 50 termites; another, of a juvenile, contained 7 large caterpillars. Plant food includes many *Ziziphus* berries (Tunisia, Western Sahara), also *Salvadora* berries, small seeds and grain. Flock readily becomes tame and bold if fed regularly at desert camp, and eats scraps of bread dough, dead moths, and small bits of meat.

Breeding Habits. Nests solitarily, pair (if monogamous) helped by other members of group. Of 6 nests, Sudan, 5 had pair and 1 helper, one had pair and 2 helpers; all helpers were immature birds (Goodman and Atta 1987). No obvious courtship behaviour known; ♂ copulated with ♀ squatting in crown of palm.

NEST: loosely woven platform of coarse, dry grass, thin twigs, rushes and bulrush *Typha* leaves (long grass leaves sometimes going whole way round), with deep, scantily lined cup in middle. Sited in *Ziziphus, Balanites* or other thorn tree or bush, mainly 1–2 m above ground, in brushwood or in crown of *Phoenix* palm.

EGGS: 3–6, usually 4–5, av. (n = 24 clutches, Algeria, Tunisia) 4·3. Smooth, glossy; subelliptical; immaculate turquoise. SIZE: (n = 28) 22·4–25·8 × 16·9–19·5 (24·4 × 18·2).

LAYING DATES: Morocco, (nest-building Jan–May; juveniles with gape-flanges Nov); Algeria, Nov–Apr; Tunisia, Mar–July; Libya, Feb; Mali, May–Oct; Mauritania Apr, Aug; Niger, (adults feeding young June, July); Chad, Aug (breeds during wet season); Gebel Elba (SE Egypt/NE Sudan), Mar–Apr.

DEVELOPMENT AND CARE OF YOUNG: ♀ on nest (Tunisia) was fed by 3 other birds, which whistled when arriving with food item. Young depart from nest with tails half-grown and before they can fly, and soon move some distance away to be attended by adults. In distraction display near fledglings, adult hopped to within few m of observer then away again, holding short, round wings up over back and uttering descending trills (Bundy and Morgan 1969).

BREEDING SUCCESS/SURVIVAL: all nestlings in one nest eaten by Fan-tailed Raven *Corvus rhipidurus*.

Reference
Cramp, S. and Perrins, C.M. (1993).

Turdoides rubiginosus (Rüppell). Rufous Chatterer. Cratérope rubigineux.

Plate 4 (Opp. p. 31)

Crateropus rubiginosus Rüppell, 1845. Syst. Uebers. Vög. Nord-ost.-Afr., p. 47; Shoa, Ethiopia.

Forms a superspecies with *T. fulvus*.

Range and Status. Endemic resident, E Africa. Ethiopia, frequent to common in Rift Valley and NE Ethiopia lowlands nearly to Djibouti border, frequent in S, east to 42–43°E; up to at least 850 m. Sudan, common in SE, north to 07°N and west to 30°E. Uganda, common in N, south to Chua, Karamoja and L. Opeta. Kenya, quite common throughout although absent from much of E; less common in W and in highlands, and absent above 1500 m. Somalia, locally very common, Jubba valley, lower Webi Shabeelle and coastal plains, and extreme S. Tanzania, south along coastal lowlands to Dar es Salaam and Soga and inland to Kidugallo, Kingolwira and Kilimanjaro lowlands, with another population south of L. Victoria, from Mwanza and Speke Gulf to Shinyanga, Mkalama and Wembere (98 localities: N. Baker, pers. comm.).

Description. *T. r. rubiginosus* (Rüppell): SE Sudan, Ethiopia except SE, N Uganda, and Kenya except E. ADULT ♂: upperparts rather dark, plain olivaceous brown, forehead and forecrown with puce tinge, fine glossy shafts and dark grey shaft streaks. Tail strongly graduated, light brown above, grey below. Lores, ear-coverts and sides of neck olivaceous grey-brown; cheeks and sides of chin and throat buffish brown, tinged rufous. Chin pale buff, grading into rufous-buff on throat; breast and flanks rufous-brown, the former with very fine whitish streaks; belly, thighs and undertail-coverts rufous-buff. Remiges dark brown with broad olivaceous brown outer fringe; alula and upperwing-coverts dark olivaceous brown. Underwing-coverts and axillaries tawny

Turdoides rubiginosus

buff; inner borders of flight feather undersides pale tawny-buff. Bill decurved, pale horn above, yellowish or whitish brown below; eyes light yellow to yellowish brown; legs pale grey to light brown or brownish flesh. Sexes alike. SIZE: (10 ♂♂, 10 ♀♀) wing, ♂ 85–90 (87·0), ♀ 79–88 (85·0); tail, ♂ 105–118 (110), ♀ 97–118 (105); bill, ♂ 20–22 (20·5), ♀ 20–21 (20·2); tarsus, ♂ 29–30 (29·4), ♀ 29–30 (29·5). WEIGHT: Kenya and Uganda, ♂ (n = 11) 33–49 (40·0), ♀ (n = 7) 36–47 (41·6).

IMMATURE: juvenile like adult, but paler, more rufous-brown above; eyes greyish brown. Recently fledged young has dark brown eyes and yellow gape flanges.

T. r. sharpii (Ogilvie-Grant): SE Ethiopia (Dolo, Unsi) and adjacently in Somalia. Like *rubiginosus* but larger: wing 91–96·

T. r. heuglini (Sharpe): S Somalia, SE Kenya (inland to Voi and Taveta) and NE Tanzania (inland to Mkomazi Game Res. and Kilimanjaro). Darker above than *rubiginosus*, top of head and mantle with distinct streaks and shiny blackish shafts; underparts deeper rufous-brown. WEIGHT: E Kenya, ♂ (n = 2) 46, 50, ♀ (n = 4) 40–50 (43·5), unsexed (n = 36) 36–48 (42·9).

T. r. emini (Reichenow): N central Tanzania. Like *heuglini*, but feathers of forehead and forecrown with minute pale grey tips, hindneck and mantle without distinct streaks and underparts a richer shade.

Field Characters. Length 19–20 cm. A common dry-country babbler of lowland thickets, uniformly coloured brown above and *cinnamon-rufous* below (these colours darker in race *heuglini*), except for a few pale streaks on forehead and breast; *eye pale yellow*, bill yellow-horn. Lacks the extensive scaling, streaking and mottling of larger and darker *Turdoides* babblers, and voice completely different. Can only be confused with the drably-coloured Scaly Chatterer *T. aylmeri*, which in Kenya and Tanzania is local and much less common, with specialized habitat (dry *Commiphora* scrub); for details see that species.

Voice. Tape-recorded (B, C, GREG, HOR, McVIC, MOY, ROC, STJ). Common call a shrill descending 'tschyeerss', repeated many times at intervals of 2–3 s; also various chattering, growling and squealing calls, a soft 'queer' and a longer quavering whistle (Zimmerman *et al.* 1996).

General Habits. Inhabits thickets, scrub and bushes in dry, grassy *Commiphora* and *Acacia* woodlands; also thick coastal scrub with baobab trees; in Yavello hills, Ethiopia, extends from thorn-acacia country into juniper woods, keeping to thick, matted growth (Benson 1946b). Lives in cohesive, family flocks of 5–7 birds. Flock is group-territorial. In one study size of one flock varied little; a flock of 6 colour-ringed birds (one juvenile) had increased to 7 birds 2 years later and the 7 included the former juvenile and at least 2 other ringed adults (Huels 1982). Adjacent flocks commonly have noisy confrontations at territorial borders, all year. All 5 adult birds in a flock preened each other and preened 2 fledglings. Flight whirring and laboured, with gliding. Flock keeps in dense cover for much of time, and birds fly low down from bush to bush in follow-my-leader fashion. Forages communally in leaf litter on ground, in shade of bushes. Flock of 4 adults and 2 juveniles once behaved excitedly, interpreted as possibly anting using ants attracted to small dead snake (Harvey 1974). Flock moved through scrub at dusk, close to and on ground, birds gathered in close-knit group on patch of sand and preened energetically, shuffling around as though dust bathing and occasionally calling in chorus. An adult hopped forward carrying 15 cm dead snake, dropped it and preened itself ecstatically over it (attracting 4 Arrow-marked Babblers *T. jardineii* to watch from adjacent bushes).

Food. Insects, including termites.

Breeding Habits.
NEST: open cup made of plant stems (including *Cadaba farinosa*), leaves and rush-like grasses, scantily lined with fine grass stems, or lined with coarse twigs on thick foundation of broad leaves, some still green; outside of nest rather untidy. One nest sited 3 m above ground, in leafy tree; one in dense tangle of *Cissus* and other creepers at base of small tree, 50 cm above ground (Duckworth *et al.* 1992); one on large branch amongst thick foliage, 2 m up in tree; another in dense mass of *Bougainvillea*. Ext. diam. 125, diam. of cup 75, ext. depth 67, depth of cup 50.

EGGS: 2–3. Laid at daily intervals. Immaculate bright blue, with smooth, glossy surface and thin shell. SIZE: (n = 1) 21·3 × 16·5.

LAYING DATES: Ethiopia, Aug. E Africa: Region D, Feb–Apr, June–July, Region E, May–June, Sept, Nov–Dec.

INCUBATION: eggs incubated by ♀ (only?), sitting close.

DEVELOPMENT AND CARE OF YOUNG: 'nestlings fed communally' (nest record card); all adults in flock feed fledglings (and once a fledgling Jacobin Cuckoo *Oxylophus jacobinus*).

BREEDING SUCCESS/SURVIVAL: parasitized by Jacobin Cuckoo (Somalia, Kenya) (Huels 1982). One nest with 3 babbler eggs contained 1 cuckoo egg, another 2 cuckoo eggs. Nest torn apart by Baboons *Papio anubis* or Vervet Monkeys *Cercopithecus aethiops*.

Reference
Huels, T.R. (1982).

Turdoides aylmeri (Shelley). Scaly Chatterer. Cratérope ardoisé.

Argya aylmeri Shelley, 1885. Ibis, p. 404, pl. 11, fig. 1; British Somaliland.

Plate 4
(Opp. p. 31)

Range and Status. Endemic resident, from NW Somalia to NE Tanzania. Somalia, very common but patchily distributed in NW (where locally one of the commonest birds at 300–1000 m, from Ber and Burao to Hargeisa and Ged Gedibtaleh), along Jubba valley and in coastal lowlands from equator north to about Raas Cabbad. Ethiopia, uncommon, to east and south of SE Highlands. Kenya, uncommon and very local, up to 1500 m, mainly in semi-arid terrain on plateau north, east and south of central highlands, also in Indunumara Mts. Tanzania, similarly uncommon and local in NE as mapped; records south of 06°S may be identification errors (N. Baker, pers. comm.).

Turdoides aylmeri

Description. *T. a. aylmeri* (Shelley): Somalia and E Ethiopia. ADULT ♂: forehead and crown tawny brown, flecked with small buff feather tips; rest of upperparts paler greyish brown. Tail long, narrow, strongly graduated, feathers with *c.* 28 structural bars, greyish brown above, glossy brown below. Lores pale buff; cheeks, ear-coverts and sides of neck brown with some buffish mottling; bare skin around and behind eye pale blue. Chin and throat scaly, feathers dark brown with broad pale buff fringes; breast scaly, feathers grey-brown with buff fringes; belly, flanks, thighs and undertail-coverts warm buff, flanks darker and greyer. Remiges and upperwing-coverts greyish brown. Underwing-coverts and axillaries sandy buff; broad inner borders of flight feather undersides pale buff. Bill decurved, yellowish brown to yellowish or pinkish white; eyes yellow or yellowish white, surrounded by oval patch of pale bluish-grey bare skin; legs pale brown to whitish flesh. Sexes alike. SIZE: (10 ♂♂, 6 ♀♀) wing, ♂ 70–79 (74·8), ♀ 69–75 (72·0); tail, ♂ 96–122 (114), ♀ 99–120 (112); bill, ♂ 20–23 (21·4), ♀ 20–21 (20·7); tarsus, ♂ 26–28 (26·9), ♀ 25–26 (25·7).

IMMATURE: juvenile browner, throat and breast lacking scaly markings; bill and eyes brown.

T. a. boranensis (Benson): S Ethiopia (Boran to upper Shabeelle), and N Kenya from South Horr to Barsaloi. A little darker above than nominate race, underparts slightly darker, more ochreous. A poor race, barely distinguishable from following ones.

T. a. kenianus (Jackson) (includes '*loveridgei*'): SE Kenya (Meru, Embu, Kitui, Amboseli, Tsavo, Taru, Taveta) and NE Tanzania (Moshi, North Pare foothills, Kihurio, Pangani R.). Browner above than nominate race; deeper, more brownish buff below. WEIGHT: ♂ (n = 3) 35–42 (38·7), unsexed (n = 12) 31–40 (34·5).

T. a. mentalis (Reichenow): interior of NE Tanzania (Lossogonoi, Lolbene, L. Natron, Dodoma, E Singida, Mesembe; and, perhaps this race, Olorgesaillie and Magadi in S Kenya border country). Like nominate race, but slightly darker above and darker below.

Field Characters. Length 21–23 cm. A slim dry-country babbler with long, graduated tail; more drab than partly sympatric Rufous Chatterer *T. rubiginosus*. Upperparts vary from ashy brown (nominate race) to medium brown (*mentalis*) or rich brown (*kenianus*); top of head markedly darker and browner (grey races) or richer (*kenianus*). Underparts vary from buff to cinnamon-brown; all races have *dark throat* and breast feathers with *pale fringes* giving *scaly* effect; belly plain. Rufous Chatterer has rufous-cinnamon underparts without any scaling. From behind, *kenianus* appears richer brown above than Rufous Chatterer, but it never has bright reddish cinnamon underparts. Both species have horn-yellow bill and pale yellow eye, but eye of Scaly is surrounded by patch of bare *blue-grey skin*.

Voice. Tape-recorded (B). Said to make an odd, rather feeble 'squeaking wood-screw' sound, varied by a thin, high-pitched chatter and broken whistling (Fuggles–Couchman and Elliott 1946). Foraging flock constantly gives thin, squeaky cries.

General Habits. In N Somalia inhabits thick clumps of foliage and undergrowth on banks of seasonal watercourses, tall thorn brakes enclosing cultivated and fallow fields, and dense, stunted euphorbia vegetation in arid country. In Ethiopia, thorn scrub, well out on plains away from hills. In Kenya and Tanzania, thickets in arid woodland and bushland. Occurs in groups of 6–7 birds; very active; flock constantly utters thin, squeaky cries.

Food. Not known.

Breeding Habits.
NEST: only 1 found, not described; sited 1·5 m up in clump of bushes and euphorbias.
EGGS: 2 (one clutch).
LAYING DATES: Ethiopia (Ginir), Apr, (Yavello: eggs yolking in ovary, Aug; ♂♂ with large gonads Mar, Aug).

Reference
Archer, G. and Godman, E.M. (1961).

TIMALIIDAE

Genus *Phyllanthus* Lesson

A single polytypic species of forest babbler, endemic, poorly known; similar in most respects to *Turdoides*, and might prove to be sister clade to and even congeneric with that genus. Rather smaller than most *Turdoides* spp., shorter-tailed, and legs and feet more powerful. In hand, feels as strong as and more robust than larger savanna woodland *Turdoides* sp. such as *T. plebejus*. In these respects it forms one end of forest-scrub-deserticolous babbler series, desert *Turdoides* ('*Argya*') spp. being slim, long-tailed and relatively weak-legged. Plumage mainly maroon (colour adumbrated in vinaceous cast of e.g. *T. jardineii*), head feathers frosted grey (adumbrated in several *Turdoides* spp.), bill and legs pale greenish yellow. Year-round flocks, flock size, vocal choruses, undergrowth habitat, evident sedentariness, polytypy, and what can be inferred about foraging behaviour, are all typical of *Turdoides*.

Plate 4 ***Phyllanthus atripennis*** **(Swainson). Capuchin Babbler. Cratérope capucin.**
(Opp. p. 31)
Crateropus atripennis Swainson, 1837. Birds W. Africa, 1, p. 278; western Africa, restricted to Senegal (Sclater).

Range and Status. Endemic resident, forests from Gambia to W Togo and in NE Zaïre and just into Uganda; rare and local, Nigeria to Central African Republic. Gambia, rare, but parties of 4 and 6 birds recorded so surely resident (rather than vagrant: Jensen and Kirkeby 1980, Gore 1990), and group of 6–7 near Sanyang for at least 21 months, probably breeding, also groups near Gunjur (Wacher 1993). Senegal, extreme SW only, quite common near Oussouye, flock once at Ziguinchor. Guinea-Bissau, frequent near coast. Guinea, frequent on Kounounkan Massif (Hayman *et al.* 1995). Mali, rare, extreme W and SW Mts Mandingues (Kényéba, 12°50′N, 11°14′W; Faléa, 12°16′N, 11°17′W). Sierra Leone, frequent in Kilima area and Freetown Peninsula, and several records elsewhere. Liberia, not uncommon along coast and in N highlands (Gatter 1997). Ivory Coast, frequent in forest at Mt Nimba and Taï and in gallery forests at Comoé and Maraoué in N, otherwise scarce or patchy – records from Abidjan, Lamto and Agnibilekrou. Ghana, widespread and not uncommon. Togo, not uncommon; records from Misahöhe, Bismarckburg, Dzogbegan, Koniouhou and Kpalimé. Benin, rare, Bétérou area (Claffey (1995). Nigeria, frequent in area in SW bounded by Lagos, Badagri, Abeokuta and Ibadan; also in forest outlier at Kagoro; once Ilorin. Cameroon, records at Abonando (05°54′N, 09°07′E) and Nkongsamba. Central African Republic, old record from Bangui and more recent ones from Lobaye Préfecture and Botambi (04°12′N, 18°30′E), also 'Oubangui' and common in Ouossi R. area near Baroua, about 05°20′N, 24°20′E (Friedmann 1978) and records from Rafaï (04°58′N, 23°56′E) (Germain and Cornet 1994) and on Mbomou R. on Zaïre border. Zaïre, widespread in Uele, Aruwimi and Ituri districts, from between Bondo, Aketi and Buta to Banalia, Avakubi, Bafwabaka, Gamangui, Medje, Niangara and Semliki. Uganda, quite common at 700–900 m in Bwamba lowlands, record from Bwindi (Impenetrable) Forest at 1700 m.

Phyllanthus atripennis

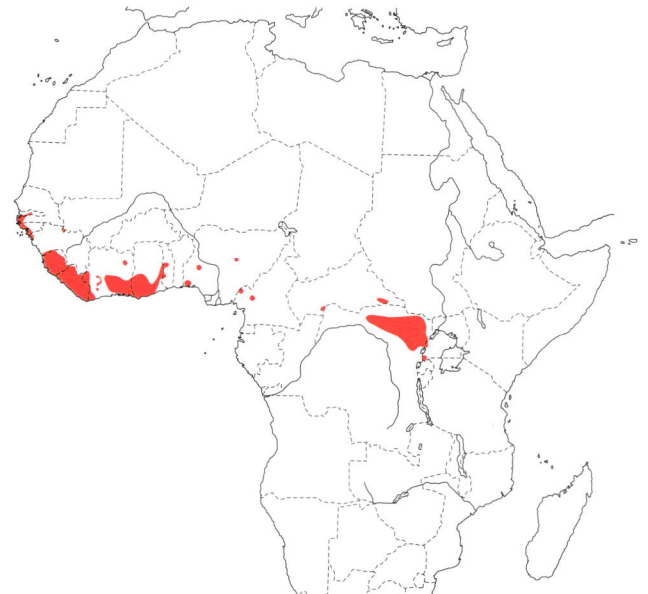

Description. *P. a. atripennis* (Swainson): Gambia to Ivory Coast. ADULT ♂: lores black; hood silvery grey, darker grey around eyes, at base of bill and on forehead and crown, the feathers with blackish grey centres and silver-grey fringes (fringes narrow on forehead and crown, broad on hindneck, sides of neck, throat and breast). Hood sharply demarcated. Parts behind hood (mantle, lower sides of neck, lower breast) glossy black or blackish mahogany, merging to maroon or rich dark vinaceous chestnut-brown on scapulars, back, rump, uppertail-coverts, flanks, belly and undertail-coverts. Tail blackish brown above and below, shiny below; rounded and graduated. Wings very dark brown, above and below. Bill pale yellow, bright cream or whitish with green tinge and small dusky area around nostril, or pale bluish green tinged yellow between nostrils; eyes clear brown or dark red-brown; legs and feet strong, pale blue-grey, greenish grey, or yellowish pale olive-green; claws whitish or pale greenish blue. Sexes alike. SIZE: (8 ♂♂, 7 ♀♀, Mt Nimba, Liberia) wing, ♂ av. 117·3 ± 4·3, ♀ 119·3 ± 3·0; tail, ♂ 79·9 ± 3·8, ♀ 81·3 ± 3·3; bill, ♂ 26·0 ± 0·5, ♀ 25·6 ± 0·5; also wing, ♂ (n = 4) 112–122·5 (118), ♀ (n = 2) 117, 118; tail, ♂ (n = 3) 84–90 (87·7), ♀ (n = 2) 86·5, 88·5; bill to skull, ♂ (n = 3) 34–38·5 (36·5), ♀ (n = 2) 37, 38. WEIGHT: (Mt Nimba) ♂ (n = 8) 83·9 ± 6·0, ♀ (n = 7) 86·4 ± 6·3.

IMMATURE: like adult but cutting edges of mandibles towards base of bill deep yellow, eyes dark brown, eyelids greenish yellow, feet pale greyish green.

P. a. haynesi (Sharpe): Ivory Coast to S Nigeria. Forehead, crown, nape and hindneck uniform dull blackish or slate grey (with sides of these parts as in nominate race); chin and throat either as in *atripennis* or rich maroon (not silvered blackish); upper breast rich mahogany (not silvery); mantle bright maroon (not black).

P. a. bohndorffi (Sharpe): NE Zaïre. Face (forehead, lores, forecheek and sides of chin) black, crown to hindneck and sides of neck silver-grey, mantle maroon, whole underparts maroon. Overall a clearer, brighter vinaceous rufous above and below than *atripennis* (which looks sooty in comparison). Smaller. SIZE: unsexed, wing (n = 4) 105–111·5 (108), tail (n = 4) 71–79 (77·3), bill to skull (n = 2) 35·5, 37. WEIGHT: (Central African Republic) ♂ (n = 3) 80–90 (84·7), ♀ (n = 3) 86–94 (88·7); (Nigeria, unsexed, n = 5) av. 87·4; (Uganda) ♂ (n = 7) 75–90 (79·6), ♀ (n = 3) 86–96 (92·0).

Field Characters. Length 24 cm. A stout, short-tailed, round-winged babbler the size of a *Turdus* thrush; flocks skulk in dense undergrowth at forest edge and beside streams within forest; usually first detected by their distinctive *hoarse calls*. *Dark maroon body* appears blackish in the field except in very good light, but *silvery grey head* with black lores and *pale yellow bill* diagnostic. Races differ: crown black or grey, mantle blackish or maroon, foreneck grey or maroon.

Voice. Tape-recorded (104, B, CHA, LEM, MAC, MOR, SALA). Contact calls from feeding flock low and hoarse, 'kawk', 'kark', 'kup', 'kop', 'kewp', 'kerk'; notes given singly, with pauses between each, but in flock may sound like continuous low-intensity conversation. Calls have grating quality of *Turdoides* babblers, and reminded Chapin (1953) of Dusky Babbler *Turdoides tenebrosus*, but notes brief and much lower in pitch. In chorus, excited cackling mixed with grating trills, 'krrrrrrr ...'; when flock disturbed, calls 'accelerate into piercing squeaks' (Field 1974). Very different call from individual is pure, loud, down-slurred whistle.

General Habits. Shade-loving, keeping mainly to thick, leafy tangles of vegetation near ground at edge of tall forest and dense, lush growth bordering old secondary forest where it abuts onto roads, paths and clearings. Inhabits rain forest borders, forest-savanna mosaic, and forested ravines; noted in undergrowth by a stream, and in low bushes on edge of rice field, in tea plantation and in cacao farm. Once flock seen well into interior of large patch of tall forest surrounded by open fields. Occasionally appears in well-timbered gardens (e.g. Ibadan, Nigeria). Lives year-round in groups of 4–12 birds, usually 5–7, noisy but secretive and most difficult to observe since flock stays constantly in thickest, darkest undergrowth. Dislikes strong light – fit bird released in bright clearing 70 m from forest edge where just mist-netted (SW Nigeria), appeared disoriented, refused to fly back and had to be re-caught twice and carried to forest. Forages on ground and in growth up to 3 or 4 m above it, like *Turdoides* babblers. Flock falls silent when much disturbed. Sometimes joins (or is nucleus to) mixed-species foraging parties. Once reported in mixed scolding party with Blackcap Babblers *Turdoides reinwardtii*, Grey-headed Bristlebills *Bleda canicapilla*, Little Greenbuls *Andropadus virens* and Black-throated Weavers *Ploceus nigricollis* (Wacher 1993). Easily netted, and snared: adults and young apt to fall into hands of local people.

Sedentary, but in SW Mali may move north in rains (Lamarche 1981).

Food. Mainly insects: cockroaches, grasshoppers, black ants, small beetles, sparsely haired caterpillars, pupa in thin cocoon, lepidopteran eggs; also millipedes, a snail, a small frog, seeds and possibly some fruit. Beakfuls of insect larvae brought to unseen nest (Wacher 1993). Insects in all and seeds in 3 of 12 stomachs.

Breeding Habits. Not known. Birds in flock seen in Gambia, June, tugging at creepers and carrying strands of dry vegetation to a place cloaked in dense herbage under sagging frond 3 m up trunk of short, creeper-laden oil palm *Elaeis guineensis* inside forest thicket (Wacher 1993). Group appeared to drive tree squirrel *Heliosciurus* sp. away from site but, 13 months later, tolerated pair of African Goshawks *Accipiter tachiro* nesting overhead.

LAYING DATES: Gambia, (carrying nest material June and, at same site, carrying insect larvae Nov); Liberia, (breeds June–Aug: large gonads June–July–Aug, wing-moult well advanced Apr, plumage fresh all summer, worn Nov–Dec); Mali, (said by local hunters to breed all year); Zaïre (Ituri), probably breeds all year (adults in breeding condition Jan–Feb and Sept; adults with fledgling cuckoo Oct; juveniles in dry and wet seasons: Chapin 1953).

BREEDING SUCCESS/SURVIVAL: strong circumstantial evidence of parasitization by African Striped Cuckoo *Oxylophus levaillantii* – fledgling cuckoo once collected from within party of Capuchin Babblers.

References
Bannerman, D.A. (1936).
Chapin, J.P. (1953).
Colston, P.R. and Curry-Lindahl, K. (1986).
Wacher, T. (1993).

Genus *Lioptilus* Bonaparte

Single monotypic species, endemic to temperate South African forest, size of smallest *Kupeornis* and outwardly quite like that genus but relatively short-billed, short-legged and short-toed, with weak hind toe nail; soft parts pink, cap black. Frugivorous; melodious song: both characters quite unlike the *Kupeornis/Phyllanthus/Turdoides* assemblage of babblers. Nest an open cup, eggs whitish; neither particularly like those of typical babblers where known. Distinctive feature, shared with *K. rufocinctus*, is narrow half-ring of black feathers below eye, but we do not share Hall and Moreau's (1970) confidence that *Lioptilus* and *Kupeornis* are thereby closely related. Affinities remain uncertain. 'In behaviour, ecology and size it has more in common with bulbuls than with most Afrotropical babblers' (T.B. Oatley *in* Harrison *et al.* 1997).

Plate 1
(Opp. p. 14)

Lioptilus nigricapillus (Vieillot). Bush Blackcap. Lioptile à calotte noire.

Turdus nigricapillus Vieillot, 1818. Nouv. Dict. Hist. Nat., nouv. éd., 20, p. 256; south of Africa, restricted to Bruintjes Hoogte, Somerset East.

Range and Status. Endemic resident, nomad and altitudinal migrant, South Africa and Swaziland. Transvaal, scarce and localized along Escarpment from Soutpansberg to Kangwane and in SE Highveld from Ermelo to Wakkerstroom and Josefsdal. Known from W Swaziland but status uncertain. Orange Free State, locally fairly common in Ficksburg area and at 5 other localities along E borders. Natal, quite widespread in W (recorded from 23% of 166 squares, Cyrus and Robson 1980); good localities include Ferncliffe Forest, Giant's Castle Game Res., Monk's Cowl Nat. Res. and Goodhope (Sani Pass) (Bennett and Herbert 1995); frequent at 1080–1380 m, commoner at 1380–1850 m (Vincent 1951). Lesotho, not recorded, but known from Orange Free State bank of Caledon R. (near Wepener, also Ficksburg) which forms border between Lesotho and OFS. Cape Prov., locally fairly common in forest isolates in E, west to Bosberg and kloofs of Baviaans R., southwest to Uitenhage and east to E Griqualand.

Description. ADULT ♂: lores, forehead, crown, nape and hindneck black; mantle, scapulars, back, rump and uppertail-coverts olive-green, darkest and greyish on mantle, clearer on rump. Tail dark brown, feathers fringed olive-green at base. Upper chin black (between rami of lower mandible), lower chin, throat, breast, cheeks, ear-coverts and sides of neck grey, the colour merging to white in midline of belly and to light greyish olive-brown on flanks, thighs and vent; undertail-coverts buffy grey or whitish. Primaries, secondaries and tertials with dark brown inner vanes and olivaceous brown outer vanes, outer primaries narrowly edged pale grey; rest of upperwing same olive as back; underside of remiges shiny grey with pale inner edges, greater underwing-coverts and axillaries yellowish buff, lesser underwing-coverts olive. Bill coral pink, salmon pink shading to buffy cream at tip, pale orange, or (immature?) sepia with some pink-orange on lower mandible and around nostrils; eyes reddish brown, chestnut or dark sepia; bare ring around eye pink; legs and feet dark pink. Sexes alike. SIZE (unsexed, n = 17): wing 76–88 (81); tail 68–83; bill 12–15; tarsus 22–25 (Maclean 1993).

IMMATURE: like adult but cap dark brown and less well delineated, throat, sides of neck, and breast brownish white (not clear pale grey), and bill paler pink and with dusky tip.

NESTLING: unknown.

Lioptilus nigricapillus

Field Characters. Length 18–20 cm. Sparrow-sized, with olive-brown upperparts, glossy *jet black cap* reaching to below eye and onto hindneck, coral *pink* to pale *orange bill*, dull pink legs; underparts pale grey except for whitish belly and light brown wash on flanks plump (**A**). At close range shows pale pink eye-ring surrounding red eye. Immature like adult but duller and browner, with dull pink bill.

Voice. Tape-recorded ((75, 88, 91–99, B, F, GIB, GIL). Varied, powerful song contains rich whistles and shorter connective notes, 'cheep-ta-woy, cheep-cheep-ta-woy', 'cho-woy, cheep-cheep-cho, cho-wee-cho'; the 'woy' sound is rich and melodious, other notes also tuneful; another variant, 3–4 upslurred 'wheep's followed by 3–4 ringing 'woy's – 'wheep-wheep-wheep, woywoywoywoy'. Another bird had some 'warm-up' notes followed by rich whistles, 'taw-tee-wip, trit-trit, wawdilay-wawdilay-wawdi'; the 'wawdilay' is fast, with tone and suppressed power of Nightingale *Luscinia megarhynchos*. Song often compared with Common Bulbul *Pycnonotus barbatus*; some notes similar, but songs of *Pycnonotus* bulbuls have a casual quality with less depth and richness. Alarm, a fairly loud, guttural 'burgg' (Maclean 1993).

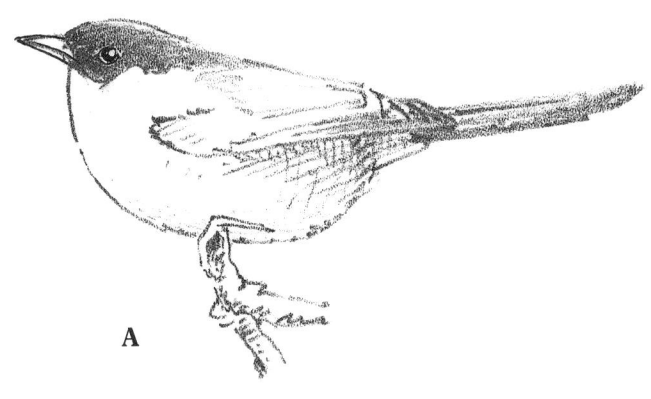

A

General Habits. Inhabits edges of montane forests and wooded valleys, ravines and kloofs at borders of Drakensberg and other montane evergreen forest, occurring largely in adjacent thickets of *Buddleia* and on *Leucosidea* scrub-covered hillsides in mistbelt. Occurs singly or in pairs or small flocks. Usually silent, creeping about unobtrusively in middle strata at edge of upland forest or low down in bushes. Movements slow and deliberate. Flight direct, rather undulating. Tame, inquisitive – responds well to 'pishing' (Maclean 1993). May be territorial outside breeding season, since tape-recordings call birds up near Natal coast where only a winter visitor (D. P. Cyrus *in* Ginn *et al.* 1989).

Partial altitudinal migrant from forests on Drakensberg Mts to lower mistbelt forests and occasionally to river valley bushveld (T.B. Oatley *in* Harrison *et al.* 1997). In Transvaal a localized breeding resident, but recorded only Sept–June; non-breeding visitor to coastal Natal. One appeared in a Johannesburg garden in June, some 300 km from its normal range and habitat of mistbelt and montane forest; it kept to ferns and shrubs near a small fishpond where it often drank (G. Selfe, *Africa Birds & Birding* **4**, 4, 1999, 29).

Food. Fruits and berries.

Breeding Habits. Poorly known, but evidently a solitary nester, monogamous and strongly territorial (reacts to tape-recording of call, which may stimulate song). Sings continuously for periods of 15 min or more, mainly in Nov–Jan (Vincent 1951, D. P. Cyrus *in* Ginn *et al.* 1993).

NEST: neat open cup made of moss and small twigs and lined with rootlets and thin bits of bark. Placed 1–2 m above ground in fork in tree by a clearing, often where ground boggy.

EGGS: 2. Dull white, streaked and mottled with slate grey and light brown, mainly at large end. SIZE: (n = 3) 23·0–24·3 × 16·5–17·0 (23·4 × 16·8).

LAYING DATES: South Africa, Nov–Jan (Transvaal Dec).

References
Ginn, P.J. *et al.* (1993).
Maclean, G.L. (1993).

Genus *Kupeornis* Serle

Robust, acrobatic, montane equatorial forest babblers, with chestnut and white or rufous, brown and black plumage, dark cap, pale eyes and bills, short, curved wings, short, thrush-like bills, rather short, rounded tails and rather short, strong legs and feet with long, strongly curved hind claw. Insectivorous, foraging among large masses of forest epiphytes, often hanging head down. Crown feathers erectile to form short crest. Poorly known, but in one species tail constantly moved up and down, and in display spread and held vertically above back. Occur year-round in monospecific flocks, which give chattering chorus like *Turdoides* babblers. Flight very like *Turdoides*. One species dimorphic and variable. Plumage of *K. gilberti* like that of the larger *Phyllanthus atripennis*; and even before *K. gilberti* was discovered the small *K. rufocinctus* had been placed in *Phyllanthus* (Verheyen 1947). (*Phyllanthus* and *Turdoides* are closely allied.)

Endemic. 3 species, one in Cameroon montane and 2, a superspecies, along or near Albertine Rift.

68 TIMALIIDAE

Kupeornis chapini superspecies

1 *K. chapini*
2 *K. rufocinctus*

Plate 1   *Kupeornis gilberti* Serle. White-throated Mountain Babbler. Phyllanthe à gorge blanche.
(Opp. p. 14)

Kupeornis gilberti Serle, 1949. Bull. Br. Orn. Club 69, p. 50; Mt Kupé.

Range and Status. Endemic resident, SE Nigeria and SW Cameroon. In Nigeria common on Obudu Plateau at about 1520 m. In Cameroon common on Mt Kupé at 1370–2030 m but on W flank down to 950 m, and near Dikume Balue in Rumpi Hills at 1100–1700 m, at Foto near Dschang at 1670 m, on Mt Nlonako at 1600–1800 m and on Bakossi Mts at 1100–1600 m (C. G. R. Bowden, pers. comm.). On Mt Bakossi abundant around L. Edib, with perhaps several thousand birds (Dowsett-Lemaire and Dowsett, pers. comm.).

Description. ADULT ♂: forehead, crown, nape and sides of neck dark mahogany or rich dark rufous-brown, the colour merging via hindneck to dark olivaceous brown on mantle, scapulars and back; rump dark rufescent brown, uppertail-coverts dark rufous. Tail blackish brown, feathers with dark rufous edges at base. Lores, cheeks and ear-coverts white, heavily tinged with very pale orange or straw-colour; often odd dark rufous feathers mixed in, and upper ear-coverts all dark rufous in some birds. 4 birds, all ♀♀, out of 45 examined have a sharply demarcated white superciliary stripe 2–4 mm wide, usually broken into a loral and a post-ocular part. Chin, throat and upper breast snowy white, lower breast, belly, flanks, thighs, vent and undertail-coverts dark rufous, or sometimes olive-brown ('olive-brown' birds also have warmer brown mantle and more russet on rump and uppertail-coverts than normal birds, evidently a true dimorphism, not related to gender or wear: Serle 1954). Remiges with blackish brown inner webs and dark rufescent brown outer webs; rest of upperwing dark rufescent brown; underside of remiges shiny black distally and brown proximally; underwing-coverts and axillaries dark brown. Bill grey, mixed dusky brown and greyish, palest clear brown, pale brown with darker culmen, creamy brown, or mainly yellowish with grey-brown culmen; gape soft

Kupeornis gilberti

and yellow; eyes clear light grey, bluish white, grey-white or grey; legs and feet greenish grey. Sexes alike, but of the 2 morphs most ♂♂ are 'rufous-brown' and most ♀♀ 'olive-brown' (Serle 1954). SIZE (13 ♂♂, 10 ♀♀): wing, ♂ 101–115 (109), ♀ 99–113 (106); tail, ♂ 68–77 (72·1), ♀ 69–80 (73·3); bill to feathers, ♂ 17–18

(17·5), ♀ 17–19 (17·3), bill to skull, ♂♀ 20–23 (21·4); tarsus, ♂ 32–34 (32·6), ♀ 31–33 (32·3). Also (19 ♂♂, 8 ♀♀, partly same series of skins as above): wing, ♂ 108–119 (112), ♀ 107–113 (100); tail, ♂ 70–79 (73·4), ♀ 69–80 (72·75); bill to feathers, ♂ 17–18 (17·6), ♀ 17–19 (16·5); tarsus, ♂ 31–34 (32·4), ♀ 32–33 (32·1) (Serle 1954).

IMMATURE: in museum series of 45 skins, nearly half have irregularly mixed brown and white feathers in chin and throat and 2, thought to be immature, have chin and throat all brown but otherwise resemble adults; 2 have all-white chin and forethroat and almost all brown lower throat and upper breast. Some sexually mature individuals (large gonads) have brown throat spots (Serle 1954).

Field Characters. Length 21·5 cm. A plump, active babbler moving in noisy flocks through forest canopy. Chorus of loud throaty calls coming from tree tops is often first indication of birds' presence (Serle 1950). Unmistakable when seen from below: startlingly *white bib and ear-coverts* contrast with chestnut head and body; in good view of upperparts, dark chestnut cap demarcated sharply from olive back, *tawny rump* separates back from dark brown tail. Variable: sometimes with white eyebrow or post-ocular streak and with bib blotched with brown or all brown. Movements in tree *Parus* tit- or *Sitta* nuthatch-like, but flight and voice like *Turdoides* or *Phyllanthus* babbler. Almost always associates with Grey-headed Greenbuls *Phyllastrephus poliocephalus*.

Voice. Tape-recorded (104, B, CHA, LEM, RODE). Contact calls of birds in flock a series of grating, hoarse chuckling and chattering notes reminiscent of Capuchin Babbler *Phyllanthus*, 'chaak', 'chup', 'chukka', 'twee-cha-cha', 'chichichi' and longer series, 'kakakakaka ...', 'cherchercher ...', punctuated at intervals by louder, explosive '*chack!*' or *chuck!*' See sonagrams in Dowsett-Lemaire and Dowsett (1990). Flock often combines these notes into group chatter.

General Habits. A bird of the canopy and upper stratum of primary montane forest, often where it borders hillside grassland, but sometimes occurs in mid and ground strata and in mature secondary forest. Moss forest; frequents upper boles and branches with epiphytes, and shuns thick tangles of vines and dense herbage. Lives year-round in party of 5–10 (av. 6·9) and up to 13 birds. Larger flocks invariably include other species, which babbler leads. Very active and vocal.

Forages mainly by probing for insects (90% of foraging actions: C.F.C. Hesse, pers. comm.) on or within mosses and other epiphytes and in crevices in bark. Continually on the move, hopping along and flying between branches, moving now in canopy, now dropping 15 to 20 m to lower but still lofty branches. Forages on branches (58% of sites), trunks (38%) and twigs (4%); at heights above ground of <5 m (20% of occasions), 5–10 m (25%) and >10 m (55%) (Hesse 1995). Flock progresses through forest quite slowly despite its activity, but travels faster, without feeding, when moving towards new feeding area. Associates constantly in lively and clamorous parties with Grey-headed Greenbuls, and at times with Tullberg's Woodpeckers *Campethera tullbergi*, Elliot's Woodpeckers *Dendropicos elliotii*, White-bellied Crested Flycatchers *Elminia albiventris*, Fork-tailed Drongos *Dicrurus adsimilis*, Waller's Starlings *Onychognathus walleri* (Serle 1950, 1954), Pink-footed Puffback *Dryoscopus angolensis* and Dark-backed Weaver *Ploceus bicolor* (Stuart and Jensen 1986). Occurrences of other birds in White-throated Mountain Babbler flocks were: Grey-headed Greenbul in 62% of flocks, Dark-backed Weaver *Ploceus bicolor* in 44%, woodpeckers 35%, Mountain Greenbul *Andropadus tephrolaemus* 22%, and Cameroon Olive Greenbul *P. poensis* 18% (Hesse 1995). A flock regularly came to bathe in rain water among aerial roots of fallen fig tree, 1·5 m above ground (F. Dowsett-Lemaire, pers. comm.).

Food. Large, hard insects.

Breeding Habits. No nest yet found.
BREEDING INDICATIONS: Nigeria, young just out of nest, June; Cameroon, large yolking egg in oviduct, June, juveniles with unossified skulls, Nov–Jan, and juvenile-plumage birds July–Jan (Mt Kupé).
DEVELOPMENT AND CARE OF YOUNG: fledglings out of nest beg, calling querulously, with drooped and quivering wings, when being fed by parents.

References
Collar, N.J. and Stuart, S.N. (1985).
Serle, W. (1949, 1950, 1954).

Kupeornis rufocinctus (Rothschild). Red-collared Babbler. Phyllanthe à collier roux.

Plate 1
(Opp. p. 14)

Lioptilus rufocinctus Rothschild, 1908. Bull. Br. Orn. Club 23, p. 6; Rugege Forest, SW Rwanda.

Forms a superspecies with *K. chapini*.

Range and Status. Endemic resident, E Zaïre from Ruzizi Valley to Mt Kabobo (north of Kaliémié (Albertville), L. Tanganyika), and SW Rwanda (Nyungwe Forest). Common throughout Nyungwe Forest (1700–2700 m), less so in E, commonest in very wet forest at 2200–2500 m; 4 parties of 6–8 birds regularly encountered on 2 km of forest trail and seemed to have home ranges of 20–25 ha (including clearings: Dowsett-Lemaire 1970). (2 localities plotted by Hall and Moreau 1970, just south of L. Edward and on equator northwest of Lubero, are likely to refer to Chapin's Babbler *K. chapini*.)

Description. ADULT ♂: lores, forehead, crown and nape black, slightly glossy; hindneck, sides of neck and ear-coverts rufous,

Kupeornis rufocinctus

sharply demarcated from the black but merging into colours behind; mantle, scapulars and back greyish olive-brown; rump and uppertail-coverts bright rufous-orange. Tail blackish brown, feathers with olive edges at base. Cheeks dusky, chin buffy rufous, throat and upper breast rufous-orange (yellower than cheeks), lower breast olive-brown, rufous or tinged rufous in middle, belly, flanks, thighs and vent same olive-brown as back, undertail-coverts bright rufous-orange. Remiges dark brown, rest of wing olive-brown, tertials and greater coverts structurally banded (dark and darker bands alternating every 1–2 mm). 1 specimen has indistinct orange tips to greater primary and greater coverts. Underside of remiges shiny black-brown, greater underwing coverts rufous, lesser underwing coverts and axillaries olive-brown. Bill pale horn or whitish brown, dusky around nostrils; eyes yellowish white or pale straw yellow; legs and feet light grey. Sexes alike. SIZE: (5 ♂♂, 5 ♀♀, Prigogine 1971) wing, ♂ 103–111 (106), ♀ 105–111 (107); tail, ♂ 78–85 (81), ♀ 80–90 (82.8); bill, ♂ 18–19 (18.5), ♀ 17.5–19.5 (18.7); tarsus, ♂ 27–29 (28.0), ♀ 25–28 (26.5); also (2 ♂♂) wing, 99, 100.5; tail 67, 100; bill to feathers 15, 15.5; bill to skull 18, 20; tarsus 29, 29.5.

IMMATURE: like adult but eyes dark brown.
NESTLING: not known.

Field Characters. Length 19–20 cm. A sociable, noisy bark-gleaning babbler, usually high up in trees; flicks tail up and down. Distinctively marked: *orange-rufous* neck, upper breast and undertail-coverts contrast with dark body, black cap and yellow eye and bill; when feeding bird is hanging upside-down, orange-rufous hind collar and rump contrast with dark cap, back and tail. Wings uniform dark brown. Unlike any bird in its range except Chapin's Babbler *L. chapini* (q.v.).

Voice. Tape-recorded (104, B, F, LEM). Contact calls from feeding flock hoarse and grating, 'kakak', 'cherr', 'jajajut', 'jaa-gut', 'kukukuk', 'dsuk-kakak' and variants, also described as 'kerr-ker' (Dowsett-Lemaire 1990, and see sonagrams in Dowsett-Lemaire and Dowsett 1990). Group often joins in *Turdoides*-style chorus. Very similar to voice of White-throated Mountain Babbler *K. gilberti*.

General Habits. Inhabits montane forest and bamboo forest at 1690–2770 m. Keeps quite high in forest but sometimes comes down into understorey, 2–4 m from ground, in low, dense, mossy forest (Dowsett-Lemaire 1990). Lives year-round in cohesive flock, of 3–8 birds in breeding season and at other times up to 15 birds. Very active and vocal; constantly uttering a harsh chattering, often in chorus. Not shy. Forages by searching acrobatically for insects in crevices – almost exclusively in fissures in bark and on or within large masses of epiphytes like mosses, ferns and orchids; often hangs head down or upside down, and pecks and probes with some force, recalling nuthatch *Sitta* sp. (Vande weghe 1988). Continually on the move, hopping along and flying between branches, the flock soon moving away to new foraging tree. Constantly moves tail up and down. Aggressive interactions between members of a flock are frequent: birds raise crown feathers, chase and supplant each other. 'During active display the tail is spread and erected vertically over the back' (nature of display not studied: Vande weghe 1988). Frequently accompanies bark-probing insectivores, especially White-headed Wood Hoopoes *Phoeniculus bollei*, also *Andropadus* bulbuls, woodpeckers, weavers and insectivorous squirrels *Funisciurus boehmi* (Dowsett-Lemaire 1990). Flies only for short distance, fluttering wing beats alternating with glides when tail spread and drooped, much like *Turdoides* babblers. Adults with juveniles vigilant, stopping stiffly and watching alertly whenever accompanying bulbul or squirrel calls. Juvenile calls like adults' but more rasping. Flock emits rasping chatter with slight beat, lasting at least 2 s.

In each of 4 parties of 3 adults one bird always moved and explored ahead of other 2. In 2 parties each of 6 adults and 2 immature birds, 2 adults dominated: after spell of quiet foraging they flew to another tree, hopped and looked around, then called up rest of party with 'kerr, ker' notes. On arrival whole party gave brief chorus (like *Turdoides* babblers: Dowsett-Lemaire 1990).

Food. Insects; occasionally takes small fruits of *Urera hypselodendron*.

Breeding Habits. Barely known.
BREEDING INDICATIONS: Rwanda, carrying nest materials Apr–May and copulating May–July (Vande weghe 1988); 8 parties of birds all with young, Sept–Dec, indicating nesting July–Sept (Dowsett-Lemaire 1990); fluffy juv., Jan, hatched probably from Nov egg; Zaïre, clutches of eggs Apr, Sept and Oct (but eggs and nest not described: Prigogine 1971) and large gonads, Sept and Nov.
DEVELOPMENT AND CARE OF YOUNG: of 2 immature birds, Rwanda, Oct, one fed itself and the other still begged, following behind adult which sometimes fed it, sometimes refused it. It perched next to 3 resting, preening adults and sat stiff and upright whilst one adult preened its neck.

References
Dowsett-Lemaire, F. (1990).
Vande weghe, J.P. (1988).

Kupeornis chapini Schouteden. Chapin's Babbler. Phyllanthe de Chapin.

Plate 1
(Opp. p. 14)

Kupeornis chapini Schouteden, 1949. Rev. Zool. Afr. 42, p. 344; Mongbwalu, 01°57′N, 30°02′E, Belgian Congo.

Forms a superspecies with *K. rufocinctus*.

Range and Status. Endemic resident, forests of E Zaïre from L. Albert (Kilo) to L. Edward (Lutunguru, Kivu – 0°29′S, 28°47′E), Mt Nyombwe at 1200–1660 m and Itombwe highlands at 1180–1540 m.

Kupeornis chapini

Description. *L. c. chapini* Schouteden: Zaïre: L. Albert to L. Edward. ADULT ♂: forehead, crown, nape and hindneck dark mahogany or rich, dark orange-brown; sides of forehead and of crown, also lores and cheeks, paler orange; around eye, barely discernible superciliary stripe, and forepart of ear-coverts, rufescent brown; hind part of ear-coverts and sides of neck orange-brown; mantle, scapulars, back and upper rump olive-brown; lower rump and uppertail-coverts bright foxy rufous. Tail blackish brown, feathers with rufescent olive edges at base. Chin buff, throat pale orange, breast orange-brown (colour merging into that of throat but delineated from belly), belly, flanks, thighs and vent greyish olive-brown more or less heavily tinged with rufous mainly in middle; undertail-coverts dark rufous. Primaries, secondaries and tertials with dark brown inner vanes and olive-brown outer webs, those of primaries edged with rufous, making long rufescent panel in closed wing; rest of wing dark olive-brown with rufescent tinge. Underside of remiges shiny black-brown, underwing-coverts rufous, axillaries olive-brown. Bill with upper mandible brownish and lower mandible paler horn-brown; eyes dark; legs and feet dark brown. Sexes alike. SIZE: (6 ♂♂, 1 ♀, Prigogine 1964) wing 92–100, tail 66·5–72, bill to skull 16·5–18, tarsus 24–27; also (1 ♂, 1 ♀): wing, ♂ 97, ♀ 96·5; tail, ♂ 67, ♀ 69; bill to feathers, ♂ 12·5, ♀ 15; bill to skull, ♂ 17, ♀ 18; tarsus, ♂ 27, ♀ 26·5. Also: (unsexed, n = 3) wing, 93–97 (95), tail 65–71 (67) (Chapin 1953).
    IMMATURE and NESTLING: unknown.
    K. c. nyombensis (Prigogine): Mt Nyombe (at Butokolo, 02°42′S, 28°16′E), Kivu. Crown duller and less orange-brown than in nominate race, sides of head clear grey not rufescent brown, upperside greyer, less olivaceous, and throat paler. Intermediate in most characters between *chapini* and *kalindei*. SIZE: ♂ (n = 3) wing 95–98, tail 67·5–71, tarsus 23·5–26, bill to skull 17–18.
    K. c. kalindei (Prigogine): SW part of Itombwe highlands, near Kikoko, Mwenge, Kiliza (Kilizo) R. (about 03°42′S, 28°10′E) and Kitongo R. (03°46′S, 28°11′E). Crown darker and less reddish than in *nyombensis*, sides of head darker grey, hindneck less red, back darker and less greyish; edges of primaries and secondaries brighter rufous, undertail-coverts brighter rufous. SIZE: (6 ♂♂, 6 ♀♀, Prigogine 1971) wing, ♂ 101–105 (102), ♀ 98–103 (100), tail, ♂ 68·5–75·5 (71·8), ♀ 67–72 (69·2); (2 ♂♂, 2 ♀♀, Prigogine 1964) wing 85–102, tail 67–72, bill to skull 16–18, tarsus 25–26.

Habits. Almost nothing appears to be known about this rare bird, whose races were described from small series collected 35–50 years ago. Occurs in parties of about 12 under crowns of large trees, at much lower altitudes than closely allied Red-collared Babbler (Chapin's Babbler under and Red-collared over 1700 m). Joins mixed-species foraging flocks. Clutches in Jan, Feb and Dec (but eggs and nest not described: Prigogine 1971); ♂ with large gonads, May, and fledgling mid Jan, probably from Dec egg (Prigogine 1964, 1971).

Field Characters. Length 18 cm. A smaller and duller version of Red-collared Babbler *K. rufocinctus*, from which it differs as follows: cap chestnut, rufous of throat and upper breast paler and not extending onto sides of neck and ear-coverts, undertail-coverts dark chestnut, duller rufous hind collar contrasts little with cap and back, rump chestnut, tail dark brown. *Chestnut* wing edges form *panel on dark brown wing*. Ranges of the 2 come very close west of N end of L. Tanganyika – they may just be sympatric.

Voice. Unknown.

TIMALIIDAE

Genus *Horizorhinus* Oberholser

A small babbler or babbler-like bird, endemic to Príncipe Island. Olive-brown above with grey cap; white below with olive breast band and flanks. Conspicuous curved rictal bristles at gape; many forward-pointing bristles on chin. Plumage soft; rump feathers thick and soft. Rectrices somewhat pointed at ends and so tip of tail finchlike (**A**). Overall quite like *Kupeornis*, ethologically and morphologically (de Naurois 1994), but bill longer, legs less robust, and feet weaker.

Endemic; monotypic.

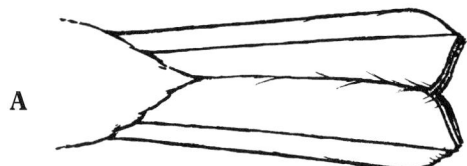

Plate 1
(Opp. p. 14)

Horizorhinus dohrni (Hartlaub). Dohrn's Thrush-Babbler. Cratérope de Príncipe.

Cuphopterus dohrni Hartlaub, 1866. Proc. Zool. Soc. Lond., p. 326; Príncipe Island.

Range and Status. Endemic resident, Príncipe I., where widespread and abundant in cocoa plantations and forest regrowth (Jones and Tye 1988). Abundant in all disturbed and edge habitats: plantations, farmland, gardens, bushy country and savanna, at all altitudes from sea level to top of Mt Pico Papagaio at 630 m (Atkinson *et al.* 1991).

Description. ADULT ♂: forehead, crown, nape, hindneck, sides of neck and upper mantle dark grey; tiny patch of white featherlets behind each nostril; lores, cheeks and ear-coverts darker grey than crown – lores almost black; lower mantle, scapulars, back, rump and uppertail-coverts greyish olive-brown, greyer on mantle, slightly brighter on rump. Tail dark brown, at base feathers fringed olive. Chin and throat white, sharply demarcated from adjacent colour; upper breast olive-brown, in band 5–10 mm deep in centre and 20 at sides of breast; lower breast, belly and vent white, flanks and thighs olive, undertail-coverts white. In breeding season flanks and belly washed yellow, almost saffron, the colour disappearing rapidly after death (de Naurois 1994). Wings uniformly brown, outer primaries with very narrow pale outer edges. Underside of remiges shiny dark grey, inner edges pale or whitish; underwing-coverts and axillaries white. Bill horn with pale yellow edges or dark brown upper mandible and horn brown lower mandible; eyes brown; legs and feet pinkish brown with orange soles. Sexes alike. Size (3 ♂♂, 3 ♀♀): wing, ♂ 68–69 (68·3), ♀ 67·5–69·5 (68·3); tail, ♂ 50–54 (52·2), ♀ 55·5–59 (56·5); bill to feathers, ♂ all 13, ♀ 13–14 (13·3); bill to skull, ♂ 17·5–19 (18·5), ♀ 12–13 (12·3); tarsus, ♂ 26–28 (26·7), ♀ 25–27 (26). WEIGHT: unsexed (n = ?) 17–22.

IMMATURE and NESTLING: not described.

Field Characters. Length 15 cm. The most abundant of the 7 passerines on Príncipe I. after the weaver *P. princeps*; the only warbler-like bird – about size and shape of Wood Warbler *Phylloscopus sibilatrix*. Grey-brown upperparts, breast-band and stripe on flanks, otherwise white below, often tinged yellow; white forms inverted 'V' on sides of

Horizorhinus dohrni

neck below ear-coverts. Loud, attractive song one of the characteristic bird sounds of Príncipe.

Voice. Tape-recorded (104, B, GUL, JOPJ, SIN, TYE). Beautiful loud song like Nightingale *Luscinia megarhynchos*, of clear, ringing notes, lasting 3–6 s. Usually starts with 2 down-slurred whistles, first one longer, followed by a central section of repeated double notes, ending with a faster series of notes on same pitch, the final ones often upslurred: 'cheeeoo, heoo, ti-charr-ti-char-ti-char, ti-cher-ti-cher-ti-cher, wawawawawa, wee-wee'; 'cheeeoo, woo, tsee-cha-tsee-cha-tsee-cha, tee-weeoo-tee-weeoo, wawa, wee-wee-whooee-whooee'. Final section sometimes long and trailing away. Displays same spirit of extempore inventiveness as Nightingale; no 2 songs exactly the same. Also has a long, anxious, scolding chatter call, varying in length, form and intensity. This may be given alone, as an alarm, or be interrupted sometimes by a few notes of song, the whole performance a 'repeated, 'inward', creaking note, 'chee', accelerating into a trill, then another sequence of loud melodious notes, then a fast trill, high-pitched and mechanical-sounding, duration 2–3 s, like song of Striped Kingfisher *Halcyon chelicuti*' (Snow 1950). Other calls include 'wit', 'weet' or 'tuik', repeated every 2–3 s.

General Habits. Occurs at all levels in all vegetation in all woody habitats; mainly in understorey of dense bushy growth, bases of crowns of taller trees, and low amongst lianas, where moves with agility amongst vertical stems but can be very secretive. Inhabits mainly dense forest, but comes into open growth of abandoned clearings to feed in *Trema* shrubs and canopy of large parasol trees *Musanga*; keeps in shade. Common, less so in coconut plantations in S; the commonest bird species in upper valley of R. Papagaio.

Usually in pairs, but often in larger parties, of about 6–10 birds. Extremely active; pairs and parties lively and vocal, with much trilling, singing, bill-snapping and wing-trembling by birds in foraging parties. Flock forages cohesively from tree to tree, moving mainly at low and mid levels, sometimes low down through bushwood. Gleans small insects from foliage and makes short sallies upwards from a slanting branch to take insects from undersides of leaves above. Explores trunks and foliage of lower branches of *Erythrina* trees; inspects undersides of small branches and twigs. Makes quite long hops from branch to branch, wagging whole rear end of body from side to side; locomotory movements often accompanied by marked balancing twitches of tail, making bird look nervous. When actively foraging, stance mainly horizontal. Sometimes comes down to rocks, but otherwise avoids the ground. Sometimes foraging pair can be silent – perhaps when feeding young (Christy and Clarke 1998).

Food. Small invertebrates including caterpillars, beetles and snails; also berries, fruit pulp and seeds (de Naurois 1994).

Breeding Habits. Solitary nester; monogamous (possibly co-operative breeder).

NEST: unlined, thin, shallow bowl of dry twigs or grass, surrounded by loose dry leaves. Placed in trees or thick bushes, *c.* 1–3 m from ground.

EGGS: 2–3. Dull white ground, stippled with brown spots forming band at large end. SIZE: *c.* 25×16.

LAYING DATES: Sept–Jan, also evidently June–July (de Naurois 1994).

References
Atkinson, P. *et al.* (1991).
Christy, P. and Clarke, W.V. (1998).
Jones, P.J. and Tye, A. (1988).
de Naurois, R. (1994).
Snow, D.W. (1950).

Genus *Parophasma* Reichenow

Single monotypic species: a distinctive Ethiopian/Eritrean endemic, sooty grey, with short, rather stiff, whitish feathers on forehead, black lores and rufous belly and undertail-coverts. Soft parts black; bill rather short and fine. Wing short and very rounded; legs and feet moderately robust; shape and proportions much like those of same-sized *Sylvia* warbler. A noted songster, song often compared with that of Nightingale *Luscinia megarhynchos*. Affinities uncertain; generally agreed to be a timaliid, and several recent authorities have concurred with Hall and Moreau's (1970) opinion that it may be related with *Kupeornis*. Hall and Moreau noted that rufous belly and undertail-coverts are like those of the *K. rufocinctus* superspecies, and that grey in plumage and black edging before or below eye recall *Lioptilus nigricapillus*. Occurs sometimes in small flocks, but whether they are transient family parties or year-round cohesive social groups as in *Kupeornis*, *Phyllanthus* and *Turdoides*, remains to be discovered. Fine song is very un-babblerlike.

Parophasma galinieri (Guérin-Méneville). Abyssinian Catbird. Phyllanthe de Galinier.

Plate 1 (Opp. p. 14)

Parophasma Galinieri Guérin-Méneville, 1843. Rev. Zool. [Paris], 6, p. 162; Ethiopia.

Range and Status. Endemic resident, Ethiopia and Eritrea. Ethiopia, widespread and frequent to common or at least locally common, in W Highlands north to Simien Mts and in SE Highlands. Eritrea, occurs along E side of central plateau from about Adigrat (Ethiopia) to near coast northeast of Asmara.

Description. ADULT ♂: forehead white, lores black extending in crescent around front of eye, forecrown pale grey, hindcrown, rest of head, neck to uppertail-coverts and breast and flanks brownish grey, with chin slightly paler and ear-coverts ochre-tinged. Tail dark brown above, dark grey with white feather-shafts below, with *c.* 15 regular structural bands in vanes. Lower breast often whitish in midline, belly, hind part of flanks, vent and undertail-

Parophasma galinieri

coverts bright foxy orange, thighs grey. Wing dark brown. Remiges dark brown, primaries fringed grey except for emarginated portions of P5–P7 which are narrowly white-edged. P5–6 longest, P7 2, P8 6, P9 15 and P10 36 shorter; P5 emarginated 24, P6 28 and P7 28 from tip. Upperwing-coverts dark grey, but alula and greater primary coverts shiny black with matt black outer edges. Underside of remiges shiny grey with whitish inner edges, lesser underwing-coverts white and dark grey, others pale fawn. Bill black; eyes crimson; legs and feet sepia, soles whitish or pale brownish. Sexes alike. SIZE: wing, ♂ (n = 6) 82–85 (83·8), ♀ (n = 4) 83–86 (85·0); tail, ♂ (n = 5) 70–84·5 (77·0), ♀ (n = 4) 71–75 (73·2); bill to feathers, ♂ (n = 3) 12·5–14·5 (13·3), ♀ (n = 2) 13, 13; bill to skull, ♂ (n = 5) 17–18 (17·2), ♀ (n = 4) 16–19 (17·8); tarsus, ♂ (n = 3) 27–29 (27·8), ♀ (n = 3) 26·5–28 (27·4). WEIGHT: 1, unsexed, 21·5.

IMMATURE: like adult but greys paler and less sooty and belly and undertail-coverts paler rufous.

NESTLING: featherless at 4–6 days.

Field Characters. Length 18–19 cm. Uniform grey, wings and tail a bit darker; lores black, forehead paling to whitish at base of bill; lower belly and undertail-coverts bright tawny. Endemic to Ethiopia and Eritrea, and cannot be confused with any bird there.

Voice. Tape-recorded (B, 109). Song a rich warbling trill lasting 3–4 s, becoming progressively louder and sharper and decelerating at the end; has same feeling of suppressed power as Nightingale *Luscinia megarhynchos*. ♀ answers with low dry churr.

General Habits. Inhabits dense tangles of vegetation and bushy thickets with vines, in juniper and *Podocarpus* forests, mainly at 2425–3000 m but ranging down to 1800 m and up to 3500 m (Urban 1987); also clumps of bamboo. Such vegetation often confined to steep-sided gullies and rocky slopes (Fishpool *et al.* 1996). Characteristic of bushy ground stratum in open *Podocarpus* forest (Dorst and Roux 1973). Commonly occurs in spacious, well-timbered gardens of plateau cities. Lives in pairs and parties of up to 8 birds. ♂ sings intensely in rainy season; loud, clear, ringing call given whilst bird stretches neck skyward and holds wings half out; ♀ answers with purring or churring note (Mackworth-Praed and Grant 1960, Urban 1987).

Food. Mainly small fruits, including juniper berries. Abyssinian Catbird takes notably large share of berry and small fruit content of *Podocarpus* forest (Dorst and Roux 1973).

Breeding Habits. Territorial (Friedmann 1937).
NEST: one was small, frail, thin, cup-like structure, made of fine plant stems, placed loosely on top of 2–3 small branches and vines, 5 m up *Hypericum lanceolatum* tree (Urban *et al.* 1970). Cup 70 across and 30 deep.
EGGS: 2. Pale flesh colour, uniformly covered with fine flesh marks and a few chestnut spots (Urban 1987).
LAYING DATES: Eritrea/Ethiopia May, July (and various indications Feb–June) (Urban and Brown 1971).
DEVELOPMENT AND CARE OF YOUNG: young have short grey feathers at 9–11 days; well feathered but unable to fly at 14–16 days.
BREEDING SUCCESS/SURVIVAL: 1 survived for 6 years (Urban 1975).

Genus *Panurus* Koch

A single species of paradoxornithid babbler, adapted to life in Palearctic reedbeds *Phragmites*. Bill short, yellow, culmen arched, tip uncinate. Nostrils with operculum covered with bristles. Rictal bristles feebly developed. Wing rather short but somewhat pointed, P10 rudimentary, a tiny stiff feather shorter than its covert; P6–P8 longest and about equal. Tail long, strongly graduated; rectrices narrow, tip pointed; undertail-coverts long, as long as T6. Tarsus strong, scutellate in front. Plumage soft. Large cloacal protuberance, forming a copulatory organ when everted (Birkhead and Hoi 1994). Stomach lining becomes plate-like in winter, 1 g heavier than in summer, and filled with *c.* 600 uniform-sized bits of grit. Sexually dimorphic, ♂ with broad black moustache, moustachial feathers elongated. Juvenile very distinctive, like adult ♀ but back with large black patch, outer rectrices pied, and P10 long – half length of P9. Nest an open cup built onto reed stems, made of dead reed or sedge *Scirpus* leaves, lined by ♂ only with reed flower heads, so soft that an egg sometimes becomes buried (Witherby *et al.* 1938). Eggs white, finely scribbled.

Insectivorous in summer; eats mainly reed seeds in winter. Pinging voice carries well through reedbed. Adept at straddling space between 2 swaying reed stems. Direct flight, through or just above reeds. Sedentary, nomadic and partial migrant (Cramp and Perrins 1993). Mating system monogamous, with ♂ and ♀ rearing several broods in succession, but ♀ regularly initiates extra-pair copulations (Birkhead and Hoi 1994).

Strong resemblance in plumage colours and pattern, bill, tail and leg morphology and other characters to group of bamboo *Bambusa* and reed-dwelling Oriental babblers, particularly *Paradoxornis* (18 species). Group is distinctive and composes its own subfamily Paradoxornithinae, often raised to rank of family (Paradoxornithidae). *Panurus* is only W Palearctic representative or offshoot of *Paradoxornis*.

1 species; rare vagrant to NW Africa.

Panurus biarmicus (Linnaeus). Bearded Reedling. Panure à moustaches.

Plate 7
(Opp. p. 94)

[*Parus*] *biarmicus* Linnaeus, 1758. Syst. Nat., ed. 10, 1, p. 190; Europe, restricted to Holstein by Hartert 1907.

Range and Status. W Europe (where range now fragmented and peripheral, including E Ireland, E England, central and E Spain, Denmark and Gulf of Riga to Greece and central Turkey) to China. About 250,000 pairs in Europe (Tucker and Heath 1994). Vagrant south to N Africa and Kuwait.

Accidental in Morocco (1, Fez, Mar 1929; singles Oued Massa estuary, Apr 1987 and Jan 1991) and Algeria (3, L. Fetzara, Apr 1911).

Description. *P. b. biarmicus* (Linnaeus): Europe and Transcaucasia. (Only race in Africa). ADULT ♂: forehead, crown, hindneck and sides of neck pale bluish grey; mantle, back, rump, uppertail-coverts and most of tail rufous-chestnut, tips of T3 and T4 greyish, tips and outer webs of T5 and T6 pale grey and bases dark grey or blackish. Remiges and greater primary coverts dark greyish brown; outer vanes of P9–P6 and their coverts white. Outer vanes of P10 and P5–P1, secondaries, and remaining primary coverts bright rufous. Outer 2 tertial feathers with black outer vane and shaft, whole of inner vanes white, creamy or pale pinkish; innermost tertial mainly white; all tertials with outer vanes broadly fringed rufous. Greater and median upperwing-coverts black, narrowly or broadly fringed rufous. Lesser upperwing-coverts mixture of pale grey, buff and cream. Alula feathers black with white outer edges. Underwing-coverts and axillaries buff-white. Lores, cheeks and moustachial area black, feathers rather long and pointed, forming large, crescentic moustache. Chin and throat white, breast white with greyish tinge in centre and pinkish one at sides, merging into rufous flanks and cream-buff belly. Undertail-coverts black. Bill yellowish or orange-horn; eyes yellow with narrow surround of black skin; legs and feet black. ADULT ♀: like ♂ but lacks crescentic black moustache, blue-grey on head, and black undertail-coverts. Forehead, crown, and hindneck yellowish grey-brown; lores dusky; cheeks, ear-coverts, sides of neck, mantle, back, rump, uppertail-coverts and central tail feathers less rufous, more yellowish brown than in ♂; crown often with small black spots or streaks, sometimes heavily streaked; lower mantle and back often lightly black-streaked and sometimes heavily so; rump usually with fine black streaks. Chin and throat buffy white; breast browner than in ♂, less grey- and pink-tinged; flanks buff-rufous; undertail-coverts yellowish buff. Bill and eyes not so bright yellow. SIZE: wing, ♂ (n = 49) 58–65 (60·8), ♀ (n = 45) 57–61 (59·2); tail, ♂ (n = 49) 72–87 (79·9), ♀ (n = 42) 67–78 (74·4); bill to skull, ♂ (n = 28) 10·2–11·7 (10·8), ♀ (n = 37) 9·9–11·2 (10·6); tarsus, ♂ (n = 27) 20·0–21·9 (21·1), ♀ (n = 37) 19·4–21·6 (20·4). WEIGHT (Netherlands, Mar–May) ♂ (n = 8) 15·0–18·0 (16·2), ♀ (n = 24) 13·2–16·0 (14·4); monthly averages of >1200 ♂♂ vary from 14·7 (Netherlands, July) to 16·3 (France, autumn) and 18·9 (Germany, winter) (Cramp and Perrins 1993).

IMMATURE: deep, somewhat dingy, yellowish buff, except for wings, tail, and large oblong black patch on back. In tail, T1 yellow-buff, T2 mainly yellow-buff, T3–T6 black with pale whitish buff tips and outer vanes. Wings like adult ♂ but rufous parts yellower, primary coverts, alula and outer greater coverts black, and remaining coverts black with rufous fringes. Lores blackish; soft parts duller.

Field Characters. Length 12·5 cm. Any twanging '*ping*' calls in N African reedbed are likely to be from this rare vagrant. Small, rather short-winged and long-tailed, tail strongly graduated (**A**); ♂ *rufous* with blue-grey cap and large *black moustache*, easily seen in flying bird, and (easy to see in bird momentarily at rest) *black and white* marks in *rufous wings* and *black* undertail-coverts. ♀ much plainer, but has same pied-and-rufous wings and graduated tail. Immature bird dingy yellowish with large oblong or *ragged black patch on back*, wings like adult's but blacker, tail blackish at sides.

Voice. Tape-recorded (62–73, 93, 105). Main call, usually given when pair or family party fly through reeds, is vibrant, explosive, clipped or ringing twangy monosyllable, 'ping' or 'tschin', ' ching' or 'pzing'; also a harder 'tjick' or 'tschick', which speeds into twittering from excited flock.

General Habits. Confined to large reedbeds *Phragmites* with stands of reedmace *Typha*, though in hard winters moves into mixed reed and sedge *Carex* beds near to surrounding lakeside woody tangles. In pairs and small family parties, active, restless, vocal. Flight direct, whirring and tail-trailing, like tiny pheasant *Phasianus* sp. (Cramp and Perrins 1993); through or just above reed-heads. Agile, hops up reed stem and often straddles between stems; quite *Parus* tit-like when foraging on reed flowering spikes, pecking sideways and sometimes hanging upside down. Forages on all parts of reeds, mainly flowering heads but often low down on stems, sometimes on wet, tussocky ground at edge of reedbed; moves along stems blown down on surface of water, taking insects on them and from or even a few mm under water surface. When perched on stem, uses bill or foot to pull adjacent reed-head closer. Flies out from reed to catch midges. Hops and runs quickly on ground; scratches ground like hen and makes vey rapid small hops both forwards and backwards, soil flying in all directions (Cramp and Perrins 1993). Flips over sodden reed leaves on mud.

Largely sedentary, but erupts out of NW European stronghold, the Netherlands, every few years, particularly in harsh winters, many birds reaching e.g. Britain, Gironde (France) and Switzerland.

Food. Great variety of small invertebrates in summer, mainly small but up to size of damsel flies, dragonflies and stoneflies, also caterpillars, spiders and snails. In winter almost exclusively seeds (see *Panurus* diagnosis for condition of stomach then), of Polygonaceae, Chenopodiaceae, Gramineae including reeds, rushes Juncaceae, *Typha*, *Epilobium* and *Urtica*.

A

References
Birkhead, T.R. and Hoi, H. (1994).
Cramp, S. and Perrins, C.M. (1993).
Witherby, H.F. *et al.* (1938).

Family AEGITHALIDAE: long-tailed tits

Tiny birds; plumage copious and soft; mostly in pale greys, white, browns, sooty and pink, with white and blackish marks on head; sexes more or less alike. Bill conical, extremely short; narrow, bare ring around eye; wing short, rounded, P10 less than half length of P9; tail very long, strongly graduated, feathers narrow. Legs slender, scutes in front usually fused into single lamina; foot weak but hind claw long and quite strong. Insectivorous; rather sedentary; one species a vertical migrant, another subject to irruptive movements. Highly sociable. At least some species are co-operative breeders. Best diagnostic feature is nest, a long-lasting upright oval bag, large for size of bird, with small side entrance, thickly and closely made of moss, lichen, spider web and up to 2000 small feathers. Several characters suggest distant alliance with Timaliidae rather than (for instance) Paridae.

8–10 species in 3 genera: *Aegithalos* Palearctic, *Psaltriparus* American and *Psaltria* Javan.

Genus *Aegithalos* Hermann

Characters those of the family. 6 species, lowland and alpine forests, Europe to Himalayas, China and Japan; one accidental in NW Africa.

Plate 7 (Opp. p. 94)

Aegithalos caudatus (Linnaeus). Long-tailed Tit. Mésange à longue queue.

Parus caudatus Linnaeus, 1758. Syst. Nat., ed. 10, p. 190; Sweden.

Range and Status. Whole of Europe south to Gibraltar, Sicily and S Turkey; central latitudes of Asia east to Japan, south in China to *c.* 30°N; resident, partial and irruptive migrant.

Vagrant; flock of 60, Jebel Ressas, Tunisia, in Dec 1957; 1 record Morocco (Cramp and Perrins 1993).

Description. *A. c. irbii* (Sharpe and Dresser): S Portugal, S Spain, Corsica. (Races of vagrants to Africa unknown, probably *irbii*, *italiae* (Italy) or *siculus* (Sicily)). ADULT ♂: forehead and crown in centre white with dusky streaks, and at sides black, forming wide black stripe from eye to nape; nape blackish, hindneck, sides of neck, mantle and back grey-black, scapulars greyish pink, rump and uppertail-coverts black. Tail black, outer 2 feathers mainly white. Chin and throat white, streaked with pale grey and pinkish, breast white streaked with pink, belly whitish, flanks mainly pink, thighs whitish, vent and undertail-coverts pink. Primaries and secondaries black, outer vanes narrowly edged white, tertials black, outer vanes broadly edged white, wing coverts mainly black, underside of remiges dark grey, underwing-coverts and axillaries pinkish white. Bill black; ring of bare skin around eye red, eyes dark hazel; legs and feet black-brown. Sexes alike. SIZE: wing, ♂ (n = 10) 56·5–60·5 (58·5), ♀ (n = 7) 55·0–58·5 (56·8); tail, ♂ (n = 8) 72·5–80·5 (76·3), ♀ (n = 6) 71·5–79·0 (75·4); bill to nostrils, ♂ (n = 10) 3·7–4·8 (4·3); ♀ (n = 6) 4·0–4·4 (4·2); bill to skull, ♂ (n = 10) 7·0–8·3 (7·4), ♀ (n = 6) 6·9–7·3 (7·1); tarsus, ♂ (n = 8) 15·7–17·0 (16·4), ♀ (n = 6) 15·2–16·9 (16·0). WEIGHT: (*A. c. europaeus*, n = 206): subset averages from 7·7 to 9·0 (Cramp and Perrins 1993).
IMMATURE: forehead white, forecrown white in centre, brown at sides; hindcrown, nape, sides of neck, upper ear-coverts, cheeks and around eye brown; rest of upperparts including wings and tail mainly dark brown but scapulars and tertials and edges of inner secondaries whitish and T5–6 mainly white. Underparts white, flanks and undertail-coverts pinkish.

Field Characters. Length 14 cm. A very small bird, narrow tail conspicuously longer than body, with tiny bill and distinctive combination of pink, dingy white and blackish in plumage; head can appear white with broad black stripe behind eye. Active, vocal, usually in loose flock moving through trees and hedges.

Voice. Tape-recorded (62–73, 93, 105, B). Contact call, given freely when flock foraging and flying through trees, a soft, almost metallic 'pit' or 'pt', also a high, thin 'zee zee zee' and a rippling or trilling 'tsirrrrrup'.

General Habits. Inhabits deciduous woods, open thickets, undergrowth, tall hedges, and scrub and spaced trees bordering rides, roads and marshy ground; likes oak *Quercus*, ash *Fraxinus*, sycamore *Acer*, hazel *Corylus avellana* and alder *Alnus glutinosa*; sometimes in ornamental trees in gardens and parks, and conifer plantations. Usually keeps at about 2–5 m above gound, sometimes higher, rarely coming near ground. Occurs in loose flocks of 6–15; always on the move, birds in flock foraging mainly >1 m apart and flying 1 or 2 at a time in a somewhat follow-my-leader fashion between leafy but open trees. Flight rather weak and laboured. Often forms mixed foraging parties with tits *Parus* spp. Flock roosts at night huddled on twigs near tree trunk, birds facing in same direction; in severe weather clusters in bunch of up to 50 birds.

Sedentary, but dispersive (seldom moving more than 100 km), and irruptive in hard autumns following population build-up.

Food. Insect eggs and larvae, small adult insects, small spiders.

References
Cramp, S. and Perrins, C.M. (1993).
Gaston, A.J. (1973).
Witherby, H.F. *et al.* (1938).

Family PARIDAE: tits and chickadees

Small, arboreal, gregarious, insectivorous birds. Bill rather short, sharp-pointed, quite strong, culmen curved to almost straight, varying in shape from conical and almost finch-like to slender and warbler-like. Nostrils small, round, concealed below dense, antrorse plumules. Rictal bristles inconspicuous, sometimes obsolete. Tongue truncated, tipped with strong bristles. Wings rounded, P10 longer than primary-coverts but much less than half length of P9; P8–P5 generally longest and about equal. Tail rounded, of 12 feathers. Leg and foot strong, front of tarsus scutellate. Some species crested. Plumage soft and copious, mainly pied and grey in African species, yellow, olive, brown and black in Palearctic ones, and pied and brown in American ones (chickadees); many tits have glossy black crown, black bib and white cheeks. Essentially sedentary; sociable, vocal; calls generically similar yet individually varied, notes repetitive and complex-looking in sonagrams (Harrap 1996); songs derived from calls. Eat wide variety of vegetable matter, mainly that available in trees and deciduous and coniferous woods, also variety of mainly small and soft-bodied insects; young fed largely on caterpillars. Acrobatic in foraging, clinging sideways and upside down; often stand on plant food (fruits, buds, nuts, bird-table scraps) with feet together and hammer at it in tiny gap between feet. Several species store food (Sherry 1989). Nest in holes, generally in wood.

One large genus, *Parus* (Palearctic, Nearctic, Oriental and Afrotropical) and 2 monotypic genera closely related to it (*Sylviparus* and *Melanochlora*, Himalayas and montane SE Asia).

Plate 5

Plate 6

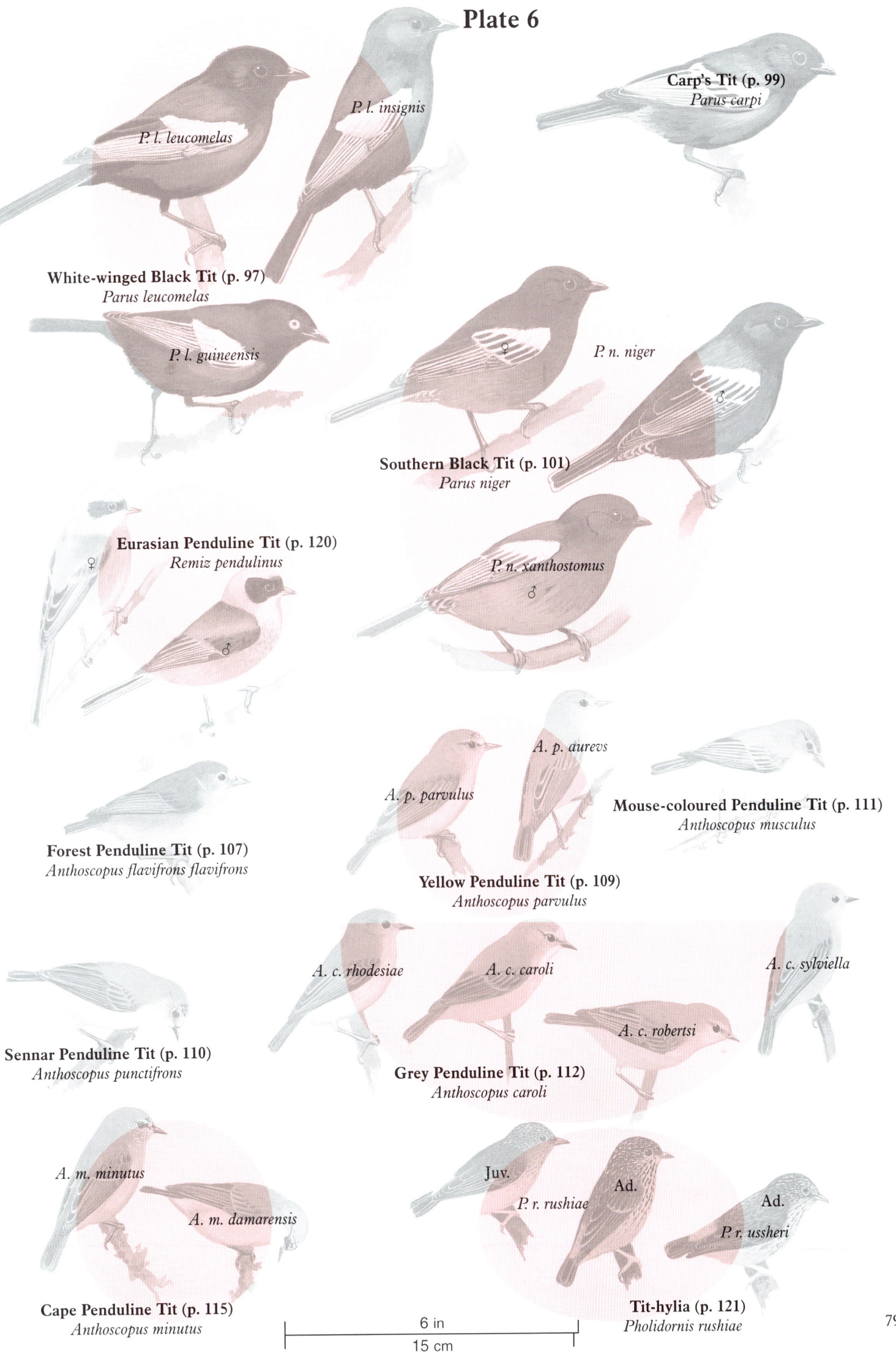

Genus *Parus* Linnaeus

Characters those of Family Paridae. 55 species (Harrap and Quinn 1996), or 51 (Sibley and Monroe 1990, excluding penduline-tits which Sibley and Monroe treat as a subfamily in Paridae); Holarctic, Oriental and Afrotropical. 17 in Africa: 3 Palearctic species resident in Maghreb, a European species accidental in Maghreb, and 13 endemic tropical species.

Several subgenera have been recognized, including *Melaniparus* which supposedly embraces all Afrotropical species (Hailman 1989). Hall and Moreau (1970) recognized a single independent species, *P. funereus*, of lowland and intermediate rain forest canopy, and allocated all of the remainder to 3 superspecies: grey tits, black tits, and 2 part-rufous tits (*P. rufiventris*, *P. fringillinus*). A grey tit, *P. afer sens lat.* is now widely seen to comprise 3 separate species (*P. griseiventris, P. cinerascens* and *P. afer sens str.*) and a black tit, *P. leucomelas-P. niger sens lat.*, is also 3 species (*P. leucomelas, P. carpi* and *P. niger sens str.*). In each trio the species are parapatric and ecologically segregated, with limited marginal overlap between *P. cinerascens* and *P. griseiventris*, and between *P. leucomelas* and *P. niger*. We regard each trio as a superspecies (see specific TAXONOMIC NOTES).

Of the 13 tropical African tits, the Red-throated Tit *P. fringillinus* is an independent species of no obvious affinity. The remaining 12 comprise 5 'grey' tits and 7 'black' ones. Although brightly coloured, the grey, green and yellow European Great Tit *P. major*, is another 'grey' tit. All except the Albertine montane *P. fasciiventer* have conspicuous white cheek patches; and at least the 2 end-of-chain species (*P. major, P. afer*) have alarm calls with strong structural similarities in sonagrams (Dowsett and Dowsett-Lemaire 1993a, Cramp and Perrins 1993). The distinctive *P. fasciiventer* may not in fact be a 'grey' tit at all. *P. thruppi* of E Africa, its white cheek patch enclosed by black, resembles *P. major* as much as it does the *P. afer* superspecies; but its alarm calls are very like those of *P. afer*, and we follow Short *et al.* (1990), Sibley and Monroe (1990) and Dowsett and Dowsett-Lemaire (1993a) in admitting *P. thruppi* to the *P. afer* superspecies.

Only one 'black' tit is all-black (i.e. ♂; ♀ is grey) – the rain forest *P. funereus*, a distinctive species with short, broad-based bill, red eyes, and sexual dimorphism. The others are black with white in wing (*P. leucomelas/P. carpi/P. niger* superspecies), in wing and on belly (*P. albiventris*), on mantle (*P. leuconotus* of Ethiopia), or white in wing with isabelline belly (*P. rufiventris*). *P. leucomelas* and *P. rufiventris* both have yellow-eyed and brown-eyed races, but are unlikely to be closely related since they are sympatric. Dowsett and Dowsett-Lemaire (1993a) show with sonagrams that the alarm call of *P. carpi* is much more like that of *P. niger* than of *P. leucomelas* and conclude that the first 2 are conspecific. Yet the calls are not identical; one taxon is sexually monomorphic, the other dimorphic; we therefore retain *carpi* as a full species.

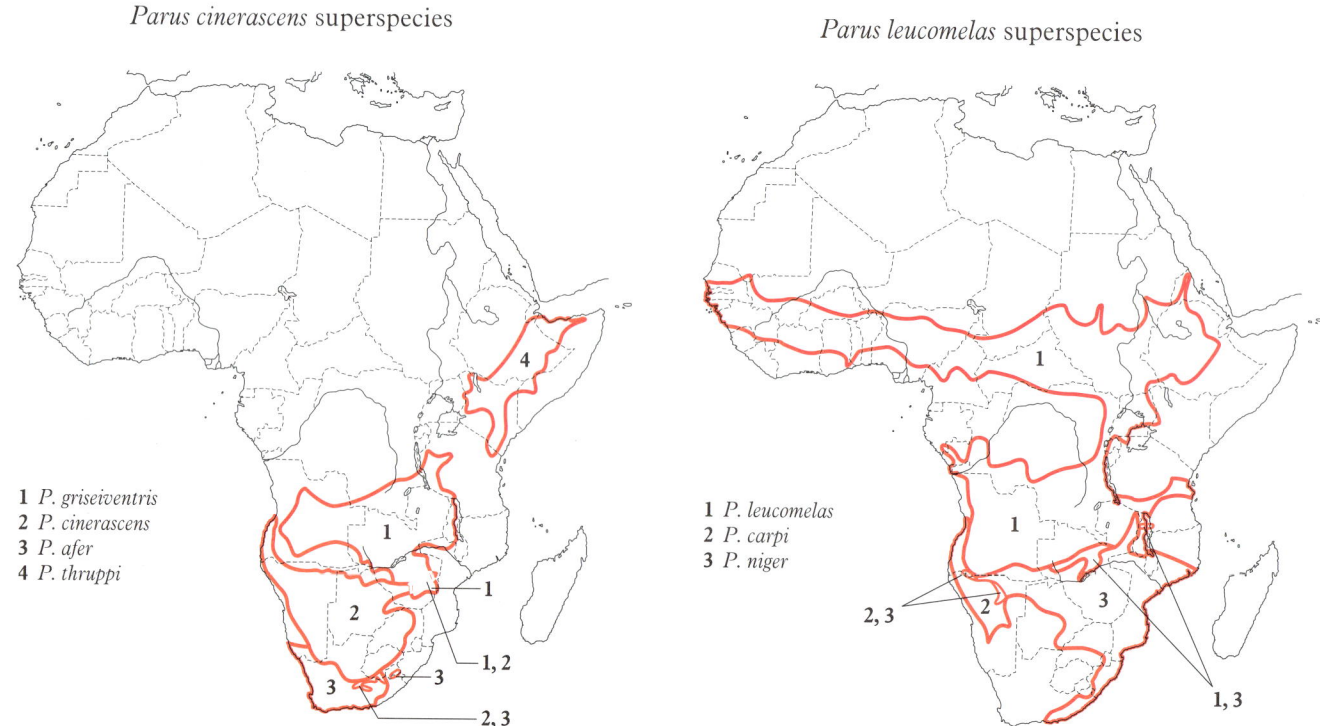

Parus cinerascens superspecies

1 *P. griseiventris*
2 *P. cinerascens*
3 *P. afer*
4 *P. thruppi*

Parus leucomelas superspecies

1 *P. leucomelas*
2 *P. carpi*
3 *P. niger*

Parus fasciiventer Reichenow. Stripe-breasted Tit. Mésange à ventre strié.

Parus fasciiventer Reichenow, 1893. Orn. Monats., 1, p. 31; Ruwenzori.

Range and Status. Endemic resident, mountains of Albertine Rift from 01°N to 05°S. E Zaïre: frequent in Rwenzori and Virunga Mts north and northwest of L. Edward, at 2000–2800 m and occasionally up to 3385 m; Mitumba Mts, and Kivu Volcanoes where fairly common at 3385 m and reported up to 3815 m; Mt Kandashomwa; quite common and widespread in Itombwe Mts and other highlands west of Ruzizi Valley and Uvira and northwest of Baraka, and Mt Kabobo. SW Uganda: locally abundant at 1800–3400 m, Rwenzoris and Bwindi-Impenetrable Forest, Kigezi. SW Rwanda: Nyungwe Forest, rare as low as 1900 m, widespread above 2100 m, common above 2300 m. NW Burundi in Teza forests and Rwegura.

Description. *P. f. fasciiventer* Reichenow: Zaïre except Itombwe and Mt Kabobo; Uganda, Rwanda and Burundi. ADULT ♂: top of head black with slight bluish gloss, grading to sooty brown on hindneck; mantle to scapulars, rump and uppertail-coverts greyish brown. Tail blackish brown, T1–T5 with narrow white fringe at tips, T6 with broad white fringe along outer edge and around tip. Lores, supercilium, cheeks, ear-coverts and sides of neck to chin, throat and centre of upper breast brownish black, blacker with slight gloss on centre of throat, browner on sides of neck and along sides of breast bib. Broad black mid-ventral stripe from lower breast to upper belly. Rest of underparts including sides of breast buffish white. Flight feathers blackish with narrow greyish white outer edges, extending around tips on secondaries, and becoming broader and whiter near base on primaries; tertials blackish with broad greyish white fringe. Alula blackish, fringed white; larger upperwing-coverts blackish, primary coverts edged buffish white, greater and median coverts edged and broadly tipped white; lesser coverts sooty black, broadly fringed greyish brown. Axillaries, underwing-coverts and inner borders on underside of flight feathers white, tinged buff. Bill black; eyes dark brown or hazel to reddish brown; legs grey, slate grey or blue–grey. ADULT ♀: like adult ♂, but hindneck, sides of face and sides of upper breast browner, black on cap and centre of throat more restricted. SIZE (8 ♂♂, 4 ♀♀): wing, ♂ 77–82 (78·8), ♀ 75–80 (76·8); tail, ♂ 58–63 (61·3), ♀ 57–60 (58·5); bill, ♂ 11–12 (11·4), ♀ all 11; tarsus, ♂ 19–20 (19·1), ♀ 18–19 (18·5). WEIGHT: 1 ♂ 14, 2 ♀♀ 14, 15 (Impenetrable Forest, Uganda, S. Keith pers. comm.).

IMMATURE: juvenile like adult but paler brown on sides of head, neck and breast bib; black less glossy, confined to cap and narrow band from chin to centre of upper breast. Upperparts browner, pale wing feather markings buffier; underparts buffier, and dark mid-ventral stripe narrower.

NESTLING: not described.

P. f. tanganjicae Reichenow: Itombwe Mts, Zaïre. Both sexes resemble ♀ of nominate race, with sides of head and neck and sides of breast bib brownish, contrasting more with black cap and mottled blackish centre of throat; underparts more strongly washed buff. SIZE: ♂ (n = 9) wing 76–82 (79·1), tail 57–61 (59·2).

P. f. kaboboensis Prigogine: Mt Kabobo, E Zaïre. Like *tanganjicae*, but upperparts purer grey, underparts darker and greyer. SIZE: ♂ (n = 6) wing 77–82 (79·3), tail 57–61 (59·3).

Field Characters. Length 14 cm. Best identified by *range*; the only tit in montane forest in the Albertine Rift. Plumage, with *all-dark* head and breast, reminiscent of Cinnamon-breasted Tit *P. pallidiventris*, but belly white with *black stripe* down centre. At border of montane and

Parus fasciiventer

intermediate forest may meet Dusky Tit *P. funereus*, which is entirely sooty black.

Voice. Tape-recorded (104, B, C, LEM). Territorial calls (also used in contact) are combination of high notes (sibilant 'ps' or dry 'ti') and low ones, buzzy 'dzerr' or grating 'grrr': 'ps-ps-der-der-ps-der', 'tititi-grr-ti-grr', 'grr-ti-grr-ti-grr'; 'ti' may also be run into grating trill, 'titititirrrrrrr'. Also known are a frog-like 'crud-y' (almost 'ribbit'), a rapid, thin clinking 'chit-lilit' and a very deep, powerful, full 'chur-chur-chur-chur' (Harrap and Quinn 1996). Song a 'tea-cher' motif recalling Great Tit *P. major*.

General Habits. Inhabits clearings in and edges of montane forest, woodland undergrowth in bamboos, *Hagenia* woods and tree heath. Does not penetrate understorey of closed forest (Dowsett-Lemaire 1990).

Occurs in pairs or more commonly family parties, foraging through trees often in company of other species, including Montane White-eyes *Zosterops poliogaster*, Cardinal Woodpeckers *Dendropicos fuscescens* and Dusky Tits *P. funereus*. Probes foliage, dead vegetation and bark in canopy of smaller trees, descending to within 1–2 m of ground (Dowsett-Lemaire 1990). Confiding and approachable. One bird roosted in same place on 4 consecutive nights, underneath thick clump or tent of dead, rust-coloured, undehisced *Hagenia* leaves; it remained dry despite lengthy rain (Davies 1993).

Sedentary.

Food. Nestlings fed mainly on caterpillars (C.M. Perrins, pers. comm.).

Breeding Habits. Monogamous, territorial, solitary nester. One family party with 2–3 dependent young also included 2 independent immatures, so species probably multi-brooded (Dowsett-Lemaire 1990). 4–5 birds sometimes seen around a nest box where pair was nesting; once ♀ entered box with nesting material whilst 2 birds (♂♂?) perched on top of box and a fourth bird was nearby (C.M. Perrins, pers. comm. 1998).

NEST: only 3 known, made of dry lichens, lined with snake skins (in nest boxes, Bwindi Forest, Uganda: C.M. Perrins, pers. comm.).

EGGS: 3 (1 clutch; also broods of 1 and 2 chicks). White, speckled with reddish spots (C.M. Perrins, pers. comm.).

LAYING DATES: Uganda, (Bwindi) Feb, Mar, July, (Rwenzori Mts) (nestling, Dec); Zaïre (adult attending at old woodpecker nest hole in stub of tree, June; ♂♂ in breeding conditiom, July), Mt Kabobo (inferred to lay from Dec, Prigogine 1960), Itombwe (inferred from gonad condition to breed May–July, Prigogine 1971).

DEVELOPMENT AND CARE OF YOUNG: at 5–6 days, young brooded by ♀ for most of time and fed mainly by ♂ (in 135 min period ♂ fed brood 17 times and ♀ 4 times); ♂ brings food to nest, or calls ♀ out of nest and gives her food to take to young (C.M. Perrins, pers. comm.).

Plate 5 (Opp. p. 78) *Parus griseiventris* Reichenow. Miombo Grey Tit. Mésange à ventre gris.

Parus griseiventris Reichenow, 1882. J. Orn., p. 210; Kakoma, Tabora district, Tanganyika

Forms a superspecies with *P. afer*, *P. cinerascens* and *P. thruppi*.

Range and Status. Endemic resident. In *Brachystegia* woodland: central Angolan highlands; Zambia (absent from extreme W and Luangwa valley); SE Zaïre (in Lualaba, Haut-Luapula, Tanganika, S Kwilu and SE Shaba districts); central and N Malaŵi (west of Shire R.); SW Tanzania, north to Kibondo and central Tabora district; Zimbabwe, central (north to Zambezi escarpment) and E highlands (500–1600 m; from Inyanga Nat. Park south to Chimanimani and Chipinga Mts); Mozambique in NE Tete Prov. Frequent to common throughout most of range, but local and uncommon in Tanzania. Occurs above 915 m on central plateau and at 500–1600 m in E highlands of Zimbabwe, at 1220–1900 m in Zaïre and 975–1950 m in Zambia. In Zambian miombo woodland, 14 birds per 100 ha (Alerstam and Ulfstrand 1977).

Description. ADULT ♂: forehead to nape black, faintly glossy; mantle and back grey; small dull white patch (sometimes absent) on upper mantle, separating cap from back; rump, uppertail-coverts grey to sooty black; tail black, feathers with pale grey tips, and pale grey outer edges to T4, T5 and T6. Cheeks, ear-coverts and sides of neck and breast pale buff-white; throat, upper belly, and midline of lower belly sooty black; rest of breast, belly and flanks off-white washed with buff; undertail-coverts grey, broadly tipped white. Wings black, with narrow white fringes to primaries and secondaries; wing coverts, tertials black to blackish brown, finely tipped and narrowly fringed white, underwing dull white. Bill black; eyes grey to brown; legs grey-brown to black and feet black with pale yellow soles. ADULT ♀: like ♂, but slightly duller throat and cap. SIZE (9 ♂♂, 9 ♀♀): wing, ♂ 74–88 (77·9), ♀ 74–77 (75·7); tail, ♂ 51–59·5 (54·9), ♀ 50–57 (52·9); bill, ♂ 10·8–12·1 (11·4), ♀ 10·6–12·3 (11·4); tarsus (sex?) 18·0–20·0 (Maclean 1993).

IMMATURE: like ♀, but less white on fringes and tips of tail. Fringes of primaries and secondaries pale yellow-buff. Black throat patch slightly duller and not extending to lower belly.

Field Characters. Length 14 cm. A black, white and grey tit endemic to miombo woodland. Stockier and *paler-looking* than Ashy Tit *P. cinerascens*, with smaller black bib,

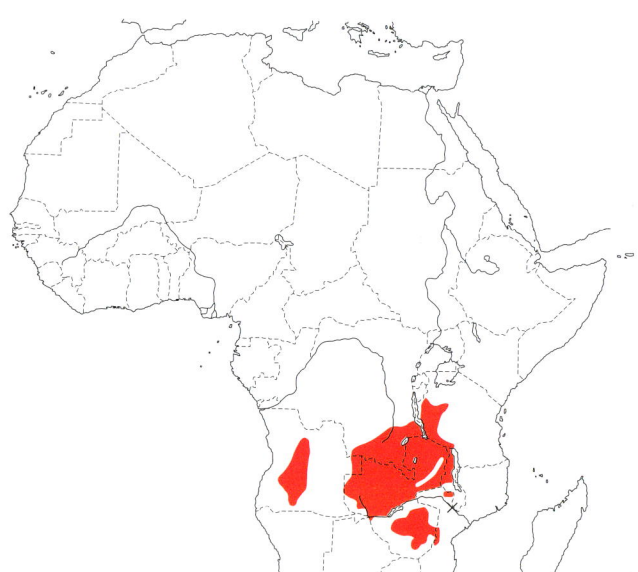

Parus griseiventris

white underparts faintly washed greyish buff, *distinct white bar on wings* formed by tips to median coverts, but *cheeks darker* (light brown rather than white); cap and bib sooty black, with little or no gloss. Ecologically separated from Ashy Tit in areas of potential overlap (Victoria Falls and Inyanga Park, Zimbabwe, and along E edge of Benguela escarpment, W Angola), Ashy inhabiting acacia steppe. Does not meet either Southern Grey Tit *P. afer* or Acacia Tit *P. thruppi*. Loud, grating call distinctive.

Voice. Tape-recorded (86, 91–99, 104, B, F, STJ). Common call a grating, rather low-pitched 'jrrr-jirrr-jirrr-jirrr-jirrr', with or without an introductory 'tit', 'plit' or more sibilant 'tsi', 'tsi-tsi-jerr-jerr-jerr'. Other calls include a less harsh 'cheerrr' and a down-slurred 'tsirrrrrr'. Song, various whistled phrases, e.g. 'toowee-toowee, toowee-teeyoo, toowee-teeyoo', 'tsi-trreeeoo-tweetoo' or 'trit, twee-too-you, twee-too you'.

General Habits. Frequents mature *Brachystegia* and (in E Zimbabwe highlands) *Uapaca* woodlands. In Zimbabwe, absent from most poor woodlands lacking distinct canopy (Irwin 1959). In Zambia, also occurs in mavunda and *Julbernardia* forest. In pairs and small groups; birds call often as they move quickly through trees. Sympatric with Rufous-bellied Tit *P. rufiventris* throughout much of range; tends to feed on larger branches and trunks of trees, where Rufous-bellied Tit forages on smaller twigs and branches in canopy (Alerstam and Ulfstrand 1977). Often in mixed-species foraging parties. Resident.

Food. Caterpillars, insect larvae, spiders.

Breeding Habits.
NEST: soft, thick pad of mammal fur, feathers and plant material; in tree holes, as well as old woodpecker and barbet holes, and occasionally fence posts, metal pipes and termite hills (of 24 Zimbabwe nests, 13 in holes in trees, 9 in pipes/fence poles and 2 in termite hills).
EGGS: 3–5. Oval; dull white, with reddish brown speckling. SIZE: ($n = 4$, Zaïre) $17 \cdot 1 – 18 \cdot 3 \times 13 \cdot 6 – 14 \cdot 2$; ($n = 19$, 4 clutches, Zimbabwe) $18 – 19 \cdot 5 \times 11 – 14$ ($18 \cdot 5 \times 13 \cdot 4$).
LAYING DATES: Angola, Oct; Zaïre, Sept; Zambia, Aug–Sept; Malaŵi, Sept–Oct; Zimbabwe, Aug–Dec.
BREEDING SUCCESS/SURVIVAL: reported to be frequent host of Lesser Honeyguide *Indicator minor* (Mackworth-Praed and Grant 1960).

References
Alerstam. T and Ulfstrand, S. (1977).
Clancey, P.A. (1996).
Harrap, S. and Quinn, D. (1996).

Parus cinerascens Vieillot. Ashy Tit. Mésange cendrée.

Plate 5 (Opp. p. 78)

Parus afer cinerascens Vieillot, 1818. Nouv. Dict. Hist. Nat. 20, p. 316; Palla's Drift.

Forms a superspecies with *P. afer*, *P. griseiventris* and *P. thruppi*.

Range and Status. Endemic resident, common to frequent in acacia savanna in S Angola (north along coast to Benguela area), Namibia (except SW and extreme N and NE), Botswana (very common in W and S, sparse in far N and NE), W and central Zimbabwe (north to Inyanga; spreading with clearing of miombo woodland) and South Africa (Harrison *et al.* 1997). Common; about 1 pair per 50 ha in acacia thornveld in Transvaal.

Description. *P. c. cinerascens* Vieillot: Namibia (except perhaps extreme NW), Botswana, Zimbabwe, Cape Prov., South Africa. ADULT ♂: forehead to nape glossy black; mantle and back grey; small white patch on upper mantle, separating cap from back; rump grey, uppertail-coverts greyish black; tail black, feathers all with white tips, white outer edges to T6 and variably to T5. White patch extends from lower mandible to cheeks, ear-coverts, sides of neck; chin, throat, most of breast and midline of belly dull glossy black; rest of breast and flanks grey, belly buffy grey – the greys paler and creamier in broad band next to black ventral stripe; undertail-coverts dull black, broadly fringed grey-white. Wings grey-black (fresh) to brown (worn, during breeding), feathers with narrow pale white to buff outer edges except when worn; tertials and wing coverts black, narrowly fringed dull white; underwing dark grey, tipped white. Bill black; eyes brown to dark brown; legs and feet dark blue-grey to black. Sexes alike. SIZE: wing, ♂ ($n = 14$) $75–83 \cdot 5$ ($78 \cdot 4$), ♀ ($n = 8$) $73–78$ ($75 \cdot 1$); tail, ♂ ($n = 10$) $52 \cdot 5–61$ ($56 \cdot 7$), ♀ ($n = 6$) $54–61 \cdot 5$ ($56 \cdot 9$); bill, ♂ ($n = 11$) $12 \cdot 5–15$ ($13 \cdot 6$), ♀ ($n = 6$) $13–14 \cdot 2$ ($13 \cdot 8$); tarsus, ♂ ($n = 5$) $21 \cdot 2–23 \cdot 4$ ($22 \cdot 6$), ♀ ($n = 3$) $22 \cdot 2–22 \cdot 5$ ($22 \cdot 3$). WEIGHT: (3 ♂♂, 3 ♀♀, Windhoek, Namibia, while feeding young, Jan–Mar) ♂ $20 \cdot 0–21 \cdot 7$ ($20 \cdot 8$), ♀ $18 \cdot 8–20 \cdot 6$ ($19 \cdot 9$); unsexed ($n = 3$) $18 \cdot 5–20 \cdot 1$ ($19 \cdot 4$).

Parus cinerascens

P. c. orphnus Clancey: Transvaal and Orange Free State. Darker above and flanks darker grey than *cinerascens*; underparts nearly lacking pale band next to black midline strip.

P. c. benguelae Hall and Traylor: SW Angola and perhaps extreme NW Namibia (Kaokoland). Paler, plainer and slightly smaller than *cinerascens*. SIZE: (2♂♂, 2♀♀, Angola): wing, ♂ 75·5, 76, ♀ 69, 72; tail, ♂ 57, 57, ♀ 52·5, 54·5; bill, ♂ 14·1, 14·6, ♀ 14, 14·1.

IMMATURE: like adult but cap duller black and mantle and back slightly browner; black on throat dull, not glossy.

TAXONOMIC NOTE: the 4 species here united in a superspecies, following Hall and Moreau (1970), are often treated as composing a single species (Clancey 1996). *P. cinerascens* is essentially parapatric with *P. griseiventris* and *P. afer* (with small zones of overlap with both: Harrison *et al.* 1997 and see superspecies map), whilst *P. thruppi* is allopatric. *P. thruppi* and *P. cinerascens* inhabit acacia; the intervening *P. griseiventris* inhabits miombo and the southern *P. afer* scrub.

Field Characters. Length 14·5–15 cm. Meets Miombo Tit *P. griseiventris* in Zimbabwe and W Angola but avoids miombo woodland, although they may occasionally be in the same bird party. *Underparts darker grey* than Miombo Tit, but *face patch pure white* contrasting with glossier black cap and bib; white edges to wing-coverts do not form bar. Overlaps with Southern Grey Tit *P. afer* in Namibia and NE Cape, where ecological separation not complete, although Ashy prefers acacia scrub and Southern Grey Tit Karoo scrub; further distinguished from Southern Grey by *grey back* and *flanks, white nape patch* and *white edges* to wing-feathers (Southern Grey has brownish back and wings, buffy flanks and nape patch). Loud ringing song distinctive.

Voice. Tape-recorded (88, 91–99, B, CHA). Song a liquid trill, variable in pitch and speed, 'tlu-tlu-tlu-tlu-tlu-tlu ...', sometimes beginning or ending with up-slurred notes, 'tutututututututu-tuwee-tuwee-tuwee'; 'tuwee' may be given separately, sometimes in 3 syllables, 'tu-*wee*-u'; may be heard at a great distance during the breeding season. Other calls include a down-slurred 'peeoo-peeoo-peeoo-peeoo', a dry chatter, 'jejejejejejeje' and a thinner, squeakier 'pstrrrrrrr'. For other calls see Harrap and Quinn (1996).

General Habits. Inhabits open acacia savanna and sparse Kalahari woodland throughout most of range, but deciduous woodland in W Transvaal. Largely confined to watercourses in Orange Free State. Along fringes of Namib desert in Namibia inhabits desolate acacia scrub, preferring taller vegetation along dry riverbeds. At elevations of 275–1220 m in Namibia, 915–1370 m in South Africa, and typically above 1200 m in Zimbabwe. Pairs move restlessly through trees and bushes, often joining bird parties (rarely in company of Miombo Grey Tit) in winter. Occasionally roosts at night in nests of Sociable Weaver *Philetairus socius* and Brown-throated Sand Martin *Riparia paludicola* (Maclean 1993), and in nestboxes. In central Namibia sympatric with Carp's Tit *P. carpi* and both species nest in same acacia-dominated habitat. However, Ashy tends to forage in poorer scrub and Carp's in better-developed acacia along watercourses. In central Namibia, sometimes forages on ground. Moults in Mar–Apr (Namibia); may moult when breeding during years with late rains.

Food. Mainly lepidopteran caterpillars and pupae, also beetles, moths and spiders. In Namibia, nestling diet almost exclusively lepidopteran larvae and pupae.

Breeding Habits. Monogamous, helpers suspected at some nests (W.R.J. Dean *in* Ginn *et al.* 1989), but no helping observed over 3 breeding seasons in central Namibia (pers. obs.). Territorial; ♂ typically sings from top of tall tree, and continues to sing vigorously throughout incubation period and at least into early nestling stage.

NEST: thick cup of plant material, mammal hair and feathers; in hole in tree, abandoned woodpecker or barbet hole, also in pipes and poles; generally 2–4 m above ground.

EGGS: 3–6 (Namibia), laid at 1-day intervals; Namibia mean (21 clutches) 4·6. Oval; dull white with brownish red flecks concentrated at broad end. SIZE: (n = 20, Namibia) 17–19 × 12–14 (18·5 × 13·8); (n = 11, S Africa) 17·8–19·7 × 13·6–14·2 (18·8 × 13·9).

LAYING DATES: Namibia, Oct–Mar (highly dependent on local rainfall); Botswana, Apr, Nov–Dec; Zimbabwe, Sept–Nov; South Africa, Oct–Dec.

INCUBATION: begins with 3rd egg or later; by ♀ only. Period: 14–15 days. ♂ feeds ♀ during egg laying and incubation; he approaches nest hole with food, calling softly and rapidly shaking wings; ♀ appears and shakes wings as she accepts food from him. She may then resume incubation or follow ♂ away on foraging trip. Rate of feeding by ♂ varies little during course of incubation (about 1 feed per h). On average, ♀ incubates 45·5 min per h.

DEVELOPMENT AND CARE OF YOUNG: weight of newly-hatched nestlings 1·2–1·6 g. Young brooded by ♀ until 6–9 days old. Both parents remove faecal sacs. Sheaths of primary feathers begin opening 12 days after nestlings hatch. Young leave nest 20–22 days after hatching. ♂ brings most food during first week, typically passing it to the brooding ♀ either at nest hole or just outside nest; thereafter both parents contribute equally, each bringing 1–1·5 food items per nestling per h.

BREEDING SUCCESS/SURVIVAL: in Namibia (and probably in other arid areas), breeding success dependent on rainfall. Of 24 Namibian nests, 12 failed (6 being abandoned, half during hot, dry weather) prior to hatching, and 12 fledged at least one young.

References
Clancey, P.A. (1996).
Ginn, P.J. *et. al.* (1989).
Hall, B.P. and Traylor, M.A. (1959).
Irwin, M.P.S. (1959).
Wiggins, D.A. (in press a, b).

Parus afer Gmelin. Southern Grey Tit. Mésange grise australe.

Parus afer Gmelin, 1789. Syst. Nat. 1, pt. 2, p. 1010; Cape of Good Hope.

Forms a superspecies with *P. cinerascens*, *P. griseiventris* and *P. thruppi*.

Plate 5
(Opp. p. 78)

Range and Status. Endemic resident of Karoo and sparse woodlands of extreme SW Namibia (north to Aus) and W South Africa (Cape Province, to Upington in NE and Port Elizabeth in SE; S Orange Free State). Reported disjunct population in central and E Namibia (e.g. Maclean 1993) requires verification (R. Simmons, pers. comm., D. Wiggins, pers. obs.). Upland Lesotho from Marakabei and Maseru east to Sani Pass; population in Lesotho (formerly thought to be Ashy Tit *P. cinerascens*) estimated to be 1000–10,000 birds (Bonde 1983, Osborne and Tigar 1990). Recorded in Natal before 1970, but not during atlassing work in 1970s (Cyrus and Robson 1980). Sparsely distributed throughout range, but in SW Cape 'fairly common' between Cape Town – Stellenbosch and Olifants R. (Hockey *et al*. 1989).

Parus afer

Description. *P. a. afer* Gmelin: Namibia, and Cape Prov. east to 24°E. ADULT ♂: forehead, crown, lores, upper cheek and side of neck and upper mantle black with slight bluish gloss; patch on centre of hindneck and border of upper mantle buffish white; rest of mantle to scapulars, rump and shorter uppertail-coverts dark ashy brown; longer uppertail-coverts blackish brown. Tail blackish brown; tips of T1–T2 with narrow buffish white fringes, and of T3–T6 with broader fringes; T6 also with broad buffish white outer edge. Broad band from base of bill across cheek and ear-coverts to side of neck buffish white. Lower cheeks to chin and throat black, with faint bluish gloss, extending as large bib to lower sides of neck and to ventral point on lower breast, and linked to ventral stripe on upper belly; sides of breast and flanks buffish brown; breast adjacent to black bib, belly and undertail-coverts pale buff; thighs buffish brown. Flight feathers dark grey-brown, outer webs narrowly edged pale greyish buff (except emarginated parts of primaries). Tertials dark grey-brown more broadly edged greyish buff, tips fringed buffish white. Alula and larger upperwing-coverts dark grey-brown, alula and primary coverts finely edged, greater and median coverts more broadly edged and tipped greyish buff; lesser coverts ashy brown. Axillaries and underwing-coverts ashy brown, the latter tipped light buff. Bill black; eyes dark brown or blackish; legs blue-grey or grey to blackish. ADULT ♀: like adult ♂, but cap less glossy, more sooty black; throat brownish black without gloss, bib on average slightly more restricted, and ventral stripe narrower. SIZE (10 ♂♂ 10 ♀♀): wing, ♂ 72–80 (75·5), ♀ 68–75 (71·6); tail, ♂ 55–62 (59·0), ♀ 53–59 (55·6); bill, ♂ 14–16 (14·8), ♀ 13–14 (13·6); tarsus, ♂ 20–22 (21·2), ♀ 19–21 (19·9). WEIGHT: (n = 14) 17·3–22·4 (20·2).

IMMATURE: juvenile like adult ♀ but crown slightly duller and less white on tips of tail feathers; 1st year bird separable from older adult by incomplete moult making contrast between old greater primary coverts, primaries and secondaries and fresh greater coverts and tertials.

P. a. arens Clancey: E Cape Prov., S Orange Free State, Lesotho, Natal. Upperparts browner and flanks and sides of breast darker and buffier than in *afer*, and stripe on upper belly much wider, c. 20 mm wide; size the same (Clancey 1996).

Field Characters. Length 13 cm. Endemic to southern Africa. Overlaps with Ashy Tit *P. cinerascens* in Namibia and NE Cape, but it prefers poorer, stunted scrub, while Ashy occurs in trees and bushes in well-developed acacia. *Browner above* than Ashy Tit, wings almost plain (few pale feather edges); nape patch buff, not white, and underparts *buffy brown* rather than grey; tail shorter. Voice also distinctive.

Voice. Tape-recorded (22, 39, 58, 75, 88, 91–99, B, F). Song, variable repertoire of sweet, mellow whistles; takes a phrase and repeats it, 'we-toolee-too, we-toolee-too', 'see-tyew-tyew, see tyew-tyew', 'tse-woo-woo, tse-woo-woo, tse-woo-woo', 'tuwee-tuwer-tuwer, tuwee-tuwer-tuwer', 'tuwee-tuloo, tuwee-tuloo', 'tsee-way-way, tsee-way'; apparently lacks trills of Ashy Tit (Harrap and Quinn 1996). Also gives 'chick-a-dee' type scolds, 'tsick-jer-jer-jer-jer-jer' or 'tsick-tsick-tsick-jer-jer'. Alarm a harsh 'chrrr'.

General Habits. Inhabits arid Karoo and fynbos vegetation in rocky, open country, as well as scrub in agricultural areas; in SW Cape rolling, low-lying country with remnant patches of renosterveld (small- and broad-leaved shrubs including *Elytropappus rhinocerotis*); bushy sandveld. South of Kimberley lives in open *Tarchonanthus/Acacia* scrub. Occurs at 400–945 m in Little Namaqualand in NW Cape Province, and up to 2745 m in Lesotho mountains. Forages on tree trunks, branches and among twigs and foliage: investigates bark crevices; hangs upside down to hammer at prey on underside of branch; pecks repeatedly at food item held under one foot; energetically taps acacia pods, particularly camelthorn *A. erioloba* (pods often dislodged and fall to the ground, noise alerting observer to bird's presence), and tears pod open to extract parasitic insect; often in mixed foraging flocks – in about one-third of mixed bird parties near Kimberley – and hunts with crombecs *Sylvietta*, Black-chested Prinias *Prinia flavicans*, Rufous-eared Warblers *Malcorus pectoralis* and Cape Penduline Tits *Anthoscopus minutus* (W.R.J. Dean *in* Ginn *et al.* 1989). Moves restlessly in pairs or family groups of 3–6 birds, calling frequently when foraging. Spends much time prying into bark and woody niches. Flight bounding. Sedentary.

Food. Insects: mainly caterpillars, also wasps and beetles.

Breeding Habits. Poorly known. Suspected to have helpers at some nests (Ginn *et al.* 1989).

NEST: thick cup of grass, hair, and feathers sited in holes in rocks, banks, trees (scarce in favoured habitat), fence posts, culverts, also down pipes and in abandoned buildings.

EGGS: 2–5; S Africa mean (6 clutches) 3·3. Oval; flat white, with reddish or grey speckling. SIZE: (n = 3, S Africa) 18·9–19·5 × 14·5–14·6 (19·2 × 14·6) (Maclean 1993).

LAYING DATES: South Africa, Aug–Mar (only July–Oct in SW Cape region).

INCUBATION: period *c.* 14 days.

References
Clancey, P.A. (1996).
Ginn, P.J. *et al.* (1989).
Harrap, S. and Quinn, D. (1996).
Maclean, G.L. (1993).

Plate 5
(Opp. p. 78)

Parus thruppi Shelley. Northern Grey Tit; Acacia Tit. Mésange somalienne.

Parus thruppi Shelley, 1885. Ibis p. 406; south of Burao, Somaliland.

Forms a superspecies with *P. afer*, *P. cinerascens* and *P. griseiventris*.

Range and Status. Endemic resident in *Acacia* scrub in E Uganda (Moroto area), Kenya (generally below 1000 m, absent in E, far W, and SW; isolated records from NW Lake Turkana region), extreme NE Tanzania (below 1500 m, to lowlands southeast of Mt Kilimanjaro), S and E Ethiopia (up to 2000 m), and Somalia (915–2100 m; uncommon in NW east to 49°E; local and uncommon in S, south of 05°S). Uncommon throughout most of range, frequent in S Ethiopia.

Description. *P. t. thruppi* Shelley: Somalia, Ethiopia. Top of head to hindneck glossy black with bluish tinge; narrow collar across border of upper mantle white, tinged cinnamon–buff; rest of mantle to scapulars and rump ashy grey; shorter uppertail-coverts dark grey, longer feathers black. Tail black; T2–T6 with narrow white fringe around tip, T4–T5 also with narrow white outer edge and T6 with broad white outer edge. Lore and around eye pale buff, merging with large buff-tinged white patch from base of bill through cheek and ear-coverts to side of neck. Chin and throat black, extending ventrally to rear of upper breast to form bib, and these linked to black cap by narrow black band across rear of side of neck; bib connects with black ventral stripe on lower breast and upper belly. Central breast and belly otherwise pale cinnamon–buff; sides of breast and flanks ashy grey; undertail-coverts white, tinged buff. Flight feathers blackish, with narrow whitish outer edge except on emarginated parts of outer primaries; secondaries also with narrow white fringe around tips. Tertials blackish, with broad buffish white fringe along outer edge and around tip. Alula and larger wing-coverts blackish; alula and primary coverts narrowly edged and tipped whitish, greater and median coverts fringed and broadly tipped white to form 2 prominent wing-bars. Lesser coverts ashy grey. Axillaries ashy grey; underwing-coverts buffish white. Bill black; eyes dark brown to reddish brown; legs dark grey to blackish. Sexes alike, but ♀ has slightly smaller and duller black bib. SIZE (10 ♂♂, 9 ♀♀): wing, ♂ 64–71 (67·2), ♀ 62–68 (63·9); tail, ♂ 45–51 (48·9), ♀ 44–50 (47·2); bill, ♂ 11–12 (11·6), ♀ 10–11 (10·9); tarsus, ♂ 18–19 (18·2), ♀ 17–18 (17·7). WEIGHT: unsexed (n = 1, Uganda) 12.

IMMATURE: like adult ♀, but slightly duller overall with no gloss on crown, back brownish grey, less white on tips and outer

Parus thruppi

fringes of T6, ventral stripe narrower and less distinct; 1st year adult separable from older birds by having mixture of old and fresh wing-coverts, tertials, secondaries and occasionally primaries. All juvenile tail feathers retained in 1st year adult.

P. t. barakae Jackson: SW Somalia, central and S Kenya, NE Tanzania, E Uganda. Very similar to nominate *thruppi* but has slightly paler mantle, reduced or absent nuchal spot, and flanks generally paler grey. SIZE (3 ♂♂, 2 ♀♀): wing, ♂ 63·5–68 (65·7), ♀ 61·5–63 (62·3); tail ♂ 49–50 (49·5), ♀ 46–48 (47); bill ♂ 11·2–11·7 (11·5), ♀ 10·7.

Field Characters. Length 11·5 cm. Smallest of the grey tits, confined to acacia country of NE Africa. Does not overlap with any of the others, and thus readily identified by plumage combination, unique in NE Africa, of *large white patch* on face from lores to cheeks surrounded by *black cap*, *bib* and *band* on sides of neck; black on breast continues as line down centre of belly. Buff line separates black cap from bill, unlike other grey tits.

Voice. Tape-recorded (B, 109, ZIM). Song, 3–7 clear, sweet liquid whistles, 'tuwee, tuwee, tuwee ...'; up-slurred, with emphasis on second syllable. Calls very similar to harsh series produced by Southern Grey Tit *P. afer*, but in some the timbre is definitely mellower, with almost disyllabic notes, 'tsi-tsi-tchuerr-tchuerr-tchuerr ...' (Dowsett and Dowsett-Lemaire 1993). Contact calls a buzzy 'chya-chya' or 'tsi-tsa-char' and a quiet hoarse 'chet, chet, chet' or 'chut'chut'chut'. In annoyance or agitation gives a 'chick-a-dee' type, grating and nasal, 'ssss, jaa-jawaa-jawaa-jawaa', or rapid gong-like 'tsi-tsi-jojojojojojojojo'.

General Habits. Frequents arid acacia woodland (and *Commiphora* scrub in Kenya), especially along watercourses. Travels in pairs and small groups, often in mixed-species foraging flocks.

Food. Insects: mainly caterpillars, but also wasps, beetles.

Breeding Habits. Barely known. Territorial; ♂ sings from tops of trees (Zimmerman *et al.* 1996).
  NEST: in hole in tree; 2 were 2·5 and 7 m above ground.
  EGGS: oval; white with small reddish-brown speckling and heavy blotching at broad end.
  LAYING DATES: Ethiopia, Feb–Apr (♀ with egg Apr, young in nest April); Somalia, Mar–June.
  DEVELOPMENT AND CARE OF YOUNG: unknown.
  BREEDING SUCCESS/SURVIVAL: nest in dead tree in Somalia robbed by snake.

References
Archer, G.F and Godman, E.M. (1961).
Harrap, S. and Quinn, D. (1996).

Parus major Linnaeus. Great Tit. Mésange charbonnière.

Plate 5
(Opp. p. 78)

Parus major Linnaeus, 1758. Syst. Nat., ed. 10, p. 189; Europe.

Range and Status. Palearctic and Oriental Regions, north to N Norway and Sakhalin I., south to NW Africa, Egypt (N Sinai), Sri Lanka and Indonesia. Resident, but eruptive in N of range.

Morocco, not uncommon in Mamora, widespread but sparse in the Rif, Moyen Atlas, Zaërs, Jbel Tazzeka, central plateau, Haut Atlas, Souss and Doukkala; occurs at Aoufour, Mamora and Marrakech; from coast up to 2000 m, also in palm groves at Errachidia. Algeria, widespread from Tell to Aurès; rare in Atlas Saharien (Ksours Mts up to 1700 m); scarce in Oranais (Tlemcen, Saïda, M'Sila, Moulay Ismael), Mitijda and Réghaïa. Tunisia, scarce, from Kroumirie to Le Kef, east to Sedjenane; records at Ousselat, Pont du Fahs, Mateur and L. Ichkeul.

Description. *P. m. excelsus* Buvry: Morocco to Tunisia. ADULT ♂: whole head including chin glossy black, except for large white patch from corner of bill to cheeks and ear-coverts, and for small whitish patch on nape and greenish hindneck. Mantle and scapulars olive-green, back, rump and uppertail-coverts bluish or greenish grey. Tail greyish blue, T6 largely white. Throat and broad, ragged line from breast to belly black; sides of breast and of belly, and flanks and thighs, yellow; undertail-coverts white, with blackish streak in midline. Wing feathers blackish, tertials quite broadly fringed white, secondaries and primaries narrowly edged blue-grey, greater coverts with outer vanes mainly greyish blue and both vanes white-tipped. Underwing-coverts and axillaries white or yellow-white. Bill black with greyish cutting edges, eyes dark brown; legs and feet blue-grey. Sexes alike, but ♀'s crown less glossy, and black band below cheeks narrower. SIZE: wing, ♂ (n = 62) 74–84 (77·0), ♀ (n = 22) 72–77 (74·2); tail, ♂ (n = 9) 59–68 (62·8), ♀ (n = 10) 56–66 (60·0); bill to skull, ♂ (n = 39) 12·9–14·5 (13·6), ♀ (n = 9) 12·8–13·9 (13·2); tarsus, ♂ (n = 9) 20·5–22·2 (21·3), ♀ (n = 9) 19·8–22·2 (20·7). WEIGHT: no data; (*P. m. major*, France, Apr–Sept) ♂ (n = 20) 16·5–20·0 (18·0), ♀ (n = 15) 14·0–22·0 (17·5).

IMMATURE: like adult but black parts dull, white cheek patch and white in wings creamy, and lacks black line under cheek patch joining blackish bib and black sides of neck; juvenile similar but crown washed with olive, cheeks yellowish, and chin and midline of breast and belly dark grey.

NESTLING: naked, pink, with plentiful dark grey down on head and upper back. Mouth orange, gapes pale yellow.

Field Characters. Length 14 cm. A large, chunky tit identified by solid black cap, white cheek patch, single broad white bar on wing, broad black bib continuing as *black line down centre of yellow* underparts. Juveniles have yellow cheek and nape patches, black line on underparts still present though less distinct. Caution: birds in rear view show small pale patch on nape which may cause confusion with Coal Tit *P. ater*.

Voice. Tape-recorded (62–73, 93, 105, B). Song of 2–3 notes, first highest, 'tea-cher, tea-cher, tea-cher' or 'pity-wee, pity-wee, pity-wee'; calls include a Chaffinch-like 'pink, pink', a dry 'char' and a churring scold.

General Habits. Inhabits woods in lowlands and mountains up to treeline, orchards, farmland and gardens; in Morocco mainly cork oaks *Quercus suber*, holm oaks *Q. ilex*, olives, *Argania spinosa*, and palm groves; in Tunisia mainly cork oaks, holm oaks and cedar *Cedrus atlantica*.

Keeps in pairs and, in winter, flocks of 6 birds or more. Spends most of time in trees, foraging amongst foliage and on small and large limbs; quite often comes to ground; readily attracted to bird feeders. Active, agile and rather restless (though less so than Blue Tit *P. caeruleus*, and more skulking than Blue Tit). Forages by gleaning; holds small fruit under foot, removes seed(s) and discards pulp; regularly opens foil-topped milk bottles (Europe); can use bill and feet together in feeding (e.g. pulls up hanging string with nut attached to end, standing on string loop by loop until nut in reach); only uncommonly feeds by hanging with feet under twigs, buds and inflorescences; beats insect against branch; snips butterfly wings off before eating it. Moves by hopping, jumping, clambering, sidling, creeping and making very short flights; flight fluent, with burst of loose wing-beats and momentary wing-closing, direct and level over short distances, erratic and undulating if of more than *c*. 30 m. Usually tame and confiding. Gently flicks wings and tail (Cramp and Perrins 1993).

Sedentary, but seems to occur in Réghaïa only in June–Feb (Ledant *et al.* 1981).

Food. Insects, particularly small caterpillars and beetles; also spiders; small seeds and fruits in winter. Huge body of knowledge on diets and foraging biology (and breeding biology) in Europe (Cramp and Perrins 1993), but little is known about diet in Africa.

Breeding Habits. Monogamous, territorial. Territory size (Israel, n = 33) 0.6–3.6 (1.5) ha. ♂ uses song display, singing from various heights in trees and shrubs, demonstratively turning in different directions, white outer tail feathers flashing. Threatens rival with folded wing-tips slightly raised (**A**) or, intensely, with wings half open and drooping and tail fanned (**B**), and uses Head-up posture (**C**). Copulation preceded by wing-shivering by both birds.

NEST: foundation made mainly of moss, with some dry grasses, cup thickly lined with wool, hair and sometimes feathers; placed in hole in tree or wall; readily uses nest boxes, in woods and around houses. Built by ♀.

EGGS: 5–12, usually 7–8. Sub-elliptical, smooth, not glossy. White, variably speckled or spotted with reddish brown on violet-grey under-markings, mainly around broad end. SIZE (n = 52) max. 20.0 × 13.5, min. 16.0 × 13.0; also (*P. m. major*, France, n = 137) 15.9–18.8 × 12.5–13.9 (17.45 × 13.5).

LAYING DATES: NW Africa, Mar–May (Mar–Apr in lowlands, Apr–May in highlands).

INCUBATION: by ♀ only. Period 12–15 days, av. (England) 13.7 days.

DEVELOPMENT AND CARE OF YOUNG: young brooded for first few days by ♀; later cared for and fed by both parents. Nestling period 16–22 (18.9) days in Europe but av. in Turkey (n = 133) 21.9 days.

BREEDING SUCCESS/SURVIVAL: success in Mamora, Morocco (3 nests with 24 eggs) 44% and in Ras el Ma, Morocco (4 nests with 28 eggs) 86%. In Europe, heavy predation by weasels *Mustela nivalis* and for that and other reasons success generally rather low. Greatest longevity 15 years.

References
Baouab, R.E. (1981, 1983).
Cramp, S. and Perrins, C. (1993).
Thévenot, M. *et al.* (1982).

A   B   C

Parus fringillinus Fischer and Reichenow. Red-throated Tit. Mésange à gorge rousse.   Plate 5

Parus fringillinus Fischer and Reichenow, 1884. J. f. Orn., 32, p. 56; foot of Mt Meru, Arusha, Tanganyika.   (Opp. p. 78)

Range and Status. Endemic resident, Kenya and Tanzania. Uncommon and local in SW Kenya, mainly between 1000 and 1600 m, but widespread in Masailand and fairly common in Mara Game Res. and Loita plains, Serengeti and Arusha Nat. Park. Extends northwest to Narok and northeast to Nairobi Nat. Park, where frequent in Athi R. Hippo Pool area (Turner 1977); east to Machakos and Amboseli; and southeast to Oloitokitok District. In Tanzania from Mwanza southeast through Masai steppe to Tarangire Game Res., Hanang, Kibaya and Dodoma.

Description. ADULT ♂: forehead, lores and around eye to chin, throat, cheeks, ear-coverts and broad band around hindneck cinnamon-rufous; crown to nape dark grey, feathers fringed brownish; mantle to shorter uppertail-coverts brownish grey; longer uppertail-coverts blackish. Tail feathers black, tips fringed white; T5–T6 also with white outer edges (narrow on T5). Upper breast and sides of breast cinnamon-buff, merging into ashy grey on belly and flanks; thighs and undertail-coverts buffish white. Remiges dark grey-brown; primaries and secondaries narrowly edged pale buff, tertials with outer edges and tips broadly fringed white. Alula and larger upperwing-coverts dark grey-brown; alula finely edged buffish white, primary coverts finely edged buff, greater and median coverts broadly edged and tipped white, forming 2 prominent wing bars; lesser coverts brownish grey. Axillaries pale greyish buff; underwing-coverts buffish white; inner borders of underside of flight feathers buffish white. Bill greyish or blue-grey with paler cutting edges and dark tip to upper mandible; eyes dark brown; legs dark grey, dark blue-grey or blackish. ADULT ♀: similar to adult ♂, but rufous on head and neck paler and duller. SIZE (4 ♂♂, 3 ♀♀): wing, ♂ 70–75 (73·5), ♀ 69–70 (69·7); tail, ♂ 49–51 (49·8), ♀ 45–47 (46·0); bill, all 12; tarsus, ♂ 18–19 (18·8), ♀ all 18.
IMMATURE: juvenile similar to adult ♀ but browner (less grey) on crown and upperparts.
NESTLING: unknown.

Field Characters. Length 11·5–12 cm. Endemic to *acacia country* of S Kenya and N Tanzania. *Rufous face, neck and breast* contrast with *dark cap* and black-and-white wings. Only other tit in its range is White-bellied Tit *P. albiventris*, which has all-black head and breast and white belly.

Voice. Tape-recorded (B, GREG, McVIC, PEA). Commonest calls a hard 'jik' or 'chick' (which may become a subdued, conversational 'chick-chick, chick-chick-chick ...' given in long series: Harrap and Quinn 1995), and a grating 'drrrrrrr' or 'drdrdrdrdrdr'; often in combination, 'tsik-a-drdrdrdr'. Possibly the same is a (contact?) call lasting 1.2 s described as 'chick-tuc-de-de-de-de-de-de-de-de-de-de-de-de-de' (sonagram in Halliman 1989). Other calls: lilting 'dadaweet-dadaweet-dadaweet-dadaweet', persistent 'see-er, see-er, see-er', buzzy 'bzee-zee-zee', thin 'tsi' or 'tsisisisisi', harsh 'zwi-zwi-zwi', low, conversational 'prr', hard rattling trill, and occasionally a few high-pitched sweet whistles.

General Habits. Inhabits stands of *Acacia xanthophloea* and similar woods in bushy grassland in medium rainfall regions. Occurs in pairs, foraging sometimes in mixed species flocks. Sedentary.

Parus fringillinus

Food. Unknown.

Breeding Habits. Probably monogamous, territorial, solitary nester.
NEST: one was in cavity behind bark of *Balanites aegyptiaca* tree; lined with fibres and plenty of down; tree was in area surrounded by occupied huts in middle of a veterinary station (Moreau and Moreau 1937).
EGGS: 3. White, with pale red-brown spots and speckles, with some intra-clutch variation. SIZE: (n = 3, one clutch) 16·6–18·3 × 13·2–14·1 (17·2 × 13·6).
LAYING DATES: E Africa, Region C, Feb, Aug; Region D Jan, Apr, Sept.
INCUBATION: ♀ sat so tight that she was lifted off eggs by observer; after eggs were measured she was replaced and resumed incubating.

Plate 5
(Opp. p. 78)

Parus funereus **(J. and E. Verreaux). Dusky Tit. Mésange enfumée.**

Melanoparus funereus J. and E. Verreaux, 1885. J. f. Orn., 3, p. 104; Gabon.

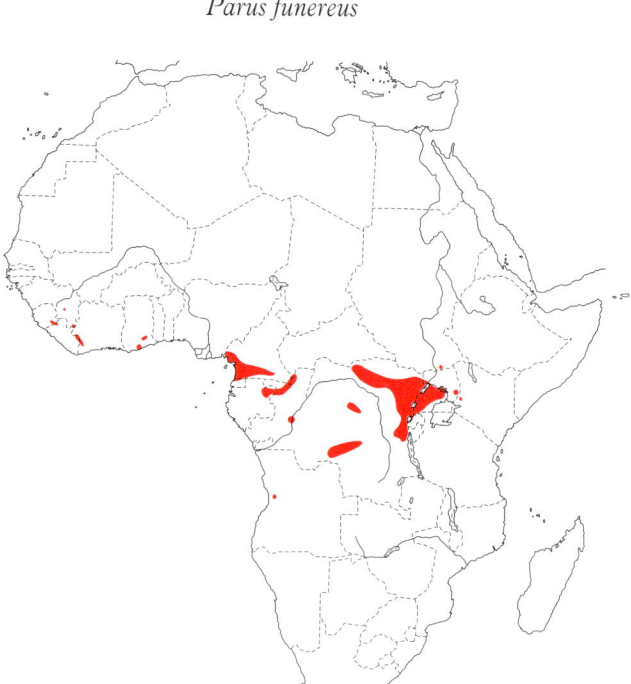

Parus funereus

Range and Status. Endemic resident, rain forest. Guinea, uncommon, above 800 m on Ziama Massif (Halleux 1994). Sierra Leone, Gola Forest. Liberia, R. Gbe, Voinjama, Ziatown, Zwedru, Kpelle Forest, and not uncommon on Mt Nimba up to about 800 m (Gatter 1997). Ivory Coast, rare: known from Mt Nimba and Taï Nat. Park; in Taï and Gola, recorded only in unlogged forest (Gartshore *et al.* 1995). Ghana, rare: Kakum Nat. Park (Helsens 1996), Worobong and Atewa Forest Reserves and old record from Elmina (Macdonald and Taylor 1977). Cameroon, scarce, in S, northwest to Korup Nat. Park (Rodewald *et al.* 1994). Equatorial Guinea (Fa 1990). Gabon, in NE on Ivondo R. at M'Passa, Bélinga and between Makokou and Bélinga, also Mékambo, and Olindi (60 km southeast of Makokou: Sargeant 1993); recorded regularly at M'Passa. Congo, widespread in Odzala and Nouabalé-Ndoki Nat. Parks. Zaïre, said to range from Gabon 'eastward across the whole lower Guinea forest to Uganda' (Chapin 1954) but, in the absence of definition we map it conservatively; widespread in NE and E as shown, mainly below 1550 m, north to Koloka (03°12'N, 24°28'E) and Medje (02°26'N, 27°17'E); ascends higher in hills west of L. Albert; records from southeast of Boende (Equateur Prov.) and several in Kasai Occidental Prov.; also between Kwamouth and Bolobo on lower Congo River. Central African Republic, 1 at Bayanga (02°54'N 16°23'E, Dzanga-Sangha Rainforest Res.: Green and Carroll 1991). Sudan, uncommon in secondary growth up to 2000 m in Imatong Mts. Uganda, uncommon and local, at 900–1700 m, from Bunyoro (Budongo, Bugoma) to SW corner (Lutoto, Kalinzu, Maramagambo, Bwindi-Impenetrable Forest), east to Mabira For. and R. Sezibwa. Kenya, confined to Kakamega and Nandi Forest where fairly common; formerly on Mt Elgon. Rwanda, common in dense forest in W, up to 2100 m, less common up to 2500 m (Uwinka). Angola, at Gabela and Conde (Cuanza Sul).

Density of about 25–30 birds per km² at M'Passa, Gabon, and of 4 groups per 4 km² at Ikessi, Congo.

Description. *P. f. funereus* (Verreaux): Guinea to Zaïre and Kenya. ADULT ♂: head and upperparts black, with slight oily green gloss, strongest on crown, feathers of upperparts with slightly greyer border when fresh. Tail sooty black. Chin to belly dull blackish, flanks, thighs and undertail-coverts sooty grey. Remiges black, P4–P7 finely edged off-white along proximal part of emarginations, inner primaries narrowly edged greyish, secondaries and tertials more broadly edged grey; alula and upperwing-coverts black, fringed grey. Axillaries dark sooty grey; underwing-coverts grey, with paler fringes; inner borders of underside of flight feathers white. Bill black; nostril surrounded by raised membrane (not apparent in skins; Bates 1909); eyes scarlet; legs grey or blue-grey to blackish. ADULT ♀: greyer than ♂; upperparts greyish black, feathers of crown fringed paler brownish grey (looking scaly), those of mantle edged with grey when fresh to give faint mottling; sides of head and neck and underparts ash grey, browner on sides of head, chin and throat, darker on sides of breast and flanks; eyes orange-red. SIZE (10 ♂♂, 8 ♀♀): wing, ♂ 85–92 (87·2), ♀ 81–86 (84·0); tail, ♂ 54–59 (57·3), ♀ 54–57 (55·6); bill, ♂ 11–12 (11·8), ♀ 12–13 (12·1); tarsus, ♂ 18–20 (18·9), ♀ 18–19 (18·6). WEIGHT: (Liberia) ♂ (n = 6) av. 25·6, s.d. 3·0; ♀ (n = 5) av. 25·2, s.d. 1·7; imm. ♂ (n = 5) av. 25·8, s.d. 1·4; 1 imm. ♀ 23·3; unsexed (n = ?) 22–29.

IMMATURE: juvenile like adult ♂ but duller, underparts and sides of head tinged brownish; greater and median coverts with triangular white spot at tip; eyes brown.

P. f. gabela Traylor: Angola. In ♂, throat and breast dull blackish slate; ♀ paler, more bluish slate below than in nominate race. Smaller: wing (2 ♂♂) 86, 86, (2 ♀♀) 77, 84.

Field Characters. Length 13–14 cm. The only tit in *lowland* and intermediate *rain forest*, readily distinguished by *entirely black* (♂) or *dark grey* (♀) plumage; eye *red*. Immature has 2 rows of white spots on wing. At borders of intermediate and montane forest in Albertine Rift meets Stripe-breasted Tit *P. fasciiventer*, which has pale belly and much white in wing. Noisy, often announcing the approach of a mixed-species flock of which it is a member.

Voice. Tape-recorded (32, 68, 104, B, C, ERA, FOR, KEI, LEM, STJ, ZIM). Typical call a sweet, ringing, slow 'here, see-here, see-here' (Kenya), clear but less ringing 'pyew', 'ss-pyew' or 'fui-tsiu' (Rwanda) or more abrupt 'tyomp', 'tyop' or 'see-tyop' (Liberia); the second note is down-slurred and lower than the first. Other calls include ringing 'wee-hooey', drawn-out, down-slurred 'peeeeyoo', a bubbling 'tyuyuyut', sharp 'wit' and thin 'seee' and a brief trill, 'teerrrr'; frequently given are a buzzy 'zit-zit' or 'see-zit' and nasal 'joey-joey', all with a distinctly drongo-like quality. Song a long, conversational medley of liquid whistles, twangy drongo notes, high, sibilant 'sisisi', rolled trills, burry notes and little chatters, occasionally with imitations of other birds. Also, 2 measured, buzzy notes, the first up-slurred, the second down-slurred, 'zzweee ... zweerrr', called 'song' (Chappuis, in press).

General Habits. Inhabits upper strata of dense primary lowland rain forest and treetops in secondary forest. When in mixed-species foraging flocks also visits other habitats such as isolated trees in forest clearings and forest outskirts (Colston and Curry-Lindahl 1986), and once seen in forest-grassland mosaic (Mt Nimba, Liberia). In Gabon occurs in very old plantations, borders of forest roads, and around margins of forest clearings, keeping to canopy and descending no nearer to ground than 20 m. Occasionally in parasol trees *Musanga* and fruiting fig *Ficus*. In Congo occurs in open-canopy forest and at N boundary of Zaïrean lowland forest occurs in small forest outliers and gallery woods. Coffee forest and gallery forests in Angola.

Occurs in bands of 3–15 birds, larger parties evidently consisting of 2 or 3 family groups. Typically in groups of 6–8. Party keeps to home range of 12–15 ha. Territory defended by singing; usually only one bird in flock sings, but several do so when 2 parties meet. Song like that of a drongo or like whistled notes of Fraser's Forest-Flycatcher *Fraseria ocreata* (Brosset and Erard 1986), which may be in fact imitated since song is a medley of mimicry, calls and churrs (Dowsett-Lemaire 1990). Vocal, with wide variety of calls. In Liberia nearly always occurs in mixed bird parties where it is generally one of the more conspicuous species; in Congo a noisy and very active leader of mixed bird parties (Dowsett-Lemaire and Dowsett 1998). Forages along upper boughs of tall forest trees, in manner typical of *Parus* tits: active and exploratory, hopping along and around branches, clinging acrobatically to probe bark, pecking at clumps of dead leaves, tearing apart rolled-up leaves to extract leaf-rolling caterpillar within, taking short flights within part of tree, and creeping through foliage. Tears at caterpillar and eats it piecemeal (Bates 1909). Bill is short, and wider at base than in other tits; besides caterpillars, bird appears to eat more hard insects than do its congeners.

Sedentary.

Food. Insects: caterpillars, beetles, microlepidopteran moths, orthopterans (Brosset and Erard 1986). 1 or more small caterpillars, mostly hairless, in each of 9 stomachs, and other insects (one a beetle) in 5 of them (Zaïre: Chapin 1954).

Breeding Habits.
NEST: only one found was in hole in dead tree, 15 m above ground, in primary forest; hole lined with lichens, thin stems and downy pappus (Prigogine 1972).
EGGS: 3. Creamy white, heavily blotched with red-brown. SIZE: (n = 3) 17·5–18·0 × 14·5–14·6 (17·8 × 14·6).
LAYING DATES: Liberia, (large gonads Nov); Ghana, (adults feeding fledgling Apr, adult with begging juv. May); Zaïre, (adults in breeding condition Mar, June, Sept); Uganda (fledglings, Oct).
DEVELOPMENT AND CARE OF YOUNG: group of birds typically contains 2 juveniles that are still begging for food 3 months after leaving nest (Dowsett-Lemaire and Dowsett 1998).

References
Brosset, A. and Erard, C. (1986).
Chapin, J.P. (1954).

Parus albiventris Shelley. White-bellied Tit. Mésange à ventre blanc.

Plate 5
(Opp. p. 78)

Parus albiventris Shelley, 1881. Ibis, p. 116; Ugogo, northern Tanganyika Territory.

Range and Status. Endemic resident, montane areas of Cameroon/Nigeria and montane and lowland ones in E Africa. Nigeria, sight records on Obudu Plateau. Cameroon, known from Mt Manenguba, Bamboutos and Banso Mts, Bamenda, Ndu, Tchabal Mbabo and N edge of Adamawa Plateau. SE Sudan, common above 1600 m in Dongotona Mts. Uganda, confined to NE (Karamoja) and E (W Elgon). Kenya, fairly common and widespread in W and S, in foothills and highlands from 1200 to 3400 m, north to Mt Elgon and Laikipia Escarpment and southeast nearly to coast; forest-island populations further north on Matthews Range, Ndotos, Mts Kulal and Nyiru. Formerly occurred regularly in one area in Tsavo East (Lewis and Pomeroy 1989). Tanzania, record on SW shore of L. Victoria; widespread from NE (Serengeti Nat. Park, Lake Manyara Nat. Park, Mkomazi Game Res.) south to Dodoma and Ruaha Nat. Park, southwest to Rukwa and Ufipa, and west to Tabora and Mwanza (mapped from 54 localities: N.E. Baker, pers. comm.).

Parus albiventris

Description. ADULT ♂: upperparts glossy black, tinged violet-blue. Tail black, feather tips fringed white, narrowly on T1–T3, more broadly on T4–T5; T6 with broad white outer web border and broad white area at tip. Lores, cheeks, ear-coverts and sides of neck black, tinged brownish. Chin, throat, upper breast and foreflanks black, with faint violet-blue gloss; feathers of lower border of black breast tipped white, especially at sides. Lower breast and hind-flanks to undertail-coverts white; thighs sooty black. Remiges black; outer primaries narrowly edged white proximal to emarginations, inner primaries and outer secondaries narrowly edged pale buff, inner secondaries more broadly edged whitish; outer edge of outermost tertial broadly white and of central tertial variably white; pale edges of secondaries and tertials extended narrowly around tips. Alula blackish, finely fringed white; primary coverts blackish, narrowly edged buff; greater coverts black, edged and broadly tipped white (fringes broader on inner feathers); median coverts white; lesser coverts black, larger feathers broadly tipped white. White on upperwing-coverts forms large white patch on closed wing; that on tertial and inner secondary edges forms a longitudinal stripe. Axillaries black; lesser underwing-coverts dark grey, rest of underwing-coverts white; inner borders of underside of flight feathers white. Bill black; eyes dark brown or blackish; legs blue-grey to blackish. ADULT ♀: similar to adult ♂, but duller, less glossy above, sides of head browner; throat and breast lacks gloss, typically greyer. SIZE (10 ♂♂, 10 ♀♀): wing, ♂ 79–85 (83·4), ♀ 74–82 (77·8); tail, ♂ 62–70 (66·2), ♀ 61–69 (64·5); bill, ♂ 12–13 (12·3), ♀ 12; tarsus, ♂ 19–20 (19·3), ♀ 19–20 (19·2).

IMMATURE: juvenile sooty brown above, chin to upper breast dark ashy brown; lower underparts buffish white; pale wing feather edges and tips tinged buff or (very young bird) yellow.

TAXONOMIC NOTE: although 2000 km apart, the 2 populations are identical. In SE Kenya, between Taveta and coast, birds are small (*P. a. 'curtus'*: wing, ♂ 75–77, ♀ 72–76) but otherwise identical to western populations. Almost exactly parapatric with White-winged Black Tit *P. leucomelas*: the 2 are separated altitudinally but seem to occur close together in W Kenya and Ufipa Plateau, SW Tanzania) (*contra* Sibley and Monroe 1990). May be consuperspecific with *P. leucomelas* (Short *et al*. 1990).

Field Characters. Length 14–14·5 cm. A black tit with *dark eye*, *white* lower breast and belly, much white in wing (almost solid on wing-coverts) and white edges and tip to tail. In Cameroon, W Kenya and SW Tanzania (Ufipa) closely approaches range of White-winged Black Tit, which is wholly black except for white in wing, and race *guineensis* in Cameroon and Kenya (sometimes considered a separate species, White-shouldered Tit) has *yellow eye* and almost no white in tail.

Voice. Tape-recorded (68, 104, B, C, HOR, McVIC, NOR, STJ). Song, given infrequently, rendered as 'chee-er-weeoo, chee-er-wheeo' (Zimmerman *et al*. 1996) or a repeated, warbling 'pee-pee-purr' (Harrap and Quinn 1996). Calls include a 'chick-a-dee' type, high thin notes followed by 3–6 low buzzy ones, 'tss-tss-tcher-tcher-tcher' or 'pt'tsi-tsi cher-cher-cher-cher'. Some calls not unlike those of Great Tit *P. major*, 'tch-ir tch-ir tch-ir' (van Someren 1956, presumably referring to the 'teach-er' call of Great Tit). Contact calls 'twach-twach-twach' or 'chrip-twsich-twach', also a sharp 'tss', 'tsee' or 'tss-tss-tss-tee;' alarm a harsh 'chirrr'.

General Habits. Inhabits edges of montane evergreen forest, gallery forest along river banks and streams, open woodland, orchard bush, scattered copses and trees in upland savanna and on high mountainsides; also shrubs in acacia woods, and long established gardens with indigenous trees. Keeps mainly near tree tops, just below canopy. Usually in pairs, sometimes in small family parties, often in mixed-species foraging flock. Active and restless; creeps and hops up, down, around and under large and small branches and twigs, searching fissures in bark, clumps of lichen, curled-up leaves, and clumps of epiphytes such as ferns and orchids (van Someren 1956). Most movements are lively and rapid; small flock flits from branch to branch, often flying only 1–2 m, intermittently giving small contact calls. At night roosts in tree holes, knot-holes, under sheet of dislodged bark, or in crevice between 2 upright, close-set trunks. Flock moving to roost is noisy and fussily takes time to settle down; several birds roost in same hole, foraging until last minute, and going in and out of hole until settled, and having last look out before disappearing from view. Most roost sites 13–26 m above ground; some as low as 1–2 m. Roost holes are mere depressions or recesses, or may be true, deep hole (van Someren 1956). Some roost holes later used for nesting, and *vice versa*.

Sedentary.

Food. Spiders and insects fed to nestlings, mainly caterpillars (hairless ones), some quite large; once a long-horned grasshopper.

Breeding Habits. Monogamous, territorial, solitary nester.

NEST: a scant pad of bark fibres, hair, fur and beard-lichen covering all or part of base of tree hole. Bird chooses rot hole in trunk or at end of broken branch, in rotting crevice where parallel trunks touch, and, frequently, holes and fissures in gnarled trunk of olive tree; sometimes old barbet's or woodpecker's nest hole. Size and shape of hole entrance varies considerably, but a narrow cleft just wide enough for tit to enter is preferred (van Someren 1956). One hole was 45 cm deep, and pair of birds prepared it for 10 days before eggs laid.

EGGS: 3–5. White, with sparse maroon speckles mainly at large end. SIZE: no data.

LAYING DATES: E Africa, Region C, Oct, Nov, Jan; Region D (18 records), all months except Oct, 6 records in Mar–Apr and 4 in Dec–Jan, suggesting breeding mainly in rainy seasons.

INCUBATION: in one nest 2 young hatched on 12th day.

DEVELOPMENT AND CARE OF YOUNG: when young 4 days old, adult ♂ once brought food item (a large green caterpillar), alighted 2 m from nest hole, quivered wings rapidly, called, flew to tree trunk well above nest then spiralled down into top of narrow nest crevice; ♀ flew straight to crevice, entered, called 'chir-chr-chrrr', and after 90 s reappeared carrying faecal sac which she dropped in flight several m away. ♀ made 7 more visits before ♂ returned. ♀ brought large, flat, hairless larva which the young rejected; she flew away with it and dropped it (when it was eaten by a White-eyed Slaty Flycatcher *Melaenornis fischeri*). ♂ sometimes entered this nest by a slit at the side. In 4 h ♀ made 15 provisioning visits and ♂ 8. When chicks 7 days old, parents made 37 visits in 5 h; most food items (spiders, caterpillars, a grasshopper) were brought one at a time; sometimes 3–4 caterpillars were brought at once. Young seen out of nest, in adjacent trees, 20 days after hatching when just able to fly but unable yet to follow parents around. At dusk that day the parents took 15 min to coax brood back into nest crevice for the night; ♀ entered with young and ♂ squeezed in 30 min later (van Someren 1956).

References
Harrap, S. (1996).
van Someren, V.G. L. (1956).

Parus leuconotus Guérin-Méneville. White-backed Black Tit. Mésange à dos blanc.

Parus leuconotus Guérin-Méneville, 1843. Rev. Zool. [Paris], 6, p. 162; Abyssinia.

Plate 5
(Opp. p. 78)

Range and Status. Endemic resident, highlands of Ethiopia and adjacently into Eritrea. In Eritrea, formerly frequent around Senafe and Guna Guna in mountains close to Ethiopian border (but no recent records: Smith 1957). In Ethiopia, frequent to common and widespread in highlands at about 3000 m, from Tigray and Simen to Harrar, Arsi (Arussi), Balé Mts, Shoa, Djamdjam and Djimma (Jima). As common in 1989 as prior to 1977 (Ash and Gullick 1989).

Description. ADULT ♂: top of head to sides of upper mantle glossy black, tinged violet–blue; rest of mantle to back and inner scapulars pale brownish white, feathers with blackish bases; outer scapulars, rump and uppertail-coverts black. Tail black, T4–T5 with faint whitish tips, T6 with fine white outer fringe and tip. Lores, supercilium, cheeks and ear-coverts sooty black. Chin, throat, sides of neck and breast black with faint violet–blue gloss; rest of underparts sooty black, undertail-coverts sometimes with whitish tips. Upperwing feathers sooty black; primaries (proximal to emarginations), secondaries, alula and primary coverts finely edged glossy blue-black, tertials and rest of coverts edged and tipped glossy blue-black. Axillaries and underwing-coverts sooty black; inner borders of underside of flight feathers whitish. Bill black, rather small; eyes brown; legs dark slate to black. Sexes alike. SIZE (10 ♂♂, 7 ♀♀): wing, ♂ 74–80 (76·9), ♀ 72–79 (74·8); tail, ♂ 55–62 (58·1), ♀ 55–60 (57·7); bill, ♂ 12–13 (12·4), ♀ 11–13 (12·3); tarsus, ♂ 19–20 (19·2), ♀ 19–20 (19·2). WEIGHT: (2, unsexed) 16·6, 16·8.

IMMATURE: juvenile similar to adult, but head and underparts duller and browner with little gloss; mantle with dusky brown mottling, buffish white practically confined to feather tips.

Field Characters. Length 13·5 cm. Endemic to highlands of Ethiopia and Eritrea. *All black* except for *white patch on back*. Only other tit in Ethiopia is White-winged Black Tit *P. leucomelas*, which lives at lower levels and is readily separated by large white patch on wing and no white on back.

Voice. Tape-recorded (109, BOR, MAC). Song motifs, usually repeated several times, 'pit-tyo-whee'; 'whee-dodo'; liquid 'whee-whee-whee-whee'; 'pi-ja-pi-dyow' (3 short notes and a hollow down-slurred one): and a 'duet' by 2 birds, down-slurred 'hay-ho' answered by upslurred 'chop-wee'; also noted: 'tchu, plit-kli chueee'u', repeated with little variation, either singly or in couplets, for well over a minute, at *c*. 2-second intervals; the 'tchu' very sparrow-like (Harrap 1996). In recording involving at least 2 birds, one gave rather mellow, bulbul-like song, repeated in short series 1–6 times, 'tchip-pi'kiu't' or 'plit, tchiu-p'du', answered by another bird with a 'chick-a-dee' call, 'tseeh-cherr-cherr', the initial note having a distinctive rising tone, and sometimes given alone (N. Borrow *in* Harrap 1996). Calls include buzzy 'bee-bzz-bzz'; for further vocalizations see Harrap and Quinn (1996), also sonagrams in Harrap (1996).

General Habits. Poorly known. Inhabits bushy wooded valleys and gorges, juniper, bamboo and giant heath forests, mainly at 2830–3140 m. Occurs in pairs and small parties, down to 1850 m (at Mensa, N Ethiopia) and up to about 3400 m (Amba Ras, Desfayes 1975). Presumed to be sedentary, though ringed bird recovered 223 km away (Ash 1994). No nests found. ♀ with oviduct egg, late Dec (Friedmann 1937). 1 bird lived 8 years (Ash 1994).

Parus leuconotus

Plate 7

Plate 8

Plate 5
(Opp. p. 78)

Parus rufiventris Bocage. Rufous-bellied Tit. Mésange à ventre cannelle.

Parus rufiventris Bocage, 1877. J. Sci. Nat. Lisboa 6, p. 161; Caconda, Angola.

Range and Status. Endemic resident in *Brachystegia* woodland. S Congo (in 'savanna bush') along Congo R., W (in Kinshasa area), S (southern Kasai Occidental) and SE Zaïre (Tshikapa area through S Katanga district), central and E Angola (from Huila and southern Cuando–Cubango in south, to Malanje district in west), extreme NE Namibia (*Baikiaea* woodland in Kaudom area and W Caprivi), extreme NW Botswana (a few records along Okavango R.: Harrap and Quinn 1996; 1 near Mohembo: *Babbler* 35, 1999, 37), Zambia (except S Southern Prov., E Barotse Prov., and Luangwa and Middle Zambezi valleys), Malaŵi (except extreme S), Zimbabwe (highlands from Harare to Inyanga Nat. Park and Chimanimani Mts), Tanzania (mapped from 90 records: N.E. Baker, pers. comm., 1998); N-central Mozambique (north of Zambezi). Common in Zaïre and Zambia, uncommon and local throughout rest of range.

Parus rufiventris

Description. *P. r. rufiventris* (Bocage): Angola, W Zaïre, NE Namibia, W and central Zambia. ADULT ♂: forehead and crown glossy black; entire head, neck and throat black, mantle and back slate grey, upper breast black fringed with grey ventrally, lower breast, belly, and undertail-coverts rich chestnut, rump and uppertail-coverts blackish grey. Tail black, tipped white (T5–T6), with white outer edges to T6 and T5. Wings brownish black with white-edged median and greater coverts. Tertials black, tipped white. Wing coverts blackish grey, tipped and fringed white. Primaries fringed white at base, forming white flash (Harrap and Quinn 1996). Underwing-coverts white. Bill black; eyes light yellow to sulphur-yellow; legs and feet grey to blue-grey. Sexes alike. SIZE: wing, ♂ (n = 5) 80–87 (84); tail, ♂ (n = 4) 63–67 (65·1); bill, ♂ (n = 4) 12·1–13·3 (12·6); tarsus, 18–20.

IMMATURE: like adult but duller overall, with greyish crown, greyish brown upper mantle and throat, brown wing with pale yellow fringes to feathers, brown tail, and pale chestnut breast, belly and flanks. Eyes black, grey or brownish grey, becoming yellow some time after first moult.

P. r. masukuensis Shelley: extreme SE Zaïre, E Zambia, Malaŵi. Like *rufiventris*, but upperparts generally paler, breast paler grey, and lower breast and belly paler pinkish cinnamon. Eyes pale yellow to brown. SIZE: wing, ♂ (n = 8) 80·5–87·5 (83·3), ♀ (n = 5) 77–80·5 (78·8); tail, ♂ (n = 7) 62–70 (66·9), ♀ (n = 3) 60·5–66 (63); bill, ♂ (n = 7) 11·9–13 (12·4), ♀ (n = 3), 11–11·8 (11·4).

P. r. pallidiventris Reichenow: Tanzania, N Mozambique, Zimbabwe and perhaps extreme E Zambia. Like *masukuensis*, but with paler grey breast and with pale, washed-out cinnamon on lower breast and belly. Eyes dark brown. SIZE (10 ♂♂, 6 ♀♀): wing, ♂ 79–86 (82·9), ♀ 74–77 (75·9); tail, ♂ 61–68 (64·9), ♀ 56–60·5 (58·2); bill, ♂ 11·2–12·9 (12·1), ♀ 11–11·9 (11·6).

TAXONOMIC NOTE: Sibley and Monroe (1990) split *P. pallidiventris* from *P. rufiventris* because of differences in eye colour (yellow in *rufiventris*, dark brown in *pallidiventris*, and light brown in 'hybrids' – Hall and Moreau 1970) and belly shade. However, belly shade appears to vary clinically (from pale cinnamon in Mozambique and Zimbabwe to rust in Zambia and Malaŵi to deep chestnut in Angola, Zaïre and Namibia). Dowsett and Dowsett–Lemaire (1993a), question the use of eye colour as a species isolating mechanism and correctly, in our view, question the validity of a supposed lack of interbreeding between 'hybrid' forms in E Zambia and Zimbabwe. We therefore do not follow Harrap and Quinn (1996), who treat *P. rufiventris* and *P. pallidiventris* as separate species. Brown-eyed birds may be juveniles (K. Hustler, pers. comm.).

Field Characters. Length 14–15 cm. Readily recognized by unique combination of *black head and breast* and *rufous belly*; eye yellow in nominate *rufiventris* and *masukuensis*, dark in *pallidiventris*, which has pale cinnamon belly and is sometimes considered a separate species, Cinnamon-breasted Tit; latter distinguished from sympatric White-bellied Tit *P. albiventris* by *grey back and breast* and *white wing-edgings* not forming solid patch. Range in Tanzania abuts (but does not overlap) that of Red-throated Tit *P. fringillinus*, which has rufous face and neck contrasting with black cap.

Voice. Tape-recorded (86, 91–99, 104, B, F, CART, MOY, WALK). Wide variety of grating, buzzy and sibilant notes, 'tsit', 'jaaa', 'tsit-terit', 'chizzeeya', 'jer-jer-tup', 'tsi-cher, tsi-cher', often combining to give 'chick-a-dee' calls: 'tsitsi-jaa-jaa-jaa-jaa', 'jit-jer-jer-jer-jer'; also more liquid, song-like 'weetlu-chu, weetlu-chu, weetlu-chu', 'worra-pee-chu, worra-pee-chu', 'tser-la-wee, tser-la-wee, tser-la-wee'. Alarm, harsh, grating 'churrr-churrr-churrr' or 'za-za-za-za'.

General Habits. Frequents *Brachystegia* woodlands, and *Burkea* (W Zambia), *Baikiaea* (NE Namibia), and *Uapaca* (E Zimbabwe). In pairs and small groups. Remains high in trees, often foraging on leaves and twigs near canopy. Foraging niche similar to that of Southern Black Tit *P. niger*, although the 2 species do not occur in same habitat.

Occasionally joins bird parties along with Miombo Grey Tit *P. griseiventris*. Resident, though reported to be locally migratory (Mackworth–Praed and Grant 1973).

Food. Insects, especially caterpillars.

Breeding Habits. Poorly known. A report of a group of 4 birds inspecting nest holes in miombo woodland in Angola suggests co-operative breeding.
  NEST: a soft pad of plant matter, fur, and feathers in natural tree hole or in old woodpecker or barbet hole.
  EGGS: 2–5. Oval; cream–white with reddish brown speckling and dark grey markings at the broad end. SIZE: (n = 12, Zimbabwe) 15·8–19·5 × 12·8–14·1 (17·2 × 13·4).
  LAYING DATES: Angola, Sept–Nov; Zaïre, Sept–Oct; Malaŵi, Oct; Zambia, Sept–Nov; Tanzania, Nov; Zimbabwe, Sept–Dec.

References
Alerstam, T. and Ulfstrand, S. (1977).
Dowsett, F. and Dowsett–Lemaire, F. (1993).
Harrap, S. and Quinn, D. (1996).

Parus leucomelas Rüppell. **White-winged Black Tit; White-shouldered Black Tit; Northern Black Tit. Mésange à épaulettes.**

Plate 6
(Opp. p. 79)

Parus leucomelas Rüppell, 1840. Neue Wirbelt., Vögel, p. 100; Halei Prov., Temben, Ethiopia.

Forms a superspecies with *P. carpi* and *P. niger*.

Range and Status. Endemic resident, woodland to about 14°N, 15°S and 43°E, outside rain forest.
  Common to frequent in mesic woodlands across northern tropics, south to woodland–savanna/rain–forest interface. Mauritania, scarce visitor in rainy season to S Guidimaka and upper Senegal R. valley. Senegal, widespread and quite common in S half of country, north to about 14° or 14°30′N. Gambia, rare on upper Gambia R., scarce to frequent on lower river. Common in Guinea-Bissau. In N Sierra Leone in Karine and Port Loko districts, Yana, Bintumane, Tingi Mts, Kamasigi and Banda Karafaya. Liberia (Gatter 1997). Widespread throughout Guinea. Mali, quite common in S, records north to Mopti (14°30′N). Ivory Coast, common, south to Lamto. Burkina Faso, known only from Arli-Pendjari Nat. Park, where frequent. Ghana, quite common, south through Volta Region to Ho (06°37′N), approaching coast in woods around inselbergs on Accra Plains, but absent from coastal thicket zone. Togo, south to Bismarckburg and Ebéva. Nigeria, frequent to common, north to Zaria, Sokoto Prov. and Falgore and frequent at Gaya (11°55′N), and south to Ilorin in SW and Enugu in SE. Niger, a few records in extreme SW ('W' Nat. Park, Sabongari); once Tillabéry (14°13′N). Cameroon, widespread, north to Yagoua (10°20′N) and south to Kounden, Bafia, Kombetiko and Nachtigal (04°21′N). Reported occurrence in Equatorial Guinea requires confirmation (Dowsett 1993). Chad, frequent in S soudanian zone, recorded north to Zakouma (about 10°55′N). Central African Republic, east to at least 20°E where common in Bamingui-Bangoran Nat. Park; also in extreme E. Sudan, one race fairly common south of line between Jebel Marra and Sobat R., another fairly common in Nile Province. Eritrea, occurs sparsely everywhere except Dancalia and coastal plains below 300 m. Ethiopia, frequent to common in Rift Valley, and occurs in adjacent lowlands of W and SE Highland Regions, also in S and presumed to occur in W Ethiopia. NE Zaïre, common in drainage of Uele and Bomakani rivers and to east and north, west to about Bondo. Uganda,

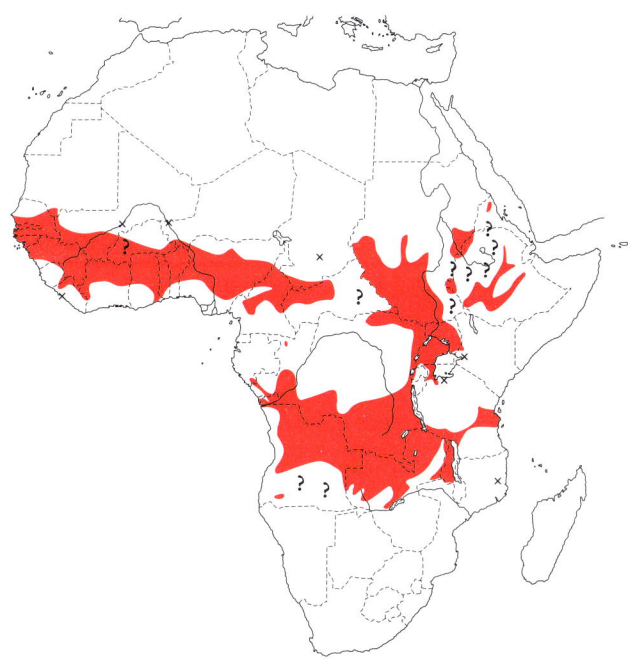

Parus leucomelas

common and widespread except in NE and extreme E. Kenya, records from Mt Elgon, Kapenguria, Saiwa Swamp, Bungoma, Kibingei and Nyando Valley; early record from about Kerio R. and one (only one south of Equator) from east of Kavirondo Gulf (Hall and Moreau 1970).
  In southern tropics occurs from coast to coast, south to 16°S. Gabon, confined to extreme S. Congo, Odzala Nat. Park; Brazzaville, Djambala and elsewhere in S savannas. Zaïre, scarce or local in E along Albertine Rift foothills but common in Semliki Valley, widespread in SW and whole of SE, common at least locally, in some areas not so common (Vincent 1949). Frequent in W Rwanda and W Burundi. Tanzania, in NW a record from Nyarumbugu

and probably frequent in Ngara distict; also recorded at Mwanza; in S, highlands from Ufipa and Njombe plateaus to Udzungwas and records at Kidugallo, Bagamoyo, Dar es Salaam and near Nguru hills. Angola, from Zaïre and Zambia border west to Malanje, Cuanza Norte and Cabinda and south to N Huila. Zambia, common in N half, south to Mazabuka, Chunga, Mankoya, Kalabo, South Lueti R., and in Eastern Prov. south to Chipata and Chadiza; up to 2060 m on Nyika Plateau. Malaŵi, from Viphya Mts to Dzonze Mt, at 900–1700 m. In Mozambique given as occurring in N (White 1963), and a single record from near Namapa (Hall and Moreau 1970).

Description. *P. l. leucomelas* Rüppell: central and SE highlands of Ethiopia. ADULT ♂: entire head and body glossy black, with violet–blue sheen. Tail glossy black. Flight feathers sooty black, primaries narrowly edged white (proximal to emarginations), secondaries edged white with fringe continuing narrowly around tip, white edge broad on S4–S6. Tertials black, outermost with narrow white outer edge. Alula and primary coverts black, the latter narrowly edged white; greater and median coverts with concealed blackish base and very broad white outer edge and tip; lesser coverts black. White on upperwing-coverts forms large patch on closed wing, and white secondary edges form broad longitudinal panel. Axillaries black; greater primary underwing-coverts whitish, rest of underwing-coverts black; inner borders of underside of flight feathers whitish. Bill black; eyes dark brown; legs black or very dark bluish grey. ADULT ♀: like adult ♂ but underparts less glossy. SIZE (10 ♂♂, 6 ♀♀): wing, ♂ 85–91 (87·7), ♀ 80–84 (82·2); tail, ♂ 68–76 (71·6), ♀ 66–71 (68·3); bill, 12–13 (12·3), ♀ 12; tarsus, ♂ 19–21 (20·3), ♀ 18–20 (18·8); also (21 ♂♂, 7 ♀♀): wing, ♂ 85–92, ♀ 80–94; tail, ♂ 63–70, ♀ 61–67; bill ♂ ♀ 13–14·5 (Hall 1960). WEIGHT: unsexed, (Ethiopia, n = 4) 21–24 (22·7) (J.S.Ash, pers. comm.).

IMMATURE: juvenile has gloss confined to a little on cap and mantle; sides of head and underparts sooty brown; white wing feather edges and tips tinged buff; fledgling has white parts of wing washed yellowish.

NESTLING: bill black with light greenish tomia, gape yellowish white, eyes grey at first, feet purplish grey, claws black.

P. l. guineensis Shelley (including '*camerunensis*' and '*purpurascens*'): N tropics from Senegal to Sudan, NE Zaïre, Uganda south to Equator, and W Kenya; birds of SW Ethiopia (near L. Turkana), NW Ethiopia and N Eritrea probably also this race, but eye-colour data are lacking. Like nominate race but eyes yellow or yellowish white, and underwing coverts mainly white (smaller inner feathers variably black). Small; wing, ♂ (W Africa, n = 10) 77–82 (79·2). Birds of central/SE Uganda ('*purpurascens*') and of Cameroon ('*camerunensis*') slightly larger. Wing, ♂, (W Africa, n = 30) 74–82, (Sudan, n = 16) 75–83, (Cameroon, n = 7) 78–87, (N Uganda and NE Zaïre, n = 10) 79–83, (central and SE Uganda, n = 12) 79–85, (SW Ethiopia, n = 6) 82–87 (Hall, 1960). WEIGHT: (Ghana, n = 10) 13·1–17·5 (16·1).

P. l. insignis Cabanis: S tropics, north to E Zaïre (to Rwenzori area) and SW Uganda (to Ankole, Toro and Masaka). Eyes dark brown as in nominate race, but sheen greenish blue, underwing-coverts all-white and edge to outer tail feathers white. Large; wing, ♂, (Zambia and Malaŵi) 88–95 (90·6); birds of Gabon/NW Angola and of SW Uganda/E Zaïre somewhat smaller. Wing, ♂, (Gabon/NW Angola, n = 12) 81–88, (rest of Angola, S Zaïre, Zambia, S Tanzania and Malaŵi, n = 34) 86–95, (SW Uganda/E Zaïre, n = 9) 83–89 (Hall, 1960).

TAXONOMIC NOTE: yellow-eyed and brown-eyed races are treated as separate species by Sibley and Monroe (1990) and Harrap and Quinn (1996), *P. guineensis* and *P. leucomelas*, composing a superspecies, but the occurrence of yellow-eyed and brown-eyed birds in the same foraging flock in Uganda is more likely a case of intergradation (White 1963) or of age differences, than an indication of specific difference (Dowsett and Dowsett-Lemaire 1993a). *P. leucomelas* has been treated by many authorities as conspecific with *P. niger*, even though Benson *et al.* (1959) showed that they are biologically distinct species. They were treated as forming a superspecies by Hall and Moreau (1970) (with *P. albiventris* and *P. leuconotus*) and by Short *et al.* (1990) (with *P. albiventris*); but now that it is realized how widely they overlap in central Zambia (R.J. Dowsett pers. comm. and Zambia Bird Atlas, in prep.), they are best regarded as independent species.

Field Characters. Length 16 cm. *All black*, with noticeable *gloss*, except for large *white patch on wing-coverts* (conspicuous in flight) and *white edges* to rest of wing-feathers; eye pale (race *guineensis*) or dark (other races); edges and tip of tail with some white but much less than in Southern Black Tit *P. niger*. Parapatric White-bellied Tit *P. albiventris* told by white belly. Very similar to Southern Black Tit and Carp's Tit *P. carpi*. Where it meets Southern Black Tit in Zambia they are separated ecologically, White-winged occurring in scattered bushes in open country at higher altitudes, Southern Black in mopane and other dry woodland (Harrap and Quinn 1996). Plumage darker than Southern Black Tit, especially belly and undertail-coverts which are grey in Southern Black; white shoulder patch more solid than in Southern Black, which shows black in most greater coverts, and white edges to secondaries and tertials broader, forming pale panel. Juvenile White-winged is dark sooty brown without any gloss; juvenile Southern Black is much paler, with white undertail-coverts and more white in tail. For differences from Carp's Tit, see that species.

Voice. Tape-recorded (68, 86, 104, 108, B, BRU, CART, LEM, MOR, MOY, SMIT, WATSN, ZIM). In southern race *insignis*, typical call (alarm?) a short, harsh, rather deep trill, 'ch-dzrrrrrrr-ya', 'chut-d'r'r'r'du' or 'chit-z'z'z'z'rr' (see sonagrams in Dowsett and Dowsett–Lemaire 1993a and Harrap 1996). This is considered by Harrap (1996) to be a 'chick-a-dee' call, *contra* Harrap and Quinn (1996). A possible variant 'chick-a-dee' call is a combination of a short, high-pitched note and 1–2 low frequency ones, 'chit-zorrrr-zorrrr', 'chit-zorrrr' or 'chut-tzi'orrr' (sonagrams in Harrap 1996). Also gives a 'teach-er' call reminiscent of Great Tit *P. major*, 'pss-tyoo-yoo, pss-tyoo-yoo', and a shrill, explosive 'chut' or 'chit'. Song a sweet whistled, unhurried phrase, 'wheee-to-trrrr-tooee', described as 'recalling lark or chat and not at all tit-like, perhaps more mournful than song of Southern Black Tit' (Harrap and Quinn 1996).

Western race *guineensis* has a wide vocabulary, see sonagrams in Harrap (1996). Song, sweet whistled phrases, sometimes of 3–4 notes, 'way-teetuwer-teetuwer', 'tee-tooyoo, tee-tooyoo, tee-tooyoo', 'tee-tyoo-puwer, tee-tyoo-puwer', sometimes 2 notes, 'tyiyi-tyoowi-tyoowi-tyoowi-tyoowi', or 'teacher'-like 'tee-tew, tee-tew'; whistles can be combined with grating noises in the same phrase, 'teekrolo-zzheeyi'. 2 long sequences of complex vocalizations recorded by C. Chappuis are described as 'chant en duo avec notes vibrées spécifiques'. At first glance these are quite unlike the utterances of any Holarctic tit, but may in fact be elaborate 'gargles' similar to those given by many tits in the subgenus *Poecile* (Harrap 1996). Calls include a sweet, upslurred 'toowee', a nasal, finch-like rising 'dzwee' or 'dzwee-zee', which in excitement or alarm is accelerated into a slightly higher-pitched rapid crescendo, a very thin, silvery 'sisisi-pee' and a lower-pitched, slightly grating 'ruwi-uwi' (Harrap 1996). Also has harsh grating 'zizizizi-zhaa-zhaa-zhaa-zhaa' and 'dzi-jraa-jraa', and a 3 syllable 'Reykjavik, Reykjavik'.

General Habits. In W Africa inhabits mature guinean and soudanian savanna woodland; shuns areas degraded by farming; in Darfur (Sudan) keeps to thick bushes on plains and in dry watercourses. In Eritrea, upland juniper woods, lowland riparian thickets and acacia woodland; in E Africa, wooded and bushy grassland, well-treed farmland and gardens up to 2000 m, also, in S Uganda, thickly wooded and forested country; in Angola, coffee forest and *Brachystegia* woodland; and in Zambia open country with scattered bushes, often at edge of grassland on plains or damp, grassy ground. In Malaŵi *Brachystegia*-grassland ecotone and montane grassland with scattered bushes.

Occurs in pairs and family parties of 3–4 and up to 6 birds, foraging often in mixed-species flock, including e.g. *Sylvietta*, *Eremomela* and *Hyliota* warblers and Spotted Creepers *Salpornis spilonota*. Restless, vocal, not particularly shy. Flight short and undulating. Foraging party moves quite quickly, bird moving constantly along twigs and through leafy bushes and small trees, diligently searching for insects, tapping and probing bark and inflorescences, exploring acrobatically, inspecting vertical surfaces by clinging sideways, reaching across with bill, hanging momentarily upside down, then flying with rest of party to adjacent tree; birds often calling as they fly. Occasionally makes very short flight after insect on the wing.

Sedentary, but said to be a wet-season (summer) visitor in Mauritania.

Food. Largely caterpillars. Caterpillars repeatedly brought to nest (Gambia), and small hairless ones in 4 adult stomachs (Zaïre) and a beetle grub in one. Party once seen feeding on custard-apple fruit *Annona*.

Breeding Habits. Poorly known; evidently a monogamous, territorial, solitary nester. ♂ said to rise in air from top of bush when it sings (Bannerman 1948).

NEST: soft pad of chaffed grass with a few bits of lichen, in knot-hole in tree-stump or stump of branch, or behind tiny fissure in side of trunk, 1–4 m above ground. Tree-stump may stand in open. One hole was 5 × 6 cm diam. and 60 cm deep (Vincent 1949).

EGGS: 3–4. White or cream, heavily spotted with brown or red-brown, with grey undermarkings towards large end. SIZE: (n = 3) 19·2–20·2 × 14·0–14·2 (19·8 × 14·1).

LAYING DATES: Gambia, (courtship, and bird entering tree hole, Jan); Nigeria, (nestlings, Feb, Apr; adult carrying food, Mar); Ghana, (breeding condition Jan–Feb, dependent young Apr, July); E Africa, Region B May, July, Region C Dec; NE Zaïre, Mar–Apr, Shaba, Oct–Nov; Angola, Aug–Oct; Zambia, Sept–Dec; Malaŵi Sept–Nov.

INCUBATION: by both sexes.

References
Hall, B.P. (1960).
Harrap, S. and Quinn, D. (1996).

Parus carpi Macdonald and Hall. Carp's Tit. Mésange de Carp.

Plate 6 (Opp. p. 79)

Parus niger carpi Macdonald and Hall, 1957. Ann. Transvaal Mus. 23, p. 32; Warmquelle, Kaokoland, South West Africa.

Forms a superspecies with *P. leucomelas* and *P. niger*.

Range and Status. Endemic resident. Frequent to common in mopane and acacia woodland. S Angola (west of and below Escarpment) north to Sumbe (Novo Redondo, 11°S) in Cuanza Sul Prov., south along coast to Namibe (Mossamedes), and east to Quilengues and Sá da Bandeira in Huila Prov. Found in NW Namibia (Kaokoland) east of the coastal strip, south along fringes of Namib desert to Hardap Dam and Mariental, and east to Etosha Nat. Park, Otavi Mts, and Waterberg plateau.

Description. ADULT ♂: entire plumage glossy bluish black, with white median coverts, most greater coverts white-edged, and thin, white outer edges to primaries and secondaries. Wings appear dark brown when worn (breeding), otherwise dull black. Tertials with wide white outer edges. Tail black with white tips and white outer edges (T6 and variably T5). Undertail-coverts and vent black, occasionally with narrow white fringes. Bill black; eyes dark brown; legs and feet greyish black. ADULT ♀: like ♂, but with duller black plumage around head and back, and brownish face. SIZE: wing, ♂ (n = 7) 76·5–82 (79·8), ♀ (n = 8) 72–77 (75·1); tail, ♂ (n = 4) 65–68 (66·5), ♀ (n = 5) 59–63 (61·2); bill, ♂ (n = 3) 11·4–12·2 (11·7), ♀ (n = 2) 10·6–11·6 (11·1); tarsus, ♂ (n = 7) 21·8–23·3 (22·5), ♀ (n = 8) 20·5–22·6 (21·4). WEIGHT: (7 ♂♂, 8 ♀♀, Windhoek, Namibia, while feeding young, Jan–Mar) ♂ 18·7–20·3 (19·4), ♀ 17·2–19·4 (18·2).

IMMATURE: resembles adult ♀, but duller overall. Recently fledged young has yellow at base of bill and pale yellow fringes to primaries and secondaries. 1st year breeding birds said to be separable by contrast between retained (brownish) and newly (black) moulted secondaries (Harrap and Quinn 1996).

TAXONOMIC NOTE: originally described as a subspecies of *P. niger*, and subsequently of *P. leucomelas* (Clancey 1972), this form is now generally treated as a full species, based on its voice (but see Dowsett and Dowsett–Lemaire 1993a), lack of sexual dimorphism, and habitat (e.g. Clancey 1995).

Field Characters. Length 15 cm. Confined to Angola and Namibia, where it abuts ranges of both Southern Black Tit *P. niger* and White-winged Black Tit *P. leucomelas*. Smaller than both, with shorter tail and smaller bill; appears slightly square-headed with full nape (Harrap and Quinn 1996). Differs from adjacent race *xanthostomus* of Southern

Parus carpi

Black Tit in having *more white in wing*: median and inner greater coverts solid white, and another white patch on secondaries and tertials. Carp's is otherwise almost wholly black, including belly and vent (Southern Black has paler belly, grey vent), undertail-coverts black or with narrow white tips (broadly white-tipped in Southern Black), and with less white on outer tail feathers. Carp's has less pronounced sexual dimorphism: ♀ overall duller than ♂, face and throat appearing browner in good light. Immature Carp's is sooty brown, much darker than immature Southern Black. Has *less white in wing* than White-winged Black Tit; outer 4–5 greater coverts show dark centres, whereas these are white in White-winged Black, producing a larger solid white area on wing. Assuming that all field identifications are correct, overlaps with Southern Black Tit in Cunene R. valley between 13° and 14°E, and on Waterberg Plateau between Otjiwarongo, Tsumeb and Grootfontein (Harrison *et al.* 1997). However, 'we have yet to see a Southern Black Tit on Waterberg Plateau' (D. Wiggins, pers. comm.).

Voice. Tape-recorded (91). Voice 'typically that of *P. niger*' (Dowsett and Dowsett-Lemaire 1993a). Normal call (alarm?) a buzzy, grating 'jer-jer-jer-jer ...' often preceded by single or double 'tsi', 'tsi-tsi-jer-jer-jer-jer ...', i.e. a 'chick-a-dee' call similar to that of Southern Black Tit; also gives a medley of whistled and squeaky notes. Other calls include bulbul-like 'churia-churia-churia', a clear, whistled 'piu-piu-piu-piu-piu' or 'witch-a, witch-a, witch-a ...', and a squeaky, slurred 'sii-u, si-u, si-u' (Harrap 1996). Song of Carp's Tit said to be very 'quiet' or 'soft', in stark contrast to the loud ringing song of Ashy Tit *P. cinerascens* (D. Wiggins, pers. comm.).

General Habits. Occurs in mopane woodland in N-central Namibia and S Angola, thickets in coastal grasslands of SW Angola, and acacia scrub and woodland in central Namibia. Recorded at 1220–1830 m in Angola, and normally above 1000 m in Namibia. Generally keeps to large, mature trees, particularly along dry watercourses. Pairs remain together during winter (Namibia). When not breeding, often in small groups (up to 6 birds: unclear whether family groups or temporary coalitions of unrelated adults). While foraging, pairs and family groups maintain vocal contact with short, buzzy calls. In central Namibia, usually forages in large trees along river-courses; also in adjacent acacia scrub and among rocks on cliff faces. In successive years in NW Namibia, ♀ roosted at night in horizontal metal pipe 2 m above ground; ♂ accompanied her to roost site, but left area after she entered pipe. Mainly sedentary, but disappears from some breeding areas and occurs in suburban gardens in winter (R.E. Simmons *in* Harrison *et al.* 1997).

Food. Lepidopteran larvae, insect pupae, termites, spiders.

Breeding Habits. Monogamous, helpers reported to occur at some nests (e.g. C.F. Clinning *in* Ginn *et al.* 1989) with helpers taking no part in nest building or incubation, but feeding fledglings. However, observations of colour-ringed individuals in central Namibia over 3 years gave no indication of helpers there (pers. obs.). Prior to and during egg-laying, ♀ follows ♂ on foraging trips, foraging on her own, but also soliciting food from ♂. ♂ also courtship feeds ♀ at nest during egg-laying and incubation.

NEST: thick cup of plant material, grasses, hair, snake skins and feathers; in natural hole in tree, old woodpecker or barbet hole, and in open pipe or nest box. Built by ♀ alone. One Namibian nest was in iron road-notice pole 50 mm in diam., with nest cup 2 m below entrance hole.

EGGS: 2–5; (Namibia, 12 clutches, av. 4·1); normally laid at 1-day intervals. Oval; white to pale pinkish buff with reddish brown speckling, mainly at broad end. SIZE: (n = 9, Namibia) 16–18 × 13–14 (17·6 × 13·8)

LAYING DATES: breeding highly dependent on rainfall: Namibia, Oct–Apr, usually Jan, also May and July.

INCUBATION: begins with last or penultimate egg; by ♀ alone. ♂ provides ♀ with food; when arriving with food, ♂ calls ♀ off nest with soft, buzzy call or with quiet song; after luring incubating ♀ out of hole he passes food to her. Courtship feeding is sometimes at nest hole, but usually several m away; both sexes rapidly quiver wings, like begging juveniles; generally 1–2 courtship-feeds per h, rate remaining constant throughout incubation period. In first 4 days, ♀ may spend considerable periods off the nest foraging. Thereafter, ♀ spends 42–50 min per h in nest hole. Period: *c.* 13–15 days.

DEVELOPMENT AND CARE OF YOUNG: for first week after young hatch, ♂ brings majority of food, often passing it to ♀ outside nest hole. ♀ broods young 45 min per h during first 6–8 days. Thereafter, parents contribute equally to feeding young. Combined feeding rate of parents increasing only slightly during nestling stage, from 1·5 to 2·0 feeds per nestling per h. In central Namibia, over 90% of food items were caterpillars and other insect larvae. Young increase in weight from 1·2–1·4 g at hatching to 17–20 g at 18 days. Both parents remove faecal sacs. Young remain close to nest for first day, then each day move further from nest site, remaining in loose flock. They are fed by parents for at least 3–4 days (and probably for much longer) after fledging.

BREEDING SUCCESS/SURVIVAL: in Namibia, breeding success largely dependent on rainfall. Parents may abandon nest (eggs or young) if insufficient rain falls following clutch initiation. In Namibia, of 14 nesting attempts, young nestlings disappeared from 4 nests, and 9 produced at least one fledgling. 3 of the 5 nest failures were due to apparent desertions during hot, dry weather.

References
Clancey, P.A. (1972, 1995).
Hall, B.P. (1960).
Harrap, S. and Quinn, D. (1996).
Wiggins, D.A. (in press a, b).

Parus niger Vieillot. Southern Black Tit. Mésange noire australe.

Plate 6
(Opp. p. 79)

Parus niger Vieillot, 1818. Nouv. Dict. Hist. Nat. 20, p. 325; Sunday River, Cape Province.

Forms a superspecies with *P. carpi* and *P. leucomelas*.

Range and Status. Endemic resident. Frequent to common in broad-leaved woodland in SE Angola (Cuando Cubango Prov.), NE Namibia (Caprivi and west to about Okavango and Caprivi districts, where common), N and E Botswana (south to Kanye), at low elevations in S Zambia (north to Senanga, Nangweshi, and Lupuka in Western Prov., throughout Southern Prov., north to Chunga and Chilanga in Central Prov.) and E Zambia (north to 10°45'N in Eastern Prov.), locally in Songea, SW Tanzania (Harrap and Quinn 1996), S and NW Malaŵi (Shire R. valley north and west to Kasache, and from Kasuni to Katumbi, plentiful only in mopane and acacia woodlands); Zimbabwe (except E highlands); central and S Mozambique (common in Tete district and on Manica plateau); Swaziland and SE South Africa (Transvaal, Natal and E Cape). Density of *c.* 45 birds in 400 ha, N Transvaal (Tarboton 1981); est. population *c.* 500,000 (*Afr. Birds & Birding* 4 (2) 1999, 54).

Description. *P. n. niger* (Vieillot): E. Cape Prov., Natal, SE Swaziland and extreme S Mozambique. ADULT ♂: entire head, mantle, back and uppertail-coverts glossy black; tail black, feathers with white tips, T6 and variably T5 with white outer edges. Chin, throat, and breast dull black, merging to sooty black-grey on belly. Undertail-coverts sooty grey with white tips. Wings dark blackish brown with white leading edges to primaries, secondaries and tertials, white median coverts and white-edged greater coverts. Bill black; eyes brown; legs and feet black. ADULT ♀: like ♂, but much duller overall; upper breast dull black, merging to sooty black-grey on lower breast and belly. Tail black, feathers with white tips, T6 and variably T5 with white outer edges. Undertail-coverts sooty grey with white tips. Wing blackish brown with white leading edges to primaries, secondaries and tertials, white median coverts, and white-edged greater coverts. Bill black; eyes brown; legs and feet black. SIZE (large series: Clancey 1972): wing, ♂ 82–85, ♀ 76·5–81; tail, ♂ 69–73·5, ♀ 64·5–71; bill ♂ 12–13, ♀ 11·5–13, tarsus (n = 53, sex and race not specified) 17–21 (Maclean 1993). WEIGHT (11 ♂♂, 11 ♀♀, race not specified): ♂ 19·7–26·0 (21·9), ♀ 17·2–24·8 (21·1) (Maclean 1993).

IMMATURE: like adult ♀ but with duller, brownish grey breast and belly.

P. n. ravidus Clancey: Zambia east of about 28°10E, Malaŵi, Zimbabwe except NW and SW, Mozambique (except lower Limpopo). Extensive white in secondary coverts. SIZE (large

Parus niger

series: Clancey 1972): wing, ♂ 82–88, ♀ 80–82; tail, ♂ 70–75, ♀ 70·5–75; bill, ♂ 12·5–13·5, ♀ 11·5–13.

P. n. xanthostomus Shelley: Angola, Namibia, Botswana, Zambia west of about 28°10'E. Poorly differentiated from *niger*, but larger, and ♂ and ♀ have slightly paler underparts (sides of breast and flanks). ♀ with slightly darker underparts that ♀ *xanthostomus*. SIZE: wing, ♂ 85–89, ♀ 80–84; tail, ♂ 72·5–78, ♀ 70–77; bill, ♂ 12·5–13, ♀ 11·5–13.

Field Characters. Length 16 cm. A black tit with white wing-coverts. Very similar to Carp's and White-winged Black Tits *P. carpi* and *P. leucomelas*; in overlap zones shows *less white in wing* than either, with only a small shoulder patch (median coverts) solid white, the greater coverts all having black centres. Also differs from both in having *barred undertail-coverts* (black and white in ♂, grey and white in ♀), and is overall more drably-coloured, less glossy, ♂ with a touch of grey on lower belly, ♀ with much greyer underparts; immature sooty brown, paler below. For detailed comparison with Carp's Tit, see that species.

Voice. Tape-recorded (58, 75, 86, 88, 91–99, B, F, LEM, WALK). Noisy, maintaining a continual conversational calling among members of group (Harrap and Quinn 1996). Common call a buzzy, nasal 'jee-jee-jee-jee-jee ...' or grating 'jrrr-jrrr-jrrr ...', often becoming a 'chick-a-dee' type, 'ts-ts-ts-ts-jee-jee-jee-jee-jee ...', 'tsit'chaaa' or 'tsit-zzzzzzzz' (sonagrams in Harrap 1996). Also has a wide variety of whistled calls (songs?), 'peeoo-pu-peeoo-pu-peeoo-pu-peeoo', 'cheep-wah, cheep-wah', 'wah-pew, wah-pew', 'whee-u-wee' (or perhaps 'where-are-we?'), 'willo-willo-willo-willo ...'; for further variations see Harrap (1996).

General Habits. Prefers broad-leaved woodland although largely absent from *Brachystegia* woodland, where it is replaced by Rufous-bellied Tit *P. rufiventris*. Occurs in mopane and mixed deciduous woods, camelthorn and other thornveld types, plantations and well-wooded suburbs. In Zambia keeps to thinner, drier types of miombo. Generally below 1070 m. Forages in parties of up to 6 birds in middle and upper levels of trees and bushes. Noisy, members of flock continuously in vocal contact. Often joins mixed-species flocks in winter. Often taps on *Combretum* pods and opens those that contain insect larvae; forages on branches and trunks; sometimes tears off bits of lichen and bark. Foraging flocks (family groups) may travel 10 km in one day (Harrap and Quinn 1996). Roosts in tree holes.

Food. Mainly lepidopteran caterpillars and pupae, also various insect larvae, beetles, moths and spiders.

Breeding Habits. Co-operative, monogamous, group-territorial breeder; helpers (young ♂♂) at some nests (58% of nests in Transvaal: Tarboton 1981). Although helpers are presumed to be offspring of the pair they mainly help, recently fledged young may switch territorial groups during territorial skirmishes. Primary ♂ courtship-feeds ♀ for up to 7 weeks before egg laying, and prevents helper ♂♂ from feeding her until incubation begins. 40% of ♂♂ fail to find a mate (Tarboton 1981). Territories range in size from 25 to 48 ha and are defended all year by all members of breeding group. Pair rarely uses same nest site in successive years.

NEST: thick cup of plant material, mammal hair and feathers, sited in natural hole in tree e.g. *Baikiaea plurijunga*, abandoned woodpecker or barbet hole, open pipe or nest box. Of 87 nests in Zimbabwe, 60 in holes in trees, 16 in metal pipes, and 11 in nest boxes. Typically 2–4 m above ground (2–5 in South Africa); in Zimbabwe, 77 nests 0.31–6.15 (av. 2.4) m above ground. Built by ♀.

EGGS: 2–6; southern Africa mean (39 clutches) 3.6; clutch not known to exceed 3 eggs in Malaŵi (Harrap and Quinn 1996). Oval; white with reddish brown speckling, mainly at broad end. SIZE: (n = 82, southern Africa) 16.6–20.7 × 13.0–14.6 (18.0 × 13.9); (n = 18, Zimbabwe) 17.8–19.5 × 13.5–15.0 (18.9 × 14.3).

LAYING DATES: normally just before rains. Zambia, Sept–Nov; Malaŵi, Oct–Nov; Zimbabwe, Aug–Dec; Mozambique, Oct; Botswana, Sept–Dec; South Africa, Oct–Dec.

INCUBATION: begins with 3rd or 4th egg; by ♀ only. Period: 14–15 days; primary ♂ and helper ♂♂ bring food to incubating ♀.

DEVELOPMENT AND CARE OF YOUNG: young hatch asynchronously over period of 12–48 h (Zimbabwe); brooded by ♀ until 6–8 days old; both parents remove faecal sacs; young leave nest 20–24 days after hatching, taking 1–2 days (Zimbabwe). Fed by parents and helpers for up to 40 days after fledging.

BREEDING SUCCESS/SURVIVAL: in South Africa, 13 breeding groups (13 ♀♀, 22 ♂♂) produced a total of 16 fledged young (range 0–3 young per group) during one season. Pairs with helpers raise more young. In the same study population, 5 of 10 ringed adult ♂♂ and 3 of 6 adult ♀♀ survived to the following breeding season.

References
Clancey, P.A. (1972).
Harrap, S. and Quinn, D. (1996).
Tarboton, W.R. (1981).

Plate 7
(Opp. p. 94)

Parus cristatus Linnaeus. Crested Tit. Mésange huppée.

Parus cristatus Linnaeus, 1758. Syst. Nat., ed. 10, p. 189; Europe, restricted to Sweden by Hartert 1905.

Range and Status. Palearctic, restricted to Europe, where resident and sedentary from Iberian Peninsula (south to Gibraltar), north-east to Finland, south to Greece, and east to Urals. Isolated population in Scotland. Has expanded range in north in last hundred years. Vagrant to Morocco, Moldavia, Caucasus, Lapland, Kazakhstan and Siberia.

Morocco, very rare vagrant (no details: Etchécopar and Hüe 1967. No subsequent records).

Description. *P. c. weigoldi* Tratz: Portugal and S Spain (race of Moroccan vagrants unknown, presumed to be this one). ADULT ♂: forehead frosted, feathers blackish with narrow white fringes; forecrown feathers similar but, being larger, more black shows; hindcrown feathers elongated, black with white fringes; hindneck dusky brown; lores, malar region, cheeks, ear-coverts and sides of neck white, ear-coverts speckled with black; narrow black line from eye backwards, joining black comma-shape at hind edge of ear-coverts; chin and upper throat black, forming neat bib, with narrow black line running backwards from it around neck to nape. Mantle to rump, tail and wings uniformly warm brown. Lower throat, breast and centre of belly white, sides of breast, flanks and undertail-coverts warm buff. Bill black; eyes dark brown; legs and feet black. Sexes alike, though crest of ♀ av.

shorter than that of ♂. SIZE: wing, ♂ (n = 6, Portugal) 61–65·5, 1 ♀ 57·5; (*P. c. cristatus*), wing, ♂ (n = 17) 64–67 (65·5), ♀ (n = 14) 60–64 (62·0); tail, ♂ (n = 13) 47–52 (49·8), ♀ (n = 13) 45–51 (48·1); bill to skull, ♂ (n = 13) 10·7–11·7 (11·2), ♀ (n = 11) 10·6–11·8 (11·4); tarsus, ♂ (n = 13) 18·2–19·3 (18·8), ♀ (n = 12) 17·6–18·9 (18·1). WEIGHT (*P. c. cristatus* and *P. c. mitratus*, Romania): ♂ (n = 46) 60–68 (64·0), ♀ (n = 43) 48–56 (51·6).

IMMATURE: like adult but blacks tinged brownish, crown feathers shorter and less pointed, and flanks paler buff.

Field Characters. Length 11·5 cm. Unmistakable – the only tit in its range with *crest*. Crest always erect; crown blackish, sides of head whitish, narrow black line around hind end of ear-coverts. Below neck, rather like Coal Tit *P. ater*, which, however, has 2 rows of white spots on wing.

Voice. Tape-recorded (62–73. 93, B). Commonest call a purring, tremolo trill, 'burrurrlt, burrurrlt' or 'brrrrrr, t'brrrrrr'. Bird lacks the vocal variety of most congeners.

General Habits. In S Spain inhabits mainly cork oak *Quercus suber* woodland with shrub layer and low, rotten stumps (for nesting); elsewhere, in birch *Betula pubescens* or (N Europe) pure, mature pine forest, e.g. *Pinus scoticus*; also scrub, bushy heath, juniper, orchards and parks. Breeds up to treeline. Solitary, in pairs, or groups of up to 6 birds; confiding or aloof. Forages in canopy amongst twigs, leaves and pine needles, and on larger branches and knobbly treetrunks; in winter, feeds in shrub layer, saplings, heather, and on tussocky ground under cover. Agile and energetic, probing vegetation, constantly on the move, hanging upside down from tips of twigs, and searching amongst lichens. Caches food, wedging larvae or seeds into bark crevice. Visits bird tables. Readily uses nest boxes. Sedentary.

Food. Small insects, spiders, invertebrate eggs, seeds.

References
Cramp, S. and Perrins, C.M. (1993).
Etchécopar, R.D. and Hüe, F. (1967).
Harrap, S. and Quinn, D. (1996).

Parus ater Linnaeus. Coal Tit. Mésange noire.

Plate 7
(Opp. p. 94)

Parus ater Linnaeus, 1758. Syst. Nat., ed. 10, p. 190; Europe, restricted to Sweden by Hartert 1905.

Parus ater

Range and Status. Palearctic, from NW Africa, Spain, Ireland and Sweden to Greece, Cyprus, Turkey and N Iran, north to *c*. 66°N, east to S Korea, Japan and Taiwan. Resident, but in N half of range strongly eruptive.

Resident in NW Africa, but some evidence of autumn dispersal and of limited vertical migration. Single ringing recovery of N European bird in Morocco, so that many migrants, not yet distinguished from local residents, might reach Africa in winter. Morocco, widespread and locally common at 1000–2500 m in Moyen Atlas and Haut Atlas Mts; common in winter down to Jaaba, scarce down to Marrakech, with records on Rif coast at Jnan en Nich. Algeria, locally common from about Ouarsenis region east to Mostaganem near Tunisian border; west of 05°E, only in some seaward-facing slopes of Hauts-Plateaux; east of 05°E, in Hauts Plateaux Constantinois and Tell à l'Aurès and down to coast; inland to Djelfa in Atlas Saharien.

Description. *P. a. atlas* Meade-Waldo: Morocco. ADULT ♂: forehead, crown and sides of neck glossy black; nape and hindneck white; mantle, scapulars, back, rump and uppertail-coverts bluish grey, tinged olive; tail dull black, feathers narrowly edged with olive-grey. Cheeks and ear-coverts white. Chin and throat black; breast and centre of belly white, flanks warm buff, undertail-coverts buff. Wing feathers blackish, all except alula with olivaceous blue-grey outer edges, and median and greater coverts with white tips. Underwing-coverts and axillaries buffy white. Bill black; eyes dark brown; legs and feet bluish grey. Sexes alike, though black bib of ♀ narrower than ♂'s. SIZE: wing, ♂ (n = 38) 65–74 (68·2), ♀ (n = 20) 63–70 (65·7); bill to skull, ♂ and ♀ (n = 53), av. 11·7; tarsus, ♂♀ (n = 41), av. 18·6. Tail (*P. a. ater*), ♂ (n = 11) 43–49 (45·5), ♀ (n = 10) 43–48 (45·1). WEIGHT: 1 ♂ 11·8, 1 ♀ 11·6.

IMMATURE: juvenile like adult but duller, white parts creamy or yellowish, flanks paler than in adult.

NESTLING: long, dense tufts of grey down on crown, some short grey down on shoulders and back; mouth bright yellow or orange, gapes pale yellow.

P. a. ledouci Malherbe: Algeria, Tunisia. Like *atlas* but upperparts greener, white patches on cheeks and nape tinged yellowish, flanks paler. Population in Atlas Saharien Mts intermediate between *atlas* and *ledouci*.

P. a. ater Linnaeus: Europe. Once, Morocco. Upperparts less green-tinged and flanks paler buff than in *atlas*; some calls different (Cramp and Perrins 1993); smaller. Wing, ♂ (n = 11) 61–66 (63·3), ♀ (n = 10) 58–63 (60·7). Weight (Norway): ♂♀ (n = 176) 7·9–10·7 (9·2); (Turkey), ♂♀ (n = 76) av. 9·7.

Field Characters. Length 11·5 cm. A small, thin-billed tit of coniferous woods, with white cheek patch and black bib like larger Great Tit *P. major*, distinguished by buffy white underparts without black line down centre, 2 rows of white spots on grey wing, and extensive *white patch on nape* and hindcrown (not always easy to see); song higher-pitched and thinner.

Voice. Tape-recorded (see Cramp and Perrins 1993). High-pitched, pure, sweet; main call a plaintive 'pui' or 'tsuee' and sweet 'swee-pi'; also 'pipi', and a thin 'sisisi' very like Goldcrest *Regulus regulus*; alarm, a churring or trilling scold (both races – quite different from alarm of European races). ♂ song a 2–3-syllable motif repeated up to 8 times, usually 'peechoo-peechoo-peechoo' but also 'tu-wa-chi', emphatic 'ti-t'wa', rapid 'chi-chi-chi', or slurred 'sit'tui-sit'tui-sit'tui'. ♀ song quieter, with motif given only once or twice.

General Habits. Inhabits montane woods, mainly evergreen: cedars and evergreen oaks up to 2000 m, junipers up to 2200 m, xen oaks, cork oaks, hollies, pines, thuyas, cyprusses *Tetraclinis articulata* and jujubes *Ziziphus mauritiana* in winter. Indifferent to human habitation, enters ornamental gardens and suburban habitats in winter if they contain suitable conifers. Occurs in pairs in summer; solitary, in pairs or roving parties of *c.* 6 birds in winter. Foraging and other habits barely studied in Africa but known in great detail in Europe (Cramp and Perrins 1993). Arboreal; in Spain feeds in trees by hanging upside-down from tuft of pine needles or clump of leaves (56% of 141 observations), gleaning (33%), hovering (6%), clinging onto vertical trunks or branches (5%) and, rarely, taking sluggish insect in flight. Holds large food item under foot, dismembering an insect with bill or stabbing at seed. Extracts seeds from conifer cones by alighting on cone and hanging head-down, or sometimes hovering (if cone fully open). Forages at all levels of tall and short trees, usually higher than 2 m above ground, and far more on outside of tree than within it.

Mainly sedentary, but a few birds move from mountains down to sea-level in Morocco (see above). Bird ringed in Germany recovered in Morocco, 2500 km away.

Food. Very small insects and spiders; some plant material. In Europe many aphids, and locally can take caterpillars to exclusion of all else. Tends to eat more spiders in pine forests where caterpillars sparse. Outside breeding season, if insects hard to find, takes wide variety of small seeds and fruits (listed by Cramp and Perrins 1993).

Breeding Habits. Solitary nester; monogamous; not conspicuously territorial. Very well known in Europe; scores of nests found in Africa though no detailed biological studies so far made there. 2nd broods common in N Europe but not in Africa.

NEST: cup made of moss, lined with hair and wool and sometimes a few feathers, sited in hole in tree, rotten tree stump or on ground amongst tree roots; less often in crevice in rock or hole in masonry. Built by ♀ only.

EGGS: 4–7 (av 5·7) in Africa; up to nearly 10 in N Europe. Subelliptical, smooth, glossy, white, finely speckled with red-brown, spots sometimes forming band around large end. SIZE (*P. a. ater*, central Europe, n = 102) 13·7–16·7 × 10·7–12·7 (15·3 × 11·9); more oval in Turkey, (n = 43), av. 15·6 × 11·45 (Cramp and Perrins 1993).

LAYING DATES: Morocco and Algeria, Apr–May, a few June. Av. 1st-clutch dates: Morocco, Moyen Atlas Mts at 1830 m, 6th May; Algeria, Grande Kabylie Mts at 915 m, 10th Apr, Aurès Mts at 1830 m, 1st May.

INCUBATION: by ♀ only. Period (Europe): 14–16 days.

DEVELOPMENT AND CARE OF YOUNG: young brooded by ♀ only, fed by ♂ and ♀. Nestling period (Europe) 18–20 (19) days.

BREEDING SUCCESS/SURVIVAL: in Europe, clutch size close to number of young that are usually reared if nest not attacked by predator (data in Cramp and Perrins 1993).

References
Berndt, R. and Jürgens, R. (1977).
Cramp, S. and Perrins, C. (1993).
Harrap, S. and Quinn, D. (1996).
Heim de Balzac, H. and Mayaud, N. (1962).
Snow, D.W. (1952).

Plate 7
(Opp. p. 94)

Parus caeruleus **Linnaeus. Blue Tit. Mésange bleue.**

Parus caeruleus Linnaeus, 1758. Syst. Nat., ed. 10, p. 190; Sweden.

Range and Status. W Palearctic, from 67°N in Norway and N end of Baltic Sea to Britain, Ireland, whole of central and S Europe, NW Africa, Canary Is, Mediterranean islands except Cyprus, N Libya, Turkey, Lebanon, Caucasus, Iraq and N Iran. Mainly resident, but eruptive in N; migrants have reached Spain from Germany and Russia (1480 and 2200 km away).

Endemic races in NW Africa are resident. Morocco, wooded mountains (Rif, Tazzeka, Moyen Atlas, Jbel Ayachi, Haut Atlas up to 1600 m on Timenkar Plateau and Jbel Afirouane and 1800 m at Imlil; on south-facing mountainsides south to Tizi-n-Tichka and Aït Ben Haddou) and plains (Sidi-Bou-Rhaba, Zaërs, Mamora); uncommon on Mediterranean coast, with records at Jnan en Nich, Bou Ahmed, Cap des Trois Fourches, El Jebha and Cala Idris; frequent in parks and gardens of Casablanca and Rabat. Algeria, widespread from coast up to treeline and inland to the most northerly oases; locally quite common at Chréa, Djurdjura and Aurès up to 1600 m and in the Constantinois, Grande Kabyle, Oranie, Ksours

Parus caeruleus

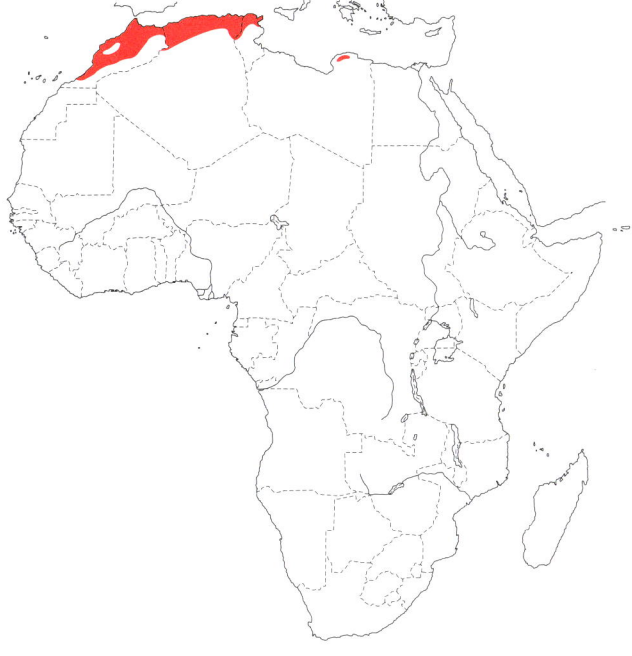

Mts up to 1800 m, breeding south to oasis at Baniane. Tunisia, common and widespread in N south to central plateau, uncommon and local in W south to Tozeur and Nefta; commonest in Kroumirie, and coastal areas south to El Djem and Mahdia. Libya, race endemic to Jebel Akhdar, around Wadi el Kuf, Merg and Barce.

Description. *P. c. ultramarinus* Bonaparte: Morocco to Tunisia. ADULT ♂: forehead white, crown ultramarine-blue (feathers black with blue tips), surrounded by neat, narrow white line from forehead to nape; upper hindneck blackish; lower hindneck, mantle, scapulars, back, rump and uppertail-coverts greyish ultramarine. Tail mainly greyish blue. Neat, narrow black line from bill through lores and eye to upper hindneck. Cheeks, ear-coverts, moustachial and malar areas white. Chin and upper throat black, sharply demarcated from lower throat and running as narrow black line below large white cheek patch. Lower throat, breast, sides, belly, flanks, thighs, vent and undertail-coverts bright lemon-yellow, except for irregular, usually rather patchy black line in middle of lower breast and upper belly. Wing feathers black, all broadly edged with blue (primaries, secondaries) or ultramarine (greater coverts, primary coverts); greater coverts broadly tipped with white, primary coverts narrowly tipped white, tertials broadly tipped with off-white or greyish white. Underwing-coverts and axillaries pale sulphur yellow. Bill slaty; eyes dark hazel-brown; legs and feet slate-blue. Sexes alike, though blues possibly slightly less intense on ♀. SIZE: (coastal regions) wing, ♂ (n = 80) av. 62·7, ♀ (n = 41) av. 60·6. bill, sexes combined (n = 120) av. 6·8; montane populations larger – wing av. 1·9–2·0 and bill av. 0·1–0·6 mm longer (Cramp and Perrins 1993). Supplementary measurements (*P. c. caeruleus*, Europe): wing, ♂ (n = 129) 65–71 (67·5), ♀ (n = 90) 62–67 (65·0); tail, ♂ (n = 62) 49–54 (51·5); bill to skull, ♂ (n = 37) 8·4–9·6 (9·1); tarsus, ♂ (n = 34) 16·5–17·8 (17·6). WEIGHT: (Algeria), ♂ (n = 5) av. 10·9, ♀ (n = 3) av. 10·5; (Morocco) unsexed (n = 14) av. 10·7.
IMMATURE: juvenile pale yellow where adult white, dark brown where adult black and dusky olive where adult ultramarine-blue (but outer edges of primaries, secondaries and rectrices bluish); lacks black chin, upper throat, and line under cheeks (those parts being same yellow as rest of underparts).
NESTLING: naked, pink, tuft of long grey down on forecrown and another on nape; bill and gape yellowish; mouth yellow.

P. c. cyrenaicae Hartert: Libya. Like *ultramarinus* but less white on forehead, upperparts duller blue and underparts less pure a yellow. Smaller: wing, ♂ (n = 5) 56–61 (58·4), ♀ (n = 5) 56–68 (57·1).

Field Characters. Length 11·5 cm. A small tit with *blue cap, wings and tail*; black stripe through eye separates white cheek patch from white stripe over eye and around crown; underparts yellow. Juvenile has brownish upperparts and yellow cheek patch but has striped face pattern of adult.

Voice. Tape-recorded (62–73, 93, 105, 106, B). Song in North Africa said to be different from European birds (Cramp and Perrins 1993); typical song a repeated phrase of 2–4 rather metallic notes, 'tizee tizee tizee ...' or 'tee-zay-brre tee-zay brre ...', harder than Coal Tit *P. ater* and not so ringing as Great Tit *P. major* (Snow 1952); sometimes with introductory 'see koo-zree koo-zree' (see sonagrams in Cramp and Perrins 1993). Scolding alarm calls similar to European birds, but North African birds have variety of other sharper calls, similar to some song-phrases, 'tsee-tsi-brree tsee-tsi-brree', 'chi-chichiwee' and 'pichoo', with distinctive metallic quality (Snow 1952).

General Habits. Inhabits gardens, thickets, cork oak *Quercus suber* forests, evergreen oaks *Q. ilex*, xen oaks *Q. faginea*, pines and cedars; palm groves in oases. Occurs in pairs, family parties, and small flocks in winter, sometimes of a dozen or more birds in hard weather. Forages mainly amongst foliage, gleaning leaves and twigs for caterpillars, restlessly exploring in canopy and outer branches, on trunk and large boughs, sometimes coming momentarily to ground. Confiding, inquisitive; probes and pries with bill; perky, acrobatic and agile, commonly hangs sideways or upside-down when searching for food or dealing with it, e.g. on flowering heads of small shrubs, at bird-feeders, and at twiggy tips of branches. Feeds at all levels in vegetation, mainly at 2–4 m above ground. Whether singly or in flocks, seldom in one place for more than a few s, moving on in short flights. Flight fast and direct, rather erratic, fluttering and bouncing over greater distances. Vocal. Constantly flicks and twitches wings and tail.

Sedentary, though post-breeding dispersal may take birds some distance from suitable cover (Algeria: Heim de Balsac and Mayaud 1962).

Food. Insects and spiders, with fruits and seeds in winter; seeds, nuts and fat at bird-feeders. In Mamora and Ras El Ma, Morocco, young fed mainly with caterpillars (47–79%) and spiders and spider eggs (7–23·5%); also small beetles and many other undetermined small prey (Baouab *et al.* 1986).

Breeding Habits. Monogamous, territorial. Details of pair-bonding behaviour, Dance-display and courtship-feeding well known in Europe (Cramp and Perrins 1993) but little studied in Africa. In Moroccan plains, replacement clutches are commonplace but 2nd broods are rare (Baouab *et al.* 1986).

NEST: in tree-hole, nest-box or hole in masonry or wall; a pad of moss with bits of dead grass and other plant materials, cup lined with hair, wool, feathers, fine dry grass and bits of fibrous bark. Built by ♀ alone, taking 5–12 days from start to laying of 1st egg; she continues to add fresh plant material during incubation and after eggs start to hatch.

EGGS: 4–10; in Tunisia and Algeria below 1500 m (n = 35 clutches) av. 7·4; in cork oak woods, lowland Morocco, 5–9 eggs, av. in 3 years (n = 40 clutches) 6·3, 6·9 and 7·2 eggs; in xen oak woods, Moyen Atlas (Morocco) at 1600 m, 4–10 eggs, av. in 3 years (n = 34 clutches) 5·7, 6·7 and 7·8 eggs (Baouab *et al.* 1986). Sub-elliptical, not glossy; white, with small reddish spots mainly around large end. SIZE: (n = 38 eggs) max. 16 × 12, min. 14 × 11·5; also (*P. c. caeruleus*, France, n = 172) 14·5–17·9 × 11·2–12·8 (16·0 × 12·0).

LAYING DATES: NW Africa, Apr–May; in Moroccan lowlands mainly mid Apr and in Moyen Atlas mainly late May (Baouab *et al.* 1986).

INCUBATION: by ♀ only. Period 13–16 (14·2) days (Europe).

DEVELOPMENT AND CARE OF YOUNG: brooded only by ♀; cared for and fed by both parents. Sigmoidal growth curves (graphed in Baouab *et al.* 1986), young fledging at 12 g after 17–18 days.

BREEDING SUCCESS/SURVIVAL: excluding predation, 50–85% success in cork oak woodland, Morocco, and 80–92% success in xen oaks, Moyen Atlas, Morocco (Baouab *et al.* 1986, Cramp and Perrins 1993); numerous data, Europe.

References
Baouab, R.E. *et al.* (1986).
Cramp, S. and Perrins, C.M. (1993).

Family REMIZIDAE: penduline tits

Small or tiny arboreal birds; bill short, straight, sharp-pointed, with straight culmen; wing rather short and rounded; 10 primaries; tail short with 10 rectrices; legs rather short (length varies), feet small but strong. Sexes and ages alike in plumage (*Anthoscopus*, *Pholidornis*), or juveniles much plainer than adults (*Remiz*). Plain olive, grey or brown above, white, yellowish or buff below with darker wings and tail and speckled forehead (*Anthoscopus*); or streaky dark brown and yellow (*Pholidornis*), or with chestnut back and black head or grey head with black mask (*Remiz*). The most diagnostic characters are behavioural ones: feed by probing, inserting closed bill into vegetable matter then opening mandibles to pry it apart (*Anthoscopus*, *Remiz*; ?*Pholidornis*); very agile climbers in reeds, bushes and tree foliage, often clinging upside down; build rain-proof, felt-like or kapok-like pendent nest with entrance high or low in side, in *Anthoscopus* with conspicuous false, blind entrance below concealed, self-closing true entrance. At least one species clamps food (caterpillar) with foot (and tears it piecemeal with bill).

Excluded from Remizidae are the Himalayan/Oriental Fire-capped Tit *Cephalopyrus flammiceps* which has characters of both Paridae and Remizidae, and the Verdin *Auriparus flaviceps* of SW North America, which may be a certhiid (Sibley and Ahlquist 1990).

3 genera: 2 in Africa and 1 in Palearctic.

Genus *Anthoscopus* Cabanis

For characters, see under family Remizidae. Cranium not wholly pneumatized. Use foot to open true nest entrance. Voices high-pitched, rhythmic, mechanical buzzes and trills. The species are all allopatric or parapatric, and are placed in a single superspecies by Hall and Moreau (1970), Sibley and Monroe (1990, who include a 7th species, '*A. sylviella*', separated from *A. caroli*) and Harrap and Quinn (1996). The one rain forest species, *A. flavifrons*, is saturated green above, has only vestigial speckles on forehead, and is relatively large-billed; we treat it as an independent species. A semicircle of 4 species, *A. punctifrons*, *A. musculus*, *A. caroli* and *A. minutus*, differ a little clinally but are basically very alike and we regard them as a superspecies. *A. parvulus* is almost uniformly pale green; with no great conviction we retain it also in Hall and Moreau's superspecies.

The remarkable architecture and properties of the nest have excited much interest and, formerly, wild conjecture (e.g. Barrow (1801) *An Account of Travels into the Interior of Southern Afric in the Years 1797 and 1798*). Yet nest biology of only one species is at all well known: *A. minutus*, an admirable study by Skead (1959). Some statements culled from the literature about other species in the following review, in conflict with Skead's (e.g. nest closure), are likely to be inaccurate.

Endemic. 6 species.

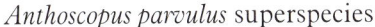

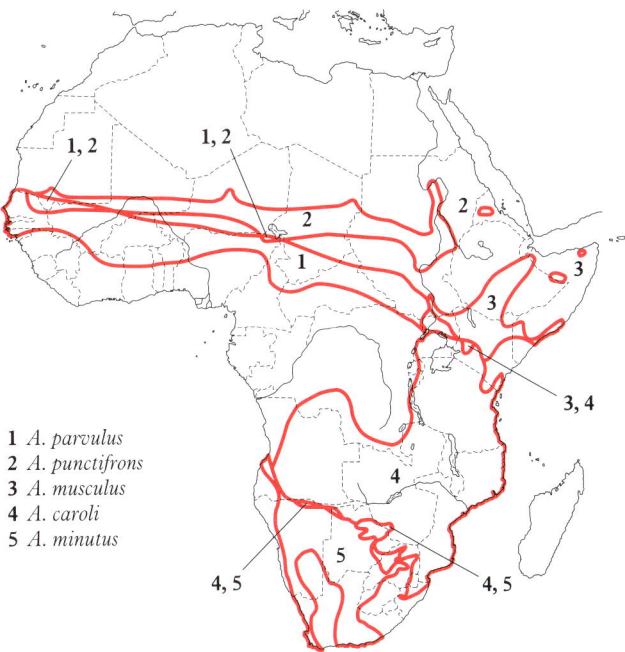

1 *A. parvulus*
2 *A. punctifrons*
3 *A. musculus*
4 *A. caroli*
5 *A. minutus*

Anthoscopus flavifrons (Cassin). Forest Penduline Tit. Yellow-fronted Penduline Tit. Rémiz à front jaune.

Plate 6
(Opp. p. 79)

Aegithalus flavifrons Cassin, 1855. Proc. Acad. Nat. Sci. Philadelphia, 7, p. 325; Moonda River, Gabon.

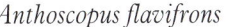

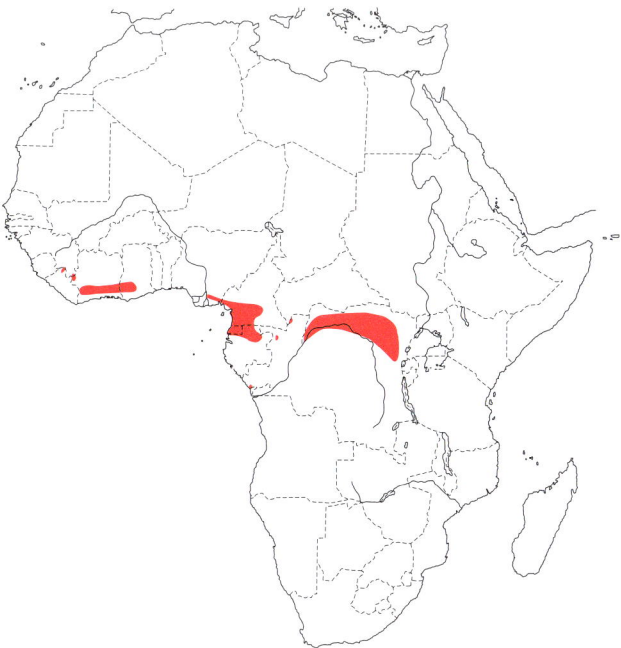

Range and Status. Endemic resident, lowland rain forest, from Liberia to E Zaïre; records sparse, but species probably commoner and more widespread than sightings suggest. Liberia, several records on Mt Nimba and N Lofa county (Gatter 1997). Ivory Coast, rare: Taï (encountered twice in 53 days), Fresco, Yapo forests. Ghana, Goaso (where 'may be fairly common, as I noticed several old nests': Lowe 1937) and Kakum Forest Res. (Macdonald 1980). Nigeria, old records at Itu and Umuagwu in SE. Cameroon, said to range throughout forest zone (Louette 1981), though hardly any published locality records (Lolodorf, Elat, Aboulou). Mbini, sight record. Gabon, rarely seen (Bélinga, M'Bes, Makokou, M'Passa, Mondah cstuary) but density thought to be in order of 3–5 pairs per km^2. Congo, Odzala and Nouabalé-Ndoki Nat. Parks, and in SW Goumina and Béna (Kouilou). Zaïre, widespread in Equateur and Haute-Zaïre Provs, and through Ituri to c. 02° S in Kivu (Nyamiringe, Bungakiri) (Harrap and Quinn 1996).

Description. *A. f. flavifrons* Cassin: S Nigeria to Gabon and NE Zaïre. ADULT ♂: forehead greenish yellow, sometimes with faint blackish feather tips; rest of upperparts dark olive-green. Tail blackish, feathers edged dark olive-green. Lores, ear-coverts and sides of neck dark olive-green streaked yellowish; cheeks and chin paler olive, feathers with yellow tips and shaft streaks; throat and breast olive, throat mottled and breast streaked with yellowish green; belly and flanks greenish yellow, streaked olive; undertail-coverts dusky olive with yellow fringes. Upperwing blackish, primaries (proximal to emarginations) and secondaries edged olive-yellow, tertials and coverts fringed olive-green. Axillaries, underwing-coverts and inner borders of flight feathers whitish, tinged yellow. Bill black, lower mandible and cutting edges light grey to bluish white; eyes brown; legs grey or blue-grey. Sexes alike. SIZE (5 ♂♂, 3 ♀♀): wing, ♂ 52–58 (55·2), ♀ 56–57 (56·7); tail, ♂ 27–30 (28·8), ♀ 27–29 (28); bill, ♂ 9–11·5 (10·3), ♀ 10–11 (10·3).

IMMATURE: juvenile like adult, but forehead band narrower and dull buff; upperparts duller and greyer; throat whiter, less mottled.

A. f. waldroni Bannerman: Liberia to Ghana. Only 1 specimen available: brighter and more yellowish green above; breast paler and yellower.

A. f. ruthae Chapin: E Zaïre (W Kivu). Differs from nominate race in having forehead tawny buff, cheeks washed cinnamon and throat cinnamon buff; rest of underparts off-white, washed pale yellowish buff, with brownish patches on breast sides.

Field Characters. Length 9 cm. The only penduline tit confined to *forest*, where hard to see well since it lives in canopy and mid-levels. With olive-green upperparts and yellowish underparts it might be taken for a small sunbird, but has *short, stubby bill*. Yellow forehead only visible at close range. Yellow Penduline Tit *A. parvulus* is only in savanna. Abuts range of Grey Penduline Tit *A. caroli*, a woodland bird, in Congo and Zaïre; distinguished by olive-yellow, rather than creamy white, underparts.

Voice. Tape-recorded (68, 104, B, CHA, ERA). Various extremely high and thin tit-like calls, 'tseet', 'tit', 'tsitsit', often accelerating into 'titititititititi', almost a trill. Also transcribed as an emphatic 'pshit', high-pitched and almost cricket-like, easily overlooked, and a clearer 'tsi', accelerating into a shrill, trilling 'tsi-si-si-si-si ...' (Harrap and Quinn 1996). Said to be rather silent; no vocalization heard during 6-week stay at Mt Nimba, Liberia (pers. obs.), but this may well be because 'les vocalisations aigues passent inaperçues dans l'ambiance forestière' (Brosset and Erard 1986).

General Habits. Inhabits flowering trees in primary lowland rain forest, also in old secondary growth, treefall gaps in forest, mature plantations and old secondary bush and scattered trees in adjacent fallow ground. Keeps mainly to canopy but occasionally comes into lower strata. Lives in pairs or parties of 3–5 birds; once a flock of at least 30, scattered singly or in groups of 2–3, foraging amongst debris of old red weaver-ant *Oecophylla* nests in low foliage, Ghana (Macdonald 1980). Very mobile, probably working large area in search of flowers. Once seen feeding in flowers of *Bombax* tree in company with numerous sunbirds (Lowe 1937); seen investigating inflorescences of *Harungana* tree, with Little Green Sunbirds *Anthreptes seimundi*; also watched climbing through twigs of large strangling fig *Ficus*, knocking down small fruits whilst apparently searching for insects, and a flock of 5 explored branches of small trees and a mass of mistletoe *Loranthus* (Chapin 1954). Also rummages systematically into old nests of social spiders. Bird forces entrance to its own nest (see below) by pushing bill then head into closed slit at tip of entrance tube (Brosset and Erard 1986), and by grasping lower lip of entrance with one foot whilst clinging to side of nest below with other (Chapin 1954). Dead adult once found in nest, outside presumptive breeding season, so birds probably roost in nest then. Birds preen each other. Sedentary.

Food. Small insects, and small pulpy fruits (Chapin 1954).

Breeding Habits. 3–4 birds watched coming and going between nest and forest canopy (Brosset and Erard 1986); 3 birds, all ♂♂ with enlarged gonads, taken at nest or in nest tree on same day (Chapin 1954); so species probably a co-operative breeder.

NEST: bag, suspended from twig, with self-closing short-tubular entrance sloping down from top and large elliptical depression or false 'entrance' in side of nest below the real one; made of soft brown vegetable down or kapok, worked into close-woven, supple, thick, wool- or felt-like tissue. Wall of bag is especially thick and resilient at false entrance. Aperture of entrance tube, spout or protrusion closes automatically to a horizonal slit whether adult bird(s) in or out of nest, and thought to be completely impenetrable, as is also substance of bag itself, to predatory ants (Brosset and Erard 1986). Sited 5–10 m above ground; once nest was being built in a leafless tree; another was in *Bombax* ceiba tree (and it seems likely that ovary hairs of *Bombax* are one source of birds' kapok for nests). Cotton from cotton *Gossypium* crops also used for nest fabric, in one nest almost exclusively (Uele: Chapin 1954). Old nests extremely durable and long-lasting.

EGGS: unknown.

LAYING DATES: Nigeria, (breeding condition Nov); Cameroon, (fresh nests, Dec); Gabon, (fresh nests in Oct and Mar); Congo, (3 birds chasing each other over piece of downy nest material, Oct); Zaïre, (N Equateur, nest with young, July; nest-building, Medje, Oct).

References
Brosset, A. and Erard, C. (1986).
Harrap, S. and Quinn, D. (1996).

Anthoscopus parvulus (Heuglin). Yellow Penduline Tit; West African Penduline Tit. Rémiz à ventre jaune.

Plate 6 (Opp. p. 79)

Aethithalus ? parvulus Heuglin, 1865. J. f. Orn., 12, p. 260; Bongo, Bahr el Ghazal, Sudan.

Forms a superspecies with *A. punctifrons*, *A. musculus*, *A. caroli*, and *A. minutus*.

Range and Status. Endemic resident, Senegal to NE Zaïre. Many field workers have commented that species is easily overlooked, and it may be much commoner and more uniformly distributed than following remarks suggest. Mauritania, Senegal R. valley and S Guidimaka. Senegal, widespread at least peripherally, but scarce; records at Bogué, Fatick, Bandia and Saint-Lewis. Gambia, scarce lower river and rare upper river. Guinea-Bissau. Mali, locally common, Boucle du Baoulé and from Bamako north to Mopti and about 15°N. Ivory Coast, occurs locally in N, south to about 09°N. Ghana, frequent at Mole and Tumu and widespread in savannas in between. Togo, sight record at Lomé on coast. Burkina Faso, Fada-N'Gourma region (Harrap and Quinn 1996) and Arli-Pendjari Nat. Park. Nigeria, uncommon, throughout N soudanian zone (Sokoto to Potiskum and L. Chad); records south to Zaria, where frequent, and Kainji Lake Nat. Park. Cameroon, only record, Bénoué Nat. Park at 08°06′N, 13°52′E. Chad, scarce, north to Chari-Baguirmi district and Abéché. Central African Republic, records at Bozoum and on Bamingui River. Sudan, uncommon, S Kordofan and to south and east. Zaïre, records in extreme NE from Mauda, Dungu, Aba and Mahagi.

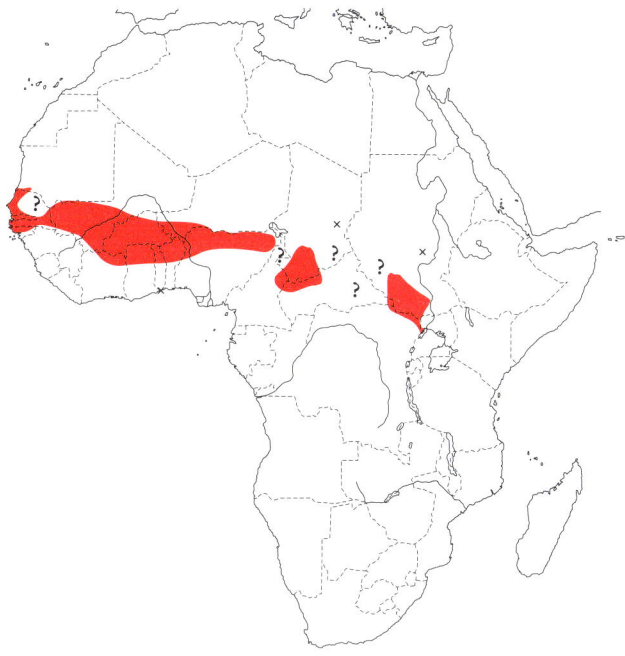

Anthoscopus parvulus

Description. *A. p. senegalensis* (Grote) (includes '*citrinus*'): Senegal to Central African Republic, intergrading with nominate race. ADULT ♂: forehead and forecrown yellow, spotted with blackish brown; rest of upperparts, including sides of neck, yellowish olive-green, brighter yellow on rump and uppertail-coverts. Tail blackish brown, feathers fringed olive-yellow, tipped pale buff. Lores dark olive-grey; short stripe behind eye dark olive-grey, and above this a narrow yellow superciliary stripe connecting with yellow forehead. Ear-coverts and cheeks to chin, throat and rest of underparts rich yellow with faint olive tinge. Upperwing dark brown; primaries (proximal to emarginations) and primary coverts narrowly edged buffish white, alula finely edged olive-yellow; secondaries more broadly edged olive-yellow; tertials and greater coverts fringed olive-yellow, the latter more whitish at tips; median and lesser coverts broadly fringed yellowish olive-green. Axillaries pale yellow; underwing-coverts and inner borders on undersides of flight feathers pale creamy buff. Bill blackish or dark grey with whitish or bluish lower mandible and cutting edges; eyes brown or dark brown; legs dark slate grey or blue-grey. Sexes alike. SIZE (2 ♂♂, 5 ♀♀): wing, ♂ 50, 48, ♀ 46–53 (48·8); tail, ♂ 28, 28, ♀ 26–31 (28·0); bill, ♂ 9, 9, ♀ 8·5–9·5 (8·8); tarsus, ♂ 11·5, 12, ♀ 11·5–13 (12·0).
IMMATURE: juvenile duller yellow on forehead; yellow underparts paler, less bright.
A. p. aureus Bannerman: N Ghana. Duller than *senegalensis*, forehead olive–yellow, contrasting less with crown and with few spots; paler yellow below, tinged olive on throat and upper breast.
A. p. parvulus (Heuglin): Chad to Sudan and NE Zaïre. Upperparts duller than in *senegalensis*, more olive-green, with contrasting yellow forehead; yellow underparts duller and paler. Intergrades with *senegalensis*.

Field Characters. Length 8 cm. A *mainly yellow* bird of semi-arid savanna north of the equator. Yellow-green above, forehead *brighter* yellow with tiny black spots; wings and tail *dark* with conspicuous *white feather-edgings*. Abuts range of Sennar Penduline Tit *A. punctifrons* which lives in drier country to the north; readily distinguished from it by yellow rather than creamy underparts. Forest Penduline Tit *A. flavifrons* has yellow underparts but is confined to forest.

Voice. Tape-recorded (104, 108, B, MOR). Song, rapid insistent 'ska-ska-ska-ska-ska-ska-ska' or double 'pichee-pichee-pichee ...'; 6–12 notes lasting 1–2 s; dull and tuneless, not whistled. Contact call a thin, quiet 'si, slii-li-lii', vaguely reminiscent of Long-tailed Tit *Aegithalos caudatus*; other calls include a high-pitched but full and slightly hoarse 'bzee-bzee-bzee-bzee', and a deep, buzzing, rhythmical phrase of 3–8 notes, 'chura-chura-chura' or 'duza-duza-duza', also a short 'ch, ch, ch' (Harrap and Quinn 1996).

General Habits. Inhabits semi-arid soudanian and N guinean savannas, keeping to dry and riparian woodlands and open areas with spaced shade trees, as well as stands of *Acacia nilotica* and acacia scrub in open sandy or short-grass (sometimes rank grass) areas. At Zaria, near S limit in Nigeria, occurs in light woodland, plantations including fuelwood reserves with plentiful secondary growth of native trees, and shrubs only 2 m tall anywhere in degraded, deforested and overgrazed countryside. Occurs in pairs, parties of 3–6 and occasionally up to 25 birds. Active, mobile, rather silent. Forages often in mixed-species flocks, which include white-eyes *Zosterops* and warblers *Eremomela*. Flight buoyant. Resident and possibly partial migrant; in Mauritania thought to move northward in rains; in Gambia described as a wet-season visitor to lower river (Jensen and Kirkeby 1980); in Mole Nat. Park, near south of range in Ghana, species fairly common in wet season (May–Sept) but not seen at end of dry season (Apr) (Taylor and Macdonald 1978); not thought to be migratory in Nigeria (Elgood *et al.* 1973).

Food. Insects: 2 observers have noted small caterpillars.

Breeding Habits.
NEST: (only one described: Serle 1943) a beautiful and elaborate pendent pouch made of white pappus 'so closely woven as to have almost the texture and toughness of surgical lint'. Upper end closely knit to suspending twig. Shape ellipsoidal, ext. 115 from top to bottom, 60 from side to side, 67 from front to back. Entrance passage slopes down from top of nest and is 40 mm long; roof and floor of passage 40 broad, 'in apposition in their whole extent' and needing to be separated by bird entering or leaving nest chamber. Walls of nest uniformly 2 mm thick but floor (of nest, not passage) 15 thick. In places, outside surface thrown up into ridges. Sited 5·5 m above ground, suspended from one of highest sprigs of a young, soft-leaved tree, in arid country with thorn and soft-leaved shrubs (Serle 1943).
EGGS: 2. Ovate or oval, thin-shelled, glossless dead white.
SIZE: (n = 2) 12·9 × 9·2, 13·1 × 9·6.
LAYING DATES: Gambia, Mar; Mali (Bamako), Feb–Mar; Ghana, (♀ had finished laying, Jan); Nigeria (Argungu) June.

References
Harrap, S. and Quinn, D. (1996).
Serle, W. (1943).

Plate 6 (Opp. p. 79)

Anthoscopus punctifrons (Sundevall). Sennar Penduline Tit; Sudan Penduline Tit. Rémiz du Soudan.

Aegithalus punctifrons Sundevall, 1850. Kongl. Svensk. Vet. Akad., 7, p. 129; Sennar.

Forms a superspecies with *A. parvulus*, *A. musculus*, *A. caroli* and *A. minutus*.

Range and Status. Endemic resident, S borders of Sahara: Mauritania to Eritrea. Mauritania, very local but abundant, from Senegal valley north to about 17°N but to 17°30′N at Tagant. Senegal, in NE west to Richard-Toll (where breeds, and evidently overlaps with Yellow Penduline Tit *A. parvulus*) and south at least to 15°S. Mali, widespread but local in sahel zone between 15°N and 17°30′N; locally abundant in woods around lakes and boreholes. Niger, records (most in 1920s) from Tahoua region, Takoukout, Farak and Bagzans Mts (17°45′N, 08°45′E). Nigeria, confined to sahel zone in extreme NE; rare – records from Logomani and near Mongonu and Gashegar. Cameroon, rare, in extreme N: Waza Nat. Park and S border of L. Chad. Chad, quite common in Kanem region (northeast of L. Chad) and throughout sahel and soudanian savannas (Salvan 1967); in Ouadi Rime-Ouadi Achim Faunal Res., present in all major wadis including Wadi Sofaya, 15°55′N (Newby 1980). Sudan, fairly common through Darfur east to about 30°E, and near Nile north to about El Kirbekan and south to L. No, also widespread in Blue Nile Province. Eritrea, uncommon in W and centre, below 1075 m.

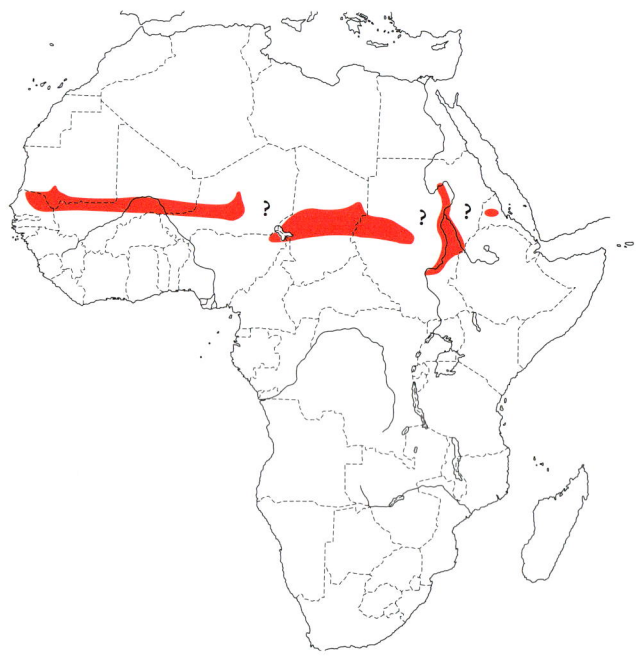

Anthoscopus punctifrons

Description. ADULT ♂: forehead and forecrown yellow, variably spotted with blackish brown; rest of crown, nape and hindneck pale olive-green; mantle and scapulars to uppertail-coverts pale greenish grey. Tail dark brown, feathers fringed pale buff. Dusky stripe across lore, and short, narrow dusky stripe behind eye; narrow yellow supraloral stripe extending back to above eye; ear-coverts, cheeks and sides of neck buffish white. Chin and throat off-white; rest of underparts washed with buff, more strongly from lower breast and flanks to undertail-coverts. Upperwing brown; primaries, alula and primary coverts narrowly edged greyish buff; secondaries more broadly edged olive-yellow; tertials and greater coverts fringed olive-yellow and tipped pale buff; median and lesser coverts broadly fringed greenish grey. Axillaries, underwing-coverts and inner borders of flight feather undersides buffish white. Bill blackish or slate grey, base of lower mandible and cutting edges pale grey; eyes light hazel to blackish brown; legs slate or blue-grey. Sexes alike. SIZE (10 ♂♂, 10 ♀♀): wing, ♂ 50–53 (51·6), ♀ 48–52 (50·3); tail, ♂ 27–31 (28·7), ♀ 27–30 (29·0); bill, ♂ 8–9·5 (8·8), ♀ 8·5–9·5 (9·0).

IMMATURE: juvenile like adult but upperparts slightly duller and greyer, fringes of wing feathers less olive.

Field Characters. Length 8·5 cm. A pale bird of the Sahel, told from Mouse-coloured Penduline Tit *A. musculus*, which it meets in E and SE Sudan, by *greenish upperparts* and more conspicuously spotted forehead, and from Yellow Penduline Tit *A. parvulus*, which lives in less arid country to the south, by *creamy white belly*. Forehead brighter than top of head, pale yellow with black spots; rump paler than back and buffier. Dark brown wings and tail contrast with paler back and underparts but pale feather edgings are less conspicuous than in Yellow Penduline Tit.

Voice. Tape-recorded (68, 104, B, CHA, MOR). Song (?) a buzzy 'dzzeewi-dzzeewi ...' or high, thin whistled 'seewi-seewi ...' repeated 6–9 times, second bird following or partly accompanying with a hard, dry rattle lasting 2–3 s. Same or similar calls transcribed as 'a high, thin, plaintive "tsui-tsui-tsui-tsui", an even thinner "tsee-tsee-tsee ...", and a rhythmic, mechanical, buzzing "bizur, bizur, bizur, bizur"' (Harrap and Quinn 1996); also said to give a short 'tsit'.

General Habits. Inhabits sahel zone, semi-desert or desert steppe with acacia and *Balanites* scrub in wadis, near wells and boreholes, and, towards southern border of range in Chad, soudanian savanna woodlands including gallery woods along permanent and seasonal watercourses.

Food. Not known. When actively exploring acacia foliage, birds thought to be foraging for small insects.

Breeding Habits. Barely known; nests found but nest and eggs not described.

NEST: said to be like nest of Eurasian Penduline Tit *Remiz pendulinus*, but larger. One nest contained 2 young.

LAYING DATES: Mauritania, (breeding season June–Sept); Senegal (Richard-Toll), June–July (and bird ready to lay, last day July); Mali (Goundam), July–Aug; Nigeria (Logomani), July; Chad, (nest with young, Sept; in Ouadi Rime breeds in wet season, i.e. July–Sept); Sudan, Feb, Mar, Sept; Eritrea (Af Abet, ♂♂ with enlarged gonads, Apr).

Anthoscopus musculus (Hartlaub). Mouse-coloured Penduline Tit. Rémiz souris.

Plate 6
(Opp. p. 79)

Aegithalus musculus Hartlaub, 1882; Orn. Centralbl., 7, p. 91; Lado, Upper Nile, Sudan.

Forms a superspecies with *A. parvulus*, *A. punctifrons*, *A. caroli* and *A. minutus*.

Range and Status. Endemic resident, E Africa. Ethiopia, frequent in hot lowlands below 450 m in Rift Valley and S and SE Ethiopia. Somalia, rare, <15 records, mainly in S. Sudan, single record in Kassala district, but fairly common in S between Bor and foothills of Dongotona Mts. Uganda, uncommon at 400–1600 m from Kidepo Valley Nat. Park to Moroto and Elgon. Kenya, uncommon in arid and semi-arid country, widespread north of SW highlands and west of L. Turkana, southeast of L. Turkana through Marsabit to Isiolo and Tana R. drainage, with a few scattered records to northeast including Huri hills and Wajir district; frequent in Samburu region; uncommon in Rift Valley (Bogoria, Baringo, Magadi), widespread in W Tsavo but rare in Tsavo East Nat. Park, mainly in north (Lack 1985). Just into Tanzania, in lowlands between Mt Kilimanjaro, Arusha and North Pare Mts (occurs up to 900 m in North Pares).

Description. ADULT ♂: forehead, crown, upperparts, and sides of neck, greyish olive. Tail dark grey brown, feathers narrowly fringed pale greyish buff. Stripe through lore and upper ear-coverts dark grey-brown. Some birds with indistinct pale buff superciliary stripe. Rest of ear-coverts and cheeks buffish white. Underparts whitish, washed pale buff on flanks, belly and undertail-coverts. Upperwing dark grey-brown; primaries (proximal to emarginations), secondaries, alula and primary coverts narrowly edged pale buff; tertials and greater coverts fringed pale greyish buff; median and lesser coverts broadly fringed greyish buff. Axillaries, underwing-coverts and inner borders on undersides of flight feathers buffish white. Bill black, base and cutting

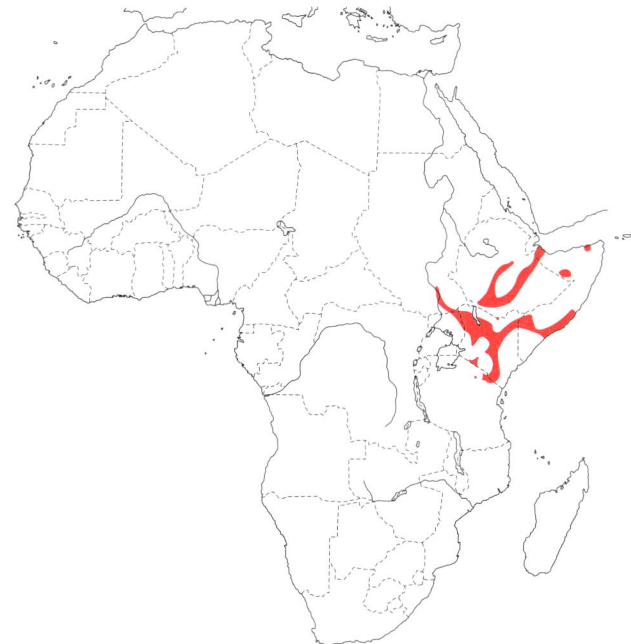

Anthoscopus musculus

edges pale greyish or yellowish; eyes brown; legs slate-grey to blackish. Sexes alike. SIZE (10 ♂♂, 10 ♀♀): wing, ♂ 46–52 (49·3), ♀ 46–51 (48·0); tail, ♂ 25–29 (27·1), ♀ 23–28 (25·9); bill, ♂ 8–9 (8·7), ♀ 8–9 (8·6); tarsus, ♂ 13–14 (13·4), ♀ 12–14 (12·9).

IMMATURE: juvenile resembles adult but upperparts and edges of primaries and secondaries a warmer brown.

Field Characters. Length 6·5–8·5 cm. The penduline tit of NE Africa; range abuts those of Yellow Penduline Tit *A. parvulus* in Nile Valley near Mongalla, Sudan, and Grey Penduline Tit *A. caroli* in NE Uganda and S Kenya (breeding within 50–100 km of each other in Kerio Valley, and have wandered into each other's range: Wilson and Wilson 1994). Distinguished from Yellow Penduline Tit by *white underparts*; from Sennar Penduline Tit by *drab*, mouse-coloured *upperparts*, and from Grey Penduline Tit by *grey* top of head continuing *to bill* (no pale band on forehead). More washed-out and colourless than all races of Grey Penduline Tit, whose underparts and forehead vary from yellowish to buff and cinnamon. Tiny size, conical bill and short tail should distinguish it from similar nondescript warblers, e.g. Yellow-vented Eremomela *Eremomela flavicrissalis* which has longer legs and finer bill (yellow on vent can be hard to see), or Buff-bellied Warbler *Phyllolais pulchella* (buff underparts, longer, white-sided tail).

Voice. Tape-recorded (B, GREG, McVIC, PEAR). 2 main song motifs, often alternating with each other (different birds?), a hard rattling trill with a sibilant overtone, lasting 1–2 s, and a light ringing 'si-clee-si-clee-si-clee ...' or 'see-cli-see-cli-see-cli ...', with quality of European Goldfinch *Carduelis carduelis*; also a thin 'si-tsi-tsi-tsi-tsi-tsi'; for further renditions see Harrap and Quinn (1996). Contact call, often given in flight, 'tit, tit' or 'stit, stit'; also a harsh 'jeea'.

General Habits. Inhabits arid acacia woodland and wooded grassland and thick stands of semi-arid thorn scrub. Habitat around wells at Bohetleh, Somalia, was such dense thorn bush that human entry was almost impossible without cutting a track (Archer and Godman 1961).

Food. Unknown.

Breeding Habits. Solitary nester.
NEST: pendent pouch or bag with closable entrance at end of short tube near top, almost identical to nests of Grey Penduline Tit *A. caroli* and Cape Penduline Tit *A. minutus* (Benson 1946b); one made of fibres from plants of families Compositae and Asclepiadaceae, another of dead white *Acacia* flowers. Sited 1·5, 2, 4·5 and 6 m above ground, quite unconcealed, in thorn bush or tree.
EGGS: 4 (single clutch; another nest contained 2 hatchlings: Harrap and Quinn 1996). Pure white. SIZE: (n = 3, from 1 clutch) 13·2 × 9·2, 13·5 × 9·5, 13·0 × 10·0.
LAYING DATES: Sudan, Dec; Ethiopia, Feb, Mar, Sept, Oct; Somalia, Nov; Kenya Apr or May.
INCUBATION: by ♀ (and ♂?); in one nest, ♀ came several times into exit spout to view observer, then closed entrance aperture from inside (and presumably resumed incubating); later, entire nest with contained ♀ and eggs was collected, the ♀ apparently not attempting to escape.

Reference
Harrap, S. and Quinn, D. (1996).

Plate 6
(Opp. p. 79)

Anthoscopus caroli (Sharpe). Grey Penduline Tit; African Penduline Tit. Rémiz de Carol.

Aegithalus caroli Sharpe, 1871. Ibis, p. 415; Ovaquenyama, Damaraland, South West Africa.

Forms a superspecies with *A. parvulus*, *A. punctifrons*, *A. musculus* and *A. minutus*.

Range and Status. Endemic resident, mainly southern tropics. Uganda, widespread but uncommon in S half, from Lango district southwest to Toro and southeast to Elgon; absent from extreme SW? E Zaïre (Kivu): Butembo (08′N, 29°17′E). Kenya, uncommon, mainly at 1000–2000 m but up to 2200 m, in semi-arid parts of SW, scarce in more humid parts of highlands, evidently absent from Suna-Kericho-Mau-Narok-Loita Plains region. Another population in SE coastal lowlands with records at Mombasa, Taru and Samburu just inland from Mombasa, and north of lower Tana River; this population thought to have disappeared in recent years (Harrap and Quinn 1996). Rwanda, Burundi, and NW Tanzania: in Kagera R. system (Rwanda/Tanzania border, and Kagera district of Tanzania partly along Uganda border), Akanyaru R. system (E Rwanda/E Burundi border), Ruvuvu and Kumoso basins (NE Burundi); east in Tanzania to Nyarumbugu and Usambiro (02°59′S, 32°30′E). Absent from much of west half of Tanzania, but known from Kigoma in W and Ufipa Plateau and Mbeya-Rungwe district in SW; widespread in SE south of 09°S (Songea, and S Nachingwea District) and in E (west to Dodoma) and NE (Kidugallo, Korogwe, Naberera and Longido, to Kenya border except Lushoto District and Mkomazi Game Res.) (map data: N. Baker, pers. comm.).

Congo, uncommon in Odzala Nat. Park (Dowsett-Lemaire and Dowsett 1998). Zaïre: Butembo (see above); Lukolela-Bolobo-Kwamouth stretch of Congo. R. valley; widespread in Kasai Occidental; Shaba Prov. north to about 10°S; likely to range continuously between Lukolela, Angola border and S Shaba; E Shaba from Marungu north to about R. Lukuga. Angola: NW Malanje south thrugh E Cuanza Sul to plateau where widespread above 1500 m, south to Cunene District and east up Cubango valley; record(s) from Dundo, NE Lunda Norte; not known to occur in Moxico, although likely to be in Zambeze salient in E. Namibia, extreme N Ovamboland, Kavango and Caprivi; recorded at Windhoek. Zambia, almost through-

Anthoscopus caroli

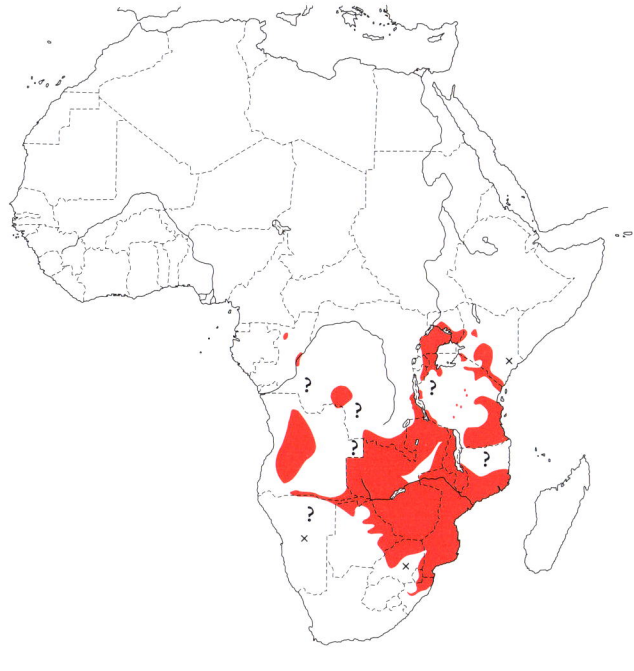

out; absent from Luangwa and most of Middle Zambezi valleys. Malaŵi: widespread and common at 750–1550 m; somewhat less common below 750 m, but recorded down to 80 m. Mozambique, widespread south of Lurio R. and probably north of it. Botswana, uncommon to quite common in N and E; in E, may extend as far west as Orapa only in good rainfall years. Zimbabwe, frequent throughout, absent only from E highlands (though present in Chipinga uplands). Transvaal, uncommon, local in bushveld but widespread in lowveld. Swaziland, except in W. Natal, in N Zululand from littoral plain southwestward; west to Ladysmith and southwest to about Estcourt.

Description. *A. c. caroli* (Sharpe): N Namibia, S and SW Angola (central and S Huila and Cubango), N and E Botswana and SW Zambia (Barotse, Southern and Central Provs). ADULT ♂: forehead pale cinnamon-buff, rest of upperparts and sides of neck light grey, tinged olive-green. Tail dark grey-brown, feathers narrowly fringed pale buff. Dark blackish stripe through lore and upper ear-coverts, and above this a narrow pale cinnamon-buff superciliary stripe, extending to just behind eye; rest of ear-coverts and cheeks pale cinnamon-buff. Chin to upper breast whitish, faintly tinged buff, merging through buff lower breast to deeper tawny-buff on belly, rear flanks and undertail-coverts. Upperwing dark grey-brown; primaries narrowly edged pale greyish buff proximal to emarginations, secondaries, tertials, alula and primary coverts edged pale olive-grey; greater coverts fringed pale greyish buff, whiter on tips, forming a diffuse wingbar; median and lesser coverts broadly fringed light grey (dark centres concealed). Axillaries, underwing-coverts and inner borders on underside of flight feathers whitish. Bill dark grey with pale cutting edges; eyes dark brown or blackish; legs slate-grey to blackish. Sexes alike. SIZE (10 ♂♂, 10 ♀♀): wing, ♂ 51–55 (52·5), ♀ 52–55 (53·6); tail, ♂ 28–31 (29·7), ♀ 28–31 (30·1); bill, ♂ 8–9·5 (8·7), ♀ 8·5–9 (8·7); tarsus, ♂ 13–14 (13·5), ♀ 13–14 (13·4).

IMMATURE: juvenile more brownish on head and upperparts; breast mottled dusky; lower underparts paler buff.

A. c. hellmayri Roberts: E and S Zimbabwe, S Mozambique (north to Inhambane district and inland to Save R. and SW Manica Sofala), South Africa (N and E Transvaal and N Natal) and E Swaziland. Poorly marked. Like nominate *caroli* but throat and breast more strongly tinged buff; tawny-buff rear flanks, belly and undertail-coverts averaging darker. WEIGHT: Zimbabwe, 1 ♂ 6·5, 2 ♀♀ 6·2, 6·9.

A. c. robertsi Haagner (including '*taruensis*'): Mozambique north of Inhambane district, Malaŵi, SE Zambia (E Province plateau) and interior E Tanzania and SE Kenya. Similar to *caroli*, but more olive-grey above; sides of head pale yellowish buff (rather than cinnamon-buff); throat and breast whiter, tinged yellowish; buff lower underparts paler, tinged more yellowish. Birds of NE Tanzania and SE Kenya ('*taruensis*') average slightly smaller; wing, ♂ (n = ?) 49–51.

A. c. winterbottomi (White): NW Zambia (Northwestern and Western Provinces south to *c.* 14°) and adjacent S Zaïre (S Katanga). Poorly marked, intermediate between nominate *caroli* and *rhodesiae*. Resembles *robertsi* but upperparts slightly more olive.

A. c. roccatii Salvadori: Uganda, NE Zaïre (Butembo in N Kivu), Rwanda, Burundi, extreme NW Tanzania and W Kenya borders (Kongelai and Kapenguria to Bungoma). Differs from all previous races in having upperparts greyish olive; forehead and superciliary stripe pale yellow; sides of head, chin and throat tinged grey, rest of underparts pale yellow, deeper and buffier on vent and undertail-coverts.

A. c. ansorgei Hartert: Angola (except S and SW) to SW and W Zaïre; intergrades with *rhodesiae* in SW Katanga. Upperparts bright olive-green, forehead and superciliary stripe yellow; sides of head and underparts whitish, rear of flanks, belly and undertail-coverts tinged yellow; upperwing and tail feathers dark grey, edged olive-green, greater coverts tipped yellowish.

A. c. rhodesiae Sclater: SE Zaïre (SE Katanga from Lubumbashi north to Marangu Mts and Tembwe) to NE Zambia (south to Luwinga and Kasama, east to Mbala) and SW Tanzania (Ufipa Plateau); intergrades with *winterbottomi* in S Katanga and E Zambia. Differs from *ansorgei* in having forehead paler yellow, olive-green upperparts duller, sides of head and chin to breast greyish white, and rear of underparts pale cinnamon-buff; secondaries, tertials, greater coverts and tail feathers edged pale olive-yellow.

A. c. pallescens Ulfstrand: W Tanzania (Kigoma, Kungwe–Mahari). Like *rhodesiae* but paler, more greyish green above; greyish white underparts washed yellow, undertail-coverts tinged buff.

A. c. sylviella Reichenow: S-central Kenya east of rift (Murang'a and Kitui south to Kajiado, Simba and Voi) and central Tanzania (Longido south to Dodoma, Iringa and Mbeya-Rungwe district); west of *robertsi*. Similar to *caroli*, with grey upperparts (olive tones faint or absent), but whole underparts tawny-buff, paler on chin and throat; forehead pale buff to deep tawny-buff; sides of head pale buff.

A. c. sharpei Hartert: SW Kenya and N Tanzania east and south of L. Victoria (Kakamega and Nyanza to Lolgorien, Mara Game Res., Serengeti and Usambiro); probably this race in W-central Kenya (L. Baringo to Kerio valley and Nakuru district). Poorly marked. Like *sylviella*, but darker, more cinnamon below, forehead dull cinnamon.

A. c. rankinei Irwin: NE borders of Zimbabwe (Middle Zambezi and Lower Mazoe rivers to Humani Ranch). Differs from *hellmayri* in having upperparts dark slate grey, without olive tinge; forehead and ear-coverts whitish, not buff; tawny-buff lower underparts paler.

TAXONOMIC NOTE: the 11 subspecies fall into 5 groups (Harrap and Quinn 1996): *caroli*, *hellmayri*, *robertsi* and *winterbottomi* of eastern and southern Africa are mainly grey and buff; *roccatii* is an isolated E-central African olive-and-yellow form; *ansorgei*, *rhode-*

siae and *pallescens*, from humid parts of central Africa, are mainly green and yellow; *sylviella* and *sharpei* of E Africa are grey and tawny-buff; and *rankinei* (NE Zimbabwe) is grey-backed, quite distinct from adjacent *hellmayri* and *robertsi*.

Vast area of apparent absence in Tanzania is surrounded by complex of races. *A. c. pallescens* is endemic to Kigoma and Mahari in W; clockwise from it are *roccatii*, *sylviella*, *robertsi* and *rhodesiae*; taxonomy and range need further research; *sylviella* seems to be 'interior' race and may prove to fill out at least N parts of 'absent' area.

Field Characters. Length 8–9 cm. Widespread in central and southern Africa, plumage varying considerably according to race. Overlaps with Mouse-coloured Penduline Tit *A. musculus* in Kenya and N Tanzania, where races *robertsi* and *sylviella* are distinguished by *pale forehead, buff to cinnamon underparts*, brightest on lower belly, and *sylviella* also by buff ear-coverts; typically occupies moister habitats. Nominate race meets Cape Penduline Tit *A. minutus* in S Angola, N Namibia, N and E Botswana, SW Zimbabwe and W Transvaal, but the 2 are usually segregated by habitat; *greyer*, with buff forehead, plain buff face *without speckling* on dark lores, white throat and breast and yellowish buff belly and vent. Short, conical bill distinguishes it from eremomelas and other warblers.

Voice. Tape-recorded (86, 88, 91–99, 104, B, F, GREG). Song a long up-slurred note preceded by 2 short ones, either buzzy, 'chi-chi-dzwizzz' or thin and lisping, 'ti-ti-ssssweee' or 'tsi-tsi-weee', repeated 3–4 times; some songs lack introductory notes, 'tsweee, tsweee, tsweee', 'chizzizzi, chizzizzi, chizzizzi' or a high, thin 'tsee-weee, tsee-weee, tsee-weee'; sometimes has a dry, *Prionops*-like quality, 'dzzizzyaa' or 'dzzazzyaa'. A squeaky, disyllabic, whistled motif, 'tschwee, tschwee ...' given as location call by 'lost' birds separated from flock (Vincent 1934, 1935). Other calls include buzzy 'jaa', thin 'jip', a hard trill, 'trrrrri' and a high-pitched, sibilant trill, 'see-see-see ...', gradually dying away (Ulfstrand 1960). Calls may vary by subspecies, see Zimmerman *et al.* (1996).

General Habits. Inhabits medium rainfall, lowland savanna woodland, but whether microphyllous thorn or broad-leaved types varies across range. In Odzala, Congo, in *Hymenocardia* woodland. In E Africa occurs in woodland, wooded grassland and forest-edge situations from coast up to 2200 m (in Kenya 72% of records at 1000–2000 m), in semi-arid, sub-humid and humid regimes. In Zambia mainly in *Brachystegia spiciformis* and *Julbernardia globiflora* (miombo), but in SW in *Baikiaea* woods (mutemba). In Zimbabwe has strong preference for *Acacia* which it inhabits especially in absence of Cape Penduline Tit, but in Chibi district (S-central Zimbabwe), where common, is confined to Mountain 'Acacia' *Brachystegia glaucescens*. Inhabits miombo woodland canopy on Mashonaland Plateau, Zimbabwe, but on Charama Plateau and Chirisa Game Res., on Kalahari Sands, avoids miombo and keeps to drier savanna woods in valley bottoms (Irwin 1981). Local use of various habitats complicated by seasonal movements.

Occurs in pairs and family parties of up to 7 birds. Keeps mainly to tree canopy but sometimes in low shrubs; active, restless, movements very rapid; hard to find and re-find. Forages among foliage, around leaf buds, in clusters of flowers and in fruiting trees and bushs. Birds of pair or in flock keep in contact by calling frequently. Joins mixed-species foraging flocks, e.g. in W Tanzania feeds with Yellow-bellied Hyliotas *Hyliota flavigaster*, Green-capped Eremomelas *Eremomela scotops*, African Dusky Flycatchers *Muscicapa adustus* and Spotted Creepers *Salpornis spilonotus*. Seen in flock with Cape Penduline Tit in Botswana. Flight dipping, light and airy; birds in flock tend to fly between trees one-by-one. Roosts gregariously in its nest, even when it contains eggs (Mackworth-Praed and Grant 1963), and once 12 birds roosted in old nest of Spectacled Weaver *Ploceus ocularis* (Clancey 1962). Lips of nest are closed by bird on leaving by tweaking with bill – bill marks said to be sometimes visible on felt at entrance slit of sealed entrance spout (but better-known Cape Penduline Tit, *q.v.*, closes entrance in another way); entrance is also closed by entering bird from inside. To open entrance from outside, bird alights on side of nest, holds on with one foot, grasps lower lip of entrance with other foot and pulls it down to open it (Ginn *et al.* 1989).

First prebasic moult incomplete. Second and subsequent prebasic moults complete; primary and rectrix moult regular, secondary moult irregular; some contour-feathers moult during breeding season; first-year birds moult after adults have finished moulting (Austin 1978).

Food. Small insects; ticks (photo in Ginn *et al.* 1989); said to eat small fruits.

Breeding Habits.
NEST: one of the wonders of the bird world (Milstein 1975); very like better-described nest of Cape Penduline Tit (Skead 1959); pendent, pocket-, ball- or pear-shaped bag made of woolly vegetable down, e.g. wild cotton, often putty-coloured or greyish white, densely felted, with soft, smooth surface and texture of ordinary blanket, but resistant to deformation rather than floppy – springs back ino shape after being lightly pressed (Vincent 1949); fabric so strongly woven that nest can be torn by person only with difficulty. Entrance at side, just above half way up, through floppy, pinched-together 25 mm long projection which opens to 25 mm in diameter (**A**). Immediately below entrance spout, partly obscured when spout flops over it, is shallow concavity, the false entrance. When entrance is closed, spout does not flop but projects outwards and cavity below is not concealed. Height *c.* 125, diam. *c.* 75. Entrance spout wall *c.* 2 mm thick; rear wall of egg chamber *c.* 9 mm thick; floor 30 thick, and front wall of egg chamber, immediately below false entrance, *c.* 40 thick; chamber is thereby toward top of nest and is shape of head of driver golf club (**B**). Nest firmly attached by top to twig or twigs, which are usually embedded in nest fabric. Sited 2 m above ground in acacia veld but up to 7 m high or in canopy in miombo woods; attached to fork in twigs near end of branch; not concealed in foliage. Nest is repaired and given constant attention during breeding season (Ginn *et al.* 1989).

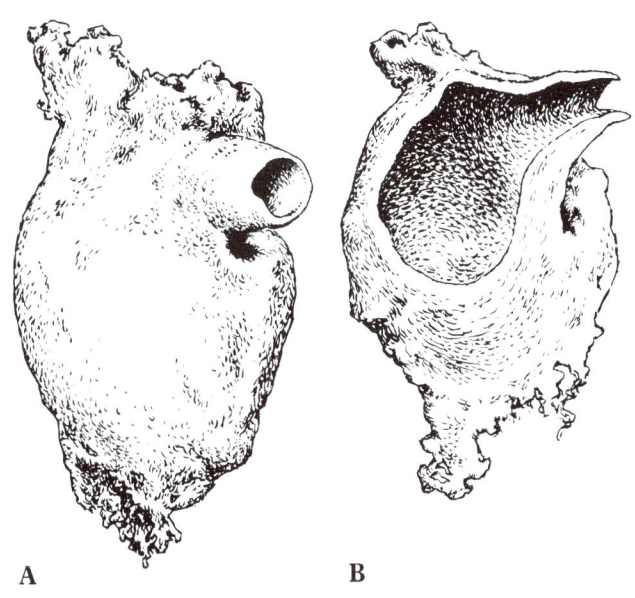

EGGS: 3–6 (av. 4·4). Chalky white, with overlay of dirty cream colour giving shell a fair gloss (Vincent 1949). SIZE: (n = 67) 13·3–16·6 × 8·5–10·5 (14·2 × 9·5).

LAYING DATES: mainly in rainy season. Congo, (nest-building Feb); E Africa, Region A, Dec, Region B, Apr–June and Oct; Region C, Dec, Feb, Region D, Sept, Dec; Tanzania, (pair copulating by recently completed nest, Dec); Zambia, Sept–Nov; Zimbabwe, Aug–Feb, mainly Sept–Nov (79 out of 88 records); Botswana, Nov; Transvaal, Oct–Dec, mainly Oct; Natal, Sept–Dec.

INCUBATION: bird incubating hard-set eggs flew out of nest on observer's approach, perched nearby, and preened unconcernedly while nest was examined.

DEVELOPMENT AND CARE OF YOUNG: young fed by both parents. Entrance spout may be left open for up to 5 min when parents bringing food to chicks frequently (Ginn et al. 1989). One young bird in well-feathered brood waits in entrance tube, its head poking out, for parent's return with food (Vincent 1949).

References
Austin, G.T. (1978).
Benson, C.W. (1955).
Ginn, P. J. et al. (1989).
Harrap, S. and Quinn, D. (1996).
Milstein, P. le S. (1975).

Anthoscopus minutus (Shaw and Nodder). Cape Penduline Tit. Rémiz minute. Plate 6

Sylvia minuta Shaw and Nodder, 1812. Nat. Misc., 23, pl. 997, ex Levaillant 1802, Oiseaux d'Afrique, 3, pl. 134, fig. 1; Heerenlogement, Cape Province, South Africa. (Opp. p. 79)

Forms a superspecies with *A. parvulus*, *A. punctifrons*, *A. musculus* and *A. caroli*.

Range and Status. Endemic resident in regions with <250 mm annual rainfall, and somewhat nomadic. W Angola and W Zimbabwe to S Cape Prov., but absent from a broad belt about longitude 19°E, south of Tropic of Capricorn (i.e. SE Namibia and W-central Cape Prov.). Mapped from Harrison et al. (1997).

Angola, known only along arid coast between Benguela and Huxe, and at Mucope, NW Cunene. Namibia, sparse throughout N half above about 1000 m, and occurs in corridor east of Namib Desert from inland of Walvis Bay to Orange R. east to about 18°E. Zimbabwe, restricted to area from Nkai and Nyamandhlovu to Matopo Hills, Figtree and Inseze; at Dzivanini Pan, Hwange Nat. Park; and a record further north, near Zambezi R. (Hall and Moreau 1970). Botswana, uncommon to fairly common, but absent from central and N areas as mapped. Transvaal, locally common in bushveld region, between about 24°30' and 26°S and 27°30' and 30°E, east to Burgersfort; scarce north of 24°S, east to Messina; very local in W Highveld; almost absent south of 26°S except in extreme SW. Orange Free State, uncommon in extreme W and rare elsewhere, with records near Bloemfontein (where has bred), near Parys and Marquard and between Bethulie and Aliwal Noord. Cape Prov., uncommon in SW Cape, west of 20°E, from Orange R. to Cape of Good Hope and Cape Agulhas; frequent, locally common in bushveld, in central Cape,

Anthoscopus minutus

mainly between 23° and 25°E. Density: flocks of 3 and 5 birds can forage only 300 m apart (Skead 1959).

Description. *A. m. damarensis* Reichenow: Angola, N Namibia, Botswana except SW, Zimbabwe and Transvaal. ADULT ♂: broad black band across forehead at base of bill to lores; behind it a broad whitish band across forehead (feathers black with white tips; forehead often looks mottled), turning back as a short, deep, whitish superciliary stripe; narrow, speckled, dark grey band across forecrown (dark feather bases showing through) behind the white; rest of crown, nape, hindneck and sides of neck mid grey. Narrow black stripe behind eye; fore ear-coverts and cheeks to chin whitish, mottled with grey; rear ear-coverts and sides of neck grey, concolorous with hindneck. Mantle, scapulars and back greyish green, becoming greener on rump and greenish yellow on uppertail-coverts. Tail dark grey-brown, feathers fringed pale greyish buff. Throat pale yellow, merging into deeper yellow on rest of underparts. Thighs buffish white. Upperwing feathers dark grey-brown; primaries (proximal to emarginations) and secondaries edged pale greyish or whitish buff, more broadly on inner secondaries; primary coverts and alula narrowly fringed pale greyish buff, tertials and greater coverts more broadly so; lesser coverts broadly tipped mid grey (dark feather centres concealed). Axillaries, underwing-coverts and borders of inner webs of flight feathers creamy white. Bill black or blackish, with pale cutting edges; eyes yellowish brown to dark brown; legs blackish, grey, cobalt blue, blue-grey or grey-green. Sexes alike. SIZE (9 ♂♂, 9 ♀♀): wing, ♂ 48–52 (50·1), ♀ 49–53 (51·0); tail, ♂ 32–37 (33·8), ♀ 33–37 (34·4); bill, ♂ 8·5–9·5 (8·7), ♀ 8–9 (8·4); tarsus, ♂ 13–15 (13·8), ♀ 13–15 (13·9). WEIGHT: (n = 4) 6·9–8·6 (7·5).

IMMATURE: juvenile similar to adult, but browner above and slightly paler and duller yellow below; fringes of wing and tail feathers washed yellow.

NESTLING: naked, skin orange, gapes yellow.

A. m. minutus (Shaw and Nodder): areas with <120 mm rainfall in W Namibia (south of about 23°S), W Cape Prov., SW Botswana and adjacently in N Cape, NW Orange Free State, and just into S Transvaal (Hamanskraal: Clancey 1997). Darker and more brownish above; duller, more buffish (less yellow) below.

A. m. gigi Winterbottom: Cape Prov. east of 20°E and SW Orange Free State; intergrades with *minutus* along lower Vaal R. (Clancey 1997). Upperparts darker than *minutus*, dark greyish olive with rump ochreaceous olive; chin and throat light greyish fawn with sides darker than in other races; rest of underparts more olivaceous than in *minutus*.

Field Characters. Length 8–9 cm. This and the last species are the smallest birds in southern Africa, told from short-tailed eremomelas and crombecs by short, conical bills. Marginally overlaps Grey Penduline Tit *A. caroli* in N and E Botswana and W Transvaal, and comes close to it in S Angola, N Namibia and SW Zimbabwe, but they are normally segregated by habitat (Penry 1994, Tarboton *et al.* 1987); overlap may involve both vagrants and residents. Distinguished from Grey Penduline Tit by *black speckles on forehead coalescing into black line above bill*, short white supercilium, and *black line through eye* from lores to just behind eye; further, northern race *damarensis* also has yellow underparts, southern races *minutus* and *gigi* have buffy underparts but brown or grey-brown upperparts and some light speckling on face and throat.

Voice. Tape-recorded (88, 91–99, B, F, LUT). No structured song; calls simple and unmusical; the following are detailed by Skead (1959): contact call a single, quiet, bell-like 'tillink' or 'tillilink'. Assembly call (when flock about to move to new foraging area) a loud, raspy 'zizzit'. Courtship-flight call an abrupt 'zwayt'. High-pitched, plaintive 'tswee, tswee ...', 'chawee, chawee ...' or 'tsiwizza, tsiwizza ...', repeated 6–8 times from prominent perch or from inside bush, is given in excitement, or as advertisement during breeding season, or to assemble flock in pre-roost gatherings. Aggression within flock accompanied by quiet, stuttering 'ch-ch-ch. ...' Other calls (context?) recorded by Gillard (88) and Gibbon (91) include high, thin 'weess' repeated 5–6 times, an up-slurred 'pitooi' or 'tsooi' repeated 3–4 times, a clipped 'chi-chu' or 'chi-chucher' and a brief 'tsit' or 'tsi-tit'.

General Habits. Inhabits arid tree and scrub savanna and sparse acacia woodland on desert edge; in central Kalahari (Botswana) in sub-desert areas with only isolated thorn bushes; also in acacia savanna and secondary thorn bushland (Zimbabwe), and strandveld and karroo fynbos (SW Cape). In strandveld, keeps mainly to islands of remaining natural woody vegetation on steep, stony kopjes. Especially partial to *Acacia karroo*. Arboreal; occasionally forages down close to ground, but very rarely alights on ground except to snatch piece of nesting material.

Occurs in pairs or flocks of 3–4, sometimes up to 8, occasionally up to 21 birds. Flock is close-knit, probably a family party (though unrelated or only distantly-related birds known sometimes to roost in same nest). Flocks occur all year except in period between nest-building and departure of first brood. Restless, active, tame and approachable. Forages in bushes and canopy of thorn trees by moving continually in hops and short flights within canopy, acrobatically investigating foliage, buds, flowers, twigs and the trunk. Peers under leaves, swings beneath twigs, and pecks at galls, and obtains much of its insect food from acacia leaves (caterpillars), from spider's webs among acacia leaflets, and from stem galls and rough bark (Skead 1959). Caterpillar is held beneath one foot and torn piecemeal. Birds in flock call frequently to maintain contact. Flock tends to keep to itself, but outside breeding season occasionally forages with Grey Penduline Tits *A. caroli*. (e.g. Makgadikgadi Pan, Botswana: Ginn *et al.* 1989), Southern Black Tits *Parus niger*, Chinspot Batises *Batis molitor*, or with small warblers and white-eyes *Zosterops*. Short flights between trees are bouncy, jerky and dashing, birds in small flock tending to follow one by one. When party on the move there are occasional lulls when 2 birds sit touching and alternately preen each other's head. When flock ready to return from foraging area and go to roost in nest, one or more birds give distinctive assembly call, loud 'zizzit', and party moves away. Phrase of repeated 'tswee' notes uttered during excited interactions, bird leaning well back on perch; phrases accompanied by much wing-flicking and some chasing.

Pair roosts at night in nest; young of first brood sleep in nest with parents throughout incubation and nestling periods of second brood; second brood continues to roost in nest at night with parents and first brood: once for 3–4 months after breeding (Skead 1959). One pair slept in nest for 6 months. Highest count was 18 birds roosting in nest (nest capacity only 207 cm^3), and 3 birds in the flock of 21 had to roost elsewhere (Skead 1959). Occasionally roosts in ball nest of other species, e.g. Masked Weaver *Ploceus velatus* (Courtenay-Latimer and Clancey 1961). Flock approaches nest tree a little before sunset, sits preening on surrounding bushes, then birds enter nest one by one, some often leaving just before dark to sleep elsewhere. The 18 birds roosting in one nest had not all been reared there, so birds evidently use nests somewhat indiscriminately. Roosting birds generally close nest entrance from within, but sometimes leave it open. When ambient shade temperature is 16 or 35°C, temperature inside nest is 3·5 or 2·3°C higher, respectively (Skead 1959).

Post-juvenile moult incomplete: includes tail, secondaries, tertials and greater coverts; post-breeding moult (Oct–June) complete; post-juvenile moult rather later in season than post-breeding moult (Austin 1978).

Resident, but sightings in unusual locations and occasionally outside supposed breeding range suggest that some birds wander widely.

Food. Small insects, insect eggs, and berries (Maclean 1993).

Breeding Habits. Solitary nester, strictly monogamous (i.e. nesting and rearing of young conducted by same 2 birds). Double-brooded, second clutch laid about a month after first brood leaves nest; may be serially multiple-brooded. Not a co-operative breeder, but young of first and second broods fail to disperse for many months, so flock of up to 21 birds builds up, as many sleeping in nest at night as it can accommodate. Only courtship behaviour described is flight, whilst nest being built by pair and a few days before egg-laying starts: instead of pair flying from nest more or less side by side, one bird follows the other and repeatedly puts on spurt to catch up, giving abrupt 'zwayt' note. Not obviously territorial. Aggressive interactions rare and only half-hearted (bird advancing to another, gaping bill widely, and calling 'ch-ch-ch...'). Family party ranges up to about 1·5 km from nest (Skead 1959).

NEST: globular or oval bag, with quite long entrance spout and below it a thick-rimmed, deep concavity, the false entrance. Fabric of nest is felt-like, composed predominantly of hairy seeds of kapokbush *Eriocephalus africanus* or (where kapokbush not available) of cotton, silky pappi of Asclepiadaceae and Compositae, cottony fibres of seeds of camphor wood tree *Tarchonanthus camphoratus*, hair of Cape Hare *Lepus saxatilis* and domestic animals, particularly sheep, and cobweb, often of social spider *Stegodyphus*. One nest contained material including old nylon from nearby rubbish dump. Spiders' white silk egg bags often built into outside of walls. Nowadays most nests contain cotton or sheep's wool, even in remote parts of birds' range. Most nests white when newly built, but soon weather to grey or brown; nests made of wool from karakul sheep are blackish. Nest suspended from drooping twig or built around mass of twigs or small upright branches of thorn tree or spiny shrub, commonly *Scutia myrtina* or *Lycium campanulata*, at height above ground of 0·3–5 m, generally 2–3 m. Nest usually conspicuous but often (in tall thorn tree) practically inaccessible. One nest built upon strand of fence wire. Ext. height 130–150, ext. diam. 70–95, int. height 70–110, int. diam. of brood chamber 50–66, wall thickness 5 at top and 20 at base; spout 20–60 long, 20–30 across and 8 deep, thinner-walled, softer and more deformable than rest of nest, and spout projects somewhat downward; false entrance rim 30–40 wide and 20 deep. Nest weight 16–21 g.

Nest constructed by both sexes. Birds start by winding strands of spider web around twig fork to act as anchor, then build a flimsy vertical ring of spider web, 35 mm diam., hanging from it; bird perches in ring like a building weaver *Ploceus*. At 3rd day base of ring is much thickened and walls are starting (**A**). Walls are extended backwards and forwards to form back and front walls of nest; at this stage bag can be thin-walled and lacy, the material only loosely applied (Plowes 1951). Bird brings large wad of plant down, carried in tip of beak, dumps it on nest, seizes strand and pulls it across nest surface, jabbing its ends into the mass, repeating the action until wad dispersed after about 5 min (Skead 1959). Birds of pair often arrive and leave together; on arrival, one starts work, the other dumps its wad or passes it to first bird then waits nearby. Sometimes second bird dumps wad on twig near nest and waits to use it, or leaves it on twig and flies away. Several twigs near nest often have wad dumps on them.

After 10 days bag is completed, birds constantly picking up, placing, jabbing, teasing, pulling and pummelling to make walls matted and thick, but nest still looks lacy and lacks entrance spout (**B**). Opening where spout to be added is 13 wide and 22 deep. Spout takes *c.* 2–5 more days to complete. At one nest work began on false entrance on 17th day, work on thickening walls continuing at same time. At this stage spider web still being added to outside surface, bird scuttling over nest drawing threads back and forth; but it spends more time working inside, entering with wad, constantly jabbing from inside, or leaning far out of opening to place wad outside. Some days earlier, birds sheltered in nest during a rainstorm and encased themselves by closing entrance vertically, from inside, not horizontally. Effect of building false entrance was to deform real entrance from a round opening to a horizontal slit. Nest material is not woven in-and-out, but felt-like nature results from constant teasing and jabbing: bird takes a few downy strands in tip of bill, pulls and pulls at them until they are teased out, then sweeps head from side to side to lay threads over the mass, repeatedly jabbing at them. Some threads are given tiny twists. Nest completed by 22nd day (when first egg was laid), after much work on false entrance and its rim. Birds continue to add material during incubation period and into nestling period, but only to inside of nest. Inside of chamber and spout always smoother than outside. Completed nest is usually (not always) waterproof, with thick floor and impenetrable but deformable walls, neat false entrance with thick rim (**C**). Nest is strongly attached to twigs, felted kapok sometimes extending a few cm along them (**C**). Nest resists rain downpour of 50 mm; it is highly durable and may become more so as time passes (Skead 1959).

Parent arriving at nest with food for young is said sometimes to mislead any watching nest predator by half

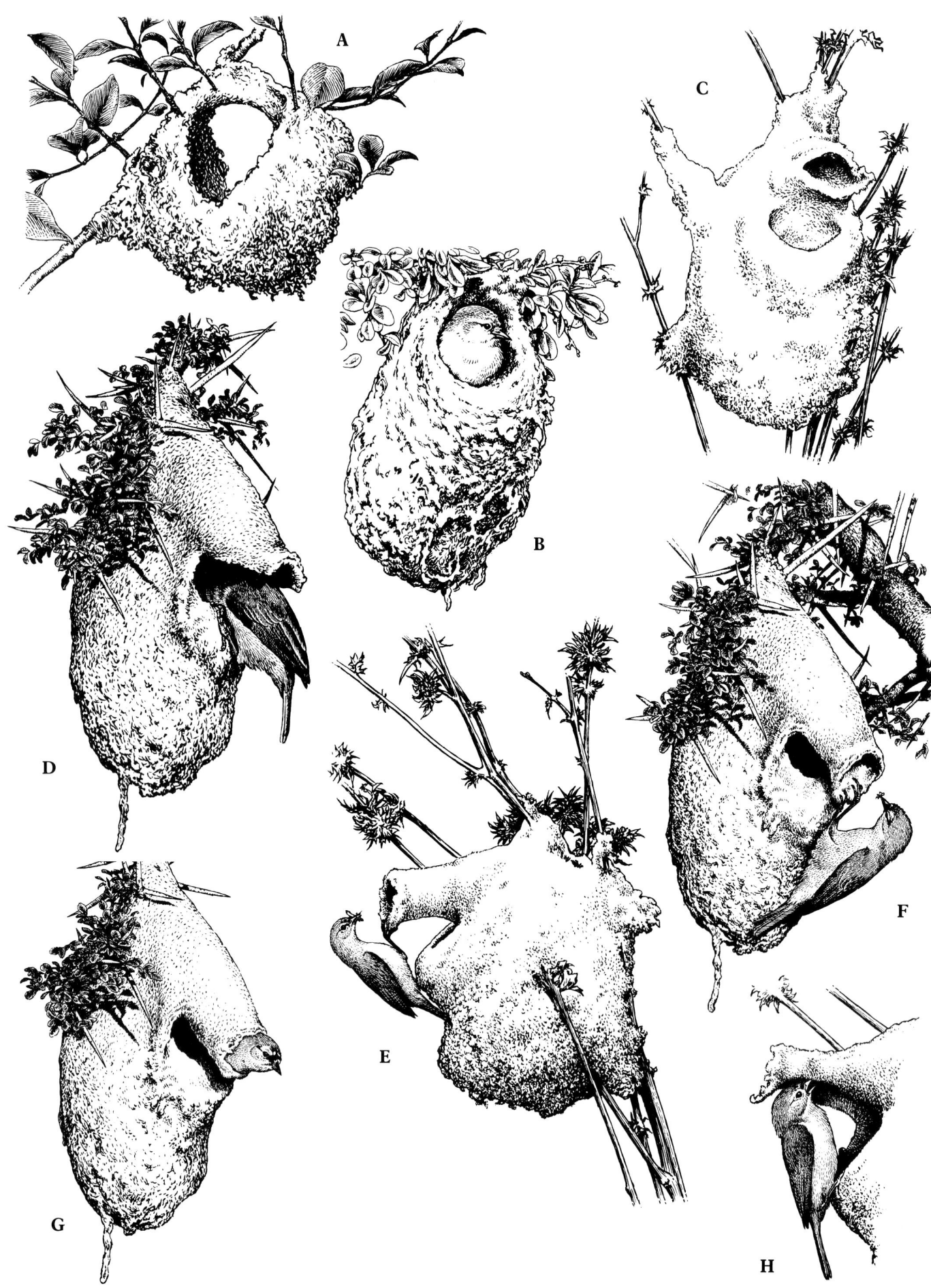

entering false entrance (**D**), then withdrawing if coast is clear (Chadwick 1983).

Generally bird arrives at nest by alighting on wide rim of false entrance, then grasping lower lip of entrance spout with bill or one foot (**E**) and pulling lip down to open horizontal slit of true entrance; it then grasps lip with both feet (**F**), levers itself into spout, and turns around to close first the corners then middle of opening; thereafter shaking movements of nest indicate that bird is closing tube along its length by pulling floor up against ceiling: 'floor and ceiling are lined with coarse cobweb so they stick together like strips of velcro' (Steyn 1996). On leaving nest, bird merely pushes its way along tube (**G**), swings down, perches on rim of false entrance, jabs underside of spout upward with forehead and opened bill (**H**), using extremely rapid movements that have prompted some observers to state, incorrectly, that bird pinches entrance closed with bill. Jabbing and prodding are directed along whole underside of spout, bird leaning well into concavity of false entrance for the purpose; it also prods sides and bottom of concavity, repeatedly and decisively, perhaps to maintain shape of false entrance and bottom of spout. False entrance may deceive predators, but may also function primarily as strong platform for bird opening and closing nest spout. Bird arriving at nest often perches on lip of false entrance and opens then closes real entrance up to 6 times before opening it again and finally entering nest (Skead 1959). Juveniles often have trouble in opening spout from outside and only do so after much trial and error.

EGGS: 4–8 (once 10, 3 times 12 eggs, perhaps laid by 2 ♀♀), av. (n = 15 clutches) 4·9 (Maclean 1993). Laid at 24 h intervals. White. SIZE: (n = 38) 13·3–14·9 × 9·3–11·1 (14·1 × 9·8).

LAYING DATES: Namibia, Jan–Feb; Botswana, Sept, Nov–Jan; Zimbabwe, Oct, Feb; Transvaal, Oct–Jan; SW Cape, July–Oct, E Cape, June–Dec. In the most arid regions, thought to breed opportunistically after rain, e.g. Aug–Mar in Kalahari (Maclean 1993).

INCUBATION: period (13–)15 days. Both sexes incubate; spells usually of 30 min or less. No change-over ceremony. Both birds often bring nest material, passing it to incubating bird or entering with it. Incubating bird constantly tinkers with nest fabric. At one nest each egg in turn, after it was laid, was covered with thin, soft layer of nest material; thereby whole clutch was covered, although sometimes observer also found the clutch uncovered; at another nest, clutch was never covered. Observer inserted finger under incubating bird to count eggs; bird did not try to escape but came up to the entrance from inside 'and closed it for me' (Skead 1959).

DEVELOPMENT AND CARE OF YOUNG: young are fed by both parents equally. Period 22–24 days. Entrance kept closed after each entry and exit for most of period. Near end of nestling period, young move increasingly from brood chamber of nest into spout, to be fed there by adult. Young in nest beg noisily at first, but later receive food quietly. After leaving nest, young return to it to roost at night with their parents, even when second clutch is being incubated; by day they stay with parents in family party for considerable time. At first fledged young follow parents, importuning with repeated, quiet 'zittery' note but without wing-shivering.

BREEDING SUCCESS/SURVIVAL: nests particularly prone to be robbed of their kapok by sunbirds, in SW Cape particularly by Malachite Sunbirds *Nectarinia famosa* and Southern Double-collared Sunbirds *N. chalybea*: nest can be dismantled by pair of sunbirds as fast as penduline tits build it. Tits also harrassed by sunbirds nesting nearby, to point of tits abandoning nesting attempt (Uys 1966). Tits deserted when nest pulled apart by Streaky-headed Canary *Serinus gularis*. Nests sometimes infested by moth larvae which may destroy them. Out of 16 eggs in 3 clutches, 5 were infertile (Skead 1959).

References

Clancey, P.A. (1997).
Harrap, S. and Quinn, D. (1996).
Skead, C.J. (1959).
Steyn, P. (1996).
Uys, C.J. (1966).

Genus *Remiz* Jarocki

A single, Palearctic superspecies of remizid, differing from *Anthoscopus* in greater size, chestnut mantle, black or grey and black head, and nest lacking false entrance. Migratory. Eats insects and seeds; inhabits shrubs and reeds. Often congenerized with African penduline tits (in genus *Remiz*).

4 species (Harrap and Quinn 1996), 1 vagrant to NW Africa.

Plate 6
(Opp. p. 79)

Remiz pendulinus (Linnaeus). Eurasian Penduline Tit. Rémiz penduline.

Motacilla pendulinus Linnaeus, 1758. Syst. Nat., ed. 10, p. 189; Poland, Lithuania, Hungary and Italy.

Forms a superspecies with Asiatic *R. macronyx, R. coronatus* and *R. consobrinus*.

Range and Status. W Palearctic: Spain, Sicily, Crete and Turkey, north to *c.* 56°N (further north in Sweden, Finland and Urals), east to Uzbekistan, Tadzhikistan, Aral Sea and L. Balkhash. Migratory, wintering south to Gibraltar, Sicily, Cyprus, N Sinai, Iraq and Iran. A visible migrant in S Turkey; marked passage through Israel. Rare in Egypt. Vagrant to Britain, Morocco, S Sinai, NW Saudi Arabia, Oman, etc. Wintering and breeding populations increased rapidly in SW Iberian Peninsula in 1980s.

Morocco, 3 at Settet (33°04′N, 07°37′W), Nov 1976 to Feb 1977; 4 at Larache, Jan 1988; 1 at Merja Zerga, Feb 1989, and 3 there Jan 1990; 2, Merja Barga, Dec 1993; 3, Marais du Bas Loukkos, Jan 1995. Egypt, rare vagrant: several singletons, flock of 20 in Suez area in Feb–Mar 1982 and small numbers there 2 and 3 years later; one seen and several heard at Bir Beida, Quseir, Red Sea coast, Mar 1985.

Remiz pendulinus

Description. *R. p. pendulinus* Linnaeus (only race in Africa): W Europe. ADULT ♂: head and neck pale grey, forehead black and rusty, lores and ear-coverts black, forming large, well-defined mask. Mantle and back chestnut, merging to tawny buff on scapulars, rump and uppertail-coverts. Tail feathers black with narrow pale grey or whitish margins. Chin and throat white; breast pale buff, dappled with reddish brown, middle of belly white, sides of belly, flanks and undertail-coverts buffy, thighs rufous. Primaries and secondaries black, narrowly edged whitish; tertials black, with broad buff edges proximally and narrow whitish ones distally; greater coverts black with russet outer edges and buff tips; primary coverts black with narrow buff borders; alula black; lesser coverts chestnut. Bill dark grey with blackish culmen; eyes dark brown; legs and feet dark slate grey. ADULT ♀: like ♂ but forehead pale grey, mask blackish and not so well defined; chestnut and russet parts paler and browner; and breast pale buff. SIZE (14 ♂♂, 13 ♀♀): wing, ♂ 55-59 (57·0), 54-59 (55·9); tail, ♂ 45-49 (46·8), ♀ 42-47 (45·0); bill to skull, ♂ 10·1–12·2 (11·1), ♀ 10·7–11·8 (11·1); tarsus, ♂ 14·3–15·9 (15·0), ♀ 14·3–15·3 (14·7). WEIGHT (Cyprus): ♂ (n = 22) 8·4–10·2 (9·2), ♀ (n = 18) 8·8–10·4 (9·4).

IMMATURE: like adult ♀ but head plain, lacking mask, and mantle, scapulars and back uniform greyish brown; bill pale.

Field Characters. Length 11 cm. A small, fine-billed, compact and lively bird of waterside vegetation, ♂ unmistakable with pale grey head and neck, broad black mask, chestnut mantle and variegated wings. ♀ less strongly marked, mask smaller. Juvenile difficult to identify unless with adult: can resemble Common Linnet *Carduelis cannabina* or even a *Sylvia* warbler: plain and pale with small black eye, small, sharp, conical bill, dusky tail, and variegated wings, greater coverts forming black and russet patch with buffy wingbar (formed by greater covert tips).

Voice. Tape-recorded (62–73, 93, B). Common contact call a thin, soft, plaintive 'tsee' or 'seeou', like Reed Bunting *Emberiza schoeniclus* (similar range and habitats); also utters *Parus*-like 'tsi-tsi-tsi'. Song a trill recalling European Goldfinch *Carduelis carduelis* or European Greenfinch *C. chloris*.

General Habits. Inhabits tamarisks, willows and other small trees, also reeds, reedmace and tall trees such as poplars, by water: marshes, ditches, rivers, lakes and estuaries. In pairs or, usually, small rather loose flocks of 6–12 birds. Agile, erratic, flighty. Forages in reeds, shrubs and outermost twigs of trees, clinging *Parus*-like among flowers and seed-heads, often upside down; explores crevices in bark, regularly gleans reed stems and pokes into reedmace heads; stands on large insect and dismembers it by tugging with bill.

Food. Insect larvae, also eggs and imagos; spiders; small seeds and other plant material.

References
Cramp, S. and Perrins, C.M. (1993).
Goodman, S.M. and Meininger, P.L. (1989).
Harrap, S. and Quinn, D. (1996).

Genus *Pholidornis* Hartlaub

Single tiny species of rain forest remizid; bill slender, sharp-pointed, culmen slightly decurved; base of lower mandible yellow; eye colour differs sexually; legs and feet yellow; stumpy-tailed; sexes alike, juvenile similar, foreparts heavily striped with black on whitish, rump and belly yellow, wings and tail blackish. Nest a felt-like ball with entrance low down on one side; no false entrance (*cf. Anthoscopus*); usually pendent, sometimes built into mass of leafy twigs. Trilling song. Tit-hylia *P. rushiae* has been placed in no less than 7 families, but voice, social behaviour, flight pattern, foraging behaviours, nest characteristics, body size and juvenile plumages all resemble *Anthoscopus* (Vernon and Dean 1975). Several recent field observers are agreed that in the field the species is typically remizid (e.g. Fishpool 1993, C.J. Vernon pers. comm. via W.R.J. Dean 1973 and J. Heigham pers. comm. 1986). Recently-described advertising song thought to indicate affinities with warblers (Sylviidae) (Dowsett-Lemaire and Dowsett 1991) but is actually very like those of some other remizids (Harrap and Quinn 1996).

Endemic; 1 species.

Pholidornis rushiae (Cassin). Tit-hylia; Tit-weaver. Mésangette rayée; Astrild-mésange.

Plate 6 (Opp. p. 79)

Dicaeum rushiae Cassin, 1855. Proc. Acad. Nat. Sci. Philadelphia, p. 325; Moonda River, Gabon.

Range and Status. Endemic resident. Easily overlooked, and probably far commoner and more widespread than records suggest. Sierra Leone, known from Yele, Yonnibanna, Kwendu, Sandaru and Gola Forest. SW Guinea, Macenta and Mt Nimba. Liberia, known from Nannu Kru and Du River, and quite common on Mt Nimba up to 1200 m. Ivory Coast, records at Taï, Man, Gagnoa, Lamto, Abidjan, Adiopodoumé, Bingerville and Yapo Forest (where frequent: Demey and Fishpool 1991). Ghana, not uncommon along S scarp of Volta basin (Kumasi, Atewa For. Res., Tafo, Aburi) and in Bia and Kakum Nat. Parks, and occasional at Cape Coast. Togo, 3 at Balla, 1990. Not yet known from Benin. Nigeria, locally not uncommon in SW (Ife to near Lagos and Bendel State: race *ussheri*?) and in SE in Umuagwu-Calabar area (*rushiae*) and probably continuously distributed in between. Bioko, scarce: known from between Luba and Moka at 750 m and Malabo (Pérez del Val *et al.* 1994). Cameroon, many records throughout forested region; fairly common around Victoria. Mbini (Fa 1990). Gabon, on Moonda River, and 4 birds once at M'Passa in 1975 and 4 present there in 1985. Central African Republic, single bird, Botambi (Germain and Cornet 1994). Congo, common in open canopy in Nouabalé-Ndoki Nat. Park and occurs in tall secondary forest at Mbandza in Odzala Nat. Park (Dowsett-Lemaire 1997a,b), and records at Goumina and Béna in Kouilou basin (Dowsett-Lemaire and Dowsett 1991). Angola: N'Dalatando, Bologongo, Quiculungo, and not uncommon at Salazar in Cuanza Norte (W.R.J. Dean, pers. comm.). Zaïre, evidently widespread in 4 peripheral areas but not yet recorded in NW or in vast forested regions of central and SE Zaïre: in middle Congo River from Ingende and Mbandaka to Keseki, Kunungu and Bokalakala; in Uele and Ituri drainage (Bajwabaka, Bambesa, Wamba, Mambasa, Bunia, Zobia, Irumu); central Kivu Prov.; and S Kasai Occidental, from about Mkumbi to Tshikapa and Kananga. Uganda: Budongo Forest, and Kifu Forest, Mabira Forest and N Buganda near Kampala (Fishpool 1993).

Pholidornis rushiae

Density of 5–10 territories per km^2 (Kouilou basin, Congo: Dowsett-Lemaire and Dowsett 1991).

Description. *P. r. ussheri* Reichenow: Sierra Leone to Togo (this race in SW Nigeria?). ADULT ♂: forehead buff, minutely speckled with blackish; crown blackish, finely streaked with buff; hindneck and sides of neck blackish-and-buff speckled; mantle and scapulars dull dark brown with bright olive streaks (i.e. feather edges); back, rump and uppertail-coverts bright yellow-olive. Tail dark brown, feathers fringed yellow-olive proximally on outer vanes. Chin, throat and breast greyish buffy white, evenly covered with neat, small, narrow, dark brown streaks, short on throat, longer on breast; belly, flanks, thighs, vent and undertail-coverts greenish yellow. Remiges dark brown, narrowly fringed with bright yellow-olive on entire vanes; upperwing coverts dark brown, narrowly fringed yellow-olive, underside of remiges shiny dark grey, axillaries and underwing-coverts cream with yellow wash. Bill black with base of lower mandible yellow-orange; eyes dark orange, orange-red, red, or orange-brown; legs and feet bright waxy yellow. Sexes alike except for eye colour: eyes of ♀ light grey, dark grey, grey-brown, or creamy. SIZE (9 ♂♂, 7 ♀♀): wing, ♂ 45·2 ± 1·3, ♀ 44·4 ± 0·8; tail, ♂ 17·3 ± 0·4, ♀ 16·9 ± 0·5; bill, ♂ 8·2 ± 0·3, ♀ all 8 (Colston and Curry-Lindahl 1986); also (n = ?) wing, ♂ 42·5–51, ♀ 42–49, tail, ♂ 23–25, ♀ 25–26, bill 9·2–11·4, tarsus 11–13 (Harrap and Quinn 1996). WEIGHT: ♂ (n = 9) 5·0 ± 0·4, ♀ (n = 7) 5·3 ± 0·2, 1 unsexed 4·8.

IMMATURE: juvenile like adult but head dark brown with pale greyish feather fringes, mantle olivaceous brown, wings brown with all feathers fringed olive-brown or dull olive-yellow; lores and ring around eye buffy or rusty brown; chin to breast pale yellowish grey, dark-streaked but less boldly so than in adult; flanks, belly and undertail-coverts dull greyish yellow. Eyes grey-brown. One fledgling has crown blackish, mantle dark olivaceous brown with feathers lacking pale edges; rump greener, less bright yellow, than in adult; chin to breast unstreaked pale grey, belly and flanks pale yellowish. WEIGHT: 1 unsexed 4·6, 1 ♀ 5·3.

NESTLING: unknown.

P. r. rushiae Cassin: SE Nigeria, SW Cameroon, Gabon. Like *ussheri* but streaks on chin to breast darker, broader, and more profuse.

P. r. bedfordi Ogilvie-Grant: Bioko Island. Like *rushiae* but streaks on chin to breast even broader; belly and flanks softly marked with diffuse broad dark streaks.

P. r. denti Ogilvie-Grant: SE Cameroon to Uganda and Angola. Very like *ussheri* but streaks on chin to breast sparser, softer, slightly broader and darker; belly brighter yellow. WEIGHT: 1 ♀ 6·0.

Field Characters. Length 7·5 cm. The smallest bird in Africa, with the possible exception of the African Piculet *Sasia africana*. Plump shape, short tail and quick foraging actions remind one of a Goldcrest *Regulus regulus*. Streaked foreparts combined with yellow rump and belly render it unmistakable. Bright yellow legs and feet noticeable in the field (Harrap and Quinn 1996).

Voice. Tape-recorded (104, B, CHA, ERA, LEM). Advertising song 'consists of 2 clear trills, 'puipuipui-tjitjitjitjit-jitu', the second faster than the first, and lasts *c*. 1·5 s. It is sometimes preceded by 2 grating notes, 'ruirui', and then lasts nearly 2 s. 'Reasonably loud, far-carrying and stereotyped ... more characteristic of a sylviid than a remizid ...' (Dowsett and Dowsett–Lemaire 1991). Tape playback from ground does not bring singers down from canopy. Calls include short 'ptu' or 'ptiu' and high-pitched 'psii' like anxiety call of Tree Pipit *Anthus trivialis*, and sometimes 'psii-ptc' (Brosset and Erard 1986); begging young give shrill, rapid 'tsi-tsi-tsi ...'.

General Habits. Inhabits secondary evergreen forest and edges of primary forest, large trees such as silk cotton trees isolated in clearings and gardens, coffee forest, cassava plantations, oil-bean trees, balsa plantations; occasionally occurs in leafless trees and degraded scrub. Often in 'spring' trees which are budding flush of new leaves. In Liberia in treefall gaps, along wooded river banks, clearings in depleted high forest, abandoned farmland, and cocoa and coffee plantations with large shade trees, from coast up to 1200 m (Gatter 1997). In Ivory Coast occurs near houses but in Ghana (Kumasi) tends to avoid their vicinity. Occurs in pairs and groups of 3–7, keeping mainly at 5–15 m above ground in upper and middle canopy of giant trees; sometimes in lower canopy; nests much lower down than it feeds. Gleans leaves, explores leaf petioles, nodes and stems, and pokes at vines and clumps of mistletoe; hammers at surface of twig, like a tiny woodpecker; acrobatic when foraging, hanging upside down under twig to investigate inflorescences and buds; moves about rapidly, creeps through foliage, clings to vines; flock of 7 birds gathered at one point high in tree where they seemed to have found plenty of (scale) insects on bark, but otherwise foraged in a loose group (W.R.J. Dean, pers. comm.). Spends *c.* 65% of time on undersides of branches up to at least 30 mm diam., mainly 5–10 mm (Gatter 1997). Often flicks wings. Flight when sustained is rapid and direct. 'Does not associate with other species' (Serle 1965) although at flowers will feed near sunbirds *Cinnyris*. Roosts gregariously (up to 5 birds) in nest after breeding season. Allopreening, and bathing against wet leaves, recorded (R. Farmer, pers. comm.).

A

Food. Insects and seeds; mainly scale insects, Coccidae (Nigeria); of 24 stomachs, 19 contained insects and 5 contained seeds (Serle 1965).

Breeding Habits. Territorial, solitary nester, territory defended by singing whilst on the move; evidently monogamous (Liberia: Bannerman 1939). May be cooperative breeder: nest under construction (Ghana) occupied by 4 birds (though only one when incubation commenced); and 2 juveniles fed in quick succession by 4 out of 5 adults (Angola: Vernon and Dean 1975).

NEST: ball, with overhanging entrance spout at one side, the entrance narrow and facing vertically downward (**A**). Built of downy vegetable material (one nest composed entirely of 'rubber-seed' down, *Funtumia elastica*, a common apocynaceous tree in regrowth forest; another nest made of down from exotic balsa tree *Ochroma lagopus*, worked into a thick, felt-like fabric. Very strongly bound by short extentions of the fabric onto thorny vine or in tangle of long thin vines hanging from tree foliage; or a loose mass placed amongst twigs and leaves at end of lateral bough (Serle 1965), and once reported built into old weaver's nest. Sited 3–20 m above ground. Nest large for size of bird, *c*. 150 in diameter.

EGGS: 2. Pure white. SIZE: (n = 2, Nigeria) 13·5 × 10·0, 12·5 × 10·5.

LAYING DATES: Liberia, Nov–Apr (n = 16 of which 11 in Jan–Mar: Gatter 1997); Ghana, (nest building Apr, dependent fledglings Nov–Jan); Nigeria, (adults feeding fledged young, Dec); Gabon, (dependent young Mar); Congo, (sings less in Nov–Dec than in Sept–Oct, and not at all in Jan; adults chasing each other over nest material, Oct); Angola, (adult feeding fledglings Aug, building nest Oct; breeding condition Nov).

DEVELOPMENT AND CARE OF YOUNG: young out of nest importune adults with low, sibilant notes and quivering of wings.

BREEDING SUCCESS/SURVIVAL: bird became entangled in spider's web (Bates 1909); unsuspended balsa-down nest blown to ground (Serle 1965).

References

Bannerman, D.A. (1949).
Dowsett, R.J. and Dowsett-Lemaire, F. (1991).
Gatter, W. (1997).
Serle, W. (1965).
Vernon, C.J. and Dean, W.R.J. (1975).

Family SITTIDAE: nuthatches and Wallcreeper

1 large genus (*Sitta*) and 1 monotypic one (*Tichodroma*) of tree- or rock-haunting Holarctic and tropical Asian birds. Bill as long as head, culmen straight, bill strong and pointed (*Sitta*) or longer than head and slightly decurved (*Tichodroma*); nostrils round, partly covered by feathers; rictal bristles short, soft; 10 primaries, P10 short, wing tip rounded; flight strong, direct or undulating; tail short, broad, square-tipped, often with white corners (both genera); tarsus scutellate. Plumage blue-grey, *Sitta* with eye-stripe and rufous in underparts (and some Oriental spp. blue-black or bright blue, with yellow or vermilion bills), *Tichodroma* with carmine in wing; sexes alike or nearly so (*Sitta*) or a little different (*Tichodroma*); juveniles like adults. Eat small insects (*Tichodroma*) or insects, seeds and nuts (*Sitta*); nuts stored; hard items wedged in bark and hammered with bill; 1 Nearctic *Sitta* sp. uses bark flake as tool in feeding. Very agile in climbing rocks (*Tichodroma*) and trees or rocks (*Sitta*). Nest a simple cup in tree hole (*Sitta*), hole entrance daubed with mud by some spp. to reduce size, and protected by daubing with resin or noxious insect fluids. 2 Palearctic *Sitta* spp. (rock nuthatches) build cup nest in rock fissure, protecting entrance with bottle-shaped outer shell of mud; *Tichodroma* has cup nest inside heap of boulders or deep in wet rock cleft. Sedentary, or vertical migrant (*Tichodroma*).

Conventionally the 2 genera have been placed in separate families, Sittidae and Tichodromadidae, but comfort, reproductive, locomotory and other behaviours are alike and familial merger was recommended by Löhrl (1975); DNA–DNA hybridizations show close relationship (Sibley *et al.* 1988, Sibley and Ahlquist 1990). Wallcreeper *Tichodroma muraria* perhaps best regarded as a rock nuthatch extremely adapted to harsh montane environment (and Spotted Creeper *Salpornis spilonota* (*q.v.*) may yet prove to be a tropical nuthatch adapted to tree-creeping life).

2 genera; up to 25 species; 1 superspecies Holarctic, 3 other species Nearctic, remainder Palearctic and Oriental.

Genus *Sitta* Linnaeus

For characters, see family diagnosis. 24 species, being 4 superspecies and 7 independent species (Sibley and Monroe 1990). Holarctic, and Oriental, to Philippines. 2 species in N Africa: 1 endemic to Djebel Babor, Mediterranean Algeria, the other a Palearctic nuthatch which breeds in Morocco.

Plate 7
(Opp. p. 94)

Sitta ledanti Vielliard. Algerian Nuthatch. Sittelle kabyle.

Sitta ledanti Vielliard, 1976. Alauda 44, pp. 351–352; Djebel Babor, Algeria.

Forms a superspecies with *S. whiteheadi* (Corsica), *S. krueperi* (Asia Minor), *S. yunnanensis* (SW China), *S. villosa* (China, Korea) and *S. canadensis* (N America).

Sitta ledanti

Range and Status. Endemic resident, confined to 4 areas in Petite Kabylie Mts, NE Algeria, within 3–45 km of coast; at 1450–1900 m in Djebel Babor, Tamentout and Djimla Forests and at 350–1120 m in Guerrouch Forest in Taza Nat. Park. Discovered in 1975. Some 80 pairs on Djebel Babor and perhaps 360 birds in Taza Nat. Park (where 91 birds in 800 ha surveyed in 1989). Forest covers 1300 ha on Djebel Babor (only 250 ha of which is optimum habitat for nuthatches), 3200 ha in Guerrouch (was nearly 8600 ha in 1955), 9500 ha in Tamentout and 1000 ha in Djimla; together, the 4 forests may harbour a summer population of 1200 birds (Harrap and Quinn 1996).

Description. ADULT ♂ (breeding): forehead and forecrown black, rest of upperparts bluish grey. Long white superciliary stripe, from nostril almost to side of neck. Central tail feathers bluish grey, T2–T6 black with narrow bluish grey tips, T3–T6 with subterminal white patches (only on outer vane of T3); white patch largest on T6. Lores and upper part of ear-coverts black, forming long narrow line from gape to side of neck. Chin white, throat and undertail-coverts white tinged buff, rest of underparts uniform orange-buff or salmon-cream. Cheeks and lower part of ear-coverts mottled grey and white. Primaries and secondaries black, secondaries and inner primaries narrowly edged with white, bases of primaries and outer vanes of secondaries fringed bluish grey. ADULT ♂ (non-breeding): forehead and forecrown greyish black, supercilium not so white, blackish stripe through eye mottled with white and less distinct, upperparts bluer grey. Bill greyish black; eyes brown-black; legs and feet grey. ADULT ♀ (breeding): like breeding ♂ but black patch on forehead smaller, duller and less well demarcated. ADULT ♀ (non-breeding): forehead patch greyer (feathers black, with grey tips that wear off during winter). SIZE (3 ♂♂, 3 ♀♀): wing, ♂ 80–83 (81.3), ♀ 77–81 (79.0); bill, ♂ 16–17.5 (16.7), ♀ 16–17 (16.5); tarsus, ♂ 19–22 (20.7), ♀ 19–21 (20.0). WEIGHT: 1 ♂ 18.0, 1 ♀ 16.5.
IMMATURE: literature descriptions conflicting – forehead black, or grey; plumage otherwise like adult; possible that juvenile ♂ has black forehead and juvenile ♀ has grey one (Cramp and Perrins 1993).
NESTLING: not described.

Field Characters. Length 13.5 cm. Endemic to a small area in Algeria, where Eurasian Nuthatch *Sitta europaea* is not present. In remote possibility that vagrant Eurasian Nuthatch might reach Algeria, Algerian Nuthatch identified by *prominent white eyebrow* above *dark stripe* through eye and variable amount of black on forehead.

Voice. Tape-recorded (105, 106, B, C, GARD, VIEL). Details from Harrap and Quinn (1996): call in flight and whilst foraging a quiet, soft, nasal 'kna' or inquisitive 'quuwee' (like 'diuwee' call of European Greenfinch *Carduelis chloris*). Territorial defence and excitement call a harsh, rasping 'vschrr vschrr' or 'schrr, schrr-schrr' recalling Eurasian Jay *Garrulus glandarius*, also an agitated, loud, even 'chwa-chwa-chwa'. Another call a nasal, inquisitive 'du-wa du-wa' or 'qu-wa-di-wa'. Song a stereotyped repetition of 7–12 motifs like the 'du-wa du-wa' call, either nasal or fluty, 'quair-di, quair-di, quair-di ...' or 'verdi, verdi, verdi ...' or a faster 'du-wid-du-wid-du-wid ...'

or (even faster) 'vid-vid-vid-vid ...', lasting 2–4 s and rising slightly in pitch.

General Habits. On Djebel Babor inhabits cool montane forest with pure stands as well as mixtures of Atlas oak *Quercus afares*, Algerian chestnut-leaved oak *Q. faginea*, Canary oak *Q. canariensis*, cork oak *Q. suber* (lower down), Atlas cedar *Cedrus atlantica* and Algerian fir *Abies numidica*; trees covered with epiphytic mosses and lichens, and birds, since they appear to be resident, evidently tolerate 4 m depth of snow in winter; in Guerrouch Forest, mainly oaks *Q. afares*, *Q. faginea* and *Q. suber*, with understorey of alder, cherry, willow, ash and maple (Harrap and Quinn 1996).

Occurs in pairs, which defend winter territories and are strongly territorial in summer, ♂ and ♀ defending territorial boundary. Forages in outer branches and twigs of oaks and on trunks and branches of mossy and lichen-covered oaks, cedars and firs; searches undersides of branches, and flowers; knocks off pieces of bark by hammering and prising with bill and searches for insects on exposed tissue; eats seeds of firs, maples, oaks and cedars, and caches seeds in lichen-covered branches and thick cushions of epiphytic mosses, retrieving them weeks or months later. Of 123 observations, Mar, 45% were of birds foraging on oaks, 32% on cedars and 22% on firs; 59% on trunks and 41% on branches and twigs; of 138 observations, June, 90% on oaks, 6% on firs, 4% on cedars. Seldom comes lower than 2–3 m above ground; once seen eating seeds on ground. Sometimes joins mixed-species foraging flocks.

Food. Insects (imagos, larvae, eggs), spiders, seeds of *Quercus faginea*, *Acer obtusatum*, *Cedrus atlantica* and *Abies numidica*. Young fed on earwigs, noctuid and geometrid moth caterpillars, beetles, spiders and seeds. 104 feeds consisted of geometrid larvae (57%), conifer seeds up to 10 mm long (19%), noctuid larvae (6%), beetles (4%), spiders and earwigs (2%) and unidentified arthropods (13%).

Breeding Habits. Solitary nester, evidently monogamous and territorial. Territory size *c.* 3·3 ha. No information on courtship. Single-brooded (?).

NEST: in tree hole; material includes wood chips, bristles of Wild Boar *Sus scrofa*, dead leaves, and feathers. Bird excavates its own cavity or appropriates old nest hole of woodpecker; hole 35–65 mm across; single record of hole size being reduced to diam. of *c.* 40 by plastering with mixture of clay and rotten wood (in manner of much better known Eurasian Nuthatch *S. europaea*); 2 holes 150 and 200 deep. Nest holes sited in dead or dying tree, usually in branch, at height of 4–15 m (n = 12) above ground, av. (n = 5, another sample) 8·5 m. Bird prefers huge old isolated tree; of 21 nests, 14 in firs *A. numidica*, 6 in cedars *C. atlantica*, and 1 in oak *Q. faginea*.

EGGS: not known. (Brood sizes recorded of 2 and 4 fledglings.)

LAYING DATES: about May (when ground can still be under deep snow); in one year 5 broods fledged in 6–8th July; in another, fledging started before 18th June, completed by 25th June; another brood just hatched on 18th June.

INCUBATION: thought to be by ♀ only (since ♂ has no incubation patch).

DEVELOPMENT AND CARE OF YOUNG: young fed by both parents. One pair with territory in pure stand of cedar flew long distances to obtain food from oak trees for young. Nestling period about 22–25 days.

References
Bellatreche, M. (1991).
Cramp, S. and Perrins, C.M. (1993).
Gatter, W. and Mattes, H. (1979).
Harrap, S. and Quinn, D. (1996).
Ledant, J.P. (1978, 1981).
Ledant, J.P. and Jacobs, P. (1977).

Sitta europaea Linnaeus. Eurasian Nuthatch. Sittelle torchepot.

Plate 7
(Opp. p. 94)

Sitta europaea Linnaeus, 1758. Syst. Nat., ed. 10, 1, p. 115; Europae, Americae: restricted to Sweden by Hartert 1905.

Forms a superspecies with Oriental *S. nagaensis*, *S. cashmirensis* and *S. castanea*.

Range and Status. Palearctic and Oriental Regions, from Morocco, Portugal, Spain (breeding south to Gibraltar), Britain and S Norway to Cambodia, Taiwan, Japan and Kamchatka; north to about 65°N, south to Nilgiri Hills, India. Sedentary, occasionally eruptive.

Morocco, widespread and locally frequent in Moyen Atlas Mts (El Hajeb, Azrou, Aïn Leuh, Khénifra, Col du Zad, Timahdit, Michliffen, Ifrane, El Harcha, Tifourhaline, Aguelmos, El Khab, Itzer, Aguelmane Aït Tchou, Oued Serrou, Aguelmane Azigza, Ras el Mar); Jbel Tasseka; and the Rif (Jbel Tidighine, Talassemtane, Ketama, Rhafsai, Taounate, Bab Lahcen, Jbel Tisirene); scarce in Haut Atlas near Zaouïa Ahansal on Ighil-n-Aït Tamajjout, at 2150–2350 m, and from near Midelt to Aït-Tamlil near Télouet, south to Jbel Anrhomer (31°20'N, 07°00'W).

Description. *S. e. atlas* Lynes: Morocco. ADULT ♂: narrow black line across front of forehead joining lores and continuing backwards through eyes to end on sides of neck; forehead, crown, nape, hindneck, mantle, scapulars, wings, back, rump, uppertail-coverts and central pair of tail-feathers uniform bluish grey. Rectrices T2–T4 black, with blue-grey tips 2–3 mm wide; T5 and T6 with black bases, grey tips, and broad white subterminal bars; outer vane of T6 largely white. Short,

Sitta europaea

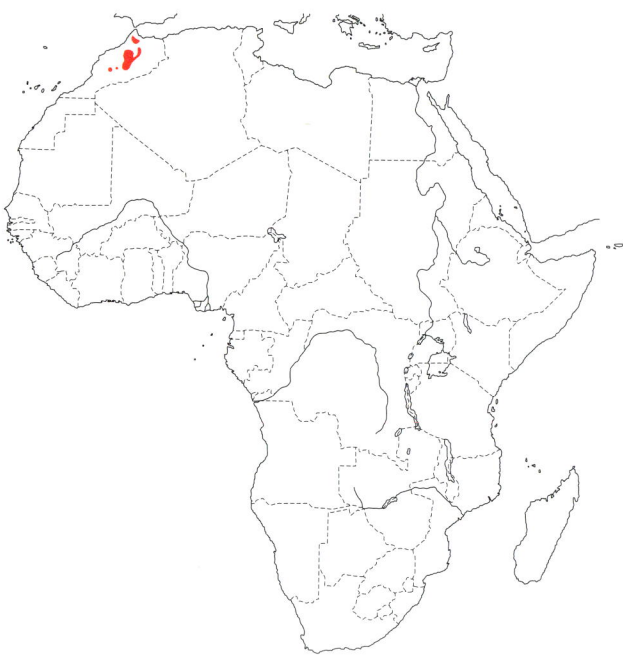

indistinct, pale superciliary stripe. Cheeks white, speckled blackish. Chin and upper throat white, merging to buff on lower throat; breast, belly and flanks quite deep buff, hind part of flanks with c. 4 jagged rufous stripes; undertail-coverts rufous with white centres. Underwing strongly patterned: coverts and axillaries grey with dark rufous-buff tips; inner under primary coverts black, longer outer ones white. Bill bluish slate-grey, base of lower mandible bluish white, sometimes tinged pink; eyes dark brown; legs and feet greenish or greyish yellow-brown. Sexes alike, but eye-stripe of ♀ tends to be brownish black and less well defined than in ♂, and inner under primary coverts greyish black. SIZE: wing, ♂ (n = 5) 84–91 (87.4), bill to skull, ♂ (n = 5) 20.3–21.6 (20.9); (*S. e. hispaniensis*, Spain, 11–13 ♂♂, 13–15 ♀♀): wing, ♂ 81–87 (84.6), ♀ 80–85 (82.6); tail, ♂ 41–46 (43.3), ♀ 40–44 (41.6); bill to skull, ♂ 18.8–20.3 (19.6), ♀ 18.3–20.0 (19.1); tarsus, ♂ 18.7–20.3 (19.4), ♀ 18.2–20.1 (19.2); middle toe with claw (*atlas*, *caesia* and *cisalpina*, n = 30) 19–22 (20.3). WEIGHT: (*S. e. cisalpina*, Italy and Sicily), ♂ (n = 5) 21.1–23.6 (22.7), ♀ (n = 4) 19.0–30.5 (23.1) (Cramp and Perrins 1993).

IMMATURE: like adult but narrow black mask brownish and less well defined, upperparts scaled by feather tips being very narrowly pale, and rufous stripes on flanks less conspicuous. Juvenile the same but base of bill pale blue, legs and feet pale greenish or yellowish flesh-brown.

NESTLING: naked but for scanty dark grey down, short on crown, longer on shoulders and middle of back; skin pink, gapes ivory white.

TAXONOMIC NOTE: African race *atlas* is sometimes united with *hispaniensis* (Iberia except N) (Vaurie 1959, Cramp and Perrins 1993), though darker underparts, longer bill, and voice of *atlas* make it distinct.

Field Characters. Length 14 cm. The only nuthatch in its range; blue-grey above, buff below, *long black stripe* through eye, *short tail* (pied when spread, but markings usually not easy to see); quite vocal; creeps jerkily on tree trunks and boughs.

Voice. Tape-recorded (62–73, 93, 105, B). Song loud, variable, from a fast, stuttered 'chu-tchu-tchu-tchu...' (up to 19 'tchu's per s) to a slower, deliberate 'pee-pee-pee-pee...' (2–3 per s), lasting for 1–3 s. Also (in Morocco, not in Europe) repeated whistles, falling then rising in pitch (Cramp and Perrins 1993). Calls, given freely, also variable; commonest a loud, ringing 'chwit-chwit' or 'chwit-chwit-chwit'; also shrill 'tsit' or 'tseet', often repeated, and grating 'tsirrp' of excitement or alarm. For range of calls and sonagrams see Cramp and Perrins (1993).

General Habits. Inhabits mature, mixed broad-leaved and coniferous woods at 1750–2350 m in Moyen Atlas and Haut Atlas Mts, mainly cedars, xen oaks, cork oaks, holm oaks and pines. Solitary or in pairs; vocal, though can be difficult to locate even when calls known, since birds move freely, often fly 100–200 m within forest and across open spaces, and when foraging on a tree they often disappear from view on other side of trunk or branch. Flight fast, slightly undulating or not undulating, without any gliding; with round wings, short tail and compact form, bird looks broad-ended when wings open and spear-like when wings closed. Gait on tree-trunk a short, jerky jump; works its way upwards in large, fissured tree-trunk when foraging or moves horizontally underneath large, outward-reaching bough, but excitable and flighty, and less methodical in working full length of trunk than Short-toed Treecreeper *Certhia brachydactyla*. Sometimes works its way downwards; uncommonly, hangs upside-down from twigs, and forages on ground. In spring opens buds, looking for larvae. Often stops and hammers quite violently at morsel in bark crevice, clinging with splayed legs and toes, head and bill moving through arc of 10 mm 2–3 times per s. Removes pieces of bark, searching for insects behind. Wedges seeds and large and hard insects into bark crevice, and smashes them by hammering with bill. In autumn, caches seeds. Foraging height 2–8 m. In Europe comes to bird-tables (where can be confiding) and hammers nuts and seeds, while other birds such as tits and finches keep their distance. Highly sedentary (Enoksson 1998).

Food. In Europe invertebrates and, mainly in autumn and winter, seeds. Feeds on sap exuded from maple, birch, poplar and ash (Cramp and Perrins 1993).

Breeding Habits. Well known in Europe; not known to differ in Africa though barely studied there. Solitary nester; monogamous and territorial.

NEST: in natural tree-hole or old woodpecker hole; bird prefers old holes where tree growth has created short entrance tunnel (Cramp and Perrins 1993); entrance hole often narrowed by plastering with mud, and nest hole walls often plastered inside, with small cracks stopped up with plaster made of mud and wood chips. Nest itself is heap of bark flakes (usually of conifer, often of *Pinus sylvestris* in N Europe) on foundation of wood chippings and bits of soft, rotten wood. One nest cavity, Morocco, consisted of mud roof at angle of 45° over top of broken-off tree stump. 40% of nests (England) in oak tree holes. Height of holes (Germany, n = 12) 1.9–19.4 (11.1) m. Diam. of natural cavities (Netherlands, n = 12) 100–240 (162) mm and (n = 12) 40–80 (55) mm, the latter sample reduced after plastering to 27–30 (28) (Cramp and Perrins 1993). Tree-hole plastered and nest built almost entirely by ♀, regularly accompanied by ♂.

EGGS: 6–10 (8·3, n = 6 clutches, Morocco). Laid daily, in early morning. Subelliptical, smooth, slightly glossy; white, lightly speckled with red-brown. SIZE: (n = 50, Morocco) 17·0–21·0 × 13·5–15·8.

LAYING DATES: Morocco, late Apr to late May.

INCUBATION: by ♀. Period (Germany, n = 26) 13–18 (14·8 days).

DEVELOPMENT AND CARE OF YOUNG: young cared for and fed by both parents. Nestling period (Germany, n = 19) 18–28 (24) days.

BREEDING SUCCESS/SURVIVAL: high (many European data, Cramp and Perrins 1993).

References
Cramp, S. and Perrins, C.M. (1993).
Enoksson, B. (1998).
Harrap, S. and Quinn, D. (1996).

Genus *Tichodroma* Illiger

For characters, see family diagnosis. Single, highly distinctive species of Palearctic montane bird, usually placed in its own family Tichodromadidae, although increasingly seen as an aberrant member of nuthatch family Sittidae. Accidental in N Africa.

Tichodroma muraria (Linnaeus). Wallcreeper. Tichodrome échelette.

Plate 7
(Opp. p. 94)

Certhia muraria Linnaeus, 1766. Syst. Nat., ed. 12, p. 184; southern Europe.

Range and Status. Mountains in S Europe (Cantábrica, Pyrenees, Alps, Apennines, Carpathians, Balkans), Corsica, S Turkey and Transcaucasia, to Himalayas, central and E Asia. Resident, altitudinal and short-range migrant; occurs regularly in winter in only a few W Palearctic localities outside breeding range, including SE Spain, much of Italy, S Turkey and in some years Israel. Resident. Vagrant Portugal, Gibraltar, NW Africa, Britain, etc.

Very rare winter visitor, Morocco and Algeria, up to 1950s; none recently (though 2 records in Gibraltar in 1970s). Old record in Egypt rejected (Goodman and Meininger 1989).

Description. *T. m. muraria* (Linnaeus) (only race in Africa): S Europe to N Iran. ADULT ♂: forehead, crown, hindneck, sides of neck, mantle, scapulars and back grey; rump and uppertail-coverts greyish black. Tail black, feathers narrowly tipped pale grey, T6 with white outer edge and T5 and T6 with small white corners. Lores, cheeks, chin, throat and breast black. Lower breast, belly, flanks and thighs grey, rather darker than upperparts; undertail-coverts grey, tipped whitish. Wings: upperwing-coverts mainly carmine; primaries black, P9–P6 with large white oval near tip of inner vane; all primaries except P10 with more or less concealed white patch near base of inner vane and with outer vanes proximally carmine; secondaries black, with narrow whitish tips, and outer vanes proximally carmine; tertials black with narrow whitish margins. Underwing-coverts and axillaries pale red-pink. Bill black; eyes black-brown; legs and feet brown-black or grey-black. Adult ♂ in worn plumage has crown paler grey and mantle almost white. ADULT ♀: like ♂ but lores grey, ear-coverts and sides of neck pale grey, whitish eye-ring, chin and breast white, throat whitish usually with mottled dark grey patch in middle. SIZE: wing, ♂ (n = 23) 95–104 (99·9), ♀ (n = 15) 94–101 (98·0); tail, ♂ (n = 21) 51–57 (54·1), ♀ (n = 12) 48–58 (52·6); bill to skull, ♂ (n = 13) 29·7–38·8 (33·0), ♀ (n = 10) 30·0–38·6 (34·0); tarsus, ♂ (n = 19) 21·8–23·9 (23·0), ♀ (n = 10) 21·3–23·7 (22·7). WEIGHT: (n = 7, unsexed, in captivity) av. 17·3 (Oct) and 19·3 (July).

IMMATURE: whole head with chin, throat and neck very pale grey (crown sometimes brownish), which gradually darkens backwards via mantle and breast to mid grey or dark grey on uppertail-coverts and hind part of flanks. Wings, tail and undertail-coverts same as adult ♂. Juvenile has shorter bill with yellowish base and cutting edges.

Field Characters. Length 16·5 cm. Unmistakable in all plumages, a long-billed, long-toed, grey bird creeping on precipitous rock faces, with largely *crimson*, very round-tipped wings with large *white spots*.

Voice. Tape-recorded (62–73, 93, 105, B). Contact call a quiet, brief, chirping whistle 'tseeoo, tschju'; also a thin 'tuee'; in aggressive encounters bird utters whistling 'chuit dweeoo', rising or falling in pitch (Harrap and Quinn 1996). Song, unlikely ever to be heard in Africa, is ascending series of 4–5 clear piping whistles.

General Habits. Inhabits cliffs, rock faces and precipitous boulder-strewn slopes in limestone mountains; gorges, quarries, often near waterfalls; in winter also sea cliffs, rocky riverbeds, clay banks and cliffs, road cuttings, bridges, ruins, and grand buildings like cathedrals and castles. Occurs in pairs in summer, generally solitary in winter, ♂ and ♀ vigorously defending a feeding territory; sometimes occurs in small (and rarely in quite large), loose groups. Territorial disputes can involve 2 birds fighting in air. Forages on rock surfaces, and in summer sometimes along stream bed and boulders and rarely on trees; in winter quite often forages along stream beds, earth banks and roadsides. Forages by creeping over rocks, moving with jerky hops, in any direction and plane, moving head sinuously, constantly flicking wings, clinging with spindly toes but, unlike Short-toed Treecreeper *Certhia brachydactyla*, not using tail as prop. Flight erratic, skipping, with exaggerated wingbeats, like butterfly or Hoopoe *Upupa epops*. Sometimes flycatches; may disappear into deep cleft or hole. Roosts solitarily, in rock crevice.

Food. Small invertebrates.

References
Cramp, S. and Perrins, C.M (1993).
Harrap, S. and Quinn, D. (1996).
Löhrl, H. (1975).

Family SALPORNITHIDAE: Spotted Creeper

A small arboreal bird with cryptic plumage, sunbird-like decurved bill and long toes. Bill laterally compressed; nostrils longitudinal, near base of bill; no rictal bristles; tongue long, narrow, horny, with *c.* 5 bristles at tip. Skull of some adults has unpneumatized, transparent cranial window. 10 primaries, P10 less than half length of P9; wing long and pointed, tip reaching to within 10–12 mm of tip of tail; tarsus thick, short, with transverse scales in front; foot large; toes long and spindly; claws long, sharp, strongly curved, hind claw especially so; tail square-ended or slightly forked, full, with 12 feathers. Plumage long and soft, cryptic, patterned in browns and greys with white spots; sexes alike; young like adults. Sedentary but may wander widely. Climbs tree trunks more like nuthatch *Sitta* sp. than treecreeper *Certhia* sp., creeping fast but jerkily upward, around trunk or sometimes downward, or along underside of bough with belly almost brushing bark. Insectivorous. Nest a deep open cup, bound with spider web, outside decorated so nest blends in with branch. Eggs turquoise, maculate.

Conventionally the Spotted Creeper is placed in its own family, but it has been put in Sittidae (e.g. Mayr and Amadon 1951) and some recent authorities have placed it in Certhiidae. DNA–DNA hybridization data indicate that *Salpornis* is certhiine, i.e. belongs to the cluster of families embracing treecreepers, wrens Troglodytidae and nuthatches Sittidae (Sibley and Monroe 1990). *Salpornis* seems closer to *Tichodroma* (Sittidae) than to *Certhia* (Certhiidae), but is so distinctive that we retain it in its own family.

1 genus and species: Africa and India.

Genus *Salpornis* Gray

Characters those of the family.

Salpornis spilonotus **(Franklin). Spotted Creeper. Grimpereau tacheté.**

Certhia spilonota Franklin, 1831. Proc. Zool. Soc., London, 1830, p. 125; between Calcutta and Benares.

Range and Status. Africa and India (foothills of Himalayas and plains to Delhi, Bombay to Bihar and Orissa).

Resident, woodlands from Gambia to Ethiopia and Kenya and from SW Tanzania to Angola and Zimbabwe. Senegal, once, Niokolo-Koba Nat. Park. Gambia, 3 records: Bondali, Keneba and Abuko Nat. Reserve. Guinea-Bissau, 3 or more records. Guinea, old record from Kouyéyé, S Fouta Djallon. Sierra Leone, no recent records; formerly not uncommon in N, and common in Tembikunda region, including Nerekoro (770 m); also Bintumane Peak, Momoria Bedala, and near E border with Guinea. Ivory Coast, uncommon in N, including Comoé, Odienné and Tingrela (Mali border). Burkina Faso (Dowsett 1993). Ghana, uncommon: at least 9 localities;

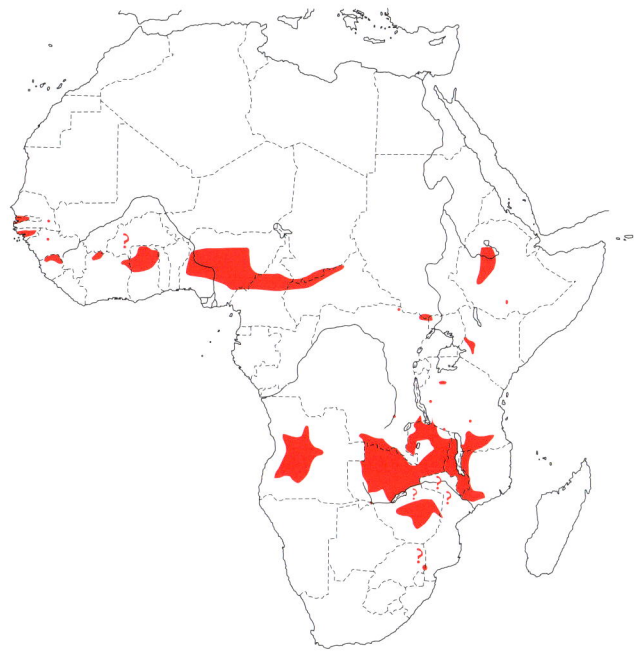

Salpornis spilonota

seen regularly at Mole (Grimes 1987). Togo, rare: single record, Naboulgou. Nigeria, uncommon, from Zaria, Falgore and Kari south to Shaki, Enugu, Serti and Obudu (where frequent), uncommon in Lake Kainji Nat. Park and Gashaka-Gumti Game Res., not known from Yankari Nat. Park. Cameroon, known from a few localitis on Adamawa Plateau and S Benue Plain. Chad, scarce, north to Azoum/Chari confluence. Central African Republic, a few records near Chad border, east to Manovo-Gounda-Saint Floris Nat. Park. Sudan, rare and local in S, from Tambura to Kajo Kaji. Single record from near point where Central African Republic, Sudan and Zaïre meet. NE Zaïre, only in NE Uele District, record from Aba and frequent at Faradje and Garamba. Uganda, in extreme NW in West Nile Prov. near Zaïre border, and in extreme E, on NW slopes of Mt Elgon. Kenya, local and uncommon at 1000–2300 m in W, from Elgon to Kapenguria, Kitale, Tambach, Londiani and Kaimosi. Ethiopia, uncommon to locally frequent or common in W Highlands, in woodland mainly above 1500 m, from Dangila to Kaffa and Alghe; occurs in SE Highlands.

In S tropics, occurs in SW Tanzania (Mpanda) and in SE from Songea to Liwale; thence widespread to Angola and Zimbabwe. In Angola confined to climax *Brachystegia* woodland on central plateau from S Cuanza Sul and W Malanje to N Huila (where common), W Huila and NW Moxico; rare in Congo basin (Dean *et al*. 1988). S and SE Zaïre: frequent south of 10°S in SW Shaba; Upemba Nat. Park; and Marungu Highlands (Kasoko, Kampia, 1250–1380 m). Zambia, widespread almost throughout, usually quite common, but sparse in drier types of woodland; recorded in 71% of country's 391 map squares (R.J. Dowsett, pers. comm.); absent west of Zambezi in Balovale, Kalabo and Barotse districts; absent from much of L. Bangweulu basin, and from parts of Nchelenge, N Mporokoso and Isoka. Botswana, once, Kasane (Randall 1994). Malaŵi, widespread, from N Nyika Plateau to L. Malaŵi littoral and east of Rift in Blantyre, Zomba, Mangochi and Phirilongwe districts. Zimbabwe, Zambezi Escarpment and Mashonaland Plateau, west to Kalahari Sand miombo woods on Charama Plateau, Malimasindi and Chirisa Game Reserve, south to Selukwe, Mt Buhwa and Great Zimbabwe, and east to Rusape and Inyazura. Mozambique, uplands of W Niassa and N Zambezia Provs, common west of Kirk range, rare farther east; records at Revue and Vila de Manica on Manica Platform; and in S sight record at Bangu Gorge on South African Kruger Park border. Transvaal, sight record inside Kruger Nat. Park, given without supporting detail, and another, Sabi, 'many years ago' (Skead 1967).

Description. *S. s. salvadori* Bocage (includes '*rovumae*'): Uganda (Mt Elgon), Kenya and S tropics except Zimbabwe. ADULT ♂: forehead and crown to scapulars, back, rump and uppertail-coverts dark chocolate brown, heavily spotted with white or whitish. Each feather on head is dark brown with white tear-drop down middle, not quite reaching tip; each one on back has dark fringe, white subterminal oval mark of *c*. 2–3 × 4–5 mm, then a broad dark band, then a white band, then a grey fluffy base. The 'white' is often cream or pale pinkish buff; these pale shades accentuated where they overlie each other. Attractive result is heavy but irregular spotting of variable pale shades, white or buff, on chocolate background. Pale spots small and teardrop-shaped on head, large, round, more profuse and sometimes coalescing towards tail. Tail blackish brown, each feather strikingly marked with 6 pairs of white patches, of which proximal 2 pairs concealed below coverts; white patch-pairs of T1 to T6 all aligned, so spread tail is dark brown with 4 neat white visible bands; even central pair (T1) banded; distal pair of white patches, one on each web, is at 'corner' of feather tip and does not reach shaft; most other patches reach shaft; T1 further variegated by areas of grey next to white patches on outer web. Lores and ear-coverts dark brown; superciliary stripe and stripe below eye (cheeks) whitish (but stripes not well defined). Underparts from chin to flanks and vent buff or greyish white, lightly or heavily scalloped with dark brown crescents; missing feather reveals dark brown central oblong of underlying feather, so moulting bird has several dark oblongs amongst its finer dark crescents; crescents most profuse on breast; oblong spots sometimes more profuse than crescents on flanks and belly. Undertail-coverts whitish with blackish bands. Wings dark brown, heavily spotted with white, cream or pale buff. P9 with plain outer web; other primaries with 4 visible white spots on outer webs, the spots on adjacent feathers not aligned; inner primaries, and secondaries and tertials, white-tipped; all remiges with regular pale blobs along edge of inner webs – 4–5 visible on underside of wing, scalariform in closed wing but forming regular white rows in open wing. Upperwing-coverts like back feathers. Underwing-coverts buff, speckled blackish. Upper mandible dark sepia, lower mandible flesh-pink with sepia cutting edge and grey-white underside, gape whitish, inside mouth dull gamboge yellow with greenish patches; eyes dark brown; legs and feet brownish slate grey or horn-brown or greyish. SIZE: wing, ♂ (n = 40) 89–100 (94·1), ♀ (n = 28) 87·0–98·5 (92·1); (3 ♂♂, 3 ♀♀) wing, ♂ 92–97 (94·0), ♀ 90–95 (92·3); tail, ♂ 44–52 (47·7), ♀ 44–56 (50·3), T1 up to 3·5 mm shorter than T6; bill to feathers, ♂ 17–19 (18·0), bill to skull, ♂ 22·5–23·0 (22·7), ♀ 22–23 (22·5); tarsus, ♂ 15·5–16·5 (16·0); middle toe with claw, ♂ 19·5–20·5 (20·0). WEIGHT: 1 unsexed 16. (Nominate race, India, 1 unsexed, 14).

IMMATURE: like adult but underparts uniform, brindled grey. Fledgling just out of nest also like adult but paler and with yellow gape (Howland 1988); legs and claws almost full-sized but bill, tail and wing short (one with bill to feathers 10·5, toes 18, tail 26, wing 66).

NESTLING: skin dark grey or black.

S. s. emini Hartlaub: Gambia to NE Zaïre and NW Uganda. Like *salvadori* but upperparts slightly less heavily white-spotted, therefore looking blacker brown; white marks in tail smaller and greyish; underparts greyer, dark-barred rather than scalloped. SIZE (3 ♂♂, 3 ♀♀): wing, ♂ 90–94·5 (91·3), ♀ 90·5–95·5 (92·3); tail, ♂ 47–49 (47·3), ♀ 50–53 (50·3); bill to skull, ♂ 22·5–25·5 (24·3), ♀ 23–24 (23·7).

S. s. erlangeri Neumann: Ethiopia. Like *salvadori* but upperparts less heavily spotted (i.e. like *emini*) with the whites decidedly buff-tinged on crown, mantle, back and upperwing coverts; tail like *salvadori*; underparts buffier, even rufescent, with numerous round, cream spots each with short blackish bar in front. Wing, ♂ (n = 3) 90–94 (92), ♀ (n = 4) 90–96 (93).

S. s. xylodromus Clancey: Zimbabwe. Like *salvadori* but ground colour of upperparts sooty black (rather than dark brown), feather barring white (not vinaceous buff) and spot at feather tips snowier; tail more sharply black and white; underparts paler than in *salvadori*. Wing, ♂ (n = 16) 90–100 (94·8), ♀ (n = 7) 91–97 (93·3).

Field Characters. Length 15 cm. Barred and spotted plumage with long white supercilium, long curved bill and bark-gleaning habits render it unmistakable within its range. Smaller Short-toed Treecreeper *Certhia brachydactyla*, with unmarked underparts, occurs only north of the Sahara.

Voice. Tape-recorded (86, 91–99, 104, B, F, CHA, HOR, LEM). Song extremely high-pitched, thin and sibilant, like treecreeper *Certhia* or sunbird *Cinnyris*; somewhat variable in length, form and accent, sometimes with up to 10 notes: '*tsee*-wee, tsi-*weee*, tsi-*weee*, tsi-*weee*, tseee', 'tse-tsi-tsi-*weee*, tseee, tseee, tseee, tseee', 'tsi-tsi-*weee*, tsi-tsi-*weee*, tseee, tseee', 'si-*sweee*-y, se-*weee*-y, si-*sweee*-y, si-*sweee*-y'. Indian birds said to give 'feeble, whistling song of the volume and timbre of a sunbird's, 'chichichichiu-chi-chiu-chiu-chiu', lasting *c*. 5 s. (Ali and Ripley 1973). Bird recorded by C. Chappuis in Benin had a very different song, sweet and warbling and much lower in pitch. Call (less common than song) a sharp 'keck' repeated 5–6 times, reminiscent of coot *Fulica* (Vincent 1936). Other calls include tit-like 'tseee', 'see' and 'see-it'. Juvenile begging call a repeated high, very thin 'seep' or 'sip'.

General Habits. Inhabits mature but open *Isoberlinia* woodland in W Africa, evergreen juniper forest in Ethiopia, and climax *Brachystegia* (miombo) woodland in S tropics. Also: recently burnt scrub and edge of dense forest (Gambia), rough-barked, fire-resisting trees in orchard-bush savanna (SE Nigeria), plantations where ground layer burnt over (Chad), open wooded habitats especially with acacias (Kenya), in Zambia *Cryptosepalum* forest (mavunda, restricted to S Mwinilunga and a small area northwest of L. Bangweulu), in Malawi *Brachystegia*/riparian forest ecotone, exceptionally in evergreen forest up to 5 km from nearest *Brachystegia*, in open stands of *Terminalia sericea*, once in *Eucalyptus*; once on maize stalk. Forages alike in large, majestic trees and lower growth of stunted trees. Occasionally occurs in town parks and large gardens, even to nest (in musasa trees *Brachystegia spiciformis* in Harare and Marondera, Zimbabwe).

Mature, undisturbed *Isoberlinia* and *Brachystegia* woodlands look much alike, but bird's habitat in Ethiopia, inside evergreen forest, is strikingly different (Benson 1946b).

Singly or in pairs; commonly in mixed-species foraging flocks (when seen at all – many observers remark on difficulty in finding Spotted Creepers, which may be both cryptic and shy, and are enigmatically scarce over much of range; they also hide from observer around back of tree trunk: Ginn *et al.* 1989). However, pair in Mole Nat. Park, Ghana, landed in small tree under which observers were standing and fed unconcernedly for several min. (S. Keith and C.G.R. Bowden, pers. comm.). In some areas seems always to occur in mixed flocks, of e.g. *Eremomela*, *Sylvietta* and *Hyliota* spp., Brubrus *Nilaus afer* and Olive-headed Weavers *Ploceus olivaceiceps*; even joins large mixed flock moving through trees near nest. Forages by creeping, clambering, fluttering and running jerkily but quite rapidly, in stages, up tree trunk or large limb, stopping to examine fissures and crevices in bark; on reaching upper branches, flies down to foot of nearby tree. Sometimes flies more distantly ('half a mile' once claimed); flight dipping or undulating, distinctly woodpecker-like, with rapidly fluttering wings. Alights by tumbling onto treetrunk with wings momentarily open and flapping, rather like quail *Coturnix* sp. pitching into grass. Described as going into a glide when flying from nest, and using gliding action between trees when cautiously returning to nest (Skead 1967). Remains still ('freezes') if in danger. Actions in tree rapid, decisive and agile, more like Wallcreeper *Tichodroma muraria* or nuthatch *Sitta* sp. than treecreeper *Certhia* sp.; works branches forwards and back again, clings upside down, and moves up or around trunk in series of restless hops, sometimes with wings fluttering rapidly; tends to edge sideways and creep round and round trunk or bough, often taking bird in upward spiral (Skead 1967); hugs tree trunk so legs difficult to see, though large feet noticeable (Randall 1994).

Usually considered sedentary, but sometimes found well away from known range (Skead 1967, Tree 1987 and 1992, Randall 1994), so must disperse or wander widely (in keeping with its long wings).

Food. Insects, including small beetles, beetle larvae, winged ants and other small hymenopterans, moths, bugs and caterpillars; also spiders. 'Cicadas', reported in stomachs (Bannerman 1948), seem a most unlikely food item and observer may have meant bugs.

Breeding Habits. Solitary nester; monogamous.

NEST: neat, high-walled, open cup made of flower stalks, leaf petioles, rootlets, thin stems and lichens, bound together with spider web or cotton-like fibres. One nest (**A**) made almost entirely of lichens and spider web (Steyn 1974). Lined with spider web, spider egg bags, insect cocoons and soft plant down. Outside surface heavily decorated with bits of lichen, bark chips, caterpillar droppings and vegetable down including seeds with hairy pappi, perfectly camouflaging nest against branch and making it practically invisible from below. Nest fabric rather loose, soft and pliant but strong; very like nest of

Brubru or batis *Batis* sp. (though larger). Nest placed in Y-shaped fork or on horizontal bough and often against vertical branch, 3–12 m above ground; bound to substrate with spider web. 3 nests in fork of musasa tree and 1 on short horizontal kink in musasa branch (Zimbabwe: Masterson 1970). Ext. diam. *c*. 60–70, int. diam. 45, int. depth 45, wall 12 thick. Nest usually about same diam. as branch it is built upon. Nest generally part-shaded by light foliage above it. Nest is added to throughout incubation period, so sides become more upright (Skead 1967). Same tree may be used for nesting in successive seasons.

EGGS: 2–3, usually 3 in S tropics. Pale green or turquoise-blue, zoned with lavender, heavily speckled all over with small dark brown or black spots, with a few brown and grey smudges towards blunt end. SIZE: (n = 9) 16·2–19·5 × 12·0–13·8 (17·9 × 13·3).

LAYING DATES: Ghana, Jan; Nigeria, (chasing, singing and soliciting, May); Ethiopia, (recently fledged young, May); E Africa: Region A, Dec, Region C, Nov; Angola, Aug; Zaïre (Lubumbashi), (young in nest Oct); Zambia, Sept (7 records); Malaŵi, Sept (3 records; and fresh nest Oct); Zimbabwe, Aug (3 records), Sept (26) and Oct (7).

INCUBATION: by ♀ only (Skead 1967) or by both sexes (Ali and Ripley 1983); ♀ fed on nest by ♂. Bird disturbed from incubating by photographer soon returned to tree, alighting half way up, ran up trunk and hopped onto nest in single quick movement (Steyn 1974); also described as hopping along nest branch, weaving its way rapidly over last few m before settling on nest (Ginn *et al.* 1989). Confiding; at 2 nests, incubating bird ignored nearby person 1 m away and settled onto nest with bill pointing up at a 60–75° angle (**A**). Bird appears to sit very tight and normally with bill pointing up at 45°. One bird sang loud and long whilst incubating (Masterson 1970); 'song' from nest perhaps better interpreted as contact call, high-pitched and twittering, since mate forages seldom out of earshot (Ginn *et al.* 1989).

DEVELOPMENT AND CARE OF YOUNG: at a few days of age when still quite tailless, head, dorsal midline, rump, and sides of breast have feathers 1–3 mm long, buff speckled with black (i.e. like adult); plenty of bare skin shows; primaries and secondaries in long grey sheaths (at this age bill to feathers 7 mm; and no tail). Both parents brood young (Skead 1967). Adult carrying food approaches nest indirectly, via 2–3 trees, by creeping up tree then flying down to foot of next one. Parent removes faecal sac and drops it away from nest (Steyn 1996). Adult distraction display (Masterson 1970): at one nest parent brooded 3 well-grown young; when disturbed (virtually chased away) by person, it flew in faltering manner down to foot of

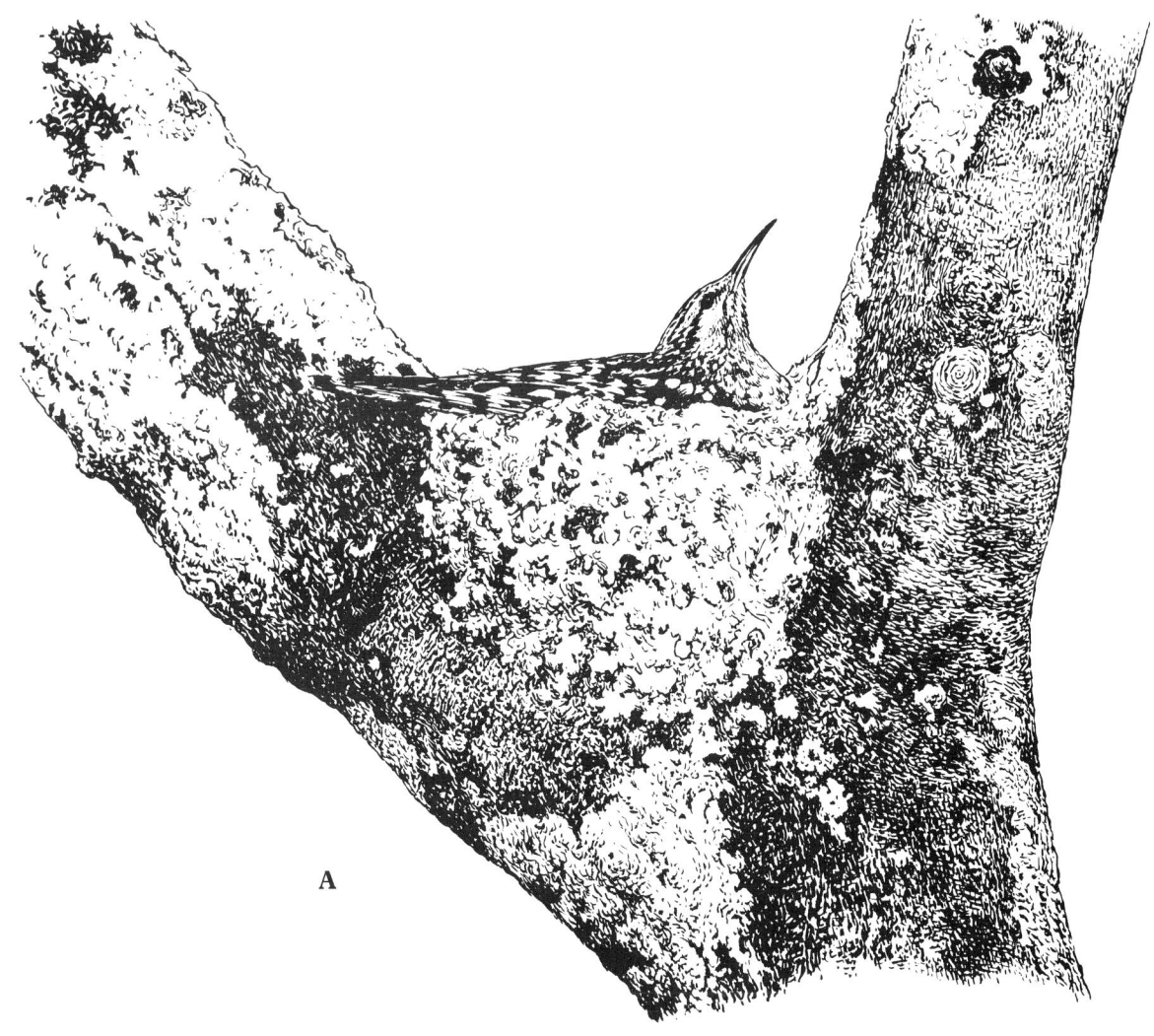

A

nearby tree trunk, where first one wing then the other momentarily flopped down; then bird crept into head-down position with both wings hanging loosely out from its back. Adult care of single fledgling (Howland 1988): juvenile in *Brachystegia boehmii* tree begged by quivering wings and calling; parent flew away, into next tree, and foraged; then juvenile flew to parent only when latter called; juvenile was immediately fed, then juvenile followed very close behind parent as latter foraged, its bill touching adult's tail tip; during observation lasting 1 h, juvenile followed parent to new foraging tree only when summoned and it moved up and down tree only when following immediately behind parent. When left alone, juvenile moved (crept) little or not at all. Parent twice brought food item to it, rather than summoning it to another tree.

References
Ali, S. and Ripley, S.D. (1973).
Clancey, P.A. (1975).
Ginn, P.J. et al. (1989).
Masterson, A. (1970).
Skead, C.J. (1967).
Steyn, P. (1974, 1996).
Vincent, J. (1936).

Family CERTHIIDAE: treecreepers

Small or very small arboreal birds with cryptic plumage, thin decurved bills and long toes. Bill laterally compressed, needle-like but curved; nostrils longitudinal, near base of bill; rictal bristles vestigial; tongue long, narrow, horny, with *c*. 5 bristles at tip. 10 primaries, P10 < half length of P9; wing short and rounded with P5–P7 longest and about equal; tarsus short, with either scutellate or booted lamina in front; outer toe long, middle toe longer, inner toe short or quite long; claws rather long and thin but sharp and strongly curved, hind claw especially so; tail strongly graduated, with 12 stiff feathers. Plumage long and soft, richly patterned in browns, greys and white; sexes alike; young like adults but more spotted. Sedentary. Lives of treecreepers bound to tree trunks and large limbs which are used for foraging, roosting and nesting purposes. Climb with feet wide apart, creeping jerkily upward on trunk or along underside of bough with belly almost brushing bark and tail pressed against tree, used like woodpecker's tail. Insectivorous. Nest a deformed saucer, built into crevice behind bark; eggs white with corona of reddish spots.

In Sibley and Monroe's taxonomy (1990), Certhiidae embraces wrens (Troglodytinae), gnatwrens and gnatcatchers (Polioptilinae), treecreepers (Certhiinae, Certhiini) and Afro-Indian Spotted Creeper *Salpornis spilonotus* (Certhiinae, Salpornithini). Having already adopted the conservative approach of separating wrens as family Troglodytidae (Vol. IV, p. 382), we adhere to traditional scheme and restrict Certhiidae to treecreepers *Certhia*.

1 genus, of 6 species in Holarctic and Orient.

Genus *Certhia* Linnaeus

Characters those of the family. 6 species, N America, Palearctic including NW Africa, and Orient; 2 in W Palearctic, of which 1 breeds in NW Africa.

Plate 7
(Opp. p. 94)

Certhia brachydactyla C. L. Brehm. Short-toed Treecreeper. Grimpereau des jardins.

Certhia brachydactyla C. L. Brehm, 1820. Beitr. Vogelkunde, 1, p. 570; Roda valley, Thuringia, Germany.

Forms a superspecies with *C. americana* (N America).

Range and Status. Endemic to W and S Europe and NW Africa. In Europe, Iberian Peninsula to Denmark, Poland, Hungary, Balkan Peninsula, Sicily, Crete, Turkey, also W Georgia.

N Morocco to NW Tunisia. In Morocco, widespread from sea level up to 2000 m in Rif, Tidighine, Talassemtane and Tazzeka Mts, Moyen Atlas (Ifrane, Col du Zad, Itzer, Kerrouchène, Azigza, Al Hamman, An Leuh,

Certhia brachydactyla

Azrou), and Haut Atlas (Toubkal, Imlil, Tizi Oussen, Zaouïa Ahansal); record from Meski (Errachidia). Algeria, locally frequent in N within 200 km of coast, from Tlemcen Mts to Mitidja Plain and El Harrach (Alger), Oued Sebaou and Medjerda, commoner in E than in W; also in Mts des Ksours; from sea level up to 2000 m. Tunisia, scarce, confined to NW, from Kroumirie south to Le Kef; formerly along Algerian border south to Tozeur, but no records there in 1970s (Thomsen and Jacobsen 1979).

No density data from Africa; in S France, 26 pairs per km^2 in oak *Quercus ilex* woods and 94 pairs per km^2 in *Q. pubescens* (Cramp and Perrins 1993).

Description. *C. b. mauritanica* Witherby: NW Africa. ADULT ♂: feathers of forehead, crown, nape and hindneck brown, with narrow whitish shaft streaks and blackish borders; long greyish or brownish white superciliary stripe; feathers of mantle, back and scapulars the same as crown, but centres as well as shafts whitish, and only tips blackish; feathers of rump and uppertail-coverts cinnamon or rufous-brown, whitish along shafts. Tail brown, shafts pale rufous-brown, vanes darker next to shaft than at edges. Lores black; ear-coverts dark brown, mottled with whitish buff. Chin and throat greyish white; breast greyish white, buffy at sides; belly white, grading to buff on flanks; thighs greyish; undertail-coverts pale rufous, tipped whitish. Wing variegated in grey-browns, black, white and buff, in complex, almost dead-leaf, pattern; portrayed by Harrap and Quinn (1996, p. 190) where well described: Lesser coverts brown, tipped orange-buff. Median and greater coverts black-brown, median coverts tipped orange-buff and centred whitish, greater coverts fringed and broadly tipped pale buff to rufous-brown, sometimes with a pale shaft streak. Alula and primary coverts black-brown, tipped pale buff, largest feather of alula often with complete pale fringe on outer web. Inner web of tertials mid-light brown, outer web dark brown, black from centre towards tip, with narrow buff-white fringe and broad white tip. Primaries and secondaries mid brown, P4/P5–P10 and secondaries with broad pale buff band across feather (only on outer web P4/P5), bordered on each side by a blackish brown band, fringed orange-buff distally of these bands (especially on the secondaries) and tipped off-white. Underwing-coverts and axillaries white. Bill dark brown; eyes dark brown; legs and feet pale brown. Sexes alike. SIZE: wing, ♂ (Morocco, n = 10) 62–70 (65·5), ♂ (Algeria and Tunisia, n = 10) 60–68 (64·2); bill to skull, ♂ (Morocco, n = 10) 17·5–21 (19·4), ♂ (Algeria and Tunisia, n = 10) 17–22 (20·1); supplementary measurements, *C. b. brachydactyla*, Spain, 8 ♂♂, 3 ♀♀: wing, ♂ 62–66 (63·4), ♀ 60–63 (61·5); tail, ♂ 56–65·5 (59·8), ♀ 59·5–60·5 (60·0); bill to skull, ♂ 17·5–19·1 (18·4), ♀ 16·0–16·2 (16·1): tarsus, ♂ 14·8–16·3 (15·5). WEIGHT: (*C. b. dorotheae*, Cyprus, unsexed, n = 19) 7–9·5 (8·5).

IMMATURE: like adult, but pale feather centres in ear-coverts and feathers of mantle, back and scapulars larger; superciliary stripe greyer, mottled, and less distinct; and breast and flanks faintly mottled.

NESTLING: plentiful long, grey-black down, on head only. Mouth orange-yellow; no tongue spots; gape pale yellow; bill brown; legs and feet very pale brown.

Field Characters. Length 12·5 cm. Only passerine bird in Africa that habitually uses its tail as a prop; creeps up tree-trunks and around branches; easily overlooked. Streaky brown upperparts, whitish underparts, variegated brown and buffy wings, thin decurved bill and spiky tail make identification easy (since Eurasian Treecreeper *C. familiaris* does not occur in Africa).

Voice. Tape-recorded (62–73, 93, 105, 106, B). Song in Europe high-pitched, buzzy and rapid, 'tsee-tsee-didiree-too-see-too-tsee' and variations, sometimes a shorter 'tee-tee-doodawy-tee'; call a distinct '*ting!*', like tit *Parus*, very different from high thin call of Eurasian Treecreeper. Voice of North African birds lower pitched than those from France, and song distinctly different; some units common to both populations, but some of *mauritanica* not recognized as conspecific by European birds.

General Habits. Inhabits lowland and mountain woods with pine *Pinus halepensis*, cedar *Cedrus atlantica*, cork oak *Quercus suber*, holm oak *Q. ilex* and other trees with rough bark, including poplars; rarely palms; avoids areas where tree growth stunted and with annual rainfall below 400 mm. In Algeria commoner in cedars than in cork oaks. In Europe associated with broad-leaved trees, particularly oaks, occurring in woods, forest edges, parks, orchards and copses with plenty of undergrowth and in gardens with mature trees. Mainly solitary. Thoroughly arboreal; forages by creeping mouse-like up trunk and large limbs, keeping mainly where bark deeply fissured; often moves spirally, with short, hopping gait and splayed toes, and tail depressed against substrate and used as a prop. Occasionally gleans small twigs and foliage. Active, quiet, easily overlooked, often confiding, but tends to disappear to far side of trunk when being observed. Tends to work its way upwards in large tree, then to fly down to bole of next large tree. Momentarily moves sideways around vertical trunk; gleans underneath large horizontal bough, but never progresses head-downwards. Flight distinctive, with bursts of fluttering beats, bird side-slipping or tumbling just before alighting. Often joins foraging parties of tits *Parus* and *Aegithalos*, Firecrests

Regulus ignicapillus, Eurasian Nuthatches *Sitta europaea* and Chiffchaffs *Phylloscopus collybita*. In S Spain in winter, forages mainly on holm oaks, at av. height above ground of 1·8 m, or 1·5 m when in mixed-species foraging flock; of 3750 observations, birds spent 41% of time on large branches, 34% on trunks, 21% on small branches, 2% on twigs, and 2% on leaves, the ground, in shrubs and in the air (Cramp and Perrins 1993). In cold conditions roosts communally in dense cluster of 5–8 and sometimes up to 20 birds, usually on building, birds clinging under eaves, sometimes in tree; in cluster, bodies are in centre and tails point outwards (**A**), and cluster of 20 birds may be so tight that only 14 tails are visible; birds arrive at roost one by one, over period of 30 min (Cramp and Perrins 1993).

Sedentary; no evidence of immigration from Europe into Morocco, but increase in abundance in Strait of Gibraltar area from Sept suggests possible small-scale passage (Cramp and Perrins 1993).

Food. Small invertebrates, mainly insect larvae and pupae and spiders.

Breeding Habits. Moderately well known in Europe; not known to differ in Africa though barely studied there. Solitary nester; apparently monogamous; territorial. One instance of bigamy (Isenmann and Guillosson 1993). ♂ displays by singing, stationary in tree or whilst climbing it; conflict between rivals involves pursuit in trees, flying attacks, and aerial fighting during slow descents, as well as song duels near territorial boundaries (Cramp and Perrins 1993). Little by way of postural courtship displays known. Threatens by gaping. Courtship-feeds.

NEST: in deep crevice, often behind piece of loose bark, sometimes in masonry, woodpile, thick mass of ivy or old woodpecker hole; untidy foundation, often large and irregular, made of twigs, fibres, conifer needles, grass, bits of bark and artifacts, lined with moss, lichen, cocoons, hair, rootlets and feathers. Size (Germany, n = 9): int. diam. 40–60 (44), depth 35–45 (41). Foundation built by ♂; usually 2–3 foundations built, one chosen and lined by ♀.

EGGS: 4–6 (n = 5, Morocco), av. 5·2. Sub-elliptical, smooth, matt, white with red-brown or red-purple spots and blotches (sometimes very faint), mainly around large end. SIZE: (n = 50, Morocco) 17·0–21·0 × 13·5–17·5 (18·9 × 14·3).

INCUBATION: by ♀ only; period 13–14 days.

DEVELOPMENT AND CARE OF YOUNG: young fed and cared for equally by both parents. Nestling period (S France, n = 6) av. 17·3 days.

BREEDING SUCCESS/SURVIVAL: in S France 8–10% of eggs (2 sites) lost to predators, 8–16% lost to rain, 0–11% infertile; at the 2 sites together 57% of eggs produced fledglings.

References
Cramp, S. and Perrins, C.M. (1993).
Etchécopar, R.D. and Hüe, F. (1967).
Harrap, S. and Quinn, D. (1996).
Isenmann, P. and Guillosson, J.-Y. (1993).
Isenmann, P. *et al.* (1986).

A

Family NECTARINIIDAE: sunbirds and spiderhunters

A well-defined, mainly paleotropical family of small or very small songbirds, strongly associated with trees in flower. Feed upon insects, spiders and nectar, and somewhat convergent with other nectarivorous birds such as sugarbirds Promeropidae, flowerpeckers Dicaeidae, honey-eaters Meliphagidae and hummingbirds Trochilidae. Sexes alike in size and colour (spiderhunters – Oriental *Arachnothera* – and a few sunbirds) or more or less strongly colour-dimorphic (most sunbirds). Spiderhunters and most sunbird ♀♀ olive above and streaky or mottled grey, buff, whitish or yellowish below; breeding sunbird ♂♂ strongly patterned and brightly coloured. ♂ colours are mainly structural, i.e. iridescent or metallic, strikingly so, in black-background greens, violets and blues; breast in many ♂♂ pigmentary (non-glossy) scarlet or maroon, feather tips sometimes structural blue; belly pigmentary black, yellow, green or white; many species with conspicuous tuft of pigmentary yellow or orange feathers at side of lower breast ('pectoral tufts'), concealed by carpal bend of wing or conspicuously erect in front of it. Non-breeding ♂ plumage somewhat like ♀; ♂♂ of many species have eclipse plumage – like ♀ only for duration of wing-moult.

Bill sharp-pointed, slender but hard and quite strong, varying in length and curvature, in sunbirds from short (about length of rest of head) and only slightly decurved or almost straight, to short (not much longer than head) and strongly decurved, to twice length of head and more or less strongly decurved, and in spiderhunters bill about 3 times length of rest of head and decurved, though less so than in some sunbirds. Base of bill curved, or more or less straight: a character of importance in defining sunbird genera. Maxilla and mandible same length, close-fitting and hence similarly decurved; radius of curvature rather uniform (in *Arachnothera* about 65–105 mm), or proximal third of bill slightly decurved and distal two-thirds more markedly decurved (radius of curvature only 20 mm in some sunbirds and *c.* 10 in one). Edges of mandibles very finely serrated near tip. Bill does not open very wide. Tongue long and tubular, in cross-section U-shaped proximally and ∞-shaped distally; tip of tongue bifurcate or trifurcate, each fork with spiral ridges and microscopically branched at very tip; hyoid arms elastic, long, arising by glottis, passing up occiput and inserted (as in woodpeckers Picidae) above line between eyes; tongue flicks in and out through short distance when birds feeding on nectar and sometimes for few moments afterwards, bill seemingly closed; tip of tongue occasionally protrudes momentarily from 'closed' bill tip at other times. Nostrils oval, operculate. No rictal bristles. Wings rather short and rounded; 10 primaries, P10 25–35% length of P7–P8. Flight strong and direct; birds often hover momentarily when feeding. Tail of 12 feathers, very short (*Arachnothera*) or short or medium (sunbirds), usually rather round-ended, sometimes graduated, with or without white signal-marks on outer rectrices, ♂♂ of *Aethopyga* and some *Nectarinia* spp. with long central pair of rectrices. Leg short, tarsus slender, hard, and scutellate, toes short but sharp-clawed and with quite strong grip.

Arboreal, feeding mainly at flowers and inflorescences, at outer edges of open-canopy woodland trees and in well-lit uppermost part of canopy of continuous forest; many species also feed low down, on flowering shrubs and upright flowering spikes or hanging inflorescences. It involves quite acrobatic postures, but sunbirds are not as acrobatic as tits (Paridae, Remizidae) although they are at least as active. Most species also more or less 'insectivorous' (spiders, small insects) and many eat small fruits and fruit pulp. Faeces expelled in strong, liquid jet. Vocal, mostly with simple calls and quite loud, euphonic ♂ territorial songs, often based on repetition of chirping call notes. Monogamous. Nest, built only by ♀, an untidy-looking pendent bag with plain or porched entrance at side; nests of some species look rather like weavers' nests (Ploceidae) but are more cryptic and fabric is softer; other nests can be almost felted, like nests of penduline tits *Anthoscopus*. Eggs pale grey, bluish or pale mauve, sparingly or quite heavily blotched, or in a few species almost immaculate. Nestling naked or covered in grey down; inside of mouth yellow, orange or red.

In the classification of Sibley and Monroe (1990), Nectariniidae embraces 2 subfamilies, Promeropinae (sugarbirds) and Nectariniinae, and the latter comprises 2 tribes, Dicaeini (flowerpeckers) and Nectariniini. We regard sugarbirds and flowerpeckers as separate families, although we recognize that flowerpeckers resemble sunbirds/spiderhunters in many regards and probably constitute their sister clade. Nectariniidae *sens. str.* is biologically and morphologically uniform (as is Dicaeidae), and has been reduced by most recent authorities to only 5 genera, following Delacour (1944) (with removal of Malagasy false-sunbirds *Neodrepanis* and recognition of Purple-naped Sunbird as a distinct, monotypic genus *Hypogramma*).

Rand (1967) recognized 116 species in 5 genera: *Anthreptes* (17 species), *Hypogramma* (1), *Nectarinia* (74), *Aethopyga* (14) and *Arachnothera* (10). Wolters (1977) admitted 97 species in 25 genera. In their Tribe Nectariniini Sibley and Monroe (1990) recognized the same 5 genera as Rand's, with 2 more species in *Anthreptes* and 5 more in *Nectarinia*, totalling 123 species. One new species, of *Aethopyga*, has since been discovered (Kennedy *et al.* 1997).

Two-thirds of all sunbirds are Afrotopical. *Arachnothera*, *Aethopyga* (yellow-rumped sunbirds) and *Hypogramma hypogrammicum* are Oriental; *Anthreptes* and the huge genus *Nectarinia* are mainly African. Sunbird taxonomy has been reviewed by M.P.S. Irwin (in press), using morphological and biological characters. Impressed by revealed differences as much as by resemblances, we adopt his classification. We recognize 81 species in Africa (including one, the Tohâ Sunbird, newly discovered and not yet formally named), in 10 genera: *Anthreptes*, *Nectarinia* (*sens. str.*, comprising only 7 species), *Deleornis*, *Anabathmis*, *Dreptes*, *Anthobaphes*, *Cyanomitra*, *Chalcomitra*, *Hedydipna* and *Cinnyris*. All genera are endemic except *Anthreptes* and *Cinnyris*.

Genus *Anthreptes* Swainson

A heterogeneous, taxonomically isolated and primitive group of unspecialized, morphologically conservative sunbirds, somewhat warbler-like in habits. Bill relatively short, heavy, almost straight or moderately decurved; basal third lacks curvature, which is only apparent distally (Irwin 1993); mandibles finely serrated, nasal operculum naked; outermost primary less than half length of next; tail square, central feathers never elongated. Sexes alike or moderately to strongly dimorphic. 2 unrelated species lack iridescent plumage, others have iridescent parts replaced by grey, brown or blackish brown. Reduction in iridescent plumage may be secondary; some dull forms, particularly forest ones, thought to be descended from iridescent or bright-coloured ones. Some species insectivorous, others also eat fruits; a few species rarely visit flowers, others take nectar opportunistically.

14 species: 10 in Africa, 4 in Oriental Region (Nepal and Bangladesh to Philippines and Indonesia). In Africa 3 species in savanna (1 in dry savanna), the rest in forest; none wholly montane. *A. longuemarei*, *A. neglectus*, and *A. orientalis* are long winged and large bodied and form a superspecies, closely related to Oriental *A. malacensis*, *A. singalensis* and *A. rhodolaema*. Hall and Moreau (1970) included the small *A. aurantium* and *A. gabonicus* as divergent members of the same superspecies; but while *aurantium* may be close, we think *gabonicus* is not – each is restricted to forested rivers and mangroves, where they probably evolved independently. The forest *A. rectirostris* and *A. rubritorques* form a superspecies. *A. reichenowi* is closely related to Oriental *A. simplex* (Irwin 1995). *A. seimundi* is difficult to place in any sequence and has no clear affinities with the others; because of its long bill, dull plumage and similarly plumaged sexes, Delacour (1944) suggested that it was the most primitive member of the *Nectarinia* complex but this we cannot accept. Lastly, *A. anchietae*, one of the most brilliantly coloured African sunbirds, endemic to miombo woodland, shows very little sexual dimorphism, and may be unrelated to the others; we retain it in *Anthreptes incertae sedis*.

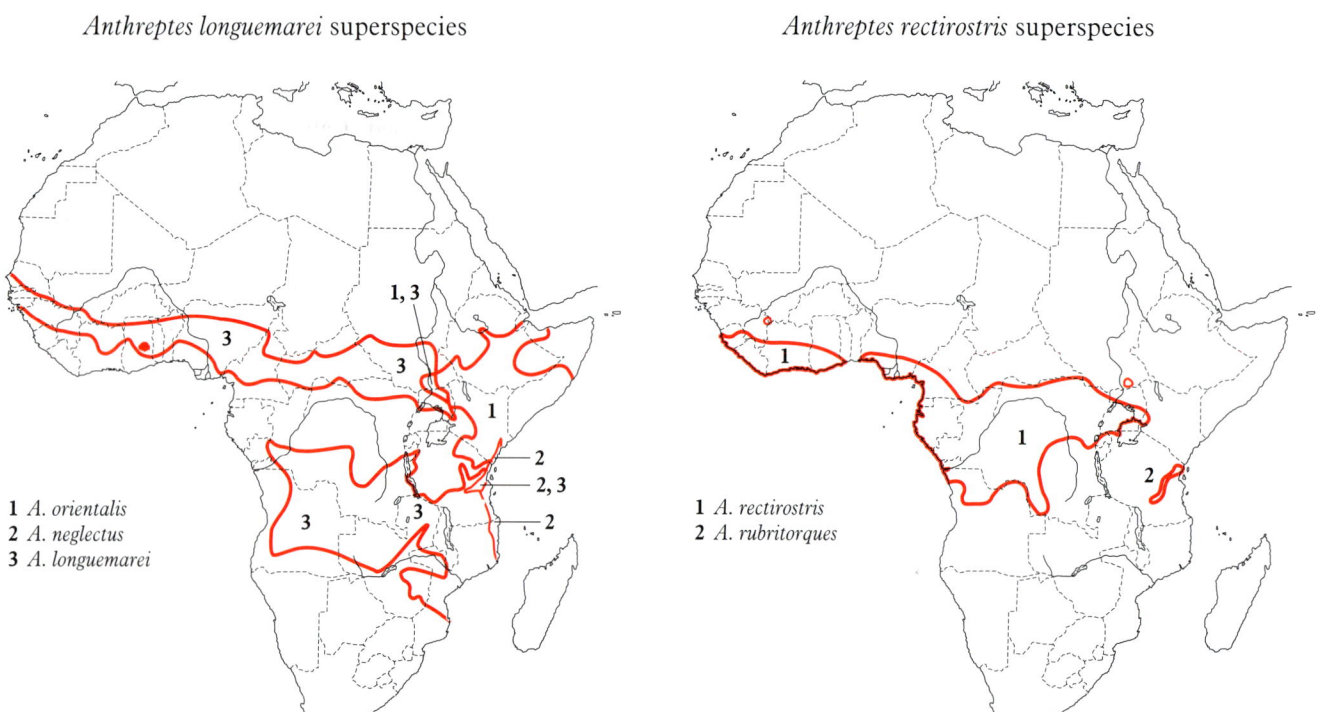

Anthreptes longuemarei superspecies

1 *A. orientalis*
2 *A. neglectus*
3 *A. longuemarei*

Anthreptes rectirostris superspecies

1 *A. rectirostris*
2 *A. rubritorques*

Anthreptes neglectus Neumann. Uluguru Violet-backed Sunbird. Souimanga violet des Ulugurus. Plate 8
Anthreptes longuemarei neglectus Neumann, 1922. Orn. Monatsb., 30, p. 13; Uluguru Mountains, Tanganyika Territory. (Opp. p. 95)

Forms a superspecies with *A. longuemarei* and *A. orientalis*.

Range and Status. Endemic resident. In Kenya rare, only 6 records; known from 2 tiny forest patches on lower Tana R., Makeri West and Kitere; not seen since 1962 in Makeri West, which is now very damaged, but a pair seen recently in Kitere, which is 18 ha in extent but has only 4–5 ha of tall forest suitable for this species (Butynski 1994); also Buda and Jadini Forests, inland to Shimba Hills. Tanzanian lowlands (44 recent sight records: N. Baker, pers. comm.) from Pande, Kazimzumbwe, Kurasini and Kiwengoma Forests, Nguhi, Soga, Kidugallo and Ngerengere R. to Mikindani; eastern arc forests to 1400 m in East and West Usambara Mts, Nguru, Ukaguru and Uluguru Mts (possibly to 1800 m), Rubeho Mts (Usagara), Udzungwa Scarp Forests (Mwanihana, Magombera, Chita), Ndundulu and Nyumbanitu Mts. N Mozambique at Lurio R. and Netia. Locally common to uncommon; probably declining with forest clearance. Classed as near-threatened by Collar and Stuart (1985).

Description. ADULT ♂: forehead and crown reddish violet, hind crown and nape sooty black; mantle, back and scapulars reddish violet (without bluish gloss), rump metallic bluish green, uppertail-coverts metallic reddish violet; tail metallic violet, blackish below, tips of outermost feathers off-white. Lores, cheeks and ear-coverts dusky-brown, chin metallic reddish violet, throat creamy white, breast, sides of neck, belly and flanks isabelline, washed buff, greyer at sides, thighs off-white, undertail-coverts buffy white. Wings dark brown, outer webs of primaries and secondaries margined greenish olive, greater wing-coverts margined greenish olive, shoulder and median wing-coverts metallic green; underwing-coverts and axillaries off-white, pectoral tufts bright yellow or orange. Bill dark sepia to blackish; eyes dark brown; legs and feet black. ADULT ♀: like ♂ above, metallic reddish violet, but centre of forehead dusky blackish brown. Chin, throat, sides of neck, breast and flanks dusky grey with obscure off-white or buffish streaks; belly bright yellow, streaked olive at sides, undertail-coverts and thighs yellow. SIZE: wing, ♂ (n = 10) 68–75 (71·3), ♀ (n = 6) 63–66 (64·8); tail, ♂ (n = 10) 50–54 (52·6); bill, ♂ (n = 10) 15·5–18 (17·2); tarsus, ♂ (n = 10) 16·5–18·5 (17·4). WEIGHT: ♂ (n = 3) all 10, ♀ (n = 2) 9,9.

IMMATURE: mantle and back sooty black, washed with violet, uppertail-coverts bluish violet. A well-developed off-white supercilium from above lores to well behind eye; belly to undertail-coverts lemon yellow. Shoulder metallic green.

NESTLING: unknown.

TAXONOMIC NOTE: birds from Tana R. (Kenya) appear smaller; yellow on belly of ♀♀ paler, more restricted; they may represent an undescribed race (Keith 1968).

Field Characters. Length 11·8–12·2 cm. The forest counterpart of Eastern and Western Violet-backed Sunbirds *A. orientalis* and *A. longuemarei*. Both sexes have violet crown and back separated by dark brown hind-collar, blue-green rump, violet uppertail-coverts, deep purple-blue tail; ♂ has variable underparts, off-white to grey-brown or isabelline, with violet chin and yellow pectoral tufts; ♀ lacks white supercilium and has chin to upper belly grey, lower belly yellow. Meets Eastern Violet-backed Sunbird along Tana R., Kenya, although ecologically segregated; both sexes most clearly differentiated from it by *grey-brown*, rather than pure white, *underparts*, and ♀ also by lack of white supercilium (Butynski 1994). These same characters differentiate it from Western Violet-backed Sunbird where their habitats meet in coastal Tanzania.

Anthreptes neglectus

Voice. Not tape-recorded. Call, a 'typical sunbird sharp 'tsssp' given 1–4 times in succession ... sometimes followed immediately by a melodious 1-s long warbling trill' (Butynski 1994). Also said to give loud 'sweep-sweep-sweep' or 'seep-sureep, sureep ...', persistently uttered (Zimmerman *et al.* 1996).

General Habits. Inhabits coastal forest and nearby moist woodland, ascending into mid-altitude forest in mountains; also secondary forest and shade trees in tea plantations (perhaps only on feeding excursions), riparian forest and adjacent moist bush. Common tree species in Kitere Forest, Kenya, are *Alangium salvifolium*, *Cordia ghoetzii*, *Ficus sycamorus*, *Oxystigma msoo*, *Pachystela msolo*, *Sorindela obtusifoliata*, and lower vegetation is dominated by *Polysphaeria multiflora* and *P. reclinata*; phoenix palms *Phoenix reclinata* on the edge of the forest are avoided (Butynski 1994). Forages mainly in middle levels, sometimes higher and occasionally near ground.

Occurs in pairs or family parties, usually <10, occasionally in larger concentrations. Frequently joins mixed-species parties. In Kitere Forest a pair was present in a party of at least 18 species (for list see Butynski 1994); they searched actively for insects in dense foliage in outer branches of trees, usually within a few m of each other. ♀ once preened ♂ briefly as they sat side by side on a twig. In 3 h this party moved about 120 m through tall evergreen forest (tree height 10–30 m) (Butynski 1994). Not yet seen to feed among flowers.

Food. Insects; fine sand recorded from stomach. Nectar?

Breeding Habits. Monogamous; territorial.

NEST: domed oval, with side-top entrance, untidily constructed of moss, lichens and grasses, with trail of material hanging below; lined with white plant down; suspended from twig at forest edge 8 m above ground.

LAYING DATES: E Africa: Region D, Dec, Region C (feeding recently fledged young Dec).

DEVELOPMENT AND CARE OF YOUNG: recently fledged young fed by ♂, undoubtedly also by ♀.

Reference
Butynski, T. (1994).

Plate 8 (Opp. p. 95)

Anthreptes longuemarei (Lesson). Western Violet-backed Sunbird. Souimanga violet.

Cinnyris longuemarei Lesson, 1831. Illustr. Zool., pl. 23; 'Senegambia superior' = Senegal.

Forms a superspecies with *A. orientalis* and *A. neglectus*.

Range and Status. Endemic resident. Guinea savannas from S Senegal, Gambia and Guinea to Sierra Leone, N Ivory Coast south to Lamto (but mainly at 08–10°N) and SW Mali (borders of Guinea and Mandingo Mts), S Burkina Faso, N Ghana (rare, known only from Mole and Ejura); N Togo south to *c*. 08°N; NW Benin (Arli Nat. Park) and Nigeria, uncommon north of Niger and Benue rivers to Zaria and Kari, formerly south of Benue at Enugu but no recent records; N Cameroon (north of forest belt), SW Chad, Central African Republic north to Bamingui–Bangoran and Manovo–Gounda–Saint Floris Nat. Parks; Sudan (west of 32°E, to 10°30'N), and Zaïre (north of forest zone) to Uele R. and west of L. Albert; Uganda from W Nile and Acholi south to Bwamba, Nakasongola and Soroti, and scattered records from L. Victoria basin; W Kenya, scarce and local in Muhoroni and Kapenguria Districts; formerly Bungoma (Zimmerman *et al*. 1996); southern savannas from SE Gabon (Leconi and Franceville), SE Congo along Congo R., S Zaïre in Kasai and Shaba north to Manyema and west of L. Tanganyika to about Uvira; central plateau of Angola (south to S Bie and N Huila), Zambia (except extreme SW and Luangwa and Zambezi valleys). Tanzania as mapped, from Nyarumbugu, Kibondo and Bugarama south to Ufipa (Sumbawanga), and in S and E from Songea to Nandembo and coastal lowlands north to Kidugallo and Dar es Salaam (N.E. Baker, pers. comm.); SE Burundi; N Mozambique and Malaŵi, coastal lowlands to south of Save R. to Vilanculos (Rumbacaca) and plateau of NE Zimbabwe south to Mtao Forest Reserve and in E to Chipinga Uplands. Locally common to uncommon.

Description. *A. l. angolensis* Neumann; Gabon, S Zaïre and Angola to W Tanzania, NW Mozambique and Malaŵi west of Rift. ADULT ♂: forehead and crown metallic violet, nape blackish brown, mantle and back metallic violet (often glossed bluish), rump sometimes with metallic blue or green gloss. Uppertail-coverts metallic violet, tail dark with metallic violet sheen, blackish with faint metallic sheen below. Lores, cheeks and ear-coverts blackish brown, chin metallic violet, throat, breast and belly to flanks, thighs and undertail-coverts off-white, strongly washed buff; pectoral tufts orange-yellow. Primaries and second-

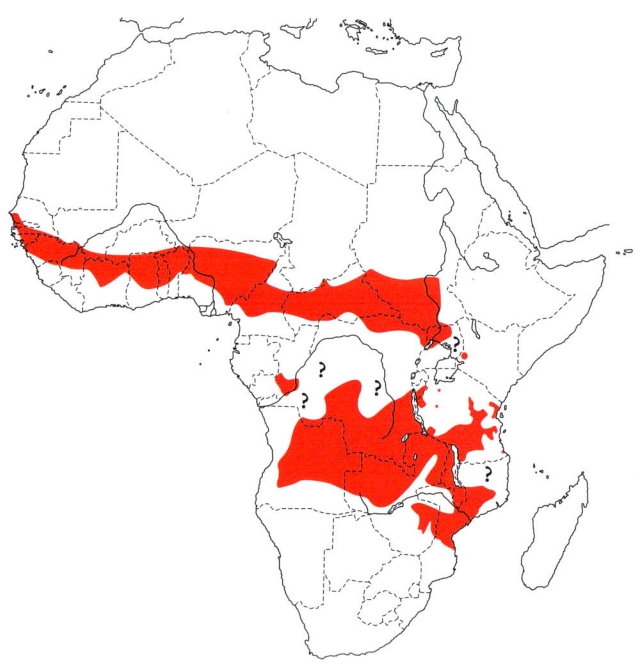

Anthreptes longuemarei

aries dark brown, outer webs fringed yellowish, ends of secondaries with metallic greenish sheen, primary and secondary-coverts with metallic greenish wash on outer webs, scapulars metallic violet, lesser wing-coverts and shoulder metallic blue; underwing-coverts and axillaries off-white. Bill blackish to blackish brown, lower mandible sepia; eyes dark brown; legs and feet black. ADULT ♀: upperparts greyish brown, washed olive on mantle and back, uppertail-coverts metallic violet. Amount of metallic violet on each tail feather reduced outwardly, T5–6 plain brown or almost so, T6 with variable pale to off-white ends. Lores, cheeks and ear-coverts brown; a well-marked whitish supercilium from above lores to well behind eye, a short white line immediately below; chin, throat and breast dull off-white with slight greyish streaking, lower belly, flanks and thighs to undertail-coverts bright lemon yellow. Wings grey brown, outer webs of primaries and secondaries tinged olive. SIZE (44 ♂♂, 24 ♀♀): wing, ♂ 74–82 (78·6), ♀ 66–72 (69·2); tail, ♂ 49–57 (53·9), ♀ 43–49 (46·0); bill, ♂ 16–19 (17·1), ♀ 15·5–18 (16·7); tarsus, ♂ 16·5–18 (17·0), ♀ 16·5–17 (16·8). WEIGHT: Zaïre, ♂ (n = 5) 12–13 (12·6), ♀ (n = 10) 10–13 (11·3); Zambia, 1 ♂ 11·5.

IMMATURE: uniform grey-brown above, tail violet-blue; a black loral spot before eye; entire underparts pale yellow. Young bird still under parental care washed olive above, paler yellow below.

NESTLING: undescribed.

A. l. longuemarei (Lesson) (including '*haussarum*'): Senegal to N Zaïre, Sudan and Kenya. ♂ with lesser wing-coverts and shoulder feathers with only trace of green edging (sometimes absent); whiter, less washed buff below.

A. l. nyassae Neumann: coastal Tanzania to Malaŵi (east of Rift), E Mozambique and Zimbabwe. ♂ purer violet above than *angolensis*, lacking occasional greenish wash; underparts more cleanly white with no buffy wash, pectoral tufts lemon-yellow; reduced yellowish wash on outer webs of primaries and secondaries. ♀ darker brown above than *angolensis*, lacking olive wash; below, breast more extensively grey with whitish streaking, paler yellow of belly confined to centre, and flanks greyer. WEIGHT: Zimbabwe and S Mozambique, ♂ (n = 15) 10·7–13 (11·7), ♀ (n = 10) 9·8–12·2 (10·8).

Field Characters. Length 12·5–13 cm. ♂, with *violet chin* and *upperparts* (which can look black in poor light) and white underparts, unlike any sunbird in its range except in E Africa. Meets Eastern Violet-backed Sunbird *A. orientalis* in S Sudan, N Uganda and W Kenya but they are usually segregated ecologically, Western preferring moister habitat. Both have blue shoulder spot and yellow pectoral tufts, but lower back and rump are blue-green in Eastern, violet in Western (sometimes with a little blue or green gloss). ♀♀ more readily distinguished: both have brown upperparts and white supercilium, but Western has *yellow belly* and *violet tail*, Eastern has white belly and dark blue tail. Immatures have different underparts, entirely pale yellow in Western, dull white in Eastern (with just a trace of yellow in centre of belly). For differences from Uluguru Violet-backed Sunbird *A. neglectus*, see that species. Immature Variable Sunbird *Cinnyris venusta* is similarly yellow below, but much smaller, with curved bill and blue, not violet tail.

Voice. Tape-recorded (86, 91–99, 104, B, GIB, HOR, McVIC, MOY). Song a high, thin but liquid twitter, sometimes ending in brief 'chi-preea-chi'. Variety of other calls include grating, 4-note 'ti-claa-teee-tee' and hard dry 'chit'; alarm, 'skee'.

General Habits. Inhabits wooded guinea savannas, particularly along streams, also thicker vegetation around inselbergs and gallery forest fringes. In Gambia, open savanna woodlands and regenerating areas where young saplings have begun to re-establish; also occurs on periphery of *Avicennia* mangroves, occasionally in forest thickets (Barlow *et al.* 1997). In moister southern savannas restricted to richer miombo woodland, scarce or absent in drier miombo types; sometimes wanders into vegetation types nearby, even acacia in Mozambique (Hanmer 1976). In W Kenya, gardens, cultivation and forest borders (Zimmerman *et al.* 1996).

Generally in pairs or family parties, but larger numbers concentrate at particular flowering trees or other sources of nectar. Frequent member of mixed bird parties. Moves rapidly from one tree to another; feeds like warbler, searching assiduously among leaves and twigs for insects; searches under bark in upside down or other acrobatic positions; flies out after insect or hawks briefly in air. Much attracted to seasonally flowering trees (particularly *Erythrina* spp. and *Faurea speciosa*), also *Spathodia*, *Loranthus*, *Bauhinia*, *Aloe cameronii*, *Leonotis leonura*, *Tecomaria capensis*, *Parinari macrophyllum* and baobab. Punctures bases of long flowers to feed on the nectar (Ginn *et al.* 1989). Spends most of its time in canopy, but freely leaves it to feed on particular plants at lower levels, like *Leonotis* (Tree 1990). ♂ allopreens ♀ briefly; ♂ bathed in small stream, preened in bush nearby (pers. obs.). Flight direct, buoyant, with floating quality.

In miombo woodland-forest mosaic of coastal Tanzania meets Uluguru Violet-backed Sunbird *A. neglectus*, but they must be largely mutually exclusive ecologically there; in interior acacia woodland, well separated ecologically from Eastern Violet-backed Sunbird *A. orientalis* and not in immediate contact there.

Resident except for very local wandering in search of food or flowering plants; thought to move in Mali and possibly a migrant (Lamarche 1981), but resident elsewhere in W Africa; no clear evidence of movement in Zambia (Aspinwall 1983).

Food. Predominantly insects (caterpillars, beetles, Hymenoptera and small flies), spiders and nectar; also small fruits (frequently) and hard seeds as large as small pea.

Breeding Habits. Monogamous; territorial. Display hardly known; ♂ drives other birds away from vicinity of nest; ♂ once drooped wings and spread tail in front of 2 ♀♀ (Chapin 1954).

NEST: oval, bag-like structure, with side-top entrance, woven of fine grass, vegetable fibres, small dead leaves, shredded papery bark and fine silky material, bound together with cobwebs (**A**); exterior decorated with dry leaves, pieces of bark or twiglets, but lacks trail of material hanging beneath it; lined with vegetable down. Structure varies from compact to flimsy, loose and untidy. Ext. depth 135–230; ext. diam. 102, depth below entrance 40–52, int. diam. 39–45. Placed 4–22 m (usually 6–10 m) above ground, slung from end of branch, either exposed or well hidden in mass of tangled matter on periphery of tree, such as clump of dry seed pods or cluster of cobwebs of social spiders; also in association with aggressive arboreal tailor ants *Oecophylla* (Wells 1966). Once placed in a creeper growing against a building (Ginn *et al.* 1989). Built by ♀ only, frequently accompanied by ♂, mostly in early morning and late afternoon; she often calls when arriving with material and again when departing. Takes from 15 days to 1 month to build; nest-lining started 5 days before 1st egg laid.

EGGS: 2 (occasionally 1); ovate, pale grey or smoky buff, marked with hairline (mostly longitudinal) streaks of sepia; also greenish white or pale buff, with scribbled markings of very deep brown or black. SIZE: (n = 14) 18·1–20·2 × 12–13·3 (19·1 × 12·8).

A

LAYING DATES: Gambia, Feb–May, Oct; Ghana, Feb; Togo, (juv. Mar, nest-building Feb–Mar); Nigeria, Feb, Nov; Zaïre, Feb, Sept (gonads enlarged Mar); Sudan, Feb; E Africa: Region C, Mar; Malaŵi, Sept (nest-building Apr, Nov, gonads enlarged Sept); Zambia, Feb (1), July (1), Sept (4), Dec (1) (gonads enlarged Aug, Oct, Mar); Zimbabwe, Aug–Dec (Aug (6), Sept (5), Oct (6), Nov (2), Dec (1)).

INCUBATION: by ♀ only.

DEVELOPMENT AND CARE OF YOUNG: both sexes bring food to nest and to fledged young; in one instance food was mainly green caterpillars. Adults with young roosted side by side 4 m up in tree (Skead 1967).

BREEDING SUCCESS/SURVIVAL: ♀ while feeding in miombo woodland unsuccessfully attacked by Souza's Shrike *Lanius souzae* (Medland 1991).

Reference
Skead, C. J. (1967).

Plate 8 (Opp. p. 95) *Anthreptes orientalis* Hartlaub. Eastern Violet-backed Sunbird. Souimanga violet oriental.

Anthreptes orientalis Hartlaub, 1880. J. Orn., 28, p. 213; Lado, on Bahr-el-Jebel (White Nile), Sudan.

Forms a superspecies with *A. longuemarei* and *A. neglectus*.

Range and Status. Endemic resident. SE Sudan north to 06°N along Nile and to 07°N along Ethiopian border: common. Central and S Ethiopia and Rift Valley (absent from highlands); Somalia, fairly common south of 07°N, less so in NW, west of 46°E; NE Uganda in Karamoja and scattered records from Fatiko, Masindi and Soroti; in Kenya widespread north and east of highlands (except in some dry areas in NE), including Kongelai, the Kerio Valley, lakes Baringo and Bogoria, Isiolo, Meru, Kitui, the Tsavo region and Garissa, and south of highlands in Olorgesaile–L. Magadi area; absent coastal plain south of Galana River. Dry interior of Tanzania from Kenya border west and south to Shinyanga, Singida, Dodoma, Kilosa and Masai Steppe, also S-central Tanzania as mapped (based mainly on sight records, some likely to be of *A. longuemarei* (N.E. Baker, pers. comm.).

Description. ADULT ♂: forehead, crown, nape, mantle and upper back metallic bluish violet, lower mantle and rump metallic green. Uppertail-coverts metallic violet; tail glossy violet-blue, some variable white tipping on inner webs of T 5–6, outer web of T 6 dull brown. Lores and chin metallic violet, throat, breast and belly to undertail-coverts pure white, flanks washed grey, thighs mottled white and dusky brown; pectoral tufts sulphur yellow. Primaries, secondaries, primary and secondary coverts dusky brown, outer webs of outermost primaries narrowly margined yellowish white, inner secondaries with dull violet sheen, median wing-coverts and scapulars metallic violet, lesser wing-coverts and shoulder tipped metallic green; underwing-coverts and axillaries white. Bill black; eyes dark brown; legs and feet black.
ADULT ♀: entire upperparts warm brown except for violet-blue uppertail-coverts and tail; T6 duller, outer web brownish. Lores, line through eye and ear-coverts warm brown, a well-marked off-white supercilium from above lores to behind eye; underparts from chin and throat to undertail-coverts and thighs off-white, belly and flanks faintly washed yellowish (bleaching to off-white with wear). SIZE (33 ♂♂, 26 ♀♀): wing, ♂ 63–71 (67·0), ♀ 58–64 (61·3); tail, ♂ 44–52 (49·0), ♀ 40–46 (43·0); bill, ♂ 15–17·5 (16·5), ♀ 15–16·5 (15·8); tarsus, ♂ 16–17 (16·5), ♀ 16–16·5 (16·2). WEIGHT: ♂ (n = 15) 8–12 (10·4), ♀ (n = 9) 8–12 (10·2).

Anthreptes orientalis

IMMATURE: like ♀ but underparts from chin to undertail-coverts dull white faintly washed yellowish or buff, brightest on centre of belly.

Field Characters. Length 11·5–12·5 cm. Replaces Western Violet-backed Sunbird *A. longuemarei* in dry country of eastern Africa. ♂♂ very similar but Eastern has conspicuous *blue-green rump*; ♀ Eastern has *white belly* and *dark blue tail*, ♀ Western yellow belly and violet tail. Immature Eastern is dull white below, while immature Western is pale yellow. For differences from Uluguru Violet-backed Sunbird *A. neglectus*, see that species.

Voice. Tape-recorded (GREG, McVIC)). ♂ gives nasal 'chwee' or 'tswee-tswee' (Zimmerman *et al.* 1996).

General Habits. Inhabits acacia woodland and bush in arid and semi-arid country, where often very local, usually in vicinity of water; also thickets and dry riparian cover along watercourses; on Mt Kulal (Kenya) frequents clumps of juniper *Juniperus* at forest edge, surrounded by flowering *Leonotis* (Diamond and Keith 1980).

Occurs in pairs or family parties. Favours acacia canopy (particularly flat-topped trees), moving rapidly from tree to tree; frequently joins mixed bird parties. Searches for insects in warbler-like manner in fine-leaved acacias, gleans twigs and bark, occasionally flycatches. Nectar comprises 54% of diet (51% from trees, 49% from bushes), insects 46% (88% gleaned from leaves, 12% in air, 14% taken in flight (Kenya: Lack 1985)). Frequently visits flowering *Loranthus* and *Leonotis*, but not attracted to flowers in Somalia (Ash and Miskell 1998).

Food. Arthropods, principally insects; nectar.

Breeding Habits. Monogamous; territorial.
NEST: compact oval, with side-top entrance and small overhanging porch, neatly made of fine grass and down felted together; usually placed low down (often in association with wasps' nests), in small thorn tree, suspended from end of small branch.
EGGS: 1–2; greenish grey, closely speckled with darker grey, brown or black. SIZE: (n = 5) 17–19 × 12–12·5 (17·9 × 12·3).
LAYING DATES: Sudan, Nov–Mar; Ethiopia, Apr–June; Somalia, Apr–June; E Africa: Region C, Jan; Region D, Feb 2, Apr 2, May 5, June 1, July 1, Aug 1, Oct 2, Nov 4, Dec 4 (most Region D records are in rains, 7/22 Apr–May, 8/22 Nov–Dec).

Anthreptes aurantium Verreaux and Verreaux. Violet-tailed Sunbird. Souimanga à queue violette. Plate 8 (Opp. p. 95)

Anthreptes aurantium J. and E. Verreaux, 1851. Rev. Mag. Zool. (Paris), ser. 2, 3, p. 417; Gabon.

Range and Status. Endemic resident. Forested rivers. S Cameroon north to Korup Nat. Park and Mt Cameroon; Gabon; Congo north to Nouabalé-Ndoki Nat. Park; SW Central African Republic (Lobaye Préfecture); Zaïre, rivers of Congo basin to Ubangi, Uele and Ituri, north to Niangara, in south to Kasai (Kananga) and Ulindi R; NE Angola in N Lunda (Dundo and Kasai R). Single record from Nigeria is an error (Elgood *et al.* 1994). Common. 1 pair or family party per 500–800 m of river frontage on Ivindo R., Gabon (Brosset and Erard 1986).

Description. ADULT ♂: forehead, crown, nape and mantle to uppertail-coverts metallic bluish purple, lower mantle and rump with green sheen; tail bluish purple. Lores, cheeks, ear-coverts, chin and upper throat metallic blue; lower throat to belly, flanks, thighs and undertail-coverts dirty off-white, variably washed buff. Primaries and secondaries dusky brown, wing-coverts and scapulars metallic green; underwing-coverts and axillaries dull off-white; pectoral tufts vivid orange. Bill black, greyer at base; eyes dark brown; legs and feet dark grey. ADULT ♀: forehead, crown, nape and back to uppertail-coverts metallic bluish, lower mantle and uppertail-coverts with green sheen, tail metallic blue-green. Cheeks buffy white, well-marked whitish supercilium from above dusky loral spot to well behind eye, dark line through eye, area directly behind eye metallic bluish, ear-coverts buffy; chin, throat, sides of neck and breast buffy white, belly, flanks undertail-coverts and thighs bright yellowish. Primaries, secondaries and greater wing-coverts dusky, margined olivaceous on outer webs; lesser wing-coverts edged metallic green. SIZE (12 ♂♂, 6 ♀♀): wing, ♂ 63–68 (65·9), ♀ 59–62 (60·3); tail, ♂ 44–49 (46·3), ♀ 39–43 (40·6); ♂ 16–17·5 (16·5), ♀ 16–17 (16·3); tarsus, ♂ 15·5–17 (16·3); ♀ 15·5–16·5 (16·0).
IMMATURE: olivaceous-brown above, tail washed dull metallic green; sides of head olive, pale yellowish supercilium, dark line through eye; underparts dull olivaceous green or yellowish.
NESTLING: undescribed.

Plate 9

Anthreptes aurantium

TAXONOMIC NOTE: ♂♂ from Gabon and lower Congo R. are more heavily washed with buff on the breast and have rather more violaceous lustre on the throat than those from Ituri (Chapin 1954); some adult ♂♂ have a white, not yellow belly (Dowsett-Lemaire and Dowsett 1991). Variation in this species needs further study.

Field Characters. Length 11·5–12·7 cm. ♂ with iridescent *blue*, *violet* and *green upperparts* and throat, orange pectoral tufts and ochre-buff underparts unmistakable; ♀ mainly *iridescent green* above with violet-blue tail, broad white supercilium, dark patch through eye and white ear-coverts; throat and breast white, belly yellow. Riverine forest habitat separates it ecologically from Western Violet-backed Sunbird *A. longuemarei*, a woodland bird, where their ranges meet in W, S and N Zaïre. ♂ Western Violet-backed is violet above and on throat, has narrow blue rump, yellow pectoral tufts; ♀ brown above with violet rump and tail.

Voice. Not tape-recorded. Brief calls given by pair in flight high-pitched and not distinctive (Christy and Clarke 1994). Said to be 'rather silent' by Chapin (1954).

General Habits. Restricted to larger, forest-fringed rivers, river banks and gallery forests and even rather narrow strips of trees; also swamp forest, edges of salt-pans (Congo), thickets and seasonally flooded forest; rarely mangroves.

Forages in pairs or family parties. Searches for insects in leafy vegetation; once seen poking bill into base of *Globimetula* flower (perhaps for nectar) and visits flowering vine *Mussaenda erythrophylla*. In Gabon feeds mainly on berries gathered in low vegetation a short distance from water; never seen at flowers (Brosset and Erard 1986). Behaviour less nervous than other sunbirds.

From known localities on or near the coast in Cameroon and Gabon, Violet-tailed must occur alongside Brown Sunbird *A. gabonicus* with the latter perhaps more or less restricted to mangroves.

Food. 8 of 11 stomachs contained up to 4–5 hairless caterpillars, 5 others contained small fruit or their seeds. Also eats chrysalids, winged ants, small spiders, seeds, possibly nectar. Partial to *Cissus* fruit. In Gabon seen to eat mainly berries and fruit; in Congo, fruit of *Macaranga assas* (Dowsett-Lemaire, 1996).

Breeding Habits. Monogamous; territorial.
NEST: pear-shaped oval, with side-top entrance overhung by short porch, made principally of dead leaves, strips of brown fibre and black, thread-like fungi, strengthened with fine silky material or cobwebs, with strands of loose vegetable matter trailing below; interior lined with white plant down; suspended from small-leaved tree (e.g. *Cynometra alexandri*), growing far out over water, 1–3 m above surface or sometimes close to where branch touches water; also in tree growing out of water. Nest sometimes incorporates web-masses of social spiders *Agelena*, beside which it is frequently placed.
EGGS: 1–2, usually 2. Elongated, dull light blue or grey-blue, with long scrawls of purplish black markings, darkest and thickest at blunt end, forming ring, other surfaces somewhat clouded. SIZE: (n = 2) 18·0–19·2 × 12·0–13·1 (18·6 × 12·6).
LAYING DATES: Gabon, Apr, Nov (♂ in breeding condition Feb, pair feeding young early Jan, late Mar, nests under construction Jan, Sept); probably breeds all year except in the dry season (June–Aug); Congo, Nov–Dec; Cameroon, Mar; Zaïre, (gonads enlarged Feb–Mar, nest-building Apr–June, but probably breeds all year: Chapin 1954).
DEVELOPMENT AND CARE OF YOUNG: young fed by both sexes, partly on fruit, also small insects and spiders.

References
Brosset, A. and Erard, C. (1986).
Chapin, J.P. (1954).

Anthreptes gabonicus (Hartlaub). Brown Sunbird. Souimanga brun.   Plate 8

Nectarinia gabonica Hartlaub, 1861. J. Orn., 9, p. 13; Gabon.   (Opp. p. 95)

Range and Status. Endemic resident in mangroves on coast and forested rivers inland. S Senegal, W Gambia, Sierra Leone, Guinea; Liberia, abundant on coast, common along rivers and larger creeks (Gatter 1997); Ivory Coast (to 09°50′N), coastal Ghana, occasionally inland to Opintin Gallery Forest and Kete Kratchi; SW Burkina Faso; S Nigeria on coast and north in gallery forest to Iseyin, Kainji and Pandam Wildlife Park; Cameroon north to Korup Nat. Park at 05°00′N, 09°00′E; Gabon inland to Ivindo R., M'Passa; Congo to Goumira at 04°08′S, 12°07′E, extreme SW Zaïre (Banana) and N Angola (Cabinda). Common to abundant. Density in Liberia, 6–15 birds per km on mangrove creeks.

Description. ADULT ♂: forehead, crown and whole of upperparts grey-brown, lower back and rump with faint olive wash. Uppertail-coverts and tail grey-brown, faintly washed olive at sides; all but 2 central pairs of tail feathers with white tips, *c.* 8 mm on T 6, becoming smaller inwardly. Grey-brown area in front of eye, dark loral streak running through eye, supercilium and line below eye white, ear-coverts grey, cheeks dull white. Chin and throat whitish grey, breast and sides of neck dark grey, flanks lighter; belly greyish with faint olive wash extending to undertail-coverts, thighs greyish. Wings grey-brown, primaries and secondaries blackish, outer webs washed olive, shoulder, underwing-coverts and axillaries white. Bill black; eyes dark red or reddish brown; legs and feet black. Sexes alike. Birds in worn plumage are greyer, particularly below. SIZE (11 ♂♂, 7 ♀♀): wing, ♂ 56–60 (57·8), ♀ 53–57 (54·8); tail, ♂ 35–40 (37·3), ♀ 32–39 (34·1); bill, ♂ 13·5–14·5 (14·2), ♀ 13·5–15 (14·2); tarsus, ♂ 14·5–15·5 (15·0), ♀ 13·5–15 (14·3).
IMMATURE: like adult but with more olive wash above and on wings; belly and flanks to undertail-coverts washed yellowish.
NESTLING: naked, skin black.

Field Characters. Length 9·5–10·8 cm. Well-named the Brown Sunbird. Dull greyish brown above, wings plain brown with white carpal edge, underparts pale grey. Readily identified by *double white line* from bill *above* and *below eye* enclosing dark streak through eye. Waterside habitat and warbler-like habits also help identification.

Voice. Tape-recorded (104, B, CHA). Song, given from dead bough at top of tall tree by riverbank, 'tser-tser-tsew-tsi-tsi-tsi-tsi-tsi-tseuuur'; latter part of song has twittering quality common to many sunbirds; first part, 'tser-tser-tsew' sometimes uttered on its own (R. A. Cheke, pers. comm.). Soft 'tsurp-tseeep-tseeep' given in flight; another call, a very high and thin 'tsit' or 'tseeet'. Also said to give regular conversational 'wit-wit-squee-witter-witter', the 'squee' slightly rising in pitch (Barlow *et al.* 1997); and see R.A. Cheke, *Malimbus* 21, 1999, 51.

General Habits. On coast restricted to mangroves, in Gambia both *Avicennia* and *Rhizophora*, swamp thickets and lagoons with palms and other trees nearby; occasionally visits adjacent woodland (Gambia); occurs inland in low shrubs and plants along larger rivers where trees overhang water, less often in farmland and gardens. Feeds

Anthreptes gabonicus

singly or in pairs and family parties; unobtrusive, warbler-like, keeping to thicker cover or in canopy; does not usually associate with other sunbirds. Constantly on the move, with rapid energetic movements; makes quick whirring flights, one bird following shortly behind another, over and through vegetation (Gambia: Barlow *et al.* 1997). Gleans insects from leaves, lianas and small twigs; not known to visit flowers but may do so. Forages low, near water surface, rarely above 3 m (Liberia: Gatter 1997).

Food. Small insects.

Breeding Habits. Monogamous; territorial.
NEST: domed, with side-top entrance, overhung by very short porch; a tough outer wall of green moss, held together by spider webs, decorated with dry grass, a few dead leaves, fine vegetable fibres and tiny rootlets; lacks trailing 'tail' below; bulk of interior a felted mass of vegetable down, especially thick at base, innermost lining a few shreds of fine grass and 1–2 feathers; suspended from dead branch or small twig over water, several m from bank, 0·3–1·5 m above surface or high water mark; 1 nest in a garden (Liberia).
EGGS: 2. Ovate, slender, smooth, without gloss; brown or grey; washed violet, heavily overlain with clouds of fine, delicate, close linear marks of softer shades of grey and brown, sometimes almost obscuring surface. SIZE: (n = 3) 16·8–18 × 12·4–12·7 (17·3 × 12·5).

LAYING DATES: Gambia, Dec–Mar (nest building Aug); Sierra Leone, Feb–Mar, July–Sept; Liberia, Oct–Mar; Nigeria, Jan, and at Sapele nests in dry season; Cameroon, Jan (2), Feb (1), Mar (1); Gabon, June; Zaïre, Sept.
INCUBATION: by ♀.

Reference
Serle, W. (1949).

Plate 8
(Opp. p. 95)

Anthreptes rectirostris (Shaw). Green Sunbird. Souimanga à bec droit.

Certhia rectirostris Shaw, 1811–12. General Zool., 8, p. 246; Ashanti, Gold Coast.

Forms a superspecies with *A. rubritorques*.

Range and Status. Endemic resident. Guinea (Conakry, borders of Mt Nimba), Sierra Leone, SW Mali (Kangaba); Liberia, not uncommon to locally common; Ivory Coast (to *c.* 09°N), S Ghana, Benin and S Nigeria (north to Akure and Ife), Bioko; S Cameroon north to Rumpi Hills, Manenguba and Mt Kupé, up to 1250 m; Gabon, Congo north to Nouabalé-Ndoki Nat. Park, SW Central African Republic (Lobaye Préfecture), SE Sudan (foothills of Imatong Mts), Zaïre in Congo basin, mainly fringes, south to Lutunguru at 1550 m; also in Kasai and Shaba (Kasaji); S and W Uganda between 700 and 2150 m (Budongo and Bugoma to Bwindi (Impenetrable) and Malibigambo Forests, east to Entebbe, Kifu, Mabira, Mbale and Mt Elgon), W Kenya above 1000 m (Mt Elgon, Sio R., Nandi and Kakamega Forests, south to Kaimosi, Kericho and Sotik), NW Tanzania (Minziro Forest) and N Angola (Cabinda south to Cuanza Norte and N Lunda at Capaia). Common to local and uncommon; density in Gabon (M'Passa) 3–5 pairs per km^2 (Brosset and Erard 1986).

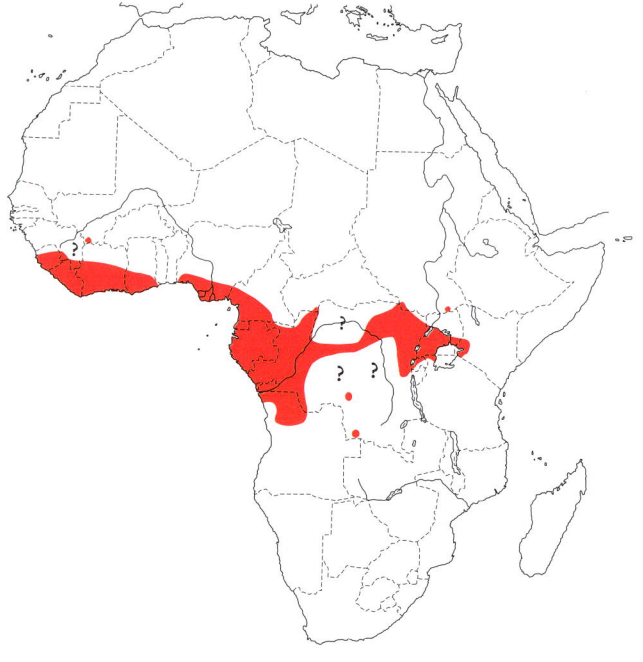

Anthreptes rectirostris

Description. *A. r. tephrolaema* (Jardine and Fraser) (including 'elgonense'): Nigeria to Kenya and Angola. ADULT ♂: forehead, crown, nape, mantle and back metallic green; rump olive-yellow, uppertail-coverts metallic green, tipped olive; tail feathers blackish brown with some metallic green sheen and olive edges. Lores and cheeks metallic green, ear-coverts tinged golden green; chin and upper throat dusky grey, lower throat and upper breast metallic green, base of feathers dusky grey; below this a narrow orange breast-band; lower breast dusky grey, pectoral tufts bright yellow, flanks and belly to undertail-coverts dull yellowish, thighs dusky olive. Primaries and secondaries dusky brown, secondaries and secondary coverts edged olivaceous, wing-coverts and scapulars metallic green, shoulder dusky grey; underwing-coverts and axillaries white. Bill black; eyes red-brown to deep brick red; legs and feet black. ADULT ♀: forehead, crown, nape, mantle and back dusky olive, rump and uppertail-coverts deep olive. Lores, cheeks and ear-coverts dusky olive, a paler olive area before and just above eye. Tail dusky brown, edged olive-green. Chin, throat and whole of underparts olive-green, dullest on throat, brightening on breast to undertail-coverts. Wings deep olive. SIZE (22 ♂♂, 23 ♀♀): wing, ♂ 55–59 (57), ♀ 53–57 (55.0); tail, ♂ 31–33.5 (31.8), ♀ 28.5–32.5 (30.2); bill, ♂ 12–15 (14.1), ♀ 13.5–15 (14.1); tarsus, ♂ 13.5–14.5 (14.2), ♀ 13.5–14.5 (14.0). WEIGHT: ♂ (n = 34) 10–12 (11.3), ♀ (n = 16) 7–12 (9.9).

IMMATURE: ♂ olivaceous above, dull green sheen on forehead, back and mantle; chin and throat greyish, remainder of underparts to flanks and undertail-coverts bright yellow; shoulder, underwing-coverts and axillaries yellowish. ♀ lacks green sheen on upperparts; chin, throat and rest of underparts yellowish.

JUVENILE: at fledging resembles immature, but with slight yellowish superciliary and some yellowish spotting on wing-coverts; bill black except at base, an orange-red tubercle at sides of mouth.

NESTLING: unknown.

A. r. rectirostris (Shaw): Guinea and Sierra Leone to Benin. ♂ with chin and throat yellow, metallic green on lower throat and upper breast less extensive. ♀ with belly to flanks and undertail-coverts brighter yellow.

TAXONOMIC NOTE: *A. rectirostris* and *tephrolaema* have sometimes been treated as specifically distinct (Chapin 1954, Zimmerman 1972) but are now generally regarded as conspecific following White (1963) who also included *A. rubritorques* as conspecific, here regarded as forming a superspecies. *A. 'pujoli'* Berlioz has been shown by Erard (1979) to be based on a fledgling. *A. r. elgonense* van Someren has been proposed for larger birds from W Kenya (♂♂ wings 59–64) but is not recognized here.

Field Characters. Length 9–10 cm. A very small canopy sunbird with short, straight bill. ♂ has *golden green* upperparts, throat and breast, chin and upper throat yellow in W Africa (nominate race), grey elsewhere. *Dull orange band* across lower breast, narrow and not easy to see at a distance; pectoral tufts *yellow*. ♀ dull olive-green above, yellowish olive below, dull yellow on belly, hint of a pale supercilium; at close range some birds show grey chin, some green iridescence on shoulder and on a few back feathers, but most lack these features; with drab plumage and short, straight bill, it could be mistaken for small warbler. Immature like ♀ but with a little dark mottling on throat and breast. Collard Sunbird *Nectarinia collaris* is also small but bill slightly longer and decurved, and ♂ has chin and throat iridescent green, like breast, and rest of underparts bright yellow; ♀ is entirely bright yellow below.

Voice. Tape-recorded (104, B, CHA, ERA, LEM, SALA). Song variable but follows same pattern: starts with single notes, speeds up into trill in the middle and ends with more single notes, e.g. 'chik ... chik,chik,tss-tss-tss-chchchchchchch-tsee-tsee-chi-chi-chik'; notes are strong, without the light twittering quality of many sunbirds. Song lasts 2–3·5 s; av. of 14 songs, 2·8 s; intervals between songs 1·5–5 s; song bouts last for a minute or more. Songs more or less connected during intervals by irregular 'chik' notes. Another song type is shorter, with chattery notes in the middle and thin ones at either end. Also has a loud, pure, 'peee' and down-slurred 'peeew', often alternating: 'peee, pee-peeew, peee-peeeew, peee-peee-peeeew ... '.

General Habits. Inhabits forest edges and clearings, entering forest where canopy broken and along tracks; also dry swamp forest, regenerating growth, second growth, gallery forest and thickets; cocoa and coffee plantations; farms (Cameroon). Keeps mainly to tops of tall trees, and seldom below 20 m, but sometimes descends to lower levels at edges of gallery forest and small clumps of trees, even down to bushes in savanna (Gabon: Christy and Clarke 1994).

Occurs singly or in pairs or family parties; in Gabon, in territorial groups of 3–7. Feeds like warbler, searching twigs (usually <1 cm in diam.) and leaves (especially undersides) and clumps of hanging lichens; association with flowers needs confirmation. In Congo feeds together with Collared Sunbird *Hedydipna collaris* without obvious ecological separation, and sings mainly from medium heights (10–20 m) (Dowsett-Lemaire and Dowsett 1991).

Considerable seasonal movement apparent but no evidence for migration.

Food. Insects: black flying termites, small caterpillars, minute larvae; spiders; variety of fruits (often red or orange), including *Macaranga* and *Xylopia* fruits up to 4 mm diam., and in Congo fruits of *Alchornea cordifolia*, *Ochthocoosmus africanus* and *Macaranga assass* (Dowsett-Lemaire 1996); small hard seeds; large seed size of pea swallowed whole. Nectar?

Breeding Habits. Territorial; in Gabon territorial groups of 3–7 suggest it may be a cooperative breeder (Brosset and Erard 1986).

NEST: oval, with side-top entrance, lacking porch (**A**); made of green moss, held together with spider web, decorated with bits of stems and other materials, with trail

A

of vegetable matter hanging below; lined with soft brown material (perhaps from ferns) or white vegetable down and a few rootlets. Ext. depth 130, ext. diam. 100, entrance 40. Attached to end of twig, overhanging branch or leafy vine, 2–10 m above ground, usually at forest edge or beside river.

EGGS: 2; ovate, ashy grey with slight violet wash, irregularly marked with dark lines and spots.

LAYING DATES: Liberia, Sept–Dec (Sept (1), Oct (2), Dec (1)), (immatures Jan–Apr); Ghana, Dec; Nigeria, June; Cameroon, Jan–Feb, Apr–Dec; Gabon, (juvs Mar); Congo, Feb, Sept–Oct; Zaïre, July and probably all year in lowlands; Itombwe, Jan–Apr, July; E Africa: Region B, Aug; Angola, breeding condition Feb.

INCUBATION: by ♀ only.

DEVELOPMENT AND CARE OF YOUNG: young fed by both parents; in Gabon, group of 7 young fed by several adults (Brosset and Erard 1986).

Plate 8
(Opp. p. 95)

Anthreptes rubritorques Reichenow. Banded Green Sunbird. Souimanga à col rouge.

Anthreptes rubritorques Reichenow, 1905. Orn. Monatsb., 13, p. 181; Mlalo, Usambara Mountains.

Forms a superspecies with *A. rectirostris*.

Range and Status. Resident, endemic to E Tanzania in mid-altitude eastern arc forests at 750–1600 m, rarely down to 250 m in lowlands. Locally common in E and W Usambaras, uncommon in Ngurus and Ulugurus, local and uncommon in Udzungwa Scarp Forests (W Kilombero, Magombera and Mwanihana) and Ndundulu Mts. Threatened by forest destruction; only *c.* 200 km² of suitable habitat remains in Usambaras (where densities may be high); elsewhere habitat probably very small and fragmented.

Description. ADULT ♂: forehead, crown, cheeks, ear-coverts, nape, mantle and back to uppertail-coverts metallic green; tail green with metallic purple sheen. Chin, throat, sides of neck and upper breast grey, narrow red breast band, bright orange pectoral tufts; lower breast and flanks grey, centre of belly to thighs and undertail-coverts yellowish. Primaries and secondaries dusky brown, margined olive-green on outer webs, secondary coverts edged dull metallic green, median and lesser coverts and scapulars metallic green; underwing-coverts and axillaries off-white. Bill black; eyes red-brown; legs and feet black. ADULT ♀: forehead, crown and entire upperparts dull metallic green, tail feathers dusky, edged olivaceous. Cheeks, lores, ear-coverts, chin and throat off-white, remainder of underparts dull olive, brightest in centre of belly. SIZE (6 ♂♂, 1 ♀): wing, ♂ 59–62 (60·6), ♀ 56; tail, ♂ 34–36 (35·1), ♀ 31; bill, ♂ 13–15 (14), ♀ 13·5; tarsus, ♂ 13·5–14 (13·8), ♀ 13·5. WEIGHT: 2 ♂♂ 10·5, 10·6.
IMMATURE: (fledgling almost ready to fly) upperparts wholly blackish, washed olive-green; underparts uniform yellowish olive; gape dull yellow, inside of mouth orange-red.
NESTLING: undescribed.

Field Characters. 8·5–9 cm. A tiny, short-tailed, short-billed canopy sunbird. ♂ has *grey underparts* with conspicuous *red breastband* and *yellow* pectoral tufts; upperparts shining green. ♀ olive above and dull greyish yellow below. Occurs in E Usambaras with similar-sized Amani Sunbird *Hedydipna pallidigaster*, which may also be in canopy; ♂ Amani has shining blue-green upperparts and breast, red pectoral tufts and white belly; ♀ Amani is pure white below, grey above. Eastern Double-collared Sunbird *Cinnyris mediocris* is larger and longer-tailed, with curved bill; ♂ has green throat and red breastband, ♀ is olive-brown.

Voice. Not tape-recorded. ♂ said to have repeated chirp of remarkable carrying power, the same note repeated several times, sometimes accelerating into a song, also a disyllabic 'thk-eeer'; ♀ generally silent (Moreau and Moreau 1937).

General Habits. Frequents edges and canopy of mainly mid-altitude forest, but found nesting as low as 300 m in E Usambaras (Evans 1997). Often in highest tree tops where difficult to detect. Adapts to secondary habitats, including plantations, if some forest trees survive; leaves forest to feed on suitable trees outside it, such as *Casearia*

Anthreptes rubritorques

battiscombei, and visits flowering *Erythrina* spp.; forages on *Usnea* lichens.

Usually in pairs or family parties, also in larger groups of 10–60 (Stuart and Hutton 1977); often joins mixed bird parties. Moves around erratically in search of suitable fruiting or flowering trees. ♂ sings from topmost perch for hours at a time, day after day between June and Jan. ♀ seen to bathe in gutter of building (Moreau and Moreau 1937).

Food. Insects, nectar; also fruits and berries, including little green *Macaranga* berries, *Rubus* fruits. Young fed on bright cotton-stainer bug *Dysdercus* larvae (generally considered distasteful) (Moreau and Moreau 1937).

Breeding Habits. Monogamous; territorial.
NEST: oval, with side-top entrance just below point of attachment; extremely neat, lacking tail of material trailing below; bulk of nest made of *Usnea* lichen, also some spider webs, exterior decorated with a few *Casuarina* needles, interior (including roof), lined with soft vegetable down. Ext. depth 150, ext. diam. 75, entrance diam. 45. Situated at forest edge (once in *Widdringtonia*), also frequently in exotics planted nearby; placed high up, suspended from end of thin branch. One was placed 15 m above ground in crown of leafless tree over main road through forest (E Usambaras: Evans 1997). ♀ alone builds, ♂ often in attendance.
EGGS: probably 2 (2 young in nest).

LAYING DATES: Jan (carrying nest material Jan, July, Sept–Oct, Nov).

INCUBATION: by ♀ only.

DEVELOPMENT AND CARE OF YOUNG: young fed by ♀; food brought at 3 min intervals when chicks almost fledged.

References
Collar, N.J. and Stuart, S.N. (1985).
Moreau, R.E. and Moreau, W.M. (1937).
Stuart, S.N. and Hutton, J.M. (1977).

Anthreptes reichenowi Gunning. Plain-backed Sunbird; Blue-throated Sunbird. Souimanga de Reichenow.

Plate 8
(Opp. p. 95)

Anthreptes reichenowi Gunning, 1909. Ann. Transvaal Mus., 1, p. 173; Mzimbiti, near Beira, Portuguese East Africa.

NOMENCLATURAL NOTE: not to be confused with the sunbirds *Nectarinia reichenowi* or *Cinnyris reichenowi*.

Range and Status. Endemic resident. Kenya, forested coastal lowlands lower Tana R. (Wema), Sokoke–Arabuko Forest, Diani and Jadini Forests, and Inland to Mrima and Shimba Hills. Coastal Tanzania from Moa, Tanga and Muheza to foothills of E Usambara Mts (below 500 m) and Pande Forest (Dar es Salaam); also Mikindani and inland on Rondo Plateau; locally common; abundant at some localities in E Usambaras (Evans 1997). Central Mozambique on lower Zambezi R. (Lacerdónia and Mopeia) and widely distributed south of there from Inhaminga and Inhamitanga through Chiniziua, Dondo, Beira, Vilanculos and Massinga to Inharrime, Zandamela, Chimonzo and Macia. Occurs irregularly (perhaps seasonally) inland to E Zimbabwe: Bvumba (1580 m), Mutare, Huroni R., Chipinga Uplands, Save and Mwenezi rivers and Chinhoyi (Manyame R.). South Africa, ♂ near Punda Maria Camp, Kruger Nat. Park, 1993 (Hockey *et al.* 1996); unconfirmed reports from NE Transvaal (Tzaneen) and N Zululand.

Description. *A. r. reichenowi* Gunning: Mozambique, Zimbabwe and South Africa. ADULT ♂: forehead dark metallic blue (feathers with greenish bases), crown, nape, mantle, back and rump to uppertail-coverts and tail olive-green. Lores, cheeks and ear-coverts to sides of neck olive-green; chin, throat and centre of upper breast dark metallic blue, sides of breast, belly and flanks to thighs and undertail-coverts olive-yellow, brightest in centre of belly; pectoral tufts lemon yellow. Primaries and secondaries dusky brown, margined olive-green on outer webs, upper wing-coverts and scapulars olive-green, shoulder yellow; underwing-coverts and axillaries pale yellowish. Bill blackish horn, blue-grey at base; eyes dark brown; legs and feet blue-grey. ADULT ♀: like ♂ but wholly uniform olive-green above; yellowish below, dullest on chin and throat, brightening on centre of belly. SIZE (21 ♂♂, 13 ♀♀): wing, ♂ 53–57 (55·2), ♀ 50–54 (51·8); tail, ♂ 36–41 (38·2), ♀ 34–38 (35·6); bill, ♂ 16–17·5 (16·6), ♀ 14–17 (15·4); tarsus, ♂ 13·5–16 (14·9), ♀ 15·5–16·5 (15·9). WEIGHT: (Mozambique) ♂ (n = 11) 6·7–8·8 (7·8), ♀ (n = 7) 6·5–7·8 (7·1).

IMMATURE: like adult ♀, but duller olive-green above, lighter yellow below; bill yellowish horn. Immature ♂ moults from greenish plumage into adult-like dress but metallic chin and throat are duller and smaller, forehead with admixture of green and blue feathers. Juvenile just out of nest (tail 26) indistinguishable from immature except for fluffier plumage.

NESTLING: undescribed.

A. r. yokanae Hartert: Kenya and Tanzania. ♂ has chin and throat sooty black with narrow bluish fringes to feathers of upper throat, merging into purplish blue on lower throat and upper breast; centre of belly pale greenish yellow. WEIGHT: ♂ (n = 5) 5–10 (7·2), ♀ (n = 8) 6–10 (7·2).

Anthreptes reichenowi

Field Characters. Length 10–11·5 cm. A very small short-billed bird of E coastal forest. ♂ has *dark blue chin* and *throat* patch and yellow underparts; from below, *yellow border* to throat patch *from breast up to eye* distinguishes it from ♂ Collared Sunbird *Hedydipna collaris*, which is entirely green from chin to breast. Upperparts of both ♂ and ♀ plain olive (shining green in Collared). Imm. ♂♂ of Variable and White-bellied Sunbirds *Cinnyris venusta* and *C. talatala*, with dark throat patches, can resemble ♂ Plain-backed, but they both have small green shoulder patches and lack blue-black forehead. Occurs with Amani Sunbird *H. pallidigaster* in Kenya/Tanzania; ♂ Amani has blue-green head and back, white belly, ♀ Amani is grey above, white below. Olive upperparts and yellow underparts of ♀ Plain-backed Sunbird give it a superficial resemblance to white-eye *Zosterops*, but it lacks white around eye and has thinner, slightly decurved bill.

Voice. Tape-recorded (32, 91–99, B, F, CHA, GREG, KEI, McVIC, STA). Very pretty song starts with high, thin notes and goes down the scale; lower notes become increasingly sweet and warbler-like, and at the end may accelerate into a trill. Notes often divided into 3 or more different groups given at different speeds, with a general tendency to accelerate, like Willow Warbler *Phylloscopus trochilus*, and the mellow 'tooee-tooee-tooee' notes in the middle also have a Willow Warbler quality. Songs last 2–4 s, sometimes 5 or 6 s; intervals between songs a few s. Calls include a nasal 'zlui' and a hard 'tik-tik'. Alarm, 'wee-wee-wee', 'tew-tew-tew' or a complaining 'eea-eeea-eeea' (Zimmerman *et al*. 1996).

General Habits. Inhabits lowland and semi-deciduous forest, *Afzelia–Brachystegia* (and rarely *Cynometra*) woodland, second growth and thickets; in south of range prefers forest edges, heavily foliaged trees and riparian forest in dry savanna, sometimes far inland; in E Usambaras at high densities, even in degraded forest.

Occasionally occurs singly, but usually feeds in pairs or family parties; frequently joins mixed bird parties. Forages at all levels from thickets to mid-stratum and canopy; in thickets both in deep shade and in open sunlit places. Searches actively for arthropods like warbler or white-eye, making quick sallies for insects both within cover or in the open; feeds industriously in flowers, flitting quickly from one flower cluster to another; frequents thick canopy of *Diospyros* and *Mimusops*, also flowering *Kigelia* trees and *Loranthus* plants which are popular with other sunbirds. Attracted to feathery inflorescences of *Albizzia*, perhaps by the abundant insects. No evidence of major dependence on nectar. Sings mainly from canopy (Sokoke Forest: S. Keith, pers. comm.). Mobs African Barred Owlet *Glaucidium capense*.

Sedentary or with very local movements in E Africa; in southern part of range possibly breeding visitor in interior, in riparian fringing forest and thickets in savanna (Irwin 1995a).

Food. Arthropods: termites, minute insect larvae; stomach contents 80% spiders by bulk (Williams 1951); nectar requires confirmation.

Breeding Habits. Monogamous; territorial. Displaying ♂ pursues ♀ in flight; after several min ♀ alights on bare hanging vine or leafless sapling, ♂ settles immediately above, flicks wings, fans pectoral tufts, utters brief snatches of song; ♀ quivers wings, flies off, ♂ continues pursuit.

NEST: oval, with side-top entrance, a well-defined porch at top; bulk of material compromises shredded grass stems, fine slivers of bark, small twigs and leaves, flimsily bound with cobwebs, sometimes decorated with spider or insect cocoons; another nest was built of skeletal leaves in last stages of disintegration; nests lack trail of material hanging below; size 115 × 65, entrance 34. One in Tanzania was a pouch apparently formed of dead leaf fragments, silk and many horsehair-like black threads (fungal mycelia?) attached along about 5 cm of side branch of liana (rather than hanging by a point or loop), placed *c*. 5 m above ground in opening several m across in area of low degraded forest (Evans 1997).

EGGS: 2–3, usually 2; pointed ovals, white, without gloss, evenly spotted with reddish brown and mauve with a few underlying freckles, forming a distinct band around larger end. SIZE: (n = 6) 15–16·2 × 10·4–11·4 (15·7 × 10·9).

LAYING DATES: E Africa: Tanzania, Nov, Region D, Mar; Region E, Apr–Nov (1 record for each month); Mozambique, Oct–Nov (juveniles under parental care June, Nov–Dec); Zimbabwe, Oct (feeding recently fledged young Nov).

References
Skead, C.J. (1967).
Williams, J.G. (1951, 1953).

Plate 8 (Opp. p. 95)

Anthreptes seimundi (Ogilvie-Grant). Little Green Sunbird. Souimange de Seimund.

Cinnyris seimundi Ogilvie-Grant, 1908. Bull. Br. Orn. Club, 23, p. 19; Banterbari, Fernando Po.

Range and Status. Endemic resident. W Guinea (Kounounkan, rare: Hayman *et al*. 1995), Sierra Leone (Bo and Kenema), Liberia, uncommon; widespread but uncommon Ivory Coast, to 09°N; S Ghana (uncommon), S Togo (only 2 records but probably overlooked: Cheke and Walsh 1996), S Nigeria from Lagos and Ipake to Umuagwu and Oban; uncommon, Bioko, S Cameroon (north to Kumba and Korup Nat. Park); Gabon except SE, S Congo and in N in Odzala and Nouabalé–Ndoki Nat. Parks. S Central African Republic (Lobaye Préfecture), and pair recently collected in extreme SW (P. Sweet, pers. comm.). Apparently absent W Zaïre between 15° and 20°30′E; present central and NE Zaïre from Semliki valley south to SW Kasai and Itombwe at 1130 m; S Sudan (Bengengai and Talanga Forests); forests of Uganda at 700–1400 m (Budongo, Bugoma, Maramagambo, Bwindi-Impenetrable and Sango Bay), wandering east to Entebbe (once) and Kifu; NW Rwanda at 1750–1850 m (Bururi valley); N Angola in Cuanza Norte (Ndalla Tando) and N Lunda (Rio Cassai). Common to local and uncommon.

Description. *A. s. traylori* (Wolters): Nigeria to Sudan, Uganda and Angola. ADULT ♂: forehead, crown, nape, mantle and back to uppertail-coverts uniform dark green. Tail feathers green, edged yellowish. Lores, cheeks and ear-coverts olive; a pale yellow area around eye, but no supercilium; chin, throat, breast and belly to flanks and undertail-coverts uniform pale yellow, thighs greenish yellow. Wings dark brown, primaries broadly and secondaries narrowly edged green, inner webs of inner primaries and secondaries broadly bordered yellow; upperwing-coverts and scapulars green, shoulder pale yellow; underwing-coverts and axillaries whitish olive. Bill black or dark brown, dull yellow below; eyes brown; legs and feet dark grey or black. Sexes alike. SIZE (22 ♂♂, 19 ♀♀): wing, ♂ 51–56 (53·5), ♀ 47–52 (49·5); tail,

Anthreptes seimundi

♂ 26–30 (28·2), ♀ 23–28 (25·8); bill, ♂ 15–16·5 (15·7), ♀ 14–15·5 (14·9); tarsus, ♂ 14·5–16 (15·4), ♀ 14–16·5 (15·0). WEIGHT: ♂ (n = 16) 6·5–11 (7·6), ♀ (n = 5) 5–6·5 (5·7).

IMMATURE: duller green above, greyer on throat and breast, less yellowish on belly and flanks.
NESTLING: undescribed.

A. s. kruensis (Bannerman): Guinea and Sierra Leone to Ghana and Togo. Poorly differentiated; chin to breast and flanks tinged greyer, yellow restricted to lower breast and belly.

A. s. seimundi (Ogilvie-Grant): Bioko. Poorly differentiated; slightly clearer green above, brighter yellow below.

Field Characters. Length 9–10 cm. Tiny and short-tailed. Olive-green above, with *narrow, pale, broken* eye-ring; olive-yellow below, yellowest on belly. Very difficult to separate in the field from Bates's Sunbird *Cinnyris batesi*; distinguished by its more horizontal stance, straighter bill, brighter yellow and more uniform underparts, slightly longer tail. Looks like a miniature Fraser's Sunbird *D. fraseri*, while Bates's is more like a small Western Olive Sunbird *Cyanomitra obscura* (Brosset and Erard 1986). Also said to be brighter olive-green above and yellower below than Bates's Sunbird (Rodewald *et al.* 1994). W African race *A. s. kruensis* is duller, but Bates's Sunbird does not occur in its range.

Voice. Tape-recorded (104, B, CHA, KEI, MAC). Song a very thin, high-pitched, insect-like 'pssss' or 'pssssssup', given at irregular intervals.

General Habits. Inhabits forest edge, gallery forest, clearings and regenerating growth, particularly where there are dense tangles and thickets; coffee plantations where trees remain. In Liberia, logged forest, older secondary forest and forest-grassland mosaic (Gatter 1997). In Gabon, abandoned plantations, and the final stages of forest regrowth where pure stands of parasol trees *Musanga* have disappeared; not in true primary forest (Brosset and Erard 1986). In Congo, fairly open forest, forest edge, canopy and gaps (Dowsett-Lemaire 1996). All levels from canopy to thickets.

Occurs in pairs or family parties, monospecific groups of up to 8, i.e. does not generally associate with other sunbirds. Sometimes joins bands of insectivores. Active, restless; moves wings very rapidly while foraging. Searches foliage for insects and other invertebrates, especially in outermost parts of canopy. Frequents fruiting trees, and also visits flowers, but probably more to obtain minute insects than nectar (Brosset and Erard 1986).

Food. Principally insects, including small beetles and caterpillars; small spiders, fig seeds, fruits of *Urera hypsilodrendron*; some nectar, in Gabon especially of *Loranthus*.

Breeding Habits. Monogamous; territorial.
NEST: oval, with side-top entrance; compact, without any loose material trailing below; lined with fine plant down; suspended from end of branch or vine.
EGGS: 2. Without gloss, cream-coloured, almost covered by fine brown freckling. SIZE (n = 2): 11–11·5 × 14·5–15 (11·25 × 14·7).
LAYING DATES: Liberia, July (dependent young Feb, Apr); Cameroon, Apr–May, July, Nov–Dec; Gabon, Nov; Congo, Oct; Zaïre, Feb, Sept (but probably extended); E Africa: Region B, (building nest Aug).
INCUBATION: by ♀ only.

Anthreptes anchietae (Bocage). Anchieta's Sunbird; Red-and-blue Sunbird. Souimanga d'Anchieta. Plate 9 (Opp. p. 142)

Nectarinia anchietae Barbosa du Bocage, 1878. J. Sci. Math. Phys. Nat. Lisboa, 6, p. 208: Caconda, Huila, Angola.

Range and Status. Endemic resident. Moister miombo woodlands of southern savannas from central plateau of Angola in NW Huila (Caconda) and Huambo (Mts Moco and Soque), N Bie and Lunda (Cacolo, Cafunfo and Cuango north to Cacumbi), SW Zaïre (Dilolo); Zambia north of 15°S from Salujinga, Mwinilunga and Mankoya to Kabwe, Mkushi, Serenje and the Muchinga Escarpment, north to Mpika, Kasama, Luwingu, Kawambwa, Mporokoso, Mbala and Isoka, and west of Luangwa Valley south to Chipata and Chadiza. NW Mozambique (Furancungo) and Malaŵi (west of Rift) to 1800 m from Dedza northwards; local but not uncommon in SW Tanzania (Tunduma, Tukuyu and 5 localities in Kasanga, Ufipa plateau and L. Rukwa region: N.E. Baker, pers. comm.).

Anthreptes anchietae

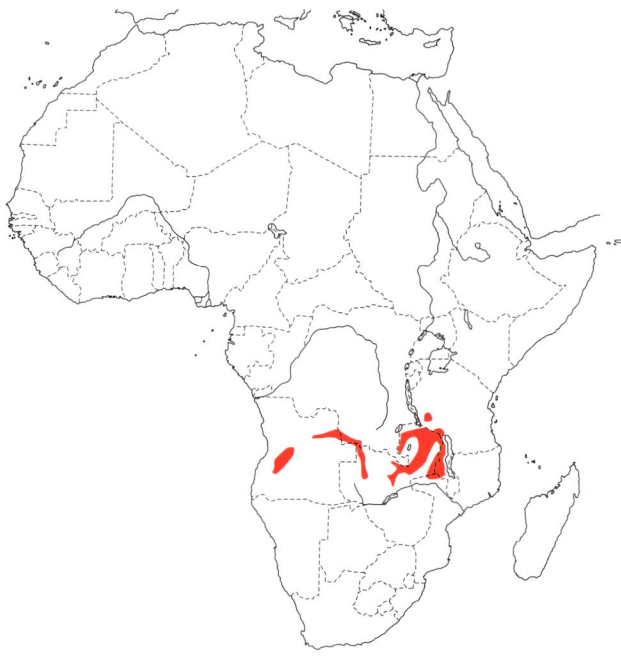

Locally common but frequently sparse or patchily distributed; absent from some likely areas.

Description. ADULT ♂: forehead and crown dark metallic blue; hind-crown, nape, and mantle to uppertail-coverts earth-brown; tail uniform earth-brown, faintly glossed above, brown below. Lores dull metallic blue, area behind eye, ear-coverts and sides of neck earth-brown; chin, throat and upper breast dark metallic blue, sides of breast bright yellow, centre of lower breast to upper belly vivid orange-red, lower belly, flanks and thighs light greyish brown, undertail-coverts vivid orange-red; pectoral tufts bright yellow (uniform with sides of breast). Primaries, secondaries, wing-coverts and scapulars earth-brown, margin of outer webs of primaries buff; shoulder dusky brown, an obscure orange-yellow spot on bend of wing; underwing-coverts and axillaries buffy brown. Bill black; eyes dark brown; legs and feet black. ADULT ♀: similar to ♂, but dark metallic blue restricted to forehead (not reaching crown) and lores dusky brown. Chin and throat dull brown, feathers narrowly tipped dark metallic blue, blue developed more extensively on upper breast, but with brownish feather bases showing through; sides of breast paler yellow, faintly washed olive, orange-yellow on centre of belly reduced in extent, lower belly and flanks washed olive-yellow, not contrasting with yellow at sides of breast; shoulder pale buffy white. SIZE (42 ♂♂, 16 ♀♀): wing, ♂ 62–67 (64·3), ♀ 59–64 (61·6); tail, ♂ 40–46 (42·6), ♀ 38–43 (39·7); bill, ♂ 15·5–18 (16·5), ♀ 15–16·5 (15·8); tarsus, ♂ 15·5–16·5 (16·0), ♀ 15·5–16 (15·8).

IMMATURE: like adult ♀ above, but forehead and rest of upperparts uniform; chin and throat to upper breast dusky olive, lower breast and centre of belly and flanks to undertail-coverts dull olive.

NESTLING: undescribed.

Field Characters. Length 11·5–12·7 cm. A miombo sunbird with a combination of colours unique in Africa – sooty brown above, forehead, crown and chin to breast metallic dark purple-blue, broad *vermilion-red line* down centre of *breast* and upper belly and on *undertail-coverts;* sides of breast *yellow*, lower belly grey. Sexes similar.

Voice. Tape-recorded (86, 102, 104, B, CART, CHA). Songs include a repeated thin, tinkling round with quality of European Goldfinch *Carduelis carduelis*, and louder, stronger, purer notes, 'tsee-pew-pew', the 'pew' sometimes continuing, separated by a quick bubbly jumble, 'pew ... werdlyiddly-pew ... werdlyiddly-pew ... '. Call, a rising 4-note phrase, 'wer-tee-ti-ti', repeated in long series.

General Habits. Restricted to well-developed miombo (*Brachystegia*) woodland, but also occurs in poor or degraded woodland, particularly where there are rocks (Dowsett 1977), though the reason for this is unclear.

Feeds singly, in pairs or small family parties; frequently joins mixed bird parties. Moves actively about in canopy, restlessly flying from one tree to another. Much less warbler-like than congeners (showing strong resemblance to sunbirds of genus *Cinnyris*), perhaps not primarily insectivorous. Visits *Protea* bushes growing on ecotone between woodland and dambo grassland; favours flowering *Tecoma* and *Protea*. Everywhere sympatric with Western Violet-backed Sunbird *A. longuemarei*; occurs in same mixed bird parties but without clear ecological distinction although behaviour is different.

Undergoes some erratic seasonal movements in Zambia when it becomes locally common, but no evidence of migration.

Food. Nectar; insects; fruits of *Ochthocosmus africanus* and *Macaranga assas* (Dowsett-Lemaire 1996).

Breeding Habits. Monogamous; territorial.
NEST: oval, with side-top entrance, a compact structure woven with fine twigs, bits of *Protea* flowers and hairy ovaries of *Faurea saligna*, bound together with rachids of pinnate leaf of *Indigofera* and awns and glumes of *Loudetia superba*, decorated outside with protea seeds; interior lined with white felted plant down or brown protea seed pappae; placed 1·3–6·5 m above ground in low protea bush, higher *Monotes* tree or tall *Pterocarpus angolensis*, suspended from outer branch.
EGGS: 2 (occasionally 1); smooth, without gloss; off-white or bluish white, with black spots and bunting-like scrawls of sepia and underlying grey, mostly in zone around larger end. SIZE: (n = 2) 17·5–18 × 11·5 (17·8 × 11·5).
LAYING DATES: Gambia, Apr–May, Malaŵi, Nov–Dec.

Genus *Deleornis* Wolters

A single species, without close relatives; formerly placed in *Anthreptes*. Large; tail short, not quite square, outermost feather (T6) 4–7 mm shorter than the others. Bill shape most like *Anthreptes* but proportionately heavier, almost straight, lower mandible completely so (lower mandible in *Anthreptes* always partly curved), largely yellowish or horn-coloured; feet pale; legs and feet strong. Plumage non-metallic, uniform bright green; sexes alike except for orange-yellow pectoral tufts of ♂; E populations *axillaris* have grey head and may prove specifically distinct.

Insectivorous; not known to visit flowers. Occurs in lowland forest in groups with mixed flocks of insectivores, and may be a nucleus species for their formation; its behaviour is essentially that of a warbler (Sylviidae) rather than sunbird. The nest is undescribed.

Deleornis fraseri Jardine and Selby. Fraser's Sunbird; Grey-headed Sunbird; Scarlet-tufted Sunbird. Souimanga de Fraser.

Plate 8 (Opp. p. 95)

Anthreptes fraseri Jardine and Selby, 1843. Illust. Orn., (new series), pl. 52 with text; Fernando Po.

Range and Status. Endemic resident. Guinea (Kounounkan Massif, Mt Nimba and Macenta), Sierra Leone, SW Mali (Baoulé, Bougouni and Sankarani), Liberia (common), Ivory Coast to 09°N, S Ghana, locally not uncommon, S Togo, only 2 records: Badou and Misahöe. Uncommon SW Nigeria from Lagos north to Gambari and Benin; common in SE; Bioko; S Cameroon (north to Mamfe, Mts Kupé and Nlonako) east to SW Central African Republic (Mbaiki), south through Gabon and Congo; through Congo basin in Zaïre north to Niangara, Ubangi and Uele R., in east to Lutunguru at 1550 m and Itombwe at 1360 m, south to Kasai (Basongo), and single record from Katanga (Hall and Moreau 1970). S and W Uganda (Murchison Falls Nat. Park, Budongo and Bwamba to Maramagambo, Lugalamba and Impenetrable (Bwindi) Forests, east to Sezibwa R, Kifu, Mabira, Buvuma Is., and W Elgon); single record from NW Tanzania at Bukoba; NW Angola (Cabinda, south to Cuanza Norte at Canzele). Generally common to abundant, local in a few areas. Density in Gabon (M'Passa) 18–22 pairs per km^2 (Brosset and Erard 1986).

Deleornis fraseri

Description. *D. f. cameroonensis* Bannerman: Nigeria to Congo, Angola and W Zaïre (Mayombe). ADULT ♂: forehead, crown, nape, mantle, back, rump and uppertail-coverts bright green; tail feathers brown, margined olive-green on outer webs. Cheeks, lores and ear-coverts bright green, ring of pale yellow feathers around eye; underparts bright yellowish green, paler than upperparts, pectoral tufts bright orange-red, yellowish at base. Primaries and secondaries dusky brown, margined golden olive-green on outer webs, inner webs pale yellowish; upper wing-coverts and scapulars bright green, underwing-coverts and axillaries whitish, washed pale olive, becoming yellowish at carpal joint. Upper mandible dark grey, lower mandible light pinkish grey; eyes red-brown to dull red; legs and feet grey, undersides of toes yellowish, claws yellowish horn. ADULT ♀: like ♂ but lacking pectoral tufts. SIZE (20 ♂♂, 20 ♀♀): wing, ♂ 65–74 (69·4), ♀ 58–63 (59·9); tail, ♂ 42–56 (49·6), ♀ 37–42 (39·6); bill, ♂ 19–21·5 (19·8), ♀ 16·5–18·5 (17·5); tarsus, ♂ 15·5–17·4 (16·2), ♀ 15–16·5 (15·9). WEIGHT: (3 unsexed) 11·5–15·3 (12·9).

IMMATURE: like adult but duller olive-green above, paler below; bill darker.

NESTLING: undescribed.

D. f. idius Oberholser: Guinea and Sierra Leone to Ghana and Togo. Darker green above than *cameroonensis*, tail darker brown, less greenish; bill darker brown; legs olive-green; size smaller: (9 ♂♂, 11 ♀♀) wing, ♂ 59–66 (62·7), ♀ 55–60 (57·2); tail, ♂ 39–45 (43·1), ♀ 37–41 (39·0); bill, ♂ 15·5–18 (16·8), ♀ 15–17·5 (16·3). WEIGHT: 2 ♂♂ 10, 11·3, 2 ♀♀ 11·2, 12·3.

D. f. fraseri Jardine and Selby: Bioko. Poorly differentiated from *cameroonensis*: brighter on underparts; size larger: (5 ♂♂, 5 ♀♀) wing, ♂ 71–78 (74·6), ♀ 63–72 (68·2).

D. f. axillaris (Reichenow): Congo basin from about 16°E and Ubangi R to Uganda. ADULT ♂: forehead, crown and nape, cheeks and ear-coverts dark grey; chin and throat light grey (contrasting with head). ADULT ♀: green-headed (as *cameroonensis*); immature ♂♂ are green-headed, like ♀. WEIGHT: ♂ (n = 28) 11–15 (12·3), ♀ (n = 16) 8·3–13 (10·4).

TAXONOMIC NOTE: green-headed *cameroonensis* and grey-headed *axillaris* appear to meet without integradation in Zaïre, and they may represent incipient species (Hall and Moreau 1970). Wolters (1979–1980) and Sibley and Monroe (1990) treated them as specifically distinct. However, until more is known about how they behave where they meet, we follow Short *et al*. (1990) and Dowsett and Forbes-Watson (1993) in regarding them as conspecific.

Field Characters. Length 11·5–12·7 cm. A warbler-like canopy sunbird with plain green plumage and rather long, straight bill; ♂ is the only lowland forest sunbird to have *orange pectoral tufts*. W races have green heads concolorous with black and breast; from Zaïre eastwards head is grey, contrasting with olive-yellow underparts. Western birds easily confused with Grey Longbill *Macrosphenus concolor* (Demey and Fishpool 1994); they are best distinguished by conspicuous *pale yellow eye-ring* and dark legs, while the longbill lacks head markings and has pale legs; these marks are easier to see than bill differences (Fraser's Sunbird has upper mandible dark and slightly curved, lower mandible pale horn and almost straight; bill of Grey Longbill straight and dark with pale base).

Voice. Tape-recorded (104, B, CHA, ERA, KEI, LEM). Song, lasting 2–3 s, 4–6 high-pitched notes, thin, squeaky and tuneless, given at even, measured rate; most are on same pitch, although 1–2 may be lower: 'tseep-tseep-tseep-tsup-tseep'. 5–10 s intervals between songs. Frequent contact call between pair members, very high, thin 'psi'; also a nasal up-slurred 'jewy'.

General Habits. Inhabits primary and old secondary forest; in Gabon, also regenerating forest in final stages of transition from abandoned agriculture; elsewhere, small forest patches and narrow gallery forest. Lowlands, up to 1100 m in Liberia, up to 1550 m in E Zaïre (Lutunguru) and Uganda. Frequents forest interior at all levels from undergrowth to canopy (2–30 m) but chiefly in middle levels, at 15–25 m. Avoids sunlit clearings, also park-like forest (Congo).

Forages in pairs or ?family parties of up to 8 birds. Regularly joins mixed flocks of insectivores, and may be nucleus for their formation. Often associates with Chestnut-capped Flycatcher *Erythrocercus mccallii*. Restless and noisy. Behaves more like a warbler than a sunbird: runs up vine or along small branch and explores undersides of leaves closely, gleaning insects; when entering clumps of leaves, arches and spreads wings and sways from side to side (Brosset and Erard 1986), presumably to flush insects. Rarely seen on flowers (Liberia).

Food. Insects only in 7 stomachs; small beetles, orthopterans, butterflies, termite alates, caterpillars, ant pupae; spiders (plentifully); small fruits and unidentified seeds.

Breeding Habits. Monogamous; territorial; some birds spend their lives in the same territory. Actively defends territory against congeners, and confrontations can be aggressive. In threat display, body is held straight while head and tail are pointed forward, wings held half open and drooped, tail feathers spread, pectoral tufts fluffed out; the beak is wide open, showing orange-red inside of mouth, and bird gives series of sharp cries (Brosset and Erard 1986). One was retrapped in same spot 3 times within 11 years, another was retrapped 8 years 5 months after being ringed (Brosset and Erard 1986).

NEST: not described. Nest-building observed at 9, 13 and 15 m above ground on outermost twigs of understorey trees (Gatter 1997).

LAYING DATES: Liberia, (dependent young Nov, Jan, breeding condition June, Aug); Ghana, Oct (breeding condition June); Nigeria, breeding condition Apr, Nov (juvs with unossified skulls Jan, Apr, June–Aug, Oct); Cameroon, (juv Mar, breeding condition Mar, Oct); Gabon, (juvs Mar, Nov); Congo. Jan (1), Oct (4), Nov (1); Zaïre, lowlands July, Oct (probably breeds most of year); Itombwe, (breeding condition May, Sept); Angola, (breeding condition Feb, Aug–Sept); E Africa: Region B, (breeding condition July).

DEVELOPMENT AND CARE OF YOUNG: young fed by both sexes.

BREEDING SUCCESS/SURVIVAL: 1 bird lived at least 10 years 11 months.

Reference
Brosset, A. and Erard, C. (1986).

Genus *Anabathmis* Reichenow

Small- to medium-sized sunbirds with iridescent violet-blue chin and throat (and crown in one species); back olive; tail quite long, evenly graduated and rather wedge-shaped; all rectrices except T1 broadly pale-tipped. ♂♂ 5–10% larger than ♀♀. Bill quite robust and decurved, curvature starting at skull. Mainland species (*A. reichenbachii*, a waterside bird, range centred on Gulf of Guinea coast) without sexual dimorphism, with yellow pectoral tufts. 2 insular species (which have presumably derived from ancestral *reichenbachii*), *A. hartlaubii* (Príncipe) and *A. newtonii* (São Tomé) a little sexually dimorphic, without pectoral tufts. *A. hartlaubii* markedly smaller than other 2. Feed on nectar and small insects; mainland species reported also to hunt insects in manner of *Elminia* flycatcher. Nest suspended.

In plumage broadly resembles *Cyanomitra*, and *A. reichenbachii* has quite strong plumage resemblance to *C. verticalis*; but bill profiles of the 2 genera differ significantly (M.P.S. Irwin, pers. comm.).

Endemic. 3 species. Surprisingly, genus seems to be absent from Bioko Island.

Anabathmis reichenbachii (Hartlaub). Reichenbach's Sunbird. Souimanga de Reichenbach. Plate 9 (Opp. p. 142)

Nectarinia reichenbachii Hartlaub, 1857. Syst. Orn. Westafr., p. 50; Gabon.

Anabathmis reichenbachii

Range and Status. Endemic resident, rare west of Nigeria and east of 20°E, common on coast from W Cameroon to 05°S, and probably widespread and locally common but overlooked between 10°E and 20°E.

Liberia, rare; records at 5 coastal localities (Gatter 1997). Ivory Coast, sight records in Parc Nat. d'Azagny, on Bandama R. at Grand-Lahou, Dabou, Abidjan (Eccles 1985, Demey 1986), Assinie and Eloka; fairly common at Grand Bassam (Balchin 1988); very common in swamp bush in Azagny (Demey and Fishpool 1991). Ghana, records from Elmina, Hwini R. near Takoradi and (2 birds) mouth of Volta River. Togo, rare (Cheke and Walsh 1996). Nigeria, not uncommon, very locally, from Lagos to Niger Delta; the commonest sunbird along tributaries of Ossa R. near Okomu, and along Jamieson R. near Sapoba (Elgood *et al*. 1994). Cameroon, locally frequent to abundant between Idenau and Ikona, frequent along coast to Campo on Mbini border, frequent along Ja River, and records elsewhere north to Yabassi (north of Douala) and south to Aboulou (Gabon border); also a record north of Adamawa Plateau (on upper Benue R.?) (Hall and Moreau 1970). Equatorial Guinea (Mbini), Bata. Gabon, occurs mainly along coast, notably at Libreville, and along banks and islands in Ogooué, Ivindo and other larger rivers, and in forest but never more than 500 m from water. Congo, locally common in Odzala Nat. Park and noted at Ndoki and Mbéli in Nouabalé-Ndoki Nat. Park (Dowsett-Lemaire 1997a, b). Angola, known from Cabinda, and once taken at L. Carumbo, N Lunda. Zaïre, scarce and little-known; on lower Congo R., tributaries near Mbandaka, Rubi R. (Buta) and Aruwimi R. (Panga) in Haut Zaïre, and in Itombwe only a single record, at Kamituga.

Density of a dozen pairs on Chimpanzee I., lower Ivindo R., Gabon.

Description. ADULT ♂: top and sides of head to chin, throat and centre of upper breast dark metallic blue, tinged violet; rear and sides of neck to mantle, back and scapulars dark olive-green with mottling due to darker brown feather centres; rump and uppertail-coverts olive-yellow. Tail graduated, T6 11–16 shorter than T1, blackish brown, feathers edged olive-yellow basally and tipped pale grey, broadly so on T5–T6. Lores blackish. Sides of breast, lower breast, upper flanks and upper belly grey; lower belly, lower flanks, thighs and undertail-coverts olive-yellow; pectoral tufts vary from spectrum-yellow (most birds) to bright orange-yellow (Cane and Carter 1988). Upperwing blackish brown; primaries edged olive-yellow, secondaries and tertials edged olive-green, all upperwing-coverts fringed olive-green. Axillaries and underwing-coverts white; inner borders of undersides of flight feathers pale buff. Bill black; eyes dark brown; legs black. Sexes alike, but ♂ larger and longer-tailed than ♀, and pectoral tufts of ♀ usually a little paler spectrum-yellow. SIZE (10 ♂♂, 10 ♀♀): wing, ♂ 58–61 (59·1), ♀ 54–57 (55·9); tail, ♂ 47–50 (48·8), ♀ 40–44 (42·2); bill, ♂ 17–19 (18·2), ♀ 17–19 (18·0); tarsus, ♂ 16–18 (17·5), ♀ 16–17 (16·4). WEIGHT: 1 ♂ 9·8.

IMMATURE: juvenile has top and sides of head to throat and centre of breast dark greyish brown, merging into olive-brown upperparts and yellowish olive underparts, latter with some dusky mottling. Immature after post-juvenile moult has upperparts, wings and tail like adult, but top of head dark olive-green, mottled brownish black; narrow yellow superciliary stripe; sides of head dark olive-green, mottled yellowish; cheeks and underparts wholly olive-yellow, chin and throat barred dusky, sides and flanks washed olive-green.

NESTLING: late nestling has very fluffy plumage, upperparts olive, throat olivaceous grey, breast and belly olive-yellow inclining to lemon-yellow.

Field Characters. Length 12–14 cm. A sunbird of coastal habitats and riverine vegetation in lowland forest, very active and easily observable; usually located by voice – calls and sings constantly (Christy and Clarke 1994). Smaller and shorter-billed than Green-headed Sunbird *Cyanomitra verticalis*, with which it often occurs, with greenish steel-blue head and throat; from below, *longish* *white-tipped tail* and *vivid yellow* splash on belly and undertail-coverts distinctive (Serle 1965). Sexes alike, both with yellow pectoral tufts. Immature has brown crown, yellow underparts (no grey), chin to breast heavily blotched brown. ♂ Green-headed has iridescent green head, blue throat and upper breast and creamy pectoral tufts, but rest of underparts entirely grey (no yellow on lower belly) and tail square. In ♂ Blue-throated Brown Sunbird *C. cyanolaema*, blue throat and crown are separated by broad area of brown on face and sides of neck; rest of body is uniform brown except for creamy pectoral tufts; ♀ Blue-throated quite different, with brown crown, white stripes on face, whitish underparts with blotchy brown throat and breast. See also Cameroon Sunbird *C. oritis*.

Voice. Tape-recorded (104, B CHA, LEM). Song long and bustling, a medley of high sweet whistles and thin squeaky sounds, with repeated lilting 'chippity-hee' and 'chippity-chuju'; somewhat reminiscent of European Goldfinch *Carduelis carduelis*. Shorter and more formal version an excited 'chee-ti-pity' repeated 3–4 times; sometimes ends in some down-slurred notes, 'chee-ti-pity, chee-ti-pity, teep-chewp-chyoo'. Call a rather melancholy whistle, 'tewy-tewy', also rendered 'tchui tchui tchu-ih, tchu-ih, tiyi, tiyie' (Christy and Clarke 1994); said to be quite unlike any other sunbird in lowland Cameroon (Serle 1965), but said to be practically identical to Green-headed Sunbird (which also occurs in lowland Cameroon) (Christy and Clarke 1994).

General Habits. Confined to immediate vicinity of water: marshes, swamps, and coastal lagoons, but particularly on islands in and along banks of rivers, where it feeds and breeds in vegetation overhanging water. In Kouilou Basin, Congo, inhabits the more open sections of low bush including *Alchornea*, *Drepanocarpus* and *Mimosa*, and thickets with *Alstonia*, *Ficus congensis*, *Nauclea pobeguini* and other small trees overhanging water; builds nests in shrub *Mimosa pigra* which seldom grows higher than 1 m and in heads of papyrus *Cyperus papyrus* up to 3 m above water; utilizes similar situations near coast, especially *Alchornea* thickets over water (Dowsett-Lemaire and Dowsett 1991). Inland in Congo, lives in wet thickets on edges of salt-pans and marshes and in open riverine thickets. In Liberia bird netted in mangrove interspersed with a few thorny shrubs, within 50 m of ocean on one side and 25 m of a coconut plantation on the other. In Ivory Coast inhabits scattered trees and shrubs in gardens and in grassy areas near forest. In Nigeria occurs in swamp forest (between coastal mangrove and rainforest), and timbered and scrub-covered banks of lowermost reaches of rivers. In Cameroon, quite common in open places within a few hundred m of the sea with flowering trees and shrubs, gardens and abandoned cultivation. In Gabon, occurs in coastal vegetation, coastal town gardens, large trees overhanging edges of larger rivers, river islands, canopy of forest up to 500 m from rivers, and gallery forests (but never in adjacent savanna: Christy and Clarke 1994).

Very active. Feeds largely at blossoms in W Cameroon (Serle 1965) but in Gabon never feeds at flowers; instead, visibly hunts insects in leafy trees, mainly between 5 and 30 m above ground, 'effectue en séries de brusques retournements et des pirouettes', with tail fanned and wings drooping, recalling African Blue Flycatcher *Elminia longicauda* (Brosset and Erard 1986) – foraging behaviour unique amongst African mainland sunbirds. However, in Congo Reichenbach's Sunbird was never seen foraging in that manner; rather, it foraged only within a few m of ground and probed a variety of flowers for nectar (Dowsett-Lemaire and Dowsett 1991). Old report of a bird, with 2 Splendid Sunbirds *Cinnyris coccinigaster*, making short, circular flights from low tree on Lagos beach, Nigeria, as if catching flying insects (Bannerman 1948).

Keeps in pairs; sometimes singly or family parties; uncommonly, joins mixed-species foraging flocks; in Liberia seen with other sunbirds in flowering tree, including Collared *Hedydipna collaris*, Western Olive *Cyanomitra olivacea* and Olive-bellied *Cinnyris chloropygia*; often in company of Green-headed Sunbirds. Flight undulating. Calls and sings constantly; sings all year. High-pitched, jingling song delivered with persistence and vigour from prominent perch such as top of coconut palm *Cocos nucifera* (Serle 1965), uttered with head up, body erect, wings drooped and tail spread (Christy and Clarke 1994). Roosts in tall weeds.

Food. Insects and evidently nectar. Feeds at blossoms, at least in west of range, 'apparently sucking nectar; yet insects were found in every one of over 30 stomachs' (Serle 1965). In Congo 'readily probes flowers for nectar', of *Psychotria djumaensis*, *Stipularia africana* and mistletoes *Englerina gabonensis* and *Globimetula braunii* (Dowsett-Lemaire and Dowsett 1991). Said to feed at flowers of coconut palm (Chapin 1954).

Breeding Habits. Solitary nester, evidently monogamous and territorial.

NEST: suspended pouch with side entrance, composed of fine strips of plantain leaves, grass and vegetable fibres; not lined.

EGGS: 1. Not glossy. One egg nearly immaculate, light chocolate-brown with fine dusting of dark chocolate-brown over broad end; another greyish buff, similarly with fine dusting of deeper greyish buff over broad end. SIZE: (n = 2) 17·5 × 12·5, 17·5 × 13·0.

LAYING DATES: Liberia, (immature Sept); Cameroon June (3), Sept (1), Oct (2), Nov (1), Dec (1) (Serle 1981), (and fledgling being fed by adults, May; nest almost certainly with young, Oct; immature birds in May–July); Gabon, Nov, Feb. Angola, (hypertrophying gonads, Mar); Congo, Sept (2), Oct (2), Dec (2) (F. Dowsett-Lemaire 1997 and pers. comm.).

INCUBATION: by ♀.

References
Brosset, A. and Erard, C. (1986).
Dowsett-Lemaire, F. and Dowsett, R.J. (1991).
Serle, W. (1965).

Anabathmis hartlaubii (Hartlaub). Príncipe Sunbird. Souimanga de Hartlaub. Plate 9

Nectarinia hartlaubii Hartlaub (ex Verreaux mss), 1857. Syst. Orn. Westafr., p. 50; 'Angola' (= Príncipe, Dohrn, 1866, Proc. Zool. Soc. London p. 326). (Opp. p. 142)

Range and Status. Endemic resident, Príncipe Island, uncommon in some habitats but common to abundant in others; possibly absent from montane forest.

Description. ADULT ♂: top and sides of head, neck, mantle to rump and scapulars dark olive; long uppertail-coverts blackish with steel-green fringes. Tail graduated, T6 12–18 mm shorter than T1, blackish, glossed slightly bluish above; T2–T4 narrowly tipped and T5–T6 broadly tipped buffish white. Lower cheeks to chin, throat and upper breast dark steel-blue with violet reflections. Sides of breast and lower breast dark olive, grading into paler yellowish olive on flanks, belly, thighs and undertail-coverts. Upperwing dark brown, flight feathers edged olive, tertials, alula and coverts fringed olive. Axillaries and underwing-coverts white; inner borders on undersides of flight feathers buffish white. Bill black; eye colour unrecorded; legs black. ADULT ♀: differs from adult ♂ in having chin mottled whitish and dark olive, throat and breast olive, mottled dusky, merging into paler olive-yellow underparts streaked with dusky olive. SIZE (10 ♂♂, 8 ♀♀): wing, ♂ 60–65 (62·5), ♀ 57·5–60 (58·4); tail, ♂ 49–56 (54·2), ♀ 42–47 (45·3); bill, ♂ 23–24 (23·5), ♀ 19–21 (20·0); tarsus, ♂ 19–21 (20·0), ♀ 17·5–19 (18·4). WEIGHT: 1 unsexed 10·5.
  IMMATURE: like adult ♀, but chin to upper breast entirely dark grey-brown. Immature ♂ like adult ♂ but parts which become metallic in adult are dull sooty grey.
  NESTLING: unknown.

Field Characters. Length ♂ 14, ♀ 13 cm. Endemic to Príncipe, and the only sunbird on the island. Dark olive-green above, tail blackish with blue gloss; throat and upper breast dark violet steel-blue, whitish pectoral tufts, rest of underparts brownish olive becoming yellower on belly; whitish underwing peeps out at bend of wing on perched bird. Chin and throat of ♀ mottled light and dark.

Voice. Tape-recorded (104, B, CHR, JOPJ). Call, a loud, clear, up-slurred 'too-wee', repeated insistently; also a variety of very high-pitched, thin hissing notes, 'psit-sit', 'tissss' or 'tssit'. Song rapid, dry and lively, 'tik-e-tik, tik-e-tik, tik-e-tik, ee-tik' or a ringing 'tiu-tiu-hiut-tiu-tiu-hiut' (Christy and Clarke 1998).

General Habits. Inhabits primary and secondary forest, cocoa and coconut plantations, banana groves, farms and gardens; much commoner in forest regrowth and plantations than in cultivated areas and gardens. In forest keeps to lower strata; out of forest at all levels in trees, often the canopy. Usually in pairs but also in foraging parties of 3–10. Active and lively. Often feeds at flowers, especially of coconut palms, but spends at least as much time in gleaning twigs and leaves as in attending blossoms (Snow 1950). Probes into ends of broken twigs (Atkinson *et al.* 1991). Occasionally comes to ground. Vocal, more so in late afternoon than earlier in day; ♂ in pair and ♂♂ in parties equally vocal, even in season when birds evidently not breeding; both sexes sing, with 20 notes in 15 s.

Anabathmis hartlaubii

Food. Insects in stomachs (Snow 1950). Nectar, small invertebrates, and perhaps fruit pulp (de Naurois 1994). Feeds at flowers of *Erythrina*, palms and bananas (Christy and Clarke 1998).

Breeding Habits.
  NEST: oval pouch suspended from one or more twigs, made of palm-bark fibres, plant strips and dry leaves rather loosely woven together, with small entrance at one side below small porch, and insubstantial 'beard' below; thickly lined with soft filaments of flowers, cotton and other fine plant materials. Ext. height 170–200, ext. diam. *c.* 40, entrance diam. *c.* 35 (Bannerman 1948). Placed 1–4 m up in bush or tree, well concealed, and hangs more freely than nest of São Tomé Sunbird *A. newtonii* (de Naurois 1994).
  EGGS: 2. White ground, almost obliterated by numerous reddish spots.
  LAYING DATES: breeding season (Sept)–Feb or Oct–(Mar) (de Naurois 1994); juveniles in all months (J.G. Keulemans *in* Bannerman 1948 and Atkinson *et al.* 1991).

References
Atkinson, P., Peet, N. and Alexander, J. (1991).
Bannerman, D.A. (1948).
Christy, P. and Clarke, W.V. (1998).

Plate 11

Plate 12

Plate 9
(Opp. p. 142)

Anabathmis newtonii (Bocage). Newton's Sunbird. Souimanga de Newton.

Cinnyris newtonii Barboza du Bocage, 1887. J. Sci. Math. Phys. Nat. Lisboa, ser. 1, 11 (1886), p. 250; S. Thomé.

Range and Status. Endemic resident, São Tomé Island, where frequent to abundant in variety of wooded habitats.

Description. ADULT ♂: top and sides of head, neck, mantle to rump and scapulars dark olive brown; long uppertail-coverts sooty black, tips faintly glossed green. Tail graduated, T6 6–9 mm shorter than T1, blackish brown with faint violet gloss above; T2–T4 tipped whitish, T5–T6 more broadly tipped with whitish extending along edge of outer web. Lower cheeks to chin and throat dark metallic greenish blue with violet reflections, grading to violet-purple on upper breast. Lower breast yellow, merging with pale creamy-olive flanks, belly and undertail-coverts. Upperwing dark brown; flight feathers edged dark olive; tertials, alula and coverts fringed dark olive, lesser coverts darker and faintly glossy. Axillaries and underwing-coverts whitish; inner borders on undersides of flight feathers buffish white. Bill black; eyes dark chestnut; legs black. ADULT ♀: differs from ♂ in having chin to upper breast olive-grey, barred dusky; lower breast creamy olive, streaked greyish. Bare parts as in adult ♂. SIZE (9 ♂♂, 7 ♀♀): wing, ♂ 52–57 (54·3), ♀ 48·5–50 (49·0); tail, ♂ 35–42 (39·1), ♀ 28–35·5 (32·5); bill, ♂ 17–18 (17·6), ♀ 16–18 (16·8); tarsus, ♂ 17–19 (18·0), ♀ 15·5–17·5 (16·5). WEIGHT: 1 ♂ 7, 1 ♀ 6.

IMMATURE: ♂ like adult ♀ but slightly darker and browner above, less green; belly somewhat paler.

NESTLING: unknown.

Field Characters. Length 10–11·5 cm. Endemic to São Tomé and the only small sunbird on the island. Olive above, ♂ with throat and breast dark metallic greenish- to violet-blue, belly yellow; ♀ similar but throat and breast dark olive with indistinct pale barring.

Voice. Tape-recorded (104, B, ALEX, CHR, TYE). Song rapid, lively, high-pitched, rather explosive and slightly grating, ending abruptly (Christy and Clarke 1998). Calls include an insistent 2-note 'see-tzee' or 'see-tsew', second note lower, a loud, buzzy 'bzueeh', a repeated hard 'jit-jit-jit ...' and various little 'sips' and chatters; for further calls see Christy and Clarke (1998). Begging call of young a continuous nasal 'bjuit-bjuit-bjui'.

General Habits. Inhabits steep forested hillsides, edges and interior of forests, plantations, dry savanna woodland, and gardens; common in habitats with tall trees, including towns, cocoa plantations, forest and woodland; abundant in high altitude coffee plantations at Monte Café and Zampalma in centre of island and in overgrown plantations on W coast between Lemba R. and Binda (Jones and Tye 1988). Sea-level to at least 1800 m. Uncommon in tree-lined avenues in São Tomé Town. Some observers have found Newton's Sunbird only locally common, suggesting that it may move seasonally between open country and forest or distantly between stands of flowering trees. In Apr, none in gardens despite profusion of flowers there (Eccles 1988).

In pairs and family parties; often in parties of up to 10; gathers at flowering *Erythrina* tree in loose aggregations of up to 30 birds (Atkinson *et al.* 1991); fond of banana plants; flocks with other species, including São Tomé Speirops

Anabathmis newtonii

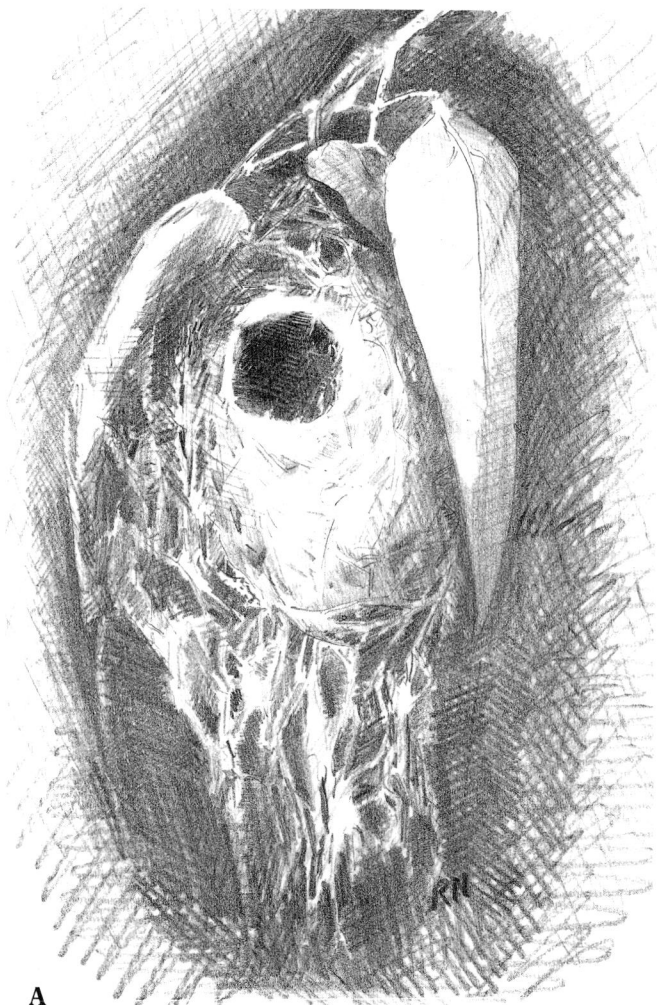

A

Speirops lugubris and São Tomé White-eye *Zosterops ficedulinus* and others (Christy and Clarke 1998). Gleans insects from leaves and feeds high in trees that are in flower (Atkinson *et al.* 1991); acrobatic, inspecting undersides of leaves. Very active; sings frequently; often 4–5 ♂♂ together sing loudly, almost in chorus, perching close together and bowing excitedly at one another; posture erect, opened wings vibrating to show the white underwing-coverts, bill open, body turning from side to side (Snow 1950, Christy and Clarke 1998).

Food. Nectar; small invertebrates including beetles and spiders (de Naurois 1994). Takes nectar of *Leea tinctoria*; attracted to flowers of *Cestrum levigatum*, *Rubus*, and especially *Canna* and *Erythrina* (Christy and Clarke 1998).

Breeding Habits. Solitary nester; possibly co-operative breeder – 4 birds including 1 ♂ once seen exploring an old nest.

NEST: suspended pocket, made of mosses, leaves and plant fibres, with entrance high on one side under short porch, and 'beard' of plant stems and fibres (**A**); lined thickly with wool, down and feathers (Bannerman 1948, de Naurois 1994). Hangs from outer branch of tree or from twigs near edge of cacao or other shrub, often close to bank or wall.

EGGS: 2. White, whitish or greyish white, covered in small violet spots, sometimes confluent. Size (n = 2) 15 × 11.

LAYING DATES: Aug–Jan (and juveniles in June and July).

References
Atkinson, P., Peet, N. and Alexander, J. (1991).
Christy, P. and Clarke, W.V. (1998).
de Naurois, R. (1994).

Genus *Dreptes* Reichenow

Monotypic (*D. thomensis*); the world's largest sunbird, endemic to São Tomé Island. Bill proportionately long (up to 39 mm), deep and decurved. Tail long, steeply graduated, all rectrices except T1 white-ended. Plumage mainly sooty brown with vestigial iridescence on head and chest; lower belly and undertail-coverts yellowish olive; no pectoral tufts; sexes alike. Forages partly by bark-probing, to greater extent than in any other sunbird, and diet may include fruit pulp.

Appears to be derivative of *Anabathmis*: in overt features *D. thomensis* is to *Anabathmis newtonii* (the other endemic São Tomé sunbird) as *A. newtonii* is to *A. hartlaubii*, and it surely represents an earlier invasion of São Tomé by the stock that gave rise to *A. newtonii* there.

Dreptes thomensis (Bocage). Giant Sunbird. Souimanga de São Tomé.

Nectarinia thomensis Barboza du Bocage, 1889. J. Sci. Math. Phys. Math. Lisboa, ser. 2, 1, p. 143; St Miguel, St Thomé.

Plate 9
(Opp. p. 142)

Range and Status. Endemic resident, primary forest on São Tomé Island. Common in primary forest and forest scrub along Xufexufe R. in SW and Ana Chaves R. in centre of island; occurs (or used to occur) on coast in SW; common in mid-altitude and high altitude forest at Lagoa Amélia (1500 m, 10 km north of Ana Chaves R., 8 km from NW coast), but survival there is threatened by encroaching settlement (already within 500 m of forested Lagoa Amélia Crater in 1990); evidently now absent from mature secondary forest along Quija R. (3 km from Xufexufe R.) where occurred in 1920s, despite area having been left undisturbed for a long time (Atkinson *et al.* 1991), but can easily be overlooked (Jones and Tye 1988).

Distribution in lowland forest is extremely patchy; birds concentrate to exploit flowering plants in shrubby growth along streams; 7 birds foraged in area of 0.25 ha (Atkinson *et al.* 1991).

Forests on lower slopes of Pico (2040 m) had almost completely disappeared 50 years ago (Snow 1950). Inroads have been made into what was formerly one of the larger stretches of forest, around Lagoa Amélia. Further clearance of primary forest will seriously endanger this remarkable sunbird (Atkinson *et al.* 1991).

Description. ADULT ♂: upperparts sooty brown, feather tips glossed dark violet-blue. Tail quite long, strongly graduated, T6 28–48 shorter than T1, blackish brown, upperside glossed steel blue; T4–T5 narrowly tipped and T6 broadly tipped white. Underparts sooty brown, narrow tips to feathers of chin to upper throat, sides of breast and lower breast with slight bluish gloss; lower belly and lower flanks yellowish olive-green; undertail-coverts paler olive-yellow. Upperwing blackish brown, tertials and coverts fringed dark glossy violet-blue. Axillaries and underwing-coverts pale greyish buff; inner borders on undersides of flight feathers narrowly tawny-buff. Bill black, eyes red, legs and feet black. Sexes alike. SIZE (11 ♂♂, 10 ♀♀): wing, ♂ 88–94

Dreptes thomensis

(91·7), ♀ 79–84 (81·9); tail, ♂ 84–95 (89·0), ♀ 72–76 (74·4); bill, ♂ 36–42·5 (40·6), ♀ 31–36 (32·9); tarsus, ♂ 28–31 (29·5), ♀ 26–28 (26·3). WEIGHT: ♂ (n = 1) 26·2, ♀ (n = 4) 18·0–18·9 (Atkinson *et al.* 1991).
IMMATURE and NESTLING: unknown.

Field Characters. Length, ♂ 20–23, ♀ 18–19 cm. A huge blackish sunbird endemic to São Tomé; nothing remotely like it on the island or in Africa. Bill very long and curved, indistinct steel-blue wash on head and upperparts, underparts dark sooty brown, lower belly and undertail-coverts yellowish. Tail graduated, blue-black, with white tips to outer feathers.

Voice. Tape-recorded (104, B, CHR, JOPJ). Territorial song a series of rich, warbled but rather abbreviated notes connected by little hisses and chips; overall effect is pleasant but a little disjointed; rendered 'huèt-tsip-tsuit – huèt-tsip-tsueep – huèt-tsip-tsueep' (Christy and Clarke 1998). Birds assembled in 'lek' give drongo-like calls, beginning with grating, rattling 'tic-tic-tic-tic-tic-tic', then a loud, explosive series of 'tsi-tsu-huee – tsi-tsiu-tsiu' or 'huit-huit-rruit-rruit', very emphatic; in late evening one of the noisiest birds in the SW forests (Christy and Clarke 1998). Contact calls by pair foraging at nectar of flowers, a softer 'tu-huit, tu-ut'; for other calls see Christy and Clarke (1998).

General Habits. Inhabits undisturbed primary forest, with *Craterispermum*, *Canthium*, *Syzygium* and *Ficus*, feeding in and below canopy and at blossoms in ground layer. Occasionally visits cultivation up to 1 km from forest. Active and lively. Commonest methods of foraging are probing into flowers and gleaning and hover-gleaning undersides of leaves (Atkinson *et al.* 1991); also feeds by hopping or creeping along branches and digging bill into

A

bark, using it as a probe (Atkinson *et al.* 1991, Sargeant 1994) – 'the whole manner of feeding more like a creeper than a sunbird' (Snow 1950); probes into epiphytes and banana inflorescences. Vocal; strong, strident calls usually given in flight (but silent near nest). Sings all day; several birds assemble in 'a sort of lek' (Christy and Clarke 1998) and sing strongly.

Food. Nectar, insects and fruit pulp (de Naurois 1994).

Breeding Habits. Solitary nester, may be polygynous (twice as many ♀♀ as ♂♂ noted at 3 sites: Atkinson *et al.* 1991); apparently territorial, but pair territory may be little more than surroundings of nest.
NEST: suspended pouch (**A**), made of moss and plant fibres, with side entrance below small porch, and long hanging 'beard' of lengthy plant fibres (de Naurois 1994). Sited at very end of long, hanging branch, sometimes of bamboo, 4–10 m above ground.
EGGS: 2. Rather long; white, with small reddish spots.
LAYING DATES: Sept–Jan.

References
Atkinson, P., Peet, N. and Alexander, J. (1991).
Christy, P. and Clarke, W.V. (1998).
Collar, N.J. and Stuart, S.N. (1985).
de Naurois, R. (1983, 1994).

Genus *Anthobaphes* Cabanis

Monotypic: a fynbos *Erica* and *Protea* specialist in SW Cape. Small; fine-billed. ♂ long-tailed, tail strongly graduated with T6 *c.* 14 mm shorter than T2 and T1 *c.* 28 longer than T2; with iridescent green head and violet upper breast; no pectoral tufts; lower breast orange, belly yellow. ♀ olive, with orange wash on belly. Placed in *Cinnyris* by Delacour (1944). Allied superspecifically with *Cyanomitra olivacea* by Hall and Moreau (1970), though habitats and voices are quite different and that alliance not supported by Clancey (1986, 1987), Dowsett and Dowsett-Lemaire (1993) or Short *et al.* (1990). Perhaps closer to *Anabathmis* than to *Cyanomitra* or *Cinnyris* (Irwin, in press).

Anthobaphes violacea (Linnaeus). Orange-breasted Sunbird. Souimanga orangé.

Plate 12
(Opp. p. 159)

Certhia violacea Linnaeus, 1766. Syst. Nat., ed. 12, 1, p. 188; Cape of Good Hope.

Anthobaphes violacea

Range and Status. Endemic resident, SW Cape Prov., South Africa., from Vanrhynsdorp and Cape Peninsula to Port Elizabeth, but on W coast only a vagrant away from mountains. Adversely affected by fires in and clearance of its specialized macchia, *Protea/Erica*, habitat, and locally less common than formerly (Fraser and Crowe 1990). Density of up to 70 birds in 100 m^2 patch of *Erica gilva* near Cape Point.

Description. ADULT ♂: head and neck to upper mantle and upper scapulars bright metallic green; rest of upperparts non-iridescent olive green. Tail graduated, T1 long and narrow, projecting 20–25 mm; dark brown, feathers edged olive-green. Lores blackish. Chin and upper throat iridescent green, concolorous with head, grading to broad violet band across upper breast; pectoral tufts bright yellow; lower breast orange, duller with olive tinge at sides; belly yellow; flanks and thighs olive-green; undertail-coverts orange-yellow. Flight feathers dark brown, edged yellowish olive-green; tertials, alula, primary coverts, greater coverts and median coverts dark brown, fringed olive-green; lesser coverts metallic green. Axillaries and underwing-coverts pale olive-green, flight feather inner borders grey-buff. Bill black; eyes dark brown; legs black. ADULT ♀: upperparts olive-green, top of head to ear-coverts and sides of neck browner, crown with faint darker mottling. Tail slightly graduated, dark brown, feathers edged olive-yellow. Underparts pale olive-green, faintly mottled on cheeks, chin and throat, yellower on centre of breast, belly and undertail-coverts. Upperwing feathers dark brown, flight feathers edged olive-yellow, tertials, alula and coverts all fringed olive-green. Axillaries and underwing-coverts olive-yellow; inner borders of flight feathers grey-buff. SIZE (10 ♂♂, 5 ♀♀): wing, ♂ 53–57 (55·6), ♀ 51–53 (52·2); tail, ♂ 71–80 (76·0), ♀ 45–52 (48·4); bill, ♂ 24–25 (24·5), ♀ 21–23 (22·4); tarsus, ♂ 16–18 (16·9), ♀ 16. WEIGHT: ♂ (n = 15) 9–11·3 (10), ♀ (n = 3) 8·6 = 9·7 (9·1) (Maclean 1993).
IMMATURE: juvenile like ♀, but rather browner above, duskier from chin to upper breast. In immature ♂, adult plumage appears first on breast and lesser coverts.
NESTLING: not described.

Field Characters. Length ♂ 14·5–16·5, ♀ 12·5–13·5. Endemic to W Cape Province, South Africa. Gorgeous ♂, with long tail, green head, purple breast-band and orange-yellow belly, unmistakable in its range. ♀ very uniform, olive-green above, warm olive-yellow below with orange wash on belly, no pale supercilium or malar stripe; somewhat larger ♀ Malachite Sunbird *Nectarinia famosa* has prominent malar stripe, dark blotching on yellow underparts; ♀♀ of Lesser and Greater Double-collared Sunbirds *Cinnyris chalybea* and *C. afra* are dull grey-brown; ♀ Amethyst Sunbird *Chalcomitra amethystina* is brown above and streaky below.

Voice. Tape-recorded (88, 91–99, B, BBC, F, LEM, WALK). Common contact/warning call a harsh, down-slurred 'dzeeu' or 'dzeeu-dzeeu', also described as a soft tinny-and-wheezy 'sshraynk' (Skead 1967). Song of 2 types, (1) a short, rapid, low-key rambling jumble of tuneless tinny and scratchy notes; and (2) a slower, disjointed, more conversational medley of different notes, mixed with little tinny ones and frequent 'jeeeu' call, 'chup-chup, chippy-chippy, jeeeu, tin-tin, chewit, trrrrrr, jeeeu, wippit, titi, turrt ... '. Spells may run for 50–60 s, sometimes up to 3 min, with pauses of 3–4 s between repetitions (Skead 1967). When chasing ♀, ♂ calls 'ke-ke-ke...'; alarm call of ♀ a quiet 'eet-eet' (Skead 1967).

General Habits. A bird almost exclusively of fynbos (Fraser *et al.* 1989): open macchia scrub, low, aromatic, woody heaths, proteas and other shrubs interspersed with small trees; from sea level to 1200 m. A specialist strongly associated with *Erica* spp.; occurs also on shrubby hillsides, in flower gardens, and wherever *Protea*, *Leucospermum* and *Mimetes* spp. are flowering, and even in strandveld (perhaps in response to burning of mature mountain fynbos: Fraser and McMahon 1992b). Favours damp or swampy places and forages on swarming gnats above sedges and reedbeds.

Often quite elusive in bush and veld, but commonly tame and confiding in gardens, parks and suburbs. Solitary, in pairs and family parties, and loose foraging assemblies of up to 20 in one tree and 50 or even 100 in a 40 × 40 m patch of flowering vegetation, often with Lesser Double-collared Sunbirds *C. chalybea* and others. Birds in such an assembly sing, chase and squabble, but the assembly is not entirely adventitious and incoherent since it progresses across a bushy hillside as a body, birds sometimes concentrated, sometimes dispersed. In nonbreeding season small parties of ♂♂ occur but most parties are mixed; same 2 birds breed in successive years though ♂ and ♀ separate outside breeding season. Adult ♂ chases mate, other Orange-breasted Sunbirds and other sunbird spp. throughout year. Perched ♂ threatens by thrusting head forwards and down with feathers sleeked, wings drooping, slightly opened and sometimes quivered, tail pointing downward, and pectoral tufts raised. When about to give chase, ♂ raises short head feathers (particularly crest ones), which then look black not shiny. When more aggressive, ♂ also jerks tail to one side, and sweeps head and bill tip in semi-circular up-and-over motion from one side to the other. Birds chase in zigzags, one following close behind the other, usually low over vegetation; lasts for only a few s; adversary resumes foraging generally at a distance, but sometimes both birds return to bush where aggression started and forage without further chasing. If adversary is not intimidated, aggressor may sing rather than give chase. Chasing may involve small group of ♂♂, which can appear almost oblivious of surroundings when in full pursuit (Skead 1967). Chases Lesser Double-collared and Malachite Sunbirds *N. famosa* and other birds up to size of Cape Robin-Chat *Cossypha caffra*; reacts to alarm calls of Cape Sugarbird *Promerops cafer* and often displaced by sugarbirds. Thought to reduce competition with Cape Sugarbird by feeding lower on flowering spike of *Mimetes hirtus* (Collins 1983).

Takes nectar of *Protea lepidocarpodendron* by clinging to side of inflorescence and thrusting bill between tight-fitting bracts; takes nectar of *Mimetes hirtus* and *M. hartogi* (Proteaceae) by moving to side of flower-head and probing into florets. Feeds at deep cup-shaped inflorescence of *P. mellifera* by hopping down into cup. Stands on top of pincushion flower-head of *Leucospermum* and probes into it with bill, or probes it from adjacent twig. Perches below pendent flowers of *Liparia spherica* and probes upwards into them. Makes 2 or 3 rapid, hard jabs to pierce small hole in base of tough corolla of *Watsonia tabularis*, to reach nectar which cannot be reached by probing down the long corolla tube. One bird twice visited and probed into bunch of artificial flowers (Skead 1967). Frequently probes into leaf clusters at end of branches of *Protea* spp., presumably seeking insects. Mean duration of bout of feeding at *Leucospermum conocarpodendron* 11 s (39 ♂ bouts) and 10 s (18 ♀ bouts), with av. 8 and 6 probes per bout respectively (Wooller 1982). Commonly hawks insects, e.g. mass of midges above reeds, by perching on top of vegetation and flying near-vertically upward for *c.* 2·5 m, momentarily hovering and snatching up insect then diving straight back, often to same perch. Often feeds on the ground, on lawns, hopping and probing grass for spiders and insects (Spottiswoode 1993, Cohen and Winter 1993).

Lone ♂ or ♀ loafs, preening, or in relaxed, hunched posture with head sunk in shoulders, bill slightly raised, plumage fluffed, body sunk onto perch, yet alert and eyes beady. Vocal, ♂♂ commonly and ♀♀ sometimes singing, and both sexes often giving wheezy, tinny 'sshraynk, sshraynk' call. 4 or 5 ♂♂, foraging adjacently, may call repeatedly then join in an aerial skirmish. Bathes in dew and rain drops, particularly in leaf rosettes of *Leucodendron* and *Protea*, by spreading wings and tail and nestling body down, wriggling from side to side until plumage thoroughly wetted, when bird moves to open perch and preens vigorously.

Some birds sedentary, but most disperse and wander locally after breeding. Locally, birds track the flowering of food plants (M.W. Fraser *in* Harrison *et al.* 1997), making movements mostly unpredictable but in some places regular; e.g. appears in early Feb on slopes west of Knysna Lagoon, breeds, and disappears about Sept; descends to lower ground (and enters Claremont suburbs) in Feb–Apr (P. Steyn, *Promerops* 238, 1999, 18). Ringed bird seen to move 4 km and 320 m vertically between Kirstenbosch and Table Mt; another moved 8 km (Fraser *et al.* 1989). In Feb 1964 population returned there even though local food plant *Erica speciosa* had not yet bloomed (Broekhuysen 1963). Restricted to winter rainfall region. Bird collected in 1907 near Amatole Forests (King William's Town) must have been a vagrant (Skead 1967).

Food. Nectar and some insects. Feeds at most nectar-bearing plant species but mainly at heaths *Erica* spp. (of which 520 endemic to S Cape), especially those with long, tubular corollas: *E. mammosa*, *E. plukenetti*, *E. phylicaeifolia*, *E. speciosa*. (May be main pollinating agent of some with protruding stamens, e.g. *E. coccinea*). Commonly feeds at flowers of proteas, e.g. *P. lepidocarpodendron*, *P. mellifera*, *Mimetes hirtus* and *M. hartogi*; also *Leucospermum* spp., *Watsonia tabularis*, *Lobostemon fruticosum*, *Lobelia pinifolia* (Fraser and McMahon 1993), *Leonotis leonurus*, *Liparia spherica*, and many exotics, especially *Eucalyptus* spp, also e.g. *Agapanthus*, *Hedera helix*. Adults' insect food not studied but includes midges and spiders (see above). Wide variety of small arthropods fed to nestlings; 56% of items spiders and 42% insects, mainly grasshoppers, small cockroaches and flies, also bugs, 'beetles, ants, midges Chironomidae, hover-flies Syrphidae, flies Muscidae, butterflies, moths and caterpillars; once a mole cricket 32 mm long (swallowed by nestling with great difficulty); also nectar and bits of vegetation' (Broekhuysen 1963).

Breeding Habits. Solitary nester; monogamous. Territorial, or at least high frequency of disputes, counter-singing and chasing between neighbouring ♂♂ gives impression of territoriality, though nature of territory needs more study. ♂ perches prominently in, and chases away intruders from, area of up to 0·2 ha; pair spends most of time in nesting area, leaving it to forage and collect nest materials (Broekhuysen 1963). Some ♂♂ occupy or at least periodically visit their territories outside breeding season; others do not. In lead up to breeding, ♂ becomes more aggressive, size of defended territory diminishes, and ♀ is admitted and courted later, just before nest building. One marked pair occupied same territory for 3 breeding seasons (years); the ♂ with a new ♀ nested there in 4th year, and occupied territory alone in 5th year. Breeding pair tolerates incursions by neighbours towards end of nesting period. Double- or multiple-brooded; tends to return to same territory in successive years. ♀ may use same nest a second time but usually builds anew; a marked ♀ built 6 nests in May–Oct (young fledged from 2 of them); another built 6 in Apr–Nov (young fledged from one); another built 4 in May–Nov (young fledged from 3).

Displays tend to be given out of breeding season (see General Habits) as well as prelude to it or within it. 2 ♂♂, perhaps at territorial boundary, once perched facing each other c. 15 cm apart, one constantly chattering with crest raised and wings slightly drooped, other swaying slowly from side to side, with crest flat and wings drooped. They then grappled, fell to ground, broke apart and flew away. Another ♂, with body erect and plumage compressed, touched ♀'s bill or forehead, 2 days before egg laying (courtship? allopreening?). A perched ♂, evidently alone, repeatedly spread wings, erected pectoral tufts and sang, with tail jerking up and down. ♀ responds to courting ♂ by adopting fledgling's begging posture, crouching with wings half open and quivering, tail raised a little, and head and bill either pointing downward or raised obliquely (Broekhuysen 1963).

NEST: neat spherical or oval ball, thick-walled, with entrance in middle of side or just above middle, without porch, without 'beard' below, not suspended but built into foliage or microphyllous twiggy mass (**A**). Compactly built of great number of rootlets 12–15 long or dry heath twigs, bound with spider web. Fine grass stems sometimes mixed in, particularly around entrance. Dome and walls lined, and floor densely lined, usually with brown fluff and petals of *Protea* spp., often with downy seeds of *Eriocephalus*. Much of lining is forced deeply into rootlets or twigs. Some nests have only a few rootlets or twigs and bulk of nest is 'lining' rather than 'wall'. Decoration on outside of nest almost always absent; rarely a feather, piece of green vegetation or spider's egg cocoon added to exterior. Size variable: height ext. 85–120, int. 50–80; width ext. 64–90; front to back ext. 75–100, int. 45–55; entrance diam. 20 × 40 (Skead 1967). Nests markedly oriented with entrance to southeast-by-east, and lesser orientation to northwest; height above ground (n = 29) 0·1–1·2 m (0·62 m) (Williams 1993b).

Built by ♀. ♂ accompanies her on all journeys between nest and material collection sites (sometimes near to nest, usually not), and perches near nest or collection site,

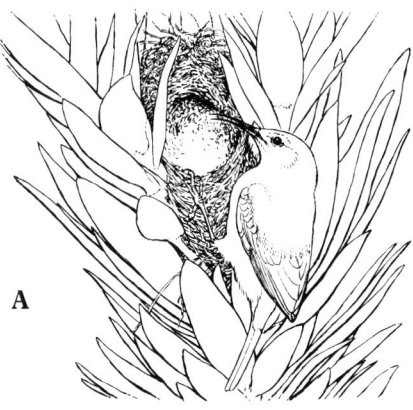

singing. ♀ brings several bills-full of material and dumps it perfunctorily on nest; about every tenth visit, she stays to work material for 1–11 min (generally 2 min) by poking and pulling it with bill and compressing it with pushing movements of breast and sides, with wings half open and legs splayed. Building mainly in forenoon; activity depressed by wet weather. Nest built in 5–18 days. Apparently a lapse of 1–8 (av. 3) days between end of nest-lining and start of egg-laying (Broekhuysen 1963).

EGGS: 1–2, av. 1·69. Laid within 2 h after sunrise; 2nd egg laid 24 h later. ♀ may sit in nest for long spells on day before 1st egg laid. Eggs white, whitish, cream or greyish green, marked with grey undermarkings overlain with grey-brown or sepia dots, streaks, clouding and hair-lines, uniformly over shell or concentrated as ring at broad end. SIZE: (n = 122) 14·9–18·2 × 10·8–13·5 (16·5 × 12·4) (Maclean 1993).

LAYING DATES: SW Cape Prov., Feb–Nov, mainly June–July (Apr, May and Aug together 2/3rds of June/July peak; only a few clutches in Feb–Mar and Oct–Nov; in Hottentots Holland Mts breeds av. 2 months earlier (Apr–July) than in Cape Peninsula only 45 km to west (June–Sept)

INCUBATION: by ♀ only, starting as soon as last egg laid. ♂ stays nearby, singing frequently from prominent perch. Attentiveness (min per h in nest) averages 39 min in daylight hours, lowest (33 min) at 16h00–17h00, thereafter rising until dusk when ♀ settles in nest 10–20 min after sunset, for the night; first absence at 07h25, 20 min before sunrise; absences av. 6·7 min per h (Williams 1993a). When brooding, she occasionally utters short burst of high-pitched, warbling song, evidently in response to ♂'s song, and may then leave nest for a spell (Skead 1967). ♀ incubates facing entrance, and tail almost invariably becomes bent; it stays bent for most of time until young have flown. Period: 14–15 days, i.e. 14·5 days, 14 days 5 h, 14 ± 1 days, and 15 days 19 h. 3 eggs hatched between evening and early morning; 7 hatched in afternoon.

DEVELOPMENT AND CARE OF YOUNG: immediately upon hatching, young fed by ♂ and ♀; feeding rate per h remains constant throughout day, ♂'s rather lower than ♀'s. However, only ♀ broods young, for 1st week at constant rate of 20–30 min per daytime h, rather less (though still regularly) later on (Skead 1967). If ♀ is on nest when ♂ arrives with food, she comes off and waits nearby, or flies away to forage. When ♂ leaves she enters nest giving 'brooding' song (see Incubation). During first 5 days, faecal sacs are swallowed by parents, or a few carried away. After 5 days, sacs always carried away and dropped, by each parent at every 5–6th nest visit. Nestling period given as 15–22 (av. c. 19) days. Young leave nest with bills c. 5 mm shorter than adult length. For 5–15 days they return at dusk to sleep in nest, for first 5–7 evenings accompanied by ♀ who coaxes them back by flying ahead, calling 'seep, seep', going in and out of nest, and finally entering to sleep with them. Once ♀ and young roosted at night in a previous brood's nest. Young remain partly dependent on parents for about 3 weeks (Skead 1967).

BREEDING SUCCESS/SURVIVAL: of 93 nests, clutches hatched in 65 (70%), and 28 clutches (30%) either infertile or taken by predators or destroyed by mice which appropriated nests. Of 44 other nests with young, young fledged from 32 (72%). Of 164 eggs laid 65% hatched; of 78 nestlings 68% fledged; of 142 eggs laid 56 young fledged (overall success 39%: Broekhuysen 1963). Adult birds taken by Southern Boubous *Laniarius ferrugineus* and Rufous-chested Sparrowhawks *Accipiter rufiventris*. Lives for at least 5 years. Parasitized by Klaas's Cuckoo *Chrysococcyx klaas*.

References
Broekhuysen, G.J. (1963).
Harrison, J.A. *et al.* (1997).
Skead, C.J. (1967).
Williams, J.B. (1993a, b).
Wooller, R.D. (1982).

Genus *Cyanomitra* Reichenbach

8 quite large sunbirds, of evergreen lowland or montane forest and mature, dense humid savanna woodland; rather slim; bill quite long and curved, rather heavy; ♂♂ of 5 species with head and breast all or mainly iridescent blue-green, mantle olive (orange-olive in one, dark brown in another), pale yellow pectoral tufts, belly grey or yellow-olive; ♀♀ similar, but iridescent areas smaller or iridescence vestigial, and pectoral tufts absent. 3 further species have sexes alike; one (*C. veroxii*) with vestigial iridescence and scarlet pectoral tufts; 2 (*C. obscura*, *C. olivacea*) without iridesence, with olive plumage and yellow pectoral tufts (absent in ♀♀ of some races), ♂ larger and longer-tailed than ♀. 'Tck, tck' calls, wide range of other calls; warbling songs. Some species have lekking mating system. Nest a bulky, cryptic, bag made of vegetable matter, with conspicuous porch above side entrance and long beard hanging below; nest of 1 species 1 m long; spider web not used in construction.

One W-and-central African lowland superspecies (*C. verticalis*/*C. bannermani*); another superspecies (*C. obscura*/*C. olivacea*) is very widespread; 4 independent species, *C. cyanolaema* (rainforest zone), *C. oritis* (Cameroon montane), *C. alinae* (E Zaïre montane) and *C. veroxii* (coastal lowlands from Somalia to South Africa).

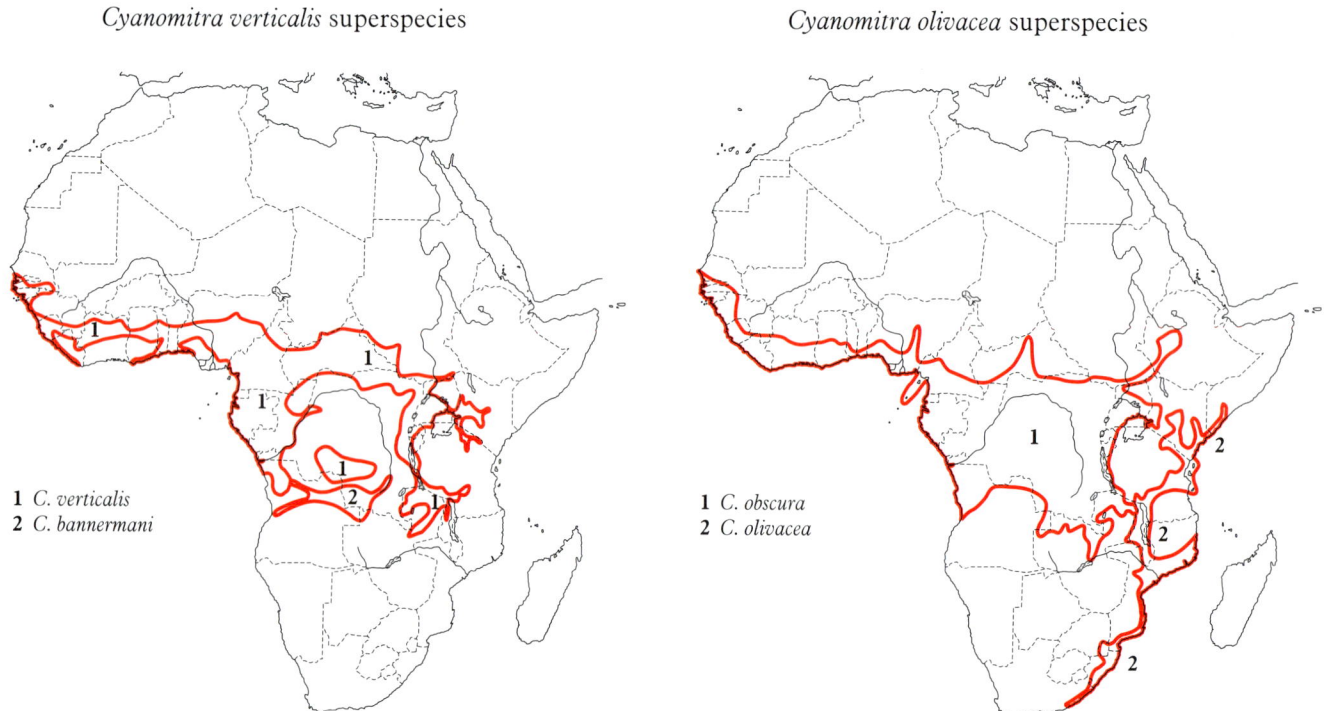

Cyanomitra verticalis superspecies

1 *C. verticalis*
2 *C. bannermani*

Cyanomitra olivacea superspecies

1 *C. obscura*
2 *C. olivacea*

Cyanomitra verticalis (Latham). Green-headed Sunbird. Souimanga à tête verte.

Certhia verticalis Latham, 1790. Index Orn., 1, p. 298; Africa (restricted to Senegal by Sclater and Mackworth-Praed, 1918, Ibis, p. 622).

Plate 10

(Opp. p. 143)

Forms a superspecies with *C. bannermani*.

Range and Status. Endemic resident, from Gambia to Kenya, south to N Angola and NE Zambia; partial migrant near N borders of W African range. Frequent in mature savanna woodlands surrounding rain forest zone. Gambia, frequent near coast, rare inland. Sénégal, rather uncommon in SW, quite common in SE, in Niokolo-Koba Nat. Park and Kédougou area; in dry season not north of 14°N, but moves to 15°N (Dakar) in rains (July). Widespread in Guinea-Bissau, frequent in W and S Guinea. Throughout Sierra Leone. Liberian records peripheral – along coast, in NW and NE (Gatter 1997). Ivory Coast to Nigeria, commonest in S Guinean savanna woodland zone, abundance decreasing sharply at rainforest boundary in S and markedly towards soudanian savanna boundary in N. In Ivory Coast not north of 09°N. Mali, records in wet season at Kangaba (11°56′N, 08°25′W), May, and Siby, June (Lamarche 1981). Burkina Faso, recorded in S (Dowsett 1993). Togo, along coast, and north to Landa-Pozanda (09°31′N). Benin, recorded. Nigeria, common in coastal mangrove and in S Guinean woodlands north of rain forest, less common north to about 11°N (Zaria) but still frequent in thick woodland at Danbagudu (10°N), Anara and Dunbi (10°50′N) (Fry 1975). Cameroon, very common and widespread in SW, frequent across middle belt on Adamawa Plateau; range limits near Poli, Djohong and Kombetiko. Central African Republic, sparse records north to Bamingui-Bangoran Nat. Park (Green 1983) and Manovo-Gounda-Saint Floris Nat. Park where breeds (Carroll 1988); south to Dzanga Res. (Green and Carroll 1991).

Sudan, common in SW and S in mature woodland and forest-edge habitats up to 3000 m. NE and E Zaïre, common and widespread; very common in Itombwe. In E Africa more a highland bird, fairly common at 1200–2400 m. In Uganda north and east to Victoria Nile and West Nile; also on Mt Lonyili in NE. In Kenya typically above 1700 m, sometimes down to 1500 m, throughout W and central highlands, from Mt Elgon east to E flanks of Mt Kenya. Rwanda, widespread in N, and in S occurs on W margin of Nyungwe forest hills up to 1900 m, and widespread in open forest in E Nyungwe and at Gisovu, up to 2500 m. Known from W Burundi. Tanzania, in N on Grumeti R. in Serengeti Nat. Park and small populations at Oldeani and Kilimanjaro; in W on Mt Mahari and highlands in NW Rukwa and W Kigoma Provs; and in S on Ufipa Plateau, Rungwe and Poroto Mts and Dabaga Highlands. Zambia, in Northern Prov. along Muchinga Escarpment to Musense, Lusiwashi R. and Mkushi R.; absent from Isangano Game Res. region; occurs from S Luapula and E Kawambwa through Mporokoso to Mbala and Isoka Districts, and on Nyika Plateau at 2000–2150 m; appears to replace Western Olive Sunbird *N. olivacea* between Mpika and Mkushi Districts and to be replaced by

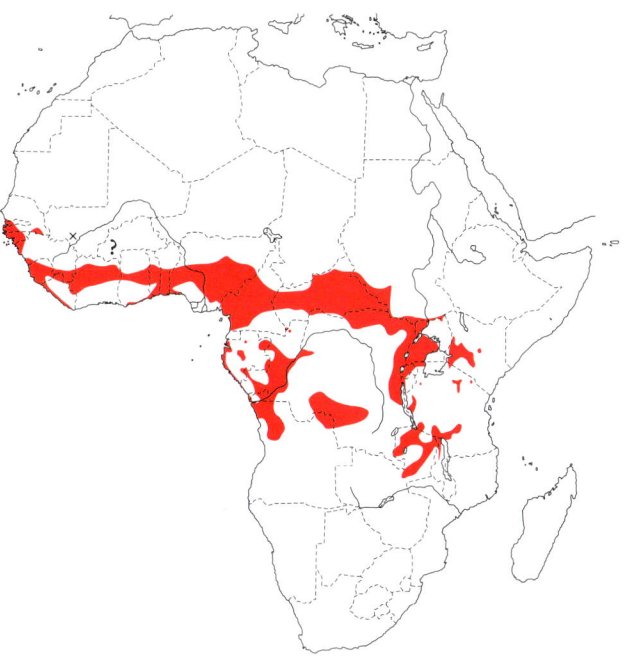

Cyanomitra verticalis

it at Kankonto, N Serenje. Malaŵi, Misuku Hills, E side of Mafinga Mts, Nyika Plateau and E Mzimba District, in riparian forest up to 1550 m and common in evergreen forest up to 2050 m.

Recorded in Mbini (Rand *et al.* 1959), and a subspecies is endemic and quite common along Gabon–Congo–Cabinda coast to south, to mouth of Congo River. Along Ogooué R. in Rés. de la Lopé, central Gabon; frequent and widespread in Ivindo Basin in NE Gabon, and in Odzala Nat. Park, adjacently in Congo. Elsewhere in Congo a single record in Nouabalé-Ndoki Nat. Park, and widespread south of equator. In Zaïre (other than NE and E – see above) occurs along Congo River south of equator, and widespread in Kasai Occidental, S Kasai Oriental and NW Shaba. Angola: Cabinda, Zaire Prov., Cuanza Norte, NW Malanje and NE Lunda Norte Province.

Density of *c.* 12 pairs per km^2 at M'Passa and between 15 and 20 pairs per km^2 at Bélinga, Gabon (Brosset and Erard 1986).

Description. *C. v. verticalis* (Latham): Sénégal to Cameroon, where intergrades with *bohndorffi* (Serle 1950, 1965). ADULT ♂: top and sides of head to hindneck and sides of neck metallic green with blue reflections; rest of upperparts non-iridescent olive-yellow. Tail graduated, dark grey-brown, T6 and tip of T5 slightly paler; outer webs of T1–T5 broadly edged olive-yellow. Lores sooty black. Chin to upper breast metallic greenish blue with violet reflections, merging with greener sides of head; pectoral tufts pale yellow; underparts otherwise dark ash-grey, paler on centre of belly and undertail-coverts. Upperwing dark grey brown; primaries edged olive-yellow proximal to emarginations, secondaries and tertials edged yellowish olive-green; alula and primary coverts narrowly fringed olive-yellow, rest of upperwing-coverts more broadly fringed yellowish olive-green. Axillaries and underwing-coverts ash-grey; inner borders of undersides of flight feathers pale buff. Bill black; eyes brown; legs dark greyish or black. ADULT ♀: differs from adult ♂ in having entire underparts, including chin and throat, pale ash-grey; no yellow pectoral tufts. Top and sides of head metallic green (less bluish). SIZE (10 ♂♂, 10 ♀♀): wing, ♂ 60–67 (64·9), ♀ 60–65 (61·9); tail, ♂ 43–47 (46·1), ♀ 40–43 (41·6); bill, ♂ 24–28 (26·5), ♀ 24–27 (24·8); tarsus, ♂ 17–19 (17·9), ♀ 17–18 (17·3). WEIGHT: Liberia, ♂ (n = 5) 10·4–13·5 (12·5), 1 ♀ 10·7; Ghana, ♂ (n = 3) 11·8–13·7 (12·9), 1 ♀ 13·5.

IMMATURE: juvenile has entire upperparts, including top and sides of head, dull yellowish olive. Chin and throat sooty grey; rest of underparts yellowish olive-grey, streaked yellow on centre of breast and belly; typically with narrow yellower band across upper breast, below throat patch. Birds in SW Cameroon have grey on head darker and more extensive than on those further west (Serle 1965). Wings and tail as in adult.

NESTLING: not described.

C. v. bohndorffi (Reichenow): Cameroon to Central African Republic, south to N Angola and Kasai, S Zaïre (but not along coast between Mbini and Congo River mouth). Mantle to uppertail-coverts slightly darker and greener than in nominate race, underparts darker grey. Slightly larger: wing, ♂ (n = 10) 64–70 (66·8); bill to skull, ♂ (n = 9) 26–29 (27·4), ♀ (n = 4) 24–27·5 (25·9). WEIGHT: Cameroon, 1 (sex?) 13·8, Angola, 1 ♂ 14·5.

C. v. cyanocephala (Shaw): coastal Gabon to mouth of Congo River. Mantle to uppertail-coverts greener than in *bohndorffi*; throat and upper breast glossed more violet, less steel-blue; underparts of ♂ more sooty grey.

C. v. viridisplendens (Reichenow): east of 25°/27°E: Sudan and NE Zaïre to Kenya and N Malaŵi. ♂ greener (less bluish) on chin to upper breast, without violet reflections; top of head bright metallic green without blue tinge. Mantle to uppertail-coverts duller and greener than in *bohndorffi*; underparts as in *bohndorffi*, but flanks usually tinged green in both sexes. Large: wing, ♂ (n = 10) 65–70 (67·8). WEIGHT: Kenya, ♂ (n = 33) 13–16·5 (14·9), ♀ (n = 23) 12–16 (13·5); Malaŵi, ♂ (n = 11) 12·9–15·4 (14·0), ♀ (n = 5) 11·8–13·8 (12·7) (Dowsett-Lemaire 1989b).

Field Characters. Length 13–14·5 cm. A widespread sunbird of forest and woodland, with bright olive-green upperparts and grey underparts. ♂ has iridescent *head and breast*, colour varying from green head and blue-green throat in E of range (*viridisplendens*) to blue-green head and blue-violet throat in W (nominate race), but colour also varies within each race. ♀ has *green confined to top* and sides of head, underparts entirely grey. Pectoral tufts creamy yellow in ♂, whiter in ♀. Immature has top of head and chin to throat sooty black, separated by narrow yellow band from dull olive underparts. ♂ most resembles Blue-headed Sunbird *C. alinae* of Albertine Rift Mts; they overlap altitudinally but apparently avoid each other; Blue-headed (both sexes) has violet throat, brown-orange tone to green upperparts, blackish belly with yellow wash on lower belly. ♂ Green-headed Sunbird said to look blue-headed in the field and easily mistaken for Blue-throated Brown Sunbird *C. cyanolaema* (Zimmerman *et al.* 1996) but this can only be true on a casual glance from below, when shared features of blue throat, grey underparts and creamy pectoral tufts are visible; on side view, Blue-throated Brown Sunbird seen to have brown face separating blue throat from blue crown, and brown, not green, upperparts.

Voice. Tape-recorded (102, 104, B, C, F, BRU, ERA, GREG, MOY). Song a brief (*c.* 1 s) sharp metallic chatter all on one pitch; intervals between songs filled with irregular light 'tip' notes. Another song described as prolonged twittering preceded by rapid 'chip' notes (Zimmerman *et al.* 1996). Said to mimic calls of other species (Mackworth-Praed and Grant 1973). Calls include a sharp 'tsip' or louder 'tsyip', also 'whew', 'tzit, tzit' and a plaintive mewing 'chiuwee' or 'tseea-wee, cheea-weu' (Zimmerman *et al.* 1996), and a 'tchouee' very like that of Reichenbach's Sunbird *Anabathmis reichenbachii* (Christy and Clarke 1994). Alarm a harsh 'chaa' or 'pss-chaa' with rather tit-like quality.

General Habits. Inhabits forest-savanna mosaic, mature *Isoberlinia* savanna woodland, evergreen forest and bamboo forest edges and clearings, coastal scrub, mangrove, gardens, cultivation, banana plantations, riverine forest, secondary forest, well-wooded watercourses, and gallery forest in savanna woodland. In NE Gabon inhabits primary forest as well as secondary formations. Occurs solitarily, in pairs and family parties, sometimes joining mixed foraging flocks, in mid levels of trees and in top of forest canopy and crowns of tallest emergents about 50 m high, but also comes down to within 2 m of ground (Brosset and Erard 1986). Gleans foliage for spiders and insects, catches insects on the wing in short chase, sometimes hovers when gleaning, and takes insects from flowers (Dowsett-Lemaire 1983), feeding at them from in front. Not known to pierce corollas in Malaŵi (Dowsett-Lemaire 1989b) but in Kenya often pierces base of long tubular corollas (Zimmerman *et al.* 1996), doubtless to take nectar; forages in canopy, liana tangles and thickets in clearings (Dowsett-Lemaire 1990).

Food. Thought to be more insectivorous than nectarivorous. Insects and 'one large hard seed' in stomachs (Liberia). 3 stomachs in Uele, Zaïre, contained only small spiders, and one only small insects; a ♀ was taken carrying 8 small spiders in bill to her young (Chapin 1954). Insects in all 30 stomachs examined in Cameroon (Serle 1965). Drinks oil-palm *Elaeis guineensis* sap (from which palm wine made) (Morel and Morel 1990); often feeds at flowers of pawpaw *Carica papaya*, banana and (Gabon) *Baikiaea*. Nectar obtained from relatively unspecialized flowers, in Malaŵi of *Bersama abyssinica*, *Bridelia brideliifolia*, *Halleria lucida*, *Syzygium cordatum*, *Canthium gueinzii*, *Clematis brachiata*, *Clerodendrum johnstonii*, *Dendrophthoe pendens*, *Phragmanthera usuiensis*, *Tapinanthus sansibarensis*, *T. subulatus*, *Hypericum revolutum*, *Leonotis decadonta*, *L. mollissima* and *Tecomaria capensis* (this list 'certainly incomplete': Dowsett-Lemaire 1989b), in Rwanda of *Alangium*, *Clematis*, *Ixora*, *Syzygium*, *P. usuiensis*, *T. brunneus* and *T. constrictiflorus*, and in Congo of *Symphonia globulifera* (F. Dowsett-Lemaire, pers. comm.). In NE Gabon feeds at inflorescences of several wild and cultivated plant species; seen eating fruit of *Xylopia aethiopica* and capturing small orthopterans, caterpillars and spiders (Brosset and Erard 1986).

Breeding Habits. Territorial, solitary nester. Assemblies of up to 30 ♂♂ (rather like those of Blue-throated Brown Sunbird, *q.v.*) observed in one particular fruiting *Xylopia aethiopica* tree in NE Gabon in Feb (Brosset and Erard 1986): each ♂ occupied a sector 2–4 m in diam. and actively defended it against other ♂♂ by singing and moving rapidly at its perch in an intimidating posture with pectoral tufts standing out, tail raised and bill pointing at adversary. ♂'s behaviour to any ♀ showing interest was quite different: ♂ lifted head, lowered tail and made series of rapid wing-flaps. Pair occupies same territory for at least 3 seasons (Malaŵi, Dowsett 1985); territory is often single forest patch or forest-island, 11 such patch territories being of 0·5–10 ha one year (10 being clustered together) and several of them being occupied the next year (Dowsett-Lemaire 1983). Does not defend clumped flowers persistently when breeding (as some other sunbirds do: Dowsett-Lemaire 1989b).

NEST: bulky, untidy, suspended globe, like large nest of Western Olive Sunbird, with side entrance and a ragged, wispy beard up to 30 cm long hanging below entrance and under floor of nest chamber. Made of dead leaves and bits of dry bark and fibres (including those of banana plants) bound with thin torn-off strips of fibrous bark, lined with fine fibres and grass. Sometimes nest roof slightly overhangs entrance. Hung from spiny liana or small tree or bush, sometimes well screened behind foliage, at heights of 1·5–12 m above ground; 4 nests in man-made structures (grass huts etc.: T. Butynski, pers. comm.).

EGGS: 2. Pale pink, sparingly marked with small spots and short lines of deep brown on top of pale lilac-grey blotches. Size: (n = 2) 18 × 13·5, 18·5 × 13·5, and (n = 2) av. 19 × 14.

LAYING DATES: Liberia, (gonads enlarged and birds in fresh plumage, Feb–Mar); Nigeria, (nest-building Mar, adult carrying food, Aug); Cameroon, Apr (and pair in breeding condition, May); Gabon, Nov-Dec (and ♀ collecting spider web, Mar); Zaïre (Uele), (breeds in May–Aug; enlarged gonads in Jan); (Itombwe), (various evidences Mar, Apr, July, Oct, Dec; and juveniles in all months, so thought to breed year-round); E Africa, Region B, Mar–Oct (23 clutches, mainly in long rains in Apr and between rains in July–Sept); Zambia, Jan, Mar, May (and large gonads Apr and Oct); Malaŵi, Dec–Feb and Apr.

DEVELOPMENT AND CARE OF YOUNG: young fed by both parents.

References
Brosset, A. and Erard, C. (1986).
Dowsett-Lemaire, F. (1989b).
Serle, W. (1965).

Cyanomitra bannermani Grant and Mackworth-Praed. Bannerman's Sunbird. Souimanga de Bannerman.

Plate 10 (Opp. p. 143)

Cyanomitra verticalis bannermani Grant and Mackworth-Praed, 1943. Bull. Br. Orn. Club, 63, p. 63; Kayoyo, southern Belgian Congo.

Forms a superspecies with *C. verticalis*.

Range and Status. Endemic resident, about 10°S in Angola, Zaïre and NW Zambia. Angola, single specimens from Malanje and Luacano (Traylor 1962), near border between Cuanza Sul and Benguela Districts, and between Lunda Norte and Lunda Sul near Cacolo (Hall and Moreau 1970). Zaïre, not uncommon in 'Upper Katanga' (= W Shaba, east to Lualaba R. valley) (Chapin 1954). Zambia, common in N Mwinilunga up to at least 1260 m, south to Mundwiji Plain and Matonchi (Benson *et al.* 1971).

Description. ADULT ♂: top and sides of head to hindneck and sides of neck metallic greenish blue; rest of upperparts olive-green. Tail graduated, dark grey-brown, tip and outer web of T6, tip of T5 and narrow tips of T3–T4 paler grey-brown; outer webs of T1–T5 edged yellowish olive-green. Lores black. Chin, throat and upper breast metallic greenish blue, merging with sides of head; throat and upper breast tinged violet; pectoral tufts pale yellow; rest of underparts ash-grey, darker on lower breast and upper flanks. Upperwing dark grey-brown; primaries edged dull yellowish, secondaries and tertials edged yellowish olive-green; alula and primary coverts fringed yellowish olive-green; rest of coverts broadly fringed olive-green. Axillaries, underwing-coverts and inner borders of flight feather undersides pale grey. Bill black; eyes dark brown; legs black. ADULT ♀: top and sides of head to hindneck and sides of neck dark grey-brown with only

Cyanomitra bannermani

faint iridescent bluish tips to feathers; throat and breast dark grey, not iridescent; otherwise like ♂. SIZE (4 ♂♂, 5 ♀♀): wing, ♂ 69–70 (70·0), ♀ 64·5–66 (65·2); tail, ♂ 48–51 (49·5), ♀ (no data): bill, ♂ 21–24 (22·5), ♀ 21·5–23·5 (22·75); tarsus, ♂ 18–20 (18·3), ♀ (no data). WEIGHT: 1 ♀ 14·1.

IMMATURE and NESTLING: not known.

TAXONOMIC NOTE: '*Nectarinia sororia*' Ripley 1960 is now known to be ♀ *C. bannermani* (R.A. Cheke, pers. comm. 1998).

Field Characters. Length 12·5–13 cm. Very similar to Green-headed Sunbird *C. verticalis* but with restricted range; they apparently meet in NE Angola. Slightly smaller and bill a trifle shorter and less curved. Head and breast of ♂ duller and bluer than average ♂ of local Green-headed Sunbird *C. v. viridisplendens* but yet within colour range of that species; blue on breast less extensive. ♀ more distinctive, with top and sides of head mainly grey brown, blue iridescence confined to tips of a few feathers; however, not all ♀♀ of Green-headed Sunbird have solid green head tops – in some the green colour is spotty. In most places the 2 can be told apart by range.

Voice. Tape-recorded (B, CART). Call a nasal, up-slurred 'djoowi'.

General Habits. Inhabits edges and middle strata of moist evergreen forest, small and isolated swamp forests and gallery relics, and penetrates short distance into adjacent miombo woodland (Zambia); strips of heavy woods along streams, and fringes of riparian cover (Zaïre). Habits have barely been reported but appear not to differ from those of Green-headed Sunbird *N. verticalis*.

Food. One bird had eaten a small and a large spider and 30 small winged termites ?*Pseudacanthoptermes* (Oatley 1969).

Breeding Habits.

NEST: only one found (Colebrook-Robjent 1990). Suspended, broad and bulky bag, with porch above entrance; made of rather long, coarse grass stems, many extraneous to construction of walls proper and hanging well below nest to form 'beard'; roof made of fine, black fibrous stems and porch of fine, twisted stems. Spider silk not used. Nest lined with soft, fine, twisted stems. Outside adorned with curled strips of banana bark on each side and large leaves, mainly at back. Ext. height, from point of attachment on vine to base of nest, 180, with beard a further 160; ext. diam. 100 across and 90 from front to back, with 'extraneous' material extending width to 190; entrance diam. 42; lower lip of entrance 84 from base of chamber; int. height *c*. 90, int. width *c*. 40. Nest was sited near ground in narrow gallery forest with abandoned banana grove nearby, and suspended from drooping, leafy vine 1·2 m above sluggish stream, well hidden in and deeply shaded by overhanging vegetation.

EGGS: 2. Regular ovals, without gloss. Warm, pink-brown ground, densely marked all over with irregular, dark purplish brown streaks, giving general smudgy puce or dull claret appearance. SIZE: (n = 2) 19·6 × 13·9 and 18·3 × 13·6. Weight: (both eggs together, with small embryos) 3·9 g.

LAYING DATES: Zambia, Sept (and gonads active Mar–Sept).

INCUBATION: ♀ once found on eggs at night.

Reference
Colebrook-Robjent, J.F.R. (1990).

Plate 10 (Opp. p. 143)

Cyanomitra cyanolaema **(Jardine and Fraser). Blue-throated Brown Sunbird. Souimanga à gorge bleue.**

Nectarinia cyanolaemus Jardine and Fraser, 1851. Contrib. Orn., p. 154; Clarence, Fernando Po.

Range and Status. Endemic resident, rainforest zone of W Africa and Congo Basin. Guinea, records only at Nzérékoré (07°45'N, 08°49'W) in 1930s and 1950s. Sierra Leone, uncommon in Freetown Peninsula (Field 1974); old records from Kailahun and Sandaru (Konno district). Liberia, uncommon to common; old records from Stampfli in W and Gedetabo in E; common on Mt Nimba. Ivory Coast, abundant, north locally as far as Lamto and Daloa; throughout Taï Nat. Park and in forests at Téné, Yapo, Irobo and Mopri. Ghana, not uncommon in forest, e.g. Kakum, Tafo and Mpraeso, and in wet season extends north into woodland savanna at Tamale and Mole (Harvey and Harrison 1970). Togo, single record, Djodji. Benin, old record, requires confirmation (Dowsett 1993). Nigeria, not uncommon, north to Ibadan, Ife and Mamu Forest. Bioko, uncommon (or uncommonly noted): Pico Basilé, Luba to Moka, Malabo, Ureca; up to 700 m. Cameroon, frequent in S, north to about 6°N. Central African Republic, 2 ♂♂ collected near Baroua (Friedmann 1978). Gabon and Congo, common and widespread throughout (in Congo north at least to Nouabalé-Ndoki Nat. Park). Angola, Cabinda, Zaire Prov., Cuanza Norte and N Lunda Provs. Zaïre, common and widespread, almost throughout, but apparently absent from N and SE. Record from near junction of Central African Republic, Zaïre and Sudan (upper Bomu R. or Bengengai Forest?: Hall and Moreau 1970). Rwanda, Nyakabingo Valley, Nyungwe, at 1900–2000 m (Dowsett-Lemaire 1990). Uganda, locally common in W and S at 700–1700 m: West Nile, Budongo, Bugoma, Bwamba, Lutoto, Kalinzu, Maramagambo and Bwindi Forests, east along N shore of L. Victoria to Jinja and Buvuma I. (Kenya, single old record, now rejected, in Kakamega Forest.) Tanzania, frequent in Gombe Stream Game Reserve.

Description. *C. c. octaviae* (Amadon): Ghana to Uganda and Angola. ADULT ♂: forehead and forecrown dark metallic blue;

Cyanomitra cyanolaema

rest of upperparts dark brown with slight greenish tinge. Tail dark brown, tinged greenish above, T5–T6 with paler tips. Lores sooty brown; ear-coverts, cheeks and sides of neck dark brown, concolorous with upperparts. Chin to throat dark metallic blue. Upper breast brown, grading to pale greyish brown on flanks, belly, thighs and undertail-coverts; pectoral tufts pale yellow. Upperwing feathers dark brown. Axillaries, underwing-coverts and inner borders of undersides of flight feathers pale brownish grey. Bill black; eyes dark brown; legs black. ADULT ♀: top of head dark brown, streaked olive; hindneck to mantle and scapulars yellowish olive-green, mottled dark brown; back, wings and uppertail-coverts plain yellowish olive-green. Tail dark brown, feathers tipped pale greyish, broadly on T5–T6, and edged olive-yellow. Narrow whitish superciliary stripe; lores and ear-coverts dark brown; cheeks and sides of neck paler grey-brown. Chin and throat whitish with variable brown suffusion and mottling; breast whitish, tinged olive-yellow, with diffuse dusky mottling and streaking; belly, flanks, thighs and undertail-coverts pale olive-yellow, streaked greyish. Upperwing feathers dark brown, flight feathers edged olive-yellow, tertials and greater, median and lesser coverts broadly fringed olive-yellow. Axillaries white; underwing-coverts white, tinged yellow near bend of wing; inner borders of undersides of flight feathers greyish white. Bare parts as in adult ♂. SIZE (10 ♂♂, 10 ♀♀): wing, ♂ 65–73 (68·4), ♀ 63–67 (65·0); tail, ♂ 50–61 (54·9), ♀ 44–50 (46·7); bill, ♂ 22–26 (24·0), ♀ 22–24 (23·2); tarsus, ♂ 16–17 (16·6), ♀ 16–17 (16·5). Wings in Ituri (♂♀, n = 9) 64–71 but in Nigeria 63–69 (Chapin 1954). WEIGHT: Uganda, ♂ (n = 16) 14·5–20 (16·9), ♀ (n = 12) 13·5–20 (17·0); 2 ♂♂ 15, 16; 1 (sex?), Cameroon, 16·0.

IMMATURE: juvenile similar to adult ♀, but upperparts more uniform dark olive-brown; chin to breast dusky brown.

NESTLING: not described.

C. c. cyanolaema Jardine: Bioko. Like *octaviae*, but ♀ greener on upperparts and wing edgings. Wing averages *c.* 3 mm longer than in Nigerian population of *octaviae*.

C. c. magnirostrata (Bates): Sierra Leone to Ivory Coast. (Since there is a break in the range of the species between Ghana and Nigeria, Ghanaian birds, usually ascribed to *octaviae*, are likely to be *magnirostrata*.) Like *octaviae*, but bill slightly longer and slightly less decurved: ♂ (n = 10) 24–27 (25·7). WEIGHT: Liberia, ♂ (n = 12) 13·9–18·8 (15·3), ♀ (n = 7) 14–16·6 (15·0).

Field Characters. Length 14–15 cm. A rather dark sunbird of lowland forest canopy; iridescence dull and only visible in good light. Field impression is big, rangy and long-billed (D. Roberson, pers. comm.). ♂ sooty brown with small patch on chin and throat and round patch on forecrown iridescent blue-green; pectoral tufts creamy. Dark head contrasts with lighter upperparts (Christy and Clarke 1994). Told from ♂ Green-headed Sunbird by *brown* upperparts and much *smaller* area of iridescence on throat, separated from *blue* crown patch by brown face. ♀ even more distinctive: no iridescence on dark head, *white stripes* above and below eye, underparts *white* with *brown blotches*, vent yellow. Has shorter bill and longer tail than Blue-headed Sunbird *C. alinae*, which it meets in transition forest of Albertine Rift mountains. Blue-headed has entire head violet-blue and orange-brown upperparts.

Voice. Tape-recorded (104, B, CHA, ERA, LEM). Song very brief (<1 s, usually 0·5 s), a descending trill of 4–7 high, sharp metallic notes, varying in speed, notes sounding either well-spaced or run together; same individual sings both speeds, often alternately; given in flight or when moving around in canopy (Christy and Clarke 1994). Quality of song not unlike that of Western Olive Sunbird *C. obscura*. Calls include a sharp sibilant 'tsit' and a rapid but irregular series of 'chip' notes.

General Habits. Occupies all types of forest at all strata, but particularly characteristic of tall, dense lowland forest, where usually the commonest sunbird species in canopy. In tall primary lowland forest, keeps mainly in general canopy and in tops of emergent trees, but freely comes down to forage in wind-fallen trees, in clearings and at forest edges. Inhabits old, regenerating and mature secondary growth near lowland and montane forests, up to about 1200 m (Liberia), 1350 m (Rutshuru Valley, E Zaïre), 1700 m (Uganda, Kenya) and 2000 m (Rwanda); abundant in coastal woodlands and coconut palm groves (Ivory Coast); occurs in gallery forests in savanna country, perhaps seasonally (Aug, Ghana); moss forest and de-graded forest (Bioko); occasionally in gardens; often visits flowering trees and shrubs but by no means dependent upon them; also forages commonly at epiphytes and vines (Liberia) and in fruiting trees. Perches on twigs diam. (n = 40) 5–10 mm (Gatter 1997). Flowers visited include bright red ones, where bird often gets forehead dusted with yellow pollen; forehead also tends to harbour mites. Defends parts of the canopy against conspecifics and congeners (Dowsett-Lemaire and Dowsett 1991). Takes insects by gleaning and hover-gleaning; plucks spiders from their webs (Gatter 1997).

Active and noisy; constantly on the move. Lives in pairs and family parties; regularly joins mixed flocks of insectivores. Territorial in flowers of forest canopy (Gabon: D. Roberson, pers. comm.). Assemblies of singing ♂♂ occur at start of rainy season (Gabon): generally 4–7 birds but once 40–45, in a forest emergent tree, often on dead branches, the birds quite close to one another.

Assembles sometimes in flowering or fruiting *Xylopia* tree (Brosset and Erard 1986).

Food. Insects, small spiders, hard seeds like grape pips, and what appear to be fragments of flowers in stomachs (Bates 1909); birds seen taking caterpillars and spiders. Seeds from small, soft fruits, and arillate seeds from hard capsules, e.g. *Xylopia aethiopica*. Eats fruits of *Macaranga assas*, *M. barteri* and *Musanga cecropioides* (Dowsett-Lemaire 1996). Seen feeding in *Caesalpinia* and *Erythrina* flowers (Nigeria) and *Albizia gummifera* (Rwanda). Eats fruits of *Tetrorchidium disymostemon* and attends flowers of *Combretum platypterum* (F. Dowsett-Lemaire, pers. comm.), *Pentadesma*, *Maranthes* and *Anthocleista* (Gabon: Christy and Clarke 1994).

Breeding Habits. Solitary nester, monogamous and territorial. Assemblies of singing ♂♂ (see above) have all the elements of a lek: sites are quite small and appear to be used constantly (M'Passa, Gabon, Brosset and Erard 1986); birds hop excitedly from branch to branch and sing aggressively at each other with pectoral tufts exposed, breast inflated, head thrust forward, bill wide open, wings drooping, tail depressed and fanned. ♀♀ regularly visit periphery of assembly area, to watch but not to participate nor copulate.

NEST: an astonishing construction up to 1 m long, looking like a mass of vegetable débris caught up at end of liana (Chapin 1954, Brosset 1974, Field 1974), or rope-like with a bulge half way down, an untidy jumble of dead leaves, short dry twigs, weed stems and plant litter, held together by tangled, interlacing threads of *Marasmius* and spider web, with trailing ends at sides and below (**A**). Bird seems to select twig with *Marasmius* fungus growing out of it, and fungal threads (hyphae) may continue to grow after twig is in wall of nest (Bates 1909). Somewhere in the middle or two-thirds of the way down is a typical sunbird nest, a deep cavity, lined with fine shreds of bark or stuffed with mammal hair (particularly of Gorillas *Gorilla gorilla* in Gabon), and with a little of the same material projecting above oval entrance. Suspended from side of thorny shrub, end of hanging vine, often thorny, or of spiny rattan palm frond, between 1·8 and 12 m above ground in clearing or windfall gap, often over a path or open spot in forest, or overhanging edge of motor road or river.

EGGS: 2 (n = 5 clutches). Rather long; not glossy. Colours vary: pink with large, contrasting reddish spots; violaceous grey, heavily spotted with dark brown; cream, nearly covered by obscure dark purplish grey with heavy dark brown speckles and flecks; or buff, almost obscured by dense mottling of several shades of dark brown. SIZE: (n = 4) 18–20 × 13–14 (19·4 × 13·75).

LAYING DATES: Liberia, (nestlings being fed Dec and Feb–Mar; enlarged gonads, Aug–Oct); Nigeria, (fledgling Aug, ♀ at nest Nov); Cameroon, June–July; Gabon, Dec–Apr; Zaïre (Ituri), (evidence that it breeds all year).

INCUBATION: by ♀. Period: 14 days.

Reference
Brosset, A. and Erard, C. (1986).

Plate 10
(Opp. p. 143)

Cyanomitra oritis (Reichenow). Cameroon Sunbird. Souimanga à tête bleue.

Cinnyris oritis Reichenow, 1892. J. Orn., 40, p. 191; Buea, Mt Cameroon.

Range and Status. Endemic resident, Cameroon montane area. Nigeria, not uncommon resident on Obudu Plateau and frequent in Gashaka-Gumti Nat. Park (07°20'N, 11°35'E on Mambilla Plateau). Cameroon, common on Mt Cameroon at 570–2300 m and (another race) highlands to the north and northeast: Dikome Balue in Rumpi Hills at 1000–1700 m, Mt Nlonako at 1650–1700 m, Foto, Fotabong, Mt Kupé, Mt Manenguba, Bamenda, L. Bambili, L. Bambulue and Bafut-Ngemba Forest Res., Sabga Pass, forest patches between Mbengwi and Tinachong at 1700–1800 m, Bamboutos Mts at 2000–2600 m, Ndu, Mt Oku, Mt Lefo and Banso Hills (but probably not

Cyanomitra oritis

Tibati plateau as earlier claimed: Stuart and Jensen 1986). Bioko, frequent (the commonest sunbird between 1200 and 1600 m) on Pico Basilé at 1200–2000 m, at Moka (1300 m), and in S highlands from 800 m into Caldera of Luba, above Ureca and up to 1800 m around Biao Lake (Pérez del Val *et al.* 1994). Reported in Equatorial Guinea (Fa 1990), almost certainly in error (Dowsett 1993). On Mt Cameroon abundance increases with altitude and, with records down to 570–670 m in June–Sept, species may be a partial vertical migrant (Serle 1964).

Description. *C. o. bansoensis* (Bannerman): W Cameroon highlands (not Mt Cameroon). ADULT ♂: top and sides of head metallic bluish green; hindneck, sides of neck and rest of upperparts non-iridescent olive-green. Tail graduated, dark grey-brown; outer web and tip of T6 and narrow tips of T1–T5 paler, more olive; both webs of T1 and outer webs of T2–T5 edged dark olive-green. Lores black. Chin, throat and centre of upper breast metallic blue, tinged violet, merging with greener sides of head; pectoral tufts pale yellow; otherwise sides of breast and rest of underparts olive-yellow. Upperwing dark grey-brown; primaries narrowly edged yellowish olive-green, secondaries and tertials more broadly edged olive-green; primary coverts finely edged and greater, median and lesser coverts broadly fringed olive-green. Axillaries pale yellow; underwing-coverts creamy; inner borders on undersides of flight feathers buffish white. Bill black; eyes dark brown; legs black. ADULT ♀: like adult ♂, but glossy blue on upper breast slightly more restricted and less violet tinged; lacks yellow pectoral tufts. SIZE (10 ♂♂, 10 ♀♀): wing, ♂ 58–63 (61·5), ♀ 51–60 (55·8); tail, ♂ 39–48 (42·7), ♀ 35–38 (36·6); bill, ♂ 27–29 (27·9), ♀ 25–28 (26·2); tarsus, ♂ 18–21 (19·5), ♀ 17–20 (18·6). WEIGHT: ♂ (n = 73) 9·9–13·3 (12·0), ♀ (n = 41) 8·7–11·9 (10·4) (C. Bowden, pers. comm.).
IMMATURE: juvenile differs from adult in having top and sides of head dull olive; chin to upper breast sooty grey; rest of underparts dark olive-green, streaked yellowish on centre of breast and belly.
NESTLING: not known.

C. o. oritis (Reichenow): Mt Cameroon. Like *bansoensis*, but top and sides of head bluer (less green), chin to upper breast metallic violet-blue. Bill and wing longer but tail shorter than in *bansoensis*: wing, ♂ (n = 3) 62–65 (63·3), 1 ♀ 57; tail, ♂ (n = 3) 38–40 (39), 1 ♀ 34; bill, ♂ (n = 10) 31–33 (31·8).

C. o. poensis (Alexander): Bioko. Metallic feathers of top and sides of head and chin to upper breast dull steely green, which colour extends to hindneck, sides of neck and further onto breast.

TAXONOMIC NOTE: regarded by Hall and Moreau (1970) and several later authors as belonging to the *C. verticalis*/*C. bannermani* superspecies. It has considerable altitudinal overlap with *C. verticalis*, and we consider its plumage differences too great for inclusion in the superspecies.

Field Characters. Length 11·5–13 cm. Endemic to Bioko and highlands of Cameroon and E Nigeria. Bill long and curved. Top and sides of head iridescent green (*bansoensis*) or blue (other races), chin to breast violet-blue. Sexes alike except only ♂ has pale yellow pectoral tufts. A highland species, the only blue-headed sunbird present in most of its range, usually altitudinally separated from Green-headed and Blue-throated Brown Sunbirds *C. verticalis* and *C. cyanolaema*, but may meet them around the 1000 m level; distinguished from both by *purple* throat and *olive-yellow* (not grey) underparts.

Voice. Tape-recorded (104, B, CHA, LEM, PER). Song a rapid, metallic jingle lasting *c*. 2 s, ending with 3–4 distinct 'jee' notes followed by a swirl, 'jee-jee-jee-tiddliddle'; often preceded by a soft 'tip'. 2 other song types described (Serle 1954): a single note repeated vehemently without pause for a min or more, the pitch altering slightly every few s; and a very sweet, sustained warbling subsong audible at range of only a few feet. Calls include a 3-note 'titi-seee', emphasis on last note, a soft, unmusical 'tick tick tick' given by both sexes, and (alarm?) a nasal 'jeep'.

General Habits. Inhabits undergrowth in interior of forest, especially along streams, also forest clearings and edges. Solitary or in pairs, but a tree in flower sometimes attracts large numbers (Serle 1950). An unobtrusive, insectivorous sunbird, normally rather quiet but with a range of vocalizations including 3 song types.

Food. Mainly insects (Serle 1965), also nectar and small bits of plant material. Regularly visits flowers in understorey, of *Cephaelis mannii* and other Rubiaceae, and *Leea guineensis* (Leeaceae) (F. Dowsett-Lemaire, pers. comm.).

Breeding Habits. Solitary nester, evidently monogamous.
NEST: only 2 known. One (Serle 1965) was a loose, bulky, pear-shaped, suspended bag with a 'beard' of dry grasses trailing down from underside and with entrance at one side overhung by a porch 50 mm long. Nest made of broad, dry grass blades bound together with rootlets, fine tendrils and moss, and lined with fine fibres and, at bottom, some plant down; much fresh green moss around point of attachment. Hung from end of pendent spray of fern, 1 m above ground; well concealed behind screen of ferns drooping over rocky bank by a stream. Other nest was almost identical in construction and site (Stuart and Jensen 1986).

EGGS: 2 (in one nest; 1 nestling in other). Ovate, smooth, without gloss; whole surface so thickly and finely speckled with brown as to be almost uniform, but some obscure darker mottling about large end (Serle 1965).

LAYING DATES: Nigeria, (adults feeding juveniles Apr); Cameroon, Nov, Mar (and breeding indications Nov, Dec, Jan, Mar, Apr and July).

References
Stuart, S.N. and Jensen, F.P. (1986).
Serle, W. (1965).

Plate 10
(Opp. p. 143)

Cyanomitra alinae Jackson. Blue-headed Sunbird. Souimanga d'Aline.

Cyanomitra alinae Jackson, 1904. Bull. Br. Orn. Club, 14, p. 94; Ruwenzori.

Range and Status. Endemic resident, Albertine Rift mountains from Lendu Plateau to Marungu Highlands. In Zaïre, ranges in mountains of E borders, from Mt Wago and Aboro in Lendu Plateau northwest of L. Albert, Rwenzori Mts across Zaïre–Uganda frontier, Volcano region across Zaïre–Uganda–Rwanda borders, mountains west of L. Edward and L. Kivu, Itombwe Plateau, Mt Kabobo and Marungu Highlands. Uganda, quite common in Rwenzoris, Impenetrable-Bwindi Forest and Ankole. Rwanda, common throughout Nyungwe (except for part of Crête Zaïre-Nil), commonest in W (Dowsett-Lemaire 1990), and on Volcanoes and Gishwati mountains in NW. Burundi, abundant in forested massifs from Rwanda frontier south to Mt Teza and Bugarama (03°20′S, 29°32′E) (Gaugris et al. 1981). Range altitudes: Lendu Plateau 1750–2400 m, Rwenzoris 1680–3280 m, west of L. Edward 1550–2700 m, Bwindi Forest 1600–2400 m, Volcanoes 2300–2500 m, west of L. Kivu 1700–2550 m, Nyungwe 2300–2400 m, Itombwe 1320–2340 m, Kabobo 1980–2480 m, and Marungu 1660–1710 m (Prigogine 1975).

Reaches highest density where main food plant *Balthasaria* is common: 23 birds caught along 220 m of nets at 2350–2500 m at Uwinka, Rwanda (Dowsett-Lemaire 1990).

Description. *C. a. alinae* Jackson: Uganda and NW Rwanda. ADULT ♂: top and sides of head to hindneck and sides of neck deep metallic blue-green with violet reflections; mantle and back bright orange-washed green; scapulars, rump and uppertail-coverts olive-green. Tail slightly graduated, blackish brown, tips of T5–T6 paler, more olive; feather outer webs edged olive-green. Lores sooty black. Chin to upper breast metallic violet-blue, merging with greener sides of head; pectoral tufts yellow; lower breast, upper flanks and belly dark sooty brown, merging with dull olive-green lower flanks, thighs and undertail-coverts. Median and lesser coverts bright metallic green; upperwing otherwise blackish brown, primaries edged olive-yellow, secondaries and tertials edged olive-green, alula, primary coverts and greater coverts fringed olive-green. Axillaries, underwing-coverts and inner borders of flight feather undersides greyish white. Bill black; eye reddish brown to dark chestnut; legs black or olive-black. ADULT ♀: like adult ♂, but crown glossy green (lacking blue lustre), and lacks pectoral tufts. SIZE (10 ♂♂, 4 ♀♀): wing, ♂ 59–68 (64·0), ♀ 62–64 (62·8); tail, ♂ 37–46 (42·5), ♀ 27–30 (28·8); bill, ♂ 28–31 (29·7), ♀ 27–30 (28·8); tarsus, ♂ 18–19 (18·6), ♀ 18–19 (18·5). WEIGHT: ♂ (n = 5) 12–14·5 (13·5), ♀ (n = 8) 10–13 (11·3).

IMMATURE: juvenile has entire upperparts dull olive-green, with orange tinge on mantle; below, sooty grey with blackish barring from chin to upper breast, grading to dull greenish grey on lower flanks, belly and undertail-coverts.

Cyanomitra alinae

NESTLING: unknown.

C. a. derooi Prigogine: NE Zaïre, west of L. Albert and L. Edward, from 02°N to 00°30′S. ♂ with crown darker green than in nominate race, less bluish; chin to upper breast blue, not violet-blue; tail (n = 26, west of L. Edward) 3·0–3·5 shorter than in *alinae* (Prigogine 1975).

C. a. tanganjicae (Reichenow): E Zaïre from 01°30′S (L. Kivu) to 04°S, SW Rwanda and W Burundi. ♂ crown lustrous bottle-green, not blue-green; thereby less difference between the sexes than in nominate race. Weight: ♂ (n = 23) 11–16 (13·5), ♀ (n = 23) 10–15 (11·7).

C. a. kaboboensis Prigogine: Mt Kabobo, Zaïre. ♂ with crown and breast blue with green reflections; belly dusky grey; large. Size (9 ♂♂, 4 ♀♀): wing, ♂ 65–69 (67·4), ♀ 58–62 (60·8); tail, ♂ 46–49·5 (47·5), ♀ 36–40 (38·5); bill, ♂ 31–33 (32·2), ♀ 31·5–32 (31·6).

C. a. marungensis Prigogine: Marungu Mts, Zaïre. Belly clear grey (not dark sooty brown); bill 3–4 mm shorter and less robust than in *kaboboensis*.

TAXONOMIC NOTE: regarded by Hall and Moreau (1970) and several later authors as belonging to the *C. verticalis/C. bannermani* superspecies. It has considerable sympatry and altitudinal overlap with *C. verticalis*. A curious, short-tailed ♂ bird from Kiliza,

Itombwe, may be a hybrid *alinae* × *verticalis* (Prigogine 1978). We think its plumage differences are too great for its inclusion in the superspecies.

Field Characters. Length 12·5–14 cm. Endemic to mts of Albertine Rift, where it is the only blue-headed sunbird in montane forest. In transition forest meets Green-headed and Blue-throated Brown Sunbirds *C. verticalis* and *C. cyanolaema*; longer-billed and shorter-tailed, with strong *brownish-orange tinge* to green upperparts; throat blue-violet, rest of underparts *sooty* black with yellowish wash on flanks and undertail-coverts. Sexes alike, both with pale yellow pectoral tufts.

Voice. Tape-recorded (104, B, LEM). Territorial advertising song a far-carrying series of rapid metallic chips going up and down the scale at rate of 8–9 notes/s for 3–6 s or more, one song joined to the next by just a brief hesitation (see sonagram in Dowsett-Lemaire 1990). A hastened, high-pitched subsong is used in aggressive encounters and in courtship; main call notes are very high 'ssee' or 'tsee', reminiscent of *Cryptospiza* (Dowsett-Lemaire 1990).

General Habits. Inhabits fairly dense primary, secondary, gallery and riverine forests, above 1350 m, keeping mainly to understorey, also to edges. Solitary or in pairs. Joins mixed foraging parties outside breeding season. Especially fond of nectar of *Balthasaria schliebenii* (flowers orange-pink, with long tubular corolla, like aloe), and in Nyungwe (Rwanda) distributions of bird and tree nearly coincide (Dowsett-Lemaire 1990). Uses long bill to probe down tubular corollae from perching position and from brief hovering flight in front of blossom. Feeds at flowers of vine *Canarina eminii* only by hovering. Not observed to 'cheat' by pecking through base of corolla (as some short-billed sunbirds do). Overlaps altitudinally with Green-headed Sunbird *C. verticalis*, e.g. in Bwindi at 1400–2100 m (Uganda) and Nyungwe (Rwanda); in Nyungwe the 2 species seem to avoid each other (Dowsett-Lemaire and Dowsett 1990); appears to replace Green-headed Sunbird in riparian forest on Lufuko R. in Marungu Highlands (Dowsett and Prigogine 1974).

Food. Wide variety of small insects and spiders, and nectar of *Balthasaria schliebenii* (Theaceae), *Phragmanthera usuiensis*, *Tapinanthus brunneus*, *T. constrictiflorus* (Loranthaceae) and *Canarina eminii* (Campanulaceae), also *Brilliantaisia*, *Impatiens niamniamensis*, *Ixora burundensis*, *Lobelia gibberoa*, *Pseudosabicea*, *Symphonia* (Nyungwe, Rwanda: Dowsett-Lemaire 1990) and *Erythrina*.

Breeding Habits. Solitary nester. Territorial bird defends flowering *Balthasaria* tree. Territory advertised by singing, often in early morning; a conspecific or other sunbird approaching singer closely is chased and pursued away from tree, pursuer giving fast, high-pitched subsong.

NEST: a suspended pouch made of green moss, lichen, herbs, dry leaves (of bamboo at high altitudes), and rootlets; lined with thin stems and fine plant down, some from flowering heads of Compositae. Entrance at one side, with pronounced porch over it. One nest was made of *Usnea* lichen, inflorescences of *Panicum* (Graminae) and *Thalictrum rhynchocarpum* (Ranunculaceae) and lined with pappi of *Gynura vitellina* (Compositae) and a few small feathers. Ext. height 165–200, ext. width 85–100 and from front to back 75, entrance diam. 30; from top of nest to centre of entrance, 60–80 (Prigogine 1971, 1972). Sited *c.* 2 m above ground in bush or tree.

EGGS: 1 (in Itombwe) or 2 (Kivu). Grey or greyish white ground, very finely stippled with grey or brownish red spots and blotches, darker and more numerous at broad end; egg colour and markings may vary geographically (Prigogine 1972). SIZE: (*alinae*, n = 2) 17·3 × 13·0, 17·8 × 13·1; (*tanganjicae*, n = 2) unexpectedly large: 20·2 × 13·2, 20·2 × 14·6.

LAYING DATES: Zaïre, Mt Kabobo (gonads mature Oct–Nov), Itombwe (gonads mature or nests found Jan–Feb, Apr–June and Oct; juveniles May, June, Aug: Prigogine 1971, 1984). Rwanda, (primary moult in Aug–Sept and fresh plumage in Sept–Oct indicate breeding soon thereafter).

References
Dowsett-Lemaire, F. (1990).
Prigogine, A. (1971, 1975).

Cyanomitra obscura (Jardine). Western Olive Sunbird. Souimanga olivâtre de l'Ouest.

Plate 10
(Opp. p. 143)

N.[*ectarinia*] *obscurus* Jardine, 1843. Nectariniadae, in Nat. Library, 13, p. 253; Fernando Po.

Forms a superspecies with *C. olivacea*.

Range and Status. Endemic resident and migrant (Senegal), forests and dense woodlands from Senegal to Ethiopia and Kenya west of Rift Valley, on Bioko and Príncipe Is, and from Angola to Zambia north and west of Luangwa Valley. Outlying population on Zimbabwe–Mozambique border. Common to abundant except as noted.

Senegal, scarce or locally frequent: Dakar, Mbour, Ziguinchor, Oussouye, Niokolo-Koba Nat. Park and Kédougou. Unknown in Gambia. Guinea-Bissau, formerly frequent at Gunnal. Guinea, Fouta Djalon; Conakry, Foulayah and Kolenté (Demey 1995); frequent on Kounounkan Massif (Hayman *et al.* 1995); Macenta, Nzérékoré, Mt Nimba. Sierra Leone, widespread at least south of line from Freetown to Sefadu. Liberia, widespread and frequent to common. S Ivory Coast, S Ghana. Togo, north to 06°45′; Nigeria, north to nearly 10°N (Kainji, Kagoro). Cameroon, widespread in S, north to Ngoumé (05°30′N). Bioko, frequent from sea-level to 800 m; higher than that scarce – up to 1000 m on Pico Basilé and 1200 m at Moka. Príncipe. Central African Republic, uncommon, Manovo-

Cyanomitra obscura

Gounda-Saint Floris Nat. Park, Lobaye and Haute Sangha Préfs. Mbini, Gabon, Congo and Zaïre; in Ivindo Basin (Gabon) the most abundant bird species after Yellow-whiskered Greenbul *Andropadus latirostris*; Itombwe at 790–1750 m. Angola in Cabinda and Cuanza Norte and occurs in N Malanje, N Lunda Norte, and along escarpment to S Benguela; probably also in Zaïre District. Sudan, up to 3000 m from Bengengai to Didinga Mts and Boma Hills. Uganda, widespread north and east to Bugoma, Budongo, Masindi, Sezibwa R. and Busoga (Britton 1980). Kenya, widespread and locally common up to 2300 m and less common up to 3000 m; only occurs west of Rift Valley, from Mt Elgon, Kapenguria and Cherangani Hills to Kakamega, Sotik, Mau Narok, Lolgorien and Migori River. Rwanda, common locally; in SW Nyungwe Forest up to 2000 m and in NW regular up to 2300 m and occasional up to 2500 m (Uwinka). Burundi. Tanzania, local in W: Ukerewe I., Bukoba, Kibondo, Gombe Stream Game Res., Mt Mahari and Ufipa Plateau. Zambia, frequent and quite widespread in N up to 1500 m as mapped (R.J. Dowsett pers. comm), very common in mavunda forest; in S only in E Lusaka District (*lowei*). Zimbabwe, isolated race *sclateri* occurs at 350–2000 m in moist, seaward-facing slopes of Inyanga highlands south to Melsetter, Chimanimani Mts and Chipinga Uplands. Mozambique, Vumba, Chimanimani Mts, Búzi R. drainage, Chikamboge valley, Kurumadzi R., Manica Platform and Espungabera (Clancey 1996).

Density of about 4 birds per ha, NE Gabon.

Description. *C. o. cephaelis* (Bates): S Nigeria and probably Benin to N Angola and Congo Basin, intergrading with *ragazzii* in E Zaïre. ADULT ♂: forehead and crown dark olive-green, a little paler at sides, making faint superciliary stripe; hindneck to uppertail-coverts slightly brighter. Tail feathers dark brown, edged yellowish green, more broadly near base. Lores dark grey-brown, paler grey in front, cheek feathers and ear-coverts dark grey-brown with pale olive or yellowish centres, making them speckled. Chin, throat, breast, belly and undertail-coverts olivaceous pale grey, tinged yellowish on throat; flanks slightly greyer; conspicuous bright yellow pectoral tufts. Median and lesser coverts dark olive-green, rest of wing dark grey-brown, primaries narrowly and secondaries, tertials and greater coverts broadly edged yellowish green, alula and primary coverts finely edged olive-green. Underwing coverts and inner borders of remiges pale buff; axillaries pale olive-yellow. Bill black, base of lower mandible pale, almost whitish; eyes dark sepia; legs black or dark brown. Sexes alike except that ♀ lacks yellow pectoral tufts. SIZE (6 ♂♂, 7 ♀♀): wing, ♂ 63·0–66·5 (65·2), ♀ 56·5–58·0 (57·5); tail, ♂ 47·5–56·5 (51·5), ♀ 39·5–43·0 (41·3); bill, ♂ 26–28 (27·1), ♀ 24–26 (25·0); tarsus, ♂ 16–18 (17·0), ♀ 16–17 (16·1). WEIGHT: Cameroon, ♂ (n = 19) 10·0–12·8 (11·3), ♀ (n = 10) 7·8–11·5 (9·6); Gabon, ♂ (n = 5) 10·0–11·0 (10·5), ♀ (n = 4) 8·6–9·8 (9·2).

IMMATURE: juvenile like adult but crown a browner olive; chin whitish, throat to undertail-coverts strongly washed yellow.

NESTLING: gapes deep orange.

C. o. guineensis (Bannerman): Senegal to Togo, intergrading with *cephaelis*. Like *cephaelis* but lower mandible all black. Slightly darker and greyer above, more greyish below. Smaller: wing, ♂ (n = 10) 54–61 (57·9). Weight: Liberia, ♂ (n = 11) 8·0–10·8 (9·4), ♀ (n = 9) 7·7–9·7 (8·4).

C. o. obscura Jardine: Bioko and Príncipe. Like *cephaelis*, but brighter green above and on wing edgings, still paler greyish white below. This and next 3 races poorly differentiated.

C. o. ragazzii (Salvadori): SW Ethiopia. Underparts rather darker than in *cephaelis* and bill longer. Wing, ♂ (n = ?) 66–73, ♀ (n = 2) 59, 65.

C. o. vincenti (Grant and Mackworth-Praed): S Sudan, Uganda, W Kenya, W Tanzania, E Zaïre; intergrades with *cephaelis* and perhaps with *ragazzii*. Like *ragazzii* but larger. Wing, ♂ (n = 10, Kenya, Uganda) 64–74 (68·1), ♀ (n = 2) 57–59. Weight: Kenya, ♂ (n = 30) 9–12·8 (10·9), ♀ (n = 27) 8·1–11 (9·6).

C. o. lowei (Vincent): Zambia, and probably this race adjacently in S Zaïre. Like *cephaelis*, but underparts slightly darker and more uniform though paler in midline. Slightly larger: wing, ♂ (n = 15) 64–68 (66·2), ♀ (n = 15) 56–62 (59·1) (Clancey 1978).

C. o. sclateri (Vincent): Zimbabwe–Mozambique border mts. Upperparts dull olive-green, underparts pale yellowish olive-buff, feathers of chin to breast with dusky fringes. T5 and T6 usually tipped pale buffy olive. Juvenile with throat and breast olivaceous. Wing, ♂ (n = 15) 63–67·5 (65·0), ♀ (n = 10) 57–60 (58·1) (Clancey 1978). Weight: ♂ (n = 20) 9·2–12·4 (10·5), ♀ (n = 19) 8·0–10·5 (9·4).

TAXONOMIC NOTE: see under *C. olivacea*.

Field Characters. Length, ♂ 12·5–15 cm, ♀ 11·5–12·5 cm. This and its close relative the Eastern Olive Sunbird *C. olivacea* are the most characteristic and common sunbirds in forest throughout Africa; their distinctive song of well-spaced notes going down the scale and then up, frequently given, is the best indicator of their presence in the dark forest interior. Plumage details often hard to see; appears uniform olive except in good light, where dark green upperparts seen to differ from dull olive-grey underparts, head darker, throat paler, some races with slight yellow wash on throat and belly, wing edgings with greenish wash. *Long curved bill* and *yellow pectoral tufts* of ♂ distinguish it from similar drab forest sunbirds; ♀ lacks pectoral tufts, bests told by bill, voice and unmarked olive plumage. Caution: yellow pectoral tufts often concealed in perched birds. Has longer tail and bill than similar but much smaller Bates's Sunbird *Cinnyris batesi*, and inhabits lower levels in forest, not canopy; for further differences, see that species.

Voice. Tape-recorded (86, 104, KEI). Song starts hesitantly, sometimes in an undertone, then proceeds with series of pure, clear, well-separated notes, accelerating slightly, sometimes on same pitch but more often going down the scale and then up, perhaps more than once depending on length of song; ending may tail off in a fuzzy jumble of notes: 'chip ... chip, chip-choop-choppy-chippy-chuchewp'; 'chop-chip-chop-chop-poppachichichi'; 'zi ... zi ... tsip, chip-chop-chip-chop-chichichi-chip ... '; may last up to 15 s. In song battle between rival ♂♂ a rapid, bustling version is given with individual notes barely discernible, rising and falling in regular motion, lasting up to 20 s. Gives a 'midday call' from perch, a continuous series of evenly spaced chip notes, either on one pitch at rate of 2 per s (Ghana, C. Chappuis) or alternating high and low notes, 'weep ... wup ... weep ... wup ...' at rate of 1 per s (Liberia, S. Keith). Calls include a harsh 'jaa-jaa-jaa ...' at rate of 5 per s, a less excited, slower 'der-der-der ... ' (4 per s), an irregular, hard 'tic', and an up-slurred 'cooee'.

General Habits. Inhabits gloomy undergrowth in mature and regenerating lowland evergreen forest up to 3000 m, riverine forest, forest edges and clearings, *Cryptosepalum* forest (Zambia), dense mesic woodland, bushy thickets, banana groves, coffee forest, bamboo forest, lower limits of moss forest, open coastal bush, forest–shrub–grassland mosaic, gallery forests in savanna, edges of mangrove swamps, and gardens. Occurs at all levels in forest from undergrowth to canopy, but commonest within 6 m of ground. Occasionally enters open houses.

Lives singly; ♂ often chases ♀, when they can appear to be bonded, but ♂ and ♀ do not form permanent pair bonds; often accompanied by 1–2 juveniles; ♂♂ gather in leks; large aggregations of both sexes together sometimes occur in single, large flowering tree. Regular member of insectivorous mixed-species foraging flocks: seen in 76% of 228 flocks (Gabon). Inconspicuous, keeping mainly in dense foliage, but sometimes bird shows itself boldly. Forages by searching foliage using acrobatic, tit-like postures; hovers to probe flowers, and flycatches. When foraging, nervous and active, constantly in motion; flicks wings; pecks into clusters of dead leaves, curled-up living leaves, and debris, explores thin, intertwining stems of leafy vines, pokes around in bases of Marantaceae in undergrowth, and inspects red terrestrial fruits of *Aframomum* (Zingiberaceae) (Brosset and Erard 1986). Once seen eating small red berries from clusters in roadside tree attended also by Bates's *Cinnyris batesi*, Green-throated *Chalcomitra rubescens*, Collared *Hedydipna collaris* and Little Green Sunbirds *Anthreptes seimundi*, all arriving in twos and threes, feeding on berries for a few min then departing, and amounting to 'great numbers' (Serle 1954). Displaced from flowers of *Tecoma stans* by honeybee *Apis mellifera* which flies in zigzags at bird's face and scares it off (Akinpelu 1989). In dull, overcast weather feeds in open all day; when aloes in flower, feeds on them in the open irrespective of weather (Hanmer and Chadder 1997). Defends flowers of *Symphonia globulifera* for days on end, each of several birds in one tree top controlling part of canopy; does the same in crown of *Syzygium congolense* tree (SE Cameroon), territorial birds dividing flowering crown into portions, each defended against conspecies, Collared Sunbirds and any other sunbirds (F. Dowsett-Lemaire, pers. comm.).

Sedentary, but all Senegal records in Dec–Mar and bird appears to be a non-breeding visitor there. Most pairs completely sedentary in Gabon, where one ringed bird moved 2·5 km and another 3 km.

Food. Insects, spiders, berries and nectar. One of the most insectivorous sunbirds (Brosset and Erard 1986, Dowsett-Lemaire 1990). Seen catching flies, spiders and small caterpillars. Of 28 stomachs in Cameroon, insects in 24, insects and seeds in 2, and fruit pulp in 2; in Guinea (Mt Nimba), flying termites, spiders and small yellow seeds; in 6, Zaïre, small berries, seeds, insects and a spider; in Cameroon, tiny fragments of flowers; termites, small hemipterans, spiders, nectar and banana fruit pulp (Príncipe I.); elsewhere commonly small fruits or seeds from them. Can swallow fruits of diam. *c.* 5 mm; droppings contained intact seeds of parasol tree (Brossett and Erard 1986). Eats fruits of *Macaranga assas*, *M. barteri* and *Musanga cecropioides* and interested in fruits of *Xylopia aethiopica* (Dowsett-Lemaire 1996). Feeds at nectar-bearing flowers: race *cephaelis* largely at flowers of *Cephaelis mannii* (Rubiaceae), evidently as much on spiders as nectar or pollen. Takes nectar from *Tecoma stans*, a favourite flower in Nigeria (Akinpelu 1989), and in Congo from *Syzygium congolense* and *Symphonia globulifera* (F. Dowsett-Lemaire, pers. comm.); feeds at flowers of *Oxyanthus troupinii* in forest understorey in Rwanda (Dowsett-Lemaire 1990).

Breeding Habits. Lekking mating system; solitary nester. At beginning of rains (Gabon), 7–8 ♂♂ assemble in a twiggy but open part of foliage, and actively sing whilst hopping around pursuing each other in area of 2–3 m². When singing, pectoral tufts conspicuous, throat inflated, head held high, wings droop, and tail held up at 45° angle (Brosset and Erard 1986).

NEST: tidy, oval suspended purse; side entrance with porch above and untidy 'beard', which hangs from below entrance rather than under bowl; made of dry leaves, fine fibres, herb stems, green moss, twiglets, bound together (and bound at point of attachment) with strands of *Marasmius* and spider web; lined with a little plant down and fine fibres. Size (n = 2) height 190–220, breadth 75–90, from front to back 75–90, entrance diam. 30 and 40 × 50; entrance 80 below top of nest. Sited between 0·3 and 2·5 m above ground; usually at tip of drooping branch or at end of cut liana in undergrowth; often in open situation; sometimes hung from root projecting from bank. Nest built by ♀ only.

EGGS: 2 (once 1, once 3). 2nd egg laid 24 h after 1st. Shells of 3 types: (a) pale green, mottled with yellowish brown, with rounded, diffuse, dark brown spots and blotches, (b) greyish white, clouded with pale lilac-grey, with small dark brown spots and dashes scattered thinly over surface, and (c) greenish white, finely and densely clouded at large end with yellowish brown and dull grey blotches (Bannerman 1948). In Gabon type (b) is usual, other types rare or unknown (Brosset and Erard 1986). SIZE: (n = 19, Zaïre, Itombwe) 15·8–18·0 × 11·9–13·2 (17·4 × 12·5).

LAYING DATES: Ghana, (fledglings May–July); Nigeria, Jan (and oviduct egg Apr); W Cameroon, Apr–Aug; S Cameroon, (birds in breeding condition all months); Príncipe, Nov–Jan (–Apr?); Gabon, all year (61 nests; fledglings), mainly June–Sept; Angola, Oct–Mar; Sudan, June–Aug; Zaïre (Ituri), (various indications, all months), (Itombwe), July–Apr; Uganda, Feb–Apr, June, Sept, Dec; Zambia, July, Sept, Nov (and gonads active Jan–Apr); Zimbabwe, Sept–Mar (45 clutches), 68% in Nov–Dec.

INCUBATION: by ♀ alone, starting when 2nd egg laid.

DEVELOPMENT AND CARE OF YOUNG: young fed by both sexes. Out of nest, may stay with parents for long time, since adults and young often taken 'simultanément dans le même filet' (Brosset and Erard 1986). Skin of tibio-tarsal joint bright pink-orange up to 6 months of age, then pink-yellow for a year, then yellow or dull yellowish brown for a further year (Hanmer and Chadder 1997).

BREEDING SUCCESS/SURVIVAL: parasitized by Emerald Cuckoo *Chrysococcyx cupreus* (Gabon, Zaïre) and once by a honeyguide (sp. indet.: Indicatoridae). Of 23 nests, Gabon, 16 destroyed by predators, 1 parasitized by cuckoo and 1 by honeyguide. At least 2 instances of Western Olive Sunbird being caught in spider's web (whether fatally, not recorded). Of 326 birds netted, 15 recaptured after 2–5 years (2 birds at 61 months) (Brosset and Erard 1986). Sex ratio skewed (?): twice as many ♂♂ as ♀♀ mistnetted (Bowden 1986). Longevity records of over 5·5 years (3 birds), 6, 6 and 7 years (Hanmer and Chaddon 1997) and 5–>6 years (n = 13) (Hanmer 1997).

References
Brosset, A. and Erard, C. (1986).
Chapin, J.P. (1954).
Clancey, P.A. (1993).

Plate 10
(Opp. p. 143)

Cyanomitra olivacea (A. Smith). Eastern Olive Sunbird. Souimanga olivâtre de l'Est.

Cinnyris olivaceus A. Smith, 1840. Ill. Zool. South Africa, text to pl. 57, footnote; Kafirland to Port Natal.

Forms a superspecies with *C. obscura*.

Range and Status. Endemic resident and local nomad, forests, dense woodlands and scrub from Kenya east of Rift Valley and S Somalia to E Cape, South Africa; inland to Malaŵi and just into SE Zambia; Pemba, Zanzibar and Mafia Is.

Somalia, locally very common in lower Jubba valley; in Boni Forest. Kenya, in central highlands east of Rift Valley, east to Nyambenis, south to Kiambu, also Mt Endau, Chyulu Hills, Taita Hills and Mt Kasigau; coastal lowlands from Somali border, Witu and lower Tana R. upstream to Bura and Garissa, south to coastal Tanzania. N Tanzania from Loliondo, Monduli, Arusha Nat. Park, Mt Meru, Kilimanjaro, North and South Pare Mts and Usambaras south through eastern arc mountains from Ngurus, Ukagurus and Ulugurus through Udzungwas and Iringa to Mt Rungwe and west to 33°E, also Matengo highlands, Songea, Selous Game Res. and between Nachingwea, lower Mbekuru valley and Lindi (N.E. Baker, pers. comm.). Zambia, east of Luangwa Valley only, at Chipata and on Mt Mwanda. Malaŵi, locally common, from lakeshore (Karonga to Chintheche, common between Nkhata Bay and Chintheche) up to nearly 1850 m in Misuku Hills; south to Malaŵi Hill. Mozambique, north of Zambezi in Zambézia and Nampula Districts from Zambezi valley and Malaŵi border to coast north to Pemba (13°S); south of Zambezi to Beira; Mt Gorongoza; and coastal lowlands from Inhambane south to border; occurs on Inhaca and Bazaruto Is. Transvaal, scarce in mountains east of Barberton and south of Malelane, west to Kangwane. Most of Swaziland. Natal, common in coastal lowlands, inland up to about 600 m. Coastal E Cape Prov., south to East London.

Density of 5–7 pairs per ha around Lake St Lucia, Natal (Skead 1967).

Cyanomitra olivacea

Description. *C. o. olivacea* (Smith): coastal lowlands from south of Save R. (Mozambique) to East London (E South Africa), inland to Swaziland. ADULT ♂: upperparts dark olive-green, feathers of top of head and neck with dark grey-brown fringes. Tail graduated, dark grey-brown; feathers edged olive-green, more broadly near base, and tipped pale greyish, broadly on T4–T6;

T5–T6 with whole outer web paler grey. Lores, cheeks and ear-coverts dark grey-brown, the latter flecked olive-yellow. Chin to upper breast yellowish olive-green, chin and sometimes throat with faint dusky barring; pectoral tufts yellow; rest of underparts pale olive, greyer on sides and flanks. Median and lesser coverts dark olive-green; rest of upperwing dark grey-brown, primaries narrowly edged yellowish olive-green, secondaries, tertials and greater coverts broadly edged olive-green, alula and primary coverts finely edged olive-green. Axillaries pale olive-yellow; underwing-coverts and inner borders on undersides of flight feathers pale buff. Bill black; eyes dark sepia; legs black or dark brown. Sexes alike. SIZE: (10 ♂♂, 10 ♀♀): wing, ♂ 61–70 (65·1), ♀ 59–64 (61·2); tail, ♂ 49–62 (53·9), ♀ 47–53 (50·1); bill, ♂ 27–29 (28·1), ♀ 27–28 (27·7); tarsus, ♂ 16–18 (17·1), ♀ 16–17 (16·3). WEIGHT: ♂ (n = 126) 9·2–14·0 (12·6), ♀ (n = 112) 7·8–11·8 (9·8), unsexed (n = 192) 9·0–14·7 (11·8) (Maclean 1993).

IMMATURE: juvenile like adult but upperparts brighter olive-green, and rusty yellow tinge on throat.

NESTLING: not described.

C. o. olivacina Peters: NE Zululand and coastal S Mozambique (up to 350 m, Sul do Save, Manica e Sofala, offshore islands). Very like *olivacea* but slightly paler above and below; smaller and bill finer. Wing, ♂ (n = 6) 58–62 (59·9); bill 22–25.

C. o. alfredi (Vincent) (includes '*intercalans*'): S Tanzania, Malaŵi, extreme SE Zambia, and Mozambique south to Beira. Lighter, brighter green above than nominate race, top of head less greyish; paler greyish olive below. Wing, ♂ (n = 5) 62–64 (63·0), unsexed (n = 30, Udzungwa) 55–66 (60·4). Weight: (unsexed, n = 159, Udzungwa) 8·3–10·7 (9·5) (L.A. Hansen, pers. comm.).

C. o. changamwensis (Mearns): S Somalia and coastal Kenya (inland to Kasigau and Taita) and NE Tanzania (coast, Usambaras and South Pare Mts); Mafia I. Like *alfredi* but still greener above; central underparts tinged somewhat yellower. Smaller: wing, ♂ (n = 10) 56–63 (60·3). Weight: S Somalia, ♂ (n = 54) 6·8–9·1 (7·7), ♀ (n = 49) 6·2–8·7 (7·1); Kenya, ♂ (n = 26) 7·5–9·9 (8·4), ♀ (n = 21) 7·0–9·2 (7·8).

C. o. granti (Vincent): Zanzibar and Pemba Is. Like *changamwensis* but ♀ lacks pectoral tufts. Small: wing, ♂ (n = 7) 57–61 (59·7); bill shorter and straighter, ♂ (n = 7) 20–23 (20·6).

C. o. neglecta (Neumann): S-central Kenya, NE Tanzania (North Pare Mts, Kilimanjaro, Meru, Monduli, Arusha). Like *alfredi*, but slightly darker above, top of head greyer.

TAXONOMIC NOTE: *C. olivacea* and *C. obscura* were formerly regarded as a single species. We follow Clancey (1993) in recognizing 2 parapatric allospecies, ♀♀ of the eastern *C. olivacea* with yellow pectoral tufts and ♀♀ of the western *C. obscura* without. 2 isolated populations are anomalous: *sclateri* (Zimbabwe–Mozambique border) and *granti* (Zanzibar and Pemba Is), being geographically close to *C. olivacea* but with ♀♀ like *C. obscura*. Clancey (1993) placed both in *N. obscura*. We think it likely that at least one, *granti*, is more closely allied with *C. olivacea* (mainland *changamwensis* only a few km away) than with *C. obscura*, despite the evidence of ♀ pectoral tufts to the contrary, and so place it in *C. olivacea*. We leave *sclateri* in *C. obscura*.

Field Characters. Length, ♂ 12·5–15 cm, ♀ 11·5–12·5 cm. This and its close relative the Western Olive Sunbird *C. obscura* are the most characteristic and common sunbirds in forests throughout Africa; their distinctive song of well-spaced notes going down the scale and then up, frequently given, is the best indicator of their presence in the dark forest interior. Plumage details often hard to see; appears uniform olive except in good light, where dark green upperparts seen to differ from dull olive-grey underparts, head darker, throat paler, some races with slight yellow wash on throat and belly, wing edgings with greenish wash. *Long curved bill* and *yellow pectoral tufts* (both sexes) distinguish it from similar drab forest sunbirds. Caution: yellow pectoral tufts may be concealed in perched birds. Can only be confused with Mouse-coloured Sunbird *C. veroxii*, often in same coastal habitat, but Mouse-coloured is grey, not olive, and has scarlet pectoral tufts. Voice diagnostic.

Voice. Tape-recorded (88, 91–99, KEI, LIV, STAN). Characteristic song of well-spaced, short, sharp, clear notes; speed irregular, but usually starts hesitantly, accelerates down the scale, then either slows up at the bottom and gets faster towards the end, or continues at the same speed and trails off at the end: 'tip ... chip, chip-chip-chichii-chop-chop-chop-chiichi-chip'; 'tsip-tsitsip-chep-chep-chop-chichop-chop-chip'; or may descend and not come up again, ending with marked 'chop, chop'. Lasts 2·5–5 s, but many may be joined without a break to produce lengthy performance; one bout lasted nearly 2 min (Gedi, Kenya: S. Keith). Normal break between songs 5–10 s, interspersed with dry 'jit' or 'jit-jit'. Several ♂♂ may gather and sing together in chorus, a long, continuous babble of notes rising and falling like waves; notes are sharp, without the tinny or sibilant quality of other sunbirds, and sometimes song slows to allow interpolation of regular 'chip' notes. Has same 'midday call' as Western Olive Sunbird, continuous 'wip-wip-wip-wip ... ' at rate of 2/s (Gedi, Kenya). What is probably the same call described by Skead (1967) as 'phit, zeet' or 'woot, zeet', repeated monotonously at 1-s intervals for sessions of an hour or more at all times of day. Alarm scold a rasping 'jet-jet-jet ... '; other calls include a sharp, wooden 'tuk, tuk, tuk' uttered on the move or in flight, a high-pitched frog-like piping 'tsweep-tsweep-tsweep' while foraging, a feeble, piping, but attractively fluty 'tseeng-tseeng-tseeng-tseeng ... ' and a rapid chittery stutter, 'ch, ch, ch, ch, ch ... ' in mild excitement while dashing through bush (Skead 1967).

General Habits. Inhabits undergrowth in mature and regenerating lowland evergreen forest up to 1850 m (wandering up to 2150 m in Malaŵi: Dowsett-Lemaire 1989b), riverine forest, bushy thickets, open coastal bush, dense dune forests and adjacent thornveld, thorn hedges, banana and eucalypt plantations, edges of mangrove, and town parks and gardens; in Mozambique commonest in sapling undergrowth with *Achyrospermum* and *Macrorungia* (Swynnerton 1908). Occasionally enters open houses.

Solitary or in pairs and groups of 4–5, birds in vocal but not necessarily visual contact. However, large aggregations can occur in single, large flowering tree: 30 in a *Jacaranda*, and up to 100 in a *Eugenia* all day for the 5 weeks it was in bloom. Singing assemblies of 10–20 birds in breeding season (see below). Joins mixed-species foraging flocks. Aggressive towards other sunbirds such as Southern Double-collared *Cinnyris chalybea* and Greater Double-collared *C. afra*. Active, alert and noisy; convulsively flicks wings; attacks its image in window-panes; birds chase each other in zigzagging flight. Often calls loudly and persistently.

Forages at all levels from forest canopy to floor. In thick cover in N Mozambique almost invariably keeps very low down, darting into shrub next to observer, poking about confidingly, then darting away. In Malaŵi, birds regularly netted in forest at 07h00–08h30 and 15h30–17h00, but not in between unless sky is overcast (D.B. Hanmer, pers.

comm.). In Zululand feeds mainly at flowers at edge of forest, and pollinates giant strelitzias. Forages by hovering at blossoms but mainly by moving through leafy herbage, peering at and gleaning leaves and twigs, and probing into tangles of thin vines and accumulations of dead leaves, also into masses of old cobwebs on buildings and debris caught up in them. Feeds at flowering stem of *Leonotis mollissima* by perching sideways on stem and, without changing grasp, probing into each flower in succession until half way around stem, when grasps stem anew and shifts attention to flowers on other side. Takes short flights from bushtop perch to catch termites on the wing. Eats many spiders, hunting them out with methodical deliberation.

Sedentary, but prone to be 'here today, gone tomorrow', and frequency of appearance in gardens and open plantations some distance from usual leafy, forest habitat suggests that it wanders outside breeding season.

Food. Berries, nectar, spiders and insects. Eats small berries of *Trema orientalis* (= *guineensis*) (Natal, Skead 1967) and *Antidesma venosum* (Ginn *et al.* 1989). Feeds at nectar-bearing flowers of *Leonotis mollissima*, *Achyrospermum carvalhi*, *Macrorungia pubinervia*, *Loranthus*, *Burchellia*, *Erythrina*, *Hibiscus tiliaceus*, *Ipomoea*, *Strelitzia nicolai* (a favourite), fuchsias and particularly *Jacaranda* and *Schotia brachypetala*. In Malaŵi many Eastern Olive Sunbirds caught in nets set amongst aloes, *Tecomaria*, *Cestrum*, *Holmskioldia*, cannas, salvias and proteas, which are thought to attract them (D.B. Hanmer, pers. comm.). Adept at catching spiders.

Breeding Habits. Solitary-nesting, polygynous, territorial. 'Groups of 10–20 collect and sing amongst themselves' in treetop (Skead 1967); territorial ♂♂ regularly assemble in tree or bush and give loud, lilting song 'as each bird tries to perch higher than the others' (S.K. Frost *in* Ginn *et al.* 1989), which suggests a lekking mating system. Dominant ♂ establishes territory around favoured flowering plants, and relentlessly chases all other sunbirds away; it advertises with persistent, frog-like 'phweep-phweep' call; arriving ♀ is allowed to visit flowers only after displaying or copulating; ♂ copulates with several ♀♀, which make their nests in his territory (S.K. Frost *in* Ginn *et al.* 1989).

NEST: somewhat elongated, pear-shaped, suspended bag, with side entrance, straggly porch projecting sideways above entrance (sometimes quite long, making nest comma-shaped rather than pear-shaped), and untidy beard hanging below entrance; made of fine dry grass, plant fibres, small twigs, moss, lichen, leaves (in S Mozambique often of *Behnia reticulata*), thin dry stems (often of *Thalictrum rhynchocarpum*) and plant down (e.g. of liana *Oncinotis chirindica*) loosely bound and held together with spider web; lined with down from seeds, fine tendrils and strands of grass, and feathers. One nest was made in part from thin, dry, wiry fibres taken from nest of Thick-billed Weaver *Amblyospiza albifrons*. Some nests made mainly of moss; others of dry leaves. In all nests, dry leaves built into fabric at base to support the cup. Nest's attachment to twig or leaf often made of moss. 'Beard' composed of grass blades, twigs, roots, moss, leaves and strips of bark; both attachment and 'beard' vary in length, bulk and construction. Ext. height 90–130 (usually 90–100), width *c.* 80, front to back *c.* 70, entrance diam. *c.* 40. Attachment up to 175 long and 'beard' up to 260, so entire construction (attachment, nest proper and 'beard') can be nearly 0·5 m long (Swynnerton 1908). Sited in undergrowth, only 45–180 cm above the ground, hanging loose from drooping twig of small-leaved shrub such as *Sclerochiton harveyanum* (S Mozambique) or from tip of *Dracaena* leaf (Swynnerton 1908); also on wires or light fittings in buildings (Maclean 1993). Built by ♀ only.

EGGS: 2, rarely 3. White, greyish white, bluish or buff, marked with dark brown scrolls, spots, speckles and marbling on underlying grey. SIZE: (n = 31) 16·8–20·1 × 11·7–13·6 (17·9 × 12·8).

LAYING DATES: E Africa, Region D, Sept and Nov–Mar, Region E, Apr–Aug and Oct; Malaŵi, Sept, Nov–Jan; Mozambique, Nov; Transvaal, Aug–Mar; Natal, Sept–Mar (mainly Oct–Dec); E Cape, Sept–Jan.

DEVELOPMENT AND CARE OF YOUNG: young fed by ♀ (♂ defends nest area).

BREEDING SUCCESS/SURVIVAL: parasitized by Emerald Cuckoo *Chrysococcyx cupreus* (Malaŵi) and possibly by Klaas's Cuckoo *C. klaas* (South Africa). 4 birds lived at least 7 years (Ziika, Uganda: Dranzoa 1997).

References
Ginn, P.J. *et al.* (1989).
Skead, C.J. (1967).
Swynnerton, C.F.M. (1908).

Plate 10 (Opp. p. 143) *Cyanomitra veroxii* (A. Smith). Grey Sunbird; Mouse-coloured Sunbird. Souimanga murin.

Cinnyris veroxii A. Smith, 1831. South African Quart. J., ser. 1, no. 5, p. 13; Cafferland (= E Cape Province, Sclater, 1930, Syst. Av. Aethiop., p. 702).

Range and Status. Endemic resident, E African coastal lowlands from S Somalia to S Cape.

Somalia, rare: records along coast south of 2°15'N, and extends at least 240 km up Jubba R. Kenya, scarce and local, all along coast; inland along lower Tana R. for about 50 km, and single record well inland at Nzaui Hill (01°54'S, 37°32'E), at 1200 m. Tanzania, record inland at Kidugallo, otherwise confined to coastal strip within a few km of the sea, and a race is endemic to Zanzibar. Small coastal breeding population (and breeding record) SE Tanzania (N.E. Baker, pers. comm.). In S Mozambique extends about 250 km up Zambezi valley to Umquasi. Very sparse on Manica e Sofala littoral, present on Bazaruto and Inhaca Is, local on Sul do Save littoral; in Mozambique and

Cyanomitra veroxii

Swaziland occurs in foothills of Lebombo Range. Malaŵi, 2 records, Nchalo (Hanmer 1981). South Africa, fairly common along whole Natal littoral, reaching 70–80 km inland along thickly wooded valleys, and along E Cape littoral south to Algoa Bay (and, formerly?, to about Knysna: Hall and Moreau 1970).

Description. *C. v. veroxii* (Smith): South Africa. ADULT ♂: upperparts dark olive-grey, feathers fringed oily green on forehead and crown, more broadly glossed oily green elsewhere. Tail dark grey-brown, slightly glossy above; feathers narrowly edged olive-green, T6 and tip of T5 paler grey-brown. Lores dark olive-grey; cheeks and ear-coverts olive-grey. Chin, throat and underparts pale grey, slightly darker across breast; pectoral tufts scarlet. Upperwing dark grey-brown, remiges narrowly edged pale brown, greater coverts fringed pale grey-brown, alula more blackish, median and lesser coverts fringed oily bluish green. Axillaries, underwing-coverts and inner borders on undersides of flight feathers pale grey-buff. Bill black; eyes dark brown; legs black. Sexes alike; one ♀ with 2 yellow feathers each side of red pectoral tuft (Hanmer 1979). SIZE (6 ♂♂, 8 ♀♀): wing, ♂ 64–68 (65·5), ♀ 57–62 (59·0); tail, ♂ 50–56 (52·3), ♀ 50–54 (51·9); bill, ♂ 26–28 (26·5), ♀ 23–26 (25·1); tarsus, ♂ 16–18 (16·8), ♀ 15–17 (16·0). WEIGHT: ♂ (n = 8) 9·4–11·5 (10·4), ♀ (n = 10) 8·6–11·7 (9·7), unsexed (n = 20) 9·5–12·6 (11·1) (Maclean 1993).

IMMATURE: juvenile like adult but underparts yellowish at first, later more olivaceous. Gape bright orange 1–2 months after leaving nest, and still fairly bright orange a year later (Hanmer 1981)

NESTLING: not known.

C. v. fischeri (Reichenow): Somalia to Mozambique. Upperparts paler and greyer, less glossy than in nominate race; underparts paler grey. Smaller; wing, ♂ (n = 6) 60–63 (61·0). WEIGHT: Somalia, unsexed (n = 1) 8·7; Kenya, ♂ (n = 4) 9·3–10 (9·8), ♀ (n = 5) 7·8–8·6 (8·0); Tanzania, 1 ♂ 9·0; Malaŵi, 1 ♀ 8·6, 3 weighings of imm. ♀ 8·6, 8·9, 9·1 (Hanmer 1981).

C. v. zanzibarica (Grote): Zanzibar. Like *fischeri*, but underside still whiter. Small; wing, ♂ (n = 6) 59–61 (59·8).

Field Characters. Length 11·5–14 cm. A dull greyish sunbird confined to coastal eastern Africa; mousy plumage almost featureless, relieved only by *scarlet pectoral tufts*. In good light, dull bluish sheen visible on crown, back and wing-shoulders. Sexes alike. Shares habitats with similar-sized Olive Sunbird *C. olivacea*, which is easily recognized by olive plumage and yellow pectoral tufts.

Voice. Tape-recorded (9, 16, 58, 75, 88, 91–99, B, C, F, STA). Variable song of clear, sharp, metallic notes, of 2 general types: (a) 4–6 deliberate, well-spaced notes descending the scale, sometimes accelerating, often with 1–2 low introductory notes, '(za-tsa) tyee-tyeee-cha-choop', '(tutu) tsee-tsee-tya-tya-tyop-tyoop', '(tsu)-tyee-tyee-chop-chop-tachop'; or (b) faster, almost a trill, 'tyutyutyutyu-tyutyu-tyup', '(tsa) tyee-chachachachachacha', 'chichichi-chichichichachachoop', either on one pitch, or descending the scale, or undulating up and down. Continuously on the move, when it gives chatters of various speeds interspersed with clear 5-note songs (Haagner 1964). 2 birds once sang at the same time for half an hour; calls included repeated 'zzik', 'chip, chop, chop' and a loud, richly contralto 'chip-choy-choy'; not known if this was a pair or 2 ♂♂ (Skead 1967). Subdued subsong of protracted tippering given by 2 ♂♂ chasing each other (Skead 1967). Calls include 'zit', 'zaa' a loud 'dzip' and a ringing 'tzeep-tjip-cha'; alarm an agitated twittering; for further calls see Skead (1967).

General Habits. Inhabits coastal lowland bush, dune scrub, woods, mangrove, riverine forest, gardens, evergreen forest tops and edges, also dry *Euphorbia* thickets (Mozambique) and *Acacia* thornveld (South Africa). In forest keeps to tops of trees and leafy parts of more open areas.

Occurs singly in non-breeding season and travels widely each day, visiting small patches of flowering plants. Shy. Sings all year, a bold, attractive song that carries far (S.K. Frost *in* Ginn *et al.* 1989). When calling, ♂ perches near showy flowering plant, flicks wings and shows red pectoral tufts; if ♀ visits the blossoms she is chased until becoming submissive. Prone to make sudden dash at congener, chasing it momentarily at speed. ♂ has a second courtship call, see below. Sings noisily and protractedly from forest tree-top, but silent for long periods when can easily be overlooked. Sometimes feeds low down in vegetation, for instance at garden shrubs in flower. To probe into individual flowers in large, disc-like inflorescence of *Scadoxus puniceus*, bird stands on top of flowering head and leans down, dusting its feet with pollen as it does so (S.K. Frost *op. cit.*). Forages restlessly, moving rapidly from flower to flower and probing quickly. Often probes into bunches of mistletoe *Viscum*. Hawks flying insects, and gleans spiders by hovering at webs (Maclean 1993). Flight across open spaces fast and erratic. Resident; however 'one day a forest will be alive with song....the next and subsequent days will be silent at the same time of day' (Skead 1967), so this rather secretive species may be subject to minor local movements between dense forests. Moves seasonally between forest and thornveld (Maclean 1993).

Food. Nectar and small invertebrates (both found in stomachs: usually more insects than nectar). Feeds at mistletoe *Loranthus*, giant strelitzia *Strelitzia nicolai*, *Schotia*, *Erythrina*, *Leonotis leonurus*, tree-fuschia *Halleria lucida*, snakelily *Scadoxus puniceus* and aloes. Eats spiders, spider eggs, small insect imagos and caterpillars.

Breeding Habits. Solitary nester. Territorial?: 3 breeding ♂♂ were 200–300 m apart at edge of thickly-forested ravine (Skead 1967), but said not to be territorial (S.F. Frost *op. cit.*). 2 noisy birds in tree tops gave song display for 25 min : with last 3 syllables of 'zzik-zzik-zzik-zzik-zzik-chip, choy, choy' song, calling bird thrust head and neck upwards and forwards, with outer wing stiffly lowered and half fanned; when bird continued song with stuttering or trilling phrase it thrust head forward and up and vibrated lower mandible, but did not fan wings (**A**) (Skead 1967). Display of each bird seemed to be directed at the other.

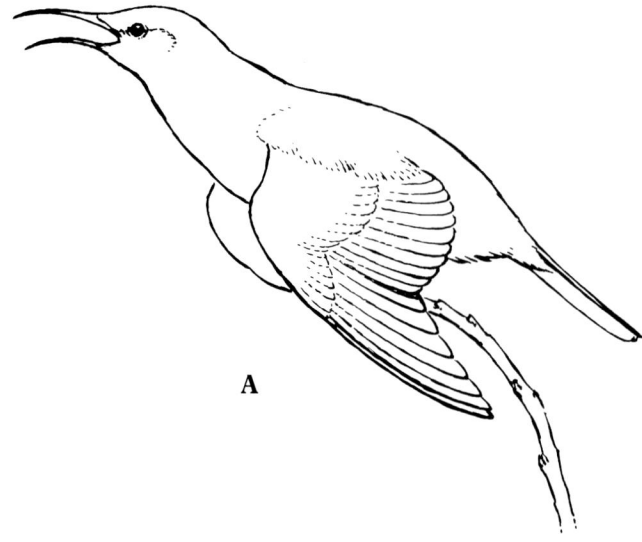

NEST: suspended purse or bag, with side entrance hidden below long, overhanging porch. Nest quite large; sometimes oval and compact but usually attenuated, with long neck below point of suspension. Tidily built, at least inside, but outside generally decorated with leaves and twiglets dangling from bits of spider web, often forming 'beard' hanging below, and can then look dishevelled. Made of black *Marasmius* hyphae, sometimes exclusively, or with strips of bark fibres, small rootlets, rotted twiglets and dead leaves introduced into *Marasmius* weave. Occasionally *Marasmius* absent. Hardly any spider web in fabric of nest, but some is applied to outside surfaces. Lined with thin layer of dead leaves at bottom, and upon it a layer of dead grass inflorescences up to 10 mm thick, run up insides of nest to ceiling where 3 mm thick. Porch droops over entrance. Ext. height up to 230, including neck 110 long; width 75, from front to back 80; entrance 40 × 30; porch 85 long, 60 wide, 10 thick.

In Kenya Grey Sunbird nests almost entirely in buildings: nests made hanging from horizontal wire in room, 2 nests attached to dead creeper hanging from house roof, one in a garage, 3 on wires in bedrooms and verandahs, one on a creeper tendril in an office porch (Britton and Britton 1977). In South Africa nests are 2–6 m, usually 4–6 m, above ground, occasionally in cow-stall or disused hut, but usually outside, under overhanging bank of rock or earth or beneath thick, leafy canopy of tree. Nest hangs from long, drooping twig, vine or piece of wire; same site is used year after year; each year fresh nest made, often next to previous year's old one.

EGGS: 1–3 (av. of 45 South African clutches 2·2). Immaculate; buffish, brown, chocolate, often slightly darker at large end. SIZE: (n = 24) 16·5–19·6 × 11·9–13·2 (18·1 × 12·5).

LAYING DATES: Kenya, May–July (10 clutches), Sept and Nov–Dec (4 clutches); Mozambique, 'from about Oct' (Clancey 1971); Natal, Sept–Jan (mainly Nov); Cape Prov., Dec–Jan. In Kenya breeds mainly in the rains, and predilection for nesting in buildings is probably to prevent nest becoming waterlogged in protracted rain (Brown and Britton 1980).

BREEDING SUCCESS/SURVIVAL: suffers much nest predation from boubous *Laniarius*, monkeys and snakes (S.K. Frost *op. cit.*).

References
Ginn, P.J. *et al.* (1989).
Maclean, G.L. (1993).
Skead, C.J. (1967).

Genus *Chalcomitra* Reichenbach

7 large sunbirds, long-bodied, square-tailed, bills long, relatively heavy and quite strongly decurved. 6 species sexually dimorphic, ♂♂ with mainly velvety brown-black plumage, iridescence only on cap, uppertail-coverts, chin-to-breast and lesser wing-coverts; without pectoral tufts except for *C. fuliginosa*; ♀♀ without iridescence and with underparts heavily striped or mottled with blackish. Nest neat, well-knit, strong; decorated outside; materials usually including spider web; generally with porch and beard.

C. adelberti and *C. rubescens* (W African and Congo Basin forests) form a superspecies; *C. fuliginosa* (open habitats on W African coast) and *C. amethystina* (open woodland, E and southern Africa), which occasionally hybridize, form another;

and *C. senegalensis* (widespread south of Sahara, in savanna woodlands) and *C. hunteri* (parapatric with it in E Africa) compose a third superspecies. The seventh species, *C. balfouri*, endemic to Socotra I., is smaller, monomorphic, without iridescence, and is almost certainly derivative of ancestral *C. senegalensis/C. hunteri* stock. It is also held by some to link *Chalcomitra* with other Indian Ocean endemics (Hall and Moreau 1970). The new Tohâ Sunbird, from Djibouti, may prove to belong to this genus.

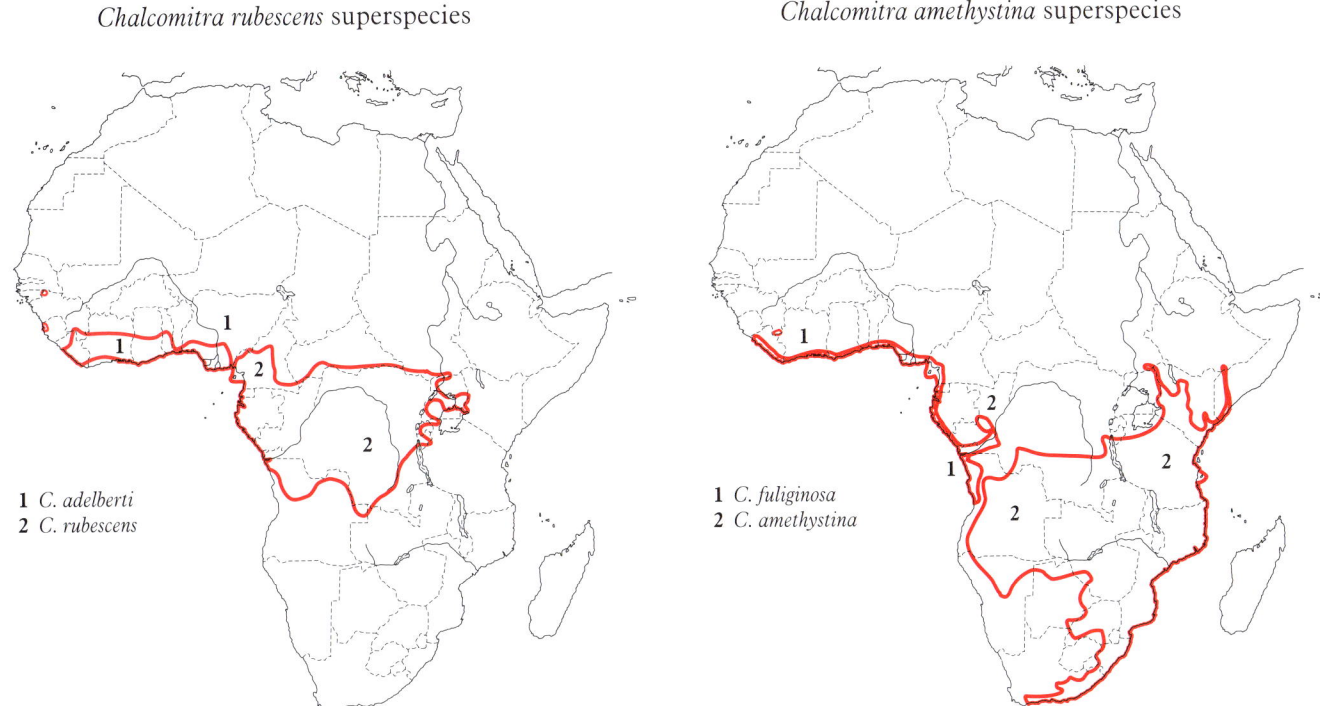

Chalcomitra rubescens superspecies

1 *C. adelberti*
2 *C. rubescens*

Chalcomitra amethystina superspecies

1 *C. fuliginosa*
2 *C. amethystina*

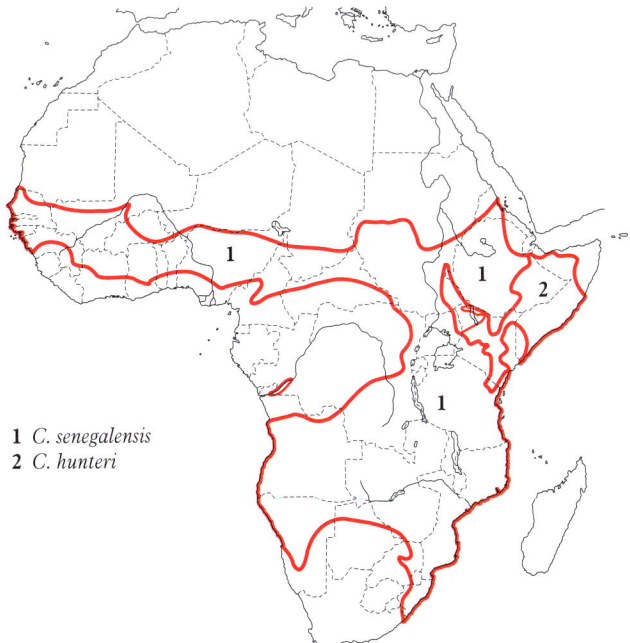

Chalcomitra senegalensis superspecies

1 *C. senegalensis*
2 *C. hunteri*

Plate 11
(Opp. p. 158)

Chalcomitra adelberti (Gervais). Buff-throated Sunbird. Souimanga à gorge rousse.

Cinnyris adelberti Gervais, 1835. Mag. Zool. (Paris), 3 (1833), cl. 2, pl. 19; 'Senegambia' (in error?).

Forms a superspecies with *C. rubescens*.

Range and Status. Endemic resident. Guinea, known from Koundara region in extreme NW, and in SE frequent at Macenta (Morel and Morel 1988, Halleux 1994). Sierra Leone, known from Mt Aureol (Freetown) and York Pass. Liberia, widespread, uncommon to locally common, from coast to N highlands. Ivory Coast, frequent but local, mainly between 06–08°N, south to Grand Bassam and north once to Korhogo; uncommon in Yapo Forest. Ghana, widespread and not uncommon, Akwapim Hills, S scarp forests, Western, Central and Ashanti Regions; outside forest at Tafo and Kumasi, Hohoe (07°10′N), Amedzofe and Legon (Grimes 1987). Togo, common from coast north to Déguingué (08°05′N). Benin, recorded but not well known. Nigeria, not uncommon, from coast north to Ibadan, Ife, Ankpa, Onitsha and Ogoja; once Nindam For. Res. (09°32′N); occurs east on coast to Calabar. Cameroon, single record at Kribi (Turner and Forbes-Watson 1979) and common at Ekok near Nigerian border (Hopkins 1998).

Description. *C. a. adelberti* (Gervais): Sierra Leone to Ghana. ADULT ♂: forehead and forecrown metallic emerald green; hindcrown and nape to upper mantle blackish brown, grading to dark brown on scapulars, lower mantle, back and rump, and dark rufous-brown on uppertail-coverts. Tail blackish brown. Lores, around eye and ear-coverts sooty black, merging with slightly browner hindcrown. Chin sooty brown, bordered by broad metallic emerald green moustachial stripe. Throat pale creamy buff, bordered posteriorly by narrow sooty black band across upper breast; below this, rest of underparts bright chestnut. Lesser and median coverts light orange-chestnut; remaining upperwing feathers dark brown, flight feathers with bronzy gloss above and below and narrow buff edges, tertials and greater coverts more broadly edged rich buff. Axillaries and underwing-coverts chestnut. Bill black; eyes dark brown; legs black. ADULT ♀: top of head to cheeks and ear-coverts olive brown; rest of upperparts olive-brown, greener on uppertail-coverts. Tail dark brown, feather tips fringed whitish, more broadly on T4–T6. Underparts pale buff, whiter on throat, more olive-yellow on breast and belly; lower throat, breast and flanks broadly streaked grey-brown. Lesser and median coverts olive-brown; rest of upperwing dark brown, flight feathers edged olive, primary coverts and greater coverts fringed buffy brown. Axillaries and underwing-coverts pale olive-yellow; inner borders on undersides of flight feathers greyish white. Bare parts as adult ♂. SIZE (10 ♂♂, 6 ♀♀): wing, ♂ 60–67 (63·2), ♀ 58–63 (58·8); tail, ♂ 34–39 (37·0), ♀ 32–36 (34·0); bill, ♂ 18–21 (18·9), ♀ 18–20 (19·3); tarsus, ♂ 14–16 (15·0), ♀ 14–15 (14·8). WEIGHT: Liberia, ♂ (n = 7) 8·9–11 (10·1), ♀ (n = 4) 8·3–10·1 (9·4), imm. ♂ (n = 7) 9·1 ± 0·6.

IMMATURE: juvenile like adult ♀, but cheeks, chin and throat sooty grey; breast and flanks more heavily spotted and appearing barred (rather than streaked). Immature ♂ like adult ♀ but has blackish lores, blackish chin, buff throat and very narrow blackish band or line across upper breast.

NESTLING: unknown.

C. a. eboensis Jardine: Togo to SE Nigeria. ♂ differs from nominate race in having underparts darker chestnut; lesser and median coverts and edges of flight feathers dark brown, concolorous with back. ♀ slightly less yellow below.

Chalcomitra adelberti

Field Characters. Length 11·5–12·5 cm. ♂ unique, with *brilliant green* forehead and crown, bright *yellow-buff* throat and breast and rich *chestnut* belly. ♀ yellowish white below, yellower on belly, with narrow brown streaks; very like ♀ Green-throated Sunbird *C. rubescens*, whose range it approaches but does not meet in E Nigeria, but distinguished by *lack of pale supercilium*.

Voice. Tape-recorded (104, B, CHA). Song brief (1 s), thin, high-pitched, 4–6 up-down notes, clear, quite sweet, not sibilant, 'see-sipit-si-tyew', 'si-tew-sipit-si-tyew', 'tyu, si-soo-sit'. Typical call, given from treetop, a loud, clear, down-slurred 'tyeew' prefaced by soft 'pi' or 'p', audible at close range, 'pi-tyeew' or 'p'tyeew'.

General Habits. Inhabits primary and secondary lowland rain forest, forest edges and clearings; often visits flowering trees in clearings, residential areas and cultivated country outside forest; likes tall trees and forages often in canopy, but also frequently visits flowers within 2 m of ground in gardens and compounds. In SE Nigeria, quite common and widespread in intensively farmed, deforested country, where partial to tall trees near water (Serle 1957), and in orchard bush, old farms in forest zone, gallery forests and small patches of high bush (Marchant 1953).

Occasionally solitary but generally in pairs; up to 10 ♂♂ in full breeding plumage can forage together without remarkable aggression; often in mixed bird parties. Very active, particularly in the hottest, sunniest hours; readily approachable. Many gathered in a lone *Malacantha alnifolia* tree (Sapotaceae, a very small-flowered plant) (Bannerman 1948). Forages at flowers of trees, vines, orchids and other epiphytes, often with other sunbirds, e.g. Superb Sunbird *C. superba*. Visits cassava fields and gardens far from forest. At one locality, Nigeria, often observed hovering in front of hanging ants' nests high in trees, apparently picking ants off (Bannerman 1948). Takes insects by hovering beneath leaves and twigs and gleaning them. Sustained flight is high and undulating.

In Liberia, migrates south for the dry season (Gatter 1997). In Ghana, formerly quite common around Kumasi in Jan–Mar and absent in Apr–Dec (Bannerman 1948); occurs outside forest at various localities only in the wet season (May–Aug), probably only a local movement of a few km rather than distant migration. Single record well north of forest zone in Nigeria, in Nindam in Feb, was ascribed to 'blossom vagrancy' (Dyer *et al.* 1986).

Food. Insects and nectar. Stomach contents insects, including small caterpillars (Liberia: Colston and Curry-Lindahl 1986). 'Nine of those collected had eaten insects, but from field observation it seems to be chiefly a nectar eater, and five stomach contents that were tested tasted strongly of sugar' (Nigeria: Serle 1957). Feeds in *Bombax*, *Malacantha alnifolia* and flame tree *Spathodea campanulata* (Bignoniaceae).

Breeding Habits. Solitary nester. ♂ once seen in 'butterfly' display flight, singing passionately (Serle 1957). A nest recently found at Kakum, Ghana, not yet described (D. Moyer, S. Keith, pers. comm.). Eggs not known.

LAYING DATES: Liberia (Mt Nimba), (various indications May–June and Aug); Ghana, (nest-building Dec); Nigeria, Oct–Mar.

References
Bannerman, D.A. (1948).
Gatter, W. (1997).

Chalcomitra rubescens (Vieillot). Green-throated Sunbird. Souimanga à gorge verte.   Plate 11
Cinnyris rubescens Vieillot, 1819. Nouv. Dict. Hist. Nat., nouv. éd., 31, p. 506; Kingdom of Congo and Cacongo.   (Opp. p. 158)

Forms a superspecies with *C. adelberti*.

Range and Status. Endemic resident, Congo Basin: E Nigeria and Cameroon to W Kenya, south to NW Zambia. Nigeria (Serle 1957), several seen on Kam R., SW Gashaka-Gumti Game Res. (Green 1990), also at Tara (07°01′N, 10°53′E), Anape (06°26′N, 09°23′E) and Afi River For. Res. (06°19′N, 08°59′E) (Hopkins *et al.* 1999). Bioko I., frequent and widespread: Pico Basilé up to 250 m, Malabo, Riaba, Musola, at 350 m from Luba and Ureca to Moka, and up to 1200 m on Mt Moka. Cameroon, quite common in forest zone from Mt Cameroon to north-east and east and in gallery forests in savanna, north to Tibati and Kounden; Korup Nat. Park, 1 record (Green and Rodewald 1996); abundant at Victoria and Kumba – one of the commonest sunbirds in flower gardens. Central African Republic, recorded from Mboko, Bangui, Géringou and Mbaiki (Germain and Cornet 1994), Manovo-Gounda-Saint Floris Nat. Park (error?) and Lobaye Préf. (Carroll 1988); frequent in Dzanga Res. Sudan, uncommon, close to Zaïre border, from 27°E to Imatongs. Mbini. Gabon, frequent and widespread in NE and probably throughout. Congo, occasional in Odzala and Nouabalé-Ndoki Nat. Parks, occasional records in centre and SE, and rare (2 localities) in Mayombe. Zaïre, widespread and often common throughout Congo R. basin, but absent from SE Katanga, and towards range boundaries in N and SW Zaïre restricted to outlying forest patches and heavy gallery forest; common in Itombwe at 720–1840 m. Uganda, fairly common and widespread in W and S at 700–1800 m, from West Nile, Budongo, Masindi and Bugoma to Ankole and Kigezi, east at least to Jinja and Mbale, with sight record from Soroti (Britton 1980). Kenya, locally common in Nandi and Kakamega Forests; formerly on upper Yala R.

Chalcomitra rubescens

Extreme NW Tanzania (4 localities including Bukoba and Kibondo, N. E. Baker, pers. comm.). W Rwanda and NW Burundi. Angola, Cabinda, Zaïre and Cuanza Norte, to N Malanje and N Lunda. Zambia, known only from N Mwinilunga Prov. at Mbulungu–Zambezi confluence, Lisombo Stream and near Mwinilunga town.

Estimated density of 5–6 pairs per km^2, NE Gabon.

Description. *C. r. rubescens* (Vieillot): range of species except Bioko and Cameroon from Mamfe to Bamenda. ADULT ♂ (breeding): forehead and forecrown metallic emerald green, merging with glossy violet-purple band across hindcrown; rest of upperparts sooty brown with faint bronzy tinge. Tail blackish. Lores dull black; cheeks and ear-coverts to sides of neck sooty brown. Chin to upper breast metallic golden green, bordered laterally by emerald green moustachial stripes and posteriorly by narrow glossy violet band; rest of underparts sooty black. Upperwing sooty brown, concolorous with upperparts. Axillaries and underwing-coverts sooty black. Bill black; eyes blackish or dark brown; legs black. ADULT ♂ (non-breeding): in contrast to fresh, dark breeding plumage, winter ♂♂ are worn and bleached (Serle 1950). ADULT ♀: upperparts rich dark brown with olivaceous tinge. Tail dark brown; T6 paler brown, with broad whitish inner web tip; other rectrices with narrow whitish tips. Narrow buffy superciliary stripe. Lores to cheeks and ear-coverts dark brown. Underparts pale buff, tinged olive-yellow from breast to undertail-coverts; chin to upper throat suffused or finely mottled greyish, lower throat, breast and flanks boldly streaked dark grey-brown. Upperwing dark brown, primaries narrowly edged pale brown, secondaries edged olive-brown, tertials and coverts fringed pale brown. Axillaries pale olive-yellow; underwing-coverts pale olive-yellow with dusky bases; inner borders on undersides of flight feathers pale grey-buff. Bare parts as adult ♂.
SIZE (10 ♂♂, 10 ♀♀): wing, ♂ 65–71 (68·5), ♀ 58–62 (60·5); tail, ♂ 43–48 (45·0), ♀ 34–38 (36·4); bill, ♂ 21–23 (21·4), ♀ 20–22 (20·9); tarsus, ♂ 15–17 (15·6), ♀ 15–16 (15·5). WEIGHT: (Uganda), ♂ (n = 15) 9–13 (11·2), 2 ♀♀ 10, 11; (Kenya) 2 ♂♂ 8, 14; 2 ♀♀ 6·5, 9.
IMMATURE: juvenile like adult ♀, but chin and throat dark grey-brown, separated from olive-brown ear-coverts by buffish white moustachial stripe; underparts more boldly mottled. Immature ♂ like juvenile, but chin and upper breast metallic green with bordering purple band.
NESTLING: unknown.
C. r. crossensis Serle: Nigeria (see above) and Cameroon (Mamfe and to west of Bamenda). ♂ like nominate race but throat and breast dark chocolate brown, concolorous with rest of underparts. ♀ not known.
C. r. stangerii Jardine: Bioko. ♂ differs from nominate race in having chin to upper breast emerald green. ♀ buffish white rather than yellowish below; chin and throat densely barred dark brown, breast and flanks more heavily streaked.
TAXONOMIC NOTE: the close (consuperspecific) relationship between *C. rubescens* and *C. adelberti* is shown by the intermediate population *crossensis* (Serle 1963, Louette 1982), even though the last is still known only from 2 adult ♂♂.

Field Characters. Length 11·5–14 cm. A rather large sunbird appearing entirely black in poor light. Overlaps with 2 similarly dark sunbirds, Amethyst *C. amethystina* and Carmelite *C. fuliginosa*, but meets them only at forest edge; Green-throated is the only one inside forest. Can be mistaken for Copper Sunbird *Cinnyris cuprea*, but posture *upright* (Hopkins *et al.* 1999). ♂ distinguished from both by shiny *golden-green patch* on throat and upper breast, bordered below by narrow violet or maroon band, and lack of iridescent uppertail-coverts or shoulder patch; *no pectoral tufts;* chocolate body uniformly darker than Carmelite Sunbird. Forehead iridescent green bordered behind by *broad violet band* on crown. ♀ has dark brown upperparts (pale in ♀ Carmelite), pale yellow underparts with heavy dark streaking; told from similar-looking ♀ Buff-throated Sunbird *C. alberti* by *pale* supercilium; very similar to ♀ Amethyst Sunbird but darker above and more heavily streaked below. ♀ Blue-throated Brown Sunbird *Cyanomitra cyanolaema* is spotted rather than streaked below and has pale stripes either side of dark mask.

Voice. Tape-recorded (102, 104, B, C, ERA, GREG, HOR, McVIC). Call, incorporated into song, a loud, pure, downslurred 'chewp'; during song it is given at different speeds, either at irregular intervals, or at steady rate of 5 per s, sometimes increasing to a dry chattery trill lasting 5 s; interpolated among the 'chewp's are little 2–3 s bursts of softer rapid notes, 'chippity-chippity-chiddle-chip', including some silvery tinkling notes reminiscent of European Goldfinch *Carduelis carduelis*. Also said to give soft 'tui' regularly repeated on song posts (Christy and Clarke 1994).

General Habits. Inhabits montane shrubby heath and grassland, lowland cocoa and rubber plantations, broken forest canopy, open forest and forest edges, tall trees in gallery forest, large clearings around villages, patches of cultivation surrounded by forest, fallow ground, second growth, town parks and village gardens. Forages in canopy and middle strata, 15–30 m above ground, seldom coming to within 5 m of ground. ♂ sings from perch near top of dead or damaged tree.
Occurs singly, in pairs and often in company of other sunbirds, associating freely with them, although at least at some seasons also extremely aggressive (F. Dowsett-Lemaire, pers. comm.). Common in Entebbe Botanic Gardens, Aug–Nov; 5 ♂♂ once seen chasing each other in an *Erythrina* tree. Gleans insect prey in foliage and hovers persistently about tree trunks. Often visits flowers, including some low-down ones, e.g. of banana and *Leonotis nepetifolia*. ♂ defends patches of flowers of Loranthaceae and *Dalbergia* against conspecifics (except its mate) and other sunbirds; territorial ♂ tolerates presence of mate but no other sunbird (F. Dowsett-Lemaire, pers. comm.). Once seen eating small red berries from clusters in roadside tree attended also by Bates's *Cinnyris batesi*, Western Olive *Cyanomitra obscura*, Collared *Hedydipna collaris* and Little Green Sunbirds *Anthreptes seimundi*.
Zambian records fall in Sept–mid Oct, so species may be a visitor there from Katanga.

Food. Small insects (caterpillars, ants) and spiders; berries, and presumed to eat pollen and nectar when visiting flowers of banana, *Leonotis nepetifolia*, etc. (Brosset and Erard 1986). Feeds at flowers of *Erythrina abyssinica*, *Dalbergia* and Loranthaceae. 8 out of 9 stomachs contained insect sclerites; 2 contained small caterpillars, 2 >30 ants and 2, 5 spiders. One contained a hard round seed and another a disc of rubber 2·5 × 4·5 mm in size (probably coagulated latex accidentally ingested when bird pierced long tubular corollae, as in several other sunbirds) (Chapin 1954).

Breeding Habits. Solitary nester.
NEST: suspended, long-oval, domed pouch, very neat and unusually strong, with or without a 'beard' below nest. One made of fine shreds of vegetable material and pieces of dry leaves; decorated with bits of bark, lichen, and caterpillar droppings; lined with plant down. Another made of *Usnea*

and encrusting lichens, mossy rootlets, hair from tree-fern *Cyathea*, and decorated with fine asclepiad down. A third made of *Usnea*, gossamer, and leaf skeletons, thickly lined inside with vegetable down, and decorated outside with grey-green lichen and bits of bark; small, circular entrance near upper pole of nest, overhung by short porch. Suspended by spider web binding near tip of drooping branch; one nest 6 m above ground. Height 200 (plus 'beard' hanging below nest, 50), breadth 70, entrance diam. 25 (entrance 110 from top of nest).

EGGS: 2. Creamy white, almost entirely covered with longitudinal streaks of lavender-grey and dark brown. SIZE: (n = 4, Cameroon) 17·0–17·7 × 11·2–12·5 (17·45 × 12·2), (n = 2, Itombwe) 18·9–19·0 × 12·1–12·4.

LAYING DATES: Cameroon, July, Aug, Oct; Gabon, (sings Sept–Feb; ♀ carrying material, Sept); Zaïre (Ituri: gonads enlarged Dec, Feb, Mar), (Elila R., Kivu: breeding condition Aug, Oct), Itombwe, Mar (and various indications Apr, May, Nov: probably breeds all year); E Africa, Region B, May, June, Aug, Oct–Dec; Zambia, (gonads active Oct).

DEVELOPMENT AND CARE OF YOUNG: a nest with newly hatched young was attended only by ♀, though ♂ was nearby in same tree.

References
Hopkins, M. T. E. *et al*. (1999)
Serle, W. (1963, 1965).

Chalcomitra fuliginosa (Shaw). Carmelite Sunbird. Souimanga carmélite.

Plate 11
(Opp. p. 158)

C[*erthia*] *fuliginosa* Shaw, 1811–1812. General. Zool. 8, p. 222; Malimba, Portuguese Congo.

Forms a superspecies with *C. amethystina*.

Range and Status. Endemic resident, along coast from Sierra Leone to NW Angola, and lower R. Congo. Sierra Leone, sight record (G.D. Field in Dowsett 1993). Liberia, rare to locally common along coast; common in Monrovia; also inland at Ganta, Yekepa (Mt Nimba) and Wologizi. Ivory Coast, locally common on coast from San Pedro (06°37′W) to Assini (03°15′W). Ghana, rare ; 100 years ago frequent on coast from mouth of Volta R. eastward to Togo with 8 birds collected on lower Volga, but only one record since, near Elmina (Grimes 1987, Helsens 1996). Togo and Benin, rare or extinct; 9 birds collected in 1910–1911, some at Porto-Novo (Benin); reportedly rare in Togo in 1923, but no records in either country since then (Cheke and Walsh 1996). Nigeria, common, from Badagri (Benin border) to Opobo (east of Bonny, Niger R. delta), and common at Port Harcourt; record 50 km inland at Abeokuta (Elgood *et al*. 1994). Cameroon, not uncommon near sea at Victoria (Serle 1965), and records along coast elsewhere. Several records in Mbini and Gabon coastal lowlands. Congo, common along coast – the most numerous sunbird there (Dowsett-Lemaire and Dowsett 1991), and inland along and up to about 100 km away from lower Congo R. to Brazzaville (and doubtless to Ngabe, 'opposite' Kwamouth in Zaïre: see below). Angola, widespread in Cabinda and frequent from Zaïre to Cuanza Norte and Cuanza Sul (Gabela and near to Sumbe): Chiela, Sassa-Zau, Malembo, Landana, and Luali-Chiluango confluence (Cabinda) and Chinchoxo, Canzele, Muxima, Carmona, Uige and Salazar (W.R.J. Dean, pers. comm.). Zaïre, along Congo R. valley from mouth to Boma (where common), Kinshasa, Stanley Pool and Kwamouth, and a few records up to about 100 km south of Congo R. (Louette 1989). Specimens thought to be hybrids between *C. fuliginosa* and *C. amethystina* obtained in Nkiene (Zaïre, 04°10′S, 15°55′E) and N'gabe (Congo, 03°08′S, 16°10′E) (Louette 1989) and at a locality *c*. 100 km northwest of N'gabe (Hall and Moreau 1970).

Chalcomitra fuliginosa

Description. *C. f. aurea* (Lesson): Liberia to Cameroon and coastal Gabon. ADULT ♂: forehead and forecrown glossy violet; hindcrown to neck and upper mantle light tawny brown, grading to rich dark brown on scapulars, back and rump; uppertail-coverts metallic purple, in some tipped greenish. Hindcrown commonly bleaches with wear to pale greyish buff, mantle to pale buffish brown. Tail blackish brown. Lores black; ear-coverts and cheeks brown, merging with more buffish crown and nape. Chin and throat metallic violet-purple; pectoral tufts pale yellow; rest of underparts rich dark brown. Leading lesser coverts glossy purple; upperwing otherwise blackish brown. Axillaries and underwing-coverts blackish brown. Bill black; eye dark brown; legs black. ADULT ♀: upperparts ochreous brown, paler, more buffy, on neck and upper mantle; long uppertail-coverts dark brown. Tail blackish brown, paler on T5–T6; tips of T1–T3 narrowly fringed white, of T4–T6 with broader white area on inner web. Lores blackish brown; cheeks and ear coverts dark brown. Chin and throat rich dark brown, feathers fringed buff; bordered by creamy buff moustachial stripes. Rest of underparts pale buffy yellow, breast and flanks mottled rich dark brown, undertail-coverts with dark brown centres. Upperwing blackish brown; primaries narrowly edged yellowish, secondaries edged pale buff, tertials and wing coverts fringed pale buff. Axillaries pale yellowish; underwing-coverts buffy yellow with dark brown bases; inner borders on underside of flight feathers pale grey-buff. Bare parts as adult ♂. SIZE (10 ♂♂, 8 ♀♀): wing, ♂ 67–72 (70·2), ♀ 62–67 (64·3); tail, ♂ 42–46 (43·5), ♀ 37–41 (39·0); bill, ♂ 22–26 (24·6), ♀ 22–25 (23·3); tarsus, ♂ 16–18 (17·0), ♀ 17. WEIGHT: 1 ♀ 12·2.

IMMATURE: juvenile ♂ resembles ♀, but cheeks to chin and throat sooty brown (no moustachial stripes); breast deeply mottled sooty brown.

NESTLING: unknown.

C. f. fuliginosa (Shaw): lower Congo R. to W Angola. ♂ darker brown above, glossy areas slightly more extensive on throat and crown. WEIGHT: Angola (Cabinda), ♂ (n = 1) 11·6.

TAXONOMIC NOTE: hybridizes with and forms a superspecies with *C. amethystina* but not with *C. rubescens*, even though range of the last is to some extent intervening (Louette 1989, *contra* Hall and Moreau 1970).

Field Characters. Length 12·5–15 cm. ♂ mainly dark chocolate-brown; the only sunbird of this type in most of W Africa, but from Cameroon east meets 2 similar species, Green-throated and Amethyst Sunbirds *C. rubescens* and *C. amethystina*, although it remains mainly a coastal bird. ♂ differs most conspicuously from these 2 in *bright yellow* pectoral tufts, also in light brown hind-neck and mantle which become *conspicuously pale* in worn plumage; also has velvety black area around base of bill, including chin and lores, *small* dull blue-violet patch on forehead, *small* amethyst throat patch with golden-green reflections, amethyst uppertail-coverts and shoulder patch. ♀ much lighter than ♀♀ of the other 2, pale brown above, especially on hind-neck and mantle, pale feather edgings on brown wing; below, brown throat patch bordered by white malar stripe, underparts pale buff, yellower on belly, with dark blotches on breast. Immature like ♀ but without pale malar stripe.

Voice. Tape-recorded (104, B, C, CHA). Gives an irregular series of short (0·5 s) dry, descending trills (possibly territorial song: Chappuis, in press); call, a sharp 'tsit'.

General Habits. Inhabits maritime plains, open coastal areas, mangroves, sublittoral thickets, gardens, coffee forest, forest clearings and disturbed, secondary forest (but avoids primary forest or only just penetrates its very edges). In Congo feeds mainly along forest edges but nests in more open places, e.g. plantations or isolated, small *Annona senegalensis* bushes. Feeds at blossom by perching on or hovering in front of it; bird once seen progressing on ground below a *Russellia* shrub by hopping, then reaching up with neck extended to suck at a drooping bloom (Serle 1965). Forages in groups of up to 8 on flowers and insects (Gatter 1997).

Occurs commonly in and around Lagos, Nigeria, from late July to late Oct, when plumage becomes markedly bleached; disappears in late Oct (Bannerman 1948) but probably moves only a few km to nest along coast.

Food. Strongly attracted to *Moringa pterygosperma*, banana and *Hibiscus* flowers in Liberia, flamboyant trees, canna lilies and coconut palms in Nigeria (Bannerman 1948) and *Russellia* flowers in Cameroon (Serle 1965). Large insects sucked dry and then discarded (Gatter 1997).

Breeding Habits. Solitary nester.

NEST: suspended, well-knit, domed pouch with circular entrance in one side surmounted by a porch and with short 'beard' hanging below. Made of plant fibres, skeletonized leaves from rubber tree, and a little moss, bound together with spider web and fine black fibres; decorated outside with bits of bark; profusely lined with plant down. One nest was suspended by spider web and plant fibres from extremity of drooping branch of rubber tree in a plantation; another sited in avocado tree near tidal mangrove creek; others right out in the open in an isolated shrub; one suspended from end of wire hanging from telegraph pole, another in low bush over water next to nest of Reichenbach's Sunbird *Anabathmis reichenbachii* (Dowsett-Lemaire and Dowsett 1991).

EGGS: 2. Long ovals, with smooth, rather glossy surface. Brownish white ground, largely covered with streaks, spots and longitudinal blotches in shades of brown with similar lilac-grey shell marks; marks form a well-defined cap at large end (Serle 1950). SIZE: (n = 4) 19·0–21·7 × 13·1–14·6 (20·1 × 13·75).

LAYING DATES: Liberia, (juvs and imms Oct); Nigeria, breeds about Sept–Mar (nest with young early Oct, nest Nov, accompanied juveniles Oct, Nov, nest being built late Jan, copulation Mar: Elgood *et al.* 1994); Cameroon, Sept–Nov (1 clutch Sept, 3 clutches Oct, 3 Nov); Zaïre (Kinshasa), (gonads enlarged Dec).

References
Bannerman, D.A. (1948).
Serle, W. (1950, 1954, 1965).

Chalcomitra amethystina (Shaw). Amethyst Sunbird; Black Sunbird. Souimanga améthyste.

Certhia amethystina Shaw, 1811–1812. General Zool., 8, p. 195; Cape of Good Hope.

Forms a superspecies with *C. fuliginosa*.

Range and Status. Endemic resident and partial migrant, S Congo to SW Angola, east to eastern seaboard, north to S Sudan and S Somalia, south to E South Africa.

Gabon, unpublished record (Dowsett 1993). Congo, common in Léfini Res. and record at Djambala. Specimens thought to be hybrids between *C. amethystina* and *C. fuliginosa* obtained in Nkiene (Zaïre, 04°10′S, 15°55′E) and N'gabe (Congo, 03°08′S, 16°10′E) (Louette 1989) and at a locality *c.* 100 km northwest of N'gabe (Hall and Moreau 1970). Angola, widespread on central plateau, west to escarpment, south to about Chibia, north to Guissama Nat. Park, eastern limits unclear. Zaïre, E Bandundu west to Gungu (05°44′S, 19°19′E on Kwilu R.), Kasai and Katanga, north to S Kivu; just into S Burundi (Schouteden 1966a). Ethiopia, only in extreme SE, frequent around Dolo. Somalia, uncommon, whole length of Jubba Valley. Sudan, restricted to Didinga Mts with single record from Imatong Mts. Uganda, Kidepo Valley Nat. Park and Moroto to Elgon. Kenya, common from coast up to 2200 m in medium rainfall regions, mainly west of 37°E and south of 02°N; record at Mandera, on Kenya/Somalia/Ethiopia border, and widespread resident with some local wandering; from Mt Loima, Kitale, Kabarnet and Tugen Hills, Horr Valley and Mt Kulal, Marsabit, Ndoto Mts, Mt Uraguess, Laikipia, lower Tana R. valley, Lamu I. and Manda I., to Tanzania border (Short *et al.* 1990). Tanzania, widespread in NE near Kenya border, south to Ngorongoro, Usambara Mts and coast to mouth of Rufiji R. and Mafia I. South of Rufiji R. thought to be migratory, being common in Iringa uplands and Songea only seasonally, with records in SE Tanzania at Nandembo, Masasi and Mikindani. Poorly known in W Tanzania, in Kigoma and W Tabora, and on Ufipa Plateau. Zambia, throughout except for parts of upper and lower Luangwa Valley. Malaŵi, common throughout except for L. Malaŵi littoral and Shire Valley where scarce; mainly at 615–1540 m. Mozambique, widespread in Nampula, Zambezia and from lower Zambezi to Beira, also in N and W Tete; also occurs in Lebombo Mts (Mt Meponduie, Namaacha, Goba and Pequeños Limombos) and widespread in S, north to Inhambane. Ranges in Namibia, Botswana, and Zimbabwe, South Africa and Swaziland (where common), mapped by Harrison *et al.* (1997). Not in Lesotho. Widespread in Zimbabwe, locally abundant, including up to 1800 m on seaward-facing slopes of E mountains.

Description. *C. a. kirkii* (Shelley) (includes '*doggetti*'): Sudan, Uganda, Kenya (except E) and Tanzania (except NE) to S and E Zambia, Malaŵi, Zimbabwe and central Mozambique. ADULT ♂: forehead and crown metallic emerald green; rest of upperparts and sides of head and neck sooty black. Tail blackish brown. Moustachial stripe, chin and throat metallic pinkish purple; rest of underparts sooty black. Leading lesser coverts metallic purple; upperwing feathers otherwise blackish brown, edged blacker when fresh. Axillaries and underwing-coverts sooty black. Bill black; eyes brown; legs black. ADULT ♀: upperparts olivaceous

Chalcomitra amethystina

For migrations and possible non-breeding areas, see text

grey-brown; long uppertail-coverts dark brown. Tail blackish brown; T2–T3 narrowly tipped greyish white, T4–T5 more broadly so, T6 paler with narrow whitish outer edge. Narrow superciliary stripe creamy white; lores, cheeks and ear-coverts dark grey-brown. Underparts pale grey-buff, tinged olive-yellow, throat, breast and flanks diffusely streaked grey-brown. Lesser and median coverts as upperparts; rest of upperwing dark grey-brown, primaries edged buffy white, secondaries edged olive-buff, tertials and coverts fringed buff. Axillaries and underwing-coverts pale buffy yellow; inner borders on undersides of flight feathers grey-buff. Bare parts as adult ♂. SIZE (10 ♂♂, 10 ♀♀): wing, ♂ 65–68 (66·6), ♀ 60–64 (61·8); tail, ♂ 42–46 (43·8), ♀ 36–40 (38·3); bill, ♂ 25–26 (25·5), ♀ 23–25 (23·5); tarsus, ♂ 16–17 (16·3), ♀ 15–16 (15·7). WEIGHT: Kenya, ♂ (n = 4) 10–11·8 (10·9), ♀ (n = 11) 10–11·6 (10·7); Zimbabwe, ♂ (n = 23) 9·2–12·8 (11), ♀ (n = 11) 8·3–11·4 (9·8).

IMMATURE: juvenile like adult ♀, but chin and throat dusky or blackish, bordered by pale moustachial stripe; breast more boldly spotted, appearing barred (rather than streaked). Immature ♂ like adult ♀ but chin and throat metallic pinkish purple, lores blackish and some variable black feathering on breast, upperparts and wings (in South Africa retained until over 12 months old).

NESTLING: hatches naked and blind, with egg tooth, skin pale orange; later, mouth bright orange and gapes pale yellow.

C. a. kalkcreuthi (Cabanis): S Somalia, coastal lowlands Kenya and N Tanzania. Like *kirkii*, but ♀ has superciliary stripe more pronounced and throat paler. Slightly smaller: wing (10 ♂♂) 62–65 (63·4).

C. a. deminuta (Cabanis): Zaïre to Angola, N and W Zambia, N Namibia and N Botswana. Hybridizes with *C. fuliginosa* (Louette 1989). Like *kirkii*, but ♂ has long uppertail-coverts glossy purple. Slightly larger: ♂ (n = 10) wing 67–71 (69·3), bill 24–28 (26·4).

C. a. amethystina (Shaw): South Africa (S and E Cape to W Natal). ♂ like *kirkii*, but with glossy purple uppertail-coverts. ♀ has ground colour of underparts paler, less yellow, than *kirkii*, but chin and throat dusky or sooty black with contrasting buffy white moustachial streak, and breast more heavily marked, barred rather than streaked. Larger; ♂ (n = 10) wing 71–75 (73·5), bill 27–32 (29·7).

C. a. adjuncta Clancey: South Africa (N Natal, Transvaal) and Swaziland to S Mozambique. Like nominate *amethystina* but ♀ slightly paler and greyer above and less heavily marked below.

Field Characters. Length 13–14·5 cm. A dark sunbird of open wooded country, widespread in E and S Africa and unlike any other sunbird there, but in W and N of range meets 2 similar sunbirds at forest edge, Green-throated *C. rubescens* and Carmelite *C. fuliginosa*. ♂ distinguished by *metallic green* forehead and crown (not backed by violet band as in Green-throated; Carmelite has small violet patch on forehead), *broad* round bronzy purple-pink throat patch (Carmelite has similar but much smaller patch), small violet and blue shoulder patch (absent in Green-throated); rose-violet uppertail-coverts in southern birds and in race that meets Carmelite in Congo, absent from most birds in E Africa. *No pectoral tufts*. ♀ similar to ♀ Green-throated but paler above, less heavily streaked below. However, intensity of streaking on underparts very variable, at least in southern Africa, and is completely lacking in some birds, giving the impression of a ♀ Miombo Double-collared Sunbird *Cinnyris manoensis* with a yellow wash on the underparts; Amethyst distinguished by having some degree of white in the tail (Tree 1991). Juv. ♂ like ♀ but with blackish throat. Immature ♂ assumes purplish throat before rest of dark plumage and is almost identical to Marico Sunbird *C. mariquensis* at same stage; breast of Amethyst *usually* more strongly marked and appearing barred rather than streaked, and bill more strongly decurved (Tree 1991). All birds other than adult ♂♂ have full supercilium, lack white on alula and outer primary coverts which is present in ♀ and immature Scarlet-chested Sunbird *C. senegalensis* (Tree 1991).

Voice. Tape-recorded (16, 86, 88, 91–99, 104, B, C, F, KEI, LEM, STA). Song a rapid, continuous bubbling and twittering lasting several min without a break; muted, almost like a subsong, without the sharp quality of e.g. Variable Sunbird *Cinnyris venusta*; with short breaks, singing spell may last up to an hour (Skead 1967). This song can be heard at any time of year and is often given when ♂ is alone, and is not necessarily stimulated by either the need for territory advertisement or the proximity of a ♀; also given by immature ♂, and soft warbled version by ♀ (Skead 1967). Another version, while still quite rapid, is slow enough to hear individual calls and little trills, 'chip-chipper-chewy, tyip-tyew-tyip, warty-witty trrrrrrr', joined together with little silvery notes. Common call is a series of clear, sharp, measured notes given at rate of about 2 per s, either monotonous repetition, 'chip-chu-chip-chu-chip-chu ... ' or more varied, 'chip-chyew-chiju-chip ... '. This call (song?) heard most often at dawn but may be repeated at intervals during the day; calling spells may last up to 30 min (Skead 1967). Another characteristic note is a single, abrupt 'tschiek', with variants 'chist', 'choot' and 'poyt'; ♂ pursuing another bird gives plaintive 'ssweeek-ssweeek' (Skead 1967). Alarm a hard, coarse 'jit-jit-jit-jit ... '.

General Habits. In E Africa inhabits woodlands, gardens and edges of forest, up to about 2000 m; and in dry, lowland areas confined to riverine and other dense woods. In Zambia in open *Brachystegia* and other woodland with flowering trees and shrubs; in Zimbabwe in *Brachystegia* and *Baikiaea* woodland, sometimes abundant in flowering *Baikiaea plurijuga* woods; avoids foraging under closed canopy; occurs on crowns of trees emergent from forest (e.g. flowering *Erythrina*); aloe-covered hillsides, proteas, town gardens.

Equally often singly or in pairs; sometimes also in small, excitable single-sex or mixed groups (for instance, of an adult ♂, 2 adult ♀♀, 3 sub-adult ♂♂ and a juvenile). Active, restless, conspicuous though often retiring, quite aggressive at times (less so than Malachite Sunbird *Nectarinia famosa*) but forages amongst flowering plants alongside other sunbirds (Collared *Hedydipna collaris*, Greater Double-collared *Cinnyris afra*, Lesser Double-collared *C. chalybea*, Malachite) with only limited chasing now and then; tends to dominate smaller species but to be dominated by Greater Double-collared. Both sexes attack their images in mirrors and window-panes, when said to warble quietly and sway from side to side. In conspecific groups, social chasing and jostling occurs year-round, accompanied by stuttering chatter 'chee, chee, chee, chee'. Frequent display by adult ♂ directed at hunger-calling juveniles in flowering tree: ♂, also uttering hunger-call ('sssweeek, sssweeek, sssweeek ... '), flies at juvenile in slow, butterfly-like flight, half planing, causing juvenile to flee, or else ♂ and juvenile fight and flutter to ground interlocked. Once young bird broke away, flew to bush and resumed calling, and adult perched 10 cm away facing juvenile and uttered burst of chippering warbles; juvenile reacted by leaning back, quivering wings, raising neck feathers, and swaying slowly from side to side, holding position for long time at end of each sway (Skead 1967). Also in party, adult ♂ often planes or parachutes down from high perch towards conspecifics, with wings held well up and quivering rather than beating, tail inclined with tip above level of back, bird uttering stuttering, plaintive 'sssweeek' calls (**A**). Another frequent display by adult ♂ is sudden deceleration at end of dashing flight at conspecific, bird 'applying air brakes' at the last moment, with body drawn up vertically and wings flapping jerkily fore and aft (**B**).

Has barely-perceptible habit of gently bobbing head rhythmically up and down (**C**) whilst quietly perching between activities (Skead 1967) (perhaps working tongue without protruding it). Flight erratic, speedy, and un-certain in duration; jinks on the instant to right or left and makes sideways swoops down and round.

Takes nectar by perching below, in front of, or some-times behind blossom, and occasionally by hovering before it, and thrusting bill and head into corolla tube. Probes into juicy flesh of fallen orange through cracks in skin or holes made by bulbuls, barbets or starlings. In Mombasa (Kenya) seen drinking sap flowing from young shoots of

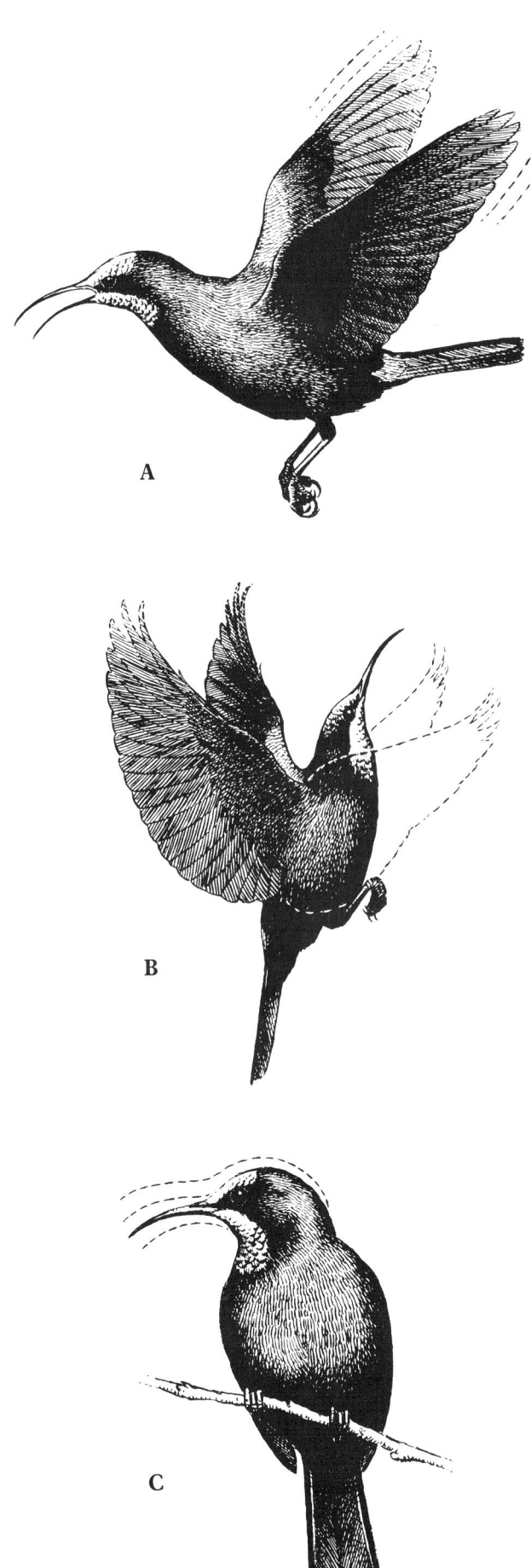

coconut palm (pierced for commercial collection). Picks spiders out of webs by hovering and stabbing; takes spiders from webs between tree branches, amongst aloe leaves, under rocks, and in eaves of steadings. Sometimes hovers to take spider out of flower; spider may be carried in bill to perch and there mandibulated before being swallowed whole. Commonly catches insects, especially termites, in flight, in late afternoon and evening; termites also taken from ground at moment of emergence from crevice in soil; termite wings generally rubbed off against hard perch, then bird swallows body whole (Skead 1967). Once fed persistently by flying from fence into swarm of hovering flies, fluttering momentarily, snatching a fly, and returning with it to fence to eat it. Sometimes flies afar (often calling 'tschiet') and lone bird can occur quite a distance from usual habitat.

Roves widely: local population often vanishes, to reappear suddenly some weeks later at same place (that even applies in leafy suburbs, where nectar-bearing plants more constantly in flower than in the countryside). In addition to such essentially local wandering, Amethyst Sunbird is a partial migrant, decreasing in abundance in about June–Sept in NE Namibia and Botswana and decreasing about Mar–May in E South Africa (A.J. Tree in Harrison *et al.* 1997). It is a breeding visitor from Sept or Oct to Kalahari Sand woodlands in NW Zimbabwe and in Aug–Apr in Kariba Basin; at Great Zimbabwe arrives in Sept, breeds, disperses, and disappears in Dec; at Mutare arrives in mid June and is the commonest sunbird from late July; at Chirawanoo occurs erratically in all months but a marked increase from Jan to July and decrease thereafter (Tree 1990). ♂ ringed in Zambia recovered 115 km away, 4 months later, and a recovery of 115 km in SE Cape, South Africa. Common in Iringa and Songea, Tanzania, Apr–Oct. In S Mozambique thought to be a non-breeding visitor (Clancey 1996). In South Africa an albino ♀ remained in same garden for 2 years, and other, similar examples of site fidelity have been inferred.

Food. Nectar, many spiders and insects. Nectar of *Strelitzia augusta*, *S. reginae*, *Schotia*, *Bauhinia*, *Poinsettia*, *Erythrina*, *Watsonia*, *Cestrum*, *Crotalaria australis*, *Mina lobata*, many species of aloes, *Salvia*, *Hibiscus*, *Tecomaria*, *Canna*, *Protea*, *Kniphofia*, *Royena*, *Brunsfelsia*, *Petraea*, *Cassia*, *Tithonia tagetiflora*, *Leonotis ocymifolia* and *Baikiaea plurijuga* (*Babbler* 35, 1999, 56) and *Eucalyptus*. Birds strongly attracted to *Loranthus*, *Syzygium* and *Dalbergia nitidula* (Tree 1990). Spiders evidently form large part of diet (see above). Small insects, taken from ground and on the wing, include termites and flies; insects taken in foliage include aphids. Birds feeding at blossoms may take arthropods as well as nectar: 'bills withdrawn from aloe corolla-tubes ... small black objects at their tips ... necessitated much manipulation of the long, thin whippy tongue before swallowing' (Skead 1967). Orange juice and coconut palm sap (see above).

Breeding Habits. Solitary nester, evidently monogamous, although some old observations, repeated below, might suggest a promiscuous mating system. ♂ defends immediate area around nest and pair uses same home range year after year, but not known whether ♂ or pair is territorial in conventional sense, since pairs seem never to nest near each other. In prelude to breeding, ♂ pursues ♀ in high-speed chases, at frequency where he hinders her gathering of nest material. With ♂ only a few cm behind ♀, the pair dash and weave with side-to-side swerves through trees, zooming and diving in the open 'and changing direction with astonishing right-angled suddenness, the male always maintaining station astern so perfectly that he is never caught unawares by a female's unexpected diversionary jink. He gives the impression of being in tow. Sharp clicking sounds emanate from him during the chase, apparently as wing-snaps' (Skead 1967). Sometimes both birds hover momentarily in mid chase, twittering excitedly; occasionally ♂ flies at ♀ then air-brakes with body held vertically (**B**). With short breaks a bout of chasing can last 30 min, then pair perches together and ♂ quivers wings (like a soliciting juvenile).

Another pre-breeding behaviour is vent-pecking. ♂ and ♀ perch together, calling excitedly; ♂ bobs up and down then bends forward and pecks ♀'s vent several times. Once a small black object was seen in tip of ♂'s bill, interpreted as faecal pellet (Skead 1967) (though it may have been a cloacal plug?). Once, ♂ flew from perch to hover above perching ♀, who crouched; another ♀ flew toward the pair but was ignored; first ♀ then displayed by tensing body, stretching head and neck far forward, outstretching one wing, warbling, fanning tail to off-centre position, and swaying body slowly from side to side. ♂ paid no attention and there was no copulation. ♂♂ occasionally breed when in eclipse or immature plumage; a ♀ was attended by 'immature' ♂ throughout 2 successive nestings (Skead 1967).

Double broods are common and multiple broods occur. Although new nests are built year after year in a particular site, old nest may be used for second and third broods in a season. One ♀ laid 3 clutches in same nest in one season, and used nest again in following year. One ♀ built on a clothes line, starting first nest each year at nearly the same date (12th, 16th, 17th Sept).

NEST: neat, oval or pear-shaped, suspended, domed pouch, narrow at top, where side entrance situated; strongly and compactly made of fine plant fibres, chaffed grass blades, small leaves, down, wool, hair, fibres, bark, and grass-stems (often of *Eragrostris*) bent right round so that seed-heads finish off at the top and form a porch over entrance; sometimes no porch (Serle 1943); quite often a beard under nest, of cobweb, stringy lichen and dead leaves; plant materials in nest walls bound with spider web; walls thick and compact, outside decorated with pieces of lichen and a few tiny dead leaves, bits of bark, clusters of tiny reddish seeds, wood shavings or quite large, white wood splinters; inside with smooth, solid, felt-like lining of woolly white, yellowish or reddish plant down, sometimes from *Tarconanthus* seeds (Compositae) or asclepiads. One nest was lined and another decorated externally with red petals.

Attached to down-pointing tip of twiggy branch, often incorporating a few living, hanging leaves into back of nest; usually somewhat exposed, sometimes in leafy cover among branches and not readily seen; occasionally suspended from stem of climbing plant trailing under tree foliage, in which case nest material usually twisted up climber for several cm; also occasionally attached to bare wires, clothes lines and ropes, around country and town houses. Often sited by a road, pathway or other gap in woods; often in clump of shrubs growing on termite mound; once in house verandah. Most nests loose-knit and easily blown down, others appear flimsy but in fact hang in exposed position for more than a year. Nests of *kirkii* in Zimbabwe consistently placed low down, at *c.* 2 m from ground, in sapling or open, thinly-branched shrub, e.g. among scattered young *Brachystegia* and *Berlinia* trees near *Eucalyptus* plantation; also consistently smaller than nests of *amethystina* and *deminuta*: height of body of nest (excluding 'beard') (n = 6) 25–40 less than in nominate race and width at base 12·5–25 less (Vincent 1949). Height above ground: 2–6 m, mostly 2·5–4 m; once 1 m (above water). Brachystegias, acacias, mangroves, tamarinds, pines, firs, eucalypts and wattles commonly used. External size (n = 6): height 120–171 (150), width 53–89 (76), front to back 64–72 (66); entrance (n = 2) 50 × 55, 40 × 35; also (n = 1, *kalkcreuthi*) int. height of cavity 78, entrance diam. 28, depth of floor below lower lip of entrance 43 (Serle 1943).

Built by ♀ only. While she works, ♂ sings incessantly from favourite perch in tree near nest and from perch near ♀'s collecting point. All ♀'s outward and homeward flights accompanied by much chasing and interference by ♂. Sometimes ♀ performs hovering, fluttering display on slow-beating wings during her flights between nest and collection point, and ♂ hovers just above her. Nest is begun by ♀ winding spider web around twig; lichen is pushed onto or into web; more spider web and lichen added successively, to form small oblong wad; into it ♀ thrusts her bill and head, parting the mass to make hole which will become nest entrance; then more material is added to bottom of wad, under hole, to make growing nest signet-ring-shaped; thereafter nest cavity and entrance take shape by ♀'s regular entry and compression of material within; nest becomes loosely-knit shell, mainly of cobweb and lichen; then walls are thickened and porch of grass seed-heads is built above entrance. ♀ snips off dead seed heads with her bill, either perching on porch or hovering in front of it. ♀ often brings single thread of spider silk at a time, and winds it into position with weaving action of her head. 'Nest material is easily collected and plentiful, but females consistently travel further than they need, egged on by their ever-importunate males' (Skead 1967). Later, nest is lined; during that stage, ♀ may sleep in nest at night, and kneads lining evidently with her feet. 3 nests took 33, 35 and 36 days to build.

EGGS: almost invariably 2 (once 1, rarely 3, once 4). Laid on successive days. Ovate; shell smooth and slightly glossy or may be slightly rough. Eggs of nominate race white, cream, greenish grey, pale green or light grey, thinly covered in short, light brown and ash-grey lengthwise streaks and smudges, markings heavier at large end; or with black or blackish brown smudges and small grey clouding and streaky spots in irregular band at large end. Eggs of *deminuta* either light fawn, thickly covered with fine, streaky, sepia and grey mottling making faint band around large end, or deep creamy buff, with short twirly pencilling and a few black or dark brown streaks mixed with grey, streaky spots. SIZE: *C. a. amethystina* (n = 36) 18·4–23·2 × 12·6–13·5 (20·8 × 13·1), and (n = 6) 18·1–19·0 × 12·0–13·4 (18·7 × 12·8); *C. a. deminuta* (n = 11) 16·8–19·0 × 12·3–13·5 (17·8 × 12·8); *C. a. kirkii* (n = 6) 17·6–19·5 × 11·9–12·9 (18·3 × 12·4).

LAYING DATES: Angola, May (and enlarged gonads, Oct–Dec); Zaïre, (Lubumbashi), Sept–early Nov; E Africa, Region A, Aug, Region D, Mar–Jan, mainly Mar–Apr and Oct–Dec, Region E, Mar, May–Sept, Nov; Zambia, Aug–Mar, mainly Sept and Feb–Mar; Somalia, (♀ on nest, Aug); Malaŵi, Sept–Feb and Apr, mainly Sept–Oct; Zimbabwe, Aug–Mar (146 records), mainly Sept (38%), Oct (41%) and Nov (12%); Transvaal, Aug–Mar; Natal, Aug–Mar and June; E Cape, July–Apr.

INCUBATION: by ♀; sits deep in nest, bill tip sometimes showing behind entrance; she sleeps in nest at night during incubation period (and ♂ sleeps in nearby tree). ♀ easily alarmed and flees nest during first few days, but later sits more steadily and is less easily put off. ♂ spends much time perching prominently nearby, singing and chasing away other sunbirds, and from time to time flies to nest making chippering call, when he and ♀ fly off, doubtless to forage. Period: (n = 3) 13 ± 1, 14 ± 1, 18 ± 1 days.

DEVELOPMENT AND CARE OF YOUNG: (single nestling:) bill 6 long on day 3, 9 on day 5, 10 on day 7, 12·5 on day 10, 16 on day 13, 19 on day 16, much shorter than adult's when chick left nest on day 19; day 4, wing quills emerging; day 5, skin purplish above, reddish orange below, mouth bright orange, chick becoming active; day 6, most quills broken open, except on forehead and chin; day 7, eyes 75% open; day 8, maxilla (previously shorter) now same length as mandible; days 8–10, chick almost fully feathered; day 10, maxilla longer than mandible; day 13, ring of bright yellow feathers around cloaca; days 14–16, quills on forehead (previously unopened, close-packed and shield-like) beginning to open.

Young fed mainly by ♀; ♂ brings the occasional spider, and also often visits nest without bringing food. Period (n = 2) given as 14 and 18 days. After leaving nest the 2 young keep close together, persistently giving penetrating and far-carrying hunger calls and following parents around. If danger threatens, parents show much consternation and incessantly make stuttering alarm calls; young may continue to give hunger calls, though less noisily. Young out of nest fed by both parents and quiver wings rapidly when fed; they also quiver wings when sitting waiting, and even when probing flowers on their own.

BREEDING SUCCESS/SURVIVAL: exposed nests often blown down by wind (Vincent 1949, Skead 1967), though ♀ may continue to brood if nest only dislodged to hang askew. Parasitized by Klaas's Cuckoo *Chrysococcyx klaas* and Emerald Cuckoo *C. cupreus* about equally, in Malaŵi, Zimbabwe, South Africa and probably throughout sunbird's range. Amethyst Sunbirds seen mobbing Jacobin Cuckoo *Oxylophus jacobinus* and imm. Red-chested Cuckoo *Cuculus solitarius*, and also once a Lesser Honeyguide *Indicator minor*, though none is known to brood-parasitize them (M.P.S. Irwin in Fry *et al.* 1988). Sometimes carries ticks *Hyalomma aegyptium*. Ringed birds over 5, 6·5 and 7 years old (Hanmer and Chaddon 1997).

References
Harrison, J.A. *et al.* (1997).
Skead, C.J. (1953, 1967).
Tree, A.J. (1990).
Vincent, A.W. (1949).

Chalcomitra senegalensis (Linnaeus). Scarlet-chested Sunbird. Souimanga à poitrine rouge. Plate 11 (Opp. p. 158)

Certhia senegalensis Linnaeus, 1766. Syst. Nat., ed. 12, 1, p. 186; Senegal.

Forms a superspecies with *C. hunteri*.

Range and Status. Endemic resident and partial migrant, subsaharan Africa except for rain forest zone and arid parts of E and SW Africa. Mauritania, frequent in extreme S, in S Guidimaka and upper Senegal R. valley. Senegal, common in S, frequent near coast in NW, absent from NE. Gambia, the commonest sunbird; uncommon on coast, common to locally abundant inland. Guinea-Bissau. Guinea, Fouta Djallon in NW, and recent records from Conakry coast, Kankan and Beyla areas. Mali, occurs north to *c.* 16°N, abundant at about 13°30′N, from west to east. Sierra Leone and Liberia, no certain records. Ivory Coast, frequent throughout guinean and soudanian zones. Ghana, commonest sunbird north of forest zone, generally abundant; resident in S on Shai Hills and other inselbergs on Accra Plain; record from Cape Coast (Grimes 1987). Burkina Faso, common and widespread (Holyoak and Seddon 1989); resident in Ouagadougou area; resident, frequent, Pendjari and Arli Nat. Parks, across border into Benin; also adjacently in extreme SW Niger, north to *c.* 12°30′N. Togo, abundant, south to 07°32′N. Nigeria, common, south to Igbetti, Ilorin, Okene, Serti and Enugu, with records at Ife and Lagos; north to Niger border, but not in Sahel zone in NE. Cameroon, frequent, south to edge of rain forest zone (Nachtigal, Efok). S Chad. Central African Republic, known from Vakaga Préf., Manovo-Gounda-Saint Floris Nat. Park and Bamingui-Bangoran Nat. Park (where common) in NW and N; also records at Bayanga and Bai Bakalonga in Dzanga Res. in extreme SW (Green and Carroll 1991), though not found in adjacent Nouabalé-Ndoki Nat. Park in Congo (Dowsett-Lemaire 1997b). Sudan, common in S half of country (but in SE only in Boma Hills) and in Jebel Marra uplands, central Darfur. Eritrea, at 1200–2460 m altitude, north to Anseba; absent from NW and eastern escarpment. Ethiopia,

Chalcomitra senegalensis

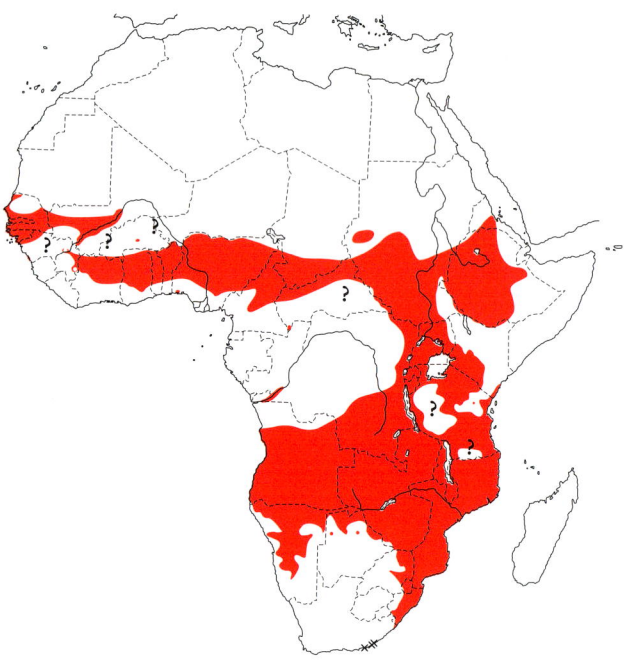

common to abundant, but absent in E. Absent from Djibouti and Somalia. Kenya, Ethiopian race just into N Kenya at Sololo; highlands race common in central areas and SW, mainly over 1000 m; coastal race local, below 500 m. Tanzania, widespread in N and E, absent from much of NE as shown, and no acceptable records in vast area south of L. Victoria nor in small area in S (N. E. Baker, pers. comm.); abundant resident, Zanzibar; uncommon non-breeding visitor, Pemba. Zaïre, race *acik* very common in NE; *lamperti* up to 1700 m in E; *gutturalis* widely distributed in Katanga in SE; and *saturatior* in Kasai Occidental and near Congo R. downstream from Kinshasa. Angola, common and widespread south of *c*. 08°N. even in arid SW (Iona Nat. Park and along Cunene R.). Throughout Rwanda, Burundi, Zambia, Malaŵi and Mozambique. Throughout Zimbabwe except near Shashe R. upstream of Tuli. Namibia and Botswana, as mapped (Harrison *et al.* 1997). Transvaal, only in lowveld and base of escarpment. Natal, common in NE, frequent in SE, absent from W. Not in Lesotho. Possibly just into Cape Prov. west of Port Edward (Natal); vagrant, East London and Port Alfred.

May have increased in abundance with fragmentation of miombo woodland (Irwin 1981). Density of 3 pairs in 5·5 ha on shores of L. St Lucia, Natal.

Description. *C. s. senegalensis* (Linnaeus): Senegal to Nigeria, intergrading with *acik* in N Cameroon. ADULT ♂: forehead and crown metallic green; sides of head, neck and rest of upperparts sooty black. Tail blackish brown. Moustachial area, chin and upper throat metallic green; lower throat and breast scarlet, with many metallic blue bars formed by subapical feather bands; rest of underparts sooty black. Lesser coverts blackish; rest of upperwing dark brown. Axillaries and underwing-coverts sooty black. Bill black; eyes dark brown; legs black. ADULT ♀: upperparts including sides of head dark brown, tinged olivaceous. Tail dark brown; T6 paler brown with white fringe to tip of inner web. Chin and throat dark brown, narrowly fringed paler brown; rest of underparts olive-yellow, breast densely mottled and flanks heavily streaked dark brown; undertail-coverts pale yellowish with dark centres. Upperwing dark brown, primaries and secondaries edged olive-brown; tertials and greater coverts fringed and tipped pale brown; outer webs of primary coverts and alula broadly fringed whitish; outermost lesser coverts also fringed white. Axillaries olive-yellow; underwing-coverts olive-yellow with dark brown bases; inner borders on undersides of flight feathers buffy brown. Bare parts as adult ♂. SIZE (10 ♂♂, 9 ♀♀): wing, ♂ 66–69 (67·2), ♀ 59–65 (61·6); tail, ♂ 42–50 (45·7), ♀ 37–44 (40·2); bill, ♂ 22–23 (22·3), ♀ 20–24 (21·6); tarsus, ♂ 16–17 (16·3), ♀ 15–16 (15·5). WEIGHT: Ghana, ♂ (n = 16) 9·6–11·9 (10·3), ♀ (n = 7) 6·8–9·8 (8·4).

IMMATURE: juvenile similar to adult ♀, but more uniform sooty grey from lores to chin, throat and upper breast; rest of underparts olive-yellow, lower breast, flanks and belly more boldly mottled and barred blackish. Immature ♂ has metallic green chin and upper throat and scarlet upper breast with metallic blue bars; otherwise like juvenile.

NESTLING: feathered nestling has cream gapes, olive back, dark-mottled throat and greenish yellow underparts.

C. s. acik Hartmann: N Cameroon to Central African Republic, W and S Sudan (except where replaced by *lamperti*), NE Zaïre (Uele) and NW Uganda. Like nominate race, but ♂ with slightly lighter and purer scarlet breast patch (little blue barring visible).

C. s. cruentata Rüppell: Eritrea, Ethiopia and SE Sudan (Boma Hills). ♂ differs from other races in having chin and throat sooty black, bordered by metallic green moustachial stripe; breast patch with prominent blue barring; purple patch on leading lesser coverts small or absent. ♀ similar to *gutturalis*, but buffish white moustachial stripe more prominent; primary coverts and alula lack white edges. Large; wing (10 ♂♂) 72–78 (74·6).

C. s. lamperti (Reichenow) (includes '*aequatorialis*'): E Zaïre, Rwanda, Burundi, Uganda (except NW), SW and central Kenya, S Sudan (Imatong and Leboni) and Tanzania. ♂ like *acik* with little blue barring, but breast patch slightly deeper red. ♀ with throat as in nominate race, but breast with more distinct olive yellow streaking. Larger; ♂ (n = 10) wing 69–76 (72·2), bill 26–29 (27·5). WEIGHT: Kenya, ♂ (n = 39) 11·4–16 (13·9), ♀ (n = 12) 12·1–14 (13·3).

C. s. gutturalis (Linnaeus) (including '*inaestimata*'): SE Kenya to Tanzania, Malaŵi, E Zambia, E and S Zimbabwe, Mozambique, E Botswana and E South Africa. ♂ as in *lamperti*, but leading lesser coverts metallic purple. ♀ with distinct pale buff moustachial stripe; pale fringing on feathers of chin and throat more pronounced; breast with more pale streaking, less densely mottled. Southern birds large: Zimbabwe/Malaŵi, ♂ (n = 10) wing 75–80 (77·8), bill 28–31 (29·3). A cline of decreasing size towards E African coast: E Tanzania ('*inaestimata*'), ♂ (n = 10) wing 67–74 (69·8), bill 24–27 (25·3). WEIGHT: Zimbabwe, ♂ (n = 2) 13·7, 14·2, ♀ (n = 3) 11·2–13·8 (12·3); Malaŵi, monthly means, Mar–Dec, ♂ (n = 94) 12·5–13·6, ♀ (n = 118) 11·3–12·2 (Hanmer 1981); (mainly *gutturalis*), ♂ (n = 38) 11·5–16·4 (14·9), ♀ (n = 17) 10·2–15·3 (13·1) (Maclean 1993).

C. s. saturatior (Reichenow): Angola, W Zambia, W and S Zaïre, N and central Namibia, N Botswana and NW Zimbabwe. Like *gutturalis*, with purple patch on lesser coverts, but ♂ with slightly darker red breast patch with more prominent blue barring. ♀ with chin to breast darker, densely mottled. Large; wing (10 ♂♂) 74–78 (75·5).

Field Characters. Length 13–15 cm. One of the commonest and most familiar sunbirds in Africa, with long, curved bill. ♂ sooty black, wings and tail browner; forehead, crown, chin and upper throat iridescent green, lower throat and *breast scarlet*; no pectoral tufts or eclipse plumage; small violet shoulder patch in race *gutturalis*.

Subterminal blue band on breast feathers visible at close range in E and S Africa, more readily visible in W African race where scarlet tips are narrower, giving an overall more purplish, less scarlet effect. ♀ has upperparts and face brown with *no* superciliary stripe, separated from dark throat by pale malar stripe; underparts pale dull yellowish, heavily mottled on belly, becoming barred on breast, throat feathers dark brown with pale tips. Juvenile like ♀ but with throat uniformly dark, underparts yellower with bolder blackish mottling and barring; immature ♂ has some green and red on throat and breast. ♂ Amethyst Sunbird *C. amethystina* is also blackish with green forehead but has small round golden-pink throat patch, violet uppertail-coverts. ♀♀ of Amethyst and Green-throated *C. rubescens* Sunbirds have pale superciliary stripes and are streaked rather than mottled or barred below. Parapatric with Hunter's Sunbird *C. hunteri* and separated from it by differing habitat preferences even where they may nearly meet as wanderers; however, the 2 may overlap in wooded grassland on lower Tana R. (Lewis and Pomeroy 1989) and along their interface between equator and Tanzanian border (Hall and Moreau 1970); for further differences, see that species.

Voice. Tape-recorded (10, 16, 22, 66, 75, 86, 88, 91–99, 104, NIV). Territorial song a series of clear, loud, measured notes, audible for 80 m or more, given at rate of 2 per s or 5 per 2 s, 'chip, tyewp, choop, chip', 'chip, tyewp, chip, tyewp, chichu, chip', 'chichip, cheep, chew, cheep, chew'; typically 4–7 notes in a phrase, but one phrase follows another after pause of *c*. 4 s, often less, and song period may last for an hour; subadult ♂♂ also sing vigorously (Skead 1967). Song variants include a lilting 'widdly-widdly-widdly ... ', 'weekly-wip, wip-weedly-wip' and 'ti-purri-chip, ti-purri-chip, chichichichi'. Subsong, given while other individuals present, a continuous fast, low-key, liquid babble. Call a clear, double 'pi-tee'; harsh aggression/alarm call, 'ju-ju-ju-ju-ju ... ', 'jaa-jaa-jaa ... ' or faster 'jijijijijiji'; also said to give a 'rounded, fluty and abrupt 'tjoyp'' (Skead 1967). Hunger call of fledgling a high-pitched, plaintive 'weeping' note.

General Habits. Inhabits light woodland and hot, open country, farmland, riverine and dry acacias, *Brachystegia*, proteas, orchard bush, shrubs, parkland, coastal bushveld and wooded, flowery gardens, formal and informal, in the country, suburbs and towns. In Eritrea, mainly in rich *Stercularia* woodlands and riparian tamarinds; not much in acacia.

Singly and in pairs, active even through heat of day, noisy and conspicuous. Loose flocks may aggregate at flowering tree (such as *Erythrina*, *Schotia*, *Bauhinia*), mixing freely with other birds: in South Africa particularly Amethyst Sunbird *Chalcomitra amethystina*, White-bellied *Cinnyris talatala*, Lesser Double-collared *C. chalybea* and Violet-backed *Anthreptes longuemarei*, and white-eyes. Aggressive, constantly chasing away other sunbirds foraging close to it, and sometimes chasing other birds such as Brubrus *Nilaus afer*. Also aggressive in defence of territory, at least the ♂: pair occupied territory in Zimbabwe for 4 years, but when ♂ disappeared, neighbouring Amethyst and Lesser Double-collared Sunbirds quickly entered and remained in his territory. ♀ with 2 young out of nest was unable to expel intruders, nor was new ♂ who joined her 2 months later (Skead 1967). Attacks its image in window-panes.

Restless; feeds for several s at a blossom, moves quickly to adjacent or distant one in same plant, calls, interacts with conspecific or other sunbird, makes several flights through or around open branches, feeds again, then suddenly makes lengthy flight to distant trees or disappears from view. May then often fly quite high. Flight fast, dashing but jinking, with sudden changes in direction and altitude. Forages at blossoms by hovering in front or, more commonly, perching at side, leaning and probing sideways or upside-down. Sometimes pokes bill into several flowers in a panicle, by leaning in different directions without changing grip on perch. Pierces bases of flowers with tubular corollas too long for sunbird's bill, then perches to suck nectar. Hovers in front of spider webs to pluck spider out; often feeds from spider webs on woodwork and stonework around buildings. Flies momentarily down to ground to take ants; gleans leaves; hawks flying termites in darting flights from perch. Water-bathes; one bird seen many times to immerse head then fly to perch in tree and rub forehead, lores and bill on rough bark. Sings from prominent twig, pivoting from right to left, with bill open, tongue visible and throat vibrating and glistening. Moult, see Hanmer (1981) (200 birds, Malaŵi).

Mainly resident, but nomadic and migratory near some edges of range. Visitor to Mauritania in June–Oct, rarely Nov. Burkina Faso, commoner in Dec–Feb than in other months. Some suggestion that it may be partial migrant in Sudan, Ethiopia and Tanzania. No evidence of seasonality in Eritrea. In Kenya, annual influxes into Ruaraka–Kahawa–Ruiru area in each dry season (Feb–Apr and Sept–Nov), augmenting resident population (Lewis and Pomeroy 1989). In Zimbabwe an influx onto plateau, normally in Mar–July; in drought year of 1987, a major irruption into Zimbabwe in June–Sept, including some *saturatior*, perhaps from Namibia (Tree 1990). In South Africa visits Pietermaritzburg mainly during summer months, and thought partially to vacate hot coastal lowlands for the cooler midlands during summer (Skead 1967). Only a few birds return to same place on successive years (Tree 1990); in S Zimbabwe, however, pairs often occupy territories constantly for up to at least 4 years. Ringed bird moved 360 km (Harrison *et al.* 1997). Higher reporting rates in Namibia and Botswana in Sept–Mar than Apr–Aug, and in E southern Africa in June–Aug than Sept–May, suggest an east–west movement there (Harrison *et al.* 1997).

Food. Nectar, spiders and insects. Feeds at flowers (presumably on nectar) of *Erythrina*, *Leonotis*, *Loranthus*, *Aloe*, *Tithonia*, *Schotia*, *Kniphofia*, *Crotalaria*, *Callistemon* and *Canna*. Attracted to large red blossoms, e.g. of silk cotton and coral trees in Gambia, where also feeds at blooms of parasitic *Tapinanthus bangwensis* (Barlow *et al.* 1997). Commonest in Burkina Faso when *Bombax costatum* and *Combretum panisulatum* in flower (Thonnérieux *et al.* 1989).

Takes nectar from (and pollinates) large orange flowers of *Phragmanthera dshallensis*, a mistletoe parasitic on *Acacia* (Gill and Wolf 1975). Around Zaria, Nigeria, relies mainly on nectar of ornithophilous trees *Tapinanthus globiferus*, *Erythrina senegalensis*, *Bombax costatum*, *Albizia coriaria* and *A. zygia*; feeds much less commonly on bat- and insect-pollinated flowers of *Parkia clappertonia*, *Butyrospermum paradoxum*, *Daniellia oliveri*, *Vitex doniana*, *Cassia sanguinea*, *C. sieberiana*, *Carissa edulis*, *Tacazzea apiculata* and *Macrosphaera longistyla*; regularly takes nectar of exotics *Spathodea campanulata*, *Euphorbia pulcherrima*, *Caesalpinia pulcherrima*, *Thevetia peruviana* and *Citrus limon*, and rarely feeds at *Bauhinia*, *Delonix regia*, *Albizia lebek*, *Gmelina arborea* and *Hippeastrum equestre* (Pettet 1977).

Many spiders; also ants, grubs, caterpillars, crickets and flying termites (Skead 1967). 4 stomachs, Zaïre, contained small spiders and small insects including a leaf-hopper and tiny caterpillar; 5 out of 7 stomachs also contained pea-sized ball of rubbery latex from Ceará rubber trees *Manihot glaziovii* (Euphorbiaceae), thought to be ingested when bird feeds at blossoms (Chapin 1954). Small arthropods fed to nestlings: white moths, caterpillars, black beetles, flies and spiders, and ♀ seen from hide to feed nectar to young (van Someren 1956).

Breeding Habits. Solitary nester, evidently monogamous; territorial. Territory advertised by ♂ singing from prominent perch. Courtship involves ♂ swaying from side to side and pivoting body when perched immediately in front of ♀ (Skead 1967). Commonly double-brooded, sometimes treble-brooded (e.g. Nigeria: Brown 1948). 3 broods once raised over period of 5·5 months. Builds new nest for each brood.

NEST: suspended, oval, fig-shaped or pear-shaped bag, rough and ragged, with side-top, porched entrance, consistently more bulky than nests of e.g. Amethyst Sunbird or Lesser Double-collared Sunbird, never decorated with lichen (Vincent 1949, though lichen listed by Skead 1967), but always with brownish skeletal dead leaves. Composed of fine, fibrous, chaffed grass, mixed with small bits of skeletal leaves (mainly towards outside of walls), old cobweb litter, sometimes some soft, dry leaf petioles and a few bits of woolly plant down; yellowish grass stems around entrance and forming (and projecting from) the porch; decorated outside with dead leaves, grass seeds, wool, string, feathers, scraps of paper, caterpillar droppings or a few clusters of tiny reddish brown seeds (Vincent 1949, Chapin 1954, Skead 1967); lined with woolly, silvery plant down and, rarely, hair or a few feathers. Materials vary regionally; in E Uganda nests often cottony (van Someren 1956). SIZE (n = 2): height av. 132, width av. 75, from front to back av. 80, entrance diam. av. 35.

Nest built entirely by ♀, who is chased and harried by ♂ whenever possible, particularly on flights between nest and material-collecting site. When ♀ is collecting material or working at nest, ♂ perches prominently nearby and sings. Sometimes ♀ drives ♂ away from nest. Nest takes 3–6 days to build (n = 3). One nest construction was commenced in early morning; 2 days later ♀ was collecting material to thicken walls (taking round trips of 15–120 s) and inside nest was pressing round and round to shape it; next day lining was added, several loads being dumped on floor of nest, then every 4–6th journey ♀ worked lining into place; nest was completed 2 days after that. Sometimes uses kapok from fallen nest of penduline tit *Anthoscopus* sp.

Sited in borders of woodland near open ground and often near water, sometimes over it; commonly in thicket of bushes and small trees growing on *Macrotermes* termite mound. Commonly also in thorn trees with thick clusters of heavy cobweb (when nest hard to see). Built in thorn or other tree or shrub at height of 1·5–6 m from ground, nest usually in plain view on outside of tree, but sometimes hidden amongst foliage. Often uses *Brachystegia*, flamboyants, *Jacaranda*, *Colophospermum*, *Grevillea*, *Eucalyptus* and wattle trees; sometimes uses leafless dead tree. Suspended at top of nest or attached by back of top, to projecting, down-sloping twig at or near tip of branch; when embedded in thick cluster of spider web, nest may be securely held and need no further support. Unusual nest in Kenya was enclosed in living spider nest, and the spiders may have enveloped the sunbird's nest originally built adjacently (van Someren 1956). One nest was secured by spider web and fibres to flat surface of tall, upright cactus stem. Another was a double-storeyed nest hung from an iron bar under a bridge. Nest quite frequently placed close to hornets', wasps' or spiders' nest (Vincent 1949, Skead 1967, Pakenham 1979); one was inside mass of cobweb in *Acacia* tree, another inside web of spider *Eresus purcelli*. Occasionally builds nest hung from flex near house, or under a thatched roof or in a porch or in dense brambles; often uses telephone wires. Heights are 1·5–9 m from ground, usually 4·5–9 m. Pair generally uses same tree year after year, or adjacent tree or shrub. Pair built in *Jacaranda* one year and in *Opuntia* the next. A ♀ started to build new nest only 11 days after young had left old one.

EGGS: 2, rarely 3, sometimes only 1 near equator (Chapin 1954). Colour and markings variable; usually ground cream, tinged greenish or slightly pinkish, sometimes white, marked in lengthways streaks of light sepia-brown or pale grey; sometimes speckled on open ground, or almost immaculate except for even, close mottling which can form ring around large end. SIZE: (*C. s. gutturalis*, n = 40) 17·8–20·5 × 12·2–14·0 (19·2 × 13·2).

LAYING DATES: Gambia, (Apr, May), June, July, Sept, Dec; Ghana, May, indications Mar and June–Aug or Sept; Togo, Aug; Nigeria, Feb–Oct; Sudan (Darfur), new nests July–Aug, eggs only in Sept; Eritrea, Aug; Ethiopia, Apr–Nov; E Africa, Region A, Feb–May, July, Sept–Oct; Region B, all months, especially June–July, uncommon Aug and Oct–Dec; Region C, Jan–June, Aug–Oct; Region D, all months, mainly Feb–June, especially Mar; Region E, Apr–July, Sept–Oct. Zaïre, Feb–May in NE, Aug–Oct in SE (near Lubumbashi). Angola, Mar (and active gonads Aug–Sept); Namibia (Damaraland), Nov–Mar; Botswana, Aug–Dec; Zimbabwe, July–Apr (30% of 396 clutches in Sept, 39% in Oct, 15% in Nov, 11% in Dec–Jan); Mozambique, Sept–Mar; Transvaal, June and Aug–Dec, mainly (60% of nests) Oct–Nov; Natal, Aug–Jan.

INCUBATION: by ♀ only. ♀ sleeps in nest at night. Period (n = 2) 13·5 ± 0·5 and 15 ± 1 days; also 13–14 days.

DEVELOPMENT AND CARE OF YOUNG: young fed by ♂ and ♀, mainly ♀. Period: 15–19 or 20 days and *c*. 17 days. After leaving nest, young stay near it for a few days; once they kept within 20 m of it; they return to nest to sleep for *c*. 4 nights; thereafter young roost with parents in thick foliage of tree; young fed by both parents; one fledgling took nectar by itself 9 days after leaving nest; young (of 3rd brood) fed by parents for at least 8 weeks.

BREEDING SUCCESS/SURVIVAL: killed by striking windows, and young often taken by Fiscal Shrikes *Lanius collaris*. Brood of nestlings taken by boomslang *Dispholidus typus*. Parasitized by Klaas's Cuckoo *Chrysococcyx klaas*, African Emerald Cuckoo *C. cupreus* and possibly Diederik Cuckoo *C. caprius* (Colebrook-Robjent 1984). Long-lived: 3 ♂♂ retrapped after 8, 9·5 and at least 11 years (Hanmer 1989).

References

Harrison, J.A. *et al.* (1997).
Pettet, A. (1977).
Skead, C.J. (1967).
van Someren, V.G.L. (1956).
Vincent, A.W. (1949).

Chalcomitra hunteri (Shelley). Hunter's Sunbird. Souimanga de Hunter.

Cinnyris hunteri Shelley, 1889. Proc. Zool. Soc. London, p. 365, pl. 41, fig. 2; Useri River, Kilimanjaro.

Plate 11
(Opp. p. 158)

Forms a superspecies with *C. senegalensis*.

Range and Status. Endemic resident, E Africa. Sudan, Elemi Triangle only, at 500 m. Ethiopia, uncommon, only east of *c*. 42°E, mainly below 1500 m. Somalia, fairly common and widespread, but absent in NE, SW and possibly W-centre; scarce in N, ranging north to about Burao–Oadweina–Hargeisa line. Kenya, uncommon, in N in regions shown, up to 1200 m; southwest border of range defined by highland contours and, although Hunter's and consuperspecific Scarlet-chested Sunbird *C. senegalensis* are mapped here as overlapping (Lewis and Pomeroy 1989) they are separated altitudinally, with Hunter's in semi-arid lowlands, and in fact may have no geographical overlap at all (D. Pearson, pers. comm.); absent from much of L. Turkana basin and from parts of E and coastal Kenya. Mt Kulal to Kerio Valley, L. Baringo, L. Bogoria, Samburu Game Res., Shaba Game Res., and E lowlands to Tanzanian border; common at Garissa. Uganda, Moroto. Tanzania, Mkomazi Game Res., Lembeni and Same. Common in Tsavo East Nat. Park, Kenya: density of 2 (Jan) to 13 (Aug) birds per 10 ha and av. over 12 months of 6·25 birds (Lack 1985).

Description. ADULT ♂: forehead and crown metallic green; lower rump and uppertail-coverts metallic purple; sides of head, neck and rest of upperparts velvety black. Tail blackish. Chin and upper throat black, bordered by metallic emerald green moustachial stripe. Lower throat and breast scarlet; feathers with largely hidden metallic subapical blue and greenish bands, but blue bars visible on lower throat and along upper breast sides. Leading lesser coverts metallic purple; upperwing otherwise blackish brown. Axillaries and underwing-coverts sooty black. Bill black; eyes brown; legs black. ADULT ♀: upperparts greyish brown; long uppertail-coverts darker brown. Tail dark grey-brown; T6 paler brown, tips of T5 and T6 fringed whitish. Cheeks and ear-coverts greyish brown, lores rather darker. Chin and throat dark greyish brown, barred pale buff, bordered by narrow pale buff moustachial stripe; rest of underparts buffish white, breast and flanks heavily mottled dark greyish brown. Upperwing dark grey-brown, primaries narrowly edged creamy white, tertials and greater coverts fringed pale brown, primary coverts and alula edged buffy white, median and lesser coverts broadly tipped greyish brown. Axillaries buffy brown; underwing-coverts greyish brown tipped buffy white; inner borders on undersides of

Chalcomitra hunteri

flight feathers greyish white. Bare parts as in adult ♂. SIZE (10 ♂♂, 5 ♀♀): wing, ♂ 67–73 (70·0), ♀ 59–64 (62·4); tail, ♂ 46–51 (48·4), ♀ 38–42 (39·9); bill, ♂ 23–28 (25·8), ♀ 25–27 (26·2); tarsus, ♂ 16–18 (17·2), ♀ 16–18 (16·6). WEIGHT: Kenya, ♂ (n = 12) 10–14 (11·9), 1 ♀ 10.

IMMATURE: juvenile like adult ♀ but lacks moustachial stripe; lores to chin and throat more uniform dark brownish grey; underparts more boldly and darkly mottled. Immature ♂ like juvenile, but chin and upper throat black, with green moustachial stripe, and lower throat and breast red with some blue barring.

TAXONOMIC NOTE: 2 races, *hunteri* and *siccata*, recognized by Clancey (1986).

Field Characters. Length 12·8–14·5 cm. Very similar to Scarlet-chested Sunbird *C. senegalensis* which it replaces in dry country of NE Africa. The 2 are usually ecologically separated, but could meet in wooded grassland on lower Tana R. and along their interface between equator and Tanzanian border. ♂ blacker than ♂ Scarlet-chested, with brown wing and green crown and forehead, but differs in reddish *violet rump*, uppertail-coverts and *shoulder* patch; throat black, bordered by narrow *green stripe*. ♀ somewhat paler brown above than ♀ Scarlet-chested, underparts off-white, not tinged yellow, but with same mottling and barring; chin and throat generally paler than Scarlet-chested but become darker as pale feather edges wear off. Some juveniles have light yellow, not brown throat; immature and ♀ have dark throat.

Voice. Tape-recorded (B, GREG, McVIC). Calls very similar to those of Scarlet-chested Sunbird. Usual note a loud, repeated 'tew'; also a harsher scolding 'tchew-tchew-tchew' or 'tchi-tchi-tchi-tchi'. More song-like is a series of clear distinct notes, 'tew-tew-tew-tew-tew-tee-tee-tee-tee' (Zimmerman *et al.* 1996)

General Habits. Inhabits *Acacia* and *Commiphora* scrub in woody and bushy semi-arid grassland at 50–1200 m, mainly below 1000 m. In lower Tana R. valley inhabits *Dobera-Albizia* parkland, and range may overlap there with Scarlet-chested Sunbird. Generally separated from Scarlet-chested geographically, altitudinally, and by preference of Hunter's for arid and of Scarlet-chested for mesic habitats, but the 2 may also overlap as local wanderers near common border. Occurs in stands of flowering aloes, acacias in flower, and particularly attracted to flowering *Delonix elata* trees.

Shier than most sunbirds, restless and flighty. General behaviour and interactions with other bird species, so far as known, are very like those of Scarlet-chested Sunbird. Feeding ecology studied in Tsavo East Nat. Park, Kenya: hovers in front of, and perches next to or just above, a tree blossom evidently to take nectar; gleans leaves and twigs for insects, and captures them one at a time in the air by snatching from a perch or flying out for a metre or two; also feeds by probing ripe *Commiphora* fruits.

Food. Of 422 feeds, 10 were on *Commiphora* fruit, 77% of nectar and 21% of insects. Of nectar feeds, 28% were from bushes, 35% from *Delonix elata* and 37% from flowers of other trees; of insect prey, 43% were gleaned from leaves, 36% from twigs, 21% from the air. 67% of insects were caught by birds at perch and 34% by birds in flight (Lack 1985).

Breeding Habits. Solitary nester.

NEST: 3 described (Serle 1943, Archer and Godman 1961, Lack 1976). Oval bag made of fine grasses, rootlets, partly decayed dead leaves, fragments of bark and paper, and small feathers, compacted and strengthened with binding of spider silk, covered with pieces of lichen; with side entrance above middle of nest, overhung by porch 24 long; below nest a 'beard' 200 mm long. Profusely lined with feathers (many of Vulturine Guineafowl *Acryllium vulturinum*), on floor, walls and ceiling of cavity. Length 126, greatest breadth 64, from front to back 78; entrance diam. 34; int. depth below lower margin of opening, 47. Suspensory stalk, of grasses and gossamer, constructed around a 150 long, downward-pointing, straight thorn (tip of which was in body of nest) 125 long, from horizontal branch 1·3–2 m up in a thorn tree. One nest was attached to flex of electric light in an inhabited building; length *c.* 200: stalk and beard together *c.* 80, body of nest *c.* 120. Another built in *Platycelyphium voense* tree (Leguminosae).

Built by ♀, generally accompanied by ♂ as she flew to and from nest. ♀ added feathers to lining long after incubation had commenced.

EGGS: 1–2. 2nd laid 1 day after 1st. Oval, with smooth, lustreless surface. White or pale olive-grey, minutely speckled with dark grey, particularly around broad end, or ground largely obscured by streaks and longitudinal blotches of pastel browns and greys (like streaked type of egg of Scarlet-chested Sunbird). SIZE (n = 3) 19·0–19·5 × 11·6–13·0 (19·2 × 12·5).

LAYING DATES: Somalia, July, (and nestlings May, in N and S); Kenya, Oct; Tsavo East, nests Nov–Feb (Lack 1976).

INCUBATION: by ♀, the ♂ usually staying nearby.

References
Clancey, P.A. (1986).
Lack, P. (1976).
Serle, W. (1943).

Plate 11 (Opp. p. 158)

Chalcomitra balfouri (Sclater and Hartlaub). Socotra Sunbird. Souimanga de Socotra.

Cinnyris balfouri Sclater and Hartlaub, 1881. Proc. Zool. Soc. London, p. 169, pl. 15, fig. 2; Socotra.

Range and Status. Endemic to Socotra, where a common resident in most areas with trees and abundant in densely wooded places, from sea-level to at least 1370 m. Thought not to be at risk; habitat is overgrazed and overbrowsed by livestock, but bird persists even where vegetation sparse. Up to 50 can be seen in a day (Porter and Martins 1993).

Description. ADULT ♂: top of head to hindneck dull dark brown, streaked pale buff or greyish white; rest of upperparts dark brown, mantle and scapulars with diffuse pale streaking. Tail slightly graduated, blackish, with slight green gloss above; feathers narrowly fringed buffish white; T5–T6 with broad white border to distal part of inner web, and T6 with wholly white outer web. Lores and around eyes sooty blackish; ear-coverts dark brown; narrow black moustachial stripe; white sub-moustachial stripe broadening backwards to cheeks and leading to white band around throat and sides of breast. Malar stripe black, bordering dark grey chin and upper throat; lower throat and breast blackish brown, white feather fringes forming prominent scaly barring on upper breast and spotting on lower breast; pectoral tufts bright yellow; rest of underparts whitish, feathers of flanks and thighs

Chalcomitra balfouri

showing a few dark centres. Upperwing dark brown, primaries finely edged white; secondaries edged buffish yellow, tertials edged warm buff to grey-brown; alula and upperwing-coverts edged buff. Axillaries white; underwing-coverts white with dark bases; inner borders on undersides of flight feathers greyish white. Bill black; eyes reddish brown or orange-brown; legs black. ADULT ♀ like adult ♂, but lacks pectoral tufts; however, adults near same nest (photos in Porter and Stone 1996) differ – [♂] with white sub-moustachial stripe and clear-cut dark chin and spotted lower breast, [♀] with barred sub-moustachial stripe and less clean looking chin and lower breast. SIZE: (12 ♂♂, 6 ♀♀): wing, ♂ 63–67 (65·6), ♀ 58–60 (59·2); tail, ♂ 47–53 (49·6), ♀ 42–46 (44·1); bill, ♂ 22–25·5 (23·1), ♀ 20–22 (20·7); tarsus, ♂ 20–21 (20·9), ♀ 17–20 (18·3). WEIGHT: (n = 33, ♂ and ♀) 8–15 (Ripley and Bond 1966); 1, unsexed, 10·5.

IMMATURE: juvenile similar to adult but eyes brown, lower mandible pale yellowish, legs and feet dark grey with pale soles. Uncertain whether pectoral tufts present. Young bird thought to attain adult plumage at first moult (Ogilvie-Grant and Forbes 1903).

NESTLING: unknown.

TAXONOMIC NOTE: a distinctive sunbird not obviously closely related to any African species. Closest ally may be *N. dussumieri* of the Seychelles (Hall and Moreau 1970); or, in the view of M.P.S. Irwin (*in litt.* 1997) it may be derivative of the *Chalcomitra* group, in which case its closest relative may be *C. hunteri* in Somalia and Kenya.

Field Characters. Length, ♂ 12·5–14 cm, ♀ 11·5–12 cm. A rather large, robust sunbird with stout bill and thick legs. Plumage highly distinctive: head and upperparts striped brown and white, blackish throat patch separated from face by white stripe, underparts white barred black on breast, with yellow pectoral tufts. Endemic to Socotra, where it is the only sunbird.

Voice. Tape-recorded (DAV). Song, usually delivered in short bursts, sometimes longer, is a series of quick jangling notes fairly typical of sunbirds (Showler and Davidson 1996); said to mimic calls of other birds, especially Socotra Warbler *Incana incana* (Ogilvie-Grant and Forbes 1903). Main call a squeaky, strident 'zii' or 'zee'; alarm or territorial dispute call a repeated harsh grating 'tchee-up' or 'tchee' (Showler and Davidson 1996).

General Habits. Inhabits wadis and other wooded areas where dominant trees include *Rhus thyrsiflora*, box *Buxus hildebrandtii*, *Carphalea obovata* and *Sterculia* sp., also common on open, rocky hillsides with sparse trees (*Adenium obesum*, *Euphorbia arbuscula*), and occurs on mountains with box scrub (Showler and Davidson 1996). Absent from bare limestone plateaux and probably absent from sparsely-vegetated Noged Plain (although occurs in thicker cover on Kallansiya Plains). Lives singly or in pairs, and in family parties after breeding. Noisy and vivacious, constantly on the move. Does not often visit flowers, foraging among branches of trees and shrubs, picking insects off leaves and twigs and occasionally catching one in flight or taking one from ground. Paired birds inseparable, except when ♂ display-sings from prominent perch, with yellow pectoral tufts spread widely (Ripley and Bond 1966). Resident.

Food. Insects including small cicadas and small spiders; seeds and small fruits. Most commonly taken seeds are large, black, oval ones. Observed eating fruits of *Euphorbia* sp. (bird had great difficulty in swallowing them but managed eventually after much head-stretching and gasping: Ripley and Bond 1966). Seen at flowers of *Calotropis procera*, perhaps taking nectar.

Breeding Habits. Indulges in noisy territorial and sexual chases, birds pursuing each other in fast, manoeuvring flight with noisy wing-beats.

NEST: suspended (usually) or unsuspended, domed, pear-shaped purse made from fine, loosely woven grasses and cobwebs, lined with white, silky or woolly plant material and possibly goat hairs (lining in one nest thick, in another patchy); ext. height *c*. 80, ext. width 55 × 60; with large, oval entrance at one side, 40–60 high and 20–40 across; nest usually well concealed amongst creepers or twiggy branches in tree. One sited 2·5 m high in *E. arbuscula* tree on rocky hillside; others in shrubs at base of low cliffs. An unsuspended nest was built 2·5 m above ground in top of bare, twiggy tree; nest small, neat and frail, supported at top and sides by twigs incorporated into walls; large side entrance topped by short, indefinite porch made of plant down; dead leaves and a spider egg-case attached to walls, disguising nest (Ripley and Bond 1966).

EGGS: 3. Not described.

LAYING DATES: Mar. Juveniles in early Feb and late Mar, nest from which young had flown, mid Feb, nest with young early May: all indicate eggs early Jan to early Apr.

DEVELOPMENT AND CARE OF YOUNG: nest with three 4-day-old young was visited by parent(s) 5 times in 75 min and adults twice removed faecal sac; parents foraged within 80 m of nest, called often and the ♂ occasionally

Not illustrated.

sang. Juvenile on leaving nest, 8 g; stomach contained only insects (Ripley and Bond 1966). Fledglings solicit food from parents with wing-quivering and querulous calls.

References
Ogilvie-Grant, W.R. and Forbes, H.O. (1903).
Ripley, S.D. and Bond, G.M. (1966).
Showler, D.A. and Davidson, P. (1996).

'Tôha Sunbird'

Welch and Welch (1986, 1998) have reported upon 3 yellow-crowned birds seen together in Djibouti in 1985, thought to be a ♂, a ♀ and a well-grown young. After examination of 8 photos of the '♀', M.P.S. Irwin regards it as impossible to reach a conclusion about the birds' identity (pers. comm. Jan 1998), R.A. Cheke suggests that they might be pollen-dusted Marico Sunbirds *Cinnyris mariquensis osiris* (pers. comm. Dec. 1997), and we consider that they might represent a new race of Shining Sunbird *C. habessinica* or a new race or allospecies of *C. mariquensis*. That the yellow of the crowns could have been caused by pollen is most unlikely, since the ♂'s was not matt but a little glossy and the amount and distribution of yellow on all 3 birds appeared identical (Welch and Welch 1998; we concur).

Range and Status. Seen in Wadi Tôha (11°48′N, 42°45′E) in E foothills of Goda Massif, Djibouti, on 24 Nov 1985.

Description. ♂: crown bright, slightly metallic yellow-green; remaining upperparts, tail and wings like '♀'; chin, throat and upper breast bright metallic green with narrow black band below, separating green from white belly; overall slightly darker than '♀', and bird may have been in non-breeding plumage. '♀': forehead greyish brown (not yellowish?); crown pale yellow or yellow-green, especially hindcrown, the colour reaching well back and to the sides and rather sharply demarcated from greyish brown hindneck; crown less metallic than in ♂; lores blackish; cheeks, ear-coverts, mantle and scapulars uniform dull greyish brown; tail blackish; distinct, long, pale yellow moustachial streak separating grey-brown cheeks from dappled brownish chin and throat; rest of underparts pale brown or grey-brown, darker on belly and flanks than on breast, all feathers pale-fringed (or underparts dirty white dappled with grey-brown), undertail-coverts slightly paler and with distinctly dark tips; wings slightly darker than rest of upperparts; pale tips to median and primary coverts, forming 2 pale wing-bars; outer primaries (and tertials?) with pale margins, forming distinct longitudinal pale stripe in closed wing; alula pale brown. Bill, eyes, legs and feet black. 'Immature': like '♀' but a distinct superciliary stripe, underparts paler, undertail-coverts without blackish tips. [Descriptions of ♂ and 'imm.' from Welch and Welch (1998), that of '♀' mainly from photos.]

Habitat. Secondary forest at 180 m, mainly of *Acacia mellifera* and *Rhigozum somalense* scrub, with numerous taller *Acacia seyal* trees (habitat of Djibouti Green-winged Pytilia *Pytilia melba flavicaudata*, endemic in the same area).

References
Welch, G. R. and Welch, H. J. (1986, 1998).

'Tôha Sunbird'

Genus *Nectarinia* Illiger

7 large, long-tailed sunbirds, mostly alpine or montane (1 also in lowlands); bill medium to long, rather straight or decurved or scimitar-like; ♂♂ uniformly iridescent above and below (5 coppery black, 2 green), without contrasting patterns (but *N. reichenowi* with golden yellow in wings and tail); pectoral tufts only in 2 (yellow in *N. famosa*, red in *N. johnstoni*); tail square or slightly graduated, central rectrices greatly elongated. ♀♀ dark above with much yellow and olive in plumage, only *N. johnstoni* with pectoral tufts; tail blackish, somewhat iridescent. No superspecies. Feeding specialized; species in this genus exploit morphologically-specialized inflorescences (particularly of non-arborescent plants), eating mainly nectar, obtained from limited spectrum of flowers with curved tubular corollae fitted by the various bill shapes. Also eat small invertebrates in flowers; some species feed whilst hovering in front of flower.

Nectarinia bocagei Shelley. Bocage's Sunbird. Souimanga de Bocage.

Nectarinia bocagii Shelley, 1879. Monog. Nectariniidae, p. 21, pl. 6, fig. 2; Angola: Caconda.

Plate 12
(Opp. p. 159)

Nectarinia bocagei

Range and Status. Endemic resident, Angola and SW Zaïre. W highlands in W Huila (mountains of Caconda) above 1500 m, N Huila, Huambo and Bié Provs of Angola, northeast to north of Cambundi-Catembo (= Nova Gaia), Malanje; also extreme NE Lunda Norte (Hall and Moreau 1970) and adjacently in Zaïre: records near Gungu (05°43′S, 19°20′E) on middle Kwilu River at 800 m, further south at 1000 m, and in Kasai R. drainage basin in SW Kasai Occidental.

Description. ADULT ♂ (breeding): head, neck and rest of upperparts glossy violet with bluish green reflections. Tail graduated, T1 long and narrow, projecting *c*. 35 mm; black, feathers narrowly edged glossy violet. Lores black. Chin to upper breast glossy violet-green, concolorous with head; lower breast to flanks and undertail-coverts sooty black. Lesser coverts glossy violet; rest of upperwing feathers sooty black, tertials, greater coverts and median coverts edged glossy violet-black. Axillaries and underwing-coverts black. Bill black; eyes dark brown; legs black. ADULT ♂ (non-breeding): after nesting in Zaïre moults into greyish eclipse plumage, looking like adult ♀, but retaining long central rectrices; eclipse plumage has not been described in Angola (Lippens and Wille 1976). ADULT ♀: top of head and upperparts light olive, rump paler and greener. Tail graduated, blackish brown; T4–T6 with narrow pale olive fringe toward tip of inner web, T5 also with narrow pale olive outer edge, T6 with whole outer web pale brown. Lores blackish brown; cheeks and ear-coverts dark olive-brown. Chin and throat olive-yellow; rest of underparts olive-yellow with diffuse dusky streaking, sides of breast more dusky olive. Upperwing dark greyish brown; primaries and secondaries edged pale olive; tertials and greater coverts fringed pale olive-brown, lighter on tips; alula and primary coverts narrowly fringed pale olive-brown; median and lesser coverts broadly fringed olive. Axillaries olive-yellow; underwing-coverts greyish, tipped pale yellow; inner borders of flight feather undersides pale grey-buff. Bare parts as adult ♂.
SIZE: (2 ♂♂, 6 ♀♀): wing, ♂ 74, 75, ♀ 67–69 (68·5); tail, ♂ 92, 94, ♀ 52–55 (53·7); bill, ♂ 26, ♀ 24–26 (24·3); tarsus, ♂ 17, 18, ♀ 17–18 (17·3).

IMMATURE: juvenile has upperparts and tail very like adult ♀, chin feathers grey with pale tips, breast grey, rest of underparts pale yellow with irregular grey mottling; primaries with broader and brighter yellow-white edges than in ♀, forming pale panel; underside of flight feathers like adult ♀ but underwing-coverts and axillaries pale grey. Immature moulting into adult dress: upperparts like adult ♀ but with a few violet feathers on scapulars, ear-coverts and throat; chin and throat dusky, rest of underparts including thighs blotchy blackish with yellow line down centre, broadening onto lower belly and undertail-coverts; upperwing coverts blackish, flight feathers pale-edged like adult ♀, underwing mainly dark brown like adult ♂, with a few yellow feathers in axillaries.

TAXONOMIC NOTE: the Zaïrean and Angolan populations appear to have quite different habitats. The former has an eclipse plumage and the latter may not. They may represent different taxa.

Field Characters. Length, ♂ 18–19 cm, ♀ 13·5 cm. ♂ mainly black, with inconspicuous dull bronzy violet reflections on head, breast, upperparts and wing-shoulders; *the only* long-tailed black sunbird in its restricted range in Angola/SW Zaïre; does not come close to Purple-breasted Sunbird *N. purpureiventris*. ♀ has black lores and *pale-tipped blue-black* tail, pale outer web of outer tail-feathers, *unstreaked* yellow underparts; larger and longer-billed than ♀ Copper Sunbird *Cinnyris cuprea* and *lacks pale supercilium*; darker (olive-yellow) on throat and brighter yellow on belly.

Voice. Not tape-recorded. Said to give strident calls, 'wisp-wisp' or (♂) 'trik-trik' (Lippens and Wille 1976).

General Habits. Inhabits *Brachystegia* woodland and montane evergreen forest edges in Angola, and deforested boggy land covered with scrub on plateaux between Kwilu and Kwango in Zaïre. In dry season (SW Zaïre) occurs in small flocks travelling between sparse flowering shrubs in marshes; in wet season, courtship involves rapid, zigzag aerial pursuits (Lippens and Wille 1976).

Food. Nectar.

Breeding Habits. Solitary nester.
NEST: 2 nests found. Quite large and thick, made of moss, lined with thick bed of raffia fibres. Sited in small raffia palms in a marsh.
EGGS: 2 (2 clutches), spotted, grey.
LAYING DATES: Zaïre, Oct, Jan; Angola, (♂♂ in breeding condition, Jan).

Reference
Lippens, L. and Wille, H. (1976).

Plate 12
(Opp. p. 159)

Nectarinia purpureiventris (Reichenow). Purple-breasted Sunbird. Souimanga à ventre pourpre.

Cinnyris purpureiventris Reichenow, 1893. Orn. Monatsb., 1, p. 61; Migere, Mpororo (SW Uganda).

Range and Status. Endemic resident, northern Albertine Rift mountains. Zaïre, widely distributed and locally numerous: mountains southwest of L. Albert enclosing Semliki valley, Rwenzoris at 1500–2400 m, mountains west of L. Edward, Kivu mountains, Idjwi I. in L. Kivu, Itombwe and mountains west of Baraka and Fizi in SW Kivu Province. In Itombwe, common or locally abundant above 1900 m. Uganda, locally common at 1550–2450 m in Kalinzu and Bwindi-Impenetrable Forests and Rwenzoris. Rwanda, widespread in W; throughout Nyungwe, up to 2500 m. Burundi, probably widespread in W: known from Lua and Rukwa valleys, Rwegura, Teza and Bugarama.

Nectarinia purpureiventris

Description. ADULT ♂ (breeding): forehead to lores dull black; top and sides of head otherwise glossy black with coppery purple reflections; hindneck to upper mantle metallic green, merging through bands of black and glossy purple into coppery lower scapulars, back, rump and short uppertail-coverts; long uppertail-coverts coppery green. Tail graduated, T1 very long and narrow, projecting 70–115 mm; blackish, feathers edged coppery green. Lower cheeks to chin and throat black, glossed bluish green; upper breast coppery green, merging into narrow metallic bluish green band. Below this, flanks and belly glossy purple with blue reflections; vent sooty black; undertail-coverts glossy purple. Lesser coverts metallic purplish blue; median coverts bronzy purple; rest of upperwing blackish, tertials and greater coverts broadly edged coppery green. Axillaries and underwing-coverts black. Bill black; eyes dark brown; legs black. ADULT ♂ (non-breeding): wings, tail and back to uppertail-coverts as in breeding ♂; otherwise same as adult ♀, but some metallic feathers retained on underparts and mantle. ADULT ♀: forehead to hindneck light greyish brown, dark feather centres showing through to produce barring on top of head; rest of upperparts dull olive-green, slightly yellower on uppertail-coverts. Tail graduated, T1 projecting *c.* 5 mm; dark grey-brown, tips of T3–T5 and broad tip and outer web of T6 paler grey; both webs of T1 and outer webs of T2–T5 fringed olive-green. Indistinct, narrow, brownish white supraloral line. Lores dark greyish brown; cheeks, ear-coverts and sides of neck paler greyish brown. Chin to upper breast greyish white, barred dusky on centre of throat and breast; rest of underparts pale greyish, tinged olive-yellow, most strongly across breast. Median and lesser coverts dull olive-green; rest of upperwing dark grey-brown, primaries (proximal to emarginations) and secondaries edged olive-green, alula and primary coverts finely fringed olive-green, outer edges and tips of tertials and greater coverts more broadly fringed olive-green. Axillaries greyish olive; underwing-coverts pale greyish, tinged olive-green; inner borders of flight feather undersides pale buff. Bare parts as adult ♂. SIZE (9 ♂♂, 8 ♀♀): wing, ♂ 63–72 (66·7), ♀ 56–62 (59·4); tail, ♂ 115–167 (139), ♀ 48–63 (59·4); bill, ♂ 20–23 (22·0), ♀ 19–22 (20·0); tarsus, ♂ 15–17 (16·2), ♀ 15–16 (15·6). WEIGHT: ♂ (n = 32) 10–13·5 (11·7), ♀ (n = 9) 10–12·5 (11·2).

IMMATURE: at about 4 months young ♂♂ become heavily spotted with dull black on foreneck; first metallic feathers to appear are purple lesser coverts, followed by scattered metallic feathers on trunk and greater coverts; middle tail feathers distinctly lengthened, and grow to adult length before head acquires metallic feathers. Breeding plumage (and hypertrophied gonads) acquired at 6–7 months (Chapin 1959).

NESTLING: unknown.

Field Characters. Length, ♂ 20–24 cm, ♀ 13–14 cm. *Bill rather short and straight*. Gorgeous ♂ is the *only long-tailed purple* sunbird in its range (nowhere near either Bocage's or Tacazze Sunbirds *N. bocagei* or *N. tacazze*). Overlaps with Bronzy Sunbird *N. kilimensis*, which is bronzy green with black belly and no pink, violet or purple. ♀ has graduated tail, blackish mask with no pale supercilium, olive-green upperparts, dull yellowish green underparts with brighter band across breast and white throat; ♀ Bronzy is browner above with pale supercilium and streaked yellow underparts. Juv. ♂ like ♀ with a few pink tips to scapulars, blackish throat patch broadening onto breast with a few pink, blue and green feather tips, indistinct dark barring across breast and upper belly.

Voice. Tape-recorded (104, B, C, LEM). Typical song a continuous twitter of calls, a mixture of dry rattles and squeaky notes (Dowsett–Lemaire 1990, and see sonagram there); squeaky calls have husky quality, 'zay-zay-zit' or 'zay-zay-zit-zit-zit-zit'.

General Habits. Inhabits tall flowering trees in montane forest edges, clearings and glades. In Nyungwe, Rwanda, closely tied to flowering *Albizia gummifera* and *Symphonia globulifera* trees, numbers of birds correlated with abundance of flowers at which they feed (Dowsett-Lemaire 1990). At edges of forest, comes down to feed at flowers in herb and shrub layers. In crown of *Symphonia* tree or *Albizia* tree, single ♂ or ♀ attempts to defend a portion of the tree top with its contained flowers against intrusion by conspecifics and other sunbirds such as Collared *Hedydipna collaris*, Blue-headed *Cyanomitra alinae*, Northern and

Greater Double-collared *Cinnyris preussi* and *C. afra*, and Regal Sunbird *C. regia*. At other times a dozen Purple-breasted Sunbirds may forage unaggressively in upper boughs of *S. globulifera* tree, with white-eyes *Zosterops* spp. and 8 other species of sunbirds (Chapin 1959). A foraging/defending bird spends much time singing. Commonly also flycatches, upward and outward from canopy (Dowsett-Lemaire 1990).

Resident, but moves locally and altitudinally according to tree flowering seasons. At Uwinka, 2350–2500 m in Nyungwe, Rwanda, *Symphonia* flowering is over in Dec–Jan and Purple-breasted Sunbirds disappear but are common at *Albizia* and *Symphonia* trees still flowering at lower altitudes of 1750–2000 m.

Food. Feeds, presumably on nectar, at flowers of *Albizia gummifera* (Leguminosae) and hog plum *Symphonia globulifera* (Guttiferae) (Rwanda, see above), the last being extremely attractive (Chapin 1954, 1959). The strongly decurved bill seems ideal for probing the deep, curved corollae of these flowers (Dowsett-Lemaire 1990). Also feeds at flowers of *Lobelia gibberoa*. At different altitudes some *Symphonia* trees are in flower at all times of year, which promotes the evident reliance of this sunbird upon it. Bulk of food in stomachs is winged insects, mainly very small hymenopterans and dipterans, including tiny flies *Bradysia* (Sciaridae); also leaf-hoppers, and minute spiders, but never beetles (Chapin 1959).

Breeding Habits. Solitary nester, evidently monogamous and territorial. In the run up to nesting, ♂ and ♀ keep together, ♂ chasing and courting ♀ (Dowsett-Lemaire 1990).

NEST: suspended purse made of mosses and lichens (*Usnea, Peltigera*) and inflorescence stalks of *Thalictrum rhynchocarpum* and grass *Panicum*, with side-top entrance and sometimes lichen 'beard' hanging from base up to 170 long. Decorated with foliaceous lichens and a few grass-tops, and lined with pappus down of *Gynura vitellina* and several feathers. Measurements (n = 3): ext. height 110–125, ext. width 65–75, ext. depth (front to back) 65–70, entrance diam. 30 × 35. Attached to branch of small tree or bush, e.g. *Maesa* and *Macaranga*, usually 2–3 m above ground, but also at *c*. 11 m and *c*. 25 m, the last in drapery of *Usnea* lichen beneath large bough. One was built by ♀, later by ♂, inside mass of epiphytic moss attached to underside of horizontal *Alangium* branch 16 m above ground (Dowsett-Lemaire 1990).

EGGS: 1–2, usually 2 (Prigogine 1972). Clear grey ground, with small olive-grey spots, numerous at wide end, forming a dark band. SIZE: (n = 1) 19·4 × 13·1.

LAYING DATES: Zaïre (Kivu), Apr, May, July, Nov (Chapin 1959), (Itombwe), Apr–Aug, Nov (other indications: Feb–Aug and Oct–Nov: Prigogine 1984); Rwanda, Jan–Feb.

INCUBATION: by ♀, sitting with head and bill showing at nest entrance. When ♀ takes a break from nest, ♂ often returns with her, hops about in branches near nest, and sometimes clings to side of nest to look in (Chapin 1959).

References
Chapin, J.P. (1954, 1959).
Dowsett-Lemaire, F. (1990).

Nectarinia tacazze (Stanley). Tacazze Sunbird. Souimanga tacazze.

Plate 12
(Opp. p. 159)

Certhia tacazze Stanley, 1815. In Salt, Voyage Abyssinia, app. 4, p. 58; Abyssinia.

Range and Status. Endemic resident and vertical migrant, E African mountains from Eritrea to NE Tanzania. Eritrea, central highlands between Ethiopian border, Asmara and Keren, above about 1850 m; abundant at 3200 m along border. Ethiopia, common and widespread, up to 4000 m, seldom below 1800 m. Sudan, seasonally very common on forest edges and in mountain meadows with *Lobelia*, above 1800 m in Imatong, Dongotona and Didinga Mts. Uganda, common in mountains in N and E: Morongole, Lonyili, Moroto, Kadam and Elgon. Kenya, frequent to common at 1800–4200 m, from Mt Elgon to Cheranganis, Mau, Aberdares, Mt Kenya and Nyambenis, with outliers to the north on forest islands of Mt Kulal, Mt Nyiru and Ndoto Mts. Tanzania, Mt Kilimanjaro, Mt Meru, above 2150 m in Crater Highlands and at 2900 m on Mt Hanang; single (breeding) record in NW (N. E. Baker, pers. comm.).

Density of *c*. 20 pairs breeding in 0·2 ha in garden at Mau Narok, Kenya, reported to Brown and Britton (1980), must be quite exceptional.

Description. *N. t. tacazze* (Stanley): Eritrea and Ethiopia. ADULT ♂ (breeding): head, hindneck and mantle bronzy green, tinged purple, merging into bronzy purple on scapulars, sides of neck, rump and uppertail-coverts. Tail graduated, with T1 long and narrow, projecting 25–50 mm; sooty black, feathers narrowly edged glossy violet-black. Lores black. Chin to upper breast bronzy green, merging with head, grading to bronzy purple on lower breast; belly and flanks to undertail-coverts sooty black. Median and lesser coverts metallic purple; upperwing otherwise sooty black. Axillaries and underwing-coverts sooty black. Bill black; eyes brown; legs black. ADULT ♂ (non-breeding): wings and tail as in adult ♂ (breeding), but otherwise like adult ♀ (see below); metallic purple feathers on rump and uppertail-coverts, and a few on breast. (A breeding population monitored year-round at Mau Narok, Kenya, with 2 breeding peaks per year, evidently lacked non-breeding or eclipse dress in the ♂♂: P.H.B. Sessions *in litt.* to Brown and Britton 1980). ADULT ♀: upperparts greyish olive-brown. Tail graduated, T1 pointed, projecting *c*. 5 mm, dark brown, T3–T4 with narrow whitish fringe at tip, T5 with broad whitish tip, T6 with outer web and distal part of inner web buffish white. Prominent superciliary stripe buffish white, extending to rear of ear-coverts; lores to ear-coverts dark greyish olive. Chin to throat pale greyish buff, bordered by broad whitish

Nectarinia tacazze

moustachial stripe; breast and flanks greyish olive with diffuse mottling; belly and undertail-coverts pale olive-yellow. Upperwing dark brown, remiges edged pale olive, alula and coverts fringed olive-brown. Axillaries pale olive-yellow; underwing-coverts dark greyish, tipped olive-yellow; inner borders of undersides of flight feathers buff. Bare parts as adult ♂. SIZE (10 ♂♂, 10 ♀♀): wing, ♂ 77–83 (79·8), ♀ 67–73 (70·5); tail, ♂ 98–106 (103), ♀ 48–56 (54·0); bill, 30–34 (32·5), ♀ 28–31 (29·7); tarsus, ♂ 18–20 (19·3), ♀ 17–19 (18·0). WEIGHT: (Ethiopia) ♂ (n = 29) 12·3–18·6 (15·7), ♀ (n = 19) 10·0–16·2 (14·1) (Urban 1975 and J.S. Ash, pers. comm.).

IMMATURE: juvenile like adult ♀, but greyer above; yellower below with centre of throat blackish. In immature ♂, metallic feathers appear first on rump and uppertail-coverts, and in coverts near bend of wing, together with bronzy feathers on throat.

NESTLING: not described.

N. t. jacksoni Neumann: Sudan to Tanzania. ♂ with head and neck to mantle more bronzy; chin to upper breast tinged more purplish, and whole of lower breast metallic purple. Central tail feathers longer, projecting 45–60 mm. Some immatures reported to have chin bright orange (dusted with pollen?) but chin usually blackish (C. Jackson, pers. comm.). WEIGHT: Kenya, ♂ (n = 14) 12–17·9 (15·9), ♀ (n = 8) 12·5–14·5 (13·4).

Field Characters. Length ♂ 20–23 cm, ♀ 15–16·5 cm. ♂ is the *only* long-tailed purple sunbird in E Africa; however, if only the bronzy green head is in sunlight and the rest appears dark, could be mistaken for Bronzy Sunbird *N. kilimensis*. ♀ has dark mask highlighted by white supercilium and malar stripe; ♀ of *N. t. jacksoni* meets nominate ♀ Bronzy in Kenya, distinguished by *dull olive-grey underparts* with just a tinge of yellow on belly (♀ Bronzy much yellower); ♀ of nominate *tacazze* is brighter below, with pale throat and pale yellow belly, and in Ethiopia meets Malachite Sunbird *N. famosa*; ♀ Malachite also has yellow belly but has olive throat, indistinct malar stripe and supercilium.

Voice. Tape-recorded (C, KEI, McVIC). Song a lengthy sibilant sputtering twitter interspersed with a few louder notes but otherwise little fluctuation in pitch, 'sweetsius-witterr *tseu* seet-swirursittii, tsit-tsit-tsit-chitichitichiti ... '; less frequent is slower 'tew tew tew tew tew' (Zimmerman *et al.* 1996).

General Habits. Inhabits highland eucalyptus and acacia woods (Ethiopia, Eritrea), montane forest edges and meadows with *Lobelia* (Sudan); in E Africa forest edges and glades, bamboo, grassland, shrubby heath, riverine woods, gardens and cultivation (Britton 1980). Usually singly or in pairs, but aggregations of hundreds occur at stands of flowering trees in non-breeding season, especially flowering eucalyptus. A noisy and quarrelsome sunbird, that tends to displace smaller congeners from food plants. In Kenya Highlands appears to eat more insects than do other sunbirds, and often hawks for flying insects (C. Jackson, pers. comm.).

Resident and vertical migrant. Disappears from mountains in S Sudan when *Lobelia* no longer in flower. In E Africa present all year at high elevations (as on Elgon, Mau Narok, Naro Moru and Mt Meru), but only a non-breeding visitor below 2000–2100 m in cool, wet season (June–Aug in W Kenya). Has then occurred down to 1650 m, in Nairobi. Ringed ♂ seen repeatedly on same hectare, Ethiopia, at all times of year for 6 years, and ♀ stayed put for 1 year (Urban 1975).

Food. Nectar: *Lobelia*, canna lilies, banana flowers, flowering thistles, aloes, *Leonotis*, *Kniphofia*, *Callistemon* and *Eucalyptus*. Visits sugar-water feeders in gardens. Young in nest given mainly moths, a few small insects (ants, beetles?) and very few large spiders; adult sometimes visits nest with no visible insect but seems to feed the young – presumably on nectar (C. Jackson, pers. comm.). Also insects (see above). Captive nesting ♀♀, England, took the large number of spiders given to them and a few crickets, grasshoppers, caterpillars and moths (Martin 1976, Coles 1978).

Breeding Habits. Solitary nester, evidently monogamous and territorial. ♂ stays close to nest when ♀ is nesting, driving off all other sunbirds and especially conspecifics (C. Jackson, pers. comm.). ♂ once alighted on back of Fiscal Shrike *Lanius collaris* (a nest predator). ♂ of pair in captivity, England, extremely aggressive to nesting ♀; copulation described as rape (Martin 1976, Coles 1978).

NEST: suspended, pear-shaped purse or bag with side-top entrance, composed largely of spider web and vegetable fibres, grey lichen, dry grass, leaves and bits of string; decorated outside with patches of yellow bark, lichen, wool and leaves, and thickly lined with feathers including those of domestic hens and guineafowl (Cheesman and Sclater 1936–1936), wool or cotton-wool. 3 nests made of *Usnea* lichen, one with an inner layer of fine grass and bits of bamboo-sheaths (C. Jackson, pers. comm.). Lower rim of entrance thickened somewhat. Sometimes a 'beard' of lichen hangs below nest. In E Africa placed mainly on forest edge in tree, at 1·5–10 m above ground; once in rose bush on house wall. Suspended from thin branch, sometimes at very tip; once on a juniper bough 4 m over water; others in cedars, black wattles *Acacia mearnsii*, high in *Podocarpus* tree, high up in dense *Eucalyptus* foliage, and in creepers on house verandah. Nest used only once, but sometimes twice if food plentiful, permitting 2nd clutch to be laid soon after 1st brood has fledged (C. Jackson, pers. comm.).

Built by ♀ only, once bringing material every 30 s. ♂ visits nests every 10–30 min, but only to inspect the work. Nest completed in 3–8 days.

EGGS: 1. Greenish blue with dark brown splashes and faint pinkish spots. SIZE: (n = 5) 19·0–20·75 × 14·0–15·0 (20·2 × 14·35).

LAYING DATES: Ethiopia, May, July, Aug–Nov (various indications Apr, June); E Africa, Region A, Feb–Aug, Oct–Dec (24 clutches, mainly Apr and Nov–Dec); Region D, Nov–Feb, May–July, Sept (16 clutches, no clear-cut breeding season, other records suggest mainly May–June and Nov–Jan). Other indications: Region A, 2 nests being lined mid Nov, Region D, adults feeding young Aug.

INCUBATION: by ♀ alone, facing out of entrance with head just showing. Period: 14–17 days.

DEVELOPMENT AND CARE OF YOUNG: young fed by both parents, mainly ♀; in a 7·3 h period ♂ fed 50 times, ♀ 61 times; visits were clumped, with a frequent feeding followed by gap of *c.* 40 min. ♀ removes faecal sacs; as nestling grows, ♀ waits at nest entrance for faecal sac, but leaves after *c.* 30 s if none is produced. ♀ also collects feathers or down from nestling (C. Jackson, pers. comm.). In last 3 days before young due to leave nest, ♀ enters nest which can be seen to vibrate violently for several min – thought to be young exercising wings. Nestling period *c.* 19 days. Soon after young left nest, ♀ visited it several times with food despite the young being in tree only 10 m away where ♂ was feeding them (C. Jackson, pers. comm.). Young return to roost in nest for following 6 days.

BREEDING SUCCESS/SURVIVAL: of 5 nests with known outcome, young reared successfully in 4, and 5th deserted after egg hatched. May be parasitized by Klaas's Cuckoo *Chrysococcyx klaas*, since singing cuckoo is chased away. Main predator (Mau Narok, Kenya) is Fiscal Shrike (C. Jackson, pers. comm.). ♂ lived for at least 8 years, Ethiopia (Urban 1975).

Nectarinia kilimensis Shelley. Bronzy Sunbird; Bronze Sunbird. Souimanga bronzé.

Plate 12
(Opp. p. 159)

Nectarinia kilimensis Shelley, 1884. Proc. Zool. Soc. London, p. 555; Kilimanjaro.

Range and Status. Endemic resident, W-central Angola, and E and SE African highlands from Uganda and Kenya to Chimanimani Mts along the Zimbabwe/Mozambique border. Angola, S Cuanza Sul to NW Bié and extreme N Huila; also Duque de Braganza Falls, NW Malanje Province. Zaïre, frequent to common (the commonest large montane sunbird): Lendu and Djalasinda highlands near Mahagi northwest of L. Albert, Semliki, Lubero highlands west of L. Edward, Kivu mts, Itombwe, Kabobo and Marungu. Descends to 1200 m at NW end of L. Victoria (where breeds) and in Semliki Valley and perhaps even lower around N end of L. Tanganyika; ascends to 1850 m in W Rwenzori, 1240 m west of L. Edward, 2770 m on Mt Sabinyo in volcano region, at least 1880 m (where breeds) in Itombwe, and 2050 m at Pepa (where common) in Marungu Highlands. Uganda, common in SW, uncommon north to L. Albert and L. Kyoga; occurs around whole N end of L. Victoria, commoner west than east of Sezibwa R., to Bukoba (Tanzania) and Kavirondo Gulf (Kenya). Kenya, common at 1200–2100 m and erratic and scarce at 2100–2800 m, throughout SW except for S Nyanza, and in Chyulus. Rwanda and Burundi, widespread on all higher ground. Tanzania, as mapped (8 breeding and >90 sight records, N. E. Baker, pers. comm.): in NW in Kagera and NW Kigoma Prov. (Kasulu, Kibondo); in NE, Kilimanjaro, Meru, and Mbulu uplands; in SW, Ufipa Plateau, Mt Rungwe; and in S, from Kipengere Range (Njombe Highlands) to Uvidunda Mts. Malaŵi, Nyika Plateau south to Tambo, Mwanza Prov., in hills above 1380 m (occasionally descends to 1230 m). Zimbabwe, up to 2200 m in Inyanga Highlands, through Stapleford and the

Nectarinia kilimensis

Vumba to Melsetter, Banti, Chimanimani Mts, Chipinga Uplands (Mt Selinda); record at Rusape; breeds down to 1150 m and descends to 1000 m. Mozambique, headwaters of Pungwe R. north to about 17°30'S, south to Mafusi and Chimanimanis, not below 1375 m.

Plate 13

Plate 14

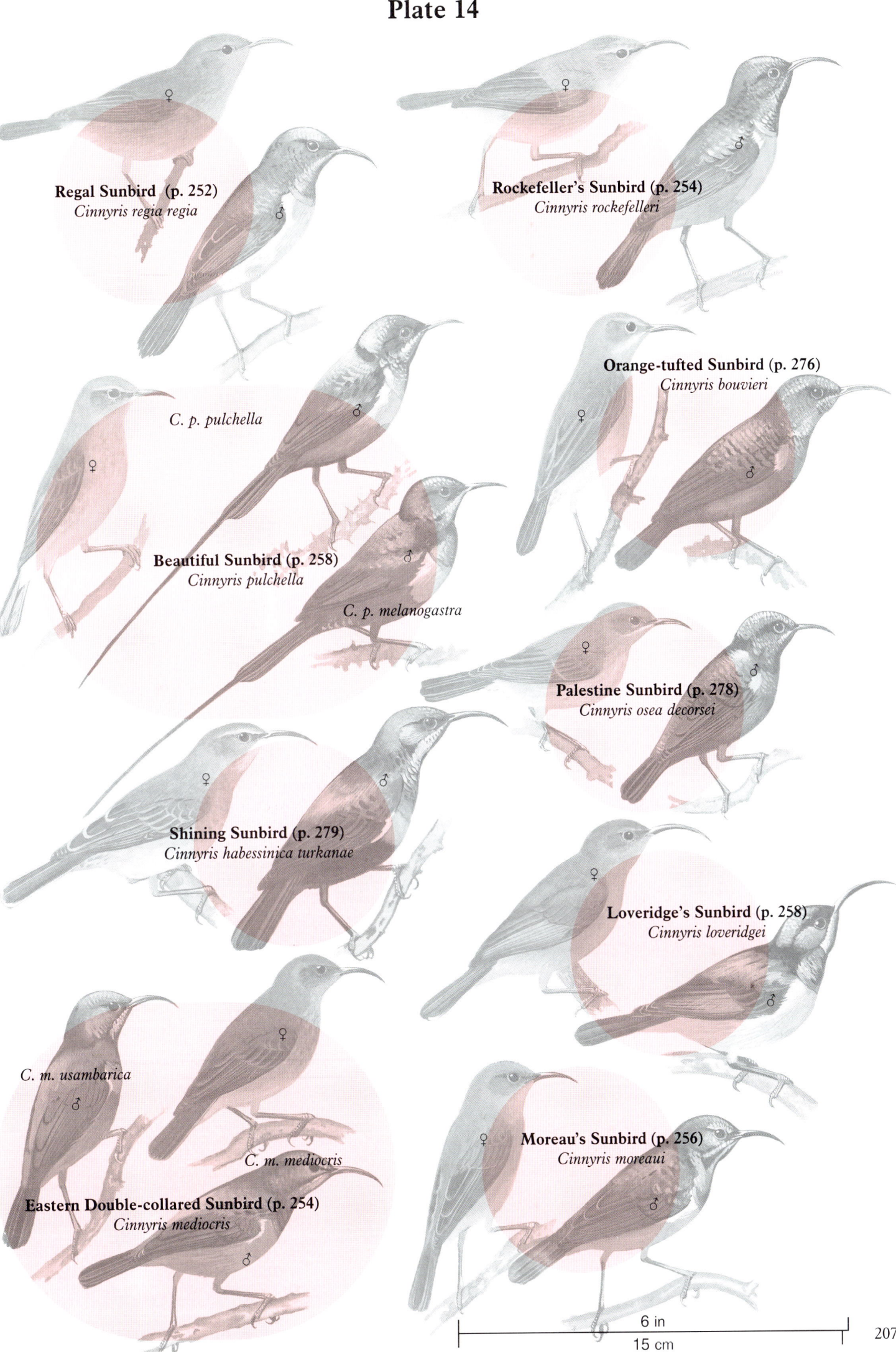

Regal Sunbird (p. 252)
Cinnyris regia regia

Rockefeller's Sunbird (p. 254)
Cinnyris rockefelleri

Orange-tufted Sunbird (p. 276)
Cinnyris bouvieri

C. p. pulchella

Beautiful Sunbird (p. 258)
Cinnyris pulchella

C. p. melanogastra

Palestine Sunbird (p. 278)
Cinnyris osea decorsei

Shining Sunbird (p. 279)
Cinnyris habessinica turkanae

Loveridge's Sunbird (p. 258)
Cinnyris loveridgei

C. m. usambarica

C. m. mediocris

Eastern Double-collared Sunbird (p. 254)
Cinnyris mediocris

Moreau's Sunbird (p. 256)
Cinnyris moreaui

6 in
15 cm

207

Description. *N. k. kilimensis* Shelley: Uganda, Kenya and NE Tanzania to SE Zaïre. ADULT ♂: head, neck and rest of upperparts metallic bronzy green, with strong coppery reflections on back, rump and uppertail-coverts. Tail blackish brown, T1 narrow and elongated, projecting 53–68 mm. Lores blackish. Chin to upper breast bronzy green like head, upper breast with coppery tinge; underparts backward from lower breast sooty black. Median and lesser coverts metallic coppery green; rest of upperwing blackish brown. Axillaries and underwing-coverts sooty black. Bill black; eyes black; legs black. ♂ does not have eclipse plumage. ADULT ♀: upperparts dull olive, darker feather centres on top of head and side of neck giving faintly streaked effect. Tail dark brown, T1 projecting c. 5 mm; tips of T2–T3 fringed narrowly, those of T4–T5 more broadly, brownish white; inner web of T6 with brownish white distal fringe, outer web with large brownish white area near tip and outer edge brownish white. Narrow superciliary stripe buff, extending to rear of ear-coverts. Lores to below eye and ear-coverts dark olive-brown. Cheeks to chin and throat buffish white, streaked dusky brown. Breast to belly and flanks olive-yellow, streaked olive-brown except on centre of belly; undertail-coverts pale buffish yellow. Upperwing dark brown; primaries (proximal to emarginations) and secondaries edged olive-yellow, tertials, alula and primary coverts narrowly edged yellowish buff, rest of upperwing-coverts more broadly fringed buff. Axillaries olive-yellow; underwing-coverts buffish white with dark greyish bases; inner borders of flight feather undersides pale grey-buff. Bare parts as adult ♂. SIZE (10 ♂♂, 10 ♂♂): wing, ♂ 73–79 (77·0), ♀ 67–69 (67·5); tail, ♂ 115–135 (125), ♀ 54–59 (55·5); bill to feathers, ♂ 27–31 (29·0), ♀ 27–29 (28·1); tarsus, ♂ 18–20 (19·2), ♀ 18–19 (18·4). WEIGHT: Kenya, ♂ (n = 108) 14–19·9 (16·7), ♀ (n = 68) 13·3–16·7 (14·6).

IMMATURE: juvenile similar to adult ♀, but more uniform olive-green above; less streaked below, chin to throat dark greyish, bordered by pale yellowish moustachial stripe.

NESTLING: naked, blind, skin dark brownish, bill short and soft, gape pads thick and yellow.

N. k. arturi Sclater: SW Tanzania to Zimbabwe. ♂ has head, neck and chin to throat coppery green, grading to coppery purple on upperparts and upper breast (strongly purple on median and lesser coverts, rump and uppertail-coverts). Smaller; wing, ♂ (n = 10) 71–75 (72·7); bill to skull, ♂ (n = 13) 26·5–31·0 (28·2), ♀ (n = 7) 25·0–29·0 (27·1). WEIGHT: Malaŵi, ♂ (n = 17) 14·8–18·6 (16·3), ♀ (n = 6) 12·7–15·5 (14·0) (Dowsett-Lemaire 1989b); Zimbabwe, ♂ (n = 1) 18·0.

N. k. gadowi Bocage: Angola. ♂ purer, deeper metallic green than nominate race on head and throat; upperparts and breast less coppery. ♀ rather greyer above. Same size as nominate race.

Field Characters. Length, ♂ 21·5–23 cm, ♀ 14–15 cm. Bill shorter and more curved than Tacazze or Malachite Sunbirds *N. tacazze* and *N. famosa*. ♂ has bronzy green head and back, becoming coppery on rump, wing-shoulder and lower breast; otherwise black, and looks all blackish in poor light. Lacks any purple in plumage, but could be mistaken for Tacazze Sunbird if only head seen. ♀ has dark mask and pale supercilium like ♀ Tacazze but throat white so *lacks contrasting malar stripe;* underparts *yellow with dark streaking*. Most resembles ♀ Malachite Sunbird but brighter yellow below and streaked rather than mottled. Central tail feathers slightly elongated, pale tips to others. Juv. ♂ like ♀ but with indistinct supercilium, dark mottled throat, yellow underparts mottled rather than streaked.

Voice. Tape-recorded (10, 16, 91–99, 104, B, C, F, LEM). Characteristic call a loud, nasal 'jer-jooey', second note rising; in one variation second note high and pure 'dzu-wee'; this call frequently incorporated into song, 'jer-jooey, jer-jooey-jooey, jer-jooey-jooey, tik-tik, jer, tik, jer-jooey', running into purer sharp notes, 'tyup, tyup-tyup-tyeek-tyeek'; a shorter version is 'jer, ju-jer, wip-wip'. Excited song from one of 2 displaying ♂♂ (Gibbon 1991) a continuous rapid, bustling medley of chattery, tinny and sibilant notes lasting >30 s; similar but slower song (context?) a disjointed series of chips, chatters and jingles, including a section of tinny babbling given in an undertone, 'chip-chip, chi-chee, chi-chee, pssst, tiddliddliddle, chippity ... '. Song of ♀, 'tippy-chippity, tsi-tsi, tippy-chippity ... ' interrupted by call note 'jee-jweeu', second note downslurred; ♂ has a similar song, including the repeated phrase 'pi-chappy-chop'. ♂ sings strident 'peo-view' near nest while ♀ incubating. ♂ chasing intruder gives twittering 'chee-wit-chee chee-wit-wit'; sharp little calls made by chick when ♀ arrives at nest (van Someren 1956). Territorial ♀ has insistent 'psew-psew-seep' or 'tsew-*eep*' (Zimmerman *et al.* 1996). Another call is a high-pitched, rather faint 'si-tsi-tsu'.

General Habits. Inhabits open montane areas with profusion of flowering trees and shrubs such as *Erythrina*, aloes and *Leonotis*, edges of montane evergreen forest, clearings, secondary growth, bushy grassland, highland savannas, heath, open grassland with scattered bushes, bracken-briar, gardens and farmland. ♂♂ widely recognized as being exceptionally intolerant territorially (e.g. Simpson 1971) and ♀♀ as unusually industrious at the nest. In Zimbabwe, ♂♂ regularly pursue Malachite Sunbirds, and increase of Bronzy Sunbirds held responsible for decrease of Variable Sunbirds *Cinnyris venusta* at one homestead (Meikle 1985). Elsewhere displaces Blue-headed Sunbirds *Cyanomitra alinae*. ♂ spots a conspecific rival, chases after it, twittering loudly, and having driven it off he perches on top of vegetation, calls loudly, swings body from side to side and spasmodically fans tail (van Someren 1956). Feeds on nectar by hovering or fluttering momentarily in front of flower and by perching next to and leaning into it; but mainly by alighting just below inflorescence at end of almost upright stalk, standing upright and pushing bill deep into corolla. Frequently punctures base of corolla of flowers that are too long for bill to reach nectar, such as fuchsias. Gleans leaves and twigs industriously for tiny spiders and minute insects, flitting from tree to tree. ♀ hunting for young alights on twig, scrutinizes foliage overhead, moves head up and down jerkily, and either reaches up to take insect or flutters below leaf to dart at prey. Flutters in front of spider web. Anting behaviour known (Stewart and Stewart 1964).

Sedentary; but on Nyika Plateau, Malaŵi, disperses locally after breeding to forage on forest edges and in river valleys, mainly at flowers of *Syzygium cordatum* and mistletoes, occasionally wandering into adjacent protea grassland.

Food. Nectar and small invertebrates. Commonly feeds at flowers of *Erythrina abyssinica*; also *Aloe*, *Kniphofia*, *Lobelia*, *Tephrosia*, *Pentas*, *Triumfetta*, *Canna*, *Fuchsia*, *Balthasaria*, *Leonotis* and *Syzygium*. The last is a favourite in Malaŵi (*S. cordatum*) and Rwanda (*S. guineense*), as is *Leonotis*

(Dowsett-Lemaire 1989b, 1990); other flowers in Malaŵi, all seen to be probed for nectar, are of forest trees *Bridelia brideliifolia*, *Halleria lucida*, mistletoes *Phragmanthera usuiensis*, *Tapinanthus sansibarensis*, *T. subulatus*, forest edge or secondary growth plants *Impatiens gomphophylla*, *Lobelia gibberoa*, *L. mildbraedii*, *Leonotis decadonta*, *L. mollissima*, *L. pole-evansii*, *Pentas schimperana*, *Tecomaria capensis*, rocky grassland species *Protea madiensis*, *P. petiolaris*, *P. welwitschii*, and upper woodland *Protea angolensis* and *P. gaguedi*. *Erythrina abyssinica* is favourite nectar flower (and insect tree) around towns and villages; especially likes nectar of suikerbos *P. repens*, and with cultivation of large protea garden in Zimbabwe Bronzy Sunbird numbers increased dramatically (Meikle 1985).

Breeding Habits. Solitary nester, monogamous and territorial. In Kivu, pair maintains fixed territory for several years and pair-bond persists indefinitely (Chapin 1978). One territory, around an ever-flowering *Erythrina abyssinica* tree, was only c. 45 m in diameter. ♂ defends territory pugnaciously against other sunbirds, large weavers such as *Ploceus cucullatus*, and African Paradise-Flycatchers *Terpsiphone viridis*. ♀ is less aggressive, but once seen to chase a Least Honeyguide *Indicator exilis* near nest. Territories and their food plants studied in Malaŵi (Dowsett-Lemaire 1988, 1989b): in Feb Bronzy Sunbirds spend their time feeding, chasing others, calling and displaying; somewhat less aggressive in Mar, enabling territories to be defined and ♀♀ to build nests. One territory, of 0·4 ha with c. 450 *Leonotis* stems, had initially been fought over by 3 ♂♂ and 2 ♀♀; another was disputed by 4 ♂♂ and 3 ♀♀, with 1 pair eventually nesting there; a pair in another of 0·3 ha with c. 250 *Leonotis mollissima* stems produced 1 nestling fed in Apr when most flowers had withered; another of 0·1 ha with 130 *Leonotis* stems contained a nest which, with the territory, was abandoned. A pair defended a territory of 0·2 ha with c. 140 food-plant stems but did not attempt to breed. Renewed activity in June when *L. decadonta* started flowering. At low latitudes breeds almost continuously – see below. Nesting is extremely sensitive to human interference. One pair (♂ colour-ringed) built c. 15 nests in a territory in 4 years (Chapin 1959).

NEST: rather large, elongated, suspended bag or purse, with side entrance and porch above it; made of grasses and fibres or shreds of bark, copiously interlaced with spider web, sometimes with downy pappus of Compositae incorporated into fabric. Elaborate yet untidy, with vegetable strands left hanging. Although suspended, nest is not freely movable, since neck of nest is thickly built around the support and small living leaves are often worked into roof or body of nest (van Someren 1956). Attachment to supporting branch can be weak, however, and some nests soon fall down (Zimbabwe: Meikle 1985). Porch is made of fine fibres and seedless grass-heads, latter placed with stem in body of nest and feathery ends projecting out over entrance. Threshold of entrance, under porch, is thickened and bound with spider web. Favoured plant down for lining nest is from *Leleshwa* and *Clematis*; the down is felted into compact bed, 25 mm thick on floor of nest. One nest made of pine needles and lined with plant down, obtained from 100 m away (Jackson 1970). Size of 1 nest 120 × 90. Suspended quite conspicuously from tip of stem sloping up or from twiggy branch, under large leaves or mass of foliage of tree, 0·8–13 m above ground. Nest built only by ♀, taking up to 2–3 weeks; but one nest made in 5 days after first one fell down. Nest-building is industrious, but lining is added in leisurely and spasmodic way.

EGGS: 1 (rarely, 2). Pale cream or bluish white, sparsely or heavily spotted with lilac or grey-brown. SIZE c. 21·0 × 13·5.

LAYING DATES: Zaïre (L. Kivu and Tshibati, 2000 m, Kivu), (breeds every month: Chapin 1954), (Rwenzori), ('ready to breed in early Nov': Chapin 1954), (Itombwe), Apr, Sept, (Marungu), (enlarged gonads, Feb). E Africa, Region A, Mar, June–Aug, Nov–Dec, Region B, Feb–July, Oct–Dec Region C, Oct, Region D, all months: (n = 60 clutches) Jan 4 clutches, Feb 2, Mar 9, Apr 15, May 8, June 5, July 3, Aug 1, Sept 4, Oct 4, Nov 1, Dec 4; Malaŵi, Feb, Mar, June, July; Zimbabwe, Sept–May (half of 16 clutches in Oct–Dec).

In Uganda there are 2 main breeding seasons: Feb–Aug and Oct–Dec, and Bronzy Sunbird is variously seasonal in S tropics. However in Kivu, although the species nests in all months (observed at 2 localities: Chapin 1978), a pair nests 4–5 times a year, taking 40–50 days from nest-building to fledging each time, and has an annual non-breeding or resting season of about 2 months. Resting seasons of adjacent territorial pairs do not necessarily coincide.

INCUBATION: by ♀ only, who spends much time outside nest (Kivu), but when incubating (Kenya) sits close, deep down in nest, with bill tip just above lip of entrance; if uneasy, she lifts head slightly to look out. ♂ tends to stay near nest, singing strident 'peo-view', and chivvies ♀ when she comes out 4–5 times daily to feed. Period: 14–16 days.

DEVELOPMENT AND CARE OF YOUNG: the single young brooded only by ♀ and fed mainly by ♀. She attends assiduously, coming to nest every few min with spiders, small larvae and nectar, and clinging to front of nest to feed chick through entrance (van Someren 1956). Later, ♂ makes spasmodic and desultory provisioning visits (Chapin 1978). One chick was brooded for 53 min in 2 h and fed 12 times by ♀ and 3 times by ♂; 3 days later it was fed 5 times in 60 min by ♀ and 7 times by ♂; both parents brought insects on all trips (Dowsett-Lemaire 1988). Nestling defaecation: ♀ arrives at nest, clings below entrance, leans in and prods nestling, which backs up to entrance from inside and excretes faecal sac; ♀ takes it in bill and flies away with it. Period: 14–16 days, once 22 days. Bill still short when chick leaves nest. Young hesitates for a long time before leaving nest for first time, then flies only very short distance. ♀ shepherds it back into nest at dusk for 5–6 days (once only 2 days) and roosts in nest with it (♂ roosts nearby under foliage). During day, young follows ♀ around, quivering wings, and soon starts probing *Erythrina* blossoms and gleaning spiders and small insects, but continues to be fed by ♀ for some time, even after she has started to make new nest for next brood. Young able to drink nectar at 4 days after leaving nest, and independent of parents at 3–4 weeks. ♂ largely ignores fledgling, or occasionally flies aggressively at it, when it topples over and hangs upside down under perch (Chapin 1978).

BREEDING SUCCESS/SURVIVAL: ringed birds over 6, 7, 7, 7·5 and 12·5 years old (Hanmer and Chadder 1997). Parasitized by Klaas's Cuckoo *Chrysococcyx klaas* (12 records, E Africa). A ♀ Bronzy Sunbird once seen to remove and drop an egg from her nest – almost certainly a Klaas's Cuckoo egg (Chapin 1978).

References
Chapin, R.T. (1978).
Dowsett-Lemaire, F. (1989b).
van Someren, V.G.L. (1956).

Plate 12 *Nectarinia reichenowi* (Fischer). **Golden-winged Sunbird. Souimanga à ailes dorées.**
(Opp. p. 159) *Drepanorhynchus reichenowi* Fischer, 1884. J. Orn., 32, p. 56; Lake Naivasha.

NOMENCLATURAL NOTE: not to be confused with sunbirds *Anthreptes reichenowi* or *Cinnyris reichenowi*.

Range and Status. Endemic resident and partial (mainly vertical) migrant, E African mountains. Kenya, widespread and locally common at 1200–3400 m, breeding above 1600 m but mainly a cool weather non-breeding visitor below 1800 m, spreading to L. Victoria basin in Mar–Sept and Ruaraka–Kahawa–Ruiru areas near Nairobi in Apr–June, with records at Baringo in Apr, Olorgesaille in June and Nov and Ngulia in Nov. One population endemic to montane forest islands in N Kenya: Mt Kulal at 1800–2300 m, Mt Nyiru at 2200–2400 m, Matthews Range and Mt Uraguess; another, widespread in highlands from Chyulu Hills to Mt Kenya, Aberdares, Nyahururu, Mau Plateau, Nandi and Mt Elgon. On Mau Plateau may be resident above 2100 m (breeding Nov–Jan), but reported to be migrant to 3000 m in Mau Narok in Aug–June (breeding Oct–Apr). Uganda, breeds on Mt Elgon, otherwise erratic visitor in SE, west to Ankole and Masaka. Tanzania, Crater Highlands, Mbulu, Arusha Nat. Park, Mt Meru, Mt Kilimanjaro, and (less commonly) North Pare and South Pare Mts and West Usambaras. Zaïre, quite common and widespread in Itombwe above 2400 m, localities including Mt Kandashomwa, Lubuku, L. Lungwe, Kitoga, Mt Mohi and Rurambo.

Nectarinia reichenowi

Description. *N. r. reichenowi* (Fischer): Uganda, Kenya and Tanzania. ADULT ♂ (breeding): head, neck and rest of upperparts black, with strong bronzy gold reflections from crown to back, and more coppery reflections on rump and uppertail-coverts. Tail graduated, T1 very long and narrow, projecting 45–70 mm; blackish, T1 edged bright golden yellow on both webs, T2–T6 on outer web, more broadly near base, narrowly near tip. Lores velvety black. Chin to upper breast black with strong coppery gold reflections; rest of underparts sooty black. Median and lesser coverts coppery gold; rest of upperwing black, primaries edged bright golden yellow proximal to emarginations, secondaries and tertials more broadly edged, forming large golden patch on closed wing; alula, primary coverts and greater coverts edged bright olive-green. Axillaries and underwing-coverts sooty black. Bill black; eyes dark brown; legs black. ADULT ♂ (non-breeding): whole trunk and head velvety black; wings and tail like breeding ♂. ADULT ♀: forehead and crown dark brown, feathers fringed olive-green; rest of upperparts dull olive-green with diffuse brownish mottling. Tail graduated, T1 projecting 5–10 mm; dark brown, feathers edged yellow, more broadly so towards base. Lores and ear-coverts dark brown. Cheeks to chin and throat pale olive-yellow; rest of underparts deeper olive-yellow with greyish cast, centre of breast and belly mottled dusky. Upperwing dark brown; primaries edged golden yellow, secondaries and tertials broadly edged olive-yellow; alula, primary coverts and greater coverts edged bright olive-green; median and lesser coverts broadly fringed dark olive-green. Axillaries pale olive-yellow; underwing-coverts sooty brown, tipped pale olive-yellow; inner borders of flight feather undersides grey-buff. Bare parts like adult ♂. SIZE (10 ♂♂, 10 ♀♀): wing, ♂ 79–84 (81·6), ♀ 67–74 (69·6); tail, ♂ 115–140 (130), ♀ 52–67 (54·8); bill, 28–31 (28·8), ♀ 25–29 (26·8); tarsus, ♂ 16–18 (17·5), ♀ 16–17 (16·7). WEIGHT: Kenya, ♂ (n = 61) 12·8–17·5 (15·7), ♀ (n = 23) 11–15·9 (13·8); also Kenya, ♂ (n = 10) av. 14·3, ♀ (n = 11) av. 14·35 (Cunningham-van Someren 1976).

IMMATURE: juvenile like adult ♀, but cheeks and chin to throat sooty black, rest of underparts olive-green, barred black; in some birds, entire underparts and sides of head blackish; wing and tail feathers edged more saffron-yellow.

NESTLING: not described.

N. r. shelleyae Prigogine: E Zaïre. Poorly defined: bill less curved; ♀ with crown feathers edged greyish (not greenish like mantle).

N. r. lathburyi Williams: N Kenya (Mt Kulal to Mt Uraguess). Poorly defined: bill more strongly curved than in other races; ♂ with more coppery reflections; smaller: wing, ♂ 75–80.

Field Characters. Length, ♂ up to 24 cm, ♀ 14–15 cm. An E African highland sunbird with strongly decurved bill and *golden wing patches* and tail feathers in both sexes. ♂ has elongated central tail feathers; ♀ has yellow underparts with dark streaking and from below (only) might be mistaken for ♀ Bronzy Sunbird *N. kilimensis*.

Voice. Tape-recorded (104, CHA). Song a tinny, jingling babble given in series of short bursts, also described as a 'prolonged twittering warble interspersed with high chi-chi-chi ... phrases' (Zimmerman *et al*. 1996); a high, sharp, metallic chatter lasting 1–2 s is probably also a song. Calls include a dry, rattling 'chachachacha ... ' and a nasal 'jwee' like shorter version of Bronzy Sunbird call; also said to give rapid 'chuk-chi-chi-chek', insistent 'cher-cher-cher' and a single 'tweep' (Zimmerman *et al*. 1996).

General Habits. Inhabits edges of montane evergreen forest, clearings, and edges of bamboo forest, extending into cultivated areas and gardens. Solitary, in pairs (♂ and ♀, or ♀ with juvenile), sometimes assembles in great abundance as at yellow-flowered 'tall, laburnum-like shubs' at Limuru, Kenya (Jackson and Sclater 1938). Does not join mixed foraging flocks in Itombwe, Zaïre (Prigogine 1971), but near Naivasha, Kenya, commonly feeds together with Malachite, Bronzy and Variable Sunbirds *N. famosa*, *N. kilimensis* and *C. venusta* (Gill and Wolf 1979). 2000 sunbirds of various species, largely Golden-winged, occupied 50 ha patch of *Leonotis nepetifolia* for protracted period, establishing individual feeding territories, often close to or adjoining territories occupied by Scarlet-chested Sunbirds *C. senegalensis* and Bronzy Sunbirds (Gill and Wolf 1975a); in that area (Hell's Gate, Kenya) Golden-winged is closely associated with flowering *L. nepetifolia* and feeds only infrequently at other flowers. Territory size 5–12 m^2 in Apr, 12–47 m^2 in Mar and in another year 10–27 m^2 in July; most contain 500–2500 *L. nepetifolia* inflorescences, some >6000. Numerous birds fail to establish feeding territory, and territorial ♂ loses 18% of nectar resource to roaming non-territorial Golden-winged Sunbirds and 17% to Malachite and Variable Sunbirds; territorial ♀ loses 44% in total (Gill and Wolf 1979). Feeds mainly at open mistletoe *Phragmanthera dshallensis* flowers but also commonly at not-yet-open flowers which have slits in their corolla tubes (Gill and Wolf 1975b).

Resident, but makes regular and extensive migrations from high breeding grounds to lower elevations, occurring down to 1200 m in Mar–Sept. Bird ringed near Nairobi, May, found 65 km away and nearly 1000 m higher, on South Kinangop Plateau in Aug; another ringed on South Kinangop Plateau, Mar, was recovered 100 km away in Machakos (Britton 1980).

Food. Nectar, especially of *Leonotis nepetifolia*, *L. mollissima*, *Phragmanthera dshallensis*, *Erythrina abyssinica* and *Crotalaria agatiflora*, and insects: beetles, flies, hymenopterans and caterpillars in stomachs.

Breeding Habits. Solitary nester, monogamous and territorial.

NEST: suspended, globular bag with side entrance but little or no porch above, made of fine grass blades, slender flower stalks, *Usnea* lichen and other plant materials sometimes including thistledown, mixed with spider web, and lined with plant down. Sited 1–4 m above ground, in thick patch of *Salvia* or other shrub or at side of an isolated bush. Ext. height 190, ext. width 70, entrance diam. 30, top of entrance 40 below top of nest

EGGS: 1. Whitish or pale grey, heavily but irregularly spotted with greyish brown, forming a patch at or band around large end. SIZE: (n = 1) 20·5 × 14·2.

LAYING DATES: E Africa, Region A, Aug–Mar (n = 15 clutches, peaking in Nov–Dec; birds avoid wetter, cooler months); Region D, Jan, Apr, June, July, Sept (n = 11 clutches, 5 below 2000 m in long rains in Apr and June, 6 in high altitudes avoiding rains); Zaïre (Itombwe), May.

References
Gill, F.B. and Wolf, L.L. (1975a, b, 1977, 1978, 1979).
Pyke, G.H. (1979).

Nectarinia johnstoni Shelley. Scarlet-tufted Malachite Sunbird. Souimanga de Johnston.

Plate 12
(Opp. p. 159)

Nectarinia johnstoni Shelley, 1885. Proc. Zool. Soc. London, p. 227, pl. 14; Kilimanjaro.

Range and Status. Endemic resident, E African and S Tanzania/Zambia/Malaŵi highlands. Zaïre/Uganda, alpine zone on Rwenzoris, on Ugandan side at 3850–4460 m; on Mt Stanley down to 3880 m, common at 4150 m, less so at 4280 m, with a few birds up to 4550 m (including some by Stanley Glacier at 4510 m: Chapin 1954). Zaïre/Rwanda/SW Uganda, Virunga range; Mt Karisimbi (on border), common above 3690 m, but scarce towards bleak and windy summit at 4507 m. Zaïre, on Mt Nyamaragira (Nyamlagira) at 2615 m, Kivu Volcanoes, and Itombwe: Mt Kandashomwa at 2680–2710 m, Mt Muhi (Mohi) at 3020–3090 m, Muusi, L. Lungwe – very common throughout above 2300 m. Not known from Burundi. Kenya, common to abundant at 3000–4500 m on Mt Kenya and in Aberdares; on Mt Kenya descends to 3385 m on W slopes, to 3080 m in E and to 2770 m in NE (Williams 1951). Record from Menengai Crater at 2150 m. Tanzania, at 3400 m on Olosirwa in Crater Highlands and common at 3000–4500 m on Mts Meru and Kilimanjaro; record in North Pare Mts – single bird at 2150 m, doubtless a wanderer, 60 km from Kilimanjaro; in S, records from Njombe Highlands (Hall and Moreau 1970) and Makonde, Livingstone Mts, at 2000 m (Britton 1980). Zambia/Malaŵi, Nyika Plateau, above 2000 m; also (Malaŵi) Nyankhowa Mt (at 2000 m: near S end of Nyika Plateau).

Nectarinia johnstoni

Description. *N. j. dartmouthi* Ogilvie-Grant: E Zaïre, SW Uganda and NW Rwanda. ADULT ♂ (breeding): forehead to lores black; crown, sides of head and neck black with shiny green reflections; mantle to back and scapulars metallic bluish green, grading to blue on rump, and violet-blue on long uppertail-coverts. Tail black, T1 very long and narrow, projecting 75–115 mm; base of T1 (both webs) and outer web of T2–T6 narrowly edged shiny blue-green. Chin to upper breast metallic green, grading to blue on lower breast and violet-blue on upper belly and flanks; pectoral tufts scarlet; lower belly, thighs and undertail-coverts sooty black. Median and lesser coverts metallic green; rest of upperwing sooty black, inner feather of alula fringed bright blue-green, greater coverts edged iridescent blue. Axillaries and underwing-coverts black. Bill black; eyes brown or hazel; legs black. ADULT ♂ (non-breeding): head and body brownish, wings and tail as in breeding ♂. ADULT ♀: upperparts sooty brown. Tail blackish brown with slight bluish gloss above; T6 paler with narrow brownish white outer edge and tip. Narrow, indistinct pale superciliary stripe; lores to ear-coverts and sides of neck sooty brown. Chin to breast dusky brown, with indistinct mottling, especially on chin; chin and throat bordered by paler, narrow moustachial streak; flanks and undertail-coverts dull brown; belly pale buff; pectoral tufts orange-red. Upperwing blackish brown, fresh tertials and greater coverts narrowly fringed pale brown. Axillaries and underwing-coverts sooty brown; inner borders on undersides of flight feathers narrowly buffy brown. Bare parts as adult ♂. SIZE (8 ♂♂, 4 ♀♀): wing, ♂ 82–88 (84·4), ♀ 74–77 (75·3); tail, ♂ 130–162 (144), ♀ 46–50 (48·4); bill, ♂ 26–30 (28·1), ♀ 25–27 (26·3); tarsus, ♂ 19–21 (20·0), ♀ 19–21 (20·0). WEIGHT: ♂ (n = 3) 14–17 (15·7).

IMMATURE: juvenile like adult ♀ but lacks red pectoral tufts.
NESTLING: not described.

N. j. johnstoni Shelley: Kenya and N Tanzania. Breeding ♂ with upperparts purer metallic green than ♂ *dartmouthi*, blue confined to uppertail-coverts; lower underparts bluish green without violet tinge. Size similar to *dartmouthi*, but bill longer, ♂ (n = 6) 33–36 (34·3); T1 projection longer (up to 135 mm). ♀ less sooty above than *dartmouthi*, pale marking on chin and throat more prominent, belly whiter; pectoral tufts more orange. Non-breeding ♂ like ♀, but darker brown; wings and tail like breeding ♂; metallic green feathers also on back, uppertail-coverts and lower scapulars, and a few on head and breast. Weight: 2 ♂♂ 14, 16.

N. j. itombwensis Prigogine: E Zaïre (Itombwe Mts). ♂ like ♂ *johnstoni* but mantle, back, throat, breast and belly more bluish green. ♀ slightly paler than ♀ *johnstoni*. Smaller: wing, ♂ (n = 24) 75–79·5 (77·8), ♀ (n = 11) 67·5–73·5 (70·5); tail, ♀ (n = 10) 43·5–48·5 (46·4); bill, ♂ (n = 22) 27·5–31 (29·1), ♀ (n = 10) 27·5–29 (28·1).

N. j. nyikensis Delacour: Zambia and Malaŵi (Nyika Plateau) and S Tanzania (Livingstone Mts). ♂ like ♂ *johnstoni* but upperparts of some birds more bronzy; underparts slightly greener (less bluish on lower breast). ♀ and non-breeding ♂ same as *itombwensis*. Small: wing, ♂ (n = 9) 75–80 (77·5), ♀ (n = 3) 67·5–70 (68·8); tail, ♀ (n = 3) 43–47 (45·0); bill, ♂ (n = 7) 26–28 (27·0), ♀ (n = 3) 24·5–26·5 (25·7).

Field Characters. Length, ♂ up to 29 cm, ♀ 14–15 cm. A *moorland* species except in Malaŵi, and often the only sunbird present; meets Malachite Sunbird *N. famosa* on Nyika Plateau, but apparently ecologically separated. ♂ metallic iridescent green, appearing *darker* in the field than Malachite Sunbird since those parts of plumage not reflecting light appear black (Malachite less iridescent, more uniform); race *dartmouthi* (Albertine Rift mts) much bluer green, belly and uppertail-coverts almost purple. Pectoral tufts *scarlet* in ♂, *orange-red* in ♀ (yellow in ♂ Malachite, absent in ♀); ♀ *uniform brown*, without pale underparts or face markings of ♀ Malachite or Bronzy *N. kilimensis* Sunbirds; tail black, square-tipped, without white edges. ♀ Tacazze Sunbird *N. tacazze* is also dark below but has pronounced pale supercilium and malar stripe separated by dark 'mask'.

Voice. Tape-recorded (102, B(BBC), F, KEI, LEM). Song faster, thinner and more scratchy than that of Malachite Sunbird (see sonagrams in Dowsett-Lemaire 1988). In Kenya, starts hesitantly with 3–4 accelerating introductory notes, running into a rapid sibilant trill, 'tseet ... tseet, tseet-tsoo tsssrrrrrrr', lasting 4–6 s. Song in Malaŵi (race *nyikensis*) more simple and shorter, 1·3–2 s, consisting of an accelerated repetition of dry 'tserrep tserrep tserrep' notes following a few detached metallic 'tsec' notes (Dowsett-Lemaire 1988).

General Habits. Inhabits misty afro-alpine moorland with *Senecio* (groundsel) trees, and giant lobelias *Lobelia*, also living and dead bamboo, open *Hagenia* forest, and alpine meadow with *Hypericum* trees. Nests in tussock grass and (below 4100 m) *Erica* shrubs and giant lobelia and groundsel-tree inflorescences (Zimmerman *et al.* 1996). In Malaŵi (Nyika Plateau, too low for alpine vegetation) establishes territories in bushland of *Philippia* (= *Erica*), *Kotschya* and *Protea* scrub, where main food plant is *Protea welwitschii* on rocky outcrops and *Protea madiensis* in bushy grassland (Dowsett-Lemaire 1989b). Fearless of man. Active all day – as active in cold fog as in sunshine – but not as restless as many other sunbirds. Forages by perching on giant lobelia or in leafy crown of groundsel-tree, clinging with ease onto inflorescences and probing drooping bracts (where masses of small weevils congregate: Chapin 1954). Frequently flycatches, flying up from top of bush into swarms of midges or to catch passing blue butterflies; see also under Development and Care of Young, below. Often forages in company of other sunbirds: Ludwig's Double-collared *Cinnyris ludovicensis*, Tacazze, Bronzy, Golden-winged *N. reichenowi* and Eastern Olive *Cyanomitra olivacea*. Dominates Malachite Sunbird, aggressively displacing it.

Mainly resident, but on Nyika Plateau, Malawi, a local migrant, absent in May–June. Menengai Crater (Kenya) and North Pare (Tanzania) records (see above) strongly suggest considerable displacements.

Food. Nectar and small arthropods. Commonest plants visited on Mt Kenya are *Lobelia keniensis* and *L. telekii*, then *Senecio brassica*, *S. keniodendron* and *Protea kilimandscharica* (Coe 1961). In Malawi, feeds mainly at *Protea* flowers (*P. madiensis*, *P. petiolaris*, *P. welwitschii* and *P. angolensis*), secondarily at *Leonotis* (*L. decadonta*, *L. mollissima*, *L. pole-evansii*), also at *Tecomaria capensis* (Dowsett-Lemaire 1989b). Occasionally feeds at yellow flowers of *Hypericum* sp. and red flowers of *Hypericum keniensis* (Zaïre). Insects, taken from groundsel-trees and giant lobelias (Zimmerman *et al.* 1996); many tiny flies in stomachs (Chapin 1954, Coe 1961). On Mt Kenya eats many midges Chironomidae and lycaenid (blue) butterfly *Harpendireus aequatorialis*, the latter commonly fed to nestlings (Coe 1961).

Breeding Habits. Solitary nester, monogamous, territorial. On Mt Kenya territorial boundaries formed by *Senecio* and *Lobelia* stumps, used as song posts; territory can be only c. 70 m² in area, several territories often packed together, but other territories much larger, with ill-defined boundaries (Coe 1961). In Malawi ♂♂ arrive on Nyika Plateau in July; ♂ begins to advertise territory in July–Aug, singing from prominent perch around midday and in afternoon, chasing other sunbirds away, especially Malachite Sunbirds and feeding away from but not inside territory; by Oct, pair remains permanently in territory (one of 1 ha, 1 of 0.5 ha), feeding on insects and on newly-opened flowers of *Protea welwitschii* and *P. petiolaris*; pair and young leave territory by mid Feb, disperse into neighbouring valleys to feed at *Leonotis mollissima* flowers; pair deserts territory a few days after young have become independent (Dowsett-Lemaire 1989b). Single-brooded.

In courtship, ♂ approaches ♀ and perches close to her, usually slightly higher than her; ♂ raises bill slightly, puffs feathers out, holds tail curved up over back, and raises half-opened wings. Uttering high-pitched 'psurr-psurr-psurr' call, he turns his body by slowly pivoting to one side (Coe 1961). Whilst ♀ nesting, ♂ divides time between singing, displaying, chasing intruders, diving at ♀ and occasionally feeding her, when she begs like a nestling, with vibrating wings (Dowsett-Lemaire 1988).

NEST: domed oval with side entrance: one made entirely of white down from everlasting flowers, 0.5 m above ground, firmly woven into small *Erica* bush; another made largely of beard-lichen mixed with pappus from Compositae, brownish plant fibres or rootlets (Chapin 1954). Nest lined with soft materials, including fur of alpine rodents and hyraxes and, commonly, down from leaves of *Senecio brassica* (Coe 1961). Ext. height 140, ext. width c. 120. Nests also sited in centre of clump of tussock grass, often in dead inflorescences of *Lobelia* and dead leaf rosette of *Senecio keniodendron*, also in giant heathers, 1.8 and 2 m above ground deep in *Philippia* bush, and on triple fork in *Hagenia* tree.

EGGS: 1 (E Africa), 1–2 (Malawi). Eggs laid as soon as nest completed. White, streaked with pinkish brown, mainly at broad end. SIZE: (n = 1) 19–20 × 12.

LAYING DATES: Zaïre (Rwenzori), (♀ carrying beakful of tiny insects, Nov; ♂♂ with enlarged gonads, late Nov to early Jan; nest Dec); Zaïre/Rwanda, (courtship, Mar; large gonads and nest-building, June); Uganda, Nov–Dec; Kenya, July–Aug (mainly), also Sept–Oct, Dec–Feb; Zaïre (Itombwe), (juveniles Apr, May, Oct); Malawi, Oct and Dec–Jan.

INCUBATION: period (n = 1 nest) 22 days; both chicks hatched and left nest on same day (Dowsett-Lemaire 1988).

DEVELOPMENT AND CARE OF YOUNG: young brooded and fed in nest entirely by ♀ (2 nests, Malawi); ♀ fed young 5 times per h on day 1 and day 9 and 12 times per h on day 15 (after each visit, ♀ fed herself at flowers of *Protea welwitschii*, then hunted for insects and spiders, often hovering and hawking low between bushes and above tussock grass clumps: Dowsett-Lemaire 1988). Young brooded at first for 10 min per h. After leaving nest, young fed by ♂ or (another year) ♀ only, for 14 days, during which time they often take nectar on their own from *Protea* flowers (Dowsett-Lemaire 1988); ♂ keeps watch on prominent tree, sometimes flying aggressively at young, which react with begging posture and 'seep seep' calls.

References
Coe, M.J. (1961).
Dowsett-Lemaire, F. (1988, 1989b).
Prigogine, A. (1977).
Williams, J.G. (1951).

Nectarinia famosa (Linnaeus). Malachite Sunbird. Souimanga malachite. **Plate 12**
Certhia famosa Linnaeus, 1766. Syst. Nat., ed. 12, 1, p. 187; Cape of Good Hope. **(Opp. p. 159)**

Range and Status. Endemic resident and partial migrant, highlands from Ethiopia to Zimbabwe/Mozambique border, and South Africa. Ethiopia, uncommon, mainly above 2400 m, apparently subject to local movements. Sudan, Imatong Mts only. Zaïre, widespread in Rift Valley mountains from Lendu/Mahagi Highlands and Rwenzori to Kivu, Itombwe and Marungu Highlands; above 1200 m, common above 1800 m, but uncommon in Marungu

Nectarinia famosa

(Lufuko R., Ketendwe, Kasiki); up to 1690 m on Rwenzoris and 2920 on Mt Muhavura. Rwanda, several records in SW; in Nyungwe restricted to marshy areas above 2200 m. Burundi, highlands in NW. Uganda, Morongole and Lonyili Mts in NE, Mt Elgon in E and up to 2580 m in Kigezi District in SW. Kenya, frequent and widespread, locally common, mainly at 1800–3000 m; in W and central highlands, from Mt Elgon and Cheranganis to Mau, Aberdares and Mt Kenya; also on Mt Nyiru and Chyulu Range. Breeds up to 2700 m on Mau Narok and 3400 m on Mt Elgon and occurs seasonally as low as 600 m (near Njoro). Tanzania, common in Crater Highlands; on Mt Meru, Mt Kilimanjaro, and W Usambaras (down to 1650 m); occurs in Kungwe-Mahale Highlands in W, and widespread from Ukagurus through Iringa and Udzungwa Highlands to mountains around N end of L. Malaŵi, Ufipa Plateau and Matengo Highlands in S (70 records, N. Baker, pers. comm.). Zambia, only on Nyika Plateau and Mafinga Mts, above 1850 m. Malaŵi, widespread on Nyika Plateau and in Viphya Mts. Zimbabwe, patchily distributed but can be locally very common, in Inyanga Highlands to Stapleford, Vumba, Banti and Melsetter at 1500–2200 m and Chimanimani Mts down to 1200 m. Mozambique, known from Mafusi, upper Buzi R. drainage, Chimanimanis, and headwaters of Pungwe R. Transvaal, common in E Highveld and along crest of Escarpment, west to Magaliesberg, Waterberg and Suikerbosrand. Orange Free State, fairly common east of 26°E and in SW. Swaziland, in W only. Natal, common in mountainous country in W; Cape Prov., common, but absent from centre and N, and increasingly local and scarce towards E coast. Lesotho, throughout. Namibia, restricted to Orange R. valley.

Densities in South Africa, 3·5 birds per ha (*Protea roupelliae*, Transvaal, De Swardt 1993), 2·3 birds per ha (mountain fynbos) and 1·4 birds per ha (mixed alien shrubs with *Acacia cyclops* and *Eucalyptus lehmanii* (M.W. Fraser *in* Harrison *et al.* 1997), 0·5 birds per ha (shrubland, Fraser 1989) and 0·04 birds per ha (coastal renosterbosveld, Winterbottom 1968).

Description. *N. f. famosa* (Linnaeus) (includes '*major*' of Lesotho): South Africa to E Zimbabwe and W Mozambique. ADULT ♂ (breeding): head, neck, mantle, back and scapulars golden- or bronzy green with dull lustre; rump and uppertail-coverts emerald-green. Tail blackish, upperside with faint blue gloss, T1 very long and narrow, projecting 60–80 mm beyond the rest; base of T1 with broad green fringe on both webs, T2–T6 with narrow green outer edge. Lores black. Underparts green, merging with sides of head and neck, tinged golden across throat and upper breast, bluer on belly, flanks and undertail-coverts; pectoral tufts yellow. Median and lesser coverts bright golden green; rest of upperwing sooty black, tertials and flight feather edges tinged glossy blue, tertial outer webs and greater coverts fringed bright green and alula finely so. Axillaries and underwing-coverts sooty black. Bill black; eyes dark brown; legs black. ADULT ♂ (non-breeding): head and upperparts like adult ♀ (see below), but rump and uppertail-coverts bright green; underparts yellow, spotted and barred bright green and blackish, especially on breast; wings and tail as adult ♂ (breeding). ADULT ♀: upperparts greyish brown, feathers of forehead and forecrown with dark centres and paler fringes. Tail dark brown, T1 projecting up to 2–3 mm; T4–T6 with narrow whitish terminal fringe on inner web, T6 outer web white with pale brownish tip. Narrow superciliary stripe buffish white; lores and below eye to ear-coverts dark greyish brown. Chin to upper breast greyish brown with diffuse mottling, chin and throat bordered by buffish white moustachial stripe; sides of breast and flanks greyish brown, tinged yellow; lower breast to belly dull greyish yellow; undertail-coverts greyish white, streaked dusky brown. Upperwing dark brown; primaries and secondaries edged pale brown, tertials with pale brown fringes and paler grey tips; alula and primary coverts finely fringed pale brown, greater, median and lesser coverts broadly so. Axillaries and underwing-coverts whitish with pale grey bases; inner borders of undersides of flight feathers grey-buff. Bare parts as in adult ♂. SIZE (10 ♂♂, 10 ♀♀): wing, ♂ 76–80 (78·4), ♀ 68–73 (70·2); tail, ♂ 118–135 (126), ♀ 45–53 (48·0); bill, ♂ 33–36 (35·0), ♀ 30–34 (32·3); tarsus, ♂ 16–18 (17·4), ♀ 16–18 (16·5). WEIGHT: ♂ (n = 24, Zimbabwe) 13·0–18·7 (15·5), (n = 24, South Africa) 15·5–21·0 (17·0). ♀ (n = 14, Zimbabwe) 11·5–17·5 (13·9), (n = 27, South Africa) 12·0–16·2 (14·7) (Maclean 1993).

IMMATURE: juvenile like adult ♀, but upperparts greener and underparts yellower, especially breast, belly and flanks; moustachial stripe yellow. Juvenile ♂ sometimes has chin and throat blackish. In immature ♂, bright green appears first on lesser and median coverts, breast and rump.

N. f. cupreonitens Shelley: Ethiopia to N Malaŵi. ♂ with lower breast to belly bluer (less green) than in nominate race. Smaller; (10 ♂♂): wing, 70–75 (72·4), bill 29–31 (30·3). WEIGHT: Kenya, ♂ (n = 78) 12·5–16·6 (13·9), ♀ (n = 71) 9·1–14·5 (12·1).

Field Characters. Length, (nominate race) ♂ up to 27 cm, ♀ 15 cm, (*cupreonitens*) ♂ 24 cm, ♀ 13–14 cm. ♂ uniform bright golden green, bluer on belly and rump, with *yellow pectoral tufts* and elongated central tail feathers. In N part of range overlaps broadly with Scarlet-tufted Malachite Sunbird *N. johnstoni* but altitudinally separated (Scarlet-tufted at higher altitudes) except on Nyika Plateau, Malaŵi, where they may be ecologically segregated. ♂ Scarlet-tufted is darker (and bluer in race *dartmouthi*), with scarlet pectoral tufts, ♀ is brown below with orange-red pectoral tufts. ♀ Malachite has *pale yellow* malar stripe broadening onto cheeks, contrasting with dark mottling on breast and throat; darker, duller and longer-billed than ♀ Bronzy Sunbird *N. kilimensis*, yellowish olive below, whiter on belly, and *unstreaked*; outer tail feathers edged white, conspicuous in flight. Similar-sized ♀ Tacazze Sunbird, which it meets in E Africa, is mainly grey below.

Voice. Tape-recorded (16, 86, 88, 91–99, 104, B, F, LEM, MEY, WALK). Typical call a loud, clear 'chip', often repeated 7–8 times in irregular series. 2-part advertisement song, given from prominent perch, begins with measured series of 9–11 (sometimes 4–5) 'chip' and other notes, often accelerating, and ends with bustling thin, high-pitched sibilant notes, 'chip ... chip ... chip, chip, chip-chip-chip-chip-chi-chi-syoo-see-suwee'; 'tseeep ... tsee, tseee, tsu-tsu-tsu-tizitizitizitzi'; 'chip ... chop, chip, chopchop-seechopeechopeecho'; 'chop-chop-chip-chu-chee-seepissy-seepissysee'; average songs lasts 4–5 s, shorter versions 2–2·5 s. Song in Malaŵi shorter, 1·5–2·5 s, described as a 'pleasant motif of fine whistles, often with an alternation of higher and lower-pitched notes, sounding like 'pesui pesui pesui' and introduced by call notes similar to the contact call, a fluid 'tseeu, tseeu'' (Dowsett-Lemaire 1988). In courtship display ♂ gives high-pitched rapid medley of thin notes interrupted at irregular intervals with 'chop' or 'chip'; generalized songs include a loud, excited, rapid warble given by full- or eclipse-plumaged ♂ from perch or in flight; and a subdued warble, audible from only 10–30 m, uttered by bird under cover in centre of bush, lasting for 1–2 min with breaks of 1–2 s (Skead 1967). Another common call is plaintive, weak, but far-carrying 'sssseep', often repeated, with pauses, for long periods; alarm a plaintive, repetitive trilling, or a more sedate, repetitive 'tjoep, tjoep, tjoep ... '; ♂ gives whinnying cry when chasing another bird in flight (Skead 1967).

General Habits. In most of range inhabits open montane grassland scattered with flowering shrubs; also bamboo and forest edges and glades, but does not penetrate forest. In Nyungwe, Rwanda, marshy areas with *Kniphofia princiae*, *Lobelia mildbraedii* and *Hagenia*. In Malaŵi occurs mainly amongst *Protea* bushes in montane grassland at 1600–2500 m and in Zimbabwe occupies scrub-covered hillsides with flowering herbs and shrubs, particularly *Erythrina* and *Protea*. In South Africa inhabits open country, often arid, thorny and bleak, from sea level up to 2900 m (in Drakensbergs), and occurs in bare country of Oranjemund (Namibia) and Richtersveld Mts, and in macchia and karoo vegetation; commonest in alpine grassland and fynbos; strongly associated with main food plant *Lycium* spp. (Solanaceae), in scrub by dry, lowland riverbeds and in thornveld. Aloe-clad hillsides and proteaceous vegetation (W Cape) and thorn-bushveld of dry, karroid valley-bush in E Cape and Transkei. Freely enters farm and suburban gardens, parks, nurseries and town streets lined with flowering shrubs and trees.

Occurs singly, in pairs, family parties, in small, sometimes single-sex groups of 2–4 birds in eclipse plumage, and loose, unco-ordinated aggregations of up to about 40, seasonally attracted to stand of flowering plants; once c. 100 in 1·2 ha of flowering *Cotyledon macrantha* (Niven 1968). Usually (not always) rather wary of people, but dashingly pugnacious towards its own species, other sunbirds and many other birds up to size of Hoopoe *Upupa epops* and Laughing Dove *Streptopelia senegalensis*, both in and out of breeding season; once even a passing Egyptian Goose *Alopochen aegyptiacus* was chased. Often suddenly breaks off activity (foraging, singing) to chase and occasionally fight with others. Flight fast and strong, often high and distant. ♂ flies 100 m across aloe-clad hillside to attack another ♂, the attack generally accompanied by loud, whinnying call. 2 fighting ♂♂ may rise high in air, seemingly clawing each other breast to breast (Skead 1967), sometimes interlocking and falling to ground together. Known to chase Karoo Prinia *Prinia maculosa*, Cape Sparrow *Passer melanurus*, often Fiscal Shrike *Lanius collaris* (sunbird so close behind and above shrike that it can seem almost on its back), Cape Canary *Serinus canicollis* (chase becoming series of high-speed circlings), Common Starling *Sturnus vulgaris* (grabbing starling's tail in flight and then grounding the bird) and foraging flock of Rock Martins *Hirundo fuligula*. Chasing flight remarkably swift: a Pearl-breasted Swallow *Hirundo dimidiata* 'was hard pressed to draw away on a straight flight' (Skead 1967), and ♂ chased a White-rumped Swift *Apus caffer* for c. 10 s and touched it. ♀ Malachite not aggressive. Pair of sunbirds once invaded territory of pair of Cape Robin-Chats *Cossypha caffra*, resulting in protracted chasing by robin-chats, and ♂ Malachite and a robin-chat repeatedly grappling on ground (outcome not recorded). ♂ sometimes ousted, e.g. once by 2 ♂♂ Orange-breasted Sunbirds *Anthobaphes violacea*, and a resident ♂ once displaced by arrival of pair of Gurney's Sugarbirds *Promerops gurneyi*. ♂ and ♀ Malachite attack their reflections in window panes.

2 displays appear to be agonistic although context and function not studied. ♂ may rise high in air then descend in fast vertical dive. Common display among loose aggregations, e.g. on hillsides where aloes are flowering, is by ♂ making long, gliding, zigzagging flight downhill, singing usually all the way, either with wings held straight up and almost motionless, or with slow, regular, loose-jointed flapping, in both cases with bill and tail raised at

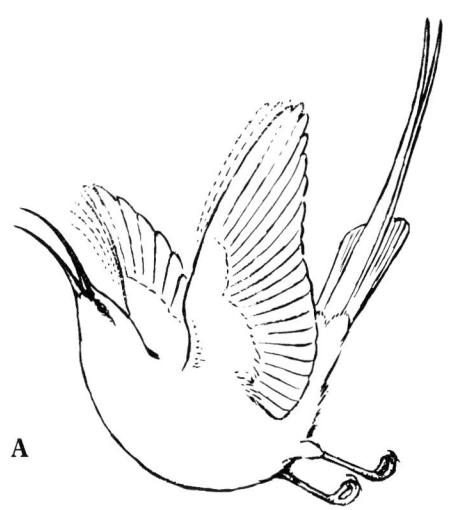

A

steep angle so that profile of body is shallow U shape (**A**) (Vincent 1949, Skead 1967). See also ♂ breeding display (below).

Flight strong, direct or with quick changes in direction and altitude, but less jinking and unpredictable than in many smaller sunbirds. Ground speed of flying bird once timed from car at *c.* 55 km per h.

Forages at large variety of flowers, mainly by perching and probing. Feeds on *Leucospermum* 'pincushions' by probing into all parts in leisurely manner. ♂ feeds at massed florets of *Mimetes hartogii* (Proteaceae) similarly, perching on top and repeatedly probing florets with bill; ♀, however, often clings to side of inflorescence to probe it. Visits giant protea *Protea cynaroides* (to long corolla of which Malachite Sunbird's long bill seems adapted) more frequently than any other sunbird or even sugarbirds *Promerops* spp. Takes nectar from *Leonotis nepetifolia* in Kenya, commonly feeding together with Bronzy, Golden-winged *N. reichenowi* and Variable Sunbirds *C. venusta* (Gill and Wolf 1978). Study of foraging strategies showed: 1st flower probed in a cluster is used as a test of how much nectar the remaining flowers may have; territorial bird feeds preferentially at inflorescences not visited for a long time; and flight distance to next flower depends on nectar 'reward' of previous flower (Gill and Wolf 1977). Sometimes sucks nectar whilst hovering; once probed 5 *Tecomaria* flowers in cluster, hovering in same position with only head noticeably turning. Always probes tubular corollae directly – does not pierce their bases. Commonly feeds by hawking for flying insects; bird darts up from bush-top perch, adroitly snaps insect up in bill, and returns, sometimes half gliding, to same or different perch; many Malachites hawking in small area of vegetation present distinctive 'jumping jacks' spectacle. Flies across open space between bushes to seize an insect in flight, often jerkily with brief hovering. Frequently interpolates spell of hawking into period of feeding at blossoms. Sometimes forages low in vegetation, only 25–50 cm above ground (see Food), fluttering to dislodge insects including moths, snapped up as they take flight. ♂ once foraged on ground, on flowers of parasite *Hyobanche* 2–3 cm above sand surface (Ryan 1998). One ♀ usually fed on aloe nectar just after visiting young in nest, then flew up to 500 m away to collect insects for them (Wolf and Wolf 1976).

Bathes in foliage wetted by rain and in dew on tops of *Protea* and other bushes; uses bird baths, and deliberately flies through spray of water from garden-bed irrigator or hand-held hose-pipe. Occasionally comes to ground, where stands with tail raised (**B**).

Sometimes completely sedentary, but most populations subject to erratic, poorly understood movements, presumably but not convincingly driven by flowering times of nectar-bearing plants. In some places local movements are annual, regular and predictable, e.g. on Cape Flats (SW Cape Prov.); species is summer visitor arriving in Dec and slowly declining in abundance thereafter. Seasonal mapping in South Africa suggests movement out of drier parts of W Cape in Nov–Apr and into SE Cape in July–Dec (Harrison *et al.* 1997). Nomadic (Craig and Hulley 1994) and partly altitudinal migrant (Johnson and Maclean 1994): regular displacements amount to short-distance

migration; highland population '*major*', breeding at up to 3000 m in Lesotho, winters in Natal down to sea level and in E Orange Free State. Pair observed for 5 years at Inyanga, 2300 m, Zimbabwe, arrived regularly in last week Aug, reared 2–3 broods, and departed in mid Apr. Occurs in Bvumba Botanical Gardens only in Apr–Oct (Hanmer and Chadder 1997). Ringed bird recovered 160 km away.

Food. Nectar and insects. Nectar especially of *Lycium* (at least 3 species), many *Protea* spp. including *P. lacticolor*, *P. roupelliae*, *P. madiensis*, *P. petiolaris*, *P. welwitschii*, *P. angolensis*, *P. gaguedi* and *P. cynaroides*, aloes including *Aloe arborescens*, *A. graminicola*, *A. mzimbana*, *A. nuttii*, *A. ferox* and *A. striata*, *Agave* sp., and numerous native and exotic garden plants: *Mimetes hartogii*, *Hypericum revolutum*, *Tecomaria capensis*, *Impatiens gomphophylla*, *Phygelius capensis*, *Chelonia*, *Nicotiana*, *Opuntia*, *Strelitzia*, *Erica mammosa*, *Erythrina lysistemon*, *Leonotis mollissima*, *L. nepetifolia*, *L. decadonta*, *L. pole-evansii*, *L. leonurus*, *Lobelia aberdarica*, *L. gibberoa*, *Halleria lucida*, *Jacaranda*, *Leucospermum conocarpum*, *Sutherlandia frutescens*, *Cotyledon macrantha*, *C. orbiculata*, *Kniphofia*, *Gladiolus*, *Watsonia tabularis*, *Crocosmia*, *Sarcocolla formosa*, *Melianthus villosus*, *Buddleia salviifolia*, *Greyia sutherlandi*, *Hyobanche*, *Chasmanthe*, *Digitalis* and *Vinca*.

Arthropods in stomachs include pupae, small flies and small spiders; 16 stomachs from 3450 m on Mt Kameligon, Kenya, contained 145 items, mainly bibionid fly *Dilophus erythraea* and other flies, also hymenopterans, Psocoptera, Neuroptera, Hemiptera, a beetle, and spiders (Cheke 1971). Young fed mainly or entirely on small arthropods: moths, beetles, caterpillars and spiders, but no evidence of regurgitative feeding with nectar (Wolf and Wolf 1976, contrary to van Someren 1956); sight records of adults catching beetles, flies, and small and large moths to take to young; and small lizards (Stark and Sclater 1900). Insects for young are caught by bird hovering around weeds and grasses or plucking them from leaves and flowering heads (Wolf and Wolf 1976).

Breeding Habits. Solitary nester, monogamous, territorial. Breeding territory may be small (0·08 ha); defended by ♂ perching prominently and singing, and chasing out other Malachite Sunbirds and small birds in general. ♂ also chases and constantly harasses his mate throughout nest-building and incubation periods. However, ♂ of pair with nest and eggs in 50 × 20 m patch of flowering *Protea lacticolor* appeared to tolerate invasion of patch by 10–20 foraging Malachite Sunbirds (Skead 1967). Near Gilgil, 2300 m, Kenya, territories generally centred on flowering *Aloe graminicola* plants, clumps of which are defended by ♂ and (to lesser extent) ♀; boundaries and shape of territory change during nesting season, due partly to presence of dominant sunbird spp (Wolf and Wolf 1976). On Nyika Plateau, Zambia/Malaŵi, at about 2150 m, ♂ and ♀ each defends its own territory in early morning and late afternoon (in June, when probably not breeding: Dowsett-Lemaire 1989b). At Hell's Gate, Kenya, ♂ Malachite Sunbirds, mostly non-territorial wanderers, always chased away from *Leonotis*-based territories of Golden-winged Sunbird and Bronzy Sunbird. Territory is essentially a feeding one, and needs to embrace e.g. 800 *Aloe graminicola* flowers to support the pair. Nest is generally but not always built in territory; one nest 30 m from territory (Wolf and Wolf 1976). Interspecifically territorial with Tacazze Sunbird *N. tacazze* and Eastern Double-collared Sunbird *N. mediocris* in *Protea* and heather moorland above 3450 m in Cherangani Mts, Kenya (Cheke 1971).

In courtship display, perched ♂ sings or subsings with head raised, bill pointing up and wings raised high, half spread and vibrating, and yellow pectoral tufts raised; tail may be cocked (**C**); in a variant, ♂, making chippering calls, crouches beside perched ♀, his feathers fluffed out, head sunk, tail fanned, and wings quivering. In common pre-copulatory display, conspicuously-perched ♂ utters emphatic whistles, flicking part-open, drooping wings at each note; voice changes to rapid warble and wing-flicks change to a rapid vibrating; tail sometimes cocked up sharply, or depressed, or wagged up and down; he then flies up with rapidly beating wings, hovers momentarily above perched ♀, alights on her back and they copulate. The display and copulation sometimes repeated immediately. ♂ seen to peck at ♀'s cloaca immediately before copulating; on another occasion ♂ pecked ♀'s cloaca 4–5 times, flew away, returned and cloaca-pecked again (Skead 1967). Generally 2–3 broods per breeding season. 2nd brood occasionally raised in 1st nest, but usually new nest built for each brood. After seasonal absence pair commonly returns to same site: records for 3, 3, 6 and 10 successive years.

NEST: a rather ragged, domed, oval, globular or somewhat pear-shaped bag, with large side or side-top entrance, sturdily attached at top or top-back to several twigs and living leaves, or occasionally suspended by narrow attachment between twig and apex of nest. Built of coarse, dry grass intermingled with small twigs, fibres, rootlets, feathers and sometimes a little wool or leaves, and lined with feathers, fine grass, hair and a little plant down; or (Kenya) dry grasses and plant down from thistle *Cirsium*, bound with spider web. One nest lined with 385 domestic turkey feathers; others lined with kapok taken from nests of Cape Penduline Tit *Anthoscopus minutus*. Occasionally decorated with moss or lichen. A small porch of dead grass seed-heads above entrance; seeding grass can extend around most of entrance. Some nests lack porch, and some have a beard hanging below. Entrance hole nearly always faces into nest-supporting tree. Size: height ext. 140, int. 50, width 45, from front to back ext. 90, int. 65, entrance diam. 49.

Sited at tip of branch or in outer foliage of bush or small tree in sheltered place, such as hillside hollow, incised gully or erosion cutting, commonly near water or where ground is marshy. Sometimes built into cluster of leaves, e.g. in *Protea* bush. Nest may be built over stream or small pool, or sited where considerable drop below it – under road or railway culvert or often at side of shrub or attached to creeper jutting out from high bank of erosion gully or quarry. Uncommonly, nest attached to wire, in clump of tall, thick-foliaged flowering plant, in weedy waterside growth, in mass of driftwood caught up in tree, amongst similarly snagged mass of dead leaves, or to pendent reeds or rushes on bank of stream (Vincent 1949). Nest sited in wide variety of native and introduced plants (listed by Skead 1967), mainly between 1 and 2 m above ground, but once base of nest only 15 cm above water, and nest once in an overhanging cliff-side bush 25 m from ground. Entrances of 8 nests, Kenya, all faced away from prevailing wind (Wolf and Wolf 1976).

Only ♀ builds, though site may be selected by ♀ and ♂ together. She obtains much material whilst hovering, at places up to 100 m from nest site. ♂ accompanies ♀ back and forth whilst she is building and often seems to harry her (van Someren 1956, Skead 1967). ♂ sits near nest when ♀ adds material and sometimes when she is absent; he sings, occasionally hovers by nest or clings to entrance (Wolf and Wolf 1976). From near start of building, when nest only at signet-ring stage, to completion except for lining, can take only 2–4 days; adding lining takes 3 days more; but nest-building often takes 1·5–4 weeks (10, 16 and 16 days, Kenya; 20, 24, 26 and 30 days, South Africa). 2nd nest started within 6 days of young leaving 1st nest,

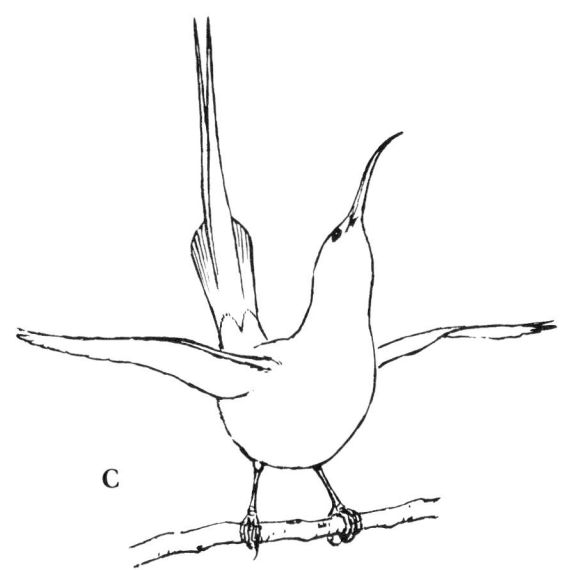

and 3rd nest (same ♀) started 11 days after young left 2nd nest (Wolf and Wolf 1976); 2nd clutch started 5 weeks after 1st clutch laid (Skead 1967).

EGGS: 1–3, usually 2 throughout range, although C/1 common near equator. Laid on successive days. Rather long; shell smooth and glossy, sometimes dull. Ground colour light brown, cream, greenish buff, fawn, olive-white, rarely white; marked with grey, olive or dark brown spots, streaks, dusting and clouding, mainly at broad end, sometimes completely obscuring ground colour. SIZE: (n = 85) 17·9–21·0 × 13·0–15·0 (19·6 × 13·6).

LAYING DATES: Ethiopia, July–Sept; E Africa, Region A, Nov–Dec, Region D, May–Dec (mainly July–Oct, after the long rains); Zaïre (Itombwe), Apr–July (and juveniles in Apr, Aug and Sept); Zambia, June; Zimbabwe, Aug–Mar; Malawî, Nov, Apr, July, Aug, with evidence of breeding season Mar–Aug below 2150 m and Oct–Feb above 2150 m (Benson and Benson 1977); Transvaal, Aug–Feb, mainly Nov; Orange Free State, Aug–Jan; Natal, Oct–Apr; Cape, Aug–Jan in E and May–Nov in SW; Lesotho, Nov–Jan.

INCUBATION: in a clutch of 2, both eggs hatched on same day. Incubated by ♀ only, sitting closely, spending 44–70% of daylight hours on nest. ♂ perches nearby and defends area. ♀ leaves nest with some regularity in early morning and again in late afternoon, accompanied by ♂ as she flies off. Sometimes ♂ appears to call incubating ♀ away from nest, and may cling to edge of nest, calling, until she vacates, when ♂ flies behind her. ♂ sometimes follows or chases ♀ back to nest. Period (n = 3) 13 ± 1 days.

DEVELOPMENT AND CARE OF YOUNG: for first few days young fed by ♀ only, thereafter by ♀ and ♂. ♀ of one pair made 83% of feeding visits, ♀ of another 76%, and ♀ of a 3rd 40%. 4–12 feeds given to each nestling per h, av. 7·2 (Wolf and Wolf 1976). Young brooded 7 times per h, reducing to once per h in 2nd week. One ♂ fed and reared brood unaided after ♀ had been killed, and continued to feed young out of nest for 3–4 weeks. Parent brings larger food items one at a time, but smaller, softer ones in a mass held in bill or concealed from observer's view inside closed bill. Both sexes remove faecal sacs, at rate of up to 1·7 sacs per young per h. Nestling period (n = 3) 15–17 days; in a brood of 2, both chicks left nest on same day (Haig 1987). After leaving nest, young remain nearby and are fed by both parents, about twice per young per h when 2–4 days out of nest. Young tend to follow parents around, and beg using repetitive, feeble hunger call, quivering wings, and opening mouth wide when food offered. By 4th day young take nectar themselves, e.g. from aloe flowers inside parental territory; by 6th day young appear to be as proficient as their parents in nectar feeding, and by 11th day are fully independent. Parents induce young to return to nest to sleep for up to 14 nights. Young once chased away by parents 18 days after they had left nest; another young chased away by ♂ 11 days after it left nest, on day that ♀ started to build another nest.

BREEDING SUCCESS/SURVIVAL: 12+ birds died of strychnine poisoning (source not traced) in 10 days in Transkei, 1954. Young broods thought to be sometimes killed by ants. Parasitized by Red-chested Cuckoo *Cuculus solitarius* and Klaas's Cuckoo *Chrysococcyx klaas*. Feather mites: *Proctophyllodes legaci* (Cheke 1978).

References
Chapin, J.P. (1954).
Cheke, R.A. (1971).
Harrison, J.A. *et al.* (1997).
Skead, C.J. (1967).
van Someren, V.G.L. (1956).
Taylor, J.S. (1946).
Vincent, A.W. (1949).
Wolf, L.L. and Wolf, J.S. (1976).

Genus *Hedydipna* Cabanis

4 small or very small sunbirds with short, stubby, only slightly decurved bills, broad at base; bill curvature commences at skull (Irwin, in press). In 3 species (*H. collaris* and a superspecies, *H. platura*/*H. metallica*) ♂♂ have upperparts, head and breast iridescent green; no pectoral tufts; belly yellow; central pair of rectrices (T1) very long and narrow in *platura* and *metallica*. ♀ *collaris* like ♂ but throat and breast non-iridescent yellow; ♀ *platura* and *metallica* non-iridescent olive and yellow, central tail feathers not elongated. ♂ of fourth species (*H. pallidigaster*) dull iridescent blue above, with orange pectoral tufts and white belly; ♀ non-iridescent grey and whitish, without pectoral tufts. Eat insects and small berries, but not much nectar.

H. collaris is one of the most widespread sunbirds in evergreen woodland; the *H. platura*/*H. metallica* superspecies is widespread in arid parts of N tropics (*H. metallica* north to Nile delta). *H. pallidigaster* is restricted to E coastal forests.

Hedydipna platura superspecies

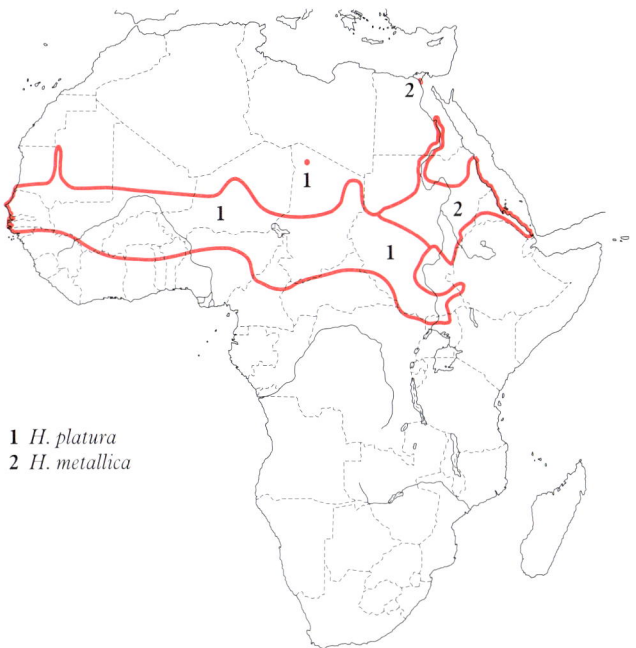

1 *H. platura*
2 *H. metallica*

Hedydipna collaris (Vieillot). Collared Sunbird. Souimanga à collier.

Plate 9
(Opp. p. 142)

Nectarinia collaris Vieillot, 1819. Nouv. Dict. Hist. Nat., nouv. éd., 31, p. 502; Gamtoos (Cape Province).

Range and Status. Endemic, resident. One of the most widespread sunbirds, common to abundant except where noted; throughout forested Africa south of 05–09°N; absent from drier parts of SW and S Africa.

Senegal, records in Dec–Jan near Dakar, common in SW and in extreme SE. Gambia, frequent in Abuko Nature Res., Sukuta woods and other woods in W, otherwise rare. W Guinea-Bissau. Mali, June records from Naréna and Kourémalé. Burkina Faso, rare (Dowsett 1993). Guinea, a few records between Conakry and Kindia; abundant in Macenta region. Ivory Coast, S half except in denser rain forests and drier savannas. Ghana, in secondary forest, forest-edge habitats and coastal thickets to W edge of Accra plains at Abokobi, on Akwapim hills and at well-wooded inselbergs; uncommon in riverine forest just north of forest zone, north to Ejura. Togo, from coast north to Tinkiro (08°05'N). Benin, frequent in SE. Nigeria, from coast to just north of forest belt in Ibadan and Nsukka; sparse in gallery forests north to Kainji, Kafanchan and Pandam. Bioko. S Cameroon, north to Bafia, Garoua-Boulai, and gallery forest in Benue floodplain. Central African Republic, Haute Sangha Préf. (uncommon in Dzanga Reserves), Lobaye Préf., and Bangui area. Zaïre, absent from extreme N, otherwise widespread although not certainly recorded in central regions, as mapped. Extreme S Sudan: Bengengai, Maridi, Yambio and Tambura (*somereni*), and Imatong, Dongotona and Didinga Mts (*garguensis*); a third race (*djamdjamensis*) is likely to occur in extreme SE Sudan between Naita and Todenyang. Uganda, north to Murchison Falls Nat. Park, Lango and Mt Elgon,

Hedydipna collaris

also extreme NE in Kidepo Valley Nat. Park. Kenya, up to 2800 m, with outlying populations on Marsabit, Mt Nyiru and Mt Kulal (*garguensis*); records from near Lokichokio (probably *somereni*), near Mt Labur, Ileret and Jibisa (doubtless *djamdjamensis*) and along Daua R. in extreme NE (*djamdjamensis*). S Ethiopian records mainly from wooded valleys of Omo, Alghe, Sayan, Daua, Wenz and Webi Shabeelle rivers. Somalia in lower Jubba and Webi Shabeelle valleys. Rwanda; Burundi at 800–2600 m, at least as far south as Bururi; probably absent from Rusizi Plain and perhaps from SE border area. Tanzania, 3 intergrading races occur around peripherally; widespread in E half as shown, sparse in centre and W (N. E. Baker, pers. comm.); Zanzibar, Mafia and records on Pemba.

Gabon, Congo. Zaïre, peripherally widespread except in far N, but (as in Tanzania) may be absent from vast central region; up to 2350 m (on Mt Kandashomwa). Angola, Bengo (Quissama Nat. Park), plateau of Uige, Cuanza Norte, escarpment of Cuanza Sul, N-central Malanje Provinces, and Luando Res. (N Bié); in coffee forest at Salazar; probably more widespread than mapped. Zambia except in SW. Botswana in Okavango and along Linyanti R, sparse along Chobe R., frequent at Shashe R./Limpopo R. confluence. Namibia, recorded in Caprivi Strip. Zimbabwe, peripheral, in fringing forests along all main river valleys; west of Victoria Falls; in Middle Zambezi Valley; in extreme NE up to 700 m; eastern slopes of E highlands from Inyanga to Lusitu-Haroni and Chipinga Uplands, sparse above 1200 m but common below; up to 1500 m in Vumba and 1700 m in Inyangani massif; in Sabi Valley up to Hot Springs; throughout SE lowveld and in Limpopo Valley west to Sentinel Ranch; sparse population in Great Zimbabwe. Malawi at low levels (though inexplicably absent from Nkhata Bay District); much less common at higher altitudes, up to 1850 m. Mozambique, absent from most of Gaza and Inhambane Provs except for coastal lowlands, and possibly from N (east of L. Malawi). Transvaal, in Limpopo Valley west to Breslau, Escarpment, Soutpansberg and Blouberg, and riparian forests in lowveld valleys. Natal, only in coastal half, up to about 900 m; coastal E Cape Province.

Description. *H. c. collaris* (Vieillot): E South Africa (S and E Cape and Natal to S and W Zululand). ADULT ♂: head, neck and rest of upperparts metallic green, with brassy tinge on ear coverts and on back and rump, purer green on uppertail-coverts. Tail blackish, upperside with slight dark green gloss, feathers edged bright metallic green, more broadly near base. Lores sooty brown. Chin and throat to upper edge of breast metallic green, concolorous with head, bordered posteriorly by very narrow violet band; pectoral tufts bright lemon yellow; rest of underparts dull yellow, flanks tinged olive-grey. Lesser and median coverts bright metallic green. Rest of upperwing blackish brown; primaries narrowly edged olive-green; secondaries, alula and primary coverts finely edged, tertials and greater coverts more broadly edged metallic green. Axillaries and underwing-coverts pale yellow; inner borders of flight feather undersides buffish white. Bill black; mouth pink in centre, orange-yellow at sides; eyes dark brown or blackish; legs black. ADULT ♀: differs from adult ♂ in having cheeks to chin pale olive-grey, merging through yellowish olive-grey throat and upper breast with rest of yellow underparts. SIZE (10 ♂♂, 10 ♀♀): wing, ♂ 50–55 (52·9), ♀ 48–52 (49·4); tail, ♂ 33–41 (36·2), ♀ 32–36 (33·9); bill, ♂ 14–16 (14·9), ♀ 13–15 (14·7); tarsus, ♂ 16–17 (16·3), ♀ 16–17 (16·0). WEIGHT: ♂ (n = 167) 6·3–9·4 (7·8), ♀ (n = 113) 5·8–9·7 (7·0), also ♂ (n = 58) 6·9–11·0 (8·2), ♀ (n = 51) 6·3–9·7 (7·5).
IMMATURE: juvenile like adult ♀.

NESTLING: bill light grey, becoming black in a few days; gape swollen, bright yellow.

H. c. zambesiana (Shelley) (including '*zuluensis*', '*chobiensis*' and '*beverleyae*'): NE South Africa (N and E Zululand, E Transvaal, Limpopo valley), SE lowland Zimbabwe, Zambezi valley and S Zambia west to Caprivi, N Botswana, NE Namibia and SE Angola, and lowland Mozambique to S Malawi and SE Tanzania. Differs from nominate race in having all remiges edged olive-yellow with little or no metallic green; slightly brassier green above; below, less olive wash on flanks, violet breast band broader. Wing, ♂ (Malawi and Tanzania, n = 10) 49–53 (51·1); birds from Namibia/Angola border ('*chobiensis*') slightly larger, ♂ (n = 6) 55·5–55 (Clancey 1980). WEIGHT: Malawi, ♂ (n = 7) 6·0–7·1, ♀ (n = 3) 6·0–6·8 (6·3).

H. c. patersoni Irwin: highlands of E Zimbabwe and adjacent Mozambique. Like *zambesiana*, but sides strongly washed olive-grey. This and the following races lack metallic green edges to remiges and greater coverts. WEIGHT: ♂ (n = 10) 7–9·3 (8·3), ♀ (n = 5) 6·5–8·3 (7·4).

H. c. elachior Mearns: Zanzibar, and coastal N Tanzania to S Somalia inland to Kilosa, Moshi and foot of Kenya highlands. Like *zambesiana*, but violet breast band narrower; slightly paler yellow below, especially in ♀. Intergrades inland with darker forms. Wing, ♂ (n = 10) 49–55 (51·2). WEIGHT: Somalia, ♂ (n = 9) 5·3–6·9 (6·2), ♀ (n = 8) 5·4–6·5 (5·9); Kenya, ♂ (n = 4) 6·9–7·6 (7·2), ♀ (n = 2) 6·3, 8·5.

H. c. djamdjamensis Benson: S Ethiopia (Alghe and Sayan rivers area). Like *zambesiana* with broad violet breast band, but brighter yellow below; ♀ yellower on throat and upper breast.

H. c. garguensis Mearns: S Sudan east of Nile, Uganda, W Kenya, Rwanda, Burundi, E Zaïre (Kivu and Katanga), W Tanzania, SE Zaïre (Katanga) and Zambia north of *zambesiana*. Intergrades with *elachior* and *zambesiana*. Deep yellow below, but flanks more dusky olive; ♀ with throat to upper breast darker olive-grey. Larger than *zambesiana*: wing, ♂ (n = 10) 54–57 (55·3). WEIGHT: Kenya, ♂ (n = 3) 8·0–9·3 (8·6), ♀ (n = 5) 7·3–9 (7·8); W Zambia, ♂ (n = 6) 6·2–7·9 (7·0).

H. c. somereni Chapin: Niger R. delta in S Nigeria, to N Angola, Zaïre (except Katanga) and Nile R. in Sudan. Paler, duller yellow below than any of the above, flanks dusky olive. ♀ with throat pale olive-grey. Wing, ♂ (n = 10) 52–55 (53·7); bill, ♂ (n = 10) 15–16 (15·6).

H. c. hypodila Jardine: Bioko. Like *somereni*, but bill larger, ♂ (n = 6) 17–18 (17·6).

H. c. subcollaris Hartlaub: Senegal to Niger R. delta in S Nigeria. Rich yellow below with orange tinge, little olive on flanks and narrow violet band; typically brighter emerald green above than other races. ♀ with pale buffish throat, and metallic green mottling on sides of throat and sides of neck. Juvenile distinctive: head feathers grey, tipped olive, forehead brighter olive, crown and nape greyer; long yellow streak above and a paler one below eye; back, wings and uppertail-coverts bright olive; scapulars slightly glossy; tail brown, feathers edged olive; lores grey; cheeks and sides of throat yellow-olive; chin, centre of throat and upper breast very pale grey, colour merging into clear pale yellow on belly; lower mandible pinkish horn (Field 1971). Rather small: wing, ♂ (n = 10) 47–52 (50·0). Weight: 10 ♂♂ av. 7·4 ± 0·8, 10 ♀♀ av. 7·0 ± 0·7; juveniles: 5 ♂♂ 7·1 ± 0·8, 3 ♀♀ 6·6–7·8 (7·1) (Colston and Curry-Lindahl 1986).

Field Characters. Length 10–10·5 cm. Small, with short, almost straight bill; no pectoral tufts. ♂ has bright iridescent metallic green upperparts, throat and upper breast, separated from yellow underparts by narrow purple band; ♀ has similar green upperparts and sides of head but is entirely yellow below, sometimes with olive wash on throat. Underpart colour (both sexes) varies from orange-yellow in W Africa to pale yellow in coastal E Africa.

Yellow-bellied races of Variable Sunbird *C. venusta* have longer, curved bill, bluer upperparts, orange or yellow pectoral tufts, dark, mainly purple throat and breast extending further towards belly, separated from belly by dark band; ♀♀ of Variable (all races) are brown above. ♂ Green Sunbird *Anthreptes rectirostris* has green upperparts and breast but grey belly, yellow pectoral tufts. For differences from Pygmy and Nile Valley Sunbirds *Hedydipna platura* and *H. metallica*, see those species.

Voice. Tape-recorded (16, 32, 58, 75, 86, 88, 91, 104, B, C, F, KEI, STA). Song, 1–2 pure or buzzy introductory notes followed by 3–8 pure up-slurred or down-slurred whistles, either on same pitch, 'jeeu-jeeu, tui-tui-tui-tui-tui ... ' or slightly descending, 'sip-sip, tsew-tsew-tsew-tsew ... '; sometimes 8–14 notes on same pitch with no introduction, 'tyup-tyup-tyup-tyup ... '. Notes are measured and individually discernible, given at *c.* 4 per s. Said to have a tinny, rolling duet, 'chippery chippery'; ♂ has quiet, throaty warbling subsong; ♀ also has a 'full song', given even when she is on the nest (Skead 1967). Contact calls include a light twittering (perhaps the 'more confiding 'chi-rrreee'' of Skead 1967), a repeated tinny 'sloo, wee-sloo, wee- ...' and a shrill 'chip' repeated at irregular intervals and speed; other calls include buzzy 'jee', 'dzit' and 'julie'; low, hard 'jup' and a high 'see-say' or 'spit-see'; alarm a grating 'djewy-djewy ... ', also a thin 'sweee'.

General Habits. Inhabits secondary forest canopy, edges and clearings, coastal scrub, riverine forest, moist savanna woodlands, and gardens. In equatorial forests of Zaïre it lives mainly along borders of clearings and in second growth, but in open areas inhabits densest available tree growth: in Ituri, thickets and old farmlands, but does not visit flowering plants around houses. In Liberia, wide range of woody habitats from coast up to 1600 m; in primary forest keeps mainly to treefall gaps but also visits canopy of mid-strata trees in logged and open forests. In Gabon, all forested habitats, from gallery forest to primary forest interior, from canopy down to 2 m above ground, secondary forest with large trees, fallow farmland with tall bushes, and abandoned plantations. Penetrates primary forest mainly along roadsides and clearings, or where tree growth smashed by tornadoes. Common along pathways in montane forest. In Zimbabwe, keeps to creeper tangles and thickets in canopy, mid-stratum and at edges of lowveld and mid-altitude evergreen forest and riverine fringing forests (Irwin 1981). In Natal, restricted to thicker woodland in coastal evergreen bush, dry thornveld, sourveld, and mistbelt evergreen forest biomes. Visits gardens of dune-forest cottages in E Cape and will even enter rooms to suck nectar of picked flowers.

Active, alert and restless. Keeps in thick vegetation at heights above ground of 1–25 m, mainly 4–15 m, foraging mainly in leafy cover and amongst tangles of lianas, by creeping, gleaning living leaves and twigs, poking into spider webs and caterpillar tents in curled-up dead leaves, hanging upside down to peer under leaves, bending down to examine below leaves without swivelling right round, probing shallow flowers, occasionally hovering in front of flower (in the absence of a suitable perch next to it), or giving short chase to insect in flight amongst branches or in open gap between trees. From a perch, slits down length or along basal half of long, tubular corolla of flowers to obtain nectar (e.g. of *Burchellia bubalina* and *Helleria lucida*), by means of a quick jab and downward tear with the sharp bill. Sometimes every flower on a bush is slit open, and bird returns again and again to drink nectar from the same slit corollas day after day (Skead 1962).

Nearly always in pairs, sometimes in family parties of 3–5 birds, and as member of mixed sunbird parties: once seen eating small red berries from clusters in roadside tree attended also by Bates's *Cinnyris batesi*, Western Olive *Cyanomitra obscura*, Green-throated *C. rubescens* and Little Green Sunbirds *Anthreptes seimundi*, all arriving in twos and threes, feeding on berries for a few min then departing, and amounting to 'great numbers' (Serle 1954). In South Africa an almost invariable constituent of mixed foraging flocks in forest, typically with batis *Batis capensis*, shrike *Dryoscopus cubla*, white-eye *Zosterops pallidus*, and warbler *Camaroptera brachyura*. Uncommonly, quarrelsome party of 4 adults occurs, e.g. ♂ and 3 ♀♀, at favourite nectar flowers. Inquisitive and excitable; always ready to investigate unusual noise or occurrence; responds well to 'pishing', but also readily alarmed and frightened away. Flicks wings when alarmed. Readily investigates reflections in windows. Flight fast, dashing and direct. Pair evidently occupies territory all year (Skead 1962).

Food. Small red berries (Cameroon: see above); red or blackish berries 7–8 mm in diam. of *Macaranga, Rauwolfia* and other trees. Arillate seeds of *Tetrorchidium didymostemon*. Of 6 stomachs, Zaïre, 3 had various small fruits, 4 had insects, 2 had hairless caterpillars and 1 large termites. Feeds at flowers of *Stereospermum* sp. and *Berlinia grandiflora* (Togo, in flower Feb–Mar: Cheke and Walsh 1996), but at blossoms may be foraging more for insects than nectar (Brosset and Erard 1986). In Kenya takes much less nectar than most other sunbirds, but visits flowers of forest trees such as *Albizia, Erythrina* and *Loranthus*, also of herbs and succulent aloes, probably to take nectar as well as small insects (van Someren 1956). In South Africa eats small insects, spiders, snails, small berries of *Trema orientalis* and *Chrysanthemoides monilifera*, seeds, and nectar of *Schotia, Erythrina, Catunaregam* (= *Xeromphis*: Rubiaceae), *Burchellia bubalina* and *Halleria lucida*. Nestlings given spiders, aphids, moth caterpillars and small flies (van Someren 1956).

Breeding Habits. Solitary nester, monogamous, territorial. Breeds at any season but mainly in the drier months; slender evidence of occasional second broods (Skead 1962).

NEST: suspended pyriform or oval purse with side or side-top entrance overhung by substantial porch (**A**); flimsy, untidy and loosely knit; composed of fine and very fine strips of plant fibres including some grass blades and bark, tendrils, dead leaves, leaf-mould, twigs and rootlets, bound with spider silk, sometimes without much of a 'beard' below but usually with a loose, untidy one up to 75 mm long; decorated outside with lichenous material, caterpillar excrement, and small plant stems carrying dried-out flowers or seeds, and lined purely with kapok,

Plate 15

Plate 16

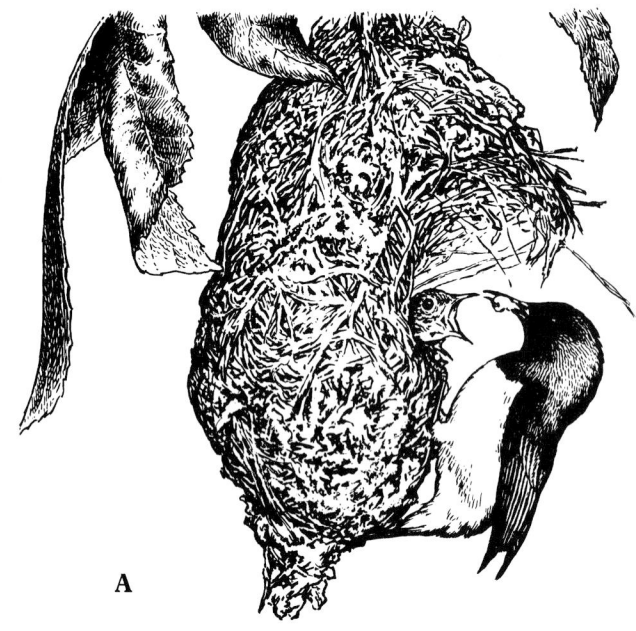

A

fine plant down, horsehair, rootlets or wiry plant fibres, with a few feathers; porch often made of grass seed-heads. Nest often made largely of spider and caterpillar silk, bark fibres, fine tendrils and skeleton leaves, decorated with frass from beetle workings, spider cocoons and shrivelled leaves. Width 76–90; height (without 'beard') c. 90; entrance diam. 25–30.

Sited 1·5–8 m above ground, suspended from drooping branch of leafy tree or under twigs or a small branch, well concealed amongst thick green herbage, once below silken canopy of social spider *Agelena consociata*, and often overhanging roadway or gully. Often in outer foliage of shrub growing in stream bed or inside forest (E Cape). In Kenya most nests placed high up (3–12 m), under curving creeper or drooping branch, often thorny, sometimes near wasps' nest, at edge of forest (van Someren 1956).

Built mainly by ♀, taking 2–6 days; once nest completely built between 10h00 one day and afternoon of the next. She starts by binding spider web round and round chosen twig until there is enough to allow retention of dead leaves and other dry matter, which ♀ hurriedly jabs into and onto the web. More spider web is then added until a ball-shaped structure results, flimsy and open enough to allow light through. While ♀ flies back and forth gathering and working material, ♂ chases her vigorously and speedily, calling, causing deviations and even violent evasive action by the ♀. When ♀ is out of sight ♂ waits near nest, singing loudly. ♂ sometimes helps to build. At least 5 nests may be sited successively in exactly the same place, over 4 years (van Someren 1956). One nest started as a grass ring hanging down under forked twig; porch was added last, after body of nest had been completed (Hanmer and Manson 1990).

EGGS: (1)–2 (low latitudes), 1–3 (Malaŵi, where av. of 1·6 eggs in 7 clutches), or 1–4 in South Africa (n = 112 clutches, av. 2·2 eggs). Eggs laid on consecutive days. Pale pink, off-white, cream, green or blue; spotted, streaked, mottled, scrawled or blotched with black, reddish brown, grey or mauve, mainly in band around large end. SIZE: (n = 86, South Africa) 14·5–17·8 × 10·3–12·0 (15·9 × 11·2).

LAYING DATES: Gambia, (pair feeding fledged young Oct); Senegal (Casamance) Dec; Liberia, June, Sept, (and fledglings and dependent young Nov–Apr); Ghana, July, Nov (and nest-building Mar, Apr, June, Oct and dependent young Nov, Dec, Feb, Apr, July, Aug); Togo, (dependent young Mar, Sept); Benin, Dec; Nigeria, Sept–Oct, Dec–Jan, Apr, July; Gabon, Aug, Nov–Mar; Ethiopia, Mar (and indications Oct, Feb); Somalia, (nestlings Dec); Zambia, Aug (3 clutches), Sept (6), Dec–Jan (2) and Mar–Apr (2 clutches); Malaŵi, Sept, Oct, Dec–May; Zimbabwe, Sept (7 clutches), Oct (8), Nov (12), Dec (6) and Mar–Apr (2 clutches); Mozambique, Sept–Jan; Transvaal, Nov–Dec, Feb–Mar, July; Natal, July–Feb, mainly Oct–Dec. E Cape, breeds throughout year, mainly in Sept–Jan (Skead 1962).

INCUBATION: usually by ♀ only; at some nests ♂ may take minor role (van Someren 1956). ♀ incubates for periods of 15–53 min, leaving to feed for periods of 2–25 min. Period: 12–14 days, usually 12–13.

DEVELOPMENT AND CARE OF YOUNG: young fed by both parents. Period c. 17 days. Young out of nest keep close to parents (if 2 young, one per parent) and depend on them for up to 24 days. At first they give almost constant, plaintive 'tseep' and carry wings half open and drooping, flicking tips open and shut, and quivering wings rapidly and calling vehemently when parent approaches with food. In race *subcollaris*, metallic feathers begin to appear on shoulders between 2 and 3 weeks after juvenile leaves nest (Field 1971).

BREEDING SUCCESS/SURVIVAL: lives for at least 6, 7, 8 and once 11 years (survival data in Hanmer 1989 and Hanmer and Manson 1992). Occasionally parasitized by Klaas's Cuckoo *Chrysococcyx klaas*.

References
Brosset, A. and Erard, C. (1986).
Hanmer, D.B. and Manson, A.J. (1992).
Irwin, M.P.S. (1981).
Manson, A.J. (1990).
Skead, C.J. (1962).
van Someren, V.G.L. (1956).

Hedydipna platura (Vieillot). Pygmy Sunbird. Souimanga pygmée.

Cinnyris platurus Vieillot, 1819. Nouv. Dict. Hist. Nat., nouv. éd., 31, p. 501; Senegal.

Plate 9
(Opp. p. 142)

Forms a superspecies with *H. metallica*.

Range and Status. Endemic, resident and intra-African migrant, sahelian and soudanian savannas from Senegal to Uganda. Common almost throughout. Mauritania, south of 17°N except for Adrar hills (20°30′N) and Tagant hills. Senegal, common, but scarce in N. Gambia, common on coast, scarce on lower river, uncommon on middle and upper river. Guinea-Bissau, recorded (Bannerman 1948). Guinea, only in Fouta-Djalon district. Sierra Leone, old specimens known, but not recorded by recent observers. Mali, widespread north to 17°N; abundant in sahelian zone and along inland delta of R. Niger. Ivory Coast, scarce dry season visitor in far N: Comoé, Ferkessédougou, Boundiali. Ghana, not uncommon visitor to N: Tumu, Bolgatanga and Mole. Togo, not uncommon visitor to soudanian and N Guinean savannas, south to Aledjo. Benin, frequent in Arli-Pendjari Nat. Park. Nigeria, south in dry season to Igbetti, Enugu and Serti (but rare south of 10°N except east of 10°E). Niger, common in summer wet season north to Agadès and Aïr Massif, Cameroon, in Benue Plain and far N; in dry season south to Adamawa Plateau. Chad, in sahelian zone; Ouadi Rime – Ouadi Achim Faunal Res. north to Ouadi Haddat; Ennedi; and recorded in S Tibesti. Central African Republic, resident Bamingui-Bangoran Nat. Park and Manovo-Gounda-Saint Floris Nat. Park. Sudan, locally common, east to about 30°E and in SE, breeding in N and S of range. Zaïre, frequent in NE Uele district (Faradje, Aba, Dungu). Uganda, frequent in NW (West Nile) and in some years common in semi-arid plateau country north of 02°N; records south to Soroti and Kaburoron. Kenya, vagrant to Turkana, Lodwar, Kapedo and Baringo. Ethiopia, status uncertain: may occur in SW and NW.

Description. ADULT ♂ (breeding): head, neck, and mantle to rump and scapulars metallic green with bronzy tinge; uppertail-coverts metallic violet. Tail black, upperside with slight blue-green gloss; T1 very long and very narrow, broadened at extreme tip, projecting 50–80 mm; tips of T2–T6 with narrow whitish fringe. Lores blackish. Chin to upper breast metallic green, concolorous with head; in some birds, traces of violet-blue feather tips along lower border of green upper breast, but no distinct band; rest of underparts yellow. Lesser and median wing-coverts metallic green; rest of upperwing blackish brown, primaries and secondaries narrowly edged pale greyish; tertials, greater coverts and primary coverts narrowly fringed metallic green. Axillaries and underwing-coverts sooty grey; inner border of undersides of flight feathers pale grey-brown. Bill black; eyes dark brown to hazel; legs black. ADULT ♂ (non-breeding): resembles adult ♀ (see below), but upperparts slightly greyer, and wings as adult ♂ (breeding) with bright green median and lesser coverts and sooty grey underwing-coverts. T1 short and buff-fringed, T2–T5 blackish, slightly glossy. ADULT ♀: upperparts greyish olive, slightly greener on rump. Central tail feathers dark grey-brown fringed buffish grey; T2–T5 blackish with faint blue gloss, fringed whitish at tip, T5 with whitish spot at tip of inner web, T6 with larger inner web spot and fringe along outer web brownish white. Faint, narrow superciliary stripe yellowish; lores dark grey-brown; cheeks, ear-coverts and sides of neck greyish

Hedydipna platura

Breeding resident north of dashed lines, breeding visitor south of them.

olive. Chin and throat pale yellow, grading to deeper yellow on breast, belly and flanks; undertail-coverts pale yellow. Flight feathers dark grey-brown, edged buff; alula and primary coverts blackish brown; tertials and greater coverts grey-brown, fringed buff, lighter at tips; median and lesser coverts olive-grey. Axillaries pale yellow; underwing-coverts and inner border of undersides of flight feathers pale buff. Bare parts as adult ♂. SIZE (10 ♂♂, 10 ♀♀): wing, ♂ 57–60 (58·5), ♀ 53–56 (64·6); tail, ♂ 91–113 (99·7), ♀ 32–34 (33·2); bill, ♂ 11–13 (12·2), ♀ 11–12 (11·8); tarsus, ♂ 13–14 (13·7), ♀ 13–14 (13·5). WEIGHT: ♂, Niger and Senegal (late Jan–Feb, n = 5) 6–7 (6·5), Nigeria (n = 10) 6·0–7·5 (6·9); ♀, Senegal (late Jan–Feb, n = 2) 6, 6, Nigeria (n = 5) 6–7 (6·5); unsexed, Nigeria (Dec–May, n = 25) 5·2–7·3 (6·6).

IMMATURE: juvenile like adult ♀; more greyish white (less yellow) below.

NESTLING: not described.

Field Characters. Length, ♂ with fully grown tail 17–17·5 cm, ♀ 9 cm. Tiny, with short, straight bill. Breeding ♂ iridescent golden green with long central tail feathers, violet uppertail-coverts, and golden yellow lower breast and belly. ♀ pale brown above and pale yellow below, with indistinct pale superciliary; tail blue-black with white corners. Range meets that of Collared Sunbird *H. collaris* but they are ecologically separated, latter being confined to forest. ♂ without long tail feathers resembles ♂ Collared but latter has blue-green uppertail-coverts, yellow belly without orange tones and purple band dividing belly from breast. Evidently overlaps with very similar Nile Valley Sunbird *H. metallica* in N Southern Kordofan, about 12–13°N and 30–31°E, about Sungikai and Rashad. Breeding

♂♂ are same size and have similar long tails and orange-yellow bellies, but Pygmy is golden-green (not blue-green), lacks purple band between breast and belly, and violet lacks bluish cast and is confined to uppertail-coverts, not extending to rump and lower back. ♀♀ extremely similar; ♀ Pygmy tends to have a less pronounced pale supercilium, but they are probably not safely distinguishable in the field.

Voice. Tape-recorded (104, 108, B, BRU, CHA, MOR). Common call a nasal 'dooy'. Bustling song, variable in length and content, incorporates this call with silvery trills and chips, e.g. 'dooy-tititiii-trrrrrrrr-titi-dooy-surrr-dooy-dooy-dooy-dooy'. Other calls include a 5-note 'wa-dee-dee-dee-*chew*', last note loudest, and a thin, hissing 'tsi-tsee-tsi-tsi-tsi'. 'Cheek' or 'cheek-cheek' directed by ♂ at nest-building ♀ (Serle 1940). Ticking noise reported by Bannerman (1948).

General Habits. Inhabits acacia woodland and semi-arid thorn scrub, open grassy and wooded savannas, orchard-bush, parks and large gardens. In Mauritania associated with mistletoe *Loranthus* parasitic on various *Acacia* spp. (Lamarche 1988). Occurs in pairs, and after breeding, in family parties; gathers in loose pre-migratory flocks and 7 or more may concentrate in copse of flowering trees. Rather shy. If disturbed may fly several hundred m.

Resident in most of its range, but in S a breeding visitor in dry season (about Nov–Mar). Well-known as a migrant even where present all year; migrations poorly understood but they, and dry-season breeding, thought to relate to the flowering of many savanna shrubs preceding the rains (Bates 1927). In W Africa a breeding dry season visitor to S: in Ghana, Oct–Apr at Tumu, Nov–Jan at Bolgatanga and Aug–Apr at Mole, and in Togo Nov–Apr. In Niger, present in Oct–Apr in extreme SW (Dosso area); recorded June–Sept near Tillabéri, also Dec; resident in S Niger; about 16°N, occurs in Feb–Nov, but further north in Aïr occurs frequently year-round. Nigeria: in NW (Sokoto) present Apr–Aug, in NE (Malamfatori) present year round and marked passage in late Sept/late Oct; in Zaria (11°N) present Oct–May but routine bird-netting did not reveal strong spring passage like that of some other sunbirds there; in Borgu (10°N) arrives in mid Oct/mid Nov, breeds in Dec–Jan, and departs in late Apr (Elgood *et al.* 1973); in Kafanchan/Nasarawa area (09°N), present late Nov to mid Mar; on arrival birds unpaired and gonads quiescent; nest-building from mid Dec, egg-laying from late Dec to early Mar, and birds then gathered in pre-migratory flocks (Serle 1940). Strongly migratory in Sudan, but in all parts occurs year-round and breeds in N and in extreme S. However, a breeding visitor adjacently in NE Uele district, Zaïre, and NW Uganda, mainly in Dec–Mar but with a few reports in Sept–Nov. Vagrant to Kenya in Sept, Nov and June.

Moult (obvious in ♂♂ which have eclipse plumage after breeding) varies with breeders' latitude (Bates 1934) and schedule difficult to understand because of partial migration; ♂♂ of central Nigerian population (Serle 1940) evidently moult into eclipse plumage after breeding (Dec–Mar) and emigrating (Mar), and moult back into breeding plumage (in non-breeding grounds well to north) probably in Sept–Oct; in W Sudan ♂♂ are in eclipse in June–Aug and acquire breeding dress in Oct–Nov.

Food. Ants, caterpillars, spiders, nectar (*Acacia*, *Calotropis*, *Loranthus*, *Bombax*), pollen, stamens and small petals (Chapin 1954, Lamarche 1981, Cramp and Perrins 1993).

Breeding Habits. Monogamous; territorial. Singing ♂ droops wings and cocks tail (Chapin 1954), evidently much like Nile Valley Sunbird, *q.v.* ♂ sings from perch, Nov–Mar (Nigeria, Serle 1940). Double-brooded, using same nest; once 2nd brood fledged 34 days after 1st (Nigeria, Skinner 1969).

NEST: roughly oval bag or purse, upper half of back wall closely attached by spider web and fine plant fibres to twig for length of 55–70 mm; entrance at one side, 22–27 high × 22–23 wide, overhung by small porch *c.* 25 long; nest lacks hanging 'beard' below; lined with white plant-down, scantily on walls and roof, thickly on floor; lining is strengthened externally and encapsulated with fine grasses, vegetable fibres, spider and caterpillar silk, tiny leaves and a few small feathers, all compactly inter-woven; outside of walls with silvery-grey leaves, brown seeds and caterpillar droppings added, helping to conceal nest (Serle 1940, Chapin 1954). N Nigerian nests lined with silky down from seeds of *Calotropis procera* (Skinner 1969). 4 nests: ext. height 85–100, ext. width 45–62, int. depth below entrance 24–50. Another nest 100 high, and an entrance 30 wide. Sited 2–4 m high near edge of thorny *Acacia* tree or *Citrus* shrub; sometimes nest suspended from bare branch in tree. Built by ♀ alone; in 40 min observation, at stage when attachment, roof, walls and entrance were completed but floor was unfinished and lining was still to be added, she visited nest every 90 s with bill-full of material, uttered 'cheek', and remained in nest for *c.* 30 s. Each approach was in same way: she alit on lower branches of bush, ascended bush with successive hops, and entered nest; departure was by dropping down to lower branches before flying away. When ♀ inside nest, ♂ perched very close and sang vigorously but seemed to pay no attention to ♀ as she came and went. From this stage nest took 2 more days to complete, then downy lining was added, and clutch laid a few days later (Serle 1940).

EGGS: 1–2. Slightly glossy; shell very fragile; immaculate white. SIZE: (n = 13) 14·2–16·1 × 10·3–11·6 (15·0 × 10·8).

LAYING DATES: Senegal, Feb; Gambia, Feb–Mar, Sept; Mali, said to breed Mar–Sept although supporting evidence not provided (Lamarche 1981); N Nigeria, Dec–Apr; Niger, Dec; Zaïre, Feb; N Cameroon, Feb. Breeds at height of dry season. In north, may have a more protracted breeding season but no certain evidence to that end, nor that it ever breeds in rainy season (Mar–Oct in N Guinean zone, May/June–Sept in soudanian zone and July–Aug in N sahel zone).

INCUBATION: by ♀, who seldom leaves nest during incubation period (Skinner 1969).

DEVELOPMENT AND CARE OF YOUNG: young fed by both parents, mainly ♀; intervals between ♂ feeding visits 23 min and between ♀ visits 13 min; ♂ approaches nest more cautiously than does ♀, calling several times from nearby perch on arrival and departure; ♀ seems to clean nest; from 5 days, young can reach out of nest entrance and are fed by adults clinging to outside, next to it (Skinner 1969). Young brooded at night by ♀, from sunset until well after dawn. Nestling period: 12–15 days (Skinner 1969).

References
Chapin, J.P. (1954).
Cramp, S. and Perrins, C. M. (1993).
Elgood, J. H. *et al.* (1973).
Serle, W. (1940).
Skinner, N.J. (1969).

Hedydipna metallica Lichtenstein. Nile Valley Sunbird. Souimanga du Nil.

Plate 9
(Opp. p. 142)

Nectarinia metallica Lichtenstein, 1823. Verz. Doubl. Zool. Mus. Berlin, p. 15; Dongola, Sudan.

Forms a superspecies with *H. platura*.

Range and Status. Sudan to Somalia and SW Arabia (north to between Mecca and Rabigh, east to Dhofar, Oman).

Resident, intra-African migrant and probably migrant from Arabia. Egypt, common breeding resident in Nile valley from Qena to Aswan; and uncommon non-breeding visitor in Faiyum and most of Nile valley; has spread north this century; frequent winter visitor north to Cairo, rare in winter north to Nile Delta, Alexandria and Suez Canal area, and first bred in Cairo in 1986. Sudan, frequent and widespread from Msidob Hills region, NE Darfur, east to Nile, Blue Nile and Red Sea hills; record from near Chad border, northeast of Ennedi Massif. Eritrea, frequent to common, mainly below 1200 m, in central plateau and coastal hills; western limits uncertain. Ethiopia, frequent to common in E Welo, Adal-Esa, and Rer-Ali foothills southeast of Harer. Djibouti, very common throughout, especially where acacia scrub occurs at low altitudes; 22 counted one day (Arta, Doralé, Ambouli), 43 another (Obock, Waddi) and 54 another (As Eyla) (Welch *et al.* 1986). Somalia, common, locally abundant, visitor to whole of N, not south of 9°N.

Density of 20–30 pairs in breeding season on Kitchener's Island, Aswan, Egypt; 3 active nests in one acacia tree near Aswan (Goodman and Meininger 1989).

Hedydipna metallica

Description. ADULT ♂ (breeding): head, neck, and mantle to back and scapulars metallic green; rump and uppertail-coverts metallic violet-blue. Tail blackish, upperside glossed violet, especially long narrow T1 which projects 40–70 mm; tips of T5–T6 usually fringed whitish. Lores blackish. Chin, throat and upper breast iridescent green, concolorous with head, bordered on breast by narrow glossy violet band; rest of underparts yellow, paler on undertail-coverts. Lesser and median coverts metallic green; rest of upperwing blackish brown, primaries and secondaries edged pale grey-buff; tertials and greater coverts narrowly fringed metallic green. Axillaries and underwing-coverts dark grey-brown; inner borders of undersides of flight feathers pale grey-buff. Bill black; eyes dark brown or blackish brown; legs black. ADULT ♂ (non-breeding): similar to adult ♀ (see below), but wings as in adult ♂ (breeding), with bright bronzy-green lesser and median coverts and dark greyish underwing-coverts; T1 short and buff fringed, but T2–T6 blackish with slight gloss; dark stripe down centre of chin and throat glossed greenish; some birds with a few violet feathers on rump, and green feathers on head. ADULT ♀: upperparts pale greyish olive, greener on rump. Central tail feathers dark grey-brown, fringed buffish white, T2–T5 blackish with narrow white fringe at tip, T6 with large spot at tip of inner web and narrow fringe along outer web whitish. Narrow, creamy superciliary stripe; lores grey-brown; ear-coverts and sides of neck pale greyish olive. Chin and throat creamy, grading to pale yellow on breast, belly and flanks; undertail-coverts creamy. Flight feathers dark grey-brown, edged pale buff; alula and primary coverts dark grey-brown; tertials and greater coverts grey-brown, broadly fringed pale buff; median and lesser coverts pale olive-grey. Axillaries pale yellow; underwing-coverts and inner borders of undersides of flight feathers buffish white. Bare parts as adult ♂.
SIZE (10 ♂♂, 10 ♀♀): wing, ♂ 55–60 (56·8), ♀ 53–56 (54·6); tail, ♂ 87–106 (96·2). ♀ 34–37 (35·3); bill, ♂ 12–14 (13·3), ♀ 12–14 (12·8); tarsus, ♂ 14–16 (14·9), ♀ 14–15 (14·7). WEIGHT: N Chad, ♂ (n = 3) 7–7·5 (7·2), ♀ (n = 1) 7.

IMMATURE: juvenile similar to adult ♀, but greener above when fresh.
NESTLING: unknown.

Field Characters. Length, ♂ with fully grown tail 17–17·5 cm, ♀ 10 cm. Breeding ♂ with long tail feathers very like Pygmy Sunbird *H. platura q.v.*, which it overlaps in Sudan; distinguished by blue-green (not golden green) upperparts and breast, extensive blue-violet patch on lower back, rump and uppertail-coverts (violet confined to uppertail-coverts in Pygmy), and violet band separating breast from belly. ♂ without long tail streamers very like a slim, short-billed version of ♂ Collared Sunbird *H. collaris* but they are ecologically separated, Nile Valley being in dry country, Collared in forest; additionally, Nile Valley has extensive violet patch on lower back and rump, while rump of Collared is blue-green concolorous with black. ♀♀ of Nile Valley and Pygmy Sunbirds not safely distinguishable in the field; ♀ Collared is bright green above like ♂.

Voice. Tape-recorded (HOL). Song soft, fine and high-pitched, with thin silvery trilling and hissing notes, 'pruiit-prruiit-pruitt-tiriririri-tiririri' (Hollom *et al.* 1988). Also gives quiet musical twittering like subsong of *Sylvia* warbler, a rapid succession of squeaky, chattering and trilled notes, the squeaky notes sounding like cork rubbed jerkily on glass (E. K. Dunn *in* Cramp and Perrins 1993). Contact-alarm, grating, peevish 'pee', 'pee-ee' or 'pee-e-ee' with penetrating nasal quality (P.A.D. Hollom *in* Cramp and Perrins 1993). Other calls include thin, hoarse 'veeii-veeii-veeii', 'ptscheeiii', repeated 'cheeit-cheeit' and 'tee-weee' with upward inflexion on last note (Hollom *et al.* 1988).

General Habits. Inhabits hot, arid acacia country, wooded river valleys, *Dichrostachys* scrub in wadis, grassy areas with sand and scrub, and low, bushy growth in subdesert steppe. Especially attracted to *Acacia nilotica* trees, for foraging and nesting. In gardens visits variety of flowering plants, including low, herbaceous annuals. Occurs mainly in pairs, and in non-breeding season forms loose aggregations of 5–10 birds, sometimes more when concentrated at good feeding trees, e.g. once 30 in single flowering *Bombax* tree (Luxor, Egypt), 50 in acacias (Somalia) and 70 in one tree (Arabia). Active, restless, constantly flicking wings or tail open and shut. ♂ sings from treetop perch, a vigorous and eye-catching performance, flicking wings and flirting half-spread tail, head and throat moving with strong song delivery, and body constantly swivelling so that iridescent plumage shimmers in the sun.

Forages throughout heat of day. Drinks nectar by hanging on to inflorescence or twiggy flower stem, leaning down or across or reaching up, and inserting bill into flower to reach nectaries; sometimes feeds at 2 or 3 adjacent flowers in same inflorescence, in succession. Forehead often dusted with yellow pollen. Known to pierce base of corolla to reach nectar of trumpet-shaped flowers whose corolla is longer than bird's bill, though piercing behaviour not described. Drinks for 2–4 s per flower. Sometimes hovers momentarily in front of blossom, drinking on the wing. Spends 75% of foraging time in searching for and extracting nectar (Cairo: Greaves and Tregenza 1937). Spends rest of time in hunting for insects, mainly at different sites from nectar sources: gleans leaves, shoots and stems rather than flowers, but does take occasional small insects from corollas. Sometimes hovers to pick insect from leaf and makes brief pursuit of flying insect. One bird tends to forage at same trees, visiting them in nearly same sequence, on successive days. Associates at feeding sites with Shining Sunbirds *Cinnyris habessinica*, Chiffchaffs *Phylloscopus collybita* and Graceful Warblers *Prinia gracilis*. Regularly chased away or into interior of canopy by Little Green Bee-eaters *Merops orientalis* hawking for insects at edge of tree (Luxor: Howe 1987).

Mainly non-breeding visitor to Cairo, Oct to May, but year-round resident in S Egypt. Appears to be resident in much of Sudan but probably everywhere subject to partial migratory or dispersive movements; has distinct migratory movements in Red Sea hills (Nikolaus 1987). Locally migratory in Eritrea, arriving in coastal plains for the flowering of *Acacia* spp. in Feb; birds disappear in June, probably moving to E foothills of central plateau (Smith 1957). Migratory movements southward detected along Red Sea coast in Sept–Oct. Common non-breeding visitor to E Ethiopia and Somalia, arriving – probably from Arabia – in last few days of Sept and leaving from Feb to May (latest, 24th May). Pre-migratory flocks occur on N Somali coastal plain from Feb, and northward passage noted in hinterland of Saylac up to May.

Food. Nectar of *Calotropis procera*, *Acacia nilotica*, other acacias, *Rhazya stricta*, *Ziziphus*, *Adhatoda*, *Tecoma*, *Citrus*, *Jacaranda*, *Carica*, *Nasturtium*, *Bignonia*, *Petunia*, *Loranthus*, *Capparis* and *Faidherba*; also small invertebrates (Cramp and Perrins 1993). Insects include aphids (a favourite), taken from crucifers and stems and leaves of *Duranta* and *Prunus*. Visits flowers (probably for nectar, perhaps for insects) of *Clarkia*, *Antirrhinum*, *Hibiscus*, *Bauhinia*, *Riglia*, *Salvia*, *Althaea* and *Bombax*, and eats parts of flowers of *Acacia nilotica* (Greaves and Tregenza 1937).

Breeding Habits. Monogamous, territorial. Song display (described in captive bird) involves sudden increase in duration of ♂ songs; singer drooped wings, sang and on last syllable spread wings, ruffled head plumage, raised tail almost vertically, pivoted from side to side in series of vigorous, rapid quarter-turns, then the usual 'tschai-tschahi-ttscha-tschi' song with added 'tatai-taiti' gave way to long, jumbled series of chattering and purring sounds; song display followed by copulation (Cramp and Perrins 1993). Courtship of nesting pair (Arabia) involves fast chasing flights with whirring wings. Probably double-brooded.

NEST: oval purse with unporched entrance high on one side, made of plant fibres, plant down, rootlets, flower calyxes, seeds, dead leaves, spider web and a few small feathers; lined with plant down and small feathers, finely worked together; lining thin opposite entrance, thick on floor. Dimensions (2 nests): height 90–100, width 50, circumference 140–170, entrance hole 25 across, with lower edge 54 above floor of nest. Sited resting on fork or (usually) suspended from twig of bush or small tree, generally 1·5–3 m above ground, but can be up to 5 m. Built by both sexes; captive pair took 12 days (Cramp and Perrins 1993).

EGGS: 2–3, possibly sometimes 4. Sub-elliptical; smooth, glossy; white (palest pink when fresh), broad end finely but faintly stippled with rufous on top of larger grey marks. SIZE: (n = 12) 15·0–19·0 × 11·0–12·0 (16·9 × 11·5). WEIGHT: *c.* 1·13 g.

LAYING DATES: Egypt, Mar–Sept; Sudan, Jan–Nov (various indications: Hogg 1950); Ethiopia and Eritrea, Mar (and various indications Jan–July).

References
Archer, G. and Godman, E.M. (1961).
Cramp, S. and Perrins, C.M. (1993).
Greaves, R.H. and Tregenza, L.A. (1937).

Hedydipna pallidigastra (Sclater and Moreau). Amani Sunbird. Souimanga d'Amani. Plate 9

Anthreptes pallidigaster Sclater and Moreau, 1935. Bull. Br. Orn. Club, 56, p. 17; Sigi Valley, 4 miles east of Amani, Tanganyika Territory. (Opp. p. 142)

Hedydipna pallidigastra

Range and Status. Endemic resident. SE coastal Kenya (Sokoke–Arabuko Forest); NE Tanzania, 8 localities in and near East Usambara Mts (foothills to 900 m, uncommon) and Udzungwa Scarp (Ndundulu Mts, 1500–1550 m) (N. E. Baker, pers. comm.). In the Usambaras, rarity may be due to competition with the widespread and adaptable Collared Sunbird *H. collaris* with which it interacts, and which is dominant in riparian growth and forest edge (Stuart and Hutton 1977). Sokoke population estimated between 2900 and 4700 pairs in 70 km^2 of suitable habitat (Britton and Britton 1978). Threatened by habitat destruction; only 320 km^2 of suitable habitat remains in Usambaras. Density in optimum habitat in Sokoke, 1 pair per 1·5–2·4 ha.

Description. ADULT ♂: entire head to breast and upper back dull metallic bluish green with purplish wash, lower back and rump blackish, centre of throat purple, greener at sides. Uppertail-coverts and tail dull metallic blue. Lower breast, belly, flanks and thighs to undertail-coverts dull greyish white, pectoral tufts scarlet-vermilion. Wings blackish brown, median and lesser coverts and scapulars to shoulder metallic blue, washed violet; underwing-coverts and axillaries silky white. Bill black; eyes dark brown; legs and feet black. ADULT ♀: upperparts dark grey with some dull metallic sheen. Uppertail-coverts and tail dull metallic blue with greenish sheen. Lores and cheeks blackish, whitish loral spot and pale eyestripe to just behind eye, ear-coverts off-white; chin, throat and whole of underparts to undertail-coverts off-white. Wings dark brown, median and lesser coverts, scapulars and shoulders with dull metallic sheen. SIZE (8 ♂♂, 3 ♀♀): wing, ♂ 50–53 (52·1), ♀ 49–51 (50·0); tail, ♂ 31–33 (31·7), ♀ 29–30 (29·6); bill, ♂ 12–14 (13·2), ♀ 12–14 (13·0); tarsus, ♂♀ 12–13 (12·6). WEIGHT: ♂ (n = 4) 6–7·2 (6·8), ♀ (n = 3) 6·6–8 (7·2).
  IMMATURE: like ♀ but lacks metallic sheen.
  NESTLING: undescribed.
  TAXONOMIC NOTE: ♂♂ from Ndundulu Mts are brighter metallic blue, less greenish and may represent an undescribed race (Dinesen *et al*. 1993).

Field Characters. Length 8·4–9 cm. A tiny treetop species. ♂, with *shining blue-green head and back* and *white belly*, unlike anything in its very limited range (♂ Collared Sunbird has yellow belly, yellower green head and back); ♀ differs from ♀ Plain-backed Sunbird *Anthreptes reichenowi* in having grey back and white underparts rather than olive back and yellowish underparts.

Voice. Tape-recorded (32, F, CART, HOR, McVIC). Song a continuous bustling series of high-pitched, squeaky notes on one pitch; several birds may sing together. Call, 'seeeet-seeeet'.

General Habits. Inhabits miombo (*Brachystegia*) woodland in Sokoke Forest, especially between 5–15 m in more open canopy; wanders into forest habitats outside. In East Usambaras occurs at various levels on forest edges, especially in more open, secondary growth; in Ndundulu Mts apparently in forest interior, both in canopy at 35 m and among flowers at 7 m.

Feeds singly, in pairs or family parties, often with other sunbirds; also joins mixed species bird parties, e.g. in Ndundulus, where party consisted of woodpeckers, bulbuls, warblers, drongos, weavers and Eastern Olive Sunbird *Cyanomitra olivacea* (Dinesen *et al*. 1993). In Sokoke occurs in less than 20% of miombo bird parties. Restless, continually on move, searching leaves for arthropods, visiting flowers, trees and plants. At Amani visits largely coincide with flowering of favoured trees and shrubs; groups of up to 60 recorded in monospecific flocks (not merely concentrating at suitable feeding places), attracted to flowering *Erythrina* (18 birds in a single tree) and *Loranthus*. Utters loud 'seer-seer' call-note, followed by twittering note, as it flies from one tree to another.

Somewhat nomadic at Amani, where local movements associated with flowering times of particular trees; no evidence of regular migration.

Food. Arthropods: small insects (1 stomach contained 20 caterpillars 5–8 mm long) and small spiders. No trace of nectar in 5 stomachs (Moreau and Moreau 1937), but nectar almost certainly taken.

Breeding Habits. Monogamous; territorial.

NEST: thin-walled oval, with side-top entrance, surmounted by a sizeable porch of lichen; made of very fine, grass-like fibres, interwoven with plant down, the whole enshrouded by *Usnea* lichen (closely resembling nest of Collared Sunbird), with trail of materials hanging below (nests measure 400 mm long, including 'tail'). Placed 7–14 m up at end of *Brachystegia spiciformis* branch; nest resembles lichen on branch and is thus not easily seen.

EGGS: 3. Oval, beige, heavily marked with brown (especially at broad end), covered with numerous but loosely scattered very dark brown (almost black) spots and some chocolate brown markings up to 2 mm long (not streaks). SIZE: (n = 3) 15·7–16 × 11–11·1 (15·8 × 11·0). Each weighed 1 g (to nearest 0·1).

LAYING DATES: E Africa: Region E, May–June, Sept–Oct, Dec (one for each month).

INCUBATION: by ♀ only

DEVELOPMENT AND CARE OF YOUNG: both sexes enter nest but whether to build or feed young not known; both sexes feed fully fledged young.

References
Britton, P.L. and Britton, H.A. (1977).
Collar, N.J. and Stuart, S.N. (1985).
Stuart, S.N. and Hutton, J. (1977).

Genus *Cinnyris* Cuvier

Nearly 50 species of medium-sized sunbirds (a few large, several small) with short to medium-length, fine bill, more or less strongly decurved. ♂♂ brightly coloured, with forepart of body (head, mantle, upperwing coverts, upper breast) strongly iridescent, generally green, lower breast iridescent purple or non-iridescent scarlet in a deep band, often a narrow band of iridescent blue between upper and lower breast; rump and tail black, strongly glossed with green, blue or violet; wings and belly generally matt black. Yellow or orange pectoral tufts in ♂♂ of most but not all species. Tail usually short and square, or a little graduated in a few species, and with greatly elongated, narrow central feathers (T1) in a scattering of species not necessarily closely allied within the genus. ♀♀ dull coloured, not iridescent, yellowish, olive or brownish, underparts plain or lightly streaked, without pectoral tufts.

Mainly sedentary, yet may also be subject to wide dispersal (probably juveniles) and have repeatedly colonized W Indian Ocean islands and penetrated to subtropical and temperate latitudes and to tropical montane habitats. Some northern forms are partial migrants. Occupy wide spectrum of woody habitats, where the genus seems to be differentiating and evolving as rapidly as any birds in the Afrotropics. Eat small invertebrates, nectar and some small fruits. Nest a small, suspended globe with side entrance.

Systematically complex; there has been little agreement about the numbers of subspecies or the limits and relationships of species and superspecies. Different arrangements have been proposed by Delacour (1944), Rand (1967), Hall and Moreau (1970), Wolters (1977), Sibley and Monroe (1990) and Dowsett and Forbes-Watson (1993). Modern studies in molecular genetics and numerical taxonomy are sorely needed. In the meantime we adopt the recent revision by Irwin (1999); we recognize 38 species in Africa (and there are a further 6 in Madagascar, Seychelles and Comoros, all with African affinities, and 5 in the Oriental Region), which Irwin places in six groups, as follows:

Double-collared sunbirds, 3 superspecies and 7 independent species – *chloropygia, minulla, manoensis/chalybea, neergaardi, stuhlmanni/ludovicensis/afra, reichenowi, regia, rockefelleri, mediocris/moreaui/loveridgei, pulchella*;

Purple-banded sunbirds, 4 superspecies and 4 independent species – *shelleyi/mariquensis, congensis/erythrocerca, nectarinioides, bifasciata/tsavoensis, chalcomelas/pembae, bouvieri, osea, habessinica*;

White-bellied sunbirds – *oustaleti/talatala, venusta, fusca*;

Maroon sunbirds – *johannae, superba, coccinigastra, rufipennis*;

Green-and-olive sunbirds – *batesi, ursulae*; and

Copper Sunbird – *cuprea*.

Cinnyris 231

Cinnyris manoensis superspecies

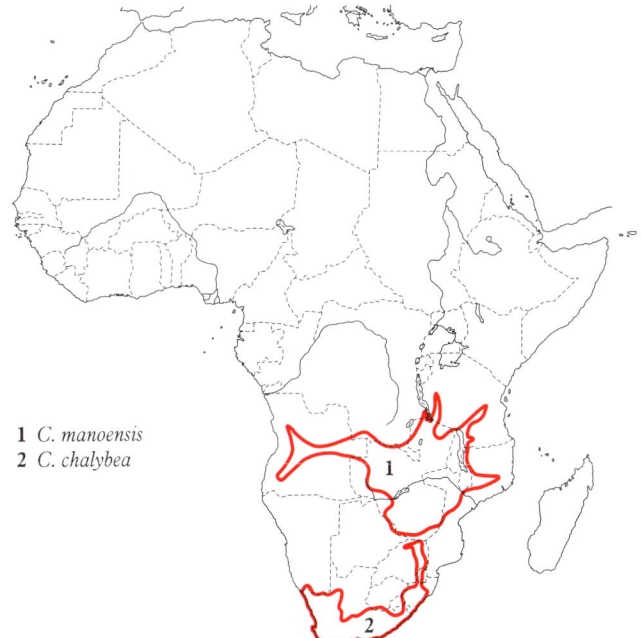

1 *C. manoensis*
2 *C. chalybea*

Cinnyris afra superspecies

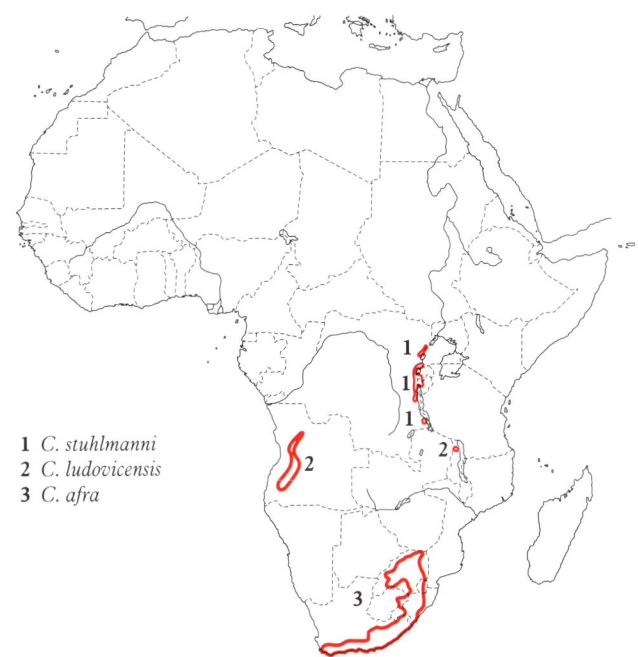

1 *C. stuhlmanni*
2 *C. ludovicensis*
3 *C. afra*

Cinnyris mediocris superspecies

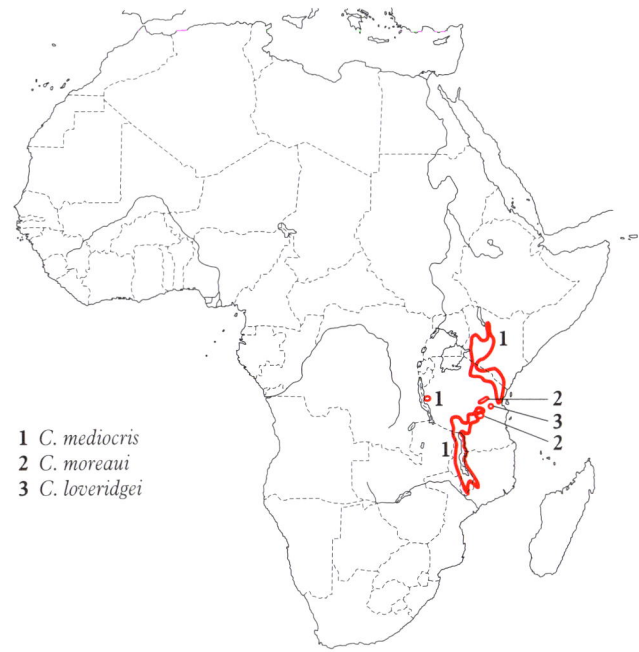

1 *C. mediocris*
2 *C. moreaui*
3 *C. loveridgei*

Cinnyris mariquensis superspecies

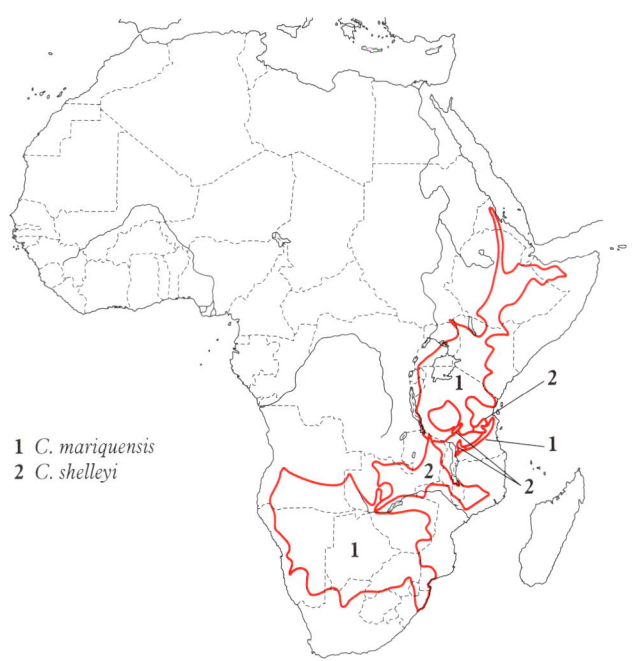

1 *C. mariquensis*
2 *C. shelleyi*

Cinnyris erythrocerca superspecies

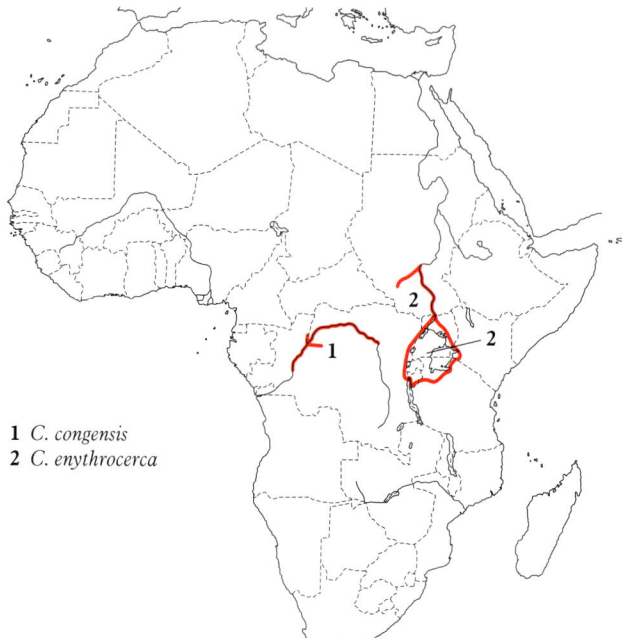

1 *C. congensis*
2 *C. enythrocerca*

Cinnyris bifasciata superspecies

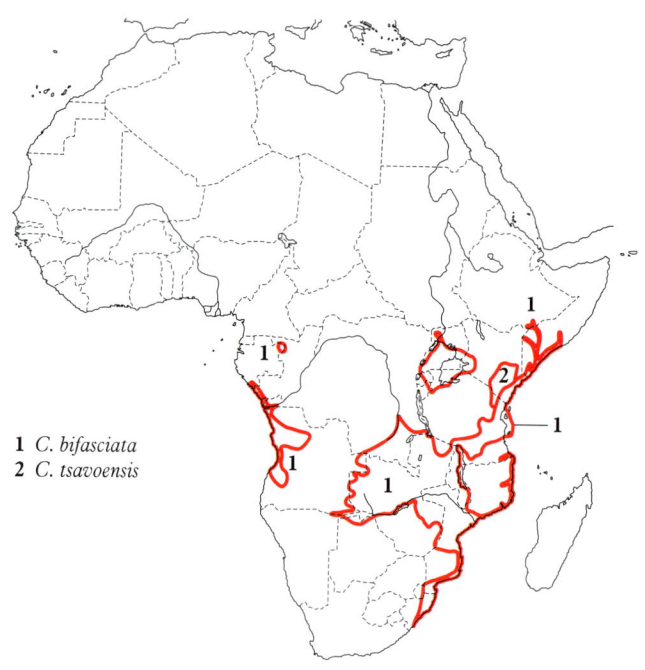

1 *C. bifasciata*
2 *C. tsavoensis*

Cinnyris chalcomelas superspecies

1 *C. chalcomelas*
2 *C. pembae*

Cinnyris talatala superspecies

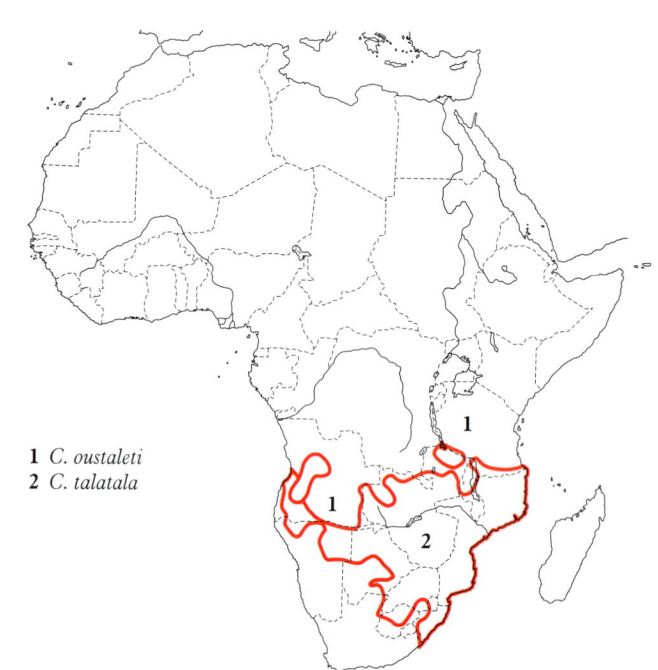

1 *C. oustaleti*
2 *C. talatala*

Cinnyris chloropygia (Jardine). Olive-bellied Sunbird. Souimanga à ventre olive.

Plate 13
(Opp. p. 206)

Nectarinia chloropygia Jardine, 1842. Ann. Mag. Nat. Hist., 10, p. 188; Niger River.

Range and Status. Endemic resident, forest edges from Senegal to Ethiopia and Angola to Kenya. Mali, 2 birds collected south of Bamako said to be this species. Senegal, several observations in Basse-Casamance and Niokolo Koba Nat. Parks and around Saloum and Mbour, but none yet in Gambia. Guinea-Bissau, in S only. Guinea, records from Fouta-Djalon, Kindia region (rare or local), Conakry area, and near Sierra Leone border (scarce in Kounounkan, frequent in Macenta). Sierra Leone, common on coastal plain and inland except in uplands north of about 09°N. Liberia, common to abundant almost throughout; up to 1200 m on Mt Nimba. Ivory Coast, abundant throughout 'to nearly 10°N' (Thiollay 1985, though here mapped north only to known localities). S Ghana. Togo, not uncommon, north to Kara (09°33′N). Benin, occurs, doubtless widespread in S, though actual abundance ill known. Nigeria, common from coast to N limits of derived savanna (Ibadan in SW, Enugu in SE); local in gallery forests further north (Ilorin, Pandam). Cameroon, very common in SW, uncommon north of 06°N to Tibati; once near Garoua. Chad, single record Sarh, Mar. Very common throughout Gabon, Congo, and Zaïre except Katanga, though abundant in Itombwe up to 1100 m and probably reaches 2000 m (Prigogine 1971). Angola, in Cabinda, N Malanje, Cuanza Norte (quite common at Salazar and N'Dala Tando) and along escarpment of Cuanza Sul to Gabela, and Cundo region of N Lunda Norte (W.R.J. Dean, pers. comm.). In Central African Republic known from Lobaye Préf., Bangui and Bamingui areas (Carroll 1988), and common in Haute Sangha Préf. and Dzanga Reserves (Green and Carroll 1991). S Sudan, rare, only on Aloma Plateau and in Yambio and Bengengai. Uganda, common at 700–1750 m in W and S, from Murchison Falls Nat. Park to Rwenzoris and Impenetrable Forest, east to shores of L. Victoria from Sango Bay to Jinja and Kenya border, north to Mbale. Reported from Rwanda but absent from (for instance) Nyungwe. In Burundi, occurs around lakes Cyohoha, Rwihinda and Rweru in Bugesera district, and abundant in W, west of 29°30′E, from Musigati to Nyanza-Lac (Gaugris et al. 1981). Kenya, locally common at 1150–1550 m, in Mumias, Busia and Kakamega Districts, Ukwala, Ng'iya, lower Yala R., Sotik (formerly), Pala and Rapogi (Britton 1980, Zimmerman et al. 1996). Tanzania, Kagera and N Kigoma, from Bukoba to Biharamulo and Kibondo; also Mahari Mts (N. E. Baker, pers. comm.). A poorly-known outlying population near Binescho, Kefa (Kaffa) in W Highlands of Ethiopia.

Density in Gabon estimated at 5–7 birds per 10 ha.

Description. *C. c. chloropygia* Jardine (includes '*insularis*' and '*luehderi*'): SE Nigeria, Bioko and Cameroon to lower Congo basin and NW Angola, east to Ubangi and middle Congo rivers. ADULT ♂: head, neck and rest of upperparts metallic green with golden tinge, uppertail-coverts more turquoise. Tail blackish, upperside with faint blue gloss; feathers narrowly edged bright metallic blue when fresh. Lores blackish. Chin to upper breast metallic green, concolorous with head, often tinged bluish along lower border; below this a scarlet band (10–15 mm deep) across lower breast; pectoral tufts bright yellow or orange–yellow; rest of underparts bright olive-green, paler on centre of lower belly. (Belly and undertail-coverts subject to marked change in colour, with wear, from bright olive-green to brown with only trace of olive wash: Serle 1965.) Lesser and median coverts metallic green; upperwing otherwise blackish brown, alula, primary coverts and greater coverts with narrow metallic green fringes. Axillaries and underwing-coverts dark olive-brown; inner borders of flight feather undersides buffy brown. Bill dark brown or black; eyes dark brown; legs black. ADULT ♀: upperparts olive-green. Tail blackish brown, upperside faintly glossed; T6 and distal part of T5 paler brown. Narrow superciliary stripe buffy white; lores dusky olive; cheeks and ear-coverts olive-green. Underparts olive-yellow, paler on chin and throat; mottled light olive-grey on throat and breast and tinged greyish on flanks. Lesser and median coverts olive green; upperwing otherwise dark grey-brown, remiges edged olive-yellow, greater coverts fringed olive. Axillaries and underwing-coverts pale yellow; inner borders of flight feather undersides grey-buff. Bare parts as adult ♂. SIZE (10 ♂♂, 10 ♀♀): wing, ♂ 47–51 (48·2), ♀ 43–46 (44·6); tail, ♂ 30–36 (33·3), ♀ 23–29 (26·1); bill, ♂ 18–21 (19·5), ♀ 17–20 (18·5); tarsus, ♂ 15–16 (15·5), ♀ 13–14 (13·9). WEIGHT: Angola, ♂ (n = 1) 7·1.

IMMATURE: juvenile similar to adult ♀, but upperparts darker and browner; chin and throat dusky grey, separated from dark ear-coverts by olive-yellow moustachial stripe; upper breast sometimes barred dusky grey.

NESTLING: not described.

C. c. kempi (Ogilvie-Grant): Senegal to SW Nigeria. Like nominate race, but ♂ has olive lower underparts lighter and greener; breast band more orange-scarlet, rather narrower. Slightly smaller; ♂ (n = 10) wing 46–49 (47·4), bill, 17–19 (18·2). WEIGHT: Liberia, ♂ (n = 5) 4·7–5·5 (5·1), ♀ (n = 1) 5·0, Ghana, 1 ♀ 5·0.

C. c. orphogaster (Reichenow) (includes '*bineschensis*'): NE Angola and S and E Zaïre to Sudan, Ethiopia, Uganda, Kenya and Tanzania. ♂ differs from nominate race in having lower underparts darker olive–brown. Larger; ♂ (n = 10) wing 52–56 (53·3), bill 20–22 (21). WEIGHT: Uganda, ♂ (n = 34) 5·5–8 (6·3), ♀ (n = 12) 5–7·5 (6·1); Kenya, ♂ (n = 3) 6·3–6·7 (6·5), 2 ♀♀ 5·8, 6·0.

Field Characters. Length 10·5–11 cm. A lowland forest species, very active and easy to see. ♂ of small W African race *kempi* has golden-green head and upperparts, red breast-band *c.* 1·25 cm broad *not separated by blue or violet breast-band from green throat*, yellow pectoral tufts, yellowish olive belly. ♀ dark greenish olive above with narrow pale supercilium, blue-black tail; below, *bright yellow centre of breast and belly*, paler on throat, washed olive on flanks, light streaking on throat, breast and flanks. Only other double-collared sunbird in lowland forest (and the only one of any kind west of Cameroon) is Tiny Sunbird *C. minulla*, which is smaller (8–10 cm) and shorter-billed; ♂ has lower throat blue and red breast-band mixed with blue feathers, belly duller, less yellow; ♀ duller below and more heavily streaked. ♂ of larger central African race *orphogaster* lacks golden tinge to upperparts which are more bluish, especially on uppertail-coverts (but uppertail coverts *not strongly contrasting* as in Northern and Eastern Double-collared Sunbirds *C. reichenowi* and *C. mediocris*); feather tips of lower throat variably bluish where they border red breast-band but never forming a wide blue band as in Eastern; *belly dark olive-brown*; ♀ has pale supercilium, *whitish chin and throat*, pale yellow underparts streaked on breast and flanks. Replaced by Northern Double-collared Sunbird in highlands of Cameroon and central and E Africa, but meets it at lower levels; ♂ Northern has contrasting violet uppertail-coverts and violet breast-band and uppertail-coverts. ♀♀ of both Northern and Eastern have dark throats and are darker below, olive rather than yellow except on belly; Northern is unstreaked, Eastern lightly mottled. Just overlaps Miombo Double-collared Sunbird *C. manoensis* in Angola, ♂ of which has brown rump, purple band separating throat from breast, ♀ wholly grey.

Voice. Tape-recorded (104, B, C, F, ERA, GRI, KEI, LEM). Song begins with some thin introductory notes, followed either by a rapid, rather liquid, sibilant trill, 'tseep, tseep, see-see-see tissrrrrrrrr', 'tseep, tsee-tsi-tsi-tsi tisssrrrrrrr' or shorter 'tseep, sisisisi tissrrrrr', the whole lasting 3–4·5 s; or by 7–8 individually separable notes, 'tseep, sisisisi che-che-che-che-che-che-che-che', lasting 2·5 s; louder, lower-pitched and with more trills than song of Tiny Sunbird (F. Dowsett-Lemaire, pers. comm.). Fighting ♂ gives continuous rapid chattering, trilling and thin sibilant notes (Chappuis, in press); call, a plaintive 'tyeep-tyeep-tyeep ...'; alarm a rapid, hard scolding 'dzidzidzidzidzidzi ...'. Call while foraging, a sharp 'tsip' or 'tseep' at irregular intervals.

General Habits. Inhabits lowland forest edges and clearings, secondary and gallery forests, small patches of woodland in savanna, woods around inselbergs, cultivation, farms, coffee forest and plantations, swampy woods, riverine growth, mangroves, thickets, scrub, nurseries, small-holdings, gardens and town parks. Generally in low growth, foraging 0·5–3·0 m above ground in *Hibiscus* hedges, tangles of *Ipomoea*, and shrubs, e.g. *Cogniauxia podolaena* (Gabon).

In pairs, small ♂ parties, and mixed-sex aggregations at flowering shrub or tree, often mixing freely with other sunbird species. Sometimes pugnacious and territorial, but evidently not when blossoms are sparse or conversely very abundant. Interspecifically territorial with Tiny Sunbird and the 2 interact such that local population densities seem to reciprocate throughout year. Interacts territorially at blossoms with hawk moths Sphingidae (Brosset and Erard 1986). However, breeding pair studied feeding on nectar of *Hibiscus rosasinensis* did not interact with skipper, white and swallowtail butterflies competing for nectar on same shrub (Prendergast 1983). Active all day, sprightly, mobile, rather silent for much of time, feeding mainly at flowers by perching on them and probing or by hovering in front, also by moving through foliage and gleaning leaves and twigs. Pierces bases of tubular corollae too long to be probed, e.g. *Canna*. Pecks at and pokes around spiders' webs. Especially prone to feed at flowers of herbs, shrubs and creepers immediately around houses. Significant preference for pink over red flowers of *Hibiscus rosasinensis*, especially by ♀ of a nesting pair (Ivory Coast: Prendergast 1983). Often confiding; known to enter houses to investigate cut flowers.

Most W African records well to north of N borders of rainforest zone are undated; dated ones are Mar–May, which suggests a tendency to disperse or migrate northward in early wet season.

Food. Nectar and small spiders and insects; some fruits. Often feeds at inconspicuous greenish flowers of *Manihot utilissima* and at showy flowers of *Hibiscus* and other Malvaceae, *Spathodea nilotica* and *Canna*. Particularly favours *Cogniauxia* (Cucurbitaceae), *Ipomea* and bindweeds (Convolvulaceae), *Solanum* and guava (Gabon). Probes coffee and rubiaceous flowers (Dean *et al.* 1988). Tiny bits of flowers, also sometimes snail shells fragments and grains of sand and quartz eaten as gastroliths. 5 stomachs contained small spiders, small beetles and tiny caterpillars (Chapin 1954). Eats arillate seeds of *Ochthocosmus africanus* (Dowsett-Lemaire 1996).

Breeding Habits. Solitary-nesting, evidently monogamous and territorial.

NEST: suspended, domed purse with side entrance overhung by untidy porch, made of pliant vegetable fibres, bits of dead leaves and lichen, bound with spider web, raggedly decorated or disguised with bits of bark and palm leaf-base fibres, and lined with soft plant down (like that from *Funtumia* rubber tree). Height 120–130, entrance diam. 20. Nest hung from tip of leafy branch or liana, 1·6–8 m above ground, usually (6 out of 8 nests, Gabon) between 1·6 and 2·0 m; in interior of tree or shrub or on outside, when nest entrance faces outward; often sited in abandoned cultivation.

EGGS: 2 (once a completed clutch of 1 egg, Gabon; once 3 eggs, Ghana). Rather long or pointed ovals, more or less glossy. Pale grey, pale blue, bluish white or greenish white, with bold, dark grey or dark brown spots, clouds and longitudinal scrolls or wreath of confluent blotches around large end and with scattered minute black dots. One clutch glossy immaculate white (Serle 1954). SIZE: (n = 6, Cameroon) 13·6–15·6 × 9·6–11·3 (14·3 × 10·5).

LAYING DATES: Sierra Leone, Sept–Nov; Liberia, Nov, Feb (and dependent young Sept–Apr); Ghana, (various evidences Feb–Apr, June and Sept–Oct); Nigeria, (indications nearly all months, particularly May–Oct); W Cameroon, June, Aug–Jan and Mar (mainly Dec), S Cameroon, Mar-June and Nov–Dec (mainly Apr); Gabon, Oct–Feb and once (in exceptional year) Aug; Zaïre (Medje), Sept–June, (Uele), (various indications of breeding in driest season in Feb and in wettest in Sept–Nov), (Itombwe), Jan, June, Aug–Nov (and indications that breeds throughout year); E Africa, Region B, Feb, Apr–Nov (records spread rather uniformly); Ethiopia, July; Angola, Dec.

INCUBATION: by ♀ only.

DEVELOPMENT AND CARE OF YOUNG: young fed by both parents.

References
Bannerman, D.A. (1948).
Brosset, A. and Erard, C. (1986).
Chapin, J.P. (1954).

Cinnyris minulla Reichenow. Tiny Sunbird. Souimanga minule.

Plate 13
(Opp. p. 206)

Cinnyris minullus Reichenow, 1899. Orn. Monatsb., 7, p. 170; Jaunde, Cameroon.

Range and Status. Endemic resident, W African and Congo Basin rain forests. Status difficult to ascertain owing to difficulty in separating this species from the common *C. chloropygia*.

Sierra Leone, Gola Forest (Allport *et al.* 1989). Liberia, not uncommon as mapped (Gatter 1997). Ivory Coast, known from Abidjan, Yapo Forest (where rare) and Korhogo (Thiollay 1985, Demey and Fishpool 1994). Ghana, 2 19th-century records; current status uncertain (Grimes 1987). Nigeria, specimens from Badagri, Lagos, Warri and Degema and unconfirmed sight records in Lagos area. Cameroon, widespread in S, common at least locally (Bates 1911, Serle 1965); from sea level at Victoria up to 1070 m on Mt Kupé and 1385 m on Dikume Balue. Bioko, widespread and not rare; 14 specimens collected in last 70 years (Malabo, Luba, Riaba, Bonyoma); recently netted at Ureca and at 1000 m on Pico Basilié. Mbini, recorded (Fa 1990). Gabon, widespread and common in NE and Réserve de la Lopé, known also in W and S. Congo, in Odzala Nat. Park (Andzoyi), Nouabalé-Ndoki Nat. Park (Bomassa), and in Mayombe where common around Dimonika, scarce further west (Goumina). Zaïre, known from lower Congo R. upstream to Kunungu, Kasai (Macaco), Lualaba R. in W Kivu, Ituri, Semliki and Itombwe forests; frequent near Irumu; rather rare in Itombwe. Uganda, at 700–1700 m at N end of Rwenzoris: common at Bwamba, scarce east to Kibale.

Density of up to 2–3 birds per ha (Gabon).

Description. ADULT ♂: head, neck and rest of upperparts metallic green with bronzy tinge, more turquoise on uppertail-coverts. Tail rather short, blackish, upperside with faint blue gloss. Chin to upper breast metallic green like head, more bluish green along lower border; below this a dull orange-red band across lower breast (10–15 mm deep), with dusky and bright blue barring due to subapical feather bands; pectoral tufts yellow; rest of underparts olive. Lesser and median coverts metallic green; upperwing otherwise blackish brown, greater coverts narrowly edged metallic green. Axillaries and underwing coverts white or whitish (grey in *C. chloropygia*); inner borders on undersides of flight feathers greyish white. Bill black; eyes brown; legs black. ADULT ♀: upperparts olive green. Tail blackish brown, upperside with faint gloss; T6 and distal part of T5 paler brown. Narrow superciliary stripe olive-yellow; lores dusky olive; cheeks and ear-coverts olive-green. Chin to upper breast and flanks dusky olive-green, chin and throat flecked whitish; rest of underparts olive-yellow. Upperwing dark brown, remiges edged olive-yellow, greater coverts fringed olive, median and lesser coverts broadly tipped olive-green. Axillaries, underwing-coverts and flight feather undersides as adult ♂. Bare parts as adult ♂. SIZE (10 ♂♂, 7 ♀♀): wing, ♂ 47–51 (48), ♀ 43–46 (45); tail, ♂ 28–29 (28·4), ♀ 24–26 (25); bill, ♂ 15–18 (16·5), ♀ 16–17 (16·5); tarsus, ♂ 13–15 (14·1), ♀ 13–15 (13·9). WEIGHT: Liberia, 1 ♂ 6, 1 ♀ 4·5; Cameroon, 1 ♀ 6·5; Uganda, ♀ (n = 6) 5·5–6·5 (5·8) and a ♂ given as 3.

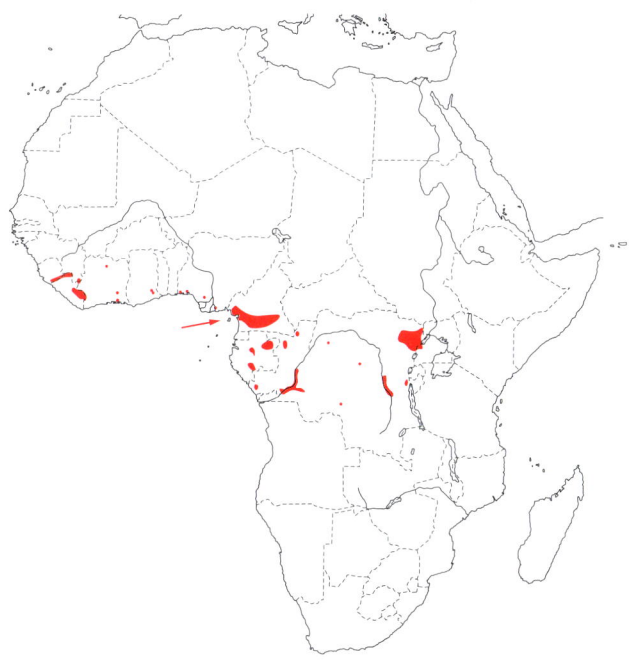

Cinnyris minulla

IMMATURE: resembles adult ♀, but chin and throat dusky grey.
NESTLING: naked; inside of mouth orange, tongue unmarked, swollen gapes whitish (Bates 1911).

Field Characters. Length 8–10 cm. A lowland forest bird, occurring regularly with the larger and longer-tailed but otherwise very similar Olive-bellied Sunbird *C. chloropygia*. *Bill structure* is helpful; both sexes of Tiny have an appreciably shorter and less curved bill, the lower mandible appearing almost straight except for the very tip, whereas it is equally curved over most of its length in Olive-bellied (Demey and Fishpool 1994). In ♂ of Tiny, metallic green areas said to appear visibly bluer in good light (Christy and Clarke 1994). Well-marked *blue-violet band* separates throat from red breast-band (mainly absent in Olive-bellied, although in nominate *C. c. chloropygia* throat feathers grade into blue where they overlap red band, forming narrow line); red breast-band appears darker in the field, being partly crossed by blue bars and flecks as feather bases show through; these are very obvious in skins but said to be 'virtually impossible to see in the field' (Demey and Fishpool 1994) but 'possible to see in good light' (Gartshore 1989). ♀ Tiny duller and less yellow below than *C. c. kempi* but similar to other races of Olive-bellied; best distinguished by small size, short tail and straight bill. While they can occur together their habitat preferences differ, Tiny being mainly a forest bird while Olive-bellied is more catholic in its choice of habitats (Demey and Fishpool 1994).

Voice. Tape-recorded (104, B, C, CHA, ERA, LEM). Song of 2 main types, either beginning with 1–4 thin 'tit' notes followed a short burst of individually separable notes, 'tit ... tit ... tsee-tsee-tsoo-tsee-tsitsitsitsu', 'tit ... tit ... ti ... ti ... ti*tsee* tippy-ti-tsee-titititi', the whole lasting 2–3 s; or prefaced by a nasal up-slurred 'zhwee' and then a rapid trilling chatter, sometimes slowing at end into individually audible notes, lasting 3 s. Higher pitched than song of Olive-bellied Sunbird (F. Dowsett-Lemaire, pers. comm.), but so similar that each species is attracted by playback of the other's song, 'ce qui est assez étonnant pour deux espèces vivant dans le même milieu et la même région' (Chappuis, in press); specific recognition apparently achieved by calls, which differ (but are not described). Song also described as a series of 7–10 shrill notes, 'suisui sui sui sui' (Christy and Clarke 1994).

General Habits. Inhabits primary and old secondary forests, especially discontinuous formations like old plantations (cocoa plantations in Bioko), long-abandoned fallow land, clearings, and forest with pioneering trees such as *Harungana*, *Trema* and *Musanga* (Gabon), and old regeneration near villages, sometimes in open canopy, also in gardens (Congo). More dependent on forest than Olive-bellied Sunbird. Seen (and netted) usually singly. Foraging and social habits not obviously different from those of much better-known Olive-bellied Sunbird, and the 2 are interspecifically territorial, and interact such that local population densities seem to reciprocate throughout year (Brosset and Erard 1986). Forages by alighting on and probing flowers, and by gleaning leaves; sometimes feeds alongside other sunbird species. Sings frequently during breeding season from tops of tall trees in gallery forest (Christy and Clarke 1994).

Food. Field observations indicate mixed diet of pollen, nectar and small arthropods (Brosset and Erard 1986). Only small insects in stomachs.

Breeding Habits. Solitary nester, evidently monogamous and territorial. 2 nests found in successive years in same place on same date – probably same pair of birds.

NEST: small and exquisite oval, suspended, domed purse with porched side entrance, like nest of Olive-bellied Sunbird but without a 'beard' and darker, made of different materials (Cameroon: Bates 1911). Built mainly of 'fine, dark-colored rootlets, probably from some epiphytic plant' (Chapin 1954) or fine black rootlets and moss (Bates 1911), covered largely with small, flat bits of bark or flat pieces of grey-green lichen with a few grass seed-heads or dry palm flowers and bits of dead leaves, bound together with *Marasmius* fungal filaments and spider silk, with projecting porch of *Marasmius* filaments above entrance (Brosset and Erard 1986). Softly lined with kapok or slender vegetable fibres and a little brown vegetable down. Dimensions: (1 nest) height 130, breadth 75, entrance diam. 25, from top of nest to centre of entrance hole 90. Sited 1·5–3·0 m above ground at edge of forest, hung from tip of leafy branch or thin twig or once from 2 epiphytic fern fronds, with entrance facing away from densest trees.

EGGS: 1–2, av. (n = 5) 1·9 (Gabon). 2nd egg laid 24 h after 1st. Rather long ovals, described both as glossy and devoid of gloss; ground pale blue, bluish white or greenish white, with bold violet-brown or dull dark brown spots and dark lilac-grey blotches, scattered evenly or forming zone around large end. SIZE: (n = 2, Cameroon) 14·5–15·5 × 10, (n = 2, Irumu, Zaïre) 13·8–14·1 × 10·3–10·8, (n = 4, Itombwe, Zaïre) one pair 13·8–14·0 × 10·4–10·5, other pair 14·5–15·0 × 10·5–10·9.

LAYING DATES: Cameroon, Apr–June, Oct; Gabon, Sept, Nov, (and nestlings Jan); Zaïre (Medje), June; (Itombwe), Mar, Nov, (Irumu), (♂♂ in breeding condition and ♀♀ nest building Sept).

INCUBATION: by ♀ only.

DEVELOPMENT AND CARE OF YOUNG: young fed by both parents.

References
Bates, G.L. (1911).
Brosset, A. and Erard, C. (1986).
Chapin, J.P. (1954).
Prigogine, A. (1971).

Cinnyris manoensis Reichenow. Miombo Double-collared Sunbird. Souimanga du Miombo. Plate 13
(Opp. p. 206)

Cinnyris manoensis Reichenow, 1907. Orn. Monatsb., 15, p. 200; Missale, west of L. Nyasa.

Forms a superspecies with *C. chalybea*.

Range and Status. Endemic resident, *Brachystegia* woodland in southern tropics. Angola, central plateau from W and N Huila to Huambo, N Bihe, and N Moxico to border with Zambia. Zaïre, north to 10°S in SW Katanga where scarce; Lubumbashi and presumably eastward to Zambian border; and in Marungu (Pande, Sambwe, Matafali: 1875–1915 m). Zambia, restricted to *Brachystegia*, nowhere common; absent from Nchelenge, Samfya, N Mpika, Luangwa Valley, Kabompo except N (south to Mayau), Balovale, Kalabo, and Barotse (except for record in Mankoya at 15°S, 23°50′E); occurs south to Kafue Gorge, Choma and Kalomo; up to 1850 m in Mafinga Mts. Tanzania, local and uncommon at 1000–1400 m from Busondo (05°20′S, 30°24′E) to Tukuyu and from there to Iringa, and E Mikumi Nat. Park; also widespread in Songea. Malaŵi, from Mulanje and Zobue northward, on both sides of Rift; not uncommon at 920–1690 m, and up to 2150 m on Mt Mulanje. Mozambique, uplands in W Niassa Prov, and in Nampula Prov. east to about Nampula; south of Zambezi, known from above 750 m in Mafusi, Buzi R. drainage, Chimanimani Mts (where the commonest sunbird in *Brachystegia*), the Vumba and Mt Gorongoza. Botswana, 2 old and 1–2 recent records near Francistown. Zimbabwe, widespread on central plateau, absent from gusu woodland (*Baikiaea*) in NW (Hwange, Chete and W Chizarira Nat. Parks: Worsley 1983), from Matabeleland south of about Gwanda, from Gonarezhou, and from NE (area along Mozambique border); frequent to common; large concentrations can occur in winter in *Acacia* woodland, and sometimes abundant in suburban gardens (Irwin 1981, Tree 1990).

Seems to have benefited from fragmentation and opening up of miombo woodland (A.J. Tree *in* Harrison *et al.* 1997).

Description. *C. m. manoensis* (Reichenow): Zimbabwe, Mozambique (except Mt Gorongoza), E Zambia (east of Luangwa valley), S Malaŵi and SE Tanzania (W Songea). ADULT ♂: head, neck, mantle, back and scapulars metallic green with bronzy reflections; rump and smaller uppertail-coverts pale olive-brown; longer uppertail-coverts violet-blue. Tail blackish brown, faintly glossed above; T6 paler brown with narrow greyish white outer web. Lores blackish. Chin to upper breast bronzy green, concolorous with head, bordered by narrow (*c.* 2 mm deep) metallic band, blue with violet intrusions; below this a bright scarlet band (*c.* 10 mm deep) across lower breast; pectoral tufts yellow; rest of underparts olive–buff. Lesser coverts bronzy green; rest of upperwing (including median coverts) grey–brown, flight feathers and greater coverts edged buffy brown. Axillaries and underwing-coverts whitish; inner borders of undersides of flight feathers pale grey. Bill black; eyes dark brown; legs black. ADULT ♀: upperparts greyish brown, rump tinged olivaceous. Tail dark brown, faintly glossed above; T6 paler brown with brownish white outer web, T5 with brownish white tip. Faint, narrow superciliary stripe buffy white; lores dusky brown; cheeks and ear-coverts greyish brown. Chin to breast and upper flanks buffy grey, flecked whitish; lower flanks, belly and undertail-

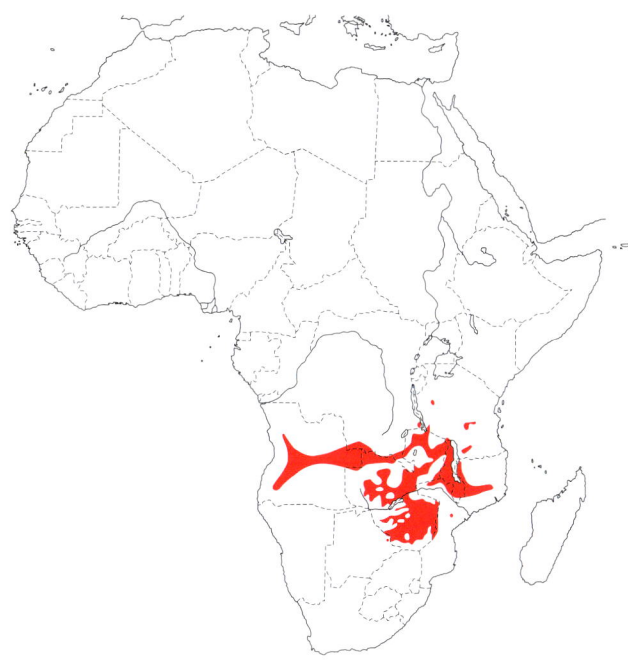

Cinnyris manoensis

coverts paler buffy grey, tinged yellow medially. Upperwing feathers greyish brown, flight feathers edged pale buffy brown. Axillaries whitish, tinged yellow; underwing-coverts dull white; flight feather undersides as adult ♂. Bare parts as adult ♂. SIZE (10 ♂♂, 10 ♀♀): wing, ♂ 61–67 (63·0), ♀ 57–61 (58·8); tail, ♂ 42–47 (45·2), ♀ 38–42 (40); bill, ♂ 23–25 (24·3), ♀ 20–24 (22·1); tarsus, ♂ 16–18 (16·9), ♀ 16–17 (16·5). WEIGHT: Zimbabwe, ♂ (n = 15) 8·4–12·8 (9·8), ♀ (n = 8) 7·4–9·3 (8·4).

IMMATURE: juvenile resembles adult ♀.

NESTLING: unknown.

C. m. pintoi Wolters: Angola, Zambia (except E borders), Zaïre, N Malaŵi, and Tanzania except Songea. ♂ differs from *manoensis* in having large uppertail-coverts plain olive-brown or tipped bronzy green (not violet); lower underparts slightly paler, centre of belly and vent tinged yellow. Bill shorter: ♂ (n = 10) 19–21 (20).

C. m. amicorum Clancey: Mt Gorongoza, Mozambique. ♂ greener (less bronzy) above than nominate *manoensis*, with upper breast band more violet (less blue) and lower underparts darker. ♀ darker above; duskier, more brownish below. Larger, with longer bill: (3 ♂♂) wing 63–67 (64·3), tail 45–47 (45·8), bill to feathers 24–26 (24·5) (Clancey and Irwin 1978).

Field Characters. Length 11·5–13 cm. Endemic to miombo woodland. ♂ has green upperparts and red breast-band bordered above by narrow blue band; ♂ of nominate race in Zimbabwe has blue-violet uppertail-coverts and is almost identical to Lesser Double-collared Sunbird *C. chalybea* except for broader olive rump, but their ranges are separated by the Limpopo Valley so this is not a field problem. Overlaps in Angola with Ludwig's Double-collared Sunbird *C. ludovicensis*, where ♂ of local race *C. m. pintoi* lacks violet on uppertail-coverts, and instead has *broad olive* area from lower *back to uppertail-coverts*, with no green iridescence except on tips of uppertail-coverts; also distinguished by narrower and duller red breast-band. Occurs in same habitats as Shelley's Sunbird *C. shelleyi*; ♂ distinguished by *olive* (not black) *belly*, ♀ by *lack of streaking* on underparts. ♀ with grey underparts and small yellow patch on mid-belly can be confused with grey-bellied form of ♀ White-bellied Sunbird *C. talatala* (Tree 1990) but is larger; freshly-fledged young and some ♀♀ have yellowish wash on much of underparts (Tree 1990), and might be mistaken for the smaller ♀ Variable Sunbird *C. venusta*. In Malaŵi just meets Eastern Double-collared Sunbird *C. mediocris*; for differences, see that species.

Voice. Tape-recorded (86, 91–99, F, GIB, LEM, STJ). Song (slow version), a jumble of rapid yet individually audible notes with up-down quality, 'chewy-chichwit-chichi-cheedo-chuji-chwit-cheedit ...' lasting 6–8 s; or a shorter (2·5–5 s), faster version more like that of Lesser Double-collared Sunbird, a sibilant chatter, individual notes barely audible, beginning with single 'jet' or 'tee-tss-jet' and ending with 'jet' or hard 'trrt'. Also gives a hard, descending trill lasting 1 s, and high, thin 'tit' or 'tsit'.

General Habits. Inhabits mainly miombo (*Brachystegia*) woodland and in some areas restricted to it; also mavunda forest (*Cryptosepalum*) in Zambia, bushland in Tanzania, and flowery suburban gardens as well as montane vegetation in Malaŵi and Zimbabwe. Occurs up to 2200 m. In woodland feeds in canopy. Concentrates in acacia savannas in Zimbabwe when parasite *Loranthus* is flowering; in gardens much attracted to *Bauhinia*. Often hawks for small flying insects and gleans spiders from verandah hanging flower baskets (Hanmer and Chaddon 1997).

Spends much time chasing other sunbirds, e.g. White-bellied, from nectar sources e.g. *Aloe* spp. (A.J. Tree *in* Ginn *et al.* 1989). Resident but (as with most sunbirds) subject to wandering, with rapid turnover of individuals at a given locality (Tree 1990). Visits gardens seasonally: mainly in June-Aug. Ringed bird moved 26 km.

Food. Nectar and presumably some insects obtained at flowers of *Tecomaria capense, Aloe* spp, *Leonotis, Kniphofia, Loranthus, Bauhinia* and variety of garden flowering plants, (Tree 1990); spiders and airborne insects (see above).

Breeding Habits. Solitary nester, monogamous and probably territorial.

NEST: oval, pear-shaped or ball-shaped, thickly lined or (if thinly lined) grassy outer structure is thick (**A**). Cobweb used as binding on the outside, sometimes profusely, giving nest greyish appearance. One beautiful nest was thickly and neatly built of pure white plant down, with only a slight covering of fine, soft grass (Vincent 1949). Often attached to mass of cobwebs, at height of 2–4·5 m (once 6 m) above ground, in thorn veld and open woodland with good undergrowth of shrubs and small trees, and in bushes along a low bank or dry stream bed; sited out in the open at edge of bush or as often in middle near top of open-

A

growing evergreen bush; nest attached to twigs embedded in structure at top and top-back of dome (Vincent 1949).

EGGS: 1–3, av. (n = 173) 1·91 in Zimbabwe and av. 1·25 in Malaŵi where C/3 unknown. SIZE: (n = 14) 16·1–17·9 × 11·2–12·0 (16·7 × 11·8).

LAYING DATES: Zambia, Aug, Oct, Nov (and gonads active Mar and Sept); Malaŵi, May–June, July–Nov, mainly June and Sept; Mozambique, May; Zimbabwe, all months, mainly Sept–Oct (56% of 195 records; 28% in Aug and Nov, only 4% in Jan–Apr).

BREEDING SUCCESS/SURVIVAL: adult occasionally killed by accumulating ball of coagulated latex in stomach, ingested when probing rubber plant *Manihot galziovii* for nectar. Ringed birds over 3, 5, 6, 7 and 8 years old (Hanmer and Chadder 1997, Hanmer 1997).

References
Harrison, J.A. *et al.* (1997).
Tree, A.J. (1990).
Vincent, A.W. (1949).

Cinnyris chalybea (Linnaeus). Southern Double-collared Sunbird; Lesser Double-collared Sunbird. Souimanga chalybée.

Certhia chalybea Linnaeus, 1766. Syst. Nat., ed. 12, 1, p. 186; Cape of Good Hope.

Plate 13
(Opp. p. 206)

Forms a superspecies with *C. manoensis*.

Range and Status. Endemic resident, South Africa, NW Swaziland and just into S Namibia. Fairly common and locally common from Namaqualand through Karoo to E Cape and Natal. A separate population in E Transvaal, to Zoutpansberg. Natural habitat reduced in extent by growth of agriculture and towns near coast; however, has recently invaded Cape Town gardens and become much commoner in Cape Peninsula with replacement of native vegetation with exotic *Eucalyptus lehmannii* (Fraser and Crowe 1990). Has also become commoner in W Swaziland by occupying alien trees (Parker 1994).

Description. *C. c. chalybea* (Linnaeus): South Africa (mountains of SW and W Cape). ADULT ♂ (breeding): head, neck, mantle, back and scapulars metallic green with golden tinge; rump olive with metallic green feathers admixed; uppertail-coverts steely blue. Tail dark blackish brown with slight bluish gloss; T6 and distal part of T5 paler brown, T6 with narrow brownish white outer edge. Lores black. Chin to upper breast metallic green, concolorous with head, separated by very narrow (2 mm deep) steel-blue band from broader scarlet band (c. 10 mm deep) across lower breast; pectoral tufts yellow; rest of lower breast dusky olive-grey; flanks, belly, thighs and undertail-coverts paler, olive-grey with yellow tinge. Lesser and median coverts metallic green; upperwing otherwise dark brown, primaries narrowly edged olive-yellow, tertials and greater coverts fringed olive when fresh. Axillaries and underwing-coverts olive-grey; inner borders of undersides of flight feathers greyish. Bill black; eyes dark brown; legs black. ADULT ♂ (non-breeding): wings and tail as in adult ♂ (breeding); metallic feathers usually retained on mantle, back and uppertail-coverts, and a few on chin, throat and head; also some red feathers on lower breast; otherwise as adult ♀ (Martin 1983). ADULT ♀: top and sides of head greyish brown; rest of upperparts greyish brown with olivaceous wash, slightly paler on rump. Tail dark grey–brown; T6 paler with narrow whitish outer edge and whitish terminal fringe on inner web; also a whitish terminal fringe on inner web of T5 (and sometimes T4). Underparts buffy brown; throat and upper breast with faint dusky streaks, centre of lower breast and belly tinged greenish yellow. Lesser and median coverts same as upperparts; upperwing otherwise dark grey-brown, primaries narrowly edged pale buff, secondaries, tertials, alula and greater coverts edged light brown. Axillaries and underwing-coverts whitish, tinged yellow; inner borders on undersides of flight feathers pale greyish. Bare parts as adult ♂. SIZE (10 ♂♂, 10 ♀♀): wing, ♂ 55–59 (56·6), ♀ 49–54 (51·6); tail, ♂ 43–47 (45), ♀ 35–39 (37·1); bill, ♂ 21–23 (21·9), ♀ 18–20 (19·3); tarsus, ♂ 16–18 (16·8), ♀ 15–17 (15·8). WEIGHT: South Africa (all races combined), ♂ (n = 26) 6–9·5 (8·6), ♀ (n = 21) 6–9·5 (7·3) (Maclean 1985).
  IMMATURE: resembles adult ♀.
  NESTLING: hatches naked with skin dark pink, bill short, light horn with dark tip and white egg-tooth, and gapes white.
  C. c. albilateralis Winterbottom: South Africa (Cape Prov. from arid W coast and Orange R. valley (where it just enters Namibia) to Bushmanland and through Karoo to Great Fish R. valley). Similar to nominate race, but adult ♂ has breast band lighter (more orange-scarlet) and narrower (c. 7–8 mm); lower underparts slightly paler and greyer with less yellow tinge; rump more often plain olive. ♀ slightly paler, less olive above; more greyish olive–buff (less yellowish) below.

Cinnyris chalybea

C. c. subalaris (Reichenow) (including '*capricornensis*'): South Africa (E Cape to Natal and N Transvaal). Similar to nominate race, but adult ♂ has lower underparts darker olive; ♀ darker and greener above, and tinged more greenish below. Bill longer: ♂ (n = 8) 24–26 (24·8).

Field Characters. Length 11–13 cm. Endemic to southern Africa south of the Limpopo, and does not overlap with the very similar Miombo Double-collared Sunbird *C. manoensis*. Coexists with Greater Double-collared Sunbird *C. afra*, of which it is a smaller version, with much shorter bill. ♂ further differs from Greater by much narrower red breast-band; ♀♀ are almost identical except for bill and size. ♀ smaller than ♀ Orange-breasted Sunbird *Anthobaphes violacea* and grey below, not yellow.

Voice. Tape-recorded (16, 58, 75, 88, 91–99, 104, B, C, F, STA). Song exceptionally thin and high-pitched, a rapid burst of silvery tinkling and swizzling notes, including repetition of 'weeta-weeta-weeta ...'; some songs include a few brief pure sweet notes among the jumble of sibilant ones; before and between songs adds typical call notes, grating 'jee-jee-jee', hard 'tip' or sharp 'tsip'. Song bursts typically 3–5 s, sometimes up to 8 s, often with very short break before the next song; singing spells may last 15 min or more. Variations include 'harder, more metallic, staccato, stuttering and rolling with noticeable changes in tempo during each bout of song', and an abbreviated version said to bear strong resemblance to song of Fairy Flycatcher *Stenostira scita* (Skead 1967). ♂ on song post often gives plaintive, long-drawn, wispy 'sssteeeeu' or 'ssssweeeee', and similar call given while foraging; ♀ gives quiet subsong in between bouts of nest-building (Skead 1967).

General Habits. Great range of habitats in southern Africa: arid strandveld in NW Cape, macchia in SW Cape, coastal and inland forests, sparsely vegetated mountainsides with *Protea* spp., dry karroid valley-bushveld near rivers in E Cape; common wherever stretches of scrub remain in intensively cultivated areas; erosion gullies and dry watercourses with small trees and scattered bushy growth, gardens, scrubby sea-shores and dune vegetation, and amongst flowering eucalypts. Occurs in karoo and false karoo in SW Cape but generally avoids *Acacia karroo* in E Cape and Transkei (Skead 1967).

Noisy, somewhat aggressive to its own kind and to other sunbirds, mixing freely with them although dominated by Amethyst *Chalcomitra amethystina*, Dusky *Cinnyris fusca* and Greater Double-collared Sunbirds. Forages alongside Collared Sunbirds *Hedydipna collaris* on flowering heads of *Halleria lucida* and *Erythrina* without very much chasing. However, at any time of year 2 ♂ Lesser Double-collared Sunbirds will fight, clinging breast-to-breast with interlocked feet and falling to ground. One ♂ broke with another lying on back on ground and stabbed it with bill, then both flew away. Forms large, off-season foraging assemblies with Orange-breasted Sunbirds in SW Cape (Sundays River).

Lively and fast-flying, dashing from one food plant to another, interacting with other sunbirds and e.g. Wahlberg's Honeybirds *Prodotiscus regulus*; longer flights have sudden swerves with audible 'frrrt frrrt' of wings. In gardens and near people cautious but confiding. Spends much time hovering at flowers for nectar, and hovers to pick small insects and spiders out of leaves and cobwebs. Regularly pierces bases of *Canna*, *Gladiolus* and *Antholyza* flowers and of others with long corollas. With some flowers, bird inserts bill between calyx (sepals) and tubular corolla. When feeding at discoid inflorescence of *Protea lactiflora*, clings to side and probes in amongst upper bracts where they are loose. Sometimes treats *Protea* branch-end leaf clusters in same way as *Protea* inflorescences, visiting each in turn, perching across branch-tip with a leg each side and probing into rosette of young leaves, or hovering in front of branch-tip and probing in flight (**A**). Once visited flower after flower of (nectarless) *Pelargonium inquinans*, systematically touching base of each corolla-tube with bill (Skead 1967). Sips nectar from *Microloma sagittatum* flowers, when pollen parcels become attached to tongue, are carried in mouth and detached to fertilize next flower visited (Pauw 1998). Takes insects by gleaning, probing, and to large extent by hawking, dashing up from plant-top perch at steep angle and returning to perch. Bathes by flying back and forth through spray of automatic and hand-held hose-pipe.

Probably sedentary in many localities, e.g. E Cape forests. Bird known to stay in single spot for 4 years. Moves seasonally for short distances between breeding and

non-breeding areas, e.g. in SW Cape between Strandfontein/Somerset West (May–Dec) and Kirstenbosch/Plumstead (Dec–May, in parties). One ringed bird moved 18 km, another 25. Ringing evidence suggests that young birds disperse some distance and do not return to natal areas. Routine counts at Cape Flats show 2 peaks of frequency, in Dec (dispersion from nesting areas) and Apr (returning to breed). In S Cape, suddenly becomes common at Knysna in June, when aloes begin to flower.

Food. Nectar, insects, some spiders. Nectar of *Lachenalia pendula*, *Salvia aurea*, *Halleria lucida*, *Erythrina*, *Protea lacticolor*, *Cotyledon orbiculata*, *Schotia*, *Cadaba*, *Lycium*, *Tecomaria*, *Bauhinia*, *Loranthus*, *Phygelius capensis*, *Albuca canadensis*, *Nicotiana*, *Virgilia*, *Eucalyptus*, *Lobostemon fruticosum*, *Carissa*, *Cissus juttae*, *Aloe humilis*, also *Leonotis oxymifolia* (Fraser and Wheeler 1991) and *Microloma sagittatum* (Pauw 1998). Once seen feeding at orchid *Satyrium odorum*, landing on outermost inflorescence, probing lowermost flowers first, working its way upward to younger flowers, then repeating process at another inflorescence (Rebelo 1987). Probes into ripe fruits of figs *Ficus*. Eats much insect food, including larvae, small beetles, flies, gnats and midges, and winged ants; also spiders. Probably this species recorded perching on and fluttering by grass heads to eat seeds one by one (Berrington 1997). Of offered sugar solutions, prefers sucrose to fructose/hexose mixture and prefers latter to glucose (Jackson *et al.* 1998).

Breeding Habits. Solitary nester, monogamous, seasonally territorial. Territory maintained by ♂ singing at single prominent song post close to nest; adjacent territories often seem not to abut and boundaries generally ill-defined. ♂ does not commonly flash yellow pectoral tufts (Schmidt 1964). Once 5 breeding pairs along 675 m of roadside vegetation (Strandfontein). Pair returns each year to same area to breed. An aberrant ♂ (with white rump) bred in July–Oct in 3 successive years, all nests being within 25 m diam. circle (Strandfontein; birds absent Nov–June). Year-round chasing, excitability and local wandering disguise nature and duration of breeding territoriality. ♂ chases Southern Double-collared Sunbirds other than mate out of territory, yet sometimes seems to

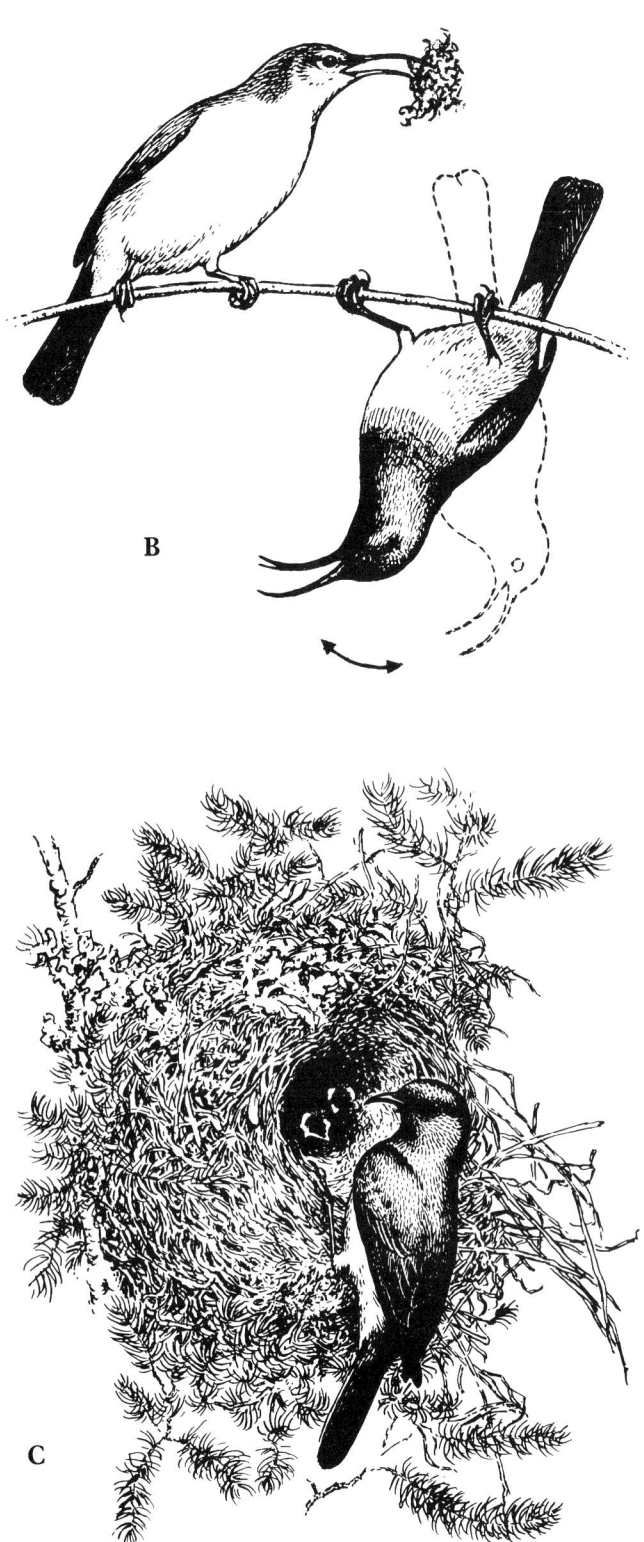

tree-top, ♀ hopped aimlessly around amongst branches below then flew to join him; facing each other, they bowed stiffly in unison then came jerkily to very upright stance; they bowed and rose repeatedly for *c.* 10 s, then copulated (Vincent 1949). Cloaca-pecking (by ♂ of ♀) is common, silent or accompanied by ♂'s soft warbling; either 1–2 abrupt pecks or a series; sometimes but not always followed by copulation; ♂ flies at ♀ and pecks her cloaca in variety of nesting situations throughout breeding cycle, from courtship and nest-building to post-fledging times (Schmidt 1964). ♂ once seen to peck cloaca of juvenile he was feeding. Some (most?) pairs double-brooded. New nest habitually built for each clutch, though exceptionally 2nd brood known to be reared in same nest (once 2nd clutch laid within 17 days of departure of 1st brood).

NEST: rather delicate, domed, oval or pear-shaped, with side-top entrance sometimes overhung by porch (**C**); made of fine, dry grass, plant fibres and pieces of *Galium tomentosum*, with woolly plant down and feathers, often with soft tendrils of *Clematis*, spider cocoons, pieces of string or paper or (in sheep districts) Merino wool on outside (and once strips of plastic: Korff 1997). Lined with feathers (often contour and down feathers of poultry) and sometimes with woolly plant down with a few feathers; one nest contained 189 poultry feathers, collected 200 m away, placed in layers on nest floor, inside dome, pushed into insides of walls, and around entrance; feathers quite absent from a few nests; occasionally nest made of kapok, and often almost entirely of *Usnea* (which binds onto itself making additional spider web binding unnecessary so it is not used). *Usnea* nests very occasionally have 100 mm beard dangling below. Situated in variety of shrubs and small trees; often evergreens, also tall, fleshy-leaved plants along streams, and isolated wattle tree or one of a small clump; at heights from ground, in favoured habitat of dense low scrub, of 12 cm to 3 m, mostly 0·5–1·25 m, or in well-wooded regions at 0·5–9·5 m. Sited in fork and attached to surrounding twigs and stems, or supported between twigs in bush, or firmly attached to foliage, or at end of branch out in the open, though most nests are screened behind outer foliage. Only uncommonly is nest suspended, with narrow attachment between dome and tip of twig (Vincent 1949). Often uses *Metalasia*, *Acacia cyclops*, *Melianthus*, *Elytropappus*, *Olea*, *Sideroxylon* and *Helichrysum*. Of 50 nests at Strandfontein, 38% in *Euclea racemosa*, 24% in *Salvia aurea*, 22% in *Acacia cyclops*, rest in *Senecio halimifolius*, *Chrysanthemoides monilifera* and *Osyris abyssinica* (Schmidt 1964). Pair once took over nest of Karoo Prinia *Prinia maculosa* and relined it. Exceptionally, suspends nest from wire. Size: (n = 20) height ext. 90–130, int. 100, width ext. 60–75 below entrance and 60–90 above it, from front to back ext. 65–90, incl. porch 90–120, int. (entrance to back) 45–70, entrance diam. 23 × 30.

tolerate protracted incursions by a pair. Main sexual display seems to be ♂ singing strongly, directed at nearby ♀; once ♂ swung forward, gripping perch, until upside-down, continuing to sing at ♀ holding nest material (**B**), then righted himself and pair flew away. Another ♂ and ♀ perched facing each other and rapidly pivoted to left and right like clockwork for 5 min (Skead 1967). In pre-copulatory display, ♂ sang strongly for several min. from

Nest built by ♀ alone, accompanied by ♂ as she flies to and fro gathering material and building; whilst she builds, ♂ perches nearby, singing energetically. Replacement nests often built near earlier, failed one, using material taken from it (Vincent 1949, Schmidt 1964). First nests take 25–30 days to build (period barely affected by weather), replacement and second-brood nests *c.* 9–16 days. If eggs soon robbed but nest undamaged, ♀ may lay replacement clutch in it. Several ♀♀ sleep in nest when building nearly completed (Schmidt 1964).

EGGS: 2–3; av. of 55 clutches in W Cape 2·3. Laid on successive days (1st egg once between 10h00 and 11h40). Cream, grey, greyish or greenish white; marked with black, dark brown and dark and pale grey spots, mottles, clouds and occasional hair-lines, usually obscuring most of ground colour, more so at large end. SIZE: (n = 40) 14·6–18·3 × 10·7–12·8 (16·3 × 11·6).

LAYING DATES: Transvaal, June, Oct; Natal, Oct–Dec; E Cape, July–Dec; SW Cape, Apr–Nov (mainly July–Sept).

INCUBATION: 2nd egg hatches 1 day after 1st one. Incubation by ♀ only. ♀ sleeps in nest at night. Period (n = 6) 14–15 ±1 days, also once 13 ± 1 days and 16 days ± 10 h.

DEVELOPMENT AND CARE OF YOUNG: quills emerge at day 4; eyes fully open at day 8; feathers break out (belly first) at days 9–10. First-hatched nestling noticeably larger than 2nd until days 9–10 and tends to remain larger during entire development (Schmidt 1964). Brooded by ♀ for a few days and by night for whole nestling period; young fed by ♀ (mainly) and ♂; both parents (mainly ♀) remove faecal sacs; sac is dropped in flight or from perch c. 100 m away. 10-day-old nestling said to pick up its own faecal sacs and pass them to parent (Skead 1967). Nestling period: (n = 5) 16–17 days, once 15 days ± 5 h, once 18–19 days. Young leave nest unassisted by parents, and return, led by parents, to sleep in nest for 3–9 nights, sometimes with ♀. Young stay near nest for some days, following parents about and begging with quivering wings and skizzing calls, even after they have started to feed for themselves. Young keep with parents for up to 27 days after leaving nest (once including 13 day period after ♀ started to incubate 2nd clutch).

BREEDING SUCCESS/SURVIVAL: 17 young hatched and 14 of them flew from 35 nests with 68 eggs at Strandfontein, Cape Prov. (20% success: Schmidt 1964). From 47 eggs in 22 nests only 9 young survived to fly (Skead 1967). In 6 nestings in 3 years, pair (♂ marked) laid 13 eggs, hatched 10, from which 6 young flew. Nests destroyed by Grey-headed Bush-Shrike *Malaconotus blanchoti* and Pied Starling *Spreo bicolor*. Eggs often robbed by Striped Mouse *Rhabdomys pumilio*. Chicks mutilated by arboreal ants *Crematogaster*. Adult once killed by Brown-hooded Kingfisher *Halcyon albiventris*. Parasitized by Klaas's Cuckoo *Chrysococcyx klaas* (Follett 1990).

References
Lloyd, P. and Craig, A.J.F.K. (1989).
Lotz, C.N. and Nicolson, S.W. (1998).
Martin, R. (1983).
Schmidt, R.K. (1964).
Skead, C.J. (1967).
Vincent, A.W. (1949).

Plate 13
(Opp. p. 206)

Cinnyris neergaardi Grant. Neergaard's Sunbird. Souimanga de Neergaard.

Cinnyris neergaardi Grant, 1908. Bull. Br. Orn. Club, 21, p. 93; Coguno, Inhambane District, Portuguese East Africa.

Range and Status. Endemic resident in SE Mozambique and N Natal, and a local migrant in Natal and extreme S Mozambique. Mozambique, one population common in mixed woodlands mainly between 34° and 35° E from Save R. and 48 km north of its mouth, south to about 24°30′ S and formerly to Xinavane (25°02′S, 32°47′E), and another in coastal sand forest in extreme S. Thought to occur on some off-shore islands (Clancey 1971), at least formerly. Natal, uncommon, from Ndumu (50 km inland) and Maputo Elephant Reserve near Mozambique border, south through Mkuzi and littoral flats around Lake St Lucia to just south of Richards Bay; mainly in coastal sand forest. Range in the south has contracted owing to destruction of natural forests. In the north slash-and-burn agriculture has destroyed its coastal sand forest habitat, though species survives well (probably >5000 birds) in woods further inland (Parker 1999).

Cinnyris neergaardi

Description. ADULT ♂: head, neck, mantle to shorter uppertail-coverts, and scapulars metallic green; long uppertail-coverts deep metallic blue. Tail blackish. Chin to upper breast metallic green like head, bordered by narrow band of metallic violet and below it a narrow metallic blue one; below those a broad red band (c. 13 mm deep) across lower breast; pectoral tufts pale yellow; rest of underparts sooty brown. Lesser and median coverts metallic green; rest of upperwing blackish brown, greater coverts edged green. Axillaries and underwing-coverts dark grey; inner borders of flight feather undersides grey-buff. Bill black; eyes dark brown; legs black. No eclipse plumage (Ginn *et al.* 1989, but *cf.* Clancey 1964). ADULT ♀: upperparts greyish brown, tinged olivaceous on rump. Tail blackish brown; T6 paler brown, especially outer web, T5 and T6 with whitish terminal fringe on inner web. Narrow superciliary stripe buffy white; lores, cheeks and ear-coverts dark grey-brown. Underparts grey-buff, paler and tinged yellow on centre of lower breast and belly. Lesser and median coverts like upperparts; rest of upperwing dark grey-brown, primaries narrowly edged greyish white, secondaries, tertials and greater coverts more broadly edged grey-buff. Axillaries and underwing-coverts dull white; inner borders of flight feather undersides greyish white. Bare parts as adult ♂. SIZE (11 ♂♂, 4 ♀♀): wing, ♂ 52–56 (54·2), ♀ 49–51 (49·8); tail, ♂ 35–38 (36·6), ♀ 30–33 (31·0); bill, ♂ 17–17·5 (17·4), ♀ 15–16 (15·8); tarsus, ♂ 16–17 (16·4), ♀ 14–15·5 (14·7). WEIGHT: ♂ (n = 7) 6·2–7·1 (6·5), ♀ (n = 2) 5·6, 5·9.

IMMATURE: juvenile said to resemble adult ♀.

Field Characters. Length 10–11 cm. A small, short-billed sunbird with a restricted range in S Mozambique and N Zululand; does not overlap with similar-looking Shelley's Sunbird *C. shelleyi* or Southern Double-collared Sunbird *C. chalybea*. ♂ differs from sympatric and somewhat larger Purple-banded Sunbird *C. bifasciata*, which also has black belly, by *yellow pectoral tufts* and *red breast-band*; ♀ differs from ♀ Purple-banded by paler, *unstreaked* underparts.

Voice. Tape-recorded (88, 91–99, F, GIB, GIL). Song a slightly descending chatter of 10–20 individually audible, evenly spaced high, sharp notes given at rates varying from 4 per s to 6 per s; between and before songs gives irregular 'tsip' call notes, 'tsip ... tsip ... tsip ... chichichichichichi-chichi ...'; song may also begin with loud 'chee' (see sonagram in Maclean 1993).

General Habits. Inhabits dense, dry, coastal bush, sand forest, dune forest, and acacia and mixed woodlands inland. At Panda, Mozambique, feeds along with Amethyst Sunbirds *Chalcomitra amethystina* in flowering trees left standing in native villages, and in scrubby bushes and undergrowth along streams running through *Brachystegia* woodland (Clancey 1971). Easily seen foraging at canopy edge; at Ndumu, Natal, seen feeding at low aloes in thorn thicket (Clancey 1964). Subject to local migration in S Mozambique where occurs only in austral summer (Clancey 1971, Harrison *et al.* 1997), perhaps between coast and the interior to the north.

Food. Nectar of flowering shrubs and trees, including *Schotia capitata*, *Syzygium cordatum* and aloes, and small insects and spiders.

Breeding Habits. Solitary nester; probably monogamous and territorial.

NEST: several found, but the only one described was a suspended, rather compact pouch, with side or side-top entrance and porch or hood projecting above it, made of vegetable down, decorated outside with pieces of lichen and insect larval exoskeletons, placed 6 m above ground near top of thorn tree (Skead 1967). Another nest (photo in Steyn 1996) was built into, and well concealed behind, a

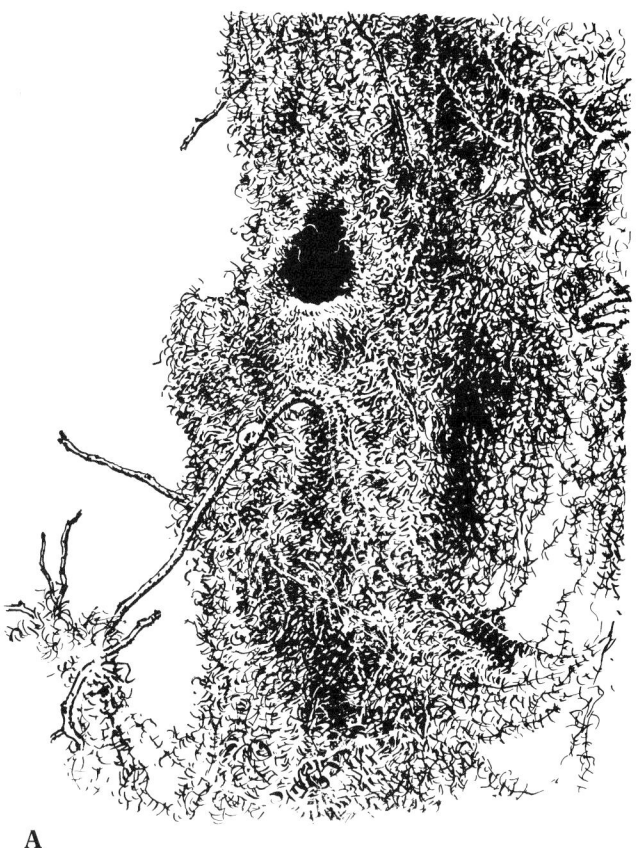

A

mass of *Usnea* lichen festooning a few well-separated slender twigs (**A**). Built by ♀; ♂ accompanies her while she collects down for lining.

EGGS: 2 (n = 8 clutches). Slate-blue, with well dispersed, dark grey spots. SIZE: (n = 14) 15·4–17·5 × 10·4–12·3 (16·2 × 11·1) (Maclean 1993).

LAYING DATES: Mozambique, July, Oct, Nov; Natal, July, Oct, Nov; breeds from Sept to Jan (Ginn *et al.* 1989).

References
Maclean, G.L. (1993).
Parker, V. (1999).
Skead, C.J. (1967).

NECTARINIIDAE

Plate 13
(Opp. p. 206)

Cinnyris stuhlmanni Reichenow. Rwenzori Double-collared Sunbird. Souimanga du Ruwenzori.

Cinnyris stuhlmanni Reichenow, 1893. Orn. Monatsb., 1, p. 61; West Ruwenzori.

Forms a superspecies with *C. afra* and *C. ludovicensis*.

Range and Status. Endemic resident, E Zaïre, SW Uganda, W Rwanda and NW Burundi, in separate populations on Rwenzori Mts (Zaïre/Uganda border), mountains west of L. Edward (Zaïre), Virunga Range (Zaïre/Rwanda border) and Kigezi (Uganda), mountains west of L. Kivu (Zaïre), Nyungwe Forest (Rwanda) and mountains northeast of Bujumbura (Burundi), Itombwe Mts (Zaïre) and Mt Kabobo (Zaïre). Altitudinal range: Rwenzoris 2020–4070 m, Mt Tshiaberimu (west of L. Edward) 2000–2720 m, Virunga 2250–3475 m, Nyungwe 2000–2600 m, Mt Kahuzi (southwest of L. Kivu) 1980–2280 m, Itombwe 1910–3240 m, Mt Kabobo 2450–2500 m.

Description. *C. s. stuhlmanni* (Reichenow): Rwenzoris. ADULT ♂: head, neck, mantle to rump, and scapulars metallic emerald green; uppertail-coverts metallic violet-blue. Tail slightly graduated (T1 8–13 longer than T6), blackish brown, upperside faintly glossed. Chin to upper breast metallic green, concolorous with head; bordered by a narrow (c. 3 mm) metallic violet band; below this a very broad (16–22 mm) scarlet band with basal brown showing as mottling through the red; pectoral tufts yellow; rest of underparts deep olive (darker than in *C. afra*). Lesser and median coverts metallic green; rest of upperwing dark brown, remiges, alula and greater coverts edged olive-brown. Axillaries, underwing-coverts and inner borders on undersides of flight feathers grey. Bill black; eyes dark brown; legs black. ADULT ♀: top of head to cheeks and ear-coverts dark olive, forehead and crown greyer, with faint dusky streaks; rest of upperparts dark olive-green. Tail blackish brown; T6 and distal part of T5 pale brown, these feathers with narrow outer edge and terminal fringe on inner web greyish white. Tail somewhat graduated. Lores, cheeks and ear-coverts dusky olive. Underparts greyish olive, chin to breast darker with dusky feather bases showing as faint barring, centre of belly paler and tinged yellower. Lesser and median coverts as upperparts; rest of upperwing dark brown, edged paler olive brown, narrow primary edges tinged golden. Axillaries and underwing-coverts white, tinged yellowish; inner borders on undersides of flight feathers greyish white. SIZE (8 ♂♂, 3 ♀♀): wing, ♂ 64–69 (67·1), ♀ 58–61 (59·7); tail, ♂ 53·5–58·5 (55·6), ♀ 43–47·5 (45·0); bill to skull, ♂ 25·5–28 (26·6), ♀ 22·5–23·5 (23·0) (Prigogine 1979); tarsus, ♂ (n = 4) 19–20 (19·5), 1 ♀ 19. WEIGHT: ♂ (n = 7) 10–11·5 (10·9), ♀ (n = 4) 9–10·5 (10·0).

C. s. chapini (Prigogine): mountains west of L. Edward (Zaïre). Like *stuhlmanni* but breast band dark red, belly dark grey, and bill slightly shorter. Bill to skull, ♂ (n = 22) 22·5–24·5 (23·6), ♀ (n = 3) 22–23·5 (22·8).

C. s. graueri (Neumann): Kigezi (Uganda) and Virunga (Zaïre, Rwanda). Like *stuhlmanni* but breast band more brick-red, belly more grey-buff; wing av. 1·6 mm shorter and bill much shorter; weighs much less. Size: wing, ♂ (n = 12) 63·5–67·5 (65·5); bill to skull, ♂ (n = 12) 19·5–21 (20·3), ♀ (n = 10) 18–20·5 (19·1). Weight: ♂ (n = 6) 6–8 (7·0), ♀ (n = 7) 5–7 (6·0). ♂ has metallic green areas less golden tinged, more bluish; uppertail-coverts and upper breast band more invaded with violet; larger than *whytei* but bill relatively smaller.

C. s. schubotzi (Reichenow): mountains west of L. Kivu (Zaïre), Nyungwe Forest (Rwanda) and range northeast of Bujumbura (Burundi). ♂ like *chapini*, ♀ with throat and upper breast much darker grey than ♀♀ of other races. Bill length same as *chapini*.

C. s. prigoginei (Macdonald): Marungu Highlands, SE Zaïre. Like *stuhlmanni* but small: ♂ (n = 4) wing 59·5–64 (62·8), tail 46–

Cinnyris stuhlmanni

50 (47·9), bill 20–22 (21·6). Like *C. ludovicensis whytei*, q.v., but ♂ has metallic green areas less golden-tinged and upper breast band and uppertail-coverts more violet-tinged (as in *stuhlmanni*); size like *whytei* but bill longer.

TAXONOMIC NOTE: see note under *C. ludovicensis*. *C. s. prigoginei* may be a race of it, or may be a separate species.

Field Characters. Length 12·5–14 cm. A moorland sunbird of the Albertine Rift mountains. Very like Greater Double-collared Sunbird *C. afra* of South Africa, with similar large body but shorter and straighter bill. Usually identifiable by habitat and altitude, but at lower levels overlaps with Northern Double-collared Sunbird *C. reichenowi*, of which it is a larger version, with longer bill and tail; ♂ has same violet uppertail-coverts and breast-band separating throat from broad red breast-band, and differs only in lighter belly; ♀ is somewhat paler above than ♀ Northern, with grey rather than olive throat.

Voice. Tape-recorded (104, LEM). Song jerky and rather short (2·5–3 s), a bustling jumble of notes, rather harder, less tinny or ringing than other double-collared sunbirds, with 'jit' or 'jeep' calls between songs; another call is a long thin 'tseeeep'.

General Habits. Occurs wherever trees and shrubs are in flower, in upper edge of bamboo zone and in tree heath zone; in vine-clad thickets and *Hagenia* trees. In Nyungwe, rarely below 2200 or 2300 m; mainly in marshes in E and *Erica-Philippia* zone on mountain ridges; occurs locally on edges of forest, and penetrates short way into forest to feed at flowers of *Balthasarea* and *Symphonia* (Dowsett-Lemaire 1990). Solitary and in pairs, feeding at flowers by probing corollae and sometimes piercing their bases (e.g. *Balthasarea* both probed and pierced); flighty and shy. Regularly chased from *Balthasarea* flowers by Blue-headed Sunbird *Cyanomitra alinae* and from *Symphonia* flowers by Purple-breasted Sunbird *C. purpeiventris*.

Food. Very small spiders and insects, and nectar of *Balthasarea schliebenii* (Theaceae), *Symphonia globulifera* (Clusiaceae), *Hypericum revolutum*, *Crotalaria agatiflora*, *Kniphofia princiae*, *Lobelia giberroa* and *L. mildbraedii* (Dowsett-Lemaire 1990).

Breeding Habits. Solitary nester, evidently monogamous and territorial. ♂ displays in front of ♀ by hopping about with wings drooping and quivering and yellow pectoral tufts raised and spread sideways.

NEST: domed oval with side entrance, fixed to ends of tree heath branches and placed inside festooning moss (or lichen?), constructed of the same moss and thus well camouflaged (Chapin 1954). Sited 1–5 m above ground.

EGGS: 1. Dark olive, freckled with darker shade so thickly as to be rather uniform. SIZE (n = 1) 19 × 13.

LAYING DATES: Uganda, Sept, Dec, Jan: Zaïre (Itombwe), (gonads large in Apr–May).

References
Chapin, J.P. (1954).
Clancey, P.A. and Irwin, M.P.S. (1978).
Dowsett-Lemaire, F. (1990).
Prigogine, A. (1979).

Cinnyris ludovicensis (Bocage). Ludwig's Double-collared Sunbird. Souimanga de Ludwig.

Plate 13 (Opp. p. 206)

Nectarinia ludovicensis Barboza du Bocage, 1870. J. Sci. Math. Phys. Nat. Lisboa, 2 (1868), p. 41; Biballa, Mossamedes, Angola.

Forms a superspecies with *C. afra* and *C. stuhlmanni*.

Range and Status. Endemic resident, highlands of W Angola, Marungu Highlands in E Zaïre, and Nyika Plateau, Zambia/Malaŵi. Angola, from central Huila Prov. and Namibe through Huambo to Cuanza Sul and extreme W Malanje at Pungo Andongo (W.R.J. Dean, pers. comm.). Marungu, known from Matafali, Sambwe, Pande, Kasiki and Lufuko R., at 1860–2280 m; common along Lufuko R., though population is endangered since Marungu Highlands are severely affected by timber felling, overgrazing and erosion (Collar and Stuart 1985). Nyika, very common above 1900 m.

Description. *C. l. ludovicensis* Bocage: highlands of W Angola. ADULT ♂: head, neck, mantle to rump, and scapulars metallic green, tinged golden; uppertail-coverts metallic blue, with slight violet tinge (bluer than in *C. afra*). Tail blackish, upperside faintly glossed bluish. Lores black. Chin to upper breast metallic green, concolorous with head; separated by very narrow metallic violet-blue band (*c.* 2 mm deep) from broad scarlet or orange-scarlet band (17–23 mm) across lower breast. Pectoral tufts yellow; rest of underparts deep olive. Lesser and median coverts metallic green; upperwing otherwise dark brown, remiges narrowly edged olive-brown. Axillaries and underwing-coverts pale grey; inner borders of flight feather undersides grey-buff. Bill black; eyes dark brown; legs black. ADULT ♀: upperparts rather dark greyish brown, blacker on long uppertail-coverts. Tail blackish brown; T6 and distal part of T5 paler brown, both with greyish white fringe at tip of inner web. Narrow superciliary stripe pale buff; lores blackish brown; cheeks and ear-coverts greyish brown. Chin grey-buff; throat and upper breast brownish grey, faintly mottled and flecked buffish; rest of underparts brownish grey, centre of lower breast and belly paler and tinged olive-yellow. Upperwing feathers dark grey-brown, edged paler olive-brown, narrow primary edges with more golden tinge. Axillaries and underwing-coverts pale grey; inner borders of flight feather undersides greyish white. Bare parts as adult ♂. SIZE (7 ♂♂, 7 ♀♀): wing, ♂ 64–66 (64·9), ♀ 56–59 (57·6); tail, ♂ 45–54 (50·5), ♀ 35–43 (39·5); bill, ♂ 19–22 (20·5), ♀ 17–19 (18·1); tarsus, ♂ 17–19 (18·1), ♀ 16–18 (17·1).

Cinnyris ludovicensis

C. l. whytei (Benson): Nyika Plateau, N Malaŵi, and adjacent E Zambia. Adult ♂ differs from nominate race in having upper breast band rather wider (*c.* 3 mm) and more steely blue; scarlet band slightly darker and narrower (12–15 mm) with some dusky bars showing through the red. Adult ♀ darker than nominate race on chin and throat; underparts duller, tinged greener (less yellow) medially. Smaller than *ludovicensis* but tail relatively longer. SIZE (10 ♂♂): wing 60–65 (61·5), tail 50–53 (51·1), bill 18–21 (19·3). Weight: ♂ (n = 29) 7·1–10·8 (9·3), ♀ (n = 13) 7·4–9·0 (8·1) (Dowsett-Lemaire 1989b).

TAXONOMIC NOTE: *C. ludovicensis* and *C. stuhlmanni* were removed from *C. afra* and recognized as separate species by

Clancey and Irwin (1978); *C. ludovicensis* comprised 5 races, *graueri*, *chapini*, *prigoginei*, *whytei* and *ludovicensis* and *C. stuhlmanni* was monotypic. Prigogine (1979) examined far more material of equatorial taxa and concluded that *graueri* and *chapini* (and *schubotzi*, not discussed by Clancey and Irwin), are best transferred from the former *C. ludovicensis* to *C. stuhlmanni*. Benson and Prigogine (1981) raised *prigoginei* to the rank of full species. In this montane superspecies complex Sibley and Monroe (1990) recognize only the 3 taxa *C. [l.] ludovicensis*, *C. [l.] prigoginei* and *C. [l.] stuhlmanni*. On the basis of field experience, Dowsett and Dowsett-Lemaire (1993) find that *afra*, *whytei*, *graueri* and *prigoginei* are vocally and behaviourally indistinguishable, so Dowsett and Forbes-Watson (1993) recognize only the single species *C. afra*. We take a middle course through these divergent opinions and follow the taxonomy of Prigogine (1979), whilst acknowledging that *C. ludovicensis* as here constituted may be unnatural.

Field Characters. Length 11·5–12·5 cm. A sunbird with a restricted range in S-central Africa. Plumage similar to Greater Double-collared Sunbird *C. afra* but body much smaller; bill moderately long and curved. Overlaps with similar-sized Miombo Double-collared Sunbird *C. manoensis*; distinguished by *much broader*, *more scarlet* breast-band and *blue* uppertail-coverts. On Nyika Plateau (Malaŵi) meets race *fuelleborni* of Eastern Double-collared Sunbird *C. mediocris* at forest edge, though generally preferring montane scrub and grassland; slightly larger, bill a little shorter and less curved; ♂ distinguished by *blue* uppertail-coverts and narrow breast-band (these are violet in this race of Eastern), broader red breast-band, brown (not yellow-olive) belly; ♀ grey-brown above (♀ Eastern olive-green); voices different (see Dowsett–Lemaire 1988).

Voice. Tape-recorded (86, LEM, STJ). Song of *C. l. whytei* (Nyika) a bustling jumble of high, thin notes, beginning with several short 'tip' notes and breaking into 'pitsy-pitsy' alternating with chattering 'chu-chu-chu' or faster 'chichi-chichi' and more 'tip' notes, with an occasional down-slurred 'pi-chew', lasting 4–6 s, 'ti, ti, titi pitsy-pitsy, chu-chu-chu pitsy-pitsy titititipitsy, pichew, chichichichi ...'; shorter version, usually lasting from 2 to 2·5 s, described as a sharp, dry, nervous strophe, often preceded by one or two sharp 'tsic' sounds; alarm/aggression call a fast series of 'tchep tchep tchep tchep ...' (Dowsett–Lemaire 1988, and see sonograms there).

General Habits. Inhabits gallery forest in Angola, edges of riparian forest along Lufupo R., Marungu Highlands, and thick tangles and cereal fields near streams (Chapin 1954, Dowsett and Prigogine 1974), and on Nyika Plateau scrub and secondary growth 1·5–4 m tall, forest edges (but never enters forest) and flowering plants in open grassland (Dowsett-Lemaire 1988).

Solitary or in pairs; active, mobile and conspicuous. Feeds by probing down tubular corollae of flowers, and by piercing base of corolla to obtain nectar from flower species with corollae longer than 20–25 mm; of the flowers at which sunbird feeds in Malaŵi (see Food), *Tapinanthus sansibarensis*, *T. subulatus*, *Gladiolus natalensis* and *Kniphofia grantii* are habitually pierced and *Tecomaria capensis*, *Aloe mzimbana* and *Kniphofia linearifolia* both pierced and probed. (No strong colour preference: most flower species are white, orange or red, remainder are yellow, pink, purple, blue or greenish.) One individual pierced 45 *Aloe mzimbana* flowers and probed 5. Pierced corollae (e.g. of mistletoe *Englerina*) may be revisited. Main food plants on Nyika Plateau, Malaŵi, at least in breeding season, are *Tecomaria capensis* then *Hypericum revolutum* and the mistletoe *Englerina inaequilatera*.

Food. Nectar and insects. In Malaŵi nectar of: *Bersama abyssinica*, *Halleria lucida*, *Syzygium cordatum*, *Canthium gueinzii*, *Dendrophthoe pendens*, *Englerina inaequilatera*, *Phragmanthera usuiensis*, *Tapinanthus* (2 spp), *Clerodendrum quadrangulatum*, *Crotalaria goetzei*, *Hypericum revolutum*, *Impatiens gomphophylla*, *Leonotis* (3 spp), *Lobelia giberroa*, *Pentas schimperana*, *Tecomaria capensis*, *Aloe mzimbana*, *Borreria dibrachiata*, *Chlorophytum engleri*, *Geniosporum rotundifolium*, *Gladiolus natalensis*, *Kniphofia* (2 spp) and *Protea* (3 spp) (Dowsett-Lemaire 1989b). Young fed on bill-full mass of small insects.

Breeding Habits. Solitary nester, monogamous and territorial. Territory (0·1–0·5 ha in extent) established to encompass plants that provide plentiful nectar or attract insects: mainly (Nyika Plateau, Malaŵi) *Kotschya africana*, *K. recurvifolia*, *Tecomaria*, *Philippia benguelensis*, *Hypericum*, *Protea* spp, *Leonotis* spp and *Buddleia salviifolia*. ♂ remains in same territory all year (without adopting eclipse plumage in non-breeding season). At borders of or outside territory chased by sunbirds *C. mediocris*, *Cyanomitra verticalis*, *Nectarinia kilimensis*, *N. johnstoni* and *N. famosa*, but chases and dominates *Cinnyris venusta*.

2–3 broods per season (*whytei*); one ♀ laid eggs in Apr, June and Aug; another laid 2nd clutch (1 egg) 1–2 days after 1st brood (1 young) left nest; the same ♀ built in same tree in the following year (Dowsett-Lemaire 1988, 1989b).

NEST: in Malaŵi, nests built on outside of *Hagenia abyssinica* tree, suspended from leaf rachis, in *Buddleia salviifolia* tree, in *Philippia benguelensis* tree where successive nests suspended from thin, arched branches, in clump of *Rhus longipes* and in dense foliage of *Anthospermum*; height above ground (n = 9) 1·5–5·0 (1·9) m. Some nests sited above rocks. 2nd and 3rd nests may be made entirely of material made from earlier one(s). Built solely by ♀.

EGGS: 1 (n = 6, Malaŵi)

LAYING DATES: Angola, (large gonads, Oct–Dec); Malaŵi, Feb–Aug, mainly Feb and Apr.

INCUBATION: by ♀ only. ♂ sings from prominent perches, follows ♀ to and away from nest, and chases conspecifics out of territory.

DEVELOPMENT AND CARE OF YOUNG: young brooded by ♀ only; she feeds it, ♂ either feeding it from 1st day also or taking a minor (10%) role. ♂'s role increased after young leaves nest: one ♂ took sole charge of young for at least 19 days; another ♂ was feeding new chick in 2nd nest whilst single young of earlier brood begged nearby (Dowsett-Lemaire 1988).

References
Dowsett-Lemaire, F. (1988, 1989b).

Cinnyris afra (Linnaeus). Greater Double-collared Sunbird. Souimanga à plastron rouge.

Certhia afra Linnaeus, 1766. Syst. Nat., ed. 12, 1, p. 186; Cape of Good Hope.

Forms a superspecies with *C. stuhlmanni* and *C. ludovicensis*.

Range and Status. Endemic resident or partial migrant, South Africa and Swaziland. Locally common in hilly parts of Transvaal, in Escarpment and Soutpansberg, less common in Magaliesberg, Waterberg and Strydpoort Mts and vagrant in E Lowveld (Tarboton *et al.* 1987, Harrison *et al.* 1997); in rest of range common and widespread.

Description. *C. a. saliens* Clancey: E Cape and Natal to Transvaal and W Swaziland. ADULT ♂: head, neck, mantle, back and scapulars metallic green with slight bronzy tinge; rump usually plain olive with a few metallic green feathers; uppertail-coverts metallic violet-blue. Tail blackish, upperside with slight bluish gloss; T6 paler near tip and on outer web. Lores black. Chin to upper breast metallic green, concolorous with head; separated by very narrow metallic violet-blue band (*c.* 2 mm deep) from broad scarlet band (15–20 mm) across lower breast; pectoral tufts yellow; rest of underparts light olive. Lesser coverts metallic green; median coverts dark brown with a few green terminal fringes; rest of upperwing dark brown, primaries narrowly edged pale buff. Axillaries and underwing coverts light olive-grey; inner borders of undersides of flight feathers pale grey-buff. Bill black; eyes brown; legs black. ADULT ♀: crown, cheeks, ear-coverts and rest of upperparts greyish brown, tinged olive; long uppertail-coverts dark grey-brown. Tail dark grey-brown; distal part of T6 and tip of T5 paler brown. Lores dusky brown. Chin to upper breast grey-buff, exposed darker feather centres forming faint dusky barring, slightly paler where bordering moustachial stripes; rest of underparts paler grey-buff, centre of lower breast and belly tinged yellowish. Primaries, secondaries, alula and primary coverts dark grey-brown, narrowly edged buffy brown, palest on primaries; tertials, greater coverts, median coverts and lesser coverts greyish brown. Axillaries pale olive; underwing-coverts whitish with olive tinge; flight feather undersides as in adult ♂. Bare parts as adult ♂. SIZE (10 ♂♂, 10 ♀♀): wing, ♂ 67–70 (68·9), ♀ 60–64 (61·4); tail, ♂ 54–56 (54·4), ♀ 45–48 (45·6); bill, ♂ 30–32 (31·0), ♀ 27–29 (27·7); tarsus, ♂ 18–19 (18·6), ♀ 17–19 (17·8). WEIGHT: ♂ (n = 3) 11·3–12·7 (12·1), ♀ (n = 1) 9·8; (subspecies?) ♂ (n = 10) 11·0–13·1 (12·2), ♀ (n = 6) 8·1–11·3 (9·8) (Maclean 1993).

IMMATURE: juvenile like adult ♀ but slightly darker on chin, throat and breast.

NESTLING: not described.

C. a afra (Linnaeus): SW and S Cape. Adult ♂ typically has red breast band darker than in *saliens*, lower underparts rather darker olive. Slightly smaller; wing (12 ♂♂) 62–68 (65·1) (Clancey and Irwin 1978).

TAXONOMIC NOTE: see note under *C. ludovicensis*.

Field Characters. Length 14–15 cm. A large double-collared sunbird confined to South Africa/Swaziland. Distinguished from the only other double-collared sunbird in its range, Southern Double-collared *C. chalybea*, by large size, long bill and (♂) *very broad* (up to 2·5 cm) *bright red* breast-band; blue of uppertail-coverts and narrow band above red breast-band has strong violet tinge, while in Southerns these tend to be more blue. Caution: young ♂ Greater in first moult has narrow red breast-band at first, broadening with age (Skead 1967). ♀♀ are identical except for size and bill.

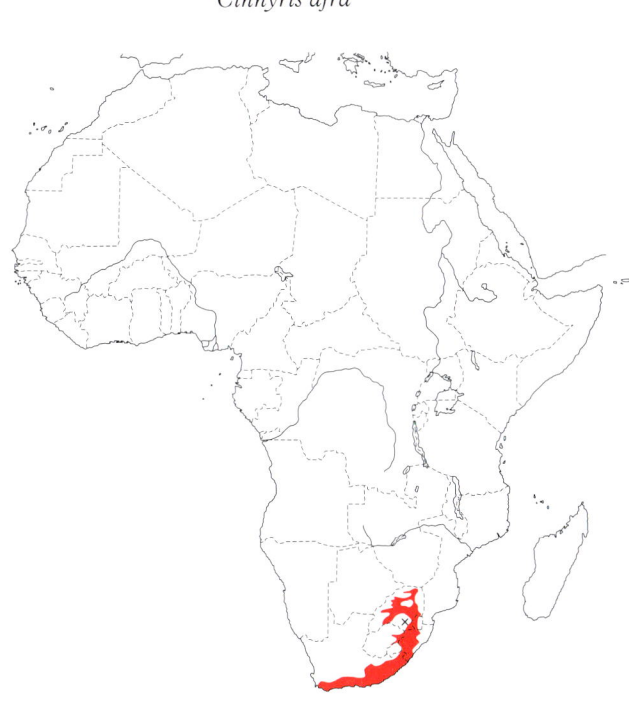

Cinnyris afra

Voice. Tape-recorded (16, 20, 88, 91–99, B, F, STA). Song often starts with a single husky, abrupt 'zhyeet' or 'jee-jee', and is followed by a long (up to 30 s) bustling medley of silvery tinkling and sibilant notes mixed with scratchy ones and intermittent 'chup' calls, sometimes with an interlude of dry 'dzi-dzi-dzi-dzi ...' notes; notes louder than song of Lesser Double-collared Sunbird and more clearly enunciated, less run together (Skead 1967); an irregularly-spaced pause between songs. Both sexes have subsong of quiet continuous warbling, usually given from inside bush. Most typical call a harsh, rasping 'sskert', used casually at any time (Skead 1967); duller versions include 'tsig, tsig' or 'chay, ing – chay, ing'. ♂ gives rapid, high-pitched stuttering 'ch, ch, ch, cher-rrreee' when chasing another bird. Other calls include a long-drawn, whistle-like 'tssweeee', a sharp 'wit-wit-wit' and a loud, clear 'peee'; alarm a series of machine-gun-like stutters, 'ss, ss, ss ...' (Skead 1967).

General Habits. Prefers mesic habitats with tall scrub and trees in hilly and coastal regions. Inhabits open bushveld, coastal grassveld, karroid valley bushveld in coastal plains, bushy dune forest, montane and protea savanna, gardens and parks. Occurs to edges of evergreen forest but does not penetrate it.

Lives singly, in pairs, family parties of 3 or 4, and loose assemblies of 5–10 birds at a flowering tree. In a *Tecoma* hedge, clump of aloes or lone, large food tree such as *Schotia* or *Erythrina*, assembly mixes with other sunbirds, white-eyes *Zosterops* spp. and other nectarivorous and small insectivorous birds; such mixed foraging aggregations subject to constant sunbird chasing, bickering and displacements, though on occasions will tolerate its own species and other sunbirds feeding only a few cm away. Tends to dominate Amethyst *Chalcomitra amethystina* and Lesser Double-collared Sunbirds. Harries another sunbird by dashing at it within tree, perching and lunging, but mainly in short, fast, aerial chase, the 2 twisting, swerving and jinking, the chaser only a few cm behind. Both sometimes return after 3–5 s to patch of flowers where chase started, but usually one bird breaks away and only the other returns. Non-chasing flight fast and strong, with slightly jerky wing action, and if much further than about 5 m, always with sudden changes of direction and often of altitude.

Takes nectar mainly by perching above flower, leaning down and using bill to move tubular corolla to convenient angle, then inserting bill, or perching under flower and reaching up to probe corolla. Sometimes inserts bill for up to 3 s whilst hovering momentarily in front of flower. One bird sometimes visits same flower several times in rapid succession, probing it each time. Examination of a *Cotyledon* flower showed that it still contained plenty of nectar immediately after 2 feeding visits by *C. afra*. Not known to pierce bases of long corollas. However, once seen puncturing unopened flower buds of *Erythrina*. Also once recorded pressing tip of bill against calyx of *Schotia* flowers, hard enough to leave mark (Skead 1967). During drought, a ♀ once punctured numerous grapes in several bunches to suck juice. Once rumoured to puncture pineapples in commercial crops, but pineapple research institute demonstrated that sunbirds were only probing flowers for nectar.

Hunts for insects by systematically probing into leaf axils and poking at curled edges of dead leaves, gleaning leafless dry twigs of deciduous tree or trailing green stems of roses, and probing into leaf clusters of air plant *Ananas* and into dense foliage of *Tecomaria* hedge (as well as probing *Tecomaria* flowers). Seizes small insect in tip of bill, withdraws bill from vegetation, briefly mandibulates insect then tosses it back into throat. Some evidence that bird uses hard perch to adjust position of prey in bill tip. Sometimes, when probing flower probably to drink nectar (e.g. aloe), bird withdraws bill with tip of tongue protruding and small insect stuck to it; such a morsel (fortuitous or deliberate prey?) is mandibulated and eaten. Larger insects, e.g. katydid or moth, seized as encountered, and may be worked against woody perch to remove legs or wings. If insect accidentally dropped, bird instantly flies down after it and may catch it before it falls to ground. If it reaches the ground it is eaten there. Bird hunts for spiders by moving from cobweb to cobweb, whether in open spaces between bushes or under eaves or in confined spaces between cactus spines or in aloe leaf clusters, hovering by each web and plucking spider from centre, or perching by small cobweb mass and winkling spider out. Spiders also removed from rolled up, web-filled leaves and crevices.

♂ persistently chasing a competing/rival ♂ may lead to both birds perching close and singing, then one swaying body rapidly from side to side or bobbing up and down, often silently, while other, dominant, ♂ continues singing but with body erect and bill pointing acutely upward. Immature bird and adult ♀ similarly sometimes pivot in front of dominating adult ♂. In ♂/♂ conflict, subdominant bird once swung forward on perch until hanging upside down. Solitary ♂, with no other sunbirds around, occasionally subsings and sways gently from side to side for up to 30 min (Skead 1967). In another agonistic display by ♂, directed at conspecifics or other birds (e.g. Cape Canary *Serinus canicollis* and Streaky-headed Seed-eater *S. gularis*), ♂ utters rapid, warbling song, swaying and bobbing with pectoral tufts erect and tail fanned; after then chasing other bird away, ♂ has been seen to hang upside down. Rarely, a ♀ chases. The 2 birds in a chase can appear almost oblivious of person or flight obstruction. ♂ Greater Double-collared and Amethyst Sunbirds once fought breast-to-breast with grappled legs and fell, interlocked, to ground.

Bathes in foliage wetted by dew or rain; freely uses garden bird baths. Commonly hovers momentarily in front of window panes, sometimes repeatedly for days or weeks (once over 2 years).

Resident, but a marked fall in abundance in E Cape in Oct–Mar (Craig and Simon 1991, Craig and Hulley 1994), when probably a partial migration to higher ground (Harrison *et al.* 1997).

Food. Nectar, insects and spiders. Nectar of *Erica, Aloe, Protea, Schotia, Erythrina, Cotyledon, Gasteria, Tecomaria, Strelitzia, Canna, Hibiscus*, pineapple, etc. Probes into ripe fig fruits and once into grapes (see above). Small beetles, flies, ants, scale insects, moths and caterpillars commonly taken, and many spiders.

Breeding Habits. Solitary nester; monogamous; pair remained together for 6 years (F. Dowsett-Lemaire, pers. comm.). Defends territory restricted to within a few m of nest. Rears up to 3 broods in quick succession, building new nest for each clutch, often within a few m of previous one. Territorial or courting ♂ sings loudly and repeatedly from prominent perch. If a ♀ arrives, ♂ dashes at her, breaks into warbling song or stuttered calling, sways slowly and tensely from side to side whilst singing or, with pectoral tufts erect, bows toward her. If ♀ warbles, ♂ sings more strongly with head back, bill pointing skyward, and (usually) tail fanned (**A**). He may repeatedly erect or flash pectoral tufts; she may sway from side to side or half-chase him. Variants are: (a) ♂ bobs deeply and slowly and ♀ sways from side to side; (b) ♂ and ♀ stretch body up with plumage ruffled, wings half-open and quivering, and tail turned to one side and fanned, and ♀ sways, then, with ♂ edging closer, falls to hang from perch upside down, rights herself, and hangs upside down again; and (c) ♂ and ♀ stretch up, flutter wings, fan sideways-turned tail, then bow low to each other (**B**) (Skead 1967). ♂ then flies jerkily from bush to bush with body held almost vertically, chest out, head back but bill pointing forward, the jerky flight produced by very brief bursts of fast wing beats (seemingly

more forwards/backwards than up/down) alternating with equally brief wing rests; ♀ may do the same. ♂ commonly pecks ♀'s cloaca, mainly during nest-building and egg-laying stages; e.g. after aerial chase ♀ alights, ♂ alights beside her and pecks or tries to peck her cloaca a few times. Regularly triple-brooded.

NEST: compact and neat, or sometimes rough and untidy, long-oval domed bag (**C**), firmly attached by top and back to substrate vegetation, or (very occasionally) suspended; side-top entrance with porch above it. Constructed of mixture of dead grasses, bark shreds, wool, cottony materials, fur, feathers, rootlets, lichens, small twigs, lumps of leaf-mould, and bits of string and rag, all bound with lengths of spider web; the whole attached to substrate by spider web. Porch made of 60–80 dry grass seed-heads, commonly *Eragrostis* or *Panicum*, 80–90 long, with stems thrust into nest dome fabric and seed-heads projecting outward. Nest decorated outside with dead leaves in size up to 30 × 60, small fruiting or seeding plant heads, and sometimes small sloughed snake skins. Lined with feathers, sometimes with some fur. 3 nests were lined with: 539 feathers; 445 feathers (of poultry, Helmeted Guineafowl *Numida meleagris*, Hadada Ibis *Bostrychia hagedash* and Cape Glossy Starling *Lamprotornis nitens*); and 89 feathers and some Cape Hare *Lepus capensis* fur.

Another nest was composed of: 539 feathers, 538 dry grass blades, 136 dead leaves, 42 twigs up to 25 long, 6 drupes of *Scutia myrtina*, 4 pieces of paper, 4 pieces of lichen (1 *Usnea*), spider webs, and a mass of rotting leaf-mould. A few nests are flimsy; one had only a token covering over the lining; another was built onto *Opuntia* cactus leaf, such that flat, solid leaf itself formed rear wall. Size: (n = ?) height 127–150, width 60–85, from front to back 60–83, entrance diam. 30–38, porch overhang 35–60 (Skead 1967). Nest built into and securely attached to cluster of leaves near top of shrub or tree, particularly *Euclea*, *Acacia* and *Schotia*, at height from ground of 1·25–6·5 m, mostly 2–3·5 m.

Built by ♀. Mate-guarding ♂ perches prominently nearby and sings repeatedly. When she flies off for more material he accompanies her, perching and singing strongly whilst she works. ♂ harasses her during flights to and from nest, constantly chasing and attacking, causing her to deviate often from straightest flight path. Most material gathered quite near nest but some, especially lining materials, several hundred m away. Other birds' nests sometimes used as source of materials; sunbird robbed nest of African Paradise-Flycatchers *Terpsiphone viridis* as fast as it was being built (Skead 1967). Collects spider web with curious weaving motion of head and swaying of body. ♀ once collected very short down from stem of *Rhoicissus cuneiformis* by repeatedly hovering and plucking. Collection round trip typically lasts *c.* 10 min.

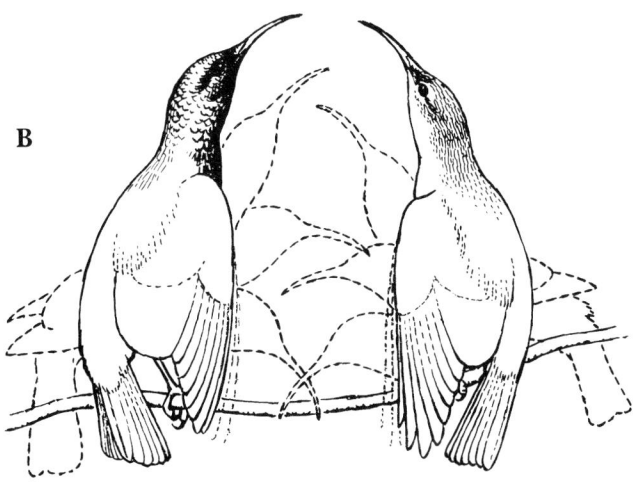

At start of nest, wisps of spider web are twisted around twig, then a few grass blades or a little leaf mould placed on web with stabbing of bill, then more and more web and vegetation added alternately until a large, loosely-formed mass accumulates. ♀ begins to force an entry at one side, using bill and feet to form a deepening indentation. ♀ anchors growing nest by pulling cobweb ends from outside of nest and fixing them to foliage. When nest has become hollow ball, porch is built over entrance, outside decorated, then lining added; during that stage ♀ starts to sleep in nest at night, for about a week before first egg laid. 2 nests took 10 and 24 days to build.

EGGS: 2; laid a day apart. White, greenish white or grey, liberally marked with olive, brown or dark grey spots, mottles and streaks, uniformly over shell so as almost to obliterate ground colour, or occasionally concentrated at broad end. SIZE: (n = 10) 17·0–19·5 × 11·8–13·1 (18·4 × 12·4).

LAYING DATES: Transvaal, July–Feb; Natal, June–Jan; E Cape, all months, mainly Oct-Nov.

INCUBATION: by ♀. ♂ stays nearby, singing and chasing intruders. When ♀ leaves nest from time to time, ♂ follows her closely wherever she goes, chasing her and trying to peck her cloaca (Skead 1967). Period (n = 1) 15 ± 1 days, (n = 1) 14 days.

DEVELOPMENT AND CARE OF YOUNG: young fed mainly by ♀, partly by ♂. ♂ still spends much time singing from favoured, prominent perch near nest and guarding area around. ♀ flies directly to nest with food, feeds young, perhaps broods them briefly, then departs. Even if ♂ returns to near nest with food, he may perch and sing before delivering it. ♂ never broods chicks. ♀ sometimes arrives without evident insect food in bill, but inserts bill into chicks' gullets up to 11 times per visit, presumably delivering nectar. Many caterpillars and moths fed to young. Nestling period 13–15 days (Harebottle 1999). Out of nest, the 2 young stay together and follow parents around, each being fed by them once every 2–4 min; later, they mingle freely with mixed sunbird assemblies at flowering plants, still being fed by parents but beginning to forage for themselves.

BREEDING SUCCESS/SURVIVAL: commonly parasitized by Klaas's Cuckoo *Chrysococcyx klaas*. Pair survived for at least 6 years (see above).

References
Benson, C.W. and Prigogine, A. (1981).
Clancey, P.A. and Irwin, M.P.S. (1978).
Lloyd, P. and Craig, A.J.F.K. (1989).
Skead, C.J. (1954, 1967).

Plate 13 (Opp. p. 206) *Cinnyris reichenowi* Sharpe. Northern Double-collared Sunbird. Souimanga de Preuss.

Cinnyris reichenowi Sharpe, 1891. Ibis, p. 444; Sotik, East Africa.

NOMENCLATURAL NOTE: formerly named *Nectarinia preussi*; not to be confused with Golden-winged Sunbird *Nectarinia reichenowi* (nor with Plain-backed Sunbird *Anthreptes reichenowi*).

Range and Status. Endemic resident and vertical migrant, with distant W and E African populations. Nigeria, abundant on Obudu and Mambilla Plateaux. Bioko, common on Pico Basilié at 1200–2900 m (seasonally down to 400 m) and S Highlands at 800–1800 m (Caldera de Luba, Ureca, Biao L.). Cameroon, common, in several more or less isolated populations: Mt Cameroon at 950 to nearly 3000 m (seasonally down to 600 m), and the commonest bird there at 1700–3000 m in Dec (Stuart 1986); down to only 240 m at Kumba; Dikome Balue at 1200–1700 m, Mt Kupé at 1950–2200 m, Mt Manenguba and environs (Essosong, Muambong) at 1950–2200 m, Bamenda-Banso Highlands at 1200–2200 m, Bamiléké Plateau, Bamenda area, Mt Oku-Kumbo abundant at 2100–2500 m, Tchabal Mbabo at 1450–1650 m, Adamawa Plateau at 800–1600 m (Meiganga and Tello where abundant, Yoko, Mbakaou near Tibati), and hills around Yaoundé (Louette 1980) and Nkolondom; also Bafut-Ngemba Forest Res., Sabga Pass, Mbengi and Tinachong forests (Stuart 1986). Central African Republic, 2 records Sarki (06°57'N, 15°25'E) (Germain and Cornet 1994).

Zaïre, from northwest of L. Albert to Itombwe; up to 2210 m on Mt Aburo, L. Albert, and numerous a little to the south at Djugu; in Rwenzoris common at 1540–1850 m and occurs up to 2460 m; Kivu Volcanoes; Itombwe, very

Cinnyris reichenowi

common and widespread at 1300–2450 m. NW Burundi. Rwanda, throughout W half, numerous below 2200 m, frequent up to 2400 m, local at 2450 m (in Nyungwe). Sudan, very common above 1800 m in Dongatona Mts and Didinga Hills. Uganda, in NE on Mt Morongole at 2100–2700 m, in SW in Rwenzoris, Kigezi and Ankole at 1250–2450 m, and in E on Mt Elgon. Kenya, widespread and locally common at 1700–2800 m (exceptionally up to 3400 m) from Mt Elgon, Elgeyu, Maralal, Mt Uraguess, Mt Kenya and Nyambenis to Nandi, Sotik, Eldama Ravine, Subukia, and Ngong (Britton 1980); numerous at Elgon, Kapenguria, North Nandi, Naro Moru, Limuru and in suburbs of Nairobi (Zimmerman *et al.* 1996).

Description. *C. r. preussi* (Reichenow) (includes '*parvirostris*' and '*genderuensis*'): Nigeria, Cameroon, Bioko and W Central African Republic. ADULT ♂: head, neck, mantle to rump and scapulars metallic green; uppertail-coverts violet or violet-blue. Tail blackish, upperside with faint blue gloss. Lores black. Chin to upper breast metallic green, concolorous with head, and bordered by a line of metallic violet-blue; below it a very broad red band (15–20 mm deep) on lower breast; rest of underparts sooty brown. Lesser and median coverts metallic green; upperwing otherwise blackish, flight feathers, alula, primary coverts and greater coverts edged bright olive-brown. Axillaries and underwing-coverts dusky grey. Bill black; eyes brown; legs black. ADULT ♀: top of head to cheeks and ear-coverts, and rest of upperparts, dark olive-green. Tail blackish, faintly glossed above; distal part of T6 and T5 paler brown. Lores dusky olive. Underparts greyish olive-green, paler and yellower on centre of lower breast and belly. Lesser and median coverts dark olive-green; upperwing otherwise blackish brown, remiges and greater coverts edged olive-brown with rich golden tinge. Axillaries pale yellow; underwing-coverts whitish, tinged yellowish green; inner borders of undersides of flight feathers pale greyish. Bare parts as adult ♂. SIZE (10 ♂♂, 10 ♀♀): wing, ♂ 56–60 (58·3), ♀ 52–55 (53·4); tail, ♂ 40–46 (42·6), ♀ 33–37 (35·8); bill, (Bioko) ♂ (n = 13) 16·0–17·5 (16·9), ♀ (n = 4) 15·0–16·5 (15·9), (Mt Cameroon) ♂ (n = 36) 18–20 (18·1), ♀ (n = 15) 16–19 (17·4), (Adamawa, Cameroon) ♂ (n = 8) 15·5–16·5 (16·0), ♀ (n = 3) 14–15 (14·5); tarsus, ♂ 16–18 (17·0), ♀ 16–17 (16·6).
IMMATURE: juvenile like adult ♀, but slightly darker and browner above; duskier on throat and underparts.
NESTLING: not described.
C. r. reichenowi Sharpe (= '*kikuyuensis*' (Mearns 1915) and includes '*shellyae*'): E Zaïre, SE Sudan, Uganda and Kenya. ♂ has lower underparts deep olive-brown (slightly lighter than in nominate race). Small: (Zaïre, 10 ♂♂, 10 ♀♀, Louette 1980) wing, ♂ 50·5–54 (53·0), ♀ 47·5–50·0 (48·8), bill, ♂ 14–15 (14·7), ♀ 13·5–14·5 (14·0). WEIGHT: Kenya, ♂ (n = 27) 5·2–8·0 (6·7), ♀ (n = 9) 5·3–8·1 (6·5).
TAXONOMIC NOTE: *C. r. preussi* includes several isolated populations, varying clinally from *preussi sens. str.* (Mt Cameroon) with long bill and wide olive-yellow fringes on most wing feathers to '*genderuensis*' (rest of range in Cameroon) with intermediate or short bill and intermediate or narrow fringes. Nigerian populations are intermediate; Bioko birds '*parvirostris*' are like '*genderuensis*' despite their proximity to *preussi sens. str*. All of these birds seem subject to vertical migration and horizontal displacement, rendering interpretation of >200 study specimens complex (Louette 1980). In E Africa bill length increases clinally from west to east by 1 mm, and curvature increases also.
C. reichenowi is so like *C. stuhlmanni*/*C. ludovicensis*/*C. afra* that it could well be placed in their superspecies, were it not for vertical overlap between the lower *C. reichenowi* and the higher *C. stuhlmanni* of 230–750 m on at least 5 mountain ranges (Prigogine 1980b, Table 18).

Field Characters. Length 11–11·5 cm. A double-collared sunbird of medium elevations in equatorial Africa from Cameroon to Kenya, with relatively short bill and tail. ♂ has *violet* uppertail-coverts and narrow breast-band and *broad* (up to 19 mm) *red breast-band*; belly olive-brown. ♀ has plain face (no pale supercilium) and is dull olive below, yellowish only in centre of belly. At lower elevations replaced by (and meets) Olive-bellied Sunbird *C. chloropygia*; ♂ Olive-bellied lacks violet breast-band and uppertail-coverts and has narrower (12·5 mm) red breast-band; ♀ has narrow pale supercilium and is mainly yellow below with streaked breast. In mts of Albertine Rift occurs with Rwenzori Double-collared Sunbird *C. stuhlmanni* in broad (up to 750 m) overlap zone; looks like a miniature version of it, since ♂ Rwenzori also has violet uppertail-coverts and narrow violet breast-band separating throat from broad red breast-band; the only plumage difference is the darker belly of Northern Double-collared. ♀ Northern is slightly darker above than ♀ Rwenzori and throat more olive than grey but otherwise identical; best told by smaller size and shorter tail and bill. Very similar to Eastern Double-collared. ♀ Northern is slightly darker above than ♀ Rwenzori and throat more olive than grey but otherwise identical; best told by smaller size and shorter tail and bill. Very similar to Eastern Double-collared Sunbird *C. mediocris* which it overlaps broadly in W Kenya highlands; has shorter bill and tail, and ♂ is a different green, blue-green rather than golden-green, with violet rather than blue uppertail-coverts (often hard to see) and breast-band, red breast-band broader (18 *vs* 12 mm), belly olive-brown *vs* yellowish olive. ♀♀ almost identical except for bill and tail, but throat of Northern usually paler.

Voice. Tape-recorded (104, B, CHA, LEM). Song a rapid, scrambling jingle of sibilant, buzzy and tinny notes; said to be frequently repeated and sung with great vehemence (Serle 1950); often prefaced by a single 'jit', then a pause of 1 s before song begins; sometimes begins as a jingle and changes into a hard chatter, 'chachachacha ...'; songs often last 3–4 s, but shorter snatches only 2 or even 1 s. Song in Kenya starts with 2 well-separated notes followed by a short, fast sizzling twitter and ending with more distinct lower notes: 'tsip ... tsweet ... sususususrisrisri-tsew-tsutsu-tsu' (Zimmerman *et al.* 1996). Calls a harsh 'jijijijiji' and a hard 'chick' or 'chip-chip'; alarm, a long, thin 'seeeeep'.

General Habits. In Nigeria inhabits patches of montane forest and nearby shrubby grassland; Cameroon, forested ravines and open mountainsides with scattered trees and shrubs, uncommon in tall, mature montane rain-forest and common in open mountainside forest with natural clearings; Bioko, moss forest, lichen forest and montane heathland, and cocoa plantations after breeding; Zaïre, near lower edges of montane forests, and forest clearings in Ituri; in Itombwe, primary and secondary forest and gallery forests; in E Africa occupies edges of highland forest, bamboo, riverine woods, secondary growth and gardens. Forages in forest canopy (Rwanda) and at all levels in clearings but avoids closed understorey.

Singly or in pairs, or flocks outside breeding season. Joins mixed foraging parties. Very active and aggressive all year; flock sometimes stops foraging to chase away other sunbirds, e.g. Olive-bellied and Cameroon Sunbird *Cyanomitra oritis*. Only territorial ♂ sings, usually at a perch, sometimes in flight. Often associates with other birds, e.g. Orange-tufted Sunbird *Cinnyris bouvieri*, Cameroon Sunbird, Black-capped Speirops *Speirops lugubris*, African Blue Flycatcher *Elminia longicauda*, Oriole Finch *Linurgus olivaceus* and Yellow Bishop *Euplectes capensis*.

Cameroon population moults in Mar–Apr, immature birds acquiring breeding plumage by July; metallic feathers appear first on bend of wing, immature non-metallic ones persisting longest on crown, mantle and back. ♀ may moult later than ♂; in tail T1 moulted first not last (Eyckerman and Cuvelier 1982).

After breeding in W Cameroon in the dry season, many birds move by c. 800–1000 m downwards for the rains, amounting to regular invasion in July–Aug in Kumba (altitude 240 m) and even at Victoria on the coast; Kumba vacated in early Sept, when birds thought to go to Mt Cameroon or Mt Manenguba (Serle 1950, 1965). Common around Saxenhof at 520–615 m in mid June to mid Aug, also thought to be non-breeding vertical migrants (Serle 1965).

Food. Nectar and insects including caterpillars and grubs (Serle 1950), hard-shelled insects and hairless caterpillars (Chapin 1954). In Rwanda, flowers probed for nectar include *Alangium*, *Albizia*, *Brillantaisia*, *Englerina woodfordioides*, *Impatiens niamniamensis*, *Mitragyna*, *Oxyanthus troupinii*, *Pavetta*, *Pseudosabicea*, *Psychotria*, *Rhynchostigma*, *Symphonia*, *Syzygium*, *Triumfetta cordifolia* and *Virectaria* (Dowsett-Lemaire 1990). In Nigeria visits grassland flower *Lobelia columnaris* near forest edge (F. Dowsett-Lemaire, pers. comm.).

Breeding Habits. Solitary nester, evidently monogamous and territorial. In Cameroon parties start to disperse in Sept; ♂ occupies territory, defending it by singing and displaying; after nesting in Sept–Feb territory is deserted and birds aggregate into flocks (Serle 1954).

NEST: in Cameroon a suspended, domed oval with side entrance overhung by short porch; bulky and loosely knit, made of dry moss, *Usnea* lichen, strips of grass, slender dry flower stalks and cobweb, without any 'beard' below; lined with thick, felted layer of plant down and a few feathers, contained within inconspicuous layer of fine grass stems. In E Africa nest is globular and generally lacks a porch. In W Africa nest very well concealed in moss-draped branches of forest tree; one in a *Convolvulus* creeper hanging from roof of shed. In E Africa usually situated under a leaf: one attached to *Brillantaisia* leaf petiole, entrance and whole nest hidden behind the leaf; one on underside of terminal leaf high in tall *Markhamia platycalyx* tree (this nest described as a 'tight ball'); one under 2 large leaves high in tall *Polyscias fulva* tree (T. Butynski, pers. comm.). 1·5–15 m above ground (n = 8, av. 7·2 m). Built only by ♀.

EGGS: 2 in Cameroon, 1 in Zaïre and E Africa. Ovate, pointed, shell smooth and glossless. Wine-red or claret-brown, spotted, speckled and streaked with shades of dark red and purple-brown; the markings profuse, indistinct, and tending to merge into ground colour. SIZE (n = 2) 18·3 × 12·0, 17·9 × 11·9 (Serle 1950).

LAYING DATES: Nigeria, (nestlings Nov, building nest Dec, oviducal egg May); Cameroon, Oct–Jan (and replacement clutches probably in Mar–Apr); Rwanda, (indications in Nov and Jan–Feb); Zaïre (Itombwe), Apr–June and Dec (and juveniles Jan, Feb, Sept); E Africa, Region A, Dec, (and occupied nests Sept–Nov), Region B, Aug, Nov, Dec.

References
Clancey, P.A. and Irwin, M.P.S. (1978).
Louette, M. (1980).
Serle, W. (1950).

Plate 14 (Opp. p. 207) *Cinnyris regia* Reichenow. Regal Sunbird. Souimanga royal.

Cinnyris regia Reichenow, 1893. Orn. Monatsb., 1, p. 32; central Africa (restricted to W Ruanda by Gyldenstolpe, 1924, Kongl. Svensk. Vet.-Akad. Handl., 1, p. 94, amended to Ruwenzori by Schouteden, 1937, Rev. Zool. Bot. Africa, 30, p. 166.

Range and Status. Endemic resident, Albertine Rift mountains from Rwenzori to near Baraka and Mt Kabobo (Zaïre, W side of L. Tanganyika) and from SW Uganda to Burundi with an isolate on Mt Kungwe and Mahale (Mahari) Highlands, Tanzania.

Zaïre, numerous in Rwenzoris, at 1850–2770 m (near lower limit of heath zone), recorded up to nearly 3100 m; common on Kivu mountains at 1500–2500 m, up to 2850 m on Mt Niragongo and 3000 m on Mt Karisimbi; widespread in Itombwe Highlands. Uganda, locally common at 1900–3000 m in Rwenzoris and 1550–2500 m in Kigezi. Rwanda, in Nyungwe montane forests numerous above 2200 m or 2300 m, frequent at 2000–2200 m, and local down to 1800 or 1700 m in Tangaro Valley (Dowsett-Lemaire 1990). Burundi, known from Rwegura, Teza, Ijenda and Mt Heha. Tanzania, locally common above 1800 m in Mahale Mts.

Description. *C. r. regia* (Reichenow) (includes '*kivuensis*'): range of species except Mahale Highlands. ADULT ♂: head, neck, mantle to rump, and scapulars metallic emerald green; uppertail-coverts glossy violet. Tail graduated, feathers rather narrow, T1 projecting c. 3 mm; blackish, with faint violet gloss above; T6 paler brown. Lores blackish. Chin, throat and upper breast metallic green like head, bordered by narrow glossy violet upper

Cinnyris regia

breast band; below this, centre of lower breast and belly scarlet or orange-scarlet, sides of breast and flanks yellow; vent and thighs olive-green; undertail-coverts orange-red. Lesser and median coverts metallic green; rest of upperwing blackish brown, remiges broadly edged olive-green; alula, primary coverts and greater coverts fringed olive-green. Axillaries and underwing-coverts dull white, tinged olive; inner borders of flight feather undersides greyish. Bill black; eyes dark brown or hazel; legs black. ADULT ♀: top of head to cheeks and ear-coverts, and rest of upperparts, dark olive-green, upperside faintly glossed; T6 paler brown, T5 and T6 fringed buffy white at tip of inner web. Lores dark grey-brown. Chin and throat dusky olive-green, with variable, rather narrow yellow streaking; breast dusky olive-green; rest of underparts lighter, yellowish olive-green. Lesser and median coverts olive-green; rest of upperwing dark brown, edged yellowish olive. Axillaries pale olive; underwing-coverts whitish, tinged yellow; inner borders of undersides of remiges pale greyish. Bare parts same as adult ♂. SIZE (10 ♂♂, 6 ♀♀): wing, ♂ 52–56 (54·8); ♀ 48–53 (50·2); tail, ♂ 45–51 (49·1), ♀ 32–38 (35); bill, ♂ 19–20 (19·5), ♀ 18–19 (18·3); tarsus, ♂ 16–17 (16·7), ♀ 15–17 (16·4). Birds from L. Edward to Mt Kabobo ('*kivuensis*') have bills av. 2 mm shorter (Chapin 1954). WEIGHT: ♂ (n = 34) 5–8 (7·0), ♀ (n = 12) 6–7 (6·2).

IMMATURE and NESTLING: unknown.

C. r. anderseni (Williams): Mahale Highlands, W Tanzania. ♂ has red underparts less intense, more orange than in nominate race, extending only to upper belly; lower belly olive green. Bill shorter: ♂ (n = 2) 17, 18.

Field Characters. Length, ♂ 11·5 cm, ♀ 10 cm. Endemic to mountains of Albertine Rift, where it overlaps with Stuhlmann's, Northern Double-collared and Rockefeller's Sunbirds *C. stuhlmanni*, *C. reichenowi* and *C. rockefelleri*. Bill (both sexes) shorter than Rockefeller's or Stuhlmann's, longer than Northern Double-collared. ♂ distinguished by *golden-green* head and upperparts (blue-green in Stuhlmann's and Reichenow's), *bright yellow* sides of breast and flanks, *narrow red* breast-band complete just below *violet* band, then *continuing broadly down* centre of breast and belly only; lower belly olive, undertail-coverts *orange-red*. ♀ greener above and yellower below than Stuhlmann's or Northern Double-collared; similar to ♀ Rockefeller's but lacks metallic fringes to mantle feathers and has less distinct eye-stripe.

Voice. Tape-recorded (104, B, C, LEM). Rapid swizzling, tinkling song shorter than other double-collared sunbirds, lasting only 1–2 s; notes span wide frequency (3–8 kHz) (Dowsett-Lemaire 1990). Calls, buzzy 'djer' and 'dzit' and sharper 'dit' or 'didit'; on take-off, dry 'tchic-tchic' or 'thickathic', similar to that of other double-collared sunbirds (Dowsett-Lemaire 1990).

General Habits. Occurs in montane evergreen forest, where forages at all levels in canopy and open understorey, but at lower altitudes keeps more to secondary woody growth at the edges; also in forest glades, secondary scrub and bamboo (Uganda).

Food. Nectar of *Englerina woodfordioides*, *Triumfetta cordifolia*, *Virectaria*, *Albizia*, *Impatiens niamniamensis*, *I. gesneroides*, *I. stuhlmanni*, *Pavetta*, *Oxyanthus troupinii*, *Symphonia*, *Syzygium*, *Canthium*, *Lobelia giberroa*, *Ipomoea* and *Sericostachys* (Dowsett-Lemaire 1990).

Breeding Habits. Solitary nester, evidently monogamous and territorial.

NEST: suspended, domed pouch, oval, rather small, securely attached to tip of fern frond or tip of stem of bamboo, *Syzygium* or *Hagenia*. Made of plant fibres, shreds of grass, tendrils and moss, with thick lining of plant down, pappus and feathers; sometimes bound with spider web. One built from moss, remains of *Panicum* inflorescences, half-decomposed leaves of ?*Dracaena*, and tree-fern roots, and lined with feathery down of composite *Gynura vitellina* and hairs of fern *Cyathea*, with a few turaco feathers mixed in. Another lined with down from *Sericostachys*. Nest not suspended by any narrow neck; instead, top of dome is built around branch or twig (T. Butynski, pers. comm.). Ext. height 160, ext. width 90 × 85, entrance diam. 35; from centre of entrance to top of nest, 70. Sited 4 m high above forest path, or out on a small limb 4 m up in wattle copse, or c. 25 cm from tip of leaf 125 cm long, 5 m up in tall *Polyscias fulva* tree, or 2·5 m high near top of shrub in dense *Mimulopsis* grove (T. Butynski, pers. comm.).

EGGS: 1. Clear grey, with dark stippling making broad end appear dark brownish grey. Size: (n = 1) 17·1 × 12·2.

LAYING DATES: Zaïre (Itombwe), Apr–May; Uganda, (occupied nests Oct, Apr, May, nestlings June: T. Butynski, pers. comm.); Rwanda, (nest-building Dec).

Reference
Dowsett-Lemaire, F. (1990).

Plate 14
(Opp. p. 207)

Cinnyris rockefelleri Chapin. Rockefeller's Sunbird. Souimanga de Rockefeller.

Cinnyris rockefelleri Chapin, 1932. Amer. Mus. Novit., 570, p. 16; Mt Kandashomwa, west of Ruzizi Valley, Belgian Congo.

Range and Status. Endemic resident, afroalpine zone of Albertine Rift mountains, E Zaïre. Mt Kahuzi and Mt Kabushwa (west of L. Kivu, at 3000–3200 m). N Itombwe highland plateau at 2050–3300 m; known from Mt Kandashomwa (03°04′S, 28°59′E) and, very near it, L. Lungwe, Mt Mohi and Muusi, also (35 km southwest of Kandashomwa) Ngussa and Nzombe. Sight record of a ♂ on Mt Karisimbi, north of L. Kivu, Zaïre (A. Forbes-Watson *in* Collar and Stuart 1985) and another of a ♂ in bamboo zone west of Nshili, 2400–2500 m, Rwanda (J.P. Vande weghe *in* Dowsett-Lemaire 1990) (not found again in 3 days' searching at Nshili: Dowsett-Lemaire 1990); sight record requiring confirmation of 2 birds at Rwegura, Burundi (Gaugris *et al.* 1981). Locally common, but entire range covers area of only about 250 km², partly in Kahuzi-Biega Nat. Park. 10–20 years ago the species was not thought to be at risk unless forest destruction were to accelerate; the present status of these high altitude forests is unknown.

Description. ADULT ♂: head, neck, mantle to rump and scapulars metallic green; uppertail-coverts glossy violet-blue. Tail blackish above with faint gloss. Chin to upper breast metallic green like head, bordered by narrow glossy violet band; below this, lower breast and upper belly bright scarlet; small pectoral tufts yellow; lower belly, flanks and thighs deep olive-green; undertail-coverts scarlet. Lesser and median coverts metallic green; rest of upperwing blackish brown, remiges edged olive-green; secondaries with protruding tip to outer vanes. Underwing-coverts whitish, tinged olive. Bill black; eyes very dark brown; legs and feet black. No eclipse plumage known. ADULT ♀: forehead and crown mottled dark olive and black (feathers blackish brown, margined with olive-green); distinct, pale eyestripe from bill to 3 mm behind eye; rest of upperparts dark olive-green; lores black; chin and throat bright yellowish olive-green, breast duller and darker olive-green, belly and undertail-coverts bright olive-green with yellowish tinge; primaries and secondaries dark brown, secondaries with protruding tip to outer vanes, upperwing coverts olive-green; underwing coverts white with yellowish edges, undersides of remiges grey-brown with pale inner edges (R.A. Cheke, pers. comm.). SIZE: wing, ♂ (n = 10) 56–58 (57·2), ♀ (n = 3) 51–52 (51·3); tail, ♂ (n = 10) 36–45 (39·5), ♀ (n = 2) 30, 34; bill, ♂ (n = 9) 20–23·5 (22·4), ♀ (n = 2) 22, 22; tarsus, ♂ (n = 9) 18–20 (18·8), ♀ (n = 2) 17, 17 (R.A. Cheke, pers. comm.). WEIGHT: ♂ (n = 3) 5·2–5·5 (5·3).
IMMATURE and NESTLING: unknown.

Cinnyris rockefelleri

Field Characters. Length 12 cm. Rare, known only from a few Albertine Rift mountains. Co-exists with Regal Sunbird *C. regia*; bill longer (both sexes); ♂ has same narrow violet breast-band but below that *entirely scarlet*, including undertail-coverts, except for yellow pectoral tufts and olive lower belly; ♀ similar to ♀ Regal but has metallic fringes to mantle feathers and more distinct eye-stripe.

Voice. Unknown.

Habits. Inhabits primary forest, gallery forest and thickets along streams in bamboo forest, at 2050–3300 m; also in heather. Solitary or in pairs, noisy and active, foraging at mid levels in trees. Joins mixed-species foraging flocks. Gonads enlarged in May.

Reference
Collar, N.J. and Stuart, S.N. (1985).

Plate 14
(Opp. p. 207)

Cinnyris mediocris Shelley. Eastern Double-collared Sunbird. Souimanga du Kilimandjaro.

Cinnyris mediocris Shelley, 1885. Proc. Zool. Soc. London, p. 228; Kilimanjaro.

Forms a superspecies with *C. moreaui* and *C. loveridgei*.

Range and Status. Endemic resident, E and SE African mountains between L. Turkana and R. Zambezi. In E Africa, common at 1800–3700 m, rather less common down to 1400 m, and wanders or migrates down to 900 m. Kenya, Mt Kulal, Mt Nyiru, Ndoto Mts, Mt Uraguess, common in Cheranganis, Mau, Aberdares and on Mt Kenya, once on

Cinnyris mediocris

Mt Elgon; also in Subugu Hills, Chyulu Range and Taita Hills. Tanzania, Loliondo area south to Crater Highlands and Mt Hanang, and east to Meru, Kilimanjaro, N and S Pares, W and E Usambaras; also Mahari Mts and towards Karobwa in W, and in S in Iringa Hills, Mbeya, Kipengere Range and hills in W Ruvuma Province (N.E. Baker, pers. comm.). Zambia, Mafinga Mts and Nyika Plateau. Malaŵi, mainly above 1550 m, on Nyika Plateau, west of Rift south to Mwanza/Ntcheu district boundary (but not on Dzalanyama Mts nor Mt Ntchisi which are too low), and east of Rift only on Mulanje, Zomba and Malosa Mts and the summit of Chiradzulu; recorded in Zomba and Dedza and down to 700 m on Mulanje. NW Mozambique, known from Unango Mts, Mt Namuli and Mt Chiperone.

Density of 2 territorial pairs per ha (larger patches of forest, Nyika Plateau, Malaŵi: Dowsett-Lemaire 1983).

Description. *C. m. mediocris* Shelley: Kenya and Tanzania southeast to N Pare Mts. ADULT ♂: head, neck, mantle to rump, and scapulars metallic green with golden tinge; uppertail-coverts blue. Tail slightly graduated, blackish brown, faintly glossy above; T6 and distal part of T5 pale grey-brown, both with whitish terminal fringe to inner web, T6 also with narrow whitish outer edge. Lores black. Chin to upper breast metallic green, concolorous with head, and bordered by line of blue and violet feathers; below this a scarlet band across lower breast (10–13 mm deep at sides, often narrower in centre); pectoral tufts yellow; rest of underparts light olive. Lesser and median coverts metallic green; upperwing otherwise dark grey-brown, flight feathers and greater coverts edged buffy brown. Axillaries pale olive; underwing-coverts whitish; inner borders of undersides of flight feathers pale grey. Bill black; eyes dark brown or black; legs black. ADULT ♀: top of head to cheeks and ear-coverts dark olive-green; rest of upperparts dark olive-green. Tail slightly graduated, grey-brown, faintly glossy above; T5 and T6 with buffy white fringe at tip of inner web, T6 also with narrow buffy white outer edge. Lores dark olive-grey. Chin and throat olive-grey flecked yellowish; rest of underparts olive-yellow, greyer on breast and flanks. Lesser and median coverts dark olive; rest of upperwing dark brown, flight feathers and greater coverts fringed olive-green. Axillaries olive-yellow; underwing-coverts white, tinged yellow; underside of flight feathers same as adult ♂. Bare parts like adult ♂. SIZE (10 ♂♂, 10 ♀♀): wing, ♂ 53–57 (55.7), ♀ 48–52 (50.3); tail, ♂ 42–50 (46.2), ♀ 34–41 (37.8); bill to feathers, ♂ 19–21 (20.4), ♀ 17–19 (18.1); tarsus, ♂ 17–19 (17.9), ♀ 16–17 (16.6). WEIGHT: (Udzungwa Mts, Tanzania) ♂ (n = 54) 7.1–9.6 (8.7), ♀ (n = 38) 6.4–8.9 (7.6) (L.A. Hansen, pers. comm.); also ♂ (n = 33) 5.2–8.3 (6.9), ♀ (n = 17) 5.2–7.5 (6.6).

IMMATURE: juvenile similar to adult ♀, but chin and throat darker and greyer, underparts darker, more greenish.

NESTLING: not described.

C. m. usambarica (Grote): NE Tanzania: S Pare Mts and Usambaras (and Chyulu and Taita Hills in Kenya? – van Someren 1939, Zimmerman *et al.* 1996). ♂ differs from nominate race in having uppertail-coverts violet-blue, and metallic green areas tinged bronzy; scarlet band narrower (6–10 deep), and lower breast below band dark greyish olive; lower underparts darker olive-green. ♀ slightly greener above than nominate race; brighter, more greenish below, throat more distinctly streaked.

C. m. fuelleborni (Reichenow): S Tanzania to N Malaŵi (Karonga) and NE Zambia. ♂ like *usambarica*, but red band much broader (12–16 deep) and lower underparts deep golden-olive. ♀ darker than *usambarica*, underparts generally yellowish olive-green with greyer chin to breast and flanks; wing edgings more yellowish olive. Bill to skull: ♂ (n = 11) 21.0–24.5 (23.4), ♀ (n = 9) 20.0–22.5 (21.2). WEIGHT: ♂ (n = 42) 7.6–9.8 (8.8), ♀ (n = 46) 6.9–9.2 (7.9) (Dowsett-Lemaire 1989b).

C. m. bensoni (Williams): Malaŵi (except extreme N) and N Mozambique. Like *fuelleborni*, but ♂ darker, more brownish olive on lower underparts.

Field Characters. Length 11–12 cm. A common sunbird of E highlands from Kenya to Mozambique. Does not live on same mountains as Loveridge's and Moreau's Sunbirds *C. loveridgei* and *C. moreaui*, and in most places is the only double-collared sunbird present. Range overlaps that of Northern Double-collared Sunbird *C. reichenowi* above 1800 m in Kenya, but only rarely (Kieni Forest, Aberdares) are the 2 species found together (Zimmerman *et al.* 1996). Bill and tail longer than in Northern (both sexes); ♂ of sympatric race has *golden green* (not blue-green) *head and upperparts, blue* (not violet) *uppertail-coverts and breast-band*, narrower red breast-band and paler, yellowish olive belly. ♀♀ very similar but Eastern rather darker. Range abuts that of Olive-bellied Sunbird *C. chloropygia* in W Kenya but the 2 are altitudinally separated, Olive-bellied below 1550 m, Eastern above 1850 m. ♂ Olive-bellied has no contrastingly-coloured uppertail-coverts, only a trace of a blue breast-band, and darker belly; ♀ is yellowish below. On the Nyika Plateau, Malaŵi, meets Ludwig's Sunbird *C. ludovicensis* at forest edges; smaller than Ludwig's but bill longer and more curved; ♂ of local race *fuelleborni* has same golden green upperparts, but in reversal from northern populations, uppertail-coverts and breast-band are violet (blue in Ludwig's); belly yellowish olive (not plain brown), red breast-band narrower; ♀ Eastern has olive-yellow (not grey-brown) belly. In Malaŵi also just meets Miombo Double-collared Sunbird *C. manoensis*. ♂ Miombo has uppertail-coverts olive with green tips (not violet); ♀ Miombo lacks any olive tones to plumage, and is dull grey-brown, paler below, with hint of pale yellow on belly.

Voice. Tape-recorded (86, 102, B, C, F, HOR, KEI, LEM, PARK). Song a rapid, forceful, insistent, bustling medley of 4 main sounds, a high piercing 'teece', low, brief 'ja', grating 'tsrrt' and faster 'peety-peety-peety': 'peety-ja-teece-ja-teece-ja tsrrt-tsrrt-ja-peety-peety-peety-ja', or 'teece-ja-teece-ja-teece-ja-peety-peety-peety-ja-tsrrt-tsrrt-ja-teece-teece-ja-peety-peety-peety'. Songs last 1–3 s, often with brief (as little as 5 s) pause between songs. The overall effect is thin, sibilant and tinkling. In some songs notes are so close together they become an indecipherable chatter, 'tiddlyiddlyiddlyiddly ...'. Songs may be prefaced by irregularly spaced call notes, 'tseet ... tsee ... too-tiddlyid-dlyiddly ...'. Calls include a long-drawn high, thin 'seeeeee', 'tsip' and a harsh 'jack' or 'jick'.

General Habits. In E Africa inhabits edges of montane evergreen forest, bamboo forest and tree-heaths; also open alpine grassland, fern-covered hillsides and gardens. Favours areas with plenty of red-hot pokers *Kniphofia*. On Nyika Plateau, Malaŵi, occurs widely in all types of forests including small patches, and forages from canopy to lower stratum of understorey and at edges of forest. Takes nectar from variety of flowers by probing from hovering or perching stance, and regularly takes nectar from mistletoes *Tapinanthus sansibarensis* and *T. subulatus* (Loranthaceae) by piercing base of the corollas, which are longer than birds' bills.

Sedentary, but in Kenya descends to lower elevations in cool weather. In Malaŵi (Nyika) disperses up to 750 m (Dowsett 1985); few altitudinal movements noticed at Zomba or Mulanje (F. Dowsett-Lemaire, pers. comm.).

Food. Nectar and insects (Dowsett-Lemaire 1983). On Nyika Plateau, N Malaŵi, feeds at flowers of *Bersama abyssinica*, *Bridelia brideliifolia*, *Halleria lucida*, *Syzygium cordatum*, *Canthium gueinzii*, *Clematis brachiata*, *Clerodendrum johnstonii*, *C. quadrangulatum*, *Ipomoea involucrata*, *Brillantaisia ulugurica*, *Hypoestes aristata*, *Pavella subumbellata*, *Dendrophthoe pendens*, *Englerina inaequilatera*, *Phragmanthera usuiensis*, *Tapinanthus*, *Crotalaria goetzei*, *Hypericum revolutum*, *Impatiens gomphophylla*, *Leonotis decadonta*, *L. mollissima*, *Lobelia giberroa*, *Pentas schimperana* and *Tecomaria capensis*. About half of these are insect-pollinated open-cup or short-tubular flowers (Dowsett-Lemaire 1989b). Feeds at *Kniphofia* flowers (Kenya).

Breeding Habits. Solitary nester, monogamous. Strongly territorial for much of year. Does not organize territory around particular flowering plants, but defends a small forest island or a small patch of continuous forest (Dowsett-Lemaire 1989b). Consistently dominates slightly heavier Ludwig's Double-collared Sunbird *C. ludovicensis*, but gives way to Green-headed Sunbird *Cyanomitra verticalis* (Malaŵi: Dowsett-Lemaire 1989b). On Nyika, Malaŵi, pair occupies forest patch of just under 0·2 ha, maintaining it as nesting and feeding territory, but sometimes crossing open land to adjacent forest patch to feed, usually being chased away again by territorial pair there. Pair can remain in territory for at least 6 breeding seasons (Dowsett 1985).

NEST: suspended, domed pouch with side entrance, made of fine vegetation and lined with soft white seed down. Built by ♀. One sited at tip of hanging conifer branch, 2 m above ground at edge of an overgrown conifer plantation (Williams 1978).

EGGS: 1–2. Pale greenish white, covered with finely fused greyish brown marks, almost obliterating ground colour. SIZE: (n = 2) 16·5 × 12·0, 17·5 × 12·0.

LAYING DATES: E Africa, Region A, Feb–May and Sept–Nov; Region D, May–June, Sept–Jan, mainly Sept–Dec (14 out of 19 clutches), i.e. before and in short rains (all long-rainy-season records, May–June, are from below 2000 m: Brown and Britton 1980). Zambia, (hypertrophied gonads, May). Malaŵi, Apr–Oct, mainly July–Aug (26 out of 41 clutches: season thought to be determined by large number of forest edge shrubs such as *Buddleia*, *Gnidia*, *Hypericum* and *Kotschya* that flower and attract insect food in the dry season: Dowsett and Dowsett-Lemaire 1984).

INCUBATION: by ♀; at start of incubation period, ♂ sings persistently from nearby perch (one 8 m away). ♂ once chased away ♀ Variable Sunbird *C. venusta* (but ignored singing ♂ Variable); however, ♀ Variable persisted and managed to 'accidentally parasitize' the Eastern Double-collared Sunbirds' nest by adding an egg (so that clutch was eventually of 3 eggs: Williams 1978).

BREEDING SUCCESS/SURVIVAL: on Nyika, Malaŵi, can live at least 6·1 years; av. annual mortality 51–57% (♂♂) and 76–79% (♀♀) (Dowsett 1985).

References
Benson, C.W. and Prigogine, A. (1981).
Dowsett-Lemaire, F. (1983).

Plate 14
(Opp. p. 207)

Cinnyris moreaui (Sclater). Moreau's Sunbird. Souimanga de Moreau.

Cinnyris mediocris moreaui W.L. Sclater, 1933. Ibis, p. 214; Maskati, Nguru Range, east-central Tanganyika.

Forms a superspecies with *C. mediocris* and *C. loveridgei*.

Range and Status. Endemic resident, S Eastern Arc Mts, Tanzania. Known from Ngurus, Ukagurus, Mt Kiboriani (west of Ukagurus), Uvidundas and Mwanihana Forest, E escarpment of Udzungwa Mts. Common in Mwanihana Forest at 1500–1850 m (Stuart *et al.* 1987, Evans and Anderson 1993a); abundant in some, if not all, parts of range (Stuart and van der Willigen 1980). Abundant above 1300 m in Mamiwa Kisara Forest Res., Ukagurus at 06°21'S, 38°47'E (Evans and Anderson 1993b).

High density indicated by capture rate of 27–28 birds per 10,000 mistnet m per h. In no immediate danger in Ukagurus where, however, the remaining 10,000 ha of

Cinnyris moreaui

natural forest are under intensive pressure from felling for firewood and clearance for agriculture; status of Near-Threatened is merited (Evans and Anderson 1993b).

Description. ADULT: ♂: head, neck, mantle, back and scapulars bronzy metallic green; rump non-metallic olive-green; uppertail-coverts glossy violet. Tail slightly graduated, blackish brown, upperside faintly glossed; T6 paler brown. Chin to upper breast bronzy green like head, bordered by very narrow band (c. 2 mm deep), glossy blue with violet reflections; below this a scarlet band (10–15 mm) on centre of lower breast, but sides of breast yellow; rest of underparts deep olive-green. Lesser and median coverts bronzy green; rest of upperwing blackish brown, remiges, alula, primary coverts and greater coverts edged olive-yellow. Axillaries and underwing-coverts pale olive-yellow; inner borders of flight feather undersides greyish. Bill black; eyes dark brown; legs black. No eclipse plumage. ADULT: ♀: top and sides of head dark greyish olive, duskier on lores; sides of neck and rest of upperparts dark olive-green. Tail blackish brown; T6 paler brown; feathers narrowly edged olive-yellow when fresh. Underparts yellowish olive-green, chin and throat greyer, breast and flanks darker and greener. Lesser and median coverts like upperparts; otherwise, upperwing blackish brown, remiges and greater coverts edged yellowish olive. Axillaries, underwing-coverts and undersides of flight feathers as adult ♂. Bare parts as adult ♂. SIZE (20 ♂♂, 15 ♀♀, Evans and Anderson 1993): wing, ♂ 53–59 (56.0), ♀ 50–53 (51.5); tail, ♂ 29–43 (39.2), ♀ 32–36 (34.1); bill, ♂ 19.7–26.4 (23.7), ♀ 19.2–24.8 (22.1); tarsus, ♂ 17.6–20.0 (18.0), ♀ 16.9–19.0 (17.8). Also (4 ♂♂, 5 ♀♀): wing, ♂ 55–58 (56), ♀ 50–54 (51.4); tail, ♂ 37–40 (38.8), ♀ 32–36 (34); bill, ♂ 23–25 (24), ♀ 20–22 (20.8); tarsus, ♂ 17–18 (17.8), ♀ 17–18 (17.3). WEIGHT: ♂ (n = 20) 7.6–9.6 (8.3), ♀ (n = 15) 6.8–9.1 (8.1) (Evans and Anderson 1993); ♂ (n = 36) 6.8–9.6 (8.6), ♀ (n = 28) 5.1–10.4 (7.9) (Ukagurus, N.E. Baker, pers. comm.).

IMMATURE: juvenile ♂: head, nape, mantle, rump and uppertail-coverts dusky greyish olive; chin, throat, upper breast, flanks and undertail-coverts dusky yellowish olive, brighter and yellower on lower breast and middle of belly; wings dusky brown, feathers margined warm olive-brown; tail feathers blackish with narrow olive margins. Immature ♂ like adult ♂, with a few olive-green feathers amongst metallic ones on crown. Immature ♀ like adult ♀, but upperparts with less pronounced metallic green fringes to feathers and underparts duller. First adult ♂ plumage differs from older ♂♂ in red breast patch being duller and smaller, metallic fringes to feathers on upperparts narrower (upperparts look scaly with non-metallic feather bases showing), and belly duskier (Williams 1950).

TAXONOMIC NOTE: intermediate between *C. mediocris* and *C. loveridgei* in plumage, and can be regarded as a subspecies of either of them (Hall and Moreau 1970). *C. moreaui* and *C. loveridgei* are each endemic to an adjacent mountain and together they sunder the range of *C. mediocris*. Stuart and van der Willigen (1980) present evidence that *moreaui* originated by hybridization between *mediocris* and *loveridgei* and is now a stabilized hybrid species. Validity of form in Udzungwas has been questioned with finding that *C. mediocris* ♂♂ netted at Mufindi have plumage variation 'well into the concept of *C. moreaui*' (Jensen and Brøgger-Jensen 1992).

Field Characters. Length 11.5 cm. ♂ like Eastern Double-collared Sunbird *C. mediocris* but with broad yellow pectoral tufts which extend towards centre of breast, reducing breast-band to a central red patch; belly yellowish olive. ♀ olive-yellow below, throat greyer. Easily identified by range – it is the only double-collared sunbird on the mountains where it occurs.

Voice. Not tape-recorded. Song appears to be identical to that of Eastern Double-collared Sunbird (Stuart and van der Willigen 1980); described as 'typical of sunbirds – a rapid, almost musical cascade of high-pitched notes' (Evans and Anderson 1993b).

General Habits. Inhabits interior of primary montane forest, typically in canopy, and in bushes at sides of roads through forest. Sometimes joins mixed-species flocks.

Breeding Habits. Solitary nester, evidently monogamous and territorial.

NEST: one described as a small, domed purse, made entirely of dry grasses, without embellishment, attached to upright stem of small shrub, 2 m above ground, with entrance near top (Fuggles-Couchman 1986); others were domed construction suspended in low vegetation beside dirt road through forest; one 3 m up, suspended from bracken overhanging a cutting, another 10 m up in tall tree. ♀, building one nest, repeatedly gathered silk at nearby insects' nest of leaves bound together with silk (Evans and Anderson 1993b).

LAYING DATES: Ukagurus, Aug–Sept (occupied nests and other indications), and Dec (2 half-grown nestlings, late Dec.)

References
Evans, T.D. and Anderson, G.Q.A. (1993b).
Stuart, S.N. and van der Willigen, T. (1980).
Williams, J.G. (1950).

Plate 14 (Opp. p. 207)

Cinnyris loveridgei Hartert. Loveridge's Sunbird. Souimanga de Loveridge.

Cinnyris loveridgei Hartert, 1922. Bull. Br. Orn. Club, 42, p. 49; Uluguru Mts, Tanganyika.

Forms a superspecies with *C. mediocris* and *C. moreaui*.

Range and Status. Endemic resident, Uluguru Mts, Tanzania. Fairly common around Bunduki at 800–2000 m (Turner 1977); the most abundant species there (in 1963: S. Keith, pers. comm.).

Description. ADULT: ♂: head, neck, mantle to rump and scapulars metallic green; uppertail-coverts violet. Tail slightly graduated, blackish, faintly glossed above. Lores black. Chin to upper breast metallic green like head, bordered below by narrow band of metallic violet *c*. 3 mm deep; below this, centre of lower breast orange-red, merging with golden olive at sides and on flanks, belly and undertail coverts; pectoral tufts yellow. Lesser and median coverts metallic green; upperwing otherwise blackish, secondaries, tertials, primary coverts and greater coverts edged golden olive, forming distinct wing panel. Axillaries olive-yellow; underwing-coverts pale olive-grey; inner border of underside of flight feathers greyish. Bill black; eyes brown or dark brown; legs black. ADULT: ♀: top of head to cheeks and ear-coverts dark grey, tinged olive; sides of neck and rest of upperparts dark olive-green, tinged grey. Tail blackish brown; T6 paler brown. Lores blackish. Chin greyish olive, throat dusky olive-green, merging into olive-yellow on rest of underparts. Lesser and median coverts dark olive-green; upperwing otherwise dark grey brown, primaries narrowly edged buffish, secondaries and tertials broadly edged golden olive, primary coverts and secondary coverts fringed olive-green. Axillaries and underwing-coverts whitish, tinged yellow; inner borders on undersides of flight feathers pale greyish. Bare parts as adult ♂. SIZE: wing, ♂ (n = 9) 55–61 (58), ♀ (n = 3) 55–57 (56); tail, ♂ (n = 9) 38–43 (40·1); ♀ (n = 3) 35–37 (36); bill, ♂ (n = 7) 25–27 (25·9), ♀ (n = 3) 23–24 (23·7); tarsus, ♂ (n = 7) 19–21 (19·6), ♀ (n = 3) 18–19 (18·3). WEIGHT: ♂ (n = 6) 9–12 (10·8), ♀ (n = 8) 8·5–10 (9·3).

IMMATURE and NESTLING: unknown.

Field Characters. Length 11·5–12·5 cm. Endemic to Uluguru Mts of Tanzania, where it is the only double-collared sunbird. Bill long and curved, like Eastern Double-collared Sunbird *C. mediocris*. ♂ has head, throat and upperparts golden green to blue-green, uppertail-coverts and breast-band violet; broad yellow pectoral tufts, orange-red patch in centre of belly, rest of underparts olive-yellow; ♀ has top and sides of head olive-grey, rest of upperparts greenish olive with few iridescent green feathers; below, olive-yellow, greyer on throat.

Cinnyris loveridgei

Voice. Tape-recorded (B, F, CHA, STJ). Swizzling song similar to that of Eastern Double-collared Sunbird but faster and almost formless, no individual notes discernible. Call a harsh 'pzit'.

General Habits. Inhabits interior, edge and immediate vicinity of primary and secondary montane forest; appears to be adaptable – remains fairly common despite encroachment of its forest habitat by native cultivation (Turner 1977).

Breeding Habits.
NEST: a rather long bag made of grasses, moss and dead leaves, lined with fine rootlets and hair-like fungal mycelia; sited amongst hanging moss and very well camouflaged (Mackworth-Praed and Grant 1960).
LAYING DATES: breeds Sept–Feb.

Plate 14 (Opp. p. 207)

Cinnyris pulchella (Linnaeus). Beautiful Sunbird. Souimanga à longue queue.

Certhia pulchella Linnaeus, 1766. Syst. Nat., ed. 12, 1, p. 187; Senegal.

Range and Status. Common to abundant endemic resident and migrant, sahelian, soudanian and guinean savannas from Senegal to Eritrea, Ethiopia and Kenya, south in E Africa to Tanzania.

Northern limits: Mauritania, common in Senegal R. valley and Lac d'Aleg, moving in the rains to 17°N (e.g. 60 km north of Rosso: Gee 1984) and as far north as between Tidjikdja and Atar, i.e. nearly 20°N. Mali, common and

Cinnyris pulchella

widespread to 17°N. Niger, abundant summer visitor north to about 15°N (Tahoa, Tanout), seasonally frequent to abundant north to Agadez and central Aïr (18–20°N: see Migrations paragraph below). Chad, common throughout S half, north to Fada and Ennedi; in bushy wadis in Ouadi Rime – Ouadi Achim Faunal Res. at least 20 times as abundant as the only other sunbird, Pygmy Sunbird *Hedydipna platura*. Sudan, common and widespread north to Jebel Marra, N Kordofan, on Nile to about Atbara, and in E to about Kassala. Eritrea, very common in S and W, up to about 1380 m north to about Algena.

Southern limits: widespread in Guinea-Bissau; Guinea, Fouta Djalon, with records from Conakry coast. Sierra Leone frequent on estuary of Little Scarcies R. and adjoining coast. Ivory Coast, quite common south to 09°N, scarce south of Bouaké, vagrant to Abidjan (Mar). Ghana, common south to Mole, records from near Tamale. Togo, common in north, south to Péwa. Nigeria, seasonally abundant in north, south to Lokoja. Cameroon, south to Touroua and Koum. Central African Republic, frequent in N, south to Bamingui-Bangoran Nat. Park. Absent from Western Equatoria, Sudan. Northern race *pulchella* just enters Zaïre along NW shores of L. Albert (Mahagi Port, Mswa). Uganda, common in N, south through Bunyoro to Ankole in W, and in E to Teso and L. Bisina and Debasien Game Reserve. Rwanda, pair at Kamitumba, on Uganda border in NE Akagera Nat. Park (Vande weghe 1984). Kenya, nominate race common below 1300 m from Lokichokio and Turkana District south to Kongelai Escarpment and Kerio Valley, and penetrates in Rift Valley south to L. Baringo and L. Bogoria.

In east, ranges in Ethiopia east to Adal-Esa and lowlands north, east and south of Harar, but does not enter Djibouti or Somalia.

Distinctive southern race *melanogastra* ranges in interior of Tanzania (Pangani R., Arusha, Serengeti, Dodoma, N Iringa, Shinyanga, Katavi Plain Game Res., Rukwa, Ufipa) and in S Kenya from S Uaso Nyiro R., Magadi, Kajiado, Amboseli and Tsavo to Kitui and upper Tana R., and in L. Victoria basin.

Description. *C. p. pulchella* (Linnaeus) (includes '*aegra*' and '*lucidipectus*'): Senegal to Uganda and NW Kenya. ADULT: ♂ (breeding): head, neck and rest of upperparts metallic golden green, more bluish on uppertail-coverts. Tail blackish brown, with faint bluish gloss above; feathers fringed bright bluish green; T1 very long and narrow, projecting 45–70 mm. Lores black. Chin and throat metallic golden green, concolorous with head, merging into narrow greenish blue band across upper breast. Below this, centre of breast scarlet, bordered by broad yellow sides; flanks and belly metallic green; thighs and vent sooty black; undertail-coverts greenish blue. Median and lesser coverts metallic green; upperwing otherwise dark brown. Axillaries and underwing-coverts sooty brown. Bill black; eyes brownish black; legs black. ADULT ♂ (non-breeding): wings and long central tail feathers as adult ♂ (breeding); patch on centre of chin and throat and a few scattered feathers on rump and uppertail-coverts metallic green; otherwise like adult ♀. ADULT ♀: upperparts olive-brown. Tail feathers blackish with faint blue gloss; T5 with broad whitish fringe at tip, T6 with tip and outer edge brownish white. Narrow superciliary stripe pale yellow; lores and ear-coverts dark olive-brown. Cheeks to chin and throat creamy; rest of underparts pale yellow with faint dusky mottling on breast and flanks. Upperwing dark brown; flight feathers edged pale olive, tertials, alula and coverts fringed pale olive-brown, lighter on tips of greater coverts. Axillaries and underwing coverts pale yellow, marked with dusky; inner borders on underside of flight feathers buffish white. Bare parts as adult ♂. SIZE (10 ♂♂, 10 ♀♀): wing, ♂ 57–60 (58.4), ♀ 47–56 (52.2); tail, ♂ 78–108 (93.2), ♀ 33–41 (36.0); bill, ♂ 17–19 (18.2), ♀ 16–18 (17.2); tarsus, ♂ 14–16 (14.9), ♀ 14–16 (14.6). WEIGHT: Ghana, ♂ (n = 12) 5.6–10.2 (7.1), ♀ (n = 7) 5.6–7.3 (6.6); Nigeria (Zaria), ♂ (n = 17, Apr–May) 6.1–8.2 (7.3), ♀ (n = 18, Apr–May) 5.9–7.0 (6.5), unsexed (n = 16, Nov–Feb) 5.3–6.7 (6.1); ♂♂ significantly (12%) heavier than ♀♀ (Fry 1971); Chad, unsexed (n = 10) 5–9 (6.5).

IMMATURE: juvenile like adult ♀, but centre of chin and throat dusky grey, bordered by creamy moustachial stripe. Immature ♂ develops metallic green first on throat and lesser coverts.

NESTLING: unknown.

C. p. melanogastra Fischer and Reichenow: W and central Kenya south through dry interior of Tanzania. ♂ has red rectangle on breast band slightly longer, and belly and flanks entirely black; undertail-coverts tipped metallic greenish blue. Larger: wing, ♂ (n = 10) 61–65 (62.4); central tail feathers shorter, projecting 28–45 mm. WEIGHT: Kenya, ♂ (n = 35) 7.5–9.5 (8.5), ♀ (n = 3) 7.1–9.7 (8.1).

TAXONOMIC NOTE: *melanogastra* and nominate *pulchella* may prove to be specifically separate, as suggested by Chapin (1954). There is some overlap in their ranges, though that seems to be due to displacements or migrations (by *melanogastra*), rather than being a breeding overlap.

Field Characters. Length, ♂ 14–17.5 cm, ♀ 8.5–11 cm. Body small, ♂ with elongated central tail feathers. Breeding ♂ golden green above with *brassy pink sheen;* large *red patch* in centre of breast and *broad yellow patches* either side; belly wholly black in race *melanogastra* (Kenya/Tanzania), iridescent green with black undertail-coverts elsewhere; non-breeding ♂ like ♀ but retains black wings and long tail feathers. ♀ has dark lores and pale stripe from above eye backwards, underparts creamy yellow, with indistinct streaks on breast and flanks in *melanogastra*; juv. ♂ like ♀ but with dark throat patch which becomes green with age. In Black-bellied Sunbird *N. nectarinioides*, which it meets in Kenya/NE Tanzania, ♂ has complete red breast-band, with only small yellow pectoral tufts (nominate race) or no yellow at all (*erlangeri*), ♀ has only a hint of supercilium, paler yellow underparts streaked on breast and flanks. ♀♀ of Pygmy and Nile Valley Sunbirds *Hedydipna platura* and *H. metallica* have brighter yellow underparts, short bills.

Voice. Tape-recorded (104, 108, GREG, HOR, McVIC, MOR, PAY, STJ). Song a continuous jumble of rather dry, unmusical notes with cyclical quality, seeming to rise and fall, groups of chattery notes connected by thin high-pitched ones, 'tli-tli-pss-chip-bochiko-choppy-chippy-cheepy-pee-ss-ss-chop-choppy-chippy-peechu-pss-pss ...'; not rapid, notes individually audible. Scolding call a hard 'je-je-je-je ...'. Other calls include annoyance (loud 'tseu-tseu-tseu-tseu') and foraging call (weak 'chip ... chip') (Zimmerman *et al.* 1996).

General Habits. Inhabits wide variety of open woodlands and bush, farmland and gardens; and stands of *Acacia, Ziziphus* and other thorny trees and scrub in desert wadis. On Sierra Leone coast, mangrove scrub, village gardens, and raised sandy beaches where vegetation is mostly small bushes and creepers.

Occurs singly or (usually) in pairs. An active, mobile, conspicuous and often confiding sunbird that commonly lives in village and suburban gardens in drier, lowland parts of N tropics. Interacts with other sunbird spp. but no species 'pecking order' has been remarked. Forages in and around foliage at edges of trees, often quite low down in bushes, within 1 m of ground; feeds from blossoms by perching on flower or flowering head and probing downwards or sideways, or briefly hovering in front.

Across subsaharan Africa resident and migrant, moving more or less north and south. What happens everywhere best illustrated by situation in Niger and Nigeria, where it occurs from 08°N to about 19°N, breeding at southern limit (Lokoja, 07°48'N) and at or near northern limits (Timia, 18°07'N); breeds in wet season, which is protracted in south of range, about Feb–Oct, and short in north of range, about June–Aug. It is resident (but not known whether individuals are ever sedentary) in guinean and soudanian savanna woodlands, i.e. from 09–10° to 11–12°N (vegetation-zone latitudinal limits vary longitudinally). At boundary between S guinean and the more southerly derived-savanna belts (boundaries actually increasingly difficult to define, owing to vast deforestation and agricultural changes in last 25 years), Beautiful Sunbird is both resident (e.g. in well-watered Lokoja area) and a non-breeding dry season visitor (Ilorin, Jebba, Feb–Mar). Although it is resident in mid latitudes, many birds migrate northward in Mar–Apr and southward in Sept–Oct, e.g. at Zaria (11°N) where 100 netted in 24 spring days, without retraps (Fry 1971). North of soudanian woodlands, in sahel zone and deep into Sahara (at least far from watered regions), it is a spring immigrant, breeding wet season visitor, and autumn emigrant; e.g. at Malamfatori (13°37'N in NE Nigerian sahel) visits in Mar–Oct, abundant on passage at same times as at Zaria. Presumed to migrate by day; yet migrations have been detected, not visually, but only by routine netting. 228 birds ringed at Malamfatori in spring and 184 in autumn 1968 (more or less constant netting effort); 2 autumn birds were retraps marked in spring (Elgood *et al.* 1973). Abundant visitor June–Oct in cultivated S sahel zone in Niger, and abundant in pastoral N sahel zone in Apr–Aug and Oct–Dec. Pattern in Niger affected by 2 major geographical features: River Niger in SW and Aïr Massif in NW; it is year-round resident in Niger valley and floodplain and particularly abundant at height of dry season in Jan; and is frequent or common in Aïr at least in Dec–Mar and June–Sept (Giraudoux *et al.* 1988) or is restricted to period Mar–Oct (Newby *et al.* 1987).

Food. Nectar and insects. In central Chad feeds to great extent on nectar of *Tapinanthus globiferus*, a parasite of jujube tree *Ziziphus mauritiana* and acacias (Arabic name 'tuyr al-annaba' means *Tapinanthus*-bird); also feeds on variety of small insects attracted to aromatic jujube flowers (Newby 1980). Attracted to *Bombax, Acacia* and *Loranthus*. At Zaria, Nigeria, feeds on nectar of *Albizia coriaria, Jacaranda mimosifolia, Hibiscus sinoornata* and *Gmelina arborea* (Pettet 1977). Feeds regularly at flowers of *Moringa pterygospermum* (Nigeria) and *Calotropis procera* (Eritrea).

Breeding Habits. Solitary nester, monogamous and territorial. Seasonally territorial and aggressive; ♂ sings at prominent perch on top of bush or small tree, jerkily flicking wings up, fanning tail and switching it from side to side. In N Nigeria (Sokoto) newly-arrived ♂♂, having already moulted from eclipse into breeding plumage, establish territories in Apr by singing vehemently and chasing away rivals; yet ♀♀ do not lay until mid June (Serle 1943). Whether single or multiple-brooded, and whether a pair might breed at successive migration staging areas in 1 year, is unknown.

NEST: suspended, compact, domed pouch with side entrance, but only vestigial porch above entrance and 'beard' below nest; porch and 'beard' sometimes absent. Constructed out of fine plant stems, bits of bark, fibres, tiny twigs, leaves, plant heads, and sometimes a few feathers, bound together with spider web; profusely lined with feathers and a little plant down. Spider web is particularly abundant around entrance and at point of attachment of nest to twig. Outside decorated with a few bits of bark, lichen and snake skin. Size (n = 1): ext. height 128 (including suspensory stem of 35 mm), breadth 52, from front to back 53, int. depth below foot of entrance 35, entrance diam. 26 (Serle 1943). Suspended near end of branch or twig toward outside of tree or shrub; height above ground 1·5–5·5 m; direction in which entrance points (whether into or away from tree) seems random. In Chad, nest generally sited in jujube and acacia tree; elsewhere commonly uses *Citrus, Bauhinia reticulata* and tamarind trees.

EGGS: 1–2; of 25 clutches, N and S Nigeria, only 2 were C/1 (complete clutches). Long ovate; slightly glossy or without gloss; shell thin and fragile. Ground pale brown, pale grey or pale green (Nigeria) or bluish (E Africa); markings, in brown, grey and green-grey, vary considerably from blotched to streaked types (Nigeria) and some eggs finely speckled all over (E Africa); grey and erythristic types also known; blotches usually sparse and uneven; streaks evenly distributed and profuse; blotched eggs generally also have scrawls and hair lines. SIZE: (n = 32) 16·0–17·0 × 9·9–11·6 (16·3 × 10·7).

LAYING DATES: Mauritania, June–Oct; Mali, July–Oct; Senegal (Richard-Toll), June–Oct; Sierra Leone, (coastal population may nest in Jan, when ♂♂ singing and displaying and gonads enlarged: Bannerman 1948); Ghana (Mole), July–Sept; Nigeria (Sokoto), June–Sept, (Lokoja), Oct, Feb; Niger, June, (Timia) and June–July (Zinder); Chad, (nests with young July, late Aug); Eritrea, July (and probably May–Aug); E Africa, Region A, Feb–July and Sept–Dec, mainly Apr–May, Region B, Dec–Mar, May–Sept, mainly Aug–Sept, Region C, Nov–Mar, mainly Dec–Jan, Region D, Jan–Sept, mainly Mar–May (especially May).

INCUBATION: by ♀ only (although both sexes commonly occur at the nest).

References
Bannerman, D.A. (1948).
Serle, W. (1943).

Cinnyris shelleyi Alexander. Shelley's Sunbird. Souimanga de Shelley.

Plate 15
(Opp. p. 222)

Cinnyris shelleyi Alexander, 1899. Bull. Br. Orn. Club, 8, p. 54; Zambezi River, 60 miles below Kafue-Zambezi confluence.

Forms a superspecies with *C. mariquensis*.

Cinnyris shelleyi

Range and Status. Endemic resident, SE Africa. Zambia, sparse; uncommon in mutemwa woodland (*Baikiaea*) at Livingstone, and elsewhere restricted to miombo woodland (*Brachystegia*): records from Kachalola, Chadiza, Muliro; middle Zambezi valley; Mpika, Muchinga Escarpment; Isoka; Ndola, Luanshya; Solwezi Mazabuka, Choma, Kalomo; virtually absent from Luangwa Valley (Benson *et al.* 1971, R.J. Dowsett, pers. comm.). SE Katanga, Zaïre: single record in salient east of Ndola (Zambia) (Hall and Moreau 1970). S-central Tanzania (N. Baker, pers. comm.): NW edge of Iringa highlands from east of Mbeya to about Idadi (just east of Ruaha Nat. Park); W and E sides of Selous Game Res. between 08° and 09°S; S edge of Iringa highlands, just southeast of Njombe (many records); uplands of W Ruvuma District but not Matengo hills (many records between 10°30′ and 11°S and 35° and 35°30′E); and N Morogoro Prov. from Mikumi Nat. Park to Magogoni and Ruvu valley, north to foothills of Uluguru and Kingolwira – local and rather uncommon, at 500–1200 m. Malaŵi, sparse, from Edingeni on Zambian border and L. Malaŵi littoral (Nkhotakota, Salima, and E shore at Makanjila) to Blantyre. Mozambique, occurs north of Zambezi south to Boror region and south of Zambezi in W Tete Prov., but does not occur in S Mozambique (Parker 1999) as previously claimed (Hall and Moreau 1970). Zimbabwe, record at Mana Pools.

Description. *C. s. shelleyi* Alexander: range of species except N Morogoro Prov., E Tanzania. ADULT ♂ (breeding): head, neck and rest of upperparts metallic bronzy green, more emerald green on uppertail-coverts. Tail blackish, slightly glossy above, T1 finely edged bright metallic green. Lores blackish. Chin to upper breast metallic green like head, bordered by narrow blue band, and below this a broad scarlet band (10–15 mm deep) across lower breast; flanks, belly and undertail-coverts blackish brown. Lesser and median coverts metallic green; rest of upperwing blackish brown, primaries and secondaries finely edged olive-yellow when fresh. Axillaries and underwing-coverts sooty brown. Bill black; eyes dark sepia; legs black. ADULT ♂ (non-breeding): wings and tail as adult ♂ (breeding); colourful feathers otherwise present mainly on chin to breast (including much of red band) and uppertail-coverts; rest of upperparts, sides of head and breast, and lower underparts like adult ♀, but olive-brown feathers of head and mantle with variable metallic green tipping. ADULT ♀: upperparts greyish olive-brown. Tail blackish brown; T6 paler brown with broader whitish tip, especially on inner web, T5 with narrow whitish terminal fringe. Supercilium narrow and inconspicuous, buffy white; lores dusky; cheeks and ear-coverts dark grey-brown. Underparts pale buff, breast to belly and flanks tinged olive-yellow; chin to throat mottled and breast and flanks heavily streaked dark grey-brown; undertail-coverts with dark centres. Upperwing dark grey-brown, primaries narrowly edged buffy white, secondaries edged light olive, tertials and coverts fringed pale buff when fresh. Axillaries and underwing-coverts buffy white with dusky bases; inner borders on undersides of flight feathers greyish white. Bare parts as adult ♂. SIZE (10 ♂♂, 3 ♀♀): wing, ♂ 62–68 (64·1), ♀ 58–60 (59·0); tail, ♂ 40–45 (42·9), ♀ 38–39 (38·7); bill, ♂ 21–23 (21·9), ♀ 21–22 (21·3); tarsus, ♂ 15–16 (15·9), ♀ 15–16 (15·7).

IMMATURE: juvenile similar to adult ♀ but yellower below; chin to throat sooty grey or blackish, bordered by olive-yellow moustachial stripe; immature ♂ with scarlet pectoral band.

C. s. hofmanni Reichenow: E Tanzania (N Morogoro Prov., see above). Distinctively different, and has been regarded as a separate species. ♂ like ♂ *shelleyi* but flanks, belly and under-tail-coverts black; ♀ with upperparts dull green, tail blackish, supercilium narrow but distinct, yellowish, chin and throat dusky, breast with irregular orange-red band, rest of underparts pale yellow, feathers of breast and flanks with dusky centres. Small and short-billed: wing, ♂ (n = 3) 58–61 (59·0), ♀ (n = ?) 56–57; tail, ♂ (n = ?) 35–38, ♀ (n = ?) 31–34; bill, ♂ (n = 4) 18–19 (18·5), 1 ♀ 19.

Field Characters. Length (*shelleyi*) 12–13 cm, (*hofmanni*) 9–9.5 cm. A miombo sunbird which co-exists with Miombo Double-collared Sunbird *C. manoensis*. ♂ readily identified by combination of *scarlet* breast-band and *black* belly; wings and tail also black (brown in Miombo), and golden green upperparts grade into bluish green uppertail-coverts (Miombo has grey-brown rump and violet upper-tail-coverts). ♀ told from ♀ Miombo by brown streaks or blotches on yellow underparts (underparts unmarked in ♀ Miombo). Meets larger and longer-billed Marico Sunbird *C. mariquensis*; ♂ Marico easily told by purple and maroon breast-bands, but ♀♀ quite similar; ♀ Shelley's is greyer above and less yellow below. Immature like ♀ but throat dark grey bordered by pale moustachial stripe. Does not overlap with Neergaard's Sunbird *C. neergaardi*.

Voice. Tape-recorded (86, 91–99, B, F, GIB, GIL, STJ). Insignificant song is rapid, high, thin and tuneless, a jumble of hissing and liquid notes interspersed with hard 'chup', 'tissssllllllchuptssssllllchup-chup-llllllll1tssssss...', lasting up to 8 s. Call/scold a repeated hard, grating 'jik-jik-jik ...' or faster 'jijijijijiji ...'; also gives a very high, thin 'tsss-tsss-tsss' or 'seep-seep'.

General Habits. Nominate race almost restricted to *Brachystegia* woodland, but has been found also in *Baikiaea* and *Acacia* (Irwin 1981). Resident, though possible that birds in Zambezi valley may have moved down from adjacent plateau country (Benson *et al.* 1971). Race *hofmanni* inhabits woodland, bushland, gardens and thick, riverine scrub.

Food. Feeds at flowers of *Loranthus* (nectar favoured by many sunbirds), and in gardens at *Tecomaria*, *Holmskioldia* and pride-of-India; said to avoid aloe flowers (Brooke 1964); takes spiders, flies and small lepidopterans (Skead 1967).

Breeding Habits. Solitary nester; double brooded.

NEST: one nest was a fairly compact, suspended pouch *c.* 120 high, with side entrance *c.* 30 in diam., made of leaves, leaf ribs, bits of bark, grass and other vegetable matter, bound together with spider web, and lined with domestic chicken feathers; floor of nest *c.* 30 thick, so eggs were lying *c.* 50 below bottom of entrance. Sited 3 m up in small pride-of-India tree in garden. After the young flew, ♀ laid 2nd clutch within 14 days, having pulled out (then presumably replaced) the lining (Brooke 1964).

EGGS: 1–2. 1st clutch (in nest described above) pale drab with a few fine specks of purple-drab and complete purple-drab zone 3·5 mm wide around broad end; 2nd clutch pale drab with many fine specks of purple-drab all over (Brooke 1964). SIZE: (n = 4) 17·0–17·5 × 11·5–12·0 (17·25 × 11·75).

LAYING DATES: Zambia, Sept-Oct; Malaŵi, ('nest built, apparently in August, ... two broods were reared': Brooke 1964).

References
Brooke, R.K. (1964).
Irwin, M.P.S. (1981).
Skead, C.J. (1967).

Plate 15 (Opp. p. 262)

Cinnyris mariquensis A. Smith. Marico Sunbird. Souimanga de Mariqua.

Cinnyris mariquensis A. Smith, 1836. Report Exped. Centr. Africa, p. 53; north of Kurrichaine, W Transvaal.

Forms a superspecies with *C. shelleyi*.

Range and Status. Endemic resident, E and southern Africa.

Eritrea, uncommon, from Anseba to Senafe at 920–1840 m. Ethiopia, frequent to locally abundant in central highlands. S Sudan, only in Didinga Mts, uncommon, up to 2000 m. Uganda, Karamoja and Teso in E; central areas, southwest to Ankole, northwest to north of L. Kwania, and southeast probably to Kenya border. Kenya, northern race uncommon: Lapurr Mts, Moyale and Tabaka; Mt Kulal, Mt Nyiru, Ndoto foothills; and from Kapenguria and Kerio Valley to Maralal, Laikipia, Samburu Game Res. and Meru Nat. Park; southern race common and widespread: Busia, Mumias and Kakamega Districts south to Tanzania border and east to Lolgorien, Mara, Nairobi, Thika, Athi River, Machakos and Sultan Hamud Districts (Zimmerman *et al.* 1996); at 800–2000 m, mainly 1000–1500 m; old sight records below 500 m in Tsavo East Nat. Park and lower Tana R. (not mapped). Tanzania, uncommon and local but may be considerably more widespread than mapped: in NW at Ngara and Kibondo; along shore of L. Victoria from Kenya border to Mwanza; in NE east to Arusha and Pare Mts with a record on coast; in SW in Kitavi Plain Game Res.; in interior Tabora district and N Dodoma District; and in S in Mbeya and Njombe

Cinnyris mariquensis

highlands, Iringa and Songea. N Zambia, widespread in N Isoka and NE Mbala Districts and a record from Mansa District.

Angola, sparse and local in most of S half of country west to Huila. SW Zambia, frequent and widespread, north to Balovale and Kafue Nat. Park, east to N Kalomo District. Namibia, Botswana, South Africa and Swaziland, sparse to very common and widespread as mapped; may well prove to occur in those parts of NE Namibia where shown absent (Harrison *et al.* 1997). In Zimbabwe very local, but common around Bulawayo; very sparse in Sabi Valley, only between Birchenough Bridge and Devuli R. to Dakati Pan; commoner in Limpopo Valley to Beit Bridge, Siyanje and near Mateke Hills; absent from SE lowveld and from much woodland on Kalahari Sand (Irwin 1981). Mozambique, common, as mapped (Parker 1999). Commonest in Okavango, NW Botswana and in SE Botswana and NW Transvaal; max. density in N Kalahari of 1–2 birds per ha and av. density of 1 bird per 2·4 ha (Herremans 1992). Absent from E Transvaal escarpment.

Description. *C. m. mariquensis* (Smith): southern Africa south of 10°S, except for parts occupied by *lucens* and *ovamboensis*. ADULT ♂: head, neck and rest of upperparts metallic green with bronzy tinge, uppertail-coverts more bluish green, sometimes blue. Tail blackish brown; T1–T4 edged bright metallic green. Chin to upper breast metallic green, concolorous with head, bordered below by narrow band of metallic violet-blue; below this a broader non-metallic band of purple-maroon (8–10 mm deep) with a few violet or bluish bars formed by metallic subapical feather bands; rest of underparts sooty black, feathers immediately below maroon band tipped with slight greenish gloss. Lesser and median coverts metallic green; rest of upperwing dark brown, faintly glossed when fresh, especially on tertials. Axillaries and underwing-coverts sooty black. Bill black; eyes dark brown; legs black. ADULT ♀: upperparts olivaceous grey-brown; large uppertail-coverts blackish brown. Tail blackish brown, upperside faintly glossed; T6 paler, with broad greyish white spot at tip of inner web and narrow greyish white outer edge, T5 with narrow greyish white fringe at tip of inner web. Narrow supercilium buffy white; cheeks and ear-coverts greyish brown. Underparts pale grey-buff, throat and breast streaked dark brown, flanks more diffusely so, centre of belly paler and yellower; undertail-coverts buffy white with dark brown centres. Lesser and median coverts greyish brown; rest of upperwing dark grey-brown, primaries edged pale buff, secondaries and tertials more broadly edged olive-brown, tertial tips and fringes of primary coverts and greater coverts buff; alula fringed buffy white. Axillaries buffy white; underwing-coverts buffy white with dark bases; inner borders on underside of flight feathers greyish white. Bare parts same as adult ♂. SIZE (10 ♂♂, 7 ♀♀): wing, ♂ 66–71 (68·2), ♀ 61–64 (63); tail, ♂ 45–51 (47·5), ♀ 41–43 (42·1); bill, ♂ 23–25 (23·7), ♀ 22–24 (22·8); tarsus, ♂ 17–19 (18·1), ♀ 17–18 (17·4). WEIGHT: Transvaal, ♂ (n = 14) 11·3–14·1 (12·3), ♀ (n = 11) 10·3–12·3 (11·1).

IMMATURE: juvenile similar to adult ♀, but chin and throat dark grey, bordered by long buffy white moustachial stripes; lores darker; underparts tinged more olive-yellow.

NESTLING: not described.

C. m. lucens (Clancey): SE Zimbabwe, N and E Transvaal, S Mozambique, E Swaziland, Natal. Like *mariquensis* but 4–5% smaller: wing, ♂ (n = 18) 63–67 (65·2), ♀ (n = 9) 58–62 (60·4) (Clancey 1973a).

C. m. ovamboensis (Reichenow): Ovamboland, N Namibia. Like *mariquensis* but upperparts, face and chin to breast less bronzy (as in *osiris*) and belly browner; slightly smaller than *mariquensis* (Clancey 1979).

C. m. suahelica (Reichenow): central Uganda, W and central Kenya, Rwanda, Tanzania and NE Zambia. Similar to nominate race, but ♂ has lower underparts sooty grey rather than blackish; ♀ yellower below. Bill shorter: ♂ (n = 10) 20–22 (21·3). WEIGHT: Kenya, ♂ (n = 64) 9·1–13·3 (11·7), ♀ (n = 7) 8·3–11·3 (10·1).

C. m. osiris Finsch: Eritrea, Ethiopia and Somalia to N Kenya (south to equator), E Uganda west to E Teso, and SE Sudan. ♂ differs from *suahelica* in having glossy violet breast band broader, but lower maroon band darker and narrower; lower underparts blacker (as in nominate race); upperparts purer green (less bronzy). ♀ less yellowish below than *suahelica*, more densely mottled on throat and breast.

Field Characters. Length 11·5–14 cm. Dark and rather stocky; widespread and common from Ethiopia to South Africa. ♂ readily distinguished from Shelley's *C. shelleyi* and other scarlet-breasted sunbirds by *narrow violet-blue* and *broad deep maroon* breast-bands; belly, wings and tail black; most similar to Purple-banded Sunbird *C. bifasciata* but maroon breast-band broader (12–13 mm *vs* 7–10 mm) and violet band narrower. Both sexes larger and longer-billed than Purple-banded, ♀ brighter yellow below, with heavier dark streaking. Immature ♂ has dark throat patch bordered by pale moustachial stripe; almost identical to imm. ♂ Amethyst Sunbird *Chalcomitra amethystina*; breast *usually* less marked, streaked rather than barred, and bill straighter, but these are the most difficult sunbirds to identify in the hand and impossible in the field (Tree 1991). ♀ similar to some ♀♀ of Amethyst but throat *mottled* grey, mottling radiating out to breast, flanks and belly (Tree 1990).

Voice. Tape-recorded (9, 16, 75, 86, 88, 91–99, B (BBC), C, F, NIV). Common call a hard, sharp 'jik-jik-jik …', 'co-ji-ji-ji-…' or 'chip-chip', sometimes accelerating into a hurried chatter, 'jijijijijiji …'. Song of 2 types: (1) a rapid swizzle of high, thin, sibilant and liquid tinkling notes, almost canary-like, typically starting with and interrupted by call and other notes, 'chip-tsip-chop-chop-pssssssstle-chup-ssslllllll-chirp-chewy-ssssssstllllll-trrr-trrr-sissssssl-jik-jik-jik ...'; swizzle can also be given on its own, lasting c. 2 s; (2) starts with some sharp chip notes which accelerate like bouncing ball (cf. Copper Sunbird *C. cuprea*) and end with brief swizzle, 'tsip-tsip-chap-chap-chap-ta-ta-tatatatata-swizzllll'; swizzle may be omitted, 'chip-chip-chap-chap-chap-chap-titi'; this song heard especially in early morning (Haagner 1964). Said to imitate other bird calls (Maclean 1993); song session may last several min (Skead 1967). Also calls a very high, hissing 'tsssssss'.

General Habits. Inhabits well-wooded, park-like acacia savanna grassland with isolated clumps of acacias, combretums and aloes, open bush and broad-leaved woodland containing *Acacia* spp (especially *A. robusta* in Zimbabwe), and woods with plenty of thorny vegetation fringing larger rivers; in Okavango occurs mainly in swamp-fringing forest. Reporting rates in various southern African vegetation types: Okavango 38%, Northern Kalahari 22%, Arid Woodland 19%, Moist Woodland 18%, Central Kalahari 15%, Mopane 14.5%, Namibian Escarpment 3%, Miombo 3%, Southern Kalahari 2% (Harrison *et al.* 1997). Sometimes wanders into suburban gardens.

Solitary or in pairs; sometimes in small aggregations foraging in flowering tree, and mixes with Scarlet-chested and White-bellied Sunbirds *Chalcomitra senegalensis* and *Cinnyris talatala* and others. At most times, however, very aggressive; once pursued a Cape White-eye *Zosterops pallidus* for c. 200 m; Diederik Cuckoos *Chrysococcyx caprius* attacked and driven away. Moves through foliage to glean insects, and may hang upside down to do so. Hawks passing insects by flying upward from perch with erratic twists, audibly snapping insect up, then returning to same or another perch. Hovers in front of grass seed heads to pick small insects out. Hovers briefly in front of flowers to take nectar, but generally takes that food by perching on top of flower or inflorescence and probing downwards.

Mainly sedentary. Thought to be subject to seasonal movements in e.g. Eritrea, E Africa and S Mozambique (Smith 1957, Benson 1982, Clancey 1996). Evidently quits Kalahari in driest season and times of drought (Herremans 1992), and numbers increase in Zimbabwe at those times (M. Herremans *in* Harrison *et al.* 1997).

Food. Nectar, insects and many spiders. Favours flowers of *Kigelia*, *Erythrina*, *Aloe* (**A**), *Thevetia*, *Leonotis*, *Schotia*, *Grevillea*, *Crotolaria*, *Jacaranda*, *Callistemon*, *Loranthus* (parasites of, e.g. *Acacia nigrescens*) and *Cadabe termitaria*. Insects include flies, termites taken in flight, small moths, wasps and caterpillars.

Breeding Habits. Solitary nester, evidently monogamous and – to judge by ♂'s aggressiveness – territorial, though no close observations have been made.

NEST: compact, pear-shaped, domed purse, with circular side-top entrance. Nest is not so much suspended loosely beneath twig by narrow neck as built around twig, so that twig or twigs pass through dome, rest of nest being below

A

but firmly attached to them. Thick-walled, built of soft, hair-thickness grass bound together with cobweb, the outside sometimes smoothed over with cobweb; decorated with fragments of bark, pieces of *Eucalyptus* resin, fibres, feathers and small clusters of tiny reddish seeds or furry seed capsules; inner part of nest walls made of woolly plant down; thickly and raggedly lined with small francolin's feathers or goose down, which once bulged through walls like split cushion (Vincent 1949). Size (1 nest): ext. height 128, breadth 64, horizontal circumference 228, entrance diam. 39; top of entrance c. 26 from top of nest. Sited at tip of branch on outside of bush or tree (nest 2–8 m above ground, usually 2–4 m), usually conspicuously, sometimes nest very well concealed; generally in top half of tree, often in bare or leafy canopy. One suspended over standing water; another built 1 m from a wasps' nest. Uses native trees, also exotics e.g. pines, eucalypts and jacarandas. Pair often returns to same place to nest in successive years, and once the same nest was used twice in a season.

Built by ♀ alone. She does not build weaver-like circular frame, but makes a long, slim hank of white down suspended below twig, starts to camouflage it on 2nd day, then forms entrance and cavity by shaking, jerking and trampling (Milstein 1963). ♂ accompanies her in flights to and from nest (Vincent 1949) and was once seen carrying material; whilst ♀ is working at nest ♂ perches nearby and sings almost continuously. 1 nest took 12 days from commencement to laying of 1st egg.

EGGS: 2; rarely 3. Ground colour pale cream, greyish white or pale greenish white; marked with dense streaks and smears, blotches, speckles and occasional scrolls of brown, black, grey or olive; markings either evenly spread over whole shell, or form band around broad end. SIZE: (n = 20) 16·8–20·0 × 11·0–13·6 (18·4 × 12·1).

LAYING DATES: Eritrea, (gonads active Mar, June, July); Ethiopia, May; E Africa, Region B, Feb, Apr–Sept and Dec, Region C, Apr, June–July, Sept–Feb; Region D, Mar, May; Zambia, Sept, Oct, Feb; Zimbabwe, Sept–Dec (Sept 4 clutches, Oct 19, Nov 7, Dec 3); Mozambique, July; Transvaal, Aug–Feb, mainly Sept–Nov; N Natal, Oct–Nov.

INCUBATION: by ♀ only, sitting deep in nest with only tip of bill showing at rim of entrance; incubation begins after 2nd egg laid. Period (n = 1) 13·5–14·5 days.

DEVELOPMENT AND CARE OF YOUNG: ♀ took food to nest every 2 min, ♂ accompanying/pursuing her but not himself feeding young. A single young that had left nest a few days earlier, returned to nest to sleep at night with both parents.

BREEDING SUCCESS/SURVIVAL: parasitized by Klaas's Cuckoo *Chrysococcyx klaas*. One nest torn apart by Red-faced Mousebirds *Urocolius indicus*.

References
Harrison, J.A. *et al.* (1997).
Milstein, P. le S. (1963).
Skead, C.J. (1967).
Vincent, A.W. (1949).

Cinnyris congensis (van Oort). Congo Sunbird. Souimanga du Congo.

Plate 15
(Opp. p. 222)

Nectarinia congensis van Oort, 1910. Orn. Monatsb., 18, p. 54; Boma, lower Congo; corrected to Irebu by Chapin, 1954, Bull. Amer. Mus. Nat. Hist., 75B, p. 268.

Forms a superspecies with *C. erythrocerca*.

Range and Status. Endemic resident, quite common on forested banks of Congo River from equator above Kisangani downstream to about Lukolela, with records on lower Oubangui R. near Impfondo and on lower Busira-Momboyo R.; record between Léfini and Ngo, Congo (Dowsett-Lemaire 1997c). Type locality was 'Boma' on Congo River much further downstream, near mouth, but was doubted and corrected by Chapin (1954) to Irebu; the new Léfini locality suggests that species may yet prove to range to Boma.

Description. ADULT ♂: head, neck and rest of upperparts metallic green, bluer on uppertail-coverts. Tail graduated, T1 narrow and elongated, projecting *c*. 70 mm; sooty black, feathers narrowly edged metallic blue-green. Lores black. Chin to upper breast metallic green like head, bordered by line of metallic blue; below this, a broad scarlet band 15 mm deep across lower breast; rest of underparts sooty black. Lesser and median coverts metallic green; upperwing otherwise sooty black. Axillaries and under-wing-coverts sooty black. Bill black; eyes dark brown; legs black. ADULT ♀: upperparts dull dark brownish olive, lores blackish, short, narrow pale line from above to behind eye. Tail feathers blackish brown with dull greenish sheen, especially on edges, graduated, central feathers 3–4 longer than T2, *c*. 10 longer than T6; central feathers with or without very narrow pale tips, the rest with pale brownish white tips to inner vanes increasing in extent outwardly to 6–7 mm on outer feather, which also has distal half of outer web pale brown, darkening proximally. Chin grey, upper throat with variable amount of dark mottling, lower throat heavily mottled blackish brown, forming patch; pale line from base of lower mandible down side of neck, separating dark throat from olive face; rest of underparts pale yellow, brightest in centre of belly, mottled and streaked on breast and flanks; undertail-coverts yellowish white. Bill and legs blackish. SIZE: wing, ♂ (n = 15) 62–66 (63·7), ♀ (n = 4) 56–59 (57·25); tail (T1), ♂ (n = 12) 110–125 (115), 2 ♀♀ 41, 43, (T2) ♂ (n = 6) 45–50·5 (47·7); bill, ♂ (n = 7) 21–22·5 (21·7), 2 ♀♀ 18·5, 20; tarsus, ♂ (n = 6) 16–17 (16·3), 2 ♀♀ 16, 16·5.

Cinnyris congensis

Field Characters. Length, ♂ 18–20 cm, ♀ 11·5 cm. A specialized sunbird confined to forested banks of Congo and adjacent rivers. ♂ with blue-green back grading to blue-purple rump, scarlet breast-band separated from throat by narrow blue line, no pectoral tufts, black belly and long central tail feathers distinctive; the only other double-collared sunbird in the region is Olive-bellied *C. chloropygia*, ♂ of which lacks long tail feathers and has olive belly, yellow pectoral tufts. ♀ is brownish olive above, has dark throat bordered by pale malar streak, pale yellow underparts streaked dark on lower belly and undertail-coverts; ♀ Olive-bellied has entire underparts pale yellow, with streaked breast. ♀ somewhat resembles immature ♂ Green-throated Sunbird *Chalcomitra rubescens*; latter is larger and longer-billed and browner above, with solid black throat, underparts white, not yellow.

Voice. Unknown.

General Habits. Forest edge, forested banks of great rivers, clearings around forest villages and trading posts, and small forests on river islands. Seems not to move more than a few hundred m from the riverside. Seen feeding at large red flowers of flamboyant tree *Poinciana regia*.

Food. 9 stomachs contained tiny spiders but no insects. May also feed on nectar.

Breeding Habits. Unknown. A pair seen by a nest hanging from bush 2 m above water; inferred from gonad maturation to breed year round (Chapin 1954).

Reference
Chapin, J.P. (1954).

Plate 15 (Opp. p. 222) *Cinnyris erythrocerca* (Hartlaub). Red-chested Sunbird. Souimanga à ceinture rouge.

Nectarinia erythrocerca Hartlaub, 1857. Syst. Orn. Westafr., p. 270; no locality; White Nile, south of 08°N, designated by Sclater, *ibid.*, following Heuglin, *ibid.*

Forms a superspecies with *C. congensis*.

Range and Status. Endemic resident. Sudan, R. Nile from Malakal and L. No to Uganda border, and R. Sue south of Wau and probably elsewhere in S Bahr al Ghazal. Uganda, locally common in W and S, scarce in Teso and most northern regions, absent from Karamoja District; record from Sudan/Uganda border (S Didinga Mts/N Kidepo Nat. Park) (Hall and Moreau 1970); common to abundant around Kampala, and in mid-1980s penetrated city centre, occurring in gardens with swimming pools (Carswell 1986). Common on shores and islands of L. Victoria in Uganda, Kenya and Tanzania. Kenya, fairly common on lake shore and rarely far from it, but occurs east to Ahero, Ruma Nat. Park and Migori (Zimmerman *et al.* 1996). Zaïre, common along shores of L. Albert, L. Edward and L. Kivu, ascending streams up to 1970 m. Rwanda, common throughout. Burundi, only in N. Tanzania, shores of L. Victoria, east to Mara and south to Mababma, and in W Kagera and extreme NW Kigoma Districts to Kahama and Kagera Swamp.

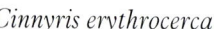

Cinnyris erythrocerca

Description. ADULT ♂: forehead, crown and sides of head metallic green, merging with bluer green nape and sides of neck; hindneck and mantle to scapulars and rump metallic bluish green, shot with violet-blue; uppertail-coverts violet-blue. Tail graduated, T1 long and narrow, projecting 18–25 mm; blackish brown, glossed slightly bluish above, feathers edged bright bluish green. Lores sooty black. Chin and throat metallic green, concolorous with sides of head, and merging into narrow violet-blue band across upper breast. Broad non-iridescent band across lower breast deep red; flanks and belly to undertail-coverts sooty black. Median and lesser coverts metallic greenish blue; rest of upperwing blackish brown, fresh secondary edges, tertials, alula, primary coverts and greater coverts with faint bluish gloss. Axillaries and underwing-coverts sooty black. Bill black; eyes dark hazel or dark brown; legs black. ADULT ♀: upperparts olive-brown, dark feather centres on forehead and forecrown giving mottled effect. Tail blackish brown, with slight bluish gloss above when fresh; T4–T5 with narrow brownish white fringe near tip of inner web, T6 with broad fringe near tip of inner web and narrow fringe at tip and along outer edge of outer web. Narrow, inconspicuous superciliary stripe yellowish buff; lores and below eye to ear-coverts dark olive-brown. Chin and throat dark greyish with narrow olive-yellow barring, bordered by broad yellowish buff moustachial stripe; breast pale olive-yellow, mottled dark olive-grey; flanks and belly pale olive-yellow with diffuse greyish streaking; thighs dark olive; undertail-coverts pale creamy buff with dark olive-brown centres. Upperwing blackish brown; flight feathers edged olive-yellow to form noticeable wing-panel in fresh plumage, tertials, alula and coverts fringed olive-brown. Axillaries olive-yellow; underwing-coverts pale olive-yellow with dusky centres; inner borders of flight feather undersides greyish white. Bare parts as adult ♂. SIZE (10 ♂♂, 10 ♀♀): wing, ♂ 62–67

(63·7), ♀ 54–61 (57·0); tail, ♂ 65–77 (70·5), ♀ 41–46 (43·8); bill, ♂ 19–21 (19·7), ♀ 17–20 (18·3); tarsus, ♂ 16–17 (16·4), ♀ 15–16 (15·5). WEIGHT: Uganda, ♂ (n = 56) 8·2–10 (9·2), ♀ (n = 51) 7·4–9·5 (8·4).

IMMATURE: juvenile like ♀, but upperparts greener, flight feather edges more yellowish; chin and throat more uniform sooty black, yellowish moustachial stripe prominent.

NESTLING: unknown.

Field Characters. Length, ♂ 13·5–16 cm, ♀ 11–12 cm. ♂ has head, throat and upperparts iridescent *blue-green*, becoming *purple* on rump; narrow *violet* band separates throat from broad *dark red, non-iridescent* breast-band; belly black; central tail-feathers extend only *about 2 cm* beyond rest of tail. ♀ *has blue-black graduated, pale-tipped* tail; pale yellow primary edges form wing panel when fresh; chin and throat with heavy dark streaks bordered by indistinct pale malar stripe, rest of underparts pale yellow with heavy streaking. Juv ♂ like ♀ but with black throat bordered by conspicuous white malar stripe, breast and belly with dark mottling. ♂ of black-bellied race of Beautiful Sunbird *C. pulchella* has golden green head and upperparts, scarlet breast-band flanked by yellow patches, ♀ has square tail, faintly streaked yellow underparts. Does not overlap Black-bellied Sunbird *C. nectarinioides*.

Voice. Tape-recorded (C, GREG, KEI, McVIC). Song a short thin twitter on same pitch, 'tsi-si-sip-see-see-swee' or 'tsi-tsi-tsi-tsi-tsi-tsi-tsi-tsip' (Zimmerman *et al.* 1996); call/scold a hard, loud 'jik' given in long series, 4–5 notes at a time in irregular tempo, sometimes speeded up.

General Habits. Inhabits waterside, swamp and streamside vegetation, bushy low cover, and gardens. Usually in pairs, conspicuous, active and pugnacious. Forages low amongst flowering shrubs and garden annuals, but sometimes in large tree, e.g. flowering tamarind. Sedentary.

Food. Nectar, insects and spiders. Spiders twice as common in stomachs as insects, latter include beetles (most minute) and flies.

Breeding Habits. Solitary nester, evidently monogamous and territorial.

NEST: rather small purse, suspended from outer twig or bush or small tree, often over water; made of variety of vegetable materials including grasses and bark fibres, bound together with spider web; decorated outside with dead leaves and seed heads; lined with plant down and feathers (Chapin 1954).

EGGS: 1–2, usually 1. Bluish white, with streaks and spots of grey and grey-brown; some eggs dull grey, not much spotted.

LAYING DATES: E Africa, Region B, all months; of 80 records, 19 in Apr, 10 each in May, June and July, 6 in Aug–Oct, 9 in Nov–Dec, 8 in Jan–Feb and 8 in Mar; Region C, Feb.

Reference
Chapin, J.P. (1954).

Cinnyris nectarinioides Richmond. Black-bellied Sunbird. Souimanga nectarin.

Plate 15
(Opp. p. 222)

Cinnyris nectarinioides Richmond, 1897. Auk, p. 158; plains east of Kilimanjaro.

Range and Status. Endemic resident, mainly in E Kenya. Ethiopia, sight record at L. Abaya (Beals 1966), and must occur in extreme SE on Ethiopian bank of R. Daua. Somalia, rare: 7 old records in upper R. Jubba valley. Kenya, locally common at 50–1300 m: Daua R. in extreme NE, Wajir, N Uaso Nyiro R., most of Tana R., Kitui, Kiboko, Tsavo West and Tsavo East Nat. Parks (where common along Athi-Galana, Tiva and Tsavo Rivers), Taveta Plains. Tanzania, plains east of Kilimanjaro and Mkomazi Game Res.

Description. *C. n. nectarinioides* (Richmond): interior E Kenya to NE Tanzania. ADULT ♂: head, neck and upperparts metallic green, more golden on head, more bluish on uppertail-coverts. Tail black with slight bluish gloss above, T1 narrow and elongated, projecting 16–22 mm; bases of feathers edged bright blue. Lores black. Chin and throat golden green, concolorous with head; upper breast greener, merging with narrow violet-blue band; broad band across lower breast orange-red; pectoral tufts yellow; rest of underparts dark sooty brown. Median and lesser coverts metallic green; upperwing otherwise blackish, secondaries, tertials, alula and larger coverts with faint blue gloss. Axillaries and underwing-coverts sooty brown. Bill blackish; eyes dark brown or blackish; legs black. ADULT ♀: upperparts olive-brown, forehead and crown with faint darker streaks. Tail blackish brown, T6 paler and with broad whitish fringe at tip

Cinnyris nectarinioides

of inner web. Narrow superciliary stripe pale yellowish buff; lores to ear-coverts dark olive-brown; sides of neck as upperparts; cheeks pale yellowish buff. Underparts pale olive-yellow, chin to breast and flanks finely streaked olive-grey. Upperwing dark brown, flight feathers finely edged olive-yellow, tertials, alula and coverts fringed olive-brown. Axillaries and underwing-coverts pale olive-yellow; inner borders of flight feather undersides pale buff. Bare parts as adult ♂. SIZE (10 ♂♂, 4 ♀♀): wing, ♂ 51–55 (53·8), ♀ 47–49 (48·0); tail, ♂ 54–64 (55·4), ♀ 31–33 (32·0); bill, ♂ 17–19 (18·0), ♀ 17–18 (17·3); tarsus, ♂ 14–15 (14·4), ♀ 13–15 (14·0). WEIGHT: ♂ (n = 5) 4–6 (5·0), 1 ♀ 4.

IMMATURE: juvenile similar to adult ♀, but flight feather edges more yellowish; chin and throat uniform dusky grey, bordered by pale yellowish moustachial stripe; breast mottled rather than streaked.

C. n. erlangeri Reichenow: Somalia, (SE Ethiopia?) and NE Kenya from Daua R. to Uaso Nyiro R. where intergrades with nominate race. ♂ with breast band deeper red (not orange); lacks pectoral tufts.

Field Characters. Length 12–13·5 cm. A small sunbird with restricted range in dry areas of E Africa. ♂ has golden green head and upperparts, said to have brassy pink sheen, lacking in Beautiful Sunbird *C. pulchella*, by Zimmerman *et al.* (1996), but in skins at any rate the reverse is the case; black belly and central tail feathers extending about an inch beyond rest of tail; ♂ of nominate race (from central Kenya south) has *complete orange-red* breast-band and *yellow* pectoral tufts, and meets black-bellied race of Beautiful Sunbird, which has incomplete red band in centre of breast and broad yellow patches at sides; northern race *erlangeri* has *pure red* breast-band and *no pectoral tufts*, does not quite meet nominate Beautiful Sunbird in NW Kenya; latter would be readily identified by green belly as well as yellow breast patches. ♀ pale yellow below with *thin dark streaks* on breast and flanks, trace of red or orange on breast; ♀ of Tsavo Purple-banded Sunbird *C. tsavoensis* is whiter and more heavily streaked below. Juv. ♂ has blackish throat bordered by pale malar stripes.

Voice. Tape-recorded (B, GREG, McVIC, STJ). Song a 2-part, thin 'tsit-tsit-tsit-tsit-tsitsereetsereet' or 'tsit-tsit, sit-sreet, sit-sreet' (Zimmerman *et al.* 1996).

General Habits. Inhabits riverine acacias, courses of larger rivers, and watered gardens in dry thornbush regions. Solitary, in pairs or family parties. Forages high in crowns of riverside trees.

Food. Spiders, insects, nectar. Of 16 stomachs (various localities) 15 contained spiders, 3 contained insects (flies, caterpillars), and at least 3 contained nectar. Feeds at red flowers of *Loranthus* parasitic on *Acacia* trees, and forages in flowering baobab trees.

Breeding Habits. Solitary nester.
NEST: suspended, domed purse with side entrance and 'beard' below, built in low tree or shrub.
EGGS: not described.
LAYING DATES: E Africa, Region D, May–Aug.

Reference
Zimmerman, D.A. *et al.* (1996).

Plate 15
(Opp. p. 222)

Cinnyris bifasciata (Shaw). **Purple-banded Sunbird; Little Purple-banded Sunbird. Souimanga bifascié.**

Certhia bifasciata Shaw, 1811–1812. General Zool., 8, p. 198; Malimba, Portuguese Congo.

Forms a superspecies with *C. tsavoensis*.

Range and Status. Endemic resident and partial migrant, S Gabon to Angola, and W Zambia to S Somalia and Natal. Gabon, Congo and Cabinda, common resident along coast; uncommon and irregular as a migrant in NE Gabon; extends up Congo R. from mouth to about 14°E. Angola, Zaire Prov., Bengo, Cuanza R. and N Malanje east to about border of Lunda Norte, south to Benguela and S Benguela escarpment (where the commonest sunbird), NW Huila south to Quilengues, also NE Moxico.

Range in Somalia not clearly understood, owing to confusion with *C. tsavoensis* (Ash and Miskell 1983, 1998); range shown in Somalia is that of both birds; one of them just enters Ethiopia at Dolo. Uganda, scattered records in West Nile Prov., Masindi, Rwenzori Nat. Park, Kigezi, south of L. Kyoga, and Teso (where not uncommon). Kenya and N Tanzania, in W in Lake Victoria basin north to Siaya and Uasin Gishu, east to Ahero, Migori, Lolgorien and Mara Game Res., and south to Grumeti R. and W Serengeti Nat. Park. Not uncommon around Siaya. In NW Tanzania at Kibondo and Nyarambugu and in S L. Victoria on Ukerewe I. and around Mwanza; numerous records in E and a few in S Tanzania, as mapped (N. Baker, pers. comm.). Scattered records in central Kenya. Frequent to locally common along Kenya coast, north to Malindi, Manda I. and Boni Forest; south to Shimoni and abundant on offshore Mpunguti I.; record from inhospitable Kisite I.; frequent inland to about E border of Tsavo East Nat. Park. In NE Tanzania occurs south on mainland to about Mligasi R. and on Zanzibar (where common) and Mafia Island. Record from Kilosa district (Hall and Moreau 1970). In S Tanzania occurs near Mikindani and 100 km inland, and may be much more widespread than shown (Britton 1980).

Zaïre, widespread in Katanga, north in Lualaba valley to 06°S. Zambia, common and widespread in N half, uncommon in S half and Eastern Prov., but so far found only in the areas shown. Present but sparse in central Luangwa Valley. Namibia, frequent in Caprivi Strip.

Cinnyris bifasciata

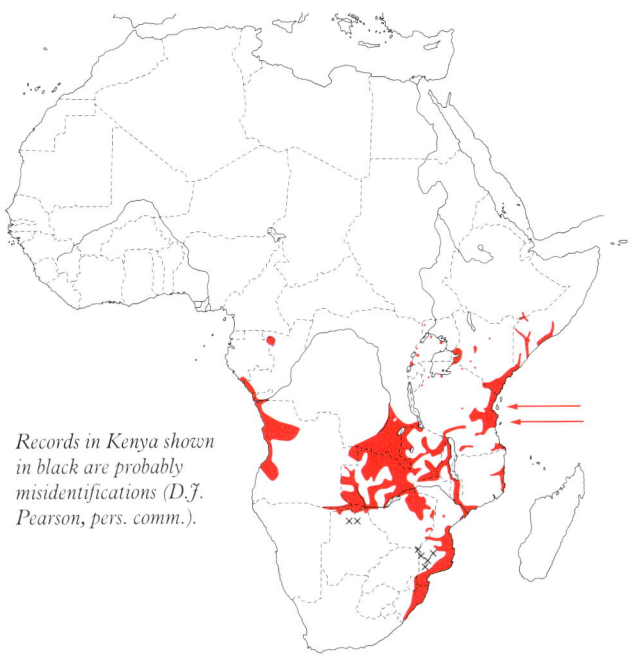

Records in Kenya shown in black are probably misidentifications (D.J. Pearson, pers. comm.).

Botswana, 4 records (Penry 1994). Zimbabwe, local and patchily distributed as shown; in E Highlands restricted to moist, east-facing slopes above Honde and Nyamkwawara Valleys up to 750 m and to Haroni-Lusitu confluence at 350 m (Irwin 1981). Malaŵi, occurs along whole of L. Malaŵi littoral, common between Songwe and Karonga and between Nkhata Bay and Chintheche; all along Shire Valley; in the hills known at 680–1230 m from Mwanza District, Zomba, Malaŵi Hill (Kasingale), Mcocha and Edingeni. Mozambique, very common in coastal lowlands south of 20°S and hinterland west to Mt Gorongoza and Save Valley to Zimbabwe border. Transvaal, status uncertain because of confusion with *C. mariquensis* but may occur just north of NE Swaziland where a pair nested at Malelane in 1986. Natal, locally common in E but rare south of Durban.

Description. *C. b. bifasciata* (Shaw): Gabon to lower Congo R. and W Angola. ADULT ♂ (breeding): head, neck and rest of upperparts metallic green with bronzy tinge. Tail blackish brown with faint bluish gloss above. Chin, throat and upper breast metallic green, concolorous with head, merging through turquoise into narrow band of glossy violet; below this, on lower breast, a dark reddish maroon band (6–8 mm deep), suffused with violet and with a few metallic violet bars; rest of underparts sooty black. Lesser and median coverts metallic green; upperwing otherwise black. Axillaries and underwing-coverts sooty black. Bill black; eyes dark brown; legs black. ADULT ♂ (non-breeding): wings and tail like adult ♂ (breeding); metallic feathers otherwise confined mainly to centre of throat and breast, and to back, rump and uppertail-coverts; variable black feathering below and a few metallic green feather tips on head and mantle. ADULT ♀: upperparts greyish olive-brown, large uppertail-coverts darker brown. Tail dark grey-brown, faintly glossed above. Narrow supercilium pale buff; lores, cheeks and ear-coverts grey-brown. Chin to upper breast pale buff, breast narrowly and faintly streaked grey–brown; rest of underparts pale buffy yellow, flanks with diffuse darker streaking. Upperwing dark grey-brown; remiges edged olive; tertials, tips and fringes of alula, primary coverts and greater coverts buffy brown; median and lesser coverts broadly tipped greyish olive-brown. Axillaries and underwing-coverts creamy white; inner borders on undersides of flight feathers greyish white. Bare parts as adult ♂. SIZE (10 ♂♂, 10 ♀♀): wing, ♂ 55–59 (57.3), ♀ 51–54 (52.4); tail, ♂ 37–40 (38.1), ♀ 31–34 (33); bill, ♂ 20–22 (20.7), ♀ 18–20 (19.6); tarsus, ♂ 16–17 (16.7), ♀ 15–17 (16.1).

IMMATURE: juvenile similar to adult ♀, but rather darker above; below, chin to upper breast sooty brown, bordered by pale moustachial stripe; breast and flanks mottled blackish brown. Mouth gapes bright orange up to 5 months; some 9-month-old birds in full breeding plumage still have bright yellow gapes; gapes thought to blacken by 20 months (Hanmer 1981).

NESTLING: not described.

C. b. microrhyncha (Shelley): coastal Kenya and Tanzania (and this race in W and central Kenya?). Breeding ♂ brighter green, less bronzy, above than nominate race; maroon breastband broader (c. 8 mm), tinged more purple, and upper glossy violet band also broader. ♀ less olive tinged above than nominate race; less yellow below, and more sharply streaked; on some, throat mottled greyish, contrasting with pale moustachial stripes. Wing and tail length similar to nominate race but bill shorter: ♂ (n = 18, E Tanzania), tail 33–40 (37.0), bill 16.5–20 (18.5) (Clancey and Williams 1957). WEIGHT: Kenya coast, ♂ (n = 6) 6.5–7.2 (7.0), ♀ (n = 3) 6–6.5 (6.2).

C. b. strophium Clancey and Williams: Uganda, E and S Zaïre and S Tanzania to South Africa. Like *microrhyncha*, but glossy purple breast band broader, and lower purple–maroon band also broader (10–12 mm); lower abdomen more coal black. ♀ usually darker and greener than in *microrhyncha*, underparts greyer. Tail longer than in *microrhyncha*, wing and bill lengths similar: ♂ (n = 26, S Mozambique), tail 40–45 (41.8), bill 17–19.5 (18.4); population in SW Rwenzori has particularly long tail, av. 43.0 (Clancey and Williams 1957). WEIGHT: W Kenya, ♂ (n = 12) 6.5–8.1 (7.1), ♀ (n = 8) 6–6.5 (6.3); Malaŵi, ♂ (n = 40) 6.4–8.9 (7.9) monthly means varying from 7.4 in Apr to 8.3 in Aug-Jan, ♀ (n = 13) 6.0–7.9 (6.8) (Hanmer 1981); southern Africa, ♂ (n = 73) 6.0–8.9 (7.6), ♀ (n = 22) 6–8 (6.8) (Maclean 1993).

Field Characters. Length 9.5–11 cm. Smaller and slimmer than Marico Sunbird *C. mariquensis*, bill *shorter* and *less decurved*. Dark-plumage breeding ♂ appears metallic blue-green and black at a distance (Zimmerman *et al.* 1996). Breast-bands of ♂ differ racially but *purple more extensive* than Marico, merging into maroon band which is darker and narrower (7–10 *vs* 12–13 mm) and often suffused with purple; in Marico the division between broad maroon and narrow violet bands is usually more clear-cut; race *microrhyncha* (Kenya/Tanzania) has narrow violet band below the maroon and is almost identical in plumage to race *suahelica* of Marico except upperparts bluer, less golden green, belly jet black, not greyish black. ♀ similar to ♀ Marico but underparts paler yellow and less streaked; throat usually white but some individuals of *microrhyncha* have mottled grey throat contrasting with pale moustachial stripe; young ♂ has sooty brown throat and upper breast, dark mottling on breast and flanks. In South Africa (and possibly elsewhere) habitat preference can also be helpful (though both species can occur together), Marico preferring dry thornbush and Purple-banded trees and riverine bush (Skead 1967). Smaller and shorter-billed than Violet-breasted Sunbird, which has broad purple band with no maroon. For differences from Tsavo Purple-banded Sunbird *C. tsavoensis*, see that species.

Plate 17

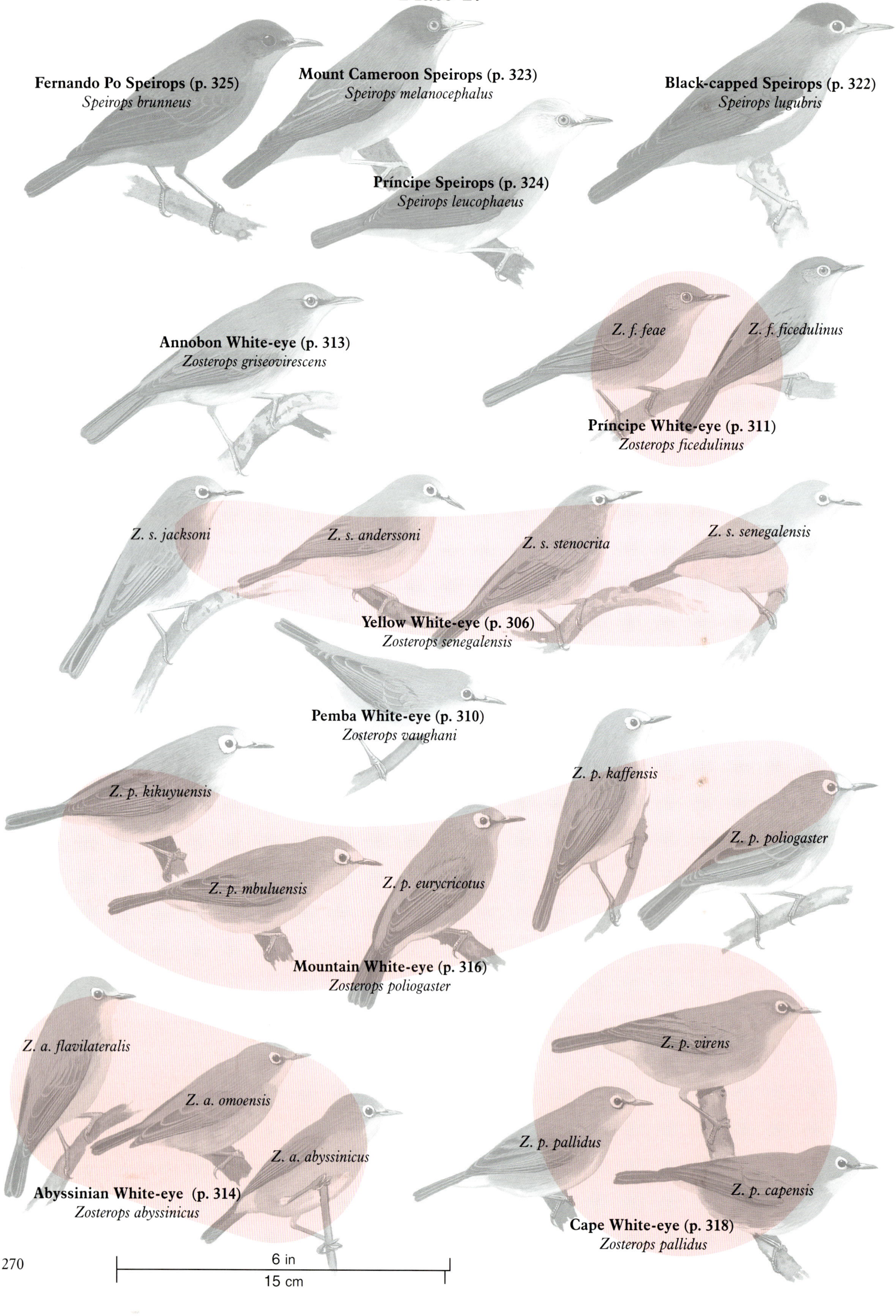

Fernando Po Speirops (p. 325)
Speirops brunneus

Mount Cameroon Speirops (p. 323)
Speirops melanocephalus

Black-capped Speirops (p. 322)
Speirops lugubris

Príncipe Speirops (p. 324)
Speirops leucophaeus

Annobon White-eye (p. 313)
Zosterops griseovirescens

Z. f. feae   *Z. f. ficedulinus*

Príncipe White-eye (p. 311)
Zosterops ficedulinus

Z. s. jacksoni   *Z. s. anderssoni*   *Z. s. stenocrita*   *Z. s. senegalensis*

Yellow White-eye (p. 306)
Zosterops senegalensis

Pemba White-eye (p. 310)
Zosterops vaughani

Z. p. kikuyuensis   *Z. p. kaffensis*

Z. p. mbuluensis   *Z. p. eurycricotus*   *Z. p. poliogaster*

Mountain White-eye (p. 316)
Zosterops poliogaster

Z. a. flavilateralis   *Z. p. virens*

Z. a. omoensis

Z. a. abyssinicus   *Z. p. pallidus*

Abyssinian White-eye (p. 314)
Zosterops abyssinicus

Z. p. capensis

Cape White-eye (p. 318)
Zosterops pallidus

6 in / 15 cm

Plate 18

Voice. Tape-recorded (58, 75, 86, 88, 91–99, B, C, F). Song a rolling metallic chatter with sibilant overtones, typically starting slowly with introductory chip calls and accelerating, 'chip, chip, chip-chippy-tsurrrrrr-titi-trrrrrrr-ssrrrrrrr-sit-tsit'; variable: chatter may undulate down the scale and up, or may end with bubbling 'tsibbly-tsi-tsi-tsi', or turn into a jumble of blurred, thin squeaky notes. Songs last 3–4 s; sometimes given in almost continuous series, barely perceptible breaks giving it a stop-go quality. Calls include distinctive 4-syllable 'tsikit-y-dik', buzzy trill 'brrrz' and annoyance chatter 'chi-chi-chi-chi ...'.

General Habits. Range is markedly but by no means exclusively coastal and riverine. In Kouilou Basin, Congo, inhabits coastal and sublittoral thickets, hardly penetrating tall forest; in Angola mainly in open woodland and acacia scrub; in E Kenya inhabits coastal bush, thickets, mangrove and suburban gardens; in Zambia commonest along edges of moist evergreen forest, also on edges of riparian forest, in *Cryptosepalum* forest in Balovale, *Terminalia* and camelthorn woods in dry country in S Senanga, and dense thickets; in Mozambique and South Africa, scrub thickets, riverine bushes, coastal bush, mangrove (e.g. on Santa Carolina I., where one of the commonest sunbirds), acacia veld, thickets, borders of woodland in more open ground where plenty of shrubs and small trees, edges of evergreen forest, gardens and cultivated land. On the whole avoids tall, mature, dense savanna woodland, though forages along edges of thick woods and forest.

Occurs singly and in pairs. Moves widely, seasonally or unpredictably in search of flowering food plants, where it can gather in large flocks and forages alongside other sunbirds including Scarlet-chested *Chalcomitra senegalensis* and White-bellied *Cinnyris talatala*. Very restless, moving rapidly from flower to flower; aggressive to other sunbirds. Foraging techniques include searching for spiders in webs under eaves of houses. Given to a good deal of chasing, mainly of ♀♀ by ♂♂. Moult schedule as in other sunbirds (Hanmer 1981).

Mainly resident; little or no good evidence of movement in e.g. Zambia (Aspinwall 1979). In Zimbabwe clearly undertakes seasonal movements; in Dichwe Forest at 1150 m a breeding visitor between July and Nov–Dec, absent in Jan–June. Birds seen in NE Gabon are probably non-breeding migrants. In Nchalo, S Malaŵi, a non-breeding visitor, adults and juveniles arriving in late Nov and departing in June (–Aug); 2 birds retrapped in following year, suggesting migratory fidelity (Hanmer 1981). In S Mozambique thought to be an abundant visitor to lower Zambezi Valley, coastal plain and offshore islands in the dry season (Clancey 1996, but see Parker 1999).

Food. Nectar and small arthropods. Feeds especially at flowers of acacias, mistletoes, *Mimusops caffra* and *Syzygium cordatum*. Small flies and winged termites in stomachs. 'A pair obtained at the nest were carrying small seeds to the young one' (Vincent 1949). Recently-fledged young given spiders.

A

Breeding Habits. Solitary nester, presumably monogamous.

NEST: suspended, oval or pear-shaped bag with round side-top entrance; small, neat and copiously decorated (**A**); often with wispy beard dangling below. Made of dry, chaffed grass and fine, delicate plant fibres mixed with spider web and some woolly plant down, the outside entirely covered with bits of lichen, bits of dead leaves, white chips of wood, caterpillar droppings, and clusters of tiny reddish seed; thinly lined with feathers or, uncommonly, soft vegetable down including asclepiad seed hairs. Height *c.* 100, width *c.* 70. Situated in clumps of small trees, bushes and climbers growing thickly on sides of large *Macrotermes* termite mounds (Lubumbashi, Zaïre), over water, by roads, and over dry river-beds (South Africa). One nest 15 cm from nest of wasps. Nest placed at end of branch, suspended from tip of twig, or end of bamboo stem curving out of clump of thorn trees (Vincent 1949), on climber and in thorn tree (Skead 1967); mostly 3–4.5 m from ground.

EGGS: 1–2 (Zaïre), 2 (southern Africa). Slightly elongated; shell smooth and slightly glossy. Slate, white, khaki or purplish grey; lightly marked with black, dark grey or purplish brown or pale purple spots, streaks, splashes and lengthwise smears, often overlying each other and concentrated at large end (Skead 1967). SIZE: (n = 20) 15.0–17.8 × 10.4–11.8 (16.0 × 11.2).

LAYING DATES: Angola, May, Oct; E Africa (following records may include some of *C. tsavoensis*), Region B, May,

Aug, Region E, Mar–May, July, Oct–Nov; Zambia, Aug–Nov and Jan, mainly Sept–Oct; Malawî, May–Aug; Zimbabwe, Sept–Dec and Mar, mainly Sept–Nov; Mozambique, Sept–Dec; Transvaal, Sept (once); Natal, Oct–Nov.

DEVELOPMENT AND CARE OF YOUNG: young fed by both parents.

BREEDING SUCCESS/SURVIVAL: longevity (2 ♂♂) 7·5 and 8·5 years (Hanmer 1989).

References
Clancey, P.A. and Williams, J.G. (1957).
Hanmer, D.B. (1981).
Skead, C.J. (1967).
Vincent, A.W. (1949).

Cinnyris tsavoensis (van Someren). Tsavo Purple-banded Sunbird. Souimanga bifascié du Tsavo. **Plate 15 (Opp. p. 222)**

Cinnyris bifasciatus tsavoensis van Someren, 1922. Novit. Zool., 29, p. 106; Tsavo.

Forms a superspecies with *C. bifasciata*.

Range and Status. Endemic resident, S Somalia to NE Tanzania. Range in Somalia not clearly understood, owing to confusion with *C. bifasciata microrhyncha* (Ash and Miskell 1983, 1998); range shown in Somalia is that of both birds; one of them just enters Ethiopia at Dolo. In Kenya, occurs in dry E from about equator to Simba, Kibwezi, Mtito Andei and Maungu, including Tsavo East Nat. Park, and Tsavo West Nat. Park. In NE Tanzania, from North Pare Mts to Same, Mkomazi Game Res., Korogwe and Handeni.

Cinnyris tsavoensis

Description. ADULT ♂: head, neck and upperparts metallic green, more bluish green on rump and uppertail-coverts, head and neck with slight bronzy tinge. Tail blackish, slightly glossy above, feathers narrowly edged metallic green when fresh. Chin, throat and upper breast metallic green, merging through bluish green into narrow band of glossy violet *c.* 2 mm deep with bluish intrusions; below this, on lower breast, a narrow maroon band 3–5 mm deep; rest of underparts black. Lesser and median coverts metallic green; upperwing otherwise black. Axillaries and underwing-coverts sooty black. Bill black; eyes black; legs black. ADULT ♀: upperparts greyish olive. Tail dark grey-brown above, faintly glossed; T6 with whitish tip and narrow white edge along outer edge, T5 and T4 narrowly tipped whitish. Narrow supercilium buffish white; lores, cheeks and ear-coverts grey-brown. Underparts pale buffish white, breast lightly streaked grey-brown, lower breast to undertail-coverts tinged yellow. Upperwing dark grey brown; remiges, alula and greater coverts edged olive, median and lesser coverts broadly tipped greyish olive. Axillaries and underwing-coverts buffish white, mottled grey-brown; inner borders of undersides of flight feathers buffish white. Bare parts as adult ♂. SIZE (9 ♂♂, 4 ♀♀): wing, ♂ 54–57 (55·2), ♀ 48–51 (49·0); tail, ♂ 38–41 (40·7), ♀ 33–35 (34·3); bill, ♂ 16–17 (16·8), ♀ 16–17 (16·5); tarsus, ♂ 14–15 (14·6), ♀ 14–15 (14·5). WEIGHT: SE Kenya, ♂ (n = 10) 6·3–7·6 (7·0), ♀ (n = 5) 6·0–6·8 (6·5).

IMMATURE: juvenile similar to adult ♀, but chin and throat dark grey brown, bordered by pale moustachial stripes, underparts more boldly barred dark brown.

TAXONOMIC NOTE: status of taxon *tsavoensis* not resolved. It was formerly thought to be partly sympatric with *C. bifasciata microrhyncha*; that now appears unlikely, although non-breeding examples of *bifasciata* do occur in breeding range of *tsavoensis*. The latter is distinguished from *bifasciata* by having a much narrower maroon breast band and by apparent absence of eclipse plumage; but *tsavoensis* is also noteworthy for its range bisecting that of *bifasciata* (in Kenya).

Field Characters. Length 9·5–11 cm. Almost identical to Purple-banded Sunbird *C. bifasciata*, except that ♂ differs in maroon *breast-band* being either *very narrow* (3–5 mm), or present *only at sides*, or even *lacking*. The 2 species are parapatric, so *range* can be used in identification except at borders, also habitat preference, Tsavo inhabiting dry *Commiphora* and acacia bush and scrub, Purple-banded preferring moist coastal bush and thickets, mangroves and suburban gardens (Zimmerman *et al.* 1996). Overlaps with larger Violet-breasted Sunbird *C. chalcomelas* in Tsavo region of Kenya; bill shorter and more deeply curved; ♂ Tsavo has *narrow purple breast-band*; ♀ Tsavo has pronounced *white malar streak* contrasting with grey throat.

Voice. Tape-recorded (GIB, McVIC). Song a rapid, sputtering 'tsusitiseesee, chuchiti-tsi-tsi-tsi-tsi-sitisee-see-see-see-chitisee ...' and 'sitisee-see-see tseu-tseu-tseu chiti-tisiti-see-swee', sometimes shortened to only the last few notes (Zimmerman *et al.* 1996).

General Habits. Inhabits dry *Commiphora* and *Acacia* wooded savannas and scrub. Habits studied in Tsavo East Nat. Park, Kenya (as '*C. bifasciata*', before separation of *C. tsavoensis* recognized: Lack 1985): resident, almost confined to woodland, but occurring also in other thick habitats in N of Tsavo. Eats both nectar and insects; foraging habits not remarkably different from those of closest allies. Of 178 feeds 82% were of nectar obtained mainly in trees including *Delonix elata*, also in bushes, and 18% were of insects. 70% of insects caught by bird in flight and 30% by bird gleaning them from twigs.

Food. Nectar and insects: see above.

Breeding Habits.

NEST: only one described (Serle 1943); suspended, elliptical purse with side entrance in upper half, without porch, with 'beard' below 30 mm long; compactly made (though less compact or elaborate than nests of most other sunbirds) of fine strips of wood fibre, grasses, dried acacia-buds, bark and tiny twigs, bound together with gossamer, the fragments of bark chiefly adhering to the outside. No feathers inside nest which appeared to be unlined though clutch already laid. Sited 2 m from ground near tip of slender shoot of thorny *Acacia* tree; nest entrance faced away from tree; suspended below branch which was incorporated into fabric of nest roof. Size: height (excl. 'beard') 150, greatest breadth 50, from front to back max. 53, int. height of cavity 58, depth of floor below lower lip of entrance, 30.

EGGS: 1 (proved to be only 1 by examination of ♀ caught on nest: Serle 1943). Ovate, smooth, glossless; palest grey, delicately and beautifully marked with slightly darker grey suffusions, blotches and spots, mainly around large end, with a few scattered fine grey-brown speckles. SIZE: (n = 1) 15·4 × 10·6.

LAYING DATES: E Africa, Region D, Jan, Mar-Dec; Kenya (Garissa), Oct.

References
Clancey, P.A. and Williams, J.G. (1957).
Lack, P. (1985).
Serle, W. (1943).

Plate 15 (Opp. p. 222) *Cinnyris chalcomelas* Reichenow. Violet-breasted Sunbird. Souimanga à poitrine violette.

Cinnyris chalcomelas Reichenow, 1905. Vög. Afr., 3, p. 482; Kismaju, S Somaliland.

Forms a superspecies with *C. pembae* (and Malagasy *C. notata*?).

Range and Status. Endemic resident, Somalia and Kenya. Somalia, fairly common but local south of 03°N in lower Jubba R. and Webi Shabeelle R. valleys and adjacent lowlands up to about 70 km away. In Boni Forest on Kenyan border. Kenya, uncommon and locally fairly common in coastal lowlands from Somali border and Kiunga to Tana R. delta, Watamu and Kilifi and inland to Ijara, Bura and Galana Ranch; several old records between Kilifi and Tsavo and Tanzania border (Hall and Moreau 1970), which probably refer to Purple-banded Sunbird *C. bifasciata* (D. Pearson, pers. comm.). Record from Moyale, probably also a misidentification.

Cinnyris chalcomelas

Description. ADULT ♂: head, neck and rest of upperparts metallic emerald green. Tail black, upperside with bluish gloss. Lores blackish. Chin and throat metallic bluish green, merging with greener side of head, and grading through blue to broad glossy violet band on upper breast 7–8 mm deep. Rest of underparts black. Lesser and median coverts metallic green; upperwing otherwise blackish, alula and lesser primary coverts fringed metallic green. Axillaries black; underwing-coverts black apart from glossy green marginal coverts. Bill black; eyes dark brown; legs black. ADULT ♀: upperparts greyish brown, tinged olivaceous; large uppertail-coverts dark grey-brown. Tail blackish brown, faintly glossed above; T6 with broad greyish white fringe at tip of inner web, and very narrow whitish outer edge, other feathers often narrowly fringed whitish at tip. Supercilium buffy white; lores dusky brown; cheeks and ear-coverts greyish brown. Underparts buffy white, tinged grey on chin and upper throat and faintly streaked greyish on lower throat, breast and flanks; slightly yellow on centre of breast and belly. Upperwing dark grey-brown, primaries edged greyish white and secondaries grey-buff, tertials, alula, primary coverts and greater coverts fringed buff, median and lesser coverts broadly tipped greyish brown. Axillaries buffy white; underwing-coverts dusky brown, tipped buffy white; inner borders on undersides of flight feathers buffy white. Bare parts as adult ♂. SIZE (4 ♂♂, 5 ♀♀): wing, ♂ 60–63 (61·8), ♀ 55–58 (56·6); tail, ♂ 40–43 (41), ♀ 32–38 (33·2); bill, ♂ 18–20 (19·3), ♀ 18–19·5 (18·7); tarsus, ♂ 16–17 (16·5), ♀ 14·5–16 (15·2) WEIGHT: ♂ (n = 4) 7–9 (8·5), 1 ♀ 7.

IMMATURE and NESTLING: not described.

TAXONOMIC NOTE: formerly known as a race of *C. pembae*. Plumage, size and voice differences between mainland *chalcomelas* and insular *pembae* were emphasized by Pakenham (1979), Archer and Turner (1993) and Zimmerman *et al.* (1996), and we follow those authors in treating *C. chalcomelas* and *C. pembae* as separate species.

Field Characters. Length 11–12 cm. Meets both Purple-banded and Tsavo Purple-banded Sunbirds *C. bifasciata* and *C. tsavoensis*; larger than both, with *longer bill* and *shorter tail*. ♂ similar above but has *broad purple* breast-band *without a maroon* band below it; throat greenish blue rather than golden green. ♂ Tsavo Purple-banded can lack maroon breast-band but its purple band is very narrow. ♀ pale grey to very pale yellow below, *plain or with only a hint* of streaking; browner and lighter above than ♀ Tsavo, with more pronounced pale supercilium. Juv. ♂ has black throat.

Voice. Tape-recorded (McVIC). Calls while foraging a mixture of dry 'chut' and 'chip' notes and hard trills, the latter often turning into a brief (1–2 s) swizzling song, 'chrrrrrrsssswwwizzzzlllle'. Another song lacks swizzling quality, a variable, much slower series of individually audible notes, sharp and clear, often ending with a buzzy note, 'chee-per-chichi-woo-per-chichi-chee-dzurr'; a third type (subsong?) is low-key, protracted, jerky and conversational, mixing clear notes like those of previous song, 'chipper-chee-chee', with dry rattles, hard 'tip', clear 'tewp', short sweet warbles (imitation of other species?), clear trills, machine gun-like 'ratatat' and bursts of rapid swizzling; uneven rambling, subdued quality reminiscent of Barn Swallow *Hirundo rustica*.

General Habits. Inhabits coastal shrub-savanna and grassy thickets. Forages rather less actively, with more deliberate movements, than Tsavo Purple-banded Sunbird, in shrubs and small flowering trees; perches quietly for long periods (Zimmerman *et al.* 1996). Occurrences inland may be seasonal (Kenya: Zimmerman *et al.* 1996). Occurs singly or in small parties; once 100 at *Acacia* flowers (Afgoye, Somalia, Feb: Ash and Miskell 1998).

Food. Not known; doubtless nectar and small insects.

Breeding Habits.
NEST: not described. Built by ♀ only (Ash and Miskell 1998).
LAYING DATES: Somalia, (nestlings, Dec; nest-building Apr, July, Oct) (Ash and Miskell 1998). Evidence of breeding at Ijara, Kenya (Percy *et al.* 1953).

Cinnyris pembae Reichenow. Pemba Sunbird. Souimanga de Pemba.

Plate 15
(Opp. p. 222)

Cinnyris pembae Reichenow, 1905. Orn. Monatsb., 13, p. 180; Pemba Island.

Forms a superspecies with *C. chalcomelas* (and Malagasy *C. notata*?).

Range and Status. Endemic resident, Pemba Island, where abundant throughout.

Description. ADULT ♂: head, neck and rest of upperparts metallic green or bluish green, usually bluer on uppertail-coverts. Tail black, feathers narrowly edged metallic green (T1 on both webs). Lores blackish; chin and throat metallic bluish green like head, merging with glossy violet-purple band on upper breast 4–7 mm deep; rest of underparts black. Median coverts and largest outer lesser coverts metallic green, merging with glossy violet-purple smaller and inner lesser coverts; greater coverts finely edged metallic green; upperwing otherwise black. Axillaries and underwing-coverts black. Bill black; eyes dark brown; legs black. ADULT ♀: upperparts olivaceous grey-brown. Tail dark grey-brown, faintly glossed above; T3–T4 with narrow whitish terminal fringe, T5 with broader fringe at tip of inner web, T6 with larger whitish spot at tip of inner web and narrow whitish outer edge. Supercilium pale buff; lores dusky brown; cheeks and ear-coverts greyish brown. Underparts plain creamy white, throat contrasting with side of head; tinged yellower on belly. Lesser and median coverts grey-brown; upperwing otherwise dark grey-brown, remiges edged olive-brown, alula, primary coverts and secondary coverts fringed grey-buff. Axillaries and underwing-coverts creamy white; inner borders on undersides of flight feathers greyish white. SIZE (8 ♂♂, 2 ♀♀): wing, ♂ 52–54 (53), ♀ 48, 50; tail, ♂ 31–36 (33·5), ♀ 29, 30; bill, ♂ 16–18 (17·3), ♀ 16, 18; tarsus, ♂ 14–16 (15), ♀ 15, 16.

Cinnyris pembae

IMMATURE: juvenile like adult ♀, but chin and throat barred dark grey, separated from cheeks by broad buffy white moustachial stripe; breast and flanks mottled dark grey. Immature ♂ acquires metallic plumage first on throat and upper breast.

NESTLING: not known.

TAXONOMIC NOTE: based partly on the findings of Clancey and Williams (1957), *pembae* was treated as conspecific with *C. notata* (Comoros, Madagascar) and *C. chalcomelas* by White (1963) and with *chalcomelas* by Hall and Moreau (1970), Wolters (1982), Short *et al.* (1990) and Dowsett and Forbes-Watson (1993). Wolters (1982) placed *pembae* and *notata* in subgenus *Angaladiana* (sequenced between *C. (Maricornis) bifasciatus* and *C. (Cinnyris) coccinigaster*). ♂ *notata* resembles a giant, long-billed *pembae*, and ♀ *notata* (heavily streaked below) is more like ♀ *C. johannae/C. coccinigaster* than ♀ *pembae* (Hall and Moreau 1970), so *pembae* and *notata* are best kept as separate species (Dowsett and Dowsett-Lemaire 1993). Plumage, size and voice differences between insular *pembae* and mainland *chalcomelas* were emphasized by Pakenham (1979), Archer and Turner (1993) and Zimmerman *et al.* (1996). Predicting that the morphological differences between the 2 taxa would obviate interbreeding were they ever to meet, we follow those authors in treating *pembae* and *chalcomelas* as separate species.

Field Characters. Length 9·5–10 cm. Endemic to Pemba Is., where it is the only 'purple-banded' sunbird; similar species occur on Zanzibar and the mainland. ♂ like smaller version of Violet-breasted Sunbird *C. chalcomelas*, with wing-coverts violet and blue rather than green; ♀ more distinct, white below with yellowish belly, not grey.

Voice. Tape-recorded (B, McVIC). Call note loud and distinctive, a frequently-repeated 'tslink-tslink-tslink' or high-pitched 'ssweek', given continuously, sometimes for several min, from bare exposed branches of tall trees; totally unlike any call of Violet-breasted Sunbird (Archer and Turner 1993; Zimmerman *et al.* 1996). In flight, a repeated 'tsik' or 'tsik-kik-kik'; bird in mist net gave 'jik' or 'juk' of alarm.

General Habits. Widespread throughout Pemba, in towns, villages, the countryside and offshore islands including small ones. Feeds at flowers, presumably upon nectar, and widely attracted to single scarlet flowers of *Hibiscus* cultivars (Vaughan 1930). Also feeds on pulpy, sweet, white berries of shrub *Flueggia virosa* (Euphorbiaceae), perching and jabbing at them with bill to extract small segments which are swallowed whole, or even plucking and swallowing whole small berries (Archer and Parker 1993).

Food. Nectar and small fruits (see above).

Breeding Habits. Evidently monogamous and territorial. Breeding ♂♂ pugnacious towards each other. Sings only in breeding season.

NEST: a suspended bag or purse with side entrance, typical of sunbirds in shape, but rather thick-walled; well lined, often with fluffy down from seeds of grass *Imperator cylindrica*. Sited 1–2·5 m above ground, suspended from outer twigs of shrub, sometimes partly concealed amongst foliage.

EGGS: 2. Ground colour greenish white, almost obscured by streaks and marbling with various shades of ashy brown.

LAYING DATES: May–Jan, mainly May–Nov. Post-breeding moult Dec–Apr.

References
Pakenham, R.H.W. (1979).
Vaughan, J.H. (1930).

Plate 14 (Opp. p. 207) *Cinnyris bouvieri* Shelley. Orange-tufted Sunbird. Souimanga de Bouvier.

Cinnyris bouvieri Shelley, 1877. Monog. Nectariniidae, p. 227, pl. 70; Landana, Cabinda.

Forms a superspecies with *C. osea* and Oriental *C. asiatica*.

Range and Status. Endemic resident (and migrant?), with an apparently fragmented range in savanna woodlands encircling Congo forest block. Nigeria, common on Obudu and Mambilla Plateaux. Cameroon, frequent at about 1350–2050 m on Bamenda-Banso highlands (Bamboutos Mts); Mt Manenguba, Adamawa Plateau, Mieri and Yaoundé. Central African Republic, records from Nana-Mambéré region in W, Bangui, Bangui R. near 20°E, and Rafaï (04°56′N, 23°55′E). Equatorial Guinea, listed by Fa (1990). NE Zaïre, reported in 1950s from Pawa (where common), Niangara (frequent) and Aba (1 bird, 7 km from Sudan border). Uganda, generally uncommon but seasonally numerous: at 700–1800 m around NE corner of L. Albert, Bugoma Forest, Bwamba, E slopes of Rwenzori Mts, Kibale and Kalinzu, east to Kampala, Entebbe, Mabira and about Ikubwe. Kenya, only in Busia District on Ugandan border and at about 1700 m in Kakamega Forest, where frequent records since discovery in 1965. Rwanda included in range given by White (1963), probably in error.

Southern tropics: Gabon, frequent (many collected) at Tchibanga in SW and records of vagrants in NE. Zaïre (south of equator), record from near Congo border at *c*. 13°E (Hall and Moreau 1970), locally frequent in Cabinda (Angola); not found in Kouilou basin (Congo, near Cabinda) by Dowsett-Lemaire and Dowsett (1991) but several noted in SE Kouilou at Gungu, Gudi and Katenga by Lippens and Wille (1976); reported from Port Nganciu (03°17′S, 16°11′E); records in Kasai Occidental (Kayembe Mukulu and Kananga); W Katanga, and Sandoa on L. Tanganyika. Angola: frequent at Landana, Cabinda; and Cuanza Norte, N Malanje and W Lunda Norte Provs. Zambia, known only from Zambezi Rapids, N Mwinilunga and Mwombezhi River, Solwezi (Benson and Irwin 1966).

Cinnyris bouvieri

Description. ADULT ♂: forehead and forecrown metallic violet-purple; hindcrown, sides of head, neck and rest of upperparts metallic bronzy green, more bluish on uppertail-coverts. Tail blackish, with slight violet gloss above; T1–T5 edged metallic blue-green. Lores to upper chin black; lower chin glossy dark blue, merging with green side of head and with green throat. On upper breast, metallic green grades through narrow line of turquoise to purple band *c.* 5 mm deep; below this a non-metallic dark maroon band; pectoral tufts bright orange, tipped yellow; rest of underparts sooty brown. Lesser and median coverts metallic green; upperwing otherwise dark brown. Axillaries and underwing-coverts sooty brown. Bill black; eyes brown; legs black. ADULT ♀: upperparts olivaceous brown. Tail blackish brown with faint blue gloss above; T6 with outer web and tip of inner web pale brown, outer edge narrowly whitish; T4–T5 with pale brown terminal fringe. Superciliary stripe pale buff; lores dusky brown; cheeks and ear-coverts olive-brown. Underparts pale grey-buff, tinged olive-yellow on breast and belly, more greyish on flanks and thighs; fine mottling on centre of chin and throat often coalescing to form grey patch; breast with faint grey mottling; undertail-coverts grey-buff with dark brown centres. Lesser and median coverts olivaceous brown; upperwing otherwise dark brown, flight feathers edged olive. Axillaries and underwing-coverts pale olive–yellow; inner borders on underside of flight feathers pale greyish. Bare parts as adult ♂. SIZE (10 ♂♂, 10 ♀♀): wing, ♂ 55–59 (57·5), ♀ 52–54 (53·5); tail, ♂ 36–42 (39·5), ♀ 29–36 (33·9); bill, ♂ 19–22 (20·9), ♀ 18–20 (19·2); tarsus, ♂ 16–18 (16·8), ♀ 15–17 (15·6). Also (17 ♂♂, 12 ♀♀): wing, ♂ 54–60 (57·6), ♀ 52–55 (53·5); tail, ♂ 36–41 (38·6), ♀ 32–36 (33·5); bill, ♂ 20–22 (21·2), ♀ 19–20·5 (19·6) (Benson and Irwin 1966). WEIGHT: Kenya, ♂ (n = 7) 7–10 (8·5), Nigeria, 1 ♂ 10·0.
IMMATURE and NESTLING: not described.

Field Characters. Length 10–12 cm. Bill long and lightly curved. ♂ differs from Marico and Purple-banded Sunbirds *C. mariquensis* and *C. bifasciata* by *purple forecrown, blue chin* and *orange* and yellow pectoral tufts. Violet and maroon breast-bands narrower (latter hard to see), and head and upperparts shiny golden green, without blue tinge. ♀ very undistinguished, dull olive above, yellow-olive below with faint streaking and mottling, more marked on throat; extremely hard to tell from ♀ Bates's Sunbird *C. batesi* and even museum skins have been misidentified (Benson and Irwin 1966); best characters are *dark mottled throat* bordered by indistinct pale malar stripe and *long*, relatively *straight bill*. Tail feathers glossy black without the olive tips and margins characteristic of ♀ Bates's; outermost pair paler unglossy brown tipped whitish and with whitish margins to outer webs (Benson and Irwin 1966).

Voice. Tape-recorded (102, 104, B, F, CHA, McVIC). Song begins with 1–2 introductory notes and continues as regular series of up–down notes on one pitch, 'tsit ... tsi, see-chu-see-chu-see-chu-see-chu ...', lasting 4–6 s, each note individually audible; sometimes starts with dry chattery scolding trill, 'trrrrrrr-see-chu-see-chu ...', or 'see-chu' part may be interrupted by softer trills. Calls include a low 'cheep' or 'chip-ip' and a loud 'tchew' (Zimmerman *et al.* 1996).

General Habits. Prefers recently modified habitats such as plantations and deforested areas with plenty secondary growth; also edges of moist evergreen forest, adjacent rich *Brachystegia* woodland; margins of riverine forest; glades; savanna woods and open tall-grass country with trees and bushes. In Cameroon, lives in more open parts of woods in hill savanna country, second-growth shrubbery and orchard bush, and occurs (uniquely among sunbirds there) far away from woody growth, on open hillsides with short grass, bracken, thistles, and herbaceous flowering plants (Serle 1950). At Niangara (Uele, Zaïre), feeds in Ceará rubber trees when in flower in May–June, and at Pawa (Uele) on large thistle-like plants, legumes and *Erythrina*. Mainly solitary; sometimes with other sunbird species. Tends to forage low down; favours thistle-like *Acanthus* flowers.

Too poorly known for movements to be understood; usually thought to be sedentary although, like many other sunbirds, subject to local movements; appearances and disappearances doubtless governed by favourite-tree flowering seasons; however, also said to be an erratic Afrotropical migrant from S Gabon to NE Gabon, where noted in Apr and Oct (Brosset and Erard 1986).

Food. When foraging at blossoms (see above) may take nectar, but only food found in 5 stomachs were very small insects including tiny beetles, in 2 stomachs mixed with coagulated rubber (Chapin 1954).

Breeding Habits. Solitary nester.
NEST: domed purse, made of fine grasses and spider web, lined with vegetable down; suspended in bush or from tall grass stem (Lippens and Wille 1976) or, only 70 cm from ground, from thistle stem (Serle 1950).
EGGS: not described.
LAYING DATES: Cameroon, Oct; Zaïre (Uele), (inferred to breed in late rains: Chapin 1954). Zambia, Mar.

Reference
Benson, C.W. and Irwin, M.P.S. (1966).

Plate 14
(Opp. p. 207)

Cinnyris osea Bonaparte. Palestine Sunbird; Orange-tufted Sunbird. Souimanga de Palestine.

Cinnyris osea Bonaparte, 1856. Compt. Rend. Acad. Sci. Paris, 42, p. 765; 'plaines de Jéricho', Palestine.

Forms a superspecies with *C. bouvieri* and Oriental *C. asiatica*.

Range and Status. Resident and partial migrant, S Lebanon (formerly), Israel, W Jordan, Palestine, Sinai, W Saudi Arabia, Yemen, S Oman; migrant, Afrotropics.

Chad, not recorded since the African subspecies *decorsei* was described from L. Chad in 1904. Cameroon, bird collected at Tello (07°10′N, 14°18′E), Oct 1974. Central African Republic, 1 north of Ndélé, N Bamingui-Bangoran Prov., 1948; 2 ♂♂, Bozo (05°10′N, 18°30′E), Sept 1976 (Germain and Cornet 1994). Zaïre, not infrequent at Faradje, Uele, and known also from Dungu; all records in Oct–Mar and bird is 'undoubtedly migratory', almost certainly breeding in Jan–Feb (evidence from gonads: Chapin 1954). Sudan, common breeding visitor to Jebel Marra, Darfur, at 2000–2900 m in Nov–Mar/Apr, breeding in Nov–Jan; rare in S Sudan, Oct–Mar, between Bengengai, Tambura and Kajo Kaji, thought not to breed (Nikolaus 1989), although breeding status is presumably the same as that of population in neighbouring Uele.

Levant and Arabian nominate race, not yet certainly recorded in Africa, is abundant resident and partial migrant in Israel, breeding all months, mainly Feb–Aug; some migrants winter in Israel coastal plains; obvious broad-front diurnal migrant in Oct–Nov at Eilat, perhaps to Sinai, where much commoner in winter (up to 1000 m) than in summer (Shirihai 1996). In N Sinai, resident from Rafah west to El Arish (Baha el Din 1985).

Where Jebel Marra and S Sudan/NE Zaïrean populations go between Mar and Oct is not known. Birds found in Chad, Cameroon and Central African Republic may be migrants, from Darfur?; they may even be migrants from Levant, but, if so, have been racially misidentified.

Description. *C. o. decorsei* (Oustalet): only race in Africa. ADULT ♂ (breeding): forehead and forecrown dark metallic blue; hindcrown, sides of head, neck and rest of upperparts bronzy green; uppertail-coverts turquoise or blue. Tail blackish, upperside glossed violet; outer webs of T1–T5 edged metallic blue-green. Lores black. Chin dark metallic blue; throat greenish blue, merging with bronzy side of head, and with blue upper breast which grades to violet-blue next to lower breast; pectoral tufts orange; rest of underparts sooty black. Lesser and median coverts bronzy green; rest of upperwing dark grey-brown. Axillaries and underwing-coverts blackish brown; inner borders on undersides of flight feathers pale grey. Bill black; eyes black-brown; legs black. ADULT ♂ (non-breeding): wings and tail same as breeding ♂; retains yellow pectoral tufts and metallic feathering on rump, uppertail-coverts, chin, throat and breast; upperparts with scattered metallic green feather tips, underparts only patchily black; otherwise like adult ♀. ADULT ♀: uppertail-coverts black glossed with metallic green, rest of upperparts olivaceous grey-brown. Tail blackish above, slightly glossy, dark grey-brown below; T6 with outer web and tip of inner web pale brown; inner web of T5 with pale tip. Lores dusky brown; cheeks and ear-coverts grey-brown. Chin to breast and flanks grey-buff, separated from cheeks by rather paler moustachial stripe; lower breast to belly paler, tinged yellow; undertail-coverts greyish white with dark brown centres. Primaries and secondaries dark grey-brown, edged olive-brown; rest of upperwing olivaceous grey-brown.

Cinnyris osea

Axillaries pale buffy yellow; underwing-coverts buffy white; undersides of flight feathers like adult ♂. SIZE (10 ♂♂, 5 ♀♀): wing, ♂ 50–56 (52·9), ♀ 47–52 (49); tail, ♂ 31–37 (35·2), ♀ 31–33 (31·8); bill, ♂ 16–18 (16·8), ♀ 16–18 (16·6); tarsus, ♂ 14–16 (14·9), ♀ 13–15 (14). Weight unknown (weight of nominate race, Israel: 8 ♂♂ av. 7·6, 12 ♀♀ av. 6·8)

IMMATURE: juvenile similar to adult ♀, but yellower below.
NESTLING: not described.
TAXONOMIC NOTE: *C. o. osea* has blue-green sheen on lower breast and upper belly, bill to skull averages 2·4 mm longer, and wing and tail are 2–3 mm longer than in *decorsei*. Wing measurements of ♂ and ♀ specimens from Jebel Marra (Chapin 1954) are almost exactly the same as in Nat. Hist. Mus. (London) series of *decorsei*, but were said by Chapin to be almost equal to those of *C. o. osea*. Further study is needed.

Field Characters. Length 8–9·5 cm. African race *decorsei* short-tailed and *tiny*; smaller and shorter-billed than Orange-tufted Sunbird *C. bouvieri*. ♂ looks very *dark*, with black wings and belly, purple-blue forehead and rump, blue-green upperparts and throat; has orange and yellow pectoral tufts like Orange-tufted but differs in *broad purple breast-band*. Does not overlap Shining Sunbird *C. habessinica*, which would be readily identified by red breast-band and larger size. ♀ is duller and less yellow below than ♀ Orange-tufted, without mottling or streaking, and has *glossy blue-green* uppertail-coverts.

Voice. Tape-recorded in Palearctic only (B, HOL). Many renderings of song of nominate *osea*: '2–3 introductory notes followed by fast, lower-pitched flourish'; '1st 2 notes slow, remainder rapid and lower-pitched, 'tsi-tsi-tser-tser-

tser-tser-tser', like Blue Tit *Parus caeruleus*'; '2 loud sweet whistled notes, then rapid hard musical trill, 'tweet-tweet-chee-e-e-e-e-e-', trill lower pitched but inflected up then down'; 'fast high-pitched trilling, 'dy-vy-vy-vy-vy-vy', rising 'tveeit-tveeit-tveeit' or 'veet-tji veet-tji veetji' accelerating and often ending in trill like European Serin *Serinus serinus*' (Cramp and Perrins 1993). Calls include thin 'ftift', European Robin-like 'tiiu', hard 'tskak' and loud, sharp 'te-*veeit*, te-*veeit*', second note stressed and rising (Porter *et al.* 1996). For further vocalizations see Cramp and Perrins (1993).

General Habits. In Africa inhabits desert gardens and vegetated wadis, open savannas, bushy grassland above 2000 m and mature woodland below that level. In Levant, dry, open, grassy places with scattered trees and shrubs, coastal gardens, orchards, flowering shrubs and creepers along lanes, rocky and bushed grassland, bushy river banks, cultivated valleys, rock-strewn hillsides, juniper forest on mountain summits, and cypress groves up to 1500 m. Mainly solitary, in pairs, or loose flocks on migration and (up to 15 birds) in food trees. Sips nectar by hovering in front of flower or perching beside it. Perforates base of long corollae with bill, making small hole through which nectar is taken (in Israel *Convolvulus*, *Hibiscus*, *Malvaviscus*). Tame and approachable (in Israel).

Mainly resident; for migratory and dispersive movements, see above.

Food. Invertebrates, nectar, and some other material. 2 stomachs, Uele, contained a small caterpillar and other tiny insects. Flowers visited for nectar and perhaps also insects and spiders (W Palearctic, Cramp and Perrins 1993, 35 species listed) include *Aloe*, *Yucca*, *Echinops*, *Cordia*, *Acacia*, *Capparis*, *Lonicera*, *Bauhinia*, *Oleander*, *Nicotiana*, *Robinia*, *Cytisus*, *Lupinus*, *Punica* and *Citrus*, as well as variety of exotic herbs, shrubs and trees. Said to feed on juices of small fruits and on date fruits. Often feeds at flowers of *Loranthus acaciae* (a semi-parasite of *Acacia* and *Ziziphus*) (Goodman and Meininger 1989). Also eats small seeds, and insect larvae up to 5 mm long. Insects and spiders fed to nestlings, but not nectar (Markman *et al.* 1999).

Breeding Habits. Barely known in Africa, well known elsewhere. Solitary nester, monogamous and territorial. Pair bond maintained all year. Territory size (Tel-Aviv, Israel, n = 19) 0·5–14·3 ha. ♂ establishing territory sings from treetop or wires; rivals sometimes engage in song-duels, perching near one another until one flees, other pursues (Israel). ♂ performs soliciting-display by erecting orange pectoral tufts, bobbing body with neck stretched up and head held high, drooping wings and spreading tail (Goldstein and Yom-Tov 1988). In few days before ♀ starts to lay, ♂ especially vigilant and chases out intruding ♂♂ whilst also attempting to guard his mate from them.

NEST: untidy, suspended, pear-shaped purse with side entrance and small porch over it and trailing 'beard' at base; in Darfur made of pieces of dry thistle leaves and lined with thistledown, and sited in low bush just above the ground. Elsewhere made of thin stems, rootlets, dead leaves, plant down and bits of bark, bound with spider web, wool and hair, and lined with feathers, wool, down, leaves and bits of paper. Built only by ♀, taking av. 8·4 days (Israel). Nest placed at tip of drooping branch of sheltered bush or tree, often close to wall of house (Israel).

EGGS: 2 (Darfur), 1–3 (Israel). Laid at daily intervals, in early morning. Sub-elliptical; smooth, glossy; white, blurred and clouded with fine spots of pale grey and grey-brown, mainly in zone around large end. SIZE: 15×11; (nominate race, n = 8) $15·2–16·2 \times 10·5–11·4$ ($15·4 \times 11·2$). WEIGHT: (nominate race, n = 22) av. 0·99.

LAYING DATES: Sudan (Darfur), Nov, Dec.

INCUBATION: by ♀ only, beginning with penultimate egg. Period, Israel (n = 22) av. $13·1 \pm 0·6$ days.

DEVELOPMENT AND CARE OF YOUNG:: young brooded by ♀ only, but fed by ♂ and ♀ in and out of nest. However, ♂'s main role is to guard territory, and he leaves ♀ to do most of the provisioning (Israel). Indulges in infanticide – intruding ♂ occasionally manages to peck at and kill territorial pair's well-grown nestlings; he may also interfere with ♀'s attempts to feed her young. It is thought that intruding ♂ kills young as a means of securing rival's territory and mate (Goldstein *et al.* 1986). Nestling period av. 15·6 days (n = 25). Fledglings perch together on branches and are fed by parents every 2–4 min for 4 or 5 days after leaving nest. Young return to nest with ♀ to sleep for 7–10 nights after first departure. Young independent 1–2 weeks after leaving nest.

BREEDING SUCCESS/SURVIVAL: in Israel, of 100 eggs in 41 clutches, 67% hatched and 47% fledged (Cramp and Perrins 1993).

References
Cramp, S. and Perrins, C.M. (1993).
Goldstein, H. (1988).
Goldstein, H. and Yom-Tov, Y. (1988).
Markman, S. *et al.* (1998, 1999).

Cinnyris habessinica (Ehrenberg). Shining Sunbird. Souimanga brillant.

Nectarinia (Cinnyris) habessinica Ehrenberg, 1828. *In* Hemprich and Ehrenberg, Symp. Phys. Av., folio a, pl. 4; Eilet in Eritrea.

Plate 14
(Opp. p. 207)

Range and Status. NE Africa and SW Arabia, east to Dhofar, Oman.

Resident, Red Sea coast south of 22°N to Kenya. Egypt/Sudan, locally common in Gebel Elba region, e.g. Gebel Shellal and Wadi Kansisrob. Sudan, common in Red Sea hills; records in SE Kassala Prov. and near Naita in extreme SE. Eritrea, common throughout, up to 1700 m, but absent from offshore islands and absent from coastal

Cinnyris habessinica

plains in summer. Ethiopia, common in NE, Rift Valley and S, up to 1500 and probably 1800 m. Djibouti, common at low elevations (Randa, Colonie, Garrab, Arta, Forêt du Day, Mabla Mts, Yokobi to Gaggadé). Somalia, common north of 07°N, mainly at 900–1250 m but up to 1550 m in Golis Mts and 2000 m at summit of Mt Wagar, and very common towards Gulf of Aden coast (though avoids the arid coastline itself, and avoids environs of coastal towns), and a few old records on upper Jubba River. Uganda, Mt Moroto (Matthews *et al.* 1997). Kenya, uncommon and local at 400–1000 m in arid and semi-arid parts of N, west to Kozibiri R. (W Turkana District), south to foothills of Kongelai escarpment, Kito Pass, Baringo, Isiolo and Meru Nat. Park.

Description. *C. h. habessinica* Hemprich and Ehrenburg (including '*altera*'): NE Sudan (Gebel Elba), Eritrea, Ethiopia south to Yavello and Harar, and N Somalia. ADULT ♂: forehead and crown glossy purple; sides of head, neck and rest of upperparts metallic green with golden tinge, more bluish green on uppertail-coverts. Tail black, with faint violet-blue gloss above, T1 with narrow metallic green edges. Chin to upper breast metallic golden green, concolorous with side of head, bordered below by narrow bluish green band, and below that a broader scarlet band (*c.* 10 mm deep) across lower breast; feathers immediately below scarlet band tipped dark glossy blue; pectoral tufts yellow; rest of underparts black. Lesser and median coverts metallic golden green; rest of upperwing blackish with slight violet-blue tinge; tertials, greater coverts and primary coverts finely fringed metallic green when fresh. Axillaries and underwing-coverts black. Bill black; eyes dark brown; legs black. ADULT ♀: upperparts rather pale grey-brown; long uppertail-coverts dark grey-brown. Tail blackish brown, faintly glossed above; T6 paler brown distally, with large brownish white spot at tip of inner web and narrow brownish white outer edge; a broad pale fringe also at tip of T5 inner web, and a narrower fringe on T4. Lores dark greyish; cheeks and ear-coverts grey-brown. Chin greyish white; throat, breast, flanks and thighs buffy grey with faint paler buff mottling on lower throat and breast; belly pale buff, tinged yellowish; undertail-coverts greyish white with dark brown centres. Lesser and median coverts grey-brown like upperparts; upperwing otherwise dark grey-brown, primaries edged pale grey-buff, secondaries and tertials edged buff, alula, primary coverts and greater coverts fringed buff. Axillaries pale grey-buff; underwing-coverts greyish white with dusky bases; inner borders of undersides of flight feathers pale grey-buff. Bare parts as in adult ♂. SIZE (10 ♂♂, 10 ♀♀, Ethiopia): wing, ♂ 67–71 (68·5), ♀ 58–62 (59·6); tail, ♂ 45–51 (47·9), ♀ 37–42 (39·6); bill, ♂ 21–24 (22·7), ♀ 19–23 (20·8); tarsus, ♂ 26–27 (26·4), ♀ 15–16 (15·3). Sudan birds slightly smaller: ♂ (n = 10) wing 65–68 (66·6), bill 20–23 (21·8). N Somalia birds ('*altera*') slightly larger: ♂ (n = 10) wing 68–71 (69·6), bill 23–26 (24·6). WEIGHT: Kenya, ♂ (n = 8) 9–11·5 (10·0), ♀ (n = 13) 7–11 (9·1).

IMMATURE: juvenile like adult ♀ above; below, chin to upper breast sooty grey, or sooty with pale mottling, bordered by whitish moustachial stripes; rest of underparts darker than adult ♀, with some pale buff feather fringes and mottling.

C. h. turkanae (van Someren): SE Sudan, Kenya, SW Somalia and S Ethiopia north to Arsi (= Arussi). Like nominate race, but ♂ has breast band broader, more orange–red, and lacks glossy blue feather tips below this band.

Field Characters. Length 12–14 cm. A sunbird of NE Africa, distinguished from similar-looking double-collared sunbirds by arid habitat and larger size; ♂ also by *black belly* and *coppery-purple* iridescence on crown. ♀ *uniform pale brownish grey* with whitish supercilium and throat; larger than ♀ Beautiful and Black-bellied Sunbirds *C. pulchella* and *C. nectarinioides*, with longer and straighter bill, no yellow on underparts; similar-sized ♀ Marico Sunbird *C. mariquensis* has streaked underparts with yellowish wash. Juv. ♂ like ♀ with black throat.

Voice. Tape-recorded in Palearctic only (B). Song fast, fluty, trilling and whirring, 'tuu-tuu-tuu-tuu-vita-vita-vita-du-du-du-du', often ending in long trill like Winter Wren *Troglodytes troglodytes*; subsong a fast whispering warble (Porter *et al.* 1996). Usual call in Africa sharp 'spik-spik' or 'speek-speek' (Zimmerman *et al.* 1996); calls in Middle East include hard 'dzit' and 'chewit-chewit', also fast, dry 'tje-tje-tje-tje' (Porter *et al.* 1996).

General Habits. Inhabits green vegetation in wadis, valleys and low mountain passes, acacia savanna and dense, arid thorn scrub. Solitary and in pairs; sometimes in large aggregations (up to 75 at one locality in S Yemen on 1 day). In Arabia visits gardens and cultivation, at least seasonally. Active, mobile; feeds (evidently on nectar) by perching next to blossom and probing it, and by hovering in front of it.

Resident, but subject to local movements in accordance with flowering times of preferred food trees; common in inner end of Wadi Kansisrob, Gebel Elba, Egypt/Sudan border, in Feb–Mar when *Delonix elata* in flower, but disappears by Apr when trees are no longer in bloom (Goodman and Meininger 1989).

Food. Forages in flowering Malvaceae and tree *Delonix elata* in Gebel Elba, and in *Acacia, Capparis* and *Stereospermum* trees and shrubs in Eritrea. Particularly attracted to aloes (Somalia).

Breeding Habits. Solitary nester, evidently monogamous and territorial. Thought to be double-brooded.

NEST: suspended, domed pouch with side entrance, sometimes but not always with slight porch above; made almost entirely of cobweb (spider web, insect cocoons). Hangs from near tip of branch, on underside; not concealed, though readily overlooked since it resembles a mass of spiders' silk.

EGGS: usually 1, occasionally 2. Pyriform; ground colour greenish white, with ring of lavender spots and smudges around broad end. SIZE: (n = ?) 18–19 × 12–13.

LAYING DATES: Egypt/Sudan (Gebel Elba), Dec–Jan; Eritrea, Jan (and various evidences, Dec–Mar); Ethiopia, Apr (and probably Feb, and bird in Aug 'appeared to be on the point of breeding': Benson 1946b); Somalia, Feb–May; Kenya, (nest-building, Dec).

Reference
Archer, G. and Godman, E.M. (1961).

Cinnyris oustaleti (Bocage). Oustalet's Sunbird. Souimanga d'Oustalet.   **Plate 16**
Nectarinia oustaleti Barboza du Bocage, 1878. J. Sci. Math. Phys. Nat. Lisboa, 6, p. 254; Caconda, Huila, Angola.   **(Opp. p. 223)**

Forms a superspecies with *C. talatala*.

Range and Status. Miombo endemic with 2 disjunct populations, in Angola and near Zambia/Tanzanian border. Angola, on western plateau, from Cuanza Sul to Huambo, N Bié and N and W Huila Provinces. Zambia, widespread but rather uncommon (and easily overlooked), in Mbala, E Mporokoso, NE Kasama, N Chinsali and NW Isoka districts of Northern Province. Tanzania, 2 or 3 Zambian records are within a few km of border; twice at Kasesya (08°40′S, 31°28′E) (Moyer 1983), and extraordinary record of ♂ collected at Kigoma in 1961 (Dillingham 1984). Malaŵi, single record from Zambian border (Aspinwall 1989). Not yet recorded in Zaïre, though likely to occur in extreme E Katanga.

Description. *C. o. oustaleti* Bocage: Angola. ADULT ♂ (breeding): forehead and forecrown metallic blue, merging into metallic green of hindcrown, neck, mantle to rump and scapulars; uppertail-coverts metallic blue. Tail blackish, upperside with slight blue gloss; outer webs of T2–T5 and both webs of T1 edged bright metallic blue; T6 browner with whitish tip and outer web. Lores black; sides of head and neck glossy green, concolorous with upperparts. Chin glossy blue, merging into bluish green throat and broad glossy purple band across upper breast, with admixed maroon feathers along posterior border; below this a narrow sooty brown band; pectoral tufts orange and yellow; underparts behind lower breast white. Lesser and median coverts metallic green; rest of upperwing blackish brown, narrowly edged pale brown. Axillaries, underwing-coverts and inner borders on undersides of flight feathers greyish white. Bill black; eyes dark brown; legs black. ADULT ♂ (non-breeding): wings and tail as in adult ♂ (breeding); otherwise, metallic feathering mainly confined to centre of throat and breast and rump to uppertail-coverts; rest of head and body as in adult ♀. ADULT ♀: upperparts dark greyish brown; longer uppertail-coverts blacker, tips glossed greenish. Tail blackish brown, faintly glossed blue-green above; feathers narrowly edged metallic green; T5 with narrow white inner web tip, T6 with broad whitish tip and outer web. Narrow superciliary stripe pale buff; lores, cheeks and ear-coverts dark greyish brown. Underparts grey-buff, paler on belly and undertail-coverts; chin to breast and flanks streaked grey-brown. Upperwing dark grey-brown, flight feathers narrowly edged buff. Axillaries and underwing-coverts whitish; inner borders of flight feather undersides greyish white. SIZE (13 ♂♂, 6 ♀♀) wing, ♂ 54–60 (56·2), ♀ 51–55 (52·4); tail, ♂ 33–42 (37·7), ♀ 31–36 (33·8);

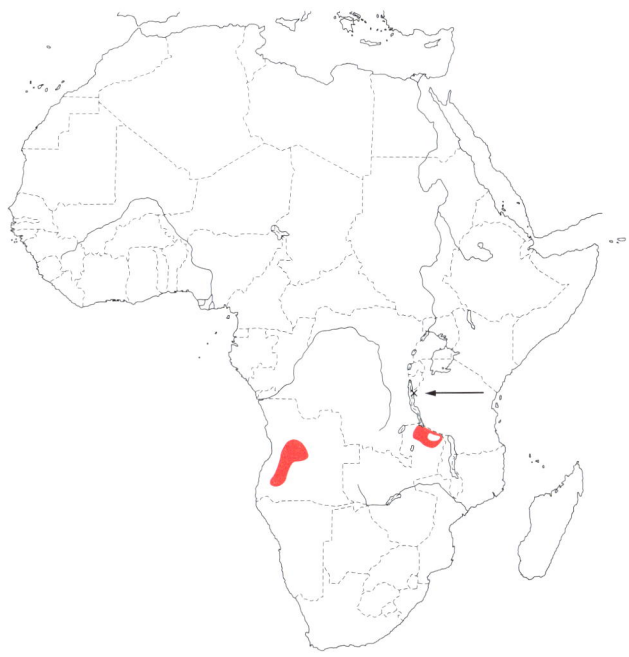

Cinnyris oustaleti

bill, ♂ 20–22 (20·6), ♀ 18–21 (19·7); tarsus, ♂ 14–16 (15·4), ♀ 14–15 (14·8).

IMMATURE: juvenile like adult ♀, but chin to breast greyish, lower flanks and belly tinged yellow.

C. o. rhodesiae Benson: NE Zambia. Smaller, with shorter bill; wing (3 ♂♂) 52–55 (53·3), bill (4 ♂♂) 18–19 (18·3). ♀ paler on belly.

TAXONOMIC NOTE: provides the link between black-bellied (*C. bouvieri*, *C. osea*, *C. habessinicus*) and white- or yellow-bellied sunbirds (*C. talatala*, *C. venusta*). ♂♂ of *oustaleti* and *talatala* are white-bellied, but close relationshiop between *oustaleti* and *bouvieri* is suggested by blue gloss on backs, orange (not yellow) pectoral tufts, and narrow maroon breast bands (none in *talatala*). 'These intermediate characters and the location of the two isolated populations of *oustaleti* suggest that it is a species derived from hybridization between *talatala* and *bouvieri*' (Hall and Moreau 1970).

Field Characters. Length 10–11·5 cm. Very similar to closely-related White-bellied Sunbird *C. talatala*; their ranges abut but do not overlap in N Zambia. Most easily located by its distinctive call as it forages in low, dense vegetation. Bill shorter than White-bellied, creating the impression of a smaller bird; bill shape also different, fairly straight, only curving down towards the tip, whereas bill of White-bellied more or less evenly curved; bill of ♀ even shorter, almost like *Anthreptes* (Beel 1993). Gloss on head of ♂ Oustalet's greener, centre of throat bluer; maroon breast-band broader and clearer; pectoral tufts orange and yellow (only yellow in White-bellied), but these can be very hard to see and considered useless as a field character by Beel (1993). ♀ extremely similar to ♀ White-bellied but upperparts a little darker, chin to breast darker with indistinct streaks.

Voice. Tape-recorded (MOY). Foraging call, given by both sexes, a soft, short, high-pitched 'tzzip', different from any call of White-bellied Sunbird (Beel 1993); also gives a harder ticking call reminiscent of Amethyst or Scarlet-chested Sunbird *Chalcomitra amethystina* and *C. senegalensis*.

General Habits. Inhabits *Brachystegia* woodland (miombo) and has especial liking for extremely degraded woods – a very dense scrub sometimes only 1 m tall. Solitary or in small groups, once foraging on flowers in harvested maize fields >100 m away from nearest miombo (scrub) woodland (Beel 1993). Does utilize larger trees in better developed woodland where, however, generally keeps in scrub very low down, near ground. Sings from a prominent perch. Sedentary.

Food. Said to feed mainly on insects when visiting flowers of trees and shrubs (Mackworth-Praed and Grant 1963).

Breeding Habits. A few nests have been found but nidification not described.
LAYING DATES: Angola, Oct–Feb; Zambia, May.

Plate 16
(Opp. p. 223)

Cinnyris talatala A. Smith. White-bellied Sunbird. Souimanga à ventre blanc.

Cinnyris talatala A. Smith, 1836. Report Exped. Centr. Africa., p. 53; between Orange River and Kurrichaine.

Forms a superspecies with *C. oustaleti*.

Range and Status. Endemic resident, southern Africa. Tanzania, below 1000 m, from Ruhuhu Valley in S Njombe, through Songea and Masasi to coast at Mikindani and Mtwara. Malaŵi, on L. Malaŵi littoral in Karonga District from Ngerenge to Vintukutu; possibly in W Rumphi District; reappears further south in Nkhotakota District, frequent from Benga and Makanjila to Shire Valley, where common; at 630 m at Chididi, Nsanje District; sight record at 1100 m at Bunda, Lilongwe District. Zaïre, known only from Mopala in extreme SE Katanga, though likely to be widespread where mapped, since many records on all sides in Zambia. In Zambia common and widespread in S, thinning out northward; north to Mpika, Milambo and Fwaka in NE, Ndola, Luanshya, Ntambu, Mayau, Ndubeni and 13°30′S on Lungwebungu R. Angola, in S and W Huila and adjacently in Namibe Prov., along escarpment and landward side of coastal plains to S and coastal Benguela; recorded in SE near Zambia border (Hall and Moreau 1970). In Namibia, Botswana and South Africa ranges further to southwest (as mapped here) than formerly realized (Harrison *et al.* 1997). May be expanding range, e.g. with opening up of miombo woodlands in Zimbabwe (A.J. Tree *in* Harrison *et al.* 1997). Common in Zimbabwe and NE South Africa; very common in Mozambique, where probably over 2 million birds (Parker 1999). Density of 1 bird per 3 ha in *Acacia* and 1 per 10 ha in broad-leaved woodland, N South Africa.

Description. *C. t. talatala* (Smith) (includes 'anderssoni': Clancey 1967): range of species except Natal and S Transvaal. ADULT ♂ (breeding): head, neck and rest of upperparts metallic green, long uppertail-coverts bluer green. Tail blackish, upperside with slight

Cinnyris talatala

blue gloss; T1–T5 edged metallic blue; T6 browner with inner web tip fringed whitish. Lores black. Throat metallic green, concolorous with head, merging with glossy blue chin and with broad purple band on upper breast, the latter bordered below by very narrow band of dull blackish brown; pectoral tufts yellow; rest of underparts white. Lesser and median coverts metallic green; rest of upperwing blackish brown, flight feathers edged pale buff when fresh, primary coverts, greater coverts and alula finely edged metallic green. Axillaries and underwing-coverts whitish; inner borders on undersides of flight feathers greyish white. Bill black; eyes dark sepia; legs black. ADULT ♂ (non-breeding): wings and tail as in breeding ♂, but metallic feathers otherwise mainly confined to patch on centre of chin to upper breast and on back to upper tail-coverts; rest of head and body as adult ♀. ADULT ♀: upperparts greyish brown with olivaceous tinge, tips of uppertail-coverts glossed metallic green. Tail blackish, upperside with faint bluish green gloss; T1–T5 narrowly edged metallic green; T5 with narrow white tip to inner web, T6 with larger white tip to inner web and pale outer web. Superciliary stripe pale buff; lores and ear-coverts dark greyish brown. Underparts pale buff, chin to breast streaked greyish; undertail-coverts dull white. Upperwing dark brown; when fresh, primaries edged pale buff, secondaries edged olive, primary and greater coverts edged buff. Axillaries and underwing-coverts whitish; underside of flight feathers as adult ♂. Bare parts as adult ♂. SIZE (10 ♂♂, 10 ♀♀): wing, ♂ 52–55 (53·4), ♀ 49–53 (50·6); tail, ♂ 34–37 (35·5), ♀ 27–35 (30·9); bill, ♂ 20–23 (21·2), ♀ 18–21 (19·6); tarsus, ♂ 15–17 (16·1), ♀ 14–16 (15·2). Population in middle Zambezi Valley and eastward smaller (wing, ♂, n = 20, 51–56, av. 54·1) than that to southwest (e.g. Zimbabwe, ♂, n = 10, wing 57–58, av. 58·0) (Clancey 1967). WEIGHT: Zimbabwe, ♂ (n = 2) 6·7, 8·0, ♀ (n = 1) 6·0; Transvaal, ♂ (n = 3) 7·5–9·2 (8·1); also (both races, southern Africa) ♂ (n = 163) 6–10 (8·2), ♀ (n = 79) 4·8–10·5 (7·1) (Maclean 1993).

IMMATURE: juvenile like adult ♀, but rather greyer on throat and lacks gloss on uppertail-coverts. In immature ♂, metallic feathers appear first on centre of throat.

NESTLING: hatches naked, skin dark orange, browner on head and back; feathered nestling has olive-grey upperparts, dark grey-brown wings and tail, and yellow underparts.

C. t. aresta Clancey: Natal to S Transvaal highveld. ♂ like nominate ♂ but lower breast band even blacker, and deeper (10–15 mm) at sides, plastron reddish violet (violet in *talatala*); flanks slightly streaky. ♀ not always separable from ♀ *talatala* but tends to be darker above and deeper yellow below (Clancey 1967).

Field Characters. Length 10–12·5 cm. ♂ has upperparts and sides of head and neck iridescent green, uppertail-coverts blue; chin black, throat blue-green; broad violet breast-band separated from white belly by dusky brown band; yellow pectoral tufts. Remarkably like white-bellied race of Variable Sunbird *Cinnyris venusta albiventris*, but sympatric race *falkensteini* has yellow belly, as does ♂ Collared Sunbird *Hedydipna collaris*. ♀ light greyish brown above, typically off-white below, paler on belly, with indistinct streaking on breast; extremely like ♀ Dusky Sunbird *C. fusca*, which lacks streaking but is probably not separable in the field (Maclean 1993). Belly colour of ♀ variable; birds with grey belly and yellowish patch on midbelly extremely similar to ♀ and immature Miombo Double-collared Sunbird *C. manoensis* (Tree 1991) but smaller; ♀♀ sometimes washed pale yellow below, looking very like ♀ Variable Sunbird; distinguished by lack of white tips to all but outer tail feathers. Freshly fledged juvs have greenish wash on upperparts, yellow supercilium, greenish yellow throat and light yellow underparts (Tree 1991). Moulting ♂ with metallic blue throat and metallic green wing shoulder can be mistaken for both Variable Sunbird in same plumage and adult ♂ Plain-backed Sunbird (Hustler 1985). For differences from Oustalet's Sunbird *C. oustaleti*, see that species.

Voice. Tape-recorded (9, 16, 17, 22, 51, 75, 86, 88, 91–99, 100, 104, LEM, MAC, POO). Common call a strident, husky, nasal 'dzoowi-dzoowi-dzoowi ...'; this is frequently incorporated as first part of song, followed by 1–2 high sibilant notes and then a trill, 'dzoowi-dzoowi-tsi-tsee-srrrrrrrr'; the trill varies in speed and may be hard, 'ddrrrrr' or more often liquid, 'tutututu ...', or sibilant, 'tississississ'; instead of a trill there may be a jumble of high, thin, liquid notes, but they are always preceded by some single notes, 'sip-sip', 'tsi, tsee', with or without the 'dzoowi': 'sip, sip, swee-swee-swee-swee-swee-swee', 'tsee, tsip, tsusususususu'. Songs are short, with only a short pause between each during song session. Sings almost all day in the bushveld (Haagner 1964). ♀ has been heard singing a song not unlike ♂'s (Skead 1967). Some notes have been likened to those of other species, but are not necessarily imitations; however, the sunbird was heard to imitate a Streaky-headed Seedeater *Serinus gularis* when the two were feeding close to each other (M.B. Markus *in* Skead 1967). Alarm a loud 'chak-chak-chak' or hard 'chet'; flight call a thin 'zit-zit'.

General Habits. Inhabits acacia savanna, keeping to denser stands of thorn trees in long grass with a good amount of low, thorny growth; in Zambia in *Acacia*, mopane (*Colophospermum*) and mutemwa (*Baikiaea*) woodlands and poorer types of miombo (*Brachystegia*), but absent from richer miombo; common in riverine forest and dense scrub. Out of breeding season invades gardens and parks in villages and cities. Abundance of *C. talatala* and Southern Double-collared Sunbird *C. chalybea* tends to be reciprocal, and in absence of *C. chalybea* from Mozambique *C. talatala* occupies greater range of habitats there. Habitat near Pietermaritzburg, Natal, where species common, is acacia savanna ecotone between impenetrable scrub and open grassland dominated by *Acacia karroo*, *Ehretia rigida* and *Aloe candelabrum* (Earlé 1982).

Occurs singly, in pairs and family parties, and in loose and sometimes large, quite dense aggregations. Lively and active, foraging out in the open, hovering to feed from flowers, often mixing freely with Copper *C. cuprea*, Scarlet-chested *Chalcomitra senegalensis*, Variable, Southern Double-collared *C. chalybea*, Dusky, Marico Sunbirds *C. mariquensis* (which dominate the White-bellied) and with other birds. Highly vocal; sings intermittently for 30 min from inside bush (Maclean 1993). Flight rapid and jerky. Once seen probing mud and drinking oozing water.

In N and E of range mainly resident, but mobile and evidently subject locally to seasonal and altitudinal movements; any movements are, however, more inferential than proven (Aspinwall 1979). Nomadic and partially migrant in W and S of range; may quit Namibia in the dry season (Tree 1990), and migration visible to northeast in Chobe R. floodplain at Ngoma Bridge in Apr (Borello 1992). Extremely abundant on Chobe R. when *Combretum mossambicense* blooming in Sept (e.g. 107 birds in 20 ha), perhaps migrants (Herremans 1992). Much commoner in Bulawayo, Zimbabwe, in Apr–July than rest of year, evidently attracted to flowering *Tecoma*, *Bauhinia* and *Callistemon*, and again in Sept when *Jacaranda* blooms. Non-breeding visitor to Pretoria city suburbs and parks

(Transvaal), consistently present only in Mar–June; scarce in July–Dec, when breeds in *Acacia* thornveld surrounding city; becoming commoner in Jan–Feb. Race *aresta*, breeding in South Africa, thought to winter to north and east (Clancey 1967). In Mozambique, population in Sul do Save lowlands thought to be augmented in southern winter by birds from interior uplands (Clancey 1971). No evidence of seasonality in Swaziland (Parker 1994), but 400 counted (at once?) during netting activities in 0·25 ha at Umhlatusi Bridge, Natal, suggests a migratory concentration (Skead 1967).

Food. Nectar and small arthropods. Nectar of *Loranthus* (birds strongly attracted to flowering mistletoes *Loranthus* parasitic on thorn trees), *Schotia, Acacia, Eucalyptus, Mina lobata, Strelitzia, Bunsfelsia, Salvia, Aloe* spp., *Cestrum, Bauhinia, Jacaranda, Kigelia, Watsonia, Petrea, Canna, Protea* spp., *Poinsettia, Hibiscus, Cotyledon, Kniphofia, Tecoma, Callistemon, Combretum mossambicense, Agapanthus, Japonica, Grewia, Tithonia tagetiflora* and *Leonotis leonurus* (Skead 1969). Thought to glean *Ficus* and other tree foliage and flowers for small spiders and insects; readily gleans rose bushes for aphids. Brooding ♀ seen to eat ants on her nest and to feed some of them to young beneath her (Skead 1967); nestlings also given aphids, small grasshoppers and moths.

Breeding Habits. Solitary nester, monogamous, territorial. ♂ relies mainly on singing for territorial display, but unlikely that definite territorial boundaries are ever established (Earlé 1982). In open *Acacia* veld where neighbouring pairs at least 80 m apart, ♂♂ sing over whole area and seldom approach each other (Earlé 1982). ♂ sings often, near nest, only during nest construction and early-incubation stages, and flashes yellow pectoral tufts only in presence of mate. In evident display, ♀ picked up piece of material, flew to top of tree where ♂ was singing and dropped it, repeating the performance (Earlé 1982). After copulating with ♀, ♂ once gave warbling song 'whilst flying with seemingly half speed, cupped wingbeats in a semi-circle' of *c*. 5 m radius before returning to perch (Medland 1992). Only one instance of double-broodedness known.

NEST: oval or pear-shaped bag, suspended low down in vegetation, with side-top entrance (**A**); composed of soft, chaffed grass with very fine stems, usually *Eragrostris*, which project stiffly above entrance to form a small porch; grass mixed with bits of dead leaves and (mainly towards or on outside) whole dead leaves; bound outside with spider web and cocoon silk, and more or less covered with dead leaves; occasionally decorated with small pieces of bark; lower half of chamber lined with woolly plant down including thistle pappi, or fine, soft grass with a few feathers; sometimes lining not woolly but consists only of fluff from seeding grass (Vincent 1949); one nest lined with sheep's wool. Untidy and loosely knit, compared to nests of Greater Double-collared Sunbird *C. afra* and Amethyst Sunbird *Chalcomitra amethystina*.

Height *c*. 125, width 64–77, entrance diam. 19–38. Situated 0·25–3 m, mostly 0·6–1·5 m, above ground; attached through top and back to tip of branch of *Acacia*

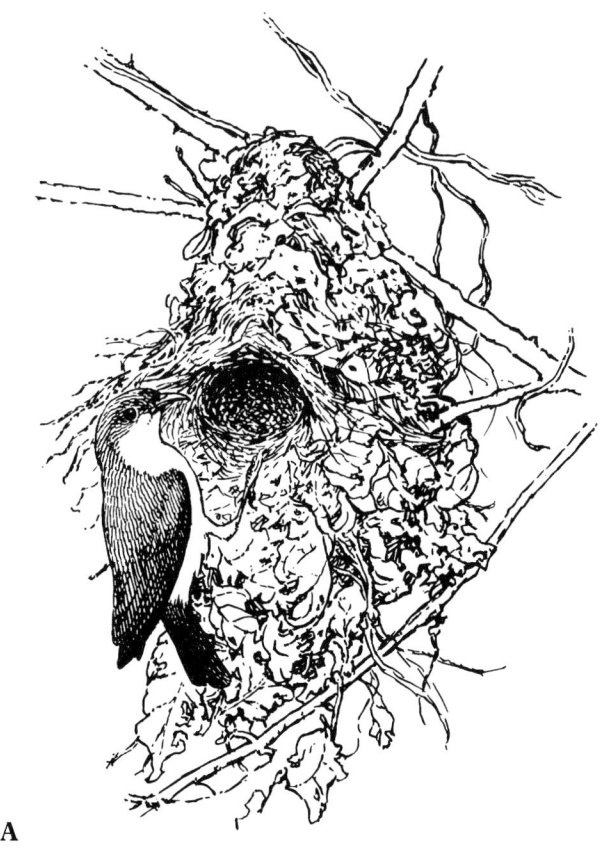

A

tree or thorny shrub and quite often also embedded in mass of spiders' web. Of 54 nests 40 hung free, 11 were attached both at top and back, 2 at top and side and 1 at top and front (Earlé 1982). Where several large, whitish masses of spider web occur close together, sunbird's nest built into one is very hard to see, sheets and streamers of cobweb anchoring it to twigs. Nest sometimes built onto thistle, often against hard *Opuntia* cactus, suspended from vine, in isolated shrub on garden lawn, and in streamside bushes, reeds, and once on a wire fence. Bird favours plants with thorns at least 15 mm long. One nest was made of materials taken from old mouse nest in thistle; another was close to nest of hornets *Belanogaster*. Built by ♀ only, taking 5–8 (av. 6) days, duration being affected by the weather (Earlé 1982); ♂ stays nearby, singing at prominent perches, and sometimes accompanying ♀ back and forth when she collects nest material. ♀ uses same site but builds new nest in successive years. When nest well advanced ♀ sleeps in it.

EGGS: 1–3 (av. 2·0, Natal: 5 × c/1, 43 × c/2, 4 × c/3). Laid before 07h30, 1–5 days after completion of nest. White, cream, greenish cream to greyish; marked with speckles all over, or minute scrolls and dots, or smudges which are sometimes comma-shaped; markings often concentrated at broad end; some eggs almost immaculate; colour of marks slate, fawn, reddish, mauve, sepia, grey or black. In a study in Natal eggs were of 2 types: (a) white with purplish grey and olive speckles mainly at large end, and (b) indistinctly speckled in pale fawn and light grey; of 74 eggs, 44 were type (a) and 30 of type (b), ♀♀ laying one type consistently (Earlé 1982). SIZE: (n = 167) 14·3–18·0 × 10·5–12·9 (16·2 × 11·3).

LAYING DATES: Malaŵi, Jan, Apr, May, July, Sept; Zambia, Sept–Oct; Angola, Nov–Dec; Botswana, mainly Sept–Dec (18 records), also Jan, Mar and July (3 records); Zimbabwe, mainly Sept–Oct (81 records), also July–Aug and Dec–Jan (14 records) and Mar (2 records); Transvaal, July–Mar, mainly Sept–Nov; Natal, July–Feb, mainly Aug–Dec, especially Sept then Oct (Earlé 1982).

INCUBATION: by ♀ only, starting after 2nd egg laid. ♂ seldom near nest during incubation. In Natal study, ♀♀ fed from 09h00 to 11h00 and incubated from midday to 17h00. Period (n = 2) 13 ± 1 days (Buchanan and Steyn 1964, Earlé 1982).

DEVELOPMENT AND CARE OF YOUNG:: on day 2 eggtooth still present; eye slits formed; pterylae visible as dark patches. Days 4–5: eyes begin to open; primary sheaths 2–3 mm long. Days 9–10: primaries 2–4 mm out of sheaths, rectrices breaking open; gape orange-yellow; no mouth spots. Young fed by both parents, mainly ♀, the ♂ often flying back and forth with her between nest and foraging sites but not necessarily himself gathering food. ♀ arriving with food gives quiet 'tzick' call. ♀ broods young chicks at night. Faecal sacs carried away by both sexes. When about to leave nest, young slightly heavier than adult ♀. Nestling period (n = 1) 14–15 days. Young do (Buchanan and Steyn 1964) or do not (Earlé 1982) return to nest to sleep. Young out of nest fed by both parents, mainly ♀. One young was moulting into adult plumage 7 months after fledging.

BREEDING SUCCESS/SURVIVAL: in Natal, overall success rate over 2 seasons was 21·7%: 0·6 fledglings reared per nesting attempt and 1·9 per successful nesting attempt. Losses were due to bad weather (wind and rain) and predators. Nests more successful in *Acacia* than in *Opuntia*, and unsuccessful if as high as 3 m above ground (Earlé 1982). Longevity 10 years (1 ♂) and 10·5 years (1 ♀; Hanmer 1989).

References
Clancey, P.A. (1967).
Earlé, R.A. (1982).
Skead, C.J. (1967).
Vincent, A.W. (1949).

Cinnyris venusta (Shaw and Nodder). Variable Sunbird; Yellow-bellied Sunbird. Souimanga à ventre jaune.

Plate 16 (Opp. p. 223)

Certhia venusta Shaw and Nodder, 1799. Nat. Misc., 10, pl. 369; Sierra Leone.

Range and Status. Endemic resident and migrant, Senegal to Somalia, south to Zimbabwe.

Nominate race: Senegal, frequent, Dakar to Casamance, records in E and SE. Gambia, locally common on coast, uncommon inland. Guinea-Bissau, widespread. Mali, quite common and widespread in Boucle du Baoulé, Baoulé-Banifing and in S Mandingo Mts. Guinea, scarce in N and NE, uncommon on Conakry coast and inland to Kolenté (Demey 1995) but common on Conakry Peninsula and on Kakoulima Hill 50 km to northeast, common on Kounounkan Massif (Hayman *et al.* 1995), and occasional on Ziama Massif and in SE. Sierra Leone, not in Kilimi or Outamba regions in NW but plentiful south of that in N half of country and common around Freetown, but absent from S half. Liberia, common in coastal belt and in N highlands. On Mt Nimba frequent at 1000–1650 m (Liberia) and very abundant at upper limit of forest at 1600–1700 m (Ivory Coast). In remainder of Ivory Coast local or uncommon from coast north to 09°20′N. Ghana, on coast at Denu, uncommon resident in woodland savanna just to north of forest zone, uncommon in Mole but very common in Walewale (Mar). Togo, common resident at coast, occasional at forest edge, and records north to 10°21′N. Benin, records from coast north to 09°12′N. Burkina Faso, rare (Dowsett 1993). Nigeria, widespread at coast in SW, and north of forest zone and west of 10°E, north to Nindam (frequent) Zaria (common in spring) and Kano (rare); in E Nigeria, frequent and widespread north to at least Serti (07°30′N) and south to Obrubra on Cross R.; absent from Maiduguri region but

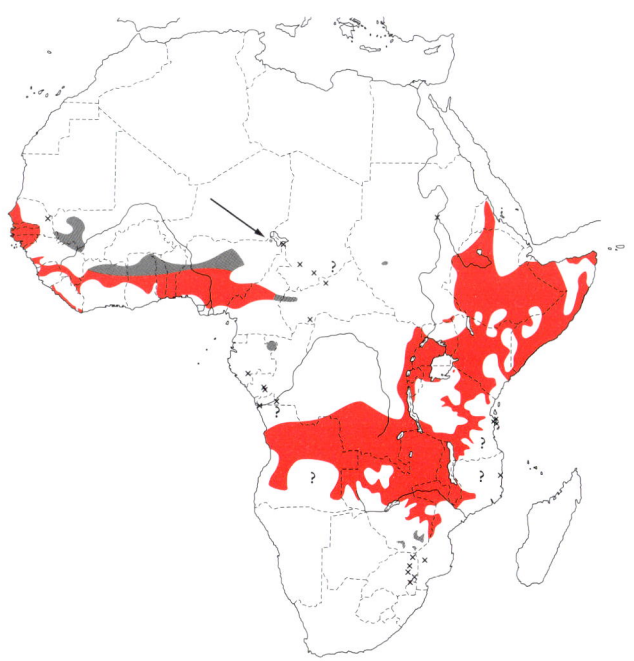

Cinnyris venusta

occurs uncommonly at Malamfatori on W shore of L. Chad. Cameroon, occurs across Adamawa Plateau and in far N, but perhaps not between. Chad, rare: records from Bousso, Sarh and in Salamat Préf. Central African

Plate 19

Republic, rains visitor (June–July) around Bozoum in W (Bannerman 1948); elsewhere 2 sight records, Lobaye Préf. and Bamingui-Bangoran Nat. Park (Green 1983, Carroll 1988). Gabon, scarce and irregular, Mar–Oct, in NE.

Sudan, records from Kordofan and Khartoum, also from E near Ethiopian border, and Boma Hills; another race common in Didinga and Imatong Mts, up to 2770 m. Eritrea, very common above 1230 m (at least west of 40°E). Ethiopia, widespread at all altitudes between 900 and about 1500 m, but few reports in W and SE and very few in NE (Hall and Moreau 1970, Urban and Brown 1971). Not in Djibouti. Somalia, common, almost throughout, the numerous records falling within areas shown.

Uganda, common and widespread north to Bunyoro, Teso and Elgon. Kenya, common and widespread, especially in W and central highlands; another race (*albiventris*) in arid N and NE Kenya, from east of L. Turkana to Marsabit and Isiolo Districts, lower Tana R. and coastal lowlands from Somali border to Lamu; abundant on Manda I.; probably more widespread than shown. Tanzania, in NE common from about 35°E (south to Dodoma) to Kilimanjaro, Pare Mts and W Usambaras; common in Kagera District – also in Rwanda and W Burundi – and W Kigoma, with a few records south of L. Victoria; widespread in uplands from S end of L. Tanganyika to N half of L. Malawî and from Njombe to Nguru, vagrants reaching coast at Dar es Salaam (map: N.E. Baker, pers. comm.).

Congo, 2 records in Mayombe (Dowsett-Lemaire and Dowsett 1989). Zaïre, widespread in E; Lendu Plateau at summit of Mt Avu at 1850 m, Nioko, Bunia and Bogoro; in Rwenzoris very common at 2150–2460 m; highlands west of L. Edward; on lower slopes of Kivu Volcanoes, mainly above 1200 m; Itombwe (Kamituga, Luiko, Mt Kandashomwa) at 1910–1730 m. In S Zaïre occurs north to Kabinda (06°08′S, 24°27′E); widespread, frequent but localized in Katanga. Angola, central Huila (south to Bicuari Nat. Park), Cuanza Norte, Lunda, N Moxico. Zambia, widespread in N and Eastern Prov., sparse in NE end of Luangwa Valley but common in SW end, common on Middle Zambezi but does not occur upstream of Chirundu (R.J. Dowsett, pers. comm.). Zimbabwe, more widespread (Harrison *et al.* 1997) than previously thought (Irwin 1981); has extended range from the east into wooded suburbs, and is now one of the commonest sunbirds in Harare (Tree 1990). Mozambique, Niassa Prov., common in Tete District, and W Manica e Sofala east to Mt Gorongoza; once in Sul do Save (Parker 1999). Transvaal, 5 records: Kruger Park, Barberton, Crocodile R., Sandringham and Punda Milia.

Description. *C. v. venusta* (Shaw and Nodder): Senegal to Central African Republic. ADULT ♂ (breeding): forehead and forecrown glossy violet, merging with glossy greenish hindcrown; mantle to rump and scapulars metallic bluish green; uppertail-coverts metallic blue. Tail blackish, upperside with slight blue gloss; feathers narrowly edged bright metallic blue when fresh, T5–T6 with whitish fringe at tip of inner web. Lores blackish. Sides of head and neck to throat metallic green, concolorous with upperparts, merging into glossy purple chin and broad purple band across upper breast, the latter bordered posteriorly by very narrow band of dull black; pectoral tufts orange and yellow; rest of underparts pale yellow. Lesser and median coverts metallic green; rest of upperwing blackish brown, flight feathers narrowly edged pale brown, tertials faintly glossed greenish, primary coverts, greater coverts and alula with fine metallic green edges when fresh. Axillaries pale yellow; underwing-coverts creamy; inner borders of flight feather undersides greyish white. Bill black; eyes brown; legs black. ADULT ♂ (non-breeding): wings and tail as in adult ♂ (breeding), but metallic feathers otherwise mainly confined to patch on centre of chin, throat and upper breast, and to back, rump and uppertail-coverts; rest of head and body as adult ♀; eclipse plumage thought to occur in some but not all populations (Lane 1996). ADULT ♀: upperparts greyish olive-brown, greener on rump. Tail blackish, slightly glossy above; feathers narrowly edged metallic blue; T2–T6 tipped whitish, T5–T6 broadly so on inner web; T6 outer web brownish white. Faint, narrow supercilary stripe buffish white; lores, cheeks and ear-coverts dark greyish brown. Chin, throat and upper breast buffish white, breast with grey streaking; rest of underparts pale yellow, sides and flanks suffused greyish, undertail-coverts more creamy. Lesser and median coverts greyish olive; rest of upperwing blackish brown, primaries edged olive-yellow, tertials edged olive, greater coverts fringed pale buff. Axillaries pale yellowish; underwing-coverts creamy; inner borders of flight feather undersides greyish white. Bare parts as adult ♂. SIZE (10 ♂♂, 10 ♀♀): wing, ♂ 48–53 (50·7), ♀ 45–53 (47·5); tail, ♂ 32–37 (34·4), ♀ 28–35 (30·4); bill, ♂ 17–19 (18·0), ♀ 16–18 (16·9); tarsus, ♂ 14–16 (15·1), ♀ 13–15 (14·1). WEIGHT: Liberia, ♂ (n = 5) 5·7–7·9 (6·8), ♀ (n = 1) 5·8; N Nigeria, ♂ (n = 14, Apr–May) 5·0–6·5 (5·9), ♂ (n = 6, July) 6·1–6·8 (6·4), ♀ (n = 5, May–June) 5·1–6·0 (5·3) (Fry 1971).

IMMATURE: juvenile like adult ♀ but more yellowish below; juv ♂ has blackish patch on chin and throat. In immature ♂ metallic feathers appear first on throat, lesser coverts and uppertail coverts.

NESTLING: not described.

C. v. fazoglensis Heuglin: Eritrea, N and W Ethiopia and Sudan except mts in S. Like nominate race, but ♂ deeper yellow below. Larger; wing (10 ♂♂) 53–56 (53·3). WEIGHT: (1 ♂, Ethiopia) 6·8.

C. v. falkensteini (Fischer and Reichenow) (includes '*kuanzae*' and '*niassae*'): Gabon, Congo, W and S Zaïre, Angola, Zambia, Malawî, Zimbabwe, central Mozambique to Tanzania, Kenya (except NE), and S Sudan in Imatong and Didinga Mts. Intergrades with *igneiventris*, *fazoglensis* and *albiventris*. ♂ bluer green above than nominate race; throat violet-blue; underparts deeper yellow; ♀ more yellow on belly than nominate race. Slightly larger; wing (10 ♂♂) 50–57 (53·8). In Zimbabwe, Middle Zambezi population smaller than E Highlands one: ♂ wing av. 49·5 and 52·8, ♂ bill av. 19·4 and 21·4 (Irwin 1981). WEIGHT: Kenya, ♂ (n = 48) 6·1–8·8 (7·6), ♀ (n = 19) 5·5–7·9 (7·0); Zimbabwe, ♂ (n = 27) 5·4–9·0 (7·0), ♀ (n = 15) 5·3–8·5 (6·7); Malawî, monthly ranges Sept–Dec, ♂ (n = 59) 6·7–7·3, ♀ (n = 53) 6·0–6·4 (Hanmer 1981).

C. v. igneiventris (Reichenow): E Zaïre, Rwanda and Uganda, intergrading with *falkensteini* in extreme W Kenya. Like *falkensteini*, but ♂ has lower breast and upper belly washed red, and pectoral tufts scarlet.

C. v. albiventris Strickland: SE and S Ethiopia to Somalia and N and E Kenya, intergrading with adjacent yellow-bellied forms. Like *falkensteini*, but ♂ has white (not yellow) underparts; ♀ greyish brown (not olive) above, whitish below with faint grey streaking on throat and breast. Large; wing (10 ♂♂) 54–58 (55·5).

Field Characters. Length 9·5–11 cm. A common and widespread sunbird through much of Africa; plumage, especially belly colour, very variable. ♂ above iridescent green (W Africa) to greenish blue (elsewhere), with blue uppertail-coverts, *violet* forehead; chin and upper throat *dark violet*, lower throat green (W Africa) or blue, *breast* *broadly purple-violet*, separated by variable-width black band from belly; pectoral tufts orange or yellow. Belly colour varies from pale yellow (W Africa: nominate race) to bright orange (Albertine Rift: *igneiventris*), rich yellow or orange-yellow (E and S Africa: *falkensteini*) or white (Horn of Africa: *albiventris*). ♂ distinguished from sympatric double-collared sunbirds by lack of scarlet breast-band, and race *falkensteini* from sympatric White-bellied Sunbird *C. talatala* by yellow belly. ♀ told from similar-sized ♀♀ of other species by combination of olive tone to brown upperparts, *blue-black* tail and *unstreaked pale yellow* underparts. Juvenile ♂ like ♀ but with small black throat patch; this becomes larger and iridescent purple with age, when green also appears on wings and rump. Moulting ♂ with metallic blue throat and metallic green wing shoulder can be confused with both White-bellied Sunbird in same plumage and adult ♂ Plain-backed Sunbird *Anthreptes reichenowi* (Hustler 1985).

Voice. Tape-recorded (16, 86, 88, 91–99, 104, B, C, F, KEI, NOR). In variable 1·5–2·5 s advertising song, 2–7 introductory notes are followed by a rattling chatter; preliminary notes may be harsh, like alarm, dry and hard, thin and high-pitched or some combination of these; chatter variable in speed, faster versions being a trill, in slower ones individual notes audible; speed sometimes changes during song, and song may trail off a bit and slow towards end; pitch usually constant but may change slightly: 'chewy-chewy-chatatata-tatatatata', 'tsee-sip-titi-chatatatatatata', 'chup, tatiti-turrrrrrrtatatatata', 'chup, chuppychup, chutsee-chititititititi'; slower versions, 'chew-titichew-cha-cha-cha-cha-cha' or 'cha-cha-cha-cha-cha-tititsee-cha-cha-cha-cha'. Lower-volume rolling song, characterized by Chappuis (in press) as 'de combat', is rapid, undulating jumble of thin, high-pitched squeaky and chattering notes, with or without sibilant, tinkling overtone, continuing in stop-start bursts for 20 s or more. Alarm a loud, scolding 'djoowi-djoowi-djoowi ...'; other calls include scolding 'chew-*tseep*' or 'chew-tsew-*eep*' (Zimmerman *et al*. 1996), 3-syllable 'tsiu-tse-tse' and clicking sounds in flight (Skead 1967).

General Habits. In W Africa inhabits dry, open country with scattered trees, forest-grassland mosaic, savanna woodland, gardens, coastal scrub, fringing forests, wooded ravines, small copses of open trees, low shrubs, farmland almost devoid of trees, and grassland at 1150 m (Cameroon). In Liberia occurs in coconut plantations, mangroves and on inselbergs in S, and in N at 1400 m in secondary savanna of *Pteridium* ferns and Solanaceae. In eastern Africa bushes and scrub in moorland, light woodland, edges of eucalyptus and juniper woods (Eritrea), cultivation, gardens and bushy woodland (Kenya) and open spaces with bracken (Itombwe, Zaïre); absent from closed-canopy woodland and drier areas; near Lubumbashi confined to open, grassy areas with streams and marshy ground with flowering plants. In Angola mainly in *Brachystegia* woodland. In southern Africa favours mesic vegetation and rank, herbaceous growth, mainly between 300 and 2300 m: *Burkea* savanna and second-growth and wind-stunted *Brachystegia*, bracken-briar, lowland riverine scrub, highland herbaceous scrub, sheltered valleys incised into hillsides, also exposed, *Protea*-covered hills; edges of small patches of evergreen forest. In Umtali district, Zimbabwe, common to numerous in luxuriant, flowering undergrowth in sheltered folds on steep S-facing hillsides, up to 2300 m, from Inyanga to Melsetter in bushy hollows sheltered in lee of boulders or evergreen forest.

Usually in pairs, year-round. Lively, energetic, never still; pair forages often close to ground, making short or very short flights in and between flowering small trees and low bushes, from flower-stem to flower-stem, bird clinging to thinnest stems to probe the flower, even though its weight bends stem right over, when continues feeding head downwards. Alights at flowers and immediately starts to probe them, working systematically through inflorescence, moving jerkily, often looking up and calling, then suddenly flitting to next flower head; on leaving tree may fly quite far, with rapid beats and slightly dipping flight. Gleans small insects and larvae from plant stems and leaves; makes short, fluttering flights straight upwards from perch to seize slow-flying small insects (Vincent 1949). Sudden and erratic movements and frequent tumbling chases make bird hard to watch, though its progress in foliage can be followed from frequently-repeated, low, ticking contact call and occasional, drawn-out 'cheer, cheer' alarm call; but often feeds out in open. Caterpillars beaten against perch before ingestion. Sings from treetop or similar high, exposed perch, with tail spread, bird switching body from side to side. ♂ freely entered a house in Karen, Kenya, and frenziedly attacked its image in mirrors and windows over period of 5 months (Sheppard 1988).

Moults primaries centrifugally in sequence 1–10, secondaries in sequence (counting centripetally) 8, 1, 9, 2, 7, 3, 4, 5, 6, and rectrices either 1– 6 or 1, 2/3, 4, 5/6 (Hanmer 1981). In Malaŵi ♂♂ moult *c*. 20% of head, neck, mantle and breast feathers and upperwing-coverts soon after fledging, replacing it with metallic green feathers, and assume 50–80% of adult plumage at end of 1st year; they assume full adult plumage at 3rd moult at end of 2nd year. One multiple-retrap bird was in 95% adult dress in Jan, 100% in Feb, 90% in Mar and Apr, and the next year 100% in Mar, slightly less than 100% in May, and moulting from eclipse into adult plumage in Nov (Lane 1996).

Mainly resident; perhaps sedentary at low latitudes, but partly migratory in W Africa (and probably elsewhere). Mali, only present May–June and Nov–Dec when quite common; present all year in Gambia but commonest Mar–June. In Ghana occurs on coast (Denu) only in May–July, not in Jan–Apr; very common in N (Walewale) in Mar. In Nigeria resident on coast near Lagos but absent from much of country south of the great rivers in Apr–Oct (Elgood *et al*. 1973, 1994); in Nindam Forest Res. (09°32′N, 08°30′E) a dry-season visitor, Oct–Mar (Dyer *et al*. 1986); rare in Yankari Game Res. (about 09°45′N, 10°30′E) – sight records in Feb–Mar only (Crick and Marshall 1981); well-marked passage at Zaria (11°02′N, 07°43′E) from late Apr to early July, with ♂♂ far more in evidence than ♀♀, and birds again common there in Sept (Fry 1971); further north, rare in Kano State – 4 records, Kano (12°N, 08°30′E), all June–July, and undated records at Maiduguri (see above). Occurs around Bozom, W Central African

Republic, only in June–July. Thought to be resident in E Africa though probably in part a vertical migrant. In Zimbabwe much local wandering, but sedentary in newly settled areas, e.g. Harare (A.J. Tree in Harrison et al. 1997); wandering may amount to migration south of 20°S in May–Sept, when vagrants appear in Transvaal.

Food. Nectar, spiders and small insects including caterpillars. Feeds at flowers of *Leonotis*, *Salvia*, *Erythrina*, *Kniphofia*, *Loranthus*, *Combretum paniculatum* and *Protea* spp. in South Africa, *Petria nobilis*, *Bryophyllum*, *Salvia*, *Fuschia* and *Grewia similis* in Kenya and *Albizia coriaria*, *Thevetia peruviana*, *Pedilanthus tithymaloides* and *Jacaranda mimosifolia* in Nigeria. Moths, very small flies, caterpillars and spiders fed to nestlings, but bulk of food for them thought to be nectar (van Someren 1956).

Breeding Habits. Solitary nester, monogamous. Presumably territorial; ♂♂ not aggressive to each other except when breeding. In courtship ♂ perches on topmost twig of small tree, stands upright, and pivots from side to side with wings held slightly open, breast feathers fluffed out and orange pectoral tufts showing; he repeatedly utters short, rippling song, with throat expanded and vibrating. ♂ makes sudden flight towards ♀, who may be quietly feeding but puts head down and opens and shuts bill (van Someren 1956). In Kenya, double-brooded in Apr–June season and single-brooded in Oct–Dec (van Someren 1956); pair (♂ ringed) nested twice, nests 3 months and 2 m apart (Cunningham–van Someren 1976). In Malawi breeding may also be bimodal (Hanmer 1981).

NEST: suspended oval bag, sometimes narrower at top and more pear-shaped, with circular side-top entrance. Constructed (not as compactly as most sunbirds' nests) of very fine, soft, dry grass stems and blades, often intermingled with dry reed strips and broad, coarse grass blades, bits of pith, rootlets, fibres, small withered leaves and flower heads; bound outside with spider web; lined (usually thinly in Zimbabwe, usually thickly and warmly in W and E Africa) with silvery thistle down, *Clematis* seed hairs or other white, woolly vegetable down, to which a few small feathers occasionally added; lining is scant inside walls and ceiling (Vincent 1949, van Someren 1956, W. Serle in Bannerman 1948). Some nests in southern Africa, most or all nests in W and E Africa, have small porch over entrance, formed from projecting stiff grass stems or sometimes from one or a few living leaves of plant from which nest hung. Outside of nest not decorated (Vincent 1936) or may be decorated with bits of spider cocoons and dry composite flowers (van Someren 1956); often camouflaged with bits of fibre dangling from sides and bottom, and sometimes attached to adjacent leaves by long strands of spider web. Nest height (n = 3+) 115–130, width 58–76, from front to back (ext.) 70, entrance diam. 25–35 horizontally and 25–30 vertically, int. depth of cup below entrance 35. Appearance generally brownish yellow; nest like that of White-bellied Sunbird *C. talatala* (which has, however, far more dead leaves built into it).

Nest sited in areas where much grass and undergrowth has been burnt out, leaving thin woodland with scattered shrubs and only a few patches of long grass still standing. Built in tall, dry weed or small sapling growing amidst grass or dense clump of shrubs. Attached across full width of top or top-back, to branching stem or projecting twig or tall bracken frond (and once onto top of stunted maize plant) 60–150 cm from ground; assimilates well with dry, yellowish-brown growth in which placed; nest often appropriated by mouse, when birds build second nest (Vincent 1949). Nest built behind tip and not *at* tip of stem, and generally sheltered below leaves. Construction of nest is mainly by ♀; ♂ accompanies ♀ when she flies from nest to collect material, she gathers a billful, ♂ flies with her back to nest and then helps to work the material in. Building takes 10–20 days.

EGGS: 1–2 (Zimbabwe) or 2 (Nigeria, Zaïre, E Africa). 2nd egg laid sometimes 1 day but usually 2 days after 1st (van Someren 1956). Ovate; not glossy; white, pale cream or greyish white, closely and evenly freckled or with fine, olive dusting with some underlying grey at large end, sometimes forming cloudy brown ring or cap. SIZE: *C. v. falkensteini* (n = 19) 13·9–16·8 × 10·4–11·7 (15·4 × 11·2) (Vincent 1949) and (n = 35) 13·9–17·3 × 10·4–12·1 (15·0 × 11·3) (Skead 1967); *C. v. venusta* (n = 2, Nigeria) 14·5–14·6 × 10·5–10·8.

LAYING DATES: Gambia, Nov, Mar; Liberia (Mt Nimba), Dec (and juveniles Jan–Apr). Sierra Leone, Oct; Togo, July (Glidje, 06°15′N); Nigeria, Oct–Jan (and juvs on coast May–July); Sudan, (mts in S) Nov–Jan; Eritrea, June (nest-building July, Sept; ♂♂ in breeding plumage Apr–Sept); Ethiopia, May–Oct (*fazoqlensis*), Apr–May (*albiventris*); Zaïre (Itombwe) June (to Aug), (Lubumbashi) Aug–Sept; E Africa, Region A, Oct; Region B, Feb–Apr, June, Oct, Dec; Region C, Jan–June, Aug, Nov, mainly Jan–May; Region D, Jan–June, Aug–Dec (116 records, mainly Apr–June and Dec); Region E, May (Brown and Britton 1980). Angola, Nov (and gonads enlarged Aug); Zambia, Apr, July–Oct (and gonads active Mar and May); Zimbabwe, Feb–Dec, mainly Mar, May–June and Sept; Mozambique (Vumba), Apr–Sept.

INCUBATION: ♀ sits close when clutch completed, leaving to forage in early morning and late afternoon; in between, she may be fed by ♂ (van Someren 1956). Period (1 nest, Kenya) 14 ± 1 days.

DEVELOPMENT AND CARE OF YOUNG:: ♀ broods young almost constantly for 1st 2 days; thereafter young fed by both sexes equally (van Someren 1956); pair once seen flycatching from shrubbery and carrying 4–5 insects at a time to nest (Vincent 1949). Period: (1 nest, Kenya) 18 ± 1 days.

BREEDING SUCCESS/SURVIVAL: 'low elevation of the nests renders these birds very liable to attention from predators such as mongooses, rats, egg-eating snakes, and bush shrikes' (van Someren 1956). Parasitized by Klaas's Cuckoo *Chrysococcyx klaas* (Kenya, Ethiopia). Longevity records of >5 – >6·5 and 8 years (n = 10; Hanmer 1989, 1997, Hanmer and Chadder 1997).

References
Bannerman, D.A. (1948).
Cheke, R.A. (1976).
van Someren, V.G.L. (1956).
Vincent, A.W. (1936, 1949).

Cinnyris fusca Vieillot. Dusky Sunbird. Souimanga fuligineux.

Plate 16
(Opp. p. 223)

Cinnyris fuscus Vieillot, 1819. Nouv. Dict. Hist. Nat., nouv. éd., 31, p. 506; Great Namaqualand.

Cinnyris fusca

Range and Status. Endemic, irruptive nomad, W Angola to S Cape Province. Angola, arid coastal plain south of Benguela; common in gardens in Namibe (Moçâmedes), and the commonest sunbird in Iona Nat. Park in SW Namibe Prov.; recorded in W Cunene Province. Namibia, common and widespread, from near Nkurenkuru on Cubango R. in NE, west to Skeleton Coast Nat. Park, south to Walvis Bay and Orange R. valley west to mouth, and east to Kalahari Gemsbok Nat. Park. Botswana, uncommon to locally quite common in Gemsbok Nat. Park and in Molopo valley (border with South Africa) east to near Werda. Particularly common in arid regions where Acanthaceae, aloes, exotic *Nicotiana* and other flowering shrubs grow along watercourses and under Escarpment; follows major river valleys across Namib Desert.

South Africa, common in some areas, e.g. Namaqualand and Great Karoo, mostly uncommon, from Atlantic coast east to 26–27°E in Cape Prov. and Orange Free State; Graaff-Reinet, Cradock, Somerset East, rare near Port Elizabeth. Nomadic, breeding in arid regions in any month and thereafter sometimes having to move distantly to avoid drought. Rare annual visitor into SW Cape, usually Aug–Mar. Mainly in Oliphants R. valley, but major irruptions can occur, as in July–Aug 1978, when reached (and nested near) Cape Town. Absent from Little Karoo, but 3 pairs bred in Anysberg Nature Reserve in June 1989 following good rains (Martin *et al.* 1989). In Orange Free State occurs mainly in false karoo veld in S, and rare in W northeast to Sandveld Nature Reserve. Irruptive vagrant in Transvaal: up to 14 at Barberspan, June–Aug 1970 (Milstein 1975). (Old Botswana record 'near Transvaal border' mentioned in Tarboton *et al.* 1987 was actually from Tschabong in SW Botswana.)

Density of 2–3 ♂♂ per 100 m of linear *Nicotiana* growth in drainage lines in NW Namibia, or 1–2 birds per 200 m where *Nicotiana* absent; once a concentration of 'thousands' in 1 km of *Nicotiana* near Keetmanshoop (Hoesch and Niethammer 1940).

Description. ADULT ♂ (breeding): upperparts grey-brown, variably tipped dull black with bronzy green and purplish reflections; long uppertail-coverts blackish, glossed violet-blue. Tail blackish, upperside with slight blue gloss; T3–T6 with narrow whitish terminal fringe. Lores blackish; cheeks and ear-coverts blackish brown. Chin to upper breast dull black with green and purple reflections; pectoral tufts bright orange or orange-yellow; lower breast to upper belly and upper flanks sooty brown; lower belly and lower flanks to undertail-coverts off-white. Lesser and median coverts dull black with green reflections; rest of upperwing grey-brown, flight feathers and greater coverts prominently edged pale grey-buff. Axillaries and underwing-coverts sooty brown. Bill black; eyes dark grey-brown; legs black. ADULT ♂ (non-breeding): upperparts and sides of head largely (in some almost entirely) grey-brown. Below, black restricted to centre of throat and breast; sides of throat and breast greyish white, forming stripe connecting with whitish underparts. Wings and tail as in breeding ♂. ADULT ♀: upperparts grey-brown. Tail blackish, upperside with faint blue-green gloss; T3–T6 with narrow whitish terminal fringe. Lores, cheeks and ear-coverts dark grey-brown. Chin to breast and flanks pale grey, merging with creamy white belly and undertail-coverts. Upperwing grey-brown, flight feathers and greater coverts edged pale buff. Axillaries and underwing-coverts grey-brown; inner borders on undersides of flight feathers grey-buff. Bare parts as in adult ♂. SIZE (10 ♂♂, 10 ♀♀): wing, ♂ 56–61 (58·6), ♀ 50–54 (52·5); tail, ♂ 38–43 (40·2), ♀ 30–35 (32·7); bill, ♂ 21–24 (22·4), ♀ 18–21 (19·8); tarsus, ♂ 17–19 (18·0), ♀ 15–17 (16·2). WEIGHT: 1 ♂ 10·0, 2 unsexed 8, 8·.

IMMATURE: juvenile similar to adult ♀, but more olive above, tinged yellowish below.

NESTLING: hatches naked, blind, skin blackish olive; gape yellow, bill rather long; still like that at 2–3 days (photo, Jensen and Clinning 1974), but at 5 days quills well developed and eyes starting to open.

TAXONOMIC NOTE: placed by Wolters (1982) between *venusta* and *talatala* in genus *Arachnechthra*; we consider this species to be an arid-zone derivative of *C. cuprea*, although the 2 are now too divergent to be a superspecies (*cf.* Hall and Moreau 1970).

Field Characters. Length 11–12·5 cm. Breeding ♂ looks *blackish* except for *white belly*; pale edges to wing-feathers and *orange* pectoral tufts. ♀, eclipse ♂ and immature ♂ dark grey-brown above, white below; immature has black on centre of throat, ♂ in eclipse has broader black area on centre of throat and breast, rest of underparts often patchy black and white. ♀ purer white below than ♀ White-bellied Sunbird *C. talatala*, with no streaks on breast, but 'probably not safely separable in field' (Maclean 1993).

Voice. Tape-recorded (16, 88, 91–99, B, C, F, LIV, NIV, ROC, STA). Common call a shrill 1–2 s chatter, slightly descending the scale, with extra notes at beginning and end, 'tsi-trrrrrrrrrrr-jt' or 'tsirrrrrrrr-zut-zut-zut'; sometimes slower, with individual notes audible, 'tsit-chi-chi-chi-chi-chi ...'. Song a low-key, conversational medley of short musical trills at different speeds and pitches mixed with chuckles, sweet notes, nasal 'jay-jay-jay' and dry 'dzut-dzut', lasting 3–4 s. Flight call a double 'ji-dit'.

General Habits. Inhabits semidesert with scattered shrubs and trees along watercourses and at foot of rocky outcrops; desolate desert with the odd shrub, and sand dunes on Namib coast with only the most meagre growth. Usually in pairs, sometimes solitary, or in flocks of up to 20, sometimes foraging in company of other sunbirds and white-eyes *Zosterops*. Restless, aggressive; forages actively even in mid-day heat, probing from a perch or hovering at flowers and gleaning tree foliage. Catches small insects in flight by hawking up into air from perch. Longer flights between trees fast and straight, not quite so jinking and erratic as in other sunbirds. Rarely, drinks from water pool.

♂ moults into nuptial plumage in Oct–Nov (NW Namibia), breeds, then moults into eclipse plumage from Apr–July onward; if rainfall is early and heavy, ♂♂ may breed whilst still moulting into nuptial plumage (Hoesch and Niethammer 1940).

Nomadic. In W Namibia occurs mainly as non-breeding visitor below, or west of, 80 mm isohyet; shows considerable local movement, and subject to major irruptive immigration into extremes of range (see above).

Food. Nectar and insects. Feeds at flowers of *Aloe dichotoma* in Richtersveld, also probably *A. marlothii*, *A. littoralis*, *A. zebrina*, *A. grandidentata*, *A. hereroensis* and *A. gariepensis* in Kalahari and along Orange River – all nectar-rich plants (Skead 1967) – and *A. asperifolia* and *Drosanthemum luderitzii* in NW Namibia (Williams *et al.* 1986); also *Lycium campanulatum*, *Cadaba*, and *Leonotis oxymifolia* (Fraser and Wheeler 1991). Other sought-after food sources are *Loranthus*, *Psilocaudon*, various garden flowers (*Montbretia*, *Canna*, *Oleander*, *Hibiscus*, *Lantana*) and, particularly, alien *Nicotiana glauca*. Insects (including moths) may feature in diet to greater extent than in most other sunbirds, especially when Dusky Sunbird is breeding (Skead 1967).

Breeding Habits. Solitary nester, evidently monogamous and territorial. Courtship behaviour includes vigorous singing, aerial display, and aggressive maintenance of territory. Play-back of song elicits aggressive reaction; one ♂ attacked the play-back tape-recorder. ♂ sings and moves around in nesting season in well-defined territory, of 0·5–1·0 ha (in high-density population at Spitskoppe, 22°S, 15°E, Namibia); territory and nest-site within it remain rather constant from year to year (Jensen and Clinning 1974). Flight, when ♂ excited by presence of another, moth-like and fluttering (Skead 1967). Double-brooded (in a good year); pair re-nests quickly if nest lost.

NEST: suspended, oval pouch with side-top entrance, made of grasses, plant fibres, dry leaves and bark, bound with spider web, lined with fine silky or downy seed fibres and animal hair, e.g. of Gemsbok *Oryx oryx* (in NW Namibia: Williams *et al.* 1986), and decorated with a few shreds, once of tissue paper, bound onto surface with spider web. Ext. height *c.* 125, ext. diam *c.* 75, int. diam. 40–50 (Jensen and Clinning 1974). Attached to supporting plant by back, and sited 0·1–1·65 m above ground (av. 0·7 m) in bushes of 11 spp. named by Williams *et al.* (1986); quite conspicuously placed, or occasionally hidden; often along banks of streams or dry water courses. In Cradock district, Cape Prov., often builds nest against leaves of *Opuntia* (Skead 1967). Nest built by ♀ alone, taking 6 days. The 'preliminary nest' of Skead (1967) is a misinterpretation (Jensen and Clinning 1974).

EGGS: 2–3 (av. 2·6, n = 31 clutches, NW Namibia). Eggs laid on consecutive days, beginning 4–6 days after completion of nest. White, mottled, clouded and spotted with greys and browns, mainly at thick end. SIZE: (n = 15) 14·7–16·2 × 10·3–11·6 (15·5 × 10·9), (n = 7) 15·0–15·9 × 0·7–10·9 (15·4 × 10·8), and (n = 25) 14·1–16·1 × 10·6–11·6 (15·4 × 11·0); egg-size varies within clutch, particularly in clutch of 3 (Williams *et al.* 1986).

LAYING DATES: Namibia (Skeleton Nat. Park), Sept–June, (Windhoek) mid Sept to early Apr (mainly Feb–Mar), (Pro-Namib lowlands below escarpment) Nov–June (mainly Feb–Apr), (Tses near Keetmanshoop) late Dec to early Mar; breeds in months with most reliable rainfall (Feb, Mar); breeds early in well-watered gardens in Windhoek. South Africa, mainly Aug–Mar, but thought to breed at any time after rain (Maclean 1993).

INCUBATION: by ♀ only. Incubation starts sometimes with 1st egg, usually with 2nd egg, occasionally with 3rd egg (Jensen and Clinning 1974). Period 12–13 days; eggs hatch asynchronously.

DEVELOPMENT AND CARE OF YOUNG: growth very rapid after eyes open. Young brooded by ♀ but fed by both parents; ♀ removes faecal sacs. Young spend at least 13–15 days in nest, then may leave nest for short flights for a further 8 days, and return to nest to roost at night for 5 more days (Benseler 1970). Fledgling has underparts strongly washed with yellow, and variable amounts of black on throat (Jensen and Clinning 1974). Both parents feed young out of nest, for 2–3 weeks.

BREEDING SUCCESS/SURVIVAL: 7 out of 64 nests (11%) in Pro-Namib parasitized by Klaas's Cuckoo *Chrysococcyx klaas*, whose eggs closely resemble those of Dusky Sunbird (Jensen and Clinning 1974). 2 features of breeding biology (within-clutch variation of egg size, and asynchronous hatching) promote a size hierarchy in the brood; when food in short supply the 1–2 smaller chicks starve to death. Such 'brood reduction' occurs early in at least 25% of nests, and by fledging time may operate in most or even nearly all broods (Williams *et al.* 1986).

References
Harrison, J.A. *et al.* (1997).
Jensen, R.A.C. and Clinning, C. (1974).
Skead, C.J. (1967).
Williams, A.J. *et al.* (1986).

Cinnyris johannae Verreaux and Verreaux. Johanna's Sunbird. Souimanga de Johanna.

Cinnyris johannae J. and E. Verreaux, 1851. Rev. Mag. Zool. (Paris), sér. 2, 3, p. 514 (= 314); Gabon.

Plate 16
(Opp. p. 223)

Cinnyris johannae

Range and Status. Endemic resident, Sierra Leone to E Zaïre. Mali, rare: 3 records around Kangaba (Lamarche 1981; requires validation, in view of bird's rain forest habitat and similarity to *C. superba*). Guinea, rare; Quinandou, Ziama Massif (Macenta) (Halleux 1994). Sierra Leone, common in Freetown Peninsula (Field 1974), east to Bo and Sulima R. Liberia, widespread from coast inland in N to Belle Yella region and Mt Nimba (where frequent). Ivory Coast, uncommon, Nimba to Taï Nat. Park, Maroué and Ayamé Dam, with a record in N at Dabakala; not uncommon in Yapo Forest. Ghana, rare: 19th-century records from Fumsu, Prahsu, Aburi, 'Fanti' and 'Denkera', and several recent ones from Tafo and Subri River Forest Res. (Grimes 1987) and Atewa Hills. Togo, record at Klouto near Ghana border now doubted (Cheke and Walsh 1996). Benin, reported only at Abomey.

Nigeria, sight record of ♀ near Lagos in 1951 and ♂ near Mambilla Plateau in 1985 (Gray 1986). Cameroon, rare in W (Kumba, Dikume Balue – 1230 m in Rumpi Hills) but widespread and locally frequent in S, south of 04°N and west of 14°E. Mbini. Gabon, in Rés. de la Lopé, and widespread in NE but not very common. Congo, common and widespread in Odzala Nat. Park and in all forest types in Nouabalé-Ndoki Nat. Park; occurs in Léfini Rés.; rare in sublittoral forests near Koubotchi and local in Mayombe, particularly near Makaba. Angola, known only from Landana, Cabinda. Zaïre, uncommon: known from about 12 equatorial localities, from 3 localities between Tshikapa and NE Angola border in Kasai Occidental, and from Itombwe. Uncommon in Itombwe, at 840–1530 m (Butokolo, Kamituga, Mulungu, Mwana, Itula).

Density roughly 3–4 pairs per km^2 (NE Gabon).

Description. *C. j. fasciata* Jardine and Fraser: Sierra Leone to Benin. ADULT ♂: head, neck and rest of upperparts metallic golden green. Tail black, feathers glossed violet-blue above, and narrowly edged metallic green when fresh. Lores blackish. Chin and upper throat golden green, concolorous with head, merging with a broad glossy violet band across lower throat and upper breast; pectoral tufts lemon yellow; lower breast, upper belly and upper flanks non-iridescent dark red; lower belly, lower flanks, thighs and undertail-coverts sooty black. Lesser and median coverts metallic green; rest of upperwing feathers sooty black. Axillaries and underwing-coverts sooty black. Bill black; eyes dark brown; legs black. ADULT ♀: upperparts dark olive. Tail feathers blackish brown, tipped brownish white, more broadly on T5–T6. Narrow superciliary stripe creamy; lores, ear-coverts and sides of neck dark olive. Cheeks to chin and throat creamy, breast to flanks, belly and undertail-coverts pale olive-yellow; entire underparts sharply streaked dark olive-grey, most heavily on breast and flanks. Upperwing feathers dark greyish brown; primaries and secondaries narrowly edged olive; tertials and greater, median and lesser coverts narrowly fringed pale olive. Axillaries creamy white; underwing-coverts creamy white with dark shaft streaks; inner borders of flight feather undersides greyish white. Bare parts as adult ♂. SIZE (10 ♂♂, 6 ♀♀): wing, ♂ 62–67 (65·1), ♀ 60–65 (62·0); tail, ♂ 35–37 (35·7), ♀ 31–35 (32·0); bill, ♂ 31–33 (31·8), ♀ 30–32 (30·7); tarsus, ♂ 16–18 (16·7), ♀ 16–17 (16·5). WEIGHT: Liberia, ♂ (n = 11) 12·6–14·7 (13·7), ♀ (n = 3) 11–15 (12·4).

IMMATURE: juvenile like adult ♀, but greyish streaking on underparts denser, less sharp; sides of throat, breast and flanks generally dusky.

C. j. johannae (Verreaux): Cameroon to Zaïre. ♂ slightly darker red below than *fasciata*, ♀ darker and browner above. Larger: wing, ♂ (n = 7) 70–72 (71·6); bill, ♂ (n = 7) 35–37 (35·7).

Field Characters. Length 12·5–14 cm. Long-billed and short-tailed, giving a top-heavy appearance. ♂ has iridescent green head and upperparts, *purple* breast, *dark crimson* belly and *yellow* pectoral tufts (lacking in Superb Sunbird *C. superba*); further distinguished from Superb by smaller size and *green* chin and throat. Unique ♀ has white chin and throat, rest of underparts *yellow*, completely covered by *long, heavy dark streaks*; unlike anything in Africa; striping reminiscent of Asian spider-hunters *Arachnothera*.

Voice. Tape-recorded (104, B, BRU, CHA, KEI, LEM, SAR). Song given by 2 birds together, both perched and in flight, a rapid, long, rambling, undulating string of squeaky and chattery notes and a few melodious ones, almost whistles (imitation of other species?). Calls: a loud, descending, ringing but rather plaintive 'peeeurr', shrill 'pwee-pwee-pwee ...' and a brief, thin 'pssst' or 'pssit'. Alarm, harsh screech, 'chraaa'. Call on take-off or in flight, 'pit-pit', very typical (F. Dowsett-Lemaire, pers. comm.)

General Habits. Inhabits canopy, edges and clearings in primary lowland rain forest and old secondary forest, gallery forest, cocoa plantations, and (often in Freetown, uncommonly in NE Gabon) gardens and large flowering trees around villages and in town suburbs in the forest zone. Occupies upper and middle strata of forest; seen mainly along tracks and in clearings (Demey and Fishpool 1994).

In pairs, also parties of 3–5 birds, mainly ♂♂ in breeding plumage. ♂♂ sometimes gather in larger flocks, high in the canopy, like the evident lekking aggregations of ♂♂ of Western Olive, Blue-headed and Blue-throated Brown Sunbirds *Cyanomitra obscura*, *C. verticalis* and *C. cyanolaema* (Brosset and Erard 1986). Very active and mobile, visiting flowers high in trees and lianas, constantly moving on, and inspecting masses of *Usnea* and other filamentous lichens covering large limbs. Feeds at flowers of wide range of trees, shrubs and epiphytes; searches for small arthropods in epiphytes and on thin branches; makes short sallies after insects above the canopy (Gatter 1997). Joins mixed foraging bird parties.

Sedentary. All Mali records (correctly ascribed to this species?) May–June.

Food. Spiders and caterpillars in stomachs (Mt Nimba), spiders and a 'sweet taste' (i.e. nectar) in stomachs at Bitye, Cameroon (Bates 1911). Eats fruits of *Macaranga assas* and *M. barteri* (Dowsett-Lemaire 1996). Seen feeding at flowers of several species of Loranthaceae and of *Syzygium congolense* (Congo: F. Dowsett-Lemaire, pers. comm.).

Breeding Habits. Solitary nester. Intensive display (not described) in Dec and Mar, Liberia (Gatter 1997).

NEST: one nest, *c.* 22 cm long, was built of green moss and a little lichen, held together by black fungal hyphae, and attached to midrib of a pinnate palm frond (Chapin 1954). Another attributed to this species (though 'in savanna country far from the evergreen forest') was a large, loosely-built, untidy, suspended pouch at tip of a drooping palm frond 2·5 m above ground; made of palm leaf-base fibres, with thin 'beard' underneath; decorated with a few dead leaves; lined with palm fibres less soft and flexible than those of body of nest (Holman 1949). One inaccessible nest suspended from end of liana, 10 m above sloping ground, under foliage of large tree (Brosset and Erard 1986). A fourth undescribed (Demey and Fishpool 1994). 3 nests, in mature and depleted forest and at forest edge, at heights of 18, 20 and 35 m above ground, were 25–30 cm long; one 12 m above ground, in savanna (Gatter 1997).

EGGS: (from the suspect nest, see above) 2, pointed ovals, smooth and glossy, very pale blue, boldly spotted with red-brown mainly at large end. SIZE 20·5 × 14·0, 20·5 × 15·0.

LAYING DATES: Liberia, (nests being built Mar, May, June; feeding visits to nest, May: Gatter 1997); Ivory Coast (Yapo), May (and Taï: collecting nest material Jan, Feb); Ghana, June (? – Grimes 1987); Gabon, (nest building Nov); Congo, (nest building Jan); Zaïre (Itombwe), (various indications of breeding Feb–July: Prigogine 1971).

Plate 16
(Opp. p. 223)

Cinnyris superba Shaw. Superb Sunbird. Souimanga superbe.

Certhia superba Shaw, 1811–1812. General Zool., 8, p. 193; Malimba (Portuguese Congo).

Range and Status. Endemic resident and partial migrant, forests and clearings from Sierra Leone and S Mali to W Kenya, south to W Angola and S Zaïre.

Mali, rare: records from Bougoni, Sikasso and Kangaba. Guinea, uncommon on Ziama Massif (near Macenta) (Halleux 1994); Sierra Leone, old sight records at Freetown; not uncommon at Daru; uncommon at Firestone Plantation, common at Ganta; frequent in E: Buedu, Kwendu, Sefadu, Bossu and Gandahun. Liberia, widespread, not uncommon in S, common in N; occurs up to 1200 m on Mt Nimba, but less frequent than *C. johannae*; may be absent from SE and from much of centre of the country (Gatter 1997). Ivory Coast, uncommon (but commoner than *C. johannae*), 'throughout forest zone', north to Sipilou and Comoé (Thiollay 1985); however, absent from Taï Forest, rare in Yapo Forest (where *C. johannae* not uncommon: Demey and Fishpool 1994), and may not occur south of line between the two. Ghana, not uncommon in SW and S, north to Sunyani, but absent from wet evergreen forest west of Cape Coast (such as remains there), rare in Bia Nat. Park, and absent north of line from Sunyani to S end of L. Volta. Togo, not uncommon in narrow band in S (Nyivé, Tététou, Tokpli: Cheke and Walsh 1996). Burkina Faso/Benin, 1–2 sight

Cinnyris superba

records in Arli/Pendjari Nat. Parks; known from SW Benin. Nigeria, not uncommon in S, from Lagos to Benin and Sapele, north to Abeokuta, Ibadan and Ife; rare east of R. Niger, records only at Umuagwu (Owerri) and Enugu.

Cameroon, Mbini and Gabon: frequent and widespread as shown; fairly common in Cameroon, west to Kumba. Congo, frequent in Odzala and Nouabalé-Ndoki Nat. Parks, and in Kouilou Basin west to Béna and Koubotchi. Zaïre, frequent and widespread, but absent in NW (north of 02°N, west of 25°E), in Ituri and Uele occurs north only to Dungu (although a record also at Zemio (Semio) on Bomu R. at Central African Republic border), absent south of line between Bolobo and L. Edward except for broad swath from S Bandundu and S Kasai Oriental to Itombwe (where common); absent from most of Katanga, but a record in SW Central African Republic: records from Landjia, Mboko and Pisa (Germain and Cornet 1994). Angola: Cabinda, Cuanza Norte, N Malanje, N Lunda, south along escarpment to Chingoroi, Huila. Uganda, locally common in W and S at 700–1700 m, from Murchison Falls Nat Park, Budongo, Masindi and Bugoma to Maramagambo and Impenetrable Forests, east to Malabigambo, Sezibwa R., Entebbe, Kampala, Kifu, Mabira and Buvuma Island (Britton 1980). Kenya, 1 old and 3 recent records in Mumias and Busia Districts (Zimmerman *et al.* 1996). Tanzania, record from Bukoba.

Description. *C. s. superba* (Shaw): Cameroon (Serle 1965) to Zaïre (Kasai, Katanga, Itombwe) and Angola. ADULT ♂: forehead to nape metallic greenish blue with violet reflections, forming well demarcated cap; rest of upperparts including rear and sides of neck glossy golden green. Tail sooty black. Lores blackish; supercilium, cheeks and ear-coverts bronzy green. Chin, throat and upper breast metallic violet with dark red and glittering blue reflections; lower breast, upper flanks and upper belly deep non-iridescent maroon; lower belly, lower flanks and thighs sooty black; undertail-coverts sooty, tipped maroon. Lesser and median coverts metallic green; rest of upperwing feathers sooty black. Axillaries and underwing-coverts sooty black. Bill black; eyes dark brown to blackish; legs black. ADULT ♀: upperparts light olive-green with faint diffuse mottling, especially on top of head. Tail dark grey-brown, feathers tipped paler brown, broadly on T5–T6, and edged olive-yellow. Prominent long superciliary stripe yellowish; lores and ear-coverts dark olive-green. Cheeks and chin to breast pale olive-yellow, mottled dusky olive-green, some with orange barring across breast; belly and flanks deeper olive-yellow, streaked olive-green; undertail-coverts orange-yellow. Flight feathers dark grey-brown, edged olive-yellow; tertials and upperwing-coverts grey-brown fringed light olive-green. Axillaries and underwing-coverts creamy; inner borders of flight feather undersides pale grey. Bare parts as adult ♂. SIZE (10 ♂♂, 10 ♀♀): wing, ♂ 72–79 (74·8), ♀ 67–71 (69·7); tail, ♂ 45–48 (46·7), ♀ 40–47 (42·7); bill, ♂ 34–37 (35·3), ♀ 33–37 (35·3); tarsus, ♂ 19–20 (19·8), ♀ 19–20 (19·8). WEIGHT: Angola, ♂ (n = 1) 18, ♀ (n = 1) 15·4.
  IMMATURE: juvenile like adult ♀.
  NESTLING: not described.
  C. s. buvuma (van Someren): N and NE Zaïre to W Kenya. Purer (less golden) green above than nominate race and slightly darker maroon below. WEIGHT: ♂ (n = 5) 17–19 (18·2), ♀ (n = 3) 17–20·6 (18·2).
  C. s. nigeriae Rand and Traylor: Nigeria. Like nominate race, but slightly brighter red below.
  C. s. ashantiensis (Bannerman): Sierra Leone to Togo. Slightly darker maroon below than nominate race. Smaller: wing, ♂ (n = 10) 67–75 (70·6); bill, ♂ (n = 10) 30–32 (31·2). WEIGHT: Liberia, ♂ (n = 2) 13·8, 14·5, ♀ (n = 1) 13·2.

Field Characters. Length 14–16·5 cm. Large, with long bill and short tail. ♂ like larger version of Johanna's Sunbird *C. johannae*; differs in *metallic blue* top of head, *blue-purple* chin to upper breast, belly deep maroon rather than scarlet, *no pectoral tufts* (yellow tufts in Johanna's). ♀ has long, pale supercilium separated from pale throat by dark mask; underparts *unstriped*, uniform olive-yellow except for pale orange undertail-coverts.

Voice. Tape-recorded (104, B, C, CHA, ERA). Song/call, 3 sharp notes, 'chip-chewp-chip', repeated with briefest gap; sometimes just 'chip-chewp', or with additional 'tip' at the end. In a less emphatic version, notes accelerate and fade away, 'chichew, chichew, chchichew, chichichi ...'. ♂ also gives long series of finch-like notes from elevated post, 'pink-pink-pink' and 'huit-huit' ♀ also calls from same post, 'pee-up, pee-up, pee-up ...' (Christy and Clarke 1994). Call, a slightly nasal 'tew'.

General Habits. Inhabits forest canopy and lower strata, clearings, forest edges, coffee forest, gallery forests and other dense vegetation along watercourses; in Liberia in logged and secondary forest, farmland with plenty of large trees, savanna woodland, gallery forest, mature mangroves and villages, from coast up to 1400 m (Gatter 1997). In E Zaïre occurs in wooded lowlands (not primary forests) and more open plateaux, e.g. west of L. Albert at 1385 m, but not in montane forests; also in secondary forests and fallow ground (Itombwe). In Ghana and Togo usually seen at tops of large trees, e.g. *Berlinia grandiflora* or flowering *Bombax*, often in company of other sunbird species, and can be plentiful in open spots in forest where the sun penetrates and there are plenty of flowers. In Rés. de la Lopé, Gabon, often explores dead trees, apparently looking for spiders. Enters and sometimes nests in cultivated areas, e.g. gardens on forested lower slopes of Kupé Mt (Cameroon).

Versatile when foraging, with tit-like actions, clinging to twigs in all positions, sometimes head down. Sometimes comes low to forage in shrubs, peck at bases of banana leaves or forage at banana flowers and fruit buds of pawpaw *Carica papaya* (Bannerman 1948). Generally in pairs, sometimes singly. Joins mixed foraging flocks. Rather timid. Hovers at flowers, inserting bill as it does so. Hunts arthropods by flitting about in bushes and by searching tree trunks in slightly creeper-like manner. Makes short aerial sallies after small insects, rather like flycatcher. Flight fast and powerful, jerky, undulating, and with irregular zigzagging.

All records north of 10°N fall in Apr–June.

Food. Insects including ants, spiders, nectar, seeds. Insects and a few small green seeds in 4 stomachs, Liberia. Only spiders in stomachs in Sierra Leone. Eats arillate fruits of *Ochthocosmus africanus* (Dowsett-Lemaire 1996). Especially fond of banana flowers.

Breeding Habits. Solitary nester. ♂ of pair 'making mating actions' sang and 'danced with outstretched wings on a large limb' (Bannerman 1932).

NEST: suspended, ragged pouch with side entrance protected by untidy porch; a long wispy 'beard' hangs below nest, which is made of soft fibres from inner surface of bark, or of *Usnea* and fibres, or of banana bark fibres and fine dry grasses; decorated with a few loosely-attached dead leaves, bits of moss and lichen; lined scantily with pappus on floor, or lined copiously. Length *c.* 22–30 cm, excluding 'beard' which can be 30 cm long or more. Nest sited in open, hanging from twig at outside of shrub or tree, 2–10 m (usually about 4 m) above ground. One nest made only a few cm from occupied hornets' nest (Serle 1954). Built by ♀; ♂ follows her on flights to and from nest and often chases her.

EGGS: 1 (Ghana, Cameroon) and 2 (E Zaïre). Pyriform; smooth and somewhat glossy. Creamy white or bluish white, strongly marked with a few bold blotches and spots of inky black and underlying grey in zone around large end (rest of shell only sparsely dotted), or with dots and freckles of pale and dark grey or purplish black. SIZE: (n = 2) 20·0–21·5 × 14·0–15·0.

LAYING DATES: Sierra Leone, (enlarged gonads, Feb–Mar); Liberia, (large gonads Feb, June, July, Nov; dependent young Nov); Ghana, Feb, Nov (nest building Oct, Nov, imm. Aug); Cameroon, Feb, July; Congo, Apr; Uganda, Mar–May, July–Sept; Zaïre (Ituri), (gonad and other evidence that it breeds at least from July to Dec), (Itombwe), (enlarged gonads Sept–Jan and June; juvs May, July, Sept–Dec).

References
Bannerman, D.A. (1948).
Chapin, J.P. (1954).

Plate 16
(Opp. p. 223)

Cinnyris coccinigastra (Latham). Splendid Sunbird. Souimanga éclatant.

Certhia coccinigastra Latham, 1801. Gen. Synop. Birds, Suppl. 2, p. 35; Africa, restricted to Senegal by Grote, 1924, Orn. Monatsb., 32, p. 71.

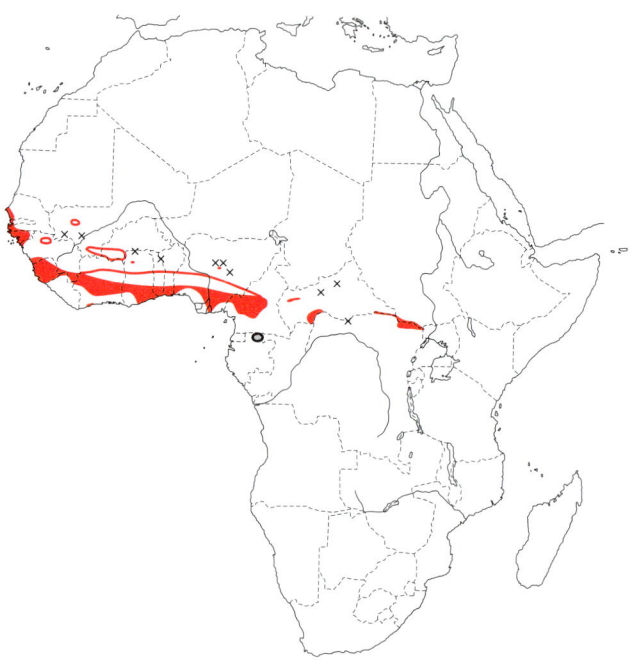

Cinnyris coccinigastra

Range and Status. Endemic resident and migrant, Senegal to NW Uganda. Senegal, locally common in coastal gardens from Cap-Vert and Mbour and on lower Casamance R. (Oussouye). Gambia, generally frequent and locally common on lower river, uncommon on middle and upper river. Guinea-Bissau, many records throughout (Hall and Moreau 1970). Guinea, Fouta-Djalon, Conakry coast to Kindia and Kounounkan (frequent), Beyla area (regular, Oct–Nov: Walsh 1987), Macenta (frequent). Sierra Leone, common in Freetown Peninsula and Kenema, frequent in Kilima area and widespread elsewhere. Liberia, only in N: once up to 13 per day, all ♂♂, at Voinjama in Dec–Jan (Gatter 1997); uncommon on Mt Nimba and at Yekepa. Mali, frequent in Boucle du Baoulé Nat. Park and further south: Bougouni, Sikasso, Kangaba and Bafing-Makana. Burkina Faso, quite common in SW, at least in Dec–Jan. Ivory Coast, very common in dense woodlands in open country from coast (San Pedro area) north to 09°N; less common north of 09°N. Ghana, common outside forest from Accra coast north to Ejura and Wenchi; further north, frequent or uncommon and local (Kintampo, Mole) and rare at Tumu (10°55′N; Oct, Jan). Togo, common and widespread in coastal region, extending north to at least Tantigou (10°51′N). Benin, known (Brunel 1958). Nigeria, common (at least seasonally) from coast at Lagos and northern borders of forest zone north in the wet season to Zonkwa and Kwal (09°58′N); an outlying population in Nindam For., where frequent, July–Feb only (Dyer *et al.* 1986); thought to breed little further north than Abeokuta, Ubiaja and Enugu, all near southern edge of range; abundant in SE Nigeria north of forest (numbers evidently not fluctuating seasonally), north to Obudu and Mambilla Plateaus. Cameroon, common along northern edge of forest, in forest-savanna mosaic, in montane area in W (up to 1540 m on Mt Cameroon), and on Adamawa Plateau (Louette 1981). Not found in Mbini or Congo. Gabon, accidental or invasive non-breeding visitor to NE in Sept–Nov (Brosset and Erard 1986). Said to be old records from Gabon and Congo coasts (Chapin 1954). Central African Republic, known in Ouham-Pendé district in W, Dzanga Res. (rare), Lobaye Préf. and Bangui Préf. in S, between Batangafo and

Ndélé in N, and in Uele and Mbomou (Bomu) river valleys along Zaïre border in E. Zaïre, in Uele District frequent at Niangara and Dungu and known at Nzoro, Madrapili's and Aba. S Sudan, only in Yei and Yambio areas (Nikolaus 1989). Uganda, rare in or vagrant to extreme NW.

Density of 4–5 territorial ♂♂ in audible singing distance of each other, and of 14 singing ♂♂ per km², Ghana coast (Grimes 1974, Payne 1978).

Description. ADULT ♂: forehead and crown iridescent purple, grading through glossy violet-blue on nape to greenish blue on rear and sides of neck and upper mantle and glossy green on lower mantle, back and scapulars; rump and uppertail-coverts metallic greenish blue to violet-blue, the latter long, almost covering tail. Tail black, feathers narrowly edged metallic greenish blue. Lores black; ear-coverts and cheeks to chin, throat and upper breast iridescent purple, concolorous with crown, merging with a broad breast band of ruby red and bright blue reflections; pectoral tuft, usually concealed, of long, pale yellow feathers; belly, flanks and thighs sooty black; undertail-coverts glossy blue. Lesser and median coverts metallic green; rest of upperwing-feathers sooty black. Axillaries and underwing-coverts sooty black. Bill black; eyes dark brown; legs black. ADULT ♀: upperparts olive. Tail dark brown, inner webs of T5–T6 with whitish terminal fringe. Narrow superciliary stripe creamy; lores, cheeks and ear-coverts dark olive. Chin and throat creamy white, barred dusky; rest of underparts pale olive-yellow, breast and flanks streaked dusky olive. Flight feathers dark brown, edged pale olive; rest of upperwing dark brown, tertials and greater, median and lesser coverts broadly fringed pale olive. Axillaries pale olive-yellow; underwing-coverts dusky brown, tipped pale olive-yellow; inner borders of flight feather undersides pale grey-buff. Bare parts as adult ♂. Plumage changes greatly with wear, olive upperparts turning brown and underparts becoming darker as yellowish feather tips abrade (Serle 1957). SIZE (10 ♂♂, 10 ♀♀): wing, ♂ 70–73 (71·1), ♀ 61–67 (64·3); tail, ♂ 41–46 (43·2), ♀ 36–38 (37·0); bill, ♂ 24–27 (25·7), ♀ 23–26 (24·4); tarsus, ♂ 16–18 (17·1), ♀ 16–17 (16·6). WEIGHT: Liberia, ♂ (n = 2) 13·3, 15·3.

IMMATURE: juvenile dark olive-brown above; sides of face and chin to upper breast densely mottled or wholly sooty grey; rest of underparts olive-yellow, breast and flanks mottled dark grey; subject to wear in same way as adult ♀. In immature ♂, glossy plumage appears first on chin to upper breast.

NESTLING: unknown.

TAXONOMIC NOTE: in Ivory Coast some ♂♂ have mantle and back glossy yellowish green, breast band with greenish blue reflections, and undertail-coverts greenish blue; their taxonomic status needs reviewing.

Field Characters. Length 13·5–14·5 cm. Comparatively long-billed and short-tailed but less markedly so than sympatric Superb and Johanna's Sunbirds *C. superba* and *C. johannae*. ♂ distinguished by unique combination of *violet-purple* head, throat and upper breast and *red and purple* lower breast, separated by black belly from blue undertail-coverts; yellow pectoral tufts shared with Johanna's but lacking in Superb. ♀ also different, with variable *dark barring* on *white throat*, light streaking on breast and flanks, lower belly unmarked yellow (underparts heavily streaked in Johanna's, plain in Superb). Juvs. have dark throat patch, more solid and extensive in ♂. ♀ and juvs larger and longer-billed than Red-chested Sunbird *C. erythrocerca*, which they just meet in NE Zaïre; not unlike ♀ and juv. Scarlet-chested Sunbird *Chalcomitra senegalensis* but bill shorter and less curved, underparts much yellower.

Voice. Tape-recorded (104, B, CHA, GRI, PAY). Song a descending series of 6–9 pure, down-slurred, well-spaced whistles, first 2 faster and higher-pitched, 'tewpi-teew-teew-teew-teew-teew', lasting 2–3 s; another form with double notes, 'pititew-tew-pitew-pitew-pitew'. Descriptions of the song by Serle (1956, 1957), as 'curious staccato song of 8 notes: *'Oh ... what-a ... splendid ... bird-I-am'* ' and 'Song of small compass and indifferent tone but considerable power ... 8–9 clearly separated notes' are misleading and fail to do it justice. A different type of song described by Chapin (1954) as 'a protracted rambling warble of no great musical virtue'. Call single sharp 'tyip', faster and repeated in flight, 'tyipipipipip ...'; alarm a typical sunbird grating scold, 'djew-djew-djew ...'.

General Habits. Inhabits mesic guinean wooded savannas and woodland with oil palms *Elaeis guineensis*, and copses, plantations, farm-bush, forest-edge, secondary growth, gardens and residential areas, scrub, rubber plantations, and some wooded coasts. On Mt Nimba open country with scattered trees, bushy growth, forest-grassland mosaic and gardens. In W Cameroon, orchard bush and tree-dotted grassland in hill country at 860–1540 m. In parts of range seems to be commoner in large flowering trees in residential than in non-residential areas.

Singly or (usually) in pairs; 12 together in tree, visiting its magnolia-like flowers (Guinea, Oct), and groups of 12 or more attracted to flowering *Berlinia grandiflora* trees (Togo: Landa-Pozanda Mar, Mo May–June) (Cheke and Walsh 1996). 'Great numbers' gather at flowers of *Bombax*, and mix freely with at least 7 other sunbird species there (Bannerman 1948). Strongly attracted also to drooping inflorescences of Ceará rubber shrubs *Manihot glaziovii* and to flowers of pawpaw *Carica papaya*. Punctures corollae of *Manihot* flowers (and in so doing accumulates indigestible rubbery latex in stomach). Hovers by flowers and leaves, and flits around blossoms of succulents, middle-sized shrubs and at tops of large trees; sometimes moves half hidden in foliage, gleaning insects; probes trunks of oil palms and pawpaws for invertebrates, and takes sap exuding from cuts made in oil palms by palm-wine makers. Once seen catching termites on the wing as they emerged from hole in ground, the sunbirds in company of swifts, swallows and weavers (Bannerman 1948). Often perches out in open, and 4 ♂♂ may perch close together on topmost twigs of tall tree, crouching forward, bowing and advancing at each other until they break into an aerial chase (Field 1974). Sings seasonally, or in SE Nigeria all year; song delivered from conspicuous tree-top perch, also sometimes in flight. Local populations have distinct song dialects, differing in rhythm (time intervals between strophes); dialect differences are maintained for at least 3 years, even across boundary only 50 m wide; there are no habitat correlations and no geographical trends (Grimes 1974, Payne 1978).

Resident in Gambia but commonest in Mar–June. Occurs in Mali only at beginning of rainy season, in May–June; yet in Burkina Faso occurs in Nov–Feb (once May), mainly Dec–Jan (Lamarche 1981, Thonnérieux *et al.* 1989). Said to be commonest in Sierra Leone from Apr to about Aug, and at one locality to appear in May and disappear in Dec

(Bannerman 1948). In N Nigeria, resident at New Bussa but on Foge I. only in Apr–Dec; Nindam July–Feb; records at Kishi, Igboho, Ago-Are, Kafanchan and Zonkwa are all June–July. No clear pattern of migration (nor of breeding) emerges, but birds seem to disperse or migrate northward with the rains in May–July (Elgood *et al.* 1973), birds returning to south in Sept–Nov and some overshooting (non-breeding visitor to NE Gabon then).

Food. Insects, spiders, nectar and seeds. Insects in 36 stomachs and seeds in 2 (Serle 1957); nectar, tiny black bees, an ant and a spider – and balls of Ceará rubber up to 100 mm across – in 8 other stomachs (Chapin 1954). Feeds at flowers of *Carica, Manihot, Bombax, Berlinia* and *Parkia* (see above).

Breeding Habits. Solitary nester. Territorial; territories not studied, but ♂ 'song is used to establish and maintain a territory' (Grimes 1974). In Jan–Oct (Ghana coast) ♂ starts to sing at dawn then sings often until late afternoon, using one or a few favoured song perches. Courtship: ♂, after long, high, erratic flight during which he sang, alighted on branch 10 m above ground; ♀ alighted 1–2 m away and perched with body held stiffly, tail depressed, wings drooping and neck stretched upward and forward. She maintained the pose for several s, then shuffled along branch towards ♂ who flew off (Serle 1957).

NEST: rather large, suspended pouch, with side entrance and a substantial porch above it, underside of nest bowl compactly and neatly finished, without any hanging 'beard'. Made of fine grass stems, bits of dry leaves, palm fibres, and guava bark scales, bound together with white down from various plant seeds, and thickly lined with the same down. Nest well concealed, hanging from branch in dense mass of foliage, 2–3 m from ground; one nest built 25 cm from ants' nest in leaf cluster. Built by ♀ only; ♂ accompanies her on material-collecting flights and remains near nest, singing, when she works. At early stage of construction nest is a suspended vertical ring of plant fibres. One nest built desultorily in 30 days; 7 days after eggs were taken by a predator, ♀ repaired nest and laid replacement clutch; 3 months after young fledged, another nest (same pair of birds?) was built using material from the earlier one.

EGGS: 1–2. Shell smooth and glossless. Two types: (a) blotched – ground blotched and suffused with dark brown and grey, chiefly as a cap at large end, with irregular scrawls of same shades; narrow end half of egg practically immaculate; (b) streaked – ground pale grey, profusely and evenly sprinkled with streaks and speckles of greys and browns (Serle 1950). SIZE: (n = 2) 18.5×15.3, 19.0×13.1.

LAYING DATES: Senegal (Casamance), Sept–Oct; Gambia, July–Oct; Sierra Leone, Sept-Oct; Ghana, Mar, Apr, June, Sept, Oct, Dec (and nest-building Dec–Jan and Mar–Apr and dependent young Mar–Nov) – breeds all year; Togo, (nest building Mar, Apr, Aug, singing Mar, July); Nigeria (Abeokuta), Mar, May, (Ovim), Sept, (Enugu), (sings all year, display flight Mar) (and indications most months, mainly Mar–June); Cameroon, (nest building Jan); Zaïre (Uele), (gonads enlarging Jan; imm. May).

INCUBATION: by ♀ only.

References
Bannerman, D.A. (1948).
Grimes, L.G. (1974).

Plate 16
(Opp. p. 223)

Cinnyris rufipennis Jensen. Rufous-winged Sunbird. Souimanga à ailes rousses.

Nectarinia rufipennis Jensen, 1983. Ibis, 125, p. 447; Mwanihana Forest Reserve, Udzungwa Mts, Tanzania.

Range and Status. Endemic resident, Mwanihana Forest, 07°45′S, 36°35′E, and Ndundulu Mts, 07°45′S, 36°29′E, Tanzania (E escarpment of Udzungwa Mts). Mwanihana Forest is in Udzungwa Mountains Nat. Park and Ndundulu in West Kilombero Scarp Forest Reserve. In Mwanihana Forest (c. 50 km^2 in extent: Jensen 1985) ranges from 600 to 1700 m; uncommon below 1000 m, quite common at 1000–1500 m, and most numerous at 1500–1700 m. In Ndundulus ranges from 1350 to 1600 m (Dinesen *et al.* 1993).

Description. ADULT ♂: top and sides of head, neck and rest of upperparts metallic violet with bluish green reflections. Tail blackish, upperside with slight violet gloss. Chin and throat glossy bronze, bordered by band of metallic violet c. 4 mm deep across upper breast; below this a broad band of chestnut c. 10 mm deep; pectoral tufts pale yellow; rest of underparts dark olive-grey, merging into olive-green on lower belly and thighs. Lesser and median coverts metallic violet; rest of upperwing black, remiges edged cinnamon-rufous, broadly so on secondaries and tertials, forming a prominent wing-panel. Underwing-coverts pale grey. Bill black; eyes dark brown; legs black. ADULT ♀: forehead to nape greyish olive, merging to olive-green on mantle, back, rump and scapulars, and brighter olive-green on uppertail-coverts; feathers on top of head, mantle, back and scapulars with blackish centres, giving strongly mottled appearance; feather tips on mantle, back and uppertail coverts with slight bluish green gloss. Tail blackish, feathers slightly glossed above and with narrow bluish green outer edges; T6 with outer edge and tip pale grey-brown; also a pale terminal fringe on T5 and (more narrowly) T4. Throat yellowish olive-green; rest of underparts olive-green, paler and yellower on centre of breast and belly, blackish feather centres giving heavy spotting throughout. Upperwing blackish brown; remiges edged tawny, broadly so on secondaries and tertials to form bright wing-panel; tertial tips and greater coverts fringed grey-buff; median and lesser coverts broadly tipped olive-green. Underwing-coverts grey. Bare parts as adult ♂. SIZE (1 ♂, 1 ♀): wing, ♂ 57, ♀ 55; tail, ♂ 43, ♀ 38; bill, ♂ 23.5, ♀ 21.5; tarsus, ♂ 16, ♀ 15. WEIGHT: ♂ (n = 1) 10, ♀ (n = 1) 8.7.

IMMATURE and NESTLING: unknown.

TAXONOMIC NOTE: ♂ shows some resemblance to double-collared sunbirds of the *C. chalybea* complex but ♀ shows only distant resemblances to them (Jensen 1985). True affinities remain obscure (Jensen 1983).

Cinnyris rufipennis

Field Characters. A very distinctive sunbird with an extremely restricted range in E Tanzania. Both sexes have a rufous wing-patch, conspicuous in the field. ♂ has crown, face and upperparts iridescent violet-blue, triangular bronze throat patch, chestnut breast-band and yellow pectoral tufts; ♀ has olive upperparts and pale yellow underparts with profuse spotting and streaking. Both plumages unique among African sunbirds.

Voice. Tape-recorded (MOY). Song high-pitched and trilling, not unlike Eastern Double-collared Sunbird *C. mediocris* but less 'electrical' (Jensen 1983); often accompanied by loud, high-pitched chirping from ♀. Very vocal. Commonest call of both sexes, a repeated loud, sharp, clear 'tyew' or 'chow', sometimes immediately followed by buzzy 'zew' or 'zee' (from second bird?), 'cha-zew'; also common is a short, dry rattle, 'tiddit', sometimes lengthened into a trill. Birds foraging among flowers give thin 'see-it' and other squeaking and fizzing noises; in flight, a short 'drep-drep' (Stuart *et al.* 1987).

General Habits. Lives in interior of wet montane forest with lichens, mosses and epiphytes growing plentifully on the trees, also in glades and in disturbed forest; forages at 2–8 m from ground, at red flowers of undergrowth shrub *Achyrospermum carvalhi*; keeps mainly to light gaps with dense undergrowth and flowering plants. Occurs singly or in pairs, foraging mainly at 2–7 m above forest floor, but sometimes comes within 0·5 m of ground, and ascends to 30 m to join mixed-species flock. Feeds at blossoms and gleans leaves. Pair, seen from time to time over 2 days, stayed in area of 1 ha, ♂ and ♀ always close together. Aggressive to other sunbirds, particularly Moreau's Sunbird *C. moreaui* (which commonly fed at same flowering shrubs) and locally abundant Eastern Olive Sunbird *Cyanomitra olivacea*, chasing and being chased by the former and often chasing the latter (Stuart *et al.* 1987). Also feeds alongside 3 other sunbird species (Dinesen *et al.* 1993).

Food. Feeds at flowers of wide range of low, flowering trees and shrubs in forest, including *Achyrospermum radicans* (Labiatae) where thought to suck nectar, and gleans leaves, presumably for small arthropods.

Breeding Habits. ♀ observed feeding juvenile, Jan.

References
Collar, N.J. and Stuart, S.N. (1985).
Dinesen, L. *et al.* (1993).
Jensen, F.P. (1983).

Cinnyris batesi (Ogilvie-Grant). Bates's Sunbird. Souimanga de Bates.

Plate 16
(Opp. p. 223)

Cinnyris batesi Ogilvie-Grant, 1908. Bull. Br. Orn. Club, 23, p. 19; Ja River, Cameroons, and Camma River, Gaboon.

Range and Status. Endemic resident, W Africa and peripherally in Congo basin, from Ivory Coast to Cabinda, E Zaïre and Zambia. Tiny treetop sunbird, easily overlooked, difficult to distinguish, so likely to be commoner and more widespread than records indicate. Ivory Coast, known from Lamto, Gagnoa and Taï. Ghana, known only from Jukwa, where seen occasionally. Nigeria, uncommon, throughout forest zone: records at Lagos, Ibadan, Benin, Ughelli, Warri, between Idah and Nsukka (north of forest zone), Degema and Umuagwu. Cameroon, uncommon throughout forest zone in SW (Victoria, Kumba, Dikume Balue, Eseka, Mieri, Messea, Assobam, Bitye, Ja R., Yukaduma), but can appear at fruiting tree in abundance. Bioko, frequent (Eisentraut 1968). Equatorial Guinea and Gabon, locally common. Congo, 3 observations in Odzala Nat. Park (Andzoyi, Ikaka, Lékoli: Dowsett-Lemaire 1997a). Angola, known from Landana and Pangamongo in Cabinda. Zaïre, 8 or more records in 3 areas in E and S as mapped. Zambia, N Mwinilunga, very near Zaïre and Angola borders: Salujinga, Source of the Zambezi, R. Lisombo and R. Sakeji.

Density thought to be *c.* 30 birds per km^2 near Makokou and *c.* 45–60 per km^2 on hillside forest near Bélinga, Gabon (Brosset and Erard 1986).

Description. ADULT ♂: upperparts dark olive-green. Tail blackish; feathers edged olive-green, more broadly near base; T3–T5 with dark olive tip, T6 with broader olive tip and olive outer edge. Narrow superciliary stripe olive-white; lores, cheeks and ear-coverts dark olive-green; chin, throat and upper breast olive-grey; sides of breast and flanks pale olive-grey; centre of breast, belly

Cinnyris batesi

and undertail-coverts pale olive-yellow. Median and lesser coverts dark olive-green; rest of upperwing blackish, primaries and secondaries edged olive-green; tertials, alula, primary coverts and greater coverts fringed dark olive-green. Axillaries and underwing-coverts creamy white. Bill black; eyes brown; legs black. Sexes alike. SIZE (10 ♂♂, 10 ♀♀): wing, ♂ 49–52 (50·3), ♀ 46–49 (47·3); tail, ♂ 27–31 (28·4), ♀ 23–26 (24·5); bill, ♂ 16–18 (16·4), ♀ 15–16 (15·5); tarsus, ♂ 13–14 (13·7), ♀ 13–14 (13·5). Bioko population slightly larger, but shorter-billed: (6 ♂♂, 3 ♀♀) wing, ♂ 52·5–54 (53·3), ♀ 48·5–50·5 (49·7); tail, ♂ 30–31 (30·4), ♀ 26–29 (27·2); bill, ♂ 14–16 (15·3), ♀ 14–15 (14·3).

IMMATURE: juvenile like adult.

NESTLING: mouth and tongue unmarked, uniform orange (Bates 1911); late nestling almost exactly like adult but thin feathers on belly and undertail-coverts more lemon yellow; bill horn colour; eyes grey; legs and feet grey (Bannerman 1948).

TAXONOMIC NOTE: placed with *Anthreptes seimundi* in genus *Paradeleornis* by Wolters (1977); the 2 species are very alike except in bill shape. Treated with *A. seimundi* for convenience by Hall and Moreau (1970), who thought that the 2 are not necessarily closely related.

Field Characters. Length 9–10 cm. A dull treetop sunbird, difficult to identify, especially at a distance. Looks like a small version of Western Olive Sunbird *Cyanomitra obscura* with a much shorter bill (only about half the length). Best field character distinguishing it from Olive Sunbird is *short, dark tail* (Christy and Clarke 1994). Dark olive-green above, greyish olive below, yellower on belly; no pectoral tufts; sexes alike. Olive Sunbird has duller and greyer underparts, almost without yellow, and pale throat; ♂ has yellow pectoral tufts, ♀ lacks them. Also, pectoral tufts can often be concealed by closed wing, so a bird *not* showing pectoral tufts could be either Bates's or Olive. ♀ Olive has brighter yellow-green edgings to flight feathers, producing wing panel on perched bird. The 2 species occupy different habitats, Bates's in canopy, Olive in understorey, and only come in contact in open areas on ridges and at edges of tracks and clearings (Christy and Clarke 1994). Curved bill with *orange base* distinguishes it from Little Green Sunbird *Anthreptes seimundi*, which is brighter green above and much brighter yellow below, and in good light has discernible pale eye-ring; Bates's perches upright, like Olive Sunbird, whereas Little Green has more horizontal stance (Gartshore 1989). Extremely similar to ♀ Orange-tufted Sunbird *C. bouvieri*; for differences, see that species.

Voice. Tape-recorded (102, 104, B, CHA, STJ). Song, high, thin 'tsip' given in uneven series, mixed with lower 'chut' or 'chut-ut'.

General Habits. Inhabits primary forest and old, mature second growth, edges of forest clearings, trees and shrubs in clearings; keeps largely to crowns of tallest, emergent trees, and has predilection for large inflorescences at tops of trees, but comes nearer to ground to nest. Forages out in sunshine at blossoms at edge of forest and in clearings, and moves around in twigs and branches covered in moss and epiphytes. Often prises into clumps of filamentous lichens. Occurs singly but mainly in loose aggregations of 6–10 individuals and as member of mixed parties of insectivores and sunbirds. Once seen eating small red berries from clusters in roadside tree attended also by Western Olive, Green-throated *Chalcomitra rubescens*, Collared *Hedydipna collaris* and Little Green Sunbird all arriving in twos and threes, feeding on berries for a few min then departing, and amounting to 'great numbers' (Serle 1954).

Food. Small invertebrates including spiders, and products of flowers including nectar (though behaves more like an insectivore than a nectarivore – Brosset and Erard 1986). Rare not to find insects in stomachs (Benson and Irwin 1966). Small red berries (see above); occasionally feeds on fruit (Benson and Irwin 1966), including capsules 3–4 mm in diam. of *Macaranga assas* (gape width is 4 mm: Dowsett-Lemaire 1996).

Breeding Habits.

NEST: suspended bag, loosely made of dry leaves, moss and spider web, and lined with fine down; entrance at side 20 mm diam. Unusual among sunbirds in having no long fibres used in construction and no loose ends hanging down (Bates 1911). Sited in thickets. 3 were suspended 2–3 m from ground, from electric wiring hanging down from roof in ruined encampment building.

EGGS: 2 (5 clutches or broods); 3 single nestlings may have been from clutches of 1. Rather short, somewhat pointed oval, without any gloss; pale pink, mottled all over with darker, greyish pink; small spots and irregular deep brown marks scattered overall, some of the marks smeared yellowish brown at their edges. Size: (n = 1) 15 × 11.

LAYING DATES: Nigeria, (birds in breeding condition Nov, Feb, Mar); (Cameroon: clutch and brood dates not recorded); Gabon, (3 nests, Feb–Mar, each with a chick); Zambia, (gonads active May).

References
Bates, G.L. (1911).
Brosset, A. and Erard, C. (1986).

Cinnyris ursulae (Alexander). Ursula's Sunbird. Souimanga d'Ursula.

Cyanomitra ursulae Alexander, 1903. Bull. Br. Orn. Club 13, p. 38; Mt St Ysabel, Fernando Po.

Plate 16
(Opp. p. 223)

Range and Status. Endemic resident, mountains of W Cameroon and Bioko. Record in Equatorial Guinea (Fa 1990) requires confirmation. In Cameroon locally common, known on Mt Cameroon at 950–1250 m, Buea at 950 m (Grote 1948), common on Mt Kupé at 950–2050 m (the summit) and on Mt Bakossi down to 1100 m, Mt Nlonako at 1650–1800 m, Nyasoso, forest near Dschang, frequent at Dikume Balue at 1000–1700 m, Rumpi Hills, and north to Foto at 1600–1700 m. In Bioko occurs in primary montane forest at 1000–1200 m in S Highlands (Moka Valley, Caldera de Luba) and on Pico Basilié (Pérez del Val *et al.* 1994). Has a circumscribed distribution and narrow altitudinal range, but not seriously at risk (Collar and Stuart 1985).

Cinnyris ursulae

Description. ADULT ♂: forehead to crown dark olive-grey with paler feather fringes, faintly washed with iridescent blue, merging with bright olive-green of rest of upperparts. Tail blackish brown; feathers fringed olive-green, T4–T6 tipped greyish. Lores dark grey; ear-coverts and cheeks olive-grey; area around eye smoke grey. Chin to breast, flanks and upper belly grey, merging with pale greyish green lower belly, thighs and undertail-coverts; tips of long flank feathers pale olive-yellow; pectoral tufts flaming orange or orange-red. Median and lesser coverts olive-green; rest of upperwing blackish, but flight feathers edged bright olive-green, tertials and greater coverts broadly fringed olive-green, and primary coverts and alula finely fringed olive-green. Axillaries and underwing-coverts white to pale creamy; inner borders of undersides of flight feathers whitish. Bill brown; eyes hazel; legs brown. Sexes alike. SIZE (10 ♂♂, 10 ♀♀): wing, ♂ 47–49 (48·1), ♀ 44–48 (46·3); tail, ♂ 25–28 (26·7), ♀ 23–26 (24·9); bill, ♂ 18–20 (19·4), ♀ 17–19 (18·0); tarsus, ♂ 14–15 (14·5), ♀ 14–15 (14·2).

IMMATURE: juvenile similar to adult, but no iridescent wash on crown chin and throat barred dark grey and whitish, bordered by yellowish moustachial stripes; top and side of head dark olive; breast to undertail-coverts dull yellowish green. No pectoral tufts.

NESTLING: unknown.

TAXONOMIC NOTE: a taxonomically isolated species which has been placed in its own genus *Haplocinnyrys* (Wolters 1977). Hall and Moreau (1970) allied it 'very tentatively' at superspecies level with *C. venusta*, on the grounds of parapatry, unsuspended nests, and slight plumage resemblance. Its taxonomic isolation is confirmed by Stuart and Jensen (1986).

Field Characters. Length 9·5–10 cm. A Cameroon highlands/Bioko endemic, with curved bill and short tail. Bright olive-green above, forehead and forecrown grey-blue; chin and throat light grey-brown, rest of underparts grey except for yellow undertail-coverts; pectoral tufts orange. In the field, dark chin and throat contrast sharply with paler breast and belly (Serle 1954). Sexes alike. Nothing resembling it in its limited range.

Voice. Tape-recorded (104, B, CHA, ROD). Territorial song simple, 5 or so pure, measured notes on one pitch, 'tee-tee-tee-tee-tee' or 'tewp-tewp-tewp-tewp...'; several birds together give high-pitched, thin, bubbly subsong. Both sexes constantly call a very soft, high-pitched 'tsit-tsit' (Stuart 1986).

General Habits. Inhabits primary forest, forests with some clearance taking place (Collar and Stuart 1985), moss forest, and most types of mature second growth at all levels from tree tops to undershrubs (Serle 1954) and at intermediate elevations. In Bioko, moss forest is floristically varied and Ursula's Sunbirds were not found in less varied, altitudinally lower and higher types of forest (Wells 1968). Reported both to avoid sunlit clearings and to enter them to feed. Occurs singly, in pairs, and in mixed sunbird parties. Very active, wide-ranging, diligently searching leaves and small branches and occasionally hovering in front of blossoms. Sedentary.

Food. Stomachs contain insects and occasional spiders, seeds and sugar (Serle 1954). Often probes white flowers for nectar, including *Rauvolfia vomitaria*, several species of Rubiaceae and one of Flacourtiaceae (F. Dowsett-Lemaire, pers. comm.).

Breeding Habits.

NEST: 2 were similarly constructed (Serle 1954): not bound to branch by spider web and not suspended; outer wall is loose, mossy envelope which blends perfectly with mossy surroundings; inside envelope is a small, compact, domed pouch made of well-integrated shreds of moss with some spider web and rootlets intermixed, with a scanty lining of white down; pouch has circular entrance at one side, with small porch of moss projecting above it. No 'beard' below nest. Sited at 3·7 and 4·3 m above ground, one in moss enveloping a liana, and one attached by the moss of its external wall to slender upright shoot of an undershrub. One nest in a large ball of moss by a mossy trunk, 4 m up in small tree (F. Dowsett-Lemaire, pers. comm.).

EGGS: 1, 2 (2 completed clutches). Ovate, immaculate white, tinged palest brown. SIZE: (n = 2) 15·2 × 11·2, 14·9 × 11·3.

LAYING DATES: Cameroon, clutches Jan, Mar (Serle 1981), (♀ with oviduct eggs, late Jan; juvenile with unossified skull late Feb; nestlings being fed, Mar). Breeds in middle of dry season.

DEVELOPMENT AND CARE OF YOUNG: young fed by both parents; brooding sessions last 30–35 min (F. Dowsett-Lemaire, pers. comm.).

References
Serle, W. (1954).
Stuart, S.N. and Jensen, F. P. (1986).

Plate 12
(Opp. p. 159)

Cinnyris cuprea (Shaw). Copper Sunbird; Coppery Sunbird. Souimanga cuivré.

Certhia cuprea Shaw, 1811–1812. General Zool., 8, p. 201; Malimba, Portuguese Congo.

Range and Status. Endemic resident and migrant in mesic savanna woodlands from Senegal R. to Ethiopia, and Uganda to Angola and Zambezi R., penetrating neither rain forest zone nor arid savannas. In N tropics resident in southern parts of range shown but a breeding wet-season visitor towards northern boundaries. Mauritania, scarce in lower Senegal valley, Aug-Oct. In Senegal, Gambia and Guinea-Bissau frequent and resident near coast; uncommon in E Senegal. Mali, quite common south of line between Bamako and Kayes. Common and widespread in Sierra Leone. Liberia, 7 records on coast; N Lofa; frequent on Mt Nimba. In N half of Ivory Coast much commoner south than north of 9°N. Throughout Ghana and Togo north to Tumu (11°N) and Mango (10°20′N), but north of about 09°20′N only a wet season visitor. In Nigeria, Cameroon and Central African Republic frequent to common as a wet season visitor towards northern limits of guinean savanna woodland zone and resident in S towards border of forest zone. Common and widespread in N Zaïre and Uganda. Uncommon in SW Sudan, north along Nile and Sue rivers to Shamba and Gogrial, east to Imatongs; a few records north of 10°N and in SE near Ethiopian border. Ethiopia, locally frequent in W, below 1800 m. W Kenya, frequent to common north to Kitale and east to Muhoroni, rare further north to about Kachagalau; occurs south to Tanzania border (and south in N Tanzania to Musoma). Widespread in SW Tanzania, on Ufipa Plateau and in Iringa uplands east to Udzungwas; also in Mahari highlands (N.E. Baker, pers. comm.).

Common along Gabon coast and occurs in Rés. de la Lopé and probably E Gabon where shown, although not yet certainly recorded. Common and widespread in Congo north to Odzala Nat. Park; single record in Nouabalé-Ndoki Nat. Park. Widespread in Angolan hinterland, absent from dry coastal and southern regions; doubtless occurs in corridor in NW Angola and SW Zaïre with question marks on map. In Zaïre, very common in grasslands bordering lower Congo R.; common and widespread in N and very common in Uele in NE; throughout Kasai and Katanga, uncommon around Lubumbashi. Almost throughout Zambia, but largely absent from Luangwa Valley. Sparse in Malaŵi, but frequent in Lilongwe and seasonally the commonest sunbird in S Shire Valley. Mozambique: only in Zambezi Valley, where

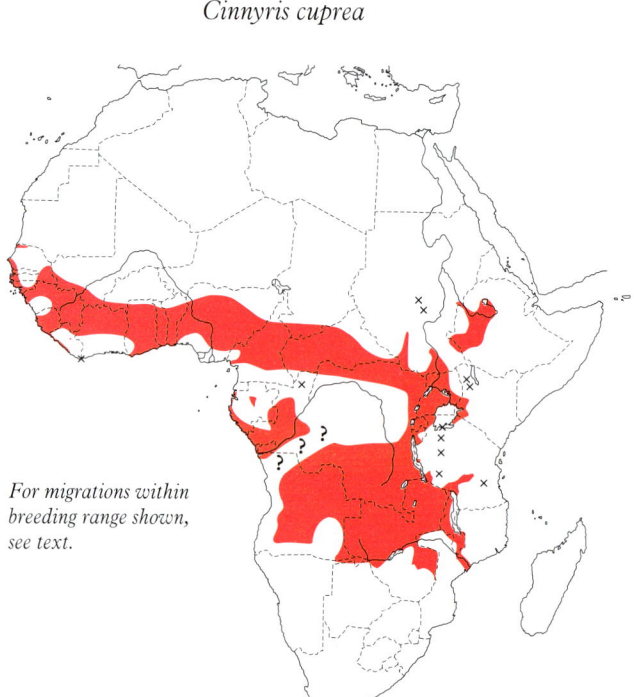

Cinnyris cuprea

For migrations within breeding range shown, see text.

locally common. Zimbabwe, uncommon: parts of Mashonaland, gardens in Harare, Dichwe Forest, Mangula, Mazoe Valley; west of Victoria Falls; middle Zambezi, upstream to Chirundu; record at Robins Camp in Hwange Nat. Park; in E Highlands known from 1300–1600 m on Nyawutari R., Mumeni R., Banti and Musapa Mt.

Density of 30–35 pairs per km^2 (Odzala Nat. Park, Congo).

Description. *C. c. cuprea* (Shaw): Senegal to Cameroon and Congo R. mouth, east to Ethiopia, N and E Zaïre, Uganda, Kenya and NW Tanzania. ADULT ♂ (breeding): head to neck and mantle shining coppery bronze with purple and greenish reflections, grading to glossy purple on larger scapulars, back, rump and uppertail-coverts. Tail black, slightly glossed above. Chin and throat coppery bronze, concolorous with head, grading to coppery purple on upper breast; underparts from lower breast backward black. Lesser and median coverts coppery or glossy purple; upperwing otherwise blackish. Axillaries and underwing-coverts sooty black. Bill black; eyes brown; legs black. ADULT ♂ (non-breeding): wings and tail as breeding ♂. Undertail-coverts black, and glossy purple retained on back, rump and uppertail-coverts; usually also much black feathering on belly and some glossy purple on breast. Otherwise as adult ♀. ADULT ♀: upperparts olivaceous grey-brown. Tail blackish brown; T6 pale brown on outer web and distally on inner, narrowly fringed whitish; T1–T4 often fringed whitish at tip, T5 sometimes broadly so. Narrow supercilium yellowish buff. Lores, cheeks and ear-coverts dark olive-brown. Chin and throat pale buff, rest of underparts pale buffy yellow; throat and breast faintly streaked and sides and flanks suffused greyish. Lesser and median coverts greyish brown; rest of upperwing dark grey-brown, primaries edged buffy yellow and secondaries olive-brown, alula, primary coverts and greater coverts fringed buff. Axillaries pale buffy yellow; underwing-coverts olive-buff with dusky bases; inner borders on undersides of flight feathers pale grey-buff. Bare parts as adult ♂. SIZE (10 ♂♂, 10 ♀♀, Nigeria and Cameroon): wing, ♂ 56–62 (59·6), ♀ 51–54 (52); tail, ♂ 41–48 (45·4), ♀ 32–38 (35·8); bill, ♂ 19–23 (21), ♀ 17–19 (17·8); tarsus, ♂ 15–16 (15·5), ♀ 14–15 (14·5). A cline of increasing size from west to east of range; wing, ♂, (n = 10, Senegal) 56–59 (57·3), (n = 10, NW Tanzania) 59–67 (62·7). WEIGHT: Liberia, ♂ (n = 5) 8·1 ± 0·4; Ghana, ♂ (n = 5) 6·8–7·6 (7·2), ♀ (n = 5) 5·5–6·8 (6·1); Nigeria, ♂ (n = 20) 7·8–8·4 (7·0), ♀ (n = 13) 6·9–7·8 (6·0); Kenya, ♂ (n = 63) 7·5–11 (9·0), ♀ (n = 10) 7·2–8·9 (8·1).

IMMATURE: juvenile like adult ♀, but ♂ has dusky throat.

NESTLING: at 2–3 days skin dark, whitish down on crown, palate orange, gape yellow, eyes closed, bill and legs blackish mauve (Howells 1971).

C. c. chalcea Hartlaub: Angola, and central-southern Africa south of 04°S. ♂ slightly more greenish bronze (less coppery) than nominate race on top of head and mantle. Slightly larger; ♂ (n = 10) wing 62–66 (64·2), tail, 47–52 (48·4), bill 21–23 (21·7). WEIGHT: Malaŵi (Nchalo), ♂ (n = 70) 8·4–11·4 (9·8), ♀ (n = 48) 7·3–10·2 (8·8) (Hanmer 1981).

Field Characters. Length 11·5–13 cm. Bill rather straight; tail relatively longer than similarly sized small sunbirds. ♂ appears dark in the field; shining coppery bronze with varying amount of pink and purple reflections, especially on wing-coverts and from lower back to rump; belly, wings and tail black; in eclipse plumage retains coppery gloss on wing-coverts and uppertail-coverts and may have dark blotches or even a dark midline on yellow underparts. ♀ has dark lores and patch behind eye, buff supercilium to behind eye, pale corners to tail; pale dull yellow underparts, brighter on belly; breast with a few faint streaks (looks plain in the field). Juv. ♂ like ♀ with dusky throat. ♂ looks like a small, 'tail-less' Tacazze *Nectarinia tacazze* or Bronzy *N. kilimensis* Sunbird. ♀ larger than ♀ Purple-banded Sunbird *C. bifasciata* and yellower below; similar to individuals of Purple-banded with unstreaked breasts, when told by dark lores and pale supercilium. ♀ also resembles a number of other ♀♀, which differ as follows: Olive-bellied Sunbird *C. chloropygia* is darker and greener above, lacks face markings; Northern Double-collared Sunbird *C. reichenowi* is darker above and greyer below and has shorter bill; Variable Sunbird *C. venusta* has brighter yellow belly, no supercilium; Orange-tufted Sunbird *C. bouvieri* is smaller and darker with proportionately longer bill and dark throat.

Voice. Tape-recorded (86, 91–99, 104, B, C, F, BRU, GRI, LEM, MOY). Song with very characteristic rhythm: starts hesitantly with a variable number of sharp 'tip's which accelerate in 'bouncing ball' fashion and end in a rapid swizzle, 'tip ... tip, tip, tiptiptip-titititiii-swizzllllllll', with or without an additional 'tip' at the end; length *c*. 5–6 s. Alarm scold a grating 'jid-jid-jid ... ' or a longer, querulous 'jaaa-jaaa', also incorporated into a different song type in which a jumble of swizzles, rattles, chips and other notes are given in short bursts punctuated by irregular series of 'jid's. Call, a sharp 'tyip' at irregular intervals, and said to give a hoarse 'tsit-chit' in flight between flowers (Zimmerman *et al.* 1996).

General Habits. Inhabits open savanna woodland, fallow fields, cultivation, forest edges, regenerating scrub with flowering weeds along streams and rivers, thickets and gardens. Attracted to *Baikiaea* flowers in gallery forest and small forest patches in Rés. de la Lopé, Gabon. Around Lubumbashi, Zaïre, occurs in and around dambos, i.e. open, waterlogged, marshy ground with very long, dense grass with a few stunted trees, in close-growing, unbroken woodland. Freely enters rural and town gardens and leafy, flowered suburbs. Occurs singly and in pairs, and forages seasonally in loose aggregations in stand of flowering trees or one large tree. Mixes with other sunbirds. Hovers in front of large flowers to feed at them; regularly and frequently pierces bases of flowers with long corollas to reach nectar (in some patches of vegetation most flowers have puncture wounds); probes some flowers, e.g. *Gloriosa superba*, *Brachystegia spiciformis* and *B. boehmii* for insects, not nectar (Howells 1971, Tree 1990); gleans outer foliage for small invertebrates; forages near windows and under house eaves for insects and spiders; hawks insects (especially termites) in flight by flying up and outwards in short sallies from top of bush or tree. Longer flight strong but erratic. Pair spent much time foraging around aloes and canna lilies not in bloom, and reeds and a sapling (*Juniperus procera*: Manson and Manson 1981). Bathes on dew-covered vegetation.

Mainly resident. A wet-season breeding visitor to northern parts of range in W Africa: Mauritania, Aug–Oct; Ghana, resident north to 09°20′N but around Tumu, 11°N, only in June-Sept; Togo, visitor in far N, May–Sept. In Nigeria visits Borgu region mid Apr to late Oct, and in Zaria ♂♂ arrive in late Apr, are common by 1st week May, and depart in 2nd–4th weeks Oct (Elgood *et al.* 1973). In Malaŵi a non-breeding visitor to Nchalo, adults arriving in late Apr and immatures from Aug, all leaving in Dec (Hanmer 1981). In Zimbabwe no longer thought to be a summer visitor to Mashonaland, but is highly nomadic (Tree 1990).

Food. Mainly insects, also nectar. In Nigeria nectar of *Tapinanthus*, *Thevetia*, *Jacaranda mimosifolia*, *Hibiscus* and *Ceiba* (Pettet 1977); in southern Africa nectar of *Jacaranda acutifolia*, *Bougainvillea*, *Pentas*, *Salvia*, *Caryopleris odorata*, *Zinnia*, *Cuphea miniata*, *Hibiscus*, *Marhamia*, *Callistemon viminalis*, *Tithonia rotundifolia*, *Tecomaria capense*, *Leonotis leonura*, *Kniphofia*, *Calliandra*, *Aloe cameronii*, *Acrocarpus*, *Bauhinia petersiana*, *Combretum mossambicense*, peaches, granadillas and *Eucalyptus* (Tree 1990). Eats spiders, caterpillars and flying termites. Young fed on nectar (parent's throat seen pulsating) and arthropods including aphids, flies, ants, leaf-hoppers, caterpillars and small spiders (Chapin 1954, Howells 1971). Often ingests latex of Ceará rubber shrub *Manihot glaziovii*, whose flowers are pierced for nectar; latex forms rubbery ball in stomach, up to 7 × 9 mm, and rubber is neither disgorged nor digested for several months (Chapin 1954).

Breeding Habits. Solitary nester, monogamous and territorial. One territory was small but well defined; ♂ aggressiveness peaks during nest building, incubation and early part of nestling periods; if bird of any species enters territory, ♂ immediately adopts aggressive posture, intensifying 'keek' and 'tzck' calls, then mobbing and even striking intruder, calling all the time, and raising then sleeking feathers rapidly whilst at perch (Howells 1971). Birds are allowed to fly through territory, but any bird alighting in it is chased away, e.g. cuckoo *Chrysococcyx caprius*, weavers *Ploceus velatus* and *P. cucullatus*, sparrow *Passer domesticus*, bulbul *Pycnonotus barbatus*, sunbird *Chalcomitra senegalensis* and thrush *Turdus libonyanus*. Fiscal Shrikes *Lanius collaris* attacked but not displaced (Howells 1971).

Courtship display leading up to copulation lasts c. 14 s; copulation lasts (n = 3) 3–7 (5·3) s. In display, ♂ flies in and perches 30 cm from ♀, then advances towards her. ♀ rocks from side to side, with tail spread and twisted towards ♂ as he shifts sides. With legs bent, wings drooped and half opened, ♂ starts to shiver and advances right up to ♀, who increases rate of rocking, sings, and may rapidly vibrate tail through small arc up and down; ♂ jumps onto her back and pair may flutter to ground, ♀ preening after copulating. In a 2nd type of display, ♂ perches prominently on twig, raises body vertically, with wings opened and raised high above head, tail slightly spread, just above the horizontal, legs straight and bill open (**A**) (Howells 1971).

A

Distraction display of ♀ disturbed from nest involves 'flap-flap movements similar to those of the Pin-tailed Widow *Vidua macroura*' (Howells 1971).

NEST: fairly neat and compact suspended oval or roughly pear-shaped bag, with round side-top entrance surmounted by small porch, and loose, ragged beard hanging below nest; made of strips of woody fibres, broad fibrous dry grasses, bits of dead bamboo leaf, thin plant stems, grass inflorescences, leaf-mould or soft, chaffed grass, mixed with greyish, old cobweb-litter, stripped spider cocoons and some plant down, sometimes decorated with bits of bark which can be quite large, wood chips and pieces of lichen; lined, moderately or thickly, with silvery thistle pappi and other whitish or yellowish plant down which becomes matted, and sometimes a few small feathers. Porch often made of dead flowering grasses, with seed-heads pointing outwards (Vincent 1967). Of 2 nests in Zimbabwe one had a small, stiff sill at bottom of entrance and the other did not (Howells 1971). Suspender is made of twists of plant fibre with a spider web binding, often for some small length up twig from which nest is hung (Vincent 1949). Size: ext. height 125–160, also 150–160 incl. attachment and 120–125 excl. attachment, width 65–90, from front to back 60–80, entrance 36–50 high and 35–41 wide, hood projection 35–108 (Skead 1967, Howells 1971). Built by ♀ alone.

Situated in woods around dambos (see above) or in isolated shrub in almost treeless, sometimes marshy grassland, and in thick clumps of woody vegetation growing around large termite hills; also sometimes in largely cleared ground around habitations. Sited at tip of twig or end of drooping branch, under cover of foliage; usually at 60–150 cm from ground; one in an *Acrocarpus* shrub at 2·8 m above ground.

EGGS: 1–2, generally 2. Evidently 2nd egg laid 24 h after 1st (Howells 1971). Ground colour pale or deep cream, sometimes brownish cream; scattered with dark brown streaky spots and lengthwise smears, with patchy, pale brown clouding. SIZE: (n = 9) 15·6–17·0 × 11·0–12·2 (16·5 × 11·8) (Vincent 1949).

LAYING DATES: Senegal, Sept–Oct; Gambia, May–June, Sept–Oct; Mali (Bamako), June; Ghana, May–July; Togo, July–Aug; Nigeria (SW and Jos Plateau), Aug–Oct; Sudan, July–Aug; E Africa, Region A, Sept, Region B, Mar–Sept and Nov, mainly Apr–July, Region C, Dec; Angola, Nov; Zaïre (Faradje) July–Oct, (Niangara) May–June, (Lubumbashi), Feb–Apr; Zambia, Jan–Mar and June; Malaŵi, Jan–Mar; Zimbabwe, Dec–Feb.

INCUBATION: thought to begin on day after clutch completed; by ♀ only; period at least 13·5 days (Howells 1971).

DEVELOPMENT AND CARE OF YOUNG: feathers emerge from sheaths between days 8 and 13. By day 11 young large enough for bill to rest on nest entrance sill; head dark with feathers in sheath, gape pale yellow, back olive, palate orange, tarsus brown, bill barely curved. Young brooded by ♀ for first few days, and fed by both parents, mainly ♀. In 6 h of observation on 5th day, ♂ fed young 10 times and ♀ 15 times. She approaches nest in cautious, circumspect way, through foliage or shrub, or making short direct flight, and at one nest in Zambia with small young ♀ made frequent visits of 30–120 s each. ♀ arriving at nest with single item, e.g. a spider or caterpillar, always thrusts bill deep down throat of one chick, withdraws the morsel and feeds it to the other chick. Such 'bluff' feeding invariably stops both young from giving loud begging calls (Howells 1971). Parent sometimes brings several items at once (Chapin 1954).

♀ sleeps with brood at night, her head projecting from entrance. ♂ spends much time on guard in nearby tree, driving away any other sunbird except another ♀ of his species or young from an earlier brood; but one breeding pair tended to keep away from nest when young of earlier brood was in area (Skead 1967). ♀ removes faecal sacs, flying away with sac in direction other than her usual one. Nestling period: (n = 2) 15 and 16 days. At nest in lone tree, the 2 young had no option but to make maiden flight down into grass, c. 10 m away; they were later fed near there by both parents. At another nest the 2 young left on successive days, so nestling period of each one was 15 days (Howells 1971).

BREEDING SUCCESS/SURVIVAL: parasitized by Klaas's Cuckoo *Chrysococcyx klaas* (Angola). Longevity at least 8 years (Hanmer 1989).

References
Chapin, J. P. (1954).
Howells, W.W. (1971).
Manson, C. and Manson, A. (1981).
Skead, C.J. (1967).
Vincent, A.W. (1949).

Family ZOSTEROPIDAE: white-eyes

A remarkably homogeneous, Old World family of small (mostly 8–16 g), gregarious, arboreal insectivores and capillary suckers of sugary plant and insect secretions. Nearly 100 spp.; family dominated by genus *Zosterops* with *c.* 73 species; remaining 23 spp. are in 12 genera, several monotypic, many insular, few of which are well-known morphologically or biologically. Distinguishing characters of the family are thus those of its principal genus and include a brush-tipped tongue and in most genera a white feathery eye-ring; see following *Zosterops* diagnosis. The large (30 g) *Hypocryptadius cinnamomeus* of Philippines may not be a zosteropid. Mostly highly sedentary, yet contradictorily a strong tendency to disperse to considerable distances in flocks and to invade islands. *Z. lateralis* became commonest land bird in Tasmania after invading it in 1850s from Australia, 2000 km away. More than half of African species are offshore island endemics. Only rarely are more than 2 spp. sympatric. Many islands have had double invasions of white-eyes and one, Norfolk I., has been invaded 3 times. Island forms tend to lose yellow colour and white eye-ring, and to become large (e.g. African *Speirops*).

200–250 taxa are recognized, with plumage variation within narrow limits. Greatest likenesses are often between geographically distant taxa; for instance, *Zosterops griseovirescens* of Pagalu I., W Africa, is strikingly similar to *Z. natalis* of Christmas I., Indian Ocean; E African *Z. abyssinicus flavilateralis* closely resembles Australian *Z. lutea*. Systematics notoriously difficult at species and generic levels. Despite the analyses of Stresemann (1931), Moreau (1957), Mees (1957–1969), Clancey (1966, 1967), Hall and Moreau (1970), Gill (1973) and Sibley and Monroe (1990), affinities of numerous taxa remain inscrutable, and assertions in the present work about the composition of superspecies and species should be taken with caution.

13 genera with *c.* 96 species. Whole Afrotropical Region, Oriental Region north to Hokkaido, Japan, and Australasia south to Campbell I. south of New Zealand. 2 genera in Africa, one being endemic.

Genus *Zosterops* Vigors and Horsfield

Small, gregarious, arboreal, yellow-green, insect-, berry- and nectar-eaters with comb-tipped tongues; mostly sedentary, yet prone to invade islands and to speciate there. Remarkable uniformity in structure, plumage and behaviour. Sexes alike. Bill black, sharp-pointed, thin, slightly decurved. Tongue short, fleshy, only 3 times as long as broad, slightly pleated in section, extensile, bifid, with each half further divided distally (but none of the 3 divisions very obvious, the distal 4 parts of the tongue holding together), with distal third horny, consisting of single row of long, quite stiff, parallel filaments, comb-like rather than brush-like, upper surface of each filament concave. Palate has depression into which tongue fits, with groove at each side. Comb sucks up liquid foods by capillarity. Plumage olive above, yellow or greyish below, flanks rufescent in some spp.; continental spp. tend to be yellow-green, insular ones to be grey-brown; ring of minute, snow-white featherlets around eye; eye-ring appears only after young have left nest. Breadth of ring varies between species; ring absent in a few, brown in one, large and exaggerated in several. Nostrils exposed, lidded. No rictal bristles. Wing short, rounded, P6–8 longest and about equal, P10 vestigial. In some spp., wing-length increases by c. 0·7 mm with every 300 m of increasing altitude. Legs and feet rather short and weak; dark grey. About 3300 feathers – many for a passerine. Nest a neat, open, deep cup, lodged or slung in bush. Highly vocal, but voices weak, rather simple, quite far-carrying, without much variation between species, though variation, such as it is, is biologically crucial. Make clattering noises with bill.

A Cameroon highlands population, *stenocricotus*, has a distinctive voice and so it is a candidate for specific separation from *Z. senegalensis* (compare the recent recognition of 3 species of chiffchaffs in place of *Phylloscopus collybita sens. lat.*: see Vol. V, p. 355). The S Sudan montane population *gerhardi* has been placed in *Z. poliogaster*; we think it is a race of *Z. senegalensis*; the taxonomic status of a population in NE Zaïre (Medje) requires clarification. South African and Namibian populations, which Clancey (1980) regarded as a single polytypic species, *Z. pallidus*, have been treated by others as 3 species; we follow Clancey, though the 'Orange River' white-eyes (*pallidus*, *sundevalli*) are an incipient species, differing vocally and in plumage from other populations. Kenyan and Tanzanian isolates of the montane *Z. poliogaster*, which we treat as a single species, have also been regarded by others as distinct species. See the various Taxonomic Notes.

73 species (comprising 12 superspecies with 36 spp., and 37 independent spp.: Sibley and Monroe 1990) including one recently extinct. 7 in Africa (6 endemic and one also in Arabia); *Z. senegalensis* and *Z. vaughani* compose a superspecies, and so do *Z. ficedulinus* and *Z. griseovirescens*.

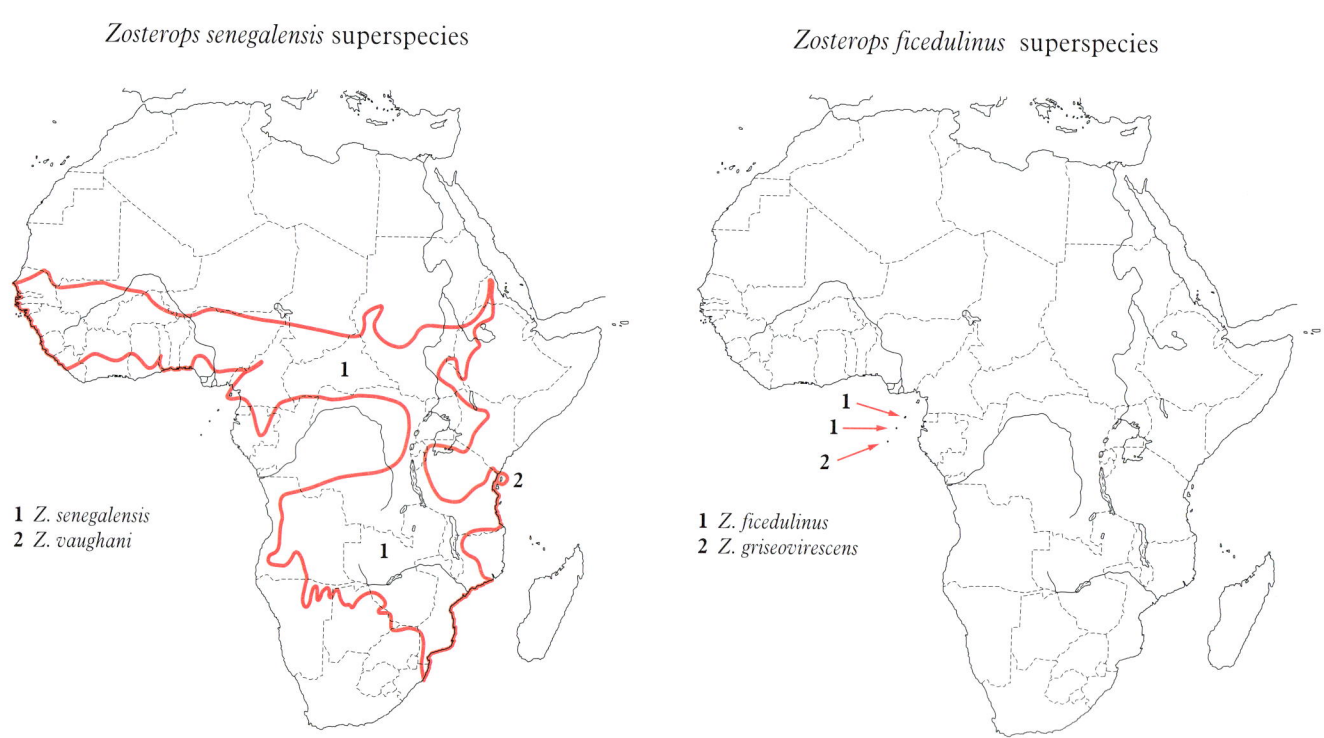

Zosterops senegalensis superspecies

1 *Z. senegalensis*
2 *Z. vaughani*

Zosterops ficedulinus superspecies

1 *Z. ficedulinus*
2 *Z. griseovirescens*

Plate 17
(Opp. p. 270)

***Zosterops senegalensis* Bonaparte. Yellow White-eye. Zostérops jaune.**

Zosterops senegalensis Bonaparte, 1850. Consp. Av., 1, p. 399; Senegal.

Forms a superspecies with *Z. vaughani*.

Range and Status. Endemic resident, mesic savanna woodlands. Common. Range limits: Mauritania, local, S Guidimaka only; Senegal, north to 14°N and Cap-Vert but common only very locally; Gambia, frequent in Lower River Div., scarce in North Bank Div. (Barlow *et al.* 1997); Mali, common in Mandingo Mts and quite common north to 16°N; Burkina Faso, common north of Ouahigouya to Mali border though apparently not to north of Ouagadou-

Zosterops senegalensis

gou; Niger, only in extreme SW, where not common. Liberia, Ivory Coast, Ghana and Nigeria: common south into derived savannas bordering rain forest zone, nearly reaching coast in Togo and Benin and at coast in SW Nigeria. Bioko. Cameroon, throughout except SE. Gabon, frequent in Rés. de la Lopé; 1, Bélinga, Aug 1983. Central African Republic, common in Bamingui-Bangoran Nat. Park, uncommon in Vakaga district in N, few observations elsewhere though probably common south to Congo and Zaïre borders. Sudan, north in W to about 10°N and also frequent in Darfur; east of 30°E north to 12°N. Ethiopia, frequent or locally common in W lowlands, status in W Highlands unclear. Eritrea, no recent records but old records from about 1200 m on Anseba. Kenya, highlands at 1100–3400 m, from Mt Elgon, Cheranganis, Laikipia, Mt Nyiru, Ndotos, Matthews Range and Mt Marsabit to Mau Forest, Naivasha, Loita and Nguruman Hills. Tanzania, Loliondo, Serengeti, Mangola (L. Eyasi), Ukerewe Island, Bukoba; more widespread in Kagera, Kigoma, W Tabora, Rukwa and W Mbeya south to Malaŵi; Iringa Highlands, Ulugurus, Ukagurus, Ngurus and Usambaras; Pugu Hills, Utete, Lindi, Masasi, Nachingwea, Kilimarondo; and Songea. Range in Tanzania shown here is based on large number of sight records, and some in W and centre may be misidentifications (N.E. Baker, pers. comm.; see under *Z. abyssinica*). Zaïre, widespread, frequent, or locally common, in savannas in N, NE, E and SE. Angola, absent from N Uige, coastal lowlands, S Huila and Cunene. Zambia, almost throughout, common, but scarce in low-lying areas in Luangwa Valley and SW, and may be absent east of L. Mweru. NE Namibia, N and NE Botswana, common in Zimbabwe except in parts of NW, SW and S where scarce or absent. South Africa, restricted to extreme N Transvaal, in valleys of Limpopo, Levuvhu and Mutale Rivers, and to NE KwaZulu-Natal, where frequent. Mozambique, throughout, except in N.

Description. *Z. s. senegalensis* Bonaparte: Senegal to NW Ethiopia and N Eritrea; intergrades with next 3 forms in humid areas of W Africa, NE Zaïre and central Uganda. ADULT ♂: upperparts greenish yellow, more strongly yellow on rump and uppertail-coverts, broadly yellow on forehead and above eye. Tail brown, feathers with narrow yellowish outer edges. Narrow white eye-ring (c. 0·5 mm broad); narrow blackish band through lores to eye. Ear-coverts and sides of neck yellow, tinged greenish; cheeks to chin, throat and entire underparts bright yellow. Remiges dark brown, edged greenish yellow except on emarginated parts of primaries; alulua dark brown; primary coverts dark brown, fringed greenish; rest of upperwing-coverts greenish yellow. Axillaries yellow; underwing-coverts and inner borders of flight feathers creamy white. Upper mandible dark horn to blackish, lower mandible paler; eyes brown; legs grey or blue-grey. Sexes alike. SIZE (10 ♂♂, 10 ♀♀): wing, ♂ 53–58 (55·5), ♀ 52–57 (54·9); tail, ♂ 32–40 (36·8), ♀ 33–38 (35·5); bill, ♂ 11–12 (11·2), ♀ 11–12 (11·3); tarsus, ♂ 14–16 (15·1), ♀ 14–16 (15·1). WEIGHT: Ghana, unsexed (n = 2) 8·1, 8·7; Nigeria, unsexed (n = 17, Oct–Mar) 7·5–9·0 (8·1), (n = 8, Apr–June) 8·0–9·0 (8·4).

IMMATURE like adult.

NESTLING: hatches naked except for patch of whitish down above eye; bill, inside mouth and tarsus yellow; gape flanges creamy.

Z. s. demeryi Buttikofer: Sierra Leone to Ivory Coast, south of range of nominate race. Greener than it on upperparts and side of head; yellow forehead band narrower and supraloral stripe narrow; yellow underparts slightly paler, sides of breast and flanks greenish. WEIGHT: (Liberia), ♂ (n = 5) 6·8–9·2 (8·2).

Z. s. stenocricotus Reichenow: SE Nigeria (Obudu Plateau) and Cameroon (north to Bamenda), Bioko, N Gabon; lowland and montane forest and thickets. (A similar population at Medje, NE Zaïre.) Similar to previous race, but more richly coloured; upperparts darker green with broader, more contrasting yellow forehead and broad supraloral stripe; sides of breast and flanks darker green, contrasting with otherwise bright yellow underparts. Montane populations slightly larger: wing, ♂ (n = 10, Cameroon highlands) 55–60 (57·9).

Z. s. stuhlmanni Reichenow: NW Tanzania to Uganda, intergrading with nominate race at about 1°N, with *jacksoni* in W Kenya and with *reichenowi* in Ruwenzoris and Kivu. Like *demeryi*, but yellowish green upperparts with slight brownish or cinnamon tinge; forehead and supraloral area yellower than in *demeryi*, sides of breast and flanks olive-green, but contrasting less with yellow ventral areas than in *stenocricotus*. Slightly larger than *demeryi*: wing, ♂ (n = 10) 56–60 (57·7).

Z. s. gerhardi (Elzen and König): S Sudan (Imatong, Didinga and Dongatona Mts) and NE Uganda border (Mt Lonyili). Similar to *stuhlmanni* but lacks brownish tinge on upperparts. Paler, less rich green above than *stierlingi*.

Z. s. toroensis Reichenow: NE Zaïre to W Uganda; lowland forest in Semliki and Ituri region, probably intergrading with Medje population of *stenocricotus* (see above). Like *stuhlmanni*, but smaller (wing 51–56, White 1963), and lacks brownish tinge to upperparts.

Z. s. reichenowi Dubois: E Zaïre, highland forests northwest of L. Tanganyika and southwest of L. Kivu. Upperparts dark green with little yellow on forehead; underparts mainly greenish, with yellow confined to chin, throat and centre of belly. Large: wing 59–64 (White 1963).

Z. s. jacksoni Neumann: highlands of Kenya and Loliondo in N Tanzania. Rich dark green above like *stenocricotus*, with well-defined yellow forehead; below, green sides and flanks contrast with bright yellow chin, throat, centre of breast and belly. Eye-ring better defined (1–2 mm wide). Large: wing, ♂ (n = 8) 60–64 (61·9). WEIGHT: (W Kenya, unsexed, n = 60, mainly this race) 9·8–13·5 (11·7).

Z. s. stierlingi Reichenow: highlands of E and S Tanzania, Malaŵi, E Zambia and NW and W Mozambique. Rich green upperparts less dark than in *jacksoni*; less yellow on forehead, supraloral stripe narrower; yellower below, with slightly less green on sides and flanks. Eye-ring well defined. Smaller than *jacksoni*: wing, ♂ (n = 10) 55–60 (57·4). WEIGHT: N Malawi, ♂ (n = 2) 10·3, 11·3, ♀ (n = 4) 10·2–11·6 (11·1); NE Tanzania (this race?) unsexed (n = 14) 8·0–11·4 (10·3) (L. A. Hansen, pers. comm.).

Z. s. kasaicus Chapin: central Zaïre (Kasai) to NE Angola. Dull green above with no yellow on forehead; sides and flanks olive-green. Eye-ring narrow. Small: wing, ♂♀, 51–56 (White 1963).

Z. s. heinrichi Meise: NW Angola. Slightly brighter than last with yellow forehead. Larger: wing, ♂♀, 54–60 (White 1963).

Z. s. quanzae de Schauensee: central Angola (Malanje to central highlands), integrating with last 2 forms and with *anderssoni*. Upperparts and sides of head entirely green; below, breast and flanks olive-green, yellowish chin, throat and belly also tinged green. Rather large: wing, ♂ (n = 2) 59, 66, ♀ (n = 2) 62, 63.

Z. s. anderssoni Shelley: savannas in S and E Angola, Zambia, SE Zaïre (Shaba), SW Tanzania, Malawi and W Mozambique (Tete) south to N Namibia, N and NE Botswana, Zimbabwe plateau and E Transvaal lowveld. Generally paler than adjacent forms of more humid or montane areas with which it integrates widely. Like nominate race, with yellowish rump and uppertail-coverts and broadly yellow forehead, but upperparts and sides of head yellowish green; bright yellow below, but sides of breast tinged green. Larger than nominate race: wing, ♂ (n = 10) 55–63 (59·2).

Z. s. tongensis Roberts: lowlands in NE Natal (E Zululand), Mozambique and SE Zimbabwe. Like *anderssoni*, but upperparts slightly duller and greener; duller yellow below, tinged more olive on sides.

TAXONOMIC NOTE: *gerhardi* was described by van der Elzen and Koenig (1983) as a race of *Z. poliogaster*, but in view of its close resemblance to *demeryi*, *stuhlmanni* and *stierlingi* we place it in *Z. senegalensis*. The nearest yellow-and-green races of *poliogaster* (*kaffensis* and *kikuyuensis*) are larger and greener, with broader eye-rings and broad yellow foreheads. On the basis of song *stenocricotus* seems to be specifically distinct (Chappuis, in press), especially since songs of other races of *senegalensis* differ little between W, E and southern Africa.

Field Characters. Length 11·5 cm. The common green and yellow white-eye over most of Africa, and the only one in its range in W and much of central Africa; confusion with other white-eyes only where it meets other species in Ethiopia–N Tanzania and in extreme S of range. Green above, *bright yellow below*; many races also green on flanks and in diffuse band across breast; forehead usually with small amount of yellow, eye-ring fairly narrow, black loral line continues under eye. In Ethiopia inhabits mainly non-montane western regions; distinguished from Abyssinian White-eye *Z. abyssinicus* by yellow belly (belly greyish white in races *Z. a. abyssinicus* and *omoensis*); W Ethiopian race of Mountain White-eye *Z. poliogaster kaffensis* has broad white eye-ring, darker green upperparts and green sides of breast and flanks. In Kenya/Tanzania told from yellow-bellied face *flavilateralis* of Abyssinian White-eye by darker green upperparts, brighter yellow underparts, black lores and yellow forehead band. In W Kenya approaches range of Mountain White-eye, which is confined to montane forest; race *kikuyuensis* of Mountain distinguished by very broad eye-ring and more extensive yellow on forehead, extending onto forecrown. Marginally sympatric with Cape White-eye *Z. pallidus* north of Otavi, Namibia, in E Botswana, extreme NE Transvaal and in NE KwaZulu–Natal; in latter area occurs in open woodland while Cape White-eye inhabits evergreen forest and coastal bush. Southern races *anderssoni* and *tongensis* distinguished from Cape White-eye by yellower upperparts and bright yellow underparts, with just a hint of green on sides; for further differences, see that species.

Voice. Tape-recorded (9, 86, 88, 91–99, 104, B, C, F, CHA). Song a bustling series of short, sweet notes with a slightly burry quality in W and E Africa (more pure in southern Africa); often starts with a few rather hesitant introductory notes, e.g. 'tit-chup-wheeyoo' or 'tree-turri-weeeu-teu', then breaks into song which may last up to 8 s (typically 3–5); alternating high and low notes impart a lilting quality, 'tew-tew-ti-tew-ti-tew-tewi-titi-tew ... '; next song begins after a gap of about 5 s. In the early morning 2 birds will sing against each other for long spells, taking turn and turn about perhaps 100–150 m apart, one starting as the other finishes (South Africa: Skead 1967). Contact calls include a single long, down-slurred 'teeeyer', a shorter, pure 'tew' and a tinny, twittering trill lasting about 1 s. ♂ said to have subdued sub-song not unlike song of Willow Warbler *Phylloscopus trochilus* (Skead 1967). Birds in Cameroon highlands (race *stenocricotus*), recorded by C. Chappuis, have a very different song. The notes are pure, without the burry quality found in other races; most have a typical white-eye sweetness. Closest to other white-eyes are the slower songs with 10–15 warbled notes, 'tyoo-tyee-tyootowee-tyootowee-tyootu'; many songs are fast, almost a trill or chatter, either a brief (1·5 s) trill of liquid notes, or longer (2–2·5 s) with a few introductory notes and characteristic ending, 'teetuwee-teetuwee-chippychippy-chippychippychippy-chee-chu' or 'tsee-chay-chippychip-pychippychippychippy-chee'; because of their speed they do not have the same sweet quality as slower songs, and in extreme examples may sound almost tuneless. Bird sings from top of tall forest tree.

General Habits. Arboreal, inhabiting canopy of *Isoberlinia*, *Brachystegia*, *Baikiaea* and *Uapaca* woodland, riverine bushes, wooded swamps, and woods and thorn scrub; lives in forest in several regions (e.g. Gabon, Kenya); forages in wide variety of woody growth; commonly visits and nests in cultivated land, small-holdings, rural and urban gardens, parks and boulevards in towns, and plantations especially of flowering *Eucalyptus*. In Liberia occurs in forest-grassland mosaic, clearings and gardens; on Mt Nimba common in secondary growth along ridges up to 1600 m, and occurs in open primary forest on steep rocky outcrops at 1000–1400 m (Gatter 1997) but not in interior of forest (S. Keith, pers. comm.). Keeps in dense foliage, usually in upper half of tree, foraging at sides by gleaning smaller twigs and leaves. Active, acrobatic and versatile, clinging even to weak stems and twiglets in great variety of postures, supported by the feet alone, hovering to pick insects from otherwise inaccessible leaves and occasionally making short sally to snatch airborne insect. Diligently explores twigs and foliage, peering at undersides, hanging sideways or upside down; gleans aphids from stems and caterpillars from leaves, and catches small moths and termites in flight. Occasionally comes down to ground, where progresses for a metre or two by hopping, 'but can walk' (Skead 1967). Bird puts bill into crevice then opens it, using considerable force to prise edges of crevice apart, then peers inside; it uses that method to examine bark of twigs and rolled-up leaves for food, and to open up flowers. Explores tiny indentations in bark, leaf axils and flowers by rapidly flicking brush-tipped tongue out to distance of 1–2 mm (behaviour hard to see since bill tip is concealed in crevice: Kunkel 1975). Tongue may be tactile but also draws water, exudates and nutrient liquids into mouth by capillarity.

Outside breeding season occurs in pairs or small, straggling parties, usually of 4–10 birds (forest, Kivu, Zaïre), sometimes up to 20, occasionally 50 (farms and towns, southern Africa). Occurs commonly in mixed-species foraging flocks – in 17–20% of such flocks in Zambia; mixes with sunbirds, Brubrus *Nilaus afer* and warblers including Green-capped Eremomela *E. scotops*. Singletons, pairs or trios leave and enter small white-eye flock, whose composition thereby alters constantly through course of day. Likes to bathe, visiting road puddles in small flocks, splashing in them and drinking from them; in damp forest can obtain plenty of fluids by lapping with tongue whilst foraging, and there visits ground to drink only occasionally. Sleeps at night 7 m up in tree, 2–3 birds huddled together; same spot may be used for several nights.

Within flock, birds keep in twos which are sometimes mated pairs but are generally 'clumping alliances' (Kunkel 1975): the 2 birds are temporary partners which forage and fly together within flock, perch side by side and preen each other. There is constant high level of aggression in flock and membership of clumping alliance often changes. Clumps of 2 (sometimes 3) birds stop foraging in order to allopreen many times a day; allopreening is sublimated aggression, and appeasement and courtship behaviours are derived from it (Kunkel 1975). Never overtly fights, but constant quarrelling in flock by pecking neck (which can turn into allopreening session) or, rarely, the crown, and by supplanting. In latter, dominant bird pursues another in short flights within tree, displacing it from several perches in succession. Supplanting is commonplace, and when several dominants are supplanting at once the flock appears just to be moving or foraging through woodland more actively than usual.

Threatens by sleeking plumage, crouching and opening bill widely. If unsure whether to pursue or flee, perched bird threatens by pivoting slowly from side to side through *c.* 30° without perceptible pause at the turn; body held stiffly and bill points down at *c.* 45° (**A**); utters no sound whilst pivoting but calls loudly between spells of pivoting. A commoner aggressive posture is 'wing dropping' (Kunkel 1975), perched bird holding body stiffly with bill pointing down, wrists held away from body, flight feathers loosely drooping to leg level, and wings quivering in short bursts (**B**). Bird challenges another whilst foraging by snapping its bill. Submission showed by bird freezing into immobility then fluffing crown feathers and sometimes opening bill (**C**), or fluffing crown and back feathers, or at high intensity fluffing whole plumage and half closing eyes so that white eye ring becomes an oval (**D**).

Resident, often sedentary (Irwin 1981), but outside breeding season may move locally (Vernon 1985); in

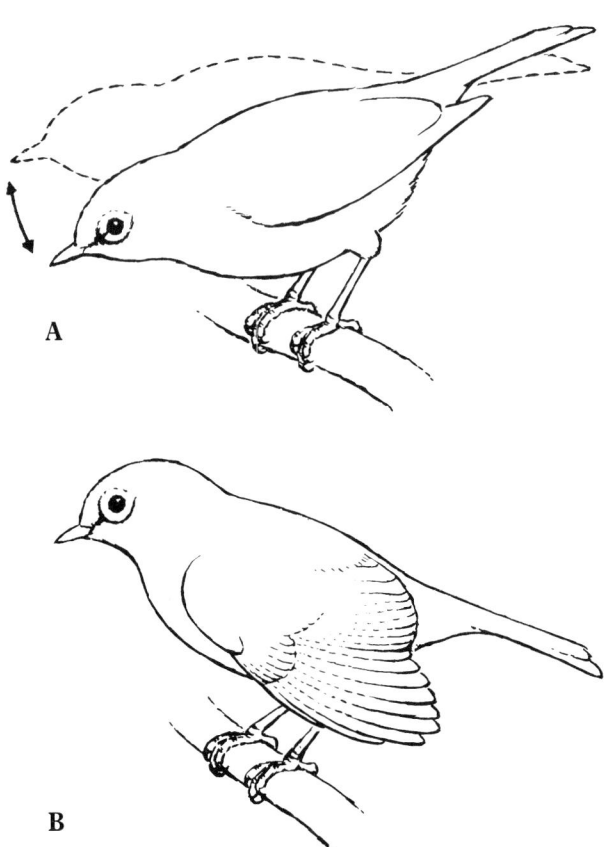

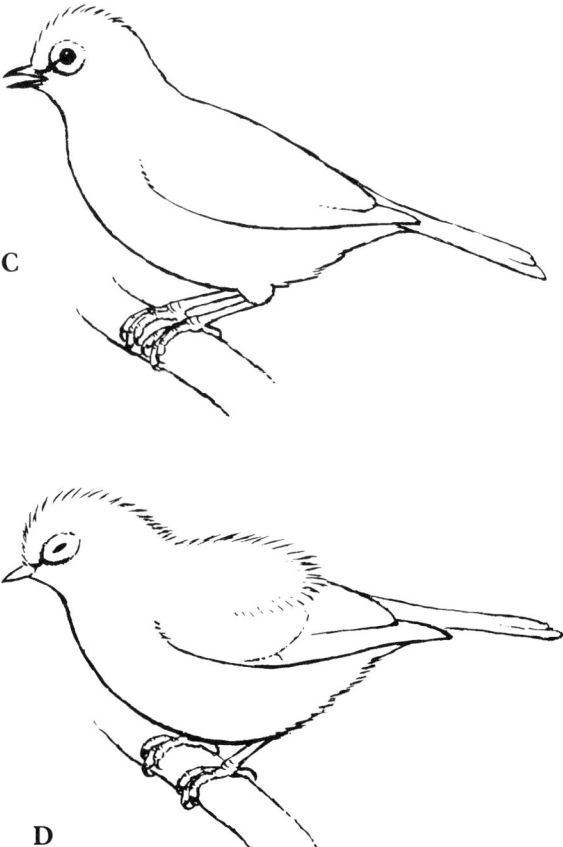

Zimbabwe a slight increase in reporting rate in winter – commonest in July (Harrison *et al.* 1997).

Food. Staple food is small arthropods; also eats nectar, juices, berries, other fruits and occasionally small seeds (Kunkel 1975). Wild and cultivated figs including *Ficus burtdavyi*; fruits of *Cussonia* spp.; flesh of soft cultivated fruits such as grapes. Eats fruit of *Tetrorchidium didymostemon* (Euphorbiaceae), Sierra Leone. Captive birds take a wide variety of offered foods: porridge, tomatoes, bananas and tinned peaches. Probes *Eucalyptus* flowers and thought to take their nectar. Insects include aphids, small moths, flies and termites, and many caterpillars.

Breeding Habits. Monogamous, solitary nester. In courtship, perched ♂ stands high, fluffs crown feathers (or more plumage at higher intensity), half opens wings horizontally and quivers their tips (**E**). Rarely, wings are fully opened. ♀ flies away, or reacts by pecking towards ♂'s neck and body in a half-hearted and hurried manner; if ♂ accepts ♀'s advance the pecking becomes allopreening and pair formation has occurred (Kunkel 1975).

E

NEST: small, neat, regular, compact but thinly-made open cup, delicately built of small strips of soft, dry grass-blades and chaffed grass, bound with a little spider-web outside, lined with tiny twiglets and on top of them a layer of hair-thin dry grass; attached by spider web to supporting twiglets and live leaves at the rim and base, or placed in vertical or horizontal fork of twigs. Some nests decorated outside with spider egg-cocoons. Nest is lightweight and flimsy-looking though holds together well, with open weaving and thin, translucent walls. Size (n = 3): ext. diam. 52–55, int. diam. 40–43, ext. depth 28–36, int. depth 24–27. Sited at tip of branch, or near top and middle of thick-foliaged shrub, very well concealed, 1·2–3·0 m from ground (Vincent 1949).

EGGS: 1–4; C/1 and C/4 are rare; av. of 17 clutches, South Africa, 2·8. Not glossy. Immaculate white, bluish-white, blue or turquoise. SIZE: (n = 3, Nigeria) 15·0–15·7 × 10·6–12·0 (15·5 × 11·5), (n = 45, South Africa) 13·2–16·9 × 11·3–12·0 (15·3 × 11·7).

LAYING DATES: Senegal, June–Nov, mainly Sept; Gambia, Dec–Mar, June–Sept; Mali, June; Ghana, (dependent young, Jan, June); Nigeria, Nov–July; Sudan, Dec–Mar, June–Sept; E Africa, Region A, Feb, May, Sept–Dec, Region B, Nov–July, Sept (mainly Apr–July), Region C, Nov–Jan, Apr, Aug, Region D, Feb–Mar; Zaïre (Lubumbashi), Aug–Oct; Angola, Sept; Zambia, Aug–Feb; Zimbabwe, Aug–Jan (mainly Sept–Oct); Zululand, Aug–Feb (mainly Sept–Oct).

INCUBATION: by both sexes, mainly ♀ (Vincent 1949); period 11 days.

DEVELOPMENT AND CARE OF YOUNG: young fed by both parents; very prone to jump out of nest prematurely if disturbed by person (Skead 1967). Period: about 14 days. Young out of nest solicit food from parents by quivering wings and giving shrill call like parental contact call.

BREEDING SUCCESS/SURVIVAL: parasitized by Klaas's Cuckoo *Chrysococcyx klaas* (once, Ghana) and Green-backed Honeybird *Prodotiscus zambeziae*. White-eyes are principal hosts of *P. zambeziae*. Longevity of 5–>7 years (n = 5, E Zimbabwe highlands).

References
Bannerman, D.A. (1948).
Chapin, J.P. (1954).
Kunkel, P. (1975).
Skead, C.J. (1967).
Vincent, A.W. (1949).

Plate 17 (Opp. p. 270) *Zosterops vaughani* Bannerman. Pemba White-eye. Zostérops de Pemba.

Zosterops vaughani Bannerman, 1924. Bull. Br. Orn. Club 44, p. 41; Pemba.

Forms a superspecies with *Z. senegalensis*.

Range and Status. Endemic resident, Pemba Island and adjacent coral islets, where abundant and one of the commonest birds in all situations.

Description. ADULT ♂: upperparts yellowish green; broad yellow band across forehead extends to above lore and eye, merging into greenish crown. Tail feathers blackish brown, outer webs narrowly edged greenish towards base, central pair yellowish green. Narrow white eye-ring, *c.* 1 mm wide. Narrow black stripe through lore, broadening and extending around front lower part of eye-ring. Ear-coverts, cheeks and sides of neck greenish yellow; underparts yellow, washed greenish on sides and flanks. Flight feathers blackish brown, narrowly edged yellowish; tertials brown, more broadly edged greenish yellow; alula blackish brown, inner webs tinged green; primary coverts brown, fringed green; rest of upperwing-coverts yellowish green. Axillaries pale yellow; underwing-coverts whitish, tinged yellow. Bill black, base of lower mandible blue-grey; eyes dark brown or blackish; legs blackish or slate grey. Sexes alike. SIZE (9 ♂♂, 6 ♀♀): wing, ♂

Zosterops vaughani

50–55 (52·3); ♀ 49–53 (51·8); tail, ♂ 33–38 (34·9); ♀ 31–37 (34·5); bill, ♂ 12–13 (12·9), ♀ 12–14 (13·0); tarsus, ♂ 15–16 (15·5), ♀ 15–16·5 (15·6).

IMMATURE: juvenile like adult.

TAXONOMIC NOTE: treated as a subspecies of *Z. senegalensis* by Hall and Moreau (1970), Britton (1980), Short *et al*. (1990) and Dowsett and Dowsett-Lemaire (1993a) and as a separate species by Mackworth-Praed and Grant (1960), White (1963), Paynter (1967), Morony *et al*. (1975), Sibley and Monroe (1990) and Zimmerman *et al*. (1996). This white-eye is clearly an insular representative of the mainland *Z. senegalensis*, and we believe that its distinctive appearance and voice mark it as of specific rank.

Field Characters. Length 10–10·5 cm. The only white-eye on Pemba. All yellow, with prominent black lores and very narrow eye-ring.

Voice. Tape-recorded (B, MELL). Song rich and sweet, fuller than that of Yellow White-eye *Z. senegalensis*, lasting 5–8 s. Also described as high and sweet, shorter than Yellow White-eye, 'seweet-sureeteet-twerila-eeta-eet' or 'weet, su-weeet-see-sur-seeiwee-see', first note distinct, others often run together in a brief warbling (Zimmerman *et al*. 1996). Also gives a low liquid twittering trill, lasting 0·5 s.

General Habits. Inhabits gardens and all types of woody vegetation: mangrove forest, Ngezi and Mwitu Mkuu Forests, thickets on coral rag, bushland, and copses and hedges in grassland. Occurs in pairs and parties of up to 20 or more birds. Very confiding. Forages in foliage, presumably searching for edible vegetable matter and gleaning leaves for insects; however, few observations on record. Sedentary.

Food. Unknown.

Breeding Habits. A solitary nester, presumed to be monogamous.

NEST: small open cup made of fine grass or palm fibres, placed 1–3 m up in branches of bush or small tree; no further details known.

EGGS: 2. Immaculate pale blue. SIZE: not known.

LAYING DATES: (fully fledged young in early Oct; breeds mainly Oct–Dec, also Mar and July: Mackworth-Praed and Grant 1960); thought to breed Aug–Mar, and moults in Jan–Feb (Pakenham 1979)).

References
Mackworth-Praed, C.W. and Grant, C.H.B. (1960).
Packenham, R.H.W. (1979).

Zosterops ficedulinus Hartlaub. Príncipe White-eye. Zostérops becfigue.

Plate 17
(Opp. p. 270)

Zosterops ficedulina Hartlaub, 1866. Proc. Zool. Soc., p. 327; Principe.

Forms a superspecies with *Z. griseovirescens*.

Range and Status. Endemic resident, São Tomé and Príncipe Islands.

On São Tomé common in 1920s, uncommon in 1970s, common to abundant near Monte Café, Nova Moça, Bombaím and north of Binda in 1987 and commonest at middle and high altitudes except on W coast (Jones and Tye 1988), and frequent to common in savannas in N (Harrison and Steele 1989). In 1990 found to be rare: only a few at Nova Moça, Lagoa Amélia and Santa Catarina, and during 3 months' fieldwork found only at Lagoa Amélia, Bombaím, Santa Catarina and Rio Ana Chaves and not at Monte Café (Atkinson *et al*. 1991, Nadler 1993). More recently, found in E at Quimpo; between Santo António de Mussacavú and Porto Alegre; quite common in central massif up to 1600 m and its N and E flanks (Fortunato, Lagoa Amélia, Bom Sucesso, Bombaím, Formosa Pequeno) quite common and in SW, along rivers An Chaves and Io Grande (Christy and Clarke 1998).

Formerly fairly common on Príncipe but had become rare by 1920s; not seen in 1970s or 1980s (de Naurois 1983, Atkinson *et al*. 1991), though it may survive in the southwest (Jones and Tye 1988); also survives in central

Zosterops ficedulinus

massif where confined to primary forest (and does not enter open country as it does on São Tomé) – flock recently on R. Papagaio (Christy and Clarke 1998).

Description. *Z. f. ficedulinus* Hartlaub: Príncipe. ADULT ♂: upperparts dark olive-green with brownish wash, top of head darker brown, rump and uppertail-coverts greener. Tail feathers blackish brown, edged olive-green. Black band through lore, and narrow buffish white supraloral line. Narrow white eye-ring. Ear-coverts and cheeks greyish white, mottled dark olive-grey; sides of neck dark olive-brown. Underparts whitish, tinged buffish yellow, throat and upper breast with faint dusky streaks, sides of breast and flanks washed olive-brown. Flight feathers blackish brown, edged olive-green; tertials dark brown, tinged green; primary coverts and alula dark brown; rest of upperwing-coverts dark olive-green. Axillaries buffish white; underwing-coverts white. Bill brownish horn; eyes very pale brown; legs and feet brownish grey. Sexes alike. SIZE (1 ♂♂, 3 ♀♀): wing, ♂ 52, ♀ 52–56 (53·7); tail, ♂ 38, ♀ 38–39 (38·3); bill, ♂ 13, ♀ 12–13 (12·7); tarsus, ♂ 18, ♀ 18.
  IMMATURE and NESTLING: unknown.
  Z. f. feae Salvadori: São Tomé. Slightly darker above, more uniform olive-green; greyer below on throat, breast and flanks. SIZE (6 ♂♂, 4 ♀♀): wing, ♂ 52–55 (53·7), ♀ 53–55 (54·6); tail, ♂ 33–36 (35·5), ♀ 36–38 (36·8); bill, ♂ 12–13 (12·3), ♀ 11–13 (12·3), tarsus, ♂ 16–17 (16·3), ♀ 16–17 (16·8).

Field Characters. Length 10·5 cm. Rather small; olive-green above, brighter on rump, white below tinged yellowish; broad and conspicuous white eye-ring contrasts with dark green head and black on lores and under eye. The only green white-eye on São Tomé and Príncipe. Each of these islands has a brown white-eye *Speirops*; for comparisons, see those species.

Voice. Tape-recorded (B, ALEX, Jo PJ, MAC). Dawn song monotonous and insistent, astonishingly powerful for the size of the bird, a rapid series of 3–6 (usually 5) distinct notes with burry quality, 'ptirrr ptirrr ptirrr ptirrr ptirrrrr', last note accentuated (Christy and Clarke 1998); given from same song post in canopy for up to 15 min at a time, with 1 s intervals between songs; 3 birds may countersing without leaving their posts. A more complex type of song is given during the day, mixing burry and tinny notes with musical ones. Contact calls, 'prrrip, prrrip ...'; in flight a weak, high pitched 'plink-plink-plink ...' (Christy and Clarke 1998). Birds in foraging flock call continuously, mixing little chippering notes with thin buzzes and a dry, tuneless 'tyup' or 'tyew' (L.R. Macaulay, pers. comm.).

General Habits. Inhabits dense primary and degraded forests from sea level up to 1000 m and patches of dry woodland and isolated large trees in savanna on São Tomé; and on Príncipe formerly occurred in uninhabited forests and canopy of tallest trees in plantations in the hilly interior. Lives in pairs, family parties, and flocks of up to 20 birds, often mixing with Black-capped Speirops *Speirops lugubris*, Príncipe Speirops *S. leucophaeus*, São Tomé Prinia *Prinia molleri*, São Tomé Paradise-Flycatcher *Terpsiphone atrochalybeia* and Newton's Sunbird *Anabathmis newtonii*. Gleans insects from leaves, twigs, under branches and in growths of lichen; habits said to be much the same as those of Black-capped Speirops. Forages at mid levels in forest and in the canopy. Vocal, giving contact calls whilst foraging and in flight; at certain times of day can be very vocal. Uses favourite song posts in canopy; sings particularly from 17h00 until dusk. Sedentary on São Tomé, though may move locally.

Food. Insects and berries.

Breeding Habits.
  NEST: a neat, open cup slung under forked twigs.
  EGGS: 3–5.
  LAYING DATES: (breeds from Sept onward; fledglings in Feb).

References
Atkinson, P. *et al.* (1991).
Christy, P. and Clarke, W.V. (1998).
Collar, N.J. and Stuart, S.N (1985).
Jones, P.J. and Tye, A. (1988).

Zosterops griseovirescens Bocage. Annobon White-eye. Zostérops d'Annobon.

Zosterops griseovirescens Bocage, 1893. J. Sci. Math. Phys. Nat. Lisboa, ser. 2, 3, p. 18; Annobon.

Plate 17
(Opp. p. 270)

Forms a superspecies with *Z. ficedulinus*.

Range and Status. Endemic resident, Pagalu (Annobon) Island. Common and ubiquitous from sea level up to highest peak (655 m). Island's area is 1750 ha; density guessed at 2 pairs or flock of 5 birds per ha, so population may be between 7000 and 9000 birds.

Description. ADULT ♂: entire upperparts uniform olive-green, but sides of crown paler, and in some birds a narrow, creamy pale green superciliary line; a conspicuous white eye-ring of feathers, broken before and often behind eye, with upper part broader than lower part (**A**); lores black; dusky line below white eye-ring and behind eye. Tail: central feathers olive-green, others blackish, broadly fringed with olive-green on outer webs; all shiny very dark grey below. Ear-coverts mainly olive-green, slightly streaked greyish, merging with dusky under-eye line; sides of neck olive-green; chin and throat very pale sulphur-yellow, merging behind into olive-green ear-coverts and below into buff breast, greyer at sides; flanks pale tawny-buff; midline of belly, thighs, vent and undertail-coverts very pale sulphur-yellow. Primaries and secondaries blackish brown, edged olive-green; tertials brown, broadly edged olive-green; alula blackish brown, upperwing coverts olive-green; axillaries and underwing coverts white. Bill blackish brown or horn-brown, with most of lower mandible pinkish yellow-brown; eyes light brown; legs and feet brownish grey. Sexes alike. SIZE (10 ♂♂, 1 ♀): wing, ♂ 61–68 (63·0), ♀ 60; tail, ♂ 46–51 (49·3), ♀ 49; bill, ♂ 14–16 (14·9), ♀ 15; tarsus, ♂ 19–21 (20·2), ♀ 21. WEIGHT: (n = 2, unsexed) 11, 11.
IMMATURE: like adult but slightly yellower below.

Field Characters. Length 12 cm. The only white-eye on Pagalu. Fairly large, olive-green above, buffy below, with white eye and black line across lores to under and behind eye. Unmistakable – the only other resident passerine on Pagalu is a paradise-flycatcher.

Voice. Not tape-recorded. Song a pleasant, quiet warble of indefinitely jumbled notes derived from the flight call. Flight call a sharp 'plic' or 'tsip', singly or repeated 2–4 times in quick succession and run together, 'plic plic-plic-pic' so that small flock calling together can sound like twittering finches; bird perching in foliage gives a trilling chur, less than 1 s in duration, very like chur of sympatric Red-bellied Paradise-Flycatcher *Terpsiphone rufiventer*.

A

Zosterops griseovirescens

General Habits. Inhabits all types of woody vegetation, i.e. mist forest, dry *Steganthus/Lannea* forest, savanna with baobabs *Adansonia digitata*, tamarinds *Tamarindus indica* and silk-cotton trees *Ceiba pentandra*, regrowth woods with bananas, coffee, cocoa and yellow plum trees *Spondias mombin*, dense oil-palm *Elaeis guineensis* plantations, thick plantations of mango *Mangifera indica* and small shrubs and spindly young trees around grassy, cultivated patches with cassava *Manihot utilissima*, *Citrus* and other bushes. Keeps mainly but not exclusively to lower strata in primary and regrowth forest, and usually about 2–5 m high in trees in Santo Antonio (the only village). Seeks dappled shade; in forest, keeps to more open parts and avoids densest shade. Sometimes carries wing tips slightly away from flanks (**A**). In pairs and flocks of up to 10 birds, actively gleaning leaves, petioles, twigs and branches and probing into inflorescences, buds and fruiting heads at all heights from the smallest bush to crowns of tallest emergent trees. Flock readily flies from tree to tree across open spaces, birds following each other on same flight path 1–2 at a time. Sedentary.

Food. Small insects including ants, and small seeds.

Breeding Habits. Evidently a monogamous, solitary breeder, not conspicuously territorial.

NEST: a small, neat, closely woven and quite robust cup, made of fine grass and vegetable fibres, built amongst fine twigs in small *Acacia* tree, *Manihot* bush, or 'leafless' cactus *Rhipsalis* hanging on rock face in shady, bouldery ground.

EGGS: 2; immaculate white. SIZE: *c.* 18 × 13·5.

LAYING DATES: Oct–Nov (n = 5), Dec (i.e. 'nesting in Dec': Bannerman 1915) and Feb (i.e. fledglings common in early Apr: Harrison 1990).

References
Fry, C.H. (1961).
Harrison, M.J.S. (1990).

Plate 17
(Opp. p. 270)

Zosterops abyssinicus Guérin-Méneville. Abyssinian White-eye; White-breasted White-eye. Zostérops à flancs jaunes.

Zosterops abyssinica Guérin-Méneville, 1843. Rev. Zool., 6, p. 162; Abyssinia.

Range and Status. Africa and SW Arabia.

Resident in E Africa from 19°N in Red Sea Hills, Sudan, south to NE Tanzania and from about 35°E in Kenya east to Socotra. Common in Sudan, below 1650 m. Common at middle altitudes, between about 300 and 1800 m in Eritrea, Ethiopia and Kenya; absent from Red Sea coastal plains and scarce on much of Eritrean plateau. Somalia, common in N and locally common in S. Socotra, rather common and widespread from sea level up to 850 m, Kenya, locally common from sea level up to 1800 m, in N along Ethiopian border, and on lower slopes from S end of L. Turkana through central Kenyan highlands, around and east of Rift Valley, to coastal lowlands from Somali border south to Ngomeni; on lower Tana R. up to Baomo; west of Rift Valley known only from Kongelai Escarpment and north of Cherangani Hills. Tanzania, from Kenyan border to Mbulumbulu, Tarangire Nat. Park, Arusha, Same and Mkomazi Game Res.; map based largely on sight records, but separation of *Z. abyssinicus* and *Z. senegalensis* in the field 'is a huge problem, especially in forested areas such as Arusha and the central highlands, *abyssinicus* being a common non-breeding visitor to forest-edge habitats around the central plateau dry-country breeding range' (N.E. Baker, pers. comm.).

Zosterops abyssinicus

Description. *Z. a. flavilateralis* Reichenow: E, central and S Kenya to N and NE Tanzania. ADULT ♂: upperparts rather pale yellowish green, yellower on uppertail-coverts, and grading into narrow band across base of upper mandible. Tail dark brown, feathers with narrow yellowish green outer edges. White eye-ring narrow (<1 mm wide); narrow blackish brown line through lore and around lower front margin of eye-ring; yellow supraloral stripe continuous with yellow above bill; sides of head and neck greenish yellow continuous with upperparts, cheeks flecked yellow. Underparts yellow, tinged greenish on sides and flanks. Remiges dark brown, primaries narrowly edged green proximal to emarginations, secondaries and tertials more broadly edged yellowish green; alula blackish brown; primary coverts dark brown, fringed green; rest of upperwing-coverts yellowish green. Axillaries pale yellow; underwing-coverts and inner borders of underside of flight feathers whitish. Bill black; eyes brown or grey-brown; legs grey, blue-grey or blackish. Sexes alike. SIZE (10 ♂♂, 10 ♀♀): wing, ♂ 53–58 (55·6), ♀ 53–59 (54·9); tail, ♂ 36–42 (38·4), ♀ 36–42 (37·9); bill, ♂ 10–12 (11·2), ♀ 10–12 (11·0); tarsus, ♂ 15–17 (15·6), ♀ 15–16 (15·3). WEIGHT: (n = 16, unsexed, Tsavo, SE Kenya) 7·2–9·0 (7·8).

IMMATURE: juvenile resembles adult.

NESTLING: hatches tiny, naked, pink, with prominent yellow gapes.

Z. a. jubaensis Erlanger: thorn savannas of S Ethiopia to N Kenya south to Kulal, and S Somalia. Like *flavilateralis* but yellowish green upperparts tinged greyish; yellow underparts slightly less bright.

Z. a. abyssinicus Guérin-Méneville: NE Sudan, Eritrea and N and central Ethiopia. Differs from previous 2 races in having sides and lower part of breast to flanks pale grey, belly and thighs greyish white; chin to throat and centre of upper breast pale yellow; undertail-coverts pale yellow; upperparts greyish green, paler and duller than in *jubaensis*. Bill horn brown above, pinkish brown below; eyes brown to light red-brown; legs brown or flesh.

Z. a. socotranus Neumann: Socotra and N Somalia. Like *abyssinicus*, but bill blackish; throat brighter yellow, sides and underparts darker grey. WEIGHT: (n = 4) 9·8–12·0 (10·8).

Z. a. omoensis Neumann: W Ethiopia (L. Tana to Omo valley). Similar to *abyssinicus*, with brown bill and legs, but upperparts light yellowish green; throat brighter yellow and grey underparts tinged buff.

TAXONOMIC NOTE: habits described by van Someren (1956) under the name *Z. senegalensis fricki* (= *Z. a. flavilateralis*: Paynter 1967).

Field Characters. Length 10·2 cm. A small white-eye with *narrow* white eye-ring, *no yellow band* on forehead and mainly *yellow lores* (black only at base of lower mandible and below the eye-ring). In Ethiopia usually below 1800 m (Urban and Brown 1971), and distinguished from Yellow and Mountain White-eyes *Z. senegalensis* and *Z. poliogaster* by whitish breast and belly. In Kenya/N Tanzania race *flavilateralis* has yellow underparts like Yellow White-eye but they are less bright, 'pale powdery canary yellow' (Zimmerman *et al*, 1996) and lack any green, and upperparts paler yellow-green. Considerable geographic overlap with Mountain White-eye but the 2 are usually altitudinally separated, Abyssinian being at lower elevations and in mostly non-forest habitats. Abyssinian meets race *kikuyuensis* of Mountain White-eye in Nairobi suburbs, where identified by pale upperparts, lack of green on underparts and narrow eye-ring; told from race *Z. p. sylvanus*, which it meets in Taita Hills, by yellow, not grey, underparts.

Voice. Tape-recorded (B, HOR, McVIC, PEA). Song like other white-eyes, sweet and lilting, with burry quality. Calls, plaintive 'teeyu', 'tew' or 'tyew-tip' and sharper, sibilant 'pseeyip'. Buzzing and twittering call notes said to be given repeatedly by foraging groups (Zimmerman *et al.* 1996).

General Habits. Inhabits broad-leaved and thorn woodland, bushland, forest edge, copses and thickets; a familiar bird in parks and gardens; in Somalia often by streams and amongst olive trees or acacias and euphorbias, and also strongly associated with fruiting figs and junipers; in Socotra, submontane woodland and lowland woods with *Adenium obesum* and *Ziziphus*. Lives typically in bands of 12–20 birds, near Nairobi, Kenya, working the canopy of tall bush where *Vernonia*, attractive to aphids, grows. Quick, restless, very confiding. Fond of water-bathing; up to 30 birds at a time come to bird-bath or small puddle; flock bathes in wet, leafy vegetation, then preens vigorously. Flock forages by gleaning leaves and twigs for small insects, and birds seem to be almost 'trap-line' feeders, following same course around home range each day and appearing at the same time on successive days at a given tree (van Someren 1956). Birds feeding at flowers often have foreheads dusted yellow or cinnamon-colour with pollen (e.g. flock of 30 foraging at *Acanthus arboreus* flowers in Yemen: Harvey 1993). Flock flies between trees in ones and twos at a time, birds following each other on same flight path and calling. In foliage, flock splits into pairs which perch and forage close together. Roosts at night in trees, 2 birds perching huddled close together. In heat of noon, flock sits in rows along sheltered twigs, dozing, loafing, warbling and preening. Flock dissociates into separate pairs of birds shortly before breeding begins.

Mainly sedentary, but partial migrant in N Tanzania – occurs all year in and around Arusha, but in other areas vacates forest-edge habitats in Oct–Nov (N.E. Baker, pers. comm.).

Food. Aphids and other small insects, including caterpillars 10–38 cm long; mainly small moth larvae given to nestlings; tiny long-horned grasshopper nymphs given to young nestlings were too large and were rejected.

Breeding Habits. Monogamous, solitary nester. Flocks split up into pairs as breeding season approaches. One bird of a pair feeds the other. Nest often built in same spot season after season; 3 nests may be close together; a small bamboo clump once contained 3 nests and 2 of Mountain White-eyes. No apparent territoriality.

NEST: small, open cup or hammock, quite deep, firmly attached to and slung between 2 horizontal arms of forked twig or between 2 parallel thin petioles, so that rim of nest is at same level as supports. Constructed of very fine strips of bark from creepers, bound together with quantities of spider web and often decorated outside with spider cocoons. Sited 1–3 up in bush or small tree. Nest built by ♂ and ♀, taking about 7–8 days; some very fine vegetable fibres added after 1st or 2nd eggs hatch.

EGGS: 2–4, usually 3. Immaculate pale blue. SIZE: *c.* 14 × 10.

LAYING DATES: Sudan, Mar, July; Ethiopia, May–June; E Africa, Region D, Feb–June, mainly Apr–May; Socotra, (♂♂ singing Nov, juvs Dec–Jan).

INCUBATION: by ♂ and ♀. Incubating bird sits very tight and can be touched by observer without it being dislodged. Period: 11 days.

DEVELOPMENT AND CARE OF YOUNG: young fed by both parents, continuously from dawn to dusk but easing off for 2 h over noon; both parents often at nest together, one at each side, but one waits its turn to give food to young. Pair brought 58 feeds in 150 min. Nestling period *c.* 14 days, or more in poor weather. Fledgling has lax plumage, olive upperparts, wispy white down at hind end of superciliary stripe, black eye, thick pale yellow gape, pinkish bill with upper mandible darkening, pale yellow chin, grey underparts and black remiges (photo, Oman Bird News 21, 1998, opp. p. 5). Fledglings accompany and are fed by parents for some days.

BREEDING SUCCESS/SURVIVAL: few clutches successful – eggs and young heavily predated by shrikes, rats, mongooses and snakes; also parasitized by Green-backed Honeybird *Prodotiscus zambeziae*, which replaces up to 2 white-eye eggs one at a time with its own; white-eyes are principal host of Green-backed Honeybird.

References
Archer, G. and Godman, E.M. (1961).
van Someren, V.G.L. (1956).

Plate 17
(Opp. p. 270)

Zosterops poliogaster Heuglin. Mountain White-eye. Zostérops alticole.

Zosterops poliogastra Heuglin, 1861. Ibis, p. 357; highlands of Abyssinia.

Range and Status. Endemic resident, E African highlands from Eritrea to NE Tanzania. Eritrea, early records from Marebquellen, Asmara and Keren; occurs around Faghena at 1690 m. Ethiopia, common in evergreen forests at 1380–3230 m. Sudan, very common above 1800 m in Imatong, Dongotona and Didinga Mts, and on or very close to Uganda border in S Imatongs. Kenya, widespread and locally common or abundant at 1500–3400 m on Mt Kulal, central Kenyan highlands from Meru District, Mt Kenya, Ol Doinyo Erok (Namanga Hill) and Aberdares to Nairobi, and Chyulus, Taita Hills (Brooks *et al.* 1998) and Mt Kasigau. Tanzania, Crater Highlands, Mbulu and Mt Hanang to Mt Ketumbeine, Longido, Mt Meru and Arusha Nat. Park, Lolkissale (Ol Doinyo Kisale), Essimingor plateau, Kilimanjaro, North Pare Mts and Lossogonoi Plateau. In South Pare Mts, perhaps several thousand in Chomme Forest Res., and present in Mwala Forest, Kwizu Forest and Chambogo Catchment Forest Res. (Collar *et al.* 1994); this population may prove to be a separate species (see below).

Zosterops poliogaster

Description. *Z. p. kikuyuensis* Sharpe: Kenya highlands from Mt Kenya and Aberdares to Nairobi. ADULT ♂: upperparts rich green with well marked yellow forehead and forecrown, extending as yellow patch above lores and eye. Tail dark brown, feathers with narrow green outer edges. Very broad white eye-ring (*c.* 3 mm wide); narrow black line through lore and around lower front edge of eye-ring. Cheeks, ear-coverts and sides of neck green. Chin and throat bright yellow, demarcated from green sides of head; centre of breast and belly to undertail-coverts yellow; sides of breast and flanks dark green. Remiges blackish brown, primaries narrowly edged yellowish green proximal to emarginations, secondaries more broadly edged green, tertials with outer webs mainly green; alula blackish brown; primary coverts dark brown, fringed greenish; greater, median and lesser coverts green. Axillaries pale yellow; underwing coverts white, tinged yellow; inner borders on underside of flight feathers whitish. Bill black; eyes brown or hazel; legs slate grey to pale grey. Sexes alike. SIZE (10 ♂♂, 10 ♀♀): wing, ♂ 60–64 (61·6), ♀ 59–62 (60·8); tail, ♂ 44–50 (47·1), ♀ 44–49 (46·4); bill, ♂ 12–14 (13·1), ♀ 12–14 (12·9); tarsus, ♂ 17–19 (18·3), ♀ 17·5–19 (18·3). WEIGHT: (unsexed, n = 20) 9·0–14·0 (11·0).

IMMATURE: like adult.
NESTLING: not described.

Z. p. kaffensis Neumann: W and SW Ethiopia (south of L. Tana, west of Omo R.). Similar to *kikuyuensis*, but yellow forehead band extends back only to between eyes; white eye-ring much narrower (1–2 mm wide); green upperparts, side of breast and flanks less dark; yellow chin and throat less well demarcated from green sides of head. Size similar to *kikuyuensis*.

Z. p. poliogaster Heuglin: Eritrea, N, central, S and E Ethiopia south to Addis Ababa and east to Harar; also Yavello. Differs from preceding races in having sides of breast and band across upper breast pale grey, flanks greyish buff, and centre of breast and belly white, sometimes with yellow tinge; throat, thighs and undertail-coverts yellow. Head, eye-ring and upperparts like *kaffensis*. Size similar: wing, ♂ (n = 10) 60–66 (63·8).

Z. p. kulalensis Williams: Mt Kulal, N Kenya. Like *poliogaster*, but grey breast and flanks slightly darker, the latter without buff tinge; mid-ventral stripe yellowish white; throat tinged greenish, merging more with green on cheeks.

Z. p. mbuluensis Sclater and Moreau: SE Kenya (Ol Doinyo Erok and Chyulu Hills) and N Tanzania (N Pares, Longido, Ketumbeine, Crater and Mbulu Highlands and Hanang). Similar to *kikuyuensis*, but yellow forehead narrower, merging into yellowish green crown; eye-ring less broad (*c.* 2 mm); yellower below, green sides less dark.

Z. p. eurycricotus Fischer and Reichenow: N Tanzania – Kilimanjaro, Mt Meru, Essimingor, Lossogonoi, Lolkissale. Top and sides of head and upperparts rich green apart from yellowish supraloral stripe. Throat greenish yellow, merging with green cheeks and breast; rest of underparts mainly green, with yellowish belly and undertail-coverts. White eye-ring very broad, as in *kikuyuensis*.

Z. p. winifredae Moreau: NE Tanzania – S Pare Mts. A grey-bellied form similar to *kulalensis*, but darker green above, forehead less yellow; below, sides darker, more blue-grey, mid-ventral stripe greyish white, yellow undertail-coverts tinged green. Smaller: wing, ♂ (n = 4) 57–59 (58·3), 1 ♀ 58.

Z. p. silvanus Peters and Loveridge: SE Kenya – Taita Hills. White eye-ring very broad (*c.* 4 mm). Top and sides of head and upperparts wholly green as in *eurycricotus*, but breast to belly and flanks grey, still darker than in *winifredae* and without whitish central stripe; chin to throat and undertail-coverts greenish yellow. Rather small: wing, ♂ (n = 10) 56–60 (58·0), ♀ (n = 2) 56, 58.

TAXONOMIC NOTE: habits described by van Someren (1956) under the name *Z. (virens) kikuyuensis*. The above are isolated montane forms, and several have been regarded as separate species. In keeping them all conspecific we take the view of most recent authorities, although *winifredae* ('South Pare White-eye') was given species rank by Collar *et al.* (1994), and the distinctive grey-bellied *silvanus* ('Taita White-eye') and *poliogaster* with *kulalensis* may yet warrant specific recognition also.

Field Characters. Length 11–12 cm. A *montane forest* white-eye of E Africa, with very variable plumage. Most races have *very broad white* eye-ring; throat and undertail-coverts yellow, rest of underparts grey, yellow or green; yellow band on forehead varies from broad to vestigial; lores black. Birds with grey or mostly green underparts readily identifiable. Near Nairobi, Kenya, yellow-bellied race *kikuyuensis* meets Abyssinian White-eye *Z. abyssinicus*; separated from it by broad eye-ring, black lores, darker green upperparts and green flanks. Combination of eye-ring width, underpart colour and habitat will identify E African white-eyes, but range is also a useful tool; most montane islands and forest patches in Kenya and N Tanzania have only one white-eye species present, and a careful reading of the ranges will indicate which one lives in each.

Voice. Tape-recorded (CHA, KEI). Song of Taita Hills race *sylvanus* starts in undertone (*c.* 2 s) and gets louder, becoming a jumbled cadence of soft, sweet, rather plaintive notes, many down-slurred, 'feeurr', alternating with short 'ti' or 'tee'; song lasts 4–6 s and has same slightly burry quality as Yellow White-eye *Z. senegalensis*. Distinctive foraging calls of this race said to include a clear, slightly querulous but not buzzing 'kweer-a-ree-ree', 'ree-tree, ter-ree-tee' or 'kwera-kwee-kwee-kwee' (Zimmerman *et al.* 1996). Flock of foraging birds also gave a continuous string of little chips and chirps in an undertone; at intervals one or more birds would break into regular 4–5 s song (S. Keith, pers. comm.). Song of race *kikuyuensis* described as 'zhree zhree zhri zhree zhri zhree zhri zhew' (Zimmerman *et al.* 1996); contact calls while foraging a rather plaintive, burry, down-slurred 'wheerr-tu' interspersed with constant, bubbling 'tu-tu-tu ...' notes, also described as 'whii-tu-tu-her-tu' or 'whii-tew' (van Someren 1956). Moving flock gives typical white-eye buzzing or twittering.

General Habits. Inhabits eucalypt and conifer woods in Eritrea, montane forest, forest edges, plantations and bushy gardens in Kenya and Tanzania; in saplings, canopy of small trees in middle strata of forest, creepers and tall undergrowth; likes forest where growth is not too thick. Forages mainly at 1–6 m above ground. In pairs, family parties and small flocks; often in mixed species foraging flocks. Around Nairobi, Kenya, particularly fond of *Vernonia* trees, and sidles down slender pendent twigs to get at small fruits amongst terminal leaflets. Forages by moving actively in little hops and short flights, searching among creepers, tops and lower sides of leaves, poking into crevices and clumps of moss, lichen, leaf debris caught up in branches and curled-up leaves. Shows little fear of man. At night flock roosts in tree, birds sitting in pairs; 2 birds of pair huddle close together to sleep, but pairs tend to be well separated in one part of the tree. Sedentary.

Food. Small berries and invertebrates, especially aphids. Larger fruits include bits of figs and juicy pulp of pawpaws *Carica papaya*. Small spiders and many tiny caterpillars fed to nestlings.

Breeding Habits. Monogamous, solitary nester. Flocks split up into pairs as breeding season approaches. No apparent territoriality, and 2 nests may be close together: a small bamboo clump once contained 2 nests and 3 of Abyssinian White-eye *Z. abyssinicus*. In Nairobi, Kenya, double-brooded in long season (when a third nesting sometimes attempted if first 2 have failed) and some pairs (same as long-season ones?) nest in short season (van Someren 1956). Nest often built in same spot season after season.

NEST: a small deep cup, neat inside but straggly outside, made of moss, beard-lichen, fine bark fibres and spider web, all firmly worked together and bound with more spider web along rim; lined with fine grass- or bark-fibres; outside sometimes decorated with green moss and spider cocoons. Sited in small-leaved bushes and trees, slung between arms of horizontal fork, in angle between an upright stem and a long-stalked leaf, or between stems of a twisting creeper; well-protected from rain and sun by overhanging foliage. SIZE: diam. at rim *c.* 45, int. depth 35.

Nest built by ♂ and ♀, taking 7–10 days; one end of material is worked around arm of twig, then the other end is made fast to opposite arm; long strands of moss laid on the sling, not evenly looped; later, bird works them into a U-shape and then gradually into a bowl.

EGGS: 2–3. Immaculate blue or white. SIZE: (n = ?) *c.* 17–18 × 12.

LAYING DATES: Sudan, Nov–Apr; Ethiopia, Apr–Dec (39 clutches), mainly Apr (18 clutches), Aug–Sept (5) and Nov–Dec (5); E Africa, Region D, Feb–June, mainly (15 out of 21 clutches) Apr–May.

INCUBATION: by ♂ and ♀; sits very tight and will allow branch to be bent down by person inspecting nest, without budging; birds living in gardens and accustomed to people can be stroked whilst incubating.

DEVELOPMENT AND CARE OF YOUNG: in Nairobi, 'the parents only brood the chicks at intervals and at night after the first two days' (van Someren 1956, doubtless meaning intervals by day and brooded at night only for the first 2 nights). One nest in a large-leaved *Datura* was exposed to full sun when the leaves wilted at noon; a parent stood above the 3 chicks rather than sitting on them, and stood on side of nest to cast shadow on chicks. Parents very attentive, bringing food every few min from surrounding vegetation and seldom going far from nest. Parent searches leaves very thoroughly, and as soon as 1 or 2 items are found parent brings them directly to nest, not creeping to it in a roundabout way. Young fed entirely on small, soft invertebrates, never on berries. Nestling period: usually 15 days. At first young can fly only very short distance; they lie up in leafy vegetation and wait for parents to feed them. At night parents sleep with fledglings, huddling close, one on either side of them.

BREEDING SUCCESS/SURVIVAL: few clutches successful – shrikes and tree-rats take a heavy toll of eggs and young; also parasitized by Green-backed Honeybird *Prodotiscus zambesiae*, which replaces up to 2 white-eye eggs one at a time with its own (Nairobi: van Someren 1956). White-eyes are principal host of Green-backed Honeybird.

References
van Someren, V.G.L. (1956).
Zimmerman, D.A. *et al.* (1996).

Plate 17 (Opp. p. 270) *Zosterops pallidus* Swainson. Cape White-eye. Zostérops du Cap.

Zosterops pallida Swainson, 1838. Anim. Menag., p. 295; east of Prieska, Cape Province.

Range and Status. Endemic resident, Namibia, South Africa, Lesotho, Swaziland, and SE Botswana and extreme S Mozambique (locally common in Lebombo Mts: Namaacha, Mt Meponduine and Goba frontier). Very common in South Africa, almost throughout. Locally common in S Namibia, scarce and irregular in much of N. In Botswana very common in Gaborone, common in Lobatse and Kanye, elsewhere sparse and uncommon.

Description. *Z. p. virens* Sundevall: E Cape Prov. from near East London to Natal, E Orange Free State and SE Transvaal; Swaziland and SW Mozambique. ADULT ♂: upperparts olive-green, usually with narrow yellow forehead band. Tail blackish brown, feathers with narrow green outer edges. Narrow white eye-ring (c. 1 mm wide); blackish line across lore, broadening around lower front margin of eye-ring; yellow supraloral stripe continuous with yellow forehead; sides of head and neck green, cheeks streaked with yellow. Chin and throat yellow; upper breast, sides of breast and flanks olive-green; centre of breast, belly, thighs and undertail-coverts yellow, tinged green. Remiges blackish brown, primaries edged green (emarginations narrowly edged pale brown), secondaries and tertials more broadly edged green; alula blackish brown; primary coverts dark brown, fringed green; rest of upperwing-coverts olive-green. Axillaries pale yellow; underwing-coverts and inner borders of underside of flight feathers whitish. Bill blackish brown, greyer at base of lower mandible; eyes dark brown or greyish brown; tarsus brownish grey or blue-grey. Sexes alike. SIZE (10 ♂♂, 10 ♀♀): wing, ♂ 61–66 (63.2), ♀ 60–64 (62.0); tail, ♂ 45–50 (47.6), ♀ 43–48 (46.5); bill, ♂ 13–15 (13.7), ♀ 12–14 (13.5); tarsus, ♂ 17–19 (18.0), ♀ 17–19 (17.8). WEIGHT: (all subspecies together) (n = 1516, unsexed) 8–20 (11.1), ♂ (n = 20) 7.7–14.6 (10.3), ♀ (n = 31) 8.4–14.0 (11.8) (Maclean 1993).

IMMATURE: juvenile like adult, but duller; white eye-ring develops at about 5 weeks.

NESTLING: hatches naked and blind; skin bright orange, gape orange, bill pale orange, mouth pink, 3 bifid plumules above each eye; remiges emerge from day 3.

Z. p. caniviridis Clancey: Transvaal highveld to SE and E Botswana. Like *virens*, but green upperparts lighter, less saturated; underparts yellower, with olive-green more restricted on sides of breast and flanks. WEIGHT: (n = 22, unsexed, Heidelberg, S Transvaal) 10–13 (11.5).

Z. p. atmorii Sharpe: inland S Cape Prov., from about Sutherland and Calvinia in Karoo east to high interior of E Cape, Transkei, Lesotho and adjacent Orange Free State and W Natal; also coastal E Cape near Port Elizabeth. Head, upperparts and eye-ring similar to *virens*, but yellow restricted below to chin, throat and undertail-coverts, with upper breast, sides and flanks buffish grey and centre of lower breast and belly whitish.

Z. p. capensis. Sundevall: W Cape Province from S Little Namaqualand to Cape Town, east through mountains to Gamtoos River. Like *atmorii*, but upperparts darker and more olive, underparts darker grey; usually lacks yellow forehead and supraloral line.

Z. p. pallidus Swainson: S Namibia from Damaraland highlands and Namib edge to Orange R., and Cape Prov. north of previous 2 races, east to Kenhardt District and Asbestos Mts. Differs from all preceding races in having sides of breast and flanks tawny-buff, centre of breast to belly buffish white and a narrow olive-grey band across upper breast; pale yellow restricted below to chin, throat and undertail-coverts; olive-green upperparts much paler and duller than in *virens*, edges of wing feathers paler and

Zosterops pallidus

yellower; yellow supraloral stripe extends over base of upper mandible. Smaller: wing, ♂ (n = 10) 54–61 (57.2).

Z. p. sundevalli Hartlaub: catchment of Upper Orange and Vaal rivers and tributaries in NE Cape Province, Orange Free State and SW Transvaal. Like *pallidus* but slightly larger: wing, ♂ (n = 5) 58–62 (60.0); upperparts rather greener (like *caniviridis*). Weight: (n = 18, unsexed, Kimberley) 8.4–10.5 (9.3).

TAXONOMIC NOTE: yellow-and-green-bellied *Z. virens* ('Green White-eye'), grey-bellied *Z. capensis* ('Cape White-eye') and white-and-buff-bellied *Z. pallidus* ('Pale White-eye') were formerly treated as separate species. In W Cape Prov. *capensis* and *pallidus* meet and behave as good species, but interbreeding occurs between *Z. p. [capensis] atmorii* and *Z. p. [virens] virens* in upland W Natal and SE Orange Free State (Moreau 1957); and a form first described as *Z. vaalensis*, from SW Transvaal and W Orange Free State, has been shown to be an intergrade between *Z. p. [virens] caniviridis* and *Z. pallidus sundevalli* (Clancey 1966). We follow Clancey (1980b) in regarding the 3 colour types as racial groupings within one polytypic species, *Z. pallidus*.

Field Characters. Length 10–13 cm. Plumage very variable. Most races darker and greener above than Yellow White-eye *Z. senegalensis*; below, throat and undertail-coverts always yellow, breast and belly white, grey, green or yellowish. Nominate race, which just meets Yellow White-eye in Namibia, has white belly and tawny-buff flanks; birds in Zululand (race *virens*) have olive-green upperparts and yellow underparts washed olive-green on breast, belly and flanks. Race *caniviridis* in E Botswana and NE Transvaal most similar to Yellow White-eye, with pale green back and yellow underparts with olive-green restricted to sides of breast and flanks; best identified by green (rather than mainly yellow) head, also by greyish green vent (Sinclair *et al.* 1993).

Voice. Tape-recorded (11, 22, 42, 58, 75, 88, 91–99, B, F, LEM, STAN). Rambling daytime song begins with short, wispy phrase of sibilant twittering, often overlooked because it is so subdued (Skead 1967), continues with louder but still low-key song or subsong, a disjointed series of chips, chirps and whistles reminiscent of Garden Warbler *Sylvia borin* or Blackcap *S. atricapilla*, including imitations of other species, and ends by breaking into a 'regular' white-eye song of sweet, slightly burry warbles and whistles. Songs vary in length from 7 to 40 s; one singing bout lasted over 5 min; species imitated include robins, bulbuls, canaries, sunbirds, glossy starlings, Eastern Black-headed Oriole *Oriolus larvatus*, Fork-tailed Drongo *Dicrurus adsimilis*, Greater Striped Swallow *Hirundo cucullata* and even the harder notes of Crowned Plover *Vanellus coronatus* (Skead 1967). 'Regular' white-eye song often given alone, and is very similar to Yellow White-eye in tone and timbre but generally slower in pace, though this is variable, some being about the same speed. One song with a robin-like whistle in the middle lasted 6 s and could be rendered 'chup, chipway, chipway-cheeo; wur-lee, cheeu-chu-cheeu-chu-chayee-chuchoo-cheeo'. Dawn song said to be similar to daytime song but shorter, 6–10 s, and lacking the subdued twitterings at the start (Skead 1967). Contact calls in flock include short 'pee', longer, down-slurred 'peeeu', double 'peeu-pew', up-slurred 'tu-tsee' and a hard 'pik' or 'tip'. Excitement call used by birds flying in follow-my-leader fashion from bush to bush, a trilling 'krrreee'; variant perhaps used by 2 birds together in excitable mood includes a tripping stutter, 'k, r, r, r, r, eeee' (Skead 1967); alarm a harder and louder version of the same 'krrreee', slower and rather less tinny than similar call of Yellow White-eye.

Voice of 'Orange River White-eye' (races *pallidus* and *sundevalli*), considered by Skead (1967) to constitute a different species, said to differ from other races, being more subdued or muted and a little deeper in tone, and according to one observer more staccato. The rolling notes are less variable, 'pee, tee, trr-per tee, ti-per, tee, ti-per, tee, trr, perr tee ti' or 'pitee, tee, tee ...', each phrase more stereotyped and repetitive, also softer and huskier, without the ringing quality of other races, also less protracted. Song phrases begin with a quiet stammering opening gambit before breaking into full sound. Phrases last for 7–10 s, sometimes up to 30 s, with 2–5 s pause between each. Contact call a quiet 'kree', 'kree-kree' or 'kirree', lower pitched than other races; birds flying from one bush to another give an unmusical 'chrrt, chrrt, chrrt' interspersed with spluttering notes.

General Habits. In SW Cape, occurs in open, hilly country with scattered stands of trees, on wooded mountain slopes, and in bushy growth, wooded gardens and borders of dense wattle groves on Cape Flats. In Orange R. and Vaal R. systems inhabits large trees and riverine thorn scrub, and favours native and alien willows *Salix capensis* and *S. babylonica*; gardens, parks and tree-lined streets. In E Cape, tall forest, bushy kloofs, thick or sometimes thin foliage in canopy of coastal and montane evergreen forest, riverine scrub, acacia veld and drier scrub-veld. Likes flower gardens with hedges, ornamental and fruit trees. Occurs in dune scrub, protea and heath macchia, woodland, acacia thickets, plantations of eucalypts, oak, pines, poplars and wattles. Ranges from sea level up to 2770 m.

Arboreal, in pairs and small, straggling parties year-round; usually keeps within screen of dense foliage; agile, restless, rather elusive but can be confiding. Foraging flock will approach observer quite closely, but retreats to other side of tree if itself approached by observer. Outside breeding season in flocks of up to 100. Flight between trees quite rapid, straight and level, with bursts of flaps but without undulation.

In winter, pairs waken well before dawn, may stay huddled together for 15 min; bird rouses itself, stretches wings and legs, then, still in darkness, flies to place up to 400 m away where pairs and single birds arrive from all sides to assemble; a particular tall tree such as a *Eucalyptus* is used as an assembly point, sometimes for only a week before a new tree is chosen. When a large flock has assembled it moves away, still well before dawn, to start foraging in surrounding vegetation; as day develops, single large flock breaks up into several small flocks which forage independently of each other, and by late afternoon most (not all) small flocks have disintegrated into dispersed pairs. Small flocks wander widely, with composition constantly changing as pairs or odd birds leave or join from other parties. When one tree has been worked over, a bird flies to next tree and remainder soon straggle after it one or two at a time. From time to time during day, flock settles to rest in tree, some birds dozing; bird preens itself, or 2 often huddle side-by-side on thin branch and preen each other in turn. First, one bird sits with legs bent and only feet visible, eyes closed, body hunched and plumage ruffled; other bird nibbles its crown and around eyes, ears, chin and neck (and occasionally other parts of body) (**A**); the bird being preened, with eyes closed, turns its head as if to assist in the nibbling process, which it soon reciprocates.

Periods of foraging alternate with spells of social excitement: birds flick wings and tails, several make

A

dashing flights between bushes and chase each other making small clicking noises (with wings?), and flock moves through foliage for a few moments without feeding. 2 birds scrap by fluttering together in mid-air, slowly falling earthwards but soon separating. Pairs and parties coalesce into large flock at sunset, from which pairs disperse to roosting sites. At roost, 2 birds sleep sitting in same direction, on small branch in tree, with heads tucked into their backs, plumage fluffed out, and bodies so close together that they can look like a single bird. There are no particular pre-roosting vocalizations.

Highly vocal. In shrub, flock gives quiet contact calls, changing to loud ones when one bird leaves to fly to adjacent tree, followed by rest of party generally one at a time. Some birds sing regularly before dawn, starting about 45 min before sunrise. Song is given from well up in tall tree; singers 100–300 m apart; even if both birds of a pair are in a tree, only one sings. Dawn singing, from Sept to Apr, lasts *c.* 20 min. No bird has a fixed song post. Birds sing when the need arises wherever they happen to be, and instead of rising to a prominent perch they keep under cover. When singing at their best they are often on the move and foraging. Once heard mimicking calls of African Paradise-Flycatcher *Terpsiphone viridis* and Cape Bulbul *Pycnonotus capensis*.

Bird in party forages by systematically exploring vegetation, probing into bark crevices and bases of flower and leaf buds, peering under leaves, twigs and branches, often hanging upside down from both feet or sometimes only one foot (**B**); takes insects by gleaning twigs and under leaves; makes brief flight out from cover in pursuit of flying insect, and momentarily hovers or flutters to peck object out of curled leaf of spider's web; hovering accompanied by quiet clicking sounds which may not be vocal. From time to time forages down to within 2 m of ground or even lower in brushwood, and occasionally comes to ground to take an insect or to retrieve a dropped item of food, or to take flesh and juices from e.g. fallen oranges; also to eat sand as gastroliths. Known to enter orange hollowed out by other birds. Pulls small, unripe figs from their stems, and pecks pieces from larger, ripe figs by pulling and twisting with bill. Visits flower gardens and soft fruit orchards; punctures corolla tube of *Watsonia* flower and makes long slits in tubular corolla of *Tecomaria* flower, in order to reach nectar. Feeds on small soft fruits by pecking and taking small amount of flesh from each of many fruits in turn; since the fruits, still ripening on the tree, are then prone to rot, Cape White-eye has been widely regarded as a pest; one fruit farmer claimed to have shot 500 birds with an air-gun in one season (Skead 1967). Also feeds at sticky exudates in bark of oaks and some fruit trees, and searches out small berry-like fruits which can be consumed whole. Eats surprisingly large insects and feeds them to nestlings. Removes eggs from spider cocoons by probing. Nibbles on some (not all) large insects to soften them before ingestion.

Bathes gregariously in bird bath, at edge of pool, in wet foliage and at garden tap. Flock first comes down to low bushes surrounding water, then a few birds at a time fly to water to drink and plunge into shallows, shaking bodies and flapping wings. Up to 35 bathe together, but usually much fewer, especially if waterside space is limited. Hangs upside down under tap, and pecks at drips. Plunges heavily into dew- and rain-soaked leaves, then flies to adjacent perch to preen. Bird once seen in relaxed stance, keeping still for 2 min with wings spread over leaves by its perch, may have been sunning. Flock readily mobs owl disturbed during daytime, but birds otherwise unaggressive. One of the few African birds known to use anting behaviour (Calahan 1981): after wet weather, seizes stingless but poisonous Pugnacious Ant *Anoplolepis custodiens* in bill and strokes it along underside of flight feathers; a captive bird 'anted' both wings alternately, first 3 weeks after fledging, vigorously from 5 weeks, using mealworms *Tenebrio molitor* given to it to eat, and later stingless but poisonous ants *Camponotus rufoglaucus* and especially *A. custodiens*; it once 'anted' with small stink bug (Pentatomidae), then ate it.

Annual moult in Feb–Mar; about May, feeds on nectar of aloes in Grahamstown Botanical Gardens and forehead feathers become matted with nectar and pollen; forehead feathers are then moulted (Craig and Hulley 1996).

Generally sedentary, but variation in reporting rates suggest immigration into arid interior of Karoo and Namaqualand during austral winter and some emigration in spring. Ringing shows that some birds move up to 164 km (Harrison *et al.* 1997).

Food. Nectar, fruit pulp, small soft or hard fruits eaten entire, exudates, fleshy petals and sepals; aphids, aphid honeydew, scale insects, larger insects, spiders and their eggs, and sugar and jam from garden tables. Nectar of *Aloe ferox*, *A. pluridens*, *Royena pubescens*, *Scutia myrtina*, *Watsonia*, *Tecomaria*, *Poinsettia*, *Erythrina*, *Callistemon* and grevilleas. Takes nectar mainly during early winter (Grahamstown). Captives prefer sucrose to glucose or fructose solutions and reject xylose solutions (Franke *et al.* 1998). Fruits of fig trees *Ficus capensis*, *F. petersii* and *F. pretoriae*, and of

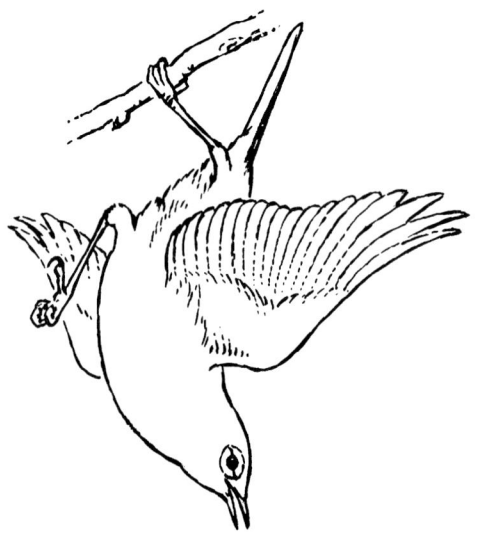

B

Kiggelaria africana, Olea capensis, Cotoneaster, Pyracantha, Lantana, Schinus molle, apricots, peaches, mulberries, grapes, plums, pears, blackberries and commercial figs. Fond of hard, dry *Rhus* berries. Visits citrus orchards; cannot break through skin of oranges but feeds on juices in any natural cracks in fruit and in rotting ones on ground. Pecks holes in *Olea capensis* fruits which are only 8 mm in diam., and swallows whole the rather smaller fruits of pepper tree *S. molle* (one bird ate 13 fruits).

Eats large numbers of aphids, systematically gleaning rose bushes, peach trees, broad beans and other plants susceptible to aphid infestation. Takes scale insects from twigs of oak trees. Makes short hawking flights to catch flying termites. Seizes moths, katydids, mantises, stick-insects up to *c.* 50 mm long, caterpillars, beetles, and spiders as encountered.

Breeding Habits. Solitary nester. Monogamous? Appears to be completely devoid of territoriality; does not have fixed song posts. No courtship behaviour observed.

NEST: a small, neat, delicate, deep cup, made of fine tendrils, pieces of lichen *Usnea barbata*, long, thin, pliable plant stems (e.g. of *Asparagus*), a few grass seed-heads and fibrous, chaffed grass mixed with some plant down and spider cocoon silk; thinly lined with plant down and one or two small feathers, sometimes with a little springy material like horsehair wound around on top; or made of thin rootlets and stems (e.g. of *Clematis*) and rather solidly lined with woolly, white vegetable down; bound together and to twigs with copious amounts of spider web. Some nests decorated outside with spider cocoons or dense masses of old spider silk. Nest delicate-looking but durable. One nest was made of pieces of string and thread with cobweb, and plant-down lining topped with coir from a doormat. Sited mainly in small shrub (*Erica, Metalasia, Salvia, Randia*), sometimes only 60 cm from ground, usually at 1–6 m, occasionally up to 9 m high. Always well concealed in foliage. Size: ext. width 65–70, ext. depth 50–63, int. width 37–45, int. depth 41–51.

Nest slung by rim between horizontal arms of forked twig or built into upright or horizontal fork of twig, or in dense, twiggy growth, wedged in quite tightly and half-suspended by short loops of material at the rim. Well hidden in foliage, towards end of branch of a tree, or in a creeper; favourite site is end of branch of leafy sapling with thin branches, in shade of larger tree; at height above ground of 1.6–5.5 m but seldom lower than 2.5 m; commonly sited in small evergreens, also in deciduous trees in hillside ravines and along mountain streams. One nest was sited in hanging spray of branches a metre above soil in a gulley.

Built by both sexes, one rather more than the other, sometimes using material from an earlier nest. One nest built in 5.5–6 days, another in about 9–11 days.

EGGS: 2–4, usually 2 in W Cape and 3 in E Cape; in Cape Peninsula, 50 C/2 and 27 C/3. Once, 4 eggs. Laid before 08h00 on consecutive days. Quite glossy; immaculate white (*Z. p. capensis*), pale blue or green-blue (other races). SIZE: (n = 140) 14.6–19.1 × 10.9–13.6 (16.8 × 12.3).

LAYING DATES: Botswana, Oct–Nov and Mar; SW Cape (Cape Town), mainly Sept–Dec, Cape Prov. generally, Aug–Apr; Natal, July–Mar; Transvaal, June–Feb.

INCUBATION: begins when clutch complete; by both sexes, sitting very tightly. Bird flushed from eggs disappears into foliage, reappears giving alarm call which summons mate, and the 2 may sit together and displacement-allopreen. Nest seldom left unattended; change-over simple: bird arrives, incubating one leaves, and mate hops into nest. Incubation period: 10 days and 17 h ± 5.5 h; measurements of 11–12 days are thought to be unnaturally high (Skead 1967).

DEVELOPMENT AND CARE OF YOUNG: clutch hatches over period of about 12 h. Quills of wing coverts and alula emerge on day 4, and of humerals, back, rump, flanks, breast, belly and crurals on day 5. Eyes open on days 5–7. Quills of frons, chin and ear-coverts emerge, and remex quills start to break open, on day 6. Most other quills are opening on day 8.

Young brooded and fed by both parents. Brooding almost continuous for first 4 days, thereafter wanes, but young still brooded up to time they leave nest. Parent broods for 1–24 min, spells broken by periods of 1–14 min when parent comes off nest but stays nearby and returns for a few very brief spells. If one bird is brooding and mate arrives with food, it leaves, then mate leaves, so nest unattended. Young fed on insects, small fruits and pieces taken from larger fruits. Av. of 16 feeding visits per h at nest in Cape Peninsula, not changing as brood grew, though feeds became more clumped, with longer intervals between bouts. Nestling period 12–13 days. Nestlings very prone to leave nests prematurely. When not disturbed, they leave nest and scramble into surrounding branches, where they bunch together waiting silently to be fed, and give hunger-calls, quiver wings and beg open-mouthed when a parent approaches. Between feeds they may allopreen. Parents forage together. Young able to fly at 12 days, with records at 10 and (once) 8 days.

Young birds out of nest ball close together on a perch, and when parent appears with food they break apart, fly to parent, perch, quiver wings and utter weak, stuttering hunger call; after being fed, they huddle into ball again (Skead 1967).

BREEDING SUCCESS/SURVIVAL: only 50–57% of nestlings survive to fledging. Young eaten by Fiscal Shrikes *Lanius collaris*, Southern Boubous *Laniarius ferrugineus*, and boomslangs *Dispholidus typus*. Adults found entangled in spider web, and impaled on cactus thorns. Greatest elapsed time between ringing and recovery 95 months. Average adult annual mortality rate estimated at 35% (Prŷs-Jones 1985).

References
Clancey, P.A. (1966).
Harrison, J.A. *et al.* (1997).
Maclean, G.L. (1993).
Skead, C.J. (1967).
Skead, C.J. and Ranger, G.A. (1958).
Vincent, A.W. (1949).

Genus *Speirops* Reichenbach

Mainly montane white-eyes or zosteropids like *Zosterops* in gross morphology and biology, but substantially larger, bill relatively stouter, bill and legs whitish in some spp., plumage grey-brown, lacking yellow or green, with tendency to dark cap and to reduction of white eye-ring.

Restricted to Gulf of Guinea islands except Pagulu (Annobon) and to Mt Cameroon. The 2 most distant populations, on São Tomé and Mt Cameroon, are the most alike and are generally treated as conspecific. Following describer Gray (1862) and Wolters (1979–1980), and swayed by the strikingly different head-on appearances shown below, we treat these 2 as separate species in a superspecies whilst recognizing that their voices are very similar and mutual affinities very close. The 2 intervening species, on Príncipe and Bioko (Fernando Po) are dissimilar. *Speirops* may have originated on São Tomé, and subsequently colonized the other 3 areas; because suitable habitat on Príncipe and Bioko must always have been extremely circumscribed, populations there are small, allowing faster evolution than in the large-habitat, numerous populations on São Tomé and Mt Cameroon, a theory that explains why the superspecies is divided geographically by the species of Príncipe and Bioko (Jensen and Stuart 1986).

Endemic; 4 species – a superspecies and 2 independent species.

Speirops melanocephalus superspecies

Plate 17
(Opp. p. 270)

Speirops lugubris (Hartlaub). Black-capped Speirops. Zostérops de São Tomé.

Zosterops lugubris Hartlaub, 1848. Rev. Zool., p. 109; São Tomé.

Forms a superspecies with *S. melanocephalus*.

Range and Status. Endemic resident, São Tomé. Common to abundant throughout island wherever there are tall trees; perhaps more numerous in S than N and at high than low altitudes. Formerly occurred on the 5 km² Ilhéu das Rolas but evidently now extinct there.

Description. ADULT ♂: thin buffy line behind nostrils from culmen to lores, lores and 2 mm wide ring of feathers around eye white (**A**), forehead, crown and nape matt black, hindneck dusky olivaceous grey, sides of neck olive-grey, mantle, scapulars, upperwing-coverts, back, rump and uppertail-coverts dark greyish olive. Tail blackish, feathers with olive fringes to bases of outer vanes. Chin and upper throat olivaceous grey with indistinct whitish spots, lower throat to flanks and belly greyish or buffy olive, thighs whitish, undertail-coverts dull yellowish olive. Primaries dark brown narrowly edged pale olive, secondaries and tertials dark brown, fringed olive, underside of remiges shiny grey, shafts white, underwing-coverts cream-white. Bill yellow-brown, culmen darker near tip; eyes brown or pale red-brown; legs and feet flesh. Sexes alike. SIZE (5 ♂♂, 2 ♀♀): wing, ♂ 73–79 (76·4), ♀ 70·5, 78; tail, ♂ 44–49 (46·6), ♀ 45, 46·5; bill to feathers, ♂ 12·5–13·5 (13·0), ♀ 11, 12; bill to skull, ♂ 16–17 (16·4), ♀ 15, 16·5; tarsus, ♂ 22·5–24 (23·2), ♀ 22·5, 22·5.
IMMATURE and NESTLING: not known.

Field Characters. Length 13·5–15 cm. A large white-eye with *black cap* and broad white eye-ring; olive-brown above, grey-brown below, belly and undertail-coverts washed cinnamon. Overall appearance resembles ♂ Blackcap *Sylvia atricapilla*. From below, *pure white* underwing and concealed *white breast patch* conspicuous in flight; this is the only noticeable feature on what can appear, in poor light, as a uniformly dark bird (Snow 1950). The only other white-eye on São Tomé is the Príncipe White-eye *Zosterops ficedulinus*, which is small, green above and pale yellow below, with black on lores and under eye.

Voice. Tape-recorded (104, B, ALEX, CHA, GUL, JO PJ, MAC, TYE). Territorial ♂ has beautiful whistled song,

A

remarkably powerful for the size of the bird, harmonious, musical and lively, reminiscent of a sylviid (Christy and Clarke 1998). Contact calls, given almost continuously, a short, dry trill, without the whinnying or tinny quality of similar calls of *Zosterops* spp.; sharp 'tyip', 'chap' or 'tyi-tyip', and low twittering. Other calls: high-pitched but soft 'tuc, tuc-tuc-tuc, tuc, tuc-tuc-tuc', often given in flight by birds following each other; also in flight a dry rattle, 'tieuktieuktieuktieuktieuk'; alarm a nervous 'trrrrriiiirrr' (Christy and Clarke 1998).

General Habits. Occurs wherever there are tall trees: humid primary forest, patches of dry forest in savanna, around lakes and along watercourses, shady understorey of cocoa and coffee plantations and forest regrowth at all altitudes. In pairs and (usually) parties of up to 10–12 or occasionally 25 birds; forages in mixed flocks with Príncipe White-eye *Zosterops ficedulinus*, Giant Weaver *Ploceus grandis*, São Tomé Paradise-Flycatcher *Terpsiphone atrochalybeia*, Newton's Sunbird *Anabathmis newtonii* and São Tomé Oriole *Oriolus crassirostris*. Often raises crown feathers. Not shy. Keeps in contact with rest of flock by frequent or almost constant quiet calling; call given very often in flight; imitates other birds including São Tomé Seedeater *Serinus rufubrunnea*, São Tomé Oriole and Emerald Cuckoo *Chrysococcyx cupreus*; often gives loud song, though song not long sustained from one place. Gleans leaves and twigs for insects, taking caterpillars from twigs and from under leaves, feeding mainly in lower strata at 5–10 m above ground; sometimes comes down to within 1 m of soil. Inspects leaf bases, buds and flowers. Perches on petioles of large leaves, inspecting their undersides and reaching from them to inspect inflorescences. Searches for insects also amongst herbaceous plants, on walls of old buildings, and in dense bushes (Christy and Clarke 1998). Eats small fruits entire and pecks small pieces off large fruits, e.g. avocados. Party moves from tree to tree in leisurely, haphazard manner. Sedentary.

Food. Small berries including those of *Cestrum levigatum* and other vegetable matter (including avocado fruits, and probably seeds); insects including caterpillars. Thought to take nectar from *Erythrina* flowers (Christy and Clarke 1998).

Breeding Habits.
NEST: deep, open cup, slung between arms of forked twig, fixed firmly at points of suspension; made of fine vegetable fibres or of twigs of even thickness, giving nest uniform appearance; delicate, though quite thick-walled. One nest 4 m up in lower branches of *Cinchona* tree, concealed amongst large leaves (Christy and Clarke 1998).
EGGS: 2–3. White, immaculate or very finely speckled with grey. SIZE: (n = 2) 20 × 13, 18 × 14.
LAYING DATES: Apr–June, (nest building from late Dec to mid Jan).
INCUBATION: when a bird is incubating it sits so deep in nest that only its tail and wing tips can be seen.

References
Atkinson, P. *et al.* (1991).
Christy, P. and Clarke, W.V. (1998).
de Naurois, R. (1994).
Snow, D.W. (1950).

Speirops melanocephalus (Gray). Mount Cameroon Speirops. Zostérops du mont Cameroun.

Plate 17
(Opp. p. 270)

Zosterops melanocephalus Gray, 1862. Ann. Mag. Nat. Hist. (3) 10, p. 444; Cameroon Mt.

Forms a superspecies with *S. lugubris*.

Range and Status. Endemic resident, restricted to Mt Cameroon, where common between 1820 and 3000 m; about 15 birds a day can generally be seen.

Description. ADULT ♂: forehead and lores greyish white, crown, nape, hindneck, cheeks, ear-coverts and sides of neck very dark brown, slightly glossy, grading into the colours behind; mantle, scapulars, back and uppertail-coverts grey-brown, rump slightly greyer. Tail dark brown above, blackish below. Chin and upper throat greyish white, a narrow patch defined by dark cheeks and malar area (**A**); lower throat merging into buffy grey breast; belly pale buffy grey, flanks buff, thighs whitish, vent and undertail-coverts pale grey. Remiges dark brown, all except outer primaries fringed pale grey, whitish, or olivaceous grey-brown, tertials widely so. Upperwing-coverts grey-brown, underside of remiges dark grey, underwing-coverts and axillaries silky white. Bill white, sometimes tinged pink near tip, or yellowish; eyes clear pale brown, creamier towards edge; or yellow or brownish white or grey; legs and feet dead white, flesh, creamy white, pink-white or pale greyish flesh. Sexes alike. SIZE (5 ♂♂, 5 ♀♀): wing, ♂ 60–64.5 (62.1), ♀ 60.5–65 (62.7); tail, ♂ 38.5–43 (42.0), ♀ 38–40.5 (40.1); bill to feathers, ♂ 11.5–12.5 (12.0), ♀ 10.5–12 (11.7); bill to skull, ♂ 15–16 (15.4), ♀ 13.5–15 (14.2); tarsus, ♂ 21–22 (21.5), ♀ 20–23 (20.8); also, averages of 8 unsexed: wing 62.5, tail 44.7, bill 11.7 (Eisentraut 1968). WEIGHT: (n = 8, unsexed) 9–12.5 (10.9).
IMMATURE and NESTLING: not described.

Field Characters. Length 13 cm. Endemic to Mt Cameroon. Similar to Príncipe White-eye *S. lugubris*, with general resemblance to ♂ Blackcap *Sylvia atricapilla*, but blackish cap has brown tinge, eye-ring very narrow, white

B

line over base of lower mandible, brown upperparts lack olive tone, underparts paler, chin and centre of throat white. Underwing pale but not conspicuously white. Strong, dagger-like white or very pale bill, white frontal patch and pale legs are excellent field marks (Serle 1954, Grimes 1971); black cap contrasts strongly with pale grey face. Only other white-eye on Mt Cameroon is green-and-yellow Yellow White-eye *Zosterops senegalensis*.

Voice. Tape-recorded (104, CHA). Song remarkably powerful for the size of the bird (Chappuis, in press), a lilting phrase of staccato, up-down, tuneless notes lasting 2–2·5 s, 'chet-chi-chop-po, chet-chi-chop-po, chi-chu-chu', 'chet-chi-chop-po, chet-chi-chop-po, po-chi-chop', 'chet-chi-chop-po, chet-chi-chop-po, po-chi-chu-chi-chop'. Staccato quality of song very similar to that of Príncipe Speirops *S. leucophaeus*. The song has also been described as 'a series of 7–8 rich, sweet notes, first rising and then falling in pitch, sounding very like a Blackcap' *Sylvia atricapilla* (Serle 1954), but this does not fit at all with the song recorded by Chappuis and transcribed above, which is neither rich nor sweet, so perhaps there are 2 song-types. A 'rattling call' described by Grimes (1971), and Serle (1954) heard a flock giving a 'chorus of little calls very much like those of the Yellow White-eye'.

General Habits. Inhabits open parts of forest; at lower altitudinal limits occurs only in clearings, and at highest limits feeds in scattered trees and shrubs in windswept grassland; avoids dense areas of closed canopy; often at edge of forest where it borders grassland; in copses, shrubby gullies, and isolated stands of trees and bushes in thickets. Keeps to middle strata and canopy; active and sprightly. Occurs singly, in pairs and parties of 5–15 birds, often in mixed foraging flocks with White-bellied Crested Flycatchers *Elminia albiventris*, Northern Double-collared Sunbirds *Cinnyris reichenowi* and Yellow White-eyes *Zosterops senegalensis*. Investigates mossy holes and branches, pecks at leaves and twigs, and tears to pieces fresh flowers and buds, sometimes hanging upside down. Flocks sometimes silent but tend to call in chorus.

Resident with some vertical movement; in Dec seen between 1850 and 2800 m, in Jan seen rather less commonly at 1950–2150 m (Stuart and Jensen 1986).

Food. Insects, berries and fragments of vegetable matter.

Breeding Habits.
LAYING DATES: (gonads enlarged in Nov–Dec and Mar; cloaca enlarged Dec). Nothing further known.

References
Eisentraut, M. (1968).
Stuart, S.N. and Jensen, F.P. (1986).

Plate 17 (Opp. p. 270) *Speirops leucophaeus* (Hartlaub). Príncipe Speirops. Zostérops de Príncipe.

Parina leucophaea Hartlaub, 1857. Syst. Orn. Westafr., p. 71; Príncipe.

Range and Status. Endemic resident, Príncipe Island. Frequent to common but somewhat local; described as abundant at all altitudes in 1970s (de Naurois 1983) and may have declined since then (Jones and Tye 1988); perhaps vulnerable to plantation development and use of pesticides (Atkinson *et al.* 1991).

Description. ADULT ♂ : white above nostrils; forehead, crown and nape pale grey, even paler at sides, forming poorly defined superciliary stripe; sides of neck white, hindneck and mantle grey, scapulars, upperwing-coverts, back, rump and uppertail-coverts dark grey. Tail dark brown. Lores grey, cheeks and ear-coverts whitish. Chin and throat greyish white, breast and flanks soft pale grey, belly grey, patchily white or creamy in centre, thighs greyish white, undertail-coverts pale grey. Primaries dark brown, secondaries and tertials dark brown with narrow olive-grey fringes to outer vanes; underside of remiges shiny dark grey, feathers with whitish inner edges, underwing-coverts greyish white. Upper mandible dark grey, lower mandible whitish; eyes golden brown; legs and feet pearl grey, soles yellow. Sexes alike. SIZE (4 ♂♂, 4 ♀♀): wing, ♂ 67–75 (71·8), ♀ 68–70 (69·2); tail ♂

Speirops leucophaeus

44–48 (46·1), ♀ 45–48 (46·7); bill to skull, ♂ 13·5–15·5 (14·5), ♀ 13–15 (14·2); bill to feathers, ♂ and ♀ 10–11; tarsus, ♂ 19·5–21 (20·4), ♀ 19·5–20·5 (20·0).

IMMATURE and NESTLING: unknown.

Field Characters. Length 12·5–14 cm. A pale bird, with *whitish* face and throat (**A**), grey crown, greyish white underparts, *white thighs, undertail-coverts* and *flank patch*; black lores, very narrow white eye-ring. Unmistakable and common, with warbler-like actions. Only other white-eye on Príncipe is Príncipe White-eye *Zosterops ficedulinus*, which is smaller, green above and pale yellow below, with conspicuous white eye-ring and black lores

Voice. Tape-recorded (104, B, C, JO, PJ, MAC, TYE). Song (or call?) a repeated phrase of hard, rattling, tuneless notes, 'chee-putu-tsi-chee-putu-tsi-chee', totally unlike typical *Zosterops*. Contact calls from feeding flock a long trill, 'trrrrrrrr-ruuuuu' or 'trrrriiiii', a high-pitched, dry, fast 'tric-tric-tric' or 'truc-truc-truc' and a continuous 'tictic-tictictictic...'; flight call a rapid 'tuctuctuctuctuctuc'; the true song is poorly known (Christy and Clarke 1998). Also gives brief, dry single notes, 'chip', 'tsit' and 'titit'.

General Habits. Commonest in forest regrowth, trees and bushes in farmland, and shady cocoa and coffee plantations under the shade of large *Erythrina* trees; occurs in primary forest, where keeps mainly to middle strata, but also in canopy and undergrowth; often *c.* 8 m above ground. Lives in pairs, family parties, and groups of up to *c.* 15 birds; members of flock keep rather close and move off together. Restless and mobile; moves amongst twigs and foliage in very lively manner, gleaning leaves, inspecting their undersides, moving acrobatically and sometimes hanging upside down, much like tit *Parus*, constantly raising crown feathers a little and flicking wings together then closing them tightly. Pokes about in clusters of curled, dead leaves; makes small leaps, stopping suddenly to search undersides of branches; puts head into large flowers, where thought to take nectar (Christy and Clarke 1998).

Food. Insects, spiders, berries, seeds and other vegetable matter.

Breeding Habits. Poorly known.
NEST: delicate openly-woven cup slung between 2 twigs; one made of fine grasses with some fine twigs or petioles, another of dry twigs and moss; attached to twigs with moth-cocoon silk. One sited 3 m above ground under large leaves in *Acalypha* shrub in garden hedge. Size (n = 1); diam. 90, depth 60.
EGGS: 2; white. SIZE: (n = 1) 19 × 16.
LAYING DATES: Sept, (said to hatch mainly in June–July; nest-building Jan; plumage freshly moulted in Feb–Mar).

References
Atkinson, P. *et al.* (1991).
Christy, P. and Clarke, W.V. (1998).
Bannerman, D.A. (1914, 1948).
Snow, D.W. (1950).

Speirops brunneus Salvadori. Fernando Po Speirops. Zostérops de Fernando Po.

Speirops brunnea Salvadori, 1903. Boll. Mus. Torino 18, no. 442, p. 1; Fernando Po.

Plate 17
(Opp. p. 270)

Range and Status. Endemic resident, Mt Malabo (Pico de Santa Isabel), Bioko, at 1900–2800 m; seems to be absent from mountains in S of island although there is a small area of suitable lichen-forest habitat there. Quite common on Mt Malabo, where lichen forest is only about 50 km^2 in extent, and the commonest bird in its habitat there (T.M. Butynski and S.H. Koster, pers. comm.). 14 birds collected at 2100 m in 1966. Collar and Stuart (1985) gave it 'Rare' status and pointed out that a survey is needed to assess status and determine where conservation action is needed to protect this, the only endemic bird species on Bioko.

Description. ADULT ♂: forehead and lores dark brown, narrow line at base of upper mandible paler, crown blackish brown, nape and hindneck reddish brown, cheeks and ear-coverts greyish brown, mantle, scapulars, back and uppertail-coverts brown. Tail dark brown above, blackish below. Chin and upper throat greyish brown; lower throat merging into brown breast, slightly paler than back; rest of underparts brown, a little paler on belly and undertail-coverts. Remiges blackish brown, edged rusty. Upperwing-coverts brown, underside of remiges dark grey-brown. Bill said to be dark horn; eye colour uncertain; legs and feet dusky brown. Sexes alike. SIZE: (n = 14, 11 ♂♂ and 3 ♀♀ combined) wing 63–69 (65), tail 51–54·5 (52·3), bill to feathers 13–14·5 (13·8). WEIGHT: (n = 14, 11 ♂♂ and 3 ♀♀ combined) 14–18·5 (16·1) (Eisentraut 1968).

Speirops brunneus

IMMATURE and NESTLING: not known.

Field Characters. Length 13–13.5 cm. Endemic to Bioko. Uniform brown, with fulvous rump and underparts; cap darker, no white eye-ring; reminiscent of Brown Parisoma *Parisoma lugens* of eastern Africa. Only other white-eye on the island is green-and-yellow Yellow White-eye *Zosterops senegalensis*.

Voice. Not tape-recorded. Said to give frequent soft 'peep's and rapid 2–5 note twitters while foraging and squeaky 'tweet's in flight and when foraging; also a loud 'chirp-chirp'; excitedly responds to person 'pishing' by giving very rapid trill of *c.* 1 s duration (T. M. Butynski and S. H. Koster, pers. comm.).

General Habits. Inhabits fairly open lichen forest (not moss forest: Wells 1968, Collar and Stuart 1985), montane heathland scrub and tree savanna, and forest with *Syzygium, Schefflera, Hypericum* and *Pittosporum*; keeps to open areas and edges of clearings, preferring to keep out of sunshine, in moderate leafy cover. In groups of 3–30 birds, foraging 0.5–6.5 m above ground; very confiding; foraging behaviour much like *Zosterops* spp. Spends much time feeding in *Pittosporum* and *Hypericum*; commonly associates with Green Longtails *Urolais epichlora*; also with Oriole-Finches *Linurgus olivaceus* and Mountain Greenbuls *Andropadus tephrolaemus* (T.M. Butynski and S.H. Koster, pers. comm.).

Food. Insects including 10–25 mm long hairless caterpillars, and berries.

Breeding Habits. Birds in breeding condition in Oct and Dec; nothing further known.

References
Collar, N.J. and Stuart, S.N. (1985).
Eisentraut, M. (1973).
Pérez del Val, J. *et al.* (1994).
Wells, D.R. (1968).

Family PROMEROPIDAE: sugarbirds

A family with a single genus of 2 species of obscure affinities, with lives centred on *Protea* shrubs and the exploitation of their nectar. Somewhat resemble huge sunbirds: same proportions, long, decurved bill, and rather spindly black legs; and territorial, courtship and dispersive behaviours quite like those of the sympatric *Protea*-specializing sunbird *Anthobaphes violacea*. However, lack spectral colouration, and tail long and steeply graduated, with central 3 pairs of rectrices of ♂♂ long and slender (**C**; very long in *P. cafer*). Plumage warm brown and drab greyish white, with olivaceous rump and yellow undertail-coverts; slight sexual colour dimorphism. Tongue long, protrusible, with brush tip, not bifurcate, distally almost tubular, proximally trough-shaped in section. Maxilla with sub-apical hairline groove; maxilla and mandible without serrated edges. External nares elongate, operculate, not feathered. Forehead and forecrown feathers lanceolate and quite stiff. Eyes rather small. Wing-tip somewhat rounded; 10 primaries, in ♂♂ inner vanes of P6 (*P. gurneyi*) and P5–7 (*P. cafer*, **A**) with large bulge or lobe, responsible for 'frrrt-frrrt-frrrt' sounds in aerial displays (De Swardt 1992). Tibia long (**B**), tarsometatarsus short, hind toe with claw shorter than tarsus. Tail with 12 feathers, central ones often disarrayed by breeze and fluttery in flight.

Nectarivorous – feed at blossoms by perching and probing; do not regularly hover or probe into tubular corollas; also glean foliage for insects and hawk for them; voice tinny, twangy, creaky, clicking and sibilant; song prolonged, given from top and interior of bush; territorial, less aggressive than e.g. most sunbirds; territories can be so tight-packed as to compose a colony; several territorial and dispersive behaviours much like sunbirds'; remarkable aerial tail-whipping, wing-fripping display; daily regime of active foraging alternating with long spells of resting deep in vegetation. Nest a quite neat, open, twiggy and fibrous cup, with 2 linings, one of fluffy *Protea* seed hairs, one of wiry stems. 2 eggs, buffy, dark-blotched and scrawled. Hatchling downy. Only ♀ builds nest and incubates (with unexplained aperiodic nocturnal absences); ♂ takes lesser part than ♀ in rearing chicks.

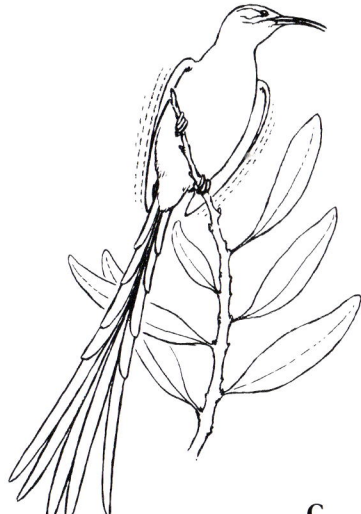

Sugarbirds are generally placed in their own family but, on the basis of one or another cluster of anatomical or biochemical characters, have also been placed in starlings Sturnidae (Sibley and Ahlquist 1974), thrushes Turdidae (Olson and Ames 1984), honey-eaters Meliphagidae (Bock 1985), and sunbirds Nectariniidae. DNA–DNA hybridization data strongly suggest that *Promerops* is a nectariniid (Sibley and Ahlquist 1985), and Sibley and Monroe (1990) place it in that family as subfamily Promeropinae. Some additional evidence for affinity with meliphagids, an Australasian family, is in skeletal character states (Farquhar *et al.* 1996). Sugarbirds, sunbirds, honey-eaters and some starlings seem to have co-evolved with flowering woody plants, in respect of nectar and pollination biology, and are morphologically convergent, often strongly so, which obscures their phylogeny and mutual affinities.

Endemic to southern Africa; a single superspecies with 2 species, one in Cape Prov., other from Natal to E Zimbabwe.

Genus *Promerops* Brisson

Characters those of the family.

Promerops cafer superspecies

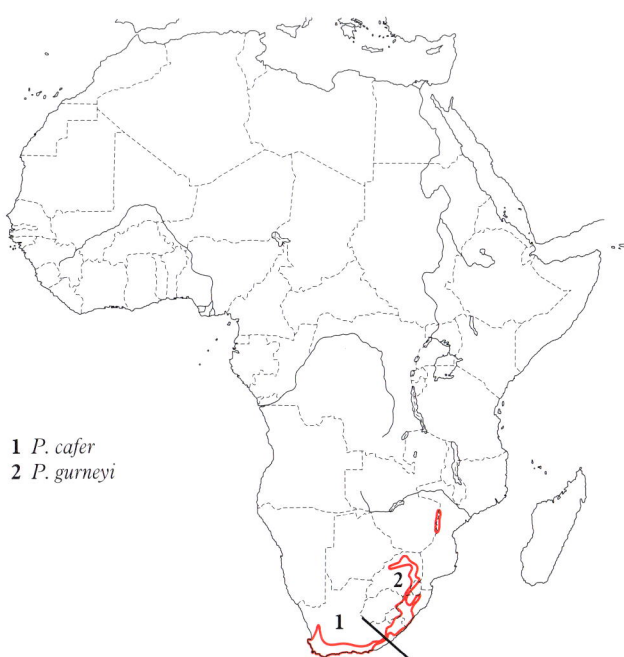

1 *P. cafer*
2 *P. gurneyi*

Plate 20
(Opp. p. 287)

Promerops cafer (Linnaeus). Cape Sugarbird. Promérops du Cap.

Merops cafer Linnaeus, 1758. Syst. Nat., ed. 10, p. 117; Cape of Good Hope.

Forms a superspecies with *P. gurneyi*.

Promerops cafer

Range and Status. Endemic resident, South Africa. Saldanha, Cape Peninsula and Cape Agulhas north to Cedarberg and between Vanrhynsdorp and Calvinia and east, between coast and Witteberge, Groot Swartberge and Grootrivierhoogte ranges, to Port Elizabeth, Grahamstown, King William's Town and Buffalo R. near East London. Record from Jagersfontein, Orange Free State (De Swardt and Buys 1992). Throughout range of *c.* 800 km from W to E Cape, almost entirely dependent on fynbos *Protea* spp. Much protea veld has been replaced by agriculture, stock-keeping, housing, industrial and other developments, or displaced by frequent scrub burning or invading aliens such as the proteaceous Australian shrub *Hakea gibbosa* (Skead 1967), and Cape Sugarbirds are bound to have been adversely affected thereby in recent decades. However, present range (Harrison *et al.* 1997) broadly like that 30 years ago. Still occurred in 1960s in greatly disturbed strip of macchia with *Protea repens* and *P. cynaroides* on crests of Suurberg range near Grahamstown, E Cape, where proteas were abundant a century ago; and may still hang on further east in Hogsback and Amatole Mts where natural plant communities now almost completely extinguished by fire and cattle. Remains common, though increasingly localized with densities of up to 20 pairs per km^2 (i.e. about 50 pairs in c. 2·6 km^2 in van Staden's Pass Wild Flower Res., Port Elizabeth); 'hundreds' of birds above Hout Bay, Cape Peninsula (Skead 1967); much less common northeast of Port Elizabeth, where bushveld and forest take over from western macchia, in Suurberg, Winterberg and Amatole Mts to Stutterheim; however, density of about 22 birds in 4 acres (i.e. 14 birds per ha) of cultivated proteas, Grahamstown, and 4 breeding pairs (with 2 of Gurney's Sugarbird *P. gurneyi*) in 4 ha (i.e. 3 sugarbirds per ha) in Pirie Mts. Density of only 4 birds per 100 ha of strandveld, Olifantsbos (Fraser and McMahon 1992b). Pressures on *Protea* fynbos and Cape Sugarbirds continue to increase: destruction of coastal lowlands by holiday-home development, agriculture, agroforestry, water-impoundment, spreading of alien plants, and too-frequent burning (Martin and Mortimer 1991, M.W. Fraser *in* Harrison *et al.* 1997).

Description. ADULT ♂ (breeding): forehead straw-coloured, feathers lanceolate, short, with yellowish shafts and dusky barbs; crown blackish brown and buffy, feathers lanceolate and short, with buff edges; nape blackish brown, hindneck, sides of neck and mantle brown with blackish smudges, back olive with diffuse blackish streaks, rump and uppertail-coverts olive-green. Tail blackish distally, greyer proximally, T1–T3 very long and ribbonlike, T5 sometimes with small whitish spot at tip. Lores, cheeks and ear-coverts olivaceous brown, moustachial stripe buff, malar stripe black. Chin and throat buff, in narrow area enclosed by malar stripe, breast warm rufescent brown dappled with more-or-less crescentic buff feather tips, upper belly white or very pale buff with warm brown streaks in centre and broad, dark brown or blackish streaks at sides, lower belly whitish, flanks buff, heavily but diffusely streaked dark brown, thighs warm brown, vent and hind part of flanks bright yellow, undertail-coverts pale yellow with dusky centres. Remiges dark brown, upperwing-coverts dark brown narrowly bordered with olive, underside of remiges shiny greyish brown, underwing-coverts dark brown. Bill black; eyes very dark hazel-brown or wine red-brown – almost black; legs and feet black. Partial albino once seen (Broekhuysen 1971). Sexes alike but ♀ shorter-tailed. SIZE: wing, ♂ (n = 67) 86–101 (93·5), ♀ (n = 46) 77–84 (82·7); tail, ♂ (n = 59) 121–360 (259), ♀ (n = 40) 85–188 (125); bill, ♂ (n = 52) 28·3–34·7 (31·4), ♀ (n = 40) 28·0–32·2 (30·1); tarsus, unsexed (n = 11) 21–24 (Maclean 1993). WEIGHT: ♂ (n = 57) 30·5–43·5 (37·5), ♀ (n = 46) 26–39 (32·2).

IMMATURE: juvenile soon after fledging has dark grey back (grey-brown in adult); breast band absent; crown darker than in adult. Immature like adult but yellow on vent region paler and smaller in extent; streaks on flanks smaller and less bold, or even lacking; brown breast and pale brown belly not sharply demarcated; base of bill pinkish.

NESTLING: hatchling blind, with skin pale pink, gapes pale yellow, bill almost straight, paler yellow than gapes, with blackish tip and white egg tooth, mouth pale pink (becoming brighter pink later), legs very pale pink, claws yellow (later becoming grey); body covered with long, pale grey-brown down, on capital, humeral, femoral, alar, posterior dorsal and posterior lateroventral tracts. Later, when wing and trunk feathers growing, eyes dark grey, and down, now relatively short and dense, restricted to cap and 'bottom' (paintings: McMahon and Fraser 1988). Neossoptile down persists on crown until after chick leaves nest. Weight at hatching (n = ?) 2·9–4·8 (3·6) g. Feathered nestling: see below.

Field Characters. Length, ♂ 37–44 cm, ♀ 24–29 cm. *Very long, floppy brown tail*, streaked plumage and *yellow vent* of this protea endemic prevent confusion with any sunbird; can only be mistaken for Gurney's Sugarbird *P. gurneyi* which it overlaps in Pirie and Hogsback Mts, E Cape (see below). Distinguished from Gurney's Sugarbird by indistinct dull brown *breast band*, variable in colour intensity and sometimes with rufous tinge but never the dark, rich chestnut-red of Gurney's, and by *pale creamy* forehead and forecrown; dark malar stripe more pronounced than Gurney's. Tail of breeding ♂ always far longer than Gurney's, but in ♀ and moulting ♂ it is about the same length. Immature has short tail, brown breast and pale brown belly not sharply demarcated, and lacks yellow under tail and streaks on flanks.

Voice. Tape-recorded (20, 88, 91–99, B, BBC, F, PAY, WALK). Song, used to advertise territory, a jumble of harsh, grating, scratchy notes with liquid ones, with frequent interjection of hard 'chit': 'tschaak-tschayli-chitchit', 'tchit-tschaluwit-tscheeluwo-chu', 'tschaak-tscha-witchi-chut'; also give a longer rambling medley of the 'tschaak', 'cheeliwee', 'chit', 'chup' and other notes, some of which have been likened to the squeaking of a rusty gate hinge; some notes said to have twangy quality not unlike those of Fork-tailed Drongo *Dicrurus adsimilis*, but this not evident on tapes by Gillard or Gibbon (above). Song types vary considerably; for further transcriptions, see Skead (1967). Threat song similar but more intense and faster. ♀ also sings, sometimes for 3–4 min, a soft sibilant song, shorter than ♂'s (Skead 1967). Calls variable, including a loud, tinny 'tcheenk, tcheenk', and a rapid, repeated 'skwidge, skwidge' or 'skeedge, skeedge', into which are introduced from time to time a 'sit-wotty-geenkle' or 'tcheekarik, rik karik' ending with a slurred, drongo- or starling-like nasal sizzling (Skead 1967). During breeding season both sexes sit giving long series of 'clack-clack-clack' or 'tsit-tsit-tsit' notes, which may be preceded by a run of excitable, more musical notes, 'tirry, tirry, tirry ...'; at dusk ♂♂ chatter to each other with gabbling-chippering calls from exposed protea flowers (Skead 1967). Mild alarm, high-pitched 'tweet-tweet-tweet', delivered with wings flicking excitedly, becoming faster and more explosive as anxiety increases; even more intense is a single harsh, wheezing note and a 'cloth-tearing' sibilant 'ssssssrrrr' or 'sssrrr, sssrrr, sssrrr' (Skead 1967).

General Habits. Endemic to fynbos or macchia: mainly narrow-leaved shrubs, aromatic, burn-resistant, thick, low, growth largely of heaths Ericaceae, 'rushes' Restionaceae and, especially, woody shrubs Proteaceae. The lives of sugarbirds revolve around proteas, which provide nectar and insect food, nest sites and materials, and shelter; breeding season is closely correlated with winter flowering, and sugarbirds are major protea pollinators (Burger *et al.* 1976, Collins 1983a, Mostert *et al.* 1980).

Inhabits protea hillsides, and flat ground where proteas may be only subdominant. Penetrates otherwise alien habitats, like gardens, Restionaceous Tussock Marsh in Cape Peninsula, Strandveld (to feed on sisal *Agave sisalana*: Richardson 1990), and Karroid Brokenveld and Valley Bushveld northwest of Grahamstown, where small patches of sugarbush/suikerbos *Protea repens* (= *P. mellifera*) occur on rock outcrops. Known or presumed to forage at large variety of proteas, including the many mentioned in this account. Outside breeding season feeds at aloes, agaves and *Eucalyptus* (see Food). Around Tsitsikamma, S Cape, lives in very dense, 3–3.5 m tall *Berzelia-Protea* community. Near Port Elizabeth mainly in dense growth of *P. neriifolia*. In Collingham Towers district (E Cape) confined to thinly-scattered *P. cynaroides* on S-facing slopes of escarpment. At 1200–1400 m in Amatole Mts near Alice, Mt Kempt near King William's Town and Dohne Peak near Stutterheim, occurs in very dense growth of *P. lacticolor*, *Cliffortia*, *Metalasia* and *Halleria lucida*. Occurs in parks, botanical gardens and large rural gardens and nurseries. Roams widely, and out of breeding season re-appears at a particular locality from time to time, staying for only a few days.

At E extremity of range, in Hogsback hills above Keiskamma valley and in Pirie Mts above King William's Town, E Cape, overlaps tiny population of Gurney's Sugarbirds (or used to: much *Protea lactifolia* forest habitat there destroyed by fire in 1962), and evidently interspecifically territorial (Skead 1964).

Solitary or in pairs or family parties; 'solitary' birds may in fact have unseen mate nearby. Usually conspicuous, but in non-breeding season easily overlooked. Disperses locally after breeding, often away from *Protea* locations and from lowlands into hills, and becomes more dependent for nectar on agaves, aloes, spider-gums, other eucalypts etc., and may then occur in mixed-sex parties of 4–12.

Forages at 'flowers' (showy, composite inflorescences) of proteas by perching on top, generally facing into any breeze so that tail is not blown awry, or straddling small inflorescence, leaning down to distance that accords with location of nectaries, and sucking nectar or probing between stamens or bracts. Perches on *P. multibracteata* flower and probes far side of mass of stamens (**A**). *P.*

A

lepidocarpodendron has more cylindrical flower head with densely-packed florets; sugarbird perches on top, leans far down far side and probes through bracts from the outside (**B**). Perhaps a sexual difference, ♂♂ tending to probe florets of *P. lepidocarpodendron* from within the bracts and ♀♀ from outside them (Seiler and Robello 1987). *P. neriifolia* has long, densely-packed florets, and bird's head, reaching down between outer florets and enclosing bracts (**C**) creates visible bulge (**C**, arrow). Front half of bird disappears from view, reaching into deep cup of sugarbush *P. repens* (**D**). Often flutters wings to balance when standing on flower. Hawks flying insects by dashing up from bush top to snatch one out of air, diving steeply back to perch. Probes clusters of young leaves at shoot tips of proteas, evidently for small insects. Larger insect is held in tip of bill and beaten against woody perch.

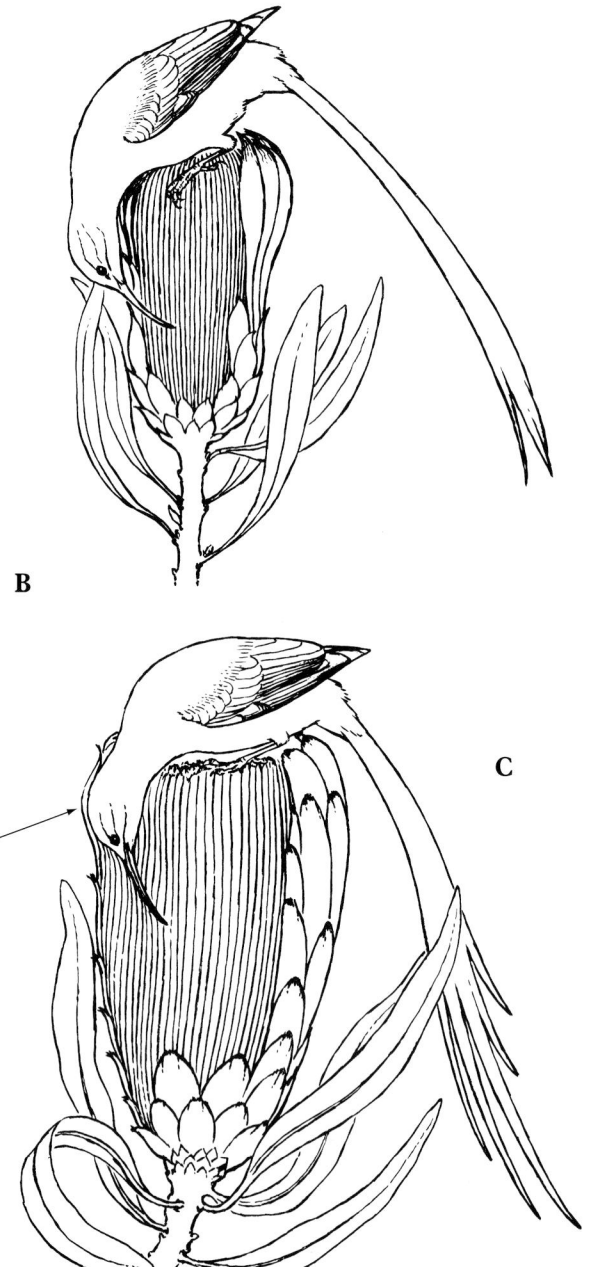

Active year-round, ♂♂ often chasing each other and ♀♀ occasionally being chased away. ♂ gapes widely if another sugarbird approaches too close, and utters 'rusty-hinge' notes breaking into chittering. ♂ sings jangling, drongo-like territorial advertisement song at one or a few favoured bush-top vantage points, mainly in early morning and late afternoon; he sings also, often for half-hour or more, from within depths of protea bush, and sings on moonlit nights in breeding season, several ♂♂ counter-singing and almost chorusing (Skead 1967). ♀ occasionally subsings, and ♂♂ threaten each other by using advertisement song at increased intensity.

♂ spends much time perched conspicuously on bush-top, with long tail often blown sideways or up over his back. ♂'s tail incoherent – long feathers hang separately, and bird can look scissor-tailed (**E**). Also spends much time within thick foliage of protea bushes, bird diving down from bushtop perch, and later shooting up out of thickest foliage and back to perch on top, with speed and precision and a little upward flip-jump both on leaving and returning to bush top. Day characteristically punctuated with irregular, alternating periods of active feeding, socializing and loafing, involving many birds synchronously. Lively and active early in morning, for 2 h, then suddenly all birds become silent and disappear to rest or loaf. Rests deep in protea bush, for 5 min or much longer, dozing and preening. Wipes bill on perch; scratches head with foot indirectly. Bill-wiping and head-scratching both thought to be used as displacement activities in certain social interactions. Relaxes with head sunk into shoulders and bill held a little upward (**F**); when alert, head held higher (**E**). Expels faeces in strong liquid jet. Flight strong, direct, without undulation and without apparent slowing or braking before alighting. Bathes in dew- or rain-wetted foliage, often of proteas, by making series of plunges onto leaves, repeatedly dipping head into the wet; flutters wings and splashes plumage. Breeding ♂♂ gather to mob a ground predator, e.g. mongoose *Herpestes pulverulenta*, hovering and calling 2 m above it. Breeding pair mobs person, flying at him and veering aside at last moment. ♂♂ and ♀♀ dive into vegetation if sparrowhawk passes, e.g. *Accipiter rufiventris*, and remain there for up to 30 min.

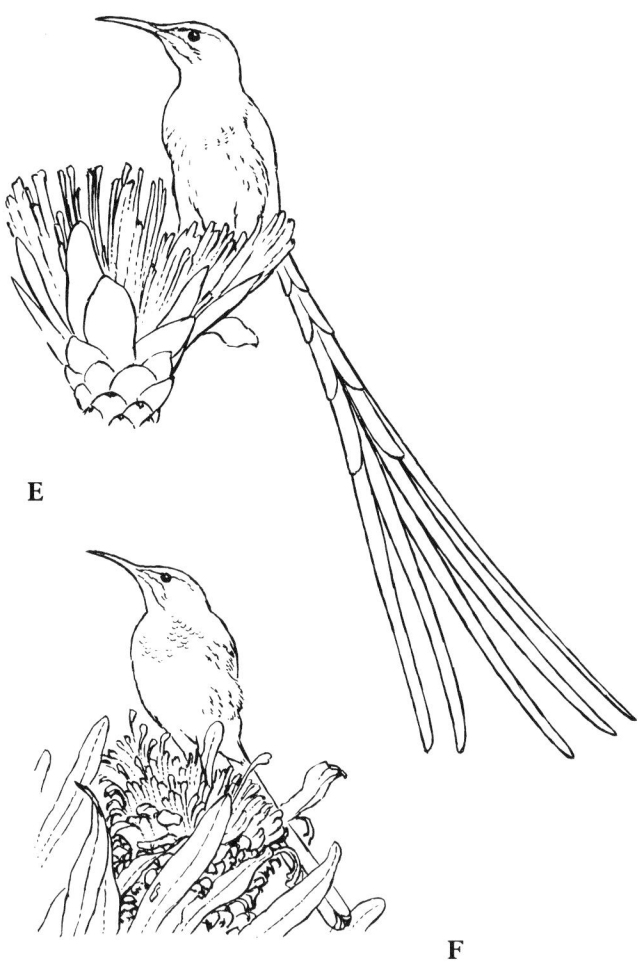

Sleeps at night in lowermost branches of protea thicket, descending at dusk after a short period of noisy socializing, birds perching conspicuously, calling and chasing. In some localities roosts communally, in flock of several hundred birds (Steyn 1996).

Resident, but sedentary (pairs present all year) hardly anywhere – only in a few large gardens and commercial *Protea* nurseries with enough species of proteas and leucospermums to supply birds' needs all year; rather, a regular short-distance migrant, moving in response to flowering of main food plants. Tracks flowering of *Leucospermum conocarpodendron* on coastal plains in Sept–Mar and of *Protea repens* on hillsides in May–Aug (Harrison *et al.* 1997); 10 birds moved 27–44 km and one 160 km (Fraser *et al.* 1989, Fraser and McMahon 1992a, Oschadleus and Fraser 1988). Population arrives locally at much the same time each year (though at different times, one or a few weeks apart, at different localities only some 20 km distant from each other: e.g. earlier on University of Cape Town campus than in Kirstenbosch Botanic Gardens or Hout Bay). At several localities in SW Cape first arrivals in late Feb/early Mar, most birds appearing from mid Mar to late Apr until a breeding population of several hundred occupies the protea slopes (Skead 1967); breeds; then population declines in July–Aug to about 10% in Aug and none in Nov. Arrives at one particular patch of *Protea repens* in 1st week Apr and at one of *P. lepidocarpodendron*/ *P. incompta* when not yet in flower, in June, ♂♂ taking up same station, singing even from same twigs, as in previous year (A. Handel-Hamer *in* Skead 1967). Also moves in response to fire, avoiding burnt vegetation until it has regrown (M.W. Fraser *in* Harrison 1997). Perhaps less migratory in east than in west of range: some present all year in Amatole Mts and around Grahamstown; various proteas flower for longer in summer-rainfall E Cape than in winter-rainfall areas further west.

Food. Nectar and small insects and spiders. Feeds at, and often takes nectar copiously from, flowers of *Protea repens*, *P. lacticolor*, *P. multibracteata*, *P. lepidocarpodendron*, *P. neriifolia*, *P. barbigera*, *P. cynaroides*, *P. incompta*, *P. longifolia*, *Leucospermum conocarpum*, *Mimetes hartogii*, *M. hirtus* (which it pollinates – Collins 1983b), *Erica* spp., *Watsonia tubularis*, *Aloe* spp., agaves including sisal *A. sisalana*, *Tecomaria*, *Kniphofia*, *Eucalyptus lehmannii* and *Halleria lucida*. Commonly takes arthropods from protea leaf-clusters and florets (flowers) and other flowers, some species of which may be visited more for their insect than their nectar content (Skead 1967). Exceptionally, feeds at *Oldenburgia grandis* (Rebelo 1987). Animal food not studied but thought to be substantial and said to include spiders and small beetles. Insects brought to young in nest mainly cockroaches, also many hymenopterans or flies, and some moths, beetles and neuropterans; also a few spiders.

Breeding Habits. Loosely colonial, monogamous. Strongly territorial, ♂ establishing and defending breeding territory by singing, chasing, and performing a visually-compelling aerial advertisement display. Most small birds (robins, thrushes, weavers, white-eyes, waxbills) up to size of Laughing Dove *Streptopelia senegalensis* tolerated, but Sugarbird reacts to e.g. Fiscal Shrike *Lanius collaris* and Tropical Boubou *Laniarius aethiopicus* by giving territorial call and attempting to drive it off. ♂♂ territorially active soon after arrival on breeding grounds (i.e. territorial activity starts in Mar near Cape Town), and display until season wanes (late July near Cape Town). One ♂ appeared in 3 successive years in same small protea patch and sang from same vantage points. Territory size varies greatly, from *c*. 0·2–0·4 ha in sparsely-populated areas down to perhaps only 75 m^2 where population dense (a Cape Town protea-and-heath garden of 120 × 30 m supported one breeding pair in some years, 2 in others). ♂ advertises and patrols large territory and pair does not leave it often nor for long periods nor do they go far away; ♂ always returns immediately an intruder appears. He flies directly at intruder which is usually dislodged; if not the territorial ♂ perches near it singing excitedly and flicking wings jerkily.

In dense 'colonies', nests may be only 10 m apart and density of birds may promote social assemblies. Sometimes several ♂♂ and a few ♀♀ gather on one or a few bushes, ignoring established territorial boundaries. Both sexes have social display (function?), perching with wings held out vertically at sides of body, quivering. Both sexes, but mainly ♂, also have flight version: in flight towards song post or opponent, bird suddenly checks, draws body into almost vertical position, and hovers for a few moments on rapidly beating wings, uttering 'tsk, tsk, tsk ...' notes. Even in smallest territory any intruder is aggressively chased away, across territorial border into adjacent territories, whose owners may join in until up to 6 ♂♂ are chasing; ♀♀ emerge from their nests to watch from bush tops. Sharp increase in frequency of territorial incursions by strange ♂♂ during time when territorial ♀ thought to be most receptive, at which time her mate guards her (Seiler and Prŷs-Jones 1989). ♂♂ with the longest tails shown to be the most sought after by ♀♀; ♀ mated with long-tailed ♂ more likely to have second brood than ♀ paired with average ♂.

Daily, characteristic lulls in activity are broken by chittering calls deep within protea bushes; then first one ♂ appears from within bush, then others, perching on it and keeping up continuous clacking calls (whose function evidently is to make ♀♀ come from interior of bushes to have a look around from top); then ♂♂, one or more at a time, perform elaborate flight display (**G**) (Skead 1967). ♂ flies up from bushes giving flow of soft, sibilant, rolling 'ssssssrrr' notes; with feathers puffed out, body tensed and moving jerkily in time with flapping, bird makes 'frrrt, frrrt, frrrt ...' sound, and jerks tail up and down faster and faster until the long, pliant tail feathers are lashing to and fro under body and above back. Display flight can be shallow parabola over no great distance, or a pronounced, high curve with bird rising at steep angle to 9–16 m, then giving twangy notes and flying down again fast but on slowing wing beats and with a dipping action, finally dashing head-long back into bushes. When ♂ is displaying at another, which may respond by fluffing out plumage, a third sometimes alights on displayer's perch and will then be driven off in charging flight, with roles reversed and the pursued doing the pursuing out of its own territory. Several adjacent territorial ♂♂ may display at same time, then activity gradually diminishes and all birds retreat silently into the bushes for a spell. Where breeding concentrations are low and territories not adjacent, few aerial displays are performed. 2 territories in Pirie Mts were adjacent to 2 of Gurney's Sugarbirds, but Cape Sugarbirds displayed to Gurney's only half-heartedly on low-looped flights over the proteas or on short horizontal flips (Skead 1967); one ♂, whose mate was collecting nest material in bush, flew vertically up from top of the bush, hovered for 1–2 s with tail whipping up and down, then plunged headlong down with wings half open, body jerking from side to side, tail feathers fluttering audibly, and yellow undertail-coverts flashing.

Copulation thought to occur in concealment. ♂ twice seen attempting to copulate with ♀ on nest, both times driven off by ♀. Habitually double-brooded; one ♀ laid 2 clutches a year for 3 successive years

NEST: fairly deep, open cup, neat inside, untidy outside, solidly constructed of fine, brittle heath and *Cliffortia* twiglets, thin dry plant stems and coarse dead grass mixed with bits of bracken, rootlets, pine needles, woolly plant-down and *Protea* seed down. Thickly lined with coarse but soft and fluffy brown *Protea* seed down, overlain neatly with dark, thin, wiry fibres mainly of *Leptocarpus paniculatus* and *Thamnochortus fruticosus* (Restionaceae), *Serruria* (Proteaceae) and *Helichrysum* (Compositae) (Vincent 1949). Size (n = 77): ext. diam. 100–180 (135), ext. depth 60–190 (110), int. diam. of cup 55–80 (70), depth of cup 40–70 (60); variation in ext. depth arises from nests sited in deep forks needing substantial foundation. Some nests have little or no plant down; a very few are slight, made almost entirely of coarse grass stems and lined with fine grass (Vincent 1949) and may disintegrate before young fledge. Situation is macchia slope covered with

dense growth of proteas and heaths, where nest generally in sheltered hollow or gully with thick undergrowth on sides, or dense mass of *Berzelia lanuginosa*. Nest is occasionally in isolated bush on open, grassy hillside; usually concealed below screen of foliage, sometimes open to the sky (Vincent 1949); sited either in middle of bush or side away from prevailing wind (Steyn 1996). Built in small, thick *Protea* bush in dense undergrowth (57% of 90 nests) or *Erica, Rhus, Metalasia, Cliffortia, Stoebe, Maytenus* (*Gymnosporia*), *Leucodendron argenteum, Acacia cyclops*, bramble, bracken, clump of restoniaceous 'rushes', in weeds on marshy ground, *Watsonia*, dense young pine, or on fork of *B. lanuginosa* or cross-twigs of spiny tree *Hakea gibbosa*. Height above ground 0·25–2·4 m (rarely, higher), 56% at 0·9–1·6 m, av 1·2. Once a nest built on ground.

♀ thought to choose site. Nest built by ♀; ♂ makes infrequent visits whilst nest being built and occasionally brings a few bits of nest material. Construction: ♀ deposits a few dead twigs loosely on chosen site, then finer material. Settling onto it, with head and neck thrust forward and wrists held slightly out from side of breast, she presses nest base down with body, rises, changes her position, and presses down again; she does not use bill to rearrange twigs any further. Lining added in 2 stages. First stage lasts 2–3 days: she collects brown fluffy down from *Protea* seeds and dumps several billfuls on nest, and later compresses them into place with her body by crouching as before. Second, ♀ collects wiry, fibrous materials and makes them into layer covering the down. Building is mainly in 08h00–10h00, and activity is not noticeably diminished in poor, rainy weather. One nest took only 72 h to build and 2 eggs were laid within 5 days of completion. 4 nests took 5–6 days to build and 5 took 7–10 (av. 8·0) days. 2nd-brood nest once started day after brood left 1st nest. Replacement nests can be built rapidly: when 2 broods of young eaten in nests by ants, one ♀ started to build new nest 2 days later and another ♀ had nearly completed her replacement nest after 6 days. Same ♀ may build successive nests in different shrubs. In 9 nests, 1st egg laid 1–5·5 (av. 2·7) days after nest completed.

EGGS: 2–3; nearly always 2; clutches of 3 are rare; 7–8% of nests reportedly contain only 1 egg, but clutches perhaps incomplete. 1st egg laid by 10h00, 2nd egg 24–33 h later (n = 11), once 48 h later. Ground colour cream to pale buff or pinkish white; marked with grey, brown, purple-black or glossy black blotches, scrawls, spots and zigzags, sparingly or liberally over shell or forming ring around wide end. SIZE: (n =100) 21·4–25·0 × 16·5–18·7 (23·4 × 17·5).

LAYING DATES: Cape Peninsula, mid Apr to late May, occasionally Mar and June–Aug; Knysna and Port Elizabeth (fewer data), also Mar–Aug; George, Dec. Analysis of 200 records shows that breeding peak in Cape Peninsula is 1 month later than in Hottentots Holland Mts 50 km away, where dominant protea is early-flowering *P. neriifolia* (absent from Cape Peninsula, where *P. lepidocarpodendron*, with long season, dominates).

INCUBATION: by ♀ only, sitting closely, with long tail over side of nest, starting as soon as 2nd egg laid. ♂ sometimes visits nest but never settles in it. In 4 nests ♀♀ covered eggs for 22–59 (av. 42) min per daytime hour (Broekhuysen 1959). 2nd clutches are incubated less than 1st ones. ♀ incubates more around mid-day than in early mornings and late afternoons, when she tends to leave nest (and territory) to feed. She settles onto nest for the night *c.* 15 min after sunset, and leaves nest *c.* 30 min before sunrise. She also leaves nest for 2–3·5 hours at a time during the night, at any time between 20h00 and 07h00, not every night, but once every few nights, quite irregularly; during her night absences air temperature was 10·5–15·5°C (Broekhuysen 1959). Incubation period 16–18 days, usually 17; once 16 days ± 22 min.

♀ disturbed from nest often perches nearby with head drawn into shoulders, bill horizontal, wings drooping and tail hanging straight down. ♀ returning to nest to find a person near is said to assume a brooding pose on branch, squatting down with breast feathers fluffed out. When perched away from nest with eggs or young, ♀ indicates intention of returning to it by quivering or flapping wings, with neck held obliquely and bill pointing down (**C** on p. 327).

DEVELOPMENT AND CARE OF YOUNG: eggs hatch mainly before 10h00 or in afternoon at 15h00–16h00. Eggshells are carried away and dropped. Eyes open from day 5 to day 8; maxilla darkens at about day 5, and by day 9 bill starts to look curved rather than straight and gapes now very bright yellow; feathers breaking open from day 11; nestling doubles its weight by days 3–4; growth rate affected adversely by poor weather, but rates of 1st and 2nd chicks in a brood are similar. ♀ broods young during day, reducing gradually until 6th day, and broods them at night for 11–16 nights. Young fed by both parents, ♂ starting on 2nd day after eggs hatch. Feeding visit made every 10 min on average (more in mornings and evenings than in middle of day). Most ♂♂ feed young at only one-third of their mate's rate. Food given to very young nestlings thought to be nectar, since adults' tongues visibly work in and out of tip of bill before and after bill is inserted into nestlings' gullets. Insects, mostly found in territory, are brought within a few days. For day or two before they fly from nest, young clamber about on adjacent leaves and twigs, often returning to nest. ♀ hovers over head of person near nest, calling agitatedly, and may then fly to the ground and hop or creep slowly away (Skead 1967). Nestling period (n = 6) 17–21 (19·1) days.

Faecal sacs removed by both parents, at about every 4th visit. Young bird backs up to nest rim and excretes sac without parent prodding it; parent seizes sac as it emerges and flies away with it always to the same bush, outside territory, where parent alights and wipes sac onto bush so that it soon becomes encrusted with dry sacs. (Once both parents used same bush one year, and another bush in next year.)

2 young of a brood may leave nest on same day or a day apart. Young away from nest fed by both parents. Chick begs by calling, holding up bill for food, and slightly quivering wings (not the exaggerated quivering of begging sunbirds, for instance). Young keep to cover of macchia scrub, and still beg for food 18 days after leaving nest (once at time when mother was incubating next clutch; the same young was chased by mother 3 days later but was still present 2 days after that).

Plate 22

Orange-breasted Bush-Shrike (p. 406)
Malaconotus sulfureopectus

Olive Bush-Shrike (p. 402)
Malaconotus olivaceus

Bocage's Bush-Shrike (p. 404)
Malaconotus bocagei

Many-coloured Bush-Shrike (p. 396)
Malaconotus multicolor

Black-fronted Bush-Shrike (p. 400)
Malaconotus nigrifrons

BREEDING SUCCESS/SURVIVAL: of 55 eggs in 28 nests, 14 did not hatch, and of the 41 chicks 23 died and only 18 survived to fly (overall success rate of 33%). Chicks killed by Argentine ants *Iridimyrmex humilis*, house snakes *Lamprophis ornatus*, and rain and cold.

References
Broekhuysen, G.J. (1959).
Fraser, M.W. *in* Harrison, J.A. *et al.* (1997).
Henderson, K. and Cherry, M. (1998).
Seiler, H.W. and Prŷs-Jones, R.P. (1989).
Seiler, H.W. and Rebelo, A.G. (1987).
Skead, C.J. (1964, 1967).

Plate 20 *Promerops gurneyi* Verreaux. Gurney's Sugarbird. Promérops de Gurney.
(Opp. p. 287)

Promerops gurneyi Verreaux, 1871. Proc. Zool. Soc. Lond., p. 135; Natal.

Forms a superspecies with *P. cafer*.

Range and Status. Endemic resident, confined to *Protea* veld, mainly *P. roupelliae*, on mountains from Zimbabwe/Mozambique border to E Cape. Zimbabwe, local and generally uncommon at 1100–2300 m (scarce below 1500 m) in Inyanga Highlands to Stapleford, Vumba, Banti, Melsetter and Chimanimani Mts; not in Chipinga Uplands. Mozambique, mountains of Manica e Sofala, close to Zimbabwe border; probably occurs in headwaters of Pungwe R. (Clancey 1996); not below 1220 m. Transvaal, locally common on Soutpansberg, Blouberg, higher parts of Waterberg, Strydpoortberg and E Highveld; appears to be absent from *Protea caffra* veld near Bryanston. NW Swaziland. Natal, from Drakensberg Range east nearly to Pietermaritzburg, breeding east to Richmond (Nevill 1987), and reaching coast at Cape Prov. border. Orange Free State, scarce to locally common close to NW Natal border. Absent from Lesotho. E Cape Prov., very local southwest to Pirie Mts, where meets and used just to overlap range of Cape Sugarbird *P. cafer*.

Thought to be only a few thousand birds in Zimbabwe (Irwin 1981). Must have been much commoner formerly than at present, since much of its habitat, in regions like Natal where bird still occurs sparsely, has been destroyed for banana plantation or sugercane culture. There is little doubt that the Pondoland coastal strip was once a stronghold; there has probably been a winter reduction in abundance in Giant's Castle and Royal Natal Nat. Parks (Skead 1967). On the other hand, has extended range with the multiplication of *Protea* nurseries (Cyrus and Robson 1980). Mean density of 4·6 birds per ha and maximum of 20 per ha in optimum *Protea roupelliae* woodland near Lydenburg, Transvaal (De Swardt 1993).

Main food plant, *P. roupelliae*, is less fire resistant than most proteas, and is destroyed by too-frequent burning and by a hot, late-summer fire. Range of Gurney's Sugarbird has contracted with destruction of its habitat by burning and its conservation depends on *P. roupelliae* conservation (De Swardt 1993).

At S extremity of range, in Hogsback hills above Keiskamma valley and in Pirie Mts above King William's Town, E Cape, a tiny population occurs with Cape Sugarbirds. 3 breeding pairs of Gurney's Sugarbirds discovered in 1962, in tall, dense *Protea lactifolia* forest, with evidence that a population had been in region 30 and 55 years earlier. Social and ecological interactions of the 2

Promerops gurneyi

species never studied; their Hogsback habitat destroyed by fire in 1964 when a fire-belt was made to protect pine plantations (Skead 1967).

Description. *P. g. gurneyi* Verreaux: South Africa. ADULT ♂ (breeding): forehead and crown rufous-brown, the feathers lanceolate, long, with disarticulated vanes, nape and hindneck grey, feathers lanceolate, each minutely tipped buff, sides of neck grey, mantle grey with diffuse dusky streaks, scapulars, and back olive with diffuse dark streaks, rump and uppertail-coverts bright yellowish olive, unstreaked. Tail very dark grey, strongly graduated, the central feathers ribbonlike, T4–T6 with narrow white outer edge and tip. T2 about 5, T3 30, T4 60, T5 83 and T6 111 shorter than T1. Lores and ear-coverts greyish brown, moustachial stripe greyish white, malar stripe (not always discernible) grey, chin and upper throat white merging to rufous on upper breast; lower breast rufous, the feathers pale tipped; upper belly white with some rufous streaks at front and long, distinct black streaks at sides, lower belly white, flanks white with long black streaks, thighs grey-brown, vent and undertail-coverts bright yellow. Remiges blackish, secondaries and tertials with narrow grey-white edge to outer vanes, greater primary-coverts blackish, with dull olive edges; underwing-coverts blackish. Bill black, mouth dark pink, tongue blue; eyes dark brown, eyelid sooty black; legs and feet black or dark slate, soles light grey. Sexes alike but ♀ a little smaller, and much shorter-tailed. SIZE (62 ♂♂, 80 ♀♀): wing, ♂ 86–101 (94·5), ♀ 79–96 (87·4); tail, ♂ 94–186 (150), ♀ 88–136 (112); bill, ♂ 25·8–31·9 (28·9), ♀ 25·2–30·6 (27·7); tarsus, ♂ 19·4–23·8 (21·7), ♀ 17·0–25·3 (20·8). Also, bill to feathers, ♂ (n = 97) 26·6–35·1 (29·0), ♀ (n = 115), 25·0–30·6 (27·7); bill to skull, ♂ (n = 97) 30·0–38·1 (34·1), ♀ (n = 115) 28·9–35·4 (32·5). WEIGHT: ♂ (n = 62) 30·0–42·8 (32·5) (De Swardt 1992).

IMMATURE: juvenile has yellow gape for several months after fledging; crown and breast brown; undertail-coverts brownish yellow. Immature, breast rufous with greenish tinge; somewhat later crown becomes rufous-brown; secondaries edged brownish; undertail-coverts duller yellow than in adult, with greenish tinge.

NESTLING: hatches blind, pink, with pale grey down on head, back and wings; at day 5 body covered with blackish down and wings very downy; gape bright yellow.

P. g. ardens Friedmann: Zimbabwe and Mozambique. Breast of ♂ considerably brighter and darker than in ♂ of nominate race; cheeks darker; upperparts darker – feather centres blackish; rump and uppertail-coverts greener, less yellowish; flanks streakier; tail darker above and below. Same size as nominate race.

Field Characters. Length, ♂ 25–29 cm, ♀ 23 cm. Most surely distinguished from Cape Sugarbird *P. cafer* by *chestnut-red* forehead and crown (creamy and brown in Cape); also by *solid* chestnut-red breast-band (breast-band usually dull brown and indistinct in Cape but can have rufous tinge). Tail of ♂ about the same length as that of ♀, never greatly elongated as in ♂ Cape. Voice less grating.

Voice. Tape-recorded (88, 91–99, B, F, CHA, GIB, GIL, STJ). Song faster and higher-pitched than Cape Sugarbird, more muted, less harsh and scratchy, but with the same rambling, disjointed, squeaky quality; short, stuttering notes are interspersed with hurried, rather liquid jumbles, churrs and hard chatters, 'chip-wip-woop-woodlyoodly-erchup-chip-ip-trrrrrr-chip-chop-cheedlyeedlyup-wip-chip-pleetututu-wup-chip-chachachachacha ...'. For further transcriptions, see Skead (1967) Song bouts can last 5 min or more, and a full session, with breaks, as long as 20 min; singer may be answered by nearby ♂ or even by Cape Sugarbird in areas of overlap (Skead 1967). Some notes said to sound like Fork-tailed Drongo (not evident on tapes by Gillard and Gibbon, above), and one ♂ sang 'a persistent rolling warble irregular in tempo like a Willow Warbler *Phylloscopus trochilus* but husky and sibilant and without any truly musical notes' (Skead 1967). Commonest calls are a single loud 'clack' or brisk 'chit'; these are run into a series when bird is excited, sometimes preceded by burst of 'tirry-tirry-tirry ...', increasing in tempo until it becomes a repetitive 'chit, chit'; this both serves as advertisement and indicates apprehension (Skead 1967). Alarm, a 'cloth-tearing' sound or a harsh, burring 'skirrrrt'; acute alarm, when person near nest, rapidly-stuttered 'tree, tree, tree ...'; when chasing mammal or another bird, 'chiddly, chiddly, chick, chick, chick' (Skead (1967).

General Habits. Inhabits Alpine and Sour Grassland vegetation types, and grassland bordering Afromontane forests. In Zimbabwe, scrub-covered hillsides, particularly on drier western slopes of mountains with abundant flowering shrubs including *Protea*, *Strelitzia* and *Erythrina*; common in *Brachystegia–Protea–Phillipia* woods east of Melsetter, and in Pungwe valley occurs in dense riverine bush surrounded by woody grassland with a few protea bushes. In Drakensberg Mts, Natal, favours tree-proteas, *Protea roupelliae* and *P. multibracteata*, in light forest beside streams and along slopes of steep-sided valleys, and marginal scrub of *Leucosidea sericea*, *Buddleia salviifolia* and *Myrsine africana*, also stands of *Greyia sutherlandii* and *Aloe arborescens* (Skead 1967). When foraging along forest margins, constantly returns to patches of *P. roupelliae* (De Swardt and Louw 1994). In Richmond, Natal, feeds at *Leucospermum*, *Kniphofia* and aloe blossoms. Inhabits *P. lacticolor* near King William's Town, E Cape; probably *P. roupelliae* and *P. gaguedi* (=*P. abyssinica*) in Swaziland, *P. caffra* in Transvaal. Occurs in orchard-like tree communities near Kokstad, E Griqualand, Natal. Enters small-town gardens and eucalypt plantations.

Occurs singly, in pairs, and parties of up to 10. Perches prominently on top of protea, foraging actively at flowers, flicking wings, then makes flip-jump (like a cormorant diving) to dash down into interior of bushes. Later, emerges at speed, flying straight up and out and making flip-turn before alighting back on bush top. Nectar-foraging behaviour appears, from general description, to be very like that of much better-known Cape Sugarbird. Hawks adroitly for flying insects, by flying fast up from bushtop perch, snatching up prey in bill-tip with audible snap, and darting down to perch again, 'making a peculiar rapid somersault back to perch' (Nevill 1987). Also hawks horizontally, for both slow- and fast-flying insects (Skead 1963). Remains hidden and silent deep in thick protea scrub for lengthy spells, then all birds emerge to forage at about same time. Dislikes wind and remains in cover for longer on windy than on calm days. Flight straight and direct or slightly undulating, with quite rapid wing beats, usually at 3–5 m above vegetation. Noisy, restless; said to be shier and less approachable than congener. Territorial; drives other sugarbirds and e.g. weavers *Ploceus* away by flying fast at them on rapidly-beating wings; veers away at last moment from larger bird like Red-winged Starling *Onychognathus morio*; when nesting, tolerates Common Kestrel *Falco tinnunculus* nearby; may tolerate Malachite Sunbird *Nectarinia famosa* feeding on same bush but not at same flower.

Moults primaries descendantly, during breeding season, and secondaries ascendantly (S1–S8) as well as tail without seasonal pattern (De Swardt 1992).

Resident and short-distance, vertical migrant, e.g. in Transvaal, where moves 7–10 km into suburban gardens in Lydenburg in Apr–Sept (De Swardt 1989, 1991); appears to be considerable movement in Natal from Drakensberg Mts to Midlands (Byrne Valley, Seven Oaks, Karkloof, Pine-town) where occurs in Apr–Sept, occasionally later; once Durban, Aug; noticeably commoner in Richmond in winter than summer. Juveniles disperse a few km from parental territory, and whole populations displaced locally by destructive burning (De Swardt 1993).

Food. Nectar, small insects and spiders. Feeds mainly at flowers of proteas, including *P. roupelliae*, *P. lacticolor*, *P. gaguedi* and *P. multibracteata*; likes *Greyia sutherlandii*, *G. radlkoferi*, aloes including *A. arborescens*, *Watsonia*, *Halleria lucida* and *Erythrina lysistemon*; also *Leucospermum spp.*, *Kniphofia*, *Callistemon* and *Eucalyptus*. Animal food includes spiders and small flies; many insects fed to nestlings, including small beetles and quite large mantids.

Breeding Habits. Solitary nester, monogamous, territorial. Nests occur 100–200 m apart (Natal Drakensberg); 2 nests 50 m apart (E Cape, Pirie Mts) in Apr and June were likely to have been 1st (robbed) and replacement nests of same pair. 2 adjacent breeding pairs in Pirie Mts occupied separate favourite areas for at least 6 months, and kept separate from an adjacent pair of Cape Sugarbirds: the 3 ♂♂ of the 2 species sang counter to each other, and the Gurney's ♂♂ sometimes chased Cape pair away from their patch and sometimes tolerated them; all 3 ♂♂ occasionally wandered together, foraging about the protea thickets, or wandered to other proteas 400 m away (Skead 1967). Dashes back and forth above proteas in presumptive territory, often planing for a few m, and disappears into dense cover, but tends to be less active and more restrained than Cape Sugarbird. ♂ of breeding pair studied in Richmond, Natal, spent much time perched on tall tree top above incubating ♀, guarding her, diving at great speed towards and aggressively chasing away other sugarbirds, sunbirds, white-eyes, bulbuls, sparrows, and even Fiscal Shrikes *Lanius collaris* (but ignoring weavers and Common Mynas *Acridotheres tristis*). ♂ chased his ♀ every time she left nest; he also responded vigorously to voice playback.

Courtship display consists mainly of bouts of rapid chasing, up to time when eggs hatch. Chase starts with ♂ flying up into air from his tree-top perch, as if chasing flying insect, then suddenly 'somersaulting' and diving headlong down towards ♀, when dazzling chase ensues with subdued chittering sounds (Nevill 1987). Sometimes flies horizontally for short distances with jerky, wing-chopping action, or with rapid fluttering. Occasionally displays by flying slowly upward, at steep angle or almost vertically, flicking tail up and down and giving soft, husky 'zikky, zikky' call, hangs momentarily at top of parabola *c*. 9 m above bushes, then flutters down at steep angle, singing, ending with quick dive onto an exposed perch (Skead 1967: display seemed to be given mainly in response to ♂ Cape Sugarbird entering territory). Double-brooded (E Cape). ♀ in nest-building period (Richmond, Natal) crouched low on branch in nest tree, with wings half extended and quivering, bird making soft chittering sound when ♂ arrived to feed her; ♂ and ♀ chased each other at speed through nest tree, *Leucodendron argenteum* (Nevill 1987).

NEST: shallow cup made of grasses, herb stems, small twigs and lengths of fibre taken from below *Protea* bark; once with a few soft leaves of *Buddleia salviifolia* amongst the twigs; may be robust, solidly built and tidy, or loose, flimsy and untidy; in either event, lining, of dense brown *Protea* seed fluff, sometimes with fine dry grass, is generally neat, compact, and built up to rim, but sometimes laid thinly only half way to rim. Neater, outside, than nest of *P. cafer*. One nest made of pliable twigs 110–220 mm long of *Cliffortia linearifolia*; the twiglets were bent to the circle of the nest and overlapped each other rather than criss-crossing, sticking to each other by the small, protruding leaf-buds; foundation of nest was slivers of dead *Protea* underbark *c*. 60 mm long; nest lined with single layer of wiry, dry grass stems or rolled-up leaves 150–200 long, like coir (Skead 1963). SIZE: (n=3) ext. diam. 100–120, ext. depth 45–60, int. diam. 50–60, int. depth 40.

Sited generally in *Protea* bush or tree, in thicket of the *Protea* species on which bird also feeds; mainly in *P. roupelliae*, also *P. multibracteata* and *P. lacticolor*, occasionally *Cliffortia linearifolia*; on a 2- or 3-pronged fork, in smaller branches, or on a terminal cluster of twigs. Some nests supported more by surrounding tough, springy foliage than by branches on which built. Height above ground *c*. 1.7 in one study, 3–4 m in another.

Built by ♀ alone, her mate perching nearby; one nest took 9–10 days to complete.

EGGS: 1–2, usually 2, laid on successive days. Ground colour creamy or buffy, marked with brown or deep purple speckles, scrawls and blotches, all over shell or in ring around broad end. SIZE: (n=16) 20.9–23.4 × 16.0–17.5 (22.5 × 16.7).

LAYING DATES: Zimbabwe, July, Oct, Apr; Transvaal, Nov–Feb, mainly (90% of 20 nests) in Dec; Natal, June–Feb. Breeding in Transvaal coincides with flowering of main food plant, *Protea roupelliae*, peaking at Lydenburg in Nov–Jan (De Swardt 1991, De Swardt and Bothma 1991).

INCUBATION: by ♀, sitting closely for greater part of day; one day ♀ left nest only twice, for 3 min in morning and 4 min in afternoon. ♂ perches on tree top above, guarding ♀, and at night sleeps deep in nesting tree, evidently close to nest. Period (n=1) 16–17 days.

DEVELOPMENT AND CARE OF YOUNG: eyes fully open by day 9, chick fully feathered by day 16, when chin and throat pure white with black malar stripe. Weight 15 g on day 9 and 25 g on day 14 (De Swardt and Bothma 1991). Young fed by both parents, on both nectar and insects. When parent arrives with food, they stand on heels with bills wide open (De Swardt and Bothma 1991). Parents fly away with faecal sacs and drop them *c*. 15 m away. At one nest parents brought several bleached millipede segments (Skead 1963). Nestling period (n=1) 23 days. Young fed by both parents for at least 2 weeks after leaving nest.

BREEDING SUCCESS/SURVIVAL: one nest destroyed in gale. Ringed adult bird survived for 4 years 5 months (De Swardt 1993). Blood parasites (*Leucocytozoon*, microfilaria, *Haemoproteus*) found in half of 74 birds sampled (Bennett and De Swardt 1989).

References
De Swardt, D.H. (1991, 1992, 1993).
De Swardt, D.H. and Bothma, N. (1991).
De Swardt, D.H. and Louw, S. (1994).
Nevill, H. (1987).

Family LANIIDAE: true shrikes

Called 'true' shrikes, to distinguish them from bush-shrikes (Malaconotidae), helmet-shrikes (Prionopidae), and other hook-billed 'shrikes' worldwide which have been associated with *Lanius* shrikes in the past. Several African species called fiscals, 'from their rapacity, which no revenue-officer could exceed' (Newton 1893–96).

Medium-sized songbirds (15–50 cm, 20–100 g), with rather short, laterally compressed and robust bill, upper mandible sharply hook-tipped with a notch and tomial tooth behind tip, lower mandible with corresponding incurvation or small indentation near tip; bite powerful, masseter muscles large, hence shrikes are large-headed; nostrils rounded, without operculum, more or less concealed by bristly antrorse feathers; distinct rictal bristles; wings not very long, P10 up to half length of P9, P9–P7 longest and usually about equal; tail quite long and graduated to very long; foot moderately strong, with strong grip powered by muscle arrangement unique amongst passerines (Raikow *et al.* 1980). ♂♂ with handsome plumage, mainly pied, often rich brown; ♀♀ and juveniles strikingly different from ♂♂ in some species, being brown and heavily barred, and very similar in others, but then ♀ often differs from ♂ in colour of flanks (the colour patch half-obscured when wings folded). Arboreal but readily come to ground; predatory; solitary nesters; territorial; several species co-operative breeders; other biological characters more or less those of main genus, *Lanius*, *q.v.* (the minor laniid genera being not so well known).

4 genera: one (*Lanius*) well represented in Africa and 3 endemic there. 2 small genera, *Corvinella* (W Africa) and *Urolestes* (SE Africa) seem to be quite closely related to *Lanius*. They are allopatric or parapatric (in SW Kenya), have similarities such as long tail and social organization, and it has become fashionable to congenerize them, in *Corvinella*. In the field they seem to us to be very different sorts of birds in respect of plumage, voice and behaviour and we revert to the former generic separation.

Eurocephalus, with a single superspecies in E and southern Africa, is generally placed in family Prionopidae, which it resembles in co-operative breeding, shallow-wingbeat-and-gliding flight, and plush forehead feathers. However, several laniid shrikes are also co-operative breeders, and *Eurocephalus* resembles Laniidae in many morphological and behavioural characters (Harris and Arnott 1988) including flank patches, in osteology (Olson 1989) and DNA (Sibley and Ahlquist 1990).

'**Shrikes**'. Analysis of over 90 morphological, biochemical, behavioural and plumage characters of shrike-like birds led T. Harris to classify an enlarged family Laniidae into 3 subfamilies: true shrikes Laniinae, helmet-shrikes Prionopinae and bush-shrikes Malaconotinae (Harris and Arnott 1988). The first 2 have the same composition as our Laniidae and Prionopidae, and Malaconotinae embraces our Platysteiridae (Vol. V, p. 548) and Malaconotidae; bush-shrikes and helmet-shrikes are thought to be sister groups, distantly allied with true shrikes and more distantly with crows, family Corvidae. DNA-hybridization studies (Sibley and Ahlquist 1990, Sibley and Monroe 1990) suggest even closer relationship of these 'shrikes', amongst themselves, and with such groups as shrike-tits, shrike-thrushes, cuckoo-shrikes and vanga-shrikes. Sibley and Monroe (1990) retain Laniidae (true shrikes only) and Corvidae as separate but related families and divide Corvidae into 7 subfamilies: 3 small and non-African, one large and non-African (Pachycephalinae, including shrike-tits and shrike-thrushes), and Corvinae, Dicrurinae and Malaconotinae. Their Corvinae embraces 4 tribes (crows, birds-of-paradise, wood-swallows, and cuckoo-shrikes with orioles); Dicrurinae embraces 3 (drongos, monarchs (Vol. V, p. 508) and fantails); and Malaconotinae consists of tribes Malaconotini (bush-shrikes) and Vangini (helmet-shrikes (Africa), vanga-shrikes (Madagascar) and platysteirines (Africa), the last including shrike-flycatchers, White-tailed Shrike, batises and wattle-eyes).

Genus *Lanius* Linnaeus

True shrikes, length 15–25 cm, weight 15–70 g; ♂ and ♀ alike, or plumages moderately dimorphic; ♂♂ striking, in grey, black and white, with strongly-contrasting black mask (ear-coverts, usually lores, often forehead) or hood, wings and tail usually strongly pied, large parts of plumage (mantle, rump, tail or flanks) often pink- or red-brown; ♀♀ and immature stages usually much duller, browner versions of adult ♂, with upper- and underparts heavily scaled with thin crescents and vermiculations and distinctive pattern in tertials and greater upperwing-coverts, each feather typically with brown centre, white border and bold black line between. Bill black (with yellowish or bluish base in juveniles, and yellowish in adult *L. meridionalis*).

Sit-and-wait predators, flying from elevated perch to snatch insect, often large and hard, or small vertebrate, mainly on ground. Kill vertebrates by striking back of head with bill and using tomial teeth to disarticulate cervical vertebrae (Cade 1995). Use feet to hold down small prey on perch whilst tearing at it with bill, and to carry large prey item in flight. Occasionally use foot to transfer morsel to bill. Large prey impaled on thorn or wedged in cleft branch, and cached there or dismembered whilst impaled. Dismembering process is quite violent, bird flexing legs and jerking wings with each tug by the bill. Lardering habit common in some species, rare in others; cached items generally eaten within 9 days but often neglected. In flight, carry prey in bill or in feet, and some shrikes can carry up to 1·3 times their own weight (Yosef 1993). Can devenom wasps and bees by squeezing abdomen and rubbing its tip onto perch, like bee-eater *Merops*. Swing and 'wind' (rotate) tail. Not very vocal; songs rather weak and unstructured. High-latitude species strongly migratory, wintering in tropics, most tropical ones resident. Nest a simple cup, made of vegetable matter, sited in bush or tree or (one species) often on ground. Solitary nesters, territorial; one species a co-operative breeder.

26–27 species, Africa, Palearctic and Oriental Regions, one east to New Guinea and 2 in Nearctic (one endemic there).

15 species in Africa. 9 are endemic and 6 are Palearctic visitors (2 also breeding in N Africa). Of the 15, we cannot discern close African relatives of the Palearctic *L. senator* or *L. nubicus*. The other 13 appear to fall into 4 groups, partly on basis of flank-patch colour (as shown by Grimes 1979a, ♀♀ of *L. dorsalis*, *L. cabanisi*, *L. excubitoroides*, *L. mackinnoni* and most but not all races of *L. collaris* have a usually-concealed maroon patch and ♂♂ have white flanks): (1) small pied shrikes, most with narrow tails and (♀♀) maroon flanks: the highly polytypic, widespread subsaharan *L. collaris* and its insular allospecies *L. newtoni*; the E Africa species-pair *L. somalicus* and *L. dorsalis*; *L. cabanisi* (E Africa) and *L. mackinnoni*; (2) large grey shrikes with black foreheads and maroon or white flanks: the Palearctic *L. minor* and tropical *L. excubitoroides*; (3) large grey shrikes with white foreheads: the Holarctic *L. excubitor* and Saharosindian *L. meridionalis* (a superspecies with the Holarctic *L. exubitor*, which does not occur in Africa); and (4) small red-and-grey shrikes: the Palearctic *L. isabellinus/L. collurio* superspecies and African species *L. gubernator* and *L. souzae*.

Lanius collaris superspecies

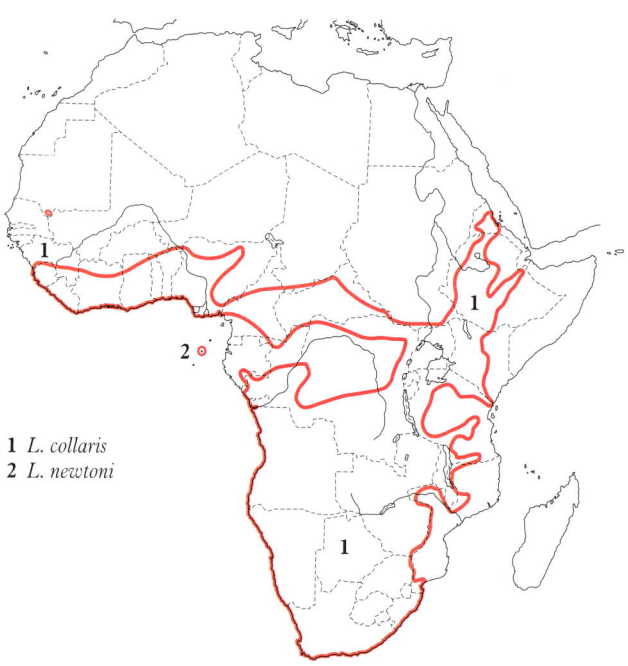

1 *L. collaris*
2 *L. newtoni*

Plate 18   *Lanius collaris* Linnaeus. Fiscal Shrike. Pie-grièche fiscale.

(Opp. p. 271)   *Lanius collaris* Linnaeus, 1766. Syst. Nat., ed. 12, 1, p. 135; Cape of Good Hope.

Forms a superspecies with *L. newtoni*.

Range and Status. Endemic resident, partially migrant in driest parts of range. A lowland bird in a spectrum of open-country habitats; in W Africa commonest near seaboard; but montane east of 30°E.

Mauritania, uncommon visitor to extreme S (Guidimaka). Not in Senegal or Gambia. Mali, doubtful record in W (de Bie and Morgan 1989); records at San and Tominian (Lefranc and Worfolk 1997). Guinea, scarce in Kounounkan, abundant in Macenta. Sierra Leone and Liberia, common and widespread in cleared areas and open farmland; has become much commoner in Nimba area with wholesale clearance of forest for opencast mining.

Lanius collaris

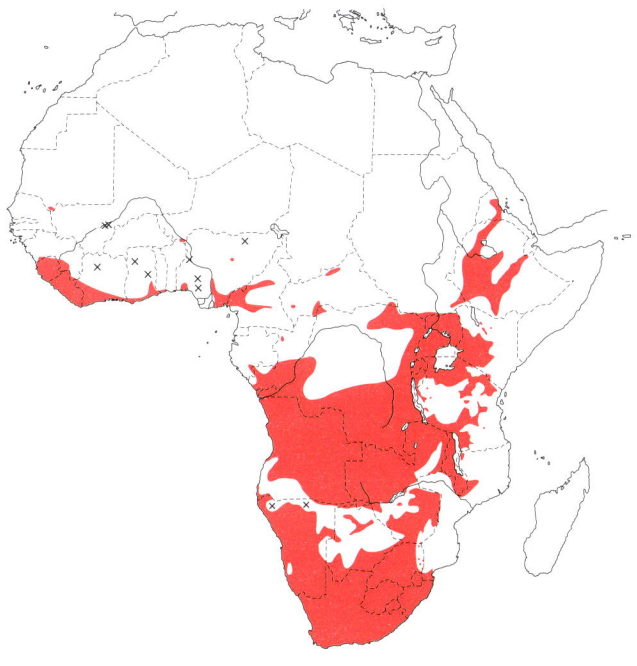

Ivory Coast, locally common in suburban gardens in S, uncommon and very local further north, to 09°30′N. Ghana, mainly on coast and in areas of cleared forest inland (Akropong, Tafo, Korforidua, Dixcove); locally north to Kete Kratchi and Mole. Togo, locally common, e.g. Lomé region and along coast near Anécho and Badou. Nigeria, a suburban bird, locally not uncommon; common at Lagos, at least formerly, and locally common around most towns in SE, e.g. Enugu, Calabar, Port Harcourt, and Obudu Town; in north, records in Kainji Lake Nat. Park (Luma), Jos and Hadejia, and frequent on Mambilla Plateau. Niger, several records in 'W' Nat. Park. Cameroon, scarce and local, from Bamenda to Adamawa Plateau and forest edge situations in S-central Cameroon; twice in N (Scholte *et al.* 1999). Central African Republic, frequent southeast of Carnot and in Mbaïki-Bangui-Damara area in S (Germain and Cornet 1994), and occurs in Manovo-Gounda-Saint Floris Nat. Park and Bamingui area in N (Carroll 1988); also in extreme E. Congo, widespread in Odzala Nat. Park; common from Léfini Rés. towards coast though absent from Kouilou Basin. In SE Gabon frequent in savannas of Ngounié R. and Nyanga R., around Mouila and Tchibanga (Y.M. de Martin de Viviès, pers. comm.).

Eritrea, very common in centre, up to 1230 m. Ethiopia, frequent to abundant in all highland areas; higher altitudinal limit unknown, but rare below 1200 m. In E Africa occurs at 500–3350 m, local and uncommon in lowlands, common in highlands. Uganda throughout, in 3 races: *smithii* in W, *humeralis* in E and *capelli* in SW (Kigezi). Kenya, common at 1400–3000 m, widespread in W and central highlands, south to Serengeti Nat. Park, Crater and Mbulu Highlands; resident on Marsabit, Nyiru, Kulal, Ndotos and Matthews Range; in Chyulu and Taita Hills. Tanzania, common in Arusha District and highlands from Mt Meru to S Pares and E Usambaras; widespread in L. Victoria Basin and in W, east to Tabora; in Eastern Arc Mts and highlands around N end of L. Malaŵi, also in S in Matengo Highlands and Nandembo. Frequent in NE Zaïre and common throughout much of E, S and W. Common throughout Angola, but in arid SW absent except for coastal population of *aridicolus*. Frequent to common throughout Zambia, but absent from Luangwa Valley and scarce on Middle Zambezi and in parts of S and W. Malaŵi, sparse, up to 1530 m, but at 1850–2100 m on Viphya and Mafinga Mts and quite common up to 2450 m on Nyika Plateau. N Mozambique, few records, in W.

Common to very common in South Africa, common in S half of Namibia but becoming gradually scarcer to north and absent from near Angola border; scarce and unpredictable in Botswana north of 24°S, but very common at Jwaneng, Phitsane Molopo and in W Molopo Valley; absent from E Transvaal and SE Zimbabwe lowveld (vagrant, Kruger Nat. Park, in austral winter), quite common in NE of central Zimbabwe Plateau but rare or absent in NW of the country; altitudinal range in Zimbabwe, 950–2200 m. Widespread in Lesotho and Swaziland. Southern African populations (*subcoronatus*, *collaris*, *pyrrhostictus*) appear to be sundered from those to the north. In S Mozambique, frequent in Zimbabwe frontier highlands, and uncommon but widespread in S Sul do Save, in Lebombo Mts and on coast north to Incomati R. mouth.

In Liberia, rare 100 years ago but densities now of 4–6 pairs per 10 ha in outskirts of larger towns, 20 per km^2 in coastal savannas and 3 per km^2 near Wologizi in N (Gatter 1997). About 4 pairs per km^2 in Odzala Nat. Park, Congo; but much higher densities in urban areas in E and southern Africa. In E Africa has spread to become a common bird of residential suburbs and even city centres (e.g. Nairobi). In southern Africa has spread into open areas where overhead lines and wires provide lookout perches at the right elevation; has undoubtedly become more numerous in wake of settlement and development (V. Parker *in* Harrison *et al.* 1997).

Description. *L. c. collaris* Linnaeus (includes '*predator*'): S Namibia, South Africa east to S Transvaal highveld, E Free State, Natal and W Swaziland. Intergrades with contiguous races. ADULT ♂: forehead, crown, hindneck, mantle, lores, upper cheeks, ear-coverts and sides of neck black with greyish brown cast, sharply delineated. Black lower mantle merges through dark grey back with dark bluish grey rump and paler bluish grey or greyish white uppertail-coverts. Scapulars white, sharply delineated. Tail black, narrow, strongly graduated, with white sides: T1 black, T6 white with black at very base, T2–T5 with increasingly large white tips. Entire underparts greyish white, lower breast to belly and flanks finely (almost imperceptibly) vermiculated with pale grey (greyer in W Cape than E Cape). Wings brownish black, bases of primaries white, the white concealed except on P9–P4 where it forms conspicuous patch. Underside of remiges shiny dark slate, underside of tail white and slate. Bill black; eyes dark brown; legs and feet black. Sexes alike. ADULT ♀: like ♂ but blacks duller, and small, diffuse, variable, chestnut patch on flank. SIZE: wing, ♂♀ (n = 14) 93–103 (98.9); tail ♂♀ (n = 10) 100–113; bill to feathers ♂♀ (n = 14) 16–25; tarsus ♂♀ (n = 10) 25–30. WEIGHT: ♂ (n = 84) 24.7–51.0 (41.8), ♀ (n = 69) 27.7–52.0 (39.8).

IMMATURE: forehead to back pale brown, heavily and evenly banded with blackish brown; ear-coverts the same but a richer brown; rump same as back but paler, and dusky and whitish bands a little wider; uppertail-coverts blackish, broadly margined with warm brown; tail dark brown where adult black; scapular feathers white, margined blackish; remiges and upperwing-coverts dark brown, broadly margined with warm, buffy brown; chin and throat white or off-white, remaining underparts off-white with c. 15 irregular brown or blackish brown bands, wider-spaced and less definite on belly and flanks than on breast; basal part of lower mandible pinkish.

NESTLING: hatches naked, blind, flesh-brown or greyish pink, prominent egg tooth, yellow mouth, without mouth spots, and yellow gapes. Peculiar distribution of neossoptiles: only on abdominal, caudal and alar tracts (Markus 1972). Feathered nestling: brownish grey above, scapulars whitish, chin to belly softly and finely spotted or evenly, softly vermiculated with fine grey and whitish bands, bill black with pinkish cutting edges and rather narrow, pale yellow or pinkish gapes, inside of mouth orange-yellow, eyes black.

L. c. smithii (Fraser): Guinea to Sudan. Like nominate race but black upperparts glossier and bluer, lacking brownish cast; rump whitish, uppertail-coverts grey; only distal half of T6 white (basal half black); tertials and inner secondaries white-tipped; underparts pure white; brown flank patch of ♀ vestigial or absent.

L. c. humeralis Stanley: Eritrea, Ethiopia, whole of L. Victoria Basin, to S Tanzania. ♂ and ♀ like *collaris* but upperparts blacker and underparts almost pure white; T5 without white tip. WEIGHT (Kenya): ♂ (n = 14) 32–50 (37·4), ♀ (n = 14) 20–45 (33·6).

L. c. marwitzi Reichenow: NE and S-central Tanzania, from Mpwapwa and Ukagurus to Mahenge, Njombe, Mt Rungwe and Tukuyu. Like *humeralis* but a thin white superciliary line from forehead to hind end of ear-coverts, and rump and uppertail-coverts dark grey; ♀ (always?) lacks brown patch on flank.

L. c. capelli (Bocage): S Congo, Angola, Zambia, Malaŵi, Katanga, Burundi, Rwanda, SW Uganda. Like *humeralis* but less white on T5 and T6, and brown flank patch of ♀ vestigial or absent. WEIGHT (Uganda): ♂ (n = 4) 33–40·5 (37·1), 1 ♀ 33.

L. c. pyrrhostictus Holub and Pelzeln: like *humeralis* but upperparts slaty (not pure) black and underparts tinged grey.

L. c. subcoronatus Smith: Namibia north of S Great Namaqualand (except for NW coast), N Cape Prov., W Free State, W Transvaal, Botswana except N, SW Zimbabwe. See NOTE below. Like *collaris* but thin white superciliary line from forehead to above hind end of ear-coverts; rump grey but uppertail-coverts whitish. WEIGHT (unsexed, n = 4): 36–42 (38·75).

L. c. aridicolus Clancey: desolate coastal dunes in NW Namibia and SW Angola. Like *subcoronatus* but less deep blackish slate above, purer silky white below, and with more white in tail.

TAXONOMIC NOTE: applying the Recognition Species Concept that employs specific mate-recognition characters, Harris (1995) speculates that Fiscal Shrike populations with red-flanked ♀♀ may prove to be specifically separable from those with white-flanked ♀♀. The white-browed *marwitzi* is a distinctive, montane bird which Short *et al.* (1990) treat as a 'megasubspecies' and Lefranc and Worfolk (1997) as a separate species (in part to encourage research on this very poorly known population). If *marwitzi* is a good species, *subcoronatus* might prove to be as well, but a 500 km wide area of introgression between *subcoronatus* at Vanzylsrus and *collaris* at Brandfort, N Cape, militates otherwise (Maclean and Maclean 1976). Here we treat the whole complex conservatively.

Field Characters. Length 21–23 cm. The most familiar shrike in subsaharan Africa, commonly seen perching on fences, telephone wires and garden trees. Black and white, with long, narrow, graduated tail. *Back black*, with *broad white scapulars forming 'V'*; rump grey to whitish (varies with race); small white spot on black wing; tail black with white sides and tip. Top of head black to below eye except in races *marwitzi* (Tanzania) and *subcoronatus* (SW Africa), which have narrow white supercilium. ♀ has variable amount of chestnut on flanks. Immature ash-brown above and off-white below, with fine dark bars, pale scapular patch duller but well-marked. Other fiscals have grey backs; immature darker and more heavily barred than immature Woodchat and Masked Shrikes *L. senator* and *L. nubicus*, which have crescentic and more widely-spaced barring on whiter underparts.

Voice. Tape-recorded (10, 22, 58, 86, 88, 91–99, 104, F, KEI). Territorial call, also used as alarm and as part of song, a double screech, 'djaaaa-djaaaa' or 'djaaaa-dju'. Song very variable: typically a disjointed medley of short clear whistles, scratchy notes and imitations of other birds; also, e.g. 3 low buzzes followed by 4 double whistles, 'zzzz-zzzz-zzzz ... tiss-weeo, tiss-weeo, tiss-weeo, tiss-weeo', or a conversational 'toyo-tiddlywit-twayee' followed by loud, tearing 'tazzz-tazzz'. Race *marwitzi* in S Tanzania gave a hesitant 'way-wyo' or 'way-wyo-dii' at irregular intervals from perch (pers. obs.). A series of mournful double whistles, 'tyooo-yooo' or 'tu-wheew' used as song in E Africa but apparently not elsewhere (Chappuis, in press); occasional variant, a louder, sharper, higher-pitched 'twee-yee'. When one bird gives alarm screech, its mate may accompany it with low stuttering subsong or soft piping notes. For further details see Harris and Arnott (1988) and Harris (1995).

General Habits. Inhabits open areas with short grass and scattered shrubs. In Guinea and Ivory Coast frequents recently burnt and overgrazed wooded savanna grassland, farmland, well-cleared land with scattered bushes and trees, suburban gardens, roadsides, wasteland around towns and town parks in the forest zone; in Ghana, coastal thickets, forest clearings, bushy margins of marshes, and suburbs (e.g. of Accra); SE Congo, wooded and rocky grassland; further east and south, grassy plains (but a habitat requirement is exposed perches such as telephone wires, fences or scattered shrubs), bush savanna, broad river valleys with cultivation and thickets, acacia bushes along drainage lines in semi-desert, wooded hillsides, alpine grassland, farmyards, roadsides, gardens and the edges of plantations.

Bold, conspicuous, confiding, territorial, and quite vocal and aggressive. Solitary or in pairs, keeping mainly in territory year-round, perching conspicuously on wire or tree. Uses lookout perches 1–10 m above ground, usually 3–5 m; there it keeps alert and watches, almost motionless but sometimes moving tail slowly up and down, then drops at shallow angle to ground to alight and seize insect, usually c. 15 m away, sometimes up to 25 m away. Young bird feeds on ground more than adult, hopping after very small insects. Eats small item after perfunctory treatment on ground, then returns to same or another lookout perch. Carries larger prey item in bill back to perch, and kills it there by nibbling and crushing it in bill and beating it hard against perch, when insect legs, beetle elytae and butterfly wings may fall to ground before remains are swallowed. Still larger prey – vertebrates and large insects – may be carried back to perch in feet rather than in bill; at

perch, shrike often stands on prey with one foot and tears at it forcefully with bill, or it may impale it on long thorn, wood splinter, sisal leaf or fence-wire barb, returning later on to eat it (**A**). A locust was wedged between stem and leaf of *Hyparrhenia* grass (Y.M. de Martin de Viviès, pers. comm.). Hops on ground, for less than a metre. On returning to elevated perch often wags tail quite vigorously from side to side. Sometimes gleans insects in foliage, often momentarily hovering, and hawks small insects in flight. Known to plunge-dive into shallow puddle and catch tadpoles. In gardens, can become a bold, fearless predator of small birds at bird-baths and bird-tables, and known to kill cage birds on verandahs; but finds small birds difficult to catch – in Ghana seen to stoop several hundred times at sparrows, wagtails and weavers feeding on ground, without a single success (Macdonald 1980). Regularly uses larders (Ghana, Kenya) – less so in southern Africa. In South Africa a Fiscal swooped onto House Sparrow *Passer domesticus*, held it down under one foot, bit into neck and shook it; the sparrow, weight 24 g, died within a min. and was carried in shrike's bill to tree and impaled through neck onto a thorn; later in day it was moved higher up tree and there eaten (Roos and Roos 1988).

Flight direct, without undulation, often low down, with rather whirring wingbeats. Scratches indirectly. 2 birds sometimes preen each other. Flicks wings (Smith-Symms 1999). Pair defends territory all year, using 'terj-terj' call which is often answered by neighbouring territory owners; territory size 0·4–18 ha; but occasionally 2 nests only 10 m apart, and 2 nesting pairs have been said to share a territory. Regurgitates small, blackish pellets.

Young birds (Kenya) start to moult at 3 months and finish at 4 months; adult undergoes complete moult in Aug–Dec and finishes moulting before breeding commences. Young birds (Ghana) start to moult at 2–3 months and finish at age 5–7 months; adult moults in Sept–Nov, when most have finished breeding. In southern Africa juveniles and adults moult mainly in Jan–Feb.

Records in Mauritania are all Nov–Dec; northern Ghana in wet season (about June–Aug), Niger all in Dec–Feb, N Nigeria (Hadejia) in Aug. In Kenya, some non-breeding birds thought to move into arid areas in wet season. However, numbers in Botswana increase in austral winter (Penry 1994, Herremans 1994). An example of *subcoronatus* in Cape of Good Hope Nat. Res. may have flown *c.* 400 km (but a fledgling there, evidently *subcoronatus*, had not!).

Most ringing recoveries within 20 km, but one 36 km away (E Cape) and another 110 km away (N Transvaal).

Food. Mainly insects: crickets, grasshoppers, mantids, locusts, beetles, butterflies, moths, damselflies, termites, and ants; 5–30 mm long (Macdonald 1980). Also spiders, millipedes, and a few vertebrates: small frogs and reptiles (snakes, skinks, agamas up to 180 long, chameleons), birds (sparrows, estrildids, wagtails, nestlings, nidifugous young including 30-g chicks of Helmeted Guineafowl *Numida meleagris*; rarely, larger birds up to size of Speckled Mousebird *Colius striatus* and Laughing Dove *Streptopelia senegalensis*), bats *Pipistrellus kuhli* (Kenya: Schwan and Hikes 1979) and small rodents; occasionally seeds and food scraps from dog bowls; once evidently figs (Roos and Roos 1988). Of 35 museum specimens which had eaten 'insects', food was specified as grasshoppers in 8, locusts in one and beetles in one. Nestlings fed on noctuid larvae, moths, grasshoppers, mantids, crickets, spiders, small skinks, tadpoles, scarabaeid and melolonthid beetles, small nestling warblers and typhlopid snakes (van Someren 1956).

Breeding Habits. Solitary nester, monogamous (Zack and Ligon 1985a). Highly territorial, defending territory from intruders with characteristic vocal and visual behaviour (which not described: Macdonald 1980); has strong aversion to Woodland Kingfishers *Halcyon senegalensis* in or near territory. Interspecifically territorial with Grey-backed Fiscal *L. excubitoroides*, Kenya, pair-territories of Fiscal Shrike forming patchwork with group-territories of Grey-backed Fiscal (Zack and Ligon 1985a). Ringed ♂ known to stay in its territory for 5 years; but ♂♂ establish new territory up to 1·2 km away quite frequently, following collapse of pair-bond (Zack 1986). Territory size in Ghana: (n = 9) 0·41–0·78 (0·59) ha.

Unmated ♂ perches conspicuously, fluffs out white scapulars, and repeatedly gives 'jujuut' call. Once pair is formed, ♂ feeds ♀ and ♀ may sometimes feed ♂, and they sing in duet. Courtship feeding is followed by zigzag display flight. At the nest an ecstatic display involves a perched bird gyrating with fanned tail held up vertically, head lifted up and twisted from side to side; display accompanied by whistles and mimickings. ♂ offers ♀ food items and bits of nest material (Toschi 1950) and ♀ solicits copulation by making 'naaa' call and quivering wings (Harris and Arnott 1988). ♂ sees off rival by flying at it and striking it with foot, or alighting on it with both feet and pecking vigorously; the rivals may then spar in the air, interlocked, or one bird seizes the other's wing and they both fall to the ground (van Someren 1956).

Regularly replaces failed clutches, using same nest several times or building anew, often using material from old nest; usually double-brooded, often triple-brooded. ♀ becomes lethargic before egg-laying, staying in thick cover, with plumage fluffed and rump and flanks exposed (Harris and Arnott 1988).

NEST: a deep, thick-walled, bulky cup, made of twigs, leafy herbs (e.g. *Helichrysum* in southern Africa), flower-heads (e.g. of *Ocimum* and labiates), bark fibres, moss and bents of grass, sometimes with spider web and cocoons and feathers and often with man-made materials added; lined with fine grass heads and fine rootlets. Size: ext. diam. 100–180 (130), int. diam av. 75, ext. depth 65–120 (95), int. depth av. 50. Sited typically in vertical fork against trunk or horizontal fork on a lateral branch, in *Acacia*, *Pyracantha* or other thorny tree or briar, about 3 m high (n = 789, 0·6–6·0 m high: Harris and Arnott 1988; once *c*. 15 m high). Nest height in Ghana bimodal: 20 nests at 1·5–2·4 m and 5 at 3·7–4·6 m (Macdonald 1980). Nest site selected by ♀. Nest can be built in only 2–5 days, by both sexes but mainly by ♀. Occasionally new nest built on top of old one. Some pairs build 2–3 nests before laying, the ♀ destroying the first 1–2.

EGGS: 1–6, usually 3–4 (southern Africa, n = 958; 5 eggs rare, 6 very rare); 1–4 in Zambia and Malaŵi (3 c/1, 17 c/2, 41 c/3 and 5 c/4); 1–3 in Ghana (of 30 clutches, 18 of c/3; av. 2·5); mainly c/3 in Kenya. Pointed or elongate ovals. Cream or pale green, spotted with browns, greys and yellow-olive, mainly around large end. Larger clutches tend to be of smaller eggs (Cooper 1971a). SIZE: (n = 630) 19·6–28·5 × 15·6–19·9 (23·5 × 17·7).

LAYING DATES: Guinea (Macenta), Mar–May; Liberia, Feb–Mar; Ghana, Dec–Oct, mainly Jan–Feb and June–July; Togo, Feb, Apr, July; Nigeria, Mar–May, July, Nov–Dec; Eritrea, Apr–June; Ethiopia, Apr–July; E Africa, Region A, Jan–June, Aug, Oct, Dec (mainly Apr–May), Region B, Feb–June, Aug, Nov; Region C, Sept–Nov, Jan, Region D, Sept–July (n = 191 clutches) with distinct peaks in Apr–May (32%) and Nov–Dec (30%); Zambia (n = 103), Aug–Dec, mainly Sept–Oct, and records in Jan, Apr and June; Malaŵi, Aug–Jan, mainly Oct; Zimbabwe, June–Jan (n = 454), 15% Aug, 35% Sept, 23% Oct, 17% Nov, once Apr; Botswana, Aug–Sept; Transvaal (n = 354), June–Jan and Mar, 63% in Sept–Oct, 34% in Aug and Nov–Dec; Natal, June–Jan; S Mozambique, Oct–Nov.

INCUBATION: by ♀ only, fed on nest by mate. She may start to incubate before clutch completed. Period: 12–16·5 days, usually 14–15.

DEVELOPMENT AND CARE OF YOUNG: eyes open at 5 days; young fully feathered at 15 days. Sometimes a marked disparity in size and weight between nestlings in same brood (Cooper 1971b). Weight increases sigmoidally, from *c*. 4 g at day 1 to *c*. 36 on day 17 (Marshall and Cooper 1969). Eggshells eaten by adults or carried away. Young fed by both parents, mainly by ♂ for first few days when ♀ broods chicks, later mainly by ♀. ♂ feeds brood himself, or gives food to ♀ who passes it to chick. Both parents eat faecal sacs, or carry them away. Nestlings are unusually noisy. Parent stands over nestlings with wings half open, to shade them (**B**). Period: 17–21 (*c*. 19) days. Young out of nest fed and attended by both parents; by this time legs and bill flesh-pink and gapes salmon-pink; they start uttering discordant, warbling subsong within a few days; and begin to feed themselves after 3 weeks and become independent at 5–7 weeks, although remaining with parents in territory for 4 months in southern Africa, 5 months in E Africa and 5–7 months in W Africa. Young from 2 or occasionally 3 previous broods may be present in parental territory at the same time (Macdonald 1980).

BREEDING SUCCESS/SURVIVAL: very high nest-failure rate, and pairs estimated to start 13 clutches in an 8-month season; in Ghana pair successfully raised av. 2·2 broods per

B

8-month season, though nest-success rate only 16% in Ghana and 15% in Kenya (Macdonald 1980, Zack 1986). Av. intervals between successive broods (Ghana) 3·7 months. Using artificial nests, pairs successfully raised 2–5 broods per year for 4 years (Butticker 1960). Parasitized by Jacobin Cuckoo *Oxylophus jacobinus* (*c*. 10% of all southern African records of Jacobin Cuckoo nest parasitization involve Fiscal Shrike); 1–2% of Fiscal Shrike nests are parasitized (South Africa), the cuckoo being reared with rest of shrike brood. Rarely parasitized by African Cuckoo *Cuculus gularis* (is one of its very few known hosts). Eggs taken by egg-eating snake *Dasypeltis scaber* (and the 53 cm snake had then evidently been killed by parent shrikes). Many shrikes killed by traffic, because of widespread habit of hunting from roadside wires and alighting on busy roads to take insects knocked down by vehicles. Adult annual survival 39% (Kenya: Zack 1986). ♂ bred annually for 9 years (Smith-Symms 1999).

References
Cooper, J. (1971a).
Dittami, J.P. and Knauer, B. (1986).
Harris, T. and Arnott, T. (1988).
Lefranc, N. and Worfolk, T. (1997).
Macdonald, M.A. (1980).
Marshall, B.E. and Cooper, J. (1969).
Toschi, A, (1950).
Zack, S. (1986).
Zack, S. and Ligon, J. D. (1985a).

Lanius newtoni Bocage. São Tomé Fiscal. Pie-grièche de São Tomé.

Plate 18
(Opp. p. 271)

Lanius (Fiscus) Newtoni Barboza du Bocage, 1891. J. Sci. Math. Phys. Nat. Lisboa (2), 2, 6, p. 79; São Tomé.

Forms a superspecies with *L. collaris*.

Range and Status. Endemic resident, São Tomé. Discovered in 1888; 100 years ago known from São Miguel and Rio Quija in SW and from 2 untraceable localities in SE; 13 specimens collected in 1928, one at Io Grande in SE and 12 in 4 weeks at Roça Jou and Rio Quija in SW, up to 1060 m. Thought to be extinct by Günther and Feiler (1985). Rediscovered in 1990 and several birds seen since, along Rio Xufexufe and Rio Quija in SW and in central São Tomé, mostly below 700 m. Total population likely to run into hundreds (Lefranc and Worfolk 1997). From near SW coast up to 700 m on Formosa Pequeno (Christy and Clarke 1998).

Description. ADULT ♂: upperparts from forehead to rump glossy black, also lores, narrow line under eye, ear-coverts and sides of neck. Uppertail-coverts blackish grey or slate-grey. Scapulars pure white. Tail black, narrow and strongly graduated, T4 with white tip, T5 with larger white tip and T6 all white. Underparts uniformly pale yellow or yellow-orange (which soon fades to white in museum specimens). Wings black, tertials very narrowly white-tipped, and bases of middle primaries white, usually concealed, sometimes showing as small white patch. Bill black; eyes dark brown; feet and legs greyish brown. Sexes alike, or ♀ like ♂ but upperparts may have brownish wash. SIZE: wing, ♂ (n = 10) 91·5–96 (93), ♀ (n = 5) 87–96 (89·5); tail, ♂ (n = 10) 110–117·5 (114), ♀ (n = 2) 100, 115; bill to feathers, ♂ (n = 7) 13–14·5 (13·8), ♀ (n = 4) 12–13 (12·5); tarsus, sexes combined (n = 14) 23–26 (Lefranc and Worfolk 1997).

IMMATURE: a young bird seen in Aug had black bill, brown upperparts, slightly greyer on head, with short, thin, buffish superciliary stripe; scapulars tawny, forming V-mark on back; tail dark brown; wings brown with coverts buff-tipped; underparts tawny-orange. Juvenile seen in Jan had horn-brown bill, upperparts brown, finely vermiculated, throat yellowish, remaining underparts tawny yellowish, barred darker (Lefranc and Worfolk 1997).

Field Characters. Length 21–23 cm. The only African laniid inhabiting forest interior. Similar to Fiscal Shrike *L. collaris* but rump black, underparts pale yellow. Juvenile looks like juvenile Fiscal Shrike, i.e. brownish, with barred upperparts, but underparts tawny-orange. No other shrike is resident on São Tomé, although there is a single record of a juvenile migrant Red-backed Shrike *L. collurio* (Christy and Clarke 1998).

Voice. Tape-recorded (104, CHR). Song of 2 types: (a) series of 10–11 regularly spaced rather plaintive notes on same pitch, given at rate of c. 1 per s, 'pew ... pew ... pew ...'; (b) faster series of somewhat metallic notes, 'tsink, tsink, tsink, tsink'. A bird once uttered 250 notes without interruption. Young bird once sang repeatedly a 3–4 note 'tieu-tieu-tieu', like adult song but more nasal. A netted adult gave squawks and a scolding, churring sound while being handled (Lefranc and Worfolk 1997).

Lanius newtoni

General Habits. Inhabits dense, closed-canopy primary lowland forest with rather open undergrowth and boulder-strewn ground. Seems to keep to lower and middle storeys; rather quiet and unobtrusive, 'probably skulking in low bushes'; once seen foraging by hopping between boulders and on rocks in dried-up stream; another pursued an insect by hopping along a mossy tree limb. Perches 3–5 m high in forest trees, often next to tree-fall gap. Very upright stance. Flies down to ground or to base of large tree, seizes insect, and returns to perch to deal with it. Inspects tufts of lichens and mosses on branches. Bird once seen trying to catch passing flying insect. 'Tiu' or 'tiûh' call is territorial, given mainly in morning, bird calling from one perch then flying to others, calling at each (Christy and Clarke 1998). Flight somewhat more graceful and less heavy-looking than in most congeners. Sometimes readily approachable.

Food. Insects, including small beetles.

Breeding Habits. Unknown. Some ♂♂ with hypertrophied gonads, Dec. Sings in Dec, Jan and Aug. Juveniles seen Jan and Aug.

References
Atkinson, P. *et al.* (1991).
Christy, P. and Clarke, W.V. (1998).
Collar, N.J. and Stuart, S.N. (1985).
Lefranc, N. and Worfolk, T. (1997).
Sargeant, D. (1992).

Plate 18
(Opp. p. 271)

Lanius dorsalis Cabanis. Taita Fiscal. Pie-grièche des Teita.

Lanius (Fiscus) dorsalis Cabanis, 1878. J. Orn., 26, pp. 205, 225; Ndi, Teita.

Range and Status. Endemic resident, E Africa, subject to wandering, amounting to migration in Somalia. Ethiopia, uncommon to frequent in S, south of 05°N (Safford *et al.* 1993). Somalia, common and widespread in lowland south, in bushland in and near valleys of Webi Shabeelle and Jubba R., between the rivers, west of Jubba R., and along coastal plain south to Kenya border. Sudan, fairly common in open *Acacia* savanna in extreme SE. Uganda, quite common in NE from Kidepo Valley Nat. Park to Moroto. Kenya, common and widespread, mainly below 1000 m, occasional up to 1500 m, north and east of central highlands; absent in W, local to south of Nairobi (Olorgesailie, L. Magadi, L. Natron), and absent from SE, south of Malindi; status uncertain in some under-worked arid areas in E and N. Tanzania, drier parts of Serengeti Plains from Ndutu to L. Eyasi; Lake Manyara Nat. Park, Tarangire Nat. Park, Arusha and Kilimanjaro lowlands, Masai Steppe, and Mkomazi Game Res.; record from Dar es Salaam.

Description. ADULT ♂: forehead, crown, nape, hindneck, upper mantle, lores, line below eye, ear-coverts and sides of neck all black, sharply delineated from chin and throat, and quite well delineated from lower mantle. Lower mantle and back pale bluish grey, back slightly paler. Scapulars white, and rump and uppertail-coverts white or whitish; sharply delineated. Tail black, well rounded or somewhat graduated, outer feathers with shortfall of *c.* 20–25; T1–T3 all black, T4 and T5 with white tips, T6 with white outer vane and white tip to inner vane. Entire underparts white. Wing glossy black, most primaries with white bases, forming white patch *c.* 20 long. Axillaries and underwing-coverts black, undersides of remiges slaty, primaries with white bases. Underside of tail brownish black and white. Bill black; eyes dark brown; legs and feet dark brown or yellowish brown. ADULT ♀: like ♂ but a half-concealed chestnut patch on flank. SIZE (unsexed, n = 4): wing 98–103 (100); tail 88–100 (92.7); bill to feathers 16.5–17.5; tarsus 25–28. WEIGHT: ♂ (n = 3) 54–58 (56.3), 1 ♀ 56.

IMMATURE: like adult but black feathers of forehead, crown, hindneck and mantle tipped pale grey, forming scallops; lores and ear-coverts black, unmarked, forming mask; scapulars narrowly tipped black; upperwing-coverts and tertials black, with broad white surround with a narrow black line running along it (from rhachis to edge of outer vane: black, white, black then white); bill black with some yellowish horn on base of lower mandible. Juvenile: forehead, crown, hindneck, mantle and back grey-brown, finely vermiculated in blackish brown; rump and uppertail coverts pale buff, similarly vermiculated. Lores, line below eye and ear-coverts dark brown, forming a mask. Tail dark brown, feathers tipped and narrowly edged whitish buff, T1–T3 with broad blackish subterminal bars. Underparts off-white, with evenly spaced pale brown bars, fine on breast and sides of face, more widely spaced on flanks. Wing black, variegated with warm buff and off-white: upperwing coverts, tertials and outer secondaries with black centres, mainly buff outer webs, pale buff or off-white tips and edges, and black subterminal bar. Bill black with yellowish brown base.

Field Characters. Length 20–21 cm. A compact shrike of dry open bush and savanna in E Africa, readily distinguished from Fiscal Shrike *L. collaris* by *pale grey back* and

Lanius dorsalis

short tail; rump whitish, wings solid black except for white primary patch; ♀ has some chestnut on flanks. Extremely similar to Somali Fiscal *L. somalicus*, which is same size and shape; differs in *lack of white tips* to secondaries, less white around edges of tail, and closed undertail *black with white spots* towards tip (all white in Somali Fiscal: Lefranc and Worfolk 1997). Juvenile dark grey above with dark barring, white below with a little barring; juvenile Somali is similarly barred but brown above, buffy white below, undertail white. Partly ecologically segregated from Somali Fiscal, preferring more wooded, less arid habitats, but they overlap in Kenya north of Marsabit, S Ethiopia, and Somalia in coastal areas around 2°S and probably elsewhere (see Ash and Miskell 1998).

Voice. Tape-recorded (B, F, McVIC, STJ). Song a quaint mixture of churrs, hollow sounds and ticking notes, 'chwaaa pikereek chrrrrrrr yook pikerchik ... skyaa, week kiook-tiureek tik ...' (Zimmerman *et al.* 1996). Also said to give a flute-like whistle, low chuckling, and a typical harsh grating *Lanius* alarm (Lefranc and Worfolk 1997).

General Habits. Inhabits hot, arid open bush and lightly wooded grassland in arid and semi-arid lowlands, up to 1500 m. Solitary outside breeding season. Shy. Perches prominently on shrub or small tree, and flies down to ground to pounce on prey. Of 135 prey captures (Tsavo East, Kenya: Lack 1985), 83% on ground and ground herbage, 8% in the air (flying insects), and 5% from leaves in bushes; bird perches at av. 2.6 m high, flies out av. 14 m to prey on ground, and to height av. 4·9 m for airborne insects, and makes av. 1·7 forays every 10 min. Flight direct, straight, usually low down, without undulation. Sedentary, but prone to wander (e.g. as far from known breeding range as Dar es Salaam).

Food. Insects: grasshoppers in all of 5 stomachs and beetles in one; also eats spiders, rodents up to size of rat, lizards and small snakes (Lack 1985).

Breeding Habits. Solitary nester, evidently monogamous and territorial.

NEST: a rather untidy, thick-walled open cup, made of grass bents and small twigs; sited on fork inside thorn bush.
EGGS: 3–4.
LAYING DATES: Somalia, May; E Africa, Region C, Jan, June, Region D, Mar–June, Dec.

References
Lack, P. (1985).
Lefranc, N. and Worfolk, T. (1997).
Zimmerman, D. A. et al. (1996).

Lanius somalicus Hartlaub. Somali Fiscal. Pie-grièche de Somalie.

Plate 18
(Opp. p. 271)

Lanius somalicus Hartlaub, 1859. *In* Heuglin, Ibis, p. 342; Bender Gam, Ker-Singeli-Somals country, Red Sea.

Lanius somalicus

Range and Status. Endemic resident, Horn of Africa. Djibouti, record north of Tadjoura Gulf (Hall and Moreau 1970) and one between Djibouti and Arta (Welch and Welch 1984b). Ethiopia, uncommon to frequent in NE, Rift Valley, S and SE. Sudan, rare or under-recorded, in extreme SE. Somalia, common and widespread in N half, from Djibouti border to Cape Guardafui, south to 05°N, and common on coastal plain south to Muodisho (Mogadishu) and about Marka; from sea-level up to 2000 m on summit of Mt Wagar. Kenya, uncommon and local, resident at 400–1000 m: L. Turkana basin, east and southeast to Turbi, Marsabit, Kapedo and Laisamis; occasionally wandering south to Baringo, Wamba, Merti and Isiolo (Zimmerman *et al.* 1996); record near Lokichokio.

Description. ADULT ♂: forehead, crown, nape, hindneck, mantle, lores, area under eye, ear-coverts and sides of neck black, sharply delineated from chin and throat; black of mantle merges into mid grey of back. Scapulars white; rump and uppertail-coverts white, sometimes with a few greyish feathers. Tail black, graduated, with white sides, outer feathers with shortfall of *c.* 20–25; T1 black, T2 black or with tiny white mark at tip, T3 with small white patch at tip, T4 with large white patch at tip, T5 mostly white, inner vane with black base, and black shaft streak narrowing towards tip, T6 white. Entire underparts white. Wing glossy black, but most primaries with white bases, forming white patch *c.* 30 long, and tertials and inner secondaries broadly white-tipped. Underside of tail more white than black, axillaries black and underside of wing slate with white primary patch. Bill black; eyes dark brown; legs and feet black. ADULT ♀: like ♂ but axillaries pale brownish grey.
SIZE (unsexed, n = 4): wing 98–103 (100); tail 88–100 (92·7); bill to feathers 17; tarsus 25–28. WEIGHT: ♂ (n = 3) 54–58 (56·3), 1 ♀ 56.
IMMATURE: juvenile with forehead, crown, hindneck, mantle and back grey-brown, finely vermiculated in blackish brown; rump and uppertail-coverts buffy white, similarly vermiculated. Scapulars whitish, feather tips buff. Mask (lores, area below eye and ear-coverts) dark brown. Tail dark brown, feathers tipped and narrowly edged whitish buff. Underparts off-white, with evenly spaced pale brown bars, fine on breast and sides of face, more widely spaced on rump and flanks; throat and middle of belly white. Wing black, variegated with warm buff and off-white: upperwing coverts, tertials and outer secondaries with black centres, buff outer webs, white or whitish edges and white tips, and narrow black subterminal line. Bill black with yellowish brown base.
NESTLING: not described.

Field Characters. Length 20–20·5 cm. A short-tailed fiscal of arid country in the Horn of Africa. Very similar to Taita Fiscal *L. dorsalis*, in flight showing white primary patch on black wing, white scapular patches beside grey back, and white rump; differs in having *broad white tips* to secondaries, prominent as bird swoops up to perch (Zimmerman *et al.* 1996), broader white edges to tail, and *wholly* white undertail when tail closed (Lefranc and Worfolk 1997). ♀ lacks chestnut on flanks. Juvenile barred like juvenile Taita Fiscal but brown above, not dark grey. Prefers more arid habitat than Taita Fiscal; for areas of overlap see that species.

Voice. Tape-recorded (B). Song composed of short, variable phrases, 'bur-ur-ur, bit-it-it ...'; common call a low churr of alarm (Zimmerman *et al.* 1996).

General Habits. Inhabits sandy wastes, plains with only a few scattered shrubs, wadis, sub-desert steppe and almost pure desert; in Somalia occurs on arid hillsides and in sandy, stony maritime plain far from nearest surface water. Occurs singly and in pairs; perches on topmost twig of bush or on outermost branch of thin thorn tree, and often on telephone wires. Flight direct, low down, rather weak and whirring, seldom very far. Aggressively territorial; flies out from perch to chase or attack passing birds, especially crows and ravens, also rollers, cuckoos and some hawks (Archer and Godman 1961). Range overlaps that of Taita Fiscal in NW Kenya and marginally in SE Somalia, the 2 separated by habitat (Somali Fiscal in less dense woody growth and rather more arid conditions: Ash and Miskell 1998) and, where pairs meet, mutually exclusive by territorial aggression. Keeps watch from lookout perch for prey on ground; flies down to alight next to insect, grabs it in bill and returns to perch to deal with it. Hard items like beetles are squeezed several times, the elytrae falling off or being torn off and, with regurgitated pellets, forming characteristic litter under favourite bush. Impales small birds on thorns.

Sedentary, but prone to wander, especially southwards in Kenya to northern periphery of central highlands, occurring exceptionally up to 1500 m (Lefranc and Worfolk 1997).

Food. Insects such as grasshoppers and mantises, and small birds.

Breeding Habits. Solitary nester, evidently monogamous and territorial.

NEST: shallow, open cup, composed of thorn twigs and silky or fibrous plant materials, sited in small bush only about a metre above ground or in centre of low-growing thorn tree, 1·8 m above ground.

EGGS: 3–4, usually 4. Broad but tapering. Cream or greenish white, heavily spotted with light sienna-brown or umber-brown on underlying ash-grey markings that may coalesce to form irregular nebula at either end. SIZE: (n = 12), av. 26·2 × 19·0 (Archer and Godman 1961).

LAYING DATES: Ethiopia, Mar, May–Aug, Nov; Somalia, Jan (2 clutches), Feb (1), Apr (5), May (2), June (6), July (2), Nov (1), Dec (3), tends to lay earlier near coast than in interior; Kenya, May, Nov.

References
Archer, G. and Godman, E.M. (1961).
Ash, J.S. and Miskell, J.E. (1998).

Plate 18 (Opp. p. 271) *Lanius cabanisi* Hartert. Long-tailed Fiscal. Pie-grièche à longue queue.

Lanius cabanisi Hartert, 1906. Novit. Zool., 13, p. 404, new name for *Lanius caudatus* Cabanis 1868, J. Orn., 16, p. 412; Mombasa.

Range and Status. Endemic resident, S Somalia to Tanzania. Somalia, common near coast from Kenya border (Boni Forest) and in low ground along lower and middle Webi Shabeelle valleys to about 03°N in Jubba Valley (where only a few records). Kenya, common from coast up to 1600 m in highlands, mainly east of Rift Valley, though also many records west of it; Samburu Game Res., Isiolo, Meru Nat. Park, lowlands south-east and south of Mt Kenya, Thika, Nairobi, Kitui, Namanga, Amboseli, Tsavo West Nat. Park, Tsavo East Nat. Park except northeast third (and, where does occur in Tsavo East, much rarer than Taita Fiscal *L. dorsalis*), Galana R., lower Tana R. and coastal plains up to Lamu and Kiunga; in Rift Valley from west of Ngong to Nkito Hills, and west to Mara Game Res., Sotik area and about Kisii-Suna road. Tanzania, from Kenya border southwest to L. Manyara Nat. Park, Tarangire Nat. Park, Mkomazi Game Res., Masai Steppe, Kilosa, Morogoro, Kidugallo, N half of Selous Game Res., and outlying population in Usangu Flats (N.E. Baker, pers. comm).

Description. ADULT ♂: forehead, crown, nape, hindneck, upper mantle, lores, below eye, ear-coverts and sides of neck black, sharply defined from chin and throat, grading into dark grey on

Lanius cabanisi

lower mantle, mid grey on back and scapulars, and white on rump and uppertail-coverts. Uppertail-coverts sometimes slightly barred. Tail black (with white base concealed by white uppertail- and undertail-coverts), long and strongly graduated (T6 with shortfall of 55–65 mm), outer feathers with tiny white mark at tip, in fresh plumage, which abrades away. Entire underparts white. Wing glossy black, primaries except P10 with white bases, forming white patch. Underside of wing mainly white, axillaries black; underside of tail brownish black and white. Bill robust, black; eyes dark brown; legs and feet black. ADULT ♀: like ♂ but a part-concealed, diffuse chestnut patch on flank. SIZE (♂♀, n = 4): wing 105–118 (112); tail 150–180; bill to feathers 17–20; tarsus 28–32. WEIGHT: ♂ (n = 3) 75–79 (77·3), ♀ (n = 2) 70, 74.

IMMATURE: forehead, crown, hindneck, mantle and back grey-brown, finely barred or vermiculated in blackish brown; grey-brown grades to buff on rump and buffy white on uppertail-coverts, rump and coverts being barred with well-spaced, zigzag, thin dark brown lines. Blackish mask (lores, upper cheeks, ear-coverts). Tail blackish brown, T3 to T6 with small buff mark at tip. Wing slaty with white patch in primaries; greater coverts, tertials and outer secondaries with narrow buff edge to outer webs and broad buff tips, the buff with narrow black line on its inner side; outer secondaries buff tipped; median primary-coverts white tipped; median coverts grey-brown with double dark lines at tips. Underparts off-white or buffy white, with well spaced pale brown bars on sides of breast and flanks. Bill black with greyish horn base.

NESTLING: not described.

Field Characters. Length 26–31 cm. A large, conspicuous, noisy and sociable shrike of dry country, with *dark grey back*, white patches on black wings and white rump, distinguished by *long, rounded black tail* with almost no white; ♀ has chestnut flanks. Smaller Fiscal Shrike *L. collaris* of higher ground and damper habitats has black back, white scapulars and shorter, white-sided tail. Juvenile grey-brown above with fine barring, black mask and white rump; readily identified by long black tail.

Voice. Tape-recorded (B, C, F, CHA, GREG, HOR, McVIC, NOR). Calls include chattering and scolding, a mellow whistle and a harsh 'chit-er-row' (Zimmerman *et al.* 1996); chorus starts with a 'cha cha raa' (Lefranc and Worfolk 1997).

General Habits. Inhabits thornbush and open *Acacia* country with scattered trees, grass and cultivation. In Tsavo East Nat. Park occurs in bushy savannas, but always in lusher areas near watercourses or damp ground, and not on open, dry plains (Lack 1985). Occurs in pairs, each pair spending most of daytime foraging in its territory; several times a day, mainly in early morning and late afternoon, 2–3 pairs congregate at top of a thorn tree and indulge in noisy and conspicuous group display (**A**). Perching rather erectly, with wings down and slightly open, each bird vigorously swings its long tail up and down, from side to side, in a figure of eight, and fans and cocks it over back, whilst the group calls 'cha cha raa' then 'chit-er-rowe, chit er row ...' in loud chorus. Forages from prominent, exposed perch on top or side of small tree, where ground can be scanned. On seeing prey, bird flies steeply down to ground, seizes it, and returns to perch to deal with it. A very large grasshopper was secured on ground only with difficulty, and carried in bill back to tree perch; there the shrike bit its head several times then tore off and discarded insect's head, back legs and wings; prey was passed to and fro sideways in bill, then carried and given to nestling (van Someren 1956). Of 81 prey items in Tsavo East, Kenya, 90% taken from ground and ground herbage, and 9% from leaves in bushes; the only *Lanius* shrike there that appears never to catch flying insects (Lack 1985). Remarkable association with Red-billed Buffalo-Weaver *Bubalornis niger* (Brosset 1989): a single shrike and a flock of buffalo-weavers accompanied each other for 90 min, buffalo-weavers foraging on seeds on ground and evidently using shrike perched on bushtop as a sentinel (much as Desert Warbler *Sylvia nana* uses wheatears *Oenanthe* spp.: Vol. V, p. 395); shrike repeatedly caught insects disturbed by buffalo-weavers; when a predator approached the birds all scattered, but soon re-assembled.

Sedentary so far as known, but most occurrences south of 05°S may be of wanderers (Lefranc and Worfolk 1997).

Food. Insects including grasshoppers (some very large) and beetles, and small vertebrates: lizards, snakes and

Plate 23

Plate 24

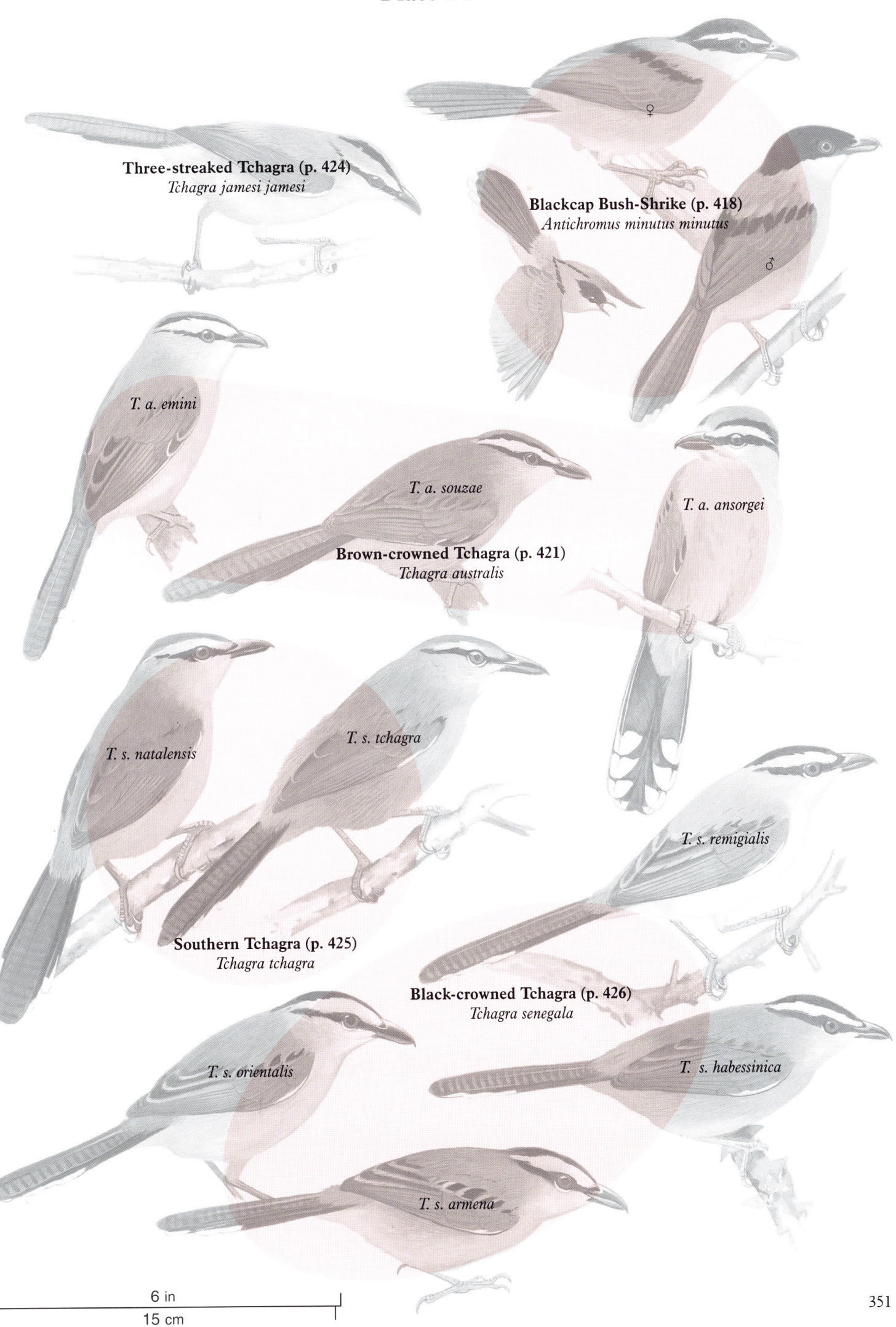

young birds (van Someren 1956). Also fruits of *Salvadora persica* (4 records, Tsavo East, Kenya, Lack 1985).

Breeding Habits. Solitary nester, evidently monogamous and territorial. Group displays are most eye-catching just before breeding begins, but no sure evidence that species (poorly studied as it is) is a co-operative breeder.

NEST: rather large, thick-walled, open cup, made of rootlets, twiglets, bits of grass, bark fibres and grass roots, and lined with fine grass roots; spider web often worked around rim of cup, and sometimes felted mass of spider web is added as lining. Sited 1·5–2·5 m high inside thorn bush.

EGGS: 3–4, usually 3. Putty-coloured, darker at large end, covered with faint yellowish brown and grey under-marks with sepia and dark brown spots on top. Size: (n = ?) c. 25 × 19.

LAYING DATES: Somalia, Apr, May, Sept; E Africa, Region D, Sept–Oct, Dec–May, mainly Apr–May and Sept–Dec, Region E, Aug, Oct.

INCUBATION: begins at or near time first egg laid. Period: 13–14 days.

DEVELOPMENT AND CARE OF YOUNG: young of a brood vary in size. Period: 16–18 days. Young leave nest when hardly able to fly, and take shelter in thorn tree for some time, later perching on top with parents.

References
Brosset, A. (1989).
van Someren, V. G. L. (1956).

Plate 19 (Opp. p. 286) *Lanius mackinnoni* Sharpe. Mackinnon's Shrike. Pie-grièche de Mackinnon.

Lanius mackinnoni Sharpe, 1891. Ibis, p. 444, 596, pl. 13; Kikuyu (= Bugemaia, Kavirondo, Kenya).

Range and Status. Endemic resident, W Cameroon to NW Angola, S and NE Zaïre to W Kenya. Nigeria, frequent on Obudu Plateau. Cameroon, fairly common in open country at 580–2000 m: Essosong at 770–930 m, Bamenda, Mt Manenguba at 2000 m, Mt Kupé, Saxenhof, Buea, Dikume Balue, and east to about 13°E. Gabon, common and widespread in N and centre, perhaps absent from SW and E; record Yaoundé (Quantrill and Quantrill 1998). Mayombe country (SW Congo/Angola [Cabinda]/W Zaïre): common, close to sea level (Chapin 1954), though not found in Kouilou Basin (Congo) by Dowsett-Lemaire and Dowsett (1991). Angola, locally common, N Cuanza Norte and at Gabela, Cuanza Sul (W.R.J. Dean, pers. comm.). Zaïre, records in Kasai Oriental and W Katanga (Hall and Moreau 1970); locally frequent in Haut-Zaïre and Kivu (known from over 20 localities in clearings in wooded and forested country up to 2500 m, but absent from savanna woodlands in far NE); very common in Itombwe, at 870–2000 m (Prigogine 1971). Uganda, widespread but uncommon, 700–2200 m, Budongo, Bwamba, Ankole, Kigezi east to Jinja and near Mt Elgon. Kenya, uncommon, 1500–2000 m, Nandi and Kakamega districts to Kericho, Sotik, Kilgoris, Saiwa Nat. Park and Mara R., and (at least formerly) west to Kenya border including Elgeyu and Mt Elgon. Rwanda, widespread, but absent from SE. Burundi, widespread but sparse, at 1260–2200 m: known from Lua valley, Rwegura, Teza, Bugarama, Ijenda, Bururi, Bugesera, Kanzigiri valley, Mukenke, Kisanze, Gashoho, Ngozi, Kirundo, Karuzi, Ruvubu valley, Kigamba and Musigati (Gaugris *et al.* 1981). Tanzania, 4 recent records in extreme NW, 2 in W Karagwe, 2 near Bukoba; may occur also on SE shores of L. Victoria (N.E. Baker, pers. comm.).

Description. ADULT ♂: forehead white in 2-mm band by bill; forecrown, crown, hindneck, mantle and back dark grey, narrow but clearly defined white superciliary stripe above black mask, scapulars white, rump and uppertail-coverts grey (paler than back). Tail black, graduated, T6 with shortfall of 40–45, T2–T6 broadly tipped with white. Mask (lores, ear-coverts) black.

Lanius mackinnoni

Underparts white or off-white with faint buffy tinge. Wings brownish black. Underwing coverts mottled grey and white, axillaries grey, undersides of flight feathers silver-grey; underside of tail blackish brown with white corners. Bill black; eyes dark brown with outer greyish ring; legs and feet black, soles yellowish grey. ADULT ♀: like ♂ but for a large and distinct but hidden rufous-chestnut patch on flank. SIZE: wing, (16 ♂♂ and 16 ♀♀ combined, W Africa) 84–91, (6 ♂♂ and 6 ♀♀ combined, N Ituri, Zaïre) 81–87, (n = 15, sexes combined, E Zaïre highlands) 84–92; tail, ♂ (n = 16) 97–104, ♀ (n = 16) 92–104; bill, ♂♀ (n = 32) 15–16; tarsus, ♂♀ (n = 32) 24–25. WEIGHT: ♂ (n = 9, E Africa) 30–35 (33·3), ♀ (n = 10, E Africa) 30–38 (35·3).

IMMATURE: upperparts grey-brown, finely barred with black on top of head and with more widely-spaced black bars on mantle, back and rump; uppertail-coverts black-tipped; scapulars white, feathers buff-tipped with dark subterminal bar; tail brownish black, all feathers narrowly tipped pale buff and with blackish subterminal bar; mask brownish black; underparts buffy off-white, finely banded or vermiculated on sides of breast and flanks; wings blackish, all feathers narrowly buff-grey edged; juvenile ♀ lacks rufous flank patch.

NESTLING: (at stage when tail half grown), upperpart feathers with buffy brown tip and subterminal narrow black bar; underparts buffy, breast and flanks indistinctly barred with blackish grey; bill dusky, corners of mouth yellow, iris grey with blackish inner rim, feet light grey.

TAXONOMIC NOTE: no constant differences are evident between the W African and E-central African populations; since they appear to be sedentary, they are likely to be in contact via undiscovered corridor populations in Zaïre. Relationships within genus are uncertain; *L. mackinnoni* thought by Brosset and Erard (1986) to form a superspecies with *L. excubitor*, but heavily barred juvenile, and rufous flank-patch of adult ♀, seem to ally it more with the *L. collaris* group of shrikes.

Field Characters. Length 20–21 cm. A shrike of *forest edge* in humid equatorial Africa. Grey above with black mask, *white supercilium*, white scapulars, and black wings with *no white* primary patch; narrow, graduated black tail with white corners. ♀ has chestnut flank patch. Juvenile has dark mask and pale scapulars of adult and is heavily barred above and below. Shares some habitats with Fiscal Shrike *L. collaris*, but latter has black head and back (and sympatric races lack white supercilium). Adult Lesser Grey Shrike *L. minor* lacks white supercilium and scapulars but has white primary patch; juvenile pale grey-brown above, almost unbarred below.

Voice. Tape-recorded (104, B, C, F, GREG, GRI, HOR, McVIC, STJ). Song a long, rambling, disjointed medley of chips, squeaky notes, sunbird-like 'whit-whit', buzzy 'jit', scratchy 'jew-rrrree' with watch-winding quality, little running warbles and trills, and imitations of other birds, especially bulbuls Pycnonotidae. Calls include a low, musical 'chickarea' and a low churr while young are being fed (Lefranc and Worfolk 1997).

General Habits. On Obudu Plateau, Nigeria, inhabits edges of montane forest and immediate environs of houses and hotels; in Gabon, around habitation, in bushy grassland, gardens, parks, and recent clearings in primary forest; in Itombwe, E Zaïre, in secondary forest, on fallow land, bushy open places and gardens; in Kenya shrubby grassland, borders of cultivation and edges of forest. Keeps to middle and lower levels of vegetation; solitary or in pairs. Behaviour recalls Great Grey Shrike *L. excubitor* (Brosset and Erard 1986). A sentinel feeder, perching in exposed places: dead branches, ends of projecting lateral branches, and tops of bushes and small trees; seizes most prey items in ground herbage and on ground; also takes insects at edges of clumps of leaves, and occasionally in flight; regularly exploits hatches of winged termites. Impales prey, generally on *Citrus* spines (M'Passa, Gabon) – usually animals too large to be eaten whole; 'larder' often holds geckos and small nestling birds. Returns later to tear impaled prey piece by piece, by tugging with bill. Lives in pairs, a pair keeping in territory year-round, the 2 birds often quite far apart. Twice seen following raiding swarms of *Dorylus* ants on ground, perching 2–5 m above them. Territory 1–6 ha in extent: size determined by amount of bushy cover (territory very large if few bushes and if bushy canopy too closed). Locally, territorially exclusive with Fiscal Shrike *L. collaris*. Can be pugnacious, driving other birds away from territory. Rather silent; song includes mimicry of other birds, notably bulbuls (a perfect mimic of Common Bulbul *Pycnonotus barbatus gabonensis*, and of 'kuiu' call of Speckled Mousebird *Colius striatus*).

Adults sedentary, juveniles and immatures subject to wandering (Gabon).

Food. Large insects including grasshoppers and beetles, small insects including termites, bugs and caterpillars; and small vertebrates, mostly geckos, also frogs and nestling estrildids and once a sunbird (*Cyanomitra verticalis*) (Bannerman 1939, Brosset and Erard 1986). Of 11 stomachs, grasshoppers in 5, beetles in 5 and caterpillars in 1.

Breeding Habits. Solitary nester, monogamous, territorial.

NEST: like that of Great Grey Shrike; a bulky cup, made of rootlets, twigs and plant stems, and lined with fine fibres and once 'silk' from maize cobs; built in thorny bushes such as *Citrus* (Gabon) or placed rather low down in trees; one in tree in grassy hollow on mountainside at 2000 m in Cameroon (Serle 1950). Size of 1 nest: ext. diam. 110, int. diam. 66, ext. depth 75, int. depth 50. One pair raised 2 broods in same season (Dec, Apr) in Gabon.

EGGS: 2–3. Creamy white or buff when fresh, dirty grey when well incubated, irregularly spotted with numerous yellowish brown and violet or lilac-grey speckles. SIZE: (n=5) 22.1–23.5 × 17.5–18.0 (23.0 × 17.65).

LAYING DATES: Cameroon, Jan, Feb–Mar and Aug (i.e. nestlings Mar–Apr and Sept), Oct or Nov (i.e. short-tailed juvenile with parents), Dec (i.e. ♀ with ruptured ovarian follicles) (Serle 1965); Gabon (M'Passa), (sings, displays and copulates in Aug–Apr; flying young being fed by parents, mid Sept to end of May); Angola, Mar–Apr; NE Zaïre, (carrying nest material, Apr; short-tailed flying young in May–July and Sept); E Zaïre (Itombwe), Aug, Oct, Jan (and juveniles July, Sept–Oct); E Africa, Region B, Feb–Aug, mainly Feb–Apr (10 out of 15 records).

INCUBATION: evidently by ♂ as well as ♀ (Bates 1911).

References
Bannerman, D.A. (1939).
Brosset, A. and Erard, C. (1986).
Lefranc, N. and Worfolk, T. (1997).

Plate 18
(Opp. p. 271)

Lanius nubicus Lichtenstein. Masked Shrike; Nubian Shrike. Pie-grièche masquée.

Lanius nubicus Lichtenstein, 1823. Verz. Doubl. Zool. Mus. Berlin, p. 47; Nubia.

Range and Status. SE Europe (Greece, Turkey, Cyprus, Levant) and SW Asia, winters in Africa, and a very few in SW Arabia. Very rare vagrant in NW Africa, Spain, France and Malta, also Sweden and Finland.

Palearctic migrant, common in autumn, winter and spring between about 15°N and 11–12°N, from Chad to Eritrea and N Ethiopia.

Vagrant on coast of Algeria (1, Apr) and Libya (2–3, Mar). Mauritania, uncommon in SE, in N and middle sahel zone (Ayoun el Atrouss, Boïbou, Néma) in Aug–Apr. Mali, uncommon, only in sahel zone in E, in Aug–Apr. Niger, vagrant: Zinder, 1927, and between Birni-Nkonni and Niamey in 1984. Nigeria, uncommon, sahelian zone in NE, near L. Chad, between Nguru and Bama, Molai and Pompomarii, Oct–Apr. Cameroon, rare: Waza area, 2 records. Chad, frequent at Ndjamena, Nov–Feb; uncommon in Ennedi; very common wintering in E, and abundant on autumn and spring passage at Abéché, mid Oct to mid Nov and in Feb, departing by mid Apr; in Ouadi Rimé-Ouadi Achim Faunal Res. common in winter, early Oct to late Mar. Sudan, seasonally common, mainly north of 11°N; main wintering grounds are west of Nile, between 12° and 14°30′N; probably ranges much further west into Kordofan than shown, and even into Darfur; occurs in all months (and may possibly breed: Hogg *et al.* 1984); west of Nile, uncommon to frequent, June–Aug, common Sept, frequent Oct–Apr; in Nile Valley, commonest (in autumn) from late Aug to mid Oct, but commoner in spring – very common during whole of Mar; very common autumn migrant from late Aug to middle third Sept in Red Sea Hills, and at Khor Arba'at near coast composes 1·6% of all passerine migrants (Nikolaus 1983). Ethiopia, mainly below 1500 m, frequent mid Aug to mid May and frequent to common, Sept–Apr (once July) in NW and NE lowlands and Rift Valley; *c.* 5 birds seen daily (Ash 1980); rare in W Highlands; one or a few records in S (though Urban and Brown 1971 and Ash 1980 imply that bird is common and widespread in S); limits on map somewhat speculative. Djibouti, noted (Laurent 1990). Eritrea, common winter visitor, Sept–May, mainly below 1540 m, but as high as 2460 m on passage in Cohaita; common in Dancalia; rare, July (1-year-old birds that had probably summered); not recorded in Dahlac Archipelago (though does occur in Farasan Is, Saudi Arabia, in spring: Felemban 1995). Somalia, regular but uncommon around Hargeisa (Clarke 1985); records from 6 other localities in NW. Zaïre, old record L. Albert. Kenya, vagrant, almost annual at L. Baringo; single records L. Kanyaboli and L. Naivasha. Egypt, scarce in winter in extreme S (Meinertzhagen 1930); only recent midwinter (Dec, Jan) records are from Ismailiya, Fayid and El Kabrit; uncommon passage migrant from late Aug to mid Oct (rare, late July, late Oct); common passage migrant mid Mar to late Apr (uncommon, late Feb, late June); commonest in last third of Mar; autumn passage concentrated in Nile Valley and E delta to Sinai; in spring

Lanius nubicus

occurs almost throughout Nile Valley and Eastern Desert (rare at Bahig and only 2 records in Western Desert), but is commonest in S Nile Valley and along Red Sea coast (Goodman and Meininger 1989).

Description. ADULT ♂: forehead and superciliary stripe white, the stripe triangular, broad at front, tapering to rear; crown, hindneck, lores, ear-coverts, sides of neck, mantle, back, rump and uppertail-coverts glossy black. Tail rather narrow, graduated, black with white sides: T1–T3 black, T4 with narrow white outer edge, widening towards tip, inner vane usually with white wedge at tip, T5 mainly white, with long black wedge along shaft and inner vane, T6 almost entirely white. Scapulars white. Chin, throat, centre of breast, vent and undertail-coverts white; sides of breast and of belly yellowish buff, grading to pale or medium rufous on flanks. Wings black, but base of all primaries white, forming long white crescent in extended wing. Underwing coverts white, but greater under primary coverts dull black; axillaries white. Underside of closed tail white. Bill black; eyes dark brown; legs and feet black. ADULT ♀: crown dull black (not glossy) and rest of upperparts dark grey; white forehead-and-supercilium patch not as well defined as in ♂, and white scapular patch a lot less well defined. Tail as in ♂. Mask brownish black with some buffy mottling. Underparts as in ♂ but slightly paler and with less rufous on flanks. Wings as in ♂ but brownish black, nearly all feathers with narrow greyish outer edges. Bill black with brownish or greyish tinge, grading to greyish at base. SIZE: wing, ♂ (n = 48) 87–95 (90·7), ♀ (n = 22) 87–96 (91·1); tail, ♂ (n = 31) 83–92 (86·6), ♀ (n = 15) 82–90 (86·3); bill to skull, ♂ (n = 51) 16·7–19·9 (18·2), ♀ (n = 23) 17·0–19·3 (18·2); tarsus, ♂ (n = 18) 22·0–23·6 (22·7), ♀ (n = 16) 21·8–23·7 (22·8) (Cramp and Perrins 1993). WEIGHT: migrants, (Chad, unsexed, n = 20) 20–37 (23), (one, Kenya) 21; also Sinai (n = 15, autumn) av. 22·6 and (n = 7, spring) av. 21·0; and residents, Cyprus (n = 16, Mar–Apr) 14·5–23·1, (n = 11, June–Aug) 23–30.

IMMATURE: forehead, crown and hindneck brown, heavily mottled with whitish; lores and ear-coverts dusky, but no dark mask and no pale superciliary stripe; usually a small whitish mark above and behind eye. Mantle to uppertail-coverts brown, sometimes with rufescent tinge, mottled whitish or pale greyish, but less mottled than on head. Tail dark brown with white sides. Chin, centre of throat, and most of midline of breast and belly, creamy white; sides of throat, breast and belly, and flanks, greyish cream, quite heavily freckled with brown. Undertail-coverts cream, tipped blackish. White or whitish feathers in scapulars form irregular patch or line. Wings brown, remiges dark brown, all feathers with narrow buff outer edge and tips, less narrow on tertials; small white mark shows at base of primaries; buff-edged innermost secondaries and tertials can form quite conspicuous buffy block in tight-closed wing. Bill grey with blackish tip.

Field Characters. Length 17 cm. A smallish, slender shrike, black above from crown to tail, broad *white forehead and eyebrow*, narrow black mask, *orange patch on flanks*; ♀ dark grey-brown above, with duller flank patch. When flying away shows same white scapular and wing patches as Woodchat Shrike *L. senator*, but *rump black*, tail longer and more slender. Looks small-headed compared to Woodchat, bill narrower, and often perches with tail slightly cocked and waved up and down, unlike Woodchat (Porter *et al.* 1996). Barred juvenile very undistinguished, can even look like *Melaenornis* flycatcher (C. H. Fry, pers. comm.); like juvenile Woodchat but grey above, not brown, some white already showing on forehead and eyebrow, and *rump pale*. Flight variously described as light, weak or graceful, less sustained than other shrikes, with fewer undulations.

Voice. Tape-recorded (62–73, 104, 110, B, F). Song a hurried and sustained jumble of twittering and scratchy notes, lasting up to 1 min; often described as warbling but it lacks any sweet or melodic tones, and sounds instead like dry, grating songs of *Hippolais*, *Acrocephalus* or *Sylvia* warblers, or at times like twittering of Barn Swallow *Hirundo rustica*. Largely silent in Africa, although occasionally gives subsong; a quiet 'crret crret' given while foraging; other calls include hard, chattering 'chek-chek', harsh 'keer keer keer' during chase, and 'krrr' like that of Woodchat in alarm.

General Habits. Inhabits low, hot acacia country, thorn scrub, wadis, sub-desert steppe, wooded oases and palm groves; in Eritrea almost any woodland, but mainly *Acacia* and thorn scrub, also junipers on passage; near L. Langano, Ethiopia, orchards, *Eucalyptus* plantations, gardens and resorts. Solitary in Africa, though 2–3 birds commonly in sight of each other in Egypt (and in Cyprus up to 7, occasionally 40, concentrate on migration). Strongly territorial in Sudan; migrants defend strictly delineated, temporary territory of 830–2250 m^2, for a few hours or days, mid Aug to mid Oct, and for shorter periods in spring (Egypt, Simmons 1951, 1954; also Cyprus).

Hunts by keeping watch from station on side of shrub or small tree, wire fence or telephone wire; swoops down onto prey, usually on ground, sometimes in grass or bush foliage. Takes flying insect after rapid, twisting chase, and often hovers above long grass, searching for prey (Kenya, Stevenson 1983). Seen carrying Lesser Whitethroat *Sylvia curruca* (half the shrike's weight) in bill and tearing at it at perch (Israel: Watson 1967); and seen to attack Lesser Whitethroats, Blackcap *S. atricapilla*, Willow Warbler *Phylloscopus trochilus*, Collared Flycatcher *Ficedula albicollis*, Barn Swallow *Hirundo rustica* and Common Redstart *Phoenicurus phoenicurus* (Jordan: Ferguson-Lees 1967); some of these were pursued very aggressively, struck, knocked down and killed. Impales insects, larvae and probably birds on thorns, palm spines and barbed wire (Egypt); doubtless it was this shrike species, the only one present, that had impaled a Lesser Whitethroat and, astonishingly, a Little Swift *Apus affinis* on palm spines (Jordan: Ferguson-Lees 1967). Eats small prey on ground, or flies back to previous or new perch in small tree. Holds prey under one foot and tears at it; seen to take insect from bill with foot, pull off wings and swallow wings and body (Cyprus, Cramp and Perrins 1993).

At perch frequently waves and swings tail in U-shaped arc. Sometimes cocks tail when giving contact-alarm call. Occasionally gives subsong in Africa. Makes harsh but quiet 'crret crret' whilst foraging (Kenya, Stevenson 1983). Trapped migrant feigned incapacity, lying on side with wing extended and clenched feet drawn into belly feathers (Egypt).

Main migrations in Sept–Oct and Mar–Apr in Egypt; winters in Oct–Feb in Chad and Sept–May in Eritrea (further details by country in Range and Status). Nile Valley is main migration route into Africa, and there may be a loop migration with spring departures tending to be further east (Zink 1975, Cramp and Perrins 1993). In any event, directions between breeding range in SE Europe and SW Asia and wintering range in Africa suggest that, whilst many migrants in Egypt and Sudan doubtless move N/S, those wintering west of Chad have moved NE/SW or ENE/WSW, and those in Eritrea and Ethiopia have moved NE/SW or NNE/SSW. Ringing recovery from K'ok'a, 08°27'N, 39°06'E, Ethiopia, Apr 1972 to Kuwait, May 1973. Imm. ringed Abéché in Oct killed in Tripoli, Libya, in Mar, 4.4 years later. Bird ringed Israel found in Sudan (Shirihai 1996). Bird ringed in E Chad found at sea off Israel and one ringed near Addis Ababa, Ethiopia, found in N Iraq (R.J. Dowsett, pers. comm.).

Food. Not studied in Africa. Eats mainly grasshoppers, beetles, and seasonally, exhausted small migrant passerines, particularly Lesser Whitethroats. Also dragonflies (Saudi Arabia), crickets, moths and ants, and small lizards (Cyprus).

References
Cramp, S. and Perrins, C.M. (1993).
Ferguson-Lees, I.J. (1967).
Lefranc, N. and Worfolk, T. (1997).
Simmons, K.E.L. (1954).
Stevenson, T. (1983).

Plate 19
(Opp. p. 286)

Lanius excubitoroides Prévost and Des Murs. Grey-backed Fiscal. Pie-grièche à dos gris.

Lanius excubitoroides Prévost and Des Murs 1847. *In* Lefebvre, Voyage en Abyssinie, 6, p. 99 (emended to *excubitorius* on p. 170 and pl. 8, because of *excubitorides* Swainson 1831: but *cf. excubitorides* and *excubitoroides*); Abyssinie et Nubie (= White Nile, Neumann, Ibis, 1927, 506–507).

Range and Status. Endemic resident and partial migrant at some borders of range; SE Mauritania to Ethiopia, south to S Tanzania. Mauritania, 2 records, localities not given (Lamarche 1988). Mali, quite common, not south of 15°N, in sahel and Sahara. Not in Burkina Faso or Niger (*cf.* Lefranc and Worfolk 1997). L. Chad basin: scarce and local in Nigeria, records from Malamfatori, Logomani and Chingurmi-Duguma; Cameroon records from Yagoua and Maroua; Chad, known from flood-plains of R. Chari and R. Logone. Chad, also in E at Am Timan and in Zakouma Nat. Park (Salvan 1967). Central African Republic, scarce and local, Manovo-Gounda-Saint Floris Nat. Park and Vakaga Préf. (Bretagnolle 1993). Sudan, frequent to common, Darfur, north in Nile valley to about Kosti, widespread south of 11°N except in SW, where absent; abundant in Bahr al Ghazal, along R. Jur. Ethiopia, frequent to abundant, in W lowlands and highlands north to 12°N (L. Tana), Rift Valley, SE Highlands, and S Ethiopia east to about 41° E; rare above 1800 m. Uganda, common throughout, except extreme NW. Kenya, locally common at 600–1900 m (and record from Mau Narok at 3000 m); 3 races, *intercedens* from Mt Elgon and Kapenguria to Kisumu, Kendu Bay and Ruma Nat. Park, *excubitoroides* in Kerio and Rift valleys from Baringo District to Naivasha, Longonot and Laikipia Plateau (where only occasional), and *boehmi* in Serengeti Nat. Park, Mara Game Res. and Loita Hills. Zaïre, numerous on dry plains near L. Albert and L. Edward; also on plains north of L. Kivu; record from Babuye country west of Fizi, SE Kivu Prov. (Hall and Moreau 1970). Uncommon in Rwanda and Burundi. Tanzania, widespread from West Lake, Mwanza and Serengeti Nat. Park east to Seronera Valley, to Ufipa, Rukwa, Chimala, Usangu and Songea; commonest in 3 areas – N Karagwe and N Biharamulo Districts in extreme NW near Uganda border, Serengeti, and Usangu Flats (N.E. Baker, pers. comm.).

Density of 0·40–0·45 birds per ha, Uganda, on 81·3 ha of Makerere Univ. campus (Banage 1969) and of 0·25–0·31 birds per ha, Kenya, in study site of 256 ha with 16 pairs, groups and group-territories (Zack 1986).

Description. *L. e. excubitoroides* Prévost and Des Murs: Mauritania and Mali, L. Chad basin, Chad, Sudan, NE Zaïre, Uganda except SW, Kenya (see above). ADULT ♂: deep band across forehead, lores, just above eyes, ear-coverts and in some birds line down side of neck, black; crown, nape, hindneck, mantle, scapulars, back, rump and uppertail-coverts pale bluish grey, paling nearly to white above eyes and ear-coverts and on rump and uppertail-coverts. Tail long, rounded, black, with proximal halves of outer feathers white; all feathers with tiny white tip; underside of tail black and white; T6 shortfall 35–40 but T5 shortfall <20 mm. Underparts white. Wing black but for large white patch formed by white bases to all primaries except P10. Bill black; eyes dark brown; legs and feet black. ADULT ♀: like ♂, but concealed rufous or chestnut patch on flanks. SIZE: wing, ♂ (n = 8) 104–119 (113), ♀ (n = 4) 116–120 (115); tail, ♂ (n = 6) 130–160 (139), ♀ (n = 4) 133–142 (137); bill to feathers, (sexes combined, n = 10) 16–17 (16·6); tarsus, (sexes combined, n = 10) 28–32 (29·25). WEIGHT (Kenya): ♂ (n = 42) 47–63·5 (55·3), ♀ (n = 26) 46·5–59 (52·2).

IMMATURE: like adult but greys darker and browner, blacks browner and less intense, whites less snowy; lores and ear-coverts dusky, forming a mask; forehead, crown, hindneck, mantle and scapulars grey-brown, barred blackish brown, the bars close together on head, wider apart on back; rump and uppertail-coverts zigzag-barred with pale brown, the bars thin and well separated; chin, throat and centre of belly off-white, lower cheeks, sides of breast and flanks lightly marked with narrow brown bars. Primaries brown-black with white marks as in adult; upperwing-coverts, tertials and inner secondaries dark brown, with broad buff edges and tips, and thin black line parallel to edge but within the buff border; the penultimate black line double near tip of greater coverts. Bill brownish black, base of lower mandible yellowish horn.

NESTLING: not described.

L. e. intercedens Neumann: Ethiopia (western limits uncertain) and W Kenya (see above). Like nominate race but larger: wing 116–126. WEIGHT: 1 ♀ 53.

L. e. boehmi Reichenow: SW Uganda (Kigezi and Ankole to Masaka), SW Kenya, Zaïre (Rutshuru Plain), Tanzania. Greys darker than in *intercedens*.

Lanius excubitoroides

Field Characters. Length 24–26 cm. Large, with long, broad tail. Upperparts pale grey, broad black mask from forehead across face and down *side of neck*; wings black with white primary patch, but no white scapulars; tail black with broad *white sides* to basal half. Readily distinguished from other fiscals and masked shrikes by behaviour: sociable and noisy, chorussing and chattering in groups from fence, telephone wire or other prominent perch while waving wings and wagging tails. Silent individuals easily told by large size and diagnostic tail pattern.

Voice. Tape-recorded (36, 40, 104, 109, B, C, F, GREG, STJ, ZIM). Calls given in group chorus have a clanging quality, brassy but not quite musical, with grating undertone, 'chewy ... chwayoo ... chewy ... chyoo ... chayoo'. Threat call a metallic 'kyoir-l'; display against intruder, 'kyoir-l, kyoi, kyoooh'; breeding ♂ gives brief song during nuptial feeding while ♀ utters loud 'pssh' just before being fed (Lefranc and Worfolk 1997). Loud harsh alarm given when hawk spotted (S. Zack *in* Lefranc and Worfolk 1997). Also gives a shrill, harsh 'jee-jee-jee-jee ...' accompanied by a few brief 'chup' and 'chip' notes.

General Habits. Inhabits sub-desert steppe (W Africa), wooded grassland, open woodland, mesic acacia country, gardens and cultivation (E Africa). Strongly associated with yellow-barked acacia *Acacia xanthophloea*, and commonest where it dominates and forms nearly monospecific stands with understorey of grass *Cynodon plectostachys* (Zack and Ligon 1985a). Perennial shrubs cover 3–23% of area of group territories (Zack and Ligon 1985b).

In pairs or, often, conspicuous, noisy groups or 'rallies' on wires and tops or outer branches on trees, which chatter, lift and half open wings and vigorously wave tail or fan tail and raise it a little; studied in Uganda (Banage 1969) and Kenya (Zack 1986). Pair or (generally) group occupies territory year-round, foraging and nesting in it. Often confiding. Group size up to 20 birds, mode 5, mean 5·3. Group consists of breeding pair and adult-plumaged helpers-at-the-nest (which are ♂♂ and ♀♀ in equal numbers; ♂♂ help at nest more than ♀♀), but pair is often not accompanied by its helpers. Indeed birds forage essentially solitarily, ♂ and ♀ not far from each other, helpers scattered throughout group territory. Forages by scanning open ground near treetop perch 2–4 m high and flying down to seize prey on ground, giving it single peck and quickly returning to tree perch. Sometimes flies up from bush to catch ants and termites on the wing. Impales caterpillars on thorns. Known to perch near gardener, seemingly to prey on insects he disturbs. Plunders nests in colony of Red-billed Queleas *Quelea quelea*. Rather silent, except when 'rallying' in territorial display; bird advertises by perching conspicuously on treetop; patrols boundary of territory and chases intruders; no evident contact calls; within group, adults not aggressive to each other (but juveniles are: see below). Flight direct, with rapid wing beats, ending in upward glide into tree. Roosts at night gregariously in thick acacia tree.

Birds gather only at territorial rallies (which occur commonly throughout year but are commonest just before nesting begins: about 12–15 rallies per day), or when patrolling territorial boundary (by flying along it, calling: Banage 1969), or where prey is concentrated, or when provisioning nestlings. Site fidelity high, and groups are rather stable, though several birds can leave a group about the same time and ♀♀ leave group to fill vacant breeding positions elsewhere. In a rally, all group members gather at a tall tree, sing, rock back and forth, raise wings (whilst singing) and spread tails; they frequently change perches. Between rallies, birds perch close to each other and preen themselves (never each other). 2 groups often rally at same time, several m apart. Birds of a party, having discovered an intruder from neighbouring group-territory, jump *c.* 30 cm up from perch with rapid wing-flapping, landing in same place with wings outstretched and tail fanned, and then swing tail as when rallying. Such display at intruder can last 30 min, and intruder usually departs without fighting; but if mated pair of birds intrude they may fight with territorial group; 2 adjacent groups sometimes scrap in flight, 2 birds grappling in the air (Banage 1969).

Resident, mainly sedentary; but disappears from Mali in rains (about July–Aug); both Mauritanian records are in dry season. Nominate race formerly thought to be a visitor in Kenya. Some birds certainly wander quite widely.

Food. Insects, small frogs or toads, nestling songbirds. Insects include caterpillars of moths *Cossus*, and driver ants *Dorylus*, both eaten extensively (Kenya, Zack 1986). Grasshoppers in majority of stomachs, beetles in some.

Breeding Habits. Solitary, monogamous, territorial, co-operative breeder. Territory, of group, (n = 7) 9–34 ha, larger ones being of poorer food quality (Banage 1969). Interspecifically territorial with Fiscal Shrike *L. collaris*, Kenya, group-territories of Grey-backed Fiscal forming patchwork with pair-territories of Fiscal Shrike (Zack and Ligon 1985a). Before nest-building starts, breeding pair isolates itself from other members of group. ♂ often feeds ♀, briefly singing as he does so; ♀ solicits food by drooping wings, flattening body and making demanding 'pssh' call. Copulation often immediately follows courtship feeding, and lasts 2–4 s.

NEST: compact, rather untidy cup, built of twigs (mainly acacia twigs with protruding thorns) and rootlets, lined with fine grasses and small feathers, placed on horizontal branch 1–>10 m above ground, only in *Acacia xanthophloea* trees (Kenya), near centre or peripherally on branch. Nest built by ♀; ♂ brings material to nest but does not help build it. Helpers rarely gather nest material. Never re-uses nest, but takes material from old nest to make new one; also takes material from nests of acacia rat *Thallomys paedulcus*, Superb Starling *Lamprotornis superbus* and Wattled Starling *Creatophora cinerea*. Eggs laid within 1–2 days of nest being completed, once 2 weeks later. Predatory birds, e.g. Long-crested Eagle *Lophaetus occipitalis*, mobbed by group which gives alarm calls, only when nesting under way.

EGGS: 2–4, mode 3, av. (n = 17) 2·8. Laid on successive days. Pale yellowish grey, with a few brown and grey spots. Size: av. 25 × 19.

LAYING DATES: Mali, Feb–Mar; Cameroon, (♀♀ about to lay, Apr; Bates 1927); Sudan, Apr–May, Aug; Ethiopia, Apr–June; E Africa, Region A, Mar–Apr, Nov, Region B, Feb–Dec (n = 61, mainly Mar–June but many Oct–Feb and July), Region D, Jan, June–Aug, Nov.

INCUBATION: by ♀ (not helpers). ♀ fed on nest av. 4·2 times per h, mainly by ♂, also by helpers, bringing single items at a time. Period: 13–15 days.

DEVELOPMENT AND CARE OF YOUNG: brood generally hatches on same day; once 3rd nestling hatched day after first 2. Young brooded by ♀ (not helpers), who for first few days is fed on nest by ♂ and all helpers. Average number of provisioning visits to nest per h, 2·9 when nestlings up to 5 days old (mainly by ♂ breeder, also by ♂ helpers) and 7·4 when nestlings older than 5 days (mainly by ♂ breeder, also by ♀ breeder). Weight at age 10 days (n = 11) 28·5–44·5 (36·6 ± 4·4). Period c. 20 days, though some fledglings stay in or next to nest for a few days longer. Recently fledged birds keep inside dense acacia foliage, coming out to be fed; they are completely dependent on parents and helpers for food for 2 weeks, except that they glean a few small insects for themselves near perches. Can fly up to 25 m within a week of leaving nest, pursuing adults and begging with wings drooping and 'pssh' calls; adult often sings just before food item is transferred. Start to make predatory forays after 2 weeks, but inefficiently, bird staying on ground for several min pecking and probing at small items. 50 days after leaving nest, young forage as efficiently as adults. Rufous flanks of ♀♀ first evident 50 days after fledging. First elements of adult song at 75 days, when young bird can warble for several min. Juveniles are aggressive to each other, supplanting each other at perches, pecking and gaping. Young do not become helpers until at least 6 months old.

BREEDING SUCCESS/SURVIVAL: in Naivasha study, of 55 nests only 8 successful; 30 failed at egg stage and 17 at nestling stage; one nest branch snapped; 6 nests probably raided by large predator; Gabar Goshawk *Micronisus gabar* took nestlings from one nest; at another nest, nestlings became caught or impaled on acacia thorns.

References
Banage, W.B. (1969).
Lefranc, N. and Worfolk, T. (1997).
Zack, S. and Ligon, D. (1985a, b).
Zack, S. (1986).

Plate 19 *Lanius minor* Gmelin. Lesser Grey Shrike. Pie-grièche à poitrine rose.
(Opp. p. 286)
Lanius minor Gmelin, 1788. Syst. Nat. 1, p. 308; Italy.

Range and Status. Europe and W central Asia, from NE Spain, S France, Italy, Hungary and Poland to Russian Altai, south to N Greece, Turkey, Iran and Tien Shan. Winters in Africa.

Palearctic migrant. Winters in extreme S Angola, Namibia (except Namib desert and SW); Botswana (from Nata southwards); Zimbabwe (south of 20°S, mainly west of 30°E); South Africa, Transvaal (mainly N and W), Orange Free State (mainly NW), N Cape Prov. (W to Bushmanland), N Natal (S to Richmond), exceptionally S to East London, Staytlerville and Montagu; E Swaziland; and extreme S Mozambique. Common and widespread in arid acacia thornveld in Kalahari basin, where c. 85% of world population winters (Herremans 1998); also in central and N Namibia, and Limpopo catchment; frequent to uncommon elsewhere. Rare winter records further north, in Gabon, N and SE Zaïre, Rwanda, S Malaŵi and Zambia.

On southward passage, common to abundant in Egypt (west to Libyan border), N and central Sudan (especially Nile valley and Red Sea coast), E Chad, and Eritrea. Rare or vagrant further west, in Mauritania, Mali (Niger delta), Libya (west to Tripoli and Fezzan), Niger (Zinder), Nigeria (once, Lagos), W Chad (Ndjamena), Cameroon (Waza) and Príncipe I. Rare but regular in coastal Gabon. Scarce in S Sudan, Ethiopia, Somalia, Kenya, Uganda, Rwanda, Tanzania and Malaŵi, but frequent to common in Zaïre (mainly E and S), Angola, Zambia and Zimbabwe (Mainly W). On northward migration, locally common to abundant in Zimbabwe, Zambia, Malaŵi, Zaïre (especially NE), Tanzania, Rwanda, Uganda, Kenya, Ethiopia (Rift Valley eastwards), Somalia (mainly N), Djibouti and coastal Eritrea; but no spring records from Chad or Sudan, and scarce in Egypt (mainly Red Sea coast). Spring vagrant

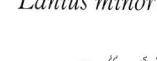

Mauritania (Atar, Chinguetti, Tidjikja: Lamarche 1988) and Mali (Niger delta: Lamarche 1981). Birds in Niger (June), Uganda (July), Zambia (June) and NE Zaïre (June and July) may have been oversummering.

Wintering density: in Kalahari thornveld, 1 bird per 13·3 ha; in central Namibian Khomas Hochland, 1 per 10·3 ha; in human-disturbed arid woodland, 1 per 10·2 ha (Upper Limpopo catchment), 1 per 14 ha (N Namibia)

(Herremans 1998). World population, estimated from wintering survey in southern Africa, 6·1 million (Herremans 1998).

Description. ADULT ♂ (breeding): forehead and lores black, extending as broad mask above and below eye to ear-coverts. Crown to uppertail-coverts, sides of neck and scapulars light grey. Central tail feathers (T1) black, T2 black with white at base and narrow white tip, T3–T4 with basal half to two-thirds white and broad white tips, T5 white with black subterminal patch on inner web, T6 all white. Underparts creamy white, tinged vinous pink on breast and flanks. Flight feathers, tertials and upperwing-coverts black, primaries with large white patch at base (extending 12–19 mm beyond primary coverts in closed wing). Underwing-coverts and axillaries white. Wing rather long, tip pointed: P8 longest, P7 3–6, P6 10–14, P5 15–20, P1 39–44 shorter; P9 2–4 shorter than P8 (= P7 or between P7 and P8); P10 very small, about equal to primary coverts; outer webs of P7 and P8 emarginated. Bill black, or grey with darker culmen and tip; eyes brown; legs slate-grey or blackish. ADULT ♀ (breeding): like breeding adult ♂, but black forehead band narrower, sometimes with grey feathers; mask tinged brown; underparts less pink; black subterminal areas on T4 and T5 usually larger. ADULT ♂ and ♀ (non-breeding): like breeding adult but forehead grey with variable black mottling. SIZE (10 ♂♂, 10 ♀♀, Europe and Asia): wing, ♂ 115–128 (120), ♀ 114–128 (119); tail, ♂ 86–102 (91·4), ♀ 84–97 (90·4); bill, ♂ 21–22 (21·4), ♀ 20–22 (21·3); tarsus, ♂ 25–28 (26·6), ♀ 25–27 (25·7). WEIGHT: ages and sexes combined: NE Sudan (n = 29, Aug–Sept) 28–52 (36·6); NE Chad (n = 2, Sept) 36, 41; Zaïre (n = 2, Oct) 41, 46, (n = 1, Mar) 47; Kenya (n = 5), Apr) 36–45 (42·4); Zambia (n = 4, Nov) 44–47 (45·8), (n = 4, Mar–Apr) 45–60 (50·1); Zimbabwe (n = 4, Feb–Apr) 41–51 (46·4); South Africa, Namibia and Botswana (n = 11, Nov–Mar) 29–63 (47·4).

Complete moult in Africa, Dec–Mar; partial moult July–Sept.

IMMATURE: first-winter bird before moult differs from adult in having upperparts, including forehead, brownish grey, often with juvenile barring on forehead, crown and mantle; black band from lores to ear-coverts brownish black; underparts creamy-white; black areas of wings and tail tinged brown, flight feathers, tertials and wing coverts fringed whitish at tips; bill horn-brown, paler at base.

Field Characters. Length 20–22 cm. Medium-sized, grey above with black wings and black mask which extends in spring birds as broad band across forehead; white underparts tinged pink. Rounded black tail shows white sides, and white bases to primaries form prominent short wing-bar in flight. Autumn adults have grey forehead with black mottling; first winter birds are browner grey above, with white-tipped wing feathers, and lack any black on forehead. Distinguished from Southern Grey Shrike *L. meridionalis* by shorter bill, shorter, less graduated tail, longer and more pointed wings; black wing not separated from back by white area (all plumages), and breeding adult has single white wing panel on otherwise all-black wing (no white wing-tips). Southern Grey lacks black forehead band, and most races have white superciliary stripe, whitish outer scapulars forming patch between wing and back, and white tips to secondaries show in flight.

Voice. Tape-recorded (62–73, 88, 91–99, 105, 110, B, F, McVIC). Song a long, bustling medley of shrill chips, chirps and twitters including harsh shrike-like grating and twanging drongo-like sounds as well as musical whistles; for transcription see Maclean (1993); lower-pitched and usually louder and more strident than song of Red-backed Shrike *L. collurio*. Much mimicry of other birds, including call of Red-chested Cuckoo *Cuculus solitarius*, song of Eurasian Blackbird *Turdus merula*, churr of Blue Tit *Parus caeruleus* and crowing of rooster; for further examples see Cramp and Perrins (1993). In Namibia in winter, song typically quiet, but by mid Mar became harsher and louder (Sauer and Sauer 1960). Most frequent and characteristic call is 'kerrib kerrib', often used in territorial defence, including against humans (Lefranc and Worfolk 1997). Other calls include 'bojit' and 'boy-ee-chip', also harsh 'jaa' or 'grrraaa,' and lower, more nasal 'jer'. Alarm, a grating, discordant 'tchak' or 'tshek-tshek ...'; also 'geer-geer' when excited, and a rattling 'tr-tr-tr-trrr' when attacking predators or defending territory. For complete catalogue of calls see Cramp and Perrins (1993).

General Habits. Inhabits arid acacia savanna, especially where bushes and tall trees alternate with grassy areas; also cleared bushland and edges of cultivation. On passage, occurs in wide range of open habitats with scattered trees or wires for perching, sometimes in more lush situations: cultivated country, grassland, thornbush, open woodland, lakeshores and edges of swamps; mainly at low to medium altitudes, but occasionally >2000 m in Malawi and >3000 m in Kenya.

Usually solitary and territorial in winter quarters, but forms loose associations upon arrival and before departure. Small parties of up to 10 or more occur on migration; in Zambia (Ndola) 80–120 in *c*. 50 ha of thickets during peak Nov passage (Cramp and Perrins 1993). Wary and difficult to approach, much less confiding in Africa than Red-backed Shrike (Dowsett 1971). Perches upright on bush, tree-top, fence or power line, typically higher than Red-backed. Wintering birds defend territories, calling and chasing intruders. During disputes, bird swings and flicks tail and calls excitedly (Harris and Arnott 1988). In Kalahari, forages from exposed perch, seeking shade only during hottest hours of day. Flies down to take insect from ground, often 15 m or more from perch; occasionally makes sallies to catch flying prey; not known to hover in Africa. In Kenya, 80% of prey taken from ground, 17% in air, 2% from bushes (Lack 1985). Flight fast and fluent with little undulation. Sings frequently in southern Africa, mainly during Mar–Apr, just before departure (Harris and Arnott 1988).

A classical loop migrant, taking more westerly course through Africa and Middle East in autumn than in spring (Dowsett 1971). Western breeding populations initially head southeast or south through Greece and Aegean; eastern ones head west of south, those from central Asia crossing Iran and Arabia. Enters Africa in autumn on a narrow front across Egypt and Red Sea coast; scarce in Libya. Heavy migration through Egypt and N and central Sudan in early Aug–early Oct (mainly late Aug–mid Sept); lesser numbers through Eritrea, late Aug–late Sept. Passes quickly through W Sudan (Darfur) and E Chad (mainly Oct) but few records further west. Main migration to southern Africa occurs west of Nile, with many records from E and S Zaïre (mid Oct–mid Nov), but few in

Ethiopia or Kenya. Scarcity in S Sudan and Uganda presumed to be due in part to overflying. Apparently crosses upper Congo forest. Scarce in autumn in Malaŵi, but common in Zambia (mainly centre and W, late Oct–early Dec); passage through Zimbabwe mainly west of 30°E. Reaches Botswana in mid Oct, with main arrival in late Nov, and in N Cape Prov. in Dec.

Northward migration begins in mid Mar. Wintering area mainly vacated during early Apr, late birds in mid Apr, exceptionally early May. Route through S-central Africa more easterly than in autumn: in Zimbabwe east of 30°E; strong movement through Malaŵi, late Mar–late Apr, but only small numbers in W Zambia. Heavy passage in NE Zaïre all Apr, when widespread in Uganda and Rwanda; absent from Sudan in spring. Strong passage through Tanzania, Kenya (mainly 5–25 Apr), S and E Ethiopia, coastal Eritrea, Djibouti, and N Somalia (mid Apr–early May). Main departure from Africa is across Gulf of Aden and S Red Sea.

A bird ringed W Kenya (Apr) was recovered in Greece (Aug).

Food. Mainly beetles, also grasshoppers, bugs, Lepidoptera and Diptera; occasionally lizards and small rodents.

References
Cramp, S. and Perrins, C.M. (1993).
Dowsett, R.J. (1971).
Harris, T. and Arnott, G. (1988).
Harrison, J.A. et al. (1997).
Herremans, M. (1998).

Plate 19
(Opp. p. 286)

Lanius meridionalis Temminck. Southern Grey Shrike. Pie-grièche méridionale.

Lanius meridionalis Temminck, 1820. Man. D'Orn., ed. 2, 1, p. 143; 'Italie, la dalmatie, le midi de la France'.

Forms a superspecies with Holarctic *L. exubitor*.

Range and Status. N Africa, SW Europe, Levant and Arabia to S and central Asia. African, European and S Asian populations resident or not highly migratory; central Asian populations partially migratory, wintering in SW Asia and NE Africa.

Resident and local migrant; also winter visitor from Arabia and central Asia. Breeds from N African coast to sahel zone, including Saharan massifs, wadis and oases, and Nile Valley; also N Ethiopia to N Somalia, and Socotra. Frequent to common in Morocco and Algeria south to Atlas ranges (avoiding high mountains), in N, central and coastal Tunisia, and coastal Libya (N Tripolitania and N Cyrenaica); widespread, often common, throughout W Sahara, avoiding only bare desert; local in SW Libya in Fezzan oases; locally common in Egypt (Nile valley, NE desert and Red Sea coast, mainly on passage in delta). Widespread and fairly common in sahel zone: in Senegal (north of 15°30′N, wintering only); Mali (14°30′ to 17°N); N Burkina Faso; Niger (north to Aïr, Ténéré and Kouar); NE Nigerian borders; central Chad; and W, central and NE Sudan (north of 11°N); rare Gambia (Dec–Apr); once N Ghana (Tono); rare N Central African Republic (Monovo–Gounda–Saint Floris Nat. Park). Frequent N and central Ethiopia (south to Gondar and Awash); locally common Eritrea; frequent Djibouti and N Somalia (south to 09°N); common on Socotra, from coast up to 1300 m. Residents in coastal Sudan, Eritrea and Somalia apparently augmented by wintering migrants.

Wintering central Asian migrants frequent to locally common in E Sudan (Nile Valley and Red Sea coast), Eritrea, Djibouti and Ethiopia (mainly Rift Valley, south to c. 07°N); uncommon NW Somalia; vagrant Egypt (once, Quseir), SE Ethiopia (once, near Gerlogubi) and NW Kenyan border (once, Kibish, Feb 1988). Breeding density of up to 5 pairs in 4 ha Algerian date-palm grove.

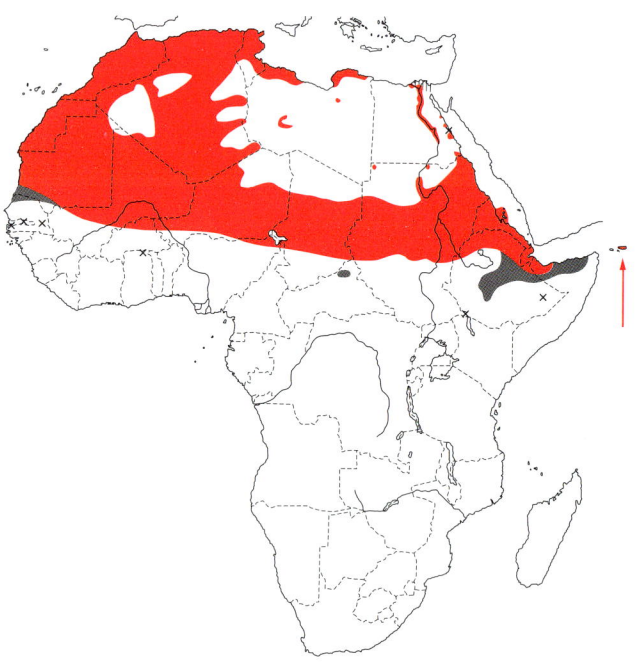

Lanius meridionalis

Description. *L. m. elegans* Swainson: N Sahara from Mauritania to Egypt, south to Ahaggar, Tibesti and Port Sudan. ADULT ♂: forehead to inner scapulars, back and uppertail-coverts light grey, usually slightly paler on rump and uppertail-coverts; outer and lower scapulars white. Tail rather long, strongly graduated (T1–T6, 25–30 mm); T1 black, T2–T3 black with white tips (broad on T3), T4 white with basal two-thirds of inner web black, T5–T6 white with black along basal two-thirds of shaft; appears white from below. Narrow whitish stripe above lores and eye. Broad black band through lores and below eye to ear-coverts, forming contrasting mask. Underparts greyish white, whitest on lower cheek to chin and throat, and belly to undertail-coverts, greyer on sides of breast. Primaries blackish; inner feathers tipped white, P1–P2 broadly so; basal half of P2–P9, and of inner web of P1, white, forming large white patch on closed wing which extends 20–25 mm beyond primary coverts; secondaries black with white inner webs, narrow white outer edges and broad white tips (*c.* 4 mm); tertials black, tipped white, broadly on outer 2 feathers; alula black, narrowly tipped greyish white; primary, greater and median coverts black; lesser coverts light grey. Axillaries, under-wing-coverts and underside of flight feather bases white. Bill black; eyes dark brown; legs slate. Wing-tip bluntly pointed; P8 longest; P7 0–2, P6 2–5, P5 9–13 shorter; P9 6–11 shorter than P8 (falls between P5 and P6); P6–P9 emarginated on outer web. Sexes alike. Size (10 ♂♂, 10 ♀♀): wing, ♂ 101–112 (106), ♀ 101–109 (105); tail, ♂ 102–112 (105), ♀ 102–111 (105); bill, ♂ 21–24 (22·3), ♀ 21–25 (22·6); tarsus, ♂ 29–32 (30·3), ♀ 29–31 (30·0). WEIGHT: Algeria and Niger, ♂ (n = 4) 49–55 (52·8), ♀ (n = 2) 50, 51.

IMMATURE: juvenile pale grey-buff above, rump and outer and lower scapulars paler and buffier; mask pale brown; wings as in adult, but tertials black with broad pale buff tips, greater coverts dark brown with broad buff-brown edges and tips, median and lesser coverts pale buff-brown; tail as in adult, but browner black, feathers more pointed. Bill dusky brown; eyes brownish; legs blue-grey. First-winter like adult but duller above, often with faint grey bars on sides of breast; many juvenile wing feathers retained.

NESTLING: not described.

L. m. algeriensis Lesson (includes '*dodsoni*'): Morocco, N Algeria, N Tunisia and coastal Libya, to N Sahara borders; and along Atlantic coast to W Mauritania. Intergrades with *elegans*. Darker grey above than *elegans*. Greyer below with whitish confined to chin and cheek; white patch at base of primaries smaller (12–20 mm beyond primary coverts); white on secondaries and tertials confined to terminal fringe, secondary inner webs black with pale greyish bases. Less white in outer tail feathers: T4 black apart from white tip and distal part of outer web; T5 with basal half of inner web black, extending distally in narrow wedge along shaft.

L. m. meridionalis Temminck: breeds SW Europe; a few winter N Morocco and N Algeria. Similar to *algeriensis*, but still darker above; narrow white stripe above lores and eye more prominent; underparts tinged vinous-pink.

L. m. leucopygos Hemprich and Ehrenberg: S Sahara from Mali through Burkina Faso, Niger (north to Aïr), N Nigeria and S and central Chad to W and central Sudan (Darfur, Kordofan, Blue and White Nile Provs north to Dongola). Similar to *elegans*, but slightly paler grey above, rump whiter, contrasting more with back. Smaller: wing, ♂ (n = 10) 94–104 (100), bill, ♂ (n = 10) 19–23 (21) (Vaurie 1959). WEIGHT: Chad, unsexed, (n = 8) 43–46 (44) (Salvan 1967–69).

L. m. aucheri Bonaparte: Red Sea coasts of Egypt and Sudan to Eritrea, N Ethiopia and N Somalia; resident and migrant visitor from Arabia. Intergrades with *elegans* in E Egypt and NE Sudan. Slightly darker above than *elegans*, slightly greyer below; narrow black frontal band, but lacks pale line above lores and eye; secondaries black with white tips; T4 and T5 with less white than in *elegans*, T5 with basal half to two-thirds of inner web black. Size same as *elegans*: wing, ♂ (n = 10) 104–112 (108). WEIGHT: NE Sudan, Aug–Oct, unsexed (mainly this race, n = 14) 37–54 (46·9).

L. m. buryi Lorenz and Hellmeyer: breeds Yemen; scarce visitor central Ethiopia and Djibouti. Similar to *aucheri*, with black frontal band, but much darker above, with less white on scapulars; much greyer below. Still less white in outer tail feathers: T5 black except tip of inner web and distal border of outer web; T6 with black inner web base extending along shaft, base of tail appears black from below. Slightly smaller than *aucheri*: wing, ♂ (n = 10) 101–109 (104).

L. m. uncinatus Sclater and Hartlaub: Socotra. Like *aucheri*, but darker above with less white on scapulars, slightly greyer below. Smaller, but bill longer: wing, ♂ (n = 14) 97–104 (101), ♀ (n = 17) 92–103 (98·5); bill, ♂ (n = 5) 24–26 (25·4).

L. m. jebelmarrae Lynes: Darfur (W Sudan), at 2000 m and above. Colour, frontal band and tail pattern like *aucheri*, but inner webs of secondaries mainly white. Smaller: wing, ♂ (n = 3) all 99, unsexed (n = 1) 96; bill, ♂ (n = 3) 20–22 (21).

L. m. pallidirostris Cassin: breeds central Asia; winters E and central Sudan to NW Somalia; vagrant Egypt and NW Kenya. Paler, more silvery grey above than *elegans*, with narrow white supercilium; lacks frontal band; lores sometimes whitish; underparts tinged pink when fresh. Inner webs of secondaries and T5 mainly white (as in *elegans*). ♂ (n = 10) 110–117 (113); tail shorter, ♂ (n = 10) 96–103 (100). Wing slightly more pointed (P8 longest, P9 usually longer than P6) and P6 outer web scarcely emarginated.

Field Characters. Length 24–25 cm. A fairly large grey-backed shrike with longer bill than Lesser Grey Shrike *L. minor*, shorter, more rounded wings and longer, more graduated tail. *Forehead grey* or with very narrow black line at base of bill (*aucheri*) (broadly black in Lesser Grey). Grey of upperparts varies from rather dark (*algeriensis*) to medium (*aucheri*) and pale (Saharan races and Asian *pallidirostris*); pale races also have narrow white line above black mask, pale rump and more white in wing; *pallidirostris* has pale bill, lores often white. Further differs from Lesser Grey in *white* scapular patch, evident both at rest and in flight, prominent *white tips* to secondaries, and sides of tail more broadly white near tip. Juveniles (all races) are brownish grey above, with brown mask and buff-tipped tertials and wing-coverts.

Voice. Tape-recorded (105, 110, B, CHA). Song a rambling medley of 2- and 3-note phrases, each brief and with a 'chopped off' feeling, with pauses between each, 'plok-kik ... ker-li ... pulili ... pujee ... kerlyerdi ... jee-wee ... pulooo ... '; harsh and melodious notes, short 'kik's, plopping noises and ringing calls mingle with some mimicry in rather leisurely and quiet performance. Calls include a jay-like screech, almost hissing 'skaaaaaa' and nasal 'gwaaaay' like Grey Lourie *Corythaixoides concolor*. For complete catalogue of songs and calls see text and sonagrams in Cramp and Perrins (1993).

General Habits. Inhabits arid open plains and plateaux with scattered thorny bushes, with or without large trees. In N Africa, scrublands below 1100 m; desert or semi-desert areas with sandy soils and patches of thorny *Ziziphus* (Lefranc and Worfolk 1997); Saharan oases and wadis, and desert with sparse dwarf shrubs; massifs up to 2000 m on Hoggar plateaux; absent from bare rock and sand desert. In sahel zone, in semi-desert with small thorny trees, especially *Acacia tortilis* and *Balanites* (Bannerman 1953). In Eritrea, *aucheri* winters in drier acacia up to 1300 m, *pallidirostris* on more arid sandy plains with clumps of *Sueda* (Smith 1957).

Usually solitary, or in pairs or family groups. Desert races rather sedentary: many ♂♂ appear to remain in territories all year. Wary, but desert birds sometimes fairly tame. Perches conspicuously on bush top, waving and spreading tail; usually 2–3 m above ground, often >5 m on

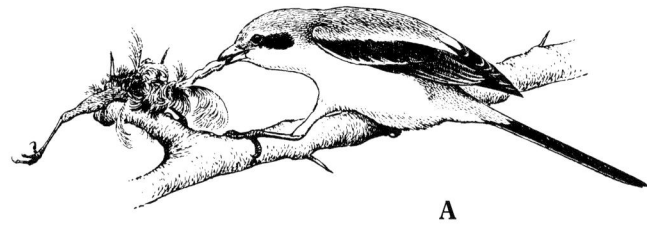

trees or wires. Hunts from vantage point. Takes large insects in air by flying vertically upwards or pursuing them directly. More distant prey caught after swooping undulating flight, and sometimes brief hovering. Pounces on invertebrates on ground below perch. Sometimes forages by hopping on ground, like a thrush. Prey eaten at once, taken to perch for dismembering (**A**), or stored; large item held under foot and eaten piecemeal (Cramp and Perrins 1993). Impales or wedges beetles and small vertebrates in caches in bushes. One, lacking upper mandible, once tackled a small snake (Oman, J. Eriksen, pers. comm.) (**B**). Flight fast and powerful, low between bushes, quite high and undulating over longer distances.

Most populations sedentary, but many *pallidirostris* are long-distance migrants. Small scale, mainly short-distance movements occur in some desert races and in *meridionalis*, which is regular on Gibraltar, Sept–Nov, and may reach Morocco; 3 records on N Algerian coast (Ledant *et al.* 1981). *L. m. elegans* regularly winters in Lower Senegal valley (where now known to breed). In Tunisia, northernmost *algeriensis* disperse (to semi-desert areas?) after breeding; *leucopygos* has occurred in Dec–Apr west to Gambia and south to N Ghana. Numbers of *aucheri* near Red Sea coast increase in Oct–Mar by migration from Arabia (perhaps further north); and *buryi* from Yemen sometimes reaches central Ethiopia and Djibouti. Central Asian *pallidirostris* reaches E Sudan, Eritrea and Ethiopia in Sept–Oct, departing by Mar.

Food. Small reptiles (also mammals and passerines) and large arthropods (insects, larvae, arachnids). Desert races take many small lizards, snakes and tenebrionid beetles; gerbils and jerboas may also be taken. 40 pellets, Libya, contained only beetles: 65% Tenebrionidae (*Adesmia, Pimelia, Akis, Blaps*) and 35% Buprestidae (*Julodis onopordi*) (Massa 1999). Grasshoppers, centipedes and small lizards in stomachs, Socotra. In S France, commonest insects taken are Coleoptera, Orthoptera, Lepidoptera and Hymenoptera (Lefranc and Worfolk 1997). Nestling diet mainly insects; in S Spain, lizards, large beetles, mole crickets, Diptera, Odonata, Lepidoptera, spiders and whole shrews (Cramp and Perrins 1993).

Breeding Habits. Monogamous, but in Negev (Israel) some ♂♂ simultaneously polygamous (Yosef and Pinshow 1988). Territorial; territory *c.* 10 ha in Sahara (*elegans*, Lefranc and Worfolk 1997), 10–25 ha in Spain. Vocal and demonstrative along territory boundaries early in breeding season. Rival ♂♂ give harsh calls, accompanied by wing-shivering. Sings mainly in mornings and evenings. Soft 'hui hui' contact calls uttered by both birds of pair in very upright posture. Circular or figure-of-8 morning display flights frequently performed early in breeding season.

NEST: bulky cup of twigs, plant stems and grasses, lined with rootlets, fibres, hair, sometimes wool or feathers; ext. diam. 170–270, int. diam. 85–110, int. depth 55–75. Usually sited 0·5–1·5 m high in thorny bush or small prickly tree, but occasionally up to 5 m, e.g. in palm.

EGGS: 3–7, usually 4–6. NW Africa (*algeriensis*, n = 72 clutches) 4–7 (4·9); W Sahara (*elegans*, n = 51 clutches) 3–6 (4·8). Laid daily. Buff or greenish white, heavily blotched and spotted with olive, purplish grey or brown. SIZE: *algeriensis* (n = 53) 25–29 × 18–20 (26·6 × 19·6); *elegans* (n = 60) 25–28 × 18–21 (26·3 × 19·6).

LAYING DATES: NW Africa, Feb–June, usually 2 broods or more; Mali June–Oct; Egypt, Jan–June; Sudan, Feb–May, Sept–Oct; Eritrea, Dec–Mar; Socotra, Nov–Dec.

INCUBATION: mainly by ♀. Period: 17–18 days.

DEVELOPMENT AND CARE OF YOUNG: young fed and cared for by both parents; brooded by ♀ for *c.* 8 days after hatching, when most food brought by ♂; fledging period 18–19 days; young typically leave nest before they can fly; in Spain, period of dependence *c.* 39 days.

BREEDING SUCCESS/SURVIVAL: in SW Spain, of 136 eggs, 67% hatched, 49% produced fledged young (3·1 per pair) (Cruz Solis and Lope Rebollo 1985).

References
Cramp, S. and Perrins, C.M. (1993).
Cruz Solis, C. de la and Lope Rebollo, F. de (1985).
Lefranc, N. and Worfolk, T. (1997).

Lanius isabellinus Hemprich and Ehrenberg. Isabelline Shrike; Red-tailed Shrike. Pie-grièche isabelle.

Plate 19 (Opp. p. 286)

L. isabellinus Hemprich and Ehrenberg, 1833. Symb. Phys. Avium, fol. E; Kanfunda, Arabia.

Forms a superspecies with *L. collurio* and Asian *L. cristatus*.

Range and Status. Central and SW Asia, from S Kazakhstan and Mongolia to Iran, Afghanistan, NW and N China. Winters in Africa, SW Asia and India.

Palearctic migrant. Winters commonly NE Africa, in Sudan (N to Darfur, Khartoum and Red Sea Hills), Eritrea, Ethiopia, Djibouti, Somalia, Kenya, NE Tanzania (south to Dar-es-Salaam and Mpanda), extreme NE Zaïre (Uele, south to Avakubi) and Uganda (mainly N and E); also central and S Chad, and sparingly N Cameroon (north of 11°N) and Nigeria (mainly N, but south to Kainji and Onitsha); scarce in Rwanda, and in Mali (14°30′–18°N), S Mauritania, N Zambia (Dowsett *et al.* 1999) and Malaŵi (Nyika Plateau; once, Livingstonia). Locally common to abundant on both passages in Sudan, Eritrea, Ethiopia, Djibouti, and central and SE Kenya, and on spring passage E Chad; uncommon E Egypt.

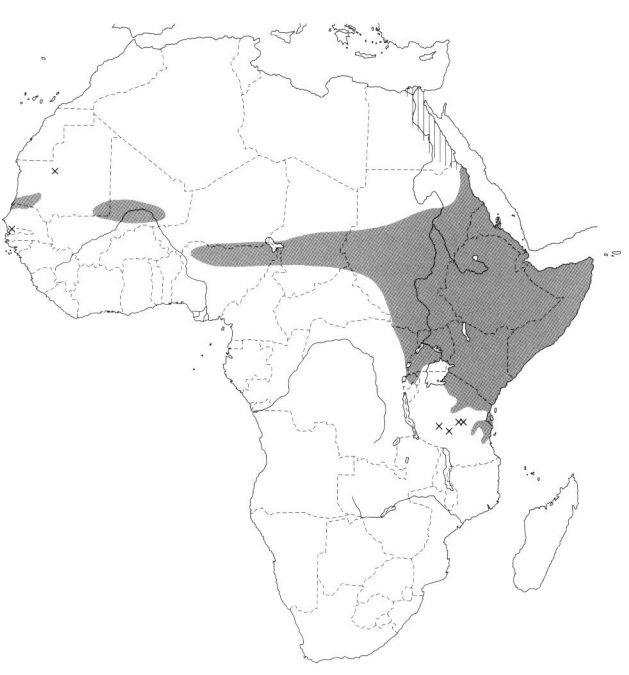

Lanius isabellinus

Description. *L. i. phoenicuroides* Schalow ('Red-tailed Shrike'): S Kazakhstan to Iran and Afghanistan; winters from Ethiopia and Somalia to Kenya and NE Tanzania; also in S Arabia. ADULT ♂: highly variable; typically, forehead and crown rufous- or tawny-brown, nape, mantle, scapulars and back brown, rump and uppertail-coverts bright rufous. Central tail-feathers rufous-brown above, the rest paler rufous; underside of tail rufous-cinnamon. Conspicuous narrow white superciliary stripe, and black band through lores, below eye and on ear-coverts. Underparts white, tinged pale creamy pink on centre of breast and sides of belly, deeper vinous pink on sides of breast and upper flanks. Flight feathers blackish brown, bases of primaries white, forming exposed band 5–10 mm wide; primary coverts blackish brown; tertials, greater coverts and median coverts blackish brown with buff-brown fringe along outer edge and tip; lesser coverts brown like mantle and scapulars. Underwing-coverts and axillaries creamy white. Some birds (variant '*karelini*') paler, drab grey-brown above without rufous crown, underparts less pinkish. Wing tip slightly more rounded than in Red-backed Shrike *L. collurio*: P7–P8 longest, P6 1–4, P5 6–11, P1 22–26 shorter; P9 5–9 shorter than P6); P10 very small, 4–11 longer than primary coverts; P6–P8 emarginated on outer web. Bill blackish brown to pale brown, paler at base, often black in spring; eyes brown; legs dark grey or blackish. ADULT ♀: variable; upperparts typically brown with rufous tinge to crown; tail much like adult ♂; superciliary stripe buffish; fore part of lore, and sometimes whole lore and feathers round eye pale, buff; patch on ear-coverts dark brown. Underparts whitish; cheeks, sides of neck, breast and flanks with narrow brown bars, faint and restricted in some birds. Wing like adult ♂, but patch at base of primaries smaller, usually tinged buff. SIZE (10 ♂♂, 10 ♀♀, central Asia): wing, ♂ 91–97 (92·9), ♀ 89–98 (92·5); tail, ♂ 74–86 (80·6), ♀ 78–82 (80·1); bill, ♂ 18–21 (19·0), ♀ 18–20 (18·8); tarsus, ♂ 24–26 (24·8), ♀ 23–25 (24·2). WEIGHT: Kenya (mainly this race); ♂ (n = 163, Nov–Dec) 22–30 (26·1), (n = 9, Mar–Apr) 25–32 (28·6). This race and *isabellinus*: NE Sudan, (n = 18, Sept–Oct) 23–36 (28·9); Ethiopia, (n = 42, Sept–Mar) 23–35 (27·8), (n = 7, Apr–May) 28–38 (32·2); E Chad, (n = 1, Apr) 41. Complete moult in winter quarters, Dec–Feb or Mar; partial post-nuptial moult in breeding area, July–Aug.

IMMATURE: first winter bird before moulting is like adult ♀, but in upperparts fresh brown feathers usually with faint barring; barring on cheeks, breast and flanks usually more distinct and extensive. Tail like adult ♀, but often paler and duller rufous. Wing like adult ♀, but pale primary patch usually hidden or absent; inner primaries, secondaries and primary coverts with broader buffish edges; tertials and greater coverts with blackish subterminal fringing line.

L. i. isabellinus Hemprich and Ehrenberg: Mongolia; winters throughout African range of species, mainly in Chad, Sudan, Ethiopia and N Somalia; also in S Arabia. ADULT ♂ isabelline-grey above, including crown; superciliary stripe creamy; underparts isabelline, tinged pinkish on side of breast and flank. Resembles *phoenicuroides* in having rufous rump and tail, black face mask, blackish brown flight feathers contrasting with body, and well developed white wing patch; size similar. ADULT ♀ like ♀ *phoenicuroides*, but upperparts usually more isabelline-yellow and upperparts more greyish or sandy brown.

TAXONOMIC NOTE: sometimes considered to form single species with *L. collurio* and Brown Shrike *L. cristatus* of central and E Siberia (Voous 1960). However, *isabellinus* overlaps with *cristatus* without interbreeding so these 2 cannot be conspecific. *L. collurio* hybridizes locally with *isabellinus* (race *phoenicuroides*) in NE Iran (Vaurie 1955), Tarbagatay area and Zayasn basin (Panow 1983), and these 2 are often treated as conspecific and separate from *cristatus* (Segmann 1930, Vaurie 1955, 1959); however, contact zone is narrow and intergradation apparently secondary. *L. collurio* and *isabellinus* are recognized here as separate species, following Olivier (1944), Stresemann and Stresemann (1972), Voous (1979), Stepanyan (1990) and Cramp and Perrins (1993).

The type corresponds with Mongolian population called '*speculigerus*' by Stegmann (1930), Vaurie (1959), Panow (1983) and Cramp and Perrins (1993), not with NW Chinese race called *'isabellinus'* by Vaurie (Pearson, 2000); *'speculigerus'* is thus a synonym of *isabellinus*.

Field Characters. Length 16·5–18 cm. Similar to Red-backed Shrike *L. collurio* in size, structure and behaviour; slightly longer tail more often cocked. Paler overall, especially underparts, and ♂♂ of both races distinguished by bright rufous rump and tail contrasting with sandy back; crown of ♂ *phoenicuroides* rufous (sandy grey in *isabellinus*); white supercilium above black mask conspicuous (faint or absent in Red-backed). Wings of both sexes dark with white patch on primaries and pale feather edges (wing of Red-backed mainly rufous). ♀ and immature similar to Red-backed but sandy brown upperparts contrast more with rufous tail and show little barring, even in young birds, barring on underparts finer and less extensive; underside of tail rufous-cinnamon, not grey. Tail of ♀ and imm. Red-backed brown or with rufous tinge above but greyish below. ♀ Isabelline usually shows buffish white patch on primaries, absent in Red-backed. However, some 1st year *phoenicuroides* can be very similar to 1st year Red-backed (Porter *et al.* 1996). For wing formula differences, see Red-backed Shrike.

Voice. Tape-recorded (105, B, KRJ, PEA). Quality and timbre of songs and calls different from Red-backed Shrike (Chappuis, in press). Song a subdued, continuous medley of high-pitched trills and chatters, with a few musical whistles as well as harsher strangled notes; more liquid and less grating than Red-backed Shrike; suggests a *Sylvia* or *Hippolais* warbler. Harsh threat and alarm calls include 'chack' or 'chaza', repeated when threatened, 'cha-cha-cha-cha ...'; also 'gree-gree ...' and down-slurred 'jeeda'.

General Habits. Frequents open country: thornbush, dry woodland, grassland with scattered bushes, scrub at edges of cultivation; also marshy areas and lake edges with small acacias; mainly below 1500 m, but up to 2000 m in Kenya Rift Valley. Usually winters in drier, less leafy habitat than Red-backed Shrike *L. collurio*.

Solitary; migrants sometimes in loose groups. Often with Red-backed Shrikes on autumn passage (E Kenya). May remain in winter territory for weeks at a time. In S Kenya, 6 birds were retrapped where ringed, in a later year.

Perches on bush or low tree, more conspicuously than Red-backed Shrike. Pounces on prey, usually after gliding a few m from perch; pursues insects in flight, and gleans from vegetation. In Kenya, 73% of prey taken from ground, 11% from twigs and leaves, 9% in the air (Lack 1985). Flight, carriage and movements same as Red-backed Shrike, but tail more often held cocked or sideways and its motions tend to be more exaggerated. In Kenya, frequently sings in Feb to early Apr.

L. i. phoenicuroides migrates southwest in autumn, crossing Iran and Iraq in early-mid Sept; *isabellinus* is slightly later, first heading west through breeding range of *phoenicuroides*. Both races winter in S Arabia, but most birds continue to NE Africa, reaching central Sudan, Ethiopia and N Somalia in late Sept–early Oct. Many *phoenicuroides* winter further south, arriving in Kenya and NE Tanzania during Nov. Pronounced migration occurs through SE Kenya in early-mid Nov. *L. i. isabellinus* is main race wintering in Nile valley, arriving in S Sudan in Nov; it reaches Chad and NE Zaïre in Oct–Nov and a few reach N Nigeria from early Dec; uncommon in Kenya. Stragglers west to Senegal are probably this race.

Both races have full moult in winter quarters, some completing by Jan, others not until Mar. Northward migration begins in Mar. Birds leave Zaïre and Sudan by mid Apr. Main passage occurs through E Kenya in late Mar–early Apr (last birds in mid Apr); and through Eritrea, Djibouti, E Ethiopia and N Somalia in late Mar–mid Apr, with stragglers to early May.

2 birds ringed Kenya (Nov and Dec) were recovered Kuwait (both mid Apr).

Food. Adult and larval insects, especially beetles, also grasshoppers, dragonflies, ant-lions, bugs and termites; other invertebrates; small lizards, frogs and tadpoles, and tired migrant birds in oases.

References
Cramp, S. and Perrins, C. M. (1993).
Pearson, D. J. (1979).

Plate 19
(Opp. p. 286)

Lanius collurio **Linnaeus. Red-backed Shrike. Pie-grièche écorcheur.**

Lanius collurio Linnaeus, 1758. Syst. Nat. (10th ed.), p. 94; Sweden.

Forms a superspecies with *L. isabellinus* and *L. cristatus* (Asia).

Range and Status. Europe from Baltic to N Spain, Italy, Turkey, Caucasus and NW Iran; also W Siberia and N Kazakhstan. Winters in Africa.

Palearctic migrant. Winters in southern Africa, from SE Kenya (including coast), E and S Tanzania, Zambia, and central and S Angola, to Namibia (except arid W and SW), Botswana, Zimbabwe, Mozambique, Swaziland (mainly E), and South Africa: Transvaal, Natal (mainly N), Orange Free State (mainly NW), Cape Province (mainly N Cape, but frequent south to Rooiberg and Beaufort West, and in E Cape west to Clarkson, rare in S and SW Cape). Common to abundant in semi-arid acacia savanna from Malaŵi and S Zambia southward, with strongholds in central Kalahari, Limpopo basin, Transvaal and Swaziland lowveld, Lebombo Mts to Zululand (Bruderer and Bruderer 1993, Harrison 1997); more local in north of range. Rare winter records north of equator: S Uganda (Feb), S Sudan (Bahr el Ghazal, Jan), Egypt (Rafa, Djebel Uweinat, Dec), Morocco (Agadir Jan, Saidia Feb; J. R. D. Vernon, pers. comm.). and Mali (Sahara, Jan; Lamarche 1981).

Lanius collurio

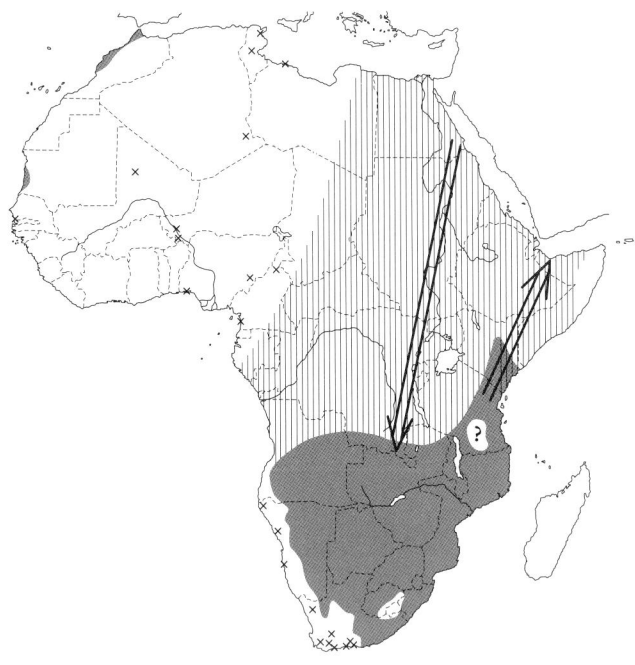

On southward passage, frequent to common in E Libya (east of Benghazi), and common to abundant in Egypt, Sudan and E Chad; frequent to common in Eritrea and Ethiopia (mainly W and central). Common to very abundant in Zaïre (mainly E), Rwanda, Uganda (especially W), W Tanzania and Zambia; scarce S Somalia, but frequent to common in central and E Kenya and E Tanzania. Rare in autumn in W Morocco and W Mauritania (Nouakchott); rare or vagrant in Algeria (once Djanet), W Libya (Tripoli), Gambia, SW Niger, Nigeria (Lagos, Obudu) and Cameroon (Garoua, Kribi); uncommon NE Gabon.

On northward migration, common in Tanzania, common or locally abundant in central and E Kenya, Ethiopia east of Rift, S Eritrea, Djibouti and N Somalia; frequent in Uganda, W Kenya, and central Ethiopia. Scarce in Egypt (mainly E desert and Upper Nile); no spring records from Sudan. Vagrant Nigeria (once Obudu, Apr), Tunisia (twice, May), Mauritania (once, Mar, Tanoudert) and Morocco (twice, Apr, Mdiq and Ouezzane).

Wintering density in thornveld, Botswana and Transvaal, 5–15 birds per 10 ha, reaching 2–8 per ha (Herremans 1993, Bruderer and Bruderer 1994). Probably >50,000 winter in S Mozambique (Parker 1999).

Description. *L. c. collurio* Linnaeus: Europe to Urals; winters throughout African range of species. ADULT ♂: crown to nape and upper mantle grey, lower mantle, scapulars and back bright rufous chestnut, rump and uppertail-coverts grey, often tinged with chestnut. Central tail feathers (T1) black, T2 similar with some white at base, T3–T5 black with basal half to two thirds white (shaft black), T6 white with black shaft and black distal third with white edge and tip; all tail feathers with narrow white tips when fresh. Black band across lores, and above and below eye to broad patch on ear-coverts, extending also as narrow band across base of upper mandible; sides of neck grey. Cheeks, chin and throat white or pinkish white, breast and flanks pale vinous pink, grading to whitish in centre of belly and undertail-coverts. Primaries, secondaries and primary coverts blackish brown, P2–6 often with white at base, occasionally visible as short patch beyond primary coverts. Tertials and greater coverts blackish brown with broad rufous fringes; rest of upperwing-coverts rufous with darker centres. Underwing-coverts and axillaries white with dusky centres. Wingtip rather pointed: P8 longest, P7 1–3 shorter, P6 5–9, P5 10–14, P1 22–30; P9 3–9 shorter than P8, falling between P6 and P7 or about equal to P6 (sometimes up to 2 mm shorter than P6); P10 very small, 1 mm shorter to 7 mm longer than primary coverts; P7–P8 emarginated on outer webs. Bill blackish with horn-brown or greyish base, becoming all-black in spring; eyes dark brown; legs brownish grey to black. ADULT ♀: upperparts variable; typically, forehead and crown rufous-brown, hindneck and sides of neck brownish grey, mantle, scapulars and back rufous-brown, rump and uppertail-coverts brownish grey; nape and rump sometimes all brown, whole top of head sometimes greyish; brown of upperparts varies from deep rufous to dull olive-brown. Sometimes some narrow barring on forehead and crown, and on back and scapulars. Tail dark brown above (often tinged rufous), grey-brown below, inner webs and tips of T2–T6 narrowly edged whitish. Narrow creamy white superciliary stripe; lore and feathers around eye creamy buff, ear-coverts rufous-brown. Underparts cream, whiter on chin and from mid belly to undertail-coverts. Feathers of lower cheeks, sides of neck, breast and flanks with narrow blackish subterminal bars. Flight feathers, tertials and wing-coverts like adult ♂, but fringes of tertials and wing coverts browner. Bare parts as in ♂, but bill browner. SIZE (10 ♂♂, 10 ♀♀, Europe): wing, ♂ 90–98 (94.4), ♀ 91–98 (93.7); tail, ♂ 73–80 (76.5), ♀ 72–78 (74.1); bill, ♂ 17–20 (18.2), ♀ 17–20 (18.6); tarsus, ♂ 23–25 (23.9), ♀ 23–25 (23.6). WEIGHT: ♂, Kenya (n = 92, Nov–Dec) 22–35 (26.9); ♀, Kenya (n = 85, Nov–Dec) 21–35 (25.9). Ages and sexes combined: NW Egypt (n = 40, Sept–Oct) 22–34 (28.0); NE Sudan (n = 107, Aug–Oct) 18–28 (23.3); Ethiopia (n = 27, Oct–Mar) 21–32 (26.2), (n = 10, Apr–May) 23–43 (32.2); Uganda (n = 66, Oct–Dec) 21–38 (30.2); Kenya (n = 649, Nov–Dec) 21–35 (26.3), (n = 105, Mar–Apr) 22–40 (27.6); Zambia (n = 44, Oct–Apr) 22–33 (26.9); South Africa (n = 22, Nov–Mar) 24–34 (27.8).

Complete moult in Africa, Dec–Mar.

IMMATURE: first-winter bird before moulting like adult ♀, but upperparts with blackish barring, hindneck and rump less grey. Upperparts vary from deep rufous brown to grey-brown. Tail often more rufous-tinged above, but grey-brown below, as in adult ♀. Tertials and upperwing-coverts with buff fringes and, usually, contrasting black subterminal bars.

L. c. pallidifrons Johansen: W Siberia and N Kazakhstan: paler on upperparts than *collurio*, ♂ with paler grey-white forehead, ♀ with slightly more rufous upperparts and greyer hindneck.

L. c. kobylini Buturlin: Crimea and Caucasus to Turkey and NW Iran; overlaps widely in Africa with nominate race. Variable, often inseparable from nominate race. Typical adult ♂ has chestnut of mantle and scapulars duller and less extensive, sometimes brownish; underparts darker vinous red.

TAXONOMIC NOTE: see under *L. isabellinus* for relationship with that species and Brown Shrike *L. cristatus*.

Field Characters. Length 16.5–18 cm. A rather small, compact, short-billed shrike. Adult ♂, with grey crown, mantle and rump, black face mask, rufous back and wings and black tail with white sides to base, can only be mistaken for ♂ Emin's Shrike *L. gubernator*; differs in grey rump (rufous in Emin's), pearly grey (*vs* duller grey) crown and hind neck, pinkish white (not rich cinnamon) underparts. ♀ has crescentic barring below, ♀ Emin's is unbarred and resembles ♂; immatures similar but Emin's has tawny sides. ♀ and immature Red-backed similar to Isabelline Shrike *L. isabellinus* but usually darker and warmer brown above and more boldly barred, brown tail often has rufous wash above, but this contrasts less with back colour; underside of tail grey. In the hand, P9 is longer than, equal to or slightly shorter than P6 (much shorter than P6 in Isabelline Shrike), and outer web of P6 lacks emargination.

Voice. Tape-recorded (73, 86, 88, 91–99, 105, 110, B, F, LEM, WALK, ZIM). Song often lengthy, a hurried, bustling mixture of shrill pure notes, grating churrs, brief chirps, high thin dry trills and tuneless chatter, with much mimicry, usually subdued in Africa, sometimes loud on breeding grounds. Also gives hollow 'glogloglo-weee' and 'gwer-gwolek'. Contact/alarm call, hard 'jit-jit', sometimes given in rapid series; other scolding calls include harsh 'jeea', 'jer', 'zeet' and repeated 'jizzy-jizzy ...' or 'dzzooy-dzooy ...'. For further vocalizations see Cramp and Perrins (1993).

General Habits. Inhabits dry thornbush, acacia woodland, moist secondary thickets, bushy grassland, and forest edges and clearings. Mainly below 1500 m, but regularly up to above 2500 m on passage through Kenya. In E Africa, ranges into greener, higher country than Isabelline Shrike *L. isabellinus*, although the 2 occur together in dry bushlands on migration. In southern Africa, winters mainly in medium dense thornbush, and *Acacia* and pockets of scrub in adjacent broad-leaved woodland; also fallow land and edges of cultivation (Harrison 1997). In Transvaal, mainly in open bush (10–50% cover) with patchy herbaceous layer; prefers low scrub (1–3 m) to higher bushes and trees (Bruderer and Bruderer 1994). ♂♂ prefer more open habitats than ♀♀, with fewer and smaller trees (Bruderer and Bruderer 1994, Herremans 1997). Typically in less arid, more thickly bushed country than Lesser Grey Shrike *L. minor*.

Usually solitary, but forms loose groups on migration, e.g. *c*. 60 in mid Nov. along 1 km of shore of L. Victoria (Entebbe, Uganda). Occupies territory for months in winter quarters, defending it most aggressively upon arrival and again shortly before departure. ♂♂ dominate ♀♀, perching and behaving more conspicuously (Herremans 1997). Fidelity to wintering and passage sites, demonstrated by recaptures of ringed birds (Malaŵi, Botswana, South Africa). Defends feeding station during transient migration stopovers.

Often unobtrusive, spending long periods within leafy cover. Passage birds in Chad prefer shadier, less prominent perches than wintering Isabelline Shrikes (Salvan 1968–69). Forages by keeping watch from bush, fence, or lower branch of tree. In Transvaal perches 1·5–2 m high in open places (Bruderer and Bruderer 1994). Most prey taken from ground after short glide and pounce, or by dropping directly from perch; also pursues and catches flying insects. In Kenya, 81% of items taken from ground and herb layer, 14% from air and 3% from bush foliage (Lack 1985). Prey occasionally impaled for storage or dismemberment on twig or thorn (less often in Africa than on breeding grounds). Flight low and direct between bushes, bird swooping up to new perch; markedly undulating over longer distances (Cramp and Perrins 1993). When excited, gives chacking calls, flicks tail and swings it sideways, partially spread. Sings at low intensity, mainly just before departure from winter quarters.

A loop migrant; passage through Middle East and Africa more easterly in spring than in autumn (Moreau 1961, Schüz, 1971, Zink 1975). Birds from W and central Europe head southeast in autumn through Greece and Aegean; those from E Europe take a more southerly heading; and birds from Siberia migrate southwest through Kazakhstan and Iran. Main African arrival is in Egypt and NE Sudan, minor one in Eritrea. Southward passage is heavy through N, central and W Sudan, E Chad, E Zaïre, Rwanda and Uganda, but sparse and more local in Ethiopia, Kenya (mainly E) and S Somalia. First birds reach Egypt in early Aug, but main passage here and in NE Sudan is in late Aug–late Sept; onward passage through W Sudan and E Chad mainly late Sept–late Oct. Most birds remain for some weeks in Sudan. First arrivals in S Uganda in mid Oct, with main passage there in late Oct–late Nov and through SE Kenya in Nov–early Dec. Adults migrate through NE and E Africa *c*. 2 weeks earlier than young birds. Onward migration south of equator is rapid, and high pre-departure weights in Uganda in Nov indicate long direct flights (Pearson 1970). Earliest birds reach Malaŵi in early Oct, and Zimbabwe and South Africa in late Oct; main influx into southern Africa mid Nov–early Dec, with marked passage in N Botswana.

Full moult takes place in winter quarters, late Jan–Mar. Northward migration begins in southern Africa in mid-late Mar; region vacated in early Apr, with last birds in mid Apr (once, Botswana, 3 May). Strong passage through Zimbabwe in early Apr, Zambia, Malaŵi and Tanzania in early–mid Apr, and Kenya in Apr; west of L. Victoria only light passage. Strong passage through E Ethiopia, S Eritrea and N Somalia in mid Apr–early May, but no spring records in Sudan and few in E Egypt. Main return route to Palearctic is across S Red Sea and Gulf of Aden.

>50 autumn recoveries of ringed birds in Egypt, mostly from Germany, Scandinavia and Poland, also from Belgium, Switzerland, Hungary, Yugoslavia and Turkey. Recoveries from Belgium to S Zaïre (1, Nov); Denmark to Tanzania (1, Apr); Germany to S Zaïre (2, Nov), Malaŵi (4, Mar–Apr) and Zimbabwe (1, Jan); Austria to Tanzania (1, Nov); Sweden to E Zaïre (1, Nov); Finland to N Sudan (1, Nov); and Hungary to Tanzania (1, Apr). A bird ringed S Zaïre (Dec) was recovered Germany; 1 ringed Zambia, in Czechoslovakia; 1 ringed Transvaal, in Yemen (May); and 1 ringed Kenya (Nov), in Georgia.

Food. Mainly insects, sometimes large; a few small vertebrates. Beetles, Hymenoptera and Orthoptera (which are main prey species in Europe – Cramp and Perrins 1993), winged termites; frogs, snakes and some birds. In E Sudan in autumn Marsh Warblers *Acrocephalus palustris* and other weakened migrants (Nikolaus 1990).

References
Bruderer, B. and Bruderer, H. (1994).
Cramp, S. and Perrins, C.M. (1993).
Harrison, J. A. *et al.* (1997).

Lanius gubernator Hartlaub. Emin's Shrike. Pie-grièche à dos roux.

Lanius gubernator Hartlaub, 1882. Orn. Centralb., 7, p. 91; Central Africa, = Langomeri, Uganda, Shelley, 1912.

Plate 19
(Opp. p. 286)

Range and Status. Endemic resident, N tropics. Mali, uncommon: Téfoulet and Tamalet lakes in inland delta of Niger R., and Séno plain. Map on p. 368. Ivory Coast, uncommon (9 records) S Comoé Nat. Park (Demey and Fishpool 1991). Ghana, rare: dry savannas at Tumu (2 records) and Mole east to Sekwi, Gambaga and Salaga; evidently not infrequent in Gambaga area 100 years ago. Nigeria, rare, only 7 or 8 localities; recorded regularly at Kainji over 3 years and in nearby Kainji Lake Nat. Park, 2 collected Kururuku (Benin Prov., locality no longer identifiable, but 'in grass woodland') and one at Gajibo (L. Chad), records between Kabba and Lokoja, Taboru, Yankari Nat. Park (1 old, 1 recent record) and Serti. Cameroon, rare or uncommon, Adamawa Plateau and S Benue Plain; collected at Yagoua, Rei Buba and between Tibati and Ngaoundéré. Central African Republic, uncommon Lobaye Préf. and Manovo-Gounda-Saint Floris Nat. Park (Carroll 1988). Sudan, uncommon, in S, from Yei and Lado to Kajo Kaji. NE Zaïre, uncommon to locally frequent around Aba and Faradje, and known from Dungu, Vassaka, Kuterma, Nzoro and Garamba Nat. Park. Uganda, thinly distributed at 600–1500 m from West Nile to Kidepo Valley Nat. Park, south to Acholi and Murchison Falls Nat. Park.

Description. ADULT ♂: forehead grey, but narrow black line across base of bill, joining black lores, ear-coverts and upper cheeks to form mask; crown bluish grey, whiter in narrow line where it meets black mask; nape and upper mantle grey; lower mantle, scapulars, back, rump and uppertail-coverts bright, deep rufous. Tail brownish black, T1 edged rufous, T2–T5 edged and tipped white, white tip tiny on T2, 2–3 mm deep on T5; T6 with white outer vane, and inner vane mainly brown, white at base, with 3 mm wide white mark at tip and diagonal black-brown subterminal line between the white and brown. Chin, throat and lower cheek white; breast, flanks and sides of belly pale rufous, warm ochre or dark rufous (more orange than back, which is mahogany shade); centre of belly whitish, undertail-coverts greybrown, edged and tipped whitish. Wings dark brown, inner tertials dark rufous, inner secondaries and greater coverts broadly fringed rufous, P1–P5 with white bases, making patch 8–11 mm long. Underside of remiges and tail mainly grey, underwing-coverts and axillaries mainly white. Bill black; 4 rictal bristles; eyes dark brown; legs and feet dark grey. ADULT ♀: like ♂ but less strongly coloured. Mask brownish black (not black) and narrower than on ♂, at lores and base of bill; forehead pale brownish grey, crown, hindneck and mantle brownish grey; extent of grey on mantle varies – whole mantle and back brownish grey, or lower mantle and back brownish rufous. Tail as in ♂, though rufous parts less bright and blacks brownish, and T6 without white on inner vane; underparts distinctly paler than in ♂; wings as in ♂ but rufous parts less bright and black parts decidedly brownish. SIZE (♂ ♀, n = 11): wing, ♂ 77–84; ♀ 78–84; tail, ♂ 60–66; ♀ 64–67; bill to feathers 13–14; tarsus, ♂ 23–25, ♀ 20–23.

IMMATURE: forehead, crown, hindneck and mantle brownish grey with rufescent tinge, rather finely barred with dull black; scapulars, back, rump and uppertail-coverts rufescent brown, similarly barred, but bars on rump very thin, zigzag-shaped, and widely spaced; later, only back and rump barred, and head and mantle plain. Tail dark brown, T6 with outer vane pale rufous.

Underparts pale buffy or creamy off-white, deeper buff on breast and flanks; moustachial region, sides of breast and flanks scaled with narrow blackish crescents. Wings brown; upperwing coverts with whitish tips and blackish subterminal bar; tertials and inner secondaries edged with warm buff and tipped whitish. Some birds with only crown grey, mantle to rump and wing coverts dark rufous, with still a good deal of banding on all of upperparts and flanks.

Juvenile tiger-banded above (forehead to uppertail-coverts, wing coverts and edges of secondaries and tertials): each feather grey or buffy grey with evenly-spaced dark brown bands. Pale superciliary stripe; mask dark brown, diffuse, mottled with rufous; wing without white patch at base of primaries; tail plain, without white at tips and edges; underparts buffy white, freckled and unevenly marked with wavy grey or dark brown bands, usually very thin; breast and flanks rufescent.

NESTLING: not described.

TAXONOMIC NOTE: has been allied superspecifically with *L. souzae* (Hall and Moreau 1970, Sibley and Monroe 1990), but Palearctic *L. collurio*, which *L. gubernator* quite strongly resembles, makes a better candidate.

Field Characters. Length 14·5–16 cm. A small shrike; ♂ differs from ♂ Red-backed Shrike *L. collurio* in broad black mask continuing across forehead, bordered above by whitish line, dull grey (not pearly grey) crown and hind neck, rufous rump as well as back (rump grey in Red-backed), rich cinnamon flanks and paler cinnamon breastband (underparts pinkish white in Red-backed). Tail less black, has light rufous wash. White patch on wing shows in flight. ♀ like ♂ but no black on forehead, upperparts duller, greyish tawny; lacks any barring above or below. Immature brown above with black barring, like immatures of Red-backed and Isabelline *L. isabellinus* shrikes, but has tawny flanks; throat white but rest of underparts barred black from breast to undertail-coverts. Some immatures are grey with little banding and can look very like Lesser Whitethroat *Sylvia curruca*.

Voice. Not tape-recorded. A 'pleasant call' noted by Emin (Jackson and Sclater 1938); ♂ utters a low hissing note (H.F. Marshall *in* Bannerman 1939). Seen frequently but no call heard (Chapin 1954). In Nigeria, various twitterings and whistles, together with low harsh 'zut zut, chuz-zoo-wit'; low 'chark, chark' as bird landed in tree (Lefranc and Worfolk 1997).

General Habits. Inhabits abandoned fields and old village sites in open bushy grassland with mature woodland. Solitary and in pairs or family parties. Forages by keeping watch from perch on bush, fence, hedge or telephone wire. Habits very poorly known, but early observers remarked that it behaves much like Red-backed Shrike *L. collurio*; said to take most of its food from the ground, and occasionally to catch small insect in aerial sortie.

Resident, though all Mali records Aug–Nov.

Food. Insects: beetles, mantises and small grasshoppers in stomachs.

368 LANIIDAE

Breeding Habits. Nest and eggs unknown.
LAYING DATES: Nigeria (Kainji), (pair with young bird just out of nest, June); Zaïre, (gonad evidence of breeding Mar–Apr).

Reference
Lefranc, N. and Worfolk, T. (1997).

Plate 19 (Opp. p. 286) *Lanius souzae* Bocage. Souza's Shrike. Pie-grièche de Souza.

Lanius souzae Barboza du Bocage, 1878, J. Sci. Math. Phys. Nat. Lisboa, 6, p. 213; Caconda, Angola.

Range and Status. Endemic resident, southern tropics. Gabon, pair at Lékoni (Sargeant 1993). Congo, local from Zanaga area, Djambala and Léfini Res. east and southwards. Zaïre, Kinshasa and Stanley Pool area; uncommon in S Kasai and Katanga, northeast to Marungu Highlands; rather uncommon around Lubumbashi. Rwanda, rare, known only from area near Rusumo Falls. Not certainly known from Burundi. Tanzania, uncommon and local, in NW from L. Burigi to Kibondo and Mpanda, also in S in Songea. Angola, central plateau from N Huila and Bié to Cuanza Sul, Malanje and Zaïre and Zambia borders; record in S in Cubango Valley. Zambia, widespread and locally common; recorded in 46% of squares; absent from Luangwa Valley. Malaŵi, sparse, at 770–1540 m, almost throughout, except for Mwanza, Chikwawa and Nsanje districts. Mozambique, restricted to Tete District. Rare in NE Namibia and N Botswana (Kavango and Chobe Rivers, Mipacha, Nkasa Island, Tobera and near Popa Falls); once W Namibia (Spitzkopje, near Karibib).

Description. *L. s. tacitus* Clancey: SE Angola (Cubango) to S Zambia west of Luangwa Valley, Malaŵi and W Mozambique. ADULT ♂: forehead grey, paling to whitish by bill, crown grey, paling to whitish at sides, forming superciliary stripe above black mask; hindneck and upper mantle darker grey, merging into olivaceous russet-brown on lower mantle; scapulars snow-white; rump and uppertail-coverts grey or brownish grey. Tail narrow, graduated, T6 with 20–35 mm shortfall; T1 and T2 proximally dark brown, feathers edged with russet-brown and finely barred with black, distally dark brown, unbarred; T3 brown with white tip 8 long at shaft, T4 brown with similar white tip but most of inner vane also white, T5 and T6 almost wholly white. Lores and ear-coverts black, forming sharply-demarcated mask. Chin and throat white; breast, flanks, belly and undertail-coverts pale grey or greyish white. Wings mainly olivaceous russet-brown but primaries and outer secondaries blackish, upperwing-coverts, inner secondaries and tertials regularly banded with blackish. Underside of flight feathers pale grey, underwing-coverts and axillaries whitish. Underside of tail white or buffy white (concealed feathers in middle blackish). Bill black; eyes dark brown; legs and feet black. ADULT ♀: like ♂ but mask dull black (not jet black), tail dark-banded further towards tip, and flanks warm rufescent buff. SIZE (22 ♂♂, 24 ♀♀): wing, ♂ 78–87 (81·3), ♀ 78–85 (81·6); tail, 72–87; bill 12·5–14; tarsus 19–23. WEIGHT: ♂ (n = 12) 21–30 (27), ♀ (n = 11) 22–30 (25·7), 1 unsexed 33.
IMMATURE: forehead, crown, hindneck and upper mantle buff-brown, finely banded with dark brown; mask dark brown; lower mantle to uppertail-coverts rufous-brown, dark-barred, the feathers tipped blackish; scapulars whitish, the feathers tipped blackish; tail rufous-brown, finely and evenly banded with blackish; each tail feather with narrow, U-shaped black line 3 mm from tip, arms of U 1 mm from edges of vanes half way down feather; chin, middle of belly and undertail-coverts whitish, cheeks and rest of underparts whitish, densely barred or vermiculated with brown; wings more rufescent than in adult, not barred, but greater coverts and tertials with narrow whitish edges and tips and blackish line just inward of the white; primaries and secondaries finely edged and tipped with buff.
NESTLING: not described.

L. s. souzae Bocage: Congo, Zaïre, Angola except S, N Zambia (?), W Tanzania (?); boundary between *souzae* and *tacitus* poorly known. Like *tacitus* but mantle less olivaceous, wings paler, and ♀ with less rufous wash on flanks than in ♀ *tacitus*. Perhaps slightly larger: (♂♀, n = 25) wing 81–90, tail 76–90. WEIGHT: (♂♀, n = 23, Angola) 21–30 (26·5).

L. s. burigi Chapin: E Rwanda, NW Tanzania, also Zambia east of Luangwa Valley (plateau of Eastern Prov.). Racial affinities of E Zambia (and Malaŵi) populations require re-examination. Like nominate race but less vermiculated and less rufous above, especially on tertials; grey on head and mantle supposedly more brownish.

Field Characters. Length 16·5–17·5 cm. A small woodland shrike of S-central Africa with black mask, grey head, white supercilium, white above bill, brown back, black barring on brown wings; somewhat like ♀ Red-backed Shrike *L. collurio* but with prominent *white scapular patch*, no barring on white underparts; ♀ has buffy flank patch.

Barred juvenile has black mask and pale shoulder patch, back and wings brown as in adult.

Voice. Tape-recorded (86, 91–99, B, F, CART, GIB, STJ). Territorial call a long (*c.* 1 s), loud, burry, ringing, down-slurred 'beeeeeer'; other calls include a tearing double 'tschaaa-zhaaaa', low 'tzzer' or 'tzzzzzick' of alarm, and chattering and harsh notes.

General Habits. Inhabits climax miombo woodlands in S part of range in Angola, but not necessarily associated with miombo in N (e.g. in Quissama Nat. Park); in woodlands of Congo R. basin in Angola, commonest in open *Uapaca* woods along drainage lines. In Rwanda, confined to *Pericopsis* savanna (vande Weghe 1981). In Malaŵi, interior or quite open miombo woodland. Forages singly or in pairs (birds well separated), perching 2–4 m high, searching ground, then flying down rapidly to seize insect on ground; also takes beetles from base of treetrunk. Uses feet like a hawk, standing on prey whilst using bill to tear off beetle's elytrae and head. Described both as very confiding, and shy and silent, disappearing when disturbed into upper foliage of trees. Sedentary, but occurrence of vagrant in W Namibia suggests that some birds also subject to distant movements.

Food. Insects including 'Christmas beetles'; spiders; one record of vigorous attack on a small bird (Violet-backed Sunbird *Anthreptes longuemarei*), which escaped after a few sec (Medland 1991).

Breeding Habits. Solitary, monogamous nester.
NEST: a rather deep cup, thick-walled, round and neat, more like helmet-shrike nest *Prionops* or Brubru *Nilaus afer* than usual *Lanius* nest. Made of pliant, soft petioles, short stems and grasses, the outside consisting of greyish, woolly plant down, bits of lichen, and coarse cobweb, which binds everything together and secures and moulds nest to substrate; lined with network of yellowish creeper tendrils and fine, soft grasses and roots (**A**). Placed in strong central fork against main stem, or on 5-cm thick, more-or-less horizontal branch away from centre, in tall shrub or small, thickly-leaved tree, 2·5–6·5 m from ground, near top of tree, in more open parts of woodland. One nest 4 m high in *Uapaca kirkiana* tree was on straight, horizontal, lichen-covered branch, 2·5 m from centre of tree, and resembled a perfectly-blending node or swelling on the limb; another high in a leafless *Dalbergiella nyassae* tree was where 2 branches crossed, looked like a slight thickening in the wood, and was invisible from below (Took 1966).

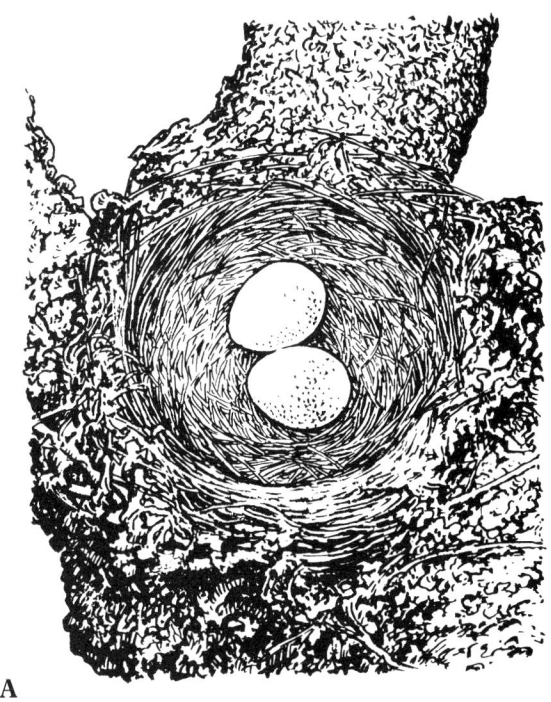

A

EGGS: 2–4, av. (n = 66 clutches, Zambia and Malaŵi) 2·64; c/4 rare. Cream or greenish white, densely speckled and with some larger spots, mostly in broad, ragged band around large half, sometimes around centre or small half, in light ash-grey and pale olive-green; with scattered specks over rest of surface. SIZE: (n = 13) 19·0–21·9 × 15·9–17·0 (20·8 × 16·3).

LAYING DATES: Zaïre, Sept–Nov; Angola, Aug–Sept; Zambia, Aug–Dec (48% of 85 clutches in Oct, 47% in Oct–Nov); Malaŵi, Sept–Oct, Dec, mainly Oct.

INCUBATION: by ♀ only (Vincent 1949). Incubating ♀ uttered low 'tzzzzzick' note several times on approach of mate carrying large spider; it left nest and flew to adjacent tree where, still calling and with much wing fluttering, it was fed; the ♂ departed and the ♀ preened then returned to incubating; it showed no fear on human inspection.

BREEDING SUCCESS/SURVIVAL: a brood taken possibly by tree lizards (Took 1966).

References
Clancey, P.A. (1970).
Dowsett, R.J. (1996).
Lefranc, N. and Worfolk, T. (1997).
Took, J.M.E. (1966).
Maclean, G.L. (1993).
Vincent, A.W. (1949).

Lanius senator **Linnaeus. Woodchat Shrike. Pie-grièche à tête rousse.** **Plate 18**

Lanius senator Linnaeus, 1758. Syst. Nat., ed. 10, p. 94; Indiis (error, type locality restricted to the Rhine, Hartert 1907). **(Opp. p. 271)**

Range and Status. NW Africa; Europe north to France, Poland and Romania; Turkey, Levant and Caucasus to Iran. Winters in Africa and S Arabia.

Breeds N Africa. Morocco, widespread, locally common south to High Atlas and Anti-Atlas Mts; N Algeria, widespread from coast to northernmost oases, common in

Lanius senator

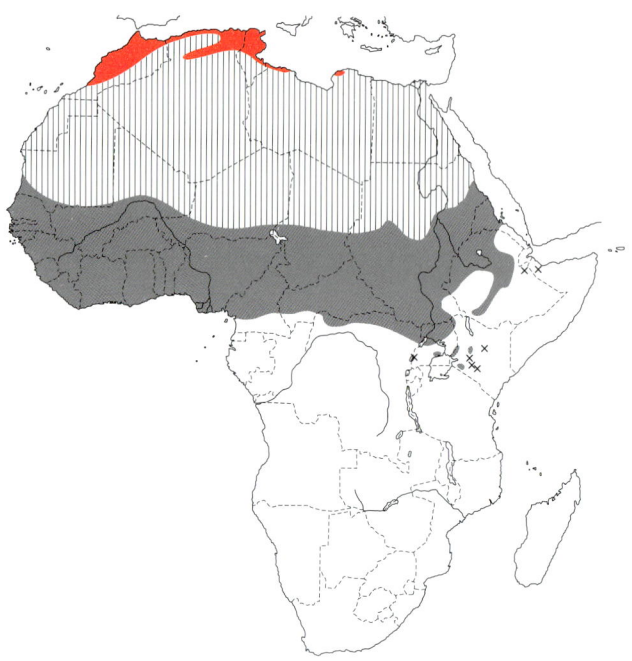

Tell, especially E, absent from Hauts-Plateaux; Tunisia, frequent to common south to desert borders and Gulf of Gabès, extending to NW Libya (coastal Tripolitania), east to Misurata; and locally NE Libya (Jebel Akhdar).

Winters south of Sahara: from SW Mauritania, Senegal, Gambia, Mali (from Niger Delta and N limit of sahel), S Niger, (from about 14°N), central Chad, central and NE Sudan and W Eritrea, south to Liberia, Ivory Coast, Ghana, Nigeria, N Cameroon, Central African Republic (except SW), S Sudan, N Zaïre border, Uganda, W Kenya, and Ethiopia (N, W and Rift Valley). Widespread and common in sahelian and soudanian savannas; more local in guinea savannas, reaching Ghanaian coastal plains and S Nigeria; fairly common in N Liberia; frequent to common N Zaïre in Ubangi and Uele grasslands, and Uganda south to L. Albert and Karamoja; scarce S Uganda and W Kenya (south to Mara Game Res., east to Samburu and Nairobi); rare NW Somalia (2 old records). A few winter in S Algerian oases.

On southward passage, frequent N African coast, Nile Valley and Saharan oases; common to abundant on Atlantic coasts of Morocco and Mauritania and Red Sea coasts of Sudan and Eritrea. In spring, common to abundant along whole N African coast (12 times as numerous as in autumn in Egypt). Much commoner than in autumn in N Saharan oases. Marked autumn and spring passages S Saharan borders and sahel zone.

Wintering densities of 1·5 birds per km^2 on natural dry steppe, Mauritania (Brown 1982); 3–10 birds per km^2 in *Mimosa* savanna, Senegal (Morel and Roux 1966); 17·3 birds per km^2 in Gambia, Jan (Cawkell and Moreau 1963); 1–4 birds per km^2 in savanna at Wologizi, Liberia (Gatter 1997). In Togo, birds spaced less than 200 m apart in baobab country.

Description. *L. s. senator* Linnaeus: Europe south to Pyrenees, mainland Italy, Sicily, Greece and W Turkey; also E Libya (Cyrenaica); winters south of Sahara, east to W Sudan (Darfur) and NE Zaïre (Uele). ADULT ♂ (breeding): forehead black; crown to hindneck and upper mantle rufous-chestnut; lower mantle, inner scapulars and back black or greyish black (feather tips fringed rufous when fresh; outer and lower scapulars white, forming prominent large patch; rump dark grey; uppertail-coverts creamy white. In tail T1 black; T2–T5 black with narrow white tips (5–10 mm on T5) and white bases (5–10 on T2, increasing to almost half of T4 and over half of T5); T6 white, inner web usually with black or greyish subterminal band 10–20 deep. Mask around eye and ear-coverts black, continuous with black forehead and extended as broad blackish band down side of neck. Variable whitish patch on lores and side of forehead, sometimes small or mottled with black; and a small creamy white streak backward from upper rear corner of eye. Entire underparts white, tinged pinkish buff or creamy buff when fresh, especially on breast and flanks. Flight feathers and primary coverts black; tips of secondaries and inner primaries narrowly fringed whitish; primary bases creamy white, forming small patch on closed wing. Tertials and greater coverts black, edged and tipped pale buff when fresh; rest of upperwing-coverts creamy white. Bill black; eyes dark brown or grey brown; legs black. Wing-tip rather pointed: P8 longest, P7 0–2 shorter, P6 2–7, P5 8–14, P1 24–31; P9 5–10 shorter than P8, falling between P5 and P6 or equal to P6; P10 very small, 3–9 longer than primary coverts; P7–P8 emarginated on outer web. ADULT ♂ (non-breeding): like adult ♂ (breeding) but black forehead and mask often mottled with rufous; lower cheeks, sides of breast and flanks often with black scaling. Bill blue-grey or flesh-grey at base. ADULT ♀ (breeding): variable, some like adult ♂, with black forehead and black band through eye, but usually with browner back, others with forehead and eyestripe mottled buff or brown (not well demarcated from buff patch on lores or from rufous crown), lower cheeks, breast and flanks often barred brownish, black of tail and upperwing tinged brown. Bare parts like adult ♂ (breeding). ADULT ♀ (non-breeding): like adult ♀ (breeding) but forehead and area around eye mainly creamy buff; ear-coverts dark brown with buffy mottling. SIZE: wing, ♂ 98–103 (100), ♀ 96–102 (98·6); tail, ♂ 71–77 (75·9), ♀ 72–79 (74·5); bill, ♂ 19–20·5 (19·8), ♀ 18·5–20 (19·4); tarsus, 24–25 (24·3), ♀ 23·5–25 (24·3). WEIGHT: SE Morocco, spring (may include some *rutilans*), ♂ (n = 26) 23–35 (29·2), ♀ (n = 15) 25–41 (31·2); Nigeria, spring (presumed mainly this race), sexes combined, Dec–Mar (n = 4) 27·5–30 (28·6), Apr (n = 10) 29–46 (38). Complete moult usually starts in breeding area, completed in Africa; partial pre-breeding moult in Africa, late winter.

IMMATURE: juvenile closely barred blackish brown and buffy white above, outer and lower scapulars buffy white with fine blackish brown submarginal bars; lores pale buff with dusky speckling; ear-coverts grey-brown, mottled blackish and streaked pale buff; underparts buffy white with dark grey barring. Tail dark brown; T1–T3 fringed rufous-buff along tip and outer web, T4–T5 with broader creamy buff tip, T6 with broad tip (>5 mm) and broad outer edge creamy buff. Upperwing dark grey-brown; flight feather tips fringed pale buff, white patch at primary bases tinged buff, less strongly defined than in adult; tertials and greater coverts broadly edged and tipped tawny buff, median coverts buffy white with blackish submarginal bars, lesser coverts tipped pale buff. Bill greyish flesh with dark culmen. Juvenile plumage replaced gradually during autumn and winter. First-winter a variable mixture of juvenile and non-breeding ♀-type head and body feathers, many of the latter with black subterminal bar; new feathers of forehead and lores buff, those of crown to upper mantle pale chestnut.

NESTLING: naked, or with a little short sparse white down on underparts.

L. s. rutilans Temminck: Iberia, Morocco to NW Libya; winters Senegal, presumably elsewhere in W Africa. Like *senator* but smaller: wing (10 ♂♂) 91–96 (94·1).

L. s. badius Hartlaub: Balearic Is., Corsica and Sardinia; winters Liberia to Cameroon and Chad, mainly south of *senator*. Differs from nominate race in lacking white patch at base of primaries (or patch largely hidden by coverts); black forehead band narrower; uppertail-coverts less white.

L. s. niloticus (Bonaparte): Cyprus, Levant and E Asia Minor to Iran; winters W Sudan and NE Zaïre to Ethiopia, Eritrea and W Kenya. Same size as *senator*, but differs at all ages in having extensive white on all tail feather bases (14–35 mm on T1); underparts whiter, white patch on primaries typically larger; ♀ often without black on face, even in breeding plumage. WEIGHT: NE Sudan, Aug–Oct, sexes combined (n = 154) 19·5–32 (24·9).

Field Characters. Length 18–19 cm. A stocky, medium-sized, rather large-headed black and white shrike. Adult ♂ (all races) has broad *rufous patch* from crown onto hind neck, surrounded by *black* forehead, mask and mantle; in flight, white patches on scapulars and primaries and broad white rump contrast with black back, wings and tail (white primary patch lacking in race *badius*; white of rump extends onto base of tail in *niloticus*); a little white on lores or around eye, varying with race. ♀ similar but duller, with more white or buff on face. Juvenile has dark barring on head, upperparts and back; distinguished from juv. Red-backed Shrike *L. collurio* and Lesser Grey Shrike *L. minor* by *pale scapular* patch, from Red-backed by dark brown tail and lack of rufous tone to upperparts and wings, from Lesser Grey by more heavily barred underparts and throat and less distinct eye patch, and from juv. Masked Shrike *L. nubicus* by brown tone to head, mask and back, *pale rump* and shorter, square tail. First autumn birds begin to acquire rufous nape feathers by Sept.

Voice. Tape-recorded (62–73, 105, 110, B, F, PAY). Song long and varied, with many imitations; chatters, whistles, churrs, shrill piping, chuckles and occasional sweet warbles, given in continuous series of short phrases separated by brief intervals, in typical *Lanius*-like jerky delivery. Common call a harsh, grating rattle. For complete catalogue of songs and calls, see Cramp and Perrins (1993).

General Habits. Breeds in open woodland, and open grassy scrubland or cultivated country with tall bushes or trees and tracts of bare soil; typically on plains or low hills, but regularly up to 1200 m and exceptionally at c. 1900 m in Moroccan High Atlas. Favours holm oak *Quercus ilex* forests which are poor, sparse and open, also open forests of cork oak *Q. suber*. Winters in acacia steppe, open bushed grassland, and cultivated country with scattered trees; favours greener acacia near water, on floodplains and swamp and lake margins; also forest clearings in Liberia and Ghana. Pauses in saharan oases on passage.

Usually solitary and territorial in winter. In E Nigeria, territories several 100 m across, defended for weeks at a time; some territories in Senegal appear to hold pairs (Cramp and Perrins 1993). Spring passage birds hold temporary territories in Egypt (Simmons 1951). Often conspicuous and easy to approach. Hunts from exposed lookout such as bare branch, post, fence or telephone wire. Commonly perches higher than Red-backed Shrike *L. collurio*, more erect, with tail drooped down. Spends much time in trees where it may stay concealed within canopy (Cramp and Perrins 1993).

Drops or glides onto ground prey or makes sallies after flying insects. Takes 65–80% of prey on ground. Rarely hovers. In Senegal, 20% of pounces are from <1 m, 48% from 1–3 m, 22% from >3 m (Tye 1984). Rarely returns to original perch with prey. Crushes small insect in bill; dismembers larger one, holding it down with foot (Cramp and Perrins 1993). De-stings wasps and bees by crushing abdomen tip in bill. Occasionally impales prey items on twig or thorn. Flight direct and dashing with upward sweep to perch; often flies for up to 100 m when disturbed. Tail in constant motion when perched, with side to side swinging movements.

Western races (*senator, rutilans, badius*) migrate on broad front across N Africa and Sahara. In autumn, some ringing evidence for initial southwest heading of W and central European birds, while records from Levant suggest some eastern *L. s. senator* move east of south (Cramp and Perrins 1993). Wintering area of *badius* lies due south of breeding range. Autumn migration through Mediterranean and Egypt extends from late July to early Oct (peak mid Aug–mid Sept); small numbers on N African coast suggest overflying. Heavy movement, however, along Moroccan Atlantic coast, mid Aug to late Sept. Arrives Senegal from mid-July (peak after mid Aug), with juveniles c. 2 weeks later (Morel and Roux 1966). Passage in N Niger Aug–Oct. Arrives Mali from Aug (mainly Sept–Oct), N Nigeria from mid Sept, Ghana early Oct. In Liberia, reaches NW in early Dec, SW forest bloc mid–Dec to early Feb; highest numbers Feb; departs Mar, latest date 15 Apr (Gatter 1997). Reaches guinea savannas of Ivory Coast and S Nigeria later, from Nov. *L. s. niloticus* migrates south to southwest, most crossing Red Sea en route to NE African winter quarters. Strong passage on Sudan coast, late Aug–early Oct (peak early–mid Sept), Eritrea from early Sept, and W Sudan (Darfur) (this race and *senator*) late Sept to late Oct. Reaches central Ethiopia from Sept, S Sudan and N Uganda from Oct.

Northward migration protracted. Begins in Senegal in late Feb, and noted Mar–Apr in Gambia, Sierra Leone, Mali, Niger and Chad. Volume of passage in Eritrea about same as in autumn. Most birds vacate southern wintering areas in Mar–early Apr and sahel zone in Apr–early May. Late records in Uganda mid Apr, NE Zaïre late Apr, Eritrea mid May, N Ghana late May, Senegal mid May. Ringing shows that some W and central European birds use more easterly route in spring than autumn (Zink 1975). N African passage much heavier than in autumn, especially in NE Morocco to Tunisia, early Mar–early June (mainly Apr). Migrants conspicuous at coast and oases on N edge of Sahara, but NW African mountains appear to be overflown. In Tunisia, where this was commonest spring migrant trapped at Gabès (Castan 1954–55), adults preceded first-year birds; weights c. 5 g less after crossing Sahara than on European breeding grounds. Breeding birds occupy NW African territories from end of Mar; pairs widespread by mid Apr. Stragglers remain in desert to mid June.

Birds ringed in SW Germany and Switzerland recovered in autumn in W Morocco. Over 3000 ringed in spring in Tunisia gave recoveries in same year in Italy and (1 bird) Albania; 2 retrapped at same site in autumn so some birds use same route in both seasons. Adult *badius* ringed Algeria (Apr) recovered in Mallorca (July); and a bird ringed N Nigeria (Oct) found in N Italy (Aug).

Food. Mainly insects (often large) and other invertebrates, occasionally small vertebrates. Beetles and Orthoptera are main prey in Europe; Hymenoptera, Lepidoptera (adult and larvae), Hemiptera, Diptera and Odonata are also taken, as well as spiders, earthworms and sometimes snails (Cramp and Perrins 1993, Lefranc and Worfolk 1997). In E Sudan, Aug–Sept, takes small passerine migrants such as Marsh Warbler *Acrocephalus palustris*, weakened by migration (Nikolaus 1990).

Breeding Habits. Monogamous and territorial; territory defended by both sexes, used for courtship and nesting (Cramp and Perrins 1993). Territories av. 8 ha in S Germany, 2·2 ha in S Spain; in Maghreb, territories often in groups, 3–4 within 10 ha, neighbouring nests usually 200 m apart (sometimes only 30–50 m). Some birds are apparently paired on arrival, begin nest-building immediately and lay within a few days. Unpaired birds engage in more calling, display and courtship feeding (Cramp and Perrins 1993). In display, ♂ often stands erect facing ♀, plumage sleeked except for ruffled forecrown. He may nod rapidly, raising and lowering head while singing continuously. ♂ and ♀, or ♂ alone, may bob, lowering bill while bending legs, then rapidly stretching to full height (Cramp and Perrins 1993). Often 2 broods.

NEST: strong cup of leafy plant material and roots, lined with rootless, wool, hair, fine roots, cobweb, moss and lichen; ext. diam. 120–150, int. diam. 70–85; int. depth 50–75. Site chosen by ♂, usually in tree on thick horizontal branch; in Algeria, mostly above 4–5 m. Built by both sexes, usually in 4–6 days.

EGGS: 5–6, occasionally 7; NW Africa (n = 106), 4–8 (4·9). Laid daily. Pale olive-green (occasionally sandy, yellowish or reddish), with brown or olive speckles and blotches, usually concentrated at broad end. Reddish eggs (with red-brown or purple speckles) comprise 10% in E Morocco (Heim de Balzac and Mayaud 1962). SIZE (n = 200) 22·5–26 × 16·5–19 (23 × 17).

LAYING DATES: NW Africa, late Apr–late May.

INCUBATION: by ♀ (rarely by ♂), beginning with penultimate or last egg; period 14–15 (16) days.

DEVELOPMENT AND CARE OF YOUNG: young cared for and fed by both parents. Brooded by ♀ for first few days; food then brought by ♂, and passed to chicks by ♀. Young fledge after 15–18 days. Family remains loosely together until migration.

BREEDING SUCCESS/SURVIVAL: of 533 eggs in German study, 69% hatched, 42% fledged; successful pairs averaged 3·4 young, all pairs 2·3 young (Ullrich 1971).

References
Cramp, S. and Perrins, C.M. (1993).
Ullrich, B. (1971).

Genus *Corvinella* Lesson

Monotypic; endemic to N tropics, southeast to L. Victoria. A large, paedomorphic *Lanius* (i.e. plumage of adult suggestive of immature *Lanius* shrike); bill yellow; plumage brown, dark-streaked and vermiculated; black mask; red-brown primaries; lanceolate feathers on crown and back; uppertail-coverts and flank feathers long, and flanks with usually-concealed colour patch, greyish rufous in ♂, maroon in ♀; sexes otherwise alike; long, strongly graduated tail with 12 narrow feathers. Juvenile very like adult though more barred than streaked. Very vocal; co-operative breeder with many helpers; in year-round social flocks of a dozen birds.

Plate 20 *Corvinella corvina* (Lesson). Yellow-billed Shrike; Long-tailed Shrike. Corvinelle à bec jaune.
(Opp. p. 287) *Lanius corvinus* Shaw, 1809. Gen. Zool., 7, p. 337; no type locality (= Senegal, Lesson 1831).

Range and Status. Endemic resident, and partial migrant near some boundaries, S Mauritania to W Kenya. Mauritania, S sahel zone and R. Senegal valley; thought to breed, but for most of year large flocks wander widely. Senegal and Gambia, throughout, but since early 1970s local or absent north of 15°30′N; common in S. Guinea-Bissau (recorded, Frade and Bacelar 1959). Mali, quite common, in erratic flocks, north to 14° N, and to 15° near to Delta although not in it; in S, resident at Ban Markala and very common on Bani R. in Oct–May. Burkina Faso, frequent all year to north of Ouahigouya and in Ouagadougou area, and occurs in E and SW, south to 11°N (Holyoak and Seddon 1989). Niger, common resident in extreme SW; in Niger R. valley known from Saga in

Corvinella corvina

Jan–Aug, also Niamey and Torodi; in S of country records from near Tillabéry, Maradi-Tanout region, and all year at Boboye and Dosso.

Guinea, frequent and widespread south to Wassou, Mambia, Koba, Kolenté, Kankan, and coast near Sierra Leone border; in E, south to Beyla. Sierra Leone, status poorly known; uncommon in Scarcies district. Ivory Coast, widespread resident north of Bouaké, and in dry season non-breeding visitor in southern guinean savannas south to Toumodi. Ghana, locally common resident in N (Mole, Bole, Gambaga), south through Kete Kratchi to Keta and Accra Plains; rare at Tumu; absent from Winneba and Cape Coast. Togo and Benin, common and widespread south to about 07°N. Nigeria, locally common resident in soudanian and guinean savannas; common around Kano, but not in Maiduguri district or sahel zone in NE; frequent around Zaria, common further south around Kaduna; very common in Yankari Nat. Park; in SW occurs south to Ibadan and Abeokuta where frequent, and occasionally Lagos (where none seen since 1960s); in SE, common in Tivland, very common on Obudu Plateau, rare as far south as Enugu. Cameroon, Adamawa Plateau (south to Galim), northern limits unclear; probably not as far north as Waza Nat. Park, although just across Chad border occurs to 12°N near Ndjamena. Chad, common resident in soudanian zone, moving north to 16°N during rains (Salvan 1969): erratic wet season visitor to Ouadi Rime-Ouadi Achim Faunal Res., north to Ouadi Achim. Central African Republic, in N a common resident in Manovo-Gounda-Saint Floris Nat. Park and Vakaga area; in W a record near Bozoum; in S, claimed occurrences in Lobaye Préf. and Bangui area need substantiating. Sudan, common breeding resident on Jebel Marra, rare in N Kordofan, fairly common south of 09°N, in corridor through N Bahr al Ghazal Prov. and from El Buheyrat Prov. to Equatoria east of 30°E. Zaïre, local in extreme NE, in Upper Uele: parties encountered around Garamba (now Garamba Nat. Park) every 5–10 days (Chapin 1954); from Garamba, west to Yambio (Sudan) and south to Nzoro. Uganda, generally common at medium elevations south to Murchison Falls Nat. Park, Lango, central Mengo, Soroti and Elgon; up to 2200 m. Kenya, west of Lokichokio; local and uncommon on NE flank of Mt Elgon (Suam R., from Kanyakwat to Kongelai and Kacheliba); uncommon at Soy, Kibigori and Awasi; formerly more widespread in areas now under cultivation (Kitale, Nandi, Lumbwa, Muhoroni, Sotik: Zimmerman *et al.* 1996).

Density (in 18 more-or-less contiguous group-territories, Legon, Ghana) 0·4–1·5 (0·8) birds per ha.

Description. *C. c. togoensis* Neumann: Guinea (except N) to Nigeria (except NW), Cameroon, Central African Republic, Chad and Darfur, Sudan. ADULT ♂ (breeding): forehead to upper back warm brown, with copious fine, short, black streaks; scapulars pale rufous with grey centres and black shaft-streaks, lower back to uppertail-coverts greyish brown with a few long dusky streaks. Sides of crown and of nape unstreaked pale rufescent buff, forming indistinct superciliary stripe. Lores, above and below eye, and ear-coverts, brownish black. T1 and T3 dark brown with buff tips and dusky penultimate bar, T4 to T6 successively paler buffy brown, with irregular penultimate black bar well behind tip. Chin and throat buffy white, breast buff with fine blackish streaks, belly pale buff, almost unstreaked, flanks grey, or grey and buff, with patch of silky, fluffy pinkish rufous feathers at front; thighs, vent and undertail-coverts buff. Flank feathers, 40–50 mm long, have dark grey base, pinkish cinnamon-rufous centre, and dark grey distal third with faint rufous barring near tip. Primaries proximally rufous on both vanes, distally blackish, buff-tipped, P10 with no rufous on outer vane and P9 with only a little. Secondaries and tertials dark brown, with rufous-buff edges and tips and thin blackish line around each feather inside the buff; greater primary-coverts dark brown; alula and all other upperwing coverts dark brown with rufous edges and tips; underside of remiges dark grey (where blackish dorsally) and silvery rufous (where rufous dorsally), underwing-coverts grey and pale rufous. Bill chrome-yellow; eyes hazel-brown, rim of eyelids chrome-yellow; legs and feet pale greenish or dirty yellowish grey. ADULT ♀: like ♂ but flank feathers long and fluffy, with dark grey base, maroon middle, and white distal third, sometimes tinged cinnamon, with faint grey barring near tip; shafts black, making long black streaks; forms conspicuous coin-sized patch if overlying feathers are parted. SIZE and WEIGHT: (see *corvina* and *affinis*; all races same size).

IMMATURE: juvenile has upperparts not streaked but freckled (on head) or scalloped (on back) with black; scapulars with conspicuous blackish chevrons; T3–T6 with warm buff edge well delineated from narrow penultimate band and grey-brown centre. Uppertail-coverts short and, like rump, banded dusky and rufescent buff. Breast and flanks covered with diffuse dusky and pale buff bars. During moult into immature plumage, both sexes have tips of flank feathers pale cinnamon-rufous; at 11–12 weeks of age the sexually dimorphic flank feather colours develop and can show even in the field (Grimes 1979a).

NESTLING: not described.

C. c. corvina Shaw: S Mauritania, Senegal, Gambia, Guinea-Bissau, N Guinea, Mali, N Burkina Faso, Niger, NW Nigeria; interface between *corvina* and *togoensis* ill-defined and may be clinal. Juvenile more rufescent above than juvenile *togoensis* and far more prominently banded; adult paedomorphic (has retained essentially juvenile plumage): mantle streaked, darker than *togoensis*, scapulars heavily black-banded on rufescent buff, tertials and secondaries and coverts broadly rufous-edged; breast and flanks banded, not streaked. SIZE (7 ♂♂, 6 ♀♀): wing, ♂ 120–129, ♀ 118–123; tail, ♂ 170–180, ♀ 163–176; bill to feathers, ♂ 16.0–18.5, ♀ 16.0–18.5; tarsus, ♂ 30–32, ♀ 29–31. WEIGHT: (unsexed, n = 11, Ghana) 58–80 (65.4).

C. c. caliginosa Friedmann and Bowen: Bahr al Ghazal, Sudan. Like *togoensis* but greyer above, with streaking more diffuse, and greyer below; may be slightly larger.

C. c. affinis Hartlaub (includes '*chapini*'): S Sudan, NE Zaïre, Uganda, W Kenya. Upperparts more greyish, less rufescent than in other races, and smaller area of rufous in primaries. SIZE (15 ♂♂, 19 ♀♀): wing, ♂ 113–128, ♀ 117–123; tail, ♂ 160–183, ♀ 160–186; bill to feathers, ♂ 16–18, ♀ 15–18; tarsus, ♂ 30–32, ♀ 30–31.

Field Characters. Length 30–32 cm. A large grey-brown shrike, streaked dark above and below, with long graduated tail and *yellow bill*, pale supercilium above dark mask; *rufous primaries* conspicuous in flight. Juvenile mottled and barred where adult streaked. *Sociable* and noisy, frequently displaying and breaking into chorus of grating calls like *Turdoides* babblers.

Voice. Tape-recorded (104, 108, B, BRU, GRI, MEES, MOR, PAY). Song, given by individual alone or in groups of up to 25, a series of rather scratchy, tuneless single and double notes, shrill but not unpleasant, 'tschweeya, choochit, chwit, tschuwee ... ', or rolling 'trrrreeeo'; sometimes harsher, more screechy 'jeeew-chrrit' or 'scrrreee', or double 'pitch-pitch'; also imitates other birds, e.g. Ringed Plover *Charadrius hiaticula*, Ruddy Turnstone *Arenaria interpres* and Snowy-headed Robin-Chat *Cossypha niveicapilla* (see Chappuis, in press). Group display accompanied by warbling note (Grimes 1980). Alarm a short, harsh 'chee-zhee-zhee', becoming harsher and louder when other birds join in chorus; alarm calls are of 4 types, differing with situation (see sonagrams in Grimes 1980); those in response to aerial predator are higher-pitched. Constantly repeated 'scis-scis' has given the bird the local name of Scissor-bird in Uganda (Lefranc and Worfolk 1997); also said to give rippling and parrot-like calls when moving in group. For calls associated with nesting, see Grimes (1980).

General Habits. Inhabits dry, shrubby savanna woodlands, bushy pasture, acacia savanna, cultivation and riverine woods; in Mali *Pterocarpus lucens* woodland; ubiquitous in all woodland types in Yankari Nat. Park, Nigeria; abundant in Legon University Campus and elsewhere in Accra (Ghana) in civic, private and botanic gardens, bushes around gamefields, and park-like grassy areas with spaced trees and shrubs and scattered buildings. Hunts from bare branch on side of leafless or leafy tree, from short, charred tree-stump, telephone wire or fence. Feeds essentially solitarily, but often attracted to abundant food source where other members of group are gathering, and feeds communally at hatches of ants and termites on the ground (Grimes 1980). Catches most prey items one by one by keeping watch from perch, then flying down to ground and pouncing; or pounces after short, low flight over feeding area. Takes large ant *Paltothyreus tarsatus* back to tree perch and rubs it along branch before swallowing it. In poor visibility feeds entirely on ground, systematically turning leaf litter over. Pulls earthworms from burrows and takes them from soil surface. Makes aerial sallies after small insects in flight.

Almost always in tightly-structured flock or group of 6–25 (av. 12) birds, which stays in and defends a territory all year (Grimes 1980). Group comprises a breeding pair, adult helpers (which help to defend territory, raise alarm against predators, build a succession of nests, feed the nesting ♀, and feed nestlings and fledglings) and young birds. Size and composition of group changes with mortality and recruitment and when a subordinate rises to sexual dominance; one group varied between 8 and 13 birds in 3 years. ♀♀ usually predominate; typical group comprises 3 ♂♂ and 5 ♀♀, but one was of 9 ♂♂ and 4 ♀♀. Territory size (n = 18) 10.6–27.1 (16.1) ha. No correlation between group size and territory size. Quarrelsome, group often driving small birds away and scolding arboreal squirrels, and frequently entering upon noisy disputes with flock in adjacent territory. Disputes precipitated when one or more birds of adjacent territory trespass or when 2 groups meet at common border. Disputes or rallies take place in high places: tallest trees along border between 2 territories or, in absence of trees, on flat roof of 4-storey block of flats or horizontal aerial on top of radio mast (Legon, Ghana). In dispute, birds either hop to and fro between branches uttering distinctive warble (the increasing noise attracting other members of social group, including incubating ♀♀, until all are participating), or bow towards nearest neighbours, rubbing bills on branches between bowing sessions, and oscillating tails slowly up and down and from side to side, with body plumage puffed out and flank patches showing. Occasionally a 3rd neighbouring group, not initially involved, joins in the display. Display often lasts 10–15 min. Display tree may change hands several times during a dispute; 2 birds sometimes fight, falling with feet and wings interlocked, separating before reaching ground.

Flight rather feeble, not very sustained; flies low, and glides upwards into tree for last few m. On approach of person or ground predator, gives alarm call and others in group quickly take it up, calls becoming harsher and louder. Alarm call for aerial predator (Shikra *Accipiter badius* or Pied Crow *Corvus albus*) is higher pitched, and all shrikes immediately seek cover. Flank patches, generally concealed, are exposed during group territorial disputes, in displays at nest, and when bird is preening. Flock roosts together all year in same tree, often *Fagara xanthoxyloides* or *Bambusa vulgaris* (Legon, Ghana); before and during flight to roost, bird utters 'dispute-warble', stopping abruptly a few moments after it enters roost. Moults completely, soon after breeding.

Resident; but in Mauritania local and irregular, birds moving widely, sometimes in flocks of 150–200, in search of grasshopper or locust concentrations (Lamarche 1988); similarly erratic towards northern limit of range in Mali. Partial migrant in Chad, moving to north in wet season, arriving at Abéché in early or mid July, and departing to south with drying of surface waters in Nov and Dec; sporadic up to 16°N in Ouadi Rime-Ouadi Achim Faunal Res. where common in 1975 but scarce in next few years, perhaps owing to desertification (Newby 1980). May be summer visitor rather than a resident in N Darfur (Lefranc and Worfolk 1997).

Food. Wingless ants and termites taken on ground and winged ones taken in flight; grasshoppers, mantises, caterpillars; spiders, slugs, earthworms and small lizards. Known to attack fledgling passerines (*Uraeginthus* sp.). Longhorn beetle in a stomach.

Breeding Habits. Solitary nester, monogamous, territorial, co-operative breeder. Social group contains a mated pair and non-breeding helpers (mostly their own progeny); it changes size and composition during course of year by birth into it, death, and some adults leaving and entering. Among 18 groups, over 3-year period, 7 ♂♂ and 9 ♀♀ changed groups locally (av. distance, 600 m), twice in small single-sex parties (Grimes 1980). During nest construction, incubation and brooding phases, ♀ calls from nest and other members of group respond by flying to nest with food item; ♀ calls more loudly as group member approaches, her call changing to one like nestling's begging.

NEST: substantial, rather loosely-constructed, open cup, made of small and quite large twigs, rootlets, grass stems and sometimes leaves, and lined with fine roots, pliable tendrils, rootlets, fine dry grass and vegetable fibres. Sited in vertical fork or upon a cluster of twigs, generally concealed in thick foliage in bush or tree in open situation but occasionally in leafless tree, nest 2–10 m high (60% are 3–6 m high), frequently near human habitation; often in shea-butter *Vitellaria paradoxa*, locust-bean *Parkia biglobosa* and flame trees *Delonyx regia* (N Nigeria). 20 tree species used in Legon, Ghana, commonest being *Fagara xanthoxyloides* and *Bambusa vulgaris*. Some (all?) ♂ and ♀ helpers assist pair in nest building. At first fine twigs are arranged in neat circle, then base built, then walls, then cup lining added; ♂ and ♀ shape the developing nest by crouching in it and rotating body. Nest built in 4 weeks; if clutch fails, replacement nest built (sometimes in same tree) in 3–4 days. Occasionally same nest used for 2 successive broods. Some groups of birds build several nests in quick succession before ♀ lays in one; others build only single nest. One group contained 2 breeding ♀♀, on nests 80 m apart. Of 94 nests built by 5 groups, eggs were laid in 75 (Grimes 1980). Some replacement clutches laid only 10 days after loss of 1st clutch. One group had 2 nests in same tree, one with fresh eggs, other with a large nestling.

EGGS: 2–6, mode 4, av. <4 in some years, >4 in others (Legon, Ghana), overall av. (by 3-month period, 5 years) 3·68–3·94. Ovate; unglossed or slightly glossy, smooth-shelled and fragile; colour variable, commonly cream with irregular light yellow-brown spots on numerous underlying grey or lilac speckles, mainly at large end; others pale greenish white, yellow-buff or deep buff, with grey and yellow-brown spots, speckles and occasionally scrawls forming irregular zone around broad end. SIZE (n = 49) 23·2–26·4 × 17·0–19·0 (25·1 × 18·2); also (n = 36 eggs laid by single ♀ in 9 clutches) 22·3–25·0 × 17·7–19·4 (22·4 × 18·3). WEIGHT: (n = 8, a few days before hatching) 3·84–4·35 (4·12).

LAYING DATES: Mauritania, (July, unconfirmed); Senegal, Richard-Toll, June–Aug, Oussouye, (young in July–Aug); Gambia, Mar–July (breeds Mar–Oct (Barlow *et al.* 1997), probably including breeding evidence in Aug–Oct other than eggs; breeds mainly in wet season); Burkina Faso, (young, Aug); Niger, July; Ghana, Legon, Dec–Oct (mainly Feb–Apr, at start of rains and into rainy season), Pong Tamale, Apr, (Mole, juvs July); Togo, June (and dependent young May–Aug); Nigeria, Jan–Oct, and at Kano, near northern limit, Apr–June; Sudan, Darfur, July, Bahr el Jebel, (fledglings Jan, Feb); E Africa, Region A, Jan–Feb, May, Region B, May.

INCUBATION: incubating ♀ gives characteristic call, audible 100 m away (see above), and other members of group respond by bringing her food; she is fed on nest on av. 2·8 to 9·5 times per hour. Period (from laying to hatching of last egg, n = 17): 15–18 days, mode (n = 9) 17 days.

DEVELOPMENT AND CARE OF YOUNG: eyes partly open by day 7, fully open and young bird alert by day 12. Faecal sacs removed by all members of adult group and either swallowed or dropped away from nest. Young close-brooded, by ♀ parent only, for first 8–9 days. Brooding ♀ calls from nest; all members of group respond, one at a time, by bringing food to her, which she passes to the young. In her absence, other members of group generally reluctant to feed young directly. Breeding ♀ brings some food to nestlings during brooding period, but ceases to provision them after day 12. From days 8–9, other group members start feeding young directly. As adult flies to nest with food it calls, and nestlings rotate in nest to face incoming bird. Average per-nestling provisioning rate increases from 1–2 per h at first to 12 per h by day 14. Brood tends to be fed in bursts of activity. Nestling period 19±1 days. On leaving nest, young cannot fly and flutter helplessly to ground, where they scramble up and cling to suitable cover with ease; they are closely attended by members of group and guided to cover. For a few days young make no attempt to escape when person approaches, and can be picked up; after 4–5 days they can fly short distances. All members of brood are attended by all members of group, more or less indiscriminately. Young independent in 7th week after hatching; they continue begging after that, but with little success; well integrated into group by week 10, when they use alarm calls, roost with adults, and participate in communal displays; they can feed fledglings themselves at 14 weeks and feed nestlings at 24 weeks.

BREEDING SUCCESS/SURVIVAL: of 366 eggs (Legon, Ghana) only 57% hatched; most clutches lost during rain storms and high winds; of 159 nestlings, only 44% fledged successfully; overall success rate, 25%. 44% of nestlings were alive one year later, so survival rate from hatching to 1-year-old is 11%. Annual mortality 24% in 2nd year, 41% in 3rd year, 23% in 4th year and 60% in 5th year. At end of 6-year study, no young birds that hatched during the period had yet begun to breed (Grimes 1980). Of 160 nests, one was parasitized by African Cuckoo *Cuculus gularis*, which evidently destroyed hosts' clutch (Grimes 1979b).

References
Bannerman, D.A. (1939).
Chapin, J.P. (1954).
Grimes, L.G. (1979a, 1979b, 1980).
Lefranc, N. and Worfolk, T. (1997).
Serle, W. (1940).

Genus *Urolestes* Reichenow

Monotypic; endemic to southern tropics, north to L. Victoria. Large, with very long tail. Bill black. Plumage black, with white scapulars, pied wing, and flanks black in ♂ but white in ♀; white rump feathers may droop and make flanks appear white in both sexes. Sexes otherwise alike. Head drongo-like: forehead with stiff feathers strongly curved backwards, feathers of head, neck, mantle, throat and breast lanceolate; long rictal bristles; bristles and stiff narrow feathers point forwards to cover nostrils. Breast feathers long, somewhat lanceolate. Juvenile brown where adult black, but feathers lack the penultimate blackish bars that characterize juveniles of *Lanius* and *Corvinella*. Vocal; duets. Co-operative breeder with only a few helpers; in year-round flocks of up to 12 birds. Not known to impale prey.

The single species *melanoleuca* is often placed in *Corvinella*, and Hall and Moreau (1970) even combined *Corvinella corvina* and *melanoleuca* into a superspecies. In our view differences are too great for acceptable accommodation in a single genus.

Plate 20
(Opp. p. 287)

Urolestes melanoleucus (Jardine). Magpie Shrike; Long-tailed Shrike. Corvinelle noir et blanc.

Lanius melanoleucus Jardine, 1831. Edinburgh J. Nat. Geogr. Sci., 3, p. 209; Orange River.

Range and Status. Endemic resident, southern tropics, in 2 populations, SW Kenya to S Tanzania, and Angola to S Malaŵi and Natal. Kenya, a few pairs resident in E Mara Game Res. near Siana Springs; formerly more widespread in Masai Mara and ranged towards Kisii and Magadi. Tanzania, locally common in Serengeti, Lake Manyara Nat. Park and Tarangire Nat. Park; record in Kilimanjaro District; more widespread and commoner from Shinyanga to Dodoma District, through central Tabora, E Rukwa District and corridor through central Mbeya District to N end of L. Nyasa.

Angola, in S Huambo Prov. (Hall and Moreau 1970), throughout Huila Prov., in W Cunene and W Cuando Cubango (probably throughout Cunene and Cubango). Zambia, in S, local and uncommon: Southern Prov. and Lower Kafue Basin from Musa R. downstream to Namwala and Kafue Bridge; north to Lusaka, wandering to Kitwe and Kapiri Mposhi; also much of Sesheke and Barotse Districts, north to Sioma and Nangweshi. N Mozambique, from Zambezi northward near Malaŵi border to about L. Amaramba. Namibia, Botswana, Zimbabwe, N Cape Prov., Transvaal, E Swaziland and N Natal as mapped (Harrison *et al.* 1997: mapping thought to be accurate since bird is so conspicuous). Rather sparse in NE Namibia; frequent to very common in N and SE Botswana. In Zimbabwe commonest in SW, absent from large parts of N and eastern highlands, rather sparse and local over much of central plateau. Transvaal, widespread or local on highveld, common in lowveld and Kruger Nat. Park. Orange Free State, isolated population in Koppies-Edenville area, and known from Clocolan and near Bloemfontein. S Mozambique, rather uncommon and local, confined to Maputo and Gaza Districts.

Has not adapted itself successfully to man-made habitats, and has become less common in, or has disappeared from, areas heavily modified by man (V. Parker *in* Harrison *et al.* 1997).

Urolestes melanoleucus

Description. *U. m. melanoleucus* (Jardine) (includes '*angolensis*'): Angola, S Zambia, N Mozambique, NE Namibia, N and E Botswana, Zimbabwe except SE, Transvaal except E, Orange Free State. ADULT ♂: entirely black, with bluish gloss, except for: white scapulars, lower back and rump (rump usually palest grey rather than white), tertials and secondaries tipped white, and primaries black with lower third white across whole feather, but P10 without white on outer vane and P9 with much shorter white patch on outer than inner vane. Rectrices with tiny white spot at tip, or sometimes a white tip 1 mm deep. Rump plumage soft and voluminous; scapular feathers lax. Bill black; inside of mouth deep pink; eyes dark brown; legs and feet black. Adult ♀: like ♂ but flanks creamy white (feathers long, lanceolate, lowermost with black lower and white upper webs). An albino had black soft parts and completely white plumage (Watson and Watson 1983). SIZE: wing, ♂ (n = 10) 137–145 (140), ♀ (n = 10) 137–145 (140); bill to feathers, (sexes combined, n = 42) 16·5–20·5; tarsus, (sexes combined, n = 42) 31·0–35·5; tail, ♂ (n = 25) 225–350, ♀ (n = 17) 215–340. WEIGHT: ♂ (n = 22) 55–98 (82·3), ♀ (n = 12) 71–96 (82·4).

IMMATURE: dark brown where adult is black; feathers of breast, belly and thighs buff tipped; whitish tips to secondaries more extensive; narrow whitish tips to tertials and inner secondaries; some soft, white, incoherent feathers in flanks, even of ♂♂, and flanks can look entirely white. No feathers are lanceolate (see generic diagnosis). In this plumage tail may be as long as adult's. Iris dark grey-brown. Fledgling like immature bird, tail <70 long, uppertail-coverts buff-tipped. Gape yellow; bill horn.

NESTLING: naked, blind, without mouth spots. Later pinkish, with thin, dry, wrinkled skin, greyish over feather tracts, bill cream, with thick, cream gapes (Steyn 1996).

U. m. expressa Clancey: SE Zimbabwe, E Transvaal, S Mozambique, E Swaziland, Zululand. Like *melanoleucus* but rump greyer, larger white tips to inner primaries, secondaries and tertials. Smaller: wing, ♂ (n = 20) 127–137·5 (133), ♀ (n = 15) 125–137 (132); tail, sexes combined (n = 10) 273–353 (305).

U. m. aequatorialis Reichenow: E Africa. Like *melanoleucus* but much shorter-tailed (up to only 260 mm), with smaller white patches at tips of tertials and secondaries, and deeper black on throat and breast. (One aberrant adult ♂ skin has mixed pale brown and dark brown feathers where plumage is normally black.) SIZE (5 ♂♂, 5 ♀♀): wing, ♂ 127–137 (133), ♀ 127–136 (130); bill to skull, ♂ 21–25·5 (23·4), ♀ 22–24·5 (23·2).

Field Characters. Length 34·5–50 cm. A large shrike whose *very long black tail* might suggest a widowbird *Euplectes* at first glance. Mainly black; *pied appearance in flight* produced by prominent white scapulars forming 'V' on back, white patch on primaries and white tips to secondaries; ♀ has white patch on flanks. Immature dark brown with some pale barring. Sociable and noisy, groups often perching conspicuously.

Voice. Tape-recorded (15, 36, 58, 75, 86, 88, 91–99, B, F). Song a variety of loud, clear, liquid whistles of 1–3 syllables, generally down-slurred at end, 'teeeyoo', 'tooeeyoo', 'tuweer', 'hewp', 'teeliyoo', 'tuweet-tuweer', also a shriller 'heee-aw' and thin 'tsleeet'; given by individual, as duet or in chorus. Duet antiphonal, ♀ giving higher-pitched whistles. Other calls include scolding 'taaa' or 'taaaya', low, harsh 'jwaak', buzzy 'tzit-tzit', and a squealing noise like silk cloth being torn; 'tchzzrrr' indicates alarm or aggression.

General Habits. Inhabits short-grass savanna woodlands and open, park-like, bushy and wooded country with spaced, tall acacias, short, grazed grass and bare ground; sometimes in broad-leaved woods in lowveld; in Botswana commonest in moist areas and river valleys and vleis with acacia trees, though also in semidesert, well away from surface water. In South African lowveld, occasionally enters town parks with lawns and native *Acacia* trees. In Orange Free State reported regularly only in tall *Acacia erioloba* veld.

Solitary or in pairs, but generally in cohesive group of 3–6 birds, often up to 11 in austral summer and 19 in winter. Perches conspicuously on top or on outer branch of bush or small tree. Noisy; rather shy. Flicks and jerks tail, particularly as bird calls. Flight fast and horizontal, usually short-distance and rather low down; slightly undulating if at treetop level; most flights short, ending in brief upward glide onto tree perch. Birds in group tend to leave tree one by one, following each other into next tree. Group occasionally flocks with other birds including White-headed Buffalo-Weavers *Dinemellia dinemelli* (Lefranc and Worfolk 1997). Hunts from exposed station on tree or sometimes fence or power-line, perching with body rather horizontal, occasionally turning head and bowing in flight-intention movement; scans ground under and near station, flies down and seizes prey, or hops around on ground searching for small insects. Sometimes hawks insect in flight, and searches or gleans branches, twigs and leaf clusters, and known to hang upside down by one foot whilst tearing with bill at prey held in other foot (Harris and Arnott 1988). Carries prey back to perch in bill, then stands on it with one foot. Not known to allopreen. Scratches over wing.

Group occupies home range or territory, once of 70 ha. Group members keep in contact with whistles and 'tchzooo' calls, and from time to time one or a few birds initiate a group display or rally which quickly draws rest of group together; all birds then rally by whistling noisily, perching with head lowered and bill pointing down and frenziedly bowing in synchrony, with wings raised and held half open above back, and tail jerking up and down with each bow (**A**). Sometimes rallies on ground, birds more or less in a circle, bowing with wings and tail raised. 2 groups from neighbouring home ranges may sit in adjacent trees and display at each other, or merge into single large and aggressive rally. Give 'tchzzrrr' alarm calls when chasing each other, and when mobbing and dive-bombing ground predator and perched hawk. Group said to roost at night with birds 1 m apart, not huddled up.

A

Resident, mainly sedentary, but moves temporarily into burnt areas, and subject to distant wandering in dry season near borders of range in N Namibia and central Zambia.

Food. Mainly insects, including grasshoppers, mantises and large grubs. Also ticks, millipedes, small reptiles, mice, fruit and fresh and rotting meat (Harris and Arnott 1988).

Breeding Habits. Solitary, territorial nester; probably a co-operative, monogamous breeder. Pair defends territory of *c.* 3 ha, within group's home range. ♂ advertises by perching conspicuously on treetop near nest, with body very upright, scapular and rump feathers fluffed out, repeating 'teelooo' whistles for long periods. Pair sings antiphonally, in duet, ♂ giving 'teelooo' and ♀ higher 'tleeeu' whistles (Harris and Arnott 1988). Apparently some, not all, group members are tolerated in territory, others being chased away aggressively with 'tchzzrrr' calls. ♂ courtship-feeds ♀; they duet throughout nesting season; ♀ solicits with loud, nasal 'tsseeeer' call, crouching with wings drooping and fluttering. Second broods probably quite commonplace.

NEST: untidy, bulky, strongly-built cup, ragged on the outside, made of very thin bents or coarse grass stems, thin, thorny twigs and dry grass and often dry asparagus stems *Protoasparagus*; not lined (**B**). Sited in thorn tree, usually acacia with formidable thorns, often near a stream, 1·5–12 m up (n = 121), av. *c.* 4 m high; nest open to view but inaccessible because of thorns. Size: (1 nest) ext. diam. 165–180, int. diam. 100, ext. depth 160, int. depth 57 (Vincent 1949). ♂ and ♀ thought to construct and shape nest, not helped by other group members.

EGGS: 2–6 (mainly 3–5). Buff or yellowish, speckled with reddish browns and greys, mainly at broad end. SIZE: (n = 166) 23·1–29·9 × 18·3–20·7 (26·8 × 19·7).

LAYING DATES: E Africa, Region C, Feb, Region D, Feb–Mar; Angola, Oct (and enlarged gonads Nov); Zambia, Oct–Mar (13 out of 18 records in Oct–Nov); Botswana, Sept–Feb; Zimbabwe, Aug–Feb (n = 145), 41% in Oct,

25% in Nov, 27% in Sept and Dec; Transvaal, Aug–Feb (n = 82), 76% in Oct–Dec.

INCUBATION: by ♀, fed on nest by ♂, also apparently by only certain members of the social group. Incubating ♀ begs for food using loud and distinctive 'tzzeeeer' call.

DEVELOPMENT AND CARE OF YOUNG: nestlings fed and cared for by parents and some (at least 3) helpers. They use noisy 'tzzeeeer' call almost indistinguishable from mother's begging call. Young stay in nest for at least 15 days (n = 1). Young clamber about in nest tree before taking their first flights, and later become a part of the social group (Harris and Arnott 1988). First brood known to feed nestlings of second brood.

References
Harris, T. and Arnott, G. (1988).
Lefranc, N. and Worfolk, T. (1997).
Vincent, A.W. (1949).

Genus *Eurocephalus* Smith

Quite large, heavy, shrikes with pied and buffy plumage, forehead plush with short, stiff feathers, snow-white crown, black mask, black sides of neck and black bill; wings rather long and pointed; tail moderately long but not graduated. Sexes identical; flank patch present in one species, absent in the other. Juveniles markedly different from adults – plumage not barred, bills yellow; juveniles of one species black-capped and of other species dusky-faced. Noisy, gregarious; co-operative breeders. Fly with shallow wingbeats and glide with wings held in V. Occasionally use foot to transfer morsel to bill. Not known to impale prey. Remarkable nest, a deep cup made of grasses and spider web. Probably often bigamous (see under *E. rueppellii*). For remarks on systematic position, see family Laniidae diagnosis.

Endemic; 2 species, forming a superspecies.

Eurocephalus rueppelli superspecies

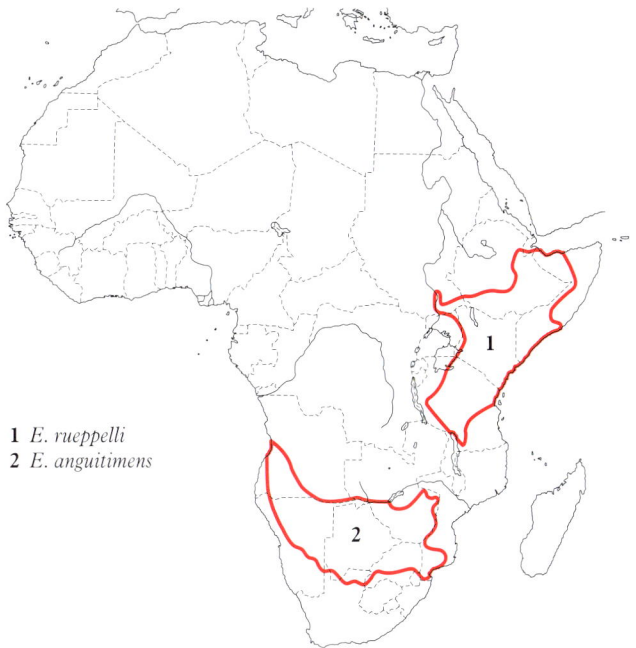

1 *E. rueppelli*
2 *E. anguitimens*

Eurocephalus rueppelli Bonaparte. Northern White-crowned Shrike; White-rumped Shrike. Eurocéphale de Rüppell.

Plate 20
(Opp. p. 287)

Eurocephalus rueppelli Bonaparte, 1853. Rev. Mag. Zool. (Paris), sér. 2, 5, p. 440; White Nile, Shoa.

Forms a superspecies with *E. anguitimens*.

Range and Status. Endemic resident, E Africa. Common in SE Sudan. Ethiopia, frequent to common in S, Rift Valley and E. Somalia, common in NW and S (south of 03°N), sparse in centre; mainly at 900–1550 m. Uganda, only in NE (Karamoja, NE Teso and Gulu). Kenya, common and widespread from sea-level up to 1600 m; very common in Samburu, Meru, Amboseli and Tsavo (Zimmerman *et al.* 1996); sparse or absent near to L. Victoria, east of L. Turkana, in NE Kenya, and in coastal lowlands south of Malindi (though many old records on that coast). Tanzania, very common in Mkomazi Game Res., widespread in NE between L. Victoria and coast, in W south to Mambali, Wembere and Ukerewe Is.; very common in Serengeti, Lake Manyara and Tarangire Nat. Parks; in interior occurs south to Mikumi and Ruaha Nat. Parks, Buhoro Flats and Njombe (but not quite as far south as Matengo Highlands), thence west to Rukwa and Mpanda.

Description. ADULT ♂: forehead and crown white, or white with greyish wash, nape and hindneck white, mantle, scapulars and back dark brown, rump and uppertail-coverts white. Tail blackish brown. Lores, cheeks, ear-coverts, sides of neck and sides of throat dark brown, forming large, oblong patch and a widely-interrupted hindneck collar. Chin white, throat and upper breast

Eurocephalus rueppelli

white in centre, lower breast and upper belly creamy white in centre, sides of breast pale brown (sometimes patchily), lower belly, vent and undertail-coverts white, forepart of flanks dark brown in large patch more or less concealed beneath wing, mid part white or pale buffy, hindpart of flanks black, forming large, well-defined, visible patch. Primaries, secondaries and tertials, greater and median primary coverts and alula dark brown; greater, median and lesser coverts mid brown; underside of remiges silvery grey, inner borders of primaries proximally white, underwing-coverts and axillaries pale greyish buff. Bill black; eyes dark brown; legs and feet dark grey or dark brown. Sexes alike. SIZE (♂♀, n = 41, Ethiopia and Kenya): wing 118–134 (131); tail 86–103; bill to feathers 16–18; tarsus 21–23·5. Sudan birds smaller (wing, n = 27, av. 123·5: Friedmann 1937). WEIGHT: ♂ (n = 12) 39–70 (48·4), ♀ (n = 10) 37–64 (52·9).

IMMATURE: juvenile (c. 2 weeks fledged) has forehead, crown, hindneck, lores, cheeks and malar region dark brown; ear-coverts, sides of neck, chin and entire throat white or buffy white. Flight feathers and all upperwing-coverts narrowly edged and tipped pale rufescent brown. Plumage of entire underparts very soft and fluffy; breast pale grey (feathers with pale grey bases, wide whitish tips, and narrow mid grey band between); forepart of flanks grey-brown, belly snowy white. Bill pale orange-pink, eyes brown, legs grey.

Later, full-size immature moults through various plumages (A); in one, like juvenile, with falcon-like moustachial patch, orange bill, still quite fluffy underparts, and grey-brown breast and flanks; in another, whole cap blackish, broad white hindcollar, dark horn-brown bill, and no 'moustache'; in another, head and neck white except for dark brown forehead, forecrown, bill, lores and separate patch on hind ear-coverts.

NESTLING: when nearly ready to leave nest, feathers of lores and over eye olive, wing-coverts and feathers of back edged and tipped with pale brown; gape bright yellow; skin of chin, visible under short white feathers, bright chrome-yellow; bill blue-grey; legs mealy white (Moreau and Moreau 1939).

TAXONOMIC NOTE: recognition of 5 subspecies was once advocated, but they have been whittled down to the point where species is regarded as monotypic, if individually rather variable (Lefranc and Worfolk 1997).

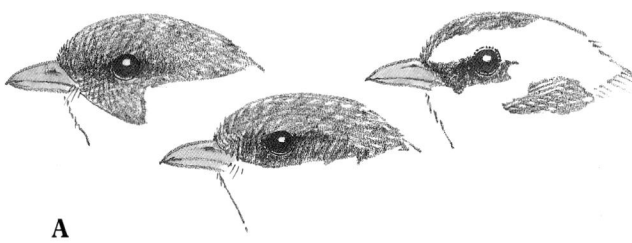

A

Field Characters. Length 19–21 cm. A bulky shrike with large head and rather short tail, often on prominent perch by roadside or in dry woodland. Noisy and sociable. *Head white*, with *black line* through eye broadening onto ear-coverts and side of neck; whitish below with brown patch on flanks; in flight, *white rump* contrasts with dark brown back, wings and tail. Immature has *dark crown* to below eye, brown band across breast and *yellowish bill*. Distinctive butterfly-like flight interspersed with long glides aids identification at a distance.

Voice. Tape-recorded (B, C, F, GREG, McVIC, NOR). Calls include harsh 'kaak-kaak' or 'weeyer-wok, weeyer-wok'; gives similar notes in series, 'chrrk, wirk-wirk, yerk-yerk, wuk-wuk, yerk ...' or 'yerk yeck-yeck-yeck'; juvenile gives sharp, piercing 'skeet' (Zimmerman et al. 1996).

General Habits. Inhabits dry *Acacia-* and *Commiphora-*dominated woodlands; sometimes in gardens, tall hedges and windbreaks, cultivation with plenty of trees and bushes left standing, and parkland, near human habitation; in N Somalia in acacia-aloe country, often perching in tall mimosa trees in wadis. Occurs in flocks of 3–4 birds (Kenya) or 4–8 (Somalia); noisy, confiding, restless, birds perching conspicuously on tops of bushes and sides of trees, flock progressing from one tree to another with rather weak-looking flight. Flight distinctive, with butterfly-like, shallow, quivering or fluttering wingbeats interspersed with frequent, unsteady-looking long glides with wings uplifted to form angle of c. 100–120° between them; bird has been likened to tiny harrier *Circus* sp.

Spends much time on ground, hopping about energetically. Hunts from exposed, elevated perch, and can stay at one perch for extended periods; takes prey mainly on ground (80% of 213 items, Tsavo East, Kenya); also airborne (15%) and, uncommonly, from leaves in bushes (1%); av. hunting perch height 3·6 m and av. distance flown 9·6 m (70 observations); flying insects taken at av. height of 7·6 m above ground (17 observations); av. feeding rate, 10 items in 3·1 min (Lack 1985). Sometimes several members of group feed on ground or in the air together. Sedentary.

Food. Insects, including considerable diversity of butterflies caught by one nesting pair. Locust nymphs fed to nestlings. 10 stomachs contained beetles (in 6), a centipede (in 1), ants (in 1), grasshoppers and mantises (in 2); a scarab and a cerambycid were both 25 mm long. Known to eat berries.

Breeding Habits. Poorly known, but evidently a monogamous, group-territorial, co-operative breeder. Nestlings in some nests attended by 2 adults only, but an incubating ♀ attended by 3 adults (Moreau and Moreau 1939), and 4 adults repeatedly fed 3 full-grown fledglings – sometimes all 7 birds together on branch (Plumb 1978). Thought to be double-brooded (N Somalia: Archer and Godman 1961).

NEST: very neat, well-constructed, usually yellow or yellowish, thick-walled, shallow cup with deep base that fills in the spaces between supporting twigs and branches; made of very thin grass stems, plant fibres, rootlets and seed down and lined with fine fibres; smoothly bound and secured to branch with copious spider web; sited on thick horizontal branch or on horizontal twiggy fork near tip of branch, 6–9 m up in *Acacia* tree by seasonal watercourse or near open water. Some nests greyish, made largely of spider web, on foundation of moss, so closely matted with silk as to resemble felt or even clay or a small cheese. Usually cryptic, but common yellow colour can make some nests stand out against green branches. Size: ext. diam. 120, int. diam. 75.

EGGS: 3–6; av. 3·3 (n = 6 clutches or broods, Somalia). Once 7 eggs, Somalia; in Tanzania 7 eggs in one nest were of 2 distinct types and 'it is apparently not unusual, at least in Uganda, for two or more females to share a nest'

(Moreau and Moreau 1939); since Southern White-crowned Shrikes also have apparent double clutches (see below), the superspecies may be frequently bigamous. Broad and tapering; ivory-white, mottled and blotched with sienna-brown and with strong secondary markings of violet-grey. Size: (n = ?) 25–28 × 21.

LAYING DATES: Sudan, Aug, Dec; Ethiopia, Mar–May, Oct; Somalia, Mar–June; E Africa, Region A, May, Region B, Mar, Region C, Mar, Nov, Region D, Feb–Apr, June–Aug, Nov–Dec.

BREEDING SUCCESS/SURVIVAL: has been preyed upon by Tawny Eagle *Aquila rapax* and Bateleur *Terathopius ecaudatus*.

References
Lefranc, N. and Worfolk, T. (1997).
Moreau, R.E. and Moreau, W.M. (1939).
Plumb, W.J. (1978).

Eurocephalus anguitimens Smith. Southern White-crowned Shrike; White-crowned Shrike. Eurocéphale à couronne blanche.

Plate 20
(Opp. p. 287)

Eurocephalus anguitimens Smith, 1836. Rep. Exped. Centr. Afr., App., p. 52; 'between Latakoo and the Tropic'.

Forms a superspecies with *E. rueppelli*.

Range and Status. Endemic resident, southern Africa. In Angola extends north along coast to Luanda and in interior to Quilengues and Caconda. Frequent in Caprivi Strip, Namibia, and far N of Botswana, yet not known in Zambia. Botswana, common to very common, except in W-centre and SW where absent. Zimbabwe, except much of N and E. In remainder of range uncommon to fairly common in most parts including central Namibia and N Cape Prov., commonest in Limpopo Valley and in Botswana/Transvaal border country. Absent from Transvaal highveld and Escarpment areas, widespread but local in bushveld and lowveld. Scarce in Mozambique: records from Maputo district, Olifants R., Goba, Mapulanguene, Guijá, Caniçado, Vista Alegre, Cabeça do Elefante, Pequenos Libombos, Mapi, Pafúri and along Kruger Nat. Park border. NE Swaziland. Population in S Mozambique probably >5000 birds (Parker 1999).

Description. *E. a. anguitimens* Smith: Angola to about 29°E in SE Africa. ADULT ♂: forehead and crown white, or white with greyish wash, lores and large patch from eye to whole sides of neck blackish brown, nape and hindneck white, mantle, scapulars, back, rump and uppertail-coverts pale warm grey-brown, uniform and with almost pinkish tinge when fresh, but most feathers in most birds broadly tipped greyish white. Tail dark brown. Lores, cheeks, ear-coverts, sides of neck and sides of throat dark brown, forming large, oblong patch and a widely-interrupted hindneck collar. Chin white, throat, breast and upper belly creamy white, lower belly, flanks, vent and undertail-coverts light greyish buff-brown, thighs a little darker. Primaries, secondaries and tertials, greater and median primary coverts and alula dark brown; greater, median and lesser coverts greyish brown; underside of flight feathers silvery grey, inner borders of primaries proximally white, underwing-coverts and axillaries pale greyish buff. Bill black; eyes dark brown; legs and feet blackish brown or light brown. Sexes alike. SIZE (sexes combined, n = 11): wing 130–143 (136); tail 100–116; bill 15–20; tarsus 22–26. WEIGHT: ♂ (n = 13) 55·7–76·5 (68), ♀ (n = 9) 59–84 (70·1), unsexed (n = 3) 51–70 (63·5).

IMMATURE: head and upper mantle white or off-white, faintly mottled and barred, except for greyish dusky on lores, immediately around and behind eye, in narrow line between forehead and bill, and on cheek, moustachial and malar region; ear-coverts dusky, sometimes mottled with buff. Lower mantle, scapulars, back, rump and uppertail-coverts unmarked warm grey-brown.

Eurocephalus anguitimens

Tail black. Chin and throat off-white, breast and undertail-coverts mainly off-white, belly and flanks pale greyish buff. Wings dark brown, inner vanes of primaries tipped buff, inner secondaries, tertials and all upperwing-coverts broadly tipped and narrowly edged with buff. Bill yellow or yellowish horn.

NESTLING: hatches blind and naked. At *c*. 5–10 days (eyes half open, first feathers just breaking open), has pinkish brown bill, thick yellowish gapes, a conspicuous patch of white down or featherlets behind ear (photo in Ginn *et al.* 1989).

E. a. niveus Clancey: S Mozambique, and Zimbabwe and Transvaal west to about 29°E. Feathers of mantle and scapulars broadly fringed white; throat and breast pure white (not cream-white).

Field Characters. Length 23–25 cm. A large, broad-winged and highly visible shrike, often in noisy groups in treetops. *Head white*, with *black line* through eye broadening onto side of neck; underparts white shading to ash brown on flanks and belly; in flight, dark brown wings and tail contrast with pale brown back and rump. Immature has white head and neck, dark patch on face, mottled crown and yellow bill. Characteristic flap-and-glide flight with wings in a shallow dihedral.

Voice. Tape-recorded (88, 91–99, B, C, F, CHA, PAR, STJ, WALK). Common call, repeated shrill and squeaky 'kee-keeya', 'wee-beeya', 'weeyatu' or 'chiapet'; often given by groups, and interspersed with short, dry notes, 'chup', 'chit', 'chat-chat-chat-chat', which may accelerate into quarrelsome chattering trill. Greeting call a stuttering 'kida-kida-kida' (Lefranc and Worfolk 1997).

General Habits. Inhabits dry, deciduous woodlands, parkland with spaced, tall trees and sparse intervening ground cover, and riverine woods, bushes and thickets; in Botswana semi-desert thorn savanna on Kalahari sand; never in woodland with closed canopy; in Zimbabwe acacia and other dry savanna woodlands, and edges of woods where there is a fringe of *Terminalia*; penetrates well into miombo *Brachystegia*.

Occurs sometimes singly or in pairs, but generally in small, socially cohesive parties of 4–8 or 5–10 birds; aggregations of up to 20 birds in austral winter. Conspicuous and noisy, often perching in party on tops of small trees or large ones like baobabs. Flight straight and slow, wingbeats shallow and fluttering, interspersed with frequent, unsteady-looking glides with wings held stiffly up in shallow V. On alighting, bird often flicks tail then squats down with only toes showing, and flicks tail again just before flying on with rest of group. Very vocal, birds in group calling and answering each other when perched, and group almost always chattering for duration of flight from one tree to another. Birds in a party may perch in tree several m apart, but 2, thought to be a mated pair and socially dominant, often perch close together and preen each other (Harris and Arnott 1988). Group commonly uses bleating 'kaee' contact call and repeated 'k-*kaeeke*'; stuttering 'kida-kida-kida' call is used when 2 group members approach each other and when adult approaches immature bird with food. In frequent greeting display droops wings or sometimes holds them half open, raised a little above back. Bird occasionally flies about above treetops, repeatedly calling 'k-*kaeeke*' as if to re-establish contact with lost group. Erects crown feathers when excited. Group may become nucleus of mixed species foraging flock (R.M. Harwin *in* Ginn *et al.* 1989).

Group appears to occupy home range (one, in winter, of *c*. 200 ha). In what may be a territorial display, members of group perch conspicuously around treetop, flick tail, shuffle wings, and call in chorus, using variety of bleating, stuttering notes. Threatens by adopting horizontal stance, raising crown and neck feathers, and giving low growl with bill wide open. Group raises alarm noisily, one bird uttering loud, nasal 'kaeeer' call and others joining in with chorus of 'keekeekee...' calls. Readily mobs and dive-bombs owls, hawks and ground predators. Roosts gregariously in thick foliage or dense thicket, sometimes using same site for months on end; members of group going to roost chase and jostle each other noisily, then settle onto a branch, make chortling and purring noises, preen themselves and each other, then huddle together and go to sleep, not all facing same way; in middle of night they give noisy alarm calls if disturbed by calling owl. In the morning, birds at roost chase each other and call raucously, then move off to forage; once 5 birds hung at end of thin, drooping branch, plucking leaves off and making soft chortling sounds (Harris and Arnott 1988).

Hunts from exposed, elevated perch, on small tree top or outer branch of large tree, scanning ground below. Members of group widely separated as they forage, and group slowly progresses by hindmost birds continuously leap-frogging up to 100 m to the front. From time to time group reforms and rests in tree. Takes most food from the ground, by flying down, alighting and seizing insect with bill; returns to perch, carrying it in bill, or eats it immediately on ground and stays hopping around for a few moments. Often accompanies *Tockus* hornbills on the ground, catching small insects they disturb. Occasionally makes short sally after flying insect, and gleans leaves and branches for small insects, even hanging upside down by one foot whilst tearing at prey held in other foot (Harris and Arnott 1988). At perch, bird habitually holds down prey too large to be swallowed whole under one foot and tears at it with bill, eating it piecemeal; often uses foot to transfer piece to bill.

Sedentary, but in drought years subject to distant wandering. In N Namibia and N Transvaal, reporting rate considerably higher in Apr–May to Aug–Sept than in austral summer (Harrison *et al.* 1997).

Food. Invertebrates, said to be caterpillars in particular; beetles; millipedes; one record of berry-eating.

Breeding Habits. Solitary nester, evidently monogamous. Group-territorial, co-operative breeder. In evident courtship display, 2 birds flew slowly, one somewhat above the other, with exaggerated wingbeats, then glided slowly down with wings held in steep V above backs (Harris and Arnott 1988). Copulation once observed at nest, during construction, the ♂ ignoring another ♀ soliciting inside nest.

NEST: very neat, exactly circular, deep cup with strong, thick, compact, perpendicular walls; made of dry grasses and occasionally bits of bark, bound together with copious amounts of spider web, lined with grass stems or sometimes hair or feathers; secured with spider web onto slender horizontal or drooping branch (often a dead one) near tip, 4–6 m up in tree dominating surrounding growth, generally an *Acacia*, once a *Parinarium mobola*, once a *Commiphora merkeri*. Spider web gives nest silvery finish, helping it to blend with substrate, though nest not otherwise concealed. Size: ext. diam. 115–120, int. diam. 90, ext. depth 65–75, int. depth 45. Placed on branch usually of same diameter as nest; 2 well-grown chicks crouching down in nest help the concealment (**A**, a nest containing feathered young!). Nest built on very slender forking twig can be hemispherical (**B**); in such nests twig fork is embedded into fabric. Built by ♂ and ♀, arriving in

turn with material, sometimes helped by 1–2 other members of group.

EGGS: 2–5, usually 3–4 (once 8, once 10, doubtless each laid by 2 ♀♀: Vincent 1949). White or cream, sparsely blotched and spotted with browns and greys, mainly at wide end. SIZE: (n = 94) 23·0–30·0 × 17·0–22·8 (27·2 × 21·3).

LAYING DATES: Angola, (nest being built, Oct; enlarged gonads Nov); Botswana, Nov–Jan; Zimbabwe, Sept–Jan, mainly Oct (51% of 154 records) and Nov (33%); Transvaal, Oct–Dec.

INCUBATION: mainly by ♀, sometimes by helpers for short periods. Period: c. 20 (?) days.

DEVELOPMENT AND CARE OF YOUNG: young fed by most or all members of group, at one nest by 6 adults; they remove some faecal sacs, but young also defaecate over side of nest. Before and after fledging, young beg with loud 'skea-skea-skea...' calls. Nestling period not known. Fledglings fed by all members of group.

References
Clancey, P.A. (1965).
Dorka, V. (1975).
Harris, T. and Arnott, G. (1988).
Lefranc, N. and Worfolk, T. (1997).
Vincent, A.W. (1949).

Family MALACONOTIDAE: bush-shrikes

Small to very large, pied or colourful, hooked-billed insectivores. Endemic. For relationships to true shrikes Laniidae and to other shrike-like birds, see 'Shrikes' in family Laniidae diagnosis. 46 species, in 7 genera; *Malaconotus* and *Telophorus* are probably closely related, and the tchagras *Antichromus* and *Tchagra* may be allied with *Telophorus*; *Dryoscopus*, black, white and buff birds with marked sexual dichromatism, would appear to be a near relative of *Laniarius*, which are mostly black, yellow and scarlet, sexually monochromatic birds with reciprocal duetting developed to a remarkable extent. *Nilaus*, a single species, shares some plumage characters with *Dryoscopus* and *Laniarius* and has others which are strongly reminiscent of batises, Family Platysteiridae (see Vol. V, pp. 548 and 574); its nest is very like batis nests, and it seems to link Malaconotidae with Platysteiridae – and perhaps with Laniidae too, for *Nilaus* has chestnut flanks, a character of *Lanius*.

Genus *Malaconotus* Swainson

Small, medium and large bush-shrikes with beautiful plumage – bright, harmonious colours in simple, striking patterns. Upperparts green and grey, usually with pale spots in wings and tail, underparts white, green or brilliant yellow more or less suffused with orange and red. Bill black, upper mandible notched; nasal and rictal bristles; bill small in smaller species (which are often separated as genus *Chlorophoneus*) and massive in larger species, hook-tipped and deeply notched;

eyes yellow, grey, dark brown or red-brown; legs and feet bluish grey. Plumage soft and rather full. Outer and middle toes fused in some species (*M. blanchoti*), separate in others (*M. sulfureopectus*).

Arboreal, in canopy and undergrowth. Predatory, taking small and quite large insects and (larger species) some small vertebrates. Stand on large prey item and dismember it by tearing with bill. Nest a quite neat open cup or rather untidy bowl or platform. Voices in simple rhythm, mainly clear, ringing and bell-like, or pure, haunting whistles.

The remarkable parallelism in plumage characters between 2 of the small and small-billed '*Chlorophoneus*' bush-shrikes and most or all of the larger, massive-billed *Malaconotus sens. str.* ones has been analysed by Prigogine (1953), Moreau and Southern (1958) and Hall *et al.* (1966). Each group has a single, monomorphic species in savanna woodlands (*M. blanchoti, M. sulfureopectus*) which quite strongly resemble each other in plumage. The other 5 species of *Malaconotus sens. str.* occupy montane or lowland evergreen forests and are monomorphic and monotypic or, in the case of *M. cruentus*, variable and almost diphasic; 4 are large and one (*M. alius*) mid-sized; most have fragmented or circumscribed ranges. 2 species of small *Malaconotus* ('*Chlorophoneus*') bush-shrikes, *M. multicolor/M. nigrifrons*, form a highly polymorphic superspecies in rain forest, the main variants having breast and belly respectively (a) red and red, (b) black and yellow, (c) black and red, (d) black and olive, (e) buff and buff, (f) yellow and yellow, (g) yellow and olive, and (h) orange and yellow (Moreau and Southern 1958, Plate 1).

1–3 variants or morphs of the Many-coloured Bush-Shrike superspecies *M. multicolor/M. nigrifrons* occur at a given locality, in proportions that vary geographically (Moreau and Southern 1958, Fig. 2). Some strongly resemble the species of large *Malaconotus sens. str.* in the same forests: in particular, morph (f) resembles *M. monteiri* and 'yellow' *M. cruentus* and morph (h) resembles 'red' *M. cruentus*. Morph (e), mainly of E Africa, resembles the small *M. olivaceus* of S and SE Africa. Morph (g) somewhat resembles the medium-sized *M. alius*. Hall *et al.* (1966) concluded that the large, monomorphic species of *Malaconotus sens. str.* derive directly from a polymorphic ancestor. None of these bush-shrikes has been well studied in the field, certainly not the forest species, most of which are rare. Other taxonomic complications are mentioned in Taxonomic Notes under *M. cruentus, M. multicolor* and *M. olivaceus*.

M. cruentus is closely related to *M. blanchoti, M. lagdeni* and *M. monteiri* (which compose a superspecies, from which we exclude *M. cruentus* because of its widespread sympatry with *M. lagdeni*). The 4 together appear to be a sister group of the *M. multicolor* superspecies, and their parallelism may be due less to mimicry (for instance) than to affinity and the evolution of the same colours in a given region under the same selective pressures operating there. It has also been proposed that *M. monteiri* is a colour phase of *M. gladiator* (Williams 1998).

Endemic. 12 species, including 2 superspecies (*blanchoti/lagdeni/monteiri*; *multicolor/nigrifrons*). Of the 7 'independent' species the closest relatives of 3 are hard to discern, but *bocagei* is close to *sulfureopectus* (juvenile plumage, voice, tendril-nests, habits), *sulfureopectus* may be close to *blanchoti* which it resembles, and *olivaceus* to *multicolor*.

Malaconotus blanchoti superspecies

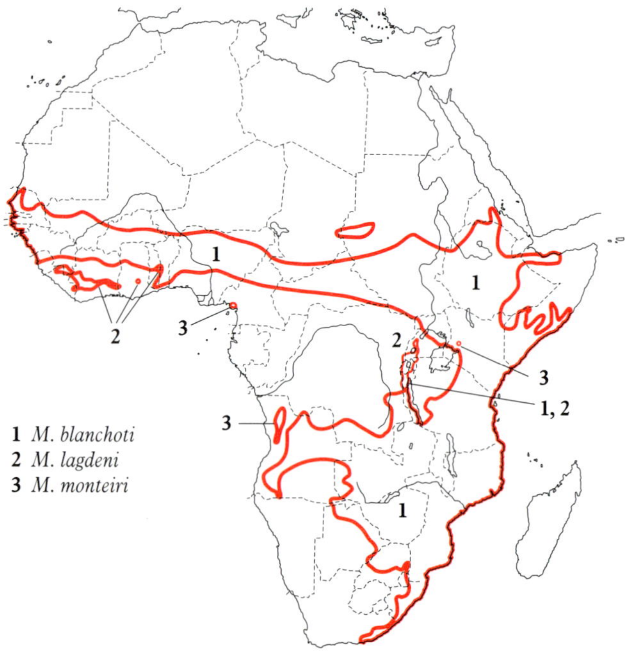

1 *M. blanchoti*
2 *M. lagdeni*
3 *M. monteiri*

Malaconotus multicolor superspecies

1 *M. multicolor*
2 *M. nigrifrons*

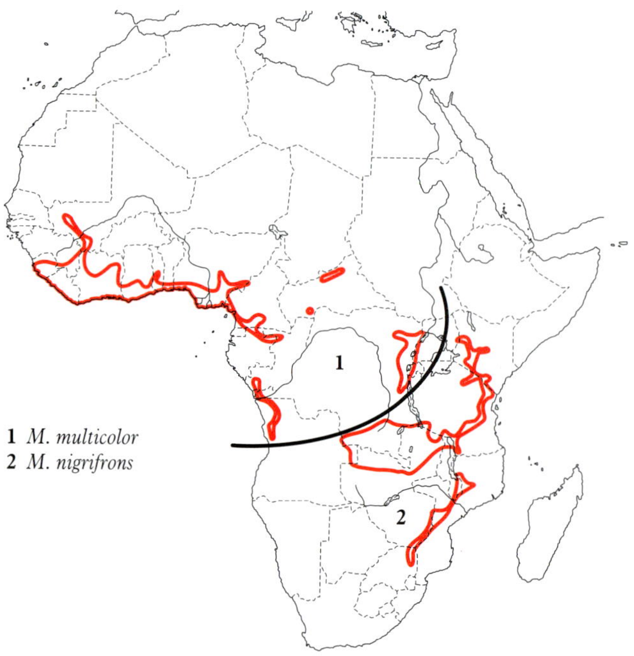

Malaconotus cruentus (Lesson). Fiery-breasted Bush-Shrike. Gladiateur ensanglanté.

Plate 21 (Opp. p. 334)

Vanga cruenta Lesson, 1830. Cent. Zool., p. 198, pl. 65; Cape of Good Hope (i.e. Cape Coast, Gold Coast, Neumann, 1899, J. Orn., 47, p. 389.).

Malaconotus cruentus

Range and Status. Endemic resident. Sierra Leone, records from Sefadu, Bo, and in NE. Guinea, uncommon, Ziama Massif (Halleux 1994). Liberia, frequent south of 07°N as mapped and locally common in N; sea level up to 1500 m; common on Mt Nimba at 800–1500 m (Gatter 1997). Ivory Coast, common in Danane-Nimba-Taï region and widespread between Abidjan, Bouaké and Maraoué (though not in Yapo Forest). Ghana, widespread and not uncommon in forest zone, but absent from wetter forests of SW; occurs in Cape Coast and forests north of it, Chama, Mampong (Ashanti), Amedzofe, Akropong and Tafo. Togo, scarce (several records) at Misahöhe near Ghana border. Not known in Benin. Nigeria, not uncommon in SW (Lagos, Ilaro, Ibadan, Ife), scarce in delta from about Ewu and Ahoada and in SE also at Ikom. Cameroon, throughout forest zone in SW. Gabon, common and widespread. Congo, occurs in Odzala Nat. Park, records at Mombongo and Ndoki in Nouabalé-Ndoki Nat. Park, where evidently rare; in Kouilou Basin known from Dimonika and may occur between Goumina and Kananga. Central African Republic, records in Lobaye Préfecture. Zaïre, several old records from Angumu near Buta, 02°48′N, 24°47′E, so species may occur all across N Zaïre; local and uncommon, or more likely common but overlooked, in E from Semliki to Kivu, and fairly common in Itombwe Highlands, at 1400–1900 m (Butokolo, Kakanda, Kamituga, Kanyaa, Kikolo, Kiloboze and Lumbokwe). Uganda, common at 700 m in Bwamba lowlands.

Description. ADULT ♂: forehead pale grey in centre, merging to white towards nostrils and lores; crown grey; white ring around eye, confluent with white lores; nape, hindneck, sides of neck, ear-coverts and cheeks grey; mantle bluish grey; back, rump and uppertail-coverts olive-green. Tail olive-green, with yellow tip and well-defined black subterminal band; yellow tip 7 mm deep on T1 to 11 deep on T6, black band *c*. 12 deep on all feathers; tail rounded, T6 12 shorter than T1. Many birds in Cameroon and some in Sierra Leone have chin orange-yellow, throat and breast brilliant orange-red, upper belly orange, lower belly, flanks, thighs and undertail-coverts brilliant yellow; in SW Cameroon breast may be yellow (Williams 1998); most birds in Gabon and Zaïre have chin and throat brilliant orange, breast bright reddish orange, middle of belly and undertail-coverts brilliant yellow, sides of belly orange, and flanks pale yellow, tinged greenish near rump; birds from NE Liberia (Mt Nimba) have orange suffusion on yellow underparts less extensive and decidedly paler. Underside of tail like upperside but all colours duller and shafts yellow (upperside of feather shafts dark brown). Wings mainly olive-green; alula and P1 blackish; other remiges with blackish inner vanes, blackish tips to outer vanes, and green outer fringes; secondaries tipped yellow, on outer vane of outer secondaries and both vanes of inner ones and tertials, where yellow tips 6 mm deep. Exposed tertials mainly black and yellow, making series of conspicuous yellow spots in closed wing. Underside of remiges shiny grey, inner edges pale yellow, underwing-coverts and axillaries bright sulphur yellow; featherlets on leading edge of wing yellow and orange. Bill black; 3 nasal and 5 rictal bristles on each side; eyes pale blue-grey or greyish white; legs and feet grey, blue-grey or dark blue. SIZE: 'cruentus' (32 ♂♂, 13 ♀♀), wing, ♂ 100–116 (111), ♀ 105–113 (109), bill, ♂ 29–33 (31·6), ♀ 29–31 (29·8); 'adolfifriederici' (12 ♂♂, 4 ♀♀), wing, ♂ 101–114 (105), ♀ 101–107 (103·5); tail, ♂ 90–99 (96·2), ♀ 91–98 (94·3); bill, ♂ 28–30 (28·8), ♀ 27–29 (27·5). WEIGHT: (Bwamba, Uganda) ♂ (n = 7) 68–79·5 (72·1), 1 ♀ 75.

IMMATURE: fledgling like adult but forehead to hindneck and ear-coverts warm buffy brown with grey feather bases showing through, cheeks grey, underparts very fluffy, chin and throat white, breast greyish pale yellow, rest pale yellow, bright on undertail-coverts. Black band in tail narrow. Bill small, black; gape whitish; eyes and legs blue. Later, when more than half grown, forehead to hindneck still buff, throat becoming orange, breast still greyish yellow, greater coverts diffusely tipped orange-brown, and black-and-yellow pattern in tertials conspicuous but duller than in adult. Immature dark olivaceous brownish grey above, rather uniform from forehead to tail, but with white mark above each nostril, tail grading to black subterminally, feathers with pale yellow tips, and tertials black with large, conspicuous, pale yellow tips; chin, throat and breast red (in E Zaïre), upper belly red or buffy grey, lower belly and flanks buffy grey, undertail-coverts buff (Prigogine 1986).

NESTLING: not known.

TAXONOMIC NOTE: 5 races have been described, the main 3 being *cruentus* (Sierra Leone to Nigeria), *gabonensis* (Cameroon, Gabon) and *adolfifriederici* (Zaïre), based on the extent and depth of red or orange suffusion on breast, and slightly decreasing size, from west to east (e.g. Mayr and Greenway 1960, Prigogine 1986). However, most populations vary in themselves in amount of yellow/orange/red in underparts; for instance, about half of long series of birds from Kumba, Cameroon, are markedly yellower than the orange or orange-red other half (the variation not gender-related). White (1962) did not recognize subspecies and nor do we, pending better knowledge. Series from Mt Nimba, Liberia, is uniform, with small area of pale orange on breast, and may need to be formally named.

Field Characters. Length 23–25·5 cm. Hard to see in spite of its bright colours, and the location of the calls is hard to pin-point (Christy and Clarke 1994). Much larger than Many-coloured Bush-Shrike *M. multicolor*, with huge bill. Underparts red-orange or orange-yellow, but lacks black mask, and face from bill to behind eye is white. Yellow phase could be confused with Lagden's Bush-Shrike *M. lagdeni*; best character is *broad black* subterminal band on *yellow-tipped tail* (F. Dowsett-Lemaire, pers. comm.). Differs from Grey-headed Bush-Shrike *M. blanchoti* in red underparts, black and yellow tail tip and barring on wings, and grey eye (yellow in Grey-headed).

Voice. Tape-recorded (104, B, F, BRU, CHA, LEM). Song, given from high in forest canopy, a series of 4–10 short, soft, mellow notes, 'hoh' or a higher 'hah', reminiscent of Hairy-breasted Barbet *Tricholema hirsutum*, given at speeds varying from 7 notes per 6 s to 10 notes per 16 s, often repeated after a short interval; 1–2 other birds perched lower down answered 'kick-ik-ik' and gave another still more rasping note (Chapin 1954). Characteristic call, often the first indication of its presence, a hard accelerating trill ending in sharp notes, 'krrrièèèk; another call, sometimes given in flight, an excited, rolling 'rrrourourou' (Christy and Clarke 1994). It also emits a curious soft chattering 'song' intermittently while foraging (Bannerman 1939). Clicks bill (F. Dowsett-Lemaire, pers. comm.).

General Habits. In Liberia inhabits primary forest along rivers, forests bordering inundation zones, semi-evergreen forest with open canopy, riverine forest, secondary forest, abandoned cultivation and (Mt Nimba) ridge forest (Gatter 1997). In Ghana and Nigeria, undergrowth of secondary forest, forest edge, and well-wooded gardens. In Gabon, gallery forest, 3–5-year-old regenerating woods in forest clearings and thick cover in old, unmaintained plantations; often seen at edges of forest paths; likes tangles of creepers and lianas where they form dense curtains; in primary forest mainly where many trees knocked down by tornados. In N Congo inhabits closed, dry-land forest, and canopy of open forest, moving in all strata (Dowsett-Lemaire 1997a).

Lives in pairs, but ♂ and ♀ often separated whilst foraging and keep in contact by calling. In secondary growth forages mainly between 2 and 15 m above ground but in semi-evergreen forest keeps to canopy, at 25–40 m above ground (F. Dowsett-Lemaire, pers. comm.), moving through foliage amongst lianas and tree branches; occasionally higher than 15 m; rarely comes to ground; pair has home range of 6–10 ha under more or less closed canopy (Gabon: Brosset and Erard 1986). Stands on large prey item and dismembers it with bill. In abandoned forest farmland, Liberia, keeps much more to the understorey than it does in riverine forest. Joins mixed-species foraging flocks. Sedentary.

Food. Insects; seen to take grasshoppers, mantises, cicadas and caterpillars. Captive birds accept small vertebrates and pieces of meat. Beetles (many), grasshoppers, a cricket, caterpillars, moth pupae, a *Charaxes* butterfly and small frogs in stomachs.

Breeding Habits.
NEST: one, Cameroon, a shallow bowl, loosely built of dry vines, small twigs and leaf petioles, with the top and lining of black rootlets, sited in *Afromomum* thicket (Bates 1911); one, Gabon, made of large dead twigs, lined with vegetable fibres and dead leaves; sited on fork 4 m up in cocoa tree in unkempt plantation. ♀ finds material 10–15 m from nest site; when she has collected a large billful she droops and vibrates wings like a begging juvenile, ♂ approaches her silently and they move to nest where ♀ builds; construction spells last *c*. 20 min, in mornings before it gets hot, each collecting and building trip taking 1·5–2 min (Brosset and Erard 1986).

EGGS: one clutch of 3. Slightly glossy; pale pink, spotted and blotched with rich maroon and pale purple, mainly in zone around large end. SIZE: (n = 3) 28·0–28·5 × 20·5 (Bates 1911).

LAYING DATES: Sierra Leone, (fledgling Aug); Liberia, (sings mainly Sept–Mar; courtship and copulation, Oct; large gonads Nov; dependent young Nov: Colston and Curry-Lindahl 1986); Cameroon, (♀♀ with yolking eggs, June, Nov); Gabon, (calls all year but particularly in Aug–Feb; nest being built Feb); Zaïre, Itombwe, (gonads indicated recent laying, May, and size suggests breeding Mar–July and Oct–Nov: Prigogine 1971).

References
Brosset, A. and Erard, C. (1986).
Gatter, W. (1997).
Hall, B.P. *et al* (1966).
Prigogine, A. (1986).
Serle, W. (1952b, 1954).

Plate 21
(Opp. p. 334)

Malaconotus monteiri **(Sharpe). Monteiro's Bush-Shrike. Gladiateur de Monteiro.**

Laniarius monteiri Sharpe, 1870. Proc. Zool. Soc. London, p. 148, pl. 13, fig. 1; Angola (Donde R., Loando: Gadow, 1883, Cat. Birds Brit. Mus., 8, p. 157).

Forms a superspecies with *M. lagdeni* and *M. blanchoti*.

Range and Status. Endemic resident, known from 6 specimens collected between 1870 and 1954 in Angolan escarpment zone, an old specimen and a recent sighting on Mt Cameroon, and a specimen (now lost) from Kenya. Angola: 1 collected on R. Dande, near Luanda, 1870; 2 in early 1900s at Dondo and Ndalo Tando; and 2 in 1954 at

Malaconotus monteiri

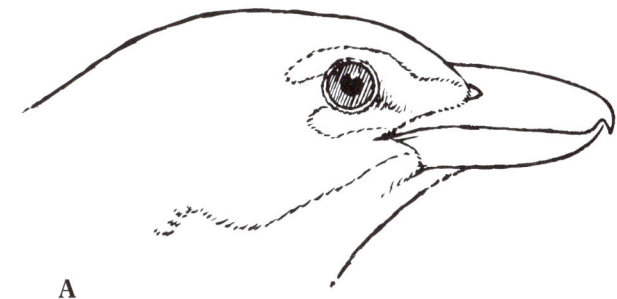

A

Mucoso (near Dondo) and Gabela; undated specimen from 'Bucaso' (perhaps Buçaco, 250 km south of Gabela) (Collar and Stuart 1985). Cameroon: one collected in 1894 from Buea, Mt Cameroon, at 1000 m (Reichenow 1894, Prigogine 1984); must be very rare, Mt Cameroon now being well worked (Stuart 1986). Bird with characters of *M. monteiri* seen in same tree as a Green-breasted Bush-Shrike *M. gladiator* on Mt Kupé (Anon. 1992, Andrews 1994), and an unconfirmed report of one heard and glimpsed in 1997 on Max's Trail, Mt Cameroon, at 1450 m (Demey 1997); 1–2 seen at 1250 m at L. Edib, Bakossi Mts, 1997 (Williams 1998). Kenya: single bird (see TAXONOMIC NOTE) said to have been obtained in Kakamega Forest (van Someren 1932). Zaïre, a bird attributed to *monteiri* collected at Kanzenze (Kansenze), 01°31'S, 29°23'E., Kivu, in 1905; regarding *monteiri* as a race of *M. blanchoti*, Chapin (1953) gave range as 'southern Congo, apparently as far as the western Katanga and Mwinilunga in Northern Rhodesia, possibly to Lake Moero'. Namibia, bird attributed to *monteiri* taken at Okombambi, Cunene valley, in 1963; 'probably occurs further east along course of the Okavango R.' but they 'may be distinguishable from true *monteiri*' (Clancey 1980a).

Description. *M. m. monteiri* (Sharpe): Angola. ADULT ♂ (breeding): forehead, crown, nape, hindneck and sides of neck uniform dark grey, except for well-demarcated white superciliary stripe, confluent with white lores and patch below eye (making C-shaped white mark on grey head – **A**). Mantle, scapulars, back and rump bright olive-green; uppertail-coverts bright olive-green with narrow yellow tips. Tail rounded, T6 with 16 mm shortfall; bright olive-green, T3–T6 tipped pale yellow, tips 3–5 long. Underparts brilliant yellow, breast yellow more or less suffused with tawny orange; flanks olive-green, largely concealed beneath spreading yellow belly feathers. Wing mainly olive-green, P10 black, P9–P4 emarginated, P9 46 from tip, P5 *c.* 38 and P4 *c.* 20 from tip; outer vanes proximal to point of emargination are olive-green and distal to it yellow, then black towards tip; P3–P1 with outer vanes all olive-green; secondaries and tertials with olive-green outer vanes, broad yellow tips, and green-tinged black inner vanes; alula and greater primary coverts olive-green; all other upperwing coverts olive-green with broad yellow tips. Underside of remiges shiny dark grey, with broad pale yellow inner edges and bases of all remiges except outermost primaries. Underside of tail olive-grey. Bill black; eyes greenish ochre or bluish grey; legs and feet slate-blue. Sexes alike. SIZE: (1 ♂, 1 ♀): wing, ♂ 114·5, ♀ 113; tail, ♂ 109, ♀ 104; bill to feathers, ♂ 28, ♀ 26·5, bill to skull, ♂ 32, ♀ 30·5, depth, ♂ 13, ♀ 13; tarsus, ♂ 35, ♀ 31·5; probably the same 2 specimens, with a 2nd ♀, measured by Hall *et al.* (1966) as: wing, ♂ 117, ♀ 109, 115; bill, ♂ 33, ♀ 31, 34. (3 ♂♂, 2 ♀♀, Prigogine 1984): wing, ♂ 117–119 (118), ♀ 110, 114; tail, ♂ 109–112 (110), ♀ 101, 107; bill to feathers, ♂ (n = 2) 32, 33, ♀ 31, 34; tarsus, ♂ 35–36 (35·7), ♀ 34·5, 36.

IMMATURE and NESTLING: unknown.

M. m. perspicillatus (Reichenow): Mt Cameroon. Like nominate race but bill more massive (photo of single known specimen in Prigogine 1984: bill to feathers 34).

TAXONOMIC NOTE: status poorly understood; in plumage (but not eye colour) some specimens are practically indistinguishable from *M. blanchoti* and taxon was generally treated as a subspecies of it until Hall *et al.* (1966) showed that *monteiri* is at least as close to the red-breasted *M. cruentus* (if, by analogy with the polytypic *M. multicolor*, yellow *vs* red colours 'may be dismissed as trivial'). *M. monteiri* resembles *M. cruentus* in iris colour and is thought to share the same habitat (forest, as opposed to savanna woodland of *M. blanchoti*). The Kenyan specimen, now lost, resembled both *M. m. monteiri* and *M. blanchoti catharoxanthus* (Zimmerman *et al.* 1996). Hall *et al.* (1966) found the proposition that *monteiri* is merely a phase of *M. cruentus* difficult to uphold; nonetheless, it remains quite a possibility.

Williams (1998) suggests another possibility: that *monteiri* is a colour phase of *M. gladiator*. Their voices seem to be identical and in Cameroon eye colour, habitat and altitudinal range are alike. However, *gladiator* appears to be a larger and heavier-billed bird than *monteiri*, there are differences in plumage, and their ranges would appear to be largely separate.

Field Characters. Length 25–26·5 cm. Plumage almost identical to Grey-headed Bush-Shrike *M. blanchoti* but eye grey, not yellow, more white on side of head, and breast yellow with only a tinge of rufous. Inhabits forest (Grey-headed in savanna woodland). Sympatric with Green-breasted Bush-Shrike *M. gladiator* in Cameroon and may be a yellow phase of it (see above). Larger and thicker-billed than yellow phase of Many-coloured Bush-Shrike *M. multicolor*; generally at higher elevations, and lacks black on face. Could be confused with orange- or yellow-breasted forms of Fiery-breasted Bush-Shrike *M. cruentus* (which on Mt Kupé, Cameroon, occurs lower down, in farmland and second growth), but *cruentus* lacks yellow tips to wing-coverts and has smaller yellow tips to tertials (Williams 1998).

Voice. Not tape-recorded. Call of bird duetting with Green-breasted Bush-Shrike described as a mournful whistle of 5 notes repeated 5, not 3, times, without inflections at end (Andrews 1994); or voice indistinguishable from that of Green-breasted (Williams 1998).

General Habits. Evergreen or gallery forests occur or used to occur in all localities where species has been found and, in absence of collectors' data, it is presumed to inhabit them. One in same tree as a Green-breasted Bush-Shrike duetted with it and rattled bill, throwing head back and moving it sharply downward with each note (Andrews 1994).

Nothing else known.

Food. Unknown.

Breeding Habits. Unknown.

References
Andrews, S.M. (1994).
Collar, N.J. and Stuart, S.N. (1985).
Collar, N.J. et al. (1994).
Hall, B.P. et al. (1966).
Moreau, R.E. and Southern, H.N. (1958).
Prigogine, A. (1984).
Williams, E. (1998).

Plate 21 (Opp. p. 334) *Malaconotus blanchoti* Stephens. Grey-headed Bush-Shrike. Gladiateur de Blanchot.

Malaconotus blanchoti Stephens, 1826. Gen. Zool., 13, pt. 2, p. 161; Senegal.

Forms a superspecies with *M. lagdeni* and *M. monteiri*.

Range and Status. Endemic resident, uncommon to fairly common in savanna woodlands throughout subsaharan Africa except for the Horn, Congo Basin and the SW.

Senegal, frequent in S, rare north of 14°N. Mali, uncommon in S in wet season; northern limits uncertain, but occurs in Boucle du Baoulé Biosphere Reserve. Burkina Faso, rare in Pendjari and Arli Nat. Parks (across border with Benin); frequent around Ouagadougou (Thonnérieux et al. 1989). Niger, uncommon in W Nat. Park.

Sierra Leone, records at Kamasigi, Freetown and Waterloo; implied to be widespread in N (Field 1974). Ivory Coast, known from 9 localities between Bouaké, Comoé and Odienné. Ghana, not uncommon, south to borders of forest in W and to Achimota, Legon and Cape Coast. Togo, common from Dapaon in N to Moretan and the coast. Benin, south at least to Bétérou. Nigeria, uncommon to frequent, north to Sokoto, Kano and Potiskum, south to Meko, possibly Ibadan, and to just south of Benue valley. Cameroon, records in Benue Plain and Mandara, but likely to be widespread in N, if uncommon. Chad, records only at Bali and Agan marsh. Central African Republic, uncommon in Bamingui-Bangoran and Manovo-Gounda-Saint Floris Nat. Parks; a few other records near W, N and E borders (Hall and Moreau 1970); not in NE (Vakaga). Sudan, widely distributed but uncommon in SW and S and a few records between 11° and 13°N. Ethiopia, frequent in all principal regions, up to at least 1800 m. Eritrea, scarce, in Mareb and Tacazze valleys with records in southern plateau. Somalia, scarce in N, up to 1650 m, commoner in S, in coastal lowlands and river valleys (Ash and Miskell 1998).

NE Zaïre, fairly common in Uele District, along upper Oubangui/Boma, south to Faradje and Dungu, east to near Mahagi. Uganda, widespread but uncommon, southwest to Mengo and Bunyoro. Kenya, widespread but uncommon

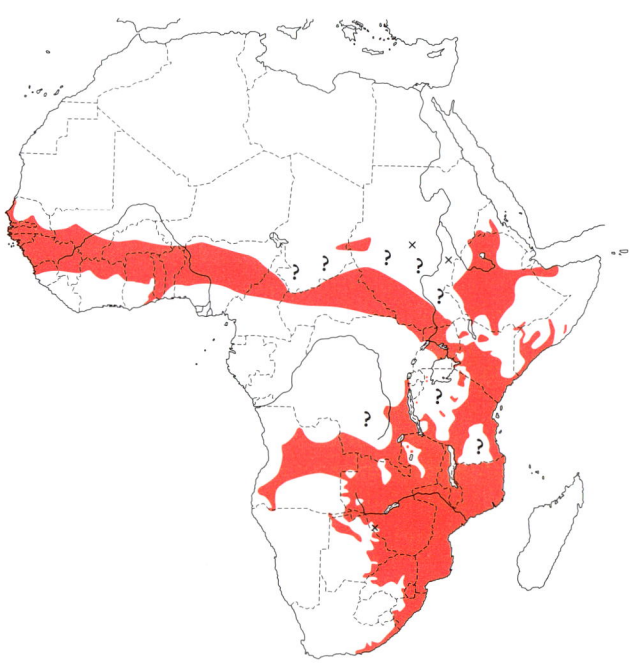

Malaconotus blanchoti

in southwestern half of country, with isolates in N (Ileret; between Moyale and Danissa Hills; in Daua valley and near Mandera), west of L. Turkana; east of L. Turkana centred on Marsabit Nat. Park; and in SW from Lolgorien to Nguruman Hills; up to 3000 m on Mt Elgon.

E Zaïre, from Ruzizi Plain and Manyema grasslands to Katanga and Marungu. Known from Rwanda (L. Kivumba, L. Birengero: Vande weghe 1974) and NW Burundi. Tanzania, Kenyan Nguruman population extends to Loliondo and central Serengeti Nat. Park; widespread in

NE, centre and SE (c. 130 records, N. Baker, pers. comm.), but scarce west of 35°E. Angola, throughout much of interior west to Malanje, Cuanza Sul, Huila and NE Namibe, but evidently absent from NE (most of Lunda Norte) and S (Cuando Cubango). Zambia, almost throughout; known from 81% of squares; sparse in W, to the Mashi at Lupuka (R.J. Dowsett, pers. comm.). Malaŵi, widespread but sparse; below 1500 (perhaps 1300) m. Zimbabwe and Mozambique, throughout, up to 1600 m. Namibia, frequent in extreme NE (E Caprivi Strip); records in W Caprivi Strip and 1 in Cunene valley. Botswana, sparse to locally very common, as on Shashe R. to its confluence and in E from Francistown to Martin's Drift. Transvaal, Swaziland and Cape Prov. as mapped (Harrison *et al.* 1997).

Density of <1 pair per 200 ha, Transvaal (Tarboton *et al.* 1987).

Description. *M. b. hypopyrrhus* Hartlaub: SE Africa south to about Durban, west to Rwanda and Southern and Central Provs of Zambia. ADULT ♂ (breeding): forehead, crown, nape, hindneck, sides of neck, ear-coverts and cheeks grey; lores and narrow eye-ring white; mantle to rump and uppertail-coverts and tail olive-green, tail tipped (sometimes obscurely) with pale yellow, up to 3 mm deep on T5–T6. Chin, throat and belly brilliant yellow, breast brownish orange, more so at sides, colour merging into yellow of throat and merging more gradually with yellow of hind part of flanks and of belly. Amount and depth of orange on breast varies considerably in this and some other races. Thighs olive-green and yellow, vent and undertail-coverts yellow. Underside of tail greyish olive, tip pale yellow, shafts yellow. In wing P10 blackish, other remiges blackish with outer vanes olive-green proximal to point of emargination and edged yellow distal to it; secondaries narrowly tipped pale yellow, tertials tipped pale yellow forming series of spots up to 6 mm long; wing coverts olive-green, greater coverts and outer median coverts broadly tipped pale yellow in triangular or axe-head shape; underside of remiges pale silvery grey, inner vanes of all but outermost primaries edged pale yellow; underwing-coverts and axillaries sulphur yellow; leading edge of wing bright yellow. (One bird examined has no yellow or green in plumage: upperparts bluish grey with white spots, underparts pure white with pale rufous breast. A somewhat similar bird seen by Donnelly (1978).) Bill black or black-brown, robust, strongly hooked, but upper mandible notch blunter than in large forest-dwelling congeners, 2 nasal bristles and 4 rictal bristles on each side; eyes orange or light yellow (may vary geographically); legs and feet bluish grey. Sexes alike. SIZE (sexes combined): wing (n = 30) 107–121 (114); tail 106–117; bill 26–31; tarsus 29–35 (Maclean 1993). WEIGHT: ♂ (n = 18) 72·9–83·9 (77·7), ♀ (n = 17) 65·0–99·2 (75·8).

IMMATURE: like adult but mottled brown on head, underparts paler yellow, bill light grey, eyes yellow (in areas where adult eye orange) or pale yellow (where adult yellow).

NESTLING: entirely naked; bill fleshy grey, dark brown at tip, gape yellowish white, tongue and palate orange-yellow (Serle 1940); eyes take at least 13 days to open (Grobler 1979). At age 7 days, feathers c. 5 mm long on wings and in line down back (Phillips 1979).

M. b. approximans Cabanis: Ethiopia south and east of Rift Valley, Somalia, Kenya, intergrading with *catharoxanthus* in W Kenya and with *hypopyrrhus* in N Tanzania. (All races intergrade at common boundaries.) Like *hypopyrrhus* but tawny orange on underparts deeper in shade and more extensive, forming gorget enclosing yellow throat and extending well onto flanks. WEIGHT: ♂ (n = 8) 70–89 (76·25), 1 ♀ 79.

M. b. interpositus Hartert: Angola, Zaïre (Katanga), W Zambia (Mporokoso; North-Western Prov.). Like *hypopyrrhus* but breast with less brown-orange suffusion; variable – some birds yellow below with only light wash of orange across breast.

M. b. catharoxanthus Neumann: N Cameroon to Ethiopia north and west of Rift Valley, Eritrea, Uganda and W Kenya. Underparts generally uniform bright yellow – no orange wash on breast although sides of breast sometimes with brown wash.

M. b. blanchoti Stephens: Senegal to Cameroon. Like *hypopyrrhus* but wing and leg longer. SIZE (11 ♂♂, 6 ♀♀): wing, ♂ 117–132, ♀ 118–131; tail, ♂ 108–121, ♀ 103–116; bill, ♂ 25–31, ♀ 29–31; tarsus, ♂ 37–41, ♀ 35–37.

M. b. extremus Clancey: SE Cape between about 25° and 30°E. Like *hypopyrrhus*, but head-top, nape and hindneck darker (less bluish) grey, rest of upperparts and wings and tail darker green (less yellowish) (Clancey 1982).

M. b. citrinipectus Clancey: Namibia: N Ovamboland to Cunene valley. Like *hypopyrrhus*, but forethroat and breast lemon-yellow not cadmium-yellow (Clancey 1982).

Field Characters. Length 23–26 cm. A large, heavy-billed bush-shrike of wooded *savanna* and acacia thickets (Lagden's and Monteiro's Bush-Shrikes *M. lagdeni* and *M. monteiri* are forest birds). Differs from Lagden's in *white lores, yellow eye,* paler grey head, *no black on wing* (wing-coverts have only indistinct pale yellow tips) and *yellow throat* (orange wash on breast only). Plumage extremely similar to Monteiro's, which distinguished by grey eye and white above and below eye. Immature paler yellow below, with barring and mottling on grey head. Size, lack of black on face, and call distinguish it from smaller Orange-breasted, Many-coloured and Olive Bush-Shrikes *M. sulfureopectus, M. multicolor* and *M. olivaceus*. Orange-breasted further distinguished by yellow supercilium, smaller bill and dark eyes.

Voice. Tape-recorded (10, 15, 75, 86, 88, 91–99, 104, 109). Song a single long (0·5–1 s), mournful, ventriloquial whistle all on one pitch, soft yet penetrating, 'whoooooo' or 'hawwwwww', with slightly quavering quality, repeated at rate of 1 note every 1–3 s for lengthy periods; often given as antiphonal duet. Some notes are shorter and have falsetto ending, 'whoooo-up' or 'hawwww-ip'; others are both grating and ringing, 'wraaaaaaanh', or they may be thinner or squeakier, 'wheeu... wheeu...' and may be preceded by soft notes audible only at close range; sometimes accompanied by a variety of clicks, ticks or clinks (Zimmerman *et al.* 1996). Also gives a hard rattle followed by a short, low 'haw', or by 3-note whistle 'hee-hah-haw' (more than one bird?). Incubating bird utters soft 'pheeo, pheeo, pheeo'; alarm at nest a loud, hoarse 'skwaark, skwaark, skwaark' and loud, harsh, explosive 'squot tot squot tot', 'squok-squok' or 'ptut, ptut'. During courtship one (or both?) birds gave 'a rattle followed by a very short whistle, with one or several whistles sounded for each rattle' (Mundy and Cook 1972). Also makes 'tch-tch-tch' sounds and soft, querulous mewings. ♀ solicits with quivering wings and soft, buzzing 'zzhoreer-zzhoreer' call.

General Habits. Inhabits thick woody growth and bushland in open savanna woodland, keeping to tangled growth, creepers, branch masses and dense foliage within a few m of ground (includes canopy of small trees): microphyllous, thorny woodland, shrubby pasture, dense riparian woods (in Eritrea especially with tamarind trees) riverine forest and thickets. In NW Nigeria inhabits old woodland, dry or wet, with very thick undergrowth and dominant trees and palms: *Khaya senegalensis*, *Ceiba pentandra*, *Borassus*, *Raphia* and *Elaeis*. In NE Zaïre occurs in plantations of Ceará rubber trees and in Rwanda in galleries of *Acacia polyacantha*; resident in eucalyptus plantations in Swaziland; occurs near neem *Azadirachta* in Nigeria; on the whole, though, rare in exotic vegetation. In Zimbabwe typically in *Brachystegia* woodland, commonest in riparian fringing forest with tall acacia trees, sometimes within lowland evergreen forest; uncommon in semi-arid savannas; in Somalia has predilection for acacias but lives in all types of woodland, including fragmented forests, riverine woods and scattered trees. Everywhere can occur in large gardens and parks, and sometimes nests close to houses.

Solitary or in pairs and sometimes small parties (of up to 6 birds – probably families). Wary, agile, inconspicuous. Forages at all levels in vegetation *c*. 12 m tall, up to canopy; occasionally descends to ground (e.g. to take worm or swarming termites); hops and bounds nimbly, in crouched posture, along and between branches and through twigs, diligently checking leaves for insects, peering about and even twisting head upside down. May pay more attention to branches than to leaves when foraging. Tends to work its way upward in a tree. Sometimes forages in leafless shrubs. Occasionally hawks insects in flight, fluttering up to catch one with audible snap of bill. A nest predator of doves, e.g. Laughing Dove *Streptopelia senegalensis* (Malzy 1962, Chadwick 1984). Attempts to catch small bird by pursuing it through vegetation. Appears to kill and dismember prey at favourite or special places, from where prey is taken to mate on nest (Harris and Arnott 1988). A shrike ate a Yellow-eyed Canary *Serinus mozambicus* by wedging it firmly between 2 vertical twigs in a mopane tree and tugging with bill until a piece came away, the pieces being swallowed instantly regardless of size; when remains of canary came adrift and fell to ground the shrike dived after it and retrieved it (Phillips 1979); killed a gecko by beating it against a branch, then ate it piecemeal. To eat a large moth, shrike pulled its wings off, beat it against ground and swallowed it whole. Small prey item held under one foot whilst being torn apart. Bird described hopping purposefully into house, flying up to cluster of mason wasps on ceiling and eating them all, then flying to another cluster on verandah and quickly eating them too (Hornby 1973). Caches food in larders (Harris and Arnott 1988). Flies for only short distances; rapid wingbeats then a short glide and swoop up into foliage of adjacent tree. In display sometimes frips wings in flight. Occasionally mobbed by small birds (and chases large ones), yet joins mixed-species foraging flocks (southern Africa, in winter).

May be territorial all year, with home range of *c*. 200 ha. That, and breeding territory (see below) defended mainly by calling. The spooky call (Afrikaans name is 'Spookvoël') can be given up to 50 times; bird calls with body sleeked and inclined, tail hanging vertically down, and head bowing downward with each call. Territorial threat signalled by loud, sharp bill snaps, like hedge-clippers, followed by low whistle. Pair often calls in duet during territorial disputes. In low-intensity alarm or threat, bird adopts hunched, crouching posture with back feathers raised, calls, opens bill and sways head menacingly from side to side (Harris and Arnott 1988). In intense alarm the explosive 'squok-squok' call is often given in dive-bombing flight.

Resident, probably mainly sedentary, but in Mali moves northward with the rains; arrives Bamako Mar and departs Oct, arrives Ban Markala May and departs Oct (Lamarche 1981). The most northerly records in Nigeria tend to be in rains. Vagrant to Cape Coast, Ghana, only in dry season. Some suggestion, but no clear evidence, for movements in southern Africa.

Food. Large arthropods and small vertebrates: locusts, grasshoppers, mantises, bees, wasps (particularly eumenid mason wasps), beetles, termites, dragonflies, caterpillars, moths and worms; chameleons, geckos and other lizards, small snakes, frogs, rodents, bats, birds and their eggs and nestlings. One snake was 'two feet long' (*c*. 60 cm: Serle 1940); another, 60–70 cm long, was dragged alive for 6 m and 20 min then left (Curtis 1987).

Breeding Habits. Solitary nester, monogamous, territorial. Breeding pair occupies territory of *c*. 50 ha, advertised by singing and defended by chasing other birds away (Harris and Arnott 1988). Courtship involves one bird chasing the other and frequently raising its head; courtship activities very conspicuous (Mundy and Cook 1972). In (rare?) display flight, ♂ frips wings and fans tail, interspersed with short glides when it utters whistle call (Harris and Arnott 1988). Nesting pair may chase away such birds as dove *Streptopelia* and Pied Crow *Corvus albus*. Once 2 shrikes chased a crow when a 3rd was brooding young (Wilkinson 1978). A pair built 2nd nest 50 m away and laid clutch 14 days after 1st nest robbed (Phillips 1979). Single brooded.

NEST: shallow, loosely knit, rather untidy, circular platform or bowl made of stout and slender twigs (sometimes black) and lined with tendrils, dry leaves, bits of soft or strong grass and fibrous roots. Some twigs 5 mm in diam. and 25 cm long. One nest, ext. diam. *c*. 170, int. diam. *c*. 70; another, ext. diam. *c*. 150, ext. depth *c*. 90, int. depth *c*. 43; another, int. diam *c*. 90. Nest (n = 125) placed 1·2–12 m (usually about 4 m) high, near top of small deciduous tree with spreading branches, e.g. *Cassia abbrevita*, *Ficus*, *Strychnos spinosa* or *Parinarium mobola*, either out on side branch or in fork in upright stem but always where there are many spreading twigs to support and screen it (Vincent 1949). Nest often sited in mistletoe *Loranthus*. Bird sometimes uses old nest of dove, sparrowhawk or Grey Go-away Bird *Corythaixoides concolor*. Nest (or at least the lining) built by ♀ and ♂; ♂ once seen collecting twigs and passing them to ♀.

EGGS: 2–4, usually 3 (Zambia and Malaŵi, 8 clutches of 2 eggs, 32 of 3 and 4 of 4; av. in southern Africa 2·9). Ovate; smooth, slightly glossy; pinkish white, lilac, pale lilac-brown, cream or pinkish buff, blotched, clouded or finely spotted especially around large end with chestnut and purplish brown on lavender shell markings. SIZE: (Nigeria, n = 6) 29·9–30·2 × 19·6–21·1 (29·9 × 20·1), (Zaïre, n = 4)

27·1–32·0 × 19·6–22·4 (29·5 × 20·9); (southern Africa, n = 69) 26·5–32·1 × 19·3–21·8 (29·4 × 20·6).

LAYING DATES: Gambia, June–July (and nestlings, Nov); Ghana, Jan, May, (fledgling Feb, Apr–May and July); Nigeria, Feb, Mar, June, Sept; Sudan, Mar; Ethiopia, June; Eritrea, (active gonads, May); Somalia, (nestlings, Mar); E Africa, Region B, Sept, Region C, Oct–Nov, Region D, July, Oct–Dec; Zaïre (Lubumbashi), Sept, Oct; Angola, Aug–Sept; Zambia, Aug–Feb, mainly (82% of 44 clutches) Sept–Oct; Malaŵi, Sept–Jan, Apr–May, mainly Oct–Nov; Botswana, (chicks near fledging mid Dec: Skinner 1998); Zimbabwe, Sept–Feb, Apr, July (of 145 clutches 44% in Oct, 25% in Sept and 21% in Nov); Transvaal, Aug–Jan; Natal, Sept–Jan.

INCUBATION: only (?) by ♀, sitting very tight; period c. 15–17 days.

DEVELOPMENT AND CARE OF YOUNG: half-grown nestling has forehead, crown, nape, hindneck and sides of neck brownish grey, all feathers with buff tips; lores, eye-ring and superciliary stripe buffy white, forming a long whitish 'mask', quite unlike adult; rest of upperparts olive-green; entire underparts pale yellow, feathers very fluffy; wing like adult but median and lesser coverts yellow-tipped. At one nest all 3 eggs hatched at intervals on same day; 4 days later, one bird arrived in nest tree with part of a Yellow-eyed Canary and gave it to mate, who took it to nest, tore it into small pieces and fed young (Phillips 1979). Young fed by both ♂ and ♀; adult seldom approaches nest directly, but spends several min beating large prey item (mantis, lizard) against branch and mandibulating it before wing-quivering, then going to nest in cautious, halting manner; when parent arrives with food, item is invariably given to brooding mate, which then gives it whole to one of the young or tears it to pieces to feed whole brood (Wilkinson 1978). Later, adults arrive with food independently of each other, but often 1st will not approach nest to deliver food until 2nd arrives in vicinity; adult commonly adopts soliciting posture, with body held horizontally and wings quivering, before flying to nest with food; one adult similarly solicits before approaching its mate near nest; brooding bird may also quiver wings rapidly before accepting food item from mate (Wilkinson 1978). Up to time of fledging one parent broods young at night and in early afternoon (to shade them?). Nestling period c. 20–21 days. Once one bird of pair roosted in a mango tree near its nest tree for all or most of nestling period. One nestling made first flight from nest at 09h20 and alighted in lower branches of adjacent tree; parent accompanied it, disappeared to find food, then made soliciting young follow it to safety of branch much higher up, when parent fed it (Phillips 1979). Young are attended by both parents before and after leaving nest, and can remain with them for up to a year.

BREEDING SUCCESS/SURVIVAL: in clutch of 4 eggs no more than 3 seem ever to hatch; at one nest the 3 young at 14 days were far too big for nest and so were 2 young at 20 days; 1–2 young sometimes fall from nest and die (Phillips 1979); however, 3 young often do fledge successfully. Brood taken by snake (Somalia).

References
Chapin, J.P. (1954).
Clancey, P.A. (1982).
Harris, T. and Arnott, G. (1988).
Phillips, R.L. (1979).
Serle, W. (1940).
Vincent, A.W. (1949).
Wilkinson, R. (1978).

Malaconotus lagdeni (Sharpe). Lagden's Bush-Shrike. Gladiateur de Lagden.

Laniarius lagdeni Sharpe, 1884. Proc. Zool. Soc. London, p. 54, pl. 5; Ashantee (Ghana).

Forms a superspecies with *M. blanchoti* and *M. monteiri*.

Range and Status. Endemic resident, with W and E African rain forest populations. Sierra Leone, Gola Forest. Liberia, uncommon but widely distributed, found in all well-surveyed forests: Nimba, Yekepa, Yéalé, Wologizi, S and N Gola Nat. Forest, 6 localities near Zwedru, and Grebo Nat. Forest (Gatter 1997). Ivory Coast, only in Taï, Mopri and Yapo Forests; recorded in 4 out of 5 localities in Taï Nat. Park and in 1 out of 5 localities outside the Park (Gartshore *et al.* 1995); frequent in Yapo (Demey and Fishpool 1994). Ghana, 1, near Kumasi, 1884; possibly this species seen in canopy in Nini-Suhien Nat. Park in late 1980s (Dutson and Branscombe 1990). Togo, sight record on Pagala to Ghana road, 1990. Zaïre, Rwenzoris, Kivu Volcanoes at about 2400 m (Kibumba, Bitashimwa, Kamatembe, Mt Karisimbi), and south to Itombwe highlands; uncommon in Itombwe, 03–04°S, at 1390–2710 m: Kiliza, Kitongo, Luiko, Lungwe L., Mayamoto, Mianga and Mikenge. Uganda, scarce, at 2200–2800 m in Rwenzoris and Bwindi-Impenetrable Forest, though may be locally quite common in latter. Rwanda, uncommon in Gishwati Forest; rare in wetter Nyungwe Forest where seen twice in 4·5 months, at Uwinka at 2500 m and to the south at 2300 m; another record near Mt Bigugu.

Density locally up to 0·5–1·5 pairs per km^2 in Liberia, where estimated to be 6000 pairs (Gatter 1997). Given Rare status in ICBP/IUCN Red Data Book (Collar and Stuart 1985).

Description. *M. l. lagdeni* (Sharpe): Liberia to Ghana. ADULT ♂: forehead, crown, nape, hindneck and sides of neck uniform dark grey; mantle, scapulars, back, rump and uppertail-coverts bright olive-green. Tail bright olive-green, T3–T6 tipped pale yellow; tail rounded, T6 15–17 shorter than T1. Underparts brilliant yellow, throat and upper breast strongly suffused with orange which gradually pales to yellow on lower breast; flanks olive-

Malaconotus lagdeni

green, largely concealed beneath spreading yellow belly feathers. Primaries blackish on inner vanes, and on outer vanes P9–10 except at base, others with increasing amounts of olive-green on outer vanes, secondaries with outer vanes almost wholly olive-green, inner vanes blackish, tertials olive-green with 2 mm deep yellow tips and wide subterminal black band; alula and greater primary coverts black; all other upperwing coverts black with broad yellow tips. Underside of flight feathers shiny dark grey, with broad pale yellow inner edges and bases to all except outermost primaries; underwing-coverts and axillaries bright yellow. Bill black, 2 nasal and 4 rictal bristles on each side (**A**); eyes light grey tinged with bluish; legs and feet blue or slate-blue. Sexes alike. SIZE (1 ♂, 1 unsexed, Mt Nimba, Liberia): wing, ♂ 115, unsexed 108; tail, ♂ 100, unsexed 96; bill, ♂ 35, unsexed 31. WEIGHT: 1 ♂ 97, 1 unsexed 96.

IMMATURE: juvenile has entire upperparts from forehead to wings and tail, also cheeks, dark greyish brown, tertials narrowly tipped grey and with broad blackish subterminal band; chin and throat white, breast white with greyish tinge or faint grey dapples in centre, greyer at sides; upper belly white in middle, grey at sides, merging to warm or rufescent buff on lower belly, hind part of flanks, thighs and undertail-coverts; bill pale grey-brown or horn at first, blackish later (Prigogine 1986). Subadult like adult but with whitish streaks or spots on throat and middle of belly, breast lightly barred, and bill black (Lippens and Wille 1976).

NESTLING: not known.

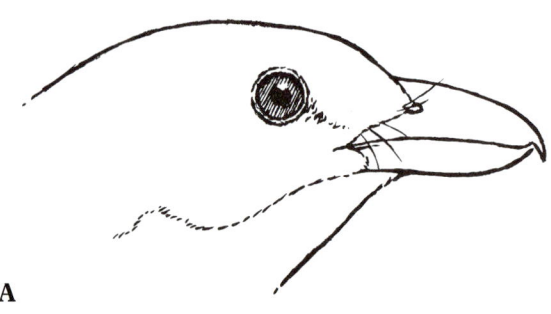

A

M. l. centralis Neumann: Zaïre and Uganda. Like nominate race but orange suffusion on throat and upper breast usually much less heavy; lower breast and belly slightly paler yellow and washed with olive; however, in Bwindi-Impenetrable Forest throat and upper breast 'bright orange-yellow' (Bennun 1985). SIZE (1 ♂, 1 ♀): wing, ♂ 110·5, ♀ 110·5; tail, ♂ 106, ♀ 109; bill to feathers, ♂ 26, ♀ 27, bill to skull, ♂ 31, ♀ 31, depth at ant. border of nares, ♂ 12·3, ♀ 12·2; tarsus, ♂ 33·5, ♀ 33·5; also (5 immature ♀♀): wing 110–114 (112), tail 104–111 (107), bill to feathers 27–32·5 (28·1) (Prigogine 1986). WEIGHT: 1 ♂, Uganda, 84.

Field Characters. Length 24·5–25·5 cm. A forest version of Grey-headed Bush-Shrike *M. blanchoti*. Differs in having wing-coverts and inner secondaries with broad black and narrower yellow barring, no white on lores, chin, throat and centre of breast orange-yellow, flanks green; eyes light grey (yellow in Grey-headed). Differs from yellow-phase Fiery-breasted Bush-Shrike *M. cruentus* by yellow-green tail without black subterminal band. Immature similar but chin to upper belly white. Montane forest in Albertine Rift and lowland forest in W Africa (Grey-headed in savanna woodland).

Voice. Tape-recorded (104, B, DEM, FISP, GAR, GUL, LEM). Song very similar to Grey-headed Bush-Shrike, a single, long (0·5–0·7 s) mournful note, 'whoooooo' or 'hawwwww', with quavering quality, repeated at intervals of 3–5 s; sometimes falsetto at the end, 'whoooo-up'; see sonagrams in Dowsett-Lemaire (1990). Other song types include double mournful 'hoop ... hooooo', second note longer and a tone lower than the first; 'wip ... hawww;, second note moderately quavering, and low, soft, 'hoh ... hoh ... hoh ... hoh', notes at rate of *c.* 1 per s, like that of Fiery-breasted Bush-Shrike *M. cruentus* but lower-pitched; typically 4 'hoh's but number varies (Demey and Fishpool 1994). Confusion with Fiery-breasted unlikely because these shorter whistles are never produced in isolation but only by second bird of pair when mate is already giving the long whistles (R. Demey, pers. comm. F. Dowsett-Lemaire). Adult and juvenile foraging together both repeatedly gave harsh, grating 'chaarr, chaarr' (Bennun 1985).

General Habits. Inhabits middle strata and canopy of evergreen rain forest, lowland in W Africa and montane in Zaïre, Uganda and Rwanda; commonest in undisturbed forest but occurs in lightly logged and occasionally in heavily logged forests, in dense creepers, and open parts of forest canopy (Liberia), and in much-disturbed forest in Uganda. In Itombwe, Zaïre, occurs in primary and gallery forests mainly over 1900 m, occasionally down to 1400 m, keeping near edges of forest, around clearings, in secondary growth and in open grassy woods at higher elevations; elsewhere in Rift Valley mountains occurs up to 2500 m. In Liberia at 100–700 m. Usually in pairs, sometimes in groups of 3 or 4; often in mixed-species foraging flocks of insectivorous birds (in *all* sightings in Liberia a member of bird parties, often in company of Many-coloured Bush-Shrikes *M. multicolor*: Gatter 1997). Once seen in a low thicket, among small birds mobbing a genet; rarely below 5 m high, usually between 10 m and 30 m high (mode, 15–20 m).

Forages mainly by gleaning branches for invertebrates, but also predatory, taking small vertebrates; small birds warn of its approach; searches branches 5–30 mm thick (Gatter 1997). Adult and juvenile foraging together flew from branch to branch, bouncing rapidly along the boughs after alighting (Bennun 1985). Beats large insect (e.g. katydid) vigorously against perch. Flight heavy, with flailing, noisy wingbeats. Sedentary.

Food. Insects and a small frog in stomachs. In Liberia seen catching 10-cm long *Agama* lizard and trying to catch juvenile Blue-throated Brown Sunbird *Cyanomitra cyanolaema*. In Uganda seen with probable katydid (Orthoptera, Tettigonidae).

Breeding Habits.
NEST: only one seen was a bulky bowl of dry leaves and bracken (**B**), placed 3·5 m up 'in the forks' of a small tree in dense secondary growth in an abandoned clearing (Chapin 1954). 2 other nests were hidden in leafy regrowth on top of broken-off tree *c.* 6 m high; nest built by ♂ and ♀, approaching it carrying long, thin, barely visible fibres and singing in duet as they did so (Demey and Fishpool 1994).
EGGS: 2 (i.e. one in nest, another in ♀'s oviduct). Eggs dull light grey with specks and spots of dark brown, slightly tinged with purplish, the spots dense around large end (Chapin 1954). SIZE: (n = 1) 30·2 × 20·9.
LAYING DATES: Liberia, (just-fledged young accompanied by parents, Aug; immatures in Oct and Dec); Ivory Coast, (singing Oct–Jan and once in July; nest-building in Nov and Dec; bird carrying food Oct); Zaïre, Rwenzori, Dec (and Itombwe, large gonads Jan, juvs in July, Oct and Nov).

B

DEVELOPMENT AND CARE OF YOUNG: 2 young, *c.* 10 days out of nest, accompanied by 2 adults (Liberia) and for next 25 days each adult guided one of them (Gatter 1997).

References
Bennun, L.A. (1985).
Chapin, J.P. (1954).
Dowsett-Lemaire, F. (1990).
Gatter, W. (1997).
Prigogine, A. (1971, 1984, 1986).

Malaconotus gladiator (Reichenow). Green-breasted Bush-Shrike. Gladiateur à poitrine verte. **Plate 21**
Laniarius gladiator Reichenow, 1892. J. Orn., 40, p. 441; Buea, Mt Cameroon. **(Opp. p. 334)**

Range and Status. Endemic resident, SE Nigeria and W Cameroon. Nigeria, only on Obudu Plateau, where 1 collected in 1961 and a few subsequent sight records. Cameroon, Rumpi Hills (1300 m, and Dikume Balue, 1520 m), Bamenda-Banso Highlands (near Bambulue, 2080 m, Bali-Ngemba Forest Res. at 1700 m, in forest patches between Mbengwi and Tinachong, and Mt Oku at 2200–2300 m), and Mt Cameroon (southern slopes above Bonenza, 950–1350 m and above Buea, 1375–1800 m), Mt Kupé at 1100–1950 m, Mt Nlonako at 1400–1600 m, and common in Bakossi Mts at 1100–1400 m. Treated as Rare in ICBP/IUCN Red Data Book (Collar and Stuart 1985); common in mid-altitude forest on Mt Bakossi, where at least 6 territorial birds or pairs in <1 km² of forest/grassland mosaic near Kodmin, and also common near L. Edib; 'there is little doubt that the Bakossi Mts hold the single most important population of the species' (Dowsett-Lemaire and Dowsett 1998).

Description. ADULT ♂: forehead, crown, nape, hindneck, sides of neck and upper mantle dark grey, remainder of plumage entirely uniform olive-green, brighter below than above; but alula feathers, primary coverts, greater coverts, tertials, secondaries and inner primaries narrowly tipped yellowish olive; P10 black;

Malaconotus gladiator

P9–P6 black with outer vanes olive-green proximal to point of emargination and black, narrowly edged green, distal to it; P5–P1 with mainly olive-green outer vanes and blackish inner ones. Tail feather shafts dark brown above, pale yellow below. Underside of remiges shiny grey, merging to strong pale yellow on inner edges of all of them except outer few primaries; underwing-coverts and axillaries olive-green, marginal coverts on leading edge of wing yellow. Bill black; eyes light grey or greyish white; legs and feet grey or blue-grey. Sexes alike. SIZE: wing, ♂ (n = 3) 119–121 (120), 2 unsexed 117, 120; tail, ♂ (n = 3) 110–114 (113), 2 unsexed 113, 114; bill to feathers, ♂ (n = 3) 28–30 (28·7), bill to skull, ♂ (n = 3) 32–34 (33·3), depth ♂ (n = 3) 13–14 (13·3); tarsus ♂ (n = 3) 34–36 (35).

IMMATURE and NESTLING: undescribed.

TAXONOMIC NOTE: some evidence suggests that Monteiro's Bush-Shrike *M. monteiri* may be a colour phase of *M. gladiator*: see Note under *M. monteiri*.

Field Characters. Length 28 cm. *Huge*, the largest bush-shrike; *uniform dark green* except for grey top of head. Sympatric with Monteiro's Bush-Shrike *M. monteiri*; their calls are identical but Monteiro's has yellow underparts.

Voice. Tape-recorded (104, B, GAR, LEM, STU, WILL). Repertoire very similar to that of Grey-headed Bush-Shrike *M. blanchoti*. Song, audible for at least 1 km, 3–10 long (1–1·5 s), drawn-out, mournful piping whistles, with very brief intervals, at rate of 1 whistle per 1·5–2 s. Some whistles end with a falsetto note. Whistles also may be shorter and given at a faster rate, see sonograms in Williams (1998). Harsh grating call is same length and given at same rate as whistles. Birds in Bamenda Highlands have thinner, more muffled whistle than those further south (Stuart 1986). Alarm a harsh 'chit-chipaa'.

General Habits. Inhabits canopy of gloomy primary and secondary montane evergreen forests at 950–2250 m. Seen also: in grove of tree ferns and old second growth; high up in foliage along mountain forest stream; in thick clump of secondary growth near cultivated clearing; and exceptionally in forest underplanted with subsistence crops. On Mt Bakossi strictly confined to canopy, though readily brought to edges of clearings by voice playback (Dowsett-Lemaire and Dowsett 1998). Occurs singly or in pairs, mainly in canopy, sometimes down to 10 m, once only 2 m above forest floor. Can readily be brought close to observer if call imitated, when bird becomes agitated and occasionally responds by drumming bill on branch (Stuart 1986). Bird annoyed by observer imitating it does not descend from canopy but glides into another high tree, then calls (Dowsett-Lemaire and Dowsett 1998).

Food. Insects including large locusts.

Breeding Habits. Unknown.

LAYING DATES: Cameroon, (gonads slightly hypertrophied Dec but not in May or July).

References
Collar, N.J. *et al.* (1994).
Dowsett-Lemaire, F. and Dowsett, R.J. (1998).
Serle, W. (1954).
Stuart, S.N. (1986).
Williams, E. (1998).

Plate 21 (Opp. p. 334) *Malaconotus alius* Friedmann. Uluguru Bush-Shrike. Gladiateur des Ulugurus.

Malaconotus alius Friedmann, 1927. Proc. New Engl. Zoöl. Club., 10, p. 5; Bagilo, Uluguru Mts, Tanganyika Territory.

Range and Status. Endemic resident, Uluguru Mts, Tanzania. Map on facing page. Restricted to altitudinal range of 1300–2100 m in area of about 250 km² (see Ulugurus map in Stuart and Jensen 1985). Rare or very elusive; known from *c.* 15 specimens taken in 1926 (when discovered) and between 1952 and 1962, mostly at Bagilo (1800 m in N Ulugurus), and from sight records near Bunduki, on both sides of Mt Lupanga, and on W scarp of Lukwangule Plateau. No records at all between 1927 and 1948 nor between 1962 and its rediscovery in 1981. Critically threatened; forest habitat is <120 km² in extent; in 1993 only 3–4 territories were found in 2 months of study (Collar *et al.* 1994).

Description. ADULT ♂: forehead, crown, nape, hindneck and sides of neck glossy black, sharply defined all around; remaining upperparts including wings and tail uniform olive-green (concealed parts of remiges blackish). Chin and throat bright yellow, breast yellow with green tinge, merging to olive-green at sides, belly yellow in midline, merging to olive-green at sides; flanks and thighs olive-green; vent and undertail-coverts greenish yellow; underparts below breast entirely green in some birds. Underside of remiges, underwing-coverts and axillaries bright yellow; underside of tail olivaceous grey. Bill black; eyes brown; legs and feet bluish grey. Sexes alike. SIZE (1 ♂): wing 105, tail 99, bill to feathers 26, bill to skull 31·5, depth 12, tarsus 31.

IMMATURE: undescribed. One recently identified as such in the field presumably differed from adults in same way as in congeners.

NESTLING: unknown.

Field Characters. Length 23–24 cm. Endemic to Uluguru Mts, Tanzania. Entire top of head to cheeks and nape *glossy black*. Dark green above, *rich yellow below*, washed green on flanks; sometimes only throat yellow, rest of underparts green. Huge bill further distinguishes it from yellow phase of much smaller Black-fronted Bush-Shrike *M. nigrifrons* which is more orange in colour and has yellow tips to underside of feathers.

Voice. Tape-recorded (B, CART, SVEN). Voice loud and far-carrying, 'tok-tok-tyeew' or 'hyewp-tyeew'; main component, clear down-slurred 'tyeew', can also be given alone. Very different from voice of Black-fronted Bush-Shrike, *q.v.*

General Habits. Inhabits canopy of montane forest, where keeps to densest foliage and extremely hard to see. Once seen in mixed-species foraging flock; 'often in company with the Grey-headed Bush-Shrike' *M. blanchoti* (Mackworth-Praed and Grant 1960).

Food. Unknown.

Breeding Habits. Unknown. Immature bird seen in July.

References
Collar, N.J. *et al.* (1994).
Friedmann, H. and Stager, K.F. (1964).

Malaconotus kupeensis (Serle). Mount Kupé Bush-Shrike. Gladiateur du Kupé.

Plate 21 (Opp. p. 334)

Chlorophoneus kupeensis Serle, 1951. Bull. Br. Orn. Club, 71, p. 41; Kupé Mountain, Kumba Division, British Cameroon.

Range and Status. Endemic resident, restricted to forest on Mt Kupé and neighbouring Mt Bakossi (04°45′N, 09°40′E), SW Cameroon. Uncommon on Mt Kupé, at 950–1550 m. Discovered in 1949 when 2 birds collected at Nyassoso, 04°45′N, 09°40′E; in 1951 3 more collected together near Nyassoso. Extent of forest there was only *c.* 21 km^2 in late 1950s. Extensive search by ICBP Cameroon Montane Forest Survey in 1984, on Mt Kupé and adjacent Mt Nlonako, failed to rediscover it; however, rediscovered in 1989 at 1220–1310 m and several have been seen on Kupé since then, and one mist-netted (photo, *Bull. Afr. Bird Club* 2, 1995, p. 31). In Mar–May 1997 1–4 birds seen 6 times (*Bull. Afr. Bird Club* 4, 1997, p. 142), but possible that there are no more than 7 pairs present near Max's and Shrike Trails; at 930–1450 m. Discovered on Bakossi in 1992, several seen there since at 1150–1200 m near L. Edib and towards Messaka; Bakossi population might be only *c.* 50 pairs; might also prove to occur on Mwenzekong and Ekomane Mts and unexplored lower, S slopes of Mt Manenguba and Mt Nlonako, but species best treated as critically endangered (Dowsett-Lemaire and Dowsett 1998).

Description. ADULT ♂: forehead black, crown grey, paler at sides above eye; nape, hindneck, sides of neck and mantle grey; back, rump, uppertail-coverts, wings and tail bright olive-green. P10 and inner vanes of remiges blackish; outer primaries fringed yellowish olive at base. Chin and throat white; line of 3–4 dark maroon spots between throat and breast (line 15 mm wide, 2–4 deep); remaining underparts grey, except for olive-green hind part of flanks, olive thighs, bright yellow patch on lower belly, and bright yellow vent and undertail-coverts. Bill black, with 4 long rictal bristles on each side; inside mouth yellow; eyes violet; legs and feet grey. ADULT ♀: like ♂ but, instead of row of small spots, one large round spot *c.* 6 mm across in midline between throat and breast, dark maroon (photo, *Bull. Afr. Bird Club* 2, 1995, p. 31) or metallic purplish black (Serle 1951); lower belly and undertail-coverts pale greenish yellow, not bright yellow. SIZE (1 ♂, 2 ♀♀): wing, ♂ 97, ♀ 94·5, 99; tail, ♂ 81·5, ♀ 78·5, 81; bill to feathers, ♂ 21·5, ♀ 21·5, 23; bill to skull, ♂ 27·5, ♀ 26; tarsus, ♂ 29·5, ♀ 29·5; 1 imm.: wing 98; tail 80; bill 22; tarsus 28.
IMMATURE: like ♂ but front and sides of forehead tinged green, sides of neck and sides of throat tinged green, grey of upperparts flecked with olive-green, chin and middle of throat flecked with pale yellow; narrow line of orangish spots where ♂ maroon; patchy green wash on breast; and larger area of olive on flanks.

Field Characters. Length 17·5 cm. Black forehead and mask, grey head and mantle and green upperparts, as in Many-coloured Bush-Shrike *M. multicolor*, but *white throat* contrasts with *blue-grey underparts*. Only other small bush-shrike near its range, Bocage's *M. bocagei*, has blackish upperparts, white supercilium and all-white underparts.

Voice. Tape-recorded (LEM). Call by ♂ as it joined bird party a loud, babbler-like introductory chatter, 'thec-thec, kh-kh-kh' followed by 3–4 'tchraaa-tchraaa-tchraaa', at rate of 2 'tchraaa' per s. Responds vigorously to playback with faster introductory chatter and louder and louder song (up to 29 'tchraaa' in a row) (Dowsett-Lemaire 1999). Also described is a distinctive call of 3 whistles of different pitch (Serle 1951); whistles are clearly detached, ascend in pitch, and sound somewhat out of tune (I. Faucher, pers. comm. to F. Dowsett-Lemaire). Song consisting of 'whistles at three different pitches repeated slowly and sometimes intermittently in an irregular order', heard on Mt Kupé in 1984, may have been of this species but the singer was not seen (Stuart and Jensen 1986). This song is apparently rarely given, but one bird called persistently for

2 days (I. Faucher, pers. comm. to F. Dowsett-Lemaire). Also apparently uncommon is the quiet, continuous insect-like grating mentioned by Bowden and Andrews (1994).

General Habits. Inhabits primary montane forest; party of 3 birds was 6–9 m high in smaller trees in primary forest on steep mountainside. Most sightings on Mt Kupé at 2–10 m and on Mt Bakossi at >10 m above ground, but 2 birds seen feeding on insects high in canopy. Usually seen singly, sometimes in pairs, twice 3 together. One bird joined a mixed-species foraging flock of Grey-headed Greenbuls *Phyllastrephus poliocephalus* and Xavier's Greenbuls *P. xavieri*, White-throated Mountain Babblers *Kupeornis gilberti*, Elliot's Woodpeckers *Dendropicos elliotti*, Black-capped Woodland-Warblers *Phylloscopus herberti*, Green Hylias *Hylia prasina*, Buff-throated Apalises *Apalis rufogularis*, Yellow-bellied Wattle-eyes *Diaphorophyia concreta*, Red-bellied Paradise-Flycatchers *Terpsiphone rufiventer*, Pink-footed Puffbacks *Dryoscopus angolensis* and Dark-backed Weavers *Ploceus bicolor*. It responded to playback of its song by flying over observer, noisily snapping or fripping its wings in 4–6 bouts and singing repeatedly with fast chatter followed by up to 29 'tchrraa's at rate of 3 per sec. In next 2 hours same bird was called up with playbacks several times, once from a distant hillside, taking nearly 10 s to arrive near tape-recorder. Although generally silent, bird sang persistently for 2 days in Mar 1998. 2 birds once heard calling repeatedly to each other for >10 min, and song thought to be territorial (Dowsett-Lemaire 1999).

Food. Insects in 3 stomachs.

Breeding Habits. Unknown.

References
Collar, N.J. *et al.* (1994).
Dowsett-Lemaire, F. (1999).
Dowsett-Lemaire, F. and Dowsett, R.J. (1998).
Serle, W. (1951).
Stuart, S.N. and Jensen, F.P. (1986).

Plate 22
(Opp. p. 335)

Malaconotus multicolor Gray. Many-coloured Bush-Shrike. Gladiateur multicolore.

Laniarius multicolor Gray, 1845. Gen. Birds, 1, (1849), p. 229, pl. 72; type locality restricted to Accra, Bannerman, 1939, Birds Trop. West Africa, 5, p. 428.

Forms a superspecies with *M. nigrifrons*.

Range and Status. Endemic resident, distributed patchily from SW Mali to E Zaïre and NW Angola.

Guinea, frequent on Ziama Massif. Mali, wet-season visitor in S (Sikasso, Bougomi, Sélingué, Bamako, north to Baoulé where frequent). Sierra Leone, frequent in S, localities including Freetown Peninsula, Bo, York Pass, Kenewa and Maltotaka; doubtless widespread. Liberia, frequent to common, 60 localities as mapped (Gatter 1997); up to 1100 m on Mt Nimba. Ivory Coast, scarce, Nimba and Täi to Lamto, Yapo, Maraoué and Comoé. Ghana, formerly not uncommon, recorded at Accra, Cape Coast and Mampong (Ashanti), Kumasi, Makessim, interior of 'Fantee' and Agomé Tongwe; only recent records at Akropong (Akwapim) and Amedzofe. Togo, uncommon: Ounabé, Dzogbégan, Balla, and several recent records at Misahöhe. Not known in Benin. Nigeria, in SW at Ilaro, Ibadan, Ife and Ikom; known from Kagoro, and Gashaga-Gumti Nat. Park. Cameroon, not uncommon at Kumba (where 14 birds collected in 1940s) and Nguti, Korup Nat. Park; also Baro and a few other localities: Ekona, Mt Cameroon at 400–1000 m, Limbe (Victoria), and in S on Dja River. Central African Republic, recorded in Lobaye Préf., Manovo-Gounda-Saint Floris Nat. Park and Bamingui-Bangoran Nat. Park. Mbini, recorded. Gabon, frequent between Makokou and Mékambo; not yet found in Rés. de la Lopé. Congo, in N only in Odzala Nat. Park where very local (Lango, Ikolo) (Dowsett-Lemaire 1997a); in SW in Kouilou Basin at Col de Bamba, Dimonika and Koubotchi. Angola, rare, 5+ records from Cuanza Norte and Escarpment in Cuanza Sul.

Malaconotus multicolor

E Zaïre, frequent west of L. Albert, from Nepoko R. and Medje via Mawambi to Djugu, Kilo, Semliki Valley and Beni in Ituri; another race frequent to common in Itombwe Highlands at 1160–2190 m; frequent between

Ituri and Itombwe (though uncertain where racial boundary falls). Rwanda, scarce in W Nyungwe Forest at 2500 m, increasingly common further down, especially below 2000 m. Not in Burundi. Uganda, Kibale (*batesi*) and Bwindi-Impenetrable Forest (*graueri*).

Density, estimated from calls at Zwedru, Liberia, 3–9 pairs per km^2 (Gatter 1997).

Description. *M. m. multicolor* (Grey): Sierra Leone to W Cameroon. 4 phases: Yellow, Scarlet, Black/Yellow and Black/Red. ADULT ♂: *Yellow Phase* forehead black, continuous with black lores and black band below eye to ear-coverts and lower sides of neck; hindcrown to upper mantle grey, paling to white on forecrown and supercilium; amount and brightness of white variable – forecrown and supercilium sometimes grey; rest of upperparts green. Tail feathers green, grading to blackish distally and with broad yellow tips, *c*. 10 mm deep on T1 increasing to *c*. 15 deep on T6; underside of tail blackish. Cheeks to chin, throat and upper breast bright orange-yellow, rest of underparts bright yellow. Flight feathers dark brown with green outer vanes, secondaries and inner primaries tipped yellow; borders of inner vanes bright yellow. Tertials green, fringed yellow at tips; alula and upperwing-coverts green. Underwing-coverts and axillaries bright yellow. Bill black; eyes blue-green, bright violet or purple; tarsus bluish grey to brownish grey. *Scarlet Phase* differs in having chin to lower breast scarlet, rest of underparts yellow, often with orange tinge; tips of tail feathers orange-yellow. *Black/Yellow Phase* with chin to upper breast black, continuous with black face mask, sharply demarcated from bright yellow on rest of underparts. *Black/Red Phase* like Black/Yellow phase but belly, flanks and undertail-coverts red. All phases: W Cameroon population darker-tailed than *multicolor* further west. ADULT ♀: all phases differ from adult ♂♂ in having forehead and lores to ear-coverts grey (not black); faint greyish white superciliary stripe behind eye; tail feathers greenish above and below, with yellow tips. *Yellow Phase* bright yellow below but lacks orange tinge on throat and breast. *Scarlet Phase* more orange-red below than in ♂, the colour more restricted, to throat and upper breast. SIZE (10 ♂♂, 10 ♀♀): wing, ♂ 95–102 (97·8), ♀ 91–98 (94·6); tail, ♂ 86–93 (90·1), ♀ 83–93 (89·2); bill, ♂ 21–24 (22·9), ♀ 21–24 (22·6); tarsus, ♂ 25–27 (26·2), ♀ 25–27 (26·2). WEIGHT (Liberia): ♂ (n = 17) 43–61 (50·9), ♀ (n = 3) 46, 52, 54.

IMMATURE: ♂ and ♀ like adult ♀ but retain yellow-tipped juv. wing-coverts. Juvenile has top of head to upper mantle grey with dark grey barring, rest of upperparts green with yellowish barring; forehead to lores and ear-coverts grey; indistinct paler supercilium; chin to upper breast, flanks and thighs barred yellowish and dark grey, lower breast to undertail-coverts yellow; tail like adult ♀; wing like adult, but alula fringed yellow and all upperwing-coverts tipped yellow. Barred juvenile plumage replaced by unbarred immature plumage probably rapidly (D. J. Pearson, pers. comm.).

NESTLING: not described.

M. m. batesi (Sharpe): S Cameroon, Gabon, NW Angola, Central African Republic, NE Zaïre (Ituri) and Uganda (Kibale). 2 phases: Yellow and Scarlet. ♂ with forecrown and sides of crown always white, and more distinctly so than in whitest birds of nominate race. Tail feathers black above with yellow tips. In *Scarlet Phase*, entire underparts often scarlet, tail feathers tipped reddish yellow. ♀ differs from ♀ *multicolor* in having whiter forecrown and superciliary stripe. WEIGHT: ♂ (n = 5) 38–41 (38·9), ♀ (n = 3) 36–39 (37·7).

M. m. graueri (Hartert): montane forests of E Zaïre (Kivu to Itombwe), SW Uganda (Bwindi) and Rwanda (Nyungwe). 3 phases: Yellow, Scarlet and Buff. ♂: white on head like *multicolor*, tail black (not green) like *batesi*. Scarlet darker and extends further onto belly than in *multicolor*. Differs further from *multicolor* and *batesi* in having *Buff Phase*: whole underparts pale cinnamon-buff.

TAXONOMIC NOTE: variation and polymorphism of this species and its allospecies *M. nigrifrons* were studied by Moreau and Southern (1958) and Hall *et al.* (1966); little improvement of knowledge since then. Certain characters other than the phasic ones vary geographically (e.g. amount of white on forecrown, blackness of green in tail, amount of yellow in tail tip, degree of sexual dimorphism) and provide basis for recognizing conventional subspecies in both members of the superspecies; but those taxa are defined also, in part, by phasic characters (accounted for by 3 pairs of alleles in a polygenic genome: Hall *et al.* 1966). It prompts the question, If only Yellow birds occur in one region, only Red in another, and both Red and Yellow in a third, freely interbreeding, are 1, 2 or 3 taxa involved? There are 5 major phases (or 7, counting Black/Yellow, Black/Red and Black/Green separately) and a further 6 quantifiable variants, as well as intermediate individuals scattered randomly (Moreau and Southern 1958). In most regions 2 or 3 phases occur together and interbreed; but only Red Phase occurs in Katanga (Zaïre), Zambia, Zimbabwe and Transvaal (*M. nigrifrons*), and evidently only Yellow in NW Angola (*M. multicolor*) and parts of E highlands of Kenya (*M. nigrifrons*).

Were a Yellow Angola ♂ ever to meet a Red Zambia ♀, only the 2 birds concerned could decide whether to mate, i.e. whether or not they belonged to the same species. Aside from the complexity of phases, we are persuaded that the *multicolor* and *nigrifrons* taxa are separate, 'good' species: *multicolor* races have blacker tails with more prominent yellow tips, are more sexually dimorphic, and juveniles have barred upperparts (usually plain in juvenile *nigrifrons* races).

REGIONAL PHASE PREVALENCE: Sierra Leone to Togo (n = 50) Scarlet 45%, Yellow 10%, Black/Yellow 6%, Black/Red 4%; Nigeria and W Cameroon (n = 82), Scarlet 44%, Yellow 46%, Black/Yellow 8·5%, Black/Red 1·5%; rest of Cameroon and Gabon (n = 36), Scarlet 67%, Yellow 33%; NW Angola (n = 7), Yellow 100%; Zaïre (Ituri) (n = 49), Scarlet 73%, Yellow 27%; Zaïre (west of L. Edward) (n = 47), Scarlet 23%, Yellow 64%, Buff 13%; Zaïre (northwest of L. Tanganyika) (n = 35), Scarlet 6%, Yellow 74%, Buff 20% (Moreau and Southern 1958). W Cameroon (Kumba, Ekona) (n = 15), Scarlet 46%, Yellow 27%, Black/Yellow 27% (Serle 1950).

Field Characters. Length 18–19 cm. A *forest* species. Upperpart colours consistent – olive-green with grey crown and mantle and *black* mask from forehead through eye onto ear-coverts; in some races black band below eye only, white forecrown extends around eye and onto face; underpart colours variable, yellow, orange, red or cinnamon-buff; ♂ occasionally has black breast. Much smaller and shorter-billed than Lagden's and Fiery-breasted Bush-Shrikes *M. lagdeni* and *M. cruentus*, which have heavy bills and no black on face; nonetheless may be hard to distinguish from Lagden's, and in Taï Nat. Park, Ivory Coast, some ♀♀ have same colours as Lagden's (Gartshore 1989). In mixed vegetation or at forest edge may meet Orange-breasted Bush-Shrike *M. sulfureopectus*, which lacks black mask and has yellow supercilium, yellow underparts with orange on breast only. Immature Many-coloured lacks black on face and is yellowish below, with dark barring in juveniles. Mellow calls very different from ringing songs of other bush-shrikes.

Voice. Tape-recorded: (104, B, ERA, HUG, LEM, MAC, ROD). Common call in W Africa (Liberia, Ivory Coast, Cameroon) single or double low whistles on same pitch 3–4

Plate 25

Plate 26

s apart, 'hoh ... hoh ...' or 'hohoh ... hohoh ...'; bird easily imitated and attracted by human; ♀ may answer with coughing note, 'hoh-gah'. In 2-note call from S Congo (Mayombe), second note is 4 tones higher than first (F. Dowsett-Lemaire, pers. comm). In Rwanda, normal song is a single 'woh' repeated at 2–3 s intervals; also gives a fast 'fufufufufufufu', usually uttered only once (Dowsett-Lemaire 1990).

General Habits. Inhabits matted tangles of creepers and lianas festooning the canopy and under-canopy of primary lowland and montane rain forest. Also occurs frequently in old secondary forest, forest-grassland mosaic, forest remnants, gallery forest; in Gabon, inhabits parts of forest with dense bushy growth in abandoned slash-and-burn farm patches and villages; also dense vegetation near occupied human habitation, and thick secondary growth in clearings; in Congo (Odzala) in swamp forest.

Lives in pairs, in Gabon evidently territorial but in parties of 4–8 birds in Itombwe. Shy, skulking, unobtrusive. Forages between 8 and 30 m from ground (90% of observations, Liberia: Gatter 1997) or at 2–10 m (Gabon, Brosset and Erard 1986); perches in and stalks through very dense creepers; sometimes in open canopy; uses perches of 3–40 mm diam.; feeds amongst tangles of lianas, epiphytes and on mossy limbs of large trees, hopping and creeping about in crouching posture, gleaning from trunks, branches, matted creepers and thick foliage, gradually working its way up tree, then flying to the next. Occasionally hawks insects in air. Often a member of mixed-species foraging flocks, in Itombwe with drongos, orioles, cuckoo-shrikes, bulbuls, batises, weavers, and *Tamiscus* and *Funisciurus* squirrels (Prigogine 1971).

Sedentary, but a wet-season visitor in May–June to Mali.

Food. Insects including orthopterans, mantises, stick insects, bugs, beetles, ants, bees, small and large wasps, moths and caterpillars.

Breeding Habits. Nothing known other than indications of season.
LAYING DATES: Liberia (Zwedru, Wologizi), (courtship and copulations Sept–Oct; large gonads Oct–Dec and Mar; independent young Sept, Nov–Jan and Apr); Gabon, (most vocal Oct–Mar); Angola, (active gonads Apr); Zaïre (Itombwe), (various evidences, Jan, Mar–June, Sept).

References
Brosset, A. and Erard, C. (1986).
Gatter, W. (1997).
Serle, W. (1950).

Plate 22 (Opp. p. 335) *Malaconotus nigrifrons* (Reichenow). Black-fronted Bush-Shrike. Gladiateur à front noir.

Laniarius nigrifrons Reichenow, 1896. Orn. Monats., p. 95; Marangu.

Forms a superspecies with *M. multicolor*.

Range and Status. Endemic resident and partial vertical migrant, Kenya to NW Zambia and NE Transvaal.

Uganda, Mt Elgon. Kenya, Mt Elgon, Saiwa Nat. Park, Cheranganis, Elgeyu-Maraquet, Tugen Hills, Kericho, formerly Nairobi; Mau, Trans-Mara, Lolgorien, Meru, Embu, Aberdares and Taita Hills (Zimmerman *et al.* 1996). Tanzania, in NE from Marang Forest (L. Manyara) to Arusha Nat. Park and Kilimanjaro, N and S Pare Mts and Usambaras, mainly at 900–2200 m, south in apparently isolated populations to Ulugurus, Ufipa Plateau, Mbeya, Tatanda, Njombe, Matengo Highlands and Rondo Plateau.

SE Zaïre, throughout Shaba (Katanga), evidently uncommon and local, at 1250–1400 m. Zambia, frequent and widespread in N half, as mapped (R.J. Dowsett, pers. comm.), common in *Cryptosepalum* forest, in Northern and Luapula Provs east to Nyika, and Western and North Western Provs. This population presumed to extend into E Angola as shown. Malaŵi, in N, Misuku Hills and on Nyika Plateau around 2000 m; in SE, east of Rift Valley in Mangochi, Blantyre, Chiradzulu, Mulanje and Thyolo Districts. Mozambique, Mt Namuli, and occurs locally in Chimanimani Mts along Zimbabwe frontier and on Mt Gorongoza. Zimbabwe, eastern slopes of Inyanga Highlands up to 2000 m, Pungwe Falls, the Vumba, Chimanimani Mts at about 1800 m, Honde Valley at 750 m, Lusitu-

Malaconotus nigrifrons

Haroni confluence, and Mt Selinda (Chirinda Forest). Transvaal, Escarpment region, from Mariepskop, Serala, Woodbush and Duiwelskloof to Soutpansberg (Hangklip to Entabeni); a few records in Punda Milia/Pafuri region in N Kruger Nat. Park, where birds (if not misidentified – T. Harris, pers. comm.) presumably vagrant altitudinal migrants.

Density: on Nyika Plateau, Zambia/Malaŵi, pairs occupy all forest patches of 7·5 ha or more; higher up, territorial pairs occupied 6 out of 10 patches of 6·5–8·5 ha; only 1 pair in each of 2 patches of 10 and 10·5 ha, but 2 pairs in a patch of 12 ha (Dowsett-Lemaire 1983).

Description. *M. n. nigrifrons* (Reichenow): montane forests of Kenya, Tanzania and N Malaŵi. ♂♂ in 4 phases: Yellow, Red, Buff and Black/Green, ♀♀ in 3 phases: Yellow, Red and Buff. ADULT ♂: *Yellow Phase* narrow black band across forehead, continuous with black lores and black mask below eye to ear-coverts; crown to upper mantle and sides of neck grey; rest of upperparts green. Tail feathers green, T3–T6 with yellow tips c. 4 mm deep. Cheeks and underparts yellow, tinged golden on breast and greenish on belly and undertail-coverts; flanks greenish with faint darker barring. Tertials, alula and upperwing-coverts green. Underwing-coverts and axillaries bright yellow. Bill black; eyes dull crimson, maroon, red-brown, claret or bright red; legs and feet bluish grey to slate-grey. In Songea, S Tanzania, Yellow Phase ♂♂ (and ♀♀) have underparts duller yellow, washed coppery. On Ufipa Plateau, SW Tanzania, Yellow Phase ♂♂ (and ♀♀) have golden-brown breast. *Red Phase* differs only in having chin to upper breast orange-red. *Buff Phase* underparts pale cinnamon-buff, whiter on chin and throat, tinged greenish grey on flanks and undertail-coverts. *Black/Green Phase* (restricted to Usambara Mts) chin to upper breast black, continuous with black face mask and merging into green on rest of underparts. ADULT ♀: all phases quite like adult ♂♂, with black face mask, though it is duller than in ♂♂ and greyer behind eye; underparts duller than in ♂♂; eyes paler crimson than in ♂. *Yellow Phase* has underparts greenish. *Red Phase* red on chin, throat and upper breast less extensive and shade less intense than in ♂. *Buff Phase* underparts duller and browner than in ♂. SIZE (10 ♂♂, 10 ♀♀): wing, ♂ 90–94 (91·4), ♀ 86–93 (88·6); tail, ♂ 82–95 (86·9), ♀ 82–89 (84·7); bill, ♂ 20–23 (20·6), ♀ 19–22 (20·1); tarsus, ♂ 23–26 (24·6), ♀ 23–25 (23·9). WEIGHT: (Kenya, Tanzania) ♂ (n = 6) 28–39·7 (34·9), 3 ♀♀ 25, 40·2, 42·5, 3 imm. 33, 34, 37; (Malaŵi, Nyika), unsexed (n = 9) av. 32·4.

IMMATURE: juvenile (all phases) has upperparts and tail like adult; tips of secondaries, tertials, alula and upperwing-coverts with narrow whitish fringes; *Yellow Phase* underparts yellow, with close greenish barring on breast, flanks and undertail-coverts; *Buff Phase* underparts greyish white with dark grey barring, tinged greenish on flanks. Juvenile plumage appears to be replaced directly by adult plumage, without any intermediary imm. stage (cf. *M. multicolor*: D.J. Pearson, pers. comm.).

NESTLING: not described.

M. n. manningi Shelley: SE Zaïre (Katanga) and Zambia (and E Angola?). 1 phase: *Red*. Chin to upper breast typically duller red than in red *nigrifrons*, and yellow tail-feather tips larger. Some ♂♂ very like Red Phase ♀ *nigrifrons*; some ♀♀ like copper-tinged Yellow Phase ♂♂ of Songea population of *nigrifrons*. In Upemba region, Zaïre, red is often darker and extends further onto belly (such birds sometimes termed *Scarlet Phase*). WEIGHT: ♂ (n = 10) 33–40 (36·5), ♀ (n = 5) 87–95 (91).

M. n. sandgroundi (Bangs): SE Malaŵi, Mozambique, Zimbabwe and Transvaal. 2 phases: Red and Black/Green (latter restricted to Mt Namuli, Mozambique). *Red Phase* like Red *nigrifrons* but underparts duller, more orange-brown. *Black/Green Phase* like Black/Green *nigrifrons*.

TAXONOMIC NOTE: see under *M. multicolor*.

REGIONAL PHASE PREVALENCE: Zaïre (Upemba) (n = 18), Scarlet 39%, Red 61%; Zambia (n = 24), Red 100%; Malaŵi and Transvaal (n = 45), Red 100% (also 2 Buff, Nyika); Kenya (Mt Kenya) and Tanzania (Kilimanjaro) (n = 72), Red 43%, Yellow 53%, Buff 4%; E Kenya Highlands (other than Mt Kenya and Taita Hills) (n = 20), Yellow 100%; W Kenya Highlands (n = 32) Red 19%, Yellow 81%; SE Kenya (Taita Hills) and E Tanzania (Pare and Usambara Mts) (n = 70), Yellow 53%, Buff 43%, Black/Green 4%; Tanzania (Uluguru) (n = 29), Yellow 72%, Buff 28%; Tanzania (S Highlands incl. Songea) (n = 46), Yellow 47%, Buff 53% (Moreau and Southern 1958). Values from northwest of L. Tanganyika are at variance with those of Prigogine (1971) for Itombwe (n = 69): Yellow 64%, Buff 29%, Red 7%.

Field Characters. Length 18–19 cm. A *forest* species. Upperpart colours consistent – olive-green with grey crown and mantle and *black* mask from forehead through eye onto ear-coverts – but underpart colours variable, yellow, orange, red or buff. ♂ with black breast occurs in Usambaras (Tanzania) and Mt Namuli (Mozambique). Where it meets Olive Bush-Shrike *M. olivaceus* in Malaŵi and Transvaal, only orange form of Black-fronted occurs, and only grey phase of Olive. In mixed vegetation or at forest edge may meet Orange-breasted Bush-Shrike *M. sulfureopectus*, which lacks black mask and has yellow supercilium, yellow underparts with orange on breast only. Immature Black-fronted lacks black on face and is yellowish below, with dark barring in juveniles. Mellow calls very different from ringing songs of other bush-shrikes.

Voice. Tape-recorded (32, 86, 88, 91–99, 104, B, C, F, CART, KEI, LEM, SVEN). Songs and calls variable; commonest is liquid, melodius, soft, almost 'fuzzy' 2-note 'woo-hah', second note an interval of a third or a fourth higher; ♂ may call alone, or ♀ accompanies with dry scold, 'skaaaa'. In southern Africa 'woo-hah' may be repeated 8–12 times at intervals of 3–5 s. Second note may be down-slurred, 'who-heeuw', or up-slurred, 'who-hoooeee'; second note is not pure but has grating quality. Other versions of duet: 'woh-woh-woh' answered by rasping 'screeee' or clicking notes, and single 'hoh' or 'wah' followed by long 'tick-zzzzhaaaa' or clicking 'titik' or 'clitiklik'. ♂ may do both parts of 'duet', e.g. 'who-nyaaanh' (F. Dowsett-Lemaire, pers. comm.). Other calls: higher-pitched double 'wop ... wop', sometimes answered by ♀ with nasal 'gaaanh'; 3 longer 'human' whistles on same pitch; 'hoopoe call', 3 fast hollow notes 'hohoho' repeated in series at rate of 3 per 8 s; a curious single long snoring note that is both grating and ringing; and a lower-pitched nasal snore that lacks the ringing quality, 'rraaanh'. Also makes clicking sounds, 'klitik' or 'click-clack', like a well-oiled lock shutting, sometimes followed by a scold, 'klitik-waaa'; occasionally triple 'klitikik'. In territorial interactions, high-pitched 'zzreee', deeper 'gruk-gruk' and bill-snapping. Alarm, raucous, scolding 'rraa-kaaa-kaaa', or bill-snapping followed by low 'grrr' or 'zor'. Young beg with high-pitched 'seeeeep-seeeeep'.

General Habits. Inhabits canopy of evergreen forest: *Cryptosepalum* forest in Zambia, various types of lowland forest in Zimbabwe and Mozambique, montane and escarpment forests in S Africa; also thickets bordering marshes. Keeps to thick tangles of creepers and mossy branches with many epiphytes. Sometimes occurs alongside Olive Bush-Shrikes *M. olivaceus*; appears to respond to some of its calls and to be aggressive to it (Harris and Arnott 1988). When calling, bird adopts inclined-upright posture, bowing slightly with each call. 2 ♂♂ often counter-sing, using same call, probably at territorial boundary; ♂ moves about excitedly, bowing, swaying body and flicking wings and tail, making short, wing-fripping flights; ♀ may join in by duetting with discordant, tearing, drawn-out 'zzrrreeet-zzrrreeet' calls (T. Harris, pers. comm.). Flight rather heavy, with rapid, shallow wingbeats, sometimes accompanied by fripping sound. Alarm signalled vocally and by bill clicking.

Mainly sedentary. In Uluguru Mts, Tanzania, occurs mainly at 1200–2200 m, but records in Kibungo and Kimboza forests at 300–400 m in June and July point to vertical migration; also occurs seasonally down to 300 m in E Usambaras. Possibly regular dispersal of 60–100 km in Malaŵi. May be an altitudinal migrant in Zimbabwe and to Kruger Nat. Park, Transvaal.

Food. Grasshoppers and caterpillars commonest in stomachs.

Breeding Habits. Solitary nester, evidently monogamous and territorial. Courtship barely known, perhaps no different from territorial interactions (see above) except that the bowing, swaying, calling, wing- and tail-flicking and short-flight wing-fripping are between ♂ and ♀. Pair territory size given as 3 and 6–10 ha; territory maintained all year, defended by calling (T. Harris, pers. comm.).

NEST: one was a flimsy, shallow saucer or flat platform made of twining dry twigs and tendrils, lined with lichen *Usnea barbata*; ext. diam. 95, int. diam. 45, ext. depth 57, int. depth 8 (Benson and Benson 1947). 1 or 2 other nests described are similar (Harris and Arnott 1988); well concealed, in creepers or dense woody and leafy vegetation 20–30 m up in forest tree. Both sexes thought to participate in nest building (Harris and Arnott 1988).

EGGS: 2 (3 clutches). Pale greenish, heavily, evenly and uniformly smeared with elongate streaks of dark brown and light chocolate, on underlying blue-grey, mauve and French grey; the dark marks mainly at large end, forming a cap. SIZE: (n = 2) 22·4 × 17·0, 22·3 × 17·3.

LAYING DATES: Zaïre (Katanga), July–Aug and Oct; Zambia, Mar and Oct (and active gonads Nov–Jan); Malaŵi, Oct–Nov; Mozambique, (Oct)-Jan; Zimbabwe, Nov, Feb.

INCUBATION: at one nest ♀ was incubating (Benson and Benson 1947).

DEVELOPMENT AND CARE OF YOUNG: both parents thought to feed young (Harris and Arnott 1988).

References
Benson, C.W. and Benson, F.M. (1947).
Harris, T. and Arnott, G. (1988).

Plate 22 (Opp. p. 335) *Malaconotus olivaceus* (Shaw). Olive Bush-Shrike. Gladiateur olive.

Lanius olivaceus Shaw, 1809. Gen. Zool., 7, pt. 2, p. 330; Algoa Bay, South Africa; ex. Levaillant, Sclater, 1930, Syst. Av. Aethiop., p. 634.

Range and Status. Endemic resident, southern Africa from S Malaŵi to E and S South Africa. Malaŵi, west of Rift Valley from Chongoni(?) and Mt Dedza to boundary between Ntcheu and Mwanza districts, and east of Rift from Malosa (Zomba) to Mulanje District; record at Fort Lister. Zimbabwe, recent record in N (Harrison *et al.* 1997); frequent in E in Inyanga Highlands up to 2300 m (uncommon below 1300 m) and south to Chimanimani Mts, Melsetter and Chipinga Uplands; in austral winter occurs down to Honde Valley and Lusitu-Haroni confluence. Mozambique, frequent near Zimbabwe frontier and on Mt Gorongoza at 900–1200 m; also in upper Búzi R. drainage and the Vumba; another race uncommon along Sul do Save littoral: Inhaca Is, Maputo, Chimonzo, Macia, Mapinhanhe; Mt Meponduine; Lebombo Mts, along E boundary of Kruger Nat. Park (South Africa). South Africa and Swaziland, uncommon to frequent and locally common, as mapped (Harrison *et al.* 1997).

Estimated density of 1 pair per 1·5 ha in E Transvaal forests.

Description. *M. o. olivaceus* (Shaw) (includes '*taylori*' and '*rubiginosus*'): E South Africa (E Transvaal to E Cape), Swaziland and S Mozambique. ADULT ♂: 2 colour phases. (1) *Grey Phase*

Malaconotus olivaceus

('ruddy form'): forehead to upper mantle bluish grey, rest of upperparts olive-green. Tail: T1–T3 blackish with olive-green outer edges; T4 blackish with distal part of outer vane dark green and broad (c. 7 mm) yellow tip; T5–T6 green on outer vane, inner vane black at base, yellow at tip. Lores white; narrow whitish superciliary stripe extending back to side of neck; broad black patch across ear-coverts extending back to side of neck, and narrowly around upper front of eye between lore and superciliary stripe. Cheeks and chin to breast light creamy buff, breast buffy rufous; belly white, undertail-coverts greyish white; flanks pale grey or olive sometimes with c. 5 faint dark bars; thighs olive. Flight feathers blackish brown, outer vanes broadly fringed olive-green, emarginated parts of outer primaries edged whitish; primary coverts blackish with dark green outer fringe and tip; rest of upperwing-coverts, tertials and alula wholly dark green. Underwing-coverts and axillaries pale yellow, coverts along leading edge of wing grey; large yellow area on underside of flight feathers formed by broad borders of inner vanes. Bill black, sometimes with greyish base to lower mandible; eyes dark brown to chestnut-red or purple-brown; legs bluish grey. (2) *Green Phase* ('olive form'): differs from Grey Phase in having whole upperparts including crown bright olive-green; lores and superciliary stripe greenish yellow, cheeks and chin to breast light orange; belly and undertail-coverts greenish yellow, and flanks bright green. ADULT ♀: Grey Phase like Grey ♂ but ear-coverts to side of neck grey like crown: superciliary stripe absent or vestigial; breast paler; barring on flanks slightly more apparent. All tail feathers green, outer 2 pairs with narrow yellow fringe to tip and inner web. *Green Phase* like Green ♂ but ear-coverts grey, not black; supercilium inconspicuous; flanks with more extensive dark green barring; tail feathers as in Grey ♀. SIZE (10 ♂♂, 10 ♀♀): wing, ♂ 80–86 (83·2), ♀ 81–85 (83·0); tail, ♂ 76–81 (79·3), ♀ 77–83 (79·9); bill, ♂ 20–22 (21·4), ♀ 20–24 (22·1); tarsus, ♂ 25–26 (25·6), ♀ 24–26 (25·0). WEIGHT: ♂ (n = 103) 31·1–38·8 (34·5), ♀ (n = 56) 24·6–43·6 (32·7).

IMMATURE: *Grey Phase* juvenile dull olive-green above, greyer on head; whitish below, chin to upper breast barred dusky, lower breast and flanks barred dark green; tail as in adult ♀; wing as in adult but tips of secondaries, tertials and upperwing-coverts fringed buffish white. *Green Phase* juvenile similar but brighter olive-green above, and greenish yellow below with dark green barring. Primary coverts barred. Bill horn-brown.

NESTLING: naked, skin dull orange-brown; at c. 10 days eyes open, black, area around eye dark grey, skin of throat pink, mouth orange, gapes bright yellow (photo in Steyn 1996).

M. o. makawa (Benson): S Malaŵi west of Shire Valley, E Zimbabwe, N Transvaal. Like Grey Phase of nominate race, but brighter green above; bill larger.

M. o. bertrandi (Shelley): S Malaŵi east of Shire Valley. Like *makawa* but ♂ has indistinct supercilium; and greener tail, as in ♀, with blackish confined to basal part of inner feathers, and yellow to narrow fringes on T5–T6.

M. o. interfluvius Clancey: Mozambique except littoral. Like *olivaceus* but breast paler and less yellow-tinged, and flanks pale olivaceous grey, faintly barred; bill less heavy than in *makawa* and *bertrandi*.

M. o. vitorum Clancey: Mozambique coast from Save R. to Algoa Bay. Like *interfluvius* but white superciliary stripe more prominent, and mask dark grey rather than black.

TAXONOMIC NOTE: only the nominate race has phases; from Zimbabwe to Natal only Grey Phase occurs, in E Cape both phases, mainly Green, and in S Cape only Green Phase. Clancey (1975) proposed that *M. o. olivaceus* may not be simply dimorphic but that 2 formerly allopatric races may be in process of intergradation. The 2 phases have been known to pair and nest together.

Field Characters. Length 17·5–19 cm. 2 forms: green phase has upperparts *including head* olive-green in both sexes (crown and mantle grey in Orange-breasted and Black-fronted Bush-Shrikes *M. sulfureopectus* and *M. nigrifrons*). Grey phase has top of head and mantle grey, *breast* and upper belly *rufous-buff* (no yellow on underparts). ♂♂ of both forms have *black mask* from eye across ear-coverts onto side of neck, with yellow or white line above it; ♀♀ lack mask. *Lores white or yellow* in both sexes (black in Orange-breasted and Black-fronted Bush-Shrikes). Immatures like ♀♀ but barred olive-brown below. Olive and Black-fronted Bush-Shrikes occur together in mountains from S Malaŵi to N Transvaal; in these areas only grey phase of Olive occurs and only orange form of Many-coloured. Immatures like ♀♀ but barred olive-brown below. Green phase immature distinguished from immature Orange-breasted Bush-Shrike by: (i) top of head olive-green (not grey); (ii) upperparts unbarred; (iii) tail plain olive, without pale edging (yellow-green with pale edging in Orange-breasted); (iv) throat and breast gingery orange with narrow dark bars (instead of yellow-orange with broader dark bars); (v) belly darker yellow.

Voice. Tape-recorded (11, 15, 58, 75, 88, 91–99, B, C, F). Song, 5–10 ringing whistled notes in a wide variety of combinations. Typically, first 1–2 notes are higher in pitch and shorter; songs may be slow and measured, 'hi-hee-hay-kway-kway-kway-kway-kway-kway'; 'hee-hoo-hoo-hoo-hoo-hoo-hoo-hoo'; 'wik-wik-weee, weee, weee'; a more hurried 'pickwick-koi-koi-koi-koi-koi-koi', or faster, almost a trill, 'wickokokokokoko'; some are descending, 'hi-hi-tew-tew-tew-tew-tew', each 'tew' being lower than the last, some ascending, 'ki-ki-ki-ki-heeee' (very like song of Orange-breasted Bush-Shrike). Other variations include a downslurred 'tew-tew-tew-tew-tew-tew', slow and measured, and a louder, faster and more insistent 'week-week-week-week-week-week-week-week'. Seldom duets, but ♀ sometimes joins in with a few tearing scolds, 'zzrrreee-zzrrreee' or 'zizzzz'. Short-range contact or anxiety call a soft, guttural 'grrrr'; ♂♂ in territorial interactions give rasping and guttural notes accompanied by rapid bill-snapping. Fledglings solicit with high-pitched 'seee-seee'.

General Habits. Inhabits canopy and middle stratum of coastal and montane evergreen forests up to mist-belt in mist forest; also matted evergreen dune forest, forest edges, riverine woodland and thickly wooded valleys, and dense patches of thornveld; in Zimbabwe occurs in *Philippia* heath, thickets, wattle plantations and isolated forest patches, up to 2300 m in formerly open country; in Chirinda Forest in thickets on forest edge at 900–1200 m.

Solitary or in pairs. Secretive, even when singing; shy but inquisitive. Agile, moving through vegetation rapidly and silently, peering and darting at prey, often making short 'fripping' flights, and gliding quite fast between trees. Forages by gleaning for invertebrates from leaves and wood, hopping and bounding amongst twigs and small branches; mainly in canopy but often in mid levels, occasionally coming down to low tangles at forest edge; frequently joins mixed species foraging flocks. Vocal; calls all year but mainly in austral spring; ♂♂ often counter-singing with each other, using same calls; when calling, bird has upright stance with tail slightly fanned. ♂ and ♀ occasionally sing in duet (see above). In territorial interactions, ♂ threatens by adopting horizontal stance, bowing, swaying body from side to side, flicking tail, and making discordant tearing notes and piping whistles; ♂ also snaps or clicks bill, gives guttural notes, excitedly bows and sways, flicks tail and makes short 'fripping' flights (T. Harris, pers. comm.). Appears to respond to calls of Black-fronted Bush-Shrike *M. nigrifrons*.

Mainly sedentary, but a vertical migrant in E Zimbabwe where it occurs down to 750 m in Honde Valley and 350 m at Lusitu-Haroni confluence in austral winter, and in S Mozambique some birds, found near Transvaal border, thought to be migrants from escarpment forests in Transvaal.

Food. Insects: Hymenoptera including bees, hornets and ants; beetles, grasshoppers, mantises; also spiders; some fruit, including figs.

Breeding Habits. Solitary nester, evidently monogamous and territorial. For calling behaviour of ♂, see above. Single-brooded. Known to nest whilst still in barred juvenile plumage.

NEST: a flimsy, untidy, shallow saucer made of long, wiry twigs, slender pieces of grass, fine *Asparagus* stems, roots, aerial roots and tendrils of pencil-lead thickness; tendrils and rootlets become finer towards centre of nest; cup quite thickly lined with rootlets and fine materials, e.g. terminal sprays of thornless wild asparagus and reddish midribs of small pinnate leaves. The whole effect is like nest of dove and eggs are often visible from below. Built in fork or across main branch of *c.* 10 mm diam. and smaller branches or twigs on either side; from 0·6 to 6 m above ground, usually (n = 15) 3 m. Int. diam. 95, ext. diam. 130, int. depth 32, ext. depth 85–90. Well concealed amongst foliage in bush or tree, sometimes in thorny tree, sapling, acacia, creeper, *Philippia* or *Woodringtonia*. Nest built by both sexes.

EGGS: 1–2, av. (n = 11) 1·6. Pale greenish white, heavily smeared and streaked with violet-grey and light red-brown. SIZE (n = 18) 19·4–23·0 × 14·3–17·0 (21·6 × 15·9).

LAYING DATES: Malaŵi, Oct–Nov; Zimbabwe, Oct–Jan, mainly Nov; Mozambique Nov–Dec; Transvaal, Nov–Feb; Natal, Sept–Dec, E Cape, Dec–Jan.

INCUBATION: by ♂ and ♀. Nest relief signalled by ♂ with piping whistles, which either summon ♀ to nest or indicate that ♂ is approaching nest to make a change-over. Incubation period (n = 1) 18 days (T. Harris, pers. comm.).

DEVELOPMENT AND CARE OF YOUNG: ♂ seen brooding young before sunset several times and was always brooding again next morning; he left on approach of ♀ who fed young; ♀ never seen to brood (T. Harris, pers. comm.); ♀ once swallowed faecal sac; parents always approach nest indirectly and stealthily. Young fed by both parents, initially by regurgitation. Nestling period (n = 2) 16–17 days.

References
Harris, T. and Arnott, G. (1988).
Harrison, J.A. *et al.* (1997).
Maclean, G.L. (1993).
Vincent, A.W. (1949).

Plate 22 (Opp. p. 335) *Malaconotus bocagei* (Reichenow). Bocage's Bush-Shrike; Grey-green Bush-Shrike. Gladiateur à front blanc.

Laniarius bocagei Reichenow, 1894. Orn. Monatsb., 2, p. 125; Jaunde, Cameroon.

Range and Status. Endemic resident, borders of Congo Basin. Cameroon, in W, records from coast north-east to Mt Nlonako, Limbe (Victoria), Saxenhof, Ekona, Kumba, Dikume Balue, Nyassoso, Bamali; frequent in S, around Yaoundé (Quantrill and Quantrill 1998) and Bitye on Ja River. Rio Muni, known from Bebai on R. Benito. Gabon, locally common, e.g. at Makokou; absent from Rés. de la Lopé. Congo, Mayombe, frequent around Ganda Sundi, and record near Bomassa, Odzala Nat. Park. Angola, scarce in Cabinda, Cuanza Norte and Malanje; frequent around Camabatela; south to Canhoca and Ndalo Tando. Zaïre, a few records in S Kasai Oriental; Haute Zaïre, from about upper Lulu R. to Garamba Nat. Park and Virunga, south to Itombwe Highlands at 1100–1670 m (Kamituga, Mwana, Nyabisanda). Central African Republic, unsubstantiated record in Manovo-Gounda-Saint Floris Nat. Park (Carroll 1988). Uganda, locally common in S and W at 1100–2200 m, from Bugoma and Kibale to Ankole and Kigezi, also Entebbe to Jinja and about Bugiri. Kenya, uncommon in Nandi Forest and Kakamega Forest; formerly at Kericho and Nyarondo. Formerly in Rwanda, now extinct (Dowsett 1993).

Description. *M. b. bocagei* (Reichenow) (includes '*ansorgei*'): Cameroon to Angola. ADULT ♂: forehead, lores and sides of crown white, forming long, narrow superciliary stripe; rest of crown jet black; nape, hindneck and sides of neck jet black; mantle, back, rump and uppertail-coverts dark slaty grey. Tail black, graduated or strongly rounded, T3–T6 with small white tips. Small patch just in front of eye, upper cheeks, ear-coverts, and area behind ear-coverts adjoining sides of neck, jet black; chin and throat white, breast to flanks and undertail-coverts white with ochre-buff tinge, usually most distinct on breast and flanks; breast can be buff. Scapulars and upperwing-coverts dark slaty grey, primaries, secondaries and tertials black with grey outer borders. Underwing-coverts and axillaries white; underside of tail blackish. Bill black; eyes dark red-brown; legs and feet slate-grey. ADULT ♀: like ♂, but tail greyish, not quite so black; eyes dark brown (not red-brown). SIZE (7 ♂♂, 8 ♀♀): wing, ♂ 80–83, ♀ 76–82; tail, ♂ 65–71, ♀ 63–70; bill to feathers, ♂ 14–15, ♀ 13–14.

IMMATURE: juvenile with upperparts grey, finely barred dull blackish, pale superciliary stripe and lores, blackish mask, tail

Malaconotus bocagei

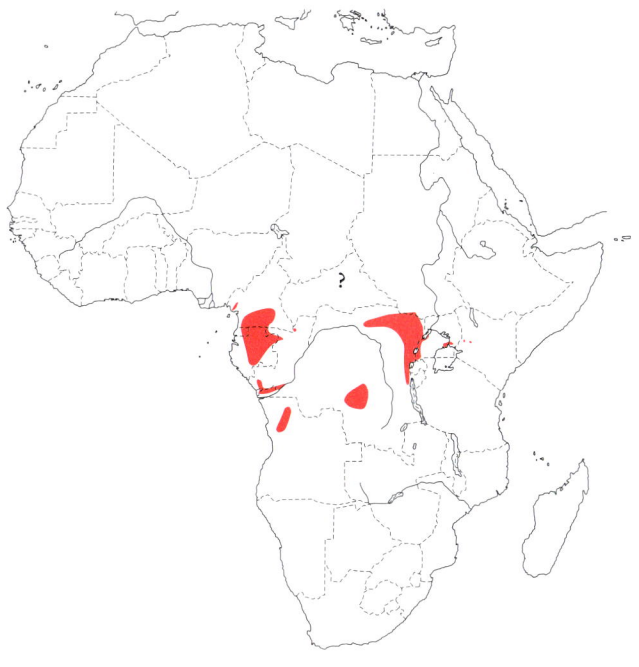

grey, wings grey, wing-coverts broadly tipped with pale yellow, each with subterminal black bar, secondaries tipped with buff, underparts dirty white, breast to undertail-coverts dirty white or yellowish white finely vermiculated with faint dark grey bars; mouth orange, tongue blackish at base (Bates 1911). Subadult differs from adult in less boldly pied appearance; crown, nape, ear-coverts and sides of neck dull blackish; tail dull brownish black, underparts greyish white with less buff suffusion than in adult, upperwing-coverts and secondaries with small white tips, and flight feathers edged olive.

M. b. jacksoni (Sharpe) (includes 'andaryae', a single juvenile bird from Entebbe, possibly a hybrid with *M. sulfureopectus*): Zaïre (Uele, Kasai), Uganda, W Kenya. ♂ and ♀ exactly like ♂ *bocagei* (i.e. tail of ♀ black, not dark grey). WEIGHT: ♂ (n = 3) 27–28 (27·3), ♀ (n = 6) 22–28 (26·5), 1 unsexed 29.

Field Characters. Length 16–16·5 cm. A small, pied bush-shrike of forest canopy; blackish above, white below with faint buff wash. Looks rather like a puffback *Dryoscopus*; distinguished by prominent *white* lores and eyebrow over *black mask*, lack of white in wings and *ringing call*. Immature duller than adult, with white spots in wings.

Voice. Tape-recorded (104, B, C, BEL, ERA, KEI, LEM, ZIM). Song of ♂ very variable, e.g. an up-slurred '(tip)-tooi-*tweeee*', sharp, clear and ringing; a more clipped 'tewp-tewp-tewp'; a 2-note 'po-*hee*' in series of 4, 'po-*hee*, po-*hee* ... po-*hee*, po-*hee*'; a faster 6–7 note 'tyo-tyo-tyo-tyo-tyo-tyo-tyop', clipped, not ringing. One version, 'kuli-kuli-kuli-kuli-kuli-kuli-kuli', surprisingly like song of African Yellow Warbler *Chloropeta natalensis* (F. Dowsett-Lemaire, pers. comm.). ♂ and ♀ perform synchronized duet, also variable, e.g. ♂: loud 'doi-doi-doi-doi' (usually 4 notes, sometimes 6–7), clear but not ringing, quality reminiscent of Gorgeous Bush-Shrike *Telophorus viridis*; immediately following song ♀ replies with 2-note scold, 'zerr-wick'; ♂: liquid trill, ♀: nasal 'nyaaaa'; ♂: long, down-slurred '*peee-yoh*', ♀: 'wizzzz'; trio also recorded (C. Chappuis, in press): ♂: liquid 'tiplee, tiplee', bird 2, 'nyaaaa', bird 3, 'wizzzz'.

General Habits. In Gabon inhabits heavily modified, tall secondary forests with many strata and dominant swathes of mimosaceous plants; quite open wooded country with large-crowned trees like *Albizia* and *Pentaclethra* scattered irregularly amongst bushes and shrubs; park-like gardens around large houses; cocoa and coffee plantations with closed canopy and plenty of Mimosaceae; and thickets with parasol trees along the borders of cultivation (Brosset and Erard 1986). In Angola occurs in gallery forest in *Brachystegia* woodland, and coffee forest (at Salazar); in Mayombe, Congo, has been seen in flamboyant trees. In Uele, Zaïre, in gallery forests, and in Ituri old second growth near borders of clearings. In E Africa keeps mainly to forest canopy; sometimes in sub-canopy and forest-edge saplings.

Lives in pairs in home range or territory of 2–3 ha (sometimes less), defined and advertised by ♂ singing and pair duetting. Forages by gleaning for insects in foliage, at 8–20 m above ground, and sometimes in the tops of very tall trees, moving through thick foliage slowly and deliberately; skulks in trees very much in the manner of Red-eyed Puffback *Dryoscopus senegalensis*. Exploits hatches of flying termites.

Sedentary.

Food. Seen taking berries, beetles, orthopterans and caterpillars; mainly insects in stomachs, often including caterpillars; only caterpillars in 6 out of 7 stomachs (Chapin 1954).

Breeding Habits.

NEST: shallow open cup, one made entirely of dry, interwoven vine tendrils (Bannerman 1939); another exactly the same, 'composed exclusively of strong spiral tendrils of some vine, which intertwined so as to give great strength' (Chapin 1954).

EGGS: 2. Greenish white or light green, thickly spotted, streaked, smeared and blotched all over with dull dark brown or olive-brown and grey. Size: (n = 2) 21·0 × 15·5, 21·9 × 16·0

LAYING DATES: Cameroon, Bitye, Sept; Gabon, (sings all year, but particularly in Nov–Feb); Zaïre, Avakubi, Aug; Uganda, Feb, July, Aug.

References
Brosset, A. and Erard, C. (1986).
Chapin, J.P. (1954).

Plate 22
(Opp. p. 335)

Malaconotus sulfureopectus Lesson. Orange-breasted Bush-Shrike. Gladiateur soufré.

Lanius (Tchagra) sulfureopectus Lesson, 1931. Traité d'Orn., p. 373; designated type locality Senegal, Neumann, 1899, J. Orn., 47, p. 395.

Malaconotus sulfureopectus

Range and Status. Endemic resident, possibly a partial migrant in W Africa, frequent in savanna woodlands throughout subsaharan Africa except for the Horn, Congo Basin and the SW.

Senegal, uncommon, north to Dakar, Mbour and Bandia in W but barely north to 14°N in E. Gambia, frequent, commonest near coast. Mali, north to Mopti. Burkina Faso, north in E to Faga valley. Niger, recorded only in 'W' Nat. Park (Nov). Nigeria, north to Sokoto, Katsina, Maiduguri, Gwoza and about 11°45′N on Komadugu Gana. Southern limits in W Africa are at derived savanna/rain forest interface, i.e. through middle of Sierra Leone and Ivory Coast; absent from Liberia and SW Ghana; reaches coast from SE Ghana to extreme SW Nigeria. Said to be in Gabon (White 1962), though no record traced; not in Congo.

Chad, record from Ndjamena, otherwise occurs only south of 09°N. Central African Republic, north to Bamingui-Bangoran Nat. Park, where uncommon to locally frequent, and Vakaga, where uncommon. Sudan, fairly common in better woodlands in SW and SE; records north to Jebel Marra and En Nahud area. Ethiopia, frequent to common, mainly below 1800 m, in Rift Valley, SW and SE; records north to L. Tana area.

Zaïre, frequent in Uele, evidently scarce in Kasai (record from Mérode), frequent in Manyema, Katanga and Marungu. Throughout W and S Uganda, but appears to be absent from NE. Kenya, widespread in S, absent from arid regions in N and E except coast and some localities with dense riverine or hill-foot acacias. Somalia, restricted to extreme S (Boni Forest), and Jubba-Webi Shabeelle triangle, and Shabeelle valley north of Muodisho (Mogadishu). Occurs in N Rwanda and S Burundi (Bururi Forest), but status in both countries and in adjacent parts of Zaïre unclear. Tanzania, common in E, sparse in W, not recently seen in centre, as mapped (N. Baker, pers. comm.). Angola, throughout interior except in highlands; along coast from Cabinda and mouth of R. Congo to about Benguela; frequent in Cunene Valley. Zambia, almost throughout, as mapped (R.J. Dowsett, pers. comm.). Namibia, Botswana, Zimbabwe and South Africa as mapped (Harrison *et al.* 1997).

Description. *M. s. sulfureopectus* Lesson: Senegal to Gabon, N Zaïre and W Uganda. ADULT ♂: forehead and forecrown bright yellow, continuous with yellow superciliary stripe extending back to above middle of ear-coverts. Hindcrown to mantle and anterior scapulars light bluish grey; rear scapulars and back to uppertail-coverts green. Tail feathers green; T2–T6 with yellow tips, *c.* 3 mm wide on T2 increasing to *c.* 8 mm on T6 and extending on T3–T6 as fringe along inner vane. Lores black; ear-coverts black, grading to dark grey on sides of neck. Cheeks and chin to throat bright yellow; upper breast orange; rest of underparts bright yellow. Flight feathers dark brown, primaries with green base to outer vane and yellow margin near tip, secondaries with green outer vanes; tertials green; tips of secondaries and tertials with narrow yellow fringe; alula and upperwing-coverts green, greater coverts tipped yellow. Underwing-coverts and axillaries yellow; underside of flight feathers extensively yellow, due to broad yellow borders of inner vanes. Bill black; eyes dark brown; mouth black; legs dark grey or blue-grey. Sexes alike, but orange on breast of ♀ less intense. SIZE (10 ♂♂, 10 ♀♀): wing, ♂ 87–95 (92·2), ♀ 87–97 (90·4); tail, ♂ 85–95 (90·9), ♀ 86–97 (90·8); bill, ♂ 20–21 (20·2), ♀ 19–21 (19·8); tarsus, ♂ 24–27 (25·1), ♀ 24–26 (25·0). WEIGHT: Ghana, ♂ (n = 3) 30–34 (31·7), ♀ (n = 2) 28, 30, unsexed (n = 9) 26–32 (29·9).

IMMATURE: juvenile lacks yellow, blue-grey and black on head and orange on breast; crown, hindneck, sides of neck, mantle and back barred grey and whitish; superciliary stripe pale buff; all upperparts behind head with small whitish or pale yellow spots; uppertail-coverts barred with pale yellow; underparts pale yellowish buff, breast, upper belly, flanks and thighs barred with whitish and dark greyish, barring soon becoming only dusky; wings and tail like adult, but upperwing-coverts tipped yellowish white. Eyes light brown. Immature like adult but forehead olive not yellow, superciliary stripe whitish, lores and ear-coverts pale grey, upperwing-coverts spotted with yellowish, underparts pale yellow; upper breast and rump with vestigial dusky bars. Eyes dark brown.

NESTLING: naked, blind, bill slate-grey at tip, with yellowish cutting edges, yellow base to mandible, and bright yellow gape; mouth spots. At about 1 week, when eyes starting to open, skin of top and sides of head grey, skin of chin, throat and neck pink (photo in Ginn *et al.* 1989).

M. s. similis Smith (includes '*terminus*' which is slightly smaller, wing av. 89): S Sudan, Ethiopia, E Zaïre, E Uganda and Kenya to southern Africa. Ear-coverts grey, blackish only near eye. WEIGHT: Uganda, ♂ (n = 8) 20–32 (25·25), ♀ (n = 10) 22–27 (24·0), unsexed (n = 5) 24–28 (25·6); Kenya, unsexed (n = 4) 23–33 (27·0); southern Africa, ♂ (n = 31) 24·0–30·5 (27·6), ♀ (n = 18) 19·5–28·4 (26·2).

Field Characters. Length 16·5–17·5 cm. A bird of acacia savanna, often in canopy, where it can be hard to spot; usually first located by its ringing song. Identified by *orange breast-band* on yellow underparts and *yellow superciliary stripe* contrasting with black lores and dark grey ear-coverts. Distinguished from green phase of Olive Bush-Shrike *M. olivaceus* by *grey* crown and mantle; ♂ Olive Bush-Shrike has yellow lores and yellow line over black mask from eye down side of neck. Resembles orange-breasted form of Many-coloured Bush-Shrike *M. multicolor* of evergreen forest, which has grey crown and black mask but no yellow eyeline; the 2 are also ecologically separated. Somewhat like small version of Grey-headed Bush-Shrike *M. blanchoti*, which has white lores and grey crown and face with no yellow. Immature lacks supercilium and resembles immature Many-coloured Bush-Shrike but has paler grey head and paler yellow underparts. Can also be confused with immature green phase Olive Bush-Shrike; for differences see that species.

Voice. Tape-recorded (15, 47, 75, 86, 88, 91–99, 104, B, C, F, KEI, LEM). Song ringing and ventriloquial. Variable; 2 main forms are: (1) 3–4 short notes and 1 long one, 'wi-hi-hi-hi-wheeee', or first note lower, 'woo-hi-hi-heee', sometimes descending, 'hi-hi-hu-hu-yooo; (2) faster, usually 4–9 notes (but up to 15) all on one pitch, given at rate of 5 per s, 'wi-wi-wi-wi-wi-wi-wi ...'. Speed varies; faster versions higher in pitch. Antiphonal duet recorded, ♀ giving soft, husky, discordant 'zrrrt', 'skizzz' or 'tzzrrik-tzzrrik' immediately after ♂ song, 'wo-hi-hi-hi-heee-zrrrrt'. Different dialects in W Africa: 4-note 'put-wit-too-weee' and 2-note 'hoop ... heeee', each repeated in series (Senegal); clear, whistled 3-note 'hooey-hooey-hooey', also repeated in series (Ivory Coast) (see Chappuis, in press). Also gives a snoring scold, 'chizzzzip' and clicking sounds. Alarm/anxiety calls, a soft, low 'chrrr' and 'tu-tu-tu-tu-tu' (T. Harris, pers. comm.).

General Habits. Inhabits wide range of types of woody vegetation, mainly rather low and with dense foliage: canopy of acacia woods especially along streams and rivers; *Erythrina* bushland, scrub, gallery forest and mixed riparian groves, forest edge, bushveld, clumps of bushes and shrubby pasture, thornveld, and thickets growing on termitaria; *Brachystegia* woodland where not occupied by Grey-headed Bush-Shrike; canopy and mid stratum of *Cryptosepalum* forest (Zambia); interior of lowland forests (Mozambique, E Zimbabwe); coastal savanna bush (Ghana, Kenya), and woods around foot of inselbergs (N tropics). In E Africa mainly below 1500 m, but sparsely up to 3000 m.

Occurs singly and in pairs. Rather skulking, but readily flies short distances between bushes and sometimes sings half-exposed to view in bush top. Flies low down between trees or over tops of bushes; wing beats rather shallow, often accompanied by fripping sound, flight ending in glide and a swerve as bird enters foliage. Forages by hopping strongly and rapidly amongst branches and leafy twigs, making short flights within tree, pausing and peering about, and darting quickly at an insect; works its way up tree to top, then glides down to base of canopy of nearby tree. A bird, persistently eating honeybees at flowering shrub, sat very still, snatched at each one as it passed, and in every case flew into adjacent tree with bee in its bill and beat it against branch (Steyn 1967). Often joins mixed-species foraging flocks. Sings from within upper canopy of prominent tree or from near top of thick bush; when singing, stretches and inclines body and points bill up. Responds well to person 'pishing'. Threatens rival by adopting sleek horizontal posture, giving drawn-out call followed by bill-clicking and scratchy notes; in more intense threat, also makes short, jerky, wing-fripping flights (T. Harris, pers. comm.). Scratches head indirectly.

Generally sedentary (Harrison *et al.* 1997), but thought to be influx of birds on Ghana coast (Legon) in dry season (Grimes 1987), and in Nigeria the most northerly records tend to be in wet season, May–July (Elgood *et al.* 1973); seasonal changes in abundance at Bamako, Mali; around Ouagadougou, Burkina Faso, appears only between late Mar and mid Sept (Thonnérieux *et al.* 1989); widespread resident in Togo but may also be partial migrant there (Cheke and Walsh 1996).

Food. Small insects: bees, wasps and other hymenopterans, caterpillars, mantises, beetles and grubs. Of 4 stomachs (Zaïre), caterpillars in 2 and other insects in 2; of 14 stomachs (Uganda), caterpillars in 10, termites in 1, beetles in 2 and other insects in 4.

Breeding Habits. Monogamous, solitary nester; territorial. Probably generally single-brooded, though 2nd broods known, and once a 2nd nest built for the purpose (T. Harris, pers. comm.).

NEST: flimsy platform made of grass, petioles and fine dry twigs, sometimes with some cobweb, with depression in centre lined with fine rootlets and creeper tendrils. Structure is openly-worked and dove-like; eggs can be seen from below. Sited usually near site of previous season's nest, in multiple fork in lateral branches of thorn tree or amongst thick foliage or creeper tangles, although nest sometimes built before tree is in leaf. Generally <4 m above ground; can be up to 10 m; 66 nests were from 1·8 to 9 m above ground. Built by ♂ and ♀.

EGGS: 1–3 (n = 33) av. 1·8. Dull (not glossy) light or greenish white, heavily streaked, blotched and spotted with various shades of brown and grey, mainly around wide end. SIZE: (n = 51) 19·5–24·0 × 14·0–18·0 (22·0 × 16·1).

LAYING DATES: Gambia, (juv. in Jan); Mali, breeds Mar–June; Ghana, May, July (and juvs in July–Aug); Nigeria, May; Sudan, Aug–Sept; Ethiopia, Apr, May; E Africa, Region A, June, Region B, May; Zaïre (Lubumbashi), mid Sept to mid Oct; Angola, (enlarged gonads Nov, Jan); Zambia, Oct–Dec, Feb; Malaŵi, Aug, Oct–Feb; Botswana, Oct, Dec–Jan; Zimbabwe, Sept–Apr, mainly Oct–Dec, records June and July; South Africa, Transvaal, Sept–Dec, mainly Oct–Nov, records Feb–Mar, Natal, Sept–Jan, mainly Oct–Nov, E Cape, Oct–Dec.

INCUBATION: by ♂ and ♀. ♂ calls on nest, probably to initiate change-overs. Period: 13–14 days.

DEVELOPMENT AND CARE OF YOUNG: young fed by ♂ and ♀; they give repeated, high-pitched 'seeeep-seeeep' begging call with ventriloquial quality (T. Harris, pers. comm.). Period: (n = 2) 12, 14 days.

BREEDING SUCCESS/SURVIVAL: adults retrapped after 5, 6 and 6·5 years (Hanmer 1983). Nestling once taken by boomslang *Dispholidus* sp., and an adult taken by Wahlberg's Eagle *Aquila wahlbergi* (Harris and Arnott 1988).

References
Clancey, P.A. (1959).
Harris, T. and Arnott, G. (1988).

Genus *Telophorus* Swainson

A genus of 4 small, semi-terrestrial bush-shrikes in eastern and southern Africa forming a parapatric series, characterized by well-demarcated black gorget and contrasting throat and breast in brilliant carmine or yellow. The geographically central species, *T. dohertyi*, is montane, and separates 2 discrete populations of the lowland *T. viridis* to the west and east. They compose a superspecies – small-billed green birds with red throat and without light corners to the tail. Another species, the well-known Bokmakierie *T. zeylonus*, inhabits open and arid bushlands to the south; it is larger, green above, yellow and black below, with white in the tail. The distinctive Rosy-patched Shrike *T. cruentus* occupies arid bushlands to the north, and is similarly large and with white in the tail, but brown above and with red and black below.

T. cruentus was placed in *Tchagra* by Mayr and Greenway (1960), White (1962) and others, and in its own genus *Rhodophoneus* by Hall and Moreau (1970), Short *et al.* (1990), Dowsett and Forbes-Watson (1993) and others, usually sequenced between *Tchagra* and *Laniarius*. It resembles *Tchagra* in respect of its white-cornered, graduated tail, but little else. 'The suggestion – which we cannot accept – that it is closest to *Malaconotus (Telophorus) zeylonus* is credited by Sibley and Monroe (1990) to A.C. Kemp (pers. comm.), but it perhaps originated with Clancey (1986), where the use of the name *Telophorus cruentus* for the Rosy-patched Shrike was clearly a slip of the pen for *R. cruentus* (through confusion with the similarly-named *Malaconotus cruentus*)' (Dowsett and Dowsett-Lemaire 1993). In fact Clancey *was* discussing the Rosy-patched Shrike and, quite independently, we have arrived at the same conclusion as Clancey's. Repression of yellow-green colours, whether pigmentary or non-iridescent structural ones, is commonplace in birds of sub-desert steppe and arid, microphyllous savanna woodlands (Baker and Parker 1979). Make the white parts of the Rosy-patched Shrike yellow and the brown parts green, and not only is its placement with the other black-gorgetted *Telophorus* bush-shrikes vindicated but a striking resemblance to red-patched *T. zeylonus* emerges, even down to tail shape and pattern. Habits and voices of *T. zeylonus* and *T. cruentus* are also similar.

Telephorus is a small bush-shrike radiation, and links the arboreal *Malaconotus* and the semi-terrestrial *Tchagra*.

Endemic; 4 species, 2 independent and 2 in a superspecies.

Telophorus viridis superspecies

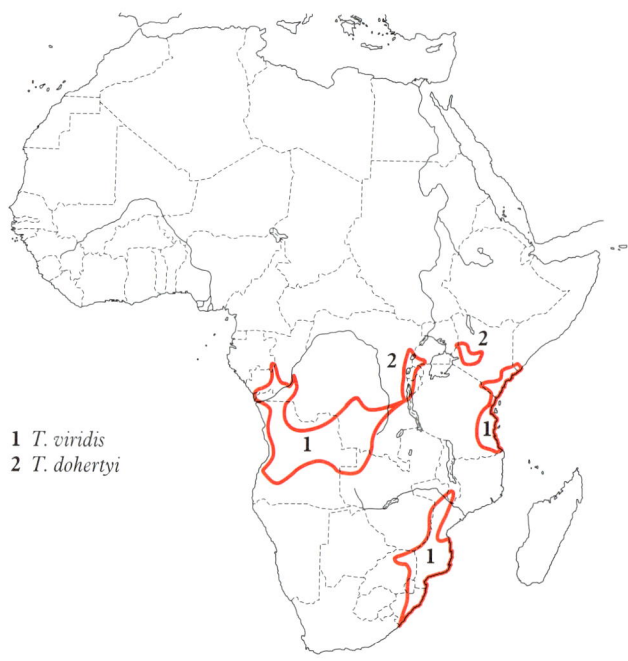

1 *T. viridis*
2 *T. dohertyi*

Telophorus viridis (Vieillot). Gorgeous Bush-Shrike. Gladiateur vert.

Plate 23

Laniarius viridis Vieillot, 1817. Nouv. Dict. Hist. Nat., nouv. Ed., 13, p. 300; Malimbe, Congo area (= Loango Coast, Portuguese Congo).

(Opp. p. 350)

Forms a superspecies with *T. dohertyi*.

Range and Status. Endemic resident. (a) *viridis* group: SE Gabon (sight record and tape recording from near Leconi, 01°30′S, 14°25′E (Macaulay and Sinclair 1999); extreme S Congo (Pointe-Indienne, Diosso, Diosso-Mpinde road); Angola (Cabinda and from Cuanza Norte and Malanje south along escarpment and up to 2000 m in montane forest patches, to N Huila and mouth of Cubal R. in Benguela; occasional in dry forest at Kisama Nat. Park; also NE Moxico along Zambezi R); W Zaïre east to Kwamouth, and SE in Kasai and Katanga (Lusambo, Kabinda, Kinda), probably continuous with populations in E Angola and Zambia; single occurrence at Baraka, L. Tanganyika (Schouteden 1969; specimen correctly identified – M. Louette, pers. comm); although not in nearby Itombwe. NW Zambia in Mwinilunga and N Kabompo (very common). (b) *quadricolor* group: extreme S Somalia; local and uncommon in coastal Kenya south from Boni Forest (pers. obs.) and Lamu, inland along lower Tana R. and on Mt Endau at 1200 m and Mt Kasigau, and Shimba Hills Nat. Park; mainly below 600 m; formerly Chyulu, Taita and Sagala Hills and Tsavo East Nat. Park (Voi R.), but no recent records (Zimmerman *et al*. 1996, Brooks *et al*. 1998). Coastal Tanzania south to Mikindani, inland to North Pare Mts at 1100–2000 m, foothills of Uluguru Mts, Liwale and Rondo Plateau; formerly E Usambara foothills (Zimmerman *et al*. 1996, and not listed by Evans 1997). S Malaŵi in lower Shire valley north to Lengwe, wherever dense thicket still remains, also at Chididi at 550 m. E highlands of Zimbabwe up to 1500 m along Mozambique border from foothills of the Vumba to Chimanimani Mts, Mt Selinda and along Buzi R. in Chipinga Uplands, reappearing at Chikwarakwara on Limpopo R; local and nowhere common. Mozambique on Manica Platform and highlands along Zimbabwe border, from Nyakafura in Tete District south to Espungabera and Buzi R.; apparently absent from littoral north of Save R. (Clancey 1971); widespread south of Save R., especially along coast, also in W on borders of Kruger Nat. Park and in Lebombo Mts. In Transvaal restricted to base of Escarpment, the Soutpansberg–Blouberg and N lowveld (Punda Milia–Pafuri), locally common; also Machadodorp and Steelport. Swaziland; locally common in coastal lowlands of Natal, penetrating up thickly-wooded valleys to *c*. 600 m; (extreme E Cape?).

Description. *T. v. quadricolor* Cassin: E Cape to Swaziland and N Transvaal. ADULT ♂: upperparts entirely green. Central tail feathers greenish, grading to blackish distally; rest of tail feathers black, T2–T5 usually with greenish base to outer web. Broad yellowish superciliary stripe, sometimes tinged orange or red, extending to rear of eye, and sometimes narrowly across forehead above base of bill; ear-coverts and sides of neck green. Black band through lores, below eye, and narrowly along sides of throat to join broad black gorget across lower throat and upper breast; cheeks to chin and upper throat orange-red. Below gorget, a

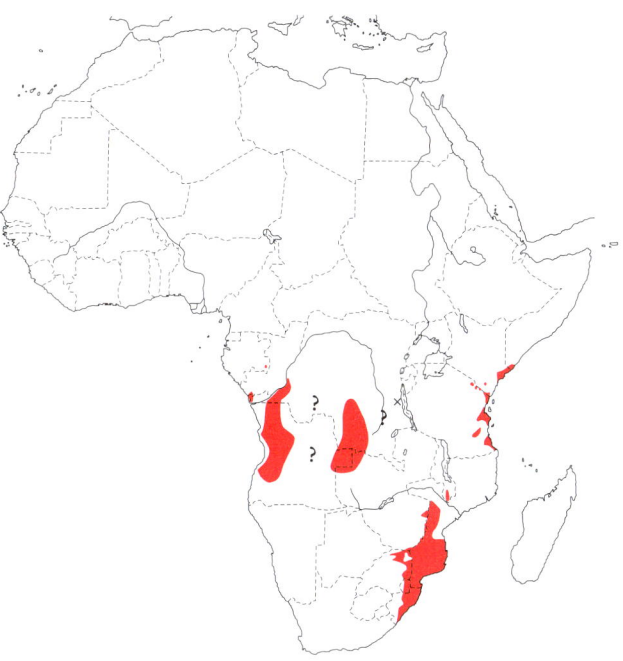

Telophorus viridis

narrow orange-red band across centre of upper breast; lower breast and belly greenish yellow; sides of breast, flanks and thighs green; undertail-coverts yellow, washed variably with orange-red. Primaries and secondaries dark brown with green outer webs, except on outer primaries where emarginated parts edged pale brownish; inner web borders pale grey. Tertials, alula and upperwing-coverts green. Underwing-coverts and axillaries green. Bill black; eyes dark brown; legs bluish grey to slaty brown. ADULT ♀: differs from adult ♂ in having almost no black on lores or around sides of throat, and narrower, broken black breast band; chin and throat more orange, with yellow feather bases showing through as mottling or flecking, chin sometimes wholly yellow; tail feathers all greenish. SIZE (10 ♂♂, 6 ♀♀): wing, ♂ 78–84 (81·3), ♀ 77–83 (77·8); tail, ♂ 81–87 (84·5), ♀ 83–86 (84·2); bill, ♂ 20–23 (21·7), ♀ 20–22 (20·8); tarsus, ♂ 26–28 (27·3), ♀ 26–27 (26·8). WEIGHT: (combined *quadricolor* and *quartus*) ♂ (n = 19) 29·6–40·5 (38), ♀ (n = 7) 31–39·5 (35·7).

IMMATURE: duller green above, yellowish supercilium poorly defined; tail feathers greenish with pale brown inner webs; chin to throat and belly to undertail-coverts greenish yellow, breast and flanks greener, the former with fine barring; wing like adult, but secondaries, tertials, alula, and primary and greater coverts with narrow fringe and blackish subterminal line. Bill dark brown above, flesh below.

NESTLING: hatches blind and naked, with mouth spots.

T. v. quartus Clancey: Mozambique, Zimbabwe and Malaŵi. Upperparts brighter green, superciliary stripe brighter yellow, yellow of underparts tinged brownish. WEIGHT: Zimbabwe, ♂ (n = 5) 29·6–40·5 (36·5), ♀ (n = 2) 37·5, 37·8.

T. v. nigricauda Clarke: Kenya and Tanzania. In ♂, central tail feathers blacker, less green; reddish on centre of breast often more extensive. WEIGHT: (unsexed, Kenya, n = 5) 36–44 (38·4).

T. v. viridis (Vieillot): Angola, Zaïre, Zambia. Upperparts darker green, forehead saffron yellow; underparts below gorget bright green, with chestnut-brown mid-ventral stripe linked to deep chestnut undertail-coverts. Tertials, upper- and underwing-coverts and axillaries dark green. ♀ has centre of belly brown, tail black tinged green at base. SIZE (10 ♂♂, 6 ♀♀): wing, ♂ 82–90 (85·3), ♀ 82–88 (84·7); tail, ♂ 79–86 (83·9), ♀ 80–87 (83·8); bill, ♂ 22–25 (22·8), ♀ 21–23 (22·0); tarsus, ♂ 28–31 (29·3), ♀ 28–29 (28·3).

TAXONOMIC NOTE: *viridis* and *quadricolor*, at least 800 km apart, were treated as species composing a superspecies by Hall and Moreau (1970), Britton (1980), Clancey (1980), Sibley and Monroe (1990) and Maclean (1993). The principal difference is that underparts below the black gorget are green and red-brown in *viridis*, yellow and red in *quadricolor*. However, their voices, though varied, are very similar (Benson and Irwin 1965, sonagrams in Dowsett and Dowsett-Lemaire 1980). We follow Dowsett and Dowsett-Lemaire (1980), Short *et al.* (1990) and Dowsett and Forbes–Watson (1993) in treating them as conspecific.

Field Characters. Length 18 cm. Readily distinguished by unique *colour combination* from all birds except Doherty's Bush-Shrike, a montane bird which it does not overlap except possibly in S Kivu, Zaïre; there, race *viridis* can be told from Doherty's by green underparts below black gorget (yellow in Doherty's), superciliary stripe (lacking in Doherty's) and narrow yellow rather than broad red forehead. Immature Sulphur-breasted, Many-coloured and Black-fronted Bush-Shrikes *Malaconotus sulfureopectus*, *M. multicolor* and *M. nigrifrons* are also yellow below but have grey head and upper back. Shy and seldom seen but readily identified by distinctive loud 3- or 4-note call from undergrowth, quite unlike that of any other bush-shrike.

Voice. Tape-recorded: *viridis* (32, 86, 104, B, F, BUL, LEM, MAC); *quadricolor* (7, 33, 50, 58, 75, 88, 91–99, B, C, F, KEI). Song loud, liquid and far-carrying, very uniform throughout extensive range, though with both individual and geographic variation: 1–3 short 'ko' or 'kok' followed by 1–2 upslurred 'kowick' or 'kowee'; compare Keith's recordings of nominate *viridis* from Zambia (32), either 'kok-kok-kokoick', 'kok-kok-kokoick-kokoick' or 'kok-kokoick-kokoick' with numerous recordings of the *quadricolor* group, e.g. Gibbon (91), who gives 3 versions, a slow 'ko-ko-whoeee', a more emphatic 'ko-koweet-koweet' and a faster 'ko-kowee-kowee'. Bird at Mtunzini, Natal, (KEI) had last notes rather less strident, 'ko-koway-koway'. Birds in Sokoke Forest, Kenya (KEI) sing a little faster, and start with 'wick', e.g. 'wick-kowick-kowick'. The 'wick' component is the loudest, and at a distance is all one hears. Song lasts *c*. 1 s, and intervals between songs vary from 2 to 5 s; bird in Kenya sang for several min on end, averaging 20 songs per min; other birds answered it in the distance. Song sometimes preceded by 'series of whispered notes, audible only when very close, followed by a small puffing sound ending in a snap' (da Rosa Pinto and Lamm 1953), also described as a faint 'klink', and followed by a 'brief rattling clicking sound' (Maclean 1993). Duetting poorly developed, but ♀ sometimes gives tearing 'zzrrreer' in response to ♂'s whistle. Contact call, low, guttural 'grr-grr ...'; alarm louder, a croaking or rasping call, somewhat crow-like or frog-like.

General Habits. In Angola inhabits isolated patches of high montane forest, coffee forest (common) and occasionally dry forest; avoids closed rain forest (Heinrich 1958). In Zambia confined to *Cryptosepalum* forest (Benson and Irwin 1965). Shrubbery of gallery forests and clearings in more wooded areas (Zaïre). Undergrowth of forest and woodland, thickets (Kenya); in Sokoke, *Afzelia*, *Brachystegia* and *Cynometra* forest. Zimbabwe, semi-evergreen thickets on edges of mid-altitude or lowland evergreen forest, and in riparian fringing or in groundwater forest; tall riparian *Ficus sycamorus* forest on Limpopo R. Dense thickets, liana tangles and 'jungle' on edge of forest and riverbanks (Mozambique). Dense tangled bush or secondary forest with thorny creepers and brambles, often along drainage lines or on hillsides (Transvaal). Dune forest, thickets in quite dry woodland, even tangles in degraded areas (Natal). Usually keeps to thick growth in understorey; sometimes well up in trees but always concealed among leaves.

Usually in pairs; ♀ keeps to thicker cover than ♂. Shy, skulking in dense undergrowth, but will respond readily to playback of song. Largely terrestrial. Forages in low bushes, gleaning trunks, branches and leaves, or hopping along the ground, turning over leaves and forest litter. When foraging in bush or creeper tends to work its way upwards, peering about and darting at prey, occasionally hawking insects in the air. Agile, moving about silently and swiftly in crouched horizontal posture. Flight heavy, with short, rapid wing-beats; seldom flies far. When excited makes 'fripping' sounds with its wings (Harris and Arnott 1988). Said to attack small birds but not seen to do so by da Rosa Pinto and Lamm (1953). Sings from canopy, in upright posture with head held up; as it sings, red throat is expanded, parting feathers to show their yellow bases (Harris and Arnott 1988). Sings all year, with peak at start of breeding season.

Not known to be a migrant, but immature once caught away from normal range at Ngulia Lodge, Kenya. In drier habitats in southern Africa it appears to undergo local movements in winter (Harris and Arnott 1988).

Food. Insects and their larvae, including sphinx moths, caterpillars and beetles; spiders.

Breeding Habits. Monogamous, territorial. Territory size usually <2 ha. One individual known to have remained in same territory for at least 6·5 years. ♂ advertises and defends territory by repeated singing; ♂♂ often countersing, in unison or antiphonally, sometimes perching <1 m apart; singer bows excitedly, jerks body from side to side and flicks wings and tail. Threat signalled by faster rate of calling, usually followed by rapid bill snapping, during which ♂ adopts more horizontal posture, with neck stretched out (Harris and Arnott 1988). ♀ accompanies these interactions with quiet low-pitched 'zzrrr' and guttural 'grr-grr ...'. In presumed courtship display, ♀ bowed in front of ♂ as he sang and snapped bill (Harris and Arnott 1988). Single-brooded.

NEST: rather skimpy saucer-like structure of twigs, roots, stalks and grasses, lined with dry leaf stalks or creeper tendrils, 'so loosely put together you can see through it'

(Priest 1936); one made entirely of creeper tendrils (Vincent 1949); ext. diam. *c*. 10 cm; placed 0·6–1·6 m above ground, sometimes among some dried twigs or sticks, in fork of low, dense bush, thicket beside stream, in tangled creeper, clump of low matted thorn bushes at edge of woodland or in small evergreen shrub amongst thorn scrub; well hidden.

EGGS: 2 (13 clutches: Maclean 1993); also 3 (Priest 1936). Pale whitish or greenish blue, spotted, blotched and streaked with pale brown, grey-brown and purplish grey, most heavily at the large end, sometimes in well-defined zone; one egg had a 'pencilling or two of jet black' (Vincent 1949). SIZE: (n = 10) 22·4–24·6 × 15·9–18·3 (23·3 × 16·9).

LAYING DATES: Angola, Mar–May (late in rains), juv. July; Tanzania (enlarged gonads Oct); Malaŵi, Apr; Zimbabwe, Oct–Feb (Oct 3, Nov 5, Dec 4, Jan 1, Feb 1 clutches); Mozambique, Nov–Feb; Transvaal, Oct–Dec; Natal, Oct–Nov.

INCUBATION: mainly by ♀, but ♂ takes over when she is disturbed.

DEVELOPMENT AND CARE OF YOUNG: black gorget and red throat appear in *c*. 60 days; immature attains adult plumage at one year. Chicks brooded by ♀; ♂ brings food, also sings nearby. Young remain with parents for at least 4 months after leaving nest, maintaining contact with call slightly higher-pitched than contact call of adults.

BREEDING SUCCESS/SURVIVAL: ringed bird lived for at least 6·5 years. Chick once taken by Shikra *Accipiter badius*.

Reference
Harris, T and Arnott, G. (1988).

Telophorus dohertyi (Rothschild). Doherty's Bush-Shrike. Gladiateur de Doherty.

Plate 23 (Opp. p. 350)

Laniarius dohertyi Rothschild, 1901. Bull. Br. Orn. Club 11, p.52; Kikuyu Escarpment, Kenya.

Forms a superspecies with *T. viridis*.

Range and Status. Endemic resident. Highlands of E Zaïre from Butembo to Mt Kabobo; fairly common in Kivu; absent from Lendu Plateau and Rwenzoris. W Rwanda (common) and W Burundi south to Bururi Forest; SW Uganda in Kigezi. Widespread and locally common in highlands of W and central Kenya on Mt Elgon and Cheranganis, Mau, Nyandarua (Aberdare) Mts and Mt Kenya; old records from Sotik, Kericho, Limuru, and Kiambu (Zimmerman *et al.* 1996).

Description. ADULT ♂: forehead and forecrown crimson, rest of upperparts green. Tail black. Ear-coverts and sides of neck green. Black band through lores, below eye, and around side of throat linked with broad black gorget across lower throat and upper breast; cheeks to chin and upper throat bright crimson. Below gorget, centre of breast and belly bright yellow, sides of breast, flanks and tibial feathers green, undertail-coverts crimson. Rare yellow phase lacks red; forehead, throat and undertail-coverts yellow. Primaries and secondaries dark brown; outer webs green except on outer primaries where emarginated parts edged pale brown; inner borders pale brown. Tertials, alula and upperwing-coverts green. Underwing-coverts and axillaries green. Eyes medium to dark brown; bill black; legs grey or bluish grey. ADULT ♀: like ♂ but has faint, narrow olive-green edges to tail feathers. SIZE (9 ♂♂, 1 ♀): wing, ♂ 78–81 (79·9), ♀ 79; tail, ♂ 77–85 (80·2), ♀ 80; bill, ♂ 21–23 (22·1), ♀ 21; tarsus, ♂ 28–30 (29·2), ♀ 28. WEIGHT: 2 ♂♂ 36, 40.

IMMATURE: olive-yellow above, narrowly barred blackish; tail greyish olive; chin to belly yellow, breast and flanks tinged olive and barred greyish; undertail-coverts pinkish with faint dark barring; wings dull green.

Field Characters. A *montane* bush-shrike with *red* forehead and throat, broad *black* breast-band and *yellow belly*. Rare yellow morph has forehead and throat yellow. Could be confused only with Gorgeous Bush-Shrike *T. viridis*. Does not come close to *T. v. nigricauda* in Kenya, from which it would be distinguished by broad red forehead, lack of

Telophorus dohertyi

orange-yellow supercilium and purer yellow belly without red or orange. Overlaps at Baraka, Zaïre, with *T. v. viridis*, which has green and brown belly. Calls of the 2 species are similar but not identical.

Voice. Tape-recorded (32, 104, B, C, F, GREG, HOR, LEM). Song quite variable, loud liquid notes given in brief phrases, e.g. 'wok-wok-week-week', an upslurred 'koo-wee, koo-wee, koo-wee', a liquid 'koi-kwer' or 'koi-kwodi', 'chop-o-chop-o-chop', triple 'wheep-wheep-wheep' or a slow trill, 'kikoikoikoikoi'; takes a phrase and repeats it several times, then switches to a different one. Also rendered 'quit-quit-quit-work' and 'quip!' followed by rising whistle, 'whee-u ...' (Chapin 1954), a prolonged 'fou-ic' (Dowsett-Lemaire 1990) or 'weeo-weeo-weeo-weeo-werk' (Zimmerman *et al.* 1996). Alarm, low rasping notes, 'krrerr' and harsh 'tchrak, tchrak'.

General Habits. Inhabits low, dense thickets of scrub and bracken in clearings and at edges of primary and secondary forest; likes dense tangles in secondary growth and nearby moist thickets; also thick bushes among bamboo. Altitudinal range 1500–2700 m in Zaïre, 1600–3350 m in Kenya.

Lives in pairs. Normally secretive, keeping well under cover, but sometimes confiding and easily seen, e.g. at The Ark in Aberdare Nat. Park (Kenya), where it feeds on ground in and near shrubbery (Zimmerman *et al.* 1996).

Food. Insects, including beetles.

Breeding Habits. Almost unknown. Nest and eggs not described.
BREEDING INDICATIONS: Zaïre (breeding condition May–July, juv. Nov); Rwanda, Aug (juv. Oct); E Africa: Region A May–June (juvs Dec–Mar); Uganda, (carrying nest material, Dec: T.M. Butynski, pers. comm.).

Plate 23 *Telophorus zeylonus* **(Linnaeus). Bokmakierie. Gladiateur bacbakiri.**
(Opp. p. 350) *Turdus zeylonus* Linnaeus, 1766. Syst. Nat., ed. 12, 1, p. 297; Cape of Good Hope.

Range and Status. Endemic resident, SW Africa from Angola and Zimbabwe to S Cape. Angola, arid coastal plain from Lobito Bay to Cunene R.; occasional in Iona Nat. Park; record in extreme SE Namibe Province. Botswana, uncommon in SW from Nossob Camp to Bokspits and Khuis, possibly commoner in austral winter; sparse resident in extreme SE. Zimbabwe and Mozambique; a population at 1350–2100 m in Chimanimani Mts, estimated at 400 birds (Irwin 1981), mainly in Mozambique east to 33°07'E, only a few pairs in Zimbabwe. Namibia, South Africa, Swaziland and Lesotho as mapped (Harrison *et al.* 1997); common and widespread in South Africa; in Transvaal, restricted to highveld and escarpment region north to about Ohrigstad and in bushveld north to about Rust der Winter. Density of 87 breeding pairs in 50 km^2, Bloemfontein, Free State (Kopij 1999). Common throughout Lesotho at 1970–2550 m; population estimated at 10,000–100,000 birds (Osborne and Tigar 1990).

Telophorus zeylonus

Description. *T. z. zeylonus* Linnaeus: S Cape Prov., central and E Orange Free State, Lesotho, Natal, Swaziland and Transvaal highveld. ADULT ♂: forehead, crown, nape, hindneck and ear-coverts grey, merging into bright olive-green of mantle; sides of forehead and well-demarcated superciliary stripe (back to just behind eye) bright yellow. Mantle, scapulars, back, rump and uppertail-coverts olive-green. Tail: T1 dark olive-green, T2–T6 black with bright yellow tips, progressively larger from T2 to T6. Lores black. Chin and throat bright yellow, bordered by black band 3–5 mm wide at side of throat and 12–17 mm wide in midline; breast, belly, flanks and thighs yellowish olive-green; undertail-coverts yellow. Wing uniformly olive-green; underwing-coverts greyish yellow. Bill black, palate black; eyes mauve-brown; legs and feet blue-grey. Sexes alike, though ♀'s gorget may average narrower than ♂'s. SIZE: wing, ♂ (n = 12) 95–102 (98·7), ♀ (n = 12) 92–101 (97·8); tail, ♂ (n = 6) 91–104 (97), ♀ (n = 3) 89–97 (94·3), sexes combined (n = 25) 92–105; bill to feathers, ♂ (n = 6) 25–29 (26·8), ♀ (n = 3) 26–27 (26·3), sexes combined (n = 25) 21–26·5; tarsus, ♂ (n = 13) 32–36, ♀ (n = 12) 29–34. WEIGHT: ♂ (n = 12) 57–71·3 (65·1), ♀ (n = 11) 48–76·1 (61·2), unsexed (n = 19) 52–76 (66·6) (Maclean 1993).
IMMATURE: forehead, crown, nape, hindneck and sides of neck dark greyish olive-green, lacking yellow stripe of adult; rest of upperparts like adult except that dark parts of tail feathers are brownish slate, not black; lores and ear-coverts dark grey, chin and throat pale grey, merging into yellowish olive breast and yellowish or greyish olive belly, flanks, thighs and undertail-coverts; breast and flanks finely and inconspicuously barred or vermiculated with blackish. Bill horn, with paler base; eyes greyish brown; legs and feet blue-grey.
NESTLING: hatches naked and blind; mouth black (T. Harris pers. comm.); at about 1 week skin of head dark grey and of chin and throat pink; bill grey, pale yellow gapes.
T. z. thermophilus Clancey: N Cape Prov. including arid W Cape coast, W Orange Free State, W Transvaal, Botswana, and Namibia north to about 21°S. Paler than *zeylonus*, particularly on crown and nape; mantle greyer; no greenish tinge on sides of breast and flanks. Wing, ♂ (n = 10) 92·5–101·5 (97·2), ♀ (n = 8) 89–99·5 (94·3).

T. z. phanus (Hartert): NW Namibia and Kaokoveld, and Angola. Like *thermophilus* but paler; flanks chalky white; same size, but bill slightly heavier (up to 30 mm).

T. z. restrictus Irwin: Chimanimani Mts, E Zimbabwe and W Mozambique. Darker than *zeylonus*, particularly crown and nape, and with heavy grey wash on flanks; somewhat shorter winged and longer tailed.

Field Characters. Length 22–24 cm. A common open-country bird, endemic to southern Africa, with *yellow underparts* and distinctive *duetting song*; distinguished from all bush-shrikes in range except Gorgeous *T. viridis* by *broad black breast-band*. Range abuts that of Gorgeous in Natal and E Zimbabwe, but they are ecologically separated; Gorgeous skulks in thick coastal and riverine undergrowth, and has red throat and different voice. Flight rather slow and laboured, with diagnostic *yellow-tipped, fanned tail*. Immature is greyish yellow below, lacks black lores and breast-band and yellow supercilium but has yellow tips to tail feathers.

Voice. Tape-recorded (11, 15, 20, 24, 75, 88, 91–99, B, C, F). Song loud, liquid and far-carrying, very variable; almost always a duet, e.g. Bird A, liquid trill reminiscent of Doherty's Bush-Shrike *T. dohertyi*, 'wik-wik-wik-wik-wik ... ', Bird B, high-pitched 'wree-wree-wree'; Bird A, a hollow 'hohoho' given at 2-s intervals while Bird B gives continuous 'kooee-kooee-kooee-kooee ...'; Bird A gives low vibrant trill while bird B says 'see-here, see-here, see-here'; Bird A sings 'kwee-kwee-kwee-kwee ...' while bird B gives high-pitched ringing 'hihihihihihihi'. Other phrases incorporated include 'wee-hoo, wee-hoo, wee-hoo', 'wick-wick-wick', 'widdly-widdly', a down-slurred, snoring 'prrreeeooo', an upslurred 'hoo-hee, hoo-hee, hoo-hee ...', and of course the onomatopoeic 'bok-bok-bokkie' or 'bok-ma-koori' and even 'bok-ma-kiri'. Either bird may change song phrase in mid-song; immatures often sing flat or off-key (Maclean 1993). Alarm, harsh 'krrrr'; warning call, knocking 'tok-tok-tok'.

General Habits. Requires bushes, clumps and thickets surrounded by plenty of quite open ground: in Namibia inhabits arid, boulder-strewn mountain-sides with scattered bushes, and dunes with sparse scrub; in South Africa, thickets in open grassland, bushy and rocky hillsides, broken thornveld on stony ground with aloes and euphorbias, scrub in hillside ravines, plantations, farmyards, gardens, tamarisk *Tamarix usneoides*, shrubby growth along dry water-courses, dense *Protea* scrub, coastal fynbos and red dunes; in Mozambique, steep boulder screes with *Philippia* heath or *Hymenodyction* scrub and borders of forested ravines where rugged quartzite and schist massifs fold.

Nearly always in pairs. Alert, shy and retiring, though can become confiding around houses. Territorial; often advertises, throughout year, by perching on top of low bush, boulder, small tree, fence, post or rooftop and, with body inclined, head stretched, bill open and pointing upward at 45° so that yellow throat is displayed, uttering ringing 'bokmakierie' or 'bocbakiri' calls, either solo or pair calling in duet. In duet, ♂ leads and ♀ responds, within same bush or from afar. In elaborate territorial threat display, bird tenses, jerkily bows body, flicks wings and fans tail, whilst making bill-clicking, croaking and tearing sounds. Territorial ♂ vigorously attacked a twice-life-sized Bokmakierie image painted on a ceramic tile (Kok 1999). Territory or home range larger in open than bushy country, and larger out of breeding season than in it. At disputed territorial boundary, 2 pairs give threat display on ground or in bushes, with chasing, posturing and counter-singing. Calls vary geographically.

Spends much time on ground, hopping and running with head drawn in and tail depressed and held just above ground surface. Hop-runs with long, swift strides; when taking flight from ground, usually takes a few rapid hops first. Flight rather slow and laboured; flies only a few m high, with fast, shallow wing-beats and tail generally fanned; dives straight into cover. When flushed, flies as close to ground as possible. Catches prey mainly on ground, bird often chasing it by running in zigzags; large prey generally beaten and chewed on ground, then carried in bill to a perch hidden in foliage. Tends to hunt in clump of grass, bushes or small trees, moving on ground, searching bases of vegetation, sometimes jumping up to seize prey out of easy reach, and hopping about in shrub searching stems and leaves, gradually working its way upward; then flies down to base of another clump a few m away. Usually wipes caterpillar on ground or wedges it in a grass fork and scrapes it with bill before eating it. Occasionally flies up from bush to catch insect in flight. Preens, concealed from view in foliage. Scratches indirectly. 2 birds of pair roost separately or together in dense bush.

Sedentary.

Food. Mainly invertebrates: longhorn grasshoppers, stick-insects, mantises, hairless and hairy caterpillars, moths, grubs, glow-worms, spiders; once a centipede; less commonly geckos, young chameleons, small snakes, and frogs; rarely takes small birds, up to size of prinias and white-eyes; occasionally takes a few seeds and small fruits.

Breeding Habits. Solitary nester, monogamous and territorial. Breeding territory *c*. 5 ha in extent in suburban habitats, larger in arid areas. Defends territory vigorously. After young have left, adults move into larger territory, which may be an extension of breeding one. Courtship display seems the same as territorial threat display (see above). Unmated ♂ displays and sings, sometimes uttering entire antiphonal 'duet' on its own; true duetting frequent once ♀ arrives and pair bond is being established. Solo singing is commoner than duetting during second half of breeding season. In Bloemfontein, vocal activity peaks in Oct and Jan (so bird may be double-brooded: Kopij 1999).

NEST: shallow, bulky bowl, quite neat and compact, made of dry grass, wet stems (e.g. of *Helichrysum*), herbs, fine twigs, roots, rootlets, sometimes with woolly plant down, leaves and man-made materials; outside sometimes decorated with asparagus fern; lined with fine grass and rootlets; sited on or near ground under matted vegetation, or up to 2 m high on fork of branch in dense foliage in bush, hedge or tree; 441 nests in SW Cape 0–6 m high. In

Plate 27

Plate 28

Bloemfontein, nests mainly in olive trees *Olea europea*, also in *Thuja* and *Calistemon* trees, at heights above ground (n = 7) of 1.0–1.8 (1.3) m (Kopij 1999). Constructed by ♂ and ♀; sometimes they build several nests before settling on one.

EGGS: 2–6, mode 3, mean (n = 54) 3. Bright, clear pale blue or greenish blue, sometimes mid blue or dark blue, sparingly spotted and very sparsely blotched with red-brown and violet-grey; large end sometimes richly and boldly spotted. SIZE: (n = 194) 22.9–29.3 × 16.5–20.8 (25.6 × 19.3).

LAYING DATES: Zimbabwe, (breeding condition Aug–Sept); Namibia, (oviduct egg Apr); Transvaal, May–Mar, mainly Aug–Nov; Natal, Sept–Mar; E Cape, July–Mar; SW Cape, July–Oct.

INCUBATION: by ♂ and ♀ by day, usually ♀ at night. Period (n = 21) 14–19 (16) days. At change-over, incoming bird either alights on ground near nest and hops up through bush to it, or alights near nest and hops through branches to it, and waits for sitting bird to depart (T. Harris, pers. comm.).

DEVELOPMENT AND CARE OF YOUNG: at about 3 weeks young almost fully feathered, plumage like immature, but large area of bare, grey skin around eye, and other areas of skin showing between feather tracts are pale yellow and bright pink; gapes yellowish white, inside mouth bright yellow (Ginn *et al.* 1989). Young brooded and fed by ♂ and ♀; parent carries more than one food item in bill at a time. Parents take shifts at feeding and guarding. Both remove faecal sacs, dropping them some distance away. Period (n = 3) 17–19 (18) days. Feathers appear from day 6; yellow throat patch begins to show at age 70 days (Robinson 1953) and young begin to duet, imperfectly, at about that time.

BREEDING SUCCESS/SURVIVAL: eggs and chicks taken by predators including snakes, mongooses, Fiscal Shrike *Lanius collaris* and probably Southern Boubou *Laniarius ferrugineus* (T. Harris, pers. comm.). Ringed adults survived for 6.5 and 8 years. Breeding density may be adversely affected by Fiscal Shrike abundance (Bloemfontein, Kopij 1999).

References
Clancey, P.A. (1960).
Harris, T. and Arnott, G. (1988).
Irwin, M.P.S. (1968).
Jackson, H.D. (1972).
Kopij, G. (1999).
Robinson, C. St C. (1953).
Vincent, A.W. (1949).

Plate 23 (Opp. p. 350) *Telophorus cruentus* (Ehrenberg). Rosy-patched Shrike. Gladiateur à croupion rose.

Lanius cruentus Ehrenberg, 1828. In Hemprich and Ehrenberg, Symb. Phys. sig. c (1833), pl. 3 (1828); Arkiko near Massawa.

Range and Status. Endemic resident, E Africa, from NE Sudan to Tanzania. Egypt/Sudan border, locally common resident around Jebel Elba. Sudan, *cruentus* common in Red Sea Hills south of 19°N; *kordofanicus* known only from 3 birds collected in Ogayeh Wells near El Obeid, Kordofan, in 1902 and one seen near Wadi el Melik (Nikolaus 1987); and *hilgerti* uncommon in extreme SE. Eritrea, below 300 m in coastal plains and Danakil and on Baraka R. south to Karkabat. Frequent in lowland E and S Ethiopia, up to 1800 m. Djibouti, common to abundant on coastal plain (>12 in 20 km, between Doralé and Loyada, seen in a day). Somalia, very common in NW, up to 1970 m, common elsewhere south to about 02°N; probably more widespread than shown. Not in Uganda. In Kenya, possibly more widespread east of 36°E than shown; uncommon or locally frequent, from near coast (lower Tana R.) up to 1600 m but scarce above 1300 m: L. Turkana, Marsabit, Moyale, Mandera, Wajir, Samburu Game Res., Shaba Game Res., Meru Nat. Park, Kora, Garissa and Bura; another race, *cathemagmenus*, from Mosiro and Olorgesailie to Amboseli, Kiboko, Tsavo West and Tsavo East Nat. Parks. Tanzania, L. Natron to lowlands north of Arusha, Ardai Plains, Tarangire Nat. Park, Mkomazi Game Res. and Masai Steppe south to Kondoa and Kilosa.

Description. *T. c. cruentus* Ehrenberg: Sudan, Eritrea and N Ethiopia. ADULT ♂ (breeding): forehead and crown light grey-brown, tinged pink, forehead whitish at sides, behind nostrils; lores and narrow, short superciliary stripe cream-white; nape, hindneck, sides of neck, mantle and scapulars pale grey-brown;

Telophorus cruentus

long fluffy back feathers pink-carmine; rump, uppertail-coverts and central tail feathers like mantle. Tail long, graduated, T1 with *c.* 18 structural bars, T2 brownish black, T3–T6 the same with white ends *c.* 13 long on T3, 23 on T4, 31 on T5 and 28 on T6; white in tail often earth-stained buff or pale rufous. Ear-coverts light pink-tinged grey above grading to whitish towards cheeks and throat. Chin whitish; throat and upper breast pink-carmine in large patch surrounded by creamy white or palest buff, the patch separated from or just confluent with an oblong or irregularly diamond-shaped patch of same colour on lower breast. Large grey-brown patch at side of breast, confluent with sides of neck. Belly silky white in midline, buff at sides; flanks buff, thighs buffy brown, vent buffy, undertail-coverts pale buff. Underside of tail black and white. Wings light brown, most feathers narrowly edged pale buff; leading edge of wing from wrist to below alula conspicuously pale buff; underside of flight feathers pale silvery brown, underwing-coverts and axillaries buff. Bill grey-black, slate or black; quite heavy and strong, hooked and notched; eyes dark brown or russet; legs and feet lead grey or greyish horn. ADULT ♀: like ♂ but, in place of pink-carmine patch on throat and upper breast, a wide black gorget extending as malar stripe up to corner of mouth, surrounding a large white patch on chin and upper throat; below gorget, a ragged pink-carmine patch as in ♂, but narrower and shorter. SIZE (5 ♂♂, 5 ♀♀): wing, ♂ 87–97 (93.1), ♀ 87–94.5 (91.8); tail, ♂ 106–109 (107), ♀ 100–110 (107); bill to feathers, ♂ 19.5–20.5 (20.0), ♀ 17–21 (20.2); bill to skull, ♂ 22–24 (23.2); tarsus, ♂ 32.5–34.5 (33.6), ♀ 31–32.5 (31.7).

IMMATURE: upperparts exactly like adult; ♂ with chin, throat and middle of breast white, lower throat and upper breast with a few black feather-bases showing through, lower breast with a scattering of carmine feathers, large grey-brown patch at side of breast, middle of belly white, feathers silkily fluffy; ♀ underparts same as in immature ♂ but indistinct black gorget beginning to show, at first as thick black malar stripes and a black dot in centre of breast, soon joined together.

NESTLING: not known.

T. c. kordofanicus Sclater and Mackworth-Praed: W Kordofan, Sudan. Like nominate race but browns much paler and rather greyer; ear-coverts white tinged with brown, grading to white on cheeks and throat (Sclater and Mackworth-Praed 1918).

T. c. hilgerti Neumann: E and S Ethiopia, Somalia, N and E Kenya. Like nominate race but entire upperparts darker, forehead to mantle and scapulars strongly tinged carmine, lores pale, supercilium and grey ear-coverts in quite strong contrast to pink-grey forehead, crown and hindneck; rump tinged carmine; in ♂ chin carmine (not white) and carmine patch(es) from chin to upper belly stronger in shade and more discrete in shape. WEIGHT: ♂ (n = 5) 40–62 (54), 1 ♀ 50.

T. c. cathemagmenus Reichenow: S Kenya, NE Tanzania. ♀ like ♀ *hilgerti* but with smaller pink-carmine patch on lower breast; ♂ like ♂ *hilgerti* but has black gorget like ♀ – chin and throat carmine, malar stripe and breast black, carmine patch on lower breast as in ♂ *hilgerti*. T2 with small buff or white tip in some birds. Juvenile like adult, with pink-carmine back, but lacks black gorget and any suggestion of carmine in underparts; cap brown, finely spotted buff; mantle and scapular feathers with narrow buff ends and dark brown subterminal bar; flight feathers and wing-coverts with buff tips and edges (tips of greater coverts 3–4 mm long), inner secondaries and tertials with dark brown subterminal line and warm buff fringes all around; T1 with rufous-buff tip and dark subterminal band. Underparts silky whitish, suffused with warm buff on breast, flanks and undertail-coverts and with quite large rufous patch at side of breast. Adult tail longer (?) than in *cruentus*: ♂ (n = 5) 109–114 (111), ♀ (n = 5) 103–109 (107).

Field Characters. Length 22–23.5 cm. A dry-country bush-shrike, often seen running along the ground. Unmistakable; broad white corners on tail of bird flying away suggest a tchagra, but readily told from tchagras by *pink rump* and pinkish tan upperparts with no red patch in wing. ♂ has *large pink patch* on throat and breast ending in point on lower breast; in southern race *cathemagmenus* the pink is interrupted by a *black gorget*; ♀♀ of all races have white throat, black gorget with pink patch below it.

Voice. Tape-recorded (B). Duet or antiphonal singing by members of a pair is a loud, piercing, somewhat slurred 'twee-u, twee-u ... ' or 'tswee-ur, tswee-ur, repeated indefinitely (Zimmerman *et al.* 1996).

General Habits. Inhabits wadis with extensive acacia groves and dry, open, low thorn-scrub country in E Sudan; hot coastal plains with *Acacia* and thorn scrub, and lava fields in Eritrea; and arid bush and semi-desert with scattered shrubs, mainly below 1300 m, in Kenya. Solitary or in pairs, sometimes in small parties (N Somalia); in small trees, a bush or shrub, but avoids thickets. Decidedly terrestrial, spending much time on open or almost bare

A

ground near to cover, progressing by hops and bounds and foraging amongst surface litter. Runs over patch of bare ground where bird might be vulnerable; can run very fast 'and if one is winged it is almost impossible to catch it among the bushes' (Archer and Godman 1961). Flies very low down over surface of ground. Restless, can be conspicuous and confiding, but likes to keep out of observer's sight. 2–3 birds call monotonously in duet for long periods from bushtop, perching out in open or with only head visible, facing each other, head and bill pointing steeply up, body rather attenuated and wings drooping somewhat (**A**). 2 pairs may duet together, the ♂♂ only a few cm apart (M.W. Woodcock, pers. comm.). Birds in small party chase each other around a large stone, with wings outstretched and tail fanned, and perch on dead bough, bobbing up and down and giving 2-note nasal, tinny call. During such a social display birds seem to be almost oblivious of an observer.

Sedentary, but in Red Sea Hills, Sudan, said to be common only seasonally. The occurrence of *hilgerti* in or very near range of *cathemagmenus* in Tsavo, Kenya (Hall and Moreau 1970) suggests that some non-breeding individuals may wander.

Food. Insects; beetles in 3 stomachs, grasshoppers in one; also 'fruits' in one.

Breeding Habits. Solitary nester, habits poorly known.
NEST: untidy, shallow cup or platform, made of vegetable fibres, twigs and dry grass stalks or grass leaves, unlined, sited 1–3 m high in small shrub (one covered with creeper *Cissus rotundifolia*); nest may be in topmost foliage, e.g. of *Suaeda fruticosa* bush; most nests placed low down in thorn bush.

EGGS: 2–3. Rather long. Green-blue, entire surface closely streaked and spotted, more boldly at the large end, with sienna or warm umber-brown on a few grey blotches. SIZE: (n = ?) 24–25 × 16·5–18 (24·6 × 17·0).

LAYING DATES: Egypt/Sudan border (Jebel Elba), (adult carrying faecal sac Mar, fledglings June); Sudan (Red Sea Hills), Oct–Nov; Eritrea, Jan–Mar (and young Dec–Feb); Ethiopia, Nov–May; N Somalia, mainly June, also Jan, Apr and July; E Africa, Region D, Apr–June, Nov.

Reference
Archer, G. and Godman, E.M. (1961).

Genus *Antichromus* Richmond

Endemic. A single, polytypic series, closely allied with *Tchagra* and often included in it. However, bill shorter and stouter, tail shorter with only vestigial white tips, and sexes dissimilar: ♀ has typical *Tchagra* head pattern with white eyebrow which ♂ lacks. Inhabits rank, marshy grassland. Nest deeper and bulkier than in *Tchagra*, usually decorated with snakeskin.

Plate 24 (Opp. p. 351)

Antichromus minutus (Hartlaub). **Blackcap Bush-Shrike; Marsh Tchagra. Tchagra des marais.**

Telephonus minutus Hartlaub, 1858. Proc. Zool. Soc. London, p. 292; Ashantee.

Range and Status. Endemic; resident (and probably local migrant) in tall moist grass and rank herbage; range extensive but discontinuous, from Sierra Leone to Ethiopia, Angola and Zimbabwe. Generally local, usually rather uncommon within restricted habitat. N Sierra Leone, Guinea (Macenta Pref., common), N Liberia (Mt Nimba), S Ivory Coast (Lamto, Sipilou and Maraoué), Ghana (Cape Coast, Deno and along Volta drainage to Kintampo and Mole), Togo (Mango south to Misahöhe), Nigeria (Abeokuta and Ibadan east to Enugu and Obudu and north to Zaria, Yankari and Aliya) and W and S Cameroon (Adamawa Plateau to Sanaga drainage and Dja R.); S Gabon (Tchibanga, Mouila); Congo (Odzala Nat. Park (common), Lefini Rés., Djambala, Brazzaville); Angola (Cabinda) and W Zaïre (Lower Congo R.); N, E and S Zaïre (Oubangi and Uele Rivers, N Ituri and Lendu plateau through Ruzizi valley and Baraka to W shore of L. Tanganyika, Katanga and Kasai). Angola in central plateau from Huambo and N Bié, north to S Cuanza Norte and Malanje, east to N Moxico. Sudan north to Wau and along Nile to L. No, also Boma Hills and in N on Atbara R.; Ethiopia (W edge of highlands from R. Dinder to Blue Nile and Omo Valley, in south on Ganale Dorya R.); Uganda (except NE), Rwanda, Burundi, W Kenya (Saiwa and Kapenguria to Mumias, Kakamega, L. Victoria basin and S Nyanza, wandering to the Mara and to Seregeti in N Tanzania) and NW Tanzania (W Lake to Kigoma, Tabora and Mwanza). Locally common S Uganda, Rwanda and Burundi. Apparently now absent from Nairobi–Thika–Murang'a area of central Kenya. Coastal and SE Tanzania (E Usambaras, Dar-es-Salaam and Morogoro to Iringa and Songea), but no longer in coastal Kenya; SW Tanzania

Antichromus minutus

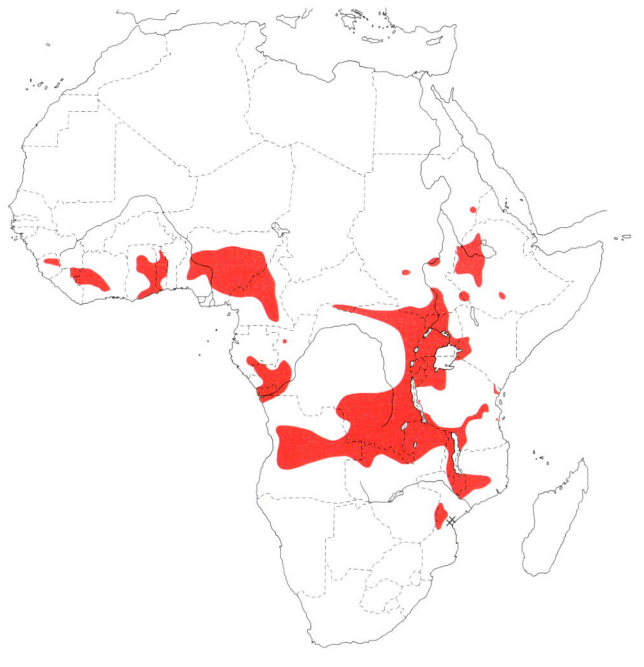

(Mahari Mt to Ufipa); Zambia (NW borders south to Mwinilunga, Solwezi and Kabompo, widespread in N and Luapala Prov., south to Mansa and Mpika, Nyika Plateau); Malaŵi (Nyika, L. Malaŵi littoral, Southern Region); Zimbabwe (local and uncommon in E highlands, Honde Valley, Melsetter, lower Pungwe R., formerly Chipinga Uplands); Mozambique (uncommon, very local, in north on central plateau and Mt Tumbini, in west near mountainous border with Zimbabwe, Upper Zona and Karumadzi rivers, Upper Buzi drainage and Vumba, Mt Gorongoza). Presumed non-breeding birds near Mozambique coast (Beira district) May and July.

Description. *A. m. minutus* (Hartlaub): Sierra Leone to Lower Congo R., N and E Zaïre south to NW shore of L. Tanganyika, Sudan, Ethiopia, Uganda, W Kenya and NW Tanzania. ADULT ♂: forehead, crown and nape to lores and below eye glossy black. Hindneck, sides of neck and mantle to rump tawny brown, tinged rufous across lower mantle, more greyish on rump; uppertail-coverts black, tipped tawny buff. Scapulars black, innermost with tawny inner web, outermost with rufous outer web and tip. Tail feathers blackish; tip of T1 with narrow buff fringe, T2–T6 with broader buff tips increasing in width toward shorter outer feathers (2 mm wide on T2, c. 5 mm on T5 and T6), T6 also buff along outer edge and on distal part of outer web. Cheeks and ear-coverts to chin and throat buffish white, merging with rich tawny buff of rest of underparts, somewhat paler on belly. Primaries blackish brown, base and distal edge of outer webs rufous; secondaries blackish brown with broad rufous outer fringe; tertials rufous with blackish brown shaft streaks; alula and upperwing-coverts rufous; marginal coverts pale tawny buff; broad inner borders form large tawny-buff area on underside of flight feathers. Bill black; eyes dull pink to pale red or reddish brown; tarsus light grey to slate-grey. ADULT ♀: differs from adult ♂ in having broad whitish superciliary stripe from above lores to side of nape; below that a black stripe extends back behind eye to join black nape. SIZE (10 ♂♂, 10 ♀♀): wing, ♂ 73–77 (74·3), ♀ 72–78 (75·0); tail, ♂ 76–86 (80·9), ♀ 78–86 (82·3); bill, ♂ 23–25 (24·1), ♀ 23–25 (23·7); tarsus, ♂ 25–27 (26·4), ♀ 25–27 (26·2). WEIGHT: Liberia, ♂ (n = 2) 25, 32·3, ♀ (n = 3) 30·3–34·6 (33·0); Ghana, ♂ (n = 1) 30, ♀ (n = 1) 32; Kenya, ♂ (n = 11) 33–38 (34·1), ♀ (n = 10) 30–37 (34·7).

IMMATURE: juvenile like adult ♀, but generally more buffish; supercilium pale buff; top of head blackish brown with pale buff mottling and tawny-buff feather fringes; a few short blackish streaks on mantle; blackish scapulars mottled buff. Bill pale horn, lighter at base; eyes grey-brown.

NESTLING: not described.

A. m. anchietae (Bocage): Angola, S Zaïre (Kasai, Katanga, meeting nominate race in Manyema northwest of L. Tanganyika), N Zambia, SW Tanzania (Ufipa to Mahari Mt) and N Malaŵi. Lacks black scapular patches; hindneck, sides of neck, mantle, back and scapulars slightly darker than in nominate race, rufous-brown rather than tawny. Eyes pale purplish or pink. Slightly larger: wing, ♂ (n = 7) 74–80 (77·0).

A. m. reichenowi (Neumann) (includes '*remotus*'): E and SE Tanzania, S Malaŵi, E Zimbabwe and Mozambique. Like *anchietae*, but paler, less richly coloured above and below, black crown patch separated from brown mantle by buffy white hindneck; white tail tips separated from black tail surfaces by grey-shaded area. Smaller than *anchietae*: wing, ♂ (n = 5, Tanzania) 71–77 (74·6).

TAXONOMIC NOTE: White (1962) retained *anchietae* as a subspecies of *A. minutus* (*Tchagra minuta*) but the 2 were treated by Sibley and Monroe (1990) as allospecies (separate species in a superspecies). We follow Dowsett and Dowsett-Lemaire (1993, p. 372) in keeping them conspecific.

Field Characters. Length 15–19 cm. A small, short-tailed, rather dumpy bush-shrike of damp places, with short, thick bill; more brightly coloured than tchagras, with *reddish back* as well as wings, *rich buffy underparts* and *red eyes*. Nominate race (W Africa to Tanzania) has *broad black stripe on scapulars* visible on closed wing, and from behind forming a black 'V'. ♂ has *solid black cap* without white eyestripe; ♀ has white supercilium and black line through eye. Tail solid black with white tips to all feathers, outer pair edged white (central feathers brown without white tips in tchagras). Juvenile has centre of crown pale.

Voice. Tape-recorded (91–99, 104, B, C, F, CART, McVIC). Territorial song, given in short flight, preceded by muffled wing-fripping, 3–6 pleasant, whistled notes, 'tuwee-twer-tuweet' 'tuwee-til-weuu', 'chee-chu-chweeoo', 'chee-woh-chiwee', 'chi-chuwee', 'chi-chuweeo', 'chwee-chwee-chweeoo'; also rendered 'tewao, tuwaro' (Zimmerman *et al.* 1996), or 'today or tomorrow' (Chapin 1954). Tone has been likened to that of Common Bulbul *Pycnonotus barbatus*, but it is rather fuller. Often a duet, ♀ replying with nasal 'cherrruu'. Territorial interactions accompanied by discordant 'tzzerrr-tzzerrr ... ' and 'tzik-tzik ... '. Scold a harsh 'tchaaa-tchaaa-tchaaa ...' or 'tchi-tchaaa', sometimes answered by harsh bleat from ♀; other alarm notes 'kiop', 'klock' and 'tchup' (Zimmerman *et al.* 1996).

General Habits. Inhabits tall rank grass near water with scattered bushes, moist grassy hollows and valley bottoms, edges of swamps and papyrus beds, damp forest edges and regenerating scrub on abandoned cultivation; tall grass and reeds along streams in southern Africa; bracken-briar at higher elevations in Zambia and Malaŵi; humid grasslands dominated by *Pennisetum* (Guinea); in Congo (Odzala) common in dambos, especially patches of *Jardinea* and *Clappertonia*, also local in tall grass in wooded grassland (Dowsett-Lemaire 1997). Cotton and coffee plantations in Ethiopia; extends into maize cultivation. Mainly at low to medium altitude, but up to 2000 m in E Africa, 2150 m on Nyika plateau, 1800 m in Cameroon highlands and 1450 m in Zimbabwe highlands.

Solitary or in pairs, in territory of *c.* 1 ha. Skulks low down in dense cover, but bolder than Black-crowned and Brown-crowned Tchagras *T. senegala* and *T. australis*, and perches for long periods on tall grass stem or small bush below 4–5 m, especially in morning and evening; posture upright with tail slightly depressed. When disturbed either creeps away or flits short distance before dropping into cover. Forages low down, often clinging to reeds or grass stems; most prey gleaned from stems and from trunks, branches and leaves of bushes; hops on ground where it takes some prey; occasionally hawks insects in air (Harris and Arnott 1988). Flies with rapid, shallow wingbeats; wings make 'fripping' sound during short excited flights. Threat to territory evokes flight display in ♂, in which he checks in level flight, rises up with muffled wing-fripping; at top of flight raises crest and head and gives slow whistling 'twee-twer-tuweet' or 'tweetitweu' call, then glides down to perch with tail fanned (Harris and Arnott 1988). ♀ may accompany ♂ in duet, giving nasal 'cherruu'; she occasionally gives less elaborate flight song display, giving this call a few times. Aggressive territorial interactions consist of repeated song flight displays combined with 'tzzerr tzeu' and 'tzik tzik' calls; when perched, birds move about excitedly, bowing, jerking body from side to side, flicking wings and cocking tails. ♀ joins interactions in duet.

In W Africa, apparently migrates during dry season; most northern records are of breeding birds during the rains. A non-breeding visitor at Mozambique coast.

Food. Large insects such as grasshoppers, beetles, dragonflies and Hemiptera. 7 stomachs, Zaïre, contained only insects; grasshoppers, present in 6, outnumbered beetles and other insects (Chapin 1954).

Breeding Habits. Little known. Solitary nester, apparently monogamous and territorial. In Nigeria pairing takes place during rains. Courtship involves duetting and repeated flight song displays with rump feathers fluffed; ♂ chases ♀, both sexes making nasal mewing calls. Both sexes involved in nest-building.

NEST: a deep cup, bulky but neat and rounded; made of tendrils, stems and rootlets, bound with a little spider web; some nests lined with fine pieces of grass (South Africa) or with a little moss, vegetable down and an odd feather (Tanzania); often decorated with snake skin; ext. diam. *c.* 90, ext. depth 75; int. diam. 70–75, int. depth 50–55; usually placed less than 1 m from ground, in fork of small bush well-hidden by long grass, or bound to supporting stems with cobweb; in S Sudan in fork of tall composite among elephant grass, at 1·8 m from ground; usually on border of marsh or in forest-edge secondary growth.

EGGS: 2–3; white with spots and blotches of warm brown and underlying greyish shell marks, mainly at large end (NE Zaïre, Nigeria); white or pale cream with small spots and specks of reddish or purplish brown and ash grey, densest at large end (Katanga, *anchietae*). SIZE: (Nigeria and NE Zaïre, *minutus*) 22·5–23·5 × 17–17·5, (Katanga, *anchietae*, n = 4) 23–24 × 16·5–17 (23·3 × 16·6), (Zimbabwe, *reichenowi*, n = 12) 19·5–26·5 × 15·5–19 (22·5–17).

LAYING DATES: mainly during rains; Nigeria, June–July; Sudan, Oct; E Africa, Region A, May–June, Region B, Mar–Apr, Sept; N Zaïre (Uele), July–Oct; S Zaïre (Katanga), Sept–Mar; Zambia, Nov–Dec; Malaŵi, Nov; Zimbabwe, Nov–Mar; Mozambique, Jan–Mar.

INCUBATION: by both sexes.

Reference
Harris, T. and Arnott, G. (1988).

Genus *Tchagra* Lesson

Medium sized to large bush-shrikes with strong notched and hooked bill, stout tarsi, and rather long, graduated tail. Plumage distinctive, with dark-streaked head, brown or grey-brown back, chestnut edged wing feathers and white tips to all except central tail feathers. Sexes alike. Skulking; inhabit dense low thickets and tangles, and prey on insects on or near ground. Loud whistling and songs given from bush-top or in display flight. Nest a shallow thin-walled cup, hidden low down in bush or small tree.

Nearly endemic; 4 species, one also in S Arabia. We follow Hall and Moreau (1970) and Sibley and Monroe (1990) in treating *T. australis*, *T. jamesoni* and *T. tchagra* as composing a superspecies. However, there are areas of marginal sympatry and some authorities, e.g. Dowsett and Forbes-Watson (1993) make them independent species.

Tchagra australis superspecies

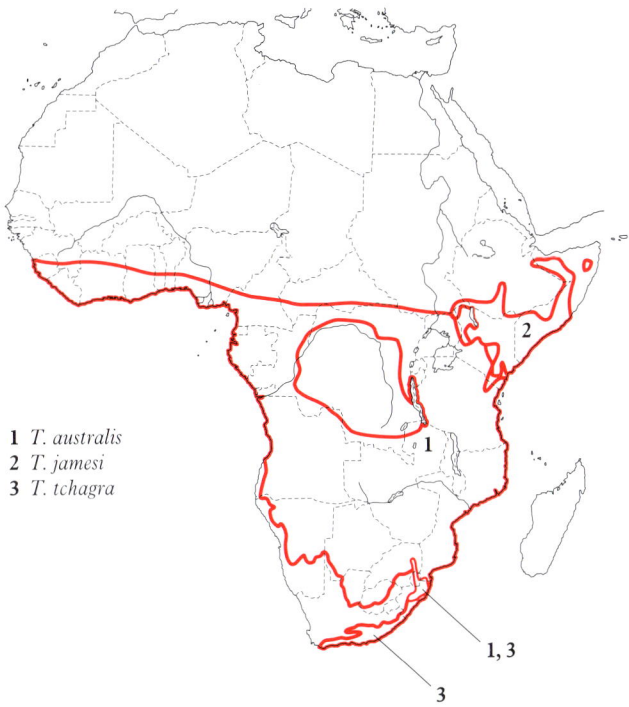

1 *T. australis*
2 *T. jamesi*
3 *T. tchagra*

Tchagra australis (Smith). Brown-crowned Tchagra. Tchagra à tête brune.

Malaconotus australis Smith, 1836. Rep. Exped. C. Afr., p. 44; north of Kurrichane.

Plate 24
(Opp. p. 351)

Forms a superspecies with *T. jamesi* and *T. tchagra*.

Range and Status. Endemic resident. In W Africa, locally common from forest zone to N guinea savannas: Sierra Leone, Liberia as mapped, S Guinea (Macenta), Ivory Coast (scarce north of 8°N), S Ghana (Cape Coast and Volta north to Kete Kratchi), central and S Togo to S Nigeria (north to Ibadan and Obudu), S Cameroon and S Central African Republic (Lobaye, Bangui); also Gabon to Angola (Cabinda) and Lower Congo R. Locally common in central and E Africa from N and E Zaïre (Lower and Upper Uele, Ituri, base of Rwenzoris, Kivu and Manyema) to Sudan border (Bengengai, Maridi, Amadi and southern mountains), Uganda, W, central and S Kenya (north to Cheranganis, Maralal and Lower Tana R.), Rwanda and Burundi. Minimal range in Tanzania as shown (N.E. Baker, pers. comm.). Widespread and locally common in southern Africa: Angola (except dry SW coast), S Zaïre (S Kasai, Katanga), Zambia (scarce and local in N and E), Malaŵi, N Namibia (east of desert, south to N Great Namaqualand), Botswana (except extreme SW), Zimbabwe, Mozambique, E Swaziland and N South Africa (Transvaal, N Cape, W Orange Free State, NE Natal). Overlaps widely with *T. senegala*, but generally in moister habitats; the commoner species in forest zone, and ranges to higher altitude.

Tchagra australis

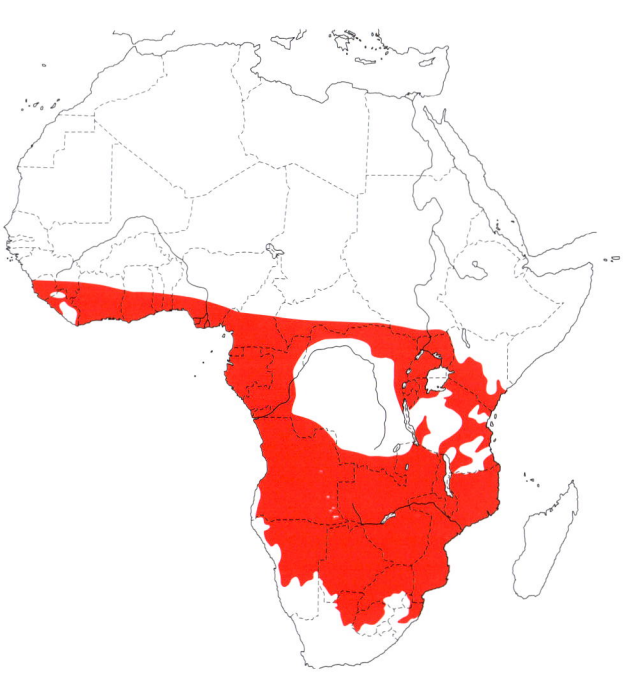

In central Transvaal, density of 1 pair per 20 ha in acacia woodland, 1 pair per 25 ha in broad-leaved woodland (Tarboton *et al.* 1987). In S Mozambique, densities of <5–8 birds per 100 ha, according to woodland type (Parker 1999).

Description. *T. a. emini* (Reichenow) (including '*frater*'): SE Nigeria (Calabar) to lower Congo R., and to N and E Zaïre, S Sudan, Uganda, W and central Kenya, Rwanda and NW Tanzania (to W shore of L. Victoria). ADULT ♂: forehead to nape brown; crown with narrow black edges extending back and broadening along sides of nape; prominent pale buff superciliary stripe extending back to side of nape, bordered below by black stripe through lores and behind eye; hindneck, mantle and inner scapulars tawny brown; rump and uppertail-coverts greyish brown; outer scapulars rufous with blackish brown centres. Central tail feathers olive-brown, with narrow, evenly spaced, dark greyish barring; rest of tail feathers tipped white (white tips 1–2 mm deep on T2 increasing to *c*. 9 on T5 and *c*. 11 on inner web of T6), white tip of T6 extending as narrow wedge along outer edge. Ear-coverts and sides of neck brown; cheeks greyish buff. Chin to belly greyish white with buff tinge; darker grey on breast and flanks, tinged brownish; undertail-coverts pale buff. Remiges blackish brown; outer webs of primaries rufous, outer webs of secondaries rufous at base, fringed rufous distally, tertials broadly fringed rufous; alula dark brown, largest feather edged whitish on outer web; primary coverts dark brown, edged and tipped rufous; outer greater coverts rufous, inners dark brown with rufous fringes; median and lesser coverts rufous-chestnut; coverts at bend of wing white. Axillaries buff; underwing-coverts buff, outermost small feathers white; large cinnamon-buff area on underside of flight feathers, formed by pale inner borders. Bill black; eyes brown or dark purplish brown with pale inner rim; legs slate grey or blue-grey. Sexes alike. SIZE (10 ♂♂, 10 ♀♀): wing, ♂ 73–79 (76·3), ♀ 71–77 (74·7); tail, ♂ 81–87 (84·1), ♀ 83–87 (85·5); bill, ♂ 20–28 (21·4), ♀ 20–22 (21·1); tarsus, ♂ 26–28 (26·8), ♀ 25–28 (26·1). WEIGHT: Kenya and Uganda, ♂ (n = 6) 35–40 (35·6), ♀ (n = 4) 31–44 (36·3); W Kenya, unsexed (n = 19) 35–42 (37·0).

IMMATURE: juvenile like adult but head pattern less distinct; more grey-buff below, throat and breast faintly mottled; tail feathers tipped pale buff; bill horn-brown, eyes greyish sepia.

NESTLING: not described.

T. a. ussheri (Sharpe): Sierre Leone to SW Nigeria. Like *emini* but slightly paler brown, less tawny, above; clearer grey, less buffish, below.

T. a. minor (Reichenow) (includes '*littoralis*' and '*congener*'): SE Kenya, Tanzania (except NW), Zambia (except Ndola and extreme SW), Malaŵi, N Zimbabwe and Mozambique (south to Save R.). Like *emini* but slightly paler brown above; whiter below, less grey on breast and sides. WEIGHT: coastal Kenya, ♂ (n = 3) 25–31 (28·7), ♀ (n = 2) 30, unsexed (n = 3) 31–34 (32·2); Malaŵi, unsexed (n = 18) 27–36 (31·0); N Mozambique, unsexed (n = 28) 29–39 (33·7). Zimbabwe, ♂ (n = 9) 30·6–45·8 (34·9), ♀ (n = 3) 29·6–32·9 (31·2).

T. a. souzae (Bocage): central Angolan plateau (from N Huila, E Benguela and S Cuanza Sul) north perhaps to lower Congo, east through S Zaïre to Upper Katanga and N Zambia (Ndola). Richer brown above than *emini* and darker grey below; differs from all other races in having brown centres of inner scapulars and tertials merging with rufous fringes.

T. a. ansorgei (Neumann): W Angola (central and S Huila and E Mossamedes, north along coast and escarpment to Luanda and S Cuanza Norte). Similar to *emini* above, but deep tawny buff below.

T. a. bocagei da Rosa Pinto: SE Angola (central Cuando-Cubango). Like *ansorgei* but upperparts darker. Large: wing, ♂ (n = 5) 76–82, ♀ (n = 6) 76–81; tail, ♂ (n = 5) 93–104, ♀ (n = 6) 94–104.

T. a. rhodesiensis (Roberts): NE Namibia along Okavango east to Caprivi, NW Botswana, SE Angola (S Cuando-Cubango), extreme SW Zambia. Paler and colder brown above than *emini*; tawny-buff below, tinged greyish on breast and sides.

T. a. damarensis (Reichenow): Namibia (except NE) to extreme SW Angola, most of Botswana, SW Zimbabwe (W and NW Matabeleland) and N South Africa (N Cape, W Orange Free State, dry W Transvaal). Lighter and greyer above than *rhodesiensis*; pale tawny-buff below with little grey on breast and sides. Slightly larger: wing, ♂ (n = 6) 76–81 (78·5), ♀ (n = 4) 76–80 (77·5).

T. a. australis (Smith): SE Zimbabwe (Sabi valley and SE lowveld), NE South Africa (Transvaal except extreme dry west, N Zululand), S Mozambique (Sul do Save) and E Swaziland. Slightly darker above than *rhodesiensis*; more dusky, greyish buff below, with paler chin and throat. WEIGHT: Transvaal, unsexed (n = 15) 29–34 (32·0).

Field Characters. Length 17·5 cm. Readily identified as a tchagra by bright chestnut wings and graduated black tail with white tips to all but central feathers, conspicuous in flight. Somewhat smaller, with shorter bill and tail, than Black-crowned Tchagra *T. senegala*, with which it is widely sympatric; rather shyer, preferring damper habitat where they meet, but mainly distinguished by *brown crown* and different voice. Juvenile like adult but with duller head stripes, buffy underparts, brown bill. Very similar to Southern Tchagra *T. tchagra* but paler and slimmer overall with smaller bill; crown dull brown (not russet), supercilium whiter, and *black line* separates crown from supercilium. For differences from Three-streaked Tchagra *T. jamesi* see that species.

Voice. Tape-recorded (22, 58, 86, 88, 91–99, 104, B, C, F, LEM). In E and southern Africa flight song, preceded by wing-snapping, is a series of 8–15 descending double whistles, becoming progressively longer, loudest in the middle and dying away at the end, 'chi-chi-chu-cheeyu-cheeyu-cheeyu ... ', 'chi-cheeya-cheeya-chooyu-chooyu-chooyu ...' or more liquid 'wee-woo-wee-woo-wee-woo ...'. Tone variable; notes may be wheezy rather than whistled. Aggressive song from perch, harsh 'jwee-jwee-jwee-jweeu-jweeu-jweeu-jweeu'. Other songs include varied conversational medley of buzzy notes and trills interspersed with a warbled, oriole-like 'tooeeoo'. Sometimes duets during flight song, ♀ giving soft, nasal 'cheru-cheru ...'. Contact calls from feeding group, 'jwee', 'jojo', 'jip', 'jeejee', 'toptop', a hard trill, a hoarse 'jaw' and a rasping song variant 'ji-jee, chweeju-chweeju-chweeju'. Calls from cover include short 'chuck', low 'cheerk' and a whistled 'quweeo'. For further variations see Zimmerman *et al.* (1996). In Gabon and Congo flight song very different; one version starts with high thin note which turns into a hard trill and ends with a rapid series of 'chup's, decelerating and descending the scale as it progresses, 'wheeeeeeeee-rrrrrrrrrrrrupupupchupchupchupchupchup ... '; another version starts with a decelerating rattle and ends with a dozen or so down-slurred whistles, 'ratatatatatatatatatata-tata-chyoo-chyoo-chyoo-chyoo ... '; both last about 6 s. Second version may omit whistles or singer may give nightjar-like steady churr on one pitch.

General Habits. Occurs in low thickets in a variety of bushed and wooded habitats with coarse grass; thick low secondary bush; forest-savanna edge; abandoned forest clearings; gallery forest in N Zaïre; in Liberia, scrub with Solanaceae and abandoned farms with very early second growth (Gatter 1997); old cultivations and gardens. Mainly at low or medium altitudes, but ranges to 1600 m on Mt Nimba; above 2000 m in Kenya; up to 2250 m in E Zaïre in bracken thicket, elephant grass and bushes at edge of bamboo; and at 1800 m in Malaŵi in bracken-briar. In Zambia, less confined to *Brachystegia* than is *T. senegala*, preferring miombo edges and underbrush. In Zimbabwe, in acacia or other woodland with grass and thorny cover; avoids miombo, preferring nearby valleys with rank forest-edge secondary growth. In South Africa, habitats range from deciduous broad-leaved and acacia woodland, mopane and dry riparian forest to bush and thornveld. Usually in denser bush than *T. senegala*; however, it extends into more arid areas than the larger species in the west. Where it overlaps marginally with *T. tchagra* it occupies acacia and mixed woodland, with *T. tchagra* confined to forest edge and broad-leaved thickets (Harrison *et al*. 1997).

Occurs singly, in pairs, and occasionally family parties. Less arboreal than *T. senegala*, skulking and more retiring; seldom perches for long in the open. A reluctant flier, usually glimpsed on ground, or moving through grass or low bush in crouched posture. Forages mainly on ground beneath bushes, running about and hopping as it searches plant bases; also gleans prey from branches and leaves of bushes and occasionally takes aerial insects (Harris and Arnott 1988); commonly in mixed species foraging parties. Flight low and jerky, bird fanning tail as it dives for cover at base of bush.

♂ defends territory with repeated flight song displays. Bird zig-zags up small bush or tree to top, makes steep upward flight with loud burst of wingbeats to a few m above vegetation, erects crest and raises head, then calls as it glides jerkily down, often fanning tail to expose white terminal spots (Harris and Arnott 1988); may land 20–60 m from take-off point. Flight song sometimes preceded by song from bush-top or tree, accompanied by posturing and tail swinging. Sometimes several birds display excitedly together, flying low from bush to bush, tails flicking and cocked, with 'chuk-chuk' and 'pechew' calls and repeated wing fripping.

Food. Mainly insects and their larvae; some small vertebrates in E Africa. Includes grasshoppers, caterpillars, beetles, whole mantis egg-cases. In Gabon, young fed with grasshoppers, mantids, spiders, beetles and caterpillars (Brosset and Erard 1986). In Kenya, hatchlings fed with small moth larvae, also nymphal grasshoppers (legs removed); at 1 week they received small crickets, moth larvae and mantid nymphs; at 2 weeks noctuid moths, geometer larvae, black weevils and a hairless *Leggada* mouse, pulped and disembowelled (van Someren 1956).

Breeding Habits. Solitary nester, monogamous, territorial. ♂ defends territory as small as 4 ha by repeated flight song displays. Responds to territorial threat by loud wing-fripping, and aggressive harsh calls, whistles and chukking; moves about excitedly, swinging body from side to side, bowing, and cocking and flicking tail (Harris and Arnott 1988). ♂ courts ♀ with repeated flight song displays, with rump fluffed out and tail fanned as he descends; chases ♀ from tree to tree; postures in front of her uttering variety of calls, bowing, stretching, pointing bill upwards and fanning tail (Harris and Arnott 1988). Nest built by ♂ and ♀, taking *c*. 7 days. Apparently single brooded.

NEST: shallow, thin-walled cup of rootlets, fine twigs, leaf stems and coarse grass, usually lined with finer grass and rootlets, sometimes with feathers, and bound on outside with spider's web; int. diam. 60–75, int. depth 25–30; placed about 1 m high (0·2–3·6 m, southern Africa, n = 166; Harris and Arnott 1988) in fork of bush or small tree, or supported in thick scrub; usually well concealed.

EGGS: 2–4; southern Africa (n = 72) 2–4 (2·4), E Africa (n = 26) 2–3 (2·3). White or pinkish white with small blotches or streaky spots of dark brown and grey or violet, concentrated at thick end. SIZE: southern Africa (n = 133) 19·4–25·3 × 14·9–18·3 (21·7 × 16·3) (Maclean 1993); Cameroon (n = 2) 22–25 × 16·5–18; Zaïre (Katanga) (n = 2) 23 × 16·5–17.

LAYING DATES: Ghana, May–June; Cameroon, Mar, Aug–Oct; Angola (*souzae*) Sept–Oct; Sudan, Dec; E. Africa, Region A, Aug, Sept, Region B, Dec, Feb, Region C, Oct–June (mainly Nov–Mar), Region D, Dec–Oct (mainly Feb–June), Region E, Apr; Gabon, Aug–Feb; Zaïre (Katanga), Oct; Zambia/Malaŵi, Sept–Mar; Zimbabwe, Sept–Feb (mainly Oct–Dec), Mozambique, Sept–Nov; Transvaal, Sept–Apr (mainly Oct–Dec); Natal, Oct–Nov.

INCUBATION: by ♂ and ♀, mainly by ♀, who is fed on nest by ♂; often begins with first egg. ♂ continues to perform display flights; they are given as he flies towards nest and appear to signal nest relief (Harris and Arnott 1988). Period: 13–16 days.

DEVELOPMENT AND CARE OF YOUNG: young fed by both parents, remain in nest about 15–16 days; young remain in family parties in South Africa for at least 5 months.

BREEDING SUCCESS/SURVIVAL: nests destroyed by rats, mongooses, snakes and ants (van Someren 1956), young taken by African Hawk-Eagle *Hieraaetus spilogaster*. Longevity, Malaŵi, >6·5 years (Hanmer 1983).

Reference
Harris, T. and Arnott, G. (1988).

Plate 24
(Opp. p. 351)

Tchagra jamesi (Shelley). Three-streaked Tchagra. Tchagra de James.

Telephorus jamesi Shelley, 1885. Ibis, p. 403, pl. 10, fig. 2; Goolis mountains, Somaliland.

Forms a superspecies with *T. australis* and *T. tchagra*.

Range and Status. Endemic resident in arid and semi-arid bushland of E Africa. Extreme SE Sudan, fairly common Natoporoputh Hills. S Ethiopia, Boran, Sidamo. Somalia, locally common and widespread (except most of NE and coastal areas north of 04°N). Kenya, locally common north and east of highlands, south to Kerio Valley, L. Bogoria, Samburu, Meru and Tsavo, and extending to extreme NE Tanzania (Mkomazi); reaches coast from Lamu northward, including Lamu and Manda Is.; local in Turkana, confined to Sudan border areas, but extending to NE Uganda (Moroto).

Sympatric with *T. senegala* along Juba R., in Tsavo and on Kenya coast; replaces *T. australis* in more arid habitat, although a specimen from W of Lodwar, NW Kenya, is apparently a hybrid between *jamesi* and *australis*.

Tchagra jamesi

Description. *T. j. jamesi* (Shelley): range of species except area occupied by *mandana*. ADULT ♂: top of head greyish brown with narrow black central stripe from base of bill to nape; sides of head pale greyish brown; black stripe through lore and behind eye above ear-coverts, and above this a narrow greyish white superciliary stripe which contrasts little with crown. Rest of upperparts greyish brown, mantle and scapulars with olive tinge. Central tail feathers greyish brown with fine darker barring; remaining feathers blackish, T2 with small triangular white spot at tip, T3–T6 with large white tips, increasing toward shorter outer feathers, especially on outer web (*c.* 6 mm deep on T3, *c.* 13 on inner web of T6), and that on T6 extending as narrow wedge along outer edge. Chin and throat to sides of neck buffish white; underparts grey-buff, darker on upper breast and flanks, whiter on belly. Primaries and secondaries dark brown, base of primary outer webs and entire outer webs of secondaries rufous; tertials brown; alula feathers dark brown, outer webs white; primary coverts brown, outer webs rufous; greater coverts rufous, innermost with brown centres; median and lesser coverts rufous; coverts along bend of wing white. Underwing-coverts and axillaries pale grey; inner borders of undersides of remiges pale grey-brown. Bill black; eyes dark brown with ring of silvery dots surrounding pupil; legs slate, blue-grey or olive-green, soles white. Sexes alike. SIZE (10 ♂♂, 10 ♀♀): wing, ♂ 67–74 (70.9), ♀ 67–72 (69.8); tail, ♂ 82–92 (86.3), ♀ 79–86 (81.3); bill, ♂ 19–21 (20.2), ♀ 19–21 (20.1); tarsus, ♂ 23–26 (24.4), ♀ 23–25 (24.0). WEIGHT: Kenya, ♂ (n = 5) 22–30 (27), ♀ (n = 2) 26, 27, unsexed (n = 5) 24–32 (27.5).

IMMATURE: juvenile like adult but head stripes blackish brown, sides of crown paler, more buffish, white tail feather tips tinged buff. Bill slate above, horn with slate tip below.

NESTLING: not described.

T. j. mandana (Neumann): Lamu and Manda Is and adjacent Kenya coast west to Witu and Tana R. delta. Slightly paler; upperparts more sandy, less ashy, belly whiter; black crown stripe wider.

Field Characters. Length 16.5 cm. A smallish, slim tchagra of dry NE Africa. Head *ashy grey* concolorous with back and breast, with *3 black streaks*, one through each eye from lores to side of neck and one over centre of crown from bill to hind neck. Lacks prominent pale supercilium of Black-crowned and Brown-crowned Tchagras, *T. senegala* and *T. australis*. Range overlaps broadly with that of Black-crowned in Ethiopia and Somalia, and marginally in Kenya; latter is larger, with solid black cap. In Kenya also meets Brown-crowned, which is somewhat larger and darker, with black line separating pale supercilium from brown crown, no central black crown stripe.

Voice. Tape-recorded (CHA, HRS, McVIC). Song in display flight, preceded by mechanical wing-fripping, similar to that of Brown-crowned Tchagra, descending the scale, an emphatic series of downslurred whistles, 'wi-weo-weo-weo-weo' or 'chweeo-chweeo-chweeo'. Protests with a nasal scold, 'chuwaa' or 'chwaa-chwaa', and 'cherraa-cherraa' (Zimmerman *et al.* 1996).

General Habits. Inhabits low thickets in arid and semi-arid bushland and *Acacia-Commiphora* woodland, dense low thorn scrub, semi-desert savanna, 'the thickest wind-shorn bush and aloe bush areas' (Jackson and Sclater 1938: race *mandana*); mainly at low altitudes; up to 1100 m in plateau country east of Kenya highlands, below 1200 m in Ethiopia.

Usually solitary or in pairs. Shy and skulking; creeps in low cover, but may forage up to 4 m high in low acacia trees. Flight jerky, white tipped tail conspicuous; bird dives back quickly into cover. Flight display, similar to that of *T. australis*, consists of fluttering upward climb accompanied by mechanical wing 'fripping', followed by gliding descent with descending whistled calls.

Food. Insects; a cockroach and mantis egg cluster with protective coating largely intact (Lack and Quicke 1978).

Breeding Habits. Solitary nester, monogamous, territorial.

NEST: small flat cup of dry grass stems, twigs and fibres, lined with rootlets; placed low in thorn bush, usually fairly conspicuous.

EGGS: 2–3, glossy white with dark brown and violet spots and blotches, chiefly at larger end; SIZE *c.* 23 × 15.

LAYING DATES: S Sudan, Dec; Somalia, Mar–Apr; S Ethiopia, Mar–July (mainly Apr–May); Kenya coast, May.

Tchagra tchagra (Vieillot). Southern Tchagra. Tchagra du Cap.

Plate 24 (Opp. p. 351)

Thamnophilus tchagra Vieillot, 1816. Nouv. Dict. Hist., 3, p. 317; Gamtoos River.

Forms a superspecies with *T. australis* and *T. jamesi*.

Range and Status. Endemic resident. SW Cape, west to Hermanus and Robertson and north to Beaufort West, including Little Karoo, where common, and Great Karoo. S and E Cape, but sparse in Transkei and Pondoland. Natal, sparse and local along coast and inland up to about 600 m, ranging north to Lebombo Range at Ingwavuma, Swaziland, to SE Transvaal (Graskop), and S Mozambique (Inhaca I.: Parker, 1999).

Range overlaps that of Brown-crowned Tchagra *T. australis* in N Natal, but it occupies broad-leaved thickets and forest edge, with *T. australis* in *Acacia* and woodland (Harrison *et al.* 1997).

Tchagra tchagra

Description. *T. t. tchagra* (Vieillot): SW and S Cape Province east to Uitenhage. ADULT ♂: forehead to nape rich dark brown; rest of upperparts dark olive-brown, tinged greyer on rump. Central tail feathers dark olive-grey with regular narrow dusky barring; rest of tail feathers blackish with white tips (*c.* 3 mm deep on T2 increasing to 9 on T5 and 11 on inner web of T6), white wedge along outer edge of T6. Broad superciliary stripe extending back to side of nape, greyish and diffuse in front of eye, whiter and sharply formed behind eye. Below this, lores and stripe behind eye black; ear-coverts and cheeks pale grey-brown; sides of neck grey-brown. Chin and throat greyish white, rest of underparts grey, darker and tinged olive on sides of breast and flanks. Remiges dark olive-brown; outer webs of primaries fringed dark rufous, outer webs of secondaries with rufous tinge; alula dark brown, outer webs fringed dark rufous; primary coverts olive-brown with rufous outer edges; greater coverts olive-brown, outers with broad rufous outer fringes; median and lesser coverts deep rufous; coverts along carpal edge greyish white. Underwing-coverts and axillaries dusky grey-brown. Bill black; eyes dark chestnut-brown; legs slate grey or blue-grey. Sexes alike. SIZE: wing, ♂ (n = 9) 82–87·5 (84·3), ♀ (n = 13) 79–87 (82·2); tail, ♂ (n = 4) 91–97 (94·3), ♀ (n = 8) 88–99 (90·4); bill, ♂ (n = 9) 28–34 (31·3), ♀ (n = 13) 28–32 (30·1); tarsus, ♂ (n = 2) 28, 29, ♀ (n = 2) 28, 29. WEIGHT: unsexed (n = 13) 42–54 (46·8).

IMMATURE: juvenile tinged more olive below; tail feathers tipped buff.

NESTLING: not described.

T. t. caffrariae Quickelberge: Fish R. in SE Cape to Natal border. Crown more reddish than nominate race; superciliary stripe more buffish; flanks, thighs and undertail-coverts washed more olive; rest of underparts slightly lighter. Size as nominate race, but bill shorter: wing, ♂ (n = 14) 79·5–84·5 (82); bill, ♂ (n = 13) 26·5–31 (28·7). WEIGHT: unsexed (n = 2) 51, 54.

T. t. natalensis (Reichenow): Natal to W Swaziland and SE Transvaal. Crown still redder than in *caffrariae*, upperparts brighter; underparts much lighter. Slightly smaller: wing, ♂ (n = 8) 76·5–81 (79·5), bill, ♂ (n = 7) 26–28·5 (27·4).

Field Characters. Length 20–22 cm. A large tchagra, often skulking and hard to see. Similar to Brown-crowned Tchagra *T. australis* but more thickset, with longer, heavier bill; overall rather darker, greyer below, with *reddish wash on crown* and duller supercilium, white over eye, but becoming buffy on lores and behind eye; best distinguished by *lack of black line* separating pale eyebrow from crown. Also overlaps with Black-crowned Tchagra *T. senegala*, which has black crown, whiter underparts, different voice, and is bolder and more easily seen.

Voice. Tape-recorded (11, 58, 75, 88, 91–99, B, F, STA). Flight call, preceded by wing fripping, often starts as trill, followed by short notes and then progressively longer ones, all on one pitch, 'trrrrrr-itititititi-tyi-tyi-tyi-tewp-tewp-tewp-tewp …', seeming to decelerate as it proceeds; usually 15–20, sometimes 30 notes. In slower version, no preliminary trill, goes right into loud whistles with just a few introductory notes, 'tyio-tyi-wheeeoo-wheeeoo-wheeeoo-wheeeoo-hooee-hooee-hooee-tooiyup-tooiyup-tooiyup-tooiyup-toooiyup …'. After flight ♂ often gives discordant, tearing 'tzzerrr-tzzerrr …', and similar call used by ♀ in duet. In aggressive interactions, harsh 'tzik', 'chok', 'tzzerrr' and more complex purring and tearing sounds. Alarm, a searing 'neeeaaa' reminiscent of puffback *Dryoscopus*, also a rattling 'kikikiki ...' and a clucking 'chok-chok'. Call originally described by Levaillant as 'tcha-tcha-tcha-gra', whence the name.

General Habits. Occurs in dense coastal bush, thorny stream-side scrub, fringes of coastal dune forest in clearings and *Lantana* tangles, brush piles around rural cultivation, thickets of exotic *Acacia cyclops*, and Natal and E Cape thornveld; at northern limit, in bracken and scrub at interface between montane grassland and forest.

Usually solitary or in pairs. Keeps mainly to thick cover, visible briefly as it flies low from one thicket to the next. Forages on ground where it hops and runs about, turning over debris; also gleans prey from lower branches and stems while creeping in dense low vegetation (Harris and Arnott 1988). Flight brief, not strong, with rapid wingbeats and tail spread. Perched posture similar to *T. senegala*, with head and bill up. Aggressive interactions involve excited wing fripping, body bowing, side-to-side movements and tail cocking and flicking (Harris and Arnott 1988). Performs low display flight over bushes, accompanied by wing fripping, with tail broadly fanned, head up and crest raised; calls after gliding down to new perch, sometimes in flight.

Food. Insects and their larvae; berries; a few molluscs.

Breeding Habits. Solitary, monogamous, territorial. Courtship involves display flight by ♂ with grey rump fluffed out; excited posturing with tail fanned and body bowed and stretched repeatedly; pair contact maintained by discordant 'tzzerrr-' call and by duetting (Harris and Arnott 1988).

NEST: a shallow cup of fibres, twigs and rootlets, lined with finer rootlets and sometimes hair; height about 1 m (0.3–3 m; n = 41; Harris and Arnott 1988); concealed in fork in thicket or matted bush.

EGGS: 2–3, av. 2.2 (n = 17); white, spotted and scrolled with dark red-brown and purplish grey. SIZE (n = 19): 23–26.5 × 17–19.5 (24.3–18.4).

LAYING DATES: Cape Prov., Aug–Mar (mainly Sept–Nov); Natal, Sept–Dec.

INCUBATION: period (n = 2 clutches) 15.5–16 days.

DEVELOPMENT AND CARE OF YOUNG: young fed by both parents; leave nest after 13–15 days. One nest appeared to be parasitized by Jacobin Cuckoo *Oxylophus jacobinus*.

Reference
Harris, T. and Arnott, G. (1988).

Plate 24 *Tchagra senegala* (Linnaeus). **Black-crowned Tchagra. Tchagra à tête noire.**
(Opp. p. 351)
Lanius senegalus Linnaeus, 1766. Syst. Nat., ed. 12, 1, p. 137; Senegal.

Range and Status. Africa and S Arabia. Resident, with some local movements.

Local and uncommon in NW Africa: W Morocco, Tangier to Oued Sous, inland to Atlas foothills; coastal NE Morocco and Algeria, Berkane to Bejaia; coastal N Tunisia, Kroumirie east to Cap Bon where more common; Libya, bird displaying Al'Aziziya (Cramp and Perrins 1993).

Widespread south of 15°N, except for Congo basin and arid parts of Horn and SW Africa. In W Africa, widespread and common to abundant in savanna habitats north to limit of sahel belt: c. 17°N in Mauritania and Mali, 15–16°N in Chad and W Sudan and reaching Atbara (18°N) on the Nile. Perhaps a rains visitor to northernmost breeding areas; scarce and local in forested areas. Ranges up to 1900 m in Cameroon highlands, 1900 m in Kenya and 3000 m in Ethiopia. Absent in Kenya from dry country north of 02°N, and in Somalia confined to extreme NW, N highlands and riverine habitat south of 03°N. In southern Africa, widespread and common as mapped.

Density, Morocco, in *Thyula* scrub 8 pairs per km^2, in macchia 2.7 pairs per km^2, in woodland 1.3 pairs per km^2 (Thévenot 1982); in central Transvaal woodland, 1 pair per 25 ha (Tarboton *et al.* 1987); in Congo (Odzala Nat. Park), c. 4 pairs per km^2 in wooded grassland (Dowsett-Lemaire 1997a).

Description. *T. s. armena* (Oberholzer) (includes '*camerunensis*', '*sudanensis*' and '*rufofusca*'): S Cameroon to N Zaïre, SW Sudan (Bahr el Ghazal), Uganda, Kenya (except coast) and Tanzania (except E lowlands) south to central Angola, Zambia (except extreme SW), N Zimbabwe, W Mozambique (Tete) and Malaŵi.
ADULT ♂: forehead to nape black; broad buff superciliary stripe

extending back to side of nape, more tawny behind eye; bordered below by black lore and black stripe behind eye; hindneck, mantle and inner scapulars tawny brown, grading to grey on rump; uppertail-coverts greyish brown, darker on larger feathers; outer scapulars blackish brown with broad rufous fringes. Central tail feathers dark greyish brown with narrow, evenly spaced, darker barring; remainder black with broad white tips (c. 5 mm wide on T2 increasing to c. 10 mm on T5 and c. 18 mm on inner web of T6), white on T6 extending narrowly along outer edge. Upper ear-coverts and sides of neck tawny brown; lower ear-coverts and cheeks grey-brown. Underparts light grey, whiter on chin, centre of throat, belly and undertail-coverts, darker on sides of breast and flanks. Remiges blackish brown; primary outer webs rufous, secondary outer webs rufous at base and with broad rufous distal fringe, tertials broadly fringed rufous (innermost on outer web only), forming large conspicuous rufous patch on closed wing. Alula dark brown, fringed rufous; primary coverts brown, outer edge and tip rufous; outer greater coverts rufous, inners blackish with broad rufous fringe; median and lesser coverts rufous; feathers along bend of wing white. Underwing-coverts and axillaries pale cinnamon, outermost small coverts white; cinnamon-buff area on underside of flight feathers, formed by broad pale borders of inner webs. Bill black; eyes dark brown or dark blue-grey; legs grey. Sexes alike. SIZE (10 ♂♂, 10 ♀♀, Kenya and Tanzania): wing, ♂ 84–90 (86·5), ♀ 81–88 (85·6); tail, ♂ 87–102 (94·6), ♀ 85–97 (94·7); bill, ♂ 23–25 (24·0), ♀ 23–26 (24·5); tarsus, ♂ 29–31 (30·3), ♀ 29–32 (30·3). WEIGHT: Kenya, unsexed (n = 28) 42–53 (49·7); Zambia, unsexed (n = 4) 55–64 (60·7).

IMMATURE: juvenile similar to adult, but crown mixed brown and blackish, eye-line dark brown; breast and flanks tinged brownish; white tips to tail feathers tinged buff; bill horn.

NESTLING: not described.

T. s. cucullata (Temminck): Morocco to Tunisia. Darker, greyer brown above than *armena*; darker grey-buff on cheeks and ear-coverts; darker and greyer below; underwing-coverts dark cinnamon. Larger and longer-tailed. SIZE (5 ♂♂, 2 ♀♀): wing, ♂ 89–96 (92·2), ♀ 89, 90; tail, ♂ 110–127 (118), ♀ 107, 112.

T. s. senegala (Linnaeus) (includes '*pallida*'): Senegal and Sierra Leone east through S Mali (north to Mopti), Ghana and Nigeria (north to c. 12°N) to S Chad and Central African Republic. Similar to *armena*, but slightly paler above and whiter below, superciliary stripe paler buff. Size similar.

T. s. notha (Reichenow) (includes '*timbuktana*'): desert border from Mali (Timbuktu) to N Nigeria (Sokoto, Kano, Bornu); intergrades with *remigialis* around L. Chad. Paler than *senegala*, rather greyish brown above; white below with pale grey breast and flanks; underwing-coverts white. Larger and longer-tailed: wing, ♂ (n = 2) 92, 91, 1 ♀ 90; tail, ♂ (n = 2) 111, 112, 1 ♀ 111.

T. s. remigialis (Hartlaub and Finsch): central Chad to W and central Sudan (Darfur, Kordofan and Nile Valley). Very pale tawny brown above; lighter rufous on wings; white below with buffish wash on breast and flanks; underwing-coverts white. Slightly larger than nominate race, and longer-tailed: wing, ♂ (n = 9) 84–92 (88·2); tail, ♂ (n = 9) 105–111 (107).

T. s. habessinica (Ehrenberg): Ethiopia, Eritrea, Djibouti, NW Somalia, and S Sudan west to Imatong Mts. Similar in size to *armena*, but darker and greyer above and below, underwing coverts greyish, superciliary stripe narrower.

T. s. warsangliensis Clarke: Warsangli, N Somalia. Similar to *habessinica*, but still darker and greyer above, cheeks greyer. Smaller: wing, ♂ (n = 3) 76–79 (78·0).

T. s. orientalis Cabanis (includes '*mozambica*' and '*confusa*'): S Somalia, coastal Kenya, E Tanzania lowlands, Mozambique (except Tete), Zimbabwe (except N and NW), Swaziland and South Africa (E Transvaal, Natal, E Cape). Poorly differentiated; slightly paler below than *armena*, with rather browner (less grey) cheeks and ear-coverts. WEIGHT: N Mozambique, unsexed (n = 6) 48–53 (51·2).

T. s. kalahari (Roberts): S Angola, N Namibia, SW Zambia (W and S Borotse), NW Zimbabwe (Hwange Nat. Park to Victoria Falls), Botswana and N South Africa (W Transvaal). Similar to *orientalis*, but slightly paler and greyer above and whiter below; superciliary stripe paler buff.

Field Characters. Length 19·5–23 cm. A large, heavy-billed, long-tailed tchagra whose lilting whistled song is characteristic of the African bush. Overlaps with all other tchagra species; tends to be bolder and more conspicuous, and told from all by combination of *black crown* and *long buffy white eyebrow* extending onto side of neck. Crowns of the other 3 species are various shades of brown, that of Three-streaked *T. jamesi* with black line through centre, and that of Brown-crowned *T. australis* with lateral black line bordering pale eyebrow; all have different voice from Black-crowned. Juvenile Black-crowned has brown crown, often mottled with black, but lacks lateral black bordering line. ♀ Blackcap Bush-Shrike *Antichromus minutus* has black crown and white eyebrow which is short and closed off by black behind eye. It is also smaller, has more uniform rufous back (nominate race with black 'V'), buffy underparts and red eye, and prefers marshy habitats.

Voice. Tape-recorded (5, 86, 88, 91–99, 101, 105, B, C, F). Song, from perch and in flight, a series of rich, lilting, warbled whistles, much sweeter than other tchagras, slower and more deliberate, becoming progressively slower and lower in pitch, 'wheeya, heeeea, hyoooee', 'whee-hoo, hweeya-hweeya-hoo, hweeya-hweeya-hoo', 'whit-tu, hweeyu hweee, hooey', 'whee, hooee, heeyoo, hyoo, hooey'. Often a duet, ♀ joining in either before, during or after ♂'s song with a hard, rattling, descending trill. In other duets one bird chatters and the other gives a single, loud slurred whistle, 'trrrrr-hoo*eee*you-trrrrr', 'tatatata-wr*eee*-tatatatata', 'chchchch-hu*wee-yee*-chchchch'.

In Morocco, ♂ has 3 main songs (R. Dawson, pers. comm.). (1) Territorial song, 1–3 whistles followed by usually 3 phrases each of 3 notes, each phrase descending and successive phrases descending: 'di-dee-de, dieu deeu dewu, dieu deeu dewu, dieu deeu dewu'. (2) Similar song but not descending; in each phrase first 2 notes on same pitch, 3rd note lower-pitched and almost disyllabic. (3) Sonorous, mournful, descending series of whistles like slowed-down song of distant Green Woodpecker *Picus viridis*; usually 4–6 notes, sometimes up to 10. ♀ utters only type (3) song. Strong local dialects; in Morocco populations near Agadir, in Oued Massa/Sidi Rabat (80 km south of Agadir) and at Tifrit (40 km north-east of Agadir), all have substantially different songs.

Aggressive interactions and territorial threats include counter-singing and excited and complex duetting, bubbling and tearing sounds, whooping whistles, 'trrrreeeeo' and clucking alarm calls. Mild alarm, a slowly repeated 'chuk ... chuk ... chuk ... ' and rattling 'krrrrr ... ', when more intense a harsh, tearing 'kzzzzrrr ... '. Clear, liquid 'chu-tu-woi' also said to be an alarm call (Zimmerman *et al.* 1996).

General Habits. Uses a wide variety of open grassy habitats with bushes or thickets and low trees. In Morocco and Algeria, scrubby semi-desert to dry open forest, plantations and gardens; prefers *Thyula* scrub, dense macchia, bushy slopes and ravines at base of mountains with wild olives, *Cistus* and holm oaks *Quercus ilex*. South of Sahara, ranges from sparse sahelian thorn scrub to moist woodland thickets in guinea and derived savannas, occurring in farmland and suburban gardens; absent from rain forest. In arid habitat in Senegal, partial to denser thickets e.g. *Salvadora*. In E Africa, in bushland and dry woodland thickets, riverine forest, neglected cultivation and gardens; from sea level up to 1900 m. In Ethiopia, occurs in forested habitats – montane olive-juniper-*Podocarpus* up to 3000 m and lowland riverine *Ficus-Acacia* as well as broad-leaved and acacia savannas, thorn bush and semi-desert scrub. In southern Africa, in light deciduous woodland with underlying rank grass or sparse cover; especially in *Brachystegia* in Zambia and Zimbabwe; absent from areas above 1400 m. Varied habitats in South Africa and Botswana include broad-leaved woodland, mopane *Colophospermum mopane*, bushveld, overgrowth in eucalypt plantations, and Natal coastal thicket; less common in thornveld and more arid scrub.

Occurs singly, in pairs or in small family groups after breeding; sometimes joins mixed feeding parties. Perches freely in small trees and bushes, and more arboreal than *T. australis*. Alert and secretive; often retires to heart of bush when alarmed, or escapes through low cover or by running along ground. Feeds mainly on ground, where it hops and creeps in grass and around bases of bushes and trees. Searches slowly and carefully, flicking debris aside with bill; jumps to take insects from low vegetation, and pulls apart dung to find prey (Cramp and Perrins 1993). Also gleans from trunks, branches and leaves, and occasionally takes aerial prey. Hammers aestivating snails off branches and eats them on ground; carefully wipes large hairy caterpillars on ground or low branch before eating them (Morocco: R. Dawson, pers. comm.). Regurgitates pellets: a fresh one 13 × 8·5 mm, weight 0·25 g. Gait well developed with hopping and leaping in cover and hopping and loping run on ground (Cramp and Perrins 1993). Flight low and hurried, usually from bush to bush, bursts of fluttery wingbeats followed by a glide; tail fanned on alighting (Harris and Arnott 1988). Closed tail moves in figure-of-8 when bird perched (G. Morel, pers. comm.). In alarm adopts horizontal posture, moves about excitedly, flicking tail and bowing, and giving discordant calls.

♂ sings from exposed perch or in song-flight, sometimes from ground. From perch adopts upright posture with head thrown back. In song-flight climbs steeply from perch to c. 10–15 m with series of wing 'frips', then ruffles crown, raises head and begins singing at start of stepped downward glides to another perch up to 60 m away (Harris and Arnott 1988). ♀ joins ♂ in duet, often zig-zagging up through bush to perch upright just below him. She may initiate duet, and occasionally joins him in combined display flight (Cramp and Perrins 1993). During territorial interactions rival ♂♂ conduct vigorous song-duels on ground or in bush, which develop into excited complex duetting by pair members (Harris and Arnott 1988). Dominance established by how tall (stretched up) bird can make itself and how vocal it is (R. Dawson, pers. comm.).

Food. Invertebrates, mostly insects: grasshoppers up to 5 cm, beetles, caterpillars, termites, mantis nymphs and beetle larvae; also small amphibians and reptiles, fruits (berries of nightshade *Lycium*: SW Arabia), and (Morocco) snails. 6 crops (Algeria) contained mainly beetles, larvae and lizards (Payne 1948). Diet of young in E Africa included moth larvae, nymphal grasshoppers and other Orthoptera, beetles, spiders and tadpoles (van Someren 1956). Scorpion fed to fledgling (Morocco: R. Dawson, pers. comm.).

Breeding Habits. Solitary nester, monogamous, territorial. ♂ defends territory of c. 4 ha, used for courtship and nesting for most of year. Pairs perform elaborate duet, both ♂ and ♀ in steeply upright posture with tail variously cocked, rotated and lowered, sometimes also fanned (Harris and Arnott 1988). Sometimes several ♂♂ may chase a single ♀. ♂ sings while ♀ brooding. Apparently single-brooded.

NEST: a compact shallow cup of rootlets, fine twigs, and tendrils, lined with finer rootlets and grass stems, rim sometimes bound with cobweb; ext. diam. 90–120, ext. depth 50–100, int. diam. 70–88, int. depth 25–45; thin enough to see eggs from below; usually 0·5–2 m high, occasionally up to 5 m (southern Africa); in fork or on horizontal branch of low thick bush or small tree. Nest built by ♂ and ♀ in just over 7 days.

EGGS: 2–3, rarely 4; E Africa (n = 33 clutches) 2–3 (av. 2·2), Nigeria (n = 10 clutches) 2; southern Africa (n = 100 clutches) 1–4 (2·5); NW Africa 2–3. White or pinkish white, spotted and blotched reddish and purplish brown, with lavender-grey undermarkings, mainly in zone at large end, markings sometimes in form of streaks and bunting-like scrawls. SIZE: (NW Africa, n = 7) 23–24 × 17·5–18; (Ethiopia, n = 7) 21–27 × 18–19·5 (24·2 × 18·4); (Nigeria, n = 18) 22–25 × 17–18 (23·8 × 17·5); (Zaïre, n = 22) 22–26 × 17·5–19 (24·0 × 17·9); (South Africa, n = 184) 22–27·5 × 17–19·5 (24·6 × 18·0). Weight (Morocco): 4·1 g.

LAYING DATES: NW Africa, Apr–June; Mauritania, Aug; Senegal, July; Gambia, Oct–Nov; Ghana, Apr–June in S, July–Sept in N; Nigeria, Feb–Sept (mostly July–Aug); Sudan, Aug–Nov (*remigialis*); Ethiopia, Apr–July; E and S Zaïre, Sept–Nov, during rains; E Africa, all months, with local laying peaks in early rains; Angola, Aug–Oct, in late dry season; Zambia/Malaŵi, Sept–May (mainly Oct–Nov); Zimbabwe, Aug–Apr (mainly Sept–Dec); Transvaal, Sept–Jan.

INCUBATION: by both sexes, mainly by ♀, who sits very tight in later stages. When ♀ sitting, ♂ flies to bottom of nest tree and works way up to nest, or sometimes alights in top of tree and moves down; ♂ and ♀ exchange calls for up to 2 min before ♀ leaves; she continues calling for several min whilst preening, scratching and stretching in nearby bush (R. Dawson, pers. comm.). Period 12–15 days.

DEVELOPMENT AND CARE OF YOUNG: young fed by both parents, leave nest after c. 16 days; family groups remain together most of austral winter (South Africa).

References
Cramp, S. and Perrins, C. M. (1993).
Harris, T. and Arnott, G. (1988).

Genus *Dryoscopus* Boie

6 species of arboreal bush-shrikes that share some characters with *Laniarius*: pied colouration in ♂♂, thick, fluffy lower back and rump plumage that can be erected into a puff (especially in ♂♂ of *Dryoscopus*), short thick plumage extending forwards from forehead to cover nostrils or at least hind border (culmen itself is bare, deep into forehead), arboreal habits, and wing fripping. Both genera show range of stoutness of bills. ♀ *D. gambensis* resembles juv. *L. luehderi* in appearance. The genera differ in *Dryoscopus* spp. being small, more or less strongly sexually dimorphic, relatively weak-legged, keeping mainly to forest and woodland canopy (*Laniarius* largely terrestrial), and in being less vocal, without much duetting (antiphonal duetting strongly developed in *Laniarius*); rump white in *Dryoscopus* ♂♂ (pied in *Laniarius*). Most *Dryoscopus* spp. have relatively weak bills, not strongly hook-tipped, shorter than head, with ridged culmen (**A**: *D. senegalensis*); however, bill of *D. sabini* is long and heavy, with culmen swollen and smooth, not ridged (**B**).

Foliage- and wood-gleaning predators, mainly of insects; territorial year-round, sedentary. Extravagant 'puff-ball' territorial and courtship displays. Nest an open cup, on or in tree fork, neater, deeper and more compact than *Laniarius* nests, in several species draped with lichen or whitish cobweb. Nest built by ♀; most incubation by ♀; other nest duties shared.

D. sabini and *D. angolensis* appear not to be quite as closely allied to the remaining 4 species, *gambensis*, *pringlii*, *senegalensis* and *cubla*, as the 4 are amongst themselves. Hall and Moreau (1970) and Sibley and Monroe (1990) treated the 4 as composing a superspecies. Short *et al.* (1990) and Dowsett and Forbes-Watson (1993) excluded *pringlii* (because of morphological differences and sympatry with *gambensis* and *cubla*) and *senegalensis* (because of sympatry with *gambensis*). The 4 are closely allied, ♂♂ (but not ♀♀) of *gambensis* and *cubla* being very alike, *gambensis* and *pringlii* being alike in plumage but different in size and habitat, and *senegalensis* and *D. cubla affinis* very similar except in bill size. Sympatry is at most marginal, sympatric species segregating by habitat and probably by means of territorial exclusion in borderline habitats. In our view the closest relationships are between *gambensis* and *pringlii* and between *senegalensis* and *cubla*. We treat them as 2 superspecies.

Endemic. 6 species: 2 independent and 2 superspecies.

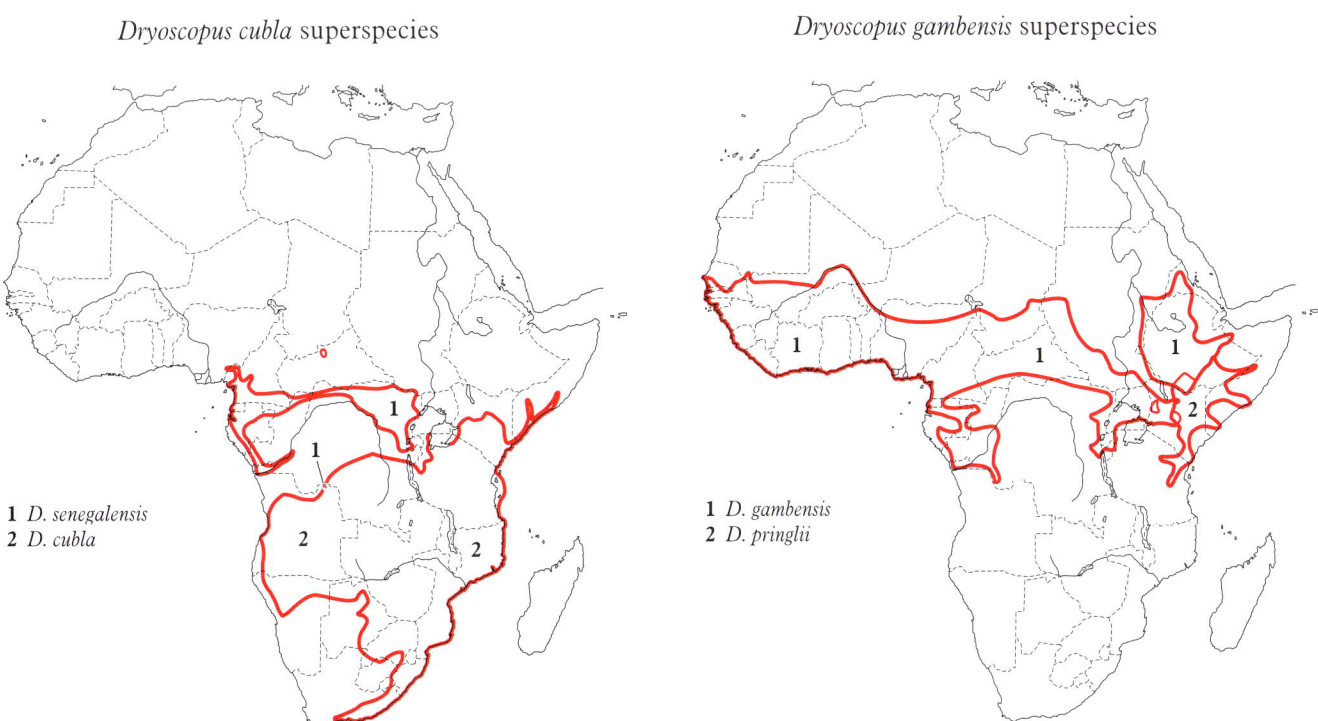

Plate 25
(Opp. p. 398)

Dryoscopus sabini (Gray). Sabine's Puffback. Cubla à gros bec.

Thamnophilus sabini Gray, 1831. Zool. Misc. 1, p. 7; Sierra Leone (see Gray, 1837, Mag. Nat. Hist., 1, n.s., p. 487).

Range and Status. Endemic resident, tall rainforests from Sierra Leone to N Angola and NE Zaïre. Sierra Leone, uncommon; records from Bo, near Matotaka, and in Sugarloaf Forest; logged and primary forest in Gola Forest; 8 birds seen in 2 years in 1930s. Liberia, not uncommon, 35 localities from coast at Cape Mount and Harbel to Nimba highlands and Wologizi range (Gatter 1997). Ivory Coast, known from 8 tall forests from Täi and Ayamé to Lamto, Yapo and Daloa; quite frequent in Täi Nat. Park and in Yapo. Ghana, not uncommon, from Sunyani, Kumasi and Tesano to forests in extreme SW, including Kakum, and on coast at Axim, Cape Coast and Accra; in 1870s was said to be very common in some areas. Togo, 1 record, Gonobé R. at 07°34′N, 0°50′E (Cheke and Walsh 1996). Nigeria, uncommon, from near Lagos to Ibadan, Ife, Benin and Umuagwu. Cameroon, frequent at Kumba, common in Rumpi Hills Forest Res., uncommon in Korup Nat. Park; records from Mbonge and southeast to Dja (Ja) River. Gabon and Mbini, uncommon, south to about 01°S. Congo, common in Nouabalé-Ndoki Nat. Park, frequent in Odzala Nat. Park. Central African Republic, known from Haut Sanga Préf. (Carroll 1988), where 1 seen more recently (Green and Carroll 1991). Angola, forests of Cabinda, and known from upper Cuango R., Lunda Norte. Zaïre, rare; known from at least 7 localities in NE and Ituri including Avakubi, Ngayu and Mawambi, 2 localities on equator (Mbandaka and upper Maringa R.), and at least 5 localities in Kasai including Kabambaie, Tshikapa and Ngombe.

Estimated densities of 2–5 pairs per km^2 (SE Liberia: Gatter 1997) and 8–10 pairs per km^2 (M'Passa, Gabon: Brosset and Erard 1986).

Description. *D. s. sabini* (Gray): Sierra Leone to Nigeria. ADULT ♂: forehead, crown, nape, lores, very narrow line under eye, hindneck, sides of neck, mantle and scapulars black with dark blue gloss; back and rump white, plumage long, very dense and fluffy; uppertail-coverts and tail black with dark blue gloss. Entire underparts white. Wings black, less glossy and less bluish than mantle; remiges curved, wing very rounded; underwing-coverts and axillaries white, underside of remiges slaty black. Underside of tail glossy black. Bill black, upper mandible pale bluish except for black tip and cutting edges, and robust, the culmen flat-convex and wide between nostrils; eyes dark brown; legs and feet blue-grey. ADULT ♀: forehead, crown, nape and hindneck mid grey, merging into olive-grey on mantle, warm rufescent brown on back and scapulars, and rufous on rump, tail and wings. Back and rump plumage thick and fluffy, though less so than in ♂. Pale grey or whitish above eye, joining with narrow white ring around eye; in some birds a short, narrow, pale line above lores. Lores grey; ear-coverts grey-brown with thin whitish streaks (feather shafts). Chin, throat, breast, flanks, sides of belly and undertail-coverts uniform pale rufous. Centre of belly and vent snow-white. Underwing-coverts and axillaries pale rufous. Undersides of remiges dark grey and of rectrices pale brown. SIZE: wing, ♂ (n = 10) 80–87 (83.3), ♀ (n = 9) 78–85 (81.6); tail, ♂ (n = 7) 67–75 (71.9), ♀ (n = 7) 72.5–86 (74.2); bill, ♂ (n = 3) 28–30 (29), ♀ (n = 3) all 25; bill depth at nostrils, ♂ (n = 4) 8.0–8.9 (8.4), ♀ (n = 3) 7.4–8.6 (8.1). WEIGHT: (Liberia) ♂ (n = 3) 31–44 (38.5), ♀ 32.4–39.2 (35.8), 2 imm. ♂♂ 41.4, 42.8.

Dryoscopus sabini

IMMATURE: ♂ like adult ♀; young ♂ moulting into adult plumage replaces grey and rufous feathers on upperparts with black ones irregularly, and has blotchy mix of grey and black feathers on head, rufous and black feathers on back, tail and wings, and rufous and white patches on underparts.

NESTLING: not described.

D. s. melanoleucus (J. and E. Verreaux): Cameroon to Angola and Zaïre. Tail shorter: ♂ (n = 5) 56–68.5 (61.2), ♀ (n = 6) 58.5–65 (62.3); and tail of ♀ slightly darker than in ♀ *sabini*. Cline of increasing wing length from 80–86 in Cameroon to 85–94 in Zaïre.

Field Characters. Length 18–20 cm. ♂ like Red-eyed Puffback *D. senegalensis* but larger, with longer bill and tail; *glossy black*, including *wings*, with *pure white* lower back, rump and underparts; eye dark, not red. ♂ Pink-footed Puffback *D. angolensis* has grey back and wings and is pale grey below; ♂ Northern Puffback *D. gambensis* has grey rump and patterned wing (feathers edged white). ♀ brighter than ♀ Pink-footed or Northern Puffback: top of head grey but upperparts and tail *tawny*, underparts bright cinnamon; rufous wings lack white feather edgings. Where it occurs with Red-eyed Puffback, inhabits forest interior while Red-eyed occurs at the edges.

Voice. Tape-recorded (104, B, BRU, CHA, HUG, KEI, LEM). Song, given from canopy, reminiscent of Yellow Longbill *Macrosphenus flavicans*, a series of 11–14 notes descending the scale, short and clipped at first and becoming longer, 'pip-pip-peep-peep-pee-pee-peee-peee ...'; notes high, thin and clear, some with ringing quality; speed varies from 2 notes per s to 3 notes in 2 s.

Variety of harsh calls, including descending 'cheeerrrr' lasting up to 2 s, prolonged (5–7 s) dry rattle on one pitch, grating 'dzhup' or shorter 'jip' and 'djip', double 'tuk-rrraaa' or 'di-rrrrraaaaa'. In apparent duet, one bird gives hard rattle and second joins in before first rattle has ended; in aggression a grating, up-slurred 'jooway' or 'joowy'.

General Habits. Inhabits canopy of high primary forest, logged and secondary forests, primary forest on islets in rivers, edges of clearings, and suburban gardens with tall timber; in Odzala, Congo, in riparian forest and less commonly in swamp forest.

In pairs, usually seen as members of mixed-species foraging flocks; forages by gleaning conspicuously on twigs and on and under leaves and vines of 3–20 mm diam. in outermost branches; examines leaves closely for prey; keeps at heights above ground of 20–45 m (Liberia), mode 30–35 m (Gatter 1997); often in canopy of emergent trees and in curtains of hanging lianas; hardly ever descends into middle strata (Gabon). Probably territorial. Makes fripping noises with wings.

Food. Insects, including grasshoppers, beetles, moths, large caterpillars, and termites.

Breeding Habits. Unknown, except for indications of breeding season. Courtship display not described, but no reason to suppose it differs from that of congeners.
LAYING DATES: Liberia, (enlarging gonads June, Aug–Sept, Nov; courtship display Aug, Oct–Nov); Ghana, (courtship display, July); Nigeria, (courtship display, June, enlarged gonads, Oct); Gabon, (family parties at end of short dry season and in following wet season).

Reference
Brosset, A. and Erard, C. (1986).

Dryoscopus angolensis Hartlaub. Pink-footed Puffback. Cubla à pieds roses.

Plate 25 (Opp. p. 398)

Dryoscopus angolensis Hartlaub, 1860. Proc. Zool. Soc. London, p. 111; Bembe, Loanda Province.

Range and Status. Endemic resident with 3 main centres (W Cameroon, NW Angola, and Rift Valley mountains) and various montane outliers.

Nigeria, 2 records on Obudu Plateau. Cameroon, known at sea level at Limbe (Victoria), uncommon on Mt Cameroon at 700–1280 m, fairly common on Mt Kupé at 760–1950 m, common to abundant on Mt Nlonako at 1200–1750 m and at Dikume Balue, Rumpi Hills, at 1000–1700 m, and quite common in Bali-Ngemba Forest Res. at 1700–1800 m; also known from Bafang, Ngaoundéré, Yoko, Yaoundé and Lolodorf. Gabon, ♀ collected Mouila in 1984 (identified by C. Erard; Y.-M. de Martin de Viviès, pers. comm.). S Congo and W Zaïre, not uncommon in Mayombe (Col de Bamba, Temvo, Makaia Ntete), inland to Thysville. Angola, locally common in Zaïre, Cuanza Norte and Malanje Provs and at Gabela, Cuanza Sul. Central African Republic, record(s) in Manovo-Gounda-Saint Floris Nat. Park (Carroll 1988) require confirmation. Sudan, only in Imatong Mts, uncommon at 2400 m. E Zaïre, widespread from Arebi in N to Baraka in S; common in Itombwe Highlands, at 1170–1830 m, and near Baraka at 2200–2300 m. Uganda, common in Impenetrable Forest, elsewhere local and uncommon at 900–2500 m from Kibale to Malabigambo; also on Mt Elgon. Kenya, confined to Kakamega and Nandi Forests where uncommon; an old record from Mt Elgon at 2400 m; also reported from Tugen Hills. Rwanda, up to 2300 m in W Nyungwe Forest. Not in Burundi (Dowsett and Dowsett-Lemaire 1993), though seemingly 1 record (Hall and Moreau 1970). W Tanzania, Minziro Forest in N (N.E. Baker, pers. comm.) and Mt Kungwe in W.

Description. *D. a. angolensis* Hartlaub: (includes '*boydi*'). E Nigeria, Cameroon, Gabon, Congo, W Zaïre, Angola. ADULT ♂: forehead, crown, nape, hindneck, sides of neck and upper mantle black with dark blue gloss; lores, upper cheeks and ear-coverts

Dryoscopus angolensis

dull greyish black. Lower mantle dark grey; scapulars, back and rump mid grey to pale grey (feathers mid grey with white bases that can show through); uppertail-coverts dark grey. Tail very dark brown, feathers structurally barred blackish. Chin and throat greyish white, grading into pale grey on breast and flanks; belly almost white in middle; thighs, vent and undertail-coverts pale grey. Wings uniform dark greyish brown. Underside of flight feathers shiny dark grey; underwing-coverts and axillaries pale grey or whitish. Underside of tail shiny dark grey. Bill black; eyes dark brown, edge of sclerotic membrane bright cobalt-blue or violet-blue, rim of eyelids dusky, tinged dark red (Chapin 1954); legs and feet pale lavender-pink, lilac-brown, pink-violet or lilac-grey with grey claws. ADULT ♀: forehead, crown, hindneck, lores, upper cheeks, upper ear-coverts, sides of neck and upper mantle uniform bluish grey, merging into olive-brown on lower mantle and scapulars; back and rump olivaceous grey; uppertail-coverts olive-brown. Tail greyish brown. Lower cheeks and lower ear-coverts rufous-buff, slightly streaky. Chin, throat, breast and upper flanks uniform light orange or rufous; lower flanks and thighs pale rufescent brown; belly white, greyish white or cream-white; undertail-coverts whitish or pale rufous. Underwing-coverts and axillaries bright rufous; underside of flight feathers glossy dark grey, with inner border of feathers proximally pale rufous. Underside of tail ochreous brown. SIZE: wing, ♂ (n = 8) 79–83 (80·75), ♀ (n = 13) 75–83 (79·5); tail, ♂ (n = 8) 63–65 (63·75), ♀ (n = 12) 63–68 (64·0); bill, ♂ (n = 8) 18–21 (19·25), ♀ (n = 8) 18–22 (19·6); tarsus, ♂ (n = 8) 22–24 (23·25), ♀ (n = 9) 23–24 (23·4).

IMMATURE and NESTLING: not known.

D. a. nandensis (Sharpe): S Sudan, NE and E Zaïre, Uganda, W Kenya, W Rwanda. ♂ like ♂ *angolensis* but forehead, crown, nape, lores, upper cheeks and upper ear-coverts dull blackish grey, grading through hindneck into glossy bluish black on upper mantle. ♀ does not differ from ♀ *angolensis*. WEIGHT: ♂ (n = 8) 32–37 (34·9), ♀ (n = 3), 24, 33, 35.

D. a. kungwensis. Moreau: W Tanzania: Kungwe Mahale (Mahare) Mt. Like *nandensis* but ♂ with greyer crown and neck.

TAXONOMIC NOTE: black upper mantle is sharply delineated from dark grey lower mantle in Angolan and some Mt Cameroon birds but intergrades in other Mt Cameroon birds and ones from e.g. Dikume Balue (Cameroon). Together with a possible size cline (smaller in N of range, greater in S?) it might be the basis for resurrecting race *boydi*.

Field Characters. Length 15–15·5 cm. A forest puffback, often high up in trees. ♂ has solid *black crown* and *dark eye* (♂ Northern Puffback *D. gambensis* has red eye; Bocage's Bush-Shrike *Malaconotus bocagei* has prominent white supercilium); mantle black but back and rump grey; *underparts pale grey*, not pure white as in Sabine's and Red-eyed Puffbacks *D. sabini* and *D. senegalensis*; wings plain grey, without white feather edgings. ♀ has *pinkish cinnamon* underparts, brighter than ♀ Northern Puffback, and lacks white wing-edgings; similar below to ♀ Sabine's Puffback but shorter, with smaller bill; above grey-brown, not tawny.

Voice. Tape-recorded (104, B, C, F, GREG, KAEST, LEM, McVIC). A variety of harsh, clipped 'chukking' calls, 'jack', 'djik', 'jaa', 'chit-tup', 'jyoop-jyoop', 'jrrrit', and a dry, hard trill lasting *c.* 1 s; calls in chorus may indicate alarm; also an emphatic sharp, grating 'cheow' repeated at an uneven pace up to 30 times.

General Habits. Inhabits mature montane forest, old second growth, also young, scrubby, secondary growth (Rumpi Hills, Cameroon: Stuart 1986) and edges and interior of coffee forest (Angola: W.R.J. Dean, pers. comm.). Usually in pairs but sometimes singly; keeps mainly in foliage high up in trees although often comes down to mid levels of forest (Cameroon, Itombwe, Zaïre), hopping and gleaning foliage on large sub-canopy branches, usually silently, much like large warblers (Dowsett-Lemaire 1990). Commonly joins mixed-species foraging flocks. Quite vocal, both sexes using the harsh, churring call (Serle 1950, Stuart 1986).

Records on Cameroon coast suggest that part of Mt Cameroon population may move to low levels in non-breeding season.

Food. Insects, mainly beetles, also grasshoppers and caterpillars.

Breeding Habits. Barely known.

NEST: ♀ seen carrying large clump of green moss to presumed nest between 8 and 13 m above ground in Impenetrable Forest, Uganda (T. M. Butynski, pers. comm.).

LAYING DATES: Cameroon, (breeding condition, Apr); Zaïre, Itombwe, (various evidences, Jan, Apr–June, Nov), Uganda, (nest-building, Sept).

References
Chapin, J. P. (1954).
Prigogine, A. (1971).
Stuart, S.N. (1986).

Plate 25 *Dryoscopus senegalensis* **(Hartlaub). Red-eyed Puffback. Cubla à oeil rouge.**
(Opp. p. 398)
Sigelus senegalensis Hartlaub, 1857. Syst. Orn. Westafr. [Bremen], p. 112; 'Senegal', probably = Gabon.

Forms a superspecies with *D. cubla*.

Range and Status. Endemic resident, secondary forest from Nigeria to Uganda and Central African Republic to N Angola.

Nigeria, 3 records: Oshogbo and Obudu (1961, 1980) (Elgood *et al.* 1994). Cameroon, very common at Kumba; record at Bali (Bamenda Highlands), widespread in S (Bitye, Efulen, Bonga – where given as abundant in 1920s, Olounou, Aboulou, Eseka, Mang, Djaposten). Mbini. Gabon, widespread in N from Ogooué (Ogobai) and Moonda (Mondah) rivers east at least to M'Passa, and throughout Rés. de la Lopé; records at Mimongo, Kango, Cape Esterias and Mbigou; frequent around Mouila (Y.-M. de Martin de Viviès, pers. comm.). Congo, common in more humid parts of Odzala Nat. Park (Mbomo-Mbandza), common Nouabalé-Ndoki Nat. Park (Bomassa region); widespread in Mayombe (Sibiti, Brazzaville, Mah). Central African Republic, frequent in Bamingui-Bangoran Nat. Park. Zaïre, widespread in NE (Uele), the population possibly confluent with that in NW Congo; common in Ituri (Bosobangi, Mawambi) and Semliki; also occurs west of Bukavu, and in W Zaïre in region of lower Congo R. valley, downstream from Kwamouth. Uganda, restricted to Bwamba lowlands. Rwanda, pair seen in Shava Valley (Dowsett-Lemaire 1990). Angola, known from Cabinda and Dundo, Lunda (W. R. J. Dean, pers. comm.).

Density of 2–4 pairs per 10 ha, N Gabon (Brosset and Erard 1986).

Dryoscopus senegalensis

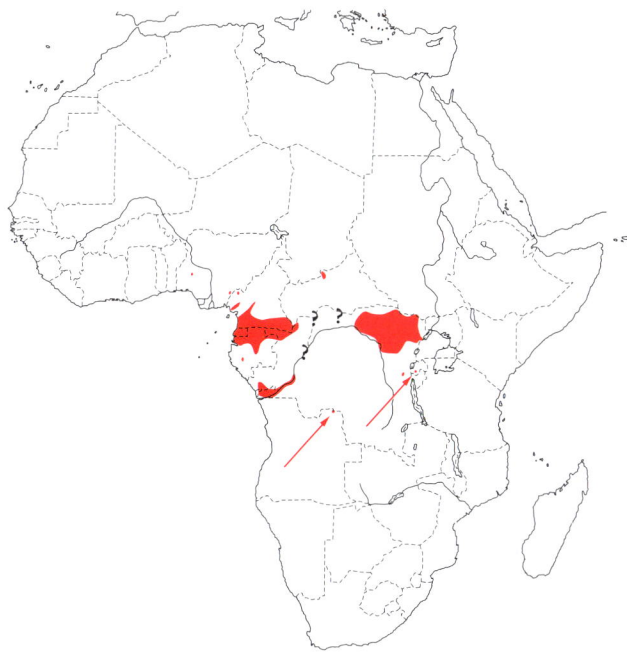

A

Description. ADULT ♂: forehead, crown, nape, lores, cheeks, ear-coverts, hindneck, sides of neck, mantle and scapulars black, glossed dark blue; back and rump white; uppertail-coverts glossy bluish black. Tail glossy black. Underparts uniform white with cream wash; patches can look greyish where dark grey feather bases ill-concealed beneath white feather ends. Chin with many filoplumes 5–8 mm long. Wings black, feathers glossy at edges; underside of flight feathers greyish black; underwing-coverts and axillaries white. Underside of tail black. An albino known. Bill black; eyes orange-red; legs and feet grey. ADULT ♀: like ♂ but a white stripe from nostril to above eye, lores black, upper cheeks white not black; lower mantle, scapulars, wings and tail dark greyish brown, or mantle sometimes glossy black; back and rump smoky mid grey, or white. Mantle and rump shades variable; of 29 ♀♀ examined, 4 (Ja R., Cameroon) have white rump (5 from Ja R. have grey rumps), 3 (Efulen, Cameroon) have white and grey rumps (2 from Efulen have grey rumps), and all 4 from E Zaïre white or pale grey rumps; remainder are intermediate. 2 ♀♀ from Kumba (Cameroon) have grey mantles and 2 have black mantles; most Ja R. ♀♀ have black mantles; most Victoria (Cameroon) ♀♀ have grey mantles; others are intermediate. SIZE (17 ♂♂, 15 ♀♀): wing, ♂ 73–83 (81·6), ♀ 72–82 (76·3); tail, ♂ 57–66 (62·5), ♀ 58–68 (60·2); bill, ♂ and ♀ 18–19 (18·7); tarsus, ♂ 22–23 (22·4), ♀ 21–22·5 (21·8) WEIGHT: 1 ♂ 34, 1 unsexed 30.

IMMATURE: assuming skins correctly sexed, imm. ♂ exactly like adult ♀. Juvenile ♂, almost fully fledged, like adult ♂ in colour but plumage fluffy; no sign of barring or pale tips to e.g. wing coverts (Bates 1911).

NESTLING: not known.

Field Characters. Length 15–18 cm. A noisy and demonstrative bird of lowland forest edge that cannot pass unobserved (Christy and Clarke 1994). *Glossy black, including wings,* with *pure white* lower back, rump and underparts; smaller than Sabine's Puffback *D. sabini*, with shorter bill and tail, and conspicuous *red eye*. ♀ like ♂ except for *white stripe* from nostrils to above eye; duller above, rump sometimes greyish; readily distinguished from ♀♀ of sympatric puffbacks by white, not cinnamon, underparts, and from Northern Puffback *D. gambensis* by lack of white in wing.

Voice. Tape-recorded (104, B, C, F, LEM, SALA, ZIM). Song from perch of 2 types, a long (1·5–2 s) rolling snore, loudest in the middle, beginning with a tearing sound, 'tzzzzrrrreeeeawrrrrr', ending with a single 'jit' (second bird?) or just dying away; or a shorter, upslurred double 'djit-djurrreeee' or 'djit-jijurrreeeeya'; ♀ joins in with softer 'tsap-tsap-tsap-tsap ...' or (more often) scolding 'chrraaay' (Christy and Clarke 1994). Call, given in flight or while foraging, a loud, high-pitched, down-slurred 'chyow' repeated on same pitch up to 20 times at rate of 2 per s. ♂ also gives a fast, dry, *Malaconotus*-like rattle.

General Habits. Inhabits secondary forest, large thickets, forest clearings and gallery forests; not so common in very open forest. Keeps chiefly in canopy of tall, isolated trees, also in smaller trees in old clearings. At M'Passa, Gabon, commonest in thin woods with open spaces and in isolated trees left in clearings; also in cacao and avocado plantations, and in clearings in primary forest (but not in primary forest itself: Brosset and Erard 1986); in Rés. de la Lopé, gallery forest and small forest patches; penetrates primary forest but only a short distance from the edge; occupies mainly middle and upper levels of vegetation but sometimes lower levels at forest edge (Christy and Clarke 1994).

Occurs in pairs and family parties of 3–4 birds. Often in mixed-species foraging flocks. Vocal, but stays out of sight. Forages at 10–25 m above ground, sometimes higher, rarely lower. Strongly territorial; ♂ regularly traverses territory, singing, fripping wings as he flies from tree to tree, quite often raising white puff on rump (**A**). ♂ often aggressive towards Bocage's Bush-Shrike *Malaconotus bocagei*, attacking it; once a ♂ thought to have misidentified a Bocage's as a ♀ Puffback and repelled it (Brosset and Erard 1986).

Food. Prey includes beetles, hemipterans, orthopterans and their eggs, dictyopterans, termites, large caterpillars, a spider and a tiny frog (Brosset and Erard 1986, Chapin 1954).

Breeding Habits. Courtship seems to be like ♂'s territorial patrols, with wing-fripping flights and rump feathers raised into puff ball (which also occur during nest construction and incubation phases: Brosset and Erard 1986).

NEST: a small, compact cup, evidently like nest of Black-and-white Flycatcher *Bias musicus*, made of fine rootlets and vegetable fibres held together with spider web, the outside uniformly and entirely covered with pieces of lichen. Placed 4–10 m above ground (av. 10 m), fairly visible from ground, on multiple horizontal fork, near end (sometimes middle) of branch. Built by ♀, accompanied by displaying ♂ (Brosset and Erard 1986).

EGGS: 2. White, with a contrasting band of reddish speckles around large end.

LAYING DATES: Cameroon, Feb, Nov (and nestling Aug); Gabon, Dec–Mar; NE Zaïre, (in breeding condition throughout year; nestling, Sept: Chapin 1954).

INCUBATION: by ♀ only. Period: 16 days.
DEVELOPMENT AND CARE OF YOUNG: young fed by both parents. ♀ broods young and ♂ passes food items to her to give to them; both parents feed young on independent short nest visits (F. Dowsett-Lemaire, pers. comm.). Fledglings remain with parents practically until their next nesting.

References
Bannerman, D.A. (1939).
Brosset, A. and Erard, C. (1986).

Plate 25
(Opp. p. 398)

Dryoscopus cubla (Shaw). **Black-backed Puffback. Cubla boule-de-neige.**

Lanius cubla Shaw, 1809. Gen. Zool., 7, pt. 2, p. 328; Knysna, *ex* Levaillant.

Forms a superspecies with *D. senegalensis*.

Range and Status. Endemic resident, E African and southern African mesic woodlands. Somalia, fairly common in lower Jubba and lower Shabeelle valleys and in extreme south; north to 03°30′N. Kenya, widespread, frequent to common, from sea level up to 2200 m, almost entirely south of equator. Tanzania, as mapped (N. Baker, pers. comm.), including Minziro Forest in NW, Zinza area (S end of L. Victoria), probably more of Kigoma and Tabora than shown, and Zanzibar and Mafia. Zaïre, Katanga; Kasai and S Kivu north to about 04°S. Rwanda, occurs in W and 'atteint certainement l'extrême sud de l'Uganda' (Vande weghe 1981). Also recorded W Burundi. Zambia, throughout (98% of 301 squares, R. J. Dowsett, pers. comm.). Zimbabwe, Malawi and Mozambique, throughout. Angola, common in most parts, absent from NW and SW. Namibia, Botswana, Transvaal and E Cape, frequent to common, as mapped (Harrison *et al.* 1997); throughout Swaziland and Natal. Orange Free State, local in montane forests along Natal border; records from Vredefort district. Density of 1 pair per 42 ha in broad-leaved woodland in South Africa, and in S Mozambique of <5 birds per 100 ha of mopane, 7 of acacia, 21 of miombo and 11 per 100 ha of other broad-leaved woodlands (Parker 1999).

Dryoscopus cubla

Description. *D. c. cubla* (Shaw): South Africa. ADULT ♂: forehead, lores, narrow line under eye, ear-coverts, crown, nape, hindneck, sides of neck and mantle glossy bluish black; back and rump white; uppertail-coverts and tail glossy black. Entire underparts creamy greyish white. Primaries, secondaries and tertials black, primaries narrowly and tertials broadly edged white, greater primary coverts black, narrowly edged and tipped white, greater, median and lesser coverts black, broadly edged and tipped white; underside of flight feathers shiny dark grey; underwing-coverts and axillaries white. Bill black; eyes bright orange-red to orange-yellow; legs and feet grey. Eye-colour appears to vary between montane forest and woodland populations (Harris and Arnott 1988). ADULT ♀: like ♂ but less intensely black, white parts creamy rather than snowy. Upper mandible black, lower mandible blue-grey or pale grey; palate black; eyes bright orange or yellow-orange; legs and feet blue-grey. SIZE (mainly *cubla*): wing, ♂ (n = 62) 76·5–88 (81·8), ♀ (n = 41) 75–85 (78·8); tail, ♂ (n = 36) 60–73 (66·7), ♀ (n = 36) 61–74·5 (66·7); bill, unsexed (n = 31) 17·5–21; tarsus, unsexed (n = 31) 21–24. WEIGHT: ♂ (n = 125) 19·3–36 (27·1), ♀ (n = 98) 21–30·4 (25·4) (Maclean 1993).

IMMATURE: juvenile like adult ♀ – underparts from chin to undertail-coverts dull white, feathers fluffy; later, immature has throat, breast and usually belly orange-buff (Brooke 1980). Eyes brown, bill horn; feathers of upperparts blackish grey with buff tips; underparts white, turning buff within days of young leaving nest (Harris and Arnott 1988).

NESTLING: naked, blind, skin flesh-brown above and pink below; gape and mouth yellow, with mouth spots; legs grey-brown; soon develops coat of fluffy white down.

D. c. okavangensis Roberts: NW Cape Prov. (Molopo R.) north to Namibia, S Angola, Botswana, Zambia (north to Balovale). Intergrades with *hamatus*. ♂ like ♂ *cubla* but white edges to wing feathers broader, underparts whiter, eyes red; ♀ like ♀ *cubla* but feathers of forehead and mantle edged greyish, rump pure grey rather than creamy, and lores pale buffy (not blackish). WEIGHT: 3 ♂♂ 24, 26, 27, 1 ♀ 28, 1 unsexed 25.

D. c. hamatus (Hartlaub) (includes '*chapini*'): Kenya west of Rift Valley to Kavirondo Gulf; Tanzania except NE; Zaïre, N Angola, N and W Zambia, E Zimbabwe, Malaŵi, lowland E Transvaal, N Zululand. Like ♂ *okavangensis* but scapulars broadly edged white, making large white shoulder patch, and underparts almost pure white; ♀ like ♀ *cubla* but lores whitish. Cline of increasing size from S to N. SIZE (6 ♂♂, 5 ♀♀, Tanzania): wing, ♂ 80·5–87 (84·3), ♀ 80·5–84·5 (82·5); tail, ♂ 59–63 (62·7), ♀ 60–67·5 (64·75); bill, ♂ 17–19·5 (18·2), ♀ 17·5–19 (18·3); tarsus, ♂ 23–25·5 (24·6), ♀ 22·5–24 (22·8). WEIGHT: unsexed (n = 7) 21–29 (24·0).

D. c. nairobiensis Rand: Kenya and NE Tanzania, east of Rift Valley from Archer's Post to Kilosa, east to Mt Endau, Kibwezi, Tiva R., Voi R., Maungu and Usambara Mts. ♂ with wings blacker than in ♂ *hamatus*, on account of white edges to all feathers being narrow, and white shoulder (scapular) patch less conspicuous; ♀ with lores dusky (whitish in ♀ *hamatus*). WEIGHT: ♂ (n = 5) 15–29 (21·2), ♀ (n = 4) 18–24 (20·25).

D. c. affinis (Gray): Somalia, Kenyan and Tanzanian coastal lowlands south to Rufiji R.; Zanzibar and Mafia; inland in Kenya to Bura and Shimba Hills. ♂ with all of scapulars and wings black; ♀ almost or entirely without whites in scapulars and wings, and with lores dusky. Pure *affinis* may occur only on Zanzibar and adjacent mainland coast; from Dar-es-Salaam to coastal S Somalia many birds are pure *affinis* but others, especially inland, have variable amounts of white in scapulars and wings and there *affinis* intergrades extensively with *nairobiensis* and *hamatus*. Strikingly heavier than *nairobiensis*: WEIGHT: ♂ (n = 17) 21–40 (29·9), ♀ (n = 20) 24–40 (30·0).

Field Characters. Length 15–18 cm. Common in bird parties in woodland canopy, drawing attention with its loud 'chick-weeeo' call. ♂ strikingly pied, with *black* crown and back, *pure white* rump and underparts; wings patterned black and white, *eye red*. ♀ duller, whites less pure, rump grey; in E Africa has black lores and white supraloral line; in S Africa lores white but has short white supercilium. In race *affinis* (coastal E. Africa), wings all black in both sexes. Does not meet similar-looking Red-eyed Puffback *D. senegalensis*, and is the only puffback present in most of southern Africa; closely approaches range of Northern Puffback *D. gambensis* in central Africa, though in Kenya at least they do not overlap (Zimmerman *et al.* 1996). Smaller than Northern Puffback, with more slender bill; eye blood-red (orange-red in Northern); ♂♂ similar, but ♂ Northern less clean-cut, with grey rump and scapulars, grey wash on underparts, more like ♀ Black-backed but without white supercilium. ♀ Northern completely different, with cinnamon underparts. For differences from Pringle's Puffback *D. pringlii* see that species.

Voice. Tape-recorded (14, 21, 75, 86, 88, 91–99, 104, KEI, LEM). Song in display flight a loud whistled 'dzlit-tuweeeyoo' or 'tzr-t'weeeyo'; less often with second note more pure and shrill, 'zzt-wheee'; repeated up to 10 times at rate of *c.* 1 per s. Whistles are clear but not ringing or melodious. In duets, one bird (♂?) gives 'tuweeeyo', second answers with tearing scold, 'chizzzrrrrr', loud 'cheeerrrr' or nasal 'wraanh'; sometimes 'scolding' bird goes first, immediately followed by mate, 'jeeeez-tweeeyu'; or one bird gives up-slurred 'tuwheet', answered by down-slurred 'peeuw'. Repeated whistles (songs?) from perch, upslurred 'tu-tu-weee, tu-tu-weee ...', down-slurred, ringing 'tyew-tyew-tyew ...', high, pure 'weeya-weeya-weeya ...' and lower-pitched 'tyow-tyow-tyow ...'. Courting ♂ calls 'chak-chak-chak' on the wing (Zimmerman *et al.* 1996). Other calls include hard, dry 'dip-dip-dip ...', grating 'jit-jaa' or 'jit-jut-jaa', buzzy 'zzzeeee', low 'charr' or 'charrway', and in flight, 'clip-clip-clip-clip-clip'. Sometimes combines variety of clicks, snores, chatters and whistles into conversational medley, like shrike *Lanius*.

General Habits. Lives in dense woodland; in southern Africa all types of indigenous woodland and forest, commoner in dense savanna woods than in forest or sparse woodland, occasionally in open acacia savanna, especially where modified by bush encroachment; commonly in black wattle *Acacia mearnsi* plantations; absent from other exotic timber plantations, but occurs commonly in well-wooded parks and gardens (V. Parker *in* Harrison *et al.* 1997); tall submontane forest at 2000 m in N Malaŵi.

Singly or in pairs, spending most of time in upper canopy of trees; sometimes comes low, rarely to the ground. Outside breeding season up to 15 birds, all or mostly juveniles, congregate noisily in trees, displaying rump feathers (see below), calling, wing fripping and chasing. Occasionally hawks for insects in the air. Often joins mixed-species foraging flocks. In treetops moves swiftly and silently, in horizontal posture, foraging by gleaning from foliage and wood; explores galls and pods at end of branch by momentarily hanging upside down (Harris and Arnott 1988). Territorial all year; territory *c.* 4 ha (in dense lowland woods). On Nyika Plateau, Malaŵi, at 2000 m altitude, pairs occupy all forest patches of 7·5 ha or more; at Zovotchipolo, 2200 m, 2 forests of 8·5 and 12 ha hold a pair each (Dowsett-Lemaire 1983). ♂ advertises and defends territory by calling, perched in exposed position in upright posture on top of tree, crest and throat feathers slightly raised, tail half fanned; 'k-wee-u' call may be repeated 60 times in succession. ♀ sometimes responds in duet, but duetting is poorly developed, as in all congeners, in relation to *Laniarius* bush-shrikes. ♂♂ signal territorial threat by counter-singing rapidly with whistles and tearing sounds, ♀♀ joining in with 'tzzerrr' calls. In high intensity interactions, calling speeds up and rises in pitch, then ♂ clicks bill stridently, often during descending, bouncy flight with wings whirring, head held up, tail fanned, and back and rump feathers fluffed out into dense snow-white hemisphere about size of golfball or small powder puff (Afrikaans name is Sneeubal – Snowball). On alighting, ♂ crouches with head and foreparts down, and sways slowly from side to side. ♀ clicks bill and occasionally fluffs rump feathers out (but feathers are grey and shorter than ♂'s).

Sedentary. 4 ring recoveries were within 10 km of ringing site.

Food. Diet of adults includes beetles, large numbers of caterpillars, *Salvadora* fruits and *Acacia* buds. Food items given to nestlings include butterflies, moths, moth caterpillars (Noctuidae, Saturniidae), long-horned grasshoppers, crickets, dragonflies, beetles, larvae, worms and small lizards, and once a stick insect (from which the legs had been removed) 115 mm long (van Someren 1956, Chittenden 1977).

Breeding Habits. Monogamous, territorial. ♂ sexual display seems to be just the same as territorial display (see above): short flights with bill-clicking, whirring wings, fanned tail and puffed-out rump; ♂ chases ♀, pair flying with loud wing fripping, bouncing swiftly through the branches (van Someren 1956). ♂ approaches ♀ in head-down posture, swaying body, with tail fanned, and white rump puff so erect that it almost envelops slightly drooping wings (**A**). ♀ bends down and partly opens wings; ♂ copulates, both birds keeping silence, then ♂ flies off with loud calling and wing fripping. One pair remained together for 4 seasons. Single brood per season.

A

NEST: small, neat, deep cup, made of dry grass, roots and thin strips of bark, bound with white spider web and some plant down, sometimes decorated outside with pieces of lichen, dry leaves and bark, and lined with fine wiry dry grass (**B**). Sited (n = 109) 2–15 (6) m high in tree, in upright fork. Nest usually well hidden; a sapling with nest in top was defoliated by caterpillars, exposing eggs to sun, but parents did not desert (van Someren 1956). Nest built by ♀, taking *c.* 10 days; ♂ escorts her to and from nest, calling and wing fripping; ♂ is extremely aggressive to other birds at this stage (Harris and Arnott 1988). ♀ first places spider web on selected site, puts down strip of bark, then works spider web across, fastening it to substrate. As the nest builds up in crotch of fork, spider web is bound around supports. ♂ sometimes carries material to nest, but ♀ always does the construction. She sits in nest, moulding it to shape of her breast. Finally completed nest is firmly knit and bound with spider web on the outside and bits of bark and lichen are worked in and held by cobweb, sometimes with cocoons still attached (van Someren 1956). ♀ builds mainly in early morning, making up to 20 (av. 8) visits per day with material carried in bill; spider silk is brought concealed inside mouth and at nest is slowly brought out and woven in (Fenn 1975).

EGGS: 2–4 (in Zambia, 2 c/1, 19 c/2, 21 c/3). Smooth, matt. White, pale cream or pink or rich pink, with fine brown or grey spots and streaks, often mainly around broad end. Erythristic eggs known. SIZE: (n = 52) 19·2–24·4 × 14·9–17·7 (22·2 × 15·9).

LAYING DATES: Somalia, (fledglings Oct); E Africa, Region C, Jan–Mar, Region D, Jan–May, July, Sept–Oct, Dec (mainly Apr), Region E, Sept, Dec; Zanzibar, July–Dec; Zambia, July–Dec and Mar, mainly Sept (52% of 71 records) and Oct–Nov (38%); Angola, (active gonads, Oct); Zimbabwe, all months except May, peaking in Sept–Nov; Mozambique, Oct, Mar; Botswana, Feb, Oct–Nov; Transvaal, mainly Sept–Jan (peak 1–2 months later than in Zimbabwe); Natal, Oct–Dec.

INCUBATION: mainly by ♀; ♂ relieves her for short periods. ♂ feeds ♀ on nest. On departure, ♂ said sometimes to fly away fast in zigzags, with rump puff expanded. Both ♂ and ♀ thought to initiate nest relief, by calling. Period: 13 days.

DEVELOPMENT AND CARE OF YOUNG: young brooded by ♀, who is given food by ♂ and passes it to chicks; later, young fed by both parents. Parent waits for and removes faecal sac after a feed. At age 7 days, in a 5-h period, young fed 21 times by ♂ and 12 by ♀, who also brooded them (van Someren 1956). Large caterpillars are well pulped by parent before being given to chick; large long-horned grasshoppers were held by ♂ in bill whilst ♀ bit off antennae and legs (van Someren 1956). At one nest ♀ disappeared, ♂ continued to feed the 3 chicks until nightfall but did not brood them at night; ♀ reappeared 2 days later and resumed both feeding and brooding (Chittenden 1977). Nestling period: *c.* 18 days. Young out of nest continue to beg from and are fed by parents for at least 3 weeks; juveniles appear to be tolerated in parental territory until next nesting season.

B

BREEDING SUCCESS/SURVIVAL: ringed bird recovered after 10 years (Zambia: Scott 1985). Adult killed by African Goshawk *Accipiter tachiro* and nestlings killed by *Crematogaster* ants. Occasionally parasitized by Klaas's Cuckoo *Chrysococcyx klaas*; once reported to be feeding young Emerald Cuckoo *C. cupreus* and once a young Black Cuckoo *Cuculus clamosus*.

References
Brooke, R.K. (1980).
Harris, T. and Arnott, G. (1988).
Harrison, J.A. *et al.* (1997).
van Someren, V.G.L. (1956).
Vincent, A.W. (1948, 1949).

Dryoscopus gambensis (Lichtenstein). Northern Puffback. Cubla de Gambie.

Plate 25 (Opp. p. 398)

Lanius gambensis Lichtenstein, 1823. Verz. Doubl., zool. Mus. Berlin, p. 48; Senegambia.

Forms a superspecies with *D. pringlii*.

Range and Status. Endemic resident and possible intra-African migrant. Frequent to common in N tropical woodlands, lowland in W Africa, highland in E.

Ranges north to upper Guidimaka valley in S Mauritania, great bend of R. Niger in Mali (where scarce, north to about 17°N), 'W' Nat. Park in Niger, S Lake Chad in Cameroon, Abéché in Chad, Jebel Marra foothills in W Sudan, Blue Nile valley to about Singa in E Sudan (apparently absent from central Sudan: Nikolaus 1987), and highlands to about Asmara in Eritrea. Ranges south in W Africa to northern limits of rainforest zone; in Liberia as mapped (Gatter 1997), occurring in mangrove all along coast; in Ivory Coast has colonized plantations south to Täi and Abidjan; in Ghana south through Volta Region to Accra Plains and coast; in Togo south at least to Misahöhe; in Cameroon south to *c*. 03°N. In Mbini; in Gabon (M'Passa) seen rarely, in July, 'manifestement individus de passage' (Brosset and Erard 1986); in SE Gabon at Lékoni; records at Mouila and Tchibanga (Y.-M. de Martin de Viviès, pers. comm.). Congo, pair seen in Odzala Nat. Park, status uncertain; widespread from Léfini to Cabinda (Dowsett-Lemaire 1997a,c). Angola, only in Cabinda. Zaïre, in W occurs east to 17–18°E; common in N Uele, sparse in Ituri and Kivu, frequent in lowlands in Semliki valley and up to 2150 m in Rwenzoris and 3000 m on Kivu Volcanoes, and quite common in Itombwe Highlands at 1850–2830 m. Uganda and Kenya, widespread, frequent to common at 900–2200 m; throughout Uganda except for area south of Masaka and east of Mbarara; in Kenya near Moyale (Ethiopian race), and from Marsabit, Mt Nyiru, Mt Loima, Horr Valley, Cheranganis, Kapenguria and Elgon south to Bondo, Ng'iya, Nandi, Elmenteita, Lake Nakuru Nat. Park and Laikipia Plateau; east to N Uaso Nyiro valley. Somalia, known only at Ged Deeble in NW (several records, 1918, 1958). NW Tanzania, known from Ngara near Rwanda and Kibondo near Burundi. Rwanda, Zaïre-Nil crest at 1700–2600 m; Kibungu region in SE at 1300–1750 m (Vande weghe 1981). Burundi, records in N.

Description. *D. g. gambensis* (Lichtenstein): Senegal to Cameroon (Bosum) and Gabon. ADULT ♂: forehead, lores, narrow line under eye, ear-coverts, crown, nape, hindneck, sides of neck and mantle glossy bluish black; back and rump greyish white; uppertail-coverts and tail glossy black. Whole underparts creamy greyish white. Primaries blackish brown, narrowly edged pale grey, secondaries and tertials blackish brown, edged and tipped greyish white, greater primary coverts black, very narrowly pale edged, greater, median and lesser coverts brownish black, broadly edged and tipped with greyish or buffy white; underside of flight feathers shiny dark grey; underwing-coverts and axillaries white. Bill black; eyes bright orange or red; legs and feet grey. ADULT ♀: forehead, upper cheeks, ear-coverts, crown, nape, hindneck and sides of neck mid grey; mantle, back and rump greyish brown; uppertail-coverts dark brown. Tail brownish black. Chin, throat, breast and upper flanks pale creamy rufous, merging into white in middle of belly and vent; lower flanks, thighs and undertail-coverts pale rufous washed brownish. Flight feathers dark brown,

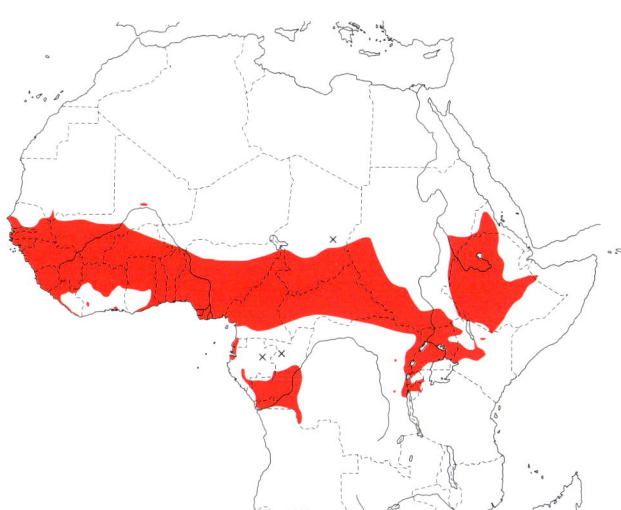

Dryoscopus gambensis

edged rufous; upperwing-coverts dark brown with grey bases, mainly fringed pale rufous or greyish white. Upper mandible black, lower mandible blue-grey or pale grey; eyes bright orange; legs and feet blue-grey. SIZE (21 ♂♂, 15 ♀♀, Bannerman 1939): wing, ♂ 87–96, ♀ 83–95; tail, ♂ 73–84, ♀ 71–81; bill, ♂ 19–23, ♀ 19–20; tarsus, ♂ 24–25, ♀ 23–24. WEIGHT: (Mt Nimba, Liberia), 1 ♂ 37·6, ♀ (n = 3) 30·7–34 (32·6), 1 imm. ♀ 30·4.

IMMATURE: ♂ and ♀ like adult ♀ but mantle feathers tipped pale rufous; flight feathers, scapulars and upperwing-coverts broadly fringed rufous; chin and throat rufous, rest of underparts white, feathers fluffy, with grey bases showing through patchily. Very young bird (Enugu, Nigeria, Sept) similar but chin and throat white, lower throat with cream wash.

NESTLING: not described.

D. g. congicus Sharpe: SW Congo, W Zaïre. ♂ like ♂ *gambensis*. ♀ deeper rufous below and flanks more uniformly rufous than in ♀ *gambensis*.

D. g. malzacii (Heuglin) (includes 'erwini'): E Cameroon, Chad and Central African Republic to Kenya; south to Itombwe, Zaïre. ♂ like *gambensis*; ♀ has forehead, crown and hindneck very dark brown (not grey) and mantle and back dark brown (not greyish brown); underparts slightly paler than in ♀ *gambensis*, and colour from chin to flanks and undertail-coverts more uniform. Immature ♂ and ♀ like adult ♀. SIZE, cline of reducing size from N to S: (S Sudan) wing, ♂ (n = 6) 89–99 (93·6), ♀ (n = 5) 88–92 (90·7); tail, ♂ (n = 7) 69·5–76 (72·4), ♀ (n = 5) 68–74 (70·8); (Rwenzori): wing, ♂ (n = 8) 87–102·5 (91·6), ♀ (n = 8) 84–89 (87·2); tail, ♂ (n = 7) 66·5–72·5 (71·5), ♀ (n = 7) 65–71·5 (68·7); (Itombwe, Zaïre – 'erwini', 11 ♂♂, 4 ♀♀) wing, ♂ 80–89 (84·3), ♀ 81–84 (81·7); tail, ♂ 70–81 (75·1), ♀ 72–77 (74·8). Race 'erwini' may differ from *malzacii* also in bill colour (black in ♂ but proximally grey in ♀) and eye colour (orange-red in ♂, orange in ♀) (Chapin 1954). WEIGHT: (*malzacii*), ♂ (n = 6) 30–37 (33·5), ♀

(n = 5) 25–40 (31·0); ('*erwini*') ♂ (n = 15) 30–35 (32·5), ♀ (n = 7) 28–33 (30·6).

D. g. erythreae Neumann: Eritrea, Ethiopia and E Sudan (Sennar). ♂ like ♂ *gambensis*. ♀ almost ♂-plumaged; upperparts differ in being dark chocolate brown where ♂ is bluish black; underparts of ♀ creamier (apparent only in series).

TAXONOMIC NOTE: whilst we believe that this species is sufficiently closely allied with *D. pringlii* to compose a superspecies with it, it is also very close to *D. cubla* (*contra* Dowsett and Forbes-Watson 1993); in E Zaïre the differences between *gambensis* ♂♂ and *cubla* ♂♂ seem no more than subspecific (Chapin 1954).

Field Characters. Length 18–19 cm. The common puffback outside the forest zone north of the equator. ♂ duller, less clear-cut black and white, than Red-eyed Puffback *D. senegalensis*, which it meets at forest edge, or Sabine's Puffback *D. sabini* of forest interior; rump grey, wings mixed grey, black and white, underparts off-white. *Orange-red* eye and *patterned* wings distinguish it from Sabine's and from Pink-footed Puffback *D. angolensis*. ♀ duller than ♀♀ of Sabine's and Pink-footed Puffbacks, brown above, *cinnamon-buff* below, with *orange* eye and *patterned* wings. Range approaches that of Black-backed Puffback *D. cubla*, which has similarly patterned wing and red eye, but ♂ has pure white rump and underparts, ♀ has dull grey-white underparts, black lores and white supraloral line. For differences from Pringle's Puffback *D. pringlii* see that species.

Voice. Tape-recorded (104, B, C, F, GREG, GRI, LEM, PAY). Contact call a single or double whistle, 'tyoop' or 'tyoo-yoo'. Alarm a grating 'ja-ja' or 'jit-jit; also said to give harsh, slow 'wrrich, wrrich, wrrich' and a rasping 'zhiuu' or 'zhrraanh' (Zimmerman *et al.* 1996).

General Habits. Inhabits dense mesic savanna woodland of nearly all types, thickets, tall trees, forest clearings, mangroves, wooded farmland and large gardens; in Liberia occurs in forest-grassland mosaic inland, up to 1500 m, secondary growth and forest edges at the coast, and in coastal and riverine mangrove; in Nigeria favours shea-butter trees *Vitellaria paradoxa* (= *Butyrospermum parkii*); in Eritrea lives up to 1540 m in wet woodlands, *Combretum*, and riparian figs and tamarinds; never in pure *Acacia* woods.

Occurs singly and in pairs. Generally retiring, secretive and easily overlooked, keeping in tree canopy and moving rapidly but stealthily amongst the branches, descending to tops of tall undergrowth. In breeding season very vocal and demonstrative, ♂♂ flying from tree to tree with white back and rump feathers puffed out to cover back and base of wings, looking white with black head and tail. Pulls appendages from insect prey, e.g. moth wings.

Sedentary, but some Gabon records may be of migrants (see above).

Food. Insects, including beetles, lepidopterans, caterpillars and orthopterans.

Breeding Habits. Biology poorly known; several nests have been found but they are very difficult to reach and examine. Evidently monogamous and territorial.

NEST: a compact cup made mainly of fine strips of bark, plastered outside with spider web and lined with fine rootlets; int. depth *c.* 38, int. diam. *c.* 51, sited in multiple fork *c.* 20 m high in tree. Another nest made of strips of bark and lined with very fine strips of bark with a little plant down; ext. depth 75, ext. width 80, int. depth 15, int. diam. 60 (A. Otim, pers. comm.).

EGGS: 2. Matt; white, speckled all over with dark grey, forming band around large end.

LAYING DATES: Mauritania, Aug; Gambia, (nests June–Sept, Dec; displaying ♂♂ Apr and July–Sept); Liberia, (displaying ♂♂ Sept–Oct and Mar–Apr); Ghana, Apr–May (and young Sept–Oct); Togo, (accompanied juvs July; song June–July); Nigeria, (various evidences Apr–June, Sept–Oct, Dec); Sudan, (Feb, June–July); Ethiopia, Apr (and various indications Dec–Mar); Zaïre, Itombwe, (various indications Apr, Aug, Oct); E Africa, Region A, Jan, Region B, Oct.

References
Bannerman, D.A. (1939).
Barlow, C. *et al.* (1997).
Prigogine, A. (1971).

Plate 25 (Opp. p. 398) *Dryoscopus pringlii* Jackson. Pringle's Puffback. Cubla de Pringle.

Dryoscopus pringlii Jackson, 1893. Bull. Br. Orn. Club., 3, p. 3; Mauungu Wilderness and between Tsavo and Kufumika Road, E Africa.

Forms a superspecies with *D. gambensis*.

Range and Status. Endemic resident, E Africa. Ethiopia, uncommon, in S and SE (Yavello, Arero, etc., and record from upper Webi Shabeelle near Kololo, about 07°40′N). Somalia, uncommon, 9 old and 7 recent records south of 03°N, one record north of 07°N (Ash and Miskell 1998). Kenya, uncommon and local below 1000 m, with small population southwest of L. Turkana in Turkwell R. valley and main range from Mandera, Moyale and Marsabit to Meru Nat. Park, Kora Nat. Res., Kitui, Mtito Andei and Tsavo Nat. Parks (Zimmerman *et al.* 1996). NE Tanzania, including L. Jipe, Mkomazi Game Res. and Lembeni, as mapped (N.E. Baker, pers. comm.).

Description. ADULT ♂: forehead, crown, nape, hindneck, sides of neck, mantle and scapulars glossy bluish black; back mid grey, rump white, uppertail-coverts glossy black. Tail black, tips to T2–

Dryoscopus pringlii

T5 greyish white, T6 with greyish white tip and outer edge 2 mm deep. Chin and throat creamy or greyish white, grading into pale grey on breast and pale brownish grey on flanks; belly white, at least in middle; thighs grey; vent and undertail-coverts whitish. Primaries dark brown with narrow grey-buff outer edges, inner secondaries and tertials dark brown with 1–1·5 mm wide whitish outer edges and tips; greater primary coverts dark brown with 1 mm wide white or whitish edges; greater and median coverts dark brown with 1–2 mm wide white or whitish edges and tips; underside of flight feathers shiny grey; underwing-coverts and axillaries white; underside of tail grey, feathers fringed whitish. Bill black, base of lower mandible pale horn; eyes dark red; legs and feet grey-black. ADULT ♀: nondescript, flycatcher-like brown. Upperparts greyish brown, lores and ear-coverts buffy whitish, narrow white ring around eye, rump whitish when brown feather tips displaced (as they often are); tail uniform brown; chin, throat, breast and flanks pale creamy or greyish buff, grading to white on belly and undertail-coverts; thighs pale brown. Upper mandible black, lower mandible whitish; eyes yellow, outer iris white; legs and feet blue-grey, soles pale greyish. SIZE (3 ♂♂, 4 ♀♀): wing, ♂ 70–73·5 (71·5), ♀ 64·5–73 (69·2); tail, ♂ 58–60·5 (59·7), ♀ 48–62 (57·75); bill, ♂ 15–16 (15·5), ♀ 13·5–16 (14·8); tarsus, ♂ 22·5–23·5 (22·0), ♀ 20–23 (21·75). WEIGHT: Kenya, 1 ♂ 19·5, 1 ♀ 13·0.

IMMATURE: not known.

NESTLING: naked and blind, skin dull purplish all over, mouth yellow, gape cream (Moreau and Moreau 1939).

TAXONOMIC NOTE: Ethiopian birds may differ subspecifically from coastal ones, having greyer underparts (Benson 1946b).

Field Characters. Length 13·5–14 cm. A puffback of dry country in NE Africa, smaller than Northern and Black-backed Puffbacks *D. gambensis* and *D. cubla*. Black and white ♂ similar to Northern, duller than Black-backed, with *grey* scapulars and lower back (only rump white), underparts washed grey-brown on breast and flanks; distinguished from both by basal half of lower mandible *white*, and white sides and tip to tail (obvious only in fresh plumage: Zimmerman *et al*. 1996). ♀ nondescript brown with pale lores and narrow white eye-ring, like *Bradornis* flycatcher but with heavier bill, more distinct white wing edgings.

Voice. Tape-recorded (B). Song short and monotonous, harsh and low-pitched; calls include a sharp repeated 'keu', which may be given by several birds, and a low-pitched churr.

General Habits. Inhabits arid and semi-arid *Acacia* and *Commiphora* bushland and woodland. An active gleaner in microphyllous foliage, often in mixed-species flocks, moving through undergrowth or low trees (Zimmerman *et al*. 1996). In Tsavo, feeds in woodland bushes, not trees – foraging height (n = 8 observations) av. 2·2 m above ground (Lack 1985).

Food. Insects including caterpillars.

Breeding Habits.

NEST: only one found (Moreau and Moreau 1939): a very neat cup moulded onto horizontal branch and anchored to vertical twig, 120 cm from ground in *Commiphora* sapling; int. depth 50, int. diam. 60.

LAYING DATES: Kenya, Nov.

DEVELOPMENT AND CARE OF YOUNG: young brooded by ♀, which sat so tightly that observers had to lift her off to see nest contents and she 'fairly edged our fingers out of the nest and settled down on the young again' (Moreau and Moreau 1939). Fed by both parents.

References
Moreau, R.E. and Moreau, W.M. (1939).
Zimmerman, D.A. *et al*. (1996).

Genus *Laniarius* Vieillot

Robust bush-shrikes, all black, or upperparts black or mainly black and underparts white, pink, red, orange or yellow. Head and wings sometimes patterned, tail and underparts always plain. In at least some species, several thin bristles, twice length of feathers, in hindneck. Plumage of back long, fluffy and very full, overlying rump; back feathers usually with half-concealed white subapical spots. Forehead feathering extends well forward to cover posterior border of nostril but not culmen. Wings round-ended; tail quite long, square-ended or round-ended but not strongly graduated. Rictal bristles weak. In some species bill thick and strong; in most species legs and feet strong. Sexes alike except in one species; juveniles usually rather unlike adults, plumage often banded, bill horn-coloured (white in 1 species); in some species where adults are almost indistinguishable, juveniles differ distinctly from each other. Eyes black, brown, yellow, orange or red. Inhabit thickets and dense, shaded, shrub and tree growth in savanna woodlands, also in mangroves, papyrus beds and reedy, woody swamps, living low down in vegetation, foraging in large part on the ground, few if any species penetrating deep into primary rain forest and only one or two sometimes visiting upper canopy of tall or even medium growth. Vocal; loud, ringing calls in simple motifs, with reciprocal duetting of the sexes highly developed; duetting well studied in one species (*L. funebris*). Monogamous. Sedentary, pair living year-round in territory defended mainly by duetting. Impaling of food has been recorded only in captive birds, of 2 species. Nest a loosely bound, twiggy, rather untidy open cup, deeper than in most malaconotids, but subject to trampling into a flattish platform. Eggs pale bluish, brown-spotted.

Taxonomy controversial, numerous taxa being parapatric, similar in appearance, but with vocal and habitat differences beginning to be understood. White (1962) recognized 9 species, Hall and Moreau (1970) 15, composing 2–3 superspecies, Sibley and Monroe (1990) 17, being 4 independent species and 4 superspecies, and Dowsett and Forbes-Watson (1993) 14 species (including the newly-described *L. liberatus*), being 9 independent species and 2 superspecies.

Dowsett and Dowsett-Lemaire (1993b) have given reasons for their taxonomic conclusions about the *poensis*, *barbarus* and *aethiopicus* complexes; in minor cases where we part company with them we give our own reasons in Taxonomic Notes in the species accounts.

Endemic; 18 species; 4 superspecies and 4 independent species: *L. leucorhynchus*/*L. poensis*/*L. fuelleborni*; *L. funebris*; *L. luehderi*/*L. brauni*/*L. amboimensis*; *L. ruficeps*; *L. liberatus*; *L. aethiopicus*/*L. ferrugineus*/*L. bicolor*/*L. turatii*; *L. barbarus*/*L. mufumbiri*/*L. erythrogaster*/*L. atrococcineus* and *L. atroflavus*.

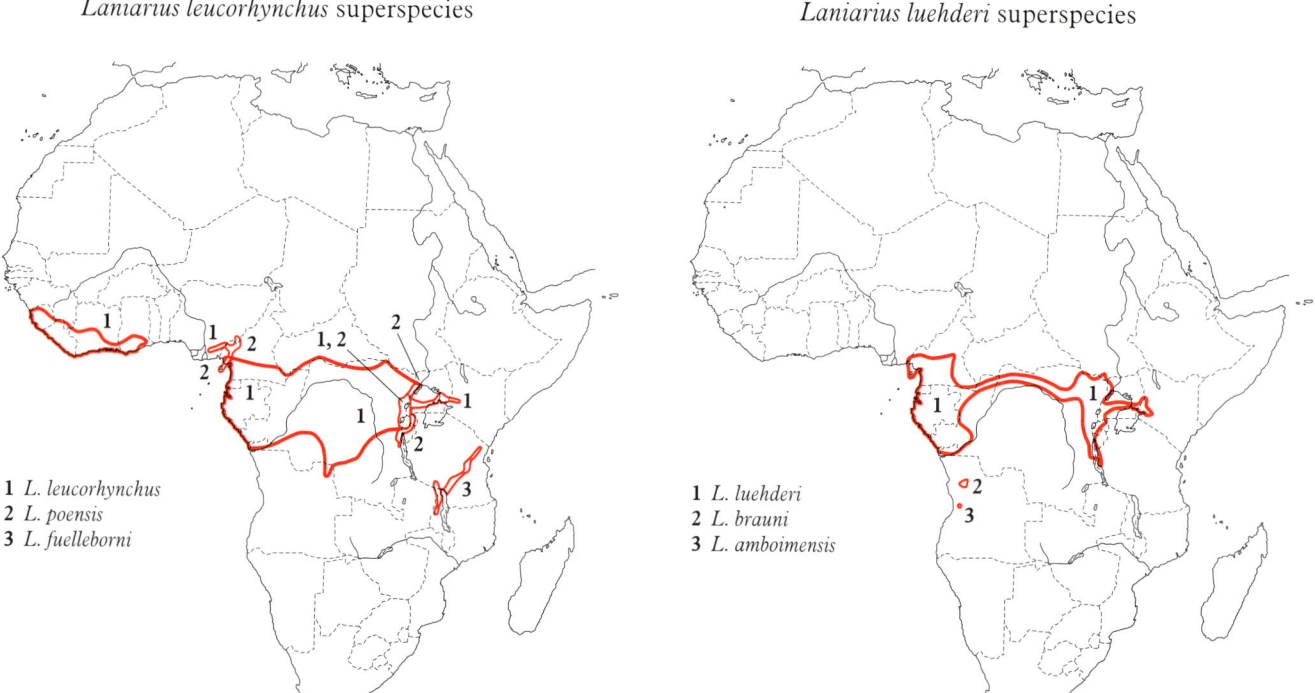

Laniarius leucorhynchus superspecies

1 *L. leucorhynchus*
2 *L. poensis*
3 *L. fuelleborni*

Laniarius luehderi superspecies

1 *L. luehderi*
2 *L. brauni*
3 *L. amboimensis*

Laniarius aethiopicus superspecies

Laniarius barbarus superspecies

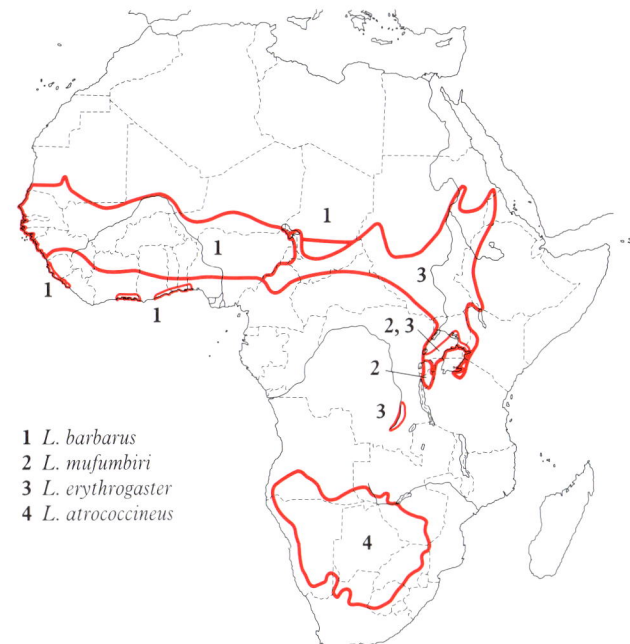

1 *L. aethiopicus*
2 *L. ferrugineus*
3 *L. bicolor*
4 *L. turatii*

1 *L. barbarus*
2 *L. mufumbiri*
3 *L. erythrogaster*
4 *L. atrococcineus*

Laniarius leucorhynchus (Hartlaub). Sooty Boubou. Gonolek fuligineux.

Plate 26
(Opp. p. 399)

Telephorus leucorhynchus Hartlaub, 1848. Rev. Mag. Zool. (Paris), 11, p. 108; Elmina, Gold Coast.

Forms a superspecies with *L. poensis* and *L. fuelleborni*.

Range and Status. Endemic resident, W African and Congo Basin rain forests. Guinea, uncommon around Ziama Massif; records near Mt Nimba. Sierra Leone, several old records from Little Scarcies R. and elsewhere west of 12°W, more recently only in southeastern quarter, west to Bo. Liberia, not uncommon along coast and in interior from Lofa County and Nimba to Zwedru, and common at Ganta; up to 900 m on Mt Nimba (Gatter 1997). Ivory Coast, not uncommon from coast north to Yapo, Lamto and Béoumi. Ghana, recorded regularly at Akropong in Akwapim; records from Cape Coast, Kakum Forest, Kumasi (where formerly common) and Tafo. Togo, sight record Fazao (Cheke and Walsh 1996). Nigeria, not uncommon around Umuagwu (Marchant 1953); 5 together near Bashu (Ash 1990). Cameroon, frequent in W, and further east known from numerous localities in Bitye, Efulen, Olounou and Zoulabot areas. Central African Republic, records in Bai Bakalonga and along Babonga R. (Dzanga Res.), in Lobaye Préf. and Bangui area, and along Mbomou R. valley at Zaïre border. Sudan, known only from Bengengai. Mbini, known to occur. Gabon, widespread and common except in S. Congo, common in Nouabalé-Ndoki Nat. Park, Odzala Nat. Park and Léfini Réserve; locally common in Mayombe, but almost absent from some areas, e.g. Goumina. Zaïre, widespread; frequent at least locally; very common around, e.g. Niangara; probably widely overlooked as it doubtless has been in much of W African rain forest zone. Angola,

Laniarius leucorhynchus

Cabinda and Dundo, Rio Luachimo, Lunda. Uganda, locally common in SW and S at 700–1400 m: Bwamba, Lugalambo, Entebbe, Kifu, Mabira, Sezibwa R. and

Busoga. Kenya, one collected in Kaimosi Forest in 1931. Rwanda, status uncertain – may be absent.

Description. ADULT ♂: entire plumage jet black, with faint violet or blue gloss; feathers of lower back and rump very full, long and soft, without any vestiges of the white spots of most congeners. Bill black, deep at base; eyes dark red-brown or very dark red (Zaïre) or given as brown-black in W Africa; legs and feet blackish or dark slaty bluish. Sexes alike. SIZE (16 ♂♂, 6 ♀♀): wing, ♂ 92–100 (97·6), ♀ 90–95 (95·4); tail, ♂ 80–86 (82·0), ♀ 77–82 (77·7); bill, ♂ 21–25 (24·3), ♀ 22–25 (24·1); tarsus, ♂ 30–32 (30·8), ♀ 28–31 (30·3). WEIGHT: (Uganda), ♂ (n = 8) 51–57 (53·5), 1 ♀ 45, 2 unsexed 44, 44; (Angola) 1 ♀ 49.

IMMATURE: plumage like adult but bill whitish to ivory white (which changes to adult black rather blotchily); several young but full-sized birds have black bills like adult, but have mantle to rump, sides of breast, and wings, dark chocolate brown with almost a maroon wash.

NESTLING: undescribed, but see **A** (from photo in Brosset and Erard 1986).

TAXONOMIC NOTE: see Note under *L. poensis*.

Field Characters. Length 21 cm. The *only* all-black bush-shrike in *lowland forest*. The scientific name, meaning 'white-billed', refers only to the immature; bill of adults is black. Range abuts that of Mountain Sooty Boubou *L. poensis* but the 2 are altitudinally separated, except perhaps seasonally on Mt Cameroon at 430 m. Adult like Mountain Sooty Boubou but slightly larger, best told by soft whistled song; white-billed immature distinctive (immature Mountain has black bill).

Voice. Tape-recorded (32, 104, B, F, CHA, ERA, LEM, SAR). Song low, hollow whistles, almost pure but with a slightly burry, 'fuzzy' quality. Number, speed and rhythm of notes very variable: a single long (1 s) 'ooooooooo' that gets louder and then fades away, with 1 s gap before next note; 3–4 note series, last 2 loudest, 'oo-oo-woo-woo', ♀ answering with low 'skaaa'; series of similar but shorter notes at rate of 2 per s, given either continuously or in groups of 4; 2-speed series of 10 notes, last 7 faster than first 3, lasting *c*. 2 s, with 1 s gap between series; very fast series of hollow notes at rate of 8 per s, either continuous or in groups of 11 notes separated by short gaps. Song dialects in S Congo the same as those in Uganda (Dowsett and Dowsett-Lemaire 1991). Faster songs could be confused with those of White-spotted Flufftail *Sarothrura pulchra* (Brosset and Erard 1986, Christy and Clarke 1994), but song of latter is louder, higher-pitched and more ringing. Whistles often given alone but may be answered by ♀ with harsh scold, either synchronously or not. Also has low clicking song of 9–10 notes, not loud or ringing, 'clicliclicli ... '. Harsh calls include double 'zzhrrrraaa-zzhrrrraaaaa', like cloth or sheet of metal being slowly and forcibly torn; scolding, scraping 'wrraa', 'chrr' or 'chrraaa', and low, dry 'tr-tr-tr-tr-tr-tr'. Short, plaintive whistles given by recently-fledged immatures waiting to be fed (F. Dowsett-Lemaire, pers. comm.).

General Habits. Inhabits impenetrable liana-covered tangles, rank vegetation, dense thickets and thick herbage in coastal shrubland, open swamp vegetation, thickets in mixed papyrus marsh (Nanga R., W Congo), low swamp forest, riparian forest, wooded islands and banks of large rivers and lakes, bushy farmland, young secondary forest growth, forest clearings and edges, gallery forest and remnant stands of forest in savanna; enters primary forest along borders of watercourses, old forest tracks, and where opened up by windfalls and storm damage (e.g. M'Passa Plateau and Rés. de la Lopé, Gabon).

Solitary or (generally) in pairs; sometimes in family parties of 4 birds, once 5. Keeps close to ground, though does forage up to 15 m above ground, sometimes even higher. Secretive, keeping in thickest tangles, moving through them with ease in small jumps. Calls often, all year. Pairs call in duet, which thought to serve both for contact and in territorial proclamation; duetting frequency increases when another pair in the vicinity. Territory seems to be as large as 6–8 ha (Brosset and Erard 1986).

Sedentary.

Food. Seen hunting or catching orthopterans, butterflies, mantises, cicadas and caterpillars; in stomachs, many beetles, several large wasps and black ants, also caterpillars, spiders and snails.

Breeding Habits. Solitary nester; territorial.

NEST: quite large, loosely structured open cup on base on intertwined twigs, rootlets and dry, thin, woody climber stems or tendrils (**A**). Sited on multiple vertical fork 2–3 m above ground, in bush in forest clearing at 900 m (Guinea), in vegetation 1 m above water, in thick ornamental tree in middle of clearing, and in young secondary growth (Gabon).

EGGS: 2. Cream-white, finely stippled with brown-grey and grey-lilac, forming band around large end; also greenish grey or greenish white, spotted with ochreous

A

or reddish brown, mainly around large end. SIZE: (n = 4) 24–25 × 17–18 (24·45 × 17·3).

LAYING DATES: Guinea (Ziama), (nest-building May); Liberia, Oct (much singing Aug–Nov, and other evidence of breeding Nov–Mar); Gabon, Dec–Jan.

INCUBATION: only by ♀.

DEVELOPMENT AND CARE OF YOUNG: young birds accompanied by both parents.

References
Bannerman, D.A. (1939).
Brosset, A. and Erard, C. (1986).
Chapin, J. (1954).

Laniarius poensis (Alexander). Mountain Sooty Boubou. Gonolek de montagne.

Plate 26
(Opp. p. 399)

Dryoscopus poensis Alexander, 1903. Bull. Br. Orn. Club, 13, p. 37; Mount St. Ysabel, Fernando Po.

Forms a superspecies with *L. leucorhynchus* and *L. fuelleborni*.

Laniarius poensis

Range and Status. Endemic resident with widely separated populations in Cameroonian montane and Rift Valley mountains. Nigeria, common on Obudu and Mambilla Plateaux and known from Gotel Mts. Cameroon, abundant on Mt Cameroon at 700–1400, common at 1400–1950 m, uncommon at 600–700 m (but once abundant at 500 m, and has occurred down to 430 m at Ekona, suggesting small-scale downward movement after breeding: Stuart 1986); abundant at Dikome Balue in Rumpi Hills at 1000–1700 m, on Mt Kupé at 1200–2000 m (where recorded down to 950 m), on Mt Nlonako at 1200–1750 m; common on Mt Manenguba at 2100–2200 m, and occurs at 2200 m on Mt Oku in Bamenda Highlands. Bioko Island, Pico Basilié at 1200–2100 m, most records at 1500–1800 m, and southern highlands at 750–1800 m.

Zaïre: mountains west of Mahagi and Kasenye, L. Albert; Rwenzoris; mountains west of L. Edward, Rutshuru and L. Kivu at 1700–3385 m; common in Itombwe Highlands at 1170–2450 m, and on Mt Kabobo. Uganda, Kigezi and Rwenzoris. Rwanda, Volcanoes, Gishwati and Nyungwe highlands; throughout Nyungwe, commoner above than below 1950 m. Burundi, only in mountains in NW.

Description. *L. p. camerunensis* Eisentraut: Cameroon and Nigeria. ADULT ♂: entirely dark glossy black, with blue reflections. Wings strongly decurved at tip and tail somewhat decurved at tip. Bill black; eyes bluish black; legs and feet black. Sexes alike. SIZE (21 ♂♂, 10 ♀♀): wing, ♂ 78–87 (82·5), ♀ 76–82 (79·4); tail, ♂ 67–76 (71·0), ♀ 65–71 (68·1); bill, ♂ 19–22 (20·5), ♀ 19·5–21 (20·1) (Eisentraut 1968); tarsus, ♂ (n = 4) 30–33·5 (31·4), ♀ (n = 4) 30–36 (33·2). WEIGHT: (Cameroon, Rumpi Hills, unsexed, n = 11) 39·3–48·9 (44·9); also ♂♀ (n = 31) 40–54 (47·3).

IMMATURE: upperparts dull black, without blue or violet gloss; underparts matt black.

L. p. poensis (Alexander): Bioko. Like *camerunensis* but smaller. SIZE (12 ♂♂, 7 ♀♀): wing, ♂ 76–80 (77·5), ♀ 73–76 (74·5); tail, ♂ 62–69 (65·7), ♀ 59·5–66 (63·5); bill, ♂ 18·5–21 (19·3), ♀ 18–19·5 (18·8). WEIGHT: ♂ (n = 8) 38–45 (41·75), ♀ (n – 7) 38–47 (41·3).

L. p. holomelas (Jackson): E Zaïre, Uganda, Rwanda, Burundi. Like *camerunensis* but less glossy. SIZE (4 ♂♂, 4 ♀♀): wing, ♂ 81–86 (83·2), ♀ 78–92 (81·0); tail, ♂ 67–79 (68·25), ♀ 67·5–68·5 (67·0). WEIGHT: 1 ♀ 38.

TAXONOMIC NOTE: *L. poensis* and *L. fuelleborni* have often been united as a single species, recently by Stuart (1986: 'no doubt' – p. 61) and Dowsett and Forbes-Watson (1993). However, they differ in adult and more so in juvenile plumages, in bill size and radically in voice. As Dowsett-Lemaire and Dowsett (1990) remark, the voice of *L. poensis*, with its duetted long, deep whistles, has nothing in common with the rhythmic, liquid song of *L. fuelleborni*. We have no hesitation in retaining them as different species, and wonder if they may actually prove to be independent rather than allospecies.

Field Characters. Length 18 cm. The all-black boubou in the mountains of Cameroon and the Albertine Rift. Altitudinally separated from the Sooty Boubou *L. leucorhynchus* of lowland forests except perhaps seasonally at 430 m on Mt Cameroon. Slightly smaller than Sooty Boubou, and immature does not have white bill, but best separated by harsh calls (see below); lacks the soft whistles of Sooty Boubou.

Voice. Tape-recorded (104, B, BUT, CHA, GRI, HOR, LEM, PER, ROD). In Cameroon has broad vocabulary, summarized by Stuart (1986) as 'loud whistles, trilling, rattling and snoring noises, squeals, harsh, grating alarm calls and duets'. Normal duet a grating 'whaa-dzu' or 'shaaaaa-tsu'; also 'wee-tchray' (probably the 'vuik-eng' of Eisentraut 1963), shorter 'weezhay'; or longer (1·5 s), 2 tearing notes, 'whaaa-chuwaaaoow'. Single calls include sharp, tearing 'whazzz', longer, nasal 'zzhaannh', hard 'clickit-clickit', a repeated drongo-like nasal, down-slurred 'jeeoonh-jeeoonh-jeeoonh ... ' (probably the loud whistled 'whee-oo' of Serle 1950) and a short (0·5 s) hollow trill.

Calls of Albertine Rift population (Rwanda and Uganda) quite different (and see comments by Dowsett-Lemaire and Dowsett 1990, p. 106). Chief component high-pitched whistles, either solo (ringing 'heeeeee' lasting 0·8 s or loud, ascending 'hooee'), or as part of duet (ringing 'heeee' answered by harsh 'ti-haarwet', or whistled 'huuuuuu' followed by snoring 'korrrrr'). Other duets: grating, toad-like 't'rrraaanh' with short 'chyow', and higher 't'zzheeeh' with sharp, abrupt 'chya'. Calls include long (1·5–2 s) tearing 'turrrraaaanh', scraping 'jaaa' and hollow 'jow-jow-jow-jow'.

General Habits. Inhabits primary montane forest edges, clearings and undergrowth where at its thickest, keeping low down in tangled vegetation with clusters of vines. Also in gallery forest (Zaïre) and bamboo (Uganda). In Cameroon keeps lower in vegetation than Yellow-breasted Boubou L. atroflavus, but in Itombwe sometimes in mid levels of the forest. Generally in pairs (which are not necessarily ♂ and ♀). Joins mixed-species foraging flocks. Follows hunting swarms of *Dorylus* ants on forest floor. Skulks in deepest shade in thickest tangles of leafy, woody vegetation; occasionally comes into view for no more than a few sec at a time; although vocal and common, very little known about it.

Sedentary.

Food. Insects include beetles, grasshoppers and caterpillars. Occasionally snails, isopod crustaceans and small frogs. Of 10 stomachs, beetles in 9 and grasshoppers in 5.

Breeding Habits. Poorly known.
NEST: the only one described was a mossy cup built in a cypress tree covered with *Gouania* creepers (Dowsett-Lemaire and Dowsett 1990).
EGGS: 2.
LAYING DATES: Cameroon, Nov–Jan and Mar (and breeding condition Oct and Apr); Zaïre, Itombwe, Apr–May (and other evidence, June–July).

References
Dowsett-Lemaire, F. and Dowsett, R.J. (1990).
Eisentraut, M. (1963, 1968, 1973).
Stuart, S.N. (1986).

Plate 26
(Opp. p. 399)

Laniarius fuelleborni (Reichenow). Fülleborn's Boubou. Gonolek de Fülleborn.

Dryoscopus fülleborni Reichenow, 1900. Orn. Monatsb., 8, p. 39; Usafua, north of L. Nyasa.

Forms a superspecies with *L. leucorhynchus* and *L. poensis*.

Range and Status. Endemic resident, eastern arc mountains, Tanzania, and N Malaŵi. Tanzania, common at 1500–2200 m in West Usambaras, common up to 2800 m in Ngurus, Ukagurus and Ulugurus (where as low as 900 m); another race in highlands in S, from Isoka, Iringa and Njombe to Udzungwa (N.E. Baker, pers. comm.). Malaŵi, Misuku Hills (local Wilindi, commoner with increasing altitude in Matipa), throughout Nyika Plateau, N Viphya (Uzumara, Chimaliro) and S Viphya Plateaux (Nthungwa, Chamambo, Kawandama), at 1170–2500 m. Zambia, restricted to Nyika Plateau at 1900–2200 m.

Description. *L. f. usambaricus* Rand (includes '*ulugurensis*'): Tanzania (Usambara, Ngurus, Ukagurus, Ulugurus). ADULT ♂: black, with bluish grey bloom, particularly on back, rump and belly; ear-coverts and tail glossy jet black; wings glossy brownish black. Wings decurved. Rump plumage very full and soft. Bill black; eyes dark brown; legs and feet black. ADULT ♀: like ♂ but head less blackish, whole plumage slightly greyer, wings brownish black, underparts with faint olivaceous or ochreous wash. SIZE (6 ♂♂, 3 ♀♀): wing, ♂ 81·5–96·5 (90·6), ♀ 82–84 (83·3); tail, ♂ 69–81 (77·1), ♀ 71·5–82 (77·2); bill, ♂ 19–22·5 (21·4), ♀ 21–22 (21·3); tarsus, ♂ 31–36 (33·0), ♀ 30–31 (30·8). WEIGHT: 3 ♂♂ 40, 42, 50; 2 ♀♀ 39, 46.
IMMATURE: like adult ♀.

Laniarius fuelleborni

L. f. fuelleborni Reichenow: Tanzania south of Uluguru Mts, Malaŵi, Zambia (Nyika). Intergrades with *usambaricus*; ♂ not quite as blackish, ♀ with belly dark blackish grey, with faint olivaceous wash, decidedly paler than belly of ♀ *usambaricus*.

IMMATURE: rather distinctive: upperparts uniform very dark olivaceous grey-brown, wings and tail browner, underparts dark greyish olive with yellowish wash on belly and, in juveniles, obscure dark olive banding.

TAXONOMIC NOTE: see Note under *L. poensis*.

Field Characters. Length 18·5–19·5 cm. The only slaty black boubou in montane forests of E Tanzania and Malaŵi. Does not overlap with Slate-coloured Boubou *L. funebris* of dry lowland habitats, nor with black form of Tropical Boubou *L. aethiopicus* which is only on coast.

Voice. Tape-recorded (32, 86, 102, B, C, F, LEM, STJ, SVEN, WOOD). Song loud, ringing, with explosive quality; often ends with loud up-slurred note, 'dududwee', 'to-wheee, to-wheee', 'pu-wheeeng'; some with 'oi' sound reminiscent of Gorgeous Bush-Shrike *Telophorus viridis*, 'co-wo-roik, co-wo-roik'. ♂ may sing alone, but duets are frequent; ♀ answers with grating trill, 'woik-woik-titrrrrrrr', 'week-week-tzzaaaa', 'dyow, dyowdyow, teeeurrr', or may sing first, 'tyorrrrr-bobobweek', 'co-heerrrr-doidoi'; sometimes ♂ whistles again after ♀'s reply, 'doidoidoi – wheez – doidoidoi', 'tyutyutyu – dzzeeeuw – tyutyutyu'. Less common is a fast liquid 5-note trill, first 2 notes lowest, 'totowiddlywit'. Variation almost endless; for further transcriptions see Zimmerman *et al.* (1996). Alarm a grating 'zi-zi-zaar' or hard clicking 'zutzutzutzut ... '.

General Habits. In Malaŵi inhabits montane forest and 3–4 m tall, dense, montane shrubland. Also forest edges. Keeps in thickets and amongst clumps of lianas. Solitary or in pairs. Forages in dense understorey, from near ground up to 16–18 m in tallest trees, also in canopy of low montane forest (8–15 m), gleaning bark and foliage. Occasionally attends ant swarms, without leaving thick cover. 1–2 birds seen following swarm of *Dorylus* ants 7 times in Usambara Mts, Tanzania, hopping as low as 30 cm above ground in weeds and vines, making short sallies to the ground, pecking liana stems and leaves, and giving variety of rattles, scolds, whistles, grunts and 'static' notes (Willis 1983). Territory size 0·6–1·0 ha (in narrow forest patches with plenty of edge habitat of thick, regenerating scrub), but larger in continuous forest, where 2–3 pairs per 10 ha; 8 patches of 8–12 ha, totalling 73 ha, contained 18 pairs (av. 4 ha per pair) (Dowsett-Lemaire 1983, 1989).

Highly sedentary; longest movement in Malaŵi (n = 7) <250 m.

Food. Invertebrates, including caterpillars, ants, grasshoppers and snails.

Breeding Habits. Solitary nester, territorial. 1 territory held for 8 seasons (Dowsett 1985).

NEST: not described.

EGGS: 1, 2 and 2 (3 clutches). Eggs not described.

LAYING DATES: Tanzania, Dec–Feb; Malaŵi, Sept–Jan, mainly Nov; Zambia, Nov–Jan (R.J. Dowsett, pers. comm.).

References
Dowsett-Lemaire, F. (1983, 1989).
Zimmerman, D.A. *et al.* (1996).

Laniarius funebris (Hartlaub). Slate-coloured Boubou. Gonolek ardoisé.

Plate 26
(Opp. p. 399)

Dryoscopus funebris Hartlaub, 1863. Proc. Zool. Soc. London, p. 105; Meninga [Unyamwesi area of Tanganyika Territory].

Range and Status. Endemic resident, E Africa. SE Sudan, fairly common southeast of Torit, and occurs between Boya Hills and NE Uganda and NW Kenya borders, also near Kibish River northwest of L. Turkana. Ethiopia, common to abundant from central Rift Valley to east in hot low country around eastern highlands and to southwest and south in Gamo Gofa and Sidamo Provs; common in arid thorn scrub at 900 m south of Mega and between Yavello, Neghelli and Giarso, in moister country at 1200 m around L. Abaya, and in evergreen scrub at 1850 m (Benson 1946b). Somalia, fairly common in NW east to 46°E and common towards Ethiopian border, and quite common in S, mainly in river valleys and near coast, north to 04°N. Uganda, only in extreme E (Karamoja) and SW (Ankole). Kenya, widespread and locally common, from sea-level up to 2200 m but mainly below 2000 m; absent east of L. Turkana and in hinterland west of it and in W from Kongelai and Kerio Valley to Kavirondo Gulf, also much of arid NE and E, and lowlands south of Malindi. Tanzania, occurs throughout much of interior as shown but absent from southeast of eastern arc mountains, and scarce in W (west of 33°E, north of 08°N: map data, N.E. Baker, pers. comm.).

Description. *L. f. funebris* (Hartlaub) (includes '*atrocoeruleus*'): NW Somalia through Ethiopia and Kenya to Tanzania. ADULT ♂: whole head, neck and throat glossy jet black, grading backwards into dark blue-grey on mantle, scapulars, back, rump and breast; long, full rump feathers with white bases and some with whitish subterminal spots; uppertail-coverts and tail jet black, belly and flanks smoky bluish grey, paler than breast; thighs and undertail-coverts bluish grey-black. Bill black; eyes dark brown; legs and feet black. Sexes alike. SIZE (5 ♂♂, 4 ♀♀): wing, ♂ 86–94 (89·6), ♀ 83–102 (89·5); tail, ♂ 81–88·5 (85·3), ♀ 84–93 (87·0); bill, ♂ 22–24·5 (22·2), ♀ 21–23·5 (21·8); tarsus, ♂ 32·5–34·5 (33·5), ♀ 33–34 (33·5). WEIGHT: ♂ (n = 18) 33–50 (41·6), ♀ (n = 15) 23–55 (37·1).

IMMATURE: upperparts dull black, forehead, hindneck, mantle to rump and upperwing-coverts with very narrow rufous barring; underparts evenly banded with dark slate-grey and buff; in underparts, buff-barred plumage replaced by adult dark bluish grey first on breast, then on sides of belly. Juvenile just out of nest: upperparts dark brown, feathers of forehead, mantle and

Laniarius funebris

upperwing-coverts with narrow rufous-buff tips; feathers of back fluffy, with rufous tip, dusky subterminal bar, buffy centre and grey base; rump tiger-banded rufous and blackish; uppertail-coverts very dark brown with narrow buff tips; short tail black; underparts buff, rather evenly banded with slate-grey, the dark bands narrow and close together on throat, becoming broader and more widely spaced backwards to undertail-coverts.

NESTLING: blind, completely naked, skin black, gape flanges conspicuously yellow.

L. f. degener Hilgert: SE Ethiopia and S Somalia, intergrading with nominate race in coastal regions and S Kenya and in Tanzania east of Kilimanjaro. Slightly paler and smaller than nominate race. WEIGHT: ♂ (n = 4) 39–40 (39·75), ♀ (n = 3) 32–37 (34·0).

Field Characters. Length 18·5–20 cm. A *uniformly blackish* bird, whose characteristic duet is often heard from thickets in dry country in E Africa, although the singers skulk under cover. The only all-black bush-shrike in most of its range (Fülleborn's Boubou *L. fuelleborni* is restricted to mountains), but in coastal Kenya and lower Shabeelle and Jubba valleys of Somalia occurs alongside black morph of Tropical Boubou *L. aethiopicus*; distinguished by smaller size, slate-grey rather than glossy black plumage and different voice. Immature barred black and brown.

Voice. Tape-recorded (5, 13, 30, 34, 64, 109, B, C, F). Song a synchronized duet, very variable; sonagrams and extensive reviews in Thorpe (1972) and Sonnenschein and Reyer (1984). Typically one bird gives 2–4 low hollow notes, answered by loud, up-slurred whistle, 'cloclo-wooeee, cloclo-wooeee'; 'woko-clerwy, woko-clerwy'; 'ko-kokoko-way'. Second bird may reply with tearing sneeze, 'woko-kitchu', or first bird give trill instead of whistle, 'wreeee-koway', or ringing trill may be answered by buzzy note, 'wreeyee-dzzrrr'. Birds often sing at the same time: vibrant snoring sound overlapping 'quick-quick', or 'quick-quick' of ♀ interrupting 'brreeee' of ♂. Either bird may start duet, and single ♂ has been seen to give complete 'duet' on its own (Thorpe 1972). Trio singing also occurs: bubbling sound from bird 'x', 'whooeeoo' from bird 'y' and 'whee' from bird 'z' (see sonagram in Thorpe 1972). Alarm a snarling 'hweerrrr' or 'cherrk'.

General Habits. Inhabits acacia woodland (Kenya) and thick bushes along watercourses (Tsavo), overgrown cultivation (Tanzania), and thorn scrub, evergreen scrub and juniper woods (Ethiopia). Keeps to bushland thickets with massed tangles of creepers, especially shrubs and small trees smothered in *Cissus rotundifolia* (Somalia).

Usually in pairs. Skulking, shy and retiring, though will show itself at edges of vegetation and even traverse open ground where accustomed to human presence; noisy. Keeps in dense vegetation, in territory of 1·5–3·5 ha, occupied year-round. Territories may be contiguous or isolated. Sometimes a third bird present in territory: once ♂ and 2 ♀♀; both ♀♀ sang (at different times) in duet with ♂. Territory defended all year by duet singing; in 3 cases of boundary transgressions, intruders were neighbouring pair whose nesting had recently failed; fighting is rare. In dry season, forages by hopping on ground in shade under woody cover, searching leaf litter; in wet season, feeds mainly amongst foliage, searching for caterpillars. Large prey item is wiped to and fro against branch before being swallowed (Sonnenschein and Reyer 1984).

Antiphonal duetting extensively studied in E Africa (Sonnenschein and Reyer 1983, 1984, Thorpe 1972, Wickler and Seibt 1979, Wickler and Sonnenschein 1989, Wickler and Lunau 1996). Pair-duet lasts 1–2 s and has 2 components, one given by ♂ and other by ♀; each component is sex-specific and is learned from a same-sex tutor in first 6–8 months of life; nonetheless, a bird is capable of producing elements that belong to the other sex's repertoire; once learned, individual repertoire remains unchanged for at least 7 years; duet can be initiated by either partner; a particular pair uses several different, standard, call-answer sequences and switches between them. At L. Baringo (Kenya), ♂♂ use 4 song types, ♀♀ only 1; one ♂ song type functions to achieve breeding synchrony; 2 others have territorial functions and their relative frequencies vary with some social correlates; 4th song type might function in mate guarding. At Seronera (Tanzania) pairs duet throughout day, mainly at 16h00–17h00 (when one pair gave 257 duets), then 17h00–18h00 (120 duets) then 10h00–11h00 (100 duets). Pair-duets and repertoires vary regionally; local dialects are transmitted from one generation to the next by social learning. Besides reacting vocally to its mate, a bird reacts vocally to same-sex rivals, whether seen or not.

Sedentary.

Food. Mainly invertebrates, including grasshoppers, beetles, caterpillars, termites, ants, bees, wasps, a small butterfly and a tick *Amblyomma variegatum* (Lack and Quicke 1978). In wet season, caterpillars form important part of diet of adults and nestlings. Seen eating *Commiphora* fruits (Lack 1985), and captive birds occasionally take fruits.

Breeding Habits. Monogamous pairs nest in permanent territories. ♂ displays by holding himself upright on extended legs, lowering head, swinging body from side to side, then bending forward, lowering wings, raising and fanning tail, and raising long, fluffy back and rump feathers into a mottled, whitish puff above back; he makes short flights between bushes, quivering wings rapidly with puff expanded (van Someren 1956). In calling display, 2 birds faced each other 30 cm apart for 2 min; presumed ♂ gave deep, churring calls and bowed with each one, answered by ♀ giving repeated, hoarse 'cough', thrusting down and forward with each note (Duckworth et al. 1992). 4 copulations were all in nesting tree soon after nest completed; ♀ crouches and quivers wings; copulation lasts a few s, then ♂ may utter 'kch' note. If clutch lost, pair usually makes a new nest; some unsuccessful pairs produce 3 clutches in a season; occasionally nest under construction is abandoned and another started elsewhere, using material taken from first nest; on day after losing first brood, one pair began to build 3 nests simultaneously (Sonnenschein and Reyer 1984).

NEST: a firmly- or sometimes loosely-made open cup, built of shreds of bark fibre stripped from dead branches or of fine rootlets and twiglets, many with remains of seed pods and flower bracts still attached; lined with delicate rootlets and dry grass blades; outer wall sometimes covered with spider web. Int. diam. 60–70, int. depth 30. One nest so poorly bound that it collapsed under weight of growing young. Nest rim often trodden down by brood, so nest changes from cup to platform shape. Placed in vertical fork of small tree or on twiggy horizontal branch of large tree growing in middle of thicket of creeper-covered bushes; usually very well hidden. Built usually by ♀ and ♂, ♂ putting in less effort than ♀ or sometimes not helping at all. Construction time 1–7 days (n = 9), first nests usually taking longer than later ones (Sonnenschein and Reyer 1984).

EGGS: 2–3 (18 clutches of 2, 3 of 3). Eggs laid on consecutive days. Pale blue, moderately glossy, finely and sparingly speckled with red-brown, pale brown or earth-brown, the speckles concentrated at large end to form a brown-grey cap with some underlying lavender spots. SIZE: (n = 2) 22 × 17, 21 × 17·5.

LAYING DATES: Ethiopia, Apr; Somalia, Apr–May (and nest building Oct); E Africa, Region B, Nov, Region C, Mar–May, Oct–Jan, Region D, Mar, May, Nov, Region E, Oct (a lengthy breeding season in Region C, Oct–May, mainly around rains, and in Region D breeds in long and short rains; at L. Baringo, Kenya, first nests found 1–3 weeks after rainfall).

INCUBATION: begins with first egg laid. One ♀ laying an egg half lifted herself out of nest and showed signs of effort. Shortly afterward, ♂ flew to nest; at first ♀ refused to leave, but change-over took place after a few min. Later, both sexes took regular turns. Av. spells, ♀ (n = 66) 40·6 min, ♂ (n = 68) 48·3 min. Nest relief is co-ordinated acoustically (Sonnenschein and Reyer 1983). At one nest ♀ drove ♂ away and prevented him from incubating (Sonnenschein and Reyer 1984). At other nests only ♀ incubates (van Someren 1956). Any shrike intruding from neighbouring territory is chased away. When person approaches silently, some ♀♀ slip away immediately, others leave only at last moment; human voices invariably cause incubating bird to leave nest. Period: (n = 4) 17 days.

DEVELOPMENT AND CARE OF YOUNG: 1 nestling hatched at 3 g, weighed 6 g on day 3, and gained weight almost linearly to 33 g at day 15. Feather tracts apparent on day 5; eyes opened on day 8; first feathers to emerge were on rump, at day 9. Young fully feathered on day 12 though wings and tail still short; it exercised by preening, stretching legs and wings, standing, and sometimes overbalancing, and begged with plaintive 'seee'. Just before young leaves nest plumage is black, every feather tipped with buff. Buff-spotted juvenile plumage takes 6–7 weeks to change to uniformly slate plumage of adult, at age *c.* 4–6 months (Sonnenschein and Reyer 1984). Eggshells, and any addled egg, carried away by parents. Young brooded for up to 7 days. Brooding ♀ with 3 young protected them from light rain by spreading wings. Brooding also serves to protect naked black chicks from the sun (Sonnenschein and Reyer 1984). At first young fed almost entirely on caterpillars, later on mixture of insects; young beg very quietly. After being fed, young produce faecal sac; in first few days parent eats sac, but later carries it away. A few days after leaving nest, a fledgling was fully capable of flight although tail still very short; it fed itself but also begged and quivered wings whenever parent approached. For next 2 weeks its calls consisted of babble of high and low sounds with no resemblance to adult duet song.

BREEDING SUCCESS/SURVIVAL: in study of 17 nests at L. Baringo, Kenya, all failed: 12 clutches and 5 broods disappeared, most nests remaining untouched; Nile monitors *Varanus niloticus* thought to be the predator responsible (Sonnenschein and Reyer 1984).

References
van Someren, V.G.L. (1956).
Sonnenschein, E. and Reyer, H.-U. (1983, 1984).
Wickler, W. and Seibt, U. (1979).
Wickler, W. and Sonnenschein, E. (1989).

Laniarius luehderi Reichenow. Lühder's Bush-Shrike. Gonolek de Lühder.

Plate 27 (Opp. p. 414)

Laniarius lühderi Reichenow, 1874. J. Orn., 22, p. 101; Cameroon Delta.

Forms a superspecies with *L. brauni* and *L. amboimensis*.

Range and Status. Endemic resident, SE Nigeria to NW Angola and S Sudan to E Zaïre and W Tanzania. Nigeria, not uncommon near Calabar. Cameroon, widespread in forest zone in SW, common (Bates 1911) or fairly common

Laniarius luehderi

in Victoria-Kumba-Dikume Balue area (Serle 1959); up to 1230 m. Bioko. Gabon, common and widespread. Congo, Odzala Nat. Park, near Mombongo in Nouabalé-Ndoki Nat. Park, and scarce in Kouilou Basin in SW, from coast to Mayombe and upstream to Stanley Pool. Angola, Cabinda, N Cuanza Norte and near Amboim (Gabela) and Conde in Cuanza Sul. (Central African Republic: given for Lobaye Préf., by Carroll 1988, record rejected by Germain 1992.) Zaïre, 'present along [rainforest's] northern edges' (Chapin 1954) though we know of no localities; widespread in NE and E, west to between Kisangani and Tshopo Falls, common between Nepoko and Bomokandi Rivers, and quite common in E up to 1850 m; south to about Kivu-Katanga border at L. Tanganyika; very common in Itombwe Highlands, at 900–2000 m. Sudan, common in Imatong Mts up to 2800 m, uncommon on Aloma Plateau. Uganda, common at 700–2400 m from Bwamba and Kibale to Ankole and Kigezi, east to Kifu and Mabira (Britton 1980). Rwanda, in western third; in Nyungwe Forest only in W, below 2050 m. Burundi, known only from Musigati and Karahe Forest at 1750–1800 m (Gaugris et al. 1981) and up to 2300 m (Vande weghe and Loiselle 1987). Kenya, locally fairly common from Mt Elgon, Kapenguria, Saiwa Nat. Park, Nandi, Kakamega, Kericho, Mau and Trans-Mara to Lolgorien, Migori R. and Olololoo Escarpment, Mara Game Reserve (Zimmerman et al. 1996). Tanzania, in W, from Gombe Stream Nat. Res. to Mahari Mts.

Density of 2–3 pairs per 10 ha of secondary growth, Gabon (Brosset and Erard 1986).

Description. *L. l. lühderi* (Reichenow): ADULT ♂: forehead and superciliary stripe pale cinnamon, merging into chestnut on crown and nape; lores, areas just above and below eye, ear-coverts, sides of neck, hindneck, mantle and scapulars glossy black with dark blue or dark green reflections, feathers with grey bases, usually concealed; lower back and rump feathers long, full and loose, overlying rump and uppertail-coverts respectively, dull black with whitish ends and whitish subterminal spot; effect is of mottled greyish rump grading to whitish uppertail-coverts. Tail black, round-ended, T1 shortfall 10–14 mm. Wing mainly black, primaries and outer secondaries with glossy outer vanes, inner secondaries, tertials and coverts matt black, but 2 secondaries with white outer vanes, and median coverts and some greater coverts broadly tipped with white. Chin, throat and breast deep cinnamon, belly, flanks, thighs and undertail-coverts white. Underside of tail black; underwing-coverts and axillaries white. Bill black; eyes dark red-brown or brown-red; legs and feet slate-blue or light blue. Sexes alike. SIZE (10 ♂♂, 6 ♀♀): wing, ♂ 85–94 (91), ♀ 81–87 (84); tail, ♂ 81–87 (82), ♀ 73–82 (74); bill, ♂ 20–23 (21·3), ♀ 19–22 (21·2); tarsus, ♂ 30–31 (30·5), ♀ 31–32 (31·1) WEIGHT: (E Africa) ♂ (n = 30) 30–49 (43·5), ♀ (n = 9) 35–45 (40·8).

IMMATURE: upperparts olive-brown with wide yellowish shoulder streak, dull rufous tail; throat and breast yellowish, rest of underparts whitish (Mackenzie 1978); or, upperparts olivaceous blackish brown with obscure, paler superciliary stripe, blackish lores and mask through eye, primaries and secondaries narrowly edged and tipped with rufescent buff, tertials rather more broadly edged and tipped buffy, and upperwing-coverts similarly tipped; chin and throat pale yellowish, merging to orange-rufous on breast and to whitish on belly, flanks and undertail-coverts; a few thin blackish bars across breast; soft parts blackish brown (Prinzinger et al. 1997). Juvenile, uppertail-coverts rufous and rest of upperparts olive-brown, all finely barred with light brown and blackish; chin and throat pale whitish grey, rest of underparts olive-yellow with fine dark bars except around vent and on undertail-coverts, latter being buff; no white in wing.

NESTLING: not described.

TAXONOMIC NOTE: see under *L. brauni*.

Field Characters. Length 17·5 cm. Common but hard to see in undergrowth of forest edge, best located by its familiar duet. *Tawny crown, throat and breast*, with *broad black mask* continuing onto back; white stripe on black wing. Adult unmistakable but immature, with brown cap and upperparts and buffy yellow underparts, might be confused with ♀ Northern Puffback *Dryoscopus gambensis*; distinguished by dark (not orange) eye, wing with single pale stripe instead of extensive white edging, and by barring (dark on underparts, buff on upperparts).

Voice. Tape-recorded (104, B, C, F, BEL, ERA, HOR, KEI, LEM, LOW, THOR). Song/contact calls a variable duet, see sonagrams in Thorpe (1972) and Dowsett-Lemaire (1990). Characteristic note of ♂ a soft, hollow, mellow 'wawh', often with slightly burry overtone, 'wrrawh', lower-pitched than whistles of Many-coloured and Black-fronted Bush-Shrikes *Malaconotus multicolor* and *M. nigrifrons*. ♀ joins in synchronously with harsh grating and snoring notes: 'wawh-kikikik', 'wawh-kik-zzha-zzha-zzha', 'wawh-kikikakaaarr'. First note may be double, 'waw-rawh-kak', or be replaced by high-pitched trill, 'rrraaauw-kakaka'; ♀ sometimes replies with excited series of cackling notes, slightly crescendo, 'wawh-kakakeke-kikiki ...'; these may also be given as an unaccompanied scold, a continuous 'kekekekekekekikikikikiki ... '. ♂

sometimes whistles alone but ♀ almost always gives some kind of reply, even if a single note, 'wawh-tuck'. A loud 'tick' followed by a snoring note, 'tick-worrrrrah', sounds like 1 bird but is probably 2.

General Habits. Inhabits thick vegetation in abandoned clearings, overgrown plantations and young secondary growth, and tall but open secondary forest near lowland towns and villages in forest zone; *Symphonia* gorge forest near Congo coast; exceptionally in primary forest (Itombwe).

Solitary or in pairs. Keeps in thick, impenetrable cover, but curious, and careful observer can draw bird momentarily into the open. Stays mainly low down in vegetation; sometimes up to mid levels. Occurs in mixed-species foraging flocks. Guttural 'crrrou' note, repeated at long, irregular intervals, uttered with marked inflation of throat immediately after decisive movement of head up and down (neck skin of ♂ thickened and distensible: Chapin 1954), or presumed ♂ bends head and neck forwards at each 'keow' utterance (Bates 1911). Pair maintains contact and defends territory by duetting; particularly prone to duet when nest is approached by person.

Sedentary.

Food. Grasshoppers, mantises, moths, caterpillars (especially those in parasol trees), termites, leaf-hoppers, beetles, spiders, 'pill-bugs' (isopod crustaceans) and tiny snails. Of 19 stomachs, E Africa, beetles in 10, beetle larvae in 1, caterpillars in 6 and grasshoppers in 6.

Breeding Habits. Solitary nester, evidently monogamous and territorial. Replacement nest once made in a palm a few m away from 1st, with clutch only 16 days after egg in 1st nest disappeared (Serle 1959).

NEST: very slight, shallow or quite deep open cup (usually shallow), loosely made of rootlets, dry weed stems or sometimes thin dry twigs and lined with fine rootlets; sited 1·8–2·5 m above ground, on a fork well inside small tree, bush or the large half-shrubby *Triumfetta* (Cameroon, Gabon); once on frond of young oil palm *Elaeis guineensis* growing amongst tangled vegetation (Nigeria).

EGGS: 1–2. 2nd egg laid 24 h after 1st (n = 3). Shape varies from ovate to long ovate; shell smooth, without gloss. Pale blue, greenish blue, cream-buff or greenish cream-white, profusely but not boldly spotted and speckled with olive brown primary and ashy or violet secondary marks, confluent and forming a cap at large end (Serle 1959). SIZE: (n = 2) 22·8 × 17·0, 24·5 × 17·4; (n = 9) 22–27 × 16–18 (25 × 17·5) (Bates 1911).

LAYING DATES: Nigeria, Sept–Oct; Cameroon, Aug–Oct, Dec; Gabon, Sept, Nov–Dec, Feb; Zaïre, Nepoko, (breeding condition Apr, July), Itombwe, (various evidences, May–July and Oct–Dec, Prigogine 1971); E Africa, Region A, June, Region B, July.

INCUBATION: by ♀ only; period *c.* 15 days.

DEVELOPMENT AND CARE OF YOUNG: young reared by both parents; nestling period *c.* 15 days.

References
Bannerman, D.A. (1939).
Bates, G.L. (1911).
Brosset, A. and Erard, C. (1986).
Chapin, J.P. (1954).
Dowsett-Lemaire, F. (1990).
Prigogine, A. (1971).
Serle, W. (1959).

Laniarius brauni (Bannerman). Braun's Bush-Shrike. Gonolek de Braun.

Plate 27
(Opp. p. 414)

Laniarius lühderi brauni Bannerman, 1939. Ibis, 14th ser., 3, p. 746; Quicolungo, Angola.

Forms a superspecies with *L. luehderi* and *L. amboimensis*.

Range and Status. Endemic resident, NW Angola, known from Quicolungo (08°29'S, 15°16'E), Canzela (within 30 km of Quicolungo), Camabatela (08°20'S, 15°26'E) and Quibaxi (08°34'S, 14°37'E), Cuanza Norte (Traylor 1962, 1963). Rare and little known. Deforestation of the northern escarpment in Cuanza Norte has proceeded steadily; given Endangered status by Collar *et al.* (1994).

Description. ADULT ♂: forehead, crown, nape and hindneck chestnut, slightly paler in front and at sides where bordering black plumage; feathers behind nostrils, lores, areas just above and below eye, ear-coverts, sides of neck, hindneck, mantle and scapulars jet black, slightly glossy; lower back and rump feathers long, full and fluffy, overlying rump, greyish black with large white subterminal spot; uppertail-coverts matt black. Tail black. Chin, throat and breast bright orange-red (area less extensive than buff area in *L. luehderi*), belly, flanks, thighs and undertail-coverts pure white. Underside of tail black; underwing coverts and axillaries white, lower flank feathers full, silky, fluffy, with dark grey bases that can show through. Wing brownish black, glossy, flight feathers becoming brown with buffy edges before being moulted; long white line in wing formed by white median coverts, large white tips to inner 2–3 greater coverts, and narrow (0·5–1·0 mm) outer edges and tips to innermost 2 secondaries. Bill black; eyes dark brown, dark brownish grey or reddish brown; legs and feet slate. Sexes alike but, in 3 skins examined, red area less extensive in 2 ♀♀ than in the ♂. SIZE (2 ♂♂, 2 ♀♀): wing, ♂ 90, 92, ♀ 91, 92; tail, ♂ 82, 86, ♀ 82, 82; bill, ♂ 21, 21, ♀ 20, 20; tarsus, ♂ 32, 33, ♀ 32, 32 (Bannerman 1939). WEIGHT: 1 ♂ 55, 2 ♀♀ 54, 54.

Laniarius brauni

TAXONOMIC NOTE: the well-known *L. luehderi*, barely-known *L. brauni* and almost unknown *L. amboimensis* would appear to be similar in most regards but differ spectacularly in colours of chin-to-breast. When first described, *brauni* and *amboimensis* were made subspecies of *L. luehderi* and subsequently nearly all authors have treated them as such, but they were regarded as incipient species by Hall and Moreau (1970) and given rank of species in the *L. luehderi* superspecies by Sibley and Monroe (1990). Treated as full species by Collar *et al.* (1994). Mindful both of current thinking about avian species concepts and limits (Haffer 1998) and of evident species limits within genus *Laniarius* as a whole, we follow Sibley and Monroe (1990) and keep the poorly-known but highly distinctive *brauni* and *amboimensis* as separate species in a superspecies.

Field Characters. Length 17·5 cm. Resembles allopatric Lühder's Bush-Shrike *L. luehderi*, with tawny crown and black mask, but differs in bright *orange-red* chin, throat and breast. Confined to a small area of forest in N. Angola, where there is no other bird like it.

Voice. Not tape-recorded. Said to be much less pleasing than that of Southern Boubou *L. ferrugineus*, and similar to that of nominate *luehderi* (Chapin 1954).

Habits. Confined to secondary and gallery rain forest region of Cuanza Norte. Said to breed in Mar. Nothing further known.

References
Bannerman, D.A. (1939).
Collar, N.J. *et al.* (1994).
Traylor, M.A. (1963).

Plate 27 *Laniarius amboimensis* (Moltoni). Gabela Bush-Shrike. Gonolek d'Amboim.
(Opp. p. 414)

Laniarius luehderi amboimensis Moltoni, 1932. Atti. Soc. Ital. Sci. Nat. Milano, 71, p. 175; Amboim.

Forms a superspecies with *L. luehderi* and *L. brauni*.

Range and Status. Endemic resident, W Angola. Known only from Amboim, near Gabela (10°56′S, 14°24′E), on escarpment in Cuanza Sul. Rare and little known; discovered in 1930, rediscovered in same area in 1960 (at Londa and Assango, da Rosa Pinto 1962) and again in 1992, when 2 pairs found in 3 days near Gabela (Collar *et al.* 1994).

Description. ADULT ♂: forehead, crown, nape and hindneck chestnut, slightly paler in front and at sides where bordering black plumage; feathers behind nostrils, lores, areas just above and below eye, ear-coverts, sides of neck, hindneck, mantle and scapulars jet black, slightly glossy; lower back and rump feathers long, full and fluffy, overlying rump, greyish black; uppertail-coverts black. Tail black. Entire underparts pure white. Underside of tail black. Wing brownish black, glossy; long white line in wing formed by white median coverts, large white tips to inner greater coverts, and narrow outer edges to innermost secondaries. Bill black; eyes dark brown; legs and feet slate. Sexes alike. SIZE (1 ♂, 2 ♀♀): wing, ♂ 93, ♀ 88, 89; tail, ♂ 85, ♀ 82, 83; bill to feathers, ♂ 19, ♀ 17, 19, bill to skull, ♂ 24, ♀ 23, 24·5; tarsus, ♂ 33, ♀ 32, 32 (da Rosa Pinto 1962).
TAXONOMIC NOTE: see under *L. brauni*.

Field Characters. Length 17·5 cm. Like allopatric Lühder's Bush-Shrike *L. luehderi*, but underparts *pure*

Laniarius amboimensis

white. Differs from Swamp Boubou *L. bicolor* by *tawny crown* and hindneck; it will also differ by call if its voice is anything like that of Lühder's Bush-Shrike.

Voice. Unknown.

General Habits. Inhabits evergreen forest. Occurs in mixed-species flocks (Collar *et al.* 1994). Nothing further known.

References
Collar, N.J. *et al* (1994).
da Rosa Pinto, A.A. (1962).

Laniarius ruficeps (Shelley). Red-naped Bush-Shrike. Gonolek à nuque rouge.

Plate 27
(Opp. p. 414)

Dryoscopus ruficeps Shelley, 1885. Ibis, p. 402, pl. 10; Somaliland (Burao).

Range and Status. Endemic resident, patchily distributed in Horn and E Africa, probably considerably more widespread than shown. Somalia, 3 subspecies: land over 1000 m around Burao, about 09°30′ to 10°N, and common at Kyal, 10 km north of Burao (Archer and Godman 1961); coastal lowlands north to 03° (and probably 05°) N; and from coast near Eyl (08°N) to hinterland of S Somalia, as far as Ethiopian and Kenyan borders. Ethiopia, uncommon, in S and SE (but formerly common around Yavello). Kenya, locally common, below 1000 m; from Garissa, Mwingi, Kitui and Mutomo Districts to Mtito Andei, Ngulia, Maungu, Taru, Lali Hills and Galana Ranch; common around Kiunga, 41°30′E, and extends south towards Boni Forest (S. Keith, pers. comm.); records in Mandera and Wajir Districts in NE.

Laniarius ruficeps

Description. *L. r. ruficeps* (Shelley): NW Somalia. ADULT ♂: forehead black, crown, nape and hindneck bright orange-rufous or rufous-red, mantle bluish grey in midline, paling towards sides of neck, upper back ash-grey, feathers of scapulars and lower back jet black with bluish gloss and large white subterminal spots; rump feathers silky and fluffy, white, with dark grey bases and narrow blackish tips; uppertail-coverts black. Tail black, T4–T6 with 10-mm-deep white patch at end and T6 also with white outer vane. Long, well-defined white superciliary stripe, from bill to side of neck, broadest above nostrils; lores black; area just above and below eye black, ear-coverts glossy black; narrow black stripe down side of neck joining ear-coverts and scapulars. Entire underparts cream-white, sides of breast and flanks washed with pale pink-buff. Wings black, with broad white longitudinal stripe formed by white median coverts, 3 inner greater coverts, and outer vanes of 3 innermost secondaries. Marginal and lesser underwing-coverts cream-white, greater under primary coverts dusky. Bill black; eyes hazel-brown; legs and feet blue-grey or slate-grey, soles white. Sexes alike except that ♀ has olive-grey back. SIZE (6 ♂♂, 4 ♀♀): wing, ♂ 78·5–79·5 (79·2), ♀ 67–77 (72·2); tail, ♂ 73–78 (77·1), ♀ 70·5–74 (72·5); bill to feathers, ♂ 17·5–20·5 (18·3), ♀ 17·5–18·5 (17·9); tarsus, ♂ 25·5–31 (28·4), ♀ 25·5–28 (26·8). WEIGHT: 1 ♂ 35, 2 ♀♀ 30, 31.

IMMATURE: juvenile has forehead to back uniform olive-grey; scapulars brown with buffy tips, giving barred appearance; primary coverts and outer greater coverts brown with well-defined, narrow buff edges and tips (otherwise wings, and tail, as adult); underparts dull white, and lacks adult's black mask – ear-coverts dusky; indistinct, pale superciliary stripe.

NESTLING: not known.

L. r. rufinuchalis (Sharpe): Ethiopia, S Somalia except coast. Like *ruficeps* but crown black and only nape and hindneck orange-rufous, and mantle brown rather than clear grey.

L. r. kismayensis (Erlanger): S Somalia coastal lowlands, and Kenya (Kiunga). Like *rufinuchalis* but mantle pale grey.

Field Characters. Length 18 cm. A black and white shrike of dry E African lowlands, with *black mask* and *pale supercilium*, distinguished by *rufous patch* on nape and hindneck, extending onto crown in Ethiopia and Somalia (forecrown black in most of Kenya). Skulks in thick bush (fiscals *Lanius* and puffbacks *Dryoscopus* are out in the open). In lusher habitats overlaps with Tropical Boubou *Laniarius aethiopicus*, which is also a skulker, but latter readily told by solid black crown and upperparts and ringing duets.

Voice. Tape-recorded (B). Calls, a continuously repeated 'kwoi-kwoi-kwoi ... ', low, whistling 'whooi-whooi', loud, harsh 'k-k-k-k-k-k' and scolding 'kwerr, kwerr'; reportedly duets like other bush-shrikes (Zimmerman *et al.* 1996).

General Habits. Inhabits dense, impenetrable thorn bush, in NW Somalia particularly *Acacia orphota*, and open thorn scrub in Garissa, Kenya. Solitary or (usually) in pairs; but may also be gregarious or loosely colonial, since 14 birds collected at Kyal, Somalia, in the course of 2 days, one in Jan, one in May (1920). Evidently keeps very much to shadiest interior of thorn thickets (Somalia). Foraging pair are generally widely separated and frequently give contact calls. Insect prey taken from ground and from lower branches in thickets, where bird hunts silently, moving rapidly but furtively. Occasionally one bird of pair climbs to top of tree and stays there in full view for a few s. Presumably sedentary.

Food. Insects including larvae in 3 stomachs.

Breeding Habits.
NEST: one was made of 'twigs and sticks' (Archer and Godman 1961).
EGGS: 3 (1 clutch). Very glossy; pale blue-green, streaked and marbled with red-brown and lilac, forming irregular cap at broad end. SIZE: (n = 3) 21 × 16, 21·5 × 16, 20·5 × 15·5.
LAYING DATES: Somalia (Kyal), May; Ethiopia, May.

References
Archer, G. and Godman, E.M. (1961).
Serle, W. (1943).
Zimmerman, D.A. et al. (1996).

Plate 27
(Opp. p. 414)

Laniarius liberatus Smith, Arctander, Fjeldså and Amir. Bulo Burti Bush-Shrike. Gonolek de Bulo Burti.

Laniarius liberatus Smith *et al.*, 1991. Ibis, 133, p. 227, pl. 1; Bulo Burti, Shabeelle R., Somalia.

Range and Status. Known from a single bird found at 140 m elevation at Bulo Berti (= Buulobarde, 03°50'N, 45°33'E), Somalia, in Aug 1988, watched from Oct, caught in Jan 1989, taken to Europe, returned to Mogadishu, Somalia, given a week's acclimatization in an open-air enclosure in acacia scrub, and released in Balcad Nat. Res. (= Balad, 02°20'N, 45°20'E) in Mar 1990 (Smith *et al.* 1991).

Description. ADULT (not sexed): upperparts from forehead, lores, cheeks, ear-coverts and sides of neck to tail and wings matt black except for: broad pale yellow superciliary stripe from bill to side of hindneck, where it can appear square-ended; hindcrown and nape mottled, feathers with pale yellow tips (Smith *et al.* 1991) or all black (Prinzinger *et al.* 1997); hindneck dark grey or black; a few back feathers with off-white tips; most rump feathers with broad white subapical spots (usually concealed). Tail almost square-ended (Prinzinger *et al.* 1997, *cf.* Smith *et al.* 1991) with small white tips to all feathers except T1; wing with tertials and secondaries dull brown-black, outer vanes of S4–S6 white-edged; inner greater coverts with broad white tips to outer vanes and next one with short narrow white border to outer vane; median wing-coverts with small white tips to outer vanes. Underparts white, but sides of throat, whole of upper breast and sides of lower breast pale yellow. Bill thick, blue-grey or blackish (bill deeper than in Tropical Boubou *L. aethiopicus*); eyes red-brown, eye-rim black; legs and feet dark grey. (Bird known to be adult, since it moulted into identical plumage whilst in captivity.) SIZE: wing 89 before moult, 92 after; bill 20·5; tarsus 34.
TAXONOMIC NOTE: from DNA analysis it was concluded that 'in all pairwise comparisons the *L. liberatus* sequence is as distinct as the other currently recognized species in this comparison. There is thus no reason to believe that the bird is a hybrid between any of the examined species or a colour morph of one.' and 'The observed number of substitutions correspond well with similar investigations ... of the variation in the cyt-b gene in birds ... supporting the conclusions drawn from morphology, that we are dealing with a well-marked *Laniarius* species.' Compared with 5 other *Laniarius* species and a *Malaconotus*, *L. liberatus* appears to be the sister-species of the closely allied *L. turatii* and *L. aethiopicus*.

Field Characters. Length 20 cm. Typical of its genus in stance (**A**), behaviour, voice and thick bill. Black upperparts and white underparts suggest Tropical Boubou *L.*

Laniarius liberatus

A

aethiopicus, with which it might occur, but distinguished by *yellow supercilium* and yellow wash on breast.

Voice. Tape-recorded (B). Low, harsh churr when flying to and from roost; single low 'chack' when feeding; in captivity, ascending whistle 'poo-eeh' (contact? See sonagrams in Prinzinger *et al.* 1997).

General Habits. Bird caught in an isolated patch of disturbed acacia thicket dominated by *A. nilotica* with a narrow band of riverine forest and scattered capparidaceous bushes, mainly *Cadaba mirabilis*, supporting a very sparse herb layer; many trees had umbrella-shaped canopies reaching the ground and supporting a wealth of climbing cucurbits, legumes and composites. Bird foraged in or below dense, tangled vegetation. Perched about 3 m high when on way to and from roost, but fed on the ground or in very low branches deep inside acacia bushes. Followed same foraging route every day, generally spending 1–2 h in same acacia before making brief, low flight to another one. Foraged rather like a *Turdus* thrush, hopping about on ground, turning over branches and leaf litter. In aviary, when bird took a lizard it carried it in its bill to a broken horizontal branch with spiky wood splinters and repeatedly beat it against the spikes until prey torn to pieces.

Food. In captivity ate crickets, cockroaches, geckos and lizards.

References
Prinzinger, R. *et al.* (1997).
Smith, E.F.G. *et al.* (1991).

(For comment on standards of description of this new species, see M. LeCroy and F. Vuilleumier 1992, *Bull. Br. Orn. Club* 112A, 191–198 and N.J. Collar, 1999, *Ibis* 141, 358–367.)

Laniarius aethiopicus (Gmelin). Tropical Boubou. Gonolek d'Abyssinie.

Plate 27 (Opp. p. 414)

Turdus aethiopicus Gmelin, 1788. Syst. Nat., 1, pt. 2, p. 824; Abyssinia.

Forms a superspecies with *L. ferrugineus*, *L. bicolor* and *L. turatii*.

Range and Status. Endemic resident, mesic woodlands in N and S tropics. Senegal, sight record from Diouloulou, Casamance (Morel and Morel 1990). Guinea, common around Beyla and Macenta; records west to Gaoual and Conakry areas. Sierra Leone, frequent in E, in Loma and Tingi Mts, south to Kono District and west in N to Kabala and Bafodia (Field 1979). Liberia, rare, records from Makona R. near Jalamai and Monrovia (Gatter 1997) and a locality in between (Hall and Moreau 1970). Ivory Coast, common north of 07°N; occurs south to Lamto and Abidjan. Ghana, uncommon in coastal belt from Axim to Accra Plains; in N, frequent at Mole and Tono Dam, otherwise known only from Wenchi, Ejura and Kete Kratchi. Togo, locally common from coast north to Landa-Pozanda and Naboulgou (Cheke and Walsh 1996). Benin, frequent north at least to Bétérou area (Claffey 1995). Nigeria, not uncommon in S guinean savannas immediately north of forest zone, and occurs north to Kaduna, Jos Plateau and Yankari Nat. Park; in suburbs of coastal towns (Lagos, Burutu, Calabar); on Mambilla Plateau. Cameroon, not uncommon north of forest zone to about 07°30'N; in S, penetrates forest zone in cleared areas and suburbs south to Gabon border. Chad, uncommon, north to Sarh and Zakouma Nat. Park. Central African Republic, frequent in Dzanga Reserves (Green and Carroll 1991), several records in mid latitudes, rare in Bamingui-Bangoran Nat. Park. NE Zaïre, frequent and widespread in woodlands and grasslands north of Congo Basin rain forest, up to 1850 m, and penetrates southward in cleared land to R. Congo. Sudan, common in SW, uncommon near Ethiopian border in E and SE. Ethiopia, common to abundant in highlands, up to 2750 m (Sen'afe) absent from hot lowlands in NE and

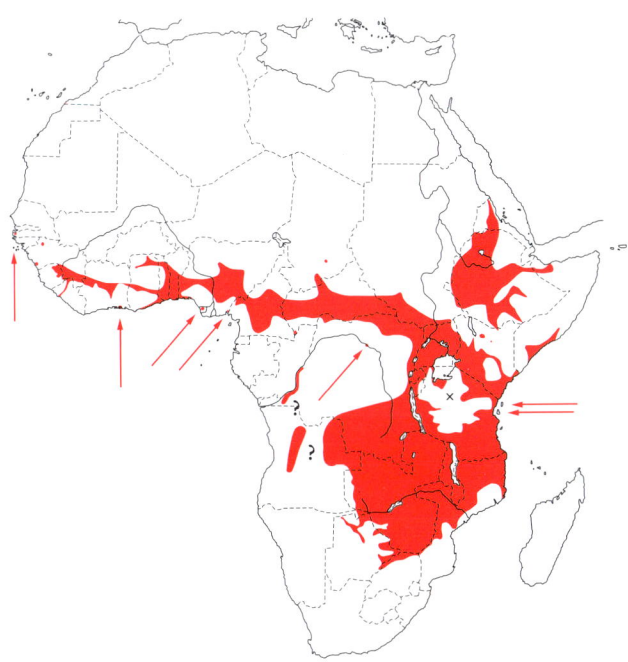

Laniarius aethiopicus

from much of Ogaden, and uncommon or absent towards Kenya border. Eritrea, frequent at all altitudes but absent from W and from dry coastal plain. Somalia, common in NW at 1540–1850 m and in SE down to sea level.

Throughout Uganda. Kenya, highlands and coastal lowlands throughout SW half of the country, north to Kulal and Marsabit, with the Ethiopian race known from

Moyale. Tanzania, in NE south to Mara region, Serengeti Nat. Park, Loliondo, Essimingor, Crater and Mbulu Highlands, Arusha, Moshi, North Pare Mts and E Usambaras; common on Pemba and Zanzibar; absent from most of interior but locally frequent in Eastern Arc Mts, Mahenge and Njombe highlands and in SW (map data: N. Baker, pers. comm.). Rwanda and Burundi, status uncertain – widespread but uncommon in W? Angola, frequent on W-central plateau between Lunda, Malanje and Huambo; known to occur in Moxico Prov. but status uncertain. Zaïre (other than NE), in W occurs east of Congo R. in Bas-Zaïre Prov. and along Congo R. between about 02° and 04°S; not certainly known from S Bandundu Prov. though probably occurs there; widespread east of 21°E in woodlands south of rain forest block, in Kasai Occidental, S Kasai Oriental, throughout Katanga and in E Kivu. Zambia, frequent to common and widespread, but uncommon in W and absent from SW west of 22°30′E. Botswana, fairly common to very common in Limpopo R. and Shashe R. drainages and at Kasane and Serondella; uncommon in Okavango and along Botletle R. (Penry 1994) – not recorded there by Harrison et al. (1997). Transvaal, riparian forest along Limpopo and Luvuvhu Rivers and tributaries in N lowveld, and Nyanda Bush in N Kruger Nat. Park (Tarboton et al. 1987). Mozambique, common and widespread north of Save R. north to Tanzanian border (Vincent 1935); absent from much of Zambezia and interior of Nampula.

Density of 55 territories (pairs) in 3·2 km^2 of linear transect along Luvuvhu R., NE Transvaal, i.e. 34 birds per km^2; estimated 10,500 birds in N Kruger Nat. Park (Hurford et al. 1996).

Description. *L. a. major* Hartlaub: W and central Africa, east to S Sudan, Uganda, Kenya to Rift Valley, Tanzania in N to Loliondo and Essimingor and in S Njombe and Mahenge, and south to NW and N Zambia and extreme N Malaŵi where intergrades with *mossambicus*. ADULT ♂: forehead, crown, cheeks, ear-coverts, hindneck, sides of neck, mantle and scapulars jet black glossed with dark blue; feathers of lower back and rump long, full and fluffy, greyish or whitish, with white subapical mark and black ends, overlying rump and uppertail-coverts respectively; uppertail-coverts glossy bluish black. Tail black above and below, slightly glossy. Entire underparts white, breast and flanks faintly tinged with salmon pink. (The pink evidently fades rapidly; in life many birds are pink, in the museum nearly all are white. See Taxonomic Note.) Wings mainly black, somewhat glossy, with long white stripe (almost straight in fully open wing, stepped and curved in closed wing); marginal and lesser coverts black, median coverts black proximally, white distally, greater coverts black but 2–3 innermost ones with white outer vanes, primary coverts and flight feathers black but 3 innermost secondaries with white outer vanes. Underwing-coverts white, axillaries white with pink tinge. Bill black; eyes deep red or dark Indian red; legs and feet bluish slate. Sexes alike. SIZE (W Africa, 12 ♂♂, 12 ♀♀): wing, ♂ 99–106, ♀ 94–103; tail, ♂ 89–100, ♀ 88–95; bill, ♂ 24–26, ♀ 22–24; tarsus, ♂ 35–37, ♀ 35–36; wing of birds in Uele, NE Zaïre, c. 10 shorter. WEIGHT: (E Africa), ♂ (n = 6) 52–62 (55·8), ♀ (n = 2) 40, 45; (Ghana) 1 unsexed 61·5.

IMMATURE: like adult but blacks duller and whites buffy; wing coverts and uppertail-coverts with buffy white tips and edges, T5–6 with white tips; juvenile has most feathers in upperparts tipped tawny buff, underparts dull whitish with breast and flanks pale brown and some dusky barring; bill horn, eyes sepia-brown.

NESTLING: naked and blind; mouth without spots; gapes pale yellowish, not particularly thick (photo, van Someren 1956).

L. a. mossambicus (Fischer and Reichenow): Zambia (except NW and N where intergrades with *major*), Malaŵi except extreme N, E Botswana, Zimbabwe except SE lowveld, and Mozambique. Like *major* but underparts strongly washed pinkish buff. SIZE (with some *limpopoensis*? – 10 ♂♂, 8 ♀♀): wing, ♂ 93–100 (96·6), ♀ 90–96 (93·2), unsexed (n = 52) 86–102 (94·7); tail, ♂ 97–105 (101), ♀ 93–100 (96), unsexed (n = 36) 90·5–105·5 (Maclean 1993); bill, ♂♀ 19–25·5 (23); tarsus, ♂♀ 30·5–37·5 (33·7). WEIGHT (with some *limpopoensis*?): ♂ (n = 26) 42·4–69·0 (52·5), ♀ (n = 51) 40·1–57·0 (47·5), unsexed (n = 51) 42·9–61·0 (49·9) (Maclean 1993).

L. a. limpopoensis (Roberts): SE Zimbabwe between Sabi, Lundi and Limpopo rivers, and N Transvaal. Like *mossambicus* but underparts buffy cream.

L. a. aethiopicus (Gmelin): E borders of Sudan; Eritrea, Ethiopia, NW Somalia, and N Kenya (Moyale). Like *major* but secondaries all black; underparts deeper pink.

L. a. erlangeri (Reichenow): lower Shabeelle and Jubba valleys, between equator and 03°N. Like *major* but secondaries and greater wing coverts all black; black parts glossed greenish rather than bluish; underparts paler pink. A fairly common morph is all black. Small; wing 80–90.

L. a. ambiguus (Madarasz): Kenya and NE Tanzania in highlands east of Rift Valley, north to Mts Kulal and Marsabit, south to Mts Meru and Kilimanjaro and Arusha and Moshi. Like *erlangeri* but larger: wing 89–103. WEIGHT: ♂ (n = 17) 47·5–60 (52·9), ♀ (n = 13) 41–60 (48·6).

L. a. sublacteus (Cassin): coastal lowlands of extreme S Somalia (Boni Forest), Kenya (inland in lower Tana valley to Bura and Garissa, also Shimba Hills Nat. Park, Mt Kasigau, Sagal Hills and Taita Hills) and NE Tanzania (inland to E Usambaras, North Pare Mts and L. Jipe) south to Zanzibar. Like *major* but wing all black. An entirely black morph occurs. WEIGHT: ♂ (n = 5) 43–50 (45·8), ♀ (n = 10) 40–55 (43·4).

TAXONOMIC NOTE: populations on plateau of W-central Angola (Mt Moco, Mt Soque, Duque de Bragança and elsewhere) have especially white underparts (see Description) and are likely to represent an undescribed taxon (Hall and Moreau 1970). For relationship with *L. ferrugineus* see Note under that species.

Field Characters. Length 19·5–25 cm. A large black-and-white bush-shrike whose ringing duets are a familiar sound throughout much of Africa. Skulks in undergrowth; fiscals *Lanius* spp. perch in the open, puffbacks *Dryoscopus* forage in treetops. Head to below eye and upperparts *glossy black*, rump grey or white; wings black, with prominent white line in most races (lacking in *sublacteus*, reduced in others). Eye brown (bright red in puffbacks). Underparts appear white in the field, but in good light show variable creamy or pinkish wash. Underpart colour is important only where Tropical meets close relatives: in W Africa, Turati's Boubou *L. turatii* (entirely pinkish buff; no white in wing); in southern Africa, Swamp Boubou *L. bicolor* (pure white) and Southern Boubou *L. ferrugineus* (rufous flanks and belly). An all-black morph occurs in Somalia and coastal Kenya and Tanzania, distinguished from Slate-coloured Boubou *L. funebris* by glossy black, not slate-grey plumage and different voice. Immature has buff spots on upperparts, off-white underparts with a little dark barring.

Voice. Tape-recorded (5, 10, 27, 36, 86, 88, 91–99, 104, 109, B, C, F, BRU, CART, KEI). Ringing bell-like song one of the great sounds of Africa, whence the popular name 'bellbird'. Voice extremely complex and varied. Normal song type is synchronized duet, but either ♂ or ♀ may whistle or scold unanswered, and sometimes more than 2 birds take part; there are also regional dialects. Common song in E and S Africa is a duet in which ♂ gives first and third notes, ♀ the second on a higher pitch, 'haw-wheee-haw' or higher-pitched 'wah-heeya-wah'; ♀ may reply with a scold rather than a whistle, 'haw-dzheeer-haw', longer, twangy 'haw-waaaangh-haw' or faster 'wah-dzh-wah', or scold may come first, 'dzheer-wah-dzheer'. Other duets: 'haw-zweeeeya', 'haw-dzz-haw-dzz-haw-dzz', 'hoo-whee-whee', 'heee-bah ... heee', 'hee-jaaawaaanh', 'waaaaang-hohohoho', 'haw-haw-whee-haw', 'haw-teeyu-haw-teeyu-haw', 'wahwah-zhraaaa-wahwah-zhreeee-wahwah', 'kyaw-hoo-kyaw-hoo-kyaw' and croaking 'hrrraw' followed by trilled 'wawawawawa'. Duets are timed so precisely they sound like 1 bird. ♂ or ♀ may omit their parts seemingly at will; in complete duet 'hoho-hah-ho', ♂ sometimes gave first 'hoho' alone, or ♀ answered but ♂ did not finish, 'hoho-hah'. In what sounded like normal duet, 'haw-haw-hah-haw', ♂ performed both parts (Chappuis, in press). Song may be triggered by soft whistle (3rd bird?), '(heee)-tyaw-zzhh-tyaw'. In 3-part song 'teeyu-woh-kzz', not known how many birds took part. In remarkable round of singing by 3 or 4 birds, series of sweet, clear ringing notes conformed almost exactly to the key of A-flat: first 2 notes, 'hoo-haw', second 3 semitones lower (D-flat, C); then 'wheee-hohohoho', second bird an octave below the first (A-flat, A-flat); next a 3-part 'wheee-haw-hohohoho' (A-flat, C, A-flat); and finally a (?4th) bird gave a harsh 'jeeeearr' accompanied at irregular intervals by 'wheee', 'haw' and 'hohohoho' (North and McChesney 1964). Unaccompanied calls include fast hollow trill of 5 notes, slower 'hoh-hoh-hoh-hoh', soft, hollow 'hoh' repeated at rate of 1 per s, a long, high-pitched ringing 'heeeeee' and a grating 'wraaaaayo'. Contact call near nest a low 'chewk' or 'chuk'. Low-level alarm, 'jit' or 'jit-jrrrr'; more intense, a loud, continuous 'kakakakakaka ... ', sometimes joined by second bird with nasal 'zhaaanh-zhaaanh'. Repertoire enormous; for extensive study of vocalizations, including those of young birds raised in captivity, see Thorpe (1972).

General Habits. Inhabits dense woody cover along riverbanks and water-courses, also thickets and unkempt hedges in farmland and gardens and dense growth on termite mounds and around bases of inselbergs and koppies. Ethiopia and Eritrea, uncommon in highland *Hagenia* forest and *Arundinaria* bamboo, common in evergreen olive-juniper-*Podocarpus* scrub and humid lowland forests, and frequent in tall grass savannas and dry *Acacia-Commiphora* bush. Occurs also in thick elephant grass, tangles of creepers, along edges of lowland forest, in thick bush in hillside ravines and on tops of hills. Sometimes well away from water, but generally in humid climates, damp vegetation and woody growth near water.

Solitary or in pairs; often seen in family parties of *c.* 5 birds. Shy and secretive, keeping inside cover, but inquisitive, even confiding, coming into the open on ground in large gardens and around game lodges. Pair remains all year in territory of *c.* 2 ha, defended by calling. Calling ♂ often perches half exposed on or near top of bush; bobs head and bows body with each note given. In territorial interactions calling ♀ often perches in bush below ♂. Paired ♂ and ♀ sing in antiphonal duets, which have considerable geographical and pair variation. Single bird can sing 'pair' duet, i.e. its own and its mate's contributions. Flight rather heavy, with rapid wing beats. Has been seen allopreening. Sleeps in dense, tangled vegetation, making 'tuk-tuk-tuk' calls before settling down to sleep (Harris and Arnott 1988).

Forages low down in woody vegetation and foliage, often coming to the ground, where hops amongst leaf litter, not flicking a leaf aside like a thrush, but rather turning a leaf or bit of bark when it sees an insect disappearing under it (van Someren 1956). Gleans trunks, branches and foliage, but probably takes more food from ground than from vegetation. On ground, stands with legs fully extended, tail raised slightly above wing tips, and body plumage drawn close. Sometimes comes out from under bush and hunts in roadside drains or, after heavy rain, in silt along path (van Someren 1956). Occasionally hawks for flying insects. A persistent raider of small birds' nests. Wedges larger prey items in a fork, and tears them apart with bill (Harris and Arnott 1988). Wild birds not known to impale prey, but a captive 'stuck pieces of ox heart onto branches in its aviary' (Sonnenschein and Reyer 1984).

Sedentary.

Food. Insects, including orthopterans, mantises, termites, many beetles and caterpillars, and other larvae and pupae; some small snails; and a few small vertebrates (chameleons, skinks, amphibians, rodents, birds' eggs and nestlings – Harris and Arnott 1988); often attacks waxbills (Vincent 1935). Occasionally takes small fruits. Young at first fed on insects including long-horned grasshoppers, mantises, *Saturnia* moths and noctuid caterpillars; later given several small tree skinks, tree-frogs, a 30-cm tree snake and several nestling waxbills (van Someren 1956).

Breeding Habits. Monogamous, territorial, solitary nester. Territories well spaced, perhaps seldom contiguous, advertised and defended by calling, and squabbles between neighbouring pairs rare or unknown (van Someren 1956). Courtship involves ♂ chasing ♀, the 2 hopping close together rapidly through foliage and branches and making short, wing-fripping flights; ♂ bobs head shallowly and, with neck extended, bows body; he gives snarling calls and flies or half-glides downwards with slow wing beats, whitish back and rump feathers fluffed out (Harris and Arnott 1988). Courting ♂ stands upright, raises head and calls 'schrang schrang', then depresses head and body whilst calling 'kit-tuu-iii'; he droops wings, fans tail, raises long, mottled, whitish back and rump feathers and 'lets them droop fanwise in a most effective manner', while moving up and down, calling and fluttering in front of mate, then copulating (van Someren 1956). Double broods on record. Mated birds with broods are aggressive and plunder nests of other small birds (Harris and Arnott 1988).

NEST: shallow, open bowl made of loosely knitted rootlets and fine twiglets, lightly bound with spider web, sparsely lined with fine rootlets; usually thin enough for eggs to be visible from below. Sited in fork generally in thorn bush, or on horizontal branch amongst climbers and epiphytes, well concealed, 1·5–3 m above ground but sometimes up to 9 m high, in tree such as *Terminalia prunioides* (Skinner 1998). Built by ♀ (mainly) and ♂. Several nests may be started, being successively plundered by pair, before ♀ lays in the definitive one (Harris and Arnott 1988). ♂ particularly vocal during nest-building period.

EGGS: 2–3. Long or short ovals, pale blue, greenish blue or greenish buff, freckled with umber-brown and lilac on pale grey undermarkings; freckles mainly in ring around large end and sometimes form a cap. SIZE: (n = 19) 22·3–26·8 × 16·7–19·1 (25·0 × 18·3).

LAYING DATES: Guinea, May; Nigeria, Feb–Mar, June; Sudan, Sept; Ethiopia, Apr–June; NW Somalia, Apr–May; Zaïre, Faradje, May, Baraka, Dec–Jan, Katanga, Sept–Dec; E Africa, Region A, Apr, June, Sept, Oct, Region B, Mar–July (mainly Apr–June, i.e. long rains), Region C, Feb, Nov, Region D, Oct, Dec–July (mainly Mar–May; breeds in wet and dry months), Region E, June, Aug; Zambia, Aug–Mar and June, mainly (94·5%, n = 92) Sept–Feb with same frequency in each month; Botswana, Oct–Dec; Zimbabwe, all months except June, rare in Mar–May and July, 5% (n = 191) in Feb and Aug, 19% Sept, 36% Oct, 22% Nov, 9% Dec and 8% Jan; Mozambique, Oct–Dec.

INCUBATION: by ♀ (mainly) and ♂, taking turns, and singing in duet at change-over, the duet being initiated by incubating bird. Period: *c*. 15 days.

DEVELOPMENT AND CARE OF YOUNG: development not described. Young fed by ♂ and ♀. Usually only one parent at nest at a time; if both away from nest and ♂ approaches it, ♀ gives call warning chicks of his coming. All or most food obtained by parents within 20 m of nest (van Someren 1956). Pair fed brood 25 times in 150 min. The 30-cm snake (see above) had its head thrust down gullet of one nestling and tail down another, and after an hour observer intervened to save them from possible death (van Someren 1956). Nestling period: *c*. 15 days. Fledglings mottled brown with horn-coloured bill; young can feed themselves by *c*. 7 weeks after leaving nest; they remain with parents for *c*. 5 months, and have most of adult vocabulary by 5 months.

BREEDING SUCCESS/SURVIVAL: 2% of nests parasitized by Black Cuckoo *Cuculus clamosus*; some pairs of shrikes appear to be repeatedly susceptible.

References
Harris, T. and Arnott, G. (1988).
Lorber, P. (1982, 1984).
van Someren, V.G.L. (1956).

Plate 27 (Opp. p. 414) *Laniarius ferrugineus* (Gmelin). Southern Boubou. Gonolek boubou.

Lanius ferrugineus Gmelin, 1788. Syst. Nat., 1, pt. 1, p. 306; Cape of Good Hope.

Forms a superspecies with *L. aethiopicus*, *L. bicolor* and *L. turatii*.

Laniarius ferrugineus

Range and Status. Endemic resident, southern Africa. Botswana, records from Ngotwane R. catchment: Metsemaswaane R., Kopong, Lobatse, Gaborone, Notwane Siding and Bokaa (Aldiss and Hunter 1985, Hunter 1988, Brewster 1996, Harrison *et al.* 1997). Zimbabwe, known only from 25 km west of Marhumbini at about 21°28′S, 32°10′E (Irwin 1981). Mozambique, Save R. valley and very common throughout Sul do Save Province. Transvaal, widespread and locally common in Bushveld, Lowveld and Escarpment regions; in Highveld, only along Vaal R., Witwatersrand, and in relict patches of forest in S. Orange Free State, record near Clocolan (Harrison *et al.* 1997). Absent from Lesotho. Natal, throughout. Cape Prov., north in E to near Aliwal North, Sterkspruit and Blikana and in W (where locally very common) to about Doring R./Olifants R. confluence. In S Mozambique, densities of 10 birds per 100 ha of mopane, 11 of miombo, 23 of acacia and 12 per 100 ha of other broad-leaved woodland (Parker 1999).

Description. *L. f. ferrugineus* (Gmelin): SW and S Cape Prov., east to Plettenberg Bay. ADULT ♂: forehead, crown, hindneck, mantle, cheeks, ear-coverts, sides of neck and scapulars glossy jet black; lower back feathers full, loose, long, each with indistinct whitish subapical spot, rump feathers tipped buff-rufous; uppertail coverts black. Tail glossy black above and below. Chin, throat and breast buffy white merging into ochreous rufous on flanks, thighs and lower belly and to rich cinnamon-rufous on hind part of flanks and undertail-coverts. Wings mainly black, somewhat glossy, with long white stripe (almost straight in fully open wing, stepped and curved in closed wing); marginal and lesser coverts black, median coverts black proximally, white distally, greater coverts black but innermost 2 with broad white outer edges, primary coverts and flight feathers black but 2 innermost secondaries with white outer edges. Underwing-coverts and axillaries cream-buff but greater under primary coverts black. Bill robust, black; eyes and eye-rim black; legs and feet slate-grey. ADULT ♀: like ♂ but duller; forehead to mantle slaty or blackish grey; greater extent of rufous on belly; eyes dark brown. SIZE: wing, (n = 9) ♂ 95–101 (99·9), ♀ (n = 10) 92–97 (94·5); tail, (n = 19, sexes combined) 91–104, av. of 6, 92·1; bill (n = 19, sexes combined) 22·5–26·0, depth 8·8–10·0; tarsus, ♂ (n = 6) 35·0–37·0 (35·9), ♀ (n = 3) 35, 36, 36·5. WEIGHT: (*ferrugineus* and *natalensis*) ♂ (n = 11) 53·0–68·8 (60·2), ♀ (n = 7) 54·1–61·9 (57·5).

IMMATURE: like adult ♀ but overall rather paler and duller, rump rufescent grey, whole underside from chin to undertail-coverts buffy, a shade darker towards rear. Juvenile similar but with variable degree of buff and dusky mottling above and rufous barring below; bill horn.

NESTLING: naked, blind; without mouth spots.

L. f. pondoensis Roberts: coastal forests of Transkei. Like *ferrugineus* but ♂ glossier black and upperparts of ♀ tinged olivaceous.

L. f. natalensis Roberts: SE Cape Prov. from Plettenburg Bay to Natal, north to Nghome Forest, W Zululand. Like *pondoensis* but ♂ whiter, less buffy, on chin, throat and breast, and flanks less richly cinnamon; upperparts of ♀ darker than in ♀ *pondoensis*.

L. f. transvaalensis Roberts: lowlands of N Zululand, E Swaziland and Transvaal (but not Limpopo Valley). Like *ferrugineus* but ♂ with lower belly paler rufous, grading insensibly into buffy white breast; ♀ with upperparts dark grey and underparts like ♂ *transvaalensis*.

L. f. tongoensis Roberts: South Africa and Mozambique: E Zululand littoral, L. St Lucia, Lebombo Range, and Sul do Save north to 20°20'S. ♂ mainly white below, with buffy flanks, thighs, belly and undertail-coverts. ♀ like ♀ *natalensis* but underparts paler. Small: wing, ♂ (n = 4) 89–95.

L. f. savensis da Rosa Pinto: Mozambique and SE Zimbabwe – E Sul do Save from 20°20'S north to Save R. valley (near Jovane and near Mavue), also Marhumbini. Like *tongensis* but uniform matt dark slate-grey, flanks tawny buff, and still smaller. Wing (1 ♀) 88, tail (1 ♀) 82. WEIGHT: ♂ (n = 4) 44·2–53·2 (50·6), ♀ (n = 5) 42·1–51·0 (44·7).

TAXONOMIC NOTE: where they meet in Limpopo R. valley in Transvaal and Mozambique, races of the northern *L. aethiopicus* and southern *L. ferrugineus* are similar and may even hybridize and intergrade (T. Harris and V. Parker *in* Harrison *et al.* 1997); but along Lundi R., Zimbabwe, *L. a. limpopoensis* overlaps the much smaller *L. f. savensis* without interbreeding, so their existence as separate species must be upheld (Irwin 1977, 1987).

Field Characters. Length 21–23 cm. The common boubou of South Africa, blackish above with white wing-stripe, creamy below becoming *orange-rufous* on lower flanks, belly and undertail-coverts. Does not overlap with Swamp Boubou *L. bicolor*, but in N of range meets Tropical Boubou *L. aethiopicus*. Their ranges abut but do not meet in Botswana; south of Limpopo R in Transvaal they are ecologically segregated, Southern favouring thickets associated with koppies, Tropical occurring in riparian woodland (Harrison *et al.* 1997). Where they occur together in S Mozambique and S Zimbabwe, Southern is considerably smaller; it also has more richly coloured underparts, rufous rather than pink, duller upperparts, slate or dark olive (especially ♀), *rump dark* (whitish in Tropical).

Voice. Tape-recorded (49, 66, 75, 88, 91–99). Extensive repertoire of synchronized duets; for full study see Thorpe (1972). Commonest are combinations of loud, ringing, up-slurred whistle 'wheee' and low hollow notes: 'hohoho-wheeeyoo', 'whee-whee-hohoho', 'hoh-tyee-tyee-tyeew-hoh', 'hooweee-kwokwo', 'haw-tooktowee-haw', 'heee-huhu-hyew', 'Marguerite-hoho'. Hollow notes may be replaced by second ringing note, 'heee-wheee'; by sweet whistle, 'tyerwi-wheee'; by harsh grating scold, 'tut-tzzhezzh-hooey-hooey'. Other calls include Brubru-like trill followed by popping, 'trrrrrrr-popopop'; 'tipoorluwee' (or 'whippoorwill'); rapid ringing 'howeeyo'; barbet-like 'pop-pop-pop-pop ...'; rasping 'tzkwrwrwr-tozzzakh'; often not clear how many birds involved.

General Habits. Inhabits dense mesic woodland, occurring in clumps of leafy bushes, tangles of creepers, coastal scrub, riverine thickets, edges of forest, stands of *Protea*, thickets around the base of koppies and on termite mounds, plantations, exotic acacias and mature gardens.

Solitary or in pairs, keeping low down in dense, woody vegetation; secretive, seldom coming into open, although responds to person 'pishing'; around habitation can become quite tame and venture into the open. Forages by leaping, hopping and creeping in rather horizontal posture through woody tangles and undergrowth, emerging briefly at edge then disappearing from view again; often comes to ground littered with dead plant matter but free of herbs, in shade under bushes, hopping around in search of food; sometimes comes into the open on edges of lawns, and visits bird tables. Bird seen removing whole egg from nest of Cape Robin *Cossypha caffra*; egg was deposited on ground 2 m away, pecked open, and contents and shell eaten; shrike then returned to nest and removed last egg. Stings of bees rubbed off against perch before insect is swallowed (Maclean 1963); hairy caterpillars vigorously rubbed against branch or in sand before being swallowed. To eat a snail, bird breaks it open by repeatedly beating it against tree trunk or branch, or snail is wedged into crevice or fork and shrike uses bill to tug flesh out (Langley 1983). Hunts for arboreal geckos by pulling bark off dead trees to find them (Langley 1983). Occasionally probes flowers, and hawks after flying insect. Captive bird regularly impaled prey (Kaumanns 1975).

Highly vocal, pair duet-singing frequently throughout day and year. ♂, when giving 'wheep-wheeo' or 'boubou' call, bobs head slightly and moves tail a little, sometimes half-opening it. ♂ whistled to advertise and defend pair-territory of *c*. 3 ha. Territorial whistle generally given from elevated perch, bird in upright stance and half exposed to view. ♀ often calls in duet with ♂; either sex may initiate duet; pair has distinctive duet pattern. Bird that has lost its mate can sing whole pair-duet by itself (Harris and Arnott 1988). In territorial interaction rival ♂♂ counter-sing, fanning tails and fluffing out white-spotted back feathers; ♂ may fly or half glide downwards with exaggerated, slow wing beats, giving snarling 'tchzananantchanana' calls with head held up (**A**) (Harris and Arnott 1988). Flight

rather heavy and slow; wing beats rapid and shallow. Scratches indirectly. Pair roosts in dense tangle of creepers, making 'tik-tik' calls before settling down. Same or similar 'tik-tik' calls also used in alarm and when plundering a nest.

Sedentary, apparently remaining in same pair-territory for life.

Food. Insects, including hairy caterpillars up to 90 long, and bees; many snails *Helix adspersa* and *Theba pisana*; occasionally birds' eggs (e.g. of Cape Robin) and nestlings (e.g. of Cape White-eye *Zosterops pallidus*: Langley 1982, 1983) and fledglings; once a mouse *Mus musculus*, and many geckos *Phyllodactylus porphyreus* and their eggs (Langley 1983); in gardens, grain, porridge and fruit (Maclean 1993).

Breeding Habits. Territorial, solitary nester; monogamous. Courtship interactions seem to be the same as interactions of rival ♂♂ (see above), birds also swaying bodies. ♂ and ♀ chase each other or rather move close together through vegetation, hopping, creeping and winding through bush, fripping wings and duet singing. Pair becomes aggressive in breeding season, chasing other birds and plundering their nests (Harris and Arnott 1988). Frequency of duetting decreases at time of egg laying. Pair destroys its own nest if much disturbed; one ♀ used the material to build another nest 150 m away (Langley 1983).

NEST: loosely-knit bowl or deep or shallow cup, made of slender twigs, roots and grasses, sometimes bound with spider web, occasionally lined with finer grass and rootlets. Twigs sometimes not used; nest can then be so flimsy that eggs visible from below. Placed 1–8 m high (usually 1–4 m) in fork in dense tree, bush, hedge or mass of creepers, or in dense foliage near end of branch, usually well concealed; occasionally in dead tree, e.g. *Acacia cyclops* in thin stand of *A. cyclops* and *A. saligna*. At one nest only ♀ observed to build, taking *c.* 6 days, though at others ♂ might help.

EGGS: 2–3 (n = 77, av. 2·5). Laid at 24 h intervals, once between 09h00 and 11h30. Pale greenish white, finely speckled with browns and greys, forming ring around large end. SIZE: (n = 90) 22·20–27·4 × 16·0–19·2 (24·6 × 18·1) (Maclean 1993).

LAYING DATES: Transvaal, Aug–Mar and May; mainly Oct (35% of records), Nov (27%) and Dec (16%); Natal, Sept–Dec, mainly Oct–Nov; SW Cape, Aug–Mar.

INCUBATION: by ♂ and ♀, starting from first egg laid. Nest relief signalled with 'houu' whistle by ♂ and 'huweyo' whistle by ♀. Period: 16–17 days.

DEVELOPMENT AND CARE OF YOUNG: eggshells removed by adult soon after young hatch, and eaten or (usually) carried away in bill. Faecal sacs removed by adult at moment young excrete them, either swallowed immediately or carried away. Young fed by both parents; in one nest one or the other constantly brooded chicks for 11 days, and both fed young thereafter (Langley 1983). Nestling period 16–17 days; at one nest 16- and 17-day-old young left nest on same morning within 2 h of each other. Fledglings do not return to nest, but stay in the area, begging with high-pitched 'pseep' calls from deep cover. At age 53–54 days, young were foraging for themselves but also being fed still by parents, though the ♂ also chased them away (Langley 1983). Young still with parents at 80 days, by which age they have adult vocabulary.

BREEDING SUCCESS/SURVIVAL: 1 ringed bird at least 11 years old. Parasitized by Black Cuckoo *Cuculus clamosus* (*c.* 2% of nests) and rarely by Jacobin Cuckoo *Oxylophus jacobinus*.

References
Harris, T. and Arnott, G. (1988).
Hunter, N. (1988).
Irwin, M.P.S. (1977, 1987).
Langley, C.H. (1982, 1983).
Maclean, G.L. (1993).
Quickelberge, C.D. (1966).

Plate 27 *Laniarius bicolor* (Verreaux). Swamp Boubou. Gonolek à ventre blanc.
(Opp. p. 414)

Dryoscopus bicolor Verreaux, 1857. In Hartlaub, Syst. Orn. Westafr., p. 112; Gaboon.

Forms a superspecies with *L. aethiopicus*, *L. ferrugineus* and *L. turatii*..

Range and Status. Endemic resident, coastal lowands from Cameroon to S Angola, and inland to Botswana. Cameroon, small population in stunted mangrove at Tiko, 04°05′N, 09°20′E in Cameroon R. delta (Serle 1965); sight record of pair near Bibundi, Mt Cameroon. Gabon, coastal records from Ogooué R. mouth, Ogooué Maritime, Fernan Vaz, Port Gentil and Nyanga Districts, and quite common around Mouila (Y.-M. de Martin de Viviès, pers. comm.). Congo, common along Kouilou Basin coast and up to 22 km inland, e.g. at Ménengué (Dowsett-Lemaire and Dowsett 1991). Angola, coast of Cabinda, and widespread west of 15°E except for arid coastal region south of

Laniarius bicolor

Benguela; probably absent from Namibe Prov. but occurs in Cunene Valley in W Cunene Prov. (and in Namibia); east of 15°E, occurs along Cuchi and Cubango valleys downstream to Caprivi Strip (Namibia); almost certainly occurs along Zambia border at 22°E, and may in fact be widespread in Cuando Cubango and S Moxico Provs. Zambia, throughout Zambezi Valley above Kapanda Bridge and Mambova (25°11′E), and west to Angola border; absent from Mwinilunga; in SW occurs along Mashi Valley but absent between Mashi and Zambezi in W Sesheke District. Botswana, common in Okavango, Linyanti and Chobe river systems; sparse at L. Ngami. Namibia, part of Cunene valley, and Cubango/Kavango and Cuando floodplains in Caprivi Strip. Zimbabwe, not definitely recorded although very probably occurs above Katombora Rapids (Irwin 1981).

Av. density in fringing gallery woodland, Okavango Swamps and Linyanti R., 1–2 birds per 10 ha (M. Herremans *in* Harrison *et al.* 1997).

Description. *L. b. sticturus* Hartlaub and Finsch: interior of S Angola, W Zambia, NE Namibia, N Botswana and extreme W Zimbabwe. ADULT ♂: entire upperparts glossy bluish black including tail and wings, but for long white stripe in closed wing; innermost greater covert with white outer and black inner vane, and next 2 are completely white; innermost 3 secondaries with 3-mm-wide white outer edges (and sometimes white tips). Cheeks and ear-coverts black. Underparts pure white. Greater under primary coverts black, and other underwing-coverts and axillaries white. Underside of tail black. Bill and palate black; eyes dark brown; legs and feet slate-blue. Sexes alike. SIZE: wing, ♂ (n = 8) 106–109 (107), ♀ (n = 6) 97–106 (101), unsexed (n = 6) 90–110 (101); tail, ♂ (n = 4) 97–106 (103), ♀ (n = 2) 101, 101, unsexed (n = 4) 95–100; bill, ♂ (n = 6) 23–25 (24.25), ♀ (n = 2) 21, 23; tarsus, ♂ (n = 6) 35.5–37.5 (36.7), ♀ (n = 2) 36, 38.5. WEIGHT: ♂ (n = 5) 47.8–56.7 (53.3), ♀ (n = 4) 43.0–58.2 (52.1).
IMMATURE: like adult but spotted with buff above and faintly barred below (Harris and Arnott 1988); outer tail feathers buff-tipped; white wing-coverts blotched with greyish brown (Serle 1965).
NESTLING: unknown.

L. b. bicolor (Verreaux): Cameroon to Gabon. Like *sticturus* but no white in secondaries; smaller. SIZE: (Cameroon, 2 ♂♂, 3 ♀♀) wing, ♂ 102, 103, ♀ 94–100 (97.7); tail, ♂ 90, 92, ♀ 75 (moulting), 85, 90; bill, ♂ 24, 24, ♀ 23–24 (23.3); tarsus, ♂ 35, 36, ♀ 35–36 (35.3).

L. b. guttatus (Hartlaub): Congo to about Sa' da Bandeira, Angola. Like *bicolor* but in some birds a single secondary feather with white outer vane; about size of *bicolor*: (unsexed, n = 3) wing 88–100 (94.3).

TAXONOMIC NOTE: *bicolor* (and *aethiopicus*) treated as subspecies of *L. ferrugineus* by White (1962) and the 3 as allospecies by Hall and Moreau (1970) and Sibley and Monroe (1990); Dowsett and Forbes-Watson (1993) regarded *L. aethiopicus* and *L. ferrugineus* as a superspecies and *L. bicolor* as an independent species. *L. bicolor* is parapatric with *L. aethiopicus* but the 2 overlap geographically (in different habitats) around the Chobe-Zambezi confluence. It is distant from the range of *L. ferrugineus* and differs from it more than it does from *L. aethiopicus*; it is larger but lighter and its bill is slender, not robust. Their voices and habitats differ, and in our view they are good species best retained as allospecies in the *L. aethiopicus* superspecies.

Field Characters. Length 23–24 cm in N, 24–25 cm in S. A black and white bush-shrike of thickets beside rivers and swamps, more often heard than seen. Very similar to Tropical Boubou *L. aethiopicus*, which it overlaps in W Zambia, Caprivi and N Botswana; distinguished by *pure white* underparts with no trace of pink. Voice simpler and less musical than Tropical Boubou.

Voice. Tape-recorded (75, 86, 88, 91–99, 104, B, C, F, LEM, NOR, SAR). Duets much less variable than Tropical Boubou. Typical in southern Africa is whistle from ♂ followed or accompanied by synchronous grating cackle from ♀, either soft but ringing 'whhawww' answered by descending 'kikikakakrrrrrr', or higher-pitched 'whheeeeww' with 'tatatrrrrr'. Whistle may be preceded by double 'werk-werk', and answered by short 'takaka' or just 'takka' – 'werk-werk-whheeee-takaka' (3 birds?). Voice in Congo rather different (F. Dowsett-Lemaire, pers. comm.); includes ringing, up-slurred 'woi-woi-woi', soft 'hop-hop', and short duets 'haw-kaka' and 'hooee-kaka'; also in Congo a repeated triple 'paw-paw-paw' with ringing quality of White-spotted Flufftail *Sarothrura pulchra* (Chappuis, in press). Various harsh scolds in alarm, 'jik-jik-tzhaaaa' and longer 'tizzh-zhizhi-zhaaaa'. Sonagrams in Thorpe (1972), Sonnenschein and Reyer (1984) and Maclean (1993).

General Habits. In Cameroon inhabits stunted mangroves with islands of drier ground surrounded by tidal shallows and soft mud; in Congo, dry thickets on coast and low secondary growth inland; in Angola, various habitats from tropical forest at Gabela to dry acacia woodland on Cunene R. (W.R.J. Dean, pers. comm.); in southern Africa restricted to major river floodplains with tall reeds, dense papyrus beds, water figs *Ficus verucculosa*, thick riverine bush, acacia thickets, riverside gardens, waterlogged swamps, marshes, floating islands in lagoons, along rivers, choked streams, and other waterways.

Solitary or in pairs; generally in densest part of vegetation, low down in papyrus or in canopy of larger riverine trees; tends to perch out in the open more often than congeners; moves in woody vegetation with inclined posture by leaping and slipping through small branches, twisting head in search for prey. Gleans from trunks, branches, reed and papyrus stems and in foliage. Often comes to ground where it is firm, dry, bare and shaded below woody growth, and forages by moving with short, bouncing hops, flicking bits of debris aside with bill, like a thrush. Flight heavy, with shallow wing beats, sometimes with wing-fripping sounds. Scratches indirectly. Roosts in thick tangled bushes, making discordant 'tik-tik-tik' calls before settling down (Harris and Arnott 1988).

Pair keeps in territory of c. 2 ha all year, defending it by calling. Frequency of calling increases in breeding season. Duetting tends to be in lengthy sequence, initiated by either sex. Pair, calling in duet, sit upright close together on exposed perch, with heads up and tails slightly fanned; ♂ bobs head as he gives 'grr' call and bows body when giving 'houuu' whistle; ♀'s tail vibrates with the force of uttering ratchet call in response to ♂. Number of strophes in ♀'s ratchet call varies individually and is greater in high intensity interactions (Harris and Arnott 1988). 'Kaw-kaw' threat call accompanied by bowing, wing flicking and tail flicking. Bird also gives twanging 'tchzenenene' calls whilst flying steeply down with exaggerated, slow wing beats, head held up, rump plumage fluffed out, and tail fanned; this behaviour has been interpreted both as threat and courtship.

Sedentary.

Food. Insects and small fruits. Only insects in 7 stomachs (Cameroon). Once a frog (Edwards 1998).

Breeding Habits. Monogamous, territorial. Courtship involves ♂ and ♀ chasing each other, keeping in tandem as they hop and bounce through vegetation, often the upper canopy of large trees; ♂ gives drawn-out whistles and glides in wake of ♀ (Harris and Arnott 1988); see also above (descending 'tchzenenene' flights).

NEST: one was flimsy saucer made of loosely-woven fine twigs; placed on horizontal fork c. 2 m up in small tree in forest fringing a reedbed; another placed at base of palm frond, 4 m from ground.

EGGS: 2. Pale greenish. SIZE: (n = 2) 23·0 × 19·8, 23·7 × 19·9.

LAYING DATES: Cameroon, (imm. with partly ossified skull, June); Gabon, (pairs singing in duet, Mouila, Oct–Jan and May: Y.-M. de Martin de Viviès, pers. comm.). Botswana, June, July, Aug–Nov, Feb–Mar (mainly Nov).

References
Harris, T. and Arnott, G. (1988).
Harrison, J.A. *et al.* (1997).
Serle, W. (1965).

Plate 27 (Opp. p. 414) *Laniarius turatii* (Verreaux). Turati's Boubou. Gonolek de Turati.

Dryoscopus turatii Verreaux, 1858. Rev. Mag. Zool., p. 304; Portuguese Guinea.

Forms a superspecies with *L. aethiopicus*, *L. ferrugineus* and *L. bicolor*.

Range and Status. Endemic resident, Guinea-Bissau to Sierra Leone. Guinea-Bissau, old records from Farim and Gunnal. Guinea, old record from Mamou, recent ones on coast and inland to Mambia, Foulayah, Koba and Kolenté (Demey 1995). Sierra Leone, common and widespread west of lines between Shenge and Farangbaia and Farangbaia and Bafodia, with records further to southeast at Bo, east of Bo and at Pujehun (Hall and Moreau 1970, Field 1979); common round Njala campus (Harkrider 1993). May be in process of extending range eastward, at least in N Sierra Leone (Field 1979).

Description. ADULT ♂: forehead to hindneck and mantle, lores, upper cheeks, ear-coverts and sides of neck black, glossed with dark blue (or dark green-blue in some lights); scapulars and back feathers glossy black, with concealed large white subterminal spots; rump feathers long, loose and fluffy, tips (overlying uppertail-coverts) white; uppertail-coverts black. Tail rather long, square-ended, black above and below. Chin, throat, breast and forepart of flanks buff-pink, merging to white on belly, hind part of flanks, thighs and undertail-coverts. Wings above entirely glossy black, below with inner coverts and axillaries and greater under primary coverts black. Bill black; eyes dark brown; legs and feet slate-grey or bluish grey. Sexes alike. SIZE (3 ♂♂, 2 ♀♀): wing, ♂ 95–100 (97·3), ♀ 89, 97; tail, ♂ 87·5–95·5 (91·3), ♀ 82·5,

Laniarius turatii

90; bill, ♂ 22·5–25 (23·8), ♀ 24, 25; tarsus, ♂ 34–37 (35·7), ♀ 31, 34·5; also (n = ?): wing, ♂ 94–106, ♀ 88–101; tail, ♂ 88–100, ♀ 83–92; bill, ♂ 21–24, ♀ 22–23; tarsus, unsexed 32–35.

IMMATURE and NESTLING: not known.

TAXONOMIC NOTE: treated as a subspecies of *L. aethiopicus* by White (1962), as an allospecies of it by Hall and Moreau (1970) and Sibley and Monroe (1990), and as an independent species by Dowsett and Forbes-Watson (1993). *L. turatii* is parapatric with *L. aethiopicus* but the 2 overlap in a tiny area in Sierra Leone, where G.D. Field's observations of lack of reaction to each other suggest that they are not closely allied. Voices differ, and so do eggshell colours. However, lacking a white wing-stripe, *L. turatii* recalls the coastal E African *L. aethiopicus sublacteus* and, with some misgivings, we retain it in the superspecies.

Field Characters. Length 20–21·5 cm. A bush-shrike confined to extreme W Africa. Largely allopatric with Tropical Boubou *L. aethiopicus*, which it meets only in a small area in N Sierra Leone; distinguished by lack of white in wing, pinkish buff breast and white belly.

Voice. Tape-recorded (104, B, DEM, MOR). Duet a pure but soft and hollow whistle from ♂, 'whawwhh' answered synchronously by ♀ with grating nasal 'dizhizhizhaaaanh'. ♀ gives similar call in aggression, 'zha-zhazhaaaaang'; alarm, a hard 'kikikikikik ... ' or 'kakakakak ... ' at variable speeds. Voice almost indistinguishable from that of Tropical Boubou (Chappuis, in press).

General Habits. Inhabits dense, creeper-festooned canopy of bushes and small trees in savanna woodland. Solitary or in pairs. Shy and retiring but vocal. Range overlaps with that of tropical Boubou *L. aethiopicus* in tiny area around Bafodia, N Sierra Leone, which was formerly forested and may have separated the 2; pair of each species twice seen in same tree without signs of mutual interest or aggression, each pair continuing on their own foraging route; one species often seen in exactly the same place previously occupied transiently by the other; Turati's less prone to mount high in tree than Tropical Boubou (Field 1979).

Sedentary.

Food. Not known – presumably insects.

Breeding Habits. Solitary nester, evidently monogamous and territorial.

NEST: a flimsily constructed bowl made of grass-stems and fine rootlets, one lined with 2 dry leaves. Sited in multiple horizontal fork in bush, *c*. 2 m above ground.

EGGS: 2. Immaculate blue. SIZE (n = 2), both 25 × 19 (Bannerman 1939).

LAYING DATES: Guinea-Bissau, May.

References
Bannerman, D.A. (1939).
Field, G.D. (1979).

Laniarius barbarus (Linnaeus). Yellow-crowned Gonolek. Gonolek de Barbarie.

Plate 26
(Opp. p. 399)

Lanius barbarus Linnaeus, 1766. Syst. Nat. ed. 12, 1, p. 137; Senegal.

Forms a superspecies with *L. mufumbiri*, *L. erythrogaster* and *L. atrococcineus*.

Range and Status. Endemic resident, Senegal to Cameroon. Gambia, frequent to common throughout. Senegal, frequent and widespread, in N seen less commonly in dry than in wet season. Mauritania, mainly in Senegal R. valley, but north to 17°N and in Tagant area to 18°N. Mali, quite common, north to 17°N; common in Boucle du Baoulé Nat. Park. Niger, common in 'W' Nat. Park, and in Niger R. valley north to Tillabéry; records between Maradi and Tanout, in Dosso, Beleyara and near Korama, and several old observations north to Tahoua; frequent in SE in Komdougou-Yobé area near L. Chad.

Guinea, poorly known; records in Koundara area in N and on Conakry coast; likely to be widespread in N half. Sierra Leone, a race is endemic to coastal areas, absolutely confined to mangrove, locally common, from near Guinea border to Bonthe (Field 1979). Liberia, 2 records, Monrovia, 1961, and nearby Congo Town lagoon, 1984 (Gatter 1997). Ivory Coast, common north of Bouaké, scarce in S guinean woodlands, common along dry coast from Azagny Nat. Park to Assinie. Ghana, widespread in riverine thickets throughout Volta region, north to Burkina Faso border at Tumu; gardens of older residential areas in main towns; and on coast confined to thickets from Elmina to Accra and Keta Plains (Grimes 1987). Togo, as in Ivory Coast, common and widespread in N,

Laniarius barbarus

Plate 29

Plate 30

uncommon in or absent from S guinean zone, and common along coast (Cheke and Walsh 1996). Benin. Nigeria, frequent and widespread in N, north to Sokoto and Kano Provs and lower Komadugu Gana valley at Malamfatori, south to Ilorin and Benue valley; absent further south except on SW coast, from Badagri to Lagos, north to Abeokuta. Cameroon, restricted to sahel zone: Maroua and Mora to Waza Nat. Park, and absent from Logone inundation zone further east. Chad, uncommon but widespread in soudanian zone (Salvan 1967); eastern limits, and to what extent it may be sympatric with Black-headed Gonolek *L. erythrogaster*, are unclear.

Description. *L. b. barbarus* (Linnaeus): range of species except Sierra Leone coast. ADULT ♂: forehead, crown, nape and hindneck olive-yellow, brighter and clearer yellow at front and sides of crown; lores, narrow line over eye, upper cheeks, ear-coverts, sides of neck, mantle and rest of upperparts including wings and tail black with dark blue gloss, some scapular feathers with concealed white mark in centre and long lower back and rump feathers with mainly concealed large white subterminal spots; uppertail-coverts glossy black. Chin, throat, breast, flanks and belly bright vermilion-red; vent, thighs and undertail-coverts buff. (Once, an adult with yellow underparts: Ndao 1989). Underside of tail black. Underwing-coverts and axillaries black. Bill black; eyes pale yellow; legs and feet slate-grey or bluish grey. Sexes alike. SIZE (14 ♂♂, 14 ♀♀): wing, ♂ 104–115 (110), ♀ 98–110 (109); tail, ♂ 98–108 (105), ♀ 90–111 (106); bill, ♂ 20–22 (21·2), ♀ 20–23 (21·4); tarsus, ♂ 32·5–36 (33·6), ♀ 33–35 (33·4). WEIGHT: (Zaria, Nigeria) (n = 7, unsexed) 44–52·5 (48·1), (imm., n = 2) 49·0, 49·5.

IMMATURE: very young bird has top of head mustard-yellow, rest of upperparts uniform dark brown, upperwing-coverts with yellowish tips; underparts ochre-yellow, sides of breast, flanks and belly with narrow, greyish black bars; later, yellow feathers replaced with scattering of red ones, starting mainly on throat and lower belly; eyes dark brown.

NESTLING: not known.

L. b. helenae Kelsall: mangroves, Sierra Leone. Like nominate subspecies but forehead, crown and hindneck rufous or orange-brown, not olive-yellow.

Field Characters. Length 23–25 cm. Strikingly coloured black and crimson; shy, usually located by abrupt duets. The only gonolek in most of W Africa; distinguished from Black-headed Gonolek *L. erythrogaster*, which it meets in NE Nigeria/N Cameroon/S Chad, by broad *yellow stripe* over crown. Eye yellow, undertail buff. Immature ochre-yellow below with some dark barring.

Voice. Tape-recorded (40, 104, 108, B, C, GRI, MEES, MOR, PAY). Duets short, sounding almost truncated, typically a single whistle from one bird, rising in the middle, accompanied by 2–3 note chatter from partner: 'kweeho-kikiki', 'hoyho-kaka', 'hyawwee-kaka', 'tyaw-titi'; first note may be trill, 'hyerrrr-dididit'; in more complex form, first bird gives 2 whistles, answered separately, 'hyaw-chichi-way-chi'. Calls may be reversed, duet starting with 'kikiki', and partners may switch roles, either one doing the whistle. Duets precisely timed; for discussions of timing and sonagrams showing variation see Grimes (1965, 1966) and Payne (1970). One bird may give 'kweeho' call alone, sometimes varying it with down and up 'heeoowy'. Contact call from both birds in deep cover a series of rapid clicking notes; in courtship flight bird gives call like winding of fishing reel (sonagrams in Grimes 1966).

General Habits. Inhabits dense, shady, woody undergrowth in mesic and dry wooded savannas; thickets on large, old termite mounds surrounded by bare soil or light grassland with only a few scattered, low shrubs; riverine bush; dense acacia woods with *Ziziphus* and other small thorn trees along watercourses, especially if plenty of woody litter on ground in shade of the trees; farmland hedges; small groves of coppiced tamarinds; a single large fig or *Parkia* tree with shrubby growth and vegetation litter on ground below; large, mature suburban gardens (Ghana), dense coastal scrub (Nigeria), and mangrove along coast, estuaries and islands in W Sierra Leone, where bird hardly ever leaves mangrove even for adjacent coastal thickets.

Solitary or, usually, in pairs. Forages low down in woody growth, moving through branches and on ground (where spends much time if undisturbed) in long, agile hops or sometimes creeping. ♂ and ♀ usually keep within a few m of each other, frequently calling whether in sight of each other or not. On ground assumes rather upright posture, with tail held horizontally or inclined a little; flicks tail with every hop. Flicks aside dry leaves and woody litter; makes short run after fleeing insect. Often feeds up to 5 m high in canopy, moving through and inspecting small branches for prey. Skulking, but less shy than some congeners; often easy to watch, foraging on ground in dappled shade or, within 1–2 m of cover, in full sun; moves boldly across irrigated hotel lawns (Gambia). Flies reluctantly and never far; flies low down over open patch, with very short glide into foot of bush.

Pair calls antiphonally or in 'reciprocal duet' (Grimes 1965, 1966, Payne 1970). One bird (the ♂: *Bull. Br. Orn. Club* 118, 1998, p. 135) calls 'whee-u' and other responds instantaneously with very short trill of 2–4 clicks; duet generally but not always initiated by 'whee-u' and lasts 0·3–0·4 s; ♂ and ♀ motifs each last 0·2–0·3 s and overlap, with response or auditory reaction time of 0·08–0·1 s. 'Whee-u' caller makes small bow with head and neck at every whistle. 8 pairs of birds used 8 different duet patterns, and any one pair uses several different forms of 'whee-u' call. Single type of duet is generally repeated 10–40 times with very regular intervals of 1·16–1·70 s between duets, then after pause of several seconds pair gives a series of another form of duet. Quite different call, a rapid series of clicks lasting 0·9 s by one bird followed after 0·2 s by clicks at different pitch, lasting 0·5 s, is delivered from deep cover; function unknown.

Sedentary, often highly so, pair keeping to same dense thicket all year.

Food. Insects, including caterpillars and large numbers of grasshoppers and locusts.

Breeding Habits. Solitary nester, poorly known, but evidently monogamous and territorial. A call, apparently part of courtship, sounds like a fisherman's reel and is given only by bird in flight, with audible wing beats and bill wide open whilst calling (Grimes 1966). What is doubtless the same flight display described as involving deep, slow, stalling wingbeats, with audible wing-snaps, accompanied by reverberant, churring 'bed-spring' call (Barlow *et al.* 1997).

NEST: open cup, built of loosely intertwined coarse plant stems, lined with fine tendrils and roots or (1 nest, made of small rootlets) unlined. Sited 1·5–4·5 m high in dense hedge or *Citrus* bush or small, soft-leaved tree, in thicket or shrubby copse. Ext. diam. at rim of cup, 100; int. diam. 74; ext. depth 54; int. depth 44.

EGGS: 2. Ovate; slightly glossy. Pale green, blue-green or grey-green, marked with reddish brown or orange-brown spots, speckles and small linear blotches on purplish grey undermarks, evenly distributed, except at broad end where marks coalesce to form cap. SIZE: (n = 5) 22·8–24·4 × 17·2–18·0 (23·3 × 17·5).

LAYING DATES: Mauritania, Aug–Sept; Mali, (breeding season given as Nov–Dec south of 13°N and Apr–Sept north of that latitude); Gambia, Jan–Sept, most activity June–Aug (Barlow *et al.* 1997). Ghana, May (and oviduct egg Apr, fledglings, early Apr); Nigeria, June–Aug (and nestling Feb and juvs Nov).

BREEDING SUCCESS/SURVIVAL: a 'yellow' adult survived at least 7 years (Ndao 1999).

References
Bannerman, D.A. (1939).
Grimes, L.G. (1965, 1966).
Payne, R.B. (1970).
Serle, W. (1943).

Laniarius mufumbiri Ogilvie-Grant. Papyrus Gonolek. Gonolek des papyrus.

Plate 26
(Opp. p. 399)

Laniarius mufumbiri Ogilvie-Grant, 1911. Bull. Br. Orn. Club, 29, p. 30; 'Mufumbiro Volcanoes', corrected to Vitshumbi (Vichumbi), L. Edward.

Forms a superspecies with *L. barbarus*, *L. erythrogaster* and *L. atrococcineus*.

Range and Status. Endemic resident, Uganda and adjacent territories. Zaïre, known only from Vitshumbi and Rutshuru, south of L. Edward. Uganda, locally common at 1100–1600 m in interior of papyrus swamps, in S and W Lango and L. Kyogo to Ankole and Kigezi, shores of L. Victoria from Masaka to Kenya. Old report from W Elgon doubtless in error. Kenya, shores of L. Victoria from Uganda to Kisumu and Kendu Bay. Widespread and locally abundant in E Rwanda west to 29°45′E and E Burundi west (in S) to 30°00′E (map in Vande weghe 1981); up to 2050 m (Rugezi); L. Bulèra, L. Luhondo, Rugezi Marsh, Mulindi, Ruvubu papyrus beds, Kayongozi, Akanyaru Marsh, Bugesera, Nyamushwaga Valley, Ndurumu Valley, Nyakijand Valley, Nyamabuye Marsh, Kinyinyi and Giofi papyrus beds, Malgarazi Basin and other places. Tanzania, in NW in middle Akagera Basin and along Ruvubu R. and in Malagarazi Basin on Burundi border to about 04°S (Vande weghe 1992).

Description. ADULT ♂: forehead, crown, nape and (in midline) hindneck dull golden yellow; lores, areas just above and below eye, ear-coverts, hindneck, sides of neck, mantle, scapulars, back, rump, uppertail-coverts, tail and most of wings black, slightly glossy; rump feathers full, soft and fluffy, with whitish tips and concealed subterminal white spots; median upperwing-coverts white with black bases; greater coverts black, innermost one with large white patch at tip of outer vane. Chin and throat red with orange tinge; breast, belly and most of flanks bright pinkish red; hind part of flanks black; thighs, vent and undertail-coverts dingy white. Underwing-coverts black. Bill black, much less deep-based and robust than bill of *L. barbarus*; inside of mouth black; eyes pale lemon-yellow; legs and feet slate-black; tarsus and toes much more slender than in *L. barbarus*, and claws longer and less strongly decurved. Sexes alike. SIZE: (3♀♀): wing 84–91·5 (88·5); tail 72–78 (75·7); bill, all 32; tarsus 22–23 (22·5); (1 unsexed) wing 87, tail 74, bill 19·5, tarsus 30. WEIGHT: 2 ♂♂ 45, 46, 1 unsexed 45, imm. ♀ 40.

Laniarius mufumbiri

IMMATURE: upperparts like adult, but forehead and crown blackish olive, becoming black on hindneck, and white spots in upperwing-coverts may include greater primary coverts; chin and throat yellowish buff, merging to buffy pale rufous-vermilion on breast and flanks; hind part of flanks dusky blackish, rest of underparts as in adult. Eyes greyish brown.

TAXONOMIC NOTE: generally treated as an allospecies of *L. barbarus*. The affinities of *L. mufumbiri* were not discussed by Dowsett and Dowsett-Lemaire (1993b), but Dowsett and Forbes-Watson (1993) placed it in the *L. barbarus* superspecies whilst sequencing *L. atrococcineus* (including *L. erythrogaster*) between *L. barbarus* and *L. mufumbiri*. In our view *L. mufumbiri* is more closely allied with *L. barbarus* than with the intervening *L. erythrogaster*, which is exactly parapatric with *L. barbarus* but is broadly sympatric with *L. mufumbiri*. Although *L. erythrogaster* and *L. mufumbiri* are sympatric, we treat them as allospecies since their habitats appear to render them mutually exclusive.

Field Characters. Length 19 cm. Colour pattern (yellow crown, black upperparts and bright red underparts) like Yellow-crowned Gonolek *L. barbarus* of W Africa, but this species is confined to papyrus swamps in the L. Victoria region. Sympatric with Black-headed Gonolek *L. erythrogaster* but largely ecologically segregated, although they occasionally meet at swamp edges; differs in *yellow crown*, broken white line on wing, orange tinge to throat, black lower flanks, and white undertail-coverts. Immature has olive crown, underparts *unbarred*, pinkish with yellow-buff throat.

Voice. Tape-recorded (104, B, FISP, GREG, KEI, LEM, McVIC). Common call a double 'pyo-pyop' with a hollow gong-like quality, given alone or answered with grating 'zeetu' or buzzy 'tzrrrr'; sometimes 4 syllables, 'pyo-pyo-pyo-pyop' answered by similar but faster call. Other duets include 'peetoo, peetoo', answered with identical call, and 'tyo-wzzz'.

General Habits. Confined to papyrus *Cyperus papyrus* swamps and beds, mainly large ones, in meandering-river valleys and along lakeshores; inhabits pure stands of papyrus, papyrus mixed with woody shrubs, *Miscanthidium* (Gramineae) papyrus, and mosaics of floating prairie-papyrus (Vande weghe 1981).

Solitary or in pairs. Skulks in thick vegetation, sometimes leaving it to make short flight over open water. Vocal. Once seen chasing a Black-headed Gonolek *L. erythrogaster* at edge of papyrus swamp. Sedentary.

Food. In one stomach myrmecine ants and beetles (Carabidae, Curculionidae, Elateridae, Lagriidae, Staphylinidae) and in another, insect fragments including beetles; in another, about 2/3rds of contents were insects (weevils and other beetles, small fly, hymenopterans, caterpillar) and bits of snail shell, and 1/3rd remnants of fruit (Chapin 1978).

Breeding Habits. Not known.

References
Britton, P.L. (1970).
Vande weghe, J.P. (1981, 1992).
Zimmerman, D.A. *et al.* (1996).

Plate 26
(Opp. p. 399)

Laniarius erythrogaster (Cretzschmar). **Black-headed Gonolek. Gonolek à ventre rouge.**

Lanius erythrogaster Cretzschmar, 1829. *In* Rüppell, Atlas (1926), p. 43, pl. 29; Kordofan, Sudan.

Forms a superspecies with *L. atrococcineus*, *L. barbarus* and *L. mufumbiri*.

Range and Status. Endemic resident, L. Chad to Eritrea and Cameroon to W Kenya and Rwanda. Nigeria, common on shores of L. Chad, south to Logomani. Cameroon, occurs in N in Logone R. inundation zone and Benue Plain, south to Adamawa Plateau (Louette 1981, Map 47). Chad, common in soudanian zone, local in sahelian zone; thought to move north in wet season to 16°N (Salvan 1968); whether and to what extent it may be sympatric with Yellow-crowned Gonolek *L. barbarus* is unclear. (They do replace each other in N Cameroon: Scholte *et al.* 1999). Central African Republic, common in Manovo-Gounda-Saint Floris Nat. Park and uncommon in Vakaga Préfecture. Sudan, common and widespread north to W Darfur and Khartoum; record near Atbara; absent from S Bahr al Ghazal and most of Western Equatoria. Ethiopia, frequent to common in NW, W and SW. Eritrea, common along all western river valleys, up to about 900 m. Zaïre, common in lowlands around L. Albert and L. Edward and in Rutshuru area; also in Lualaba R. valley, from about Kabalo to L. Kisale and Bukama. Uganda, common throughout (except perhaps extreme NW), at 600–1600 m. Kenya, around Lokichokio and N end of L. Turkana; in W, in Suam valley at Kongelai, in Marich, Nasolot Nat. Res. and S Kerio valley, with records at Marigat and L. Baringo; common in L. Victoria basin, east to Muhoruni

Laniarius erythrogaster

and Migori. E Rwanda and E Burundi (Schouteden 1966a, b). Tanzania, Mara, W Serengeti Nat. Park, N Shinyanga and Mwanza districts, and common in extreme NW (N.E. Baker, pers. comm.).

Description. ADULT ♂: forehead, crown, nape and hindneck, lores, upper cheeks, ear-coverts, sides of neck, mantle and rest of upperparts including wings and tail black with dark blue gloss, long lower back and rump feathers (overlying rump and in part uppertail-coverts respectively) with mainly concealed large white subterminal spots. Typical rump feather 45 mm long, proximal 2/3rds dark grey, radiating tip dark grey in band 7 mm deep ending proximally in black arc 1 mm deep, proximal to which is silky white patch 6–7 mm long at shaft. Some specimens, especially in Lualaba R. valley, SE Zaïre, have scattered yellow feathers on crown. Chin, throat, breast, flanks and belly bright vermilion-red; vent, thighs and undertail-coverts buff. Underside of tail black. Underwing-coverts and axillaries black. Bill black; eyes pale straw yellow; legs and feet dark lead-grey, or legs black and feet bluish black. Sexes alike. SIZE (15 ♂♂, 15 ♀♀): wing, ♂ 96–108 (106), ♀ 95–108 (101); tail, ♂ 87–112 (95·2), ♀ 90–97 (91·1); bill, ♂ 20–23 (22·0), ♀ 20–22 (21·8); tarsus, ♂ 30–35 (34·4), ♀ 30–34 (33·4). WEIGHT: (Chad, n = 6, unsexed) 42–53 (48); (Uganda) 1 ♂ 52, 1 unsexed 46·5, 3 ♀♀ 46, 49, 55.

IMMATURE: very young bird blackish above with buff tips to mantle, back and rump feathers and upperwing coverts; underparts dull yellowish buff, banded from chin to vent with narrow blackish bars; chin tinged pink at sides, breast tinged pink, with occasional vermilion feathers on flanks; eyes dark brown; in another young bird only a red wash in centre of belly.

NESTLING: not described.

TAXONOMIC NOTE: often treated as a race of either *L. barbarus* or *L. atrococcineus*. United with the latter by e.g. Dowsett and Dowsett-Lemaire (1993b), although on morphological grounds (though not vocally?) *erythrogaster* is nearer to *barbarus*. Precise parapatry of *barbarus* and *erythrogaster* in NE Nigeria and N Cameroon, and the isolated population of *barbarus* possibly sympatric with *erythrogaster* in Chad, suggest that the 2 are best kept as separate species.

Field Characters. Length 20·5–21·5 cm. A large *red and black* bush-shrike; differs from Yellow-crowned Gonolek *L. barbarus* and Papyrus Gonolek *L. mufumbiri* by *black crown*, and from Papyrus Gonolek by lack of white wing-bar (sometimes shows a few white spots), uniform red underparts, yellow-buff undertail-coverts. Underparts of immature barred black and buff (unbarred pinkish in immature Papyrus Gonolek).

Voice. Tape-recorded (10, 84, 104, B, F, BOUR, BRU, GREG, STJ). In typical duet, first bird (almost always ♂: Thorpe 1972) gives loud, hollow 'tyoyo', ♀ replies with tearing noise, 'jaaa' or 'zeeer'. Duets are extremely precisely timed; see sonagrams and discussion in Thorpe (1972). Pair also duets with hollow whistles, 'chuyo-chuyo-chyochochocho'; alarm a loud, rapid 'chk-chk-chk ... ', the notes run together in a continuous rail-like chatter (Zimmerman *et al.* 1996).

General Habits. Inhabits wooded and bushy grasslands, thick belts of woody growth along streams and lakesides, river plains subject to flooding, dense riparian scrub especially with doum palms *Hyphaene thebaica*; papyrus swamps in west of range; acacia woods, dense thickets and shrubby clumps, copses and gardens (e.g. at Entebbe, Uganda).

Solitary or in pairs. Skulks in dense, shady, woody vegetation but can become bold and conspicuous, e.g. around Entebbe; inquisitive and responsive. Pair sings in duet. Duet initiated by presumed ♂ giving 'yoick' call, immediately followed by hiss given by its mate; sonagrams in Thorpe (1963). 'Yoick' is quite variable in sonagram shape, and in duration (0·1–0·25 s); hiss varies from 0·2 to 0·35 s in duration but is constant in shape. In 8 consecutive duets by 1 pair, interval between initiator's calls were 2·5–5·6 (4·2) s and second bird responded after 125–160 ms (av. 144, S.D. 12·6 ms: Thorpe 1963). Little else is on record about habits, but they are not known to differ from those of Yellow-crowned Gonolek or Crimson-breasted Shrike *L. atrococcineus*, with each of which this species is often considered conspecific. Sedentary.

Food. Insects and some small fruits; orthopterans, caterpillars.

Breeding Habits. Seems to be a solitary, monogamous and territorial nester, like its paraspecies.

NEST: open, loosely constructed cup formed of rootlets, grass or bark fibres and lined with fine rootlets. Eggs sometimes plainly visible through nest material, from side and below. Placed in fork in middle of thick bush or in outer branch of tree (often a fig), from 4 to 8 m above ground.

EGGS: 2. Rather variable but usually somewhat pointed and glossy. Pale blue or bluish green, heavily blotched with reddish brown and grey, the marks forming distinct zone at large end. SIZE: (n = ?) 22·9–25·0 × 17·8–18·0.

LAYING DATES: Sudan, Mar, June–July; E Africa, Region A, Mar–Apr, Region B, Mar–June and Sept–Jan (mainly May–June – 22 out of 41 records).

References
Chapin, J.P. (1954).
Jackson, F.J. and Sclater, W.L. (1938).
Thorpe, W.H. (1963).

Plate 26
(Opp. p. 399)

Laniarius atrococcineus (Burchell). **Crimson-breasted Shrike. Gonolek rouge et noir.**

Lanius atrococcineus Burchell, 1822. Travels, 1, p. 387; Spuigslang Fontein (confluence of Vaal and Orange Rivers).

Forms a superspecies with *L. erythrogaster*, *L. barbarus* and *L. mufumbiri*.

Range and Status. Endemic resident, Kalahari Basin and adjoining hardveld regions: S Angola, SW Zambia and Zimbabwe, Botswana except parts of centre and N, Namibia except W and S (Namib and Karoo), N Cape Prov., NW Orange Free State and Transvaal, as mapped (Harrison *et al.* 1997). Frequent nearly everywhere, commonest in well-developed acacia woodland and bushveld and in NW part of Central Kalahari; absent from apparently suitable thornveld habitat in Transvaal lowveld. Vagrant to Kruger Nat. Park (Joubert and English 1973).

Densities of 1 pair per 4–5 ha of optimal habitat and 1 pair per 9–12.5 ha of sub-optimal habitat, Transvaal; of 2 birds per 10 ha in E Botswana and 1 bird per ha in Okavango, NW Botswana (Harrison *et al.* 1997).

Description. ADULT ♂: lores matt black, rest of head except chin and throat jet black, strongly glossy, gloss bluish in some lights; rest of upperparts, wings and tail similarly glossy black; back and rump plumage full, soft and fluffy, black with mainly concealed white blotches; tail rounded or slightly graduated, T6 sometimes with white triangular mark up to 6 mm deep at tip; closed wing with long white stripe formed by white outer lesser coverts, most median coverts, inner 3 greater coverts (of which outer 2 all white and innermost one with white outer and black inner vane), and white outer edges (2–3 mm wide) to inner 3 secondaries. Underside of tail glossy black; underwing-coverts and axillaries black. Chin to flanks and undertail-coverts brilliant crimson-red or, in a rare Yellow Phase, bright daffodil-yellow; colour of chin, throat and sides of breast sharply demarcated from black sides of head and neck. Thighs black. Birds from arid regions slightly paler. Bill black; mouth black; eyes dark violet; legs and feet black. Sexes alike. SIZE (23 ♂♂, 13 ♀♀): wing, ♂ 98–105 (100), ♀ 93–101 (97); (sexes combined) tail, 90–109, av. (n = 6 ♂♂) 98; bill 20.5–26, av. (n = 6 ♂♂) 21.2; tarsus 30–34, av. (n = 6 ♂♂) 33.0. WEIGHT: (unsexed, n = 41) 40–55 (48.3) (Maclean 1993).

IMMATURE: upperparts dull black, most feathers with narrow buff tips, wing brown where adult black, underparts mainly crimson but with large patches of feathers barred dusky black-brown and whitish. Juvenile: top of head and mantle blackish brown with fine rufescent buff bars; back, rump and lesser wing-coverts dark brown with broader pale bars (formed by buff feather tips), uppertail-coverts and tail feathers black with narrow rufous tips; chin, throat and breast evenly and finely barred with buff and blackish, belly and flanks evenly and more broadly barred likewise, undertail-coverts crimson; flight feathers dark brown, median and greater coverts buffy white with dark brown subterminal bars. As bird moults from juvenile to immature plumage, first belly and flanks become crimson, then centre of breast (leaving large brindled brown patches on sides of breast and in middle of lower breast/upper belly), then chin and throat and sides of breast and then entire underparts become crimson, except for thighs and small patch in middle of lower breast.

NESTLING: blind and completely naked, with purplish flesh-coloured skin and conspicuous yellow gapes, palate and tongue.

TAXONOMIC NOTE: *L. atrococcineus* and *L. erythrogaster* are sometimes treated as a single species, e.g. by Dowsett and Dowsett-Lemaire (1993), on the grounds that some calls are identical and the former responds to voice playback of the latter. However, in addition to the obvious wing bar and undertail-

Laniarius atrococcineus

covert colour differences, the 2 differ in iris colour and in juvenile plumage, so we treat them as separate species.

Field Characters. Length 23 cm. Strikingly coloured but more often heard than seen; shy but also very vocal. Bright crimson underparts prevent confusion with any bird in its range (does not meet Black-headed Gonolek *L. erythrogaster*). Rarely, underparts are yellow. Juvenile is dark brown above with buff barring, underparts barred black and buff; with age, barring disappears on upperparts and crimson patches appear on underparts.

Voice. Tape-recorded (42, 75, 86, 88, 91–99, B, C, F). Wide vocabulary of hollow whistles (♂) and grating or snarling notes (♀). Commonest call of ♂, used solo or in duet, 'tyotyo' or 'quiquip'; in duet ♀ intervenes with scratchy note, 'tyotyo-zilwaa-tyotyo' or 'quiquip-chui-quiquip'. In other duets, ♂ calls twice: 'pyop-zulwi-pyop', 'weeyu-jaweer-weeyu', 'chop-zer-chop', or ♀ twice, 'jawaaa-beeyu-jawaaa'; both call twice, 'beeyu-jawaaa-beeyu-jawaaa', 'tyotyo-ziluwaa-tyotyo-ziluwaa', or both once, 'chizhaanh-pyup', 'djuzhu-dorlee', 'dorlee-djeeweeya'; ♂ may give ringing trill, 'torrrr', answered by grating 'djuzhu', or dry trill 'tirrrrr' answered by tearing 'zaaaa'; either bird may call first. A high-pitched, pure 3-note whistle, 'pee-yee-yoo' is also given solo. Alarm a hard 'tikk'. For further vocalizations see Tarboton (1971) and Harris and Arnott (1988); sonagrams in Maclean (1993).

General Habits. Inhabits thornveld with scattered clumps of small trees; keeps to densest vegetation. Of trees in study site in Transvaal, 54% were *Acacia karroo*, 18% *Diospyros* spp., 10% *Rhus pyroides*, 7% *A. caffra*, and 9% *Celtis africana*, *Ziziphus* spp. and *Maytenus* spp. In Zambia in dense undergrowth of *Baikiaea* woodland.

Occurs singly or (mainly) in territorial pairs. Secretive, wary, agile and active; not always shy; appears nervous – constantly changes posture and jerks or swings tail. Territory held all year; most territories embrace intermediate tree densities (not thickets or orchard savanna). Territory size (n = 11) 2·4–7·0 (4·5) ha. Responds to pishing. Pair sings in antiphonal duet, 'qwipqwip/tzui/qwipqwip', initiated by either sex (usually the ♂). Territorial pairs countersing, and interact with excited body bowing, side-to-side movements, and tail jerking (Harris and Arnott 1988). Signals alarm with loud duetting, threat calls, and slow 'tik, tik, tik' when ground predator or perched hawk nearby.

Forages mainly on tree trunks, hopping around, inspecting rough bark and peering under loose pieces of bark; also among branches and foliage; occasionally flycatches; and commonly comes to ground, where progresses by hopping. Once attacked bat, dislodging it from behind bark; known to kill other small birds in captivity. In 22 observations, bird foraged on trunk in 11, on ground in 7, in foliage in 3 and on the wing once (Tarboton 1971), but most food taken on ground (Harris and Arnott 1988). On the ground hops with very upright posture, wing-tips pointing down (**A**) and tail horizontal and jerking up at each bounce; flicks bits of vegetable matter aside, like thrush; sometimes runs. After a few moments on ground, bird zigzags up through bush, bouncing rapidly from branch to branch (Harris and Arnott 1988). Pair invariably joins flock of Pied Babblers *Turdoides bicolor* passing through its territory, and forages opportunistically with them; also forages with Arrow-marked Babblers *T. jardineii*, twice seen with Hoopoes *Upupa epops* on ground, and attends Scimitarbill *Phoeniculus cyanomelas* probing low in trees (Tyler and Tyler 1996). Flies fairly high between patches of bushes or trees; flight heavy, with shallow wing beats.

Roosts in mid to lower stratum of thorn tree, especially *Dicrostachys cinerea* (Nylstroom, South Africa, Tarboton 1998), 1·0–4·5 (2·4) above ground; ♂ and ♀ go to roost at same time but sleep 1 to a few m apart, sitting pressed against trunk where a thin branch arises from it. In 3 years pair used 19 roost sites (*c.* 5 favoured, others used transiently); pair goes to roost usually 5–10 min before sunset, earlier on overcast days, later if some disturbance like an owl in the vicinity. In morning ♂ invariably gives one or more soft 'chop' notes just before leaving roost, ♀ responding with 'tsui' in duet; they leave roost 10 min before sunrise in austral summer and 15–20 min before in winter (Tarboton 1998).

Sedentary, with some local wandering in austral winter.

Food. Insects, including beetles, ants and caterpillars; and some small fruits. Ants a favourite prey when bird feeding on ground. Once took peanuts from bird table, husking then breaking them into small pieces (Borello 1988). Young fed with small moths, grubs, and once a 50 mm centipede.

Breeding Habits. Solitary nester, monogamous, territorial. Most territories surrounded by natural boundaries (e.g. open thicket surrounded by grassland); once, 2 nests in same open thicket divided between 2 territories. Unmated birds generally more vocal than mated pair. 2 birds, probably both ♂ ♂, once had song contest across territorial boundary for several min. ♂ in one territory often responds with same call to call of unseen bird in adjacent territory; main function of song presumed to be territorial advertisement (Tarboton 1971). ♂ advertises and defends territory mainly by calling from elevated position in crown of thorn tree (Harris and Arnott 1988). Fighting of rivals rare; once 2 birds (♂ and ♀?) confronted each other with tails jerking up and down, then flew and struck each other's breasts. Courtship involves some chasing of ♀ by ♂ in tree, the 2 bouncing and zigzagging through branches in tandem, tails jerking with each bounce; ♂ also pursues ♀ in flight, wings making fripping sounds, or ♂ gliding briefly with head held up. Copulation: ♂ alighted on branch, ♀ perched 15 cm away, ♂ hopped around her then onto her back whilst she crouched; ♂ quivered wings; copulation lasted 6–7 s. Single or (often) double-brooded, although not known how often both broods are successful; some pairs make 4 nesting attempts in a season; new nest started within 1–2 weeks of earlier one being destroyed or brood successfully fledging (Tarboton 1971, 1998). Breeding pair generally co-exists peacefully with other small birds, but can occasionally harass a pair of weavers, paradise-flycatchers or barbets.

NEST: rather untidy, poorly bound, quite bulky but weak open cup (**B**), made almost entirely of shreds of fibrous bark collected from branches and trunks of acacia trees by stripping from inside of bark peelings; many strips *c.* 75 long and 50 wide, added to nest by each bird in turn; some strips up to 250 long. Nest cup lined with rootlets, dry shreds of weed stalks and broad strips from mealie (maize) leaf. 2 birds made 6 nest-building visits in 35 min, spending 40–45 s in building on each visit. Material brought from 50–90 m away. Base of nest secured to

substrate with spider web. Cup quite neat; ext. diam. 100–125, ext. depth 50. By time young ready to fledge, cup has been trampled into flat platform. Of 32 nests, 69% in *Acacia karroo*, typically about half way up, 22% in *A. caffra*, others in *Rhus pyroides* and *Maytenus*; 63% in vertical fork, 25% in horizontal fork (usually with small lateral twigs giving further support), and 12% in sloping forks. Height above ground (n = 32) 1·1–7·6 (3·2) m. In Nylstroom, South Africa, nests sited in sort of place in which birds roost (see above), namely in vertical fork in main stem or fork between main stem and side branch, up to 3·9 m above ground (av. 2·1 m) in thorn tree (149 out of 155 nests: Tarboton 1998). New nests dark brown, fading to grey after a few weeks. Nest construction (n = 4), most intensive in early morning, takes 5 ± 1 days. Distances between consecutive nests built by one pair, 25–140 (75) m (Tarboton 1971).

EGGS: 2–3, av. (20 clutches) 2·8. Laid on consecutive days, early in morning, 1st egg being laid 5–7 days after nest building begun. Ground colour pale brown, very pale green, pale blue or white; marked with small light brown blotches or freckles with underlying slate or lilac, or with dark brown or red-brown, or with diffuse red freckles, or immaculate; marks usually form well defined cap or ring at broad end (Tarboton 1971). SIZE: (n = 69) 22·0–25·5 × 16·3–19·7 (23·7 × 17·5).

LAYING DATES: Zimbabwe, Sept–Jan, mainly Oct–Nov; Transvaal, Sept–Apr, mainly Oct–Nov; Namibia, Oct–Jan; Botswana, Aug–Jan and May, mainly Oct–Nov.

INCUBATION: begins sometimes with 1st egg laid, sometimes with 2nd or 3rd egg. On person's approach, bird slips off nest and calls until mate summoned, then the pair duets. Period 16·5 days. At one nest the 3 eggs hatched within 24 h. Eggshells removed from nest immediately, and once dropped 23 m away.

DEVELOPMENT AND CARE OF YOUNG: feather tracts apparent at 5 days; eyes opening at 8 days, when feathers breaking through sheaths on wings and back; at 10 days young feathered, some feathers 10 mm long; at 15 days red undertail-coverts apparent for first time. Tail grows rapidly during first few days after young leaves nest. At 15 days out of nest, black of tail has spread forwards to rump and red of undertail-coverts forwards to belly and flanks; at 20–25 days the spread has reached mantle and lower breast, so only head and upper breast remain in juvenile feather; at 35–40 days after fledging, bird is in adult plumage except for small brown patch on chin and throat, which soon becomes red. Young brooded continuously by parents for 5 days; no longer brooded at 11 days, even when raining. Brooding bird flies away just before arrival of relieving mate, which first feeds young and then settles to brood them. Young fed by both adults; parent often collects faecal sac after giving a feed, and swallows it or flies away with it. A single large chick fed 9 times in one hour, twice in the next hour. At 17 days young spends much time exercising, preening, stretching legs and wings and running bill through outspread feathers; when doing so it may overbalance and sit down abruptly, or suddenly fall asleep. On arrival of adult with food, young stands up, quivers wings, and gapes. Nestling period 18·5–19·5 ± 0·5 days.

Young out of nest use same squeaking 'wo-siee' note as before fledging; at first they beg from adults; by 28 days out of nest young still keep close to parents but for the most part forage independently. At that age young are remarkably quiet and unobtrusive (in marked contrast with young *Lanius* shrikes, for instance). Young roost in same trees as their parents, all sleeping in different places in the tree. Young may still be in parents' company when latter are in incubation and nestling stages of a 2nd brood.

BREEDING SUCCESS/SURVIVAL: of 34 nests built, eggs laid in 27, hatched in 13, and young fledged in 8; of 70 eggs, 30 hatched and 12 young flew; in brood of 3, 2 young often die in nest, on one occasion because of infection by mites *Ornithonyssus bursa*. Inexplicably, 19 of the 34 nests were destroyed by their owners, before or after eggs laid; one bird tore nest to pieces with bill, and jumped up and down on loosened material, demolishing nest in 10 min; nest destruction reported from study area at Olifantsfontein as well as Warmbaths and Windhoek (Tarboton 1971). Heavily parasitized by Black Cuckoo *Cuculus clamosus*. In Nylstroom breeding success abysmal: of 66 nests, 18 failed before or during egg laying, 38 failed during incubation, 2 failed during nestling stage, and of 8 broods raised one was a Black Cuckoo, so success rate only 10% of broods reared (Tarboton 1998). Adult once killed by Bateleur *Terathopius ecaudatus*.

References
Harris, T. and Arnott, G. (1988).
Tarboton, W.R. (1971, 1998).
Tyler, S.J. and Tyler, L. (1996).

Laniarius atroflavus Shelley. Yellow-breasted Boubou. Gonolek à ventre jaune. **Plate 26 (Opp. p. 399)**

Laniarius atroflavus Shelley, 1887. Proc. Zool. Soc. London, p. 124, pl. 13; Cameroon Mountain (7300 ft).

Range and Status. Endemic resident, montane E Nigeria and W Cameroon. In Nigeria confined to Obudu and Mambilla Plateaux; not uncommon on Obudu Plateau; single record in Mambilla but common on Chappal Hendu in adjoining Gashaka-Gumti Nat. Park. Cameroon: Mt Cameroon, abundant at 1500–2100 m on SE slopes, frequent to common at 700–1300 m and abundant at 1300–2300 on S slopes (Stuart 1986); common on Mt Manenguba at 1950–2250 m; and frequent and locally common in Bamenda Highlands (very abundant on Mt Oku at 2000–2900 m, also L. Bambulue, Bafut-Ngemba Forest Res., Bali-Ngemba Forest Res., Santa, Sabga Pass, Mbengwi-Tinachong at 1800 m, common in Bamboutos Mts at 2000–2600 m, Wum, Mt Lefo, Banso Mts); also Tchabal Waddi, Tchabal Mbaba (Genderu) and between there and Banyo.

Laniarius atroflavus

Description. ADULT ♂: forehead, crown, nape, hindneck, lores, upper part of cheeks, ear-coverts, sides of neck, mantle, back, tail and wings black with very dark blue or oily green gloss; upper rump feathers black with concealed large white subterminal spots; lower rump feathers broadly buff-tipped, long, overlying uppertail-coverts, which are black, glossed dark oily green. Chin and throat lemon-yellow, grading to brilliant orange-yellow on breast, flanks and sides of belly; middle of belly white with buff-pink tinge; thighs and undertail-coverts buffy white. Marginal wing coverts oily black, lesser and greater under primary coverts blackish, greater under coverts buff, axillaries lemon-yellow. Bill black; eyes dark brown; legs and feet slate-black. Sexes alike, but ♀ may be slightly larger than ♂. SIZE (6 ♂♂, 8 ♀♀): wing, ♂ 82–88 (84.75), ♀ 82–90 (87.0); tail, ♂ 65–79 (72.25), ♀ 65–77 (73.25); bill, ♂ 17–19.5 (18.6), ♀ 18–19 (18.5); tarsus, ♂ 30–32 (31.25), ♀ 29–31 (30.25). Mt Cameroon, wing (10 ♂♂) 78–84 (83.3), (7 ♀♀) 79–81 (80.25); Bamenda Highlands, wing (1 ♂) 90, (3 ♀♀) 85–87, (4 unsexed) 84–88. WEIGHT: (unsexed, n = 5) 40.1–46.9 (43.0).

IMMATURE: like adult but upperparts with dark brown wash, wings dark brown, greater primary coverts with buff tips, and T5 and T6 with buff tips up to 123 mm long.

NESTLING: not known.

TAXONOMIC NOTE: a subspecies, *craterum*, has been separated on greater size (particularly tail length), duskier underparts and greenish tinge to throat. However, the size difference, whilst good, is clinal (Mt Cameroon birds small, Mt Manenguba ones intermediate, Bamenda and Oku ones larger), and there is no colour variation except for seasonal wear (Serle 1950).

Field Characters. Length 20 cm. A striking bird; head to below eye and upperparts black except for a few white spots on rump; below golden yellow, becoming white on lower belly. Confined to highlands of E Nigeria and Cameroon, where no other bird like it.

Voice. Tape-recorded (104, B, GRI, CHA, LEM, STJ). Synchronized duets abrupt, rather harsh; whistles of ♂ do not have ringing quality of Tropical Boubou *L. aethiopicus*, and reply of ♀ short and grating, e.g. dry 'wheeyop' of ♂ answered by grinding, croaking 'djyow'. Call of ♀ follows ♂ so closely that they sound like 1 bird, 'wheeju', 'whodjaa', and they may overlap: in 'tyoho/bzz' the 'bzz' is synchronous with the 'ho'. ♀ may call first, 'djaaa-whoeee', ♂ may trill instead of whistle, 'kwarrrr-jiho', or both birds give harsh notes, 'chadow-zzheee'. Alarm a repeated hard 'jik', 'jak' or 'juk'. For further variations and sonagrams see Grimes (1976). Many calls inseparable from those of Mountain Sooty Boubou *L. poensis* (Stuart 1986).

General Habits. Inhabits dense undergrowth of clearings and secondary scrub in and near montane forest, and low undergrowth in small remnant forest patches and bamboos in ravines and on ridges. Also in topmost branches of figs *Ficus*, wind-shaped bushes on highest peaks, bramble, heath and bracken; and in tiny remnant copses and thickets around villages including those on bleakest hilltops.

Almost invariably in pairs, foraging actively in slightly more open situations, retreating into deep shade in dense vegetation if disturbed (Serle 1965). Gleans foliage and branches in undershrubs and trees, searching up to 7 m above ground (Stuart 1986); forages in small flowering shrubs *Agauria salicifolia* and *Hypericum* sp., and in gloomy woods of old trees covered with moss and lichen. Very noisy, with wide repertoire of loud whistling, swishing, rattling and harsh grating notes, 'whee-oo/chook' duets, and evidently much individual variation (Stuart 1986). Pair duets sometimes in canopy, birds often in sight of each other, invariably approaching tape-recorder on hearing playback and continuing to duet (Grimes 1976). Always on the move, restless, skulking; never flies far. Highly sedentary, pair regularly duetting and staying in same spot year after year.

Food. Insects.

Breeding Habits. Pairs nest solitarily; doubtless territorial, since that is the main function of duetting (Grimes 1976).

NEST: 'neatly woven' or 'badly constructed' rather flimsy and shallow cup made of coarse dry tendrils, lined with fine, hair-like plant materials; sited 0·6–4 m above ground, deep in undergrowth such as bushes, bracken and brambles (Eisentraut 1963).

EGGS: 2. Slightly glossy. Pale light green, sparingly flecked with small brown spots which can form cap at large end.

LAYING DATES: Cameroon, Nov–Jan and Mar (and various other evidence of breeding Nov–Mar; breeding completed by late Mar: Stuart 1986).

References
Bannerman, D.A. (1939).
Eisentraut, M. (1973).
Grimes, L.G. (1976).
Stuart, S. (1986).

Genus *Nilaus* Swainson

An anagram of *Lanius*. A single, polytypic species, the Brubru, a shrike-like but small bird with notched upper mandible and pied plumage, most races with chestnut stripe along sides; sexes similar, ♀ a duller version of ♂; juvenile plumage basically similar, dark brown and buffy rather than black and white; plumage of rump very full, lax, soft and fluffy; tail rather short, black, with white corners and sides; wings quite long, black, with white stripe in closed wing made by same white feathers as in many *Dryoscopus* and *Laniarius* bush-shrikes: median coverts, inner greater coverts, and outer edges of inner secondaries. Arboreal, insectivorous; main voice a high-pitched, fast trill ('Brubru' is onomatopoeic); pair often calls in duet; strongly territorial; resident; nest an open cup built on tree fork, bound with spider web and very well camouflaged with bits of lichen.

The nest is reminiscent of nests of *Batis* spp. (Family Platysteiridae, see Vol. 5 pp. 548, 574) and so are some plumage characters (white nape and back, fluffy rump, white wing stripe), and *Nilaus* may link Platysteiridae and Malaconotidae. Duetting, self nest destruction and parental sharing of nest duties are malaconotid, not platysteirid. Harris and Arnott (1988) combine those 2 families into a single subfamily in their Family Laniidae, sequencing *Nilaus* between *Batis* and *Malaconotus*. However, *Nilaus* seems to us more like a diminutive relation of *Dryoscopus*/ *Laniarius*, sharing pied plumage, wing stripe and fluffy rump, and we think it is much closer to them than to *Malaconotus*. Sibley and Monroe (1990) place *Nilaus* next to *Dryoscopus* in their Family Corvidae, Tribe Malaconotini (and treat platysteirines and prionopid helmet-shrikes in their following corvid Tribe Vangini).

Endemic, 1 widespread species.

Plate 20
(Opp. p. 287)

Nilaus afer (Latham). Brubru. Brubru africain.

Lanius afer Latham, 1801. Ind. Orn., Suppl., p. 19; Senegal.

Range and Status. Endemic resident and partial intra-African migrant. All subsaharan Africa south to 30°S, except W African and Congo Basin rainforest zone.

Mauritania, resident in Senegal R. valley but occurs up to 100 km north of it in rains; also in Karakoro valley. Senegal, uncommon, perhaps absent from NE. Mali, overall quite common, but unevenly distributed, north to Tombouctou (Timbuktu: 17°N); can be abundant in Bas Delta in rains. Common in Gambia. Niger, resident in 'W' Nat. Park; records Torodi, Zinder (breeding), frequent Tillia and Aïr and very common in Bagzans Mts. Nigeria, throughout the savannas north of rainforest belt, but everywhere uncommon. Chad, frequent resident in Ennedi and common in soudanian zone, but evidently only a passage migrant in between; records in Kanem; not in Ouadi Rime – Ouadi Achim Faunal Res. nor in Tibesti. Central African Republic, uncommon in Manovo-Gounda-Saint Floris Nat. Park and Vakaga area in N, and in Lobaye Préf. in S; a few records in W; hardly any records in centre or E. Also seems to be absent from SW Southern Darfur and W Bahr al Ghazal in Sudan, where otherwise fairly common as mapped, north to Northern Kordofan and 19°N in Nile Valley. Eritrea, uncommon in SW and on central plateau up to 1380 m; absent from Dancalia. Ethiopia, frequent to common and widespread in highlands and lowlands, except in N east of 38°E and between Cherch'er and Eritrea, where absent. Somalia, common and widespread as mapped.

Sierra Leone, 1 old record in extreme E. Liberia, 1 record, Cess R. (Gatter 1997). Elsewhere in W Africa not

Nilaus afer

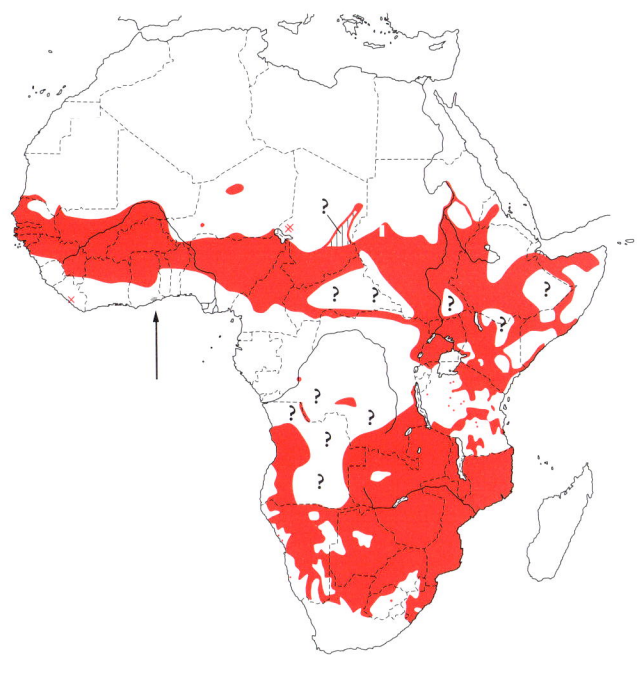

uncommon in soudanian zone and scarce in guinean zone, south to Béoumi in Ivory Coast and to Legon and inselbergs on Accra Plains in Ghana. Togo, very scarce; records from Bismarckburg, Mango (twice), Naboulgo and Pagala, all north of 08°N.

Zaïre, uncommon in acacia country in N, NE (Uele), E (S Rwenzori, N Semliki Plain, E Rutshuru Plain) and S (see racial distributions, below). Just into N Rwanda; not in Burundi. E Africa, wide-ranging but rather uncommon; rare or absent in large areas in E Kenya and in Tanzania as mapped (N.E. Baker, pers. comm.). Southern Africa, uncommon to locally very common (e.g. N Botswana); further details below. Densities of 1 pair per 30–50 ha in Transvaal, 1 bird per 9 ha in 11 woodland types in N Botswana and 1 bird per 4 ha in Okavango tall acacia (M. Herremans *in* Harrison *et al.* 1997). Probably >100,000 birds in S Mozambique (Parker 1999).

Description. *N. a. brubru* (Latham): Angolan littoral north to Luanda, SW and S Angola, Namibia, Botswana, extreme SW Zambia, W Zimbabwe from Kazuma area to Hwange Nat. Park, W Gwaai and NW Matabeleland, N Cape Prov., W Transvaal, W Free State and W Swaziland. ADULT ♂: forehead white above nostrils, black by culmen; small black loral spot in front of eye; crown and nape black; lores and superciliary stripe white; ear-coverts and sides of neck black, latter divided by rearward extension of white superciliary stripe to below nape; hindneck whitish; mantle black at sides, buffy white in midline; scapulars black; back black with large white spots and blotches in midline; rump black and white, uppertail-coverts black. Plumage of rump very full, lax, soft and fluffy, each feather with blackish tip, large white subterminal patch, and dark grey base. Tail black, T3 sometimes with white tip, T4 and T5 with small white triangle at tip and white streak on outer edge, T6 with white tip and white outer vane. Underparts white, with long, sharply defined chestnut patch from side of breast to lower flank. Primaries black with narrow white edges, P6 and P7 or sometimes P5–P8 emarginated, inner primaries with white tips; secondaries black, 3 inner ones with broad white outer edges; tertials black; primary coverts black with long white oval on inner vanes, alula and greater coverts black, 3 inner ones with very broad white or cream outer edges and tips; median coverts mainly cream-white, lesser coverts black; underside of flight feathers grey, inner borders white; underwing-coverts and axillaries white, with blackish carpal patch. Bill black, upper mandible notched; eyes dark reddish brown or dark Indian red, near crimson; legs and feet grey. ADULT ♀: like ♂ but upperparts dark brown where ♂ is black; chin, throat and upper breast with blackish streaks and speckles; sides pale rufous where ♂ is chestnut, the patch being smaller than ♂'s. SIZE (10 ♂♂, 10 ♀♀): wing, ♂ 87–90·5 (88·4), ♀ 83–91·5 (86), unsexed (n = 60) 80–91 (84); tail, ♂♀ 52–63; bill, ♂♀ 14–17·5; tarsus, ♂♀ 20–24. WEIGHT: 2 ♀♀ both 24·0, 2 ♂♂ 22, 23, (n = 4, unsexed) 23·6–24·6.

IMMATURE: top of head dark brown with small white speckles; white superciliary streak mottled with dark brown; hindneck and sides of mantle and back dark brown with broken white bars and white speckles; centre of mantle and back with large white blotches (feathers white with broad blackish subterminal band and whitish tip); rump and uppertail-coverts dark brown and whitish. Tail like adult. Ear-coverts whitish, speckled dusky; dark line from behind eye to hindneck. Underparts white with short, irregular, dark brown bars and a few rufous feathers at sides of breast and on flanks. Wings dark brown where adult black and pale buff where adult white; primary coverts with buff tips; small white speckles on lesser and median coverts; outer primaries with small white spots at tip; inner primaries, secondaries and tertials with large white or cream spots at tips. Juvenile has forehead and crown whitish with dark brown bars, buffy hind collar, mantle whitish with spaced dark brown bars, back mainly white, tail dark brown and warm buff where adult black and white respectively, underparts whiter than in immature, with widely spaced, narrow dark bars or vermiculations and without any rufous feathers at sides; wings warm, deep buff where immature pale buff.

NESTLING: naked and blind.

N. a. solivagus Clancey: NW Cape Prov., central plateau of Zimbabwe, W half of Sul do Save, Mozambique, N and E Transvaal, E Swaziland, Natal, Zululand (except coast). Black in upperparts deeper than in *brubru* and white in back tinged yellowish; underparts cream not white; chestnut stripe narrower and darker. Smaller; wing, ♂ (n = 4) 80·5–84·5 (82·9), 2 ♀♀ 82, 83. WEIGHT: (mainly *solivagus*?) ♂ (n = 32) 17–32 (23·2), ♀ (n = 26) 18–30 (22·8).

N. a. afer (Latham): N tropics from Senegal to Eritrea and Ethiopia east to Rift Valley, south to Ghana coast, about 07°N in W Cameroon and about 05°N in central Cameroon and Central African Republic, and to Chua and Lango in NW Uganda. ♂ like ♂ *brubru*, but no black loral spot in front of eye; centre of mantle and back rufous-buff not white; line from side of breast to lower flank darker chestnut and much narrower; median coverts, inner greater coverts and edges of primaries and tertials warm rufescent buff (not cream-white). ♀ like ♀ *brubru* but head, upperparts and wings warm brown where *brubru* dark brown and rich rufescent buff where *brubru* white; rump feathers white with black tips; tail like *brubru*; in underparts, narrow dark chestnut line runs from below eye, down sides of throat and breast to flanks (not just from side of breast to flanks). WEIGHT: 1 ♂ 17·0, 1 ♀ 19·0.

N. a. camerunensis Neumann: Cameroon highlands, S Cameroon, S Central African Republic, N and NE Zaïre east to N end of L. Albert. ♂ like ♂ *afer* but whites on head and underparts greyish. ♀ like ♀ *afer* but chin, throat and breast greyish white, slightly dark-speckled.

N. a. minor Sharpe (includes '*brevialatus*'): SE Sudan west to 34°E; Ethiopia, in Rift Valley and to east and south of it; Somalia; Kenya, plateau country in N and E, south to Kerio valley, Baringo, Isiolo, Tsavo and Taru; Tanzania in NE south to about 06°S on coast, west to about 35°E. Exactly like *brubru* but smaller.

SIZE (7 ♂♂, 7 ♀♀): wing, ♂ 72·5–78 (76·3), ♀ 73–76 (74·5); tail, ♂ 51·5–55 (53·9), ♀ 50–55·5 (53·7); bill, ♂ 13–15·5 (14·2); tarsus, ♂ 23–28 (25·6). WEIGHT: 2 ♂♂ 15·0, 15·5; 3 ♀♀ 13, 15, 20; 2 unsexed 15, 17.

N. a. massaicus Neumann: Zaïre on plains near L. Edward and S end of L. Albert; Uganda north to about 02°N, intergrading with *afer* in Karamoja; almost throughout Kenya, except range of *minor* in N and E; Tanzania, in N from Serengeti to Shinyanga and Maswa Game Reserve. ♂ like ♂ *afer* but light areas in upperparts white, creamy or pale rufous, not rich rufous-buff; in underparts, chestnut line wider and slightly paler than in *afer*. ♀ like ♀ *brubru*. WEIGHT: 1 ♀ 18.

N. a. nigritemporalis Reichenow (includes '*occidentalis*'): E Angola; Zaïre, in Katanga up to 1740 m and probably this race also in E Bas Zaïre and S Bandundu; intergrades with *affinis* and *brubru* in Katanga (Louette 1989); Tanzania except Kagera and NE (occupied by *massaicus* and *minor*) – north to Kibondo, Mwanza, Dodoma, Morogoro and Bagamoro; Zambia except extreme SW (occupied by *brubru*), intergrades with *solivagus* in middle Zambezi valley and known from Zimbabwe side; N Mozambique south to about 24°S. ♂ with black loral spot and white loral line above it, but white line does not continue behind eye as an eyebrow. Underparts like ♂ *afer* but band on sides rufous rather than chestnut. ♀ like ♀ *afer* but light areas in upperparts pale buff not rufous-buff, and band along side of breast and flank narrow, fragmentary, rufous, not chestnut.

N. a. miombensis Clancey: coastal Zululand, lowlands in E Sul do Save, Mozambique, and Save valley upstream to Sabi/Lundi confluence, Gulene (Gonarezhou Nat. Park) and Marhumbini, Zimbabwe; north in Mozambique to at least 24°S, but northern limits unclear. ♂ like ♂ *nigritemporalis*, ♀ with underparts pure white, unmarked on throat and breast, band along side of breast and flank reduced to chestnut streaks. WEIGHT: ♂ (n = 3) 22·6–27·1 (24·7), 1 ♀ 24·9.

N. a. affinis Bocage: Angola, escarpment and inland of it, from Duque de Bragança and Cuango to central highlands; S Zaïre in Kasai and N Katanga, east to L. Tanganyika (Mt Kabobo: Louette 1989). ♂ lacks white eyebrow (like *nigritemporalis*) and underparts pure white – completely devoid of rufous. ♀ dark brown and creamy above, greyish white below with slight buffy wash, chin to breast quite heavily streaked and speckled with black. Intergrades with *nigritemporalis* and *brubru*.

TAXONOMIC NOTE: northern tropical and southern African races are all rather similar in appearance but are separated by the distinctive *Brachystegia* races *nigritemporalis* and *affinis*. Since the last 2 races seem to intergrade at all of their boundaries with the northern *afer* and southern *brubru* groups, we treat the complex as a single species as have nearly all earlier authorities.

Field Characters. Length 12·5–15 cm. A small, short-tailed bush-shrike, common in pairs in open country. Plumage varies throughout range, but most birds have broad *white supercilium* separating *black cap* from *black eyeline* and *chestnut flanks*. Upperparts and wings black mixed with white, creamy or rufous-buff, depending on race, underparts white; flank colour varies in shade and extent, and black eyeline may be incomplete or even absent. Unlike any other shrike; colour combination suggestive of batis, but Brubru is larger and longer than any batis, has heavier bill, and lacks pale eye prominent in batises.

Voice. Tape-recorded (39, 75, 86, 88, 91–99, 104, 108, B, C, F). Song of ♂ a high-pitched trill lasting *c.* 1 s; may be fast buzz (individual notes barely discernible) or slower ringing trill, sounding like distant telephone. Sometimes (especially in southern Africa) trill is preceded by one or more clicking notes, 'tippy-tip-wrrrrreeeee' or 'wop-prrrrreeeee'. Trill given alone or in duet; ♀ answers with shrill 'weeyo' during or at end of trill, often synchronously when trills given in series; after single trill ♀ may respond with single or double 'weeyo' at varying intervals. Other calls include a hard 'yaah' and a long, rapid chatter, 'chatatatatatata ... ', and said to give a penetrating high, whistled 'wutititititi' (Zimmerman *et al.* 1996).

General Habits. Densities vary considerably but in southern Africa are greatest in pure, tall acacia woodland (often in isolated stands), mature *Brachystegia* and *Colophospermum* and other tall broad-leaved woodlands, and in mosaics of mixed woods; also occurs in patches of scrub in arid regions; in N Botswana, found in 11 out of 17 woodland types (M. Herremans *in* Harrison *et al.* 1997). Similarly in N tropics, commonest in flat-topped acacias in open, park-like country; and in desert country, can occur in small, isolated clump of trees.

Spends nearly all of its time in foliage, often near tops of emergent broad-leaved or microphyllous savanna trees; solitarily or in pairs, never gregarious, unobtrusive; pair members often forage far apart; call is far-carrying, given often, betraying presence, although calling bird often difficult to locate and seems to keep its distance from observer. In Transvaal, pair defends permanent territory; 3 territories were 33, 34 and 42 ha in extent (Tarboton 1984). Advertises territory by calling year-round; pair often involved in disputes with adjacent territorial pair, with chasing and supplanting. Silent for long periods when exploring for food (Vincent 1935). Active, gleaning foliage and branches particularly in canopy of tall *Acacia* trees, where quite acrobatic, feeding tit-like with body at all angles, sometimes head-down or completely upside down as it investigates leaf sprays, inflorescences and pods at ends of branches. Said to forage sometimes 'like a nuthatch' (Zimmerman *et al.* 1996); that comparison, which may go back to Bannerman's (1939) 'running along the branches and limbs of trees like a Nuthatch', not substantiated by Harris and Arnott (1988). Holds large prey item under one foot. Occasionally hawks insects in air. Often joins mixed-species foraging flocks, e.g. of hyliotas, eremomelas and Spotted Creepers *Salpornis spilonotus* (Vincent 1935). Flight fast and undulating, at or above tree-top level, often quite distant. ♂ sings from exposed tree-top perch, with body inclined, head raised, neck outstretched, crown and throat feathers raised, tail slightly fanned; ♀ usually responds in duet with short, high-pitched 'eeuu' notes (Harris and Arnott 1988). Once seen mobbing bird snake *Thelotornis capensis*.

Sedentary in most regions, but a partial migrant in W Africa. In S Ghana (Legon, Accra Plains), a dry-season visitor from late Oct to mid Mar. Little evidence for seasonal movement in Nigeria, where resident as far south as Enugu. Chad, occurs in Ouaddaï in July and Sept and thought to be migratory there; common resident in Ennedi to north-east but perhaps also partially migrant.

Food. Insects, including caterpillars, ants and beetles; also spiders.

Breeding Habits. Monogamous, territorial (see above). Intense territorial interactions between pairs involve counter-singing and zigzag display flights with wing-fripping and deep, rather slow wing beats; and, when birds perched, bowing and rapid side-to-side movements of body. Not known to fluff out thick, patterned rump plumage. In sexual display, pair duetted rapidly, ♂ stood (on prospective nest site?) with food in bill and gave series of high-pitched 'tuet-tuet-tuet' calls interspersed with 'prrreeee' trills, then flew to ♀ sitting with drooping wings, sidled up to her and copulated whilst repeatedly singing (Harris and Arnott 1988). Strong tendency to build and abandon early nests without laying in them (Tarboton 1984); 1 pair built and destroyed 3 nests in 4 weeks, starting 4th nest at end of that time and then laying in it. Sometimes early nest can be found intact on ground below nest tree. ♂ not known to courtship-feed ♀ (but probably does so: Harris and Arnott 1988).

NEST: small, neat, not very deep, open cup, made of small bits of fine, pliant plant materials (petioles, twigs, tendrils, pieces of bark) felted together with spider web and generally covered profusely outside with bits of foliaceous lichen, making nest highly cryptic and moulding it seamlessly to lichenous substrate (**A**). 2 nests were constructed entirely of only 3 materials: leaf petioles (451 and 550 pieces), lichen (95 and 95 pieces), and spider web (Tarboton 1984). Sited generally on 2- or 3-pronged fork on gently sloping branch of *c.* 70 mm diam., at height above ground of 2·5–9·0 (4·9) m (n = 21); at Nylsvlei, Transvaal, mainly in *Terminalia sericea* and *Burkea africana* trees. Size (n = 2): ext. diam. 65, 70, int. diam. 45, 45, int. depth 20, 20.

♂ and ♀ take equal part in nest building, collecting material and visiting nest independently. If both arrive at same time one waits until other has finished adding material (which takes 30–60 s). Bird bringing material first sits on nest then places material on outside; it taps the area repeatedly with lower mandible, rotates body through *c.* 30°, compacting the wall, then drops onto breast and uses feet to compact inner side, using treading motion and rotating body. Bird collecting leaf petioles of *Peltophorum africanum* (Leguminosae) flew with beakful to nearby spider web where it gathered web onto petioles before taking them to nest (Tarboton 1984). In one nest 1st egg laid 6 days after building started; evidently all material brought and nest built in first 2 days, then pair continued to shape and compact it up to time of laying; another nest was complete at 7 days but eggs not laid until 11 days thereafter.

EGGS: 2, rarely 3 or 4. 1st egg laid in early morning, 2nd *c.* 26–27 h later. White, greenish or greyish, heavily streaked with grey and brownish. SIZE: (n = 44) 18·6–22·3 × 14·3–16·8 (20·2 × 15·4).

LAYING DATES: Gambia, (young out of nest being fed, Nov); Mali, about June–Nov; Ghana, Jan; Nigeria, Kainji, Jan–Feb; Niger, Tillia, (♀ about to lay, May); Sudan, Apr–May, Nov; Ethiopia, Apr; Somalia, May; Zaïre, Uele, (various indications, about Jan–Feb); E Africa, Region C, Feb, Apr–June, Nov, Region D, Mar, May, July, Sept–Oct, mainly Mar and May in long rains; Zambia, Sept–Dec, Jan (mainly Sept: 65% of 34 records); Malaŵi, Aug, Sept, Mar; Botswana, Oct–Nov, Jan–Feb; Zimbabwe, Sept–Jan (142 records: Sept 48%, Oct 39%, Nov 11%); Namibia, Oct–Nov; South Africa, Transvaal, Aug–Jan, Natal, Sept–Dec.

INCUBATION: starts with 1st egg laid; clutch incubated for 98% of daytime, 54% of that time by ♂ and 46% by ♀; only ♀ incubates at night. 2 change-over patterns: when ♀ incubating (n = 20 occasions) ♂ initiates change-over by calling as he approaches, ♀ responding in duet; when ♂ incubating (n = 18 occasions), he initiates change-over by calling on nest, ♀ arriving immediately and taking over. Incubation spells (n = 28) 17–178 (42) min. Several times incubating ♀ called when Crested Barbets *Trachyphonus vaillantii* nesting in same tree approached too close, and ♂ promptly chased them away. Period: (n = 1) 19 days, eggs hatching on consecutive mornings.

DEVELOPMENT AND CARE OF YOUNG: well-feathered nestling can have crown dark brown speckled with white, and superciliary stripe and hindneck can be white (**A**); eyes rich burnt umber-brown. Young brooded by either parent continuously for first 6–7 days, then sporadically for 3–4 days; after that, only at night. Faecal sacs removed and eaten by either parent, even when nestling 21 days old; young fed by both parents. Nestling period (1 brood of 2) 21·5 and 22 days. Young out of nest stay with parents for *c.* 55 days; acquire adult plumage in 2 months; later, young still in parental territory though not in parents' company.

BREEDING SUCCESS/SURVIVAL: in one study only 3 nests out of 10 were successful. One ♂ at least 8 years old (Hanmer 1989). Nests and eggs destroyed by squirrels *Paraxerus cepapi* and fledglings by Fiscal Shrike *Lanius collaris* (Harris and Arnott 1988).

References
Harris, T. and Arnott, G. (1988).
Tarboton, W. (1984).

A

476 PYCNONOTIDAE

Family PYCNONOTIDAE: bulbuls

Bulbuls were dealt with in Vol. IV (1992); for family diagnosis, see p. 279 therein. At the time it seemed likely that 2 enigmatic genera of African birds, *Neolestes* and *Nicator*, would prove not to be bulbuls as had often been supposed previously, so discussion of them was deferred (see Vol. IV, p 352). Olson (1989) argued that monotypic *Neolestes* might be a prionopid shrike and the 3 *Nicator* species malaconotid shrikes. Recent evidence now suggests that *Neolestes* is indeed a bulbul relative, and we follow Dowsett *et al.* (1999) in placing it *incertae sedis* in Pycnonotidae. The affinities of *Nicator* continue to remain obscure (see below), and we treat it here also *incertae sedis*.

Genus *Neolestes* Cabanis

An enigmatic bird that has been allied variously with bulbuls and 'shrikes' (Malaconotidae, Laniidae or Prionopidae). Bill small, shorter, broader and more decurved than *Pycnonotus*; no nasal septum. Posterior margin of bony nostril entire, with no trace of ossification; presence of an ossified nasal margin is one of the best characters defining the Pyconotidae. The skull overall shows more affinities with bulbuls than with shrikes (Dowsett *et al.* 1999): maxillopalatines long, thin and slightly expanded at tips (as in bulbuls, unlike shrikes); palatines expanded and truncate posteriorly, lacking transpalatine processes, and descend below jugal bar, unlike shrikes; postorbital process vestigial, as in *Pycnonotus* (well developed in *Malaconotus*). Syrinx similar to that of Sombre Greenbul *Andropadus importunus* but markedly different from White Helmet-Shrike *Prionops plumatus* or Bokmakierie *Telophorus zeylonus*. Humerus has incipient accessory fossa partially undercutting head (absent in *Malaconotus*). No sexual size dimorphism as in many bulbuls (though none in *Pycnonotus*).

Plumage striking and boldly patterned, with broad, shiny black band from bill through eye and over ear-coverts onto neck and down to form broad collar across breast. (No bulbul has a plumage pattern remotely resembling this: similarity of collar to that of bush-shrikes *Telophorus* influenced earlier systematists to place *Neolestes* among them). Tuft of stiff, plush, chestnut feathers behind ear, which Olson (1989) noted was similar to condition in *Prionops*. No filoplumes (which are present in most but not all bulbuls). Immatures lack any of the fine barring present in young laniids.

Nest and eggs similar to both bulbuls and bush-shrikes; clutch size 2, normal for Pycnonotidae and Malaconotidae, while in Laniidae and Prionopidae c/4 is usual. Eats much fruit, like bulbuls but unlike shrikes; vocalizations bulbul-like (Dowsett *et al.* 1999).

DNA analysis provides strong support for placing *Neolestes* and bulbuls in a monophyletic clade. Within it *Neolestes* is basal, although its exact phylogenetic position cannot be ascertained until all pycnonotids are analysed together (Dowsett *et al.* 1999).

In spite of these biological similarities to bulbuls, the extremely divergent plumage and lack of nasal ossification prevent too close an association with them. *Neolestes* could perhaps be the primitive sister-group of the Pycnonotidae, or a member of some larger group of which the bulbuls are but one lineage (Dowsett *et al.* 1999).

Monotypic; endemic.

Plate 23
(Opp. p. 350)

Neolestes torquatus Cabanis. Black-collared Bulbul. Bulbul à collier noir.

Neolestes torquatus Cabanis, 1875. J. Orn. 23, p. 237, pl. 1, fig. 1; Chinchoxo, Loango Coast.

Range and Status. Endemic resident in lightly wooded savanna, from extreme S and SE Gabon (Tschibanga, Leconi) and Congo (widespread in Odzala Nat. Park and Léfini Reserve (Dowsett and Dowsett-Lemaire 1997); local on Batéké Plateau between Djambala and Brazzaville, rare Kouilou basin), to Angola in Cabinda and W highlands from Cuanza Norte to Lubango, where rare, also in E near borders of Zaïre and Zambia; S and SE Zaïre, and in extreme E on Idjwi I. in L. Kivu and on W shores of L Tanganyika from near Mt Kabobo to Marungu Highlands; NW Zambia south to Mwinilunga town and Mundwiji Plain. Common from sea-level to 1750 m, small numbers to 2060 m in Marungu (Dowsett and Prigogine 1974). Vagrant Rwanda. Density in Odzala Nat. Park 4–5 pairs per km^2.

Description. ADULT ♂: forehead to hindneck grey; rest of upperparts olive-green. Tail feathers dark grey-brown, edged olive-green. Broad black band through lores and around eye to ear-coverts, continuing down side of neck to form broad black gorget across upper breast. Ear-coverts buffy, cheeks to chin and throat creamy white; underparts below gorget buffy white, sides of breast and upper flanks tinged greyish, lower flanks tinged

Neolestes torquatus

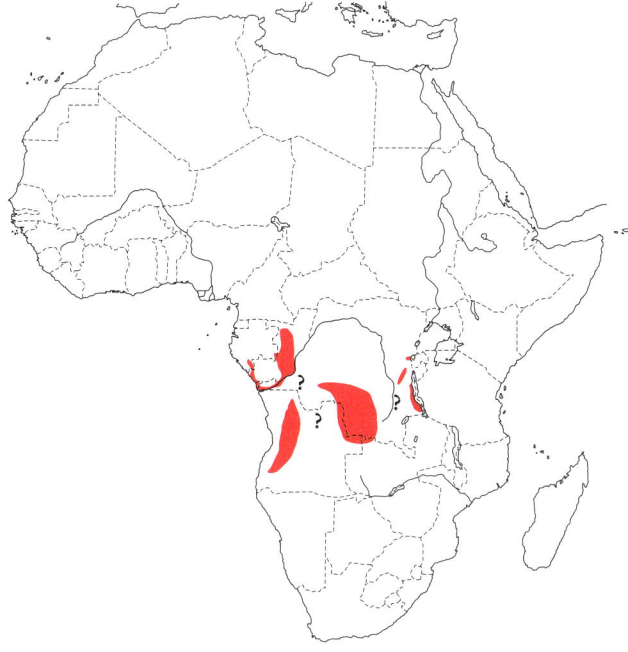

olive-green, thighs greenish yellow. Flight feathers dark grey-brown, edged bright yellowish green except emarginated parts of outer primaries; primary coverts and alula dark grey-brown, fringed bright olive-green; bend of wing bright yellow; rest of upperwing-coverts olive-green. Axillaries and underwing-coverts pale yellow; inner borders of undersides of flight feathers greyish white. Bill greenish black or dark olive-green; eyes dark brown; legs black. Sexes alike. SIZE: wing, ♂ (n = 17) 67–76 (72·9), ♀ (n = 12) 68–76 (72·7); tail, ♂ (n = 6) 61–67 (64·8), ♀ (n = 7) 60–64 (61·6); bill, ♂ (n = 6) 15–16 (15·6), ♀ (n = 7) 14–15 (14·5); tarsus, ♂ (n = 6) 20–21 (20·8), ♀ (n = 7) 20–22 (21). WEIGHT: ♂ (n = 4) 19·9–23·5 (21·8), ♀ (n = 5) 19·8–27 (23·2).

IMMATURE: like adult but duller; crown and nape greenish, a few lower mantle feathers with pale edges, giving spotted effect, breast band dull, not glossy, edges of flight feathers olive-yellow (duller, less green); scapulars and upperwing-coverts olive-brown, tipped buff, producing pale bars across greater and median coverts.

NESTLING: hatches naked, body purple, eyelids dark blue, gape yellow.

Field Characters. Length 16–17 cm. Unmistakable, with *black collar* across white breast, continuing up sides of neck and round and *through eye*. Black collar is reminiscent of *Telophorus* bush-shrikes, but underparts of *Neolestes* are white, not yellow as in Bokmakierie *T. zeylonus* (only overlaps in S Angola) or red and green (*T. quadricolor viridis*, sympatric in many areas). Unlike bush-shrikes perches boldly out in open. *Yellow* bend of wing conspicuous. Chattery bulbul voice very different from powerful ringing calls of bush-shrikes.

Voice. Tape-recorded (86, LEM). Song and call notes reminiscent of Common Bulbul *Pycnonotus barbatus*; song phrases more mellow and fluid, and call-notes 'tji-li-li' and 'dee-de-de-de-de' and nasal trill are also softer (Dowsett *et al.* 1999). When foraging gives bulbul-like twittering notes.

General Habits. Inhabits wooded grassland with 20–30% bush cover, the woody vegetation being mainly *Hymenocardia acida* and *Annona senegalensis*; avoids recently burned savannas, and feeds mainly in tall dense grassland at least 0·5 m, preferably 1–2 m tall, often *Hyparrhenia* spp. (Congo: Dowsett and Dowsett-Lemaire 1997). In Zambia, village gardens and regenerating chopped-down miombo woodland interspersed with occasional rock outcrops (Penry 1979), also overgrown cultivation, termite hill thickets, secondary miombo scrub and *Lantana* (Bowen 1983). Not associated with evergreen forest edge (Dowsett *et al.* 1999, contra Benson *et al.* 1971).

Singly, in pairs or in groups of up to 4, presumably family parties. Perches boldly on bare branches and even fence posts like shrike (Bowen 1983). Hawks for insects in air, also drops down into long grass to take them.

Food. Insects and fruit.

Breeding Habits. Solitary nester. Territorial; territories well spaced out so that one bird rarely sings within hearing range of a neighbour (Dowsett *et al.* 1997). Playback of song provokes strong reaction from both members of pair. In possible breeding display, bird 'would dart up into the air for 100 feet, in short jerky flight with noisy wing beat, and remain there within a 5-foot circumference for several minutes before descending rapidly and jerkily; this performance is accompanied by continuous twittering' (Beatty, in Rand *et al.* 1959).

NEST: frail cup, untidy and with no proper lining, wholly made of grass, or with a few plant stems, full of gaps; ext. diam. 76 × 86, int. diam. 57 × 64, depth of cup 38; placed in fairly open situation c. 1 m above ground in fork of small (1·5 m high) bush surrounded by grass 1–1·5 m high in area of scattered bushes and tall grass outside miombo scrub (Bowen 1983). Another was placed in a *Combretum* bush (by Brazzaville airport, Congo: Salvan 1972).

EGGS: 2; oval; white or pale pink, with pink and brown markings, sometimes with faint rufous zone around large end. SIZE: (n = 2) 20 × 14·4, 20·8 × 14·2.

LAYING DATES: Angola (juv. Mar); Zaïre Aug–Oct, Dec–Jan (juvs Dec–Feb); Zambia (juv. Mar).

INCUBATION: no details.

DEVELOPMENT AND CARE OF YOUNG: at day 5, nestling still blind, trace of feathering on thighs; day 7, pin feathers starting on wings; day 10, pin feathers 8 mm long and ochreous; day 13, some olive-brown on crown, buffy below, black collar developing, belly still bare, weight 9 g, wing 33·5; day 16, almost ready to leave nest; day 17, out of nest in adjacent bush, attended by adults. Young fed by both parents, mainly with fruit, occasionally insects; both parents usually visit nest at same time, at intervals of up to 30 min. On approach to nest, they perch on tops of nearby bushes, then swoop down to a dense bush near nest and fly to it (Bowen 1983); sometimes they perform a distraction display, flying down into grass and feigning injury.

References
Bowen, P.St.J. (1983).
Dowsett, R.J. *et al.* (1999).

Plate 31

Spotted Nutcracker (p. 567)
Nucifraga caryocatactes macrorhynchos

Juv. Ad.
Piapiac (p. 560)
Ptilostomus afer

Rook (p. 554)
Corvus frugilegus

Western Jackdaw (p. 555)
Corvus monedula spermologus

House Crow (p. 557)
Corvus splendens splendens

Cape Crow (p. 533)
Corvus capensis

Somali Crow (p. 544)
Corvus edithae

Pied Crow (p. 545)
Corvus albus

Carrion Crow (p. 535)
Corvus corone pallescens

Brown-necked Raven (p. 540)
Corvus ruficollis

Fan-tailed Raven (p. 548)
Corvus rhipidurus

Common Raven (p. 537)
Corvus corax tingitanus

White-necked Raven (p. 550)
Corvus albicollis

Thick-billed Raven (p. 552)
Corvus crassirostris

6 in
15 cm

Plate 32

Genus *Nicator* Hartlaub and Finsch

Bill stout, laterally compressed, with well-marked subterminal notch and prominent hook at tip of upper mandible; nostrils semilunar, with well-developed operculum and no ossified connective tissue covering the posterior margin. Metatarsi scutellate; basal phalanges of middle and outer toes completely fused, as in *Bleda*. True rictal bristles poorly developed, replaced functionally by bristly pre-orbital feathers with stiff shafts and greatly reduced barbs around front of eye; behind these and above the lores a pad of short, dense, upward-pointing feathers; yellow eye-ring in 2 species, surrounded by area of bare skin. Plumage soft, loose and abundant, feathers decomposed, with well-separated barbs; a gap in feather tract at back of neck, producing bare patch; no long filoplumes on nape; in some nape feathers the rachis and distal rami lack terminal barbules, these feathers being tipped with a number of short bristles. Upperparts and tail olive-green, with prominent yellow terminal spots on upperwing-coverts and tail feathers. Undertail-coverts yellow, as in bulbuls of the *Pycnonotus barbatus* complex. Sexual size dimorphism pronounced; ♂ *N. chloris* up to 37% heavier than ♀, ♂ *N. gularis* 50% heavier than ♀. In nestlings of 2 species, head remains unfeathered, except for 2 lateral lines on crown which merge on nape (**A**), for some time after feathers of wings and body have developed (nestling of *vireo* unknown).

Voices of 2 species (*chloris, gularis*) unique, loud and explosive, unlike either bulbuls or bush-shrikes; the voice of *vireo* is softer, and some phrases have a bulbul-like quality. Nest completely different from that of either bulbuls or bush-shrikes (Brosset and Erard 1976): flat, built of crossed twigs, like that of pigeon; often so small and rudimentary that egg falls through onto ground when nest rocked by wind; nest fragile and vulnerable to being dislodged when first built, but after a few days become glued to its support by mycelium of fungus *Marasmius*, so securely that it can be held upside down without falling off (Brosset 1974). Inhabit lowland forest; live in pairs; pair members keep some distance apart when foraging, calling to maintain contact. 2 species (*N. chloris, N. gularis*) raise tail with feathers slightly spread in territorial encounters or when disturbed by observer, and (much less often) when relaxed or feeding (F. Dowsett-Lemaire, pers. comm.).

The systematic status of *Nicator* has been the subject of much debate. Prior to 1943 it was placed in the bush-shrikes Malaconotidae. Delacour (1943) transferred it to the Pycnonotidae, derived from *Bleda* which in turn was derived from *Phyllastrephus*. Chapin (1953) agreed that it showed more resemblances to Pynonotidae, though not closely allied to *Bleda* or Asian *Setornis*; the nestling looks quite unlike a young shrike, lacking barring, while its colouration is like that of the adult, not rufous as in *Bleda*. Subsequent authors placed *Nicator* in the Pycnonotidae, e.g. Mayr and Greenway (1960), White (1962), Hall and Moreau (1970). A study of feather protein electrophoresis showed the affinities of *Nicator* to be with the Pycnonotidae rather than with *Lanius* or *Dryoscopus* (Hanotte et al. 1987); curiously, *Malaconotus* was not used in this study. DNA–DNA hybridization techniques also placed *Nicator* in Pycnonotidae (Sibley and Ahlquist 1985, 1990). However, Olson (1990) found that the Pycnonotidae can be defined by the presence of a thin sheet of bone covering the posterior margin of the nostril which is usually pierced by 1–2 small neural foramina. Since *Nicator* lacks this character he removed it from the Pycnonotidae and returned it to the Malaconotidae, probably closest to *Malaconotus*. Bush-shrike-like behavioural characters include members of pair foraging some distance apart, dropping to ground to capture insect, and chasing adversary through leaves (Brosset and Erard 1986); however, Jackson and Sclater (1938) note that bolder flight and quicker movements when searching for food are more like one of the larger forest bulbuls than the slow-moving bush-shrikes, and Short et al. (1990) believe its 'floppy disjointed movements' are bulbul-like.

Endemic. 3 species, 2 in a superspecies (*chloris/gularis*).

A

Nicator chloris superspecies

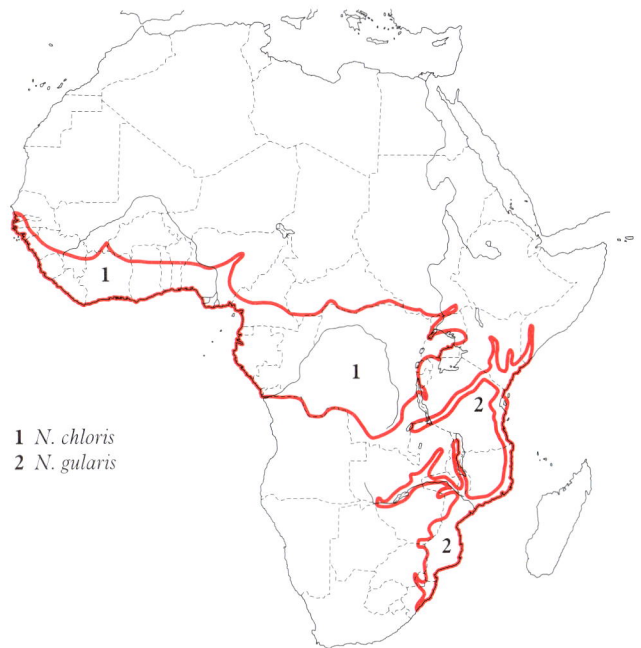

1 *N. chloris*
2 *N. gularis*

Nicator chloris (Valenciennes). Western Nicator. Bulbul nicator.

Lanius chloris Valenciennes, 1826. Dict. Sci. Nat., 40, p. 226; Galam (Senegal).

Plate 23
(Opp. p. 350)

Forms a superspecies with *N. gularis*.

Range and Status. Endemic resident. Uncommon W Gambia and SW Senegal; rare Guinea-Bissau, where noted by Monard (1940) but not seen in recent visits by Hazevoet (1996) or Rodwell (1996); SW and SE Guinea (common Macenta, lowlands up to 1000 m); Sierra Leone. Liberia, common throughout, up to 1180 m on Mt Balagizi, 1300 m on Mt Wuteve; Mali (once, Sikasso), Ivory Coast north to 09°N, common in Yapo Forest; Ghana (widespread and common), Togo (common north to *c.* 08°N), S Benin, Nigeria (common in S, also at Kagoro and on Jos Plateau). S Cameroon, abundant, north in lowlands to *c.* 05°N, also Korup Nat. Park (up to 1100 m, common), Mt Cameroon (to 700 m, uncommon), Mt Kupé (to 950 m); Mbini, Gabon (common), Congo (Nouabalé–Ndoki and Odzala Nat. Parks, Léfini Reserve, common; Kouilou Basin and Mayombe, abundant); Angola (Cabinda and in extreme N on Cuango and Luachimo rivers); Central African Republic near Bangui, and common in SE in Ouossi R. forests; Zaïre, almost throughout, south in E to Mt Kabobo and Upemba Nat. Park, common; extreme S Sudan in Imatong Mts and in SW along Zaïre border; W and S Uganda from Bwamba, Budongo, Teso and Elgon south to Bwindi (Impenetrable) and Malabigambo forests and into NW Tanzania at Minziro Forest and Bukoba; W Tanzania in Gombe and Mahali Nat. Parks (Archer 1994, Stanford and Msuya 1995). Density in Gabon about 1 bird per ha.

Description. ADULT ♂: upperparts dark green, yellowish green on rump and uppertail-coverts. Tail feathers dark green, edges tinged yellowish, more strongly so on undersides; T2–6 with increasingly large yellow tip, <1 mm wide on T2, up to *c.* 5 on T5, *c.* 7 on inner web and *c.* 10 on outer edge of T6. Lores, sides of

Nicator chloris

head and sides of neck dark green; a white spot above lores; eye-ring yellow. Chin pale grey; throat grey, yellow at sides; breast grey, greenish at sides, paling to greyish white on flanks and whitish on centre of belly; thighs greenish yellow; undertail-coverts bright yellow. Flight feathers dark brown, outer primaries with narrow bright yellow outer edges, inner primaries, secondaries and tertials with outer webs mainly green; tertials and innermost secondary with large yellow spot at tip of outer web, other secondaries with narrow yellow mark at tip. Alula dark brown with greenish outer web and pale yellow spot at tip. Large upperwing-coverts dark brown, primary coverts with greenish outer edge and yellow fringe at tip, greater, median and largest lesser coverts with large yellow spot at tip, remaining lesser coverts green, producing 3 rows of yellow spots across upperwing-coverts. Underwing-coverts and axillaries yellow; underside of flight feathers extensively yellow, due to broad bright yellow fringes on inner webs (confined to basal part of outer primaries). Bill black; eyes brown, grey-brown or chestnut-brown; legs blue-grey. ADULT ♀: like ♂ but sometimes lacks white spot above lores. SIZE (10 ♂♂, 10 ♀♀): wing, ♂ 99–111 (106), ♀ 87–95 (90·5); tail, ♂ 100–118 (110), ♀ 96–102 (98·5); bill, ♂ 22–25 (23·3), ♀ 20–24 (22·1); tarsus, ♂ 29–32 (30·6), ♀ 27–29 (28·3). WEIGHT: Liberia, ♂ (n = 6) 48–56 (52·3), ♀ (n = 6) 33–42 (38·1); Central African Republic, ♂ (n = 3) 50–58 (54), ♀ (n = 4) 40–43 (41·8); Zaïre, ♂ (n = 11) 52–67 (57), ♀ (n = 4) 47–55 (50); Uganda, ♂ (n = 26) 36–55 (49·1), ♀ (n = 17) 32–51 (43·3).

IMMATURE: slightly paler above than adult, lower breast and belly silky white rather than pale grey; tail feathers narrower and pointed; primaries yellow-tipped, outer ones more rounded; lacks feathering on face when newly fledged.

NESTLING: skin blue at hatching, which affords a remarkable camouflage: crouching, with eyes closed, it is not perceived by a potential predator as a bird (Brosset and Erard 1986). Partly feathered bird had naked skin on crown and sides of head greenish grey, that on throat grey; bill black, gape light yellow, eyes grey with black inner rim, rim of eyelids greenish yellow, feet light blue-grey (Chapin 1953).

Field Characters. Length 20–23 cm A common and characteristic bird of lowland forests, large but shy and retiring; usually located by its loud '*chok*' of alarm. Dark green upperparts and whitish underparts suggest a bulbul, but *yellow spotting* on wings distinguishes it from all but 2 bulbul species: the very rare Liberian Greenbul *Phyllastrephus leucolepis*, which is much smaller and has yellow underparts, and the Spotted Greenbul *Ixonotus guttatus*, for which it has been mistaken (Keith *et al.* 1992). Western Nicator is a sluggish undergrowth bird with a heavy shrike-like bill, and has dull white underparts and yellow undertail-coverts; Spotted Greenbul travels through forest in noisy chattering treetop parties; from below it appears almost pure white including underwing and underside of tail. Larger than Yellow-throated Nicator *N. vireo*, and has white throat and no yellow supraloral stripe.

Voice. Tape-recorded (32, 104, B, C, F, BROS, KEI, LEM, LOW). Song varied, a mixture of chuckles and clear whistles reminiscent of Nightingale *Luscinia megarhynchos* or even robin-chat *Cossypha*, but notes not quite no pure; not like a bulbul at all. Typically begins with churring or chuckling notes and ends with pure ones, 'tuk-tutuk-turrrr-*wick*', 'tuk-tichur-ti-*weety-weety-wok*', 'tok-tok-chick-chick-tok-tok *watawa tichew*', 'wik-toktok *werchu-weechu-weechu*', 'wee-tip-chup *chorra chorra chorra chorra*'; sometimes whistles with no introduction, 'weeoo *chopa*', 'wop-wop-cheep-cheep *woja*', or begins with Nightingale-like whistles '*hee-hoo-hee-hoo-hee-hoo*', followed by a hard, dry machine-gun 'tatatatata' and ending with loud '*tuck*'. Sometimes no pure whistles: 'chup-chup-chup-chup' followed by shrike-like, rolling, down-slurred snore, 'teeurrrrrrr'. Imitates other species, including Sjöstedt's Owlet *Glaucidium sjostedti*, African Pied Hornbill *Tockus fasciatus*, Yellow-throated Nicator, Yellow-bellied Wattle-eye *Diaphorophyia concreta* and Bates's Paradise-Flycatcher *Terpsiphone batesi*. Common call, given in mild alarm, a loud '*chuck*' or '*chok*'; also harsh 'keak', a longer and more subdued series of dry notes given at irregular intervals, 'tuk ... tutuk ... tuktuk ... tuk ... tutuk ... ', and a loud, insistent, complaining 'gwa-gwa-gwa-gwa ... ' at rate of 3 notes per s. Contact call between members of pair a series of clicks.

General Habits. Inhabits lowland primary and secondary forest, also transition forest up to 1400–1500 m, and logged forest and trees shading cacao plantations (Cameroon); penetrates into savanna in gallery forest and thick shrubbery. In Liberia, along waterways and logging roads and in all kinds of secondary forest, occupying dense parts of understorey and middle stratum, 5–25 m and tangled vines in treefall gaps (Gatter 1997). In Gabon frequents all levels in forest, from ground to tops of emergent trees, but generally between 8 and 25 m above ground; also in old plantations. In Nigeria (Ipake) occurs in thick, spindly sapling undergrowth in swamp forest (C.H. Fry, pers. comm.); in Bwamba Forest (Uganda) found especially along streams; in Congo (Odzala Nat. Park), absent from open areas under closed canopy (Dowsett-Lemaire 1997). Lowlands; up to 1800 m in Uganda.

Lives in pairs, but often seen singly since pair members often some distance apart. Joins mixed species flocks. Forages by creeping among vine-clad tree-trunks, in tangled creepers over smaller trees, and among lianas hanging down from giant trees; examines bunches of dead leaves, even those hanging down over water. Also seen descending to ground by small stream to capture orthopterans. Holds tail vertical when searching the underbrush. Flicks wings and wags tail. Inquisitive and vigilant; alarm call given on approach of human is a familiar forest sound. Of 11 birds ringed only 1 recaptured, after 4 months.

Moult does not interrupt breeding; adults in full moult taken on the nest (Gabon: Brosset and Erard 1986).

Sedentary, at least in NE Gabon.

Food. Insects, including caterpillars, grasshoppers, mantids and beetles; small frogs; a few berries and seeds.

Breeding Habits. Monogamous, territorial. Territory size 3–4 ha (Congo, Odzala Nat. Park: Dowsett and Dowsett-Lemaire 1997). Courting ♂ once squatted on branch, spread wings and tail, hopped about a few times and squatted again; did not call (Rand *et al.* 1959). In aggressive posture, sings and holds wings out from body and spreads tail. Chases adversary through leaves.

NEST: tiny, av. diam. 60 × 60; rudimentary, built of a few sticks which support tendrils and twigs; the latter rotting, with mycelium growing on them, soon gluing them to supporting sticks or leaves. This 'nid auto-collant' (Brosset 1974) or self-adhesive nest is so flimsy that it seems poised to fall at the slightest touch or movement of air, and in fact many are destroyed, eggs and all, before the 'glue' takes hold; nevertheless, once the nest is firmly attached it can be held upside down and will not fall off the branch (Brosset 1974). Placed 0·4–4·5 (2) m above ground among leaves in undergrowth, sometimes in a fork or on a bunch of dead leaves hanging from a branch.

EGGS: always 1 (n = 23 clutches, Gabon); reports of 2 must be due to misidentification, since the nest is only large enough to hold a single egg (Brosset and Erard 1986). Elongated; thin-shelled; ground colour variable, white to beige, pale yellowish clay or dark olive-green, spotted with dark brown, red-brown, grey and lilac, sometimes lightly, sometimes so heavily that entire surface is covered, frequently forming a cap or ring around large end. SIZE (n = 1) 31·2 × 18·0.

LAYING DATES: Sierra Leone, Apr; Liberia, Feb, Nov (immatures Feb, Apr, July–Aug); Ivory Coast, Dec; Nigeria (♀ with egg in oviduct June, breeding condition Mar, May, feeding young Oct); Cameroon, Apr–June, Aug; Gabon, all year, especially during rains and the short dry season; Angola (breeding condition Feb–Mar); Central African Republic (breeding condition June); Zaïre, Ituri and Uele, breeding coincides with rains, May–Nov; Lukolela (♂♂ in breeding condition Sept); Upemba Nat. Park (juv. Sept, breeding condition Feb, Apr–June, Nov–Dec); Itombwe, May (breeding condition Dec–Jan, June, Sept); Sudan July–Aug; E Africa: Uganda, June (breeding condition Apr), Region B Apr, June–July.

INCUBATION: by ♀ only; period (n = 3) 17–18 days. When disturbed she leaves nest and 'rodent-runs' on ground.

DEVELOPMENT AND CARE OF YOUNG: at 8–9 days chick large enough to cover the nest completely; at 10–12 days feather tips appear. Fed by both parents, on average every 30 min during 15 h of observation; prey brought to it is relatively large. During the first week one parent broods so tightly that it can often be touched by an observer; later, when disturbed, it drags itself along the ground pretending to be injured or, while perched, raises wings and spreads tail jerkily, giving shrill cries. At 16–18 days nest is abandoned completely, well before the chick can fly; it moves about on branches with agility, watched over by one parent; in case of danger parent gives alarm, and young remains motionless among the leaves, well camouflaged by its spotted appearance.

BREEDING SUCCESS/SURVIVAL: 30 nests produced 17 flying young (Gabon). 4 nests blown down by the wind just after the eggs were laid, before the mycelial growth had anchored nest.

Reference
Brosset, A. and Erard, C. (1986).

Nicator gularis Hartlaub and Finsch. Eastern Nicator. Bulbul à tête brune.

Plate 23 (Opp. p. 350)

Nicator gularis Hartlaub and Finsch, 1870. Vog. Ost-Afr., p. 360; Shupanga, Zambesi.

Forms a superspecies with *N. chloris*.

Range and Status. Endemic resident. S Somalia (lower Jubba R. and extreme S coast), Kenya, on coast (common), also inland along Tana R. to Bura, and in SE to Mt Kasigau, Sagala, Ngulia Hills, Taveta, Kibwezi and Endau; formerly on Mt Uraguess and Karissia Hills, current status uncertain (Zimmerman *et al.* 1996). Tanzania, on coast, inland in N to E Usambaras, N Pares, Mkomazi Game Res., Mt Kilimanjaro and S Moshi district, including Kahe II Forest Res. (Cordeiro *et al.* 1995); also to Kazinzumbwe Forest Res. (Mlingwa *et al.* 1993), Ulugurus and Rondo Forest Res. (Holsten *et al.* 1991), with scattered records west to Kondoa, Itigi and Rukwa; Zanzibar. Extreme SE Zaïre by L. Tanganyika near Zambian border. Locally common in N Zambia between lakes Mweru and Tanganyika, in Luangwa and Zambezi valleys and Lower Kafue Basin (up to Kafue Bridge); also west of Victoria Falls at Katombora and north to Mulanga-Buwe Pool area, W Kalomo District (and in *Baikiaea* habitats to west and north?: Dowsett-Lemaire and Dowsett 1968). Malaŵi in N from Livingstonia to Nkhata Bay, and in S throughout Shire Valley, north to L. Malaŵi, including mountains to 1350–1500 m; common. Mozambique in Zambezi valley and common from there southwards; widespread in Sul do

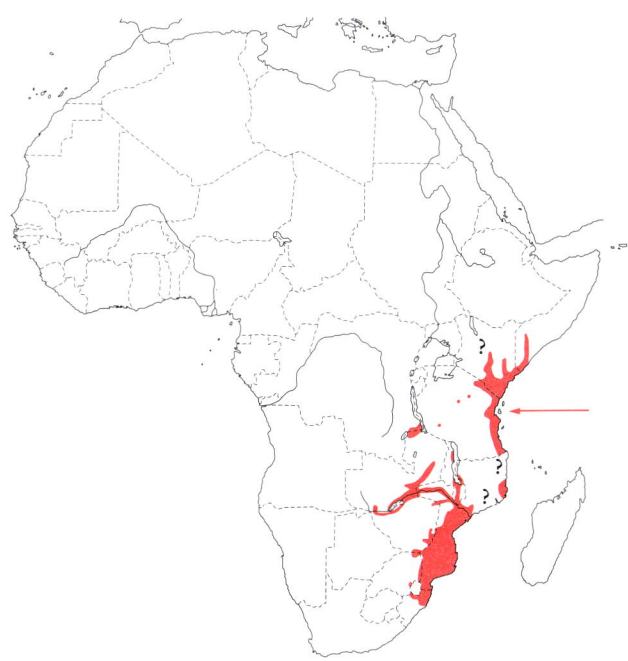

Nicator gularis

Save and not uncommon in Lebombo Mts and littoral, less common in drier interior (Clancey 1971); also known from N-central coast, and probably present along whole coastline from Tanzania to Natal. Zimbabwe in Zambezi valley, and once inland from L. Kariba at Nyamasaka (17°25'S, 28°51'E: Tree 1987); in NE along Mazoe R., and in SE along Sabi, Lundi and Limpopo rivers, Buzi R. in Chipinga Uplands and in Lusitu–Haroni area; uncommon. In Transvaal restricted to lowveld, mainly along larger rivers (Pongola, Komati, Nwanedzi, Olifants, Levuvhu, Limpopo); fairly common in N Kruger Nat. Park, scarce or rare elsewhere. E Swaziland (Lebombo Mts), NE Natal south to Umfolozi R; vagrant Durban. Population in S Mozambique probably >10,000 birds (Parker 1999).

Description. ADULT ♂: forehead and crown olive-grey; rest of upperparts dull green. Tail feathers greenish brown, edges tinged yellowish, T2–T6 with increasingly large pale yellow tips (<1 mm on T2 to *c*. 7 mm on inner web of T6), yellow on T6 extending narrowly along most of outer edge. Lores, cheeks and ear-coverts pale buffish brown, with dusky band along lower edge of ear-coverts and short dusky malar stripe; a yellow spot above lores; eye-ring yellow. Sides of neck green. Chin and throat buffish white; upper breast greyish brown, tinged green at sides, grading to greyish white on flanks and whitish on centre of belly; thighs pale greenish yellow; undertail-coverts pale yellow. Primaries and secondaries dark brown, outer primaries with narrow whitish outer edges, others with green fringe to outer web; tertials and innermost secondaries with large pale yellow tip to outer web, other secondary tips with narrow yellow fringe. Alula dark brown with narrow yellow fringe at tip; large upperwing-coverts dark brown, primary covert tips with narrow yellow fringe, greater and median covert tips with large yellowish white spot; lesser coverts green. Outer underwing-coverts bright yellow; inner underwing-coverts and axillaries pale yellow; large patch on underside of flight feathers formed by pale yellow borders of inner webs (confined to basal part of outer primaries). Bill horn or greyish brown to pale grey, cutting edges paler; eyes light brown or yellowish brown; legs slate grey to pale blue grey. ADULT ♀: like adult ♂, but often lacks yellow supraloral spot. SIZE (10 ♂♂, 10 ♀♀): wing, ♂ 105–114 (108·5), ♀ 83–99 (92·9); tail, ♂ 106–113 (109), ♀ 85–103 (95·6); bill, ♂ 24–28 (25·7), ♀ 20–23 (21·4); tarsus, ♂ 32–34 (32·6), ♀ 27–29 (27·9). WEIGHT: Kenya, ♂ (n = 18) 40–58 (50·4), ♀ (n = 7) 29–38 (33·6); Kenya and Tanzania, unsexed (n = 49) 32–57 (43·4); southern Africa, ♂ (n = 10) 50–63 (55·7), ♀ (n = 11) 33·6–41·3 (37·3).

IMMATURE: like adult but with yellow-tipped primaries and pointed outer tail feathers. Recently fledged bird has bare skin on face and neck; older fledged bird well able to fly still had bare greenish chrome skin on forehead to behind eye (so feathering evidently proceeds from the back forwards: Moreau and Moreau 1939).

NESTLING: bird with 1-cm quills on tail and wing feathers just breaking from their sheaths, has wing-coverts already well grown and with 2 rows of pale yellow spots like adult's; body plumage greenish grey rather than olive-green above, breast grey, becoming white on belly; entire head and throat greenish grey, pinker on throat, bare except for ridge of small feathers on supercilium; bill pale grey, just beginning to develop hook, gape pale yellow; tongue and inside of mouth orange-yellow, eyes dark brown with very narrow pale yellow eye-ring (Moreau and Moreau 1937).

Field Characters. Length 20–23 cm. A shy bird of dense thickets, with bulbul-like plumage (olive above, white below) but easily distinguished from all bulbuls in its range by *pale yellow spots* on wing and *loud, whistled*, un-bulbul-like *song*. Flicks wings and tail. Yellow tips to tail feathers conspicuous as bird flies away (Sinclair *et al*. 1993). Heavy bill, sluggish manner and yellow wing spots recall Grey-headed Bush-Shrike *Malaconotus blanchoti*, but latter has bright yellow underparts, long melancholy whistle. ♀ Black Cuckoo-Shrike *Campephaga flava* has greenish upperparts and much yellow on wing, but has small bill and is heavily barred below.

Voice. Tape-recorded (10, 32, 58, 75, 86, 88, 91–99, B, F, GIB, GIL, KEI). Song starts with low introductory notes, almost in an undertone, 'tik-chop-weeoo-tok', followed by a more emphatic 'trrr', then loud whistles, '*hip-to-wee-to-chip-to-weet*', or '*chip-chop-chipweooo-trrr ... chip-o-chi-weet*', '*ee-chopti-trrrr ... oputo-weeto-wheet*', '*weoo-chok ... chip-chee-turrrr*', '*chip-chop ... choy-chawidja-wachoy*'. Song lasts *c*. 3 s; interval between songs may be filled with low conversational notes such as those which precede main song, 'wee-chop-tee-to-churrr'; 14 songs and 5 intervening subsongs given in 1 min 35 s (Kenya, Sokoke Forest: pers. obs.). Rich and powerful; has been likened to Nightingale *Luscinia megarhynchos*. Somewhat ventriloqual. Mimics other species, including Moustached Green Tinkerbird *Pogoniulus leucomystax* and Eastern Black-headed Oriole *Oriolus larvatus*. Alarm a hollow, wooden 'zokk'; also gives a complaining nasal 'kare-kare-kare ... '.

General Habits. Inhabits evergreen forest and second growth, especially coastal forest and scrub; in Sokoke Forest, Kenya, common in all vegetation types, including *Afzelia*, *Brachystegia* and *Cynometra* (Britton and Zimmerman 1979); riparian forest, dense thickets, forest edge; thickets in miombo (Mozambique); dense thickets of *Bauhinia* in and on edge of *Baikiaea* forest (Zambia); moist thornveld (Natal); also suburban gardens (Tanzania, Dar-es-Salaam: Harvey and Howell 1987). All levels, from ground to canopy. Lowlands, occasionally as high as 1900 m (Kenya). Solitary or in pairs. Skulks in thick bush; easily overlooked when not calling. Rather silent in cool weather; sings increasingly with approach of rainy season, when very noisy (Mozambique: da Rosa Pinto and Lamm 1953–1960). ♂ sings from hidden perch in centre of thick bush or top of tall tree. When foraging, hops from branch to branch in search of insects, giving repeated small, nervous wing-flicks with wings half open and drooped; also forages in low vegetation and on ground among leaf litter. Regular member of mixed flocks, with e.g. Woodwards' Batis *Batis fratrum*, Square-tailed Drongo *Dicrurus ludwigii* and Dark-backed Weaver *Ploceus bicolor*, but apparently only ♀♀ join these parties, not ♂♂ (Clancey 1964). Flicks wings once or twice on perching, like cuckoo-shrike, and also flicks wings when alarmed.

Food. Insects, including caterpillars.

Breeding Habits. Solitary nester; evidently monogamous; territorial.

NEST: flat, rudimentary, dove-like; a triangle of stout twigs supporting smaller twigs and tendrils, 'a lacy affair that one would have judged incapable of supporting a bird that weighs a couple of ounces', and 'lodged precariously in a spindly bush of the forest undergrowth' (Moreau and Moreau 1937); sometimes rootlets, bamboo leaves, moss or stems and leaves of *Lantana* are incorporated; usually well concealed, but one was in plain view 1·5 m from path, in mass of thorns above a large rock (Long 1959); placed low down, *c*. 1 m above ground.

EGGS: 1–4 (2–3); slightly glossy; ground colour variable, yellowish green to white faintly tinged blue, buffy or dull pinkish cream, profusely spotted, freckled or blotched with dull purple, light brown and bright brown over grey, French grey and lilac, especially in cap or belt around large end. Said to be 'not unlike some lark eggs' (C. R.S. Pitman *in* Long 1959). SIZE: (n = 10) 24·7–27 × 16·6–19·5 (25·5 × 17·5).

LAYING DATES: E Africa: Tanzania, Dec, Region D Nov–Dec, Region E June, all in rains; Zambia (breeding condition Oct, Nov); Malaŵi Dec–Jan (gonads enlarged Apr); Mozambique, Nov–Dec (breeding condition Mar); Zimbabwe, Jan; Natal, Nov–Jan.

INCUBATION: by ♀; one sat very tightly and did not attempt to distract attention from nest as it did after young had hatched; ♂ sat 100 m away, singing at intervals, and ♀ did not call to him (Moreau and Moreau 1937). However, when observers handled hatched young, one of the adults performed injury-feigning display, dragging itself through forest floor litter, with tail spread and wings fluttering, giving low whimpering note; it kept this up for at least 15 min, approaching within 1 m of observers. That habit is so well known to local people that bird is called 'lubozi', 'loony' (Moreau and Moreau 1939).

DEVELOPMENT AND CARE OF YOUNG: a captured young became tame almost immediately, taking insects and meat from fingers within 3 h of capture. It slept with mantle feathers spread sideways, concealing wing spots (Moreau and Moreau 1939).

References
Maclean, G.L. (1993).
Moreau, R.E. and Moreau, W.M. (1937, 1939).

Nicator vireo Cabanis. Yellow-throated Nicator. Bulbul à gorge jaune.

Plate 23
(Opp. p. 350)

Nicator vireo Cabanis, 1876. J. Orn. 24, p. 333, pl. 2, fig. 2; Chinchoxo, Portuguese Congo.

Range and Status. Endemic resident. S Cameroon, north and west to Kumba, and Korup Nat. Park (Green and Rodewald 1996); Mbini; Gabon (widespread); common in N Congo in Nouabalé-Ndoki and Odzala Nat. Parks, local in S in Mayombe and Kouilou Basin; Angola in Cabinda, Cuanza Norte through Calulo to escarpment near Gabela in Cuanza Sul, and on Cuango and Kasai rivers on Zaïre border; Zaïre in SW and locally common in large area in NE, up to Semliki R.; rare at Kamituga (Kivu). W Uganda (Bwamba Forest). Density in Gabon, 1 pair per 20 ha within primary forest, 2–3 pairs per 10 ha at forest edge.

Description. ADULT ♂: upperparts dark green, forehead and crown tinged greyish. Tail feathers lighter yellowish green, T3–T6 with progressively increasing yellow tip (<1 mm wide on T3 to >5 mm wide on T6). Lores dark greenish grey; short yellow supraloral stripe to above eye, bordered above by narrow blackish line; cheeks and ear-coverts olive-grey; sides of neck green. Chin grey; throat yellow; breast and flanks grey, tinged greenish at sides, paling to whitish on belly; thighs green; undertail-coverts greenish yellow. Flight feathers blackish brown; outer primaries narrowly edged brownish white on outer web, inner primaries and secondaries fringed green, tertials and innermost secondaries with large yellow spot at tip. Alula and larger upperwing-coverts blackish brown, tips of alula and primary coverts with narrow yellow fringe, tips of greater and median coverts with large pale yellow spot; lesser coverts dark green. Underwing-coverts and axillaries yellow; underside of flight feathers with extensive broad yellow borders to inner webs. Bill blackish or dark horn above, pale grey or blue-grey below; eyes brown or grey-brown; legs grey or blue-grey. Sexes alike. SIZE (10 ♂♂, 7 ♀♀): wing, ♂ 78–82 (80·4), ♀ 71–74 (72·1); tail, ♂ 77–86 (81·7), ♀ 72–80 (75·4); bill, ♂ 19–21 (20·0), ♀ 17–18·5 (17·6); tarsus, ♂ 23–25 (24·1), ♀ 20·5–24 (22·1). WEIGHT: (Uganda) 2 ♂♂ 24, 26.

IMMATURE: like adult but forehead green, less yellow on throat.
NESTLING: unknown.

Nicator vireo

Field Characters. Length 14 cm. Similar to Western Nicator *N. chloris*, with which it is sympatric almost throughout its range, but *sparrow-sized*, with *yellow throat* and very different song. Yellow wing spots very obvious; bright yellow supraloral stripe also distinctive, but these feathers are erectile and may be concealed unless bird is excited (Brosset and Erard 1976). Secretive, quite unlike

Spotted Greenbul *Ixonotus guttatus*, which travels in noisy parties and is pure white below.

Voice. Tape-recorded (104, B, C, F, CHA, CART, HUG, LEM, STJ). Song of 8–11 notes, loud down-slurred and up-slurred whistles with soft notes in between, 'po-*tyoo*-ho-ho-ho-*wheee*, *tyoo*-ho-ho-ho' or a shorter '*tyop*-*tyop*-*whee*-ho-ho-ho-*wheeep*', with little variation. Sings for long periods from same spot with only short pauses between songs; when surprised by human, interrupts song with a loud 'krrio-krrio-tjerritjerriokrriio', and becomes silent (Heinrich 1958).

General Habits. Inhabits lowland primary forest, in dense understorey and around disturbed areas formed by trees blown down in storms; commonest in clearings in contact zone between primary and secondary forest; frequents old second growth and stages of regenerating forest on formerly cultivated ground, as soon as there are large enough masses of dense vegetation; likes low damp spots and riverside vegetation, and sometimes visits forest remnants near villages. Very fond of vine tangles at forest edge, around clearings and by roadsides; avoids pure Marantaceae (Congo, Odzala Nat. Park: Dowsett and Dowsett-Lemaire 1997). In Angola, forest remnants in coffee plantations (Heinrich 1958).

Lives in pairs, which are territorial and probably very sedentary (Gabon: Brosset and Erard 1986). One seen accompanying a band of Swamp Palm Bulbuls *Thescelocichla leucopleura* (Germain *et al.* 1973). Usually 2–8 m above ground. Sings from dense cover; exceedingly shy; answers person's imitation of song but will not come into view; when approached, moves away and starts singing again from another perch some distance away (Heinrich 1958).

Food. Insects, especially large orthopterans, also caterpillars, beetles and mantids.

Breeding Habits. Territorial; same breeding areas occupied for many consecutive years (ringing study: Brosset and Erard 1986).

NEST: only one described (Brosset and Erard 1976): built upon base of dead twigs, which form a rectangle, a tangled mass of thin, dry creeper tendrils with shallow cup in centre; ext. diam. 150, int. diam. 70; placed 0·75 m above ground on small branches and skeletonized leaves of bush in thick undergrowth at forest edge, 12 m from a cut-over area; completely exposed, not concealed at all. Larger and more elaborate than that of Western Nicator.

EGGS: 2; greyish white, finely spotted with yellow, red and grey, forming dark cap on large end.

LAYING DATES: Cameroon, Oct; Gabon, Dec–Mar, probably also at other seasons since it sings all year; Angola (fledged young Mar, juv. July, breeding condition Feb, July).

INCUBATION: by ♀, while ♂ sings nearby. Retains its suspicious nature near nest: ♂ discovered well hidden and motionless observer and kept on giving alarm call behind observer's back for a full hour (Brosset and Erard 1976). When ♀ finally returned to nest she did so from below.

DEVELOPMENT AND CARE OF YOUNG: young out of nest fed by both parents.

BREEDING SUCCESS/SURVIVAL: the only known nest was destroyed by a predator.

References
Brosset, A. and Erard, C. (1976, 1986).

Family PRIONOPIDAE: helmet-shrikes

An endemic family of small to medium-sized, intensely sociable birds with stiff bristle-like feathers on forehead, some extending forward and covering oval nostrils, and in most species a flesh-coloured, pectinate wattle around eye. Wings quite long but rounded; 10 primaries, P3–6 longest, P1 about half length of P3. Tail with 12 feathers; shorter than wing; conspicuously marked white in most species. Bill medium, fairly stout, shrike-like, hooked and notched at tip. Legs and feet fairly short but strong, claws sharp, tarsus covered by small scales on sides and front. Plumage conspicuously marked with blacks, browns, greys and whites. Sexes alike or nearly so.

In year-round intensely sociable groups of 6–30 birds, in open woodland. Feed and roost communally; breed co-operatively (or occasionally without helpers: Benson 1946, C.J. Vernon *in* Grimes 1976). Social organization based on dominance hierarchy with breeding ♀ dominant. Noisy, producing whistles, flute-like notes, duetting and communal chattering; much bill snapping. When foraging, tilts head to one side as if using auditory cues to locate prey. Nest small, neat and compact.

Systematic position controversial. Rand (1960), Hall and Moreau (1970), Raikow *et al.* (1980) and Harris and Arnott (1988) consider helmet-shrikes a subfamily of true shrikes Laniidae. Sibley and Ahlquist (1990) find helmet-shrikes only distantly related to Laniidae and include them with vangas, *Batis* and *Platysteira*, in the family Corvidae, subfamily Malaconotinae, tribe Vangini, closely allied to bush-shrikes of the tribe Malaconotini. We follow Campbell and Lack

(1985) and Short *et al.* (1990) in placing helmet-shrikes in their own family because of their distinctive tarsal scutellation, construction of compact nests and intensely gregarious, communal behaviour. For additional remarks on systematic position of helmet-shrikes, see family Laniidae diagnosis.

1 genus, 8 species, endemic to subsaharan Africa. 3 superspecies: *P. plumatus/P. poliolophus, P. caniceps/P. rufiventris, P. retzii/P. gabela.*

Genus *Prionops* Vieillot

Characters those of the family.

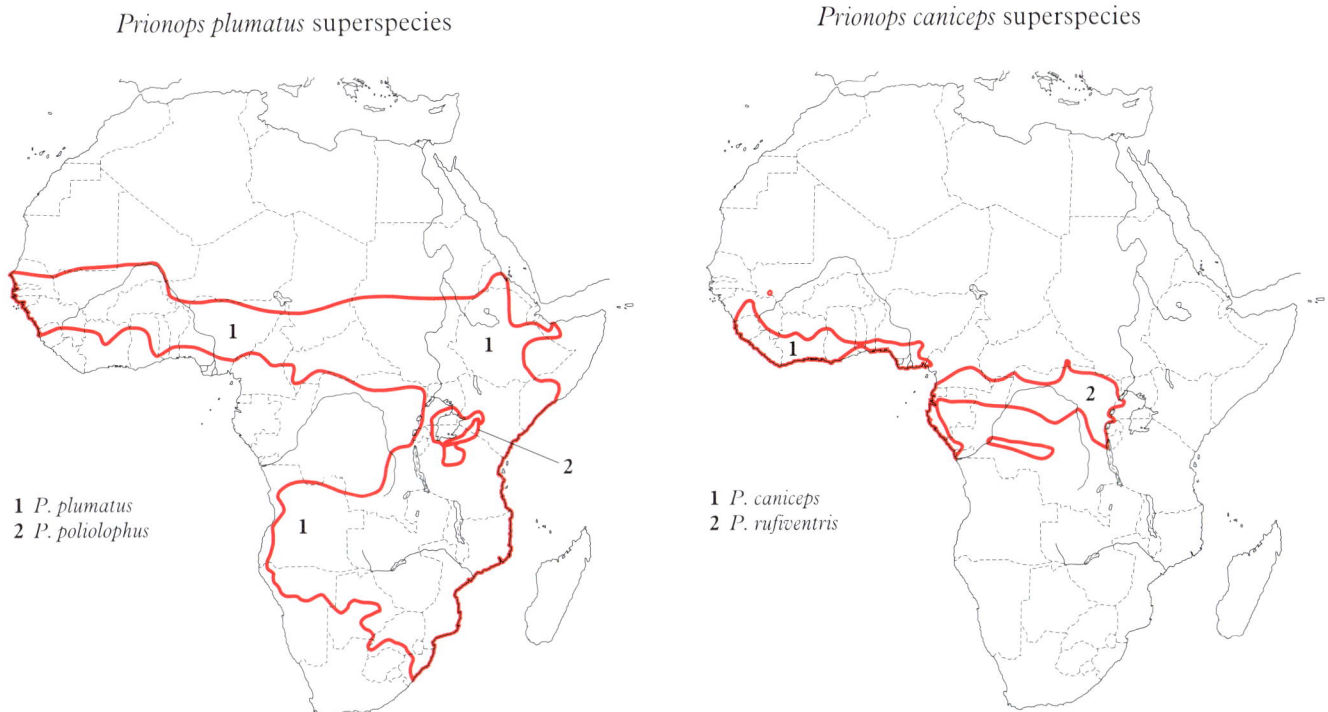

Prionops plumatus (Shaw). White Helmet-Shrike. Bagadais casqué.

Lanius plumatus (Shaw), 1809. Gen. Zool., 7, p. 292; Senegal.

Plate 28
(Opp. p. 415)

Forms a superspecies with *P. poliolophus*.

Range and Status. Endemic resident, in open woodland and thornbush from Senegal and Sierra Leone east to Somalia, south to Namibia and Natal; absent from rain forests of W and equatorial Africa. Frequent to common throughout its range except where noted below.

S Mauritania (east of Kaédi, Guidimaka), uncommon and local; Senegal and Mali south of *c.* 15°N; Gambia, frequent; Guinea Bissau; Guinea except SE; Burkina Faso, uncommon to frequent; N Sierra Leone; absent Liberia; Ivory Coast south to Lamto; Ghana south to Hohoe, Ho and Anum in Volta Region and inselbergs of Accra Plains, uncommon along coast west to Legon (Accra); Togo south to about Notsé and Kpalimé; Benin south to *c.* 07°N. SW Niger from Parc du W to Tessaona and Maradi; Nigeria from Sokoto and Kano south to Abeokuta, Ibadan, Benin, Enugu and Obudu, common in guinean zone; N and central Cameroon from *c.* 11°N south to Adamawa Plateau; uncommon S Chad, mainly south of 10°N, but north to Biltine in E; uncommon Central African Republic from Lobaye and Haute Sangha Préfectures in SW to Vakaga

Prionops plumatus

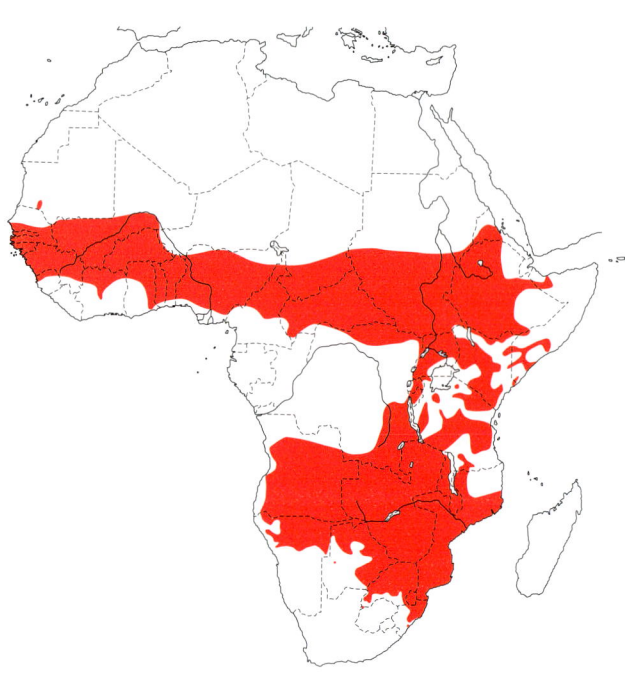

Préf. in NE. Sudan north to c. 14°N; Eritrea in highlands below 1200 m, uncommon in W plains, absent from coastal lowlands; Ethiopia except NE and SE lowlands; Somalia, uncommon to frequent, in NW east to c. 46°E and south of 05°N; Uganda except in L. Victoria basin (once Kampala); widespread Kenya but rare in NW north of Turkwell R. drainage and west of Huri Hills, and absent from L. Victoria basin and most of SW highlands.

Rwanda except W; Burundi. Tanzania, uncommon in Gombe Game Res., rare in L. Victoria basin and Serengeti Nat. Park, and absent in SE south and east of Utete and Nandembo (map data: N.E. Baker, pers. comm.). Zaïre, in NE in Uele from about Buta to L. Albert, and in central E and SE from N end L. Tanganyika and S Kivu to Katanga (including Marungu); throughout Angola from Luanda, Cuanza Norte and Malange south to Huila, Cunene and Cuando Cubango, west to escarpment in Moçamedes, but absent from Benguela coastal plain; Zambia; Malaŵi; Zimbabwe (uncommon E highlands); Mozambique, absent only in NE in Cabo Delgado and Nampula north of c. 15°S. N and NE Namibia, including Owambo, Kavango, Bushmanland and Caprivi Strip; N and E Botswana west into central Kalahari woodlands, also Takatshwane in W; Swaziland, common lowveld, uncommon elsewhere; South Africa: Transvaal, bushveld and lowveld, erratic in highlands and escarpment; Natal, common south to c. 29°S, uncommon south to Durban. Peak densities of 1 group per 10 ha estimated in South Africa (Vernon 1977). In Mozambique, 23 birds per 100 ha in acacia woodland and 45 per 100 ha in miombo (Parker 1999).

Description. *P. p. plumatus* (Shaw): Senegal to N Cameroon. ADULT ♂: lores, forehead and forecrown whitish, feathers rather stiff and straight, c. 10 mm long, forming a median frontal crest directed upward from crown and forward over basal half of maxilla. Sides of crown and central crown feathers greyish white, the latter up to 45 mm long, rather straight and soft, forming an erect median crest covering grey hindcrown and nape (**A**). Side of nape dark grey, forming bar behind whitish cheeks and ear-coverts; collar around hindneck whitish; rest of upperparts black with oily green gloss. Central tail feathers black; T2–T5 black with progressively broader white tips; distal half of T5 outer web also white; T6 wholly white. Entire underparts white, tinged greyish from chin to breast. Primaries black, middle part of inner webs white forming broken bar across spread upperwing; secondaries black, outer 5 feathers with white fringe at tip, innermost with broad white outer border and tip; tertials black, outer feather with broad white outer fringe and tip like innermost secondary, middle feather with narrow white outer fringe. Inner greater and inner median coverts white, forming, with white secondary and tertial borders, a broad longitudinal stripe on closed wing; alula and rest of upperwing-coverts black. Underwing-coverts and axillaries black; broad white bar across underside of primaries. Bill greenish black; eye-wattle yellow to orange-yellow, broad outer edge digitate; eyes greenish grey, iris with bright yellow outer ring; legs and feet orange-red. Sexes alike. SIZE (10 ♂♂, 10 ♀♀): wing, ♂ 115–124 (120), ♀ 119–129 (122); tail, ♂ 94–107 (99·4), ♀ 97–108 (102); bill, ♂ 22–25 (23·1), ♀ 21–26 (22·7); tarsus, ♂ 23–25 (23·8), ♀ 23–25 (23·7). WEIGHT: (Ghana, unsexed, n = 18) 40–49 (43·0).

IMMATURE: similar to adult, but crest on hindcrown short; top of head, cheeks, ear-coverts and hindneck collar tinged greyish buff; bar on side of nape rufous-brown; upperparts dark greyish brown with narrow buff feather fringes; wings brownish black, white areas tinged buff, alula and outer greater coverts tipped buffish white, tips of tertials finely fringed buffish white. Bill greenish brown; no eye-wattles; eyes dark brown to greenish yellow; legs ochre-yellow.

NESTLING: naked, blind, with mouth spots. Feathered nestling at 10–15 days has distinctive warm buff crown, white collar and brown upperparts.

P. p. concinnatus Sundevall (includes '*adamauae*'): central Cameroon to central and SW Sudan, NW Ethiopia, NE Zaïre and N Uganda. Like *plumatus*, also with 2 crests, but that above hindcrown shorter and bilateral, comprising feathers up to 25 mm long which curl backwards and inwards; wing-stripe narrower.

P. p. cristatus Rüppell: Eritrea, N, W and central Ethiopia, SE Sudan, E Uganda and NW Kenya. Curly-crested like *concinnatus*, but lacks white stripe on closed wing; secondaries, tertials and upperwing-coverts entirely black. WEIGHT: (Ethiopia) ♂ (n = 2) 38, 45, 1 ♀ 44; (Kenya) ♂ (n = 3) 40–46 (44), unsexed (n = 32) 41–45 (43·7).

P. p. vinaceigularis Richmond (includes '*melanopterus*'): Somalia, E and S Ethiopia, N and E Kenya, NE Tanzania. Lacks wing-stripe like *cristatus*. Short, stiff white crest on forehead and

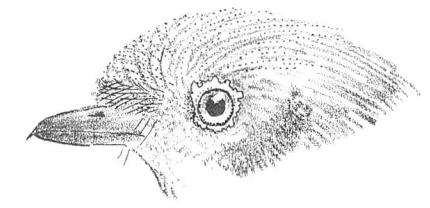

forecrown as in previous races, but white feathers of central crown shorter (c. 15 mm), stiffer, and directed backwards, not forming distinct second crest.

P. p. poliocephalus (Stanley) (includes '*angolus*' and '*talacomus*'): central and S Kenya and S Uganda, west to SE Zaïre and Angola, south to Namibia, Botswana and South Africa. White wing-stripe broad as in *plumatus*. Crest on forehead and forecrown short and stiff, but grey; central crown feathers grey, but stiff and rather short as in *vinaceigularis*, not forming second crest (**B**). Cheeks greyer than in *plumatus*, bar on side of nape blacker; more white in tail (distal half of T5 inner web and whole outer web white). Much smaller; wing (♂, n = 10) 103–111 (105). WEIGHT: (Angola) ♂ (n = 4) 35–37 (35·8), ♀ (n = 2) 35, 36; (Zaïre) 1 ♀ 41; (Zambia) unsexed (n = 21) 28–37 (32·4); (Zimbabwe) ♂ (n = 16) 27–36 (31·9), ♀ (n = 17) 30–41 (34·4); (southern Africa) ♂ (n = 44) 27–36 (32·7), ♀ (n = 41) 30–40 (34·9), unsexed (n = 25) 25–37 (29·5).

Field Characters. Length 17–25 cm. The common black-and-white helmet-shrike of open country in most of Africa. Varies racially in size, length and colour of crest and amount of white in wing, but features common to all races are pale yellow eye with broad *eye-ring* of bare *yellow* skin, and in flight, *white collar* separating grey head from black back, *white band* across base of primaries and *white* sides and tip of tail. Further, except in races *cristatus* and *vinaceigularis* of NE Africa, shows broad *white stripe* on closed wing, conspicuous also in flight. Immature lacks eye-wattle and has brown eye. Travels in small flocks; flight undulating. White-crowned Shrikes *Eurocephalus* have plain brown wings, white rump, dark tail with no white. Parapatric with Grey-crested Helmet-Shrike *P. poliolophus* in Kenya and Tanzania; for differences see that species.

Voice. Tape-recorded (55, 75, 86, 88, 91–99, 104, B, F, BRU, LEM). Variety of calls, usually given in group chorus, include loud, down-slurred double note, 'chrreeo', 'jeeup' or 'chirrow' or deeper 'chaw-dawah', whistled but with harsh overtone, and dry, buzzy 'dzurrri', 'jrrrzi' or 'dzwee', like watch being wound, and a similar but more nasal 'goy' or 'gurry'; these are interspersed with a jumble of chips, chatters, dry scolds and bill-snapping; also a soft 'dowayo', almost in an undertone. Calls differ racially, see Zimmerman *et al.* (1996). Chorussing associated with territorial display usually initiated by dominant ♀ with 'chi-chi-chi-cherow-cherow'; chorus that follows has 4 main components, 'zrrreeeu', 'tzzzr', 'kloop' and bill clicks; pair bond maintained by duetting, ♂ calling 'yuki-yuki' and ♀ 'kidoki-doki' (Harris and Arnott 1988). Chattering calls given when birds have found food or nest material. Short-range contact maintained by flight calls, 'whit-whit'; individuals that have become separated from flock utter loud 'treeu'; aggression signalled by low-pitched 'grrr' and rapid bursts of bill snapping; alarm, high-pitched 'tzzee-tzzee' (Harris and Arnott 1988).

General Habits. In W and E Africa occurs in savanna woodland, also coastal forests, open bush country, oil palm plantations (Guinea Bissau), gardens, *Terminalia* woodland (Central African Republic), *Combretum* woodland (Eritrea), riparian forests (but not in Tsavo, Kenya, where replaced by Retz's Helmet-Shrike *P. retzii*: Lack 1985), forest edges and miombo woodland (Tanzania). In southern Africa prefers deciduous broad-leafed woodland but also (especially in non-breeding season) other woodland such as acacia savanna and eucalypt plantations; also bushveld and urban areas. Mainly below 1500 m, rarely above 1800 m; in SW Kenya and NW Tanzania rare above 1200 m, where replaced by Grey-crested Helmet-Shrike.

Gregarious, in groups of 2–22 (av. 7, n = 1096: C. J. Vernon *in* Harris and Arnott 1988). Groups smallest in breeding season, reach maximum size in non-breeding season. Group structure based on dominance hierarchy. Rank is: breeding ♀ dominant, then breeding ♂; next are remaining adult ♀♀ (usually sisters of the dominant ♀), then adult ♂♂ (usually brothers of the dominant ♂); then adult offspring, then 1st year offspring, and finally fledglings. Dominant ♀ and ♂ are usually the heaviest birds, with longest eye wattles. Forms new group when up to 4 sisters from one group join up to 4 brothers of another group; the dominant sister pairs with the dominant brother. If dominant ♀ or ♂ dies, dominant status passes to next ranking sister and brother. When 5 months old, juveniles leave group and form wandering unisex groups; later they form new breeding groups. Breeding group remains together for up to 10 years (C. J. Vernon *in* Harris and Arnott 1988).

Group noisy; moves as a tight unit, with activity of individual members coordinated by calling. Flight buoyant, undulating and planing, with members of group following each other from tree to tree. Forages in all layers, but usually in lower and mid-strata, among twigs and leaves and on trunks; occasionally hangs upside down looking for food; hawks insects in air. Sometimes bird or flock drops almost perpendicularly to the ground, then returns to tree. Occasionally feeds on burnt ground, often soon after fire. In southern Africa tends to feed in vegetation in summer, on ground in winter. In Zimbabwe group forages on ground (47% of observations), in bushes (23%), in leaves (19%), on trunks (11%), on branches (5%), on twigs (1%) and in air (2%) (Vernon 1980).

Group occupies an undefended home range that may overlap with home ranges of other groups; home range becomes larger after nesting. Flock tends to travel same route daily; often active in middle of day. Progresses through woodland with birds from the back overtaking and flying to the front; some birds act as sentries while others feed. Often feeds in mixed parties with Retz's Helmet-Shrike, Eastern Black-headed Oriole *Oriolus larvatus*, Southern Black Tit *Parus niger*, Fork-tailed Drongo *Dicrurus adsimilis* and Brubru *Nilaus afer* (Vernon 1966, Garcia 1975). In Mole Nat. Park (Ghana) never forages with other species (Greig-Smith 1976). Birds roost huddled together on a horizontal branch, jostling and changing positions before settling down; flock roosts in one area for a few nights, then moves on to another. Mobs predators (including Leopard: Chapin 1954) by continuously dive-bombing and bill-snapping.

Resident, with some seasonal movements. Sierra Leone, present Dec–May; Ghana, some move at end of dry season from N (Mole Nat. Park) to S (Accra); N Nigeria (Kano), arrives at end of rains; N Central African Republic (Bamingui-Bangoran Nat. Park), present only Aug–Sept; Kenya, tends to avoid Baringo area during wetter periods (Lewis and Pomeroy 1989), *poliocephalus*, a non-breeding migrant from Tanzania in dry season (May–Sept), returning to Tanzania in rains to breed. In southern Africa, disperses (after nesting dry season) into more semi-arid and urban regions, becoming more nomadic in drought years (e.g. 1953, 1970, 1979) when erratic influxes may occur sometimes to W (Maclean 1993), or to E (Tree 1995, Brooke 1994); in winter 1992, unprecedented numbers in highveld of Swaziland: Parker 1994).

Food. Mainly small invertebrates including caterpillars, beetles, grasshoppers, praying mantises, termites, ants, butterflies, moths, flies, spiders and centipedes; also small reptiles including geckos, and small fruits. Contents of 5 stomachs (NW Zimbabwe) were 42% beetles, 30% ants, 12% grasshoppers, 16% other unidentified food (Douthwaite 1995).

Breeding Habits. Co-operative breeder; all members of a group help in constructing nest, incubating, brooding, providing for young and in removing faecal material. Some evidence (Grimes 1976) suggests that groups with several helpers of both sexes are common in dry woodland at lower elevations (200–800 m) and lower latitudes, while breeding units of 2 or with helpers of only one sex are common in moist woodland at higher elevations (800–1600 m) and higher latitudes. Sometimes double-brooded; 2–2·6 nesting attempts per season (1971–1974, Zimbabwe: Vernon 1984a).

Territory maintained only in breeding season; usually 12 ha (n = 14, 1·7–19·5). Group defends territory from rival group with members of each group facing each other while perched on adjacent trees. All sit vertically with heads and necks stretched towards the sky, then all call and snap bills. Some (all?) move their heads slowly from side to side and fan their tails; some move closer together while others fly with rapid wing-beats and body feathers fluffed out. Dominant ♀ usually starts chorus; eventually some members of one group fly at members of the other, sometimes physically touching; usually one group retreats and the other leaves there shortly thereafter (C. J. Vernon *in* Harris and Arnott 1988).

Pair bond maintained by allopreening and duetting; bond lasts for about 2·5 years (Harris and Arnott 1988). Courtship begins when established pair or dominant pair of new group displays to each other, slowly flapping wings; sometimes ♂ initiates courtship by approaching ♀ with nesting material in his bill. When about to copulate, pair moves away from group, and birds perch horizontally together next to each other, flap their wings, then ♂ mounts ♀.

NEST: a smooth, compact cup of small pieces of bark, bound together with spider webs (**C**), and lined with smaller bits of bark, rootlets, grass and lichen; ext. diam. 81–90 (84), ext. depth 35–65 (50), int. diam. 63–71 (68), int. depth 24–28 (26), on branch or in fork of tree 2–10 m above ground (mostly 3–6 m) (n = 11: Maclean 1993) or 2–9 m (*c*. 5) (n = 113: Harris and Arnott 1988). Neater, narrower and deeper than nest of Retz's Helmet-Shrike. Nests always solitary, never in neighbouring trees, and usually 50 m or more apart. Breeding pair goes from site to site, calling and making nest-building motions before selecting the nest-site. All members, but mainly the breeding pair, collect nest material and construct the nest. Upon discovering nest-material, member gives chattering call, which stimulates others to collect it. All then fly to nest where each takes turn adding material and shaping the nest. Construction takes at least 4 days; then maybe a week elapses before laying begins.

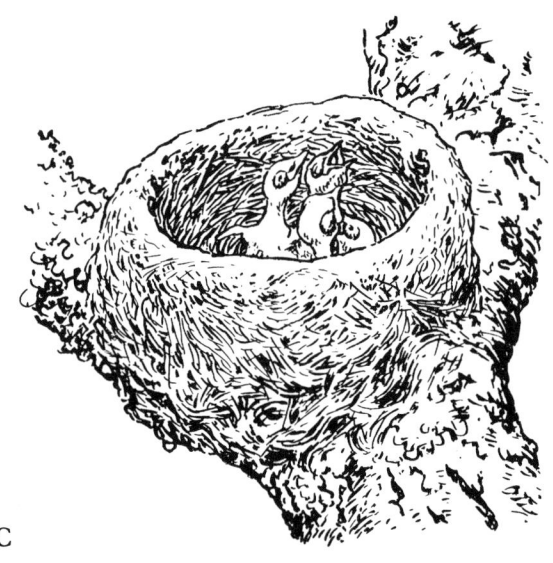

C

EGGS: 3–5, av. (109 clutches, Zimbabwe) 3·9 with 2 ♀♀ occasionally laying in same nest (8 such cases in 225 clutches, Zimbabwe: C.J. Vernon *in* Grimes 1976; up to 9 eggs in a nest, the product of 2 ♀♀: Steyn 1996). Oval, pale green, pink or buff, blotched and spotted with purple, brown and grey, mainly at thick end. SIZE: (n = 271) 18·5–23·5 × 14·5–18 (20·6 × 16).

LAYING DATES: Senegal, June, Nov; Gambia, May–Oct; Ghana, Jan (fledglings Mar, July, Aug, nestlings Oct); Togo, Nov (dependent young May); Burkina Faso July, Dec (young, Nov, Dec); Nigeria, Mar–June, Oct–Nov; Sudan, Apr–June; Ethiopia, Feb–June, Dec; Somalia (newly fledged young May, Oct); E Africa: Uganda, Mar; Tanzania, Mar (newly fledged young July); Region C, Mar, June, Oct, Nov; Region D, Feb, Aug; Zaïre, July–Jan, mostly Sept–Oct; Angola, June, Aug, Sept; Zambia, Mar, Aug–Nov, mainly Sept–Oct; Malaŵi, Sept (10), Oct (12), Nov (12), Dec (3); Mozambique, Sept–Dec, Feb; Zimbabwe, Feb (2), Apr (3), Aug (6), Sept (162), Oct (175), Nov (82), Dec (33), breeding season always started between 2 and 22 Sept, 1970–1974, (Vernon 1978); Namibia, Oct; Botswana, Oct (2), Nov (2), Dec (2), Jan (2), Feb (1); South Africa: Natal, Oct–Dec; Transvaal, Sept (3), Oct (21), Nov (17), Dec (10), Jan (1); southern Africa, Aug–Apr, mainly Sept–Oct.

D

INCUBATION: period 17 days (rarely 21), by all members of group. Incubating bird waits for group to return or summons them vocally.

DEVELOPMENT AND CARE OF YOUNG: young fledge when 17–22 days old (av. c. 20); before they can fly well, young huddle together waiting to be fed by adults. At about 21 days young birds accompany group and begin to feed on their own; fed sporadically by group until 10 weeks old; independent by 5 months; resemble adult when 1 year old.

They either remain with flock, becoming helpers during subsequent nesting season, or depart to form new group.

Young fed, brooded and guarded by all members of group, who eat faecal sacs of very small young, but carry away those of older young; occasionally nestling voids faecal sac over side of nest. 1 or more adults give chattering call when they have found food; group members then return to nest and take turns feeding nestlings. Last bird to feed young usually remains with them, brooding or remaining close to them (**D**).

BREEDING SUCCESS/SURVIVAL: parasitized by Black Cuckoo *Cuculus clamosus* and Red-chested Cuckoo *C. solitarius*. Adults and nest contents preyed upon by reptiles, mammals and birds including Bateleur *Terathopius ecaudatus*, African Harrier Hawk *Polyboroides typus* and Gabar Goshawk *Accipiter gabar*. In southern Africa starvation may contribute to nestling mortality in poor seasons (Harris and Arnott 1988). In Zimbabwe, 1970–1974, length of breeding season varied between 78 and 104 days; of 129 nests, 39 (30·2%) were successful; of 464 eggs laid, 118 (25·4%) fledged; 1·4–1·9 young fledged per group per season; 60% of 70 fledglings survived; 83% of 200 adults survived (Vernon 1978, 1984a).

References
Gill, P.J. (1939).
Greig-Smith, P.W. (1976).
Harris, T. and Arnott, G. (1988).
Vernon, C.J. (1966).

Prionops poliolophus Fisher and Reichenow. Grey-crested Helmet-Shrike. Bagadais à huppe grise. Plate 28 (Opp. p. 415)

Prionops poliolopha Fischer and Reichenow, 1884. J. Orn., 32, p. 180; Lake Naivasha, Kenya.

Forms a superspecies with *P. plumatus*.

Range and Status. Endemic resident in Kenya and Tanzania, uncommon to frequent from L. Nakuru, L. Elmenteita, L. Naivasha and Kedong Valley through Narok District, Masai Mara Game Res. and Sabaringo Valley to Loita and Nguruman Hills, Loliondo and Serengeti Nat. Park; also Mwanza (Britton 1980); range centred mainly between Narok and Serengeti. In early 1900s reported in Kigoma and Tabora regions (c. 05°S) but no records there since (Lewis 1981). Known in the Naivasha/Nakuru area until 1926, rediscovered there in Oct 1978 and recorded several times since but not known if it breeds there (Lewis 1981). Total population probably small and declining; given Vulnerable status by BirdLife International (Bennun 1994).

Description. ADULT ♂: lores, forehead and forecrown grey, feathers rather stiff, whitish tipped, forming a median frontal crest directed upward and forward to cover most of maxilla. Side of crown grey; centre of crown dark grey, feathers c. 20 mm long and rather straight, forming an erect median crest covering dark grey hindcrown and nape. Sides of nape blackish, forming bar behind pale grey cheeks and ear-coverts; broad collar around hindneck greyish white. Rest of upperparts black with greenish gloss. Inner tail feathers (T1–T4) black, with progressively

Prionops poliolophus

broader white tip (*c.* 8 mm on T1, *c.* 20 mm on T4); T5 with broad white tip extending into long narrow wedge along outer edge; T6 white with black base to inner web. Lores, sides of head, chin and throat grey; flanks greyish; rest of underparts greyish white apart from black patch at side of breast. Primaries black, innermost with whitish tips, inner webs with white central part, forming bar of broad white spots on extended upperwing. Secondaries and tertials black, outer secondaries (S1–S5) with broad white spot at tip, S6 with whole outer web white, innermost tertial with broad white outer border, middle tertial with narrow white outer edge. Alula and primary coverts black; outer greater, outer median and lesser coverts black; inner greater and inner median coverts white, forming, with inner secondary and tertial edges, broad longitudinal white stripe on closed wing. Underwing-coverts and axillaries black. Bill black; no eye-wattles; eyes bright yellow, iris with orange outer ring; legs orange-red. Sexes alike. SIZE (3 ♂♂, 3 ♀♀): wing, ♂ 125–132 (128), ♀ 126–134 (129); tail, ♂ 98–105 (101), ♀ 102–107 (105); bill, ♂ 25–26 (25·7), ♀ 23–25 (24·3); tarsus, ♂ 23–26·5 (24·8), ♀ 24–26 (25). WEIGHT: 1 ♂ 49.

IMMATURE AND NESTLING: not known.

TAXONOMIC NOTE: traditionally *poliolophus* is considered a full species because it lacks the eye-wattles of *plumatus* complex members. Except in the Loliondo district of N Tanzania where *poliolophus* mixes with flocks of *plumatus* (Williams 1963), these two species are largely allopatric and ecologically separated. Hall and Moreau (1970) have included *poliolophus* in a species group with *plumatus* and *alberti*, the latter a divergent member. We follow Short *et al.* (1990) in combining *poliolophus* and *plumatus* but not *alberti* into a superspecies.

Field Characters. Length 24–26 cm. A large helmet-shrike endemic to S Kenya and N Tanzania. Shows white stripe on closed wing, like much smaller parapatric race *poliocephalus* of White Helmet-Shrike *P. plumatus*, and resembles it in flight, but has *prominent grey crest* (**A**)(crest of White short and whitish) and *black patches* at sides of breast, yellow eye but *no eye wattle*.

Voice. Not tape-recorded. Said to have single 'chwerr' note and rolling, descending, churring phrases, such as 'chichi-cherrro', elaborated with harsh, scratchy, chattering notes and bill-snapping by several birds in group vocalizations, e.g. 'chikiki-chi-chirrrow chi-chirrro che-chiwow-cherrow chk chk skrrk chrrrk' (Zimmerman *et al.* 1996).

General Habits. Inhabits open woodland with whistling-thorn *Acacia dephanolobium* and leleshwa *Tarchonanthus*, riparian woodland with fever tree *A. xanthophloea*, *A. abyssinica* and *Protea* and wooded grassland and bushland between 1200 and 2200 m.

A

Gregarious; conspicuous and noisy; in flocks of up to 12, chattering and snapping bills as they move from tree to tree like other helmet-shrikes. Sometimes occurs in mixed parties with White Helmet-Shrike *P. plumatus* (Williams 1963).

Probably sedentary, but after nesting some foraging parties move northeast in Rift Valley from the presumed Mara–Narok breeding area to L. Nakuru Nat. Park, Longonot, Naivasha and Menengai Crater where they remain Oct–Feb.

Food. Not known.

Breeding Habits. Co-operative breeder; all members of a group help to raise young (Bennun 1994).

NEST: a cup-shaped structure composed of grass and spider webs.

EGGS: bluish green with greyish and reddish brown spots. SIZE: (n = ?) 21 × 17.

LAYING DATES: Kenya, Apr–May, (nearly fledged young May; adults feeding young July).

DEVELOPMENT AND CARE OF YOUNG: young in nest (age ?) largely brownish white.

References
Bennun, L. (1994).
Lewis, A.D. (1981).
Lewis, A.D. and Pomeroy, D. (1989).

Plate 28 (Opp. p. 415) *Prionops alberti* Schouteden. Yellow-crested Helmet-Shrike. Bagadais d'Albert.

Prionops alberti Schouteden, 1933. Rev. Zool. Bot. Afr., 24, p. 211; summit of Mt Mikeno, 4400 (= 4467) m, Kivu District.

Range and Status. Endemic resident, E Zaïre, in mountains west of Albertine Rift. Known from just north of the equator (2 plots, Hall and Moreau 1970) and 4 areas south of it (Prigogine 1985, T. Pedersen, pers. comm.): (1) Bitakongo, Kanyabisika and Pinga to Bilati (west of L. Edward); (2) Kamatemba and Virunga Nat. Park (Nyiragongo Volcano) to Butokolo (north and west of L. Kivu);

(3) Mwana to Kitongo and Kilize (Itombwe); and (4) Mt Kabobo. Unconfirmed sight record in Bwindi-Impenetrable Forest, Uganda (Keith *et al.* 1969) but has not been seen since despite extensive exploration. Status since mid-1980s not documented; probably still common in Itombwe Mts and 'perhaps west of L. Edward, as these forests are largely untouched' (T. Pedersen, pers. comm.).

Prionops alberti

A

Description. ADULT ♂: top of head lemon yellow, feathers of forehead and crown forming stiff crest, some with brownish or greyish tips, the longest up to 20 mm (**A**). Rest of head and body, including lores, cheeks, ear-coverts and hindneck, black with slight greenish gloss. Tail and wings black. Bill black; narrow eye-wattle yellow; eyes greenish yellow; legs light red or flesh-coloured, claws greyish to blackish. Sexes alike. SIZE (12 ♂♂, 14 ♀♀): wing, ♂ 130–135 (132), ♀ 130–136 (133); tail, ♂ 93–106 (97·1), ♀ 93–110 (98); bill, ♂ 21–26 (24·8), ♀ 20–26 (23·7); tarsus, ♂ 22–24 (23·3), ♀ 21–24 (22·9). WEIGHT: 1 ♂ 61, ♀ (n = 2) 61, 62.

IMMATURE: similar to adult but less glossy; feathers on head white or brownish white, some with tinge of yellow, bill light brown, eye brown.

NESTLING: not described.

Field Characters. Length 21–25 cm. Confined to mountains of Albertine Rift in Zaïre, where it lives in forest canopy. Unmistakable – all black with yellow crest (whitish in young birds). However, one's initial impression is of a starling, since the yellow crest is not obvious to the naked eye (T. Pedersen, pers. comm.).

Voice. Not tape-recorded. Usual call described as 'tlu-uk' or 'clu-uk' repeated 2–6 times (F. Hendrickx *in* Chapin 1954).

General Habits. Inhabits montane woodland including *Hagenia* and bamboo forests at 1100–2600 m, mainly 1400–1800 m, on Mt Kabobo 1560–2500 m; originally discovered in alpine area at *c.* 4400 m. At lower elevations also occurs in primary rain forest.

Social behaviour little known although reported to be similar to that of Retz's Helmet-Shrike *P. retzii* (Chapin 1954). Gregarious, in flocks of 4–6, up to 20, usually in higher levels of forest. Birds on Nyiragongo Volcano at 2300 m were in loose flock moving through the forest with some speed, often sitting on exposed branches in upper one-third of trees; not shy (T. Pedersen, pers. comm.). If one bird is wounded and cries out, all its companions turn back excitedly to see what is wrong (Chapin 1954). Sometimes feeds like a flycatcher, diving from a perch to capture its prey, then returning to its point of departure (Prigogine 1971).

Food. Insects, including small grasshoppers.

Breeding Habits. Co-operative breeder.
NEST: made of sticks, lined with small roots and lichens, in fork of tree 6–7 m from ground (Prigogine 1953).
LAYING DATES: Feb–Mar, June–July, (enlarged gonads Jan, Apr, May, Dec.).

References
Prigogine, A. (1949, 1953, 1984).

Prionops caniceps (Bonaparte). Red-billed Helmet-Shrike. Bagadais à bec rouge.

Plate 28
(Opp. p. 415)

Sigmodus caniceps Bonaparte, 1851. Consp. Av. 1, p. 365; West Africa = Boutry (Butré) River, Gold Coast.

Forms a superspecies with *P. rufiventris*.

Range and Status. Endemic resident, frequent to common. In Mali, uncommon, Mandingues Mts (Nana Kényeba, Dyoumanssana); Guinea, uncommon to frequent Fouta Djalon and Kounounkan massifs, south to Sierra Leone border, also Macenta; uncommon W and central Sierra Leone including Outamba area and Gola forest; Liberia, throughout all older forests, including Mt Nimba; Ivory Coast north to *c.* 08°N from Sipilou and Taï Nat. Park in W to southern limits of Comoé Nat. Park in E; Ghana, Kakum Forest and north to Bui Nat. Park and Kintampo and east to Atewa Forest Res. and SE scarp of Volta basin, also SE at Afegame, Tafo and Hohoe/Wli Falls/Togo

Prionops caniceps

border; Togo, throughout forest zone, also gallery forest (Mono R., 07°06′N; Amou–Oblo, 07°23′N); Benin, rare, in SE at Pobé Forest Res.; Nigeria, uncommon from near Lagos and Benin border through Niger R. delta to Cameroon border in Ikpan block of Cross River Nat. Park (Ash et al. 1989) and Calabar (Mackenzie 1979); SW Cameroon in Korup Nat. Park.

Description. *P. c. caniceps* (Bonaparte): Guinea and Mali to Togo. ADULT ♂: forehead and crown to lores and above eye pale blue-grey; rest of upperparts black with oily green gloss. Tail black, glossy above. Sides of head and neck to chin and throat black; breast and upper flanks white; lower flanks, belly, undertail-coverts and thighs pale rufous-buff. Wings black apart from centres of primary inner webs, which form broad broken white bar across extended wing. Underwing-coverts and axillaries dark grey. Sexes alike. Bill large (**A**), dark red; narrow eye-wattle orange; eyes bright yellow; legs and feet dark orange. SIZE (10 ♂♂, 10 ♀♀): wing, ♂♂ 111–119 (117), ♀ 114–121 (118); tail, ♂ 74–83 (79·6), ♀ 75–83 (80·4); bill, ♂ 23–27 (25·1), ♀ 23–27 (25·2); tarsus, ♂ 19–21 (20·2), ♂ 19–21 (20·2). WEIGHT: (Liberia) 1 ♂ 56; ♀ (n = 3) 52–62 (55·7).

IMMATURE: forehead and crown buffish white, streaked dark grey-brown on hindcrown; rest of upperparts blackish brown; streak through lore and above eye blackish brown; cheeks, sides of neck and underparts pale buff, breast whiter. Wings and tail blackish, middle part of primary inner webs white; inner secondaries, alula and primary coverts narrowly tipped buffish white; outer web of T6 tipped buff. Bill dark horn; eyes dark grey-brown; legs and feet orange.

NESTLING: no information.

P. c. harterti (Neumann): Benin to W Cameroon. Like nominate race but blue-grey of crown extends below and behind eye; buff of belly and undertail-coverts paler, less rufous.

TAXONOMIC NOTE: see under *P. rufiventris*.

Field Characters. Length 20–23 cm. Quiet and unobtrusive birds of the forest canopy, easy to overlook in spite of their conspicuous colouring. Head pale *blue-grey* (looks white in the field), contrasting with black upperparts and throat; extent of grey varies with race: ends at eye level in *caniceps*, reaches below eye in *harterti*; bill *red*. Usually seen in canopy from below, where pattern on underside is *black throat* contrasting with *pale buff* breast and belly. No similar species in its range; does not overlap with Rufous-bellied Helmet-Shrike *P. rufiventris*.

Voice. Tape-recorded (104, B, BRU, ERA, LEM, MAC). Typical call in Liberia a dry and leathery 'wrrraak, wrraak', with little carrying power (S. Keith, pers. comm.). This and similar but louder calls used in contact: harsh, grating 'waaah, waaah, waaah ... ' and 't'-waah, t'-waah, t'-waah', like cawing of crow; variations include 'chyip-to-waah' and a deeper 'tor-waah' or 'torrraaa'. Song a series of loud pure whistles, rising slightly in the middle, 'tooyoo, tooyoo, tooyoo ... ' or down-slurred 'weeyo, weeyo, weeyo'. In possible duet, 'tooyoo' answered immediately by 'waah'. Double 'kewp-kewp' like similar calls of Rufous-bellied Helmet-Shrike, *q.v.* Most typical motif of birds on Mt Kupé, Cameroon (race *harterti*) is a double 'kui-kui, kui-kui?' like that of *rufiventris* (F. Dowsett-Lemaire, pers. comm.). Aggressive calls given by group following a monkey, loud squirrel-like chattering 'chipchipchip ... ' and 'chapchapchap ... ', with bill-snapping (Chappuis, in press).

General Habits. Inhabits lowland primary forest, often in middle and canopy layers; also in mature secondary forest, dense bushy growth near streams, and tall trees at forest edges and around clearings. In Guinea noted in an old coffee plantation (Halleux 1994); in Togo, gallery forest; in Cameroon, logged forest and farms (Rodewald et al. 1994). Lowlands; up to 700 m in Guinea.

Gregarious, travelling in family (?) groups of 4–8, sometimes up to 20; often noisy, keeping in contact by calling and snapping bill. Sometimes in mixed foraging parties with cuckoo-shrikes, drongos, malimbes, and White Helmet-Shrike *P. plumatus*. In mixed parties tends to keep in own sub-group, often travelling behind the mixed flock. Forms looser flocks than White Helmet-Shrike. Flight buoyant, fluttering and fast when moving from branch to branch through canopy. Catches most prey in flight like a flycatcher; also forages by walking in a low crouch along branch and taking insect within reach on leaf or twig. Sometimes jumps after prey onto a nearby twig. Birds usually stay within 1–4·5 m of each other, with some members of group feeding, others perching and preening. Sedentary.

A

Food. Insects: caterpillars, grasshoppers, stick insects, praying mantises up to 5 cm long, cicadas and beetles; also spiders and seeds. In Ivory Coast diet included 50% adult arthropods, 30% fruits and 20% larvae (Thiollay 1973).

Breeding Habits. Probably co-operative breeder, but details not documented (Grimes 1976).
NEST: cup-shaped, compact, greyish green, 15 m up in fork of tree just below canopy (Taï, Ivory Coast: Balchin 1988).

LAYING DATES: Liberia (5 birds at nest July; independent young Jan, Nov; immatures July, Oct); Ivory Coast, Dec, Jan (fledglings Mar, juvs May–Nov); Ghana, Feb; Nigeria (feeding young Jan, Mar, flocks with young juvs Jan–Mar). Double brooded.

References
Bannermann, D.A. (1939).
Greig-Smith, P.W. (1976).
Hart, J.A. (1971).

Prionops rufiventris Bonaparte. Rufous-bellied Helmet-Shrike. Bagadais à ventre roux. Plate 28

Sigmodus rufiventris Bonaparte, 1853. Rev. Mag. Zool. (Paris), p. 441 – Mozambique; error = Gabon (see Hartlaub, 1857, Syst. Orn. Westafr., [Bremen], p. 105.)

(Opp. p. 415)

Forms a superspecies with *P. caniceps*.

Range and Status. Endemic resident, frequent to common. In Cameroon, south and east of Sanaga R. to SW Central African Republic (Dzanga-Ndoki Nat. Park, Dzanga-Sangha Rain Forest Res. and Bayanga) and N Congo (Odzala and Nouabalé–Ndoki Nat. Parks). W and N Gabon (Ogauma, Fougamou, Rabi, Moukalaba, Gamba, Makokou and M'Passa) to S Congo (Mayombe) and Angola in Cabinda (Landana). Zaïre, in centre in E Kasai and Manyema District; in NE from Kisangani to Bangala District, Uele region, Semliki Forest, south to Itombwe; and in NW and N regions along Oubangui R. on Congo and Central African borders (4 plots, Hall and Moreau 1970); W Uganda in Bwamba, Bugoma and Bwindi-Impenetrable Forest/Ishasha Gorge. Formerly Rwanda (Hall and Moreau 1970), but now extinct (J.-P. Vande weghe *in* Dowsett 1993).

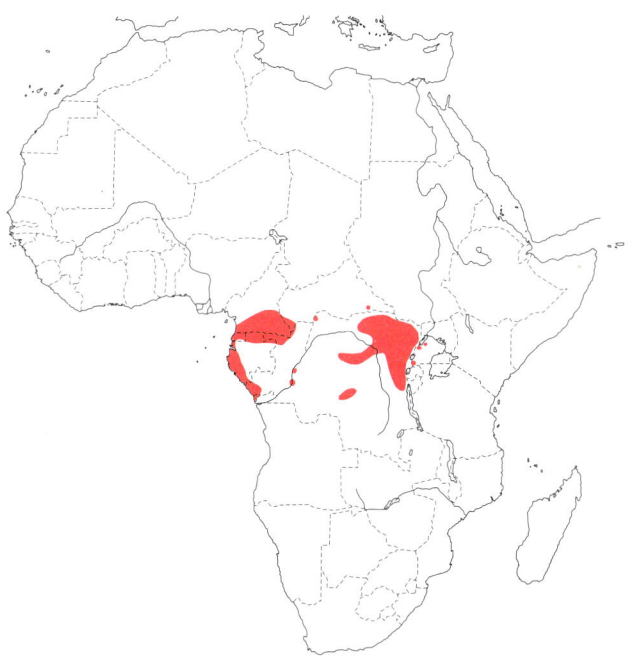

Prionops rufiventris

Description. *P. r. rufiventris* (Bonaparte): S Cameroon and Central African Republic to Congo, Cabinda and NW Zaïre. ADULT ♂: forehead and crown to ear-coverts, cheeks and chin blue-grey, paler, more whitish, on forehead, lores and around eye. Hindneck black, linked by black collar round side of neck to black throat. Rest of upperparts blackish with oily green gloss, merging with hindneck. Tail black, glossy above. Upper breast white, rest of underparts rufous. Wings black apart from broad white band across middle of inner web of each primary, this forming broken bar across extended upperwing. Underwing-coverts and axillaries blackish. Bill dark red to black; eye-wattle red; eyes yellow; legs and feet orange-red. Sexes alike. SIZE (2♂♂, 4♀♀): wing, ♂ 112, 116, ♀ 114–117 (116); tail, ♂ 83, 83, ♀ 76–80 (77.5); bill, ♂ 21, 22, ♀ 20–22 (21); tarsus, ♂ 20, 21, ♀ 20–21 (20.3).
IMMATURE: forehead and crown buffish white, crown streaked dark grey-brown; rest of upperparts blackish brown; streak through lore and above eye to upper ear-coverts and nape blackish; lower ear-coverts and cheeks buffish white; sides of neck mottled buffish and dark grey, linked in some by mottled buff and grey collar around hind-neck. Chin and throat buffish white, throat sometimes with some rufous feathers and blackish sides; upper breast whitish; lower breast pale cinnamon; rest of underparts rufous. Wings and tail blackish, primaries with white band across middle of inner web; inner secondaries, alula and primary coverts narrowly tipped buffish white, outer web of T6 tipped buffish white. Bill black; eyes dark grey; legs red.
NESTLING: no information (see *P. r. mentalis*).

P. r. mentalis (Sharpe): central and NE Zaïre to Uganda. Differs from nominate *rufiventris* in having narrower black collar, black more confined to centre of throat; whitish band across upper breast narrower, with some blue-grey anteriorly; lower breast to undertail-coverts deeper rufous. Bill dark red, lighter at tip; eye dark grey-brown with yellow outer rim and orange sclerotic membrane; bare eye rim whitish; legs orange-red. SIZE: smaller, wing, ♂ (n = 3) 108–112 (110), ♀ (n = 5) 105–119 (111). WEIGHT: (Uganda) ♂ (n = 8) 39–44 (42.4), ♀ (n = 4) 44–50 (47.3).
IMMATURE: juvenile (in nest), hind crown with 2 bare patches, later covered with black feathers which become pale blue; nape white becoming black; throat buffy white; greater and middle coverts with large white patch which becomes black. Immature, like nominate *rufiventris* except lower breast darker cinnamon; eye greyish green, becoming dark brown on inner edge; feet orange-red (Chapin 1954).

NESTLING: bill black with corners of gape pale yellow; feet dull yellow.

TAXONOMIC NOTE: the taxa *rufiventris* and *mentalis* have been treated as races of *P. caniceps* (Rand 1960, White 1962, Dowsett and Forbes-Watson 1993), as megasubspecies of it (Short *et al.* 1990), as an incipient species (Hall and Moreau 1970), and as an allospecies of *P. caniceps* by Sibley and Monroe (1990), whom we follow. Striking differences between *P. caniceps* and *P. rufiventris* are belly colour and bill size. The amount of grey on the head differs (as it does also between *P. c. caniceps* and *P. c. harterti*). Vocal evidence is inconclusive (F. Dowsett-Lemaire, pers. comm.); both species use double whistles and the 'kyop-kyop' call; the common, dry 'watch-winding' call of *P. caniceps* has not yet been heard from *P. rufiventris*; the 'chaja' call of *P. rufiventris* has not yet been heard from *P. caniceps*.

Field Characters. Length 20–23 cm. Very similar to its close relative the Red-billed Helmet-Shrike *P. caniceps* but their ranges do not overlap. A canopy bird, and when seen from below, white breast contrasts with black throat and chestnut belly. Pale blue-grey of head extends to cheeks and chin, i.e. farther down that in Red-billed Helmet-Shrike, which also lacks rufous belly. Red eye-wattle around yellow eye, instead of standing out against pale head, is sunk deep into the surrounding feathers, but red bill is very conspicuous (Christy and Clarke 1994). White on inner webs of flight feathers shows as pale band in flight. Young told from adults in flock by black line through eye. Can occur with Retz's Helmet-Shrike *P. retzii*, which has red bill but black head and underparts.

Voice. Tape-recorded (B, CHA, LEM, SAR). Noisy and garrulous. Probable song, often used to locate flock, a liquid 'tyooyoo' repeated for long periods on same pitch. Common call a strident, repeated harsh 'chaja', often answered by a lower 'chojo'; 'chaja' sometimes speeded up to 'chajajaja', like quick burst from machine gun, also a more liquid form, 'chillilli'. Other calls combine low introduction with loud whistles, 'jzzz-*yeeoo*' or 'tyaw-*weeweeyo*'. Double 'kyop-kyop, kyop-kyop ...' strongly resembles that given by Red-billed Helmet-Shrike. Other double notes include a whistle with a hoarse undertone, 'pyew-pyew', clicking 'chip-chip' and hard 'cack-cack'; also gives up-slurred 'tyaw-wee', low 'jerp' and bill-snapping. Nasal 'ouin ouin ouin' (French pronunciation) likened to electronic games (Christy and Clarke 1994); young accompanying adults give shrill 'cricricricricri'.

General Habits. Inhabits primary forest, also mature secondary forest. Often seen just beneath or in canopy. In N Congo, widespread in both closed and open canopy of dry land forests (Dowsett-Lemaire 1997). In NE Gabon, associates with canopy and crown of emergents in primary forest on slopes (Brosset and Erard 1986). In Zaïre, occurs in primary and secondary forests, often along borders of clearings and open glades up to 1450 m. In Uganda, in forests between 700–1100 m.

Lives in groups of 4–8 (sometimes 10–12), each group consisting of sub-groups of pairs and their young; birds maintain contact with constant calling, sometimes accompanied by bill snapping and noisy wingbeats. Group cohesion is greatest during the long dry season; at other times sub-groups are more likely to split off. Groups move around in large home ranges (at least 20 ha at M'Passa, Gabon: Brosset and Erard 1986), which are in effect communally defended territories. Home ranges in NE Gabon are centred around sloping sides of valleys. Forages by fluttering among leaves and around curtains of lianas, and hopping along branches. Birds never spend long in any one tree, and when one bird moves off the others usually follow in a straggling procession (van Someren and van Someren 1949). Flight buoyant and butterfly-like, with much gliding. Often joins mixed-species flocks, where it plays an important role as a catalyst. Sedentary.

Food. Insects: caterpillars, grasshoppers, cicadas, stick insects, butterflies, moths, praying mantises and beetles; also spiders, seeds and once fruits of *Musanga* and *Croton* (Brosset and Erard 1986).

Breeding Habits. Probably co-operative breeder but details not documented (Grimes 1976).

LAYING DATES: Gabon (young not yet fledged Nov, Feb; young accompanying adults May, July, Sept); Zaïre, May (enlarged gonads June, July; juvs Jan, June, Aug).

DEVELOPMENT AND CARE OF YOUNG: young that are fledged but still dependent are fed by parents and by other members of group (Brosset and Erard 1986).

References
Brosset, A. and Erard, C. (1986).
Chapin, J. P. (1954).
van Someren, V.G.L. and van Someren, G.R.C. (1949).

Plate 28
(Opp. p. 415)

Prionops retzii Wahlberg. Retz's Helmet-Shrike. Bagadais de Retz.

Prionops retzii Wahlberg, 1856. Okv. K. Vet.-Akad. Förh., 13, p. 174; ad flumen Doughe, i.e. Okavango River.

Forms a superspecies with *P. gabela*.

Range and Status. Endemic resident. Locally common to frequent except as noted. Somalia, south of 01°30'N in Juba R. valley and Boni Forest; Kenya, in SE from entire coastal lowlands inland along Tana, Athi and Galana rivers, north to Meru Nat. Park and west to Kiboko and Thika, also Tsavo Nat. Parks and Chyulu Hills; formerly in Kiambu and Nairobi District. Tanzania, widespread but scattered in E (not coast), inland to Arusha Nat. Park; Mwanza, Kibondo, Gombe Stream Game Res., N Tabora, Iringa, Ruaha Nat. Park, and Udzungwa Mts (Mwanihana and Magombere forests) (map data: N.E. Baker, pers. comm.). Angola, Cuanza Sul and Malanje, west to

Prionops retzii

Benguela and Moçamedes coastal plains and south to Namibia. SE Zaïre, to Kivu, L. Tanganyika and Katanga; throughout Zambia and Malaŵi. Mozambique and Zimbabwe, frequent to uncommon south of Limpopo R., and uncommon in NW savannas and Mashonaland Plateau in Zimbabwe. Namibia, uncommon in N Owambo, Kavango and Caprivi Strip. Botswana, in Okavango Delta and Linyanti, Chobe and Nata river systems, also in E along Shashe and Limpopo drainage. NE Swaziland, uncommon in lowveld and Lebombo Mts, once Mlumati valley. South Africa: NE Transvaal in lowveld and Limpopo Basin west to Mmabolela, also central bushveld (Hangklip near Naboomspruit; Lapalala Wilderness); and Natal south to c. 28°30′S. Density of 35 birds per 100 ha of miombo but <5 per 100 ha of other woodland types, S Mozambique (Parker 1999).

Description. *P. r. tricolor* Gray (includes 'intermedia'): W and S Tanzania, Malaŵi, S and E Zambia and Mozambique to E Transvaal and N Natal. ADULT ♂: head and neck to upper mantle glossy black, feathers of forehead and forecrown rather long and back-curved, but not forming an upstanding crest (**A**); lower mantle, scapulars, back and rump grey-brown; long uppertail-coverts black. Tail black, T1 usually with small triangular white spot at tip, T2–T6 with broad white tips (c. 8 mm long on T2 increasing to c. 27 mm on outer web of T6). Below, chin to flanks, thighs and upper belly glossy black; lower belly and undertail-coverts white. Flight feathers black, central part of primary inner webs white, forming broad broken bar across extended wing; tertials brownish black with broad grey-brown outer borders; primary coverts and alula black; outer greater coverts blackish, inner greater coverts, median and lesser coverts grey-brown, like scapulars. Underwing-coverts and axillaries grey-brown. Bill red, orange-yellow near tip; eye-wattles orange-red, broad, with digitate outer edge; eyes golden yellow; legs orange-red. Sexes alike. SIZE (10 ♂♂, 10 ♀♀): wing, ♂ 119–131 (126), ♀ 120–131 (126); tail, ♂ 87–97 (92.3), ♀ 84–98 (91.1); bill, ♂ 24–26 (25.7), ♀ 24–26 (25.3); tarsus, ♂ 23–25 (23.9), ♀ 23–25 (23.7). WEIGHT: Zimbabwe, ♂ (n = 2) 37.9, 42.5, ♀ (n = 2) 46.4, 46.7.

IMMATURE: juvenile grey-brown above, mantle feathers fringed pale buff; grey-brown below with whitish lower belly and undertail-coverts; breast and flanks of some birds barred buffish white; wings and tail as adult, but tertials and inner secondaries narrowly fringed white; alula feathers and primary coverts often tipped white; bill blackish brown; no eye-wattle; eyes dark brown; legs yellow or orange. Immature plumage at c. 6 months to 1 year: body and head uniform grey-brown apart from white lower belly and undertail-coverts; occasionally some adult black body feathers present; juvenile wing and tail feathers retained; bill dusky grey becoming yellowish at base and on lower ridge of lower mandible, later orange to red; eye-wattle ochre; eyes pale brown becoming yellow; legs dull orange, later red; black plumage of adult attained usually by 18, sometimes 24 months.

NESTLING: naked, blind, with mouth spots.

P. r. retzii Wahlberg: South Africa (N and W Transvaal) to Zimbabwe, S Zambia, Botswana, Namibia and Angola (S Huila and probably Cubango). Like *tricolor* but upperparts darker grey-brown; less white on lower belly; usually no white spot at tip of T1. Wing slightly longer: unsexed (n = ?) 131–137 (134). WEIGHT: (southern Africa) ♂ (n = 19) 38–48 (43.1), ♀ (n = 19) 38–49 (44.6), unsexed (n = 41) 38–54 (47.8).

P. r. nigricans (Neumann): Angola (except extreme S), NW Zambia and SE Zaïre. Like nominate *retzii* but upperparts even darker and greyer, less brown, less sharply demarcated on upper mantle. Large: wing, ♂ (n = 4) 133–138 (136), ♀ (n = 8) 131–140 (135). WEIGHT: (Angola) ♂ (n = 5) 45–57 (51.8), 1 ♀ 53; (Zaïre, Upemba Nat. Park) ♂ (n = 5) 46–53 (49.6), ♀ (n = 13) 50–61 (53.5) (Verheyen 1953).

P. r. graculinus Cabanis (includes 'neumanni'): S Somalia to NE Tanzania. Like *tricolor* but feathers of forehead and forecrown 15–20 mm long, and rather stiff, forming a bilateral crest (**A**). Upperparts slightly paler and browner; white inner web patches on primaries absent or (in some Tanzanian birds) reduced to small spots. Slightly smaller than *tricolor*: wing, ♂ (n = 10) 119–128 (123), ♀ (n = 10) 120–130 (124). WEIGHT: (Kenya) ♂ (n = 15) 33–55 (49.3), ♀ (n = 22) 40–60 (53.4).

Field Characters. Length 19–20.5 cm. Mainly *black*, with brown back and wing-coverts, white lower belly and undertail-coverts, and *white tail corners* conspicuous in flight. *Short black crest, red bill*, orange eye with *red eye wattle*. Immature (first year) brown with pale tips to wing feathers; no eye wattles, bill and eye brown, (second year) bill and eye-wattles orange; plumage becomes black only in third year (Maclean 1993). Chestnut-fronted Helmet-Shrike *P. scopifrons*, which it meets in coastal woodlands, lacks crest and is smaller, dark grey (not black), and has blue eye wattles, chestnut patch on forehead; bill and legs red. Second-year bird seen from below, with orange bill and legs and grey-brown plumage, can be mistaken for adult Chestnut-fronted Helmet-Shrike (see illustration in

A

Newman and Hanmer 1991); in side view readily distinguished by lack of chestnut forehead, and eye wattle, if present, orange or red, not blue.

Voice. Tape-recorded (10, 33, 75, 86, 88, 91–99, 100, 104, B, LEM). Song, also given in response to playback, 'tweeooh-tweew', oriole-like in quality, clearer than similar call of *P. scopifrons*, or a blasting 'tyeeeow, tyeeee-owp' (Short and Horne 1985). Most characteristic calls are churring and rolling, with 'watch-winding' quality, usually in chorus: up-slurred 'turrrwee', down-slurred 'weeerrrrr', up and down 'turrweeeurrr', 'turr-you-wee', sometimes followed by a low grating 'werjer – 'trrurwee, werjer-werjer', or preceded by a single 'whip', 'whip-reeer' (Maclean 1993; a duet?); duetting thought likely in maintenance of pair bond (Harris and Arnott 1988). Chorusing initiated by individual calling 'jor-rrreeea', followed immediately by duetting and chorussing from rest of group (Harris and Arnott 1988). Wide variety of other calls includes series of loud high-pitched whistles, 'tyoo-ee' or 'too-yee'; repeated double notes with 'chomping' malaconotine quality, 'tyo-wo, tyo-wo ... ', 'pi-tyoo, pi-tyoo' or 'twoo-weeyo ... '; gradually descending, low 'chaw-chaw-chaw'; a rather nightjar-like 'weeoo-wip'; and a barbet-like 'titi-wok, titi-wok, titi-wok'; also bill-snapping. Gives 'chucker' call when preferred food found; low 'grrr' signifies mild aggression at roost; alarm, 'jor-rrreeea' and 'cherr-weeu', intense alarm, sharp squeals and 'chirr' sounds; in courtship, ♂ gives subdued juvenile-like soliciting call while perching next to ♀ (Harris and Arnott 1988). For further variations, see Short and Horne (1985) and Zimmerman *et al.* (1996).

General Habits. Occurs mainly in well-developed riverine woodland and lowland forest with miombo *Brachystegia*, mopane *Colophospermum* and *Baikiaea*, from sea level to *c.* 1900 m, but usually below 1500 m; also scattered thickets and bush (coastal Tanzania), mangrove (Lamu, Kenya) and (outside breeding season) tall savanna and mixed acacia woodland. Tends to occupy moister woodland than White Helmet-Shrike *P. plumatus*.

In groups of 2–15, generally of *c.* 5 but occasionally up to 30; group members usually related. Group structure probably based on dominance hierarchy with breeding ♀ (who is heaviest) dominating her mate, the pair dominating other adults and fledglings (who weigh least). Group occupies a home range and defends territory of up to 30 ha; members perch close together on a branch, calling in chorus with tails fanned and wings lowered; all then bow back and forth (Harris and Arnott 1988). Chorusing initiated probably by one bird (dominant ♀?), others join in. Sometimes 4 or 5 approach intruder, calling; then some (all?) slowly bow with crest erect, breast feathers fluffed out, and wings held out slightly from body; then (all?) raise heads to vertical position with bill pointing skyward, and sometimes bow 2–3 times in succession (Short and Horne 1985). Also fly over intruder, singing with crest erect and body feathers fluffed.

Flight rapid, buoyant and fluttering, interspersed with gliding. Forages usually in trees, probably higher in canopy than White Helmet-Shrike. Moves from place to place along branches, among twigs and leaves and on trunks, searching for food. Hangs upside down by feet and also tilts head to one side as if listening for prey. Occasionally hawks insects in flight or captures them on ground (Zimbabwe: forages on branches (48% of observations), on trunks (25%), on twigs (22%), in leaves (2%), in air (1%), in bushes (1%), on ground (1%): Vernon 1980). Forages also in mixed flocks with White and Chestnut-fronted Helmet-Shrikes as well as orioles, drongos, tits and woodpeckers. Roosts as a group, birds huddled in a row along a branch. Mobs snakes, hawks and owls, monkeys and small carnivores, snapping bill and giving alarm calls. Occasionally performs threat display on ground with wings held straight up and mouth open.

Sedentary, but subject to dispersive movements after breeding or during droughts.

Food. Mainly insects: larvae (caterpillars up to 47 long: Oatley 1969), grasshoppers, praying mantises (up to 50 long: Oatley 1969), butterflies, moths and beetles; also spiders, a gecko 50 long and bits of egg shell.

Breeding Habits. Co-operative breeder. Solitary nest in group-defended territory. Territory large (Harrison *et al.* 1997). Nesting thought to be initiated by trees coming into leaf (southern Africa: Tarboton 1963). All members of group are involved in constructing nest, incubating and other nesting activities; occasionally a pair breeds unhelped (once in 36 nestings, Zimbabwe: C. J. Vernon *in* Grimes 1976). Sometimes double-brooded, possibly the result of a group dividing to produce 2 active ♀♀ (Harris and Arnott 1988).

Dominant pair maintains bond by allopreening and probably by duetting. ♂ initiates courtship by approaching dominant ♀ with nest material in his bill, wings drooped and half opened; ♂ (both?) then searches for a nest site, finds a place and shuffles about as if shaping a nest. Pair separate themselves from group before copulation, face each other or stand side-by-side, then slowly open and close their wings. ♂ faces ♀, slowly arches neck, fans wings and tail (as does ♀) and mounts her (Harris and Arnott 1988).

NEST: a shallow, nest cup of lichens, thin bark and grasses, bound together with spider webs (**B**); ext. diam. 100–110, ext. depth 35–50, int. diam. 70–75, int. depth 15–30. Broader, shallower and squatter with thicker walls than, and not as smoothly finished as, nest of White Helmet-Shrike. In fork or on horizontal branch *c.* 7 m (3–20 m) above ground; often in tree-fern *Pterocarpus rotundifola* (southern Africa: Harris and Arnott 1988). Built by dominant pair with help from other members of group; sometimes all share equally in building it (Tarboton 1963). All (?) fly to nest with nest material in their bills; they take turns placing material onto nest, each taking *c.* 30 s, then all fly off together. They bring spider webs first, then add grasses and bark.

EGGS: 3–5, av. (25 clutches, Zimbabwe, and 33 southern Africa) 3·4; laid on consecutive days. Oval, pale green to bluish, spotted and blotched with browns, blacks and greys, especially at blunt end. SIZE: (n = 36) 22·7–25·8 × 15·6–18·2 (24·4 × 17·1) (Harris and Arnott 1988).

B

LAYING DATES: E Africa: Kenya, Jan, Mar (enlarged gonads Nov); Region D, Mar; Region E, Feb, Sept; Angola, Aug–Oct (enlarged gonads Feb); Zaïre, Aug–Jan (enlarged gonads Mar); Zambia, Mar, Sept–Dec; Malaŵi, Sept–Oct; Mozambique, Oct–Nov; Zimbabwe, June (1), Aug (10), Sept (61), Oct (52), Nov (12), Dec (6); Namibia (feeding fledged young Aug); Botswana, Nov (2); South Africa, Sept (1), Oct (7), Nov (1), Dec (1), Mar (1).

INCUBATION: period 17 days (up to 20); by all members of group, beginning when clutch complete; incubating bird seldom left alone for more than a few min; if other members of group stay away for long time, incubating bird summons them with loud calls. No change-over ceremony.

DEVELOPMENT AND CARE OF YOUNG: fledges c. 20 days (n = 1) after hatching. Brooded, guarded, fed and attended by all members of group; at least one member stays with young most of the time; reliefs occur about every 11 min when nestlings 1–2 days old (n = 6, 2–20 min: Benson 1946). Adult stands over young at nest with wings outspread (shading young?). Young remain huddled together until an adult arrives to feed them; they depend on adults for at least 3 months after leaving nest; become independent at c. 7 months; and take 12–24 months to attain adult plumage.

BREEDING SUCCESS/SURVIVAL: parasitized by Thick-billed Cuckoo *Pachycoccyx audeberti*; 55% of 29 nests parasitized (Zimbabwe: Vernon 1984a). Cuckoo chick, within 5 days of hatching, evicts helmet-shrike eggs and chicks. Susceptible to disturbance during nest construction and early stages of incubation, birds sometimes destroying their own nests (Harris and Arnott 1988). Breeding success low (in southern Africa), probably due to cuckoo parasitism. 40 groups (Zimbabwe) reared av. 0·6 fledglings per group, with 79% of fledglings surviving to first year and 85% of adults surviving to following year (Vernon 1984a).

References
Harris, T. and Arnott, G. (1988).
Short, L.L. and Horne, J.F.M. (1985).
Tarboton, W. (1963).

Prionops gabela Rand. Gabela Helmet-Shrike. Bagadais de Gabela.

Plate 28
(Opp. p. 415)

Prionops gabela Rand, 1957. Fieldiana: Zool. (Chicago), 39, p. 43; Gabela, Angola.

Forms a superspecies with *P. retzii*.

Range and Status. Endemic resident, W Angola along and below escarpment from Chio (09°52′S, 14°23′E) to Gabela. Known from southwest of Chio (Dean 1974) and several places within 40 km of Gabela including 15 and 19 km south of it, 64 km south of Mumbondo on Gabela-Muxima road, Londa (= Conda, see Collar and Stuart 1985), Roça Cassemba and Amboim Forest. Not uncommon in Amboim Forest (da Rosa Pinto 1962).

Description. ADULT ♂: head, neck and upper breast glossy black; feathers of forehead up to 15 mm long, curling forward over basal half of bill. Mantle, scapulars, back and rump grey with brown tinge; long uppertail-coverts and tail black; T1 with narrow white edges and tip, T2–T6 with broader white tip (c. 4 mm on T2 increasing to c. 12 mm on T6). Black merges into grey on lower breast, flanks and upper belly; lower belly and undertail-coverts white. Flight feathers blackish, outer primaries with white spot on edge of outer web; tertials dark brownish grey; alula, primary coverts and outer greater coverts blackish; inner greater, median and lesser coverts grey. Underwing-coverts and axillaries dark grey. Bill red with yellowish tip, eye wattle red, broad, with outer edge digitate, markedly so above eye; eyes yellow, iris with red outer ring; legs bright orange-red. Sexes alike. SIZE (1♂, 2♀♀): wing, ♂ 112, ♀ 110, 105; tail, ♂ 71, ♀ 78, 77; bill, ♂ 22, ♀ 23, 23; tarsus, ♂ 22, ♀ 23, 20.

Prionops gabela

IMMATURE AND NESTLING: not known.

Field Characters. Length 17·0–18·0 cm. Very similar to Retz's Helmet-Shrike *P. retzii*, with red bill and eye wattles; their ranges come very close in Angola but do not overlap. Smaller, head and neck black but back and wings grey, *lower breast* and upper belly *grey*, not black; *crest feathers curl forward* over bill; white spots on wings confined to outer primaries.

Voice. Unknown.

General Habits. Inhabits dry secondary forest, tangled thickets and clearings at 300–900 mm; not in *Brachystegia* woodland where Retz's Helmet-Shrike occurs. In canopy; often in forest underplanted with coffee (Hall 1960a). In small parties of 3–5.

Food. Insects including larvae.

Breeding Habits. Nothing known except season: lays July–Aug (possibly June: Hall 1960b; enlarged gonads Sept: da Rosa Pinto 1962).

References
Collar, N.J. and Stuart, S.N. (1985).
Rand, A.L. (1957).

Plate 28 (Opp. p. 415) *Prionops scopifrons* (Peters). Chestnut-fronted Helmet-Shrike. Bagadais à front roux.

Sigmodus scopifrons Peters, 1854. J. Orn., 2, p. 422; Mozambique.

Range and Status. Endemic resident, mainly E coastal woodland, and fragmented range inland from S Somalia to N Natal; locally common. Somalia, south of 01°N between 32 and 40 km south of Kolbia; Kenya along coast (but declining), and endangered population in central Kenya in Meru and Ngaia forests, formerly also Mt Kulal, Mt Marsabit and Embu District; Tanzania along coast (but declining with reduction of coastal forests) and inland in E Usambara Mts, Pugu Hills, Uluguru Mts and Udzungwa Mts (Magombera, Mwanihana), once N Pare Mts. Mozambique, coast south to mouth of Limpopo R., common north of Save R., uncommon south of it; also Amatongas Forest, Vila Paiva and Vila Fontes regions near Zimbabwe border. Zimbabwe, uncommon, in E at Honde Valley and confluence of Rusitu-Haroni; South Africa, rare, in Ndumu Game Res. on Pongola R. near Natal/Mozambique border. Not in Malaŵi (Newman 1991). Endangered race *keniensis* inhabits forest islands, where it has declined for unknown reasons, but not because of deforestation (Lewis and Pomeroy 1989). Possibly threatened in southern Africa because its habitat is being cleared for logging and for subsistence agriculture (C. J. Vernon and V. Parker *in* Harrison *et al.* 1997).

Description. *P. s. scopifrons* (Peters): Mozambique to Zimbabwe and Tanzania, intergrading with *P. s. kirki* in coastal Tanzania (Pangani and Dar es Salaam). Forehead golden chestnut, short stiff feathers forming a well-demarcated, plush 'cushion' (**A**); a grey band across forecrown; hindcrown and nape blackish grey; rest of upperparts paler grey, tinged brownish. Tail black, T1–T3 with white spot at tip, T4–T6 with broader white tip (*c*. 5 mm on T4, *c*. 14 mm on T6 where spot extends further as narrow wedge

Prionops scopifrons

A

along outer edge). Lores, ear-coverts, cheeks and chin blackish grey like hindcrown, merging through dark grey throat and sides of neck with grey breast, flanks and upper belly; lower belly and undertail-coverts white. Primaries blackish, central part of inner webs white, forming broken band across extended wing; secondaries blackish with grey outer webs; tertials grey; alula, primary coverts and outer greater coverts blackish, inner greater, median and lesser coverts grey. Underwing-coverts and axillaries dark grey. Bill red, darker at base, orange at tip; eye-wattle dark blue, broad with many small wart-like outgrowths; bare skin behind eye mauve; eyes golden yellow; legs coral red. Sexes alike. SIZE: (7 ♂♂, 4 ♀♀): wing, ♂ 97–101 (98·8), ♀ 98–101 (99·5); tail, ♂ 74–79 (76·9), ♀ 76–80 (78·5); bill, ♂ 21–23 (22·2), ♀ 21–22 (21·8); tarsus, ♂ 19–20 (19·8), ♀ 18–19 (18·5). WEIGHT: (Mozambique) 1 ♂ 28, 1 ♀ 27; (Zimbabwe) ♂ (n = 19) 27–33 (29·1), ♀ (n = 20) 26–38 (29·7).

IMMATURE: juvenile with forehead blackish brown, rest of body dull brown spotted whitish, with white undertail-coverts; wings browner than in adult, secondaries and tertials with narrow white fringes, alula and primary coverts tipped white; bill dull brownish; eye brown. Immature similar to adult, but slightly browner, and lacks golden chestnut forehead, top of head almost uniform sooty black except for a white patch separating forehead from forecrown; retains juvenile wing feathers; bill yellowish; no eye-wattle; eyes brown becoming yellow; legs pale brownish.

NESTLING: naked, blind, with mouth spots.

P. s. kirki (Sclater): S Somalia and coastal Kenya, intergrading with *scopifrons* in coastal Tanzania. Slightly paler and browner above than nominate race, crown band paler; paler below, grey chin and throat contrasting with blackish cheeks; white spots on primary inner webs smaller. WEIGHT: (Kenya) ♂ (n = 24) 20–34 (29·2), ♀ (n = 13) 26–38 (30·3), unsexed (n = 7) 33–40 (35·0).

P. s. keniensis (van Someren): central Kenya. Like nominate race but forecrown band dark grey, scarcely contrasting with hindcrown; white on each primary reduced to small mark on inner edge. Larger; wing, ♂ (n = 2) 107, 105, ♀ (n = 2) 108, 106.

Field Characters. Length 16·5–18 cm. A small helmet-shrike of coastal forest and woodland. Looks like a washed-out Retz's Helmet-Shrike *P. retzii*, with which it often flocks; largely grey rather than black, with broad *chestnut patch* on forehead. Has similar red bill and yellow eye but *eye wattles blue*. Immature greyish brown, paler below, with white patch on crown, white spots on upperparts and wings, white barring on underparts; lacks chestnut forehead and eye-wattles; eyes brown, bill and legs orange.

Voice. Tape-recorded (91–99, B, F, GREG, HOR, HRS, McVIC). Variety of calls, many difficult to describe adequately; repertoire generally similar to that of Retz's Helmet-Shrike (Short and Horne 1985). Song, a whistled, clear 'fyew-dyew-dewt' (Short and Horne 1985). Common call, not given by *P. retzii*, is a low nasal, grating 'zhwer-zhwer-zhwer-zhwer' or 'zhway-zhway ... ', which may be the repeated 'bddddt, bddddt, ddddt', resembling aggressive trill of Lesser Honeyguide *Indicator minor*, of Short and Horne (1985); also gives a louder and higher 'jee-op, jee-op, jee-op' and a typical helmet-shrike 'watch-winding' rattle, an up-slurred 'trrreee'. Also described as whirring, gobbling and chuckling calls, 'chair-rer wit wit chirro', 'tree, tree, tree', etc. (Maclean 1993) and 'unique loud churring or chattering with a strange nasal whirring or humming quality, often accompanied by softer whistled notes' (Zimmerman *et al.* 1996). Bill snaps often accompany 'tsee-zzee-zzeee-eep' calls, or end them, as 'zzee-tsip-ip' (snap), in a gun-like burst (Short and Horne 1985). Also a sharp 'shuk!' (alarm?).

General Habits. Occurs in lowland, usually coastal evergreen and riverine forests, miombo (*Brachystegia*) woodland, liana tangles, dense bushveld and mangroves. Mostly (88% of squares, Kenya) in semi-arid and dry subhumid areas below 500 m (Lewis and Pomeroy 1989), once 1230 m (Stevenson and Pearson 1986).

Gregarious, in groups of 4–12, av. (n = 16) 7 in nesting season, up to 30 at other times. Lively and noisy; birds maintain contact and respond to each other by sight and sound. One bird of group (the dominant one?) initiates chorus singing, others then respond. Flight buoyant and undulating. Flock forages in the canopy, also on ground. Shows preference for trees with smaller leaves. Searches branches, trunks and tangles, hopping from branch to branch; sometimes hangs upside down to secure food, also flops into leaf clusters possibly to flush prey (Harris and Arnott 1988); may cling to tree trunks. When perched, assumes a horizontal, crouched position and cocks head to one side. Also catches prey on wing like a flycatcher, hovering up to 30 s in front of tree blossoms before capturing prey (Short and Horne 1985). Allopreens. Roosts socially, group huddled together on branch. Readily mobs predator, dive-bombing it and snapping bill. Tends to be more aggressive than Retz's Helmet-Shrike (Short and Horne 1985). Occasionally forages with other species including Retz's Helmet-Shrike and White Helmet-Shrike *P. plumatus*, Black-headed Oriole *Oriolus larvatus*, Green Wood-Hoopoe *Phoeniculus purpureus* and woodpeckers.

Mainly sedentary; in S of range moves out of its nesting area and becomes nomadic (Harris and Arnott 1988).

Food. Invertebrates: caterpillars (up to 40 long: Short and Horne 1985), antlions, grasshoppers, beetles and spiders; also small vertebrates, and fruits including *Ochna* berries (fed to young: Britton and Britton 1977).

Breeding Habits. Co-operative breeder with members of a group helping to construct nest, incubate and brood and care for young. Group presumably consists of breeding pair with several helpers; all help to defend territory. Density of groups per unit area probably low since each group maintains a large territory (C. J. Vernon and V. Parker *in* Harrison *et al.* 1997). In courtship 2 birds bow and spread head feathers toward each other, then one bird feeds the other. Sometimes an adult solicits, crouches, makes 'eek' notes and spreads wings in front of another for up to 1 h (Short and Horne 1985). Also a solitary bird said to wave its wings in a circle to each side in the manner of a displaying, singing ♂ Common Starling *Sturnus vulgaris* for 10 min (Short and Horne 1985).

NEST: cup of thin bark and grass, bound together with fibre, moss, lichens and spider webs; ext. diam. 62 × 65, ext. depth up to 40, int. depth 24; 4·5–21·0 (av. 12) m above ground, in fork or on horizontal branch (up to 35 thick). Built by 3–4 (possibly all) members of group. Before entering nest, each bird with nest material in its bill bows to another and spreads its wings; then one by one each moves in, sits on nest, spreads both feathers, flirts wings, turns full circle, presses body down and places bits of material into nest. Occasionally 1 individual attacks another, preventing it from getting into nest. All members of the group wait until each has entered and left nest, then all fly off together. Group may be away from nest 15 min to 2 h.

EGGS: 3, pale grey, tinged turquoise, spotted and flecked with reds, greys and blues mainly in a ring close to blunt end. SIZE: (n = 3) 19·0–20·0 × 15·2–15·3 (19·5 × 15·3).

LAYING DATES: Kenya, Jan–July peaking Apr; Tanzania, Feb; Mozambique, Oct–Nov; Zimbabwe, Oct–Dec.

INCUBATION: period unknown; begins when clutch complete; members of a group (all?) take turns, each on nest up to 3·5 h. Adult on nest sometimes fed by other members of group; no ceremony when feeding or exchanging with bird on nest.

DEVELOPMENT AND CARE OF YOUNG: fledglings beg by squatting, lowering and quivering slightly opened wings; fed by all members of group; they huddle together on horizontal branch while waiting to be fed; roost with entire group at night.

References
Britton, P.L. and Britton, H.A. (1977).
Harris, T. and Arnott, G. (1988).
Short, L.L. and Horne, J.F.M. (1985).

Family ORIOLIDAE: Old World orioles and figbirds

Highly arboreal, thrush-sized, mostly colourful, sexually dimorphic, insectivorous (mainly caterpillars) and frugivorous birds, in warmer parts of Old World. 2 genera, one large (orioles) and one small (Australasian figbirds, *Sphecotheres*, with 2 species). Figbirds are like orioles but have bare, red skin around eye and are more frugivorous; otherwise their characters, and those of the family, are as given under genus *Oriolus*. Among Afrotropical birds, orioles are thought to be most closely allied to drongos (Dicruridae), crows (Corvidae) and cuckoo-shrikes (Campephagidae) (Sibley and Ahlquist 1990). Sibley and Monroe (1990) recognize 7 subfamilies in a greatly-expanded family Corvidae; one subfamily is composed of orioles and cuckoo-shrikes in one tribe and crows, birds-of-paradise and wood-swallows in 3 more tribes, and another of drongos, monarchs and fantails.

29 species, Europe to Australia (10 independent species and *c.* 7 superspecies).

Genus *Oriolus* Linnaeus

Bill about length of head; hook-tipped; varies from long and quite slender (*O. larvatus*) to robust, quite deep and broad (*O. crassirostris*), to rather short and bulbul-like (*O. brachyrhynchus*); bill very long in some races of oriental *O. chinensis*; nares feather-free, upper part covered with membrane; fine, short rictal bristles; P10 half length of P9, P8 longest; 12 rectrices; short tarsometatarsus, rather weak foot. Plumage of ♂♂ mainly brilliant yellow, black and olive and of ♀♀ streaky olive; one species black, several blackish, 2 claret coloured, others in Oriental and Australasian Regions are dull-coloured and so strongly plumage-mimetic of certain friarbirds (*Philemon*: Meliphagidae) that oriole-friarbird pairs are almost indistinguishable.

Arboreal, keeping to tops of high trees. Feed on caterpillars, other insects, and soft fruits. Tropical spp. resident, others strongly migratory. Vocal; varied repertoire, main contact call of most species a loud, ringing 'o-ri-ole' (name may be onomatopoeic or may be from Latin *aureolus*, = golden). Voices of African orioles confusingly similar, with too much overlap in both form and timbre to provide any clues to mutual relationships. Nest hammock-like, a rather thin, open cup slung from 2–3 twigs.

O. percivali and *O. larvatus* look and sound alike, hybridize, and form a superspecies (Dowsett and Dowsett-Lemaire 1993b). Prigogine (1978) assigned *O. percivali* to a superspecies with *O. monachus* and *O. nigricollis*, the 3 closely similar and allo- or parapatric. Hall and Moreau (1970) and Short *et al.* (1990) recognized a superspecies for *O. percivali*, *O. brachyrhynchus*, *O. crassirostris*, *O. monacha* and *O. larvatus*. Sibley and Monroe (1990) placed *O. brachyrhynchus* and *O. monacha* in one superspecies and *O. percivali* and *O. larvatus* in a second. *O. brachyrhynchus* and *O. nigripennis*, in equatorial forest canopy, are similar birds with almost exactly the same ranges (at least north of equator) and overlapping habitats; they often forage together. Either could form a superspecies with *O. percivali* and *O. larvatus*. *O. nigripennis* has the best parapatric fit.

Why do some Oriental orioles exactly resemble friarbirds? Is the cuckoo-shrike *Lobotos oriolinus* (Vol. IV, p. 273) a mimic of *O. brachyrhynchus* or is it in fact an oriole? Do *O. brachyrhynchus* and *O. nigripennis* mimic each other? How rapidly do visual and vocal differences amongst orioles evolve? Pending answers to these questions, we do not claim to understand evolution in *Oriolus* or relationships between its species, and restrict superspecies recognition to the hybridizing *O. percivali*/*O. larvatus* only.

27 species, Europe to China, Africa to Australia, with several in Philippines and including one large Wallacean superspecies and several small superspecies. 9 in Africa: 8 endemic and one Palearctic, the last (*O. oriolus*) sometimes united superspecifically with subsaharan *O. auratus*.

Oriolus nigripennis J. and E. Verreaux. Black-winged Oriole. Loriot à ailes noires.

Plate 29
(Opp. p. 462)

Oriolus (Barruffius) nigripennis J. and E. Verreaux, 1855. J. Orn., p. 105; Gabon River.

Range and Status. Endemic resident, rainforest zone, S Guinea to SE Sudan, and Cameroon and Bioko south to NW Angola.

Guinea, frequent in Kounounkan area (Hayman *et al.* 1995), uncommon on Ziama Massif (Halleux 1994). Sierra Leone, known from Bo, Nerekoro, near Sefada, upper R. Rokell, and Kenewa. Liberia, uncommon, range uncertain but seems to be only well inland (Gatter 1997); scarce on Mt Nimba, up to 900 m. Ivory Coast, common throughout forest zone (commoner than Western Black-headed Oriole *O. brachyrhynchus*), and records from gallery forests north to 09°30′N. Ghana, not uncommon in forest belt, also in forest outliers north to 08°N (Ejura, Kintampo); was formerly commoner than *O. brachyrhynchus*; now abundances are similar (Grimes 1987), though in forests in extreme SW much less common than *O. brachyrhynchus* (Dutson and Branscombe 1990). Togo, not uncommon, forest zone and forest outliers north to Ayagba, Tinkoro and Déguingué (08°05′N). Not recorded from Benin. Nigeria, common in SW between Lagos coast and Ife; uncommon in S around Benin and in SE; occurs on Obudu Plateau. Cameroon, uncommon in lowland rain forest but common in forest at 1050–2150 m at Dikome Balue, Rumpi Hills, at 1200–1750 m on Mt Nlonako, uncommon at 1750–2100 m on SE slopes of Mt Cameroon, frequent at 700–1500 m on S slopes of Mt Cameroon, also at Bamenda, Mt Kupé, Bali, Bamessing and Mt Oku; scarce in Korup Nat. Park, north to Baro. Bioko. Central African Republic, in SW uncommon in Dzanga Reserves and Bangui area, and along Bomu R. in SE. Sudan, known from Bengengai and Zande, and common in Imatong Mts up to 2000 m. Uganda, only in Bwamba, where rare.

Gabon, common and widespread (though less common than *O. brachyrhynchus*). Congo, common in secondary forest and near villages in Odzala Nat. Park, local in park-like forest; rare in Nouabalé-Ndoki Nat. Park, in open forest; local and rare in Mayombe. Zaïre, frequent and widespread in NE, status in NW ill-known, not certainly known to occur south of 01–02°S except in Bas-Zaïre where frequent between Kinshasa and Angola border. Angola, occurs quite commonly in Cabinda, Cuanza Norte, W Malanje and N Cuanza Sul.

Density of 2–5 pairs per km^2 (M'Passa, Gabon).

Oriolus nigripennis

Description. ADULT ♂: hood (forehead to nape, chin to upper breast) jet black, slightly glossy; upper mantle bright yellow, lower mantle and scapulars bright yellowish olive, back yellowish green, rump and uppertail-coverts bright olivaceous yellow. Tail rounded: T1 black (sometimes with greenish base in W of range), T2–T5 black with increasing amounts of yellow distally, T6 yellow with black base. Tail below bright yellow and black. Lower breast, sides of upper breast, flanks, belly, thighs, vent and undertail-coverts intense, brilliant yellow. Primaries black with narrow whitish outer edges, outer secondaries black with broad olive-yellow outer edges, inner secondaries with black inner vanes and outer vanes almost wholly bright olive-yellow, tertials with inner vanes black, outer vanes olive-yellow, outer edges yellower, tips of both vanes narrowly whitish; greater primary-coverts black, very narrowly tipped with white. Alula black; greater and median coverts with olive-yellow outer vane and blackish inner vane; lesser coverts bright yellowish olive. Underside of remiges silvery grey, paler proximally on inner edges; marginal coverts olive, lesser underwing-coverts and axillaries bright yellow, greater under primary coverts white. Bill red-brown or pinkish brown; eyes crimson or dark red; legs and feet dull bluish grey or lead grey. Sexes alike. SIZE (25 ♂♂, 10 ♀♀): wing, ♂ 112–130 (119), ♀ 110–124; tail, ♂ 70–85 (74), ♀ 73–84; bill, ♂ 20–25 (23·5), ♀ 20–25; tarsus, ♂ 21–23, ♀ 20–22. Cline of increasing size from Sierra Leone to Angola and NE Zaïre: ♂ wing, Sierra Leone and Ghana (n = 12) 110–120 (115), Cameroon and Bioko (n = 9) 114–124 (119), Angola and NE Zaïre (n = 13) 119–131 (125); ♂ bill (n = 38) av. 2 mm longer east than west of W Cameroon (Serle 1957). WEIGHT: ♂ (n = 2) 55, 59·2, 1 ♀ 50·2.

IMMATURE: head olive-yellow, blotched with black as adult feathers grow through, or blackish; rest of upperparts olive-green; chin and throat yellow with dusky streaks (feathers with blackish centres and yellow fringes), breast dark-streaked olive-green; tail like adult, wings like adult but black parts duller and olive-green parts yellower; greater coverts narrowly yellow-tipped; bill brown or pinkish brown, eyes brown. Juvenile has olivaceous black head (at stage when wispy down still attached above eyes), mantle to rump mottled, feathers with blackish centres and yellow-olive fringes, bright yellow patch on side of upper mantle and base of side of neck; chin to breast dark-streaked yellow, belly plain yellow; wings and tail like immature.

Field Characters. Length 19–21 cm. Overlaps broadly with Western Black-headed Oriole *O. brachyrhynchus*. Main distinction is almost *entirely black primaries*, with very narrow white edges and tips probably not visible in the field; wing of Western Black-headed has extensive grey central panel formed by outer edges of greater coverts and edges of outer secondaries, and white spot at tip of primary coverts. Tail from above black (no green central feathers), with yellow tips broadening outwardly, but this is not a useful character on a treetop bird; from below, folded tails of the two are almost identical. In Gabon, at least, they occupy different habitats, Black-winged in small clumps of trees and gallery forest, Western Black-headed in deep forest, but they do meet at the forest edge (Christy and Clarke 1994).

Voice. Tape-recorded (104, B, C, F, BRU, CHA, ERA, HUG, LEM, MAC). The voices of Black-winged and Western Black-headed Orioles are extremely difficult to distinguish because of the richness and similarity of their repertoires; published works describe short and simple vocalizations and give the impression that distinction is easy, but a comparative analysis of longer phrases shows the situation is complex (Chappuis, in press). Study of 24 songs of both species shows that the voice of Black-winged is higher-pitched than that of Western Black-headed, with rapid changes in pitch and intensity, some notes merging into others and leaving fewer gaps; one of the notes is usually given more emphasis than the others; for detailed analysis, see Chappuis (in press). Calls of Black-winged often turn into duets, first bird calling a short 'tyup' or 'teeyup', answered by 'co-wah', '*wee*-hoo' or a glottal 'woo-t'l'-*wah*'. Vocabulary of Black-winged includes up-slurred 'ko-*lip*', which is perhaps the 'ou-ik' of Brosset and Erard (1986) or the 'glouik' of Christy and Clarke (1994), said by those authors to be frequently given and distinct from Western Black-headed; also: down-slurred 'tyi-woh'; double 'wah-wah', like distant barking dog; 'k'*lee*yoh'; 'woh-*wee*hoo'; 'koo-*kee*wah'; partially glottal 't'lipoo-waddliay' and 'tyoo-tyoo-ko'l'wah'; 'hee-haha, hoo-*hee*-hoo'; a measured 'ho, hee, her, koher'; deeper 'hoo-you-kyuwor'; a whistle with a fast introductory 'tityipu'; a rising series, 'tyipoo, hoo-hay-hee', and a somewhat staccato 'tyee-hippo-harbolah'. For further examples see Brosset and Erard (1986) and Christy and Clarke (1994). Said to imitate other species, including Common Greenshank *Tringa nebularia* (F. Dowsett-Lemaire, pers. comm.). Aggressive/alarm call a nasal 'tyipu-wrrraaaya', more rolling than that of Western Black-headed, with no change in pitch.

General Habits. Inhabits primary and secondary forests, also mangrove (Lagos, Nigeria). In Gabon, a bird strictly of secondary forest, copses and gallery forest, around villages and native cultivation: solanaceous and parasol-tree bush in recently-abandoned slash-and-burn patches of farmland in forest, and large trees remaining in new plantations; never penetrates primary forest (Brosset and Erard 1986). Keeps to middle and upper storeys, tends to live higher in forests than Western Black-headed Oriole: the 2 have remarkably similar distribution and status, occur together in numerous localities, and sometimes feed in same tree. In N Zaïre ranges further along gallery forests than does Western Black-headed Oriole, but in central Zaïre their habitats seem exactly alike (Chapin 1954).

In pairs and family parties of up to 4 birds. Territorial and aggressive, chasing away its own species and Western Black-headed Orioles. Does not join mixed-species flocks. Forages in thick foliage; gleans caterpillars from leaves and takes insects from undersides of leaves in hovering flight (Christy and Clarke 1994). 3 birds foraging together constantly vibrated wings and flirted tails (Bannerman 1939). Sedentary.

Food. Beetles, winged ants, caterpillars, small fruits and pulp and seeds of figs in stomachs. Of 6 stomachs (Zaïre), caterpillars in 3, other insects in 3, fruits in 2. Seen to take butterflies, small orthopterans, and fruit of *Rauvolfia*, *Ficus* and *Musanga* (Brosset and Erard 1986).

Breeding Habits. Barely known. 3 adults together, seemingly searching for a nest-site (Sierra Leone, Oct) and 3 adults together (Nigeria, Feb) (Bannerman 1939) – pairs with a helper? Singing ♂ adopts special posture, neck held erect and puffed out, tail spread (Christy and Clarke 1994). Sings all year.

NEST: one, 22 m high in *Ficus* tree, made of dry herbs and fibres with much moss and *Usnea* lichen worked into outside of wall; suspended under fork near end of branch (Brosset and Erard 1986).

EGGS: unknown.

LAYING DATES: Liberia, (large gonads Sept); Nigeria, Mar–June; Ivory Coast, (nest building Jan, full-grown young June); Ghana, Feb; Gabon, (nest-building Oct, Dec, Feb, juvs Dec); Zaïre, Niangara, (fledgling, Nov), Djugu, (juvenile, Aug); Angola, (large gonads Feb).

References
Bannerman, D.A. (1939).
Brosset, A. and Erard, C. (1986).
Chapin, J.P. (1954).
Christy, P. and Clarke, W. (1994).

Plate 29
(Opp. p. 462)

Oriolus brachyrhynchus Swainson. **Western Black-headed Oriole. Loriot à tête noire.**

Oriolus brachyrhynchus Swainson, 1837. In Jardine Nat. Lib.., Orn., 8, Birds West Africa, 2, p. 35; Sierra Leone.

Range and Status. Endemic resident, rainforest zone from Guinea to W Kenya, and Cameroon south to NW Angola. Common and widespread except as noted.

Guinea, known from between Koundara and Gaoual in NW Fouta-Djalon (Morel and Morel 1988); Ziama Massif, up to 900 m (Halleux 1994) and Nzérékoré. Sierra Leone, west to Guinea border (but not known from Kounounkan, just into Guinea). Liberia, common almost throughout, up to 1400m. Ivory Coast, throughout forest zone, and records from gallery forests north to 09°30'N. Ghana, from coast

Oriolus brachyrhynchus

north to 07°30′N: Kumasi, Akropong, Begoro, Agogo, Mpraeso, Worawora, Hohoe, Amedzofe, Tafo, Bia Nat. Park. Togo, forest zone and forest outliers north to Déguingué (08°05′N). 1st records Benin, Lama Forest, 1998 (Waltert and Mühlenberg 1999). Nigeria, uncommon in SW, rare on Lagos coast, not uncommon in forest in Benin district and Niger delta; in SE, north to Obudu Plateau. Cameroon, in rainforest zone and in larger gallery forests; on Mt Cameroon, occurs at 400–700 m on S slopes and at 900–950 m on SE slopes. Central African Republic, widespread near Congo border (though uncommon in Dzanga Res.) north to Mbaïki. Sudan, bird collected in Sakure Forest in 1952 and 1 seen in Bengengai Forest in 1986 (Hillman and Hillman 1986). Uganda, local in W and S at 700–1800 m: Impenetrable and Malabigambo Forests to Sezibwa R., Entebbe, Kifu and Mabira; also on Mts Moroto and Elgon in E. Kenya, only Kakamega.

Gabon, probably throughout; abundant around M'Passa. Congo, Odzala and Nouabalé-Ndoki Nat. Parks in all types of forest; local in Mayombe. Zaïre, Ituri, Itombwe at 620–1470 m, Kasai Occidental; not certainly known from Bandundu Prov.; in Katanga only a single record, near Angola/Zambia border (Hall and Moreau 1970), a misidentification? Angola, presumably in Cabinda; only record is at Canzele, 08°17′S, 15°11′E, Cuanza Norte.

Densities of 10–12 pairs per km^2 in primary forest and 6–8 pairs per km^2 in old secondary forest, M'Passa, Gabon, and 8–15 pairs per km^2 in high forest in Liberia.

Description. *O. b. brachyrhynchus* Swainson: west of Benin (or west of 10°E; population in Kumba, Cameroon, variable in respect of amount of yellow in upper mantle and may represent intergrades between nominate race and *laetior*: Serle 1950). ADULT ♂: hood (forehead to nape, chin to upper breast) jet black, slightly glossy; black hood sharply delimited from bright yellow upper mantle; lower mantle, scapulars and back, uniform bright yellowish olive-green, rump and uppertail-coverts bright yellowish green. Tail evenly graduated; T6 14–17 shorter than T1; T1 dark olive; T2 with proximal half olive, tip yellow (patch 6 mm long) and intervening area black; T3–T6 mainly black, with progressively less olive at base, and with bright yellow tips respectively 9, 10, 14 and 23 mm long at shaft. Underside of tail yellow and dull black. Black upper breast grades through dusky yellow in midline of lower breast (and dirty brownish yellow at sides) into clear, intense, bright yellow belly, flanks, thighs, vent and undertail-coverts. Primaries and outer secondaries blackish, narrowly edged and tipped white, inner secondaries black on inner vane, olive on outer vane, broadly edged and narrowly tipped grey, tertials with inner vane olivaceous black, outer vane olive-green, yellower at edge, both vanes paler at very tip; greater primary-coverts black, outer ones with white tips 3–4 mm long, inner ones with smaller white tips, all tips overlapping in closed wing to form conspicuous white patch 10 mm long; alula black, greater coverts greyish olive, median coverts olive-green, lesser coverts bright yellowish olive. Underside of flight feathers silvery grey, inner borders proximally white; marginal underwing-coverts olive, lesser underwing-coverts yellow, greater under primary coverts dark grey and whitish, axillaries bright yellow. Bill red-brown or pinkish brown; eyes dark carmine-red; legs and feet bluish, blue-grey or dark grey. Sexes alike but ♀ has chestnut brown, not red, eyes, and may have slightly smaller patches of yellow in tail. SIZE: (adult, 9 ♂♂, 9 ♀♀, Mt Nimba, Liberia) wing, ♂ 116.7 ± 3.5, ♀ 113.4 ± 3.2; tail, ♂ 83.2 ± 3.9, ♀ 79.5 ± 3.7; bill to feathers, ♂ 23.4 ± 0.7, ♀ 22.5 ± 1.0; (imm., 4 ♂♂, 2 ♀♀, Mt Nimba) wing, ♂ 114–120 (116), ♀ 110, 115; tail, ♂ 80–84 (81.3), ♀ 76, 83; bill to feathers, ♂ 22–24 (22.8), ♀ 22, 22 (Colston and Curry-Lindahl 1986); (8 ♂♂, 8 unsexed, mainly Sierra Leone) wing, ♂ 112–116, unsexed 110–115; tail, ♂ 76–88, unsexed 77–89; bill, ♂ 21–23, unsexed 20–22; tarsus, ♂ 21–22.5 (Bannerman 1939). WEIGHT: ♂ (n = 9) 50.7 ± 4.5, ♀ (n = 9) 47.4 ± 6.1, imm. ♂ (n = 4) 42.2–56.7 (49.2), imm. ♀ (n = 2) 43, 49.2.

IMMATURE: like adult but lacks black hood, and greater upperwing-coverts and median primary coverts tipped yellow. Forehead, crown, cheeks and hindneck dark olive, grading insensibly into brighter, yellower olive of mantle; lores yellowish and featherlets around eye yellow; chin and throat olive-green, streaked with yellow; breast and flanks olive-yellow, faintly mottled, grading into intense, bright clear yellow centre of belly. Upper mandible purplish, lower mandible dark at tip, pink-yellow at base; eyes brown or reddish brown; legs and feet grey-blue, grey-green or light blue. Juvenile has crown and cheeks yellowish olive, back mottled, feathers with dusky centres and very pale yellow tips, breast and flanks pale yellow with broad dusky streaking (Chapin 1954).

NESTLING: bill very pale pink with greyish tip, eyes brownish grey, feet light blue, claws dusky; at approximately one week, nestling mostly still naked, top of head with almost invisible, thin, pale fluffy down; back, flanks and wings feathered, feathers brown with yellow tips; tail a bright yellow tuft of feathers 8–10 mm long.

O. b. laetior (Sharpe): east of Benin or east of Nigeria (see above). Differs from nominate race in having mantle bright yellow, contrasting strongly with black hood and forming a yellow collar; rump olive-yellow (not olive); and upper breast black, sharply demarcated from bright yellow of lower breast and sides of neck; tail less graduated (T6 5–11 shorter than T1), and yellow patches larger (♂ T6 with yellow patch 31–36 long at shaft (n = 3), ♀ T6 22–32 (n = 3)).

Field Characters. Length 20.5–21.5 cm. Distinguished from Black-winged Oriole *O. nigripennis* by *pale panel and white spot* on closed wing; for further differences, see that species. Very similar to Eastern Black-headed Oriole *O.*

larvatus but their preferred habitats differ, Western in forest and Eastern in dry country, though they may meet at forest edge. Both have white edges to primaries and white spot on primary coverts, but in Western, rest of closed wing looks much more uniform, grey and green rather than black with yellow or white edgings: outer secondaries *broadly edged grey*, inner secondaries *yellow-green*; outer greater coverts grey, inner ones yellow-green (all of them grey in Eastern); in the hand, spread wing shows inner edges of inner secondaries black in Western, white in Eastern. Central tail feathers uniform yellow-green, not darkening terminally as in Eastern; from below, *tail yellow* with *black* base (all yellow in Eastern). Immature said to be very bulbul-like, and has even been confused in the hand with Golden Greenbul *Calyptocichla serina*.

Voice. Tape-recorded (104, B, C, F, BRU, ERA, HUG, KEI, LEM, SALA, ZIM). Voice very similar to that of Black-winged Oriole but rather deeper, notes of songs more clearly separated and changing in frequency more slowly. Song repertoire extremely broad; more than 50 variations assembled by Chappuis. Commonly given is a rich, mellow, fluty 'waw-hah', second note higher, sometimes shorter, 'waw-chop'. Song may be a single long, down-slurred 'tyaww', but more typically 3-syllabled: 'waw-*kwee*-hoo', 'hee-ku-waw', 'hip-oo-hooah' (last note up-slurred), '*whee*, hoochop', 'waw-wohah', 'chip-wooy-haw' or a drawn-out, measured 'wheep … wah … whaw'; a low-high-low sequence, 'waw-*hawee*-haw', is reminiscent of duet of Tropical Boubou *Laniarius aethiopicus*; brief introductory notes produce 4-5 syllable 'chopchip-wee-haw', 'chipo-tyay-wah' or partly glottal 'tiddly-waw-kaw-l'here'. For other renditions, see Brosset and Erard (1986), Christy and Clarke (1994) and Zimmerman *et al.* (1996). Aggressive/alarm call nasal but less rolling than that of Black-winged Oriole, rising in the middle and falling, 'jewi-jaaa'.

General Habits. Inhabits mature and secondary forest, forest edges and clearings (Ghana), gallery forests and savanna woodlands (Liberia), lowland forest clearings with shrubs and tall trees, forest treefall gaps, logging roads and (Mt Nimba, Nimba) ridge forest, primary and old secondary and regenerating forest in Gabon, the last if not too open; often near villages and in riparian forest. Solitary or in pairs (Dutson and Branscombe 1990), keeping mainly to canopy but occurs in middle storey; rarely descends to within 7 m of forest floor but pair has been seen chasing only 1 m above ground. Searches for small arthropods on thin branches and epiphytes, and makes sallies for flying insects from forest canopy (Gatter 1997). Often in mixed-species flocks. Tends to live not as high in forests as Black-winged Oriole *O. nigripennis*: the 2 have remarkably similar distribution and status, occur together in numerous localities, and sometimes feed in same tree. Their foraging behaviour is identical (Brosset and Erard 1986). Lives in territorial pairs, birds becoming a little separated but keeping in contact vocally. Responds to person imitating contact call and can approach him closely. Sedentary.

Food. In stomachs mainly hairy caterpillars, also smooth caterpillars, winged ants, beetles, bugs, large hard seeds, fig-like fruit pulp (Liberia); 11 stomachs (Zaïre) contained mainly caterpillars, up to 7 in each; also beetles, winged ants, a spider, and fruit with hard seeds. In Gabon seen to eat fruits and seeds of *Ficus*, *Macaranga*, *Croton* and *Allophyllus*; also large orthopterans and caterpillars. Eats capsule fruits of *Macaranga assas* and berries of *Ficus mangifera*.

Breeding Habits. Solitary nester, monogamous, territorial. Territory size 6–8 ha. 2 birds chased each other near forest floor, calling excitedly, with tail spread and depressed, June (Demey and Fishpool 1994). ♂ displaying in front of ♀ is sometimes accompanied by full-grown young bird (Brosset and Erard 1986); ♂ flexes legs, raises bill, head and hindquarters, droops wings, spreads tail, and bows rapidly to ♀ whilst singing; pair chases in and just under canopy. Singing and territorial pursuits all year (Gabon).

NEST: a cup made of dry vegetable matter with moss worked into outside (sometimes with spider web) and strands of lichen *Usnea* hanging below, slung between horizontal fork at end of small branch, about 7 m high in isolated understorey tree (Brosset and Erard 1986), and 18, 20 and 35 m high (Gatter 1997).

EGGS: 2 (Gabon). Not described; a clutch taken in Itombwe was broken.

LAYING DATES: Liberia (nest-building Mar, May, June; feeding at nest May; dependent young May; displays Dec, Mar); Ivory Coast (pair displaying, Nov; collecting nest material Apr: Gartshore *et al.* 1995); Ghana, (large gonads Apr); Nigeria, (juv. Mar); Gabon, (adult feeding young, Nov, occupied nest, Dec; singing and territorial pursuits most intense in Sept–Feb); Zaïre, Ituri, (young just out of nest, Nov); Zaïre, Itombwe, Mar (and gonad evidence of breeding Jan–Apr and June–July); E Africa, Region B, Aug.

INCUBATION: by ♀ only (Brosset and Erard 1986).

References
Bannerman, D.A. (1939).
Brosset, A. and Erard, C. (1986).
Chapin, J.P. (1954).
Prigogine, A. (1971).

Oriolus crassirostris Hartlaub. São Tomé Oriole. Loriot de São Tomé.

Plate 29
(Opp. p. 462)

Oriolus crassirostris Hartlaub, 1857. Syst. Orn. Westafr., p. 266; São Thomé.

Oriolus crassirostris

Range and Status. Endemic resident, São Tomé Island, locally common in primary forest from sea level up to 1600 m. Thought, from calls, to be locally numerous in 1940s (Snow 1950), and in 1971 thought to be commonest in NW, where density of 1–2 pairs per 25 ha (de Naurois 1984). Seems to have declined, perhaps due to prevalence of pesticide use in mid-altitude plantations in early 1970s (de Naurois 1984); formerly occurred in the extensive cocoa plantations but does not now do so. On Rolas Islet, at least formerly. Presently restricted to the more remote forests mainly in SW; recently encountered at Lagoa Amélia, Rio Ana Chaves, Rio Xufexufe, Binda, Santa Catarina, Ermelinda, Rio Quija, Nova Moca, Zampalma, Monte Mario, São Nicolau and Morro Peixe. 30 birds seen or heard in primary forest on Rio Xufexufe on 4 days in Aug and 110 at Lagoa Amélia on 3 days in Dec and Aug (Sargeant 1992). Any development of remaining forests areas may seriously endanger this oriole, which should be given 'Rare' status in Red Data Book (Atkinson *et al.* 1991).

Description. ADULT ♂: hood (whole head, neck and breast) black, slightly glossy on crown, matt on throat and breast. Of 6 ♂♂ examined, 5 (not fully mature?) have throat finely freckled with white and one of the 5 has large, slightly asymmetrical olivaceous grey patch on forehead and crown (and see under IMMATURE). Upper mantle very pale yellow with broad, diffuse olive-grey streaks; lower mantle, scapulars and back greyish olive-green; rump paler, with tips of fluffy flank feathers making whitish or pale green patch at sides of rump, or even narrow band across it; uppertail-coverts yellowish grey-green. Tail: T1 dark olive, dusky near tip; T2–T6 black with yellow tips which are 13, 16, 19, 24 and 28 mm long on inner webs of T2–T6 respectively but are considerably shorter on outer webs of T2–T5 (because of intruding long black bulges) though longer on outer web of T6. Belly, flanks and thighs white, blotched grey where dark feather bases show through; vent and undertail-coverts pale yellow. Primaries slate-black with narrow white outer edges and broad white crescent at tips; secondaries with slate inner web and blue-grey outer web with neat white edge and tip; tertials with blackish inner and greyish outer webs, all suffused with green. Greater primary-coverts black with white tips, forming conspicuous white patch in closed wing; alula black, greater, median and lesser coverts dark olivaceous grey. Underside of flight feathers silvery grey; greater under primary coverts white with blackish patches, other underwing-coverts white and grey-white. Underside of tail mainly bright yellow. Bill robust, deep and broad; dull carmine or brownish red; eyes red; legs and feet grey. ADULT ♀: like ♂ but hood dark blackish, cheeks blackish streaked with white, and chin, throat and breast buffy greyish white with heavy, diffuse, broad blackish mottles; lower edge of breast not nearly as well demarcated from belly as in ♂; belly greyish white, or buffy or dirty cream; wings blackish where ♂ black; tail with T2–T6 blackish where ♂ black, underside pale yellow and olive-green. SIZE (10 ♂♂, 3 ♀♀): wing, ♂ 122–131 (128), ♀ 122–132 (127); tail, ♂ 85–103 (97·4), ♀ 73–94 (88·3); bill to feathers, ♂ 25–28 (26·7), ♀ 25–26·5 (25·5); tarsus, ♂ 27–30 (27·4), 1 ♀ 28. WEIGHT: 1 ♂ 50, 1 ♀ 50 (de Naurois 1984).

IMMATURE: forehead, crown, nape, neck and mantle olivaceous dark grey, back and rump more olive-green, wings much browner, less colourful and less patterned, white tips of greater primary coverts less distinct than in adult and not forming a patch, greater and median coverts with buffy white outer edges and tips; chin and throat off-white in midline, mottled greyish white at sides, breast whitish, strongly marked with diffuse dark brown streaks, belly creamy white. Few skins available, but field observations suggest considerable variation in and complexity of juvenile and immature plumages (de Naurois 1984), many birds with pale streaked grey-green crown, white chin, throat and breast; one had 'large white eye-ring' (Sargeant 1992), also described by Christy and Clarke (1998).

Field Characters. Length 23–24 cm. The only oriole on São Tomé; like a washed-out mainland species, without yellow. Adult has black head, olive-green back and yellow-green rump, white underparts and hind collar; tail black with yellow tip, broader on underside. Bill noticeably thicker than other African orioles. Immature has olive-grey on top of head, upperparts like adult but duller, underparts white with a few dark breast streaks.

Voice. Tape-recorded (104, B, ALEX, CHR, JO PJ, TYE). Whistled songs slower, more drawn-out and deeper than other orioles, also richer and mellower than any of the black-headed group; felt to have a flavour that is both joyful and melancholy (de Naurois 1984). Phrases very variable, as in other species: a long, mellow, down-slurred 'tyeeow'; 'way-*whee*-ya', 'chip-aw-*hah*-aw', '(tiptip)-*wee*yoo-waw', 'hugh, ha-hoo, ko-wow', 'hik-kuway-kuwow' or an ascending 'ko-ku-*waa*yoo'. Aggression/alarm shorter than other orioles, high-pitched, harsh 'keea'.

General Habits. Inhabits primary forest on slopes and undisturbed, mature secondary forest; has been seen in cocoa plantations (de Naurois 1984) and once in savanna

(Atkinson *et al.* 1991). A rather large, relatively long-tailed oriole with long, robust bill; probably derivative of Western Black-headed Oriole *O. brachyrhynchus*. Keeps to mid storey of forest and often seen at edges, occurring singly or in pairs. Prone to call particularly in early morning, when 2–3 can be heard at same time. Singing birds all repeat same motif constantly for several min., then all suddenly change to a new motif, sung for several min (Christy and Clarke 1998). Sedentary, though may wander.

Food. Seeds and invertebrates including bugs and beetles.

Breeding Habits. Poorly known.
NEST: one was *c.* 9 m above ground, slung between 2 nearly horizontal branches of tree overhanging forest path; nest was being built by one bird, thought to be ♂, and movements included sitting in nest, pushing from inside, revolving body, and shaking wings with tail elevated (Snow 1950). Another nest, long abandoned, 11 m above ground, was suspended between 2 arms near leafy end of branch and contained rotting leaves and shreds (de Naurois 1984).
EGGS: unknown.
LAYING DATES: (nest being built mid Sept: Snow 1950; anatomical evidence of breeding Aug–Dec, also oviduct egg Sept and ♂ with well-developed gonads Feb: de Naurois 1984 [but *cf.* for same bird 'gon. non dév.', p. 126, and 'testicules très développés', p. 131]; young in early July Atkinson *et al.* 1991) and Feb (Christy and Clarke 1998).

References
Atkinson, P. *et al.* (1991).
Christy, P. and Clarke, W. (1998).
de Naurois, R. (1984).
Snow, D.W. (1950).

Plate 29
(Opp. p. 462)

Oriolus percivali Ogilvie-Grant. Mountain Oriole. Loriot de Percival.

Oriolus percivali Ogilvie-Grant, 1903. Bull. Br. Orn. Club, 14, p. 18; Kikuyu.

Forms a superspecies with *O. larvatus*.

Range and Status. Endemic resident, E African and Albertine Rift montane forests. In Uganda and Kenya widespread and fairly common at 1530–3000 m (mainly 1850–2450 m): in Ankole, Kigezi and Toro districts north to Kibale Forest; Mt Elgon (Bumasifa, Mbale, Kolosia, Bukedi, Mangiki, Kinothoa, Kiptogot; and west towards Kumi: Hall and Moreau 1970), Cherangani and Tugen Hills, Nandi, Mau and Eldama Ravine, Laikipia Plateau, Aberdares, remnant forests north of Nairobi, Thika, Mt Kenya and Nyambeni Hills; also Trans-Mara. Zaïre, near Geti, north of Semliki Valley; Rutshuru; not in Rwenzoris; in highlands west of L. Edward, Virunga and Kivu Volcanoes at 1630–2490 m, around L. Kivu and on Idjwi I., and south to Itombwe, Baraka and Mt Kabobo. Common in Itombwe, mainly at 1800–2450 m, exceptionally down to 1550 m. Rwanda, in NW, and rather common throughout Nyungwe at all altitudes; Gishwati, Bugesera. Burundi, north-east of Bujumbura; Teza; not in Bururi Forest, SW Burundi. Tanzania, only on Mt Mahari.

Oriolus percivali

Description. ADULT ♂: hood (forehead to nape, chin to upper breast) black, slightly glossy; mantle bright yellow, scapulars and back olive-green, rump and uppertail-coverts olivaceous yellow. Tail square-ended: T1 black. T2 black with small band of yellow at tip, T3–T6 black with small area of bright yellow at ends – on T6 yellow patch 23–32 mm long at shaft. Tail below bright yellow and black. Lower breast, sides of upper breast, belly, flanks, thighs, vent and undertail-coverts intense, brilliant yellow. Primaries blackish with narrow grey-white outer edges and tips, secondaries black with 3-mm-broad white outer edges and tips, tertials black with most of outer vanes and tip of inner vane on each feather bright yellow; greater primary-coverts black, 3–4 outer ones with 7–10 mm long white tips, inner ones with small white tips; white marks forming a conspicuous white patch in wing. Alula black; greater coverts black with outer vanes grey at base and broadly edged yellowish white towards tips; median coverts fringed pale yellow, with triangular black centres, lesser coverts olive. Underside of flight feathers silvery grey, paler and creamier proximally on inner edges; marginal coverts dusky olive, greater under primary coverts white. Bill brown-red or brownish pink; eyes deep carmine-red; legs and feet grey or blue-grey, claws blackish. Sexes alike. SIZE: (sexes combined, n = 202, Prigogine 1978) wing 122–141 (132 ± 4·2); tail 63–72·5 (68·1 ± 1·7); bill 21·5–27 (24·1 ± 1·1).

IMMATURE: like adult but top and sides of head streaky olivaceous blackish brown, mantle olive in centre, yellower at sides, diffusely dark-streaked; chin and throat dark olive streaked with black; breast dull yellow, softly striped with pale olive, upper breast with thin black stripes; wings and tail olivaceous, much less boldly marked in black, white and yellow than in adult; bill blackish; eyes dark brown; legs and feet light blue-grey.

NESTLING: not described.

TAXONOMIC NOTE: closely related with and often considered a race of *O. larvatus*; voices are not always distinguishable; has also been treated as a race of *O. nigripennis*. Sometimes hybridizes with *O. larvatus* (Mt Elgon, Rongai, 6 localities from Narok to Mt Kenya, Kikuyu highlands: Prigogine 1978). Introgression could increase with increasing contact as forests are cleared.

Field Characters. Length 19·5–21·5 cm. A montane forest oriole, usually ecologically separated from very similar Eastern and Western Black-headed Orioles *O. larvatus* and *O. brachyrhynchus* but sometimes meets them at forest edge. Plumage colours more contrasting than either: yellow brighter, more extensive, entire back yellow, less olive, black deeper (Zimmerman *et al.* 1996). Central tail-feathers *black*, not olive-green; from below, outer half of closed tail yellow, basal half black, as in Western (solid yellow in Eastern). Shorter-tailed than Western, much *shorter-tailed* than Eastern. More black in wings than either Eastern or Western, with less extensive white edging than in Eastern, inner secondaries black with only outer edges yellow; greater coverts mainly yellow-green, outers with a little grey; area of secondaries between greater coverts and white spot on primary coverts black, not grey. Immature only faintly streaked below dark throat (heavily streaked in Eastern: Zimmerman *et al.* 1996).

Voice. Tape-recorded (104, B, GREG, LEM). Song, short, fluid phrase of 2–4 notes, similar to that of Eastern Black-headed Oriole but higher-pitched than Western; includes a short, rather clipped, '*tyee*-woh', down-slurred 'ti-*tyao*-hoh' and longer, up-slurred 'chit-chilu*weee*' (see sonagrams in Dowsett-Lemaire 1990). Duetting common, ♂ giving loud 'weeka-ku-*weeu*', ♀ replying with higher 'weekla-wee-er' (Zimmerman *et al.* 1996).

General Habits. Inhabits montane forest, submontane woods, well-timbered hillsides, evergreen forest, remnant forests and adjacent wooded farmland and gardens in Uganda, primary, secondary and gallery forests in Itombwe, Zaïre. Singly or in pairs, often in groups of 3–4; forages in canopy, upper and middle strata of forest; often in liana tangles; sometimes descends to understorey, where snatches and flycatches insects only 3–4 m above ground; up to 6 birds may join mixed feeding flocks of insectivores (Dowsett-Lemaire 1990). Sedentary, but perhaps also locally migratory: birds seen moving through Laikipia Plateau, Kenya, in Aug (Short *et al.* 1990)

Food. Of 5 stomachs, fruits in 3, hard insects in 2 and caterpillars in 2; one caterpillar hairless, 62 mm long. Small fruits include *Urera*.

Breeding Habits. Almost nothing known.

LAYING DATES: Zaïre, Ruzizi, (very young bird in Aug), Itombwe and Baraka, (large gonads May–June, Sept; juveniles Apr, Sept–Nov); probably breeds in S of Zaïre range in Feb–Sept (Prigogine 1971, 1984). E Africa, Region B, Feb, Region D, Apr–June (2 records).

DEVELOPMENT AND CARE OF YOUNG: adult fed fledgling away from nest in bamboo zone at 2490 m, Impenetrable Forest, Uganda, late Oct; fledgling black-billed (T.M. Butynski, pers. comm.).

References
Chapin, J.P. (1954).
Dowsett-Lemaire, F. (1990).
Prigogine, A. (1971, 1978, 1984).

Oriolus larvatus Lichtenstein. Eastern Black-headed Oriole. Loriot masqué.

Oriolus larvatus Lichtenstein, 1823. Verz. Doubl., p. 20; 'Terr. Caffror', = Cape Province.

Plate 29
(Opp. p. 462)

Forms a superspecies with *O. percivali*.

Range and Status. Endemic resident, E and southern Africa. Sudan, fairly common in SE. Ethiopia, widespread and common in SW, sparse and local in SE near Kenya and Somalia borders. Somalia, 3 old records in NW; frequent to common in riverine and coastal-plain woods in S. Uganda and Kenya, common and wide-ranging except in NE of each country; up to 2300 m.

Tanzania, throughout; paucity of records west of 32°E reflects lack of observers (N. E. Baker, pers. comm.). Common to abundant in SE Zaïre, at least formerly in Katanga, though sparse around Lubumbashi; common in Marungu Highlands and adjacent shores of L. Tanganyika. Angola, almost throughout but absent from NW and perhaps NE and reaches coast only between Rio Donde and Benguela. Namibia, restricted to far N and NE; frequent in Caprivi Strip. Zambia, common throughout: in 98% of squares. Malaŵi, very common below 1550 m, occurs up to 2000 m in Mafinga Mts. Zimbabwe, widespread, sparse in dry and common in mesic woodlands; absent from W Hwange Nat. Park; only sporadic as high as 1500 m in E Highlands. Mozambique, common and widespread in S, probably but not certainly present in N. Botswana, frequent to locally very common in N and E, and common in SE west to Kanye. Transvaal, common and widespread except in treeless areas and in SW and SE; Orange Free State, records in N (Harrison *et al.* 1997); Swaziland, throughout; Natal, frequent and widespread, up to 1200 m; Cape, frequent to common in E and S, resident west to Mosselbaai and wanders west to Cape Town.

Oriolus larvatus

Breeding density of about 1 pair per 50 ha (Nylsvlei woodland, South Africa). Densities of 6 birds per 100 ha of acacia woodland, 19 of miombo, 22 of mopane and 13 birds per 100 ha of other types of broad-leaved woodlands, S Mozambique (Parker 1999). May be expanding range with urban development in e.g. E Botswana.

Description. *O. l. larvatus* Lichtenstein: E South Africa, S Mozambique and S Zimbabwe (where intergrades clinally with following race). ADULT ♂: hood (forehead to nape, chin to upper breast) black, slightly glossy; mantle bright yellow, scapulars and back olive-green, rump and uppertail-coverts olivaceous yellow. Tail square-ended: T1 olive, duskier (somewhat blackish) towards tip; T2 olive, glossy black towards tip, usually with narrow, bright yellow line around very end; T3–T6 black with bright yellow ends – yellow patches respectively 11, 18–21, 22–30 and 31–37 mm long at shaft. Tail below bright yellow and black. Lower breast, sides of upper breast, belly, flanks, thighs, vent and undertail-coverts intense, brilliant yellow. Primaries blackish with narrow grey-white outer edges and tips, secondaries black with yellowish pale grey edges 1–2 mm wide, tertials black, with longitudinal outer half of outer vane olive and a bright yellow teardrop-shaped mark at tip; greater primary-coverts black, 3–4 outer ones with 7–10 mm long white tips, inner ones with small white tips; white marks forming a conspicuous white patch c. 7–10 × 10–15 mm long in extended wing. Alula black; greater coverts black with outer vanes grey at base and broadly edged pale yellow towards tips; median coverts fringed pale yellow, with triangular black centres, lesser coverts olive. Underside of flight feathers silvery grey, paler and creamier proximally on inner edges; marginal coverts dusky olive, lesser underwing-coverts and axillaries bright yellow, greater under primary coverts white. Bill coral-red or brown-red; eyes red (alizarin-scarlet); legs and feet slate grey. Sexes alike. SIZE: wing, ♂ (n = 25) 130–143·5 (139), ♀ (n = 24) 127–142 (135); tail, ♂ (n = 25) 79–100 (89·1), ♀ (n = 24) 77·5–93·5 (87·3); tarsus, ♂ (n = 18) 21–24; bill to feathers, ♂ (n = 25) 25·5–33 (28·5), ♀ (n = 24) 24·5–32 (27·8) (see also Lawson 1962, Prigogine 1978). WEIGHT: ♂ (n = 2) 59·6, 63·3; unsexed (n = 7) 59–72·4 (67·4) (Maclean 1993).

IMMATURE: like adult but top and sides of head olivaceous blackish brown, mantle olive in centre, yellower at sides, diffusely dark-streaked; chin and throat striped yellow on black, breast striped black on yellow; bill dull black; eyes brown; legs and feet light blue-grey.

NESTLING (a few days old, as observed from 6 m away using binoculars, van Someren 1956): pink, with darker, almost mauve, bill; head mottled; upstanding pale yellow down on crown; yellow down at base of wings and on dorsal tracts back to pelvis; eyes closed (?). Later, nestling very mottled. Photo (Ginn *et al.* 1989): when eyes starting to open, crest of down on crown is grey; inside mouth carmine; no swollen gape at corner of mouth.

O. l. angolensis Neumann: range of species except for races *rolleti*, *reichenowi*, *tibicen* and *larvatus*. ♂ like *larvatus* but upperparts greener, less yellow; smaller, especially bill and tail. SIZE (10 ♂♂, 10 ♀♀): wing, ♂ 136–143 (140), ♀ 133–142 (136); tail, ♂ 84–92 (88·5), ♀ 82–91 (86·5); bill, ♂ 25·5–28 (26·6), ♀ 24·5–27 (25·6).

O. l. tibicen Lawson: coastal lowlands, S Tanzania to S Mozambique. ♂ like *angolensis* but with shorter wing and longer bill. SIZE (6 ♂♂, 5 ♀♀): wing, ♂ 130–140·5 (134), ♀ 127–133 (130); tail, ♂ 79–85·5 (83·5), ♀ 77·5–85 (82·0); bill, ♂ 27–29·5 (28·1), ♀ 26·5–29 (27·9).

O. l. reichenowi Zedlitz: coastal lowlands from Somalia to about 09°S in Tanzania. Differs from all other races in yellows of upperparts and underparts being more golden. Small. SIZE (9 ♂♂, 1 ♀): wing, ♂ 122·5–129·5 (127), ♀ 120·5; tail, ♂ 74·5–81 (77·3), ♀ 76·5; bill, ♂ 25·0–27·5 (26·1), ♀ 24·0.

O. l. rolleti Salvadori: Ethiopia, Sudan, Uganda, L. Victoria basin, Rwanda, ?Burundi, E Zaïre south to about 02°S. Like *reichenowi* but upperparts brighter yellow, less golden; larger, especially bill and tail. SIZE (8 ♂♂, 5 ♀♀): wing, ♂ 122–132·5 (127), ♀ 121–130 (125); tail, ♂ 77–89 (82·6), ♀ 76–85 (81·2); bill, ♂ 22·5–25·5 (23·6), ♀ 22–24 (23·0).

Field Characters. Length 20–21·5 cm. The common open-country black-headed oriole in E and S Africa and the only one present in most of its range, but meets other species at forest edge. Has olive-green central tail feathers, like Western and Abyssinian Black-headed Orioles *O. brachyrhynchus* and *O. monacha*, so closed tail from above is greenish, darker towards tip (uniform green in Western); when spread, rest of tail black with broad yellow corners. From below, closed tail solid yellow (yellow with black base in Western and in Mountain Oriole *O. percivali*). Inner secondaries *black with yellow edges*, outer secondaries broadly edged white, greater coverts *grey with green edges* appearing more grey than green. Pattern of secondaries gives wing a more contrasting, black and white appearance, like Mountain Oriole but unlike Western and Abyssinian Black-headed, whose wings are more uniform, mainly green and grey. Longer-tailed than Mountain; heavier and longer-winged than Western. Immature has head duller with indistinct green streaks on crown, yellow half-collar, a few black streaks on upperparts, throat streaked black and yellow, breast yellow with black streaks.

Voice. Tape-recorded (46, 58, 66, 86, 88, 91–99, 104, B, C, F, KEI, LEM). A single rather high-pitched, down-slurred whistle, short '*peeo*' or '*p'wew*', longer '*pi-weeyo*', '*wheelyo*' or '*jp'wyew*-hoo' given as call or part of duet; second bird has lower, sometimes glottal whistles, '*jip-bobblebob-wah*'; whistled calls and songs often introduced by staccato chipping note, sometimes barely audible. Wide vocabulary of whistled song phrases, typically with one syllable emphasized: '*tyup, pu-wa-hoo*', '*wree-ko-wuju*', '*tyip-po-*

wheeyyo', 'w'dew-wop', 'tyop-po*week*', 'pa-*chek*-ko-*whi*dow'; glottal component common, 'tut-t'l'*weeyo*-ho', 'uncle *Hugh*, go *far*'. Also described is 'a series of running notes, low but distinct, in the form of a song, 'tuu-ga-wak-kok wok-chu-wek', 'kik-chu-woou-ku-puwa" (van Someren 1956). Mimics other species, in subdued tones, including Jackal Buzzard *Buteo rufofuscus*, Brown-hooded Kingfisher *Halcyon albiventris*, woodpecker *Campethera*, Common Bulbul *Pycnonotus barbatus*, white-eye *Zosterops*, drongo *Dicrurus* (Vincent 1936), and robin-chats, starlings, and sunbirds (van Someren 1956). Repeated contact calls given by parents returning with food to nest: ♂ called 'plew-oo' in nest tree, answered by ♀ in forest with 'o-ko-heeut'; ♂ then changed call to 'chit-cha-noble', ♀ answered 'ki-chii-ku-paewa'. As ♀ arrived with food in bill she called 'whi-re-kerrrrr', greeted by ♂ with low 'chirr'. Aggression/alarm call a high, nasal, down-slurred 'wrrreeeaa'.

General Habits. Inhabits acacia and broad-leaved woodlands, usually in vicinity of rivers, dams and other surface waters. Juniper and thorn scrub in Ethiopia, well-timbered savannas, locally in strips and extensive areas of mangrove in Kenya and Tanzania and, where sympatric with Abyssinian Black-headed or Western Black-headed Oriole, forest edges and locally interior of forest; canopy of mavunda *Cryptosepalum* forest in Zambia, and locally in *Eucalyptus* and other plantations in Malaŵi and Zimbabwe; may enter forest in Malaŵi where Green-headed Oriole *O. chlorocephalus* is absent; all woodland types in Zimbabwe but typically in miombo *Brachystegia*, and has invaded gum plantations and larger gardens. In South Africa all woodland types, forest edges, coastal woods, and large trees in parks, gardens and farmyards; shows marked preference for tall blue gums; excludes or is excluded by European Golden Oriole *O. oriolus* (Baumann 1998).

Solitary or in pairs; forms small, loose flocks at plentiful food source. Forages mainly in canopy of larger trees; also at aloes and in figs and 'kachele' trees (figs?), sometimes with Green-headed Orioles (Vincent 1936), and immature birds join mixed-species foraging parties. May feed low down, on fruits of shrubs, and occasionally descends to ground to take caterpillars from grass. 80% of insects eaten are gleaned from leaves, at height of 6.8 ± 1.0 m (Tsavo East, Kenya: Lack 1985). Flies back to woody perch with caterpillar, and there beats and pulps it. Flight fast and forceful, with long, slight undulations and short glide up to perch. Highly vocal; calls particularly in early morning. Mimicry-babbling sometimes (always?) a part of courtship (see below). At times in breeding season remains motionless in foliage, uttering medley of subdued whistles and babbling sounds.

Sedentary, but subject to short-distance seasonal wandering (Aspinwall 1985). More prone to wander in Zimbabwe than in South Africa; one bird moved 99 km in Zimbabwe, where reporting rate markedly greater in May–Aug than in Jan–Mar; in Transvaal reporting rate greatest in Aug–Oct, least in Jan–Feb (Harrison *et al.* 1997). Can arrive in numbers to feed at aloes *Aloe* spp. (Zimbabwe).

Food. Berries, small soft fruits (*Trema*, loquats, grapes, figs, mulberries), hairless caterpillars, huge hairy caterpillars (up to 125 mm long – Vincent 1936) and small, hard insects. Nectar and pollen, particularly of *Aloe* spp. (W.R.J. Dean *in* Ginn *et al.* 1989). Of 41 items eaten (Tsavo East, Kenya: Lack 1985), 15 were insects and 26 fruits. Young fed mainly on small and large caterpillars and some small imago insects. Favourite caterpillars are of saturniid moths. Feeds mainly in indigenous trees, even where nesting in alien gums.

Breeding Habits. Solitary nester, monogamous, territorial. Courtship seen only once (Kenya, van Someren 1956): ♂ called loudly from top of tall *Brachylaena huillensis* tree to ♀ in top of high ebony *Diospyros* 10 m away; ♂ increased volume and flew in downward curve to small tree 10 m below ♀, then displayed by spreading and depressing tail, lowering head, and slightly opening wings and raising wingtips (**A**); in that attitude he slowly revolved on branch, then suddenly raised fanned tail over back, lowered wings and expanded them fully (**B**); during whole performance ♂ kept up continuous, subdued but distinct medley, mimicking other bird species (see above), the babble interspersed with 'o-ri-ole' or 'o-pi-oo' calls. After short silence ♂ repeated display but bobbed up and down with neck outstretched and held head high. ♀ flew down and alighted in front of him, ♂ tried but failed to mount her then continued displaying; ♀ flew to perch beside him and pecked and pulled at his wing; ♂ flew to high tree then sailed down with quivering wings and outspread tail to alight beside ♀. He made low, purring note; ♀ crouched with drooping wings and they copulated; they then flew far away together, both calling loudly.

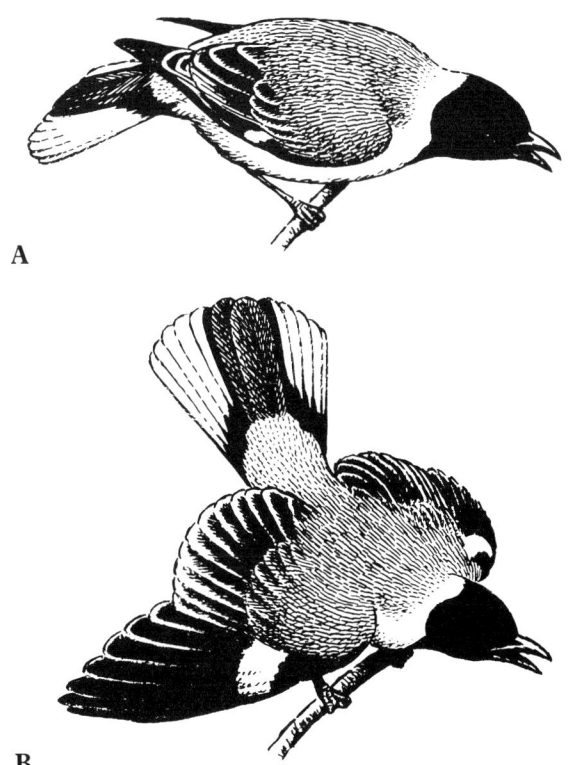

A

B

After helping ♀ feed young brood, ♂ once flew to nearby tree, fanned tail, drooped wings, and slowly and deliberately swayed from side to side.

NEST: a shallow (usually) or quite deep cradle suspended between 2–3 arms of horizontal fork, towards outside of tree or at least well away from centre, in leafy, pendent branch, often in mass of *Usnea*; made almost entirely of *Usnea* (Kenya, Zaïre) or, where none to be had, of fine fibres, moss, tendrils, grass and spider web, thinly lined with finer grass stems, creeper tendrils and rootlets (some nests unlined). Nest rather ragged and irregular, sometimes neat; loose ends of *Usnea* may straggle below (**C**);

outside may be camouflaged with bits of bark, string, cloth and acacia flower-heads; slightly smaller and shallower than nest of African Golden Oriole *O. auratus* (Vincent 1949). Firmly bound along 2 or 3 sides to supporting twigs, by mass of wiry vegetable fibres (**D**). Ext. diam. 90–100, int. diam. 65–70, int. depth 38–44. One nest described as small, open cup *on* a thin branch, in cluster of twigs (Rudnai 1994). Sited 4·5–16 m above ground, often high in *Eucalyptus* tree. In Zaïre uses tall gums, sapling gums, *Berlinia globiflora*, *Brachystegia randii*, *Parinari mobola* and *Monotes glaber* trees (Vincent 1949).

Nest built by ♀. ♂ sometimes accompanies or precedes her on material-collecting trips but generally perches in tall tree, the pair contact-calling (♂: 'chip-o-way' and 'whithoou' sometimes extended into 4- or 5-syllabic calls; ♀'s calls higher-pitched and shorter). When both birds near nest being built they make mewing notes. ♀ takes one end of long strand of lichen, twists or loops it around twig and works it into earlier loop, then she makes other end fast to other arm of branch. As cradle starts to form, ♀ gets in, stretches it with feet, adjusts loose ends, and presses breast against loosely interlaced strands. ♂ sometimes comes into nest but does not really help in building. 2 nests took 3 and 10 days to make. Replacement nest once built nearby 2 weeks after clutch stolen, then 3rd nest started 5 days after brood fell out of 2nd nest (van Someren 1956).

EGGS: 1–3, usually 2: in Zambia 2 clutches of 1, 64 of 2 and 7 of 3 eggs; average in southern Africa 2·4 (Maclean 1993). Pinkish at first, later white or creamy buff, blotched, spotted, streaked and scrolled with red-brown, olive or purplish grey, chiefly in ring around large end. SIZE: (n = 88) 26·1–32·7 × 16·7–22·3 (28·9 × 20·2).

LAYING DATES: E Africa, Region A, Mar, May–July, Region B, Jan–Feb, Apr, Region C, Nov–Jan, June, Region D, Jan, Mar–June, Sept–Nov (out of 17 records, 6 in Apr–May and 7 in Oct–Nov), Region E, May; Angola, (enlarged gonads Oct–Nov); Zaïre (Umvuma District), Oct–Nov; Zambia, Aug–Nov and Jan, mainly Sept then Oct; Malaŵi, Aug–Jan, mainly Oct; Zimbabwe, Sept–Feb – of 230 records, 46% in Oct, 23% in Nov and 21% in Sept; Botswana, Sept–Dec; Transvaal, Sept–Jan, mainly (79%) Oct–Nov; Natal and Cape, Sept–Dec.

INCUBATION: by ♀; ♂ attends in nearby tree, and occasionally brings food (once a large caterpillar) to give to incubating ♀. Period: (n = 3) 14, 14 (Kenya) and 15 days (South Africa).

DEVELOPMENT AND CARE OF YOUNG: hatchlings fed by both sexes, and ♂ also gives food to brooding ♀ who may give some to hatchlings and eat some herself (van Someren 1956). If ♂ brings a caterpillar too large for nestling, ♀ simply swallows it herself. Later, pair tend to return to nest together with food, when ♂ feeds young first then ♂ and ♀ alternate, from opposite sides of nest, until all food (caterpillars) have been given to young. ♀ rouses chicks to be fed with low 'tscher' note (van Someren 1956). Parent gives 3-note call when approaching nest with food and before feeding young; gives harsh 'piuu' call before and during feeding young, 'piuu' sometimes followed by 'whoa' call (Rudnai 1994); young respond to 3-note call by defaecating and to other calls by chirping and wing-fluttering. Parent removes faecal sac. Young fed mainly on caterpillars, small at first (many brought together in billful), large or huge later (brought singly); a large one seems to be squeezed by parent into chick's open bill like a tube of toothpaste – young described as sucking out the contents; once caterpillar skin was dropped to ground, picked up by parent and fed to nestling (Rudnai 1994). One caterpillar hung across nestling's gape; parent removed it, pulped it further, and offered it again. Parents not at all secretive in approaching nest with food, but with late nestlings of 3rd-attempt brood had become highly secretive. At age at least 2 days brood of 2 fed 5 times in 135 min and immediately prior to leaving nest fed 16 times in 180 min. One chick left nest at 09h30 when at least 16 days old, sat close to it and preened for an hour, flew to tree 8 m away 2 h later and stayed there until dusk; it was not fed often but its sibling still in nest was fed frequently; either one parent fed both young or parents fed one young each, one in and one out of nest; second nestling had left nest by 09h30 on following day, and whole family immediately moved away from the area (Rudnai 1994). Nestling period 15–18 days.

BREEDING SUCCESS/SURVIVAL: bird recovered 5 years after being ringed as adult. Clutch high in tree probably taken by Tropical Boubou *Laniarius aethiopicus* (van Someren 1956). Large nestling fell out of nest which had tilted.

References
Harrison, J.A. *et al.* (1997).
Lawson, W.J. (1962).
Maclean, G.L. (1993).
Rudnai, J. (1994).
van Someren, V.G.L. (1956).
Vincent, A.W. (1949).
Vincent, J. (1936).

Oriolus monacha (Gmelin). Abyssinian Black-headed Oriole. Loriot moine.

Plate 29
(Opp. p. 462)

Turdus monacha Gmelin, 1789. Syst. Nat., 1 (2), p. 824; Abyssinia.

Range and Status. Endemic resident, Eritrea and Ethiopia. Locally frequent or common in Eritrea and N Ethiopia; common to locally abundant in S Ethiopia (e.g. numerous at Wondo Genet and Menagesha Forest, Mt Barka). In Eritrea fairly widespread down to 950 m, but on Plateau, up to 1550 m, only in S in Guna Guna area. In Ethiopia up to 1850 m in W Highlands, W and S Ethiopia and SE Highlands; and in Rift Valley but not as far south as Nechisar Nat. Park.

Oriolus monacha

Description. *O. m. monacha* Gmelin: Eritrea and Ethiopia south to Shoa Prov., where intergrades with following race between about 10° and 08°N. ADULT ♂: hood (whole head, neck and upper breast) black, slightly glossy. Upper mantle next to hood bright yellow, rest of mantle, scapulars and back olive-yellow; rump and uppertail-coverts bright yellow. Tail: T1 pale olive or yellow-olive; T2–T6 with increasing amounts of yellow – feather bases yellow-olive, darker (almost black) next to the yellow; on T2–T6 respectively, length of yellow patch at shaft about 10, 22, 35, 35 and 40 mm, but amount of black bulge into the yellow of each vane in each feather varies, so that T5 actually yellower than T4. Lower breast, belly, flanks, thighs, vent and undertail-coverts brilliant yellow. Primaries blackish with grey or almost white outer edges; secondaries blackish with broad grey outer edges; tertials with pale grey inner vanes and yellow-olive outer vanes; greater primary coverts black with 10 mm long white tips forming conspicuous patch; alula black, greater coverts grey and yellow-olive, median and lesser coverts yellow-olive; underside of flight feathers shiny grey with inner borders almost white; underwing-coverts and axillaries yellow. Bill dull red, red-brown, sienna-brown or pink-brown; eyes brown-black or dark umber-brown; legs and feet blue-grey, slate-grey or dark olive-grey, soles dull yellow, dirty orange or pale brown (all soft-part colours from collectors' field labels). Sexes alike but ♀ may have red eyes. SIZE (5 ♂♂): wing 127–145 (136); tail 84–94.5 (90.2); bill to feathers 22–26 (23.75); tarsus 21–24 (22.8).

IMMATURE: like ♂ but forehead to hindneck brownish black, not glossy, and chin to upper breast dull brown-black, streaked and mottled with yellow. Bill black, eyes dark brown, legs and feet grey.

O. m. meneliki Blundell and Lovat (includes '*permistus*'): Ethiopia south of 08–10° N where intergrades with nominate race. Adult like nominate race except for tail; T1 and T2 entirely yellow-olive, length of yellow patch at shaft on T3–T6 respectively about 16–19, 20–22, 25 and 26 mm, T5 and T6 with olive bases but larger amount of black bordering the yellow. Immature like adult but bill black and throat with slight buff-yellow mottling; feet black with soles white or whitish; tail slightly graduated (square-ended in adult). Seems to be longer-winged but shorter-tailed than *monacha*. SIZE (8 ♂♂, 6 ♀♀): wing, ♂ 138–153 (143), ♀ 131–135.5 (133); tail, ♂ 78–91 (84.8), ♀ 84–

88·5 (86·5); bill to feathers, ♂ 22–24 (23·2), ♀ 22–24 (23·0), bill to skull, ♂ 26–28 (27·0); tarsus, ♂ and ♀ 21·5–25 (23·4).

TAXONOMIC NOTE: *O. m. meneliki* and *O. larvatus rolleti* are so similar that field confusion in S Ethiopia has made it difficult to ascribe breeding and other observations to species. Voices, and juveniles, differ, but in the hand colour pattern differences in primaries, secondaries, tertials and some rectrices must be looked for (Benson 1946b). But for those differences the 2 might be treated as composing a superspecies; however they are broadly sympatric and occur at the same altitudes at e.g. Arero and Yavello (Benson 1946b), so we treat *O. monacha* as an independent species.

Field Characters. Length 23–24 cm. A highland forest oriole endemic to Ethiopia/Eritrea. Larger than Eastern Black-headed Oriole *O. larvatus*, which it meets at forest edge. Wing pattern more uniform: outer edge of outer secondaries broadly grey, not white, forming grey panel in centre of closed wing; inner secondaries mainly yellow-green, with only inner half of innermost black (in Eastern inner secondaries black with yellow edge to outer webs); overall wing appears black, grey and green instead of black, white and yellow. Outer edge of outer greater coverts also grey (yellow-green in Eastern), increasing extent of grey panel.

Voice. Tape-recorded (109, B, MAC). Calls higher-pitched, less rich and fluty than other orioles, 'wocheelywo', 'cheelowah', 'kocheelo', 'wokachilly', 'kolyo', or in combination, 'wocholee, wocheelywo', 'wocheelywo-wollawochee'. Scold distinctive, shorter and harder than other orioles, without cat-like mewing quality, double 'grraaa-grraaa'.

General Habits. In Eritrea inhabits *Combretum* and wet or mesic woodlands on escarpments with figs and tamarinds, but never in pure acacia woods. In Ethiopia riparian and other woodlands, evergreen forest of olive and *Podocarpus* (Urban 1978) and highland juniper where it overlaps range of Eastern Black-headed Oriole *O. larvatus rolleti* in Illubabor to Shoa and NW Bale Provs. Occurs singly or in pairs, and in small flocks about June. Searches foliage and branches for insects, and hawks for insects in flycatching forays (Sim 1979). Sedentary.

Food. Searches trees 'for insects and grubs' (Sim 1979), and feeds berries to young out of nest.

Breeding Habits. Almost nothing known.
LAYING DATES: Eritrea (nest-building Aug); Ethiopia, Wondo Genet, (young out of nest being fed, June), Yavello (young out of nest, June, may have been this species or *O. larvatus*).

Reference
Benson, C.W. (1946b).

Plate 29 (Opp. p. 462)

Oriolus chlorocephalus Shelley. Green-headed Oriole. Loriot à tête verte.

Oriolus chlorocephalus Shelley, 1896. Ibis, p. 183, pl. 4; Mt Chiradzulu, Nyasaland.

Range and Status. Endemic resident, E and SE Africa. Kenya, local and uncommon in coastal forests of Arabuko-Sokoke (where appeared in late 1970s) and Diani, locally fairly common in Jadini Forest, Shimba Hills Nat. Park. Tanzania, common at 200–1200 m in Usambara Mts; very common at Amani; occurs also on Mt Lolkissale, Ngurus, Ulugurus and Udzungwas (up to 1800 m), and Rondo Plateau in SE Tanzania. Malaŵi, 5 localities: Chikala Hill, Mt Chiradzulu (where almost extinct but one seen in 1979 between Mt Chiradzulu and Lisau Hill), Soche, Thyolo and Namzadi (Nansadi), at 1000–1450 m (the last 4 localities are within 70 km of each other). In 1983 an estimated 15 pairs on Chikala Hill and 40–45 pairs in Thyolo Mt (rare above 1300 m) and nearby tea estates; density of 1 pair per 13–25 ha in Mwalantunzi Tea Estate (Dowsett-Lemaire 1989a and pers. comm.), and in Upper Thyolo 15 pairs in 6 remnant forest patches, of canopy height 20–40 m and area 1·5–20 ha – 3 pairs in one forest only 3 ha in extent (Johnston-Stewart 1982); records of wanderers at Lisau in 1895 and 1981 and on Mt Soche in 1948 and in 1970s; common in Namzadi area (Johnston-Stewart 1981). Mozambique, Mt Chiperone (though not known from Mt Mulanje, Malaŵi) and common at 950–1375 m on Mt Gorongoza (Wolters and Clancey 1969).

Description. *O. c. amani* Benson: SE Kenya, NE Tanzania.
ADULT ♂: hood (whole head, neck and upper breast) rich olive- or

Oriolus chlorocephalus

sage-green, slightly dusky on lores; mantle brilliant, intense yellow, well delineated all around; scapulars, back, rump, uppertail-coverts and much of tail rich olive-green; tail graduated (T4 2, T5 8 and T6 20 shorter than T1), T1 and T2 olive-green, T3 with yellow tip 10 long on inner vane at shaft and 20 long on outer vane, T4 with yellow tip 25 long at shaft (both vanes), T5 with distal half yellow and T6 yellow with only base of inner web green. Lower breast, belly, flanks, vent and undertail-coverts brilliant yellow, thighs yellow and olive. Primaries and secondaries with dark grey inner vanes and blue-grey outer vanes, tertials the same but heavily infused with olive-green; greater and median primary coverts bluish grey, alula bluish slate, greater, median and lesser coverts olive-green, underside of flight feathers shiny grey, with whitish inner edges proximally on P1–P6 making pale panel in closed wing; lesser underwing-coverts slate and white; other underwing-coverts olivaceous grey; axillaries yellow. Bill heavy, brown-red or brown-ochre; eyes red-brown or deep claret-red; legs and feet light bluish grey or dark slate-grey, soles yellowish. Sexes alike, but bill of ♀ brown-ochre, i.e. less red than in ♂. SIZE: wing, ♂ (n = 14) 123–144 (133), ♀ (n = 4) 123–135 (132); tail, ♂ (n = 14) 90–108 (101), ♀ (n = 4) 93–103 (99); bill to feathers, ♂ (n = 9) 25–29 (27·2), ♀ (n = 2) 22, 27; tarsus, ♂ (n = 10) 22–27 (24·6), ♀ (n = 2) 22, 23.

IMMATURE: like adult but olive-greens less intense, narrow ring of yellow featherlets around eye, mantle yellowish olive and not well demarcated, chin and throat yellow-olive, upper breast dull yellow mottled with olive, lower breast bright yellow diffusely streaked with olive, and upper belly bright yellow with small olive streaks. Secondaries quite broadly edged and narrowly tipped with white, greater primary coverts broadly tipped white, greater and median coverts narrowly tipped yellow. Bill blackish.

NESTLING: hatchling not known; late nestling has neck, cheeks, chin and throat bare; plumage of mantle and breast lax and fluffy; forehead and crown yellow-olive, mantle to rump dark olive, all feathers yellow-tipped; tail feathers not yet emerged; upperwing-coverts dark grey-olive, all yellow-tipped; breast to undertail-coverts pale yellow in centre, olive at sides.

O. c. chlorocephalus Shelley: Malaŵi and Mt Chiperone, Mozambique. Like *amani* but slightly larger and tail significantly longer: wing, ♂ (n = 12) 133–143 (137), ♀ (n = 9) 129–135 (131); tail, ♂ (n = 6) 104–114 (109), ♀ (n = 4) 102–109 (105); bill, ♂ (n = 6) 25–28 (26·5), ♀ (n = 4) 25–27 (26·3); tarsus, ♂ (n = 6) 24–26 (25·3), ♀ (n = 4) 24–25 (24·5).

O. c. speculifer Wolters and Clancey: Mt Gorongoza, Mozambique. Like nominate race, but has a conspicuous white alar speculum formed by white tips to 4 primary coverts (P4–P7); outer vanes of inner primaries and outer secondaries finely edged with white at tip. Larger than *chlorocephalus*: wing, ♂ (n = 4) 132–144 (140·5), ♀ (n = 3) 133·5–134 (134).

Field Characters. Length 21·5–24 cm. A bulky oriole, very local in eastern forest patches from Kenya to Mozambique. Visibly larger in the field than Eastern Black-headed Oriole *O. larvatus* (Johnston-Stewart 1981), and readily distinguished in reasonable light by *green* head and back separated *by broad yellow collar*. However, this is a bird of tall canopy where it remains hidden among leaves much of the time, and colour not easy to see in canopy silhouette; in poor view or at a distance head may appear black, but then back also looks black (yellow in Eastern Black-headed) and contrasting yellow collar is very conspicuous. Wing mainly grey, white speculum present only in race *speculifer* on Mt Gorongoza.

Voice. Tape-recorded (32, 91–99, B, F, HOR, KEI, LEM, McVIC, STJ). Song has same mellow, fluty tone as other orioles, somewhat more liquid and less abrupt than Eastern Black-headed: 'hweet-tyoo-*hwee*do', 'woh-jo-*koy*yo', 'wupu*chee*ya', 'co*woy*o', 'hopli-hopleero'; sometimes glottal, 'ko, kod'l'*ee*'; each utterance accompanied by quick fanning of tail (Zimmerman *et al.* 1996). Aggression/alarm, an insistent, complaining mew, louder in middle and changing in tone, 'quarreeeyaaa'.

General Habits. Inhabits high canopy in primary montane, submontane and riparian evergreen forest in Malaŵi and Mozambique, coastal forests in Kenya, and in Tanzania forests mainly below 1000 m and rich woodland and secondary growth with tall emergent trees. Usually keeps to the highest tree-tops. Occurs solitarily or in pairs; gleans or snatches insects from foliage in tall canopy, takes small fruits, and sometimes flycatches in mid-stratum (Dowsett-Lemaire 1989a). May feed in company with Eastern Black-headed Orioles *O. larvatus* and feeds in fig trees with other orioles.

Sedentary but may wander: vagrants at Lisau, Malaŵi (see above) probably from <40 km away.

Food. Berries and other fruits, large hard red seeds, flowers, particularly of *Grevillea*, and insects including caterpillars and winged imagos.

Breeding Habits. Very poorly known. In Upper Thyolo, Malaŵi, pairs form in June, calls loudly in Aug–Feb, relatively quiet in Mar–June.

NEST: deep, thick-walled pocket made of *Usnea* lichen, suspended from slender lateral fork 10 m up in forest tree on edge of clearing.

LAYING DATES: Tanzania, Amani, (advanced nestling early Nov); Malaŵi, Sept–Oct (and occupied nest Nov); Mozambique, Gorongoza, (large gonads and incubation patches, Oct).

References
Clancey, P.A. (1970).
Dowsett-Lemaire, F. (1989).
Johnston-Stewart, N.G.B. (1981, 1982, 1984).
Vincent, J. (1936).
Wolters, H.E. and Clancey, P.A. (1969).

Plate 29	***Oriolus auratus*** Vieillot. African Golden Oriole. Loriot doré.
(Opp. p. 462)	*Oriolus auratus* Vieillot, 1817. Nouv. Dict. Hist. Nat., nouv. éd., 18, p. 194; 'Africa', = Gold Coast.

Oriolus auratus

Limits of breeding range in northern and southern tropics are notional: see text

Range and Status. Endemic resident with disjunct populations in northern and southern savanna woodlands between approximately 10°N and 20°S; a post-breeding migrant into arid areas further north and south.

Northern race *auratus*. Mauritania, scarce wet season visitor (about July) into extreme south. Senegal, wet season visitor north of 14°N, common resident south of 14°N. Gambia, locally common resident. Mali, frequent resident south of 11°30′N, common in Boucle du Baoulé Res., some birds moving north to 15°N for the wet season. Burkina Faso, visitor to Ouagadougou region mid June to mid Nov, and to north of Ouahigouya, June to early Nov. Niger, breeds in extreme SW (Parc du 'W') where occurs Apr–Oct; north to Tillabéry in July–Aug. Sierra Leone, very common in N. Ghana, mainly in N (where a dry-season visitor to Mole, Oct–Apr) but occurs along coast west to Accra. Nigeria, rather uncommon to locally frequent resident in guinean savanna woodlands south to Ilaro and Enugu; mainly or entirely a visitor north to Sokoto (common July–Dec), Kano (common late June to Dec) (and once Mar, Birnin Kudu, and records Jan and July on Komadugu Gana R in NE) and Potiskum; records to south of resident range (Ibadan, Lagos, Serti) mainly in dry season, Sept–Apr (Elgood *et al.* 1973). Cameroon, north to Kalamaloué Nat. Park (Scholte *et al.* 1999). Chad, scarce; rare north of 10°30′N. Sudan, common, resident and migrant between about 6° and 10°N, dry-season visitor, Sept–May further south, and wet-season visitor north to 13°N; in Darfur only in July–Aug. Ethiopia, frequent in W lowlands (north to about 12°N) and uncommon in Rift Valley (not known in Nechisar Nat. Park) and S Ethiopia east to about 38°E; rare above 1800 m. Somalia, 6 records, Aug–Oct, mainly in Jubba valley north to 02°30′N. Savannas of NE Zaïre where breeding resident and partial migrant in N Uele and near L. Albert. Non-breeding visitor about Oct–Feb to N Uganda and W Kenya, with sight and specimen records in N Uganda south to Toro, Entebbe and Teso and in Kenya in Cheranganis, Mt Elgon, Baringo and Kapedo Districts.

Southern race *notatus*. In E Africa a non-breeding visitor, frequent, mainly Apr–Aug, throughout Uganda, in SW half of Kenya and most or all of Tanzania; much commoner in N Tanzania than in Kenya; scarce visitor to Pemba and Mafia Is, May–Dec, vagrant in Zanzibar; may breed in S Tanzania. In Zaïre common in Kivu up to 1850 m and in savannas of Katanga and Kasai, where may breed though not yet shown to do so; in Katanga scarce in Dec–Feb, uncommon in Mar–Apr and Sept–Nov, frequent to common in May–Aug. Rwanda, non-breeding visitor May–Aug (1 record Mar). Angola, frequent and widespread, north along coast to Quiçama Nat. Park and in interior to Luanda Norte; residential status uncertain but known to breed; in Malanje and Cuanza Norte a dry-season visitor in Apr–Aug. Zambia, frequent throughout, up to 1800 m; in S resident (breeding mainly Sept–Oct) and non-breeding visitor Jan–Mar; in N a partial migrant, commonest in July–Sept, scarce after Mar (Aspinwall 1978, 1980, R.J. Dowsett, pers. comm.). Malaŵi, frequent to common resident and partial migrant throughout, at 300–2000 m. Namibia, widespread visitor in Nov–May in N, uncommon in W south to Tropic of Capricorn. Botswana, sparse to frequent resident in N, common south and east of Okavango, uncommon in E, rare in SE (Gaborone, Lobatse). Commoner than wintering European Golden Orioles *O. oriolus* in Caprivi Strip, Namibia, and E Kavango, Botswana (Brown 1990, Hines 1985–87). Zimbabwe, resident and partial migrant throughout (casual up to 1600 m), commoner north than south of 20°S; in N, numbers increase steadily from late Aug to early Apr; breeds mainly in Sept–Nov; distinct decrease in abundance with partial emigration to north in Apr–Aug; many birds remain to winter in lowlands and central plateau (Irwin 1981, Harrison *et al.* 1997). W and S Mozambique, uncommon south only to Save R., mainly Sept–Mar, some resident; records Chibavava, Búzi R., Gorongoza Nat. Park, Tambarara, Mt Gorongoza, Tete and Zumbo. Transvaal, uncommon visitor, evidently non-breeding; most records in high-rainfall summers.

Av. density of 1 bird per 11 ha in 7 woodland types, N Botswana (M. Herremans *in* Harrison *et al.* 1997).

Description. *O. a. notatus* Peters: southern tropics, breeding north to *c.* 10–08°S. ADULT ♂: brilliant, intense yellow, but for: long black mask (lores, broad stripe over and narrow stripe under eye, continuing backwards above ear-coverts to taper off at side of nape), tail (which is square-ended or very slightly forked) and wing. In tail T1 blackish, T2–T6 almost entirely yellow. Wing mainly black: primaries black, inner ones with 1-mm pale yellow tips and narrow whitish outer edges; secondaries black, outer vanes tipped and edged with yellow, narrowly on outer and broadly on inner feathers, tertials black, tip of inner vanes yellow, and outer vanes broadly tipped and edged yellow; alula, greater primary and median primary coverts black, broadly tipped with yellow, other upperwing-coverts yellow; underside of remiges silvery grey. Underwing coverts and axillaries yellow. P8 longest, P9 9 and P10 50 mm shorter. Bill brown-red, brownish pink or Indian red; eyes crimson or blood red; legs and feet bluish grey, greenish slate or dark olive-green; claws blackish. ADULT ♀: forehead, crown, nape, hindneck, sides of neck, mantle, scapulars and back yellow-olive with slight dusky dappling. Tail olivaceous blackish or dark brown, all feathers yellow-tipped. Lores and around eye dusky olive. Chin, throat and breast pale yellow with soft grey streaky dappling, upper belly clear pale yellow with long narrow black streaks, lower belly and flanks pale yellow with sparse, thin blackish shaft streaks, thighs, vent and undertail-coverts bright yellow. Primaries dull brownish black with narrow whitish outer edges, secondaries and tertials blackish with outer vanes tinged olive and edged yellow; greater and median primary coverts dull blackish or greyish dark brown with well-defined, broad pale yellow outer edges and tips, greater coverts with olive or olive-blackish centres and broad yellow fringes, median and lesser coverts yellow-olive. Underside of wings as in ♂. Bill black, eyes dark brown, legs and feet slate-grey. SIZE: wing, ♂♀ 132–143, tail, ♂♀ 78–84; bill, ♂♀ 26·5–28; tarsus, ♂♀ 21–23. WEIGHT: unsexed, (n = 6) 70–79·4 (73·7).

Note: several adult-♂-plumaged skins in museum series are labelled ♀, and several immature skins, labelled ♂, have ♀ tail patterns.

IMMATURE: ♂ like adult ♀ but olive in upperparts is yellower; chin, throat and upper breast uniform pale greyish yellow, sometimes with tiny black shaft streaks (not dappled as in adult ♀); streaks on lower breast shorter and less diffuse; sides of breast bright yellow and less streaked. Immature ♀ like adult ♀. Bill dark-mottled pink-brown, eyes dark brown-red, iris shading to whitish on outer edge, feet bluish grey or greenish.

NESTLING: bill very dark brownish or purplish pink, inside mouth carmine, eyes brownish grey, legs and feet grey or green.

O. a. auratus Vieillot: northern tropics, breeding south to c. 08–06°N. ♂ like ♂ *notatus* but median and greater primary coverts have less yellow on edges and tips, primaries with even narrower white edges, secondaries and tertials with narrower yellow edges and less yellow at tips (so that closed wing blacker, less yellowish, than in *notatus*); yellow tip of T1 3–4 mm, T2 8–11, T3 20–23, T4 c. 32, T5 34–39 and T6 c. 48 mm long. ♀ like ♀ *notatus* but only T2–T6 yellow-tipped, yellow on outer vanes diminishing from c. 5 on T2 to vestige on T6, yellow on inner vanes increasing from c. 5 on T2 to 23–26 on T6. Weight: (Upemba Nat. Park, Zaïre) ♂ (n = 9) 62–72 (67), ♀ (n = 6) 60–70 (65).

Field Characters. Length 20 cm. Adult ♂ golden yellow with characteristic bulbous-shaped black line across lores and around eye, narrowing to streak behind eye. ♀ greenish yellow above with yellow rump and dark grey eye patch. In all plumages distinguished from Eurasian Golden Oriole *O. oriolus* by broad yellow edging on greater coverts, secondaries and tertials, so that (especially in adult ♂) closed wing appears more yellow than black. In flight ♂ shows whitish bases to underside of flight feathers (black in Eurasian Golden), yellow tail with black central feathers; ♀ is yellower than ♀ Eurasian, with outer half of upperwing appearing blacker in flight than inner half. Immature heavily streaked below, very similar to immature and streaked ♀ Eurasian Golden Oriole except for yellow wing-edgings; also distinguished by yellow-olive bend of wing, not contrasting with back or sides of neck, and dark postocular streak.

Voice. Tape-recorded (34, 36, 86, 88, 91–99, 104, B, C, F, KEI, LEM). Voice extremely similar to that of Eurasian Golden Oriole. Song rich and fluty, '*wee*lo-tyoo-tyoowo', '*wee*lo-lo-weelu*wee*yu', '*wee*lo-willy-*wee*yo'; displaying birds also gave higher-pitched, less tuneful phrases, 'cooly-wally-weer' and a long-drawn 'du-wee-*wee*eyer' (Gibbon 1991). A single long, down-slurred whistle, 'wheeeyoo' may be a short song or a call (Chappuis, in press). Subsong of tuneless, chattery notes possibly from immature (Chappuis, in press). Aggression/alarm, harsh 'rrraaaah', 'mwaaarr' or more drawn-out, mewing 'myaaaeeeh'; call of young, sharp 'dew-dew-dew'.

General Habits. Inhabits tall, mature, mesic, deciduous broad-leaved woodlands, forest edge, and semi-evergreen gallery woods; in southern tropics mainly in miombo *Brachystegia*, gusu *Baikiaea* and mopane *Colophospermum* woods, dense woods around hills, riparian and valley forests, and swamp-edge woods in Okavango. In N Kalahari occurs mainly in mature *Colophospermum*, *Combretum imberbe* and *Terminalia prunioides* woodlands. In Zimbabwe likes *Brachystegia randii* and especially *Berlinia globiflora* trees. Avoids open acacia savanna, but occurs in tall riverine *Acacia galpinii* in Limpopo valley, in *A. nigrescens* around Okavango, and in *A. erioloba* in mixed woods (M. Herremans in Harrison et al. 1997). Keeps mainly to canopy and upper branches of trees; often occurs at fruiting tree. Breeding habitat in Malaŵi is *Brachystegia* woodland only; in Zimbabwe nests mostly in *Berlinia* trees. Highly conspicuous and vocal in breeding season but more solitary, shy and unobtrusive out of it – less so when flocking in berry trees. Often joins mixed-species foraging flocks. Sometimes occurs alongside Eastern Black-headed Orioles *O. larvatus* and Eurasian Golden Orioles, though appears not to interact with them. Calls mainly in early morning. Flight fast and direct.

Resident and migrant. Migrations (see above) not well understood; both populations move north after breeding, some birds flying at least 1500 km; each population then travels well to the south, many birds apparently overshooting breeding ranges (limits of which are, however, poorly known). Soon before emigrating, birds in Zimbabwe put down substantial reserves of fat (Borrett 1972), suggesting that migration will be distant, perhaps rapid, and across food-poor terrain.

Food. Insects (a few hard ones, and caterpillars in 8 out of 12 stomachs, Zaïre) and small fruits including figs. Seen taking caterpillars, beetles and flies. In Nigeria stomachs contain more fruits than insects. One of very few bird species that eats fruits of exotic neem *Azadirachta indica* (Walsh 1975).

Breeding Habits. Monogamous, solitary nester, territorial. As soon as ♂ arrives on breeding grounds in Zimbabwe it claims a territory, advertised by frequent loud calling, then spends much time and energy chasing rivals away (P.J. Ginn *in* Ginn et al. 1989). Remains highly vocal until eggs hatch, calling continuously for long periods from perch near nest and in flight between trees.

largely of leaves (Nigeria). Suspended between arms of horizontal fork in slender branch near centre of tree (in contrast to Eastern Black-headed Oriole, which builds near outside of tree), and usually well hidden in foliage; sited 5–13 m above ground. Ext. diam. 85, depth c. 52–58.

EGGS: 1–3; in Zambia 10 C/1, 83 C/2, 58 C/3 (av. 2·3 eggs). Long ovate; cream or buffy pink, thinly but boldly spotted with rich chestnut-brown or black-brown and grey, suffused with pink around each brown spot; spots mainly around large end. SIZE: (n = 45, southern Africa) 26·0–32·9 × 19·2–21·8 (28·5 × 20·5), (n = 5, Nigeria) 29·0–33·2 × 19·8–21·0 (30·7 × 20·5).

LAYING DATES: Senegal, (young, July–Aug); Gambia, Apr–July; S Mali, (evidence of breeding Mar–Nov); Togo, (nest May, carrying food July, Aug); Nigeria, north of great rivers, Mar–Apr; Angola, Aug, Oct, (and large gonads Aug–Nov); Zambia (n = 183), Sept (76), Oct (98), Nov (9); Malaŵi, Oct–Dec; Botswana, Mar; Zimbabwe, Aug–Jan (n = 136; 28% Sept, 53% Oct, 15% Nov); Mozambique, Sept–Jan.

INCUBATION: nothing known.

DEVELOPMENT AND CARE OF YOUNG: young brooded only by ♀ but fed by both parents.

NEST: open cup or hammock, rather thin-walled and open in structure but nest firm and quite solid and compact at base, made of lichen *Usnea*, a few dry coarse grass-blades and tendrils, woolly plant down and a small amount of spider web, securely slung at the rim to supporting small branches on both sides by loops of material wound round with twists of woolly down and spider web (Vincent 1949) (**A**). Uses hardly any *Usnea* in Zimbabwe (Ginn *et al.* 1989). Sometimes green leaves hang over nest and are lightly attached to it by strands of spider web. One nest made

References
Bannerman, D.A. (1939).
Chapin, J.P. (1954).
Maclean, G.L. (1993).
Vincent, A.W. (1949).

Plate 29
(Opp. p. 462)

Oriolus oriolus Linnaeus. Eurasian Golden Oriole. Loriot d'Europe.

Oriolus oriolus Linnaeus, 1758. Syst. Nat. (10th ed.), p. 107; 'in Europe, Asia' (= Sweden).

Range and Status. Breeds NW Africa, Europe north to S Baltic, Siberia and N Kazakhstan to c. 95°E, Turkey and Caucasus to N Iran, and S Kazakhstan to Afghanistan and N India; winters Africa and India.

Palearctic migrant. Breeding summer visitor to Morocco (locally common south to Middle Atlas and High Atlas foothills) and N Algeria (scarce but widely recorded in Tell). In Morocco, density of 7·9 pairs per km^2 in maquis, 9·0 in woodland and 10–20 in dry *Quercus suber* forest (Thévenot 1982).

Winters from S Cameroon, S Central African Republic, Zaïre and SE Kenya south to N and central Namibia and South Africa (mainly N Cape, Transvaal and N Natal, occasionally to E and S Cape). Widespread; frequent to common N Botswana, Zimbabwe and N Transvaal, apparently scarce in Angola; mainly a passage migrant in N of wintering range. On southward passage, frequent in N Morocco (Tingitane), uncommon to rare in N Algeria (Kabylie), W Morocco and W Mauritania; frequent to common in Tunisia and Libya, common to locally abundant in Egypt, N Sudan, Eritrea and NW Somalia; common in E Chad, Ethiopia and Kenya (mainly E), frequent Uganda and Zaïre (mainly E). Common on both passages in Zambia, Malaŵi and Zimbabwe. On northward

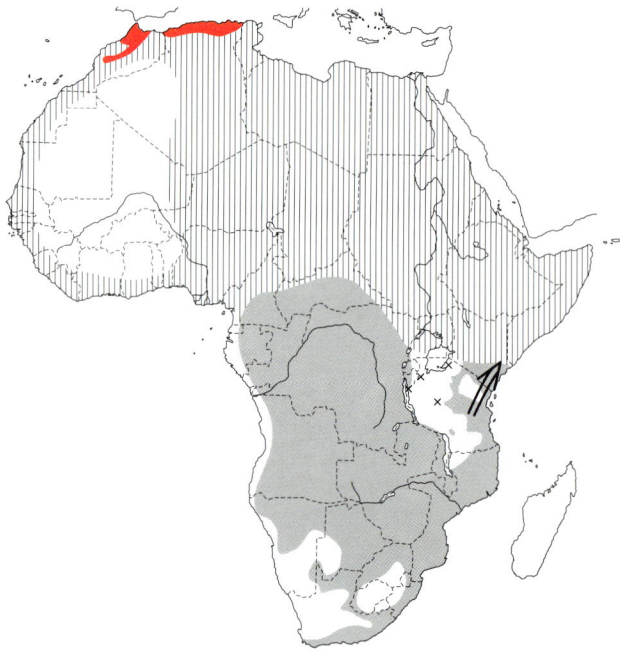

Oriolus oriolus

migration, frequent to common in Zaïre (mainly S and E), E Tanzania (map: N. Baker, pers. comm.), Uganda, Kenya (locally abundant on coast north to Malindi), Ethiopia and N Somalia, but scarce in S Somalia, Eritrea and Sudan, and absent Chad; common to abundant in Egypt, and frequent to common in rest of N Africa, including Saharan oases in Libya and E Algeria; frequent Senegal and Atlantic coasts of Mauritania and Morocco. Records from W Africa mostly from passage times: frequent in Sierra Leone and Senegal (spring only); rare E Mali (Sept, Apr), Niger (Apr), Gambia (Nov–Dec), Liberia (Mar), Ivory Coast (Oct) and Togo (Oct, Apr); uncommon Nigeria and N Cameroon.

Description. *O. o. oriolus* Linnaeus: NW Africa, Europe, W Siberia to N Kazakhstan, and Turkey to N Iran; only race in Africa. ADULT ♂: whole head and body bright yellow, apart from broad black stripe through lores, sometimes small black spot behind eye, and black outer scapulars. Central tail feathers (T1) black, often with small yellow spot at tip, others black with broad yellow tips (increasing from 10–20 mm on T2 to 25–40 on T6). Flight feathers black, outer primaries with narrow yellowish white outer borders; tertials, alula and upperwing-coverts black, primary coverts with broad yellow tips forming short wing-bar; axillaries and underwing-coverts yellow. Wing rather long and pointed: P8 longest; P7 2–5 shorter, P6 13–20, P5 23–28, P1 48–57; P9 8–13 shorter than P8, tip falling between P5 and P6; P10 small, 19–28 longer than primary coverts; P7–P8 emarginated on outer web. Bill dark red or pinkish red; eyes dark red; legs slate grey. ADULT ♀: upperparts yellowish green, yellower on lower rump and uppertail-coverts, greyish or olive-yellow on outer scapulars. Tail like adult ♂ but feather bases tinged yellowish green, central feathers mainly greenish. Lores grey. Underparts mainly greyish white, sides of breast yellowish-green, flanks and undertail-coverts yellow; throat, breast and sides of belly with diffuse grey streaks. Some (older?) birds much brighter, with crown and rump almost yellow and whole underparts yellow with only faint grey shaft streaks. Flight feathers greyish black, with greyish-white outer borders to outer primaries; primary coverts black with pale yellow tips, broad in some (older?) birds; tertials and rest of upperwing-coverts dull black or greenish black, narrowly edged greenish yellow, median coverts uniform; axillaries and underwing-coverts pale yellow. Soft parts like adult ♂, but bill and eyes often duller and browner. SIZE (10 ♂♂, 10 ♀♀, Europe): wing, ♂ 148–154 (151), ♀ 141–154 (149); tail, ♂ 85–91 (87·8), ♀ 83–92 (87·9); bill, ♂ 26–29 (27·6), ♀ 27–29 (27·4); tarsus, ♂ 22–24 (22·8), ♀ 22–24 (22·8). WEIGHT: sexes combined: Egypt (n = 51, autumn) 60–96 (79·0); NE Sudan (n = 51, Sept–Oct) 40–93 (61·6); Ethiopia (n = 14, Sept–Mar) 47–79 (59·1), (n = 3, Apr–May) 80–89 (84·2); SE Kenya (n = 16, Nov–Dec) 61–84 (73·0); Tanzania (n = 1, Feb) 67; SE Zaïre (n = 6) 62–72 (67·0); Zimbabwe, South Africa and Namibia (n = 12) 52–82 (68·3).

IMMATURE: juvenile olive-green above, yellower on rump, feathers with narrow yellowish fringes; lores dull grey; underparts whitish with grey streaks, undertail-coverts pale yellow. Tail dark greyish green, T1 with narrow yellow tip, T2–T6 with large yellow spot at tip of inner web, most extensive on T6. Wing feathers like adult ♀, but tertials and upperwing-coverts with contrasting yellow fringe at tip; and primary coverts greyer with narrower yellow tips (thus wing-bar less prominent). Bill dark grey; eyes dark brown; legs grey or blue-grey. First-winter like juvenile but plain green above with yellowish rump; whitish below with undertail-coverts, sides of breast and flanks yellow, and distinct blackish streaking; retains yellowish tips to median coverts; bill dark brown. First-summer/second-winter like first-winter but lacks yellow tips to median coverts, and yellow tips to primary coverts are broader, especially in ♂; bill dark pink, eyes reddish; (i.e. resembles typical adult ♀, but streaking on underparts always distinct, grey on breast, blackish on belly).

NESTLING: covered with short sandy or buff-white down; bill and legs flesh grey, mouth pink; no tongue spots; gape flanges whitish or yellowish white.

Field Characters. Length 21–23 cm. A large oriole with long dark wings. Adult ♂ bright yellow except for black lores, black wings with small yellow spot when folded, showing as crescent in flight, and black tail with yellow corners. ♀ and imms dull green above, yellower on rump, lores dark grey; underparts whitish with yellow sides and undertail-coverts, streaked dark; older ♀♀ yellower with fainter streaks; tail corners yellow. Same pale crescent shows on dark olive-grey wings in flight. All plumages distinguished from African Golden Oriole *O. auratus* by lack of dark streak behind eye and absence of broad yellow edgings to wing feathers, so that closed wing appears dark compared to rest of body. In ♀♀ and imms, bend of wing noticeably darker than scapulars and sides of neck (greenish like back in African Golden Oriole). In flight ♀ is darker and greener above than African Golden Oriole.

Voice. Tape-recorded (62–73, 88, 91–99, 105, B, C, F, PEA, ROC). Song a variety of loud, melodious, liquid, fluty whistles, 'weeka-laweela-weeo', 'weelo-wallo-weelyo', 'lyoo-cleeo-weelo-weelo', 'tilly-cleo-cleeoo', 'teeloo-hoh-weeyo'; similar but shorter phrases given at dawn (Chappuis, in press); given by both sexes; quieter in winter quarters than on breeding grounds. Song in NW Africa has 'slighter' quality, less rich and more hurried, 'wi-weelyo-way'; more different from songs of both European *O. oriolus* and African Golden Oriole than they are from each other (Chappuis, in press). Subsong, heard frequently in subsaharan Africa, a continuous low, starling-like chatter. A harsh, grating, drawn-out 'skaaaa', 'zhraaaz' or 'kweeer' given commonly by both sexes in alarm or excitement; often interpolated between song phrases. Other calls include a falcon-like 'jijijijijick' or 'kli-kli-kli-kli ... ' given when attacking or fleeing from predators or territorial rivals, by ♂ chasing ♀, or as contact call. Quiet antiphonal duetting reported from breeding grounds, whistling song of ♂ answered by short song-unit, 'skweeeeer' call or falcon-call from ♀. Call of young a hard, tuneless 'dew-dew'.

General Habits. Arboreal, frequenting mature leafy trees. In NW Africa, breeds in maquis, oak woodland and dry forest on plains and foothills from sea level up to 1800 m. Winters in forest edge, tree savanna, moist and semi-arid woodlands, riverine acacias, orchards and plantations; in Zimbabwe in *Brachystegia* and mopane woodland; in Botswana in patches of trees in Kalahari semi-desert. On migration, occurs in drier, more open bush and savanna, gardens, isolated fig-trees, and oases, mainly below 1500 m, but occasionally up to 2200 m in E Africa.

Shy and unobtrusive, spending much time among high dense foliage. Often solitary, but forms groups and large feeding flocks; in spring, often 50 or more together on Kenya coast, and >1000 once reported on N coast of Egypt (Goodman and Meininger 1989). Associates with other orioles, cuckoos and drongos. Feeds mainly in canopy of large trees, picking items from foliage; also catches insects in flight, and at times descends to feed on ground, moving with clumsy hops. Most flights short, between trees, direct, without undulation; over longer distances, flies fast, with long undulations and bursts of loose wing-beats (Cramp and Perrins 1993). Highly vocal in Africa, especially in Feb–Apr.

Iberian and NW African birds initially head south but their winter quarters are not known. Flight to tropics probably unbroken, for despite regular Aug–Sept passage in extreme N Morocco there are few autumn records from W Morocco, Mauritania or Mali, and none from Senegal. NW and central European birds migrate southeast or south across central and E Mediterranean, and Asian birds migrate southwest across Arabia. Heavy passage through Egypt, mainly late Aug to mid Oct, with smaller numbers in coastal Libya and Tunisia; also strong passage through N and NE Sudan, Eritrea and NW Somalia from late Aug to Oct, and E Chad from mid Sept to mid Oct. Some birds cross Zaïre forests, but main migration to southern Africa occurs between E Zaïre and Kenya from mid Oct to early Dec. First birds reach Zambia in late Sept, but main passage there is in late Oct–Nov. Early birds reach Transvaal in early Oct, but main arrival in southern Africa is in Nov.

Northward movement begins in Botswana in Feb, and most birds vacate southern Africa in Mar. Marked passage in Zambia, Zaïre and E Africa, late Mar to mid-Apr. Large numbers come to fruiting trees along Kenya coast. Latest birds remain in Zambia to early Apr, Zaïre to mid-Apr and Kenya to end of Apr. Paucity of spring records from Eritrea and Somalia suggests long onward flights from Kenya; rarity in Chad and W Sudan also suggests birds fatten at lower latitudes. Strong migration through N Africa extends further west than in autumn, suggesting a loop migration for many birds. Passage is heavy in Egypt, mainly mid Apr–May, but also marked in Tunisia, central and N Algeria and E Morocco. In W, minor passage in Sierra Leone, Senegal, Mauritania and Morocco in Apr. May and early June. NW African breeding areas occupied from mid Apr.

Birds ringed Germany (2) and Italy (1) recovered in Egypt (Sept), and birds ringed Egypt recovered Lebanon (1), Malta (1) and Greece (1). Recoveries also from E Morocco (Apr) to Belgium (1), and from Tunisia (May) to Greece (1), Rumania (1), Yugoslavia (1) and Malta (2). South of Sahara, recoveries from France to Congo (1, Oct), Holland to S Zaïre (1, Mar), Germany to S Zaïre (1, Mar) and Czech Republic to Malaŵi (1, Mar). 1 bird ringed Ethiopia (Oct) recovered in Iran (July).

Food. Berries and caterpillars; also beetles, Orthoptera, Hemiptera; seeds; occasionally lizards, birds' eggs and nestlings. In Kenya, diet comprised 75% fruits, 25% insects (Lack 1985).

Breeding Habits. Monogamous; solitary and territorial. Territories well dispersed, used for courtship, nesting and some feeding; typically 10–35 ha (Germany), but range used for feeding and gathering nest material is much larger (Cramp and Perrins 1993). On arrival, ♂ establishes territory and sings from perch in canopy of particular trees to defend and advertise it; drives off later-arriving ♂♂ and passage birds, sometimes after a long chase. ♀♀ arrive soon after ♂♂. Pair formation takes about 4–7 days. Display involves vigorous chases and vocal duets. ♂ sometimes hovers with tail spread; sings, and ♀ responds with cat-call or short antiphonal whistle (Cramp and Perrins 1993). Single brood.

NEST: shallow cup, mainly of plant fibres, grasses and rye leaves, bound with long grasses and bark fibres, and lined with fine grass, feathers, down and wool; in crown of high tree, usually suspended from horizontal fork of thin branch; ext. depth c. 90–110, ext. diam. c. 120–130, int. diam. c. 80–90; cup depth c. 60–65; site chosen by both pair members; built by ♀ in 6–12 days, but ♂ may assist in initial stages and collection of material.

EGGS: usually 3–4; white, creamy or pinkish white with scattered black spots. SIZE: (n = 737, Europe) 27–35 × 19–24 (30.4 × 20.3).

LAYING DATES: Morocco, May (and building mid-June); S Europe, May, early June.

INCUBATION: by both sexes, mainly ♀; starts with 1st or 2nd egg. Period: 15–18 (16–17) days.

DEVELOPMENT AND CARE OF YOUNG: young brooded mainly by ♀; fed partly by regurgitation; feeding and nest sanitation by both sexes; young remain in nest 13–20 days, fully fledged at 16–20 days.

BREEDING SUCCESS/SURVIVAL: of 49 eggs, SW Russia, 90% hatched and 71% produced fledged young (Cramp and Perrins 1993).

Reference
Cramp, S. and Perrins, C.M. (1993).

Family Dicruridae: drongos

Medium-sized arboreal birds, glossy black (1 species grey, 1 whitish). Head large, rather flat, sometimes crested; bill stout, somewhat hooked, notched, base covered by forward-directed feathers from forehead and lores; strong rictal bristles. Eyes usually red; bill and legs black. Tarsus short, toes strong. Wings rather long and pointed. Tail rather long, usually notched or forked, of 10 feathers; outer feathers often lyre shaped, sometimes long and twisted or with rackets. Sexes alike. Immature plumages less glossy, often pale-barred. Arboreal in forest and savanna. Perches upright. Mainly insectivorous, feeding by flycatching or pouncing. Highly aggressive, chasing hawks and crows in the air. Very vocal; give mixture of harsh discords and melodious whistles, often mimicking other species. Nest a small shallow cup, on horizontal fork near end of high branch. Largely resident (northernmost Asian populations migrant).

Affinities uncertain. Share features with bulbuls, shrikes, crows, orioles (Oriolidae) and flycatchers (Muscicapidae, Monarchidae). On the basis of DNA-hybridization studies Sibley and Monroe (1990) place them nearest to monarchs and fantails, in an enlarged family Corvidae which also embraces crows, cuckoo-shrikes, orioles and bush-shrikes.

2 genera: one, *Chaetorhynchus*, monotypic, in New Guinea, the other, *Dicrurus*, in Afrotropical, Oriental and Australasian Regions.

Genus *Dicrurus* Vieillot

Characters those of the family.

23 species: 19 in S Asia to Australia and Solomon Is, 4 in Africa (2 independent species, 1 superspecies).

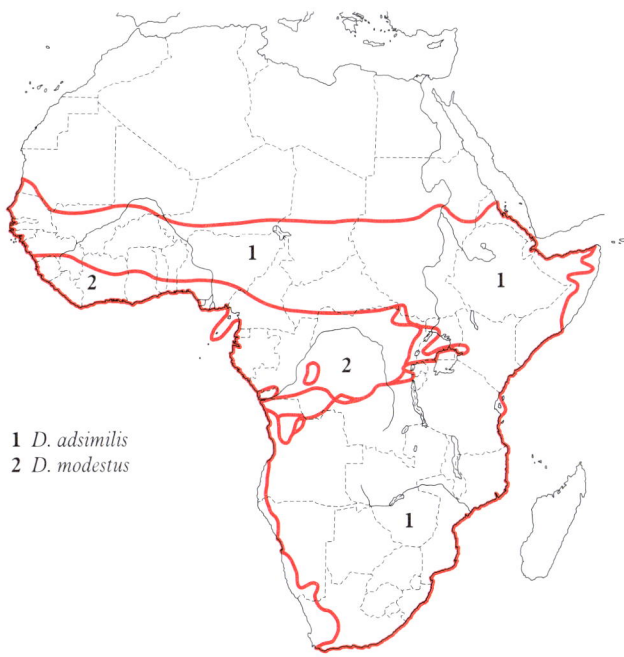

Dicrurus adsimilis superspecies

1 *D. adsimilis*
2 *D. modestus*

Plate 30
(Opp. p. 463)

Dicrurus ludwigii (Smith) Square-tailed Drongo. Drongo de Ludwig.

Edolius ludwigii Smith, 1834. S. Afr. Quart. J., 2nd ser., p. 144; Durban.

Range and Status. Endemic resident in forest and thick woodland at low to medium altitudes. Wide-ranging but local. S Senegal, common in Oussouye Region; rare in N and central Gambia; frequent Guinea Bissau, W and S Guinea, Sierra Leone and N Liberia (N Lofa County and Mt Nimba); frequent to common Ivory Coast (more numerous than *D. atripennis*), from coast north to Korhogo; local and uncommon in S Ghana, north to Yegi, sporadic in Mole Game Res.; frequent in S Togo, north to Aledjo, and in Nigeria, mainly south of great rivers, north to Zaria, Kaduna and Gashaka–Gumti Game Res.; S Cameroon north to Yoko, and in Rumpi Hills and Bamenda highlands; Central African Republic, north to Manovo–Gounda–Saint Floris Nat. Park; N Zaïre, on edge on Congo forest from Oubangi to Lendu plateau, to W Uganda in W Nile; uncommon in S Sudan on S slopes of Imatong Mts, and W Kenya in Kakamega and S Nandi Forests. E Gabon to S Congo (Mayombe) and Lower Congo R. below Kinshasa, locally common; Angola in S Cuanza Norte and adjoining NW Malanje, and escarpment forests from Chingoroi (N Huila) north to Lower Cuanza R, thence east through Huambo, N Bié–Cuando Cubango, and S Malanje to S Lunda and Moxico (Clancey 1976b). Somalia, uncommon in S forests and along Jubba R. below 02°N; E Kenya, frequent on Tana R. up to Garissa, and near coast north of Tana; highlands and lowlands of E and S Tanzania, locally common Usambaras, Ulugurus and Ngurus to Udzungwas and Pugu Hills, Mt Rungwe and Ukinga highlands, Songea to Mikindani and Lindi; Upper Katanga, S Zaïre, north to L. Mweru and NE Marungu; Zambia, locally common north of about 13°30′S, east to 31°30′E; Malaŵi, east of Rift, from Mulanje to Namweru; Zimbabwe, locally in E highlands at medium altitude; Mozambique, north and south of Limpopo, locally common, up to 1500 m on Mt Namule; E South Africa, E Transvaal highlands, Zululand, coastal Natal to E Cape, and Swaziland.

Density in primary forest NE Gabon, 7–8 pairs per km^2 (Brosset and Erard 1986).

Description. *D. l. sharpei* Oustalet: Senegal to Cameroon, Lower Congo R. and NW Angola (S Cuanza Norte and adjoining Malanje), and east to N Zaïre, S Sudan, Uganda and W Kenya. ADULT ♂: entire head and body black with slight violet-blue sheen, most marked on upperparts and breast. Tail almost square or slightly forked (T6 1–5 mm longer than T1) (**A**); black, with slight violet-blue gloss above. Flight feathers blackish brown, including inner webs. Tertials, alula and upperwing-coverts blackish brown, outer feather edges with violet-blue sheen. Underwing-coverts and axillaries blackish brown, edged glossy violet-blue. Bill black; eyes pale orange, orange-red or blood red; legs black. Sexes alike. SIZE (10 ♂♂, 10 ♀♀): wing, ♂ 105–111 (108), ♀ 104–111 (108); tail, ♂ 86–94 (90·0), ♀ 82–92 (88·5); bill, ♂ 20–23 (21·8), ♀ 20–23 (21·7); tarsus, ♂ 16–18 (17·0), ♀ 16–18 (16·8). WEIGHT: (Liberia) ♂ (n = 1) 35·5, ♀ (n = 1) 33·7; (Ghana) unsexed (n = 3) 25–29 (26·7); (Central African Republic) ♂ (n = 5) 26–32 (30·0).

IMMATURE: juvenile sooty black, speckled pale grey on mantle and breast; eyes brown. Immature like adult but feathers of belly and undertail-coverts fringed whitish; underwing-coverts tipped white.

NESTLING: naked with yellowish skin; tongue and inside of mouth deep yellow without markings.

D. l. muenzneri Reichenow: S Somalia, SE Kenya, E and S Tanzania. Tail slightly more forked, feathers more lyre-shaped than in *sharpei* (**B**). ADULT ♂ has upperparts strongly glossed bluish green; below, gloss extends down from breast over flanks and belly. Adult ♀ less glossy than ♂, underparts sooty black with blue-green lustre across breast; axillaries and underwing-coverts tipped white; smaller than ♂. SIZE (4 ♂♂, 7 ♀♀): wing, ♂ 104–112 (109), ♀ 96–105 (99·8); tail, ♂ 86–91 (88·5), ♀ 80–87·5 (83·5); bill, ♂ 21–22 (21·5), ♀ 19–21·5 (20·5) (Clancey 1976b). WEIGHT: (Tanzania) ♂ (n = 1) 25·5, unsexed (n = 22) 25–34 (29·4); (Somalia) unsexed (n = 9) 22·3–28·7 (25·8).

D. l. tephrogaster Clancey: Malaŵi, E Zimbabwe, Mozambique north of Limpopo R. Like *muenzneri*, but adult ♂ less deep black, sheen on crown and mantle slightly greener; rump and uppertail-coverts more greyish; underparts duller, lower breast, belly, sides and flanks greyer. Adult ♀ as in *muenzneri* but rump and uppertail-coverts greyer, underparts more plumbeous, especially throat and abdomen. Size similar to *muenzneri* but wing of ♂♂ shorter; wing, ♂ (n = 12) 100–107 (103), ♀ (n = 12) 95–103·5 (98·7) (Clancey 1976b).

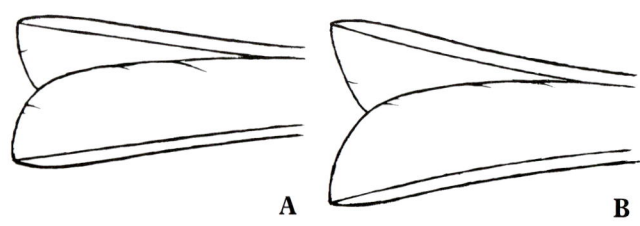

D. l. ludwigii (Smith): South Africa, Swaziland, S Mozambique. Like *muenzneri*, but blue-green sheen slightly less pronounced and less extensive below. Same size as *tephrogaster*.

D. l. saturnus Clancey: Angola (W escarpment to Moxico), Zambia, S. Zaïre (Katanga). Like *ludwigii* but even less sheen, more greenish or bronzy.

TAXONOMIC NOTE: racial treatment follows Clancey (1976b). *D. l. sharpei* lacks sexual size and plumage differences of other races, has squarer tail, and duller, more violet (rather than greenish) gloss. It may prove to warrant species status.

Field Characters. Length 18–19 cm. *Smallest* drongo, with tail *slightly notched* (square only in relation to other drongos, more distinctly forked in E and S birds); further distinguished from Fork-tailed Drongo *D. adsimilis* by *forest* habitat, and lack of pale wing-flash in flight. ♂ black with uniform dull gloss, greenish blue in E and S birds, violet in W birds. Sexes alike in W birds, but ♀♀ in E and S duller, those from Malaŵi to central Mozambique with greyish underparts. Immature like ♀ but with some whitish barring on belly and undertail-coverts. Noisy; usually heard before seen. Loud drongo calls readily separate it from similar-sized Southern Black Flycatcher *Melaenornis pammelaina* and Black Cuckoo-Shrike *Campephaga flava*; further distinguished by red eye (not dark brown), notched tail (square in Southern Black Flycatcher, rounded in Black Cuckoo-Shrike) and heavier bill, and from Black Cuckoo-Shrike by lack of yellow gape. Where range meets that of Shining Drongo *D. atripennis* in W Africa, Square-tailed inhabits secondary and riverine forest, Shining only primary forest, but they meet at mid-levels of humid forest; also meets Velvet-mantled Drongo *D. modestus* in old secondary forest and gallery forest (Chappuis, in press); distinguished from them by small size and less forked tail; for further differences see those species.

Voice. Tape-recorded (9, 51, 75, 86, 88, 91–99, 104, B, F, KEI, LEM). Repertoire extensive, with regional differences; calls of W African birds less sibilant, more muted; those from E Africa have a brassy, ringing quality. Birds from Senegal, Liberia, Ivory Coast and Gabon combine rather tuneless up-slurred and down-slurred whistles with dry buzzy notes, 'dzi-dzi-dzi-dzi-dwee', 'pew-pew-pew-zit-zit-zit', 'zer-chi-hooey'; or phrases have whistles only, 'toy-pulee', 'chi-chu-hooey'; sometimes ending with a loud '*chop*' – 'wee-wee-wi-*chop*'. Single notes may be regularly repeated, and not combined into phrases, 'pyew, pyew, pyew ... ', higher, briefer 'tyip, tyip, tyip ... ', lower 'toy, toy ... ' or 'toylu, toylu ... ', some with ringing bush-shrike quality; a long, buzzy 'watch-winding' call, 'zzzzweee' is frequent. Duet very different, quiet but rapidly delivered, a medley of short whistles with a continuum of chattery, liquid, bee-eater like notes. In W Kenya (Kakamega), typical call is a tinny 'doey-doey', interspersed with grating 'dzzit-dzzit', also 3 long rolled notes, 'prrruwcet-prrru-waaa-dzzheeeey' and a loud, clear, up-slurred '*puweet*'; song-medley, 'wit-wit-tseep-wit-dzaaa-dzaaa-prrrrt, doey-doey, dzzheeeeyaaa (screechy), peeya, trrrrrrr (watch-winding), zheeee-pwit-pwit, doey-doey'. In E Tanzania (Amani, Udzungwas) the loud '*chop*' has a hissing quality, '*tssyow*' or '*tssyoh*'; in phrases, 'tsyop-tsu-chop-wee-tsyop' and a buzzy 'tsi-tsi-za-za-tsi'; also a single 'chyoop', double 'chochop' or 'doy-doy', and a brief 'twit' extremely like the slightly drier 'chit' made by African Goshawk *Accipiter tachiro* as it flies over its territory, possibly an imitation. In Zambia common component is 'chok-ee-chok', like that of Tanzanian birds; song-phrase 'psi-jo-jwee, chewy-chewy'. In southern Africa, singers prefer an introductory chatter, often in an undertone, followed by loud '*chu-wee-ju, weeju*' or '*willy-chosso*'; calls include 'chop-suey', 'wip-choy' and 'toytoy', and nasal buzzy 'dzi dzaa dzi-dzi' and 'dzi-dzi-jaa-chipwit'.

General Habits. Inhabits forest and moist thicket, preferring edges, glades, forest patches and secondary growth; gallery forest with thick undergrowth; densely wooded S guinea savanna in W Africa; the mid-stratum of moist evergreen and riparian forest at both low and medium altitudes in Zambia and Malaŵi; primeval forest with large timber in N Mozambique. In Gabon, in lower canopy than Velvet-mantled Drongo *D. modestus*, but displaced from lower levels of primary forest by Shining Drongo *D. atripennis* (Brosset and Erard 1986). Ranges from sea level to 1000 m on Mt Nimba, Liberia, 1350 m in Cameroon, 2000 m S Sudan and S Tanzania, 1500 m S Malaŵi and on Mt Namuli, Mozambique.

Somewhat retiring and shy; usually territorial, in pairs or small family groups. Advertises presence with harsh call-notes, but less noisy and pugnacious than other African forest drongos. Darts out from bare branch to catch flying insects. Often accompanies mixed bird parties. Twitches tail sideways when perched (nominate race: Maclean 1993). Aggressive near nest, and will attack and drive off much larger birds; a pair chased a White-necked Raven *Corvus albicollis* in Tanzania, one riding on its rump and pecking at it.

Food. Large insects, especially moths, grasshoppers, mantises and beetles; exploits termite emergences (Brosset and Erard 1986).

Breeding Habits.
NEST: small, neatly-woven, thin-walled shallow bowl of lichens and dry plant stems, lined with lichen and bound with cobweb, thickly at rim; diam. *c.* 75, cup diam. 50–55, depth 25; 2–8 m high, suspended by margins from fork, usually near extremity of horizontal branch.

EGGS: 2–3; South Africa (n = 33) av. 2·6; Malaŵi (n = 7) av. 2·6. White or pinkish cream, spotted lilac and brown, mainly at large end. SIZE: southern Africa (n = 34) 20–23 × 15–16·5 (21·4 × 15·8) (Maclean 1993); Zaïre (n = 3) 20–22 × 15·5–16.

LAYING DATES: Liberia, (♀ with enlarged ovary, May); Ghana, (egg in oviduct Apr); Togo, (newly fledged juvs Mar); Gabon, Nov (newly hatched juvs Feb); Uganda, May; Zambia, Sept; Malaŵi, Sept–Oct; central Mozambique, Sept–Nov, Zimbabwe Sept–Jan; Natal, Oct–Apr.

References
Brosset, A. and Erard, C. (1986).
Clancey, P. A. (1976b).
Vincent, A. W. (1949).

Plate 30
(Opp. p. 463)

Dicrurus atripennis Swainson. Shining Drongo. Drongo de forêt.

Dicrurus atripennis Swainson, 1837; Birds of West Africa, 1, p. 256; Sierra Leone.

Dicrurus atripennis

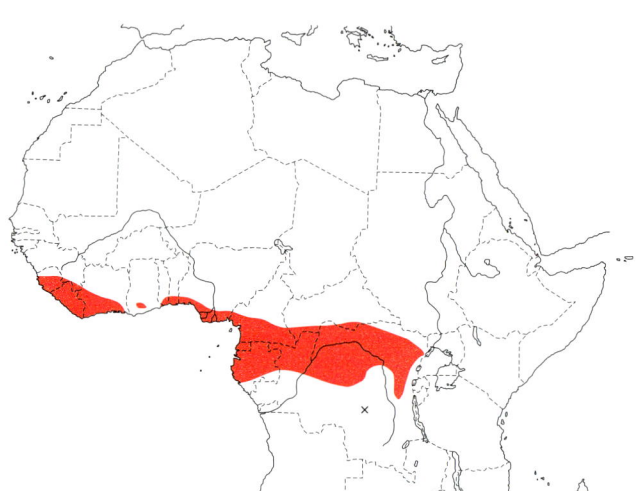

Range and Status. Endemic resident, equatorial lowland forest. Sierra Leone and Liberia to SE Guinea and S Ivory Coast, locally frequent to common; rare in S Ghana, Fumzo and 'Fanti'; rather uncommon in S Togo, north to Bismarckburg, and in forests of S Nigeria from Lagos to Ikom; locally common in S Cameroon, and most numerous forest drongo in Gabon; locally common from S Central African Republic, in Lobaye and Haut–Sangha Préf. and Mbomau Dist., to N, central and E Zaïre, from Middle Congo (Lukolela) to Upper Uele, Ituri and Semliki forests and Manyema.

Density in primary forest, Liberia, up to 10–20 pairs per km^2 (Gatter 1997).

Description. ADULT ♂: entire head and body black with blue-green gloss, strongest on upperparts, throat and breast. Stiff feathers of forehead directed upward and foreward to cover basal half of upper mandible (**A**). Tail with shallow fork (5–15 mm, T6 5–10 longer than T1); black, glossed blue-green above. Flight feathers blackish brown, including inner webs; outer edges of secondaries glossed blue-green. Tertials, alula and upperwing-coverts black with blue-green gloss. Underwing-coverts and axillaries sooty black with some bluish gloss. Bill black; eyes dark red; legs black. Adult ♀ like adult ♂, but gloss less pronounced on throat and belly. SIZE (10 ♂♂, 10 ♀♀): wing, ♂ 109–120 (114), ♀ 110–115 (113); tail, ♂ 91–111 (101), ♀ 90–102 (96·4); bill, ♂ 23–24 (23·2), ♀ 21–24 (22·5); tarsus, ♂ 16–18 (17·2), ♀ 16–17 (16·5). WEIGHT: (Liberia) ♂ (n = 7) 35–43 (39·0), ♀ (n = 9) 35–47 (38·9); (Central African Republic) ♂ (n = 1) 42.

IMMATURE: juvenile sooty black on head and body, wings and tail black with slight greenish gloss.

NESTLING: bill black with corners of mouth yellowish; eyes dark greyish brown; feet dark grey (Chapin 1954).

Field Characters. Length 21·5–24 cm. A drongo of W African primary forest, highly glossed blue-green above and below. Larger, longer-tailed and glossier than Square-tailed Drongo *D. ludwigii*, which it meets in more open areas of forest; depth of fork intermediate between that of Square-tailed and Velvet-mantled Drongo *D. modestus*. More strongly glossed than all races of Velvet-mantled or Fork-tailed Drongo *D. adsimilis*. Underside of flight feathers black like rest of wing (in Guinea forest race *atactus* of Velvet-mantled they are dark grey; in Fork-tailed Drongo of savannas strikingly pale in flight).

Voice. Tape-recorded (104, B, CHA, ERA, GAU, GRI, HUG, LEM, ROD). Wide vocabulary includes sharp calls and whistles and harsh scratchy notes: a steady 'chyip-chyip-chyip ... ' repeated up to 13 times, a lower 'chyoop-chyoop ... ', ringing double whistle, 'pee-hee' or single 'heeee'; short, shrill 'dzrip' and longer double screech, 'jreeeep-jreeeep'; buzzy rattle, 'za-za-za'; clicking 'kli-to-kli' and 'widdly-klok'. Whistled and buzzy notes given in succession, 'woy-woy, dz-jee', 'way, dz-jee, dz-jee', or whistles alone, 'woy-wit'; strung together to form 'song', 'dzirr-dzirr, way-tea, woy-tea, dzirrr'. Some songs have brassy quality, like Square-tailed Drongo, 'tsi-kway-chop', 'jizzy-chee-kwok', 'jizzy-chee-perchee-kwoyk'. Imitates other species, like Bocage's Bush-Shrike *Malaconotus bocagei*. Also gives a thrush-like chuckle, 'kewp-kewp' and 4 liquid, descending notes, 'chilly-poppa'.

General Habits. Confined to shady interior of mature, little disturbed primary and older secondary forest; typically in lowlands, but up to 1000 m in Liberia on Mts Nimba and Wuteve. Penetrates depths of Guinea forests more than other drongos. Usually at 5–25 m, below lower canopy among thick shrubs. Lively and noisy. Occurs in territorial pairs or family groups of about 4–5; often with greenbuls and flycatchers, and is conspicuous member of mixed bird parties. Sallies out after insects from loop of liana or other perch, usually below 20 m; gleans and hover-gleans. In Gabon, attracted to trails of ants *Dorylus wilwerthi*, plunging to catch falling insects, mainly in the air, from perch 5–15 m high (Willis 1983). Often chases conspecifics and other birds; flies aggressively at hawks and hornbills. Sedentary.

A

Food. Mainly insects. Stomachs of 13 birds (Zaïre) contained beetles, dragonflies, cicadas, Orthoptera, a spider and a millipede (Chapin 1954); of 19 birds (Liberia), mainly beetles, ants, grasshoppers (Colston and Curry-Lindahl 1986). Butterflies caught in air (Gabon: Christy and Clarke 1994).

Breeding Habits.
NEST: small, shallow cup slung between prongs of fork on horizontal branch; one in Cameroon was secured with cobwebs and decorated with lichens; another in Kivu (Zaïre) was attached to fork with lichens and decorated inside with fine plant stems; diam. 80.

EGGS: 2 (2 clutches, Kivu and Gabon); creamy brown with irregular reddish brown markings. SIZE (n = 2): 24·7 × 17·4, 24·5 × 17·1 (Prigogine 1961).

LAYING DATES: Liberia (enlarged gonads Aug–Dec); Nigeria (enlarged yolking ovary Oct, egg in oviduct Nov); Zaïre, Kivu, Feb, N Ituri (enlarged gonads Apr, May, Sept); Congo, Aug–Sept; Gabon Jan, Mar (nest-building Nov, young accompanying parents May).

References
Brosset, A. and Erard, C. (1986).
Gatter, W. (1997).

Dicrurus adsimilis (Bechstein). **Fork-tailed Drongo; Common Drongo. Drongo brillant.** Plate 30 (Opp. p. 463)

Corvus adsimilis Bechstein, 1794. Lathams Allgem. Uebers. Vog. 2, p. 362; Duiwenhok R., Cape Province.

Forms a superspecies with *D. modestus*.

Range and Status. Endemic resident. Common and widespread, avoiding only highlands and thick forest. In the west, throughout savanna belts, north to extreme S Mauritania (east of Rosso), inland delta of R. Niger in Mali (Niafunke, 16°N), and about 14°N in Niger; most abundant in northern guinea zone, scarce in sahel; mainly absent from forest zone, but recorded Liberia (Yekepa; Hall and Moreau 1970). Common in NE Zaïre in Uele savanna but scarcer near L. Albert. Ranges north to 14°–15°N in Chad and W Sudan (Darfur), and to c. 17°N along Nile and in N Eritrea; common at low to medium altitudes throughout S Sudan, Ethiopia, Somalia (except near E coast), E Africa (except forests in S and W Uganda and W Kenya), Rwanda, Burundi and E borders of Zaïre; common on Zanzibar but absent from Pemba I. In S Zaïre local on Lower Congo R. and in Kasai, common to abundant in Katanga; Angola, abundant north to S Cuanza Norte and Malange; Zambia, Malaŵi and Mozambique, common to abundant at lower altitudes, though scarce in richer miombo. In southern Africa, common throughout, except in Lesotho highlands and dry W Karoo, Namaqualand and Namibian coast.

Densities in Transvaal, 1 pair per 30 ha in broad-leaved woodland, 1 pair per 11 ha in *Acacia* woodland (Tarboton *et al*. 1987); in S Mozambique, respectively 18, 13 and 25 birds per 100 ha of *Acacia*, mopane and miombo woodland (Parker 1999).

Description. *D. a. fugax* (Bechstein): S Uganda and Kenya south to Tanzania, E Zambia, Malaŵi, Mozambique, Zimbabwe, NE Botswana, E Transvaal lowveld, E Swaziland, NE Natal lowlands.
ADULT ♂: head and body black; upperparts, breast and flanks with steely blue-green sheen, sides of head to chin and throat faintly glossy, belly and undertail-coverts sooty black. Tail deeply forked (15–29 mm) with T6 23–27 mm longer than T1 (**A**); black, upper surface with greenish sheen. Primaries and secondaries dark brown with inner webs pale grey-buff; tertials and alula blackish with glossy outer borders; primary coverts blackish with narrow glossy outer edge and pale grey-buff inner webs; rest of upperwing-coverts black with slight greenish gloss. Underwing-

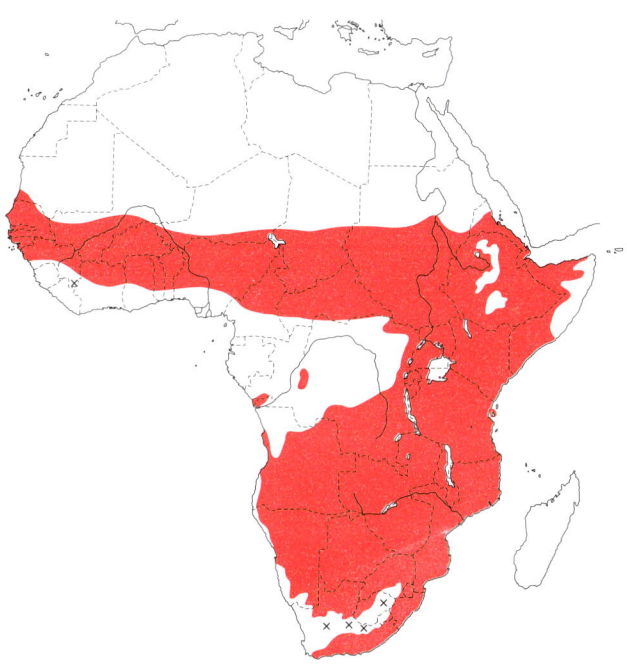

Dicrurus adsimilis

coverts and axillaries glossy black. Bill black; eyes deep red; legs black. Sexes alike. SIZE (10 ♂♂, 10 ♀♀): wing, ♂ 124–136 (129), ♀ 119–129 (122); tail, ♂ 105–115 (110), ♀ 101–111 (105); bill, ♂ 21–25 (23·2), ♀ 21–25 (23·2); tarsus, ♂ 19–21 (19·8), ♀ 18–20 (19·1). WEIGHT: unsexed: Kenya (n = 7) 39–54 (43·4).

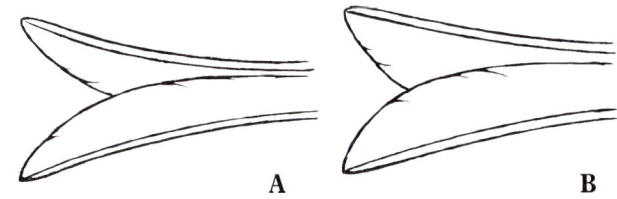

A   B

Plate 33

Plate 34

IMMATURE: like adult, but lacks gloss on underparts; lower belly feathers and undertail-coverts with whitish tips; retains juvenile wings and tail feathers. Juvenile dark brown, typically with pale buff feather fringes above and broad buff feather tips below; throat often whitish; secondaries and upperwing-coverts with fine whitish margins; underwing-coverts and axillaries tipped white; tail fork shallower; eyes brown. South African birds more uniformly sooty brown, body with little buff marking.

NESTLING: bill black, gape flanges pale yellow.

D. a. adsimilis (Bechstein): South Africa (SW, S and E Cape, Orange Free State, Natal except extreme NE, Transvaal highveld except dry W); also W Swaziland and Lesotho lowlands. Like *fugax*, with deep tail fork, but inner webs of flight feathers darker grey-brown (not pale grey-buff). Larger; wing, ♂ (n = 17) 134–143 (139), ♀ (n = 10) 132–139 (135); tail, ♂ (n = 17) 117–126 (122), ♀ (n = 10) 115–124 (119) (Clancey 1976a). WEIGHT: Transvaal (n = 5, unsexed) 48–53 (51·0).

D. a. apivorus Clancey: Angola, S Zaïre, N and W Zambia, Namibia, Botswana (except NE) and South Africa (N and NW Cape and dry W Transvaal). Large, like nominate *adsimilis*, but inner webs of flight feathers pale grey-buff, as in *fugax*; wing ♂ (n = 9, Angola) 138–143 (140) (Vaurie 1949).

D. a. divaricatus (Lichtenstein): Senegal and Guinea Bissau east to Cameroon, Sudan, N Zaïre, Ethiopia and Somalia, intergrading with *fugax* in N Uganda and Kenya. Like *fugax*, with pale inner webs of flight feathers, but tail fork shallower (8–20 mm), T6–T1 12–20 mm (B). Same size as E African birds. WEIGHT: Ghana, unsexed (n = 3) 34–40·5 (38·2); Chad, unsexed (n = 9) 37–48 (41).

TAXONOMIC NOTE: *coracinus*, *modestus* and *atactus*, often regarded as races of *adsimilis*, are treated under *D. modestus* (see Note under that species). Racial treatment of *D. adsimilis* otherwise follows Clancey (1976b).

Field Characters. Length 23–26 cm. The abundant open-country drongo of Africa, usually on conspicuous perch, showing deeply forked tail and red eye, and giving frequent harsh calls. Smaller Southern Black Flycatcher *Melaenornis pammelaina* and Black Cuckoo-Shrike *Campephaga flava* have shorter, square or rounded tails, brown eyes, smaller bills and different voices. Black body and head of adults glossed blue-green, and in perched bird primaries and secondaries appear browner than rest of wing. Underside of flight feathers light grey-brown, producing pale flash in flight. Juveniles have buffy tips to body feathers, pale gape; immatures retain pale gape and show pale-barred undertail-coverts and whitish edged lesser underwing-coverts often conspicuous at wing edge (Zimmerman *et al.* 1996), tail fork shallower than in adult. Does not normally enter forest habitat of Square-tailed Drongo *D. ludwigii*; latter is smaller, with shallow fork to tail and black underwing. For differences from Velvet-mantled Drongo *D. modestus* see that species.

Voice. Tape-recorded (22, 38, 75, 86, 88, 91–99, 104, B, KEI, LEM). Pre-dawn territorial song, pairs of rasping notes, separated by pauses, 'jwaaa-jwaaa ... jewy-jwaaa ... jeewy-jeeerr ... '. Daytime song more varied, a twangy, discordant jumble of shrill, short tuneless whistles and squeaky, grating, glottal and 'creaking hinge' notes, rather disjointed, with pauses between phrases, 'drit-drit, woy-glo-jit, drit-o, jewp-jewp, click-click, gliglaagligloo, jick-glo-jeea ... '. Certain notes delivered in synchronous duet with mate (Zimmerman *et al.* 1996). Often incorporates imitations of other species, especially raptors (e.g. Shikra *Accipiter badius*, African Goshawk *A. tachiro*, Pearl-spotted Owlet *Glaucidium perlatum*), and pure, sweet whistled notes of tchagras, thrushes and bulbuls; once a single 'koit' of Gorgeous Bush-Shrike *Telophorus viridis*. Separate calls include 'chyup', 'tjaaa', 'glick', a grating 'chidi', choked 'grit', loud, twangy 'jer-woo' and whistled 'jee-lu'. A scratchy, 3-syllabled 'my-dear-chap' given before dawn and also in daylight when attacking intruding eagle, hawk or owl (Sudan, Darfur: Lynes 1934). For further transcriptions of vocalizations see Zimmerman *et al.* (1996) and Maclean (1993).

General Habits. One of the most familiar birds of the African savanna, occupying a wide range of habitats: moist and dry woodlands; grassland and cultivation with scattered trees, wires or other perches; riverine cover in more arid country. In E Africa, equally common in semi-arid bushland, acacia woodland and moist coastal secondary habitat, and ranges into forest edge. In southern Africa, also frequents parks, farmyards and exotic plantations. In Zambia and Malaŵi, sparse in interior of richer miombo *Brachystegia* but common on edges. Avoids highlands; mainly below 1800 m in E Africa, but up to 2200 m on Mt Kenya; up to 1750 m in Malaŵi.

Bold and conspicuous; usually solitary or in pairs, sometimes in family groups or larger parties; occasionally joins mixed bird parties. Perches upright on prominent branch, post or wire; usually within a few m of ground, but sometimes high in tall tree. Periodically makes sallies to pick prey from ground or seize winged insect in twisting pursuit; often returns to same perch. Holds large insect down on perch with one foot while picking it to pieces with bill. Takes wings off butterflies before swallowing. Associates with game or livestock, catching disturbed insects. Attends grass fires, catching grasshoppers in middle of smoke. Partial to swarming ants and winged termites, and will feed until dark after evening hatch. May feed on bees in vicinity of hives. Flight from one perch to another buoyant, undulating and easy, but quick and agile when catching insects or defending territory. Pair members often perch close together, bobbing heads and bowing, while ♂ sings. Sings in duet, ♂ and ♀ alternating at rate of 12 notes (6 each) in 2·0 s (von Helversen 1980). Engages in 'aerial stunting' at dusk, rising several m almost vertically and diving steeply back to perch. Noisy and quarrelsome; may steal food from other birds. Pugnacious towards predators; chases larger raptors and crows, sometimes pursuing them to a good height, with much tumbling and twisting; also mobs owls, hornbills, shrikes and small mammals, and will attack large snakes. Kills and carries small birds (e.g. Bronze Mannikin *Lonchura cucullata*) in feet or bill. Fishes from tree perch on bank, swooping down, fluttering low over water surface, and dipping to seize prey before returning to perch. Bathes by plunge-diving from air or perch. Probes aloe flowers, presumably mainly for nectar (Oatley 1964). Vocal; often the first bird to sing at dawn and the last heard at dusk; calls throughout heat of day and sometimes during moonlit nights. ♂♂ countersing at dusk: 3 birds *c.* 100 m apart imitated down-slurred whistle of Pearl-Spotted Owlet alternately; the imitation was so perfect that observer at first thought real owls were calling (Namibia: S. Keith, pers. comm.).

Food. Mainly insects, including grasshoppers, beetles, winged termites, ants, bees and cicadas; also small birds, small fish and nectar.

Breeding Habits. Solitary nester; monogamous; territorial. Defends territory aggressively against conspecifics and potential predators. 2–3 broods per season (South Africa).

NEST: a shallow saucer, firmly woven but thin-walled, often transparent; made of rootlets, plant stems and creeper tendrils, bound and camouflaged with spider web, especially at rim, sometimes a few dry leaves and a little lichen around rim and some coarse grass at base; ext. diam. 100–115, ext. depth *c.* 50, cup diam. 75, depth 25; sited 2–10 m (usually 5–7 m) high, in small to moderate-sized soft-foliaged or thorn tree, usually attached at rim to fork near end of horizontal branch, often in the open; occasionally a more substantial bowl of plant material and sheep's wool placed on top of horizontal branch (Maclean 1993).

EGGS: Nigeria, 2–3; S Zaïre, 2–3 (usually 3); Malaŵi, 1–3, (n = 350) av. 2·6; South Africa, 2–4 (usually 3), (n = 172) av. 2·8; highly variable; usually white, cream or pinkish cream, with large freckles and small blotches, light red or reddish brown with underlying ashy or pale violet, usually mainly in zone at large end; a few eggs sparsely speckled black and charcoal-grey. SIZE (n = 336, southern Africa): 21·2–28·3 × 16·7–20·1 (24·3 × 18·2) (Maclean 1993); Nigeria 20·4–24·3 × 16–18·2, S Zaïre 20–25 × 17·1–19·2 (Chapin 1954).

LAYING DATES: N Ghana, Apr; Burkina Faso, June; Togo (carrying food, July); Nigeria, Nov–June (mainly Feb–Apr); Zaïre, (Uele) Feb–Mar, (Upemba) Aug–Oct, (Katanga) Sept–Oct; Sudan, Apr–June, Nov; Ethiopia, Mar–Sept; E Africa, Region A May, Region B Feb–June, Sept, Region C July, Sept–Dec, Region D Mar–June, Sept–Dec, Region E Jan, Apr–May, Nov–Dec, mainly in pre-rains period in Regions B and C and in short rains in Region D; Malaŵi, Sept–Nov; N Mozambique, Aug–Sept; Zimbabwe, mainly Sept–Nov; Natal, Oct–Jan.

INCUBATION: period 16–17 days.

DEVELOPMENT AND CARE OF YOUNG: young remain in nest for 17–18 days.

BREEDING SUCCESS/SURVIVAL: only known host of African Cuckoo *Cuculus gularis*. Bird retrapped after 6 years (Zambia: Dowsett 1980). Adult taken in mid-air by Red-necked Falcon *Falco chiquera* (Namibia: S. Keith, pers. comm.).

References
Chapin, J.P. (1954).
Clancey, P.A. (1976).
Maclean, G.L. (1993).
Meise, W. (1968).
Vaurie, C. (1949).
Vincent, A.W. (1949).

Dicrurus modestus Hartlaub. Velvet-mantled Drongo. Drongo modeste.

Plate 30
(Opp. p. 463)

Dicrurus modestus Hartlaub, 1849. Rev. Mag. Zool., p. 495; Príncipe.

Forms a superspecies with *D. adsimilis*.

Range and Status. Endemic resident. Widespread and locally common in Sierra Leone, Liberia, S Guinea, Ivory Coast (to *c.* 08°N) and SW Ghana (north to Kintampo and Tafo); scarce in S Togo, S Benin (Pobe) and extreme SW Nigeria. Common S Nigeria (north to Ibadan, Idah and Obudu); S Cameroon (north to Ngambe and Sanage and Kadei drainage), Bioko, Príncipe (common), SW Central African Republic (Lobaye Préf.), Gabon, Congo; common and widespread in Zaïre from Congo R. mouth north to Oubangi, Uele and Upper Bomu Rivers, south to Kasai and Manyema; extreme SW Sudan (Bengengai); forests of W and S Uganda (north to Budongo, Mabira and Elgon) and W Kenya (Kakamega); NW Angola (evergreen forest, Cuanza Norte to W Malanje and N Cuanza Sul).

Density in mature forest, Liberia, up to 10 pairs per km² (Gatter 1997).

Description. *D. a. coracinus* J. and E. Verreaux: S Nigeria (except extreme SW) to W Kenya, Angola and Bioko. ADULT ♂: head and body black; crown with blue-green sheen, outer scapulars and uppertail-coverts with greenish sheen, the latter with dull black fringes; forehead, sides of head, neck, mantle to rump and inner scapulars, and whole of underparts dull velvety black. Tail rather

Dicrurus modestus

* = Apparent *modestus* × *adsimilis* hybrids

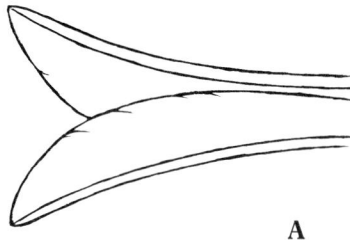
A

long, deeply forked (15–29 mm, T6 c. 20–25 longer than T1) and lyre-shaped (see *D. adsimilis*, **A**); black, upperside with steely green gloss. Flight feathers blackish brown, including inner webs; primaries narrowly edged and secondaries more broadly edged black with greenish gloss. Tertials, alula and upperwing-coverts black with steely green gloss (more bluish on lesser coverts). Underwing-coverts and axillaries sooty black. Bill black; eyes orange-red to deep scarlet; legs black. Sexes alike. SIZE (10 ♂♂, 10 ♀♀): wing, ♂ 121–132 (127), ♀ 121–133 (127); tail, ♂ 98–121 (113·5), ♀ 105–120 (113); bill, ♂ 24–27 (25·5), ♀ 23–27 (24·6); tarsus, ♂ 18–20·5 (19·6), ♀ 18–20·5 (19·4). WEIGHT: Uganda, unsexed, 45–53.

IMMATURE: juvenile sooty brown, wings and tail blackish brown with slight greenish gloss; lacks pale barring on underparts or underwing-coverts.

NESTLING: unknown.

D. m. modestus Hartlaub: Príncipe. Differs from *coracinus* in having faint violet-blue gloss on mantle and back as well as bluish gloss on crown, but sides of head and underparts unglossed; inner webs of flight feathers dark grey-brown (not blackish brown). Juvenile has pale bars on central belly and undertail-coverts and pale-tipped underwing-coverts. Larger than *coracinus*: wing, ♂ (n = 8) 130–139 (135), bill, ♂ (n = 7) 27–30 (28·3); tail relatively long, ♂ (n = 7) 117–131 (126), well-forked (12–24 mm) and lyrate.

D. m. atactus Oberholzer: Sierra Leone to S Ghana and SW Nigeria where it apparently intergrades with *coracinus* in Ondo and Benin Provinces. Upperparts (except forehead to lores) with stronger violet-blue gloss than in nominate *modestus* (greenish blue on crown); slight violet-blue gloss on sides of head and neck, breast and flanks, and tips of axillaries and underwing-coverts; wings and tail glossed steely-green as in preceding 2 forms; inner webs of flight feathers dark grey-brown. Juvenile usually with pale barring on lower belly and undertail-coverts. Same size as *coracinus*: wing, ♂ (n = 10) 118–131 (127); tail, ♂ (n = 10) 106–121 (114). Tail fork deep (14–29 mm). WEIGHT: Liberia, ♀ (n = 3) 44–48 (45·6).

TAXONOMIC NOTE: *D. m. coracinus* and *D. adsimilis* mainly behave as good species where they meet at forest edges. W African *atactus* has some intermediate characters, and apparent *coracinus/adsimilis* hybrids from Zaïre and Angola resemble this form. Accordingly, Vaurie (1949) and most subsequent authors have regarded *atactus*, *coracinus* and *modestus* as forest races of *D. adsimilis*. That view was not subscribed to by Mackworth-Praed (1942) or Meise (1968), and in view of limited hybridization in areas of (secondary?) contact with *adsimilis*, and differences in behaviour, voice and plumage, we treat *coracinus* as a subspecies of *D. modestus*. The position of *atactus* is debatable, but since it resembles nominate *modestus* closely in colour of flight feathers and body sheen, we consider it another forest form of *modestus*.

Field Characters. Length 24–28 cm. A forest species that broadly overlaps range of Shining Drongo *D. atripennis*. Larger than Shining, tail more deeply forked and characteristically fish-shaped in race *coracinus* of Lower Guinea forests. Much less glossy than Shining; *atactus* of Upper Guinea forests has subdued violet-blue (not greenish blue) gloss above, and little below, while *coracinus* has matt black underparts and velvety black back. Meets Fork-tailed Drongo *D. adsimilis* at forest edge and occasionally interbreeds with it; distinguished by darker underside of flight feathers (blackish in *coracinus*), less gloss (or none at all) on back and underparts, and more ringing and musical voice.

Voice. Tape-recorded (32, 104, B, CHA, GUL, HOR, JO PJ, KEI). Very variable repertoire, scratchy and grating notes with nasal twang combining with shrill whistles and pure notes, with similarities to (imitations of?) all 3 other species; generally much more tuneful than voice of Fork-tailed Drongo. Song of individual (*coracinus*) from Cameroon, 'dzaa-dzaa-weet-*wheep*' or 'dzaa-titi-way-tea-wait', reminiscent of Square-tailed Drongo *D. ludwigii*; duet by pair from top of dead tree at forest edge in same area a series of scratchy, tuneless phrases very like Fork-tailed Drongo; bird in Gabon sounded like Shining Drongo *D. atripennis*, a low grating 'jarr-jarr' followed by pure whistles, 'way-tea-too' 'way-tee-too-wee-*whew*'. One bird sang while taking a bath, shaking itself off between each phrase (Chappuis in press): 'tsu-tsu-*wheat*-jar-*wheat*-jar', sometimes adding 'wheet-tsee-weeyo'. Bird in Bwamba (Uganda) sang a repeated, nasal 'chichi-wew-jor-chop' with additions and variations, 'chichi-wew-jor-woo-chi', 'chip-poo-jor-wup-chip', 'chichi-wew-jor-klop, woo-titi-woo-titi-woo-titatawoo'. Birds also give pure, sweet, tchagra-like whistles, a loud rasping sound like tearing paper, hard, shrill '*chip*', up-slurred 'chip-*way*' and loud, clear, pure '*way*'. For further transcriptions see Zimmerman *et al.* (1996).

General Habits. *D. m. coracinus* inhabits forest, avoiding lower levels and thick cover; prefers canopy on edges of clearings, large trees along tracks and roadsides, and gallery forest. In Uganda and Kenya, avoids open habitat, thus seldom in contact with Fork-tailed Drongo. In Gabon and S Nigeria, where Fork-tailed Drongo absent, frequents farmland, plantations and forest-grassland mosaic with large trees. Mainly in lowlands, but ranges to 1600 m in Cameroon, E Zaïre and Kenya. On Bioko occurs in thick forest up to 1200 m, and in tops of high cotton trees among cocoa plantations near coast. On Príncipe, the nominate race occupies old secondary forest, but favours more open coffee and cocoa plantations and perches on roadside wires. In W Africa, occupies wide range of habitats: glades and clearings and high canopy in mature and secondary forest (up to 1100 m on Mt Nimba), forest-grassland mosaic, grassland with scattered trees, gardens and parkland; in high bush near villages in Sierra Leone (Bannerman 1939). In Liberia, occupies forest canopy above 25 m, thus ecologically separated from Shining Drongo *D. atripennis* (Gatter 1997).

Usually occurs singly or in small groups of 2–4; *coracinus* is rather shy and unobtrusive, foraging near openings and forest tracks or perched upright and unconcealed on tree branch 6–10 m high; *atactus* is more noisy and lively. Makes circular flights to catch larger insects in the air, swoops to pick prey from ground, and also gleans from high branches and foliage; may sally above canopy;

sometimes (*atactus*) involved in mixed-species feeding 'drives', taking prey above secondary cover disturbed by other species below (Bannerman 1939). Often almost nocturnal. Dismembers large items with bill, holding them down with foot. Pugnacious; pursues and mobs large hornbills; *atactus* reported to mob crows, herons, cranes, vultures, kites and buzzards, sometimes soaring above victim and lunging down onto its back (Bannerman 1939). Will also attack smaller birds. Pairs perch together, ♂ bowing from side to side and singing, ♀ crouched, fluttering wings.

Food. Insects, caught mainly on the wing, including grasshoppers, beetles, termites, mantids, cicadas, moths and butterflies.

Breeding Habits. Little information. Territory size, Gabon, 6–10 ha (Brosset and Erard 1986).

NEST: one a shallow cup camouflaged beneath with lichen, sited 22 m high on thin fork of small forest tree; a similar nest built mainly of black palm fibres and lichen, bound with spider web to fork *c.* 6 m high; 2 others 35 m high in fork near end of lateral branch. Liberia, nesting height 6–18 m in secondary forest, 30–40 m in primary forest (Gatter 1997).

EGGS: no information.

LAYING DATES: Liberia (nestlings June, fledglings May–June); S Ghana, Dec–Feb; S Nigeria, (May, ♀ about to lay; Nov, ♀ with discharged follicle in ovary); Cameroon, (in main dry season); Gabon, Sept; E Zaïre, Mar (breeding condition June, July, Nov).

References
Bannerman, D. A. (1939).
Brosset, A. and Erard, C. (1986).
Chapin, J. P. (1954).
Gatter, W. (1997).

Family CORVIDAE: crows, jays, magpies and others

Medium to large birds including the largest passerines. Conspicuous, bold, inquisitive and highly adaptable. Bill stout, heavy, powerful and fairly long with a small terminal hook; in *Pyrrhocorax* long and decurved. Nostrils usually round, without an operculum and covered with forward-projecting bristles; rictal bristles also present. Forehead often with stiff feathers. Wings broad; 10 primaries with P10 somewhat reduced. Tail variable in length and shape but never emarginated or forked; usually more or less rounded, often graduated; with 12 feathers, 10 in *Ptilostomus*. Legs sturdy, powerful and long; tarsus strongly scaled at the front, smooth at the rear except smooth front and rear in *Pyrrhocorax*. Feet large and strong. Plumage wholly black or black, grey, brown or white, or (jays and magpies) brightly coloured with blues and greens, also yellow and purple; often large areas with solid colours. Sexes alike or nearly so.

Occupy numerous habitats including forest, woodland, grassland, tundra, desert and cliffs; often around human habitation. Many arboreal but others terrestrial. Flight in most strong with deliberate wing-beats; more laboured-looking in shorter-winged, longer-tailed genera such as *Pica*. Some (especially *Corvus*) glide, soar, tumble and perform aerial displays. Gait a hop or gallop; many walk and run. Omnivorous; predator and scavenger; foot used to hold down food items and (*Corvus*) to carry objects. Bill used to hammer open hard food and to flick or probe food. One species (New Caledonia Crow *C. moneduloides*) uses sticks and leaves held in bill as hook-tools to obtain insects and other arthropods from holes in trees and crevices (Hunt 1996). Often drop shellfish and tortoises from height onto hard surface to break them. Sometimes cache and recover stored food. Scratch head by indirect method although *Corvus* species scratch directly. Apply ants to plumage or expose body to ants. Bathe in water in 'stand-in' manner; also sunbathe, positioning body sideways to sun with body feathers erected; not known to dust-bathe. Drink by dipping, then lifting head, and sometimes by sucking in water. Voice unspecialized with calls simple, loud and harsh. Most sedentary; some migrate.

Usually gregarious when not breeding and especially when roosting; form large roosts, regularly up to 50,000, once 328,000 (American Crow *C. brachyrhynchos*: Brady 1997). Monogamous; typically maintain large territory and nest in isolation, but some species nest in colonies. Some pair for life. Perform complex behaviours including courtship feeding when selecting a mate and forming pair-bond. Usually both sexes build bowl-shaped nest; usually ♀ incubates and broods young with ♂ nearby; both feed young; and some have non-breeding 'helpers' at nest.

Crows, jays and allies make up a well-defined family, amongst the most advanced in evolutionary terms of all bird families. Closest to bowerbirds (Ptilonorhynchidae), birds-of-paradise (Paradisaeidae), bell-magpies and allies (Cracticidae) and magpie-larks (Grallinidae). Based on their DNA–DNA work, Sibley and Ahlquist (1990) have enlarged Corvidae to encompass some 650 passerine species in 7 subfamilies including the Subfamily Corvinae. Sibley and Ahlquist further subdivide this subfamily into 4 tribes: Corvini, crows jays and allies; Paradisaeini, birds-of-paradise; Artamini, currawongs and wood-swallows; and Oriolini, Old World orioles and cuckoo-shrikes. Since aspects of this DNA-DNA classification remain controversial, we prefer to maintain Corvidae as a family that contains only crows, jays and allies.

Nearly cosmopolitan; absent from Antarctic, Arctic, New Zealand and most Pacific islands; most numerous in Northern Hemisphere. 23–25 genera and 112–117 species worldwide (Blake and Vaurie 1962, Madge and Burn 1994, Cramp and Perrins 1994 and Zimmerman *et al.* 1996). 5 genera in Africa including one endemic (*Ptilostomus*, 1 species), 3 in N Africa (*Garrulus*, 1 species; *Pica*, 1 species; and *Pyrrhocorax*, 2 species, one with isolated population in mountains of Ethiopia) and one with nearly worldwide distribution (*Corvus*, 11 species).

Genus *Corvus* Linnaeus

Large, powerful corvids. Bill strong, stout and slightly compressed with gently curved culmen; no notch. Rictal bristles long. Wing relatively long; P8–P5 longest, P10 usually about half as long as P9. Tail short to moderately long. Legs and feet strong. Plumage entirely glossy black, bill and legs black; feathers with metallic gloss; some spp. pied or with grey; some with throat feathers elongated or bifurcated at tips. Juvenile plumage similar to adult's but generally paler. One annual moult.

Ecologically adaptable, exploiting anthropogenic habitats and situations; behaviourally versatile; often claimed to be intelligent; at least one species (*C. corax*) with complex play behaviours. Omnivorous; predator and scavenger; usually on ground, sometimes in trees.

37 (Blake and Vaurie 1962) to 43 (Cramp and Perrins 1994) species; Nearctic, Palearctic, Afrotropical, Oriental and Australasian except New Zealand (where *C. frugilegus* introduced). 12 species in Africa: 4 endemic (*capensis, edithae, albicollis* and *crassirostris*), 1 introduced (*splendens*) and 4 Palearctic (1 a vagrant). 1 superspecies (*C. corax/C. ruficollis*) and 2 species groups (*C. edithae/C. albus* and *C. albicollis/C. rhipiduras/C. crassirostris*) in Africa.

C. edithae has been considered a separate species (Blair 1961, North 1962, Madge and Burn 1994, J. G. Williams, pers. comm.) and a race of *C. ruficollis* (Blake and Vaurie 1962, Hall and Moreau 1970, Goodwin 1986, Cramp and Perrins 1994). Zimmerman *et al.* (1996), while retaining it in *ruficollis*, suggest the 2 may not be conspecific. They overlap in size and in colour of feather bases (white in *edithae*, grey in *ruficollis*), for which reason Goodwin (1986) kept *edithae* in *C. ruficollis*. Far more important, in our view, is that their voices are completely different: *edithae* almost exactly like Pied Crow *C. albus* (North 1962), *ruficollis* like Common Raven *C. corax* (Meinertzhagen 1926, Cramp and Perrins 1994) though higher-pitched and less croaking. M.E.W. North suggested, and we agree, that *edithae* is a 'crow', not a 'raven'; he agreed, as we do, with C.M.A. Blair's suggestion that the bird should be named Somali Crow. *C. albus* often hybridizes with *C. edithae* but never with *C. ruficollis*, although their ranges abut all across the southern edge of the Sahara. We believe that *C. edithae* is not a race of *C. ruficollis* and may not even be closely related to it.

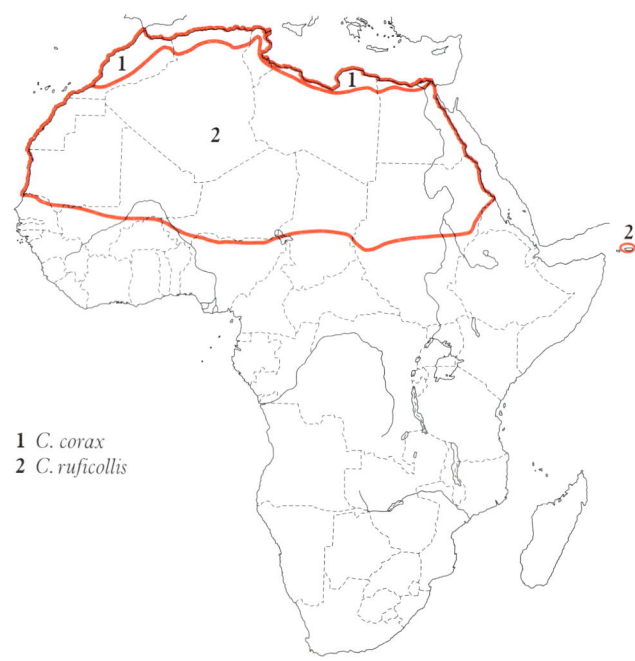

Corvus corax superspecies

1 *C. corax*
2 *C. ruficollis*

Corvus capensis Lichtenstein. Cape Crow; Cape Rook; Black Crow. Corneille du Cap.

Corvus capensis Lichtenstein, 1823. Verz. Doubl., p. 20; Cape of Good Hope.

Plate 31
(Opp. p. 478)

Range and Status. Endemic resident in 2 disjunct populations, one in NE and the other in southern Africa. Eritrea, in highlands north to *c*. 15°N; Ethiopia, SW, W Highlands and SE Highlands; SE Sudan north to 11°N; Somalia, north of 09°N, also along E coast between 05°30' and 02°N; W and central Kenya, commonest in highlands from Mt Elgon to Kerio Valley, Naro Moru and Thika, less common in N from L. Turkana basin east to Moyale; increasing in Rift Valley highlands in 20th century, probably due to large scale farming, but possibly now declining in SE (Lewis and Pomeroy 1989); SE Uganda (Busoga, Bukedi); Tanzania, in E Serengeti Plains, Ngorongoro, Monduli and Moshi.

Angola from coast of Benguela and Mossamedes east through Huila, Huambo, Bié and Moxico to Zambia and south to Namibia; Zaïre, vagrant Kwilu (Kandale and Kianza, south of Gungu: Lippens and Wille 1976); W Zambia west of Zambezi R.; record from Livingstone possibly an escape (Benson *et al*. 1971); Zimbabwe above 900 m from Kazuma Depression, Zambezi Escarpment and N Mashanoland east to Inyanga and south to Plumtree and Bulawayo; Namibia, throughout but uncommon in SE and absent E Caprivi Strip; Botswana except NE; Swaziland in middle- and high-veld; Lesotho; South Africa: Natal, mainly in higher interior grassland, scarcer along coast, absent from flats of E Zululand; Transvaal north to *c*. 23°S, vagrant in E lowveld; Orange Free State, common in E but declining due to shooting and pesticides, stragglers in central and W; Cape Province, common in S, absent dry areas in N, much of Karoo and extreme W. Possibly SW Mozambique in Lebombo mountain area (Clancey 1971) but no certain records (Parker 1999). Generally common or local, sometimes uncommon.

Description. *C. c. capensis* Lichtenstein: Angola, Zambia and Zimbabwe to South Africa. ADULT ♂: entire plumage blackish. Top of head glossy; mantle to scapulars and uppertail-coverts glossy with greenish and violet reflections. Uppertail glossed and tinged greenish. Sides of head to chin and throat black, neck and breast brownish black; chin to breast slightly glossed, the long pointed throat feathers rather lax. Belly and flanks to undertail-coverts sooty black. Upperwing feathers glossy, tinged violet and green on secondary edges, tertials and greater, median and lesser coverts; duller and more greenish on primary edges, primary coverts and alula. Axillaries and underwing-coverts sooty black. Bill black, long and rather slender; eyes brown; legs black. Sexes alike. SIZE (10 ♂♂, 8 ♀♀): wing, ♂ 301–377 (347), ♀ 293–397 (333); tail, ♂ 175–210 (202), ♀ 179–201 (184); bill, ♂ 60–73 (68·4), ♀ 62–69 (65·2); tarsus, ♂ 66–79 (73·7), ♀ 66–73 (70·8). WEIGHT: Zimbabwe, unsexed (n = 2) 410, 500; Botswana, 1 ♂ 612, 1 ♀ 490; South Africa, 1 ♂ 697.

IMMATURE: similar to adult but browner, gloss confined to wings and uppertail.

NESTLING: skin reddish orange with some grey down on feather tracts; mandible reddish orange, gape pink; eyes closed.

C. c. kordofanensis Laubmann: Sudan, Eritrea and Somalia to Uganda, Kenya and Tanzania. Smaller; wing, ♂ (n = 10) 300–330 (321). WEIGHT: Ethiopia, 1 ♀ 387.

Corvus capensis

Field Characters. Length 48–50 cm. A large but rather slim crow, with *rounded head* and *slender bill*, culmen slightly decurved; entire plumage glossy black, becoming browner when worn; immature also browner; throat feathers lax, often giving slightly shaggy look. Sympatric in N Kenya with Somali Crow *C. edithae* which is also black, sometimes with brownish head and neck; separated by slim bill, squarer tail, glossier plumage and dark bases to neck feathers.

Voice. Tape-recorded (20, 36, 38, 58, 88, 91–99, B, C, F). Normal harsh cawing call variable: short, rasping, up-slurred 'rrooak' or 'rrooah'; longer 'rroooaaa', last call of series sometimes down-slurred, 'rrooaauw'; deeper, slower 'rrawrr', sometimes with glottal ending, 'rrawwa' or 'rrawwagla'. A bubbling, glottal 'gloglogloglo' given in display, also in flight and when perched; also rendered 'kwollop, kwollop', 'kwer-kaplop' or 'gur-lalop'. Bowing call a low, gurgling 'gwurrr' followed by a sharp 'tik' (Zimmerman *et al*. 1996); often mixes repertoire in 'conversational' calling (Maclean 1993).

General Habits. Inhabits moorland, alpine meadows, acacia savanna, edges of rivers, seasonally flooded plains in semi-deserts, farmland, exotic tree plantations and along beaches. Prefers uplands to lowlands and open to urban areas; highest numbers in arable farmland. Prefers cultivated areas: in Ethiopia, 89% sightings in cultivation, 8% in high grassland, 2% in open thorn bush, 1% in urban areas, a few in dry savanna and in rocky areas (Wilson 1993). In Ethiopia at 1800–4100 m, usually 2000–2100 m. In E Africa commonest at 1200–2500 m. In W Cape (South Africa), prefers Karoid broken veld, coastal Renosterbosveld and ploughed lands; in Natal occurs down to 300 m, occasionally to coast.

Solitary or in pairs, sometimes in large flocks, up to 200 in Transvaal, 1000 in Kenya. Remains with mate on territory throughout year; strongly territorial, driving away flocks of unmated adults and immatures. Flight laboured and cumbersome; flies with deep regular wingbeats; soars less often than other crows. When chasing intruder or displaying, flies with stiffly bowed wings held well below body, quivering open-fingered wing-tips up and down in shallow and rapid beats, giving appearance of bird hovering, uttering bubbling and gargling calls. Other displays include billing (touching another bird's bill), allopreening and bowing accompanied by calls (see Voice) (bird lowers beak slowly between legs, then raises its head); also flaps wings and hops about as if dancing; flicks tail upward and raises crest upon alighting; and when defending territory puffs out throat feathers, raises crest, partly opens wings, raises tail, and gives bubbling call.

Walks easily on ground with long strides; sometimes hops. Spends 54% of time feeding on ground, 18% flying, 16% on telephone poles, 6% scavenging on road, 3% perching in trees, 2% on buildings, and 2% in other activities (Ethiopia: Wilson 1993). Forages mostly in grasslands and on ploughed fields, often following tractors and domestic stock. Digs for insects and grain with strong downward stabs of bill; also probes in and turns over droppings of large mammals in search of insects. Catches termites as they emerge from their holes; hops several times from place to place before catching a locust. Sometimes carries a maize cob several hundred m before eating it. Occasionally gleans ectoparasites from backs of large mammals and takes eggs of domestic poultry and wild birds. A well-documented 1874 instance of 'black crows' breaking 19 out of 60 Ostrich *Struthio camelus* eggs near Grahamstown, South Africa, by hovering over clutches and dropping stones (Brooke 1979). Very occasionally kills newborn lambs and sick sheep. Sometimes takes carrion, including dried flesh from dead Fur Seals (Shaughnessy and Shaughnessy 1987), but rarely travels to rubbish dumps. Known to bury scraps of food in sand, then later retrieve them (Jessnitz and Jessnitz 1986). Regurgitates indigestible matter as oval or round pellets of $c.$ 25 × 21 mm which are often below roosts and nests. Sometimes mobs large owls, lizards and snakes.

Roosts communally in non-breeding flocks, usually up to 30 or 50, occasionally up to 600, typically in stands of trees, or on telephone wires in open plains. Sometimes flies up to 20 km to roost (Vernon 1984).

Probably sedentary, although wanders widely.

Food. Variety of invertebrates and vertebrates including insects (dung beetles, grasshoppers and locusts, termites, caterpillars and other larvae), millipedes and centipedes, worms, frogs, lizards, birds' eggs and young, domestic chickens up to 450 gm, rodents; also plant material (maize and other grains, seeds, roots, *Cyperus* bulbs, peanuts and fruits and berries of especially *Scutia*, *Royena* and *Opuntia*); rarely carrion and human foods.

Breeding Habits. Solitary nester; territorial; territory size $c.$ 60 ha.; density 1 pair per 5 km^2, South Africa (Vernon 1984). Possibly pairs for life (Skead 1952); courtship behaviour not documented. Occasionally double-brooded.

NEST: large mass of sticks, twigs and sometimes wire, inner cup lined with wool, fur, cloth, dry dung, feathers and string; 1 contained 392 sticks (Skead 1952). Ext. diam. 275–375, ext. depth 200; int. diam. 200, int. depth 100; weight $c.$ 3·2 kg, of which 2·7 kg is sticks; on telephone pole, in fork near or at top of tree, in bush or aloe, on windmill and rarely on cliff edge; 2–24 m above ground. Constructed probably only by ♀ (Madge and Burn 1994); may use same nest each year (Morgan-Davies 1967).

EGGS: 1–6, usually 4, av. (n = 162 clutches, Zimbabwe, Namibia and South Africa) 3·5; narrow oval; moderately glossy; buff, buffish white or pale pink, marked with brown, deeper pink and grey. SIZE (n = 217): 40–53·5 × 27·6–34·3 (45·4 × 31·1).

LAYING DATES: Eritrea, Apr–May; Ethiopia, Feb–May; Sudan, Dec–Feb; Somalia, Mar–Dec; E Africa: Kenya, Mar–Apr, Dec; Tanzania, Nov; Region A, Mar, Nov; Region D, Mar–Apr, Oct–Dec, mainly in rains, especially short rains. Angola (breeding condition Aug); Zambia, Aug–Sept; Namibia, Feb, Oct; Zimbabwe, Aug (4), Sept (15), Oct (35), Nov (25), Dec (2); Botswana, Sept (3), Oct (19), Nov (11), Dec (14), Jan (3); Lesotho, Oct; South Africa: Natal, Aug–Nov; Transvaal, Aug (14), Sept (49), Oct (17), Nov (3), Dec (1); Orange Free State, Aug–Jan; E Cape, June–Dec, peak Aug–Nov, and SW Cape, July–Mar but mainly Aug–Dec.

INCUBATION: probably begins with first egg (Allan 1981), by both sexes but length of time each at nest not known; period 18–19 days. Bird not incubating is often near nest; no display at change-over; sometimes both leave nest for short periods.

DEVELOPMENT AND CARE OF YOUNG: young vary in size, the first 1–2 hatchlings heavier than later ones; wing quills visible by day 6, all feathers visible and eyes beginning to open by day 12, eyes half-open by day 14, eyes open by day 19, well feathered by day 24, fully grown and leave nest by days 36–39. Remain with parents for up to 6 months after leaving nest. Both parents feed and tend young; they remove eggshells and faeces up to day 5, thereafter droppings often accumulate at or near nest; young defaecate over edge of nest by day 28.

BREEDING SUCCESS/SURVIVAL: Zimbabwe and South Africa (n = 60) av. 2·7 young fledged per nest (Winterbottom 1975); sometimes parasitized by Great Spotted Cuckoo *Clamator glandarius*. In Great Karoo, half the farmers kill Cape Crows by shooting, poisoning and destroying nests.

References
Allan, D. (1981).
Madge, S. and Burn, H. (1994).
Morgan-Davies, A.M. (1967).
Skead, C.J. (1952).
Winterbottom, J.M. (1975).

Corvus corone Linnaeus. Carrion Crow. Corneille noire.

Plate 31
(Opp. p. 478)

Corvus Corone Linnaeus, 1758. Syst. Nat., ed. 10, 1, p. 105; 'Europa,' restricted to England by Hartert (1903, Vög. Pal. Fauna, 1, p. 11).

Range and Status. Europe, Middle East, Asia to Kamchatka and N China; south to Morocco, Egypt, Iraq, N Iran, N Afghanistan, Kashmir and W Tibet.

In Africa, winter visitor Morocco and resident Egypt; vagrant Tunisia and Libya. In N Morocco (Tangier Peninsula south to Asilah, Larache and Si Allal Tazi), uncommon to locally common in some years (e.g. in 1974 several flocks of up to 250 birds: Pineau and Giraud-Audine 1979). In Egypt, Nile Delta and Valley south to Aswan, Wadi el Natrun, Faiyum and along Suez Canal south to Hurghada; common, although has dramatically decreased in Delta, possibly due to use of pesticides (Mullié and Meininger 1985). In Tunisia, 1 record about 40 km northwest of Gabès; in Libya, once, 'several' birds at Tobruk. Not recorded in Algeria (Ledant *et al*. 1981).

Description. *C. c. pallescens* Madarasz: Egypt, Cyprus, S Turkey, Levant, N Iraq. ADULT ♂: head, including chin, throat and bib on centre of upper breast black, slightly glossy. Hindneck and sides of neck to mantle, scapulars, rump and shorter uppertail-coverts pale ash grey, feathers with narrow dark shaft streak; longer uppertail-coverts grey with black subterminal blotch. Tail black, slightly glossy on upperside. Sides of breast and lower breast to undertail-coverts ash grey, continuous with grey of neck and upperparts; feathers at side and lower margin of breast bib grey with black central streak. Upperwing black, slightly glossy. Axillaries grey; underwing-coverts black. Bill black; eyes dark brown to dark reddish brown; legs black. Sexes alike. SIZE (10 ♂♂, 10 ♀♀): wing, ♂ 292–331 (310), ♀ 270–314 (294); tail, ♂ 163–196 (181), ♀ 159–181 (172); bill, ♂ 53–58 (55·8), ♀ 47–54 (51·0); tarsus, ♂ 52–60 (57·7), ♀ 52–56 (54·4). WEIGHT: Egypt, 1 ♂ 295, 1 unsexed 333; this race or *sharpii*, Turkey, ♂♀ (n = 4) 450–524 (481).

IMMATURE: like adult, but head to upper breast more brownish black, tail and upperwing less glossy.

NESTLING: short, sparse, pale grey or whitish down restricted to upperparts; skin flesh-pink becoming yellowish green by 6–8 days; gape-flanges narrow, yellowish; mouth bright pink; bill brown-violet becoming greyish; legs flesh, scutes becoming grey.

C. c. sharpii Oates (includes '*sardonius*'): S Italy to Greece, Black Sea area, Caucasus, N Iran, Turkmenistan and N Kazakhstan; (probably this race) vagrant Tunisia and Libya. Like *pallescens* but a little darker grey; slightly larger: wing, (19 ♂♂) 308–345 (325).

C. c. corone Linnaeus: W and SW Europe; winters N Morocco. Differs from preceding races in having body as well as head, wings and tail entirely black, with slight bluish tinged gloss. Larger than *pallescens* with heavier bill: wing (229 ♂♂) 318–340 (331); bill (20 ♂♂) 56–61 (57·9).

TAXONOMIC NOTE: We hold the conventional view of treating this as one species comprising an all-black *corone* group and a grey-and-black *cornix* group. We follow Cramp and Perrins (1994) in recognizing African grey-and-black *sharpii* and *pallescens* and not recognizing *sardonius*. This treatment differs from that of Vaurie (1959), Blake and Vaurie (1962), Goodwin (1986) and Madge and Burn (1994) who call all African grey-and-black birds *sardonius* (including *pallescens*) (see Cramp and Perrins 1994).

The records of *cornix* from Egypt in 1911, 1923 and 1924 (Nicoll 1912, Meinertzhagen 1930) probably refers to *pallescens* (= *sardonius* in Goodman and Meininger 1989).

Corvus corone

Field Characters. Length 46–48 cm. Nominate race (winter visitor to Morocco) all black with heavy black bill feathered at base. Sleeker and less glossy than Rook *C. frugilegus*, without shaggy thighs or pale facial skin; lower angle to forehead gives *flatter profile*; in flight, tail squarer. Very similar to young Rook, which lacks pale facial skin, but culmen curved and bill stouter (bill of Rook straight and conical); bristles at bill base usually merge with bill so that culmen profile is not interrupted by a 'bump', as in young Rook. Smaller than Common Raven *C. corax* (resident in Maghreb) with less massive head and square, not wedge-shaped tail. In Egypt a grey and black race ('Hooded Crow' *C. c. pallescens*) occurs with Brown-necked Raven *C. ruficollis*, which is black with variable amount of brown on head and neck, depending on wear, and around the Suez Canal meets House Crow *C. splendens*, which has similar grey nape and hind neck but much darker back and underparts.

Voice. Tape-recorded (62, 73, 93, 105, B). Normal call a short, deep, guttural 'kaa' with vibrant or resonant quality, harsher than that of Rook; variable, also rendered 'kraarrr', 'karaa' and 'aaarr'. When uttered 2–6 times in quick succession becomes Advertising or Self-assertive call, given in territorial situations. An emphatic, angry-sounding double 'kar, kar' or 'ark, ark' given in territorial disputes. Alarm call, given when mobbing predator, a longer, harsher, flatter 'kaaaar'. For further variations see Goodwin (1986) and Cramp and Perrins (1994).

General Habits. Little known in Africa. Occurs in wide variety of habitats from farmland to forest clearings and cliffs; prefers open country with trees and human habitation and cultivation. In Egypt, common in most cultivated parts (Goodman and Meininger 1989), and rarely sympatric with Brown-necked Raven *C. ruficollis* except along ecotone between the desert and Nile basin.

Solitary or in pairs year round, also in flocks in the non-breeding season. In Morocco flocks of up to 15–50, once 250; 'small' flocks Egypt. Walks steadily on ground, occasionally hops or jumps to one side. Flight steady, powerful and direct. Rarely soars, and flies high only when on migration or when moving to and from roost. Rarely, carries stick into air, dropping and then catching it before it reaches the ground; usually less aerobatic than Common Raven *C. corax*.

Forages on ground in fields, suburbs and along shores, picking, probing and turning over stones and other objects. Scavenges; sometimes feeds in trees. Wary of man, even in urban areas. Steals food from gulls, other corvids and small raptors, chasing and forcing them to drop food. Captures small birds and bats, also insects, in flight. When killing small vertebrates or robbing nests, often does so cooperatively in pairs or small groups, one bird distracting while the others attack (Cramp and Perrins 1994). Perches on backs of domestic animals, feeding on ectoparasites. Drops mussels and nuts onto hard surfaces to break them; also pecks open nuts. Caches food under stones, wood and grass.

Social behaviour not recorded in Africa, but well documented elsewhere (see Coombs 1978 and Cramp and Perrins 1994). Often gathers in small groups and flicks wings, fans tail and raises head feathers. Not aggressive except when competing at a good food source or when defending territory from another pair, when ♂ is aggressive toward other ♂♂, ♀ toward other ♀♀. Displays include Advertising-cawing: prominently perched bird extends neck forward and upward and lowers head while calling; Forward-threat: bird holds body low and near horizontal, places head forward with bill extended and ruffles its feathers; Upright-threat: bird stands upright, raises head and points bill down; Bristle-head: bird raises head feathers; Pot-belly: bird moves neck and head upright and forward, ruffles belly feathers, tilts tail downward, then runs and hops close to intruder, and Walking: 2 ♂♂ stand or walk close to each other with bodies and tails tilted toward each other. Sometimes chases conspecific in flight, pecking and grappling with it in air and falling to ground.

Roosts usually near nest during breeding season but in non-breeding season roosts communally in trees, also on sheltered cliff ledges and occasionally on ground; frequently roosts with other corvids. Roosts often large, numbering several hundreds or (*C. c. cornix*) thousands. When travelling to roost in early or mid afternoon, gathers first on open ground or in trees, then later moves to roost; some arrive at roost after sunset. Leaves roost early, before sunrise.

In Morocco, some movements in autumn and spring across Strait of Gibraltar (Finlayson 1992) with earliest birds arriving 23 Oct and latest leaving 18 Apr (Pineau and Giraud-Audine 1976). Probably sedentary in Egypt.

Food. Not documented in Africa. Elsewhere principally invertebrates and grain; also feeds on carrion and scraps but not as often as other corvids (Goodwin 1986). Eats earthworms, molluscs, crustaceans, centipedes, spiders, insects including ectoparasites, stranded or dying fish (rarely healthy fish), frogs, small reptiles, eggs, young and injured birds; also grain, seeds, potatoes, fruits and nuts.

Breeding Habits. Poorly known in Africa. Monogamous, solitary nester; territorial; territories usually 14–49 ha but in Israel and central Asia nests may be only 100–200 m apart (Madge and Burn 1994). Courtship displays include flight-chases; bowing: either ♂ or ♀ lowers head, then jerks head upward, spreads tail and covers eyes with the nictitating membranes; soliciting: ♀ begs for food; and pre-copulatory: ♂ ruffles plumage, lowers wings, partly spreads tail and erects head feathers to form 'ears' while ♀ crouches, holds and quivers tail slightly above horizontal, tilts bill slightly upward and partly opens and droops wings. Pair allopreens, and ♂ feeds ♀ during courtship. Single brooded.

NEST: large bulky structure of sticks and twigs, in centre a cup lined with mud, dung, grass, moss and hair; av. ext. diam. 471, int. diam. 198, ext. depth 330, int. depth 126 (n = 52, Finland); usually high in tree, on telephone pole or rarely on cliff, rock, building or ground; height above ground (n = 119, Germany) 7–24 (14·8) m. Built by both sexes.

EGGS: 2–7, usually 4–5; laid every 24 hours; sub-elliptical to oval, smooth and slightly glossy; light blue to green with spots, blotches and scrawls of olive-green to blackish brown. SIZE (n = 140, SE Europe, *sharpii*): 37·5–49·3 × 26·5–34·0 (42·8 × 28·7).

LAYING DATES: Egypt, Jan–June.

INCUBATION: begins probably when second egg laid; period 17–22 (19·5) days; by ♀ with ♂ feeding her on or near nest. ♀ leaves nest for short periods when ♂ near or on nest.

DEVELOPMENT AND CARE OF YOUNG: eyes begin to open on day 8, young fledge by days 32–36, independent 3–5 weeks after fledging; leave parents and parental territory 2 months after fledging; first breed when 2 years old. ♀ young grow faster than ♂ (full details, Norway, in Rofstal 1986).

Fed and cared for by both parents; ♀ broods young and ♂ brings all or almost all food for ♀ and young for first 15 days; thereafter ♀ also searches for food for young. Occasionally young up to 2 years old help feed the incubating ♀ and nestlings and defend parental territory (3 nests, Switzerland: Richner 1990). Sometimes parents eat eggs if they fail to hatch and remove faecal material of small young; older young defaecate over sides of nest. Defends young by mobbing, dive-bombing and calling at intruders.

BREEDING SUCCESS/SURVIVAL: no information for Africa; in Israel usually 2 or 3, rarely 5 young raised per nest (Shirihai 1996). In Israel and other areas of Mediterranean region parasitized by Great Spotted Cuckoo *Clamator glandarius*.

References
Coombs, F. (1978).
Cramp, S. and Perrins, C.M. (1994).
Goodwin, D. (1986).
Shirihai, H. (1996).

Corvus corax Linnaeus. Common Raven. Grand Corbeau. Plate 31

Corvus corax Linnaeus, 1758. Syst. Nat., ed. 10, 1, p. 105; 'Europa', restricted to Sweden by Hartert (1903, Vög. Pal. Fauna, 1, p. 2). (Opp. p. 478)

Forms a superspecies with *C. ruficollis*.

Range and Status. Holarctic, south to Canary Is., N Africa, Middle East, N Pakistan, NW India, Tibet, Mongolia, N Manchuria, and Nicaragua.

Resident from Morocco to Egypt. In Morocco common from Atlantic and Mediterranean coasts east and south in mountains from Rif Mts in NE through Middle, High and Anti-Atlas to about Goulimine in SW; absent southeast of these mountains; in Algeria common north of Atlas Saharien, rare in Sahara at Daiet-Tiour (Beni Abbès), El Goléa and Mzeb. In Tunisia, frequent to common, widely distributed south to Tozeur, Gabès, Medenine, Tatahouine and Libya border; apparently absent east of Kairouan; in Libya along coastal zone south to *c.* 100 mm isohyet; common Tripolitania, status not clear along Gulf of Sirte, probably frequent to uncommon in east along Cyrenaica coast. In Egypt, rare along narrow coastal belt east to *c.* 30° E in Nile Delta.

Corvus corax

Description. *C. c. tingitanus* Irby (only race in Africa). Entire plumage black with violet and dark green gloss, strongest on head, upperparts, upper breast and outer edges of tertials, secondaries and lesser and median wing-coverts. Bases of head and body feathers light grey, palest on mantle and hindneck. Throat feathers rather long and pointed, tips often bifurcated. In worn plumage, sides of head and neck, hindneck and underparts more bronzy, almost oily olive to brown. Bill black; eyes dark brown; legs black. Sexes alike. SIZE (7 ♂♂, 4 ♀♀): wing, ♂ 362–420 (392), ♀ 376–393 (385); tail, ♂ 192–227 (214), ♀ 202–217 (209); bill, ♂ 64–70 (67.0), ♀ 62–64 (63.0); tarsus, ♂ 67–72 (69.3), ♀ 64–68 (65.4). Bill heavy, depth at centre of nostril (n = 38 ♂♀) 25–28.

IMMATURE: similar to adult but duller; head to throat sooty black, rest of body greyish black; gloss confined to wings and tail, and a little on lower mantle and scapulars. Throat feathers shorter and rounder. Eyes blue-grey.

NESTLING: covered with rather short grey-brown down on upperparts, upperwing and thigh; otherwise naked. Skin flesh pink becoming greyish flesh by 9–10 days; gape flanges pale yellow; mouth dark red or purplish; bill dusky grey becoming blue-grey with black culmen and tip; legs becoming dusky grey.

Field Characters. Length 63–65 cm. A huge glossy black crow with massive head, deep, curved bill and *long wedge-shaped tail*. Throat feathers shaggy, giving thick-necked shape in flight. Tips of primaries well spread in short flights, but in mountains may soar like raptor at considerable heights, with wings held flat and straight out and tips of primaries close together, which with long neck and tail form a 'flying cross'. *Deep croaking* calls diagnostic. Larger than Rook *C. frugilegus* or Carrion Crow *C. corone*; Rook has slightly wedge-shaped tail but narrow bill, very different call, pale face when adult; Carrion Crow has square or gently rounded tail, slimmer bill, shorter neck, tail and wings. For differences from Brown-necked Raven *C. ruficollis*, see that species.

Voice. Tape-recorded (62, 73, 93, 105, B). Common call a short, deep croak, variable, rendered 'pruk', 'kruk', 'krok', 'prok' or 'kra': a contact call between ♂ and ♀; often involves threat from non-conspecific enemy; also used in territorial advertisement and defence. 3–4 calls in rapid succession indicate alarm. Contact call a deep, low, throaty 'gro'. For complete catalogue of calls see Gwinner (1964), also Goodwin (1976) and sonagrams in Cramp and Perrins (1994).

General Habits. Little known in Africa. Occurs on mountain cliffs and massifs up to 3000 m, open plains with rocky wadis and coastal areas with cliffs; avoids extensive woodlands and rarely penetrates Sahara.

Solitary, in pairs or family groups, and outside breeding season in flocks. Pair remains in territory all year; but some pairs accompany large flock near their territory or at a food source. Large flocks occur in Morocco (e.g. de Naurois 1961); as many as 100 can soar together, up to 600 m or more above ground.

Walks easily on ground, occasionally hops; perches easily on cliffs, telephone poles and trees. Flight direct, powerful and almost effortless, with slow swooshing wing beats. Glides over long distances and performs tumbling and rolling aerobatics: dives with closed wings, then suddenly rolls over in the air. Often flies long distances in search of food; upon locating food recruits others.

Forages mainly on ground on carrion. Kills small animals by striking with bill. Can be bold around human dwellings, but shy if persecuted. Approaches carrion cautiously, sometimes making several vertical jumps before pecking at carcass. At carrion mixes with Brown-necked Ravens, gulls, vultures and raptors. Eats maggots and beetles on carrion. Robs vultures of pieces they have torn from carcass; readily attacks sick and injured animals. Holds large food object under one or both feet; sometimes carries food in flight. Follows plough, catching animals disturbed by it; hawks flying ants and catches and kills birds up to size of a partridge. Drops objects on incubating gulls, dislodging them and then feeding on exposed eggs or nestlings. Drops bones in order to crack them, and hides and stores surplus food in holes, under stones or in sand. Known to toboggan on back down snow slope (C. H. Fry, pers. comm.).

Often aggressive over food, particularly when many ravens at food source. Usually directs aggression at conspecific of same sex and displays by: Advertising: ♂ erects feathers just above each eye to form 'ears' and raises feathers of throat, neck and breast, then walks about giving a soft 'ko', while ♀ displays as above except her 'ear' tufts are less conspicuous and she gives no vocalization; Thick-necked posture (**A**): ♂ raises feathers on head and throat and holds his body horizontally, then bows several times giving 'kro' or 'krua' calls and moves toward conspecific ♂, continuing to bow and call; ♀ performing this display points bill and tail upward (giving her a U-shape) and gives squeaking calls; Forward-threat: ♂ (and ♀) retracts feathers close to body, erects 'ear' tufts, lifts wings and opens and snaps bill until display reaches high intensity and bird charges at and sometimes bites or pulls feathers of conspecific. In this display ♂ makes 'krua' call and ♀ produces squeaky sounds. Fighting may become intense, the combatants lying on the ground with claws interlocked, pecking at each other's bills but not at the eyes; Defense-threat: either sex raises plumage of back giving a

B

ruffled profile, then assumes hump-backed posture and opens bill; Bill-up: ♂ or ♀ crouches on ground or on a perch with bill raised; Head-flagging: bird perches next to a conspecific and each slowly turns its bill away from the other, then looks back at it before again turning bill away. During hostile contests keeps eyelids wide open and pupils dilated, while in friendly meetings keeps eyelids almost closed and pupils somewhat contracted. When bowing and calling may also snap bill, partly spread tail and draw nictitating membrane over eyes. Also chases conspecifics in flight, begs for food in apparent appeasement behaviour, allopreens with 'friendly' conspecific even if not paired with it and occasionally hangs upside-down (**B**), possibly in play.

Roosts communally in trees or cliffs, sometimes buildings and utility pylons, often close to good food supply; some roosting sites traditional, some used only for a few nights. Numbers at roost not recorded in Africa but up to 300 Denmark (most non-breeding birds); paired adults roost at or near nest-sites in territory throughout year. Arrives at a regular gathering place at first singly or in small groups, then a little before sunset all proceed in a flock to roost.

Probably sedentary in Africa, with some wandering by young birds.

A

Food. Carrion (afterbirths, and feeds at dead sheep, cattle, rabbits, fish and human offal). Takes centipedes, spiders, crabs and other crustaceans, earthworms, molluscs, sea-cucumbers, fish, frogs and toads, snakes, lizards, birds and

C

their eggs, various small and sometimes large mammals; seeds, fruits and berries, wheat, oats, maize, moss and seaweed.

Breeding Habits. Poorly known in Africa. Monogamous, probably pairing for life; solitary nester. Territorial: 0.7–2.5 breeding pairs per 10 ha in Strait of Gibraltar area, Spain and Morocco (Finlayson 1992). Pair-formation probably takes place in flocks of non-breeding birds. During courtship ♂ first performs Advertising display (see above), then stretches neck forward, slightly spreads and raises tail, stretches wings backward, gives 'krua' call and draws nictitating membrane over eyes. He then begins Precopulatory display (**C**), half-spreading his wings, drooping them well away from his body, spreading and raising his tail, extending his neck forward, raising his head, and pointing his bill slightly above horizontal. ♀ invites copulation by crouching, drooping and then opening and closing her wings and shaking her tail. Mating follows. They perform courtship flights, allopreen and touch each other's bills. ♂ commonly feeds ♀ by regurgitation during courtship, and later the incubating and brooding ♀ may solicit food from ♂ by begging and flapping her wings. Single-brooded but replacements possible.

NEST: a large bulky structure of twigs and branches, held together with earth, dung, grass and roots, the inner bowl lined with wool, hair, fur, grass, lichens and stems, sometimes also wire and bone; ext. diam. 702 (430–930), int. diam. 276 (200–370), ext. depth 446 (260–650) and int. depth 111 (80–160) (n = 15, Poland); usually on rocky cliff but also on or near top of tree, telephone pole, electric pylon or building. When on cliff, generally in upper third, rarely within 3 m of top or bottom and protected by overhang; when in tree, c. 9 m above ground. Built by both sexes; sometimes same nest-site and even same nest re-used, often for many years; most pairs, however, use 2 or more sites over the years. Completed 1–4 weeks before eggs laid. Both aggressively defend nest against non-breeding birds when they wander within 0.5 km of nest. Sometimes non-breeding conspecifics destroy nest, scattering contents over a wide area.

EGGS: 1–8, usually 3–6; (n = 93 clutches Morocco, Algeria, Tunisia) av. 4.1; laid every 24 h, in early morning, sometimes at intervals of 2 days; sub-elliptical, smooth and slightly glossy; light blue to greenish blue, speckled, spotted and blotched with various shades of dark brown or olive-brown. SIZE (n = 220, race *tingitanus*): 42.4–54.0 × 30.0–35.5 (47.6 × 32.5).

LAYING DATES: Morocco, Feb–June; Algeria, Feb–May; Tunisia, Mar–May; Libya, breeds but no dates recorded; Egypt, Apr.

INCUBATION: begun when 2nd egg laid; period 18–21 days; by ♀ only with ♂ feeding her at or near nest. When occasionally ♀ leaves nest briefly to stretch or preen, ♂ may substitute at the nest. ♀ turns hatching egg so that bill of hatchling always points up; helps chick hatch and cleans it; also eats egg shell.

DEVELOPMENT AND CARE OF YOUNG: dorsal feather tracts present as grey band by days 5–6; chicks beg with calls at day 6, eyes open by days 12–14, move out of nest onto nearby branch or ledge by c. day 40; fledge by c. day 41; make short glides up to 150 m from nest at days 35–45; and remain with parents 4–6 months before joining flocks of non-breeding birds. Fed and cared for by both parents; ♀ usually broods young for first 18 days; sometimes stands or crouches over young. When temperature high, ♀ and probably ♂ bring water to young; ♀ also wets her underparts and then broods young during hot days. Faecal material eaten by parents, larger young defaecate over edge of nest; older young also regurgitate pellets. Parents protect small young by crouching over them, also by actively mobbing and dropping stones on intruders.

BREEDING SUCCESS/SURVIVAL: no information for Africa; in Wales (n = 181 pairs), average of 1.9 birds fledged per nest; in Britain mortality in 1st year 37%, 2nd year 52%, and years 3–5 50%.

References
Coombs, F. (1978).
Cramp, S. and Perrins, C.M. (1994).
Goodwin, D. (1986).
Heim de Balsac, H. and Mayaud, N. (1962).
Heinrich, B. (1989).
Madge, S. and Burn, H. (1994).
Ratcliffe, D. (1997).

Plate 31
(Opp. p. 478)

Corvus ruficollis Lesson. Brown-necked Raven. Corbeau brun.

Corvus ruficollis Lesson, end of 1830 or early 1831. Traité Orn, p. 329; no locality; type locality accepted as Cape Verde Archipelago, as fixed by Hartert (1921, Vög. Pal. Fauna, 3, p. 2020).

Forms a superspecies with *C. corax*.

Range and Status. Cape Verde Archipelago, Africa north of equator, Middle East and Arabian Peninsula to W Pakistan and Transcaspia.

Resident, generally common, locally uncommon. In Morocco east of Middle Atlas, High Atlas, Anti-Atlas and Jabel Bani and south to Western Sahara; absent from these mountains and W lowlands from Agadir northward; in Algeria, from Sahara Atlas (Béjar, Laghouat, Ckaiba) southward to Mali and Niger borders, apparently extending range to north (Ledant *et al.* 1981); in Tunisia, widely distributed in southern deserts north to Feriana, Gafsa, Hadjeb el Aioun, Kairouan and Zaghouan. Widespread in Western Sahara and Mauritania, becoming uncommon *c.* 50 km from Senegal border; Senegal south to Djoudj, Rosso, Podor, Cascas, once Dakar; Gambia, vagrant (5 records from Cape St. Mary, Yundum, Mardina Bah); Mali south to *c.* 14°N; a few records from N Burkina Faso (Oursi, Markoye and Dori: Y. Thonnérieux and G. and M.-N. Balança, pers. comm.); Niger throughout except SW; Nigeria south to Sokoto, Bulatura Oases, Yo and Sahel west of L. Chad. Libya south of 100 mm isohyet in W (Tripolitania) and E (Cyrenaica) provinces, through the Fezzan and Libyan Desert (status uncertain) to central Chad at 14°N (Batha Prefecture). Egypt in W and E deserts up to 2200 m, Red Sea coast and coastal islands, Nile Valley and edge of Nile Delta; largely absent in cultivated areas and N coastal belt between Salum and Alexandria, where replaced by Common Raven *C. corax*. Sudan, south to *c.* 12°N; NW Ethiopia, vagrant (Bahar Dar, Axum). Socotra, coastal plains and mountains up to 460 m.

Description. ADULT ♂: entire plumage blackish. Forehead to nape, and lower mantle to uppertail-coverts glossed dark greenish, with violet tinge on scapulars. Hindneck and upper mantle to sides of head, sides of neck, chin and throat dark glossy brown, tinged bronzy purple, becoming duller blackish brown in worn plumage; throat feathers rather long and pointed. Rest of underparts sooty black with slight violet gloss on breast and flanks. Hidden bases of body and head feathers greyish, palest on neck and upper mantle. Tail glossy above with slight violet tinge. Secondary edges, tertials and upperwing-coverts glossed violet, primary edges less strongly so. Axillaries and underwing-coverts sooty black. Bill black; eyes dark brown; legs black. Sexes alike. SIZE (10 ♂♂, 10 ♀♀): wing, ♂ 367–429 (390), ♀ 345–404 (378); tail, ♂ 194–218 (206), ♀ 193–215 (205); bill, ♂ 66–74 (68·3), ♀ 60–67 (63·8); tarsus, ♂ 65–71 (68·7), ♀ 63–68 (65·7). Bill less heavy than in N African *C. corax*; depth across nostril (n = 35 ♂♀) 20–24. WEIGHT: Algeria, 1 ♀ 560; N Chad, 1 ♂ 580; N Niger, 1 ♂ 595, 1 ♀ 500.

IMMATURE: similar to adult, but gloss less intense, hindneck blacker (less bronzy); underparts more greyish black (browner on belly, lower flank and vent); throat feathers shorter. Eyes grey or grey-brown; base of upper mandible grey, of lower mandible yellowish; mouth yellowish flesh.

NESTLING: covered with dense, fairly long light brownish grey down; cheeks, throat and belly naked. Bare skin yellowish orange; gape flanges orange-yellow; mouth orange; bill yellowish orange with blackish culmen and tip; legs yellowish flesh-grey, scutes becoming dark grey.

TAXONOMIC NOTE: *C. ruficollis* sometimes is considered a race of *C. corax* (e.g. Ali and Ripley 1972), replacing *corax* in arid regions of N Africa, Middle East and W-Central Asia. Their ranges, however, overlap in parts of W-Central Asia and Israel and meet in N Africa, all without any intergradation. This along with different ecological requirements and differences in voice, bill size, wing formulae, tail shape and plumage colour are highly suggestive of distinct species, and we follow Vaurie (1954) and Cramp and Perrins (1994) in treating them as separate species. The Socotra population of Brown-necked Raven tends to be large, and it may form a distinct race.

Field Characters. Length 52–56 cm. Slightly smaller than Common Raven *C. corax*, which it overlaps narrowly where Sahara Desert meets mountains in N Africa, more broadly where Brown-necked has penetrated hills in Algeria; they apparently do not quite meet in NE Egypt (Goodman and Meininger 1989). Brown sheen on head and neck hard to see at a distance, although it becomes more pronounced in worn plumage, and African race *tingitanus* of Common Raven becomes dark brown in worn plumage. Structural characters are more reliable; Brown-necked has *slimmer, less strongly arched bill* and *longer, sleeker head* without shaggy throat; in flight bill often points downward, head appears thinner, wings slightly longer and narrower, and *tail shape more rounded, less graduated*, with 2 central feathers often protruding. Best distinguished by *higher-pitched* voice and

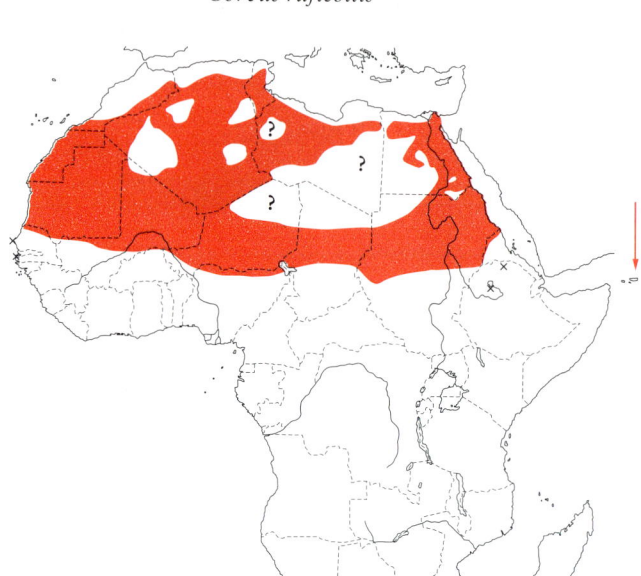

Corvus ruficollis

to some extent by habitat preference (Brown-necked prefers *Artemisia* steppe: Madge and Burn 1994).

Voice. Tape-recorded (60, 105, 109, B, CHA, HAZ, JOHN, MIL, ROC). Common call a harsh, deep very throaty croak similar in quality to Common Raven *C. corax* but more drawn-out, 'rraarrh', 'oowarrh', 'crrrarrr', 'cuworrr', 'crrra-waa', sometimes with hissing overtone, 'corrrax'; also gives a dry, up-slurred 'aarg-aarg-aarg', soft 'carr' and short, muffled 'urrk' or 'kruk-kruk-kruk', softer and less resonant than croak of Common Raven. Also described is a loose, throaty rattle or 'quiet, frog-like gargling' as bird stretched its neck and arched its head and bill downwards (Cramp and Perrins 1994; and see their sonagrams); this may be the equivalent of the 'watch-winding' call of Pied Crow *Corvus albus*, q.v.

General Habits. Inhabits primarily deserts, arid open plains and dry savanna from sea level to *c*. 2500 m; also visits rocky ledges and cliffs, oases, drinking troughs and wells, thornbush, cultivation at edges of deserts, palm groves and human settlements. In NW Africa, especially E Morocco, largely confined to *Artemisia* steppe with groups of jujube trees *Ziziphus*. Where sympatric with Common Raven, tends to avoid mountains, occurring in semi-desert and true desert.

Solitary, in pairs or flocks of 3–4, rarely over 20; paired adults usually on territory all year. Gather in groups of up to 40–50 at drinking troughs and 150 at refuse dumps or carcasses; immatures and non-breeding birds congregate in flocks of up to 1200 at communal roosts. Bold and fearless when not disturbed by man, but shy and wary where persecuted. Walks easily and methodically on ground; occasionally hops. Flight relatively slow with weak wing beats; flies quite low over feeding territory; locates food visually in flight. Wheels and soars high on thermals with storks up to 500–600 m. Performs aerobatics in groups of up to 40, chasing, spiralling upward and descending, often in pairs, in a series of tumbling dives and swoops.

Forages mainly on open ground; turns stones and digs in ground with bill for grubs; searches dung for beetles and undigested grain. Perches on camel or donkey, taking ectoparasites and pecking at cuts and sores. Tears open sacks to eat grain. Feeds co-operatively when searching for locusts, with some birds acting as beaters (Bannerman 1948); catches locusts and termites with feet in flight and eats them in flight. Catches and kills small invertebrates with bill, tearing prey into pieces before eating it. Flies along shoreline searching for dead fish, shellfish and eggs in exposed nests. Feeds on carcasses; attacks new-born lambs and gazelles. Captures exhausted migrant birds and takes eggs of Western Reef Heron *Egretta gularis* (Goodman and Storer 1987) and other birds. Feeds on fruits of date palms *Phoenix dactylifera* in the tree. Hops about streets and slaughter houses, looking for and taking refuse. Follows people travelling through desert, eating their food scraps.

Forms flocks with Fan-tailed Raven *C. rhipidurus*, Common Raven and Pied Crow. Tends to harass and attack competing vultures at carrion. Noted to attack, with Common Raven, a Short-eared Owl *Asio flammeus* in Morocco (Cramp and Perrins 1994). Once (Israel) a flock reportedly flushed and chased hare *Lepus*, attempting to strike it with their feet (Cramp and Perrins 1994).

Roosts communally on rocky ledges, trees and telephone poles and wires, in flocks of 10 or more, sometimes over 1000 (Sinai, Israel). Arrives at roost singly or in flocks; leaves at first light. Rather quiet at roosts in Morocco and Egypt but noisy just after leaving roosts in morning in Saudi Arabia (Cramp and Perrins 1994).

Resident and probably sedentary in Africa, although flocks of non-breeding birds wander in search of food. In Israel and some areas in northern part of range some seasonal movements reported (Cramp and Perrins 1994).

Food. Ground-dwelling animals; some plants and carrion. Grasshoppers, locusts, beetles, termites, ticks, spiders, snails, fish, *Agama* lizards, snakes, birds and their eggs, and small mammals. Seeds, maize, rice, and dates. Carrion of all sorts. In Western Sahara, of 50 pellets taken from roost, 65% contained fruit, 30% barley, 10% beetles, 5% crickets and 2% snails; a few also contained remnants of small lizards and seeds; and 2 stomachs of birds feeding at refuse dump contained beetles, fish remains and mammal hair and bones (Valverde 1957).

Breeding Habits. Solitary nester; territorial; monogamous; pair bond long-lasting, possibly for life. Min. distance between nests in Israel 1 or 2 km and av. distance 3·3 km (Shirihai 1996). A pair seen rolling in aerial display in Algeria.

NEST: large bulky structure of twigs and branches, lined with plant fibres, grass, feathers, hair, wool, rags and paper; usually placed in crown of tree but also on telephone pole, cliff, rocky outcrop, old building, isolated bush; in open desert, on ground among tamarisk *Tamarix* shrubs in dunes (Algeria: Cramp and Perrins 1994); on corner of chain-link fencing, concrete hut, pile of rusting oil-drums and similar discarded iron scraps and fixtures and commonly on high-tension cable pylons, 3–15 m high in harsh desert (Oman: C.H. Fry, pers. comm.); nests commonly on pylons in Libya (Massa 1999); occasionally also an old nest of Osprey *Pandion haliaetus* or buzzard *Buteo*. Size in W-Central Asia (n = 11) ext. diam. 320–600, int. diam. 200–300, ext. depth 300–390, int. depth 110–150 with lining up to 150 thick; 0·8–30 m from ground. Built by both sexes; in captivity takes 5 days. Old nest-site used for several years but not known if same pair involved.

EGGS: 1–7, usually 4–5; (n = 61 clutches NW Africa, av. 4·0); laid every 24 h in morning over 5–6 days. In NW Africa clutches in semi-desert typically 5–6, those in true desert 2–3, occasionally 4; clutch larger if laid earlier in season (Morocco, Mar (n = 11), av. 4·9, Apr (n = 14), av. 3·85: Cramp and Perrins 1994). Sub-elliptical; smooth and glossy; pale blue, spotted, streaked and blotched with olive-buff to blue-grey. SIZE (n = 140) 38·5–52·0 × 28·0–35·4 (45·3 × 31·7).

LAYING DATES: Morocco, Feb–May; Algeria, Feb–Apr; Tunisia, Mar–Apr; Libya, Apr; Egypt, Feb–Apr; Western Sahara, Jan–Mar; Mauritania, Jan–Apr, June, Sept–Dec; Mali, July–Oct; Niger, Nov–Dec; Sudan, Dec–Mar.

Plate 35

Shelley's Starling (p. 624)
Lamprotornis shelleyi

Chestnut-bellied Starling (p. 628)
Lamprotornis pulcher

Hildebrandt's Starling (p. 625)
Lamprotornis hildebrandti

Superb Starling (p. 626)
Lamprotornis superbus

Fisher's Starling (p. 632)
Lamprotornis fischeri

White-crowned Starling (p. 633)
Spreo albicapillus

Juv. Ad.
African Pied Starling (p. 634)
Spreo bicolor

Príncipe Glossy Starling (p. 609)
Lamprotornis ornatus

Splendid Glossy Starling (p. 610)
Lamprotornis splendidus splendidus

Long-tailed Glossy Starling (p. 612)
Lamprotornis caudatus caudatus

Ad. Juv.
Rüppell's Glossy Starling (p. 614)
Lamprotornis purpuropterus

Burchell's Starling (p. 618)
Lamprotornis australis

Meves's Long-tailed Starling (p. 616)
Lamprotornis mevesii mevesii

6 in
15 cm

Plate 36

INCUBATION: begins often with 1st egg; mainly by ♀; period 18–23 days. ♀ typically incubates for some time, then stands up, preens for 5 min, then either fed at nest by ♂ or flies within 200 m of nest to feed for about 3 min, rarely as long as 8 min, then returns to nest and continues incubating. ♂ usually stays in nesting area, acting as sentinel or mobbing or attacking intruding birds, including migratory kites and eagles (Harvey and Harvey 1992).

DEVELOPMENT AND CARE OF YOUNG: young hatch at intervals of 1–2 days. Young fed and tended by both parents. In W-Central Asia eyes open by day 8, wings and back partially feathered by day 15, crouch in nest and adopt threat posture by days 16–17, fledge by days 37–38 but not able to fly properly until days 42–45 (Cramp and Perrins 1994); young remain with parents for a few months thereafter (Shirihai 1996).

BREEDING SUCCESS/SURVIVAL: little information; in Morocco (n = 13 nests) av. 4·46 young fledged per nest; in dry years success rate lower, no more than 1 young per nest, due to shortage of food (Cramp and Perrins 1994). In Yemen, 2 out of 4 young killed by cats within day of leaving nest; in 2nd nest, 2 young successfully fledged.

References
Cramp, S. and Perrins, C.M. (1994).
Harvey, D. and Harvey, M. (1992).
Madge, S. and Burn, H. (1994).
de Naurois, R. (1981).
Valverde, J.A. (1957).

Plate 31
(Opp. p. 478)

Corvus edithae Phillips. Somali Crow. Corbeau d'Edith.

Corvus edithae Phillips, 1897. Bull. Br. Orn. Club, 4, p. 36; Somaliland (Hainwaina Plain).

Range and Status. Endemic resident Eritrea, Ethiopia, Somalia, Kenya and SE Sudan. Generally common, locally uncommon. Eritrea, coastal areas and Danakil desert north to about Thio. Ethiopia in NE, SE Highlands and S, rare W Highlands just west of Addis Ababa. Djibouti; Somalia except west of Juba R. and NE tip. Sudan in extreme SE; Kenya, locally in N south to *c.* 01°N (Kapedo, Laisamis, Mado Gashi and Wajir). Hybridizes with Pied Crow *C. albus* in all countries (except Djibouti) in its range, and especially in SE Highlands of Ethiopia.

Corvus edithae

Description. Adult ♂: entire plumage blackish. Forehead to nape and lower mantle to uppertail-coverts glossed dark greenish, with violet tinge on scapulars. Hindneck and upper mantle to sides of head and neck, chin and throat dark glossy brown, tinged bronzy purple, duller blackish brown when worn; throat feathers pointed. Rest of underparts sooty black. Bases of body and head feathers pale grey, those of neck and throat feathers white. Tail glossy above with slight violet tinge. Secondary edges, tertials and upperwing-coverts glossed violet. Axillaries and underwing-coverts sooty black. Bill black; eyes dark brown; legs black. Sexes alike. SIZE (9 ♂♂, 6 ♀♀): wing, ♂ 316–368 (343), ♀ 300–361 (327); tail, ♂ 163–188 (173), ♀ 150–183 (165); bill, ♂ 51–57 (52·9), ♀ 48–52 (49·5); tarsus, ♂ 55–61 (57·8), ♀ 54–58 (54·8). Depth of bill at middle of nostril (n = 22, ♂♀) 18–20. WEIGHT: Ethiopia 1 ♂ 450, 1 unsexed 435.

Field Characters. Length 46 cm. An all black crow confined to the Horn of Africa. Usually occupies drier habitats than similar black Cape Crow *C. capensis* (though both occur in deserts of N Kenya); bill stouter than Cape Crow, tail longer and wedge-shaped and plumage less glossy (brown colour on head of some adults not diagnostic since Cape Crow is browner in worn plumage). Concealed *white bases* to feathers of throat, breast and nape become evident in display or when feathers ruffled by wind (Zimmerman *et al.* 1996).

Voice. Tape-recorded (10, C, NOR). Calls extremely similar to those of Pied Crow. Normal call 'wraaa', like Pied Crow, becoming harsher and shriller when excited, sometimes higher-pitched and rather nasal, more like House Crow *C. splendens*. Also gives a soft 'waw-kah', second note higher, a husky 'tschop' and 2-part watch-winding call 'korrrrh-karrrrh'. Birds in feeding flock gave short, quick series of clucking, clicking or gobbling sounds (N Kenya: D. T. Holyoak *in* Cramp and Perrins 1994). Also described are short, metallic 'onk', 'wonk' or 'kwonk', a double 'rrawnk-rrawnk' and a flat 'yack-yack' (Zimmerman *et al.* 1996).

General Habits. Inhabits deserts, semi-deserts, arid plains, dry savannas and open thorn bush from sea level to usually *c.* 2000 m, occasionally 4000 m in Ethiopia (mean altitude

1110 m: Wilson and Balcha 1989), 3000 m in Kenya but commonest below 1000 m (Lewis and Pomeroy 1989); also cultivation and towns. Where sympatric with Pied Crow in Kenya, occurs more in desert habitats. In SE Ethiopia and other parts of Horn of Africa, has same habitat as Pied Crow and they regularly hybridize (see *C. albus* Taxonomic Note). Where sympatric with Fan-tailed Raven *C. rhipidurus*, found more in sandy wadis and coastal plains.

General behaviour not well documented. Solitary or in pairs and in flocks of up to 100 in the non-breeding season (Somalia: Ash and Miskell 1998). Walks easily; occasionally hops. Fearless and aggressive. Feeds usually on ground, often with conspecifics from which it maintains a distance of several body lengths (D.T. Holyoak *in* Cramp and Perrins 1994). Sometimes points head and bill upwards in front of a conspecific for *c*. 2 s; this causes the other bird to turn its head away and to retreat (Ethiopia: Londei 1995). Also jabs bill at competitors. Scavenges in towns and camps. Flies along shores seeking carrion. Sometimes perches on camel, taking ectoparasites. Tears open sacks to eat grain. Noted to attack Tawny Eagle *Aquila rapax* in Kenya. Wheels and soars high. During day prefers to rest on sand rather than in trees around oasis pool. Allopreens (**A**). Roosts in large numbers in palms (North Horr, Kenya: Tomlinson 1950). Probably sedentary.

Food. Ground-dwelling animals, carrion and some plants including grain; also birds' eggs, ticks and lice.

Breeding Habits. Solitary nester although sometimes in small loose colonies (Archer and Godman 1961, Zimmerman *et al*. 1996). Territories usually 1·5–3 km apart (Blair 1961). Few details of display: a pair allopreening head and neck (**A**) (N Kenya: Cramp and Perrins 1994); also exposes snow-white bases of throat, nape and breast feathers in display (Kenya: Zimmerman *et al*. 1996).

NEST: large bulky structure of sticks and twigs, lined with plant fibres, wool and feathers; usually in trees

(Archer and Godman 1961; Zimmerman *et al*. 1996); also reported on cliffs, in caves, on telephone and power poles and at top of concrete post (Ash and Miskell 1998); 4 m or more above ground. Probably built by both sexes.

EGGS: 4–5, rarely 6. Sub-elliptical; greenish blue, spotted and blotched olive-green and brown with underlying markings of violet-blue. SIZE (n = 8, Somalia): 38·5–44 × 28–30 (41·4 × 28·9).

LAYING DATES: Eritrea, Feb; Ethiopia, Feb, May–June; Somalia, Mar (2), Apr (18), May (13), June (1) with breeding commencing 3–4 weeks earlier in S; Kenya, Apr.

INCUBATION AND CARE OF YOUNG: no information.

BREEDING SUCCESS/SURVIVAL: parasitized by Great Spotted Cuckoo *Clamator glandarius*.

Reference
Archer, G. and Godman, E.M. (1961).

Corvus albus Müller. Pied Crow. Corbeau pie.

Corvus albus P. L. S. Müller, 1776. Natursyst. Suppl., p. 85; Senegal.

Plate 31
(Opp. p. 478)

Range and Status. Africa, Aldabra, Comoros and Madagascar.

Common resident almost throughout subsaharan Africa except in closed forest or arid desert; closely associated with human habitation, but also occurs in savanna. Northern limits of range: in Mauritania 17°N; in rains to 18°N at Nouakchott, Boutilimit, Tidjika and Moudjeria; Mali, *c*. 17°N; Niger, *c*. 19°N in Aïr Massif; Chad, *c*. 19°N; and Sudan, *c*. 17°N in NW to *c*. 20°N at Red Sea; Ethiopia, absent only SE lowlands; Somalia, in N only north of 09°30′N, in SE along coast south of 03°N, and Ogaden desert (square 48a: Ash and Miskell 1998) near Ethiopian border. In semi-arid to arid parts of N and NE Kenya confined to villages; in coastal areas, especially Mombasa, being replaced by House Crow *C. splendens*. Namibia, absent E arid region; Botswana, common east of 23°E, restricted to towns in W. Lesotho, absent high mountain region. South Africa, absent from arid Kalahari basin in N Cape; present in Karoo since 1950s and E Cape (Port Elizabeth, East London) since 1970s and 1980s (Anon. 1987). Also present Bioko, Zanzibar, Pemba and Mafia.

North of Sahara, vagrant Algeria (I-n-Azaoua, and once noted in extreme S but no locality given: Dupuy 1969) and Libya (Jala).

Description. ADULT ♂: head, neck, throat and centre of upper breast glossy black with slight violet and blue-green iridescence; throat feathers rather long and pointed. Broad collar across upper mantle white, continuous with white sides to neck and breast, lower breast, upper flanks and upper belly. Lower mantle and scapulars to rump and uppertail-coverts glossy black, tinged

Corvus albus

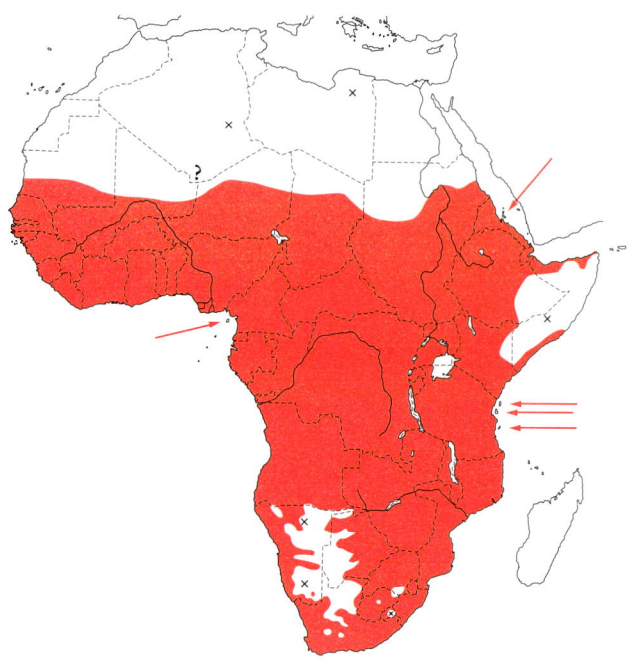

violet and blue-green. Tail black, slightly glossy above. Lower flanks, lower belly, thighs and undertail-coverts sooty black. Upperwing glossy black with greenish reflections, duller on flight feathers. Axillaries and underwing-coverts black. Bill black; eyes dark brown; legs black. Sexes alike. SIZE (11 ♂♂, 10 ♀♀): wing, ♂ 324–376 (358), ♀ 325–377 (351); tail, ♂ 178–197 (187), ♀ 181–199 (188); bill, ♂ 57–64 (60·2), ♀ 54–63 (58·8); tarsus, ♂ 57–65 (61·3), ♀ 59–65 (61·7). WEIGHT: Niger, ♀ (n = 3) 407–445 (424); Chad, ♂ (n = 2) 465, 560; Sudan, ♂ (n = 2) 587, 594, ♀ (n = 4) 487–501 (494), unsexed (n = 3) 490–594 (529); Uganda, ♂ (n = 9) av. 605 ± 20, ♀ (n = 9) av. 535 ± 10; Zimbabwe, ♀ (n = 5) 421–581 (519); Botswana, ♂ (n = 8) 474–592 (510), ♀ (n = 2) 490, 515; Southern Africa, ♀ (n = 5) 421–581 (519), unsexed (n = 6) 474–700 (556).

IMMATURE: similar to adult, but has black areas less glossy, underbody more brownish black; white areas tinged buff, border between white mantle and belly and black back and vent area less well defined.

NESTLING: naked, skin reddish pink, gonys of bill clear pink.

TAXONOMIC NOTE: speckled individuals with varying amounts of black on breast and collar are known from Sudan, Eritrea, Ethiopia and Somalia. They all occur in areas where *albus* and *edithae* overlap, and are considered hybrids between these two species (Blair 1961, Ash 1983, Madge and Burn 1994); there is evidence of actual interbreeding. Alamargot (1987), Jollie (1978) and Kleinschmidt (1906), however, suggest that these speckled/intermediate forms are not hybrids but the result of a typical 'pied' *albus* breeding with an undescribed black race of *albus*. However, we agree with Madge and Burn (1994) that the presence of 2 races (pied and dark forms) in the same highlands of Ethiopia is unlikely.

Field Characters. Length 45–50 cm. *Black and white plumage pattern unique in Afrotropics*; flight shape 'normal', i.e. wings average corvid width and length, tail average length and rounded. White-necked Raven *C. albicollis* has white half-collar across upper mantle but black underparts (breast feathers sometimes with a few white fringes), massive bill and different flight shape (short tail, broad wings). In Ethiopia and Somalia hybrids with Somali Crow *C. edithae* have white areas speckled with black. Vagrants to Palearctic differ from 'Hooded Crow' *C. corone pallescens* in having black back and belly and white, not grey, breast and hind collar.

Voice. Tape-recorded (5, 22, 66, 86, 88, 91–99, 104, B, C, F). Common (contact?) call variable, typically a strident 'wraaa' or 'kaaaa', also deeper, 'rrawrr', or shorter, up-slurred 'krow' or flatter 'raa', sometimes with almost cat-like ending, 'raaaaeow'. Another frequent call is a dry, snoring, 'watch-winding' rattle, often on 2 levels, lowest first, 'torrrrrrrh-tarrrrrrh'. Other calls include a deeper, more throaty rattle with a slight rasp to it, a muffled, nasal 'hawnng' or 'kaaaang' like an imitation of an automobile horn, a hollow 'blob-op' or 'klok-klok', a down-slurred 'tschow', flat 'ack-ack' and throaty 'glupp'; some of these are given with raised head and exaggerated movements of head and tail (Zimmerman *et al.* 1996). Displaying ♂ gives low but loud 'coow, coow' (Gwahaba 1975); in another display, one bird gives low, creaky growl, 'urrrrrkkk' answered quickly by mate's higher clicking 'tkkk' (Zimmerman *et al.* 1996).

General Habits. Inhabits grassland, scrub, savanna, open woodland, clearings in forest, sides of rivers and lakes; homesteads, farms, villages, towns, gardens, parks, plantations, rubbish dumps, sewage ponds, slaughterhouses and roadsides. Frequent at oases, but scarce or absent in other arid areas; and absent in dense forest. From sea-level up to at least 3700 m but commonest at lower elevations. In Ethiopia 81% of records are in cultivated areas, 17% in urban areas and 1% in open thorn bush; rare in acacia savanna and rocky areas (Wilson 1990). Does not mix with Brown-necked Raven where they are sympatric.

Solitary, in pairs or in small parties; sometimes up to 1000 or more at rubbish dumps and slaughterhouses, and occasionally over 3500 at roosts. Bold but wary, especially if disturbed by man. Walks easily on ground, hops quickly when moving fast. Flicks its folded wings when uneasy or after alighting. Flight laboured but powerful, with slow wing-beats; in flight plays with seedpods, sticks and stones; often soars at considerable height, regularly with vultures and Black Kite *Milvus migrans*. Up to 300 may gather in trees in morning, then they take to the air, spiralling upward on thermals, as many as 100 in the same thermal. At top of spiral, pairs form and dive in a steep glide with wings half-closed, twisting, turning and cart-wheeling as if trying to 'bomb each other' (Cooper 1969). May spiral up and dive down 2–3 times over a period of 30 minutes before alighting on ground or perch.

Forages largely on ground, occasionally in trees; takes a variety of plant and animal food and scraps of human food. In southern Africa shows a preference for vegetable food (Brooke and Grobler 1973). Feeds in the morning for first 3–4 hours until sun hot, then rests in shade until about 16h00 when again feeds actively. Holds large item under one or both feet, then breaks or tears it with bill. Often carries food in claws or beak up to a high perch to eat it there. Digs up young maize plants; sits at top of maize stalk until it breaks, then strips off kernels from cob on ground. Eats flesh of young coconuts and ground-nuts. Scavenges at refuse bins, along roadsides and train right-of-ways, with kites, vultures and Marabou Stork *Leptoptilos crumeniferus*. Kills small roosting bats, nesting African Palm Swift *Cypisurus parvus*, weavers and waxbills; also resting doves and pigeons (Brooke and Grobler 1973, Ade 1975). Occasionally takes whole nest of small bird, flying away with nest in its bill, presumably to perch and feed on nest contents (Tye 1983). Catches flying ants, locusts and small birds and bats in flight (Hamling 1953, Madge and Burn 1994). Raids heronries, taking eggs and nestlings. Follows cattle and plough, feeding on disturbed animals. Forages just behind a fire line, capturing prey as they escape the fire. Jumps up to snatch grapes. Picks ectoparasites off domestic and game mammals, either walking on their back or walking on the ground among them as they rest. Once attacked Bat-eared Fox *Otocyon megalotis* cub, but unable to lift it on wing (**A**) (Chadwick 1983). Kills injured shorebirds and sickly newborn sheep. Takes food from tables in houses and hotels. Drops stones on Ostrich *Struthio camelus* eggs, breaking them and eating their contents (Brooke and Grobler 1973). Buries scraps of food in sand under coconut palms (Tyler 1980).

Frequently mobs and harasses large birds including Osprey *Pandion haliaetus*, snake-eagles, Bateleur *Terathopius ecaudatus*, African Harrier Hawk *Polyboroides typus*, Red-necked Buzzard *Buteo auguralis*, vultures and Marabou Storks. Harasses African White-backed Vultures *Gyps africanus* and resting Marabou Storks by pulling their tail feathers. Occasionally flies at and bites large nesting owl such as Verreaux's Eagle-Owl *Bubo lacteus*, forcing it to abandon its nest (Loefler 1984).

Roosts communally in stands of trees, sometimes in large numbers in cities; several hundred in Accra (Ghana), hundreds or thousands in Kano (Nigeria: Serle 1943) and up to 3900 in the austral winter in Harare (Zimbabwe: Jones 1983). Birds at a roost start calling at about 05h00, fly around roost 06h00, then leave for feeding areas; they return to roost between 17h00 and 18h00, at first singly, then in pairs and finally in flocks in long lines. Birds returning to roost fly from one tree to another and wheel about before settling down.

Chiefly sedentary but with some migration in northern areas; also local movements. Liberia, irregularly Dec–May; in NW Central African Republic and W Sudan appears to be present only in dry season, Oct–May; and present chiefly at the same time in Sierra Leone, N Ghana (Mole), NW Niger (Niamey), N Nigeria (Malamfatori, Serti, Maiduguri) and S Chad (Abéché). Moves northward in rains (to nest?) in Mauritania (north to Nouakchott), Senegal, Mali, Nigeria (leaves Lagos in rains presumably moving northward) and N Chad. In Gambia, mainly inland in dry season in Apr–June to breed, departs for coast in rains after nesting. In Ethiopia's SE Highlands (Arussi) present from Aug to Jan–Mar and a few until June. In N Somalia, probably a visitor only (Ash and Miskell 1998). In E Africa probably sedentary, but some movements to arid regions in rains. In Zambia, local seasonal movements (Aspinwall 1985).

Food. Well documented by Brooke and Grobler (1973). Omnivorous, with vegetable matter predominant (southern Africa). Animal matter includes ectoparasites, butterflies, moths, caterpillars, crickets, grasshoppers, termites, ants, beetles (including large numbers of scarab beetle *Oryctes* larvae, Kenya: van Someren 1956), crustaceans, molluscs, fish, toads and frogs, chameleons, agama lizards, snakes, small and young birds (including doves, pigeons, cuckoo-shrikes, starlings, weavers, waxbills and domestic poultry), birds' eggs (including those of Ostriches, pelicans, herons, ducks, hawks, francolins, plovers, larks and domestic poultry), shrews, small monkeys, carrion, small fruit bats, serotine (insectivorous) bats, and rodents. Plant matter includes seeds, oil-palm nuts, cassava, potatoes, dates, tomatoes, berries, groundnuts, various grasses, sorghum, rice and maize. Also carcasses of varying sizes and human food offal.

Breeding Habits. Solitary nester, territorial. During courtship presumed ♂ faces ♀, their heads nearly touching; he raises feathers of head, neck and chest, and lifts his head; he takes a few steps, lowers head with neck fully stretched and bill pointed at ♀, half-crouches with wings lowered and makes low, but loud 'coow, coow' call (Gwahaba 1975); then he rises and may take a few steps back. He sometimes repeats this performance several times before mating with ♀, and may also step on her foot. They allopreen, touch bills and feed each other during courtship. A Pied Crow and a White-necked Raven once seen building 2 nests together; remained together for several weeks but neither nest produced eggs (Grobler 1974). Also hybridizes with Somali Crow *C. edithae* (see Taxonomic Note).

NEST: large bulky structure of sticks, twigs, roots and sometimes wire, with deep cup lined with grasses, mud, dung, string, wool and other soft material; sometimes, by weight, wire forms almost 75% of nest (Markus 1967); (n = 1) ext. diam. 500, int. diam. 200, int. depth 90; in fork near top of tall tree, on telephone pole, radio pylon, windmill platform or top of office building; 3–30 m above

ground (n = 70, Zimbabwe) av. 13·6 m (n = 33, Transvaal) av. 9·6 m; rarely on cliff ledge. Built by both sexes; often uses same site though builds new nest each year; takes 11–12 days to build. Defends nest vigorously, calling and dive-bombing at the intruder.

EGGS: 1–7, usually 4–5, (n = 140 clutches, southern Africa) av. 4·1; laid once every 24 h; elongated; smooth, slightly glossy; pale greenish to bluish green, spotted and blotched with grey, brown and olive, but less blotched than other *Corvus* eggs. SIZE (n = 74, Nigeria): 40–50 × 28·5–33 (45·0 × 30·7); (n = 187, southern Africa) 39–51 × 26·2–33 (44·7 × 30·5).

LAYING DATES: Mauritania, Aug–Oct; Senegal, June–July, Nov; Gambia, Feb–Apr, May–Sept; Mali, Aug–Oct; Burkina Faso, May–July; Niger, July–Aug; Guinea, Mar; Sierra Leone, Apr; Liberia (building Feb–Mar); Ghana, Apr–May; Togo, Apr–June; Nigeria, Mar–July; Chad, Apr–May; Cameroon, Feb–Mar, Dec; Sudan, Jan–July, Sept, Dec; Eritrea, Feb–Mar; Ethiopia, Feb–Apr; Somalia, Apr, June, Oct, Dec.

E Africa: Uganda, Jan–Feb, May–Oct; Kenya, Jan; Tanzania, Sept–Feb; Region B, Jan–Sept, Dec; Region C, Apr, Aug, Oct–Dec; Region D, Feb–Mar, Sept–Dec; Region E, Aug, Nov–Dec.

Zaïre, July–Feb; Congo, Oct; Gabon, Oct; Angola, Aug (grown young Dec); Zambia, May, Aug–Oct; Malaŵi, June, Sept–Nov; Zimbabwe, Aug (6), Sept (51), Oct (66), Nov (17), Dec (2); Botswana, Sept (4), Oct (10), Nov (2), Dec (2); Namibia, Mar, Sept–Dec (peak in Dec); Lesotho, Sept; South Africa: July–Jan (mainly Sept–Oct); Transvaal, Aug (5), Sept (18), Oct (38), Nov (8), Dec (2); Natal, Aug–Nov; Orange Free State, Nov–Dec; SW Cape, June, Aug–Nov.

INCUBATION: begins probably with 1st or 2nd egg; 75–80% of incubation by ♀, 15–20% by ♂; only ♀ incubates at night. ♀ on nest 8–68 min, usually 10–25 min at a time. Sitting ♀ may solicit food from ♂ by flapping wings; ♂ then feeds her at nest. ♂ on nest 2–8 min at a time, usually when ♀ leaves immediate vicinity. When not on nest ♂ remains nearby on ground or goes away. Either bird may join other Pied Crows as they circle overhead or as they go to roost, but nesting bird returns after a few moments. Period 18–19 days.

DEVELOPMENT AND CARE OF YOUNG: young hatch at 1–2 day intervals; covered with pin feathers by day 11, stand and flap wings by day 12, move to rim of nest by day 17, well feathered by day 23, size of adult by day 28; leave nest by days 35–45. Both parents feed young; ♂ feeds newly hatched chicks for first few days while ♀ still incubating remaining eggs; thereafter both share in feeding young. Rate of feeding 2·9 times per h first 18 or so days, 3·5 times per h next 9–10 days, 1·7 times per h for final 12 days (Lamm 1958). Early on ♂ (and later ♀) frequently probes nest with bill, presumably looking for faecal material which usually is swallowed; occasionally taken to nearby branch and dropped. Older young defaecate probably over side of nest.

BREEDING SUCCESS/SURVIVAL: apparently low in some areas. In Nigeria, of 100 eggs in 23 nests, 16 and probably 18 young fledged (Mundy and Cook 1974); in Ghana 1 nest produced 3 eggs but only 1 fledged young (Lamm 1958). Parasitized by Great Spotted Cuckoo *Clamator glandarius*: 12·7% of nests in South Africa (Payne and Payne 1967), 22% of 28 nests in Nigeria (Mundy and Cook 1971). In Nigeria, growth rate of Pied Crow chicks in nest normal until cuckoo nestlings begin to fledge; thereafter adult crows increasingly called to feed cuckoos, resulting in weight loss and death of crow chicks (Mundy and Cook 1974). In South Africa 17% of farms destroy nests (Siegfried 1963) and spread poison (Collett 1982). In Mombasa (Kenya) harassment by House Crow may depress Pied Crow numbers (Ryall 1991). Young preyed upon by Verreaux's Eagle-Owl (Brown 1970). Nestlings sometimes fall from nests in palms in Zanzibar. Parasites include nematodes, tapeworms, mites, lice, pupiparan flies, and ticks (Gwahaba 1975). 1 adult bird ringed Harare, Zimbabwe, recovered 3 years 114 days later (Irwin 1981) and 1 ringed Transvaal, recovered 9 years 4 months later (Cramp and Perrins 1994).

References
Alamargot, J. (1987).
Brooke, R.K. and Grobler, J.H. (1973).
Gwahaba, J.J. (1975).
Lamm, D.W. (1958).
Winterbottom, J.M. (1975).

Plate 31 (Opp. p. 478)

Corvus rhipidurus Hartert. **Fan-tailed Raven. Corbeau à queue courte.**

Corvus rhipidurus Hartert, 1918. Bull. Br. Orn. Club, 39, p. 21, new name for *Corvus affinis* Rüppell, 1835, Neue Wirbelt. Abyssinien, Vög., 2, p. 20, pl. 10, fig. 2; Massaua, Eritrea.

Range and Status. Africa and S Middle East (Israel, Sinai, Jordan, and Arabian Peninsula).

Resident in arid hilly and mountain areas. Mali, rare, Adrar des Ipras, also Homberi in S; NW Niger, on Aïr Massif (Mt Baggans, Mt Tamgak, Iferouane, Timia), also Zinder. Chad, Tibesti and Kapka Mts, also Ouaddaï, Melfi, Biltine and Abéché; NE Central African Republic, 1 record (Hall and Moreau 1970). Sudan, from Gebel Elba and Red Sea coast to Uganda border, Darfur and Jebel Marra, Nubia Mts and Meshra er Reque area. Eritrea, above 300 m; Djibouti, above 300 m, less common below 300 m; Ethiopia, above 600 m, uncommon below 600 m and on SE slopes of SE Highlands; Somalia, north of 09°N, also 2 records along coast south to *c.* 05°N. E Uganda, locally common from Kidepo Valley Nat. Park south to Mbale and Tororo; also Otze in NW; Kenya, locally common in N

Corvus rhipidurus

A

ranging south to Mt Elgon, Kerio Valley, Nakuru, Isiolo and upper Tana R.

Description. *C. r. rhipidurus* Hartert (only race in Africa). Entire plumage black, with oily dark green lustre on upperparts, breast, upperside of tail, flight feather edges, tertials and upperwing-coverts. Sides of head to chin and throat more brownish black, with slight bronzy gloss. Hidden feather bases of hindneck and sides of neck white, elsewhere grey. Throat feathers slightly elongated (less so than in *C. corax* or *C. ruficollis* in N Africa) but broad, with bifurcated tips. Bill black; eyes brown; legs black. Sexes alike. SIZE (10 ♂♂, 10 ♀♀): wing, ♂ 364–400 (381), ♀ 344–390 (374); tail, ♂ 154–170 (161), ♀ 146–164 (153); bill, ♂ 58–61 (59·1), ♀ 56–59 (57·8); tarsus, ♂ 65–70 (67·4), ♀ 62–67 (65·4). Bill less heavy than in *C. corax*; depth at nostril (n = 44 ♂♀) 20–24. WEIGHT: Niger, 1 ♂ 705, 2 ♀♀ 565, 610; N Chad, 1 ♂ 512, 2 ♀♀ 560, 595.

IMMATURE: like adult, but underparts duller, chin greyer; throat feathers shorter and rounder.

NESTLING: not described.

Field Characters. Length 46–47 cm. A black corvid of rocky arid areas, with short stout bill; broad wings and *extremely short tail* give distinctive shape in flight; looks almost tail-less at a distance. Wings extend well beyond tail tip at rest. Sympatric with Brown-necked Raven *C. ruficollis* and Somali Crow *C. edithae*, which have longer bills, longer, more narrowly-based wings and much longer tails (when perched, their wing-tips are about even with tail tip). In flight overhead, black underwing-coverts contrast with somewhat paler and browner flight feathers, and pale greyish soles to feet show against dark plumage (**A**) (Hollom *et al.* 1988).

Voice. Tape-recorded (5, B, C, GREG, MAC, McVIC, NOR). Normal call a croaking 'krroo', 'krroo-uk' or down-slurred 'krroo-oh'; also gives a longer, guttural, growling 'urrrrrrr' or 'rrrrrrraaa', sometimes changing in pitch, 'rrrrrraaarrrooo'. Apparent song recorded by M. E. W. North (5) includes wide variety of calls strung together loosely in lengthy series: hollow 'tok-tok', high-pitched 'wack' and 'caa', lower 'cur', longer 'waaw', brief trill 'trrrr' reminiscent of African Scops Owl *Otus senegalensis*, and less burry version of normal call, 'koo-wok'. In flight a double 'cuhcuh' given repeatedly in answer to normal call by accompanying bird (Cramp and Perrins 1994).

General Habits. Inhabits largely arid rocky hills, cliffs, crags, canyons and ravines, often near oases; ranges widely in search of food, after breeding season over moorland, bushland, grassland, cultivation, towns and villages. In Ethiopia at 300–4000 m, mainly above 1550 m. In Kenya mainly 400–1500 m but up to 2600 m. In Ethiopia 11% of observations in rocky areas, 22% in open thornbush, 9% in acacia savanna, 58% in cultivation and occasionally in towns (Wilson 1990, 1993). Where range overlaps with Brown-necked Raven and Somali Crow, Fan-tailed occurs more in rocky areas or oases rather than in desert or semi-desert country; where it overlaps White-necked Raven *C. albicollis* and Thick-billed Raven *C. crassirostris*, Fan-tailed more often in arid areas and at lower altitudes.

Solitary or in pairs; sometimes in flocks of 70 (at food, Israel: Shirihai 1996), 130 (Saudi Arabia; Cramp and Perrins 1994) and 500 (Yemen: Goodwin 1986). Excellent flier, soaring in thermals with other crows, kites, hawks, eagles and vultures. Performs group aerobatics; sometimes plays in flight, dropping and catching paper, feathers, stones and twigs as they fall. Spends large part of day flying: 71% of observations flying, 7% feeding on ground,

4% perched in trees, 10% on telephone poles, 5% scavenging on roads, 3% on buildings and on fences (Wilson 1990). Walks with rather upright carriage, often with bill open; sometimes hops, jumps to one side or leaps into air. Forages on ground, searching for small animals, carrion, scraps and grain. Perches on back of goat *Capra* or camel *Camelus* gleaning ectoparasites; actively pecks at mammal's body, pulling out tufts of hair and throwing head back a little when swallowing food (Lewis 1989). Captures locusts in mid-air with its feet and eats them in flight (Gallagher and Woodcock 1980). Once reported to capture a ping-pong ball, drop it to the ground, hammer it several times with its bill, roll it over a large stone and hit it several times with a stone 2–3 cm in diam. (presumably egg-breaking behaviour: Andersson 1989). Feeds unaggressively with other crows. Roosts communally in flocks of up to 370 in Israel (Shirihai 1996) or 1000 in Yemen (Goodwin 1986) in non-breeding season, usually in palm tree plantations but also on sheltered ledges and buildings; sometimes roosts with Brown-necked Raven. Sedentary.

Food. Invertebrates, including beetles, locusts, ectoparasites and worms; birds' eggs and young; grain, berries and other fruits, carrion and human food scraps.

Breeding Habits. Nests and eggs not recorded in Africa. Most of following details from Middle East (Cramp and Perrins 1994). Solitary nester; sometimes colonial, with 2–5 pairs on same cliff ledge and nests 50–200 m apart. Territorial entire year. Pair reported to soar almost out of sight, then plummet back to cliff; also performs pursuit-flights and vocalizes, but details lacking. Probably single-brooded.

NEST: a platform of sticks, twigs and roots, cup lined with wool, hair, cloth, fresh twigs and other soft material; on crevice or sheltered ledge usually on sheer cliff, rarely on building or in tree.

EGGS: 2–6, usually 4–5; sub-elliptical to long oval; smooth and glossy; greenish blue, blotched and speckled with various shades of brown and grey. SIZE (n = 31 Ethiopia, Somalia): 41·8–50·4 × 27·6–34·5 (45·0 × 30·7).

LAYING DATES: Niger, June; Sudan, May, June; Egypt, Mar, Apr; Eritrea, May, June; Djibouti, Mar; Ethiopia, Feb, Mar, May, June, Sept; Somalia, Apr–June; Kenya, Jan–June.

INCUBATION: presumably by ♀; period 18–20 days.

DEVELOPMENT AND CARE OF YOUNG: fledging period 35–40 days; fledged young remain with family parties but details unrecorded.

BREEDING SUCCESS/SURVIVAL: sometimes parasitized by Great Spotted Cuckoo *Clamator glandarius*.

References
Cramp, S. and Perrins, C.M. (1994).
Lewis, A.D. (1989).
Madge, S. and Burn, H. (1994).

Plate 31 (Opp. p. 478) *Corvus albicollis* Latham. White-necked Raven. Corbeau à nuque blanche.

Corvus albicollis Latham, 1970. Ind. Orn. 1, p. 151; Great Namaqualand.

Range and Status. Endemic resident in SE Africa; locally common in mountainous and hilly country throughout. Kenya, from Mt Elgon, Kerio Valley, Mt Nyiru and Ndoto Mts to Tanzania border from Nyanza district in W to Taita Hills and Voi in E; sympatric with Fan-tailed Raven *C. rhipidurus* at Mt Nyiru, Ndoto Mts, Kerio Valley and L. Baringo area; absent from coastal plains, rare in L. Victoria basin. Sudan, vagrant, near 04°N, 34°E (Hall and Moreau 1970). Uganda, from *c.* 02°N south to Rwanda border in Ankole District, rare in L. Victoria basin. W Rwanda; W Burundi; E Zaïre from Lendu Plateau and mountains west of L. Albert south to Marungu Highlands, rare in mountains of SE Katanga (Manika Plateau, Mandoko). Tanzania, as mapped (N.E. Baker, pers. comm.). Malaŵi. E Zambia, west to Central and Southern Districts at Chilanga and Livingstone. W Mozambique, from Manica and Tete Districts south along Zimbabwe border, also Lebombo Mts in extreme SW. Zimbabwe, Inyanga and Chimanimani Mts, west to Batoka Gorges, Victoria Falls and Matopos Hills. Botswana, vagrants near 25°S, 25°E (Hall and Moreau 1970) and unconfirmed record at 22°S, 27°E (Penry 1994). Namibia, uncommon visitor to lower Orange R. Lesotho; Swaziland. South Africa: Natal, absent only NE coastal plains north of L. St Lucia; E Transvaal, mainly along escarpment, vagrant in N Kruger Nat Park;

Corvus albicollis

19th century records from Pretoria, Magaliesberg and Klerksdorp suggest a wider distribution in the past (Tarboton *et al.* 1987); E and S Orange Free State; and Cape Province north to Klawer, Great Karoo, Eastern Cape and Transkei.

Description. ADULT ♂: top of head black with slight gloss. Hindneck, sides of neck and sides of head brownish black; broad white crescent across upper mantle, rest of upperparts black with oily green lustre. Tail black, with slight gloss on upperside. Chin and throat brownish black with bronzy gloss, throat feathers pointed and bifurcated; rest of underparts sooty black, feathers of upper breast sometimes fringed white, especially at sides. Upperwing black, tertials and coverts with greenish lustre. Axillaries and underwing-coverts sooty black. Bill massive, black, tipped ivory; eyes dark brown; legs black. Sexes alike. SIZE (10 ♂♂, 10 ♀♀): wing, ♂ 357–434 (412), ♀ 358–420 (403); tail, ♂ 170–194 (177), ♀ 148–182 (169); bill, ♂ 65–70 (66.9), ♀ 62–67 (63.9); tarsus, ♂ 74–80 (77.0), ♀ 70–77 (74.5). Depth of bill at middle of nostril (n = 20 ♂♀) 30–35. WEIGHT: Tanzania and Zimbabwe, ♂ (n = 2) 865, 1157, ♀ (n = 5) 762–931 (873), unsexed (n = 2) 762, 900.

IMMATURE: duller than adult, white mantle crescent with some black streaks; scattered white or white-edged feathers on neck and breast of some birds. Bill lacks ivory tip.

NESTLING: not described.

Field Characters. Length 50–56 cm. Larger than any other crow in its range, with *massive*, deep ivory-tipped bill (**A**). *Broad white crescent* across upper mantle is diagnostic at rest but hard to see in flight, when *shape* (broad wings, short broad tail, large head and bill) is more useful; calls also distinctive. Smaller Pied Crow *C. albus* has longer tail, much slimmer bill, solid white breast as well as hind collar (some White-necked Ravens, especially immatures, have scattered white feather edgings on breast). Fan-tailed Raven *C. rhipidurus* has similar broad wings but even shorter tail, lacks white and has smaller bill.

Voice. Tape-recorded (20, 86, 88, 91–99, 104, B, C, F, KEI). Normal call surprisingly high-pitched for such a large bird, a short, rolled 'krra' or 'kraa', or longer 'krrraaa'; sometimes lacks rolling quality and becomes trumpet-like blare. One bird gave series of 1, 2, or 3 rolling croaks with distinct burry quality, 'crrraauw', lasting *c.* 0.5 s, and was answered by presumed mate with similar but lower-pitched call (Taita Hills, Kenya). Soft, husky 'haa' and other low notes occasionally given. A short, somewhat metallic clattering 'cluk-cluk-cluk' accompanies bowed head in presence of presumed mate (Zimmerman *et al.* 1996).

General Habits. Inhabits primarily mountains and hills: gorges, crags, cliffs and escarpments; but ranges widely over rocky outcrops, open mountain forest, grassland, pasture, camps, settlements, villages, towns and lower open country including lakeshores. Occurs chiefly between 1000–3000 m, but visits coastal lowlands in southern Africa, and at 5800 m on Mt Kilimanjaro.

Solitary or in pairs; may congregate in large flocks of up to 150, occasionally up to 800 (Lewis and Pomeroy 1989). Excellent flier, wingbeats slower than other corvids; spends much of day soaring and gliding in thermals with other corvids, kites, hawks and vultures. Performs aerobatics, rolling, tumbling and stick-passing with conspecific, making quite loud swishing and whining sounds with wings. While in the air reported to drop a small stone from its beak and then swoop down and catch it before it reaches the ground; also reported to do a quick somersault between dropping and catching the stone (Mackie and Landman 1982). Walks with rather upright carriage; gait swaggering; also hops. Seeks food mostly on ground; holds food with its foot, and tears or hammers it with beak. Forages aggressively alongside Cape Rook *C. capensis*, Pied Crow, Fan-tailed Raven, kites and vultures. Has little fear of man when not persecuted, scavenging around houses and courtyards. 'One pair became extremely bold around camp, perching on huts and walking within 6 m of people. One stole a bar of soap and ate the whole bar; the other tested the inside of an empty tin of ravioli and consumed the remains of the tomato sauce in the bottom of the tin!' (Mt Muhavura, Uganda: S. Keith, pers. comm.). Flies along highways searching for food and usually is first to arrive at a carcass. Steals eggs and live chicks and attacks young lambs and sick sheep, starting by pecking out their eyes. Also attacks birds up to size of tern *Sterna* sp. (Baron 1981). Occasionally feeds in trees, taking insects among the foliage. Picks ectoparasites from backs of large mammals; drops small tortoises, especially *Chersine angulata* up to 8–9 cm long (Uys 1996), onto rocky ground to break their shells. Feeds on locusts on ground and in air. Roosts on cliffs, in groups of up to 40, occasionally several hundred (Winterbottom 1975).

Sedentary, partly nomadic; large flocks wander widely outside breeding season.

Food. Various small animals: caterpillars, grubs, locusts, grasshoppers, beetles, ticks and other arthropods, tortoises, lizards, snakes, small birds and mammals, eggs of domestic chickens, Verreaux's Eagle *Aquila verreauxi* and other birds, young birds including domestic chickens and geese, and small adult birds; lambs, sick sheep; also carrion including dead fish, frogs, domestic and wild mammals, birds and road kills; some plant material (maize, peanuts, fruits, berries); honeycomb; many kinds of human foods, and scraps from campsites and refuse pits.

A

Breeding Habits. Solitary nester, territorial. Pairs probably for life: 1 pair together for 14 years (Parry 1981). Allopreens; ♂ feeds ♀ during courtship; also pursues her in flight, flying over trees and around rocky hills. Other times one (or both?) flies directly upward, then turns downward at rapid speed. Bows head to presumed mate and gives a short, somewhat metallic, clattering 'cluk-cluk-cluk'. Once a Pied Crow and a White-necked Raven seen building 2 nests together; remained together for several weeks, but neither nest had eggs (Grobler 1974). Occasionally double-brooded.

NEST: large platform of twigs, sticks and branches, the central deep cup lined with grass, dried algae, hair, wool, feathers, rags and other soft material; on cliff ledge or in crevice, occasionally in tree (90% on ledges or crevices, 10% elsewhere: Winterbottom 1975). Av. height above ground (n = 18 nests, Zimbabwe) 32·0 m, (n = 11, South Africa) 11·1 m.

EGGS: 2–7, usually 4, av. (n = 21 clutches, Zimbabwe) 3·9, (n = 12, South Africa) 3·7; long oval; glossy; pale green, greenish white or bluish green, spotted and streaked with olive, grey and brown. SIZE (n = 24): 45–56·9 × 31·6–35 (51 × 33·3).

LAYING DATES: E Africa: Kenya, Oct–Dec; Uganda, Aug–Oct; Region B, Feb; Region C, Sept–Oct; Region D, July, Oct–Nov; Zaïre, Feb; Tanzania, Oct; Malaŵi, Sept–Nov; Zambia, Sept (2 young left nest Nov); Zimbabwe, Apr (1), July (1), Aug (9), Sept (109), Oct (19), Nov (6); Lesotho, Sept–Oct (adult feeding full grown young Dec); South Africa, Aug–Dec.

INCUBATION: presumably by ♀; period 19–21 days.

DEVELOPMENT AND CARE OF YOUNG: young fed mostly by ♀ by regurgitation; ♂ often accompanies ♀ to nest but rarely feeds young. Young fed 30 times in 3·5 h, each visit lasting 0·5–2 min (Uys 1966). ♀ (and ♂?) removes faecal material; young do not eject faeces from nest. Fledging period 21–28 days (Liversidge 1991); young appear to remain with parents until 1–2 months prior to next breeding season (T.M. Butynski, pers. comm.).

BREEDING SUCCESS/SURVIVAL: in Zimbabwe, in 18 nests av. 2·7 young fledged per nest; in South Africa, in 7 nests, av. was 2·1 (Winterbottom 1975). Declining in W and Central Highlands of Kenya due to eating poisoned carcasses meant for stock predators (Lewis and Pomeroy 1989); in South Africa, shot, poisoned, trapped and eggs and young destroyed (Siegfried 1963).

References
Madge, S. and Burn, H. (1994).
Uys, C.J. (1966).
Winterbottom, J.M. (1975).

Plate 31 (Opp. p. 478)

Corvus crassirostris Rüppell. Thick-billed Raven. Corbeau corbivau.

Corvus crassirostris Rüppell, 1836. Neue Wirbelt ... Abyssinien, Vög., 2, p. 19, pl. 8; Abyssinian highlands.

Range and Status. Endemic resident Eritrea and Ethiopia, wandering to Sudan and Somalia. In Eritrea, southern highlands north to Senafe, Halai and the highlands east of Asmara; also Mt Ramlu (Denkalia Desert: Smith 1957), uncommon to frequent. In Ethiopia W and SE Highlands from Eritrean border south to Sidamo Province; common to locally abundant above 1200 m, less so in Sidamo; uncommon to rare below 1200 m. Stragglers at c. 04°40'N, 36°55'E in the Chew Bahir area (Friedmann 1937), Moyale area near Kenya border (Wilson and Balcha 1989) and at c. 08°N 40'E in the Ogaden (J.S. Ash, pers. comm.); also E Sudan at Famaka and Galabat. Once NW Somalia (Ben Seilah).

Description. ADULT ♂: forehead and crown glossy black, tinged bluish. Large white patch on nape joined by short narrow white line down hindneck to small white patch on front of mantle. Rest of upperparts black with slight greenish blue and violet-blue gloss. Tail black, glossed on upperside. Lores to cheeks and ear-coverts black; sides of neck to chin and throat brownish black, throat feathers short and rounded; rest of underparts sooty black, feather tips slightly glossed. Upperwing black, with slight bluish or violet gloss on tertials and coverts. Underwing-coverts and axillaries sooty black. Bill massive, black tipped ivory, upper mandible laterally compressed, culmen ridged; eyes dark brown; legs and feet black. Sexes alike. SIZE (8 ♂♂, 2 ♀♀): wing, ♂ 430–460 (442), ♀ 438, 441; tail, ♂ 237–271 (253), ♀ 246, 255; bill, ♂ 81–88 (85·4), ♀ 84, 85, bill depth at middle of nostril (n = 4 ♂♂) 42–44; tarsus, ♂ 77–86 (81·1), ♀ 79, 82. WEIGHT: Ethiopia, 1 ♀ 1135.

IMMATURE: similar to adult but browner above, sooty brown below. Bill lacks ivory tip; ridge on upper mandible less developed.

NESTLING: naked; skin on head pink, rest of body black.

Field Characters. Length 60–64 cm. *Huge*, with *gigantic bill* (**A**) and *white patch* on nape. Looks like an outsize version of White-necked Raven *C. albicollis* but easily identified by range: endemic to Eritrea and Ethiopia, where White-necked Raven does not occur. Should they meet, Thick-billed Raven distinguished by longer, wedge-shaped tail and different voice; white patch higher up on back of head. In flight neck extends forward, which with huge bill gives it a hornbill-like appearance (Urban 1980).

Voice. Tape-recorded (109, B, MAC). Normal call a deep, pig-like grunt, 'urrrrk', 'wrrrrurrr', 'grrrarrrr' or a double, coughing 'harr-harrr', more like a mammal than a bird; described as 'a low, wheezy croak 'phlurk-phlurk', a most obscene sound' (Brown 1965), as if the bird had lost its voice and was suffering from a sore throat (Urban 1980). During courtship ceremony pair produces hoarse gurgling and choking noises.

Corvus crassirostris

General Habits. Montane; frequents alpine screes, cliffs and gorges; moorlands with giant lobelia, *Alchemilla*, tussock grass and giant heath; high grasslands; forests with St John's wort, bamboo, *Hagenia*, juniper, podocarpus and olive; lowland-subtropical-humid forests; cultivation and urban areas (Urban and Brown 1971). Occurs mainly at 1200–4100 m, preferred altitude 1500–3000 m (mean 2290 m); not below 800 m (Wilson and Balcha 1989). Associated with human habitation more than most corvids; 88% of observations in cultivation, 7% urban areas, 3% rocky areas, 2% acacia and savanna (Wilson 1990, 1993).

Solitary or in pairs; occasionally 10–20 at rubbish dumps and slaughtering places. Pairs seem to remain together throughout year. Excellent fliers and soarers, flying often together along sheer cliff-faces; sometimes performs tumbling aerobatics with 2 birds grasping feet and falling for 200 m before breaking apart. When taking off, beats wings with loud swishing sound; once airborne beats wings fairly slowly. Spends 85% of day in flight, 4% perched in trees, 4% on telephone poles and 4% scavenging on roads (Wilson 1990). Occasionally perches on fences. Forages in camps, towns, cities and countryside, scavenging for offal and carrion (including scraps in ashes of camp fires: Urban 1980), tearing at wheat, corn and other crops and searching for small animals. Digs out Giant Root-Rat *Tachyoryctes macrocephalus*; kills rodents in open grasslands. Scatters dung with its bill held almost upside-down, looking for beetle larvae and other insects. Feeds on grain, sometimes levelling whole corners of a field (Urban 1980). Hides surplus food (captive bird: Goodwin 1986). Steals bones which Lammergeier *Gypaetus barbatus* has dropped and smashed. Bold and fearless, arriving at first sign of slaughter of animals; often arrives first at carrion. Feeds with vultures, kites, eagles, Marabou Stork *Leptoptilos crumeniferus* and Hamerkop *Scopus umbretta*. To displace competitor, stretches and moves neck from side to side, then, with bill open and throat distended, slowly moves towards it and sometimes pecks it.

Mainly sedentary.

Food. Carrion, scraps of meat and human offal; grubs, beetle larvae and insects; reptiles; rodents; grain and fruit; see above.

Breeding Habits. Solitary nester; territorial. Breeding ♂ obtains a piece of food and flies to a branch where he sits and calls ♀; she arrives and flutters wings; he feeds her, then both make hoarse gurgling and choking noises. In separate display, ♂ fluffs up head feathers, stretches bill toward ♀ and produces rattles and squeaks. Sometimes double-brooded.

NEST: one was a large structure of sticks and branches, hollowed on top, the hollow lined with coarse hair, carpet fibre, bits of paper, string and other synthetic material. Ext. diam. 680–780, ext. depth 350, int. diam. 360–370, int. depth 130. Sited on 2 horizontal branches in large tree 18 m above ground; also on cliff face.

EGGS: 4 (2 clutches); oval; turquoise-blue with pale brown or reddish brown spots, especially at larger end. SIZE (n = 9): 46–57 × 34–36 (51·3 × 35·0).

LAYING DATES: Ethiopia, Dec–Apr.

DEVELOPMENT AND CARE OF YOUNG: once 4 adults attacked Verreaux's Eagle Owl *Bubo lacteus*, breaking twigs with their beaks to get close to it; each then jumped up at or flew fast past the owl, pecking at it (de Castro and de Castro 1993).

References
Alamargot, J. (1976, 1983, 1990).
Goodwin, D. (1986).
Madge, S. and Burn, H. (1994).
Urban, E.K. (1980).
Wilson, R.T. (1990).

CORVIDAE

Plate 31
(Opp. p. 478)

Corvus frugilegus Linnaeus. Rook. Corbeau freux.

Corvus frugilegus Linnaeus, 1758. Syst. Nat., ed. 10, 1, p. 105; 'Europa', restricted to Sweden by Hartert (1903, Vög. Pal. Fauna, 1, p. 13).

Range and Status. Breeds from N and central Europe across N Asia to c. 140°E in Russia, south to Turkey, NW Iran, Kazakhstan, the Tian Shan and central China. Winters N Africa, S Europe and S Asia to Middle East, NW India, S China, Korea and Japan.

Winter visitor from Palearctic. Rare and irregular Egypt late Nov–mid Mar; 3 records from Suez Canal zone (El Qantara, Ismailiya, Suez), 2 from near Cairo (Giza, El Marg) and one from Luxor. Accidental Algeria (small flock near Bone, winter 1943; group of 8 between Djelfa and Ksar el Boukhari, April 1969).

Corvus frugilegus

Description. *C. f. frugilegus* Linnaeus (only subspecies in Africa). ADULT ♂: entire plumage black. Whole head and body strongly glossed with violet-purple. Gloss of uppertail, flight feather edges, primary coverts and alula tinged more greenish, that of tertials and rest of upperwing-coverts violet-purple. Axillaries and underwing-coverts sooty black. Base of upper mandible, lores, cheek back to below eye, chin and throat bare, skin greyish white, but chin and throat with downy grey feathers in autumn. Bill rather slender, greyish black; eyes brown; legs black. Sexes alike. SIZE (10 ♂♂, 10 ♀♀): wing, ♂ 299–320 (308), ♀ 290–316 (299); tail, ♂ 162–182 (170), ♀ 157–184 (167); bill, ♂ 56–62 (59·0), ♀ 53–60 (56·0); tarsus, ♂ 49–57 (54·0), ♀ 51–57 (53·6). WEIGHT: (Britain) ♂ (n = 162) 405–560 (489), ♀ (n = 126) 325–525 (418).

IMMATURE: more sooty or greyish black, gloss confined to wings and tail; face feathered. After post-juvenile moult like adult, but face still feathered; retains less glossy juvenile flight and tail feathers.

Field Characters. Length 45–47 cm. Black, with purplish gloss. Similar to black race of Carrion Crow *C. c. corone* (winter visitor to Morocco) but plumage rather looser, with shaggy thighs, forehead steeper, giving head a peaked look, and bill greyish black, straighter, thinner and more pointed; adult further distinguished by patch of greyish white skin on face merging with pale base of bill. In flight shows narrower bill and more rounded tail. On the ground, loose belly feathers sag down thighs; Carrion Crow does not have 'trousered' profile. Juvenile lacks pale facial skin and is very similar to Carrion Crow but duller and less glossy; black nasal bristles form slight lump on bill, lacking on bill of Carrion Crow. Bill of young Rook is narrower and straighter, becoming slightly pale at base with age. Resident crows in Egypt belong to the *C. c. pallescens* group ('Hooded Crow') and are readily identified by grey body contrasting with black head and wings.

Voice. Tape-recorded (62, 73, 93, 105, B). Vocabulary outside breeding season limited, chiefly a harsh 'kaaaa', higher-pitched and more drawn-out than call of Carrion Crow, usually given singly; sometimes shorter, 'kaa'.

General Habits. In Egypt generally occurs in small flocks or singly in cultivated areas (Goodman and Meininger 1989). In Europe inhabits mainly farmland with nearby woods or lines of trees. Forages on ground in grassland, bare ploughed fields and stubble, water meadows, agricultural land and growing crops. Requires tall trees for roosting; flies considerable distance each day to feed. Gregarious, forming large flocks outside breeding season, often with Western Jackdaws *C. monedula*, and regularly roosts communally with them. Forages more successfully in large (>10 birds) than small flocks (<10 birds) (Höglund 1985). Not particularly shy, feeding on lawns in parks and large gardens (even small gardens in hard weather). Sedentary and partial migrant, northernmost birds moving south in winter.

Food. Nothing recorded for Africa. In Europe chiefly invertebrates, especially beetles and earthworms, and plant material, mainly cereal grain, also fruits and nuts; some small invertebrates (mice and voles), carrion, and scraps of all kinds.

Reference
Cramp, S. and Perrins, C.M. (1994).

Corvus monedula Linnaeus. Western Jackdaw. Choucas des tours. Plate 31

Corvus monedula Linnaeus, 1758. Syst. Nat., ed. 10, 1, p. 106; 'Europa', restricted to Sweden by Hartert (1903, Vög. Pal. Fauna, 1, p. 15). (Opp. p. 478)

Range and Status. Breeds NW Africa, Europe and across central Asia to about 100°E, south to Turkey, Israel and N Middle East, N Afghanistan, the Tian Shan and Kashmir and NW Mongolia. Winters mainly within breeding range but leaves N part of Siberia, occurring further south in Middle East, Afghanistan and NW India, and in E to Sinkiang, NW China.

Breeding resident Morocco and Algeria, occasional winter visitor from Palearctic. A colony in Tunis in the late 19th century has now disappeared, and the bird is not known in Tunisia (Thomsen and Jacobsen 1979). Reports from Egypt, perpetuated by Blake and Vaurie (1962), are almost certainly referable to the House Crow *C. splendens* (Goodman and Meininger 1989), although Western Jackdaw occurs in winter south to central Israel (Shirihai 1996). Morocco: occurs only on E side of Tingitane Peninsula, common in massifs of Beni Hozmar and l'Haouz and spreads out into lowlands to the east, also l'Ile de Leila. Absent from plains and hills on W side of the peninsula, even in non-breeding season, occurring no further east than Ksar Sehir (but once at cap Spartel: Pineau and Giraud-Audine 1979), but has recently colonized Ouezzane (Cramp and Perrins 1994). Occurs in Rif Mts, and widespread in Moyen Atlas south from Fes and Taza, west to Ouaouizert, with outlying colonies further east in Aïn Kebira above Debdou and Djebel Mahsseur on Algerian border south of Oudja. Scattered colonies also in Haut Atlas, west and south to Demnat and Massif du Toubkal south of Marrakesh. Algeria in NW from Morocco border (Tlemcen Mts) north and east to Tessala and Saïda Mts, Tiaret and Massif de l'Ouarsenis, south to Daïa and Tiffrit Mts; also along coast between Tenès and Algiers; in E a small colony near Constantine, where formerly more widespread south to mountains near Batna, and a single recent observation near Babor; a colony discovered in the 1970s at Oued Kébir near Ben Harroun was not relocated in 1989, possibly due to dam construction (Madge and Burn 1994). Vagrant NE Algeria; a few probably winter in Morocco (see below); accidental Mauritania (Nouadhibou). Locally common; most abundant in N Morocco south of Tangier (2000 roosting with 800 Cattle Egrets *Bubulcus ibis* near Tétouan: Thévenot *et al.* 1981; 2000 at mouth of Oued Negro and 1500 at Emsa: Thévenot *et al.* 1982), becoming progressively less common southward in Morocco and eastward in Algeria. Algerian populations considered endangered (Madge and Burn 1994); only 25 noted in gorge at Constantine in 1982.

Description. *C. m. spermologus* Vieillot: W Europe to Morocco and Algeria. ADULT ♂: forehead and crown glossy black with violet-purple tinge, continuous with duller black of lores and around eyes to chin and throat; feathers of chin and throat with pale grey shaft streaks. Nape to ear-coverts and rear of cheeks, hindneck and sides of neck grey. Rest of upperparts black, mantle and scapulars with slight violet gloss. Tail black, upperside slightly glossy. Breast to undertail-coverts black or greyish black, with slight gloss on upper breast. Upperwing feathers black,

Corvus monedula

secondary outer webs, tertials, and greater, median and lesser coverts with violet-purple gloss. Axillaries and underwing-coverts greyish black with glossy fringes. Bill black; eyes pale grey, pale blue-grey or whitish; legs black. Sexes alike. SIZE (10 ♂♂, 10 ♀♀): wing, ♂ 225–242 (235), ♀ 226–240 (231); tail, ♂ 127–138 (131), ♀ 124–132 (129); bill, ♂ 34–38 (35.6), ♀ 32–36 (34.5); tarsus, ♂ 41–46 (44.7), ♀ 42–46 (44.0). WEIGHT (W Europe): ♂ (n = 98) 185–280 (240), ♀ (n = 89) 179–250 (224).

IMMATURE: duller than adult; upperparts sooty black, paler areas on head less contrasting; duller and browner below, chin and throat feathers without pale shaft streaks; gloss confined to wing feather edges; eyes dark brown.

NESTLING: back, crown, thighs and upperwing covered with pale grey down; otherwise naked. Skin light pink, later greyish flesh; gape-flanges broad, pale yellow; mouth dark red or purplish pink; bill flesh or dark horn; legs pale yellowish pink, scutes becoming greyish black.

C. m. cirtensis (Rothschild and Hartert): vagrant to NE Algeria. Underparts paler than in *spermologus*, more dark ash grey; gloss of cap bluish rather than purple.

Field Characters. Length 33 cm. A small blackish crow with rounded head, short bill, grey nape and ear-coverts and pale eye. In mixed flocks with Rooks *C. frugilegus* readily distinguished in flight by smaller size, frequent 'jack' call and faster wingbeats. 'Hooded Crow' *C. corone pallescens* has grey nape but back and most of underparts also grey; larger and heavier-billed, and in Africa confined to Egypt. Sympatric Red-billed Chough and Alpine Chough *Pyrrhocorax pyrrhocorax* and *P. graculus* share rocky habitats but are all-black, with red and yellow bills respectively; in mixed flock told from Western Jackdaw by 'flapping butterfly' flight.

Voice. Tape-recorded (62, 73, 93, 105, B). Commonest vocalization is the contact call, given in flight or when perched, a high-pitched 'jack' or 'tchak', sometimes lengthened into 'jackajack'; variants include 'keak', 'k'yow', 'kia' and deeper 'kiaw' and a shorter 'tchk'. In flight, 'jack' call may be interspersed with a long-drawn 'kraare' or 'chaairr'. Predator-mobbing and alarm call a harsh 'kaaarr'. Song, given both at perch and in flight, a medley of call notes varying in loudness and inflection; its volume and variability give the listener the impression that he is listening to a flock (Goodwin 1986). For variety of other calls used at nest and during the breeding season, see Goodwin (1986) and Cramp and Perrins (1994).

General Habits. In Africa mainly a mountain bird, 1200–2000 m (up to 2600 on Massif du Toubkal); locally to coast; inhabits cliffs, rocky gorges and caves, where there are perennial streams or springs and some woods nearby; also offshore islands and sea cliffs. 2 colonies in Algeria (Daïa Mts) were in poplar trees, an unusual site (Ledant *et al.* 1981). In N Morocco leaves the massifs of Beni Hozmat and l'Haouz and spreads out onto the neighbouring plains on Mediterranean side; up to 1500 birds frequent rubbish tips in Tétouan (Pineau and Giraud-Audine 1979). Other colonies are continually occupied, birds seldom venturing more than a few km from their cliff home, even outside the breeding season (Brosset 1961).

Generally pairs for life, and pair remains basic unit within flocks. Gregarious outside breeding season, in Africa occurring with Red-billed and Alpine Choughs in the mountains and Cattle Egrets in the lowlands, in Europe with Rooks. Birds from colony at Djebel Mahsseur left cliffs at daybreak and spread out over nearby plateau; they returned to drink around 10h00, left again after noon and returned about an hour before sunset (Brosset 1961). Forages mainly on the ground, in grassland, ploughed fields, stubble, parks and other open spaces; coastal birds readily forage between tidelines (Richford 1978). Partly commensal with man, nesting in chimneys and on buildings, feeding in gardens and taking scraps from bird tables, scavenging around farmyards, towns and refuse tips. Foraging methods include some digging and probing, but more often bird picks food from surface and low vegetation; also turns over stones, lumps of earth and other small objects and scatters animal dung; also jumps up to catch flying insects. Feeds seasonally in treetops on caterpillars, beetles, fruit and acorns. Walks in a quick, jaunty manner; sometimes runs or hops. Wing-beats faster than other crows; near breeding cliffs given to 'spectacular gliding, twisting, swooping and diving aerobatics' (Goodwin 1986). Roosts communally at colony sites on cliffs; in Morocco (Tétouan) group roosts at rubbish tip with Cattle Egrets; in Europe often in trees with Rooks.

Sedentary in Africa, mainly so in central and S Europe but with some local movements in winter. Occasionally seen crossing Strait of Gibraltar both spring and autumn, so apparently winters on Moroccan side of strait in small numbers (Finlayson 1992).

Food. No data for Africa. In Europe invertebrates (especially beetles, ants, spiders, caterpillars, Diptera (adults and larvae), Orthoptera, worms and snails), plant material (mainly cereal grains, also other agricultural crops, fruits, acorns and other large seeds, and weed seeds) and some carrion; occasionally eggs and young of birds. For complete list see Cramp and Perrins (1994).

Breeding Habits. No detailed accounts for Africa. The following condensed from Goodwin (1986) and Cramp and Perrins (1994). Colonial nester; defends only nest site and immediate vicinity. Displays exhaustively described and illustrated in Lorenz (1952) and Cramp and Perrins (1994). Threat indicated by Bill-up Posture, with raised bill and sleeked plumage, and Bill-down Posture with lowered bill and raised nape feathers; in more intense situations involving disputes over ♀ or nest site, lowers head and neck, fluffs out back feathers, spreads tail and points bill either at rival or towards the ground. Pair formation occurs prior to acquisition of nest site. Heterosexual displays include allopreening, food-begging by ♀ and courtship-feeding by ♂. When partner returns to nest mate may perform Meeting-display, half spreading and drooping wings and quivering tail laterally. Single-brooded.

NEST: untidy mass of sticks and other debris, lined with fibrous bark, hair, fur, wool or rags, with a little dry earth or mud; placed in crevice among rocks or in cliffs (Africa), tree hole or other natural cavity, and in man-made structures. Often nests in house chimney; nest sometimes falls into fireplace below and birds fly around room spreading soot everywhere (pers. obs.). Built by both sexes; nest added to while birds are incubating or brooding young.

EGGS: 2–6 (usually 4–5). Sub-elliptical; smooth and glossy; pale light blue to greenish blue or bluish white, variably speckled and blotched with blackish brown, light olive, pale grey or greyish violet, blotches becoming larger toward broad end. SIZE (n = 317) 30·4–39·4 × 21·5–25·5 (35·0 × 24·7); calculated weight 11·3 g.

LAYING DATES: Morocco, April–June.

INCUBATION: by ♀ only; starts with 2nd or 3rd egg, rarely with last. Period: 16–20 (17–18) days.

DEVELOPMENT AND CARE OF YOUNG: young fed by both sexes but brooded by ♀ only. For first few days ♀ feeds young with food brought by ♂, later ♂ also feeds them directly. Young fledge at 30–35 days.

BREEDING SUCCESS/SURVIVAL: no data for Africa; in Europe, breeding success varies with habitat. In Spain, of 965 eggs, 70·3% hatched and 26·1% produced fledged young, giving 1·4 fledged young per pair (n = 184). Most egg losses due to predation by Common Raven *C. corax*, intraspecific interference, human disturbance or parasitism by Great Spotted Cuckoo *Clamator glandarius*. 24% of broods failed completely, half from death by starvation in first 5 days (mainly last chicks hatched); chicks also taken by Common Raven. On Skomer I. (Wales) av. 0·65 chicks per nest, on Welsh mainland 1·21, in deciduous woodland near Oxford 1·7.

References
Cramp, S. and Perrins, C.M. (1994).
Goodwin, D. (1986).
Heim de Balsac, H. and Mayaud, N. (1962).
Lorenz, K. (1952).
Richford, A. S. (1978).
Thévenot, M. *et al.* (1981, 1982).

Corvus splendens Vieillot. House Crow. Corbeau familier.

Corvus splendens Vieillot, 1817. Nouv. Dict. Hist. Nat., nouv. Éd., 8, p. 44; Bengal.

Plate 31
(Opp. p. 478)

Range and Status. Indian subcontinent, Sri Lanka, Maldive and Laccadive Is to S China (Yunnan). Introduced or self-established resident in Malaysia, Singapore, Iran, Kuwait, Oman, Saudi Arabia, Persian Gulf States, Yemen, Jordan, Israel, Mauritius and NE, E and S Africa. Sightings in Australia, Japan, Chile, Netherlands and Gibraltar.

Introduced first into Zanzibar in 1890. In Egypt established about 1922 in Suez city, now in Suez area (Port Said, Ismailiya, Port Taufiq) south along Red Sea coast (Ain Sukhina, Ras Gharib, Safaga) to Quseir; abundant Suez area, 800–850 in 1981 (Bijlsma and Meininger 1984), uncommon elsewhere. Sudan, established late 1930s in Port Sudan, flourishing 1941 (Kinnear 1942) and since; Eritrea, Massawa since 1971, now abundant; Assab since mid-1980s, now plentiful. Djibouti, first reported in Djibouti City in 1958, now several thousand, also Obock; Somalia, Berbera in 1988 several dozen apparently well established (Douthwaite and Miskell 1991), also 4 on Raas Caseyr (Cape Guardafui) in early 1950s (Davis 1951). Kenya, first introduced into Mombasa 1947, up to 1 million birds there 1991 (Schmidt 1996); also 50 km inland along road from Mombasa to Nairobi; common Kenya coast from about 25 km north of Malindi south to Tanzania border; several reports Nairobi, Mtito Andei and Tsavo East Nat. Park (C. Ryall pers. comm., Lack 1985, Lewis and Pomeroy 1989). Tanzania, abundant a century after introduction into Zanzibar, introduced into Dar es Salaam in 1968; now 15,000–20,000 (Schmidt 1996); also Tanga and Pemba I. (Zimmerman *et al.* 1996). Mozambique, 1 record early 1950s Bazaruto I; established population Inhaca I. since the 1960s, 2250 in 1998 (Nhancale *et al.* 1998); and a few recently in Maputo (C. Ryall, pers. comm.). South Africa, first established in Durban in 1972 (possibly 1966–67: Newman 1974) with populations now about 1000; also 1 record East London 1975; small numbers Cape Town since 1977, in 1998 *c.* 300, with flocks of 60 and 120 within 15–20 km of Cape Town airport (T. Williams, pers. comm.).

Description. *C. s. splendens* Vieillot: main race in Africa. ADULT ♂: forehead to crown, side of head (back to ear-coverts and behind eye) and chin to throat glossy black; top of head with violet-purple and greenish reflections; throat with dark greenish lustre, feathers rather long, many of them bifurcated. Nape and posterior ear-coverts to upper mantle and sides of neck ash grey, well demarcated from black on side of head but merging into black on mantle. Lower mantle and scapulars to rump and uppertail-coverts glossy black with violet-purple and greenish reflections, rump sometimes greyer. Tail black, glossed on upperside with greenish tinge. Breast ash grey, meeting grey of neck and mantle, merging into black of throat and more gradually into sooty greyish black on flanks and sides of belly, and sooty black on tibiae and undertail-coverts. Upperwing black, primary edges, primary coverts and alula glossed greenish, tertials, secondary edges and rest of coverts glossed violet-purple. Axillaries ash grey; underwing-coverts greyish black. Bill black; eyes dark brown; legs black. Sexes alike. SIZE (10 ♂♂, 10 ♀♀): wing, ♂ 253–281 (267), ♀ 240–266 (254); tail, ♂ 155–175 (166), ♀

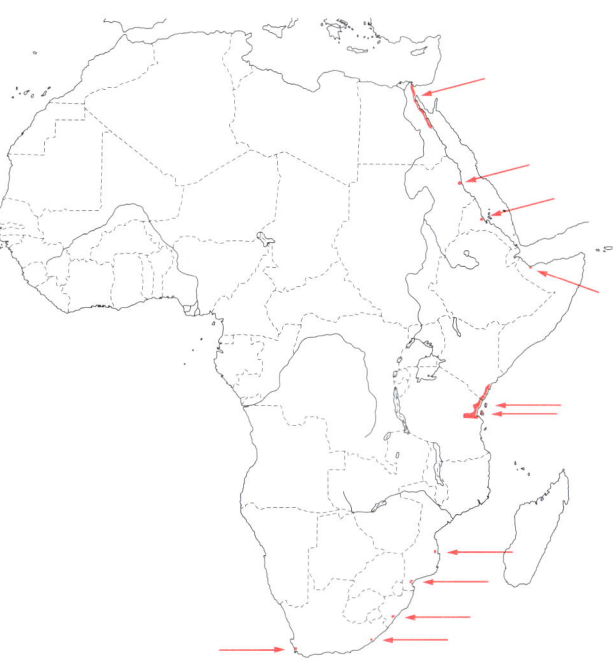

Corvus splendens

140–168 (159); bill, ♂ 49–55 (51·7), ♀ 46–52 (48·4); tarsus, ♂ 48–52 (49·5), ♀ 44–51 (47·5). WEIGHT: South Africa, ♂ (n = 2) 310, 326, ♀ (n = 5) 252–304 (av?), unsexed (n = 2) 266, 288 (Maclean 1993); Mauritius, ♂ (n = 16) 270–371 (317), ♀ (n = 10) 245–295 (270) (Feare and Mungroo 1989).

IMMATURE: similar to adult but duller, black areas more sooty, grey areas paler, gloss confined to wings and tail.

NESTLING: naked; skin light flesh, bill, legs and feet light flesh; eyes closed.

[*C. s. zugmayeri* Laubmann: probably this race in parts of NE and E Africa. Like nominate race but grey parts very milky grey contrasting strongly with iridescent black of head and wings.]

[*C. s. protegatus* Madarász: probably this race in Somalia. Like nominate race but grey distinctly darker: ash grey to slate grey.]

TAXONOMIC NOTE: racial status of introduced populations in Africa not well documented. *C. s. splendens* in Egypt, Sudan, Kenya, Tanzania, Mozambique and South Africa (Durban); probably *protegatus* in Somalia (Raas Caseyr where 4 birds arrived in ship from Sri Lanka); *zugmayeri* and *splendens* said by Cramp and Perrins (1994) to be introduced or self-established in many localities in Egypt, Red Sea and eastern Africa with populations often forming variable mixture between both races.

Field Characters. Length 40–43 cm. A *smallish*, *slim* crow with comparatively long neck and legs, most obvious in the alert position, when they give it a distinctive shape. *Grey nape and hind neck* suggest 'Hooded Crow' *C. corone pallescens* (which it meets in Egypt), but grey comes further up onto back of head and sides of neck, giving different 'look'; *grey darkens* on back and underparts, whereas on Hooded Crow these areas remain uniformly pale. Bill longer than Hooded Crow, and upper mandible more deeply curved. Hooded Crow has a heavier, more squat shape.

Voice. Tape-recorded (91–99, B, C, CHA, GIB, GREG, McVIC, NOR). Normal call, given in contact or alarm, a flat 'waaa' or 'kaaa', more nasal and 'slighter' than other crows, sounding as if it comes from a smaller bird. For catalogue of calls in Asia see Ali and Ripley (1972) and Cramp and Perrins (1994).

General Habits. Found always in or near human coastal habitation, usually where there are trees or telephone wires. Sometimes in scrub, open grassland, beaches, tidal pools and mangroves.

Gregarious, pairs or family parties joining into groups of 5–130. Tame and bold but wary. Walks with striding gait, sometimes hops; often flicks wings. Flight steady and direct. Aerobatic, sometimes twisting, turning and somersaulting with wings closed. Aggressive over food and when defending territory and nest. In threat, holds body horizontally, extends neck, ruffles throat feathers, holds wings close to body, lowers tail and flicks it each time it calls. On finding food or intruder, bird calls, and soon others arrive at food or dive at intruder. If on ground, birds often attack intruder as a group, with one bird distracting it in front while the others peck at it from behind. Sometimes mobs of up to 200 attack a bird of prey flying by, pursuing it up to 1 km.

In Mombasa feeds in early morning, loafing for rest of day. Forages largely on ground. Regularly patrols tidal pools, catching fish, taking garbage; while in flight snatches fish from water. May follow domestic stock, feeding on disturbed insects. Occasionally rides on domestic stock (**A**), semi-domesticated oryx *Oryx gazella* and eland *Taurotragus oryx*, picking off ticks and pecking at sores. Raids heron and weaver colonies, tearing holes especially in weaver nests and taking eggs and young; also attacks shorebird nests. Climbs silently in undergrowth searching for nests of small passerines. Hawks insects awkwardly; steals food from egrets, gulls, terns and kingfishers. Catches Epauletted Fruit Bats *Epomophorus wahlbergi* in trees. Kills poultry, newly born lambs, goats and calves. Eats grain and fruits; occasionally hangs upside-down in a tree to feed on fruit or flesh of coconut. Stores food at caches in sand (S. Tyler, pers. comm.) or larders among nuts of coconut palms (Angwin 1977). Scavenges in rubbish dumps, slaughterhouses, sewage and harbour areas. Steals food from street vendors, market stalls and restaurant tables. Hops onto crocodiles at crocodile farms, searching for food. Occasionally attacks monkeys and people, scratching and pecking head, neck and back.

Harasses other birds: 22 species listed in Mombasa including herons, ibises, storks, ducks, 9 raptors, 2

A

shorebirds, gulls, cuckoos, kingfishers, bee-eaters, rollers, orioles and Pied Crow *C. albus* (Ryall 1992a). Known to jump on and off backs of vultures sitting on ground. Sometimes harasses monitor lizards, cats and foxes. Aggressiveness and destruction of nests, eggs, and young in Mombasa have caused reductions in several indigenous birds including Black Kite *Milvus migrans*, mousebirds *Colius*, African Palm Swift *Cypsiurus parvus*, Collared Palm-Thrush *Cichladusa arquata*, Rufous Chatterer *Turdoides rubiginosa*, sunbirds, puffbacks *Dryoscopus*, Pied Crow and Golden Palm Weaver *Ploceus bojeri* (Ryall 1992a); has displaced Pied Crow as urban scavenger in Mombasa and Hooded Crow in Suez area.

Roosts communally in traditional tree-site, up to 800–850 in Egypt. Before roosting flock may glide and soar. Arrives at roost from up to 20 km away well before sunset. Much bustle and clamour as birds go to roost before and after nightfall. Leaves roost before sunrise in parties of 3–20.

Sedentary in Africa.

Food. Animal and plant matter and carrion: earthworms, molluscs, crustaceans, spiders, beetles, grasshoppers, locusts, termites, ticks, centipedes, fish, frogs, toads, geckos and skinks, eggs and young of domestic and wild birds up to size of whistling ducks, lambs, newborn goats and calves, sick sheep, cats, rodents, fruit bats and other small mammals. Maize and other grains, flower buds, nuts, papayas, mangos, bananas, tomatoes, flesh of young coconuts, melons, guavas and nectar. Takes carrion of all sorts including molluscs, fish, turtles and tortoises, birds and mammals; also human offal.

Breeding Habits. Chiefly monogamous but promiscuity not uncommon; bond for entire year and presumably for life in India, but for one season of about 14 weeks in Kenya (Ryall and Reid 1987). Usually a solitary nester, territorial; sometimes loosely colonial with 6–12 nests in same tree or building. Once paired, keeps close to mate. During courtship, pair allopreens and ♂ feeds ♀. When mate returns, partner performs meeting display, adopting an upright posture, flicking wing tips, lowering tail, and fluffing feathers on head; they sometimes rub bills together. Copulation lasts only a few seconds, ♀ pulling in her neck and partly opening her wings, while ♂ holds her head with his bill. Often mobbed by conspecifics while mating. Double brooded.

NEST: a large mass of twigs, bark, sticks, and wire, lined with soft plant and animal fibres; ext. diam. 250–300, int. diam. 120–150, int. depth 70–100 (India); weighs 3·5–8 kg (Mombasa); in fork near top of tall tree or in outermost branches; also top of building, construction crane, girders of bridge and derelict ship; usually 4 m or more above ground (n = 159 nests, Mombasa), av. 6·8 m. ♀ probably selects nest site, both gather material, only ♀ usually builds; takes 14–17 days.

EGGS: 3–6, usually 4–5 (n = 15 clutches, Kenya av. 3·9), laid in morning, usually at intervals of 1–2 days; short, oval; fairly glossy; pale blue-green with grey and brown blotches and speckles. SIZE (n = 225, nominate *splendens*): 30·4–44·1 × 23·0–29·1 (38·0 × 26·8).

LAYING DATES: Egypt, Mar, May, July, Nov; Sudan, Jan–Feb; Djibouti, Apr; Kenya (Mombasa), Sept (5), Oct (22), Nov (2), Dec (1), Jan (1); Tanzania, Sept–Jan; South Africa (Durban) Oct–Nov, (Cape Town) Oct (probably Aug–Nov).

INCUBATION: by both sexes but mostly by ♀; only ♀ at night. Period 15–18 days.

DEVELOPMENT AND CARE OF YOUNG: young hatch at intervals of 24–28 h; move and feed by day 1, eyes open by days 2–3, and fledge by days 27–35. Remain with parents for several weeks thereafter. Both parents feed young, eject dead nestlings from nest and fiercely protect young from predators, dive-bombing and attacking intruder on head and back and chasing it off for up to 200–300 m; attacks often carried out by several conspecifics.

BREEDING SUCCESS/SURVIVAL: in Africa (Mombasa), high breeding success rate due to lack of predators or brood parasites (Ryall and Reid 1987). Has been identified as agricultural pest; a threat to public health (by carrying human pathogens such as *Salmonella*, *Giardia*, *Entamoeba*, possibly cholera and other microorganisms: Ryall and Reid 1987); a nuisance (noise and droppings); a food thief; and a destroyer of urban wildlife. Damages TV aerials and telephone and electric wires and once reported to damage aircraft in Mombasa (Anon. 1997). Control measures now underway in Kenya, Tanzania and South Africa include trapping, poisoning and destroying nests, eggs and young.

References
Ash, J.S. (1984).
Bijlsma, R.G. and Meininger, P.L. (1984).
Cramp, S. and Perrins, C. M. (1994).
Ryall, C. (1990, 1992a, 1994).
Ryall, C. and Reid, C. (1987).

Genus *Ptilostomus* Swainson

Monotypic, endemic; a slender, black crow-like bird with tail long and graduated, 10 rectrices; bill short, high and arched; 5 rictal bristles; nasal bristles soft, upcurving, covering nostril, pointing forward and inward, each side overlapping on culmen, mixing with stiff and upstanding frontal feathers. Primaries brownish and rather translucent, like *Onychognathus* starlings. Eyes violet. Bill black, but in juvenile mainly bright pink. Forages amongst livestock; roosts and nests in palms. Molecular analysis suggests that *P. afer* is more closely allied to jays than to *Corvus* crows (Cibois and Pasquet 1999).

Plate 31
(Opp. p. 478)

Ptilostomus afer (Linnaeus). Piapiac. Piapiac africain.

Corvus afer Linnaeus, 1766. Syst. Nat., ed. 12, 1, p. 157; Senegal.

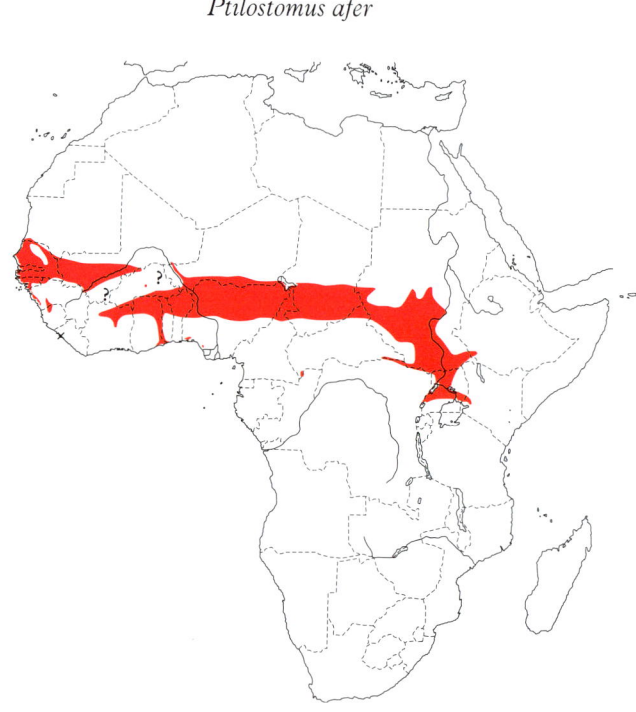

Ptilostomus afer

Range and Status. Endemic resident, N tropics from Senegal to W Kenya, sea-level to 1500 m. Mauritania, in lower Senegal R. valley from delta to Rosso and in S Guidimaka. Senegal and Gambia, frequent to common, almost throughout in S, less common and local in N. Guinea-Bissau, widespread; on some offshore islands, including Bulama. Guinea, frequent in NW in Koundara and Gaoual, flocks in NE at Niandokoro; occurs in coastal areas south to Conakry, and from Kindia (Foulayah: Demey 1995) to Kounounkan, where rare. Mali, quite common, north to at least 15°N. Sierra Leone, old records from Rokupr. Liberia, single bird, Monrovia, May–June 1988 (Gatter 1997). Ivory Coast, uncommon in N, south to 08°30'N, record Bouaké (07°40'N). Burkina Faso, uncommon at single locality north of Ouahigouya (Balança and Visscher 1997), rare and erratic near Ouagadoudou, occasional in SE (Holyoak and Seddon 1989) but common in Pendjari and Arli Nat. Parks. Ghana, common in N south to Mole, Tamale and Yendi; occurs through Volta region, Ho and Adidome to E part of Accra and Keta coastal plains. Togo, common in N, rare in S guinean savannas (records Ahepe, Ayengré, Davie, Sokodé), common on coast at Lomé and Anécho (Cheke and Walsh 1996). Benin, widespread in N, south to Bétérou (where rare resident: Claffey 1995). Nigeria, common in N, south to Ilorin and Yankari Nat. Park and just south of R. Benue; also not uncommon in coastal savannas near Badagri and common in Lagos suburbs and at Sapele. Niger, frequent along Niger floodplain (Ayorou, Niamey, Say, 'W' Nat. Park), common in S from Nigerian border north to Maradi, Tessaoua and Zinder. Cameroon, frequent in Benue Plain. Chad, locally frequent, perhaps widespread, north to L. Fitri and Batha watercourse and south to Moïssala. Central African Republic, uncommon in Dzanga Res. (Green and Carroll 1991), although absent from adjacent Nouabalé-Ndoki Nat. Park, Congo (Dowsett-Lemaire 1997b); common in Vakaga Préf. in N. Sudan, probably the commonest corvid, widespread in S, resident in Darfur north to Kas and Garsila, records north to 13°N at Zalingei and Nyertete (Wilson 1982); in Tewfikia, Bahr el Ghazal, common inside the town one year when unknown 3 years earlier (Bannerman 1948). Ethiopia, recorded in extreme SW. Zaïre, distinctly local, along NE borders, from Semio to W shores of L. Albert, Semliki and west of L. Edward. Uganda, widespread in N, south to Katwe, but absent from much of east. Kenya, scarce in extreme W (records from Busia, Alupe, Mumias, Maseno and Kisumu).

Description. ADULT ♂: entirely black, except for tail and wings. Very glossy, except for wings, tail, uppertail-coverts, belly and undertail-coverts. Forehead and lores velvety. Tail long, stiff and graduated, dark brown above and below, shafts black above and brown below. Primaries dark brown with bluish gloss on exposed tips of P1–P5; outer secondaries dark brown, not glossy, inner ones and tertials glossy black; primary coverts dark brown, other upperwing-coverts glossy black. Underside of flight feathers silvery pale grey; underwing-coverts and axillaries dark brown. Bill black; eyes intense violet-blue or violet-purple with red-brown outer rim (also described as violet-brown or red-brown with mauve outer rim); legs and feet black. Sexes alike. SIZE:

wing, ♂ (n = 20) 160–177 (169), ♀ (n = 12) 160–173 (166), pink-billed juv. ♂♀ (n = 17) 150–163; tail, ♂ (n = 20) 225–300, ♀ (n = 12) 250–283, pink-billed juv. ♂♀ (n = 17) 230–266; tail graduated – shortfall ♂ (n = 4), T2 50–65, T3 100–111, T4 117–156, T5 176–190 shorter than T1, ♀ (n = 4), T2 58–62, T3 99–108, T4 119–138, T5 164–181 shorter than T1; bill to skull (n = 4) 37·0–39·5, ♀ (n = 4) 36·0–38·0; bill to feathers, ♂ (n = 20) 32–36, ♀ (n = 12) 32–36; tarsus, ♂ (n = 20) 45–51 (48·1), ♀ (n = 12) 43·5–50 (47·4). WEIGHT: not known.

IMMATURE: like adult but bill bright pink, variably tipped black. Juveniles with pink bills known at least from Oct to June; nearly a year may pass before bill turns black.

NESTLING: not described.

Field Characters. Length *c.* 46 cm. Unmistakable; a *slender, leggy black* crow with brownish wings, primaries appearing pale brown and *translucent* in flight, and *long, stiff*, strongly graduated brown tail. Beautiful violet eye. Commonly in group of *c.* 8 birds, walking amongst cattle or flying into crown of palm. Juvenile has bright *pink bill*.

Voice. Tape-recorded (104, B, CHA, MOR). Utters loud chattering and scolding cries, 'pee-ip' repeated 2–3 times (*Piapiac* is probably onomatopoeic), rather like calls of Long-tailed Glossy Starling *Lamprotornis caudatus* but shriller; also short, metallic chirrups not unlike Jackdaw *Corvus monedula* and, when disturbed, a short, rasping 'kweer' and in alarm a harsh, scolding chatter (Barlow *et al.* 1997, Zimmerman *et al.* 1996).

General Habits. Inhabits well-grazed cattle pasture in open savanna woodlands and shrub savanna with sparse trees, stubble, cultivated and fallow fields (particularly ones that are extensive and newly harvested), lawns, playing fields, airports, and vicinity of *Borassus* and *Hyphaene* palms in wet and dry floodplains and farmland. Sometimes invades villages, 'walking about fearlessly close to the houses, sitting in rows on the corrugated iron roofs' (Bannerman 1948); sometimes nests in palms in town gardens (Nigeria).

Spends much time on ground in cohesive flocks usually of about 8 birds, sometimes 4–5 and 20–40, uncommonly 50, foraging by striding on bare earth between trampled grass or on short grass sward, birds moving more or less in the same direction, often amongst cattle, donkeys, sheep or goats, pecking at insects on ground and darting hither and thither in pursuit of insects put up by cattle. Also forages on backs of cattle 'more in the manner of an oxpecker than a crow' (Wilson 1981), searching for ectoparasites, and uses elephants likewise (Thomson 1964), though not wild ungulates (Rice 1963). Seizes termites as they emerge from ground, and makes wild hops into air as termites fly off. If disturbed on ground, flies short distance away, alights, and scolds with jerking movements of head. In breeding season often in pairs (Serle 1940), despite fact that it nests co-operatively (*c.* 8 adults and immatures attend nestlings).

Gait on ground a leggy walk; also runs and hops with agility; in trees, hops and runs nimbly along branches. Flies rather slowly with rapid, shallow wingbeats, often low down, without undulation, tail carried stiffly trailing behind, giving unbalanced appearance. Often entire flock flies from ground keeping close and alighting together 100 m away; or part of flock may break away to alight in top of tree or palm. Arrives in tree and leaves it with hardly any gliding. Tail of bird perched in tree hangs down; tail of bird on ground often held askew, dragging on earth. Noisy; birds in flock regularly break out into simultaneous calling, ducking and bobbing heads with each call. Flock of 50–60 birds once seen bickering and parading around excitedly, attracting attention of Pied Crows *Corvus albus* and Black Kites *Milvus migrans* (Giraudoux *et al.* 1988). Adults preen each other, and adults and juveniles allopreen. Flock roosts together at night, often in crown of palms, quitting roost at dawn. Flies to roost in compact flock, uttering 'pee-ip' calls. Drinks from puddles in early morning and before going to roost in evening; flock retires to shady tree and rests during midday heat.

Mainly sedentary but wanders locally; said to occur on N edge of range in Maradi, Niger, only in wet season.

Food. Bugs, beetles, termites, spiders; seeds and fruits including pericarp of *Elaeis* fruits.

Breeding Habits. Poorly known. Co-operative breeder; several adults defended a nest with young (Bannerman 1948); 4 adults and an imm. fed young just out of the nest and still unable to fly (Grimes 1976); nestlings attended by 5 adults and 3 immature birds (Grimes 1987).

NEST: compactly built of sticks, long strips of palm-frond fibres and grass stems, with some earth in foundation; in the middle a deep cup, thickly lined with palm fibres (Serle 1940); some nests made mainly of palm fibres, with a few small twigs. Int. diam. 200. Often sited at base of frond of *Hyphaene* and *Borassus* palms; once 8 m up in a baobab, wedged into fork near top of tree. Another nest 11 m from ground in top of *Elaeis* palm.

EGGS: 3–7. Ovate, smooth, quite glossy, or rough and matt. Immaculate pale blue; or very sparsely spotted and blotched with mauve-grey; or pale greenish blue, most of shell almost immaculate but at one end or the other a cap of brown blotches superimposed on large grey-lilac blotches. SIZE: (n = 30) 29·2–32·1 × 20·8–23·6 (30·45 × 21·6); also (n = 7) 27·5–30·5 × 20·5–22·0 (29·4 × 21·4).

LAYING DATES: Mauritania, July–Sept; Senegal, June; Gambia, June–July; Mali, 'breeds July–Oct'; Burkina Faso, (adults feeding flying young, June); Ghana, Apr–May (carrying nest material Mar; recently fledged young late July); Nigeria, Mar–June; Sudan, Dec, Mar–June; Uganda, 'nesting' Nov.

INCUBATION: nothing known.

DEVELOPMENT AND CARE OF YOUNG: young birds seen huddled together on branch, parents flying up from ground with food.

BREEDING SUCCESS/SURVIVAL: no data.

References
Bannerman, D.A. (1948).
Serle, W. (1940).

Genus *Pica* Brisson

A distinctive, bush-nesting, ground-feeding, food-caching corvid with strikingly pied plumage; black wings and tail strongly burnished; round nostrils, concealed below stiff bristles; no crest; 10th primary short, narrow and stiff; tail long and steeply graduated, legs quite long. Vocal. Domed stick nest.

2 species; one Palearctic, Oriental and in W Nearctic, the other confined to California, forming a superspecies.

Plate 30
(Opp. p. 463)

Pica pica (Linnaeus). Common Magpie. Pie bavarde.

Corvus Pica Linnaeus, 1758. Syst. Nat., ed. 10, 1, p. 106; 'Europa', restricted to Sweden.

Range and Status. Palearctic, from Morocco and Norway (north to 71°N) to Korea, south to Iraq, Taiwan and Hainan, with outlying population in Yemen; also in W half of N America.

Morocco, widespread but not common, from Atlantic coast (mouth of Tensift) and Essaouira, Sidi Kaouki, Sidi Bou Rhaba and El Kansera, to Sehouls, Zemmours, Zaërs, Khatouat Massif, central plateau, Moyen Atlas up to 2000 m, Jbel Tazzeka, and Haut Atlas east to Aïn Beni Mathar; up to 2300 m in Haut Atlas at Télouet. Algeria as mapped. Tunisia, rare in Bou Hedma area, perhaps now extinct; 12 breeding colonies in 1920s between Cap Bon, Medjez el Bab and Gafsa, but birds persecuted and adversely affected by woodland logging; only 4 sight records (Zaghouan, Hammanet, Kairouan, Kelbia) in 1960s and 1970s (Thomsen and Jacobsen 1979).

Densities of *c.* 3–7 breeding pairs per km^2 over much of European range (but up to 32 pairs per km^2 in N England). In NW Africa, where nests are aggregated, densities locally may be as high as 1 pair per 2·6 ha.

Description. *P. p. mauritanica* Malherbe: Morocco to Tunisia. ADULT ♂: nasal bristles, head, neck, mantle to uppertail-coverts, chin to lower breast, thighs, vent and undertail-coverts black, glossy, with blue reflections; large, triangular, cobalt-blue patch of bare skin behind eye; scapulars white, joining in front of wing with white flanks and belly; all black and white patches sharply defined; tail very long and strongly graduated, black, T1 very glossy structural green, tip iridescent bronzy violet-blue, and outer borders of T2–T6 similarly glossy green, burnished violet toward tips. Underside of tail black, slightly glossy. Primaries white, each bordered all around with sharply-defined black band; all other upperwing feathers black with strong steely blue gloss on those surfaces not concealed when wing closed. Undersides of primaries and secondaries like upsersides but duller; underwing-coverts and axillaries black. Bill black; eyes black; legs and feet greyish black, soles yellowish. Sexes alike. SIZE: wing, ♂ (n = 15) 165–172 (169·5), ♀ (n = 16) 152–165 (157); tail, ♂ (n = 14) 247–298 (268), ♀ (n = 16) 210–270 (235); bill to skull, ♂ (n = 12) 24·4–27·2 (26·0), ♀ (n = 12) 23·5–26·0 (24·3); tarsus, ♂ (n = 12) 49·2–53·9 (51·5), ♀ (n = 12) 44·6–50·2 (47·4) (Cramp and Perrins 1994). WEIGHT: 1 ♀ 180; (*P. p. pica*, SE Germany, ♂ (n = 143) 185–247 (221), ♀ (n = 136) 161–240 (191)).

IMMATURE: juvenile like adult but black parts of head and body dull, not glossy; bare patch behind eye dark grey; white in scapulars, flanks and belly tinged isabelline; tail shorter than adult's (until juv. 2–3 months old) and less brightly iridescent,

Pica pica

tail feathers 5–15 mm narrower, tips rounded (rather than square-ended as in adult).

NESTLING: naked; mouth deep pink, palate with white spines, gapes pale pink.

Field Characters. Length 45–47 cm. Unmistakable; a boldly pied corvid with very long, strongly graduated tail; in flight, primaries appear mainly white above and below, and white scapulars conspicuous (**A**). Noisy; short-winged and rather long-legged; walks in pasture; seldom gregarious.

Voice. Tape-recorded (60, 62, 73, 93, 105, B). Song a soft, warbling babble of quiet, low- and high-pitched call notes. Commonest call a harsh, staccato, rattling, far-carrying chatter, 'chacker chacker' or 'cha-cha-cha-cha-cha' or 'cha-cha-cha-cha-ch-ch-ch-ch ... ' with up to 16 'ch's at rate of 10–11 per s; also commonly a gentle, throaty, exclamatory 'tchrow, tchrow', 'tchurch' or' 'chwikoo',

uttered by a solitary bird or by 2 birds answering each other. Several birds within calling distance, scattered along a tall hedge, call back and forth uttering an almost conversational medley of different monosyllabic and disyllabic calls interspersed with dry, wooden, throaty rattles. Less commonly a high-pitched 'trirr', sharp 'keeack-kak', and shrieking 'cheerk'. These and other voices and their contexts described in Cramp and Perrins (1994).

General Habits. Inhabits open or lightly wooded country with pasture, arable land, natural grassland, hedges and scattered bushes and low trees; glades and clearings in broad-leaved and coniferous woodland; sheep and cattle fields; parkland, cultivation, gardens and suburbs with trees, shrubs and expanses of short grass. Perches freely on fence posts and buildings but not on fence or telephone wires; commonly perches on top of shrub or small tree, but seldom higher than 12 m in tall tree.

Forages to some extent in trees and shrubs but mainly on the ground; walks over short grass or almost bare earth in pasture, arable fields and rough grassland, frequently stopping and searching, then hops or walks quickly to make a stab at live prey or seize a grain or dropping; makes small jump to seize insect from tall weed; throws ground litter aside; stands on sheep's back and probes wool for ticks. Forages opportunistically, eating mainly ground invertebrates, and takes small vertebrates when it can, including birds' eggs and nestlings. In hard weather sometimes catches small bird on ground, on nest or even in flight. Stores food, mostly domestic scraps, cereal grains and animal droppings, mainly in autumn, by carrying it in buccal pouch (once 120 wheat grains in a pouch), making small hole in ground and emptying contents of pouch into it, then covering hole with leaf or debris. In field experiments with bread in SW Spain, Magpie ate small pieces *in situ* but carried away and cached nearly all pieces of 4 cm^2 or larger (Henty 1975). Most caches seem to be retrieved within 1–2 weeks.

Usually in pairs, often singly, sometimes in loose group of 3–6 adults. On ground progresses with high-stepping walk, sidling hops and brisk jumps; holds tail horizontally or upward, flicking, raising and fanning it when excited. Often sits on top of bush, tail drooping down (**B**). Flies mainly short distances, between ground and adjacent hedgetops or cover; sometimes flies for several hundred m at treetop height, alternating quite slow, regular beats of its short wings with bursts of very rapid beating, and alights after brief, unstable-looking glide. Furtive and wary where persecuted but can be confiding and cheeky where tolerated. Territorial when breeding and usually in non-breeding season also. In border disputes, aggressively flicks folded wing tips up and out (**C**) and, standing upright or bowing head, jerks tail vigorously upward (**D**).

Pair may roost at night in territory year-round, but (at least in N Europe) communal roost of up to 200 birds may form, birds sleeping in reedbed, willows or hawthorn *Crataegus*.

Sedentary; young may disperse 10–20 km; in Cyprus abundance fluctuates seasonally and in spring birds seen heading out to sea towards Turkey; no movements reported across Strait of Gibraltar.

Food. In Europe beetles and wide range of other ground invertebrates; seeds, nuts, scraps, faeces, carrion; a few small vertebrates and fruits. Huge range, listed and tabulated for adults and nestlings in Cramp and Perrins (1994), includes fish, frogs, snakes, hedgehog *Erinaceus europaeus*, hare *Lepus capensis*, birds up to size of feral pigeon *Columba livia* (all mainly as carrion), adult and larval insects, ticks, woodlice, shrimps, centipedes, earthworms, snails and bivalves; and seeds, fruits or perennating organs of pine, juniper, oak, beech, chestnut, walnut, rose, apple, prune, holly, sea buckthorn *Hippophae*, grape, olive, fig, potato, pulses, cucurbits and grasses. Composition of diet varies regionally, seasonally and individually.

Breeding Habits. Strongly territorial, generally monogamous, but ♂♂ often promiscuous. Pair defends all-purpose territory of 5–7 ha (NW Europe). In NW Africa loosely gregarious, with nests distributed patchily; in a patch nests are rather evenly spaced, on average 264 m apart (D. Gosney *in* Cramp and Perrins 1994; Etchécopar and Hüe 1967). Pair-bonding behaviour subtle, without demonstrative displays; at potential nest site ♂ and ♀ may flick wings, flirt tail, and utter varied quiet calls; ♂ courtship-feeds ♀ in few days before egg-laying (Cramp and Perrins 1994).

NEST: roofed ball, loosely made of twigs and sticks, often thorny, with 1–2 entrances; inside, a bowl of mud and dung, applied to the sticks and incorporating twigs and roots; inside bowl is a layer of finer twigs and coarse grasses, lined with finer, soft grass, leaves, feathers, hair and scraps of cloth and paper. Nests in S Spain always roofed (Alvarez and Arias de Reyna 1974); a few nests in N Europe not roofed. Placed about 2 m up in tree or on man-made structure, in low bush, under brambles, or on ground; in Morocco, placed in thorn bush, juniper and cork oak. 1 nest (Denmark) made of 598 sticks, weight 4·6 kg, size 44 × 38 × 70 cm high (Birkhead 1991). Built by ♂ and ♀, taking 1–2 weeks in S Spain and 1–8 weeks in N Europe, often close to last year's nest and using material from it; outer ball of sticks built first, then mud bowl, then linings.

EGGS: 3–10, usually 4–6 in Africa and 5–7 in Europe; av. in S Spain (n = 108) 6·1. Generally laid daily. Subelliptical, smooth, glossy; pale blue, greenish blue, light or dark olive-brown, immaculate, or with large, sharply-defined dark brown and pale grey blotches all over or concentrated around broad end or at narrow pole. SIZE: (*P. p. mauritanica*, n = 84) 30·0–38·1 × 22·0–25·2 (33·2 × 23·5); weight *c.* 9·4.

INCUBATION: starts with 2nd-3rd egg laid or (S Spain) evidently with penultimate egg. By ♀ only. Period from laying of 1st egg to hatching of 1st nestling av. 23·7 days (n = 52); incubation period 21–22 days (Cramp and Perrins 1994).

DEVELOPMENT AND CARE OF YOUNG: young brooded only by ♀ and fed by ♂ and ♀, probably mainly ♂. Nestling period 24–30 days. Young independent of parents at 70–80 days.

BREEDING SUCCESS/SURVIVAL: of nests in S Spain, 35 of those built less than 0·8 m above ground were predated and 9 of higher ones; altogether 50–70% of eggs failed to produce flying young. Main predators were rodents; also lizards, Common Buzzard *Buteo buteo*, Imperial Eagle *Aquila heliaca*, Red Kite *Milvus milvus* and Black Kite *M. migrans*. In another study in S Spain, 20% of broods parasitized by Great Spotted Cuckoos *Clamator glandarius*. Mortality in N Europe 46% in 1st year, 58% in 2nd year and 55% in 3rd–5th years (Cramp and Perrins 1994). Oldest wild bird 15 years 1 month.

References

Alvarez, F. and Arias de Reyna, L. (1974).
Arias de Reyna, L. *et al.* (1984).
Birkhead, T.R. (1991).
Birkhead, T.R. *et al.* (1986).
Cramp, S. and Perrins, C.M. (1994).

Genus *Garrulus* Brisson

Jays with predominantly pinkish, pale grey or chestnut plumage, solid black moustache or crown or solid dark blue head, wing-coverts (and in one sp. tail) banded with black and azure blue; bill black or yellow or yellow-and-blue; legs yellowish or blackish. Plumage full and soft; crown feathers somewhat elongated. Wings rather short and rounded, P5 and P6 longest. Tail square-ended in 2 spp. and rounded in 3rd. Bill short (as in most jays). Nostrils round, covered by thick, dense bristles. Voice harsh. Inhabit woodland. Cache food, specializing on acorns *Quercus* spp.

3 species; one (markedly polytypic) in Palearctic including NW Africa, one in Himalayas, one in Ryukyu Is.

Garrulus glandarius (Linnaeus). Eurasian Jay. Geai des chênes.

Plate 30
(Opp. p. 463)

Corvus glandarius Linnaeus, 1758. Syst. Nat., ed. 10, 1, p. 106; 'Europa', restricted to Sweden.

Range and Status. Palearctic, from Morocco and Norway (north to 70°N) to Japan, south to Himalayas, Burma and Taiwan. Eruptive migrant in N Europe, sedentary in south.

Sedentary in NW Africa, where 3 endemic subspecies. Uncommon, locally frequent, forested massifs in Rif, Middle and High Atlas and N Algeria from coast to hinterland at 35°N. Morocco, Jbel Tazzeka, Zaërs, Central Plateau, Beni Snassen, Jbel Mahseur, many localities in Rif, Ouezzane region, Azrou-Ifrane-Sefrou-Skoura-Boulemane-Itzer-Azigza sector of Middle Atlas, and High Atlas up to 2400 m (Marrakech, Tizi-n-Test, Timenkar, Jbel Ifirouane, Ourika). Algeria, widespread in N, north of 34°N in W and 35°N in E, but absent from Saïda-Mascara-Oued Rhiou-Tiaret quadrilateral. Tunisia, scarce in oak forests of Kroumirie south to Jendouba and pine woods south to Bou Chebka; records from Cap Bon and between Tunis and Sfax (Thomsen and Jacobsen 1979).

Breeding densities of 6 pairs per km^2 (S France), 10 pairs per km^2 (Corsica), 3 pairs per km^2 (NW Spain) and in Morocco 6 pairs per km^2 in semi-arid *Tetraclinis* scrub and rather thinner densities in maquis and woodland (Thévenot 1982).

Garrulus glandarius

Description. *G. g. whitakeri* Hartert: N Morocco (Tangier area, Rif) and extreme NW Algeria (Tlemcen area: Saïda). ADULT ♂: forehead and superciliary stripe pale vinous grey, crown and nape black with whitish and vinous-grey streaks, hindneck and sides of neck vinous-brown, mantle, scapulars and back greyish vinous-brown, rump and uppertail-coverts white. Tail distally black, and proximally black with greyish barring (concealed below uppertail-coverts), more pronounced on sides than centres of feathers. Cheeks and ear-coverts off-white. Broad black malar stripe, clearly delineated, from base of lower mandible, broadening towards sides of lower throat. Chin and throat dirty white; breast vinous, paling to greyish vinous or buffy puce on flanks; belly off white; undertail-coverts silky white. Primaries dull black, outer vanes mainly silvery pale grey; P10 black. Secondaries and outer tertials black, middle third of outer vane of most secondaries white and bases barred blue; inner tertials dark chestnut; alula, outer vanes and tips of inner vanes of primary coverts and outer vanes of outermost 5 greater coverts barred with black and bright structural azure blue; median upperwing-coverts mainly chestnut; lesser coverts vinous brown-grey; remaining upperwing-coverts black; underwing-coverts and axillaries greyish vinous-pink, undersides of remiges and rectrices glossy silvery grey. Bill brownish black; eyes pale blue-grey; legs and feet flesh-brown. Sexes alike. SIZE: wing, unsexed (n = 16) 170–190 (179).

IMMATURE: juvenile like adult but white parts of head (forehead, around eye, ear-coverts, chin and throat) sandy or buffy pale grey, and black feathering on head (crown, malar stripe) brownish and less boldly contrasted.

NESTLING: blind, naked, skin pinkish yellow, bill and legs brownish pink or mauve-grey, mouth bright pink, gapes pinkish white.

G. g. minor Verreaux: Morocco (High Atlas) and Algeria (Saharan Atlas to coast, east to Djelfa). Like *whitakeri* but forehead streaked black, hindcrown blacker, whitish superciliary stripe better defined, and area around eye buffier white. SIZE: wing, ♂ (n = 20) 158–175 (167), ♀ (n = 17) 153–167 (160); tail, ♂ (n = 6) 132–142 (137.5), ♀ (n = 9) 132–142 (136); bill to skull, ♂ (n = 5) 27.3–30.5 (29.5), ♀ (n = 8) 28.0–30.5 (28.9); bill to nostril, ♂ (n = 4) 17.6–18.9 (18.5), ♀ (n = 5) 16.8–18.3 (17.5); tarsus, ♂ (n = 5) 39.8–40.9 (40.3), ♀ (n = 7) 38.5–39.7 (39.2). WEIGHT: unknown; (*G. g. glandarius*, E Germany) ♂ (n = 107) 153–197 (172), ♀ (n = 101) 129–188 (163.5).

G. g. cervicalis Bonaparte: Tunisia and NE Algeria, west to Algiers and Médéa, south to Batna. Like *whitakeri* but forehead feathers white with blackish tips, crown solidly black, large area of white around eye and on ear-coverts and cheeks in strong contrast with black crown, large black malar stripe, and chestnut sides of neck; malar stripe reaches chestnut sides of neck and breast. SIZE (16 ♂♂, 11 ♀♀): wing, ♂ 174–190 (183), ♀ 174–182 (178); tail, ♂ 149–167 (157), ♀ 150–158 (153); bill to skull, ♂ 31.8–36.0 (33.7), ♀ 30.5–34.6 (32.5); tarsus, ♂ 40.8–45.4 (43.2), ♀ 40.8–44.0 (42.9). WEIGHT: 1 ♂ 183, 3 ♀♀ 164, 169, 171.

Field Characters. Length 34–35 cm. Unmistakable; only jay in Africa; a pinkish grey-brown bird with heavy black malar stripe, strongly pied wings with banded azure blue patch, strongly pied head (*cervicalis*) or speckled crown and greyish face (*minor*, *whitakeri*). In flight, azure carpal patch is less striking than black secondaries with broad white band across them, and large white rump patch contrasts with square black tail. More often heard than seen; commonest voice a loud screech. Often put up from quiet road verges in woods, when disappears immediately into the trees; sometimes glimpsed within woods (screech, white rump) or flying across open field (white rump; flight laborious, weak-looking, with jerky beats of flailing wings).

Voice. Tape-recorded (62, 73, 93, 105, B). By far the best-known is loud, harsh, strident, rasping screech, 'dzjay', given once, or twice in quick succession. Repertoire varied. Song (not often heard) a quiet medley of chirruping, chattering, percussive, and buzzing or wheezing sounds, with mimicked voices of other bird species. In Europe mimics calls of Common Kestrel *Falco tinnunculus*, Grey Partridge *Perdix perdix*, Common Crane *Grus grus*, Common Magpie *Pica pica*, Carrion Crow *Corvus corone* and Common Starling *Sturnus vulgaris*, and some non-bird noises. Flight-appeal call, mewing call, food-offering call, nest call, chirruping, 'kraah' call, hissing, castanet call and several others described by Cramp and Perrins (1994).

General Habits. Strongly arboreal, though often perches on derelict buildings, woodside walls and the ground. In Morocco inhabits mainly oak and ilex woods on hillsides and in elevated valleys, maquis and semi-arid *Tetraclinis* scrub; across Europe commonest in dense cover of broad-leaved woods, scrub and undergrowth and in pure or mixed woods of oak *Quercus*, beech *Fagus* and hornbeam *Carpinus*; locally in coniferous forests, also rural and urban parks, and spinneys, orchards, vineyards, gardens and farm hedgerows.

Occurs singly or in pairs. Shy and wary of man where persecuted (i.e. over much of range) but can become indifferent in city parks. When suspicious sleeks plumage but raises crown and belly feathers, and flicks tail down and, partly spreading it, up again. In spring 3–10 birds, often 5 or 6, aggregate in a treetop for 15 min or more, perching 1 or a few m apart, moving little (although in 15 min group can move randomly in hops and short flights and progress through 2–3 adjacent trees) but interacting and displaying and uttering diversity of calls, mainly chirruping, flight-appeal and 'kraah' calls, but not any screeching. In commonest display posture 1–2 birds perch erect, looking pot-bellied, and with rump feathers fluffed out making white patch highly conspicuous (**A**). Function of such 'spring gatherings' thought to be to bring unpaired birds together.

Forages solitarily along branches amongst foliage and on ground. Gleans leaves for caterpillars. Digs in leaf litter on ground with decisive swings of head and bill; scatters light soil; pokes bill into crevices, holes and rolled-up leaves and opens bill to prise material apart. Pecks at soil surface to glean cereal seeds at sowing and harvesting times. Occasionally pursues a flying insect and known to catch cicadas *Cicada orni* in flight. Digs into wasps' nests. Holds large insect under both feet and pulls it apart with bill. Robs Syrian Woodpecker *Picoides syriacus* of *Carya* nuts that the woodpecker has opened (Israel). Takes eggs and nestlings from nests of Song Thrush *Turdus philomelos*, Eurasian Blackbird *T. merula*, Pied Flycatcher *Ficedula hypoleuca* and other small bird species. In autumn can spend much time – up to 10 h per day – in collecting acorns; also takes peanuts from bird tables, and systematically collects beech nuts and sometimes hazel nuts *Corylus* to cache them. Pulls unripe acorn with cup and stalk from oak tree, or loosens ripe acorn from cup and swallows it whole; when several have been collected, it flies away to cache them in ground. Generally carries 1–3 (once 9) acorns in gullet and 1 in bill; in 1 study, av. time to collect a load of acorns and cache them 1·2 km away was 10 min; 1 bird can cache 5000 acorns during the autumn. Caches are often in clearing in a wood, or at edge. 1–2 or more acorns are pushed into soil at an angle and sometimes then hammered in with bill, then hole is filled in with sideways swipes of bill and place covered with small sticks, stones and leaves. Cached acorns are dug up in winter and spring; bird flies directly to a site and digs acorns up, even when under 40 cm of snow. Experiments show that Eurasian Jay has remarkable memory of exact cache sites. Those that are forgotten about or not needed in mild weather germinate, and Jay is major dispersal agent for oaks (Cramp and Perrins 1994).

Sedentary in Africa so far as known. Strongly migratory in NE Europe but migrants not known to reach SW Europe or Africa.

Food. In Europe invertebrates (mainly beetles and, when feeding young, caterpillars), fruits, seeds and nuts (especially acorns); spiders, woodlice, millipedes, centipedes, earthworms, snails, frogs, mice, lizards, small birds and their eggs and nestlings. Numerous insects and plants tabulated in Cramp and Perrins (1994).

Breeding Habits. Well known in Europe but not Africa. Solitary nester, monogamous with lifelong bond, territorial. ♂ courtship-feeds ♀. ♀ solicits copulation by bowing, fluffing rump feathers, spreading and arching wings, and quivering tail and wings (**B**).

NEST: rough platform of twigs (3–15 mm in diam.) with softer layer in centre made of thin twigs, roots and stalks, lined with rootlets, grass, moss and leaves. Sited in fork in tree or close to trunk on branch; sometimes in tangle of creepers or on building. Constructed by ♂ and ♀.

EGGS: 3–8, av. (n = 90, Germany) 5·7. Oval, smooth, slightly glossy; pale green, blue-green, olive, finely (occasionally densely) speckled with pale brown or grey. SIZE (*cervicalis*): (n = 52) 29·6–33·6 × 22·0–24·4 (31·3 × 22·8). Weight *c*. 8·4.

LAYING DATES: NW Africa, Apr–May (Etchécopar and Hüe 1967).

INCUBATION: by ♀, beginning with 2nd–4th egg. Period: 16–17 days.

A

DEVELOPMENT AND CARE OF YOUNG: young brooded by ♀, for 14–15 days. Fed by ♂ and ♀ about equally; at first ♂ gives food to ♀ who passes it to young. Nestling period: 21–23 days, usually 21–22 days.

BREEDING SUCCESS/SURVIVAL: success rates vary considerably; in S Russia 67% of 181 eggs hatched and 38% resulted in flying young. In some regions most nests, clutches and broods destroyed by man; others by crows *Corvus* spp., Common Magpies *Pica pica*, European Sparrowhawks *Accipiter nisus*, Common Buzzards *Buteo buteo*, squirrels *Sciurus vulgaris* and *S. carolinensis*, and dormice *Eliomys quercinus* (Cramp and Perrins 1994). Mortality in 1st year c. 61% (Europe). Oldest bird nearly 18 years.

Reference
Cramp, S. and Perrins, C.M. (1994).

Genus *Nucifraga* Brisson

2 spp. of jay-like corvids, with long, straight, sharp-tipped bills, round nostrils concealed below dense bristles, somewhat rounded wings with P5–P7 longest, square-ended tails, and plumage soft and full, white-spotted brown in one species and pale grey in the other, both with black wings, white undertail-coverts and mainly black tails. Inhabit conifer woods, feed mainly on ground, cache food, specializing on *Pinus* nuts.

1 species in Palearctic, vagrant to NW Africa, and 1 in Nearctic.

Nucifraga caryocatactes (Linnaeus). Spotted Nutcracker. Cassenoix moucheté.

Plate 31
(Opp. p. 478)

Corvus Caryocatactes Linnaeus, 1758. Syst. Nat., ed. 10, 1, p. 106; 'Europa', restricted to Sweden.

Range and Status. Palearctic, from about 01°E in Europe to Korea and Japan, south to Balkans, Tien Shan Mts, Himalayas, N Yunnan, Taiwan and Kuril Is. Mainly sedentary, but also an eruptive migrant in N and especially Siberia. 4 major and 8 minor eruptions from Siberia into W Europe in 20th century, where small populations may remain to breed (e.g. in Belgium) for some years. Spectacular day-time movements of autumn immigrants can occur, e.g. over 10,000 birds on Baltic coasts. Irruptive birds rare in Britain, S France, N Italy and Turkey and very rare in Portugal and N Africa (where race presumed to be *macrorhynchos*).

1 record NW Africa (Cramp and Perrins 1994).

Description. *N. c. macrorhynchos* C. L. Brehm: Siberia. ADULT ♂: forehead, crown, nape and hindneck very dark brown; mantle, scapulars and back umber-brown, each feather with triangular white mark at distal end of shaft, the white narrowly bordered at sides by narrow blackish line; rump and uppertail-coverts umber-brown. Tail blackish brown with slight oily green gloss, T1 narrowly white-tipped, T2–T6 with progressively larger white tips (up to 24 deep on T6). Lore and small area around eye whitish; ear-coverts, sides of neck, cheeks and breast umber-brown, heavily speckled and streaked with white; upper chin whitish, lower chin and throat dark umber-brown with small white streaks; flanks and belly like mantle and back; vent and undertail-coverts white. Wings blackish brown with small white spot. Bill blackish, greyer at base of lower mandible; mouth black; eyes brown; legs and feet black, soles yellowish. Sexes alike. SIZE: ♂ (n = 28) 183–190 (188), ♀ (n = 25) 177–189 (182); tail, ♂ (n = 24) 118–130 (124), ♀ (n = 23) 114–128 (121·5); bill to skull, ♂ (n = 25) 52·3–59·9 (55·8), ♀ (n = 25) 48·4–59·0 (53·6); tarsus, ♂ (n = 24) 39·0–43·1 (41·3), ♀ (n = 8) 38·9–42·1 (40·6). WEIGHT (Germany): ♂ (n = 7) 164–191 (175), ♀ (n = 4) 140–176 (154) (Cramp and Perrins 1994).

Field Characters. Length 32–33 cm. A mainly dark brown bird with large round-ended wings and disproportionately short, white-ended black tail; black cap, frosted whitish face, white-speckled body and white undertail-coverts. In conifer woods. Size of Eurasian Jay *Garrulus glandarius* and, like it, round-winged with laboured flight, white undertail-coverts, and voice a rasping screech, but there the resemblances end.

Voice. Tape-recorded (62, 73, 93, B). Main call far-carrying, a high-pitched, rasping 'kraak' or 'kreak', like screech of Eurasian Jay but less strident (Cramp and Perrins 1994).

General Habits. Arboreal, but feeds largely on ground. Inhabits woods of Norway spruce *Picea abies* on mountain slopes, also larch *Larix* sp. and fir *Abies alba*. Irruptive vagrants in winter in SW Europe likely to occur in stands of any species of spruce or pine *Pinus*. Slender-billed Siberian race *macrorhynchos* specializes on seeds of *Pinus sibirica* and thick-billed W Palearctic nominate race on seeds of arolla pine *P. cembra* and nuts of hazel *Corylus avellana*, so sedentary Spotted Nutcrackers tend to live in or near such trees. Migrants are more wide-ranging, however, and occur singly or in flocks in fields and woods and around habitations at all altitudes up to treeline; often very tame, coming to bird tables and foraging on ground in gardens, farmland and picnic places. Gait on ground a strong hop, with sidling jumps. Distant flight over forest steadier than Eurasian Jay's, wing beats not so jerky, with slight undulation.

Food. Hymenopterans, beetles and range of other insects; snails, earthworms and small vertebrates; but mainly (and especially in winter) seeds and nuts of pine, spruce, fir and hazel (see above); also household scraps, carrion and dung.

Reference
Cramp, S. and Perrins, C.M. (1994).

Genus *Pyrrhocorax* Tunstall

Glossy black, montane and coastal cliff corvids with red or yellow bill and legs, strongly emarginated outer primaries, rather square-ended wings, and soaring and acrobatic flight. Bill more or less decurved. Tarsus booted (very young birds show traces of scutes).

2 species, with rather similar ranges from NW Africa and W or SW Europe to Himalayas and China, one species also in Ethiopia; resident.

Plate 30
(Opp. p. 463)

Pyrrhocorax pyrrhocorax (Linnaeus). Red-billed Chough. Crave à bec rouge.

Upupa Pyrrhocorax Linnaeus, 1758. Syst. Nat., ed. 10, 1, p. 118; England.

Range and Status. Ethiopia, NW Africa, S Europe from Ireland and Portugal to Turkey, and central Asia from Altai to Mongolia, L. Baikal region and China, south to Himalayas and Kansu. Montane, but coastal in Britain and W France. Resident, with some altitudinal migration.

NW Africa. Morocco, widespread from Tangérois, Rif, and Mediterranean coast at Targa, up to 4000 m or more in Moyen Atlas and Haut Atlas Mts where often encountered in flocks of tens, sometimes in hundreds (100 Oukaïmeden, 100 Tétouan, 300 Aguelmane Azigza, 200 Tizi-n-Test, 500 Aït Ourir, hundreds at Boumalne du Dadès in winter) and once in thousands (Oukaïmeden, Oct, at about 2100 m, with mostly immature birds at lower limits and mostly adults at upper limits: Thévenot *et al.* 1982); in E Morocco known from Jbel Mahssor (a colony near Oudja), near Boumalne du Dadès, Taliouine, and near Tazenakht (Tizi-n-Bachkoum). In winter, some evidence that flocks descend to lower altitudes, with records in Anti Atlas Mts, Tizi-n-Taghatine, Tinerhir, Ouarzazate, Gorges du Dadès, Boumalne du Dadès and elsewhere.

Algeria, local; *c.* 20, Djebel Ahmar Kaddou; a colony of 500 in Tlemcen Mts in 1960 and 3–4 colonies there totalling 100 pairs in 1979; in Kherrata-Babors; 150 in Djurdjura; once near Jijel (Ledant *et al.* 1981). Tunisia,

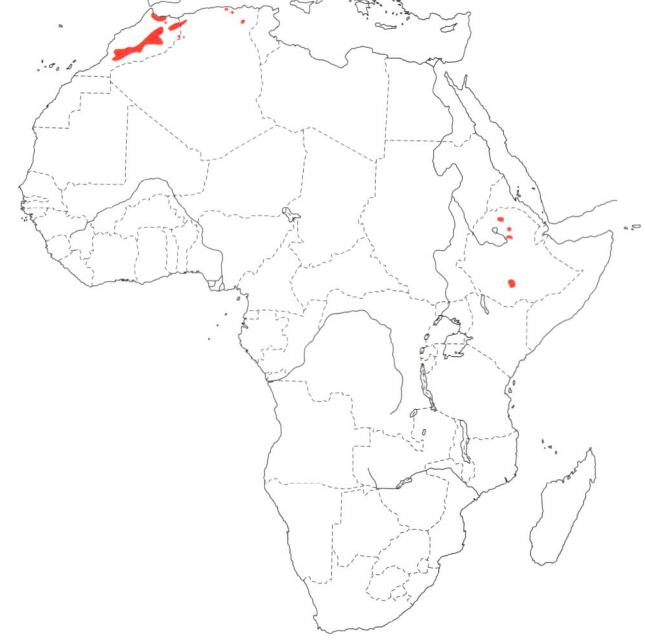

Pyrrhocorax pyrrhocorax

known to have been present in 19th century but none seen in 20th.

Ethiopia, between about 3600 m and 4530 m in: Simien Mts in N, where estimated to be 350–500 birds in Simien Mountains Nat. Park; Mt Abune Yosef, Dilanta Highlands and probably Mt Amba Farit in Welo Prov.; and Bale Mts and possibly Agarfa in S, where estimated to be 250–400 birds in Bale Mountains Nat. Park (Delestrade 1998). Total Ethiopian population estimated at 675–1300 birds. Choughs being highly sedentary, all Ethiopian populations regarded as Vulnerable, even Threatened.

Densities est. 1·9–2·8 birds per km^2 in Simien Mountains Nat. Park (179 km^2) and 1–2 birds per 10 km^2 in Bale Mountains Nat. Park (2471 km^2) excluding Harenna Forest (1700 km^2) (Delestrade 1998). In SE Spain av. 39 (max. 123) pairs per 10 km of cliff (n = 324 pairs, Blanco et al. 1991).

Description. *P. p. barbarus* Vaurie: Canary Is (La Palma), Morocco, Algeria and formerly Tunisia. ADULT ♂: black, strongly glossed with purple on head, body, lesser wing-coverts, median wing-coverts and underwing-coverts, somewhat less glossy, greenish, on rest of wings and tail. Underside of flight feathers silvery black. Bill carmine-red, orange-red, pink or crimson; eyes dark brown; legs and feet same as bill, claws black. Sexes alike. SIZE (8 ♂♂, 5 ♀♀): wing, ♂ 286–310 (297), ♀ 272–281 (274·5); tail, ♂ 135–152 (144), ♀ 128–139 (134); bill to skull, ♂ 57·9–62·8 (60·9), ♀ 54·5–57·5 (55·5); tarsus, ♂ 53·5–56·1 (55·2), ♀ 51·5–53·5 (52·7). Also (Morocco and Algeria only, 15–17 ♂♂, 4 ♀♀): wing, ♂ 290–310 (302), ♀ 275–284 (280); tail, ♂ 129–153 (145), ♀ 131–147 (138); bill to skull, ♂ 61–67 (63·2), ♀ 55–58 (57·0). WEIGHT: not known (3 other races, Switzerland, Crete, Kazakhstan: ♂ (n = 9) 258–375 (312), ♀ (n = 6) 219–300 (263)) (Cramp and Perrins 1994).

IMMATURE: like adult but wings slightly less glossy; juvenile like adult but back, rump and underparts dark grey-brown and woolly-looking, and head and rest of body brownish or greyish black, with purple-black gloss confined to narrow margin at tip of each feather; bill flesh-pink to orange, eyes dark brown; legs and feet yellowish, orange-red or reddish pink.

NESTLING: dorsal surfaces and thighs covered with dark grey down up to 20 mm long; underparts bare, red-pink; bill mauve or horn-pink, mouth orange-pink or red-pink, gapes yellow-white, legs red-pink.

P. p. baileyi Rand and Vaurie: Ethiopia. Like *barbarus* but larger though bill shorter. SIZE (7 ♂♂): wing 310–329 (319), tail 145–155 (151), bill to skull 54–65 (60·6) (Cramp and Perrins 1994).

Field Characters. Length 39–40 cm. A dashing, montane corvid with glossy black plumage and *red bill and legs*. Only other corvid in Africa with red bill is lowland Piapiac *Ptilostomus afer* (juvenile bill pink, adult black). Juvenile Red-billed Chough has dull yellowish-flesh bill and legs and, when bill still short, can be very difficult to distinguish from Alpine Chough *P. graculus*, sympatric with it in NW Africa. Choughs can be told from other corvids at a distance by broad, vulturine wings with convex trailing edge to secondaries and 5 'fingers' (emarginated primaries) at tip. Red-billed has broader wing than Alpine, shorter tail, and glossier plumage. Yellowish-billed juvenile Redbill best told from Alpine Chough only by accompanying adults; its bill is longer, and orange-yellow rather than greenish yellow. (In the hand the 2 spp. readily distinguished also by bill-base bristles: in Red-billed, bristles at base of upper and lower mandibles equal in density; in Alpine, bristles at base of upper mandible dense, but lower mandible almost bare.)

Voice. Tape-recorded (62, 73, 93, 105, B, C, CLO). Song, seldom heard, a quiet babble of warbling chitters and churrs. Contact call, given commonly throughout year, a yelping, loud, variable 'chwee-ow'. Call reported in Morocco, 'chuff' (though the name Chough is thought not to be onomatopoeic). Alarm a harsh, scolding 'ker ker ker'; mobbing call a long, loud screech 'chaaaaay'. These and other calls described with sonagrams in Cramp and Perrins (1994). Voice in Ethiopia differs substantially from that in Alps, Europe (Delestrade 1998).

General Habits. Inhabits craggy high mountain terrain with precipitous rock slopes, cliffs, escarpments, screes and boulders, and with not-too-distant alpine meadows, low afro-alpine vegetation, grassy plateaux, pasture, fields of barley, wheat and other upland crops. Occurs in open habitats between 2800 m and 4200 m in Ethiopia and mainly between 2500 m and 4100 m in Maghreb, where also winters partly in lowlands and occurs regularly on some rocky coasts where probably breeds. Does not require trees and hardly ever perches on trees, bushes or posts even where present.

Breeds in well-spaced pairs, sometimes almost colonial, and outside breeding season strongly gregarious, generally in flocks of 10–200. Flock size in Ethiopia (n = 26) 9–150 (av. 60) birds. Forages entirely on ground, in pairs or flocks, on 20–40 mm high vegetation or stubble and on burnt or tilled ground, keeping largely to interfaces between earth or vegetation and stones or bare rocks, around tussocks, and along edges of snow patches (Cramp and Perrins 1994, Delestrade 1998). Probes crevices and soil deeply, bill sometimes sunk up to its base, with quick movements at all angles, sometimes holding head parallel with and almost touching substrate; takes underground ants with up to 8 rapid jabs of slightly-opened bill, moving head from side to side or using sewing-machine action like a wader (Cowdy 1973) and tipulid larvae with sudden probe, half jump around and more bill-probing and levering (Goodwin 1986). Picks small items from surface of substrate, and flicks small stones and dry dung over in search for prey. Takes berries and caterpillars from low bush, fluttering wings to balance momentarily, and in Canary Is seen to mandibulate caterpillars and beat them against stone before eating them; however, usually swallows whole item without mandibulation, though may hold large prey under feet to treat it first. Often wipes bill against substrate after probing.

On ground walks, much like Rook *Corvus frugilegus*; also hops, runs, makes small jumps, and clambers in tight places (Cramp and Perrins 1994). Not shy. When excited, flicks wing tips and flirts tail. Commonly associates with Alpine Choughs. Flight buoyant, aerobatic and accomplished, with much solitary or social soaring, steep, graceful, sweeping dives and leisurely gliding along cliff faces or high in air, with upturned primary tips, and occasional rolls and tumbles like Common Raven *C. corax*. Vocal all year; perched bird often flicks wing tips when calling. Flies upward to avoid raven or hawk and gives alarm call; on ground, may freeze. Flying flock bunches tightly if surprised in open by Peregrine *Falco peregrinus*, and individual bird or pair hides in scrub or rock crevice. Threatens conspecific by walking up to it with neck outstretched. Roosts gregariously in rock holes and on cliff ledges; sometimes on buildings; flocks move conspicuously along cliffs to roost just before sunset (Ethiopia).

Food. In Europe mainly invertebrates: adult and larval moths, hymenopterans, flies and beetles (particularly the last 2), grasshoppers, earwigs, bugs, spiders, harvestmen, scorpions, woodlice, millipedes, centipedes, molluscs, earthworms; also small vertebrates (lizards, mice), carrion, and fruits and seeds of juniper, rowan, pear, rose, holly, fig, olive, grape, *Citrus*, *Vaccinium* and cereals and other grasses. In study in S Spain grain and wild seeds formed 70% of diet by volume in autumn (Soler 1989).

Breeding Habits. Well-known in Europe, not in Africa. Nests in solitary pairs, widely spaced (see Densities, above) or in loose, dispersed colony. Monogamous; helping 3rd adult has been reported. Not conspicuously territorial, but breeding bird often harries passing chough or other corvid. Forages gregariously, up to 2 km or more from nest. Pair often uses same nest site in successive years. Pair often visits nest site outside breeding season (Ethiopia). Courtship displays barely known, but sexual bowing display by 3 birds described in Tibet. ♂ courtship-feeds ♀ away from nest; it can be prelude to copulation (Holyoak 1972).

NEST: untidy, bulky pile of twigs, stems and roots, sometimes bound together with mud, thickly lined with wool and plant down; placed in crevice in rock or ledge in cave or on cliff (Britain) or in holes in clay cliffs (S Spain). Size: (1 nest) ext. diam. 350, int. diam. 120, ext. height 170, int. depth 75. Built mainly by ♂ at beginning, ♀ completing and lining it.

EGGS: 1–6, usually 3–5; laid at intervals of 1–3 days; subelliptical, smooth, glossy, very pale greenish grey or cream, with pale grey-lilac undermarkings and pale to dark brown spots, and well-defined, scattered or somewhat concentrated toward large end. SIZE (*P. p. barbarus*): $37 \cdot 0 \times 27 \cdot 5$, $37 \cdot 5 \times 27 \cdot 5$. WEIGHT $c.$ 14·6 g.

LAYING DATES: Morocco, (courtship Apr, nest-building May, flying young June); Algeria, (nesting Apr–June, juv. Oct).

INCUBATION: by ♀ only, starting with 1st egg laid. ♀ sometimes fed by ♂ on nest after she begs, crouching with ruffled plumage and fluttering partly opened wings; in spells of $c.$ 60 min with breaks of 5–10 min. Period: 17–18 days (Europe).

DEVELOPMENT AND CARE OF YOUNG: at 7 days, skin pink, bill mauve with whitish tip and edges, mouth pink with white spurs on palate, gape pale yellow, legs pink; near fledging, mouth redder, legs yellow or dirty orange. ♀ carries eggshells away; young brooded by ♀ for 2–3 weeks. For first 3 days ♀ feeds young by inserting her bill into throat; later, food is regurgitated deep into throat with shaking movements; after 11th day, young fed equally by ♂ and ♀ (Cramp and Perrins 1994). Both parents take faecal sacs away but only ♀ preens young.

Young beg vigorously. Nestling period 31–41 days. After leaving nest, young hide in rock crevices nearby, appearing only when parent arrives with food, and following parents and dispersing up to 1–2 km away after about a week. Family roosts at night near nest site for several weeks.

BREEDING SUCCESS/SURVIVAL: pair may not breed every year; hatching success (Britain) 2·7 young per pair (n = 194); in various studies, 42–76% of eggs laid resulted in fledged young, and productivity varied from 1·82 (Britain) to 3·7 (Italy) young per pair per year (Cramp and Perrins 1994). 9% of nests in Granada, Spain, parasitized by Great Spotted Cuckoo *Clamator glandarius*. Ravens and foxes *Vulpes vulpes* are major predators of nests and Peregrines of adults.

References
Cramp, S. and Perrins, C.M. (1994).
Delestrade, A. (1998).

Plate 30
(Opp. p. 463)

Pyrrhocorax graculus (Linnaeus). Alpine Chough. Chocard des Alpes.

Corvus Graculus Linnaeus, 1766. Syst. Nat., ed. 12, 1, p. 158; Swiss Alps.

Range and Status. Resident, with some altitudinal movements, in Morocco, NW Spain, Pyrenees, Alps, Corsica, Italy, W and S Balkans, Crete, S and E Turkey, Lebanon, Syria (Mt Hermon), Caucasus, N Iran, and central Asia from Afghanistan, Himalayas and Tien Shan Mts to Sayan Mts and W China. Vagrant north to Poland.

Morocco, widespread and often numerous (flocks of up to 250 birds) in Haut Atlas and Moyen Atlas; less common and more local in N. In Rif, records at Chechaouen, between Ceuta and Tanger, near Fez, near Taza (Jorf Lakehal, Jbel Masgout), Jbel Lakraa and Jbel Bou Halla; in Moyen Atlas at Aguelmane Sidi Ali, Col du Zad, Timahdit, Aguelmane Azigza, Ouaouizerth and source of Oum Er Rbia; in Haut Atlas at Tizi-n-Test, Seksaoua, Oukaïmeden, Jbel Toubkal at 2200–3000 m, Neltner at 2300–3500 m, Ourika at 2200 m, Tizi-n-Tichka (several localities), Zaouia Ahansal, Jbel Mourik and Imilchil.

Tunisia, 8–10 at 900 m altitude on 'La Serpentine' pass on Djebel en Nagueb, Feb and May 1998 (Maumary *et al.* 1998).

Description. *P. g. graculus* (Linnaeus) (only race in Africa): Morocco, Europe, Caucasus, Iran. ADULT ♂: black; head, body and lesser wing-coverts a little glossy. Bill yellow, golden, almost white, or orange-yellow, tinged greenish at base; mouth black; eyes dark brown; legs and feet light orange, yellowish orange or coral-red. Sexes alike. SIZE (Europe): wing, ♂ (n = 10) 270–284

Pyrrhocorax graculus

(277), ♀ (n = 7) 251–267 (259); tail, ♂ (n = 10) 161–179 (169), ♀ (n = 6) 157–165 (162); bill to skull, ♂ (n = 12) 33·2–37·2 (34·9), ♀ (n = 11) 32·2–36·4 (34·0); tarsus, ♂ (n = 13) 45·3–48·5 (46·9), ♀ (n = 11) 43·4–46·8 (45·1). WEIGHT (Switzerland): unsexed (n = 4) 188–240 (210), 1 ♂ 251·5.

IMMATURE: like adult with feathers shorter and some more pointed. Juvenile like adult but dull sooty or brownish black, without gloss; legs and feet blackish with orange marks. Very young bird has dusky orange-horn bill with darker tip and culmen, and olive-brown to dirty yellowish legs.

NESTLING: underside naked, upperparts (crown, back, upperside of wings) with short, sparse down; mouth pink.

Field Characters. Length 38 cm. A graceful, montane corvid with black plumage and yellow bill and legs. Sympatric with Red-billed Chough *P. pyrrhocorax* in Moroccan Atlas Mts where the 2 sometimes occur in mixed flocks; told from it by bill and leg colour; bill also shorter and less decurved, and wings not quite so broad and with only 4 'fingers' at tip (5–6 in Red-billed). Adult and particularly juvenile difficult to distinguish from juvenile Red-billed Chough. The 2 species not infrequently hybridize.

Voice. Tape-recorded (62, 73, 93, 105, B). Song a squeaking, chittering and churring warble. Calls a thin, musical whistle, 'chirrish', 'keerreesh' 'trree', 'sree', 'kee' or 'skweea' and a clear, sharp 'tsi-eh' or 'dschirr' (becoming a metallic whistling when given by several birds at once); also a rolled 'krrrree' and various other calls (Cramp and Perrins 1994). Threat a short 'tchiupp' and bill-snapping.

General Habits. More exclusively high-altitude than Red-billed Chough, inhabiting mountains mainly between treeline and snowline, at 2200–3900 m in Morocco (and up to 5000 m in Kashmir and 8100 m in Himalayas).

Habits known mainly from Europe. Makes substantial daily movements, in Switzerland of up to 20 km horizontally and 1600 m vertically, roosting high up at night and coming down to forage at lower levels by day, favouring alpine meadows, pasture, rocky short grass, boulder slopes, and vicinity of human habitation and tourist facilities: ski lifts, hotels, crofts and stores.

Gregarious. Forages on ground singly, in pairs or flocks, moving with shuffling walk and occasional hop. Becomes very tame, scavenging around houses and winter-sports centres. Probes crevices in rocks, walls and buildings; searches short vegetation; forages along receding snowline, opportunistically takes insects blown onto glaciers; overturns stones and dung; catches scraps in air thrown by person and known to catch insects in flight; flies into wind so that ground speed is low, and dips down to rock surface or mown meadow to take insect without alighting; occasionally feeds on berries by clinging to shrub with much flapping; takes birds' eggs and nestlings and small rodents; holds large prey under feet, with tarsus and heel on ground. Sometimes caches food items in winter. Birds in ground-foraging flock often flick wing tips up (a flight-intention behaviour) and sometimes bow. Highly vocal, all year and especially in flying flock. Flight easy and graceful; aerobatic, sailing, side-slipping, diving steeply then curving upward and riding the wind along cliff faces, soaring and coping readily with turbulent conditions over broken, barren terrain. Flock often soars around on thermal at great height. Several birds dash down, check in flight near foraging grounds, then dash about fast before alighting. Very mobile, spending much time on wing and moving considerable distances every day. One bird reported to fly 8·3 km down from roost to feeding grounds in 2·5 min, i.e. at 199 km per h (Büchel 1983). Can be quite aggressive, one bird threatening with bill-thrusting, wing-spreading and head-up postures before supplanting another; subordinate to Red-billed Chough when foraging together; evidence of a pecking-order or social ranking in a flock. Roosts communally in caves and holes and crevices in crags.

Food. In Europe mainly insects in summer, grain at harvest time and berries and scraps from around human habitation in winter. Many grasshoppers, leatherjackets (tipulid larvae) and beetles; also dragonflies, crickets, caterpillars, flies, hymenopterans, spiders, small snails, amphibians, reptiles, birds' eggs and nestlings, rodents and carrion. Plant food includes fruits and seeds of pine, juniper, rowan, olive, holly, barberry, hackberry, mulberry, grape, orange, *Vaccinium*, grasses and cereals (Cramp and Perrins 1994).

Breeding Habits. Following details mainly from Europe: monogamous; some pairs may have helpers. Not obviously territorial; in Switzerland closest nests 70 m apart; said to be loosely colonial in parts of range but confirmation needed. Pair bond maintained all year and many pairs stay together for several years, up to 8 (Büchel 1983). Up to 8 birds indulge in group-display, perching close together and posturing with head raised, bill pointing up, and plumage loose and ruffled – may function in pairing. ♂

occasionally feeds ♀ at any season and often when breeding, ♀ begging with fluttering wings, crouching, ♂ regurgitating food from throat. Copulation mainly on ground, but is suspected to take place sometimes in the air (Cramp and Perrins 1994).

NEST: bulky mass of sticks, twigs, stems, roots and moss, with neat, compact lining made of fine twigs, grasses, hair and some feathers. Size (n = 7, Alps): ext. diam. 28–50 × 30–60 (34·6 × 42·8) cm, int. diam. 12–15 (13·5) cm, ext. height 12–30 (18) cm, int. depth 50–70 (62·5) mm. Sited in crevice or on ledge in cliff, cave or sometimes in building (usually on wide beam just under roof). ♂ brings material for nest-building, which is evidently done by ♀.

EGGS: 3–5; laid at 1–2 day intervals; subelliptical, smooth, glossy; buff or very pale greenish white, with pale lilac undermarkings and pale brown to dark grey-brown blotches, either scattered evenly and rather sparsely over whole surface or concentrated at large pole (when equator to small pole can be almost immaculate). SIZE: (n = 111) 34·2–42·3 × 23·0–28·5 (38·2 × 26·8). Weight: c. 14·1.

LAYING DATES: Morocco, (nests being built Mar, Apr, May; young in early June (at 2000 m) and early July (at 2500 m)).

INCUBATION: eggshells carried away by parents. Incubation by ♀, fed on nest by ♂. Period: 18–21 days. Eggs hatch asynchronously.

DEVELOPMENT AND CARE OF YOUNG: at first young brooded by ♀, for 80% of time in first 3 days; ♂ brings food, regurgitates it, and ♀ passes items to young; later, young fed directly by both ♀ and ♂. Faecal sacs removed by ♀. Young out of the nest beg from parents and are fed by them.

BREEDING SUCCESS/SURVIVAL: success rate rather higher than in Red-billed Chough, *q.v.* (Cramp and Perrins 1994).1 ringed bird 11 years old.

References
Büchel, H.P. (1983).
Cramp, S. and Perrins, C.M. (1994).

Family STURNIDAE: starlings

A large and diverse Old World family, mostly tropical, of small to medium-sized birds (16–45 cm), robustly built with hard thrush-like, crow-like or pointed bill, quite long wings, strong legs and feet, usually short but sometimes long and graduated tail; in most species plumage vividly iridescent in glossy dark blues, greens and violets; other species brown, grey, pink, pied or snowy white. Bill and legs black or yellow; eyes blackish, crimson, red, orange, yellow or white; often coloured, bare skin around eye; some species with wattles, seasonal lappets, erectile or permanent crests, or forehead cushion of dense feathers. Sexes alike or nearly so; juveniles usually rather different from adults.

Arboreal, feeding on small fruits and (when breeding) insects, or markedly terrestrial and insectivorous. Gait on ground a walk. Flight swift and direct. Mostly resident; frugivores move locally in search of fruiting trees; some species strongly nomadic (e.g. Wattled Starling *Creatophora cinerea*), others migratory. Nearly all species markedly gregarious, feeding in flocks, flying and roosting communally. Most nest in holes in trees and often in buildings; some dig burrows in earthen banks; some build spherical or pear-shaped dry-grass or thorn-twig nests; others make pendulous nests like a weaver's (Oriental *Aplonis*) or bore holes into dead trees (Oriental *Scissirostrum*). Monogamous; breed more or less colonially and a few species co-operatively; egg-dumping and other forms of sexual 'cheating' commonplace in some well-studied species. Eggs pale bluish green with small brown speckles.

Placed by Sibley and Monroe (1990) in superfamily Muscicapoidea, comprising 4 families: Bombycillidae (waxwings), Cinclidae (dippers) Muscicapidae (thrushes, Old World flycatchers and chats) and Sturnidae (tribes Sturnini, starlings, and Mimini, thrashers). Sturnids in Africa seem to be of 2 distinct clades: (a) endemic glossy starlings, of at least 7 genera; and (b) 'true' starlings, feeding on soil insects by 'open-bill probing' or 'prying' and having immaculate bluish eggs. 'True' starlings are of Asiatic origin (*Sturnus*, *Acridotheres*); the African endemic *Creatophora* certainly belongs there and *Neocichla* and *Zavattariornis* may do so.

Of the endemic glossy starlings, there is broad agreement about the composition of *Poeoptera* and *Onychognathus*, which appear to be rather closely allied. Other generic boundaries remain controversial: White (1962), Amadon (1962), Hall and Moreau (1970), Sibley and Monroe (1990), Dowsett and Forbes-Watson (1993) and Craig (1997) all offer different schemes. Morphological and biological differences between '*Coccycolius*' (*iris*), '*Hylopsar*' (*purpureiceps*) and *Lamprotornis* are no greater than within *Onychognathus*, so we place the 2 in *Lamprotornis*. A brilliant starling, *regius*, has generally been united with the dull grey *unicolor* in *Cosmopsarus*, both being long-tailed tree-hole nesters, but Craig (1997) and Feare and Craig (1998) place the former in *Lamprotornis* and the latter, with *S. fischeri*, *albicapillus* and *bicolor*, in *Spreo*. Craig and Hartley (1985) and others have remarked on similarities between *regius* and the 4 orange-bellied but short-tailed '*Spreo*' (or '*Lamprospreo*') starlings; the '*S.*' *shelleyi*/*hildebrandti* superspecies, being tree-hole nesters, may be closer to *regius* than are the obligate or facultative ball-nesters 'S.' *superbus* and 'S.' *pulcher*. Another glossless grey starling, 'S.' *fischeri*, a short-tailed ball-nester, seems to us closer to *pulcher* than to the remaining 2 *Spreo* starlings – the large, pied, ball-nesting *albicapillus* and the large, blackish, burrow-and-eaves-nesting *bicolor*. *Lamprotornis* (sens. str.), *regius, shelleyi/hildebrandti, superbus, pulcher* and *fischeri*, in that sequence, form a graded series. Being far from certain if and where to apply generic limits in the series, we place all of these species in *Lamprotornis*, an arrangement like that of Craig (1997) and Feare and Craig (1998) although not identical with it. It creates a large and diverse genus, hard to diagnose yet (we believe) monophyletic and natural. We add *unicolor*, because of supposed affinities with *regius* and *fischeri*. It leaves only *albicapillus* and *bicolor* in the much-reduced genus *Spreo*. The remaining endemic glossy starlings are the monotypic *Speculipastor, Grafisia* and *Cinnyricinclus*, and *Pholia* with 2 species.

Some authorities sequence African genera from *Poeoptera* (supposedly the most primitive) to *Creatophora/Sturnus/ Acridotheres* (supposedly the most advanced or derived). However, Craig's cladistic tree (1997), built on character comparisons with *Sturnus* as the outgroup and lacking feather ultrastructure data for some species and skeletal data for most (A.J.F.K. Craig, pers. comm., 1998), gives derivations almost in the reverse sequence. We place the 13 genera recognized here in a modified *Poeoptera*-to-*Acridotheres* sequence.

115 species (Sibley and Monroe 1990), including one extremely endangered (*Leucopsar rothschildi* of Bali) and others extinct (*Necropsar rodericanus* and *Fregilupus varius* of Mascarene Is). 114 species (Feare and Craig 1998), including the 2 spp. of *Buphagus* which we treat as a distinct family, and excluding *Zavattariornis stresemanni* (usually thought of as a crow; we treat it as a starling). 50 in Africa, in 13 (or 12–18) genera, all endemic except *Acridotheres* (introduced) and *Sturnus*.

Genus *Poeoptera* Bonaparte

Small, long-tailed, glossy black forest-canopy starlings; bill short, moderately robust; iris grey or yellow, legs and feet rather weak; tail graduated; females with chestnut in primaries. Eat small fruits, some insects. Resident, wandering locally. Nest in hole high in tree, often in colony of *Gymnbucco* barbets. Sometimes placed in genus *Onychognathus*, but the 2 genera may not be very closely allied (Craig 1997).

Endemic. 3 species, 2 composing a superspecies.

Poeoptera stuhlmanni superspecies

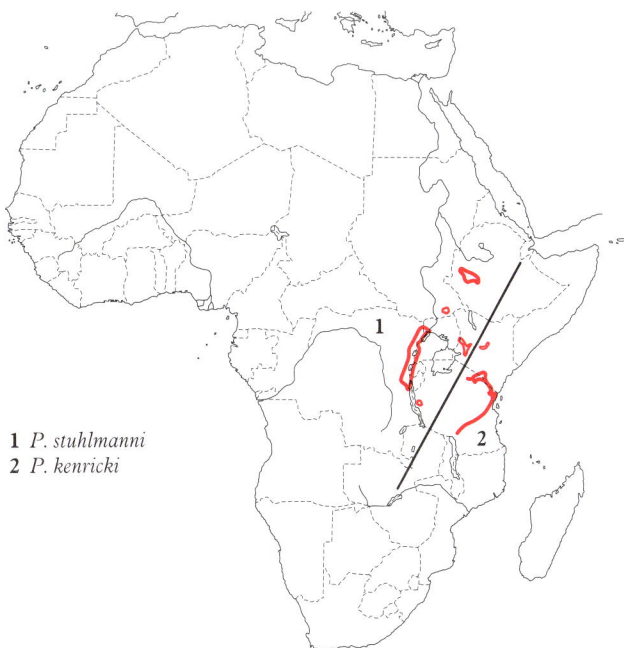

1 *P. stuhlmanni*
2 *P. kenricki*

Plate 32
(Opp. p. 479)

Poeoptera lugubris Bonaparte. Narrow-tailed Starling. Rufipenne à queue étroite.

Poeoptera lugubris Bonaparte, 1854. Compt. Rend., 38, p. 381; no locality, = Gabon (Hartlaub 1857, Syst. Orn., Westafr., p. 69).

Range and Status. Endemic resident. Sierra Leone, old record from Banda Kerafaya; rare in Gola Forest (where found only in cocoa and coffee plantations). Liberia, locally not uncommon near N and E borders (20 localities: Gatter 1997). Ivory Coast, from San Pedro to Ayame, inland to Man, Lamto, Yapo Forest and Abengourou; flock of c. 70, Bossematié Forest (Walter *et al.* 1999); uncommon in Täi Nat. Park (Gartshore *et al.* 1995). Ghana, rare: 2 old records from 'Fanti'; more recently in Subri River Forest Res. (where sporadic) and Dixcove. Togo, single record: 5 birds, Tomégbé (Cheke and Walsh 1996). Not found in Benin. Nigeria, uncommon in remnant forest from Ipake to Ife and Ado-Ekiti; in E known from Okwango and Oban. Cameroon, widespread, not uncommon. Present Bioko (Pérez del Val *et al.* 1994) and Mbini. Gabon, common and widespread, at least in parts of N; in numerous colonies of barbets *Gymnobucco* spp. (see below) between Makokou and Mékambo (Brosset and Erard 1986). Congo, Odzala Nat. Park, locally fairly common in W; also in S Congo. NW Angola: Cabinda, Cuanza Norte, along Cuango R., Carmona (Uige), Canzele and Ndala Tando. Central African Republic, single record, La Maboké (Germain and Cornet 1994). Zaïre, around Thysville in W; frequent and widespread in Kasai and Manyema and in NE from lower Uele R. to Semliki, W Rwenzori, forested foothills around L. Edward, Kivu, and south to Kamituga, where common. Uganda, in W, erratic, at 700–1700 m, often common: Bwindi-Impenetrable, Itwara, Maramagambo, Kalinzu, Kibale and Bwamba forests and Fort Portal (Britton 1980). Not in Rwanda or Burundi.

Description. ADULT ♂: entire head and body glossy violet-black. Tail long and strongly graduated, feathers narrow and pointed; black above with bluish gloss, brownish black below. Primaries and primary coverts dark grey-brown; secondaries, greater coverts and outer edges of tertials pale brown; tips of flight feathers, inner webs of tertials and primary coverts with faint violet gloss; lesser and median coverts glossy violet-black. Axillaries and small underwing coverts along carpal edge glossy violet-black; rest of underwing-coverts sooty brown. Bill black; eyes bright cadmium yellow; legs black. ADULT ♀: head, neck and underparts dusky grey, glossed bluish on crown to hindneck, and on sides of neck, breast and flanks, less strongly so on sides of head, throat and belly. Mantle to scapulars and uppertail-coverts glossy greenish blue with dark brown feather centres showing through. Tail as in adult ♂, but glossed more greenish. Primaries dark grey-brown; marked with chestnut on basal three quarters of inner webs and bases of outer webs of P1–P7; basal two thirds of inner web of P8; and basal two thirds of inner border of P9. Wings otherwise as adult ♂, but secondaries and greater coverts less pale and contrasting, lesser and median covert tips glossed greenish blue. Axillaries and underwing-coverts grey-brown. Bare parts as in adult ♂, but eye not so rich a yellow. SIZE (10 ♂♂, 10 ♀♀): wing, ♂ 88–93 (90·5), ♀ 83–89 (85·6); tail, ♂ 102–118 (111), ♀ 82–108 (95·6); bill, ♂ 18–20 (18·9), ♀ 17–19 (17·8); tarsus, ♂ 19–21 (20·0), ♀ 17·5–19·5 (18·7). WEIGHT: Liberia, ♂ (n = 4) 37·6–42·2 (40·8), ♀ (n = 3) 35·4–39 (37·5); also, ♂ (n = 13) 35–43, ♀ (n = 5) 35–39 (Feare and Craig 1998).

Poeoptera lugubris

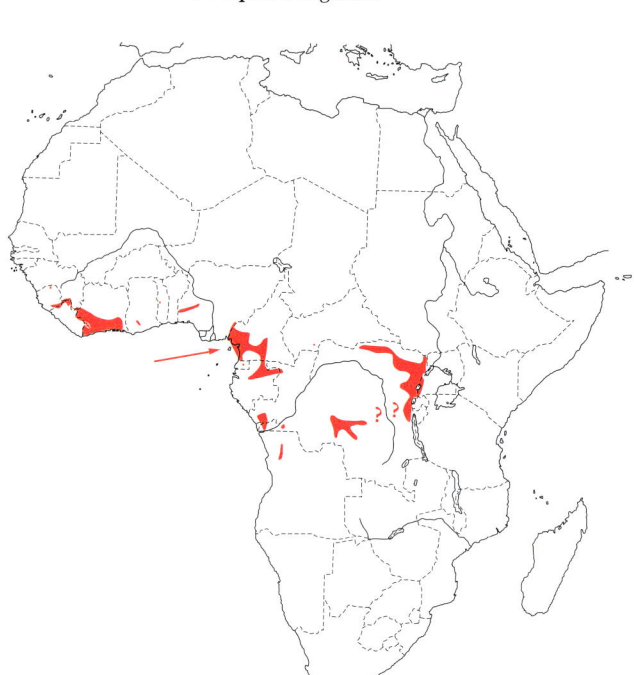

IMMATURE: juvenile like adult ♀, but gloss less intense, more greenish above, and confined below to trace on throat and breast; tail shorter.

NESTLING: not described.

TAXONOMIC NOTE: *P. l. 'major'* Neumann, E Zaïre, said to be larger, though that disclaimed by Chapin (1954); *'webbi'* Keith, W Uganda, not recognized by Friedmann and Williams (1970) or Feare and Craig (1998).

Field Characters. Length 20–24 cm A slim starling of forest treetops. In Albertine Rift occurs with Stuhlmann's Starling *P. stuhlmanni*; distinguished by *long, narrow tail*, ♂ also by rich, deep *purple body gloss* contrasting with *pale brown wing*. ♀ glossed blue-green above, brown below with slight gloss, small chestnut patch on wing.

Voice. Tape-recorded (104, B, C, BAR, CHA, KAL, KEI). Calls (contact?), given at intervals, long, pure, down-slurred whistles, 'weeeeooo' or sharper 'pyeeooow'; in between calls a soft, low grating 'zh-zh'. In group chorus, short, sweet high-pitched whistles, almost bell-like. In flight said to give a 'confused, shrill cheeping chorus' (Chapin 1954). Adults at nest give soft chirping notes.

General Habits. Inhabits canopy of mature lowland forest; in Liberia mainly at 40–50 m above ground and down to 30 m (n = 13 observations, Gatter 1997); also forest edges and clearings, and cocoa and coffee plantations. In Gabon mainly in secondary forest, plantations, and cultivated patches in forest clearings with plenty of trees left scattered about; in tall, primary forest occurs in windfall clearings.

Lives mainly in flocks, of 2–12 birds (Congo), 4–22 (Ivory Coast) or 10–30 (Zaïre), with up to 37 birds (Liberia) and several dozens (Gabon, where flocks commonest in Apr–Sept), usually seen flying fast above forest or perched at top of tall, dead tree.

Regularly chases flying termites and ants (Brosset and Erard 1986); once seen hawking insects from tree top, in company of Grey-throated Barbets *G. bonapartei* and White-headed Wood-Hoopoes *Phoeniculus bollei* (Baranga and Kalina 1991). Occasionally joins mixed-species foraging flocks (Prigogine 1971) and regularly forages with Grey-throated Barbets (Baranga and Kalina 1991). Roosts gregariously, e.g. in tree near village, Medje, NE Zaïre Jan. Flight swift, direct and purposeful, with rapidly beating wings, like small parakeet; flying flock keeps in close formation; when swooping into tree, bird fans its tail wide, each feather showing separately, like fingers (Bannerman 1948, Serle 1954). Perches in sun to preen; once came to ground to bathe in small stream (Baranga and Kalina 1991).

Food. Fruits of *Musanga*, *Rauvolfia*, *Macaranga*, arils of *Pycnanthus*; doubtless termites and ants (see above) (Brosset and Erard 1986); a large fly (Baranga and Kalina 1991); fruits and insects (Gatter 1997); only fruits found in stomachs (Liberia, Zaïre).

Breeding Habits. Nests colonially: 8–18 birds (Liberia: Gatter 1997) or 4–5 pairs (Angola), in colonies of hole-nesting Bristle-nosed Barbets *Gymnobucco peli*, Naked-faced Barbets *G. calvus* and Grey-throated Barbets in tops of dead emergent forest trees; also with *Gymnobucco* barbets in Ivory Coast, Gabon, Zaïre, Uganda and doubtless throughout starling's range. No nest hole yet found away from *Gymnobucco* nests. The nest itself has not been described, and the only eggs described were in a tree felled for the purpose (Bates 1927). At first, barbets try to drive starlings away; later in season starlings are in undisputed possession of some holes and barbets of others, the 2 species evidently living unaggressively and without apparent interactions, though barbet and starling nest holes can be only 20 cm apart and nestlings of the 2 species can see each other behind their nest entrances (Baranga and Kalina 1991). Starling flock alights on branches of adjacent trees, then pairs visit barbet holes, one bird clinging to trunk and inserting head into nest hole, the other clinging to trunk next to its mate (Bannerman 1948).

NEST: in old tree hole (often in rotten dead thick vertical limb of *Musanga* tree) made by colonial *Gymnobucco* barbets. Holes appropriated by starlings 5–8 m above ground in open middle forest stratum, and nest entrances lightly shaded; nest entrance diam. *c.* 50 mm; hole often faces southwest; In Kibale Forest, Uganda, starling/barbet mixed colonies sometimes in tree with nesting Black-and-white-casqued Hornbills *Bycanistes subcylindricus* (Baranga and Kalina 1991).

EGGS: 3 (n = 1 clutch). Somewhat glossy, pale blue-grey, sparsely and finely spotted with brown, mainly toward large end. SIZE (n = 1): *c.* 22 × 16.

LAYING DATES: Liberia, (evidently breeding, Dec–Jan); Nigeria, (nest Jan; exploring nest holes Aug); Cameroon, Jan; Gabon, (evidently breeding, Dec–Mar); Angola, (young outside nest holes, Mar); Zaïre, (Thysville), (breeding condition, Dec); (Itombwe), (various evidences Mar and June–July: Prigogine 1984); Uganda, (nestlings Mar).

INCUBATION: nesting ♀♀ have crimped tails, ♂♂ do not, suggesting only ♀♀ incubate.

DEVELOPMENT AND CARE OF YOUNG: nestlings fed by both parents, once at rate of 8 nest visits per h. Young make 'airy, rhythmic squeaking' begging calls (Baranga and Kalina 1991). Visiting adults not often seen carrying food item in bill, so they probably feed young by regurgitation. (♂ and ♀ seen exchanging food item back and forth near nest, partly by regurgitation.) Arriving adults typically perch 5 m from nest, give soft chirping note, then after 5 s fly straight to it and enter without perceptibly alighting at entrance; they feed chicks out of sight inside nest.

References
Bannerman, D.A. (1948).
Baranga, J. and Kalina, J. (1991).
Brosset, A. and Erard, C. (1986).
Serle, W. (1940).

Poeoptera stuhlmanni Stuhlmann's Starling. Rufipenne de Stuhlmann.

Plate 32
(Opp. p. 479)

Stilbopsar stuhlmanni Reichenow, 1893. Orn. Monatsb., 11, p. 31; Badjua, L. Albert.

Forms a superspecies with *P. kenricki*.

Range and Status. Endemic resident. Ethiopia, uncommon in W Highlands at 2520–2830 m. Sudan, Imatong Mts, common above 1900 m. Kenya, locally common at 1500–2600 m: Mt Elgon, Elgeyu, Tugen Hills, Nandi, Kakamega Forest and Kericho. Uganda, in Impenetrable, Maramagambo, Kalinzu and Kibale Forests. Tanzania, on Mt Mahari. Zaïre, from Lendu Highlands (west of L. Albert) through Mitumba Mts at 1540–2030 m to Itombwe Plateau at 1210–2000 m, where common. W Rwanda, throughout Nyungwe Forest, up to 2500 m. W Burundi in Rwegura, Teza, Ijenda and Bururi Forests.

Description. ADULT ♂: head and body glossy black; head, neck, mantle and chin to upper breast with strong violet-blue sheen; scapulars, back to uppertail-coverts, and lower breast to undertail-coverts more violet. Tail graduated, black, with upperside tinged violet. Lesser coverts glossy violet-black; rest of upperwing brownish black, secondaries, tertials, alula and larger wing-coverts narrowly edged, and median coverts tipped, glossy violet-

Poeoptera stuhlmanni

black. Bill black, eyes brown with yellowish outer ring; legs black. ADULT ♀: head, neck and underparts dusky grey, top of head and neck glossed blue-green, breast and flanks with fainter blue-green sheen. Mantle to scapulars and uppertail-coverts brownish black, glossed bluish. Tail brownish black. Primaries dark brown, marked with pale rufous-chestnut as follows: basal two-thirds of inner webs of P1–P7, extending across shaft near bases; basal third of inner web of P8, extending as wedge to half way along inner edge; base of inner border of P9. Rest of upperwing brownish black, median and lesser coverts tipped glossy violet-black; tertials and greater coverts narrowly edged violet-black. Axillaries and underwing-coverts greyish brown. Bare parts as adult ♂. SIZE (10 ♂♂, 8 ♀♀): wing, ♂ 101–107 (104), ♀ 95–102 (98·6); tail, ♂ 76–87 (82·4), ♀ 71–80 (83·0); bill, ♂ 18·5–20 (19·2), ♀ 17–19·5 (18·5); tarsus, ♂ 20–22 (21·3), ♀ 19·5–21 (20·3). WEIGHT: ♂ (n = 16) 38–46, ♀ (n = 4) 35–39 (Feare and Craig 1998).

IMMATURE: juvenile like adult ♀, but browner above with less bluish gloss; grey-brown below with trace of greenish gloss on breast.

Field Characters. Length 18·5–19 cm. A slim starling of highland forest. Does not overlap with similar Kenrick's Starling *P. kenricki*; told from Narrow-tailed Starling *P. lugubris*, which it meets at middle elevations of Albertine Rift mts, by much *shorter tail*, ♂ also by *blue-green* gloss and *uniform appearance* (no pale wing); ♀ darker and glossier below than ♀ Narrow-tailed, bluer above; both have similar chestnut wing-patch. Immature duller, with chestnut in wings in both sexes.

Voice. Tape-recorded (104, B, C, CHA, GREG, HOR, KEI, LEM, ZIM). Common contact call from treetop, up-slurred liquid 'toolee', 'twee', or rolled 'prrr' or 'prrr-leet'; also down-slurred 'tyewp' or more drawn-out 'pu-way-o' or 'pu-wee-yu'. Notes pure but lack musical quality of some other starlings. Whistles and chuckles strung together to form loose song, 'creep-creep-turreep-chlerreep-tleeo'; for further renditions see Zimmerman *et al.* (1996). Rolling flight call not unlike European Bee-eater *Merops apiaster*.

General Habits. Inhabits secondary and also primary montane forest, in pairs and flocks of 6–8 birds (once over 15) keeping in canopy or at mid levels in tall trees. Flock sometimes perches on top of dead emergent tree. Often gathers in large flocks to eat fruits (including yellow ones) in certain trees, but does not join mixed-species foraging flocks (Prigogine 1971). Nomadic, flocks wandering widely in search of fruiting trees, birds flying high and often perching on dead branches emerging from top of forest (Zimmerman *et al.* 1996).

Food. Small fruits, including *Maesa lanceolata*, *Polyscias fulva*, *Ilex mitis*, *Macaranga neomildbreadiana*, *Alangium chinense*, *Schefflera goetzenii*, *Olea*, *Sapium*, *Trema* and *Urera* (Dowsett-Lemaire 1990, Moermond *et al.* 1993).

Breeding Habits. Nests in holes high in trees, but nest, eggs and biology not studied (Brown 1975). Once seen prospecting an old hole of Bearded Woodpeckers *Thripias namaquus* in a tall dead tree (Ethiopia). 2 pairs present near 5 old woodpecker holes *c.* 10 m high in jacaranda tree, Kericho, Kenya; 1 pair kept on going in and out of a hole and chased away a honeyguide, and ♂ of other pair repeatedly fluttered wings, sidled up to ♀ and offered objects carried in bill (Cunningham-van Someren 1975).

LAYING DATES: Ethiopia, (Mar, Apr); Sudan, (Apr); E Africa, Region A, Dec, Region B, Oct, Feb–Mar; Zaïre (Itombwe), June (and testes enlarged May, nestling July).

Plate 32 *Poeoptera kenricki* Shelley. **Kenrick's Starling. Rufipenne de Kenrick.**

(Opp. p. 479) *Paeoptera [sic] kenricki* Shelley, 1894. Bull. Br. Orn. Club, 3, p. 42; Usambara Mts, Tanganyika.

Forms a superspecies with *P. stuhlmanni*.

Range and Status. Endemic resident and vertical migrant, E African mountains, in Kenya at 1500–2500 m on Mt Kenya (at times common around Irangi and Castle Forest near Embu), Meru Forest and Nyambeni Hills, and in Tanzania at 900–2500 m in Arusha Nat. Park, Oldeani, Mt Meru, Kilimanjaro, North Pare Mts, L. Jipe, East and West Usambaras, Ulugurus, Udzungwa and Iringa Highlands and Mdandu Forest, Njombe. (Tanzania map based on 55 plotted localities: N.E. Baker, pers. comm.). Generally common.

Poeoptera kenricki

Description. *P. k. kenricki* Shelley: Tanzania. ADULT ♂: head and body glossy black, with bronzy brown tinge; gloss less pronounced from lower breast to undertail-coverts. Tail graduated, brownish black. Lesser and median coverts glossy black; rest of upperwing brownish black. Axillaries and underwing-coverts sooty black. Bill black; eyes slate or yellow; feet black. ADULT ♀: head, neck and underparts dusky grey, tinged blue-grey on head, breast and flanks; lores blackish; upperparts brownish black, mantle to scapulars slightly glossed, rump and uppertail-coverts tipped greyish; tail brownish black. Primaries dark brown, marked with rufous-chestnut as follows: basal two thirds of inner webs of P1–P7, extending near shaft onto base of outer webs; basal third of inner web of P8, extending as wedge to half way along inner edge; and basal third of inner border of P9. Rest of upperwing as in adult ♂, but lesser and median coverts less glossy. Axillaries and underwing-coverts greyish brown. Bare parts as in adult ♂. SIZE (6 ♂♂, 2 ♀♀): wing, ♂ 99–105 (104), ♀ 99, 100; tail, ♂ 77 82 (80·0), ♀ 76, 76; bill, ♂ 19–20 (19·2), ♀ 19·5, 20; tarsus, ♂ 20·5–22 (21·6), ♀ 20, 20·5. WEIGHT: ♂ (n = 4) 46–54 ♀ (n = 5) 38 (Feare and Craig 1998).

IMMATURE: juvenile like adult ♀, but duller, more sooty on body.

P. k. bensoni van Someren: Kenya. Larger; wing, ♂ (n = 4) 105–111 (108), 1 ♀ 105.

Field Characters. Length 18·5–19 cm. A forest starling of E African highlands. Appears matt black except in good light, when *dull bronze-brown gloss* visible; ♀ dark grey below. Often flocks with Waller's Starling *Onychognathus walleri* in treetops, where it looks smaller, slimmer and narrow tailed, without purple and green gloss; no red shows in wing at rest, and in flight ♀ (only) shows small red patch (much red in wing of Waller's).

Voice. Tape-recorded (B, KEI, STJ). Call a repeated loud, monotonous 'pleep-pleep'; flocks produce a musical babbling (Zimmerman *et al.* 1996).

General Habits. Keeps to forest canopy and tops of tall trees in open country next to forest edges. Exclusively arboreal. In pairs and flocks, which wander widely. Often in company with other tree-top starling species, particularly Waller's Starling *Onychognathus walleri*, in mixed-bird parties (Turner 1977). 'Pleep' calls become a pleasing babble from a large flock.

Moves down as low as 450 m above sea level in East Usambaras (Zimmerman *et al.* 1996).

Food. Fruits, including *Ficus natalensis* and *Trema*.

Breeding Habits. Nests in old barbet and woodpecker holes high in trees, but nest, eggs and biology not studied (Turner 1977, Feare and Craig 1998). Once, in flock of 7, ♂♂ chased ♀♀ then flew to nest holes, clinging under entrances. Pair once nested in same tree as pair of Waller's Starlings *Onychognathus walleri*.

LAYING DATES: E Africa, Region D, Jan, Mar–May, June–July, Sept–Nov; mainly Oct.

Reference
Feare, C. and Craig, A. (1998).

Genus *Onychognathus* Hartlaub

Quite large starlings, attenuated, with long neck, slim body and square-ended or (several species) long, graduated tail. ♂ black with bluish, violaceous or greenish gloss, ♀ with streaky grey head; primaries largely non-iridescent chestnut or pale orange. Eyes dark red. A few spp. with tail not elongated, or with sexes alike; 2 spp. with cushion of velvet-like feathers on forehead. Petrophilous, montane and arboreal; eat mainly small fruits, also insects. Gregarious. Cup or platform nest in cavity; semi-colonial. Main voice a musical whistle.

Near-endemic to Africa; 11 species, 10 in Africa and 1 in Sinai, Israel, Jordan and W and S Arabia.

Plate 32
(Opp. p. 479)

Onychognathus walleri (Shelley). Waller's Starling. Rufipenne de Waller.

Amydrus walleri Shelley, 1880. Ibis, p. 335; Usambara Mts, Tanganyika.

Range and Status. Endemic resident. Nigeria, common on Obudu Plateau. Cameroon, common to abundant on: S slopes of Mt Cameroon at 1000–2000 m, with records up to 2300 m and down to 950 m and once only 30 m; Mt Kupé at 1200–2000 m; Mt Manenguba at 1700–2250 m; Mt Nlonako at 1250–1750 m; Rumpi Hills (Dikome Balue) at 1000–1700 m (abundance increasing with altitude); Mt Oku at 2000–2800 m; and Bamenda highlands at 1700–2000 m; also known from Mt Lefo and Kumbo (Banso, 06°12′N) (Stuart 1986). Bioko, widespread (Banterbari, Pico Joaquim, Moka, Apu R.). Mbini. Zaïre, Rift Valley mountains from west of L. Albert to Itombwe Highlands where common at 1100–2070 m. Sudan, common in Imatong and Dongotona Mts above 1800 m. Rwanda, Nyungwe Forest, up to 2500 or 2600 m. Burundi, Teza, Ijenda and Mt Bururi (Vande weghe and Loiselle 1987). Uganda, locally common at 900–3000 m in SW, wandering widely, in Kibale, Ankole and Kigezi; also on Mt Elgon. Kenya, local from Mt Elgon and Saiwa Nat. Park to Mau, Trans-Mara and Nguruman Forests; east of Rift locally common at 1600–3000 m from Karissia Hills, Mt Marsabit and Mathews Range to Aberdares, Mt Kenya, Meru Forest and Nyambenis (Zimmerman *et al.* 1996). Tanzania, widespread in NE and E Arc Mts as mapped (N.E. Baker, pers. comm.): W Serengeti, Mt Meru, Mt Kilimanjaro, Arusha Nat. Park, Pare Mts, Usambaras (with some altitudinal movement down to 300 m: Zimmerman *et al.* 1996), Udzungwa, Iringa highlands and Mbeya area. Zambia, Nyika Plateau at 1970–2200 m (R. J. Dowsett, pers. comm.). Malaŵi, Misuku Hills, Nyika and N Viphya Plateaux at 1600–2220 m (F. Dowsett-Lemaire, pers. comm.).

Density in Chowo forest, Zambia, 6 pairs in 90 ha; distribution on Nyika Plateau, Malaŵi, uneven, but overall 12 breeding pairs in 15 km² including 160 ha of forest patches; one forest patch of 10·5 ha without any pairs (Dowsett-Lemaire 1983b).

Description. *O. w. walleri* (Shelley): Kenya east of Rift, Tanzania, N Malaŵi. ADULT ♂: head and body glossy black; head to neck and throat with strong blue-green sheen; upperparts with violet sheen, and coppery green reflections on lower mantle, back and scapulars; breast to undertail-coverts glossed violet. Tail black, slightly graduated, upperside with faint greenish gloss. Primaries dark chestnut on basal two-thirds of P1–P8 and on basal half of inner web of P9; otherwise dark brown. Rest of upperwing black, edges of secondaries, tertials, greater coverts, primary coverts and alula glossed greenish, median and lesser coverts glossed violet. Axillaries and underwing-coverts black with glossy violet fringes. Bill black; eyes crimson or red-brown; legs black. ADULT ♀: head, neck, and chin to upper breast almost uniform grey, with faint bluish sheen; lores dark grey; glossy dark bluish green streaks from hindcrown to hindneck and on sides of neck; mantle to uppertail-coverts as in adult ♂. Breast to undertail-coverts grey-brown with violet sheen, strongest on sides of breast and flanks. Tail and upperwing as in adult ♂. Axillaries and underwing-coverts dark grey-brown. Bare parts as in adult ♂. SIZE (10 ♂♂, 10 ♀♀): wing, ♂ 132–140 (136), ♀ 126–134 (130); tail, ♂ 84–98

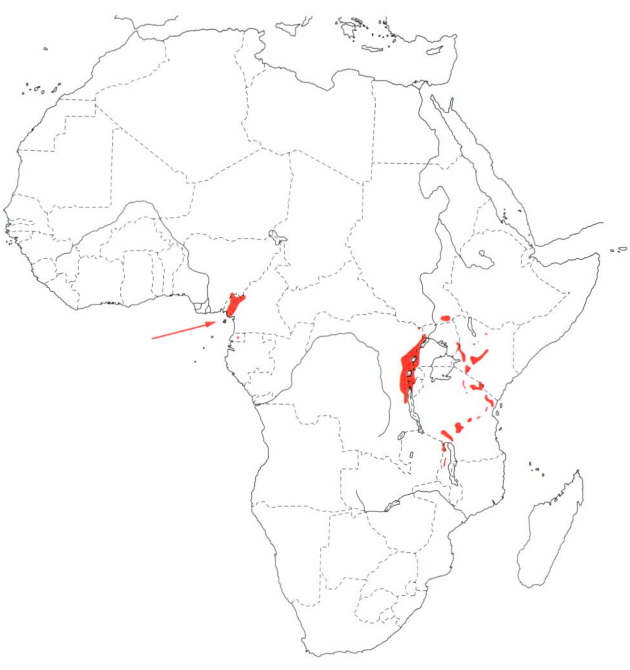

Onychognathus walleri

(92·9), ♀ 78–96 (88·0); bill, ♂ 24–27 (25·2), ♀ 22–24 (23·5); tarsus, ♂ 26–28 (27·2), ♀ 25–27 (25·8). WEIGHT: ♂ (n = 8) 73–92, 1 ♀ 93.

IMMATURE: juvenile similar to adult ♂ but plumage less lustrous, more sooty brown on head and underparts.

NESTLING: not described.

O. w. elgonensis (Sharpe): E Zaïre and adjacent Uganda, SE Sudan and Kenya west of rift. ♀ differs from nominate race in having grey more restricted on head and throat; crown to nape and lower throat with heavier dark bluish green streaks; hindneck, sides of neck and upper breast glossy violet-black. Slightly smaller: wing, ♂ (n = 4) 124–133 (127), ♀ (n = 6) 117–127 (121).

O. w. preussi Reichenow: Nigeria, Cameroon, Bioko and Mbini. Like *elgonensis* but smaller: wing, ♂ (n = 10) 113–121 (117), ♀ (n = 5) 105–112; tail, ♂ (n = 8) 86–94, ♀ (n = 5) 72–83; bill, ♂ (n = 8) 20–22, ♀ (n = 5) 18–20; tarsus, ♂ (n = 8) 24–25, ♀ (n = 5) 23–25 (Bannerman 1948).

Field Characters. Length 22·5–23 cm. A common montane forest starling with *barely-graduated, medium-length tail*. Flocks with Stuhlmann's and Kenrick's Starlings *Poeoptera stuhlmanni* and *P. kenricki*; bulkier, with green and purple reflections visible at close range, ♀ with grey head; shows much *chestnut* in wings in flight, some also when perched. ♂♂ of Stuhlmann's and Kenrick's lack chestnut, ♀♀ have small patches which do not show when perched. Slender-billed Starling *O. tenuirostris* is larger, with long graduated tail, long narrow bill, ♀ mottled grey.

Voice. Tape-recorded (32, 86, 102, 104, B, C, F, LEM, NOR). Wide variety of clear whistles, as sweet as those of Red-winged Starling *O. morio* but higher-pitched. Common contact call is loud descending 'wheeeyoo', with variations: 'woy-yeer', 'woy-yeep', 'wo-cheeoo' and sometimes a convincing human wolf-whistle, 'whee ... wheeew'. Other calls include sweet 'tew-tew', nasal 'wayaanh', 'waanyap' or 'chip, wa-eeeyanh', and flight intention call, 2 or 3-note 'poy-pyip', 'whi-chip', or 'whi-chip-chip', first note lower. Song mainly whistles but also some chips and scratchy conversational notes, often at measured intervals, 'huweep ... jerwiya ... tuwee ... jerwa ... huweep', 'heeyo ... tyoop ... tihee', 'heeyo, chippo-heepo-heeyo'; some suggest human phrases, 'Leo sleep half the day'; one or more singers repeat a motif several times before switching to a different one. Alarm a harsh, grating 'chrrra' (Dowsett-Lemaire 1983) and said to utter shrill alarm whistles as bird of prey flies overhead (Bannerman 1948). Further details in Zimmerman *et al.* (1996) and Dowsett-Lemaire (1983b, with sonagrams).

General Habits. A bird of mature, tall submontane forests, small forest patches, and gallery forests. Occurs in pairs in breeding season and in flocks out of it, often of 12–24 birds, sometimes of 50–100. Generally keeps high up in canopy, coming down to mid strata only in very windy weather. Feeds more on insects than on fruit (see below); catches insects by sallying out from treetop, probing lichen on branches, and gleaning foliage. Home range of pair 12–16 ha of forest; birds range at least 1 km away to feed. In breeding season pair often occupies a forest patch, prospects a hole and even carries nest material, yet does not breed; ♂ and ♀ call regularly from song posts on high branches, depart together to forage, return together after 20–30 min, and in general behave like nesting pair (Malaŵi: Dowsett-Lemaire 1983b). Other birds in breeding season wander in loosely-associated small flocks (once of 100 birds) and are probably unmated.

Noisy; descending whistle uttered repetitively by unmated ♂♂ at regular posts high in canopy; 2-note whistle is flight-intention call, given by birds in flock about to depart on foray, also by ♂ calling ♀ off nest. ♂ and ♀ maintain close contact by frequently exchanging variety of soft whistles; and see Incubation below. Birds in flock gathering in morning may sing together (Dowsett-Lemaire 1990). Flock invariably whistles shrilly whenever a raptor flies overhead.

Food. Fleshy fruits (especially figs) and contained seeds and insects; also spiders and occasionally chameleons *Chamaeleo goetzei* 60–70 long. Nestlings given fruit and insects (mainly flying ones), only 25–37% of feeds (n = 259) being fruit. 10 spp. of fruits eaten in breeding season but only 2 out of it, including *Measa lanceolata*, *Podocarpus latifolius*, *Allophylus abyssinicus*, *Polyscias fulva*, *Ficus natalensis* and *Afrocrania volkensii* (Malaŵi: Dowsett-Lemaire 1983b). Concentrates at fruiting *Bridelia*, *Ocotea usambarensis* and *Sapium ellipticum* trees in Rwanda and eats *Alangium*, *Harungana*, *Prunus*, *Rapanea*, *Schefflera*, *Trema* and *Urera* (Dowsett-Lemaire 1990); *Ekebergia capensis*, *Ilex mitis*, *Macaranga mildbreadiana* and *Maesa lanceolata* (Moermond *et al.* 1993); *Symphonia*, *Juniperus* and *Cinnamon zeylanicum* (Feare and Craig 1998).

Breeding Habits. Solitary nester, monogamous, pair thought to remain together all year and to return to same nest site in successive years. Remarkably unaggressive, though area within 25–30 m of nest only rarely entered by a third starling. Pair formation: ♂ calls throughout Sept (Malaŵi), and is joined by ♀; from calling perch, ♂ repeatedly flies to tree hole, 'showing' it to mate. Courtship-feeding is rare, ♂ giving ♀ insect or berries at or near nest entrance. Pair formation not necessarily followed by breeding; estimated that only 25–33% of adult population in Nyika, Malaŵi, bred one year and <20% the next (Dowsett-Lemaire 1983b). Single-brooded; once a pair renested, after 1st nest failed, using a different nest hole in same dead tree.

NEST: one a flat platform 20–30 thick, 120 in diam., made of moss and pieces of *Hagenia* bark with a few dead leaves around rim, with interior layer 90 in diam. of thin twigs, liana fibres and pieces of *Hagenia* bark (Malaŵi: Dowsett-Lemaire 1983b). ♂ and ♀ seen collecting lichen, taking it in turns at same place (Bannerman 1948). 11 nests sited in live or dead *Hagenia abyssinica* trees, mostly in holes with oval entrance c. 150 in vertical diameter; one in crack 70 wide, 4 m up in heavily-fissured bark of *Agauria salicifolia* tree in small clump over a bog; one in enlarged woodpecker hole, diam. 70–80, in vertical dead trunk in isolated clearing; one in fissure 1 m long; generally in knot-holes in vertical trunks; sometimes in irregular cracks in stumps, split trunks, or holes on underside of limbs. Height above ground: 9 nests 4–11 m, 7 nests 12–20 m. Nest evidently made by ♂ and ♀ (or at least both birds take material into hole); ♂ takes old material out of previous season's hole.

EGGS: clutch size unknown, but broods of 1–3 seen inside nest entrances.

LAYING DATES: Cameroon, (breeding condition June, Nov; courtship and collecting nest material Dec, Jan; entering tree hole, Jan); Zaïre, Itombwe, (Nov and perhaps Mar–Apr); E Africa, Region A, May, Region D, Aug–Sept; Zambia, Sept–Nov, mainly Oct; Malaŵi, Aug–Nov, mainly Sept–Oct.

INCUBATION: by ♀, starting immediately after nest has been repaired (i.e. with 1st egg laid?); ♂ whistles and sings from high branch when ♀ incubating or brooding. At one nest ♀ brought nest material for up to 12 days after starting to incubate. Eggs covered for 71% of daytime; incubating spells (n = 20) last 6–35 (19) min, but during storms can last up to 95 min; ♀'s feeding trips between incubating spells last 6–12 (8) min. ♀ relies on call by ♂ for her to start and finish incubating; if ♂ has wandered away, ♀ appears at nest entrance and waits until she hears him whistle, the signal to leave (Dowsett-Lemaire 1983b). Period: 13, 13, 15+ and 16 days.

DEVELOPMENT AND CARE OF YOUNG: ♀ removes pieces of egg shell. Small nestlings brooded only by ♀, in 1st week for 20–50% of daytime, much less in 2nd week, not in 3rd week. ♀ roosts in nest hole at night. ♂ starts bringing food to nest hole on afternoon after morning 1st egg hatches. Later, young fed by ♂ and ♀ who arrive at nest together, usually with the same type of food, and leave together or, when young still very small, ♀ remains to brood them. 51% of 251 feeds were by ♂, 49% by ♀; ♂ feeds chicks first, followed by ♀; but if pair exploits hatch of insects near nest the rhythm of strict alternation breaks down temporarily. Brood's feeding rate varies from 6 per h on day 1 to 9–12 per h at 2 weeks of age. Young have far more insects than fruit, and shortage of insects in windy weather diminishes growth and can prove fatal (Dowsett-Lemaire 1983b). From age 19–20 days, head and body well feathered and young fed at or just inside nest entrance. At 19 days wing 85 long (n = 1). Faecal sacs carried away by ♂ and ♀. Nestling period: 29, 28+ and 29–30 days (3 broods of 1); 26–27 days (Moreau and Moreau 1940). After leaving nest, juveniles never return but stay near it for 1–2 days being fed by parents, then wander away with parents.

BREEDING SUCCESS/SURVIVAL: of 22 nests 9 (41%) produced flying young and 13 failed, due to ?predation (8 nests), starvation (2), desertion (1), unknown causes (2) (Dowsett-Lemaire 1983b). Although clutch size sometimes at least 3 eggs, many nests produce only a single juvenile.

References
Bannerman, D.A. (1948).
Dowsett-Lemaire, F. (1983b, 1988, 1990).
Stuart, S.N. (1986).

Plate 33 (Opp. p. 526)

Onychognathus fulgidus Hartlaub. Chestnut-winged Starling. Rufipenne de forêt.

Onychognathus fulgidus Hartlaub, 1849. Rev. Mag. Zool. (Paris), p. 495; São Tomé.

Range and Status. Endemic resident, lowland forest. Sierra Leone, scarce in Kilima area; records from Bo. Liberia, not uncommon in NW and SE; common at Monrovia in 1920s, rare there now. Guinea, common on Ziama Massif and on Mt Nimba, records from Bossu (Bakaman). Ivory Coast, absent from much of forest zone; occurs south of it along or near coast from Dabou to Sassandra; widespread north of it to Sipilou, Maroué Nat. Park, Lamto and Comoé Nat. Park in NE; Bossematié Forest in E; absent from Taï Forest, rare in Yapo Forest. Ghana, absent from high forest in extreme SW (Dutson and Branscombe 1990), not uncommon in logged mature and secondary forest elsewhere, west and south to Bia Nat. Park and Kakum Forest. Togo, known only from several sightings at Misahöhe. Nigeria, common and widespread from coast north to Ibadan, Ife, Oyo and Enugu. Cameroon, widespread in and just outside forest zone. Bioko. São Tomé, common in forest up to 1500 m (Christy and Clarke 1998). Mbini. Gabon, widespread, not as common as other forest starlings, commonest in mountain forest as at Bélinga, Bengoué and Batouala. Congo, local in Odzala Nat. Park: Okoungou, Ikessi, Andzoyi; fairly common in Nouaboulé-Ndoki Nat. Park, common in Léfini Rés., scarce in Kouilou Basin (Dimonika, Mayombe, Mah, Ouesso, Impfondo, Col de Bamba, Béna). Angola, Cabinda and S Cuanza Norte; frequent at Salazar. Central African Republic, Lobaye Préf. (Carroll 1988), in error (Germain 1992). Zaïre, widespread as mapped; in Itombwe scarce, at 880–1470 m. Sudan, only in Bengengai Forest. Uganda, uncommon and local, at 700–1300 m: Budongo, Bugoma, Bwamba, Kibale and Mpanga, also Kifu and Mabira.

Description. *O. f. hartlaubii* Grey: Sierra Leone to W Sudan, and Bioko. ADULT ♂: head and body glossy black; head, neck and throat with blue-green sheen, rest of upperparts and underparts with violet-purple sheen, bluer on uppertail-coverts. Birds in Nigeria and Cameroon bluer on top of head than those further west, and with slight coppery green tinge on lower mantle. Tail graduated, feathers rather narrow; black, upperside glossed greenish. Primaries dark chestnut on basal three quarters of P1–P6, basal two thirds of P7–8 and basal half of inner web of P9;

Onychognathus fulgidus

otherwise blackish brown. Rest of upperwing black; secondary edges, tertials and greater coverts glossed greenish, median and lesser coverts glossed violet; secondaries, tertials and coverts with non-reflective black submarginal lines, broader and more pronounced on outer webs. Axillaries and underwing-coverts brownish black, fringed glossy violet. Bill black; eyes scarlet; legs black. ADULT ♀: like ♂ but head, neck and throat grey-brown, heavily streaked dark glossy green; lower belly browner than in ♂; eyes red to pale orange-yellow. SIZE (10 ♂♂, 10 ♀♀, west of Nigeria): wing, ♂ 119–129 (126), ♀ 118–126 (121); tail, ♂ 114–120 (118), ♀ 108–120 (113); bill, ♂ 31–35 (32·8), ♀ 30–33 (30·6); tarsus, ♂ 27–28 (27·7), ♀ 26–28 (26·9). WEIGHT: Liberia, ♂ (n = 1) 103, ♀ (n = 4) 81·4–92·8 (87·7). Birds from Nigeria and Cameroon slightly larger: wing, ♂ (n = 8) 125–134 (130).

IMMATURE: juvenile brownish black above with slight violet gloss on body, and greenish gloss on top and sides of head; sooty brown below, chin and throat mottled pale buff. Wings and tail as adult, but browner, less strongly glossed, black submarginal lines on tertials and upperwing-coverts less pronounced. Grey head plumage of adult ♀ assumed at post-juvenile moult.

NESTLING: iris dark brown, corners of mouth yellowish, bill and feet black. Down dark brown (Chapin 1954).

O. f. fulgidus Hartlaub: São Tomé. Sheen on head and neck bluer than in *hartlaubii*, forehead and crown tinged violet; lower mantle tinged coppery green. Large: wing, 2 ♂♂ 152, 168, 1 ♀ 152.

O. f. intermedius Hartert: Gabon, Congo, Zaïre and Angola. Sheen on head and neck bluish, that on forehead to crown violet; lower mantle and back with coppery green sheen; slight coppery green tinge on lower breast, upper belly and flanks. Slightly larger than *hartlaubii*: wing, ♂ (n = 7) 127–138 (133).

Field Characters. Length 28–35·5 cm. A lowland forest starling whose long bill and low sloping forehead give it a flat-headed look. ♂ black, glossed green on head and neck, purple elsewhere; ♀ duller, head and neck grey with dark streaks; at a distance looks black with grey head (Christy and Clarke 1994). Shows much red in wings in flight. Larger and bulkier than Narrow-tailed Starling *Poeoptera lugubris*, tail fairly long and graduated but not pointed. Neumann's Starling *O. neumanni* of open country has deeper, more rounded bill, ♂ without green neck gloss, head of ♀ much less streaked; Slender-billed Starling *O tenuirostris* of Albertine Rift highlands has longer tail, more slender bill, ♀ has pale tips to all feathers.

Voice. Tape-recorded (104, B, F, CART, CHA, JO PJ, TYE). Song a leisurely conversational mixture of low glottal notes and high screechy ones, 'chweep, koklo, cheewy, koklo, chway'; 'koklo, korkkwyar, kloowo, cheewy, koojo'; song also described as a repeated 'churng-chuzick' (Chapin 1954). Rather quiet in Gabon but very vocal on São Tomé (Christy and Clarke 1998), where calls include clear high down-slurred whistle, 'peeeyoo' or 'puweeeyoo'; low 'tew' often in intervals between whistles; subdued 'cor-lew' (second note up-slurred); double 'tweep-yup'; soft whistles like African Grey Parrot *Psittacus erithacus* and musical and nasal notes like Splendid Glossy Starling *Lamprotornis splendidus*. Flight call a short 'tee-yup'; calls of young nasal, plaintive 'ah-rrap, ah-rrap, teeyap', throaty 'rrouep-ahap-ahap', and when begging from adult, 'ourroap' (Christy and Clarke 1998).

General Habits. In Liberia inhabits open forests at foot of inselbergs and along large rivers, gallery forests and mature, humid savanna woodlands especially with *Borassus* palms; rare in undisturbed closed forest, but occurs along logging roads in primary forest and benefits immediately from forest destruction (Gatter 1997) by using dead emergent trees for nesting and roosting. Scarce in long-cultivated land with only a few large trees left standing. Up to 1400 m on Mt Nimba, where breeds up to at least 1000 m. On São Tomé in lowland and montane forest, forest edges, coffee and cocoa plantations with tall trees and (in S) coconut palms, also in fruiting trees around villages, particularly avocados, papayas and *Cecropia* (Christy and Clarke 1998). In Gabon mainly in hilly primary forest, also old secondary forest and cultivation around villages.

Usually 2–3 together, often in flocks of 5–10, occasionally up to 22 birds (Liberia, Feb–July, Gatter 1997). Easy to overlook, since not very vocal; keeps to canopy, and highly mobile; evidently not territorial, but ranges over wide areas in course of a day; when feeding young, returns to a fruiting tree only once or twice before disappearing in search of another one (Gabon) or spends large part of day in one tree eating fruit pulp (São Tomé). Occasionally comes down to bathe in forest stream or descends to 1–2 m above ground in cocoa plantations; rarely lands on ground. When foraging in canopy mixes freely with Purple-headed, Copper-tailed and Splendid Glossy Starlings *Lamprotornis purpureiceps*, *L. cupreocauda* and *L. splendidus*. Forages by searching in trees, running and hopping along branches and making occasional sallies after insects from branches up to 100 mm diam. (Gatter 1997). In Nigeria seen at flowers of *Bombax*, with sunbirds (Bannerman 1948).

Like most frugivores, subject to wandering, in places amounting to local, seasonal migrations, e.g. in Ivory Coast common at Lamto in Oct–June but practically absent in July–Sept, and occurs in Yapo Forest only in Oct–Jan.

Food. Fruit, including pulpy large-seeded ones and palm-nuts; fruits up to 10 mm diam., in São Tomé mainly of *Uapaca*, *Cecropia* and figs (Christy and Clarke 1998); in Gabon rather large fruits, of *Dacryodes*, *Pycnanthus*, *Guibourtia* and *Trichoscypha* (Brosset and Erard 1986); in Zaïre 10 out of 11 stomachs contained fruit pulp or large stones in fruit up to size of an olive (Chapin 1954); also eats large fruits of oil palm *Elaeis guineensis*, fruits of *Maesopsis eminii*, and seeds, carabid and other beetles, winged ants and termites.

Breeding Habits. Barely known, since nests inaccessible at great heights in trees. Pairs evidently territorial (Cameroon, Apr–July: Serle 1965).

NEST: in tree holes 15–35 m above ground, in tree and palm trunks and stumps of large dead branches; also in angle between trunk and frond-stump of palm, large masses of epiphytes, and crevices in occupied eyries of Red-necked Buzzards *Buteo auguralis* and Crowned Eagles *Stephanoaetus coronatus* (the latter once, in underside: Skorupa 1982), and (Liberia, Gabon) Vulturine Fish Eagles *Gypohierax angolensis*; nests often sited near colonies of Yellow-mantled Weavers *Ploceus tricolor* (Gatter 1997).

EGGS: not described.

LAYING DATES: Liberia, thought to breed in Nov–Apr, mainly Oct, but seen feeding young in nest in Nov, (Oct and Dec–Mar); Ivory Coast (pair in nest hole, Mar); Ghana, (copulation Mar, carrying nest material Dec (Feare and Craig 1998) and July, using tree holes Oct–Nov); Nigeria, (feeding nestlings Feb, May; carrying nest material Mar); Cameroon, May–June; São Tomé, (family parties Dec–Jan); Gabon, (oviduct egg, Aug); Zaïre, Itombwe, Nov, Mar, Apr, (Avakubi, ♀ with brood patch Sept; Niangara, nestling May; Angumu, nestling Aug).

References
Brosset, A. and Erard, C. (1986).
Christy, P. and Clarke, W.V. (1998).
Gatter, W. (1997).

Plate 32
(Opp. p. 479)

***Onychognathus morio* (Linnaeus). Red-winged Starling. Rufipenne morio.**

Turdus morio Linnaeus, 1766. Syst. Nat., 12th ed., I, p. 297; Cape of Good Hope.

Onychognathus morio

Range and Status. Endemic resident, locally nomadic. Sudan, uncommon in SE, Boma Plateau and river gorges. Ethiopia, W and SE highlands, common in south. Old records from Eritrea (Hall and Moreau 1970) were misidentified Somali Starlings *O. blythii*. E Uganda, common on Mts Elgon, Moroto and Morongole; W Kenya from L. Turkana south to E limit in Taita Hills, commonest west of rift valley, with isolated populations in forest islands of Mt Kulal, Mt Nyiru, Ndoto Mts and Matthews Range; absent from Marsabit. Tanzania, widespread and locally common from Gombe Stream Res. to Serengeti, Usambara Mts, south to L. Tanganyika, L. Malaŵi, Uluguru Mts; coastal records as at Mikindani (Britton 1980). SE Zaïre, W shores of L. Tanganyika. Zambia, local on E border from Mbala to Nyika, along Muchinga and Zambezi escarpments to gorges of Zambezi Valley; absent north and west of Lusaka and Magabuka. Malaŵi, throughout, common in montane areas. N Mozambique, local in Zambezi Valley gorges, on high central plateau (Namuli), and on isolated inselbergs as at Mirrote; also vagrant to coast (Quelimane); S Mozambique on Mt Gorongoza, along E border with Zimbabwe and South Africa to Lebombo Mts. Botswana, restricted to E hardveld, both on granite outcrops and in towns; reported from tourist lodges in Okavango basin, but not established there. Zimbabwe, common throughout, except in W sandveld south of Hwange; often in tobacco barns. South Africa, widespread and common in natural and urban habitats in Transvaal and Natal (except NE lowveld), E Free State with vagrants west to Bloemhof and Koffiefontein; N Cape, scattered records, presumably of vagrants, from Kuruman, Molopo R., Augrabies, Pofadder and O'Kiep; W Cape, common south of Klawer and Calvinia to Cape Peninsula and Agulhas, all along coast, inland absent from plains north to 33°S but local in mountainous areas near Sutherland, Beaufort West, Murraysburg and Richmond. Extensive overlap with Pale-winged Starling *O. nabouroup* in this Karoo region. Lesotho, widespread, including the highest mountain areas. Swaziland, common in highveld areas. By utilizing man-made structures, has expanded its range into regions lacking suitable natural habitat (e.g. Gaborone, Botswana).

Description. *O. m. morio* (Linnaeus) (includes '*shelleyi*'): South Africa to Kenya. ADULT ♂: entire plumage uniform glossy black except for wings: both webs of P1–9 bright rufous-chestnut with black tips. Eyes dark red; bill, legs and feet black. ADULT ♀: like ♂, but forehead, crown, nape, sides of head, chin, throat and upper breast ash grey, each feather with dark central streak. Bird with a yellow throat (T. Buchan, *Babbler* 35, 1999: 27). SIZE (50 ♂♂, 50 ♀♀): wing, ♂ 142–159 (150), ♀ 134–153 (143); tail, ♂ 120–143 (133), ♀ 109–139 (128); bill to skull, ♂ 30·0–36·5 (33·4), ♀ 28·7–34·1 (32·1); tarsus, ♂ 30·6–36·8 (34·3), ♀ 30·6–36·0 (33·0). WEIGHT: (South Africa, W Cape) ♂ (n = 27) 133–158 (143); ♀ (n = 18) 125–149 (136); (E Cape) ♂ (n = 40) 130–158 (140), ♀ (n = 60) 117–149 (127).

IMMATURE: like adult ♂, but less glossy; eyes brown, bill dark brown. Grey head feathers of ♀ first appear at 4 months.

NESTLING: skin pink, naked except for long tuft of grey down on head and in centre of back. Gape whitish.

O. m. rueppellii (Verreaux): Ethiopia. Larger and longer-tailed; has different moult cycle. SIZE (43 ♂♂, 34 ♀♀): wing, ♂ 153–169 (160), ♀ 147–160 (154); tail, ♂ 146–177 (162), ♀ 147–165 (156). WEIGHT: ♀ (n = 1) 145.

Field Characters. Length 27–31 cm. A large cliff-dwelling red-winged starling now common in cities. ♂ appears black (slight blue gloss visible at close range). Rocky habitat shared by Bristle-crowned Starling *O. salvadorii* (N Kenya and Ethiopia: tail much longer with narrow pointed tip, bristly crown feathers give head different shape), White-billed Starling *O. albirostris* (Ethiopia: white bill, short tail) and Pale-winged Starling *O. nabouroup* (SW Africa: white-wing-patch). In E Africa generally at lower elevations than Slender-billed Starling *O. tenuirostris*; Red-winged has stouter bill, shorter and less graduated tail, more melodious voice; ♀ Red-winged has grey head and neck, plain dark body, ♀ Slender-billed mottled black and grey.

Voice. Tape-recorded (75, 86, 88, 91–99, 104, B, C, F). Most characteristic vocalization is contact call, given by both sexes, a clear, rich, melodious whistle, variable in form; song consists of a series of these whistles, 'twee-twoo, wor-twee, wortiwor-twee, tyoo-yoo, weeyo-tyoo, wortiyawoo, weeto-toodliay ... '; up to 20 different phrases noted (Rowan 1955). Muted warbling given by pair close together. Alarm, harsh 'karr'. Bill-snapping, hissing and clicking sounds given by captive bird (Rowan 1971).

General Habits. Nests and roosts on cliffs and rocky outcrops, from which it makes foraging flights to well-vegetated areas. Common town bird in some parts of E Africa (Nairobi, Arusha) and in southern Africa. Avoids arid habitats; confined to montane forest islands in N Kenya, and well-vegetated mountains with permanent water in the Karoo of South Africa. Seldom occurs below 1000 m or above 2500 m in Ethiopia, yet reported up to 4000 m on Mt Elgon; in southern Africa occurs from sea level to 3000 m, but generally absent from coast north of 28°S.

Pairs associate throughout the year, often roosting at nest sites. Forms flocks at fruiting trees; during winter groups of >200 birds occur, and >1000 at roosts. Roosts in trees may be shared with Common Mynas *Acridotheres tristis* and Wattled Starlings *Creatophora cinerea*; in reedbeds with Wattled Starlings and weavers *Ploceus* spp.; once roosted with Black-shouldered Kites *Elanus caeruleus* (Brooke 1965).

Forages in trees and on ground, where it hops; does not run like more terrestrial starlings. Removes ectoparasites from domestic stock, also Cape Mountain Zebra *Equus zebra*, Eland *Taurotragus oryx*, and Impala *Aepyceros melampus*. Perches on mammals which are used as 'beaters' to disturb insects. Seen to groom Klipspringer *Oreotragus oreotragus* at several localities (Gargett 1975). When taking large food items to the nest, first breaks them up at anvil site, then feeds them to chicks. Scavenges at picnic sites, and removes insects from grilles of parked vehicles. One record of active anting from South Africa; both adult and subadult birds utilized the ant *Anoplolepis custodiens*.

Moult of wing feathers Nov–Apr in *O. m. morio*, Apr–Dec in *O. m. rueppellii*. In N Kenya, S Sudan and S Ethiopia both patterns are found, presumably as a result of local sympatry and possible intergradation of the two populations (Craig 1988a). Adults with second broods often already moulting. Juvenile birds replace body plumage only, later than the complete moult of adults.

Foraging flocks may move extensively; birds ringed during winter in South Africa subsequently recovered 60 km south, 100 km north of ringing site. In Kenya, some indication of regular seasonal movements (Lewis and Pomeroy 1989).

Food. Almost omnivorous, taking carrion, small vertebrates, arthropods and other invertebrates, fruit, nectar and human food. Eats lizards, nestling birds, intertidal molluscs, sandhoppers, crabs, scorpions, ticks, spiders, millipedes, locusts, stick insects, mantids, beetles, butterflies, cicadas, caterpillars, bees, wasps, winged ants, termite alates. Attacked and grasped Palm Swift *Cypsiurus parvus*, but not clear if this incident was predation (Mortimer 1975). Eats fruits of *Acacia cyclops, Cassine capensis, Cassytha ciliolata, Chrithagra* sp., *Chrysanthemoides menilifera, Colpoon compressum, Cussonia paniculata, Ficus* spp., *Juniperus, Lycium* sp., *Melia azaderach, Phoenix reclinata, Rhus longispina, Royena pubescens, Scutia myrtina, Trema* sp., *Trichilia emetica*. Nectar of *Agave sisalana, Aloe arborescens, A. bainesii, A. candelabrum, A. ferox, A. marlothii, A. spectabilis, Erythrina caffra, Leucospermum conocarpodendron, Mimetes fimbriifolius, Protea subvestita, Salvia africana-lutea, Schotia brachypetala*.

Breeding Habits. Monogamous; pair remains together for at least 3 successive seasons. Territorial; ♂ sings from regular song posts. Uses wing displays in threat; displaces intruders in aerial chases. Both intra- and interspecific aggression occurs frequently. Species displaced include potential nest predators or parasites: Great Spotted Cuckoo *Clamator glandarius*, Steppe Buzzard *Buteo buteo*, African Harrier-Hawk *Polyboroides typus* (often mobbed by flocks), Common Kestrel *Falco tinnunculus*, Pied Crow *Corvus albus*. Also other cliff-nesting species, e.g. Pale-winged Starling, Speckled Pigeon *Columba guinea*, Mocking Cliff-Chat *Thamnolaea cinnamomeiventris*, Cape Rock Thrush *Monticola rupestris*, and casual visitors, e.g. Red-eyed Bulbul *Pycnonotus nigricans*, Familiar Chat *Cercomela familiaris*, Cape Glossy Starling *Lamprotornis nitens* and Green Wood-Hoopoe *Phoeniculus purpureus* (Craig et al. 1991). People and other mammals approaching nest may be struck from behind by diving birds.

♂ often courtship-feeds ♀ before copulation, and also feeds her on nest during incubation. Allopreening, primarily of head area, occurs throughout the year. Pairs regularly exchange contact calls. Usually double-brooded.

NEST: saucer-shaped, with mud base, built of grass, rootlets and sticks (**A**), lined with grass, leaves, horse hair; outside diam. *c.* 30 cm, cup 12 cm; placed on horizontal ledge, either on cliff or rock or on building or other man-made structure, also on *Hyphaene* palms; at least 2 m above ground. Has nested on buildings for more than 100 years (Holub and von Pelzeln 1882). Both sexes select site, build nest in 6 days. Same nest used for second brood, and may be re-used in successive years. One nest site used annually for 35 years (Mitchell 1976).

EGGS: 2–4, av. 3·1 (75 clutches). Blue with red-brown spots. SIZE: (n = 138) 21·8–38·6 × 20·3–28·5 (33·8 × 23·2).

LAYING DATES: Ethiopia, Apr, Oct; SE Zaïre, Dec–Jan; E Africa: Kenya, Nov–Feb; Region C, Oct; Region D, Oct–Apr, also May, Aug, Sept; Zambia, Dec–Mar; Malaŵi, Oct–Dec; Botswana, Zimbabwe and South Africa, Sept–Mar.

INCUBATION: by ♀, ♂ rarely seen on nest (Broekhuysen 1951, Brown 1965). Period usually 13–14 days, but eggs have hatched after 23–25 days (Broekhuysen 1951, Rowan 1955).

A

DEVELOPMENT AND CARE OF YOUNG: by day 3, skin pinkish-grey with dark feather tracts visible; feather sheaths begin breaking through skin on days 4–5; tips of rufous primary feathers emerge on day 9; at 15 days body covered in feathers. Eyes open between days 5 and 10. Young fed by both sexes, about 10 times per hour. Nestling period 22–28 days. Young accompany and roost with adults after leaving nest; first brood driven out of territory after about 20 days, when second clutch laid; second brood tolerated for up to 6 weeks, then driven off by ♂.

BREEDING SUCCESS/SURVIVAL: adult killed by Taita Falcon *Falco fasciinucha* (Woodall 1971). Nests occasionally robbed by Pied Crow and African Harrier-Hawk, also by baboons. Sometimes parasitized by Great Spotted Cuckoo.

References
Broekhuysen, G.J. (1951).
Craig, A.J.F.K. (1988a).
Craig, A.J.F.K. and Hulley, P.E. (1992).
Craig, A.J.F.K. *et al.* (1989, 1991).
Rowan, M.K. (1955).

Plate 32
(Opp. p. 479)

Onychognathus blythii (Hartlaub). Somali Starling. Rufipenne de Blyth.

Amydrus blythii Hartlaub, 1859. J. Orn., 7, p. 32; Somaliland.

Range and Status. Endemic resident and wanderer or migrant, Eritrea, Ethiopia, Djibouti, N Somalia and Socotra Island. Eritrea, throughout, above 920 m; also in Dancalia. Ethiopia, common in central and S highlands in May–June though not in winter, above 3800 m on Bale Mts; records Mt Fantale, Awash Nat. Park, and near Ankober (Ryan and Sinclair 1998). Djibouti, common resident, mainly about 1000 m in Forêt du Day area; only records below 700 m are from coast and coastal plains in Oct. Somalia, locally abundant, south to 07°30′N, breeding at 1400–2000 m, wandering widely in non-breeding season, down to sea level; abundant on Golis Range, Mt Wagar, near Sheikh, Gan Libah and Bagan. Socotra, common and widespread from sea-level to 1100 m, commonest at 1000–1100 m.

Description. ADULT ♂: head and body glossy black; top of head with greenish blue sheen, sides of head more greenish, upperparts with dull greenish sheen, underparts tinged more violet. Tail black, rather long and graduated, feathers narrow. Primaries 1–8 chestnut, with dark brown tips and inner borders and, on P5–P8, distal quarter of outer webs; P9 dark brown with base of outer web and basal two-thirds of inner web chestnut; P10 dark brown. Rest of upperwing black, with glossy edges to secondaries, tertials, primary and greater coverts, and glossy tips to median and lesser coverts. Axillaries and underwing-coverts black. Bill black; eyes crimson; feet black. ADULT ♀: differs from adult ♂ in having head to neck and upper breast pale unglossed grey. SIZE (10 ♂♂, 10 ♀♀): wing, ♂ 165–180 (174), ♀ 155–167 (164); tail, ♂ 150–187 (175), ♀ 139–169 (162); bill, ♂ 29–32 (30·2), ♀ 28–30 (29·5); tarsus, ♂ 36–39 (37·7), ♀ 35–37 (35·8). Also (10 ♂♂, 7 ♂♂, Socotra, Porter and Martins 1996), wing ♂ 165–175 (171), ♀ 155–164 (159); tail, ♂ 172–190 (177), ♀ 155–169 (163); bill to feathers, ♂ 23–25 (24·3), ♀ 22–24 (22·6).

IMMATURE: juvenile similar to adult ♂, but less strongly glossed, more sooty, browner below and shorter-tailed.
NESTLING: not described.

Field Characters. Length 33–35·5 cm. Common in Horn of Africa and Socotra, where more slender than sympatric congeners; best told by diamond-shaped tail, with long central feathers and narrow base; broadest point of tail one-third to halfway from base (in Slender-billed Starling *O. tenuirostris* broadest point is two-thirds along from base: Ryan and Sinclair 1998). ♂ very similar to ♂ Red-winged Starling *O. morio rueppellii* but bill narrower, tail slightly longer and more graduated; like ♂ and ♀ Socotra Starling

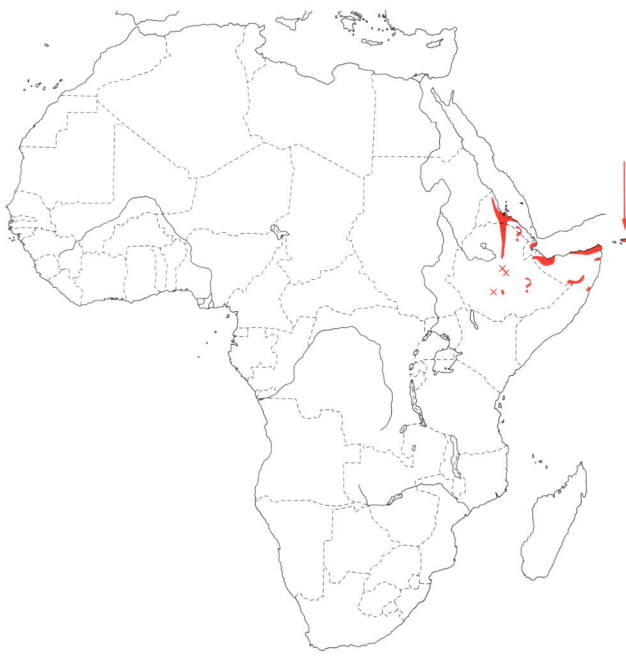

Onychognathus blythii

O. frater but slightly larger, with longer, graduated tail narrower at base, shorter, stouter bill, distinct voice (Porter and Martins 1996). ♀ Somali Starling has *solid pale grey* head and neck, becoming scaly on breast, with paler mask from lores to just behind eye (on Socotra more a ring around eye than a mask through it). Head of ♀ Red-winged darker and duller, neck with black streaks; ♀ Slender-billed is black with grey mottling; ♀ Socotra Starling is black like ♂.

Voice. Tape-recorded (DAV). Commonest call, often given in flight, a musical, high-pitched 'tleep' or 'chuit'; other calls include soft 'chee-wee' or 'chwee' and a harsh alarm note like that of Socotra Starling. Call also transcribed as 'chee-chee-chee-che-whoup' (Ogilvie-Grant and Forbes 1903). Said to utter occasionally a weak, high-pitched 'tink-ink-ink' (Ryan and Sinclair 1998).

General Habits. Inhabits mountains with junipers and rocky hills with cliffs, scarps and scattered trees and bushes; wadis and dry watercourses with figs and other trees, but occurs also in areas lacking substantial woody cover; patches of open grass; in Bale Mts, Ethiopia, large mountaintop stands of red hot pokers *Kniphofia* and huge stems of giant lobelia; near Berbera, Somalia, on sandstone bluffs overlooking settlements with gardens and springs.

Lives singly, in pairs, family parties and flocks of up to 50 (Bale Mts) and 110 birds (Socotra). Forages for fruit in trees and bushes; feeds on *Kniphofia* and *Lobelia* inflorescences by clinging to upright stems and probing individual flowers, like a sugarbird *Promerops*; shares that resource with Malachite and Tacazze Sunbirds *Nectarinia famosa* and *N. tacazze*, but does not interact with them (Ryan and Sinclair 1998). Associates with cattle on Socotra (though not reported to do so elsewhere), standing and pecking around on their backs, presumably looking for ectoparasites. Fearless of man; but flocks can be skittish, flying long distances when disturbed, often retreating to sanctuary of large cliff (Ryan and Sinclair 1998). Flight buoyant and masterful, and when bird flies far, direct and undulating, with short bursts of wing beats, long glides, and occasional abrupt changes of direction (Porter and Martins 1996). Not very vocal.

Sedentary, but after breeding wanders down to coastal plains (Eritrea, Djibouti, Somalia), and thought to be a seasonal visitor to Bale Mts at 3800 m in May–June (Ryan and Sinclair 1998).

Food. Mainly fruits, especially juniper berries (Somalia), also figs, wild olives and *Dracaena*; large black seeds. Nectar and possibly pollen of *Kniphofia* and *Lobelia* (Bale Mts, Ethiopia). Invertebrates, including grasshoppers, beetles and ticks.

Breeding Habits.

NEST: loosely-made structure composed of grass and a few feathers; sited mostly in inaccessible holes in high cliffs; also in holes in road embankments and in buildings – once in a bathroom, once 'under the outer fly of my tent on the ridge-pole and the parent birds proved extraordinarily tame and companionable' (Archer and Godman 1961); also in chimney and cavity under eaves of shed.

EGGS: 4–5. Pale bluish green speckled with Indian red. SIZE: (n = 8) 30·5–35·5 × 22–25 (33·2 × 24·1).

LAYING DATES: Eritrea, (building nest Sept; occupied nest Apr); Ethiopia, May (–June); Somalia, May (n = 6 clutches) (and nest-building Apr); Socotra, (some evidence that breeding is over by Dec: Porter and Martins 1996, but several nests being built in late Nov: Clouet *et al.* 1998).

References
Archer, G. and Godman, E.M. (1961).
Porter, R.F. and Martins, R.P. (1996).
Ryan, P.G. and Sinclair, I. (1998).

Onychognathus frater (Sclater and Hartlaub). Socotra Starling. Rufipenne de Socotra.　　　Plate 32

Amydrus frater P.L. Sclater and Hartlaub, 1881. Proc. Zool. Soc. London, p. 171; Socotra.　　(Opp. p. 479)

Range and Status. Endemic resident, Socotra I.; local and uncommon, from near sea level up to 1500 m. Vulnerable. Fairly common at Homhil and Adho Dimellus in 1903. In 1993, 41 birds found at 6 sites (compare with 538 Somali Starlings *O. blythii* at 14 sites on Socotra); thought to be fewer than 1000 birds altogether (Collar *et al.* 1994, Porter and Martins 1996).

Description. ADULT ♂: entire head and body glossy black with bluish green or bottle-green sheen. Tail black, upperside glossy, tinged greenish. Primaries 1–7 dark chestnut with dark brown tips, on P5–P7 extending to distal inner borders and distal quarter of outer webs; P8 chestnut with distal third dark brown; P9 dark brown with base of outer web and basal half of inner web chestnut; P10 small, dark brown. Rest of upperwing black; edges of secondaries, tertials, greater coverts and primary coverts, and tips of median and lesser coverts glossed bluish green. Axillaries and underwing-coverts black. Bill black; eyes dark reddish brown; feet black. Sexes alike. SIZE (5 ♂♂, 2 ♀♀): wing, ♂ 152–165 (158), ♀ 149, 155; tail, ♂ 130–142 (135), ♀ 125, 134; bill, ♂ 33–39 (36·2), ♀ 33, 35; tarsus, ♂ 37–40 (38·2), ♀ 35, 37. Also (7 ♂♂, 9 ♀♀, Ripley and Bond 1966), wing, ♂ 157–163 (159·5), ♀ 148–156 (151·5); tail, ♂ 139–146 (142), ♀ 127–136 (131); bill to feathers, ♂ 28–32 (29·2), ♀ 27–29 (27·8). WEIGHT: 1, unsexed, 100.

IMMATURE: juvenile blackish, upperparts, chin to throat and wing feather edges with slight blue-green gloss; lower breast to undertail-coverts browner.

Onychognathus frater

NESTLING: not described.

Field Characters. Length 28–33 cm. Glossy black with bluish-purple sheen (both sexes). Smaller than Somali Starling *O. blythii*, with much shorter, square-ended tail (corners sometimes slightly rounded) and longer, thinner bill, slightly decurved at tip (Porter and Martins 1996). ♀ Somali Starling readily told by grey head and neck.

Voice. Tape-recorded (DAV). Commonest call a pure, far-carrying whistle, 'tyooo', 'pseeeoo' or 'psoo' recalling European Golden Plover *Pluvialis apricaria* or Eurasian Bullfinch *Pyrrhula pyrrhula*; alarm a harsh 'scraich' (Porter and Martins 1996).

General Habits. Inhabits rocky hillsides, limestone cliffs, foothills, valleys, open plains with scattered trees, thickets, grassy uplands, wooded wadis, towns, gardens and lagoons. Occurs singly, in pairs, and flocks of up to 16 birds (Hamadiroh Plateau). Feeds on standing crop of fruits in trees and bushes; obtains pea-like legume seeds by splitting pod open. Forages in methodical way, recalling manner of *Turdus* thrush; up to 5 birds feed in cover of leafy tree for up to 20 min at a time; pair ate 2 *Ficus* fruits per min during a 5 min period (Porter and Martins 1996). Occasionally forages on ground; more closely associated with trees than is Somali Starling on Socotra. Comes down to puddles to drink. More shy (Porter and Martins 1996) or much less shy (Ogilvie-Grant and Forbes 1903) than Somali Starling.

Food. Fruits of *Ficus*, berries of *Ziziphus*, small red berries probably of *Dracaena cinnabari* (which are also fed to juveniles accompanying adults), and some legume seeds (Porter and Martins 1996); grasshoppers in stomachs, and birds seen taking grasshoppers and other insects (Ripley and Bond 1966).

Breeding Habits. Thought to be a colonial nester (Porter and Martins 1996). Nest of 1 pair only 2 m from nest of Somali Starling (Clouet *et al.* 1998).
NEST: one an untidy mass of grass and small sticks, in cavity in roof or walls of limestone cave (at 550 m on rocky slopes above Hamadiroh Plateau).
LAYING DATES: Nov (several ♀♀ incubating, Massif de Hagghiers at 750 m: Clouet *et al.* 1998), Mar (i.e. in first week of Apr many birds feeding fledglings in and out of nest).
DEVELOPMENT AND CARE OF YOUNG: several pairs were each accompanied by single young bird (Ogilvie-Grant and Forbes 1903).

Reference
Porter, R.F. and Martins, R.P. (1996).

Plate 32
(Opp. p. 479)

Onychognathus neumanni (Alexander). Neumann's Starling. Rufipenne de Neumann.

Amydrus neumanni Alexander, 1908. Bull. Br. Orn. Club, 23, p. 41; Petti, Nigeria.

Range and Status. Endemic resident, W Africa. Senegal, vagrants east of Dakar, in north near Malam; common in SW near Kédougou. S Mauritania, on Assaba Escarpment, in Tagant and Affolé. Guinea, local in N and W highlands; and adjacently in N Ivory Coast. Mali, from Satadougou and Bamako to Mopti along Niger R., locally common; also on E highlands, north of Niger R.; Burkina Faso, local in north; Niger, along R. Niger. No records from Benin, Togo or Ghana. Nigeria, most or all larger inselbergs and rocky hills from Jos Plateau northwards to Safana and Runka; absent from NE; south of Benue R. in Gashaka–Gumti Nat. Park and on Afu Hills in Benue Province (Serle 1940). Chad, in E highlands. Cameroon, common in Bamenda area, also at Yaoundé. W Central African Republic, local resident in Kare Mts; birds in NE region possibly vagrants. W Sudan, common on Jebel Marra.

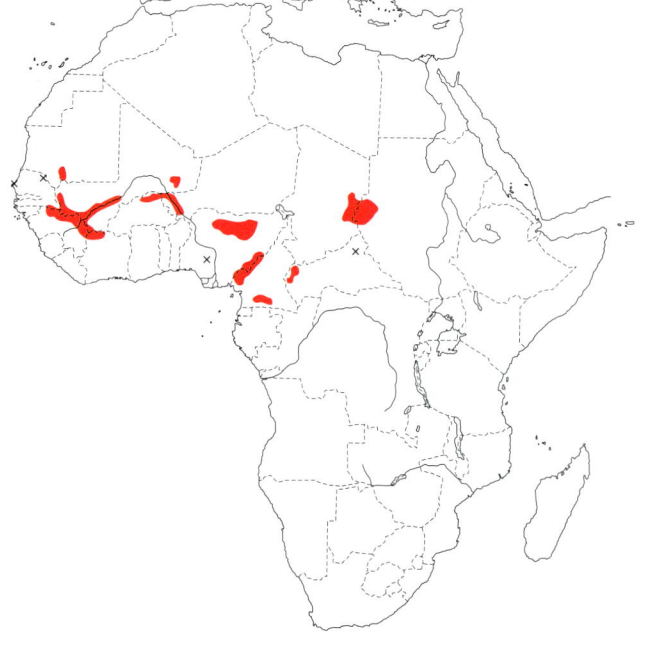

Onychognathus neumanni

Description. *O. n. neumanni* (Alexander): E Mali from 0° to Sudan. ADULT ♂: head, body and uppertail-coverts uniform glossy blue-black, with purplish sheen; greenish sheen on upperwing-coverts and rectrices. Tail strongly graduated. P2–9 chestnut on both webs with narrow black tip; P1 black, brownish on inner web. Feathers on lores point forward, covering nostrils. Bill black, eyes dark red, legs and feet black. Bill 22% deeper at nostril than bill of *O. morio*, with culmen narrow and decurved. ADULT ♀: like ♂, but forehead, crown, nape, sides of head, chin, throat and upper breast ash grey with dark streak in centre of each feather. SIZE (9 ♂♂, 6 ♀♀): wing, ♂ 155–171 (164), ♀ 156–

164 (159); tail ♂ 159–189 (173), ♀ 167–177 (171); bill to skull, ♂ 29·3–34·3 (32·9), ♀ 30·3–31·9 (31·1); tarsus, ♂ 37·2–39·8 (38·8), ♀ 35·8–37·3 (36·2).

IMMATURE: like adult ♂, but plumage lacks gloss and eyes brown.

NESTLING: undescribed.

O. n. modicus Bates: Senegal to W Mali at 03° W; apparently allopatric to nominate population, and status warrants study. Smaller and shorter-tailed, tail square, not graduated, and shorter than wing. Wing, ♂ (n = 14) 147–160 (152), ♀ (n = 10) 141–151 (145); tail, ♂ 134–155 (143), ♀ 135–145 (140). Iris described as brownish-black (Bannerman 1948). Bates (*in* Bannerman 1948) described bill as less stout and deep than in *O. n. neumanni*: ♂ *modicus* (n = 14) av. bill depth 9·2, width 6·9; ♂ *neumanni* (n = 9) av. bill depth 10·4, width 7·7.

TAXONOMIC NOTE: this species was formerly treated as a race of *O. morio*. However, their bill shapes differ considerably, and head feathering directed forwards to cover the nostrils is a feature shared only with *O. salvadorii* and *O. albirostris*. A preliminary phylogenetic analysis grouped *O. neumanni* and *O. salvadorii* as a superspecies (Craig, unpubl. data). *O. neumanni* is allopatric to all other members of the genus, and we feel that it warrants treatment as a separate species.

Field Characters. Length 25 cm. Black, with chestnut wing-patch conspicuous in flight; no other red-winged starlings occur in its range.

Voice. Tape-recorded 104, ROC. Call deeper than that of Red-winged Starling *O. morio*, an up-slurred 'kwoooy'. Musical 'too-whee-oo' reported from nesting pair (Smith 1964); strong fluted notes (Paludan 1936). Birds at rest give plaintive 2-note whistle; flight call a loud, rather plaintive twittering (Bannerman 1948). Alarm, a harsh 'kerr' or 'air, air'.

General Habits. Inhabits inselbergs, rocky outcrops and cliffs up to 2400 m in Sudan, 1800 m in Cameroon, at 500 m in Nigeria; around houses in Bamako. Groups of 20–30 birds feed in fruiting trees; bird once seen to pursue a monkey feeding in a fig tree. Large food items such as snails broken up on anvil stone. Hops on rocks, does not walk. Moult apparently takes place between Mar and Aug, which suggests that it may overlap with breeding in some populations. Apparent absence from breeding sites at Bamenda, Cameroon, in July–Aug and occurrence of flocks in grassland and savanna in Senegal, far from rocky areas (Morel 1985), suggests seasonal movements or nomadism.

Food. Figs *Ficus* spp. probably main food item; also fruit of mountain palm *Phoenix reclinata*, ants, snails, and nectar of *Bombax* trees.

Breeding Habits. Thought to be monogamous and territorial. During breeding season at Bamenda, Cameroon, birds displaced bulbuls, pipits, Black-shouldered Kites *Elanus caeruleus* and Common Kestrel *Falco tinnunculus* from their territory. ♂ made short flights, rolling and tumbling in the air while ♀ perched nearby; whistled calls given for long periods.

NEST: only one described, a simple cup of straw, placed on horizontal rock ledge. Pair in Cameroon seen carrying dry grass.

EGGS: undescribed.

BREEDING INDICATIONS: breeding activity reported Mar–Sept in Mali, July in Ivory Coast and Burkina Faso, Apr–May in Nigeria, Feb–Mar in Cameroon, July–Aug in Sudan.

INCUBATION: no information.

DEVELOPMENT AND CARE OF YOUNG: ♂ and ♀ feed young.

References
Bannerman, D.A. (1948).
Parelius, D.A. (1967).

Onychognathus salvadorii (Sharpe). Bristle-crowned Starling. Rufipenne de Salvadori.

Plate 33
(Opp. p. 526)

Galeopsar salvadorii Sharpe, 1891. Ibis, p. 241; Turquel, Suk country, Kenya.

Range and Status. Endemic resident. Ethiopia, uncommon or local, mainly below 1200 m, though nests found at 1345 m and 1400 m, in S, SE and NE. Somalia, locally common in NW, east to Gabileh, in upper Webi Shabeelle valley, and around Baydhabo (Baidoa) and Buur Haybo (Bur Hebi). Kenya, locally common at 400–1300 m from Baringo, Wajir, Isiolo, Timau, Bogoria and northeast of Tugen Hills to Marsabit, Mt Kulal, Mt Nyero, Endoto Mts (where nearly 100 birds have been seen in a day), Turkwel Gorge, Karasuk Hills, Lokitaung and Sudan and Ethiopia borders. Uganda, Mt Moroto.

Description. ADULT ♂: feathers of forehead, forecrown and lores short and stiff, forming velvety-black cushion, directed forward over basal third of bill. Otherwise, head and body glossy black with blue-green sheen. Tail very long, strongly graduated; central feathers narrow and elongated, projecting 65–90 mm beyond T2; black, upperside glossed greenish. Primaries 1–9 rufous-chestnut, with dark brown tips, the brown extending to outer web and broadly along inner border of distal quarter of P6–P7, distal third of P8 and distal half of P9; P10 small, dark brown; rest of upperwing black but outer webs of secondaries glossed greenish, primary coverts edged rufous, and tertials and greater, median and lesser coverts glossed greenish. Axillaries and underwing-coverts black with glossy blue-green fringes. Bill black; eyes dark ruby-red; legs black. ADULT ♀: like ♂, but sheen slightly greener; smaller 'cushion'; duller on head, neck and throat, with grey-brown feather bases showing through. SIZE (7 ♂♂, 5 ♀♀): wing, ♂ 152–165 (161), ♀ 153–158 (156); tail, ♂ 230–262 (240), ♀ 214–235 (226); bill, ♂ 30–35 (32·0), ♀ 30–33 (31·2); tarsus, ♂ 36–39 (37·6), ♀ 35–37 (35·6). WEIGHT: 1 ♂ 160, ♀ (n = 3) 121–150.

IMMATURE: juvenile much duller, with faint gloss (Zimmerman *et al*. 1996).

NESTLING: not known.

Field Characters. Length 39·5–42 cm. A dry-country starling of NE Africa with exceptionally *long tail* and

Onychognathus salvadorii

cushion of velvety-black feathers forming 'bump' on forehead. Glossy black with blue-green sheen and chestnut patch in wing; ♀ with some grey on head.

Voice. Tape-recorded (B, C, GREG, McVIC, NOR). Characteristic staccato whistles, 'suk-sweek' or 'su-weer', and a high, 2-syllabled whistling 'chreep-rr' often given in flight; annoyance call a harsh scolding 'schwaah' (Zimmerman *et al.* 1996).

General Habits. Inhabits trees amongst crags, cliffs and rocky gorges, often but not necessarily close to water, in arid and semi-arid bush country; common inside Baydhabo town, Somalia, and occurs in gardens in Marsabit town, Kenya. Occurs in pairs which often come together in flocks of 10–20 and up to 50 birds. Described as colonial, e.g. always *c.* 20 birds together in Gabileh, NW Somalia, and 'several small colonies breeding in the cliffs south of Ginir', Somalia (Archer and Godman 1961). Flocks wander widely between fruiting trees. Forages in leafy tree, and sometimes comes to ground to feed on fallen fruits. Perches on boulders, sand cliffs, and topmost branches of leafy or stunted thorn trees. Comes down to drink at small pool in ones and twos till pool surrounded by ring of birds chattering shrilly; drink in unison, all lowering and raising heads together (Brown 1965). Generally shy, wary and difficult to approach, though not so in Baydhabo.

Food. Fruits and seeds.

Breeding Habits.

NEST: cup of sticks, stems and grasses held together with clay; 2 in a rock passage 50–75 m inside entrance of Sof Omar caves east of Goba and one in hole in arch of stone bridge, 2 m above ground, Ethiopia (Urban *et al.* 1970). Strong circumstantial evidence that it nests colonially in holes in high sand cliffs in NW Somalia (Archer and Godman 1961). At L. Baringo, Kenya, nests alongside congeners (Feare and Craig 1998).

LAYING DATES: Ethiopia (Dire Dawa), (nestlings May), (east of Goba), (adults roosting on nests though eggs not yet laid, June); Somalia, (adults collecting food to take to inaccessible nest, May); E Africa (L. Baringo), (breeding May–June, nestlings Oct: Feare and Craig 1998).

References
Archer, G. and Godman, E.M. (1961).
Feare, C. and Craig, A. (1998).
Lewis, A. and Pomeroy, D. (1989).
Urban, E.K. *et al.* (1970).

Plate 33 *Onychognathus albirostris* (Rüppell). **White-billed Starling. Rufipenne à bec blanc.**
(Opp. p. 526)

Ptilonorhynchus (Kitta) albirostris Rüppell, 1836. Neue Wirbelt., Vögel, p. 22; Ethiopia.

Range and Status. Endemic resident, Eritrea and Ethiopia. Eritrea, above 2460 m on sheer cliffs between Adi Caieh and Senafe (only 30 km apart). Ethiopia, commonest north and west of Rift Valley; frequent to locally abundant in W Highlands, at 2000–3000 m and in Semien up to 3100 m; in Bole Valley, Addis Ababa, at 2200–2500 m; common in central Tigrai; uncommon in S Highlands, e.g. Bale Mts.

Description. ADULT ♂: head and body glossy black with violet sheen; some greenish reflections on underparts. Stiff velvety black feathers on lores and across base of upper mandible, forming small 'cushion'. Tail almost square; black above, faintly glossed; brownish black below. Primaries 1–9 chestnut, with broad brownish black tips and (on P4–P9) distal part of outer web (distal third on P9), and broadly along distal inner border; P10 small, brownish black. Rest of upperwing blackish; tertials, greater coverts, and outer webs of secondaries slightly glossed, median and lesser coverts strongly glossed violet-black. Axillaries and underwing-coverts blackish, fringed glossy violet. Bill white; eyes chestnut-brown; legs black. ADULT ♀: head to upper mantle and upper breast grey with faint bluish sheen; otherwise like adult ♂. SIZE (8 ♂♂, 1 ♀): wing, ♂ 152–168 (161), ♀ 149; tail, ♂ 110–121 (116), ♀ 107; bill, ♂ 27–29 (27·7), ♀ 27; tarsus, ♂ 31–34 (33·1), ♀ 30. WEIGHT: ♂ (n = 2) 130, 145, ♀ (n = 3) 130–135 (Feare and Craig 1998).

IMMATURE: less strongly glossed, mainly on upperparts and wing-coverts; browner below; primaries browner, chestnut pattern less strongly demarcated; primary coverts with pale chestnut distal half and dark tip. Juvenile ♂ and ♀ recently out of nest dull black with grey head; reddish remiges and white bills like adult, although some juvs have dark bill (Brown and Thorogood 1976).

NESTLING: not known.

Onychognathus albirostris

Field Characters. Length 28–30·5 cm. A chestnut-winged starling restricted to mountain cliffs in Ethiopia and Eritrea; readily identified by *white bill* and *square, medium-length tail* (sympatric congeners all have long tails). ♂ glossy dark purple, ♀ with grey head and breast.

Voice. Tape-recorded (MAC). Intermediate between sharp excited chatter of Slender-billed Starling *O. tenuirostris* and sweet whistling calls and songs of Red-winged Starling *O. morio*, but more like the latter. ♂ calls a variety of short whistles, 'chee-up', 'tu-it', sometimes low 'chut' and 'chirip'; song near nest includes repeated sharp 'kwit-kwit'. Anxiety scold near nest a harsh, grating 'charr' or 'churrr' (Brown and Thorogood 1976).

General Habits. Inhabits mountain cliffs, boulders, screes and rocky gorges. Flock often perches on slanting rock face devoid of vegetation. Perches on bushes and in trees to feed on fruits. Associated with waterfalls, and nesting sites put it into close association with White-collared Pigeon *Columba albitorques* (Smith 1957). In Bole Valley, Addis Ababa, lives alongside Red-winged and Slender-billed Starlings and keeps to vicinity of water; in central Tigrai common on rocky mountains with remnant 'olive-cedar' forest (*Olea, Juniperus*) and also in towns – much more abundant and widespread than other 2 starling spp. Frequents (and nests in) bridges, factories and other man-made structures.

Occurs singly but generally in pairs, and in flocks of *c.* 10–12 and occasionally *c.* 100 after breeding season. In morning, flock may split up into small parties and singles, which join up again into flock to roost in evening. Feeds on fruits in trees; in breeding season often makes short sally after flying insect. Flock often visits waterfall in heat of day to bathe.

Food. Fruits of figs and junipers (Eritrea). Nestlings fed mainly or entirely on insects, including grasshoppers, butterflies and hard beetles.

Breeding Habits. Same nest site occupied for 3 successive years. Another nest thought to have been used repeatedly, not rebuilt but merely added to each year. One nest taken over by Speckled Pigeon *Columba guinea*, perhaps after having been robbed. ♂ once held rootlet in bill and sang beside ♀ at edge of cereal field *c.* 1 km from nest. Many more pairs of adults seen in a small area than nests found, and likely that not all pairs breed every year (Brown and Thorogood 1976).

NEST: cup made of grass and cereal leaf blades, lined with fine plant fibres and rootlets, apparently without any mud base. Estimated dimensions, ext. diam. 150, ext. height 110–120, int. diam. 90–100, int. depth 40–50. Site nearly always inaccessible to person; in crannies high up in vertical rock faces (Eritrea); in narrow crevice amongst grass tussocks 20 m from top and 60 m from foot of sandstone cliff; 2 m above surge pool of waterfall (the nest concealed behind maidenhair fern *Adiantum* and wetted by spray); in damp hole at rear of cavern behind waterfall, 3 m above surge pool, visible from across the pool; also in hole 9 m up in vertical wall of factory; in drainage hole 3 m above concrete base of road-bridge and 50 cm in above door lintels on upper floor balcony in old Makalle palace, and behind drainpipe on outside wall of hotel. Nest thought to be made by ♀, who often stays for long time in nest hole whilst ♂ sings and postures at song-post outside. Once ♂ collected scraps of dyed thread and took them to nest hole when ♀ was inside.

EGGS: 4. Broad ovals; clear turquoise-blue with small, well-defined brown and blackish spots mainly around broad end. SIZE: (n = 4, 1 clutch) 30–33 × 20–21 (31·25 × 20·7).

LAYING DATES: Eritrea, (nest building, June); Ethiopia, Aug–Oct. Laying dates may be related to flood conditions (Brown and Thorogood 1976).

INCUBATION: evidently by ♀, fed by ♂ (once with shiny beetle). ♀ roosts in nest hole at night, ♂ roosts on ledges or trees on cliff face above.

DEVELOPMENT AND CARE OF YOUNG: young fed by both parents, mainly ♀. ♂ and ♀ generally depart together on a foray and return together, often ♀ with food visible in bill and ♂ with none; first one enters nest then the other, or sometimes both at once. ♀ sometimes enters nest hole with billfuls of small morsels, probably insects. Combined incubation and nestling periods *c.* 35 days. Young stay near nest, apparently dependent on parents, for a few days after first flight, then accompany parents to feeding grounds and form a family party. Later, young flock with adults.

BREEDING SUCCESS/SURVIVAL: of 5 nests, no eggs laid in 2, and 1 failed; 8 young fledged from 12 known and 20 'potential' eggs (Brown and Thorogood 1976).

References
Brown, L.H. and Thorogood, K.M. (1976).
Smith, K.D. (1957).

Plate 33
(Opp. p. 526)

Onychognathus tenuirostris (Rüppell). Slender-billed Starling. Rufipenne à bec fin.

Lamprotornis tenuirostris Rüppell, 1836. Neue Wirbelt., Vögel, p. 26; Ethiopia.

Range and Status. Endemic resident, E African mountains from Eritrea to Malaŵi. Eritrea, scarce and local at high altitudes in Senafe and Guna Guna areas. Ethiopia, frequent to locally abundant in W and SE Highlands. Zaïre, common at 1500–3400 m in eastern mountains and occurs up to 4250 m (Itereré, Stanley Glacier); uncommon in Itombwe Highlands, at 1980–2070 m. Uganda, locally common to abundant in W and SW: Rwenzoris and Bwindi Forest. Kenya, locally common in central highlands up to 4500 m; Elgeyu Escarpment to Mau Narok; Nyahururu (Thomson's Falls) to near Nairobi; Mt Kenya east to Meru; commonest on Aberdares and Mt Kenya; colonies at small waterfalls on Mau, Elgeyu and Kongelai Escarpments; said to be uncommon on Mt Elgon above 2700 m (Brown 1965) and reported from Kaibibich, though these records need confirmation (Lewis and Pomeroy 1989). Rwanda, widespread, but much scarcer and more localized than Waller's Starling *O. walleri*. Burundi, all western mountains: Crête Zaïre-Nil, Rwegura, Bugarama, Mt Heha, south to Bururi Forest and Musigati. Tanzania, widespread between Ngorongoro and Kilimanjaro (10 localities) though not on either of those massifs themselves; Usambaras, Nguru, Ukagurus and Iringa highlands (N.E. Baker, pers. comm.), also Njombe highlands (Hall and Moreau 1970); in Mwanihana Forest, Udzungwa Mts, at 400 m near large waterfall on Sanje R. and at 1300–1700 m (Stuart *et al.* 1987). Zambia, confined to Mafinga Mts and Nyika Plateau at 1900–2200 m (R.J. Dowsett, pers. comm.). Malaŵi, Nyika and S Viphya Plateaux, at 1200 (E scarp of S Viphya) to 2450 m (F. Dowsett-Lemaire, pers. comm.). Nyika localities: Juniper Forest Stream, Chilinda Bridge, North Rukuru Stream, Zungwara Falls, Chisanga Falls, and Chire Stream (Chowo, Zambia).

Description. *O. t. tenuirostris* (Rüppell): Eritrea and Ethiopia. ADULT ♂: head and body glossy black; top and sides of head, and neck, with greenish sheen; chin blue-green; hindneck, sides of neck, mantle to uppertail-coverts, scapulars, and throat to undertail-coverts tinged violet. Tail long, strongly graduated, central feathers narrow at tips; black, upperside with faint greenish gloss. Primaries 1–8 rufous-chestnut, with distal quarter of P1–P6 and distal third of P7–P8 blackish brown; P9 blackish brown, with rufous-chestnut on base of outer web and basal half of inner web; P10 blackish brown. Rest of upperwing black; edges of secondaries, tertials, primary and greater coverts and alula glossy, tinged greenish; tips of median and lesser coverts glossy violet-black. Axillaries and underwing-coverts black, marginal coverts fringed glossy blue-green. Thick down grows on all apteria (an adaptation against extreme cold: Chapin 1954). Bill, black, greyish at extreme tip; eyes dark brown; legs and feet black, soles yellowish. ADULT ♀: head to neck and throat blackish, with spotting produced by buffy white feather tips, top and sides of head with greenish sheen. Upperparts and upperwing as in adult ♂, but with faint scaly buff feather fringes, most prominent on rump and uppertail-coverts. Underparts blackish with scaly buff feather fringes, breast and flanks with violet sheen. Tail as in adult ♂ but shorter. Axillaries and underwing-coverts blackish brown with narrow buff fringes. Bare parts as in adult ♂. SIZE (10 ♂♂, 10 ♀♀): wing, ♂ 151–158 (155),

Onychognathus tenuirostris

♀ 140–151 (146); tail, ♂ 159–181 (173), ♀ 144–165 (155); bill, ♂ 27–32 (29.7), ♀ 28–31 (30.1); tarsus, ♂ 30–34 (32.1), ♀ 30–33 (31.4). WEIGHT: ♂ (n = 5) 127–142, 2♀♀ 113, 122.

IMMATURE: juvenile similar to adult ♂, but head and upperparts less strongly glossed; sooty brown below with trace of violet gloss on breast.

NESTLING: not described.

O. t. theresae Meinertzhagen (includes '*raymondi*', ♂♂ said to have crown purplish rather than greenish (Benson 1946b, Chapin 1954)): E Zaïre to Kenya, Tanzania and N Malaŵi. Differs from nominate race in having top and sides of head and chin glossed blue.

Field Characters. 29·5–33 cm. A montane forest starling associated with waterfalls. Slimmer than Red-winged Starling *O. morio*, with *longer*, more graduated *tail, thinner bill*; at close range ♂ glossy green on head, purple on body; ♀ looks *mottled*, black with grey-tipped feathers. In Ethiopia often with Somali Starling *O. blythii*; best told by tail shape in flight: in Slender-billed, broadest point is two-thirds from base, in Somali one-third to halfway from base, producing diamond shape (Ryan and Sinclair 1998). ♂♂ similar but ♀ Somali has pale grey head and neck, dark body without mottling.

Voice. Tape-recorded (32, 102, 104, B, C, CHA, LEM, STJ). Common call a short, sharp 'pyip' or 'psip'; song in chorus includes whistles, scratchy notes and low-key chattery babble, 'chivit, chivit, jaarrr, pyeep, tyoop, jurr-

jurr, chiveet, tyaw-tyaw, blublublublublub, pyip'; for other renditions see Zimmerman et al. (1996).

General Habits. Inhabits wet evergreen montane forests, forest edges, and open alpine moorland at up to 4600 m; requires expanse of montane forest-and-grassland mosaic; breeds near high-altitude waterfalls and rocky streams; descends to cultivated land at 1400–1500 m (Kenya, Malaŵi); on Mt Kenya and in Rwenzoris, flies down mountainsides to forage during the day and returns to Alpine zone late in afternoon to roost during night; individuals might move 3000 m altitudinally every day (Brown 1965). In pairs when breeding, otherwise in noisy flocks of 12–16 birds, often 20–30 and occasionally 50–100 (Rwanda, Malaŵi) or hundreds (Aberdares, Kenya). Active, noisy and excitable. Congregates in trees and at water, often with Somali Starling *O. blythii*; also sometimes flocks with Sharpe's Starling *Pholia sharpi*. Forages in trees and shrubs along streams, feeding on fruits; collects insects to carry to nest; on Mt Kenya seen searching giant lobelias 'for small snails' (Chapin 1954), but also visits giant lobelia flowers for nectar (Feare and Craig 1998). Descends to ground to feed on insects; forages on water weeds on wet boulders at sides of streams; sometimes catches insects in flight. Roosts gregariously on dry ledges near waterfalls or in damp caves behind curtain of falling water. Comes to roosting sites about 16h00 (Kenya); often bathes in stream water before settling in roost. Where roosting cave is also a breeding site, far more birds roost than breed, e.g. 20–30 roosting birds at Harurumwe, Kenya, but only 2 nests (Brown 1965).

Food. Fruits, also some nectar and invertebrates. In Nyungwe Forest, Rwanda, fruits of *Alangium*, *Polyscias*, *Sapium*, *Schefflera goetzenii* and small orange fruit of creeper *Urera* (Dowsett-Lemaire and Dowsett 1990). At Kalongi, Zaïre, feeds mainly on *Urera hypselodendron* (fruits in 6 out of 7 stomachs, and a 13 mm snail in the 7th); in Rwenzoris, many *Podocarpus* berries (Chapin 1954); in Aberdares, Kenya, only wild olive *Olea* fruits in stomachs but flocks feed in fruiting *Trema* trees; in Rwanda, flocks visit fruiting *Prunus africanus*, *Ocotea michelsonii*, *Ilex mitis*, *Macaranga neomildbreadiana*, *Polyscias fulva* and *Maesa lanceolata* (Moermond et al. 1993); on Nyika Plateau, Zambia/Malaŵi, fruits of *P. fulva* and *Allophylus abyssinicus*. Pupae of simuliid flies (Diptera) in stomach, Itombwe, Zaïre. Brings billfuls of insects to nestlings (Dowsett-Lemaire 1983a). Feeds amongst giant lobelias and eats small soft-shelled snails in high moorlands (Kenya: Brown 1965).

Breeding Habits. Solitary and colonial nester. Pair uses same nest site for 2 or more years in succession. Double-brooded (Lynes 1934). Displaying ♂ has high cruising flight with piping calls, ♀ perching on a tree top (Cheesman and Sclater 1936).

NEST: cup-shaped mass of mud, grasses and moss, on foundation of mud (Malaŵi: Dowsett-Lemaire 1983b). Sited on ledge, in crevice or fissure on rock face (sometimes mossy, sprayed by adjacent waterfall), in water-filled recess close to rocky cascade, and behind waterfall. Site usually moist but not very wet. One nest in crack in water-sprayed rock face, 2 m above stream; another in gloom at end of rock tunnel. Nest built by ♂ and ♀ about equally: once ♂ brought moss 3 times and ♀ 4 times in 25 min. Same site can be used for up to 16 years, nest being largely rebuilt each year, in Kenya in 10 days or up to 4 months (Oct–Feb), depending on amount of water flowing in stream (Brown 1965).

EGGS: 2–4, mostly 3, av. (n = 11 clutches) 3·1. Laid at daily intervals. Blue or white, finely spotted with red-brown all over, or sometimes with large blotches. SIZE: (n = 3) av. 32·3 × 21·9.

LAYING DATES: Ethiopia, (eggs in Dec and Jan, judging by nestlings); Zaïre, Kalongi, (breeding ♀, Dec), Baraka, (bird recently out of nest, July); E Africa, Region A, Nov, Jan, Region D, Jan–Mar, May, Aug–Sept, Nov (n = 23; mainly (n = 17) Feb–Mar); Tanzania (Njombe), Oct–Nov and at same place second clutches in Jan; Zambia, Sept, Dec; Malaŵi, Oct–Dec. Laying season thought to be timed critically with regard to rain and volume of waterfalls (Brown and Britton 1980, Lewis and Pomeroy 1989).

INCUBATION: by ♀ only, fed on nest by ♂ (or ♀ leaves nest when ♂ approaches and is fed by him, by regurgitation directly into her throat, after soliciting on nearby boulder). Period: 13 days.

DEVELOPMENT AND CARE OF YOUNG: young at first feeble and quiescent, becoming active by 3rd day. Young brooded by ♀ at night and during most of first few days; ♀ leaves for spells of 2–16 min when ♂ visits nest. Later, young fed by both parents about equally, by regurgitation, or sometimes with conspicuous billfuls of insects collected within 100 m of nest (Dowsett-Lemaire 1983a). Parents arrive at and depart from nest together, usually ♀ feeding young first, then the ♂. Frequency of feeding visits to nest falls from 2·4–2·9 per h at first to <1 per h when young ready to leave. At one nest 3 well-grown young walked out of nest crevice into full spray of adjacent waterfall; feeding lasted several min as first ♀ then ♂ regurgitated slowly, and young became completely drenched by spray.

BREEDING SUCCESS/SURVIVAL: low – in one year 12 eggs produced 5 nestlings of which only 1 survived to fly, in another year 13 eggs resulted in 6 flying young (Brown 1965).

References
Brown, L.H. (1965).
Dowsett-Lemaire, F. (1983a).
Feare, C. and Craig, A. (1998).

Plate 32
(Opp. p. 479)

Onychognathus nabouroup **(Daudin). Pale-winged Starling. Rufipenne nabouroup.**

Sturnus nabouroup Daudin, 1800. Traité d'Orn., ii, p. 308; Kamiesberg, Little Namaqualand, South Africa, ex Levaillant, pl. 89.

Onychognathus nabouroup

Range and Status. Endemic resident, SW Africa. Angola, local in arid coastal belt from Benguella to R. Cunene. Namibia, common in western and central highlands, largely absent from S Namib Desert and from area east of 19°E; vagrants to Bushmanland. Botswana, vagrants to extreme SW, adjacent to Kalahari Gemsbok Nat. Park (Penry 1994). South Africa, common from Orange R. to Olifants R. in W, to Cape Fold Mts in S, E limit on Orange R. at Zastron (Free State) in Drakensberg foothills; north of Orange R. in S Free State, N Cape to Kuruman district, occasionally to Vryburg area and Kalahari Gemsbok Nat. Park. Absent from plains in central Karoo around Brandvlei, in S Karoo between Swartberg Mts and Laingsburg. Vagrant in SW Cape to Saldanha Bay and Stellenbosch; on S coast to Mossel Bay; in SE Transvaal to Delareyville and Bloemhof Dam.

Description. ADULT ♂: entire plumage uniform glossy black, except for primary feathers. Outer web of P 6–9 rufous-brown, inner web creamy white; both webs of inner primaries white; all feathers have broad black tip, so that white wing patch appears to be in centre of wing. Bill black, eyes bright orange-yellow, legs and feet black. Sexes alike. SIZE (50 ♂♂, 42 ♀♀): wing, ♂ 129–156 (145), ♀ 128–151 (140); tail, ♂ 98–121 (108), ♀ 90–112 (104); bill to skull, ♂ 26·0–31·9 (28·8), ♀ 24·4–29·8 (27·6); tarsus, ♂ 29·8–37·6 (33·5), ♀ 29·2–37·9 (32·6). WEIGHT: (Cradock, South Africa, Nov–Mar) ♂ (n = 5) 94–122 (106), ♀ (n = 13) 80–120 (101).

  IMMATURE: plumage lacks gloss, eyes initially brown.
  NESTLING: undescribed.
  TAXONOMIC NOTE: there is some geographical variation in size with southern birds longer-winged, but we prefer to treat this species as monotypic (Craig 1988b).

Field Characters. Length 26–30 cm. Glossy black, with orange eye; large white wing-patch conspicuous in flight. When perched with only rufous outer edges of primaries visible, separated from Red-winged Starling *O. morio* by short tail and orange eye; Red-winged Starling has uniform rufous wing-patch and dark eye, and ♀ has grey head. Only other similar species in same rocky habitat is black phase of Mountain Chat *Oenanthe monticola*, which differs in having white on rump and tail. Red-billed Buffalo Weaver *Bubalornis niger* is black with small white wing-patch but has thick red bill and lives in savanna.

Voice. Tape-recorded (35, 88, 91–99, B, F, GIB, GIL). Song, given by both sexes, a sustained conversational medley of low dry notes mixed with whistles, e.g. 'chuchuchup-weepo-churr-chupweechuchu-chuchupeeeer-oo-jerjerjerjer'; also written as 'chirrup-wirri-wip-chirrup' (Maclean 1993). Whistles lack sweet quality of Red-winged Starling. A ringing, rolled 'prrreeeoo' given in flight and as a flight intention call, and sometimes incorporated into song. Alarm a harsh churr, like that of Red-winged Starling.

General Habits. A bird of rocky outcrops, escarpments and gorges in arid country, found from sea level to 2000 m. Roosts at rocky sites throughout the year, breeding pairs generally roosting at their nest site. Dependent on water; drinks and bathes regularly.

Away from nest sites occurs in small flocks; flight calls stimulate other birds to reply and to fly with calling bird. In group displays, birds gather at top of a cliff, hopping and posturing. During displays wing patches are exposed to varying degrees, apparently as signals. Displays occur most often prior to nesting, in the early morning or late afternoon.

Feeds in bushes or trees or on ground; hawks insects in flight. Removes ectoparasites from Mountain Zebras *Equus z. zebra* and *E. z. hartmannae*, and regularly grooms Klipspringers *Oreotragus oreotragus*. In some areas enters towns to forage, and feeds on commercially grown dates and other fruit.

There is a partial post-juvenile moult, while adults undergo a complete moult Nov–May, with no apparent regional variation. Moult may overlap with breeding, but wing feather replacement is slow, with only 1 or 2 remiges growing at the same time.

Food. Arthropods, fruit and nectar. Insects: grasshoppers, beetles, termites, adult and larval butterflies. Ticks. Fruits of *Boscia albitrunca*, *Cotoneaster salicifolia*, *Crotalaria stuedneri*, *Cussonia paniculata*, *Diospyros lycioides*, *Ficus* spp.,

Heeria insignis, Lycium spp., *Olea africana, Phoenix reclinata, Rhus africana* and *Ziziphus mucronata*. Nectar of *Aloe ferox*.

Breeding Habits. Monogamous; pairs apparently remain together for several seasons. Defends nest site against conspecifics and occasionally against other species; much less aggressive than Red-winged Starling. Displaces potential predators or nest parasites such as Great Spotted Cuckoo *Clamator glandarius*, Common Kestrel *Falco tinnunculus* and Pied Crow *Corvus albus*. Also displaces other cliff-nesting species, e.g. Red-winged Starling, Speckled Pigeon *Columba guinea*, Cape Rock Thrush *Monticola rupestris*, and Mocking Cliff-Chat *Thamnolaea cinnamomeiventris*, and casual visitors, e.g. Red-eyed Bulbul *Pycnonotus nigricans*, Cape Glossy Starling *Lamprotornis nitens* and Green Wood-Hoopoe *Phoeniculus purpureus*. Pale-winged Starlings displaced by Mocking Cliff-Chat and Cape Rock Thrush (Craig *et al*. 1991). Group displays (see above) possibly related to individual dominance or to territory rather than to courtship. Sometimes double-brooded.

NEST: usually deep in vertical crevice in rock; rarely on building or other site. Birds wedge sticks across crevice, supporting cup of dry grass; no mud is used. Both adults bring nesting material.

EGGS: 2–5, av. 3·4 (11 clutches). Pale greenish with reddish-brown spots and smudges. SIZE: (n = 18) 29·2–35·3 × 20·7–23·0 (31·4 × 21·7).

LAYING DATES: South Africa, Oct–Mar; Namibia, Nov–Apr; dates vary according to rainfall.

INCUBATION: by ♀ only. Period: *c*. 20 days.

DEVELOPMENT AND CARE OF YOUNG: nestling period *c*. 25 days; young fed by both parents.

BREEDING SUCCESS/SURVIVAL: parasitized by Great Spotted Cuckoo.

References
Craig, A.J.F.K. and Hulley, P.E. (1992).
Craig, A.J.F.K. *et al*. (1989, 1991).

Genus *Lamprotornis* Temminck

A rather diverse assemblage of African starlings: small to large; short- to very long-tailed; most species with matt black lores, white or yellow eyes, and entire plumage burnished green, blue and violet with black spots on wings; a few with lower breast, flanks and belly bright yellow, orange or white; 2 species non-iridescent grey or grey and white. Feed on small fruits in trees but at least as much on insects on the ground. Strongly gregarious, often in mixed-species flocks. Solitary or clustered nesters, evidently monogamous, some species with helpers at the nest; little evidence of territoriality; nest a simple grassy pad in hole in tree or elsewhere, but there may be striking geographical variation in nest situation (*L. chalybaeus*: holes in trees or banks, interstices in large birds' nests); 3 species make globular nests of grass and thorn twigs. Eggs pale blue, lightly speckled with brown. Breeding biology of most species ill-known, owing to inaccessibility of nest holes in tall trees.

Hall and Moreau (1970) recognized 15 species, 6 seeming to be particularly closely allied: *nitens, chalcurus, chalybaeus, corruscus, chloropterus* and *acuticaudus*; they treated *nitens, chalcurus* and *corruscus* as a superspecies. *L. ornatus*, endemic to São Tomé, is closely related to continental *L. splendidus* (which also lives on São Tomé). Craig (1997) and Feare and Craig (1998) recognize 19 species, placing *fischeri* and *unicolor* in *Spreo* and *purpureiceps* and *cupreocauda* as sister taxa in *Hylopsar*. They separate *fischeri* generically from *superbus* (in *Lamprotornis*) on the basis of a single character, feather melanin granule structure; but the 2 species are similar in size, proportions, stance, wing formula, overall plumage pattern, range, behaviour and breeding biology, so we return *fischeri* to *Lamprotornis*. Being uncertain about the taxonomic value of feather melanins, we also retain *purpureiceps* and *cupreocauda* in *Lamprotornis*, uniting them in a superspecies. Craig treats *mevesii* and *australis* as sister taxa; we regard *mevesii* as forming a superspecies with *caudatus* and *purpuropterus*, excluding *australis* on the grounds of its extensive sympatry with *mevesii* in N Botswana (Harrison *et al*. 1997). We admit one other superspecies, *shelleyi* and *hildebrandti*.

Endemic. 23 species: 3 superspecies and 16 independent species.

Lamprotornis purpureiceps superspecies

Lamprotornis caudatus superspecies

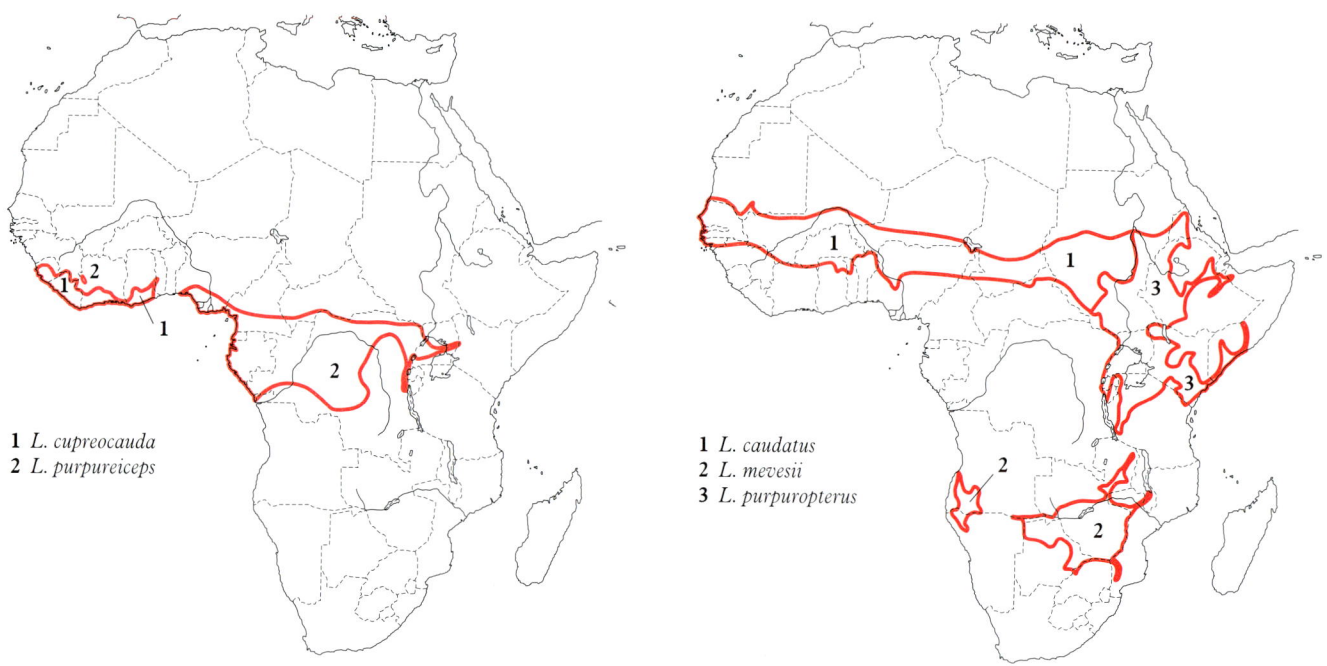

1 *L. cupreocauda*
2 *L. purpureiceps*

1 *L. caudatus*
2 *L. mevesii*
3 *L. purpuropterus*

Lamprotornis hildebrandti superspecies

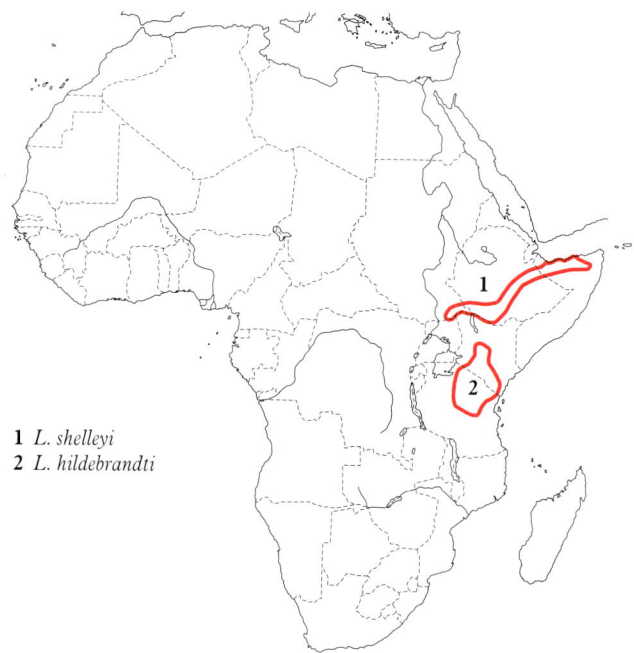

1 *L. shelleyi*
2 *L. hildebrandti*

Plate 34 *Lamprotornis cupreocauda* (**Hartlaub**). **Copper-tailed Glossy Starling. Choucador à queue bronzée.**
(Opp. p. 527) *Lamprocolius cupreocauda* Hartlaub, 1857. Syst. Orn. Westafr., p. 119; Sierra Leone, Aguapim, Gabon.

Forms a superspecies with *L. purpureiceps*.

Range and Status. Endemic resident, Sierra Leone to Ghana. Guinea, only in extreme SE (Hald-Mortensen 1971). Sierra Leone, very common in forests in SE (Collar and Stuart 1985); frequent in Kambui Hills and Nimmini Mts. Liberia, frequent to locally common throughout; commoner in years following felling and burning of forest, nesting in dead emergent or relict trees 'with consequent population decrease in intensively used older farmland and

Lamprotornis cupreocauda

savanna' (Gatter 1997); from coast up to 900 m in Wologizi Mts and Mt Nimba, groups crossing ridges up to 1300 m. Ivory Coast, common in and around Täi Nat. Park, locally frequent in semi-deciduous forest zone from San Pedro to Abidjan, north to Maroué Nat. Park and Yapo, where quite common. Ghana, one of the commonest bird species in Nini-Suhien Nat. Park and Ankasa Game Production Res. in extreme SW (Dutson and Branscombe 1990); not uncommon throughout forested parts, east to Tafo (06°15'N, 0°22'W). Togo, one in flock of Splendid Glossy Starlings *L. splendidus* in Kpété Menou (07°41'N, 0°27'E), Mar 1990 (Cheke and Walsh 1996).

2–4 pairs per km^2 in mature forest, Liberia (Gatter 1997). 2–3 pairs in *c.* 2 km^2 of cleared forest, Gola, Sierra Leone (Allport *et al.* 1989).

Description. ADULT ♂: forehead black, crown, nape and sides of neck glossy purple, hindneck purple, tinged reddish; rest of upperparts glossy greenish blue. Tail feathers blackish, outer webs of all except T6 with bronzy sheen. Lores black; ear-coverts and cheeks to chin sooty black with slight purple iridescence. Throat and upper breast glossy purple, merging into glossy blue sides of breast and band across breast, and bluish green lower breast, belly and flanks; undertail-coverts glossy purple. Flight feathers blackish brown, outer webs glossed greenish blue except for emarginated parts of outer primaries; inner webs without notches; tertials, alula and upperwing-coverts glossy greenish blue. Underwing-coverts sooty brown with bluish gloss; axillaries sooty brown. Bill black; eyes yellow; legs black. ADULT ♀ like adult ♂, but browner on lower belly. SIZE (10 ♂♂, 5 ♀♀); wing, ♂ 116–132 (122), ♀ 113–116 (114); tail, ♂ 65–77 (72·2), ♀ 65–72 (66·2); bill, ♂ 20–21 (20·4), ♀ 18–20 (19·2); tarsus, ♂ 23–24 (23·5), ♀ 22–23 (22·8). WEIGHT: (Liberia), ♂ (n = 9) 53–66 (60·7), ♀ (n = 3) 53–65 (58·3).

IMMATURE: juvenile sooty brown, crown with slight purple gloss, rest of upperparts tinged glossy green. Tail sooty brown; wings as in adult, but glossy areas greener. Only tips of feathers are metallic, dull dark brown feather bases showing through.

NESTLING: not described.

Field Characters. Length 20–21·5 cm. A small, short-tailed forest starling, similar to Purple-headed Glossy Starling *L. purpureiceps* which it replaces west of the Dahomey Gap. Glossy steel blue except for *reddish violet* head, neck and breast. Only other congener in forest is Splendid Glossy Starling *L. splendidus*, which is twice the size, glossy blue-green above and purple below, has grating voice. Lesser Blue-eared Starling *L. chloropterus* lives outside forest and is uniform blue-green. Could be confused with similar-sized Purple-throated Cuckoo-Shrike *Campephaga quiscalina* (Bannerman 1948), but latter is uniform steel-green, with purple confined to patch on throat.

Voice. Tape-recorded (104, B, CHA). Calls include short, harsh 'jeeu', scratchy 'pshili' and 'jer-yu' and ringing 'weee-iyee'; song 'very squeaky' (Bannerman 1948).

General Habits. Inhabits mature lowland forest, swamp forest, and large emergents along forest rivers (Gatter 1997); semi-deciduous forests in Ivory Coast and 'all habitats' in Täi Nat. Park. Forages at heights above ground of 10–45 m (rarely, at 5–10 m), with mode at 15–20 m and mean at *c.* 23 m, searching dense and open parts of canopy (Gatter 1997). In pairs and flocks of up to 50 birds. Keeps to topmost branches of tall trees, often isolated ones left standing on cleared ground, perching prominently on dead branches. Seen in canopy more commonly after than before 15h00 (Dutson and Branscombe 1990). Retiring, solitary, in pairs and family parties which often join mixed-species foraging flocks. Flight swift and direct. Not very vocal, and does not produce swishing sound with wings in flight. In Liberia moults remiges Jan–May, in fresh plumage June–Aug and worn plumage Aug–Nov (Gatter 1997).

Food. Insects, figs and berries (Gatter 1997); fruits of '*Rauwogia*' (= *Rauvolfia*? – Bannerman 1948).

Breeding Habits.

NEST: in hole high in standing dead tree (Gartshore *et al.* 1995); hole in dead branch 15 m up (Dutson and Branscombe 1990); hole in dead branch 15 m up (Dutson and Branscombe 1990); holes in dead trees in recently cleared forest planted with rice (Allport *et al.* 1989).

EGGS: 3 (i.e. 3 fledglings once seen being fed).

LAYING DATES: Sierra Leone, (nesting Oct and imms Dec: Allport *et al.* 1989). Liberia, (adults feeding nestlings Nov–Jan, mainly Oct; independent fledglings Oct–Nov, Jan: Gatter 1997). Ivory Coast, (active nest, Mt Kopé, Nov; adults feeding 3 fledged young, Täi, Feb: Gartshore *et al.* 1995); Ghana, (carrying nest material to tree hole, Aug).

References
Allport, G. *et al.* (1989)
Bannerman, D.A. (1948).
Colston, P.R. and Curry-Lindahl, K. (1986).
Dutson, G. and Branscombe, J. (1990).
Gartshore, M.E. *et al.* (1995).
Gatter, W. (1997).

Plate 34 (Opp. p. 527) *Lamprotornis purpureiceps* (J. and E. Verreaux). **Purple-headed Glossy Starling.** Choucador à tête pourprée.

Lamprocolius purpureiceps J. and E. Verreaux, 1851. Rev. Mag. Zool., p. 418; West Africa (= Gabon).

Forms a superspecies with *L. cupreocauda*.

Range and Status. Endemic resident. Guinea and Liberia, a few old records which doubtless refer to *L. cupreocauda* (Holgersen 1956, Gatter 1997). Ivory Coast, status uncertain: but collected Sipilou (Thiollay 1985); flock sighted Täi, perhaps also *L. cupreocauda*, since not found in Taï by Gartshore *et al.* (1995) nor in Yapo Forest by Demey and Fishpool (1994). 1st records Benin, Lama Forest, 1998 (Waltert and Mühlenberg 1999). Nigeria, uncommon, from Lagos (4 records in 7 years, in Agbara, Ijede and Tarkwa: Alexander-Marrack *et al.* 1985) and Calabar north to Gambari, Benin and Umuagwu. Cameroon, common in forest zone. Central African Republic, known only from SW, in Dzanga Res. (Bai Hokou, Ngoubounga), Bayanga and Lobaye Préfecture, not certainly recorded in S (Hall and Moreau 1970). Congo, common and widespread in Odzala and Nouabalé-Ndoki Nat. Parks, scattered records through belt in centre, and very common in Kouilou Basin and central Mayombe. Zaïre, probably much more widespread than shown, south to Kasai and Itombwe, but does not occur in N Uele; very common in Itombwe, at 860–1490 m. Uganda, common at 700–1800 m in forests in W and S: Budongo, Kibale, Bwamba, Maramagambo, Lugalambo and Bwindi east to Kifu, Entebbe, Mabira and W Elgon. Angola, confined to Cabinda.

Lamprotornis purpureiceps

Description. ADULT ♂: forehead, crown and nape deep velvety black, with slight purple or violet gloss; rest of upperparts glossy bluish green; uppertail-coverts long, metallic green, posterior ones more golden bronze with violet edges. Tail purplish black, upperside of T1 and outer webs of T2–T5 with bronzy sheen. Lores to ear-coverts black; cheeks and sides of neck to chin and throat black with slight purple gloss. Upper breast and sides of breast glossy purple, sharply demarcated from glossy bluish green lower breast, belly and flanks; undertail-coverts glossy purple. Flight feathers purplish black, outer webs glossy blue except for emarginated part of outer primaries; P10 22 mm longer than its under covert, broad, curved inwards; P7 longest; P7–P9 with shallow notch on inner web, *c.* 20–25 from feather tips; tertials, alula and upperwing-coverts glossy greenish blue. Underwing-coverts and axillaries black with blue-green gloss. Bill black; eyes dark brown; legs black. Sexes alike. SIZE (10 ♂♂, 7 ♀♀): wing, ♂ 115–129 (121), ♀ 112–119 (114); tail, ♂ 60–68 (65·2), ♀ 59–67 (63·4); bill, ♂ 20–23 (21·4), ♀ 21–22 (21·2); tarsus, ♂ 23–25 (23·5), ♀ 21–24 (22·2). WEIGHT: ♂ (n = 14) 54–79, ♀ (n = 4) 50–70 (Feare and Craig 1998).

IMMATURE: like adult but less glossy, some brown feather bases visible, middle of throat brown, sides of throat brown with feather tips metallic, inner secondaries brown. Juvenile duller and browner; underparts dark brown with slight purple gloss on upper breast and greenish one on belly.

NESTLING: naked (Chapin 1954).

Field Characters. Length 18·5–20 cm. A small *forest* starling, dark steel-blue with *purple* head and breast; recognized when perched by dumpy silhouette and in flight by *broad wings and short tail* (Christy and Clarke 1994). Only other congener in forest is Splendid Glossy Starling *L. splendidus*, which is much larger, with longish tail, blue-green upperparts and black face, loud nasal voice.

Voice. Tape-recorded (32, 104, B, C, CHA, ERA, LEM). Song from single bird rather undistinguished, a measured series of tuneless scratchy notes, interspersed with loud liquid 'pleep', given at *c.* 1·5 s intervals: 'jeep ... jyup ... jer ... pleep ... jyup ... jeeyup ... '. Variety of similar notes given in chorus, including 'psup', 'pishu', short 'peep', low 'wew', querulous 'jeejee' and 'jyu-jer'. Song also described as a ringing series of bell-like notes, 'twee-hwee-too' (Christy and Clarke 1994). Imitates other species, e.g. Common Bulbul *Pycnonotus barbatus* and Chestnut Wattle-eye *Dyaphorophyia castanea* (Chappuis, in press).

General Habits. Inhabits lowland rain forest and old second growth forest, often in clearings; gallery forest; forested islands in large rivers. In canopy and mid strata; descends into small trees and undergrowth more readily than other forest starlings, to search for small fruits. Retiring, solitary, in pairs, and in flocks of 4–5, 10–12 and 20–30, or rarely up to 50 or 100 birds (in Gabon in Apr–Sept), concentrating to feed in crown of fruiting tree and moving over forest between fruiting trees. Congregates at tree-top food source with Splendid Glossy Starlings *L. splendidus*, Chestnut-winged Starlings *Onychognathus fulgi-*

dus, large hornbills and other birds, where it is usually near bottom of dominance hierarchy because of its small size (Brosset and Erard 1986). Joins mixed-species foraging flocks of insectivores. Hawks after flying insects, like a roller (Christy and Clarke 1994). Not very vocal, and does not produce swishing sound with wings in flight.

Food. Fruits of *Rauvolfia*, *Ficus*, *Heisteria*, *Musanga*, *Pycnanthus*, *Trichoscypha*, *Morinda*, *Macaranga*, *Allophyllus*, *Polyalthia* and *Xylopia*; capsules of *Macaranga assas* and berries of *Ficus mangifera*; also 'red berries as large as acorns' and plum-like fruits, and insects: caterpillars, orthopterans, mantis nymphs, and termites (Brosset and Erard 1986). In 13 stomachs, berries and other small fruits and a small snail (Chapin 1954).

Breeding Habits. Barely known. 2 birds repeatedly entered high knot-hole in forest tree, Cameroon; one shot was a ♂ in breeding condition; 2 days later still 2 birds entered the hole, 1 carrying twig in bill (Bates 1911). Bird carrying twig entered knot-hole high in tree in clearing (Zaïre, Chapin 1954). Pair flew to 30 m high tree, Cameroon, presumed ♂ entered hole, followed by ♀ which stayed in hole for 10 min (to brood chicks?) whilst ♂ sang softly and intermittently on adjacent branch; in following week both birds continued bringing food, often arriving together (Dowsett-Lemaire 1998). Pair of birds visited broken stump *c*. 25 m above ground in *Santiria trimera* tree (Odzala, Congo). One nest in stump left by fallen lateral branch, *c*. 26 m up in tall *Alstonia boonei* tree, in semi-evergreen forest with open canopy (Lobeke Res., Cameroon: Dowsett-Lemaire 1998). Pair at hole 4 m high in dead tree in gallery forest clearing (Gabon: Christy and Clarke 1994).

NEST: in knot-hole, branch-stump or other cavity in dead branch or dead emergent tree, 4–25 m high in forest. Only nest examined, in *Musanga* tree cut down for the purpose, consisted of 'green leaves cut in small bits by the birds' (Chapin 1954).

EGGS: up to 3 (i.e. 3 nestlings in nest hole). Only egg taken (from the *Musanga* tree, see above) was blue with brown spotting. SIZE 23·5 × 18 (Chapin 1954).

LAYING DATES: Cameroon (Efulen), (♂ in breeding condition, and building, June), (Lobeke), (pair entered hole carrying food, Apr); Gabon, (Jan); Congo (Odzala), (pair prospected hole Nov–Dec); Zaïre (Avakubi-Medje area), (oviduct egg, Aug; breeding condition June–Oct), (Beni), (building, Oct), (Yokolo), June, (Itombwe), Jan–Mar and May (laying months by extrapolation: Prigogine 1971).

References
Bates, G.L. (1911).
Brosset, A. and Erard, C. (1986).
Chapin, J.P. (1954).
Christy, P. and Clarke, W. (1994).
Dowsett-Lemaire, F. (1998).
Feare, C. and Craig, A. (1998).

Lamprotornis purpureus (Müller). **Purple Glossy Starling. Choucador pourpré.**

Plate 34
(Opp. p. 527)

Turdus purpureus P.L.S. Müller, 1766. Natursyst., Suppl., p. 143; Juida (= Dahomey or Benin).

Range and Status. Endemic resident and partial migrant. Mauritania, upper Senegal valley and S Karakoro valley. Senegal, north to Linguère and Révane (15°35′N), widespread south of 15°N but uncommon in Oussouye region. Gambia, common to abundant. Guinea-Bissau. Guinea, Koundara and Gaoual regions of Fouta Djallon; record at Kankan in NE. Mali, common and widespread south of 15°N; abundant in Upper Niger and Bani valleys and Ban Markala in wet season. Ivory Coast, frequent in N guinean savannas, scarce south to Toumodi. Burkina Faso, resident north to Ouagadougou and Ouhigouya, and in E occurs north to 12°31′N. Ghana, Togo and Benin, widespread and common – the commonest *Lamprotornis* starling. Niger, resident in 'W' Nat. Park; uncommon elsewhere: records Ayorou, Tillabéry, Niamey (where frequent Oct–Feb), north of Niamey (June–July), and between Maradi and Birni-Nkonni. Nigeria, widespread north of forest zone; uncommon towards SW coast; north to Sokoto and Kano, south to Badagri and (scarce) Serti. Cameroon, common south to about 07°30′N and Adamoua Plateau. Chad, quite common, restricted to soudanian zone. Central African Republic, common in Bamingui-Bangoran and Manovo-Gounda-Saint Floris Nat. Parks and Vakaga region. Sudan, fairly common in less-disturbed woodland from Jebel

Lamprotornis purpureus

Marra foothills to Equatoria and in Sudd and Nile lowlands north to S Kordofan and about Kaka; common in extreme S. Zaïre, only in extreme NE border country, in R. Bomu valley along Central African Republic border west to about Bangassou, and in Garamba Nat. Park near Sudan border. Uganda, widespread and locally common at 600–1800 m; absent from arid and forested areas. Kenya, uncommon from N slopes of Mt Elgon, Saiwa Nat. Park and Kapenguria to Busia, Ukwala, Ng'iya, Akala, Maseno, Nyangweso and Nyarondo; formerly more widespread.

Description. *L. p. purpureus* (Müller): Senegal (Casamance) to Cameroon (where intergrades with *amethystinus*). ADULT ♂: forehead and crown glossy purple, merging through bluish nape, hindneck and sides of neck to glossy bluish green on mantle and scapulars, blue on back and violet-blue on rump; upper tail-coverts glossy violet, larger feathers purple. Inner tail feathers, T1–T3, glossy purple, T4–T5 purple with bluish tip, T6 blue with purplish inner web; tail dark brown below. Lores blackish; ear-coverts purplish black; cheeks to chin, throat and upper breast glossy purple, grading to deep glossy violet on lower breast, belly, flanks and thigh feathers; undertail-coverts violet. Flight feathers glossy blue-green; tertials, alula and primary coverts glossy greenish blue, greater and median coverts greener; lesser coverts glossy blue; tertials, greater coverts and median coverts tipped with matt blackish spots. Underwing-coverts and axillaries violet; median under primary coverts glossy blue. Inner webs of P9–P7 slightly notched (**A**); P10 very small. Bill black; eyes yellow (iris and adjacent sclera), orbit large (Bates 1924); legs black. Sexes alike. SIZE (10 ♂♂, 10 ♀♀): wing, ♂ 150–161 (155), ♀ 140–158 (149); tail, ♂ 74–91 (83·7), ♀ 76–86 (81·7); bill, ♂ 28–32 (29·8), ♀ 27–30 (28·1); tarsus, ♂ 34–38 (35·6), ♀ 33–36 (33·9). WEIGHT: 1 ♀ 83, unsexed (n = 3) 91–140 (119).

IMMATURE: juvenile brown above, feather tips glossed green on mantle and scapulars, violet on rump; head and underparts sooty brown, breast and top of head with slight violet gloss; wings like adult but duller and greener; central tail feathers glossed purple, the rest blue on outer web; eyes pale greenish yellow.

NESTLING: not described.

L. p. amethystinus Heuglin: Cameroon to W Kenya. Slightly bluer above than nominate race; more purple, less violet, below. Tail longer. SIZE: (6 ♂♂), wing 148–157 (153), tail 89–98 (93·2).

Field Characters. Length 26 cm. A bulky open-country starling with *glossy purple* head, body and tail and large orange-yellow *eye*. Long bill and *flat forehead* give head a vulturine jizz in some postures (Barlow *et al.* 1997); *broad wings and short tail* produce characteristic flight silhouette; at rest, wing-tips almost reach end of tail. Immature duller, underparts sooty with slight violet gloss, eye greyish becoming pale yellow with age.

Voice. Tape-recorded (104, B, BRU, GREG, GRI, HOR, McVIC, PAY). Single ♂ gave measured series of unmusical scratchy calls at 30 s intervals, 'jrip ... bajreep ... jip ... jreep ...'. Contact call a rising 'squee-caree'; alarm scold a hoarse, persistent 'shree' (Barlow *et al.* 1997). Babbling chorus includes shrill, clear 'tewy', grating 'jawy' and 'djwayo', and squeaky notes.

General Habits. Inhabits savanna woodlands, parkland, bushland, bushy grassland, hillsides dotted with bushes and small trees, cultivated land with plenty of woody growth, recently harvested fields, cotton fields, fruiting trees around villages and suburbs; sometimes in forest clearings (Ghana), and favours burnt areas in forest (Bannerman 1948); comes down to watered lawns; in late afternoon often comes to puddles to bathe and drink. Gregarious; in pairs when nesting, otherwise in flocks of tens or hundreds. Feeds 'on a variety of fruits in trees' (Chapin 1954); strongly attracted to flowering kapok trees *Ceiba pentandra* (Benin: Jan–Feb, Claffey 1995); perches on topmost branches of trees; sometimes hawks flying insects, and quick to exploit hatches of termites. Often comes to ground, where gait a walk, and forages there in company of other glossy starlings. Flight swift, steady and direct, without undulation. Flock of several thousand seen in evening flying in long line to roost (Mali). In Ghana, hundreds roost with other starlings in a neem *Azadirachta indica* plantation between Legon and Achimota. Flocks can be very noisy. Mobs owls.

Flocks of several hundreds in S Mali in rains, July–Sept and in N Ghana in dry season, Oct–Apr; moves south in Ivory Coast in dry season.

Food. Small fruits and insects (see above). Nestlings fed with caterpillars (Bannerman 1948). Eats *Ziziphus* drupes; picks termites and ants from ground; gathers in large numbers to eat fruits of exotic neem *Azadirachta indica* (Bannerman 1948). Flock of 20 fed in large *Ficus lecardii* tree for 4 weeks (Zaria, Nigeria); often in *Ficus* trees in Uganda.

Breeding Habits.

NEST: in hole in tree trunk or stump, under eaves, and in drainpipes (Grimes 1987). One in tree hole 4 m above ground, in passage *c.* 45 cm long, 'too narrow to admit an arm' (Bannerman 1948); one in locust-bean tree. One nest (captive birds) was pad or low cup of dried grasses, lined with fragments of leaves (Feare and Craig 1998).

EGGS: 2–3. Pale blue to dark blue, with numerous red-brown marks and paler flecks. SIZE: (n = 7) 22·3–30·5 × 18·8–23·0. Weight: 5·9.

LAYING DATES: Gambia, (breeds in end of dry season and early wet season); Ghana, (enlarged gonads, Feb; entering holes, Apr–July; nestlings, May); Nigeria, (nestlings June; flying young Apr); Sudan, (Mar–May); E Africa, Region B, (Feb, Apr).

INCUBATION: by ♀ only.

DEVELOPMENT AND CARE OF YOUNG: at one nest (captive birds) ♀ continued to bring leaf fragments in for a week after eggs hatched. Nestlings fed by both parents.

References
Bannerman, D.A. (1948).
Feare, C. and Craig, A. (1998).

A

Lamprotornis nitens (Linnaeus). Cape Glossy Starling. Choucador à épaulettes rouges.

Turdus nitens Linnaeus, 1766. Syst. Nat., 12th ed., p. 294; Cape of Good Hope.

Plate 34
(Opp. p. 527)

Lamprotornis nitens

Range and Status. Endemic resident. Gabon, old records from Ogowe (Reichenow 1903); Congo, vagrants at Djeno, pair at Lac Bleu (Dowsett and Dowsett-Lemaire 1997); Zaïre, reported from mouth of Congo R. and in SW. Angola, from Congo mouth south to Benguella, inland along lower Cuanza R. and S Cuanza Norte, in S Benguella and N, S and central Huila; locally common, sparse on coast to Mossamedes. SW Zambia, local around Chunga Pools, Cholola. Zimbabwe, absent from Zambezi Valley from L. Kariba eastwards; local on plateau around Harare, commoner in W and S, from Hwange to Gwelo and south to Sengwe. Mozambique, common south of R. Limpopo. Namibia, local in arid coastal belt, and sparse in S and SE sector; elsewhere widespread and common. Botswana, widespread throughout, especially abundant in S and SE. South Africa, widespread and locally common north of Orange R. from Augrabies and Kalahari Gemsbok Nat. Park through N Cape, Free State, Transvaal and Natal; absent from grasslands of SE Transvaal and Drakensberg area; south of Orange R., isolated population in interior of Namaqualand from Springbok south-east to Garies, and just north of Calvinia; in E Cape along coast and inland westwards to Uniondale, but absent from highland grasslands; locally common resident. Vagrants in SW Cape at George, Mossel Bay, Calitzdorp, Mooreesburg, Matjiesfontein and Saldanha. Lesotho, local in W lowlands. Swaziland, common in lowveld, uncommon on higher ground.

Description. ADULT ♂: lores black; forehead, crown, nape, chin and sides of head glossy blue; purple wash on ear-coverts. Mantle, back, rump, uppertail-coverts and tail glossy blue-green. Throat and upper breast blue-green with bluish sheen, lower breast, belly and undertail-coverts with greenish sheen; thighs bluish. Wings blue-green with small dark blue terminal spots on some of the lesser coverts; small bronzy-purple patch may show as copper bar on wrist. P 6–9 strongly indented on inner webs (**A**). Underwing blue. Bill black, eyes orange-yellow, legs and feet black. Sexes alike. A partial albino had pure white head, mantle and chest, and no iridescence in plumage. SIZE (10 ♂♂, 10 ♀♀): wing, ♂ 130–137 (134), ♀ 119–134 (128); tail, ♂ 84–93 (88·3), ♀ 77–90 (85·6); bill to skull, ♂ 24·9–29·0 (26·7), ♀ 23·7–28·5 (26·3); tarsus, ♂ 32·6–35·6 (34·3), ♀ 31·0–34·0 (32·5). Birds largest in south of range; wing, ♂ (n = 10) 132–147 (141). WEIGHT: (South Africa) ♂ (n = 8) 65·3–116 (88·5), ♀ (n = 7) 71·5–98·2 (88·7), unsexed (n = 16) 76·3–105 (85·6); (Namibia) ♂ (n = 9) 77–85 (80·9), ♀ (n = 5) 67–80 (73·8).
IMMATURE: dull greenish above with little gloss; underside appears matt black. Eyes dull grey, becoming dull yellow after about 3 months, and orange-yellow by 6 months.
NESTLING: undescribed.

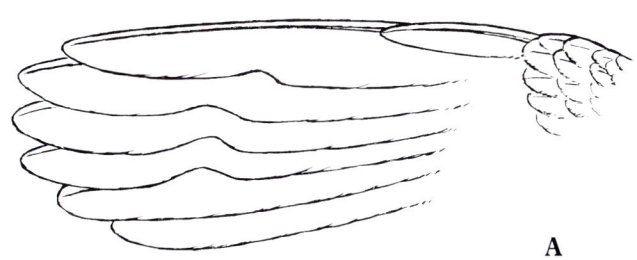

A

Field Characters. Length 22–23 cm. Commonest glossy starling over much of southern Africa. *Uniform glossy blue-green*, including head (no dark ear-coverts), flanks and belly. Most quickly identified by distinctive *rolling call note*. Wings make a loud swishing sound in flight. Often in mixed flocks with very similar Greater Blue-eared and Lesser Blue-eared Starlings *L. chalybaeus* and *L. chloropterus* which have contrasting blue-black ear-coverts, blue or magenta belly and flanks. Immature duller than adult, often appearing blackish below; eye grey, becoming dull yellow with age.

Voice. Tape recorded (17, 51, 75, 88, 91–99, 104, B, F, KON). Flight call a rolling 'turr-reee' or 'turr-rreeu', always uttered on take-off, sometimes preceded by staccato 'tup-turr-rreeu' (Maclean 1993). Song, given by both sexes all year, often in chorus, a continuous but disjointed medley of dry notes and whistles, with a casual quality; whistled 'chop-chirry', 'chu-cheeup' or 'taweeo-chip' interspersed with grating 'jewee', nasal 'dweeya-dweeya' and low conversational notes; flight calls sometimes incorporated. Whistled notes are clear but not sweet. Reported to imitate alarm calls of *Pycnonotus* bulbuls (Vernon 1973). Alarm, harsh 'kraa'.

General Habits. Inhabits savanna and riverine bush; in arid regions occurs along watercourses; has adapted to disturbed habitats such as forest margins, plantations, farmyards, parks and gardens; from sea level to 1800 m.
Mostly in flocks of 10–20 birds which forage together and share a regular roost site; sometimes forms much

larger concentrations. Groups call in chorus from trees during hot part of day, also at roost site (even after dark on moonlit nights). Feeds extensively on ground, where runs and hops; often associates with wild or domestic ungulates, feeding at their feet, sometimes perching on them to hawk insects or remove ectoparasites.

Adults have a complete moult, juveniles a partial moult, during Dec–May; no evidence of moult-breeding overlap.

Food. Insects: flies, ants, termites, beetles, wasps, caterpillars and short-horned grasshoppers. Occasionally millipedes, ticks and carrion. Bread, bone meal, and other scraps at feeding tables or picnic sites. Fruits of *Atriplex bacifera, Azima tetracantha, Diospyros pubescens, Lycium, Olea, Rhus, Scutia myrtina*; also cultivated figs, dates and grapes. Nectar of *Aloe candelabrum, A. davyana, A. ferox, A. marlothii, A. spectabilis, A. vandalenii, Boscia albitrunca, Erythrina caffra, Protea subvestita*.

Breeding Habits. Monogamous, with up to 4 helpers which are usually the offspring of the breeding pair. Territorial, and ♂ may be aggressive towards helpers at nest site. Aerial pursuits occur, but courtship behaviour not described.

NEST: pad of dry grass, twigs, feathers, dried dung and occasionally shed snakeskins, in tree hole or in man-made structure, 1–4 m above ground; rarely in hole in river bank; also in nest box, or hollow pipe. Built by both sexes. Nest site often used in successive seasons.

EGGS: 2–6, av. 2·8 (53 clutches). Light greenish blue, speckled with red. SIZE: (n = 128) 25·4–32·5 × 19·0–22·5 (28·1 × 20·0).

LAYING DATES: no breeding data for Gabon, Zaïre, or Zambia. Angola, ♀ ready to lay Oct (Traylor 1963), probable breeding Feb–Mar (Günther and Feiler 1986). Namibia and Botswana, Oct–Apr; Zimbabwe, Sept–Dec; Mozambique, Oct–Feb. South Africa, Sept–Mar (primarily Oct–Jan, also records for Apr, June, July in lowveld) in Transvaal; Oct–Feb in Natal, Sept–Feb in Free State and Cape.

INCUBATION: by ♀ only; period: 12–14 days. Helpers bring feathers to ♀ on nest.

DEVELOPMENT AND CARE OF YOUNG: young fed by both adults and helpers; nestling period *c.* 20 days.

BREEDING SUCCESS/SURVIVAL: parasitized by Great Spotted Cuckoo *Clamator glandarius*, perhaps also by Greater Honeyguide *Indicator indicator*.

References
Craig, A. (1983a, c).

Plate 34
(Opp. p. 527)

Lamprotornis chalcurus Nordmann. Bronze-tailed Glossy Starling. Choucador à queue violette.

Lamprotornis chalcura Nordmann, 1835. In Erman's Reise Naturh., Atlas, p. 8; Senegal.

Range and Status. Endemic resident and partial migrant. Status hard to determine, because of widespread confusion in the field with commoner Lesser Blue-eared Starling *L. chloropterus*. Senegal, Révane (15°35'N); quite common near Tambacounda and Oussouye and east of Kédougou. Gambia, frequent to locally common except within 30 km of coast. Guinea-Bissau. Guinea, Gaoual and Kindia regions. Mali, common south of 13°N, including Mandingo Mts and Bamako. Ivory Coast, resident or dry-season visitor in N guinean savannas, scarce in wet season. Burkina Faso, north to 10°48'N (Holyoak and Seddon 1989). Ghana, not uncommon in N where may be a dry season visitor; in S to coast, where probably resident. Togo, so far found only at 10–11°N. Benin, rare resident, Béterou (Claffey 1995). Nigeria, uncommon, breeds north to Kaduna and Sokoto; in E occurs south to Serti in dry season, where by far the commonest glossy starling species (Hall 1977c). Cameroon, uncommon in N; record near Douala. Occurs in S Chad, but range and status there uncertain. Central African Republic, widespread but uncommon in N. SW Sudan as mapped. NE Zaïre, only in extreme NE: Garamba Nat. Park and Faradje to Aba. Uganda and Kenya, uncommon in bushy grasslands at 500–1200 m in medium rainfall areas, in Uganda south to Bunyoro and Teso and east to Kidepo Nat. Park and Mt Elgon, and in Kenya to Lokichokio, Elgon-Kimiriri R.-Kitale-Kapenguria area and parts of Kerio valley; old records from Kakamega.

Lamprotornis chalcurus

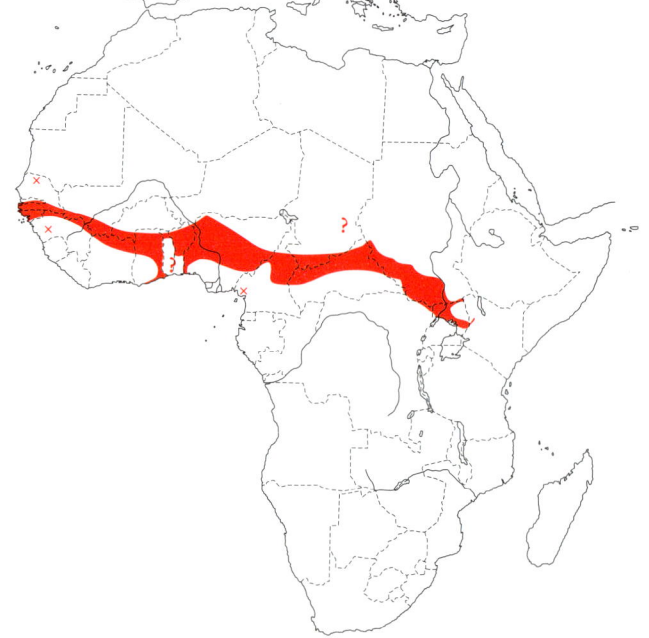

A

Description. *L. c. emini* (Neumann): E Cameroon to W Kenya. ADULT ♂: forehead to hindneck, sides of neck, mantle, scapulars and back glossy bluish green; rump and uppertail-coverts violet-blue. Inner tail feathers, T1–T2, glossy purple with bluer tip; outer ones glossy blue, tinged purple basally; all can show bronzy reflections in some lights. Tail blackish brown below. Lores black; ear-coverts black with violet gloss. Cheeks to chin and throat glossy greenish blue, grading into greener upper breast, then into violet on lower breast and glossy purple on belly and flanks; vent dark brown; undertail-coverts glossy blue. Flight feathers dark brown, upperside of outer webs glossy green; tertials, alula, primary coverts and greater coverts glossy green; median and lesser coverts bluer, inner leading lesser coverts violet-blue; secondaries, tertials, greater coverts and median coverts tipped with non-reflective black spots. Underwing-coverts and axillaries glossy purple. Inner webs of P6–P9 notched (**A**); P10 small, *c.* 6 mm longer than primary coverts. Bill black; eyes deep yellow or orange; legs black. Sexes alike. SIZE: wing, ♂ (n = 10) 131–139 (137), ♀ (n = 10) 124–143 (133) (wings of ♂♂ up to 7–8 mm longer in Uganda than in NE Zaïre: Chapin 1954); tail, ♂ (n = 10) 79–88 (84·4), ♀ (n = 10) 73–87 (82·2); bill, ♂ (n = 9) 24–27 (25·5), ♀ (n = 8) 23–26 (24·8); tarsus, ♂ (n = 10) 31–33 (31·9), ♀ (n = 8) 30–32 (30·8). WEIGHT: ♂ (n = 1) 63.

IMMATURE: juvenile dark brown above with glossy feather tips, blue-green on mantle, violet on rump; dark brown on head and underparts, with faint blue-green gloss, mainly on crown and breast. Wings and tail similar to adult, but green and purple gloss less intense, and wing feathers lack black spots.

NESTLING: unknown.

L. c. chalcurus (Nordmann): Senegal and Guinea Bissau to N Cameroon (where intergrades with *emini*). Similar to *emini*, but rump bluer, less violet. Wing longer but tail shorter: wing, ♂ (n = 9) 133–148 (142); tail, ♂ (n = 9) 75–86 (80·9).

Field Characters. Length 19·5–23 cm. A glossy blue-green starling of N savannas, generally less common than other congeners; *eye reddish orange* (often the first feature to attract attention in a mixed flock), and black wedge in front of eye creates an angry expression (Barlow *et al.* 1997). Tail short, wing-tips reaching about half-way along it at rest. Told from Greater and Lesser Blue-eared Starlings *L. chalybaeus* and *L. chloropterus* by *purple* belly, flanks and tail (tail shows bronze reflections at some angles, especially when bird flies against the light: Zimmerman *et al.* 1996), and from Purple Glossy Starling *L. purpureus* by blue-green head and neck. Violet-blue rump and uppertail-coverts contrast with blue-green back. Immature sooty brown with reduced gloss; eye grey, becoming pale yellow with age.

Voice. Tape-recorded (104, B, CHA, GREG, MOR). Characteristic call an up-down 'ju-wee-yurr', whistled but with husky quality; in song from perch individual mixed this with shrill, sharp and grating calls, 'wo-tchaap ... tchweep ... ju-wee-yurr ... chocho-weeyip ... wo-cheep ... wok-a-wee-yurr ... '. In chorus, 'wee-yurr' stands out among low-pitched chatter and squeaky babble.

General Habits. Inhabits mesic savanna bushlands and woodlands, both mature (forming deciduous forest: Claffey 1995) and degraded (into scrubby grasslands: Chapin 1954); in W Africa typically in S-guinean zone; seems to require better woodland in west than in east of range. Also cultivation and fruiting trees around villages and suburbs; comes down to watered lawns; often perches on leafless tree branches. Occurs in pairs in short breeding season and in flocks of up to 100 or more birds in rest of the year. Flighty and mobile. Feeds in trees but mainly on ground, often in company of other starling species, especially Greater Blue-eared Starling *L. chalybaeus* and Lesser Blue-eared Starling; gait on ground a walk. Calls on ground and in flight. Flight strong and direct, without undulation, with even wing beats and little or no gliding except momentarily when alighting in tree; wings make considerable swishing noise.

Mainly resident but (like most frugivores) seasonally mobile; appears to be a partial migrant towards S borders of range in W Africa: mainly a dry-season visitor in N Ivory Coast, scarce in rains (June–Sept); seemingly a dry season visitor in parts of Ghana and SE Nigeria. Present in Sudan in Sept–Apr (Nikolaus 1987).

Food. Of 6 stomachs, 4 contained insects and 2 contained berries (Chapin 1954).

Breeding Habits.

NEST: one in natural cavity 2 m up in bole of shade tree, had whole interior of cavity stuffed with grasses to depth of 20 cm, with thick layer of feathers on top, mainly from adjacent colony of Cattle Egrets *Bubulcus ibis*, clutch laid on top of the feathers several inches below lower lip of entrance (Serle 1943); another made of dry leaves and grass in bottom of cavity 4 m up in old tree stump, in high grass near swamp (Chapin 1954).

EGGS: up to 4. Long ovals, very slightly glossy; pale blue, immaculate or sparingly and finely speckled and spotted with pale orange-brown. SIZE: (n = 4, 1 clutch) 26·5–27·7 × 19·8–20·2 (27·3 × 19·85).

LAYING DATES: Senegal, (young Aug); Gambia, (young June); Nigeria, June (and entering hole Mar; juvs May); Zaïre, (Feb–Mar).

INCUBATION: by ♀ (Serle 1943).

Reference
Bannerman, D.A. (1948).

Plate 34
(Opp. p. 527)

Lamprotornis corruscus Nordmann. Black-bellied Glossy Starling. Choucador à ventre noir.

Lamprotornis corrusca Nordmann, 1835. *In* Erman's Reise Naturhist. Atlas, p. 9; Galgenbosch, Eastern Province, South Africa.

Range and Status. Endemic resident, eastern coastal belt. S Somalia, locally common in Jubba valley and on coast south from Kismayu. Kenya, common in coastal lowlands, penetrates inland in wooded habitat along Tana R. to edge of central highlands. Tanzania, common throughout coastal lowlands as mapped (N.E. Baker, pers. comm.), on Mafia I., Zanzibar and Pemba, and inland to Ngurus and E Usambaras. Mozambique, widespread in coastal lowlands and on offshore islands with some inland records; abundant at certain times of year. SE Zimbabwe, known only from the lowland forest at Haroni-Lusitu junction. E Swaziland, common in Lubombos, vagrant elsewhere. South Africa, seldom more than 50 km from coast, except in SE Transvaal where occasional records on lower reaches of Crocodile, Komati and Pongola rivers; common from Natal south to East London, but further west occurrence more sporadic; regularly reaches S Cape forests of George-Knysna area, and vagrants reported as far west as Wydgelee and Onrus in W Cape.

Lamprotornis corruscus

Description. *L. c. corruscus* Nordmann (includes '*mandanus*' and '*jombeni*'): range of species except Pemba Island. ADULT ♂: lores black; forehead, crown, nape, mantle, chin, throat, and upper breast glossy dark green; ear-coverts bluish. Back, rump and uppertail-coverts green glossed with blue; tail dark violet-blue; lower breast and belly black with bronzy sheen; thighs green with blue gloss. Alula and outer webs of primaries violet-blue, inner webs matt black; no indentations on primary feathers; wing coverts and secondaries dark blue-green. Eyes orange-yellow (may flush red when bird handled); red in breeding ♂. Bill, legs and feet black. ADULT ♀: like ♂, but lower breast and belly matt black, lacking gloss. SIZE (45 ♂♂, 35 ♀♀): wing, ♂ 104–117 (111), ♀ 99–112 (105); tail, ♂ 72–90 (82.3), ♀ 71–85 (79.0); bill to skull, ♂ 19.7–24.2 (22.1), ♀ 19.6–22.7 (21.4); tarsus, ♂ 23.3–27.3 (26.3), ♀ 23.3–26.5 (25.3). WEIGHT: (Kenya and Tanzania) ♂ (n = 12) 51.0–61.7 (56.2), ♀ (n = 10) 46.0–53.1 (49.2); (South Africa) ♂ (n = 14) 54.2–67.6 (61.6), ♀ (n = 3) 52.7, 60.0, 61.8.

L. c. vaughani (Bannerman): Pemba Island. Head glossy purple, not greenish. Larger: wing, ♂ (n = 9) 111–122 (117).

IMMATURE: uniform dull black, lacking gloss; iris grey.
NESTLING: undescribed.

Field Characters. Length 18–21 cm. A dark glossy starling of E coastal forests, smaller, slimmer and duller than others in its range; has orange-yellow eye in dark face, but lacks black spots on wing-coverts. Belly and flanks black, in ♂ with a faint purple or bronze sheen in sunlight. ♀ and immature look all black at a distance; latter has grey eyes. Flight not noisy (*contra* Maclean 1993), as primaries lack indentations. Cape Glossy Starling *L. nitens* is larger, bright green, lives in open country; immature black below but has some green gloss on upperparts.

Voice. Tape-recorded (9, 32, 58, 88, 91–99, B, F, KEI, WALK). Song a jumble of brassy, scratchy and rolling notes mixed with high-pitched liquid whistles, given in phrases of up to 2.5 s (usually 1–2 s), with pauses of 5–10 s between phrases, 'way-tito-wheeto-hwee', 'way-tipooti-wheeto-jeew', 'chuwit-chuwerrr, chop, wee-o-wit', 'rrrwee-dowit-chuwee'; during pauses gives low-pitched nasal 'jup' or 'wer-jido'. Mimics other species, e.g. Diederik Cuckoo *Chrysococcyx caprius*, Cardinal Woodpecker *Dendropicos fuscescens*, Sombre Greenbul *Andropadus importunus*, Common Bulbul *Pycnonotus barbatus*, Bleating Warbler *Camaroptera brachyura*, African Paradise Flycatcher *Terpsiphone viridis*, Black-backed Puffback *Dryoscopus cubla*, Square-tailed Drongo *Dicrurus ludwigii*, Purple-banded Sunbird *Nectarinia bifasciata* and Forest Weaver *Ploceus bicolor* (Pakenham 1936 and pers. obs.). Single calls include short 'tyup', double 'jaa-pwik' and brassy 'chee-jay-jo'. Flocks give sustained high-pitched babbling interspersed with a few liquid 'pleep's and nasal notes. Alarm a harsh 'jaaa'.

General Habits. Coastal forests and bush; favours tall, dense vegetation including lowland forest and tall *Brachystegia*; also in Afromontane forest patches at low altitude and in riverine forest; generally from sea level to 500 m and where annual rainfall >500 mm. Inland in East Africa ranges up to 1400 m, visits formerly forested areas and other habitats where fruiting trees are available. Feeds in trees, in low branches and the canopy, or on the ground. Foraging flocks most often number 15–35 birds, but flocks of >100 birds recorded in South Africa, up to 400 in Sokoke Forest, Kenya. Roosts in trees, including mangroves.

Moults from late Jan to May, throughout range from East Africa to South Africa. Immatures undergo partial moult at same time as complete moult of adults. Some birds are moulting body plumage in every month.

Seasonal movements in some areas; in Zimbabwe and inland Mozambique a summer breeding visitor only. At W limits in South Africa, occurs primarily during fruiting of *Rapanea melanophloeos*. Along E Cape coast, numbers fluctuate considerably, but some birds present in all months; also reported in all months at Dar es Salaam, and presence of birds throughout the year may obscure north-south movements along the coasts. Population on Pemba I. presumed resident.

Food. Insects: flying ants, termite alates. Small snails. Vertebrates: lizards, frogs *Hyperolius* sp. Fruits of *Acacia cyclops, Acokanthera oppositifolia, Apodytes dimidiata, Bridelia micrantha, Clausena anisata, Clerodendron myricoides, Dovyalis longispina, Ekebergia capensis, Euclea natalensis, Ficus* spp., *Grevillea* sp., *Halleria lucida, Harpephyllum caffrum, Lantana* sp., *Melia azederach, imusops caffer, Morus* sp., *Phoenix reclinata, Rapanea melanophloeos* (possibly major item in some areas), *Sapium maniamum, Sideroxylon inerme, Trema orientalis, Trichilia emetica*. Nectar of *Aloe marlothii*.

Breeding Habits. Monogamous and apparently territorial; pursuit flights may be prelude to courtship. Bill-dipping display reported (Short and Horne 1985).

NEST: in tree hole, either natural hole or nest hole of woodpecker or barbet, 2·5–6 m above ground; once in a building; lined with dry leaves, grass, feathers and mammal hair.

EGGS: 2–4, av. 2·9 (15 clutches). Pale greenish blue, rarely with brown spots. SIZE: (n = 36) 24·2–27·2 × 17·9–19·8 (25·7 × 19·0).

LAYING DATES: E Africa, Oct–Jan, also single records May and Aug; Zimbabwe, Dec; Mozambique, Nov–Dec; South Africa Oct–Jan, peak Nov–Dec.

INCUBATION: ♀ only, period unrecorded. ♂ perched near nest, singing, while ♀ incubated.

DEVELOPMENT AND CARE OF YOUNG: young fed by both sexes; nestling period unrecorded.

BREEDING SUCCESS/SURVIVAL: may be displaced from nest holes by other species, including Green Wood-Hoopoe *Phoeniculus purpureus*, Kenrick's Starling *Poeoptera kenricki*, Waller's Starling *Onychognathus walleri*, and Common Starling *Sturnus vulgaris*.

Reference
Craig, A.J.F.K. (1989).

Lamprotornis chalybaeus Ehrenberg. Greater Blue-eared Starling. Choucador à oreillons bleus.

Plate 34
(Opp. p. 527)

Lamprotornis chalybaeus Ehrenberg, 1828. Symb. Phys., folio y, pl. 10; Ambukol, Dongola District, Sudan.

Range and Status. Endemic resident. Mobile and nomadic throughout range; in W Africa resident even at northern limits at border of Sahara but also a partial migrant, a few birds wandering well to the south in the dry season.

Mauritania, resident in S Guidimaka and upper Senegal R. valley; common in Nouakchott in July–Oct 1974; frequent in S sahelian zone and along Senegal valley in 1985 wet season; has bred north to Boutilimit (17°33′N). Senegal, widespread and common north of 14°N, local and much less common further south though locally common in Ziguinchor (Sauvage and Rodwell 1998). Gambia, resident, common near coast, abundant inland. Guinea, small flock Kourémalé, record near Conakry. Mali, common and widespread between 14°30′N and 17°N, resident at least to 16°N. Ivory Coast, rare dry season visitor north of 09°N. Burkina Faso, very common north of Ouahigouya (Balança *et al.* 1997); records south to 12°N and perhaps 11°30′N (Holyoak and Seddon 1989). Ghana, uncommon dry-season visitor south to Mole and Gambaga; record Legon. Togo, scarce dry-season visitor south to Namoundjoga (10°54′N) and Tantigou (10°51′N), but also at Namoundjoga in July–Aug (wet season; Cheke and Walsh 1996). Benin, scarce south to 11°N and rare south to 10°N (Holyoak and Seddon 1989). Niger, common in S half of Niger R. valley; uncommon in N half of the valley and frequent (and resident) in country from Nigeria border north to 15°N and east to L. Chad; common breeder between 15°N and southern border of Aïr Massif (about 17°N) and of Erg du Ténéré (16°N); in Aïr, singles and

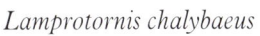

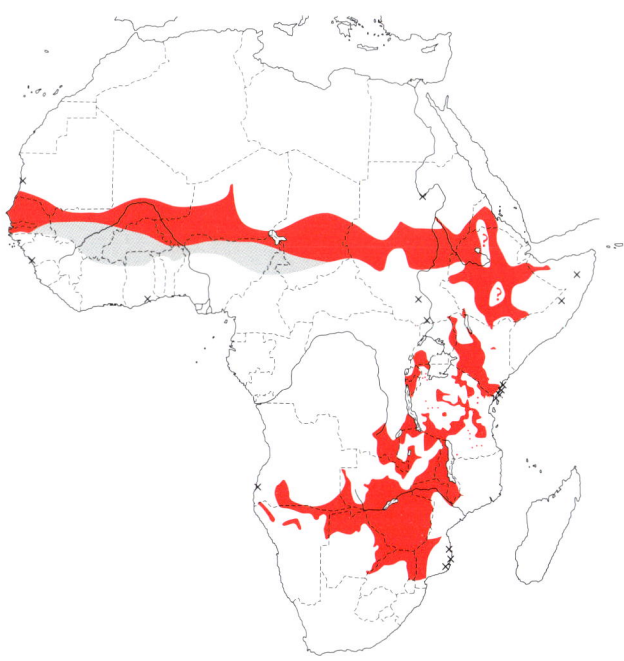

flocks occur irregularly north to Tin Telloust (18°34′N, where breeds) and Tamgak Mts (19°10′N). Nigeria, locally common resident in N Sokoto Prov. and in sahelian zone between Maiduguri and Malamfatori; records south to Zaria and Yankari. Cameroon, extreme N including Waza Nat. Park; record at Poli. Chad, very common in N soudanian zone; status in S soudanian zone unknown; local resident in sahelian zone in well wooded areas, north to Wadi Achim and Wadi Sofaya (15°50′–15°55′N). Sudan, widespread at 10–15°N, common in Darfur, records at 18°N and in El Buheyrat and W Eastern Equatoria Provs. Ethiopia, almost throughout, frequent to locally abundant; partially migrant. Eritrea, common at all altitudes, though absent from Danakil and much of coastal plain. Somalia, locally very common and ubiquitous in NW at 600–1975 m; scattered records further south.

Kenya, common in W, centre and S, north to Matthews Range; scarce and local west and north-west of L. Turkana; occurs Moyale. Tanzania, common in the minimal range mapped (N.E. Baker, pers. comm.); absent from miombo in SE. Zambia as mapped (R.J. Dowsett, pers. comm.). Uganda, north to Karamoja and Teso in E and to equator in W. Rwanda and Burundi, widespread in open country. Angola, N Cunene and S Huila Provs: Mupa Nat. Park, Mulondo, Caculuvar valley north-west to Huila-Humpata area; north to Cussece; record Namibe. Zaïre: Katanga. Zambia, absent only from parts of Northern Prov., upper Luapula valley and most of North-Western Prov.; west from Copperbelt only to Solwezi (R.J. Dowsett, pers. comm.). Malaŵi, frequent, up to 650 m, but along South Rukuru valley up to 1230 m. Mozambique, widespread in Tete District, local further south. Namibia, Botswana, Zimbabwe and Transvaal, frequent to common (R.J. Dowsett, pers. comm., Harrison *et al.* 1997).

Density of *c.* 80 (Dec) to 800 (Aug) birds in 200 km^2, Lake Nakuru Nat. Park, Kenya (Dittami 1983); in S Mozambique, 9 birds per 100 ha in acacia woodland and up to 6 in broad-leaved woodlands (Parker 1999).

Description. *L. c. cyaniventris* (Blyth): Eritrea and Ethiopia to central and W Kenya and E Uganda. ADULT ♂: top of head to mantle, scapulars and back glossy bluish green (more bluish on crown and nape); rump and uppertail-coverts violet-blue. Tail glossy blue-green above, blackish brown below. Lores blackish; ear-coverts and sides of neck deep blue. Cheeks to chin and throat glossy bluish green, upper breast bluer and merging through violet-blue on lower breast, flanks and thighs to glossy purple on belly; undertail-coverts greenish blue. Flight feathers, tertials and alula glossy blue-green, primary coverts more bluish; greater, median and lesser coverts blue-green; inner marginal coverts violet. Secondaries, tertials, greater coverts and median coverts tipped with non-reflective blackish spots. Underwing-coverts bronzy purple; axillaries violet; median and lesser under primary coverts violet-blue. P6–P9 with V-shaped notch on inner web and P10 *c.* 10 mm longer than primary coverts (**A**). Bill black; eyes variable, usually yellow, sometimes orange, occasionally white (Wilkinson 1984); legs black. Sexes alike except in size. SIZE (10 ♂♂, 10 ♀♀): wing, ♂ 143–157 (151), ♀ 129–152 (139); tail, ♂ 94–104 (97·8), ♀ 80–98 (87·2); bill, ♂ 24–26 (25·4), ♀ 22–25 (23·5); tarsus, ♂ 32–36 (33·8), ♀ 30–33 (32·0). WEIGHT: (Kenya), ♂ (n = 68) 79–106 (92·3), ♀ (n = 44) 69–91 (79·3); also ♂ (n = 191) 79–106 (median 94), ♀ (n = 109) 66–96 (median 79), imm. ♂ (n = 63) 76–104 (median 90), imm. ♀ (n = 52) 60–99 (median 78) (Dittami 1983).

IMMATURE: juvenile like adult above, but green and violet gloss less intense; dark brown below with glossy green feather tips confined to breast; sides of head dark brown. Wings and tail as adult, but wings lack blackish spots. Eyes grey-brown.

NESTLING: not described.

L. c. chalybaeus Hemprich and Ehrenberg: Senegal to E Sudan. Differs from *cyaniventris* in having head, neck, breast and upperparts greener; rump blue rather than violet-blue; marginal coverts blue; eyes yellow, occasionally orange; slightly smaller: wing, ♂ (n = 10) 142–149 (145·2). WEIGHT: (Chad, unsexed, n = 7) 71–92 (84). (Gambia, unsexed, n = 4) 77–90 (83·2) (M. King, pers. comm.).

L. c. sycobius (Hartlaub): SW Uganda to Burundi, SE Kenya, Tanzania, SE Zaïre, N and E Zambia and W and central Mozambique. Differs from *cyaniventris* in having top of head, neck and mantle rather greener, rump bluer; eyes usually orange or yellow, sometimes red (Wilkinson 1984). Smaller: wing, ♂ (n = 10) 126–143 (133).

L. c. nordmanni (Hartert and Neumann): Zimbabwe and S Zambia to Angola, S Mozambique and South Africa. Like *sycobius*, but marginal coverts coppery rather than violet; eyes orange. SIZE: wing, ♂ (n = 206) 130–156 (148), ♀ (n = 117) 121·5–146 (135). WEIGHT: (southern Africa) ♂ (n = 191) 79–106 (94), ♀ (n = 109) 66–96 (79).

Field Characters. Length 21·5–24 cm. A common open-country starling, glossy greenish blue above, bluer below and on rump; 2 rows of black spots on wing-coverts, eye golden yellow. Very similar to Lesser Blue-eared Starling *L. chloropterus*; bigger head, heavier bill and longer tail give 'rangier build' (Barlow *et al.* 1997); generally larger, but size varies by race and sex, so not very reliable; best distinguished by *broader* face mask, and in most of range by *bluer overall* colour and royal blue flanks and belly (but race *sycobius* is small and greenish, with magenta patch on belly, often mistaken for Lesser Blue-eared Starling: Zimmerman *et al.* 1996). Voices are different; nasal notes of Greater not given by Lesser. Tail longer than Bronze-tailed Glossy Starling *L. chalcurus* (wing-tips reach less than half way down tail at rest), and blue-green, not bronzy purple. Immature has dark brown face and underparts, otherwise like adult except for lack of black spots on wing.

Voice. Tape-recorded (22, 86, 88, 91–99, 104, B, C, F). Common call a hoarse, scratchy, rolled 'jawrreeya' or more nasal 'wraaanh' or 'wraanya'. Song often continuous, with disjointed quality of *Acrocephalus* warbler, a varied medley of shrill whistles, throaty coughs, buzzy churrs, hard trills, short dry notes and chuckles, woven around the characteristic nasal 'wraaa-nyaaa' or 'wraanya-wraaanh', e.g. 'pleep-chuk-tick-tewp-chuckul-plip-plip, wraaa-nyaa, chack-gurrrr-chippu-chup-chwok-chark, wraanya-wraaanh, tweep-tsirrrr-chewyjut ... '. In chorus, shrill, high-pitched babbling. Alarm, harsh 'shwarr'.

General Habits. Inhabits open savanna woodlands with quite tall and dense ground cover, bushy country, riverine forest; often in mopane and acacia woods, also in climax

A

miombo (though much less common there than e.g. Lesser Blue-eared Glossy Starling *L. chloropterus* or Sharp-tailed Starling *L. acuticaudus*). In Zambia and Malaŵi, in plateau and valley areas dominated by *Acacia*; in Chad and Botswana, in more luxuriant vegetation along larger rivers and watercourses; in Eritrea, hot foothills, dry woodland and, on mountains, open moorland with scattered bushes; in Kenya, woodland and cultivation, often around villages and towns.

Occurs in pairs when breeding and in small and large flocks in rest of year; in dry season or non-breeding season flocks commonly of 300 birds or more. Social structure flexible; 2 distinct classes: some pairs stay in territory or home range all year together with some juveniles, but in same area the large majority occupies territory only when breeding and otherwise forms wandering flocks (Malaŵi, Benson and Benson 1977; Kenya, Dittami 1983). Forages mainly on ground; gait a hop or hopping walk; often makes short run; feeds industriously, several birds working forwards in same direction across ground. Also feeds in fruiting trees. Mixes freely with congeners including Superb Starlings *L. superbus*, also Wattled Starlings *Creatophora cinerea*. Feeds on ground behind grazing sheep and goats and perches readily on back of sheep (but not slippery-coated goat) as it moves along, picking off ectoparasites from ears and forehead, balancing carefully to do so, and flying down to pick food item from ground; stays on sheep for up to *c*. 30 s (Bennun *et al.* 1990). Uses African Buffalo *Syncerus cafer*, Brindled Gnu (Blue Wildebeest) *Connochaetes taurinus* and Common Zebra *Equus quagga* as vantage points, dropping down to ground from them to forage (Koenig 1994). Drinks and bathes at puddles. Wary, but can become confiding e.g. around camps and visitor lodges, taking food scraps from tables and rubbish pits. Vocal and noisy. Flight direct, without undulation, fast, with short, even wingbeats and loud swishing noise. Roosts gregariously in trees and reedbeds. If alarmed, flock flies considerable distance before alighting. Flock suddenly arrives to cluster in tall, bare tree and after a while flies on again, as if breaking a long journey. Anting behaviour: seizes ant in bill, e.g. pugnacious ant *Anoplolepis custodiens*, and with very rapid movements rubs it on feathers of wings (above and below), breast or thighs, drops it after <2 s or eats it and seizes another (Whyte 1981).

Mainly resident, but wanders or migrates in dry season into north of Ivory Coast, Ghana, Togo and Benin (see above), also doubtless Nigeria where, however, it breeds commonly near N borders and has not been recognized hitherto as a seasonal (non-breeding) visitor further south (Elgood *et al.* 1973). In central Chad, conspicuous as daytime northward migrant at end of June, moving to arid country between 14° and 16°N and returning southward in Sept–Nov (Salvan 1967), but also resident north to range limits, with a general northward shift in the wet season (Newby 1980). No migratory movements yet detected in Sudan. In Kenya, wanders to lower altitudes after breeding, to L. Victoria and L. Turkana basins. Difficulty of distinguishing between local, seasonal dispersal and distant migration demonstrated in Lake Nakuru Nat. Park, Kenya, whence a mass exodus in Oct–Nov; very few birds remain in Dec–Jan, numbers slowly building up by factor of 10 until onset of long rains when population pairs off to breed in Apr–June, juveniles swelling population up to Oct departure (and proportion of juveniles increasing to 80% by Nov); searches of adjacent areas revealed very few birds in Dec–Jan, implying that seasonal exodus is not merely local but reflects a general, regional movement (Dittami 1983). Resident in Zimbabwe but non-ringing evidence of influx into some areas in austral winter (Irwin 1981, Harrison *et al.* 1997).

Food. Small fruits including figs, and seeds and insects. In N tropics a staple is fruit of *Boscia senegalensis* (Capparidaceae); starling eats fruits whole and defaecates the seeds, greatly accelerating germination (Tréca 1998). Also eats fruits of exotic neem *Azadirachta indica*. Takes ripe grains of 'corn, in large quantities'; after a fire, burnt grasshoppers; also beetles, ants, termites and many other such insects (Bannerman 1948, Bennun 1990). Nestlings in 10 nests, Eritrea, were all fed with locust nymphs. Campsite scraps.

Breeding Habits. Territorial, solitary nester, evidently monogamous. ♂♂ preponderate over ♀♀, amongst juveniles and adults (Dittami 1983).

NEST: bowl-shaped pad of soft chaffed grass and feathers (and sometimes coarse grass, and snake skins), topped or lined with feathers, often thickly; occasionally lined with a few dead leaves; usually in knot hole or branch stump; often in old woodpecker or barbet hole; at Nakuru, Kenya, 48 nests all in hole or cavity in dead tree or pole (Dittami 1983); in Mali commonly in empty nests of White-billed Buffalo-Weavers *Bubalornis albirostris*, Sacred Ibises *Threskiornis aethiopicus* and Abdim's Storks *Ciconia abdimii* (Lamarche 1981); in Chad (Wadi Achim) in fairly short thorn trees, usually but not always in cavities; in Eritrea in tunnels in vertical banks, often by roadsides; sometimes in cavity in building, coil of bark (an old bee hive), hollow fence post, and hollow iron fence standard open at top (nest on a bolt 0·5 m down). One nest only 1 m above ground, in *Colophospermum* trunk; another 6 m high up in dead branch.

EGGS: 2–5, usually 3–4 (southern Africa); 3–5, mainly 4 (Somalia). Pale blue or greenish blue, immaculate or lightly spotted with red-brown and greyish, the spots evenly scattered or mainly at broad end. SIZE: (n = 34) 23·4–30·7 × 18·3–20·9 (27·8 × 19·7).

LAYING DATES: Mauritania, (Aug–Oct); Senegal, June–Oct, mainly July; Gambia, (June); Mali, June–Aug; Niger, (oviduct egg June, numerous nests Aug, feeding nestlings June and July); Nigeria, (nestlings Aug); Sudan, (Apr–June); Eritrea, (10 nests with young, 1st half July); Somalia, May–June (n = 13). In E Africa, Region A, Mar–Apr, Region C, Oct–Nov, Jan, Region D, Nov–July; in Lake Nakuru Nat. Park, lays in savanna in Apr–June (mainly May) and in forest in Dec–Jan, May and July–Oct (mainly July) (Dittami 1983). Zambia, Sept–Nov (n = 7); Malaŵi, Oct–Nov; Zimbabwe, Aug–Jan, mainly Sept–Nov; Transvaal, Aug–Nov.

INCUBATION: by ♀ only. Period: 14 days.

DEVELOPMENT AND CARE OF YOUNG: iris changes from brown to yellow at 4–5 months; inside mandibles yellow until *c.* 14 months; complete body moult begins *c.* 6 weeks after hatching, flight feathers replaced in 2nd year of life (Dittami 1983). Young fed by both parents, mainly ♀. Nestling period: 23 days.

References
Bannerman, D.A. (1948).
Dittami, J. (1983, 1987).
Feare, C. and Craig, A. (1998).
Maclean, G.L. (1993).
Vincent, A.W. (1949).
Wilkinson, R. (1984).

Plate 34
(Opp. p. 527)

Lamprotornis chloropterus **Swainson. Lesser Blue-eared Starling. Choucador de Swainson.**

Lamprotornis chloropterus Swainson, 1838. Anim. Menag., p. 359; western Africa.

Range and Status. Endemic resident, nomadic and migratory toward some borders of range. The commonest glossy starling in much of its range and particularly in guinean savanna woodlands in W Africa and structurally similar woodlands in N Zimbabwe and Zambia; however, sight records continue to be bedevilled by confusion with other glossy starlings (Harrison *et al.* 1997) and the range boundaries mapped here are approximate.

In N tropics extends south everywhere to borders of lowland rain forest zone. Senegal, locally common south of 14°30′N. Sierra Leone, only in Outamba area in NW (Happel 1985). Common and widespread in Gambia, Guinea, N half of Ivory Coast, much of Ghana including coastal Accra Plains, whole of Togo and Benin, Nigeria south to Old Oyo Nat. Park in SW and Enugu and Serti in SE, Cameroon south to at least Benue Plain and Adamawa Plateau, and NE Zaïre in Upper Uele south to R. Bomokandi. In E Africa locally common up to 1600 m, in Uganda south to Bunyoro, Mengo, Teso and S Karamoja, and in W Kenya in Kerio Valley, Tugen Hills, Kongelai escarpment and foothills of Ol Doinyo Sabuk (Lewis and Pomeroy 1989, *cf.* Zimmerman *et al.* 1996); sight record at South Horr. Northern limits are: in Senegal, Thiès and about Bakel; at least 15°N in W half of Mali including Boucle du Baoulé Nat. Park, and in E half north to Tombouctou and bend of the Niger (17°N) with records from Tabereshat Wells at 17°44′N; barely known in Burkina Faso but has occurred near Mali border north of Ouahigouya as well as in Arli Nat. Park in SE; in Togo, Tantigou (10°51′N); in Nigeria north to L. Chad but in W only to S Sokoto Prov.; in Chad few positive records, north to 10–12°N; in Sudan frequent, as mapped, and seasonally common north to Khartoum (Nikolaus 1987); poorly known in Ethiopia, where populations are likely to be continuous, not isolates as shown here; in Eritrea frequent in N, at all altitudes, but absent from plateau moorland.

Tanzania, 134 records as mapped (N.E. Baker, pers. comm.); old records of wanderers north to Tanga and SE Kenya (Vanga, Mombasa and near Malindi); no recent records there. Record E Burundi. Locally common in SE Zaïre and throughout much of Zambia, Malaŵi and Zimbabwe as mapped (R.J. Dowsett, pers. comm.; Harrison *et al.* 1997). Several recent reports in Limpopo R. valley and catchment, only 1 accepted (Harrison *et al.* 1997) and 2 older ones in Kruger Nat. Park, including breeding record (Tarboton *et al.* 1987). In S Mozambique,

Lamprotornis chloropterus

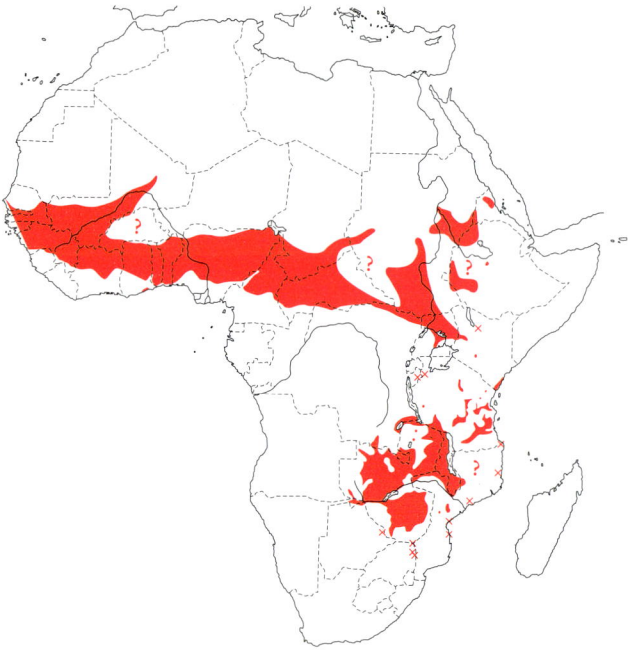

frequent in N Tete District and in Gorongoza Nat. Park, and records in coastal lowlands. In Botswana, several records north of 18°S in extreme NE, and others from Serondella and Francistown. Frequent in Caprivi Strip, NE Namibia.

Description. *L. c. chloropterus* Swainson: Senegal to Eritrea and NW Kenya. ADULT ♂: upperparts including top of head glossy green, slightly bluer on scapulars. Tail glossy green above, dark brown below. Lores blackish, tinged violet-blue; ear-coverts violet-blue. Sides of neck and cheeks to chin and throat glossy bluish green, merging through greener upper breast to glossy blue on belly and flanks; thighs blue; undertail-coverts glossy green. Flight feathers dark brown glossed green; tertials glossy bluish

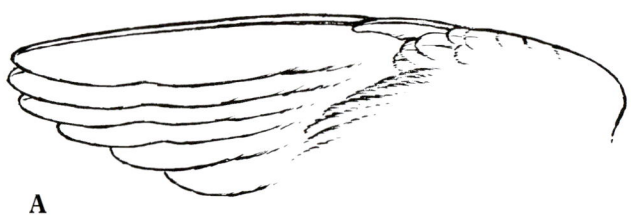

A

green; alula and primary coverts more bluish; greater, median and lesser coverts glossy green, but inner leading lesser coverts blue or violet. Secondaries, tertials, greater coverts and median coverts tipped with blackish spots. Underwing-coverts violet-blue; axillaries violet. Inner webs of outer primaries lack distinctive notches (**A**). P10 very small, *c.* 2 mm longer than primary coverts. Bill black; eyes yellow or orange-yellow; legs black. Sexes alike. SIZE (10 ♂♂, 10 ♀♀): wing, ♂ 117–126 (122), ♀ 111–121 (115); tail, ♂ 69–77 (72·2), ♀ 63–77 (68·9); bill, ♂ 23–24 (23·7), ♀ 20–23 (21·3); tarsus, ♂ 26–29 (27·8). ♀ 26–28 (27·4). WEIGHT: 1 ♂ 65, ♀ (n = 2) 57, 70; unsexed (n = 26, Gambia, Feb–Mar) 63–86 (73·1) (M. King, pers. comm.); unsexed imms (Nigeria, June, n = 5) 52·5–58 (54·8).

IMMATURE: juvenile brown above, feather tips glossy green; buffish brown below, darker on undertail-coverts; lores and ear-coverts darker brown. Wings and tail as adult, but gloss less intense, wings more bronzy green, with less conspicuous blackish spots; eyes dark.

NESTLING: not described.

L. c. elisabeth (Stresemann): SE Kenya and Tanzania to Zimbabwe. Like nominate race but lesser wing-coverts purple and juvenile darker, more tawny-brown. WEIGHT (this subspecies?), unsexed (n = 7) 52–66 (Feare and Craig 1998).

Field Characters. Length 18–20 cm. Often flocks with Greater Blue-eared Starling *L. chalybaeus*, separated by neater, more compact posture (Barlow *et al.* 1997), smaller, rounder head, narrower bill, somewhat shorter tail, *dark mask narrower and more sharply defined*; in most of range appears greener, and has magenta patch on flanks and belly, but both these features possessed by race *sycobius* of Greater Blue-eared (Tanzania to N Zambia). Smaller size sometimes obvious, but sexual size difference in both species may cause confusion, ♂ Lesser approaching ♀ Greater in size (Barlow *et al.* 1997). Eye usually duller orange-yellow and voice different, lacking nasal notes of Greater. Said to differ from Greater and from Cape Glossy Starling *L. nitens* by single row of black spots on wing (Maclean 1993), but has 2 rows of spots, see e.g. photograph in Ginn *et al.* (1989). In all 3 species the upper row of spots may be obscured, giving the impression of a single row. Immature of nominate race has dull brown head and underparts, but immature of southern *elisabeth* has diagnostic tawny underparts.

Voice. Tape-recorded (86, 88, 91–99, 104, B, C, F). Song, phrases of 4–12 measured dry churping notes mingled with a few liquid ones, 'churp, cheep, chip, chorp, peeloo, chyow', or 'chit, churrit, chick, hoowi, chee, walla, chip, yoohu'. Calls high-pitched, shrill but liquid, 'pleewee', 'pirrit', 'pyoowee' or 'weerit'; in flock, repeated rolled musical 'purrrlee'.

General Habits. Remarkably poorly documented, even for a starling. Inhabits savanna woodland, bushland and farmland in N tropics (also forest clearings in highlands of Togo) and *Brachystegia* (miombo) in S tropics, wandering widely outside breeding season when it can occur in almost all types of deciduous woodland. Occurs in pairs when breeding and in small or large flocks at other times, commonly of 10–20, often of several hundred, up to 800 birds (foraging) and 1200 (roosting). Mixes freely with other starlings, including the very similar Greater Blue-eared Starling *L. chalybaeus*; in W Africa seems to associate closely with Purple Glossy Starling *L. purpureus*; sometimes with Violet-backed Starlings *Cinnyricinclus leucogaster*. Feeds on small fruits in trees and forages for fallen fruits and insects on ground; gait a walk. Flock flies in compact body. Roosts in reedbeds, *Brachystegia spiciformis* trees, and *Acacia* and other thorn trees, sometimes with Greater Blue-eared Starlings and Wattled Starlings *Creatophora cinerea* (Feare and Craig 1998).

Resident but (like other glossy starlings) mobile and wanders widely. In addition a partial migrant; shifts northward in Ivory Coast during rains, and further north subsaharan records are mostly Aug–Dec (Tabereshat Sept, Tombouctou Oct, Ouahigouya Dec). In Nigeria, vacates Borgu area, 10°N, in dry season, Oct–Mar, when a visitor to Enugu area in SE (Wells and Walsh 1969, Serle 1957); appears at Kumbotso, Kano State, only in rains. Common visitor to Khartoum, Sudan, only in July–Oct. Appears to be seasonal in Caprivi Strip area (SW Zambia, NE Namibia, N Botswana).

Food. Feeds in *Daniella*, *Eriodendron* and *Bombax* trees and on fruits of exotic neem *Azadirachta* and probably champak *Michelia champaca* (Ghana, Bannerman 1948), and *Sterculia quinqueloba* fruits. In 11 stomachs, Zaïre (May, Oct, Nov), only beetles and other insects; in 7 stomachs, Nigeria (Oct–Mar), fruits and insects equally. 4 pairs fed young on locust nymphs ('hoppers': Eritrea). Takes carabid and other beetles, grasshoppers, caterpillars and winged termites (Feare and Craig 1998). Eats insects, seeds, fruits and flowers (Maclean 1993).

Breeding Habits. Poorly known. Solitary nester. 4 adults 'seen close to an active nest' (R. Wilkinson, p. 52 *in* Elgood *et al.* 1994); 2 entered with food, other 2 took an interest in proceedings and were probably helpers; bird leaving nest sometimes shivers tail before flying off (R. Wilkinson, pers. comm.).

NEST: mass of straw and leaves with hollow on top, sometimes quite deep. In knot hole, crevice in branch or trunk, behind loose bark, or in old woodpecker or barbet hole; once in hole in wall of fire-damaged mud building.

EGGS: 3–5. Pale blue-green, lightly spotted with red-brown and grey. SIZE: (n = 3) 26·0 × 19·2; weight 5·1 (Schönwetter and Meise 1983).

LAYING DATES: Ghana, (carrying nest material Apr, dependent young June); Nigeria, Feb–Apr; Eritrea, (June, July); E Africa, Region B, Mar, May, June; Zambia, (Sept, Oct); Malaŵi, (Oct–Dec); Zimbabwe, Sept–Oct, 1 record Nov; Mozambique, Oct.

DEVELOPMENT AND CARE OF YOUNG: parents feeding nestlings entered nest separately and together; 4–5 feeds in 90 min (R. Wilkinson, pers. comm.).

References
Feare, C. and Craig, A. (1998).
Maclean, G.L. (1993).

Plate 34
(Opp. p. 527)

Lamprotornis acuticaudus (Bocage). Sharp-tailed Starling. Choucador à queue fine.

Lamprocolius acuticaudus Barboza du Bocage, 1870. J. Sci. Math. Phys. Nat. Lisboa, 2, p. 345; Huilla, central Angola.

Range and Status. Locally common endemic resident, S-central Africa. Angola, S Cuanza Sul and Malanje to central Huila and Cubango, east to Zambian border; parts of Cunene and Cubango valleys. Namibia, NE Kavango east to Okavango valley. Botswana, rare, in NW (4 records in July–Aug and one in Mar, 1980–1994). Zambia, plateau woodland areas in N and W (mainly where Lesser Blue-eared Starling *L. chloropterus* is absent), south to Kasanka, Copperbelt, and Machili area east to Nanzhila and Dimba Dambo (R.J. Dowsett, pers. comm.). Zaïre, upper Katanga, east to Marungu (Lubenga, 1740 m) and Tembwe; common in Marungu and around Lubumbashi.

Lamprotornis acuticaudus

Description. *L. a. acuticaudus* Bocage: central Angola to Katanga, Zambia, SW Tanzania. ADULT ♂: entire upperparts, including top of head, bright glossy green. Tail long and graduated (T6 *c.* 70 shorter than T1), glossy green above, sooty brown below. Lores to ear-coverts blackish with slight blue gloss. Cheeks and sides of neck to chin, throat and upper breast glossy green, merging into glossy greenish blue on lower breast, belly and flanks; undertail-coverts glossy green. Flight feathers brown, upperside of outer webs glossy green; inner webs of P9–P7 slightly notched (**A**); tertials, alula, and greater and median coverts glossy green; lesser coverts green; inner marginal coverts purple, bordered posteriorly by line of bright blue; tertials, alula, and greater and median coverts tipped with non-reflective black spots. Underwing-coverts sooty brown; axillaries sooty brown with glossy blue-green fringes; median and lesser under primary coverts glossy green. Bill black; eyes bright orange, orange-red or scarlet, inner rim brown; legs black; soles grey. Sexes alike, but eyes of ♀ more orange (Chapin 1954). SIZE (7 ♂♂, 9 ♀♀): wing, ♂ 119–133 (123), ♀ 113–130 (123); tail, ♂ 91–101 (95·1), ♀ 82–104 (93·9); bill, ♂ 23–26 (24·1), ♀ 23–27 (24·6); tarsus, ♂ 28–31 (29·8), ♀ 25–31 (28·6). WEIGHT: ♂ (n = 4) 61·5–76 (68·2), ♀ (n = 5) 61–71·6 (67·1).
IMMATURE: like adult but lacks purple marginal coverts; retains brownish juvenile upperwing-coverts; belly feathers replaced last during post-juvenile body moult; eyes yellowish. Juvenile brown above with golden-green gloss, and buff feather fringes, especially on neck; tail bluish green; lores to ear-coverts dark brown; underparts buffish brown with dark brown scaly barring; wings as adult but more bronzy green, and upperwing-coverts with pale buff terminal fringes and dusky subterminal lines; eyes dark brown.
L. a. ecki Clancey. N Namibia and adjacent S Angola. Differs from nominate race in having upperparts, wings and tail glossy greenish blue; sides of breast and flanks deep glossy blue. Juvenile more greyish or earth brown (less buffy), feather fringes whiter.
TAXONOMIC NOTE: closely related to *L. chloropterus* (Feare and Craig 1998), but in our view not quite so closely as to compose a superspecies with it.

Field Characters. Length 21–25 cm. A woodland starling with a *wedge-shaped tail*; glossy green with dark blue mask and *red* (♂) or *orange* (♀) *eye*. In flight, underside of primaries appears pale (black in other glossy starlings: Sinclair *et al.* 1993). Immature dark grey-brown below with pale scaling; eye brown at first, becoming yellowish with age.

Voice. Tape-recorded (86, 91–99, 104, B, F, ASP, CART, STJ). Song a continuum of high-pitched warbling notes mixed with grating chuckles and churrs. Call, pleasant liquid 'puwee-o-wit'; flock chorus shrill and liquid. Birds feeding in tree gave low conversational chuckling calls, including hoarse 'skwarch', low 'chuck' and 'churrut', higher 'chark' and shrill 'chip-weer'.

General Habits. Inhabits *Brachystegia* woodland on higher ground in Angola, *Colophospermum* woods along Cunene R.; usually at edges of deciduous woods near open tree savannas, plains, floodplains and marshes; habitat in Zambia similar to that of Lesser Blue-eared Starling, though tends to be drier and more open woodland. Occurs in pairs in breeding season and in flocks of *c.* 5–50 birds out of it; sometimes flocks with Lesser and Greater Blue-eared Starlings *L. chalybaeus*. Forages on ground. Roosts gregariously, flocks performing aerial evolutions beforehand. Mainly sedentary, but said to be a dry season migrant in S Zaïre (Lippens and Wille 1976).

A

Food. Fruits, e.g. of *Diospyros kirkii* (Angola: Dean *et al.* 1988) and *D. lycioides* (Namibia: Harrison *et al.* 1997).

Breeding Habits.

NEST: pad or bowl of grass, hair and feathers, placed on floor of old barbet or woodpecker nest hole or in knot hole in tree (W.R.J. Dean *in* Ginn *et al.* 1989).

EGGS: 3 (or 4) (nest with 3 young in Angola).

LAYING DATES: Angola, Aug–Oct; Zambia, Oct; Zaïre, (enlarged gonads Aug).

References
Benson, C.W. *et al.* (1971).
Chapin, J.P. (1954).
Feare, C. and Craig, A. (1998).
Maclean, G.L. (1993).

Lamprotornis ornatus (Daudin). Príncipe Glossy Starling. Choucador de Príncipe.

Plate 35
(Opp. p. 542)

Sturnus ornatus Daudin, 1800. Traité Orn., 2, p. 309; locality unknown, = Príncipe Island.

Range and Status. Endemic resident, Príncipe I.; abundant in all wooded habitats in N and centre of island and common in hill habitats in S (Snow 1950, Atkinson *et al.* 1991, Christy and Clarke 1998).

Description. ADULT ♂: top of head to upper mantle and anterior scapulars bright glossy bluish green, slightly bluer on nape, hindneck and sides of neck; upper mantle separated by narrow violet-blue band from bronzy green lower mantle, posterior scapulars, back and rump; uppertail-coverts glossy violet-blue mixed with bright purple. Tail blackish above, with broad glossy violet-blue feather tips, confined mainly to outer webs of outer feathers; blackish brown below. Glossy uppertail-coverts and tail feather tips with dull blackish barring. Lores blackish; ear-coverts violet; cheeks to chin, throat, breast and belly dark coppery green; flanks and thighs dark glossy green with purplish tinge; undertail-coverts deep violet-blue. Primaries bronzy green above, tips and outer edges glossed purplish; secondaries blackish, outer web bronzy green with dull blackish barring; tertials bronzy green with blackish centres, purplish at base of inner webs. Alula bronzy with blue outer web; primary coverts bronzy, fringed purplish blue; greater coverts bronzy with blackish centres and purplish tips; median coverts bronzy with broad purplish blue tips; lesser coverts bright blue. Underwing-coverts and axillaries dark coppery green. Outer primaries with distinctive hook-shaped notch on inner web (**A**) (as in *L. splendidus*, *q. v.*). Wingtip rather rounded with P7 and P8 longest, and P9 *c.* 10 shorter, falling between P5 and P4; P10 rather large, *c.* 12 longer than primary coverts. Bill black; eyes white; legs black. ADULT ♀ differs from adult ♂ in having top of head and mantle greener; rump, uppertail-coverts and tail feather tips bluer (less purple). On wings, inner web of tertials and greater coverts more extensively glossed with purple; primary outer webs and greater and median coverts edged bluer; lesser coverts bluish green. SIZE (10 ♂♂, 11 ♀♀): wing, ♂ 152–160 (156), ♀ 133–150 (143); tail, ♂ 108–127 (115), ♀ 95–107 (103); bill, ♂ 27–31 (29·2), ♀ 27–29 (28); tarsus, ♂ 33–35 (33·6), ♀ 30–34 (31·8).

IMMATURE: juvenile ♂ has top of head to mantle like adult; bronze of back darker and more coppery; reduced amount of purple on tail tip. Lores and band over base of bill brown, cheeks and ear-coverts blackish with only a hint of iridescence; entire underparts blackish brown with no iridescence, feathers of chin to breast edged paler brown producing mottled and barred effect. Wing like adult but duller. ♀ like ♂ but lacks blackish mottling on underparts, chin to chest plain brown.

Field Characters. Length 28–30·5 cm. Large, almost the size of a Western Jackdaw *Corvus monedula*, appearing dark in poor light; mainly *iridescent dark bronzy green*, with reflections of blue on head and upper mantle, purple on rump and uppertail-coverts, green and purple on wing-shoulder, blue tip to tail. Forms mixed flocks with similar-sized Splendid Glossy Starling *L. splendidus*, the only other starling on Príncipe; can be confused at first glance because both have pale eye in dark mask (the first thing one notices: Christy and Clarke 1998), have some similar calls, and make swooshing sound with wings in flight, but

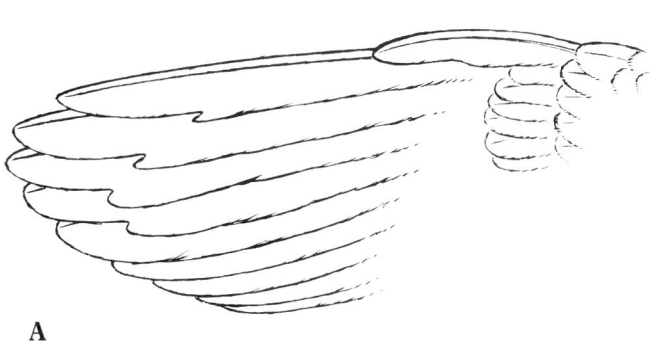

A

Príncipe Glossy Starling is much less brightly coloured; underparts look blackish in the field, whereas Splendid is purple below, green and blue above, with no bronze.

Voice. Tape-recorded (104, B, CHR, GUL, JO PJ, TYE), Commonest call a deep, grating, nasal, almost meowing 'meearrnh' or 'jwaarrnh'; likened to Red-billed Chough (Snow 1950), and sometimes followed by musical 'tu-ping' like twanging of banjo string. Also high-pitched liquid trill, sometimes alternating with low creaking brassy 'kyarnk'; rolled 'trrrrrrreeeeu' and 'turrrrweerr'; and medium-pitched nasal 'jyaaanh', like that of Splendid Glossy Starling. Song, short disjointed phrases, including characteristic hollow 'plonk', 'plip-plonk-puwit-kuchoo'. Birds building nest uttered 'pee-to-woo', the 'pee' slightly harsh (Snow 1950). Also gives fluty calls reminiscent of African Grey Parrot *Psittacus erithacus* (Christy and Clarke 1998, *q.v.* for additional vocabulary).

General Habits. In primary and old secondary forest and all types of woodland, cultivated land with scattered trees, shady coffee plantations and cocoa plantations under cover of *Erythrina*, from sea level up to at least 650 m; in large trees in town of Santo António. Generally keeps high in trees. Occurs in pairs, parties of up to 10, and from Sept in flocks of 15–30 or more birds; sometimes flocks with Splendid Starlings *L. splendidus*. Rather shy. In flight makes loud swishing noise with wings, particularly in short flights from tree to tree. Forages mainly high up in trees; occasionally comes to feed on ground. In Dec and Jan searches for caterpillars in *Erythrina* trees. Makes sallies from canopy after flying termites. Several dozen birds sometimes congregate in tree-top, calling together, audible several hundred m away (de Naurois 1994). Flocks are conspicuous in the evening, flying high to Pico Papagaio, where they may roost (Jones and Tye 1988).

Food. Large berries taken from garden shrub; drupes of *Dracaena draco*, and other fruits (de Naurois 1994); figs, fruits of *Dacryodes*; spiders, caterpillars and flying termites (Christy and Clarke 1998).

Breeding Habits.
 NEST: in hole high in dead tree or dying limb, one 11 m from ground.
 EGGS: unknown.
 LAYING DATES: Príncipe (breeds Oct–May: Keulemanns 1866; probably Sept–Dec: de Naurois 1994; carrying nest material, Sept: Snow 1950).

References
Christy, P. and Clarke, W.V. (1998).
de Naurois, R. (1994).
Snow, D.W. (1950).

Plate 35 *Lamprotornis splendidus* (Vieillot). **Splendid Glossy Starling. Choucador splendide.**
(Opp. p. 542)
T[*urdus*] *splendidus* Vieillot, 1822. Tabl. Encycl. Méth. Orn., 2, p. 653; Malimbe, Portuguese Guinea.

Range and Status. Endemic resident and migrant. Locally frequent in W and SW Senegal (Kolda to Sédhiou, Oussouye, Saloum, Dakar), W Gambia (uncommon; commonest near coast), Guinea-Bissau and Guinea (Fouta Djallon, and records in S). Sierra Leone and Liberia, common along coast, not uncommon in S hinterland and E Liberia. Ivory Coast, numerous along coast (Demey and Fishpool 1991). Ghana, frequent north and east to Ho (06°37′N). Togo, very common in forested parts of highlands and their periphery. Benin, rare, in S. Nigeria, frequent to locally abundant in forest zone; outlying populations on S edge of Jos Plateau, north to Jos town; also in Gashaka-Gumti Nat. Park. Bioko, locally frequent. Príncipe, common on all cultivated land in N and centre. Cameroon, north to N edge of Adamawa Plateau.

Entire Congo Basin except Lower Katanga (Chapin 1954) and evidently much of Kasai Oriental. Central African Republic, in W half, probably widespread. Sudan, uncommon along Zaïre border and in Boma Hills. Ethiopia, frequent to abundant in SW West Highlands, West Ethiopia north to 09–10°N, and S Rift Valley. Uganda, widespread at 700–1800 m, north to 02°N in W and 01°N in E (Bugoma, Budongo, Kabalega Falls Nat. Park, Ankole, Sango Bay and Entebbe). Kenya, uncommon on Mt Elgon and in Saiwa Nat. Park; formerly widespread at 1700–2300 m from Elgon to Nandi and Kakamega, but

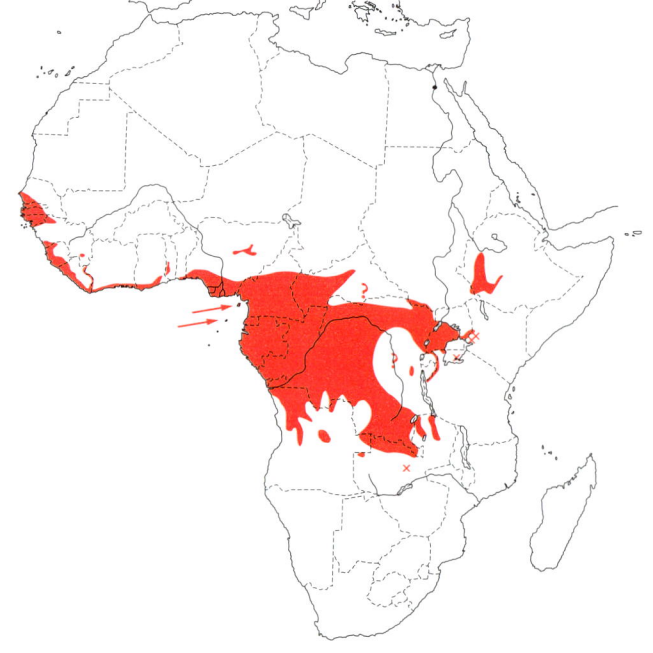

Lamprotornis splendidus

has declined drastically with deforestation (Zimmerman *et al.* 1996). Tanzania, at 1200–1500 m, in L. Victoria basin at Bukoba, Nyarumbugu and Ukerewe I. and at Ngara to Gombe Stream Game Reserve. Rwanda, L. Rwampanga; common in Akagera Nat. Park, only in Akagera valley, from Gatare to Kagitumba (Vande weghe 1974). Burundi, length of lower Ruvubu and Kayongozi valleys, and in Kumoso at Kinyinya and Giharo and abundant in July–Aug at Giofi, Nkanga and Gatanga (Gaugris *et al.* 1981). Angola, Cabinda, Zaïre, Cuanza Norte and Malanje Provs, and locally on central plateau to Cuanza Sul and N Huambo (Bailundo); also Lunda and NE Moxico. Zambia, forests in N, east to Shiwa Ngandu; Copperbelt; Mwinilunga; west to Lukolwe, Zambezi Distr., and Kambizana Stream (R.J. Dowsett, pers. comm.).

Description. *L. s. splendidus* (Vieillot): S Benin to NW Angola, east to Ethiopia and Zambia. ADULT ♂: top of head to mantle glossy blue-green (greener on forehead, bluer on neck), merging into glossy violet-blue on back and blue-green on rump and uppertail-coverts; scapulars deep glossy blue, larger feathers with blackish subterminal band. Tail violet-black, feathers with broad bright blue tips, extending on T5–T6 along much of outer web; blackish below. Lores and narrow band above base of bill black; ear-coverts black; sides of neck glossy greenish blue with metallic coppery patch. Cheeks to sides of throat and breast violet, grading to glossy purple on chin, centre of throat and upper breast and into a broad band of bronzy reflections across lower breast; belly and flanks glossy violet; thighs blue; undertail-coverts greenish blue. Primaries violet-black, tips and outer webs of outer ones bluish green and of inner ones blue; secondaries black, tips and distal part of outer webs glossed violet-blue, central part of outer webs black. Tertials black, fringed glossy blue, forming, with central part of outer webs of secondaries, a broad blackish band across closed wing. Alula and primary coverts bluish; greater coverts glossy green with narrow blue fringe and black subterminal spots; median and lesser coverts bluish green. Flight feathers blackish brown below; underwing-coverts and axillaries violet; median and lesser under primary coverts glossed bluish. Outer primaries with distinctive hooked notch on inner web (**A**). P9 about equal to P6; P10 short, *c.* 5 mm longer than primary coverts (*cf. L. ornatus*). Bill black; eyes pale creamy yellow; legs black. Birds in Cameroon and Nigeria have underparts more violet as in *chrysonotis*, with narrower bronzy band. ADULT ♀: similar to adult ♂, but more violet (less purple) below, bronzy breast band almost absent; coppery neck patches much smaller. SIZE (10 ♂♂, 10 ♀♀): wing, ♂ 151–163 (158), ♀ 137–152 (143); tail, ♂ 115–124 (120), ♀ 103–116 (110); bill, ♂ 26–29 (27·1), ♀ 24–27 (25·2); tarsus, ♂ 33–36 (34·4), ♀ 29–33 (31·6). WEIGHT: ♂ (n = 4) 119–150 (133); also (this subspecies?), ♂ (n = 47) 111–155, ♀ (n = 25) 96–139 (Feare and Craig 1998).
IMMATURE: juvenile has glossy feather tips above, greenish on head and mantle, bluish on scapulars, back and rump; underparts mostly brown; lores and forehead to ear-coverts darker brown. Wings and tail glossy as in adult, but duller, black wing feather tips and wing band less pronounced.
NESTLING: not described.

L. s. chrysonotis Swainson: Senegal to Togo. Slightly greener than nominate race on head, breast and mantle; throat and breast violet-blue rather than purple; bronzy breast band of ♂ narrower. Slightly smaller: wing, ♂ (n = 5) 142–152 (149).
L. s. lessoni (Pucheran): Bioko. Like typical nominate race above, but ♂ has upper breast more violet, bronzy band on lower breast narrower, as in Nigerian birds. Larger: wing, 2 ♂♂ 166, 164.
L. s. bailundensis (Neumann): central Angola to Zambia and SE Zaïre. Greener than nominate race on head, hindneck and mantle; lesser wing-coverts blue; violet-blue band across upper breast. WEIGHT: (Zambia), 1 ♂ 114, 1 ♀ 102.

Field Characters. Length 24–28 cm. A splendidly iridescent forest starling, blue and green above, mainly *purple below* with coppery reflections on lower breast; small bronze patch on neck, blackish area in centre of folded wing below green coverts, *broad black band* across centre of tail. Pale yellow eye stands out in black mask. *Longish broad rounded tail*, narrower at base, gives characteristic flight silhouette (Chestnut-winged Starling *Onychognathus fulgidus* has longer, graduated tail, Narrow-tailed Starling *Poeoptera lugubris* has thin pointy tail). Large size, loud nasal voice and swooshing sound made by wings in flight distinguish it from Purple-headed and Bronze-tailed Glossy Starlings *L. purpureiceps* and *L. cupreocauda*. On Príncipe flocks with the even larger Príncipe Glossy Starling *L. ornatus*; for distinctions see that species. Immature dull blackish below.

Voice. Tape-recorded (32, 102, 104, B, C, F, LEM, MAC). A collection of almost comical sounds; most distinctive is long nasal 'wrrreeyyyaanh', and variations, drawn-out 'wrro-heeyyaa-heeyyaaaanh' or shorter, more grating 'wrraadi-jow'. These can mix with explosive gong-like notes to form loose 'song', 'wrreeyyaanh ... chlonk ... wrockaggy ... chlonk ... wrrraheeyaanh'. Also gives scraping 'chyaa' or 'chyaank' and loud 'chang', and is capable of high clear musical whistles and more delicate notes, including light 'hop', reedy down-slurred 'bleew' and short 'jee'. Reported to imitate other birds and monkeys (Brosset and Erard 1986). For further renditions see Christy and Clarke (1994, 1998) and Zimmerman *et al* (1996). Calls in concert with Príncipe Glossy Starling; some calls are similar, but Splendid recognized by its 'ktuong' and nasal calls (Christy and Clarke 1998).

General Habits. Inhabits mainly primary and tall, mature secondary forests, but preferred habitat varies regionally. In Gabon, forest (mainly in canopy, sometimes descends into fruiting understorey and small trees at forest edges), also parks and gardens. In Gambia strongly associated with remnant forest patches and stands of tall trees and oil palms; occasionally in more open habitats, also in mangroves and swamp forest margins (Barlow *et al.* 1997). In Liberia, coastal and mangrove forests; savanna and cultivation. In central Nigerian plateau in wooded slopes and gullies. On Príncipe, plantations and farmland with *Erythrina*, figs and other fruiting trees. In Rwanda, gallery forests and *Acacia* galleries, in Burundi restricted to gallery forests along borders of marshes and rice-fields, and in Angola occurs in coffee forest.

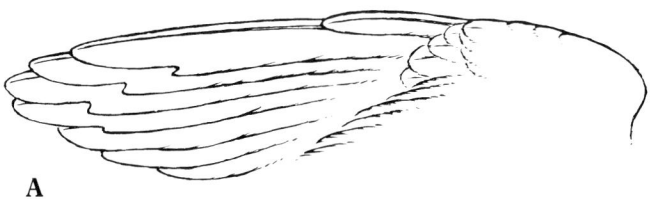

A

In pairs when nesting, when can be shy and elusive, otherwise strongly gregarious, in foraging flocks of 30, 60 and sometimes 100 (or 300, E Rwanda). Forages mainly at 30–50 m above ground (Gatter 1997). Highly mobile; not territorial. Continually on the move searching for fruiting tree, and suddenly appears in large numbers at a particular place where not seen previously for weeks or months. At a productive tree flock stays all day in crown of a favoured sheltering tree 50–150 m away; birds continually fly to and fro between food tree and shelter tree, which has very dense foliage often covered with lianas. Bird comes to food tree, swallows 2 or 3 fruits, e.g. arillate fruits of *Pycnanthus*, then flies off with another in bill to shelter tree. Seeds are regurgitated at perch (or rarely in flight) and lie below shelter tree in abundance. Flock in food tree is quiet but in shelter tree noisy, particularly during heat of day, attracting hornbills and other starling species (Brosset and Erard 1986). In flight makes loud swishing noise with wings, out of all proportion to size of bird, and sounding like 'the churning of some sternwheel steamer in the distance' (Chapin 1954). Comes to ground much less than do congeners (Gambia) but in Príncipe regularly descends to ground to feed on fallen fruits and comes down to puddles to drink. Large flock makes conspicuous daily flights high over forest between feeding and roosting localities. Roosts can be huge – tens of thousands in trees in a marsh (Makokou, Gabon), small groups dispersing up to 15–20 km away in all directions by day. Roosts in *Raphia* palm swamps (Nigeria). In daily flights between roost and feeding grounds, flock very vocal.

Mainly resident, but a breeding visitor to Kenya, especially (formerly) the Endebess-Kitale-Kapenguria area in Oct–May, to NE Rwanda in Oct–late May and to E Rwanda in Oct–Apr; and breeding visitor to Angola, Zambia and Upper Katanga (Zaïre) late July or Aug to late Nov or early Dec, presumably from further north in Zaïre (Benson *et al.* 1971, Penry 1979, Prigogine and Benson 1979). May also be migratory in parts of E Zaïre, Mayombe and Gabon (Chapin 1954). New interpretation of movements offered by findings (above) about foraging dispersion (Brosset and Erard 1986), could apply to some 'migrant' populations (Prigogine 1983).

Food. Fruit of *Pycnanthus*, *Coelocaryon* and other Myristicaceae, *Ficus*, *Rauvolfia*, *Dacryodes*, *Polyalthia*, *Trichoscypha*, *Beilschmiedia*, *Xylopia*, *Musanga*, *Maesopsis* and avocado *Persea americana*; berries of *Ficus mangifera*. Some fruit stones so large that they can only be disposed of by regurgitation. Insects, mainly large orthopterans, fed to nestlings; also ants, termites, tree snails, small lizard and small frog in stomachs. Once a lizard in adult stomach, and a lizard 10 cm long fed to nestlings.

Breeding Habits. Solitary hole nester; pair formation and (undescribed) display flights in Liberia in Sept–Oct. Many nest holes re-occupied on exactly the same dates for many years in succession (Gabon).

NEST: rough or well-made pad or bowl of dry leaf petioles, on floor of cavity in oil palm trunk, old hole of Fire-bellied Woodpecker *Dendropicos pyrrhogaster*, knot hole or stump hole, mainly 8–37 m above ground; one in hollow horizontal branch where wood made lumpy by growth of mistletoe; 3 nests 1·8 m high in hollow trunk of lone tree on island in river and others in forest clearings (Gabon); many nests in mangrove trees (Guinea-Bissau); nests in dead *Adansonia digitata* trees and dead oil palms *Elaeis guineensis* (Mouila, Gabon: Y.-M. de Martin de Viviès, pers. comm.).

EGGS: 2–3. Oval, almost without gloss, shell rough and slightly pitted. Pale greenish blue, sparingly spotted and blotched all over with pale red-brown and lilac-grey. SIZE: (n = 5) 29·0–31·5 × 21·5–24·5 (30·2 × 22·6).

LAYING DATES: Senegal, Gambia, (attending holes, June–Aug); Ghana, (Feb, Apr); Liberia, (carrying food to nest holes Dec–Jan); Togo, (nest-building Feb, entering holes Mar–Apr, feeding nestlings May); Nigeria, Dec, Feb (and building Jan); Cameroon, Feb–Mar and Aug–Oct; Gabon, Jan–Feb (♂ singing and visiting nest hole Dec; holes occupied Jan–Mar); pairs at breeding sites June–Dec (Y.-M. de Martin de Viviès, pers. comm.); Zaïre, (Lukolela, carrying nest material Dec; Kananga, young just out of nest Dec; Mbandaka, enlarged gonads July; Beni, nestling Oct); E Africa, Region A, Mar–May, Region B, Jan–May, July and Aug, mainly Mar (14 out of 21 records) (breeds in early rains); Rwanda, (Nov–Dec, Feb–Mar); Zambia, Sept–Nov.

INCUBATION: by ♀ only (♂ keeping watch nearby). Period: *c.* 18 days (in captivity).

DEVELOPMENT AND CARE OF YOUNG: young reared by both parents.

References
Brosset, A. and Erard, C. (1986).
Chapin, J.P. (1954).
Feare, C. and Craig, A. (1998).
Prigogine, A. (1983).
Prigogine, A. and Benson, C.W. (1979).
Sawyer, R.C.J. (1982).

Plate 35
(Opp. p. 542)

Lamprotornis caudatus (Müller). Long-tailed Glossy Starling. Choucador à longue queue.

Turdus caudatus P.L.S. Müller, 1776. Natursyst., Suppl., p. 144; Senegal.

Forms a superspecies with *L. purpuropterus* and *L. mevesii*.

Range and Status. Endemic resident, frequent to abundant and widespread in sahelian and soudanian savannas from Senegal to Nile. Mauritania, from Senegal valley in *Borassus* palm zone to 50–60 km to the north; Guidimaka and Karakoro; south to Guinea-Bissau and in Guinea the Kondara and Gaoual areas. Mali, abundant north to

Lamprotornis caudatus

of variant with breast and wings blue, shot with purple rather than oily green, known from Senegal, Ghana, Nigeria and Cameroon (Bannerman 1948). SIZE (10 ♂♂, 10 ♀♀): wing, ♂ 182–197 (189), ♀ 168–184 (174); tail, ♂ 290–335 (311), ♀ 256–298 (272); bill, ♂ 27–29 (27·9), ♀ 26–29 (27·1); tarsus, ♂ 40–46 (43·4), ♀ 39–44 (40·8). WEIGHT: unsexed (n = 12, Nigeria) 103–133 (121).

IMMATURE: juvenile with head sooty brown, mantle, back and rump glossy green, duller than in adult, uppertail-coverts violet and blue; underparts sooty brown, breast with some green gloss; eyes dull yellow.

NESTLING: naked except for tiny patch of grey down on vertebral mid-line and another on scapulars (Serle 1943); pink; thick white gape flanges.

Field Characters. Length up to 54 cm (tail up to 34 cm). The only long-tailed starling in W African savannas. Glossy green, with duller, bronzy head, pale yellow eye in black mask, purple belly and *extremely long* purple tail. At a distance could be mistaken for Piapiac *Ptilostomus after* but in flight wing-beats deep and laboured (shallow and rapid in Piapiac) and tail feathers lax, not stiff (Barlow *et al.* 1997). In Sudan just meets Rüppell's Starling *L. purpuropterus*, which has similar bronzy head but is otherwise all purple, and has much shorter tail. Immature has sooty brown head and underparts but tail already long.

Voice. Tape-recorded (104, 108, B, BRU, CHA, MOR). Common call a shrill, grating, slightly rolled 'chrrrrreeyo' or 'chrreelyo', probably the same as the 'harsh, grating, measured 'kchirow-kcheree', rising in pitch' of Barlow *et al.* (1997). Other calls include 'ina-yoriek' or 'élékélé', the bird's name in Hausa (Wilkinson 1988), an explosive 'skliolya', a long rasping 'chraaaaa' or 'chraaaa-ja', and hoarse, coughing 'chuc-chu-chu'; alarm a raucous shrieking. Call from bird at roost, low, husky 'guwarr-jui'; flock resting in tree makes a not unpleasant penetrating, chattering, garrulous chorus.

General Habits. Inhabits farmland, park-like savanna woodland especially where plenty of shea-butter trees *Vitellaria (Butyrospermum) paradoxa*; well-wooded residential areas and large gardens with well-spaced large trees and woody ground-litter; thorn thickets; rocky, tree-clad hills surrounded by cultivation; open riverine woods; mangroves. Occurs singly or in pairs when breeding, in flocks of 4–7 or 3–10 in late dry season, and in wet season (June–July, Kano, Nigeria) up to 20 or (Gambia) 30 birds; foraging flock scatters loosely over perhaps 0·1 ha with members in or under 2–3 trees, but is cohesive in the sense that all birds pursue same activity at about same time – foraging, resting and preening, or flying off. Flock lives in stable group, maintaining a group territory; near Kano, 3 groups held adjacent territories. However, flock also said to scatter by day and re-assemble in evening to roost communally, e.g. in mango plantation (Bannerman 1948). Very vocal, at perch and on the wing; in July large flock has much calling, chasing and displaying behaviour, thought to relate to boundary disputes. Alert and wary. Feeds on ground, hopping and walking nimbly with quite strong strides and tail held inclined; freely uses resting perches in bushes and low boughs in trees (often *Parkia*),

15°30'N, occurs to 16°30'N. Sierra Leone, recorded (G.D. Field, pers. comm. to Dowsett and Dowsett-Lemaire 1993). Ivory Coast, rare: 4 dry-season records north of 09°30'N. Ghana, uncommon resident in N in Tamale and Bolgatanga; influx of non-breeding birds at start of wet season (Grimes 1987), south to Tamale, Tumu, Gambaga and Pong Tamale. Niger, frequent in SW, and S south of about 15°N. Togo, common in N in Dapaon area (10°50'N), frequent south to Kara (09°33'N), recorded south to Bassar and Pagala (08°11'N). Benin, only in Pendjari-Arli Nat. Parks. Nigeria, throughout N though local in sahel zone in NE; south to Kainji, Okene and Kabba; a vagrant or escapee at Lagos. N Cameroon, S Chad and Sudan as mapped. Central African Republic, common in Manovo-Gounda-Saint Floris Nat. Park and N Vakaga Préf.; record, Garba (09°12'N); absent from Bamingui-Bangoran Nat. Park; 2 records in upper Kandjia valley (Hall and Moreau 1970).

Description. ADULT ♂: head bronzy green, apart from sooty black lores and ear-coverts; hindneck and sides of neck to back and scapulars glossy green, rump and uppertail-coverts glossy blue. Tail very long and strongly graduated (T1 4 times length of T6), glossy violet above with blackish barring, T2–T6 with blue edges; blackish brown below. Lower throat to upper breast glossy green, concolorous with upperparts; a narrow blue lower breast band and glossy blue flanks grade into bronzy purple belly; undertail-coverts and thighs sooty brown, glossed with blue. Flight feathers blackish, upperside with glossy green tips and outer webs; tertials, alula and upperwing-coverts glossy green; greater and median coverts with velvety black tips. Underwing-coverts and axillaries sooty brown, glossed greenish. Wing rounded, P6 and P7 longest, P9 *c.* 20 mm shorter, near length of P3; P10 large, *c.* 35% of length of P7. Bill black; eyes creamy white; legs black. Sexes alike, though in field much smaller birds (presumed ♀♀) slighter-headed and green-glossed where large ones (♂♂) are more purple-glossed (Wilkinson 1988). Specimens sometimes also in tree tops; also forages in trees, taking variety of insects and fruit (Wilkinson 1988). Flight powerful, laboured, with deep wing beats and, at close range, audible swishing sound; usually flies low down, c. 2 m above ground, and seldom flies far or above height of trees. Often visits puddle to drink, before going to roost.

Group in captivity had linear dominance hierarchy, breeding pair dominating sub-adults which dominated juveniles. Not aggressive with other bird species, except when breeding (in captivity).

Resident, but seasonal wandering amounting to immigration into N Ghana; seasonal at Tamale in Feb–Oct (at least in 1950s) and Tumu in May–Sept. In Togo penetrates south into N guinean area only in Jan–Aug. Moult protracted, from about Mar to Oct, and not well synchronized within a population.

Food. Insects (particularly important in diet when nestlings being fed) including large terrestrial ant *Messor galla*; fruits including neem *Azadirachta indica* (Wilkinson 1988), of which it is very fond (Bannerman 1948).

Breeding Habits. Semi-colonial: 'many pairs may nest in close proximity where there are groves of baobab trees' (Serle 1943); co-operative breeder. Habits poorly known in wild, but several adults seen bringing food to nest hole (Gambia), 3 adults visited tree hole (Nigeria), and a captive pair (England) bred co-operatively, parents at their 3rd nesting helped by their 2 surviving 1-year-old and 2-year-old offspring (Wilkinson 1988). Becomes aggressive when breeding. ♀ threatens conspecifics by lunging with open bill; parents and helpers in large aviary chase away doves and other birds. In the wild 2 presumed ♂♂ locked together by their feet, one on its back, other on top of it, for 5 min; observer approached to 0.5 m, birds separated but immediately re-engaged like fighting cockerels, with wings and tails spread, legs outstretched towards each other, feet grasping then interlocking; they were watched by a presumed ♀ (Wilkinson 1988). Nesting ♀ has open-billed begging display when soliciting food from sub-adult bird to pass to nestlings. ♂, approached by ♀ dominating near nest, droops wings. Captive pair bred successfully twice within 5 months.

NEST: only one described was natural hole in quite a small tree, entrance c. 100 mm diam., leading to spacious cavity extending downward, floor c. 0.6 m below lower rim of entrance hole, eggs 'on a considerable quantity of the spines of leaves which had apparently been introduced by the birds' (Shuel 1938). In hole high up in tree, often a large baobab (Serle 1943); 1 in hole 4 m up in river-bank tree.

EGGS: 3 and 4 (2 clutches) (and 2 and 3 nestlings in captive nestings). Oval; slightly glossy; immaculate pale blue; or glossy, immaculate greenish blue, or deep blue with evenly-distributed pale brown dots (Bannerman 1948). SIZE: (n = 7, 2 clutches from same nest) 27.5–28.6 × 19.75–20.6 (28.15 × 20.2) (Shuel 1938).

LAYING DATES: Senegal, (Sept–Oct); Guinea, (feeding young in nest Oct); Mali, (attending nest holes Aug–Oct); Burkina Faso, (nests occupied Aug–Oct, young flying Oct–Nov); Niger, (nest hole June); Nigeria, Kano, Sept and replacement clutch in same nest Oct (also just-hatched nestlings early Oct); Sudan, (Sept and Aug–Oct).

INCUBATION: thought to be only by ♀ (Wilkinson 1988).

DEVELOPMENT AND CARE OF YOUNG: young brooded only by ♀; faecal sacs removed only by ♀; young fed in nest by ♀, ♂ and sub-adult and juvenile helpers; ♂ and helpers wait for aggressive ♀ to leave nest before entering, otherwise she prevents them entering, supplants them or takes food from them outside nest and passes it to nestlings. Feeding rate at captive nest with 2 chicks (15 hours of observation), by ♀ av. 20.6 feeds per h (1st week), 17.2 (2nd week) and 9.4 (3rd week), by ♂ av. 1.8 per h (whole nestling period), by sub-adult helper av. 6.1 per h (whole period) and by juvenile helper av. 2.6 (whole period) (Wilkinson 1988). Nestling period, captive birds, 17 and 21–22 days. Fledglings stay hidden in cover for first few days after leaving nest; adults enticing fledglings into cover droop wings and shiver wings and tails. Adult ♀ and sub-adult helper elicit begging from recently-fledged young by pecking at its bill and crop. Captive young began to feed themselves 2 weeks after leaving nest and were still fed occasionally by parents, mainly ♀, a week later. Young leave nest with tails only half grown. Fledglings have dull, darkish yellow eyes, still noticeably darker than adult eye after 12–20 months; young birds lack gloss on head and breast for at least 5 months after they leave nest.

References
Bannerman, D.A. (1948).
Wilkinson, R. (1988).

Plate 35
(Opp. p. 542)

Lamprotornis purpuropterus Rüppell. Rüppell's Glossy Starling. Choucador de Rüppell.

Lamprotornis purpuroptera Rüppell, 1845. Syst. Uebers. Vög. Nord-ost.-Afr., p. 64; Shoa District, Ethiopia.

Forms a superspecies with *L. caudatus* and *L. mevesii*.

Range and Status. Endemic resident, eastern Africa. Sudan and Ethiopia, frequent to abundant, up to 1800 m; in S Ethiopia frequent on plains around Giarso and Gardulla but absent from plains around Yabello, Mega and Arero. Eritrea, common, up to 1540 m, east to about 39°E. Zaïre, restricted to extreme NE borders near L. Albert, L. Edward, Kagera R. and lower Ruzizi; scarce in E Kivu. Uganda, common throughout. Kenya, in NW on Turkwel R., at Lodwar and Lokichokio; Karoli Desert; Marsabit; from Lokori southward common in Rift Valley and west of

Lamprotornis purpuropterus

it; east of Rift, uncommon and local: Athi, Galana, Tana and Tiva valleys, Laikipia Plateau, Meru Nat. Park south to Tsavo, and coastal lowlands from Garsen to Somalia border. Somalia, Jubba and Shabeelle valleys south to Boni Forest; 2 old records in NW. Rwanda and Burundi, known from Gabiro, E Ruzizi, Kamonye and Luvungi. Tanzania, on Ukerewe and other islands in S L. Victoria, widespread southwest of L. Victoria and to east of it; Serengeti, Loliondo, Sonjo area; Shinyanga region; Malaragasi and Ugalla rivers; in NE in Pangani valley; occurrence near Kitavi Plain Res. in SW needs confirming.

Description. *L. p. purpuropterus* Rüppell: E Africa; Sudan and Ethiopia north to *c.* 10°N. ADULT ♂: head bronzy green apart from blackish lores and violet-brown ear-coverts; sides of neck and narrow band around nape glossy blue, merging with glossy purplish hindneck, violet-blue mantle and back, and purple rump and uppertail-coverts; scapulars glossy blue, the longest with reflective black subterminal spots. Tail long, narrow and well graduated, with T6 *c.* 60% length of T1; bronzy purple above with blackish barring, blackish brown below. Chin and throat bronzy green, bordered by narrow purplish band merging into glossy blue upper breast; lower breast more violet, belly and flanks purple with bronzy reflections; thighs and undertail-coverts sooty brown. Primaries and secondaries dark brown, upperside of outer webs glossy bluish green; tertials bluish green; alula and primary coverts bluish; rest of upperwing-coverts glossy bluish green, apart from violet-blue patch along inner leading lesser coverts. Underwing-coverts and axillaries glossy violet-blue. Wing rounded: P7 longest; P9 *c.* 17 mm shorter, about equal to P3; P10 large, *c.* 30% length of P7. Eyes creamy white; bill black; legs black. Sexes alike. SIZE: (10 ♂♂, 10 ♀♀): wing, ♂ 149–163 (155), ♀ 130–154 (141); tail, ♂ 138–163 (151), ♀ 127–155 (138); bill, ♂ 24–26 (25·1), ♀ 23–26 (24·6); tarsus, ♂ 37–39 (38·0), ♀ 34–38 (36·4). WEIGHT: (Kenya), ♂ (n = 9) 78–92 (83·7), ♀ (n = 8) 65–78 (71·6).
IMMATURE: juvenile has upperparts glossed violet-blue; underparts dark brown with faint gloss on breast; head dark brown with slight gloss on crown. Wings glossy blue-green, uppertail glossy purple. Eyes dark brown.
NESTLING: not described.
L. p. aeneocephalus Heuglin: Eritrea, Sudan and Ethiopia south to *c.* 10°N. Mantle and back more violet than in nominate race; upper breast violet rather than blue; wings glossy blue (less greenish). Tail much longer, and more strikingly graduated (T6 *c.* 40% length of T1). SIZE: (n = 9 ♂♂), wing 150–165 (161), tail 184–217 (199).

Field Characters. Length 27–31·5 cm. A glossy *dark blue and purplish* starling with bronzy head, white eye in dark mask, and *long graduated* purple tail; no spots on wing. In Sudan just meets Long-tailed Glossy Starling *L. caudatus*, which is green, with much longer tail. Does not overlap with Meves's Glossy Starling *L. mevesii*. Immature has head and underparts dark brown, shorter tail.

Voice. Tape-recorded (10, 66, 104, B, C, F, GREG, KON, PAY). Song simple, of 4–10 variable elements per strophe (Dittami 1987), incorporating long rolled whistles, shrill but pleasant, 'prrreeo-ray-po-prreerreeo', and tinny, hollow 'chaw' or 'taw': 'to-jay-dee-chaw, chaw, tutrrrraaayo', 'to-jerry-taw-taw-teerrrro'; also squeaky and grating sounds and short warbles, each note distinct and uttered in a precise manner; more musical than other members of the genus. For further renditions see Zimmerman *et al.* (1996).

General Habits. Inhabits open and well wooded savannas and verdant bushlands, in Eritrea especially with *Acacia* and riparian figs, baobabs and *Hyphaene* palms; often in pasture and cultivated areas and around habitation; usually near water. Occurs in pairs and flocks of up to 6 birds. Forages on ground, often in company of other starlings. Spends much time resting and singing on large, shady boughs low in leafy tree, especially in heat of the day; both sexes sing, often in duet, but ♀ sings less frequently than ♂. Pairs remain in territories all year, boundaries altering little or not at all; juveniles stay in parental territory for at least 2 years but are forced to periphery when pair is nesting (Dittami 1987). On alighting on ground or in tree, raises tip of one wing, keeping it folded, then sometimes raises other wing; may be comfort activity or display (Wickler 1966).

Food. Not known.

Breeding Habits. Poorly known; solitary nester, evidently monogamous, territorial, and a partly co-operative breeder.
NEST: in old woodpecker holes in *Hyphaene* palms (Eritrea), also knot holes, with rough lining of cow dung and litter. Nests in Baringo, Simba and Taita, Kenya, were in the open, on branches of baobabs and wide-spreading acacias, made of thorny twigs (Jackson and Sclater 1938). Nest of captive birds made of mud, rootlets and leaves, in a nestbox (Ezra 1933).
EGGS: 2–3. Pale blue-green, immaculate or with rusty spotting. SIZE: (n = 4) 26·0–28·4 × 20·3–20·8.
LAYING DATES: Sudan, (Oct, Dec); Eritrea, (nestlings July); Ethiopia, (Mar–Apr, June–July, Sept); Somalia, (newly flown young July); E Africa, Region A, Feb–July, Sept, Region B, Jan, Mar–May, Oct, Regions C and D, Oct–

Nov; Rwanda, (breeds in Aug–Nov and Feb–Apr). Breeds mainly in rains, but also in driest month(s) immediately before rainy season (Brown and Britton 1980); monitoring of physiological condition indicates lack of breeding seasonality (Dittami 1987).

INCUBATION: period (captive birds) *c.* 14 days.

DEVELOPMENT AND CARE OF YOUNG: 'both parents and helpers may feed the young at some nests' and 'juveniles were tolerated at the nest, and fed their younger siblings' (Kenya: J. Dittami, pers. comm. *in* Feare and Craig 1998). Period (captive birds): *c.* 25 days.

References
Dittami, J. (1987).
Feare, C. and Craig, A. (1998).

Plate 35 *Lamprotornis mevesii* (Wahlberg). Meves's Long-tailed Starling. Choucador de Meves.
(Opp. p. 542)

Juida mevesii Wahlberg, 1856. Öfv. K. Sv. Vet.-Akad. Forh., 13, p. 174; Doughe [= Okavanga River], Bechuanaland.

Forms a superspecies with *L. caudatus* and *L. purpuropterus*.

Range and Status. Endemic resident, in 2 disjunct populations mainly on major river systems. (1) Angola, along escarpment from behind Lobito and Catengue (13°02′S, 13°46′E) to Villa Arriaga and Capangombe (15°05′S, 13°08′E); another very different but adjacent race from about Tchikala-Tcholohanga in upper Cunene valley, more widespread from 15°S (Cassinga, Chipopia, Huila, Bikuar Nat. Park) southwest to Cainde (= Cahinde, 15°42′S, 13°12′E) and south and southeast to Namibian border; doubtless occurs also on Angolan side of Cubango valley east of 19°E, though yet to be formally recorded there. Namibia, locally common in NW from lower Cunene valley to extreme SW corner of Etosha Nat. Park. (2) Several records in Kavango (Cubango) valley (Angola/Namibia border) east of 19°E and in Caprivi Strip. Botswana, frequent to common on Okavango, Linyanti and Boteti river systems; sparse in NE; very common on lower Shashe R. and Limpopo R. south to Martin's drift; uncommon around Francistown, common around Nata and Tsebanana (Penry 1994). Zambia, locally abundant in middle Zambezi and Luangwa valleys, to Chama, Luano valley, Kafue Nat. Park north to 16°10′S and S Kafue Flats (Monze and Lochinvar Ranch west to Nanzhila) (R.J. Dowsett, pers. comm.). Zimbabwe, complex distribution with regionally varying ecological requirements (Irwin 1981), as mapped (Harrison *et al.* 1997). Mozambique, plentiful in Tete between Mazoe and Luenya (= Luenha) rivers; and recorded on Zambezi; local and patchy in Sul do Save in middle and upper Limpopo valley, and on Save R. near Transvaal border. Malaŵi, an isolated population in upper Shire R. valley from Monkey Bay and Mtakataka to Liwonde. South Africa, locally common in N Kruger Nat. Park, becoming progressively scarcer westward along Limpopo valley, west to Mmabolela and Ellisras (Tarboton *et al.* 1987).

Description. *L. m. mevesii* (Wahlberg): Cubango/Okavango R. (Namibia) to Luangwa R. (Zambia), Shire R. (Malaŵi), Zambezi and Limpopo Rs (Mozambique), Transvaal. ADULT ♂: top of head to mantle and scapulars glossy greenish blue or violet-blue, merging into bronzy purple on rump and glossy purple on uppertail-coverts. Tail very long and strongly graduated (T6 *c.* 40% length of T1); feathers narrow, glossy violet-blue above with blackish barring, blackish below. Lores to ear-coverts blackish;

Lamprotornis mevesii

cheeks, sides of neck, and chin to upper breast glossy blue or greenish blue (concolorous with upperparts), grading through narrow violet band on lower breast to bronzy purple on belly and flanks; thighs and undertail-coverts sooty brown, glossed purple. Primaries and secondaries blackish above, glossed greenish blue on outer webs; dark brown below; tertials, alula and upperwing-coverts glossy bluish green. Underwing-coverts and axillaries glossed violet-blue. Bill black; eyes brown with dark blue centre; legs black. Sexes alike. In Zambia, valley birds slightly smaller and redder glossed than plateau ones (Benson *et al.* 1971). SIZE (8 ♂♂, 5 ♀♀): wing, ♂ 142–152 (147), ♀ 130–140 (135); tail, ♂ 189–219 (201), ♀ 179–198 (190); bill, ♂ 19–22 (21.3), ♀ 21–22 (21.8); tarsus, ♂ 38–41 (39.3), ♀ 35–40 (37.2). WEIGHT: (Zambia), unsexed (n = 8) 74–85 (80.0); Zimbabwe, ♂ (n = 8) 77.3–95.0 (88.5), ♀ (n = 7) 65–78 (70.7); also, unsexed (n = 15) 66–85 (Feare and Craig 1998) and unsexed (n = 29) 56–77 (64.2) (Maclean 1993).

IMMATURE: juvenile like adult but feathers less glossy at base; head and underparts browner, gloss on crown fainter and greener; rump, uppertail-coverts and belly with reddish rather than bronzy iridescence.

NESTLING: naked except for dirty white down on crown, sides of neck and more abundantly on spine; gape flanges bright yellow (become duller later); bill and legs whitish.

L. m. violacior Clancey: N Namibia and SW Angola (Huila and Moçamedes). Like nominate race but top of head to mantle, sides of neck, throat and breast deep violet; wing feathers glossed violet-blue.

L. m. benguelensis Shelley: S end of W Angola escarpment (replaced by *violacior* in acacia zone). Similar in size and structure to other races, but plumage strikingly different: entire upperparts bronzy green, more coppery on scapulars; underparts dark bronzy green, tinged violet from chin to breast; rump feathers bronzy green, edged violet; tail coppery.

TAXONOMIC NOTE: the 2 Angolan races occur within 80 km of each other near Sá da Bandeira; there is no evidence that they interbreed and 'it seems probable that *benguelensis* has achieved specific status' (Hall 1960b). White (1962) treated *benguelensis*, *mevesii* and *purpuropterus* as races of *L. caudatus*. We follow Hall and Moreau (1970), and Sibley and Monroe (1990) in treating *benguelensis* as a race of *L. mevesii*, and *L. mevesii*, *L. purpuropterus* and *L. caudatus* as composing a superspecies.

Field Characters. Length 30–36 cm. *Glossy purple* with blue and bronze reflections, no spots on wing, *eye dark*. Body smaller and legs shorter than Burchell's Starling *L. australis* but graduated tail *pointed* and *much longer*, with black barring visible in good light, 'rather floppy and ribbonlike in flight' (Maclean 1993). Immature has brown head and underparts but tail is already long.

Voice. Tape-recorded (38, 86, 88, 91–99, B, C, F, DUV, LUT). Commonest calls are rolled, rather shrill and unmusical, 'trrrreeo', 'urrrraaa', 'rrrer-chweeyo' or 'urrrik'; also gives tearing 'tssssaa', high insistent 'jeeea', short 'hwik' and a glottal 't'bar'.

General Habits. Habitat is tall woodland on flat alluvial soils where ground seasonally inundated, mainly *Colophospermum mopane*, *Acacia albida*, *A. xanthophloea* and *A. tortilis*; also mature *Combretum imberbe* woods and *Adansonia digitata* with thin understorey and bare ground below; absent from *Baikiaea*, stunted *Colophospermum* and degraded woodlands. Occurs in pairs and small flocks; at some seasons roosts communally in large flocks of up to 150 birds. Spends most of time on ground, foraging on sandy earth or heavily grazed ground by walking or running and pecking, often in company of other glossy starlings. Follows elephants, feeding on the ground behind them (Kruger Nat. Park, South Africa: Dean and MacDonald 1981). Readily perches in trees, generally on large, well-shaded limbs low down; flies low down or fairly high between treetops. Before settling to roost, sometimes in reedbeds, flock performs aerial evolutions (Maclean 1993); pair with young roosts in thick foliage in tree, often waking up and calling in middle of night. Flocks break up and birds disperse a few km at start of rains (Brooke 1965).

Food. Insects including termites, carabid and scarabaeid beetles, ants, mole crickets, and psyllids gleaned from leaves; thought to be mainly insectivorous in breeding season (Dowsett 1967). Outside breeding season feeds mainly on fruits; flock seen feeding in *Acacia albida* tree, either on fruits or flowers; fruit includes fallen ripe fruit of *Diospyros mespiliformis*.

Breeding Habits. Presumably monogamous and territorial; partly co-operative breeder. One pair nested in same tree in 2 successive seasons, in different holes; pair once accompanied by 3rd adult (their young of previous year) which did not help in nest-building; they chased away a Striped Kingfisher *Halcyon chelicuti* which visited hole and mobbed a Woodland Kingfisher *H. senegalensis* and Lilac-breasted Roller *Coracias caudatus*. First breeds at *c.* 11 months. Once 3 pairs nested in one tree. Probably single brooded.

NEST: pad of dry plant fibres and twigs, often on substrate of loose and rotting wood in tree hole; lined with a few leaves; once a cup of grass with no other material. Sited in knot hole or other natural hole in near-vertical or 45° branch of e.g. *Kigelia pinnata* or *Trichilia emetica* tree and acacias, baobabs, mopanes and palms, in live or dead, whole or rotten wood; also ventilation pipe or cavity in fence post; 2–6 m above ground. One chamber 25 cm deep, 10–11 cm in diam. Tree hole partly-excavated by ♀; ♂ perches higher in tree and often calls and flies down to peer into hole after ♀ has entered. ♀ carries wood chips away in bill for *c.* 25 m, accompanied by ♂. Nest built by ♀, who may carry twigs into hole whilst still busy excavating it. ♂ seldom enters hole; once ♂ carried wood chip in, which was promptly removed by ♀.

EGGS: 3–4. Immaculate pale greenish blue. SIZE: (n = 6) 25·0–28·5 × 18·2–20·0 (27·2 × 19·5).

LAYING DATES: Angola, Apr; Botswana, Feb; Zambia, Feb–Mar (7 clutches); Zimbabwe, Nov–Apr, mainly Feb–Mar (35 out of 44 records); Malaŵi, (nest-building Dec, occupied hole Mar).

INCUBATION: no information.

DEVELOPMENT AND CARE OF YOUNG: on day 10 spinal, femoral, humeral, crural, ulnar and ventral feather tracts present; inner and outer supraorbital and occipital tracts poorly developed. Tarsus off-white, darkening blotchily to blackish. Increases in weight and tail length graphed by Dowsett (1967). Young fed by both parents and helpers (helpers being parents' offspring in previous season). By day 15 nestlings defaecate out of entrance hole. Nestling period *c.* 22 days. Fledged young dependent on parents for several weeks. Post-juvenile moult affects only belly and flank feathers (Brooke 1967a).

BREEDING SUCCESS/SURVIVAL: sometimes parasitized by Great Spotted Cuckoos *Clamator glandarius*.

References
Brooke, R.K. (1965, 1967a).
Dowsett, R.J. (1967).
Feare, C. and Craig, A. (1998).

618 STURNIDAE

Plate 35
(Opp. p. 542)

Lamprotornis australis (Smith). Burchell's Starling. Choucador de Burchell.

Megalopterus australis Smith, 1836. Rep. Exped. Centr. Afr., p. 42; Kurrichane, South Africa.

Range and Status. Endemic resident, southern Africa. SE Angola, Huila Province north to Mupo, Mulondo on Kunene R. and Cabama on R. Caculovar. SW Zambia between Zambezi and Mashi rivers north to Nangweshi and Shangombo, east to R. Machili. Namibia, locally common in central and E region north of 24°S; absent from coastal belt. Botswana, widespread and often common; absent from open plains and areas with poor tree cover. Zimbabwe, local in NW border region near Matetsi. South Africa from Kalahari Gemsbok Nat. Park through N Cape and central and E Transvaal north of 27°S; local in N Cape, sparse in N Transvaal and absent from high escarpment area; common in E Transvaal lowveld, especially Kruger Nat. Park. Mozambique, restricted to SW border adjoining Kruger Nat. Park. Swaziland, uncommon in E lowveld.

Description. ADULT ♂: forehead and crown iridescent blue-green, nape with blue to purple sheen, almost forming collar in some birds. Lores black, ear-coverts and patch below eye bronze with blue tinge around border, chin and throat blue-green. Mantle and back glossy blue-green, rump and uppertail-coverts purple; tail violet barred black. Underparts blue-green, centre of belly purple, flanks, thighs and undertail-coverts blue. Primaries and secondaries dark blue with purple sheen and dark barring; no indentations on primary feathers. Wing-coverts blue-green, with large dark blue subterminal spots on greater coverts and a bronze epaulet bordered by purple. Bill black, eyes dark brown, legs and feet black. Sexes alike. Birds with areas of white plumage recorded. SIZE: (10 ♂♂, 10 ♀♀): wing, ♂ 165–195 (182), ♀ 148–178 (167); tail, ♂ 153–178 (166), ♀ 137–160 (151); bill to skull, ♂ 26·7–29·7 (28·1), ♀ 25·5–28·6 (26·8); tarsus, ♂ 44·0–48·4 (46·6), ♀ 42·2–45·3 (44·0). WEIGHT: (Zambia, unsexed, n = 8) 73·5–84·5 (80); (Namibia) ♂ 126, 127, 138, ♀ 112, 115, 125.

IMMATURE: duller than adult, with greenish sheen on dorsal plumage, matt black ventrally, lacking any purple iridescence.
NESTLING: undescribed.

Field Characters. Length 30–34 cm. A *heavily-built* blue-green and purple starling, noisy and conspicuous; *wings broad and rounded*, flight loud and laborious. In good light, dark barring visible on primaries and tail. Larger than other glossy starlings, with *dark eye; tail shorter and broader* than Meves's Long-tailed Starling *L. mevesii*, ear-patch larger and blacker, contrasting with surrounding blue-green feathers.

Voice. Tape-recorded (72, 88, 91–99, 103, B, C, F, KEI, LUT). Song a mixture of rolling whistles, high-pitched scratchy and creaky noises and hollow notes, given in brief 2–5–note phrases, 'trrreerroo', 'tyoo-pu-joy', 'chicktaw-chweeyo', 'ti-trock-ti-troohoo', a rusty 'screeo-creedo', a deep 'tyoow-charr' and a hollow, almost oriole-like 'chopo'. Pauses between phrases variable, 1–4 s. Songs often lengthy, and flocks produce a babbling chorus, which at roost is 'almost deafening' (Maclean 1993).

General Habits. Inhabits open woodland and savanna dominated by *Acacia* and mopane trees; particularly

Lamprotornis australis

associated with camelthorn *A. erioloba* and knobthorn *A. nigrescens*; absent from *Brachystegia* woodland. Favours areas with bare ground between trees, so often in heavily grazed areas. Distribution consequently patchy, suggesting a habitat specialist. Below 500 m in E lowveld, elsewhere up to 1500 m.

Occurs in small parties, often together with other glossy starlings. Foraging flocks number up to 50 birds; winter

A

roost in Botswana contained more than 1000 birds (R. K. Brooke, pers. comm.). Roosting birds often call for extended periods. In group sings with head flung back, bill pointed upwards and wings drooped, perching low in tree, well separated from other birds; often sings on ground (A). Feeds primarily on ground, walking with long strides; scavenges at camp sites. Moult occurs mainly before breeding, but wing-moult sometimes suspended and completed after breeding.

Food. Primarily terrestrial arthropods: locusts, termites, ants, beetles, centipedes. A mouse once eaten, and birds have been caught on bal-chatri traps baited with mice (A. C. Kemp, pers. comm.). Berries and similar fleshy fruits, of e.g. *Diospyros mespiliformis*; flowers of *Acacia giraffae*; also kitchen scraps.

Breeding Habits. Monogamous; apparently territorial; no helpers.

NEST: usually in natural tree hole or old barbet or woodpecker hole, 2–7 m above ground; occasionally crevice on cliff, hole in building or other structure; lined with grass, often fresh and green, also green leaves; in nest boxes near farm, feathers, string, cloth, paper, snake skin and strips of plastic used (W. R. Tarboton, pers. comm.).

EGGS: 2–4, glossy blue to greenish blue, plain or with scattered reddish spots, mainly at thicker end. SIZE: (n = 14) 27·5–31·7 × 19·8–22·1 (29·2 × 20·5). WEIGHT: 7·9.

LAYING DATES: Zambia, Mar; Namibia and Botswana, Jan–Apr; South Africa mainly Oct–Jan, some records Sept–Mar.

INCUBATION: *c.* 15 days.

DEVELOPMENT AND CARE OF YOUNG: nestling period 20–24 days; young fed by both parents.

BREEDING SUCCESS/SURVIVAL: parasitized by Great Spotted Cuckoo *Clamator glandarius*, but young starlings may be raised successfully with single cuckoo chicks (W. R. Tarboton, pers. comm.)

Reference
Brooke, R. K. (1967b).

Lamprotornis iris (Oustalet). Emerald Starling. Choucador iris.

Plate 34 (Opp. p. 527)

Coccycolius iris Oustalet, 1879. Bull. Soc. Philom. Paris, (7), 3, p. 85; Los Islands, Portuguese Guinea, probably in error for interior of French Guinea.

Range and Status. Endemic resident. Guinea, very local, in W around Fouta Djallon, Vallée de Konkouré and Kinsam, and in E at Badala. Sierra Leone, in NW from Yana and Kabala to Bendugu; uncommon northwest of Little Scarcies R. but increasingly common eastwards; extends south to about 08°55′N in centre and 08°45′ in E; absent from mountains in N (G.D. Field, pers. comm.). Mali, 2 records, Bafing-Makana and Kényéba, May (Lamarche 1981). Ivory Coast, in central guinean savannas north to Touba and Dabakala, south to Biankouma, Marahoué Nat. Park, Bandama, Béoumi and Bouaké. Recently, wild-caught birds have commonly been offered for sale in Europe, though extent of international trade unknown. In 1981–1984 large numbers, thought to be from Guinea, were held by bird traders in Monrovia, Liberia; 12 in captivity Voinjama, N Liberia, May 1984; 1 free, Monrovia, Apr 1982 had doubtless escaped (Gatter 1997).

Description. ADULT ♂: top of head, hindneck, and mantle to scapulars and uppertail-coverts glossy emerald green. Tail bright emerald green above, dark grey-brown below (feathers glossed green above except inner borders of T2–T6, but outer edges grey-brown when worn). Chin and fore-cheeks to centre of throat, sides of neck and upper breast glossy emerald green, bordered behind by narrow line of glossy blue; lores and below eye to ear-coverts, hind-cheeks and sides of throat glossy purple. Lower breast to upper flanks, upper belly and thighs glossy purple with bronzy tinge and bluish reflections; lower belly dark brownish; lower flanks and undertail-coverts glossy emerald green. Wing glossy emerald green except for grey-brown inner borders of primaries and secondaries; primary tips grey-brown when worn. Axillaries and underwing-coverts brown, tinged purple; tips of larger coverts fringed glossy blue; marginal coverts glossy green.

Lamprotornis iris

Bill black; eyes dark brown; feet black. Sexes alike. SIZE (8 ♂♂, 7 ♀♀): wing, ♂ 104–112 (109), ♀ 102–108 (105); tail, ♂ 73–78 (74·8), ♀ 67–73 (69·3); bill, ♂ 23–26 (24·5). ♀ 21·5–25 (23·1); tarsus, ♂ 25–27 (26·1), ♀ 24–26 (25.)

IMMATURE: juvenile similar to adult, but lower breast to upper belly and upper flanks unglossed dusky brown; wings and tail glossed more bronzy green; axillaries and underwing-coverts dusky grey-brown.

Field Characters. Length 18–19 cm. A small, short-tailed starling with limited range in W African wooded savanna; *bright iridescent emerald green* with purple ear-coverts, neck patch and belly and *dark eye*; no blue tones to plumage, no black wing spots. Immature wholly brown below, mixed brown and green above, with green wings.

Voice. Tape-recorded (WILK). Calls given when taking flight a long-drawn-out 'wheeze' or wheezy 'weee' rising in pitch (G. D. Field, pers. comm.); pre-flight call, 'chewp-chewp'. Other calls, 'we we' and a rather squeaky 'wheet'. Commonest call in captivity a high-pitched 'pee-pee-pee' of alarm; harsher, more grating version used as ground predator alarm or mobbing call; also gives low-volume contact calls (R. Wilkinson, pers. comm.).

General Habits. Inhabits orchard bush and wooded and open savanna; avoids forest but occasionally at edge of gallery forest. In forest/savanna mosaic in central Sierra Leone occurs in same areas as Red-winged Starling *Onychognathus morio* and Copper-tailed Glossy Starling *L. cupreocauda* (G.D. Field, pers. comm.). Occurs singly, in pairs, and flocks of up to 10 (Marahoué Nat. Park, Ivory Coast, Dec–Feb: Demey and Fishpool 1991). In Sierra Leone occurs in non-breeding flocks of up to 50 birds: mainly immatures (Yealland 1955). Forages in trees (for fruit) and on bare, burnt-over ground (for insects).

In Ivory Coast, all records from Marahoué Nat. Park at S border of range are in Dec–Feb, and in Sierra Leone occurs towards N border of range at Yana and Kabala in Feb–Mar (Bannerman 1948), which suggests possibility of seasonal movements. Evidently resident in N Sierra Leone, though birds could move north in July–Oct (G.D. Field, pers. comm.).

Food. Small fruits and seeds, and insects including small black terrestrial ants and caterpillars (Bannerman 1948). Often eats fruit, particularly of *Ficus* and *Harungana madagascariensis* (Sierra Leone). Young fed entirely on insects taken from ground (G.D. Field, pers. comm.).

Breeding Habits. Loosely colonial; in captivity breeds co-operatively, immature birds helping their parents at the nest (Wilkinson 1996, 1997). Aggressive when breeding, once repeatedly attacking and driving off pair of Abyssinian Rollers *Coracias abyssinica* and once a Lizard Buzzard *Kaupifalco monogrammicus* (G.D. Field, pers. comm.). In captivity, conspicuously displays intention to breed by carrying single green leaf or piece of one in bill (but shy of taking it to the nest).

NEST: in hole in stump or small tree; one nest of captive pair was pad of green leaves, carried in by ♂ and ♀, with eggs laid on top.

EGGS: 4 (one clutch, G.D. Field, pers. comm.); in captivity 3, 3, 3 and 4. Pale blue, some (all?) with red-brown blotches (Bruch 1983, Robiller and Geistner 1985, Pyper 1994).

LAYING DATES: Sierra Leone, Mar (9 nests, eggs seen in one; birds carrying food to nests, early Apr (G.D. Field, pers. comm.); enlarged testes, Feb).

INCUBATION: period 13–15 days (in captivity, England, Pyper 1994).

DEVELOPMENT AND CARE OF YOUNG: in captivity, young fed by both parents and helpers; in 10 20-min watches over 12 days, 1st ♂ fed young 33 times, 2nd ♂ 30 times, and ♀ 19 times; next year, in 12 30-min watches over 10 days, previous year's '1st' ♂ did not feed young, the '2nd' ♂ brought 28 and the ♀ 12 feeds, and 2 imm. birds brought 29 and 5 feeds (Wilkinson 1996, 1997). Nestling period *c.* 21 days (in captivity, Germany, England). Young retain patches of dull plumage on underparts until 12–14 months old.

BREEDING SUCCESS/SURVIVAL: captive birds (England) lived at least 8, 10 and 14 years (S. Tonge, pers. comm.).

References
Bannerman, D.A. (1948).
Wilkinson, R. (1996, 1997).

Plate 36 *Lamprotornis unicolor* (Shelley). Ashy Starling. Spréo cendré.
(Opp. p. 543)

Cosmopsarus unicolor Shelley, 1881. Ibis, p. 116; Ugogo, Dodoma District, Tanganyika.

Range and Status. Endemic resident, interior of Tanzania. Locally common in dry country, between 1000 and 1850 m, from Mwanza, Maswa Game Res., Mangola Springs, L. Manyara and Tarangire Nat. Park south to L. Malaŵi. Eastern Arc mountains delimit range in S and SE. Quite common in Ruaha R. gorge but stops abruptly where river enters greener vegetation to east. Range delimited in SW, W and WNW by miombo woodland and in N largely by crater highlands, although it crosses the Eyasi Rift north into Maswa Game Res. (but unclear what limits range east of Tarangire (04°00′S, 36°00′E), where there is no abrupt change in habitat or altitude – N.E. Baker, pers. comm.). In Maswa Game Res. occurs in southern, drier, rocky area with baobabs *Adansonia*, absent from northern acacia woodland and bushland (D. Bygott, pers. comm.). A vagrant at L. Jipe, Kenya, in 1917 may have been on the Tanzanian side of the border (Zimmerman *et al.* 1996).

Description. ADULT ♂: upperparts brownish grey, mantle to uppertail-coverts with faint greenish gloss. Tail long, strongly graduated (T6 *c.* 40% length of T1), feathers very narrow; blackish brown, upperside with greenish gloss and narrow dark barring. Lores blackish; ear-coverts dark ashy grey; cheeks and sides of neck brownish grey, concolorous with upperparts. Entire underparts paler brownish grey. Flight feathers blackish brown, outer webs and greater coverts glossed slightly greenish; alula and primary coverts dark grey-brown; tertials and rest of upperwing-

Lamprotornis unicolor

coverts brownish grey. Underwing-coverts and axillaries dark grey-brown. Wing-tip rather rounded, with P9 much shorter than wing point, falling between P4 and P3; P10 rather large, *c.* 17 mm longer than primary coverts. Bill black; eyes creamy yellow; legs black. Sexes alike. SIZE: wing, ♂ (n = 14) 118–132 (126), ♀ (n = 7) 118–128 (122); tail, ♂ (n = 14) 146–180 (165), ♀ (n = 7) 126–174 (152); bill, ♂ (n = 9) 21–23 (22·0), ♀ (n = 4) 20–21·5 (20·6); tarsus, ♂ (n = 14) 32–36 (34·2), ♀ (n = 7) 32–35 (33·6). WEIGHT: ♂ (n = 5) 55–66 (60·2), ♀ (n = 1) 62.

IMMATURE: juvenile similar to adult, but duller ashy brown; eye grey. Fledgling like adult but has pale horn bill, pale orbital ring surrounding dark iris, and shorter tail.

Field Characters. Length 28–30 cm. Endemic to woodlands of interior Tanzania. Uniform ashy *brownish grey* with *long, narrow tail*; *white eye* contrasts with black lores. Immatures similar but bill pale. Combination of colour, shape and size highly distinctive; cannot be confused with any other bird (D. Bygott, pers. comm.).

Voice. Tape-recorded (B, C, NOR, PARK, WILK). Not very vocal. Rasping calls, 'zip-zee' or 'zhweep'; song, low conversational notes mixed with harsh ones, 'zwer, zup, zreee, zer, zukubah-zhrreeeh, zwer ... ', also 2 alternating phrases, 'wup-tirrrro, zwooey-pu, wup-tirrrro, zwooey-pu ... '. Song also rendered 'koora tcheeo chink chink', repeated twice; call, sometimes preceded by 2–3 chuckled or warbled notes, a slight, plaintive 'kuri-kiwera' with a squeaky or rusty quality, second note higher, and sometimes accompanied by light bill rattling (Fuggles-Couchman 1984). Common contact call between family members 'weu-weu' or 'kuri-kuri', given particularly when on the move. Alarm call a loud, harsh 'kaaaaa! kaaaa!' directed at snakes and other enemies. Squeaky, trilling, rusty calls given more often when in large groups than between family members (D. Bygott, pers. comm.).

General Habits. Occurs in dry savanna and acacia woodland. Conspicuous when keeping to crowns of flat-topped acacias (Fuggles-Couchman 1939, Hall and Moreau 1970). Pairs or small flocks perch in low tree tops and forage both on ground and in bushes. On ground said to hop around holding tail at sharp angle to body (Irwin 1956), but that is not characteristic. In some areas shows an attachment to baobabs *Adansonia digitata*. At Mangola Springs, where permanent water is available, pays daily visits to drink and bathe (D. Bygott, pers. comm.). Often in flocks rather than just family parties (N. E. Baker, pers. comm.).

Food. Insects: grasshoppers, mantids, dragonflies, adult antlions, termites and caterpillars; also fruits, including berries of *Cordia* and *Commiphora*, and kitchen scraps of cornmeal porridge and rice (D. Bygott, pers. comm.).

Breeding Habits. Co-operative breeder, monogamous; 1 pair remained breeding together for at least 4 years. 1–2 helpers at each of 4 nests; helpers were offspring from previous or earlier years. One bird, from larger size believed to be ♂, helped at its parents' nests in its 3rd and 4th year (D. Bygott, pers. comm.).

NEST: in tree hole (**A**); 2 at tops of doum palms *Hyphaene thebaica*, in grove in grassland (L. Manyara): one *c.* 5 m high in a broken-off palm, the other in a dead palm (J. S. S. Beesley, pers. comm.). One in Nzega District was in a baobab tree, and another in Tarangire Res. was in a hollow *Acacia spirocarpa* (Thomas 1960). Bird carried short pieces of dry grass to the latter nest. In contrast, captive birds fashioned shallow cup in a layer of peat in one corner of

A

their nest-box, and added no other nest-material. At Mangola Springs, nests were in old woodpecker or barbet holes, mostly 5–16 m high, in dead trunks or branches of fig trees *Ficus sycamorus* (D. Bygott, pers. comm.).

EGGS: up to 4 (from observed sizes of fledged broods). In captivity 3 (n = 2 clutches). Sub-elliptical; smooth, slightly glossy; pale greenish blue with several small, obscure, pale reddish brown blotches on narrow end, and uniformly marked all over with minute faint brownish speckles.

LAYING DATES: Feb–Mar, at wettest time of year (and nest-building Apr).

INCUBATION: period: *c.* 14 days, (estimated in captive birds in England).

DEVELOPMENT AND CARE OF YOUNG: both parents and 1–2 helpers fed nestlings with insects and fruits; fledged broods were of 2, 3 and 4 young (D. Bygott, pers. comm.). Nestling period (2 nestings, in captivity): in one nest with 2 chicks the 1st fledged at 26 and the 2nd at 28 days; at another nest both chicks left at 31 days.

BREEDING SUCCESS/SURVIVAL: 1 captive ♀ (England) lived >9 years (S. Tonge, pers. comm.).

References
Fuggles-Couchman, N.R. (1939).
Wilkinson, R. and McLeod, W. (1991).

Plate 36 (Opp. p. 543) *Lamprotornis regius* (Reichenow). Golden-breasted Starling. Spréo royal.

Cosmopsarus regius Reichenow, 1879. Orn. Centralb., p. 108; Massa, Tana River.

Range and Status. Endemic resident and local migrant in dry desert and thornbush. Ethiopia, uncommon to frequent in S, SE and Rift Valley. Somalia, common in NW north of 09°N, east to 47°E; south of 08°N extends to E coast and south through W-central area; widely distributed south of 03°N. Kenya, locally common resident east of Rift Valley up to 1200 m, from Mandera, Moyale and North Horr south to Samburu, Shaba, Meru, Tsavo and along Tana R. to coast north of Malindi; absent from coast south of Malindi and very arid parts of N and E, rare at Ileret. NE Tanzania from Kilimanjaro lowlands east to Mkomazi, west to Naberera (Masai Steppe); vagrants south to Kilosa.

Description. ADULT ♂: top of head glossy green, rest of upperparts glossy blue. Tail long and strongly graduated (T6 *c.* 30% length of T1), feathers narrow, bronzy green above, blackish below. Lores and ear-coverts deep blue. Cheeks and sides of neck to chin and upper throat glossy green, concolorous with top of head, and grading through blue and violet on lower throat to glossy purple crescent on centre of upper breast; sides of breast and lower breast to undertail-coverts golden-yellow. Primaries dark brown with coppery purple outer webs; secondaries violet apart from brownish inner web borders. Tertials, alula, primary coverts and greater coverts glossy violet-blue; median and lesser coverts glossy blue. Underwing-coverts and axillaries bright yellow; median and lesser under primary coverts glossy blue. Wing-tip pointed, with P7–P9 about equal; P 10 small, *c.* 4 mm longer than primary coverts. Outer primaries notched (**A**). Bill black; eyes creamy white; legs black. Sexes alike. SIZE (10 ♂♂, 9 ♀♀): wing, ♂ 132–146 (136·5), ♀ 123–133 (127); tail, ♂ 202–238 (213), ♀ 177–221 (187); bill, ♂ 21–25 (23·6), ♀ 20–24 (22·2); tarsus, ♂ 32–36 (33·3), ♀ 28–33 (30·6). WEIGHT: ♂ (n = 1) 62, ♀ (n = 1) 43.

IMMATURE: juvenile has head, neck and chin to upper breast greyish buff, darker on lores and ear-coverts; upperparts blackish brown, feathers tipped glossy bluish green; lower breast to undertail-coverts buffy yellow. Wing as in adult but rather duller; tail brown with slight bronzy green gloss. Eyes greyish.

NESTLING: unknown.

Field Characters. Length 34–35 cm. Slim and long-legged, with thin, straight bill, *very long narrow tail* and *golden*

Lamprotornis regius

underparts; rest of plumage iridescent blue, green and violet; patch of reddish violet on upper breast; eye white. Immature duller, with brown head and neck, brown eye, shorter tail. No other bird remotely like it.

Voice. Tape-recorded (CHA, GREG, McVIC, WILK). Subsong (?) a low warbling given while resting at noon; flight call 'cheeo cheeo' or 'weep-weep', rather plaintive;

A

call at nest 'quechee chee cheeo', alarm 'chiarr' (van Someren 1956). Keeps up a constant chattering, intended no doubt to hold the flock together (Archer and Godman 1961). In flight also said to give 'a whistling, chattering 'cherrrreeeeeter-cherrrree' (Zimmerman *et al.* 1996).

General Habits. Occurs in dry open bushland, thornbush and bushed grassland. Gregarious, in small parties of 4–7, rarely >12; family-sized groups, year-round, vary little in size from month to month (Huels 1981), though occurs in pairs at nesting time (van Someren 1956). Forages for insects on ground, mainly on bare patches; also searches for them in flowers and foliage; occasionally takes termites in flight. Usually shy and restless (although confiding at some safari lodges, where it feeds with other starlings and buffalo-weavers); flies low, rarely more than 2 m above ground, from bush to bush or between *Acacia* trees. Always alert and quick in movement; hops and runs on ground, briefly stopping and raising tail slightly above wings; raises tail on landing in tree. When feeding nestlings, clings to lower edge of nest-hole, supporting body with long tail like woodpecker (**B**). Rests at midday, sitting and preening, maintaining a low warbling chorus; at sunset perches on tree-tops, preening and calling (van Someren 1956). Local migration indicated by occurrence at Ileret (NW Kenya) only during the rains; numbers peak at Tsavo East from July to Sept (Lack 1985).

Food. Insects, almost entirely. In Tsavo East 76% of 394 food items were insects, of which 97% were taken on ground (Lack 1985). Termites taken both on ground and in flight; also eats fruits of *Commiphora* and *Dobera*. One stomach contained a large scarabid beetle, 130 termites, 2 ants, *Commiphora* fruit seeds, plant material and sand grains (Lack and Quicke 1978). Nestlings fed large larvae of moths and beetles; also a grasshopper and a berry (van Someren 1956).

Breeding Habits. Co-operative breeder with helpers at the nest. 5–9 adults seen near each of 5 nests observed near Kibwezi, Kenya; at one nest at least 5 adults fed young. ♀ solicits food from helpers by crouching, quivering slightly raised wings, gaping and calling (Huels 1981). Possibly double-brooded.

NEST: hole 3–6 m from ground in tree, disused nest of woodpecker or barbet, or rot hole. One entrance was diam. 40, inclined downward, >25 cm to chamber about 110 in diam.; floor of chamber with thick bed of dry grass, matted hair and feathers (van Someren 1956); other nest holes have only desiccated wood and a few feathers (Jackson and Sclater 1938) or sometimes a few roots or cast snake-skin (Archer and Godman 1961). 5 nests in Kenya were in old cavities of Nubian Woodpecker *Campethera nubica* in *Commiphora* trees (Huels 1981), and one in hole 1·5 m up in *Platycelipheum voiensis* tree.

EGGS: 4–6, laid daily. Elongated, pale greenish blue, faintly and minutely speckled with red-brown (Jackson and Sclater 1938). SIZE: (n = 7, Somalia) 24·5–28·3 × 17·5–18 (26·7 × 17·6), (n = 4, Somalia) all 27 × 17·5. (Archer and Godman 1961).

B

LAYING DATES: Somalia, Apr (and nestlings and fledglings June); E Africa: Kenya, Mar–Apr, Nov; Region D, Mar–May, Nov–Dec (during both long and short rains).

INCUBATION: by ♀ alone. ♀ adds fibrous nesting material throughout incubation period. Helpers provide ♀ with material, occasionally enter nest with material when she is absent; they also feed her at nest, feed nestlings and remove faecal sacs. Period: 14 days (captive birds in England). Bouts of incubation 15–20 min, absences 5–10 min.

DEVELOPMENT AND CARE OF YOUNG: size disparity in brood of 4, largest becoming feathered whilst smallest still almost naked; av. weight at 12 days, 45·8 g (captive birds, n = 5, Bell 1984). Total of 48 feeding visits once made in 2 h, 16 by breeding ♀ and 32 by 4 other adults (Huels 1981). Chicks reared in captivity peer out of nest at 3 weeks and fledge in fourth week (Wavertree 1930); in another captive breeding first chick left nest on day 20, second on day 22 (Risdon 1990).

BREEDING SUCCESS/SURVIVAL: 1 captive (England) lived >19 years, another >17 and a third was reportedly kept for >28 years (S. Tonge, pers. comm.).

References
Huels, T.R. (1981).
van Someren, V.G.L. (1956).

Plate 35
(Opp. p. 542)

Lamprotornis shelleyi (Sharpe). Shelley's Starling. Choucador de Shelley.

Spreo shelleyi Sharpe, 1890. Cat. Bds. Brit. Mus. 13, p. 190. Somaliland.

Forms a superspecies with *L. hildebrandti*.

Range and Status. Endemic intra-African migrant, possibly sedentary in some localities. Breeds in northernmost parts of range. In Somalia all breeding records are north of 08°N but in Ethiopia breeds at 04°30′–05°30′N (J. S. Ash *in* Lewis and Pomeroy 1989). Breeding status of birds close to or just south of this latitude on Sudan/Kenya and Ethiopia/Kenya borders remains uncertain. In Kenya, undated breeding record of large young at Simba (van Someren 1922, 1932) rejected as well south of expected breeding range; Sudan, only in SE corner, where common, and reportedly resident (Cave and Macdonald 1955) in arid acacia grassland; Ethiopia, frequent to common below 1800 m in S, SE and Rift Valley; Somalia, very common in bushland Apr–Sept, breeding south of coastal plain north of 08°N, a few scattered records below 04°N; Kenya, may be resident in NW around Lokichokio, and in NE from Mandera south to El Wak and Wajir, but in E and SE (mainly inland from coast and southwards east from South Horr) generally uncommon non-breeding visitor Aug–Mar in thorn-scrub below 1300 m, more frequent in Garissa and Tsavo East Nat. Park, occasional south to Maktau, Voi and Maunguy, and at Bamburi.

Description. ADULT ♂: top of head glossy violet; nape dark glossy green, extending as collar round sides of neck, rest of upperparts glossy violet-blue. Tail glossy violet-blue above, blackish brown below. Lores black; ear-coverts violet-black. Cheeks to chin, throat and upper breast glossy violet-blue; lower breast, flanks and belly deep maroon, thighs and undertail-coverts chestnut. Flight feathers blackish brown, primary outer webs glossed violet, secondary outer webs dark green. Tertials and greater coverts dark glossy green; alula and rest of upperwing-coverts glossy violet, median coverts with large velvety blackish subterminal spots. Underwing-coverts and axillaries chestnut; median and lesser under primary coverts glossed violet. Wing-tip rather pointed; P9 usually longer than P6 (*cf. L. hildebrandti*). Bill black; eyes orange; legs black. Sexes alike. SIZE (10 ♂♂, 10 ♀♀): wing, ♂ 113–118 (116), ♀ 105–115 (109); tail, ♂ 76–82 (80·0), ♀ 70–80 (75·4); bill, ♂ 18–21 (19·5), ♀ 18–20 (19·0); tarsus, ♂ 30–31 (30·4), ♀ 28–30 (29·1). WEIGHT: ♀ (n = 1) 45.

IMMATURE: juvenile has greyish brown top of head, lores and ear-coverts, contrasting with paler chin and throat; rest of upperparts greyish brown, slightly mottled, with practically no gloss; underparts buffish brown, more tawny on belly and undertail-coverts. Wings brown with faint green gloss, remiges and larger upperwing-coverts with pale buffish fringes; uppertail glossed bluish green. Undersides of flight feathers greyish brown. Eyes brown.

NESTLING: unknown.

Field Characters. Length 18–19 cm. Very similar to Hildebrandt's Starling *L. hildebrandti*, with similar orange-red eye and deep violet-blue head and breast, but *belly and undertail-coverts darker rufous-chestnut*, not becoming paler where belly meets breast. Immature differs from young Hildebrandt's in greyer brown head and back, latter without gloss (Zimmerman *et al*. 1996).

Lamprotornis shelleyi

Voice. Tape-recorded (CHA, GREG, McVIC). Conversational song a mixture of whistles and nasal and scratchy calls, 'jaaw, wah, dzit, wrreeyanh, juyu-juyu, wrraa, poweet ... '; 'chavat, preeuwee, chavat, chuyee-choyo, jiraa, tyoolee'. Calls include repeated nasal 'jaraanh' or 'jararaanh'.

General Habits. In Sudan and Somalia inhabits arid and semi-arid thornbush; in Sudan often associates with *Salvadora persica* bushes (Nikolaus 1987). In Kenya, in *Acacia* and *Commiphora* thorn-scrub, perching on tops of *Commiphora* trees. Generally shyer and more difficult to approach than Hildebrandt's Starling or Superb Starling. Gregarious when not breeding, usually in small parties or flocks (sometimes hundreds). In E Africa often associates with nomadic Magpie Starling *Speculipastor bicolor* (Britton 1980), and in Sudan with Superb Starling. Feeds less on ground than Hildebrandt's Starling or Superb Starling. Attracted to fruiting trees; in Kenya common along Tana River when *Salvadora* bushes in fruit. Migratory; present Somalia Apr–Sept, where it breeds, and Ethiopia mostly Mar–Nov (once Jan). In Kenya a non-breeding visitor and wanderer, normally Oct–Mar, rarely May–Aug. Movements noted Dec–Feb, Kenya (Curry-Lindahl 1981). Seasonal occupancy and movements in non-breeding areas require further investigation.

Food. Insects and fruit.

Breeding Habits. Reported not to breed communally; probably double-brooded (Archer and Godman 1961).

NEST: untidy nest of grass and feathers in hole or cleft, in black stunted thorn tree 1·5–3 m above ground. Sometimes lays directly on pulverized or chipped wood at base of hole (Somalia: Archer and Godman 1961). In Ethiopia nests in termitaria (Ash and Miskell 1998).

EGGS: 4–6; sometimes elongated; immaculate light blue or turquoise-blue, with an occasional indistinct brown freckle. SIZE: (n = 10, Somalia) 24·5–27 × 18 (25·8 × 18)

LAYING DATES: Ethiopia, Apr, 4 records (J. S. Ash *in* Lewis and Pomeroy 1989), (♀ with enlarged ovary Mar); Somalia, Apr–June (Archer and Godman 1961), May (7 clutches), June (2) (Ash and Miskell 1998).

INCUBATION: by ♀ (♂ of a captive pair sang continuously from a nearby perch).

DEVELOPMENT AND CARE OF YOUNG: 2 chicks (of 3 hatched) fledged at 22–23 days (captive breeding, England: Scamell 1964).

BREEDING SUCCESS/SURVIVAL: 1 captive (England) lived >8·5 years (S. Tonge, pers. comm.).

References
Ash, J.S. and Miskell, J.E. (1988).
Archer, G. and Godman, E.M. (1961).
Scamell, K.M. (1964).

Lamprotornis hildebrandti (Cabanis). Hildebrandt's Starling. Choucador de Hildebrandt. Plate 35 (Opp. p. 542)

Notauges hildebrandti Cabanis, 1878. J. Orn., p. 233; Ukamba.

Forms a superspecies with *L. shelleyi*.

Range and Status. Endemic resident, S Kenya and N Tanzania. Kenya, locally common from Maralal, Laikipia plateau and Thomson's Falls east to Isiolo, Meru Nat. Park and Embu, south through upper Tana R., Thika and Nairobi Nat. Park to Tanzanian border from Mara Game Res. and Loita plains to Amboseli and Tsavo East Nat. Park. Tanzania, ranges from Mwanza, Ikoma (fairly common) Serengeti Nat. Park (frequent) and Maswa Game Res. east to L. Manyara, Tarangire Nat. Park, Masailand, Arusha, North Pare Mts and Mkomazi Game Res. and south to Singida, N Kilosa and Dodoma.

Description. ADULT ♂: top of head glossy violet; nape and sides of neck dark glossy bluish green; rest of upperparts glossy violet-blue. Tail glossy violet-blue above, blackish brown below. Lores and ear-coverts black; cheeks to chin, throat and upper breast glossy violet; lower breast and flanks pale orange-buff, grading to darker chestnut on belly, thighs and undertail-coverts. Flight feathers dark brown, primary outer webs edged glossy blue, secondary outer webs glossy green. Tertials and greater coverts bronzy green, the latter with small velvety black spots; alula and rest of upperwing-coverts glossy violet, median coverts with large velvety blackish subterminal spots. Underwing-coverts and axillaries chestnut; median and lesser under primary coverts glossy violet. Wing-tip rounded; P9 usually shorter than P6. Bill black; eyes red or orange-red; legs black. Sexes alike. SIZE (10 ♂♂, 5 ♀♀): wing, ♂ (n = 10) 108–123 (118), ♀ (n = 5) 107–118 (111); tail, ♂ (n = 4) 73–79 (77·0), ♀ (n = 4) 70–78 (73·3); bill, ♂ (n = 10) 20–23 (21·6), ♀ (n = 5) 20–22 (20·8); tarsus, ♂ (n = 10) 30–35 (32·0), ♀ (n = 5) 28–32 (30·4). WEIGHT: (Kenya), unsexed (n = 12) 50–69 (55·9).

IMMATURE: juvenile has dark brown upperparts, mantle to uppertail-coverts with bluish gloss; dark brown lores and ear-coverts contrasting with pale chin; underparts pale tawny brown with diffuse darker band across throat and upper breast; wings brown with green gloss; uppertail glossed greenish blue; undersides of flight feathers dark brown; eyes brown.

NESTLING: unknown.

TAXONOMIC NOTE: closely related to *shelleyi*, with which it has been considered conspecific.

Lamprotornis hildebrandti

Field Characters. Length 19–20 cm. Widely sympatric with much commoner Superb Starling *L. superbus*; differs in *violet-blue* head, breast and back, giving darker overall appearance, *red eye*, *rufous* undertail-coverts and underwing, and *no white band* separating belly from breast. Superb Starling appears lighter and greener, has white undertail and underwing, white eye. Young Superbs lack white breast-band but have brown head and breast, brown or pale eye. Very similar to Shelley's Starling *L. shelleyi* but rufous of underparts lighter, becoming progressively pale towards breast. Immature similar to immature Shelley's but has browner head with some rufous on face, blue-green gloss on back.

Voice. Tape-recorded (5, B, C, F, CHA, GREG, HOR, STJ, WILK, ZIM). Song described as 'a curious but agreeable succession of whistles and harsh reedy notes, all in 6/8 time' (North 1958). Principal components are a harsh 'jaay-jaay', nasal 'jerzy-jerzy' and a more clipped 'jijer-jijer', interspersed with brief, clear, high-pitched whistles, 'wheep-hoo' or 'tlee-oo'; rendered 'cherraah-cherrah, squirk, kwerra-kwerra, eeeeek, querk' and 'errraaa-errraaa, turlewp, queeleree, cherrah, eeeep ... ' (Zimmerman *et al*. 1996). Also described are a guttural subsong, 'kwa-aa, kw-kweeo, kwer-kwee-er' (possibly the same as the low, warbling 'chu-er-chu-er-chu-er-cher-cher-chule' of van Someren *in* Mackworth-Praed and Grant 1960), and a frequently repeated 'chiweh-chiweh-chiweh' (Zimmerman *et al*. 1996). Flock call a whistling 'chule'; alarm, 'chu-ee'.

General Habits. Occurs in open bushland, wooded grassland, riverine woodland, acacia woodlands and around cattle enclosures and human habitation at 700–1700 (sometimes 600–2200) m. In pairs or family groups. Forages on ground. Associates with Superb Starling in same flock (Jackson and Sclater 1938). Less confiding than Superb Starling; can be rather shy and retiring (Irwin 1957). Sometimes roosts in tree hole at night. Displays of captive birds include bowing whilst lowering head to ground and stretching wings up above back; threatens other starlings by stretching head forward with bill open.

Food. Insects and seeds; probably also takes fruits seasonally.

Breeding Habits.
NEST: in hole in tree branch or trunk, lined with pad of hair or fibre. Usually 2–3 m (range 1–6 m) high. Nest trees include *Commiphora* and *Acacia* spp. Sometimes nests in old woodpecker holes, and 7 records in telegraph poles, lamp posts and fence posts.
EGGS: 3–4. Slightly glossy; white. SIZE: (n = 3) 25·5–26 × 16·8–17 (25·8 × 16·9).
LAYING DATES: E Africa: Kenya, Mar, Apr; Tanzania, Nov; Region C, Nov; Region D, Mar–May, Oct, Dec; nests in rains.
DEVELOPMENT AND CARE OF YOUNG: parasitized by Great Spotted Cuckoo *Clamator glandarius* (Ndutu, Tanzania, Geertseema 1976).
BREEDING SUCCESS/SURVIVAL: pair ejected from nest hole by Von der Decken's Hornbills *Tockus deckeni* (Moreau and Moreau 1939). 1 captive (England) lived >16, another >8 years (S. Tonge, pers. comm.).

Plate 35 (Opp. p. 542) *Lamprotornis superbus* (Rüppell). Superb Starling. Choucadour superbe.

Lamprocolius superbus Rüppell, 1845. Syst. Uebers, p. 65; Shoa.

Range and Status. Endemic resident and local migrant. SE Sudan, local migrant west of Torit, presumed resident to east; Ethiopia, common to abundant in central areas below 2000 m, absent from far N and extreme W; Somalia, abundant throughout except in far NE and parts of E coast; N and E Uganda from West Nile and Acholi east to Kidepo Valley Nat. Park and south to NE Teso, Debasien Game Res. and slopes of Mt Elgon; Kenya, common and widespread resident, absent from L. Victoria basin (except around Ahero and Ruma Nat. Park) and from coastal lowlands south of Ngomeni; Tanzania, from Seregenti Nat. Park (frequent resident), Arusha and Mkomazi Game Res. south to Mbeya and Iringa in SW, west to Wembere and Singida, and east to Kilosa and Kilangali, Morogoro.

Description. ADULT ♂: head sooty brown; upperparts including sides of neck glossy blue, scapulars and mantle to rump tinged greenish. Tail glossy blue above, dark brown below. Chin and throat sooty brown with slight greenish gloss; upper breast glossy greenish blue, and below this a narrow white band; lower breast to belly, flanks and thighs orange-rufous; undertail-coverts white. Flight feathers dark brown, outer webs glossy bluish green; primary coverts bluish; rest of upperwing-coverts and alula bluish green; broad velvety blackish tips to greater and median coverts, forming dark wing-bars. Outer primaries notched (**A**). Underwing-coverts and axillaries white. Bill black; eyes creamy or yellowish white; legs black. Sexes alike. SIZE (10 ♂♂, 10 ♀♀): wing, ♂ 119–130 (124), ♀ 112–123 (119); tail, ♂ 66–77 (70·8), ♀ 66–79 (69·9); bill, ♂ 21–23 (21·4), ♀ 20–22 (21·2); tarsus, ♂ 31–35

A

(32·8), ♀ 30–33 (31·9). WEIGHT: (Kenya), ♂ (n = 12) 63–86 (70·4), ♀ (n = 8) 56–64 (60·6), unsexed (n = 36) 52–77 (64·9).

IMMATURE: juvenile has upperparts and head to breast sooty brown, with slight greenish gloss on mantle, back and breast, bluish gloss on uppertail-coverts and stronger greenish blue gloss on scapulars. No white band below; lower breast, belly and flanks orange-buff; undertail-coverts whitish. Wings and tail as adult but duller, wing covert spots less pronounced. Eyes brown. Iridescent blue feathers of adult begin to replace the dull black plumage on breast and back 4–8 weeks after fledging; immature becomes superficially indistinguishable from adult 12–28 weeks after fledging, and irides become creamy-white after 29–46 weeks (T. Huels, pers. comm.).

NESTLING: on hatching skin black with little or no down (Lawrence 1973), bill broad and yellow; at c. 5–6 days, pink, naked except for a few hair-like feathers on head and back, blue lines of feather roots of dorsal tracts showing; thick-necked, eye slits narrowly open (**B**); at c. 14 days old, eyes fully open, colours show on developing feathers, mantle and scapulars fully feathered, quills breaking at tips of rectrices and remiges (Restall 1968).

Field Characters. Length 18–19 cm. A common and familiar bird in eastern Africa, often picking up scraps in gardens and around campsites. Plump and short-tailed; *white band* separates *light orange-rufous belly* from *deep blue breast; pale eye* conspicuous in black head. Undertail-coverts and wing-linings white. Hildebrandt's and Shelley's Starlings *L. hildebrandt* and *L. shelleyi* lack white breast-band and have rufous undertail-coverts and wing-linings. Immature lacks white breast-band and is sometimes mistaken for Hildebrandt's Starling but eye brown or off-white, not red, head and upper breast brownish black without any blue sheen, undertail-coverts white.

Voice. Tape-recorded (5, 19, 25, 66, 109, B, C, F, GOR, KON, PAR, WILK). Calls (song?) a series of loud, rather unmusical phrases, commonest of which is 'weeoo-chu', often with a rolled introduction, 'turrrreeee-chu-weoo-chu'; frequently given in chorus, when wide variety of whistles and chattery notes, including pure, down-slurred 'cheeooo', are added. Subsong when loafing in midday slower, softer and more casual, 'joy, cheer-chit, chirrichit, swee, weeo-cherry, chivee, weejito-churry, cherry-ripe, two-way-chip, jzt ... '; for other transcriptions of this song see Zimmerman *et al.* (1996) and van Someren (1956). Also mimics other birds. Alarm a long-drawn 'chiirrr'; excitement indicated by a repeated 'whit-chor-chi-vii' (van Someren 1956). When feeding, a rasping 'cherrah-cherrreet', and a similar but higher-pitched and more screeching call given in flight (Zimmerman *et al.* 1996).

B

General Habits. Widely distributed in open habitats in arid and semi-arid savanna and grassland and semi-humid areas up to 2200 m (sometimes to 3000 m), wooded grassland, lakeside and riverine woodland, and cultivated areas; also in suburban gardens, city parks and car-parks. Gregarious, often around human habitation, bold and confiding. Forages mainly on ground, where it hops and runs. When alarmed or excited assumes upright posture, with wings lowered and tail slightly raised. Intense excitement signalled by upright stance, with pupils contracted to tiny dots, bird repeatedly calling 'whit-chor-chi-vii' (van Someren 1956). ♂ observed 'eye blazing' with pupils rapidly expanding and contracting (Wragg 1967). Indulges in 'anting': picks up large camponotine ants and strokes them along under- and upperside of spread wings and on back; sometimes for c. 1 min crouches trance-like with wings spread and head on one side, looking up at sky (Fennessy and Brown 1975). In Somalia associates with White-headed Buffalo-Weaver *Dinemellia dinemelli*, Greater Blue-eared Starling *Lamprotornis chalybaeus*, and sometimes Shelley's Starling *L. shelleyi* (Archer and Godman 1961). In S Ethiopia associates with Ethiopian Bush-Crow *Zavattariornis stresemanni* and White-crowned Starling *Spreo albicapillus* (M.W. Woodcock, pers. comm.). In Kenya associates with Hildebrandt's Starling in same small flocks (Jackson and Sclater 1938); and in Tanzania with Rufous-tailed Weaver *Histurgops ruficauda* and Red-billed Buffalo-Weaver *Bubalornis niger*. Noted roosting in large congregations in nests of Rufous-tailed Weaver (Bowen 1931) and appropriates nests of Rufous-tailed Weaver and Red-billed Buffalo-Weaver for breeding (Irwin 1957). During heat of day rests in leafy tree, preening and singing softly. Mobs raptors, owls and snakes. Once 3 Superb Starlings joined 10 Anteater Chats *Myrmecocichla aethiops* and 2 Rufous Sparrows *Passer motitensis* in a 'dancing ring', function unknown (Hayes 1982). Occasionally wanders outside normal range. In Sudan found west of Torit in Aug–Jan only. In Kenya, near Kibwezi, groups maintain territories with core areas, whose boundaries were clearest Oct–May; groups wander and members interact most with other groups June–Sept when not breeding (T. Huels, pers. comm.).

Food. Mainly insects, including termites, ants, grasshoppers, mantids, beetles, larvae of moths, flies and beetles; also fruits, including figs, berries, flowers and seeds. Insects, all taken on the ground, comprised 87% of 597 food items at Tsavo East; another 10% were fruits and the remaining 3% were *Acacia* flowers (Lack 1985). Around houses and campsites eats kitchen scraps, also chicken mash scattered for domestic poultry.

Breeding Habits. A group-living co-operative breeder with extended breeding season and helpers at nests. Near Kibwezi, Kenya, groups had 1–6 breeding pairs and helpers; smallest unit was pair with 1 ♂ helper, largest had 6 pairs nesting simultaneously, 2 unpaired adults and 6 young of the year acting as helpers (T. Huels, pers. comm.). Each pair had its own nest, only one ♀ laid in each nest and she alone incubated. At 9 nests av. 12·2 individuals fed the young. Immatures may begin feeding nestlings only a month after fledgling. Unpaired adults and pair (♂ and ♀) may help. Paired ♂ may help whilst its ♀ is incubating but stop when its own chicks hatch. ♀♀ with eggs or nestlings do not help. Pairs with fledglings often feed the fledglings and nestlings of other pairs. Non-group members and recently immigrated ♀♀ are aggressively discouraged by group members from feeding young. In one flock 6 ♀♀ between them nested at least 22 times from mid Oct to late June, one of them laying 6 clutches. ♂♂ remain in natal group, ♀♀ leave when *c.* 1 year old. Monogamous with some cheating. Prior to and during egg-laying ♀ normally only solicits and copulates with mate, but may also approach, solicit and copulate with older and more dominant ♂♂. After egg-laying, ♀ ignores mate and may copulate with any dominant adult ♂ in nest vicinity. ♀♀ sometimes solicit ♂♂ to avert ♂ aggression (T. Huels, pers. comm.).

NEST: untidy, bulky, domed structure of thorny twigs with side entrance, often with tunnel approach, lined with dry grass and feathers. Usually 1·5–9 m up in thorn tree or bush; if nest not in thorny tree, bird adds barricade of thorn twigs as protection around it (van Someren 1956). Also nests in holes in cliffs or hollow trees, and once inside an old vultures' nest. In Somalia prefers to nest in hole in dead tree, often in disused woodpecker nest, 1·5–3 m from ground, where eggs laid on pulverized wood or nest lining of grass, feathers or shreds of camel matting. Also nests in bushes, building an untidy nest of bundle of grass, in places where no tree-holes available (Archer and Godman 1961). Other nest sites in Somalia, under corrugated iron roofs, on bare metal girder of a bridge, and in old nest of White-headed Buffalo-Weaver (Clarke 1985). In Tanzania nests in old nests of Rufous-tailed Weaver and Red-billed Buffalo-Weaver; several pairs may occupy nests in same tree. In Kenya, near Kibwezi, utilized old nests of White-headed Buffalo-Weaver, Red-billed Buffalo-Weaver and, once, White-browed Sparrow-Weaver *Plocepasser mahali* (T. Huels, pers. comm.). Also nests in eaves of buildings in old nests of Northern Grey-headed Sparrows *Passer griseus*. At Kibwezi, Superb Starlings did not attempt to build own nests or transport thorny twigs. Nest lined by ♂ and ♀, rarely by helpers (T. Huels, pers. comm.).

EGGS: 2–5, glossy; bright greenish blue, laid daily or at intervals of 1–2 days. SIZE: (n = 12, Somalia) 25·2–26·0 × 18·0–19·3 (25·6 × 18·7); (n = 2, Uganda) 25·3 × 18·1, 24·5 × 18·1; (n = 15, Kenya) 25·0–28·8 × 18·3–20·5 (26·7 × 18·9).

LAYING DATES: Ethiopia, Feb–Oct; Somalia, Feb–June; E Africa: Uganda, May–June; Kenya, Mar–July and Oct–Dec; Tanzania, Jan–July, Sept–Dec; Region A, May; Region C, Jan, Mar, June, July, Sept–Dec; Region D, Jan–June, Aug–Dec; Region E, Apr.

INCUBATION: normally begins with penultimate egg, occasionally delayed 1–2 days; by ♀ alone (T. Huels, pers. comm.). Period *c.* 14 days (Kenya), 13 days (captive birds, France), 12–15 days (captive birds, England). Nest material added during incubation by ♂, ♀ and helpers. ♀ not fed by ♂ or helpers (T. Huels, pers. comm.).

DEVELOPMENT AND CARE OF YOUNG: spacing in hatching often leads to discrepancy in size and development of nestlings (van Someren 1956). Brood reduction common at Kibwezi, Kenya. Nestling period: 17–21 days, normally 19 (Kenya); 21–25 days (captive birds, France, Germany and England). Young fed for 4–7 weeks after fledging by ♂, ♀ and helpers (T. Huels, pers. comm.).

BREEDING SUCCESS/SURVIVAL: following data are from T. Huels (pers. comm.): one ♀ laid 6 clutches in <9 months, resulting in fledged young. Of 64 focal nests, 22 failed during incubation and 25 during the nestling period. 17 nests fledged a total of 40 young: 2 nests fledged 1, 8 fledged 2, 6 fledged 3 and 1 fledged 4 young. Ten of these 40 fledglings disappeared within 10 days of leaving the nest; 27 survived to 1 month, 19 to 3 and 12 to 6 months after fledging. Eggs were eaten by egg-eating snakes *Dasypeltis scabra* and White-headed Buffalo-Weavers. One clutch was destroyed by a neighbouring group of Superb Starlings during a territorial conflict. Overheating resulted in the loss of 2 clutches in eaves of buildings. Gabar Goshawk *Micronisus gabar* was main predator of nestlings which were also lost to African Harrier Hawk *Polyboroides typus*, rock python *Python sebae* and children; other chicks died of malnutrition or in storms. Elsewhere in Kenya parasitized by Great Spotted Cuckoo *Clamator glandarius* (Trevor and Lack 1976). 6 captive birds (England) lived for over 12 years (3 birds), 13, 17 and 21 years (one each) (S. Tonge, pers. comm.).

References
Grimes, L. G. (1976).
van Someren, V. G. L. (1956).

Plate 35 *Lamprotornis pulcher* (Müller). Chestnut-bellied Starling. Choucador à ventre roux.
(Opp. p. 542) *Turdus pulcher* Müller, 1776. Syst. Nat., Suppl., p. 139; Senegal.

Range and Status. Endemic resident and partial local migrant, N tropical soudanian and sahelian savannas. Senegal, common around Richard-Toll, Dakar-Thiès and lower Senegal valley; Gambia, uncommon resident at Fajara and Farafenni; Mauritania, common resident in R. Senegal valley; Mali, common north of 13°N, abundant from 14°30′N to about 17°N; Guinea-Bissau, one record (Hall and Moreau 1970); Guinea, rare in Basse-Guinée (Maclaud 1906), no recent records; Ivory Coast (2 records from extreme N in dry season); Ghana, uncommon and local at Bolgatanga, one old record north of Lawra; Burkina Faso, around Ougadougou; Togo (1 record north of Dapaon and 3 records near Cinkansé); Niger, common, 13–17° N, occasionally south to Gaya and Parc 'W', and north to Tamgak Mts (19°10′N); Nigeria, locally common from Sokoto to Maiduguri and L. Chad, common around

Lamprotornis pulcher

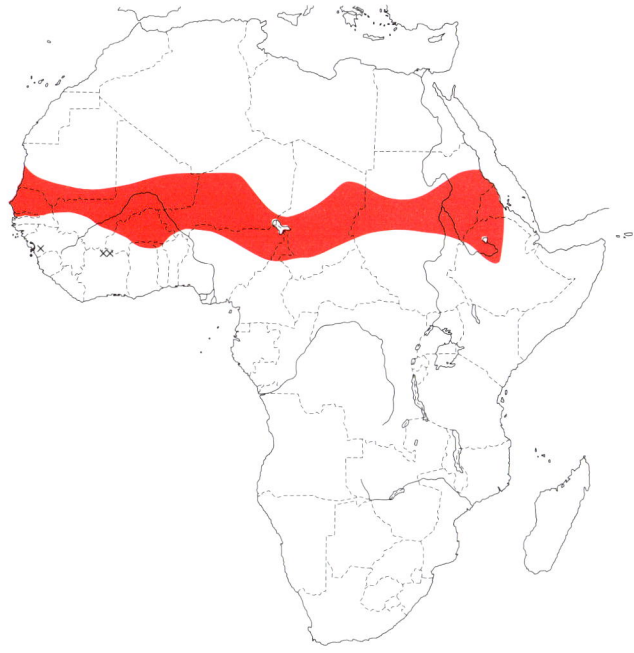

Kano, vagrant south to Zaria and Yankari; Chad, common in soudanian savanna, very common throughout the sahel, reaching north of Fada; Cameroon, in soudanian and sahel savanna north of Garoua (09°18′N); Sudan, common and locally abundant in central savanna belt in W, but has decreased greatly in E (Nikolaus 1987); Ethiopia, uncommon in W south to Addis Ababa, occurring up to 2440 m; Eritrea, frequent except near coast.

Description. ADULT ♂: top of head brown; rest of upperparts including sides of neck pale bronzy green. Tail dull glossy greenish blue above, brown below. Lores dark brown; cheeks and ear-coverts brown; chin to breast pale bronzy green, concolorous with sides of neck; belly, flanks, thighs and undertail-coverts chestnut. Flight feathers brown, outer webs glossy green; inner webs buffish white, forming large pale area on underwing. Primary coverts bluish green; tertials and rest of upperwing-coverts green. Underwing-coverts and axillaries chestnut. Bill black; eyes white or creamy white; legs black or blackish brown. Sexes alike. SIZE (10 ♂♂, 10 ♀♀): wing, ♂ 112–118 (116), ♀ 111–118 (114), tail, ♂ 64–70 (67·5), ♀ 62–70 (66·1); bill, ♂ 23–24 (23·8), ♀ 20–23 (21·7); tarsus, 31–34 (32·5), ♀ 31–34 (32·5). WEIGHT: N Nigeria, ♂ (n = 39) 59–76 (68·0), ♀ (n = 31) 60–74 (66·6); Chad, unsexed (n = 14) 51–70 (58).

IMMATURE: juvenile has top and sides of head dark brown, rest of upperparts dark brown with slight greenish gloss; chin to breast pale tawny brown; belly, flanks and undertail-coverts pale chestnut; wing feathers with slight green gloss only; uppertail glossed dull bluish. Eyes brown.

NESTLING: hatches blind, nestling has pink skin, pinkish yellow bill and yellow gape flanges.

TAXONOMIC NOTE: Hall and Moreau (1970) considered *hildebrandti* and *shelleyi* to form a superspecies together with *pulcher*. However *pulcher* is stockier and less highly glossed and we agree with Short *et al.* (1990) that it should be excluded from this superspecies.

Field Characters. Length 19–21·5 cm. A uniquely-coloured starling confined to a band of acacia savanna and steppe just south of the Sahara, with *grey-brown head*, *white eye* and *chestnut belly*; rest of plumage dull green, bluer on tail; dark green breast cleanly divided from chestnut belly; inner webs of primaries buffy white showing as *pale patch in flight*. Immature has head and underparts mainly brown, some chestnut on belly, eye brown. The only chestnut-bellied starling in its range; does not overlap with Shelley's or Hildebrandt' Starlings *L. shelleyi*, *L. hildebrandti*. Stocky; walks boldly in upright posture.

Voice. Tape-recorded (104, B, CHA, MOR, WILK). Flight calls, a high-pitched, rather mechanical, down-slurred trill, 'trrrreeeairrrr' or shorter 'trrreeeoo', and a short 'trrreee' on one pitch. Danger signals, trilling 'wheeee' and 'chewy-chewy'; response to ground predator, loud, repeated 'ch! ch! ch!' developing into mobbing call, 'churr-churr'; response to aerial predator, e.g. Red-necked Falcon *Falcon chiquera*, very high-pitched 'pee-pee'. Rapidly-repeated high-pitched calls given by helpers at the nest. Song (subsong?), rarely heard, given by adults and immatures when resting with other group members during hottest part of day, varied, low volume, rambling, including soft trilling notes, 'pee-pee' and 'wheeee' calls; does not appear to have any obvious territorial or advertisement function. Also said to give breathy 'fit-trrr, fit-trrr' (Barlow *et al.* 1997).

General Habits. Occurs in thorn scrub, acacia savanna, and open bushy country. Favours areas with mosaic of bare ground and bushy scrub, degraded savanna and farmed parkland; also in cultivation, around villages, and sometimes in towns. Wandering birds south of normal range in dry season inhabit degraded, dry, scrubby areas.

Gregarious; commonly in groups, occasionally in flocks of up to 50. Forages mainly on ground, sometimes near cattle; in loose groups with individuals scattered over large area. In Nigeria often associates with White-billed Buffalo-Weaver *Bubalornis albirostris*. In Chad sometimes breeds together with Greater Blue-eared Starling *Lamprotornis chalybaeus* (Salvan 1969). Stocky; walks boldly with upright posture (**A**); can become quite tame in spacious gardens. Searches bare ground for ants and termites, also chases

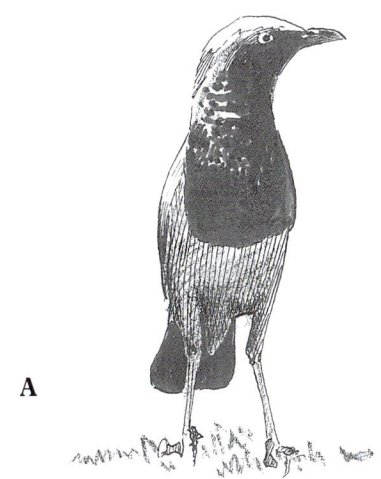

A

grasshoppers through grass. When a bird finds food it is often joined by others. Occasionally gleans insects from leaves and, rarely, clings to tree trunk and probes in bark. Feeds at fruiting trees, taking smaller berries and large fruits. Flies up to fruiting neem *Azadirachta indica*, grasps fruit in bill and hovers to pull it off. Regurgitates large neem seeds whole. One or more birds sometimes perch higher than flock in tree and act as sentinels. When a bird gives alarm calls, flock flies away in tight group, with rapid wing beats and trilling calls.

Territorial in breeding season, when group of 10–30 birds keeps to day range which barely overlaps neighbouring groups' ranges. At Kano, Nigeria, 5 groups roosted at night in area of thick acacia scrub near to but outside their day ranges; the groups kept separate at night. Group comprises 2–6 breeding pairs, non-breeding adults of both sexes, and immature birds. Exceptionally, large flocks of *c*. 50 birds may comprise several social groups. Social interactions include supplanting behaviour, proffering of sticks, stones, feathers, grasses or food (**B**); and a deep bow (**C**), often with wings raised upwards and opened butterfly-like above back (**D**). Stretches one wing and leg on same side together. Proffering, bowing, wing-stretching and leg-stretching may be associated with intra-group dominance relationships. During heat of day may loaf in leafy tree and sing.

Adults begin wing moult in Mar–Apr; some birds still in moult in Nov. Individual may take 6–7 months to complete moult leading to moult-breeding overlap (Kano, Nigeria). Post-juvenal wing moult normally commences *c*. 3 months after fledging (Wilkinson 1983).

Sedentary, but ranges quite widely when not breeding. Observation that it occurs in Senegal and Gambia only in rainy season (Curry-Lindahl 1981) not supported recently (Morel and Morel 1990), but in dry season some birds do wander south (Thiollay 1985, Walker 1965).

Food. Mainly insects and fruit. Of 79 birds examined (Kano, Nigeria) 96% had eaten ants, 92% termites, 37% flies, 32% bugs, 23% beetles and 14% grasshoppers. Termites were main food for 38%, ants for 22%, fruits (mainly *Azadirachta*, *Loranthus* and *Lantana*) for 19% and seeds (chiefly *Acacia*) for 9% of these 79 birds. *Salvadora* berries are important elsewhere. Unusual food items include a dead mouse and baby chameleon. Kitchen scraps taken close to human habitation. Nestlings and fledglings fed a diet largely of insects but also fruits including the large *Azadirachta*. Of 648 nestling food items (Kano, Nigeria) 88% were invertebrates, mainly termites and grasshoppers. Nestlings and young fed locust hoppers (Eritrea: Smith 1957).

Breeding Habits. Monogamous; co-operative breeder, with helpers at the nest. Birds breed (and live year round) in groups, each group maintaining its own feeding and breeding territory. Birds may roost outside this area and be intermittently absent from territory in middle of dry

E

F

season (Dec–Jan), when not breeding. Group comprises 2–6 breeding pairs, non-breeding adults of both sexes, and immatures. Not all pairs breed at any one time; in Nigeria most pairs attempt to breed in the long pre-rains season (Feb–June), and fewer in the short post-rains season (Sept–Nov) (Wilkinson 1982, 1983). At Kano, Nigeria, may be double or triple-brooded in the pre-rains breeding season, single-brooded in the shorter post-rains breeding season. The same nest may be used for subsequent broods or a new nest constructed. ♀ vacates nest to forage, only rarely fed on nest by ♂. Nests scattered within group territory, normally 50–300 m from nearest neighbour. Exceptionally nests may be closer; once 2 active nests were 4 m apart in same *Parkia* tree. Although ♀♀ of breeding status normally feed only their own nestlings, breeding ♂♂, non-breeding adults and immatures may act as helpers. Helpers are most evident at nests with large broods of older chicks, which may attract up to 13 helpers (**E**). Non-breeding ♂♂ help at more nests within their group than do breeding ♂♂ (which normally help only when they have no chicks of their own). Young birds of both sexes may act as helpers, preferentially assisting their own parents in their first year. ♂♂ normally remain in natal group where they help other breeders and may themselves become breeders; young ♀♀ may also help, but disperse to an adjacent or more distant group (once 8·5 km away) when about a year old. One yearling ♀ helped simultaneously at nests in its natal group and an adjacent group to which it transferred; this help was not welcomed by the breeding ♀ which vigorously chased it away from nest. Young ♂♂ may become breeders in their natal group when 2–3 years of age; one ♂ still remained a non-breeding helper until 5 years old. ♀♀ normally transfer to other groups; one ♀ became a breeder at 2 years old, another was at least 3 years old.

Courtship display not recorded. ♀ on ground once solicited, not crouching, but with body at angle of *c*. 45° with head up, wings drooped and slightly spread, wings and tail quivering. ♂ mounted briefly and then flew off.

NEST: untidy, large, round, domed structure with side entrance (**F**), mainly of dry grass, lined with feathers and sometimes snake skin; typically built in thorny trees, often *Acacia albida*, *Balanites aegyptiaca* or *Ziziphus mauretanica*, mostly at heights of *c*. 1·5–5 m. Other favoured nest sites (Kano, Nigeria) were within clumps of mistletoe *Loranthus*, especially in *Parkia* trees (6–12 m from ground) and inside large communal stick nests of White-billed Buffalo-Weavers. In Merab Valley, Ethiopia, each colony of White-billed Buffalo-Weavers had a few pairs of Chestnut-bellied Starlings as tenants (Cheesman and Sclater 1936). Sometimes nests also in *Cassia*, tamarind, *Eucalyptus* and neem trees. Nest built by ♂ and ♀.

EGGS: 3–5; ovate, somewhat pointed; glossy or sometimes almost devoid of gloss; greenish blue sparingly blotched and spotted with rust-brown and lilac. SIZE: (n = 14, Kano, Nigeria) 24·5–28·0 × 17·5–19·0 (26·4 × 18·3); (n = 7, Sokoto, Nigeria) 24·4–26·1 × 17·3–19·2 (25·4–18·0); (n = 10, Sudan) 24–25 × 17·9–18·5 (24·5 × 18·2).

LAYING DATES: Senegal, Jan, Mar–Nov; Gambia, Apr, June, July; Mali, Apr–Nov; Burkina Faso, (observed at nest in July); Niger, (nest building in Apr), July, Aug; Nigeria, Feb–June and Sept–Nov; Chad, May–July and Oct–Nov; Sudan, Jan–June and Sept–Nov; Ethiopia, Feb, Apr–June. Avoids coldest months of Dec–Jan (dry season). In Nigeria and Sudan the 2 breeding seasons avoid the wet months of July–Aug.

INCUBATION: by ♀ only. Period: *c*. 13–18 (15) days.

DEVELOPMENT AND CARE OF YOUNG: in some nests smallest chick less than half weight of largest, but in others with similar brood sizes there is little disparity. Chicks of 1 brood weighed 53, 52, 48 and 24 g, 6 days before first chick fledged. Chicks fledge together or one-by-one after *c*. 18–25 days, but if disturbed entire brood may leave nest explosively and prematurely. Both parents and up to 13 helpers care for brood; they feed nestlings, remove faecal sacs, and defend nest against predators and rare visits from members from neighbouring groups.

Where several pairs in a group breed, individual birds may help them successively or simultaneously. Provisioning rates increase with age of chicks and according to number of helpers; rate was once >60 visits per h on evening before first chick fledged. Fledglings continue to be fed by helpers for several weeks; as early as 6 weeks after fledging, a young bird may attend its parents' subsequent brood.

BREEDING SUCCESS/SURVIVAL: nests with more than 4 helpers tend to fledge more chicks than those with fewer helpers. Nestlings taken by Senegal Coucal *Centropus senegalensis*; sometimes die of mite infestation; small young chicks may starve to death. Parasitized by African Striped Cuckoo *Oxylophus levaillantii* in Mali and Gambia (Irwin 1988). At Kano, Nigeria, no records of brood parasitism at >120 nests but African Cuckoo *Cuculus gularis* and Greater Honeyguide *Indicator indicator* were chased from nest sites; also nests broken open by children and by Pied Crows *Corvus albus*, and blown down in high winds. <50% fledglings survive to 3 months, but mean annual survival of adults is high: *c*. 80% for both breeding and non-breeding ♂♂, and 90% for breeding ♀♀ (survival of non-breeding ♀♀ not calculated as it is complicated by emigration). 1 captive (England) lived >13 years (S. Tonge, pers. comm.).

References
Wilkinson, R. (1982, 1983).
Wilkinson, R. and Brown, A. E. (1984).

Plate 35
(Opp. p. 542)

Lamprotornis fischeri (Reichenow). Fischer's Starling. Spréo de Fischer.

Notauges fischeri Reichenow, 1884. J. Orn., p. 54; Pare Mts.

Range and Status. Endemic resident. Ethiopia, uncommon in SE around Dolo. Somalia, very common and widespread resident in bushland south of 05°N and west of 46°E. Kenya, common resident and wanderer in bushland and bushed grassland below 1400 m, rarely up to 1900 m; in N, east of L. Turkana, from Moyale to Mandera, south to South Horr and through eastern lowlands to Namanga, Amboseli and Tsavo, lower Tana R. and lowlands around Ijara and Garsen. NE Tanzania from lowlands around Longido and southeast of Kilimanjaro south to Tarangire Nat. Park, Masai Steppe and Mkomazi Game Reserve.

Description. ADULT ♂: top of head buffish white; rest of upperparts greyish brown. Tail brown with bronzy green sheen above, blackish brown below. Lores blackish; ear-coverts and sides of neck greyish brown. Cheeks to chin and throat pale greyish buff, grading into dark greyish brown band across upper breast; rest of underparts white apart from dark greyish brown patch on upper flanks. Flight feathers brown, outer edges tinged bronzy green; rest of upperwing feathers brown, tertials and larger coverts with bronzy green sheen, lesser coverts more greyish. Underwing-coverts and axillaries white. Bill black; eyes creamy white; legs black. Sexes alike. SIZE (11 ♂♂, 10 ♀♀): wing, ♂ 107–117 (111), ♀ 102–113 (107); tail, ♂ 62–74 (67), ♀ 60–64 (62); bill, ♂ 19–24 (22.4), ♀ 20–25 (22.1); tarsus, ♂ 29–33 (30.9), ♀ 29–31 (30.5). WEIGHT: ♂ (n = 1) 51, unsexed (n = 1) 46.
IMMATURE: juvenile similar to adult, but upperparts browner, with tawny feather fringes; lower mandible bright yellow; eyes brown.
NESTLING: unknown.

Field Characters. Length 17·5–19 cm. A *grey* starling, palest on head, becoming darker on upperparts and breast; belly white with dark patch on flanks; thin black line on lores leads to *white eye*. Immature browner with dark eye and lower mandible yellow. Wattled Starling *Creatophora cinerea* in non-breeding plumage is rather uniform brownish grey with no contrast between breast and belly, dark eye, blackish wings and tail and white rump patch conspicuous in flight.

Voice. Tape-recorded (B, C, CHA, GREG, McVIC, PEA, STJ, WILK). Wide repertoire includes rolled 'burree-

Lamprotornis fischeri

burreeo' and 'bu-burry, buburry'; high, rasping 'dzaa-zaaa' and abrupt, grating double 'dzudut', sometimes in combination, 'zaaa-dzudut'; clear up-slurred whistle, 'groy' or 'wooey'. Peculiar lilting 'rudi-roowi, dzadut, rudi-roowi, dzudut ... ' may be the 'whirligig' call of Fuggles-Couchman and Elliot (1946). Flocks keep up chatter of shrill whistles, chuckles and grating notes. See also Zimmerman *et al*. (1996).

General Habits. Usually in flocks, often foraging on ground with other starlings and buffalo-weavers (Zimmerman *et al*. 1996). In NE Tanzania noted as shy; when disturbed immediately flies up from ground into tree or bush, then flies away with some gliding. Occurs in dry thorn-bush areas, usually in pairs or small parties of *c.* 6, sometimes loosely associating with Golden-breasted Starlings *Lamprotornis regius* (Fuggles-Couchman and Elliot 1946). Flocks of up to 40 in Tsavo East, where commonest in wooded savanna; numbers peak at Tsavo in June–Aug (Lack 1985).

Food. Insects (99% of 292 food items, all taken from ground), including termites, beetles and caterpillars. Once noted eating fruits (Tsavo East Nat. Park: Lack 1985) and feeds in fruiting trees (Ukunda: Lewis and Pomeroy 1989). At 2 nests chicks were fed entirely on butterflies and caterpillars (Miskell 1977).

Breeding Habits. Co-operative breeder with nest helpers (Miskell 1977). 3 adults fed chicks at one nest; a fourth, finding youngsters sated, flew to second nest 25 m away where it fed a chick. Once 3 adults fed chicks at this other nest.

NEST: large domed sphere of coarse or soft grasses or twigs, diam. *c.* 250 mm, with side entrance above small ramp; sometimes lined with feathers. Built in thorny bush, *c.* 2 m above ground. 2 nests (Wajir, Kenya) had entrances facing south (Miskell 1977) and one had entrance facing east (Jackson and Sclater 1938).

EGGS: 3–6; blue, spotted with lilac and sepia. SIZE: (n = 10) 23·1–24·8 × 17·0–18·7 (24·1 × 17·9).

LAYING DATES: Ethiopia, Apr, May; Somalia, Mar, Apr, June; Kenya, Apr, Sept; Tanzania, Dec; E Africa, Region D, Apr, May, Sept–Nov.

References
Fuggles-Couchman, N.R. and Elliot, H.F.I. (1946).
Lack, P. (1985).
Miskell, J. (1977).

Genus *Spreo* Lesson

A genus of rather short-tailed, ground-foraging starlings, which has had an unusually varied membership. Sclater (1930) recognized 6 species, Amadon (1943) 10, White (1962) 9, Hall and Moreau (1970) 7, Wolters (1975–1982) one, Sibley and Monroe (1990) and Dowsett and Forbes-Watson (1993) 3 (*fischeri, bicolor, albicapillus*). Craig (1997) and Feare and Craig (1998) admit 4 (*fischeri, bicolor, albicapillus* and the long-tailed *unicolor*), noting that they have round, hollow, unstratified feather melanin granules (different from those of *Lamprotornis*: Craig and Hartley 1985). Of the 4, we think that *fischeri* and *unicolor* may be more closely allied respectively to some short- and long-tailed species of *Lamprotornis* than to either *albicapillus* or *bicolor*, and we retain genus *Spreo* for the last 2 species only: blackish starlings with vestigial greenish gloss, pale yellow eyes, and white belly, lower flanks and undertail-coverts. *S. albicapillus* occurs in Somalia and Ethiopia and has ball-shaped nest in thorn bush; *S. bicolor* occurs in South Africa and nests in holes usually in ground.

Endemic. 2 species.

Spreo albicapillus Blyth. White-crowned Starling. Spréo à calotte blanche.

Plate 35
(Opp. p. 542)

Spreo albicapillus Blyth, 1856. J. As. Soc. Bengal, 24, p. 301; Warsangeli.

Range and Status. Endemic resident, NE Africa. Ethiopia, frequent to common in grasslands and savanna of Rift Valley, and S and SE, at Yavello around 1230–1380 m; Djibouti, probably resident, recorded between Djibouti and Petite Douda; Somalia, very common and widespread resident, especially on inland plateau, north of 06°N, extending south to 04°N in west, from sea level on coastal Gulf of Aden to highest points on Mt Wagar. Kenya, local and uncommon to rare resident in arid (0–500 mm rainfall) bushland at 300–1000 m in N, from North Horr, Huri Hills and Moyale south to Marsabit and Dida Galgalla Desert, and in extreme NE around Ramu and Mandera.

Description. *S. a. albicapillus* Blyth: Somalia and Ethiopia to extreme NE Kenya. ADULT ♂: top of head creamy white; rest of upperparts and sides of neck brown, with glossy green sheen, mantle and scapular feathers with narrow pale fringes, rump feathers and uppertail-coverts with pale shafts and small pale tips. Tail brown above, T1 and outer webs of T2–T6 bronzy green; blackish below. Lores blackish brown; ear-coverts and cheeks brown with pale flecking. Chin and throat to upper belly and upper flanks brown with prominent buffish white shaft streaks; lower belly, lower flanks, thighs and undertail-coverts white. Flight feathers dark brown, glossed bronzy green above on outer webs, secondaries with broad buffish white outer borders, forming prominent longitudinal stripe on closed wing. Tertials, alula and primary coverts brown, glossed bronzy green; outer greater coverts and median coverts buffish white, inner feathers grey-brown with slight greenish gloss; lesser coverts bronzy with pale shaft streaks, leading feathers buffish white. Underwing-coverts and axillaries buffish white. Outer primaries notched (**A**). Bill black; eyes pale yellow; legs black. Sexes alike. SIZE (10 ♂♂, 4 ♀♀): wing, ♂ 149–162 (157), ♀ 142–157 (150·5); tail, ♂ 107–121 (116), ♀ 107–120 (114·5); bill, ♂ 25–30 (27·5), ♀ 25–28 (26·5); tarsus, ♂ 38–40 (39·1), ♀ 36–39 (37·5).

IMMATURE: like adult but bill pale yellowish horn, colour persisting longest on base of mandible. Juvenile similar to adult but cap sullied grey-brown, not so sharply demarcated from ear-

Spreo albicapillus

coverts and hindneck; upperparts paler brown, greenish gloss confined to wings and tail; underparts paler brown, shaft streaks buffier and less contrasting. Bill yellowish horn with dark brown tip; eyes dark brown.

NESTLING: not known.

S. a. horrensis Keith: N Kenya (North Horr to Maikona and Turbi). Much smaller: wing, ♂ (n = 2) 130, 137; tail, ♂ (n = 2) 89, 93; bill, ♂ (n = 1) 24; tarsus, ♂ (n = 1) 33 (Keith 1964).

Field Characters. Length 26–28 cm. A rather large starling of dry NE Africa with unique plumage combination of *white crown*, iridescent brownish green upperparts, and *brown-and-white-streaked* underparts; in flight shows *broad white stripe* on upperwing at base of primaries, white wing-linings and white undertail-coverts. Eye white. Immature similar but browner, with less white streaking on underparts, brown eye and yellow bill. Flight heavy with much gliding and fanned tail.

Voice. Tape-recorded (B, MAC). Call, a shrill up-slurred 'crrrrooooee' or down-slurred 'crrreeeeo'; also written 'tschurreeeet' or 'tchu-tchu-tsureeeeet' (Zimmerman *et al.* 1996).

General Habits. Inhabits dry stony bushland and open areas with scattered trees. Not shy, sometimes found close to villages and cultivation. Gregarious all year. In small flocks, sometimes with other ground feeding birds including White-headed Buffalo-Weavers *Dinemellia dinemelli*. Mostly feeds on or near ground; seen following camels, presumably taking insects flushed by their feet.

Food. Insects and plant matter, including caterpillars, seeds and *Salvadora* fruits.

Breeding Habits. Usually in scattered colonies, several nests grouped together in same bush.

NEST: conspicuous, ball-shaped, loosely built grass structure *c.* 150 high × 230 wide with side-entrance protruding another 70 mm; placed 1–3 mm up in thorn bush (Benson 1946).

EGGS: 4–6, usually 4–5 (once 9, probably laid by 2 ♀♀); pale or greenish blue with scattered liver-coloured spots and occasionally deep mauve undermarkings. SIZE: (n = 26, Somalia) 27–30 × 20–21 (28.6 × 20.5) (Archer and Godman 1961); also (n = 6, Somalia) 27.3–29.0 × 19.0–20.0 (28.1 × 19.7).

LAYING DATES: Ethiopia, Mar–May; Somalia, Apr (4 clutches), May (12), June (5) (Ash and Miskell 1998); Kenya, Mar/Apr (dependent young June, N edge of Dida Galgalla Desert when verdant after exceptional rains).

INCUBATION: by ♀, period >13 days (estimated from captive breeding, England: Ezra 1929).

DEVELOPMENT AND CARE OF YOUNG: one captive chick fledged at 25 days.

BREEDING SUCCESS/SURVIVAL: one nest with 2 chicks, Somalia, destroyed by Cape Crow *Corvus capensis* (Clarke 1985).

References
Archer, G. and Godman, E.M. (1961).
Benson, C.W. (1946b).

Plate 35 (Opp. p. 542) *Spreo bicolor* (Gmelin). African Pied Starling. Spréo bicolore.

Turdus bicolor Gmelin, 1789. Syst. Nat. 1, pt 2, p. 385; Cape of Good Hope, South Africa.

Range and Status. Endemic resident, South Africa, Lesotho and Swaziland. South Africa, common and widespread on inland plateau and in W and S coastal areas; sparse and local in NW Cape north of 30°S and east of 23°E; in the Transvaal absent from N and lowveld areas, vagrants in Kruger Park (Balule) and near Ofcolaco; largely absent from E coastal lowlands, with occasional records from Durban, Port Shepstone and Port St Johns; no longer found on Cape Peninsula. Lesotho, W lowlands and valleys in mountains; locally common. W Swaziland, uncommon and local.

Description. ADULT ♂: head, nape, mantle, back, uppertail-coverts, throat, breast and thighs blackish with some green sheen. Belly and undertail-coverts white. Wings and tail blackish, with

Spreo bicolor

oily green gloss. Outer primaries with shallow notch (**A**). Upper mandible black, lower mandible deep yellow with black tip (black c. 40% of mandible length); corners of mouth protrude as bright yellow gape; eyes white; legs and feet black. Sexes alike. Bird with mainly white plumage reported; not clear if it was a true albino (Van Niekerk 1996). SIZE (22 ♂♂, 17 ♀♀): wing, ♂ 144–163 (154), ♀ 142–155 (150); tail, ♂ 91–108 (100), ♀ 91–103 (97); bill to skull, ♂ 26·5–30·5 (28·9), ♀ 25·1–29·6 (27·8); tarsus, ♂ 37·6–44·4 (41·6), ♀ 38·8–42·6 (40·8). WEIGHT: (Grahamstown, South Africa, Oct–Mar) ♂ (n = 23) 94–113 (105), ♀ (n = 20) 93–112 (102).

IMMATURE: newly-fledged young dull charcoal black, lacking gloss of adult. Base of lower mandible yellowish white; gape white; eyes dark brown. Gloss in plumage appears after first moult at 4–6 months. Eyes stay brown for a year, then change to white from outer margin inwards, generally completely white at 2 years, when white on gape and lower mandible has become yellow, and yellow part of mandible has become more extensive.

NESTLING: hatchling naked and pinkish, lacking down or feathers; lining of mouth bright yellow, gape white.

Field Characters. Length 25–27 cm. A large dark brown starling with *white belly* and undertail-coverts and *pale eye*; gape and lower mandible *yellow*. Largely terrestrial, with very upright stance; a common roadside bird in South Africa. In flight the wings are broad and rounded, not pointed as in Common Starling *Sturnus vulgaris*; in worn plumage, flight feathers may appear paler than body plumage. Immature charcoal black with dark eye, gape and base of lower mandible white.

Voice. Tape-recorded (88, 91–99, B, F, GREG, LEM, ROC). Characteristic loud 'wreek-wreek' call by birds in flight or about to take off. Similar calls incorporated into song, a medley of rather liquid notes, 'weely', 'wheew', 'teeyuweep' and an upslurred 'wayp', and some harsher notes, interspersed with a low 'chup-chup' and other chuckling calls. Both sexes sing. Alarm a harsh squawk, much shriller in newly fledged birds than in adults. Nestlings give a piping begging call.

General Habits. A bird of open karoo and grassland habitats including fields, pastures and farms; also intertidal zone on coasts. Common in small rural towns, but avoids extensive built-up areas; also avoids wooded areas and very arid regions. In Natal primarily above 1200 m, and up to 2500 m in Drakensberg Mts; elsewhere down to sea level.

Occurs year-round in small flocks of 15–25, including adult breeders and their subadult offspring, which serve as helpers in cooperative breeding system. Roosts communally in reed beds or trees; >1000 birds may use a roost, sometimes with Wattled Starlings *Creatophora cinerea* and weavers *Ploceus* spp; also with Lesser Kestrels *Falco naumanni*. Forages mainly on ground, walking about and picking food from surface; may turn over dried cow pats or leaves, but does not probe soil; also perches in trees to eat fruit, hawks insects, and stands on sheep and cattle to remove ectoparasites. Follows domestic stock to catch flushed insects. On seashore catches amphipods on stranded kelp, and feeds in rocky intertidal zone. Often forages in association with Wattled Starling and sometimes with Common Starling. Flock members regularly feed each other: one bird approaches another in very upright posture with food in bill; recipient crouches and gapes (**B**). Nearby birds beg but are ignored. One bird even attempted to feed its own reflection (Jubb 1980b). Birds that feed each other are mainly subadults; allofeeding most often occurs between birds in the same breeding unit, but mates rarely allofeed (suggesting a function in social bonding, not courtship – Craig 1988c). During close-range interactions, including allofeeding, pupil size and area of white iris change inversely. Birds displaying more white may be dominant, whereas those with large pupils and less white in eyes appear to be subordinates.

Sometimes moults flight feathers when breeding, but only a few feathers grow simultaneously. Moult may start in Oct, completed by Apr; birds of the year have an incomplete moult at the same time.

Mainly sedentary, but irregular extralimital occurrences suggest nomadic habits in arid NW Cape.

Food. Terrestrial arthropods; also fruit, including figs, grapes, berries of *Scutia myrtina*, seeds; nectar of *Aloe*, *Agave* and *Sideroxylon*; small reptiles and frogs, road-killed toads (Skead 1995); also kitchen scraps. Insects include locusts, grasshoppers, beetles, ants and termites (one bird had eaten 82 termites: Kok and Van Ee 1990; another 120 ants *Anoplolepis custodiens*: Skead 1995). Intertidal amphipods. Stomach contents of 105 birds included bits of *Atriplex semibaccata*, *Convolvulus* sp., *Cotoneaster horizontalis*, *Cussonia paniculata*, *Cyperus exculentus*, *Euclea crispa*, *Ligustrum lucidum*, *Medicago sativa*, *Nicandra physaloides*, *Opuntia* sp., *Portulaca* sp., *Protasparagus laricinus*, *Sophora*

japonica and *Zea mais* (Kok and Van Ee 1990). Young in the nest fed *Lycium* berries, dragonflies, caterpillars, grasshoppers, beetles, termites, centipedes, solifugids, ticks and small geckos.

Breeding Habits. Monogamous, colonial, with nest helpers. No overt territorial behaviour known, but one pair returned to same nest site for at least 4 successive years. Pair apparently remains together all year. Aerial pursuits of ♀ by ♂ occur; may form part of courtship. Often double-brooded, second clutch 30–50 days after first.

NEST: chamber of tunnel excavated by pair in river bank, road cutting or quarry; hole in wall of building, haystack, or (rarely) tree. Tunnel *c*. 1 m long, 8 cm in diam., straight or sometimes curved. Both sexes carry in beakfuls of grass, roots, wool, muddy bits of sedge, paper, rope and snakeskin, forming pad at base of egg chamber.

EGGS: 2–6, av. 4·2 (48 clutches). Oval; glossy; bright blue-green, plain or with some small, round red spots. SIZE: (n = 131) 26·9–35·3 × 19·6–22·5 (30·3 × 21·3).

LAYING DATES: Aug–Nov in W Cape (winter rainfall region), elsewhere all months, mainly Sept–Jan, also Apr–May, following late summer rains.

INCUBATION: by ♀ only. While ♀ incubates, ♂ sings from perch with forehead and crown feathers raised, making head peaked. Incubation may start before clutch completed. Period: 14–16 days.

DEVELOPMENT AND CARE OF YOUNG: young fed by both parents and up to 7 helpers (mostly subadults and juveniles), without overt aggression; helpers queue to feed in turn; they sometimes emerge with faecal sacs, which are dropped at least 10 m from nest entrance. Helpers may feed young at 3 different nests during a single breeding season; if they are feeding at more than one nest concurrently, they deliver most food to a particular nest. Nesting adults

sometimes feed young at adjacent nests. Rectrices emerge from skin on day 6. Nestling period: 23–27 days. On leaving nest young are fed by parents and helpers for at least 14 days, and remain with natal flock. Some first breed at 1–2 years, most when >2 years old. Both sexes serve as helpers, assisting their parents and other breeding adults; commonly help in 2 successive years.

BREEDING SUCCESS/SURVIVAL: some nests parasitized by Great Spotted Cuckoo *Clamator glandarius* and Greater Honeyguide *Indicator indicator*. Nile monitor *Varanus niloticus* seen at nest tunnels; nestling removed by Lanner Falcon *Falco biarmicus* (Van Zyl 1991); in both cases the predators were mobbed by adult birds. Ringed adults have survived >6 years.

References
Craig, A.J.F.K. (1985, 1987, 1988c).
Herholdt, J.J. (1987).

Genus *Speculipastor* Reichenow

Single species of stocky glossy starling. ♂ black with slight blue-violet gloss and white breast and belly, ♀ dark grey with white breast and belly; eyes red; short, thrush-like bill, like that of *Grafisia torquata* but finer and more curved at tip; wing quite long and pointed, with white patch in primaries; tail rather short. Nomadic, gregarious; feeds in trees and on ground; nests in holes, mainly in termitaria.

Hall and Moreau (1970) showed 7 points of similarity between *S. bicolor* and *G. torquata* and thought that they might prove to be congeneric, but Craig (1997) placed *Poeoptera* and *Cinnyricinclus* between *Speculipastor* and *Grafisia* in a cladistic tree of African starling genera.

Endemic; 1 species, E Africa.

Speculipastor bicolor Reichenow. Magpie Starling; Pied Starling. Spréo pie.

Speculipastor bicolor Reichenow, 1879. Orn. Centralbl., 4, p. 108; Kipini, Kenya.

Range and Status. Endemic resident and irregular migrant, E Africa. Ethiopia, uncommon to frequent, in S and E. Somalia, widespread as shown, common to abundant, resident in S and breeding visitor to N in Apr–July. Sudan, known only from Ilemi triangle in extreme SE (Nikolaus 1987). Uganda, known only from Mt Moroto in NE. Kenya, common locally in N; post-breeding dispersal or migration south to Kerio valley, Baringo and Isiolo Districts, Tana R. valley, coastal lowlands south to Malindi and in 1978 to Mombasa, and Tsavo Nat. Park; records from Nairobi and L. Magadi. Tanzania, flocks in W North Pare and W South Pare Mts in Jan 1957 and Oct 1962. Southern breeding limit in Kenya is about 02°N; southern limits of breeding in Somalia not yet established and shown here notionally.

Description. ADULT ♂: head, neck, and mantle to scapulars and uppertail-coverts glossy black, tinged violet-blue. Tail blackish. Chin and throat to centre of upper breast glossy blue-black; sides of upper breast and lower breast to undertail-coverts creamy white. Remiges blackish brown, basal half of primaries white, forming conspicuous patch on closed wing; rest of upperwing blackish, glossed blue-black on median and lesser coverts. Axillaries white; underwing-coverts creamy white, but small feathers along carpal edge blackish. Bill black; eyes blood-red; legs black. ADULT ♀: forehead to hindneck and sides of neck buffy brown or grey-brown with dark brown mottling. Mantle to scapulars and uppertail-coverts dark brown, feather tips with violet-black sheen. Tail blackish brown. Lores dark brown; sides of head, chin and throat grey-brown or buff-brown with a little dark spotting, and below this a band of glossy violet black feather tips around centre of upper breast; sides of breast and rest of underparts white. Remiges dark brown, slightly glossed; remiges and coverts often with some tawny notching and fringing. Bill and legs like adult ♂; eyes red or orange-red. SIZE (8 ♂♂, 5 ♀♀): wing, ♂ 116–122 (119), ♀ 109–114 (112); tail, ♂ 73–76 (74·6), ♀ 71–76 (74·4); bill, ♂ 20–21 (20·8), ♀ 20–23 (21·2); tarsus, ♂ 27–29 (28·1), ♀ 27–28 (27·4). WEIGHT: ♂ (n = 2) 66, 69, ♀ (n = 2) 61, 61.

IMMATURE: juvenile almost plain dark brown above, paler on top and sides of head; chin and throat pale grey-brown, bordered by dark brown band around centre of upper breast; rest of underparts buffy white. Wings and tail dark brown, white patch on primary bases tinged buff.

NESTLING: not known.

Field Characters. Length 17–19 cm. A *pied* starling of dry lowlands in NE Africa. ♂ has upperparts, head and upper breast glossy blue-black, rest of underparts white; *eye red*. In flight, *white patch* on primaries, white underwing-coverts. ♀ brownish black above with slight gloss, throat grey-brown separated from white underparts by glossy black breast-band. Immature like ♀ but browner; throat pale brown bordered by dark brown breast-band. No other pied starling in its range; Abbott's Starling *Pholia femoralis* lives in mountain forests.

Voice. Tape-recorded (B, McVIC). Perched bird gives prolonged soft, babbling 'quereeeh quaaa kereek quak-quak, quereek suaaaa, cherak-chak-chak ... ', mixed with higher, harsher notes; lacks strident trills of Fischer's

Speculipastor bicolor

Starling *Lamprotornis fischeri*; also has shrill whistling flight call (Zimmerman *et al.* 1996).

General Habits. Inhabits semi-arid and arid thornbush in plateau country and open park-like savanna woodland with numerous termite hills (N Somalia), up to 1200 m. Occurs mainly in flocks, of up to *c.* 30 birds. Often in tops of highest trees. Forages on ground and in trees. Flocks visit thicker habitats and shun thinner ones in N Tsavo East Nat. Park.

Mainly sedentary in N Kenya and central Somalia, breeding visitor to N Somalia in Apr–July, erratic wanderer or migrant south of about 02°N. Influx into S Tsavo East Nat. Park in Nov–Dec 1976, and into NE Tanzania in Oct and Jan.

Food. Many birds were feeding on caterpillars, Huri Hills, Kenya, June, and adults fed nestlings caterpillars, South Horr, Kenya, June (Haas and Nickel 1982). Fruits, including figs and *Solanum*.

Breeding Habits. A solitary nester in N Somalia, in termite hills (Archer and Godman 1961), but solitary or colonial in Kenya. One colony with *c.* 10 pairs of birds and 2 active nest burrows in a cliff face, and another with 9 adult birds and 3 occupied burrows (Haas and Nickel (1982), the numbers suggesting that the species may be a co-operative nester.

NEST: in holes in termitaria, also in cliffs and once 3 m up in a steep river bank; in N Somalia nests invariably sited in termitaria; eggs laid on bedding of grass in chamber at end of tunnel *c.* 45 cm long (Archer and Godman 1961).

EGGS: 2–6 (in Somalia clutches of 2, 3, 3, 4, 4, 6 and 6; in Kenya 1 clutch of 5). Blue-green, profusely speckled with red-brown, tending to form zone around broad end. SIZE: (n = ?) 27 × 20.

LAYING DATES: Ethiopia, (possibly Feb–Apr); Somalia, Apr–May in S, June in N; Kenya, June. In Kenya thought to be an opportunistic breeder; nests in South Horr and Huri Hills in June 1981 were in areas made green and rich in insect life after exceptionally heavy rains (Haas and Nickel 1982).

References
Archer, G. and Godman, E.M. (1961).
Ash, J.S. and Miskell, J.E. (1998).
Haas, V. and Nickel, E. (1982).
Zimmerman, D.A. *et al.* (1996).

Genus *Grafisia* Bates

Single species of stocky glossy starling. ♂ black with slight blue-violet gloss and white breast, ♀ dark grey above with glossy blue feather tips and dark brown below; eyes yellow; thrush-like bill; wing quite long and pointed; rather short-tailed. Arboreal frugivore. For putative affinities see Genus *Speculipastor*.

Endemic; in 4 areas in N tropics.

Plate 33
(Opp. p. 526)

Grafisia torquata (Reichenow). White-collared Starling. Rufipenne à cou blanc.

Spreo torquatus Reichenow, 1909. Orn. Monatsb., 17, p. 140; Banjo, Cameroons.

Range and Status. Endemic resident in Cameroon, S Chad/N Central African Republic, and N Zaïre, rare visitor or migrant in NE Gabon and rare in N Congo. Cameroon, widespread, frequent and locally common at 860–1170 m from Ndop Plateau and valleys west of Nkambé, Wum, Banso Mts, Bum, Bambalang, Foumban, upper Noun (Nun) R. valley (Bamessing), N'gaoundaba Ranch, and Bangangte to Adamawa Plateau, east to about Mabéré and Garoua-Boulaï; north to at least Banyo, south to about Bafia and Obala (where quite common: Germain *et al.* 1973). Chad, twice collected near Sarh (Fort Archambault). Central African Republic, Ndélé and 75 km northwest of it, also 2 localities further east and in Manovo-Gounda-Saint Floris Nat. Park; in S, occurs at Mongoumba, on Ubangi R. opposite Libenge (Zaïre) and only 20 km from Congo border; also Kassai (Blancou 1974). Congo, groups of 6 and 20 in Odzala Nat. Park (Dowsett-Lemaire and Dowsett 1998). Zaïre, apparently restricted to a small area of Uele in and north of Madudu country, south to Pawa in N Ituri. Gabon, pair seen in clearing between Makokou and Mékambo and 5 days later 3 birds in clearing at M'Passa, taken to be vagrants or migrants, Mar 1975 (Brosset and Erard 1977).

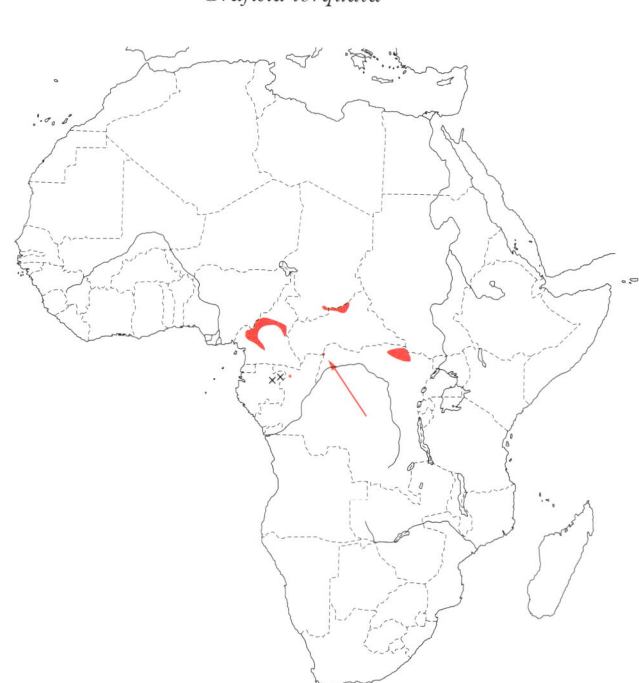

Grafisia torquata

Description. ADULT ♂: head and body glossy black with violet or dark blue sheen; broad white collar on sides of neck and across upper breast. Wings and tail black, tinged glossy violet or oily bluish green. Axillaries and underwing-coverts glossy bluish-black. Bill black; eyes yellow to orange-red; legs black. ADULT ♀: upperparts grey-brown, with spotting or mottling produced by darker feather centres; narrow glossy feather tips from violet sheen from crown to uppertail-coverts. Tail black above with violet sheen, blackish brown below. Sides of head grey-brown, ear-coverts tinged violet. Underparts grey-brown, breast to undertail-coverts lightly scaled, with pale buff feather fringes. Upperwing blackish with violet sheen. Axillaries and underwing-coverts grey-brown. Bare parts as adult ♂. SIZE (6 ♂♂, 2 ♀♀): wing, ♂ 123–133 (127), ♀ 117, 120; tail, ♂ 87–93 (89·2), ♀ 78, 83; bill, ♂ 20–22 (21·0), ♀ 20·5, 22; tarsus, ♂ 23–25 (24·0), ♀ 23, 23. WEIGHT: 1 ♂ 67, 1 ♀ 61.

IMMATURE: juvenile like adult ♀, but violet sheen less intense, confined to back, wings and tail. Sub-adult ♂ like adult ♂ but with white collar sullied dusky brown.
NESTLING: not known.

Field Characters. Length 20–23 cm. ♂ unmistakable, glossy black with broad *white breast-band*; ♀ lacks white band and is wholly grey with pale scaling and mottling, darker above with some violet gloss; both sexes have *yellow eye*. Immature like ♀ but duller.

Voice. Not tape-recorded. Call, 3 whistled notes, short and not loud; makes chirruping calls while foraging. Both sexes gave 'a somewhat weak apology for a song' (Bannerman 1948).

General Habits. In Cameroon inhabits forest-savanna mosaic and montane grasslands close to forest edge; trees around villages and farms on firm land in Noun marshes, keeping to tops of high trees and perching on topmost branches of dead trees; also said to live near high crags of barren mountains at edge of Noun R. plain. Elsewhere, rocky, almost treeless country, though not reported ever to descend out of trees to perch on ground or rocks; but roams into canopy of swamp forest to feed (Odzala Nat. Park, Congo). Occurs in pairs and commonly in flocks of 4–10 and once 20; sometimes 4 ♂♂ together without ♀♀. A flock of 6 contained 1 adult ♂; a flock of 20 contained 5 adult ♂♂. Keeps to larger trees; consorts with other fruit-eating starlings, notably Splendid Glossy Starling *Lamprotornis splendidus*; less vocal than latter. Makes short flycatching sallies from top of tree and just over the canopy. Active, climbing and hopping among branches, uttering chirruping notes. Flock of 20 birds heard 'group-singing', a rather chirpy chatter, out of breeding season (Dowsett and Dowsett-Lemaire 1998).

Resident, but for evidence of migration see above.

Food. Fruits including *Macaranga assas*, *Xylopia aethiopica* (Dowsett and Dowsett-Lemaire 1998), small wild figs and pulp of *Musanga* fruit. Sometimes eats the same berries as Splendid Glossy Starlings.

Breeding Habits. Seen exploring tree holes (Wum, Cameroon, Feb). Nothing else known.

References
Bannerman, D.A. (1948).
Blancou, L. (1974).
Chapin, J.P. (1954).
Dowsett, R.J. and Dowsett-Lemaire, F. (1998).

Genus *Pholia* Reichenow

Small starlings of highland forest canopy in E Africa, with rather broad and flat bills and pale yellow eyes; sexes alike (*P. sharpii*) or different (*P. femoralis*); glossy blue-black above, with white throat or white belly; juveniles and adult ♀ *P. femoralis* blackish or dark brown above, dark-streaked or dark-spotted below where ♂♂ white. Arboreal frugivores. Often placed with *leucogaster* in *Cinnyricinclus*, but bill shape differs somewhat, and feather melanin granule structure differs markedly (Craig and Hartley 1985).
Endemic. 2 species.

Pholia femoralis (Richmond). Abbott's Starling. Spréo d'Abbott.

Plate 33
(Opp. p. 526)

Pholidauges femoralis Richmond, 1897. Auk, p. 160; Kilimanjaro.

Range and Status. Resident, endemic to E Africa. Confined to highlands of central and SE Kenya (Kikuyu escarpment, Mt Kenya, Chyulu Hills and formerly at Molo; frequent in Taita Hills: Brooke *et al.* 1998) and NE Tanzania (Mt Meru, Ngurdoto Forest, Mt Kilimanjaro, N Pare Mts). Rare and poorly known, although frequent on Kikuyu escarpment, Kilimanjaro and Embu slopes of Mt Kenya.

Description. ADULT ♂: head and upperparts black with slight bluish gloss. Tail glossy blue-black above, blackish brown below. Chin and throat black, extending to point on centre of breast; sides of breast and rest of underparts white, apart from black patch on lower flanks; thighs black. Wings black, with slight greenish gloss on outer edges of flight feathers, tertials and upperwing-coverts. Underwing-coverts and axillaries sooty black. Bill black; eyes pale yellow; legs black. ADULT ♀: top and sides of head, sides of neck, and upperparts dark greyish brown, paler on forehead, and with paler fringes on crown and uppertail-coverts producing mottled effect. Tail dark brown. Chin and upper throat pale buff; lower throat to point on centre of breast buffish brown, with darker brown feather centres producing distinct mottling; sides of breast, upper flanks, belly and undertail-coverts pale buff with broad, sharp streaks except on centre of belly; patch on lower flanks dark brown. Upperwing dark greyish brown, median coverts and outer webs of tertials and greater coverts narrowly fringed paler tawny brown. Underwing coverts dark brown with narrow pale buff fringes. Bare parts as in adult ♂. SIZE (6 ♂♂, 1 ♀): wing, ♂ 98–103 (100), moulting ♀ 92; tail, ♂ 68–76 (71·7), ♀ 67; bill, ♂ 16–18 (16·8), ♀ 17; tarsus, ♂ 22–25 (23·2), ♀ 21.
IMMATURE AND NESTLING: not known.

Pholia femoralis

Field Characters. Length 16·5–18 cm. ♂ glossy blue-black with white belly and prominent pale eye; seen from below, bird in treetop shows black V pointing from breast down onto belly. Sympatric Sharpe's Starling *P. sharpii* is a similar size, with yellow eye, but has no black on underparts; Magpie Starling *Speculipastor bicolor* has black breast but red eye, white wing-patch in flight, lives in lowland dry country. ♀ Abbott's has almost uniform brown head and upperparts, streaked underparts; told from similar ♀ Violet-backed Starling *Cinnyricinclus leucogaster* by unstreaked upperparts, white eye, brown breast and pale belly. Immature like ♀ but with brown eye. Imm. Sharpe's Starling is dull version of adult with black spotting on white underparts.

Voice. Tape-recorded (B, McVic). Pleasant conversational song of high-pitched but quite sweet whistles interspersed with some liquid trills, churrs and squeaky notes, in brief (*c.* 2 s) phrases with intervals of 4–7 s; often starts with up-slurred whistle 'turdlee'; 'turdlee-turdlee-weelu', 'turdleedle-turdlee', 'turdlee-turdlee-widlu-trrrrrrr', 'turdlip-weedu-tur-chik'; sometimes longer, 3–4 s, merging different elements, 'turlip-surlip-toowihee-swee-weeju-weeju-weeju-churrrr'. Song may also include twangy metallic elements recalling those of Violet-backed Starling or the 'squeaking hinge' notes of Forest Weaver *Ploceus bicolor*; often a few widely-spaced soft, thin 'tseet' or 'tsuik' notes separate the louder main song phrases (Zimmerman *et al.* 1996).

General Habits. Inhabits canopy of highland forest; at 2100–2600 m in central Kenya, 1800–2500 m in Chyulus and N Tanzania. Gregarious except when breeding, in small groups or flocks of up to 40 birds; sometimes in mixed flocks with Sharpe's Starling. Shyer than Sharpe's, keeping to tops of trees, especially *Juniperus procera*. Perches on tops of dead twigs in or near canopy. ♂ sings from exposed position. Wanders widely in search of fruiting trees.

Food. Fruits, including *Cornus volkensii* (Chyulus).

Breeding Habits. Almost unknown. Nests in tree cavities. LAYING DATES: N Tanzania, Feb, Mar, Oct.

References
Taylor, P.B. and Taylor, C.A. (1988).
Zimmerman, D.A. *et al.* (1996).

Plate 33
(Opp. p. 526)

Pholia sharpii (Jackson). Sharpe's Starling. Spréo de Sharpe.

Pholidauges sharpii Jackson, 1898. Bull. Br. Orn. Club. 8, p. 22; Eldama Ravine, Kenya.

Range and Status. Resident and partial intra-African migrant, restricted to eastern Africa. Frequent S Ethiopia (SW highlands, S Rift Valley, S highlands). Fairly common S Sudan, on Imatong, Dongatona and Didinga Mts, and Mt Morongole in adjacent NE Uganda. Local and mainly uncommon throughout Kenya highlands, including Mt Marsabit, Matthews Range, Nguruman, Namanga, Chyulu and Taita Hills. NE Tanzania at Loliondo, Mbulu, Mt Meru, Arusha Nat. Park, Kilimanjaro, N Pare Mts and (rarely) W Usambaras, common Mt Meru to North Pares; W Tanzania at Mahari Mt; S Tanzania at Mt Rungwe. Frequent SW Uganda/E Zaïre, in Ruwenzoris, Bwindi Forest, mountains west of L. Albert, Mt Nyemilima west of L. Edward and Luvumba Mts northwest of Baraka; Rwanda at Nyungwe Forest.

Description. ADULT ♂: upperparts including top of head to lores and ear-coverts glossy black, hindneck to uppertail-coverts tinged violet. Tail black above with slight greenish gloss, blackish brown below. Underparts pale buff, with tawny wash on breast, richer tawny on belly, flanks, thighs and undertail-coverts; small glossy violet-black patches on sides of breast. Flight feathers dark brown, outer webs black with slight greenish gloss. Rest of upperwing black with greenish gloss, small lesser coverts glossed violet. Underwing-coverts and axillaries black. Bill black, rather short and broad; eyes golden-yellow; legs blackish. ADULT ♀: similar to adult ♂, but upperparts more sooty black, without violet gloss; less glossy on wings and tail; lacks breast patches; paler tawny on belly and undertail-coverts. SIZE (10 ♂♂, 10 ♀♀): wing, ♂ 101–109 (104), ♀ 93–102 (97·3); tail, ♂ 65–71 (67·7), ♀ 57–66 (61·6); bill, ♂ 15–17 (15·9), ♀ 14–16 (14·9); tarsus, ♂ 22–23 (22·4), ♀ 21–22 (21·7). WEIGHT: ♂ (n = 2) 40, 42. (Mt Rungwe), ♂ (n = 2) 42·5, 45, ♀ 47.

IMMATURE: juvenile like adult ♀, but less iridescent above; throat, breast and flanks with forward-directed arrow-shaped spots; eyes dark.

Field Characters. Length 15·5–17·5 cm. A small, broad-billed montane forest starling with *blue-black* upperparts

Pholia sharpii

and *buffy white* underparts, deeper buff on belly; *eye yellow*. Tail notably short and forked (M.W. Woodcock, pers. comm.). Sexes alike. Sympatric with rarer Abbott's Starling *P. femoralis* on some mountains in Kenya and Tanzania; ♂ Abbott's has black breast ending in V on belly; ♀ Abbott's is brown with streaked underparts. Immature Sharpe's is duller and browner above, with dark spots on underparts, and orange-brown eye.

Voice. Tape-recorded (104, B, C, CHA, GREG, LEM, McVIC). Call a short, sharp, high-pitched 'psink' or 'pseet', with a metallic timbre; this is also woven into song, a medley of high, thin whistles, scratchy notes, and chuckles and chatters in a nasal undertone. Songs in Rwanda lasted 1.5–2 s, accelerating at the end (sonagram in Dowsett-Lemaire 1990); song in E Africa rendered 'speenk spee- spee tsink-seresee-see cheenk seekserawn speek-speek-speek-speek' (Zimmerman *et al.* 1996).

General Habits. Inhabits highland forest, forest edges and clearings, and isolated clumps of trees in grassland. Above 1800 m in S Sudan, at 1800–2500 m in Rwanda and E Zaïre, 1700–3000 m in Kenya, but as low as 1400 m in N Tanzania. Gathers in flocks of 10–25 in trees with ripening fruit, often with Waller's Starling *Onychognathus walleri*. Pairs or small groups rest on bare emergent branches above canopy (Zimmerman *et al.* 1996). Feeds mainly in canopy but also at times very near ground. Flight swift, sometimes slightly undulating. Subject to local movements in response to tree fruiting. Occurs at top of altudinal range in Kenya Jan–Mar, during the longer, milder dry season. Mainly sedentary but apparently an intra-African migrant in Nyungwe, Rwanda (Dowsett-Lemaire 1990).

Food. Fruits: creeper *Urera hyoselodendrum* and *Rapanea pulchra* (Zaïre), *Sapium* (Kenya, Rwanda), *Bridelia*, *Ocotea usambarensis* (Rwanda), *Eugenia capensis*, *Polyscias* (Uganda).

Breeding Habits. Poorly known. Pair entered and left nest hole, together and separately, several times in 20 min; once one entered and other perched nearby, sideways on treetrunk; hole 9–10 m up unbranched trunk of large, smooth-barked tree by forest track, Mt Elgon, Apr (M.W. Woodcock, pers. comm.).

NEST: of moss or similar material, in hole usually about 10–15 m high in dead or living tree, but once as low as 3.5 m; entrance hole *c*. 5 cm. wide; nest 20–30 cm below entrance.

EGGS: 3 (1 nest Kenya), pale blue, speckled brownish.

LAYING DATES: S Sudan, Feb, Apr; W Kenya, June (and pair attending nest, Apr); N Tanzania (Arusha) Oct.

References
Beesley, J.S.S. (1972).
Dowsett-Lemaire, F. (1990).
Taylor, P.B. and Taylor, C.A. (1988).
Zimmerman, D.A. *et al*. (1996).

Genus *Cinnyricinclus* Lesson

Small, strongly gregarious and migratory starling with small bill, pointed wing (P7–P9 longest, P10 very small) with rufous in primaries; tail rather short, slightly forked; eyes dark with narrow, pale yellow outer ring. Sexes markedly different: ♂ iridescent purple with white belly; ♀ variegated brown above, dark-streaked white below. Arboreal; mainly frugivorous; nests in tree holes.

1 species, near endemic: throughout N and S tropical savanna woodlands, and in SW Arabia.

Plate 33 *Cinnyricinclus leucogaster* (Gmelin). Violet-backed Starling; Amethyst Starling; Plum-coloured
(Opp. p. 526) Starling. Spréo améthyste.

Turdus leucogaster Gmelin, 1789. Syst. Nat. 1, pt. 2, p. 819; Dahomey.

Range and Status. Africa and SW Arabia; breeding summer visitor to W Yemen and SW Saudi Arabia, wintering doubtless in Africa. Vagrant Israel (Hollom *et al.* 1988); 1 at Eilat, July–Aug 1996 (H. Shirihai, *Sandgrouse* 21, 1999: 101).

Resident and intra-African migrant, Senegal to Eritrea, Namibia and Natal. Widespread, often common to abundant; migrant flocks of thousands in Liberia. Mainly migratory, moving away from equator during N and S rainy seasons. In W Africa, *C. l. leucogaster* common from forest edges near Gulf of Guinea coast north to soudan savanna belt, but seasonal in most areas, sometimes erratic; few breeding reports. Near coast, most abundant Nov–May; in N guinean and soudanian zones, a wet season visitor, Apr–Oct, ranging north to N Senegal (Cap Verte), central Mali (Mopti), N Burkina Faso, N and NE Nigeria (Kano, Maiduguri) and S Chad (Melfi, Bousso). In S Senegal, frequent to common Basse Casamance, Apr–Aug; in Gambia, numbers vary from year to year, commonest May–July; N Senegal, mainly June–Sept. In Mali, rather common all year south of 11°N, north to 15°N during rains; breeds in south, Mar–Apr. In Burkina Faso, passage in Ouagadougou in Apr–May and Aug–Sept; May–July visitor to Yatenga, breeding June. In Liberia, common dry season visitor to coastal savannas, rare in hinterland. Seasonally abundant in Ivory Coast, mainly Nov–Mar in S (almost absent S guinea zone May–Sept), June–Aug north of 08°N. Common in Ghana: mainly an Oct–Jan dry season visitor to Cape coast and Mar–June passage migrant in N. In Nigeria, some birds all year in coastal forests, otherwise locally abundant dry season visitor south of great rivers, wet season visitor in N (except extreme NW); breeds Mar–June, near coast and in N (Kaduna, Zaria). Common in S Cameroon, Central African Republic, and N Zaïre along edge of Congo forest (south to Kasangani, Irumu, Medje and Pawa); flocks in Upper Uele in Nov. Fairly common in Sudan north to *c.* 13°N; rains visitor in central areas, Apr–Aug; passage migrant in S, Jan–Apr and Sept–Oct; no breeding records. Common in Ethiopia (mainly *arabicus* in N); seasonally abundant in Uganda north of equator, wandering to NW Kenya (Turkana) and W Tanzania (once: Mahari Mt); breeds NW Uganda (Mar). Mainly *arabicus* in Eritrea, Mar–Sept visitor to W and plateau, all year in E (at least on Mt Ghedem: Smith 1957); flocks common in Djibouti, Mar, and small flock once in early May, though species not resident there (*contra* Welch and Welch 1984b) since none at all seen in Oct to early Dec (Welch and Welch 1986); in NW Somalia, locally abundant Apr–Sept, east to 45°30'E, south to 09°30'N, and may breed. Vagrant in NW Congo (Odzala Nat. Park) and Gabon (M'Passa, Rés. de la Lopé) and known also from coast (Brosset and Erard 1986); race unknown.

C. l. verreauxi ranges from Lower Congo, S and E Zaïre (Kasai, Katanga, Kivu), Uganda and central Kenya to central and NE Namibia (on coastal escarpment to 24°S),

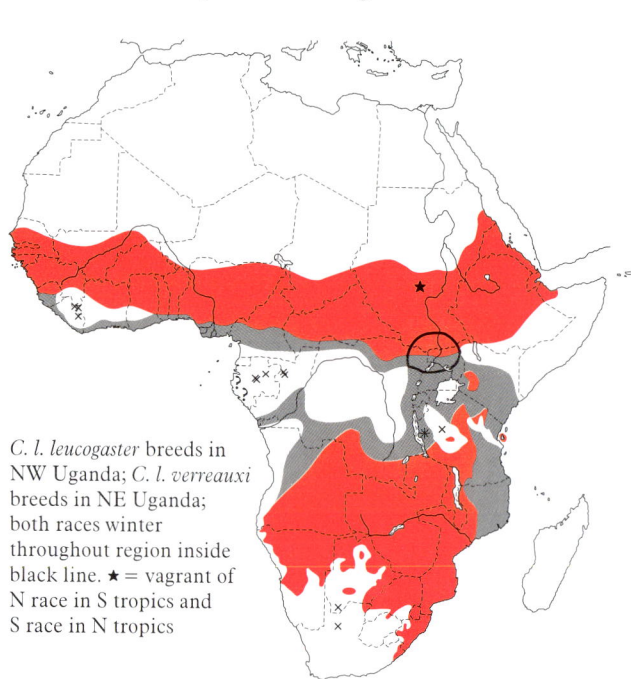

Cinnyricinclus leucogaster

C. l. leucogaster breeds in NW Uganda; *C. l. verreauxi* breeds in NE Uganda; both races winter throughout region inside black line. ★ = vagrant of N race in S tropics and S race in N tropics

N and E Botswana, and South Africa (Transvaal, Natal, coastal E Cape). Largely migratory. Common breeding visitor, SE Congo (e.g. Léfini Res.), central Angola, S Katanga, Zambia and Malaŵi southward, mainly Sept–Mar or Apr (but Apr–Oct on central Angola plateau); breeds S Kasai, many records in S and W Tanzania (Sept–Mar); a few in W and central Kenya (Jan, Mar) and Zanzibar and 1 in NE Uganda (July). As a non-breeding visitor, locally common to abundant, Mar–Sept, in E Zaïre (Kivu), throughout Tanzania, Uganda (north to Toro and Teso) and Kenya (north to 02°N, with large flocks on coast); records in NE Zaïre (Dungu) and S Sudan (Didinga Mts, Mongalla). Also mainly a non-breeding dry season visitor in SW Katanga (Kasaji area, Mar–Aug), and from SW Zambia/E Angola (Upper Zambezi) to N Botswana and NE Namibia (Okavango R.) where common Apr–Aug (Winterbottom 1959, 1966). Almost wholly a breeding migrant south of Okavango and Zambezi rivers, locally common Sept/Oct to Apr or May; very few remain in June–July in Transvaal and Zimbabwe.

Density of *c.* 30 nesting pairs in 130 ha on SW Nyika, Malaŵi (Dowsett-Lemaire 1989).

Description. *C. l. leucogaster* (Gmelin): Senegal to S Ethiopia, south to Gabon, N Zaïre, Uganda, NW Tanzania and NW Kenya. ADULT ♂: head, upperparts and chin to upper breast glossy metallic violet-purple, with blue or bronze reflections; rest of underparts white. Tail dark brown, T1 and outer webs of T2–T5 glossy purple above. Flight feathers blackish brown, secondaries with broad glossy purple outer border; tertials blackish brown, tips and outer webs glossed purple; alula and primary coverts dark brown, latter with narrow purplish fringes; rest of upperwing-coverts same as upperparts. Underwing-coverts and axillaries dark brown. Bill black, rather short; eyes dark brown with yellow outer rim; legs black. ADULT ♀: top of head and neck deep tawny buff, broadly streaked blackish brown; mantle to uppertail-coverts dark brown with tawny-buff feather fringes. Tail brown. Lores dark brown; ear-coverts dark brown, flecked tawny-buff; cheeks pale buff streaked blackish brown. Underparts buffish white, washed browner on throat and upper breast; streaked blackish brown throughout, most heavily on breast and flanks. Upperwing dark brown; outer webs of primaries narrowly edged buff; tertials and tips and outer webs of secondaries fringed buff; greater, median and lesser coverts fringed tawny-buff. Underwing-coverts and axillaries dark brown with broad tawny streaks; extensive tawny area on underwing formed by basal half of primary inner webs. Soft parts as adult ♂. SIZE (10 ♂♂, 10 ♀♀): wing, ♂ 98–105 (101), ♀ 93–100 (96·0); tail, ♂ 60–67 (64·3), ♀ 55–62 (59·7); bill, ♂ 16–17 (16·3), ♀ 16–18 (16·6); tarsus, ♂ 19–22 (20·8), ♀ 19–22 (20·1). WEIGHT: Nigeria, ♂ (n = 2) 37·5, 39·2.

IMMATURE: juvenile like adult ♀ but streaking duller and tail feathers more pointed. Juvenile body plumage (and usually tail) moulted at 2–3 months; immature then like adult ♀. Adult plumage attained at 13–14 months, after first complete moult.

NESTLING: not described.

C. l. verreauxi (Bocage): Lower Congo, S and E Zaïre, Uganda and Kenya south to Namibia, Botswana and South Africa. Like *leucogaster*, but ♂ has white base to outer web of T6. Larger: wing, ♂ (n = 10) 108–111 (110). WEIGHT: southern Africa, ♂ (n = 4) 39·5–48·2 (44·4), ♀ (n = 5) 32·7–55·6 (44·7), unsexed (n = 6) 38–48 (45·5). Kenya, ♂ (n = 4) 35–46 (40·2), ♀ (n = 1) 46.

C. l. arabicus Grant and Mackworth-Praed: N Ethiopia, E Sudan (Roseires), Eritrea, Djibouti, NW Somalia, SW Arabia. Like *leucogaster*, but adult ♀ almost plain brown above, buff feather edging greatly reduced. Larger: wing, ♂ (n = 10) 107–111 (109). Juvenile streaked above as in nominate race.

Field Characters. Length 15–17 cm. A small starling that spends most of year in small nomadic flocks searching for fruiting trees. Head and upperparts of ♂ a beautiful iridescent reddish *violet*, varying with angle of light to amethyst, plum-coloured, pink or even black; *belly white*. ♀ and juvenile brown above, white below streaked dark brown; similar to ♀ of rare Abbott's Starling *Pholia femoralis* but upperparts have *pale feather edgings* (plain brown in Abbott's), head with dark streaks, *throat and breast white* with narrow *dark streaks* and sometimes a tinge of brown on breast (breast dark brown in Abbott's, sometimes with blackish streaks); very narrow pale eye-ring visible only at close range (more prominent pale eye-ring in Abbott's). Young Sharpe's Starling *P. sharpii* has black arrow-shaped spots on white underparts but is blackish above, with red eye.

Voice. Tape-recorded (86, 88, 91–99, 104, B, C, F, CART, ROC). Song very variable: common component is tuneless chattering 'clewp-clewp', 'clep-clep' or 'chu-chu', sometimes lengthened into 'cho-chee-chi-cho-charr'; light trills, down-slurred ('trrrreew') or up-slurred ('toorrreee'); repeated, nasal 'jer-we-jerr'; 'creaking hinge' calls with ringing quality, 'tu-tueee', 'turrr-turrway', 'too-lee-loo'; thin, soft whistle, 'tsui'; the above often connected by conversational squeaky chatter. Habitual calls in W Africa (Gambia to Nigeria), sometimes given in duet, long, pure, high ringing whistles on same frequency (one bird about half an octave higher than the other), a drawn-out single note with the intense quality of a tuning fork, repeated persistently but at long intervals (Barlow *et al.* 1997; C.H. Fry, pers. comm.). In tape-recording (88) there is a remarkably good imitation of Forest Weaver *Ploceus bicolor*. Calls include harsh screeches and high buzzy scolds.

General Habits. Occurs around fruiting trees, in high fringing forest, forest clearings, woodland, wooded savanna and gardens; *Ficus* groves in Somalia; oases in Senegal; evergreen and deciduous forest, miombo and thicket in Malaŵi; bushveld and riverine forest in South Africa. From coast up to 3000 m in Kenya; breeds from 50 to 2200 m in Malaŵi. In pairs in breeding season; otherwise gregarious, forming flocks, often of ♀♀ and young birds only, or of ♂♂ only. May consort with Wattled Starlings *Creatophora cinerea* or glossy starlings *Lamprotornis* spp. Roosts in tall trees, sometimes in hundreds. Often perches conspicuously on top of tree or bush. Forages in canopy, probing lichens on branches, gleaning foliage and flycatching; also feeds on edges and amid scrub and second growth, and while hopping on ground. Sometimes hawks insects in flight. Flight swift and direct. Flicks wings like chat when alighting.

Flocks and disperses after breeding, and most populations at least partially migratory. Nominate race tends to wander north after breeding, reaching edge of sahel zone from Senegal to Sudan, May–Aug. Southward movements occur Sept–Oct, then most birds spend dry season (Nov–Feb) near guinea-savanna/forest interface before moving north to breed mainly in northern guinea savanna. Some birds remain all year in south. Some *verreauxi* populations have regular migrations over longer distances. Breeding areas south of about 20°S are almost wholly vacated in May–Aug. Comparison of moult schedules in breeding and non-breeding areas suggests that birds wintering in W Zambia/Okavango area are from South Africa and Zimbabwe, those wintering in SW Katanga from E Zambia, and those wintering in Kivu from Zimbabwe; the origin of birds wintering in Kenya is unknown (Traylor 1971). *C. l. arabicus* is a visitor to NW Somalia in May–Sept with evidence of breeding there, and to W Eritrea Mar–Sept; flocks occur in Djibouti in Mar. It is a passage migrant in S Yemen in Mar–May and Oct, thought to breed in W (Martins *et al.* 1996), and a summer visitor in SW Saudi Arabia where probably breeds; Arabian birds are presumably out of Africa, although whereabouts of *arabicus* in Oct–Feb (other than Mt Ghedem, E Eritrea: see above) remains a mystery – Kenya?

Sequence of adult moult is unusual. Main body moult takes place only after wing and tail moult complete, and moult of tertials and inner secondaries is also delayed. Migration can occur at different points in moult cycle. Some populations migrate when moult complete, others when wing and tail (but not body) finished, or only partly finished. Central Angolan birds migrate without moulting, then return early and have full moult before breeding (Traylor 1971).

Food. Fruit, mainly berries, especially when breeding; also insects. Eats mulberries and mistletoe berries in Malaŵi. 4 stomachs (Zaïre) contained berries and yellow fruit pulp, 1 also a few ants (Chapin 1954); 2 stomachs (S Cameroon) contained just berries.

Breeding Habits. Monogamous.
NEST: pad of small green leaves, in E Africa often of *Euclea keniensis*, in natural cavity in tree or stump, or in pipe; usually on foundation of grass, feathers, hair and dry cow dung. Sited 1–10 m up, entrance usually through irregular crack in firm or rotten wood, nest 25–40 cm below entrance; built by ♀, but both sexes add green leaves or lichen throughout incubation. Sometimes uses old nest-holes of Olive Woodpecker *Dendropicos griseocephalus* with entrance *c.* 4·5 cm diam.
EGGS: 2–3, sometimes 4; southern Africa (n = 35 clutches), 2–4, av. 2·6. Pale blue or greenish blue, with reddish brown and lilac spots, evenly scattered or concentrated at large end, sometimes a few small lilac blotches. SIZE: southern Africa (n = 72) 22–27 × 15·5–19 (24·5 × 17·4) (Maclean 1993); Zaïre 23–26 × 16·5–18·5 (Chapin 1954).
LAYING DATES: Gambia, (nest-building June–July, adults carrying food June, immatures with adults Oct); Sierra Leone, Mar; Burkina Faso (carrying food, June); Nigeria, Mar–June; N Uganda, Mar (*leucogaster*); NW Somali, (ovarian egg, May) (*arabicus*); Kenya, Mar (*verreauxi*); Tanzania, Nov; Zambia, Oct–Dec (mainly Oct); Malaŵi, Oct–Nov; Zimbabwe, Oct–Jan; Botswana, (nestlings, Dec); Transvaal, Sept–Feb; Natal, Oct–Feb.
INCUBATION: by ♀ only; period *c.* 12 days.
DEVELOPMENT AND CARE OF YOUNG: young remain in nest *c.* 21 days; fed by both parents.

References
Elgood, J.H. *et al.* (1973).
Traylor, M.A. (1971).

Genus *Neocichla* Sharpe

Endemic, single species of small, somewhat gregarious, brown-plumaged starling with black spot on breast, white in wing, black bill and yellow eyes and legs. Insectivorous; feeds on ground; nests in tree hole; probably a co-operative breeder but poorly known; 4 isolated populations in miombo woodlands. Skeletal characters not researched (A.J.F.K. Craig, pers. comm.), but analysis of other characters suggests placement near *Onychognathus* or *Buphagus* (Craig 1997). However, in view of its small size, yellow legs, voice and manner of foraging by moving rapidly on ground in search of insects, we think it likely that *Neocichla* will prove to be nearer to the 'true' starlings and mynahs *Sturnus*, *Creatophora* and *Acridotheres*.

Plate 36
(Opp. p. 543)

Neocichla gutturalis (Bocage). White-winged Starling. Spréo à gorge noire.

Crateropus gutturalis Barboza du Bocage, 1871. J. Sci. Math. Phys. Nat. Lisboa, 3, p. 272; Huilla, Angola.

Range and Status. Endemic resident, Angola, Tanzania, Zambia and Malaŵi, in 4 populations with strange, relict distribution. Angola, common in N and central Huila Prov. from Huambo southwest to Huila and southeast to 16°26′E on Cuvelai R. (Dean and Vernon 1988) and Cuchi R. in W Cuando Cubango (Hall and Moreau 1970); common but local between Hoque and Caluquembe and in Caconda area; 23 groups along 90 km of road between Hoque and N'gola; also around Quipungo (Dean and Vernon 1988). Zambia, isolated population in Mporokoso at several localities on E side of Mweru Marsh in Kampinda district and 2 localities further east; widespread in E and common at least locally from 10°S down E side of Luangwa valley in Lundazi sector to Kasweta and to Chipengali area of Eastern Prov. plateau (R.J. Dowsett, pers. comm.); old record from near Mozambique border (Hall and Moreau 1970). Malaŵi, confined to W Rumphi District and W Mzimba District (Manda Hill; close to Zambian border); numbers much reduced by habitat destruction, but may survive in Kasungu Nat. Park (Benson and Benson 1977). Tanzania, known from 7–8 localities in Tabora Prov., from near Tabora to about Rungwa R. (probably much more widespread further west: N.E. Baker, pers. comm.); old record from N Singida Prov. (Hall and Moreau 1970).

Density of 8 flocks totalling *c.* 40 birds in *c.* 100 ha of woodland, N'gola, Angola.

Description. *N. g. angusta* Friedmann: Tanzania, Zambia, Malaŵi. ADULT ♂: forehead to hindneck grey-brown; mantle, scapulars and back dark grey-brown with buff-brown feather fringes; rump and uppertail-coverts pale buff-brown, tinged greyer on long uppertail-coverts. Tail blackish brown, T5–T6

Neocichla gutturalis

with narrow whitish tips. Supraloral stripe pale grey; narrow black line through lores and around front of eye; cheeks, ear-coverts and sides of neck brownish grey. Chin and throat paler grey; small black bib in centre of upper breast, otherwise breast to flanks and belly cinnamon-buff, paler on undertail-coverts. Primaries blackish brown, bases of inner webs of inner feathers white. Secondaries blackish brown, S1–S2 with base of outer web white, inner web wholly white except along black shaft; S3–S4 with base and distal fringe of outer web white, base of inner web white except near shaft; S5–S6 with broad white fringe on outer web, S5 also with lateral white patch on base of inner web; forms broad longitudinal white stripe on closed wing. Tertials dusky brown, outermost fringed whitish on outer web. Primary coverts and alula blackish; greater coverts and median coverts dark brown; lesser coverts dark brown, more blackish near wrist. Axillaries and underwing-coverts cinnamon-buff. Bill black; eyes bright chrome-yellow; legs straw-yellow. Sexes alike. SIZE (3 ♂♂, 3 ♀♀): wing, ♂ 108–113 (110), ♀ 104–107 (106); tail, ♂ 83–88 (85·3), ♀ 78–85 (82·3); bill, ♂ 23–24 (23·7), ♀ 22–24 (22·7); tarsus, ♂ 30–31 (30·7), ♀ 30–31 (30·3). WEIGHT: 2 ♂♂ 64, 72, 1 ♀ 64.

IMMATURE: juvenile has top of head blackish brown; sharp cinnamon-buff fringes on dark mantle and scapular feathers. Sides of head, chin and throat streaked blackish brown, and bold blackish drop-shaped spots on upper breast, almost coalescing into patch in bib area. White bases and edges of flight feathers tinged buff; upperwing-coverts with narrow scaly buff tips. Bill yellow-ochre with blackish tip; eyes grey.

NESTLING: not known.

N. g. gutturalis (Bocage): Angola. Head and neck paler and greyer, upperparts slightly paler, more white on tips of outer tail feathers.

Field Characters. Length 21·5–24 cm. Very local in *Brachystegia* woodland. Plumage pattern unique: grey head, buffy underparts, black breast patch, dark brown back, and in flight, *white patch* on black wings; bill and lores black, eye yellow. Immature has wings like adult but pale mottling on upperparts, black streaks on underparts. Wattled Starling *Creatophora cinerea* in non-breeding plumage is plain grey with white rump and black tail, and in flight wings are two-toned black and grey.

Voice. Tape-recorded (B, LEM). Scolding calls given by birds at human presence, repeated dry, buzzy 'zzheeo', 'zzhurrrr' and 'zzhoy' and short, muted 'jik'.

General Habits. Inhabits mature *Brachystegia* woodlands with canopy height of 10–20 m and little or no undergrowth; also sometimes in *Colophospermum* woods. In pairs and flocks of 3–7 (av. of 5 birds in 8 flocks). Spends most of its time on the ground and forages there by running quickly and searching. Perches in lower branches and canopy of trees, but thought not to feed there. When flock alarmed on ground, 1 or more birds fly up into tree and call, or whole group flies away, low over the ground, to settle short distance away. Birds in flock straggle out in flight and on alighting. Flock forages after sunset until dusk then breaks up, birds flying away singly to nearby tree, preening, then flying to roost; 1 bird roosted alone in tree hole (Dean and Vernon 1988).

Food. Insects; termites in 8 out of 10 stomachs, beetles including carabids (Carabidae) and weevils (Curculionidae) in 7 (Benson and Irwin 1966).

Breeding Habits. Evidently a co-operative nester; 4 adults present at nest with young but only one brought food to it; 4 adults present at another (empty) nest; a 3rd nest had only 2 adults attending it (Dean and Vernon 1988).

NEST: one a small pad of lichens, mosses and grass c. 15 cm below entrance hole 8 m up in tree; another in knot hole, lined with hair and dry leaves.

EGGS: at least 2 (one clutch). Not glossy; smooth; white, freely speckled with reddish brown on blue-grey undermarkings (Mackworth-Praed and Grant 1963).

LAYING DATES: Angola, (empty nest with adults nearby, Aug; nestlings Sept); Zambia, Oct–Nov.

References
Dean, W.R.J. and Vernon, C.J. (1988).
Feare, C. and Craig, A. (1998).

Genus *Zavattariornis* Moltoni

Single species of starling or crow, endemic to tiny area in S Ethiopia. Body whitish grey, wings and tail black; bare blue skin in front of and behind eye; tuft of stiff bristles covering nostrils. Skull aegithognathous; bill stout, moderately long, with curved culmen; gonys rather short; nostrils circular, perforate, inferior in position (as in sturnid *Acridotheres* and corvid *Pyrrhocorax*); lachrymals free. Humerus pneumatic; tarsus scutellate; toes large and strong; number of fused vertebrae in pelvic sacrum contentious (Ripley 1955). 12 rectrices. Gait a walk. Forages on ground. Remarkable large pear-shaped nest with vertical top entrance; eggs blotched with lilac.

The putative corvid affinities of *Zavattariornis*, an enigmatic and baffling bird, are far from proven (Collar and Stuart 1985, Fishpool *et al.* 1996). Genus was assigned to Corvidae by Moltoni when first describing it and by Ripley (1955), and has stayed there for little more than reasons of convention. Corvid characters are nostril bristles, free lachrymals, single pinched foramen above ectethmoid, pneumatic humerus, tarsal scutellation, supposedly sternal morphology and perhaps voice; they are unconvincing, alone or in combination. Treated as a corvid by Goodwin (1986), with no great conviction; the bird's resemblances to Clarke's Nutcracker *Nucifraga columbiana*, *Podoces* and some Asiatic starlings taken as superficial or convergent. Sibley and Monroe (1990) placed *Zavattariornis* in Corvidae between *Pica* and *Podoces*. Mallophagan ectoparasites are not corvine and 'do not exclude possibility that *Zavattariornis* is a member of the Sturnidae' (Benson 1946). Palatal structure is not corvine at all (Lowe 1949). Bill morphology and some plumage features are strikingly similar to Asiatic ground-jays *Podoces* spp., particularly *P. panderi* (which, however, has pied wings, bill wide towards its tip, tubular nostrils, different palatal structure, and hole nest); but is *Podoces* a corvid, or a sturnid?

There are even more impressive similarities with Wattled Starling *Creatophora cinerea*: size, proportions, stance, pearl-grey and black plumage (particularly the non-breeding plumage of *Creatophora*), bare skin around eye, foraging behaviour and nest structure. Lowe (1949) thought *Zavattariornis* to be related neither to Corvidae nor to Sturnidae and placed it in its own family Zavattariornidae. Molecular and further anatomical studies are much needed. In the meantime, impressed by its resemblances to *Creatophora*, we place it in Sturnidae. Whatever its true relationships, it remains one of the most remarkable African discoveries of the 20th century.

Endemic to Ethiopia; a single monotypic species.

Plate 36 *Zavattariornis stresemanni* **Moltoni. Ethiopian Bush-Crow. Corbin de Stresemann.**
(Opp. p. 543) *Zavattariornis stresemanni* Moltoni, 1938. Orn. Monatsber., 46, p. 80; Javello, region of Borana, S Abyssinia.

Zavattariornis stresemanni

Range and Status. Endemic resident, Sidamo Prov., S Ethiopia, where restricted to area around Yabello (Yavello, Yabelo) and Mega (which are 95 km apart). Common to very common at 1230–1385 m around Yabello; known from 10 km away to the west, 15, 30 and 34 km away to north and northeast, near Arera (50 km east of Yabello), 32 km southwest of Arero, at several places between Yabello and Mega, and 15 km east and 25 and 44 km south-east of Mega (and *c.* 25 km from Kenya border). 2 birds, 'confidently identified as of this species' seen at S end of L. Tana, 700 km north of Yabello, in 1984 (N. A. Tucker, pers. comm. to Collar and Stuart 1985). Was very common around Yabello in 1940s and fairly common and widespread in 1980s; 62 birds seen in the area on 3 days in 1989 (Ash and Gullick 1989), and 312 counted in 210 km in Dec 1989–Jan 1990 on the same road transects and at same time of year as the earlier count (Hundessa 1991). Range in Yabello/Arera/Mega triangle embraces only some 4700–6000 km². Surveys in late 1989 and 1990 suggest that density remains about the same (Syvertson and Dellelegn 1991) although there may be many more birds in a greater area than formerly thought (Hundessa 1991); but the species is considered at risk from tree felling for charcoal and firewood, cattle-ranching (i.e. woodland clearance to open up cattle pasture for Didatiyarra Cattle Breeding Centre), settlement and urbanization (Mega, Metageferssa and Yabello are now Awraja capital cities, and Surupa, Dubuluk, Weib and Chewbet are urbanizing rapidly – Hundessar 1991), and fire (Collar *et al.* 1994). The 'Yabello Sanctuary' set up for this enigmatic bird and White-tailed Swallow *Hirundo megaensis* (Vol. IV, p. 182) has not been gazetted and is not actively managed (Collar *et al.* 1994), but see Anon. (1996).

Description. ADULT ♂: pale grey or whitish, with black wings and tail, and bare blue skin around eye. Forehead, forecrown, lower cheeks, chin, throat, belly, thighs and undertail-coverts creamy white, the colour merging into pale dove grey on rest of body. Underparts often and upperparts sometimes earth-stained pink-rufous. Lores, upper cheeks and long triangle behind eye and above ear naked, bright blue. Close tuft of stiff dark grey feathers point forewards from forehead, covering nostrils. Tail uniformly black with bluish gloss above, and black with brownish gloss below. Wings glossy bluish black above and blackish brown below, with lesser upperwing-coverts same grey as mantle, and with conspicuous patch of white marginal coverts on leading edge from wrist to base of primary coverts. Bill black; thickened, bare skin around eye bright blue or leaden blue and bare patch behind eye buffy; eyes dark brown; legs and feet black, soles grey or pale brown. Sexes alike. SIZE (6 ♂♂, 6 ♀♀): wing, ♂ 135–149 (142), ♀ 137–146 (139); tail, ♂ 121–124 (122), ♀ 112·5–124 (118); bill to skull, ♂ 32·5–37·5 (36·2), ♀ (30·0–33·5 (32·1); tarsus, ♂ 39·5–42·5 (41·3), 39·5–42·5 (40·6).

IMMATURE: most adult white parts and some pale grey ones buffy, especially upperwing-coverts, posterior border of scapulars, pale leading edge patch in wing, sides of breast, and back and rump; uppertail-coverts brown.

NESTLING: not described.

Field Characters. Length 30 cm. Conspicuous and unmistakable: a beautiful bluish grey bird with white face, rump and belly and solidly black wings and tail. Bill, eyes and legs blackish. Bare, bright blue skin around eye; pale, pink-buff bare patch behind eye (photos in Francis and Shirihai 1999) though patch sometimes covered by plumage (photo, *Bull. Afr. Bird Club* 3, 1996, 41). Eye skin may be thickened, making eyes look small and half-closed and bird look sleepy (Francis and Shirihai 1999); forehead profile flat and starling-like, or rounded and jay-like (**A**).

Voice. Tape-recorded (B). Call a high-pitched 'chek'.

General Habits. Inhabits park-like or orchard-like acacia savanna and short-grass savannas. Dominant trees in Yabello Sanctuary are *Acacia drepanolobium*, *A. brevispica*, *A. horrida* and *Balanites aegyptiaca* (Anon. 1996). Gregarious; in June–Feb in parties of about 6 birds and sometimes up to 30. Forages on ground by walking purposefully and pecking at surface, habitually in company of Superb Starlings *Lamprotornis superbus* and White-crowned Starlings *Spreo albicapillus* (M.W. Woodcock, pers. comm.).

Probably sedentary but locally dispersive; breeding grounds around Yabello in June are vacated by Nov and few birds present in Dec (but species also said to be present around Yabello in most months).

Food. Insects.

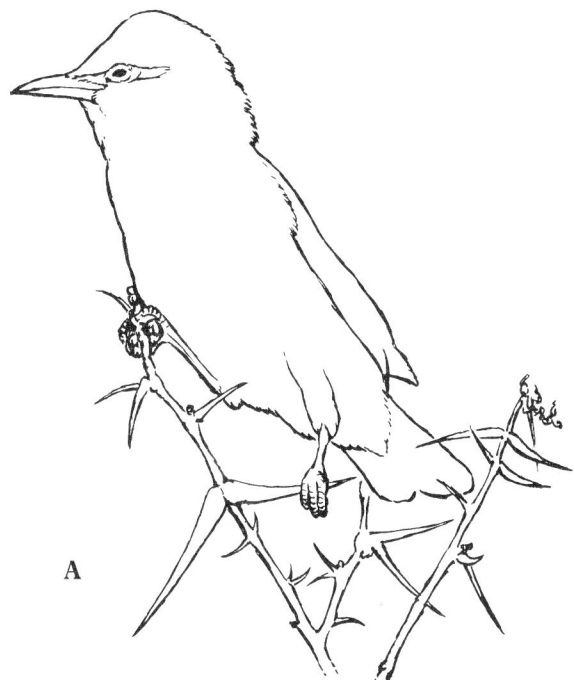

Breeding Habits. Nests solitarily. Presumed co-operative breeder: 3 adults regularly inside nest together, but no evidence from clutch to suggest that more than 1 ♀ lays.

NEST (Benson 1946): large, untidy, clumsily-made structure composed of thorny twigs 20 mm thick and *c.* 25 cm long, at first roughly globular, with entrance at top; floor lined with dry cattle dung mixed with short pieces of dry grass. Just before eggs are laid a vertical cylindrical superstructure is added at top of nest, extending and protecting the entrance, and making whole nest pear-shaped (with entrance from 'stalk' of pear to 'core', see (**B**), nest in vertical section, from Benson's record-card draw-

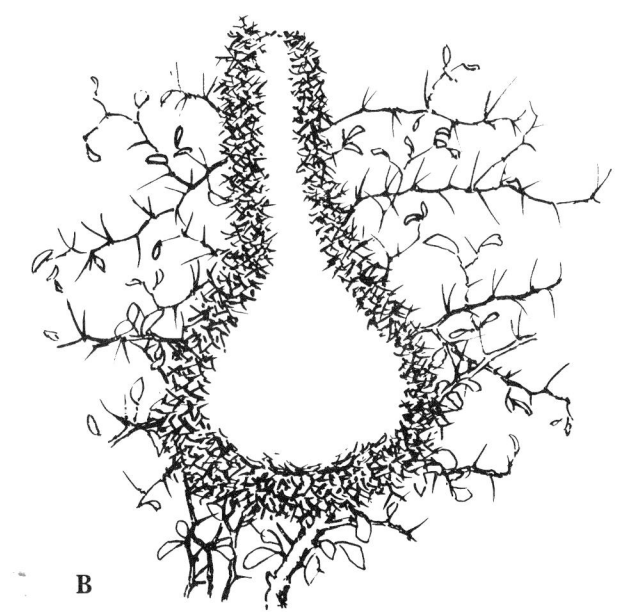

ing). Ext. diam. of globular part of nest *c.* 50 cm, int. diam. *c.* 25 cm; vertical funnel *c.* 15 cm high, int. diam. 70–75 mm, ext. diam. 25 cm. Sited in top of *Acacia* tree, *c.* 6 m above ground.

EGGS: up to 6. Smooth, slightly glossy; cream, with blotches of pale lilac, mainly forming a ring around large end. SIZE: (n = ?) av. 27·0 × 20·0.

LAYING DATES: Feb–Mar and May–June, perhaps 1st and 2nd clutches respectively.

BREEDING SUCCESS/SURVIVAL: Tawny Eagle *Aquila rapax* once seen tearing at top of a nest (J.S. Ash *in* Collar and Stuart 1985).

References
Benson, C.W. (1942, 1946).
Collar, N.J. and Stuart, S.N. (1985).
Collar, N.J. *et al.* (1994).
Francis, J. and Shirihai, H. (1999).
Hundessa, T. (1991).
Lowe, P.R. (1949).
Ripley, S.D. (1955).

Genus *Creatophora* Lesson

Single species of highly gregarious and nomadic Afrotropical starling, ♂ with distinctive grey-white plumage and in breeding season bare head with black and yellow skin and bizarre large, flat wattles; streak of bare, blue-grey skin at side of throat. Wings and tail black; ♀ and juvenile isabelline with whitish rump, very like SW Asiatic starling *Sturnus roseus*; bare skin on lores and behind eye; juvenile with bare throat-streak greenish yellow. Bill pink, eyes black-brown, legs dark flesh; thighs long, legs robust, gait a walk. Size and proportions like those of *Sturnus vulgaris* and *S. roseus*. Culmen curved, bill somewhat thrush-like, like *S. roseus*. Nest a large, untidy ball of twigs, floor covered with grass and feathers; entrance at side or near top; rather like nest of *Zavattariornis stresemanni*. Voice much like that of *S. vulgaris* and singing ♂ waves wings. Reproduction revolves largely around the exploitation of nomadic, fast-reproducing locusts and other orthopterans, as does life of *S. roseus*. Forages for insects on ground and fruits in shrubs. Feeds on soft soil by inserting and opening bill, like *S. vulgaris*; flock feeding on ground moves in one direction and 'rolls' forwards, like *S. roseus*.

Creatophora is closely related to *Sturnus* (Amadon 1943, Durrer and Villiger 1970, Sontag 1985) and particularly to *S. (Pastor) roseus* in our view, despite the pronounced plumage and head-feature differences of breeding ♂♂.

Endemic, eastern and southern Africa, wandering to W Africa and Arabia.

Plate 36
(Opp. p. 543)

Creatophora cinerea (Meuschen). Wattled Starling. Étourneau caronculé.

Rallus cinereus Meuschen, 1787. Museum Geversianum sive index rerum naturalium etc., p. 40, no. 17; [Cape of Good Hope].

Range and Status. Endemic nomad and migrant, eastern and southern Africa; rare and irregular non-breeding visitor to Yemen and S Oman; 8 in SW Madagascar, Nov 1989; 3 on Europa I., 530 km from Africa in S Mozambique Channel, May 1994; 1 on Bird I., Seychelles, July 1995–Feb 1996; 1 on Aldabra I., Sept 1998.

Gambia, 3, Cape St Mary, throughout Jan 1976 (photos confirm: Gore 1990). Cameroon, 6, Ngaoundaba, Mar 1990, and 9 there, Dec 1990 (Robertson 1992). 60, Waza Nat. Park, Feb 1992 (Robertson 1993); 1, Andirni, Nov 1998 (Scholte *et al.* 1999). Gabon, Mandji I. near Port Gentil (P. Alexander-Marrack *in* Dowsett and Dowsett-Lemaire 1993a). Congo, 1 Djéno (04°56'S, 11°57'E), Oct 1990, and *c.* 50 Djéno, July 1991 (Dowsett-Lemaire *et al.* 1993). Boundaries of breeding range in this opportunistically-breeding, mobile nomad are hard to determine; encounters marked by Xs in Central African Republic, Sudan (except SE), NE Ethiopia, NE and S-central Zaïre and W Tanzania, may be of vagrants or they may be in 'normal' range.

Eritrea, frequent on Tekeze (Tacazze) R. at Om Hajer; large nesting colony Wadi Sciotel; frequent on coastal plains. Ethiopia, common to abundant up to 2000 m and occurs up to 3000 m, in W, S, SE, NE Ethiopia and Rift

Creatophora cinerea

Valley; uncommon in N; boundaries uncertain. Djibouti, occasional small flocks (Welch and Welch 1992). Somalia, frequent resident in NW but scarce in NE, with only 6 records east of 46°E (including breeding record east of 49°E); locally abundant in S (almost entirely west of 46°E), with a breeding record on Ethiopian border, otherwise a non-breeding visitor in Aug–Dec; thousands at a roost in Afgooye (Sept) and 'myriads' on coastal plain between Raas Kaambooni and Buur Gaabo (Ash and Miskell 1998). Kenya, widespread and locally common from coast to at least 2000 m; breeds mainly between 36° and 38°E. Uganda, irregular, locally abundant in N, absent from SW. Tanzania, numerous records in corridor from NE to SW (N.E. Baker, pers. comm.). Rwanda, abundant in Kagera Nat. Park. Burundi, Ruzizi Valley. Angola, locally common in interior, along coast south of 08°S, and in S. Zaïre, erratic in Uele; specimen records from Saidi in Ituri Forest and Irumu; sporadic but can be common in Rutshuru and Ruzizi Valleys; many records in Katanga. Zambia, numerous records as shown (R.J. Dowsett, pers. comm.) but not known to breed. Malaŵi, spasmodic, uncommon, in most regions, up to 2300 m; no conclusive evidence that it ever breeds. Mozambique, uncommonly recorded in Tete, Beira district and S Sul do Save. Rest of southern Africa, highly erratic but locally common (Harrison *et al.* 1997). Population in Sul do Save varies greatly with irregular movements in and out of neighbouring regions, and is likely to exceed 10,000 birds at times (Parker 1999).

Description. ADULT ♂ (breeding): head bare, skin black on forehead, forecrown, loral area, chin and throat, and bright yellow from below eye to hindcrown and nape, bordered below by black band from lower nape to throat; naked frontal wattle and larger coronal one; large, median, bare, black, pendent wattle on chin and throat with bifurcate tip, or paired lateral wattles (**A**: greatest development of wattles); feathered hindneck, sides of neck, mantle, scapulars, back, rump and uppertail-coverts pale grey-brown with almost purplish wash; rump and uppertail-coverts generally paler than back. Tail black, glossed green. Underparts (breast to undertail-coverts) like back, slightly paler on belly. In wing, alula, primaries, secondaries and tertials black with coppery green reflections proximally on outer vanes; greater primary coverts, greater, median and lesser coverts white; undersides of flight feathers silvery grey, underwing-coverts and axillaries grey-brown with wide white fringes. Bill pale pink; eyes brown; legs and feet flesh-brown. ADULT ♂ (non-breeding): head feathered, same pale grey-brown as rest of upperparts, but bare black malar stripe *c.* 20 mm long and up to 4 wide, and bare yellow eyelids, circumorbital skin and skin behind eye; vestigial frontal and coronal wattles visible through forehead and forecrown feathering, and well before breeding season all wattles start to grow (**B**), frontal and coronal wattles usually erect but sometimes flopping over to one or both sides (**C**). Degree of wattle development varies greatly, even amongst breeding ♂♂. ♂ loses head feathers in Apr–May and replaces them in Nov–Dec (Kenya). ADULT ♀: like non-breeding ♂ though slightly duskier grey-brown, and greater primary and greater coverts blackish, not white; bill blackish around nostrils; blackish loral area and bare malar stripe conspicuous (**D**). SIZE: wing, ♂ (n = 63) 116–125 (120), ♀ (n = 50) 111–123 (118); 32 ♂♂ and ♀♀ combined: tail 61–71, bill 21–24, tarsus 27–30. WEIGHT: ♂ (n = 63) 64–85 (74·6), ♀ (n = 50) 51–83 (71·5) (Dean 1978), unsexed (n = 134) 56–92 (67·5) (Maclean 1993); also ♂ (n = 14) 74–82 (78·5), ♀ (n = 6) 61–80 (72·5) (Kok and van Zyl 1996).

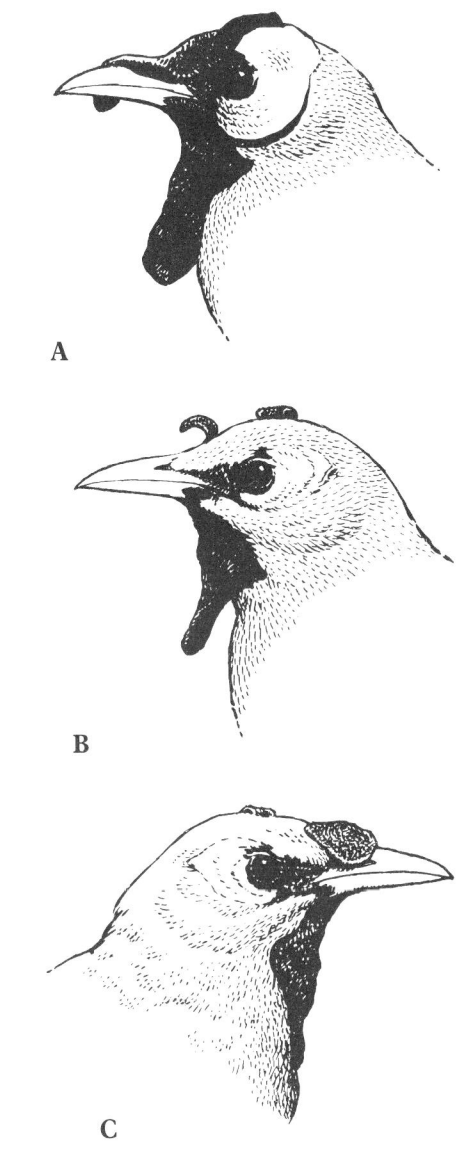

IMMATURE: subadult ♂ and ♀ like adult ♀; juvenile, rather uniform brown: forehead to upper back brown, lower back and tertials buffier and rump creamier; chin and throat buffy white, faintly streaked with darker buff, breast pale buffy brown, belly, flanks and undertail-coverts creamy; primaries, secondaries and tail mid to dark brown, not black, tertials mid to pale brown.
NESTLING: not described.

Field Characters. Length 19–22 cm. A grassland bird, often in flocks around mammals. Chunky, with short tail and pointed wings. Breeding ♂ with black and yellow facial skin and wattles unmistakable. ♀ and non-breeding ♂ light grey-brown with *pale bill*, black lores and moustachial stripe, small triangle of bare yellow skin behind eye; in flight, *white rump* contrasts with grey back and black tail; wing black and grey (♂ with small white crescent formed by tips to primary coverts). Immature like ♀ but browner. For comparison with White-winged Starling *Neocichla gutturalis* see that species.

Voice. Tape-recorded (86, 88, 91–99, 104, B, C, F). Songs long, usually lasting >15 s, and rambling, typically in 3

D

parts, beginning with faint, barely audible introduction, becoming louder in middle section and loudest at end, see sonagrams in Sontag (1985). Wide variety of sounds includes long, thin, tinny whistle, 'ssssrreeeee', deep 'kwarrr', hollow glottal 'o'clock', hissing 'tsizit', burry whistled 'wreeyoo', triple 'chochocho' and low chuckles. Each song different; may be simply a pleasant conversational babble of tinkling and gurgling notes. Said not to mimic, but a loud 'wok, wok, wok' in one song sounded like the braying of a zebra. In flock chorus, shrill babbling whistles mix with liquid 'pleep-pleep-pleep' and resonant 'boyboyboy'. Noise emanating from breeding colony is un-bird-like, a continuous succession of rodent-like high-pitched squeaks and screeches (Uys 1977).

General Habits. Inhabits wooded and bushy grassland with open ground, often sandy, and short grass sward; cattle and sheep pasture, cultivation, large-mammal country and national parks. In *Eragrostis* wooded grassland (Angola) and on seashore (South Africa).

Highly gregarious throughout year; rarely solitary or in pairs or groups of 3–10; usually flocks are of 20–30 birds, but huge aggregations of hundreds of thousands can occur (N Uganda, N Tanzania, any part of Kenya, southern Africa). Forages on ground by pecking and picking; in soft soil feeds by open-bill probing ('prying'); gait a brisk, purposeful walk; flock advances on a broad front, several birds deep, those behind constantly flying to alight ahead, so foraging party progresses in rolling, leap-frog manner (Uys 1977). Often associates with domestic stock and with larger wild mammals (e.g. White Rhino *Ceratotherium simum*, Lechwe *Kobus lechwe*, waterbuck *K. ellipsiprymnus*),

feeding almost at their feet on short grass sward, giving short chase after insect disturbed by the mammal. Forages on ground behind grazing sheep and goats and perches readily on back of sheep (but not slippery-coated goat) as it moves along, picking off ectoparasites from ears and forehead, balancing carefully to do so, and flying down to pick food item from ground (Bennun *et al.* 1990). In Masai Mara, Kenya, commonly perches on back of Brindled Gnu (Blue Wildebeest) *Connochaetes taurinus* and sometimes on zebras *Equus quagga* and *E. zebra*, bird standing alertly and occasionally feeding on ectoparasites – usually single starling on each beast, once 5 starlings together (Koenig 1994). On ground probes bill into soft soil and opens mandibles, like *Sturnus* starlings (Sontag 1979). Combines into even larger flocks with glossy starlings *Lamprotornis* spp. and Pied Starlings *Spreo bicolor*; forages with those, and with Violet-backed Starlings *Cinnyricinclus leucogaster* and Common Starlings. Shy and flighty. From large flock on ground, groups fly to perch on top of clump of shrubs then, when whole flock assembled, the entire mass of birds moves on, to settle again in tree tops a few hundred m away or to return to the ground to feed. Flight fast and direct, with even wing-beats; glides only momentarily, when about to alight. Flock dense, both on ground and in flight; large flying flock often dense and tight, birds twisting and turning in unison, like a flock of waders. A roost of thousands of birds (Somalia) was in a grapefruit orchard. Roosts in reedbeds or trees, sometimes with Red-winged Starlings *Onychognathus morio* or Cape Weavers *Ploceus capensis* (also sometimes nests alongside Cape Weaver colony). Comfort behaviours (stretching, yawning, preening, water bathing, sun bathing, anting) described by Sontag (1979). Drinks at spillage puddles. Both sexes sing, often for >15 s; ♂ song not always directed at a ♀; when uttering undirected song, perched ♂ waves wings. ♂ song has sexual connotation but ♀ song does not (Sontag 1985).

Wanders widely; in most areas quite sporadic; in some regions less unpredictable, amounting to migration, e.g. occurs in L. Victoria basin, Kenya, in June–Oct and Nairobi area in July–Oct; influx into coastal lowlands and Tsavo East Nat. Park in Oct–Nov; in Zambia scarce in Dec to Feb or Mar, common in Apr–Nov, and regarded as a migrant (Aspinwall 1986); seems to be migratory in Botswana (present Nov–June) and to be breeding visitor to Kruger Nat. Park, South Africa, in austral summer; but no true migration in southern Africa (Craig 1992). Thought to migrate between SW Arabia and (presumably) Somalia. Best described as nomadic until itinerary of individual flocks is known (Craig 1996). Ringing recovery, L. Abiata to Dukem, 138 km away, Ethiopia. Moult pattern is typical of passerine species in general; interrupted wing-moult is widespread, and seems to be associated with nomadism rather than with opportunistic breeding (Craig 1996).

Food. Locusts, grasshoppers and other orthopterans, other insects, small fruits, nectar of flowers including aloes, earthworms, snails, offal. Mainly locusts when bird breeding, largely fruits at other times; examination of nestlings shows that locust nymphs form by far the greatest part of their food (Chapin 1954); after breeding in SW Cape, flocks move to coastal macchia to feed on seeds of exotic rooikrantz *Acacia cyclops* (Uys 1977). One large breeding colony fed mainly on caterpillars of moth *Laphygma exemta* (Kenya). Mass breeding event in Etosha Nat. Park, Namibia, coincided with plague of mopane worms (caterpillars of moths *Gonimbrasia belina* and *G. tyrrhea*); nestlings fed on mopane worms, also locusts and mantises. Large colony in Free State, South Africa, fed mainly on caterpillars 20–25 mm long (Nuttall 1998). In huge colony at Windhoek, Namibia, nestlings fed mainly on armoured crickets *Hetrodes* sp. Of 108 stomachs (Cape Prov.), fruit in 42, insects and fruit in 45, insects in 19 and a lizard in 1; insects included crickets, 2 families of termites, 3 of bugs, 4 of beetles, and flies and ants (Dean 1978). Arthropods in 27 stomachs (Free State, South Africa) were 90% beetles and 10% Lepidoptera, centipedes and solifugids, and plant matter, predominantly grape vine *Vitis vinifera*, with small amount of *Pollichia campestris* (Kok and van Ee 1990). Also feeds in fruiting fig trees (Bennun *et al.* 1990) and eats fruits of *Royena*, *Azima tetracantha*, *Ziziphus mucronata*, *Ehretia rigida* and mulberry *Morus nigra*. Dependence on locusts in some areas is suggested by a nesting colony, in course of construction, being abandoned when locust 'hoppers' were destroyed (Eritrea, Smith 1957); Wattled Starling breeding certainly correlated with that of locusts (W.R.J. Dean *in* Ginn *et al.* 1989), though dependence on locusts for food perhaps less crucial than formerly believed (Maclean 1993). Eats caterpillars, bugs, spiders, earthworms and ectoparasites; mealie meal from bird tables, rubbish from tips and offal from abattoirs; and nectar of *Erythrina caffra*, *Schotia brachypetala*, *Acrocarpus fraxinifolius* and *Agave sisalana*.

Breeding Habits. Monogamous, not territorial beyond nest entrance; strongly colonial. Nesting is extremely irregular and opportunistic, dependent on verdant conditions and locust hatches after exceptionally heavy rains in normally arid regions; ♂♂ in full breeding dress, with wattles, are seldom seen – e.g. none seen at 2 breeding colonies (Craig 1996). Courting ♂ sings, lowers and spreads wing on side next to ♀, then uses frontal display, with plumage (especially white rump) fluffed out, wings held out horizontally and quivered, legs bent, head lowered, and front part of body making thrusting movements which accentuate the head markings (**E**: head here raised because ♀ perched higher than ♂). ♀ solicits using 'vulture' posture (**F**), with naked throat stripes conspicuous (Sontag 1985).

NEST: large, untidy domed mass of brittle thorny or thornless twigs and sticks with entrance, usually rather small, neat and round, at side (or occasionally near top); floor of chamber lined with dry grass and feathers. Entrance sometimes funnel-like, with thorns pointing mostly outward (Paxton and Cooper 1986). The mass of twigs is built around many living thorny branches in tree, so nest is well secured and impossible for person to pull off in one piece. Most twigs 5–10 thick, 50–100 long. Nests often clustered together into single large mass of interwoven twigs with 5 or more separate chambers. Several nests and nest clusters commonly built in one small tree, and colony of several hundred nests can cover large area of

scattered trees. Colony at L. Nakavali, S Uganda, reportedly covered 'several square miles, every bush loaded with dozens of nests' (Chapin 1954). In colony in Etosha, Namibia, c. 600 nests were in thorny *Acacia luederitzii* trees and c. 100 in large, thornless *Colophospermum mopane* trees. Also nests in *Balanites* and *A. karoo* and regularly uses *A. nigrescens* in Zimbabwe and exotic *Eucalyptus* trees in W Cape. Sited at almost any height above ground, mainly 2–3 m, often up to 8 m, in thorn bush (sometimes over water) or tree, or at top of telephone pole.

EGGS: 1–5, av. (n = 155 clutches, Namibia) 3·2 (5 × 1, 30 × 2, 57 × 3, 53 × 4, 10 × 5); 8 eggs in 1 nest probably laid by 2 ♀♀ (Paxton and Cooper 1986). In Namibia 3 eggs after mediocre rain, 4 after abundant rain (Chapin 1954). Smooth; immaculate pale blue, greenish blue or (rarely) almost white, or occasionally sparingly spotted with brown. SIZE: (n = 48) 26·4–31·9 × 18·3–21·9 (28·3 × 20·5), also (n = 45, Etosha) 25·4–31·6 × 19·1–22·4 (27·1 × 20·1); within a clutch, egg size varies considerably (Paxton and Cooper 1986).

LAYING DATES: Eritrea, (nest building July); Ethiopia, May–Aug; Somalia, May (and young being fed, June, Aug); E Africa, Region D, Apr, May, July, Dec (mainly May); southern Africa, all months, but mainly Oct–Mar (Craig

1996). Within a colony laying is highly synchronized (Paxton and Cooper 1986).

INCUBATION: by ♂ and ♀; each often sings during nest relief. Period c. 11 days.

DEVELOPMENT AND CARE OF YOUNG: young fed by ♂ and ♀, mainly between 08h00 and 10h30, each parent bringing food item every 10–15 min; sometimes parent brings several items in bill (e.g. up to 5 mopane worms). Leaves nest at 13–16 days and first flies at 19–22 days. ♂♂ do not lose head feathers in 2nd year although wattles may enlarge then; they lose feathers and enlarge wattles seasonally in 3rd and later years (Chapin 1954).

BREEDING SUCCESS/SURVIVAL: in large part connected with abundance of staple prey, i.e. locusts. '... birds were too late in commencing their nests...before their young were fully fledged the locusts began to leave ... It was easy work for the birds to follow them at first and bring back a sufficient supply of food to their nests, but as the locusts [gradually departed they] abandoned their nests [and] half-fledged young and to save themselves flew after the locust swarms...' (M. Layard, quoted by Archer and Godman 1961). In dense colony, some shrubs burdened with so many nests that, with the extra pressure of a storm, they collapse with 'an awful smash of eggs and young'. At colony in Etosha, Namibia, only predators seen eating feathered chicks clinging to outsides of nests were all birds: Secretary Bird *Sagittarius serpentarius*, Tawny Eagle *Aquila rapax*, Gabar Goshawk *Micronisus gabar* and Lanner Falcon *Falco biarmicus* (Paxton and Cooper 1986). At colony at Gweta, Botswana, 5 birds died in webs of golden orb spiders (*Babbler* 35, 1999: 56).

References
Chapin, J.P. (1954).
Craig, A.J.F.K. (1992, 1996).
Dean, W.R.J. (1978).
Feare, C. and Craig, A. (1998).
Koenig, W.D. (1994).
Liversidge, R. (1961).
Paxton, M. and Cooper, T. (1986).
Sontag, W.A. (1979, 1985).
Uys, C.J. (1977).

Genus *Sturnus* Linnaeus

Small gregarious starlings; bill straight, about length of head, shape varying from rather flat and wide with straight culmen to thrush-like with curved culmen; nostril with membrane, reached but not covered by forehead feathers; front of skull narrow, allowing eyes to rotate forwards; wings triangular, pointed but not long; P10 small, pointed, stiff, 10–20 mm long; tail rather short, square-ended, about half length of wing; plumage iridescent green-black, dark grey, grey and white or black and pink; some spp. with narrow, pointed, quite long feathers on head and breast, some crested or ruffed; sexes alike but juveniles differ markedly. Forage on ground, by walking and bill-prying: bill protractor muscles well developed and bird thrusts bill into soil and prises it open (**A**); nest colonially in holes in trees and buildings; eggs immaculate; eat insects (mainly) and fruits taken mainly from cultivated land (Feare 1984). 16 species (Sibley and Monroe 1990), mainly in SE Asia, or 4 species (Feare and Craig 1998, excluding Asiatic *Gracupica*, *Sturnia*, *Temenuchus* and *Pastor*). We retain (*Pastor*) *roseus* in *Sturnus*.

5 species, 3 in Africa: 1 (*S. unicolor*) endemic to NW Africa and Spain; 1 (its allospecies *S. vulgaris*) a common winter visitor to NW Africa, breeding across much of Palearctic and introduced to many other parts of world including South Africa; and 1 (*S. roseus*) a vagrant to NE Africa.

A

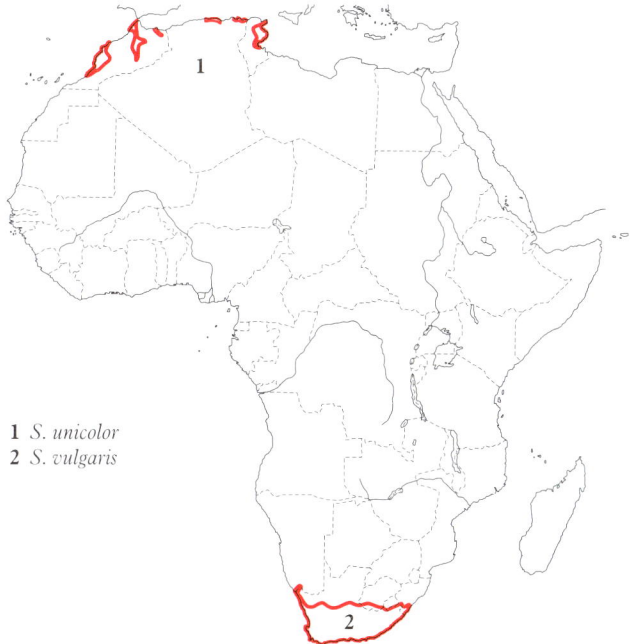

Sturnus vulgaris superspecies

1 *S. unicolor*
2 *S. vulgaris*

Sturnus vulgaris Linnaeus. Common Starling. Étourneau sansonnet.

Sturnus vulgaris Linnaeus, 1758. Syst. Nat., ed. 10, 1, p. 167; Sweden.

Plate 36
(Opp. p. 543)

Forms a superspecies with *S. unicolor*.

Range and Status. Palearctic, north to Arctic Circle, south to N Mediterranean, Iran and Indus Valley; introduced in N America, throughout, south to Baja California and Florida. Resident in W and S Europe; regular winter visitor in small numbers south to NW Africa, S Mediterranean, N and E Arabian Peninsula, Pakistan, N India and Mexico. A colonist in Iceland, Azores and some Canary Is. Introduced also to South Africa, SE Australia and New Zealand, where now a common resident.

Common winter visitor to N Africa. Mauritania, rare: 4 Cansado, 1 Cap Timirist. Morocco, common and widespread in N, south almost to desert, usually in small flocks, often in hundreds, sometimes in thousands (once 15,000 in Sidi Chiker Rés., Chichaoua, Morocco). Algeria, common to abundant, especially around olive groves, south to Sahara (records at Hassi Messaoud and Béni Abbès); thousands between Touggourt and Meghaier, 200 at Ghardaïa. Tunisia, numerous in coastal lowlands between Tunis and Kairouan, especially in Sousse, Kelbia, Kairouan, Gafsa and Nefta. Libya, uncommon to abundant around Tripoli coastal olive groves; common on Cyrenaica coast, Nov–Apr. Egypt, rare in some winters, common in others, along N coast and south to Cairo and Suez Canal; occurs in Faiyum; scarce in Nile Valley, unknown south of Luxor. Ethiopia, vagrant, old record from Dessie, W Highlands, mid June. Djibouti, vagrant (Laurent 1990). Somalia, vagrant, 1 Xawaal Barrey, 02°25'N on coast, Dec.

Sturnus vulgaris

In southern Africa common in SW Cape Prov., South Africa, where deliberately introduced to Cape Town about 1897 (by Cecil Rhodes, using probably N European birds caught in Britain in winter); reached Stellenbosch by 1910, Knysna by 1944, Addo by 1961 and Kei River by 1970 (Hockey et al. 1989); now abundant in S Cape Prov. north to about 30°S, well established in Transkei, spreading east into KwaZulu-Natal, including Durban and north to Free State; frequent in Namibia, in and around Oranjemund; numerous in Maseru, Lesotho. In W Cape Prov. occurs in roosting flocks of thousands.

Description. *S. v. vulgaris* Linnaeus: Europe; winter visitor to N Africa from Morocco to Libya; introduced in South Africa. ADULT ♂ (breeding): whole plumage black, but with such strong gloss – mainly bottle green on crown, back and wings and mainly bronze-purple on sides of neck, throat and breast – that bird never looks as black as most corvids. Many feathers, particularly on rear of flanks, retain the rufescent buff tips of non-breeding plumage, and undertail-coverts generally with buff edges, making V-shaped marks since coverts are pointed. Most feathers narrow, attenuated and sharp-pointed, except in wings and tail, and throat feathers are pointed hackles that stand out, beardlike, when bird sings. Bill yellow with pale bluish grey base; eyes dark brown; legs and feet reddish flesh-brown. ADULT ♂ (non-breeding): feathers black with green or bronzy purple gloss as in breeding plumage, but acuminate ones (on head and body) with well-defined pale tips, rufous-buff on crown, nape, hindneck, mantle, back, scapulars, rump and uppertail-coverts, or white on forehead, ear-coverts, sides of neck, lower throat, breast, flanks, belly and undertail-coverts; pale tips form tiny rufous-buff or white specks on head, neck, mantle and breast, small triangles or V-marks on back, rump, flanks and belly, and larger V-marks or U-marks on uppertail- and undertail-coverts. Tail feathers, primaries, secondaries, tertials and all upperwing-coverts similarly glossy green-black, with narrow but sharply-defined rufous-buff edges and tips. Lores, chin and upper throat black, not spotted. Underside of tail and remiges shiny dark grey; underwing-coverts and axillaries dark grey with narrow whitish edges and tips. (During non-breeding season pale feather tips and edges gradually wear away to reveal glossy unspotted breeding plumage.) Bill black; eyes, legs and feet as in breeding ♂. ADULT ♀ (breeding): like breeding ♂ but base of bill yellow like rest of bill, and eye (iris) with dark brown inner ring and pale brown outer ring. ADULT ♀ (non-breeding): like non-breeding ♂ but eyes as in breeding ♀. SIZE: (Netherlands) wing, ♂ (n = 77) 128–141 (134), ♀ (n = 47) 126–137 (131); tail, ♂ (n = 27) 60–68 (64·1), ♀ (n = 19) 58–65 (61·8); bill to skull, ♂ (n = 64) 27·8–31·2 (29·6), ♀ (n = 36) 27·0–30·5 (28·7); tarsus, ♂ (n = 58) 28·8–32·2 (30·5), ♀ (n = 50) 27·6–31·2 (29·7) (Cramp and Perrins 1994). WEIGHT: (Algeria) unsexed (n = 51) 65–87 (75·5); (South Africa) unsexed (n = ?) 55–96.

IMMATURE: in first non-breeding season at *c.* 6 months, more heavily spangled with rufous-above and white below than in following non-breeding seasons; breeding adult in first breeding season at *c.* 10 months, glossy black (with yellow bill) but usually covered with tiny buff spots on body or scales on wings. Juvenile rather uniform dark greyish brown, tail feathers, remiges and upperwing-coverts narrowly edged dark greyish fawn, ear-coverts and cheeks mottled with grey, chin and throat grey-white, rest of underparts indistinctly streaked dark brown and pale greyish; undertail-coverts dark brown with pale grey edges and tips; underside of tail and wings as in adult; bill dark horn to brownish black with paler cutting edges and base of lower mandible; eyes greyish at first, soon turning dark brown; legs and feet dusky reddish brown.

NESTLING: upperparts with grey-white down, quite thick and long, underparts naked; mouth bright yellow or orange; gapes pale yellow; blind.

S. v. purpurascens Gould: E Turkey to Causasus; winter visitor to Egypt. Adult with cap, chin, upper throat and belly glossy purple, rest of upperparts glossy green, flanks and wing-coverts bronzy purple. Same size as *vulgaris* but heavier. Weight: ♂ (n = 14, E Turkey) 80–105 (87·8), ♀ (n = 12) 67–100 (85·0).

Field Characters. Length 21–22 cm. A familiar dumpy bird with long, *sharp bill* and characteristic *waddling gait*, abundant in town and country. Low forehead gives it a rather flat-headed appearance (although singing bird may raise head feathers and fluff out neck and breast feathers). *Short tail* and *triangular pointed wings* give distinctive shape in flight. In breeding season appears black at a distance but at close range shows glossy purple and green sheen, plain below but finely spotted buff above, *bill yellow*; in winter heavily spotted white and buff all over, and bill dark. Juvenile dull greyish brown with indistinct dark markings, *pale throat* and *dark smudge in front of eye*. Noisy, gregarious and aggressive. For differences from Spotless Starling *S. unicolor*, see that species.

Voice. Tape-recorded (62–73, 76–89, 91–99, B, C, F). ♂ song lengthy, lively and varied but not loud nor far-carrying, a rambling medley of warbles, sibilant whistles, clicks and throaty chuckles, lasting up to 60 s, usually with short pauses every few s; often with strong element of mimicry – imitates many other birds, also man-made sounds such as telephone. Commonest flight call 'querrr' or 'squar'; bird taking flight often utters a quiet, soft 'prurrp'; attack calls, loud squawks and a chattering 'chackerchackerchacker'; predator-alarm a sharp 'chip' or 'spet'. Insistent begging call 'churr churr' often heard from nests under house eaves.

General Habits. Inhabits large and small suburban gardens with trees, shrubs and lawns; parks in cities and avenues of trees between highrise buildings; grassy open country, cattle pasture, small-holdings, farmyards and farmland with root crops and scattered buildings, stubble fields, arable land, hedges, berry shrubs, copses, lines of windbreak trees, orchards; perches on overhead wires. Foraging flocks in non-breeding season can occur in almost any open lowland habitat, including heaths, sewage-works, drier parts of salt marshes, rocky seashores, airfields and refuse tips.

Occurs in pairs and family parties in and just after breeding season, juveniles soon assembling into mobile flocks of a few dozens; in non-breeding season gathers into flocks of tens, thousands (Africa) or even a million (Europe). Individual distance small, and flock forages on ground with birds only 10 or 20 cm apart; huge flying flocks often very dense, birds twisting and turning together like small waders; 100 or so migrants rest more or less regularly spaced out on 10 m length in middle of powerline; roosting flocks often huge and dense, birds huddling together in cold weather in groups of 40 or so or strung out on ledge of building in 10s or 100s of birds almost touching. Flight fast, not undulating, with short glides; also has slow, fluttering, gliding flight when feeding aloft on flying insects. Freely uses elevated perches on tops and edges of trees, telephone wires and buildings, but feeds mainly on ground. Forages by striding briskly and purposefully, often making a short quick run, stopping at intervals to thrust bill into soft soil, prise mandibles apart and withdraw bill with or without prey item. Prey swallowed whole, often without perceptible break in walking pace. Also picks at soil surface, briefly chases insect, searches amongst vegetable litter and amongst stones at edge of water, gleans foliage where caterpillars abundant, hawks insects whilst slow-flying aloft, stands on sheep and pecks into wool for ectoparasites, and seizes small fruits from tips of branches with a good deal of fluttering; and feeds kleptoparasitically amongst Northern Lapwings *Vanellus vanellus*, Rooks *Corvus frugilegus* and Jackdaws *C. monedula*. Commonly forages amongst grazing ungulates and in winter with gulls *Larus* spp. and thrushes *Turdus* spp. Associates with Spotless Starlings *S. unicolor*. Quarrelsome, frequently lunging with open bill and half-open wings and squawking at conspecific or other small bird feeding near it on bird-table or ground. Nesting ♀ roosts in nest but breeding ♂♂ and non-breeding birds roost gregariously, in sugarcane fields, reedbeds, young conifer plantations, and on ledges on bridges and large buildings in city centres. On evenings in non-breeding season large, sometimes huge, flocks wheel near roost site, flying densely like small waders; at a distance they can resemble thick smoke in fast-changing swirls; at dusk flocks stream into roost site, where clamour of jostling, tight-packed birds is easily heard above din of city-centre traffic. After dawn huge roost is suddenly vacated *en masse*, in 2–4 waves at intervals of a few min, each travelling outward in all directions (visible on radar as expanding ripples) or heading in 1 direction only. In Saldanha area, W Cape, flocks of thousands fly out to sea each evening to roost on offshore islets.

Resident in South Africa; winter visitor to N Africa mainly in Oct–Mar, immigrants still arriving in Dec if weather hard in Europe. Algeria, winters Sept–Apr, abundant Oct–Mar; earliest, 7th Sept. Tunisia, rare Aug, common late Sept to Feb, scarce Mar, rare to late Apr; Libya, locally abundant Oct–Nov, uncommon Dec–Apr. Many visitors to Algeria and Tunisia are from S Poland. Juveniles from Switzerland first disperse up to 500 km to northwest into Holland and Belgium, then migrate in flocks of 50 to several hundred mainly to Algeria, presumably crossing Mediterranean on broad front, then in winter move west into Morocco and north-west to Spain (Cramp and Perrins 1994). Numerous ringing recoveries in Tunisia from E Europe: Poland, Czech Republic, Slovakia, Hungary, Yugoslavia. Bird ringed in Italy found in Tripoli, Libya; one from former USSR found in Cairo, Egypt.

Food. In Europe takes insects and plant material all year, mainly insects and larvae in spring, soft fruits in summer/autumn and seeds in autumn/winter (Cramp and Perrins 1994). Regarded as a severe pest of olive *Olea* crops in N Africa (and hunted and destroyed at roosts accordingly).

Variety of insect prey (and occasional small vertebrates) in Europe listed in Cramp and Perrins (1994). Winter diet in N Africa probably not unlike that in Israel, where many cereal grains, leaves, and olive fruits consumed. Diet in South Africa similar, so far as known, and includes larvae taken by probing turf, insects, spiders, worms, molluscs, a few small lizards, fruits, seeds, fallen cereal grains, and scraps at bird tables, around houses and in parks.

Breeding Habits. Information from South Africa supplemented with some from Europe. Solitary or loosely colonial nester, with high degree of synchrony of 1st clutches in colonies; monogamous and serially or simultaneously polygynous, with up to 60% of ♂♂ polygynous (in Belgium). Defends territory around nest, but courtship, copulation, foraging and other functions conducted elsewhere.

Courtship: ♂ chooses nest site, defends it and builds bulky nest foundation; he attracts ♀ by repeatedly singing nearby, in intense form with bill pointing up, tail down, rump and belly feathers lifted, and pointed throat feathers puffed out, accompanied by wing-waving display with wing tips flicking outward (**A**). On approach of ♀, ♂ may enter nest hole and continue singing inside; he may also show hole entrance by taking in a flower or leaf. For *c.* 5 days, from 4 days before 1st egg laid to day after, ♂ guards mate by accompanying her at all times, flying closely behind her and giving 'chackerchackerchacker' call repeatedly. Copulation, always preceded by ♂ singing, occurs in tree or on ground.

A large tree can hold several nests; nests in house eaves can be only 1 m apart; density of 170–800 nests per km² of parkland (Poland: Cramp and Perrins 1994). Regularly double-brooded; 2nd clutch 40–50 days after 1st.

NEST: bulky base made of twigs and vegetable matter, more or less filling space available in nest cavity to depth of a few cm, occasionally up to 25 cm; cup, at side of base away from entrance, made of bents of dry grass, rootlets, moss, hair, feathers and scraps of cloth or paper. Cup size (n = 11, Kazakhstan): diam. 96–153 (100), depth 68–140 (95). In South Africa nest commonly sited in cavity in house eaves; also in tree hole, crevice in rock, and on

offshore islands sometimes in bushes (A.J.F.K. Craig *in* Ginn *et al.* 1989). Built by ♂, taking c. 5 days; ♀ adds some lining to cup.

EGGS: 3–6. Subelliptical, smooth, slightly glossy, immaculate pale blue. SIZE: (South Africa, n = 68) 27·5–32·9 × 19·7–22·4 (29·6 × 21·3). Weight: *c.* 7·0 g.

LAYING DATES: South Africa, Sept–Dec.

INCUBATION: by ♀, spending nights and *c.* 71% of day (in England) on 1st clutch or longer on 2nd clutch (Feare 1984). Starts with penultimate egg. She may cover clutch with a leaf on leaving it unattended (Cramp and Perrins 1994). Period: (South Africa) 12–13 days; (Europe, n = 84) 11–15 (12·2) days.

DEVELOPMENT AND CARE OF YOUNG: ♀ broods young for 1st 5–8 days. Later, young fed by both parents, often mainly by ♀. Nestling period 20–21 days. After leaving nest, young stay near site or colony for a few days and are fed by parents, often on ground.

BREEDING SUCCESS/SURVIVAL: percentages of eggs resulting in fledglings in numerous European studies tabulated by Cramp and Perrins (1994); usually high, in 1st clutches e.g. 62–98% (Poland) and 67–81% (England) and in 2nd clutches 70% (Netherlands) and 58% (Italy). Mortality in 1st year 48% (Britain) and 68% (Czechoslovakia), in 9th year 22% and 10th year 14% (Czechoslovakia); wild birds can live at least 20 years (Cramp and Perrins 1994).

References
Cramp, S. and Perrins, C.M. (1994).
Feare, C. J. (1984).
Harrison, J.A. *et al.* (1997).
Maclean, G.L. (1993).
Quickelberge, C.D. (1972).
Winterbottom, J.M. and Liversidge, R. (1954).

Plate 36   *Sturnus unicolor* Temminck. Spotless Starling. Étourneau unicolore.
(Opp. p. 543)

Sturnus unicolor Temminck, 1820. Man. Orn., ed. 2, 1, p. 133; Sardinia.

Forms a superspecies with *S. vulgaris*.

Range and Status. Resident and short-distance migrant, endemic to W Mediterranean: Iberian Peninsula except for Pyrenees foothills; near Narbonne, France; Corsica, Sardinia, Sicily; Morocco, coastal Algeria and Tunisia. Exactly parapatric with Common Starling *S. vulgaris*, but recent spread of both species has resulted in slight marginal overlap. Vagrant Madeira, Libya, Malta and Greece.

Morocco, common but local; along Atlantic coast from Tanger to Merja Zerga and from Jorf Lasfar to mouths of Sous and Massa, common along Sous to Taroudant; common at Marrakech; Ouezzane; widespread on Plateau Central and in middle sector of Moyen Atlas foothills from Oulmes, Fes and Jbel Tazzeka (Taza) south through Dayetv Ifrah, Ifrane, Azrou, Amis, Boulemane, Timahdite, Aguelmane Sidi Ali and Col du Zad to Itzer; once Titaf (Errachidia); on Mediterranean coast from Ras Kebdana to Nador and at Beni Snassen (Oujda). Algeria, colonies near Constantine and El Arrouch and formerly at Tilremt; records L. Tonga, L. Oubeira, Azzefoun and Réghaïa. Tunisia, common in coastal lowlands between Hammanet and El Djem; scarce and local towards Cap Bon, from Kairouan to Sbeitla and Kasserine, and south to Gabès and Kettane Oasis. Once bred Maknassy. Libya, old report of bird at Zavia.

Densities of up to 45 breeding pairs per ha in *Quercus ilex* woods and up to 76 birds per km² in fallow or irrigated farmland in Spain, though most assessments much lower (e.g. 2 birds per km² in *Castanea* forest, 3·3 birds in *Q. ilex* in S Spain, 5 in mixed *Quercus*, 7 in *Juniperus* woods); in winter, up to 40 birds per km of transects in orchards and cereal fields (Cramp and Perrins 1994).

Sturnus unicolor

Description. ADULT ♂ (breeding: black, with slight oily purple gloss on head, neck, mantle, back, scapulars, upperwing-coverts, ear-coverts, throat and breast. Feathers of crown, hindneck, throat and upper breast lanceolate and long (twice length of longest head feathers of Common Starling). Tail brownish black, feathers narrowly fringed black. Wings brownish black, outer vanes of primaries, primary coverts and lesser coverts glossed greenish, those of secondaries, tertials, median and greater coverts glossed purple; primaries, secondaries, tertials and most coverts broadly tipped and fringed with velvety black. Bill pale yellow, lower mandible with pale blue base; mouth violet, palate pink; eyes dark brown; legs and feet light brown to dark brown. ADULT ♀ (breeding): like ♂ but throat feathers not so long. ADULT ♂ and ♀ (non-breeding): like breeding adult but feathers of head, neck, mantle, back, scapulars, chin, throat, breast, flanks and belly with minute pale buff tips; upperwing-coverts and undertail-coverts with warm buff edges and tips, making V-marks; bill black. SIZE: wing, ♂ (n = 20) 129–138 (133), ♀ (n = 27) 126–137 (131); tail, ♂ (n = 12) 59–68 (63.7), ♀ (n = 12) 58–67 (62.6); bill to skull, ♂ (n = 12) 27.8–31.5 (29.6), ♀ (n = 12) 27.0–30.2 (28.8); tarsus, ♂ (n = 12) 29.0–32.6 (30.7), ♀ (n = 13) 29.4–32.5 (30.7). WEIGHT: (central Spain, June–July), ♂ (n = 20) av. 95.4, ♀ (n = 12) av. 85.8; ♂♂ av. 90.4 in Aug–Oct, 93.1 in Nov, 95.8 in Dec–Jan, 92.0 in Feb and 94.1 in Mar–May (Cramp and Perrins 1994).

IMMATURE: in first non-breeding season at c. 6 months, like non-breeding adult but tail feathers less pointed. Juvenile rather uniform dark greyish brown, tail feathers, remiges and upperwing-coverts narrowly edged fawn, chin whitish, throat greyish white, upper breast, sides of breast and sides of neck streakily mottled grey-brown and whitish, lower breast and flanks dark grey-brown with faint pale streaks, belly and thighs dark brown, undertail-coverts dark brown with fawn-buff edges and tips; bill blackish, legs and feet dark brown.

NESTLING: naked, with tufts of sparse pale grey down on back and wings; after 7–8 days down becomes long, dense and dark grey (Cramp and Perrins 1994).

TAXONOMIC NOTE: sometimes treated as a race of *S. vulgaris*; however, now marginally sympatric with it (in Catalonia), without known hybridization; crosses of the 2 in captivity are infertile, and *unicolor* differs from *S. vulgaris* more than races of *S. vulgaris* differ amongst themselves. For these reasons we retain *S. unicolor* as a separate species.

Field Characters. Length 21–23 cm. Very similar to Common Starling *S. vulgaris* with similar habits. Breeding birds all-black with purplish gloss, truly spotless, lacking buffy spots and green gloss of Common Starling; also have thicker-necked profile caused by long hackles on throat and hindneck. In winter plumage pale spots are much finer so bird appears darker than Common Starling. Juvenile darker and browner, but may not be reliably separable in the field.

Voice. Tape-recorded (62–73, 93, 105, B). ♂ song lengthy, lively, varied, not far-carrying, a long (0.5 s) loud 'seeoooo' following by quiet, rambling medley of sibilant warbles, clicks and throaty gurgles, much like ♂ song of Common Starling except for the initial 'seeoooo', which may also be added to other notes and repeated in the following ramble (sonagrams in Cramp and Perrins 1994). 'Seeooo', a falling note, is often given by itself as a flight call, and in song may be preceded by rising 'seeee', the 2 sounding like a human wolf-whistle. An accomplished mimic of other birds, in North Africa (Etchécopar and Hüe 1967) and Spain: easily recognizable in song are snatches of bill-clattering of White Stork *Ciconia ciconia* and voices of Common Kestrel *Falco tinnunculus*, Common Quail *Coturnix coturnix*, Common Moorhen *Gallinula chloropus*, Little Owl *Athene noctua*, Calandra Lark *Melanocorypha calandra*, Great Tit *Parus major*, Eurasian Golden Oriole *Oriolus oriolus*, European Robin *Erithacus rubecula*, Eurasian Blackbird *Turdus merula*, Common Magpie *Pica pica* and Western Jackdaw *Corvus monedula* (Cramp and Perrins 1994). Contact call, 'gaa-haa'; threat, a rattling 'kokokokorr'; excitement, 'geee-geee-geee'; disputing at roost, a screaming 'haaa-haaa-haaa'. Also a sharp 'fiit ... fiit' and a snarling rasp.

General Habits. Occurs in open countryside in coastal plains and foothills, often around human habitation, abandoned houses, ruins, farm buildings and old wells; breeds in cedar forests in Moyen Atlas, Morocco, making lengthy daily flights to forage on open plateaux at up to 1800 m; occasionally up to 2500 m; frequents olive groves, cork oaks *Quercus suber* (isolated trees and woods), open arable land and pasture fields with holm oaks *Quercus ilex*; improved grassland, stubble fields, vineyards and rubbish tips; roosts in reeds, trees and on buildings.

Gregarious, in flocks of tens or c. 100 outside breeding season and sometimes in small flocks in breeding season. Forages mainly on the ground, by walking, from time to time pushing closed bill into soft soil then prying it open, or pushing slightly open bill into dry soil. In winter often flocks with Common Starlings and large mixed flock may 'roll' forwards on open ground, leading birds walking or running to catch prey, trailing ones constantly flying to the front. Feeds amongst cattle, and sometimes stands on head or back of cow or sheep to forage for ectoparasites. In spring feeds almost entirely on ground (Spain); in summer, 15% of feeding is in trees on caterpillars and fruit and 5% by taking insects in flight; in autumn, 88% on ground and 11% in shrubs. On ground takes mainly larvae and beetles; in air mainly crane-flies in spring and ants in autumn. Roosts communally, throughout year, in brambles, riverside willows, poplar plantations and oak woods (Spain) and coastal cliffs (Morocco); flock performs aerial evolutions before going to roost, much like those of Common Starling (Morocco, Spain: Smith 1965).

Sedentary in Africa; but flocks of c. 400 birds seen leaving Tarifa, S Spain, in early mornings in Oct, heading to Morocco, returning at dusk (Tellería 1981). (Spanish population mainly sedentary, but in winter some birds caught up in migrations of Common Starling; adults move up to 160 km and juveniles up to 700 km: Peris 1991).

Food. No African information. In Spain, mainly insects in spring and summer and fruits and seeds in autumn and winter. Food includes shrews, rats, mice, lizards, skinks, frogs, dragonflies, grasshoppers and crickets, earwigs, cockroaches, bugs; adult and larval butterflies, moths, flies, ants, wasps, bees and beetles; spiders, harvestmen, mites, scorpions, pseudoscorpions, solifugids, woodlice, millipedes, centipedes, earthworms and snails; seeds, fruits or buds of oak, elm, poplar, fig, olive, mulberry, grape, dock, strawberry, cherry, sunflower, maize and other cereals and grasses, sedges, Chenopodiaceae, Cruciferae, Leguminosae, Umbelliferae, Solanaceae and Compositae. In west-central Spain 60% of items over whole year are animals and 40% plants; of animal food (n = 8667 items), 57% beetles, 14% Hymenoptera and 12% larvae (Cramp and Perrins 1994).

Breeding Habits. Well known in Spain. Monogamous and polygamous; nests solitarily or (generally) in loose colonies. Territory *c.* 20 m in radius, around nest, serving for courtship and nesting (but pair forages up to 800 m away); colony size usually 2–8 nests, on average 70 m apart, but once 22 nests under one roof, and once 180 nest boxes (all 10 m apart) were all occupied (Cramp and Perrins 1994). Double-brooded.

NEST: mass of twigs, herbs, grass and cereal bents, with cup in middle thickly lined with grass, rootlets, leaves, feathers and flowers, placed in hole under eaves, in brickwork of old buildings, nest-boxes and other man-made sites; small minority in cedars, cork oaks, or base of large stick nest of White Stork (Thévenot *et al.* 1981) and in Spain in old woodpecker holes or other holes up to 15 m high in oak, chestnut, poplar, elm, alder, ash or pine; also in old nest of European Bee-eater *Merops apiaster*, Common Sand Martin *Riparia riparia* or Western Jackdaw. Size (n = 56, Spain): ext. diam. 180–220 (205), int. diam. 90–130 (110), ext. depth 70–180, depth of cup 40–100 (Cramp and Perrins 1994). Built by both sexes.

EGGS: 2–9, usually 3–4 in N Africa and 4–5 in Spain (where clutches larger than 7 eggs probably always due to egg-dumping). Subelliptical, smooth, slightly glossy, immaculate pale blue; eggs in 2nd clutches sometimes with tiny reddish spots. SIZE: (n = 414, central Spain) 30·1–31·2 × 20·7–21·8 (30·5 × 21·3). Weight: (n = 209) 5·0–10·0 (7·2) g.

LAYING DATES: Morocco, Apr–May (and adults commonly seen feeding newly-fledged young in May); Algeria, 1st clutches Mar–Apr, 2nd clutches May–June.

INCUBATION: by ♀, starting with 3rd–4th egg, but in one study for 20% of daytime by ♂. Period: 10·5–11·6 days (Spain) (Feare 1986).

DEVELOPMENT AND CARE OF YOUNG: weight at hatching (n = 39) 6–9 (7·1) and at fledging (day 20) 68–80 (73·7); young cared for and fed by both parents, mainly ♀. Nestling period: 18–25 days, mainly 21–22. Fledglings stay in vicinity of nest and are fed by ♂ and ♀ for *c.* 7 days, then disperse and form small juvenile flocks.

BREEDING SUCCESS/SURVIVAL: of 150 eggs (Spain), 70% hatched and 51% resulted in fledglings. Rat *Rattus rattus* is main nest predator.

References
Álvarez, S.P. (1984).
Cramp, S. and Perrins, C.M. (1994).
Feare, C. (1986).
Peris, S.J. (1980).

Plate 36 (Opp. p. 543) *Sturnus roseus* (Linnaeus). Rose-coloured Starling. Étourneau roselin.

Turdus roseus Linnaeus, 1758. Syst. Nat., ed. 10, 1, p. 170; Lapland and Switzerland.

Range and Status. Breeds patchily in Balkan Peninsula and widespread from N and E coasts of Black Sea east to *c.* 70°E in central Asia, south to 38°N. Winters in SE Arabian Peninsula, Pakistan, throughout India and in Sri Lanka. Irruptive visitor to W Europe and N Africa, rare or accidental from Finland to Seychelles; thousands on S Yugoslavian coast in May 1989, and 6500 nested in N Italy in 1875. Regular spring and autumn migrant at Eilat, Israel (Goodman and Meininger 1989).

Algeria, accidental (Etchécopar and Hüe 1967). Tunisia, 1 Tunis spring 1903. Libya, 1 Zuwatina Aug 1937, 1 Serir May 1969. Egypt, 1 Giza Aug 1864, 1 Rafa (Sinai) Aug 1917, 1 Suez Mar 1982, 1 Sharm el Sheikh (Sinai) Aug 1982 and 1 there Aug 1995, 2–4 Bahariya Aug–Sept 1983.

Description. ADULT ♂ (breeding): whole head and neck down to upper mantle and upper breast glossy purple-black, clearly delineated; crown feathers long and pointed, forming crest drooping over nape; lower mantle, back, scapulars, rump, uppertail-coverts, breast, flanks and belly pink; longest uppertail-coverts tipped black, thighs and undertail-coverts brownish black; tail and wings black, heavily glossed with dark green; undersides of tail and wings shiny dark grey. Bill pink, lower mandible with grey base; eyes bright brown; legs and feet reddish pink. Sexes alike. ADULT (non-breeding): like breeding adult but purple head covered with minute buff flecks, pinks grey, bottle-green feathers in wing and undertail-coverts very narrowly edged with buff, bill all pink (without grey base) and legs and feet brownish yellow. SIZE: wing, ♂ (n = 43) 127–139 (132), ♀ (n = 15) 125–135 (129); tail, ♂ (n = 26) 65–75 (69·5), ♀ (n = 7)

Sturnus roseus

64–74 (68·7); bill to skull, ♂ (n = 35) 22·2–26·5 (24·6), ♀ (n = 16) 22·4–25·6 (23·8); tarsus, ♂ (n = 34) 29·4–34·2 (32·1), ♀ (n = 16) 29·0–32·5 (30·9). WEIGHT: ♂ (n = 13) 71–88 (79·6), ♀ (n = 6) 60–83 (66·9).

IMMATURE: head and neck dark brown, feathers pointed and minutely tipped with buff, mantle, scapulars, back, rump and uppertail-coverts pinkish grey-brown; wings and tail black, each feather with green gloss except towards sides and tip, and narrowly edged with buff all around; breast like back, merging to pale buff on flanks and belly; undertail-coverts black with dark buff edges and tips. Juvenile overall light buffy brown with blackish brown wings and tail; forehead, crown and hindneck finely striated with dark brown, lores mid brown, narrow pale buff or whitish eye-ring, ear-coverts slightly browner than surrounding parts and finely freckled, mantle, scapulars and back plain greyish brown, rump and uppertail-coverts plain pinkish brown; sides of neck, throat and breast finely mottled dark buff and pale buff, flanks and belly softly striped with dark buff and pale buff or whitish, chin, thighs and undertail-coverts pale brown. Bill yellowish horn; eyes, legs and feet like non-breeding adult.

Field Characters. Length 21–22 cm. Adult unmistakable, even in 1st-winter plumage with blackish hood, wings and tail and buffy or pinkish grey-brown body; stocky; *pink bill*; pale legs; in flight wings triangular though less so than in Common Starling *S. vulgaris*. Juvenile (autumn; plumage sometimes retained in winter): singleton can look thrush-like (but *walks*), pale buffy with streaking on head and underparts and dark wings and tail; *bill yellow-horn*, shorter and markedly less sharp-pointed than bill of Common Starling; in flight rump paler than back, usually conspicuously so, *whitish* or pinkish.

Voice. Tape-recorded (107, B, CHA). Song, flight and other calls very like those of Common Starling *S. vulgaris*, *q.v.* Migrants wintering in Oman and vagrants in England call very little; vagrants in Africa likely also to be quiet. Main calls a harsh 'tschirr' and raucous 'kritsch'; alarm a yelping 'quilp' (Cramp and Perrins 1994).

General Habits. Inhabits semi-desert with thorn bushes and trees, open waste ground with withered grasses and dusty shrubs near human habitation, stony wadis with palms, grass tussocks, oleanders and reeds or bulrushes, open ground near industrial or brackish water, bushy ground, pasture, lawns, orchards, berry bushes, harvested cereal fields and farmland.

Singly or in small flocks in winter, in Arabia often all-juvenile flocks. Forages on ground, walking and pecking like Common Starling. Not very confiding; small flock disturbed on ground flies several hundred m away to pitch in small trees. Often feeds near grazing cattle; rarely, feeds on back of goat or sheep. Large flock may assemble at locust hatching ground and pursue hoppers, flock moving in one direction, leading birds walking briskly and often running, trailing ones constantly flying to the front, so whole flock progresses in rolling, leap-frogging manner. Occasionally flies up from ground to take insect in air. Gorges itself on fruiting mulberry bush *Morus*, flock feeding with much clambering and wing-fluttering. Drinks and bathes in road puddles and at open, shelving edges of pond. Mobile; a long-distance migrant, travelling in flocks, often by day, heading from breeding grounds mainly southeastward to wintering grounds in S Asia, but also tending to irrupt westward to S Europe. Bird ringed in Hungary recovered 4800 km away in W Pakistan 10 months later. In India lays down pre-migratory fat in spring.

Food. Mainly insects, specializing on orthopterans including grasshoppers and locusts, particularly the flightless 'hopper' instars of *Dociostaurus maroccanus* and *Calliptamus italicus*. On breeding grounds eats large amounts of grapes *Vitis* and mulberries; in winter eats insects, fruits and seeds. In Smyrna, Turkey, revered as a locust-eater in May and reviled as a grape-eater in July (Voous 1960).

References
Ali, S. and Ripley, S.D. (1972).
Cramp, S. and Perrins, C.M. (1994).

Genus *Acridotheres* Vieillot

Medium-sized Oriental starlings, plumage blackish, dark brown, grey, or pied; glossy only on head; white patches in wing, white corners to tail, several spp. with forehead crest, 1 with drooping hindcrown crest, bill yellow, yellow-brown or reddish, quite robust, shorter than in *Sturnus*, protractor muscles well developed and birds forage in part by prying or open-bill probing, like *Sturnus* (*q.v.*); eyes and legs usually yellow-orange; bare skin around eyes; legs and feet very robust; wings rather rounded. Resonant, fluty calls; eat fruits, nectar, insects, spiders; bowing displays; immaculate blue eggs.

10 species, all Oriental (Feare and Craig 1998), but one introduced in many parts of world, including South Africa.

Plate 36
(Opp. p. 543)

Acridotheres tristis (Linnaeus). Common Myna. Martin triste.

Paradisea tristis Linnaeus, 1766. Syst. Nat., ed. 12, 1, p. 167; 'Philippines', in error probably for Pondichéry, India.

Range and Status. Asia; resident and locally a partial migrant in Turkmeniya, S Kazakhstan, S Iran, Pakistan, India, Sri Lanka, Nepal and Burma. Successfully introduced or self-established in many regions, including SE Africa, Comoros, E Madagascar, Seychelles, Canaries (Gran Canaria), France (Dunquerque), W, central and SE Arabia, SE Asia from Hainan to Singapore, E and SE Australia, New Zealand (North Island), and many Pacific and Indian Ocean islands.

In South Africa abundant around Durban where introduced in about 1900 (a trade consignment of *tristoides*, with some *tristis*: Brooke 1976), and widespread and locally common in Natal; also locally common in and around Johannesburg, where separately introduced in 1930s, although 30 years passed before it became common there (Brooke *et al.* 1986); first appeared in Pretoria in 1955 and quite common there by mid 1980s; spread to Pietermaritzburg recently and first nested there in 1997; now widespread in S Transvaal and occurs in NE Free State and W Swaziland, and several records from Port Elizabeth; has bred Kimberley and Cape Town (A.J.F.K. Craig *in* Harrison *et al.* 1997). Records in Botswana (Mahalapye 1975, Gaborone 1991, Bobirwa area 1998 (Brewster 1999). In Zimbabwe, becoming established in Harare and Bulawayo (Brooke 1983, Donnelly 1982). Mozambique, 6 at Ponta Malongaue in extreme S, Nov 1997 (Parker 1999).

Acridotheres tristis

Description. *A. t. tristoides* (Hodgson): Nepal, Burma and Assam; introduced to Natal. ADULT ♂: forehead, crown and nape black with oily purple gloss (greenish in some lights); crown feathers 12–21 mm long, pointed; purple-black hood sharply delineated from dark grey neck; hindneck and sides of neck dark grey, mantle, scapulars and uppertail-coverts dark vinous grey-brown, back and rump the same but a little paler. Tail black, T1 tinged brownish, with white tip, T2–T6 with progressively larger amounts of white at ends, up to 25 deep on T6. Lores and narrow line around eye black; large triangle behind and below eye naked, skin bright yellow; lower part of cheeks and of ear-coverts black with greyish bloom; chin, throat and breast dark grey in centre, washed with vinous towards sides of throat, sides of breast and on lower breast; flanks, belly and thighs dark vinous grey-brown; vent and undertail-coverts white. Primaries black distally, white proximally, secondaries, tertials, greater, median and lesser coverts dark vinous grey-brown, alula dark vinous grey-brown with white streak in centre, primary coverts white. Underside of closed tail mainly white and of spread tail white with blackish base; underside of primaries blackish and white (sharply delineated) and of secondaries blackish; underwing-coverts and axillaries white. Bill yellow or orange-yellow, base of lower mandible dark greenish grey; eyes red-brown, mottled with white; eyelids yellow; mouth slate or palate pale dusky pink (Brooke 1976); legs and feet orange-yellow. Sexes alike. SIZE (South Africa): wing, ♂ (n = 22) 133–144 (140), ♀ (n = 23) 129–141 (134); tail, ♂ (n = 18) 74–90 (82.1), ♀ (n = 16) 74–88 (78.6); bill to skull, ♂ (n = 20) 25–31 (27.9), ♀ (n = 25) 24–29 (26.8). WEIGHT (Natal): ♂ (n = 8) 106–134 (123), ♀ (n = 8) 98–116 (109.5) (Brooke 1976).

IMMATURE: like adult but head feathers dull sooty black, not purple-glossed, hood not sharply defined, crown feathers 6–9 mm long; mantle, scapulars and wings rufescent, lacking the grey-purple bloom of adult. Juvenile like immature but back, rump and uppertail-coverts buff-brown, faintly dark-barred; white parts of tail isabelline; underparts browner than in adult, buffier on flanks; throat and breast feathers narrowly fringed buff and with narrow black shaft streaks; breast and flanks faintly dark-barred; vent and undertail-coverts off-white. Very young bird has narrow, brownish black tail feathers with fawn tips (tip *c.* 6 mm deep on T6); within 2 weeks of leaving nest, rectrices replaced by feathers like adult's but with black shafts in the white patches; adult tail feathers acquired by moult before 1st breeding season (Brooke 1976).

NESTLING: not described, but partly covered in pale grey down.
A. t. tristis (Linnaeus): India, S Iran, Pakistan, Afghanistan, S Kazakhstan; introduced to Transvaal; some birds on Natal coast are *tristis*. Slightly paler than *tristoides*, especially on throat and breast; overall less richly coloured. A little larger. SIZE (India): wing, ♂ (n = 20) 138–152 (145.5), ♀ (n = 14) 134–147 (139); tail, ♂ (n = 21) 79–92 (85.3), ♀ (n = 14) 75–85 (80.6); tarsus, ♂ (n = 24) 37.0–42.2 (39.9), ♀ (n = 17) 36.0–41.6 (38.8). WEIGHT (both races, South Africa): unsexed (n = 22) 82–138 (110).

Field Characters. Length 23–25 cm. A large, bulky, jaunty, robust suburban starling, unmistakable on the ground (yellow bill and facial skin on blackish head, strong yellow legs; walks) and in flight (flappy, blackish wings with large white patch above and larger one below; much white at corners of tail). Strident voice. Generally in small flocks.

Voice. Tape-recorded (58, 75, 88, 91–99, B, C, F, CHA). Song a sustained jumble of squawks, gurgles, liquid whistles, creaking and whining notes, each typically repeated 2–4 times: 'krr krr krr ci ri ri ri krrup chip krrup chirri chirri chirri weeu weeu ... '; begins with croaking noise, and can include some mimicry (Cramp and Perrins 1994, Maclean 1993). Call when bird taking flight a weak, querulous 'kwerrh'; contact call a strident 'chour chour kok kok kok' or liquid 'keeky-keeky-keeky'; alarm a harsh 'traaahh' or 'kharr'.

General Habits. Inhabits open areas in cities, towns and surrounding farmland. Perches in and on trees and commonly on buildings, wires and all manner of urban, industrial and agricultural substrates, but spends most of daytime on the ground, where forages by walking, looking leggy, with quite long and purposeful stride, on lawns, turf, pasture, roadsides, grassy roundabouts, pavements and in chicken runs, rubbish tips, cattle fields, stockpens, playing fields, railway yards and parks; occasionally along tideline and on boats. Sometimes bounds in long strides. Roosts in trees.

Occurs in pairs or daytime flocks of *c.* 8 birds and up to 30; sometimes solitary; huge flocks congregate at urban roosts in Durban and elsewhere. Mated ♂ and ♀ keep together by day and night. A generalized and opportunistic omnivore; scavenges by walking at a steady pace, hardly pausing to tilt head, gaze and peck at ground surface, stopping to jab opened bill into soft soil, or making running dash after a grasshopper. Forages on ground around cattle and sometimes perches on them to glean ectoparasites; does the same with sheep, goats, Burchell's Zebra *Equus quagga*, Eland *Taurotragus oryx*, Bushbuck *Tragelaphus scriptus* and Impala *Aepyceros melampus* (Dean and MacDonald 1981); follows the plough and municipal lawn-mowers. Adaptable, learning quickly; in Durban 'they have learned to feed on grasshoppers killed by lawn mowers, to the extent that starting the mower will bring the birds' (W.R.J. Dean *in* Ginn *et al.* 1989). Confident; can become very tame, approaching picnickers almost to their feet; forages unconcernedly along roadside verges within 1 m of heavy traffic. Often perch on backs of cattle, horses and antelopes, probing hair for ectoparasites. Common on ships and boats in Durban harbour, particularly when people living on board; birds perch on mastheads and booms and walk about on deck (Ginn *et al.* 1989).

Raucous, mobile and flighty. Flies fast, just above vegetation: only 50 cm high in short flight across lawn, or 5 m high, skimming treetops and low buildings, on way to roost; flies with slight undulation and rapid, even wingbeats, with only very short, veering glide on entering tree. In evening flies to roost in small flocks; roosts gregariously in large (sometimes small) tree with dense canopy, 100–200 birds gathering without aerial evolutions or much noise, though a few birds in roosting flock may call for 1 h after dusk (Oman); however, roosts in Durban can number thousands and make tremendous noise until well after nightfall (Maclean 1993). Known to roost also on buildings. Vocal all year, highly so in breeding season. Sings at perch on or near ground, bowing and nodding head (Maclean 1993). Generally not aggressive, although some supplanting behaviour occurs amongst foraging flocks, and on occasion will fight furiously.

Food. Best known in India. Mainly insects. Small vertebrates (frogs, geckos, mice); carrion; occasionally eggs and nestlings of seabirds and landbirds; wide range of larval and adult insects including bees and wasps; ticks, spiders, centipedes, millipedes, crabs; earthworms, snails; fruits of fig, papaya, neem, date, tomato, pear, apple, guava, mango, breadfruit, seeds including maize and pulses, sometimes nectar, flowers, waste food and scraps (Cramp and Perrins 1994). In Africa diet includes insects, millipedes, snails, worms, birds' eggs, small mice, frogs and lizards, and nectar, fruits and seeds (Maclean 1993).

Breeding Habits. Rather better known in India, Bangladesh, Hawaii and New Zealand than in Africa. Monogamous, with year-round, life-long bonds. ♂ and ♀ display to each other by bowing with plumage fluffed out. Nests singly or in loose colonies; somewhat territorial. 1, often 2, sometimes 3 broods. Bird once paired with a Common Starling (*Promerops* 237, 1999: 15).

NEST: untidy mass of twigs, grass bents, straw, roots, leaves, feathers, hair, bits of cloth, snakeskins, and bits of plastic, string and paper; with or without shallow cup in middle, which may or may not be lined. Placed in cavity in tree, earthen bank or building, or amongst dense mass of twisted branches, or in machinery, pipe or wall. Nest tends to fill space available; some nests can be 50 cm across and weight several kg. Built by ♂ and ♀. At first material is deposited just inside cavity entrance and may block it. Material added after eggs laid.

EGGS: in South Africa 2–6, usually 3–5, av. (n = 26) 3·9. Laid at 24 h intervals. Long ovals; glossy; immaculate mid greyish or greenish blue or turquoise. SIZE: (n = 35) 26·3–32·9 × 20·2–22·5 (29·2 × 21·5).

LAYING DATES: South Africa, all months, mainly Oct–Mar; Natal, mainly Sept–Jan; Transvaal, mainly Oct–Dec.

INCUBATION: by ♂ and ♀ by day, with *c.* 5 reliefs per h; only ♀ at night. Starts with 2nd or later egg laid. Period 13–18 days, usually 17–18 days.

DEVELOPMENT AND CARE OF YOUNG: brooded and fed by both parents by day and brooded by ♀ at night. Fed with larvae and small, soft insects for 1st 5–6 days, thereafter with worms and large, hard insects (dragonflies, beetles, bugs, butterflies), from day 10 with some fruit pulp included (e.g. mango, India). Early-hatching nestlings grow faster than later ones in same brood. Nestling period 22 days (young leaving nest before they can fly) to 35 days; young out of nest depend on parents for several weeks (Cramp and Perrins 1994). 1st generation of tail feathers are pointed and are moulted *c.* 2 weeks after bird leaves nest (Brooke 1976).

BREEDING SUCCESS/SURVIVAL: few African data. In India and Bangladesh hatching rate 92–98% and proportions of nestlings reared to eggs laid 61–76%; 1st broods more successful (75%) than 2nd (57–67%) or 3rd ones (33–40%) (Cramp and Perrins 1994).

References
Brooke, R.K. (1976).
Cramp, S. and Perrins, C.M. (1994).
Feare, C. and Craig, A. (1998).
Ginn, P.J. *et al.* (1989).
Maclean, G.L. (1993).

Family BUPHAGIDAE: oxpeckers

Semi-parasitic birds, with smooth, sleek, dark brown and buff plumage, laterally-flattened bill (red in one species, red with expanded yellow base in the other), red eyes (one species with bare yellow circumorbital skin), short, stout tarsus, toes with short, strongly curved, needle-sharp claws, and quite long, graduated tail with stiff, pointed feathers; outermost primary short and pointed. Sexes alike; juveniles have dark eyes and yellow then brown bills. Perch and nest in trees, but spend most of time clinging to hide of large even-toed and odd-toed ungulates and pachyderms, moving around all parts of them with jerky, woodpecker-like motions, using stiff tail as a prop, and scissoring or plucking with bill into fur to search out ectoparasites, wound tissue and fluid secretions. Voices hissing, buzzing and churring. Co-operative breeders. Nest placed in holes, built largely of hair taken from mammal host. Eggs starling-like.

Generally treated as a subfamily, Buphaginae, in family Sturnidae (e.g. Mayr and Greenway 1962, White 1962, Short *et al.* 1990, Feare and Craig 1998). Sibley and Monroe (1990) make the affinity even closer, on the basis of DNA–DNA hybridization data, reducing oxpeckers to the same tribe (Sturnini) as starlings. They certainly originated as starlings, and have the same feather melanin granule structure as many other starlings; oxpeckers also have the same scissoring movement of jaws and similar jaw musculature as Asian starling *Scissirostrum dubium*. However, in our opinion their distinctions from true starlings (foot and tail morphology, foraging behaviour, incipient or actual parasitism) are too great for acceptable accommodation in a single family. Association with large-mammal hosts runs deep into oxpeckers' lives: they travel with the herds, feed almost exclusively on their ectoparasites, rest and preen, court and copulate on them and may sleep on them at night, and pluck their fur and use their dung to make their nests. Recognizing that this commensal or semi-parasitic association is developed to an unique degree amongst birds worldwide, with distinctive behavioural, reproductive and morphological adaptations, we follow those authorities (e.g. Clancey 1980b, Irwin 1981) who raise the oxpeckers to family rank.

Endemic; single genus, *Buphagus*.

Genus *Buphagus* Brisson

Characters those of the family. 2 species, *B. erythrorhynchus* in eastern Africa (Eritrea to Natal) and *B. africanus* in western Africa and sparser than *B. erythrorhynchus* in eastern and southern Africa. In late 19th century *B. africanus* was widespread in Cape Prov., South Africa, but was extinct there by 1914 because of big game hunting; in 2nd half of 20th century the range and abundance of both species has been greatly reduced by the increasing use of arsenical dips and acaricides on domestic cattle. Where their ranges overlap, *B. africanus* associates with thinner-furred ungulates, particularly African Buffalo *Syncerus caffer* and *B. erythrorhynchus* with thicker-furred ones, particularly Giraffes *Giraffa camelopardalis*. Both use pachyderms, but tend to avoid mammals with very dense or long fur. Most ungulates are quite indifferent to oxpeckers, even when they cling around nostrils, eyes and ears (**A**); however Elephants *Loxodonta africana* and some ungulates, including Steinbuck *Raphicerus campestris*, reedbucks *Redunca* spp., Tsessebe *Damaliscus lunatus* and Kongoni *Alcephalus busephalus*, and in some regions Waterbuck *Kobus ellipsiprymnus*, will not tolerate oxpeckers.

Formerly much controversy amongst hunters, veterinarians and stockmen about the economic worth of oxpeckers; latterly, much interest in translocation methodology. These topics, and the birds' extraordinary lives, have spawned an unusually large literature (see Bibliography – Family Buphagidae, and Feare and Craig 1998).

Plate 36
(Opp. p. 543)

Buphagus africanus **Linnaeus. Yellow-billed Oxpecker. Piqueboeuf à bec jaune.**

Buphaga africana Linnaeus, 1766. Syst. Nat., ed. 12, 1, p. 154; Senegal.

Range and Status. Endemic resident. Still common in many national parks and wildlife areas, but numbers have declined in southern Africa and probably throughout range, because of eradication of large ungulates and widespread use of acaricide cattle-dips.

In W Africa occurs seasonally north in Mauritania to Nouakchott, in Mali to great bend of R. Niger, and Chad to Ouadi Rime-Ouadi Achim Faunal Reserve; commonest in soudanian savannas, almost exclusively in association with livestock (in virtual absence of wild ungulates in that

Buphagus africanus

zone); occurs south to N Guinea-Bissau and NW Guinea; old record from Sierra Leone; guinean savannas south to Maroué Nat. Park in Ivory Coast (and possibly in Täi Forest – see below); Benue Plain and Adamawa Plateau (where common) in Cameroon, with a forest record at Bitye; old record from Bioko (Dowsett and Dowsett-Lemaire 1993a); common in Bamingui-Bangoran Nat. Park and Varkaga area, Central African Republic, and another race in Dzanga Res. in extreme S; and SW and S Sudan. Hardly any records in Burkina Faso or Benin, but occurs in Arli-Pendjari Nat. Park and frequent near E Benin border in Nigeria (Borgu). Common in W Eritrea, from Sudan border to foot of escarpment near Cheren (= Keren, 38°25′E).

Range in Mbini, Gabon, Congo, Zaïre, Angola, Ethiopia, Uganda, Rwanda, Burundi and eastern and southern Africa now greatly fragmented, as shown (Zimmerman *et al.* 1996, N.E. Baker, pers. comm., R.J. Dowsett, pers. comm., Harrison *et al.* 1997, Clancey 1996 and other sources). Namibia, common in Caprivi (1·5 times as common as Red-billed Oxpecker *B. erythrorhynchus*). Botswana, common in Okavango, Linyanti and Chobe regions, south to Nxai Pan and Maun, otherwise sparse; once at Selebi Phikwe. Zimbabwe, scarce, but common in Hwange, Chizarira and Gonarezhou Nat. Parks and Matetsi; re-introduced successfully to Matobo Nat. Park. South Africa, became extinct about 1914; re-introduced to Hluhluwe Game Res., Zululand, in 1986, and has re-established itself in Kruger Nat. Park.

Description. *B. a. africanus* Linnaeus: range of species except Gabon, W Zaïre, Congo. ADULT ♂: head, neck, mantle and scapulars dark brown with olivaceous tinge, paling through back to buff or yellowish buff on rump and uppertail-coverts. Tail dark greyish brown, inner vanes of T3–T6 rufous; T6 shortfall 15–24 mm. Dark brown on chin and throat pales to ochreous brown on breast and warm buff on belly and flanks; thighs darker buff-brown, vent and undertail-coverts slightly paler buff than belly. Wings dark brown, P5–P8 emarginated, P7–8 longest, P6 and P9 1–2 mm shorter; underside of remiges shiny dark brown; underwing-coverts and axillaries blackish brown. Bill bright red, base of upper mandible yellow, basal half of lower mandible deep, swollen, and bright yellow; mouth blood red; eyes red; legs and feet brown. Sexes alike. SIZE (10 ♂♂, 13 ♀♀, W Africa, southern Africa): wing, ♂ 118–127, ♀ 112–131, unsexed (n = 22) 128·5–136 (131); tail, ♂ 88–98, ♀ 88–98, unsexed 88–105; bill, ♂ 14–16, ♀ 14–16, unsexed 14–18; depth of bill at nostrils, unsexed (n = 6) 7·4–8·4 (7·9), depth of lower mandible at gonys, unsexed (n = 6) 5·5–6·2 (5·7); tarsus ♂ 21–23, ♀ 20–22, unsexed 20–23. WEIGHT: 2 ♂♂ 57, 60, ♀ (n = 5) 53–71.

IMMATURE: juvenile like adult but feathers from forehead to scapulars and uppertail-coverts, also throat and upper breast, finely shadow-barred with black; upperparts lack olivaceous tinge; rump buffy; throat dark brown; underparts with slight orange tinge; at fledging, bill yellow, pale at tip, lemon-yellow at gape; later, dusky yellowish; eyes dark brown, legs and feet pale blue-grey, soles off-white. Tail short.

NESTLING: orange skin with silvery grey plumes along midline of back; prominent yellow gape flanges; mouth deep orange (Mundy and Cook 1975).

B. a. langi Chapin: Gabon, W Zaïre, Congo; race of population in Luanda area and Quissama Nat. Park (also Cambundi-Catempo), NW Angola, uncertain. Darker than nominate race, with greyer rump.

Field Characters. Length 19–23 cm. A brown and buff bird always on or around *large mammals*; tail long, stiff and pointed. The only oxpecker in W Africa, but both species occur together in E Africa, even on the same animal. Flying birds separated at a distance from Red-billed Oxpecker *B. erythrorhynchus* by *pale rump*, but Yellow-billed has distal half of bill red, and confusion easy when heads are partly obscured by an animal's fur. Base of bill *yellow and swollen*, eye red with very *narrow* yellow eye-ring. Immature like adult but finely barred black on upperparts, throat and breast; bill yellow at fledging, becoming brown; eye dark.

Voice. Tape-recorded (91–99, 104, B, F, HAZ, HOR, PAY, WALK). Long buzzy tearing and hissing calls, 'beeezzzz', 'dzzeeeeow', 'zzhaaaaa-zzhaaaa'; shorter 'dzzit', 'jeet', 'jirri' and lower 'zzhoo'; also a dry clicking trill.

General Habits. Inhabits well-wooded savannas, flood-plains and valleys with plenty of bushes and trees and substantial populations of large ungulates, or dry savannas with woody growth at least along watercourses and herds of domestic stock. In S Cameroon lives in forest, associating with small Forest Buffalo *Syncerus caffer nanus*, which inhabits swampy parts of tall rain forest. 'An apparently undescribed small oxpecker has been seen on buffaloes in the undergrowth of Taï forest' (Ivory Coast: Thiollay 1985). In Eritrea perches in thorn trees when not on ungulates.

Keeps in groups of 4–6 birds all year and in flocks of up to 20 outside breeding season. Forages on large mammals, particularly Buffalo and Giraffe *Giraffa camelopardalis*, also (in descending order of preference, Zimbabwe): White Rhino *Ceratotherium simum*, Eland *Taurotragus oryx*, Common Zebra *Equus quagga*, Sable Antelope *Hippotragus niger* (**A**), Brindled Gnu (Blue Wildebeest) *Connochaetes taurinus*,

A

Impala *Aepyceros melampus*, Common Warthog *Phacochoerus africanus* (and Greater Kudu *Tragelaphus strepsiceros*, Roan Antelope *Hippotragus equinus* and Hippo *Hippopotamus amphibius*); of domestic animals, prefers (in descending order) donkeys, horses, cows and goats (also sheep, pigs and camels). Use of Elephants *Loxodonta africana* extremely rare; only 8 records of oxpecker spp. on them, at least 4 of Yellow-billed, nearly all involving sick elephants (Mundy and Haynes 1996). Up to at least 20 birds perch on single Buffalo, well spaced out on all parts – muzzle, around eyes (**B**), ears and horns, on back and rump, neck, sides, belly, upper parts of legs, anal-genital area, but not tail. Spends about one-third of foraging time in ears, often so deep that only tail shows (P. Weeks *in* Roseneath 1977). Bird resting but not feeding sits on mammal's back and may preen. Foraging bird clings close, belly to fur, legs often hard to see, tail depressed onto mammal's hide and used as a prop; bird's movements both nimble and jerky, often sideways or backwards; uses sharp claws as hooks to cling upside down under belly of e.g. Giraffe; makes long drop downward from shoulder to foreleg; clings to upper part of walking Giraffe's foreleg or hindleg and can even feed there if Giraffe not walking too fast, but flies away if host breaks into a run. Probes and pecks towards skin; plucks quite vigorously at wound tissue; nibbles at watery secretions to drink from them.

Uses host for protection, retreating to far side if person intrudes and peering at intruder from other side of host's head, neck or trunk. If alarmed, tends to move up host to its back where several birds gather, keeping still, with foreparts raised and bill inclined at 45°, then all may fly away, calling, to alight in distant treetop. Often flies high, flock sometimes circling and returning to animal it has just left; flight can be fast and agile (Whyte *et al.* 1987). Roosts at night on host, with up to 9 birds on one animal, mainly Giraffes, also Greater Kudu and Roan Antelope (Maclean 1993); in N Nigeria roosts in dead *Acacia* trees, at least from Jan to Aug (a period encompassing the breeding season), 10–200 oxpeckers mixing with Greater and Lesser Blue-eared and Long-tailed Glossy Starlings *Lamprotornis chalybaeus*, *L. chloropterus* and *L. caudatus*; av. (n = 14 nights) 57 oxpeckers at a roost; birds very noisy when assembling at roost (Mundy and Cook 1975).

Food. Largely ticks *Amblyomma, Boophilus decoloratus* and *Rhipicephalus simus*, also semi-parasitic hippoboscid flies, lice *Damalina bovis*, fleas *Haeamopius suis* and other insects; scurf, wound tissue, bodily secretions. Estimated to eat 13,600 tick larvae or 109 engorged *B. decoloratus* ticks each day (Stutterheim *et al.* 1988). Eats a large amount of ear wax (Roseneath 1997). 8 stomachs contained ticks, (beetle) legs, a seed, blood and a good deal of hair. In Gambia birds nesting near tied cattle regularly carried billfuls of fresh red meat to nestlings (Barlow *et al.* 1997). In Senegal, regularly drinks cows' milk from receptacles left unattended by herdsmen (Ndao 1999).

Breeding Habits. Solitary nester; co-operative breeder, monogamous pair helped by 1–2 other adults. Courtship display takes place on back of large ungulate; chattering ♂ circles around ♀ with his bill open and wings outspread and drooping; copulation occurs also on ungulate's back. Single instance of viable hybridization with Red-billed Oxpecker.

NEST: untidy pad of grass, straws, feathers and much hair taken from mammal hosts; 1 nest was simple pad of white goat hair, 50 mm thick. Sited in natural hole in tree, e.g. 8 m up in large *Adansonia digitata*, or in *Acacia albida* where 1 nest was in vertical cavity, diam. 90, depth 500, in trunk (Mundy and Cook 1975), or in *Pseudolachnostylis maprouneifolia*, where chamber was *c.* 33 cm deep (Dowsett 1965). In Gambia, nests in hollow palm-log fenceposts.

EGGS: 2–3. Laid at daily intervals. White or pale blue, immaculate, or lightly or heavily spotted and blotched in various shades of red-brown and lilac-grey. SIZE: (n = 5) 23·4–26·6 × 16·6–18·0 (24·8 × 17·0). WEIGHT: *c.* 45 g.

LAYING DATES: Mauritania, June–Sept; Senegal, June–Sept, mainly Aug–Sept; Gambia, (nest-building completed in late May; fledged young late July); Sudan, May–June; E Africa, Region A, Apr–May, Region C, Dec, Feb, June, Region D, Oct, Jan, Mar; Zambia, Dec; Zimbabwe, Oct–Mar; Botswana, Oct; Transvaal, (nestlings, Dec).

B

INCUBATION: begins with first egg laid. Period at least 13 days.

DEVELOPMENT AND CARE OF YOUNG: mouth becomes scarlet by 15 days. Size at fledging (n = 1), wing 97·5, tail 47·0, bill to skull 20·0, tarsus 24·0. Young fed by both parents and their helpers; when near to fledging, young are fed noisily at tree-hole entrance. Nestling period *c.* 25 days.

References
Dowsett, R.J. (1965).
Feare, C. and Craig, A. (1998).
Hall-Martin, A.J. (1987).
Mundy, P.J. and Cook, A.W. (1975).
Stutterheim, C.J. and Brooke, R.K. (1981).
Whyte, I.J. *et al.* (1987).

Buphagus erythrorhynchus (Stanley). Red-billed Oxpecker. Piqueboeuf à bec rouge.

Plate 36
(Opp. p. 543)

Tanagra erythrorhyncha Anon. = Stanley, 1814. In Salt's Travels Abyssinia, app. 4, p. 59; Ethiopia.

Range and Status. Endemic resident, Eritrea and Somalia to Botswana and Natal, with highly fragmented range as mapped (sources including Ash and Miskell 1998, N.E. Baker, pers. comm. 1998, Tanzania; R.J. Dowsett, pers. comm. 1988, Zambia; and Harrison *et al.* 1997, Namibia, Botswana, Zimbabwe, South Africa and Lesotho). East to coast but apparently absent from most of Mozambique; common in W Somalia (breeding to *c.* 46°E) but sparse in E, east to 49°E. West to Nile in S Sudan and N Uganda, Akagera Nat. Park (Rwanda), Upemba Nat. Park (Zaïre), Caprivi Strip (Namibia), Cubango valley (Angola/Namibia) upstream to about Cuangar (and record at 14°E on Cunene R.: Hall and Moreau 1970) and *c.* 24°E in extreme N Cape Prov., South Africa.

Locally common, but much less abundant than formerly and range has contracted substantially: *cf.* Transvaal point-plotted ranges mapped by Tarboton *et al.* (1987) and Harrison *et al.* (1997). Decline due to contraction of ranges of host ungulates during 20th century, rinderpest outbreaks, and widespread use of acaricides on domestic stock. Many translocations have been made, with unsuccessful re-introduction to Matobo Nat. Park in 1975 (Grobler 1979) and, later, mainly successful ones in other areas in Zimbabwe (Rockingham-Gill 1992), and South Africa in Loskop Dam, Great Fish River Res., Addo Elephant Nat. Park and Queenstown area, E Cape (Lockwood 1995, Webster 1997). Remains common wherever large wild ungulates and untreated domestic livestock are numerous; in E Africa common and widespread up to 2500 m, but scarce and local in coastal lowlands; in Kenya much more numerous than Yellow-billed Oxpecker *B. africanus*.

Buphagus erythrorhynchus

Description. ADULT ♂: head, neck, scapulars, back, rump and uppertail-coverts uniform brown. Tail dark brown, including inner vanes (cf. Yellow-billed Oxpecker *B. africanus*); T6 shortfall 20–27 mm. Chin and throat brown, paling to light greyish buffy brown on breast and buff on belly, flanks, vent and undertail-coverts; thighs darker buff-brown. Wings dark brown, P5–P8 emarginated, P7 and P8 longest, P6 and P9 2–4 shorter. Underside of remiges shiny dark brown, underwing-coverts and axillaries blackish brown. Bill scarlet; mouth blood red; eyes red; eye-ring bright yellow, wider above than below eye (**A**); legs and feet dark brown. Sexes alike. Very slight geographical variation in tone and size. SIZE: wing, ♂ 110–128 (119), ♀ 107–121 (116); tail, ♂ (n = 59) 87–107 (96·8), ♀ 84–103 (94·6); bill, ♂ (n = 61) 16·0–18·8 (17·4), ♀ (n = 55) 15·8–18·6 (17·2); depth of bill at nostrils, unsexed (n = 6) 6·3–7·5 (6·75), depth of lower mandible at gonys, unsexed (n = 5) 2·9–4·3 (3·7); tarsus, ♂ (n = 60) 20·2–26·8 (22·2), ♀ (n = 55) 20·5–26·6 (22·1). WEIGHT: ♂ (n = 61) 45–56 (51), ♀ (n = 55) 42–59 (50·5).

IMMATURE: like adult but darker, and throat sooty brown. Bill dark olive, culmen yellowish; later, entirely dark; eyes dark brown.

NESTLING: naked except for dark grey neossoptiles 1 mm long on occipital, capital, dorsal and axillary regions; blind, immobile; skin pink, eye skin grey, bill yellow, egg-tooth white.

TAXONOMIC NOTE: several subspecies described, including *invictus* Clancey (Somalia, NE Kenya; small, pale), *angolensis* da Rosa Pinto (SE Angola, large – wing, ♂♀ (n = 14) 121–136 (127)), *scotinus* Clancey and Lawson (Panda, S Mozambique), *bestiarum* Brooke and *caffer* Grote (south of Zambezi R.); we have not been able to examine all skins; no subspecies recognized by Feare and Craig (1998).

Field Characters. Length 19–22 cm. Overlaps widely with Yellow-billed Oxpecker *B. africanus*, and has similar shape, habits and voice; distinguished by *uniform brown* upperparts (no pale rump), *all-red bill* and *broad yellow* ring around *red eye*. Immature darker than adult, bill blackish with yellow base, eye dark; told from immature Yellow-billed Oxpecker by dark rump.

Voice. Tape-recorded (47, 86, 88, 91–99, B, C, F). Long hissing, tearing calls like those of Yellow-billed Oxpecker, 'zzhaaaaa' or 'tssssaaaa'; also shorter 'jaaa', 'jaaa-yik', 't'jaaa' or 'jee' and brief 'tik' or 'tzik'.

General Habits. Inhabits broad-leaved and acacia savanna woodland and bushland, of greater diversity than does Yellow-billed Oxpecker. In Eritrea, with camels on arid salt flats and with goats on central plateau, up to 1000 m; perches readily on houses and village mosques. In Botswana often in towns, villages, kraals and cattle posts if trees nearby. Spends nearly all its time on large ungulates, favouring species with manes, especially Giraffe *Giraffa camelopardalis*; also on Greater Kudu *Tragelaphus strepsiceros*, Sable Antelope *Hippotragus niger*, Hippo *Hippopotamus amphibius*, Nyala *Tragelaphus angasi*, Impala *Aepyceros melampus*, Buffalo *Syncerus caffer*, Eland *Taurotragus oryx*, Common Zebra *Equus quagga*, White Rhino *Ceratotherium simum*, Black Rhino *Diceros bicornis*, cattle, horses, camels, donkeys, goats and pigs. Use of Elephants *Loxodonta africana* extremely rare; only 8 records of oxpecker spp. doing so including at least 2 of Red-billed, nearly all involving sick elephants (Mundy and Haynes 1996). Uses greater range of hosts than does Yellow-billed Oxpecker (Koenig 1994), listed by Feare and Craig (1998).

Occurs in pairs or parties of 4–8 birds year round; often in larger flocks, of up to 15 or 20, on herd of ungulates or even on a single Giraffe. Highly social. Sometimes mixes with Yellow-billed Oxpeckers and birds of each species can feed on same host animal (Attwell 1966) (and they have hybridized). Generally much commoner than Yellow-billed Oxpecker where both occur, despite Yellow-billed being larger and dominating Red-billed. Clambers all over head and body of host, several birds keeping well spaced on back and rump, neck, sides, belly and upper parts of legs; probes into nostrils, around eyes, ears, horns and anal-genital area. Spends about one-third of foraging time in ears, often so deep that only tail shows (P. Weeks *in* Roseneath 1977). When foraging clings close to fur, legs often hard to see, tail depressed onto host's skin or fur and used as a prop; bird's movements woodpecker-like, shuffling but nimble and jerky, often sideways or even backwards; for much of the time oxpecker's body is in vertical position with head up, but bird often moves horizontally, head sideways, and sometimes clings upside down under belly of e.g. Giraffe, using needle sharp claws to hang on; clings to upper part of walking Giraffe's foreleg or hindleg and can even feed there if Giraffe not walking too fast, but flies away if host breaks into a run; however, known to stay on galloping Greater Eland. Thrusts bill into fur, sometimes putting head on side (i.e. one eye against fur), and works mandibles with rapid nibbling action often described as scissorlike; nibbles at watery secretions; plucks quite vigorously at wound tissue and enlarges small abrasions. Inserts bill and whole head into nostrils and whole body into ears. Sometimes hawks flying insects. Drinks regularly throughout day at puddles; flock arriving to drink perches on highest vantage point near puddle, then flies down to alight at edge of water; if ungulates already drinking there, flock flies down onto them then climbs down necks and legs to reach the water.

Comes to bird tables and bird baths. Uses tail as a prop not only on host but on tree (**A**).

Uses host for protection, retreating to far side if person intrudes and peering at intruder from other side of host's head, neck or trunk. If alarmed, tends to move up host to its back where several birds gather, keeping still, with foreparts raised and bill inclined at 45°, then all may fly away, calling, to alight in distant treetop, perching upright. Often flies high, flock sometimes circling and returning to animal it has just left. Flight fast, direct, slightly undulating, and agile (Whyte *et al.* 1987). Bird resting but not feeding sits on horizontal surface like the shoulders or rump, where it may preen. Roosts at night in palms *Hyphaene natalensis* and *H. benguellensis* and communally in reedbeds; sometimes roosts on large ungulates (not Buffalo: Maclean 1993); sleeps in crevice or hole in tree or cliff face, or hole in stone or mud wall of building, under eaves, or in stone wall around paddock (van Someren 1956).

Not migratory, but highly mobile locally, moving with herd of ungulates, flying some distance between herds and water and roosting localities, and travelling afar to find new herds. In study in Kruger Nat. Park, South Africa, no marked bird found to move further than 8 km; but after being translocated birds moved 87 and 170 km back to point of capture (Lockwood 1995).

Food. Ticks (*Rhipicephalus appendiculatus*, *R. evertsi*, *R. simus*, *Boophilus annulatus*, *Amblyomma variegatum*), blood-sucking flies (Diptera, including Hippoboscidae), hairs, blood clots, wound tissue, scurf, secretions and lice. Of 58 stomachs examined, 55 contained 2291 ticks, on average 41 each, and 44 contained flies (including 2 stomachs with nothing but flies) (Moreau 1933). Eats a large amount of ear wax (Roseneath 1997). May eat leeches when foraging on hippos. Food given to nestlings 46% ticks, 19% flies and 35% hair and epidermal tissue including scurf cells (Bezuidenhout and Stutterheim 1980).

Breeding Habits. Solitary nester; monogamous, co-operative breeder. Up to 3 helpers, of both sexes; some pairs do not have helpers. All birds in a group help in selecting nest site, building nest and feeding chicks; however, only 1 ♂ and 1 ♀, the presumed parents, incubate. Not territorial, but nesting tree is defended against other oxpeckers. Home range of one breeding group was 7·0 km² and of another c. 27 km². Pre-copulatory display: ♂ and ♀, perching on back of an ungulate (or once on the ground), open bills, spread out wings and tail horizontally and move slowly around each other in circles, wings vibrating rapidly; to copulate, ♀ closes wings, pushes under ♂, crouches and lifts tail. Courtship feeding seen once, on back of giraffe, involving simple transfer of food morsel, without begging. In 'Wing-vibration' display, bird bends forward, vibrates wings and spreads tail; given by bird in front of newly-found nest hole, by bird arriving at nest to relieve incubating mate, by one or both birds when arriving simultaneously at nest, and in alarm when predator or vehicle approaches nest. In 'Open-wing' display bird bends forward and spreads wings horizontally; given by bird after nest-relief, after feeding nestlings, and when predator approaches nest. Prospecting flights to tree holes are made only from host ungulate. Double- or triple-brooded; can raise 3 broods in 176 days (Stutterheim 1982b).

NEST: pad of dung, grass and rootlets with thick layer of hair on top and quite deep depression in centre. Hair composes 93% of nest material and forms felt-like mass up to 100 mm thick. Size (n = 8): int. diam. of cup 60–100 (80), depth of cup 25–45 (30). Hair is collected from hosts, including Impala, Greater Kudu and oxen; bird grasps tuft in bill, on host's back or shoulders, and jerks it out with backward pull of head. Elephant dung and dry bits of grass are collected on ground. When starting nest, bird uses dung to fill any small holes and cracks. Built on floor of knot hole, stump hole or other tree cavity or crevice, in Kruger Nat. Park, South Africa, mainly in *Combretum imberbe*, also *Spirostachys africanus*, *Colophospermum mopane*, *Kigelia pinnata* and *Acacia nigrescens* in open parkland; nest chamber often up to 50 cm below bottom of entrance hole or slit; also in holes in metal telegraph poles (*Bee-eater* 50, 1999: 2). Chamber width av. 143 mm, depth below entrance lip av. 359, entrance hole av. 75 wide and 85 high in branch of av. circumference 1040; height above ground (n = 43) 1·2–15 (8·1) m (Stutterheim 1982b). Also sometimes in cavity amongst rocks or in hole in wall of building. Some nest holes are used repeatedly in successive years (Kenya, South Africa).

EGGS: 1–5, usually 2–3, av. (n = 85) 2·7. Laid before 09h00 daily until clutch complete. Matt; white, cream or very pale blue or pink, finely and heavily speckled with red-brown and lilac-grey, or evenly speckled with pink and maroon on top of grey and lavender. SIZE: (n = 47) 22·5–26·5 × 15·8–18·6 (24·0 × 17·2).

A

LAYING DATES: Eritrea, (nestlings Feb); Ethiopia, Jan, Mar–Apr, July–Aug; Somalia, June, Sept (and fledglings Mar; and nesting Apr); E Africa, Region A, May, Region B, Mar–Apr, June, Dec, Region C, July; Rwanda, Apr; Zambia, Dec, Feb; Zimbabwe, Nov–Dec; Botswana, Nov; Transvaal, Oct–Mar; Natal, Sept–Feb; E Cape, (building nest Nov: *Bee-eater* 50, 1999: 2).

INCUBATION: by ♂ and ♀ by day and ♀ at night. Period (n = 6) 12–13 (12·6) days.

DEVELOPMENT AND CARE OF YOUNG: chick crawls by day 4, stands upright by day 13; on day 22 a chick jumped from nest and flew to tree 20 m away; bill soon turns from yellow to olive then dark brown, starting to redden at 4 months and becoming wholly red at 7 months; adult plumage acquired at 6 months. Eggshells removed from nest cavity. Young fed by both parents and up to 3 adult helpers. 23 feeding visits per day at first, later 57 per day. Food is brought inside mouth, adult's cheeks bulging conspicuously, or sometimes in bill. After each feed, faecal sacs collected and dropped c. 30 m away. Growth curves detailed by Stutterheim (1982b). Chick hisses loudly in alarm. Nestling period: (n = 4) 26–30 days. Fledgling follows parent or helper around, begging loudly and fanning wings; makes scissoring movements with bill against branch. Fledgling just outside nest returns inside if alarmed.

BREEDING SUCCESS/SURVIVAL: of 72 eggs laid only 29 produced fledglings. Causes of losses unknown, except that Wahlberg's Eagle *Aquila wahlbergi* took a fledgling and a Ground Hornbill *Bucorvus leadbeateri* tried but failed to reach 3 nestlings (Stutterheim 1982b).

References
Feare, C. and Craig, A. (1998).
Grobler, J.H. (1980).
Moreau, R.E. (1933).
Stutterheim, C.J. (1977, 1980a, b, 1981a, b, 1982a, b).
Stutterheim, C.J. and Panagis, K. (1985a, b, 1987).
Wood, P.A. (1994).

BIBLIOGRAPHY

The bibliography is in three parts: (1) general and regional references, (2) references by family or group of genera (journal titles abbreviated) and (3) acoustic references. Together, lists (1) and (2) comprise the publications cited in the text and all other significant works consulted; titles in (1) are generally not repeated in (2), except for some family monographs. If a reference cited in the text does not appear in the appropriate systematic list (2), it will be found in (1).

General and Regional References

Ali, S. and Ripley, S.D. (1972–1974). 'Handbook of the Birds of India and Pakistan', Vols 5, 7, 9, and 10. Oxford University Press, Bombay.

Allan, D.G. and Davies, G.B. (1999). The birds (Aves) of the middle Komati River, Swaziland. *Durban Museum Novitates* **24**, 22–42.

Allport, G., Ausden, M., Hayman, P.V., Robertson, P. and Wood, P. (1989). 'The Conservation of the Birds of Gola Forest, Sierra Leone'. *ICBP Study Report* **38**, 1–104. International Council for Bird Preservation, Cambridge.

Amadon, D. (1953). Avian systematics and evolution in the Gulf of Guinea. *Bulletin of the American Museum of Natural History* **100**, 393–452.

Ames, P.L. (1975). The application of syringeal morphology to the classification of the Old World Insect Eaters (Muscicapidae). *Bonn. Zool. Beitr.* **26**, 107–134.

Archer, G. and Godman, E.M. (1961). 'The Birds of British Somaliland and the Gulf of Aden', Vol. 4. Oliver and Boyd, Edinburgh.

Ash, J.S. (1969). Spring weights of trans-Saharan migrants in Morocco. *Ibis* **111**, 1–10.

Ash, J.S. (1980). Migrational status of Palaearctic birds in Ethiopia. *Proceedings of the Fourth Pan-African Ornithological Congress*, pp. 199–208. Southern African Ornithological Society, Johannesburg.

Ash, J.S. (1983). Over fifty additions to the Somali list including two hybrids, together with notes from Ethiopia and Kenya. *Scopus* **7**, 54–79.

Ash, J.S. (1985). Midwinter observations from Djibouti. *Scopus* **9**, 43–49.

Ash, J.S. (1990). Additions to the avifauna of Nigeria, with notes on distributional changes and breeding. *Malimbus* **11**, 104–116.

Ash, J.S., Dowsett, R.J. and Dowsett-Lemaire, F. (1989). New ornithological distribution records from eastern Nigeria. *Tauraco Research Report* **1**, 13–27.

Ash, J.S. and Miskell, J.E. (1998). 'Birds of Somalia'. Pica Press, Robertsbridge.

Ash, J.S. and Sharland, R.E. (1986). 'Nigeria: Assessment of Bird Conservation Priorities'. ICBP/Nigerian Conservation Foundation, Cambridge.

Aspinwall, D.R. (1979). Bird notes from the Zambezi district, North-Western Province. *Zambian Ornithological Society Occasional Paper*, **2**, 1–60.

Atkinson, P., Peet, N. and Alexander, J. (1991). The status and conservation of the endemic bird species of São Tomé and Príncipe, West Africa. *Bird Conservation International* **1**, 255–282.

Backhurst, G.C., Britton, P.L. and Mann, C.F. (1973). The less common Palaearctic migrant birds of Kenya and Tanzania. *Journal of the East Africa Natural History Society and National Museum* **140**, 1–38.

Baker, N.E. and Hirslund, P. (1987). Minziro Forest Reserve: an ornithological note including seven additions to the Tanzanian list. *Scopus* **11**, 9–12.

Balança, G. and Visscher, M.-N. de (1997). Composition et évolution saisonnière d'un peuplement d'oiseaux au nord du Burkina Faso (nord-Yatenga). *Malimbus* **19**, 68–94.

Balchin, C.S. (1988). Recent observations of birds from the Ivory Coast. *Malimbus* **10**, 201–206.

Bannerman, D.A. (1922). The birds of southern Nigeria. *Revue Zoologique Africaine* **9**, 254–426.

Bannerman, D.A. (1936, 1939, 1948, 1951). 'The Birds of Tropical West Africa', Vols 4, 5, 6 and 8. The Crown Agents for the Colonies, London.

Barlow, C., Wacher, T. and Disley, T. (1997). 'A Field Guide to Birds of Gambia and Senegal'. Pica Press, Robertsbridge.

Barnes, J. and Bushell, B. (1989). 'Birds of the Gaborone area'. Botswana Bird Club, Gaborone.

Barreau, D., Bergier, P. and Lesne, L. (1987). 'L'Avifaune de l'Oukaimeden, 2200–3600 m (Haut Atlas, Maroc)'. *L'Oiseau et la Revue Française d'Ornithologie* **57**(4), 307–367.

Basilio, A. (1963). 'Aves de la Isla de Fernando Poo'. Editorial Coculsa, Madrid.

Bates, G.L. (1909). Field-notes on the birds of southern Kamerun, West Africa. *Ibis* **1909**, 25–74.

Bates, G.L. (1911). Further notes on birds of southern Cameroon. *Ibis* **1911**, 479–545, 581–631.

Bates, G.L. (1927). Notes on some birds of Cameroon and the Lake Chad region: their status and breeding-times. *Ibis* **1927**, 1–64.

Bates, G.L. (1930). 'Handbook of the Birds of West Africa'. John Bale, Sons and Danielsson, London.

Bates, G.L. (1934). Birds of the southern Sahara and adjoining countries in French West Africa. *Ibis* **1934**, 439–466.

Beasley, A.J. (1995). The birds of the Chimanimani Mountains. *Honeyguide* **41**, Supplement 1, 1–57.

Beesley, J.S.S. (1972). Birds of the Arusha National Park, Tanzania. *Journal of the East Africa Natural History Society and Natural History Museum* **132**, 1–30.

Beesley, J.S.S. and Irving, N.S. (1976). The status of the birds of Gaborone and its surroundings. *Botswana Notes and Records* **8**, 231–261.

Benson, C.W. (1937). Miscellaneous notes on Nyasaland birds. *Ibis* **1937**, 551–582.

Benson, C.W. (1940, 1941). Further notes on Nyasaland birds (with particular reference to those of the Northern Province). Parts III and IV. *Ibis* **1940**, 583–629; **1941**, 1–55.

Benson, C.W. (1942). Additional notes on Nyasaland birds. *Ibis* **1942**, 299–337.

Benson, C.W. (1944). Notes from Nyasaland. *Ibis* **86**, 445–480.

Benson, C.W. (1946a). A visit to the Vumba Highlands, Southern Rhodesia. *Ostrich* **17**, 280–296.

Benson, C.W. (1946b, 1948). Notes on the birds of southern Abyssinia. *Ibis* **88**, 180–205, 444–461; **90**, 325–327.

Benson, C.W. (1947). Observations from the Kota-kota district of Nyasaland. *Ibis* **89**, 553–566.

Benson, C.W. and Benson, F.M. (1947). Some breeding and other records from Nyasaland. *Ibis* **89**, 279–290.

Benson, C.W. and Benson, F.M. (1949). Notes on birds from northern Nyasaland and adjacent Tanganyika Territory. *Annals of the Transvaal Museum* **21**, 155–177.

Benson, C.W. and Benson, F.M. (1977). 'The Birds of Malawi'. Montfort Press, Limbe.

Benson, C.W., Brooke, R.K., Dowsett, R.J. and Irwin, M.P.S. (1971). 'The Birds of Zambia'. Collins, London.

Betts, F.N. (1966). Notes on some resident breeding birds of southwest Kenya. *Ibis* **108**, 513–530.

Bibby, C.J., Collar, N.J., Crosby, M.J., Heath, M.F., Imboden, C., Johnson, T.H., Long, A.J., Stattersfield, A.J. and Thirgood, S.J. (1992). 'Putting Biodiversity on the Map: Priority Areas for Global Conservation'. International Council for Bird Preservation, Cambridge.

de Bie, S. and Morgan, N. (1989). Les oiseaux de la Réserve de la Biosphère 'Boucle du Baoulé', Mali. *Malimbus* **11**, 41–60.

Black, J.G., Loiselle, B.A. and Vande weghe, J.-P. (1990). Weights and measurements of some central African birds. *Le Gerfaut* **80**, 3–11.

Bonde, K. (1993). 'Birds of Lesotho'. University of Natal Press, Pietermaritzberg.

Boyer, H.J. and Bridgeford, P.A. (1988). Birds of the Naukluft Mountains: an annotated checklist. *Madoqua* **15**, 295–314.

Bretagnolle, F. (1993). An annotated checklist of north-eastern Central African Republic. *Malimbus* **15**, 6–16.

Brewster, C. (1991). Birds of the Gumare area, northwest Botswana. *Babbler* **21/22**, 12–61.

Britton, P.L. (1970). Birds of the Balovale District of Zambia. *Ostrich* **41**, 145–190.

Britton, P.L. (Ed.) (1980). 'Birds of East Africa'. East Africa Natural History Society, Nairobi.

Britton, P.L. and Dowsett, R.J. (1969). More bird weights from Zambia. *Ostrich* **40**, 55–60.

Britton, P.L. and Zimmerman, D.A. (1979). The avifauna of Sokoke Forest, Kenya. *Journal of the East Africa Natural History Society and National Museum* **169**, 1–16.

Brooke, R.K. (1966). Distribution and breeding notes on the birds of the central frontier of Rhodesia and Mozambique. *Annals of the Natal Museum* **18**, 429–453.

Brooks, T., Lens, L., Barnes, J., Barnes, R., Kihuria, J.K. and Wilder, C. (1998). The conservation status of the forest birds of the Taita Hills, Kenya. *Bird Conservation International* **8**, 119–139.

Brosset, A. (1961). Écologie des oiseaux du Maroc oriental. *Travaux de l'Institut Scientifique Chérifien, Série Zoologie* **22**, 1–155. Université Mohammed V, Rabat.

Brosset, A. (1984). Oiseaux migrateurs européens hivernant dans la partie guinéene du Mont Nimba. *Alauda* **52**, 81–101.

Brosset, A. and Erard, C. (1986). 'Les Oiseaux des Régions Forestières du Nord-Est du Gabon', Vol. 1. Société Nationale de Protection de la Nature, Paris.

Brown, C.J. (1990). Birds of the West Caprivi Strip, Namibia. *Lanioturdus* **25**, 22–37.

Brown, C.J. (1993). The birds of Owambo, Namibia. *Madoqua* **18**, 147–161.

Brown, L.H. and Britton, P.L. (1980). 'The Breeding Seasons of East African Birds'. East Africa Natural History Society, Nairobi.

Brown, L.H., Urban, E.K. and Newman, K. (1982). 'The Birds of Africa', Vol. I. Academic Press, London.

Browne, P.W.P. (1982). Palaearctic birds wintering in southwest Mauritania: species, distributions and population estimates. *Malimbus* **4**, 69–92, 104.

Brunel, J. (1958). Observations sur les oiseaux du Bas-Dahomey. *L'Oiseau et la Revue Française d'Ornithologie* **28**, 1–38.

Brunel, J. (1982). Palearctic birds wintering in southwest Mauritania: species, distributions and population estimates. *Malimbus* **4**, 69–92, 104.

Buckley, P. and McNeilage, A. (1989). An ornithological survey of Kasyoha-Kitomi and Itwara forests, Uganda. *Scopus* **13**, 97–108.

Bundy, G. (1976). 'The Birds of Libya'. BOU Check-list 1. British Ornithologists' Union, London.

Burgess, N.D., Huxham, M.R., Mlingwa, C.O.F., Davies, S.G.F. and Cutts, C.J. (1991). Preliminary assessment of forest birds in Kiono, Pande, Kisiju and Kiwengoma coastal forests, Tanzania. *Scopus* **14**, 97–106.

Butysnki, T.M. and Kalina, J. (1993). Further additions to the known avifauna of the Impenetrable (Bwindi) Forest, southwestern Uganda (1989–1991). *Scopus* **17**, 1–7.

Campbell, B. and Lack, E. (1985). 'A Dictionary of Birds'. T. and A.D. Poyser, Calton.

Carroll, R.W. (1988). Birds of the Central African Republic. *Malimbus* **10**, 177–200.

Carswell, M. (1986). Birds of the Kampala area. *Scopus* Special Supplement **2**, 1–89.

Cave, F.O. and Macdonald, J.D. (1955). 'Birds of the Sudan'. Oliver and Boyd, Edinburgh.

Cawkell, E.M. and Moreau, R.E. (1963). Notes on birds in The Gambia. *Ibis* **105**, 156–178.

Chapin, J.P. (1953, 1954). The birds of the Belgian Congo, Vols 3 and 4. *Bulletin of the American Museum of Natural History* **75A** and **75B**.

Chapin, R.T. (1978). Brief accounts of some central African birds, based upon the journals of James Chapin. *Revue de Zoologie Africaine* **92**, 805–836.

Chappuis, C. (1974, 1975, 1978, 1979). Illustration sonore de problèmes bioacoustiques posés par les oiseaux de la zone éthiopienne. *Alauda* **42**(4), 467–500; **43**(4), 427–474; **46**(4), 327–355; **47**(3), 195–212.

Chappuis, C. (1980). List of sound-recorded Ethiopian birds. *Malimbus* **2**, 1–15, 82–98.

Chappuis, C. (1986). Revised list of sound-recorded Afrotropical birds. *Malimbus* **8**, 25–39, 79–88.

Cheesman, R.E. and Sclater, W.L. (1935–1936). On a collection of birds from north-western Abyssinia. *Ibis* **1935**, 594–622; **1936**, 163–197.

Cheke, R.A. and Walsh, J.F. (1996). 'The Birds of Togo'. BOU Check-list 14. British Ornithologists' Union, Tring.

Cheke, R.A., Walsh, J.F. and Fishpool, L.D.C. (1985). Bird records from the Republic of Niger. *Malimbus* **7**, 73–90.

Christy, P. and Clarke, W. (1994). 'Guide des Oiseaux de la Réserve de la Lopé'. ECOFAC, Libreville, Gabon.

Christy, P. and Clarke, W.V. (1998). 'Guide des Oiseaux de São Tomé et Príncipe'. ECOFAC, Libreville, Gabon.

Claffey, P.M. (1995). Notes on the avifauna of the Bétérou area, Borgou Province, Republic of Benin. *Malimbus* **17**, 63–84.

Clancey, P.A. (1964). 'The Birds of Natal and Zululand'. Oliver and Boyd, Edinburgh.

Clancey, P.A. (1971). A handlist of the birds of southern Moçambique. *Memórias do Instituto de Investigação Científica de Moçambique, Série A* **11**, 1–167.

Clancey, P.A. (1980a). On birds from the mid-Okavango Valley on the South West Africa/Angola border. *Durban Museum Novitates* **12**(9), 87–127.

Clancey, P.A. (Ed.) (1980b). 'S.A.O.S. Checklist of Southern African Birds'. Southern African Ornithological Society, Johannesburg.

Clancey, P.A. (1985). 'The Rare Birds of Southern Africa'. Winchester Press, Saxonwold.

Clancey, P.A. (1986). Endemicity in the southern African avifauna. *Durban Museum Novitates* **13**, 245–284.

Clancey, P.A. (Ed.) (1987). 'S.A.O.S. Checklist of Southern African Birds; first updating report'. Southern African Ornithological Society, Johannesburg.

Clancey, P.A. (1996). 'The Birds of Southern Mozambique'. African Bird Book Publishing, Westville, Kwazulu-Natal.

Clarke, G. (1985). Bird observations from northwest Somalia. *Scopus* **9**, 24–42.

Collar, N.J., Crosby, M.J. and Stattersfield, A.J. (1994). 'Birds to Watch 2. The World List of Threatened Birds'. BirdLife International, Cambridge.

Collar, N.J. and Stuart, S.N. (1985). 'Threatened Birds of Africa and Related Islands'. ICBP/IUCN Red Data Book, Part 1, 3rd edition. International Council for Bird Preservation, Cambridge.

Collar, N.J. and Stuart, S.N. (1988). 'Key Forests for Threatened Birds in Africa'. *ICBP Monograph* **3**, 102 pp. International Council for Bird Preservation, Cambridge.

Colston, P.R. and Curry-Lindahl, K. (1986). 'The Birds of Mount Nimba, Liberia'. British Museum (Natural History), London.

Cooper, J. (1972). A check list of the birds of the Zambezi Valley from Kariba to Zumbo. *South African Avifauna Series* **85**, 1–44. Percy Fitzpatrick Institute of African Ornithology, Cape Town.

Cordeiro, N.J. (1994). Forest birds on Mt Kilimanjaro, Tanzania. *Scopus* **17**, 65–112.

Cordeiro, N.J. (1998). A preliminary survey of the montane avifauna of Mt Nilo, East Usambaras, Tanzania. *Scopus* **20**, 1–18.

Cordeiro, N.J. and Kiure, J. (1995). An investigation of the forest avifauna in the North Pare Mountains, Tanzania. *Scopus* **19**, 9–26.

Cordeiro, N.J., Lehmberg, T. and Kiure, J. (1995). A preliminary account of the avifauna of Kahe II Forest Reserve, Tanzania. *Scopus* **19**, 1–8.

Craig, A.J.F.K. (1983). Moult in southern African passerine birds: a review. *Ostrich* **54**, 220–237.

Cramp, S. (Ed.) (1988, 1992). 'The Birds of the Western Palearctic', Vols V and VI. Oxford University Press, Oxford.

Cramp, S. and Perrins, C.M. (Eds) (1993, 1994). 'The Birds of the Western Palearctic', Vols VII, VIII and IX. Oxford University Press, Oxford.

Curry-Lindahl, K. (1981). 'Bird Migration in Africa', Vols 1 and 2. Academic Press, London.

Cyrus, D. and Robson, N. (1980). 'Bird Atlas of Natal'. University of Natal Press, Pietermaritzburg.

Dean, W.R.J. (1971). Breeding data for the birds of Natal and Zululand. *Durban Museum Novitates* **9**(6), 59–91.

Dean, W.R.J. (1974). Breeding and distributional notes on some Angolan birds. *Durban Museum Novitates* **10**(8), 109–125.

Dean, W.R.J. and Huntley, M.A. (1984). An updated list of the birds of Angola. Unpublished manuscript (available Tierberg Karoo Research Centre, Prince Albert, South Africa).

Dean, W.R.J., Huntley, M.A., Huntley, B.J. and Vernon, C.J. (1988). Notes on some birds of Angola. *Durban Museum Novitates* **14**(4), 43–92.

Dean, W.R.J. and Macdonald, I.A.W. (1981). A review of African birds feeding in association with mammals. *Ostrich* **52**, 135–155.

Dejaifve, P.-A. (1994). Contribution à l'étude de l'avifaune de la savane guinéenne du Nord-Ubangui, Zaïre. *Le Gerfaut* **84**, 63–71.

Dekeyser, P.L. (1951). Mission A. Villers au Togo et au Dahomey (1950). III. Oiseaux. *Études Dahoméennes* **5**, 47–90.

Demey, R. (1995). Notes on the birds of the coastal and Kindia areas, Guinea. *Malimbus* **17**, 85–99.

Demey, R. and Fishpool, L.D.C. (1991). Additions and annotations to the avifauna of Côte d'Ivoire. *Malimbus* **12**, 61–86.

Demey, R. and Fishpool, L.D.C. (1994). The birds of Yapo Forest, Ivory Coast. *Malimbus* **16**, 100–122.

Desfayes, M. (1975). Birds from Ethiopia. *Revue de Zoologie Africaine* **89**, 505–535.

Devillers, P. and Ouellet, H. (1993). 'Noms Français des Oiseaux du Monde'. MultiMondes, St-Foy, Québec.

Donnelly, B.G. (1985). The birds of the Matobo (formerly Matopos) National Park, Zimbabwe. *Honeyguide* **31**, 11–23.

Dowsett, R.J. (1971). The avifauna of the Makutu Plateau, Zambia. *Revue de Zoologie et de Botanique Africaines* **84**, 312–333.

Dowsett, R.J. (1985). Site-fidelity and survival rates of some montane forest birds in Malaŵi, south-central Africa. *Biotropica* **17**, 145–154.

Dowsett, R.J. (Ed.) (1989). A preliminary natural history survey of Mambilla Plateau and some lowland forests of Eastern Nigeria. *Tauraco Research Report* **1**, 1–56.

Dowsett, R.J. (Ed.) (1990). Enquête faunistique et floristique dans la Forêt de Nyungwe, Rwanda. *Tauraco Research Report* **3**, 1–140.

Dowsett, R.J. (1991). Gazetteer of zoological localities in Congo. *Tauraco Research Report* **4**, 335–340.

Dowsett, R.J. (1993). Afrotropical avifaunas: annotated country checklists. *Tauraco Research Report* **5**, 1–322.

Dowsett, R.J., Aspinwall, D.R. and Leonard, P.M. (1999). Further additions to the avifauna of Zambia. *Bulletin of the British Ornithologists' Club* **119**, 94–109.

Dowsett, R.J., Backhurst, G.C. and Oatley, T.B. (1988). Afrotropical ringing recoveries of Palearctic migrants. 1. Passerines (Turdidae to Oriolidae). *Tauraco* **1**, 29–63.

Dowsett, R.J., Christy, P. and Germain, M. (1999). Additions and corrections to the avifauna of Central African Republic. *Malimbus* **21**, 1–15.

Dowsett, R.J. and Dowsett-Lemaire, F. (1980). The systematic status of some Zambian birds. *Le Gerfaut* **70**, 151–199.

Dowsett, R.J. and Dowsett-Lemaire, F. (1984). Breeding and moult cycles of some montane forest birds in south-central Africa. *Revue d'Écologie (La Terre et La Vie)* **39**, 89–111.

Dowsett, R.J. and Dowsett-Lemaire, F. (Eds) (1991). Flore et faune du bassin du Kouilou (Congo) et leur exploitation. *Tauraco Research Report* **4**, 1–340.

Dowsett, R.J. and Dowsett-Lemaire, F. (Eds) (1993a). A contribution to the distribution and taxonomy of Afrotropical and Malagasy birds. *Tauraco Research Report* **5**, 1–389.

Dowsett, R.J. and Dowsett-Lemaire, F. (1993b). Comments on the taxonomy of some Afrotropical bird species. *Tauraco Research Report* **5**, 323–389.

Dowsett, R.J. and Dowsett-Lemaire, F. (Eds) (1997). Flore et faune du Parc National D'Odzala, Congo. *Tauraco Research Report* **6**, 1–135.

Dowsett, R.J. and Forbes-Watson, A.D. (1993). 'Checklist of Birds of the Afrotropical and Malagasy Regions'. Tauraco Press, Liège.

Dowsett, R.J., Fry, C.H. and Dowsett-Lemaire, F. (1997). 'A Bibliography of Afrotropical Birds, 1971–1990'. *Tauraco Research Report* **7**, 1–338.

Dowsett, R.J. and Prigogine, A. (1974). The avifauna of the Marungu Highlands. Hydrobiological Survey of the Lake Bangweulu Luapula River Basin 19, 1–67.

Dowsett, R.J. and Stjernstedt, R. (1973). The birds of the Mafinga Mountains. *Puku* **7**, 107–123.

Dowsett-Lemaire, F. (1983). Ecological and territorial requirements of montane forest birds on the Nyika Plateau, south-central Africa. *Le Gerfaut* **73**, 345–378.

Dowsett-Lemaire, F. (1985). Breeding productivity and the non-breeding element in some montane forest birds in Malaŵi, south-central Africa. *Biotropica* **17**, 137–144.

Dowsett-Lemaire, F. (1988). Fruit choice and seed dissemination by birds and mammals in the evergreen forest of upland Malaŵi. *Revue d'Écologie (La Terre et La Vie)* **43**, 251–285.

Dowsett-Lemaire, F. (1989). Ecological and biogeographical aspects of forest bird communities in Malaŵi. *Scopus* **13**, 1–80.

Dowsett-Lemaire, F. (1990). Eco-ethology, distribution and status of Nyungwe Forest birds (Rwanda). *Tauraco Research Report* **3**, 31–85.

Dowsett-Lemaire, F. (1997a). The avifauna of Odzala National Park, northern Congo. *Tauraco Research Report* **6**, 15–48.

Dowsett-Lemaire, F. (1997b). The avifauna of Nouabalé–Ndoki National Park, northern Congo. *Tauraco Research Report* **6**, 111–124.

Dowsett-Lemaire, F. (1997c). The avifauna of the Léfini Reserve, Téké Plateau (Congo). *Tauraco Research Report* **6**, 125–134.

Dowsett-Lemaire, F. and Dowsett, R.J. (Eds) (1989a). Enquête faunistique dans la forêt du Mayombe, et check-liste des oiseaux et des mammifères du Congo. *Tauraco Research Report* **2**, 1–51.

Dowsett-Lemaire, F. and Dowsett, R.J. (1989b). Liste commentée des oiseaux de la forêt du Mayombe (Congo). *Tauraco Research Report* **2**, 5–16.

Dowsett-Lemaire, F. and Dowsett, R.J. (1989c). Zoogeography and taxonomic relationships of the forest birds of the Cameroon Afromontane region. *Tauraco Research Report* **1**, 48–56.

Dowsett-Lemaire, F. and Dowsett, R.J. (1990). Zoogeography and taxonomic relationships of the forest birds of the Albertine Rift Afromontane region. *Tauraco Research Report* **3**, 87–109.

Dowsett-Lemaire, F. and Dowsett, R.J. (1991). The avifauna of the Kouilou basin in Congo. *Tauraco Research Report* **4**, 189–239.

Dowsett-Lemaire, F. and Dowsett, R.J. (1998a). Further additions to and deletions from the avifauna of Congo–Brazzaville. *Malimbus* **20**, 15–32.

Dowsett-Lemaire, F., Dowsett, R.J. and Bulens, P. (1993). Additions and corrections to the avifauna of Congo. *Malimbus* **15**, 68–80.

Dranzoa, C. (1997). The survival of understorey birds in the tropical rainforest of Ziika, Uganda. *Ostrich* **67**, 68–71.

Duckworth, J.W., Evans, M.I., Safford, R.J., Telfer, M.G., Timmins, R.J. and Chemere Zewdie (1992). 'A Survey of Nechisar National Park, Ethiopia'. *ICBP Study Report* **50**, 1–132. International Council for Bird Preservation, Cambridge.

Duhart, F. and Descamps, M. (1963). Notes sur l'avifaune du delta central nigérien et régions avoisinantes. *L'Oiseau et la Revue Française d'Ornithologie* **33**, No. spécial, 1–107.

Dutson, G. and Branscombe, J. (1990). 'Rainforest Birds in South-West Ghana'. *ICBP Study Report* **46**, 1–70. International Council for Bird Preservation, Cambridge.

Dyer, M., Gartshore, M.E. and Sharland, R.E. (1986). The birds of Nindam Forest Reserve, Kagoro, Nigeria. *Malimbus* **8**, 2–20.

Earlé, R. and Grobler, N.J. (1987) 'First Atlas of Bird Distribution in the Orange Free State'. National Museum, Bloemfontein.

Edington, J.M. and Edington, M.A. (1983). Habitat partitioning and antagonistic behaviour amongst the birds of a West African scrub and plantation plot. *Ibis* **125**, 74–89.

Eisentraut, M. (1963). 'Die Wirbeltiere des Kamerungebirges'. Paul Parey, Hamburg, 353 pp.

Eisentraut, M. (1968). Beitrag zür Vogelfauna von Fernando Poo und Westkamerun. *Bonner Zoologische Beiträge* **19**, 49–68.

Eisentraut, M. (1973). Die Wirbeltierfauna von Fernando Poo und Westkamerun. *Bonner Zoologische Monographien* **3**, 1–427.

Elgood, J.H., Sharland, R.E. and Ward, P. (1966). Palaearctic migrants in Nigeria. *Ibis* **108**, 84–116.

Elgood, J.H., Fry, C.H. and Dowsett, R.J. (1973). African migrants in Nigeria. *Ibis* **115**, 375–409.

Elgood, J.H., Heigham, J.B., Moore, A.M., Nason, A.M., Sharland, R.E. and Skinner, N.J. (1994). 'The Birds of Nigeria', 2nd edition. BOU Check-list 4. British Ornithologists' Union, Tring.

Elliot, C.C.H. (1972). An ornithological survey of the Kidepo National Park, northern Uganda. *Journal of the East Africa Natural History Society and National Museum* **129**, 1–31.

van den Elzen, R. and König, C. (1983). Vögel des (Süd-) Sudan: taxonomische und tiergeographische Bemerkungen. *Bonner Zoologische Beiträge* **34**, 149–196.

Erard, C. (1974). Notes faunistiques et systématiques sur quelques oiseaux d'Ethiopie. *Bonner Zoologische Beiträge* **25**, 76–86.

Etchécopar, R.D. and Hüe, F. (1967). 'The Birds of North Africa'. Oliver and Boyd, Edinburgh.

Evans, T.D. and Anderson, G.Q.A. (1993). Results of an ornithological survey in the Ukaguru and East Usambara mountains, Tanzania. *Scopus* **17**, 40–47.

Fa, J.E. (1990). 'La Conservation des Écosystèmes Forestiers de la Guinée Équatoriale'. IUCN, Gland.

Fairon, J. (1975). Contribution à l'ornithologie de l'Aïr (Niger). *Le Gerfaut* **65**, 107–134.

Farmer, R. (1979). Checklist of birds of the Ife-Ife area, Nigeria. *Malimbus* **1**, 56–64.

Feare, C. and Craig, A. (1998). 'Starlings and Mynahs'. Christopher Helm/A. & C. Black, London.

Feather, P.J. (1986). The Bulawayo garden bird survey 1973–1982. *Honeyguide* **32**, 13–33.

Field, G.D. (1974). 'Birds of Freetown Peninsula'. Fourah Bay College Bookshop, Sierra Leone.

Finlayson, C. (1992). 'Birds of the Strait of Gibralter'. T. and A.D. Poyser, London.

Frade, F. (1951). Aves coligidas pela Missão zoológica de Moçambique. *Anais da Junta das Missões Geográficas e de Investigações Coloniais, Lisboa* **6**(4), 1–220.

Frade, F. and Bacelar, A. (1959a). Catálogo das aves da Guiné portuguesa, II. *Memórias da Junta Investigações Coloniais, Lisboa* **6**, 1–294.

Frade, F. and Bacelar, A. (1959b). Catálogo das aves da Guiné portuguesa, II. *Memórias da Junta Investigações do Ultramar* **7**, 1–116.

Francis, J. and Shirihai, H. (1999). 'Ethiopia In Search of Endemic Birds'. Geometric Press, Oxford.

Frandsen, J. (1982). 'Birds of the South Western Cape'. Sable, Sloane Park, South Africa.

Friedmann, H. (1937). Birds collected by the Childs Frick Expedition to Ethiopia and Kenya Colony, Part 2. Passeres. *Bulletin of the United States National Museum* **153**, 1–506.

Friedmann, H. (1962). The Machris Expedition to Tchad, Africa: birds. *Los Angeles County Museum Contributions in Science* **59**, 1–26.

Friedmann, H. (1966). A contribution to the ornithology of Uganda. *Bulletin of the Los Angeles County Museum of Natural History* **3**, 1–55.

Friedmann, H. (1978). Results of the Lathrop Central African Expedition 1976, Ornithology. *Los Angeles County Museum Contributions in Science* **287**, 1–22.

Friedmann, H. and Loveridge, A. (1937). Notes on the ornithology of tropical East Africa. *Bulletin of the Museum of Comparative Zoology* **81**, 1–413.

Friedmann, H. and Northern, J.R. (1975). Results of the Taylor South West Africa Expedition 1972, Ornithology. *Los Angeles County Museum Contributions in Science* **266**, 1–39.

Friedmann, H. and Stager, K.E. (1964). Results of the 1964 Cheney Tanganyikan Expedition, Ornithology. *Los Angeles County Museum Contributions in Science* **84**, 1–50.

Friedmann, H. and Stager, K.E. (1967). Results of the 1966 Cheney Expedition to the Samburu District, Kenya. Ornithology. *Los Angeles County Museum Contributions in Science* **130**, 1–34.

Friedmann, H. and Stager, K.E. (1969). Results of the 1968 Avil Expedition to Mt Nyiru, Samburu District, Kenya. Ornithology. *Los Angeles County Museum Contributions in Science* **174**, 1–30.

Friedmann, H. and Williams, J.G. (1969). The birds of Sango Bay Forests, Buddu County, Masaka District, Uganda. *Los Angeles County Museum Contributions in Science* **162**, 1–48.

Friedmann, H. and Williams, J.G. (1970). The birds of the Kalinzu Forest, southeastern Ankole, Uganda. *Los Angeles County Museum Contributions in Science* **195**, 1–27.

Friedmann, H. and Williams, J.G. (1971). The birds of the lowlands of Bwamba, Toro Province, Uganda. *Los Angeles County Museum Contributions in Science* **211**, 1–70.

Fry, C.H. (1965). The birds of Zaria. IV. Residents, vagrants, and check-list (Passerines). *Bulletin of the Nigerian Ornithologists' Society* **2**, 91–102.

Fry, C.H. (1971). Migration, moult and weights of birds in northern guinea savanna in Nigeria and Ghana. *Proceedings of the Third Pan-African Ornithological Congress, Ostrich Supplement* **8**, 239–263.

Fry, C.H., Ash, J.S. and Ferguson-Lees, I.J. (1970). Spring weights of some Palearctic migrants at Lake Chad. *Ibis* **112**, 58–82.

Fry, C.H., Keith, S. and Urban, E.K. (1988). 'The Birds of Africa', Vol. III. Academic Press, London.

Fuggles-Couchman, N.R. (1953). The ornithology of Mt Hanang, in northern-central Tanganika Territory. *Ibis* **95**, 468–482.

Gartshore, M.E. (1989). 'An Avifaunal Survey of Tai National Park, Ivory Coast, 28 January–11 April 1989'. *ICBP Study Report* **39**, 1–51. International Council for Bird Preservation, Cambridge.

Gartshore, M.E., Taylor, P.D. and Francis, I.S. (1995). 'Forest birds in Côte d'Ivoire'. *ICBP Study Report* **58**, 1–79.

Gatter, W. (1987). Zugverhalten und Überwinterung von palaärktischen Vögeln in Liberia (Westafrika). *Verhandlungen der Ornithologischen Gesellschaft in Bayern* **24**, 479–508.

Gatter, W. (1997). 'Birds of Liberia'. Pica Press, Robertsbridge.

Gaugris, Y., Prigogine, A. and Vande weghe, J.-P. (1981). Additions et corrections à l'avifaune du Burundi. *Le Gerfaut* **71**, 3–39.

Gee, J.P. (1984). The birds of Mauritania. *Malimbus* **6**, 31–66.

Gee, J.P. and Heigham, J.B. (1977). Birds of Lagos, Nigeria. *Bulletin of the Nigerian Ornithologists' Society* **13**, 43–51.

Germain, M. (1992). Sur quelques données erronées concernant l'avifaune de la Lobaye, République Centrafricaine. *Malimbus* **14**, 1–6.

Germain, M. and Cornet, J.-P. (1994). Oiseaux nouveaux pour la République Centrafricaine ou dont les notifications de ce pays sont peu nombreuses. *Malimbus* **16**, 30–51.

Germain, M., Dragesco, J., Roux, F. and Garcin, H. (1973). Contribution à l'ornithologie du Sud-Cameroun. II. Passeriformes. *L'Oiseau et la Revue Française d'Ornithologie* **43**, 212–259.

Gibbon, G. (1991). 'Southern African Bird Sounds'. Southern African Birding, Durban.

Gillet, H. (1960). Observations sur l'avifaune du massif de l'Ennedi (Tchad). *L'Oiseau et la Revue Française d'Ornithologie* **30**, 99–134.

Ginn, P.J. (1976). Birds of Makgadigadi: a preliminary report. *Wagtail* **15**, 21–96.

Ginn, P.J. (1979). 'Birds of Botswana'. Chris van Rensburg, Johannesburg.

Ginn, P.J., McIlleron, W.G. and Milstein, P. le S. (1989). 'The Complete Book of Southern African Birds'. Struik Winchester, Cape Town.

Giraudoux, P. Degauquier, R., Jones, P.J., Weigel, J. and Isenmann, P. (1988). Avifaune du Niger: état des connaissances en 1986. *Malimbus* **10**(1), 1–140.

Good, A.-I. (1953). 'The Birds of French Cameroon', Part II. *Mémoires de l'Institut Français d'Afrique Noire, Série Sciences Naturelles* **3**, 1–269. Centre du Cameroun, Douala.

Goodman, S.M. and Meininger, P.L. (Eds) (1989). 'The Birds of Egypt'. Oxford University Press, Oxford.

Gore, M.E.J. (1990). 'Birds of The Gambia', 2nd revised edition. BOU Check-list 3. British Ornithologists' Union, Tring.

Green, A.A. (1983). The birds of Bamingui–Bangoran National Park, Central African Republic. *Malimbus* **5**, 17–30.

Green, A.A. (1984). Additional bird records from Bamingui–Bangoran National Park, Central African Republic. *Malimbus* **6**, 70–72.

Green, A.A. (1989). Avifauna of Yankari Reserve, Nigeria: new records and observations. *Malimbus* **11**, 61–72.

Green, A.A. (1990). The avifauna of the southern sector of the Gashaka–Gumti Game Reserve, Nigeria. *Malimbus* **12**, 31–51.

Green, A.A. and Carroll, R.W. (1991). The avifauna of Dzanga–Ndoki National Park and Dzanga–Sangha Rainforest Reserve, Central African Republic. *Malimbus* **13**, 49–66.

Green, A.A. and Sayer, J.A. (1979). The birds of Pendjari and Arli National Parks (Benin and Upper Volta). *Malimbus* **1**, 14–28.

Greig-Smith, P.W. and Davidson, N.C. (1977). Weights of West African savanna birds. *Bulletin of the British Ornithologists' Club* **97**, 96–99.

Grimes, L.G. (1976). The occurrence of cooperative breeding behaviour in African birds. *Ostrich* **47**, 1–15.

Grimes, L.G. (1987). 'The Birds of Ghana'. BOU Check-list 9. British Ornithologists' Union, London.

Grote, H. (1928). Uebersicht über die Vogelfauna des Tschadgebiets. *Journal für Ornithologie* **76**, 739–783.

Günther, R. and Feiler, A. (1986). Zur Phänologie, Ökologie und Morphologie angolanischer Vögel (Aves), II. Passeriformes.

Faunistische Abhandlungen, Staatliches Museum für Tierkunde in Dresden **14**(1), 1–29.

Gyldenstolpe, N. (1924). Zoological results of the Swedish expedition to Central Africa 1921. *Kungliga Svenska Vetenskapsakademiens Handlingar, third series*, **1**(3), 1–326.

Hall, B.P. (1960). The ecology and taxonomy of some Angola birds. *Bulletin of the British Museum of Natural History (Zoology)* **6**(7), 367–453.

Hall, B.P. and Moreau, R.E. (1962). A study of the rare birds of Africa. *Bulletin of the British Museum of Natural History (Zoology)* **8**, 313–378.

Hall, B.P. and Moreau, R.E. (1970). 'An Atlas of Speciation in African Passerine Birds'. British Museum (Natural History), London.

Hall, D.G. (1983). Birds of Mataffin, Eastern Transvaal. *Southern Birds* **10**, 1–55.

Hall, P. (1977a). Birds of the Chad basin boreholes. *Bulletin of the Nigerian Ornithologists' Society* **13**, 37–42.

Hall, P. (1977b). The birds of Maiduguri. *Bulletin of the Nigerian Ornithologists' Society* **13**, 15–36.

Hall, P. (1977c). The birds of Serti. *Bulletin of the Nigerian Ornithologists' Society* **13**, 66–79.

Halleux, D. (1994). Annotated bird list of Macenta Préfecture, Guinea. *Malimbus* **16**, 10–29.

Hanmer, D.B. (1976). Birds of the lower Zambezi (Mozambique). *Southern Birds* **2**, 1–66.

Hanmer, D.B. (1997). Bird longevity in the Eastern Highlands of Zimbabwe – drought survivors. *Safring News* **26**, 47–54.

Happel, R.E. (1985). Birds of Outamba area, northwest Sierra Leone. *Malimbus* **7**, 101–102.

Harding, D.P. and Harding, R.S.O. (1982). A preliminary checklist of birds in the Kilimi area of northwest Sierra Leone. *Malimbus* **4**, 64–68.

Harrap, S. and Quinn, D. (1995 [USA], 1996 [UK]). 'Tits, Nuthatches and Treecreepers'. Christopher Helm/A. & C. Black, London.

Harris, T. and Arnott, G. (1988). 'Shrikes of Southern Africa'. Struik Winchester, Cape Town.

Harrison, C. (1982). 'An Atlas of the Birds of the Western Palaearctic'. Collins, London.

Harrison, J.A., Allan, D.G., Underhill, L.G., Herremans, M., Tree, A.J., Parker, V. and Brown, C.J. (1997). 'The Atlas of Southern African Birds', Vol. 2: Passerines. BirdLife South Africa, Johannesburg.

Harrison, T. and Arnott, G. (1988). 'Shrikes of Southern Africa'. Struik Winchester, Cape Town.

Harvey, W.G. and Howell, K.M. (1987). Birds of the Dar es Salaam area, Tanzania. *Le Gerfaut* **77**, 205–258.

Harwin, R.M., Manson, A.J., Manson, C. and Mwadziwana, P. (1994). The birds of the Bvumba Highlands. *Honeyguide* **40**, Supplement, 1–51.

Hayman, P.V., Prangley, M., Barnett, A. and Diawara, D. (1995). The birds of the Kounounkan Massif, Guinea. *Malimbus* **17**, 53–62.

Heaton, A.M. and Heaton, A.E. (1980). The birds of Obudu, Cross River State, Nigeria. *Malimbus* **2**, 16–24.

Heigham, J.B. (1976). Birds of mid-west Nigeria. *Bulletin of the Nigerian Ornithologists' Society* **12**, 76–93.

Heim de Balsac, H. and Mayaud, N. (1962). 'Les Oiseaux du Nord-Ouest de l'Afrique'. Paul Lechevalier, Paris.

Heinrich, G. (1958). Zur Verbreitung und Lebensweise der Vögel von Angola. *Journal für Ornithologie* **99**, 322–362, 399–421.

Herremans, M. (1998). Conservation status of birds in Botswana in relation to land use. *Biological Conservation* **86**, 139–160.

Hines, C.J.H. (1985–1987). The birds of eastern Kavango, SWA/Namibia. *Journal of the South West Africa Scientific Society* **40–41**, 115–147.

Hockey, P.A.R. (1995). Rare birds in South Africa, 1991–1992. *Birding in Southern Africa* **47**(1), 14–19.

Hockey, P.A.R., Underhill, L.G., Neatherway, M. and Ryan, P.G. (1989). 'Atlas of the Birds of the Southwestern Cape'. Cape Bird Club, Cape Town.

Hodgson, C.J. (1971). The birds of the Chimanimani Mountains, Rhodesia. *South African Avifauna Series* **77**, 31 pp. Percy Fitzpatrick Institute of African Ornithology, Cape Town.

Hoesch, W. and Niethammer, G. (1940). Die Vogelwelt Deutsch-Südwestafrikas. *Journal für Ornithologie* **88**, Sonderheft, 1–404.

Hogg, P., Dare, P.J. and Rintoul, J.V. (1984). Palaearctic migrants in the central Sudan. *Ibis* **126**, 307–331.

Hollom, P.A.D., Porter, R.F., Christensen, S. and Willis, I. (1988). 'Birds of the Middle East and North Africa'. T. and A.D. Poyser, London.

Holman, F.C. (1947). Birds of the Gold Coast. *Ibis* **89**, 623–650.

Holyoak, D.T. and Seddon, M.B. (1989). Distributional notes on the birds of Burkina Faso. *Bulletin of the British Ornithologists' Club* **109**, 205–216.

Howells, W.W. (1985). The birds of the Dande Communal Lands, Middle Zambezi Valley, Zimbabwe. *Honeyguide* **31**, 26–48.

Irwin, M.P.S. (1981). 'The Birds of Zimbabwe'. Quest Publishing, Harare.

Irwin, M.P.S., Niven, P.N.F. and Winterbottom, J.M. (1969). Some birds of the lower Chobe River area, Botswana. *Arnoldia (Rhodesia)* **4**(21), 1–40.

Jackson, F.J. and Sclater, W.L. (1938). 'The Birds of Kenya Colony and the Uganda Protectorate', Vol. III. Gurney and Jackson, London.

Jackson, H.D. (1989). Weights of birds collected in the Mutare Municipal Area, Zimbabwe. *Bulletin of the British Ornithologists' Club* **109**, 100–106.

James, H.W. (1970). 'Catalogue of the Bird Eggs in the Collection of the National Museums of Rhodesia'. Queen Victoria Museum, Salisbury.

Jensen, F.P. and Brøgger-Jensen, S. (1992). The forest avifauna of the Uzungwa Mountains, Tanzania. *Scopus* **15**, 65–83.

Jensen, J.V. and Kirkeby, J. (1980). 'The Birds of The Gambia'. Aros Nature Guides, Århus, Denmark.

Johnston-Stewart, N.G.B. (1984). Evergreen forest birds in the southern third of Malaŵi. *Nyala* **10**, 99–119.

Johnston-Stewart, N.G.B. and Heigham, J.B. (1982). 'Bridging the Bird Gap'. Mountford Press, Limbe.

Jones, P.J. and Tye, A. (1988). 'A survey of the avifauna of São Tomé and Príncipe'. *ICBP Study Report* **24**, 1–64. International Council for Bird Preservation, Cambridge.

Jonsson, L. (1992). 'Birds of Europe with North Africa and the Middle East'. Christopher Helm, London.

Keith, S. and Gunn, W.W.H. (1971). 'Birds of the African Rain Forests'. *Sounds of Nature No. 9*. Federation of Ontario Naturalists, Don Mills, Ontario and the American Museum of Natural History, New York.

Keith, S., Twomey, A., Friedmann, H. and Williams, J. (1969). The avifauna of the Impenetrable Forest, Uganda. *American Museum Novitates* **2389**.

Keith, S., Urban, E.K. and Fry, C.H. (1992). 'The Birds of Africa', Vol. IV. Academic Press, London.

Kemp, A.C. (1976). The distribution and status of the birds of the Kruger National Park. *Koedoe Monograph* **2**, 1–130.

Kirwan, G.M., Martins, R.P., Morton, K.M. and Showler, D.A. (1996). The status of birds in Socotra and 'Abd Al-Kuri and the records of the OSME survey in spring 1995. *Sandgrouse* **17**, 83–101.

Kopij, G. (1997a). Birds of Bethlehem, Free State Province, South Africa. *Mirafra* **14**, 5–12.

Kopij, G. (1997b). Quantitative studies on birds of the Willem Pretorius Game Reserve, Free State Province. *Mirafra* **14**, 17–21.

Lack, P.C. (1985). The ecology of the land-birds of Tsavo East National Park, Kenya. *Scopus* **9**, 2–23, 57–96.

Lamarche, B. (1981). Liste commentée des oiseaux du Mali. Passereaux. *Malimbus* **3**, 73–102.

Lamarche, B. (1988). List commentée des oiseaux de Mauritanie. *Études Sahariennes et Ouest-Africaines* **1**(4), 1–162. Private printing, Nouakchott and Paris.

Laurent, A. (1990). 'Catalogue commenté des oiseaux de Djibouti'. Ministère du Commerce, des Transports et du Tourisme, Djibouti.

Lawson, P.C. and Edwards, J.A. (1983). Birds of Kangwane (Mswati District). *Southern Birds* **11**, 1–84.

Lawson, W.J. (1963). A contribution to the ornithology of Sul do Save, southern Moçambique. *Durban Museum Novitates* **7**(4), 73–124.

Ledant, J.-P., Jacob, J.-P., Jacobs, P., Malher, F., Ochando, B. and Roché, J. (1981). Mise à jour de l'avifaune algérienne. *Le Gerfaut* **71**, 295–398.

Lefranc, N. and Worfolk, T. (1997). 'Shrikes'. Pica Press, Sussex.

Lewis, A. and Pomeroy, D. (1989). 'A Bird Atlas of Kenya'. A.A. Balkema, Rotterdam.

Lippens, L. and Wille, H. (1976). 'Les Oiseaux du Zaïre'. Lannoo à Tielt, Belgium.

Liversidge, R. (1991). 'The Birds Around Us: Birds of the Southern African Region'. Fontein, Parklands, South Africa.

Louette, M. (1981). 'The Birds of Cameroon: an Annotated Check-List'. Paleis der Academiën, Brussels.

Louette, M. and Prévost, J. (1987). Passereaux collectés par J. Prévost au Cameroun. *Malimbus* **9**, 83–96.

Lynes, H. (1925). On the birds of north and central Darfur, with notes on the west-central Kordofan and north Nuba Provinces of British Sudan. *Ibis* **1925**, 71–131.

Lynes, H. (1934). Contribution to the ornithology of southern Tanganyika Territory. Birds of the Ubene-Uhehe highlands and Iringa uplands. *Journal für Ornithologie Supplement* **82**, 1–147.

Lynes, H. (1938). Contribution to the ornithology of the southern Congo basin. Lynes-Vincent tour of 1933–34. *Revue de Zoologie et de Botanique Africaines* **31**, 1–129.

Macdonald, J.D. (1957). 'A Contribution to the Ornithology of Western South Africa'. Trustees, British Museum, London.

Mackenzie, P. (1979). Birds of the Calabar area. *Malimbus* **1**, 47–54.

Mackworth-Praed, C.W. and Grant, C.H.B. (1960). 'Birds of Eastern and North Eastern Africa', Vol. 2. Longmans, London.

Mackworth-Praed, C.W. and Grant, C.H.B. (1963). 'Birds of the Southern Third of Africa', Vol. 2. Longmans, London.

Mackworth-Praed, C.W. and Grant, C.H.B. (1973). 'Birds of West Central and Western Africa', Vol. 2. Longmans, London.

Maclean, G.L. (1993). 'Roberts' Birds of Southern Africa', 6th edition. Trustees of the John Voelcker Bird Book Fund, Cape Town.

Madge, S. and Burn, H. (1994). 'Crows and Jays: A Guide to the Crows, Jays and Magpies of the World'. Christopher Helm/A. & C. Black, London.

Malbrant, R. (1954). Contribution à l'étude des oiseaux du Borkou-Ennedi-Tibesti. *L'Oiseau et la Revue Française d'Ornithologie* **24**, 1–47.

Malbrant, R. and Maclatchy, A. (1949, 1952). 'Faune de l'Equateur Africain Français', Vols I and II. Oiseaux. Paul Lechevalier, Paris.

Malzy, P. (1962). La faune avienne du Mali. *L'Oiseaux et la Revue Française d'Ornithologie* **32**, 1–81.

Mann, C.F. (1985). An avifaunal study in Kakamega Forest, Kenya, with particular reference to species diversity, weight and moult. *Ostrich* **56**, 236–262.

Marchant, S. (1942). Some birds of the Owerri Province, S. Nigeria. *Ibis* **1942**, 137–196.

Mayaud, N. (1988, 1990). Les oiseaux du nord-ouest de l'Afrique. Notes complémentaires. *Alauda* **56**, 113–125; **58**, 135–140, 143–148, 187–194.

Mayr, E. and Amadon, D. (1951). A classification of recent birds. *American Museum Novitates* 1496. American Museum of Natural History, New York.

Mayr, E. and Greenway, J.C. (Eds) (1960, 1962). 'Check-list of Birds of the World', Vols IX and XV. Museum of Comparative Zoology, Cambridge, MA.

Mayr, E. and Paynter, R.A. (Eds) (1964). 'Check-list of Birds of the World', Vol. X. Museum of Comparative Zoology, Cambridge, MA.

Mees, G.F. (1970). Birds of the Inyanga National Park, Rhodesia. *Zoologische Verhandelingen* **109**, 3–74.

Milstein, P. le S. (1975). The biology of Barberspan, with special reference to the avifauna. *Ostrich* Supplement **10**, 1–74.

Mlingwa, C.O.F., Huxham, M.R. and Burgess, N.D. (1993). The avifauna of Kazimzumbwe Forest Reserve, Tanzania: initial findings. *Scopus* **16**, 81–88.

Moreau, R.E. (1936). Breeding seasons of birds in East African evergreen forest. *Proceedings of the Zoological Society of London* **1936**, 631–653.

Moreau, R.E. (1938). The avifauna of the mountains along the Rift Valley in north central Tanganyika Territory (Mbulu District), Part II. *Ibis* **1938**, 1–32.

Moreau, R.E. (1940). Contributions to the ornithology of the East African islands. *Ibis* **1940**, 48–91.

Moreau, R.E. (1944). Clutch-size: a comparative study, with special reference to African birds. *Ibis* **86**, 286–347.

Moreau, R.E. (1950). The breeding seasons of African birds. *Ibis* **92**, 223–267, 419–433.

Moreau, R.E. (1972). 'The Palaearctic-African Bird Migration Systems'. Academic Press, London.

Moreau, R.E. and Moreau, W.M. (1937). Biological and other notes on some East African birds. II. *Ibis* **1937**, 321–345.

Moreau, R.E. and Moreau, W.M. (1939). Observations on some East African birds. *Ibis* **1939**, 296–323.

Moreau, R.E. and Moreau, W.M. (1940). Incubation and fledging periods of African birds. *Auk* **57**, 311–325.

Morel, G.J. and Morel, M.-Y. (1982). Dates de reproduction des oiseaux du Sénégambie. *Bonner Zoologische Beiträge* **33**(2–4), 249–268.

Morel, G.J. and Morel, M.-Y. (1988). Liste des oiseaux de Guinée. *Malimbus* **10**, 143–176.

Morel, G.J. and Morel, M.-Y. (1990). 'Les Oiseaux de Sénégambie'. Centre de l'Office de la Recherche Scientifique et Technique Outre-Mer, Paris.

Morel, G. and Roux, F. (1966). Les migrateurs paléarctiques au Sénégal. *La Terre et La Vie* **113**, 143–176.

Morony, J.J., Bock, W.J. and Farrand, J. (1975). 'Reference List of Birds of the World'. American Museum of Natural History, New York.

Moyer, D.C. (1993). A preliminary trial of territory mapping for estimating bird densities in Afromontane forest. *Proceedings of the Eighth Pan-African Ornithological Congress, Annales Musée Royal de l'Afrique Centrale (Zoologie)* **268**, 302–311.

Moyer, D.C., Lovett, J.C. and deLeyser, E.A. (1990). Birds of Ngwazi, Mufindi District, Tanzania. *Scopus* **14**, 6–13.

Nadler, T. (1993). Beiträge zur Avifauna der Insel São Tomé (Golf von Guinea). *Faunistische Abhandlungen, Staatliches Museum für Tierkunde in Dresden* **19**, 37–58.

de Naurois, R. (1969). Peuplements et cycles de reproduction des oiseaux de la côte occidentale d'Afrique. *Mémoires du Muséum National d'Histoire Naturelle, Série A, Zoologie* **56**, 9–312.

de Naurois, R. (1983). Les oiseaux reproducteurs des îles de São Tomé et Príncipe: liste systématique commentée et indications zoogéographiques. *Bonner Zoologische Beiträge* **34**, 129–148.

de Naurois, R. (1994). 'Les Oiseaux des Iles du Golfe de Guinée, As Aves das Ilhas do Golfo da Guiné'. Instituto de Investigação Científica Tropical, Lisbon.

Newby, J.E. (1980). The birds of the Ouadi Rime – Ouadi Achim Faunal Reserve: a contribution to the study of the Chadian avifauna. *Malimbus* **2**, 29–50.

Newby, J.E., Grettenberger, J. and Watkins, J. (1987). The birds of the northern Aïr, Niger. *Malimbus* **9**, 4–16.

Newman, K. (1989). 'Newman's Birds of Botswana'. Halfway House, Southern Book Publishers, Cape Town.

Newman, K. (1991). 'Newman's Birds of Southern Africa', updated edition. Halfway House, Southern Book Publishers, Cape Town.

Newman, K., Johnston-Stewart, N. and Medland, B. (1992). 'Birds of Malaŵi'. Halfway House, Southern Book Publishers, Cape Town.

Niethammer, G. (1955). Zur Vogelwelt der Ennedi-Gebirges (Französisch Äquatorial-Afrika). *Bonner Zoologische Beiträge* **6**, 29–80.

Nikolaus, G. (1979). Notes on some birds new to south Sudan. *Scopus* **3**, 68–73.

Nikolaus, G. (1987). Distribution atlas of Sudan's birds with notes on habitat and status. *Bonner Zoologische Monographien* **25**, 1–322.

Nikolaus, G. (1989). Birds of south Sudan. *Scopus* Special Supplement **3**, 1–124.

North, M.E.W. and McChesney, D.S. (1964). 'More Voices of African Birds'. Houghton Mifflin, Boston, MA.

Oatley, T.B. (1969). Bird ecology in the evergreen forests of north western Zambia. *Pukù* **5**, 141–180.

Ogilvie-Grant, W.R. and Forbes H.O. (1903). Aves. *In* 'The Natural History of Sokotra and Abd-el-Kuri' (Forbes, H.O., Ed.). H.R. Porter, London.

Olson, S.L. (1989). Preliminary systematic notes on some Old World passerines. *Rivista Italiana di Ornitologia* **59**, 183–195.

Osborne, P.E. and Tigar, B.J. (1990). The status and distribution of birds in Lesotho. Unpublished manuscript, Nature Conservation Bureau, Newbury, 336 pp.

Pakenham, R.H.W. (1979). 'The Birds of Zanzibar and Pemba'. BOU Check-list 2. British Ornithologists' Union, London.

Parker, V. (1994). 'Swaziland Bird Atlas 1985–1991'. Websters, Mbabane, Swaziland.

Paynter, R.A. (Ed.) (1967). 'Check-list of Birds of the World', Vol. XII. Museum of Comparative Zoology, Cambridge, MA.

Pearson, D.J. and Turner, D.A. (1986). The less common Palaearctic migrant birds of Uganda. *Scopus* **10**, 61–82.

Penry, H. (1994). 'Bird Atlas of Botswana'. University of Natal Press, Pietermaritzburg.

Pérez del Val, J., Fa, J.E., Castroviejo, J. and Purroy, F.J. (1994). Species richness and endemism of birds in Bioko. *Biodiversity and Conservation* **3**, 868–892.

Pineau, J. and Giraud-Audine, M. (1979). Les oiseaux de la Péninsule Tingitane. *Travaux de l'Institut Scientifique, Série Zoologie*, **38**, 1–147. Université Mohammed V, Rabat.

Porter, R.F., Christensen, S. and Schiermacker-Hansen, P. (1996). 'Field Guide to the Birds of the Middle East'. T. and A.D. Poyser, London.

Priest, C.D. (1936). 'The Birds of Southern Rhodesia', Vol. 4. William Clowes, London.

Prigogine, A. (1953). Contribution à l'étude de la faune ornithologique de la région à l'ouest du lac Edouard. *Annales du Musée Royal du Congo Belge, série in-8vo, Sciences Zoologiques* **24**, 1–114.

Prigogine, A. (1960). La faune ornithologique du Massif du Mont Kabobo. *Annales du Musée Royal du Congo Belge, série in-8vo, Sciences Zoologiques* **85**, 1–46.

Prigogine, A. (1967). La faune ornithologique de l'Ile Idjwi. *Revue de Zoologie et de Botanique Africaines* **75**, 249–274.

Prigogine, A. (1971, 1978, 1984). Les oiseaux de l'Itombwe et de son hinterland, Vols I, II and III. *Annales du Musée Royal de l'Afrique Centrale, série in-8vo, Sciences Zoologiques* **185, 223, 243**.

Prigogine, A. (1972). Nids et oeufs récoltés au Kivu. 2. (République du Zaïre). *Revue de Zoologie et de Botanique Africaines* **85**, 203–226.

Prigogine, A. (1980a). The altitudinal distribution of the avifauna in the Itombwe Forest (Zaïre). *Proceedings of the Fourth Pan-African Ornithological Congress*, pp. 169–184. Southern African Ornithological Society, Johannesburg.

Prigogine, A. (1980b). Etude de quelques contacts secondaires au Zaïre oriental. *Le Gerfaut* **70**, 305–384.

Prigogine, A. (1985). Recently recognised bird species in the Afrotropical region – a critical review. Proceedings of the International Symposium on African Vertebrates, Bonn 1985, pp. 91–114.

Prigogine, A. (1987). Disjunctions of montane forest birds in the Afrotropical region. *Bonner Zoologische Beiträge* **38**, 195–207.

Quantrill, B. and Quantrill, R. (1998). The birds of Parcours Vita, Yaoundé, Cameroon. *Malimbus* **20**, 1–14.

Quickelberge, C.D. (1989). 'Birds of the Transkei'. Durban Natural History Museum, Durban.

Rand, A.L. (1951). Birds from Liberia. *Fieldiana: Zoology* **32**(9), 561–653.

Rand, A.L., Friedmann, H. and Traylor, M.A. (1959). Birds from Gabon and Moyen Congo. *Fieldiana: Zoology* **41**(2), 221–411.

Reichenow, A. (1902–1905). 'Die Vögel Afrikas', Bands II and III. J. Neumann, Neudamm.

Reichenow, A. (1910). Über eine Vogelsammlung vom Rio Benito im Spanischen Guinea. *Mitteilungen aus dem Zoologischen Museum in Berlin* **5**, 71–87.

Richards, D.K. (1982). The birds of Conakry and Kakulima, Democratic Republic of Guinea. *Malimbus* **4**, 93–103.

Ripley, S.D. and Bond, G.M. (1966). The birds of Socotra and Abd-el-Kuri. *Smithsonian Miscellaneous Collections* **151**(7), 1–37.

Ripley, S.D. and Heinrich, G.H. (1960, 1966a). Additions to the avifauna of northern angola, I and II. *Postilla* **47**, 1–7; **95**, 1–29.

Ripley, S.D. and Heinrich, G.H. (1966b). Comments on the avifauna of Tanzania I. *Postilla* **96**, 1–45.

Roberts, A. (1935). Scientific results of the Vernay-Lang Kalahari Expedition, March to September, 1930. *Annals of the Transvaal Museum* **16**, 1–185.

Rodewald, P.G., Dejaifve, P.-A. and Green, A.A. (1994). The birds of Korup National Park and Korup Project area, Southwest Province, Cameroon. *Bird Conservation International* **4**, 1–68.

da Rosa Pinto, A.A. (1970). Um catálogo das aves do Distrito da Huila (Angola). *Memórias e Trabalhos do Instituto de Investigação Cientifica de Angola* **6**, 5–160.

da Rosa Pinto, A.A. (1972). Contribução para o estudo da avifauna do Distrito de Cabinda (Angola). *Memórias e Trabalhos do Instituto de Investigação Cientifica de Angola* **10**, 9–90.

da Rosa Pinto, A.A. (1973). Aves da colecção do Museu do Dundo. *Publicações Culturais da Companhia de Diamantes de Angola* **87**, 129–178.

da Rosa Pinto, A.A. and Lamm, D.W. (1953–1960). Contribution to the study of the ornithology of Sul do Save (Mozambique). *Memórias do Museu Dr Álvaro de Castro* **2**, 65–85; **3**, 125–159; **4**, 107–167.

Ryall, C. (1991). Avifauna of Nguuni near Mombasa, Kenya, between September 1984 and October 1987: Part I – Afrotropical species. *Scopus* **15**, 1–23.

Safford, R.J., Duckworth, J.W., Evans, M.I., Telfer, M.G., Timmins, R.J. and Zewdie, C. (1993). The birds of Nechisar National Park, Ethiopia. *Scopus* **16**, 61–80.

Salvan, J. (1968–1969). Contribution à l'étude des oiseaux du Tchad. *L'Oiseau et la Revue Française d'Ornithologie* **38**, 249–273; **39**, 38–69.

Sargeant, D.E. (1992). 'São Tomé and Príncipe'. Published by the author, Holt, UK.

Sargeant, D.E. (1993). 'A Birder's Guide to Gabon, West Africa'. Published by the author, Holt, UK.

Sargeant, D.E. (1994). Recent ornithological observations from São Tomé and Príncipe Islands. *Bulletin of the African Bird Club* **1**, 96–102.

Sauvage, A. and Rodwell, S.P. (1998). Notable observations of birds in Senegal (excluding Parc National des Oiseaux du Djoudj), 1984–1994. *Malimbus* **20**, 75–122.

Sayer, J., Harcourt, C.S. and Collins, N.M. (1992). 'The Conservation Atlas of Tropical Forests: Africa'. Simon and Schuster, New York.

Schmidl, D. (1982). 'The Birds of the Serengeti National Park, Tanzania'. BOU Check-list 5. British Ornithologists' Union, London.

Scholte, P., de Kort, S. and van Weers, M. (1999). The birds of the Waza-Logone area, Far North Province, Cameroon. *Malimbus* **21**, 16–50.

Schönwetter, M. and Meese, W. (1972, 1974–1977). 'Handbuch der Oologie', Lieferung 20–26. Akademie-Verlag, Berlin.

Schouteden, H. (1938). Oiseaux. In 'Exploration du Parc National Albert. I. Mission G.F. de Witte (1933–1935)', Fascicule 9. Institut des Parcs Nationaux du Congo Belge, Brussels.

Schouteden, H. (1954–1955). De vogels van belgisch Congo en van Ruanda-Urundi, III. *Annales du Musée Royale du Congo Belge, série in-4to, Zoologie* **IV**, fasc. 1–2.

Schouteden, H. (1957). Faune du Congo Belge et du Ruanda-Urundi, IV. Oiseaux Passereaux (1). *Annales du Musée Royal du Congo Belge, série in-8vo, Sciences Zoologiques* **57**.

Schouteden, H. (1960). Faune du Congo Belge et du Ruanda-Urundi, V. Oiseaux Passereaux (2). *Annales du Musée Royale du Congo Belge, série in-8vo, Sciences Zoologiques* **89**.

Schouteden, H. (1961). La faune ornithologique des districts de la Tshaupa et de l'Equateur. *Documentation Zoologique, Musée Royal de l'Afrique Centrale* **1**.

Schouteden, H. (1962). La faune ornithologique du territoire de Mushie. *Documentation Zoologique, Musée Royal de l'Afrique Centrale* **2**.

Schouteden, H. (1966a). La faune ornithologique du Rwanda. *Documentation Zoologique, Musée Royal de l'Afrique Centrale* **10**.

Schouteden, H. (1966b). La faune ornithologique du Burundi. *Documentation Zoologique, Musée Royal de l'Afrique Centrale* **11**, 1–81.

Sclater, W.L. (1930). 'Systema Avium Aethiopicarum', part 2. British Ornithologists' Union, London.

Sclater, W.L. and Moreau, R.E. (1932–1933). Taxonomic and field notes on some birds of north-eastern Tanganyika Territory. *Ibis* **1932**, 656–683; **1933**, 1–33, 399–439.

Serle, W. (1940). Field observations on some northern Nigerian birds, Part II. *Ibis* **1940**, 1–47.

Serle, W. (1943a). Notes on East African birds. *Ibis* **85**, 55–82.

Serle, W. (1943b). Further field observations on northern Nigerian birds. *Ibis* **85**, 413–437.

Serle, W. (1949). Notes on the birds of Sierra Leone. *Ostrich* **20**, 1–13, 70–85, 114–126.

Serle, W. (1950). A contribution to the ornithology of the British Cameroons. *Ibis* **92**, 602–638.

Serle, W. (1954). A second contribution to the ornithology of the British Cameroons. *Ibis* **96**, 47–80.

Serle, W. (1955). Miscellaneous notes on the birds of the eastern highlands of Southern Rhodesia. *Ostrich* **26**, 115–127.

Serle, W. (1957). A contribution to the ornithology of the Eastern Region of Nigeria. *Ibis* **99**, 628–685.

Serle, W. (1964). The lower altitudinal limit of montane forest birds of the Cameroon Mountain, West Africa. *Bulletin of the British Ornithologists' Club* **84**, 87–91.

Serle, W. (1965). A third contribution to the ornithology of the British Cameroons. *Ibis* **107**, 60–94.

Serle, W. (1981). The breeding seasons of birds in the lowland rainforest and in the montane forest of west Cameroon. *Ibis* **123**, 62–74.

Sharland, R.E. and Wilkinson, R. (1981). The birds of Kano State, Nigeria. *Malimbus* **3**, 7–30.

Shirihai, H. (Ed.) (1996). 'The Birds of Israel'. Academic Press, London.

Short, L.L., Horne, J.F.M. and Muringo-Gichuki, C. (1990). Annotated check-list of the birds of East Africa. *Proceedings of the Western Foundation of Vertebrate Zoology* **4**(3), 61–246.

Sibley, C.G. (1970). A comparative study of the egg-white proteins of passerine birds. *Peabody Mus. Nat. Hist. Bull.* **32**, 1–131.

Sibley, C.G. and Ahlquist, J.E. (1990). 'Phylogeny and Classification of the Birds of the World'. Yale University Press, New Haven, CN.

Sibley, C.G. and Monroe, B.L. (1990). 'Distribution and Taxonomy of Birds of the World'. Yale University Press, New Haven.

Siegfried, W.R. (1968). Ecological composition of the avifaunal community in a Stellenbosch suburb. *Ostrich* **39**, 105–129.

Siegfried, W.R. (1983). Trophic structure of some communities of fynbos birds. *Journal of South African Botany* **49**, 1–43.

Sinclair, I. and Sinclair, J. (1995). 'A Photographic Guide to Birds of Namibia'. Struik, Cape Town.

Sinclair, J.C. (1984). 'Field Guide to the Birds of Southern Africa'. Struik, Cape Town.

Sinclair, J.C., Hockey, P. and Tarboton, W. (1993, 1997). 'Birds of Southern Africa'. Struik, Cape Town.

Skead, C.J. (1967). 'Sunbirds of Southern Africa, also the Sugarbirds, the White-eyes and the Spotted Creeper'. South African Bird Book Fund, A.A. Balkema, Cape Town.

Skead, D.M. (1974). Bird weights from the Central Transvaal bushveld. *Ostrich* **45**, 189–192.

Skinner, N.J. (1995). The breeding seasons of birds in Botswana. 1: Passerine families. *Babbler* **29–30**, 9–23.

Smith, K.D. (1955). Recent records from Eritrea. *Ibis* **97**, 65–80.

Smith, K.D. (1957). An annotated check list of the birds of Eritrea. *Ibis* **99**, 307–337.

Smith, K.D. (1965). On the birds of Morocco. *Ibis* **107**, 493–526.

Smith, K.D. (1968). Spring migration through southeast Morocco. *Ibis* **110**, 452–492.

Smithers, R.H.N. (1964). 'A Check List of the Birds of the Bechuanaland Protectorate and the Caprivi Strip'. Trustees of the National Museums of Southern Rhodesia, Salisbury.

Snow, D.W. (1950). The birds of São Tomé and Príncipe in the Gulf of Guinea. *Ibis* **92**, 579–595.

van Someren, V.G.L. (1916). A list of birds collected in Uganda and British East Africa, with notes on their nesting and other habits, Part II. *Ibis* **1916**, 373–472.

van Someren, V.G.L. (1922, 1932). Notes on the birds of East Africa. *Novitates Zoologicae* **29**, 1–246; **37**, 252–380.

van Someren, V.G.L. (1939). Report on the Coryndon Museum expedition to the Chyulu Hills, Part 2. The birds of the Chyulu Hills. *Journal of the East Africa and Uganda Natural History Society* **14**, 1–2, 15–129.

van Someren, V.G.L. (1956). 'Days with Birds'. *Fieldiana: Zoology* **38**, 1–520.

van Someren, V.G.L. and van Someren, G.R.C. (1949). The birds of Bwamba. *Uganda Journal* **13**, Special Supplement, 1–111.

Steyn, P. (1996). 'Nesting Birds'. Fernwood Press, Vlaeberg.

Stuart, S.N. (1981). A comparison of the avifaunas of seven east African forest islands. *African Journal of Ecology* **19**, 133–151.

Stuart, S.N. (Ed.) (1986). 'Conservation of Cameroon Montane Forests'. International Council for Bird Preservation, Cambridge.

Stuart, S.N. and Adams, R.J. (1990). Biodiversity in sub-Saharan Africa and its islands. *Occasional Papers IUCN Species Survival Commission* **6**, 242 pp. International Union for Conservation of Nature and Natural Resources, Gland.

Stuart, S.N. and Hutton, J.M. (Eds) (1977). The avifauna of the East Usambara Mountains, Tanzania. Report on the Cambridge Ornithological Expedition to East Africa pp. 1–90.

Stuart, S.N. and Jensen, F.P. (1981). Further range extensions and other notable records of forest birds from Tanzania. *Scopus* **5**, 106–115.

Stuart, S.N. and Jensen, F.P. (1985). The avifauna of the Uluguru Mountains, Tanzania. *Le Gerfaut* **75**, 155–197.

Stuart, S.N. and Jensen, F.P. (1986). The status and ecology of montane forest bird species in western Cameroon. In 'Conservation of Cameroon Montane Forests' (Stuart, S.N., Ed.), pp. 38–105. International Council for Bird Preservation, Cambridge.

Stuart, S.N., Jensen, F.P. and Brøgger-Jensen, S. (1987). Altitudinal zonation of the avifauna in the Mwanihana and Magombera forests, eastern Tanzania. *Le Gerfaut* **77**, 165–186.

Styles, C. (1995). Notes on the bird species observed feeding on mopane worms. *Birding in Southern Africa* **47**(2), 53–54.

Svensson, L. (1992). 'Identification Guide to European Passerines'. British Trust for Ornithology, Thetford.

Tarboton, W.R., Kemp, M.I. and Kemp, A.C. (1987). 'Birds of the Transvaal'. Transvaal Museum, Pretoria.

Taylor, P.B. (1979). Palaearctic and intra-African migrants in Zambia: a report for the period May 1971 to December 1976. *Zambian Ornithological Society Occasional Paper* **1**, 1–169.

Taylor, P.B. and Taylor, C.A. (1988). The status, movements and breeding of some birds in the Kikuyu Escarpment Forest, central Kenya highlands. *Tauraco* **1**, 72–89.

Thévenot, M. (1982). Contribution à l'étude écologique des passereaux forestiers du Plateau Central et de la corniche du Moyen Atlas (Maroc). *L'Oiseau et la Revue Française d'Ornithologie* **52**(1), 21–86, 97–152.

Thévenot, M., Bergier, P. and Beaubrun, P. (1980). Compte-rendu d'ornithologie marocaine, année 1979. *Documents de l'Institut Scientifique, Université Mohammed V, Rabat* **5**.

Thévenot, M., Bergier, P. and Beaubrun, P. (1981). Compte-rendu d'ornithologie marocaine, année 1980. *Documents de l'Institut Scientifique, Université Mohammed V, Rabat* **6**.

Thévenot, M. Beaubrun, P., Baouab, R.E. and Bergier, P. (1982). Compte-rendu d'ornithologie marocaine, année 1981. *Documents de l'Institut Scientifique, Université Mohammed V, Rabat* **7**.

Thiollay, J.-M. (1971). L'avifaune de la région de Lamto (moyenne Côte d'Ivoire). *Annales, Université d'Abidjan: Série E: Ecologie* **4**(1), 5–132.

Thiollay, J.-M. (1973). Place des oiseaux dans les chaînes trophiques d'une zone préforestière en Côte d'Ivoire. *Alauda* **41**, 273–300.

Thiollay, J.-M. (1985). The birds of the Ivory Coast: status and distribution. *Malimbus* **7**, 1–59.

Thomsen, P. and Jacobsen, P. (1979). 'The Birds of Tunisia'. Nature-Travels, Copenhagen.

Thonnérieux, Y. (1988a). Commentaires sur la distribution de quelques migrateurs paléarctiques au Burkina Faso. *Le Gerfaut* **78**, 317–362.

Thonnérieux, Y. (1988b). État des connaissances sur la reproduction de l'avifaune du Burkina Faso (ex Haute-Volta). *L'Oiseau et la Revue Française d'Ornithologie* **58**(2), 120–146.

Thonnérieux, Y., Walsh, J.F. and Bortoli, L. (1989). L'avifaune de la ville de Ouagadougou et ses environs (Burkina Faso). *Malimbus* **11**, 7–40.

Traylor, M.A. (1962). 'Notes on the birds of Angola, Passeres'. *Publicações Culturais da Companhia de Diamantes de Angola* **58**.

Traylor, M.A. (1963). 'Check-list of Angolan Birds'. *Publicações Culturais da Companhia de Diamantes de Angola* **61**.

Traylor, M.A. (1965). A collection of birds from Barotseland and Bechuanaland. *Ibis* **107**, 357–384.

Traylor, M.A. and Archer, A.L. (1982). Some results of the Field Museum 1977 expedition to south Sudan. *Scopus* **6**, 5–12.

Urban, E.K. and Brown, L.H. (1971). 'A Checklist of the Birds of Ethiopia'. Haile Sellassie I University Press, Addis Ababa.

Urban, E.K., Fry, C.H. and Keith, S. (1997). 'The Birds of Africa', Vol. V. Academic Press, London.

Vande weghe, J.-P. (1973). Les périodes de nidification des oiseaux du Parc National de l'Akagera au Rwanda. *Le Gerfaut* **63**, 235–255.

Vande weghe, J.-P. (1979). The wintering and migration of Palaearctic passerines in Rwanda. *Le Gerfaut* **69**, 29–43.

Vande weghe, J.-P. (1981). L'avifaune des papyraies au Rwanda et au Burundi. *Le Gerfaut* **71**, 489–536.

Vande weghe, J.-P. and Loiselle, B.A. (1987). The bird fauna of the Bururi Forest, Burundi. *Le Gerfaut* **77**, 147–164.

Vaurie, C. (1959). 'The Birds of the Palearctic Fauna', Order Passeriformes. H.F. and G. Witherby, London.

Verheyen, R. (1953). Oiseaux. In 'Exploration du Parc National de l'Upemba. I. Mission G.F. de Witte (1946–1949)', Fascicule 19. Institut des Parc Nationaux du Congo Belge, Brussels, 687 pp.

Vernon, C.J. (1977). Birds of the Zimbabwe Ruins area. *Southern Birds* **4**, 1–50.

Vieillard, J. (1971–1972). Données biogéographiques sur l'avifaune de l'Afrique Centrale. *Alauda* **39**, 227–248; **40**, 63–92.

Vincent, A.W. (1947, 1948, 1949). On the breeding habits of some African birds. *Ibis* **89**, 163–204; **90**, 284–312; **91**, 111–139, 313–345.

Vincent, J. (1934, 1935, 1936). The birds of northern Portuguese East Africa. *Ibis* **1934**, 126–160; **1935**, 355–397, 485–529, 707–762; **1936**, 48–125.

Voous, K.H. (1960). 'Atlas of European Birds'. Nelson, London.

Voous, K.H. (1977). List of recent Holarctic bird species. Passerines. *Ibis* **119**, 223–250.

Wacher, T. (1993). Some new observations of forest birds in The Gambia. *Malimbus* **15**, 24–37.

Waltert, M. and Mühlenberg, M. (1999). Notes on the avifauna of the Noyau Central, Forêt Classée de la Lama, Republic of Benin. *Malimbus* **21**, 82–92.

Waltert, M., Yaokokore-Beibro, K.H., Mühlenberg, M. and Waitkuwait, W.E. (1999). Preliminary check-list of the birds of the Bossematié area, Ivory Coast. *Malimbus* **21**, 93–109.

Welch, G.R. and Welch, H.J. (1984a). Birds seen on an expedition to Djibouti. *Sandgrouse* **6**, 1–23.

Welch, G.R. and Welch, H.J. (1984b). 'Djibouti Expedition March 1984'. Published by the authors, Whittlesey, Cambridge.

Welch, G.R. and Welch, H.J. (1986). 'Djibouti II'. Published by the authors, Whittlesey, Cambridge.

Welch, G.R. and Welch, H.J. (1992). 'Djibouti III.' Published by the authors, Minsmere.

Welch, G.R., Welch, H.J., Coghlan, S.M. and Denton, M.L. (1986). 'Djibouti II – autumn '85'. Published by the authors, Whittlesey, Cambridge.

Wetmore, A. (1960). A classification for the birds of the world. *Smithsonian Miscellaneous Collections* **139**(11), 1–37.

White, C.M.N. (1962). 'A Revised Check List of African Shrikes, Orioles, Drongos, Starlings, Crows, Waxwings, Cuckooshrikes, Bulbuls, Accentors, Thrushes and Babblers'. Government Printer, Lusaka.

White, C.M.N. (1963). 'A Revised Check List of African Flycatchers, Tits, Tree Creepers, Sunbirds, White-eyes, Honey Eaters, Buntings, Finches, Weavers and Waxbills'. Government Printer, Lusaka.

Wilson, J.D. (1987). 'The Status and Conservation of the Montane Forest Avifauna of Mount Oku, Cameroon in 1985'. International Council for Bird Preservation, Cambridge.

Wilson, N. and Wilson, V.G. (1994). Avifauna of the southern Kerio Valley with emphasis on the area around the Kenya Fluorspar Mine site, August 1989–July 1993. *Scopus* **18**, 65–115.

Winterbottom, J.M. (1942). A contribution to the ornithology of Barotseland. *Ibis* **1942**, 337–389.

Winterbottom, J.M. (1959). Notes on the status of some birds in Northern Rhodesia. *Ostrich* **30**, 1–12.

Winterbottom, J.M. (1964). Results of the Percy Fitzpatrick Institute – Windhoek State Museum joint ornithological expeditions: Report on the birds of Game Reserve No. 2. *Cimbebasia* **9**, 1–75.

Winterbottom, J.M. (1966). Results of the Percy Fitzpatrick Institute – Windhoek State Museum joint ornithological expeditions: 5. Report on the birds of the Kaokoveld and Kunene River. *Cimbebasia* **19**, 1–71.

Winterbottom, J.M. (1971a). 'A Preliminary Check List of the Birds of South West Africa'. South West Africa Scientific Society, Windhoek.

Winterbottom, J.M. (1971b). 'Priest's Eggs of Southern African Birds'. Winchester Press, Johannesburg.

Witherby, H.F., Jourdain, F.C.R., Ticehurst, N.F. and Tucker, B.W. (1938). 'The Handbook of British Birds', Vols 1–2. H.F. and G. Witherby, London.

Wolters, H.E. (1979–1982). 'Die Vögelarten der Erde'. Paul Parey, Hamburg.

Zedlitz, O.G. (1911). Meine ornithologische Ausbeute in Nordost-Afrika. *Journal für Ornithologie* **59**, 1–92.

Zimmerman, D.A. (1972). The avifauna of the Kakamega Forest, western Kenya, including a bird population study. *Bulletin of the American Museum of Natural History* **149**, Article 3.

Zimmerman, D.A., Turner, D.A. and Pearson, D.J. (1996). 'Birds of Kenya and Northern Tanzania'. Princeton University Press, Princeton.

Zink, G. (1973, 1975, 1981, 1985). 'Der Zug europäischer Singvögel', Vols 1–4. Vogelzug, Möggingen.

Systematic References

Family PICATHARTIDAE

Allport, G.A. (1991). The status and conservation of threatened birds in the Upper Guinea Forest. *Bird Cons. Intl* **1**, 53–74.

Anon. (1995). Recent reports. *Bull. Afr. Bird Club* **2**, 123–127.

Ash, J.S. (1991). The Grey-necked Picathartes *Picathartes oreas* and Ibadan Malimbe *Malimbus ibadanensis* in Nigeria. *Bird Conserv. Intl* **1**, 93–106.

Bowden, C.G.R. (1987). The use of mist-netting for studying forest birds in Cameroon. In 'Conservation of Cameroon Montane Forests' (Stuart, S.N., Ed.), pp. 130–174. ICBP, Cambridge.

Brosset, A. (1965a). La biologie de *Picathartes orea (sic)*. *Biol. Gabonica* **1**, 101–115.

Brosset, A. (1965b). Un oiseau africain troglophile: *Picathartes orea (sic)*. *Anns Spéléol.* **2**, 425–429.

Butynski, T.M. and Koster, S.H. (1989). Grey-headed Picathartes *Picathartes oreas* found on Bioko Island (Fernando Po). *Tauraco* **1**, 186–189.

Butynski, T.M., Schaaf, C.D. and Hearn, G.W. (1996). The Grey-necked Picathartes *Picathartes oreas* on Bioko Island, Equatorial Guinea. *Ostrich* **67**, 90–93.

Cheke, R.A. (1986). The supposed occurrence of the White-necked Picathartes *Picathartes gymnocephalus* in Togo. *Bull. Br. Orn. Club* **106**, 152.

Dekker, D. (1973). Hatching the White-necked bald crow *Picathartes gymnocephalus* at Amsterdam Zoo. *Int. Zool. Yearbook* **13**, 120–121.

Delacour, J. and Amadon, D. (1951). The systematic position of *Picathartes*. *Ibis* **93**, 60–62.

Dolton, P.J. (1995). Grey-necked Picathartes: how to see one. *Bull. Afr. Bird Club* **2**, 30.
Fleig, G.M. (1993). Vestigial erectile crest in *Picathartes*. *Auk* **88**, 442.
Fotso, R.C. (1993). Contribution à l'étude du Picatharte chauve du Cameroun *Picathartes oreas*. *Proc.VIII Pan-Afr. Orn. Congr.*, 431–437.
Glanville, R.R. (1954). *Picathartes gymnocephalus* in Sierra Leone. *Ibis* **96**, 481–484.
Green, A.A. (1995). Finding Grey-necked Picathartes in Korup National Park, Cameroon. *Bull. Afr. Bird Club* **2**, 101–102.
Grimes, L.G. (1963). Some observations on *Picathartes gymnocephalus*. *Nigerian Field* **28**, 63–65.
Grimes, L.G. (1964). Some notes on the breeding of *Picathartes gymnocephalus* in Ghana. *Ibis* **106**, 258–260.
Grimes, L.G. and Darku, K. (1968). Some recent breeding records of *Picathartes gymnocephalus* in Ghana and notes on its distribution in West Africa. *Ibis* **110**, 93–99.
Grimes, L.G. and Gardiner, N. (1963). Looking for *Picathartes gymnocephalus* in Ghana. *Nigerian Field* **28**, 55–63.
Lowe, P.R. (1938). Some anatomical and other notes on the systematic position of the genus *Picathartes*, together with some remarks on the families Sturnidae and Eulabetidae. *Ibis*, **1938**, 254–269.
McKelvey, S.D. (1981). Successful hand-rearing of the White-necked picathartes *Picathartes gymnocephalus* at San Antonio Zoo. *Int. Zool. Yearbook*, **1981**, 219–223.
Moore, A. (1974). Cameroon Bare-headed Rock-fowl. *Nigerian Field* **39**, 188–190.
Moore, A. (1997). Grey-necked Picathartes. *Bull. Afr. Bird Club* **4**, 46.
Mudd, H. and Martins, R. (1996). Possible display behaviour of White-necked Picathartes. *Bull. Br. Orn. Club* **116**, 15–17.
Olson, S.L. (1979). *Picathartes* – another West African forest relict with probable Asian affinities. *Bull. Br. Orn. Club* **99**, 112–113.
Serle, W. (1952a). The affinities of the genus *Picathartes* Lesson. *Bull. Br. Orn. Club* **72**, 2–6.
Serle, W. (1952b). The Lower Guinea Bare-headed Crow. *Nigerian Field* **17**, 131–132.
Sibley, C.G. (1973). The relationships of *Picathartes*. *Bull. Br. Orn. Club* **93**, 23–25.
Thomas, J. (1991). Birds of the Korup National Park, Cameroon. *Malimbus* **13**, 11–23.
Thompson, H.S. (1993). Status of white-necked picathartes – another reason for conservaton of the peninsula forest, Sierra Leone. *Oryx* **27**, 155–158.
Thompson, H.S. and Fotso, R. (1995). Rockfowl, the genus *Picathartes*. *Bull. Afr. Bird Club* **2**, 25–28.
Tye, H. (1986). The erectile crest and other head feathering in the genus *Picathartes*. *Bull. Br. Orn. Club* **106**, 90–93.
Tye, H. (1987). Breeding biology of *Picathartes oreas*. *Gerfaut* **77**, 313–332.
Willis, E.O. (1983). Wrens, gnatwrens, rockfowl, babblers and shrikes (Troglodytidae, Polioptilidae, Picathartidae, Timaliidae and Laniidae) as ant followers. *Gerfaut* **73**, 393–404.
Wood, P. (1995). White-necked Picathartes: how to see one. *Bull. Afr. Bird Club* **2**, 29.

Family TIMALIIDAE

Chaetops

Brown, C.J. and Barnes, P.R. (1984) The birds of the Natal Alpine Belt. *Lammergeyer* **33**, 1–13.
Clancey, P.A. (1966). A catalogue of birds of the South African sub-region. *Durban Mus. Novit.* **7**, 12, 465–544.
Clancey, P.A. (1972). Miscellaneous taxonomic notes on African birds XXXIV. *Durban Mus. Novit.* **9**, 145–162.
Clinning, C.F. and Tarboton, W.R. (1972). Notes on the Damara Rockjumper, *Achaetops pycnopygius*. *Madoqua*, ser. 1, **5**, 57–61.
Craig, A. (1991). Rockjumper spotting. *Bee-eater* **42**, 17–18.
Irwin, M.P.S. (1983). The Malagasy species of Timaliidae (babblers). *Honeyguide* **116**, 26–31.
Irwin, M.P.S. (1985). *Chaetops* and the Afrotropical Timaliidae (babblers). *Honeyguide* **31**, 99–100.
Jarvis, A. and Robertson, T. (1997). Endemic birds of Namibia: evaluating their status and mapping biodiversity hotspots. Directorate of Environmental Affairs, Min. of Environment and Tourism, Windhoek, Namibia.
Martin, J. (1964). Nestlings of Rufous Rock-Jumper (*Chaetops frenatus*) being fed by two males and one female. *Ostrich* **35**, 62.
Milewski, A.V. (1975). New distributional data: 6. *Ostrich* **46**, 178.
Roberts, A. (1922). Review of the nomenclature of South African birds. *Ann. Transvaal Mus.* **8**, 187–272.
Robertson, A., Simmons, R.E., Jarvis, A.M. and Brown, C.J. (1995). Can bird atlas data be used to estimate population size? A case study using Namibian endemics. *Biol. Conserv.* **71**, 87–95.
Tait, I.C. (1948). Observations on the Orange-breasted Rockjumper. *Ostrich* **19**, 218–221.

Illadopsis, Kakamega, Pseudoalcippe, Ptyrticus

Allport, G.A., Ausden, M.J., Fishpool, L.D.C., Hayman, P.V., Robertson, P.A. and Wood, P. (1996). Identification of Illadopsises *Illadopsis* spp. in the Upper Guinea Forest. *Bull. Afr. Bird Club* **3**, 26–30.
Anon. (1995). Recent reports. *Bull. Afr. Bird Club* **2**, 123.
Aspinwall, D.R. (1971). Records from Mwinilunga District. *Bull. Zambian Orn. Soc.* **3**, 29–35.
Aspinwall, D.R. (1978). Bird notes from Northern Nchelenge District. *Bull. Zambian Orn. Soc.* **10**, 2.
Bennun, L.A. (1991). An avifaunal survey of the Trans-Mara Forest, Kenya. *Scopus* **14**, 61–72.
Bennun, L.A., Gichuki, C., Darlington, J. and Ng'weno, F. (1986). The avifauna of Ol Doinyo Orok, a forest island: initial findings. *Scopus* **10**, 83–86.
Benson, C.W. (1958). Birds from the Mwinilunga District, Northern Rhodesia. *Ibis* **100**, 281–285.
Brosset, A. and Erard, C. (1974). Note sur la reproduction des *Illadopsis* de la forêt gabonaise. *Alauda* **42**, 385–395.
Butynski, T.M. (1989). First nest record, and other notes, for the Scaly-breasted Illadopsis *Trichastoma albipectus*. *Scopus* **12**, 89–92.
Butynski, T.M. and Kalina, J. (1989). Description of the nest and eggs of the Mountain Illadopsis *Trichastoma pyrrhopterum*. *Scopus* **13**, 131–132.
Colebrook-Robjent, J.F.R. (1976). Undescribed nests and eggs of birds breeding in Zambia. *Bull. Zambian Orn. Soc.* **8**, 45–56.
Delacour, J. (1946). Les timaliines. *Oiseau et R.F.O.* **16**, 7–36.
De Roo, A. (1970). Contribution à l'ornithologie de la Republique du Togo. 2. Oiseaux récoltés par M. C. Veronese. *Rev. Zool. Bot. Afr.* **81**, 163–172.

Dowsett-Lemaire, F. and Dowsett, R.J. (1983). Notes on montane forest babblers (Timaliidae) in Malaŵi. *Nyala* **9**, 57–59.

Holyoak, D.T. and Seddon, M.B. (1990). Notes on some birds of Western Cameroon. *Malimbus* **11**, 123–127.

Mann, C.F. (1980). Notes on the avifaunas of the Kakamega and the Nandi Forests. *Scopus* **4**, 97–99.

Mann, C.F., Burton, P.J.K. and Lennerstedt, I. (1978). A reappraisal of the systematic position of *Trichastoma poliothorax* (Timaliinae, Muscicapidae). *Bull. Br. Orn. Club* **98**, 131–140.

Olson, C. (1976). Summary of field observations of birds from Begemdir and Simien Province. *Walia* **7**, 16–27.

Plumptre, A.J. and Owiunji, I. (1997). Puvel's Illadopsis *Illadopsis puveli* in Budongo Forest: a new record for East Africa. *Scopus* **19**, 114–116.

Ripley, S.D. and Beehler, B.M. (1985). A revision of the babbler genus *Trichastoma* and its allies (Aves: Timaliinae). *Ibis* **127**, 495–509.

Rodwell, S.P. (1996). Notes on the distribution and abundance of birds observed in Guinea-Bissau, 21 February to 3 April 1992. *Malimbus* **18**, 25–43.

Salewski, V. (1997). Discovery of a nest of Puvel's Akalat *Illadopsis puveli*. *Malimbus* **19**, 34–36.

Sessions, P.H.B. (1966). Notes on the birds of Lengetia Farm, Mau Narok. *J. E. Afr. Nat. Hist. Soc. Nat. Mus.* **26**, 18–48.

Stuart, S.N. and Van der Willigen, T.A. (1979). 'Report of the Cambridge Ecological Expedition to Tanzania, 1978'. Photocopy, 77 pp.

Turner, D.A. (1993). Species report. *In* East African Bird Report 1991. *Scopus* **15**, 143–163.

Vande weghe, J.-P. (1992). New Records for Uganda and Tanzania along the Rwandan and Burundian borders. *Scopus* **16**, 59–60.

Willis, E.O. (1983). Wrens, gnatwrens, rockfowl, babblers and shrikes (Troglodytidae, Polioptilidae, Picathartidae, Timaliidae and Laniidae) as ant followers. *Gerfaut* **73**, 393–404.

Turdoides, Phyllanthus

Anon. (1994). Hinde's Babbler study completed in Kianyaga. *Kenya Birds* **2**, 2–3.

Ash, J.S. (1981). A new race of the Scaly Babbler *Turdoides squamulatus* from Somalia. *Bull. Br. Orn. Club* **101**, 399–403.

Backshall, D. (1993). Submissive posture by Arrowmarked Babbler? *Babbler* **25**, 41–42.

Barbour, D. (1972). Sociability at a Pied Babbler's nest. *Honeyguide* **70**, 34–35.

Bennun, L.A. (1998). Threatened birds and rural communities: balancing the equation. Abstract 22. Int. Orn. Congr., Durban. *Ostrich* **69**, 79.

Blencowe, E.J. (1960). Hinde's Pied Babbler in Embu District. *J. E. Afr. Nat. Hist. Soc.* **23**, 248.

Bradford, H.J. (1966). On some snakes and birds. *Honeyguide* **48**, 24.

Bundy, G. and Morgan, J.H. (1969). Notes on Tripolitanian birds (part 2). *Bull. Br. Orn. Club* **89**, 151–159.

Clancey, P.A. (1958). Miscellaneous taxonomic notes on African birds XI. 4. A new race of *Turdoides jardineii* (Smith) from Sul do Save, southern Portuguese East Africa. *Durban Mus. Novit.* **5**, 123–126.

Clancey, P.A. (1974). Miscellaneous taxonomic notes on African birds XL. The *hartlaubii* subspecies-group of *Turdoides leucopygius* (Rüppell), with the characters of a new race from Botswana. *Durban Mus. Novit.* **10**, 147–150.

Clancey, P.A. (1979). Miscellaneous taxonomic notes on African birds XLV. A second southern race of *Turdoides melanops* (Hartlaub) of the Afrotropical Region. *Durban Mus. Novit.* **12**, 54–55.

Clancey, P.A. (1984). The relationship of the whiterumped babblers *Turdoides leucopygius* (Rüppell) and *T. hartlaubii* (Bocage). *Ostrich* **55**, 28–30.

Clancey, P.A. (1989). Taxonomic and distributional findings on some birds from Namibia. *Cimbebasia* **11**, 111–133.

Clark, J.E. and Clarke, J.A.C. (1985). Interactions between Arrow-marked Babblers and Striped Crested Cuckoos. *Nyala* **11**, 28–29.

Cooper, J. (1970). Arrow-marked Babblers eating loquats. *Honeyguide* **64**, 33.

Etchécopar, R.D. (1970). Extension de la zone de distribution de *Turdoides fulvus* au Maroc. *Oiseau et R.F.O.* **40**, 174–175.

Fry, C.H. (1975). The northern limits of fringing forest birds in North Central State, Nigeria. *Bull. Nigerian Orn. Soc.* **11**, 56–64.

Fry, C.H. and Hosken, J.H. (1983). Food of an Arrowmarked Babbler. *Ostrich* **54**, 178.

Fuggles-Couchman, N.R. and Elliott, H.F.I. (1946). Some records and field-notes from North-Eastern Tanganyika Territory. *Ibis* **1946**, 327–347.

Gargett, E. (1969). Unusual behaviour in the Arrow-marked Babbler. *Honeyguide* **60**, 34.

Gaston, A.J. (1977). Social behaviour within groups of Jungle Babblers (*Turdoides striatus*). *Anim. Behav.* **25**, 828–848.

Gaston, A.J. (1978). The evolution of group territorial behavior and cooperative breeding. *Am. Nat.* **112**, 1091–1100.

Ginn, P.J. (1993). Arrow-marked Babbler with nest sentry. *Honeyguide* **39**, 196–197.

Goodman, S.M. and Atta, G.A.M. (1987). The birds of southeastern Egypt. *Gerfaut* **77**, 3–41.

Harvey, W.G. (1974). Unusual behaviour of Rufous Chatterers. *E. Afr. Nat. Hist. Bull.* **1974**, 96–97.

Herremans, M. and Herremans-Tonnoeyr, D. (1995). Competition and food selection among congeneric species at a feeding table in northern Botswana. *Babbler* **29–30**, 24–27.

Huels, T.R. (1982). Co-operative feeding of conspecific and *Clamator jacobinus* young by *Turdoides rubiginosus*. *Scopus* **6**, 33–35.

Hustler, K. (1997). The status, breeding and parasitism of the White-rumped Babbler in Zimbabwe. *Honeyguide* **43**, 211–213.

Irwin, M.P.S. (1988). Order Cuculiformes. *In* 'The Birds of Africa', Vol. 3 (Fry, C.H., Keith, S. and Urban, E.K., Eds). Academic Press, London.

Jarvis, A. and Robertson, T. (1997). Endemic birds of Namibia: evaluating their status and mapping biodiversity hotspots. Directorate of Environmental Affairs, Min. of Environment and Tourism. Windhoek, Namibia.

Jones, J.M.B. (1985). Striped Crested Cuckoo parasitizing Arrow-marked Babbler. *Honeyguide* **31**, 170–171.

Jones, J.M.B. (1992). Striped Cuckoo and Arrow-marked Babbler observations. *Honeyguide* **38**, 75–76.

Lees-Smith, D.T. (1986). Composition and origins of the south-west Arabian avifauna: a preliminary analysis. *Sandgrouse* **7**, 71–92.

Lewis, A.D. (1984). Hinde's Pied Babbler *Turdoides hindei* south of Machakos, Kenya. *Scopus* **8**, 48–49.

Lindeque, M. and Kapner, J. (1993). Cooperative group defence by Pied babblers *Turdiodes (sic) bicolor* results in death of avian predator. *Ostrich* **64**, 189.

Monadjem, A. (1993). The effect of ringing nestling Arrow-marked Babblers *Turdoides jardineii* on fledging success. *Safring News* **22**, 55–56.

Monadjem, A., Owen-Smith, N. and Kemp, A.C. (1994). Position of nest, incubation period and nestling period of the Arrowmarked Babbler. *Ostrich* **65**, 341.

Monadjem, A., Owen-Smith, N. and Kemp, A.C. (1995). Aspects of the breeding biology of the Arrowmarked Babbler *Turdoides jardineii* in South Africa. *Ibis* **137**, 515–518.

Mortimer, J. (1975). Striped Crested Cuckoos and Arrow-marked Babblers. *Honeyguide* **82**, 46.

Moyer, D.C. (1982). A breeding record of the White-rumped Babbler *Turdoides leucopygius* from Tanzania. *Scopus* **6**, 103.

Njoroge, P. (1994). Hinde's Babbler survey continues. *Kenya Birds* **3**, 2–4.

Njoroge, P., Bennun, L.A. and Lens, L. (1998). Habitat use by the globally endangered Hinde's Babbler *Turdoides hindei* and its sympatric relative, the Northern Pied Babbler *T. hypoleucus*. *Biol. Conserv. Intl* **8**, 59–65.

Plumb, W.J. (1979). Observations on Hinde's Babbler *Turdoides hindei*. *Scopus* **3**, 61–67.

Reynolds, J.F. (1965). Behaviour of black-lored babbler. *E. Afr. Wildl. J.* **3**, 130.

Serle, W. (1977). The aberrant eggs of *Turdoides plebejus* in Nigeria and their relation to cooperative breeding and to victimisation by *Clamator* cuckoos. *Bull. Br. Orn. Club* **97**, 39–41.

Shaw, P. (1996). A search for Hinde's Babbler north of Embu. *Kenya Birds* **5**, 34–35.

Shuel, R. (1938). Further notes on the eggs and nesting habits of birds in Northern Nigeria (Kano Province). *Ibis* **1938**, 2, 463–479.

Turner, D. (1992). Threatened birds of Kenya 2: Hinde's Babbler. *Kenya Birds* **1**, 46–47.

Vernon, C.J. (1976). Communal feeding of nestlings by the Arrowmarked Babbler. *Ostrich* **47**, 34–136.

Witherby, H.F. (1901). An ornithological expedition to the White Nile. *Ibis* **1901**, 237–278.

Wood, B. (1989). Biometrics, iris and bill colouration, and moult of some Somali forest birds. *Bull. Br. Orn. Club* **109**, 11–22.

Zahavi, A. (1990). Arabian Babblers: the quest for social status in a cooperative breeder. *In* 'Cooperative Breeding in Birds' (Stacey, P.B. and Koenig, W.D., Eds). Cambridge Univ. Press, Cambridge.

Lioptilus, Kupeornis

Bennett, G. and Herbert, S. (1995). 'Where to see birds in Kwazulu-Natal'. Mondi Southern Birds 19, pp. 81.

Collett, J. (1982). Birds of the Cradock District. *Southern Birds* **9**, 165.

Hesse, C.F.C. (1995). Mixed-species flock-foraging behaviour of Cameroon montane forest birds. Unpublished Hons. Zoology Thesis, Aberdeen University.

Prigogine, A. (1960). Une nouvelle forme de *Lioptilus chapini* (Schouteden) de l'est du Congo belge. *Rev. Zool. Bot. Afr.* **61**, 15–18.

Prigogine, A, (1964). Un nouvel oiseau de la République du Congo. *Rev. Zool. Bot. Afr.* **70**, 401–404.

Schouteden, H. (1949). Un timaliide nouveau du Congo belge. *Rev. Zool. Bot. Afr.* **52**, 343–344.

Serle, W. (1949). A new genus and species of babbler and new races of a wood-hoopoe, swift, barbet, robin-chat, scrub-warblers and apalis from West Africa. *Bull. Br. Orn. Club* **69**, 50–56.

Serle, W. (1950). Gilbert's Babbler. *Nigerian Field* **15**, 84.

Vande weghe, J.P. (1988). The validity of *Kupeornis* Serle. *Bull. Br. Orn. Club* **108**, 54–58.

Verheyen, R. (1947). Notes sur la faune ornithologique de l'Afrique centrale. 7. Liste d'une collection d'oiseaux rares réunis à Albertville et description d'une nouveau touraco du Congo belge. *Bull. Mus. Roy. Hist. Nat. Belg.* **22**, 1–4.

Vincent, J. (1951). A note on the Bush Blackcap, *Lioptilornis nigricapillus* (Vieillot). *Ostrich* **22**, 203.

Horizorhinus, Parophasma, Panurus

Birkhead, T.R. and Hoi, H. (1994). Reproductive organ and mating strategies of the Bearded Tit *Panurus biarmicus*. *Ibis* **136**, 356–359.

Dorst, J. and Roux, F. (1973). L'avifaune des forêts de *Podocarpus* de la Province de l'Arussi, Ethiopie. *Oiseau et R.F.O.. fr. Orn.* **43**, 269–304.

Fishpool, L.D.C., Allport, G.A. and Webb, R. (1966). Photospot: Ethiopian endemics. *Bull. Afr. Bird Club* **3**, 42–43.

Poulin, B. (1998). Spatial heterogeneity and the passerine community in large reedbeds of southern France. Abstracts 22 Int. Orn. Congr., Durban. *Ostrich* **69**, 303.

Tucker, G.M. and Heath, M.F. (1994). 'Birds in Europe: Their Conservation Status'. BirdLife Intl, Cambridge.

Urban, E.K. (1975). Weights and longevity of some birds from Addis Ababa, Ethiopia. *Bull. Br. Orn. Club* **95**, 96–98.

Urban, E.K. (1987). 'Ethiopia's Endemic Birds'. Ethiopian Tourist Commission, Addis Ababa.

Urban, E.K., Brown, L.H., Buer, C.E. and Plage, G.D. (1970). Four descriptions of nesting, previously undescribed, from Ethiopia. *Bull. Br. Orn. Club* **90**, 162–164.

Families AEGITHALIDAE and PARIDAE

Alerstam, T. and Ulfstrand, S. (1977). Niches of tits *Parus* spp. in two types of African woodland. *Ibis* **119**, 521–524.

Ash, J.S. (1994). Bird-ringing recoveries from Ethiopia: II. *Scopus* **17**, 113–118.

Ash, J.S. and Gullick, T.M. (1989). The present situation regarding the endemic breeding birds of Ethiopia. *Scopus* **13**, 90–96.

Baouab, R.E. (1981). La reproduction des Mésanges bleues (*Parus caeruleus ultramarinus*) et charbonnières (*Parus major excelsius*) dans la forêt de chênes lièges de la Mamora (Maroc). C. Ecol. Anim., Fac. Sci., Univ. Rabat, 65 pp.

Baouab, R.E. (1983). Etude écologique et stratégie démographique des Mésanges *Parus caeruleus ultramarinus* Bonaparte et *Parus major excelsius* Buvry dans deux localités du Maroc. Thèse 3éme cycle, Fac. Sci., Univ. Rabat, 102 pp.

Baouab, R.E., Thévenot, M. and Aguesse, P. (1986). Dynamique des populations de la Mésange bleue en chenaies de Mamora et du Moyen Atlas. *Bull. Inst. Sci.*, Rabat, **10**, 165–183.

Benson, C.W., Irwin, M.P.S. and White, C.M.N. (1959). Some aspects of speciation in the birds of Rhodesia and Nyasaland. *Ostrich* Suppl. **3**, 397–414.

Berndt, R. and Jürgens, R. (1977). Niedersächsische Tannenmeise (*Parus ater*) als Wintergast in Nordafrika. *Vogelwarte* **29**, 65–66.

Clancey, P.A. (1958). Taxonomic notes on two species of southern African species of Paridae. *Ibis* **100**, 451–454.

Clancey, P.A. (1963). Notes, mainly systematic, on some birds from the Cape Province. *Durban Mus. Novit.* **6**, 244–264.

Clancey, P.A. (1972). The status of *Parus niger carpi* Macdonald and Hall, 1957, and a regrouping of some populations of the *Parus niger* Vieillot, sens. strict., complex. *Durban Mus. Novit.* **9**, 236–244.

Clancey, P.A. (1995). Taxonomic relationships in Namibian Black Tits *Parus* spp. *Bull. Br. Orn. Club* **115**, 181–184.

Clancey, P.A. (1996). Systematic relationships and variation in the tits *Parus afer* and *P. cinerascens* (Aves: Paridae) of the southern Afrotropics. *Durban Mus. Novit.* **21**, 37–42.

Davies, G. (1993). Bwindi bird intent on sleep. *E. Afr. Nat. Hist. Soc. Bull.* **23**, 69–70.

Gaston, A.J. (1973). The ecology and behaviour of the Long-tailed Tit. *Ibis* **115**, 330–351.

Gibb, J.A. (1950). The breeding biology of the Great and Blue Titmice. *Ibis* **92**, 507–539.

Hailman, J.P. (1989). The organization of major vocalizations in the Paridae. *Wilson Bull.* **101**, 305–343.

Hall, B.P. (1960). Variation in the African black tits, *Parus niger* and *Parus leucomelas*. *Ibis* **102**, 116–123.

Hall, B.P. and Traylor, M.A. (1959). The systematics of the African Grey Tits, *Parus afer* and *Parus griseiventris*. *Bull. Br. Orn. Club* **79**, 42–46.

Harrap. S. (1996). The vocalisations of African black tits (*Parus niger* complex). *Bull. Afr. Bird Club* **3**, 99–104.

Harrap, S. and Quinn, D. (1996). 'Tits, Nuthatches and Treecreepers'. Christopher Helm, London.

Helsens, T. (1996). New information on birds in Ghana, April 1991 to October 1993. *Malimbus* **18**, 1–9.

Irwin, M.P.S. (1959). The specific relationship of *Parus afer* and *Parus griseiventris*. *Bull. Br. Orn. Club* **79**, 46–48.

Macdonald, M.A. and Taylor, I.R. (1977). Notes on some uncommon forest birds in Ghana. *Bull. Br. Orn. Club* **97**, 116–120.

Perrins, C.M. (1965). Population fluctuations and clutch size in the great tit, *Parus major* L. *J. Anim. Ecol.* **34**, 601–647.

Sherry, F.F. (1989). Food storing in the Paridae. *Wilson Bull.* **101**, 289–304.

Snow, B.W. (1952). A contribution to the ornithology of north-west Africa. *Ibis* **94**, 473–493.

Tarboton, W.R. (1981). Cooperative breeding and group territoriality in the Black Tit. *Ostrich* **52**, 216–225.

Turner, D.A. (1977). Status and distribution of the East African endemic species. *Scopus* **1**, 2–11.

Wiggins, D.A. (in press). Comparative breeding ecology of Afrotropical and north temperate zone tits (*Parus* spp.). *J. Avian Biol.*

Wiggins, D.A. (in press). Egg size and nestling growth in tits (*Parus* spp.): a comparison of African and temperate zone species. *Biol. J. Linn. Soc.*

Family REMIZIDAE

Austin, G.T. (1978). Pattern and timing of moult in penduline tits (*Anthoscopus*). *Ostrich* **49**, 168–173.

Benson, C.W. (1955). The races of the Penduline Tit *Anthoscopus caroli* (Sharpe) in Northern Rhodesia and adjacent territory. *Rev. Zool. Bot. Afr.* **52**, 156–158.

van den Berk, V. (1991). Visible migration of Sparrowhawk *Accipiter nisus* and Penduline Tit *Remiz pendulinus* in southern Turkey. *Sandgrouse* **13**, 101–102.

Chadwick, D.H. (1983). Etosha Park – Africa's kingdom of animals. *National Geographic* **163**(3), 344–385.

Clancey, P.A. (1962). Carol's Penduline Tits roosting in weaver nest. *Ostrich* **33**, 38.

Clancey, P.A. (1997). Variation in the Cape Penduline Tit *Anthoscopus minutus* of the southern Afrotropics. *Bull. Br. Orn. Club* **117**, 52–55.

Cole, D. (1949). Notes on the Cape Penduline Tit. *Ostrich* **20**, 31–32.

Courtenay-Latimer, M. and Clancey, P.A. (1961). Penduline Tits roosting in old weaver nest. *Ostrich* **32**, 48.

Darolová, A., Hoi, H. and Schleicher, B. (1997). The effect of ectoparasite nest load on the breeding biology of the Penduline Tit *Remiz pendulinus*. *Ibis* **139**, 115–120.

Fishpool, L.D.C. (1993). New bird records from Budongo and Kifu Forests, Uganda, with an addition to the East African avifauna. *Scopus* **17**, 37–39.

Irwin, M.P.S. (1963). A new race of *Anthoscopus caroli* (Sharpe) from the Zambesi Valley. *Bull. Br. Orn. Club* **83**, 2–4.

Lawson, W.J. (1961). Comments on the geographical variation in Carol's Penduline Tit *Anthoscopus caroli* (Sharpe) in Southern Africa. *Bull. Br. Orn. Club* **81**, 149–150.

Lowe, W.P. (1937). Report on the Lowe–Waldron Expeditions to the Ashanti Forests and Northern Territories of the Gold Coast, Pt 3. *Ibis* **1937**, 830–864.

Macdonald, M.A. (1980). Further notes on uncommon forest birds in Ghana. *Bull. Br. Orn. Club* **100**, 170–172.

Marcus, M.B. (1963). Bill and leg colour of *Anthoscopus minuta* (Shaw and Nodder). *Ostrich* **34**, 170–171.

Milstein, P. le S. (1975). Observations on Penduline Tit nest structure. *Bokmakierie* **27**, 8–9.

Plowes, D.C.H. (1951). A remarkable deception. *African Wild Life* **5**, 297–299.

Skead, C.J. (1959). A study of the Cape Penduline Tit *Anthoscopus minutus minutus* (Shaw & Nodder). *Ostrich*, Suppl. **3**, 274–288.

Taylor, I.R. and Macdonald, M.A. (1978). The status of some northern Guinea savanna birds in Mole National Park, Ghana. *Bull. Nigerian Orn. Soc.* **14**, 45, 4–8.

Ulfstrand, S. (1960). A new subspecies of *Anthoscopus caroli* (Sharpe 1871) from western Tanganyika Territory. *Bull. Br. Orn. Club* **80**, 11–13.

Uys, C.J. (1966). Impressions of the Cape Penduline Tit and its nest in the south-west Cape. *Bokmakierie* **18**, 80–82.

Vernon, C.J. and Dean, W.R.J. (1975). On the systematic position of *Pholidornis rushiae*. *Bull. Br. Orn. Club* **95**, 20.

Family SITTIDAE

Bellatreche, M. (1991). Deux nouvelles localisations de la Sittelle kabyle *Sitta ledanti* en Algérie. *Oiseau et R.F.O.* **61**, 269–272.

Bellatreche, M. and Chalabi, B. (1990). Données nouvelles sur l'aire de distributon de la Sittelle kabyle *Sitta ledanti*. *Alauda* **58**, 95–97.

Enoksson, B. (1998). Winter survival in the European Nuthatch *Sitta europaea*: temperature is more important than food. Abstracts 22 Int. Orn. Congr., Durban. *Ostrich* **69**, 345.

Gatter, W. and Mattes, H. (1979). Zur Populationsgrösse und Ökologie des neuentdeckten Kabylenkleibers *Sitta ledanti* Vielliard 1976. *J. Orn.* **120**, 390–405.

Ledant, J.P. (1978). Données comparées sur la Sittelle Corse (*Sitta whiteheadi*) et sur la Sittelle kabyle (*Sitta ledanti*). *Aves* **15**, 154–157.

Ledant, J.P. (1981). Conservation et fragilité de la forêt de Babor, habitat de la Sittelle kabyle. *Aves* **18**, 1–9.

Ledant, J.P. and Jacobs, P. (1977). La Sittelle kabyle (*Sitta ledanti*): données nouvelles sur la biologie. *Aves* **14**, 233–242.

Löhrl, H. (1975) Brutverhalten und Jugendentwicklung beim Mauerläufer (*Tichodroma muraria*). *J. Orn.* **116**, 229–262.

Matthysen, E. (1998). 'The Nuthatches'. T. and A.D. Poyser, London.

Sibley, C.G., Ahlquist, J.E. and Monroe, B.L. (1988). A classification of the living birds of the world, based on DNA–DNA hybridization studies. *Auk* **105**, 409–423.

Families SALPORNITHIDAE and CERTHIIDAE

Clancey, P.A. (1975). Miscellaneous taxonomic notes on African birds XLI. The austral African races of *Salpornis spilonotus* (Franklin). *Durban Mus. Novit.* **10**, 206–208.

Howland, L.A. (1988). Interactions between adult and juvenile Spotted Creeper. *Honeyguide* **34**, 76–77.

Isenmann, P. and Guillosson, J.-Y. (1993). Simultaneous bigamy by Short-toed Treecreepers. *Brit. Birds* **86**, 371.

Isenmann, P., Cramm, P. and Clamens, A. (1986). Données sur la biologie de reproduction du Grimpereau des jardins (*Certhia brachydactyla*) en région méditerranéenne française. *Nos Oiseaux* **38**, 359–362.

Masterson, A. (1970). Notes on the Spotted Creeper. *Honeyguide* **61**, 35–37.

Randall, R.D. (1994). The Spotted Creeper *Salpornis spilonotus* new to Botswana. *Babbler* **28**, 38.

Steyn, P. (1974). A confiding creeper. *Bokmakierie* **26**, 80–82.

Tree, A.J. (1987). Recent reports. *Honeyguide* **33**, 155–160.

Tree, A.J. (1992). Recent reports. *Honeyguide* **38**, 34–40.

Family NECTARINIIDAE

Delacour, J. (1944). A revision of the family Nectariniidae (sunbirds). *Zoologica* **29**, 17–38.

Gill, F.B. and Wolf, L.L. (1979). Nectar loss by Golden-winged Sunbirds to competitors. *Auk* **96**, 448–461.

Grimes, L.L. (1974). Dialects and geographical variation in the song of the Splendid Sunbird *Nectarinia coccinigaster*. *Ibis* **116**, 314–329.

Haagner, C.H. (1964). 'Birds of Zululand, 2'. *Wildlife Series* 5. Intl. Library of African Music, Roodepoort, South Africa.

Hustler, K. (1985). Which sunbird was it? *Honeyguide* **31**, 173–174.

Irwin, M.P.S. (1993). What sunbirds belong to the genus *Anthreptes*? *Honeyguide* **39**, 211–215.

Irwin, M.P.S. (1999). The genus *Nectarinia* and the evolution and diversification of sunbirds: an Afrotropical perspective. *Honeyguide* **45**, 45–58.

Jackson, S. (1998). Avian nectarivores that breed in winter: balancing energy and water. Abstracts 22 Int. Orn. Congr., Durban. *Ostrich* **69**, 79.

Kennedy, R.S., Gonzales, P.C. and Miranda, H.C. (1997). New *Aethopyga* sunbirds (Aves: Nectariniidae) from the island of Mindanao, Philippines. *Auk* **114**, 1–10.

Rand, A.L. (1967). Family Nectariniidae. *In* 'Checklist of Birds of the World', Vol. 12 (Paynter, R.A., Ed.). Mus. Comp. Zoology, Cambridge, MA.

Stuart, S.N. and van der Willigen, T.A. (1980). Is Moreau's Sunbird *Nectarinia moreaui* a hybrid species? *Scopus* **4**, 56–58.

Tree, A.J. (1991). A ringing guide to selected species of Zimbabwean sunbirds. *Safring News* **20**, 13–20.

Wolf, L.L. and Wolf, J.S. (1976). Mating system and reproductive biology of Malachite Sunbird. *Condor* **78**, 27–39.

Wolters, H.E. (1977). Die Gattungen der Nectariniidae (Aves, Passeriformes). *Bonn. Zool. Beitr.* **28**, 82–101.

Anthreptes, Deleornis

Aspinwall, D.R. (1983). Movement analysis charts: comments on the Violet-backed Sunbird. *Zambian Orn. Soc. Newsl.* **13**, 95–96.

Butynski, T. (1994). Uluguru Violet-backed Sunbird *Anthreptes neglectus* at Tana, Kenya. *Scopus* **18**, 62–64.

Diamond, A.W. and Keith, G.S. (1980). Avifaunas of Kenya forest islands. I – Mount Kulal. *Scopus* **4**, 49–56.

Dinesen, L., Lehmberg, T., Svendsen, J.O. and Hansen, L.A. (1993). Range extensions and other notes on some restricted-range forest birds from West Kilombero in the Udzungwa Mountains, Tanzania. *Scopus* **17**, 48–59.

Dowsett, R.J. (1977). Notes on the distribution of the Red-and-Blue Sunbird. *Bull. Zambian Orn. Soc.* **9**, 64–65.

Dowsett-Lemaire, F. (1996). Avian frugivore assemblages at three small-fruited tree species in the forests of northern Congo. *Ostrich* **67**, 88–89.

Erard, C. (1979). What in reality is *Anthreptes pujoli* Berlioz? *Bull. Br. Orn. Club* **99**, 142–143.

Evans, T. (1997). Records of birds from the forests of the East Usambara lowlands, Tanzania, August 1994–February 1995. *Scopus* **19**, 92–108.

Fjeldså, J., Howell, K. and Andersen, M. (1997). An ornithological visit to the Rubeho Mountains, Tanzania. *Scopus* **19**, 73–82.

Hanmer, D.B. (1981). Mensural and moult data on nine species of sunbird from Mozambique and Malaŵi. *Ostrich* **52**, 156–178.
Hockey, P.A.R. and the Rarities Committee (1996). Rare Birds in South Africa (1993–1995). *Afr. Birds and Birding* **1**(3), 64–68.
Irwin, M.P.S. (1995a). The Plain-backed Sunbird in the Transvaal and Zimbabwe. *Honeyguide* **41**, 26–27.
Irwin, M.P.S. (1995b). On the affinities of the Plain-backed Sunbird *Anthreptes reichenowi*. *Honeyguide* **41**, 47–48.
Keith, S. (1968). Notes on birds of East Africa, including additions to the avifauna. *Amer. Mus. Novit.* **2321**, 1–15.
Krienke, W. (1941). Nesting of Nyasa Violet-backed Sunbird on Elvington farm, Beatrice district, Southern Rhodesia. *Ostrich* **12**, 24–25.
Medland, R.D. (1991). Souza's Shrike attacking Violet-backed Sunbird. *Nyala* **15**, 49.
Tree, A.J. (1990). Notes on sunbird movements and nectar sources in Zimbabwe. *Honeyguide* **36**, 171–182.
Walsh, F. (1966). A nest of the Violet-backed Sunbird. *Bull. Nigerian Orn. Soc.* **3**, 70–71.
Wells, D. (1966). The Violet-backed Sunbird nesting in Nigeria. *Bull. Nigerian Orn. Soc.* **3**, 72–74.
Williams, J.G. (1951). Notes on *Anthreptes reichenowi yokanae*. *Bull. Br. Orn. Club* **71**, 48–50.
Williams, J.G. (1953). On the nest and eggs of *Anthreptes reichenowi yokanae*. *Bull. Br. Orn. Club* **73**, 33.

Anabathmis, Dreptes

Cane, W.P. and Carter, M.F. (1988). Significant range extension for *Nectarinia reichenbachii* in West Africa. *Bull. Br. Orn. Club* **108**, 52–54.
Demey, R. (1986). Two new species for the Ivory Coast. *Malimbus* **8**, 44.
Eccles, S.D. (1985). Reichenbach's Sunbird *Nectarinia reichenbachii* new to Ivory Coast. *Malimbus* **7**, 140.

Anthobaphes

Broekhuysen, G.J. (1963). The breeding biology of the Orange-breasted Sunbird, *Anthobaphes violacea* (Linnaeus). *Ostrich* **34**, 187–234.
Cohen, C. and Winter, C. (1993). Orangebreasted Sunbirds feeding on the ground. *Promerops* **209**, 11.
Collins, B.G. (1983). Pollination of *Mimetes hirtus* (Proteaceae) by Cape Sugarbirds and Orange-breasted Sunbirds. *S. Afr. J. Bot.* **49**, 125–142.
Fraser, M. and McMahon, L. (1988). Orangebreasted Sunbird feeding from ivy flowers. *Promerops* **188**, 10.
Fraser, M. and McMahon, L. (1991). Orangebreasted sunbird feeding from *Agapanthus* flowers. *Promerops* **198**, 13.
Fraser, M. and McMahon, L. (1992b). Habitat change by Cape Sugarbirds and Orangebreasted Sunbirds in an apparent response to fire in old Mountain Fynbos. *Safring News* **21**, 51–54.
Fraser, M. and McMahon, L. (1993). Orangebreasted Sunbird feeding from *Lobelia* flowers. *Promerops* **207**, 12.
Fraser, M.W., McMahon, L., Underhill, L.G. and Underhill, G.D. (1989). Nectarivore ringing in southwestern Cape. *Safring News* **18**, 3–18.
Martin, R., Martin, J., Martin, E. and Tyler, D. (1990). Orangebreasted Sunbird in karroid area. *Promerops* **195**, 13.
Spottiswoode, C. (1993). Orangebreasted Sunbird feeding on the ground. *Promerops* **208**, 10–11.

Williams, J.B. (1993a). Energetics of incubation in free-living Orange-breasted Sunbirds in South Africa. *Condor* **95**, 115–126.
Williams, J.B. (1993b). Nest orientation of Orangebreasted Sunbirds in South Africa. *Ostrich* **64**, 40–42.
Wooller, R.D. (1982). Feeding interactions between sunbirds and sugarbirds. *Ostrich* **53**, 114–115.

Cyanomitra

Akinpelu, A.I. (1989). Competition for nectar of *Tecoma stans* flowers between Olive Sunbird (*Nectarinia olivacea*) and insects. *Malimbus* **11**, 3–6.
Bowden, C.G.R. (1986). The use of mist-netting for studying forest birds in Cameroon. In 'Conservation of Cameroon Montane Forests', (S.N. Stuart, Ed.), pp. 130–174. ICBP, Cambridge.
Britton, P.L. and Britton, H.A. (1977). Sunbirds nesting inside buildings at the Kenya coast. *Scopus* **1**, 68–70.
Brosset, A. (1974). La nidification des oiseaux en forêt gabonaise: architecture, situation des nids et predation. *Terre et Vie* **28**, 579–610.
Clancey, P.A. (1978). Miscellaneous taxonomic notes on African birds LII. On the southern and eastern races of *Nectarinia olivacea* (Smith), 1840. *Durban Mus. Novit.* **11**, 317–327.
Clancey, P.A. (1993). The status of *Nectarinia olivacea* (Smith), 1840; a unitary species or two polytypic allospecies? *Gerfaut* **82–83**, 25–29.
Colebrook-Robjent, J.F.R. (1990). The nest and eggs of Bannerman's Blue-headed Sunbird *Nectarinia bannermani*. *Bull. Br. Orn. Club* **110**, 13–14.
Dowsett-Lemaire, F. (1989b). Food plants and the annual cycle in a montane community of sunbirds (*Nectarinia* spp.) in northern Malaŵi. *Tauraco* **1**, 167–185.
Dowsett-Lemaire, F. (1996). Avian frugivore assemblages at three small-fruited tree species in the forests of northern Congo. *Ostrich* **67**, 88–89.
Fry, C.H. (1975). The northern limits of fringing forest birds in North Central State, Nigeria. *Bull. Nigerian Orn. Soc.* **11**, 56–64.
Hanmer, D.B. (1979). The Grey Sunbird *Nectarinia veroxii* in southern Malaŵi. *Bull. Br. Orn. Club* **99**, 71–72.
Hanmer, D.B. (1981). Mensural and moult data of nine species of sunbird from Moçambique and Malaŵi. *Ostrich* **52**, 156–178.
Harvey, W.G. and Harrison, I.D. (1970). The birds of the Mole Game Reserve. *Bull. Nigerian Orn. Soc.* **7**, 63–75.
Penry, E.H. (1979). The Green-headed Sunbird (*Nectarinia verticalis*) and Olive Sunird (*Nectarinia olivacea*) occurring together in southern Mansa District. *Bull. Zambian Orn. Soc.* **11**, 36.
Pettet, A. (1977). Seasonal changes in nectar-feeding by birds at Zaria, Nigeria. *Ibis* **119**, 291–308.
Prigogine, A. (1975). Etude taxonomique de *Nectarinia alinae* et description de trois nouvelles formes de la République du Zaïre. *Rev. Zool. Afr.* **89**, 455–480.
Swynnerton, C.F.M. (1908). Further notes on the birds of Gazaland. *Ibis* **1908**, 1–107.

Chalcomitra

Barnicot, N.C. (1984). Breeding the Black Sunbird *Nectarinia* (*Chalcomitra*) *amethystina amethystina*. *Avicult. Mag.* **90**, 86–87.

Brown, L.H. (1948). Notes on birds of the Kabba, Ilorin and N. Benin Provinces of Nigeria. *Ibis* **90**, 525–537.

Clancey, P.A. (1963). Notes on the South African races of the Scarlet-chested Sunbird *Nectarinia senegalensis* (Linnaeus). *Ostrich* **34**, 97–98.

Clancey, P.A. (1975). Miscellaneous taxonomic notes on African birds XLIII. On the present nominate subspecies of *Nectarinia amethystina* (Shaw). *Durban Mus. Novit.* **11**, 20–24.

Clancey, P.A. (1986). Miscellaneous taxonomic notes on African birds 67. Variation in Hunter's Sunbird *Nectarinia hunteri* (Shelley), 1889. *Durban Mus. Novit.* **14**, 7–27.

Colebrook-Robjent, J.F.R. (1984). The breeding of the Didric Cuckoo *Chrysococcyx caprius* in Zambia. *Proc. V Pan-Afr. Orn. Congr.*, 763–778.

Edwards, S.J. (1988). Unusual behaviour of Scarlet-chested Sunbird. *Honeyguide* **34**, 132.

Gill, F.B. and Wolf, L.L. (1975). Foraging strategies and energetics of East African sunbirds at mistletoe flowers. *Amer. Nat.* **109**, 491–510.

Green, A.A. and Rodewald, P.G. (1996). New bird records from Korup National Park and environs, Cameroon. *Malimbus* **18**, 122–133.

Hanmer, D.B. (1981). Mensural and moult data of nine species of sunbird from Moçambique and Malaŵi. *Ostrich* **52**, 156–178.

Hanmer, D.B. and Chadder, B. (1997). Sunbird notes from the Mutare and Bvumba areas. *Honeyguide* **43**, 220–223.

Helsens, T. (1996). New information on birds in Ghana, April 1991 to October 1993. *Malimbus* **18**, 1–9.

Hopkins, M. (1998). Buff-throated Sunbird *Nectarinia adelberti* and Fire-bellied Woodpecker *Dendropicos pyrrhogaster* in Cameroon. *Malimbus* **20**, 124–125.

Hopkins, M.T.E., Demey, R. and Barker, J.C. (1999). First documented records of Green-throated Sunbird *Nectarinia rubescens* for Nigeria, with a discussion of the distinctive race *crossensis*. *Malimbus* **21**, 57–60.

Lack, P. (1976). Nesting of Hunter's Sunbird. *E. Afr. Nat. Hist. Soc. Bull.*, **1976**, 49.

Louette, M. (1982). Allopatric species of birds approaching in western Cameroon: the *Nectarinia adelberti*, *N. rubescens* example. *Bonn. Zool. Beitr.* **33**, 303–312.

Louette, M. (1989). Additions and corrections to the avifauna of Zaïre. *Bull. Br. Orn. Club* **109**, 217–225.

Marchant, S. (1953). Notes on the birds of south-eastern Nigeria. *Ibis* **95**, 38–69.

Martins, R.P., Porter, R.F. and Stone, F. (1993). 'Preliminary Report of the OSME Survey of Southern Yemen and Socotra Spring 1993'. Orn. Soc. Middle East, Sandy, UK.

Pettet, A. (1977). Seasonal changes in nectar-feeding by birds at Zaria, Nigeria. *Ibis* **119**, 291–308.

Porter, R.F. and Martins, R. (1993). OSME in southern Yemen and Socotra. *Bull. Orn. Soc. Middle East* **31**, 1–4.

Porter, R.F. and Stone, F. (1996). An introduction to Socotra and its birds. *Sandgrouse* **17**, 73–80.

Serle, W. (1963). A new race of sunbird from West Africa. *Bull. Br. Orn. Club* **83**, 118–119.

Showler, D.A. and Davidson, P. (1996). The Socotra Sunbird *Nectarinia balfouri*. *Sandgrouse* **17**, 148–150.

Skead, C.J. (1953). A study of the Black Sunbird, *Chalcomitra a. amethystina* (Shaw). *Ostrich* **24**, 159–166.

Tree, A.J. (1990). Notes on sunbird movements and nectar sources in Zimbabwe. *Honeyguide* **36**, 171–182.

Turner, D.A. and Forbes-Watson, A. (1979). *Nectarinia adelberti* au Cameroun. *Oiseau et R.F.O.* **49**, 158.

Welch, G. and Welch, H. (1998). Mystery birds from Djibouti. *Bull. Afr. Bird Club* **5**, 46–50.

Nectarinia

Chapin, J.P. (1959). Breeding cycles of *Nectarinia purpureiventris* and some other Kivu birds. *Ostrich* Suppl. **3**, 222–229.

Chapin, R.T. (1978). Breeding behaviour of the Bronzy Sunbird *Nectarinia kilimensis*. *Bull. E. Afr. Nat. Hist. Soc.* **1978**, 52–55.

Cheke, R.A. (1971). Feeding ecology and significance of inter-specific territoriality of African montane sunbirds (Nectariniidae). *Rev. Zool. Bot. Afr.* **84**, 50–64.

Cheke, R.A. (1978). Records of birds and their parasites from the Cherangani Mountains, Kenya. *E. Afr. Wildl. J.* **16**, 61–64.

Coe, M.J. (1961). Notes on *Nectarinia johnstoni johnstoni* on Mt Kenya. *Ostrich* **32**, 101–103.

Coles, D. (1978). Breeding the Tacazze Sunbird at Padstow Bird Gardens. *Avicult. Mag.* **84**, 69–73.

Craig, A.J.F.K. and Hulley, P.E. (1994). Sunbird movements: a review, with possible models. *Ostrich* **65**, 106–110.

Cunningham-van Someren, G.R. (1976). Feeding preferences of some sunbirds. *Bull. E. Afr. Nat. Hist. Soc.* **1976**, 105–107.

Dowsett-Lemaire, F. (1988). On the breeding behaviour of three montane sunbirds *Nectarinia* spp. in northern Malaŵi. *Scopus* **11**, 79–86.

Dowsett-Lemaire, F. (1989b). Food plants and the annual cycle in a montane community of sunbirds (*Nectarinia* spp.) in northern Malaŵi. *Tauraco* **1**, 167–185.

Earlé, R.A. (1981). Weights of southern African sunbirds. *Durban Mus. Novit.* **13**, 21–40.

Evans, M.R. (1991). The size of adornments of male scarlet-tufted malachite sunbirds varies with environmental conditions, as predicted by handicap theories. *Anim. Behav.* **42**, 797–803.

Evans, M.R. and Hatchwell, B.J. (1992). An experimentl study of male adornment in the male Scarlet-tufted Malachite Sunird: II. The role of elongated tail in mate choice and experimental evidence for handicap. *Behav. Ecol. Sociobiol.* **29**, 421–427.

Evans, M.R. and Thomas, A.L.R. (1992). The aerodynamic and mechanical effects of elongated tails in the scarlet-tufted malachite sunbird: measuring the cost of a handicap. *Anim. Behav.* **43**, 337–347.

Fraser, M.W. (1989). Short-term responses of birds to fire in old Mountain Fynbos. *Ostrich* **60**, 172–182.

Gill, F.B. and Wolf, L.L.(1975a). Economics of feeding territoriality in the Golden-winged Sunbird. *Ecology* **56**, 333–345.

Gill, F.B. and Wolf, L.L. (1975b). Foraging strategies and energetics of East African sunbirds at mistletoe flowers. *Amer. Nat.* **109**, 491–510.

Gill, F.B. and Wolf, L.L. (1977). Nonrandom foraging by sunbirds in a patchy environment. *Ecology* **58**, 1284–1296.

Gill, F.B. and Wolf, L.L. (1978). Comparative feeding efficiencies of some montane sunbirds in Kenya. *Condor* **80**, 391–400.

Gill, F.B. and Wolf, L.L. (1979). Nectar loss by Golden-winged Sunbirds to competitors. *Auk* **96**, 448–461.

Haig, M. (1987). Incubation and fledging periods and post fledging behaviour of the Malachite Sunbird. *Promerops* **177**, 14–15.

Hanmer, D.B. and Chadder, B. (1997). Sunbird notes from the Mutare and Bvumba areas. *Honeyguide* **43**, 220–223.

Jackson, H.D. (1970). Unusual nest of a Bronze Sunbird. *Ostrich* **41**, 262–263.

Johnson, D.N. and Maclean, G.L. (1994). Altitudinal migration in Natal. *Ostrich* **65**, 86–94.

Löhrl, H. (1979). Tagesaktivität Nestbaumethode und Brutverhaltern des Honigsaugers *Nectarinia (Aidemonia) kilimensis* in Zentralafrika. *J. Orn.* **124**, 441–450.

Martin, R.M. (1976). The problems of breeding small nectar-feeding birds. *Avicult. Mag.* **82**, 165–168.

Meikle, J. (1985). And more on those Bronze Sunbirds. *Honeyguide* **31**, 59.

Niven, C. (1968). Mass feeding of the Malachite Sunbird *Nectarinia famosa*. *Ostrich* **39**, 40.

Prigogine, A. (1977). Populations of the Scarlet-tufted Malachite Sunbird, *Nectarinia johnstoni* Shelley, in Central Africa and description of a new subspecies from the Republic of Zaïre. *Mitt. Zool. Mus. Berlin* **53** (Suppl.), 117–125.

Pyke, G.H. (1979). The economics of territory size and time budget in the Golden-winged Sunbird. *Amer. Nat.* **114**, 131–145.

Ryan, P. (1998). Malachite Sunbird feeding on jakkalskos. *Promerops* **235**, 19.

Simpson, B. (1971). The belligerent sunbird. *Bull. East Afr. Nat. Hist. Soc.* **1971**, 163–164.

Stark, A.C. and Sclater, W.L. (1900). 'The Fauna of South Africa, Birds', Vol. 1. (Publ. not given): Cape Town.

Stewart, D.R.M. and Stewart, J. (1964). Anting behaviour in the Bronze Sunbird. *J. E. Afr. Uganda Nat. Hist. Soc.* **24**, 92.

Taylor, J.S. (1946). Notes on the Malachite Sunbird, *Nectarinia famosa*, L. *Ostrich* **17**, 254–257.

Urban, E.K. (1975). Weights and longevity of some birds from Addis Ababa, Ethiopia. *Bull. Br. Orn. Club* **95**, 96–98.

Williams, J.G. (1951). *Nectarinia johnstoni*: a revision of the species, together with data on plumages, moults and habits. *Ibis* **93**, 579–595.

Winterbottom, J.M. (1968). Bird densities in coastal renosterbosveld of the Bontebok National Park, Swellendam, Cape Province. *Koedoe* **11**, 139–144.

Wolf, L.L. and Wolf, J.S. (1976). Mating system and reproductive biology of Malachite Sunbirds. *Condor* **78**, 27–39.

Hedidypna

Britton, P.L. and Britton, H.A. (1978). Notes on the Amani Sunbird *Anthrepes pallidigaster*, including a description of the nest and eggs. *Scopus* **2**, 102–103.

Dinesen, L., Lehmberg, T., Svendsen, J.O. and Hansen, L.A. (1993). Range extensions and other notes on some restricted-range forest birds from West Kilombero in the Udzungwa Mountains, Tanzania. *Scopus* **17**, 48–59.

Field, G.D. (1971). Juvenile plumage of the Upper Guinea race of the Collared Sunbird *Anthreptes collaris subcollaris*. *Ibis* **113**, 366–367.

Greaves, R.H. and Tregenza, L.A. (1937). The Nile Valley Sunbird (*Nectarinia metallica*) in Egypt. *Oologist's Record* **17**, 79–83.

Hanmer, D.B. (1978). Nestling period in the Collared Sunbird. *Ostrich* **49**, 145.

Hanmer, D.B. (1989). The end of an era: final longevity figures for Nchalo, Malaŵi. *Safring News* **18**, 19–30.

Hanmer, D.B. and Manson, A.J. (1992). The Collared Sunbird in the Bvumba. *Honeyguide* **38**, 155–164.

Hogg, P. (1950). Some breeding records from the Anglo-Egyptian Sudan. *Ibis* **92**, 574–578.

Howe, S. (1987). The Nile Valley Sunbird in Egypt. *Bull. Orn. Soc. Middle East* **18**, 10.

Manson, A.J. (1990). Results of a ringing programme at Seldomseen, Vumba. 2. *Honeyguide* **36**, 131–141.

Skead, C.J. (1962). A study of the Collared Sunbird *Anthreptes collaris* (Vieillot). *Ostrich* **33**, 38–40.

Skinner, N.J. (1969). Notes on the breeding of the Pygmy Long-tailed Sunbird *Hedydipna platura* at Zaria. *Bull. Nigerian Orn. Soc.* **6**, 124–126.

Cinnyris

Archer, A.L. and Parker, I.S.C. (1993). Fruit-eating sunbirds. *Scopus* **17**, 60–61.

Archer, A.L. and Turner, D.A. (1993). Notes on the endemic species and some additional new birds occurring on Pemba Island, Tanzania. *Scopus* **16**, 94–98.

Aspinwall, D.R. (1979a). Movement analysis charts: Comments on Purple-banded Sunbird (Oct. 1976–Sept. 1978). *Zambian Orn. Soc. Newsl.* **9**(5), 54–55.

Aspinwall, D.R. (1979b). Comments on Southern White-bellied Sunbird (Oct. 1976–Sept. 1978). *Zambian Orn. Soc. Newsl.* **9**, 82–83.

Aspinwall, D.R. (1989). Oustalet's White-bellied Sunbird on Malaŵi/Zambia border. *Nyala* **14**, 39–40.

Baha el Din, S. (1985). The occurrence of the Palestine Sunbird (*Nectarina* [sic] *osea*) in Egypt. *Bull. Orn. Soc. Middle East* **14**, 1–2.

Bannerman, D.A. (1932). Account of the birds collected (i) by Mr. G.L. Bates on behalf of the British Museum in Sierra Leone and French Guinea; (ii) by Lt-Col. G.J. Houghton, R.A.M.C., in Sierra Leone, recently acquired by the British Museum, Part III. *Ibis* **1932**, 217–261.

Beals, E.W. (1966). Sight additions to the avifaunal list of Ethiopia. *J. E. Afr. Nat. Hist. Soc.* **25**, 227–229.

Beel, C. (1993). Some notes on Oustalet's White-bellied Sunbird. *Zambian Orn. Soc. News.* **23**, 71–72.

Benseler, A. (1970). Kleine Brutbeobachtung des Russbraunen Nektarvogels (*Cinnyris fuscus*) in meinem Garten. *Mitt. Orn. Arbeitsg. S.W.A. Sci. Soc.* **6**(1/2), 8.

Benson, C.W. (1982). Migrants in the Afrotropical Region south of the equator. *Ostrich* **53**, 31–49.

Benson, C.W. and Irwin, M.P.S. (1966). The sunbirds *Nectarinia bouvieri* and *batesi*. *Bull. Br. Orn. Club* **86**, 62–65.

Benson, C.W. and Prigogine, A. (1981). The status of *Nectarinia afra prigoginei* (Macdonald). *Gerfaut* **71**, 47–57.

Berrington, W. (1977). Seed-eating sunbird. *Bee-eater* **48**, 47.

Borello, W.D. (1992). An incidence of mass sunbird migration in northern Botswana. *Babbler* **23**, 18–21.

Brooke, R.K. (1964). Avian observations on a journey across central Africa and additional information on some of the species seen. *Ostrich* **35**, 277–292.

Brooks, D.J., Evans, M.I., Martins, R.P. and Porter, R.F. (1987). The status of birds in North Yemen and the records of the OSME Expedition in autumn 1985. *Sandgrouse* **9**, 4–66.

Buchanan, D. and Steyn, P. (1964). The incubation and nesting periods of the White-breasted Sunbird (*C. talatala* A. Smith). *Ostrich* **35**, 65–66.

Cheke, R.A. (1976). Notes on a nesting female Variable Sunbird *Nectarinia venusta*. *Bull. Br. Orn. Club* **96**, 5–8.

Clancey, P.A. (1962.) Miscellaneous taxonomic notes on African birds XIX. *Durban Mus. Novit.* **6**, 181–194.

Clancey, P.A. (1967). On variation in *Nectarinia talatala* (Smith). *Bull. Br. Orn. Club* **87**, 153–157.

Clancey, P.A. (1973a). Miscellaneous taxonomic notes on African birds XXXVII. An undescribed race of *Nectarinia mariquensis* (Smith) from the south-east African lowlands. *Durban Mus. Novit.* **10**, 12–13.

Clancey, P.A. (1973b). Miscellaneous taxonomic notes on African birds LIV. On the validity of *Cinnyris mariquensis ovamboensis* Reichenow, 1904. *Durban Mus. Novit.* **12**, 43–44.

Clancey, P.A. (1979). Miscellaneous taxonomic notes on African birds LIII. The valid subspecific criteria and ranges of the two eastern races of *Nectarina bifasciata* (Shaw). *Durban Mus. Novit.* **12**, 1–17.

Clancey, P.A. and Irwin, M.P.S. (1978). Species limits in the *Nectarinia afra/N. chalybea* complex of African double-collared sunbirds. *Durban Mus. Novit.* **11**, 331–351.

Clancey, P.A. and Willams, J.G. (1957). The systematics of the little Purple-banded Sunbird *Cinnyris bifasciatus* (Shaw), with notes on its allies. *Durban. Mus. Novit.* **5**, 27–41.

Cole, D.T. (1992). Whitebellied Sunbirds take to nesting cereously. *Babbler* **23**, 33–35.

Craig, A.J.F.K. and Simon, C. (1991). Sunbird and sugarbird seasons. *Safring News* **20**, 9–12.

Craig, A.J.F.K. and Hulley, P.E. (1994). Sunbird movements: a review, with possible models. *Ostrich* **65**, 106–110.

Crick, H.Q.P. and Marshall, P.J. (1981). The birds of Yankari Game Reserve, Nigeria: their abundance and seasonal occurrence. *Malimbus* **3**, 103–114.

Cunningham-van Someren, V.G.L. (1976). Bimodal breeding demonstrated by colour ringing at Karen. *Bull. E. Afr. Nat. Hist. Soc.* **1976**, 50.

Cyrus, D.P. (1989). Unusual plumage of Whitebellied Sunbird confusing birders? *Bokmakierie* **41**, 10.

Dillingham, I.H. (1984). The record of the Angola White-bellied Sunbird *Nectarinia oustaleti* from Kigoma in western Tanzania. *Scopus* **8**, 80.

Dowsett, R.J. (1977). Marico Sunbird in Southern Province. *Bull. Zambian Orn. Soc.* **9**, 30–31.

Dowsett-Lemaire, F. (1988). On the breeding behaviour of three montane sunbirds *Nectarinia* spp. in northern Malaŵi. *Scopus* **11**, 79–86.

Dowsett-Lemaire (1989b). Food plants and the annual cycle in a montane community of sunbirds (*Nectarinia* spp.) in northern Malaŵi. *Tauraco* **1**, 167–185.

Dowsett-Lemaire, F. (1997d). Seasonality of breeding and moult in forest and savanna birds in northern Congo. *Rev. Ecol. (Terre Vie)* **52**, 153–171.

Earlé, R.A. (1982) Aspects of the breeding biology and ecology of the White-bellied Sunbird. *Ostrich* **53**, 65–73.

Evans, T.D. and Anderson, G.Q.A. (1993b). Notes on Moreau's Sunbird *Nectarinia moreaui*. *Scopus* **17**, 63–64.

Eyckerman, R. and Cuvelier, D. (1982). The moult of some birds on Mount Cameroon. *Malimbus* **4**, 1–4.

Follett, B. (1990). Lesser Double-collared Sunbirds rearing Klaas's Cuckoo. *Promerops* **196**, 13–14.

Fraser, M. and Wheeler, G. (1991). Dusky Sunbirds at Rondevlei. *Promerops* **199**, 6.

Fraser, M.W. and Crowe, T.M. (1990). Effects of woody alien plant invasion on the birds of Mountain Fynbos in the Cape of Good Hope Nature Reserve. *South Afr. J. Zool.* **25**, 97–108.

Fuggles-Couchman, N.R. (1986). Breeding records of some Tanzanian birds. *Scopus* **10**, 20–26.

Goldstein, H. (1988). Infanticide by invading males in the Palestine Sunbird (*Nectarinia osea*). *Torgos* **7**, 31–36.

Goldstein, H. and Yom-Tov, Y. (1988). Breeding biology of the Orange-tufted Sunbird in Israel. *Ardea* **76**, 169–174.

Goldstein, H., Eisikovitz, D. and Yom-Tov, Y. (1986). Infanticide in the Palestine Sunbird (Israel). *Condor* **88**, 528–529.

Gray, H.H. (1986). Johanna's Sunbird in Nigeria. *Malimbus* **8**, 44.

Grimes, L. G. (1974). Dialects and geographical variation in the song of the Splendid Sunbird *Nectarinia coccinigaster*. *Ibis* **116**, 314–329.

Grote, H. (1948). *Chalcomitra ursulae* (Alex.) in Cameroon. *Ibis* **90**, 339.

Hanmer, D.B. (1981). Mensural and moult data of nine species of sunbird from Moçambique and Malaŵi. *Ostrich* **52**, 156–178.

Hanmer, D.B. (1989). The Nchalo Ringing Station – bird longevity and migrant return. *Nyala* **14**, 21–27.

Hanmer, D.B. and Chadder, B. (1997). Sunbird notes from the Mutare and Bvumba areas. *Honeyguide* **43**, 220–223.

Harebottle, D.M. (1999). Nestling period of the Greater Doublecollared Sunbird *Nectarinia afra*. *Bird Numbers* **8**, 22-23.

Herremans, M. (1992). Indirect evidence for the existence of movements of sunbirds in Botswana. *Babbler* **24**, 4–9.

Holman, F.C. (1949). The nest and eggs of *Cinnyris johannae*. *Ibis* **91**, 351–352.

Howells, W.W. (1971). Breeding of the Coppery Sunbird at Salisbury, Rhodesia. *Ostrich* **42**, 99–109.

Irwin, M.P.S. (1981). On the supposed occurrence of Shelley's Sunbird in southern Mozambique. *Honeyguide* **106**, 20–21.

Irwin, M.P.S. (1996). What species of double-collared sunbird occurs on Gorongosa Mountain? *Honeyguide* **42**, 246–250.

Irwin, M.P.S. and Benson, C.W. (1966). Notes on the birds of Zambia. *Arnoldia (Rhod.)* **2**, 32, 1–19.

Jackson, S., Nicolson, S.W. and Lotz, C.N. (1998). Sugar preferences and 'side bias' in Cape Sugarbirds and Lesser Double-collared Sunbirds. *Auk* **115**, 156–165.

Jensen, F.P. (1983). A new species of sunbird from Tanzania. *Ibis* **125**, 447–449.

Jensen, F.P. (1985). Something new under the sun. *Animal Kingdom* **88**(5), 18–21.

Jensen, R.A.C. and Clinning, C. (1974). Breeding biology of two cuckoos and their hosts in South West Africa. *Living Bird* **13**, 5–50.

Korff, J. von (1997). Lesser Doublecollared Sunbirds' unusual nest. *Bee-eater* **48**, 47.

Lane, S. (1996). Eclipse plumage in the Yellow-bellied Sunbird, *Nectarina* [sic] *venusta*. *Nyala* **19**, 54–55.

Lloyd, P. and Craig, A.J.F.K. (1989). Morphometrics, moult and taxonomy of the *Nectarinia afra/Nectarinia chalybea* complex of South African double-collared sunbirds. *Ann. Cape Prov. Mus. Nat. Hist.* **18**, 135–150.

Lotz, C.N. and Nicolson, S.W. (1998). A terrestrial fish: adaptations of an African sunbird for nectar feeding. Abstracts 22 Int. Orn. Congr., Durban. *Ostrich* **69**, 230.

Louette, M. (1980). The populations of *Nectarinia preussi* in the Cameroon montane area. *Proc. IV Pan-Afr. Orn. Congr.*, 9–16.

Louette, M. (1987). Additions and corrections to the avifauna of Zaïre (1). *Bull. Br. Orn. Club* **107**, 137–143.

Macdonald, J.D. (1958). Note on *Cinnyris manoensis* Reichenow. *Bull. Br. Orn. Club* **78**, 7–8.

Manson, C. and Manson, A. (1981). Notes on the Coppery Sunbird. *Honeyguide* **105**, 9–12.

Markman, S., Pinshow, B. and Wright, J. (1998). Quality of food eaten by parent sunbirds directly affects their brood-care effort. Abstracts 22 Int. Orn. Congr., Durban. *Ostrich* **69**, 196.

Markman, S., Pinshow, B. and Wright, J. (1999). Orange-tufted Sunbirds do not feed nectar to their chicks. *Auk* **116**, 257–259.

Martin, R. (1983). Possible eclipse plumage in the Lesser Doublecollared Sunbird. *Bokmakierie* **35**, 67.

Martin, R., Martin, J. and Martin, E. (1991). Lesser Doublecollared Sunbirds: new breeding areas. *Promerops* **198**, 14.

Martin, R., Martin, J., Martin, E., Pepler, D. and Tyler, D. (1989). Dusky Sunbirds breeding in the Little Karoo. *Promerops* **191**, 13–14.

Martins, R.P., Porter, R.F. and Stone, F. (1993). 'Preliminary Report of the OSME Survey of Southern Yemen and Socotra Spring 1993'. Orn. Soc. Middle East, Sandy, UK.

Matthews, R.A., Baltzer, M.C. and Howard, P.C. (1997). New bird records for Uganda, with an addition to the East African avifauna. *Scopus* **19**, 119–120.

Medland, R.D. (1992). Song-flight of White-bellied Sunbird. *Nyala* **16**, 29–30.

Milstein, P. le S. (1963). Nesting behaviour of Marico Sunbird and Greater Double-collared Sunbird. *Ostrich* **34**, 46.

Moyer, D.C. (1983). A record of the Angola White-bellied Sunbird *Nectarinia oustaleti* from Kasesya in southwestern Tanzania. *Scopus* **7**, 52.

Paterson, J. (1958). Hovering by *Nectarinia chalybea* and *N. venusta*. *Ostrich* **29**, 48.

Pauw, A. (1998). Pollen transfer on birds' tongues [*Cinnyris chalybea*]. *Afr. Birds and Birding* **3**(5), 17.

Payne, R.B. (1978). Microgeographic variation in songs of Splendid Sunbirds *Nectarinia coccinigaster*: population phenetics, habitats, and song dialects. *Behaviour* **65**, 282–308.

Percy, R.C., Percy, H.E. and Ridley, M.W. (1953). The waterholes at Ijara, Northern Province, Kenya. *J. E. Afr. Nat. Hist. Soc.* **1**(93), 2–14.

Pettet, A. (1977). Seasonal changes in nectar-feeding by birds at Zaria, Nigeria. *Ibis* **119**, 291–308.

Prendergast, H.D.V. (1983). Competition for nectar between sunbirds and butterflies. *Malimbus* **5**, 51–53.

Prigogine, A. (1979). Subspecific variation of Stuhlmann's Double-collared Sunbird, *Nectarinia stuhlmanni*, around the Albertine Rift. *Gerfaut* **69**, 225–238.

Rebelo, A.G. (1987). Sunbird feeding at *Satyrium odorum* Sond. flowers. *Ostrich* **58**, 185–186.

Schmidt, R.K. (1963). Marico Sunbird *Cinnyris mariquensis* as host of Klaas's Cuckoo *Chrysococcyx klaas*. *Ostrich* **34**, 176.

Schmidt, R.K. (1964). The Lesser Double-collared Sunbird, *C. chalybeus* (Linn.) in the south-western Cape. *Ostrich* **35**, 86–94.

Schmidt, R. (1991). Lesser Doublecollared Sunbirds nesting in gardens. *Promerops* **197**, 12.

Serle, W. (1950). Notes on the birds of south-western Nigeria. *Ibis* **92**, 84–94.

Serle, W. (1956). The Splendid Sunbird (*Cinnyris coccinigaster* (Latham)). *Nigerian Field* **21**, 78.

Sheppard, D.M. (1988). Aberrant sunbird behaviour. *Bull. E. Afr. Nat. Hist. Soc.* **18**, 6.

Skead, C.J. (1954). A study of the Larger Double-collared Sunbird, *Cinnyris a. afra* (Linnaeus). *Ostrich* **25**, 76–88.

Steyn, P. (1999). Suburban Orangebreasted Sunbirds. *Promerops* **238**, 18.

Stuart, S.N. and van der Willigen, T. (1980). Is Moreau's Sunbird *Nectarinia moreaui* a hybrid species? *Scopus* **4**, 56–58.

Tree, A.J. (1990) Notes on sunbird movements and nectar sources in Zimbabwe. *Honeyguide* **36**, 171–182.

Turner, D.A. (1977). Status and distribution of East African endemic species. *Scopus* **1**, 2–11.

Urban, E.K. (1975). Weights and longevity of some birds from Addis Ababa, Ethiopia. *Bull. Br. Orn. Club* **95**, 96–98.

Vande weghe, J.P. (1984). Further additions to the bird fauna of Rwanda. *Scopus* **8**, 60–63.

Vaughan, J.H. (1930). The birds of Zanzibar and Pemba. *Ibis* **1930**, 1–48.

Walsh, J.F. (1987). Records of birds seen in north-eastern Guinea in 1984–1985. *Malimbus* **9**, 105–122.

Wells, D.R. (1968). Zonation of bird communities on Fernando Poo. *Bull. Nigerian Orn. Soc.* **5**, 71–87.

Williams, A.A.E. (1978). Accidental parasitisation of an Eastern Double-collared Sunbird by a Variable Sunbird. *Scopus* **2**, 25–26.

Williams, A.J., Braine, S.G. and Bridgeford, P. (1986). The biology of the Dusky Sunbird in S.W.A./Namibia: a review. *Lanioturdus* **22**, 4–10.

Williams, J.G. (1950). On the status of *Cinnyris mediocris moreaui*. *Ibis* **92**, 645–647.

Winterbottom, J.M. (1965). Throat colour of the male Yellow-bellied Sunbird. *Ostrich* **35**, 239–240.

Worsley, S. (1983). The first record of the Miombo Double-collared Sunbird from Hwange National Park. *Honeyguide* **116**, 19.

Family ZOSTEROPIDAE

Bannerman, D.A. (1914). Report on the birds collected by the late Mr. Boyd Alexander (Rifle Brigade) during his last expedition to Africa. Pt 1. The birds of Prince's Island. *Ibis* **1914**, 596–631.

Bannerman, D.A. (1915). The birds of Annobon Island. *Ibis* **1915**, 227–234.

Basilio, A. (1957). 'Caza y Pesca en Annóbon'. Inst. Estud. Africanos, Madrid.

Bocage, J.V.B. du (1893). Mamíferos, aves e reptis da ilha de Anno-Bom. *J. Sci. Math. Phys. Nat. Lisboa* **2**, III, **9**, 1–46.

Broekhuysen, G.J. and Winterbottom, J.M. (1968). Breeding activity of the Cape White-eye *Zosterops virens capensis* Sundevall in the south-west Cape. *Ostrich* **39**, 163–176.

Bunning, J.L. (1985). Some notes on the Cape White-eye in the Transvaal. *Safring News* **14**, 7–12.

Calahan, B.D. (1981). Anting in the Cape White-eye. *Ostrich* **52**, 186.

Clancey, P.A. (1966). A catalogue of birds of the South African sub-region. *Durban Mus. Novit.* **7**, 13, 545–633.

Clancey, P.A. (1967). Taxonomy of the southern African *Zosterops*. *Ibis* **109**, 318–327.

Cohen, C. and Winter, D. (1992). Cape White-eye, Fiscal Shrike mimicking other birds. *Promerops* **202**, 9.

Craig, A.J.F.K. (1990). White-eyes revisited. *Safring News* **19**, 3–5.

Craig, A.J.F.K. and Hulley, P.E. (1996). Supplementary head molt in Cape White-eyes: a consequence of nectar feeding? *J. Field Orn.* **67**, 358–359.

Dowsett, R.J. (1985). Notes on white-eyes in the Cape Province and south-central Africa. *Safring News* **14**, 13–18.

Franke, E., Jackson, S. and Nicolson, S. (1998). Nectar sugar preferences and absorption in a generalist African frugivore, the Cape White-eye *Zosterops pallidus*. *Ibis* **140**, 501–506.

Fry, C.H. (1961). Notes on the birds of Annobon and other islands in the Gulf of Guinea. *Ibis* **103**, 267–276.

Gill, F.B. (1973). Intra-island variation in the Mascarene White-eye *Zosterops borbonica*. *Orn. Monogr.* **12**, 1–66.

Grimes, L. (1971). Notes on some birds seen at Buea and on Mount Cameroun. *Bull. Nigerian Orn. Soc.* **8**, 31/32, 35–41.

Happel, R. (1986). Observations of birds and other frugivores feeding at *Tetrorchidium didymostemon*. *Malimbus* **8**, 77–78.

Harrison, M.J.S. (1990). A recent survey of the birds of Pagalu (Annobon). *Malimbus* **11**, 135–143.

Harvey, D. (1993). 'Cinnamon-browed' white-eyes in Sana'a, Yemen. *Orn. Soc. Middle East Bull.* **30**, 28–29.
Hunter, N.D. (1986). An update on the status of white-eyes in Botswana. *Babbler* **12**, 14–16.
Jensen, F.P. and Stuart, S.N. (1986). The origin and evolution of the Cameroon montane forest avifauna. *In* 'Conservation of Cameroon Montane Forests' (Stuart, S.N., Ed.), pp. 28–37. ICBP, Cambridge.
Kunkel, P. (1975). Some feeding adaptations and postures of the Yellow White-eye (*Zosterops senegalensis*). *Zool. Afr.* **10**, 109–121.
Mees, G.F. (1957–1969). A systematic review of the Indo-Australian Zosteropidae. *Zool. Verh. Leiden* **35** (1957), 1–204; **50** (1961), 1–168 and **102** (1969), 1–390.
Moreau, R.E. (1957). Variation in the western Zosteropidae (Aves). *Bull. Brit. Mus. Nat. Hist.* **4**(7), 318–433.
Moreau, R.E., Perrins, M. and Hughes, J.T. (1969). Tongues of the Zosteropidae (white-eyes). *Ardea* **57**, 29–47.
Porter, R.N. (1975). Spider feeding on a Cape White-eye. *Lammergeyer* **22**, 49.
Prŷs-Jones, R.P. (1985). Movements, mortality and the annual cycle of white-eyes in southern Africa. *Safring News* **14**, 25–35.
Schmidt, R.K. (1955). Incubation and fledging periods of the Cape White-eye. *Ostrich* **26**, 157.
Skead, C.J. (1967). 'The Sunbirds of Southern Africa, also the Sugarbirds, the White-eyes and the Spotted Creeper'. A.A. Balkema, Cape Town.
Skead, C.J. and Ranger, G.A. (1958). A contribution to the biology of the Cape Province white-eyes (*Zosterops*). *Ibis* **100**, 319–333.
Stresemann, E. (1931). Die Zosteropiden der indo-australischen Region. *Mitt. Zool. Mus. Berlin* **17**, 201–238.
Vernon, C.J. (1985). Bird populations in two woodlands near Lake Kyle, Zimbabwe. *Honeyguide* **31**, 148–161.
Wells, D.R. (1968). Zonation of bird communities on Fernando Poo. *Bull. Nigerian Orn. Soc.* **5**, 71–87.
Whitelaw, D.A. (1985). Musings on the Cape White-eye in the southwestern Cape. *Safring News* **14**, 19–24.
Wolff-Metternich, G.F. and Stresemann, E. (1956). Biologische Notizen über Vögel von Fernando Poo. *J. Orn.* **97**, 274–290.

Family PROMEROPIDAE

Bennett, G.F. and De Swardt, D.H. (1989). First African record of *Leucocytozoon anellobiae* (Apicomplexa: Leucocytozoidae) in Gurney's Sugarbird *Promerops gurneyi*. *Ostrich* **60**, 171.
Bock, W.J. (1985). Relationships of the sugarbird (*Promerops*; Passeriformes, ?Meliphagidae). Proc. Int. Symp. Afr. Vert., Bonn, 348–374.
Broekhuysen, G.J. (1959). The biology of the Cape Sugarbird, *Promerops cafer* (L.). *Ostrich* Suppl. **3**, 180–221.
Broekhuysen, G.J. (1971). Partial albino sugarbird. *Ostrich* **42**, 70.
Burger, A.E., Siegfried, W.R. and Frost, P.G.H. (1976). Nest-site selection in the Cape Sugarbird. *Zool. Afr.* **11**, 127–158.
Collins, B.G. (1983a). Seasonal variations in the energetics of territorial Cape Sugarbirds. *Ostrich* **54**, 121–125.
Collins, B.G. (1983b). Pollination of *Mimetes hirtus* (Proteaceae) by Cape Sugarbirds and Orange-breasted Sunbirds. *J. S. Afr. Bot.* **49**, 125–142.
Craib, C.L. (1977). Gurney's Sugarbird. *Wits Bird Club Newssheet* **96**, 9–10.
De Swardt, D.H. (1987). Gurney's Sugarbird. *Wits Bird Club Newssheet* **139**, 17–19.
De Swardt, D.H. (1989). Some observations on the local movements of Gurney's Sugarbird in the Lydenburg area. *Safring News* **18**, 31–32.
De Swardt, D.H. (1991). The seasonal movements of Gurney's Sugarbird *Promerops gurneyi* in the Lydenburg area, Transvaal. *Ostrich* **62**, 40–44.
De Swardt, D.H. (1992). Distribution, biometrics and moult of Gurney's Sugarbird *Promerops gurneyi*. *Ostrich* **63**, 13–20.
De Swardt, D.H. (1993). Factors affecting the densities of nectarivores in *Protea roupelliae* woodland. *Ostrich* **64**, 172–177.
De Swardt, D.H. and Bothma, N. (1991). Notes on the nest of Gurney's Sugarbird in the eastern Transvaal. *Birding Southern Afr.* **43**, 2–3.
De Swardt, D.H. and Buys, P.J. (1992). Cape Sugarbird at Jagersfontein, southern Orange Free State. *Ostrich* **63**, 136.
De Swardt, D.H. and Louw, S. (1994). The diet and foraging behaviour of Gurney's Sugarbird, *Promerops gurneyi*. *Navors. Nas. Mus. Bloemfontein* **10**, 245–258.
Farquhar, M.R., Lorenz, M, Rayner, J.L. and Craig, A.J.F.K. (1996). Feather ultrastructure and skeletal morphology as taxonomic characters in African sunbirds (Nectariniidae) and sugarbirds (Promeropidae). *J. Afr. Zool.* **110**, 321–331.
Fraser, M. and McMahon, L. (1992a). Some recent ringing recoveries. *Promerops* **202**, 11.
Fraser, M.W. and McMahon, L. (1992b). Habitat change by Cape Sugarbirds and Orangebreasted Sunbirds in an apparent response to fire in old mountain fynbos. *Safring News* **21**, 51–54.
Fraser, M.W., McMahon, L., Underhill, L.G. and Underhill, G.D. (1989). Nectarivore ringing in the southwestern Cape. *Safring News* **18**, 3–18.
Friedmann, H. (1952). The long-tailed sugarbird of eastern Rhodesia. *J. Washington Acad. Sci.* **42**, 1, 31–32.
Henderson, K. and Cherry, M. (1998). Testing alternative hypotheses of provisioning in the Cape Sugarbird *Promerops cafer*. Abstracts 22 Int. Orn. Congr., Durban. *Ostrich* **69**, 253.
Jackson, S., Nicolson, S.W. and Lotz, C.N. (1998). Sugar preferences and 'side bias' in Cape Sugarbirds and Lesser Double-collared Sunbirds. *Auk* **115**, 156–165.
Martin, J. (1953). Rapid building of Cape Sugarbird. *Ostrich* **24**, 129.
Martin, R. and Mortimer, J. (1991). Fynbos: too frequent burning poses threat to Cape Sugarbirds. *Promerops* **199**, 6–8.
McMahon, L. and Fraser, W. (1988). 'A Fynbos Year'. David Philip, Cape Town.
Mostert, D.P., Siegfried, W.R. and Louw, G. N. (1980). *Protea* nectar and satellite fauna in relation to the food requirements and pollinating role of the Cape Sugarbird. *S. Afr. J. Sci.* **76**, 409–412.
Nevill, H. (1987). Gurney's Sugarbird breeding in winter. *Bokmakierie* **39**, 71–74.
Olson, S.L. and Ames, P.L. (1984). *Promerops* as a thrush, and its implications for the evolution of nectarivory in birds. *Ostrich* **56**, 213–218.
Oschadleus, D. and Fraser, M. (1988). Observations of colour-ringed Cape Sugarbirds at Kirstenbosch. *Safring News* **17**, 59–64.
Rebelo, A.G. (1987). Visits to *Oldenburgia grandis* (Thunb.) Baillon (Asteraceae) by the Cape Sugarbird *Promerops cafer*. *Ostrich* **58**, 186–187.

Richardson, D. (1990). Cape Sugarbird feeding in Sisal plants during summer. *Promerops* **193**, 9.

Seiler, H.W. and Prŷs-Jones, R.P. (1989). Mate competition, mate guarding, and unusual timing of copulations in the Cape Sugarbird (*Promerops cafer*). *Ostrich* **60**, 159–164.

Seiler, H.W. and Rebelo, A.G. (1987). A sexual difference in the Cape Sugarbird's roles as a pollinator of *Protea lepidocarpodendron*? *Ostrich* **58**, 43–45.

Sibley, C.G. and Ahlquist, J.E. (1974). The relationships of the African sugarbirds (*Promerops*). *Ostrich* **45**, 22–30.

Skead, C.J. (1964). Sugarbirds in the Amatole Mountains, King William's Town, C.P. *Ostrich* **35**, 236.

Skead, C.J. (1967). 'Sunbirds of Southern Africa also the Sugarbirds, the White-eyes and the Spotted Creeper'. A. A. Balkema, Cape Town.

Skead, D.M. (1963). Gurney's Sugarbird, *Promerops gurneyi* Verreaux, in the Natal Drakensberg. *Ostrich* **34**, 160–164.

Steyn, P. (1973). A nest of Gurney's Sugarbird. *Honeyguide* **73**, 29–31.

Steyn, P. (1997). Cape Sugarbird. *Afr. Birds and Birding* **2**(4), 27–32.

Winterbottom, J.M. (1962) Breeding season of Long-tailed Sugarbird *Promerops cafer* (L.). *Ostrich* **33**, 77.

Winterbottom, J.M. (1964). Notes on the comparative ecology of the Long-tailed Sugarbird and the Orange-breasted Sunbird. *Ostrich* **35**, 239–240.

Wooller, R.D. (1982). Feeding interactions between sunbirds and sugarbirds. *Ostrich* **53**, 114–115.

Family LANIIDAE

Aspinwall, D.R. (1984). Movement analysis charts: Comments on Fiscal Shrike. *Bull. Zambian Orn. Soc.* **14**, 130–132.

Banage, W.B. (1969). Territorial behaviour and population in the Grey-backed Fiscal Shrike. *Uganda J.* **33**, 201–208.

Beven, G. and England, M.D. (1969). The impaling of prey by shrikes. *Brit. Birds* **62**, 192–199.

Brosset, A. (1989). Un cas d'association à bénéfice mutuel, celui de la Pie-grièche *Lanius cabanisi* avec les bubalornis *Bubalornis niger*. *Rev. Ecol. (Terre et Vie)* **44**, 103–106.

Bruderer, B. and Bruderer, H. (1993a). Distribution and habitat preference of Red-backed Shrikes *Lanius collurio* in southern Africa. *Ostrich* **64**, 141–147.

Bruderer, B. and Bruderer, H. (1993b). Numbers of Red-backed Shrikes in different habitats of South Africa. *Bull. Br. Orn. Club* **114**, 192–202.

Buttiker, W. (1960). Artificial nesting devices in Southern Africa. *Ostrich* **31**, 44–45.

Cade, T.J. (1995). Shrikes as predators. *In* 'Shrikes (Laniidae) of the World: Biology and Conservation' (R. Yosef and F.E. Lohrer, Eds). *Proc. Western Found. Vert. Zool.* **6**, 1–5.

Christy, P. (1990). New records of Palearctic migrants in Gabon. *Malimbus* **11**, 117–122.

Clancey, P.A. (1955). A new geographical race of the Fiscal Shrike *Lanius collaris* from the deserts of South-West Africa and Angola. *Bull. Br. Orn. Club* **75**, 32–33.

Clancey, P.A. (1965). Variation in the White-crowned Shrike *Eurocephalus anguitimens* Smith, 1836. *Arnoldia* **1**, 1–3.

Clancey, P.A. (1970). Miscellaneous taxonomic notes on African birds XXVIII. *Durban Mus. Novit.* **8**, 325–351.

Cooper, J. (1971a). The breeding of the Fiscal Shrike in southern Africa. *Ostrich* **42**, 166–174.

Cooper, J. (1971b). Post nestling development in the Fiscal Shrike. *Ostrich* **42**, 175–178.

Cruz Solis, C. de la and Lope Rebello, F. de (1985). Reproduction de la Pie-grièche meridionale (*Lanius excubitor meridionalis*) dans le sud-ouest de la Péninsule iberique. *Gerfaut* **75**, 199–209.

Devereux, C. (1988). The Fiscal Shrike territorial imperative. *Afr. Birds and Birding* **3**(5), 52–57.

Dittami, J.P. and Knauer, B. (1986). Seasonal organisation of breeding and molting in the Fiscal Shrike (*Lanius collaris*). *J. Orn.* **127**, 79–84.

Dorka, V. (1975). Zum 'Faust'-Gebrauch beim Raubwürger *Lanius excubitor* (Laniinae) und Weizscheitelwürger *Eurocephalus anguitimens* (Prionopinae). *Anz. Orn. Ges. Bayern* **14**, 314–319.

Douglas, R. (1992). Homing instincts in the Fiscal Shrike *Lanius collaris*? *Mirafra* **9**, 31.

Dowsett, R.J. (1971). The Lesser Grey Shrike *Lanius minor* in Africa. *Ostrich* **42**, 259–270.

Dowsett, R.J. (1996). Souza's shrike *Lanius souzae*. *Zambian Orn. Soc. Newsl.* **26**, 119–120.

Felemban, H.M. (1995). Trapping of spring migrants on Qummah Island, Farasan Archipelago in the Red Sea. *Orn. Soc. Middle East Bull.* **35**, 1–13.

Ferguson-Lees, I.J. (1967). [Masked Shrikes feeding on birds.] *Brit. Birds* **60**, 303–304.

Gee, J.P. and Heigham, J.B. (1971). Red-backed Shrike at Lagos. *Bull. Nigerian Orn. Soc.* **8**, 9–10.

Grimes, L.G. (1979a). Sexual dimorphism in the Yellow-billed Shrike *Corvinella corvina* and in other African shrikes (subfamily Laniinae). *Bull. Br. Orn. Club* **99**, 33–36.

Grimes, L.G. (1979b). The Yellow-billed Shrike *Corvinella corvina*: an abnormal host of the Yellow-billed Cuckoo *Cuculus gularis*. *Bull. Br. Orn. Club* **99**, 36–38.

Grimes, LG. (1980). Observations of group behaviour and breeding biology of the Yellow-billed Shrike *Corvinella corvina*. *Ibis* **122**, 166–192.

Günther, R. and Feiler, A. (1985). Die Vögel der Insel São Tomé. *Mitt. Zool. Mus. Berlin* 61 *Suppl. Ann. Orn.* **9**, 3–28.

Hargrove, J.W., Marshall, B.E. and Mentz, D.L. (1972). Observations on the Fiscal Shrike. *Rhodesia Sci. News* **6**, 349–351.

Hargrove, J.W., Marshall, B.E. and Mentz, D.L. (1973). A simple method of trapping the Fiscal Shrike. *Safring News* **2**, 17.

Harris, T. (1995). Species recognition in the southern African population of the Fiscal Shrike (*Lanius collaris*). In 'Shrikes (Laniidae) of the World: Biology and Conservation' (R. Yosef and F.E. Lohrer, Eds). *Proc. West. Found. Vert. Zool.* **6**, 11–21.

Harris, T. and Arnott, G. (1988). 'Shrikes of Southern Africa'. Struik Winchester, Cape Town.

Herremans, M. (1993). Seasonal dynamics in subKalahari bird communities with emphasis on migrants. *In* 'Birds and the African Environment' (Wilson R.T., Ed.). *Proc. VIII Pan-Afr. Orn. Congr.*, 555–564.

Herremans, M. (1994). Fifteen years of migrant phenology records in Botswana: a summary and prospects. *Babbler* **28**, 47–68.

Herremans, M. (1997). Habitat segregation of male and female Red-backed Shrikes *Lanius collurio* and Lesser Grey Shrikes *Lanius minor* in the Kalahari basin, Botswana. *J. Avian Biol.* **28**, 240–248.

Herremans, M. (1998). Monitoring the world population of the Lesser Grey Shrike on non-breeding grounds in southern Africa. *J. Orn.* **139**, 485–493.

Herremans, M., Herremans-Tonnoeyr, D. and Borello, W.D. (1995). Non-breeding site fidelity of Red-backed Shrikes *Lanius collurio* in Botswana. *Ostrich* **66**, 145–147.

Lefranc, N. and Worfolk, T. (1997). 'Shrikes: a Guide to Shrikes of the World'. Pica Press, Sussex.

Macdonald, M.A. (1980). The ecology of the Fiscal Shrike in Ghana, and a comparison with studies from southern Africa. *Ostrich* **51**, 65–74.

Maclean, G.L. and Maclean, C.M. (1976). Extent of overlap in two races of the Fiscal Shrike. *Ostrich* **47**, 66.

Markus, M.B. (1972). Notes on the natal plumage of South African passeriform birds. *Ostrich* **43**, 17–22.

Marshall, B. (1990). Scavenging behaviour of the Fiscal Shrike. *Honeyguide* **36**, 194.

Marshall, B.E. and Cooper, J. (1969). Observation on the breeding biology of the Fiscal Shrike. *Ostrich* **40**, 141–149.

Massa, B. (1999). New and lesser known birds from Libya. *Bull. Br. Orn. Club* **119**, 129–133.

Medland, R.D. (1991). Souza's Shrike attacking Violet-backed Sunbird. *Nyala* **15**, 49.

Meinertzhagen, R. (1930). 'Nicoll's Birds of Egypt'. Hugh Rees, London.

Moreau, R.E. (1961). Problems of Mediterranean-Sahara migration. *Ibis* **103a**, 373–427, 580–623.

Newton, A. (1893–96). 'A Dictionary of Birds'. A. & C. Black, London.

Nikolaus, G. (1983). An important passerine ringing site near the Sudan Red Sea coast. *Scopus* **7**, 15–18.

Nikolaus, G. (1984). *Lanius excubitor* 'jebelmarrae'. *Bull. Br. Orn. Club* **104**, 147.

Nikolaus, G. (1990). Shrikes Laniidae feeding on Marsh Warblers *Acrocephalus palustris* during migration. *Scopus* **14**, 26–28.

Olivier, G. (1944). 'Monographie des Pies-grièches du Genre *Lanius*'. Lecerf, Rouen.

Panow, E.N. (1983). 'Die Wurger der Paläarktis'. Wittenburg Lutherstadt.

Pearson, D.J. (1970). Weights of Red-backed Shrikes on autumn passage in Uganda. *Ibis* **112**, 114–115.

Pearson, D.J. (1979). The races of the Red-tailed Shrike *Lanius isabellinus* occurring in East Africa. *Scopus* **3**, 74–78.

Pearson, D.J. (1981). Field identification of Isabelline Shrike. *Dutch Birding* **3**, 89–91.

Pearson, D.J. (2000). The races of the Isabelline Shrike *Lanius isabellinus* and their nomenclature. *Bull. Br. Orn. Club* **120**, 22–27.

Plumb, W.J. (1978). Co-operative feeding of young by White-crowned Shrikes. *Bull. E. Afr. Nat. Hist. Soc.* **1978**, 89.

Portenko, L.A. (1960). 'Ptitsy SSSR' 4. Moscow.

Raikow, R.J., Polumbo, P.J. and Borecky, S.R. (1980). Appendicular myology and relationships of the shrikes (Aves: Passeriformes: Laniidae). *Ann. Carnegie Mus.* **49**, 131–152.

Roos, L. and Roos, M. (1988). Do Fiscal Shrikes feed on their stored food? *Mirafra* **5**, 28–29.

Sauer, F. and Sauer, E. (1960). Zugvögel aus den paläarktischen und afrikanischen Region in Sudwestafrika. *Bonn Zool. Beitr.* **11**, 40–86.

Schüz, E. (1971). 'Grundiss der Vogelzugskunde'. Berlin.

Schwan, T.G. and Hikes, N. (1979). Fiscal Shrike predation on the bat *Pipistrellus kuhli* in Kenya. *Biotropica* **11**, 21.

Simmons, K.E.L. (1951). Interspecific territorialism. *Ibis* **93**, 407–413.

Simmons, K.E.L. (1954). Field notes on the behaviour of some passerines migrating through Egypt. *Ardea* **42**, 140–151.

Smith-Symms, S. (1999). Befriending fiscal shrikes. *Promerops* **238**, 19.

Spina, F., Massi, A. and Montemaggiori, A. (1994). Back from Africa: who's running ahead? Aspects of differential migration of sex and age classes in Palearctic-African spring migrants. *Ostrich* **65**, 137–150.

Stegmann, B. (1930). Über die Formen der paläarktischen Rotruken- und Rotschanzwurger und deren taxonomischen Wert. *Orn. Monatsber.* **38**, 106–118.

Stepanyan, L.S. (1990). 'Konspekt Ornithologicheskoy Fauny SSSR'. Moscow.

Stevenson, T. (1983). A second record of the Nubian Shrike *Lanius nubicus* in East Africa. *Scopus* **7**, 97–98.

Steyn, P. (1976). Protracted prelaying nest building by a Fiscal Shrike. *Ostrich* **47**, 68.

Stresemann, E. (1927) Die Wanderungen der Rotschwanzwurger (Formenkreis *Lanius cristatus*). *J. Orn.* **75**, 68–85.

Stresemann, E. and Stresemann, V. (1972). Über die Mauser in der Gruppe *Lanius isabellinus*. *J. Orn.* **113**, 60–75.

Took, J.M.E. (1966). The nest of Souza's Shrike *Lanius souzae*. *Ostrich* **37**, 155–156.

Toschi, A. (1950). Sulla biologia del *Lanius collaris humeralis* Stanley. *Lab. Zool. Appl. Caccia Suppl.* **24**, 1–136.

Tye, A. (1984). Attacks by shrikes *Lanius* spp. on wheatears *Oenanthe* spp.: competition, kleptoparasitism, or predation? *Ibis* **126**, 95–102.

Ullrich, B. (1971). Untersuchungen zur Ethiologie und Ökologie des Rotkopfwürgers (*Lanius senator*) in Südwestdeutschland im Vergleich zu Raubwürger (*Lanius excubitor*), Schwarzstirnwürger (*Lanius minor*) und Neuntöter (*Lanius collurio*). *Vogelwarte* **26**, 1-77.

Vaurie, C. (1955). Systematic notes on Palearctic birds. 17, Laniidae. *Amer. Mus. Novit.* **1752**, 1–19.

Voous, K.H. (1979). Capricious taxonomic history of Isabelline Shrike. *Brit. Birds* **72**, 573–578.

Watson, G.E. (1967). Masked Shrikes feeding on birds. *Brit. Birds* **60**, 303.

Watson, R.T. and Watson, C.R.B. (1983). A partial albino Longtailed Shrike. *Bokmakierie* **35**, 43–44.

Yosef, R. (1993). Prey transport by Loggerhead Shrike. *Condor* **95**, 231–233.

Yosef, B. and Pinshow, R. (1988). Polygyny in the Northern Shrike (*Lanius excubitor*) in Israel. *Auk* **101**, 580.

Zack, S. (1986). Breeding biology and inter-territory movements in a Fiscal Shrike population in Kenya. *Ostrich* **57**, 65–74.

Zack, S. and Ligon, J.D. (1985a). Cooperative breeding in *Lanius* shrikes. I. Habitat and demography of two sympatric species. *Auk* **102**, 754–765.

Zack, S. and Ligon, J.D. (1985b). Cooperative breeding in *Lanius* shrikes. II. Maintenance of group-living in a nonsaturated habitat. *Auk* **102**, 766–773.

Family MALACONOTIDAE

Malaconotus, Telophorus

Andrews, S.M. (1994). Rediscovery of the Monteiro's Bush-shrike *Malaconotus monteiri* in Cameroon. *Bull. Afr. Bird Club* **1**, 26–27.

Anon. (1992). Bush-shrike finds. *World Birdwatch* **14**(4), 5.

Baker, R.R. and Parker, G.A. (1979). The evolution of bird coloration. *Phil. Trans. Roy. Soc. London Ser. B*, **287**, 63–130.

Bennun, L.A. (1985). Notes on behaviour and plumage dimorphism in Lagden's Bush Shrike *Malaconotus lagdeni*. *Scopus* **9**, 111–114.

Benson, C.W. and Irwin, M.P.S. (1965). The birds of *Cryptosepalum* forests, Zambia. *Arnoldia (Rhod.)* **1**, 28, 1–12.

Bowden, C.G.R. and Andrews, M. (1994). Mount Kupe and its birds. *Bull. Afr. Bird Club* **1**, 13–16.

Chadwick, P. (1984). Grey-headed Bush Shrike taking eggs. *Honeyguide* **30**, 36.

Clancey, P.A. (1957). Geographical variation in the South African populations of *Malaconotus blanchoti* Stephens with the description of a new race. *Bull. Br. Orn. Club* **77**, 99–102.

Clancey, P.A. (1959). Miscellaneous taxonomic notes on African birds XIII. The South African races of the Orange-breasted Bush-Shrike *Malaconotus sulfureopectus* (Lesson). *Durban Mus. Novit.* **5**, 166–172.

Clancey, P.A. (1960). The races of Bokmakierie *Telophorus zeylonus* (Linnaeus), with the characters of a new form from South-West Africa. *Bull. Br. Orn. Club* **80**, 121–124.

Clancey, P.A. (1975). On the endemic birds of the Transvaal-montane forests. *Durban Mus. Novit.* **10**, 151–180.

Clancey, P.A. (1982). Miscellaneous taxonomic notes on African birds 67. On *Malaconotus blanchoti* Stephens in South West Africa. *Durban Mus. Novit.* **13**, 134–137.

Curtis, A.B. (1987). Grey-headed Bush Shrike dragging live snake along ground. *Honeyguide* **33**, 153.

Demey, R. (1997). Recent reports. *Bull. Afr. Bird Club* **4**, 142–145.

Donnelly, B.G. (1978). Albinistic Greyheaded Bush-Shrike. *Ostrich* **49**, 91.

Dowsett-Lemaire, F. (1999). First observations on the territorial song and display of the Kupé Bush Shrike *Malaconotus kupeensis*. *Malimbus* **21**, 115–117.

Dowsett-Lemaire, F. and Dowsett, R.J. (1998). Zoological survey of small mammals, birds, frogs and butterflies in the Bakossi and Kupé Mts, Cameroon. Unpublished report at BirdLife Intl, Cambridge.

Evans, T.D. (1997). Records of birds from the forests of the East Usambara lowlands, Tanzania, August 1994–February 1995. *Scopus* **19**, 92–108.

Grobler, J.H. (1979). Observations on a breeding pair of Grey-headed Bush Shrikes. *Honeyguide* **98**, 15.

Hall, B.P. (1960b). The faunistic importance of the scarp of Angola. *Ibis* **102**, 420–442.

Hall, B.P., Moreau, R.E. and Galbraith, I.C.J. (1966). Polymorphism and parallelism in the African bush-shrikes of the genus *Malaconotus* (including *Chlorophoneus*). *Ibis* **108**, 161–182.

Hanmer, D.B. (1983). Some longevity records. *Safring News* **12**, 61.

Harris, T. and Arnott, G. (1988). 'Shrikes of Southern Africa'. Struik, Cape Town.

Hornby, H.E. (1973). Shrikes and wasps. *Honeyguide* **75**, 28–29.

Irwin, M.P.S. (1968). A new race of the Bokmakierie *Telophorus zeylonus* (Linnaeus) (Aves) from the Chimanimani Mountains, Rhodesia. *Arnoldia (Rhod.)* **3**, 1–5.

Jackson, H.D. (1972). Comment on *Telophorus zeylonus restrictus* Irwin, the Chimanimani Mountain race of the Bokmakierie (Aves: Laniidae). *Arnoldia (Rhod.)* **6**, 1–5.

Kok, O.B. (1999). Kokkewiet–skyngeveg. *Mirafra* **16**(2), 23–24.

Kopij, G. (1999). Breeding ecology of the Bokmakierie *Telophorus zeylonus* in Bloemfontein. *Mirafra* **16**(1), 2–6.

MacCan, L.G. (1975). Grey-headed Bush-Shrike and doves' eggs. *Albatross* **236**, 3.

Macaulay, L.R. and Sinclair, J.C. (1999). Perrin's Bush-Shrike *Telophorus viridis*, new to Gabon. *Malimbus* **21**, 110–111.

Meise, W. (1968). Zur Speciation afrikanischer, besonders angolesischer Singvögel der Gattungen *Terpsiphone*, *Dicrurus* und *Malaconotus*. *Zool. Beitr.* **14**, 1–60.

Moreau, R.E. and Southern, H.N. (1958). Geographical variation and polymorphism in *Chlorophoneus* shrikes. *Proc. Zool. Soc. London* **130**, 301–328.

Mundy, P.J. and Cook, A.W. (1972). The birds of Sokoto, Part 2. *Bull. Nigerian Orn. Soc.* **9**, 60–76.

Ndao, B. (1999). Observations de Gonoleks de Barbarie *Laniarius barbarus* à dessous jaune Senegal. *Bull. Afr. Bird Club* **6**, 60.

Phillips, R.L. (1979). Observations on the nesting of the Grey-headed Bush-Shrike *Malaconotus blanchoti*. *Honeyguide* **98**, 31–33.

Prigogine, A. (1953). Notes sur les pie-grièches multicolores du Congo belge. *Rev. Zool. Bot. Afr.* **45**, 313–324.

Prigogine, A. (1984). Note sur deux gladiateurs (*Malaconotus*). *Gerfaut* **74**, 75–81.

Prigogine, A. (1986). Le plumage immature du gladiateur de Lagden, *Malaconotus lagdeni*, et du gladiateur ensanglanté, *Malaconotus cruentus*. *Gerfaut* **76**, 255–261.

Reichenow, A. (1894). Zur Vogelfauna von Kamerun. Zweiter Nachtrag. *J. Orn.* **42**, 29–43.

Robinson, C. St C. (1953). Notes on the breeding of the Bokmakirie (*Telophorus zeylonus*). *Ostrich* **24**, 153–158.

Schouteden, H. (1969). La faune ornithologique du Kivu II. Passereux. *Document. Zool., Mus. Roy. Afr. Centr.* **15**.

Sclater, W.L. and Mackworth-Praed, C. (1918). A list of the birds of the Anglo-Egyptian Sudan, based on the collection of Mr. A. L. Butler, Mr. A. Chapman and Capt. H. Lynes, R.N., and Major Cuthbert Christy, R.A.M.C. (T.F.). II. *Ibis* **1918**, 602–721.

Serle, W. (1951). A new species of shrike and a new race of apalis from West Africa. *Bull. Br. Orn. Club* **71**, 41–43.

Serle, W. (1952a). The polymorphic forms of *Chlorophoneus multicolor multicolor* (Gray), in the British Cameroons. *Bull. Br. Orn. Club* **72**, 26–27.

Serle, W. (1952b). Colour variation in *Malaconotus cruentus* (Lesson). *Bull. Br. Orn. Club* **72**, 27–28.

Steyn, P. (1967). Orange-breasted Bush-Shrike (*Chlorophoneus sulfureopectus*) eating bees. *Ostrich* **38**, 286.

Steyn, P. (1973). A pair of *Telophorus zeylonus* with young. *Honeyguide* **73**, 29–31.

Svendsen, J.O. and Hansen, L.A. (1995). 'Report on the Uluguru Biodiversity Survey 1993, Part A'. RSPB/BirdLife Intl.

Vande weghe, J.-P. (1974). Additions et corrections à l'avifaune du Rwanda. *Rev. Zool. Afr.* **88**, 81–98.

Williams, E. (1998). Green-breasted Bush-shrike *Malaconotus gladiator* and its relationship with Monteiro's Bush-shrike *M. monteiri*. *Bull. Afr. Bird Club* **5**, 101–104.

Wilkinson, R. (1978). Behaviour of Grey-headed Bush-shrikes at their nest. *Bull. Nigerian Orn. Soc.* **14**, 87.

Antichromus, Tchagra, Dryoscopus, Laniarius, Nilaus

Aldiss, D.T. and Hunter, N.D. (1985). A first record of the Southern Boubou in Botswana. *Babbler* **10**, 34–35.

Bannerman, D.A. (1939). A new bush-shrike from Angola. *Ibis* **1939**, 746–750.

Borello, W. (1988). Alternative diet of a Crimsonbreasted Shrike. *Babbler* **16**, 20.

Brewster, C.A. (1996). Summary of category-B records. *Babbler* **31**, 32–41.

Britton, P.L. (1970). Two new shrikes for Kenya. *Bull. Br. Orn. Club* **90**, 133–134.

Brooke, R.K. (1980). The juvenile plumage of the Southern Puffback. *Honeyguide* **101**, 22–23, 25.

Chittenden, H. (1977). Puffback Shrike. *Bokmakierie* **29**, 53–54.

Clancey, P.A. (1961). Miscellaneous taxonomic notes on African birds XVII. The southern African races of the Puffback *Dryoscopus cubla* (Shaw). *Durban Mus. Novit.* **6**, 105–118.

Clancey, P.A. (1971). Miscellaneous taxonomic notes on African birds, XXXIII. On the southern range limits of *Nilaus afer nigritemporalis* Reichenow, 1892. *Durban Mus. Novit.* **9**, 122–129.

Edwards, E.A. (1998). Swamp Boubou kills frog. *Honeyguide* **44**, 94–95.

Fenn, T. (1975). Observations on the nest-building of the Puffback Shrike. *Honeyguide* **83**, 41–42.

Field, G.D. (1979). The *Laniarius* bushshrikes in Sierra Leone. *Bull. Br. Orn. Club* **99**, 42–44.

Grimes, L.G. (1965). Antiphonal singing in *Laniarius barbarus barbarus* and the auditory reaction time. *Ibis* **107**, 101–104.

Grimes, L.G. (1966). Antiphonal singing and call notes of *Laniarius barbarus barbarus*. *Ibis* **108**, 122–126.

Grimes, L.G. (1976). The duets of *Laniarius atroflavus, Cisticola discolor* and *Bradypterus barratti. Bull. Br. Orn. Club* **96**, 113–120.

Haffer, J.H. (1998). Species concepts and species limits in ornithology. In 'Handbook of the Birds of the World', Vol. 4 (del Hoyo, J., Elliott, A. and Sargatal, J., Eds), pp. 11-24. Lynx Edicions, Barcelona.

Hall, B.P. (1954). A review of the Boubou Shrike *Laniarius ferrugineus. Ibis* **96**, 343–355.

Hanmer, D.B. (1989). The Nchalo Ringing Station – bird longevity and migrant return. *Nyala* **14**, 21–27.

Harkrider, J.R. (1993). Garden and farm-bush birds of Njala, Sierra Leone. *Malimbus* **15**, 38–46.

Harris, T. and Arnott, G. (1988). 'Shrikes of Southern Africa'. Struik Winchester, Cape Town.

Helversen, D. von (1980). Structure and function of antiphonal duets. *Acta XVII Int. Orn. Congr.*, **1**, 682–688.

Hunter, N.D. (1988). Systematics of the Southern and Tropical Boubous in Botswana. *Babbler* **16**, 7–10.

Hurford, J.L., Lombard, A.T., Kemp, A.C. and Benn, G.A. (1996). Geographical analysis of six rare bird species in the Kruger National Park, South Africa. *Bird Conserv. Intl* **6**, 117–137.

Irwin, M.P.S. (1977). Some little-known and inadequately documented Rhodesian birds. *Honeyguide* **90**, 9–15.

Irwin, M.P.S. (1987). Systematics of the Southern and Tropical Boubous in the Transvaal, Zimbabwe and southern Mozambique. *Honeyguide* **33**, 151–153.

Joubert, S.C.J. and English, M. (1973). A new bird record for the Kruger National Park. *Koedoe* **16**, 199–200.

Kaumanns, W. (1975). Haltung und Zucht von Boubous oder Flötenwürgern (*Laniarius ferrugineus*). *Gef. Welt* **99**, 209–212.

Lack, P.C. and Quicke, D.L.J. (1978). Dietary notes on some Kenyan birds. *Scopus* **2**, 86–91.

Langley, C.H. (1982). Unusual diet of the Southern Boubou Shrike. *Ostrich* **53**, 118.

Langley, C.H. (1983). Notes on the breeding and diet of the Southern Boubou. *Ostrich* **54**, 172–173.

Lorber, P. (1982). The food of the Boubou Shrike. *Honeyguide* **109**, 31.

Lorber, P. (1984). Feeding methods, mate replacement and development of the call in the Tropical Boubou. *Honeyguide* **30**, 78–79.

Louette, M. (1989). Additions and corrections to the avifauna of Zaïre (4). *Bull. Br. Orn. Club* **109**, 217–225.

Mackenzie, P. (1978). New records for Nigeria. *Bull. Nigerian Orn. Soc.* **14**, 47.

Marchant, S. (1953). Notes on the birds of southeastern Nigeria. *Ibis* **95**, 38–69.

Moltoni, E. (1932). Uccelli d'Angola raccolti da L. Fenaroli durante le spedizione 1930 Baragiola–Durini. *Atti Soc. Ital. Sci. Nat. Milan* **71**, 169–178.

Ndao, B. (1989). Au Sénégal, un Gonolek de Barbarie (*Laniarius barbarus*) à dessous jaune apparié à un sujet normal. *Malimbus* **11**, 97–98.

Payne, R.B. (1970). Temporal pattern of duetting in the Barbary Shrike *Laniarius barbarus. Ibis* **112**, 106–108.

Prinzinger, R., Becker, P., Kleim, J.-P., Schroth, W. and Schierwater, B. (1997). Der taxonomische Status von *Laniarius dubiosus* (Rchw. 1899) mit ergänzenden Daten zur Typusbeschreibung von *Laniarius liberatus*, Bulo Burti Boubou (Smith, Arctander, Fjeldså and Amir 1991). *J. Orn.* **138**, 283–289.

Prinzinger, R., Kleim, J.-P., Schroth, W. and Schierwater, B. (1997). DNA sequence analysis of mitochondrial Cyt-b and the species status of *Laniarius dubiosus* (Rchw. 1899). *J. Orn.* **138**, 291–296.

Quickelberge, C.D. (1966). A taxonomic study of the Boubou Shrike in southern Africa. *Ann. Cape Prov. Mus. (Nat. Hist.)* **5**, 117–137.

da Rosa Pinto, A.A. (1960). Endemismos ornitológicos raros de Angola. *Bol. Cult. Mus. Angola* **2**, 10–17.

da Rosa Pinto, A.A. (1962). As observações de maior destaque das expedições ornitológicas do Instituto de Investigação Científica de Angola. *Bol. Inst. Invest. Cient. Angola* **1**, 21–38.

Scott, J.A. (1985). Ringing recovery. *Zambian Orn. Soc. Newsl.* **15**, 30.

Serle, W. (1959). Some breeding records of birds at Calabar. *Nigerian Field* **24**, 45–48.

Skinner, N.J. (1996). The nest record card scheme. *Babbler* **31**, 42–46.

Smith, E.F.G., Arctander, P., Fjeldså, J. and Amir, O.G. (1991). A new species of shrike (Laniidae: *Laniarius*) from Somalia, verified by DNA sequence data from the only known individual. *Ibis* **133**, 227–235.

Sonnenschein, E. and Reyer, H.-U. (1983). Mate-guarding and other funcions of antiphonal duets in the Slate-coloured Boubou (*Laniarus funebris*). *Z. Tierpsychol.* **63**, 112–140.

Sonnenschein, E. and Reyer, H.-U. (1984). Biology of the Slate-colured Boubou and other bush shrikes. *Ostrich* **55**, 86–96.

Tarboton, W. (1971). Breeding biology of the Crimson-breasted Shrike at Olifantsfontein, Transvaal. *Ostrich* **42**, 271–290.

Tarboton, W. (1984). Breeding of the Brubru Shrike. *Ostrich* **55**, 97–101.

Tarboton, W. (1998). Scarlet skulker. *Afr. Birds and Birding* **3**(2), 31–35.

Thorpe, W. (1963). Antiphonal singing in birds as evidence for avian auditory reaction time. *Nature* (**1963**), 774–776.
Thorpe, W. (1972). 'Duetting and Antiphonal Song in Birds'. *Behaviour*, Suppl. **18**.
Tyler, S.J. and Tyler, L. (1996). Feeding associations of the Crimsonbreasted Shrike *Laniarius atrococcineus*. *Babbler* **31**, 24–25.
Vande weghe, J.-P. (1981). Additions à l'avifaune du Rwanda. *Gerfaut* **71**, 175–184.
Vande weghe, J.P. (1992). New records for Uganda and Tanzania along the Rwandan and Burundian borders. *Scopus* **16**, 59–60.
Walsh, J.F. (1987). Records of birds seen in north-eastern Guinea in 1984–1985. *Malimbus* **9**, 105–122.
Wickler, W. (1972). Aufbau und Paarspezifität des Gesangsduettes von *Laniarius funebris*. *Z. Tierpsychol*. **30**, 464–476.
Wickler, W. and Lunau, K. (1996). How do East African bush shrikes *Laniarius funebris* recognize male and female tutors during gender dialect development? *Naturwissenschaften* **83**, 579–580.
Wickler, W. and Seibt, U. (1979). Duetting: a daily routine of *Laniarius funebris*, the Slate-coloured Boubou (Aves, Laniidae). *Z. Tierpsychol*. **51**, 153–157.
Wickler, W. and Seibt, U. (1982). Song splitting in the evolution of duetting. *Z. Tierpsychol*. **59**, 27–140.
Wickler, W. and Sonnenschein, E. (1989). Ontogeny of song in captive duet-singing Slate-coloured Boubous (*Laniarius funebris*). A study in birdsong epigenesis. *Behaviour* **111**, 220–233.

Family PYCNONOTIDAE

Nicator, Neolestes

Archer, A.L. (1994). Birds observed from Kigoma to Kalambo Falls on the eastern side of Lake Tanganyika. *Scopus* **18**, 6–11.
Bowen, P. St. J. (1983). The Black-collared Bulbul *Neolestes torquatus* in Mwinilunga District and the first Zambian breeding record. *Bull. Zambian Orn. Soc*. **13–15**, 7–14.
Brosset, A. (1974). La nidification des oiseaux en forêt gabonaise: architecture, situation des nids et prédation. *Terre Vie* **28**, 579–610.
Brosset, A. and Erard, C. (1976). Première description de la nidification de quatre espèces en forêt gabonaise. 4. *Nicator vireo* Cabanis. *Alauda* **44**, 230–231.
Delacour, J. (1943). A revision of the genera and species of the Family Pycnonotidae (bulbuls). *Zoologica*, NY **28**, 17–28.
Dowsett, R.J., Olson, S.L., Roy, M.S. and Dowsett-Lemaire, F. (1999). Systematic status of the Black-collared Bulbul *Neolestes torquatus*. *Ibis* **141**, 22–28.
Dowsett-Lemaire, F. and Dowsett, R.J. (1968). White-throated Nicator in western Kalomo District. *Bull. Zambian Orn. Soc*. **10**, 76.
Green, A.A. and Rodewald, P.G. (1996). New bird records from Korup National Park and environs, Cameroon. *Malimbus* **18**, 122–133.
Hanotte, O., Knox, A.G. and Prigogine, A. (1987). The taxonomic status of *Nicator*: evidence from feather protein electrophoresis. *Biochem. Syst. Ecol*. **15**, 629–634.
Hazevoet, C.J. (1996). Birds observed in Guinea-Bissau, January 1986, with a review of current ornithological knowledge of the country. *Malimbus* **18**, 10–24.
Holsten, B., Braunlich, A. and Huxham, M. (1991). Rondo Forest Reserve, Tanzania: an ornithological note including new records of the East Coast Akalat *Sheppardia gunningi*, the Spotted Ground Thrush *Turdus fischeri* and the Rondo Green Barbet *Stactolaema olivacea woodwardi*. *Scopus* **14**, 125–128.
Long, R.C. (1959). The nest and eggs of the Nicator Shrike *Nicator gularis*. *Ostrich* **30**, 137.
Monard, A. (1940). Resultats de la mission scientifique du Dr. Monard en Guinée Portugaise 1937–1938. V. Oiseaux. *Arq. Mus. Bocage*, Lisboa **11**, 1–75.
Penry, E.H. (1979). Black-collared Bulbul (*Neolestes torquatus*) in unusual habitat. *Bull. Zambian Soc*. **11**, 34–35.
Rodwell, S.P. (1996). Notes on the distribution and abundance of birds observed in Guinea-Bissau, 21 February to 3 April 1992. *Malimbus* **18**, 25–43.
Salvan, J. (1972). Notes ornithologiques du Congo-Brazzaville. *Oiseau et R.F.O*. **42**, 241–252.
Sibley, C.G. and Ahlquist, J.E. (1985). The relationships of some groups of African birds, based on comparisons of the genetic material, DNA. *In* Schuchmann, K.-L. (Ed.) Proc. Intl Symp. Afr. Vert., Syst., Phylogeny, and Evol. Ecol., Bonn, 1984. Zoologisches Forschungsinstitut und Museum Alexander Koenig, Bonn.
Stanford, C.B. and Msuya, P. (1995). An annotated list of the birds of Gombe National Park, Tanzania. *Scopus* **19**, 38–46.
Tree, A.J. (1987). Recent reports. *Honeyguide* **33**, 68.

Family PRIONOPIDAE

Anon. (1970). White Helmet Shrike. *Bokmakierie* **22**, 72.
Bennun, L. (1994). Threatened birds of Kenya: 6. Grey-headed Helmet Shrike. *Kenya Birds* **3**, 84–87.
Benson, F.M. (1946). Field-notes from Nyasaland. *Ostrich* **17**, 297–319.
Britton, P.L. and Britton, H.A. (1971). Black-billed Barbet and Chestnut-fronted Shrike breeding in Kenya. *E. Afr. Nat. Hist. Soc. Bull*. **1971**, 126–127.
Britton, P.L. and Britton, H.A. (1977). The nest and eggs of the Chestnut-fronted Shrike *Prionops scopifrons*. *Scopus* **1**, 86.
Brooke, R.K. (1971). Breeding and breeding season notes on the birds of Mzimbiti and adjacent low-lying areas of Mozambique. *Ann. Natal Mus*. **21**, 55–69.
Brooke, R.K. (1994). Subspeciation studies and our knowledge of migration and other movements in southern African birds. *Ostrich* **65**, 49–53.
van Bruggen, A.C. (1960). Notes and observations on birds in the Transvaal, Southern Rhodesia and Portuguese East Africa. *Ostrich* **31**, 30–31.
Clancey, P.A. (1958). Miscellaneous taxonomic notes on African birds. *Durban Mus. Novit*. **5**, 117–142.

Clancey, P.A. (1960). The characters and range of the South African race of *Prionops retzii* Wahlberg. *Durban Mus. Novit.* **6**, 41–44.

Clancey, P.A. (1976). Micellaneous taxonomic notes on African birds XLV. On the validity of *Prionops talacoma* Smith, 1836. *Durban Mus. Novit.* **11**, 115–138.

Clark, J.D. (1936). The Long-crested Helmet Shrike. *Nigerian Field* **5**, 129–130.

Colebrook-Robjent, J.F.R. (1973). Some breeding records of birds in Zambia. *Zambian Mus. J.* **4**, 7–18.

Dean, W.R.J. (1987). Birds associating with fire at Nylsvley Nature Reserve, Transvaal. *Ostrich* **58**, 103–106.

Douthwaite, R.J. (1995). Occurrence and consequences of DDT residues in woodland birds following tsetse fly spraying operations in NW Zimbabwe. *J. Applied Ecol.* **32**, 727–738.

Dowsett, R.J. (1985). The conservation of tropical forest birds in central and southern Africa. *In* 'Conservation of tropical forest birds' (Diamond, A.W. and Lovejoy, T.E., Eds). *ICBP Tech. Publ.* **4**, 197–212.

Garcia, E.F.J. (1975). Ornithological observations at Chassa, Petauke District. *Bull. Zambian Orn. Soc.* **7**, 68–90.

Gill, P.J. (1939) The Long-crested Helmet Shrike *(Prionops plumata plumata)*. *Nigerian Field* **8**, 110–115.

Greig-Smith, P.W. (1976). Observations on the social behaviour of helmet-shrikes. *Bull. Nigerian Orn. Soc.* **12**, 25–30.

Hall, B.P. (1960b). The faunistic importance of the Scarp of Angola. *Ibis* **102**, 420–442.

Hart, J.A. (1971). The flocking and foraging behaviour of the West African Red-billed Shrike. *Ostrich* **42**, 294–295.

Harvey, W.G. and Harrison, I.D. (1970). The birds of the Mole Game Reserve. *Bull. Nigerian Orn. Soc.* **7**, 63–75.

Hendricks, F.L. (1946). Note sur le *Prionops alberti* Schout. *Gerfaut* **36**, 202–204.

Humphreys, C.W. (1992). White Helmet and Fiscal Shrike reaction to Lizard Buzzard. *Honeyguide* **35**, 131.

Irwin, M.P.S. (1953). Notes on some birds of Mashonaland, Southern Rhodesia. *Ostrich* **24**, 37–49.

Irwin, M.P.S. (1966). Notes on the birds of Zambia. *Arnoldia (Rhod.)* **2**(32), 1–19.

Lawson, W.J. (1964). Instability of feather pigmentation in the White Helmet Shrike *Prionops plumatus*. *Bull. Br. Orn. Club* **84**, 117–118.

Lewis, A.D. (1981). The past and present status and distribution of the Grey-crested Helmet Shrike *Prionops poliolopha*. *Scopus* **5**, 66–70.

Lewis, A.D. (1982). Further records of the Grey-crested Helmet Shrike. *Scopus* **6**, 47–48.

Lewis, A.D. (1983). A record of the Grey-crested Helmet Shrike near Naivasha, Kenya. *Scopus* **7**, 26–27.

Lorber, P. (1981). Helmet shrikes. *Honeyguide* **105**, 23.

Marchant, S (1942). Some birds of the Owerri Province. *Ibis* **1942**, 137–196.

Mayr, E. (1943). What genera belong to the Family Prionopidae? *Ibis* **85**, 216–218.

Newman, K. (1984). Chestnut-fronted Helmet Shrike *Prionops scopifrons* in Lengwe National Park. *Nyala* **10**, 121.

Newman, K. (1991). Report of Chestnut-fronted Helmet Shrikes, *Prionops scopifrons*, in Lengwe National Park, Malawi: a retraction. *Nyala* **15**, 50.

Newman, K. and Hanmer, D. (1991). Identification of helmet-shrikes. *Birding Southern Afr.* **43**, 47–50.

Payne, R.B. (1971). Duetting and chorus singing in African birds. *Ostrich*, Suppl. **9**, 125–146.

Payne, R.B. and Payne, K. (1967). Cuckoo hosts in southern Africa. *Ostrich* **38**, 135–143.

Priest, C.D. (1932). Notes on the habits of the Southern Helmet-Shrike (*Prionops poliocephala*). *Ostrich* **3**, 23–24.

Prigogine, A. (1949). Notes sur le *Prionops alberti* Schout. *Rev. Zool. Bot. Afr.* **42**, 307–321.

Prigogine, A. (1985). Conservation of the avifauna of the forests of the Albertine Rift. *In* 'Conservation of Tropical Forest Birds' (Diamond, A.W. and Lovejoy, T.E, Eds). *ICBP Tech. Publ.* **4**, 277–295.

Raikow, R.J., Polumbo, P.J. and Borecky, S.R. (1980). Appendicular myology and relationships of the shrikes (Aves: Passeriformes: Laniidae). *Ann. Carnegie Mus.* **49**, 131–152.

Rand, A.L. (1957). Two new species of birds from Angola. *Fieldiana: Zool.* **39**, 41–45.

Rand, A.L. (1960). Laniidae. *In* 'Check-list of Birds of the World', Vol. IX (Mayr, E. and Greenway, J.C., Eds.), pp. 309–365. Museum of Comparative Zoology, Cambridge.

Reynolds, J.F. (1968). Two cases of albinism in Tanzanian birds. *E. Afr. Wildife J.* **6**, 144–145.

da Rosa Pinto, A.A. (1962). As observações de maior destaque das expedições ornitológicas do Instituto de Investigação Científica de Angola. *Bol. Inst. Invest. Cient. Angola* **1**, 21–38.

Schouteden, H. (1933). Un oiseau nouveau du Parc Albert (Kivu) *Rev. Zool. Bot. Afr.* **24**, 210–212.

Schouteden, H. (1935). L'habitat du *Prionops alberti* Schout. *Rev. Zool. Bot. Afr.* **26**, 247–248.

Short, L.L. and Horne, J.F.M. (1985). Notes on some birds of the Arabuko-Sokoke Forest. *Scopus* **9**, 117–126.

Solomon, D., Solomon, S., Rawson, C. and Rawson, R. (1996). Chestnut-fronted Helmet-Shrike and others from the Honde Valley. *Honeyguide* **42**, 167–168.

Stevenson, T. and Pearson, D.J. (1986). Species report: Afrotropical and oceanic species. *Scopus* **8**, 104–114.

Tarboton, W. (1963). Breeding observations on the Red-billed Helmet-Shrike. *Bokmakierie* **15**, 1–3.

Townley, D. (1936a). Field notes on little known Southern Rhodesian birds. *Ostrich* **7**, 103–108.

Townley, D. (1936b). Some rarer Rhodesian birds. *Ostrich* **7**, 9–23.

Tree, A.J. (1995). Recent reports. *Honeyguide* **41**, 30–40.

Ulfstrand, S. and Alerstam, T. (1977). Bird communities of *Brachystegia* and *Acacia* woodlands in Zambia. *J. Orn.* **118**, 156–174.

Verheyen, R. (1947). Notes sur la faune ornithologique de l'Afrique centrale. VII. Liste d'une collection d'oiseaux rares réunie à Albertville et description d'un nouveau touraco du Congo belge. *Bull. Mus. Roy. Hist. Nat. Belg.* **23**(9), 1–4.

Vernon, C.J. (1966). Observations on the winter behaviour of the White Helmet Shrike, *Prionops plumata*. *Ostrich* **37**, 3–5.

Vernon, C.J. (1967). Some observations from the journals of K.W. Greenhow. *Ostrich* **38**, 48–49.

Vernon, C.J. (1971). Juvenile *Pachycoccyx audeberti* with *Prionops retzii*. *Ostrich* **42**, 298.

Vernon, C.J. (1978). Breeding seasons of birds in deciduous woodland in Zimbabwe, Rhodesia, from 1970 to 1974. *Ostrich* **49**, 102–115.

Vernon, C.J. (1980). Bird parties in central and South Africa. *Proc. IV Pan-Afr. Orn. Congr.*, pp. 313–325.

Vernon, C.J. (1984a). Population dynamics of birds in *Brachystegia* woodland. *Proc. V Pan-Afr. Orn. Congr.*, pp. 201–216.

Vernon, C.J. (1984b). The breeding biology of the Thick-billed Cuckoo. *Proc. V Pan-Afr. Orn. Congr.*, pp. 825–840.

Vernon, C.J. (1985). Bird populations in two woodlands near Lake Kyle, Zimbabwe. *Honeyguide* **31**, 148–161.

Williams, J.G. (1963). 'A Field Guide to the Birds of East and Central Africa'. Collins, London.

Family ORIOLIDAE

Aspinwall, D.R. (1978). Movement analysis charts – African Golden Oriole. *Zambian Orn. Soc. Newsl.* **8**, 2–3.

Aspinwall, D.R. (1980). Movement analysis charts – Comments on African Golden Oriole. *Zambian Orn. Soc. Newsl.* **10**, 118–119.

Aspinwall, D.R. (1985). Movement analysis charts – Comments on Eastern Black-headed Oriole. *Zambian Orn. Soc. Newsl.* **15**, 77–79.

Baumann, S. (1998). Where have all the orioles gone? Ecology of *Oriolus oriolus* in their resting area. Abstracts 22 Int. Orn. Congr., Durban. *Ostrich* **69**, 277.

Borrett, R.P. (1972). Fat deposition in *Oriolus auratus* in Rhodesia. *Ostrich* **43**, 64.

Clancey, P.A. (1970). On an oriole – new to the South African list. *Bokmakierie* **22**, 53–54.

Eccles, S.D. (1988). The birds of São Tomé – record of a visit, April 1987 with notes on the rediscovery of Bocage's Longbill. *Malimbus* **10**, 207–217.

Feige, K.-D. (1986). 'Der Pirol'. Neue Brehm-Bucherei 578. A. Ziemsen, Wittenberg Lutherstadt.

Gore, M.E.J. (1994). Bird records from Liberia. *Malimbus* **16**, 74–87.

Hillman, J.C. and Hillman, S.M. (1986). Notes on some unusual birds of the Bangangai area, south west Sudan. *Scopus* **10**, 29–32.

Johnston-Stewart, N.G.B. (1981). The Green-headed Oriole (*Oriolus chlorocephalus*) survives on Chiradzulu. *Nyala* **7**, 53–54.

Johnston-Stewart, N.G.B. (1982). Evergreen forest birds in Upper Thyolo. *Nyala* **8**, 69–84.

Johnston-Stewart, N.G.B. (1984). Evergreen forest birds in the southern third of Malaŵi. *Nyala* **10**, 99–119.

Lawson, W.J. (1962). On geographical variation in the Black-headed Oriole *Oriolus larvatus* Lichtenstein of Africa. *Durban Mus. Novit.* **6**(16), 195–201.

Lawson, W.J. (1969). A new name for a race of the Black-headed Oriole. *Bull. Br. Orn. Club* **89**, 16.

Morioka, H. (1986). Relationships of the drongos and the old world orioles based on the structure of the skull. Abstracts *XIX Int. Orn. Congr.*, 737.

de Naurois, R. (1984). Le Loriot endémique de l'île de São Tomé *Oriolus crassirostris* (Hartlaub) (Golfe de Guinée). *Cyanopica* **2**, 3, 121–134.

Naylor, E.G. (1975). European Golden Oriole eating a lizard. *Bull. Zambian Orn. Soc.* **7**, 28.

Prigogine, A. (1978). Le statut du Loriot de Percival, *Oriolus percivali*, et son hybridation avec *Oriolus larvatus* dans l'est africain. *Gerfaut* **68**, 253–320.

Reinsch, A. and Warncke, K. (1971). Zur Brutbiologie des Pirols (*Oriolus oriolus*). *Vogelwelt* **92**, 121–141.

Rudnai, J. (1994). Blackheaded Oriole (*Oriolus larvatus*) nesting in a Langata garden. *E. Afr. Nat. Hist. Soc. Bull.* **24**, 61–62.

Sim, L. (1979). 'Birds of Wondo Genet'. Orgut-Swedforest, Stockholm.

Spina, F., Massi, A. and Montemaggiori, A. (1994). Back from Africa: who's running ahead? Aspects of differential migration of sex and age classes in Palearctic–African spring migrants. *Ostrich* **65**, 137–150.

Tree, A.J. and de la Harpe, D.A. (1995). Ringing report for the Association 1992–1994. *Honeyguide* **41**, 256–262.

Urban, E.K. (1978). 'Ethiopia's Endemic Birds'. Ethiopian Tourist Organization, Addis Ababa.

Walsh, F. (1975). Two comments on 'The Birds of Sokoto'. *Bull. Nigerian Orn. Soc.* **11**, 87–88.

Wolters, H.E. and Clancey, P.A. (1969). A new race of Green-headed Oriole from southern Moçambique. *Bull. Br. Orn. Club* **89**, 108–109.

Family DICRURIDAE

Ballance, T.C. and Ballance, A. (1981). Fishing drongos. *Honeyguide* **106**, 32.

Clancey, P.A. (1956). Geographic variation in the southern African populations of *Dicrurus adsimilis* (Bechstein). *Bull. Br. Orn. Club* **76**, 79–85.

Clancey, P.A. (1975). Miscellaneous taxonomic notes on African birds XLIII. An undescribed race of the forest-dwelling drongo *Dicrurus ludwigii* (Smith). *Durban Mus. Novit.* **11**, 15–17.

Clancey, P.A. (1976a). Miscellaneous taxonomic notes on African birds XLIV. Further on the nominate subspecies of *Dicrurus adsimilis* (Bechstein), 1794: Duiwenhok R., Swellendam, south-western Cape. *Durban Mus. Novit.* **11**, 88–92.

Clancey, P.A. (1976b). Miscellaneous taxonomic notes on African birds XLIV. Subspeciation in the Squaretailed Drongo *Dicrurus ludwigii* (A. Smith), 1834. *Durban Mus. Novit.* **11**, 92–101.

Dowsett, R.J. (1980). Bird ringing in Zambia (1979). *Bull. Zambian Orn. Soc.* **12**, 65–69.

von Helversen, G. (1980). Structure and function of antiphonal duets. *Acta XVII Int. Orn. Congr.*, 682–688.

Mackworth-Praed, C.W. (1942). On the specific status of *Dicrurus adsimilis* (Bechstein) and *Dicrurus modestus* Hartlaub. *Bull. Br. Orn. Club* **64**, 61.

Meise, W. (1968). Zur Speciation afrikanischer, besonders angolesischer Singvögel der Gattungen *Terpsiphone*, *Dicrurus* und *Malaconotus*. *Zool. Beitr.* **14**, 1–60.

Oatley, T.B. The probing of aloe flowers by birds. *Lammergeyer* **3**, 2–8.

Prigogine, A. (1961). Nids et oeufs recoltés au Kivu (Republique du Congo). *Rev. Zool. Bot. Afr.* **66**, 359.

Tarboton, W.R. and Clinning, C.F. (1977). Nesting association between Groundscraper Thrush *Turdus litsitsirupa* and Fork-tailed Drongo *Dicrurus adsimilis*. *Madoqua* **10**, 87–79.

Tree, A.J. (1976). Fork-tailed drongos preying on Bronze Mannikin. *Honeyguide* **85**, 40.

Vaurie, C. (1949). A revision of the bird family Dicruridae. *Bull. Am. Mus. Nat. Hist.* **93**, 203–342.

Willis, E.O. (1983). Flycatchers, cotingas and drongos (Tyrannidae, Muscicapidae, Cotingidae and Dicruridae) as ant followers. *Gerfaut* **73**, 265–280.

Family CORVIDAE

Corvus

Ade, B. (1975). Pied Crow preying on Palm Swifts. *Honeyguide* **81**, 36–37.

Alamargot, J. (1976). Quelques données sur la reproduction du Corbeau corassé (*Corvus crassirostris* Rüppell). *Oiseau et R.F.O.* **46**, 73–75.

Alamargot, J. (1983). Nouveau site de nidification du Corbeau corassé (*Corvus crassirostris*). *Oiseau et R.F.O.* **53**, 393–394.

Alamargot, J. (1987). Pied Crow *Corvus albus* with atypical plumage. *Walia* **10**, 7–12.

Alamargot, J. (1990). Nids et oeufs du Corbeau corassé (*Corvus crassirostris*). *Oiseau et R.F.O.* **60**, 251–252.

Allan, D. (1981). Growth rates of nestling Black Crows. *Ostrich* **52**, 189–190.

Amadon, D. (1944). The genera of Corvidae and their relationships. *Amer. Mus. Novit.* **1251**.

Andersson, S. (1989). Tool use by the Fan-tailed Raven (*Corvus rhipidurus*). *Condor* **91**, 999.

Angwin, E. (1977). Food storage in Indian House Crows. *E. Afr. Nat. Hist. Soc. Bull.* **1977**, 131.

Anon. (1987). Early records of the Pied Crow at Port Elizabeth. *Bee-eater* **38**, 49–50.

Anon. (1997). House Crow eradication programme. *Promerops* **227**, 9.

Archer, T. (1994). Indian House Crow control in Malindi and Watamu. *Kenya Birds* **3**, 72–74.

Ash, J.S. (1984). Report of the UNEP ornithologist/ecologist on the advice to the Government of the People's Democratic Republic of Yemen on 'combatting the crow menace'. UN Environment Programme Report NEP/84/0189, 29 pp.

Aspinwall, D.R. (1985). Comments on Pied Crow. *Zambian Orn. Soc. Newsl.* **15**, 2–4.

Baron, S. (1981). Whitenecked Raven *Corvus albicollis* attacking a Common Tern *Sterna hirundo*. *Cormorant* **9**, 135.

Berruti, A. and Nichols, G. (1991). Alien birds in Southern Africa: the crow must go. *Birding Southern Afr.* **43**, 52–53, 55–57.

Bijlsma, R.G. and Meininger, P.L. (1984). Behaviour of the House Crow, *Corvus splendens*, and additional notes on its distribution. *Gerfaut* **74**, 3–13.

Blair, C.M.B. (1961). Hybridization of *Corvus albus* and *Corvus edithae* in Ethiopia. *Ibis* **103a**, 499–502.

Blake, E.R. and Vaurie, C. (1962). Family Corvidae. *In* 'Check-list of Birds of the World' (Mayr, E. and Greenway, J.C., Eds). Mus. Comp. Zool., Cambridge, MA.

Boswall, J. (1995). Communal roosting by Pied Crows in the breeding season. *Honeyguide* **41**, 112.

Brady, J.E. (1977). The too-common crow, too close for comfort. *New York Times* May 27, 1977, pp. B7, B13.

Brooke, R.K. (1979). Predation on Ostrich eggs by tool-using crows and Egyptian Vultures. *Ostrich* **48**, 257–258.

Brooke, R.K. and Grobler, J.H. (1973). Notes on the foraging, food and relationships of *Corvus albus* (Aves: Corvidae). *Arnoldia (Rhod.)* **6**, 1–13.

Brown, L.H. (1965). 'Ethiopian Episode'. Country Life, London.

Brown, L.H. (1970). 'African Birds of Prey'. Collins, London.

de Castro, J. and de Castro, M. (1993). Verreaux's Eagle Owl *Bubo lacteus* persistently attacked by Thick-billed Raven *Corvus crassirostris*. *Scopus* **17**, 62.

Chadwick, D.H. (1983). Etosha: Namibia's kingdom of animals. *Nat. Geog. Mag.* **163**, 344–385.

Clancey, P.A. (1974). The Indian House Crow in Natal. *Ostrich* **45**, 31–32.

Collett, J. (1982). Birds of the Cradock District. *Southern Birds* **9**, 65 pp.

Coombs, F. (1978). 'The Crows'. Batsford, London.

Cooper, J. (1969). Aerial evolutions in the Pied Crow (*Corvus albus*) Müller. *Honeyguide* **59**, 13–14.

Cooper, J.E. (1996). Health studies on the Indian House Crow. *Avian Pathol.* **25**, 381–386.

Critchlow, D.P. (1996). Ravens stop for a swift snack. *Nyala* **19**, 48–49.

Davis, M. (1951). Ocean vessels and the distribution of birds. *Auk* **68**, 529–530.

Dekeyser, P.-L. and Derivot, J. (1958). Étude d'un type d'oiseau ouest-africain: *Corvus albus*. Généralities-Osteologie. Initiations africaines, Dakar, Institut français d'Afrique noire 16, 5–58.

Dekeyser, P.-L. and Villiers, A. (1952). Sur une nichée du Corbeau blanc (*Corvus albus*). *Notes Afr.* **56**, 124–127.

Dorst, J. (1947). Révision systématique du genre *Corvus*. *Oiseau et R.F.O.* **17**, 44–87.

Dowsett, R.J. and Dowsett-Lemaire, F. (1978). Notes on the distribution of the White-necked Raven (*Corvus albicollis*). *Bull. Zambian Orn. Soc.* **10**, 33–34.

Drury, M.R. (1980). Pied Crow taking small bird on the wing. *Honeyguide* **102**, 33.

Dupuy, A. (1969). Catalogue d'ornithologie du Sahara Algérien. *Oiseau et R.F.O.* **39**, 140–160, 225–241.

Esterhuizen, J.R. (1994). New distribution of Black and Pied Crows. *Mirafra* **11**, 55–56.

Feare, C.J and Mungroo, Y. (1989). Notes on the House Crow *Corvus splendens* in Mauritius. *Bull. Br. Orn. Club* **109**, 199–201.

Feare, C.J. and Mungroo, Y. (1990). The status and management of the House Crow *Corvus splendens* in Mauritius. *Biol. Conserv.* **51**, 63–70.

Feuerriegel, K. (1996). Brief associations between Indian House Crows, *Corvus splendens* and antelope. *E. Afr. Nat. Hist. Soc. Bull.* **26**, 8–10.

Fry, C.H. and Smith, V.W. (1964). Lanner Falcon attacking Pied Crow. *Bull. Nigerian Orn. Soc.* **1**(2), 12–13.

Gallagher, M. and Woodcock, M. (1980). 'The Birds of Oman'. Quartert Books, London.

Goodman, S.M. and Storer, R.W. (1987). The seabirds of the Egyptian Red Sea and adjacent waters, with notes on selected Ciconuformes. *Gerfaut* **77**, 109–145.

Goodwin, D. (1986). 'Crows of the World'. University of Washington Press, Seattle.

Grobler, J.H. (1974). Pied Crow and White-necked Raven breeding attempt. *Honeyguide* **78**, 44.

Gwahaba, J.J. (1975). A contribution to the biology of the Pied Crow *Corvus albus* Müller in Uganda. *J. E. Afr. Nat. Hist. Soc. Nat. Mus.* **153**, 1–14.

Gwinner, E. (1964). Untersuchungen uber des Ausdrucks- und Sozialverhalten des Kolkraben (*Corvus corax corax* L.). *Z. Tierpsychol.* **21**, 657–748.

Hamling, H.H. (1953). Observations of the behaviour of birds in Southern Rhodesia. *Ostrich* **24**, 9–16.

Harvey, D. and Harvey, M. (1992). Breeding observations on the Brown-necked Raven in Sana'a, Republic of Yemen. *Phoenix* **9**, 16–17.

van der Heiden, J.T. (1975). Pied Crow feeding on a bat. *Honeyguide* **81**, 31.

Heinrich, B. (1989). 'Ravens In Winter'. Summit Books, New York.

Heinrich, B., Marzlutt, J.M. and Marzlutt, C.S. (1993). Common Ravens are attracted by appeasement calls of food discoverers when attacked. *Auk* **110**, 247–254.

Hofmeyr, J. (1996). House Crow. *Promerops* **226**, 13.

Höglund, J. (1985). Foraging success of Rooks *Corvus frugilegus* in mixed-species flocks of different sizes. *Orn. Fenn.* **62**, 19–22.

Holyoak, D. (1967). Breeding biology of the Corvidae. *Bird Study* **14**, 153–168.

Hunt, G.R. (1996). Manufacture and use of hook-tools by New Caledonian Crows. *Nature* **379**, 249–251.

Jennings, M.C. (1992). The House Crow *Corvus splendens* in Aden (Yemen) and an attempt at its control. *Sandgrouse* **14**, 27–33.

Jessnitz, R. and Jessnitz, J. (1986). Burying of excess food by Black Crows. *Lanioturdus* **22**, 55.

Jollie, M. (1978). Phylogeny of the species of *Corvus*. *Biologist* **60**, 73–108.

Jones, J.M.B. (1985). Black Storks, White-necked Ravens and Peregrines: an unusual train of events. *Honeyguide* **31**, 168.

Jones, M.A. (1983). The Pied Crow in Harare, Zimbabwe. *Honeyguide* **116**, 6–13.

Jones, M.A. (1984). A statistical approach to sex determination in Pied Crows. *Honeyguide* **30**, 4–8.

Kalikawe, M.C. (1990). Problem birds (Pied Crows at Palapye). *Babbler* **19**, 17–18.

Kinnear, N.B. (1942). The introduction of the Indian House-Crow into Port Sudan. *Bull. Br. Orn. Club* **62**, 55–56.

Kleinschmidt, O. (1906). Beiträge zur Vogelfauna Nordostafrikas mit besonderer Berücksichtigung der Zoogeographie. *J. Orn.* **54**, 78–99.

Lamm, D.W. (1958). A nesting study of the Pied Crow at Accra, Ghana. *Ostrich* **29**, 59–70.

Lens, L. (1996). Hitting back at House Crows. *Kenya Birds* **4**, 55–56.

Lever, C. (1987). 'Naturalized Birds of the World'. Longman and Wiley, New York.

Lewis, A.D. (1989). Notes on two ravens *Corvus* spp. in Kenya. *Scopus* **13**, 129–131.

Loefler, I.J.P. (1984). Breeding of Verreaux's Eagle Owl prevented by crows. *E. Afr. Nat. Hist. Soc. Bull.* **1984**, 105–106.

Loehr, C. (1988). The usual crow debacle. *E. Afr. Nat. Hist. Soc. Bull.* **18**, 11–12.

Londei, T. (1995). Field notes on corvids in Ethiopia *Bull. Br. Orn. Club* **115**, 164–166.

Lorenz, K. (1952). 'King Solomon's Ring'. Thomas Y. Crowell, New York.

Macdonald, I.A.W. and Macdonald, S.A. (1983). The demise of the solitary scavengers in southern Africa – the early rising crow hypothesis. Proc. Symp. Birds and Man, Johannesburg 1983, 321–335.

Mackie, C. and Landman, D. (1982). Ravens at play. *Honeyguide* **109**, 30.

Maclean, G.L. (1957). Points on the incubation of the Pied Crow, *Corvus albus*, and the Black Crow, *Corvus capensis*. *Ostrich* **28**, 178–179.

Mann, C.F. and Britton, P.L. (1972). Naturalised birds of the Kenya coast *E. Afr. Nat. Hist. Soc. Bull.* **1972**, 181–182.

Manyanza, D.N. (1989). Some observations on the Indian House Crow (*Corvus splendens*) in Dar es Salaam, Tanzania. *Gerfaut* **79**, 101–104.

Markus, M.B. (1967). Notes on the wire from two crow nests. *Ostrich* **38**, 56–57.

Marshall, B.E. (1979). On the distribution of crows in Rhodesia. *Honeyguide* **93**, 5–14.

Martin, P. (1994). An influx of Pied Crows to Port Elizabeth. *Bee-eater* **45**, 38.

Masterson, A.N.B. (1993). Changes in the behaviour of urban crows. *Honeyguide* **39**, 138–141.

Mbidde, J. (1994). Astounding encounter on the nesting habits of the Pied Crow *E. Afr. Nat. Hist. Soc. Bull.* **24**, 11–12.

McCartney, F. (1984). Bells on crow legs. *E. Afr. Nat. Hist. Soc. Bull.* **1984**, 29.

McCartney, F. (1995). Ravens with bibs? *E. Afr. Nat. Hist. Soc. Bull.* **25**, 63.

Meinertzhagen, R. (1926). Introduction to a review of the genus *Corvus*. *Novit. Zool.* **33**, 57–121.

Meinertzhagen, R. (1930). 'Nicoll's Birds of Egypt'. Hugh Rees, London.

Meininger, P.L., Mullié, W.C. and Bruun, B. (1980). The spread of the House Crow, *Corvus splendens*, with special reference to the occurrence in Egypt. *Gerfaut* **70**, 245–250.

Middlemis, E. (1958). Jakob. *Bokmakierie* **10**, 36–37.

Morgan-Davies, A.M. (1967). On the nest and eggs of the Cape Rook, *Corvus capensis* Lichtenstein in Ngorongoro, northern Tanzania. *Bull. Br. Orn. Club* **87**, 40–41.

Mullié, W.C. and Meininger, P.L. (1985). 'The decline of bird of prey populations in Egypt' ICBP Tech. Publ. 5, pp. 61–82. ICBP, Cambridge.

Mundy, P.J. (1973b). Interspecific communication between crows. *Honeyguide* **74**, 37.

Mundy, P.J. and Cook, A.W. (1971). Sokoto Province. (I) Sokoto Town and environs *Bull. Nigerian Orn. Soc.* **8**, 21–24.

Mundy, P.J. and Cook, A.W. (1974). The birds of Sokoto. Part 3: Breeding data. *Bull. Nigerian Orn. Soc.* **10**, 1–28.

Mundy, V. and Mundy, P. (1973). Jeeves – a tame Pied Crow. *Bokmakierie* **25**, 64–65.

de Naurois, R. (1961). Recherches sur l'avifaune de la Côte Atlantique du Maroc, du Détroit de Gibraltar et Iles de Mogador. *Alauda* **29**, 241–259.

de Naurois, R. (1981). Le Corbeau roux de l'Archipel du Cap Vert (*Corvus r. ruficollis* Lesson). *Bull. Inst. Fond. Afr. Noire* **43**, 202–218.

Neuby-Varty, B. (1968). Pied Crows parasitised by Great Spotted Cuckoos. *Honeyguide* **56**, 28.

Newmann, N. (1974). Indian House Crow. *Natal Wildlife* **14**, 19–20.

Nhancale, C.C., Bento, C.M. and de Boer, W.F. (1998). The impact of the House Crow *Corvus splendens* on Inhaca Island, Mozambique. *Ostrich* **69**, 443.

Nicoll, M.J. (1912). Contributions to the ornithology of Egypt. III. The birds of the Wadi Natron. *Ibis* **1912**, 405–453.

Nixon, A. (1994). Origins of Pied Crows in East Cape City centres. *Bee-eater* **45**, 38.

Nogales, M. (1994). High density and distribution patterns of a Raven *Corvus corax* population on an oceanic island (El Hierro, Canary Islands). *J. Avian Biol.* **25**, 80–84.

Nogales, M. and Hernandez, E.C. (1994). Interinsular variations in the spring and summer diet of the Raven *Corvus corax* in the Canary Islands. *Ibis* **136**, 441–447.

North, M.E.W. (1962). Vocal affinities *of Corvus corax edithae*, 'Dwarf Raven' or 'Somali Crow'. *Ibis* **104**, 431.

Oatley, T. (1973). Indian House Crow: first S. A. sightings. *Bokmakierie* **25**, 41–42.

Parry, I. (1981). White-necked Ravens. *Honeyguide* **106**, 17–19.

Payne, R.B. and Payne, K. (1967). Cuckoo hosts in Southern Africa. *Ostrich* **38**, 135–143.

Peard, S.R. (1995). White-necked Raven. *Promerops* **218**, 15.

Perlmutter, H.M. (1977). Some observations on the nesting, trapping, roosting and general behaviour of crows in Salisbury. *Honeyguide* **92**, 24–30.

Peuzhorn, B.L. (1978). Growth of a Pied Crow nestling. *Ostrich* **49**, 85–86.

Pineau, J. and Giraud-Audine, M. (1976). Notes sur les oiseaux hivernant dans l'extreme nord ouest du Maroc et sur leurs mouvements. *Alauda* **44**, 47–75.

Priest, C.D. (1939). Curious behaviour of *Corvultur albicollis*. *Ostrich* **10**, 53–54.

Ratcliffe, D. (1997). 'The Raven'. T. and A.D. Poyser, London.

Reardon, J. (1977). Pied Crow removing parasites from White Rhinoceros. *Honeyguide* **89**, 47.

Richford, A. (1978). 'The Ecology of Jackdaws on Skomer Island'. DPhil thesis, Univ. Oxford.

Richner, H. (1990). Helpers-at-the-nest in Carrion Crows *Corvus corone corone*. *Ibis* **132**, 105–108.

Rofstad, G. (1986). Growth and morphology of nestling Hooded Crows *Corvus corone cornix*, a sexually dimorphic bird species. *J. Zool., London* **208**, 299–323.

Rohloff, P. (1987). Black Crows (547) to flock or not flock. *Wits Bird Club News* **136**, 8–9.

Rudnai, J. (1996). Pied Crow (*Corvus albus*) harassing White-backed Vultures (*Gyps africanus*). *E. Afr. Nat. Hist. Soc. Bull.* **26**, 30–31.

Ryall, C. (1990). Notes on nest construction by the Indian House Crow *Corvus splendens* and other aspects of its breeding biology in Mombasa, Kenya. *Scopus* **14**, 14–16.

Ryall, C. (1991). Avifauna of Nguuni near Mombasa, Kenya, between September 1984 and October 1987: Part 1 – Afrotropical species. *Scopus* **15**, 1–23.

Ryall, C. (1992a). Predation and harassment of native bird species by the Indian House Crow *Corvus splendens*, in Mombasa, Kenya. *Scopus* **16**, 1–8.

Ryall, C. (1992b). The pest status of the Indian House Crow in Mombasa and a survey of its expansion of range in coastal Kenya. *Proc. VII Pan-Afr. Orn. Congr.*, 303–310.

Ryall, C. (1994). Recent extensions of range in the House Crow *Corvus splendens*. *Bull. Br. Orn. Club* **114**, 90–99.

Ryall, C. (1995). Additional records of range extension in the House Crow *Corvus splendens*. *Bull. Br. Orn. Club* **115**, 185–187.

Ryall, C. and Reid, C. (1987). The Indian House Crow in Mombasa. *Bokmakierie* **39**, 113–116.

Schmidt, O. (1996). Recent extensions of range in the House Crow *Corvus splendens*. *Promerops* **222**, 5–6.

Schüz, E. (1967). Ornithologischer April-Besuch in Äthiopien, besonders am Tanasee. *Stuttgart. Beitr. Naturkunde* **171**, 1–22.

Schüz, E. (1968). Ornithologischer Oktober-Besuch am Tanasee (und bei Addis Abeba), Äthiopien. *Stuttgart. Beitr. Naturkunde* **189**, 1–43.

Shaughnessy, P.D. and Shaughnessy, G.L. (1987). Birds at Wolf and Van Reenen Bays, Diamond Coast, SWA/Namibia. *Lanioturdus* **23**, 27–43.

Siegfried, W.R. (1963). A preliminary evaluation of the economic status of Corvidae and their control on sheep farms in the Great Karoo. Invest. Rep, Dept. Nature Conservation, South Africa, 16 pp.

Sinclair, J.C. (1974). Arrival of the House Crow in Natal. *Ostrich* **45**, 189.

Sinclair, M. (1992). Deaths of Pied Crows. *Kenya Birds* **1**, 11.

Skead, C.J. (1952). A study of the Black Crow *Corvus capensis*. *Ibis* **94**, 434–451.

Smalley, M.E. (1984). Predation by Pied Crows *Corvus albus* on Gambian Epauletted Fruit Bats *Epomophorus gambianus*. *Bull. Br. Orn. Club* **104**, 77–79.

Smith, K.D. (1962). Hybridizations of crows *Ibis* **104**, 259.

Steyn, P. (1965). Distribution behaviour in the Black Crow. *Bokmakierie* **17**, 16–17.

Steyn, P. (1995). Pied Crow kills Cape Weaver. *Promerops* **221**, 14.

Stutterheim, C.J. (1980). Cleaning symbiosis involving pied crows and white rhino. *Lammergeyer* **30**, 61.

Symens, P. (1994). 'Pied' Brown-necked Ravens on the Farasan Islands, Saudi Arabia. *Orn. Soc. Middle East Bull.* **32**, 26–27.

Talbot, J. (1969). Mobbing by the Black Crow. *Honeyguide* **59**, 37.

Tomlinson, W. (1950). Bird notes chiefly from the Northern Frontier District of Kenya, Part II. *J. E. Afr. Nat. Hist. Soc.* **19**(5) (89), 225–250.

Tor, J.C. (1976). Estudio ornithológico de la región de Seguiat-El-Hamra, Sahara español, en Abril de 1973. *Mus. Zool. Barcelona, Misc. Zool.* **3**, 195–208.

Tye, A. (1983). Nest predation by the Pied Crow *Corvus albus*. *Malimbus* **5**, 50.

Tyler, S.J. (1980). Notes on feeding habits of Pied and Indian House Crows *Corvus albus* and *splendens*. *Scopus* **4**, 44–45.

Urban, E.K. (1980). 'Ethiopia's Endemic Birds'. Ethiopian Tourist Commission, Addis Ababa.

Uys, C.J. (1966). At the nest of the Cape Raven. *Bokmakierie* **18**, 38–40.

Valverde, J.A. (1957). 'Aves del Sahara Español'. Instituto de Estudios Africanos, Madrid.

Vaurie, C. (1954). Systematic notes on Palearctic birds. 5: Corvidae. *Amer. Mus. Novit.* **1668** 1–23.

Vernon, C.J. (1984). Roost of Black Crows. *Bee-eater* **35**, 8–9.

Waddel, P., Thomson, M. and Thomson, P. (1991). Tortoise drop. *Promerops* **201**, 16.

Walker, S (1996). Pied Crow robbing Rock Pigeon nest. *Promerops* **222**, 12.

Watt-Pringle, S (1990). House Crow in Cape Flats. *Promerops* **192**, 8.

Wilson, R.T. (1981) The Corvidae in the Sudan Republic, with special reference to Darfur. *Afr. J. Ecol.* **19**, 285–294.

Wilson, R.T. (1982). Environmental changes in western Darfur, Sudan, over half a century and their effects on selected bird species. *Malimbus* **4**, 15–26.

Wilson, R.T. (1990). Comparative ecology of the Corvidae in Ethiopia. *J. Afr. Zool.* **104**, 593–601.

Wilson, R.T. (1993). Distribution and ecology of the African Corvidae. *Proc. VIII Pan-Afr. Orn. Congr.*, 371–378.

Wilson, R.T. and Ball, D.M. (1979). Morphometry, wing loading and food of western Darfur birds. *Bull. Br. Orn. Club* **99**, 15–20.

Wilson, R.T. and Balcha, G. (1989). Temporal and spatial ecology of the birds of Ethiopia, Order, Passeriformes, Family, Corvidae. *Walia* **12**, 30–34.

Winterbottom, J.M. (1975). Notes on the South African species of *Corvus*. *Ostrich* **46**, 236–250.

Ptilostomus, Pica, Garrulus, Nucifraga, Pyrrhocorax

Alvarez, F. and Arias de Reyna, L.M. (1974). Reproducción de la urraca (*P. pica*) en Doñana. *Doñana Acta Vert.* **1**, 77–95.

Arias de Reyna, L. M., Recuerda, P., Corvillo, M. and Cruz, A. (1984). Reproducción de la urraca (*Pica pica*) en Sierra Morena (Andalucia). *Doñana Acta Vert.* **11**, 79–92.

Birkhead, T.R. (1991). 'The Magpies'. T. and A.D. Poyser, London.

Birkhead, T.R., Eden, S.F., Clarkson, K., Goodburn, S.F. and Pellatt, J. (1986). Social organization of a population of magpies *Pica pica*. *Ardea* **74**, 59–68.

Blanco, G., Fargallo, J.A. and Cuevas, J.A. (1993). Seasonal variations in numbers and levels of activity in a communal roost of Choughs *Pyrrhocorax pyrrhocorax* in central Spain. *Avocetta* **17**, 41–44.

Büchel, H.P. (1983). Beiträge zum Sozialverhalten der Alpendohle *Pyrrhocorax pyrrhocorax*. *Orn. Beob.* **80**, 1–28.

Cibois, A. and Pasquet, E. (1999). Molecular analysis of the phylogeny of 11 genera of the Corvidae. *Ibis* **141**, 297–306.

Cowdy, S. (1973). Ants as a major food source of the Chough. *Bird Study* **20**, 117–120.

Delestrade, A. (1998). Distribution and status of the Ethiopian population of the Chough *Pyrrhocorax pyrrhocorax baileyi*. *Bull. Br. Orn. Club* **118**, 101–105.

Goodwin, D. (1986). 'Crows of the World'. Brit Mus. (Nat. Hist.), London.

Henty, C.J. (1975). Feeding and food-hiding responses of jackdaws and magpies. *Brit. Birds* **68**, 463–466.

Holyoak, D. (1972). Behaviour and ecology of the Chough and the Alpine Chough. *Bird Study* **19**, 215–227.

Maumary, L., Vallotton, L., Dutoit, V. and Fleury, Z. (1998). Chocard à bec jaune *Pyrrhocorax graculus* et Traquet isabelle *Oenanthe isabellina* en Tunisie. *Alauda* **66**, 247–250.

Rice, D.W. (1963). Birds associating with elephants and hippopotamuses. *Auk* **80**, 196–197.

Soler, M. (1989). *In* 'Choughs and Land-use in Europe' (Bignal, E. and Curtis, D.J., Eds), pp. 29–33. Scottish Chough Study Group, Edinburgh.

Thomson, A.L. (1964). Birds associated with elephants and hippopotamuses. *Auk* **81**, 436.

Wilson, R.T. (1981). The Corvidae in the Sudan Republic, with special reference to Darfur. *Afr. J. Ecol.* **19**, 285–294.

Wilson, R.T. (1982). Environmental changes in western Darfur, Sudan, over half a century and their effects on selected bird species. *Malimbus* **4**, 15–26.

Family STURNIDAE

Amadon, D. (1943). The genera of starlings and their relationships. *Amer. Mus. Novit.* **1247**, 1–16.

Amadon, D. (1962). Family Sturnidae, starlings. *In* 'Check-list of Birds of the World', Vol. XV (Mayr, E. and Greenway, J.C., Eds), pp. 75–121. Mus. Comp. Zool., Cambridge, MA.

Brooke, R.K. (1968). More on the plumages, moults and breeding seasons of southern African starlings. *Bull. Br. Orn. Club* **88**, 113–116.

Craig, A.J.F.K. (1997). A phylogeny for the African starlings. *Ostrich* **68**, 114–116.

Craig, A.J.F.K. (1983a). The timing of breeding and wing-moult of four African Sturnidae. *Ibis* **125**, 346–352.

Durrer, H. and Villiger, W. (1970). Schillerfarben der Stare (Sturnidae). *J. Orn.* **111**, 133–153.

Feare, C.J. (1984). 'The Starling'. Oxford University Press, Oxford.

Feare, C.J. and Craig, A. (1998). 'Starlings and Mynas'. Christopher Helm/A. & C. Black, London.

Poeoptera, Onychognathus

Baranga, J. and Kalina, J. (1991). Nesting association between Narrow-tailed Starlings *Poeoptera lugubris* and Grey-throated Barbets *Gymnobucco bonapartei*. *Scopus* **15**, 59–61.

Beasley, A. (1978). Red-winged starling preying on lizard. *Honeyguide* **95**, 44.

Beasley, A. (1991). Red-winged starlings: unusual nests, and preying on crabs. *Honeyguide* **37**, 17–18.

Benson, C.W. (1960). Some additions and corrections to a 'Check list of the birds of Northern Rhodesia'. *Occ. Pap. Nat. Mus. Sth. Rhod.* **24B**(3), 343–350.

Broekhuysen, G.J. (1951). Some observations on the nesting activities of the Redwing Starling, *Onychognathus morio*, and especially the feeding of the young. *Ostrich* **22**, 6–16.

Brooke, R.K. (1965). Roosting of the Black-shouldered Kite *Elanus caeruleus* (Desfontaines). *Ostrich* **36**, 43.

Brown, L.H. (1965). Redwinged Starlings of Kenya. *J. E. Afr. Nat. Hist. Soc. Nat. Mus.* **1**(110), 41–56.

Brown, L.H. (1975). Breeding of Stuhlmann's Starling and Narina's trogon. *E. Afr. Nat. Hist. Soc. Bull.* **1975**, 44–45.

Brown, L.H. and Thorogood, K.M. (1976). Ecology and breeding habits of the White-billed Starling *Onychognathus albirostris* in Tigrai, Ethiopia. *Bull. Br. Orn. Club* **96**, 60–64.

Clouet, M., Goar, J.-L. and Barrau, C. (1998). Contribution à l'étude ornithologique de l'île de Socotra. *Alauda* **66**, 235–246.

Craig, A.J.F.K. (1988a). The timing of moult, morphology, and an assessment of the races of the Redwinged Starling. *Bonn. Zool. Beitr.* **39**, 347–360.

Craig, A.J.F.K. (1988b). The status of *Onychognathus nabouroup benguellensis* (Neumann). *Bull. Br. Orn. Club* **108**, 144–147.

Craig, A.J.F.K. (1997). A phylogeny for the African starlings (Sturnidae). *Ostrich* **68**, 114–116.

Craig, A.J.F.K. and Hulley, P.E. (1992). Biogeography and sympatry of Red-winged and Pale-winged Starlings in southern Africa. *J. Afr. Zool.* **106**, 313–326.

Craig, A.J.F.K., Hulley, P.E. and Walter, G.H. (1989). Nesting of sympatric Redwinged and Palewinged Starlings. *Ostrich* **59**, 69–74.

Craig, A.J.F.K., Hulley, P.E. and Walter, G.H. (1991). The behaviour of Palewinged Starlings, and a comparison with other *Onychognathus* species. *Ostrich* **62**, 97–108.

Cunningham-van Someren, G.R. (1975). Breeding of Stuhlmann's Starling *Poeoptera stuhlmanni*. *E. Afr. Nat. Hist. Soc. Bull.* **1975**, 12–13.

Dowsett-Lemaire, F. (1983b). Studies of a breeding population of Waller's Redwinged Starlings in montane forests of south-central Africa. *Ostrich* **54**, 105–112.

Drummond, D. (1991). Red-winged starlings on impala. *Honeyguide* **37**, 16–17.

Everitt, C. (1964). Breeding the Red-winged Starling. *Avicult. Mag.* **70**, 133–135.

Fraser, M.W. (1990). Foods of Redwinged Starlings and the potential for dispersal of *Acacia cyclops* at the Cape of Good Hope Nature Reserve. *S. Afr. J. Ecol.* **1**, 73–76.

Fuggles-Couchman, N.R. (1983). On the occurrence of *Onychognathus fulgidus* the Chestnut-winged Starling in Tanzania. *Scopus* **7**, 98–99.

Gargett, V. (1975). Association between Redwinged Starlings *Onychognathus morio* and Klipspringers *Oreotragus oreotragus*. *Bull. Br. Orn. Club* **95**, 119–120.

Holub, E. and von Pelzeln, A. (1882). 'Beiträge zur Ornithologie Südafrikas'. A. Hölder, Vienna.

Lack, D. (1936). On the pugnacity at the nest of a pair of *Onychognathus walleri walleri*. *Ibis* **1936**, 821–825.

Mare, J.J. (1982). Breeding the Red-winged Starling *Onychognathus morio*. *Avicult. Mag.* **88**, 191–192.

Mitchell, C.S. (1976). Prolonged use of a nest and nest-site by Red-winged Starlings. *Honeyguide* **86**, 35.

Moermond, T.C., ka Kajondo, K., Sun, C., Kristensen, K., Munyaligogo, V., Kaplan, B.A., Graham, C. and Mvukiyumwani, J. (1993). Avian frugivory and tree visitation patterns in a Rwanda montane forest. *Proc. VIII Pan-Afr. Orn. Congr.*, 421–428.

Morel, G.J. (1985). Les oiseaux des milieux rocheux au Sénégal. *Malimbus* **7**, 115–119.

Mortimer, J. (1975). Red-winged Starlings preying on Palm Swifts. *Honeyguide* **82**, 44.

Mungure, S.A. (1973). Nest of Red-wing Starling *Onychognathus morio*. *E. Afr. Nat. Hist. Soc. Bull.* **1973**, 52.

Oatley, T. and Fraser, M. (1992). Red-ringed Redwinged Starlings. *Safring News* **21**, 43–49.

Parelius, D.A. (1967). A nest of *Onychognathus morio neumanni* in the Ivory Coast. *Bull. Nigerian Orn. Soc.* **4**, 40.

Patten, G. (1980a). Red-winged Starlings. *Wits Bird Club News* **111**, 24.

Porter, R.F. and Martins, R.P. (1996). The Socotra Starling *Onychognathus frater* and Somali Starling *O. blythii*. *Sandgrouse* **17**, 151–154.

Rowan, M.K. (1955). The breeding biology and behaviour of the Redwinged Starling *Onychognathus morio*. *Ibis* **97**, 663–705.

Rowan, M.K. (1971). Adventures of a Red-winged Starling. *Bokmakierie* **23**, 74–76.

Ryan, P.G. and Sinclair, I. (1998). Somali Starling *Onychognathus blythii* in south-central Ethiopia. *Bull. Afr. Bird Club* **5**, 56–57.

Skorupa, J.P. (1982). East African breeding records for *Cossypha cyanocampter* and *Onychognathus fulgidus*. *Scopus* **6**, 46–47.

Taylor, R.H. (1974). The use of floodlights by Redwinged Starlings for catching insects after dark. *Ostrich* **45**, 32–33.

Tilson, R.L. (1977). Palewinged Starlings and klipspringers in the Kuiseb Canyon, Namib Desert Park. *Ostrich* **48**, 110–111.

Tribe, G.D. (1991). Redwinged starlings feeding on the European wasp. *Plant Protection News* **23**, 7.

Turner, D.A. (1977). Status and distribution of the East African endemic species. *Scopus* **1**, 2–11.

Urban, E.K., Brown, L.H., Buer, C.E. and Plage, G.D. (1970). Four descriptions of nesting, previously undescribed, from Ethiopia. *Bull. Br. Orn. Club* **90**, 162–164.

Woodall, P.F. (1971). Bird notes from the northern Sengwa Gorge, Rhodesia. *Ostrich* **42**, 148–149.

Lamprotornis, Spreo

Alexander-Marrack, P.D., Aaronson, M.J., Farmer, R., Houston, W.H. and Mills, T.R. (1985). Some changes in the bird fauna of Lagos, Nigeria. *Malimbus* **7**, 121–127.

Amadon, D. (1943). The genera of starlings and their relationships. *Amer. Mus. Novit.* **1247**, 1–16.

Amadon, D. (1956). Remarks on the starlings, Family Sturnidae. *Amer. Mus. Novit.* **1803**, 1–41.

Anon. (1983). East African Bird Report 1981. *Scopus* **5**, 129–176.

Bartman, W. (1974). Eine volierenbrut des Dreifarbenglanzstars (*Lamprospreo superbus*). *Gefierd. Welte* **98**, 21–22.

Bates, G.L. (1924). On the birds collected in north-western and northern Cameroon and parts of northern Nigeria. *Ibis* **1924**, 1–45.

Bell, K. (1984). Breeding the Golden-breasted Starling *Cosmopsarus regius* at the Lincoln Park Zoo, Chicago, USA. *Avicult. Mag.* **90**, 34–35.

Bennun, L., Frere, P. and Squelch, P. (1990). Blue-eared Glossy Starlings *Lamprotornis chalybaeus* and Wattled Starlings *Creatophora cinerea* associating with livestock. *Scopus* **14**, 29–30.

Blencowe, E.J. (1963). The extension of the range of Hildebrandt's Starling, *Spreo hildebrandti* (Cabanis). *J. E. Afr. Nat. Hist. Soc.* **24**, 75.

Boetticher, H.V. (1940). Die Glanzstare Afrikas. *Anz. Orn. Ges. Bayern* **3**(3), 86–91.

Bowen, W.W. (1931). East African birds collected during the Gray African Expedition 1929. *Proc. Acad. Nat. Sci. Philadel.* **83**, 11–79.

Britton, P.L. and Britton, H. (1970). Eye colour of the Blackbellied Glossy Starling. *E. Afr. Nat. Hist. Soc. Bull.*, **1970**, 46.

Brooke, R.K. (1965). On the breeding of *Lamprotornis mevesii* (Wahlberg). *Bull. Br. Orn. Club* **85**, 139–141.

Brooke, R.K. (1967a). On the moults and breeding season of the Long-tailed Starling *Lamprotornis mevesii* (Wahlberg). *Bull. Br. Orn. Club* **87**, 2–5.

Brooke, R.K. (1967b). On the plumage (including a partial albino) moults and breeding season of *Lamprotornis australis* (Smith). *Bull. Br. Orn. Club* **87**, 60–61.

Brooke, R.K. (1968). More of the plumages, moults and breeding seasons of southern African starlings. *Bull. Br. Orn. Club* **88**, 113–116.

Brooke, R.K. (1971). An aberrant *Lamprotornis mevesii* with comments on the limits of the genus *Lamprotornis*. *Bull. Br. Orn. Club* **91**, 20–21.

Brooke, R.K., Grobler, J.H., Irwin, M.P.S. and Steyn, P. (1972). A study of the migratory eagles *Aquila nipalensis* and *A. pomarina* (Aves: Accipitridae) in southern Africa, with comparative notes on other large raptors. *Occ. Pap. Nat. Mus. Rhod.* B **5**(2), 51–114.

Bruch, K. (1983). Gelungene Zucht des Smaragdglanzstares (*Coccycolius iris*). *Trochilus* **4**, 56–57.

Chittenden, H. and Myburgh, N. (1994). Eye colour change in Blackbellied Starlings. *Birding Southern Afr.* **46**, 117.

Ciarpaglini, P. (1971). Notes on breeding uncommon birds at Clères in 1970. *Avicult. Mag.* **77**, 49–57.

Clancey, P.A. (1973). Miscellaneous taxonomic notes on African birds XXXVI. A new race of *Lamprotornis mevesii* (Wahlberg) from north-western South-west Africa and adjacent Angola. *Durban. Mus. Novit.* **9**, 279–283.

Clancey, P.A. (1974a). Miscellaneous taxonomic notes on African birds XXXIX. Comments on the range and Zambian populations of *Lamprotornis chloropterus elizabeth* (Stresemann), 1924. *Durban. Mus. Novit.* **10**, 101–102.

Clancey, P.A. (1974b). On the validity and range of *Lamprotornis corruscus mandanus* Van Someren, 1921. *Bull. Br. Orn. Club* **94**, 113–116.

Clancey, P.A. and Holliday, C.S. (1951). A systematic revision of the races of *Lamprotornis nitens* (Linnaeus) endemic to the South African subcontinent. *Ostrich* **22**, 111–116.

Cole, D.T. (1963). Cape Glossy Starling nesting in pipe. *Bokmakierie* **15**, 20.

Comins, D.M. (1966). Weights of birds recorded at the Kaffrarian Museum. *Ostrich* **37**, 64.

Craig, A.J.F.K. (1983b). A Pied Starling study. *Safring News* **12**, 8–11.

Craig, A.J.F.K. (1983c). Co-operative breeding in two African starlings, Sturnidae. *Ibis* **121**, 114–115.

Craig, A.J.F.K. (1985). The distribution of the Pied Starling, and southern African biogeography. *Ostrich* **56**, 123–131.

Craig, A.J.F.K. (1987). Co-operative breeding in the Pied Starling. *Ostrich* **58**, 176–180.

Craig, A.J.F.K. (1988c). Allofeeding and dominance in the cooperatively breeding pied starling. *Anim. Behav.* **36**, 1251–1253.

Craig, A.J.F.K. (1989). A review of the biology of the Blackbellied Starling and other African forest starlings. *Ostrich* Suppl. **14**, 17–26.

Craig, A.J.F.K. and Hartley, A.H. (1985). The arrangement and structure of feather melanin granules as a taxonomic character in African starlings. *Auk* **102**, 629–632.

Cunningham-van Someren, G.R. (1974). Sisal flowers, nectar and birds. *E. Afr. Nat. Hist. Soc. Bull.* **1974**, 104–107.

Dean, W.R.J. and Macdonald, I.A.W. (1972). *Lamprotornis australis*: a new host of *Clamator glandarius*. *Ostrich* **43**, 66.

Dittami, J.P. (1983). Notes on Blue-eared Glossy Starlings *Lamprotornis chalybaeus* at Nakuru, Kenya. *Scopus* **7**, 37–39.

Dittami, J.P. (1987). A comparison of breeding and moult cycles and life histories in two tropical starling species: the Blue-eared Glossy Starling *Lamprotornis chalybaeus* and Rüppell's Long-tailed Glossy Starling *L. purpuropterus*. *Ibis* **129**, 69–85.

Donnelly, B.G. (1966). Editorial. *Bee-eater* **17**(4), 2.

Donnelly, B.G. (1967). Editorial. *Bee-eater* **18**(1), 4.

Dorst, J. and Roux, F. (1973). L'avifaune des forêts de *Podocarpus* de la province de l'Arussi, Ethiopie. *Oiseau et R.F.O.* **43**, 269–304.

Dowsett, R.J. (1967). Breeding biology of *L. mevesii* (Wahlberg). *Bull. Br. Orn. Club* **87**, 157–164.

Dowsett-Lemaire, F. (1998). First observations on the nest of Purple-headed Starling *Lamprotornis purpureiceps*. *Malimbus* **20**, 55–56.

Durrer, H. and Villiger, W. (1970). Schillerfarben der Stare (Sturnidae). *J. Orn.* **111**, 133–153.

Every, B. (1975). Some impressions of the Karroo avifauna. *Bee-eater* **26**(1), 6–7.

Ezra, A. (1929). Breeding the White-capped Starling (*Heteropsar albicapillus*). *Avicult. Mag.* Ser. 5, **7**(8), 175–176.

Ezra, A. (1933). Breeding Rüppell's Starling (*Lamprotornis purpuropterus*). *Avicult. Mag.* **11**, 357–358.

Fennessy, R.M. and Brown, L.H. (1975). 'Birds of the African Bush'. Collins, London.

Friedmann, H. (1955). The honeyguides. *Bull. US Nat. Mus.* **208**, 1–292.

Frost, P.G.H. (1980). Fruit-frugivore interactions in a South African coastal dune forest. *Acta XVII Int. Orn. Congr.* 1179–1184.

Fuggles-Couchman, N.R. (1939). Notes on some birds of the Eastern Province of Tanganyika Territory. *Ibis* **1939**, 76–106.

Fuggles-Couchman, N.R. (1984). The distribution of, and other notes on, some birds of Tanzania. *Scopus* **8**, 1–17, 81–92.

Fuggles-Couchman, N.R. and Elliot, H.F.I. (1946). Some records and field-notes from north-eastern Tanganyika Territory. *Ibis* **88**, 327–347.

Geertseema, A.A. (1976). Great Spotted Cuckoo parasitising Hildebrandt's Starling. *Bull. E. Afr. Nat. Hist. Soc.* **1976**, 85.

Gilges, W. (1945). Notes on the birds around Richards Bay – Zululand. *Ostrich* **16**, 102–108.

Glyphis, J.P., Milton, S.J. and Siegfried, W.R. (1981). Dispersal of *Acacia cyclops* by birds. *Oecologia* **48**, 138–141.

de Grahl, W. (1982). Dreifarbenglanzstare und ihr Brutverhalten. *Gefied. Welt* **106**, 114–115.

Grant, C.H.B. and Mackworth-Praed, C.W. (1946–47). Notes on East African Birds. *Bull. Br. Orn. Club* **67**, 46–48.

Greenberg, D.A. and Colebrook-Robjent, J.F.R. (1976). First Zambian breeding record of Great Spotted Cuckoo (*Clamator glandarius*). *Bull. Zambian Orn. Soc.* **8**, 69–70.

Hald-Mortensen, P. (1971). A collection of birds from Liberia and Guinea. *Steenstrupia* **12**, 115–125.

Hall, B.P. (1960b). The faunistic importance of the Scarp of Angola. *Ibis* **102**, 420–442.

Hayes, J. (1982). Notes on bird behaviour. *Bull. E. Afr. Nat. Hist. Soc.* **1982**, 40–41.

Herholdt, J.J. (1987). Observation of the incubation and nestling period of the Pied Starling *Spreo bicolor*, with notes on nestling growth. *Mirafra* **4**, 75–77.

Herholdt, J.J. (1988). Bird weights from the Orange Free State (Part II: Passerines). *Safring News* **17**, 43–57.

Holgersen, H. (1956). On a collection of birds from Nzérékoré, French Guinea. *Sterna* **25**, 1–19.

Hopkinson, E. (1932). More additions to breeding records. *Avicult. Mag.* **10**, 319–326.

Hopson, A.J. (1964). Preliminary notes on the birds of Malamfatori, Lake Chad. *Bull. Nigerian Orn. Soc.* **1**(4), 7–15.

Huels, T.R. (1981). Cooperative breeding in the Golden-breasted Starling *Cosmopsarus regius*. *Ibis* **123**, 539–542.

Hurford, J.L., Lombard, A.T., Kemp, A.C. and Benn, G.A. (1996). Geographical analysis of six rare bird species in the Kruger National Park, South Africa. *Bird Conserv. Intl* **6**, 117–137.

Hutchinson, G.R. (1938). Breeding of Royal Starlings. *Avicult. Mag.* Ser. 5, **1**, 17–18.

Irwin, M.P.S. (1957). Some field notes on a collection of birds from Tanganyika Territory. *Ostrich* **28**, 116–122.

Irwin, M.P.S. (1988). Order Cuculiformes. *In* 'The Birds of Africa'. Vol. 3 (Fry. C.H., Keith, S. and Urban, E.K., Eds), pp. 58–104. Academic Press, London.

Jensen, R.A.C., and Jensen, M.K. (1969). On the breeding biology of southern African cuckoos. *Ostrich* **40**, 163–181.

Joubert, H.J. (1945). Starlings and others. *Ostrich* **16**, 214–215.

Joubert, E. (1972). The social organisation and associated behaviour in the Hartmann Zebra *Equus zebra hartmannae*. *Madoqua* **1**(6), 17–56.

Jubb, R.A. (1977). 'Old Wheezie'. *Diaz Diary* **43**, 4.

Jubb, R.A. (1980a). Birds feeding on ectoparasites. *Diaz Diary* **77**, 4.

Jubb, R.A. (1980b). Some window and wheel-tapping birds. *E. Cape Naturalist* **69**, 11.

Jubb, R.A. (1983). Note on the greater honeyguide. *Diaz Diary* **116**, 10–11.

Kannemeyer, M. (1951). [Nesting habits of the Cape Glossy Starling]. *Bee-eater* **2**(2), 10.

Keith, S. (1964). A new subspecies of *Spreo albicapillus* (Blyth) from Kenya. *Bull. Br. Orn. Club* **84**, 162–163.

Kemp, A.C., Kemp, M.I., Jensen, R.A.C. and Clinning, C.F. (1972). Records of brood parasitism from central South West Africa. *Ostrich* **43**, 145–148.

Keulemans, J.G. (1866). Opmerkingen over de vogels van de Kaap-Verdische Eilanden en van Prins-Eiland in de Bogt van Guinea gelegen. *Ned. Tijds. Dierk.* **3**, 374–401.

Klaptocz, A. (1913). Beitrag zur Kenntnis des Ornis Französisch Guineas. *J. Orn.* **61**, 444–455.

Koen, J.H. (1992). Medium-term fluctuations of birds and their potential food resources in the Knysna Forest. *Ostrich* **63**, 21–30.

Koenig, W.D. (1994). Two new bird-mammal associations from Kenya with comments on host use by Wattled Starlings. *Ostrich* **65**, 337–338.

Kok, O.B. and van Ee, C.A. (1990). Dieetsamestelling ven enkele voëlsoorte in die Oranje-Vrystaat en Noordwes-Kaap. 3: Lede van die Sturnidae-familie. *Mirafra* **7**(1), 18–25.

Lack, P.C. and Quicke, D.L.J. (1978). Dietary notes on some Kenyan birds. *Scopus* **2**, 86–91.

Lawrence, K.J. (1973). Breeding the Superb Spreo. *Foreign Birds* **39**(1), 8–11.

Leonard, P.M. (1998). Identification of Sharp-tailed Starling *Lamprotornis acuticaudus*. In 'Zambia Bird Report 1997', Zambian Orn. Soc., Lusaka.

Liversidge, R. (1968). Bird weights. *Ostrich* **39**, 223–227.

Martins, R.P., Bradshaw, C.G., Brown, A., Kirwan, G.M. and Porter, R.F. (1996). The status of passerines in southern Yemen and the records of the OSME survey in spring 1993. *Sandgrouse* **17**, 54–72.

McCulloch, D. (1963). Colour change in the iris of the Blackbellied Starling *Lamprocolius corruscus*. *Ostrich* **34**, 177.

Miskell, J. (1977). Cooperative feeding of young at the nest by Fischer's Starling *Spreo fischeri*. *Scopus* **1**, 87–88.

Morel, G.E. and Morel, M-Y. (1962). La reproduction des oiseaux dans une région semi-aride: la vallée du Sénégal. *Alauda* **30**, 161–203.

Neumann, O. (1944). A hitherto unnamed Glossy Starling from East Africa – *Spreo hildebrandti kelloggorum*, new subspecies. *Auk* **61**, 288–289.

Newman, K.B. (1971). Birds eating ants. *Bokmakierie* **23**, 29–31.

Newman, K.B. (1986). Identifying glossy starlings. *Bokmakierie* **38**, 84–86.

van Niekerk, D.J. (1996). Albino Pied Starling *Spreo bicolor* (Sturnidae) with some notes on albinism in the Motacillidae. *Mirafra* **13**(1), 9–10.

Nixon, A. (1992). Blackbellied Starling eating reed frogs, *Hyperolius* sp. *Bee-eater* **43**, 12–13.

Nixon, A. (1993). Food of Blackbellied Starlings. *Bee-eater* **44**, 9.

Oatley, T.B. and Skead, D.M. (1972). Nectar feeding by South African birds. *Lammergeyer* **15**, 65–74.

Odgers, J.A. (1993). More on the feeding habits of the Blackbellied Starling. *Bee-eater* **44**, 31–32.

Pakenham, R.H.W. (1936). Field-notes on the birds of Zanzibar and Pemba. *Ibis* **1936**, 249–272.

Paludan, K. (1936). Report on the birds collected during Professor O. Olufsen's expedition to French Sudan and Nigeria in the year 1927; with field notes by the collector Mr Harry Madsen. *Vidensk. Medd. Dansk Naturh. Foren.* **100**, 247–346.

Patten, G. (1980b). Ectoparasites eaten from host mammals. *Wits Bird Club News* **109**, 13.

Penry, E.H. (1979). Early and late dates for Splendid Starlings (*Lamprotornis splendidus*). *Bull. Zambian Orn. Soc.* **11**, 36–38.

Penry, E.H. (1986). A review of Sharp-tailed Glossy Starling sightings in Botswana. *Babbler* **11**, 26–27.

Penzhorn, B.L. (1981). Association between birds and mountain zebras. *Ostrich* **52**, 63–64.

Penzhorn, B.L. (1982). A partial albino Cape Glossy Starling. *Ostrich* **53**, 205.

Phillips, E.L. (1896). On birds observed in the Goolis Mountains in northern Somaliland. *Ibis* **1896**, 62–87.

Pickles, R. (1989). Burchell's Starlings breed in borehole tower. *Babbler* **18**, 40–42.

Plowes, D. (1944). Nesting of Cape Glossy Starling. *Ostrich* **15**, 70–71.

Pooley, A.G. (1967). Some miscellaneous ornithological observations from the Ndumu Game Reserve. *Ostrich* **38**, 31–32.

Prigogine, A. (1983). Contribution aux migrations de *Lamprotornis splendidus bailundensis*. *Gerfaut* **73**, 193–195.

Prigogine, A. and Benson, C.W. (1979). The mysterious movements of *Lamprotornis splendidus bailundensis*. *Gerfaut* **69**, 437–445.

Pyper, S. (1994). Breeding the Emerald Starling. *Avicult. Mag.* **100**, 35–39.

Restall, R. (1968). The Superb Spreo Starling. *Avicult. Mag.* **74**, 113–123.

Reynolds, J.F. (1968). Notes on birds observed in the vicinity of Tabora, Tanzania, with special reference to breeding data. *J. E. Afr. Nat. Hist. Soc.* **27**, 117–139.

Risdon, D. (1990). Breeding the Royal Starling. *Avicult. Mag.* **96**, 89–91.

Roberts, A. (1939). Swifts and other birds nesting in buildings. *Ostrich* **10**, 85–99.

Robiller, F. and Gerstner, R. (1985). Zucht des Smaragdglanzstars (*Coccycolius iris*). *Gefied. Welt* **109**, 158–159.

Robinson, J., Robinson, C. St. C. and Winterbottom, J.M. (1957). Notes on the birds of the Cape Agulhas region. *Ostrich* **28**, 147–163.

Rudebeck, G. (1955). Some observations at a roost of European Swallows and other birds in the south-eastern Transvaal. *Ibis* **97**, 572–580.

Sawyer, R.C.J. (1982). Breeding the Splendid Starling *Lamprocolius splendidus splendidus*. *Avicult. Mag.* **88**, 189–191.

Scamell, K.M. (1964). The breeding of the Shelley's Starling (*Spreo shelleyi* Sharpe). *Avicult. Mag.* **70**, 198–200.

Short, L.L. and Horne, J.F.M. (1985). Notes on some birds of the Arabuko-Sokoke forest. *Scopus* **9**, 117–126.

Shuel, R. (1938). Further notes on the eggs and nesting habits of birds in northern Nigeria (Kano Province). *Ibis* **1938**, 463–480.

Skead, C.J. (1968). Some bird weights (in grams), recorded at the Kaffrarian Museum, 1965–1967. *Ostrich* **39**, 268.

Skead, C.J. (1995). 'Life-history notes on East Cape birds 1940–1990', Vol. 1. Algoa Regional Services Council, Port Elizabeth.

Skead, D.M. (1966). Birds frequenting the intertidal zone of the Cape Peninsula. *Ostrich* **37**, 10–16.

Skead, D.M. (1971). Bird weights from the SA Lombard Nature Reserve, Transvaal. *Ostrich* **42**, 77–78.

de Smet, K. and van Gompel, J. (1980). Observations sur la côte sénégalaise en decembre et janvier. *Malimbus* **2**, 56–70.

Smith, V.W. (1964). Further notes on birds breeding near Vom, northern Nigeria. *Nigerian Field* **29**, 161–174.

Squire, J.E. (1977). White-crowned Starlings near Maikona. *Bull. E. Afr. Nat. Hist. Soc.* 1977, 3–4.

Stevenson, T. (1983). 'The Birds of Lake Baringo'. Sealpoint Publicity, Nairobi.

Sweijd, N. and Craig, A.J.F.K. (1991). Histological basis of age-related changes in iris color in the African Pied Starling (*Spreo bicolor*). *Auk* **108**, 53–59.

Taylor, J.S. (1936). Birds in the garden. *Ostrich* **7**, 45–48.

Taylor, J.S. (1951). Nesting habits of the Cape Glossy Starling (*Lamprocolius nitens phoenicopterus* Sw.). *Bee-eater* **2**(1), 8.

Thomas, D.K. (1960). Birds – Notes on breeding in Tanganyika: 1958–1959. *Tanganyika Notes Rec.* **55**, 225–243.

Thomsen, F. (1907). Locust birds in the Transvaal. *J. S. Afr. Orn. Union* **3**, 56–75.

Thomson, W.R. (1975). Long-tailed Starlings and Great Spotted Cuckoos at Chipinda Pools. *Honeyguide* **81**, 35–36.

Thurow, T.L. and Black, H.L. (1981). Ecology and behaviour of the Gymnogene. *Ostrich* **52**, 25–35.

Treca, B. (1998). Birds fashion savannas. Abstr. 22, Int. Orn. Congr. *Ostrich* **69**, 311–312.

Tree, A.J. (1986). Blackbellied Starling project. *Diaz Diary* **151**, 15.

Trevor, S. and Lack, P. (1976). Great Spotted Cuckoo parasitising Superb Starling. *Bull. E. Afr. Nat. Hist. Soc.* **1976**, 50.

Turner, D.A. and Forbes-Watson, A.D. (1976). Status of the White-crowned Starling *Spreo albicapillus* (Blyth) in Kenya. *Bull. Br. Orn. Club* **96**, 58.

Vande weghe, J.-P. (1974). Additions et corrections à l'avifaune du Rwanda. *Rev. Zool. Afr.*, **88**, 81–98.

Vaughan, J.H. (1930). The birds of Zanzibar and Pemba. *Ibis* **1930**, 1–48.

Vernon, C.J. (1973). Vocal imitations by southern African birds. *Ostrich* **44**, 29–30.

Vernon, C.J. (1993). Gluttonous starlings raid Natal mahogany. *Bee-eater* **44**, 9.

Walker, R.B. (1965). Two Sudan savannah birds at Zaria. *Bull. Nigerian Orn. Soc.* **5**, 22–23.

Walsh, J.F. (1986). Notes on the birds of Ivory Coast. *Malimbus* **8**, 89–93.

Walsh, J.F. (1987). Records of birds seen in north-eastern Guinea in 1984–1985. *Malimbus* **9**, 105–122.

Wavertree, Lady (1930). Further notes on the breeding of the Royal Starling and Black-winged Grackle. *Avicult. Mag.* Ser. 4, **12**, 327–328.

Webster, K. (1987). Observations on breeding at Rookwood, Queenstown district. *Bee-eater* **38**, 20.

Wells, D.R. and Walsh, F. (1969). Birds of northern and central Borgu. *Bull. Nigerian Orn. Soc.* **6**, 1–25, 63–93.

Whyte, I.J. (1981). Anting in Blue-eared Glossy Starlings. *Ostrich* **52**, 185.

Wickler, W. (1966). Flügelhochstellen als Landesignal von *Lamprotornis* (Sturnidae). *J. Orn.* **107**, 87–88.

Wilkinson, R. (1982). Social organization and communal breeding in the Chestnut-bellied Starling (*Spreo pulcher*). *Anim. Behav.* **30**, 1118–1128.

Wilkinson, R. (1983). Biannual breeding and moult-breeding overlap of the Chestnut-bellied Starling *Spreo pulcher*. *Ibis* **125**, 353–361.

Wilkinson, R. (1984). Variation in eye colour of Blue-eared Glossy Starling. *Malimbus* **6**, 2–4.

Wilkinson, R. (1988). Long-tailed Glossy Starlings *Lamprotornis caudatus* in field and aviary with observations on cooperative breeding in captivity. *Avicult. Mag.* **94**, 143–154.

Wilkinson, R. (1996). Cooperative breeding in captive Emerald Starlings *Coccycolius iris*. *Malimbus* **18**, 134–141.

Wilkinson, R. (1997). Cooperative breeding in captive Emerald Starlings *Coccycolius iris*: an update. *Malimbus* **19**, 39.

Wilkinson, R. and Brown, A.E. (1984). Effects of helpers on the feeding rates of nestlings in the Chestnut-bellied Starling *Spreo pulcher*. *J. Anim. Ecol.* **53**, 301–310.

Wilkinson, R. and McLeod, W. (1991). Breeding the Ashy Starling at Chester Zoo. *Avicult. Mag.* **97**(4), 163–166.

Wilkinson, R., McLeod, W. and Langford, D. (1993). Some observations on the breeding of African Pied Starlings at Chester Zoo. *Avicult. Mag.* **99**, 182–185.

Wilson, G.T. (1975). A second non-iridescent Longtailed Starling. *Ostrich* **46**, 185.

Winterbottom, J.M. (1975). Notes on the South African species of *Corvus*. *Ostrich* **46**, 236–250.

Woodward, R.B. and Woodward, J.D.S. (1899). 'Natal birds'. Davis, Pietermaritzburg.

Wragg, H.B. (1967). Colourful Starlings. *Foreign Birds* **33**(1), 9, 12.

Yealland, J.J. (1955). The Emerald Starling (*Coccycolius iris*). *Avicult. Mag.* **61**(6), 20.

van Zyl, A.J. (1991). Unusual falcon hunting behaviour. *Gabar* **6**, 68.

Speculipastor, Grafisia, Pholia, Cinnyricinclus, Neocichla

Benson, C.W. and Irwin, M.P.S. (1966). The *Brachystegia* avifauna. *Ostrich Suppl.* **6**, 297–321.

Beesley, J.S.S. (1972). A breeding record of Sharpe's Starling *Cinnyricinclus sharpii* in the Arusha National Park. *E. Afr. Nat. Hist. Soc. Bull.* **1972**, 12–13.

Blancou, L. (1974). *Grafisia torquata* en Afrique centrale. *Oiseau et R.F.O.* **44**, 90.

Brosset, A. and Erard, C. (1977). New faunistic records from Gabon. *Bull. Br. Orn. Club* **97**, 125–132.

Dean, W.R.J. and Vernon, C.J. (1988). Notes on the White-winged Babbling Starling *Neocichla gutturalis* in Angola. *Ostrich* **59**, 39–40.

Dowsett, R.J. (1972). Sharpe's Starliing *Cinnyricinclus sharpii* in southern Tanzania. *E. Afr. Nat. Hist. Soc. Bull.* **1972**, 56–57.

Dowsett, R.J. and Dowsett-Lemaire, F. (1998). Further additions to and deletions from the avifauna of Congo-Brazzaville. *Malimbus* **20**, 15–32.

Haas, V. and Nickel, E. (1982). Breeding of Magpie Starlings *Speculipastor bicolor* in Kenya. *Scopus* **6**, 41.

Mackworth-Praed, C.W. and Grant, C.H.B. (1950). On the migratory movements of the southern race of the Violet-backed Starling. *Ibis* **92**, 402–404.

Traylor, M.A. (1971). Moult and migration in *Cinnyricinclus leucogaster*. *J. Orn.* **112**, 1–20.

Turner, D.A. (1977). Status and distribution of the East African endemic species. *Scopus* **1**, 2–11.

Zavattariornis, Creatophora, Sturnus, Acridotheres

Álvarez, S.P. (1984). Descripcion y desarrollo del pollo del Estornino negro. *Ardeola* **31**, 3–16.

Amadon, D. (1943). The genera of starlings and their relationships. *Amer. Mus. Novit.* **1247**, 1–16.

Anon. (1962). Foods and feeding. *Lammergeyer* **2**, 66.

Anon. (= Aspinwall, D.R.) (1986). Movement analysis charts: comments on Wattled Starling. *Zambian Orn. Soc. Newsl.* **16**, 4–7.

Anon. (1996). 'Important Bird Areas of Ethiopia'. Ethiopian Wildlife and Natural history Society, Addis Ababa.

Ash, J.S. and Gullick, T.M. (1989). The present situation regarding the endemic breeding birds of Ethiopia. *Scopus* **13**, 90–96.

Bennun, L., Frere, P. and Squelch, P. (1990). Blue-eared Glossy Starlings *Lamprotornis chalybaeus* and Wattled Starlings *Creatophora cinerea* associating with livestock. *Scopus* **14**, 29–30.

Benson, C.W. (1942). A new species and ten new races from southern Abyssinia. *Bull. Br. Orn. Club* **63**, 8–19.

Brewster, C.A. (1999). Two records of Indian Myna *Acridotheres tristis* from the Bobirwa area of eastern Botswana. *Babbler* **35**, 25–26.

Brooke, R.K. (1976). Morphological notes on *Acridotheres tristis* in Natal. *Bull. Br. Orn. Club* **96**, 8–13.

Brooke, R.K. (1983). On the introduction of the Indian Myna in Harare. *Honeyguide* **116**, 15.

Brooke, R.K., Lloyd, P.H. and De Villiers, A.L. (1986). Alien and translocated terrestrial vertebrates in South Africa. In 'The Ecology and Management of Biological Invasions in Southern Africa' (Macdonald, I.A.W. *et al.*, Eds), pp. 63–74. Oxford University Press, Cape Town.

Brown, P.B. (1971). Breeding the Wattled Starling (*Creatophora carunculata*). *Avicult. Mag.* **77**, 158–159.

Corre, M. Le and Probst, J.M. (1997). Migrant and vagrant birds of Europa Island (southern Mozambique Channel). *Ostrich* **68**, 13–18.

Craib, C.L. (1971). Breeding of the European Starling at Grahamstown. *Ostrich* **42**, 145–146.

Craig, A.J.F.K. (1992). The distribution of the Wattled Starling in southern Africa. *Ostrich* **63**, 31–37.

Craig, A.J.F.K. (1996). The annual cycle of wing-moult and breeding in the Wattled Starling *Creatophora cinerea*. *Ibis* **138**, 448–454.

Dean, W.R.J. (1978). Plumage, reproductive condition and moult in non-breeding Wattled Starlings. *Ostrich* **49**, 97–101.

Donnelly, B.G. (1982). On the feral breeding of Indian Mynahs in Bulawayo. *Honeyguide* **111/112**, 53.

Durrer, H. and Villiger, W. (1970). Schillerfarben der Stare (Sturnidae). *J. Orn.* **111**, 133–153.

Feare, C.J. (1986). Behaviour of the Spotless Starling *Sturnus unicolor* Temm. during courtship and incubation. *Gerfaut* **74**, 3–11.

Fishpool, L.D.C., Allport, G.A. and Webb, R. (1996). Photospot: Ethiopian endemics. *Bull. Afr. Bird Club* **3**, 42–43.

Goodwin, D. (1986). 'Crows of the World'. Univ. Washington Press, Seattle.

Hundessa, T. (1991). Survival status review of the Ethiopian Bushcrow (*Zavattariornis stresemanni* Moltoni, 1938) in the Borana area, Ethiopia. *Walia* **13**, 9–13.

Koenig, W.D. (1994). Two new bird-mammal associations from Kenya with comments on host use by Wattled Starlings. *Ostrich* **65**, 337–338.

Kok, O.B. and van Ee, C.A. (1990). Dieetsamestelling van enkele voëlsoorte in die Oranje-Vrystaat en Noordwes-Kaap, 3: Lede van die Sturnidae-familie. *Mirafra* **7**, 18–25.

Kok, O.B. and van Zyl, J.M. (1996). Body mass of birds from central South Africa. *Ostrich* **67**, 160–162.

Liversidge, R. (1961). The Wattled Starling (*Creatophora cinerea* Meuschen). *Ann. Cape Prov. Mus.* **1**, 71–80.

Lowe, P.R. (1949). On the position of the genus *Zavattariornis*. *Ibis* **91**, 102–104.

Nuttall, R.J. (1998). Notes on flocking, feeding and breeding activity of Wattled Starlings at Sandveld Nature Reserve. *Mirafra* **15**, 31–32.

Paxton, M. and Cooper, T. (1986). Wattled Starlings breeding at Rietfontein, Etosha. *Lanioturdus* **22**, 37–40.

Peris, S.J. (1980). Biologìa del Estornino Negro (*Sturnus unicolor* Temm.). *Ardeola* **25**, 207–240.

Peris, S.J. (1991). Ringing recovery of the Spotless Starling *Sturnus unicolor* in Spain. *Ringing and Migration* **12**, 124–125.

Peris, S.J. (1998). No effects of forest spraying of Deltamatrin on the breeding Spotless Starling *Sturnus unicolor*. Abstracts 22 Int. Orn. Congr., Durban. *Ostrich* **69**, 445.

Quickelberge, C.D. (1972). Status of the European Starling at its present approximate eastern limits of spread. *Ostrich* **43**, 179–180.

Ripley, S.D. (1955). Anatomical notes on *Zavattariornis*. *Ibis* **97**, 142–145.

Robertson, I. (1992). New information on birds in Cameroon. *Bull. Br. Orn. Club* **112**, 36–42.

Robertson, I. (1993). Unusual records from Cameroon. *Malimbus* **14**, 62–63.

Sontag, W.A. (1979). Beobachtungen zum nicht-epigamen Verhalten des Lappenstars, *Creatophora cinerea* (Meuschen). *Bonn. Zool. Beitr.* **30**, 367–379.

Sontag, W.A. (1985). Song and courtship of the Wattled Starling *Creatophora cinerea*. *Malimbus* **7**, 129–135.

Sontag, W.A. (1990a). Wattled Starling *Creatophora cinerea* – a potential breeding species for Arabia. *Phoenix* **7**, 4.

Sontag, W.A. (1990b). Species, class and individual characteristics in the African Wattled Starling, *Creatophora cinerea*. *Bonn. Zool. Beitr.* **41**, 163–169.

Syvertson, P.O. and Dellelegn, Y. (1991). The status of some bird species endemic to south Ethiopia. *Scopus* **15**, 30–34.

Tellería, J.L. (1981). 'La Migracion de las Aves en el Estrecho de Gibraltar, Vol. 2: Aves no planeadoras'. Univ. Complutense, Madrid.

Uys, C.J. (1977). Notes on Wattled Starlings in the Western Cape. *Bokmakierie* **29**, 87–89.

Winterbottom, J.M. and Liversidge, R. (1954). The European Starling in the south west Cape. *Ostrich* **25**, 89–96.

Family BUPHAGIDAE

Attwell, R.I.G. (1966). Oxpeckers, and their association with mammals in Zambia. *Puku* **4**, 17–48.

Bezuidenhout, J.D. and Stutterheim, C.J. (1980). A critical evaluation of the role played by the Red-billed Oxpecker *Buphagus erythrorhynchus* in the biological control of ticks. *Onderstepoort J. Vet. Res.* **47**, 51–75.

Breitwisch, R. (1996). Oxpeckers, the genus *Buphagus*. *Bull. Afr. Bird Club* **3**, 31–33.

Brown, C.J. and Brown, S.E. (1987). Some observations on oxpeckers in eastern Caprivi, SWA/Namibia. *Lanioturdus* **22**, 74–79.

Buskirk, W.H. (1975). Substrate choice of oxpeckers. *Auk* **92**, 604–606.

Clancey, P.A. (1976). Further on subspeciation in the Red-billed Oxpecker *Buphagus erythrorhynchus*. *Bull. Br. Orn. Club* **96**, 102–105.

Cunningham-van Someren, G.R. (1984). A new race of Red-billed Oxpecker *Buphagus erythrorhynchus* from Kenya. *Bull. Br. Orn. Club* **104**, 120–121.

Dowsett, R.J. (1965). On a nest of the Yellow-billed Oxpecker *Buphagus africanus* in Zambia. *Bull. Br. Orn. Club* **85**, 133–135.

Grobler, J.H. (1979). The re-introduction of oxpeckers *Buphagus africanus* and *B. erythrorhynchus* to the Rhodes Matopos National Park, Rhodesia. *Biol. Conserv.* **15**, 51–58.

Grobler, J.H. (1980). Host selection and species preference of the Red-billed Oxpecker *Buphagus erythrorhynchus* in the Kruger National Park. *Koedoe* **23**, 89–97.

Grobler, J.H. and Charsley, G.W. (1978). Host preference of the Yellow-billed Oxpecker *Buphagus africanus* in the Rhodes Matopos National Park, Rhodesia. *S. Afr. J. Wildl. Res.* **8**, 169–170.

Hall-Martin, A.J. (1987). Range expansion of the Yellow-billed Oxpecker *Buphagus africanus* into the Kruger National Park, South Africa. *Koedoe* **30**, 121–132.

Hart, B.L., Hart, L.A. and Mooring, M.S. (1990). Differential foraging of oxpeckers on impala in comparison with sympatric antelope. *Afr. J. Ecol.* **28**, 240–249.

Hustler, K. (1987). Host preference of oxpeckers in the Hwange National Park, Zimbabwe. *Afr. J. Ecol.* **25**, 241–245.

Koenig, W.D. (1994). Host preference, behavior, and coexistence of oxpeckers. *J. Orn.* **135**, 133.

Lockwood, G. (1986). Yellowbilled Oxpeckers and lots of bull. *Bokmakierie* **38**, 73–74.

Lockwood, G. (1995). Oxpecker translocations – an SAOS success story. *Birding Southern Afr.* **47**, 131–134.

Moreau, R.E. (1933). The food of the Red-billed Oxpecker, *Buphagus erythrorhynchus* (Stanley). *Bull. Entomol. Res.* **24**, 325–335.

Mundy, P.J. (1983). The oxpeckers of Africa. *Afr. Wildl.* **37**, 110–116.
Mundy, P.J. (1992). Notes on oxpeckers. *Honeyguide* **38**, 108–112.
Mundy, P.J. and Cook, A.W. (1975). Observations of the Yellowbilled Oxpecker *Buphagus africanus* in northern Nigeria. *Ibis* **117**, 504–506.
Mundy, P.J. and Haynes, G. (1996). Oxpeckers and elephants. *Ostrich* **67**, 85–87.
Ndao, B. (1999). Le Pique-boeufs à bec jaune *Buphagus africanus* buveur de lait de vache. *Bull. Afr. Bird Club* **6**, 59.
Newman, K. (1986). Yellowbilled Oxpeckers in Zululand. *Bokmakierie* **38**, 74–75.
Olivier, R.C.D. and Laurie, W.A. (1974). Birds associating with hippopotamuses. *Auk* **91**, 169–170.
Ritchie, D. (1981). Isi-Hlalanyati Esiphuzi has returned to Natal. *Quagga* **15**, 18–20.
Rockingham-Gill, D.V. (1992). Return of the Red-billed Oxpecker in the Makonde District. *Honeyguide* **38**, 188–189.
Roseneath, W. (1997). Waxbills (oxpeckers). *BBC Wildlife* **15**, 21.
Siegfried, W.R. and Brooke, R.K. (1985). Oxpecker. *In* 'A Dictionary of Birds' (Campbell, B. and Lack, E., Eds), pp. 422–423. T. and A.D. Poyser, Calton.
Stutterheim, C.J. (1977). Dimensions of the Redbilled Oxpecker in the Kruger National Park. *Ostrich* **48**, 119–120.
Stutterheim, C.J. (1980a). Symbiont selection of Redbilled Oxpecker in the Hluhluwe-Umfolozi Game Reserve Complex. *Lammergeyer* **30**, 21–25.
Stutterheim, C.J. (1980b). Moult cycle of the Redbilled Oxpecker in the Kruger National Park *Ostrich* **51**, 107–112.
Stutterheim, C.J. (1981). The movements of a population of Redbilled Oxpeckers (*Buphagus erythrorhynchus*) in the Kruger National Park. *Koedoe* **24**, 99–107.
Stutterheim, C.J. (1982a). Past and present ecological distribution of the Red-billed Oxpecker *Buphagus erythrorhynchus* in South Africa. *S. Afr. J. Zool.* **17**, 190–196.
Stutterheim, C.J. (1982b). Breeding biology of the Redbilled Oxpecker in the Kruger National Park. *Ostrich* **53**, 79–90.
Stutterheim, C.J., Bezuidenhout, J.D. and Elliott, E.G.R. (1988). Comparative feeding behaviour and food preferences of oxpeckers (*Buphagus erythrorhynchus* and *B. africanus*) in captivity. *Ondersport J. Vet. Res.* **55**, 173–179.
Stutterheim, C.J. and Brooke, R.K. (1981). Past and present ecological distribution of the Yellowbilled Oxpecker in South Africa. *S. Afr. J. Zool.* **16**, 44–49.
Stutterheim, C.J., Mundy, P.J. and Cook, A.W. (1976). Comparison between the two species of oxpecker. *Bokmakierie* **28**, 12–14.
Stutterheim, C.J. and Panagis, K. (1985a). The status and distribution of oxpeckers (Aves: Buphaginae) in the Kavango and Caprivi, SWA/Namibia. *S. Afr. J. Zool.* **20**, 10–14.
Stutterheim, C.J. and Panagis, K. (1985b). Roosting behaviour and host selection of oxpeckers (Aves: Buphaginae) in Moremi Wildlife Reserve, Botswana, and eastern Caprivi, South West Africa. *S. Afr. J. Zool.* **20**, 235–240.
Stutterheim, C.J. and Panagis, K. (1987). Capture and transport of oxpeckers *Buphagus erythrorhynchus* and *B. africanus* from the Eastern Caprivi Strip, SWA/Namibia. *Madoqua* **15**, 251–253.
Stutterheim, C.J. and Stutterheim, I.M. (1981). A possible decline of a Redbilled Oxpecker population in the Pilansberg Complex, Bophuthatswana. *Ostrich* **52**, 56–57.
Thomson, W.R. (1982). Oxpeckers roosting on game animals. *Honeyguide* **110**, 46–47.
Watkins, B.P. and Cassidy, R.J. (1987). Evasive action taken by Waterbuck to Redbilled Oxpeckers. *Ostrich* **58**, 90.
Webster, K. (1997). Red-billed Oxpecker project. *Afr. Birds and Birding* **2**(4), 7.
Whyte, I.J., Hall-Martin, A.J., Kloppers, J.J. and Otto, J.P.A. du T. (1987). The status and distribution of the Yellowbilled Oxpecker in the Kruger National Park. *Ostrich* **58**, 88–90.
Wood, P.A. (1994). Red-billed Oxpecker survey in southern Mana Pools. *Honeyguide* **40**, 16–19.

Acoustic References

Discs and Cassettes

5. North, M.E.W. (1958). Voices of African Birds. Cornell University Press. 159 Sapsucker Woods Road, Ithaca, N.Y. 14850. One 12-inch, $33\frac{1}{3}$ r.p.m. disc. 42 species. The first African record concerned mainly with identification. Species are presented in systematic order, grouped on separate bands, and details given of circumstances, place and date of recording.

7. Haagner, C.H. (1961). Birds of the Kruger National Park. International Library of African Music, P.O. Box 138, Roodeport, near Johannesburg, South Africa. Two 7-inch, 45 r.p.m. discs, Nos XTR 17044 and XTR 27045. 31 species in systematic order following Roberts (1957; 'Birds of South Africa', Trustees of the John Voelcker Bird Book Fund, Cape Town.) and with the Roberts number; each on a separate band.

9. Haagner, C.H. (1964). Birds of Zululand. Same publisher as No. 7. Two 7-inch, 45 r.p.m. discs, Nos XTR 4 7094 and XTR 5 7095. 27 species.

10. North, M.E.W. and McChesney, D.S. (1964). More Voices of African Birds. Houghton Mifflin Co., Boston, MA, USA. One 12-inch, $33\frac{1}{3}$ r.p.m. disc. 90 species. Details of recordings are given in an accompanying booklet. These 2 discs (Nos 5 and 10) together contain the voices of 132 species and are the first major reference work for African bird voices.

11. Stannard, J. and Niven, P. (1966). Bird Songs of Amanzi. Percy Fitzpatrick Institute of African Ornithology, University of Cape Town, Rondebosch 7700, South Africa. One 12-inch $33\frac{1}{3}$ r.p.m. disc, No. ACP 524; No. 1 in 'Bird Song Series'. 37 species. On one side the birds are heard in their natural surroundings, the emphasis being on atmosphere or ambience; on the other side they are singled out and identified.

13. Hayes, C. and Hayes, J. (1966). East African Birdsong; No. 2 in *Heartbeat of Africa*, Series 1. Sapra Studios, Box 5882,

Kimathi and York Streets, Nairobi, Kenya. One 7-inch, 45 r.p.m. disc. 25 species.
14. Stannard, J. and Niven, P. (1967). Bird Song of the Forest. Percy FitzPatrick Institute (address under No. 11). One 12-inch $33\frac{1}{3}$ r.p.m. disc. GALP 1559. 32 species.
15. Walker, A. (1967). Bird Song of Southern Africa. African Music Society and International Library of African Music, Roodeport, South Africa. One 12-inch $33\frac{1}{3}$ r.p.m. disc, GALP 1501. 33 species.
16. Stannard, J. (1967). Sunbird calls and songs. One 7-inch 45 r.p.m. disc accompanying the book: 'Sunbirds of Southern Africa' by C.J. Skead. Published by A.A. Balkema, P.O. Box 3117, Cape Town, South Africa. 14 species.
17. Reucassel, R. and Pooley, A.C. (1967). Calls of the Bushveld. Published by the authors and obtainable from the Wildlife Society of South Africa. One 12-inch $33\frac{1}{3}$ r.p.m. disc, WL2; also available as a cassette. 28 species.
20. Henley, A. and Pooley, A.C. (1970). Birds of the Drakensberg. Published by the authors and obtainable from Wildlife Society of South Africa. One 12-inch $33\frac{1}{3}$ r.p.m. stereo disc, BD 100. 41 species.
21. Reucassel, R. and Adendroff, A. (1970). Nature's Melody. Published by the authors; obtainable from Wildlife Society of South Africa. One 12-inch $33\frac{1}{3}$ r.p.m. stereo disc, SWL 3. 53 species.
22. Walker, A. (1970). Garden Birds of Southern Africa. Gallo (Africa) Ltd., Johannesburg; obtainable from Wildlife Society of South Africa. One 12-inch $33\frac{1}{3}$ r.p.m. stereo disc, SGALP 1598. 40 species.
24. Roché, J.-C. (1970). L'Oiseau Musicien; The Bird as Musician, Nos 10, 11 and 12. Châteaubois, F-38350 La Mure, France. Three 7-inch, 45 r.p.m. discs, 2 species on each disc.
25. Dangerfield, G. (1970). Sounds of the Serengeti. Music for Pleasure Ltd., Astronaut House, Hounslow Road, Feltham, Middlesex, England. One 12-inch $33\frac{1}{3}$ r.p.m. stereo disc, MFP 1371. 24 species.
27. Hayes, J. (c. 1970). Bird Song of Africa. *Heartbeat of Africa*, Series 2. Sapra Studios (address under No. 13). One 7-inch, 45 r.p.m. disc. 12 species.
30. Roché, J.-C. (1971). Birds of Kenya. *Birds and Wild Beasts of Africa*, No. 1. L'Oiseau Musicien, France. One 12-inch, $33\frac{1}{3}$ r.p.m. stereo disc, G.07. 32 species.
32. Keith, G.S. and Gunn, W.W.H. (1971). Birds of the African Rain Forests. *Sounds of Nature* No. 9. Federation of Ontario Naturalists, 1262 Don Mills Road, Don Mills, Ontario M3B 2WB, Canada, and American Museum of Natural History, New York. Two 12-inch, $33\frac{1}{3}$ r.p.m. discs. 95 species. The most important reference work since the records of North (Nos 5 and 10) and the first specializing in forest birds, many of which are here published for the first time. Most species are from East Africa, some from central Africa. Species are arranged in systematic order and grouped in bands; a simple announcement of the name accompanies each species, but a lot of information is provided on the jacket.
33. Stannard, J. (1971). Bird Sounds and Songs. Fitzpatrick Institute (address under No. 11). Issued in conjunction with *Ostrich*, Supplement 9. One 7-inch, 45 r.p.m. disc, NV1. 20 species.
34. Chappuis, C. (1971). Ambiances des plaines et savanes d'Afrique orientale. *Afrique Sauvage* No. 1. One 12-inch, $33\frac{1}{3}$ r.p.m. disc, JAC 9. Edition *Jacana*, 32 rue St. Marc, 75002 Paris. 44 species.
35. Martin, R.B. (1971). Journey Across Africa. Parlophone PCSJ (D) 12.79. Obtainable from Wildlife Society of South Africa. One 12-inch, $33\frac{1}{3}$ r.p.m. disc. 34 species.
36. Ker, A. (1972). Safari 99. Equator Sound Studios Ltd., P.O. Box 30068, Nairobi, Kenya. One 12-inch, $33\frac{1}{3}$ r.p.m. disc, ESS 1001. 63 species.
38. Keibel, W.D. (1972). Wildlife of South West Africa. Wildlife Society of South Africa, P.O. Box 3508, Windhoek, Namibia. 1 cassette, 48 species.
39. Roché, J.-C. (1973a). Birds of South Africa. *Birds and Wild Beasts of Africa*, No. 2. L'Oiseau Musicien, France. One 12-inch, $33\frac{1}{3}$ r.p.m. stereo disc, G. 08. About 65 species. Seven environments are presented without commentary, created by 3 or 4 birds singing simultaneously.
40. Roché, J.-C. (1973b). Birds of West Africa – Senegal. *Birds and Wild Beasts of Africa* No. 3. L'Oiseau Musicien, France. One 12-inch, $33\frac{1}{3}$ r.p.m. disc, G. 09. 26 species.
42. Worman, D. (1974). African Birds. Soundpics Enterprises (Pty) Ltd, P.O. Box 61055, Marshalltown 2107, South Africa. One 10-inch, $33\frac{1}{3}$ r.p.m. disc, SP 002, and 16 colour slides. 16 species.
46. Anon. (1966). A Night at Treetops. Sapra Studios (address under No. 13). *Heartbeat of Africa*, Series 1, No. 3. One 17-cm 45 r.p.m. disc. About 10 species.
47. Jones, B. (1969). The Rhino Story. Wildlife Society of South Africa. P.O. Box 44189, Linden 2104, South Africa. One 17-cm 45 r.p.m. disc. 12 species.
49. Reucassel, D. (1975). Calls of the Wild. Published by the author and available from Wildlife Society of South Africa. One 30-cm, $33\frac{1}{3}$ r.p.m. disc, AV1; also available as a cassette. Comes with 32 colour slides. 17 species.
50. Hart, S. (1975). Listen to the Wild – in the Bush. EMI/Brigadiers (Pty) Ltd., South Africa. 30-cm, $33\frac{1}{3}$ r.p.m. stereo disc, Brigadiers Music LTW(W)1. 17 species.
51. Hart, S. (1975). Listen to the Wild – Among the Rocks. See No. 50. 16 species.
53. Chappuis, C. (1975). *Les Oiseaux de l'Ouest Africain*. Disc 5: Timaliidae, Pycnonotidae (first part), 32 species. Disc 6: Pycnonotidae (end), Turdidae (first part), 44 species. *Alauda*, Sound Supplement accompanying commentary in *Alauda* 43, 450–474, M.N.H.N., Laboratoire d'Ecologie, 4 Avenue du Petit Château, 91800 Brunoy, France. Two 12-inch, $33\frac{1}{3}$ r.p.m. discs. These records are a part of a series whose aim is to present all known recordings for species of a particular region, including different forms of songs and calls and geographical variation. Details of the recordings are provided in the accompanying article in *Alauda*, of which reprints may be requested when ordering the records.

These records represent a landmark in the history of African voice-recording. This is a lengthy series covering large numbers of species in great detail, and the accompanying commentaries in *Alauda* are of considerable scientific value.
58. Natal Bird Club (c. 1978). Bird Calls, Vols 1 and 2. Natal Bird Club, P.O. Box 10909, Marine Parade, Durban 4056, South Africa. 2 cassettes. 136 species presented in random order. Lengthy and numerous cuts are provided for each species.
62. Palmer, S. and Boswall, J. (1969–1972). A Field Guide to the Bird Songs of Britain and Europe. SR Records, Swedish Broadcasting Corp., 105 10 Stockholm, Sweden. Twelve 12-inch, $33\frac{1}{3}$ r.p.m. discs, RFLP 5001–5012. 530 species, nesting or accidental in Europe, mostly wintering in Africa. Presented in systematic order, on separate bands,

announced by scientific name. The most important reference work for Palearctic birds wintering in Africa.

64. Hayes, J. and Allan, J.O. Wild Africa. Andrew Crawford Productions, P.O. Box 42004, Nairobi, Kenya. One 12-inch, $33\frac{1}{3}$ r.p.m. disc, ACP 1001. 8 species.

66. Kabaya, T. (1978). Birds of the World. I: Africa. King Records Co., Japan. One 30-cm, $33\frac{1}{3}$ r.p.m. stereo disc, King Records SKS (H) 2007. 20 species.

68. Chappuis, C. (1979). *Les oiseaux de l'Ouest africain*. Sound supplement to *Alauda*. Disc 10; Sylviidae (end), Paridae, 37 species. One 12-inch, $33\frac{1}{3}$ r.p.m. mono disc, ALA 19 and 20. Commentary in *Alauda* 47, No. 3, 195–212. For details, see No. 53.

69. Walker, A. (1980). Sounds of the Zimbabwe Bush. Available from the author at 1 Northmoor Road, Oxford OXZ 6UW, England, or Queen Victoria Museum, Harare, Zimbabwe. One stereo cassette. 27 species.

72. Audio Three (1981). Bird Calls. See No. 58. Three cassettes, of which the first two are the same as those of No. 58; the third contains additional species.

73. Palmer, S. and Boswall, J. (1981). A Field Guide to the Bird Songs of Britain and Europe. 16 cassettes, RFLP 5021–5036. An updated edition of No. 62. 612 species, in boxes of 4 cassettes with commentary and list of species in each box. A first class reference collection.

75. Audio Three, Bird Calls: Bird Families, Vol. IV. 2 cassettes, 171 species. Many of these species already appear on No. 72, but here all are in systematic order.

76. Chappuis, C. (1984). *Oiseaux de France: Migrateurs et Hivernants*, Parts I and II. Obtainable from the author, 10 Vallon du Fer à Cheval, 76530 La Bouille, France. Two cassettes with booklets. Present mainly flight and contact calls, not full songs, of Palearctic birds; useful because these are the vocalizations typically made in Africa by migrants. 147 species.

86. Stjernstedt, R. (1986–1990). Bird Songs of Zambia. Distributor: Ducan Macdonald, Wildsounds, P.O. Box 9, Holt, Norfolk NR25 7AW, UK. Three mono cassettes: No. 2: Passerines, Alaudidae to Sylviidae (*Cisticola*) 108 species. No. 3: Sylviidae (*Prinia*) to Emberizidae. 140 species. One of the major collections of African bird voices. The species are presented in systematic order, often with several types of vocalization per species.

88. Gillard, L. (1987). *Southern African Bird Calls*. Gillard Bird Cassettes, P.O. Box 72059, Parkview 2122, Johannesburg, South Africa. Three cassettes of 90 min., 540 species presented in systematic order, often with several types of song and call per species. The large number of species makes this one of the most important and comprehensive collections of African bird voices so far published.

89. Chappuis, C. (1990). Sounds of Migrant and Wintering Birds, Western Europe. Obtainable from the author: 10, Vallon du Fer à Cheval, 76530 La Bouille, France. Two mono cassettes with booklet; English (and revised) version of No. 76.

91. Gibbon, G. (1991). *Southern African Bird Sounds*. Southern African Birding, P.O. Box 24106, Hillary 4024, South Africa. Six 90-min. mono cassettes in a box with booklet. 880 species presented in systematic order. The most important collection of African bird voices yet published. The quality of sound reproduction is excellent. A major reference work, even though the large number of species makes the time devoted to each one relatively short.

93. Roché, J.-C. and Couzens, D. (1985). *The Bird-Walker*. L'Oiseau Musicien, Chateaubois, Mayres-Saval 38350, La Mure, France. Three cassettes devoted to birds of Europe, with a few species from North Africa; 406 species.

99. Gibbon, G. (1995). Southern African Bird Sounds. Six CDs. Published and distributed by Southern African Birding, P.O. Box 24106, Hillary 4024, Durban. 900 species. Enlarged version of No. 91.

100. Huguet, P. and Tostain, O. (1991). Savane d'Afrique, Nature en Zambie. Distributed by Pithys International, 189, rue Grande, F-77300 Fontainebleau. Cassette and CD.

101. Strömberg, M. (1994). Moroccan Bird Songs and Calls. One cassette with booklet. Waxnäsq 44, 65341 Karstad, Sweden. 76 species.

102. Stjernstedt, R. (1996). Rare Birds of Zambia. Distributor: see no. 86. One cassette. 94 species.

103. Stjernstedt, R. (1996). Bird songs of Zambia. Addenda. Distributor: see no. 86. One cassette. 37 species.

104. Chappuis, C. (in press). Bird sounds from North-Western Africa: Vol. II. Species breeding south of the Sahara.

105. Chappuis, C. (in press). Bird sounds from North-Western Africa: Vol. I. Breeding and migratory species north of the Sahara and on the Atlantic Islands. 104 and 105 form a collection of 1450 species covering all of West and Northwest Africa.

106. Chappuis, C. (1976). Oiseaux de Corse et de Méditerranée. *Alauda* **44**, 475–503. One 30-cm, $33\frac{1}{3}$ r.p.m. mono disc. *Alauda*, 4 Avenue du Petit Château, F-91800 Brunoy, France. 61 species from the northwest perimeter of the Mediterranean.

107. Schubert, N. (1984). Stimmender Vogel VII – Vogel Stimmen Sud Ost Europas. Two 30-cm, $33\frac{1}{3}$ mono discs; Eterna 622.702 VEB. Deutsche Schallplatten, Berlin Est.

108. Smith, S. (1994). Bird recordings from The Gambia. One cassette. Available from the author: 1 Serrells Barn Cottages, Langton Matravers, Swanage, Dorset, BH19 3HX, England. 52 species.

109. Smith, S. (1996). Bird recordings from Ethiopia. One cassette. Available from the author (see no. 108). 66 species.

110. Roche, J.-C. (1993). *All the Bird Songs of Britain and Europe*. Available from *Sittelle*, Rue des Jardins, 38710 MENS, France, or *WildSounds*, P.O. Box 9, Holt, Norfolk NR25 7AW, England. Four compact discs. The CD edition is limited to 396 species since there are only 99 accessible tracks per disc; nevertheless, most birds not included are non-breeding visitors.

Principal Discs and Cassettes by Region

East Africa : Nos 5, 10, 32, 36
West Africa : Nos 53, 68, 104, 105

Southern Africa : Nos 16, 58, 75, 86, 88, 91, 99
Palearctic migrants : Nos 62, 73, 76, 89, 104, 105

Institutions with Sound Libraries

A. Audio Three, 6 Larch Road, Durban, South Africa.
B.B.C. British Broadcasting Corporation, Natural History Recording Library: through NSA.
B. BLOWS: British Library of Wildlife Sounds, now NSA.
C. Cornell University, Library of Natural Sounds, Laboratory of Ornithology, 159 Sapsucker Woods Road, Ithaca NY 14850, USA.
E. Fonoteca Zoologica, Museo de Zoologia, Parque de la Ciutadela, 08003 Barcelona, Spain.
F. Fitzpatrick Bird Communication Library, Bird Department, Transvaal Museum, P.O. Box 413, Pretoria 0001, South Africa.
N. Natal Bird Club, P.O. Box 10909, Marine Parade, Durban 4056, South Africa.
NSA National Sound Archive, Wildlife Section, 96 Euston Road, St Pancras, London NW1 2DB, UK.
S. South African Broadcasting Corporation. Library of Wildlife Sounds, P.O. Box 4559, Johannesburg 2000, South Africa.

Individual Recordists

ADE	Adendorff, G.	HAN	Hansen, L.	OSME	OSME expedition to Socotra
ALEX	Alexander-Marrack, P. D., BLOWS	HAY C	Hayes, C.	PAR	Parker, V., FITZ
		HAY J	Hayes, J., BLOWS, BBC	PARK	Parker, T., Cornell
ANO	Anon	HAZ	Hazevoet, C. J.	PAY	Payne, R. B.
ASP	Aspinwall, D. R.	HAZB	Britton, Hazel	PEA	Pearson, D. J.
ATT	Attenborough, D.	HEN	Henley, T.	PER	Perez del Val, J.
BAR	Baranga, J.	HOL	Hollom, P. A. D., NSA-BLOWS	POO	Pooley, T.
BEL	Bell, Fairfax, NSA-BLOWS			PRIN	Pringle, J. S.
BERG	Bergman, H. H.	HOR	Horne, J.	REU	Reucassel, D.
BOR	Borrow, N.	HRS	Harris, T.	ROC	Roche, J. C.
BOUR	Bourgignon, C.	HUG	Huguet, P	ROCH	Roche, J.
BROS	Brosset, A.	JOHN	Johnson, E. D. H.	RODE	Rodewald, P.
BRU	Brunel, J.	JOJ	Jones, J	SALA	Sala, A.
BUL	Bullens, P.	JO PJ	Jones, P. J.	SAR	Sargeant, D. E.
BUT	Butynski, T. M.	KAB	Kabaya, T.	SHI	Shirihai, H.
CART	Carter, C., FITZ, BLOWS	KAEST	Kaestner, P.	SIN	Sinclair, J. C.
CHA	Chappuis, C.	KAL	Kalina, J.	SMIT	Smith, S.
CHR	Christy, P.	KEI	Keith, S., Cornell	STA	Stannard, J., FITZ
CLO	Clouet, M.	KEIB	Keibel, W. D., FITZ	STJ	Stjernstedt, R., NSA-BLOWS
DAV	Davidson, P.	KER	Ker, A.	STU	Stuart, S. N.
DEM	Demey, R.	KOC	Koch, L.	SVEN	Svendsen, J. O.
DUV	Duval, C. T., NSA-BLOWS	KÖN	König, C.	THOR	Thorpe, W. H.
DYE	Dyer, M.	KRJ	Krjukov, A. P.	TOMB	Tombs, D. J.
ERA	Erard, C.	LEM	Dowsett-Lemaire, F.	TUCK	Tucker, N.
FAR	Farkas, T.	LIV	Liversidge, R.	TYE	Tye, A.
FIS	Fisher, D., BLOWS	LOS	Loskot, V. M.	VDAE	Van Daele, P.
FISP	Fishpool, L. D. C.	LOW	Low, G. C., NSA-BLOWS	VDB	Van den Berg, A.
FOR	Forbes-Watson, A. D.	LUT	Lutgens, H., BLOWS	VEP	Veprintsev, B. N.
GAR	Gartshore, M. E.	MAC	Macaulay, L. R., Cornell	VIEL	Vielliard, J.
GARD	Gardner, N.	MANN	Mannery, D.	WALK	Walker, A.
GAU	Gautier, J. P.	MAR	Martin, D. R.	WAT	Watts, D. E.
GIB	Gibbon, G.	McVIC	McVicker, R.	WATSN	Watson, C.
GIL	Gillard, L.	MEES	Mees, V., BLOWS	WHI	White, T.C.
GOR	Goriup, P.	MEY	Meyer, R. W.	WILK	Wilkinson, R.
GORD	Gordon, J.	MIL	Mild, K.	WILL	Williams, E.
GREG	Gregory, A. R.	MOR	Morel, G.	WOOD	Woodcock, M. W.
GRI	Grimes, L., NSA-BLOWS	MOY	Moyer, D., Cornell	WOR	Worman, D
GUL	Gullick, T., BLOWS	NIV	Niven, P.	ZIM	Zimmerman, D. and M.
HA	Haagner, C. H., FITZ	NOR	North, M. E. W. Cornell		

INDEXES

Plate key-page numbers in *italics*.

Scientific Names

abyssinica, Pseudoalcippe 32, *15*
abyssinicus, Zosterops 314, *270*
acaciae, Turdoides fulvus 60
acik, Chalcomitra senegalensis 194
Acridotheres 659
Acridotheres tristis 660, *543*
acuticaudus, Lamprotornis 608, *527*
adamauae, Prionops plumatus 488
adelberti, Chalcomitra 184, *158*
adjuncta, Chalcomitra amethystina 190
adolfifriederici, Malaconotus cruentus 385
adsimilis, Dicrurus 525, *463*
AEGITHALIDAE 76
Aegithalos 76
Aegithalos caudatus 76, *94*
aegra, Cinnyris pulchella 259
aeneocephalus, Lamprotornis purpuropterus 615
aequatorialis, Chalcomitra senegalensis 194
aequatorialis, Urolestes melanoleucus 377
aethiopicus, Laniarius 453, *414*
afer, Nilaus 472, *287*
afer, Parus 85, *78*
afer, Ptilostomus 560, *478*
afer, Sphenoeacus 8
affinis, Corvinella corvina 374
affinis, Dryoscopus cubla 435
affinis, Nilaus afer 474
afra, Cinnyris 247, *206*
africanus, Buphagus 662, *543*
alberti, Prionops 492, *415*
albicapillus, Spreo 633, *542*
albicollis, Corvus 549, *478*
albilateralis, Cinnyris chalybea 239
albipectus, Illadopsis 21, *15*
albirostris, Onychognathus 588, *526*
albiventris, Cinnyris venusta 288
albiventris, Parus 91, *78*
albus, Corvus 545, *478*
Alethe poliothorax 28
alfredi, Cyanomitra olivacea 179

algeriensis, Lanius meridionalis 361
alinae, Cyanomitra 174, *143*
alius, Malaconotus 394, 395, *334*
altera, Cinnyris habessinica 280
amani, Oriolus chlorocephalus 515
ambiguus, Laniarius aethiopicus 454
amboimensis, Laniarius 450, *414*
amethystina, Chalcomitra 189, *158*
amethystinus, Lamprotornis purpureus 598
amicorum, Cinnyris manoensis 237
Anabathmis 154
Anabathmis hartlaubii 157, *142*
Anabathmis newtonii 160, *142*
Anabathmis reichenbachii 155, *142*
anchietae, Anthreptes 151, *142*
anchietae, Antichromus minutus 419
andaryae, Malaconotus bocagei 405
anderseni, Cinnyris regia 253
anderssoni, Cinnyris talatala 283
anderssoni, Zosterops senegalensis 308
angolensis, Anthreptes longuemarei 138
angolensis, Buphagus erythrorhynchus 666
angolensis, Dryoscopus 431, *398*
angolensis, Oriolus larvatus 510
angolensis, Urolestes melanoleucus 376
angolus, Prionops plumatus 489
anguitimens, Eurocephalus 381, *287*
angusta, Neocichla gutturalis 644
ansorgei, Anthoscopus caroli 113
ansorgei, Malaconotus bocagei 404
ansorgei, Tchagra australis 422
Anthobaphes 163
Anthobaphes violacea 163, *159*
Anthoscopus 106
Anthoscopus caroli 112, *79*
Anthoscopus flavifrons 107, *79*
Anthoscopus minutus 115, *79*
Anthoscopus musculus 111, *79*
Anthoscopus parvulus 109, *79*
Anthreptes 136
Anthreptes anchietae 151, *142*
Anthreptes aurantium 141, *95*

Anthreptes fraseri 153
Anthreptes gabonicus 145, *95*
Anthreptes longuemarei 138, *95*
Anthreptes neglectus 137, *95*
Anthreptes orientalis 140, *95*
Anthreptes pallidigastra 229
Anthreptes rectirostris 146, *95*
Anthreptes reichenowi 149, *95*
Anthreptes rubritorques 148, *95*
Anthreptes seimundi 150, *95*
Anthscopus punctifrons 110, *79*
Antichromus 418
Antichromus minutus 418, *351*
apivorus, Dicrurus adsimilis 528
approximans, Malaconotus blanchoti 389
arabicus, Cinnyricinclus leucogaster 643
Arcanator orosthruthus 8
ardens, Promerops gurneyi 337
arens, Parus afer 85
aresta, Cinnyris talatala 283
Argya aylmeri 63
aridicolus, Lanius collaris 342
armena, Tchagra senegala 426
arturi, Nectarinia kilimensis 208
ashantiensis, Cinnyris superba 295
atactus, Dicrurus modestus 530
ater, Parus 103, *94*
ater, Turdoides hartlaubii 49
atlas, Parus ater 103
atlas, Sitta europaea 125, *126*
atmorii, Zosterops pallidus 318
atriceps, Pseudoalcippe abyssinica 32
atripennis, Dicrurus 524, *463*
atripennis, Phyllanthus 64, *31*
atrococcineus, Laniarius 468, *399*
atrocoeruleus, Laniarius funebris 445
atroflavus, Laniarius 471, *399*
aucheri, Lanius meridionalis 361
aurantium, Anthreptes 141, *95*
aurantius, Chaetops frenatus 10
auratus, Oriolus 516, *462*
aurea, Chalcomitra fuliginosa 187

aureus, Anthoscopus parvulus 109
australis, Lamprotornis 618, *542*
australis, Tchagra 421, *351*
axillaris, Deleornis fraseri 153
aylmeri, Argya 63
aylmeri, Turdoides 63, *31*

badius, Lanius senator 371
baileyi, Pyrrhocorax pyrrhocorax 569
bailundensis, Lamprotornis splendidus 611
balfouri, Chalcomitra 198, *158*
bannermani, Cyanomitra 169, *143*
bansoensis, Cyanomitra oritis 173
barakae, Parus thruppi 86
barbarus, Laniarius 461, *399*
barbarus, Pyrrhocorax pyrrhocorax 569
batesi, Cinnyris 299, *223*
batesi, Malaconotus multicolor 397
bedfordi, Pholidornis rushiae 121
benguelae, Parus cinerascens 83
benguelensis, Lamprotornis mevesii 617
bensoni, Cinnyris mediocris 255
bensoni, Poeoptera kenricki 577
bertrandi, Malaconotus olivaceus 403
bestiarium, Buphagus erythrorhynchus 666
beverleyae, Hedydipna collaris 220
biarmicus, Panurus 75, *94*
bicolor, Laniarius 458, *414*
bicolor, Speculipastor 637, *526*
bicolor, Spreo 634, *542*
bicolor, Turdoides 52, *31*
bifasciata, Cinnyris 268, *222*
billypayni, Turdoides fulvus 60
bineschensis, Cinnyris chloropygyia 234
blanchoti, Malaconotus 388, *334*
blythii, Onychognathus 584, *479*
bocagei, Illadopsis rufipennis 13
bocagei, Malaconotus 404, *335*
bocagei, Nectarinia 201, *159*
bocagei, Tchagra australis 422
boehmi, Lanius excubitoroides 356
bohndorffi, Cyanomitra verticalis 167, 168
bohndorffi, Phyllanthus atripennis 65
boranensis, Turdoides aylmeri 63
bouvieri, Cinnyris 276, *207*
boydi, Dryoscopus angolensis 431
brachydactyla, Certhia 132, *94*
brachyrhynchus, Oriolus 504, *462*
brauni, Laniarius 449, *414*
brevialatus, Nilaus afer 473
brubru, Nilaus afer 473
brunneus, Speirops 325, *270*

buchanani, Turdoides fulvus 60
BUPHAGIDAE 662
Buphagus 662
Buphagus africanus 662, *543*
Buphagus erythrorhynchus 665, *543*
burigi, Lanius souzae 368
buryi, Lanius meridionalis 361
buvuma, Cinnyris superba 295

cabanisi, Lanius 348, *271*
caeruleus, Parus 104, *94*
cafer, Promerops 328, *287*
caffer, Buphagus erythrorhynchus 666
caffrariae, Tchagra tchagra 425
caliginosa, Corvinella corvina 374
cameroonensis, Deleornis fraseri 153
camerunensis, Laniarius poensis 443
camerunensis, Nilaus afer 473
camerunensis, Parus leucomelas 98
camerunensis, Tchagra senegala 426
caniceps, Prionops 493, *415*
caniviridis, Zosterops pallidus 318
capelli, Lanius collaris 342
capensis, Corvus 533, *478*
capensis, Zosterops pallidus 318
capricornensis, Cinnyris chalybea 239
caroli, Anthoscopus 112, *79*
carolinae, Turdoides squamulatus 57
carpi, Parus 99, *79*
caryocatactes, Nucifraga 567, *478*
catharoxanthus, Malaconotus blanchoti 387, 389
cathemagmenus, Telophorus cruentus 417
caudatus, Aegithalos 76, *94*
caudatus, Lamprotornis 612, *542*
centralis, Malaconotus lagdeni 392
cephaelis, Cyanomitra obscura 176
Certhia 132
Certhia brachydactyla 132, *94*
CERTHIIDAE 132
cervicalis, Garrulus glandarius 565
Chaetops 8
Chaetops frenatus 10, *14*
Chaetops pycnopygius 8, *14*
chalcea, Cinnyris cuprea 303
chalcomelas, Cinnyris 274, *222*
Chalcomitra 182
Chalcomitra adelberti 184, *158*
Chalcomitra amethystina 189, *158*
Chalcomitra balfouri 198, *158*
Chalcomitra fuliginosa 187, *158*
Chalcomitra hunteri 197, *158*
Chalcomitra rubescens 185, *158*
Chalcomitra senegalensis 193, *158*

Chalcomitra 'Tôha Sunbird' 200
chalcurus, Lamprotornis 600, *527*
chalybaeus, Lamprotornis 603, *527*
chalybea, Cinnyris 239, *206*
changamwensis, Cyanomitra olivacea 179
chapini, Cinnyris stuhlmanni 244
chapini, Corvinella corvina 374
chapini, Dryoscopus cubla 434
chapini, Kupeornis 71, *14*
chloris, Nicator 481, *350*
chlorocephalus, Oriolus 514, *462*
chloropterus, Lamprotornis 606, *527*
chloropygyia, Cinnyris 233, *206*
chobiensis, Hedydipna collaris 220
chrysonotis, Lamprotornis splendidus 611
cinerascens, Parus 83, *78*
cinerea, Creatophora 648, *543*
cinereus, Turdoides plebejus 37
Cinnyricinclus 641
Cinnyricinclus leucogaster 642, *526*
Cinnyris 230
Cinnyris afra 247, *206*
Cinnyris batesi 299, *223*
Cinnyris bifasciata 268, *222*
Cinnyris bouvieri 276, *207*
Cinnyris chalcomelas 274, *222*
Cinnyris chalybea 239, *206*
Cinnyris chloropygyia 233, *206*
Cinnyris coccinigastra 296, *223*
Cinnyris congensis 265, *222*
Cinnyris cuprea 302, *159*
Cinnyris erythrocerca 266, *222*
Cinnyris fusca 291, *223*
Cinnyris habessinica 279, *207*
Cinnyris johannae 293, *223*
Cinnyris loveridgei 258, *207*
Cinnyris ludovicensis 245, *206*
Cinnyris manoensis 237, *206*
Cinnyris mariquensis 262, *222*
Cinnyris mediocris 254, *207*
Cinnyris minulla 235, *206*
Cinnyris moreaui 256, *207*
Cinnyris nectarinioides 267, *222*
Cinnyris neergaardi 242, *206*
Cinnyris notata 276
Cinnyris osea 278, *207*
Cinnyris oustaleti 281, *223*
Cinnyris pembae 275, *222*
Cinnyris pulchella 258, *207*
Cinnyris regia 252, *207*
Cinnyris reichenowi 250, *206*
Cinnyris rockefelleri 254, *207*
Cinnyris rufipennis 298, *223*
Cinnyris shelleyi 261, *222*

Cinnyris stuhlmanni 244, *206*
Cinnyris superba 294, *223*
Cinnyris talatala 282, *223*
Cinnyris tsavoensis 273, *222*
Cinnyris ursulae 301, *223*
Cinnyris venusta 285, *223*
cirtensis, Corvus monedula 555
citrinipectus, Malaconotus blanchoti 389
citrinus, Anthoscopus parvulus 109
clarkei, Turdoides leucopygius 47
claudei, Pseudoalcippe abyssinica 32
cleaveri, Illadopsis 23, *15*
coccinigastra, Cinnyris 296, *223*
collaris, Hedydipna 219, *142*
collaris, Lanius 340, *271*
collurio, Lanius 364, *286*
concinnatus, Prionops plumatus 488
confusa, Tchagra senegala 427
congener, Tchagra australis 422
congensis, Cinnyris 265, *222*
congicus, Dryoscopus gambensis 437
convergens, Turdoides jardineii 40
coracinus, Dicrurus modestus 529
corax, Corvus 537, *478*
corone, Corvus 535, *478*
corruscus, Lamprotornis 602, *527*
CORVIDAE 531
corvina, Corvinella 372, *287*
Corvinella 372
Corvinella corvina 372, *287*
Corvus 532
Corvus albicollis 549, *478*
Corvus albus 545, *478*
Corvus capensis 533, *478*
Corvus corax 537, *478*
Corvus corone 535, *478*
Corvus crassirostris 552, *478*
Corvus edithae 544, *478*
Corvus frugilegus 554, *478*
Corvus monedula 555, *478*
Corvus rhipidurus 548, *478*
Corvus ruficollis 540, *478*
Corvus splendens 557, *478*
Cosmopsarus, see *Lamprotornis*
crassirostris, Corvus 552, *478*
crassirostris, Oriolus 507, *462*
Creatophora 648
Creatophora cinerea 648, *543*
cristatus, Parus 102, *94*
cristatus, Prionops plumatus 488
crossensis, Chalcomitra rubescens 186
cruentata, Chalcomitra senegalensis 194
cruentus, Malaconotus 385, *334*
cruentus, Telophorus 416, *350*

cubla, Dryoscopus 434, *398*
cucullata, Tchagra senegala 427
cuprea, Cinnyris 302, *159*
cupreocauda, Lamprotornis 594, *527*
cupreonitens, Nectarinia famosa 214
curtus, Parus albiventris 92
cyaniventris, Lamprotornis chalybaeus 604
cyanocephala, Cyanomitra verticalis 168
cyanolaema, Cyanomitra 170, *143*
Cyanomitra 166
Cyanomitra alinae 174, *143*
Cyanomitra bannermani 169, *143*
Cyanomitra cyanolaema 170, *143*
Cyanomitra obscura 175, 179, *143*
Cyanomitra olivacea 178, *143*
Cyanomitra oritis 172, *143*
Cyanomitra veroxii 180, *143*
Cyanomitra verticalis 167, *143*
cyrenaicae, Parus caeruleus 105

damarensis, Anthoscopus minutus 116
damarensis, Tchagra australis 422
dartmouthi, Nectarinia johnstoni 212
decorsei, Cinnyris osea 278
degener, Laniarius funebris 446
Deleornis 153
Deleornis fraseri 153, *95*
demeryi, Zosterops senegalensis 307
deminuta, Chalcomitra amethystina 189
denti, Pholidornis rushiae 121
derooi, Cyanomitra alinae 174
DICRURIDAE 521
Dicrurus 521
Dicrurus adsimilis 525, *463*
Dicrurus atripennis 524, *463*
Dicrurus ludwigii 522, *463*
Dicrurus modestus 529, *463*
dilutior, Illadopsis fulvescens 18
distans, Illadopsis rufipennis 13
divaricatus, Dicrurus adsimilis 528
djamdjamensis, Hedydipna collaris 220
dodsoni, Lanius meridionalis 361
doggetti, Chalcomitra amethystina 189
dohertyi, Telophorus 411, *350*
dohrni, Horizorhinus 72, *14*
dorsalis, Lanius 346, *271*
Dreptes 161
Dreptes thomensis 161, *142*
Dryoscopus 429
Dryoscopus angolensis 431, *398*
Dryoscopus cubla 434, *398*
Dryoscopus gambensis 437, *398*

Dryoscopus pringlii 438, *398*
Dryoscopus sabini 430, *398*
Dryoscopus senegalensis 432, *398*

eboensis, Chalcomitra adelberti 184
ecki, Lamprotornis acuticaudus 608
edithae, Corvus 544, *478*
elachior, Hedydipna collaris 220
elegans, Lanius meridionalis 360
elgonense, Anthreptes rectirostris 146
elgonensis, Onychognathus walleri 578
elisabeth, Lamprotornis chloropterus 607
emini, Lamprotornis chalcurus 601
emini, Salpornis spilonotus 130
emini, Tchagra australis 422
emini, Turdoides jardineii 40
emini, Turdoides rubiginosus 62
erlangeri, Cinnyris nectarinioides 268
erlangeri, Laniarius aethiopicus 454
erlangeri, Salpornis spilonotus 130
erwini, Dryoscopus gambensis 437
erythreae, Dryoscopus gambensis 438
erythrocerca, Cinnyris 266, *222*
erythrogaster, Laniarius 466, *399*
erythrorhynchus, Buphagus 665, *543*
Eurocephalus 378
Eurocephalus anguitimens 381, *287*
Eurocephalus rueppellii 379, *287*
europaea, Sitta 125, *94*
eurycricotus, Zosterops poliogaster 316
excelsus, Parus major 87
excubitor, Lanius 360
excubitoroides, Lanius 356, *286*
expressa, Urolestes melanoleucus 377
extrema, Illadopsis rufipennis 13
extremus, Malaconotus blanchoti 389

falkensteini, Cinnyris venusta 288
famosa, Nectarinia 213, *159*
fasciata, Cinnyris johannae 293
fasciiventer, Parus 81, *78*
fazoglensis, Cinnyris venusta 288
feae, Zosterops ficedulinus 312
femoralis, Pholia 639, *526*
ferrugineus, Laniarius 456, *414*
ficedulinus, Zosterops 311, *270*
fischeri, Cyanomitra veroxii 181
fischeri, Lamprotornis 632, *542*
flavifrons, Anthoscopus 107, *79*
flavilateralis, Zosterops abyssinicus 314
fraseri, Anthreptes 153
fraseri, Deleornis 153, *95*
frater, Onychognathus 585, *479*

frater, Tchagra australis 422
frenatus, Chaetops 10, *14*
fricki, Zosterops abyssinicus 314
fringillinus, Parus 89, *78*
frugilegus, Corvus 554, *478*
fuelleborni, Cinnyris mediocris 255
fuelleborni, Laniarius 444, *399*
fugax, Dicrurus adsimilis 525
fulgidus, Onychognathus 580, *526*
fuliginosa, Chalcomitra 187, *158*
fulvescens, Illadopsis 17, *15*
fulvus, Turdoides 59, *31*
funebris, Laniarius 445, *399*
funereus, Parus 90, *78*
fusca, Cinnyris 291, *223*

gabela, Parus funereus 90
gabela, Prionops 499, *415*
gabonensis, Malaconotus cruentus 385
gabonicus, Anthreptes 145, *95*
gadowi, Nectarinia kilimensis 208
galinieri, Parophasma 73, *14*
gambensis, Dryoscopus 437, *398*
garguensis, Hedydipna collaris 220
Garrulus 564
Garrulus glandarius 565, *463*
genderuensis, Cinnyris reichenowi 251
gerhardi, Zosterops senegalensis 307
gigi, Anthoscopus minutus 116
gilberti, Kupeornis 68, *14*
gladiator, Malaconotus 393, *334*
glandarius, Garrulus 565, *463*
graculinus, Prionops retzii 497
graculus, Pyrrhocorax 570, *463*
Grafisia 638
Grafisia torquata 638, *526*
granti, Cyanomitra olivacea 179
graueri, Cinnyris stuhlmanni 244
graueri, Malaconotus multicolor 397
griseiventris, Parus 82, *78*
griseosquamatus, Turdoides hartlaubii 49
griseovirescens, Zosterops 313, *270*
gubernator, Lanius 367, *286*
guineensis, Cyanomitra obscura 176
guineensis, Parus leucomelas 98
gularis, Illadopsis fulvescens 18
gularis, Nicator 483, *350*
gularis, Turdoides plebejus 37
gurneyi, Promerops 336, *287*
guttatus, Laniarius bicolor 459
gutturalis, Chalcomitra senegalensis 194
gutturalis, Neocichla 644, *543*
gymnocephalus, Picathartes 1, *14*

gymnogenys, Turdoides 57, *31*

habessinica, Cinnyris 279, *207*
habessinica, Tchagra senegala 427
hamatus, Dryoscopus cubla 434
harterti, Prionops caniceps 494
harterti, Ptyrticus turdinus 35
hartlaubii, Anabathmis 157, *142*
hartlaubii, Onychognathus fulgidus 580
hartlaubii, Turdoides 49, *30*
haussarum, Anthreptes longuemarei 139
haynesi, Phyllanthus atripennis 65
Hedydipna 218
Hedydipna collaris 219, *142*
Hedydipna metallica 227, *142*
Hedydipna pallidigastra 229, *142*
Hedydipna platura 225, *142*
heinrichi, Zosterops senegalensis 308
helenae, Laniarius barbarus 464
hellmayri, Anthoscopus caroli 113
heuglini, Turdoides rubiginosus 62
hildebrandti, Lamprotornis 625, *542*
hilgerti, Telophorus cruentus 417
hindei, Turdoides 53, *31*
hispaniensis, Sitta europaea 126
hofmanni, Cinnyris shelleyi 262
holomelas, Laniarius poensis 443
Horizorhinus 72
Horizorhinus dohrni 72, *14*
horrensis, Spreo albicapillus 634
humeralis, Lanius collaris 342
hunteri, Chalcomitra 197, *158*
hypodila, Hedydipna collaris 220
hypoleucus, Turdoides 55, *31*
hypopyrrhus, Malaconotus blanchoti 389
hypostictus, Turdoides jardineii 40

iboensis, Illadopsis fulvescens 18
idius, Deleornis fraseri 153
igneiventris, Cinnyris venusta 288
Illadopsis 11
Illadopsis albipectus 21, *15*
Illadopsis cleaveri 23, *15*
Illadopsis fulvescens 17, *15*
Illadopsis puveli 26, *15*
Illadopsis pyrrhoptera 20, *15*
Illadopsis rufescens 25, *15*
Illadopsis rufipennis 12, *15*
inaestimata, Chalcomitra senegalensis 194
insignis, Parus leucomelas 98

insularis, Cinnyris chloropygyia 233
intercalans, Cyanomitra olivacea 179
intercedens, Lanius excubitoroides 356
interfluvius, Malaconotus olivaceus 403
intermedius, Onychognathus fulgidus 581
intermedius, Prionops retzii 497
interpositus, Malaconotus blanchoti 389
invictus, Buphagus erythrorhynchus 666
irbii, Aegithalos caudatus 76
iris, Lamprotornis 619, *527*
isabellinus, Lanius 363, *286*
italiae, Aegithalos caudatus 76
itombwensis, Nectarinia johnstoni 212

jacksoni, Malaconotus bocagei 405
jacksoni, Nectarinia tacazze 204
jacksoni, Zosterops senegalensis 307
jamesi, Tchagra 424, *351*
jardineii, Turdoides 39, *30*
jebelmarrae, Lanius meridionalis 361
johannae, Cinnyris 293, *223*
johnsoni, Illadopsis cleaveri 23
johnstoni, Nectarinia 211, *159*
jombeni, Lamprotornis corruscus 602
jubaensis, Turdoides squamulatus 57
jubaensis, Zosterops abyssinicus 314

kaboboensis, Cyanomitra alinae 174
kaboboensis, Parus fasciiventer 81
kaffensis, Zosterops poliogaster 316
Kakamega 27
Kakamega poliothorax 28, *15*
kalahari, Tchagra senegala 427
kalindei, Kupeornis chapini 71
kalkcreuthi, Chalcomitra amethystina 189
kaokoensis, Turdoides gymnogenys 58
kasaicus, Zosterops senegalensis 308
kempi, Cinnyris chloropygyia 233
kenianus, Turdoides aylmeri 63
keniensis, Prionops scopifrons 501
kenricki, Poeoptera 576, *479*
kikuyuensis, Cinnyris reichenowi 251
kikuyuensis, Zosterops poliogaster 316
kilimensis, Nectarinia 205, *159*
kirki, Prionops scopifrons 501
kirkii, Chalcomitra amethystina 189
kirkii, Turdoides jardineii 40
kismayensis, Laniarius ruficeps 451
kivuensis, Cinnyris regia 252

kobylini, Lanius collurio 365
kordofanensis, Corvus capensis 533
kordofanicus, Telophorus cruentus 417
kruensis, Anthreptes seimundi 151
kuanzae, Cinnyris venusta 288
kulalensis, Zosterops poliogaster 316
kungwensis, Dryoscopus angolensis 431
kupeensis, Malaconotus 395, *334*
Kupeornis 67
Kupeornis chapini 71, *14*
Kupeornis gilberti 68, *14*
Kupeornis rufocinctus 69, *14*

lacuum, Turdoides leucopygius 47
laetior, Oriolus brachyrhynchus 505
lagdeni, Malaconotus 391, *334*
lamperti, Chalcomitra senegalensis 194
Lamprocolius, see *Lamprotornis*
Lamprotornis 593
Lamprotornis acuticaudus 608, *527*
Lamprotornis australis 618, *542*
Lamprotornis caudatus 612, *542*
Lamprotornis chalcurus 600, *527*
Lamprotornis chalybaeus 603, *527*
Lamprotornis chloropterus 606, *527*
Lamprotornis corruscus 602, *527*
Lamprotornis cupreocauda 594, *527*
Lamprotornis fischeri 632, *542*
Lamprotornis hildebrandti 625, *542*
Lamprotornis iris 619, *527*
Lamprotornis mevesii 616, *542*
Lamprotornis nitens 599, *527*
Lamprotornis ornatus 609, *542*
Lamprotornis pulcher 628, *542*
Lamprotornis purpureiceps 596, *527*
Lamprotornis purpureus 597, *527*
Lamprotornis purpuropterus 614, *542*
Lamprotornis regius 622, *543*
Lamprotornis shelleyi 624, *542*
Lamprotornis splendidus 610, *542*
Lamprotornis superbus 626, *542*
Lamprotornis unicolor 620, *543*
langi, Buphagus africanus 663
Laniarius 440
Laniarius aethiopicus 453, *414*
Laniarius amboimensis 450, *414*
Laniarius atrococcineus 468, *399*
Laniarius atroflavus 471, *399*
Laniarius barbarus 461, *399*
Laniarius bicolor 458, *414*
Laniarius brauni 449, *414*
Laniarius erythrogaster 466, *399*
Laniarius ferrugineus 456, *414*
Laniarius fuelleborni 444, *399*
Laniarius funebris 445, *399*
Laniarius leucorhynchus 441, *399*

Laniarius liberatus 452, *414*
Laniarius luehderi 447, *414*
Laniarius mufumbiri 465, *399*
Laniarius poensis 443, *399*
Laniarius ruficeps 451, *414*
Laniarius turatii 460, *414*
LANIIDAE 339
Lanius 339
Lanius cabanisi 348, *271*
Lanius collaris 340, *271*
Lanius collurio 364, *286*
Lanius dorsalis 346, *271*
Lanius excubitor 360
Lanius excubitoroides 356, *286*
Lanius gubernator 367, *286*
Lanius isabellinus 363, *286*
Lanius mackinnoni 352, *286*
Lanius meridionalis 360, *286*
Lanius minor 358, *286*
Lanius newtoni 345, *271*
Lanius nubicus 354, *271*
Lanius senator 369, *271*
Lanius somalicus 347, *271*
Lanius souzae 368, *286*
larvatus, Oriolus 509, *462*
lathburyi, Nectarinia reichenowi 210
ledanti, Sitta 124, *94*
ledouci, Parus ater 103
lessoni, Lamprotornis splendidus 611
leucocephalus, Turdoides 43, *30*
leucogaster, Cinnyricinclus 642, *526*
leucomelas, Parus 97, *79*
leuconotus, Parus 93, *78*
leucophaeus, Speirops 324, *270*
leucopygius, Turdoides 46, *30*
leucopygos, Lanius meridionalis 361
leucorhynchus, Laniarius 441, *399*
liberatus, Laniarius 452, *414*
limbatus, Turdoides leucopygius 47
limpopoensis, Laniarius aethiopicus 454
Lioptilus 66
Lioptilus nigricapillus 66, *14*
littoralis, Tchagra australis 422
longuemarei, Anthreptes 138, *95*
loveridgei, Cinnyris 258, *207*
loveridgei, Turdoides aylmeri 63
lowei, Cyanomitra obscura 176
lucens, Cinnyris pulchella 263
lucidipectus, Cinnyris pulchella 259
ludovicensis, Cinnyris 245, 24, *206*
ludwigii, Dicrurus 522, *463*
luehderi, Cinnyris chloropygia 233
luehderi, Laniarius 447, *414*
lugubris, Poeoptera 574, *479*
lugubris, Speirops 322, *270*

mackinnoni, Lanius 352, *286*
macrorhynchos, Nucifraga caryocatactes 567
magnirostrata, Cyanomitra cyanolaema 171
major, Laniarius aethiopicus 454
major, Nectarinia famosa 214
major, Parus 87, *78*
major, Poeoptera lugubris 574
makawa, Malaconotus olivaceus 403
MALACONOTIDAE 383
Malaconotus 383
Malaconotus alius 394, 395, *334*
Malaconotus blanchoti 387, 388, *334*
Malaconotus bocagei 404, *335*
Malaconotus cruentus 385, *334*
Malaconotus gladiator 393, *334*
Malaconotus kupeensis 395, *334*
Malaconotus lagdeni 391, *334*
Malaconotus monteiri 386, *334*
Malaconotus multicolor 396, *335*
Malaconotus nigrifrons 400, *335*
Malaconotus olivaceus 402, *335*
Malaconotus sulfureopectus 406, *335*
maloneyana, Illadopsis fulvescens 18
malzacii, Dryoscopus gambensis 437
mandana, Tchagra jamesi 424
mandanus, Lamprotornis corruscus 602
manningi, Malaconotus nigrifrons 401
manoensis, Cinnyris 237, *206*
marchanti, Illadopsis cleaveri 23
mariquensis, Cinnyris 262, *222*
maroccanus, Turdoides fulvus 60
marungensis, Cyanomitra alinae 174
marwitzi, Lanius collaris 342
massaicus, Nilaus afer 474
masukuensis, Parus rufiventris 96
mauritanica, Certhia brachydactyla 133
mauritanica, Pica pica 562
mbuluensis, Zosterops poliogaster 316
mediocris, Cinnyris 254, *207*
melanocephalus, Speirops 323, *270*
melanogastra, Cinnyris pulchella 259
melanoleucus, Dryoscopus sabini 430
melanoleucus, Urolestes 376, *287*
Melanoparus funereus 90
melanops, Turdoides 50, *30*
melanopterus, Prionops plumatus 488
meneliki, Oriolus auratus 513
mentalis, Prionops rufiventris 495
mentalis, Turdoides aylmeri 63
meridionalis, Lanius 360, *286*
metallica, Hedydipna 227, *142*
mevesii, Lamprotornis 616, *542*

microrhyncha, Cinnyris bifasciata 269
minor, Garrulus glandarius 565
minor, Lanius 358, *286*
minor, Nilaus afer 473
minor, Tchagra australis 422
minulla, Cinnyris 235, *206*
minutus, Anthoscopus 115, *79*
minutus, Antichromus 418, *351*
miombensis, Nilaus afer 474
modestus, Dicrurus 529, *463*
modicus, Onychognathus neumanni 587
Modulatrix *stictigula* 8
monacha, Oriolus 513, *462*
monacha, Pseudoalcippe abyssinica 32
monedula, Corvus 555, *478*
monteiri, Malaconotus 386, *334*
moreaui, Cinnyris 256, *207*
morio, Onychognathus 582, *479*
mossambicus, Laniarius aethiopicus 454
mozambica, Tchagra senegala 427
muenzneri, Dicrurus ludwigii 522
mufumbiri, Laniarius 465, *399*
multicolor, Malaconotus 396, *335*
muraria, Tichodroma 127, *94*
musculus, Anthoscopus 111, *79*

nabouroup, Onychognathus 592, *479*
nairobiensis, Dryoscopus cubla 434
nandensis, Dryoscopus angolensis 431
natalensis, Laniarius ferrugineus 457
natalensis, Tchagra tchagra 425
Nectarinia 200. Species: see also
  Anthreptes, Anabathmis, Dreptes,
  Anthobaphes, Cyanomitra,
  Chalcomitra, Hedydipna, Cinnyris
Nectarinia bocagei 201, *159*
Nectarinia famosa 213, *159*
Nectarinia johnstoni 211, *159*
Nectarinia kilimensis 205, *159*
Nectarinia purpureiventris 202, *159*
Nectarinia reichenowi 210, *159*
Nectarinia rufipennis 298
Nectarinia tacazze 203, *159*
NECTARINIIDAE 135
nectarinioides, Cinnyris 267, *222*
neergaardi, Cinnyris 242, *206*
neglecta, Cyanomitra olivacea 179
neglectus, Anthreptes 137, *95*
Neocichla 644
Neocichla gutturalis 644, *543*
Neolestes 476
Neolestes torquatus 476, *350*
neumanni, Onychognathus 586, *479*

neumanni, Prionops retzii 497
newtoni, Lanius 345, *271*
newtonii, Anabathmis 160, *142*
niassae, Cinnyris venusta 288
Nicator 480
Nicator chloris 481, *350*
Nicator gularis 483, *350*
Nicator vireo 485, *350*
niger, Parus 101, *79*
nigeriae, Cinnyris superba 295
nigricans, Prionops retzii 497
nigricapillus, Lioptilus 66, *14*
nigricauda, Telophorus viridis 409
nigrifrons, Malaconotus 400, *335*
nigripennis, Oriolus 503, *462*
nigritemporalis, Nilaus afer 474
Nilaus 472
Nilaus afer 472, *287*
niloticus, Lanius senator 371
nitens, Lamprotornis 599, *527*
niveus, Eurocephalus anguitimens 381
nordmanni, Lamprotornis chalybaeus 604
notata, Cinnyris 276
notatus, Oriolus auratus 516
notha, Tchagra senegala 427
nubicus, Lanius 354, *271*
Nucifraga 567
Nucifraga caryocatactes 567, *478*
nyasae, Illadopsis pyrrhoptera 20
nyassae, Anthreptes longuemarei 139
nyikensis, Nectarinia johnstoni 212
nyombensis, Kupeornis chapini 71

obscura, Cyanomitra 175, *143*
occidentalis, Nilaus afer 474
octaviae, Cyanomitra cyanolaema 170
olivacea, Cyanomitra 178, *143*
olivaceus, Malaconotus 402, *335*
olivacina, Cyanomitra olivacea 179
omoensis, Turdoides leucopygius 47
omoensis, Zosterops abyssinicus 314
Onychognathus 577
Onychognathus albirostris 588, *526*
Onychognathus blythii 584, *479*
Onychognathus frater 585, *479*
Onychognathus fulgidus 580, *526*
Onychognathus morio 582, *479*
Onychognathus nabouroup 592, *479*
Onychognathus neumanni 586, *479*
Onychognathus salvadorii 587, *526*
Onychognathus tenuirostris 590, *526*
Onychognathus walleri 578, *479*
oreas, Picathartes 4, *14*
orientalis, Anthreptes 140, *95*

orientalis, Tchagra senegala 427
ORIOLIDAE 502
Oriolus 502
Oriolus auratus 516, *462*
Oriolus brachyrhynchus 504, *462*
Oriolus chlorocephalus 514, *462*
Oriolus crassirostris 507, *462*
Oriolus larvatus 509, *462*
Oriolus monacha 513, *462*
Oriolus nigripennis 503, *462*
Oriolus oriolus 518
Oriolus percivali 508, *462*
oriolus, Oriolus 518
oritis, Cyanomitra 172, *143*
ornatus, Lamprotornis 609, *542*
orosthruthus, Arcanator 8
orphnus, Parus cinerascens 83
orphogaster, Cinnyris chloropygyia 234
osea, Cinnyris 278, *207*
osiris, Cinnyris pulchella 263
oustaleti, Cinnyris 281, *223*
ovamboensis, Cinnyris pulchella 263

pallescens, Anthoscopus caroli 113, 114
pallescens, Corvus corone 535
pallida, Tchagra senegala 427
pallidifrons, Lanius collurio 365
pallidigastra, Hedydipna 229, *142*
pallidirostris, Lanius meridionalis 361
pallidiventris, Parus rufiventris 96
pallidus, Zosterops 318, *270*
Panurus 74
Panurus biarmicus 75, *94*
PARIDAE 77
Parophasma 73
Parophasma galinieri 73, *14*
Parus 80
Parus afer 85, *78*
Parus albiventris 91, *78*
Parus ater 103, *94*
Parus caeruleus 104, *94*
Parus carpi 99, *79*
Parus cinerascens 83, *78*
Parus cristatus 102, *94*
Parus fasciiventer 81, *78*
Parus fringillinus 89, *78*
Parus funereus 90, *78*
Parus griseiventris 82, *78*
Parus leucomelas 97, *79*
Parus leuconotus 93, *78*
Parus major 87, *78*
Parus niger 101, *79*
Parus rufiventris 96, *78*
Parus thruppi 86, *78*
parvirostris, Cinnyris reichenowi 251

parvulus, Anthoscopus 109, *79*
patersoni, Hedydipna collaris 220
pembae, Cinnyris 275, *222*
pendulinus, Remiz 120, *79*
percivali, Oriolus 508, *462*
permistus, Oriolus auratus 513
perspicillatus, Malaconotus monteiri 387
phanus, Telophorus zeylonus 413
phoenicuroides, Lanius isabellinus 363
Pholia 639
Pholia femoralis 639, *526*
Pholia sharpii 640, *526*
Pholidornis 121
Pholidornis rushiae 121, *79*
Phyllanthus 64
Phyllanthus atripennis 64, *31*
Pica 562
Pica pica 562, *463*
pica, Pica 562, *463*
Picathartes 1
Picathartes gymnocephalus 1, *14*
Picathartes oreas 4, *14*
PICATHARTIDAE 1
pintoi, Cinnyris manoensis 237
platura, Hedydipna 225, *142*
platycircus, Turdoides plebejus 37
plebejus, Turdoides 37, *30*
plumatus, Prionops 487, *415*
poensis, Cyanomitra oritis 173
poensis, Illadopsis cleaveri 23
poensis, Laniarius 443, *399*
Poeoptera 573
Poeoptera kenricki 576, *479*
Poeoptera lugubris 574, *479*
Poeoptera stuhlmanni 575, *479*
poliocephalus, Prionops plumatus 489
poliogaster, Zosterops 316, *270*
poliolophus, Prionops 491, *415*
poliothorax, Alethe 28
poliothorax, Kakamega 28, *15*
pondoensis, Laniarius ferrugineus 457
predator, Lanius collaris 341
preussi, Cinnyris reichenowi 251
preussi, Nectarinia 251
preussi, Onychognathus walleri 578
prigoginei, Cinnyris stuhlmanni 244
pringlii, Dryoscopus 438, *398*
PRIONOPIDAE 486
Prionops 487
Prionops alberti 492, *415*
Prionops caniceps 493, *415*
Prionops gabela 499, *415*
Prionops plumatus 487, *415*
Prionops poliolophus 491, *415*
Prionops retzii 496, *415*

Prionops rufiventris 495, *415*
Prionops scopifrons 500, *415*
PROMEROPIDAE 326
Promerops 327
Promerops cafer 328, *287*
Promerops gurneyi 336, *287*
protegatus, Corvus splendens 557
Pseudoalcippe 29
Pseudoalcippe abyssinica 32, *15*
Ptilostomus 560
Ptilostomus afer 560, *478*
Ptyrticus 34
Ptyrticus turdinus 34, *15*
puguensis, Illadopsis rufipennis 12, *13*
pujoli, Anthreptes 146
pulchella, Cinnyris 258, *207*
pulcher, Lamprotornis 628, *542*
punctifrons, Anthoscopus 110, *79*
purpurascens, Parus leucomelas 98
purpurascens, Sturnus vulgaris 654
purpureiceps, Lamprotornis 596, *527*
purpureiventris, Nectarinia 202, *159*
purpureus, Lamprotornis 597, *527*
purpuropterus, Lamprotornis 614, *542*
puveli, Illadopsis 26, *15*
PYCNONOTIDAE 476
pycnopygius, Chaetops 8, *14*
pycnopygius, Sphenoeacus 8
Pyrrhocorax 568
Pyrrhocorax graculus 570, *463*
Pyrrhocorax pyrrhocorax 568, *463*
pyrrhocorax, Pyrrhocorax 568, *463*
pyrrhoptera, Illadopsis 20, *15*
pyrrhostictus, Lanius collaris 342

quadricolor, Telophorus viridis 409
quanzae, Zosterops senegalensis 308
quartus, Telophorus viridis 409
querulus, Turdoides melanops 51

ragazzii, Cyanomitra obscura 176
rankinei, Anthoscopus caroli 113
ravidus, Parus niger 101
raymondi, Onychognathus tenuirostris 592
rectirostris, Anthreptes 146, *95*
regia, Cinnyris 252, *207*
regius, Lamprotornis 622, *543*
reichenbachii, Anabathmis 155, *142*
reichenowi, Anthreptes 149, *95*
reichenowi, Antichromus minutus 419
reichenowi, Cinnyris 250, *206*
reichenowi, Nectarinia 210, *159*
reichenowi, Oriolus larvatus 510

reichenowi, Zosterops senegalensis 307
reinwardtii, Turdoides 44, *30*
remigialis, Tchagra senegala 427
Remiz 119
Remiz pendulinus 120, *79*
REMIZIDAE 106
remotus, Antichromus minutus 419
restrictus, Telophorus zeylonus 413
retzii, Prionops 496, *415*
rhipidurus, Corvus 548, *478*
rhodesiae, Anthoscopus caroli 113, 114
rhodesiae, Cinnyris oustaleti 281
rhodesiensis, Tchagra australis 422
robertsi, Anthoscopus caroli 113, 114
roccatii, Anthoscopus caroli 113, 114
rockefelleri, Cinnyris 254, *207*
rolleti, Oriolus larvatus 510
roseus, Sturnus 659, *543*
rovumae, Salpornis spilonotus 129
rubescens, Chalcomitra 185, *158*
rubiginosus, Malaconotus olivaceus 402
rubiginosus, Turdoides 61, *31*
rubitorques, Anthreptes 148, *95*
rueppelli, Eurocephalus 379, *287*
rueppellii, Onychognathus morio 582
rufescens, Illadopsis 25, *15*
ruficeps, Laniarius 451, *414*
ruficollis, Corvus 540, *478*
rufinuchalis, Laniarius ruficeps 451
rufipennis, Cinnyris 298, *223*
rufipennis, Illadopsis 12, *15*
rufipennis, Nectarinia 298
rufipennis, Trichastoma 12
rufiventris, Parus 96, *78*
rufiventris, Prionops 495, *415*
rufocinctus, Kupeornis 69, *14*
rufofusca, Tchagra senegala 426
rufuensis, Turdoides hypoleucus 55
rushiae, Pholidornis 121, 122, *79*
ruthae, Anthoscopus flavifrons 108
rutilans, Lanius senator 371

sabini, Dryoscopus 430, *398*
saliens, Cinnyris afra 247
Salpornis 128
Salpornis spilonotus 128, *94*
SALPORNITHIDAE 128
salvadori, Salpornis spilonotus 129
salvadorii, Onychognathus 587, *526*
sandgroundi, Malaconotus nigrifrons 401
sardonius, Corvus corone 535
saturatior, Chalcomitra senegalensis 194
saturnus, Dicrurus ludwigii 523

savensis, Laniarius ferrugineus 457
schubotzi, Cinnyris stuhlmanni 244
sclateri, Cyanomitra obscura 176
scopifrons, Prionops 500, *415*
scotinus, Buphagus erythrorhynchus 666
seimundi, Anthreptes 150, *95*
senator, Lanius 369, *271*
senegala, Tchagra 426, *351*
senegalensis, Anthoscopus parvulus 109
senegalensis, Chalcomitra 193, *158*
senegalensis, Dryoscopus 432, *398*
senegalensis, Zosterops 306, *270*
sharpei, Anthoscopus caroli 113
sharpei, Dicrurus ludwigii 522
sharpei, Turdoides 47, *30*
sharpii, Corvus corone 535
sharpii, Pholia 640, *526*
sharpii, Turdoides rubiginosus 62
shelleyae, Nectarinia reichenowi 210
shelleyi, Cinnyris 261, *222*
shelleyi, Lamprotornis 624, *542*
shelleyi, Onychognathus morio 582
shellyae, Cinnyris reichenowi 251
siccata, Chalcomitra balfouri 197
siculus, Aegithalos caudatus 76
Sigmodus, see *Prionops*
silvanus, Zosterops poliogaster 316
similis, Malaconotus sulfureopectus 406
Sitta 124
Sitta europaea 125, *94*
Sitta ledanti 124, *94*
SITTIDAE 123
smithii, Lanius collaris 342
smithii, Turdoides leucopygius 46
socotranus, Zosterops abyssinicus 314
solivagus, Nilaus afer 473
somalicus, Lanius 347, *271*
somereni, Hedydipna collaris 220
soraria, Nectarinia 170
souzae, Lanius 368, *286*
souzae, Tchagra australis 422
spadix, Chaetops pycnopygius 8
speculifer, Oriolus chlorocephalus 515
speculigerus, Lanius isabellinus 363
Speculipastor 636
Speculipastor bicolor 637, *526*
Speirops 322
Speirops brunneus 325, *270*
Speirops leucophaeus 324, *270*
Speirops lugubris 322, *270*
Speirops melanocephalus 323, *270*
spermologus, Corvus monedula 555
Sphenoeacus afer 8
Sphenoeacus pycnopygius 8
spilonotus, Salpornis 128, *94*

splendens, Corvus 557, *478*
splendidus, Lamprotornis 610, *542*
Spreo 633
Spreo albicapillus 633, *542*
Spreo bicolor 634, *542*
squamulatus, Turdoides 56, *31*
stangerii, Chalcomitra rubescens 186
stenocricotus, Zosterops senegalensis 307
stictigula, Modulatrix 8
stictigula, Pseudoalcippe abyssinica 32
stictilaemus, Turdoides reinwardtii 44
sticturus, Laniarius bicolor 459
stierlingi, Pseudoalcippe abyssinica 32
stierlingi, Zosterops senegalensis 307
strenuipes, Illadopsis puveli 26
stresemanni, Zavattariornis 646, *543*
strophium, Cinnyris bifasciata 269
stuhlmanni, Cinnyris 244, *206*
stuhlmanni, Poeoptera 575, *479*
stuhlmanni, Zosterops senegalensis 307
STURNIDAE 572
Sturnus 652
Sturnus roseus 659, *543*
Sturnus unicolor 656, *543*
Sturnus vulgaris 653, *543*
suahelica, Cinnyris pulchella 263
subalaris, Cinnyris chalybea 239
subcollaris, Hedydipna collaris 220
subcoronatus, Lanius collaris 342
sublacteus, Laniarius aethiopicus 454
sudanensis, Tchagra senegala 426
sulfureopectus, Malaconotus 406, *335*
sundevalli, Zosterops pallidus 318
superba, Cinnyris 294, *223*
superbus, Lamprotornis 626, *542*
sycobius, Lamprotornis chalybaeus 604
sylviella, Anthoscopus caroli 113, 114

tacazze, Nectarinia 203, *159*
tacitus, Lanius souzae 368
talacomus, Prionops plumatus 489
talatala, Cinnyris 282, *223*
tamalakanae, Turdoides jardineii 40
tanganjicae, Cyanomitra alinae 174
tanganjicae, Parus fasciiventer 81
tanganjicae, Turdoides jardineii 40
taruensis, Anthoscopus caroli 113
taylori, Malaconotus olivaceus 402
Tchagra 420
Tchagra australis 421, *351*
Tchagra jamesi 424, *351*
Tchagra senegala 426, *351*
Tchagra tchagra 425, *351*
tchagra, Tchagra 425, *351*

Telophorus 408
Telophorus cruentus 416, *350*
Telophorus dohertyi 411, *350*
Telophorus viridis 409, *350*
Telophorus zeylonus 412, *350*
tenebrosus, Turdoides 45, *30*
tenuirostris, Onychognathus 590, *526*
tephrogaster, Dicrurus ludwigii 522
tephrolaema, Anthreptes rectirostris 146
terminus, Malaconotus sulfureopectus 406
theresae, Onychognathus tenuirostris 590
thermophilus, Telophorus zeylonus 412
thomensis, Dreptes 161, *142*
thruppi, Parus 86, *78*
tibicen, Oriolus larvatus 510
Tichodroma 127
Tichodroma muraria 127, *94*
TIMALIIDAE 7
timbuktana, Tchagra senegala 427
tingitanus, Corvus corax 537
togoensis, Corvinella corvina 373
tongensis, Zosterops senegalensis 308
tongoensis, Laniarius ferrugineus 457
toroensis, Zosterops senegalensis 307
torquata, Grafisia 638, *526*
torquatus, Neolestes 476, *350*
transvaalensis, Laniarius ferrugineus 457
traylori, Anthreptes seimundi 150
Trichastoma rufipennis 12
tricolor, Prionops retzii 497
tristis, Acridotheres 660, *543*
tristoides, Acridotheres tristis 660
tsavoensis, Cinnyris 273, *222*
tsumebensis, Turdoides gymnogenys 58
turatii, Laniarius 460, *414*
turdinus, Ptyrticus 34, *15*
Turdoides 35
Turdoides aylmeri 63, *31*
Turdoides bicolor 52, *31*
Turdoides fulvus 59, *31*
Turdoides gymnogenys 57, *31*
Turdoides hartlaubii 49, *30*
Turdoides hindei 53, *31*
Turdoides hypoleucus 55, *31*
Turdoides jardineii 39, *30*
Turdoides leucocephalus 43, *30*
Turdoides leucopygius 46, *30*
Turdoides melanops 50, *30*
Turdoides plebejus 37, *30*
Turdoides reinwardtii 44, *30*
Turdoides rubiginosus 61, *31*
Turdoides sharpei 47, *30*
Turdoides squamulatus 56, *31*

Turdoides tenebrosus 45, *30*
turkanae, Cinnyris habessinica 280

ugandae, Illadopsis fulvescens 18
ultramarinus, Parus caeruleus 105
ulugurensis, Laniarius fuelleborni 444
uncinatus, Lanius meridionalis 361
unicolor, Lamprotornis 620, *543*
unicolor, Sturnus 656, *543*
upembae, Ptyrticus turdinus 35
Urolestes 376
Urolestes melanoleucus 376, *287*
ursulae, Cinnyris 301, *223*
usambarica, Cinnyris mediocris 255
usambaricus, Laniarius fuelleborni 444
ussheri, Pholidornis rushiae 121
ussheri, Tchagra australis 422

vaughani, Lamprotornis corruscus 602
vaughani, Zosterops 310, *270*
venusta, Cinnyris 285, *223*
vepres, Turdoides sharpei 48
veroxii, Cyanomitra 180, *143*
verreauxi, Cinnyricinclus leucogaster 643
verticalis, Cyanomitra 167, *143*
vinaceigularis, Prionops plumatus 488
vincenti, Cyanomitra obscura 176
violacea, Anthobaphes 163, *159*
violacior, Lamprotornis mevesii 617
virens, Zosterops pallidus 318
vireo, Nicator 485, *350*
viridis, Telophorus 409, *350*
viridisplendens, Cyanomitra verticalis 168
vitorum, Malaconotus olivaceus 403
vulgaris, Sturnus 653, *543*

waldroni, Anthoscopus flavifrons 108
walleri, Onychognathus 578, *479*
warsangliensis, Tchagra senegala 427
webbi, Poeoptera lugubris 574
weigoldi, Parus cristatus 102
whitakeri, Garrulus glandarius 565
whytei, Cinnyris ludovicensis 245
winifredae, Zosterops poliogaster 316
winterbottomi, Anthoscopus caroli 113

xanthostomus, Parus niger 101
xylodromus, Salpornis spilonotus 130

yokanae, Anthreptes reichenowi 149

zambesiana, Hedydipna collaris 220
zanzibarica, Cyanomitra veroxii 181
Zavattariornis 646
Zavattariornis stresemanni 646, *543*
zeylonus, Telophorus 412, *350*
ZOSTEROPIDAE 305
Zosterops 305
Zosterops abyssinicus 314, *270*
Zosterops ficedulinus 311, *270*
Zosterops griseovirescens 313, *270*
Zosterops pallidus 318, *270*
Zosterops poliogaster 316, *270*
Zosterops senegalensis 306, *270*
Zosterops vaughani 310, *270*
zugmayeri, Corvus splendens 557
zuluensis, Hedydipna collaris 220

English Names

Abyssinian Catbird 73, *14*

Babbler, African Hill 32, *15*
Babbler, Arrow-marked 39, *30*
Babbler, Bare-cheeked 57, *31*
Babbler, Blackcap 44, *30*
Babbler, Black-faced 50, *30*
Babbler, Black-lored 47, 50, *30*
Babbler, Brown 37, *30*
Babbler, Capuchin 64, *31*
Babbler, Chapin's 71, *14*
Babbler, Dohrn's Thrush- 72, *14*
Babbler, Dusky 45, *30*
Babbler, Fulvous 59, *31*
Babbler, Grey-chested 28, *15*
Babbler, Hartlaub's 49, *30*
Babbler, Hinde's 53, *31*
Babbler, Hinde's Pied 53
Babbler, Northern Pied 55, *31*
Babbler, Pied 52
Babbler, Red-collared 69, *14*
Babbler, Scaly 56, *31*
Babbler, Southern Pied 52, *31*
Babbler, Spotted Thrush- 34, *15*
Babbler, White-headed 43, *30*
Babbler, White-rumped 46, 49, *30*
Babbler, White-throated Mountain 68, *14*
Bearded Reedling 75, *94*
Blackcap, Bush 66, *14*
Bokmakierie 412, *350*
Boubou, Fülleborn's 444, *399*
Boubou, Mountain Sooty 443, *399*
Boubou, Slate-coloured 445, *399*
Boubou, Sooty 441, *399*
Boubou, Southern 456, *414*
Boubou, Swamp 458, *414*
Boubou, Tropical 453, *414*
Boubou, Turati's 460, *414*
Boubou, Yellow-breasted 471, *399*
Brubru 472, *287*
Bulbul, Black-collared 476, *350*
Bush Blackcap 66, *14*
Bush-Crow, Ethiopian 646, *543*
Bush-Shrike, Blackcap 418, *351*
Bush-Shrike, Black-fronted 400, *335*
Bush-Shrike, Bocage's 404, *335*
Bush-Shrike, Braun's 449, *414*
Bush-Shrike, Bulo Burti 452, *414*
Bush-Shrike, Doherty's 411, *350*
Bush-Shrike, Fiery-breasted 385, *334*
Bush-Shrike, Gabela 450, *414*
Bush-Shrike, Gorgeous 409, *350*
Bush-Shrike, Green-breasted 393, *334*
Bush-Shrike, Grey-green 404
Bush-Shrike, Grey-headed 388, *334*
Bush-Shrike, Lagden's 391, *334*
Bush-Shrike, Lühder's 447, *414*
Bush-Shrike, Many-coloured 396, *335*
Bush-Shrike, Monteiro's 386, *334*
Bush-Shrike, Mount Kupé 395, *334*
Bush-Shrike, Olive 402, *335*
Bush-Shrike, Orange-breasted 406, *335*
Bush-Shrike, Red-naped 451, *414*
Bush-Shrike, Uluguru 394, *334*

Catbird, Abyssinian 73, *14*
Chatterer, Fulvous 59
Chatterer, Rufous 61, *31*
Chatterer, Scaly 63, *31*
Chough, Alpine 570, *463*
Chough, Red-billed 568, *463*

INDEXES: ENGLISH NAMES

Creeper, Spotted 128, *94*
Crow, Black 533
Crow, Cape 533, *478*
Crow, Carrion 535, *478*
Crow, House 557, *478*
Crow, Pied 545, *478*
Crow, Somali 544, *478*

Drongo, Common 525
Drongo, Fork-tailed 525, *463*
Drongo, Shining 524, *463*
Drongo, Square-tailed 522, *463*
Drongo, Velvet-mantled 529, *463*

Fiscal, Grey-backed 356, *286*
Fiscal, Long-tailed 348, *271*
Fiscal, São Tomé 345, *271*
Fiscal, Somali 347, *271*
Fiscal, Taita 346, *271*

Glossy starlings: see Starling
Gonolek, Black-headed 466, *399*
Gonolek, Papyrus 465, *399*
Gonolek, Yellow-crowned 461, *399*

Helmet-Shrike, Chestnut-fronted 500, *415*
Helmet-Shrike, Gabela 499, *415*
Helmet-Shrike, Grey-crested 491, *415*
Helmet-Shrike, Red-billed 493, *415*
Helmet-Shrike, Retz's 496, *415*
Helmet-Shrike, Rufous-bellied 495, *415*
Helmet-Shrike, White 487, *415*
Helmet-Shrike, Yellow-crested 492, *415*

Illadopsis, Blackcap 23, *15*
Illadopsis, Brown 17, *15*
Illadopsis, Grey-chested 28, *15*
Illadopsis, Mountain 20, *15*
Illadopsis, Pale-breasted 12, *15*
Illadopsis, Puvel's 26, *15*
Illadopsis, Rufous-winged 25, *15*
Illadopsis, Scaly-breasted 21, *15*

Jackdaw, Western 555, *478*
Jay, Eurasian 565, *463*

Magpie, Common 562, *463*
Myna, Common 660, *543*

Nicator, Eastern 483, *350*
Nicator, Western 481, *350*
Nicator, Yellow-throated 485, *350*
Nutcracker, Spotted 567, *478*
Nuthatch, Algerian 124, *94*
Nuthatch, Eurasian 125, *94*

Oriole, Abyssinian Black-headed 513, *462*
Oriole, African Golden 516, *462*
Oriole, Black-winged 503, *462*
Oriole, Eastern Black-headed 509, *462*
Oriole, Eurasian Golden 518, *462*
Oriole, Green-headed 514, *462*
Oriole, Mountain 508, *462*
Oriole, São Tomé 507, *462*
Oriole, Western Black-headed 504, *462*
Oxpecker, Red-billed 665, *543*
Oxpecker, Yellow-billed 662, *543*

Penduline Tit, African 112
Penduline Tit, Cape 115, *79*
Penduline Tit, Eurasian 120, *79*
Penduline Tit, Forest 107, *79*
Penduline Tit, Grey 112, *79*
Penduline Tit, Mouse-coloured 111, *79*
Penduline Tit, Sennar 110, *79*
Penduline Tit, Sudan 110
Penduline Tit, West African 109
Penduline Tit, Yellow 109, *79*
Penduline Tit, Yellow-fronted 107
Piapiac 560, *478*
Picathartes, Grey-necked 4, *14*
Picathartes, White-necked 1, *14*
Puffback, Black-backed 434, *398*
Puffback, Northern 437, *398*
Puffback, Pink-footed 431, *398*
Puffback, Pringle's 438, *398*
Puffback, Red-eyed 432, *398*
Puffback, Sabine's 430, *398*

Raven, Brown-necked 540, *478*
Raven, Common 537, *478*
Raven, Fan-tailed 548, *478*
Raven, Thick-billed 552, *478*
Raven, White-necked 550, *478*

Reedling, Bearded 75, *94*
Rockjumper 10, *14*
Rockjumper, Damara 8
Rockrunner 8, *14*
Rook 554, *478*
Rook, Cape 533

Shrike, Crimson-breasted 468, *399*
Shrike, Emin's 367, *286*
Shrike, Fiscal 340, *271*
Shrike, Isabelline 363, *286*
Shrike, Lesser Grey 358, *286*
Shrike, Long-tailed 372, *376*
Shrike, Mackinnon's 352, *286*
Shrike, Magpie 376, *287*
Shrike, Masked 354, *271*
Shrike, Northern White-crowned 379, *287*
Shrike, Nubian 354
Shrike, Red-backed 364, *286*
Shrike, Red-tailed 363
Shrike, Rosy-patched 416, *350*
Shrike, Southern Grey 360, *286*
Shrike, Southern White-crowned 381, *287*
Shrike, Souza's 368, *286*
Shrike, White-crowned 381
Shrike, White-rumped 379
Shrike, Woodchat 369, *271*
Shrike, Yellow-billed 372, *287*
Speirops, Black-capped 322, *270*
Speirops, Fernando Po 325, *270*
Speirops, Mount Cameroon 323, *270*
Speirops, Príncipe 324, *270*
Spotted Creeper 128, *94*
Starling, Abbott's 639, *526*
Starling, African Pied 634, *542*
Starling, Amethyst 642
Starling, Ashy 620, *543*
Starling, Black-bellied Glossy 602, *527*
Starling, Bristle-crowned 587, *526*
Starling, Bronze-tailed Glossy 600, *527*
Starling, Burchell's 618, *542*
Starling, Cape Glossy 599, *527*
Starling, Chestnut-bellied 628, *542*
Starling, Chestnut-winged 580, *526*
Starling, Common 653, *543*
Starling, Copper-tailed Glossy 594, *527*
Starling, Emerald 619, *527*
Starling, Fischer's 632, *542*
Starling, Golden-breasted 622, *543*

Starling, Greater Blue-eared 603, *527*
Starling, Hildebrandt's 625, *542*
Starling, Kenrick's 576, *479*
Starling, Lesser Blue-eared 606, *527*
Starling, Long-tailed Glossy 612, *542*
Starling, Magpie 637, *526*
Starling, Meves's Long-tailed 616, *542*
Starling, Narrow-tailed 574, *479*
Starling, Neumann's 586, *479*
Starling, Pale-winged 592, *479*
Starling, Pied 637
Starling, Plum-coloured 642
Starling, Príncipe Glossy 609, *542*
Starling, Purple Glossy 597, *527*
Starling, Purple-headed Glossy 596, *527*
Starling, Red-winged 582, *479*
Starling, Rose-coloured 658, *543*
Starling, Rüppell's Glossy 614, *542*
Starling, Sharpe's 640, *526*
Starling, Sharp-tailed 608, *527*
Starling, Shelley's 624, *542*
Starling, Slender-billed 590, *526*
Starling, Socotra 585, *479*
Starling, Somali 584, *479*
Starling, Splendid Glossy 610, *542*
Starling, Spotless 656, *543*
Starling, Stuhlmann's 575, *479*
Starling, Superb 626, *542*
Starling, Violet-backed 642, *526*
Starling, Waller's 578, *479*
Starling, Wattled 648, *543*
Starling, White-billed 588, *526*
Starling, White-collared 638, *526*
Starling, White-crowned 633, *542*
Starling, White-winged 644, *543*
Sugarbird, Cape 328, *287*
Sugarbird, Gurney's 336, *287*
Sunbird, Amani 229, *142*
Sunbird, Amethyst 189, *158*
Sunbird, Anchieta's 151, *142*
Sunbird, Banded Green 148, *95*
Sunbird, Bannerman's 169, *143*
Sunbird, Bates's 299, *223*
Sunbird, Beautiful 258, *207*
Sunbird, Black 189
Sunbird, Black-bellied 267, *222*
Sunbird, Blue-headed 174, *143*
Sunbird, Blue-throated 149
Sunbird, Blue-throated Brown 170, *143*
Sunbird, Bocage's 201, *159*
Sunbird, Bronze 205

Sunbird, Bronzy 205, *159*
Sunbird, Brown 145, *95*
Sunbird, Buff-throated 184, *158*
Sunbird, Cameroon 172, *143*
Sunbird, Carmelite 187, *158*
Sunbird, Collared 219, *142*
Sunbird, Congo 265, *222*
Sunbird, Copper 302, *159*
Sunbird, Coppery 302
Sunbird, Dusky 291, *223*
Sunbird, Eastern Double-collared 254, *207*
Sunbird, Eastern Olive 178, *143*
Sunbird, Eastern Violet-backed 140, *95*
Sunbird, Fraser's 153, *95*
Sunbird, Giant 161, *142*
Sunbird, Golden-winged 210, *159*
Sunbird, Greater Double-collared 247, *206*
Sunbird, Green 146, *95*
Sunbird, Green-headed 167, *143*
Sunbird, Green-throated 185, *158*
Sunbird, Grey 180, *143*
Sunbird, Grey-headed 153
Sunbird, Hunter's 197, *158*
Sunbird, Johanna's 293, *223*
Sunbird, Lesser Double-collared 239
Sunbird, Little Green 150, *95*
Sunbird, Little Purple-banded 268
Sunbird, Loveridge's 258, *207*
Sunbird, Ludwig's Double-collared 245, *206*
Sunbird, Malachite 213, *159*
Sunbird, Marico 262, *222*
Sunbird, Miomo Double-collared 237, *206*
Sunbird, Moreau's 256, *207*
Sunbird, Mouse-coloured 180
Sunbird, Neergaard's 242, *206*
Sunbird, Newton's 160, *142*
Sunbird, Nile Valley 227, *142*
Sunbird, Northern Double-collared 250, *206*
Sunbird, Olive-bellied 233 , *206*
Sunbird, Orange-breasted 163, *159*
Sunbird, Orange-tufted 276, 278, *207*
Sunbird, Oustalet's 281, *223*
Sunbird, Palestine 278, *207*
Sunbird, Pemba 275, *222*
Sunbird, Plain-backed 149, *95*
Sunbird, Príncipe 157, *142*
Sunbird, Purple-banded 268, *222*
Sunbird, Purple-breasted 202, *159*

Sunbird, Pygmy 225, *142*
Sunbird, Red-and-blue 151
Sunbird, Red-chested 266, *222*
Sunbird, Regal 252, *207*
Sunbird, Reichenbach's 155, *142*
Sunbird, Rockefeller's 254, *207*
Sunbird, Rufous-winged 298, *223*
Sunbird, Rwenzori Double-collared 244, *206*
Sunbird, Scarlet-chested 193, *158*
Sunbird, Scarlet-tufted 153
Sunbird, Scarlet-tufted Malachite 211, *159*
Sunbird, Shelley's 261, *222*
Sunbird, Shining 279, *207*
Sunbird, Socotra 198, *158*
Sunbird, Southern Double-collared 239, *206*
Sunbird, Splendid 296, *223*
Sunbird, Superb 294, *223*
Sunbird, Tacazze 203, *159*
Sunbird, Tiny 235, *206*
Sunbird, Tôha 200
Sunbird, Tsavo Purple-banded 273, *222*
Sunbird, Uluguru Violet-backed 137, *95*
Sunbird, Ursula's 301, *223*
Sunbird, Variable 285, *223*
Sunbird, Violet-breasted 274, *222*
Sunbird, Violet-tailed 141, *95*
Sunbird, Western Olive 175, *143*
Sunbird, Western Violet-backed 138, *95*
Sunbird, White-bellied 282, *223*
Sunbird, Yellow-bellied 285

Tchagra, Black-crowned 426, *351*
Tchagra, Brown-crowned 421, *351*
Tchagra, Marsh 418
Tchagra, Southern 425, *351*
Tchagra, Three-streaked 424, *351*
Thrush-Babbler, Dohrn's 72, *14*
Thrush-Babbler, Spotted 34, *15*
Tit, Acacia 86
Tit, African Penduline 112
Tit, Ashy 83, *78*
Tit, Blue 104, *94*
Tit, Cape Penduline 115, *79*
Tit, Carp's 99, *79*
Tit, Coal 103, *94*
Tit, Crested 102, *94*
Tit, Dusky 90, *78*
Tit, Eurasian Penduline 120, *79*
Tit, Forest Penduline 107, *79*

Tit, Great 87, *78*
Tit, Grey Penduline 112, *79*
Tit, Long-tailed 76, *94*
Tit, Miombo Grey 82, *78*
Tit, Mouse-coloured Penduline 111, *79*
Tit, Northern Black 97
Tit, Northern Grey 86, *78*
Tit, Red-throated 89, *78*
Tit, Rufous-bellied 96, *78*
Tit, Sennar Penduline 110, *79*
Tit, Southern Black 101, *79*
Tit, Southern Grey 85, *78*
Tit, Stripe-breasted 81, *78*
Tit, Sudan Penduline 110
Tit, West African Penduline 109
Tit, White-backed Black 93, *78*
Tit, White-bellied 91, *78*
Tit, White-shouldered Black 97
Tit, White-winged Black 97, *78*
Tit, Yellow Penduline 109, *79*
Tit, Yellow-fronted Penduline 107
Tit-hylia 121, *79*
Tit-weaver 121

Treecreeper, Short-toed 132, *94*

Wallcreeper 127, *94*
White-eye, Abyssinian 314, *270*
White-eye, Annobon 313, *270*
White-eye, Cape 318, *270*
White-eye, Mountain 316, *270*
White-eye, Pemba 310, *270*
White-eye, Príncipe 311, *270*
White-eye, White-breasted 314
White-eye, Yellow 306, *270*

French Names

Akalat à ailes rousses 25, *15*
Akalat à dos roux 34, *15*
Akalat à poitrine blanche 12, *15*
Akalat à poitrine écaillée 21, *15*
Akalat à poitrine grise 28, *15*
Akalat à tête noire 23, *15*
Akalat à tête sombre 32, *15*
Akalat brun 17, *15*
Akalat de Puvel 26, 15
Akalat montagnard 20, *15*
Astrild-mésange 121, *79*

Bagadais à bec rouge 493, *415*
Bagadais à front roux 500, *415*
Bagadais à huppe grise 491, *415*
Bagadais à ventre roux 495, *415*
Bagadais casqué 487, *415*
Bagadais d'Albert 492, *415*
Bagadais de Gabela 499, *415*
Bagadais de Retz 496, *415*
Brubru africain 472, *287*
Bulbul à collier noir 476, *350*
Bulbul à gorge jaune 485, *350*
Bulbul à tête brune 483, *350*
Bulbul nicator 481, *350*

Cassenoix moucheté 567, *478*
Chétopse à flancs roux 8, *14*
Chétopse bridé 10, *14*
Chocard des Alpes 570, *463*
Choucador à épaulettes rouges 599, *527*
Choucador à longue queue 612, *542*
Choucador à oreillons bleus 603, *527*
Choucador à queue bronzée 594, *527*
Choucador à queue fine 608 , *527*
Choucador à queue violette 600, *527*
Choucador à tête pourprée 596, *527*
Choucador à ventre noir 602, *527*
Choucador à ventre roux 628, *542*
Choucador de Burchell 618, *542*
Choucador de Hildebrandt 625, *542*
Choucador de Meves 616, *542*
Choucador de Príncipe 609, *542*
Choucador de Rüppell 614, *542*
Choucador de Shelley 624, *542*
Choucador de Swainson 606, *527*
Choucador iris 619, *527*
Choucador pourpré 597, *527*
Choucador splendide 610, *542*
Choucador superbe 626, *542*
Choucas des tours 555, *478*
Corbeau à nuque blanche 550, *478*
Corbeau à queue courte 548, *478*
Corbeau brun 540, *478*
Corbeau corbivau 552, *478*
Corbeau d'Edith 544, *478*
Corbeau familier 557, *478*
Corbeau freux 554, *478*
Corbeau pie 545, *478*
Corbin de Stresemann 646, *543*
Corneille du Cap 533, *478*
Corneille noire 535, *478*
Corvinelle à bec jaune 372, *287*
Corvinelle noir et blanc 376, *287*
Cratérope à croupion blanc 46, *30*
Cratérope à joues nues 57, *31*
Cratérope à tête blanche 43, *30*
Cratérope à tête noire 44, *30*
Cratérope ardoisé 63, *31*
Cratérope bicolore 52, *31*
Cratérope bigarré 55, *31*
Cratérope brun 37, *30*
Cratérope capucin 64, *31*
Cratérope de Hartlaub 49, *30*
Cratérope de Príncipe 72, *14*
Cratérope de Sharpe 47, *30*
Cratérope fauve 59, *31*
Cratérope fléché 39, *30*
Cratérope fuligineux 45, *30*
Cratérope maillé 56, *31*
Cratérope masqué 50, *30*
Cratérope ombré 45, *30*
Cratérope pie 55, *31*
Cratérope pie de Hinde 53, *31*
Cratérope rubigineux 61, *31*
Crave à bec rouge 568, *463*
Cubla à gros bec 430, *398*
Cubla à oeil rouge 432, *398*
Cubla à pieds roses 431, *398*
Cubla boule-de-neige 434, *398*
Cubla de Gambie 437, *398*
Cubla de Pringle 438, *398*

Drongo brillant 525, *463*
Drongo de forêt 524, *463*
Drongo de Ludwig 522, *463*
Drongo modeste 529, *463*

Étourneau caronculé 648, *543*
Étourneau roselin 658, *543*
Étourneau sansonnet 653, *543*
Étourneau unicolore 656, *543*
Eurocéphale à couronne blanche 381, *287*
Eurocéphale de Rüppell 379, *287*

Geai des chênes 565, *463*
Gladiateur à croupion rose 416, *350*
Gladiateur à front blanc 404, *335*
Gladiateur à front noir 400, *335*
Gladiateur à poitrine verte 393, *334*
Gladiateur bacbakiri 412, *350*

Gladiateur de Blanchot 388, *334*
Gladiateur de Doherty 41, *350*
Gladiateur de Lagden 391, *334*
Gladiateur de Monteiro 386, *334*
Gladiateur des Ulugurus 394, *334*
Gladiateur du Kupé 395, *334*
Gladiateur ensanglanté 385, *334*
Gladiateur multicolore 396, *335*
Gladiateur olive 402, *335*
Gladiateur soufré 406, *335*
Gladiateur vert 409, *350*
Gonolek fuligineux 441, *399*
Gonolek à nuque rouge 451, *414*
Gonolek à ventre blanc 458, *414*
Gonolek à ventre jaune 471, *399*
Gonolek à ventre rouge 466, *399*
Gonolek ardoisé 445, *399*
Gonolek boubou 456, *414*
Gonolek d'Abyssinie 453, *414*
Gonolek d'Amboim 450, *414*
Gonolek de Barbarie 461, *399*
Gonolek de Braun 449 , *414*
Gonolek de Bulo Burti 452, *414*
Gonolek de Lühder 447, *414*
Gonolek de montagne 443, *399*
Gonolek de Turati 460, *414*
Gonolek des papyrus 465, *399*
Gonolek de Fülleborn 444, *399*
Gonolek rouge et noir 468, *399*
Grand Corbeau 537, *478*
Grimpereau des jardins 132, *94*
Grimpereau tacheté 128, *94*

Lioptile à calotte noire 66, *14*
Loriot à ailes noires 503, *462*
Loriot à tête noire 504, *462*
Loriot à tête verte 514, *462*
Loriot d'Europe 518, *462*
Loriot de Percival 508, *462*
Loriot de São Tomé 507, *462*
Loriot doré 516, *462*
Loriot masqué 509, *462*
Loriot moine 513, *462*

Martin triste 660, *543*
Mésange à dos blanc 93, *78*
Mésange à épaulettes 97, *79*
Mésange à gorge rousse 89, *78*
Mésange à longue queue 76, *94*
Mésange à ventre blanc 91, *78*
Mésange à ventre cannelle 96, *78*
Mésange à ventre gris 82, *78*
Mésange à ventre strié 81, *78*
Mésange bleue 104, *94*

Mésange cendrée 83, *78*
Mésange charbonnière 87, *78*
Mésange de Carp 99, *79*
Mésange enfumée 90, *78*
Mésange grise australe 85, *78*
Mésange huppée 102, *94*
Mésange noire 103, *94*
Mésange noire australe 101, *79*
Mésange somalienne 86, *78*
Mésangette rayée 121, *79*

Panure à moustaches 75, *94*
Phyllanthe à collier roux 69, *14*
Phyllanthe à gorge blanche 68, *14*
Phyllanthe de Chapin 71, *14*
Phyllanthe de Galinier 73, *14*
Piapiac africain 560, *478*
Picatharte à cou blanc 1, *14*
Picatharte à cou gris 4, *14*
Pie bavarde 562, *463*
Pie-grièche à dos gris 356, *286*
Pie-grièche à dos roux 367, *286*
Pie-grièche à longue queue 348, *271*
Pie-grièche à poitrine rose 358, *286*
Pie-grièche à tête rousse 369, *271*
Pie-grièche de Mackinnon 352, *286*
Pie-grièche de São Tomé 345, *271*
Pie-grièche de Somalie 347, *271*
Pie-grièche de Souza 368, *286*
Pie-grièche des Teita 346, *271*
Pie-grièche écorcheur 364, *286*
Pie-grièche fiscale 340, *271*
Pie-grièche isabelle 363, *286*
Pie-grièche masquée 354, *271*
Pie-grièche méridionale 360, *286*
Piqueboeuf à bec jaune 662, *543*
Piqueboeuf à bec rouge 665, *543*
Promérops de Gurney 336, *287*
Promérops du Cap 328, *287*

Rémiz à front jaune 107, *79*
Rémiz à ventre jaune 109, *79*
Rémiz de Carol 112, *79*
Rémiz du Soudan 110, *79*
Rémiz minute 115, *79*
Rémiz penduline 120, *79*
Rémiz souris 111, *79*
Rufipenne à bec blanc 588, *526*
Rufipenne à bec fin 590, *526*
Rufipenne à cou blanc 638, *526*
Rufipenne à queue étroite 574, *479*
Rufipenne de Blyth 584, *479*
Rufipenne de forêt 580, *526*
Rufipenne de Kenrick 576, *479*

Rufipenne de Neumann 586, *479*
Rufipenne de Salvadori 587, *526*
Rufipenne de Socotra 585, *479*
Rufipenne de Stuhlmann 575, *479*
Rufipenne de Waller 578, *479*
Rufipenne morio 582, *479*
Rufipenne nabouroup 592, *479*

Sittelle kabyle 124, *94*
Sittelle torchepot 125, *94*
Souimanga à ailes dorées 210, *159*
Souimanga à ailes rousses 298, *223*
Souimanga à bec droit 146, *95*
Souimanga à ceinture rouge 266, *222*
Souimanga à col rouge 148, *95*
Souimanga à collier 219, *142*
Souimanga à gorge bleue 170, *143*
Souimanga à gorge rousse 184, *158*
Souimanga à gorge verte 185, *158*
Souimanga à longue queue 258, *207*
Souimanga à plastron rouge 247, *206*
Souimanga à poitrine rouge 193, *158*
Souimanga à poitrine violette 274, *222*
Souimanga à queue violette 141, *95*
Souimanga à tête bleue 172, *143*
Souimanga à tête verte 167, *143*
Souimanga à ventre blanc 282, *223*
Souimanga à ventre jaune 285, *223*
Souimanga à ventre olive 233, *206*
Souimanga à ventre pourpre 202, *159*
Souimanga améthyste 189, *158*
Souimanga bifascié 268, *222*
Souimanga bifascié du Tsavo 273, *222*
Souimanga brillant 279, *207*
Souimanga bronzé 205, *159*
Souimanga brun 145, *95*
Souimanga carmélite 187, *158*
Souimanga chalybée 239, *206*
Souimanga cuivré 302, *159*
Souimanga d'Aline 174, *143*
Souimanga d'Amani 229, *142*
Souimanga d'Anchieta 151, *142*
Souimanga d'Oustalet 281, *223*
Souimanga d'Ursula 301, *223*
Souimanga de Bannerman 169, *143*
Souimanga de Bates 299, *223*
Souimanga de Bocage 201, *159*
Souimanga de Bouvier 276, *207*
Souimanga de Fraser 153, *95*

Souimanga de Hartlaub 157, *142*
Souimanga de Hunter 197, *158*
Souimanga de Johanna 293, *223*
Souimanga de Johnston 211, *159*
Souimanga de Loveridge 258, *207*
Souimanga de Ludwig 245, *206*
Souimanga de Mariqua 262, *222*
Souimanga de Moreau 256, *207*
Souimanga de Neergaard 242, *206*
Souimanga de Newton 160, *142*
Souimanga de Palestine 278, *207*
Souimanga de Pemba 275, *222*
Souimanga de Preuss 250, *206*
Souimanga de Reichenbach 155, *142*
Souimanga de Reichenow 149, *95*
Souimanga de Rockefeller 254, *207*
Souimanga de São Tomé 161, *142*
Souimanga de Seimund 150, *95*
Souimanga de Shelley 261, *222*
Souimanga de Socotra 198, *158*
Souimanga du Congo 2652, *222*
Souimanga du Kilimandjaro 254, *207*
Souimanga du miombo 237, *206*
Souimanga du Nil 227, *142*
Souimanga du Ruwenzori 244, *206*
Souimanga éclatant 296, *223*
Souimanga fuligineux 291, *223*
Souimanga malachite 213, *159*
Souimanga minule 235, *206*
Souimanga murin 180, *143*
Souimanga nectarin 267, *222*
Souimanga olivâtre de l'Est 178, *143*
Souimanga olivâtre de l'Ouest 175, *143*
Souimanga orangé 163, *159*
Souimanga pygmée 225, *142*
Souimanga royal 252, *207*
Souimanga superbe 294, *223*
Souimanga tacazze 203, *159*
Souimanga violet 138, *95*
Souimanga violet des Ulugurus 137, *95*
Souimanga violet oriental 140, *95*
Spréo à calotte blanche 633, *542*
Spréo à gorge noire 644, *543*
Spréo améthyste 642, *526*
Spréo bicolore 634, *542*
Spréo cendré 620, *543*
Spréo d'Abbott 639, *526*
Spréo de Fischer 632, *542*
Spréo de Sharpe 640, *526*
Spréo pie 637, *526*
Spréo royal 622, *543*

Tchagra à tête brune 421, *351*
Tchagra à tête noire 426, *351*
Tchagra de James 424, *351*
Tchagra des marais 418, *351*
Tchagra du Cap 425, *351*
Tichodrome échelette 127, *94*

Zostérops à flancs jaunes 314, *270*
Zostérops alticole 316, *270*
Zostérops becfigue 311, *270*
Zostérops d'Annobon 313, *270*
Zostérops de Fernando Po 325, *270*
Zostérops de Pemba 310, *270*
Zostérops de Príncipe 324, *270*
Zostérops de São Tomé 322, *270*
Zostérops du Cap 319, *270*
Zostérops du mont Cameroun 323, *270*
Zostérops jaune 306, *270*